COLD SPRING HARBOR SYMPOSIA ON QUANTITATIVE BIOLOGY

VOLUME LI

Molecular Biology of *Homo sapiens*

COLD SPRING HARBOR LABORATORY
1986

COLD SPRING HARBOR SYMPOSIA ON QUANTITATIVE BIOLOGY
VOLUME LI

International Standard Book Number 0-87969-052-6 (cloth)
International Standard Book Number 0-87969-053-4 (paper)
International Standard Serial Number 0091-7451
Library of Congress Catalog Card Number 34-8174

Printed in the United States of America

COLD SPRING HARBOR SYMPOSIA ON QUANTITATIVE BIOLOGY

Founded in 1933 by
REGINALD G. HARRIS
Director of the Biological Laboratory 1924 to 1936

Previous Symposia Volumes

I (1933) Surface Phenomena
II (1934) Aspects of Growth
III (1935) Photochemical Reactions
IV (1936) Excitation Phenomena
V (1937) Internal Secretions
VI (1938) Protein Chemistry
VII (1939) Biological Oxidations
VIII (1940) Permeability and the Nature of Cell Membranes
IX (1941) Genes and Chromosomes: Structure and Organization
X (1942) The Relation of Hormones to Development
XI (1946) Heredity and Variation in Microorganisms
XII (1947) Nucleic Acids and Nucleoproteins
XIII (1948) Biological Applications of Tracer Elements
XIV (1949) Amino Acids and Proteins
XV (1950) Origin and Evolution of Man
XVI (1951) Genes and Mutations
XVII (1952) The Neuron
XVIII (1953) Viruses
XIX (1954) The Mammalian Fetus: Physiological Aspects of Development
XX (1955) Population Genetics: The Nature and Causes of Genetic Variability in Population
XXI (1956) Genetic Mechanisms: Structure and Function
XXII (1957) Population Studies: Animal Ecology and Demography
XXIII (1958) Exchange of Genetic Material: Mechanism and Consequences
XXIV (1959) Genetics and Twentieth Century Darwinism
XXV (1960) Biological Clocks
XXVI (1961) Cellular Regulatory Mechanisms
XXVII (1962) Basic Mechanisms in Animal Virus Biology
XXVIII (1963) Synthesis and Structure of Macromolecules
XXIX (1964) Human Genetics
XXX (1965) Sensory Receptors
XXXI (1966) The Genetic Code
XXXII (1967) Antibodies
XXXIII (1968) Replication of DNA in Microorganisms
XXXIV (1969) The Mechanism of Protein Synthesis
XXXV (1970) Transcription of Genetic Material
XXXVI (1971) Structure and Function of Proteins at the Three-dimensional Level
XXXVII (1972) The Mechanism of Muscle Contraction
XXXVIII (1973) Chromosome Structure and Function
XXXIX (1974) Tumor Viruses
XL (1975) The Synapse
XLI (1976) Origins of Lymphocyte Diversity
XLII (1977) Chromatin
XLIII (1978) DNA: Replication and Recombination
XLIV (1979) Viral Oncogenes
XLV (1980) Movable Genetic Elements
XLVI (1981) Organization of the Cytoplasm
XLVII (1982) Structures of DNA
XLVIII (1983) Molecular Neurobiology
XLIX (1984) Recombination at the DNA Level
L (1985) Molecular Biology of Development

All Cold Spring Harbor Laboratory publications may be ordered directly from Cold Spring Harbor Laboratory, Box 100, Cold Spring Harbor, New York 11724. Phone: 1-800-843-4388. In New York (516)367-8423.

COLD SPRING HARBOR SYMPOSIA ON QUANTITATIVE BIOLOGY

VOLUME LI

Symposium Participants

AARONSON, STUART, National Cancer Institute, NIH, Bethesda, Maryland
ALI, IQBAL U., National Institutes of Health, Bethesda, Maryland
ALT, FREDERICK, Dept. of Biochemistry and Biophysics, Columbia University, College of Physicians & Surgeons, New York
ANDERSON, FRENCH, National Institutes of Health, Bethesda, Maryland
ANDREWS, PETER, Dept. of Paleontology, British Museum, London, England
ANTONARAKIS, STYLIANOS, Dept. of Pediatrics and Genetics, Johns Hopkins University, School of Medicine, Baltimore, Maryland
ATHWAL, RAGHBIR, New Jersey Medical School, Newark
ATTARDI, GIUSEPPI, Dept. of Biology, California Institute of Technology, Pasadena
AXEL, RICHARD, Dept. of Biochemistry and Pathology, Howard Hughes Medical Institute, New York
BALAZS, IVAN, Lifecodes Corporation, Elmsford, New York
BARBACID, MARIANO, Dept. of Developmental Oncology, NCI-Frederick Cancer Research Facility, Maryland
BARBOSA, JAMES, Molecular Diagnostics, Inc., West Haven, Connecticut
BARRETT, THOMAS, University of Washington, Seattle
BATTISTUZZI, GIORGIO, Dept. of Genetics, University of Napels, Italy
BAUM, HOWARD, Lifecodes Corporation, Elmsford, New York
BEALE, A.J., Wellcome Research Laboratories, Kent, England
BEAR, J.C., Memorial University of Newfoundland, St. Johns, Canada
BEAUDET, ARTHUR, Baylor College of Medicine, Houston, Texas
BELL, JOHN, Stanford University, California
BERG, PAUL, Stanford University, California
BERKNER, KATHLEEN, Zymo Genetics, Seattle, Washington
BERKVENS, THOMAS, Dept. of Medical Biochemistry, Sylvius Laboratories, Leiden The Netherlands
BERNARDI, GEORGIO, Dept. of Molecular Genmatics, Institute Jacques Monod, Paris France
BERNSTEIN, ALAN, Dept. of M.D. Biology, Mt. Sinai Hospital Research Institute, Toronto, Canada
BERTOLOTTI, ROGER, Dept. of Molecular Genetics, CNRS, Gif-sur-Yvette, France
BHATIA, KULDEEP, National Institutes of Health, Bethesda, Maryland
BLACK, KAY, University of Texas, Dallas
BODMER, WALTER, Imperial Cancer Research Fund Laboratories, London, England
BOLLON, ARTHUR, Wadley Institutes of Molecular Medicine, Dallas, Texas
BONTHRON, DAVID, Children's Hospital, Boston, Massachusetts
BORN, WALTER, Balgrist Clinic, Zurich, Switzerland
BORRELLI, EMILIANA, Salk Institute, San Diego, California
BOTSTEIN, DAVID, Dept. of Biology, Massachusetts Institute of Technology, Cambridge
BRANDSMA, JANET, Dept. of Otolaryngology, Long Island Jewish Medical Center, New Hyde Park, New York
BROWN, MICHAEL, Dept. of Molecular Genetics, University of Texas Southwestern Medical School, Dallas
BROWN, PETER, Dept. of Biochemistry, University of Texas Health Sciences Center, Dallas
BROWNE, JEFFREY, Amgen, Thousand Oaks, California
BURGHES, ARTHUR, Dept. of Genetics, University of Montreal, Toronto, Canada
CAHILL, GEORGE, JR., Howard Hughes Institute, Boston, Massachusetts
CANAANI, DANIEL, Dept. of Biochemistry, Tel-Aviv University, Israel
CANAANI, ELI, Weizmann Institute of Science, Rehovot, Israel
CANTOR, CHARLES, Dept. of Genetics and Development, Columbia University College of Physicians & Surgeons, New York, New York
CAPECCHI, MARIO, Dept. of Biology, University of Utah, Salt Lake City
CAPRA, DONALD, J., Dept. of Microbiology, Southwestern Medical School, Dallas, Texas
CASKEY, THOMAS, C., Dept. of Molecular Genetics, Baylor College of Medicine, Houston, Texas
CATE, RICHARD, Biogen Research Corporation, Cambridge, Massachusetts
CAVENEE, WEBSTER, Dept. of Microbiology and Molecular Genetics, University of Cincinnati, Ohio
CHAKRAVARTI, ARAVINDA, University of Pittsburgh, Pennsylvania
CHAMBON, PIERRE, Institute for Chemical Biology, Faculté of Medicine, CNRS, Strasbourg, France
CHANDRA, SHARET, Indian Institute of Science, Bangalore, India
CHENG, SHIRLEY, Dept. of Neurogenetics, Massachusetts General Hospital, Boston
CHOUDARY, P., National Institutes of Health, Bethesda, Maryland

CLARK, STEPHEN, Genetics Institute, Boston, Massachusetts
COMPTON, DUANE, M.D. Anderson Hospital and Tumor Institute, Houston, Texas
COOKE, HOWARD, Dept. of Zoology, University of Edinburgh, Scotland
COOKE, ROBERT, *Newsday*, Melville, New York
COTTON, ROBIN, National Institutes of Health, Bethesda, Maryland
COX, DIANE, Hospital for Sick Children, Toronto, Canada
CRAWFORD, BRIAN D., Academic Press, Inc., New York, New York
CROCE, CARLO M., Wistar Institute, Philadelphia, Pennsylvania
DAAR, IRA O., Dept. of Human Genetics, Roswell Park Memorial Institute, Buffalo, New York
DAOUK, GHALEB H., Dept. of Biology, Massachusetts Institute of Technology, Cambridge
DAVEY, MICHAEL, National Cancer Institute, NIH, Bethesda, Maryland
DAVIE, EARL W., Dept. of Biochemistry, University of Washington, Seattle
DAVIES, KAY E., John Radcliff Hospital, Oxford, England
DE LA CHAPELLE, ALBERT, Dept. of Medical Genetics, University of Helinski, Finland
DE WET, WOUTER J., Dept. of Biochemistry, Potschefstroom University, South Africa
DEAVEN, LAWRENCE, Los Alamos National Laboratory, New Mexico
DELPECH, MARC, Institute of Molecular Pathology, Paris, France
DERYNCK, RIK, Dept. of Molecular Biology, Genentech Inc., South San Francisco, California
DETERA-WADLEIGH, SEVILLE, National Cancer Institute, NIH, Bethesda, Maryland
DONIS-KELLER, HELEN, Dept. of Human Genetics, Collaborative Research, Inc., Lexington, Massachusetts
DONOVAN-PELUSO, MARYANN, Dept. of Genetics and Development, Columbia University, New York, New York
DOOLITTLE, RUSSELL, Dept. of Chemistry, University of California, San Diego
DOWDY, STEVEN, Dept. of Medical Sciences, University of California, Irvine
DRAKE, JAMES, Academic Press (London) Limited, England
DURNAM, DIANE, Dept. of Experimental Pathology, Fred Hutchinson Cancer Research Center, Seattle, Washington
EVANS, RONALD, The Salk Institute, San Diego, California
FARBER, ROSANN, Dept. of Molecular Genetics and Biology, University of Chicago, Illinois
FELLOUS, MARC, Dept. of Human Immunogenetics, INSERM, Paris, France
FERRARA, G.B., Dept. of Immunogenetics, National Institute for Cancer Research, Genoa, Italy
FIDDES, JOHN C., California Biotechnology, Inc., Mountain View
FIERS, WALTER, Dept. of Molecular Biology, State University, Ghent, Belgium
FINTER, N.B., Wellcome Biotechnology, Kent, England
FLAVELL, RICHARD, Biogen Research Corporation, Cambridge, Massachusetts
FOWLER, MICHAEL, Dept. of Human Genetics, Roswell Park Memorial Institute, Buffalo, New York
FRANCKE, UTA, Dept. of Human Genetics, Yale University School of Medicine, New Haven, Connecticut
FRASER, G.R., Dept. of Cancer Epidemiology, Imperial Cancer Research Fund, Oxford, England
FRIEDMAN, ROBERT, S/L Health Care Ventures, New York, New York
FRIEDMAN, THEODORE, Dept. of Pediatrics, University of California, San Diego, La Jolla
FURLEY, A.J., Chester Beatty Laboratories, London, England
GEORGIEV, GEORGII, Institute of Molecular Science, USSR Academy of Sciences, Moscow
GERMAN, JAMES III, Dept. of Human Genetics, The New York Blood Center, New York
GESTELAND, RAYMOND, University of Utah, Salt Lake City
GILBERT, WALTER, Harvard University Biological Laboratories, Cambridge, Massachusetts
GIULIANO, GIOVANNI, Cold Spring Harbor Laboratory, New York
GLASER, THOMAS, Massachusetts Institute of Technology, Cambridge
GLASSBERG, JEFFREY, Lifecodes Corporation, Elmsford, New York
GODSON, NIGEL, Dept. of Biochemistry, New York University Medical Center, New York
GOEDDEL, DAVID V., Dept. of Molecular Biology, Genentech, Inc., South San Francisco, California
GOLDMAN, DAVID, National Institutes of Health, Bethesda, Maryland
GOLDSTEIN, JOSEPH L., Dept. of Molecular Genetics, University of Texas Health Sciences Center, Dallas
GOODFELLOW, PETER, Dept. of Human Molecular Genetics, Imperial Cancer Research Fund, London, England
GOOSSENS, MICHEL, Hôpital Henri Mondor, INSERM, Creteil, France
GRAY, JOSEPH, Dept. of Biomedical Science, Lawrence Livermore National Laboratory, California
GREEN, PHILIP, Collaborative Research, Inc., Lexington, Massachusetts
GREENE, WARNER C., National Cancer Institute, NIH, Bethesda, Maryland
GREGG, RONALD, Dept. of Genetics, University of Wisconsin, Madison
GROFFEN, JOHN, Oncogene Science, Inc., Mineola, New York
GRONER, YORAM, Dept. of Virology, Weizmann Institute of Science, Rehovot, Israel
GUSELLA, JAMES, Neurogenetics Laboratory, Massa-

chusetts General Hospital, Boston

HALL, JEFFREY, Public Health School, University of California, Berkeley

HAMEISTER, HORST, Dept. of Clinical Genetics, University of Ulm, Federal Republic of Germany

HAMERTON, JOHN, Dept. of Human Genetics, University of Manitoba, Canada

HARRIS, S., Dept. of Genetics, University of Leicester, England

HENNIG, ANNE K., Dept. of Clinical Pathology and Immunology, State University of New York, Syracuse

HERMAN, GAIL, Dept. of Pediatrics, Baylor College of Medicine, Houston, Texas

HESSLEWOOD, I.P., Amersham International plc, Buckinghamshire, England

HILL, ADRIAN, John Radcliffe Hospital, Oxford, England

HOBART, PETER, Pfizer Central Research, Groton, Connecticut

HOCK, RANDY, Fred Hutchinson Cancer Research Center, Seattle, Washington

HOCKFIELD, SUSAN, Dept. of Neuroanatomy, Yale University, New Haven, Connecticut

HOEIJMAKERS, J.H.J., Erasmus University, Rotterdam, The Netherlands

HONJO, TASUKU, Dept. of Medical Chemistry, Kyoto University, Japan

HOOD, LEROY, Dept. of Biology, California Institute of Technology, Pasadena

HOUSEMAN, DAVID, Cancer Research Center, Massachusetts Institute of Technology, Cambridge

HUMPHRIES, H.K., British Columbia Cancer Research Center, Vancouver, Canada

HUNG, PAUL, Weyth Laboratories, Philadelphia, Pennsylvania

IMAMURA, TAKASHI, Dept. of Human Genetics, National Institute of Genetics, Mishima City, Japan

JACKSON, CYNTHIA, Massachusetts Institute of Technology, Cambridge

JASNY, BARBARA, *Science*, Washington, D.C.

JOCHENSEN, H.G., Dept. of Medical Biochemistry, University of Leiden, The Netherlands

KALLOS, J., Dept. of Medicine, Columbia University, New York, New York

KAM, WING, Howard Hughes Medical Institute, University of California, San Francisco

KAMEN, ROBERT, Genetics Institute, Cambridge, Massachusetts

KAN, Y.W., University of California, San Francisco

KAO, FA-TEN, Dept. of Biochemistry, University of Colorado Health Sciences Center, Denver

KAPELNER, STEPHEN, Downstate Medical Center, Brooklyn, New York

KAZAZIAN, HAIG H., JR., Johns Hopkins Hospital, Baltimore, Maryland

KHAN, SOHAIB, Dept. of Anatomy and Cell Biology, University of Cincinnati College of Medicine, Ohio

KIBERSTIS, PAULA, *Cell Journal*, Cambridge, Massachusetts

KIDD, JUDITH, Dept. of Human Genetics, Yale University School of Medicine, New Haven, Connecticut

KIDD, KENNETH, Dept. of Human Genetics, Yale University School of Medicine, New Haven, Connecticut

KING, M-C., Public Health School, University of California, Berkeley

KNOLL, BRIAN, Dept. of Pathology and Laboratory Medicine, University of Texas Medical Center, Houston

KOLLIAS, GEORGE, Laboratory of Gene Structure and Expression, London, England

KOROBKO, VYACHELSLAV, Institute of Bioorganic Chemistry, USSR Academy of Sciences, Moscow

KOSHLAND, DANIEL, JR., Dept. of Biochemistry, University of California, Berkeley

KUNG, PATRICK, T Cell Sciences, Inc., Cambridge, Massachusetts

KUNKEL, LOUIS M., Dept. of Genetics, Children's Hospital, Boston, Massachusetts

LALOUEL, JEAN-M., University of Utah, Salt Lake City

LANDER, ERIC S., Whitehead Institute, Cambridge, Massachusetts

LATT, SAMUEL A., Dept. of Genetics, Children's Hospital, Boston, Massachusetts

LAU, CHRIS, Dept. of Hematology, University of California, San Francisco

LAUGHREA, M., Jewish General Hospital, Lady Davis Institute, Montreal, Canada

LAU, MARTHA, Eleanor Roosevelt Institute, Denver, Colorado

LAWN, RICHARD, Dept. of Molecular Biology, Genentech, Inc., San Francisco, California

LE BEAU, M.M., Dept. of Hematology and Oncology, University of Chicago, Illinois

LEBKOWSKI, JANE S., Applied Immune Sciences, Menlo Park, California

LEBO, ROGER, University of California, San Francisco

LEDER, PHILIP, Dept. of Microbiology and Genetics, Harvard Medical School, Boston

LEHRACH, HANS, European Molecular Biology Laboratories, Heidelberg, Federal Republic of Germany

LERMAN, LEONARD, Genetics Institute, Cambridge, Massachusetts

LEWIN, ROGER, *Science*, Washington, D.C.

LIEW, C.C., Dept. of Clinical Biochemistry, University of Toronto, Canada

LIM, BING, Dept. of Hematology and Oncology, Children's Hospital, Boston, Massachusetts

LITT, MICHAEL, Dept. of Biochemistry, Oregon Health Sciences University, Portland

LOMEDICO, PETER T., Dept. of Molecular Genetics, Hoffmann-LaRoche Inc., Nutley, New Jersey

LONG, GEORGE L., Eli Lilly & Company, Indianapolis, Indiana

LOUIE, ELAINE, Wistar Institute, Philadelphia, Pennsylvania

LUPSKI, JAMES, Dept. of Biochemistry, New York University Medical Center, New York

MACH, BERNARD, Dept. of Microbiology, University of Geneva Medical School, Switzerland
MAEDA, SHUICHIRO, Dept. of Biochemistry, Kumamoto University Medical School, Japan
MAGER, DIXIE, Terry Fox Laboratory, Vancouver Canada
MAGRAM, JEANNE, Columbia University College of Physicians & Surgeons, New York, New York
MAHLEY, ROBERT, Dept. of Pathology, University of California, San Francisco
MAK, TAK W., Ontario Cancer Institute, Toronto, Canada
MANDEL, JEAN-LOUIS, Dept. of Molecular Genetics, INSERM, Strasbourg, France
MANIATIS, THOMAS, Dept. of Biochemistry and Molecular Biology, Harvard University, Cambridge, Massachusetts
MARTIN, DAVID W., JR., Genentech, Inc., South San Francisco, California
MASON, PHILIP, Hammersmith Hospital, University of London, England
MCCLELLAND, ALAN, Molecular Therapeutics, Inc., West Haven, Connecticut
MCCLELLAND, MICHAEL, Columbia University, New York, New York
MCDONOUGH, PAUL, Dept. of Obstetrics and Gynecology, Medical College of Georgia, Augusta
MCINNES, RODERICK, Dept. of Genetics, Hospital for Sick Children, Toronto, Canada
MCKUSICK, VICTOR, Johns Hopkins Hospital, Baltimore, Maryland
MELLI, M., Scalvo Research Center, Siena, Italy
MELLON, ISABEL, Dept. of Biological Sciences, Stanford University, California
MINNA, JOHN, Navy Medical Oncology, NCI, Bethesda, Maryland
MIRZABEKOV, ANDREII, Institute of Molecular Biology, USSR Academy of Sciences, Moscow
MISKIMINS, KEITH, Yale University, New Haven, Connecticut
MORLEY, BERNARD, University of California, San Francisco, California
MOTULSKY, ARNO, Inherited Diseases Division, University of Washington, Seattle.
MULLIGAN, RICHARD, Whitehead Institute for Biomedical Research, Cambridge, Massachusetts
MULLIS, KARY B., Cetus Corporation, Emeryville, California
MURDAY, V., Imperial Cancer Research Fund, London, England
MYERS, RICHARD, Dept. of Physiology, University of California, San Francisco
NALBANTOGLU, JOSEPHINE, Imperial Cancer Research Fund, Herts, England
NEBERT, DANIEL, National Institutes of Health, Bethesda, Maryland
NEWMARK, PETER, *Nature*, London, England
NIMAN, HENRY, Dept. of Molecular Biology, Scripps Clinic and Research Foundation, La Jolla, California
NUSSENZWEIG, MICHEL, Dept. of Genetics, Harvard Medical School, Boston, Massachusetts
O'BRIEN, STEVE, NCI-Frederick Cancer Research Facility, Maryland
OKUBO, TAITI, Kitakatzuragi-Gon, Japan
OLIVERI, NANCY, Children's Hospital, Boston, Massachusetts
ORKIN, STUART, Dept. of Hematology and Oncology, Children's Hospital Medical Center, Boston, Massachusetts
PAABO, SVANTE, University of Uppsala, Sweden
PAGE, DAVID, Whitehead Institute for Biomedical Research, Cambridge, Massachusetts
PARDES, HERBERT, New York State Psychiatric Institute, New York
PATER, MARY, Faculty of Medicine, University of New Foundland, Canada
PEARSON, P.L., State University of Leiden, The Netherlands
PERSON, STANLEY, Pennsylvania State University, University Park
PETIT, CHRISTINE, Dept. of Recombinant Genetics, Institut Pasteur, Paris, France
PETTY, CAROLINE, Dana Farber Cancer Institute, Boston, Massachusetts
PINKEL, DANIEL, Dept. of Biomedical Science, Lawrence Livermore National Laboratory, California
PLACZEK, M.A., Imperial Cancer Research Fund, London, England
PREIS, PETER, University of California, San Francisco
PURRELLO, MICHELE, Memorial Sloan-Kettering Cancer Center, New York, New York
RABBITS, TERENCE, Dept. of Molecular Biology, Medical Research Council, Cambridge, England
RABBITTS, PAMELA, Ludwig Institute for Cancer Research, Cambridge, England
RABIN, DANIEL, Molecular Diagnostics, Inc., West Haven, Connecticut
RALPH, PETER, Cetus Corporation, Emeryville, California
RAYMOND, VINCENT, Dept. of Molecular Biology and Virology, Salk Institute, San Diego, California
RECZEK, PETER, Dept. of Molecular Biology, Dana-Farber Cancer Institute, Boston, Massachusetts
RIENHOFF, HUGH Y., JR., Dept. of Biochemistry, University of Washington, Seattle
ROBERTS, MICHAEL, Dept. of Biology, Yale University, New Haven, Connecticut
ROBERTSON, MIRANDA, *Nature*, Washington, D.C.
ROSSANI, MARA, Sclavo Research Center, Siena, Italy
ROYER-POKORA, BRIGITTE, Children's Hospital, Boston, Massachusetts
RUBINO, STEPHEN, Public Health Research Institute, New York, New York
RUDDLE, FRANK H., Dept. of Biology and Human Genetics, Yale University, New Haven, Connecticut
RUSSEL, DAVID W., Dept. of Molecular Genetics, University of Texas Health Science Center, Dallas
SADLER, EVAN J., Dept. of Medicine and Biochemistry, Washington University, St. Louis, Missouri

SAGER, RUTH, Dana-Farber Cancer Institute, Boston, Massachusetts
SAIKE, RANDY, Cetus Corporation, Emeryville, California
SAKAKI, YOSHIYUKI, Dept. of Genetic Information, Kyushu University, Fukuoka, Japan
SALTUS, RICHARD, *Boston Globe*, Massachusetts
SAMBROOK, JOSEPH, Dept. of Biochemistry, University of Texas Health Science Center, Dallas, Texas
SANGSTER, ROBERT, Dept. of Medical Biophysics, University of Toronto, Canada
SCHMID, CARL W., Dept. of Chemistry, University of California, Davis
SCHROEDER, HARRY, JR., University of Washington School of Medicine, Seattle
SCHWARTZ, CHARLES, Greenwood Genetic Center, South Carolina
SCHWARTZ, DAVID, Dept. of Embryology, Carnegie Institution of Washington, Baltimore, Maryland
SCOLNIK, PABLO A., Cold Spring Harbor Laboratory, New York
SEEBURG, PETER, Genentech, Inc., South San Francisco, California
SERJEANTSON, SUSAN, Dept. of Human Biology, John Curtin School of Medical Research, Canberra, Australia
SHARIK, CLYDE, Squibb Institute for Medical Research, Princeton, New Jersey
SHEN, C.-K. JAMES, Dept. of Genetics, University of California, Davis
SHIMIZU, KENJI, Dept. of Biology, Kyushu University, Fukuoka, Japan
SHOWS, THOMAS, Dept. of Human Genetics, Roswell Park Memorial Institute, Buffalo, New York
SILVER, LEE, Dept. of Biology, Princeton University, New Jersey
SIM, GEK-KEE, National Jewish Hospital, Denver, Colorado
SIMONE, P., Institute of Interdisciplinary Research, Brussels, Belgium
SINGER, MAXINE, National Cancer Institute, NIH, Bethesda, Maryland
SINISCALCO, MARCELLO, Memorial Sloan Kettering Cancer Research Center, New York, New York
SMITH, CHRIS, University of Oxford, England
SMITH, CASSANDRA, Columbia University College of Physicians & Surgeons, New York, New York
SMITH, DAVID, Health Effects Research, Dept. of Energy, Washington, D.C.
SMITHIES, OLIVER, University of Wisconsin, Madison
SOEDA, EIICHI, Institute of Physical and Chemical Research, Wako, Japan
SOMMER, STEVE, Dept. of Biochemistry and Molecular Biology, Mayo Clinic School of Medicine, Rochester, Minnesota
SORGE, JOSEPH, Scripps Clinic and Research Foundation, La Jolla, California
SPARKS, ROBERT, Dept. of Medicine, University of California Medical Center, Los Angeles
SPECTOR, DAVID L., Cold Spring Harbor Laboratory, New York
SPURR, NIGEL, Imperial Cancer Research Fund, Herts, England
STATES, J.C., Children's Hospital Research Foundation, Cincinnati, Ohio
STEGGLES, ALAN, Dept. of Biochemistry and Medical Pathology, Northeastern Ohio University, Rootstown
STEIN, JANET L., Dept. of Immunology and Medical Microbiology, University of Florida, Gainsville
STEIN, GARY S., Dept. of Biochemistry and Microbiology, University of Florida, Gainsville
STEPHINSON, JOHN, Oncogene Science, Inc., Mineola, New York
STONEKING, MARK, Dept. of Biochemistry, University of California, Berkeley
STROMINGER, JACK L., Dept. of Biochemistry and Molecular Biology, Harvard University, Cambridge, Massachusetts
STUMP, DAVID C., Dept. of Medicine, University of Vermont, Burlington
TANIGUCHI, T., Dept. of Molecular Cell Biology, Osaka University, Japan
TASSET, DIANE, University of Colorado Health Science Center, Denver
TECOTT, LAWRENCE, Dept. of Psychiatry, Stanford University Medical Center, California
THOMAS, DONALD E., Fred Hutchinson Cancer Research Center, Seattle, Washington
TODARO, GEORGE, Oncogen, Seattle, Washington
TOMITA, FUSAO, Kyowa Hakko Kogyo Company, Ltd., Tokyo, Japan
TONEGOWA, SUSUMU, Massachusetts Institute of Technology, Cambridge, Massachusetts
TONIOLO, DANIELA, International Institute of Genetics and Biophysics, Napels, Italy
TOYONAGA, BARRY, Ontario Cancer Institute, Toronto, Canada
TOYOSHIMA, KUMAO, Institute of Medical Science, University of Tokyo, Japan
TROWSDALE, JOHN, Imperial Cancer Research Fund, London, England
TSIPOURAS, PETROS, Dept. of Biochemistry, Rutgers Medical School, Piscataway, New Jersey
TSPI, ANN-PING, Syntex Research, Palo Alto, California
TSUI, LAP CHEE, Dept. of Genetics, Hospital for Sick Children, Toronto, Canada
TSUJIMOTYO, YOSHIDE, Wistar Institute, Philadelphia, Pennsylvania
ULLRICH, AXEL, Dept. of Molecular Biology, Genentech, Inc., South San Francisco, California
VANDE WOUDE, GEORGE, NCI-Frederick Cancer Research Facility, Frederick, Maryland
VARGA, JANOS M., Dept. of Molecular Immunology, National Cancer Institute, NIH, Bethesda, Maryland
VEHAR, GORDON A., Dept. of Protein Biochemistry, Genentech, Inc., South San Francisco, California
VELASCO, SUSANA, Long Island Jewish Medical Center, New Hyde Park, New York

VERMA, INDER, Dept. of Molecular Biology and Virology, Salk Institute, San Diego, California

VISSING, HENRIK, Hagedorn Research Laboratory, Gentofte, Denmark

WAGNER, ERWIN, European Molecular Biology Laboratories, Heidelberg, Federal Republic of Germany

WALLACE, BRUCE, City of Hope Medical Center, Los Angeles, California

WALTER, MICHAEL, Hospital for Sick Children, Toronto, Canada

WEISS, ANTHONY, Dept. of Biochemistry, Stanford University Medical Center, California

WEISSENBACH, JEAN, Institut Pasteur, INSERM, Paris, France

WEISSMAN, SHERMAN, Dept. of Human Genetics, Yale University School of Medicine, New Haven, Connecticut

WEXLER, NANCY, Columbia University, New York, New York

WHITE, RAYMOND, University of Utah Medical School, Salt Lake City

WIERINGA, B., Dept. of Human Genetics, Radboud Hospital, Nijmegen, The Netherlands

WIGLER, MICHAEL, Cold Spring Harbor Laboratory, New York

WILLARD, HUNTINGTON, University of Toronto, Canada

WILLIAMSON, ROBERT, Dept. of Biochemistry, University of London, England

WILSON, ALLAN C., Dept. of Biochemistry, University of California, Berkeley

WITUNSKI, MICHAEL, James S. McDonnell Foundation, St. Louis, Missouri

WOLF, STAN, Genetics Institute, Cambridge, Massachusetts

WOO, SAVIO, Dept. of Cell Biology, Baylor College of Medicine, Houston, Texas

WOOD, STEVEN, Dept. of Medical Genetics, University of British Columbia, Vancouver, Canada

WORTON, R.G., Hospital for Sick Children, Toronto, Canada

ZOGHBI, HUDA, Dept. of Pediatrics, Baylor College of Medicine, Houston, Texas

First row: A. Motulsky; R. Williamson, N. Wexler; J. Sorge
Second row: R. Sager, G.P. Georgiev; S. Aaronson
Third row: K. Davies; V. McKusick, F. Ruddle; G. Bernardi
Fourth row: D. Russell; C. Cantor, D. Botstein; M. Capecchi

First row: T. Friedmann; R. Evans; L.-C. Tsui
Second row: Wine and Cheese Party
Third row: L. Deaven; K. Berkner; T. Shows
Fourth row: Y.W. Kan; M. Van Dilla

OLIVER AND LORRAINE GRACE AUDITORIUM

Dedication Ceremony
June 1, 1986

Top (left to right): South view; Oliver R. Grace, G. Morgan Browne; Lorraine Grace
Bottom (left to right): Edward Pulling; Walter H. Page; Daniel E. Koshland, Jr.; North view

Foreword

Thirteen years marked the time between the discovery of the double helix in 1953 and the elucidation of the genetic code in 1966. A similar interval has now passed since the development by Cohen and Boyer of a simple procedure for the cloning of selective DNA fragments. The scientific advances made possible by the subsequent modification and elaboration of these original cloning procedures now amaze, stimulate, and increasingly often overwhelm us. Facts that until recently were virtually unobtainable now flow forth almost effortlessly. Most excitingly, the frenetic pace of these new discoveries, instead of marking the impending end of a glorious moment of learning, give every indication of opening up scientific frontiers that will take hundreds if not thousands of years to explore thoroughly.

This new era of enlightenment is nowhere more apparent than in our newfound ability to study ourselves at the molecular level. So it was very easy to choose the title for this our 51st Symposium. By focusing on "The Molecular Biology of *Homo sapiens,*" we would be choosing a topic that most certainly will be returned to over and over during the second 50 Symposium years. In contrast, only once in our first 50 Symposium years did our discussions focus exclusively on ourselves (Human Genetics in 1964).

Happily, this year's Symposium marked the first year of use of our new facility built expressly for the holding of meetings, the very wonderful Oliver and Lorraine Grace Auditorium. The formal dedication ceremonies occurred on June 1st, during the Symposium, with the main dedicatory speech delivered by Daniel Koshland, editor of *Science* magazine and Professor of Biochemistry at the University of California, Berkeley.

In organizing the program for this Symposium, Marcello Siniscalco and I have benefited from the thoughtful advice of many colleagues, in particular Paul Berg, Sir Walter Bodmer, David Botstein, Michael Brown, Charles Cantor, Tom Caskey, Joseph Goldstein, Lee Hood, Tom Maniatis, Joe Sambrook, and Allan Wilson. Many more potential speakers were suggested than could be included in this our most intense Symposium yet attempted. The final program comprised 123 speakers, with most of the audience of 311 staying virtually to the end. Sir Walter Bodmer, a participant of our 1964 Symposium, gave an incisive introductory talk, and Tom Caskey, a participant of our 1966 Genetic Code Symposium, gave the very able summarizing remarks. Near the conclusion, at a special session presided over by Walter Gilbert and Paul Berg, the feasibility of beginning soon a total sequencing of the human genome was discussed. This is a topic that generates diverse reactions, many of which were clearly expressed during the discussion.

The holding of this meeting required a multitude of financial backers and, in addition to our long-time Symposium supporters, the National Cancer Institute/National Institutes of Health, The National Science Foundation, and the Department of Energy, we are greatly indebted for special help provided by the Lucille P. Markey Charitable Trust, the Rita Allen Foundation, and the Banbury Fund. Help also was provided through our Corporate Sponsor Program, which provides core support for the Cold Spring Harbor Meetings Program: Abbott Laboratories, American Cyanamid Company, Amersham International plc, Becton Dickinson and Company, Cetus Corporation, Ciba-Geigy Corporation, CPC International Inc., E.I. du Pont de Nemours & Company, Eli Lilly and Company, Genentech, Inc., Genetics Institute, Hoffmann-La Roche, Inc., Monsanto Company, Pall Corporation, Pfizer Inc., Schering-Plough Corporation, Smith Kline & French Laboratories, The Upjohn Company, and Wyeth Laboratories.

Our Meetings Office Staff, Gladys Kist, Maureen Berejka, Diane Tighe, Michela McBride, and Barbara Ward, again performed at a high level, looking after the registration and housing of Symposium participants as well as making each of our participants feel thoroughly wanted. This year, her 12th at Cold Spring Harbor, marked the retirement of Gladys Kist, whose intelligent good will has been a virtually indispensable element of the Symposium mystique. The audiovisual setups were again ably provided by Herb Parsons, while the massive correspondence and telephoning needed to put together a Symposium were cheerfully performed by Andrea Stephenson. As in the past several years, our Publications Department has worked hard to ensure the publication of these books by the end of the year, and we are indebted to the efforts of Nancy Ford, Managing Director of Publications, and of editors Judy Cuddihy, Dorothy Brown, and Douglas Owen, ably assisted by Joan Ebert and Mary Cozza.

James D. Watson, Director
October 10, 1986

Contents

Part 1

Human Evolution

Drugs Made Off Human Genes

Clotting, Anti-clotting Factors

Part 2

Receptors

Human Cancer Genes

Genetics of Predisposition to Cancer

Summary

Appendix

COLD SPRING HARBOR SYMPOSIA ON QUANTITATIVE BIOLOGY

VOLUME LI

Human Genetics: The Molecular Challenge

W.F. BODMER
Imperial Cancer Research Fund, Lincoln's Inn Fields, London WC2A 3PX, England

The last Cold Spring Harbor Symposium that was wholly devoted to human genetics took place 22 years ago in 1964. The meeting was relatively small and cozy and was divided into three sections, namely, population studies, genetics of somatic cells and cells in culture, and human proteins. The main focus was on classical approaches to human and population genetics, and the proteins mentioned were predominantly hemoglobins, immunoglobulins, and glucose-6-phosphate dehydrogenase, the latter because of its use as an X-linked marker. DNA did not appear in any of the titles, hardly even by implication, and was hardly mentioned in the whole published proceedings. The main hint about the future came from studies on somatic-cell genetics, including one of the first descriptions by Littlefield (1965) of selection of mouse cell hybrids using drug resistance markers. The first description of the then *LA*, now *HLA-A*, locus was an important step in the development of the HLA system (Payne et al. 1965), which had been started a few years earlier by J.J. Van Rood and J. Dausset.

The contrast with the 1986 meeting is striking. There is surely no paper that does not mention DNA, and there are at least 40 or 50 cloned gene systems referred to, counting such complex examples as HLA and the immunoglobulins each as a single system. In some respects, this is a low number, since the actual number of gene systems cloned is now at least two to three times as many and, counting all genes individually (including those for all mammals, since a key feature of molecular genetics is the easy transition from one species to another), the total number cloned is in the hundreds. This revolution has been forged by the development of gene cloning and DNA sequencing techniques, more secure approaches to cell biology and tissue culture, and improved chemistry, including especially analytical and preparative separation techniques for small amounts of protein.

One of the most immediate applications of this "new genetics" has been the elucidation at the molecular level of some of the best known inherited diseases, including in particular the hemoglobinopathies. There are also new examples of classical diseases explained at the molecular level, such as phenylketonuria through the cloning of the phenylalanine-hydroxylase gene (see Woo et al., this volume), hemophilia and factor 8, and, most recently, color blindness and the genes for rhodopsin-like pigments (Nathans et al. 1986). New systems have been defined, most notably through the work of J.L. Goldstein and M.S. Brown on the low-density lipoprotein receptor and its association with hypercholesterolemia and early heart disease.

The second major contribution of molecular genetics is the production of proteins that have been difficult to obtain in substantial amounts by traditional approaches, mainly because of the limited availability of suitable material and the low amounts normally synthesized. Notable examples include human insulin, the interferons, human growth hormone, and, more recently, other growth factors such as interleukin-2 and certain enzymes, e.g., urokinase and tissue plasminogen activator. Added to this list, although not strictly human products, should be the vaccines that are now being produced, especially those made by inserting appropriate viral sequences into the vaccinia genome and so reproducing vaccination in the original Jennerian sense.

The challenge now is to use the new genetics to climb from the DNA level at the bottom up to the analysis of variation, whose genetics and corresponding gene products have hardly been defined. Obvious examples include cystic fibrosis and Huntington's chorea. The still greater challenge is to deal with diseases and other variations that are not simple Mendelian traits but are multifactorial, a term that basically implies that there are inherited components, but these are not well defined and may be due to a mixture of the effects of several genes and the environment. Heart disease, mental disease, cancer, and complex normal variation, including behavior and facial features as well as general physical attributes, all fall in this category. The molecular challenge, which has as its goal the characterization of the whole human genome, is to use this knowledge to unravel the genetic contributions to complex human traits.

The HLA Systems: A Model Gene Cluster

Twenty-two years ago (Payne et al. 1985) the HLA system was beginning to be defined by a combination of relatively crude serology and genetics, supported by extensive statistical analysis. A series of antigens present on white blood cells and other nucleated tissues was defined through inherited differences identified by fetal-maternal alloantisera. These are what we now call the HLA-A, -B, and -C, or class I, antigens. At the time of the second Cold Spring Harbor Symposium I attended in 1976, the Ia antigens, now referred to as HLA-D region, or class II products, had been defined, and a start had been made on the chemistry of the relevant proteins, including by that time the discovery that β_2-microglobulin was associated in a bimolecular complex with the class I products (Barnstable et al. 1977). Now, completing a trilogy of descriptions of the HLA system at Cold Spring Harbor Symposia, separated by intervals of approximately 10 years, it is possible to outline the extraordinary advances in the description of the HLA system at the DNA level. This has been a classical "top-down" as opposed to "bottom-up" approach, leading from the genetics and serology to the identification of the protein products and, through that, to the cloning of the relevant genes. A brief survey of present knowledge of the HLA system, and in particular the HLA-D region, provides a background for the rest of this paper. This demonstrates the nature of a complex gene cluster and the way that polymorphic variation can be used to analyze disease susceptibility at the population and family levels. The HLA system, which is further described in several other papers in this volume, illustrates all the detailed features that can be expected from a complex genetic region and its evolution.

The overall map of the HLA region is compared with its mouse equivalent H-2 in Figure 1, which shows the positions of the two main sets of cell-surface products, the class I or HLA-A, -B, and -C and related genes being on the telomeric side of the region on the short arm of chromosome 6 and the class II or HLA-D region being toward the centromere. In between are genes for complement components and 21-hydroxylase, the enzyme that is deficient in congenital adrenal hyperplasia. The class I region contains some 25 or more genes, about half of which are not expressed. All of these products are associated with β_2-microglobulin, which is coded by a single gene on chromosome 15. The class II region contains genes for both the α and β chains (more than 15 in all), which are combined to form a functional heterodimer. The class I and class II products function in controlling interactions between cells in the immune response and, through this, play a key role in the regulation of the immune response presumably through interaction with the T-lymphocyte antigen receptor. Sequence data show that both class I and class II products consist of closely related families of genes, which are themselves part of the immunoglobulin "superfamily." Membership of the superfamily is defined by a minimum of 20–25% homology between members of different families within the superfamily and by the sharing of common structural features such as characteristic S—S bonds and domain structures. There is, therefore, a simple evolutionary rationale for finding the class I and class II genes together in the same gene cluster. The complement genes are presumed to have entered this cluster from elsewhere at a later date and show no structural homologies whatever with the immunoglobulin superfamily. It is possible, however, to rationalize their presence in the HLA region in terms of their function. There is no similar rationale for the presence of the 21-hydroxylase genes, which may have come in as bystanders or hitchhikers with their complement gene neighbors (see, e.g., Bodmer et al. 1986).

There is a striking homology of the overall organization of the mouse and human systems, with only the mouse H-2K genes being out of line. This comparison, however, hides detailed differences in the complexity and sizes of the subregions.

Recombination hot spots and linkage disequilibrium. Another feature of the overall HLA genetic map is the identification of recombination "hot spots." These are positions where recombination is presumed to occur much more frequently than elsewhere in the region. In the mouse I region, and between *DP* and *DQ*, and *A* and *C* in the HLA region, these are identified directly by relatively high recombination fractions, i.e., between 0.5 and 1.5 or 2% as compared to the known or presumed molecular distance between the genes. In the case of *DQ*, the hot spot is identified by a relative lack of linkage disequilibrium between more or less adjacent markers.

Linkage disequilibrium, perhaps more appropriately called gametic association, is the population association that may be observed between alleles at closely linked loci (see, e.g., Cavalli-Sforza and Bodmer 1971). This association, in the absence of disturbing effects

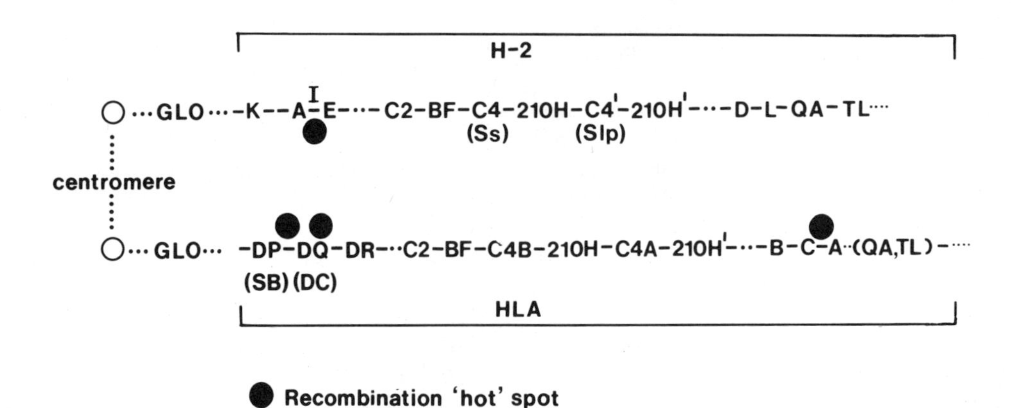

Figure 1. Schematic maps of the HLA and H-2 regions, aligned to give the maximum correspondence of the two sequences. For further details, see the text and Bodmer et al. (1986).

such as natural selection, declines eventually to zero at a rate $(1 - r)^n$, where r is the recombination fraction between the two loci and n is the number of generations. For example, if $r = 0.1\%$, the association will go down by a factor of 5 in about 15,000 generations, or for humans some 37,500 years, which is comparable in magnitude to the time of separation of the major human racial groups. If, on the other hand, $r = 0.5\%$, then the association goes down by a factor of 5 in about 320 generations or 8000 years, which is only comparable to the time since agriculture started to spread throughout Europe from the fertile crescent in the Middle East. Although some allelic variation may be very old compared to these times, perhaps even dating back a few million years or more, nevertheless it seems from patterns of recombination and linkage disequilibrium that values of r greater than 0.1% and probably more than 0.5% are needed before linkage disequilibrium becomes weak or negligible. Conversely, strong linkage disequilibrium is only likely to occur for recombination fractions less than 0.5–0.1%.

It now seems likely that recombination, at least in mammals, is mainly localized to recombinational hot spots. These are probably small regions, perhaps on the order of 1000 bp in length, where recombination occurs at least 5–10 times more frequently than elsewhere, so that 80–90% of all recombinations will occur in such hot spots. The total recombination fraction of the human genome, estimated from the number of observed chiasmata at meiosis, is about 2500 cM (1 cM = 1% recombination). If the average hot spot gives rise to a recombination fraction of 0.5%, then there would be about 5000 hot spots in the genome or about 5 in the HLA region, which is calculated to be about one thousandth of the total human genome. The average distance between such hot spots would then be about 600,000 bp. It is quite likely that 0.5% is an overestimate, since there is bound to be an observational bias favoring hot spots with higher recombination fractions. Presumably, there is a distribution of recombination fractions among hot spots, which may well be skewed toward lower values, but nevertheless have a mean of 0.1–0.2%. In this case, there would be some 10–20,000 hot spots separated by average intervals of 150–300,000 bp. On this assumption, there would be a reasonable correspondence between observed recombination fractions and physical distance as measured in nucleotide pairs, so long as distances were large compared to the average interval between hot spots. Of course, there may be further localized effects, for example, near centromeres and telomeres, and a few "very hot" spots that distort the relationship between the physical and genetic map to some extent, even on this larger scale. Differences in recombination fractions between the sexes may also influence these calculations. Nevertheless, the size of the HLA region, measured in recombination fraction terms to be about one thousandth of the total, corresponds remarkably closely with one thousandth of the total number of nucleotide pairs in the human genome, namely, $1/1000 \times 3 \times 10^9$, or 3×10^6, which is the estimate obtained from pulse gel field electrophoresis of large DNA pieces (S. Weissman, pers. comm.; Lawrance et al., this volume).

Names and functions. There is an obvious problem in naming the HLA region products, which arises because of their origin from studying genetic variation rather than a protein product. Thus, immunoglobulins and hemoglobins were named after the molecules studied, and the gene systems and names for heavy and light chains and α and β chains came later. For the HLA system, there are names for the genes, such as *HLA-A, -B,* and *-C,* but no good name for the products that, for example, for class I are formed by association with β_2-microglobulin. By analogy with the immunoglobulins, the HLA and other species major histocompatibility products could be called "histoglobulins." Then the class I products could conveniently be called histoglobulins A and the class II products could be called histoglobulins D by analogy with IgG and IgA; one could even use similar abbreviations, namely, HgA and HgD.

The functions of the two types of histoglobulins, namely, regulating the immune response, underlie their extensive polymorphism. This is presumed to have been selected for because of its association with immune response differences with respect to important pathogens. A secondary consequence of this immune response variation is susceptibility to a wide range of diseases with an immune or autoimmune etiology. This is reflected in population association of alleles of the various HLA-A, -B, -C, and -D region loci with particular diseases, including the well-known association of B27 with ankylosing spondylitis and of DR3 and DR4 with insulin-dependent juvenile onset diabetes. In many cases, these population associations probably reflect linkage disequilibrium between the marker studied and the actual genetic region determining susceptibility. For example, the first association with diabetes was with B15 and B8. This was later refined to a stronger association with DR4 and DR3, although more recent data suggest that the responsible genes are probably in the DQ subregion and in linkage disequilibrium with DR3 and DR4. Analysis of DNA level variation, most simply by restriction-fragment-length polymorphism (RFLP) but of course also by sequencing, especially indirectly using oligonucleotide probes, is being used in attempts to pinpoint the sequences responsible for the functional effects. But for this, functional studies at the cellular level are also needed (for overall recent reviews of the HLA system, see Albert et al. 1984; Immunological Reviews 1985).

The HLA-D Region

A more detailed molecular map of the HLA-D region and a comparison with its mouse counterpart are shown in Figure 2. From an initial, apparently allelic, set of determinants called DR, four subregions DP, DQ, DR, and DO/DZ have been defined. Each subre-

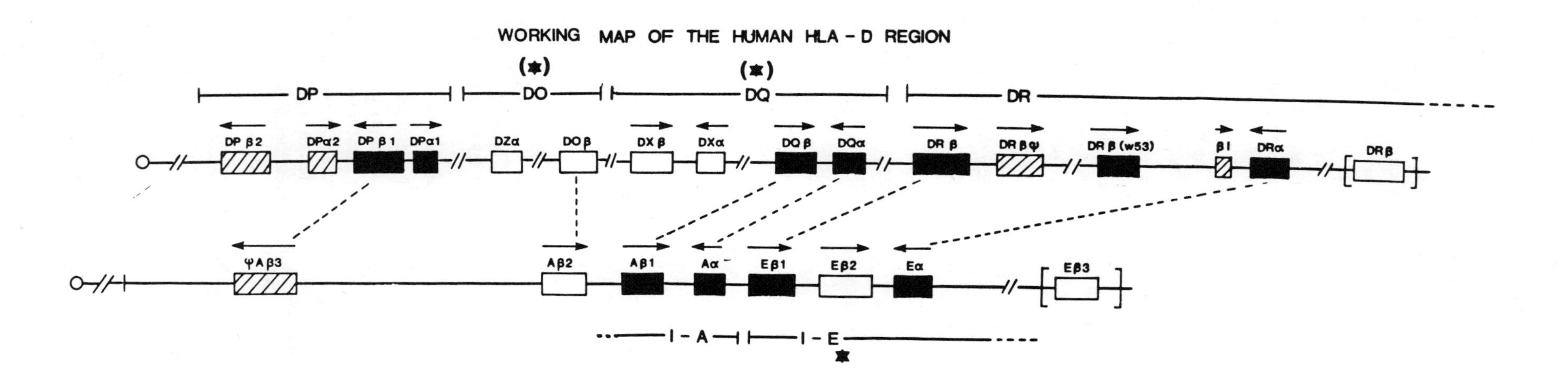

Figure 2. Schematic map of the HLA-D region (based on Trowsdale et al. 1985 and further unpublished observations). The map is compared with that of the H-2 I region. Arrows above the blocks indicating regions coding for the various gene products show the direction of transcription.

gion has its own α and β genes, although it is not yet clear whether DZα and DOβ correspond, and there are indications that there may be at least one more minor subregion. DO and DZ, as well as the DX subdivision of DQ, were clearly identified by "bottom-up" techniques using DNA cloning. Patterns of linkage disequilibrium between serological determinants and with DNA polymorphisms, together with analysis of mutants, defined the genetic map (see, e.g., Trowsdale et al. 1985). This has been strikingly confirmed by the application of pulse gel field electrophoresis (Bell et al.; Lawrance et al.; both this volume). The correspondence between classical and molecular approaches is most reassuring. The molecular data suggest a connected map of the HLA-D region that is about 10^6 bp long, with a gap of between 350 and 500 kb between *DZ*α and *DO*β (a better terminology would be to call the products α and β, and the genes, *DPA, DPB,* etc.; Bodmer et al. 1984). Duplication of pairs of α and β genes clearly seems to account for the structure of the DP, DQ, and DR subregions, although it is known that one of the DP subsets is nonfunctional, and it is not clear whether the DX genes are expressed, although the genes are potentially fully functional.

Functional diversification between the subregion products is to be expected and was presumably the basis for the original selection of duplicates to form the subregions. Evolutionary fine tuning of this sort is not uncommon and is analogous to the separation of, for example, the hemoglobin β products into embryonic, fetal, and adult forms. This functional diversification is unlikely to be clear-cut. It is most probably expressed as a tendency for certain products to deal with certain types of responses better than others. For instance, DR may tend to be better for help and DQ for suppression.

This leaves open the question as to the function of the DP products. These were originally defined only by cellular typing (which is cumbersome) but can now be identified both by RFLP and by monoclonal antibodies. One useful polymorphic monoclonal antibody that detects variation in the α chain has been produced by immunization of C3H mice with a transfected human *HLA-DP* gene expressed in mouse L cells, which are C3H-derived (Heyes et al. 1986). There is also sequence variation in the DP β-chain genes, raising the possibility for DP, as has already been suggested for DQ (Spielman et al. 1984), that there may be specific heterodimeric α/β combinations that arise in heterozygotes and may be associated with specific immune responses or disease susceptibility.

HLA and Hodgkin's Disease: Population and Family Association Analysis

The first HLA and disease association was for Hodgkin's disease, but it was very weak, with a maximum relative risk of only between 1.3 and 1.6. The significance of the association was eventually established only because it was extensively studied by many different laboratories. Although the vast majority of cases of Hodgkin's disease are sporadic, i.e., do not occur in families, a small proportion, perhaps 3% of all cases, can be found in families with two or more affected individuals, usually sibs. Within such families, HLA typing can establish whether affected pairs of sibs with Hodgkin's disease are HLA identical, share one parental HLA chromosome but not the other, or have neither of their HLA combinations in common. Combined data from a number of such family studies are shown in Table 1. The observed frequency of haplotype sharing differs markedly from the expected 1:2:1 on the assumption of Mendelian segregation for HLA and no association with Hodgkin's disease. This strong intrafamilial association, in contrast to the weak population association, is direct evidence for linkage of a gene conferring susceptibility to Hodgkin's disease with the HLA region, or most probably actually within it.

This approach to looking for linkage between a genetic marker and susceptibility to a disease works even in a situation such as Hodgkin's for which there is no evidence for familial clustering, namely, when the inherited susceptibility is determined by a gene with very poor "penetrance." If, for example, a rare dominant gene confers a 5% chance of getting a cancer, which otherwise has an incidence of only 1:10,000 as in the case of Hodgkin's disease, only $(1/2)^2 \times (1/20)^2$ = 1:1600 of sib pairs will be affected. Nevertheless, those sib pairs that are affected will share the relevant dominant gene (see Bodmer 1986). In the case of Hodgkin's disease and HLA, the question remains as to which, if any, of the HLA region products may be involved. The lack of any significant association with HLA-DR (and so DQ) or HLA-A, -B, and -C suggests either a very small effect, which is unlikely from the strong association seen in the families, or the existence of an allele at another locus not in linkage disequilibrium with these. The obvious candidate is *HLA-DP*, which is the only gene whose alleles are not in linkage disequilibrium with those for the other HLA region serological products. This suggestion can now be checked using the DP typing made possible by molecular genetic approaches.

The sib-pair approach for the detection of linkage of a marker with inherited disease susceptibility, and its generalizations, provides a powerful basis for the analysis of the genetic contribution to any complex multi-

Table 1. HLA Haplotype Sharing in Sib Pairs with Hodgkin's Disease

		No. of shared haplotypes		
		2	1	0
No. of sib pairs	Observed	16	11	5
	Expected	8	16	8

$X_2^2 = 10.7$; $p < 0.005$. Data from several studies are combined.

factorial trait. The approach will work provided (1) there are genes whose individual effect is large enough to give rise to a measurable phenotypic difference and (2) there exist enough genetic markers to be able to find one that is sufficiently closely linked to such a gene. The widespread availability of DNA clones and derived RFLPs now provides an essentially unlimited range of polymorphic markers, making realistic this classical approach to the analysis of genetic variation as first pointed out by Solomon and Bodmer (1979). The question remains, however, when one has a marker that may be 10% or only 1% recombination fraction units away from the desired "susceptibility" gene, how many functional genes are there to choose from in that distance before getting to the relevant one? In other words, what is the molecular complexity of the human genome?

Genome Complexity: The Number of Gene Clusters and Products

The fundamental question as to the number of human gene families or clusters, and expressed protein products, can now be answered within reasonable limits using available molecular data (for an earlier attempt, see Bodmer 1981). Many genes occur in related families, often clustered together in the same genetic region. The hemoglobin α- and β-chain clusters are classical examples, whereas the HLA and immunoglobulin regions appear to be at the upper limit of gene cluster complexity. These families of genes are in some cases associated together in superfamilies such as that including the immunoglobulins, histoglobulins (HLA, etc.), T-cell receptor, and certain other T-cell and immunoglobulin receptor molecules. Some gene families, such as the collagens and actins, are highly dispersed, whereas others, such as the immunoglobulins, are partially dispersed, in this case into three clusters, namely, for the heavy chains and λ and κ light chains.

Assuming, as now seems clear, that the basic genetic functional unit is the gene family or cluster, then the first question that needs to be answered is, what proportion of informational gene sequences occur in families with more than one member and what is the average number of genes per family? Orgel has suggested (see Bodmer 1983) that it may be those genes whose functions are common to prokaryotes and eukaryotes, perhaps often determining household functions, which do not occur in families and for which there may be no obvious correlation between protein domains and intron/exon organization. Complex gene clusters, such as immunoglobulins and HLA, probably evolved much later. On this basis, using the number of bacterial genes as a guide, there may be at most 1000–3000 basic genetic functions that occur singly and not in families. What then is the average size of a family or cluster? A lower limit is 2, whereas the number of coding sequences in the hemoglobin β cluster is 6 and that in the HLA region is about 50. The geometric mean of these three numbers is 8.4, which seems to be a reasonable estimate for the average size of a family. No doubt the actual size distribution will be somewhat skewed with a higher proportion of smaller families.

To complete the analysis of complexity, we need two further distributions. The first of these is the average size of gene product, considering only proteins, since functional RNAs such as tRNAs and ribosomal RNAs correspond to a relatively small proportion of the genome. RNA products that play a role in control of gene expression do not contribute directly to complexity in the sense we are considering now. Taking an average protein or polypeptide product size to be about 300 amino acids gives an average of 1000 nucleotide pairs per expressed protein product. The second distribution needed is the coding ratio, i.e., the proportion of the sequence within a cluster that codes for protein. The first well-defined example of a cluster, for hemoglobin β chains, included 6 coding sequences for products of 146 amino acids (one of which is a pseudogene) in a total region of about 60,000 bp. This gives a coding ratio of $60{,}000/6 \times 146 \times 3 = 1/23$. To apply this estimate to the whole genome, it must be assumed that the average gap between adjacent gene clusters is similar to that between coding sequences within a cluster. So far as I am aware, there is as yet no example in the human genome or, indeed, in any mammalian nuclear genome of adjacent clusters. The coding sequences within the HLA region include 6 D-region α-chain genes, 10 β-chain genes, 25 class I chain genes, and 2 C4 and 21-hydroxylase genes together with factor B and C2, for a total of 14,500 amino acids. The coding ratio is therefore approximately $3 \times 14{,}500/3 \times 10^6$ or about 1/70. Although there may still be functional gene sequences to be identified within the HLA region, so far, searches for such sequences using mRNA or cDNA probes have revealed only a pseudogene transposed from elsewhere (Trowsdale et al. 1984). However, there are clearly relatively large stretches not yet cloned, and so still potentially containing coding sequences. But, perhaps underestimates of the number of functional genes within the cluster will be compensated for by underestimating the gap between adjacent functional clusters. Let us therefore take the mean of the hemoglobin and HLA values, about 1/35, as a reasonable estimate for the coding ratio. Finally, we need to know what proportion of the total genome is not "selfish," i.e., is actually involved in functional clusters. Since, in some mammals, up to 50% of the DNA can be highly repetitive satellite DNA (Doolittle and Sapienza 1980; Orgel and Crick 1980), it seems reasonable to assume that no more than 50% of the total genome (1.5×10^9 bp) is involved in functional clusters.

Assuming 2000 nonclustered, single functional genes, these contribute only $2000 \times 1000 \times 35 = 7 \times 10^7$ bp, or about 5% of the total. The number of clusters is (total number nucleotide pairs involved in functional clusters)/(average cluster size) × (average base pairs per product) × (1/coding ratio) or $1.5 \times 10^9/8.4 \times 1000 \times 35 = 5102$. Thus, these calculations suggest a total of only about 7000 basic genetic functions, of which 2000 are single and 5000 are organized into families with an average size of just over 8.

The main sources of error in this estimate are in the family size and its distribution and in the coding ratio. It seems unlikely that either of these are out by more than a factor of 2. This suggests a maximum of 15–20,000 basic genetic functions, but with a number more likely to lie around 10,000 or less. Given the further subdivision of these functions into superfamilies, this complexity seems reasonably manageable.

It is worth noting that the estimated number of protein products is between 40,000 and 50,000, whatever the gene family size. A differentiated cell probably expresses no more than 5–10% of these, or somewhere between 2000 and 5000 proteins, which corresponds quite well to the number seen on high-resolution two-dimensional gels.

As data accumulate on gene products, their sizes, and their organization into families and genomic clusters, this information can be used to obtain more precise estimates of the number of genetic functions. Plotting the change in the estimates with time should lead to convergence to an asymptote well before the total genome is analyzed. The fact that these estimates have not changed a great deal over the last 5 or 6 years suggests that they are certainly of the right order of magnitude. It will be especially interesting to see whether there is bimodality in the distribution of the size of gene families and whether this correlates with the division into more basic household functions, as compared with other differentiated functions.

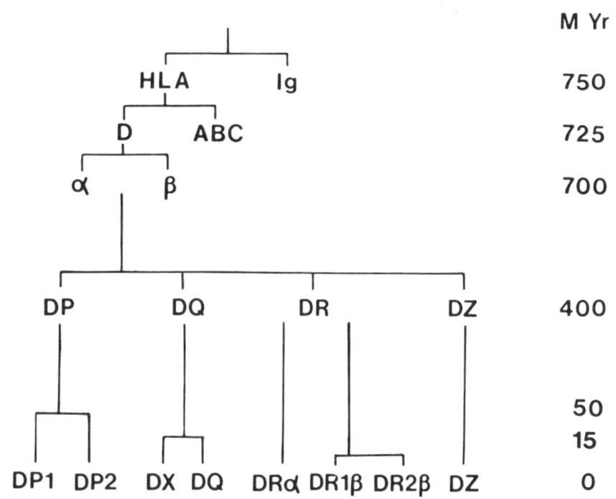

Figure 3. Evolutionary tree for the HLA-D region and associated products. HLA and Ig refer to the primordial genes for the HLA and immunoglobulin systems. D and A, B, C, α, and β refer similarly to the primordial genes for the class II and I subregions and the class II subregion chains. Approximate divergence times are given in millions of years (from Bodmer et al. 1986).

Cluster Evolution

Sequence comparisons can now provide an overall picture of the evolution of the HLA system and, in particular, the HLA-D region (see Fig. 3) (Bodmer et al. 1986). The data suggest that the HLA and immunoglobulin superfamily originated about 750 million years ago at the time of the initial evolution of the chordates, which can perhaps be thought of as the second major stage in the evolution of organismal complexity, following primitive multicellular metazoans. There appears then to have been a time of rapid diversification, during which the major features of the immune system, and so of the different members of the superfamily, evolved. The HLA system codes for about 50 different products and constitutes perhaps one fifth of the superfamily, which overall then codes for about 250 protein products. This alone could account for as much as 0.5% of the total human genome. Thus, the 250 different products divided among the members of the superfamily all presumably evolved from one ancestral gene present about 750 million years ago in a primitive chordate or its recent ancestor. It is not yet known how many such "superfamilies" there are in the human genome. The mixed-function P-450-dependent monooxygenases, the serum proteinases, and the tyrosine kinase activity containing receptor-like molecules are other examples. Even if there are as many as 100–200 such superfamilies, this would suggest that a relatively small proportion of primordial genes present 750 million years ago diversified into the present complex gene families and clusters that seem to characterize the mammalian species.

The overall molecular complexity of the mammals may therefore not be that much greater than that of their primitive ancestors. Evolutionary refinement came not so much from increasing overall complexity but from controlling it in subtler ways. In judging the quality of written English, it is not only the number of words used that counts, but also the way they are combined. That is no doubt what distinguishes my writing from Shakespeare's, although whether by comparison a scientific paper counts as a primitive chordate or just a very primitive mammal is not yet clear.

A major part of the increased functional complexity of higher organisms must be due to greater flexibility in the control of expression of combinations of genes. There is plenty of room in the mammalian genome to code for a large number of short sequences that can give rise to the RNAs proposed by J. Steitz and others that might be involved in control of gene expression (e.g., through the control of splicing; Bodmer 1981). The distinction between clustered and dispersed gene families may be related to the control strategy for a family of products. This will have been influenced by the coevolution of sequences adjacent to those coding for products, since these are presumably involved in differential control of expression in different tissues and at different times of development. A gene cluster has its own evolution, almost like a miniorganism, involving a variety of mechanisms including duplication, inversion, deletion, gene conversion, transposition, and frameshifts coupled with differential splicing of transcripts (see, e.g., Bodmer 1983; Bodmer et al. 1986).

Gene Mapping and Its Applications

The view presented above of the molecular complexity of the human genome has fundamental implications for human genetics because the basic number of types of genetic functions, say between 5,000 and 10,000, is not nearly as large as calculated previously by simply dividing the total number of base pairs by the average number needed to code for a single protein product. Thus, from an individual gene count in the millions, the complexity has been reduced up to 100-fold.

The first implication is that we may be closer to knowing something about a reasonably high proportion of the basic genetic functions than was thought to be the case until quite recently. McKusick's catalog of Mendelian inheritance in man (McKusick 1986) lists more than 3000 inherited traits, and Human Gene Mapping 8 (de la Chapelle 1985) shows that several hundred human genes or clusters have been mapped. If arbitrary DNA clones are included, the number by now is approaching 1000 (see McKusick, this volume). A second important implication is that the chance of finding a gene relevant to a given attribute may no longer be small. A guess at the relevance of a given genetic function, and some knowledge of map position, may be enough to limit the choice to a manageable number for screening. The third implication is that the genetic control of different complex attributes may overlap more than might once have been expected, because of the overlap of basic genetic functions. Thus, families of genes such as the actins, the collagens, or the P-450 monooxygenases have diversified to such an extent that they play a role in a variety of different, often apparently unrelated, physiological functions.

The obvious challenge is to map all functional genes and gene clusters, obtain their precise order, and eventually determine the complete nucleotide sequence. The genetic map is the equivalent of a dictionary, since it provides the only unique ordering of the genes. One need only think of the problem of looking up a word without a dictionary. The existence of an ordered alphabet makes the notion of a dictionary easy and obvious, but what about Chinese? My understanding is that the Chinese dictionary depends on counting strokes in symbols as an approach to systematic ordering and that the comparatively recent construction of the first comprehensive Chinese dictionary was a major achievement. A library is, of course, in some ways analogous to a dictionary. For without a unique reference system by which each book can be located by its title, author, or type of contents, the library is useless and the unreferenced book is essentially lost, especially in a large library. Thus, gene mapping is absolutely fundamental for the applications of molecular genetics to the human species and its variations, both somatically and in the germ line. Mapping genes to their chromosomes and to well-defined visible regions within them is now a relatively straightforward matter using a combination of somatic-cell genetics, DNA cloning techniques, in situ hybridization, and chromosome sorting, as documented in the recent Human Gene Mapping proceedings (see de la Chapelle 1985). It is interesting that the last Cold Spring Harbor Symposium on human genetics in 1964 just began to foreshadow these fundamental developments in the subject.

Given the ordering and location of genes on the chromosome, conventional family studies, as emphasized by Botstein et al. (1980), can provide a genetic map in terms of recombination fractions. This is especially useful for intermediate distances involving, say, recombination fractions between 5% and 25%. Ultimately, of course, mapping at the molecular level means having a defined, ordered series of overlapping DNA clones, which must be the first goal in sequencing the human genome.

The importance of gene mapping and its practical applications are often not adequately appreciated. The following are some examples illustrating the value of gene-mapping applications.

1. Gene mapping was a key to the identification of the role of the c-*abl* and c-*myc* genes, in association with specific translocations in leukemias and lymphomas. Thus, the mere fact that c-*abl* mapped to the tip of chromosome 9, and then that it was present on the Philadelphia chromosome, was more than enough to indicate that it was probably the gene actually at the junction of the translocation that contributed to the progression of chronic myelogenous leukemia. For Burkitt's lymphoma, the fact that the specific translocations involved chromosome 8 with chromosomes 14, 22, and 2, on which were located the immunoglobulin heavy-chain and λ and κ light-chain genes, respectively, was too much of a coincidence to ignore. The mapping of c-*myc* to chromosome 8 then completed the data needed to imply that it was these genes that were involved in the specific translocations in Burkitt's lymphoma.
2. The striking demonstration by Cavenee and his colleagues (1983; Cavenee and Hansen, this volume) that retinoblastoma, following Knudson's hypothesis, is associated with homo- or hemizygosity for a mutation on chromosome 13 depended entirely on gene mapping. The initial observation was that a small proportion of familial retinoblastoma cases were associated with a small visible deletion on chromosome 13. This was then generalized to families without visible deletions by using the esterase-D polymorphism, which had been assigned to chromosome 13 using somatic-cell genetic techniques. Finally, Cavenee and colleagues (1983) were able to use polymorphic DNA probes that had been assigned to chromosome 13 for their further studies.
3. The identification of linked markers for Huntington's chorea, cystic fibrosis, Duchenne muscular dystrophy, and other simply inherited diseases has depended on family linkage analysis using polymorphic DNA probes that have been mapped using somatic-cell hybrids and other techniques. Once a significant linkage with a marker is established, then

it is comparatively easy to search for closer linkage with other markers already assigned to the same chromosome or often to an identified region of the chromosome. Apart from the goal of identifying the actual genetic defect, linked markers identify high- and low-risk groups, which can be used to study factors predisposing to disease. In the case of Huntington's chorea, for example, this should help to identify earlier pointers to the disease, obviously with the aim eventually of preventing its onset following early detection.

4. Disease heterogeneity can be identified by different subgroups showing linkage to different markers. A classical example involves elliptocytosis, only one form of which was shown to be linked to the Rh blood groups (Morton 1956; see also McKusick, this volume).
5. The molecular defect of an inherited disease may sometimes be identified simply by guessing the function of the gene and knowing where the gene maps. A striking recent example is the identification of rhodopsin-related pigment genes on the part of the X chromosome where color blindness maps and through this showing how the pigment genes mutate to cause color blindness (Nathans et al. 1986).
6. Chromosome segregation in human-mouse somatic-cell hybrids provides a valuable subdivision of the genome for many aspects of its analysis. This has been used, for example, for the characterization of surface markers using monoclonal antibodies, in particular the identification of monoclonal antibodies to the EGF receptor (see, e.g., Bodmer 1982).
7. Mapping a given type of function, e.g., an integral membrane protein or cell-surface growth factor receptor, to a limited part of the genome may make a major contribution to its molecular identification. Suppose, for example, that such a function has been mapped to chromosome 21. This constitutes about 1% of the total genome and so may code for as few as 50–100 basic genetic functions. Probably at most a few percent of these are integral membrane proteins and so there may be only one or two such proteins coded by chromosome 21. Thus, identifying cDNA clones for membrane-bound messages that map to chromosome 21 may, with high probability, produce a clone for a receptor that has been mapped to this chromosome. Monoclonal antibodies to the EGF receptor were, for example, identified simply because they segregated with chromosome 7, which was known to code for the receptor. Chromosomes and their subdivisions provide well-defined segments of the genomic dictionary, just like subdivisions of the alphabet.

Given a marker that is relatively closely linked to a disease locus, such as cystic fibrosis, what is the chance of identifying the relevant functional gene for the disease? A 10% recombination fraction corresponds to about 10^7 bp, which is about 1/300 of the genome and so may code for some 20–30 functional genetic clusters. A marker just 1% recombination fraction away from the disease locus may be separated from it by only two or three functional clusters. Using the technique of pulse gel field electrophoresis (which can analyze fragments of 200–300 kb or more) for constructing overlapping restriction maps, it is now not difficult to cover regions of the genome 10^6 bp or more long. One million base pairs will be covered, on average, by 25 nonoverlapping cosmids, but for adequate mapping with overlapping cosmids, several times that number will be needed. New approaches are being developed for semiautomatic alignment of randomly chosen cosmids, e.g., by matching the sizes of simple restriction digest patterns. Such approaches will undoubtedly make a major contribution to total genome mapping (S. Brenner, pers. comm.; Poustka et al., this volume). There is, however, still an important gap in the size of the DNA piece that can be cloned. The smallest visible chromosome fragment that can be identified, e.g., in a human-mouse somatic-cell hybrid or chromosome transfer clone, is on the order of 10^7 bp long. The largest cosmid is about 40,000 bp, and so what is needed is a procedure for cloning pieces halfway between these two limits, i.e., on the order of 10^6 bp long. No doubt, approaches will be developed for cloning these, for example, as supernumery chromosomes in yeast between cloned and identified centromeric and telomeric sequences. If the calculations on the molecular complexity of the human genome are correct, then it should not be difficult to identify the functional sequences in a 10^6-bp stretch. Screening with message or cDNA libraries from a variety of tissues, such as lymphocytes, fibroblasts, epithelial cells, muscle, brain, liver, and testis (including adult and fetal sources), should give a high probability of finding the limited set of products predicted to occur in such a stretch. Once the relevant set of products has been identified, a combination of guess work as to which is likely to correspond to the function, together with straight molecular analysis, should often provide the answer. It may often be possible to identify a simple defect such as absence of message or an observable DNA rearrangement with probes from sequences for just one of the products. New techniques are also being developed to identify single base-pair changes in defined stretches of DNA (see Myers and Maniatis; Lerman et al.; both this volume).

If we had a complete dictionary and knew where a marker linked to the disease lay, it would then be possible to look up the neighboring gene clusters. In that case, a good guess at which is likely to be most relevant for the disease might well provide the answer directly, as in the case of the oncogenes mapped to chromosome 8 and 9 or the visual pigment genes mapped to the color blindness region of the X chromosome.

Analysis of Complex Multifactorial Traits

The use of linked genetic markers to analyze poorly defined inherited variation is a comparatively old idea (Kloepfer 1946), but as pointed out by Solomon and Bodmer (1979), it is one that can only now be put into practice because of the effectively unlimited availability of RFLPs using DNA clones. These authors suggested that a total of 200–300 suitably selected markers, which is what would be needed to place one at intervals of approximately 10% recombination fraction distances along the entire human genome, would satisfy the needs for an unequivocal analysis of almost any complex trait. This number, as will be discussed later, may actually be somewhat more than the minimum necessary.

Sib-pair analysis can be generalized to looking for a distortion of genetic marker segregation among any set of "affected" individuals in a pedigree whatever its complexity. The expected distribution of the marker could, for example, be determined empirically by Monte Carlo simulation of its segregation among the affected individuals in the pedigree, given identification of the presumed genotypes of the input individuals to the pedigree, i.e., parents, grandparents, etc. Incompletely identified genotypes can, to some extent, be dealt with using population frequency data on the relevant genetic markers. The power of this approach is that it makes no assumptions as to the nature of the inheritance of the trait involved, because only information on the affected individuals is taken into account. The unaffected members of the family are essentially an unresolved mixture of individuals who have the relevant genes but do not express them and those who do not have the genes. The errors involved in constructing models to attempt to regain information from such unaffected individuals are too great, in most cases, to lead to reliable results.

In a comparatively simple situation, such as for Hodgkin's disease, where there appears to be no significant disease heterogeneity, probably only one gene with a major identifiable effect, and no sporadic cases, i.e., individuals who have the disease without having the relevant gene, the sib-pair analysis and its generalizations are extremely powerful. A total of only 15–20 sib pairs can detect relatively close linkage, even when the marker is 10–15% recombination fraction units away. Complications arise with this approach if there is heterogeneity in the trait and if there are sporadic cases, perhaps because of the effects of other genes combined with the environment. Heterogeneity should be revealed by subsets of the trait showing different patterns of linkage. This is, in principle, detectable provided there are enough genetic markers available (this is solved by DNA clones and restriction enzyme polymorphisms) and there is enough relevant family data, which is mainly a matter of concerted effort. The search for linked susceptibility markers not only indicates where in the genome the relevant gene may be, but also initially helps to define the existence of a genetic contribution underlying the trait.

In pursuing this approach to the analysis of the inheritance of complex traits, there is clearly a trade-off between the number of markers studied, assuming that there can be arbitrarily many of these at known positions in the genome, and the number of families and individuals studied. Formerly, the limited availability of genetic markers placed emphasis on obtaining the maximum number of families in order to detect comparatively loose linkage. This is, however, an inefficient approach, since the ability to detect loose linkage requires a large number of individuals as soon as the recombination fraction is more than, say, 20–25%. Detection of linkage increases in efficiency greatly as the recombination fraction decreases to 10% or 15% and less. Given the now unlimited availability of genetic markers, the best strategy is to study the maximum number of markers, with the aim of finding one that is closely linked, and so limiting the number of families studied, at least in the first instance. Obviously, if there is significant expected heterogeneity, then it may become important to study more families and individuals. As pointed out by Botstein et al. (1980), emphasis must clearly be placed on using the most informative markers, i.e., those that give the highest frequency of heterozygosity. The HLA system, for example, is extraordinarily efficient, since it gives essentially 100% heterozygosity, so that all families are fully informative. The aim must therefore be to find evenly spaced, highly informative markers that span the whole genome. The logistics of DNA polymorphism analysis is simplified if the set of enzymes used can be limited to the four to six that are commonly assumed to be best for detecting polymorphism. It should also be possible to use mixtures of markers chosen carefully to identify polymorphic fragments of different molecular weights. In our laboratory, for example, we are now exploring this approach and hope that it will be possible to choose combinations of up to four or five markers that can be analyzed together in one track.

The analysis of homo- and hemizygosity in tumors, as compared with heterozygosity in matching normal tissue following Cavenee et al. (1983), requires only one marker on each chromosome if the tumor genotype is mostly due to chromosome nondisjunction, sometimes followed by reduplication. Excluding the X chromosome, this requires only 22 highly polymorphic markers, which might be analyzed with as few as five or six gel slots. One polymorphic marker per chromosome arm requires 41 markers in all. If we assume that 15% recombination on either side of a marker can be readily detected, then this already covers 50% of the genome. Twice this many (∼80) well-placed markers would therefore mean that no potential susceptibility gene is more than 15% recombination fraction units away from one of the markers. Such an analysis could be achieved with perhaps as few as 20 gel slots per individual, as long as the polymorphisms are highly selected for being informative, for their position on the chromosomes, and for a limited set of easy-to-use and informative restriction enzymes. To achieve such a set of markers probably requires the same sort of collaboration as has been practiced by the HLA workers for the last 20 years (see Albert et al. 1984), since it is unlikely that any single laboratory could hope to accumulate a good set, at least within a reasonably short period of time. This approach was first urged nearly 6 years ago (Bodmer 1981). Since then, the Human Gene Mapping Conferences have made further major contributions to the field, but there is still scope for more active collaboration and systematic exchange of probes. The CEPH family project initiated by J. Dausset and others (see White et al., this volume) makes a valuable contribution to the conventional genetic mapping of DNA markers once their chromosomal localization and subregional placements have been established using somatic-cell hybrids, in situ annealing, and other suitable techniques.

These approaches, i.e., searching systematically for linked markers to susceptibility, are now being applied to manic-depressive states, schizophrenia, heart disease, and familial clustering of relatively common cancers (e.g., of the breast, colon, or rectum). An interesting parallel to the HLA and disease associations identified through their immune etiology is the search for effects of apolipoprotein and other related genetic variation on coronary artery disease, as detected by DNA clones and restriction enzymes (see Deeb et al. this volume). Just as in the case of the HLA and disease associations, association of an apolipoprotein varient with early coronary heart disease would be due to linkage disequilibrium caused by very close linkage between the restriction enzyme variant and the functionally relevant variation in the apolipoprotein gene. Another important potential application of population marker association due to linkage disequilibrium is to the study of susceptibility to cancers due to differential metabolism of carcinogens. Thus, Ayesh et al. (1984) have shown that the genetic, recessively determined, poor metabolizers of debrisoquine to 4-hydroxy-debrisoquine occur with a much lower than expected frequency among cigarette smokers with lung cancer, than among smoking controls. Genes for the relevant cytochrome P-450 monooxygenase subfamily have been tentatively identified (Wolf 1986; Gonzalez et al., this volume) so that one can now look for RFLPs with the relevant clones. These may be associated, on the one hand, with the debrisoquine metabolism and, on the other hand, with susceptibility to lung and possibly other cancers in the face of environmental agents such as cigarette smoke. Another potential application arises from the suggestion that heterozygotes for some of the DNA repair deficiencies may have an increased susceptibility to at least certain forms of cancer (Swift et al. 1980). Once again, RFLPs identified with clones for the relevant enzymes should have a reasonable chance of associating with heterozygotes for the repair deficiencies, and thus enabling case-control studies of the association between heterozygosity and lower susceptibility to be carried out.

The key to the identification of such population associations lies in understanding the nature of linkage disequilibrium. Thus, unless the DNA-detected variation is sufficiently close, i.e., has a recombination fraction of about 0.5% or less with the actual susceptibility determining sequence, it will not show a population association. Population associations are extremely valuable because they can form the basis for screening for susceptibility when appropriate preventive measures can be taken to protect the susceptible individuals. For example, suppose a genetic marker has been found that has a 10% recombination fraction with the desired susceptibility. Then, knowing where this is on the genome, it should be possible to saturate the relevant region with further polymorphic markers and look for one that has a population association with the trait, rather than increasing the numbers of families analyzed to search for closer linkage. This is the most efficient way of finding closely linked markers, since family data are very inefficient at distinguishing between small recombination fractions such as 0.5% versus 5%. The rate of decline of linkage disequilibrium with generation number n, $(1 - r)^n$, magnifies the effect of a small recombination fraction r in population association data. Once again, the availability of an unlimited range of polymorphic DNA markers makes it possible to pursue these approaches on the basis of classical genetic notions.

Common Normal Traits

A number of classical inherited common traits have been studied over the years, including eye and hair color, hair texture, ability to roll the tongue, handedness, attachment, size and flair of the ear, PTC tasting ability, smell, short-sightedness, and pattern baldness. Just 40 years ago in a remarkably perceptive paper, Kloepfer (1946) carried out a comprehensive family analysis of many of these traits, looking for Mendelian segregation and linkage with the then known markers, mainly blood group determinants. In some cases, such as ear attachment, size, and flair, the data on inherited patterns are by no means clear-cut, although the existence of inherited tendencies seems unequivocal. In all of these situations, linkage with a DNA marker would be the first step in a proper genetic analysis. The prospects for clear-cut results would be good because these traits have been selected for showing comparatively well-defined inheritance patterns. In our laboratory, we are now devising a questionnaire to do such studies on families in which the DNA-marker analyses are in any case being done. This is the start of an approach to defining the genetic basis of normal variation.

It is well known that chromosome abnormalities, such as Down's syndrome for chromosome 21 and the fragile X syndrome, are often associated with characteristic facial and other bodily features. It is generally assumed that these associations are due to the presence on the relevant chromosomes, or chromosome region, of genes affecting these features, which are presumably not integrally involved with the mental or other abnormality associated with the chromosome defect. An analysis of the phenotypes associated with such chromosome abnormalities might provide some clues as to where one should look for polymorphic markers associated with normal differences in facial and other features. For example, is it possible that one or more genes that distinguish the characteristic Oriental and Caucasoid face are actually located on chromosome 21?

Analysis of normal human variability in facial features, character, and mental abilities is surely one of the real challenges of human genetics. Now, with the availability of polymorphic DNA markers, there is a well-defined approach. Facial features clearly show a major inherited component, as indicated by the most cursory examination of identical twin pairs. The Oxford English Dictionary defines physiognomy as "the study of the face as an index of character: art of judging character from face and form." This follows from the often intuitively assumed relationship between facial features and character. Is this an illusion or does it have some basis in fact? Facial recognition has clearly been very important in the evolution of the higher primates and has its own special region of the brain. Part of the importance of facial recognition may be mate selection, and this could well have led to a selective advantage for an association between facial features and character. Such a population association must, however, mean that the polymorphic genetic markers determining facial attributes, on the one hand, and character, on the other, remain associated, and this can only happen if these markers are closely linked and therefore in linkage disequilibrium. The DNA technology that is now available should in due course be able to answer the question as to whether indeed the genes that control facial features and certain aspects of behavior are closely linked.

The Whole Human Sequence

Knowledge of the total human genome map, with all its products defined, would, as already emphasized several times, be a revolutionary step forward in our ability to analyze the genetic contribution to all aspects of normal and abnormal variation. This knowledge alone makes it possible to analyze the genetic and functional bases of major diseases, to institute potential preventive measures, and even to consider approaches to gene therapy in selected situations. Identifying all the products, having some indications as to their function, and establishing a primary structure are entirely achievable goals. More difficult, clearly, is the structural interpretation of function, but this also will undoubtedly follow in due course.

Establishing the total human genome sequence is surely a major challenge that must be taken up worldwide. It is probably the most important, essentially technological, project of its kind that can now be identified, in terms of its potential contribution to human welfare. It is not like putting a man on the moon was when that project was first initiated and there presumably was no assurance of its success, nor is it like the present plans for putting a man on Mars, and even less like the strategic defense initiative. It is achievable, enormously worthwhile, has no defense implications, and generates no case for competition between laboratories or nations. Assuming a rate of 50,000 bases sequenced per person per year (which is probably even now conservative and will undoubtedly become more so as automated procedures are developed), 60,000 person years, or 6000 people working for 10 years, could generate the whole sequence. Calculating this as roughly 6 times the size of the Imperial Cancer Research Fund gives rise to an estimate of approximately $225 million per year, or a total cost of $2.25 billion for the project. This is perhaps equivalent to the development of the Concorde or a few Trident missiles, let alone a man on Mars. Physicists, engineers, and space and defense scientists are used to lobbying at this level for large-scale projects. Perhaps now it is the turn of the biomedical scientists, who surely have at least as good if not an even better case to lobby for additional resources for this major project.

There is, moreover, a significant difference between this and the other projects. It is no good getting a man a third or a quarter of the way to Mars. It is not even clear what the use of having a man on Mars is. However, a quarter or a third of the way along the definition of the total human genome sequence could already provide a most valuable yield of applications. The project would provide a stimulus to the development of appropriate technology, e.g., automated handling of sequencing and cloning, computer analysis and the construction and analysis of large relevant data bases, and the practical problem of the rapid and effective detection of DNA polymorphisms. Much good basic research, and development of comparative information in other species, would also undoubtedly be produced.

The first step in such a project would no doubt be to analyze the overall genome organization by obtaining a complete set of overlapping clones. This would provide the framework for the dictionary, along which the regions defining all the products could be identified, long before the complete nucleotide sequence had been obtained. The sequence of functional genetic regions may be only one tenth of the total or less and might therefore be achievable in a relatively few years time with an appropriate concerted effort. Quite apart from its applications to the analysis of normal and abnormal human variation, there would be enormous interest in unraveling the evolutionary relationships and hierarchies between gene products, both within and between species, that would be revealed by their sequences. This analysis should also help reveal the control language by which complex patterns of differential expression of genes are achieved. The sequences should form a basis for prediction of protein structures for other members of a family once one has been analyzed. Prediction of tertiary protein structure from the primary nucleotide sequence could indeed be an associated goal of the overall project. There is also a case for including the difficult problem of predicting tertiary RNA structures that may be very important for control, especially in the nucleus.

Knowledge of the total human genome sequence has

profound implications, not only for the analysis, prevention, and treatment of disease, but also for the better understanding of normal variation, and through that, hopefully making a contribution to solving the wider problems of society. These problems are still the greatest, since they impose an economic and social complexity on top of the complex biological system. But at least it must help to understand the latter.

It is my hope that there will be a worldwide consensus to pursue this most worthwhile and challenging, but achievable, goal: The complete characterization of the human genome. A not unrealistic target would be the end of this century—Project 2000.

ACKNOWLEDGMENT

I am grateful to John Trowsdale for making available the up-to-date map of the HLA-D region.

REFERENCES

Albert, E.D., M.P. Baur, and W.R. Mayr. 1984. *Histocompatibility testing 1984* (ed. E.D. Albert). Springer-Verlag, New York.

Ayesh, R., J.R. Idle, J.C. Ritchie, M.J. Crothers, and M.H. Hetzel. 1984. Metabolic oxidation phenotypes as markers for susceptibility to lung cancer. *Nature* **312:** 169.

Barnstable, C.J., E.A. Jones, W.F. Bodmer, J.G. Bodmer, B. Arce-Gomez, D. Snary, and M. Crumpton. 1977. Genetics and serology of HLA linked human Ia antigens. *Cold Spring Harbor Symp. Quant. Biol.* **41:** 443.

Bodmer, W.F. 1981. Gene clusters, genome organization and complex phenotypes. When the sequence is known, what will it mean? *Am. J. Hum. Genet.* **33:** 664.

———. 1982. Monoclonal antibodies: Their role in human genetics. In *Human genetics, part A: The unfolding genome* (ed. B. Bonné-Tamir et al.), p. 125. A.R. Liss, New York.

———. 1983. Gene clusters and genome evolution. In *Evolution from molecules to men* (ed. D.S. Bendall), p. 197. Cambridge University Press, England.

Bodmer, W.F., J. Trowsdale, J. Young, and J.G. Bodmer. 1986. Gene clusters and the evolution of the major histocompatibility system. *Philos. Trans. R. Soc. Lond. B* **312:** 303.

Bodmer, W.F., E. Albert, J.G. Bodmer, J. Dausset, F. Kissmeyer-Nielsen, W. Mayr, R. Payne, J.J. van Rood, Z. Trnka, and R.L. Walford. 1984. Nomenclature for factors of the HLA system. *Histocompatibility testing 1984* (ed. E.D. Albert), p. 4. Springer-Verlag, New York.

This reference list includes specific references needed for particular points in the text. Background references to many of the points referred to will be found in other papers in this volume. A number of the ideas discussed here are based on Bodmer (1981), which also includes a wider range of background references.

———. 1986. Genetic susceptibility to cancer. In *Accomplishments in cancer research, 1985 prize year: General Motors Cancer Research Foundation,* p. 198. J.B. Lippincott, Philadelphia.

Botstein, D., R.L. White, M. Skolnick, and R.W. Davis. 1980. Construction of a genetic linkage map in man using restriction fragment length polymorphisms. *Am. J. Hum. Genet.* **32:** 314.

Cavalli-Sforza, L.L. and W.F. Bodmer. 1971. *The genetics of human populations.* W.H. Freeman, San Francisco.

Cavenee, W.K., T.P. Dryja, R.A. Phillips, W.F. Benedict, R. Godbout, B.L. Gallie, A.L. Murphree, L.C. Strong, and R.L. White. 1983. Expression of recessive alleles by chromosomal mechanisms in retinoblastoma. *Nature* **305:** 779.

de la Chapelle, A., ed. 1985. *Human gene mapping 8.* Karger, Basel.

Doolittle, W.F. and C. Sapienza. 1980. Selfish genes, the phenotype paradigm and genome evolution. *Nature* **284:** 601.

Heyes, J., P. Austin, J. Bodmer, W.F. Bodmer, A. Madrigal, M. Mazzilli, and J. Trowsdale. 1986. Monoclonal antibodies to HLA-DP transfected mouse L cells. *Proc. Natl. Acad. Sci.* **83:** 3417.

Immunological Review. 1985. Volumes 84 and 85.

Kloepfer, H.W. 1946. An investigation of 171 possible linkage relationships in man. *Ann. Eugenics* **13:** 35.

Littlefield, J.W. 1965. The selection of hybrid mouse fibroblasts. *Cold Spring Harbor Symp. Quant. Biol.* **29:** 161.

McKusick, V.A. 1986. *Mendelian inheritance in man.* Johns Hopkins University Press, Baltimore.

Morton, N.E. 1956. The detection and estimation of linkage between the genes for elliptocytosis and the Rh blood type. *Am. J. Hum. Genet.* **8:** 80.

Nathans, J., T.P. Piantanida, R.L. Eddy, T.B. Shows, and D.S. Hogness. 1986. Molecular genetics of inherited variation in human color vision. *Science* **232:** 203.

Orgel, L.E. and F.H.C. Crick. 1980. Selfish DNA: The ultimate parasite. *Nature* **284:** 604.

Payne, R., M. Tripp, J. Weigle, W.F. Bodmer, and J.G. Bodmer. 1965. A new leukocyte isoantigen system in man. *Cold Spring Harbor Symp. Quant. Biol.* **29:** 285.

Solomon, E. and W.F. Bodmer. 1979. Evolution of sickle variant gene. *Lancet* **I:** 923.

Spielman, R.S., J. Lee, W.F. Bodmer, J.G. Bodmer, and J. Trowsdale. 1984. Six HLA-Dα chain genes on human chromosome 6: Polymorphisms and association of DCα-related sequences with DR types. *Proc. Natl. Acad. Sci.* **81:** 3461.

Swift, M., R.J. Caldwell, and C. Chase. 1980. Reassessment of cancer predisposition of Fanconi anemia heterozygotes. *J. Natl. Cancer Inst.* **65:** 863.

Trowsdale, J., A. Kelly, J. Lee, S. Carson, P. Austin, and P. Travers. 1984. Linkage map of two HLA-SBβ and two HLA-SBα-related genes: An intron in one of the SBβ genes contains a processed pseudogene. *Cell* **38:** 241.

Trowsdale, J., J.A.T. Young, A.P. Kelly, P.J. Austin, S. Carson, H. Meunier, A. So, H.A. Ehrlich, R.S. Spielman, J. Bodmer, and W.F. Bodmer. 1985. Structure, sequence and polymorphism in the HLA-D region. *Immunol. Rev.* **85:** 5.

Wolf, C.R. 1986. Cytochrome P-450: A polymorphic multigene family involved in carcinogen activation. *Trends Genet.* **2:** 209.

The Gene Map of *Homo sapiens*: Status and Prospectus

V.A. McKusick

Division of Medical Genetics, Department of Medicine, Johns Hopkins University School of Medicine, Baltimore, Maryland 21205

HISTORICAL-METHODOLOGICAL INTRODUCTION

The first gene to be mapped to a specific chromosome in man—indeed, in any mammal—was that for color blindness, deduced to be on the X chromosome by E.B. Wilson at Columbia University in 1911. A few dozen other X-linked traits (e.g., hemophilia and Duchenne muscular dystrophy) were identified by characteristic pedigree pattern during the next 57 years before the first assignment of a gene to a specific human autosome: Duffy blood group to chromosome 1 by Donahue and colleagues (1968) at Johns Hopkins University. They made the assignment by finding evidence of linkage between the Duffy locus and a heteromorphism of chromosome 1 that was segregating in a Mendelian manner in Donahue's family. Soon thereafter, haptoglobin was assigned to chromosome 16 by linkage studies in families with inherited, balanced translocations involving 16 (Robson et al. 1969) and in families with a heritable "fragile site" on 16q (Magenis et al. 1970). About the same time, Weiss and Green (1967) showed the feasibility of assigning specific genes to specific chromosomes (or regions of chromosomes) by interspecies somatic cell hybridization (SCH) (e.g., fusion of cells from mouse and man), a method of genetic study called "parasexual" by Pontecorvo and "an alternative to sex" by Haldane.

Between 1951, when the first successful autosomal assignment of a gene was achieved by the linkage method, and 1968, when the fruitful, alternative mapping method of SCH was introduced, nine autosomal linkages in man (seven pairs, one triplet, and one foursome) had been established by family studies, but the specific autosome carrying each of the linkage groups was not known. For the linkage studies in man, special methods of analysis of pedigree data were necessary because it is not possible to design matings as can be done in experimental organisms. The sib-pair method of Penrose, for example, was used by Mohr in 1951 to establish the first autosomal linkage, that between Lutheran blood group and secretor trait. The method of "lods" (log odds) was elaborated by C.A.B. Smith and by Newton E. Morton in the 1940s and 1950s. Computer methods were started about 1960, by Renwick in particular. One of the most widely used programs for linkage analysis, LIPED, was introduced by Ott (1974, 1976, 1985).

Although the potential usefulness of genetic linkage information to clinical medicine (e.g., in the recognition of the carrier and preclinical states of Mendelian disorders) may have been commented on earlier, the earliest, clearest, and most specific statement of use of the linkage principle in prenatal diagnosis is probably that by Edwards in 1956. After it had been pointed out that sexing of the unborn infant is possible by studies of sex chromatin in amniocytes, obtained by amniocentesis, Edwards (1956) suggested that, given close linkage of a testable marker, prenatal diagnosis of genetic disease in the fetus should be possible by study of amniotic material. As the concluding speaker at the Third International Birth Defects Congress at The Hague in 1969, I emphasized mapping the chromosomes of man as a great exploration for the future that was bound to have important rewards in clinical medicine (McKusick 1970). I used the close linkage of G6PD and classic hemophila (Boyer and Graham 1965) as an example of an opportunity for prenatal diagnosis. McCurdy et al. (1971) actually used this approach for hemophila. Others (e.g., Schrott et al. 1973) applied it to the prenatal diagnosis of myotonic dystrophy, using the linkage to secretor.

By SCH, mutually potentiated by family linkage studies (FLS) of the type that assigned *color blindness* and *Duffy*, a veritable explosion of information on the human gene map occurred in the 1970s. By June 1976, at least one gene had been assigned to each of man's 24 chromosomes—the 22 autosomes, the X, and the Y. The ability to identify uniquely each chromosome by its characteristic banding pattern, by methods developed by Caspersson and his colleagues (1970a,b, 1971) and by others (Patil et al. 1971; Schnedl 1971; Seabright 1971; Sumner et al. 1971), has been of fundamental importance to gene mapping. Not only could otherwise very similar chromosomes, such as those of the C group (nos. 6–12 + X), be individually identified by their stripes, even when intermingled with rodent chromosomes in the hybrid cell, but also the composition of chromosomal rearrangements could often be determined and genes could be mapped to specific bands. Banding methods applied to metaphase chromosomes

The Appendix cited throughout this paper and found at the back of this volume is a late version of the Human Gene Map newsletter, which has been prepared periodically since 1973. Figures referred to in the text are found there, and the Tables with roman numerals are there as well.

could demonstrate about 400 bands in the total karyotype; high-resolution cytogenetics as developed by Yunis (1976) and others for application to extended chromosomes in prophase or prometaphase could demonstrate more than twice that number.

In the latter half of the 1970s, the methods of molecular genetics were brought to bear on chromosome mapping—both the localization of genes to specific chromosomes or chromosome bands and molecular mapping down to the nucleotide level. Fundamental to progress in molecular genetics was the discovery, in 1970, of site-specific restriction endonucleases, restriction enzymes produced by bacteria to break down foreign DNA. These enzymes became the scalpel for dissecting the human genome. Cloning of DNA segments (genes) in *Escherichia coli* came in 1972 (Watson and Tooze 1981). Southern (1975), then in Edinburgh, devised an elegant hybridization blot method for displaying restriction fragments of DNA, now commonly called the Southern blot, capitalizing on the property of DNA to bind to nitrocellulose paper.

By the technology of recombinant DNA (Watson et al. 1983), libraries of cloned DNA fragments from the entire human genome were prepared, a well-known one being that of Maniatis (Maniatis et al. 1978); these were genomic clones. With reverse transcriptase, complementary, or copy, DNA (cDNA) clones of human genes were prepared from the corresponding messenger RNA (mRNA), first for the human (hemo)globin genes. Restriction maps were prepared of the (hemo)globin gene regions. Recombinant DNA techniques made it almost as easy to identify and isolate specific human genes as it was to electrophorese abnormal hemoglobins or determine the peptide map of hemoglobins ("fingerprinting"). In 1977, two methods of DNA sequencing (Maxam and Gilbert 1977; Sanger et al. 1977) greatly facilitated mapping to the level of individual nucleotides. In 1981, the complete nucleotide sequence of the human mitochondrial chromosome—all 16,569 bp—was published by Sanger's group (Anderson et al. 1981).

The methods of molecular biology added powerfully to the classic methods of FLS and SCH. Probes, either genomic or cDNA, provided by recombinant DNA technology, were used in combination with SCH for mapping by hybridization in solution ("Cot analysis") (e.g., α- and β-globin loci were assigned to 16p and 11p, respectively; Deisseroth et al. 1977, 1978) or by Southern blot analysis of DNA from somatic cell hybrids. A great advantage was that the gene under study did not need to be expressed in the cultured cell. Essentially, any gene for which a probe was available could be mapped.

Molecular genetics also provided new markers for FLS; as a result, such studies have enjoyed a renaissance (White et al. 1985). The new markers were not polymorphisms of the gene product but rather of the genetic material itself, i.e., variations in nucleotide sequence. Kan and Dozy (1978), using *Hpa*I (for *Hemophilus parainfluenzae*) restriction enzyme, discovered the first human DNA polymorphism, located on the 3′ ("downstream") side of the β-globin gene. They, as well as Kurnit (1979) and Solomon and Bodmer (1979), pointed out the potentially great usefulness of DNA polymorphisms as markers for linkage studies. Botstein et al. (1980) suggested that restriction-fragment-length polymorphisms[1] (RFLPs; sometimes pronounced "riflips") as revealed by Southern blots could be used for complete mapping of the human genome. The first polymorphism that was demonstrated in an arbitrary or anonymous (function-unknown) segment of DNA was that of Wyman and White (1980), which was subsequently shown (Balazs et al. 1982; DeMartinville et al. 1982) to be situated at the end of the long arm of chromosome 14 between the α-1-antitrypsin locus and the immunoglobin heavy-chain loci. Demonstration of multiple restriction polymorphisms in a segment of the genome to create a haplotype useful in FLS was the work of Kazazian, Orkin, and their colleagues at Johns Hopkins and Harvard, among others, who used the method in the DNA diagnosis of the thalassemias. The clinical usefulness of linkage using RFLPs as markers was dramatically demonstrated by the mapping of Huntington's disease by Gusella et al. (1983), of adult polycystic kidney disease by Reeders et al. (1985), and of cystic fibrosis by several groups in late 1985. In all three of these disorders, possibilities for diagnosis and understanding were opened up, as will be elaborated upon later in this review.

In recent times (mainly since 1980), direct methods of human chromosome mapping have been added to our armamentarium: (1) in situ hybridization of radiolabeled (or immunofluorescence-labeled) DNA sequences ("probes"), generated by recombinant DNA techniques, directly to spreads of chromosomes to identify the site of that piece of DNA ("gene") on the chromosome (Gerhard et al. 1981; Harper et al. 1981; Szabo and Ward 1982); and (2) fluorescence-activated sorting of chromosomes (e.g, Young et al. 1981) followed by application of molecular techniques to determine the gene content of the isolated chromosomes.

Meanwhile, fine mapping of segments of DNA up to lengths of 50–60 kb or more and down to the level of individual nucleotides has been advancing through the application of recombinant DNA technology, restriction endonucleases, and DNA sequencing. The first and perhaps the best example in man is the mapping of the 50-kb segment of the short arm of chromosome 11 that contains five genes for β-globin and the β-like globins of hemoglobins. Such fine-mapping studies revealed an "unexpected complexity of eukaryotic genes" (Watson et al. 1983).

Although the phenotype associated with most chromosomal aberrations is relatively uninformative as to the precise gene content of the chromosomes involved (Lewandowski and Yunis 1977), with the improved res-

[1]A.C. Wilson (pers. comm.) objects to the designation restriction-fragment-length polymorphism, introduced by Botstein (1980), because of confusion with one general class of mutants, i.e., length mutants as opposed to substitutions and rearrangements.

olution provided by the banding methods small deletions and other aberrations were found to be associated with specific phenotypes, especially tumors (e.g., retinoblastoma and Wilm's tumor), but also malformation syndromes such as the Prader-Willi syndrome, the Langer-Giedion syndrome, the Miller-Dieker lissencephaly syndrome, and the Beckwith-Wiedemann syndrome. The improved methods for studying the chromosomes involved in rearrangements and molecular genetic methods for demonstrating and mapping oncogenes combined to greatly further the understanding of cancer, including particularly hematologic malignancies (Mitelmann 1983; Yunis 1983).

The four commingling methodologic streams in the chromosome-mapping field—linkage, chromosomal, somatic cell hybridization, and molecular—are mutually potentiating. A combination of methods is often used, as illustrated by many examples given below. The data are cumulative; for example, data on the linkage of two loci collected from successive families (lod scores are additive), linkage data on the several loci in a stretch of chromosome, data on physical mapping provided by somatic cell hybridization, and data on the genetic map accumulated by linkage studies in families.

The explosion of information on the human gene map in the last 15 years is reflected in the numbers given in Table II (Appendix). Eight international workshops on human gene mapping[2] (HGM-1 through HGM-8) (Table I, Appendix) have been critical to the collation and validation of the information on the gene map derived from many different laboratories working with methods as diverse as lod scores in family linkage analysis on the one hand and DNA hybridization characteristics on the other.

THE STATUS OF THE HUMAN GENE MAP

Figure 1 (Appendix) presents a pictorial synopsis of the present status of the human gene map. Three levels of confidence (confirmed, provisional, or "in limbo") with which the genes have been assigned are indicated by different letter styles of the gene symbols. Gene clusters are indicated by large letters. The Key for Figure 1 (Appendix) gives not only the definition of the gene symbols (and synonymous symbols) but also information on the regional assignment and the method of assignment (see Appendix for the definition of the symbols used to designate methods).

As reflected by Figure 1 (Appendix), the chromosome that carries each of about 900 structural genes is known and many of these genes have been fairly precisely regionalized. This number represents about 47% of the well-established loci cataloged in *Mendelian Inheritance in Man* (McKusick 1986b) and 23% of all loci cataloged there. (There is an element of circularity in these figures; increasingly in the last decade, entries have been created in *Mendelian Inheritance in Man* for loci identified by somatic cell genetic or molecular genetic methods, especially if they have been mapped, even though no allelic variation had been identified.) The numbers are impressive when viewed in relation to the rather short period of time that mapping of the autosomes has been going on. In 1964, when the Cold Spring Harbor Symposium was last devoted to human genetics, not a single gene had been assigned to a specific autosome.

The number of loci that have been mapped is less than 2% of the 50,000 genes that *Homo sapiens* is thought to have as a minimum. (The known density of genes in the small segment of 11p that contains the β-globin cluster—5 genes in 50,000 bp—yields an estimate of about 300,000 genes in all, given that there are about 3 billion nucleotides[3] in the haploid human genome. The globin genes are somewhat atypically small, however [Table 1]. It may not be appropriate, further-

[2]The first workshop was organized by Frank Ruddle and held in New Haven in June 1973. The second, known as the Rotterdam Conference, was organized by Dirk Bootsma and held in The Netherlands in July 1974. The third, organized by Victor McKusick, was held in Baltimore in October 1975. The fourth, organized by John Hamerton, was held in Winnipeg in August 1977. The fifth, organized by Kare Berg, was held in Oslo in June-July 1981. The seventh, organized by Robert Sparkes, was held in Los Angeles in August 1983. The eighth, organized by Albert de la Chapelle, was held in Helsinki in August 1985. The first six were sponsored exclusively by the National Foundation-March of Dimes (now March of Dimes Birth Defects Foundation), which publishes the proceedings as part of its *Birth Defects Original Article Series*; the proceedings also appear in *Cytogenetics and Cell Genetics* (Table I, Appendix).

The published proceedings of the workships (Table I, Appendix) are revealing not only from the point of view of advancing numerology but also from the standpoint of evolving methodology. At the beginning somatic cell hybridization was the main source of information, supplemented importantly by the family linkage method. By HGM-5 in 1979, molecular genetic methods were contributing substantially, mainly in connection with somatic cell hybridization. By HGM-6 in 1981, in situ hybridization was beginning to appear on the scene (e.g, Harper et al. 1981).

The third edition of *Mendelian Inheritance in Man* (McKusick 1971) had a listing of the then-known linkages; a single page sufficed for the listing of all known autosomal assignments (three) and all known autosomal and X-linked linkage groups. Accelerating progress in mapping over the last decade is pictorially displayed in successive editions of *Mendelian Inheritance in Man* (McKusick 1986b), starting with the fourth in 1975, in the review by McKusick and Ruddle in *Science* (1977), in the review published in 1980 in *Journal of Heredity* (McKusick 1980) and in four successive versions of the *Human Gene Map Newsletter*, published every 12 to 14 months in *Clinical Genetics*, beginning in December 1982 (McKusick 1982a).

[3]Kornberg (1980) gave a value of 2.9 billion bp for the human haploid genome from estimates of the amount of DNA per cell and an average molecular weight per base pair of 660, to give the conclusion that 1 picogram (10^{-12} mg) of duplex DNA contains 9.1×10^8 bp. The DNA in single cells was measured by UV microspectrophotometry of Feulgen-stained cells (Mirsky and Ris 1951; Leuchtenberger et al. 1954), a method developed 50 years ago by Caspersson (1936). Mirsky and Ris (1951) wrote: "In a series of careful measurements by Davison and Osgood (pers. comm.) on human granulocytes and lymphocytes (from leukemic blood), the DNA per cell of the former was found to be 6.25×10^{-9} mg and that of the latter 5.84×10^{-9} mg. Our own determination, on human sperm gave 2.72×10^{-9} mg per cell, approximately one-half the value for the somatic cells." The values given by Leuchtenberger et al. (1954) were in the same range. Watson (1976) gave a somewhat larger estimate of the total number of nucleotides than did Kornberg (1980), namely 3.3 billion. Because of the difference in size of the XX and XY sex chromosome pairs, a difference of about 2% would be expected in the DNA of male and female diploid cells.

Table 1. The Size of Genes

Gene product	Genomic size (kb)	cDNA (mRNA) (kb)	No. of introns
Small			
α-globin	0.8	0.5	3
β-globin	1.5	0.6	2
Insulin	1.7	0.4	2
Apolipoprotein E	3.6	1.2	3
Parathyroid	4.2	1.0	2
Protein C	11	1.4	7
Medium			
Collagen I			
pro-α-1(I)	18	5	50
pro-α-2(I)	38	5	50
Albumin	25	2.1	14
HMG CoA reductase	25	4.2	19
Adenosine deaminase	32	1.5	11
Factor IX	34	2.8	7
LDL receptor	45	5.5	17
Large			
Phenylalanine hydroxylase	90	2.4	12
Factor VIII	186	9	25

more, to base estimates such as this on the density of genes in gene clusters.)

The variety of genes that have been mapped is as impressive as the numbers and indicates the central role of gene mapping in contemporary biomedical research. Mapped have been genes for enzymes of carbohydrate, lipid, steroid, amino acid, and nucleic acid metabolism; for hemoglobins; for serum proteins such as albumin, haptoglobin, ferritin, C-reactive protein, plasminogen, and orosomucoid; for enzymes of lysosomes, cytosol, mitochondria, and peroxisomes; for cell-surface proteins that function as receptors for hormones, growth factors, complement, viruses, and toxins, or remain with incompletely understood function being demonstrated mainly by immunologic distinctiveness (e.g., some blood groups); for histone and nonhistone chromosomal proteins; for DNA repair enzymes (e.g., DNA polymerase, α and β); for the cytosolic-nuclear receptors for hormones (e.g., the androgen receptor [mutant in the testicular feminization syndrome] and the corticosteroid receptor); for enzymes involved in the synthesis of transfer RNAs (tRNAs); for hormones such as insulin, growth hormone, ACTH, somatomammotropin (placental lactogen), and prolactin; for HLA, complement components, interferons, immunoglobins, and T-cell-antigen receptor, involved in host-defense mechanisms; for carrier proteins (e.g., transferrin), for T-cell "markers" such as T4 and T8; for cytochrome P450 enzymes; for coagulation factors and their inhibitors and activators; for growth factors, such as TCGF and EGF, and their respective receptors; for structural elements of the cell, such as spectrin, actin, myosin, desmin, and tubulin; and for structural proteins of the intracellular matrix, such as the collagens. The genetic determinants for ribosomal and U1 small nuclear RNA, and for one form of tRNA, have been mapped. More than 40 oncogenes (i.e., human DNA sequences homologous to the oncogenic nucleic acid sequence of mammalian retroviruses such as those of murine, feline, and simian sarcomas) have been assigned to specific chromosomes or chromosome regions. The homeo box genes (e.g., on chromosome 17) are examples of genetic determinants of development. In addition, pathologic phenotypes of which the biochemical basis is not yet known have been mapped (e.g., nail-patella syndrome, forms of congenital cataract and spinocerebellar ataxia, myotonic dystrophy, Huntington's disease, cystic fibrosis, and the Prader-Willi syndrome).

Gene clusters have become evident as a striking feature of the organization of the human genome. Clusters are indicated in Figure 1 (Appendix) by large letters and include the following: the three immunoglobulin clusters (on 2p, 14q, and 22q), the two (hemo) globin clusters (on 11p and 16p), the leukocyte interferon cluster (on 9p), the major histocompatibility complex (on 6p), histone complexes (on 1 and 7), growth hormone–placental lactogen complex (on 17q), the metallothionein cluster (on 16q), the myosin heavy-chain cluster (on 17p), and the β-glycopeptide hormone cluster (on 19). There is a cluster of apolipoprotein genes on 11 and a second on 19. The genes for clotting factors VII and X may be clustered (on 13q). The albumin, α-fetoprotein, and *GC* genes constitute a cluster (on 4q). The genes for arginine vasopressin and oxytocin are clustered on 20q. There is a cluster of carbonic anhydrase genes on chromosome 8. (These groupings are called *clusters* rather than *families* because the latter term is reserved for kindreds of genes that have a common ancestral origin and may or may not be syntenic; the α- and β-globin genes, for example, constitute a gene family.) Gene clustering can lead to combined deficiencies when deletion involves two or more gene loci in the cluster. A form of thalassemia in which neither δ nor β-globin chains are produced is a well-documented example; combined deficiencies of apolipoproteins A1 and C3 (MIM 23455) may be an example.

Presented in Figure 2 (Appendix) is the gene map of the mitochondrial chromosome, which is circular, like a bacterial chromosome. It carries 37 genes in all: 13 for proteins, 22 for tRNAs, and 2 for mitochondrial ribosomal RNAs. The mitochondrion has its own protein-synthesizing machinery. Obviously, most of the structural and enzymatic components of the mitochondrion are coded by genes in the nucleus. Each mitochondrion has 2 to 10 chromosomes. Whereas each nuclear chromosome is present normally in only two copies per cell, the mitochondrial chromosome is present in thousands of copies. Mutations in the mitochondrial chromosome can be expected to lead to disorders with patterns of transmission and other characteristics to some extent different from those of Mendelian disorders. The genetics is expected to be more Galtonian than Mendelian.

The Anatomy of the Human Genome

To this point I have used the customary cartographic metaphor. In reviewing the significance of the information, it may be useful to use an anatomic metaphor (McKusick 1980, 1982b). The linear arrangement of genes on our chromosomes is part of our anatomy. Knowledge of the chromosomal and genic anatomy of *Homo sapiens* has given clinical genetics (and medicine as a whole) a neo-Vesalian basis. A veritable revolution has taken place, dramatically in the practice of clinical genetics, but also generally in medicine—witness what has happened in oncology. Just as *De Humani Corporis Fabrica* of Vesalius (1543) was the basis for the physiology of Harvey (1628) and the pathology of Morgagni (1761), the information on chromosomal and genic anatomy is the foundation of our understanding and management of genetic disease in man.

The anatomic metaphor is useful in examining the significance of the mapping information because it leads naturally to a consideration of the morbid anatomy, the comparative anatomy and evolution, the functional anatomy, the developmental anatomy, and the applied anatomy of the human genome. Most of the contributions to this Symposium address one or several of these aspects. I review only selected aspects of each.

The Morbid Anatomy of the Human Genome

Figure 3 (Appendix) is a pictorial representation of the chromosomal location of mutations "causing" disorders (McKusick 1986a). Beside each chromosome are names of disorders "caused" by mutations located thereon. Many inborn errors of metabolism, such as galactosemia and phenylketonuria, have been mapped by demonstration of the location of the gene for the enzyme deficient in each. In most of these there is evidence at the protein level (and in an increasing number at the gene level, as well) that the mutation involves the structural gene for the enzyme. In other disorders, a nonenzymatic protein is known to be altered in the particular disorder, and it was the mapping of the normal gene that gave information on the location of the disease-producing mutation. Sickle cell anemia is an historic example; a recent one is familial amyloid polyneuropathy, which in several of its forms is "caused" by a mutation of the transthyretin (prealbumin) gene located on chromosome 18.

Other disorders for which the basic defect is not known have been mapped by linkage of the disease phenotype to a genetic marker that is in turn mapped, either a polymorphism of the gene product such as Rh blood group or a polymorphism of DNA (RFLPs). Traditionally, mapping has been practical almost only for autosomal dominants (and X-linked recessives, which in the male behave like dominants). However, with the increased power of the polymorphic DNA markers, recessives can also be mapped, witness cystic fibrosis, even though the heterozygote cannot be identified.

The mapping information developed in recent years has modified our classification of genetic diseases and indicates the importance of the large group of disorders that represent *somatic cell genetic diseases*. All malignant neoplasms fall into this category and some congenital malformations (e.g., aniridia) are demonstrably in that category. The locations of determinants of selected malignancies have been included in Figure 3 (Appendix) for illustrative purposes. Basically, some or most autoimmune diseases may fall into the category of somatic cell genetic diseases.

Mendelian syndromes are usually the consequence of pleiotropism of a single mutant gene. The notion that a genetic syndrome is due to the close linkage of two or more genes, each for a separate component of the syndrome, can usually be rejected as naive. Recent observations of deletions that can be seen in high-resolution karyotypes or deduced from family studies of polymorphic markers indicate that syndromes can indeed result from change in linked genes: the WAGR syndrome (11p) and the Langer-Giedion syndrome (8q) are cases in point. Although the evidence is not ironclad, in each it seems that separate components of the syndrome may occur alone. Combined deficiency of C6 and C8 (which are known to be closely linked, although the chromosomal location is not known) and of apolipoproteins A-I and C-III (close together on 11q) are cases in point. See also chronic granulomatous disease (CGD) with and without Xk deficiency and X-linked adrenal hypoplasia with and without GK deficiency.

In Figure 3 (Appendix), allelic disorders, which may be so different in phenotype as to suggest mutation in different genes, are indicated by enclosure in a box. The diversity of the phenotypes caused by mutations in the β-globin gene on 11p is a classic. Conversely, the same phenotype can, of course, be caused by mutation in different genes ("genetic heterogeneity"). Type VII Ehlers-Danlos syndrome can be caused by mutation in either the α-1 chain (Cole et al. 1986) of type I procollagen (coded by 17q) or its α-2 chain (Steinmann et al. 1980) (coded by 7q)—and there may be yet another form of Ehlers-Danlos syndrome type VII determined by mutation, not in a procollagen gene, but in the gene (not yet mapped) for the procollagen peptidase that cleaves the amino-peptide from the procollagen molecule (Lichtenstein et al. 1973). The last has been demonstrated in certain domestic animals though not yet in humans.

Some of the entities indicated in Figure 3 (Appendix) are "nondiseases"; they turn up as abnormal test values in laboratory studies and are important to know about to avoid confusion with diseases. These "nondiseases" include inborn variants of metabolism such as cystathioninuria (on chromosome 16) and pentosuria, an inborn nondisease of metabolism that has not yet been mapped. They also include abnormalities of binding by albumin, giving high levels of thyroxine or zinc without clinical evidence of intoxication.

Three entries in Figure 3 (Appendix) are infectious diseases for which the role of a single locus in susceptibility or resistance has been identified. The Duffy null gene (on chromosome 1) gives resistance to vivax malaria. As far as known, all humans are susceptible to diphtheria and poliomyelitis (Miller et al. 1974) by reason of the products of genes located on chromosomes 5 and 19, respectively. Vitamin C deficiency is a universal inborn error of metabolism in *Homo sapiens*; where the mutation is in the human genome will be known when the gene for *L*-gulonolactone oxidase is mapped (by now this gene might be only a pseudogene relic, if present at all).

Comparative Anatomy and Evolution of the Human Genome

Footprints indicating the role of gene duplication in its evolution are seen throughout the human genome—in the gene clusters and families and even in the internal structure of genes.

There is some correspondence between exons and domains of proteins. It was suggested by Gilbert (1982) that exon shuffling is a mechanism of evolution whereby exons from various sources are combined to fashion a protein of optimal characteristics for a given function. A possible rationale for introns (intervening sequences) is the opportunity they afford for recombination without disruption of the coding segments (exons).

Because of the considerable similarity in banding pattern of the chromosomes of apes and man (Yunis et al. 1980), it is not surprising that many genes that are known to be syntenic in man have been found to be syntenic in other higher primates (Lalley and McKusick 1985) when appropriate somatic cell hybridization or in situ hybridization studies are done. Furthermore, in the other primates, homologous loci have usually been found to be carried by the chromosome judged by banding pattern to be homologous. The degree of homology of synteny between mouse and man came, however, as a considerable surprise. Ohno's law (Ohno 1973), which predicts identity of the genic content of the X chromosome in all mammals, is a special case. Except perhaps for a few loci at the tip of the short arm of the X chromosome that escape lyonization, X-linkage can be expected to be conserved in all mammals. The ill effects of loss of dosage compensation would be expected to prevent movement of most genes from the X chromosome to an autosome. Because of Ohno's law, X-linked diseases in mice and other mammals (e.g., hereditary hypophosphatemia and testicular feminization) are convincing models of human X-linked diseases.

Comparative mapping has been aided greatly by molecular genetic methods. Whereas criteria for homology of the gene product such as immunologic cross-reactivity and similarities of substrate specificity were previously used, homology can be tested directly by using the same DNA probes in hybridization studies in various species.

Most would not have predicted the degree of autosomal homology of synteny that has been found between such distant relatives as man and mouse (Buckle et al. 1984; Nadeau and Taylor 1984). In the tabulation made at HGM-8 (Lalley and McKusick 1985), all the human autosomes except chromosome 13 are shown to have at least 2 loci that are also syntenic in the mouse. That gap will be filled, perhaps, when the chromosomal locations of genes *F7, F10, COL4A1, COL4A2,* and others on human 13 are known in the mouse. Human chromosome 17 has 8 loci that are all on mouse chromosome 11; the short arm of human chromosome 6 has at least 10 loci (counting all *HLA* loci as one) that are on mouse chromosome 17. Some human chromosomes bear homology in genic content to two or three mouse chromosomes. Thus, chromosome 1 of man has a distal 1p region with 6 loci homologous to loci on mouse 4, a proximal 1p region with 6 loci homologous to loci on mouse 3, and a 1q region with 4 loci homologous to loci on mouse 1. It is useful to consult such a table (Lalley and McKusick 1985) when a given locus has been mapped in mouse or other subhuman species for a guess as to where the human locus may be situated. Buckle et al. (1984) published an ingenious grid that indicates at a glance the synteny homologies between man and mouse.

An example of predicting human chromosomal assignment from findings in the mouse is the following: Because the genes coding for aminoacylase (ACY1) and for β-galactosidase-1 (GLB1) are on chromosome 3 in man and chromosome 9 in the mouse and since the structural gene for transferrin (TF) is closely linked to *ACY1* and *GLB1* in the mouse, Naylor et al. (1980) suggested that the human transferrin gene might be on chromosome 3. This was subsequently shown to be the case (Huerre et al. 1984; Yang et al. 1984). The homology did not extend to close linkage, however; in the human *TF* is on 3q, whereas *GLB1* and *ACY1* are on 3p.

An ancient tetraploidization, partial or complete, has been suggested (Comings 1972) by morphologic similarities, for example, of human chromosomes 11 and 12 and of chromosomes 21 and 22. The genic content of 11 and 12 gives some support to this idea; *LDHA* and *LDHB*, and the Harvey and Kirsten *ras* proto-oncogenes, are on 11p and 12p, respectively. *IGF1* is on 12q and *IGF2* is on 11p. Mouse chromosome 16 carries several loci that are on human chromosome 21, which is trisomic in the Down syndrome; but mouse chromosome 16 also carries the supergene for the λ light chain of immunoglobulin, which in man is on chromosome 22 (Lalley and McKusick 1985). (Other loci on human chromosome 22 are found on chromosome 15 in the mouse [Lalley and McKusick 1985].)

The coding portions (exons) of the two γ-globin genes (in the HBBC on 11p) differ in only a single codon, number 135, resulting in either alanine or glycine as the 135th amino acid. This reflects not-unexpected gene divergence after duplication, and indeed more difference might be anticipated. The finding of the same restriction polymorphism in noncoding intervening segments (introns) of these two linked genes (Jeffreys 1979) and the same nucleotide sequence (Slightom et al. 1980) may be explained by gene conversion or correction. A similar process has probably operated to preserve identity or close similarity of the two α-globin genes as well as of members of other gene clusters. Proximity of the genes involved is a necessary condition for gene conversion.

Functional Anatomy of the Human Genome

A considerable amount of information can be summarized in the following seven generalizations.

1. Although clustering of genes of similar function and common evolutional origin is a frequent finding, there is *no chromosomal aggregation of genes coding for structure and function of particular organs, such as the eye, heart, or kidney, or particular subcellular organelles, such as lysosomes or mitochondria.*

2. *The structural genes for enzymes catalyzing successive steps in a particular metabolic pathway are usually not syntenic.* The genes for at least three enzymes of galactose metabolism (GALE, GALK, and GALT), for five enzymes of the urea cycle (ARG1, ASL, ASS, CPS1, and OTC), and for eight enzymes of the tricarboxylic acid cycle (ACO1 and 2, IDH1 and 2, FH, MDH1 and 2, and CS) are known to be on separate chromosomes. On the other hand, the genes for at least four enzymes involved in glycolysis are all on chromosome 12 (TPI, CAPD, ENO1, and LDHB and probably LDHC). Glucose dehydrogenase (GDH) and 6-phosphogluconate dehydrogenase (PGD), enzymes that catalyze successive steps in the phosphogluconate pathway, are coded by linked genes on the short arm of chromosome 1, but the genes for two other enzymes of this pathway (G6PD and GAPD) are on other chromosomes. There may be a functional relationship between HPRT and PRPS, enzymes coded by genes rather closely situated on Xq. Of the nine enzymes of the purine ribonucleotide biosynthetic pathway, two are coded by genes on each of two different chromosomes: PGFT and PFGS by chromosome 14 and PAIS and PRGS by chromosome 21. OPRT and ODC, enzymes involved in successive steps of the pyrimidine synthesis pathway, are both coded by chromosome 3 and both are mutant in most cases of hereditary orotic aciduria (only ODC is mutant in a single known case). All these are not exceptions, however, because in the case of each set the enzyme activities are properties of a single multifunctional protein (D. Patterson, pers. comm.). The single bifunctional enzyme deficient in orotic aciduria is called uridylmonophosphate synthase (Patterson et al. 1983). A trifunctional enzyme that has been conserved in *Drosophila*, birds, and mammals is coded by chromosome 21. Mutants of each of the three enzymatic functions individually are known in human cells and a double mutant of the PAIS and PRGS functions are known in Chinese hamster ovary cells. It is of interest that, in the case of both the PGFT/PFGS and the PAIS/PRGS multifunctional enzymes, the reactions catalyzed are not contiguous in the metabolic chain.

There are several other examples of enzymatic functions at steps in the same metabolic processes being subserved by a single multifunctional molecule. The advantage of this arrangement is that equimolar synthesis of the two enzymatic entities is guaranteed. Linkage of the structural genes of thymidine kinase and galactokinase (on human chromosome 17) has been conserved over long evolutionary time. Coordinate function may be responsible for this (Schoen et al. 1984). It may have functional significance that *PFKP* and *HK1*, the genes for enzymes at the primary and secondary control points in the glycolytic pathway, are both on 10p.

3. *The genes determining the different subunits of a heteromeric protein are usually not syntenic.* The α and β chains of adult hemoglobin (Hb A), determined by genes on chromosome 16p and 11p, respectively, are cases in point. Other examples are shown in Table 2. The class II HLA proteins (e.g., HLA-DR), of which the α and β chains are determined by separate loci, both of which are in the major histocompatibility complex (MHC) on 6p, represent an exception to this rule of nonsynteny. Another exception is fibrinogen, of which the α, β, and γ chains are coded by chromosome 4; indeed, the genes are in the same order as the polypeptides in the fibrinogen molecules—γ-α-β (Aschbacher et al. 1985; Kant et al. 1985). Insulin and haptoglobin do not represent exceptions; in both cases the two chains are coded by a single gene with posttranslational cleavage of the proprotein into two. These are examples of proteins that are coded by a single gene but in the mature or active form consist of two subunits held together by disulfide bonds; activated PLAT and the α-γ complex of C8 are other such instances.

Table 2. Nonsynteny of Genes Coding for Subunits of Heteromeric Proteins

Coagulation factor XIII	A,B	6p, not 6p
Creatine kinase	B,M	14q,19q
Collagen, type I	α1, α2	17q,7q
Ferritin	H,L	11,19q
Glycopeptide hormones		
chorionic gonadotropin	α,β	6q,19q
follicle-stimulating hormone	α,β	6q,11p
luteinizing hormone	α,β	6q,19q
thyroid-stimulating hormone	α,β	6q,1p
Hemoglobin	α,β	16p,11p
Hexosaminidase	α,β	15q,5q
HLA-A,-B,-C	H,L	6p,15q
Immunoglobulins	H,L	14q;2q,22q
Lactate dehydrogenase	A,B	11p,12p
Phosphofructokinase, red cell	L,M	21q,1q
Platelet-derived growth factor	A,B	7,22
Protein kinase C′	α,β,γ	17,16,19
T-cell antigen receptor	$\alpha,\beta,\sigma,\gamma,\epsilon$	14q,7q,7p,11q,?

For information on mapping and other genetic aspects, see McKusick (1986b).

It must be asked whether the nonsynteny of heteromers is more than what one would expect given a random distribution as the general rule. It would appear that there is a true repulsion of the genes for the several heteromers of a protein, especially when one considers that many or even most probably originated by duplication of a common ancestral gene.

4. Whereas most heteromeric proteins are compounded of polypeptides coded by genes on different chromosomes, *some genes code for more than one polypeptide.* As just mentioned, the α and β chains of haptoglobin are coded by a single gene (on 16q) and the A and B chains of insulin by a single gene (located near the distal end of 11p). A striking example of multiple peptides from a single gene is proopiomelanocortin (on 2p); ACTH, β-endorphin, and β-melanotropin are three of the some seven peptides derived from the same gene. Tissue-specific alternative splicing of the primary RNA transcript is a method by which a single gene can code for proteins specifically suited to the differentiated function of different cell types. The calcitonin gene (on 11p) in the parafollicular cells of the thyroid codes for calcitonin but in the hypothalamus codes for calcitonin gene-related peptide (CGRP), which is read off the same primary transcript. The mechanism of this differential function is unknown.

5. *The genes determining the cytoplasmic and mitochondrial forms of a given enzyme are not syntenic.* The cytosolic (cytoplasmic or soluble) and mitochondrial forms, referred to as −1 and −2, respectively, of the following enzymes are determined by different chromosomes: ACO (9 and 22), ALDH (9 and 12), GOT (10 and 16), IDH (2 and 15), MDH (2 and 7), SOD (21 and 6), and TK (17 and 16). I know of no true exception to this rule of nonsynteny. Adenylate kinase exists in a cytosolic form (AK1) determined by a gene on 9p and in two mitochondrial forms (AK2 and AK3) determined by genes on 1p and 9q, respectively. Since the genes for AK1 and AK3 are on different arms of a long chromosome, this is probably not an exception to the rule. Similarly, ME1 (on 6q) and ME2 (on distal 6p) probably do not represent an exception. That both cytosolic and mitochondrial fumarate hydratase (FH) are determined by chromosome 1 is also not an exception: 1q carries a single structural gene for FH; posttranslational modification accounts for the electrophoretic differences in the two isozymes (Edwards and Hopkinson 1979). These observations of nonsyntenic genetic determination of cytosolic and mitochondrial isozymes are consistent with a symbiont origin of the mitochondria, with a shift of most of the mitochondrial genes from the mitochondrial chromosome to nuclear chromosomes in a random manner, and with no more homology between the mitochondrial gene and the nuclear gene for each pair of isozymes than might be expected on the basis of a very ancient origin of both from a common ancestral gene.

6. *An appreciable portion of the genome consists of functionless (unexpressed) pseudogenes*, which show similarities in nucleotide sequence to functional genes. Pseudogenes have lost critical elements necessary for transcription. Because of sequence homology to functional genes, however, they are recognized by the same DNA probes. Pseudogenes may be closely situated to the structural genes of which they are imperfect replicas (e.g., the pseudogenes in the α- and β-globin gene clusters) or may be far removed and present in many copies, as in the case of the pseudogenes of argininosuccinate synthetase (Su et al. 1984). The lack of introns in pseudogenes suggests that they are "processed genes," that is, they originated by integration of reverse transcripts of mRNA. The differentiation of functional genes from pseudogenes is aided by somatic cell hybridization; for example, the functional gene for argininosuccinate synthetase is the one demonstrated on chromosome 9 by ISH because the enzymatic function maps to chromosome 9 by SCH. (HGM-8 tentatively recognized another category of gene called "like." These are identified by in situ and other molecular hybridization methods under conditions of low stringency. The functional status of these or their relation to pseudogenes is unknown; hence the noncommittal designation.)

7. *The structural gene for a receptor and that for its ligand are usually not on the same chromosome.* Both transferrin and the transferrin receptor are coded by 3q; however, the genes are rather far apart in the 3q21 and 3q26.2 bands, respectively, and the transferrin receptor bears no sequence homology to transferrin (McClelland et al. 1984). The LDL receptor is coded by chromosome 19, as is also one of its ligands, apolipoprotein E, but the genes are on 19p and 19q, respectively. *CSF1* and *CSF1R* may be in the same band on 5q. The exceptions are, however, more numerous than the nonexceptions (see Table 3); for example, epidermal growth factor (EGF) is coded by chromosome 4, whereas the gene for its receptor is on chromosome 7.

Functional significance can, perhaps, be attached to the clustering of the various components of the MHC on 6p. These genes include not only the *HLA* loci of classes I and II, but also the determinants of certain components of the complement and alternative path-

Table 3. Sometimes the Genes for a Receptor and Its Ligand(s) Are Syntenic but Usually Not

Examples of synteny			
CSF1R(FMS)	5q	*CSF1*	5q[a]
LDLR	19p	*APOE*	19q
TFR	3q26.2	*TF*	3q21
Examples of nonsynteny			
EGFR	7p	*EGF*	4q
IFNAR	21q	*IFNA*	9p
IFNBR	21q	*IFNB*	9p
IFNGR	18	*IFNG*	12q
IGF1R	15q	*IGF1*	12q
INSR	19p	*INS*	11p
NGFR	17q	*NGFB*	1p
PDGFR	5q	*PDGFA,B*	7,22

[a]Only known example of mapping to the same band.

ways. The convertase involved in activation of C3 (gene assigned to chromosome 19) is a bimolecular complex of C4 and C2, both of which are coded by genes closely linked to *HLA-B* on chromosome 6. Properdin factor B (BF), which serves a similar role (of activating C3) in the alternative pathway, is also closely linked to *HLA-B*. (The genes for C6 and C7, linked in the dog and the marmoset, are also closely linked in man, as indicated by restriction enzyme mapping studies and by observation of combined deficiency. The genes are not on 6p, however.) No functional significance is evident for the location within the MHC of genes for 21-hydroxylase deficiency (*CAH1*) and hemochromatosis (*HFE*).

The close situation on 11p of the genes for parathyroid hormone and calcitonin (the yin and yang of calcium homeostasis) is probably happenstance and of no evolutionary or functional significance. The dissimilarity in sequence of the genes (and the peptides they determine) rules against their origin from a common ancestral gene. Possibly in favor of a functional significance of their close situation is the fact that both are on mouse chromosome 7 (P.A. Lalley, pers. comm.), which carries other genes, that are on human 11p, such as the genes for insulin, β-globin, LDH-A, and the β subunit of follicle-stimulating hormone, as well as the Harvey-*ras* oncogene.

The Developmental Anatomy of the Human Genome

The linear orientation of the cluster of β-globin genes (HBBC) on 11p appears to have ontogenetic significance. During development, the ϵ gene (at the 5′ end of the 50-kb segment) is active during the embryonic period. Later, switch occurs to the two γ genes, which are next downstream from the ϵ gene and are active during the fetal period, and then to the δ and β genes, which are active during postnatal life. The gene for α-fetoprotein, the fetal equivalent of serum albumin and a protein of diagnostic usefulness to the oncologist and medical geneticist, is closely linked to albumin on 4q. Curiously, in the mouse where the two loci are also closely linked, the postnatally active albumin gene is upstream from (i.e., on the 5′ side of) the gene for the fetal counterpart, a situation opposite to that for the non-α-globin genes of mouse and man. It turns out, however, that the gene-switching paradigm of globin ontogeny is not precisely applicable to the *AFP-ALB* system; the albumin gene is active throughout development, whereas the *AFP* gene, active in embryonic and fetal stages, is largely switched off in the postnatal period.

The genetics of differentiation, and specifically the significance of the anatomy of the human genome to morphologic development and differentiated function, are largely unknown. Among the many aspects of human biology that have been illuminated by the study of hemoglobins, this is one: the nondeletion (or heterocellular) type of hereditary persistence of fetal hemoglobin appears to result from mutation in a regulator for switch from γ- to β-globin synthesis. Tight (Old et al. 1982) and loose (Gianni et al. 1983) linkage of the mutant regulator(s) to the non-α-globin cluster has been found.

The ontogeny of the immunoglobulin-producing lymphocyte appears to be related to the anatomic orientation of the several components of the three immunoglobulin gene clusters, those for the heavy chain (on chromosome 14) and for the κ (on chromosome 2) and λ (on chromosome 22) light chains. Generation of diversity in antibodies through somatic gene rearrangements is dependent on close linkage of the *V, D, J,* and *C* genes that make up the immunoglobulin gene clusters. The developmental significance of the anatomy of the immunoglobulin genes is seen also in the case of the different genes for the *C*, or constant, part of the immunoglobulin heavy chain. Splicing of various *V, D,* and *J* genes provides diversity; the constant region gene that is closest to *D*, or diversity-generating, part of the complex is the gene activated first. Thus, production of IgM occurs early in the immune response and the switch to one or another of the constant region genes for production of IgD, IgG, IgE, and IgA ("class switch") takes place later. (Rather than representing a cluster of genes, each of the immunoglobulin-determining segments of DNA can be viewed as a single supergene in which the diversity-generating portions and those coding for the constant regions are exons.)

The rearrangements of the T-cell antigen receptor genes are another example of developmental significance of genomic anatomy. Like the immunoglobulins, the T-cell antigen receptor consists of two polypeptide chains. The α and β chains of the T-cell receptor are coded by chromosomes 14 and 7, respectively. The genes, symbolized by *TCRA* and *TCRB*, are a cluster of genes (supergene) with *V, D, J,* and *C* genes coding for constant and variable domains of the T-cell receptor molecules. The maturation of the T-cell in the thymus involves clonal rearrangement within the gene cluster to bring the constant-coding gene into contiguity with one of the variable region genes. Another *TCR* gene called γ (*TCRG*) is also situated on chromosome 7 but is on 7p, whereas *TCRB* is on 7q.

The function of the T-cell antigen receptor is to recognize antigens in combination with the individual's own MHC proteins. In the thymus, precursor lymphocytes destined to become T cells undergo a period of "thymic education." Immature T cells that respond to one or a small group of MHC proteins are allowed to propagate and continue their differentiation. The α gene (on chromosome 14) is little expressed in the immature T cell, whereas the γ and β genes (on 7p and 7q, respectively) produce a large amount of protein. Tonegawa's group (Tonegawa 1985) suggested that a switch occurs from γ-β to α-β with maturation of T cells. Obviously, since *TCRG* and *TCRA* are on separate chromosomes, the switch from γ to α is independent of anatomic proximity, unlike the β switch in hemoglobin synthesis.

The Applied Anatomy of the Human Genome

The reason that great interest has accompanied the mapping of Huntington's disease, cystic fibrosis, adult polycystic kidney disease, myotonic dystrophy, Duchenne muscular dystrophy, and other disorders is at least twofold. All of these disorders are the result of presently unknown, basic defects. For that reason, no thoroughly satisfactory diagnostic test or therapy can be designed on the basis of fundamental defect. Mapping information opens the possibility for diagnosis on the basis of linkage principle. Furthermore, it holds out hopes of determining the basic gene defect by "reverse genetics" ("chromosome walking") and using that information to devise tests for prenatal, preclinical, and carrier diagnosis. These tests may take the form of direct testing of DNA for a gene defect by a process one might call "biopsy of the human genome." Information on the basic defect may help plan methods for ameliorating the disorder even though gene therapy is not possible in the near future.

In addition to "chromosome walking," long-range mapping, and other methods for pinpointing the defect in DNA, determining the basic defect can also follow the "candidate gene" strategy. Given a protein that is a plausible candidate for the site of the basic defect, one can ask: Do the disease and the molecule map to the same area? Does the disease show linkage with a RFLP related to the cloned gene? In persons with the given disorder, is there structural abnormality of the gene for the given protein?

The Rh-linked form of elliptocytosis is probably due to mutation in the gene for protein 4.1 of the red cell membrane because they map to the same area of 1p. On the other hand, Wilson's disease (on 13q) cannot be due to mutation in the structural gene for ceruloplasmin (on 3q); nor can hemochromatosis (on 6p) be due to mutation in the structural gene for transferrin (on 3q), transferrin receptor (on 3q), ferritin light chain (on 19q), or ferritin heavy chain (on 11). (Perhaps those walking 6p in the region of the class I MHC genes will stumble on the hemochromatosis gene, which appears to be near *HLA-A* on its centromeric side.)

Even though no gross abnormality such as deletion or rearrangement can be demonstrated in the *COL1A2* gene on chromosome 7 in these cases, linkage between a *COL1A2* RFLP and osteogenesis imperfecta type IV strongly suggests that the causative mutation is in that gene (Falk et al. 1985; Grobler-Rabie et al. 1985). Phillips et al. (1981) could show that mutation in the growth hormone gene was responsible for pituitary dwarfism in some cases in which Southern blot analysis showed it to be deleted.

PROSPECTUS

Complete mapping of the human genome and complete sequencing are one and the same thing. They must go hand in hand. Nucleotide sequencing will be done within, out from, and between genes that have been localized as precisely as possible in relation to recognized landmarks, the chromosome bands, and in relation to neighboring genes. A RFLP map has properties like both a map of expressed genes and a complete nucleotide sequence. A reasonably detailed RFLP map will have great usefulness in both the mapping of expressed genes and the complete sequencing.

The potential usefulness of complete mapping/sequencing has been emphasized by some interested in birth defects (McKusick 1970) and by others interested in cancer (Dulbecco 1986). The usefulness is, in only a relatively restricted manner, indicated in the earlier section on the Applied Anatomy of the Human Genome. Great value is seen in the understanding of multifactorial disorders, a category into which most cancers fall. Dissecting out the role of individual genetic factors in disorders such as essential hypertension, atherosclerosis, mental illness, and common forms of congenital malformations promises to be valuable in the identification of unusual vulnerability and in planning preventive strategies. Characteristics that are patently genetic but presently not analyzable, such as special talents (e.g., musical and mathematical) and morphologic traits (e.g., facial characteristics, eye color, and attached/unattached earlobes), might be studied successfully, given a detailed RFLP map, for example.

The task of complete mapping/sequencing will require new techniques and improvements in existing ones. A large need, which is addressed in some of the papers in this Symposium volume, is for methods to bridge the gap between the resolution that is achieved by restriction mapping and nucleotide sequencing of overlapping cosmid clones (up to 50 or 100 kb) and that achieved with chromosome banding and linkage analysis (down to 1000 kb at best). Pulsed-field gel electrophoresis (Schwartz and Cantor 1984; Smith and Cantor 1986) is one method that can help bridge the gap. The rallying cry is for completion of total mapping/sequencing by the year 2000 or before. The magnitude of the task is indicated by the fact that the human haploid genome contains about 3.0 billion nucleotides.

Complete mapping/sequencing of the mitochondrial chromosome has been achieved. This is the goal to which mapping of the nuclear genome aspires. Anderson et al. (1981) filled three closely printed pages of *Nature* with the sequence of the mitochondrial chromosome. To print the sequence of the haploid nuclear genome of a single person in a similar manner (the nuclear genome is about 200,000 times larger) would require the equivalent of about 13 sets of the *Encyclopedia Britannica.* To print also the heterozygous variation in that individual and add the enormous range of variation between individuals will require the utmost in computer facilities. Obviously there is a large library task here and a large problem in recovering and reading the information as well as problems in the creation of indexes and concordances and devising methods for recognizing pattern similarities.

ACKNOWLEDGMENT

I am particularly indebted to Harley W. Yoder, B.A., for assistance in the recent upkeep of the Human Gene Map presented in the Appendix.

REFERENCES

Anderson, S., A.T. Bankier, B.G. Barrell, M.H.L. deBrujin, A.R. Coulson, J. Droin, I.C. Eperon, D.P. Nierlich, B.A. Roe, F. Sanger, P.H. Schrier, A.J.H. Smith, R. Staden, and I.G. Young. 1981. Sequence and organization of the human mitochondrial genome. *Nature* **290:** 457.

Aschblacher, A., K. Buetow, D. Chung, S. Walsh, and J. Murray. 1985. Linkage disequilibrium of RFLP's associated with α, β, and γ fibrinogen predict gene order on chromosome 4. *Am. J. Hum. Genet.* **37:** A186 (Abstr.).

Balazs, I., M. Purrello, P. Rubinstein, A. Alhadeff, and M. Siniscalco. 1982. Highly polymorphic DNA site D14S1 maps to the region of Burkitt lymphoma translocation and is closely linked to the heavy chain γ-1 immunoglobulin locus. *Proc. Natl. Acad. Sci.* **79:** 7395.

Botstein, D., R.L. White, M. Skolnick, and R.W. Davis. 1980. Construction of a genetic linkage map in man using restriction fragment length polymorphisms. *Am. J. Hum. Genet.* **32:** 314.

Boyer, S.H., and J.B. Graham. 1965. Linkage between the X chromosome loci for glucose-6-phosphate dehydrogenase electrophoretic variation and hemophila A. *Am. J. Hum. Genet.* **17:** 320.

Buckle, V.J., J.H. Edwards, E.P. Evans, J.A. Jonasson, M.F. Lyon, J. Peters, A.G. Searle, and N.S. Wedd. 1984. Chromosome maps of man and mouse. II. *Clin. Genet.* **26:** 1.

Caspersson, T. 1936. Ueber den chemischen Aufbau des Strukturen des Zellkernes. *Skand. Arch. Physiol.* (suppl. 8) **73:** 1.

Caspersson, T., C. Lamakka, and L. Zech. 1971. Fluorescent banding. *Hereditas* **67:** 89.

Caspersson, T., L. Zech, and C. Johansson. 1970a. Differential banding of alkylating fluorochromes in human chromosomes. *Exp. Cell Res.* **60:** 315.

Caspersson, T., L. Zech, C. Johansson, and E.J. Modest. 1970b. Quinocrine mustard fluoroscent banding. *Chromosoma* **30:** 215.

Cole, W.G., D. Chan, G.W. Chamber, I.D. Walker, and J.F. Bateman. 1986. Deletion of 24 amino acids from the pro-α-1(I) chain of type I procollagen in a patient with the Ehlers-Danlos syndrome type VII. *J. Biol. Chem.* **261:** 5496.

Comings, D.E. 1972. Evidence for ancient tetraploidy and conservation of linkage groups in mammalian chromosomes. *Nature* **238:** 455.

Deisseroth, A., A. Neinhuis, J. Lawrence, R. Giles, P. Turner, and F. Ruddle. 1978. Chromosomal localization of human β-globin gene on human chromosome 11 in somatic cell hybrids. *Proc. Natl. Acad. Sci.* **75:** 1456.

Deisseroth, A., A. Nienhuis, P. Turner, R. Velez, W.F. Anderson, F. Ruddle, J. Lawrence, R. Creagen, and R. Kucherlapati. 1977. Localization of the human α-globin structural gene to chromosome 16 in somatic cell hybrids by molecular hybridization assay. *Cell* **12:** 205.

DeMartinville, B., A.R. Wyman, R. White, and U. Francke. 1982. Assignment of the first random restriction fragment length polymorphism (RFLP) locus (D14S1) to a region of human chromosome 14. *Am. J. Hum. Genet.* **34:** 216.

Donahue, R.P., W.B. Bias, J.H. Renwick, and V.A. McKusick. 1968. Probable assignment of the Duffy blood group locus to chromosome 1 in man. *Proc. Natl. Acad. Sci.* **61:** 949.

Dulbecco, R. 1986. A turning point in cancer research: Sequencing the human genome. *Science* **231:** 1055.

Edwards, J.H. 1956. Antenatal detection of hereditary disorders (letter). *Lancet* **I:** 579.

Edwards, J.H. and D.A. Hopkinson. 1979. The genetic determination of fumarase isozymes in human tissues. *Ann. Hum. Genet.* **42:** 303.

Falk, C.T., R.C. Schwartz, F. Ramirez, and P. Tsipouras. 1985. Use of molecular haplotypes specific for the human pro-α-2(I) collagen gene in linkage analysis of the mild autosomal dominant forms of osteogenesis impertecta. *Am. J. Hum. Genet.* **38:** 269.

Gerhard, D.S., E.S. Kawasaki, F.C. Bancroft, and P. Szabo. 1981. Localization of a unique gene by direct hybridization in situ. *Proc. Natl. Acad. Sci.* **78:** 3755.

Gianni, A.M., M. Bregni, M.D. Cappellini, G. Giorelli, R. Taramelli, B. Giglioni, P. Comi, and S. Ottolenghi. 1983. A gene controlling fetal hemoglobin expression in adults is not linked to the non-α-globin cluster. *EMBO J.* **2:** 921.

Gilbert, W. 1982. DNA sequencing and gene structure (Nobel lecture). *Science* **214:** 1305.

Grobler-Rabie, A.F., G. Wallis, D.K. Brebner, P. Beighton, A.J. Bester, and C.G. Mathew. 1985. Detection of a high frequency *Rsa*I polymorphism in the human pro-α-2(I) collagen gene which is linked to an autosomal dominant form of osteogenesis imperfecta. *EMBO J.* **4:** 1745.

Gusella, J.F., N.S. Wexler, P.M. Conneally, S.L. Naylor, M.A. Anderson, R.E. Tanzi, P.C. Watkind, K. Ottina, M.R. Wallace, A.Y. Sakaguchi, A.M. Young, I. Shoulson, E. Bonilla, and J.B. Martin. 1983. A polymorphic DNA marker genetically linked to Huntington's disease. *Nature* **306:** 234.

Harper, M.E., A. Ullrich, and G.F. Saunders. 1981. Localization of the human insulin gene to the distal end of the short arm of chromosome 11. *Proc. Natl. Acad. Sci.* **78:** 4458.

Huerre, C., G. Uzan, K.H. Grzeschik, D. Weil, M. Levin, M.-C. Hors-Cayla, J. Boue, A. Kahn, and C. Junien. 1984. The structural gene for transferrin (TF) maps to 3q21-3qter. *Ann. Genet.* **27:** 5.

Jeffreys, A.J. 1979. DNA sequence variants in the $^{G}\gamma$-, $^{A}\gamma$-, δ- and β-globin genes of man. *Cell* **18:** 1.

Kan, Y.W. and A.M. Dozy. 1978. Polymorphisms of DNA sequence adjacent to human β-globin structural gene: Relationship to sickle mutation. *Proc. Natl. Acad. Sci.* **75:** 5631.

Kant, J.A., A.J. Fornace, Jr., D. Saxe, M.I. Simon, O.W. McBride, and G.R. Crabtree. 1985. Evolution and organization of the fibrinogen locus on chromosome 4: Gene duplication accompanied by transposition and inversion. *Proc. Natl. Acad. Sci.* **82:** 2344.

Kornberg, A. 1980. *DNA replication*, p. 19. W.H. Freeman, San Francisco.

Kurnit, D.E. 1979. Evolution of sickle variant gene. (Letter). *Lancet* **I:** 104.

Lalley, P.A. and V.A. McKusick. 1985. Report of the committee on comparative mapping (HGM8). *Cytogenet. Cell Genet.* **40:** 498.

Leuchtenberger, C., R. Leuchtenberger, and A.M. Davis. 1954. A microspectrophotometric study of the desoxyribose nucleic acid (DNA) content of cells of normal and malignant human tissues. *Am. J. Pathol.* **30:** 65.

Lewandowski, R.C. and J.J. Yunis. 1977. Phenotypic mapping in man. In *New chromosomal syndromes* (ed. J.J. Yunis), p. 364. Academic Press, New York.

Lichtenstein, J.R., G.R. Martin, L.D. Kohn, P.H. Byers, and V.A. McKusick. 1973. Defect in conversion of procollagen to collagen in a form of Ehlers-Danlos syndrome. *Science* **182:** 298.

Magenis, R.E., F. Hecht, and E.W. Lovrien. 1970. Heritable fragile sites on chromosome 16: Probable localization of haptoglobin locus in man. *Science* **170:** 85.

Maniatis, T., R.C. Hardison, E. Lacy, J. Lauer, C. O'Connell, D. Quon, G.K. Sim, and A. Efstratiadis. 1978. The isolation of structural genes from libraries of eucaryotic DNA. *Cell* **15:** 687.

Maxam, A.M. and W. Gilbert. 1977. A new method for sequencing DNA. *Proc. Natl. Acad. Sci.* **74:** 1258.

McClelland, A., L.C. Kuhn, and F.H. Ruddle. 1984. The human transferrin receptors gene: Genomic organization, and the complete primary structure of the receptor deduced from a DNA sequence. *Cell* **39:** 267.

McCurdy, P.R. 1971. Use of genetic linkage for the detection of female carriers of hemophilia. *N. Engl. J. Med.* **285:** 218.

McKusick, V.A. 1970. Prospects for progress. *Excerpta Med. Int. Congr. Ser.* **3:** 407.

———. 1971. *Mendelian inheritance in man: Catalogs of autosomal dominant, autosomal recessive, and X-linked phenotypes*. 3rd edition. Johns Hopkins University Press, Baltimore.

———. 1980. The anatomy of the human genome. *J. Hered.* **71:** 370.

———. 1982a. The human gene map. *Clin. Genet.* **22:** 359.

———. 1982b. The human genome through the eyes of a clinical geneticist. *Cytogenet. Cell Genet.* **32:** 7.

———. 1986a. The morbid anatomy of the human genome: A review of gene mapping in clinical medicine (first of four parts). *Medicine* **65:** 1.

———. 1986b. *Mendelian inheritance in man: Catalogs of autosomal dominant, autosomal recessive, and X-linked phenotypes*, 7th edition. Johns Hopkins University Press, Baltimore.

McKusick, V.A. and F.H. Ruddle. 1977. The status of the gene map of the human chromosomes. *Science* **396:** 390.

Miller, D.A., O.J. Miller, V.G. Dev, L. Medrano, and H. Green. 1974. Human chromosome 19 carries a poliovirus receptor gene. *Cell* **1:** 167.

Mirsky, A.E. and H. Ris. 1951. The desoxyribonucleic acid content of animal cells and its evolutionary significance. *J. Gen. Physiol.* **34:** 251.

Mitelmann, F. 1983. Catalogue of chromosome aberrations in cancer. *Cytogenet. Cell Genet.* **36:** 1.

Nadeau, J.H. and B.A. Taylor. 1984. Lengths of chromosomal segments conserved since divergence of man and mouse. *Proc. Natl. Acad. Sci.* **81:** 814.

Naylor, S.L., P.A. Lalley, R.W. Elliott, J.A. Brown, and T.B. Shows. 1980. Evidence for homologous regions of human chromosome 3 and mouse chromosome 9 predicts location of human genes. *Am. J. Hum. Genet.* **32:** 158A (Abstr.)

Ohno, S. 1973. Ancient linkage groups and frozen accidents. *Nature* **244:** 259.

Old, J.M., H. Ayyub, W.G. Wood, J.B. Clegg, and D.J. Weatherall. 1982. Linkage analysis of nondeletion hereditary persistance of fetal hemoglobin. *Science* **215:** 981.

Ott, J. 1974. Estimation of the recombination fraction in human pedigrees: Efficient computation of the likelihood for human linkage studies. *Am. J. Hum. Genet.* **26:** 588.

———. 1976. A computer program for linkage analysis of general human pedigrees. *Am. J. Hum. Genet.* **28:** 528.

———. 1985. *Analysis of human genetic linkage*. Johns Hopkins Press, Baltimore.

Patil, S.R., S. Merrick, and H.A. Lubs. 1971. Identification of each human chromosome with a modified Giemsa stain. *Science* **173:** 821.

Patterson, D., C. Jones, H. Morse, P. Rumsby, Y. Miller, and R. Davis. 1983. Structural gene coding for multifunctional protein carrying oratate phosphoribosyltransferase and OMP decarboxylase activity is located on long arm of human chromosome 3. *Somatic Cell Genet.* **9:** 359.

Phillips, J.A., III, B.L. Hjelle, P.H. Seeburg, and M. Zachmann. 1981. Molecular basis for familial isolated growth hormone deficiency. *Proc. Natl. Acad. Sci.* **78:** 6372.

Reeders, S.T., M.H. Breuning, K.E. Davies, R.D. Nicholls, A.P. Jarman, D.R. Higgs, P.C. Pearson, and D.J. Weatherall. 1985. A highly polymorphic DNA marker linked to adult polycystic kidney diease on chromosome 16. *Nature* **317:** 542.

Robson, E.B., P.E. Polani, S.J. Dart, P.A. Jacobs, and J.H. Renwick. 1969. Probable assignment of the α locus of haptoglobin to chromosome 16 in man. *Nature* **223:** 1163.

Sanger, F., S. Nicklen, and A.R. Coulson. 1977. DNA sequencing with chain-terminating inhibitors. *Proc. Natl. Acad. Sci.* **74:** 5463.

Schnedl, W. 1971. Analysis of the human karyotype using a reassociation technique. *Chromosoma* **34:** 448.

Schoen, R.C., H.C. Summers, and R.P. Wagner. 1984. Thymidine-kinase activity of cultured cells from individuals with inherited galactokinase deficiency. *Am. J. Hum. Genet.* **36:** 815.

Schwartz, D.C. and C.R. Cantor. 1984. Separation of yeast chromosome-sized DNAs by pulsed field gradient gel electrophoresis. *Cell* **37:** 67.

Schrott, H.G., L. Karp, and G.S. Omenn. 1973. Prenatal prediction in myotonic dystrophy: Guidelines for genetic counseling. *Clin. Genet.* **4:** 38.

Seabright, M. 1971. A rapid banding technique for human chromosomes. *Lancet* **II:** 971.

Slightom, J.L., A.E. Blechl, and O. Smithies. 1980. Human fetal $^{G}\gamma$- and $^{A}\gamma$-globin genes: Complete nucleotide sequences suggest that DNA can be exchanged between these duplicated genes. *Cell* **21:** 627.

Smith, C.L. and C.R. Cantor. 1986. Pulsed-field gel electrophoresis of large DNA molecules. *Nature* **319:** 701.

Solomon, E. and W.F. Bodmer. 1979. Evolution of sickle varient gene. (Letter). *Lancet* **I:** 923.

Southern, E.M. 1975. Detection of specific sequences among DNA fragments separated by gel electrophoresis. *J. Mol. Biol.* **98:** 503.

Steinmann, B., L. Tuderman, L. Peltonen, G.R. Martin, V.A. McKusick, and D.J. Prockop. 1980. Evidence for a structural mutation of procollagen type I in a patient with the Ehlers-Danlos syndrome type VII. *J. Biol. Chem.* **255:** 8887.

Su, T.-S., R.L. Nussbaum, S. Airhart, D.H. Ledbetter, T. Mohandas, W.E. O'Brien, and A.L. Beaudet. 1984. Human chromosomal assignment for 14 argininosuccinate synthetase pseudogenes: Cloned DNAs as reagents for cytogenetic analysis. *Am. J. Hum. Genet.* **36:** 954.

Sumner, A.T., H.J. Evans, and R.A. Buckland. 1971. New technique for distinguishing between human chromosomes. *Nature* **232:** 31.

Szabo, P. and D.C. Ward. 1982. What's new with hybridization *in situ*? *Trends Biochem. Sci.* **7:** 425.

Tonegawa, S. 1985. The molecules of the immune system. *Sci. Am.* **253:** 122.

Watson, J.D. 1976. *Molecular biology of the gene*, 3rd edition, p. 428. W.A. Benjamin, Menlo Park, California.

Watson, J.D. and J. Tooze. 1981. *The DNA story: A documentary history of gene cloning.* W.H. Freeman, San Francisco.

Watson, J.D., J. Tooze, and D.T. Kurtz. 1983. *Recombinant DNA: A short course.* Scientific American, New York.

Weiss, M. and H. Green. 1967. Human-mouse hybrid cell lines containing partial complements of human chromosomes and functioning human genes. *Proc. Natl. Acad. Sci.* **58:** 1104.

White, R., M. Lippert, D.T. Bishop, D. Barker, J. Berkowitz, C. Brown, P. Callahan, T. Holmes, and L. Jerominski. 1985. Construction of linkage maps with DNA markers for human chromosomes. *Nature* **313:** 101.

Wilson, E.B. 1911. The sex chromosomes. *Arch. Mikrosk. Anat. Entwicklungsmech.* **77:** 249.

Wyman, A.R. and R.L. White. 1980. A highly polymorphic locus in human DNA. *Proc. Natl. Acad. Sci.* **77:** 6754.

Yang, F., J.B. Lum, J.R. McGill, C.M. Moore, S.L. Naylor, P.H. van Bragt, W.D. Baldwin, and B.H. Bowman. 1984. Human transferrin: cDNA charcterization and chromosomal localization. *Proc. Natl. Acad. Sci.* **81:** 2752.

Young, B.D., M.A. Ferguson-Smith, R. Sillar, and E. Boyd. 1981. High-resolution analysis of human peripheral lymphocyte chromosomes by flow cytometry. *Proc. Natl. Acad. Sci.* **78:** 7727.

Yunis, J.J. 1976. High resolution of human chromosomes. *Science* **191:** 1268.

———. 1983. The chromosomal basis for human neoplasia. *Science* **221:** 227.

Yunis, J.J., J.R. Sawyer, and K. Dunham. 1980. The striking resemblance of high-resolution G-banded chromosomes of man and chimpanzee. *Science* **208:** 1145.

Construction of Human Genetic Linkage Maps: I. Progress and Perspectives

R. WHITE, M. LEPPERT, P. O'CONNELL, Y. NAKAMURA,
C. JULIER, S. WOODWARD, A. SILVA, R. WOLFF,
M. LATHROP, AND J.-M. LALOUEL
Howard Hughes Medical Institute and the Department of Human Genetics, University of Utah School of Medicine, Salt Lake City, Utah 84132

Human genetics offers opportunities to the molecular biologist through the thousands of subtle genetic variants that have been detailed by thoughtful clinicians. These variants, called mutants in other systems, can provide important insights into the workings of complex biochemical and developmental systems. Some of these insights are reflected in the dramatic progress now taking place in the analysis of developmental mutants in *Drosophila*; in the human, a good example is the light shed on cell receptor and uptake systems by Brown and Goldstein in their analysis of mutations affecting the LDL receptor pathway (Goldstein et al. 1985). The use of mutational variants to provide insight into the workings of complex biochemical and developmental pathways is a classic theme in genetics.

It is true that one of the geneticist's main tools for hypothesis testing, the construction of specific strains of known genotype, is not available in the human system. We believe, however, that a combination of an extensive system of mapped genetic markers with extended pedigrees that include large families can asymptotically overcome this limitation and provide access to the human genetic system. Within such pedigrees we may find individuals whose inferred genotype is appropriate to test our hypotheses. Such an individual becomes our test strain.

Our principal objective is to construct a set of tools suitable for mapping human genes by genetic linkage. Cloned genes, as well as genes known only by the diseases they cause when defective, are targets for mapping by this method. As the tools are refined, we expect them to have applications as well in the exploration of genetic etiology, where they will help us to determine whether certain apparently familial syndromes are indeed genetic; such information will ultimately add significantly to our understanding of human variation. In addition to increasing our ability to take advantage of human mutations for the analysis of complex systems, chromosomal linkage maps will provide a basic description of the structure of the human genome. Furthermore, a more detailed description of chromosome behavior at meiosis than we have at present should emerge.

Linkage mapping has a unique role in the localization of genes that are known only by their variants that cause disease. Such genes can be mapped by following the inheritance of the disease phenotype in families and correlating the inheritance pattern of the disease with the inheritance patterns of known markers. Importantly, no knowledge of the biochemistry of the disease or even tissue or time of expression is required for the mapping. The ultimate goal of such localization is to exploit positional information as the basis for cloning strategies. With the gene in hand, we can reasonably expect an understanding of the molecular etiology of the disorder to follow, along with the possibility of clinical intervention in the disease.

An important medical by-product of linkage mapping for many disorders is and will continue to be the increased ability of physicians to determine whether an individual carries a specific variant of a gene. For some serious diseases, it is important that the prospective parents know the diagnosis at early stages in pregnancy. The application of this by-product technology—the prenatal identification of disease-linked markers—could significantly reduce the incidence of a number of genetic disorders, as well as permit parents at risk to confidently conceive and bear children who are unaffected by the genetic disease.

Approach and Strategy for Linkage Mapping

Our approach and overall strategy was defined early on: Identify genetic markers based on DNA sequence polymorphisms (DSPs) and define their meiotic linkage relationships through family studies (Botstein et al. 1980).

Progress and findings. Family structure has proven to be important to the efficiency of map construction, in terms of information returned for the amount of effort expended. In contrast to what was originally thought, the large, extended pedigrees so useful in linkage studies with genetic diseases are almost twofold less efficient for the construction of general linkage maps than complete, three-generation families with large sibships such as the one illustrated in Figure 1. We have anecdotally ascertained and sampled more than 50 such families in Utah. DNA and lymphoblastoid cell lines have been prepared from blood samples and characterized for the standard blood group antigen and protein electromorph markers, as well as for

PEDIGREE NO. 1362

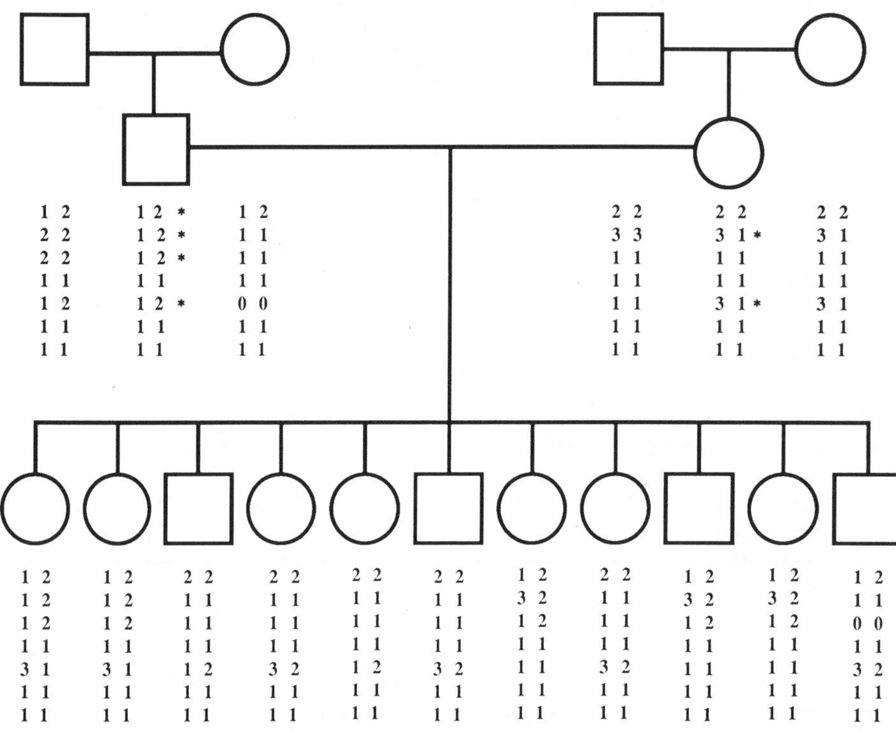

Figure 1. Pedigree of a three-generation family that is ideal for mapping studies. Genotypic data are shown for seven genetic systems on chromosome 13; heterozygosity in parents is indicated by an asterisk. (Reprinted, with permission, from White and Lalouel 1986.)

more than 100 markers consisting of arbitrary DNA sequences without known genetic function.

Because the usefulness of such family archives increases dramatically as more and more markers are characterized within them, we initially provided cell lines from some of the Utah families to the Camden Mutant Cell Respository, to make them available to other laboratories. Recently, however, Jean Dausset created the *Centre d'Etude du Polymorphisme Humain* (CEPH) in Paris, to support an international collaboration in human genotyping by providing access for many scientists to DNA samples from a specified set of 40 highly efficient and increasingly well characterized families. Many of the CEPH cell lines have been derived from the Utah pedigrees. CEPH provides the DNA free of charge to investigators willing to characterize a reasonable number of their own markers in the families. The only constraints are that the investigators must characterize all of the families in the set that are found to be informative and, at the time of publication, return the primary genotypic data on the families to the CEPH to become part of a public data base.

It was decided early on that the sample set of progeny chromosomes needed to be large enough to permit both the ordering of relatively tightly linked markers and the mapping of cloned genes that might be only poorly informative due to a low degree of DNA sequence polymorphism in their vicinity. To resolve questions of gene order with tightly linked markers, several recombinant chromosomes in at least two recombinant classes must be identified. Therefore, for resolution of gene order at the 1% recombination level, a population of between 500 and 1000 progeny chromosomes is desirable. The CEPH archive offers more than 500 progeny chromosomes.

Development of Genetic Markers

For meioses to be informative for recombination, heterozygosity at the marker loci is essential. Because we cannot construct test strains, the marker systems must have a high probability of heterozygosity in unselected individuals. Markers with multiple alleles that occur at reasonable frequencies offer the best systems in the human. Hundreds of polymorphic loci have now been identified (Willard et al. 1985). The great majority of these systems, however, consist of only two alleles, presumably due to a base pair change in a single restriction site. These systems can be improved by expanding the locus through identification of cosmid genomic clones and by testing a number of adjacent sequences for variation. When additional site polymorphisms at the locus are identified, the multiple possible haplotypes become a system of multiple alleles. These haplotype systems, however, are laborious to develop and also laborious to characterize in the families, since usually a number of different probe-enzyme systems must be run.

Hypervariable loci. Fortunately, in addition to the single base pair change that either creates or eliminates a restriction site, a second basis for DNA sequence polymorphism in the human has been identified. Polymorphic systems that give a number of alleles – several restriction-fragment-length alternatives – are associated with a number of loci: the insulin and Harvey-*ras* (Ha-*ras*) gene loci are examples. Sequencing has revealed that the basis for the variation in restriction fragment lengths is variation within the fragment in the number of short, tandem oligonucleotide repeats (Bell et al. 1982; Capon et al. 1983).

Figure 2 shows such a pattern at a new locus located just 5′ to the heavy-chain immunoglobulin *J* region. More than six alleles have been identified at this locus, and sequencing has confirmed that indeed the restriction-fragment-length variation is due to variation in number of a set of tandem repeats; in this case the repeats are 50 bp long.

The number of potential alleles at loci containing repeat sequences is therefore large, and indeed more than 10 have been found at a number of loci. Most individuals are heterozygous at these loci; furthermore, since the alleles can be characterized with only a single probe-enzyme system, a considerable increase in genotyping efficiency is achieved over the multisite haplotype approach.

Alec Jeffreys, in an insightful set of experiments, determined that scattered within the human genome were numerous highly polymorphic loci whose tandem repeats were quite closely related in sequence to the tandem oligonucleotides of the myoglobin locus (Jeffreys et al. 1985). He identified these loci using the myoglobin repeat sequence as a probe to screen phage genomic libraries.

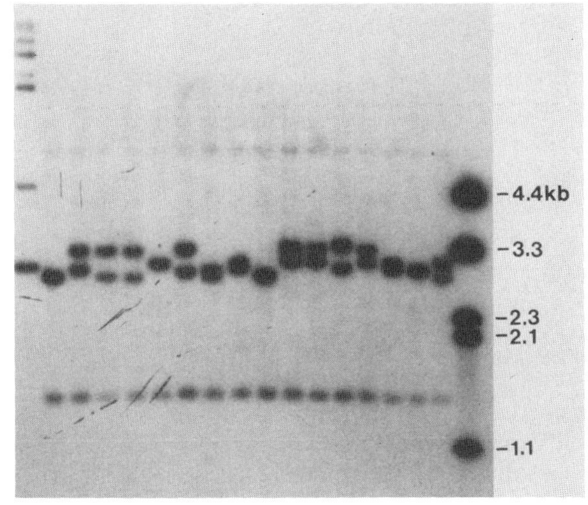

Figure 2. Polymorphism at a locus near the heavy-chain immunoglobulin *J* region, revealed on a *Bgl*II digest of 16 unrelated individuals with a 50-bp tandem repeat probe. (Reprinted, with permission, from White and Lalouel 1986.)

We have extended the Jeffreys findings by identifying a number of hypervariable regions based on sequence relationships to tandem oligomer repeats at other known hypervariable loci, such as insulin and ζ-globin. In addition, we have tested sequences, found in interesting places, that seemed reminiscent of some of the hypervariable motifs, for example, sequences from within the region of hepatitis B virus (HBV).

Table 1 summarizes the results of screening cosmid libraries with oligonucleotide probes and examining their genomic loci for polymorphisms (Y. Nakamura et al., in prep.). Two features should be noticed. First, several families of hypervariable regions exist and, therefore, a large number of hypervariable loci should be obtainable. Second, the identification of such loci by screening with oligonucleotides is an order of magnitude more efficient than identification with random clones. We have found an average frequency of more than 10% hypervariable loci as compared with the 1% frequency of hypervariable loci obtained with random clones (Braman et al. 1985).

A hypervariable "midisatellite". Although a large number of independent loci that seem to be distributed throughout the genome have been identified as hypervariable, we have also identified a single large region that seems to contain several such sequences in a clustered array that is itself hypervariable on a grand scale; perhaps the region is a "midisatellite." We have ascertained numerous clones that fall into this region both with insulin (I10) oligonucleotides and with a cosmid, Z22, that had been identified with the ζ-globin oligonucleotide and reflects a unique locus. The "midisatellite" clones reveal the complex pattern of polymorphic bands shown in Figure 3. In independent experiments, Mike Litt and his colleagues have also ascertained a clone with homology to this region (M. Litt, pers. comm.).

Analysis with pulsed-field gel electrophoresis of *Sfi*I fragments (Fig. 4) reveals that the majority of the repeat sequences in the "midisatellite" are found on a single *Sfi*I fragment that is variable in length from 250 kb to 500 kb. Almost all individuals are heterozygous for these variants. Sequencing in our hands of several subclones derived from the region has revealed only tandem repeats of a 40-mer. The repeats are not exact and do show some divergence in sequence.

Table 1. Results of Cosmid Screening

Probes	Positive clones per genome	Tested for polymorphism	Insertion-deletion polymorphism
ζ-globin	150	30	7
Insulin	100	10	3
Ha-*ras*	8	8	1
HBV (16)	150	26	6
IGJH (22)	50	4	1
Myoglobin	200	39	3
Total		117	21

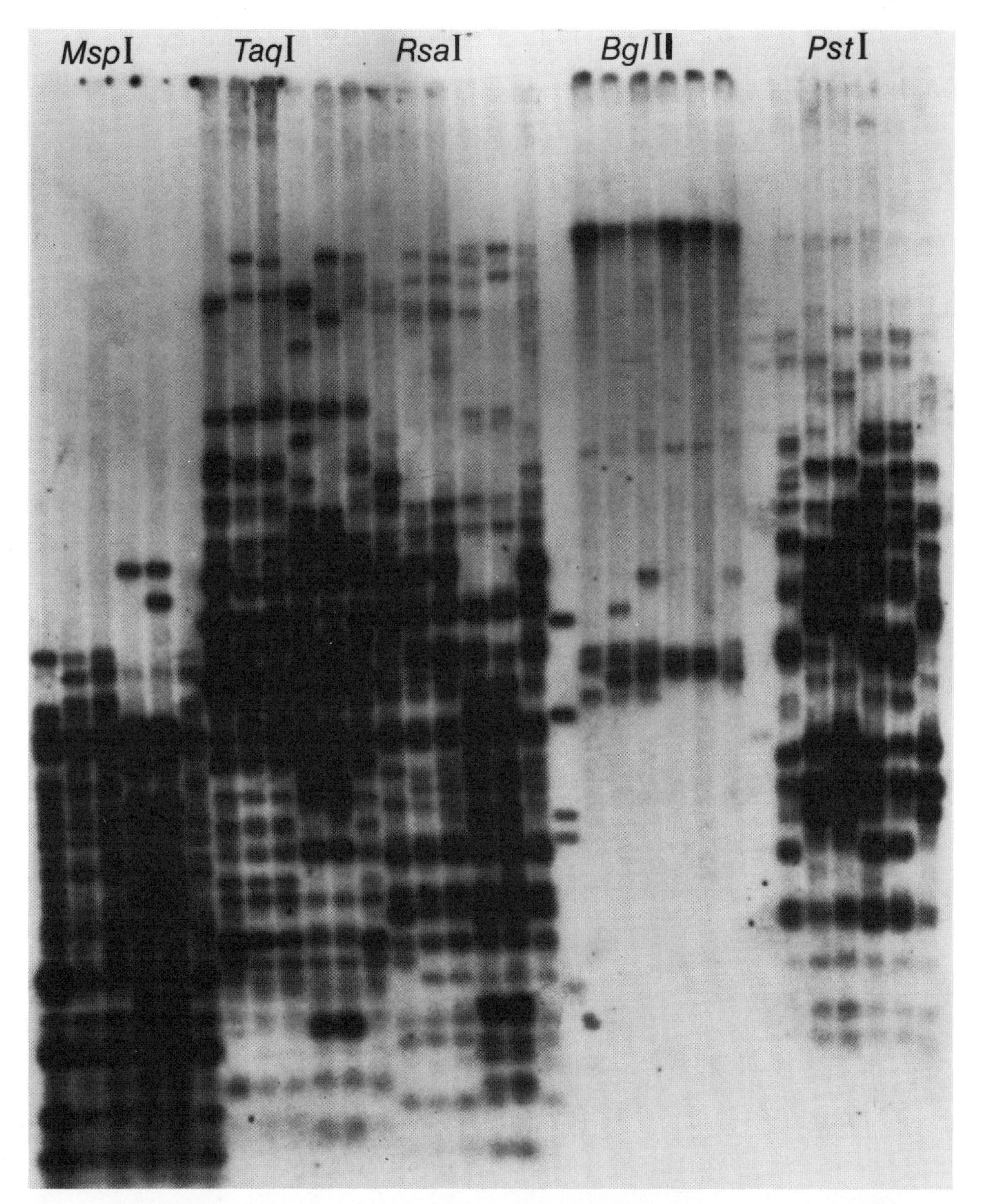

Figure 3. Hypervariable region revealed by a plasmid subclone isolated by homology to an oligomer from the insulin locus, showing polymorphism with five restriction enzymes: *Msp*I, *Taq*I, *Rsa*I, *Bgl*II, and *Pst*I.

Screening with the cosmid clones has revealed a number of single- and multiple-site polymorphic loci, as well, significantly increasing the number available to the research community. Since many of these newly defined polymorphisms have been detected by sampling with only one or two subfragments from a cosmid clone, the number of site polymorphisms at these loci, should they prove interesting, ought to be relatively easy to expand.

Our new oligonucleotide markers for hypervariable loci are available to the community of investigators seeking to map genetic diseases. Because the probes for

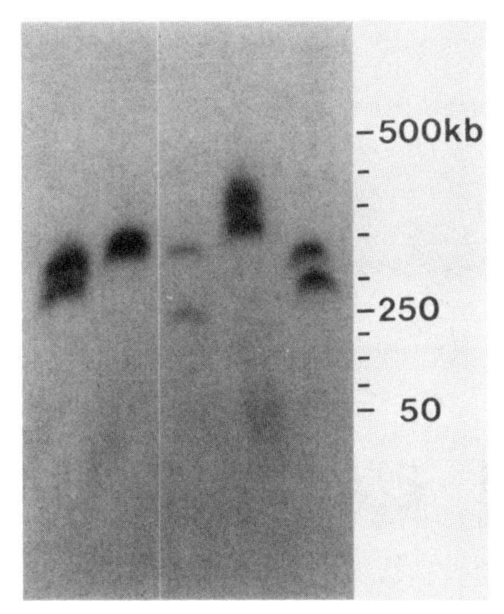

Figure 4. Pulsed-field gel pattern of *Sfi*I fragments from the "midisatellite" region, detected by probe Z22. The marker designations correspond to ligated phage λ.

these loci have been developed not only for construction of linkage maps but also for mapping of the genetic diseases outlined below, we would appreciate any use of our probes for study of these diseases being approached as a collaborative effort.

Primary Maps

Complete primary genetic linkage maps have now been constructed for several human chromosomes. Primary maps by definition have no, or few, regions with recombination frequencies greater than 0.2. Shown in Figure 5 are the maps that show the approximate chromosomal locations for sets of markers on chromosomes X, 13, and 12, with additional linkages for regions of 6p and 11p. Notice that there are no numbers on these ideograms: One should never trust a linkage map without numbers. However, in this case, for each map there are several presentations:

1. a map showing support for order (Fig. 6);
2. maps of recombination frequencies according to separate data for male and female meioses, as illustrated in Figure 7 for chromosomes 13q, 12, 6p, and 11p.

The map showing support for order is important. As pointed out by Bridges and Morgan (1923), we would expect the estimates of recombination frequency to vary somewhat as more data are obtained, but it would be much appreciated if the order were to remain invariant. Note that in some cases, support for order in Figure 6 is not strong, even with the large sample set. In general, this happens when only a small number of chromosomes are informative at the loci in question and recombinant in the appropriate intervals; the overwhelming majority of information with respect to the order of three loci comes from those meioses that are triply informative and recombinant in one or the other interval. Low support ratios indicate that the favored order could well be incorrect and that additional recombinational or physical support for order is required. In the case of the chromosome 13 map, the 5-6 order is confirmed by a translocation chromosome that separates these two loci (Leppert et al. 1986). High support ratios give good confidence that the indicated order is indeed correct, even without additional evidence.

Effect of sex on linkage mapping. One of the most striking features of the autosomal linkage maps shown here is the difference between male and female recombination frequencies. In general, recombination in female meioses is more frequent than in male meioses (see Human Gene Mapping 8, 1985). However, a notable exception is seen on the short arm of chromosome 11, where male recombination frequencies between the β-globin locus and the Ha-*ras* locus are several-fold higher than the corresponding female recombination frequencies (White et al. 1985a). To account for this finding, we suggest that certain sequences may be required for the initiation of recombination events in humans and that the active sequences are different between males and females.

One important conclusion is that we need not one map of the human genome, but three maps: a map with support for order, a map of male meiotic frequencies, and a map of female recombination frequencies.

Localization of Disease Genes

Although it would be more efficient to postpone the mapping of genetic disease loci until the normal human linkage map is completed and a reduced set of markers covering the genome at evenly spaced intervals has been developed, the exigencies of the need for markers for genetic diseases has caused us, as well as others (e.g., Gusella et al. 1983; Reeders et al. 1985), to seek linkages for genetic disease loci on a more random basis. Toward that end, we are running our better markers in a set of families that segregate unmapped genetic disease loci.

One of the major considerations guiding our choice of genetic diseases for mapping has been the availability of good family materials. For example, the pedigree segregating facioscapulohumeral muscular dystrophy (FSH) shown in Figure 8 is superb because it offers an abundance of affected individuals. We are also analyzing similarly rich pedigrees segregating the autosomal dominant form of retinitis pigmentosa.

The inherited, rare predispositions to cancer such as retinoblastoma (already mapped to 13q14) are also of major interest to molecular biologists, because these disorders likely represent lesions in genes that function early in tumorigenesis. Of particular interest to our

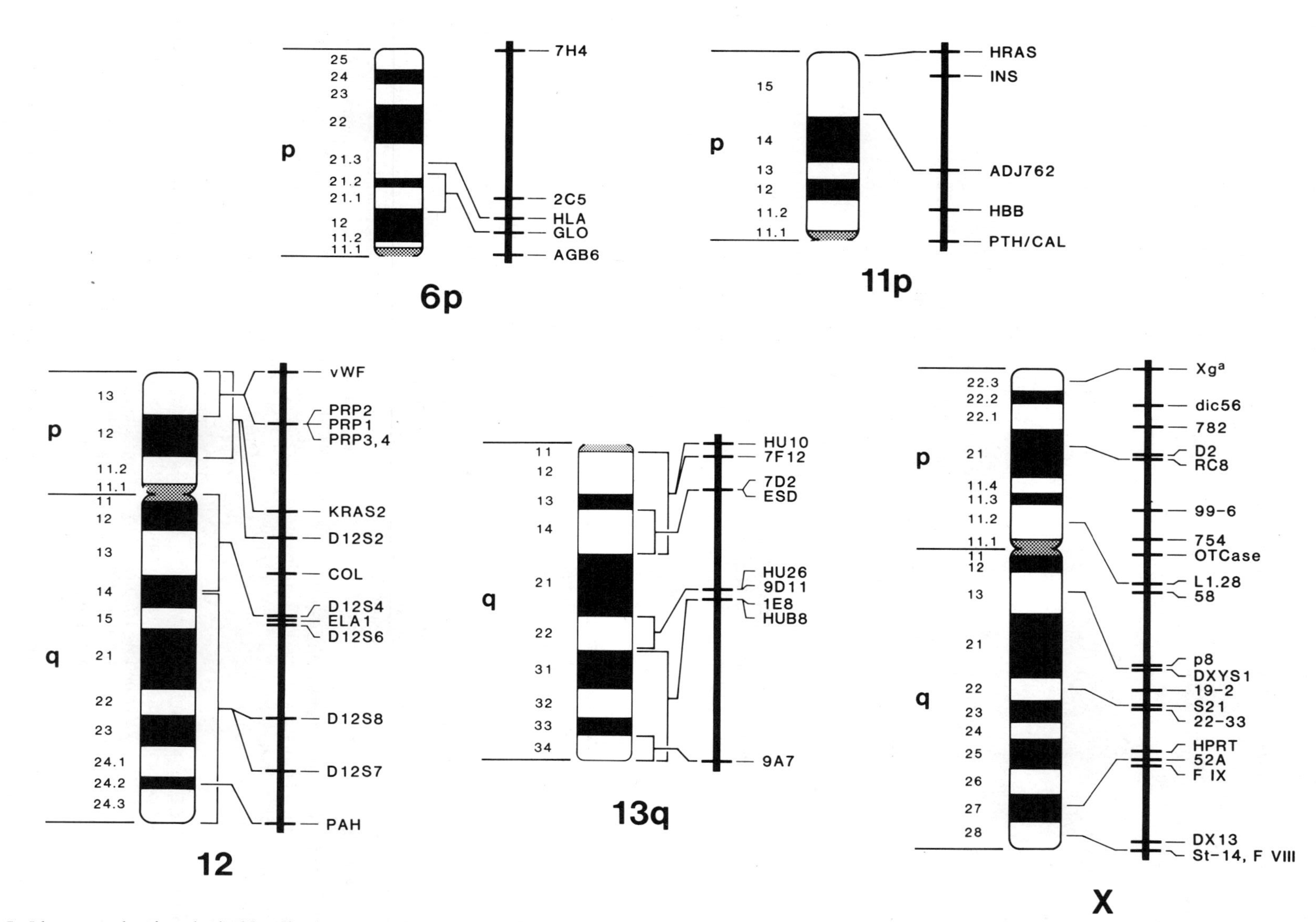

Figure 5. Ideograms showing physical localization of selected markers. The genetic linkage map to the right of each chromosome is based on recombination values for combined male and female meioses.

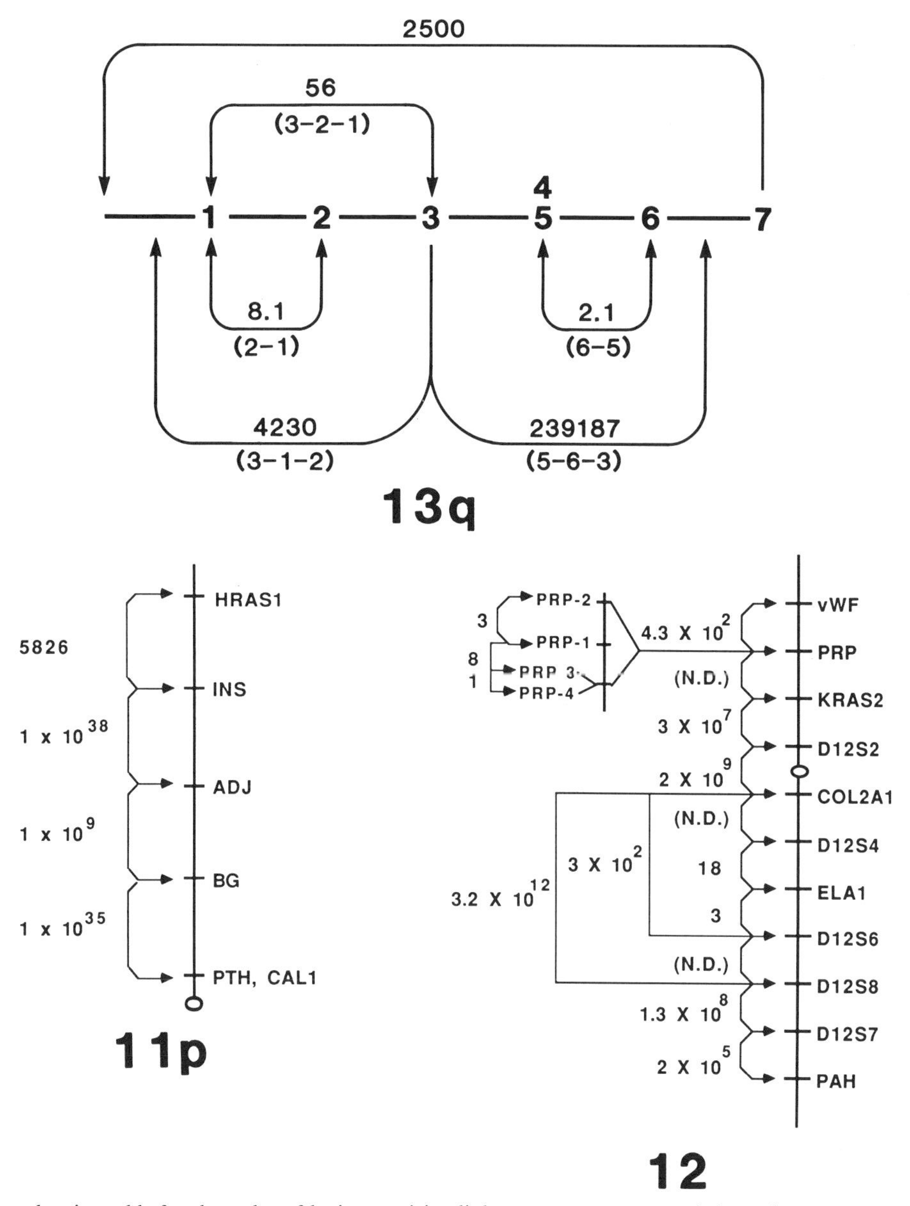

Figure 6. Maps showing odds for the order of loci comprising linkage groups on parts of three chromosomes. Locus numbers for chromosome 13q correspond to the following genetic systems: (1) HU10; (2) 7F12; (3) 7D2,ESD; (4) HU26; (5) 9D11; (6) HUB8,1E8; (7) 9A7. (N.D.) Not done. (Adapted from White and Lalouel 1986.)

group in Utah is Gardner's Syndrome, originally characterized here by Eldon Gardner (1951). This inherited cancer syndrome is believed to be a variant of familial polyposis coli and obviously involves a gene defect imbedded in the early stages of the cell's genetic pathway to colon cancer. Genetic linkage studies have eliminated as causative most of the cellular oncogenes so far identified (Barker et al. 1983; S. Woodward, unpubl.). Cell lines from several large pedigrees have been established for both Gardner's Syndrome and familial polyposis, to provide DNA for further investigation. Ataxia telangiectasia, which gives a strong predisposition to the development of lymphoma, is being studied in collaboration with Dick Gatti in an Amish kindred. Multiple endocrine neoplasia (MEN) is also a major focus of genetic linkage studies in our laboratory.

Cystic fibrosis. Following the report of linkage of the gene for cystic fibrosis (CF) to a genetic marker located on chromosome 7, we examined our repertoire of chromosome 7 markers and identified several for testing in our panel of CF families. One of these, the *met* locus, a cellular oncogene which had come to us in a collaboration with George Vandewoude and Michael Dean, proved to be tightly linked to the *CF* locus (White et al. 1985b). The results of a large collaborative effort since then indicate that initial estimates of the linkage distance were essentially correct; the *met*

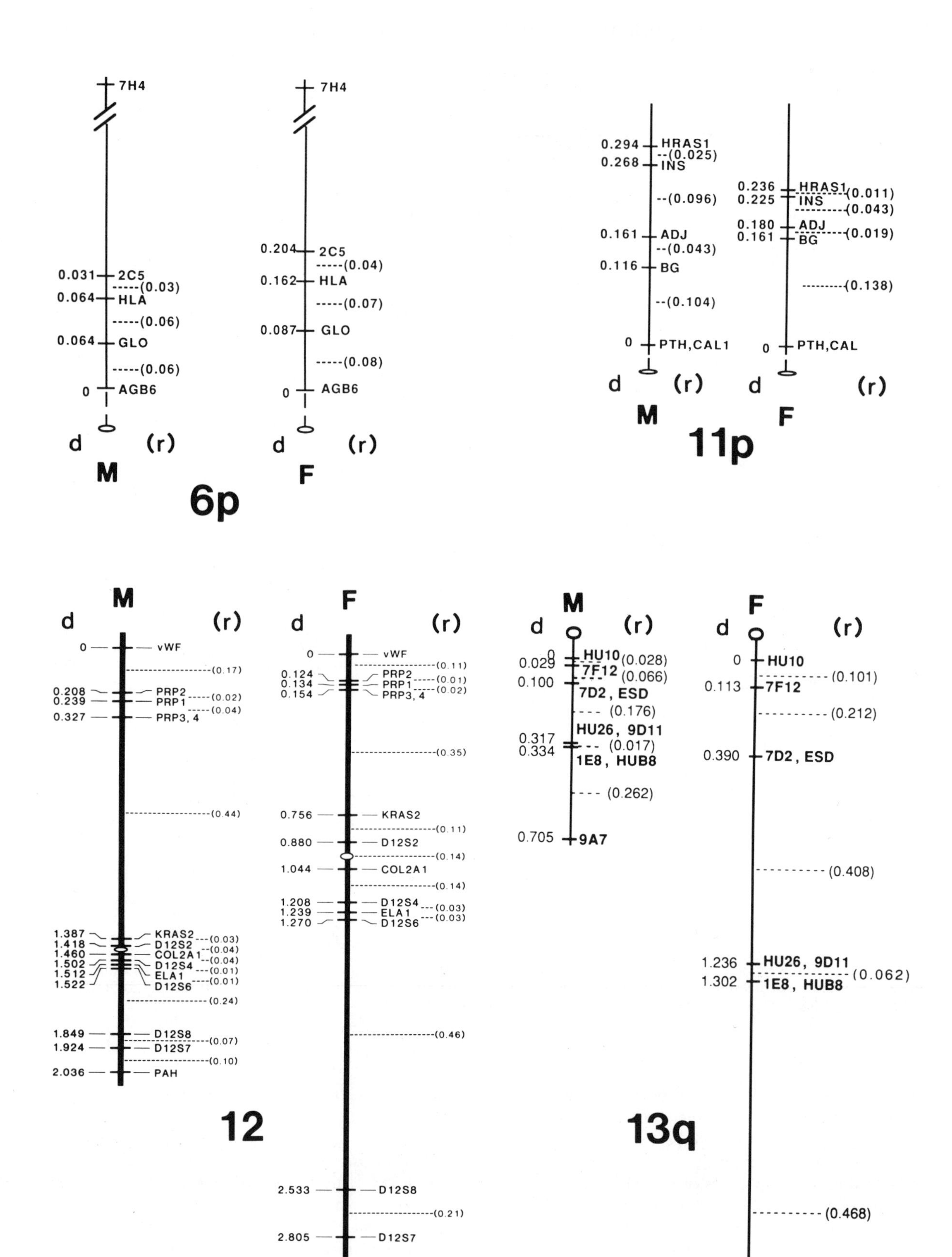

Figure 7. Genetic linkage maps showing distance (**d**) in Morgans from the mapped locus nearest the centromere, and recombination fractions (**r**) in each interval for male (**M**) and female (**F**) meioses separately. (Adapted from White and Lalouel 1986.)

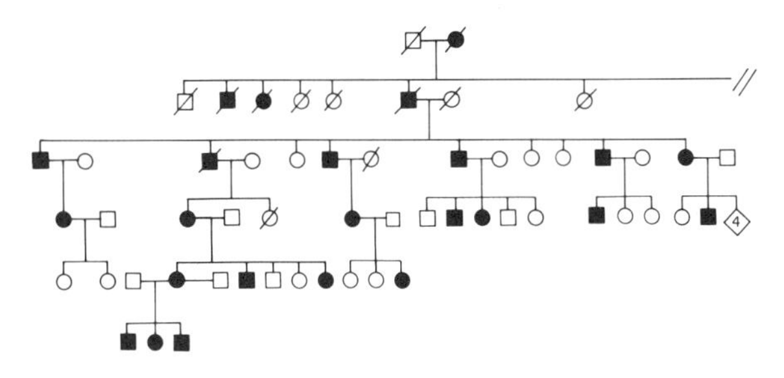

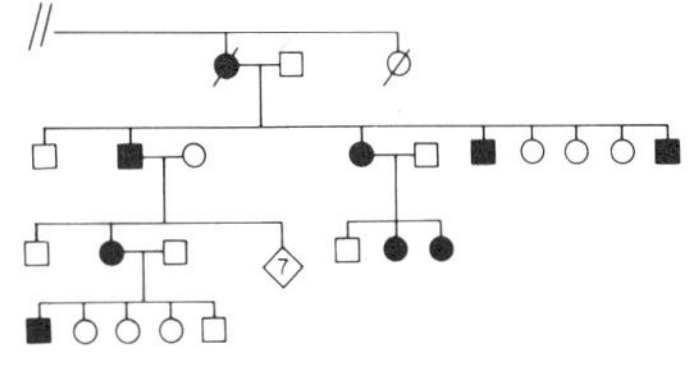

Figure 8. Pedigree segregating facioscapulohumeral muscular dystrophy (FSH). Squares represent males, circles females; diamonds indicate a number of unaffected individuals. Filled symbols denote family members affected with FSH; a line through a symbol indicates a deceased individual.

locus is likely to be located within a 1% recombination distance of the *CF* locus.

This distance is important because 1% recombination is within the range of physical distances approachable with pulsed-field gel technology for the analysis of large restriction fragments; also appropriate are available technologies that should permit covering this region with a number of clones. Figure 9 shows the *Not*I fragments identified with the *metH* clone; polymorphism in the *Not*I fragment yields a 650-kb allele and an allele more than 1,000 kb long. The latter *Not*I fragment could contain the *CF* gene, but even if not it

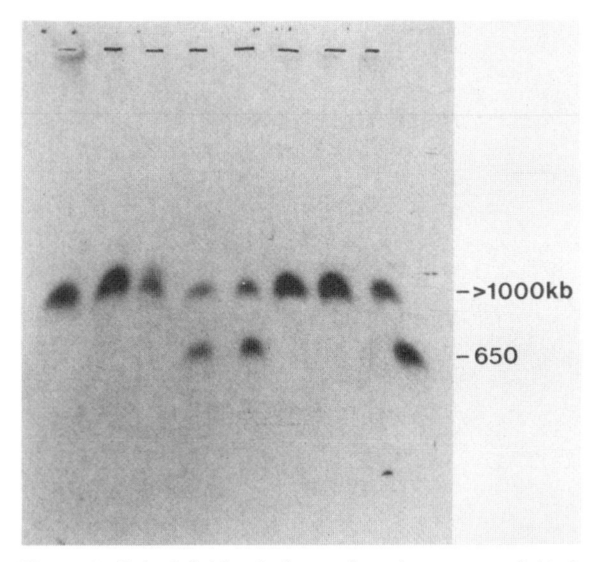

Figure 9. Pulsed-field gel electrophoretic pattern of *Not*I fragments probed with the *metH* clone.

is certainly nearby. Interestingly, the *met* gene polymorphism has been difficult to expand through cosmid isolation; the cosmids obtained have a propensity to delete the *met* gene. Meanwhile, however, as efforts to identify and characterize the *CF* gene proceed, the probes close to *CF* are already in service for prenatal diagnosis in families that are adequately informative for the markers.

ACKNOWLEDGMENTS

We gratefully acknowledge th expert technical assistance of Melanie Culver, Tom Holm, Cindy Martin, Mark Hoff, and Dora Stauffer. We also wish to thank Barbara Ogden for the collection and preparation of family material, Leslie Jerominski for the establishment and maintenance of cell lines, and Ruth Foltz for assistance in the preparation of this manuscript.

REFERENCES

Barker, D., M. McCoy, R. Weinberg, M. Goldfarb, M. Wigler, R. Burt, E. Gardner, and R. White. 1983. The test of the role of two oncogenes in inherited predisposition to colon cancer. *Mol. Biol. Med.* **1:** 199.

Bell, G.I., M.J. Selby, and W.J. Rutter. 1982. The highly polymorphic region near the human insulin gene is composed of single tandemly repeating sequences. *Nature* **295:** 31.

Botstein, D., R. White, M. Skolnick, and R. Davis. 1980. Construction of a genetic linkage map in man using restriction fragment length polymorphisms. *Am. J. Hum. Genet.* **32:** 314.

Braman, J., D. Barker, J. Schumm, R. Knowlton, and H. Donis-Keller. 1985. Characterization of very highly polymorphic RFLP probes. *Cytogenet. Cell Genet.* **40:** 589 (Abstr.).

Bridges, C.B. and T.H. Morgan. 1923. The third chromosome group of mutant characters of *Drosophila melanogaster*. *Carnegie Inst. Wash. Publ.* **327:** 1.

Capon, D.J., E.Y. Chen, A.D. Levinson, P.H. Seeburg, and D.V. Goeddel. 1983. Complete nucleotide sequences of the T24 human bladder carcinoma oncogene and its normal homologue. *Nature* **302:** 33.

Gardner, E.J. 1951. A genetic and clinical study of intestinal polyposis, a predisposing factor for carcinoma of the colon and rectum. *Am. J. Hum. Genet.* **3:** 167.

Goldstein, J.L., M.S. Brown, R.G. Anderson, D.W. Russell, and W.J. Schneider. 1985. Receptor-mediated endocytolysis: Concepts emerging from the LDL receptor system. *Annu. Rev. Cell Biol.* **1:**1.

Gusella, J.F., N.S. Wexler, P.M. Conneally, S.L. Naylor, M.A. Anderson, R.E. Tanzi, P.C. Watkins, K. Ottina, M.R. Wallace, A.Y. Sakaguchi, A.B. Young, I. Shoulson, E. Bonilla, and J.B. Martin. 1983. A polymorphic DNA marker genetically linked to Huntington's disease. *Nature* **306:** 234.

Human Gene Mapping 8. Eighth International Workshop on Human Gene Mapping, Helsinki. *Cytogenet. Cell Genet.* **40:** 1.

Jeffreys, A., V. Wilson, and S. Thein. 1985. Hypervariable "minisatellite" regions in human DNA. *Nature* **314:** 67.

Leppert, M., W. Cavenee, P. Callahan, T. Holm, P. O'Connell, K. Thompson, G.M. Lathrop, J.-M. Lalouel, and R. White. 1986. A primary genetic map of chromosome 13q. *Am. J. Hum. Genet.* **39:** 425.

Reeders, S.T., M.H. Breuning, K.E. Davies, R.D. Nicholls, A.P. Jarman, D.R. Higgs, P.L. Pearson, and D.J. Weatherall. 1985. A highly polymorphic DNA marker linked to adult polycystic kidney disease on chromosome 16. *Nature* **317:** 342.

White, R. and J.-M. Lalouel. 1986. Investigation of genetic linkage in human families. *Adv. Hum. Genet.* **16:** (in press).

White, R., M. Leppert, T. Bishop, D. Barker, J. Berkowitz, C. Brown, P. Callahan, T. Holm, and L. Jerominski. 1985a. Construction of linkage maps with DNA markers for human chromosomes. *Nature* **313:** 101.

White, R., S. Woodward, M. Leppert, P. O'Connell, Y. Nakamura, M. Hoff, J. Herbst, J.-M. Lalouel, M. Dean, and G. Vande Woude. 1985b. A closely linked genetic marker for cystic fibrosis. *Nature* **318:** 382.

Willard, H.F., M.H. Skolnick, P.L. Pearson, and J.-L. Mandel. 1985. Report of the committee on human gene mapping by recombinant techniques. *Cytogenet. Cell Genet.* **40:** 360.

Construction of Human Genetic Linkage Maps: II. Methodological Issues

J.-M. LALOUEL, G.M. LATHROP, AND R. WHITE
Howard Hughes Medical Institute and Department of Human Genetics, University of Utah Medical Center, Salt Lake City, Utah 84132

The genetic anatomy of a species consists of the description of the linear arrangement of its genes along chromosomes. Depending on the methods used in this construction, several types of maps can be distinguished. When genes are assigned to chromosomes or to chromosome subregions through somatic cell hybrids or in situ hybridization (McKusick, this volume), the map generated is strictly physical. So it is for molecular maps generated on particular genomic regions by restriction mapping of overlapping DNA sequences, and so will it be when extensive restriction maps are constructed with pulsed-field gel electrophoresis (Schwartz and Cantor 1984; Smith and Cantor, this volume).

By contrast, genetic maps constructed by analysis of the segregations at two or more loci incorporate information on both the physical distribution of loci on chromosomes and the distribution of crossing-over in all intervals considered. Because all evidence (for review, see White and Lalouel 1986) indicates that the occurrence of crossing-over is not uniform with respect to distance on the physical map, we should not expect more than a monotonic relationship between physical and genetic maps.

Our attention in this discussion bears on methodological aspects of the construction of genetic maps in humans. Together with the detection of genetic linkage and the estimation of recombination between a set of loci in appropriate familial data, the construction of genetic maps requires that the order of the genes be inferred from the observed recombination values. This enterprise presents various challenges that will call for carefully designed strategies and analytical methods. We try here to convey our interest in meeting these challenges without undue mathematical arguments, at the risk of oversimplification for the statistically minded reader.

Inference of Genetic Linkage and Estimation of Recombination

Crossing-over and recombination. Recombination results from the occurrence of an odd number of crossing-over events on a given segment of a chromatid; it follows that the primary event investigated by analysis of recombination is not directly identifiable. Rather, the density of occurrence of crossing-over in an interval, usually expressed in Morgans, can only be inferred from recombination. Yet it is only in terms of crossing-over that additivity, and therefore the definition of a distance measure, can provide the basis for generating a linear genetic map.

The design of linkage experiments. Classic linkage experiments with laboratory animals involve a mating program to control the distribution of allelic variants on parental chromosomes (Table 1). As a consequence, recombination events can be identified unambiguously in each offspring. A large number of progeny can be scored, and the observed number of recombinants can be contrasted to the theoretical proportion expected under the hypothesis of no linkage by a simple χ-square test. If this test shows significance, the recombination fraction is estimated as the ratio of the number of observed recombinants over the total number of scored progeny. A detailed account of such experimental designs is given in Bailey (1961).

Human linkage studies. In dealing with humans, one has to rely upon chance to identify families in which at least one parent is heterozygous at both loci under investigation. For a nuclear family consisting of both parents and their offspring, a situation analogous to a double backcross experiment is considered in Figure 1. It is clear that in such instances the chromosomal distribution of alleles at two loci is unknown: the heterozygous parent can be either of genotype AB/ab or genotype Ab/aB. The probability of a given genotype in the observed offspring can be expressed under each of these two alternatives; these two terms can be combined by taking into account the probability a priori of each parental allelic phase. It follows that recombination is no longer expressed as a simple function of the

Table 1. Linkage Experiments in Experimental Animals: Elaboration of a Double Backcross

P1 × P2 :	AB/AB × ab/ab			
F1 :	AB/ab			
F1 × P2 :	AB/ab × ab/ab			
F2 :	AB/ab	ab/ab	Ab/ab	aB/ab
e :	$(1-r)/2$	$(1-r)/2$	$r/2$	$r/2$
o :	$n(1)/N$	$n(2)/N$	$n(3)/N$	$n(4)/N$

Two doubly homozygous parental lines, P1 and P2, yield a first generation (F1) of heterozygotes that, when crossed with parental line P2, give segregating offspring F2. Expected proportions, e, are simple functions of recombination, r, which can be estimated directly from the observations (o), $r = [n(1) + n(2)]/N$.

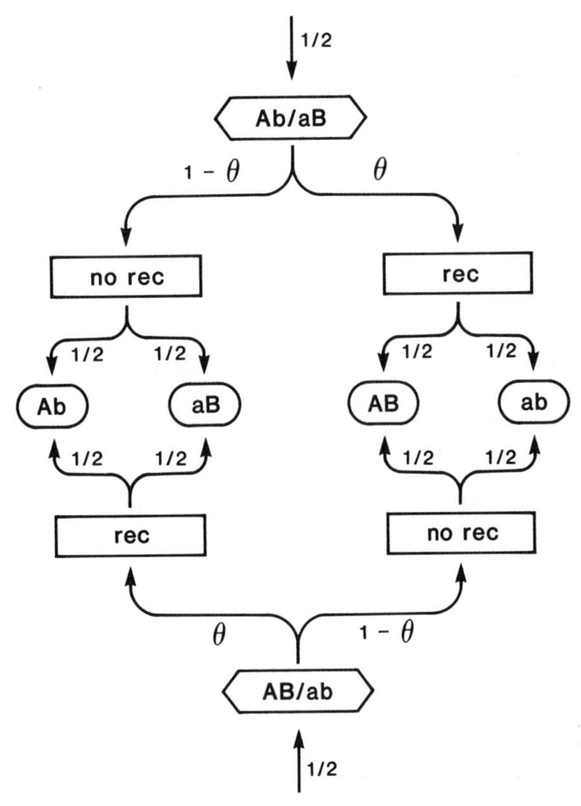

Figure 1. Probabilites of genotypes for offspring when parents are of genotypes AaBb and aabb, with allelic phase unknown. θ is the probability of recombination between the two loci.

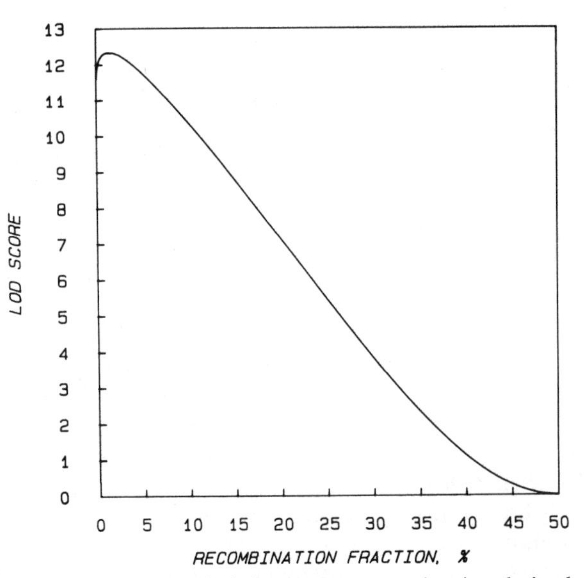

Figure 2. Example of a lod score curve assessing, in a decimal logarithmic scale, the relative odds for linkage vs. no linkage as a function of the recombination rate between two loci. In this instance, tight linkage is indicated, with a maximum lod score of 12.3 for a recombination value of 0.15.

observed number of recombinant offspring. Rather, the joint probability of all offspring tested in a sample of families is a function of the unknown rate of recombination between the two loci.

Following a general rule of statistical inference, the recombination value that has the greatest likelihood given the available observations is that value for which this function is maximum. The maximum is obtained by iteration following standard techniques. The probability obtained at this maximum can be contrasted to the value obtained at a recombination value of 0.5, providing a test of the hypothesis of independent segregation. The ratio of these two terms gives the odds in favor of linkage at the estimated recombination value, versus free recombination. In large samples, twice the logarithm of this ratio is distributed as a χ-square with one degree of freedom under the hypothesis of independent segregation. However, it has become conventional since Morton (1955) to use the decimal logarithm of this ratio in most tests of linkage in human genetics; this latter statistic, or lod score, finds its foundation in the theory of sequential sampling. It is customary to report either a lod score curve (Fig. 2) or a table of lod scores computed at conventional values of the recombination fraction. The latter affords the cumulation of linkage evidence across independent studies by summation.

The estimate of recombination obtained in a given study has an associated sampling error that must be taken into account in all applications, particularly in genetic diagnosis. The classical confidence interval based on the asymptotic standard error of the estimate is usually distributed in human linkage studies because of the limited sample sizes achieved in most studies. An alternative (Conneally et al. 1985) defines an empirical confidence region based on the observed lod score curve: Bounds on the recombination estimate are taken as those values for which the lod is decreased by one unit.

Family structures. Often, families extending over two generations are investigated in human genetics. A rare dominant condition can often be traced in a single large pedigree. The logic outlined above can be extended to handle such situations; furthermore, information accumulated over generations in extended family structures can yield knowledge of the allelic phase in doubly heterozygous parents.

This observation is relevant in determining the optimal family structure to select for a sample of reference families for genetic mapping. Although extended pedigrees are valuable for investigations of rare Mendelian disorders, the most efficient structure for mapping purposes has turned out to be a large sibship, their parents, and all four grandparents (White et al. 1985). The grandparental genotypes will often, but not always, determine the parental phase; when they do, it increases the power of such studies to detect genetic linkage. In addition, typing grandparents provides internal tests of consistency in the genetic interpretation of laboratory data. A large sample of such families was collected in Utah; a subset of these families constitutes the target of the international genetic mapping effort of the Centre d'Etude Polymorphisme Humain (CEPH) (see White et al., this volume).

Effect of sex on recombination. Experimental evidence points to significant differences between the sexes with respect to the frequency of occurrence of crossing-over in all species where this effect has been investigated; humans are no exception (Renwick 1968). Neither observations of chiasmata at meiosis in both sexes (Hulten 1974; Bojko 1985) nor cytological examination of synaptonemal complexes and recombination nodules (Bojko 1983; von Wettstein et al. 1984) have provided any clue as to the origin of this observed difference.

It was once hoped (Renwick 1969) that this difference would turn out to be constant along any chromosome segment, but the evidence is mounting that it might not be so, as summarized in the genetic maps reported by White et al. (this volume). Because differences in recombination rates between sexes reflect differences in the density of occurrence of crossing-over, it is only in terms of the latter that meaningful comparisons can be performed. By transforming recombination rates to genetic distances, the ratio of the density of occurrence of crossing-over in females over the corresponding rate in males, or map distance ratio, measures the extent of this sex effect.

Although not understood, the effect of sex needs to be taken into account in the construction of human genetic maps. For any two loci, a recombination rate needs to be estimated for each sex, and the available data will not always permit independent estimation of these parameters in each sex. This issue becomes more challenging as more loci are considered in a multilocus analysis.

Determination of Gene Order

The case of three loci. The construction of a genetic linkage map requires the determination of gene order and the estimation of genetic distances between a set of linked loci. Common sense, practical experience, and statistical reasoning have established the inefficiency of independent linkage experiments involving only pairs of loci in this context (Edwards 1982; Lathrop et al. 1984; Thompson 1984; Bishop 1985). Linkage tests with laboratory animals involve carefully designed mating programs which, as in the case of two loci, allow us to bring allelic variants into defined chromosomal configurations, as illustrated in Table 2. In a triple backcross with known allelic phases, recombination events can be scored unambiguously in the progeny. Each individual will belong to one of four possible classes. All three recombination rates can be directly computed from the observed numbers in each class, and the most likely order is obtained by defining as flanking markers the pair of loci that exhibit the highest recombination rate.

Table 2. Joint Segregation at Three Loci

		B – C	
		yes	*no*
A – B	*yes*	$p(3)$	$p(1)$
	no	$p(2)$	$p(0)$

$r(\text{A,B}) = p1 + p3$
$r(\text{B,C}) = p2 + p3$
$r(\text{A,C}) = p1 + p2$

A – B – C : $p3$ double recombinant
B – A – C : $p3$ recombination in first segment
A – C – B : $p3$ recombination in second segment

Recombination events can be described in terms of three probabilities, $p(1)$, $p(2)$, and $p(3)$. Each of these three probabilities describes different events for each of the three possible gene orders.

This approach has proven satisfactory in most instances in the laboratory, the experimenter's degree of confidence in the order indicated in his study seldom being contrasted to other possible orders. This practice finds its justification in the considerable number of progeny tested in such crosses under controlled experimental conditions, of the order of several thousands, or even tens of thousands in *Drosophila melanogaster*.

Assessing odds for alternate gene orders. In the human the situation is different for two reasons. On one hand, lack of experimental control again precludes the unambiguous determination of allelic phases in multiply heterozygous parents; multiply informative crosses, however, can be obtained if all loci are tested in a common panel of reference families, as made feasible by CEPH. On the other hand, the technology involved in preparing DNA samples and characterizing restriction-fragment-length polymorphisms (RFLPs) prohibits the consideration of the extensive sample size commonly achieved with laboratory animals; the determination of genotypes in a few hundred progeny represents a considerable effort.

If the allelic phase is not determined in some families, a difficulty anticipated for most genetic loci presently available, using all of the painfully collected information will require a generalization of the probabilistic approach used for pairs of loci to consider three or more loci. Again, recombination events are not unambiguously identified.

A consequence of the limited sample size achieved in humans is that it is no longer sufficient to report the gene order best supported by the data. Sampling errors can affect this inference, and therefore a measure must be given of our degree of confidence in that order relative to other possible orders. For a three-locus experiment, this is achieved by expressing the probability of the observations under each possible order as a function of recombination rates and contrasting these probabilities to assess the odds in favor of the most likely order against its alternatives. Such odds do not measure significance as do conventional statistical tests; however, they reflect a clearly intuitive measure of our degree of confidence in the favored gene order.

They are an integral part of the specification of a human genetic map.

Table 3. Formulation of a Three-locus Experiment When a Given Gene Order, A–B–C, Is Postulated

Event	Recombination between A,B	B,C	Probability
E(1)	yes	no	$p(1)$
E(2)	no	yes	$p(2)$
E(3)	yes	yes	$p(1,2)$

Recombination and Genetic Distance

Crossing-over and recombination. The probabilities of the four possible outcomes of a three-point linkage test have been summarized in Table 3. They involve only three independent parameters, since they must add to one. Equivalently, three recombination parameters can be defined as follows:

$$r(A,B) = p(1) + p(1,2),$$
$$r(B,C) = p(2) + p(1,2),$$
$$\text{and } r(A,C) = p(1) + p(2).$$

By substitution, $r(\mathrm{A,C}) = r(\mathrm{A,B}) + r(\mathrm{B,C}) - 2p(1,2)$, that is, the recombination rate between flanking loci, $r(\mathrm{A,C})$, can be expressed in terms of the recombination rates in each intervening segment and the probability of recombination in both intervening segments, $p(1,2)$. If not more than one crossing-over were allowed between the flanking loci A and C, then $p(1,2) = 0$, and recombination in the interval would have the additive property of a distance measure.

The situation of complete interference between crossing-overs, where the occurrence of one inhibits the occurrence of another in the whole interval, is the exception. By contrast, if crossing-over events were to occur independently along an interval, the probability of event E(1) would be $p(1) = r(\mathrm{A,B})\ [1 - r(\mathrm{B,C})]$ and that of event E(2) would be $p(2) = [1 - r(\mathrm{A,B})]\ r(\mathrm{B,C})$; therefore $r(\mathrm{A,C}) = r(\mathrm{A,B}) + r(\mathrm{B,C}) - 2\ r(\mathrm{A,B})\ r(\mathrm{B,C})$, with $p(1,2) = r(\mathrm{A,B})\ r(\mathrm{B,C})$. In the absence of interference between crossing-over events, the probability of recombination between flanking loci is smaller than the sum of the probabilities of recombination in the two intervening segments: Recombination no longer behaves as a distance measure. This remains true for any intermediate degree of interference.

To construct linear genetic maps, a number of mapping functions have been proposed (for reviews, cf. Bailey 1961 and Ott 1985) that incorporate various degrees of interference. They express genetic distance as a function of recombination in Morgans, where a Morgan is that distance over which, on average in a large number of meioses, one crossing-over occurs per strand. Some examples of the relationship predicted by such functions are given in Figure 3.

Neglecting interference. Interference in crossing-over introduces an additional degree of complexity to genetic mapping; that it seems to vary with chromosomal location (White and Lalouel 1986) renders more elusive a biologically meaningful estimation of its effect. The question of whether the incorporation of interference can significantly improve the quality of a genetic map needs to be considered carefully.

Figure 4A shows the extent of the error introduced in the estimate of recombination between loci symmetrically flanking an intervening locus when interference operating at the intermediate level characteristic of *Drosophila* is neglected. The greatest error occurs when

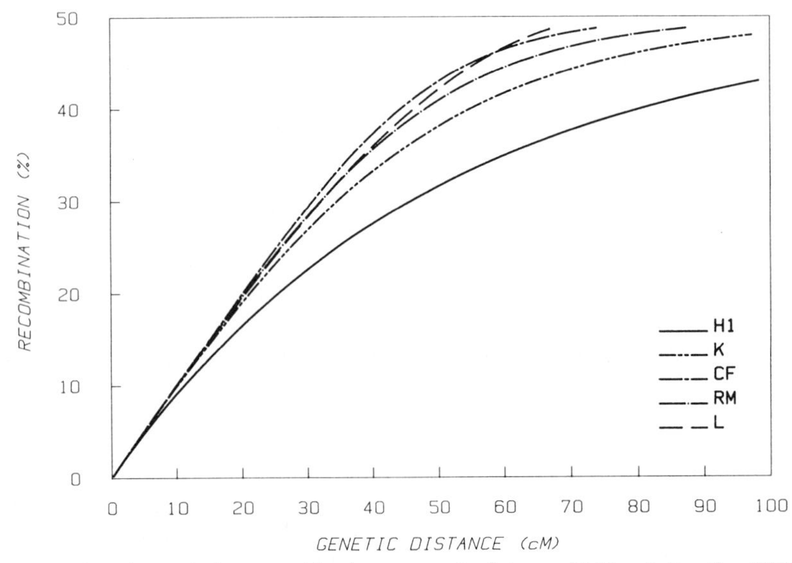

Figure 3. Various mapping functions relating recombination to genetic distance. Haldane's function (H1) assumes no interference; the degree of interference increases with the following functions: Kosambi (K), Rao and Morton (RM), Ludwig (L), and Carter and Falconer (CF). (Reprinted, with permission, from White and Lalouel 1986.)

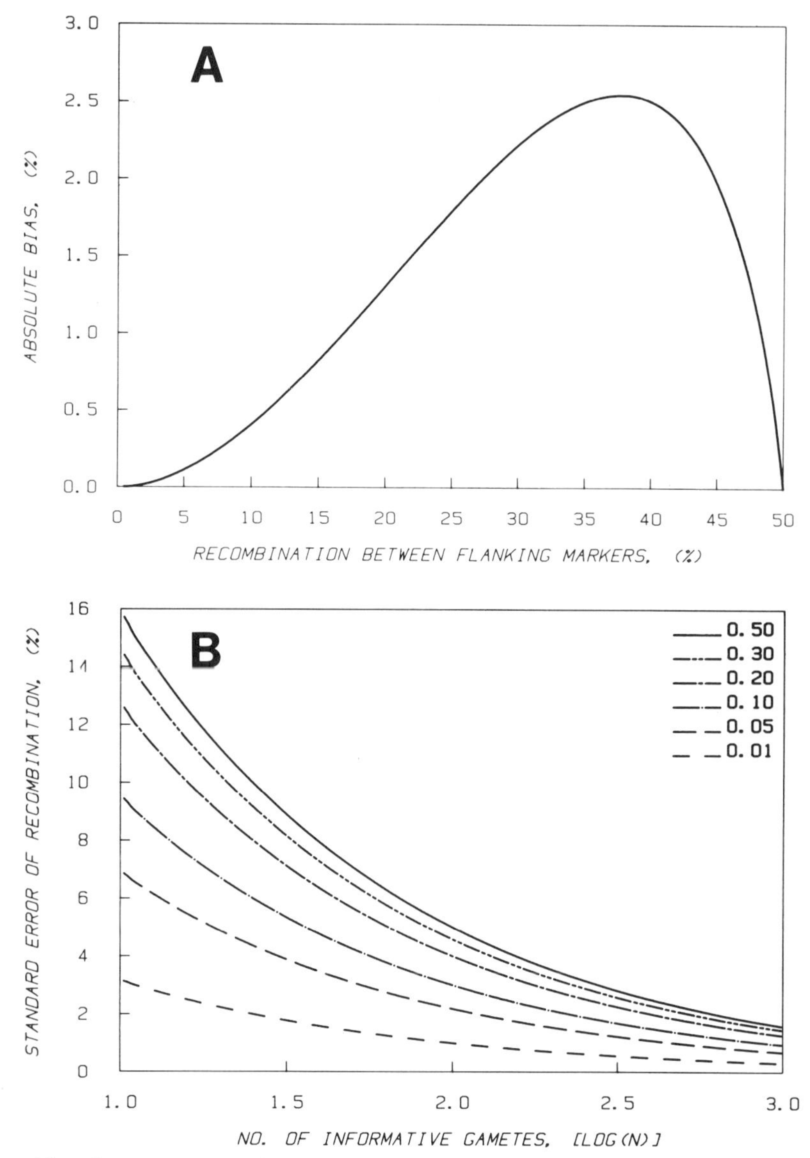

Figure 4. The effect of interference on recombination estimates. (*A*) Absolute bias in the estimate of recombination between two markers symmetrically flanking a third one when intermediate interference (Kosambi level) is neglected. (*B*) Standard error of recombination as a function of the number of informative gametes observed. (Reprinted, with permission, from White and Lalouel 1986.)

recombination between flanking loci is equal to 28%, or 0.15 in the intervening intervals; it is taken as 25.5% when interference is neglected. Considering the sampling error inherent in estimates of recombination as a function of sample size, typically on the order of about several hundred informative gametes in human applications (Fig. 4B), it appears that to neglect interference is likely to have but a small effect on the construction of human genetic maps. Because recombination in intervening segments will exert a more pronounced effect in map estimation than recombination between flanking loci, the error introduced is likely to be even smaller.

This is illustrated by an example from chromosome 11. Three very highly polymorphic loci—parathyroid hormone (*PTH*), β-globin (*HBBC*), and insulin (*INS*)—were characterized in 28 Utah families with large sibship size. Support for the order *PTH-HBBC-INS* was overwhelming. The recombination estimates in the intervening segments, assuming no interference, were 0.120 and 0.105. In view of these estimates, the large sample sizes achieved, and the very high information content of these markers, the situation was almost optimal to test for the presence of interference. As can be seen in Table 4, this test reached significance, yielding a χ-square with one degree of freedom equal to 4.79. Indeed, no double crossing-over had been observed in the data. Yet, contrasting the recombination estimates obtained for the intervening segments under each alternative, the differences introduced in the recombination estimates appear trivial.

A more rigorous statistical analysis (Lathrop et al. 1985) leads to similar conclusions: For the sample sizes

Table 4. Test of Interference for Three loci of Human Chromosome 11p

Interference	r(A,B)	r(B,C)	r(A,C)	χ-square
yes	0.124	0.105	(0.203)	
no	0.120	0.105	0.225	4.79

A, B, and C are parathyroid hormone, β-globin, and insulin loci, respectively. When no interference is assumed, r(A,C) is obtained as a function of r(A,B) and r(B,C), leaving one degree of freedom to test the hypothesis of interference.

typically achieved in linkage studies involving RFLPs, the power to estimate the effect of interference is very low. Only in ideal situations, approached in the example of Table 4, can this effect become significant.

Therefore, as is often the case in the biological sciences, it appears that for the purpose of genetic mapping, where the primary objective is to obtain good estimates of recombination between adjacent loci, interference can be neglected as a second-order effect. Parsimony is in keeping with our ignorance of the mechanism at play and provides simplicity and clarity in map construction. The probabilities of meiotic outcomes involving multiple loci can be written as simple functions of the recombination rates between intervening loci because of the independence rules that result from neglecting interference. It follows that for n loci with k alleles, the $n(n-1)/2$ recombination rates and the k^n possible meiotic outcomes can be expressed recursively in terms of only $n-1$ parameters. Simple functions relate genetic distance, d, and recombination, r, that is $d = -\ln(1-2r)/2$ and $r = [1-\exp(-2d)]/2$ (Haldane 1919). Further benefits will become apparent in a following section.

Building Initial Genetic Maps

Ordering multiple loci. Skeleton maps are being constructed from sets of loci constituting linkage groups (White et al., this volume). Once linkage between members of a group has been established, the most likely order of the loci, the odds in favor of alternate orders, and the genetic lengths of each interval need to be determined. With most available loci, the distribution of heterozygosity among parents is such that the optimal situation, in which each parental chromosome can be followed without ambiguity in the offspring, is not attained. Rather, information is distributed in a random fashion among families. To use all available information would require following the transmission of all loci jointly in families.

As the number of loci considered increases, so does the number of possible parental configurations of alleles, or haplotypes, the number of possible genotypes, and the number of possible gene orders that may need to be contemplated (Table 5). Solutions must be found to keep computations manageable.

Restricting computation to all three-point tests may not solve this problem. The number of possible three-point tests increases rapidly with the number of loci. Such tests may not use the available information with

Table 5. How the Numbers of Possible Haplotypes, Genotypes, Gene Orders, and Three-point Tests Increase as a Function of the Number of Loci and the Number of Alleles at Each Locus

Loci	Alleles/ locus	Haplotypes	Genotypes	Gene orders	Three-point tests
2	2	4	10	1	–
	4	16	136		
3	2	8	36	3	3
	4	64	2,080		
4	2	16	136	12	12
	4	256	32,896		
5	2	32	528	60	60
	4	1,024	524,800		
7	2	128	8,256	2,520	105
	4	16,384	>10 ** 8		
10	2	1,024	524,800	1,814,400	360
	4	1,048,576	>10 **11		
15	2	32,768	>10 ** 8	>10**18	455
	4	>10 ** 9	>10 **72		

maximum efficiency: Four loci can provide a better resolution of the order of two intervening loci by providing flanking loci on either side; the relative ordering of two clusters of three loci, A-B-C and D-E-F, may be best resolved by considering more than three loci jointly. Lastly, experience shows that when all three-point tests are performed, the results of these tests may be inconsistent (Lathrop et al. 1986).

The efficiency of mapping can be significantly improved in several ways:

1. The use of a reference sample yields multilocus information on segregation.
2. Optimal family structures can provide maximum linkage information for a fixed sample size (White et al. 1985).
3. Alleles and haplotypes segregating in families can be recoded without any loss of information, to reduce the set of configurations that need to be considered in a sample (Ott 1978).
4. The computation of multilocus transmission can be reduced by mathematical transformations introducing independence between subsets of loci for each family tested.
5. A strategy can be adopted that reduces the number of gene orders that need to be considered to obtain the most likely order and the odds against competing alternatives.

Likelihood calculation and factorization. The use of three-generation reference families and the assumption of no interference allow the introduction of mathematical transformations that can drastically reduce the task of computing multilocus probabilities (Lathrop et al. 1986). For three diallelic loci, there are eight possible haplotypes, 36 possible genotypes, and therefore $36 \times 36 \times 36 = 46{,}656$ possible genotype configurations in a set consisting of two parents and one of their offspring. Some of these configurations are not compatible with Mendelian inheritance and can be eliminated.

Nevertheless, a large number may yet need to be considered, and the probability in a given family will be expressed as a function of the recombination rate in each of the two intervening segments, $f[r(1),r(2)]$. However, some transformations can be found that will introduce independence in the probabilities of the events occurring in each interval; that is, the function $f[r(1),r(2)]$ can be factored into the product of two new functions, $g[r(1)]\ h[r(2)]$. As a consequence, the three-locus problem is reduced to two sets of two-locus subproblems, within each case four possible haplotypes and only $4\times4\times4=64$ possible genotypic configurations in a father-mother-child set. Computation will be eased and yet no approximation is involved; the solution obtained after factoring is exact.

A number of transformation and factorization rules have been found that can break down a multilocus problem into a number of smaller subproblems. For each gene order and each family, a particular set of transformations can be found. They involve transformations of the genotypic data and of the parameters used to describe transmission, as presented elsewhere (Lathrop et al. 1986). A summary of these rules is given in Figure 5, together with an example (Fig. 6) of an application to a family for an eight-locus situation.

Mapping strategies. Sixteen loci and more than 20 loci have been characterized for the construction of preliminary maps of chromosomes 12 (O'Connell et al. 1985) and X (Drayna and White 1985), respectively, with 10^{13} and 10^{18} possible orders of these loci. Were the computation of the full multilocus problems manageable, the number of possible orderings would remain overwhelming. Therefore, a strategy becomes necessary to investigate the network of all possible orders without having to consider each one singly. Although there is no unique solution to this problem, heuristic schemes can be designed that will yield the best supported order and assess its odds against its prime alternatives without having to consider all possibilities.

An example is provided by the preliminary map of chromosome 13, where only 9 loci were to be mapped (Leppert et al. 1986; White et al., this volume). Pairwise tests indicated no recombination between two loci in two instances, allowing us to reduce the set of loci to be mapped to 7 after these two pairs were combined into haplotypes. Moreover, pair-wise tests also revealed close linkages between loci 1 and 2 and loci 5 and 6, respectively, and loose linkage of locus 7 to all other loci, as well as confirming the poor information content of locus 4, for which a heterozygous parent was observed in only five matings. The rationale for mapping consisted of postponing the consideration of loci 4 and 7 to concentrate on the resolution of the linear order of the 5 remaining loci. Rather than perform all 60 possible 5-locus tests, we first verified through three-point tests that any of the other loci were unlikely to fall in the interval defined by each pair of tightly linked

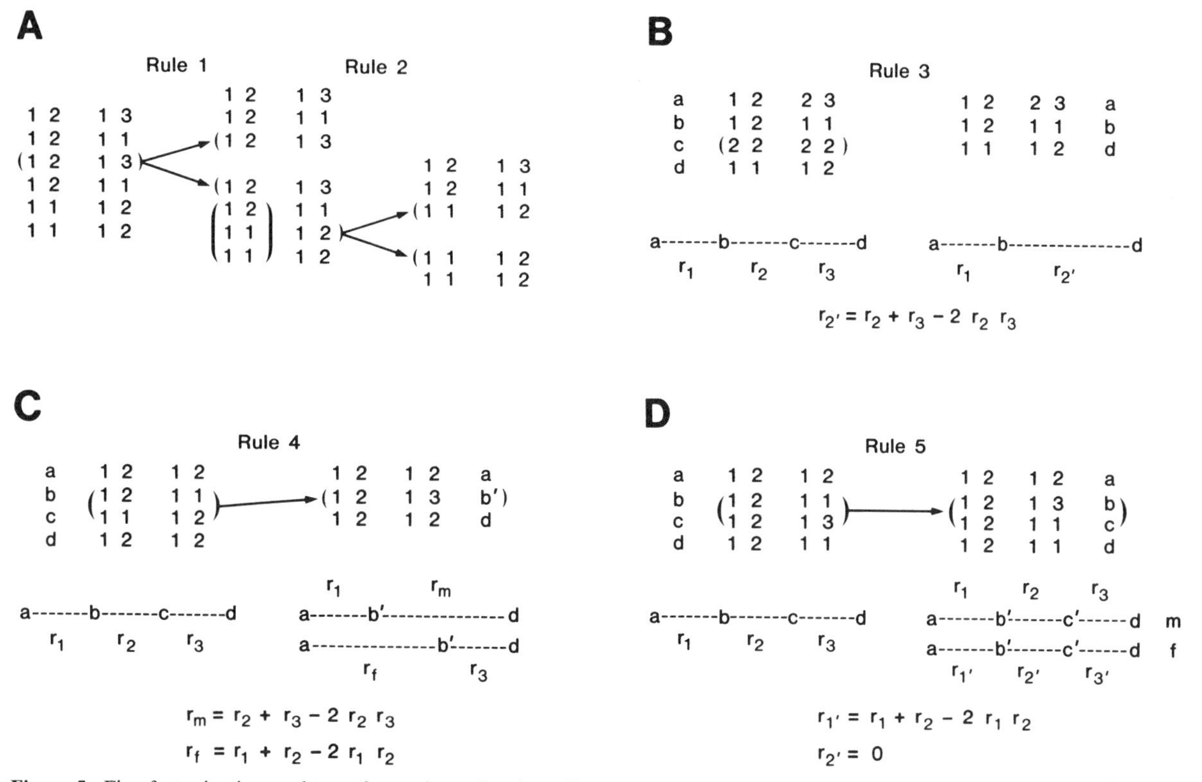

Figure 5. Five factorization and transformation rules that allow one to break down a multilocus problem into several smaller subproblems. Duplication, deletion, or permutation of genotypes at pivotal loci as a function of the observed allelic configuration in two parents and subsequent recoding of genotypes in the offspring and redefinition of recombination parameters allow considerable reduction of computational complexity.

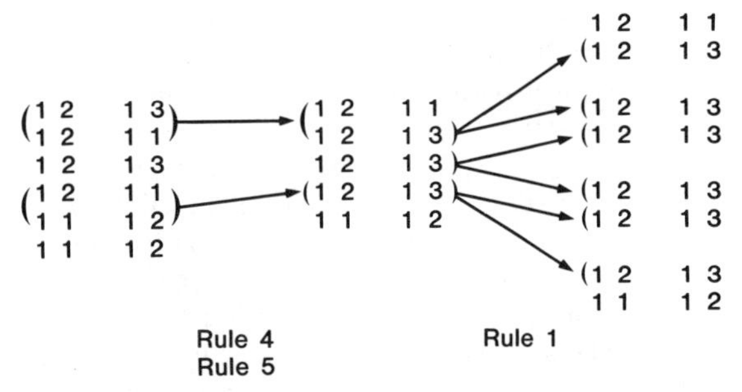

Figure 6. Example of a mating type where application of the rules summarized in Fig. 5 allow the reduction of a six-locus problem into four two-locus problems.

loci. The problem was consequently reduced to that of ordering three clusters of loci, (1,2), 3, and (5,6), involving only 12 five-point tests. The best supported order thus obtained and the odds for critical alternate orders are summarized in Figure 7.

Further developments in a general strategy to investigate the linear order of a large set of loci will yield a set of rules that can be incorporated into a production algorithm to automate the estimation of odds for alternative orders.

Adding a Locus to a Map

Once a preliminary genetic map is at hand, specifying a linear order and the recombination rates in each segment, powerful strategies can be developed to add a locus to a map or to investigate linkage between a Mendelian condition and a set of genetic markers. In both instances, greater power and efficiency can be achieved by assessing linkage between this additional locus and a whole linkage group than by considering each marker locus individually.

Location scores. Given the order and the estimated recombination rates between all loci of a linkage group, a test of linkage between an additional locus and the existing map can be performed by a method of location scores (Lathrop et al. 1984). The probabilities of joint transmission of the new locus with the members of the linkage group are expressed as a function of previously estimated recombination values as well as of a single unknown parameter describing the location of the new locus along the map. Dividing this term by the corresponding probability under the assumption of independent segregation of the new locus provides an assessment of the odds in favor of the new locus being located as assumed, as opposed to being indefinitely distant. As the location of the new locus is varied along the map, a curve is constructed that yields a continuous assessment of linkage of the new locus to all previously mapped loci (Fig. 8). If linkage is indicated within a given region of the map, tests of order or finer estimation of recombination rates in this region may be undertaken. In the absence of linkage, this approach allows exclusion of linkage within extended intervals of the genetic map (Farrall et al. 1986).

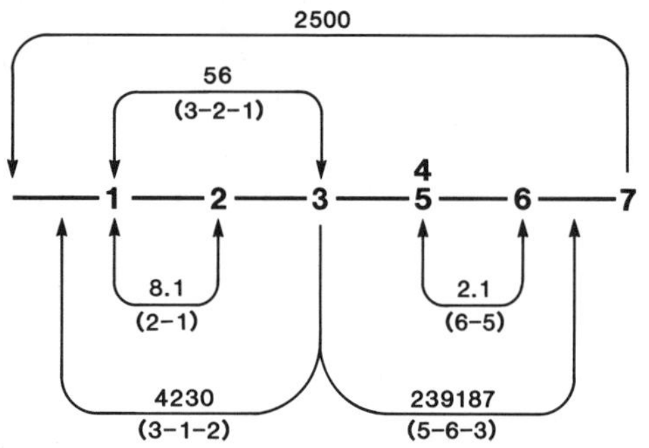

Figure 7. Odds for some critical alternate orders in a mapping problem on chromosome 13. (Reprinted, with permission, from Leppert et al. 1986.)

Chromosome distribution patterns. Another strategy (White et al. 1985) can be applied when the parental origin of alleles present in the offspring can be assigned unambiguously in the majority of informative families. Contrasting the allelic distribution at a new locus with these known parental allelic configurations provides a rapid method for assessing the approximate location of a new locus. Again, the map can thereafter be refined by reestimating recombination in the segment where the new locus is likely to reside. The efficiency of this strategy will increase as the genetic map becomes denser; this approach opens the way to a variety of rapid algorithms to assign loci to a predefined genetic map.

Applications to Clinical Disease

As the genetic map develops, it becomes an increasingly powerful way of assigning loci to chromosomes or smaller subregions. It offers unique opportunities to investigate the mode of inheritance of clinical conditions having unclear familial transmission, or to resolve etiological heterogeneity. When specific genes are presumed candidates in the etiology of a disorder, link-

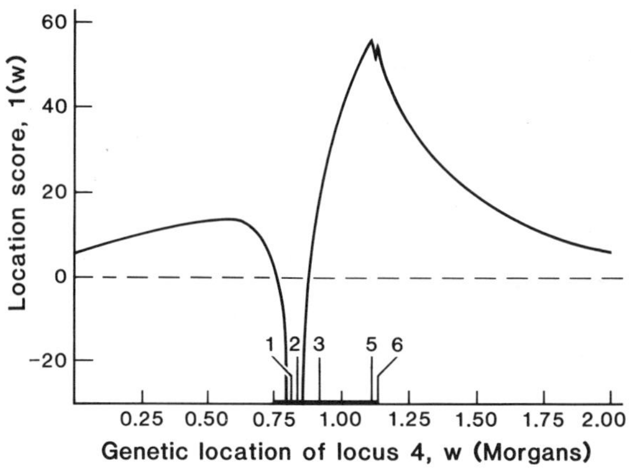

Figure 8. Relative odds on a natural logarithmic scale for linkage of an additional locus to a map of loci on chromosome 13 as a function of the location of the new locus along the map. (Reprinted, with permission, from Leppert et al. 1986.)

age provides a rapid, standard approach to confirm or refute this suspicion. Linkage with genetic loci involved in disease liability provides a valuable diagnostic tool as well as a handle for unraveling the molecular basis of a disease. Lastly, formal knowledge of the genetic and physical maps of human chromosomes is likely to yield extensive benefits of a more fundamental nature, as has been the case in all other organisms where detailed maps are at hand.

REFERENCES

Bailey, N.T.J. 1961. *Introduction to the mathematical theory of genetic linkage.* Clarendon Press, Oxford, England.

Bishop, D.T. 1985. Information content of phase-known matings for ordering genetic loci. *Genet. Epidemiol.* **2:** 349.

Bojko, M. 1983. Human meiosis. VIII. Chromosome pairing and formation of the synaptonemal complex in oocytes. *Carlsberg Res. Commun.* **48:** 457.

———. 1985. Human meiosis. IX. Crossing over and chiasma formation in oocytes. *Carlsberg. Res. Commun.* **50:** 43.

Conneally, P.M., J.H. Edwards, K.K. Kid, J.-M. Lalouel, N.E. Morton, J. Ott, and R. White. 1985. Report of the committee on methods of linkage analysis and reporting. *Cytogenet. Cell Genet.* **40:** 356.

Drayna, D. and R. White. 1985. The genetic map of the human X chromosome. *Science* **230:** 753.

Edwards, J. 1982. The use of computers. *Cytogenet. Cell Genet.* **32:** 43.

Farrall, M., P. Scambler, P. North and R. Williamson. 1986. The analysis of multiple polymorphic loci on a single human chromosome to exclude linkage to inherited disease: Cystic fibrosis and chromosome 4. *Am. J. Hum. Genet.* **38:** 75.

Haldane, J.B.S. 1919. The combination of linkage values, and the calculation of distance between the loci of linked factors. *J. Genet.* **8:** 299.

Hulten, M. 1974. Chiasma distribution at diakinesis in the normal human male. *Hereditas* **76:** 55.

Lathrop, G.M., J.-M. Lalouel, C. Julier, and J. Ott. 1984. Strategies for multilocus linkage analysis in humans. *Proc. Natl. Acad. Sci.* **81:** 3443.

———. 1985. Multilocus linkage analysis in humans: Detection of linkage and estimation of recombination. *Am. J. Hum. Genet.* **37:** 482.

Lathrop, G.M., J.-M. Lalouel, and R. White. 1986. Construction of human linkage maps: Likelihood calculations for multilocus linkage analysis. *Genet. Epidemiol.* **3:** 39.

Leppert, M., W. Cavenee, P. Callahan, T. Holm, P. O'Connell, K. Thompson, G.M. Lathrop, J.-M. Lalouel, and R. White. 1986. A primary genetic map of chromosome 13q. *Am. J. Hum. Genet.* **39:** 435.

Morton, N.E. 1955. Sequential tests for the detection of linkage. *Am. J. Hum. Genet.* **7:** 277.

O'Connell, P., M. Leppert, M. Hoff, E. Kumlin, W. Thomas, G.Y. Cai, L. Jerominski, M. Law, and R. White. 1985. A linkage map for human chromosome 12. *Cytogenet. Cell Genet.* **40:** 715.

Ott, J. 1978. A simple scheme for the analysis of HLA linkages in pedigrees. *Ann. Hum. Genet.* **42:** 225.

———. 1985. *Analysis of human genetic linkage.* Johns Hopkins University Press, Baltimore.

Renwick, J.H. 1968. Ratios of female to male recombination fractions in man. *Bull. Eur. Soc. Hum. Genet.* **2:** 7.

———. 1969. Progress in mapping human autosomes. *Br. Med. Bull.* **25:** 65.

Schwartz, D.C. and C.R. Cantor. 1984. Separation of yeast chromosome-sized DNAs by pulsed field gel electrophoresis. *Cell* **37:** 67.

Thompson, E.A. 1984. Information gain in joint linkage analysis. *IMA J. Math. Appl. Med. Biol.* **1:** 31.

von Wettstein, D., S.W. Rasmussen, and P.B. Holm. 1984. The synaptonemal complex in genetic segregation. *Annu. Rev. Genet.* **18:** 331.

White, R. and J.-M. Lalouel. 1986. Investigation of genetic linkage in human families. *Adv. Hum. Genet.* **16:** (in press).

White, R., M. Leppert, T. Bishop, D. Barker, J. Berkowitz, C. Brown, P. Callahan, T. Holm, and L. Jerominski. 1985. Construction of linkage maps with DNA markers for human chromosomes. *Nature* **313:** 101.

Mapping Complex Genetic Traits in Humans: New Methods Using a Complete RFLP Linkage Map

E.S. LANDER*†‡ AND D. BOTSTEIN†

*Whitehead Institute for Biomedical Research, Nine Cambridge Center, Cambridge, Massachusetts 02142; †Department of Biology, Massachusetts Institute of Technology, Cambridge, Massachusetts 02139; ‡Harvard University, Cambridge, Massachusetts 02138

It has been clear since the rediscovery of Mendel that humans obey laws of heredity identical with those of other organisms. The central features of Mendelism were observable in humans by following simply inherited common traits, including some diseases. However, the systematic study of human heredity using the standard concepts (complementation and recombination, tests of epistasis, etc.) has been infeasible in humans for two reasons: (1) because *Homo sapiens* is not an experimental animal that can be manipulated at will and (2) because few genetic markers had been found that were heterozygous often enough to allow random matings to be informative. The advent of recombinant DNA technology led to the suggestion (Botstein et al. 1980) that common polymorphisms in DNA sequence (conveniently observed as *r*estriction *f*ragment *l*ength *p*olymorphisms, or RFLPs) could be used as generally informative genetic markers allowing the systematic study of heredity in humans, including the construction of a true linkage map of the entire human genome.

The application of RFLP technology to simple Mendelian diseases has proceeded rapidly. RFLPs have been found closely linked to the autosomal dominant Huntington's disease (Gusella et al. 1983) and polycystic kidney disease (Reeder et al. 1985), the autosomal recessive cystic fibrosis (Knowlton et al. 1985; Tsui et al. 1985; Wainwright et al. 1985; White et al. 1985), and the X-linked recessive Duchenne muscular dystrophy (DMD) (Davies et al. 1983). In turn, closely linked DNA markers have made it possible to localize the disease genes to chromosomal regions (Gusella et al. 1983; Knowlton et al. 1985; Reeder et al. 1985; Wainwright et al. 1985; White et al. 1985), to undertake prenatal diagnosis of fetuses known to be at risk, and to begin chromosome walks in an attempt to clone the disease genes (Kunkel et al.; Worton et al.; all this volume). In addition, the RFLP linkages have shed light on the formal genetics of the disorders—showing, for example, that mutations at a single locus account for all (or almost all) cases of cystic fibrosis (Donis-Keller et al.; Tsui et al.; Williamson et al.; White et al.; all this volume) or of Huntington's disease (Gusella et al., this volume). These successes make clear that RFLPs can be found linked to any common human disease that shows simple Mendelian transmission and is caused by a single genetic locus.

Much of what we want to know about human heredity, however, concerns traits whose underlying genetics is less favorable for analysis. Some apparently identical clinical conditions can result from mutations at any one of several genes—a circumstance called *genetic heterogeneity*. Some traits have *incomplete penetrance*, with only a fraction of those carrying the appropriate mutant genotype actually displaying the trait. Conversely, some genotypes predispose individuals to a disease, but those of normal genotype may be affected as well, just at a lower rate. Environment may play an important role in the expression of the trait. In addition, *gene interactions* can occur, in which a phenotype results from the interaction of alleles at more than one locus. These complexities can underlie even the most clear-cut phenotypes.

These complex modes of inheritance are common in genetically well-studied organisms, such as bacteria, yeast, nematodes, and fruit flies, and evidence is accumulating that humans are no different. Geneticists usually surmount the problems by isolating pure-breeding stocks, each carrying a mutation at a single locus, and then arranging crosses at will. In the case of human genetics, however, we must take crosses as we find them.

Unfortunately, even with RFLPs as markers, traditional methods for genetic linkage analysis are quite inefficient at mapping genetically complex traits: A prohibitively large sample may be required before linkage can be detected or other genetic properties tested. Furthermore, traditional linkage analysis absolutely requires families with two or more affected individuals. Yet, many traits of biological or medical interest are quite rare, with most cases being sporadic. Collecting enough pedigrees with multiple affected subjects may be impossible.

It may thus appear, at first sight, that the vast majority of human heredity must remain refractory to genetic mapping, due to complexity or rarity. On the contrary, it is the main thesis of this paper that this need not be so. The important point is that studying the segregation patterns of a large number of mapped RFLPs simultaneously (rather than one at a time) can make it feasible to map many traits that are genetically complex and/or rare. We discuss below new mathematical techniques, unusual genetic resources, and clinical approaches that can potentially be brought to bear on the problem.

The RFLP Linkage Map

More-powerful methods of genetic analysis all require a linkage map of RFLPs covering the entire human genome. Such a genetic map was proposed at a time when only a few human RFLPs were known to exist, but the subsequent isolation of more than 1000 RFLPs (Willard et al. 1985; Donis-Keller et al., this volume), some as polymorphic as the HLA antigen system, in less than a decade has made clear the feasibility of the map. Indeed, construction of such a map is already well underway as part of an international collaboration (J. Dausset, pers. comm.). The eventual existence of a human RFLP map thus seems assured.

We shall assume, for simplicity, the availability of "perfect" RFLP linkage maps. By this, we mean linkage maps of RFLPs evenly spaced throughout the genome (e.g., one every 20 cM in a human genome of 3300 cM), with each RFLP so highly polymorphic that it is rarely found homozygous. Obtaining roughly even spacing presents no difficulty, although it obviously requires sorting through more than the bare minimum number of RFLPs. On the other hand, assuming that each RFLP is so highly polymorphic is unrealistic. However, if the RFLPs are informative only half the time, for instance, one can compensate by doubling the number used. (This trade-off between marker density and informativeness is accurate to a first-order approximation for family studies.) Thus, the actual number of RFLPs needed to simulate a perfect 20-cM map will vary with their informativeness.

Linkage Analysis via Likelihood Ratios

One cannot simply "count recombinants" in a human cross, because of uncertainty about the genotype of individuals or, especially, about the "phase" of markers in the parents (i.e., whether the markers are *cis* or *trans* to each other). Human linkage analysis is therefore done by the method of likelihood ratios (Fisher 1935; Haldane and Smith 1947; Morton 1955). One compares the probability that the observed data would arise under one hypothesis (e.g., linkage at 10% recombination between two markers) with the probability that it would arise under an alternative hypothesis (typically, nonlinkage). The ratio of these probabilities is called the *odds ratio* for one hypothesis relative to the other. Odds ratios from independent families can be multiplied together; when the odds become overwhelming, linkage is considered proven. By convention, one requires that the pedigree data yield an odds ratio of 1000:1 in favor of linkage over nonlinkage. (This threshold is less stringent than it may appear. A priori, the odds are 50:1 against any two loci being linked. Thus, even though the data may yield an odds ratio of 1000:1, the actual a posteriori odds in favor of linkage are only about 20:1, corresponding to the 5% confidence level.) Finding 100:1 odds against linkage is the conventional threshold for rejecting linkage. Once linkage is established, one estimates the recombination fraction as the value θ at which the likelihood ratio is largest—the so-called *maximum likelihood estimate*.

For convenience, human geneticists work with the $\log_{10}$ of the odds ratio, called the *LOD score*. LOD scores from successive pedigrees are thus added until the score grows to 3 (signifying linkage) or falls to -2 (indicating nonlinkage). Figure 1a illustrates the simple case of a phase-known meiosis with a dominant trait and a RFLP marker locus. If the loci are actually linked at 10%, then with probability 0.9 the LOD score will be $\log_{10}(9/5) = +0.25$, and with probability 0.1 it will be $\log_{10}(1/5) = -0.70$. The *expected LOD score*, or ELOD, will be $+0.16$. For the LOD score to reach the threshold of 3 required to demonstrate linkage, one would need to study about 19 ($\approx 3/0.16$) such meioses. If the loci are actually unlinked, the ELOD will be -0.22 and about 9 ($\approx -2/-0.22$) meioses will suffice to show nonlinkage. In this simple example the LOD score approach is hardly necessary, but it can be extended straightforwardly to exceedingly complicated pedigrees plagued by many uncertainties, with likelihoods of the pedigrees under competing hypotheses being calculated by computer.

Interval Mapping

With a complete RFLP linkage map of the human genome, one can test much more precise hypotheses than one can study using only unmapped marker loci. Because the hypotheses are more demanding, they are easier to prove, if correct, or to disprove, if not. The simplest example, which we call *interval mapping*, is illustrated in Figure 1b. In studying a dominant trait caused by a single genetic locus, we can compare the hypothesis that the trait maps in the middle of the interval between an adjacent pair of RFLPs with the hypothesis that it is unlinked to the interval. The hypotheses make very different predictions about the alleles we expect to see in an affected individual (see Fig. 1b). If the first hypothesis (linkage) is correct, the ELOD will be 0.22 per meiosis. Only about 13 meioses are thus needed to prove linkage, as opposed to 19 when a single unmapped marker is used. If the trait is unlinked, only 4 (as against 9) meioses are needed to show nonlinkage and allow one to move on to a new interval.

Intuitively, we can understand the power of interval mapping by thinking of two flanking markers as comprising a single "virtual" RFLP. This "virtual" RFLP is informative in meioses in which the flanking markers do not recombine and, in such meioses, it segregates as if it were very tightly linked to all loci in the interval: It will appear to "recombine" with a locus in the interval only when a double crossover occurs (the chance of which is small, even for relatively large intervals). In essence, we need to consider only two alternatives in searching for linkage: An interval is either

a. Using Single Markers

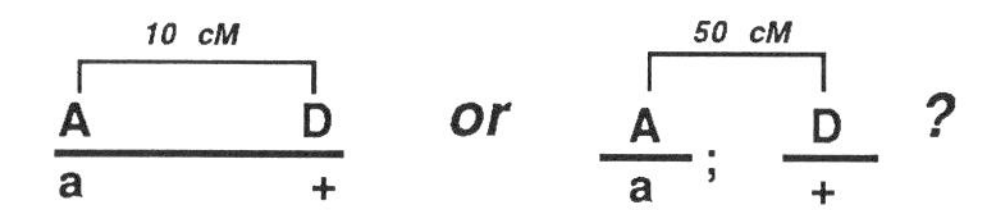

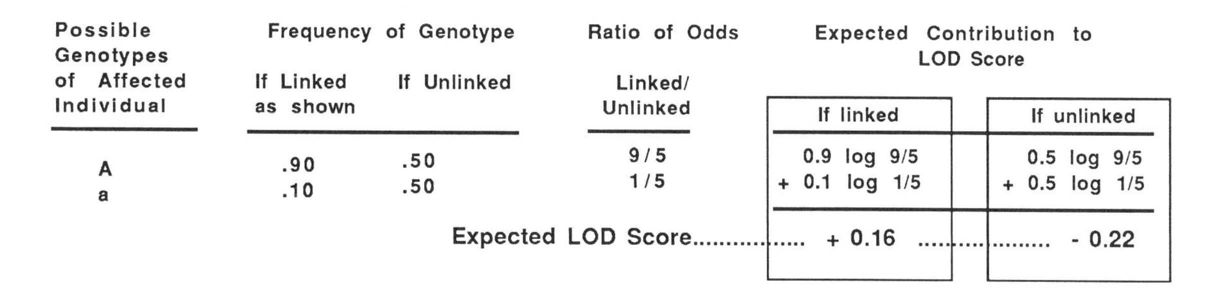

Possible Genotypes of Affected Individual	Frequency of Genotype: If Linked as shown	Frequency of Genotype: If Unlinked	Ratio of Odds Linked/Unlinked	Expected Contribution to LOD Score: If linked	Expected Contribution to LOD Score: If unlinked
A	.90	.50	9/5	0.9 log 9/5	0.5 log 9/5
a	.10	.50	1/5	+ 0.1 log 1/5	+ 0.5 log 1/5
Expected LOD Score				+ 0.16	- 0.22

b. Using Interval Mapping

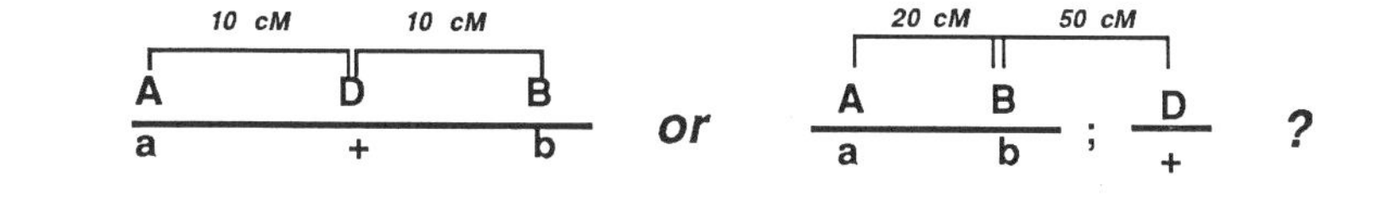

Possible Genotypes of Affected Individual	Frequency of Genotype: If Linked as shown	Frequency of Genotype: If Unlinked	Ratio of Odds Linked/Unlinked	Expected Contribution to LOD Score: If linked	Expected Contribution to LOD Score: If unlinked
A B	.81	.40	81/40	.81 log 81/40	.40 log 81/40
A b	.09	.10	9/10	.09 log 9/10	.10 log 9/10
a B	.09	.10	9/10	.09 log 9/10	.10 log 9/10
a b	.01	.40	1/40	.01 log 1/40	.40 log 1/40
Expected LOD Score				+ 0.224	- 0.520

Figure 1. Example of calculation of expected LOD score (ELOD), using a single genetic marker (*a*) and using interval mapping (*b*) in the case of a single phase-known meiosis and a dominant trait (D). In *a*, we compare the hypothesis that the trait is linked to the marker at 10 cM to the hypothesis that it is unlinked. In *b*, we compare the hypothesis that the trait lies between two markers known to be separated by 20 cM to the hypothesis that it is unlinked. (For simplicity, cM and recombination fraction are taken to be the same; no mapping function is used. In all other figures, the Haldane mapping function—corresponding to no interference—is used. Allowing for interference slightly increases the power of the methods.)

very tightly linked or else it is completely unlinked. This sharp dichotomy makes analysis easier and more powerful (even more so when genetic complications are involved, as we shall see).

In Figure 2, a and b, the number of families needed to map a simple Mendelian dominant or recessive trait using interval mapping is compared with the number needed using the traditional single-marker approach. The precise comparison depends on the spacing between consecutive RFLPs and (in the case of a recessive trait) the number of affected children per family, but one can expect to need, roughly, a sample only 40–60% as large. Clearly, interval mapping makes much more efficient use of the limiting resource, that is, families with affected members.

In planning a linkage study, it is prudent to collect more than just the expected number of families needed. Figure 2, c and d, therefore shows the sample required to ensure a 95% chance of detecting linkage, should it be present. It is striking to note (Fig. 2d) that a sample of families with three affected children, which is large enough to ensure a 95% chance of finding linkage using interval mapping, will not suffice to provide even odds of finding linkage by single-marker methods.

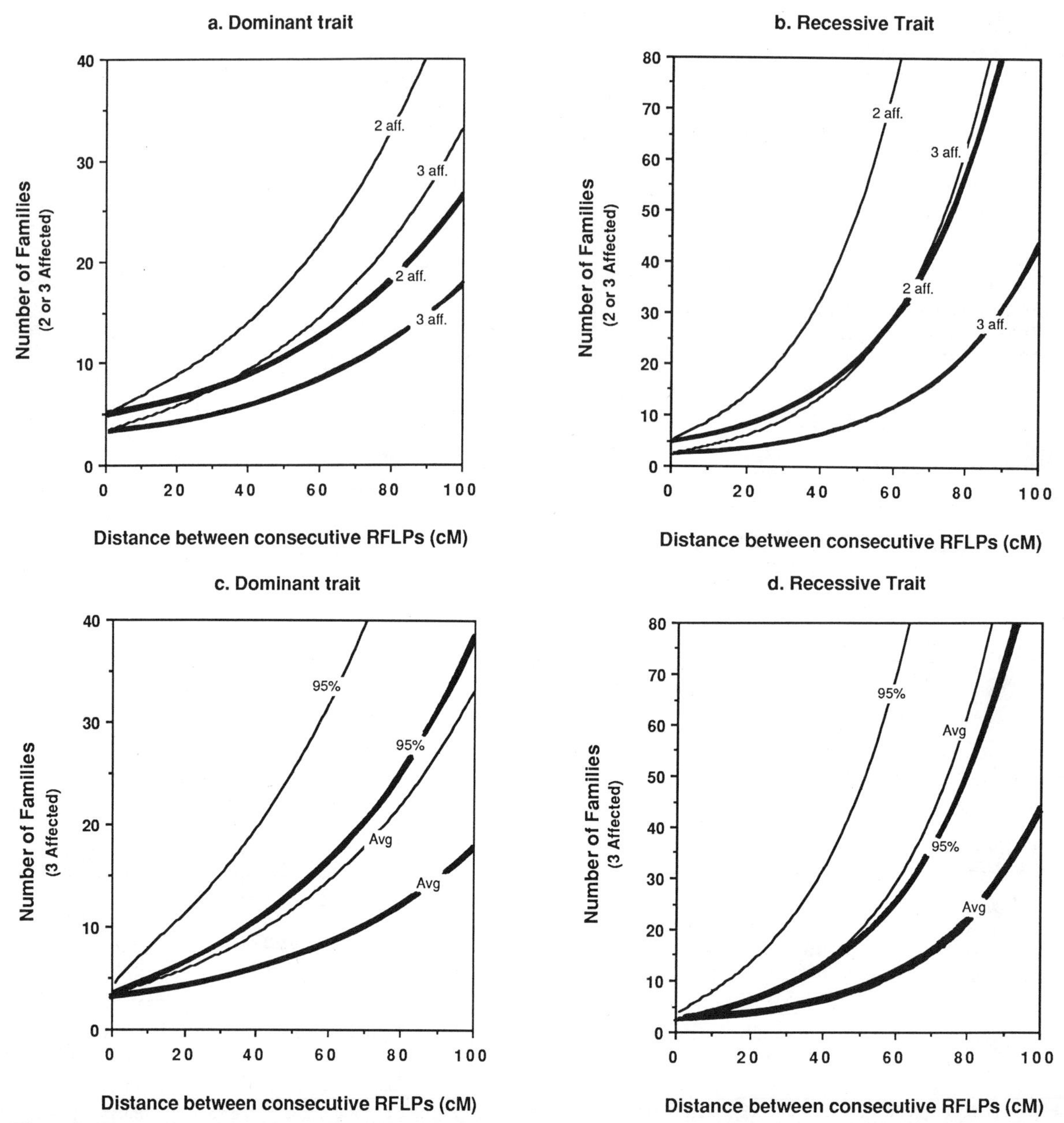

Figure 2. The number of families needed to map simple Mendelian traits with completely informative RFLPs—comparison of single-marker method (thin curves) with interval mapping (thick curves). Phase is assumed to be known for the RFLP markers and for a dominant trait, but the phase of a recessive trait is assumed to be unknown. For a dominant trait, the graphs show the expected number of meioses needed to reach a LOD score of 3.0 (*a*) and the number needed to ensure a 95% certainty of reaching a LOD of 3.0 (*c*). For a recessive trait, the graphs show the expected number of families of different types (classified by the number of affected sibs) to reach a LOD score of 3.0 (*b*) and the number needed to ensure a 95% certainty (*d*). The number of affected individuals in the family is relevant because the phase of the trait is presumed to be unknown in the recessive case; sibs thus contribute information about phase. In the phase-known dominant case, meioses can be treated independently.

Genetic Complexities

The current (sixth) edition of McKusick's (1983) catalog *Mendelian Inheritance in Man* lists 3,368 traits known or believed to mendelize to some extent. A few, like cystic fibrosis or Huntington's disease, exhibit perfect Mendelian transmission (and are now known to map to a single locus). Many, perhaps most, surely involve more complex genetics, which can greatly complicate the search for linkage. We consider the prospects and methods for mapping traits that (1) are genetically heterogeneous, (2) show incomplete penetrance, (3) amount to predispositions, or (4) involve gene interactions. More-detailed treatments will appear elsewhere (Lander and Botstein 1986 and in prep.).

Genetic heterogeneity: The method of simultaneous search. By far the most serious obstacle to linkage studies will be genetic heterogeneity, the situation of a phenotype that can be caused by mutations at any one of several loci. The classic paradigms of heterogeneity, well-studied in lower organisms, are (1) the interruption of a single biochemical pathway at any of its steps and (2) the loss or disruption of a heteromultimeric protein complex by mutation in the structural gene for any subunit, although other possibilities (including regulatory genes, posttranslational modifications, etc.) are not hard to imagine. For example, hereditary methemoglobinemia, once thought to be a homogeneous clinical entity, can be produced by mutations in either the α or β chains of hemoglobin or in NADH dehydrogenase (Stanbury et al. 1983). Elliptocytosis (Morton 1956) and Charcot-Marie-Tooth disease (Bird et al. 1983; Dyck et al. 1983; see also Ott 1985) are both known to be genetically heterogeneous, because in each case linkage to a marker has been seen in some large pedigrees but is absent in others. Xeroderma pigmentosum and ataxia telangiectasia are almost certainly genetically heterogeneous, since in vitro assays on cell fusions reveal nine and five complementation groups, respectively (Jaspers et al. 1982, 1985; Keijzer et al. 1979). In the early 1950s, two albinos married and produced normal children (Trevor-Roper 1952), confirming that albinism is heterogeneous (as had already been suspected from phenotypic distinctions and population genetics: i.e., much higher consanguinity among parents than expected for a single gene). There is also very good evidence that congenital deafness, congenital blindness, and coronary heart disease are heterogeneous in cause (Stanbury et al. 1983).

Heterogeneity has traditionally been the human geneticist's nightmare because evidence for linkage to a locus in one family will be offset by evidence against linkage in another family. Even a modest degree of heterogeneity may cause the traditional LOD score to be negative even for a marker tightly linked to one of the loci involved.[1] One solution is to concentrate only on a pedigree so large that linkage can be proven without the need for other evidence. Such situations are quite rare, as well as tedious to collect.[2]

Another solution is to allow explicitly for heterogeneity in the hypotheses to be tested, by supposing that only a certain fraction α of families are of the linked type. Such *admixture* methods (Ott 1985; Cavalli-Sforza and King 1986) have been used with single markers to detect linkage and/or test for heterogeneity. Single-marker admixture methods, however, are especially weak at detecting that a trait is heterogeneous: It is very difficult to distinguish whether (1) a marker is loosely linked to the disease-causing locus in all families or (2) it is tightly linked to the disease-causing locus in some families but unlinked in others.

We can modify the admixture method by incorporating interval mapping. Fewer families are required to detect linkage. The savings is of the same proportion as for a homogeneous trait, although it is more important here because the number of families involved is larger (Fig. 3a). Even more dramatic is the improvement in detecting the presence of heterogeneity—the number of families needed decreases from 5- to 50-fold in some typical cases (Fig. 3b). The reason is clear: Since interval mapping is essentially a test of very tight linkage (to a "virtual" RFLP), we no longer can confuse heterogeneity with loose linkage.

A complete RFLP map affords an even more powerful strategy for detecting linkage to a genetically heterogeneous trait. We call this approach *simultaneous search*, because it involves studying the segregation of several candidate loci simultaneously to see whether, as a set, they account for the transmission of the trait. From a mathematical point of view, we compare the likelihood of the hypothesis that in each family the trait cosegregates with one of the candidate loci (although which one may depend upon the family) with the likelihood of the hypothesis that the trait is unlinked to any of the loci. We examine in turn all pairs of loci (then triples, quadruples, etc.) to find a set that together explains the transmission in all of the families being studied. Because we are considering a larger number of hypotheses than usual, the threshold for the LOD score must be set higher than 3 to guard against false positives. A more detailed explanation of the mathematics appears in Lander and Botstein (1986).

From a more intuitive point of view, we are searching for a set of loci with the distinctive property that, in every family, one of the loci segregates as if it were tightly linked to the trait. Figure 4a illustrates a simplified situation in which the genome consists of nine intervals, A–I, and we have 20 isolated opportunities to observe a crossover between each interval and some dominant trait. If the trait were homogeneous, we would expect that some interval would show zero crossovers or maybe one (recall that interval mapping exhibits very tight linkage). The sample "observed" data are not consistent with this expectation. The data are consistent with the number of crossovers expected if the trait were sometimes caused by a locus in interval B and sometimes by one in interval F. However, the evidence is far from convincing. A better way to analyze the data is to ask how often the trait recombines with both of the two intervals: For the correct pair of intervals, B and F, we get the expected result of zero (Fig. 4b). Intuition suggests (and mathematics confirms) that this is more powerful evidence. This illustrates the basic principle behind simultaneous search, although in practice it is preferable to use families with several meioses rather than isolated meioses, as in the example.

We have calculated the number of families required to map heterogeneous diseases by simultaneous search (Fig. 4c). The results suggest that heterogeneous diseases caused by as many as five loci can be resolved with a feasible number of families.

[1]Accordingly, the practice of *exclusion mapping* (declaring that a trait is "excluded" from a region once a LOD score of -2 is reached) is of dubious value in the general case.

[2]Moreover, in each large pedigree one maps just one of the alternative causative genes, which may be a minor one in the population.

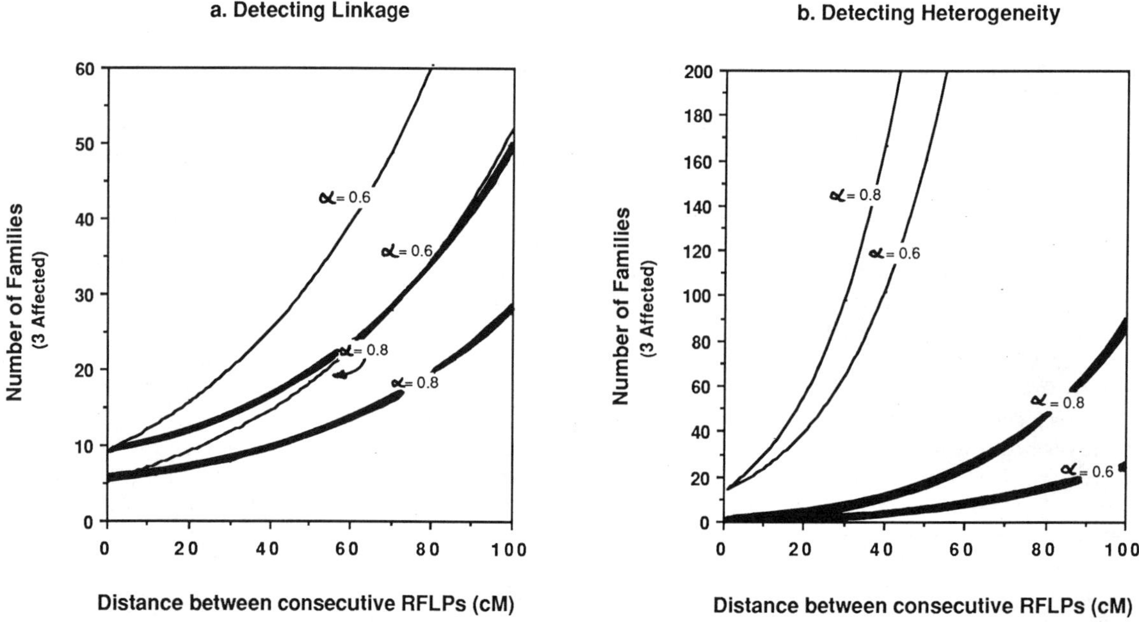

Figure 3. (*a*) The number of families with three affected sibs needed to map a heterogeneous dominant trait via admixture methods—comparison of single-marker method (thin curve) with interval mapping (thick curve). The curves pertain to a trait in which the fraction (α) of families segregate for a linked locus. (*b*) The number of families with three affected sibs needed to detect that a trait is heterogeneous via admixture methods—comparison of single-marker method (thin curve) with interval mapping (thick curve). Phases are assumed to be known.

The advantage of simultaneous search is that the set of loci that causes a trait often can be recognized before any single locus could be recognized by studying it alone. Simultaneous search offers no advantage if only one cause is a genetic locus, with other families showing the trait for nongenetic reasons, such as a virus. In such cases, admixture methods must suffice.

We should note that, although heterogeneity wreaks havoc with traditional linkage analysis, it is not even detectable by segregation analysis (the studying of pedigrees to see if they fit Mendelian patterns of transmission). No matter how many independent loci can cause a recessive trait, each family segregates the expected Mendelian 3:1 ratio. This is in sharp contrast to genetic traits showing incomplete penetrance. Incomplete penetrance disrupts the expected segregation ratios (making it hard to prove Mendelian transmission by segregation analysis), but it does not seriously complicate linkage mapping.

Incomplete penetrance: Only affected individuals contribute significantly. A genetic trait that requires additional factors to become manifest may indicate incomplete penetrance. These factors might include environment, genetic background, or chance. A simple example is retinoblastoma, in which a second somatic alteration leads to cancer in individuals carrying the dominant allele; the event is quite frequent, so penetrance is high (Macklin 1959; Cavenee et al. 1983). Wernicke-Korsakoff syndrome seems to be due to a mutation causing transketolase to bind thiamine pyrophosphate less avidly than normal, but a clinical phenotype (alcohol-induced encephalopathy) is apparent only in patients with a dietary thiamine deficiency (Blass and Gibson 1977). As this example illustrates, the degree of penetrance can vary widely with the environment and with the precise definition of the phenotype (encephalopathy vs. thiamine pyrophosphate binding). Finally, diseases with late age of onset must always be treated as incompletely penetrant.

When penetrance is incomplete, the genotype of the unaffected individuals in a sibship is in doubt. Not surprisingly, relatively little can be learned by studying the markers that they have inherited. In fact, the problem of the uncertain genotype of unaffected individuals is evident even for a recessive trait of complete penetrance. There are three possible genotypes for an unaffected sib, but only one for an affected sib. For a tightly linked marker then, the odds ratios will be 4:3 for an unaffected and 4:1 for an affected sib, making the contribution to the LOD score only one fifth as large for unaffected sibs. Even a small degree of incomplete penetrance creates a similar problem for dominant traits and further exacerbates it for recessive traits. Figure 5 shows how quickly the contribution to the LOD score plummets as the genotype of an unaffected sib becomes uncertain. Below about 80% penetrance, unaffected sibs are virtually useless. In assessing the value of a pedigree for linkage analysis then, it is usually accurate to look only at the affected sibs.[3]

Incomplete penetrance, however, does not confound

[3]It is nevertheless advisable to collect the unaffected sibs, where practical, since they may aid in determining the phases of the RFLP markers used for mapping.

a. How often does each interval segregate away from the disease?

Interval	A	B	C	D	E	F	G	H	I
Crossovers "Observed"	11	6	8	13	10	4	12	8	9
Crossovers Expected:									
if trait in B	10	0	10	10	10	10	10	10	10
if trait in F	10	10	10	10	10	0	10	10	10
if in B _or_ F	10	5	10	10	10	5	10	10	10

Total Crossovers Possible: 20

b. How often do _both_ B and F segregate away from the disease?

	A	B	C	*D*	*E*	*F*	*G*	*H*	*I*
A		3	5	*7*	*5*	*3*	*9*	*5*	*5*
B			3	*4*	*5*	**[*0*]**	*5*	*3*	*3*
C				*3*	*4*	*2*	*5*	*3*	*4*
D					*7*	*3*	*12*	*7*	*8*
E						*2*	*8*	*4*	*5*
F							*3*	*2*	*2*
G								*7*	*12*
H									*4*

c. Number of families needed to map a heterogeneous dominant trait

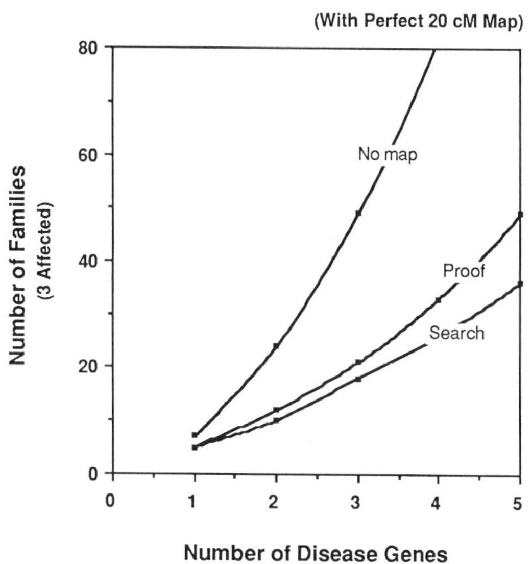

Figure 4. Illustration of the principle behind the simultaneous search method and the power of this approach for mapping heterogeneous traits. In *a*, a hypothetical genome is divided into nine intervals (A,B . . . I) flanked by completely informative RFLP markers. Twenty informative meioses are scored, resulting in the "observed" data for crossovers with a dominant disease gene shown in the second line. The number of crossovers expected is shown, under three different hypotheses about the location of the disease gene. The third hypothesis, that the disease is caused equally often by a gene in B and a gene in F, fits the data best, but the fit is not convincing. Instead, in *b*, we record the number of meioses in which the trait is seen to recombine with *both* of a pair of intervals. Here, the pair of intervals B and F stand out more markedly. The number of families with three affected sibs required to map the genes causing a heterogeneous recessive trait or a heterogeneous dominant trait is shown in *c* and *d*, respectively. In each case, the genes causing the trait are assumed to be equally frequent. The curve marked "Search" pertains to proving that the set of loci shows linkage according to simultaneous search; the curve marked "Proof" pertains to subsequently proving the involvement of each of the loci in the set. The curve marked "No map" pertains to using admixture methods alone. Phase is assumed to be known for RFLP markers and for dominant traits but not for recessive traits.

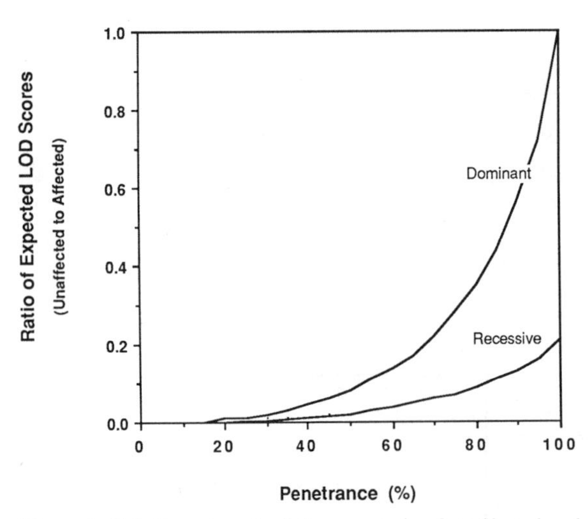

Figure 5. Relative values for linkage mapping for affected and unaffected sibs as a function of penetrance of a disease gene. Specifically, the graph shows the ratio of the ELOD for an unaffected sib to that for an affected sib, when a marker is available at 0 cM. All phases are assumed to be known.

linkage analysis in the serious way that heterogeneity does. The practical consequences are limited to the following:

1. Finding families with several affected sibs (and an affected parent, for a dominant trait) will be more difficult than if the trait were fully penetrant. A possible solution is to identify a more penetrant-related phenotype, a strategy we return to below.
2. Given the same set of pedigrees, however, a recessive trait of incomplete penetrance should not be significantly harder to map than a fully penetrant one. Unaffected sibs are of marginal value in both cases.
3. Given an affected parent, a dominant trait may require up to twice as many meioses be studied, since unaffected individuals no longer contribute. Families without an affected parent contribute less information, since the parental source of the disease gene is unknown.

Predispositions: The need to enrich for genetically caused cases. Genetic predispositions are not well understood in general, but some spectacular examples exist. Some rare families clearly segregate a dominant trait, causing at least a 100-fold increase in the risk for specific cancers (McKusick 1983). Homozygotes for a mutant allele of the gene for α_1-antitrypsin show a 20-fold higher risk of emphysema than normal (Stanbury et al. 1983), although it should be noted that most cases of emphysema are due to cigarette smoking, not α_1-antitrypsin deficiency. The relative risk of developing type-1 diabetes is 9-fold higher in individuals with HLA genotype *Dw3/Dw4*; similarly *DQ* alleles have been associated with myasthenia gravis (Stanbury et al. 1983; see also Bell et al., this volume). Genes involved in predispositions may identify steps in the pathway of pathogenesis.

Mapping a predisposition to a common disease may be difficult because of the need to discern the "genetic" families (those segregating the predisposition) from the "nongenetic" families (those with more than one affected individual just by chance). Indeed, *most* families with two affected members may be due to *non*genetic causes if the predisposition is weak, the genotype causing it is rare, or the disease is common in the general population. For example, since one in four Americans develops cancer over a lifetime, two sibs with cancer hardly suggests that the family segregates for a predisposing locus. Linkage analysis can be undertaken only if one can enrich for families with a genetic cause, amid an excess of nongenetic cases (cf. Fig. 3a concerning the admixture of causes).

The best way to enrich is to seek families with many affected individuals: If nongenetic cases occur independently in the general population, then a cluster of, for instance, four or five cases more likely will be due to inheritance rather than chance.

To illustrate the point intuitively, suppose that some families segregate a dominant trait causing a 20-fold increased risk for a disease. The chance that a child in such a family develops the disease is 10-fold higher than for a normal family (a 50% chance of inheriting the trait multiplied by a 20-fold increased risk); the chance of two children developing the disease is 100-fold higher than for a normal family, and the chance of three children in the family is 1000-fold higher.[4] Thus, even if genetically predisposed families represent only 1% of the population, they will comprise more than 90% of the families with three affected members.

The real difficulty occurs if nongenetic cases do not occur independently (e.g., due to common environmental exposure). Finding multiple cases in a family does not then ensure as great an enrichment. Because of this, in designing a linkage study, it will be important to exploit what is known about the epidemiology of the disease in order to enrich most effectively (e.g., choosing families without obvious environmental risk factors).

In many cases, it should be possible to map genes causing predispositions of at least 10-fold (calculations not shown). Weaker predispositions, however, will likely be very difficult to map.

Gene interactions. Finally, traits may result from genetic interactions between alleles at several loci. Examples abound in other organisms but have been hard to elucidate in humans. It is known, though, that α-thalassemia partly suppresses β-thalassemia. It is also suspected that at least some diseases that show partial association with HLA genotype may involve other loci as well.

The most fundamental gene interaction is a *synthetic trait*, one caused only when mutant alleles are present

[4]The exact risks also depend slightly on the number of unaffected members in the family and the incidence of the disease in the general population, but we ignore this for the sake of illustration.

at all of several loci (to be distinguished from a heterogeneous trait, which is caused by mutations at any of several loci). At some of the *component loci*, the alleles may act dominantly; at others, they may act recessively. Some may be rare; others may occur at a high frequency. The incidence of the trait in the population will reflect the product of the frequencies of the component loci.

By far the most difficult case for linkage mapping is a component locus at which the allele (mutant) frequency is very high (e.g., 50%): A large fraction of parents of affected individuals will be homozygotes at the locus but will be unaffected themselves since they lack other components. Since an affected child may have inherited either chromosome from such a parent, the trait will *not* map to the locus in such a family. Analytically, it is the same as admixture with a large proportion of families having a nongenetic cause (cf. Fig. 3a above).

On the other hand, the trait will show clear linkage to each component locus at which the allele frequency is not too high (e.g., less than 10%). Linkage analysis is relatively straightforward:

1. Recessive components involve no special considerations and require the same number of families as for a simple recessive trait (Fig. 2a).
2. Dominant components are slightly harder, since one will typically not know which parent contributed the mutant allele. The number of families needed to map a dominant trait under these circumstances is shown in Figure 6.

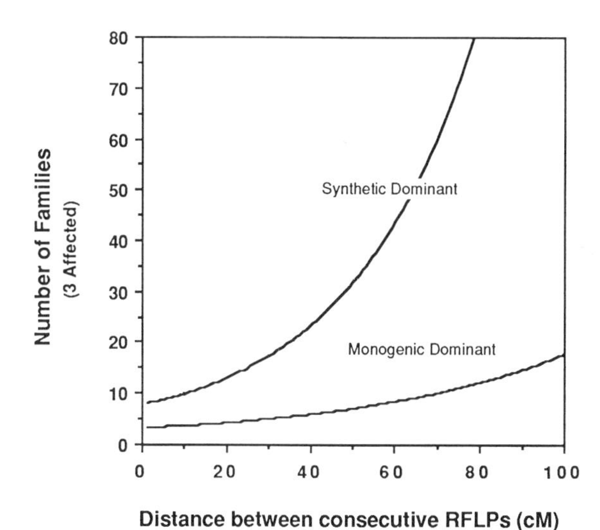

Figure 6. The number of families needed to map a dominant component allele involved in causing a synthetic trait via gene interaction. Which parent contributes the allele, as well as the phase of the allele, is assumed to be unknown. The phase of RFLP markers is assumed to be known. The allele is assumed to be present at less than 5% in the general population, so that one parent may be assumed to be a heterozygote and one parent homozygous normal (see text). For comparison, the number of families needed when a trait is caused monogenically by a dominant allele is also shown.

Component loci with intermediate allele frequency (10–50%) require special precautions and careful selection of families. For example, contrary to intuition, a very large proportion of affected individuals in a family may be *un*desirable since they may enrich for parents who are homozygotes. (We shall elaborate on these issues elsewhere.) Overall, it appears that component loci involved in synthetic interactions should be mappable if the frequency of the mutant allele at the locus is under 20%.

More-complex genetic interactions can be analyzed in a similar fashion. *Multifactorial inheritance* (which usually refers to additive interactions between a number of loci) can be modeled as involving a combination of heterogeneity and synthetic interactions. A detailed treatment of these issues, however, is beyond the scope of the present work.

Mapping without Family Studies

For many of the genetic disorders in McKusick's catalog, it will be difficult to collect enough families, each with several affected members, for traditional linkage analysis. The disorders are rare, occurring sporadically in isolated cases or small clusters.

We propose two methods for linkage analysis when family studies are not sufficient or not possible at all. *Homozygosity mapping*, appropriate for a recessive disease, exploits the unusual genetic constitution produced by inbreeding. *Disequilibrium mapping* attempts to use special features of the population genetics of genetic isolates.

Homozygosity mapping: Inbred individuals contain a surprising amount of information. In his classic study on inborn errors of metabolism, Garrod (1902) observed that an unusually high proportion of patients with alkaptonuria were progeny of consanguineous marriages. Almost immediately, Bateson (1902) proposed a Mendelian explanation: "The mating of first cousins gives exactly the conditions most likely to enable a rare, and usually recessive, character to show itself." The rarer the disease gene, the more pronounced will be the frequency of inbreds among those affected. Consanguinity was found in 82% of all cases of Brazilian type achiropody (McKusick 1983) and at least 25% of all cases of Tangier disease (Stanbury et al. 1983). By contrast, consanguineous marriages account for only about 3% of cases of cystic fibrosis, for which about 1 in 20 people is a carrier (Romeo et al. 1985).[5]

Children of first-cousin marriages provide a very sensitive assay for the location of a recessive disease gene: They will be homozygous by descent not simply at the disease locus, but throughout an entire chromo-

[5]Of course, the rate of consanguinity among parents of affected children depends on the rate of consanguineous marriages in the general population, which varies widely between different countries.

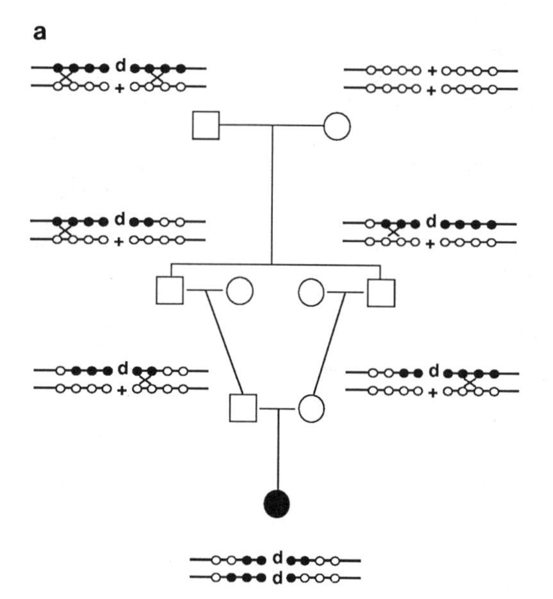

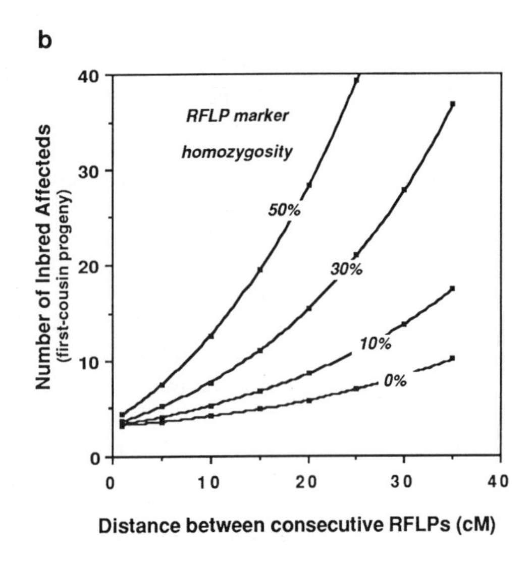

Figure 7. Illustration of the principle behind homozygosity mapping and the power of this approach for mapping a recessive trait. In *a*, the pedigree of a first-cousin marriage is given, showing the descent of marker alleles on the chromosome bearing the recessive disease allele. Homozygosity by descent for a contiguous stretch of markers in the affected child is shown. In *b*, the expected number of affected first-cousin progeny needed to map a recessive trait (i.e., to reach a LOD score of 3.0) as a function of the marker density of a RFLP map is shown. The various curves pertain to RFLPs of different degrees of informativeness: ones that are found homozygous 50%, 30%, 10%, and 0% of the time in the general population.

somal region nearby[6] (Fig. 7a). Of course, a region unrelated to the disease might become homozygous by descent in the inbred child simply by chance, but such regions will differ from child to child. Since the chance of an unrelated region becoming homozygous by descent is only 1/16 (the *coefficient of inbreeding* for a first-cousin marriage), then if a region is seen to be homozygous by descent in three unrelated inbred children, the odds are 16^3:1 (i.e., 4096:1) that it contains the disease gene. In principle then, three inbred children suffice to map a homogeneous recessive disease.

All we need to carry out this strategy, which we call *homozygosity mapping*, is an assay for whether a region is homozygous by *descent*. A RFLP map provides such an assay: We check whether all the RFLPs in a stretch of the genome are homozygous. The power of homozygosity mapping depends on the accuracy of this assay, which in turn depends on the spacing between consecutive RFLPs and their degree of polymorphism (specifically, the chance that a RFLP will be found homozygous in the general population just by chance). Using a 5-cM map of RFLPs that are homozygous 50% of the time in the population, one can map a recessive trait based on about eight inbred children (Fig. 7b). A 10-cM map of RFLPs homozygous 30% of the time, or a 20-cM map of RFLPs homozygous 10% of the time, would be equivalent. Doubling the density of the RFLPs cuts the requirement to about five children in each case.

Using homozygosity mapping with a high-quality RFLP map, a single affected child born of first cousins is roughly as informative as a nuclear family with three affected children.[7] The advantage is that, for many traits, inbred children will be easier to locate than families with three affected children. Of course, there is no need to choose between homozygosity mapping and traditional family studies; information can be combined by simply adding the LOD scores. In those cases where families with three or more affected sibs are particularly helpful (such as simultaneous search for a heterogeneous trait), inbreds should similarly prove valuable.

Children of second-cousin marriages are roughly as informative. (The probability that an unrelated region of the genome is homozygous by descent is even smaller, 1/64. On the other hand, there is a greater chance that markers near the disease locus will have recombined and/or that the disease will not be the fault of inbreeding.) For more distant consanguinity in a large population, the probability that the disease is due to homozygosity by descent (rather than random chance) falls off exponentially, making such cases much less powerful for homozygosity mapping.

Disequilibrium mapping: Recent genetic isolates may contain enough information to contribute to linkage analysis. When a fresh mutation occurs and spreads through a population, it tends to be coinherited along

[6]There is a chance that consanguinity is not the cause of the genetic disorder, but for first-cousin marriages it is small, unless the recessive allele is very common. This possibility must be included in the analysis.

[7]Two affected sibs from a first-cousin marriage are equivalent to a family with four affected sibs, and so forth.

with whatever alleles were present on the ancestral chromosome at nearby loci. Eventually, the alleles at nearby loci will be randomized by the process of recombination. But, at earlier times, one will observe *linkage disequilibrium*, an excess of one allele on chromosomes bearing the mutation. As Bodmer points out in his introduction to this volume, linkage disequilibrium should be detectable in the general population only for markers within about 0.1 cM (or about 100 kb of DNA) and then only if the mutation arose only once. Thus, searching for linkage disequilibrium in the general population will be a fruitless strategy for detecting linkage in the first instance (although disequilibrium may be useful once a locus has been narrowed down to a 1-cM interval).

There are some special situations, however, in which *disequilibrium mapping* may be possible. Some isolated populations show unusually high rates of certain genetic diseases, due to the so-called *founder effect*: The mutation was present in one of the original founders and spread by genetic drift while the population was still small. Finland is a particularly well-studied example. Congenital chloride diarrhea, congential nephrotic syndrome, cornea planta congenita recessiva, and neuronal ceroid lipofuscinosis all occur at higher rates in Finland than in the rest of the world; the clustered ancestry of cases (Fig. 8a) also supports a founder effect in isolates formed about 300 years ago (Norio et al. 1973; A. de la Chapelle, pers. comm.).

The potential power of disequilibrium mapping depends on several factors: (1) the number of chromosomes descended from the founder available for study; (2) the number of generations since the isolate was founded; (3) the spacing between mapped RFLP markers; and (4) the informativeness of the RFLP markers in the isolated population. Figure 8b illustrates the case of a recessive disease, inherited from a founder 10 generations ago, that now affects 15 individuals (thereby providing 30 chromosomes to examine for disequilibrium). One could expect to map the disease by disequilibrium mapping (i.e., attain a LOD score of 3) if one had an 8-cM map of RFLPs that were 50:50 two-allele systems, a 12-cM map of RFLPs that were three-allele systems, or a 17-cM map of RFLPs that were five-allele systems in the isolated population.

Disequilibrium mapping, alone among the methods discussed here, depends on population genetics rather than simple transmission genetics. Thus, it must be used with special care. It is crucial, for example, that the RFLP allele frequencies found in affected individuals be compared to the frequencies expected for the *isolate*, not for the general population. In passing through the bottleneck when an isolate is founded, allele systems can become either more or less polymorphic due to genetic drift. Populations that passed through very narrow bottlenecks will probably be too nearly monomorphic; HLA and other highly polymorphic systems should provide a helpful indication of this. Test cases will be needed before the power of disequilibrium mapping can be fully assessed.

How Good a Map Is Needed?

How good a RFLP map of the human genome is needed? Before answering this question, we should emphasize that we have been considering "perfect" RFLP maps, that is, ones with fully informative markers evenly spaced along the human genome. In reality, few fully informative RFLPs have been discovered, making it necessary to use many more partially informative

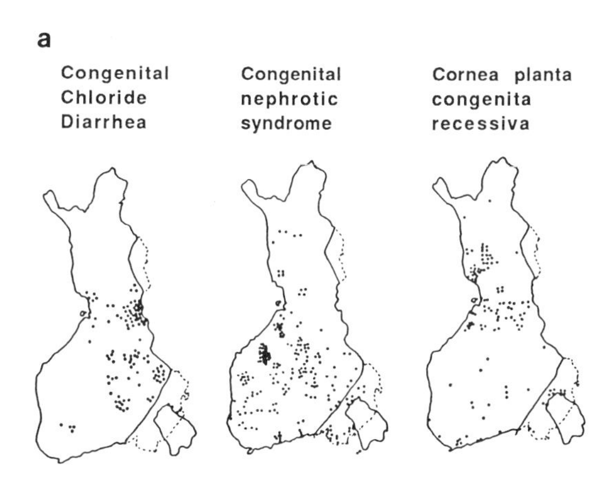

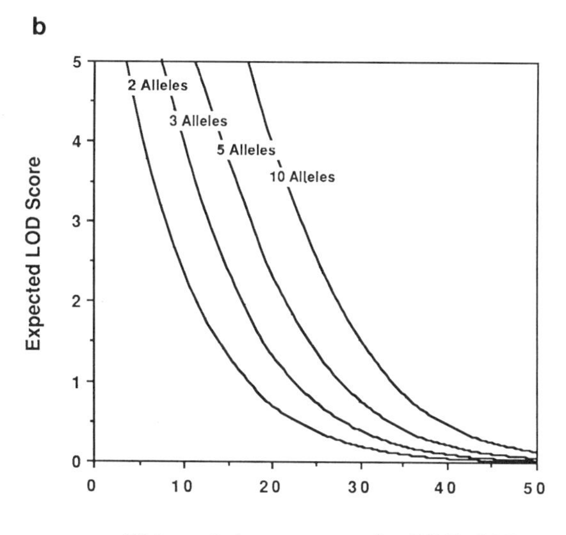

Figure 8. Disequilibrium mapping. (*a*) Maps of Finland showing geographic clustering of inherited diseases, presumably due to a founder effect. (*b*) The LOD score expected for an isolate in which 30 chromosomes, descended from an ancestor 10 generations ago, can be studied. The graph plots the LOD score against the marker density of a RFLP map. The various curves pertain to the informativeness of the markers: RFLPs consisting of 2, 3, 5, and 10 equally frequent alleles are illustrated.

RFLPs to approximate a "perfect" RFLP map of any given density.

With this caveat in mind, how good a "perfect" RFLP map of the human genome is needed? The answer varies with the application:

1. For mapping simple homogeneous Mendelian traits, a perfect 40-cM RFLP map should suffice (Fig. 2a,b). Using a perfect 20-cM map would reduce the number of families needed by about half.
2. For more complex genetic traits, using a perfect RFLP map with spacing of 20 cM or less seems advisable (Fig. 4c). This should decrease the number of families needed for these more complex cases to a more manageable level. Using a perfect 10-cM map would reduce the number needed by about half.
3. For homozygosity mapping and disequilibrium mapping, the equivalent of perfect RFLP maps of 10 cM or less are advantageous (Figs. 7b and 8b).

Thus, there are uses for RFLP linkage maps at all scales. That it is feasible to make RFLP maps of any desired resolution has been amply demonstrated by Donis-Keller and colleagues with their recent discovery and mapping of 50 RFLPs on chromosome 7, a genetic length of 140 cM (Donis-Keller et al., this volume).

Despite this achievement, the equivalent of even a perfect 40-cM RFLP map is not yet available for the vast majority of the human genome. The collection of many more RFLPs remains a major task ahead and it will be important to concentrate on finding ones that are highly informative. In fact, fewer than 200 of the 1000 reported RFLP markers come close to being fully informative (Donis-Keller et al.; White et al.; both this volume). The alternative to highly informative markers—using a much greater number of poorly informative ones—may be mathematically equivalent but is clearly inferior from the point of view of the laboratory work involved in applying the map to any specific trait.

DISCUSSION

Even though a complete RFLP map of the human genome is still some years in the future, it has become clear that such a map will be a powerful tool both for the study of human biology and the practice of clinical genetics. We have summarized above ideas and calculations that show the power of a RFLP map in mapping loci that cause, or contribute to, human traits and diseases. The main conclusions are:

1. Simple Mendelian traits can be mapped using fewer families with a RFLP map than is possible using markers one at a time.
2. In some cases, traits can be mapped without family studies at all, but only with a reasonably complete RFLP map.
3. The genetic contributions to diseases or traits of surprisingly complex etiology will be accessible to study, but, again, only with a reasonably complete RFLP map. In particular: (a) Mapping heterogeneous traits, caused by any of up to four or five different loci, should be possible, but will require a larger number of families. If more than five genes are involved, the task grows much more difficult. (b) Incomplete penetrance is not a major obstacle to linkage analysis, unless penetrance is so low that clustered cases are rare. (c) Mapping predispositions to common diseases should be possible when one can enrich the sample for families with genetic involvement. (d) It should be possible to map two or more loci that interact to cause a trait, except for those component loci at which the trait-causing allele occurs at very high frequency in the population.

Concerning the kinds of genetic resources best suited to analysis with a RFLP map, some conclusions can be drawn as well:

1. Although the traditional focus has been on very large pedigrees, families with two or more affected sibs can also suffice for linkage analysis of a simple Mendelian trait using a RFLP map. For heterogeneous traits, it is preferable to seek families with at least three affected.
2. The value of a family depends crucially on the accuracy of the diagnosis in each member asserted to be affected. Misdiagnosis can seriously hamper linkage analysis.
3. Unaffected family members contribute only slightly to linkage analysis (except for dominant traits displaying full penetrance). Nevertheless, they should be collected when feasible, because they aid in determining the phase of the RFLP markers.
4. Since the amount of DNA eventually required from each individual is not entirely predictable in advance (depending as it does on the complexity of the genetic situation), the practice of "immortalizing" peripheral blood lymphocytes with Epstein-Barr virus (Anderson and Gusella 1984) is recommended. This relatively simple procedure ensures an essentially unlimited supply of DNA and ensures that each donor has to be sampled only once.
5. Affected children of consanguineous marriages are valuable genetic resources, especially for linkage analysis of less common diseases. In collecting samples, it may be profitable to consider the many regions of the world where consanguinity is not infrequent. Also, under appropriate conditions, isolated populations showing founder effects may be useful for linkage mapping.

These are simply general comments. A more precise analysis must be performed, of course, when designing a particular study. Toward this end, we and others have written computer programs to estimate the resources needed, given assumptions about the genetics of a trait.

No matter what methods are employed, there will occur genetic situations beyond the limits of linkage analysis—because of too much heterogeneity, too high a background of nongenetic cases, too complex a genetic interaction. This will become known only in retrospect, when linkage is not detected. Certain clinical approaches may help when such problems arise or can be anticipated.

If several genes interact to produce a phenotype, the best route may be through the identification of "subphenotypes" in unaffected family members. With luck, the subphenotype may represent an effect of just one of the interacting genes and thus may be easier to map. A potential example is the photosensitivity frequently found in unaffected close relatives of epileptics (Doose et al. 1982; Botstein and Donis-Keller 1984); investigations of this subphenotype suggest that it might be advantageous to try to map it before, or at the same time as epilepsy. Similarly, clinically measurable aspects of cholesterol metabolism may provide useful subphenotypes to map in families with heart disease (M. Brown, pers. comm.). Finally, if a trait is heterogeneous but a simultaneous search turns up no candidate loci, it may be advantageous to find ways to clinically subdivide the disease into classes, each of which may be mappable.

CONCLUSION

The influence of genetic arguments in the understanding of human biology has been mainly theoretical, primarily because experimental genetics could rarely be applied directly to humans. As many of the articles in this volume make clear, the advent of RFLP mapping has begun to change this, at least in the case of simple Mendelian diseases. Our expectation is that a complete RFLP map will allow even more powerful analytic methods—methods that will draw on molecular biology, mathematics, clinical medicine, and population genetics.

By no means will complex genetic traits be easy to analyze. Our results merely offer hope that analysis is possible, at least in some cases. Despite all the obstacles, however, it should be remembered that human beings have certain unique advantages for genetics: No other organism is as sensitive to its phenotypes, nor as cooperative about reporting them. Genetic analysis by linkage methods may be able to help in the elucidation of mammalian biology, both normal and abnormal, just as it has in the understanding of biological processes in bacteria, yeast, flies, and mice.

ACKNOWLEDGMENTS

We thank David Page and Helen Donis-Keller for many helpful discussions and for their encouragement. This work was partially supported by grants from the System Development Foundation to E.S.L. and from National Institute of General Medical Sciences to D.B.

REFERENCES

Anderson, M.A. and J.F. Gusella. 1984. Use of cyclosporin A in establishing Epstein-Barr virus tranformed human lymphoblastoid cell lines. *In Vitro* **20:** 856.

Bateson, W. 1902. *Mendel's principles of heredity*. Cambridge University Press, England.

Bird, T.D., J. Ott, E.R. Giblett, P.F. Chance, S.M. Sumi, and G.H. Kraft. 1983. Genetic linkage evidence for heterogeneity in Charcot-Marie-Tooth neuropathy. *Ann. Neurol.* **14:** 679.

Blass, J.P. and G.E. Gibson. 1977. Abnormality of a thiamine-requiring enzyme in patients with Wernicke-Korsakoff syndromes. *N. Engl. J. Med.* **297:** 1367.

Botstein, D. and H. Donis-Keller. 1984. A molecular approach to defining the inherited components in epilepsy and other diseases of uncertain etiology. *Epilepsia* **25:** S150.

Botstein, D., R.L. White, M.H. Skolnick, and R.W. Davis. 1980. Construction of a genetic linkage map in man using restriction fragment length polymorphisms. *Am J. Hum. Genet.* **32:** 314.

Cavalli-Sforza, L. and M.C. King. 1986. Detecting linkage for genetically heterogeneous diseases and detecting heterogeneity with linkage data. *Am. J. Hum. Genet.* **38:** 599.

Cavenee, W.R., T.P. Dryja, R.A. Phillips, W.F. Benedict, R. GodBout, B.L. Gallie, A.L. Murphree, L.C. Strong, and R.L. White. 1983. Expression of recessive alleles by chromosome mechanisms in retinoblastoma. *Nature* **305:** 779.

Davies, K.E., P.L. Pearson, P.S. Harper, J.M. Murray, T. O'Brien, M. Sarfrazi, and R. Williamson. 1983. Linkage analysis of two cloned DNA sequences flanking the Duchenne muscular dystrophy locus on the short arm of the human X chromosome. *Nucleic Acids Res.* **11:** 2303.

Doose, H. 1982. *Genetic basis of the epilepsies* (ed. V.E. Anderson et al.), p. 2839. Raven Press, New York.

Dyck, P.J., J. Ott, S.B. Moore, C.J. Swanson, and E.H. Lambert. 1983. Linkage evidence for genetic heterogeneity among kinships with hereditary motor and sensory neuropathy type I. *Mayo Clin. Proc.* **58:** 430.

Fisher, R.A. 1935. The detection of linkage with dominant abnormalities. *Ann. Eugen.* **6:** 187.

Garrod, A.E. 1902. A study in chemical individuality. *Lancet* **II:** 1616.

Gusella, J.F., N.S. Wexler, P.M. Conneally, S.L. Naylor, M.A. Anderson, R.E. Tanzi, P.C. Watkins, K. Ottina, M.R. Wallace, A.Y. Sakaguchi, A.B. Young, I. Shoulson, E. Bonnilla, and J.B. Martin. 1983. A polymorphic DNA marker genetically linked to Huntington's disease. *Nature* **306:** 234.

Haldane, J.B.S. and C.A.B. Smith. 1947. A new estimate of the linkage between the genes for colour-blindness and haemophilia in man. *Ann. Eugen.* **14:** 10.

Jaspers, N.G.J., J. De Wit, M.R. Regulski, and D. Bootsma. 1982. Abnormal regulation of DNA replication and increased lethality in AT cells exposed to carcinogenic agents. *Cancer Res.* **42:** 335.

Jaspers, N.G.J., R.B. Painter, M.C. Paterson, C. Kidson, and T. Inoue. 1985. Complementation analysis of ataxia-telangiectasia. In *Ataxia-telangiectasia: Genetics, neuropathology, and immunology of a degenerative disease of childhood* (ed. R.A. Gatt and M. Swift), p. 147. A.R. Liss, New York.

Keijzer, W., N.G.J. Jaspers, P.J. Abrahams, A.M.R. Taylor, C.F. Arlett, H. Takebe, P.D.S. Kimmont, and D. Bootsma. 1979. A seventh complementation group in xeroderma pigmentosum. *Mutat. Res.* **62:** 183.

Knowlton, R.G., O. Cohen-Haguenauer, N. Van Cong, J. Frezal, V.A. Brown, J.C. Braman, J.W. Schumm, L.C. Tsue, M. Buchwald, and H. Donis-Keller. 1985. A polymorphic DNA marker linked to cystic fibrosis is located on chromosome 7. *Nature* **318:** 380.

Lander, E.S. and D. Botstein. 1986. Strategies for studying heterogeneous genetic traits in humans by using a linkage map of restriction fragment length polymorphisms. *Proc. Natl. Acad. Sci.* **83:** 7353.

Macklin, M.T. 1959. Inheritance of retinoblastoma in Ohio. *Arch. Ophthalmol.* **62:** 842.

McKusick, V.A. 1971. *Mendelian inheritance in man: Cata-*

logs of autosomal dominant, autosomal recessive, and X-linked phenotypes. 5th edition. Johns Hopkins University Press, Baltimore.

Morton, N. 1955. Sequential tests for the detection of linkage. *Am. J. Hum. Genet.* **7:** 277.

———. 1956. The detection and estimation of linkage between the genes for elliptocytosis and the Rh blood type. *Am. J. Hum. Genet.* **8:** 80.

Norio, R., H.R. Nevanlinna, and J. Perheentupa. 1973. Hereditary diseases in Finland. *Ann. Clin. Res.* **5:** 109.

Ott, J. 1985. *Analysis of human genetic linkage*. Johns Hopkins University Press, Baltimore.

Reeder, S.T., M.H. Breuning, K.E. Davies, R.D. Nicolls, A.P. Jarman, D.R. Higgs, P.L. Pearson, and D.J. Weatherall. 1985. A highly polymorphic DNA marker linked to adult polycystic kidney disease on chromosome 16. *Nature* **317:** 542.

Romeo, G., M. Bianco, M. Devoto, P. Menozzi, G. Mastella, A.M. Giunta, C. Micalizzi, M. Antonelli, A. Battistini, F. Santamaria, D. Castello, A. Marianelli, A.G. Marchi, A. Manca, and A. Miano. 1985. Incidence in Italy, genetic heterogeneity, and segregation analysis of cystic fibrosis. *Am. J. Hum. Genet.* **37:** 338.

Stanbury, J.B., J.B. Wyngaarden, D.S. Fredrickson, J.L. Goldstein, and M.S. Brown. 1983. *The metabolic basis of inherited disease*. McGraw Hill, New York.

Trevor-Roper, P.D. 1952. Marriage of two complete albinos with normally pigmented offspring. *Br. J. Ophthal.* **36:** 107.

Tsui, L.C., M. Buchwald, D. Barker, J.C. Braman, R.G. Knolton, J.W. Schumm, H. Eiberg, J. Mohr, D. Kennedy, N. Plesvic, M. Zsiga, D. Markiewica, G. Akots, V. Brown, C. Helms, T. Gravius, C. Parker, K. Rediker, and H. Donis-Keller. 1985. Cystic fibrosis locus defined by a genetically linked polymorphic DNA marker. *Science* **230:** 1054.

Wainwright, B.J., P.J. Scambler, J. Schmidtke, E.A. Watson, H.Y. Law, M. Farrall, H.J. Cooke, H. Eiberg, and R. Williamson. 1985. Localization of cystic fibrosis locus to human chromosome 7cen-q22. *Nature* **318:** 384.

White, R., S. Woodward, M. Leppert, P. O'Connell, M. Hoff, J. Herbst, J.M. Lalouel, M. Dean, and G. Vande Woude. 1985. A closely linked genetic marker for cystic fibrosis. *Nature* **318:** 382.

Willard, H.F., M.H. Skolnick, P.L. Pearson, and J.-L. Mandel. 1985. Report of the committee on human gene mapping by recombinant DNA techniques. *Cytogenet. Cell Genet.* **40:** 360.

Human Major Histocompatibility Complex Genes: Class I Antigens and Tumor Necrosis Factors

J.L. STROMINGER

Department of Biochemistry and Molecular Biology, Harvard University, Cambridge, Massachusetts 02138 and the Dana Farber Cancer Institute, Harvard Medical School, Boston, Massachusetts 02115

In this paper I discuss two topics relating to genes of the human major histocompatibility complex (MHC) in which new information has recently been obtained. They are (1) the use of cytotoxic T lymphocyte (CTL) variants and site-directed mutagenesis to map functional sites in class I MHC molecules and (2) the regulation of expression of class I genes by tumor necrosis factor (TNF) and the fact that TNFα and the closely related lymphotoxin (TNFβ) are themselves encoded in the MHC.

The human class II genes are encoded at the centromeric end of the human MHC and the class I genes are encoded at the telomeric end on chromosome 6 in the region of 6p21.1-.3. The elucidation of the protein structure of the human class I antigen showed that it was composed of two chains, α and β, comprising four extracellular domains, three of them in the α chain (α1, α2, and α3) and the fourth in the β chain (β_2 microglobulin, or β_2m). This domain structure was later supported by the elucidation of the class I gene structure that showed that each domain is represented by a separate exon. A physical map of the class I region of the MHC is not yet completed, but at the present time 17 different class I genes, all pseudogenes except those encoding HLA-A, -B, and -C, have been identified. The human homologs of the murine *Qa* and *Tla* genes have not yet been found (B. Koller and H. Orr, pers. comm.).

In the class II proteins the four domains are linked differently in that the α and β chains each contain two domains. In addition, only one chain of class I proteins spans the membrane whereas both chains of class II proteins span the membrane. Our knowledge of the biochemistry is not yet sufficiently sophisticated to tell us how these seemingly small differences in structure are reflected in the rather large differences in function of these two classes of molecule.

Amino Acid Polymorphism, CTL Variants, and Site-directed Mutagenesis of Class I Antigens

Amino acid polymorphism in the α1 domain of human class I antigens is rather widely distributed in a number of single, highly polymorphic residues (e.g., residues 9, 24, and 45) and one very prominent cluster (residues 63–80) (Fig. 1) (Lopez de Castro et al. 1985). In α2 also, the polymorphic regions are widely distributed. Some single, highly polymorphic residues as well as some clusters can be detected by sequence comparison. One small cluster (not especially prominent in the amino acid diversity plot) between residues 147–157 has been predicted to have an α-helical structure. From studies of CTL variants, this region appears to be a

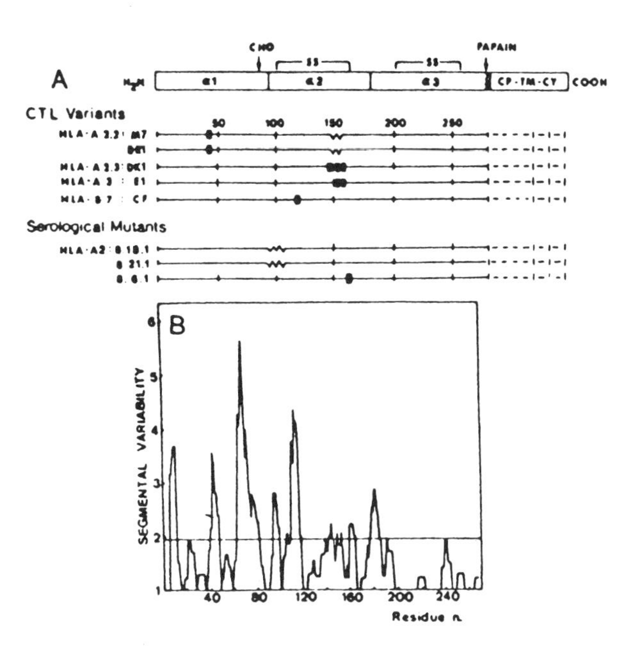

Figure 1. (*A*) Locations of amino acid substitutions in class I HLA variants and mutants. (*B*) Segmental variability analysis of class I antigens. Average variability of hexapeptide segments was computed and plotted vs. position. A reference line is shown at the average segmental variability value of residues 1–194, where the majority of polymorphic positions are located. (Reprinted, with permission, from Lopez de Castro et al. 1985.)

most important region for recognition by CTL. In the conserved domains ($\alpha3$ and β_2m) only a few conservative substitutions occur. Since the X-ray crystal structure of HLA-A2 is advancing rapidly (Bjorkman 1984; Bjorkman et al. 1985), answers to many structural questions should soon be forthcoming.

CTL variants are very useful because by mapping the locations of their amino acid changes, the question of whether the alterations occur in the polymorphic regions seen from the compilation of protein sequences can be asked. These data are summarized in Figure 1. In fact, some of the substitutions in the CTL variants occur exactly in residues of the $\alpha1$ and $\alpha2$ domains that are polymorphic in the amino acid sequences (Lopez de Castro et al. 1985). Site-directed mutagenesis is being employed to explore more precisely the effects of amino acid changes on the function of these molecules. The system used employs a double-stranded pUC plasmid into which the *HLA-A2* (or *HLA-B7*) genomic clone has been inserted. The $\alpha1$ and $\alpha2$ domains have been excised from one strand to create a gap. A mismatched oligonucleotide is annealed to the gap and the gap is filled using DNA polymerase I. The plasmid is then used to transform *E. coli* (previously made competent) and the colonies containing the mutant plasmid are then identified using the same mismatched oligonucleotide as a probe under differential-temperature washing conditions. Several cycles of transformation are necessary to further purify the mutant DNA. The presence of the appropriate mutation is also confirmed after amino acid sequencing analysis. So far a variety of mutations have been introduced into HLA-A2 between residues 62 and 86 with the following results (Santos-Aguado et al. 1987 and unpubl.).

1. When N86 (the site of the single amino-linked glycan) was changed to Q86 or D86, the resulting nonglycosylated α chain was not expressed at the surface of mouse Ltk$^-$ cells and a low level of expression was detected on human or mouse cells transfected with the human β_2m gene. However, creation of the same phenotype either by S88 to G88 (since the glycosylation signal is N-X-S) or by treatment of cells expressing the normal *HLA-A2* gene with tunicamycin resulted in close-to-normal surface expression of a nonglycosylated HLA-A2 antigen in human RD cells. These and other data suggest that N86 is essential for the interaction of the α chain with β_2m.
2. When R65 and K66 in the HLA-A2 molecule were changed to Q65 and I66 (independently or combined), the reactivity with the HLA-A2/B17 cross-reactive monoclonal antibody MA2.1 was lost, but the reactivity with other HLA-A2–specific monoclonal antibodies was retained. In addition, changing the residues nearby, H70 to Q70 or H74 to D74 (independently or combined), did not affect at all the recognition by MA2.1; its epitope appears to be relatively small. Another example is the HLA-B7–specific monoclonal antibodies MB40.2 and MB40.3, whose recognition site was lost by altering residues 176–178 in the HLA-B7 molecule, while recognition by several other monoclonal antibodies that recognize HLA-B7 was retained.
3. Some residues when changed singly resulted in molecules that were not expressed at the surface. However, expression was restored by paired changes; for examples, R65 to Q65 resulted in a molecule with a low level of expression (but inducible with γ-interferon), but R65, K66 (as in HLA-A2) to Q65, I66 (as in HLA-B7) resulted in a fully expressed molecule. The replacement K66 to I66 produces a molecule expressed at a normal level in human cells but with a low level in mouse L cells. This phenomenon may reflect the requirement at several adjacent positions for proper folding and/or association with human or mouse β_2m.

These data have shown the efficacy of site-directed mutagenesis as a powerful tool to study the relationship between structure and function of class I molecules. Thus, by continued use of this procedure, the relationship between particular amino acid substitutions at many positions and the function of the molecule can be fully explored. At the present time additional mutations are being introduced into the positions at which CTL variants of HLA-A2, -A3, -B7, and -B27 are known to differ (i.e., residues 9, 43, 78, 107, 114, 116, 152, and 156) in order to understand the effects of individual and simultaneous changes on CTL recognition.

Tumor Necrosis Factors Are Encoded by MHC Genes

The expression of class I MHC genes is markedly up-regulated by TNF (Collins et al. 1986). We have recently also found that the genes encoding TNFα and the closely related lymphotoxin (TNFβ) are encoded within the region of the human MHC (Spies et al. 1986). TNF is a powerful inducer of expression of class I MHC antigens in cells that are low expressors. However, in cells that are already expressing at relatively high levels, it cannot enhance the expression beyond a maximum limit. The enhancement of class-I-antigen gene transcription in low expressors was in the range of 50-fold to 100-fold. The time course of enhancement was relatively slow, reaching a maximum at 7 days, and thus the possibility should be considered that stimulation of cells by TNF is only the first step in this enhancement and that several intermediate factors may intervene.

In the course of these studies, the Genentech group published data using somatic cell hybrids, including some cells with translocated chromosomes, to localize *TNFα* and lymphotoxin (*TNFβ*) to the region between 6p23 and 6q12, that is, including the proximal 10–20% of the long arm and the proximal 80% of the short arm of chromosome 6 (for reference, see Nedwin et al. [1985] in Spies et al. 1986). The human MHC is located in the region of 6p21.1 to 6p21.3 (i.e., within the proximal half of the short arm of chromosome 6) and thus is clearly within the region defined for the location of the *TNF* genes. Moreover, independently, the genes for TNFα and TNFβ had been obtained on a single phage clone, linked within a segment of about 7 kb by a group at the Institute for Molecular Biology and Bioorganic Chemistry in Moscow (for reference, see Nedospasov et al. [1986] in Spies et al. 1986). We, therefore, asked the question whether these genes that regulate the expression of class I MHC antigens were themselves encoded within the human MHC; to this end we used two techniques—mapping of deletion mutants and chromosomal in situ hybridization.

Seven deletion mutants of the human MHC were available, of which four were single-haplotype deletions and three double-haplotype deletions. The location of the deletions and their breakpoints, as far as they are known, are shown in Figure 2. Southern blots of DNA from these deletion mutants were probed with the *TNFα* and *β* genes, as well as with *DRα*, *DQα*, *DPα*, and globin genes as controls. The control blots revealed the presence of full-intensity bands (e.g., globin in all the blots), half-intensity bands (e.g., *DRα*, *DQα*, or *DPα* in DNA from 8.1.6 cells), and fully deleted bands (e.g., *DRα* in 9.22.3 cells or *DQα* and *DRα* in 721.82 cells). When DNA from these seven deletion mutants was probed with the *TNFα* and *TNFβ* probes and compared with control B-cell DNA, full-intensity bands were observed for DNA from the 8.1.6 and 9.22.3 cells while half-intensity bands were observed in all of the other five deletion mutants (indicating that these genes had been deleted from one haplotype). These data lead to the conclusion that the *TNF* genes are located within the MHC in one of the possible regions shown in Figure 2. Notice that a location in the large segment of DNA comprising the class II region between the *DP* gene cluster and the *DRα* chain gene is eliminated by the deletions in 721.80 cells and 8.1.6 cells; that is, the *TNF* genes cannot be *within* the class II region. The DNA from 9.28.6 cells was most informative since this single-haplotype deletion, which contained only half-intensity *TNF* bands, localizes the *TNF* genes to within a region defined by its two deletion breakpoints, that is, just centromeric to the *DP* gene cluster and just telomeric to the *HLA-A* locus but within the remainder of the class I cluster. Several other points are noteworthy:

1. Cosmid clones covering the class III region were available. Their restriction maps, compared with the map of a cosmid containing both the *TNFα* and *β* genes, showed a location eliminated *within* the class III region. However, the regions between the class III and class II clusters or between the class III and class I clusters are the most likely locations for these genes.
2. Similarly, cosmid clusters (total of 89 cosmids containing 17 different class I genes) containing all or nearly all of the human class I genes were also negative, probably eliminating a locus *within* the class I region.
3. Murine cosmid clusters extending from the *H-2K* genes to the *IEα* gene, obtained from Dr. R. Flavell (Biogen), were similarly negative when probed under low stringency conditions in an attempt to detect a cross-hybridizing sequence.

In addition, a cosmid library from the human cell line Priess was probed with the *TNFα* gene. A 45–50-kb cosmid in which the *TNFα* and *β* genes were centrally placed on a 7-kb fragment was obtained (thus confirming the previous data obtained with a λ phage clone). The restriction map of the ends of this clone does not overlap with any of the class I, class II, and class III cosmids available. Thus, the precise localization of these genes within the human MHC awaits further experimentation. From the data available, the possible locations are (1) centromeric to the *DP* gene cluster, (2) on one or the other side of the class III cluster, or (3) within the class I region in a segment between cosmid clusters and not yet cloned.

To confirm the location of the *TNF* genes within the MHC, in situ hybridization was carried out using a *TNFα* probe. Of the grains localized to chromosome 6, 49% were found in the region 6p21.1 to 6p22, the locus of the human MHC. Their precise localization within the region must await further experimentation.

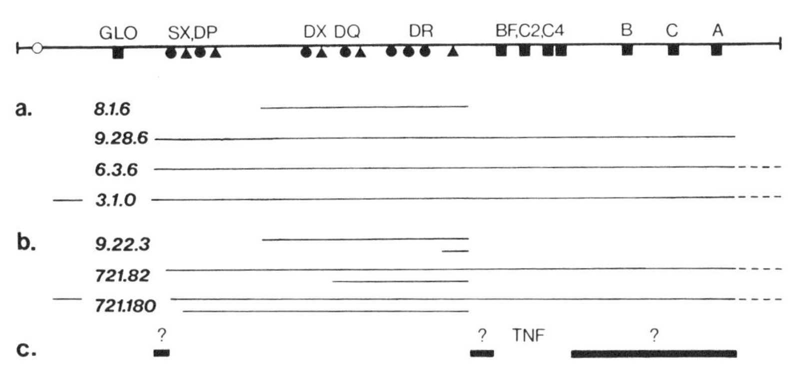

Figure 2. Linear order of genetic loci in the MHC and possible locations for the *TNF* genes (*c*) from studies of the deletion mutants listed (*a* and *b*). (Reprinted, with permission, from Spies et al. 1986.)

ACKNOWLEDGMENTS

This research was supported by National Institutes of Health grants AI-10736, AI-20182, and AM-30241.

REFERENCES

Only a few references to recent publications from our laboratory are provided. Full documentation and references to the many contributions of other laboratories can be found in the references cited.

Bjorkman, P.J. 1984. "Crystallographic studies of HLA." Ph.D. thesis, Harvard University, Cambridge, Massachusetts.

Bjorkman, P.J., J.L. Strominger, and D.C. Wiley. 1985. Crystallization and X-ray diffraction studies on the histocompatibility antigens HLA-A2 and HLA-A28 from human cell membranes. *J. Mol. Biol.* **186:** 205.

Collins, T., L.A. Lapierre, W. Fiers, J.L. Strominger, and J. Pober. 1986. Recombinant human tumor necrosis factor increases mRNA levels and surface expression of HLA-A,B, antigens in vascular endothelial cells and dermal fibroblasts *in vitro*. *Proc. Natl. Acad. Sci.* **83:** 446.

Lopez de Castro, J., J.A. Barbosa, M.S. Krangel, P.A. Biro, and J.L. Strominger. 1985. Structural analysis of the functional sites of class I antigens. *Immunol. Rev.* **85:** 149.

Santos-Aguado, J., P.A. Biro, U. Fuhrmann, J.L. Strominger, and J.A. Barbosa. 1987. Amino acid sequences in the α domain and not glycosylation are important in HLA-A2/β2-microglobulin association and cell surface expression. *Mol. Cell. Biol.* (in press).

Spies, T., C. Morton, S.A. Nedospasov, W. Fiers, D. Pious, and J.L. Strominger. 1986. The tumor necrosis factor (TNF) α and β genes are linked to the human major histocompatibility complex (MHC). *Proc. Natl. Acad. Sci.* (in press).

Polymorphism and Regulation of HLA Class II Genes of the Major Histocompatibility Complex

B. MACH, J. GORSKI, P. ROLLINI, C. BERTE, I. AMALDI, J. BERDOZ, AND C. UCLA
Department of Microbiology, University of Geneva Medical School, 1211 Geneva 4, Switzerland

Until recently, immunology had remained a controversial field for molecular biologists, with provocative and subtle theories outnumbering hard facts. In the last 10 years however, due in large part to the availability of cDNA cloning (Rougeon et al. 1975), dramatic progress in molecular immunology has clarified the genetic basis of immune diversity and of the control of the immune response. The diversity of immunoglobulins and of T-lymphocyte receptors is a somatic phenomenon, with each clone of a B or T cell exhibiting a distinct specificity. On the other hand, the diversity of major histocompatibility complex (MHC) antigens results from allelic polymorphism at the population level, with all cells of a given individual expressing the same specificity. The MHC encodes two major groups of antigens: class I, also known as transplantation antigens, and class II, involved in the control of the immune response. In man, the MHC antigens are referred to as HLA.

We first outline some of the characteristic features of MHC class II antigens and indicate their biological and medical importance. HLA class II, or Ia, antigens are highly polymorphic molecules encoded in the MHC. They consist of α and β chains forming a transmembrane, glycosylated heterodimer. Contrary to MHC class I transplantation antigens, Ia antigens are expressed on a restricted number of cell types. The key function of class II antigens is in the interaction of an antigen-presenting cell (APC) and T lymphocytes. Indeed, for the stimulation of T lymphocytes, and thus for the development of an immune response, the T-cell receptor must interact both with a foreign antigen and with the HLA class II molecule on the surface of an APC (Fig. 1). The T-cell receptor can distinguish among the different allelic products of class II genes, and in antigen presentation it only interacts with class II antigens of the same allelic specificity. T-cell stimulation by an antigen is thus genetically "restricted" to autologous or self Ia molecules (Klein 1979).

MHC class II genes also control the extent of an immune response to individual foreign antigens; they are thus referred to as immune response (*Ir*) genes (Benacerraf and McDevitt 1972). Since different alleles of class II *Ir* genes encode different effectiveness in the presentation of individual antigens, this polymorphism results in either a high or a low level of immune response and can thus be considered as a polymorphism of immune performance. An outstanding feature of the MHC class II genes is their remarkable polymorphism. It is now evident that these genes are both more numerous and more polymorphic than had been presumed (Mach et al. 1986). The most important biological significance of HLA class II polymorphism is therefore the ability it confers to the species to cope with a variety of pathogens and thus to survive as a species. An understanding of the generation of this polymorphism in evolution and its subsequent selection is therefore one of the important issues in immunology.

The different loci encoding class II genes are tightly linked, and one frequently observes a "linkage disequilibrium" in the expression of individual alleles at different loci. Finally, a medically important observation is the existence of a striking association between the susceptibility to certain important diseases, such as diabetes, and certain specific HLA class II haplotypes. In most instances, the diseases that do show this association with HLA alleles are thought to involve, directly or indirectly, an autoimmune process. This HLA class II association, if better understood, could be of important predictive value in medicine.

Gene Organization and Genetic Complexity

The *HLA-D* region contains multiple α and β chain genes organized in three related subregions, *DP, DQ,* and *DR*. The exact gene order, as predicted in Figure 2, has recently been confirmed by Southern mapping

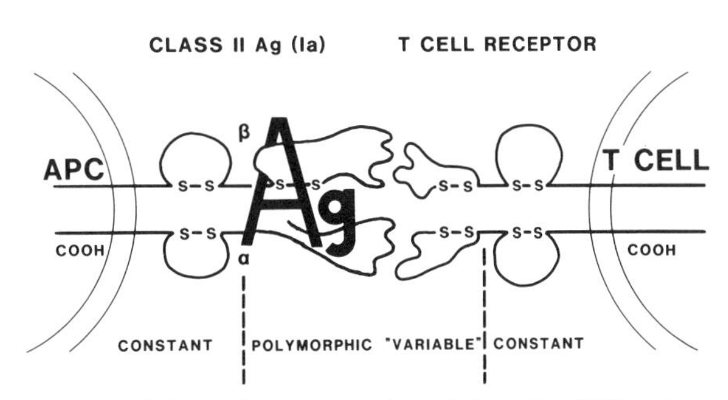

Figure 1. Schematic representation of the role of HLA class II molecules in antigen presentation to T lymphocytes. (APC) Antigen-presenting cell; (Ag) foreign antigen.

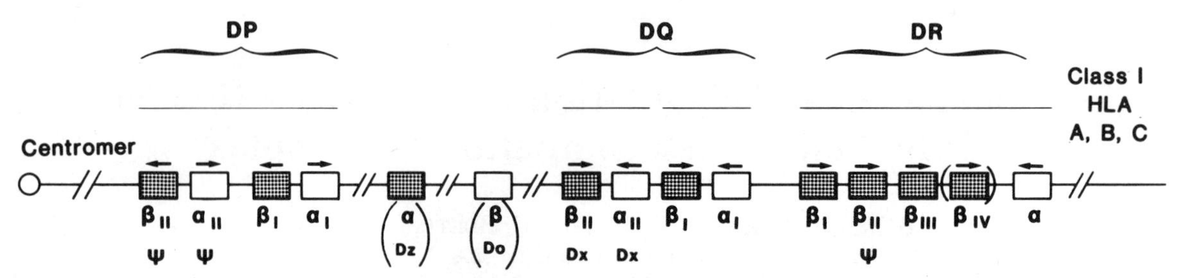

Figure 2. Map of the *HLA-D* region of the human MHC. α and β chain genes of the subregions *HLA-DP, -DQ,* and *-DR* are indicated. DQ_{II} genes are also referred to as *DX*. Pseudogenes are also indicated. Arrows indicate the direction of transcription. *Do* and *Dz* are genes related to class II genes but show a different pattern of expression.

of high-molecular-weight fragments by pulsed-field electrophoresis (Bell et al., this volume). Within each subregion, the genes from different haplotypes are quite homologous. Evidence was obtained earlier for multiple class II genes within the *DR* subregion (Long et al. 1982, 1983a; Gorski et al. 1984b), an observation that has been extended to the *DP* and *DQ* subregions (Gorski et al. 1984a; Korman et al. 1985; Rask et al. 1985; Trowsdale et al. 1985). This genetic complexity might confer a greater "immune response" potential to the individual. It is also a source or a "reservoir" of sequences for the generation of allelic diversity by mechanisms involving recombination or heteroduplex formation. Several of the multiple HLA class II genes are known to be nonfunctional or pseudogenes (Fig. 2). Therefore, although there has been an expansion of the number of class II genes to almost 20, this protein-coding potential is paradoxically not fully utilized.

The *HLA-DR* subregion encodes the most important Ia antigen in man, and the majority of T cells are stimulated by an interaction with HLA-DR antigens. In the *DR3* haplotype, this region consists of several β chain loci, which have been linked and identified as β_I, β_{II}, and β_{III} (Rollini et al. 1985). The β_{II} locus lacks the first domain exon and is a pseudogene, whereas β_I and β_{III} are both expressed into *DR* β chains. There is evidence that this DRβ gene cluster results from a relatively recent duplication event (Rollini et al. 1985). The existence, throughout the *D* region, of highly homologous genes makes the identification of individually isolated genes difficult. It is, however, essential to correlate structural or functional features of class II molecules with individual loci within the MHC. This can now be done with a knowledge of the *HLA-DR* subregion linkage map (Rollini et al. 1985; Spies et al. 1985) and a combination of structural studies and expression of individual class II genes in transfected cells (Rabourdin-Combe and Mach 1983; Gorski et al. 1985).

Nature of the *HLA-DR* Allelic Polymorphism

It is the β chain of the *DR* heterodimer that carries the polymorphism. There is evidence for the existence of groups of related alleles, referred to as "supertypic groups." Within these groups, the *DR*β genes are structurally and probably evolutionarily related. The different β chain loci are very homologous, and thus, without concomitant structural information, it is not evident if sequence or functional comparisons are truly allelic. When DNA sequences from different β chain genes are compared in a nonallelic manner, the variations are predominantly found in the first domain exon, clustered in three "hypervariable" regions (Fig. 3A). When true allelic comparisons are made with sequences for the β_I locus within two distinct supertypic groups, *DRw52* (Fig. 3B) and *DRw53* (Fig. 3D), it is observed that the polymorphic differences are predominantly found in the so-called third hypervariable segment. It is therefore this segment of the gene that is responsible for the polymorphic differences recognized by the T-cell receptor.

Three other conclusions were drawn from the analysis of β chain DNA sequences:

1. There is much more polymorphism than was suspected and, at that level, the number of HLA class II alleles in the human population is greater than had been anticipated.
2. Locus *HLA-DR*β_I is more polymorphic than locus *DR*β_{III}. The latter is characteristic of the individual supertypic groups and allelic comparisons do not show the characteristic hypervariable region (Fig. 3C).
3. One observes a "patchwork" pattern of nucleotide differences among different β gene sequences, with conservation and sharing of short blocks of sequences (Gorski and Mach 1986).

Generation of *HLA-DR* Allelic Diversity

The patchwork nature of sequence differences noted above is indicative of gene-conversion-like events, which can copy short blocks of DNA sequences from one gene to another. Figure 4 shows a comparison of DNA sequences of first domain exons of several β chain genes. In the three cases illustrated, at the position corresponding to the hypervariable region, a short segment of sequence has been copied from one gene (donor) to another (recipient), thus generating a novel sequence, that of the "converted" gene. One of these examples has been studied in more detail (Gorski and

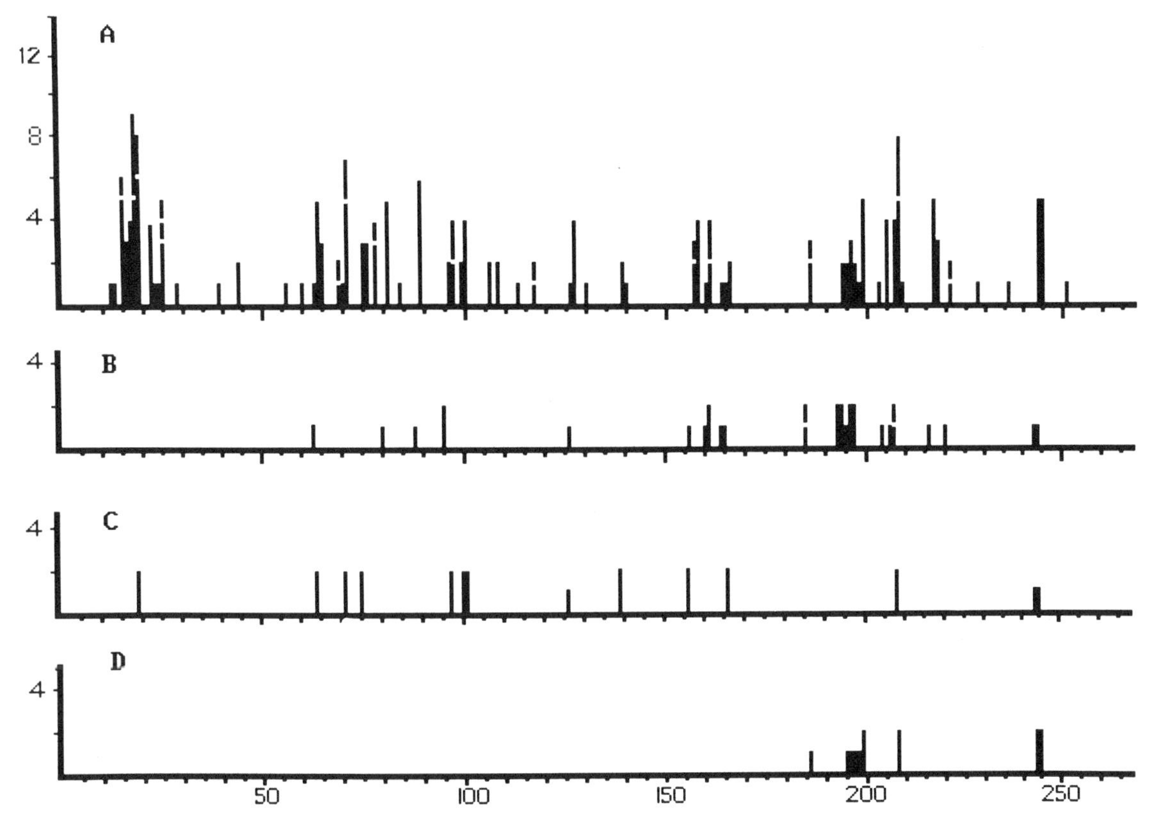

Figure 3. Variability plot of the first domain of *DR* β chain genes. (*A*) Comparison of 12 different *DR* β genes. (*B*) Comparison of 4 allelic β genes from the β_I locus of the *DRw52* supertypic group (Gorski and Mach 1986; Tieber et al. 1986). (*C*) Four sequences from the β_{III} locus of the *DRw52* supertypic group. (*D*) Comparison of 5 β_I sequences from the *DR4* specificity (Gregersen et al. 1986).

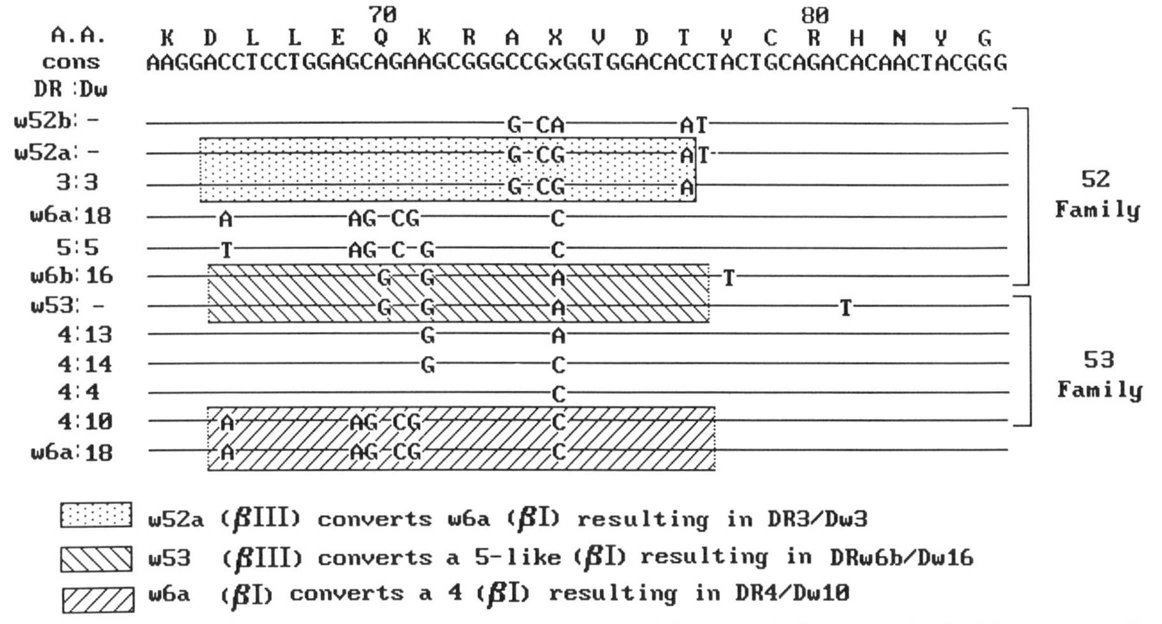

Figure 4. DNA sequence of the first domain segment of *HLA-DR* β chains. Boxes indicate genes involved in gene-conversion events, with donor, recipient, and converted gene sequences.

Mach 1986) and has revealed that the *DR3* specificity has beeen generated by gene conversion involving the $DR\beta_{III}$ and the $DR\beta_{I}$ loci of an *HLA-DRw6* haplotype as donor and recipient genes, respectively (Fig. 5, top). An important feature of gene conversion as a mechanism for the generation of polymorphism is the possibility of conserving, in the population, preselected DNA segments corresponding to relevant epitopes, while generating new variants.

Evolution of the *HLA-DR* Subregion

From the comparison of a large number of *DR* β chain genes and a knowledge of their organization, it is possible to propose a scheme that reflects the evolution of this polymorphism (Fig. 6). One of the key features is the separate pathways followed by the *DRw52* and *DRw53* lineages. These two "supertypic families" represent more than 80% of all known *DR* alleles. In the *DRw52* lineage, there is evidence for a recent duplication involving the β_{I} and β_{III} loci as well as a deletion involving the first domain of the β_{II} locus (Rollini et al. 1985). Diversification of the duplicated loci followed. At the less polymorphic β_{III} locus, this gave rise to at least two alleles, *DRw52a* and *DRw52b*. The β_{I} locus underwent more-extensive diversification, including the gene conversion events mentioned above. The *DRw53* lineage did not undergo the duplication and deletion events, but the same diversification proceeded at the β_{I} locus.

Recombination events have also taken place and, as shown on Figure 5 (bottom), we have observed that some *DR3* haplotypes have a β_{III} locus of the *DR52a* variety, while others have the β_{III} locus found in *DR5* individuals corresponding to *DR52b* (J. Gorski et al., in prep.). This second *DR3* haplotype differs therefore from the other only at the *DR* β_{III} locus.

Finally, the origin of the expressed loci is different in the two lineages (Fig. 6). The two expressed $DR\beta$ genes in *DRw52* are very related and derived from a common ancestor locus. However, in *DRw53* individuals, the two expressed $DR\beta$ genes, β_{I} and β_{III}, are quite different

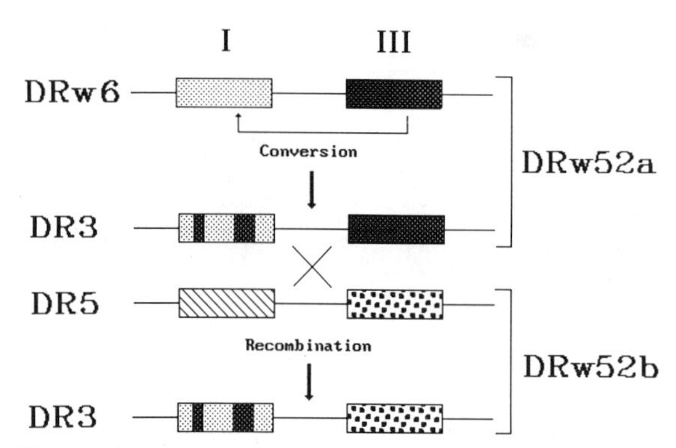

Figure 5. Schematic representation of the role of gene conversion (loci β_{III} and β_{I} of *DRw6*) and of recombination (*DR3* and *DR5* haplotypes) in the generation of two different *DR3* haplotypes (Gorski and Mach 1986 and in prep.).

and are derived from two different ancestral genes. This observation implies that different alleles in the population are not "equidistant." We have therefore suggested (Mach et al. 1986) that the nature of individual mismatch in histocompatibility typing for transplantation might be important, some mismatches (such as *DRw52/DRw53*) having more influence than others (within *DRw52* or *DRw53*). This evolutionary scheme concerns exclusively HLA gene sequences from Caucasians. The comparison of some of the specific structural features observed here with other ethnic groups will be of obvious importance. It is going to be of great interest to compare, at the DNA sequence level and across different human populations, the patterns of generation of one of the most polymorphic systems in the human genome.

Analysis of Relevant Polymorphism by DNA Typing

Since the extensive polymorphism of the MHC is functionally relevant, there is a crucial need for procedures allowing the analysis of the HLA polymorphism in a more accurate manner than the current serological assays. This concerns both matching for transplantation and the study of HLA-disease association. We had shown earlier (Wake et al. 1982) that the HLA class II polymorphism can be analyzed directly at the DNA level, comparing by hybridization the patterns of restriction fragments specific for individual class II genes such as *DR*, *DQ*, or *DP*; we had referred to this approach as "DNA typing" or HLA genotyping. This direct analysis of HLA polymorphism is now widely used.

More recently, however, some of the limitations of HLA genotyping by RFLP have become apparent, and we have presented evidence (Angelini et al. 1986) that hybridization with loci-specific and allele-specific oligonucleotide probes represents an important improvement in the analysis of HLA polymorphism. It allows the study of the phenotypically relevant regions of the polymorphic genes and allows an absolute discrimination among alleles, with a resolution of single nucleotide differences. With appropriate oligonucleotides, we have been able to screen and identify DNA from individuals of different *DR* specificities. More importantly, this oligonucleotide HLA-typing procedure allows the identification of micropolymorphism within classical HLA specificities and their "split" into multiple subtypes (Angelini et al. 1986). The procedure is well-suited for the subtle HLA typing that is very much needed for histocompatibility matching and for large-scale studies of HLA-linked diseases. One can foresee an extensive use of the oligonucleotide HLA-typing procedure, in particular if it can lead to the prediction of susceptibility to certain important HLA-linked autoimmune diseases such as insulin-dependent diabetes. Besides the beneficial impact of this new technology in predictive medicine, one might fear the possible social consequences of the misuse and abuse of such a predic-

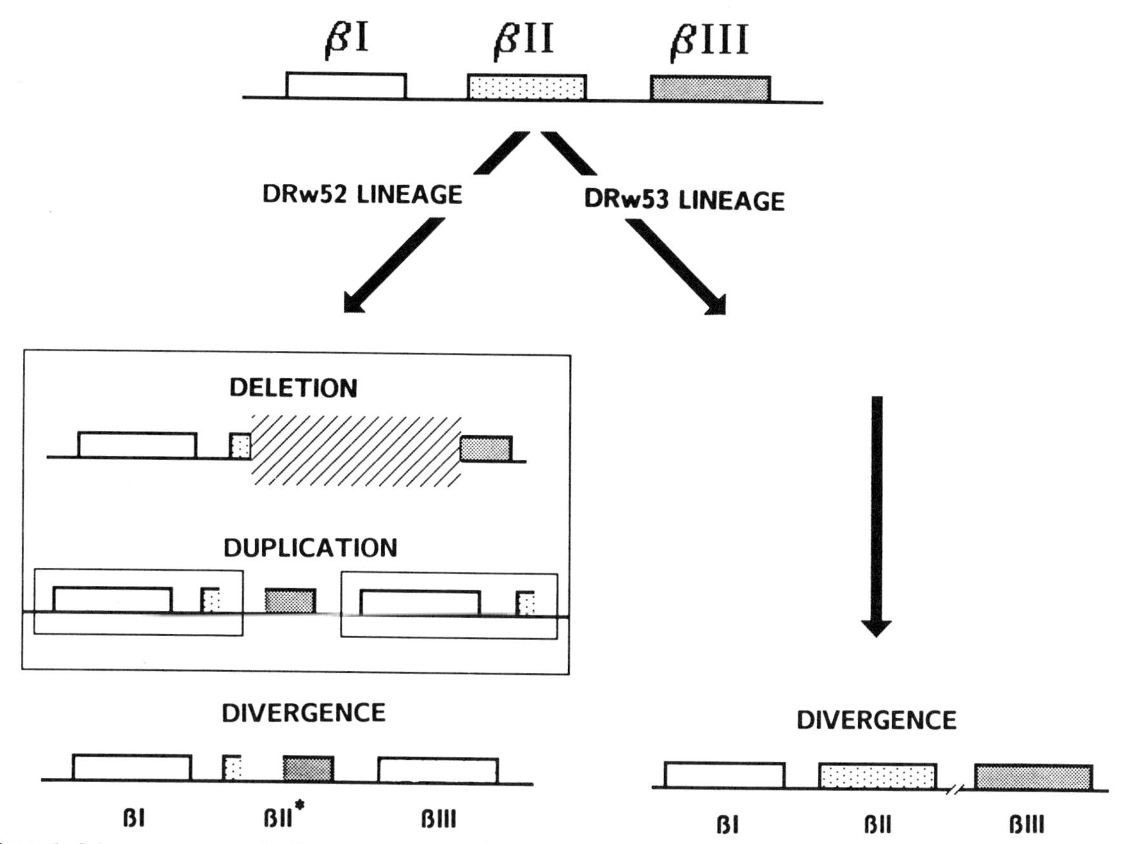

Figure 6. Scheme accounting for the generation of the DNA *DR* β chain polymorphism within the *DRw52* (*left*) and the *DRw53* (*right*) supertypic groups of *DR* haplotypes. The order of events within the box is not known. See text for details.

tive analysis, in particular in the context of insurance and employment.

Regulation of Class II Gene Expression

The expression of class II genes is tightly regulated developmentally, and in general there is a global regulation of the expression of *DR, DQ,* and *DP* α and β genes, together with the gene for the HLA-associated invariant chain (*Inv*), which is located on another chromosome (Long et al. 1983b). Induction of HLA class II genes by γ interferon also follows this global pattern involving the *DR, DQ, DP,* and *Inv* genes. We have recently shown by nuclear run-on experiments that the effect of γ interferon on class II gene expression involves induction at the transcriptional level (I. Amaldi and B. Mach, in prep.). The expression of class II genes can also be induced by other stimuli, in particular by the gene product of the *tat* gene of HTLV-I and HTLV-II retroviruses (Greene et al., this volume).

Instead of the global regulation of all class II subregions, one observes in some cases a "dissociation" of class II gene expression, generally with *HLA-DR* alone being expressed. This situation has been observed at certain stages of hemopoiesis and in certain leukemic and lymphoma cell lines. The mechanism responsible for this dissociation of expression is not known. With loci-specific probes for the individual *DR* β chain genes, we have recently observed that both locus $DR\beta_{I}$ and locus $DR\beta_{III}$ are expressed not only in B lymphocytes, but also in T-cell clones, in macrophages and in γ interferon–induced fibroblasts, in all cases with an abundance ratio of about 5 to 1 (J. Berdoz et al., in prep.).

Studies of certain congenital immunodeficiencies, characterized by an absence of class II antigens (class II SCIDs), have provided us with naturally occurring class-II-negative variants (Lisowska-Grospierre et al. 1985). In these patients, there is an absence of expression of all class II mRNAs, although the *Inv* chain mRNA is present (de Préval et al. 1985; Lisowska-Grospierre et al. 1985). The defect involves all cell types and cannot be corrected by γ interferon. Family studies have shown that the genetic defect in these patients does not map in the MHC. On the basis of these observations, we have proposed the existence of a recessive, autosomal, *trans*-active class II regulatory gene, affected in the patients with SCID (de Préval et al. 1985). It is not known if this class II regulatory gene is the same as the gene thought to be affected in in-vitro-generated, class-II-negative B-cell mutant lines (Gladstone and Pious 1980; Accolla 1983). In these two cell lines, we had observed an absence of mRNA for all class II genes (Long et al. 1984; Levine et al. 1985).

There are therefore three levels of regulation of HLA class II genes (Fig. 7A). In some cases, one observes a dissociation in the expression of individual subregions (Fig. 7A, line 1). In the case of the naturally occurring (SCIDs) or in-vitro-generated class-II-negative variants, the defect concerns class II genes but not other γ interferon–inducible genes such as the *Inv* chain gene (Fig. 7A, line 2). Finally, the effect of γ interferon, or of the *tat* gene product, involves the entire class II region, as well as the *Inv* chain gene (Fig. 7A, line 3). The most simple way to account for these observations is to postulate two levels of regulatory genes as outlined schematically on Figure 7B: (1) Upstream, a gene that is sensitive to γ interferon and controls class II and *Inv* genes, and (2) a second regulatory gene that specifically controls the expression of class II genes and that is affected in SCID patients.

Regulation of Class II Genes and HLA-linked Autoimmune Diseases

A number of important diseases show a low but significant degree of association with specific HLA haplotypes. The correlation is generally weak and is thought to involve, in addition to an MHC-linked gene, one or more additional factors such as viral infections. Heterogeneity in the diseases themselves also affects the degree of correlation. More importantly, the linkage has been reported and calculated only for the polymorphic MHC loci for which adequate typing procedures exist, initially for HLA class I and now for *HLA-DR*. It is quite possible, however, that the locus (or loci) responsible for disease susceptibility is another class II locus, such as *DQ* or *DP*, for which adequate typing has not existed.

Under normal circumstances, class II genes are only expressed in certain cell types such as macrophages, which can then function as APCs and lead to the stimulation of T lymphocytes. In a normal class II-positive cell, the relative amount of α and β chains of the different subregions, and their relative affinities is such that the class II α/β dimers formed are subregion-specific (i.e., $DR\alpha/DR\beta$ or $DQ\alpha/DQ\beta$).

In certain pathological situations, abnormal or aberrant, expression of class II antigens can take place (Bottazzo et al. 1983). We presume that this results from the local production of γ interferon and/or TNF α or β, which are produced upon viral infection. When class II antigens are induced in certain specialized endocrine cells such as the pancreas islet cells, these can then function as APCs, in association with tissue-specific autoantigens, and thus stimulate T lymphocytes. This localized T-cell proliferation initiates an autoimmune reaction, which can lead to a humoral antibody response directed against the tissue-specific antigen, but more importantly to the production of cytokines with cytotoxic activities or to the activation of antigen-specific cytotoxic T lymphocytes, which are capable of specifically destroying the HLA class II APC.

The key question has remained largely unanswered, namely, how can one explain the linkage between autoimmunity (pathology due to aberrant T-cell stimulation) and individual HLA specificities? In addition to individual variations due to exposure to environmental factors such as viral infections, why is there effective T-cell triggering only in certain haplotypes? We propose, briefly, two models that might account for this linkage, one based on micropolymorphism in the protein-coding region and the other based on allelic differences involving regulation of class II genes.

As the result of a particular polymorphism in the amino acid sequence, an Ia molecule made in certain rare haplotypes could allow a more effective presentation of some tissue-specific autoantigens to T cells. This "*Ir* gene" effect then results in an antigen-specific and therefore cell-specific T-cell response. One can easily argue that the lack of absolute correlation found with HLA typing (including that found with RFLP analysis) results from our inability to type with sufficient precision those class II loci that might be most relevant for autoimmunity, such as *DQ* and *DP*. This view implies that the DNA sequence of some class II genes (and the structure of Ia chains) will be found characteristically different in affected haplotypes (e.g., in insulin-

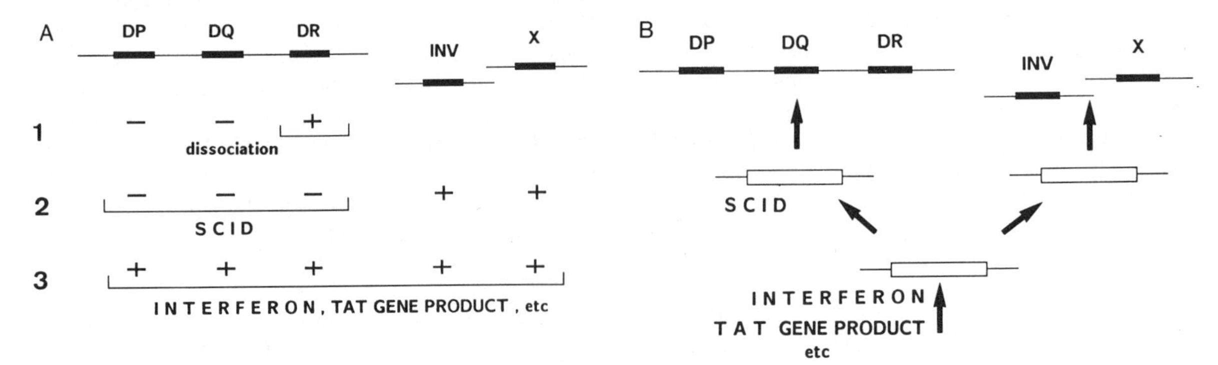

Figure 7. (*A*) The three levels of regulation of HLA class II genes. (*B*) Schematic representation of two distinct regulatory genes involved in the control of HLA class II gene expression. See text for details.

dependent diabetes [IDD] versus normal subjects). This will be settled very soon by searching for correlations, using sequence analysis and, more importantly, large-scale analysis with relevant oligonucleotides.

Alternatively, we would like to suggest that a feature of disease-prone haplotypes concerns the regulation of expression of class II genes. In such a case, allelic differences would affect promoter sequences (rather than amino acid–coding regions) and lead, in these susceptible individuals only, to an abnormal threshold of class II responsiveness. This will result in an abnormally high sensitivity to agents inducing class II genes (e.g., γ interferon and/or other cytokines such as TNF α or β) and to the induction (only in these hyperresponsive haplotypes) of some class II antigens in certain cells that are normally class-II-negative. When occurring in certain haplotypes (such as *DR3* for IDD), this aberrant expression allows an effective presentation (*Ir* gene effect) of a tissue-specific autoantigen and leads to T-cell triggering.

Polymorphism of class II gene regulation could also account for HLA-disease association on the basis of the known ability of Ia α and β chains to form "illegitimate" hybrid dimers across individual subregions (Germain and Quill 1986). In certain haplotypes, an allelic difference in the promoter of one given class II gene could result in the abnormal expression of a particular class II locus. As a consequence, an abnormal amount of that particular α or β chain will be produced and the normal balance of α and β chains made from each subregion will be upset (Fig. 8). This imbalance would result in the illegitimate pairing of α and β chains from different subregions. This could allow the synthesis, for instance, of *DX/DQ* or *DQ/DR* α/β dimers, which normally do not exist. The *DX* locus in particular does not seem to be expressed under normal conditions and might be inducible only in certain haplotypes. The newly formed hybrid Ia molecules would have, in certain cases, new specificities that could allow effective antigen presentation of tissue-specific antigens and lead to T-cell triggering.

We would like to propose therefore that HLA-associated susceptibility to autoimmune diseases is either

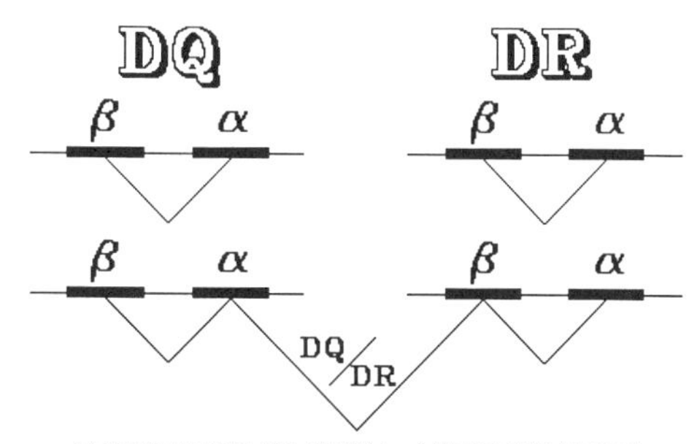

Figure 8. Normal α/β pairing within a given subregion and possible illegitimate hybrid formations. An abnormal level of expression of one of the gene products (e.g., *DQα*) would allow formation of DQ/DR illegitimate hybrid molecules.

due to an *Ir* gene effect with an allele-specific amino acid sequence at one of the class II loci or to polymorphic variations affecting the regulation of HLA class II genes (such as variations in promoter strength). This would lead to allele-specific differences in the abnormal expression of Ia antigens on Ia-negative cells. Such a polymorphism of HLA class II regulation might well account for the HLA-linked autoimmunity observed in the case of a number of important diseases.

ACKNOWLEDGMENTS

We gratefully acknowledge the technical assistance of S. Cuenod, M. Loche, C. Mattmann, and M. Zufferey. This work was supported by a grant from the Swiss National Science Foundation.

REFERENCES

Accolla, R.S. 1983. Human B cell variants immunoselected against a single Ia subset have lost expression of several Ia antigen subsets. *J. Exp. Med.* **157:** 1053.

Angelini, G., C. de Préval, J. Gorski, and B. Mach. 1986. High resolution analysis of the human HLA-DR polymorphism by hybridization with sequence-specific oligonucleotide probes. *Proc. Natl. Acad. Sci.* **83:** 4489.

Benacerraf, B. and H.O. McDevitt. 1972. Histocompatibility-linked immune response genes. *Science* **175:** 273.

Bottazzo, G.F., R. Pujol-Borrell, and T. Hanafusa. 1983. Role of aberrant HLA-DR expression and antigen presentation in induction of endocrine autoimmunity. *Lancet* **II:** 1115.

de Préval, C., B. Lisowska-Grospierre, M. Loche, C. Griscelli, and B. Mach. 1985. A *trans*-acting class II regulatory gene unlinked to the MHC controls expression of HLA class II genes. *Nature* **318:** 291.

Germain, R.N. and H. Quill. 1986. Unexpected expression of a unique mixed-isotype class II MHC molecule transfected L-cells. *Nature* **320:** 72.

Gladstone, P. and D. Pious. 1980. Identification of a *trans*-acting function regulating HLA-DR expression in a DR-negative B cell variant. *Somatic Cell Genet.* **6:** 285.

Gorski, J. and B. Mach. 1986. Polymorphism of human Ia antigens: Gene conversion between two DR β loci results in a new HLA-D/DR specificity. *Nature* **322:** 67.

Gorski, J., P. Rollini, E.O. Long, and B. Mach. 1984a. Molecular organization of the HLA-SB region of the human major histocompatibility complex and evidence for two SB β-chain genes. *Proc. Natl. Acad. Sci.* **81:** 3934.

Gorski, J., P. Rollini, E. Kawashima, E.O. Long, and B. Mach. 1984b. Complexity of the β chain genes of the HLA-D region. *UCLA Symp. Mol. Cell. Biol.* **18:** 47.

Gorski, J., R. Tosi, M. Strubin, C. Rabourdin-Combe, and B. Mach. 1985. Serological and immunochemical analysis of the products of a single HLA DR-α and DR-β chain gene expressed in a mouse cell line after DNA-mediated cotransformation reveals that the β chain carries a known supertypic specificity. *J. Exp. Med.* **162:** 105.

Gregersen, P.K., M. Shen, Q.-L. Song, P. Merryman, S. Degar, T. Seki, J. Maccari, D. Goldberg, H. Murphy, J. Schwenzer, C. Yi Wang, R.J. Winchester, G.T. Nepom, and J. Silver. 1986. Molecular diversity of HLA-DR4 haplotypes. *Proc. Natl. Acad. Sci.* **83:** 2642.

Klein, J. 1979. The major histocompatibility complex of the mouse. *Science* **203:** 516.

Korman, A.J., J.M. Boss, T. Spies, R. Sorrentino, K. Okada, and J.L. Strominger. 1985. Genetic complexity and expression of human class II histocompatibility antigens. *Immunol. Rev.* **85:** 45.

Levine, F., H.A. Erlich, B. Mach, and D. Pious. 1985. Transcriptional regulation of HLA class II and invariant chain genes. *J. Immunol.* **134:** 637.

Lisowska-Grospierre, B., D.J. Charron, C. de Préval, A. Durandy, C. Griscelli, and B. Mach. 1985. A defect in the regulation of MHC class II gene expression in HLA-DR negative lymphocytes from patients with combined immunodeficiency syndrome. *J. Clin. Invest.* **76:** 381.

Long, E.O., B. Mach, and R.S. Accolla. 1984. Ia-negative B-cell variants reveal a coordinate regulation in the transcription of the HLA class II gene family. *Immunogenetics* **19:** 349.

Long, E.O., C.T. Wake, J. Gorski, and B. Mach. 1983a. Complete sequence of an HLA-DR β chain deduced from a cDNA clone and identification of multiple non-allelic DR β chain genes. *EMBO J.* **2:** 389.

Long, E.O., C.T. Wake, M. Strubin, N. Gross, R.S. Accolla, S. Carrel, and B. Mach. 1982. Isolation of distinct cDNA clones encoding HLA-DR β chains by use of an expression assay. *Proc. Natl. Acad. Sci.* **79:** 7465.

Long, E.O., M. Strubin, C.T. Wake, N. Gross, S. Carrel, P. Goodfellow, R.S. Accolla, and B. Mach. 1983b. Isolation of cDNA clones for the p33 invariant chain associated with HLA-DR antigens. *Proc. Natl. Acad. Sci.* **80:** 5714.

Mach, B., C. de Préval, P. Rollini, C. Berte, and J. Gorski. 1986. Polymorphism and regulation of HLA class II genes. In *Proceedings of the 11th American Society of Histocompatibility and Immunogenetics Conference, 1985* (ed. B. Schacter). (In press).

Rabourdin-Combe, C. and B. Mach. 1983. Expression of HLA-DR antigens at the surface of mouse L cells cotransfected with cloned human genes. *Nature* **303:** 670.

Rask, L., K. Gustafsson, D. Larhammar, H. Ronne, and P.A. Peterson. 1985. Generation of class II antigen polymorphism. *Immunol. Rev.* **84:** 123.

Rollini, P., B. Mach, and J. Gorski. 1985. Linkage map of three HLA-DR β-chain genes: Evidence for a recent duplication event. *Proc. Natl. Acad. Sci.* **82:** 7197.

Rougeon, F., P. Kourilsky, and B. Mach. 1975. Insertion of a rabbit β globin gene into *E. coli* plasmid. *Nucleic Acids Res.* **2:** 2365.

Spies, T., R. Sorrentino, J. Boss, K. Okada, and J. Strominger. 1985. Structural organization of the DR subregion of the major histocompatibility complex. *Proc. Natl. Acad. Sci.* **82:** 5165.

Tieber, V.L., L.F. Abruzzini, D.K. Didier, B.D. Schwartz, and P. Rotwein. 1986. Complete characterization and sequence of an HLA class II DRβ chain cDNA from the DR5 haplotype. *J. Biol. Chem.* **261:** 2738.

Trowsdale, J., J.A.T. Young, A.P. Kelly, P.J. Austin, S. Carson, H. Meunier, A. So, H.A. Erlich, R.S. Spielman, J. Bodmer, and W.F. Bodmer. 1985. Structure, sequence and polymorphism in the HLA-D region. *Immunol. Rev.* **85:** 5.

Wake, C.T., E.O. Long, and B. Mach. 1982. Allelic polymorphism and complexity of the genes for HLA-DR β-chains—Direct analysis by DNA-DNA hybridization. *Nature* **300:** 372.

Molecular Biology of the Class II Region of the Human Major Histocompatibility Complex

J.I. Bell, D. Denney, L. Foster, B.S.M. Lee, D. Hardy, and H.O. McDevitt

Department of Medicine & Medical Microbiology, Stanford University School of Medicine, Stanford, California 94305

The class II region of the human major histocompatibility complex (MHC) consists of at least 14 loci clustered together on the short arm of chromosome 6 (Bell et al. 1985a). At least 9 of the loci are expressed, and they encode either the α or the β chains that associate on the surface of B cells, activated T cells, and macrophages and that are involved in the generation of a normal immune response. The class II region has been subdivided into three subregions, the *DR* subregion composed of one α and three β chains (Spies et al. 1985), the *DQ* subregion composed of two α and β chain loci (Okada et al. 1985a), and the *DP* subregion, which contains two α and two β chains (Trowsdale et al. 1984; Okada et al. 1985b). Two additional loci have been described, a *DO*β locus (Tonnelle et al. 1985) and a *DZ*α locus (Trowsdale and Kelly 1985). These two loci are both expressed, although not coordinately (Tonnelle et al. 1985). Although these loci have all been identified, no comprehensive map order of the human class II region is available. The development of a technique for studying large genomic DNA fragments (>100 kb) with pulsed-field gradient gel electrophoresis has permitted the mapping of the loci of this region.

The function of these molecules is related to their ability to present foreign antigen to T helper cells. The extensive polymorphism that exists within these genes serves to produce varying abilities to respond immunologically to specific antigenic stimuli. For this reason, these genes compose an important group of immune response genes. In addition to determining the responsiveness of an individual to distinct foreign antigenic challenges, the polymorphism in these molecules has been shown to determine, in part, susceptibility in a variety of diseases in which immune reactivity to self antigens is the primary pathophysiological process. There exist more than 40 different diseases for which associations with particular class II polymorphisms have been described (Stastny et al. 1983).

Classically, the polymorphism within this region has been defined serologically and with the mixed lymphocyte reaction (MLR). Primed lymphocyte typing has permitted the characterization of the polymorphism within the *DP* subregion (Sanchez-Perez and Shaw 1985). The use of two-dimensional gel electrophoresis, in conjunction with monoclonal antibodies, has also contributed to our knowledge of the polymorphism of this region. The development of molecular biological approaches for studying this polymorphism has recently permitted the characterization of previously undetected polymorphism and the determination of the molecular basis of the polymorphism. Two approaches have proved most fruitful: the use of restriction-fragment-length polymorphisms (RFLPs) to study variation within populations, and allelic nucleic acid sequencing of cDNA clones, which has permitted a more precise description of the polymorphism. We have used these techniques first to study the polymorphism present in the normal population and then to extend these studies to patients with diseases associated with serologically defined class II polymorphism.

METHODS

Pulsed-field Gradient Gel Electrophoresis. DNA was extracted from the lymphoblastoid cell line WT51 (DR4,4) in agarose plugs. Nuclei were prepared by centrifugation through a sucrose cushion and resuspended at 4×10^7 to 6×10^7 nuclei/ml of RSB (10 mM NaCl, 10 mM Tris [pH 7.5], 25 mM EDTA) (Hieter et al. 1981). They were embedded in an equal volume of 1.0% low-melt agarose (Bio-Rad) in RSB containing 60 μg/ml of proteinase K and cooled on ice. The nuclei were lysed in RSB with 1.0% SDS at 50°C for 16–24 hr, washed in 10 mM Tris, 10 mM EDTA, and equilibrated in restriction enzyme buffer. The DNA was digested, and the plugs were melted at 65°C and loaded into wells of a 1.0% agarose gel in 0.5% TBE. The gel was run at 300 volts for 6–12 hr. 0.5% TBE buffer was circulated cold so that the temperature was maintained at 11°C. The pulse interval was varied between 10 and 60 sec to resolve the different size bands.

After electrophoresis, the gels were stained with ethidium bromide and either irradiated at 245 nm for 2 min or treated with 0.25 M HCl for 30 min. Denaturation was performed in 0.5 M NaOH and 1.5 M NaCl for 30 min, and neutralization was performed in 0.25 M Na_2HPO_4/NaH_2PO_4 (pH 6.8) for 30 min. Southern transfer was performed by standard techniques, and the DNA fixed by UV irradiation. The filters were prehybridized and hybridized by standard techniques and probed with *DR, DQ, DP, DZ,* and *DO* probes labeled by hexamer priming to yield specific activities of 1×10^8 to 1×10^9 cpm/μg of DNA. The filters were hybridized for 16 hr at 42°C and washed in 0.1% SDS and 2× SSC at 25°C for 1 hr and in 0.1% SDS and 0.1× SSC at 65°C for 1 hr. The filters were exposed and then washed at 70°C in 5 mM Tris (pH 8.0), 0.2 mM Na_2 EDTA, 0.05% sodium pyrophosphate, 0.002% polyvinyl pyrrolidine, 0.002% bovine serum albumin, and 0.002% Ficoll (m.w. 400,000) before reprobing with other probes.

RFLP sequence analysis. Standard restriction fragment patterns were obtained from lymphoblastoid cell lines homozygous at the MHC by consanguinity. The enzymes *Bam*HI, *Bgl*II, *Eco*RI, *Hin*dIII, *Pvu*II, and *Hin*cII were all utilized. DNA was extracted by conventional techniques, digested, and electrophoresed on 0.8% agarose gels. Southern transfer was performed by standard techniques, and the filters were hybridized with a *DRβ* (2918.4) and a *DQβ* (2918.8) probe (Bell et al. 1985b). DNA from diseased individuals was isolated from whole blood, as previously described, and blotted in the same manner according to the Southern technique.

Sequence analysis was performed on cDNA clones obtained from cDNA libraries made from the homozygous consanguineous cell lines LG2 (DR1), PGF (DR2), WT87 (tb24), WT49 (DR3), WT51 (DR4), and LBF (DR7). These libraries were made using a modification of homopolymer tailing and ligated into λgt10. Screening was performed using standard techniques. Sequence analysis was done using the chain termination method with the modification of [^{35}S]dATP and salt gradient gels. Sequencing of whole cDNA inserts was performed using synthetic oligonucleotides specific for conserved regions of the alleles.

RESULTS AND DISCUSSION

Genomic Organization of the Class II Region as Determined by Using Pulsed-field Electrophoresis

Cosmid cloning studies have permitted the linking of individual loci within subregions, but to date they have failed to link the subregions and determine the overall size of the class II region (Trowsdale et al. 1984; Okada et al. 1985a; Spies et al. 1985). This has been due, in part, to the frequency of repeat sequences in this region and the apparent large size of this region relative to the mouse. Family studies have permitted the ordering of *DP* centromeric to *DR* and *DQ* (Shaw et al. 1981), but no order has been determined for either of the subregions *DR* or *DQ*, or the individual loci *DO* or *DZ*. We have chosen to apply the technique of pulsed-field gradient gel electrophoresis to study this problem (Hardy et al. 1986). This technique was originally utilized for separation of yeast and trypanosome chromosomes (Carle and Olsen 1984; Schwartz and Cantor 1984). With some modifications, we have utilized it to study large fragments of human genomic DNA. We digested high-molecular-weight DNA in agarose plugs with restriction enzymes with either 8-bp recognition sites or recognition sites with frequent CG dinucleotides. CG dinucleotides are found less commonly than expected in the human genome (Bird 1980). Digested DNA was electrophoresed using pulsed-field gradient gels and transferred using classical techniques (Southern 1975) to nylon membranes and probed with *DR, DQ, DP, DO,* and *DZ* probes. Probes for each of the subregions are readily available (Bell et al. 1985b; Tonnelle et al. 1985; Trowsdale and Kelly 1985). α and β chain probes were used sequentially to avoid any problem of cross-hybridization, although at the stringencies used, none would have been predicted. This approach has allowed us to establish the order of the subregion within this region. A number of fragments hybridized to both *DR* and *DQ* probes, establishing their linkage. These included *Sac*II 500- and 450-kb fragments, *Not*I, *Mlu*I, and *Pvu*I fragments, and a BSSHII 325-kb fragment. Digests with the enzymes BSSHII and *Sal*I resulted in fragments hybridizing to *DQ* and *DO* but not to *DR*. This established that *DQ* separated the *DOβ* locus and the *DR* subregion. *DP* and *DZ* were invariably found on the same fragment, indicating that they were very near each other. Many enzymes shared this linkage, including *Sac*II, *Cla*I, *Pvu*I, BSSHII, *Sal*I, and *Nar*I. Two fragments, a *Pvu*I and a *Cla*I fragment, had both *DP* and *DZ* as well as *DO* on them. This allowed the localization of *DO* telomeric to *DP/DZ* and centromeric to *DQ* and *DR*. Additional information was obtained from a BSSHII digest that resulted in three fragments: a 200-bp fragment hybridizing to *DO* and one of the *DQ* loci, a 325-bp fragment hybridizing to the second *DQ* locus and *DRβ*, and a 300-bp fragment hybridizing to *DRα* only. On some digests the restriction site between these two latter fragments was not cleaved, yielding a 625-kb band containing *DQ, DRβ*, and *DRα*. This series of digests allows us to place *DRα* telomeric to *DRβ*. We are confident about the linkage data using this technique because we have utilized the data from a large number of enzyme digests. This is important because of the risk of comigrating bands. All of our data fit consistently into a single map of the region, and multiple overlaps and ambiguities have been clarified by using double digests or by varying pulse times.

The complete ordering of this region is now possible if these data are combined with those available from cosmid mapping and deletion mutant studies (Fig. 1). *DRα* is telomeric to the three *DRβ* chains, and all these have been linked by cosmid clones (Rollini et al. 1985). The *DQ* subregion is centromeric to the *DR* subregion, the *DXα* and β pair being centromeric to the *DQα* and β pair (Erlich et al. 1986). *DOβ* is centromeric to *DQ*, and *DZ* is adjacent to *DP* and centromeric to *DO*. Deletion mutants have defined the location of *DZ* as being telomeric to *DP* (Erlich et al. 1986). The order of the loci is shown to scale in Figure 2.

The size of this region is of interest. Cosmid-walking data from the mouse have demonstrated the class II region in that species to be 260 kb in size (Steinmetz et al. 1984). From our study, the size of the human class

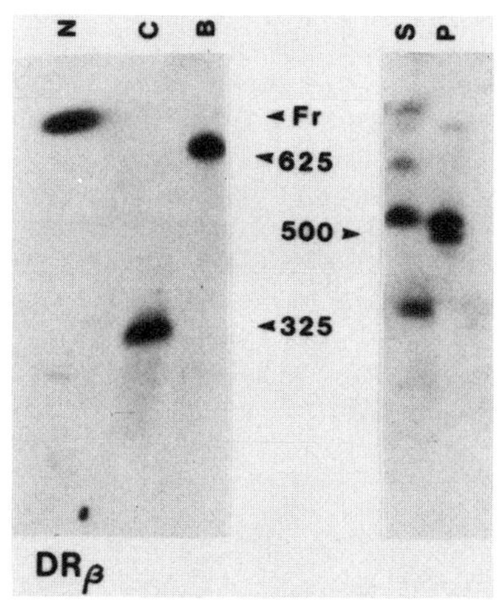

Figure 1. Example of a pulsed-field electrophoresis gel blotted and probed with *DRβ*. Enzymes used were *Not*I (N), *Cla*I (C), BSSHII (B), *Sal*I (S), *Pvu*I (P).

II region is 1100 kb. This can be determined by adding three adjacent *Cla*I fragments or three adjacent BSSHII fragments. Both these calculations lead to an estimate of 1.1 megabases. There has, therefore, been significant change in the size of this region over evolutionary time. The order of the loci, however, remains similar between man and mouse. *Aβ3* and *Aβ2* are homologous to *DPβ* and *DOβ*, while *Eβ* and *Aβ* are homologous to *DRβ* and *DQβ*, respectively. The order of these loci in the mouse is identical with that seen in man. The differences between the class II regions of the mouse and man can be explained by several events. The *DQ* loci could have been duplicated to account for the presence of the *DXα-β* pair. A deletion between *Aβ3* (*DPβ*) and *Aβ2* (*DOβ*) could account for the loss of two *DPα* loci, one *DPβ* locus, and a *DZα* locus. This would, in part, help to explain the difference in size between the mouse and man. Of interest is that the distance between *DR* and *DQ* and between *DQ* and *DP* is very similar. There is very little recombination between *DR* and *DQ*, but there is a recombination frequency of 2–3% between *DQ/DR* and *DP* (Sanchez-Perez and Shaw 1985). This difference cannot be explained by physical distance and most probably results from a recombination "hot spot" between *DQ* and *DP*.

These studies are the first to demonstrate the power of the technique of pulsed-field gradient gel electrophoresis in mapping large segments of the human genome. It fills an important role in establishing a physical map for the region that had not been successfully achieved using other techniques.

MHC Class II Polymorphism

We initiated our studies of the polymorphism within the class II region of the MHC by utilizing a set of cell lines, genetically homozygous at the MHC, through consanguinity. These lines have allowed us to study one haplotype at a time and to establish the RFLP patterns predicted for given serological specificities. They have also allowed us to identify subtypes of haplotypes defined by RFLP by comparing the patterns derived from several representatives of each serologically defined haplotype.

We have used six different restriction enzymes to study 30 different cell lines representing the *DR* types 1–9 (J.I. Bell et al., unpubl.). We have probed Southern blots of these lines with *DR* and *DQβ* chain probes. The patterns obtained with the *DRβ* probe in part reflect the standard serological specificities but more predominantly reflect those of the supertypic specificities MT2 and MT3. These patterns are similar between the haplotypes *DR3, 5,* and *6* and between the haplotypes *DR4, 7,* and *9* (see Fig. 3).

Another constant feature of the band patterns seen in the different haplotypes utilizing a *DRβ* probe is that the *DR4* and *DR7* haplotypes have more bands than the MT2-associated haplotypes or the *DR1* or *DR2* haplotypes. When these bands are compared with the band sizes predicted from cosmid cloning studies with *Eco*RI and *Hin*dIII restriction enzymes, at least two bands with each enzyme are not accountable from the predicted bands on the *DR* subregion. This suggests that in the MT3-associated haplotypes an additional *DRβ* chain may be present but has not been identified with cosmid clones.

Polymorphism beyond that previously predicted can be identified using RFLPs on these consanguineous lines. Polymorphic bands can be identified that subdivide most haplotypes using the *DRβ* probe. In particular, *DR3* can be subdivided with every enzyme into two distinct subtypes. These polymorphisms imply the presence of structural polymorphism beyond that described serologically. The *DQβ* patterns observed reveal distinct patterns for each of the *DR* haplotypes, except *DR2* and *DR6,* which are often identical. There are substantial differences between the haplotypes *DR3* and *DR7* (*DQw2*), and between *DR4* and *DR5* (*DQw3*). These patterns support the protein amino-terminal se-

Figure 2. Scale map of the human class II region showing the order of the known loci and the distance between them as determined by pulsed-field electrophoresis mapping.

	13	12	11	10	7.6	7.0	6.5	6.3	5.8	5.6	5.2	4.9	4.5	4.1	3.2	3.1	3.0	2.9	2.7	2.6	1.9	1.5	(Blot #)
DR 1																							
BVR		:	X		X		I		:	I			I					:			[I]		A23
IBW4		:	X		X		I		:	I			I					:			[I]		A9,57,67R
OOS		:	X		X		I		:	I			I					:			[I]		A9,JB2
MVL		:	X		X		I		:	I			I					:			[I]		LF3
DR 2																							
AKIBA			X	X	I	[I]	I		:			[I]	I	I		X							A9,25
IWB			X	X	I	[I]	I		:			[I]	I	I		X		:					A23,25,28,67R
JHE			X	X	I	[I]	I		:			[I]	I	I		X		:					A28
PGF			X	X	I	[I]	I		:			[I]	I	I		X		:					A9,25,57,JB2
DR 3																							
AVL				X		[I]	[I]		:		I	I	I	I			:	X				:	A9,28,LF3
WT20				X		[I]	[I]		:			I	I	I			:		X			:	A23,JB1
QBL				X		[I]	[I]		:			I	I	I			:		X			:	A28
HAR				X		[I]	[I]		:		I	I	I	I			:	X				:	A28
WT49				X		[I]	[I]		:			I	I	I			:		X			:	A28,57,JB2
DR 4																							
BSM	X		X	:				[X]	:	I		I	I	[X]	I		I	I					A10,27
WT51	X		X	:				[X]	:	I		I	I	[X]	I		I	I					A10,27,JB2
DR 5																							
JVM				X		[I]	[I]		:			I	I	I				X					A10,27,JB2
IDF				X		[I]	[I]		:			I	I	I				X					A27
THR				X		[I]	[I]		:			I	I	I				X					A10,27,57
DHI				X		[I]	[I]		:			I	I	I				X					A67R
FPF				X		[I]	[I]		:		I		I	I		:		X					A10,27,LF4
DR 6																							
APD				X		[I]	[I]		:		I		I	I		:		I					A27
WVD				X		[I]	[I]		:		I		I	I		:	:			I			A27,57
WT46				X		[I]	[I]		:		I		I	I		:	:						JB1,JB2
DR 7																							
LBF	X							[X]	:	I		I	I	[X]	I		I	I					A27,57,JB2
IBW9	X		:					[X]	:	I		I	I	[X]	I		I		I				JB1,LF4

Sizing based on A27,28

Figure 3. Class II RFLPs observed in Southern blots of homozygous consanguineous cell lines digested with *Bgl*II and probed with *DRβ*. Fragment sizes (in kilobases) are listed across the top, and band intensities are given as follows: light (:); medium (I); strong (X).

quence data, which suggest that the *DQβ* chains of each *DR* haplotype are distinct (Giles et al. 1985).

Subtypes of the haplotypes *DR2* and *DR4* are of particular interest. These two haplotypes have been shown by cellular typing to be diverse. The DR4 subtypes have been defined at the molecular level by cloning and sequencing representatives of the different MLR specificities. The differences between these are not reflected in RFLP patterns and consist of several hypervariable residues in the third variable region of the first domain.

The situation is much different for DR2 subtypes. We have studied the different DR2 subtypes by using two-dimensional gel electrophoresis and RFLPs (B.S. Morley-Lee et al., in prep.). The subtypes we have chosen to study are DR2 Dw2, DR2 Dw12, DR2 tb24, DR2 AZH, and DR2 REM. They have been probed with *DRβ*, *DQβ*, and *DQα* probes after digestion with a range of enzymes. The restriction fragment patterns clearly distinguish DR2 Dw2, DR2 Dw12, and the DR2/Dw2/Dw12 group as three distinct groups of *DR2* haplotypes for the probes *DRβ* and *DQβ*. With these probes, most of the bands are shared between *DR2* haplotypes, but some bands are unique to either Dw2,Dw12, or non-Dw2/Dw12 cell lines (Fig. 4). Two-dimensional gels of these same cell lines show similar variability with the anti-DR monoclonal antibody SG171. Two β chain spots are seen with all the lines. One spot is invariant between all the DR2 lines, but the second spot is variable between Dw2, Dw12, and non-Dw2/Dw12 lines. Variation exists between these lines at the *DQβ* locus, as well. RFLPs and two-dimensional gels both show clear differences between the same three groups of DR2 subtypes using the *DQβ* probe. Less polymorphism is seen associated with *DQα*. These studies suggest, therefore, that the *DR2* haplotype, as serologically defined, represents a supertypic family akin to that seen with MT2 and MT3 sera. Within this family, distinct structural variation occurs at one *DRβ* locus and at the *DQβ* locus. Nucleic acid sequencing studies should resolve these differences at a more precise molecular level.

Nucleic Acid Sequencing

We have utilized our collection of homozygous consanguineous lines to obtain sequence data defining the polymorphism seen between different haplotypes (see Fig. 5). This approach has allowed us to obtain sequences that may be unambiguously attributed to certain haplotypes. These studies have confirmed that in most haplotypes studied to date (*DR2, 3, 4, 5,* and *7*), there are two *DRβ* chains expressed. In the *DR1* haplotype we have identified only a single *DRβ* chain. The sequence variation between these haplotypes in the *DR* subregion indicates that one β chain is variable between all the serological haplotypes studied. The second *DRβ* chain shows strong similarity between the haplotypes sharing the same supertypic specificity MT2 or MT3. For example, a *DRβ* chain sequenced from the *DR3* haplotype is identical with one obtained from the *DR6* and *DR5* haplotypes (MT2), and a *DRβ* chain from the *DR4* haplotype is identical with one obtained from a *DR7* haplotype (MT3). Although this does not necessarily indicate that this locus is the one determining the supertypic serological reactivity, it does imply that the haplotypes associated with a common supertypic type may have evolved from a common precursor (Gorski and Mach 1986). No distinct sequence differences will differentiate *DRβ* alleles of one locus from the other. Although two-dimensional gel patterns imply that the two *DRβ* chains are expressed at approximately the same level, our experience from screening cDNA libraries is that one *DRβ* chain is present at approximately 10% of the frequency of the other.

The polymorphism in human class II molecules is similar to that described in murine α and β chains (Benoist et al. 1983; Mengle-Gaw and McDevitt 1984; Estess et al. 1986). In *Aβ* chains, the polymorphism is found between residues 8–17, 60–66, and 85–89, while in *DQβ* chains it is largely confined to the residues 52–57, 70–78, and 84–90. *Aα* chains also have the majority of their polymorphism confined to the first domain at residues 11–15, 53–59, and 69–77. *DQα* chains have variability in one long stretch between residues 45–55. The *Eα* and *DRα* chains are nonpolymorphic, but *Eβ* has three hypervariable regions, 1–13, 24–35, and 68–75, and *DRβ* has variability between 9–13, 26–38, and 70–77.

The polymorphism is abundant in the first domains of these molecules and most probably has arisen by mutation followed by selection in the regions of hypervariability. The mechanism generating the polymorphism is most probably either point mutation or gene conversion. The latter mechanism was clearly described in studies of the *bm12* mutant in the mouse (McIntyre and Seidman 1984) and has now been confirmed in human sequencing studies of *DRβ* chains (Gorski and Mach 1986). Both processes are probably involved in the generation of human class II polymorphism.

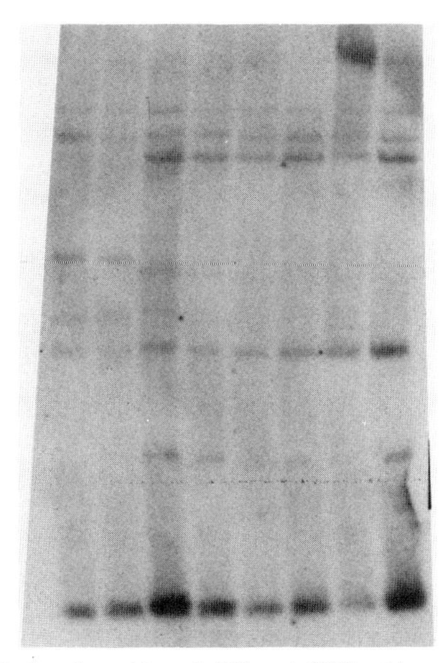

Figure 4. Southern blot of different *DR2* subtypes digested with the restriction enzyme *Pst*I and probed with a *DRβ* chain probe. Lanes from *left* to *right*: PGF (Dw2); IWB (Dw2); Akiba (Dw12); BGE (Dw12); REM (LD-5a); WT87 (tb24); FJO; and AZH. RFLPs clearly distinguish Dw2 from Dw12 and from non-Dw2/Dw12 lines.

Class II Polymorphisms and Human Autoimmune Disease

One of the goals of these studies of class II polymorphism is to identify subtypes of haplotypes and discrete sequence variations associated with diseases known to be associated with MHC alleles. RFLP studies provide a strong indication that certain class II polymorphisms not yet serologically defined but identified with restriction fragments are associated with disease. In particular, Nepom has identified a *DQβ* *Taq*I polymorphism in DR4 insulin-dependent diabetes patients that is strongly correlated with disease (Ne-

DR1	Gly Asp Thr Arg Pro Arg Phe Leu Trp Gln Leu Lys Phe Glu Cys His Phe Phe Asn Gly Thr Glu Arg Val Arg Leu Leu Glu Arg Cys Ile Tyr Asn Gln Glu Glu Ser Val Arg Phe
DR2a	--- --- --- --- --- --- --- --- --- --- Asp --- Tyr --- --- --- --- --- --- --- --- --- --- --- --- Phe --- His --- Asp --- --- --- --- --- --- Asp Leu --- ---
DR2b	--- --- --- --- --- --- --- --- Gln --- Pro --- Arg --- --- --- --- --- --- --- --- --- --- --- --- Phe --- Asp --- Tyr Phe --- --- --- --- --- --- --- --- ---
DR3a	--- --- --- --- --- --- --- --- Glu Tyr Ser Thr Ser --- --- --- --- --- --- --- --- --- --- --- --- Tyr --- Asp --- Tyr --- His --- --- --- --- Asn --- --- ---
DR3b (MT2)	--- --- --- --- --- --- --- --- Glu Leu --- --- Ser --- --- --- --- --- --- --- --- --- --- --- --- Phe --- --- --- Asn Phe His --- --- --- --- Tyr Ala --- ---
DR3b (MT2b)	--- --- --- --- --- --- --- --- Glu Leu Arg --- Ser --- --- --- --- --- --- --- --- --- --- --- --- Tyr --- Asp --- Tyr --- --- --- --- --- --- Phe Leu --- ---
DR4a	--- --- --- --- --- --- --- --- Glu --- Val Trp His --- --- --- --- --- --- --- --- --- --- --- --- Phe --- Asp --- Tyr Phe --- His --- --- --- --- Tyr --- --- ---
DR4b (MT3)	--- --- --- --- --- --- --- --- Glu --- Ala Trp Cys --- --- --- --- Leu --- --- --- --- --- --- Trp Asn --- Ile --- Tyr Ile --- --- --- --- --- --- Tyr Ala --- Tyr
DR7a	--- --- --- Gln --- --- --- --- --- --- Gly --- Tyr Lys --- --- --- --- --- --- --- --- --- --- Gln Phe --- --- --- Leu Phe --- --- --- --- --- --- Phe --- --- ---
DR7b (MT3b)	--- --- --- --- --- --- --- --- Glu Glu Ala Trp Cys --- --- --- --- --- --- --- --- --- --- --- Trp Asn --- Ile --- Tyr Ile --- --- --- --- --- --- Tyr Ala --- Tyr

DR1	Asp Ser Asp Val Gly Glu Tyr Arg Ala Val Glu Glu Leu Gly Arg Pro Asp Ala Glu Tyr Trp Asn Ser Gln Lys Asp Leu Leu Glu Gln Arg Arg Ala Ala Val Asp Thr Tyr Cys Arg His
DR2a	--- --- --- --- --- --- --- --- --- --- Thr --- --- --- --- --- --- --- --- --- --- --- --- --- --- --- Phe --- --- Asp --- --- --- --- --- --- Asn --- --- --- ---
DR2b	--- --- --- --- --- --- Phe --- --- --- Thr --- --- --- --- --- --- --- --- --- --- --- --- --- --- --- Ile --- --- --- Ala --- --- --- --- --- Asn --- --- --- ---
DR3a	--- --- --- --- --- --- Phe --- --- --- Thr --- --- --- --- --- --- --- --- --- --- --- --- --- --- --- --- --- --- --- Lys --- Gly Arg --- --- Asn --- --- --- ---
DR3b (MT2)	--- --- --- --- --- --- --- --- --- --- Arg --- --- --- --- --- --- --- --- --- --- --- --- --- --- --- --- --- --- --- Lys --- Gly Arg --- --- Asn --- --- --- ---
DR3b (MT2b)	--- --- --- --- --- --- --- --- --- --- Thr --- --- --- --- --- Val --- --- Ser --- --- --- --- --- --- --- --- --- --- Lys --- Gly Arg --- --- Asn --- --- --- ---
DR4a	--- --- --- --- --- --- --- --- --- --- Thr --- --- --- --- --- --- --- --- --- --- --- --- --- --- --- --- --- --- --- Trp --- --- --- --- --- --- --- --- --- ---
DR4b (MT3)	Asn --- --- Leu --- --- --- Gln --- --- Thr --- --- --- --- --- --- --- --- --- --- --- --- --- --- --- --- --- --- Arg --- --- --- Glu --- --- --- --- --- --- Tyr
DR7a	--- --- --- --- --- --- --- --- --- --- Thr --- --- --- --- --- Val --- --- Ser --- --- --- --- --- --- Ile --- --- Asp --- --- Gly Gln --- --- --- Val --- --- ---
DR7b (MT3)	Asn --- --- Leu --- --- --- Gln --- --- Thr --- --- --- --- --- --- --- --- --- --- --- --- --- --- --- --- --- --- Arg --- --- --- Glu --- --- --- --- --- --- Tyr

Figure 5. List of amino acid sequences of the first domain of $DR\beta$ chains showing the three major hypervariable regions.

pom et al. 1986). In addition, an RSAI *DQβ* RFLP identifies a *DQ2* subtype strongly associated with coeliac disease (Howell et al. 1986), and Festenstein et al. (1986) have utilized several *DQα*- and *β*-associated polymorphisms to study diabetic *DR4* haplotypes, identifying several haplotypes strongly associated with disease. Pemphigus vulgaris is another disease in which RFLPs strongly suggest a unique allele associated with disease. In more than 20 patients with this disease, a unique pattern of RFLPs has been found in DR4 and DR6 patients compared with DR4 and DR6 controls (L. Steinman, pers. comm.). We have studied the association between DR3 and myasthenia gravis by using subtypes of DR3 defined by homozygous lines (Bell et al. 1986). A *Hinc*II *DQβ* polymorphism was noted to be present infrequently in DR3$^+$ diabetics, coeliac patients, and patients with premature ovarian failure. The 15.0-kb band associated with this polymorphism was seen rarely in DR3$^+$ individuals and in most DR7$^+$ individuals. When myasthenia gravis patients were studied, all DR7$^+$ patients had the 15.0-kb band, as did 7 of 15 DR3$^+$ patients (Fig. 6). Other myasthenia gravis patients not having DR3 or DR7 have also now been demonstrated to have the polymorphic band despite the observation that this is almost never seen in the non-DR3$^+$/DR7$^+$ normal population. The association of this band to diseases in DR3 patients alone increases the relative risk to 32.

RFLP data would therefore indicate that polymorphisms exist that are strongly associated with disease; although they are most likely in flanking regions, they still probably are linked to coding-region polymorphisms. Of interest is the observation that all of the well-defined disease-associated RFLPs are associated with either the *DQ* or *DX* loci, and none have been associated with the *DR* loci.

Two examples of class II variation at the level of nucleic acid sequence would support the idea that such class II polymorphism may be very important in conferring susceptibility to disease on individuals. In the model of experimental myasthenia gravis, the *b* haplotype mouse is normally susceptible to disease. In mice possessing the *bm12* mutation, known to alter only three amino acid residues in the $A\beta^b$ allele, susceptibility is lost (Christadoss et al. 1985). This demonstrates that susceptibility is directly correlated with a small sequence variation in the class II $A\beta$ allele. A second example of a class II sequence variation relevant to disease is in the nonobese diabetic (NOD) mouse (H. Acha-Orbea and H.O. McDevitt, in prep.). In this mouse, only the $A\beta A\alpha$ heterodimer is expressed. The $A\alpha$ chain has a sequence identical with $A\alpha^d$ as does the second domain, the transmembrane region, the cytoplasmic domain, and the 3′ untranslated region of the $A\beta$ allele. However, the first domain, leader sequence, and 5′ untranslated region of this $A\beta$ allele are remarkably different from other $A\beta$ alleles. It differs in the first domain by 21–39 bp, leading to 10–19 variable amino acids. One region of the first domain has 5 consecutive amino acid changes between NOD and all of the other haplotypes. Because one of the susceptibility genes for NOD maps to the MHC, this striking first domain seems a likely candidate to be a disease-susceptibility allele.

The only means to establish whether similar sequence variation lies buried within the human class II region is to obtain sequences from multiple patients with disease and to compare their sequences to those found in wild-type individuals. This work is currently underway, and hopefully it will emerge that specific sequences can be used to identify alleles that carry with them a very high risk of developing certain autoimmune diseases. If so, these can be used to identify individuals at risk and will bring us a step closer to defining the structural basis for the biological sequence of events leading to disease that remains completely unexplored.

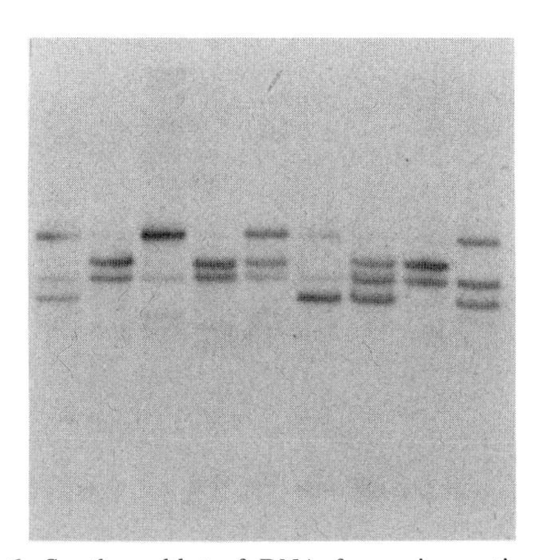

Figure 6. Southern blot of DNA from nine patients with myasthenia gravis. The upper band is 15 kb in size and represents the polymorphic band seen in DR3 and DR7 myasthenia patients.

REFERENCES

Bell, J.I., D.W. Denny, Jr., and H.O. McDevitt. 1985a. Structure and polymorphism of murine and human class II major histocompatibility antigens. *Immunol. Rev.* **84:** 51.

Bell, J.I., P. Estess, T. St. John, R. Saiki, D.L. Watling, H.A. Erlich, and H.O. McDevitt. 1985b. DNA sequence and characterization of human class II major histocompatibility complex b chains from the DR1 haplotype. *Proc. Natl. Acad. Sci.* **82:** 3405.

Bell, J.I., L. Rassenti, S. Smoot, K. Smith, C. Newby, R. Hohlfeld, K. Toyka, H.O. McDevitt, and L. Steinman. 1986. HLA-DQ_β chain polymorphism linked to myasthenia gravis. *Lancet* **I:** 1058.

Benoist, C.O., D.J. Mathis, M.R. Kanter, V.E. Williams, and H.O. McDevitt. 1983. Regions of allelic hypervariability in the murine A_α immune response gene. *Cell* **34:** 169.

Bird, A.P. 1980. DNA methylation and the frequency of CpG in animal DNA. *Nucleic Acids Res.* **8:** 1499.

Carle, G.F. and M.V. Olsen. 1984. Separation of chromosomal DNA molecules from yeast by orthogonal-field-alteration gel electrophoresis. *Nucleic Acids Res.* **12:** 5664.

Christadoss, P., J. Lindstrom, R. Melvold, and N. Talal. 1985. I-A subregion mutation prevents murine experimental autoimmune myasthenia gravis. *Immunogenetics* **21:** 33.

Erlich, H., J.S. Lee, J.L. Peterson, T. Bugawan, and R. DeMars. 1986. *Hum. Immunol.* (in press).

Estess, P., A.B. Begovich, M. Koo, P.P. Jones, and H.O. McDevitt. 1986. Sequence analysis and structure-function correlation of murine q, k, u, s, and f haplotype I-A_β cDNA clones. *Proc. Natl. Acad. Sci.* **83:** 3594.

Festenstein, H., J. Awad, G. Hitman, S. Cutbush, A. Groves, P. Cassell, W. Ollier, and J. Sachs. 1986. New HLA DNA polymorphisms associated with autoimmune diseases. *Nature* **322:** 64.

Giles, R.C., R. DeMars, C.C. Chang, and J.D. Capra. 1985. Allelic polymorphism and transassociation of molecules encoded by the HLA-DQ subregion. *Proc. Natl. Acad. Sci.* **82:** 1776.

Gorski, J. and B. Mach. 1986. Polymorphism of human Ia antigens: Gene conversion between two DR_β loci results in a new HLA-D/DR specificity. *Nature* **322:** 67.

Hardy, D.A., J.I. Bell, E.O. Long, T. Lindsten, and H.O. McDevitt. 1986. Pulse field gel electrophoresis mapping of the class II region of the human major histocompatibility complex. *Nature* **323:** 453.

Hieter, P.A., G.F. Hollis, S.J. Korsmeyer, T.A. Waldmann, and P. Leder. 1981. Clustered arrangement of immunoglobulin λ constant region genes in man. *Nature* **294:** 536.

Howell, M.D., R.K. Austin, D. Kelleher, G.T. Nepom, and M.F. Kagnoff. 1986. An HLA-D region restriction fragment length polymorphism associated with celiac disease. *J. Exp. Med.* **164:** 333.

McIntyre, K. and J. Seidman. 1984. Nucleotide sequence of mutant I-A_β^{bm12} gene is evidence for genetic exchange between mouse immune response genes. *Nature* **308:** 551.

Mengle-Gaw, L. and H.O. McDevitt. 1984. Allelic variation in the murine Ia β chain genes. *UCLA Symp. Mol. Cell. Biol.* **18:** 29.

Nepom, B.S., J. Palmer, S.J. Kim, J.A. Hansen, S.L. Holbeck, and G.T. Nepom. 1986. Specific genomic markers for the HLA-DQ subregion discriminate between $DR4^+$ insulin-dependent diabetes mellitus and $DR4^+$ seropositive juvenile rheumatoid arthritis. *J. Exp. Med.* **164:** 345.

Okada, K., J.M. Boss, H. Prentice, T. Spies, R. Mengler, C. Auffray, J. Lillie, D. Grossberger, and J.L. Strominger. 1985a. Gene organization of DC and DX subregions of the human major histocompatibility complex. *Proc. Natl. Acad. Sci.* **82:** 3410.

Okada, K., H.L. Prentice, J.M. Boss, D.J. Levy, D. Kappes, T. Spies, R. Raghupathy, R.A. Mengler, C. Auffray, and J.L. Strominger. 1985b. SB subregion of the human major histocompatibility complex: Gene organization, allelic polymorphism and expression in transformed cells. *EMBO J.* **4:** 739.

Rollini, P., B. Mach, and J. Gorski. 1985. Linkage map of three HLA-DR_β chain genes: Evidence for a recent duplication event. *Proc. Natl. Acad. Sci.* **82:** 7197.

Sanchez-Perez, M. and S. Shaw. 1985. HLA-DP: Current status. In *Human class II histocompatibility antigens* (ed. B.G. Solheim et al.), p. 83. Springer-Verlag, New York.

Schwartz, D.C. and C.R. Cantor. 1984. Separation of yeast chromosome-sized DNAs by pulsed field gradient gel electrophoresis. *Cell* **37:** 67.

Shaw, S., P. Kavathas, M.S. Pollack, D. Charmot, and C. Mawas. 1981. Family studies define a new histocompatibility locus, SB, between HLA-DR and GLO. *Nature* **293:** 745.

Southern, E.M. 1975. Detection of specific sequences among DNA fragments separated by gel electrophoresis. *J. Mol. Biol.* **98:** 503.

Spies, T., R. Sorrentino, J.M. Boss, K. Okada, and J.L. Strominger. 1985. Structural organization of the DR subregion of the human major histocompatibility complex. *Proc. Natl. Acad. Sci.* **82:** 5165.

Stastny, P., E.J. Ball, and P.J. Dry. 1983. The human immune response region (HLA-D) and disease susceptibility. *Immunol. Rev.* **70:** 113.

Steinmetz, M., M. Malissen, L. Hood, A. Orn, R.A. Maki, G.R. Dastoornikoo, D. Stephan, E. Gibb, and R. Romaniuk. 1984. Tracts of high or low sequence divergence in the mouse major histocompatibility complex. *EMBO J.* **3:** 2995.

Tonnelle, C., R. DeMars, and E.O. Long. 1985. DO_β; a new β chain in HLA-D with a distinct regulation of expression. *EMBO J.* **4:** 2839.

Trowsdale, J. and A. Kelly. 1985. The human HLA class II α chain gene DZ_α is distinct from genes in the DP, DQ and DR subregions. *EMBO J.* **4:** 2231.

Trowsdale, J., A. Kelly, J. Lee, S. Carson, P. Austin, and P. Travers. 1984. Linked map of two HLA-SB_β and two HLA-SB_α related genes: An intron in one of the SB_β genes contains a processed pseudogene. *Cell* **38:** 241.

HLA Class II RFLPs Are Haplotype-specific

S.W. SERJEANTSON, M.R.J. KOHONEN-CORISH, H. DUNCKLEY, AND M.A. REID
Department of Human Genetics, John Curtin School of Medical Research, Australian National University, Canberra, Australia

The human major histocompatibility complex (MHC) is one of the most polymorphic multigene families known in man. The MHC genes are clustered within a 2.5-cM region on chromosome 6 in a complex characterized by marked linkage disequilibrium (Bodmer and Bodmer 1978) between the class I *HLA-A, -B,* and *-C* and class II *HLA-DR* and *-DQ* loci. This disequilibrium occurs despite comparatively high estimates for recombination rates in the MHC, with calculations of 0.8% for recombination between *HLA-A* and *-B* (*HLA-A,B*) and 0.7% for recombination between *HLA-B* and *-DR* (*HLA-B,DR*) (Bodmer 1978).

In studies of HLA associations with disease, linkage disequilibrium in the MHC has often made it difficult to recognize the primary disease-susceptibility locus. For instance, the class II antigen HLA-DR3 is associated with an increased risk for insulin-dependent diabetes mellitus (IDDM), whether carried on *B8.DR3* or *B18.DR3* haplotypes (Svejgaard et al. 1980), but complement component *C4* null alleles are found on both of these haplotypes (Awdeh et al. 1983) and cannot be excluded as possible contributors to the pathogenesis of IDDM. Similarly, the class II *HLA-DQ* locus is so closely linked with *HLA-DR* that the relative contributions of these two loci to HLA-associated diseases is being sought at the molecular level.

Molecular genetic studies of the class II antigens have revealed three sets of genes, namely *HLA-DP, -DQ,* and *-DR*. There is one functional α gene at the *DR* locus and a variable number of β genes, at least one of which is a pseudogene (Böhme et al. 1985; Larhammar et al. 1985). Both *DP* and *DQ* subregions have two α and two β genes, but for both subregions only one type of product has been identified (Bodmer 1984); also one of the α and one of the β chain genes of *DP* are clearly pseudogenes (Kappes et al. 1984; Servenius et al. 1984). The β chain genes are highly polymorphic, as is *DQ*α (Spielman et al. 1984), but *DR*α and *DP*α are comparatively invariant. The serologically defined specificities DR1 to w14 are encoded by the *DR*β*1* gene, and the supertypic specificities DRw52 and w53 are encoded by the *DR*β*2* gene (Bodmer 1984). It is not known whether the DQ specificities w1, w2, and w3 are encoded by the *DQ*α or *DQ*β gene.

Restriction fragment length polymorphisms (RFLPs) of the *HLA-DR*β genes show remarkable correlation with serologically defined DR specificities (Andersson et al. 1984; Hui et al. 1985; Kohonen-Corish and Serjeantson 1986a). However, variation within DR specificity does occur, as has been described for *Taq*I *DR*β fragments of DR3, DR5, and DRw6 (Kohonen-Corish and Serjeantson 1986a,b). Southern blotting indicates that the total number of DR specificities in Caucasoids does not greatly exceed those defined previously by serological techniques, but in the *HLA-DQ* genes there is much greater variability (Spielman et al. 1984; Cohen et al. 1984a; le Gall et al. 1985) than evidenced by serology.

Heterogeneity in *DQ* and *DR* RFLPs within HLA-DR antigenic specificity has led to investigations of whether these RFLP subtypes show closer association with HLA-associated diseases than do the DR antigens themselves. It has not been established, however, whether the RFLP class II subtypes are simply more definitive markers for specific *HLA-B,DR* haplotypes than have been available to date. Here we report heterogeneous RFLPs for HLA-DR1 (*DR*α), -DR2 (*DQ*α, *DQ*β), -DR3 (*DR*α, *DR*β), -DR4 (*DQ*β), -DR5 (*DR*β), -DRw6 (*DR*α, *DR*β, *DQ*α, *DQ*β), and -DR7 (*DR*α, *DR*β, *DQ*β) and determine their HLA-B antigen linkage relationships.

MATERIALS AND METHODS

Study population. The panel of 250 subjects comprised normal, healthy, unrelated Caucasoid controls and unrelated Caucasoid patients with one of the following diseases: IDDM, multiple sclerosis (MS), systemic lupus erythematosus (SLE), or a lupus-like connective tissue disease (CTD). The panel was made up of 71 healthy control subjects, 59 IDDM patients, 24 MS patients, 46 SLE patients, and 50 CTD patients. Patients and control populations were drawn from the three Australian cities of Sydney, Melbourne, and Canberra.

HLA typing. All subjects, with the exception of MS patients, were typed in our laboratory for HLA-A, -B, -C, and -DR antigens by the standard complement-dependent microcytotoxicity method, after lymphocyte isolation over a Ficoll-Hypaque gradient. HLA-A,-B, -C, and -DR typing results for MS patients were provided by the Red Cross Blood Transfusion Service, Sydney.

DNA digestion of Southern blots. Genomic DNA was prepared for all subjects from buffy coat white cells. Approximately 10 μg of each sample was then digested with each of the following restriction enzymes: *Taq*I (40 units) for 2 hr at 65°C; *Bam*HI (70 units) overnight at 37°C; *Bgl*II (38 units) overnight at 37°C; *Eco*RV (70 units) overnight at 37°C; and *Eco*RI (40 units) overnight at 37°C; otherwise, conditions were according to manufacturers' recommendations (*Taq*I, Biolabs; *Bam*HI, *Eco*RV, and *Eco*RI, Boehringer Mannheim; *Bgl*II, Pharmacia). Samples were electrophoresed through a 0.8% agarose gel at 20V for about 16 hr in TAE-buffer (0.04 M Tris-acetate, 0.001 M EDTA), using submarine apparatus. DNA was then transferred onto Gene-Screen Plus membranes by the method of Southern (1975), according to recommendations of the manufacturer (New England Nuclear).

Probes. cDNA clones for *HLA-DRβ* and *DQβ* (Long et al. 1982), for *HLA-DQα* (Schenning et al. 1984), and for *HLA-DRα* (Stetler et al. 1982) were used. The whole plasmid containing the cDNA clone was used as a probe.

Hybridizations. Filters were prehybridized for at least 1 hr at room temperature in a solution containing 10% dextran sulfate, 0.6 M NaCl, 0.18 M Na_2HPO_4, 0.06 M EDTA, 1% sodium lauroyl sarcosine, and 50 μg/ml of sonicated, denatured salmon sperm DNA (pH 6.2) following the method of Nasmyth (1982). Hybridizations with the *HLA-DRβ*, *DRα*, *DQβ*, and *DQα* gene probes were carried out at 65°C for about 44 hr in a buffer the same as that used for prehybridization but containing 1×10^8 cpm/μg of ^{32}P-labeled DNA. For every 100 cm^2 of filter, 10 ml of solution was used with 0.1 μg of DNA.

Filters were washed in 2× SSC, 0.1% SDS for 5 min at room temperature and then consecutively at 65°C for about 30 min in each of the following solutions: 1× SSC, 0.1% SDS and 0.5× SSC, 0.1% SDS until the counts were about 5 cps.

Filters were autoradiographed with intensifying screens at −70°C using Fuji X-ray Film (FX100) for 4–10 days.

Statistical analysis. The nonrandom association between particular HLA-B,DR combinations and class II RFLPs was measured by the disequilibrium statistic Δ (Hill and Robertson 1968; Mittal et al. 1973), where Δ is the correlation coefficient between uniting gametes. When the association between gametes is random, $\Delta = 0$; the null hypothesis of $\Delta = 0$ was tested here by the χ^2 statistic. When any number in a cell of a 2×2 contingency table fell below 4, Fisher's test for calculating the exact significance value was used. The Δ values for HLA-B,DR combinations and class II RFLPs were calculated separately for healthy controls and for patients in each of the four disease groups (SLE, CTD, IDDM and MS), then examined further for heterogeneity in 2×2 contingency tables. In no instance was heterogeneity in Δ values detected between disease groups. Therefore the data for patients have been pooled with those for healthy controls for estimation of the association of *HLA-B,DR* haplotypes with particular class II RFLPs.

The HLA-B,DR antigen combinations chosen for study were those showing significant linkage disequilibrium in the Ninth Histocompatibility Workshop (Baur et al. 1985). Then, HLA-DR antigen specificities showing several HLA-B associations were examined for heterogeneity in class II RFLPs. For instance, HLA-DR3 is in known linkage disequilibrium with HLA-B8 and HLA-B18. Cells with B8 and DR3 or B18 and DR3 co-occurring were examined for nonrandom association with heterogeneous class II RFLPs.

RESULTS

HLA-DRα Linkage Relationships

The *HLA-DRα* chain gene is comparatively invariant, in the noncoding as well as the coding region, but a *Bgl*II polymorphism in the 3′ untranslated region of the gene has been described (Stetler et al. 1982). DNA digestion with *Bgl*II yields three patterns with fragment sizes of 4.5 kb, 4.4 kb, or 3.9,0.7 kb. The most commonly occurring pattern is the 4.5-kb band, which is present on 59% of healthy Caucasoid haplotypes in this series. More restricted is the 4.4-kb fragment, which occurs in some HLA-DR3 and some DRw6-positive cells. The 3.9,0.7-kb pattern is invariably associated with HLA-DR1, HLA-DRw10, and a subset of HLA-DR7. These *DRα Bgl*II fragments are haplotype-specific, as is clear from Table 1. In 203 *DRα* hybridizations, 58

Table 1. *HLA-DRα Bgl*II Linkage Relationships (Δ per 1,000)

Antigen combination	No. of occurrences[a]	4.5 kb		4.4 kb		3.9,0.7 kb	
		%	Δ	%	Δ	%	Δ
B8.DR3	58	78	−49[b]	91	109[b]	5	−13
B18.DR3	9	100	8	33	−1	0	−6
B51.DRw6	2	50	−5	100	4	50	2
Bw55.DRw6	2	100	2	100	4	0	−1
B27.DR1	4	75	−3	25	−2	100	9[b]
B35.DR1	13	69	−17	8	−12	100	28[b]
B37.DRw10	1	0	−6[b]	100	2	100	2
Bw57.DR7	7	29	−30[b]	43	1	100	15[b]
B44.DR7	12	92	4	25	−5	25	0

[a]Total number of cells tested = 203.
[b]P < .01.

cells were positive for both B8 and DR3, and in 53 cases the 4.4-kb pattern was observed. The Δ value of 0.109 is highly significant ($P < .001$) for linkage disequilibrium between the *B8.DR3* haplotype and the 4.4-kb *Bgl*II *DRα* fragment. Consequently, this haplotype is in negative linkage disequilibrium with the allelic 4.5-kb pattern, which occurs on *B18.DR3* haplotypes. The 4.4-kb pattern is seen also in the majority of HLA-DRw6-positive cells, but only in the DRw13 subset with *DQα*, *DQβ* *Taq*I arrangements that correlate with HLA-Dw18. This includes *B51.DRw6* (*w13*) but not *Bw55.DRw6* (*w14*) haplotypes. HLA-B linkage relationships with *HLA-DRw6* are not particularly strong, and the Δ values given in Table 1 for *DRw6* do not reach statistical significance.

The 3.9,0.7-kb *Bgl*II *DRα* RFLP occurred in 33 of 38 (87%) examples of HLA-DR1 and was present irrespective of whether the DR1 antigen was carried on B27- or B35-positive haplotypes, as seen in Table 1. The exceptional haplotype is *B14*(*w65*).*DR1*, which has the 4.5-kb pattern. HLA-DRw10 is serologically similar to DR1, and all examples of DRw10 in this series showed the 3.9- and 0.7-kb bands. Heterogeneity in the *DRα* RFLPs was seen in HLA-DR7 cells. The *HLA-Bw57.DR7* haplotype invariably had the 3.9,0.7-kb pattern, as shown by the significant Δ values given in Table 1, but other DR7-bearing haplotypes, such as *B44, B13, Bw50,* and *B14*(*w64*), did not.

HLA-DRβ Linkage Relationships

The correlation between *Taq*I *DRβ* fragments and DR antigen specificities has been described previously (Kohonen-Corish and Serjeantson 1986a), with unique RFLPs generated for all DR types with the exception of DR3 and DRw6. Heterogeneity in the larger *Taq*I fragments of *DRβ* occurs commonly in HLA-DR3 and HLA-DRw6. Observation of similar heterogeneity in large *Taq*I fragments associated with HLA-DR5 in Pacific Island populations has led us to propose (Kohonen-Corish and Serjeantson 1986b) that the allelic 12-kb and 10-kb *Taq*I *DRβ* fragments are probably associated with the *DRβ2* chain gene that encodes HLA-DRw52. The 12-kb or 10-kb bands are haplotype-specific, as shown in Table 2. The 12-kb fragment is in linkage disequilibrium with B18.DR3, while the 10-kb fragment occurs in B8.DR3-positive cells. Similar heterogeneity occurs in DRw6, although the Δ values are an order of magnitude less, with the 12-kb fragment occurring on Bw55.DRw6(w14) cells. The majority of *DRw6* haplotypes in Caucasoids bear the 10-kb *Taq*I *DRβ* fragment, as shown in Table 2 for B51.DRw6(w13).

HLA-DR5-positive cells usually have the 12-kb *Taq*I pattern at *DRβ* and this correlates with B18.DR5(w11), as shown in Table 2, as well as with B51.DR5(w11). In 250 cells, only three instances of the joint occurrence of B14(w65) and DR5(w12) were observed, but in all cases the 10-kb rather than the 12-kb *Taq*I fragment was evident. Allelic to DRw52 and encoded by the *DRβ2* gene is the DRw53 specificity found on *DR4, 7,* and *w9* haplotypes. HLA-DRw53 correlates with a 15-kb *Taq*I fragment and all DR4-, 7-, and w9-positive cells examined in this series had the 15-kb band. *Taq*I reveals little heterogeneity in *DR4* at *DRβ*.

In contrast to DR4, HLA-DR7 shows marked variability in *Taq*I fragments of *DRβ*, and four different DNA subtypes can be detected. All DR7 specimens share a 15-kb, 5.8-kb, and 4.2-kb fragment, but other fragments vary in size. We have yet to resolve whether the variability seen in DR7 can be attributed to RFLPs in the *DRβ3* pseudogene, but certainly there is little evidence for any serological heterogeneity in the DR7 antigen.

The HLA-DR7–associated *Taq*I patterns are haplotype-specific. The pattern designated as type DR7(I) in Table 3 occurs not only on the *B44.DR7* haplotype but also on *B14*(*w64*).*DR7* haplotypes. The other DR7 patterns are invariably seen with the HLA-B antigens given in Table 3; that is, HLA-B13, -Bw50, and -Bw57 antigens are each uniquely associated with particular *DRβ* patterns of DR7.

HLA-DQ Linkage Relationships

Each HLA-DR antigen is associated with characteristic *DQα* and *DQβ* chain gene RFLPs. The correlations of DR antigen specificities with *Taq*I fragments of *DQα* (Spielman et al. 1984; Kohonen-Corish and Serjeantson 1986a) and *DQβ* (Kohonen-Corish and Serjeantson 1986) have been described previously, as have DR correlates with *Bam*HI fragments of *DQβ* (Owerbach et al. 1983). Of course, the correlation between DR type and *DQ* RFLPs is not absolute, with rare recombinatorial events generating novel *DR,DQ* linkage arrangements.

For example, HLA-DR2 in Caucasoids is usually as-

Table 2. HLA-DRw52 *Taq*I Linkage Relationships

Antigen combination	No. of occurrences[a]	*Taq*I *DRβ* fragments			%	Δ (per 10^3)
		12 kb	10 kb	other		
B18.DR3	13	+	–	6.8	85	20[b]
B8.DR3	77	–	+	6.8	97	123[b]
B55.DRw6	2	+	–	6.8	100	4[b]
B51.DRw6	2	–	+	6.8	100	3
B18.DR5	7	+	–	6,4.2	100	13[b]
B14.DR5	3	–	+	6,4.2	100	6[b]

[a]Total number of cells tested = 250.
[b]$P < .01$.

Table 3. HLA-DR7 Heterogeneity in *DRβ* Is Haplotype-specific

Antigen combination	No. of occurrences[a]	*Taq*I *DRβ* pattern	%	Δ (per 10^3)
B44.DR7	16	I	88	27[b]
B14.DR7	9	I	89	15[b]
B57.DR7	7	II	100	14[b]
B50.DR7	3	III	100	6[b]
B13.DR7	7	IV	86	12[b]

[a]Total number tested = 250.
[b]P<.01.

sociated with the *Taq*I, *Bam*HI, and *Eco*RV *DQ* fragments designated as "DR2-like" in Table 4. This pattern occurs in 84% of healthy HLA-DR2-positive individuals in this series and equates with the cellular type, HLA-Dw2. Two other *DQ* patterns are found in association with HLA-DR2. These patterns are given in Table 4 and correspond to Dw12 and "AZH" cellular types. The three DR2-associated patterns can be distinguished using any of a number of restriction endonucleases and, as described previously (Cohen et al. 1986), have characteristic *Eco*RI *DQβ* fragments also. HLA-Dw12 occurs commonly in Asian and Oceanian populations, but in these cells a *DQα* 6.2-kb *Taq*I fragment is seen rather than the 6.8-kb fragment (Kohonen-Corish and Serjeantson 1986b), and at *DQβ* there is neither a 3.2- or 3.4-kb *Bam*HI fragment. The "DR1-like" pattern occurs in only 4% of DR2-positive healthy Caucasoids (in this series) but is commonly seen in DR2-positive IDDM patients (D. Cohen et al. 1984b; N. Cohen et al. 1986). These characteristic *DQβ* fragments occur with either 6.2-kb *Taq*I and 13.0-kb *Bam*HI or 3.8-kb, 2.9-kb *Taq*I, and 12.0-kb *Bam*HI *DQβ* fragments.

Table 5 gives linkage disequilibrium relationships in 250 cells examined for RFLPs at *DQα* and *DQβ*. The common Caucasoid haplotype *B7.DR2* was in significant linkage disequilibrium with the Dw2-like pattern. The Dw12-related RFLPs occurred in five of the cells examined and in three instances on *B5.DR2* haplotypes, or on *Bw52.DR2* when B5 subtyping was available. The "AZH"-related *DR2* pattern was not strongly linked with any HLA-B antigen, but two instances of *B16.DR2."AZH"* were observed.

The HLA-DRw6 linkage arrangements tend to overlap those already described for *DR2* but, as shown in

Table 4. *HLA-DQ* RFLPS Associated with HLA-DR2 in Caucasoids

Probe	Enzyme	DR2-like (Dw2) (kb)	DRw6-like (Dw12) (kb)	DR1-like (AZH) (kb)
DQα	*Taq*I	6.2	6.8	6.2
DQα	*Bam*HI	13.0	12.0	13.0
DQβ	*Taq*I	3.0	not 3.0 or 5.5	5.5
DQβ	*Bam*HI	3.2	NT	3.4
DQβ	*Eco*RV	5.2	NT	2.6

(NT) Not tested.

Table 5. HLA-DR2 Linkage Relationships with *DQ* RFLPs

Antigen combination	No. of occurrences[a]	*DQα,DQβ* pattern	%	Δ (per 10^3)
B7.DR2	33	Dw2	97	58[b]
B5.DR2	7	Dw12	43	6[b]
B16.DR2	4	AZH	50	4[b]

[a]Total number tested = 250.
[b]P<.01.

Table 6, can be characterized by one of several enzymes. The most common pattern in Caucasoid DRw6-positive cells is a 6.8-kb *Taq*I *DQα* band, with a 3.0-kb *Taq*I *DQβ* band. This configuration is in linkage disequilibrium with the 10-kb *Taq*I *DRβ* and 4.4-kb *Bgl*II *DRα* fragments, as shown in Table 6. This *DR,DQ* profile occurs in approximately 50% of healthy DRw6 Caucasoids and comprises a subset of HLA-DRw13-positive cells that can be defined cellularly as Dw18 (Trowsdale et al. 1985). Two alternative DRw6-associated patterns are shown in Table 6. These *DQ,DR* configurations are not always strictly maintained, with some examples of "mixing" of the three patterns. These exceptions were rare and could not be assigned to any particular HLA-B antigen. Linkage disequilibria between the three main DRw6 subtypes and HLA-B antigens known to be linked with DRw6 (i.e., HLA-B37, B51, B38, Bw60, B35, and Bw55) were examined. Δ values were low, as seen previously for DRw6 in Table 2 and, with the exception of linkage of Bw55.DRw6 with the DRw14 pattern, were not statistically significant.

Heterogeneity in HLA-DR4-associated class II RFLPs is evident in *DQβ*, where a 3.7-kb *Bam*HI fragment found in most DR5-positive cells occurs in 25% of the healthy DR4-positive individuals in this series. HLA-DR4 is in known linkage disequilibrium with HLA-B44, -Bw62, and -Bw60. As seen in Table 7, the 3.7-kb *Bam*HI fragment is in significant linkage disequilibrium with B44, whereas the allelic 12-kb band is evidenced on *Bw62* and *Bw60* haplotypes as well as on some *B44* haplotypes. In the single instance in this series of a HLA-DR5-positive cell lacking the *Bam*HI 3.7-kb *DQβ* band, genotyping revealed the haplotype as *A2.B35.Cw4.DRw11.DQ blank.TA10-negative.*

HLA-DR7-positive cells also show heterogeneity in *DQβ* fragments, which correlates with the *DRβ* RFLPs described here. The *DRβ* subtype of DR7 (type II) associated with B57 invariably shows 2.9-kb *DQβ* *Taq*I and 12-kb *Bam*HI fragments, whereas the other subtypes have 6.7-kb *Taq*I and 3.4-kb *Bam*HI bands. HLA-DRw9-positive cells that share the *DRβ* *Taq*I RFLP associated with B13.DR7, have neither the 6.7-kb or 2.9-kb *DQβ* *Taq*I bands but have instead the *Bam*HI 12-kb fragment more usually associated with DR4.

DISCUSSION

Nonrandom Recombination

The estimates of Δ values for HLA-B,DR antigen combinations and class II RFLPs show that linkage

Table 6. DRw6-associated Class II RFLPs

Probe	Enzyme	DRw13 (Dw18) (kb)	DRw13 (Dw19) (kb)	DRw14 (Dw9) (kb)
DQα	*Taq*I	6.8	6.2	3.8,2.9
	*Bam*HI	12.0	not 12.0	12.0
DQβ	*Taq*I	3.0	5.5	5.5
	*Bam*HI	3.2	not 3.2,3.4	3.4
	*Eco*RV	5.2	not 2.6,5.2[a]	2.6
DRβ	*Taq*I	10.0	10.0	12.0
DRα	*Bgl*II	4.4	4.5	4.5

[a]Some variants with 5.2 kb do occur.

disequilibrium in this region of the genome is considerably greater than previously thought. For instance, whereas the antigen HLA-DR7 is known to be linked with HLA-B44, B14, B57, B13, and Bw50, a single enzyme, *Taq*I, has revealed four different *DRβ* patterns associated with DR7 and each of these is haplotype-specific. In the absence of estimates of the physical distance between *HLA-B* and *HLA-DR*, one can only speculate that the *HLA-B,DR* region of the genome is a recombination "cold spot." It is becoming increasingly evident that meiotic recombination in the human genome is nonrandom in nature. For instance, when specific β-thalassemia mutations were correlated with background chromosome 11 DNA profiles (Kazazian et al. 1984), it was apparent that recombination within a 9-kb interval 5′ to the β-globin gene was exceedingly common and approached 30 times the average recombination rate (Chakravarti et al. 1984). This hot spot of recombination has been confirmed experimentally in yeast (Treco et al. 1985). Similarly, claims for increased recombination in a 20-kb segment at the insulin locus have been made (Chakravarti et al. 1986) and for reduced recombination at the serum albumin locus (Murray et al. 1984), probably due to its proximity to the centromere on chromosome 4. Within the MHC itself, there is also evidence, in the mouse at least, for nonuniform recombination. Steinmetz et al. (1982) examined recombination between murine class II *Aβ* and *Eα* genes, which are separated by 85 kb. The observed recombination rate of 0.1% was not very different from the expected value of 2000 kb/cM in the mouse, but all recombinants showed crossover events in a small segment of the 85-kb region and were subsequently shown to be confined to a 2-kb segment (Kobori et al. 1984).

Recombination rates may also be haplotype-specific, as is evidenced in the *t* haplotype of the mouse, where recombination suppression is due to chromatin mismatching (Silver and Artzt 1981). The presence of a similar locus in man has been proposed (Awdeh et al. 1983) to account for the marked linkage disequilibrium in the human MHC. Alternative mechanisms, however, have their proponents. As shown by Fisher (1930), if selection favors a given allele, a closely linked variant may also increase in frequency and enhance population linkage disequilibrium. Bodmer and Bodmer (1978) have argued that this phenomenon, known as "hitchhiking," is more likely to account for linkage disequilibrium in the MHC than other possible mechanisms such as migration and population admixture or a functional interrelationship between loci.

The current estimate of* HLA-B,DR *recombination is probably too high. It is clear from our estimates of Δ values for HLA-B,DR combinations and class II RFLPs that the generally accepted recombination rate (r) between *HLA-B* and *HLA-DR* could well be too high. This has been estimated as 0.7% (Bodmer 1978). Linkage disequilibrium declines between autosomal loci at a rate $(1-r)$ per generation of random mating. If $r=0.7\%$, the Δ value would decline in 10,000 years, or 400 generations, from a hypothetically high value of 0.180 to barely detectable levels of 0.006.

Recombination rates in the MHC may have been overestimated in the past as a consequence of several factors, such as technical errors in serology, biased selection of families with *HLA* recombinants for inclusion in international workshops, extrapaternity offspring, and inclusion of disease families who may have differential recombination rates. For example, the three instances of crossovers between *HLA-B* and *HLA-C* reported by Hawkins et al. (1980) in the Eighth International Workshop all occurred in families with 21-hydroxylase deficiency probands. These families could conceivably segregate for haplotypes with atypical chromosomal rearrangements, contributing to an overestimate of the recombination rate of 0.2% for *HLA-B* and *HLA-C* loci.

Table 7. *HLA-DQβ BamHI* Linkage Relationships

Antigen combination	No. of occurrences[a]	12 kb %	12 kb Δ	3.7 kb %	3.7 kb Δ
B44.DR4	24	75	23[b]	54	23[b]
Bw62.DR4	*21*	*95*	*33*[b]	29	−8
B60.DR4	14	86	18[b]	5	4

[a]Total number tested = 214.
[b]$P<.01$.

Other Haplotype-specific Markers

Many of the *HLA-B,DR* haplotypes described here as having particular class II RFLPs are known to have additional distinctive markers in the complement *C4*, *C2*, or *BF* loci. For example, *B8.DR3* haplotypes bear the *C4A* null allele, *C4A*QO*, whereas *B18.DR3* haplotypes have the *C4B* null *C4B*QO* (Awdeh et al 1983) as well as the *BF* allele *F1*. Similarly, *DR4*-positive haplotypes show differences at the *C4* locus, with *B62.DR4* linked with *C4A*3.B*3* and *B44.DR4* with *C4A*3.B*1* (Baur et al. 1985). The *DR7* RFLP subtypes that occur in conjunction with particular HLA-B antigens also have characteristic complement types. For instance, *B57.DR7* is in linkage disequilibrium with *C4*A6* in Caucasoids (O'Neill et al. 1983), whereas *BF*S1* invariably occurs in conjunction with HLA-Bw50 (Serjeantson 1985), usually on *Bw50.DR7* haplotypes. Some of the heterogeneity in *DR7* RFLPs may have been anticipated from cellular and serological studies that have shown *DR7* to be linked with *DQw3* on Bw57-positive haplotypes and linked with *DQw2* on other *DR7* haplotypes, with corresponding cellular equivalents called "Dw1" and "Dw7." In addition, B13.DR7 cells have a unique cellular reactivity known as "DB1."

Implications of Linkage Disequilibrium for Studies of HLA and Disease

Our observation that class II RFLPs are haplotype-specific in autoimmune patients and healthy controls has important implications for studies of HLA and disease. For instance, study of large numbers of IDDM patients has shown an increased frequency of both *B8.DR3* and *B18.DR3* haplotypes (Svejgaard et al. 1980); in the converse analysis, Winearls et al. (1984) have shown that DR3 is no more closely linked to *HLA-B8* in IDDM patients than in controls. Despite this, an increase in the 12-kb *Taq*I *DRβ* fragment in French IDDM patients has been claimed (Cohen-Haguenauer et al. 1985), whereas the 10-kb *Taq*I *DRβ* fragment is reportedly increased in U.S. IDDM patients (Arnheim et al. 1985). In this series, the 10-kb and 12-kb fragments were clearly in significant linkage disequilibrium with the *B8.DR3* and *B18.DR3* haplotypes, respectively, and therefore serve as new haplotype markers rather than as new markers for IDDM susceptibility. Similarly, the *DRα Bgl*II 4.4-kb fragment is found on *B8.DR3* haplotypes, as shown here, and therefore is unlikely to represent a better marker for IDDM than *DR3* alone, as preliminary studies have suggested (Erlich and Stetler 1984: Stetler et al. 1985).

The decreased frequency of the 3.7-kb *Bam*HI *DQβ* fragment in DR4-positive IDDM patients is well-documented (Owerbach et al. 1983; Cohen-Haguenauer et al. 1985). In serological analysis, Winearls et al. (1984) observed stronger linkage disequilibrium beween Bw62 and DR4 in IDDM patients than in controls. Our data permit reinterpretation of these two independent observations by showing that *Bw62.DR4* haplotypes lack the 3.7-kb *Bam*HI *DQβ* fragment and are in positive linkage disequilibrium with the allelic 12-kb fragment.

CONCLUSION

We have shown that the common serologically defined HLA-DR specificities DR1 to DR7 are associated with heterogeneous class II RFLPs. We have observed little variation in *DRβ* DNA polymorphisms for DR1, 2, and 4, but DR1 shows variation at the *DRα* locus, DR2 at *DQα* and *DQβ*, and DR4 at *DQβ*. We report for the first time four new variants in HLA-DR7 RFLPs at *DRβ* and assign *DRα* RFLPS to HLA-DRw6 subtypes. The heterogeneous RFLPs at the *DR* and *DQ* loci are in strong linkage disequilibrium with HLA-B antigens and have led us to conclude that the *HLA-B,DR* region of the genome is a recombinational "cold spot" and that currently accepted values for *HLA-B,DR* recombination are overestimated.

ACKNOWLEDGMENTS

We thank Dr. B. Mach, University of Geneva, for providing the cDNA probes for *DRβ* and *DQβ*; Dr. D. Larhammar, University of Uppsala, for the cDNA *DQα* probe; and Dr. H. Erlich, Cetus Corporation, for the cDNA *DRα* probe.

We thank our collaborators in associated studies of HLA and disease for referring specimens from patients with SLE and CTD (Dr. P. Gatenby, Clinical Immunology Research Centre, University of Sydney), with MS (Dr. G. Stewart, Clinical Immunology, Westmead Centre), and with IDDM (Professor P. Zimmet, Royal Southern Memorial Hospital, Caulfield South). We thank the Red Cross Blood Transfusion Service for providing blood specimens (Canberra Branch) and HLA-phenotype results for MS patients (Sydney Branch).

REFERENCES

Andersson, M., J. Böhme, B. Andersson, E. Möller, E. Thorsby, L. Rask, and P.A. Peterson. 1984. Genomic hybridization with class II transplantation antigen cDNA probes as a complementary technique in tissue typing. *Hum. Immunol.* **11:** 57.

Arnheim, N., C. Strange, and H. Erlich. 1985. Use of pooled DNA samples to detect linkage disequilibrium of polymorphic restriction fragments and human disease: Studies of the HLA class II loci. *Proc. Natl. Acad. Sci.* **82:** 6970.

Awdeh, Z.L., D. Raum, E.J. Yunis, and C.A. Alper. 1983. Extended HLA/complement allele haplotypes: Evidence for T/t-like complex in man. *Proc. Natl. Acad. Sci.* **80:** 259.

Baur, M.P., M. Neugebauer, and E.D. Albert. 1985. Reference tables of two-locus haplotype frequencies for all MHC marker loci. In *Histocompatibility testing 1984* (ed. E.D. Albert et al.), p. 677. Springer-Verlag, Berlin.

Bodmer, W.F. 1978. The HLA system: Introduction. *Br. Med. Bull.* **34:** 213.

———. 1984. The HLA system, 1984. In *Histocompatibility testing 1984* (ed. E.D. Albert et al.), p. 11. Springer-Verlag, Berlin.

Bodmer, W.F. and J.G. Bodmer. 1978. Evolution and function of the HLA system. *Br. Med. Bull.* **34:** 309.

Böhme, J., M. Andersson, G. Andersson, E. Möller, P.A. Peterson, and L. Rask. 1985. HLA-DRβ genes vary in number between different DR specificities, whereas the number of DQβ genes is constant. *J. Immunol.* **135:** 2149.

Chakravarti, A., S.C. Elbein, and M.A. Permutt. 1986. Evidence for increased recombination near the human insulin gene: Implication for disease association studies. *Proc. Natl. Acad. Sci.* **83:** 1045.

Chakravarti, A., K.H. Buetow, S.E. Antonarakis, P.G. Waber, C.D. Boehm, and H.H. Kazazian. 1984. Non-uniform recombination within the human β-globin gene cluster. *Am. J. Hum. Genet.* **36:** 1239.
Cohen, D., I. le Gall, A. Marcadet, M.-P. Font, J.-M. Lalouel, and J. Dausset. 1984a. Clusters of HLA class IIβ restriction fragments describe allelic series. *Proc. Natl. Acad. Sci.* **81:** 7870.
Cohen, D., O. Cohen, A. Marcadet, C. Massart, M. Lathrop, I. Deschamps, J. Hors, E. Schuller, and J. Dausset. 1984b. Class II HLA-DCβ-chain DNA restriction fragments differentiate among HLA-DR2 individuals in insulin-dependent diabetes and multiple sclerosis. *Proc. Natl. Acad. Sci.* **81:** 1774.
Cohen, N., C. Brautbar, M.-P. Font, J. Dausset, and D. Cohen. 1986. HLA-DR2-associated Dw subtypes correlate with RFLP clusters: Most DR2 IDDM patients belong to one of these clusters. *Immunogenetics* **23:** 84.
Cohen-Haguenauer, O., E. Robbins, C. Massart, M. Busson, I. Deschamps, J. Hors, J.-M. Lalouel, J. Dausset, and D. Cohen. 1985. A systematic study of HLA class II-β DNA restriction fragments in insulin-dependent diabetes mellitus. *Proc. Natl. Acad. Sci.* **82:** 3335.
Erlich, H. and D. Stetler. 1984. HLA class II DNA polymorphisms: Markers for genetic predisposition to insulin-dependent diabetes. *Banbury Rep.* **16:** 321.
Fisher, R.A. 1930. *The genetical theory of natural selection.* Oxford University Press, London.
le Gall, I., A. Marcadet, M.-P. Font, C. Auffray, J. Strominger, J.-M. Lalouel, J. Dausset, and D. Cohen. 1985. Exuberant restriction fragment length polymorphism associated with the DQα-chain gene and the DXα-chain gene. *Proc. Natl. Acad. Sci.* **82:** 5433.
Hawkins, B.R., J.A. Danilovs, and G.J. O'Neill. 1980. Analysis of recombinant families. In *Histocompatibility testing 1980* (ed. P.I. Terasaki), p. 148. UCLA Tissue Typing Laboratory, Los Angeles.
Hill, W.G. and A. Robertson. 1968. Linkage disequilibrium in finite populations. *Theor. Appl. Genet.* **38:** 226.
Hui, K., H. Festenstein, A. de Klein, G. Grosveld, and F. Grosveld. 1985. HLA-DR genotyping by restriction fragment length polymorphism analyses. *Immunogenetics* **22:** 231.
Kappes, D.J., D. Arnot, K. Okada, and J.L. Strominger. 1984. Structure and polymorphism of the HLA Class II SB light chain genes. *EMBO J.* **3:** 2985.
Kazazian, H.H., S.H. Orkin, A.F. Markham, C.R. Chapman, H. Youssoufian, and P.G. Waber. 1984. Quantification of the close association between DNA haplotypes and specific-β-thalassemia mutations in Mediterraneans. *Nature* **310:** 152.
Kobori, J.A., A. Winoto, J. McNicholas, and L. Hood. 1984. Molecular characterization of the recombination region of six murine major histocompatibility complex (MHC) I region recombinants. *J. Mol. Cell. Immunol.* **1:** 125.
Kohonen-Corish, M.R.J. and S.W. Serjeantson. 1986a. HLA-DRβ gene DNA polymorphisms revealed by *TaqI* correlate with HLA-DR specificities. *Hum. Immunol.* **15:** 263.
———. 1986b. RFLP analysis of HLA-DR and -DQ genes and their linkage relationships in the Pacific. *Am. J. Hum. Genet.* **39:** (in press).
Larhammar, D., B. Servenius, L. Rask, and P.A. Peterson. 1985. Characterization of an HLA DRβ pseudogene. *Proc. Natl. Acad. Sci.* **82:** 1475.
Long, E.O., C.T. Wake, M. Strubin, N. Gross, R.S. Accolla, S. Carrel, and B. Mach. 1982. Isolation of distinct cDNA clones encoding HLA-DRβ chains by use of an expression assay. *Proc. Natl. Acad. Sci.* **79:** 7465.
Mittal, K.K., T. Hasegawa, A. Ting, M.R. Mickey, and P.I. Terasaki. 1973. Genetic variation in the HLA-A system between Ainus, Japanese, and Caucasians. In *Histcompatibility testing 1972* (ed. J. Dausset and J. Colombani), p. 187. Munksgaard, Copenhagen.
Murray, J.C., K.A. Mills, C.M. Demopulos, S. Hornung, and A.G. Motulsky. 1984. Linkage disequilibrium and evolutionary relationships of DNA variants (restriction enzyme length polymorphisms) at the serum albumin locus. *Proc. Natl. Acad. Sci.* **81:** 3486.
Nasmyth, K.A. 1982. The regulation of yeast mating-type chromatin structure by *SIR*: An action at a distance affecting both transcription and transposition. *Cell* **30:** 567.
O'Neill, G.J., C. Nerl, and M.S. Pollack. 1983. Analysis of active and inactive complement C4 complotypes associated with subtypes of HLA-B17 in different racial groups. *Am. J. Hum. Genet.* **35:** 309.
Owerbach, D., Å. Lernmark, P. Platz, L.P. Ryder, L. Rask, P.A. Peterson, and J. Ludvigsson. 1983. HLA-D β-chain DNA endonuclease fragments differ between HLA-DR identical healthy and insulin-dependent diabetic individuals. *Nature* **303:** 815.
Schenning, L., D. Larhammar, P. Bill, K. Wiman, A.-K. Jonsson, L. Rask, and P.A. Peterson. 1984. Both α and β chains of HLA-DC class II histocompatibility antigens display extensive polymorphism in their amino terminal domains. *EMBO J.* **3:** 447.
Serjeantson, S.W. 1985. Properdin factor B. In *Histocompatibility testing 1984* (ed. E.D. Albert et al.), p. 317. Springer-Verlag, Berlin.
Servenius, B., K. Gustafsson, E. Widmark, E. Emmoth, G. Andersson, D. Larhammar, L. Rask, and P.A. Peterson. 1984. Molecular map of the human HLA-SB (HLA-DP) region and sequence of an SBα (DPα) pseudogene. *EMBO J.* **3:** 3209.
Silver, L.M. and K. Artzt. 1981. Recombination suppression of mouse *t*-haplotypes due to chromatin mismatching. *Nature* **290:** 68.
Southern, E.M. 1975. Detection of specific sequences among DNA fragments separated by gel electrophoresis. *J. Mol. Biol.* **98:** 503.
Spielman, R.S., J. Lee, W.F. Bodmer, J.G. Bodmer, and J. Trowsdale. 1984. Six HLA-D region α-chain genes on human chromosome 6: Polymorphisms and associations of DCα-related sequences with DR types. *Proc. Natl. Acad. Sci.* **81:** 3461.
Steinmetz, M., K. Minard, S. Horvath, J. McNicholas, J. Srelinger, C. Wake, E. Long, B. Mach, and L. Hood. 1982. A molecular map of the immune response region from the major histocompatibility complex of the mouse. *Nature* **300:** 35.
Stetler, D., F.C. Grumet, and H.A. Erlich. 1985. Polymorphic restriction endonuclease sites linked to the HLA-DRα gene: Localization and use as genetic markers of insulin-dependent diabetes. *Proc. Natl. Acad. Sci.* **82:** 8100.
Stetler, D., H. Das, J.H. Nunberg, R. Saiki, R. Sheng-Dong, K.B. Mullis, S.M. Weissman, and H.A. Erlich. 1982. Isolation of a cDNA clone for the human HLA-DR antigen α chain by using a synthetic oligonucleotide as a hybridization probe. *Proc. Natl. Acad. Sci.* **79:** 5966.
Svejgaard, A., P. Platz, and L.P. Ryder. 1980. Insulin-dependent Diabetes Mellitus. In *Histocompatibility testing* 1980 (ed. P.I. Terasaki), p. 638. UCLA Tissue Typing Laboratory, Los Angeles.
Treco, D., B. Thomas, and N. Arnheim. 1985. Recombination hot spot in the human β-globin gene cluster: Meiotic recombination of human DNA fragments in *Saccharomyces cerevisiae*. *Mol. Cell. Biol.* **5:** 2029.
Trowsdale, J., J.A.T. Young, A.P. Kelly, P.J. Austin, S. Carson, H. Meunier, A. So, H.A. Erlich, R.S. Spielman, J. Bodmer, and W.F. Bodmer. 1985. Structure, sequence and polymorphism in the HLA-D region. *Immunol. Rev.* **85:** 5.
Winearls, B.C., J.G. Bodmer, W.F. Bodmer, G.F. Bottazzo, J. McNally, J.I. Mann, M. Thorogood, M.A. Smith, and J.D. Baum. 1984. A family study of the association between insulin dependent diabetes mellitus autoantibodies and the HLA system. *Tissue Antigens* **24:** 234.

Identification of DNA Repair Genes in the Human Genome

J.H.J. Hoeijmakers, M. van Duin, A. Westerveld, A. Yasui, and D. Bootsma
Department of Cell Biology and Genetics, Erasmus University, 3000 DR Rotterdam, The Netherlands

The ubiquitous presence of DNA-damaging agents and the instability of certain chemical bonds in DNA have made it necessary for living organisms to develop DNA repair systems. These repair processes help prevent lesions from interfering with essential DNA functions or from converting into permanent mutations that cause cellular malfunctioning and cell death and in higher organisms contribute to malignancy and possibly aging. In view of the wide spectrum of possible lesions, a network of repair systems has evolved to cope with different types of DNA damage (for an extensive review on DNA damage and repair, see Friedberg 1985). In recent years there have been considerable advances in understanding the mechanism and genetic control of two important repair systems in *Escherichia coli*: the excision-repair pathway operating at UV-induced DNA lesions and bulky DNA adducts and the adaptive response directed toward alkylation lesions (recently reviewed by Walker 1985). Genetic studies have identified a number of genes involved in these processes. Application of recombinant DNA techniques has enabled the cloning of the genes and the subsequent purification of the corresponding proteins. The excision pathway, which constitutes part of the SOS response in *E. coli*, is mediated by the concerted action of the *uvrA*, *-B*, *-C*, and *-D* gene products. First, *uvrA* binds to the DNA. In conjunction with *uvrB* a stable complex is formed at the site of the lesion. Incision of the damaged strand (which occurs on both sides of the damage) is catalyzed in the presence of *uvrC*. The unwinding activity of the *uvrD* gene product stimulates the release of the 12- to 13-nucleotide fragment containing the lesion, after which DNA synthesis and ligation complete the excision-repair process.

Current understanding of the mechanisms of repair pathways in eukaryotes is rather limited compared with the knowledge of these pathways in *E. coli*. In the lower eukaryote *Saccharomyces cerevisiae,* more than 30 *RAD* (radiation sensitive) loci have been identified (for a review, see Haynes and Kunz 1981). These fall into three epistatic groups considered to reflect three distinct cellular responses to DNA injury. The *RAD3* epistasis group (consisting of more than 10 loci) is deficient in excision-repair of UV-induced pyrimidine dimers and cross-links. The *RAD6* group (at least 14 members) is disturbed in postreplication repair and double-strand (ds) break repair, and the *RAD52* group (composed of >10 mutants) in a process called recombinational repair. With the recent cloning of a series of genes functioning in these repair pathways (e.g., Yasui and Chevallier 1983; Adzuma et al. 1984; Naumovski et al. 1985; Nicolet et al. 1985; Reynolds et al. 1985a,b), prospects for rapid advances in our understanding of the mechanisms of DNA repair in yeast look favorable.

As far as higher eukaryotes are concerned, valuable tools for the study of repair and the isolation of genes are available in the form of naturally occurring human and laboratory-induced rodent mutants. The human mutants are cell lines derived from patients suffering from hereditary disorders such as xeroderma pigmentosum (XP), Fanconi's anemia, ataxia telangiectasia (AT), and Bloom's syndrome. These rare diseases are characterized by hypersensitivity to specific categories of damaging agents and, in general, display strongly increased incidence of neoplasia (for a review on the clinical and biochemical aspects of these heritable disorders, see Kraemer 1983).

The most extensively investigated repair syndrome is XP. Individuals with this autosomal recessive disorder clinically present extreme skin sensitivity to sun (UV) exposure. Furthermore, XP is associated with predisposition to skin cancer and frequently with neurological abnormalities. The primary defect in most XP patients resides in the excision-repair of UV-induced lesions (Cleaver 1968); a minority of XP cases (XP variants) is suggested to be deficient in a process termed "postreplication repair" (Lehmann et al. 1977). Genetic studies involving cell hybridization have disclosed the existence of nine complementation groups within the excision-deficient class of XP patients (de Weerd-Kastelein et al. 1972; Fischer et al. 1985) and at least five within excision-deficient mutants generated from Chinese hamster ovary (CHO) cells (Thompson et al. 1981, 1982; Thompson and Carrano 1983). All these mutants are unable to perform efficiently the first step postulated in the excision pathway, i.e., the incision of the damaged DNA strand. Since the relationship between the affected human and rodent loci is unknown, these data suggest the involvement of at least 9 and perhaps more than 13 genes and proteins in nucleotide excision. Considerable genetic complexity is also found for AT, in which the response to ionizing radiation is disturbed (Jaspers and Bootsma 1982; Murnane and Painter 1982). Up to now progress with respect to the isolation of the components involved in these processes is limited, due to the complexity of the systems and the experimental limitations of the organisms. Recently, we have cloned the first human gene implicated in the excision of lesions induced by UV and cross-linking agents. In this paper we summarize a molecular and

  0-87969-052-6/86 $1.00

cell biological characterization of this gene. Detailed information on some of the results presented here has appeared elsewhere (Westerveld et al. 1984; van Duin et al. 1986).

MATERIALS AND METHODS

The isolation of* ERCC-1 *cDNA clones. A human cDNA expression library was kindly provided by Dr. H. Okayama (Okayama and Berg 1983). This library was constructed from poly(A)$^+$ RNA of SV40-transformed human fibroblasts and allows for the expression of the cDNA inserts directed by the SV40 early promoter. Using colony filter hybridization (Maniatis et al. 1982), we have isolated several cDNA clones using a nick-translated, ^{32}P-labeled, 1-kb genomic *Pvu*II fragment as a probe.

Cell culture and DNA transfection. To determine the biological function of the different *ERCC-1* cDNAs, CHO 43-3B cells were transfected with each of the cDNA clones in coprecipitate with the dominant marker pSV3gptH. Culture, transfection, and selection conditions were essentially as previously described (Westerveld et al. 1984). One day prior to transfection, 5×10^5 cells were seeded in 10-cm petri dishes. After 10–14 days of selection, the cells were fixed and clones were counted.

Northern blotting and hybridization. The total RNA of HeLa cells was isolated by the LiCl procedure as described elsewhere (Auffray and Rougeon 1980). Poly(A)$^+$ RNA was obtained by two passages over oligo(dT)-cellulose and electrophoresed in 1% agarose gels containing formaldehyde. Blotting was as described by Maniatis et al. (1982) on nitrocellulose filters (Schleicher & Schuel, ph79). Filters were hybridized at 42°C in the presence of 50% formamide (1 M NaCl) to ^{32}P-labeled, nick-translated *ERCC-1* probes.

DNA sequencing. The complete *ERCC-1* cDNA sequence was determined in two directions following the base-specific chemical cleavage technique of Maxam and Gilbert (1980). To facilitate the sequence strategy, *ERCC-1* cDNA fragments were subcloned in pUC vectors as described by Maniatis et al. (1982).

RESULTS AND DISCUSSION

Molecular Cloning of *ERCC-1*

The strategy followed for the isolation of the human repair gene *ERCC-1* involved genomic DNA transfection of partially digested human DNA fragments to the excision-deficient mutant 43-3B. This mutant, isolated by Wood and Burki (1982), is very sensitive to UV irradiation as well as the cross-linking agent mitomycin C (MM-C) and falls into complementation group 2 described by Thompson et al. (1981). Prior to transfection, a dominant marker gene (*Ecogpt,* Mulligan and Berg 1981) was ligated in vitro to the 40- to 60-kb human restriction fragments to facilitate selection protocols and to tag the transferred gene. After selection for the uptake of the dominant marker gene and subsequently for UV or MM-C resistance, primary transformants were obtained with UV and MM-C survival close to the parental CHO cells. After a second round of transfection and selection using (undigested) DNA of a primary transformant, resistant secondary clones were isolated that contained only a small amount of human sequences and a few copies of the dominant marker gene physically attached to the human DNA. The linkage of *Ecogpt* to the human fragments was used to identify cosmids carrying human sequences in a cosmid library constructed from the DNA of a secondary transformant. One of the cosmids isolated in this way (cos 43-34) was able to confer UV and MM-C resistance to the 43-3B cells with a very high efficiency and apparently carried the human gene (designated *ERCC-1*) responsible for the correction (Westerveld et al. 1984).

Genetic Correction of 43-3B Cells by *ERCC-1*

After transfection *ERCC-1* induces the repair-proficient phenotype in the mutant cells as judged by all repair parameters investigated. These include: pyrimidine dimer removal (measured by the T4 endonuclease assay), UV-induced unscheduled DNA synthesis, UV and MM-C survival, mutability, and induced chromosomal aberrations (Westerveld et al. 1984; M. Zdzienicka; F. Daroudi; both pers. comm.; J. Hoeijmakers et al., unpubl.). Furthermore, correction is complementation-group-specific, since *ERCC-1* does not compensate for excision defects in CHO mutants from other complementation groups than group 2.

43-3B is not corrected in all repair end points to the wild-type level; e.g., colony-forming ability after UV irradiation or in the presence of MM-C is somewhat below that of the parental CHO line (Westerveld et al. 1984). This level is not due to a gene dose effect, since the difference between transformants and wild-type cells is not diminished when amplified *ERCC-1* gene copies are integrated into the genome of the mutant cell. A possible explanation might be that the human gene product is unable to replace fully the Chinese hamster equivalent in the excision process. Further experiments are required to decide between this and other explanations.

Molecular Characterization of *ERCC-1*

Localization of the* ERCC-1 *gene. By restriction enzyme digestion, subcloning of portions of the cos 43-34 in λ vectors, and Southern blot analysis of independent genomic 43-3B transformants, the location of *ERCC-1* could be narrowed down to a 15- to 17-kb region on cos 43-34. A provisional physical map of this region is presented in Figure 1. The isolation of unique probes from this area was hampered by the abundance

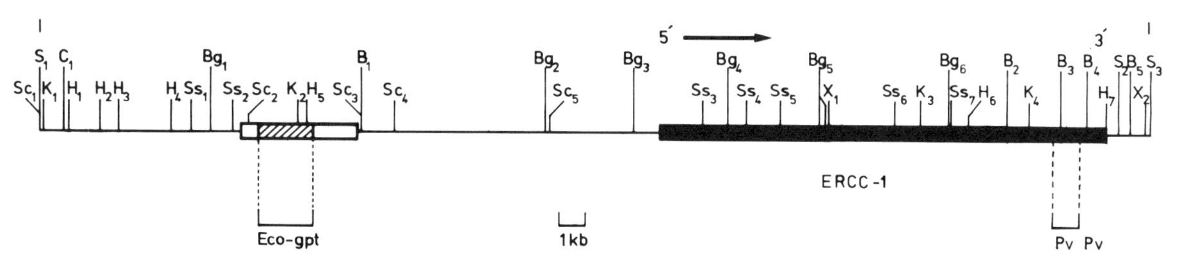

Figure 1. Physical map of the human DNA insert of cosmid 43-34. Cosmid 43-34 contains, in addition to the human excision-repair gene *ERCC-1*, the dominant marker gene *Ecogpt*, which renders transfectants resistant to mycophenolic acid and which was ligated to the human DNA prior to transfection (see text). The position of the genes and the location of the unique 1-kb *Pvu*II probe are indicated. Symbol designation for restriction enzyme cleavage sites: (B) *Bam*HI; (Bg) *Bgl*II; (C) *Cla*I; (H) *Hin*dIII; (K) *Kpn*I; (Pv) *Pvu*II; (S) *Sal*I; (Sc) *Sca*I; (Ss) *Sst*I; (X) *Xho*I. Sites S1 and S3 are from the cosmid vector. Not all *Pvu*II sites are indicated.

of repetitive elements. Nevertheless, the 1-kb *Pvu*II fragment situated at one end of the identified segment (Fig. 1) was found to be largely free of repeats. This fragment was used as probe for Southern and northern hybridizations as well as for screening cDNA libraries. Hybridization with DNA from a well-characterized panel of human/rodent hybrids demonstrated that the presence or absence of the human *ERCC-1* gene correlated well with the presence or absence of chromosome 19, indicating that this chromosome harbored *ERCC-1* (van Duin et al. 1986). Subsequent subchromosomal localization carried out by Brook et al. (1985), using various chromosome 19 translocation hybrids, assigned *ERCC-1* to band q13.2–13.3.

Expression of* ERCC-1 *and isolation of cDNA clones. To obtain information on the size of *ERCC-1* transcript(s), poly(A)$^+$ RNA of HeLa cells was analyzed by northern hybridization. Main hybridization was observed with an RNA species migrating as a relatively broad band at the position corresponding with a size of 1.0–1.1 kb. Weak hybridization was also found with a ~3-kb transcript (Fig. 2). Similar hybridization patterns were obtained with RNA isolated from human fibroblasts, keratinocytes, and blood cells. We conclude from these results that *ERCC-1* is constitutively expressed in a variety of human cell types, and that the 1.0- to 1.1-kb transcript is probably the mature mRNA. Additional evidence (J. Hoeijmakers et al., unpubl.) indicates that the ~3-kb minor transcript is most likely the result of the use of an alternative polyadenylation signal located downstream of the termination site used for the main *ERCC-1* messenger. The ~3-kb RNA species, therefore, possesses a correspondingly longer 3′ untranslated region. Finally, the constitutive level of *ERCC-1* transcripts in HeLa cells is not significantly altered by UV irradiation or MM-C treatment (J. Hoeijmakers et al., unpubl.), suggesting that this repair gene does not belong to a family of inducible genes, as in the case of the SOS response in *E. coli*.

Using the unique *Pvu*II probe from the *ERCC-1* gene region, three cDNA clones (pcD3A, pcD3B7, pcD3C) varying in size from 800 to 1000 bp were isolated from a human expression cDNA library constructed by Okayama and Berg (1983). The physical maps of the inserts are given in Figure 3. Extensive characterization, including sequence analysis, revealed that all three clones lacked different portions of the *ERCC-1* transcript. However, by combining different segments of each clone, a complete cDNA version of the *ERCC-1* transcript could be generated (pcDE, Fig. 3). Comparison of the cDNA sequence with the genomic DNA demonstrated that the gene consists of 10 exons (see Fig. 4), one of which (exon VIII, 72 bp) corresponded exactly with a 72-bp region missing in one of the cDNA

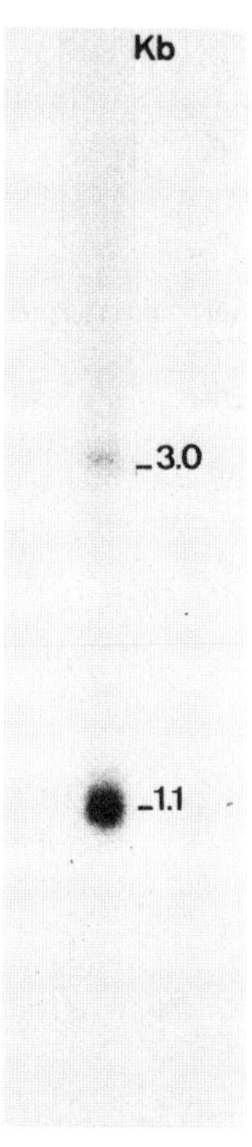

Figure 2. Northern blot analysis of poly(A)$^+$ RNA. Poly(A)$^+$RNA (20 μg) of HeLa cells was size-fractionated by agarose gel electrophoresis in the presence of formaldehyde. After transfer to nitrocellulose, *ERCC-1* mRNA was visualized by hybridization with a ^{32}P-labeled probe from the *ERCC-1* gene.

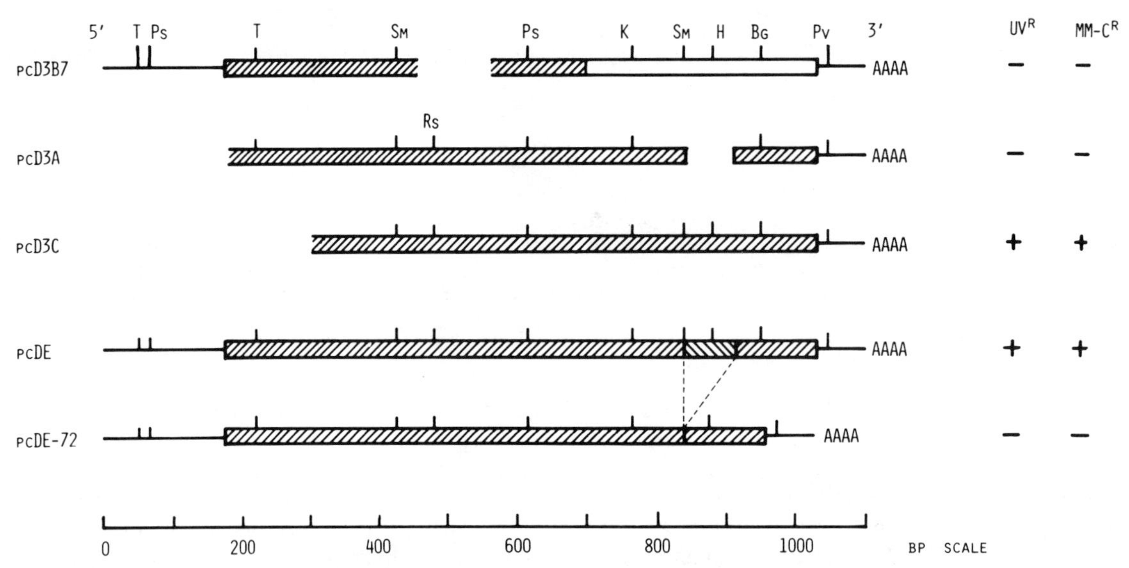

Figure 3. Physical maps of *ERCC-1* cDNAs. The three *ERCC-1* clones isolated from the human expression cDNA library using a genomic *ERCC-1* probe (Fig. 1) are depicted in the upper three rows. Clone pcD3B7, which extends most to the 5′ end has an internal deletion of 104 bp causing a frameshift in the reading frame (shaded area) and premature termination of the coding sequence. Clone pcD3A lacks an internal region of 72 bp, which appears to correspond with exon VIII of the *ERCC-1* gene. The incomplete clone pcD3C stops 161 nucleotides before the start of the open reading frame and therefore lacks 54 amino-terminal codons (a potential ATG start codon is present in the vector). See Fig. 4 for detailed information on the regions covered by these cDNAs. cDNA clone pcDE was constructed by substituting the internal *Sma*I fragment of pcD3B7 with the corresponding fragment of pcD3C. Clone pcDE-72 was obtained by ligation of the *Sma*I–*Bgl*II fragment of pcD3A to the equivalent sites in pcDE. The results of transfection of the cDNA clones to 43-3B cells are summarized in the right part of the figure; (UV^R, MM-C^R) induction of UV and MM-C resistance in 43-3B. Symbol designation for restriction enzyme cleavage sites: (Bg) *Bgl*II; (H) *Hin*dII; (K) *Kpn*I; (Ps) *Pst*I; (Pv) *Pvu*II; (Rs) *Rsa*I; (Sm) *Sma*I; (T) *Taq*I.

clones (pcD3A, Fig. 3). This finding made it very likely that this cDNA insert was derived from an alternatively spliced *ERCC-1* transcript. S1 analysis confirmed the occurrence of differential processing of *ERCC-1* precursor RNA yielding two mature mRNAs of 1.1 and 1.0 kb (see van Duin et al. 1986 for further details). Therefore a complete cDNA clone lacking the alternatively spliced 72-bp exon (pcDE-72) was also constructed as indicated in Figure 2. S1 analysis of the transcriptional start at the genome indicated that both cDNAs were complete except for the first 9–11 nucleotides.

Biological functions of the* ERCC-1 *transcripts. To examine the role of the two transcripts in the excision process, the corresponding cDNAs inserted in "Okayama" mammalian expression vectors were transfected to 43-3B mutant cells. As shown in Figure 3 the cDNA of the large transcript (pcDE) conferred resistance to UV and MM-C, in contrast to that of the small mRNA and two of the three incomplete cDNAs. This rules out the possibility that one of the full-length clones functions in the removal of UV-induced photolesions, and the other in the repair of cross-links. Apparently, the presence of the differentially spliced exon is essential for correction of the mutation in 43-3B cells. The role of the small transcript, if any, is unknown because it is not required for complementing the repair deficiency in the mutant cell.

An interesting observation is provided by the finding that the incomplete clone of pcD3C also compensates for the 43-3B defect (Fig. 3). This clone lacks the first 54 amino acids of the putative *ERCC-1* gene product. Apparently, these are not essential for correction.

Sequence analysis of the* ERCC-1 *cDNAs. The nucleotide and deduced amino acid sequence was determined for both cDNAs and is presented in Figure 4. The sequence contains open reading frames (ORFs) for largely identical polypeptides of 297 amino acids (in the case of pcDE) and 273 amino acids (for pcDE-72). The ORF is preceded by an untranslated region of 142 bp and followed by a 3′ noncoding sequence of 65 nucleotides, which contains the common polyadenylation signal (AATAAA) at the expected distance (~20 nucleotides) from the poly(A) tail. The alternatively spliced exon (Fig. 4, in italics) encodes an internal protein part of 24 amino acids, which is rather rich in threonine residues (~30%). The calculated molecular mass for the two predicted gene products is 32,562 and 29,993 daltons. The hydrophobicity value of −26 and the absence of long hydrophobic stretches suggest that the polypeptides should be water soluble and probably do not represent membrane-associated proteins. Experi-

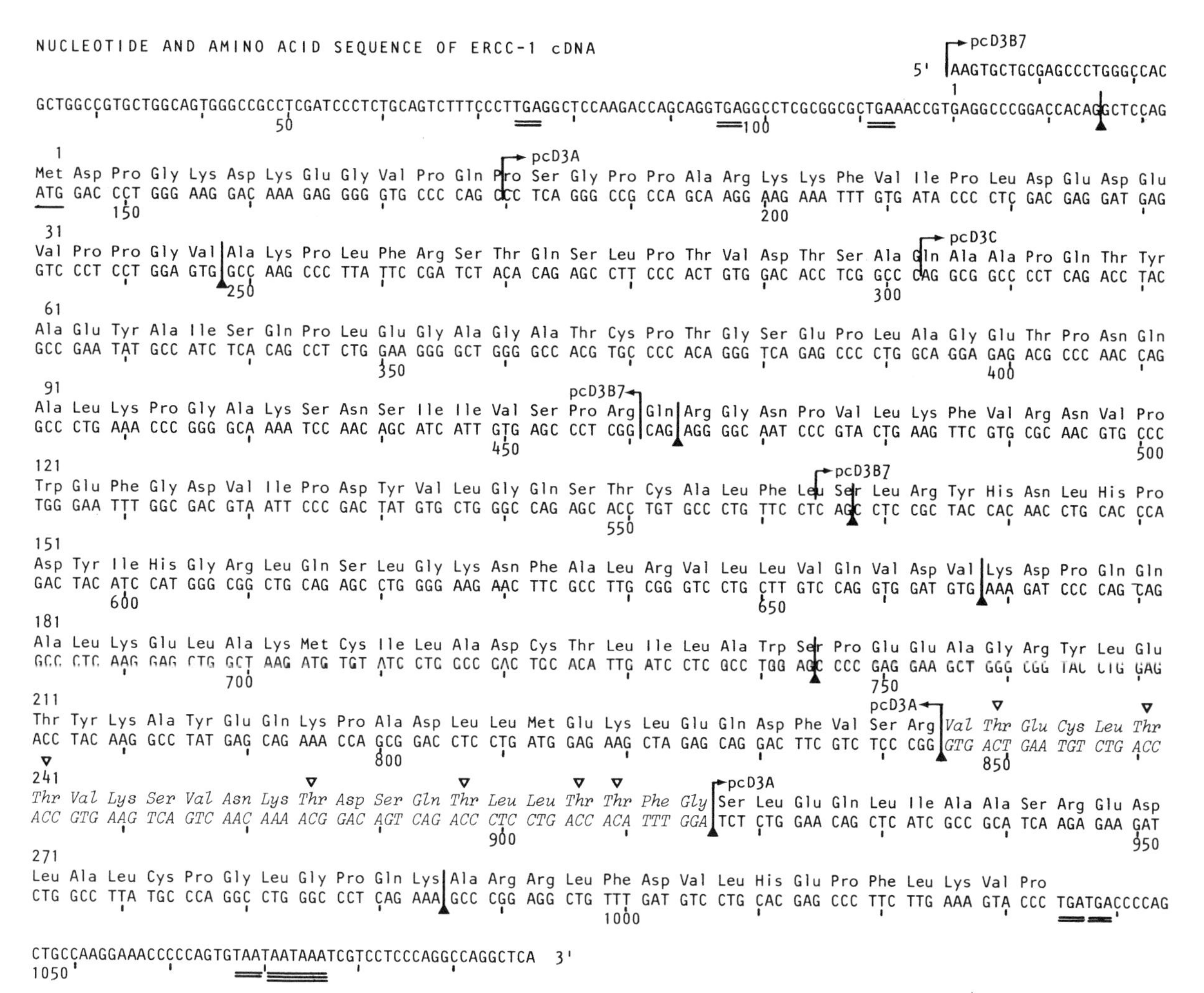

Figure 4. Nucleotide sequence of the *ERCC-1* cDNA clone pcDE and deduced amino acid sequences. The position of exon borders is indicated by ⊥. The alternatively spliced 72-bp exon absent in pcDE-72 is printed in italics. Arrows indicate the regions covered by the various cDNA clones isolated from the Okayama cDNA library. Stop codons and the polyadenylation signal AATAAA are underlined. Sequencing was done by the Maxam and Gilbert procedure.

ments to isolate *ERCC-1* gene product(s) using *E. coli* expression systems and gene amplification in CHO cells are in progress.

Homology of* ERCC-1 *gene products with other proteins. Using DIAGON software (Staden 1982), the *ERCC-1* nucleotide and predicted amino acid sequence was compared with a number of prokaryotic and yeast repair genes and proteins. Significant amino acid homology was detected with the yeast excision-repair protein *RAD10*. Figure 5A presents a possible alignment of both proteins, and quantitative data are compiled in Figure 5B. Since *RAD10* is predicted to be only 210 amino acids (Reynolds et al. 1985b), the *ERCC-1* gene product possesses an extra carboxy-terminal region of 83 amino acids, which harbors the protein segment encoded by the alternatively spliced exon. For the remainder the proteins show 26.5% identity. If substitutions of physicochemically closely related amino acids are allowed (see legend to Fig. 5 for explanation), the overall homology raises to 39%. Particularly, the carboxy-terminal 125 amino acids display a high degree of similarity (34% identical, 51% strongly homologous residues). The finding of extensive homology leads us to believe that *ERCC-1* and *RAD10* probably are evolutionarily related. This idea is strengthened by the fact that the mutant phenotypes of *RAD10* and 43-3B are very similar (Table 1). The relatively low level of homology found between the amino-terminal parts of both proteins can be explained by the indication from the pcD3C transfection experiment (Fig. 3) that the amino-terminal 54 amino acids are not required for 43-3B correction. If this part of the *ERCC-1* protein is not essential for its function, it is conceivable that it is less subject to evolutionary conservation. The main and striking difference between *RAD10* and *ERCC-1* concerns the carboxy-terminal extension of the human gene product absent in the predicted yeast homolog. In this part resides the 24-amino-acid stretch specified by the alternative exon that is shown to be essential for genetic complementation of the 43-3B mutation. It might be that *ERCC-1* has gained additional functions not

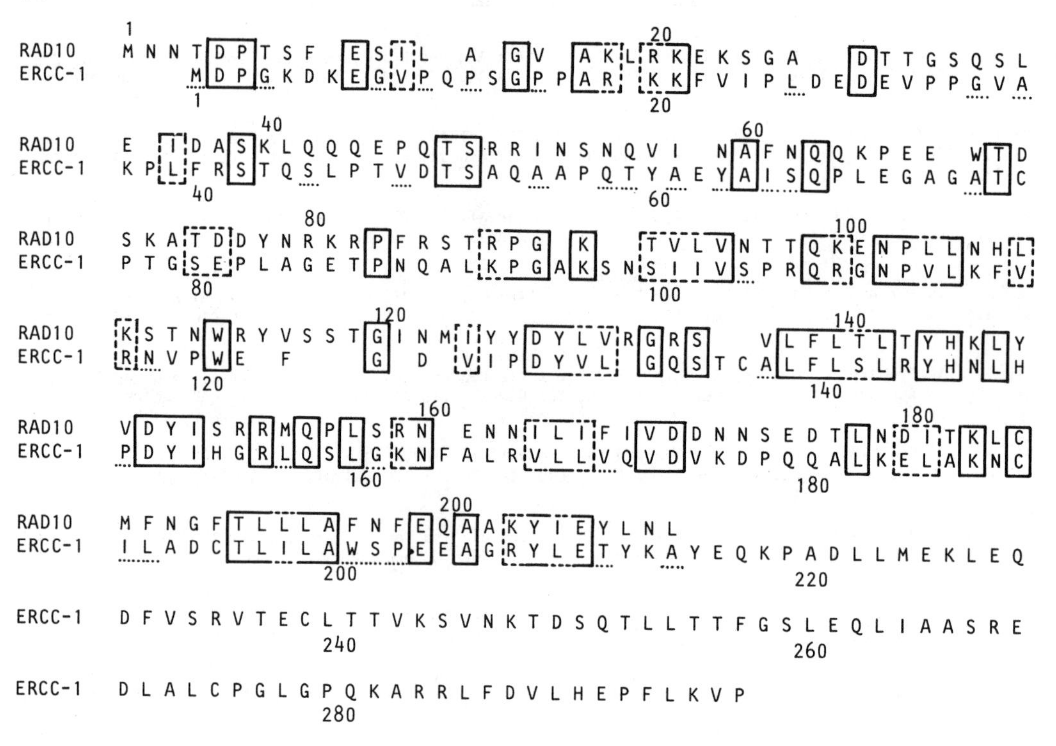

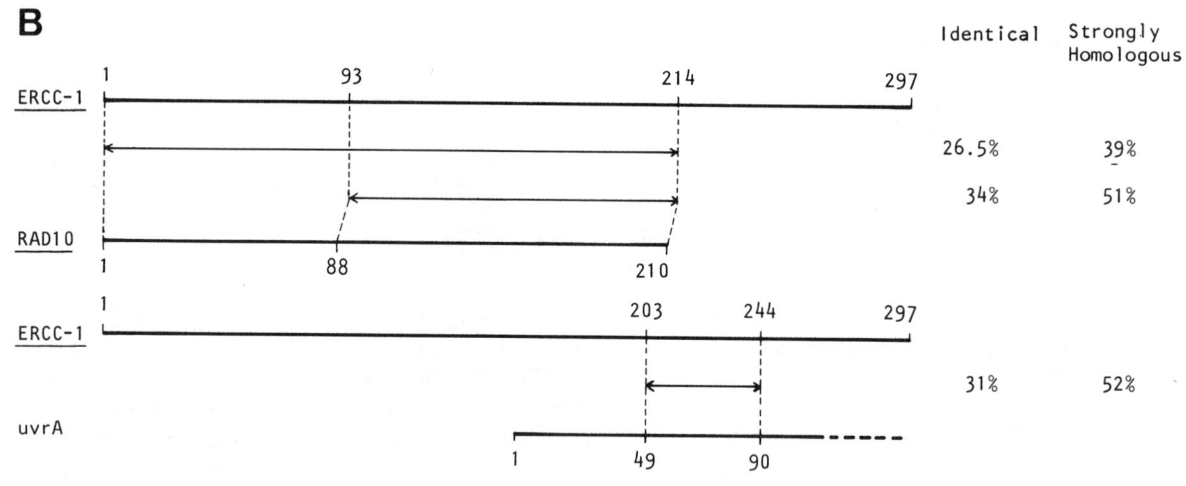

C

ERCC-1 E E A G R Y L E T Y K A Y E Q K P A D L L M E K L E Q D F V S R V T E C L T T V K S

uvrA E G Q R R Y V E S L S A Y A R Q F L S L M E K P D V D H I E G L S P A I S I E Q K S

Figure 5. Homologies of the predicted protein sequence of *ERCC-1* to that of *RAD10* and *uvrA*. (*A*) Alignment of amino acid sequence of *ERCC-1* and *RAD10*. (*B*) Schematic representation showing *ERCC-1/RAD10* and *ERCC-1/uvrA* homologies. The table on the right gives percentage of identity and close homology of the region indicated by the arrows. The numbering refers to the amino acid sequence of the respective proteins. (*C*) Alignment of amino acid sequences of part of *ERCC-1* and *uvrA*. Sequence identities are indicated by solid-line boxes; physico-chemically closely related amino acids (K,R; D,E; I,L,V) by dashed-line boxes; and weakly related amino acids (see Schwartz and Dayhoff 1978 for group classification) by dotted underlining. The standard one-letter amino acid abbreviations are used. The *RAD10* and *uvrA* sequences are from Reynolds et al. (1985b) and Husain et al. (1986), respectively.

Table 1. Phenotypic Comparison of Excision-repair Mutants CHO 43-3B and *S. cerevisiae RAD10*

			References[c]	
	CHO-43-3B[a]	*RAD10*[a]	CHO 43-3B	*RAD10*
UV sensitivity[b]	6–7×	~10×	1	3
UV-induced mutagenesis	38× (0.5–2 J/M^2)	15–50× (5–25 J/M^2)	1	4
4-NQO sensitivity[b]	5×	~4×	1	5
4-NQO-induced mutagenesis	1.5–3×	~2×	1	5
MMS sensitivity[b]	1.5–2×	1.3×	1	6
X-ray sensitivity	–	–	1	3
Excision deficient (incision$^-$)	+	+	2	7
MM-C sensitivity	>100×	unknown	1	

[a]Compared with wild type.
[b]Sensitivities compared at D_{10}.
[c]References: 1. Zdzienicka and Simons (1986); 2. Wood and Burki (1982); 3. Cox and Parry (1968); 4. Lawrence and Christensen (1976); 5. Prakash (1976); 6. Zimmermann (1968); 7. Prakash (1977).

specified by the yeast protein. However, other explanations have not been ruled out, and further research, including mutual complementation experiments using the yeast and human genes and mutants, is required to resolve this issue. Such studies have been initiated.

In addition to the extensive homology between *ERCC-1* and *RAD10*, computer analysis revealed also a regional homology between *ERCC-1* and *uvrA*, which is shown in Figures 5B and C. This concerns a stretch of 42 amino acids that exhibits 31% identity and 52% strong homology. As depicted in Figure 5B, the homologous part in *ERCC-1* overlaps just with the point where the *RAD10* homology stops. If the *uvrA*/*ERCC-1* homology represents a functional domain shared by the two proteins, it is apparently not present in a complete form in *RAD10*. As nothing is known about the function of this particular region in *uvrA*, the significance of this observation remains to be determined. Comparison of *ERCC-1* with published sequences of other repair proteins (*uvrB*, *alkA*, *phr* of *E. coli* and yeast, bacteriophage T4 *denV*, and yeast *RAD1*, *-3*, *-6*, and *-52*) did not reveal significant homologies.

To obtain additional information on possible functional properties of the *ERCC-1* gene product(s), a search was made for amino acid homology with known functional protein domains. Since nucleotide excision is anticipated to take place in the nucleus, the *ERCC-1* amino acid sequence was screened for the presence of sequence motifs resembling nuclear location signals (NLS). The most precisely defined NLS is harbored by the SV40 large T antigens and consists of a series of predominantly basic amino acids, of which Lys-128 by mutation analysis is shown to be essential (Kalderon et al. 1984a) (Fig. 6A). When inserted into other proteins, the sequence is sufficient to redirect the hybrid polypeptide to the nucleus (Kalderon et al. 1984b). A related, but not identical, sequence found at a similar position in polyoma large T exhibits the same properties (Richardson et al. 1986). Figure 6A compares a region in the amino terminus of *ERCC-1* with the two viral NLS. The high level of similarity at important positions strongly suggests that the *ERCC-1* gene product is a nuclear protein. The putative NLS region is located in the apparently nonessential amino-terminal part of the protein. The finding that its absence does not inactivate *ERCC-1* can be explained in two ways. First, the predicted size of *ERCC-1* gene product(s) is such that it can enter the nucleus also by passive diffusion (Paine et al. 1975). Although this process is probably less efficient, it might be sufficient to allow transformants to survive our UV and MM-C selection protocol. Second, the absence of a NLS might be less serious in rapidly dividing cells, such as CHO cells, in which the nuclear membrane is frequently absent due to the high rate of mitotic events.

In addition to a NLS, we have examined whether DNA binding properties could be inferred from the *ERCC-1* amino acid sequence. A well-characterized helix-turn-helix motif has been associated with DNA binding potential for a number of prokaryotic DNA binding proteins (Pabo and Sauer 1984) and recently for homeobox proteins (e.g., Desplan et al. 1985) and yeast mating-type regulatory proteins (e.g., Porter and Smith 1986). A compilation of some of these protein domains is presented in Figure 6B. Specific amino acid positions 5, 8–10, and 15 are considered to be of structural importance, whereas residues 11–13, 16, 17, and 20 (Wharton and Ptashne 1985) and possibly 14 and 19 (Laughon and Scott 1984) are thought to be involved in determining the DNA sequence specificity and hence might vary between different proteins. Comparison of *ERCC-1* with these sequences reveals that identical or related amino acids are present at the important points. Therefore this region, which appears to be the most strongly conserved part between *ERCC-1* and *RAD10*, might comprise a DNA-binding domain. However, direct proof for this hypothesis awaits experiments with the isolated *ERCC-1* protein.

Finally, a region of *ERCC-1* was identified that shared homology with the consensus sequence for ADP-ribosylation found in a family of proteins called G proteins, or guanine nucleotide-binding proteins. Transducin and the *ras* proteins are prominent members of this family. These polypeptides interact with specific receptors and transduce signals by influencing the activity of enzymes that determine the intracellular concentration of "second messengers" such as cAMP

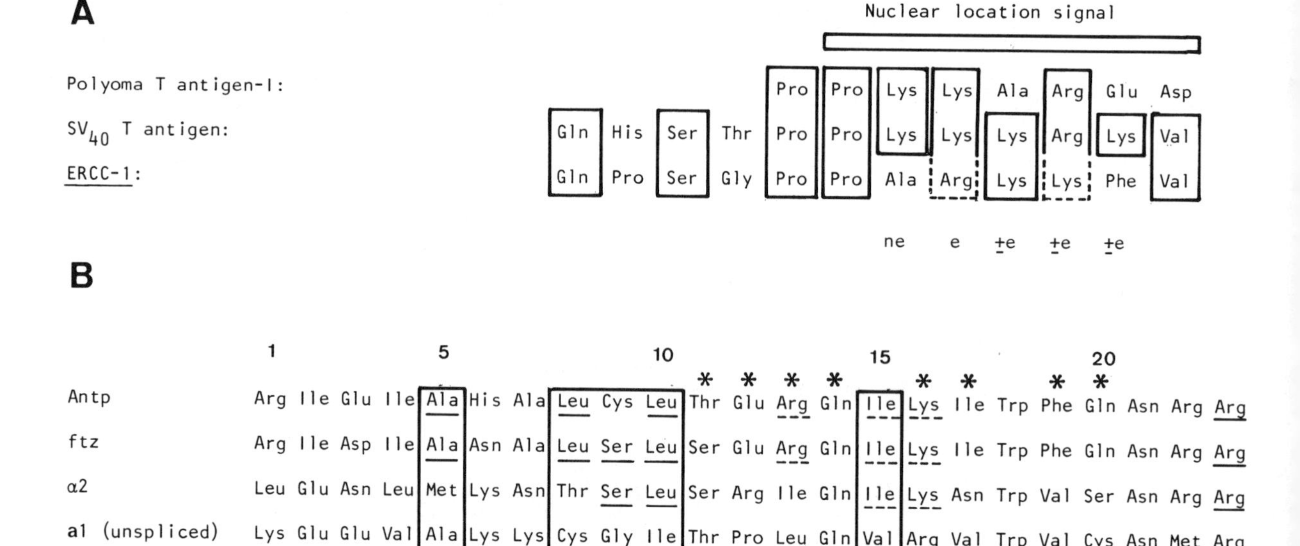

Figure 6. Homologies in amino acid sequences between the predicted *ERCC-1* gene product(s) and various functional protein domains. (*A*) Homology of part of the *ERCC-1* amino acid sequence with NLS of SV40 and polyoma T antigens. Extensive characterization of the NLS of SV40 T antigen is described by Kalderon et al. (1984a,b), and that of the polyoma T antigen by Richardson et al. (1986). Identity is indicated by solid-line boxes, and close homology by dashed-line boxes. The positions of essential (e), important (± e), nonessential (ne) amino acids are deduced from the mutation analysis of the SV40 NLS. The *ERCC-1* region shown corresponds with residues 12–23 of the *ERCC-1* amino acid sequence in Fig. 4. (*B*) Alignment of various DNA-binding protein domains with part of the *ERCC-1* amino acid sequence. The residues at positions 5, 8–10, and 15, which are considered to play an important role in determining the orientation of the two α-helices with respect to each other, are boxed; those amino acids thought to be involved in determining sequence specificity for DNA binding are indicated by an asterisk. The sequence from the *Drosophila* fushi tarazu (*ftz*) and Antennapedia (*Antp*) homeo box proteins and yeast mating-type regulatory proteins a_1 and α2 are taken from the compilation of Shepherd et al. (1984). (*C*) Homology of the *ERCC-1* amino acid sequence with the site for ADP-ribosylation of the α-subunits of transducin (Tα) and the GTP-binding stimulatory protein of adenylate cyclase (Gsα) and homologous regions of two *ras* proteins. Residues that have been retained in all proteins are boxed. Identical and closely related amino acids are indicated by solid and dashed underlining, respectively. The arrowhead points to the arginine residue that is the actual site of ADP-ribosylation by cholera toxin (Van Dop et al. 1984). The *ras* proteins appear not to be good substrates for this reaction (Beckner et al. 1985). The amino acid sequences are from Tanabe et al. (1985) and Robishaw et al. (1986). The residue numbers of the protein part shown are between the brackets.

and perhaps Ca^{++} (Lochrie et al. 1985). ADP ribosylation (e.g., by various bacterial toxins) freezes G proteins in a GTP- or GDP-bound form, thereby interfering with signal transduction. The cholera toxin site for mono-ADP-ribosylation of the α-subunit of transducin has been localized to Arg-174 (Van Dop et al. 1984) (Fig. 6C). Among the residues conserved around this site, the valine immediately following the arginine and a glutamine preceding it at some distance are most strongly conserved (Robishaw et al. 1986). Recently, it has been found that the same acceptor as used by toxins may be modified by the enzymes of host cell origin (see Ueda and Hayaishi 1985 for a recent review on ADP-ribosylation). Poly-ADP-ribosylation of specific nuclear proteins is closely related to DNA repair events; however, its exact role is not clear. Although it is uncertain whether mono-ADP-ribosyl proteins provide the initiation site for poly-ADP-ribosylation, the homology displayed by *ERCC-1* with the ADP-ribosylation sites of G proteins (Fig. 6C) is suggestive of at least mono-ADP-ribosylation. It is worth noting that the location of the putative ribosylation site is such that alternative splicing of the *ERCC-1* precursor RNA would disrupt the Arg–Val sequence (see Fig. 4). It is likely that this destroys the putative ADP-ribosyl acceptor function of this domain. Evidence for in vivo mono- and/or poly-ADP-ribosylation of one of the *ERCC-1* gene products and analysis of its function can only be obtained by direct experiments.

CONCLUDING REMARKS

We have described molecular and cell biological characterizations of the first human excision repair gene *ERCC 1*. It is clear that more work needs to be done concerning this gene, but also that many more genes implicated in the nucleotide excision process should be isolated. SV40-transformed XP fibroblasts seem to be ideal starting material for the cloning of additional (complementing) repair genes. However, so far they have not successfully been employed for this purpose, notwithstanding many attempts in different laboratories, including ours (Lehmann 1985). We think that this is due at least in part to the fact that the human cells used are very restricted in the amount of DNA that becomes integrated in their genome after genomic DNA transfections. On the average a 20-fold to 100-fold lower amount of exogenous DNA is found in such human transformants, compared for example, with CHO cells (Hoeijmakers et al. 1986). Therefore, it is not surprising that up to now the most promising results have been obtained using laboratory-induced rodent mutants. Transformants corrected by the uptake of human genes have been isolated for a number of CHO repair mutants (Rubin et al. 1983; MacInnes et al. 1984; Thompson et al. 1985), indicating that the isolation of more repair genes is underway. Another potential route to human repair genes is opened by the finding of extensive homology between the *ERCC-1* and *RAD10* gene products. This might indicate that the excision-repair system in toto is strongly conserved during eukaryotic evolution. If so, this may be utilized for the isolation of mammalian genes by virtue of homology with yeast excision-repair genes. The degree of divergence at the nucleotide level between *ERCC-1* and *RAD10* renders it unlikely that the evolutionary distance between both ends of the eukaryotic spectrum can be overcome in one step. Therefore, evolutionary intermediate "stations on the road" from yeast to mammals may have to be taken. At the same time, this study will yield valuable information on evolution of the genes investigated, and the evolutionary conservation and importance of specific parts or functional domains of the gene products. As in the case of *E. coli*, progress in the understanding of mammalian repair depends very much on the success with which these and other approaches will yield genes.

ACKNOWLEDGMENTS

We are very grateful Mrs. H. Odijk, J. van den Tol, and Mr. J. de Wit for excellent technical assistance, Mr. M. Koken and P. ten Dijke for help in some of the experiments, Dr. R. van Gorcom (Medical Biological Laboratory, TNO, Rijswijk, The Netherlands) for indispensible help with the computer, Dr. B. van Houten (University of North Carolina) for giving permission to use the *uvrA* sequence before publication, Drs. M. Zdzienicka and J. Simons for providing the characterization of the 43-3B mutant prior to publication, and Dr. H. Okayama (NIH, Bethesda, Maryland) for the generous gift of the cDNA library. Furthermore, we thank Mrs. R. Boucke for skillful typing of the manuscript and Mr. T. van Os for photography. This work was supported by FUNGO (Foundation of Medical Scientific Research in the Netherlands) and EURATOM contract number BIO-E-404-NL.

REFERENCES

Adzuma, K., T. Ogawa, and H. Ogawa. 1984. Primary structure of the RAD52 gene in *Saccharomyces cerevisiae*. *Mol. Cell. Biol.* **4:** 2735.

Auffray, C. and F. Rougeon. 1980. Purification of mouse immunoglobulin heavy chain messenger RNAs from total myeloma tumor RNA. *Eur. J. Biochem.* **107:** 303.

Beckner, S.K., S. Hattari, and T.Y. Shih. 1985. The *ras* oncogene product p21 is not a regulatory component of adenylate cyclase. *Nature* **317:** 71.

Brook, J.D., D.J. Shaw, A.L. Meredith, M. Warwood, J. Cowell, J. Scott, T.J. Knott, M. Litt, L. Bufton, and P.S. Harper. 1985. A somatic cell hybrid panel for chromosome 19 localization of human genes and RFLPs and orientation of the linkage group. *Cytogenet. Cell Genet.* **40:** 590.

Cleaver, J.E. 1968. Defective repair replication of DNA in xeroderma pigmentosum. *Nature* **218:** 652.

Cox, B.S. and J.M. Parry. 1968. The isolation, genetics and survival characteristics of ultraviolet light-sensitive mutants in yeast. *Mutat. Res.* **6:** 37.

Desplan, C., J. Theis, and P.H. O'Farell. 1985. The *Drosophila* developmental gene, *engrailed* encodes a sequence-specific DNA binding activity. *Nature* **318:** 630.

de Weerd-Kastelein, E.A., W. Keijzer, and D. Bootsma. 1972. Genetic heterogeneity of xeroderma pigmentosum demonstrated by somatic cell hybridization. *Nat. New Biol.* **238:** 80.

Fischer, E., W. Keijzer, H.W. Thielmann, O. Popanda, E. Bohnert, L. Edler, E.G. Jung, and D. Bootsma. 1985. A ninth complementation group in xeroderma pigmentosum, XP-I. *Mut. Res.* **145:** 217.

Friedberg, E.C. 1985. *DNA repair*. W.H. Freeman, San Francisco.

Haynes, R.H. and B.A. Kunz. 1981. DNA repair and mutagenesis in yeast. In *The molecular biology of the yeast* Saccharomyces: *Life cycle and inheritance*. (ed. J. Strathern et al.), p. 371. Cold Spring Harbor Laboratory, Cold Spring Harbor, New York.

Hoeijmakers, J.H.J., H. Odijk, and A. Westerveld. 1986. Differences between rodent and human cell lines in the amount of integrated DNA after transfection. *Exp. Cell Res.* (in press).

Husain, I., B. van Houten, D.C. Thomas, and A. Sancar. 1986. Sequence of *Escherichia coli uvrA* gene and protein reveal two potential ATP binding sites. *J. Biol. Chem.* **261:** 4895.

Jaspers, N.G.J. and D. Bootsma. 1982. Genetic heterogeneity in ataxia-telangiectasia studied by cell fusion. *Proc. Natl. Acad. Sci.* **79:** 2641.

Kalderon, D., W.D. Richardson, A.F. Markham, and A.E. Smith. 1984a. Sequence requirements for nuclear localization of SV40 large T-antigen. *Nature* **311:** 33.

Kalderon D., B.L. Roberts, W.D. Richardson, and A.E. Smith. 1984b. A short amino acid sequence able to specify nuclear location. *Cell* **39:** 499.

Kraemer, K.H. 1983. Heritable diseases with increased sensitivity to cellular injury. *Update: Dermatology in general medicine* (ed. T.B. Fitzpatrick et al.), p. 113. McGraw-Hill, New York.

Laughon, A. and M.P. Scott. 1984. Sequence of a *Drosophila* segmentation gene: Protein structure homology with DNA-binding proteins. *Nature* **310:** 25.

Lawrence, C.W. and R. Christensen. 1976. UV mutagenesis in radiation sensitive strains of yeast. *Genetics* **82:** 207.

Lehmann, A.R. 1985. Use of recombinant DNA techniques in cloning DNA repair genes and in the study of mutagenesis in human cells. *Mut. Res.* **150:** 61.

Lehmann, A., S. Kirk-Bell, C. Arlett, S.A. Harcourt, E.A. de Weerd-Kastelein, W. Keijzer, and P. Hall-Smith. 1977. Repair of ultraviolet light damage in a variety of human fibroblast cell strains. *Cancer Res.* **37:** 904.

Lochrie, M.A., J.B. Hurley, and M.I. Simon. 1985. Sequence of the alpha subunit of photoreceptor G protein: Homologies between transducin, *ras*, and elongation factors. *Science* **228:** 96.

MacInnes, M.A., J.D. Bingham, L.H. Thompson, and G.F. Strniste. 1984. DNA-mediated cotransfer of excision repair capacity and drug resistance into Chinese hamster ovary cell line UV-135. *Mol. Cell. Biol.* **4:** 1152.

Maniatis, T., E.F. Fritsch, and J. Sambrook. 1982. *Molecular cloning: A laboratory manual.* Cold Spring Harbor Laboratory, Cold Spring Harbor, New York.

Maxam, A.M. and W. Gilbert. 1980. Sequencing end-labeled DNA with base-specific chemical cleavages. *Methods Enzymol.* **65:** 499.

Mulligan, R.C. and P. Berg. 1981. Selection for animal cells that express the *Escherichia coli* gene coding for xanthine-guanine phosphoribosyl transferase. *Proc. Natl. Acad Sci.* **78:** 2072.

Murnane, J.P. and R.P. Painter. 1982. Complementation of the defects in DNA synthesis in irradiated and unirradiated ataxia-telangiectasia cells. *Proc. Natl. Acad. Sci.* **79:** 1960.

Naumovski, L., G. Chu, P. Berg, and E.C. Friedberg. 1985. *RAD3* gene of *Saccharomyces cerevisiae:* Nucleotide sequence of wild type and mutant alleles, transcript mapping, and aspects of gene regulation. *Mol. Cell Biol.* **5:** 17.

Nicolet, C.M., J.M. Chenevert, and E.C. Friedberg. 1985. The *RAD2* gene of *Saccharomyces cerevisiae*: Nucleotide sequence and transcript mapping. *Gene* **36:** 225.

Okayama, H. and P. Berg. 1983. A cDNA cloning vector that permits expression of cDNA inserts in mammalian cells. *Mol. Cell. Biol.* **3:** 280.

Pabo, C.O. and R.T. Sauer. 1984. Protein–DNA recognition. *Annu. Rev. Biochem.* **53:** 293.

Paine, P.L., L.C. Moore, and S.B. Horowitz. 1975. Nuclear envelope permeability. *Nature* **254:** 109.

Porter, S.D. and M. Smith. 1986. Homoeo-domain homology in yeast MATα2 is essential for repressor activity. *Nature* **320:** 766.

Prakash, L. 1976. Effect of genes controlling radiation sensitivity on chemically induced mutations in *Saccharomyces cerevisiae*. *Genetics* **83:** 285.

———. 1977. Defective thymidine dimer excision in radiation sensitive mutants *RAD10* and *RAD16* of *Saccharomyces cerevisiae*. *Mol. Gen. Genet.* **152:** 128.

Reynolds, P., S. Weber, and L. Prakash. 1985a. *RAD6* gene of *Saccharomyces cerevisiae* encodes a protein containing a tract of 13 consecutive aspartates. *Proc. Natl. Acad. Sci.* **82:** 168.

Reynolds, P., L. Prakash, D. Dumais, G. Perozzi, and S. Prakash. 1985b. Nucleotide sequence of the *RAD10* gene of *Saccharomyces cerevisiae*. *EMBO J.* **4:** 3549.

Richardson, W.D., B.L. Roberts, and A.E. Smith. 1986. Nuclear location signals in polyoma virus large T. *Cell* **44:** 77.

Robishaw, J.D., D.W. Russell, B.A. Harris, M.D. Smigel, and A.G. Gilman. 1986. Deduced primary structure of the α-subunit of the GTP-binding stimulatory protein of adenylate cyclase. *Proc. Natl. Acad. Sci.* **83:** 1251.

Rubin, J.S., A.L. Joyner, A. Bernstein, and G.F. Whitmore. 1983. Molecular identification of a human DNA repair gene following DNA-mediated gene transfer. *Nature* **306:** 206.

Schwartz, R.M. and M.D. Dayhoff. 1978. Matrices for detecting distant relationships. In *Atlas of protein sequence and structure* (ed. M.D. Dayhoff), vol. 5, suppl. 3, p. 353. National Biomedical Research Foundation, Washington, D.C.

Shepherd, J.C.W., W. McGinnes, A.E. Carrasco, E.M. De Robertis, and W.J. Gehring. 1984. Fly and frog homoeo domains show homologies with yeast mating type regulatory proteins. *Nature* **310:** 70.

Staden, R. 1982. An interactive graphics program for comparing and aligning nucleic acid and amino acid sequences. *Nucleic Acids Res.* **10:** 2951.

Tanabe, T., T. Nukada, Y. Nishikawa, K. Sugimoto, H. Suzuki, H. Takahashi, M. Noda, T. Haga, A. Ichiyama, K. Kangawa, N. Minamino, H. Matsuo, and S. Numa. 1985. Primary structure of the α-subunit of transducin and its relationship to *ras* proteins. *Nature* **315:** 242.

Thompson, L.H. and A.V. Carrano. 1983. Analysis of mammalian cell mutagenesis and DNA repair using in vitro selected CHO cell mutants. *UCLA Symp. Mol. Cell. Biol. New Ser.* **11:** 125.

Thompson, L.H., K.W. Brookman, L.E. Dillehay, C.L. Mooney, and A.V. Carrano. 1982. Hypersensitivity to mutation and sister-chromatid exchange induction in CHO cell mutants defective in incising DNA containing UV-lesions. *Somat. Cell Genet.* **8:** 759.

Thompson, L.H., D.B. Busch, K. Brookman, C.L. Mooney, and P.A. Glaser. 1981. Genetic diversity of UV-sensitive DNA-repair mutants of Chinese hamster ovary cells. *Proc. Natl. Acad. Sci.* **78:** 3734.

Thompson, L.H., K.W. Brookman, J.L. Minkler, J.C. Fuscoe, K.A. Henning, and A.V. Carrano. 1985. DNA-mediated transfer of a human DNA repair gene that controls sister chromatid exchange. *Mol. Cell. Biol.* **5:** 881.

Ueda, K. and O. Hayaishi. 1985. ADP-ribosylation. *Annu. Rev. Biochem.* **54:** 73.

Van Dop, C., M. Tsubokawa, H.R. Bourne, and J. Ramachandran. 1984. Amino acid sequence of retinal transducin at the site ADP-ribosylated by cholera toxin. *J. Biol. Chem.* **259:** 696.

van Duin, M., J. de Wit, H. Odijk, A. Westerveld, A. Yasui, M.H.M. Koken, J.H.J. Hoeijmakers, and D. Bootsma. 1986. Molecular characterization of the human excision repair gene *ERCC-1:* cDNA cloning and amino acid homology with the yeast DNA repair gene *RAD10*. *Cell* **44:** 913.

Walker, G.C. 1985. Inducible DNA repair systems. *Annu. Rev. Biochem.* **54:** 425.

Westerveld, A., J.H.J. Hoeijmakers, M. van Duin, J. de Wit, H. Odijk, A. Pastink, and D. Bootsma. 1984. Molecular cloning of a human DNA repair gene. *Nature* **310:** 425.

Wharton, R.P. and M. Ptashne. 1985. Changing the binding

specificity of a repressor by redesigning an α-helix. *Nature* **316:** 601.

Wood, R.D. and H.J. Burki. 1982. Repair capability and the cellular age response for killing and mutation induction after UV. *Mutat. Res.* **95:** 505.

Yasui, A. and M.-R. Chevallier. 1983. Cloning of photoreactivation repair gene and excision repair gene of the yeast *Saccharomyces cerevisiae. Curr. Genet.* **7:** 191.

Zdzienicka, M.Z. and J.W.I.M. Simons. 1986. Analysis of repair processes by the determination of the induction of cell killing and mutations in two repair deficient Chinese hamster ovary cell lines. *Mutat. Res.* **166:** 59.

Zimmermann, F.K. 1968. Enzyme studies on the products of mitotic gene conversion in *Saccharomyces cerevisiae. Mol. Gen. Genet.* **101:** 171.

Seven Unidentified Reading Frames of Human Mitochondrial DNA Encode Subunits of the Respiratory Chain NADH Dehydrogenase

G. ATTARDI, A. CHOMYN, R.F. DOOLITTLE,* P. MARIOTTINI, AND C.I. RAGAN†

*Division of Biology, California Institute of Technology, Pasadena, California 91125; *Department of Chemistry, University of California at San Diego, La Jolla, California 92093; †Department of Biochemistry, University of Southampton, Southampton S09 3TU, United Kingdom*

Molecular studies on the human mitochondrial genome started well before the introduction of the modern nucleic acid cloning, mapping, and sequencing technologies. The compartmentalized character of this genome, its small size, and the ease of its isolation have made it, since its discovery, an attractive object for studies of gene organization and expression in a simple eukaryotic system. As a result, about 10 years ago, mtDNA was the best characterized portion of the human genome. Its physical structure was well understood, the unique features of its transcription were known, the two rRNAs and most of the tRNAs encoded in mtDNA had been identified and mapped, and the first hints of the extraordinarily compact organization of this genome had been discovered (Attardi et al. 1976); furthermore, evidence was already available about the number of mRNAs and proteins encoded in this genome, and the probable function of some of these proteins (Costantino and Attardi 1975). It was therefore no surprise that when the dideoxy DNA sequencing technology became available in F. Sanger's laboratory in Cambridge, England, the human mitochondrial DNA was chosen as the most favorable eukaryotic nonviral system on which to test the new technology and at the same time to collect biologically important data (Anderson et al. 1981).

One of the most intriguing findings from the sequence analysis of human mtDNA was the discovery that more than half of its protein-coding capacity resides in eight reading frames (Anderson et al. 1981) (Fig. 1), which, with one exception, have no obvious amino acid sequence homology to any of the identified or unidentified reading frames (URF) of *Saccharomyces cerevisiae* (Grivell 1983) and *Schizosaccharomyces pombe* (Lang et al. 1983), although several of them were later found to occur in mtDNA of other lower eukaryotic cells (Benne 1985; Breitenberger and RajBhandary 1985; Brown et al. 1985; Burger and Werner 1985; Vahrenholz et al. 1985; de Vries et al. 1986; Simpson 1986). These reading frames were suspected very early to be functional genes, because of the evidence of their transcription into RNAs with the same structural and metabolic properties as the mRNAs of identified genes (Ojala et al. 1980; Gelfand and Attardi 1981), and because of their strong conservation in size and amino acid sequence in different mammalian species (Attardi 1985). The discovery of the mammalian mtDNA URFs, besides providing striking evidence of the individuality of the mitochondrial genomes of different organisms, created a challenge for investigators in the field. Below is an account of our recent work, which has led to the identification of the polypeptides encoded in all the human mtDNA URFs and to their functional assignment, and thus to the complete elucidation of the informational content of this genome.

RESULTS

The Human Mitochondrial Translation Products

Figure 2 (left lane) shows the electrophoretic pattern in an SDS-urea-polyacrylamide gel of the HeLa cell mitochondrial translation products, which were labeled with [^{35}S]methionine in the presence of emetine (to inhibit cytoplasmic protein synthesis) (Ching and Attardi 1982, 1985). Seventeen discrete bands of different intensities can be recognized. In an SDS-polyacrylamide gradient gel (Fig. 2, right lane), 19 discrete bands are resolved. In the two gel systems, the individual polypeptides that constitute the various bands, as identified by bidimensional fractionation (see below), are indicated by arabic numerals. In the SDS-polyacrylamide gradient gel, the various mitochondrial translation products have different relative mobilities as compared with the SDS-urea-polyacrylamide gel, as first shown for the cytochrome *c* oxidase subunits II (COII) and III (COIII) (Ching and Attardi 1982, 1985). This allowed a bidimensional fractionation of the mitochondrially synthesized polypeptides in the two gel systems and the resolution of the individual components of composite bands. A total of up to 26 components was thus reproducibly identified (Ching and Attardi 1982, 1985). The mitochondrial translation products were originally estimated to cover the range of apparent molecular weights between about 3500 and 51,000 on the basis of their electrophoretic mobilities measured relative to water-soluble marker proteins. Recent work, however, has shown that the molecular weights thus determined are underestimated by 20–50%, as

 0-87969-052-6/86 $1.00

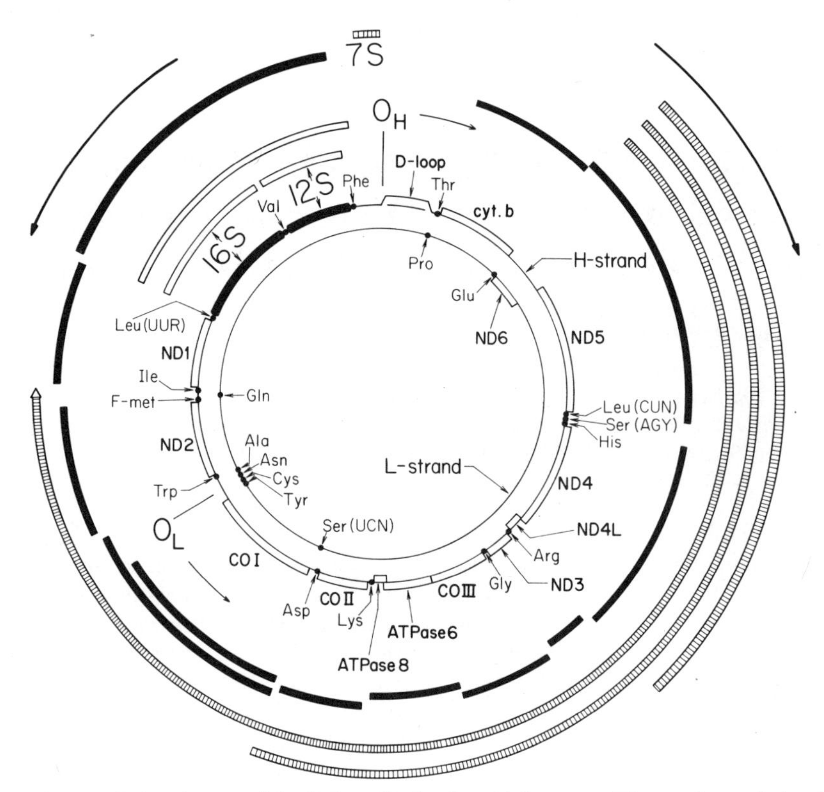

Figure 1. Genetic and transcriptional maps of the HeLa cell mitochondrial genome. The two inner circles show the positions of the two rRNA genes, as derived from mapping and RNA sequencing experiments (Crews and Attardi 1980; Ojala et al. 1980), and those of the reading frames and tRNA genes, as derived from the mtDNA sequence (Anderson et al. 1981). Mapping positions of the oligo(dT)-cellulose-bound and nonbound H-strand transcripts are indicated, respectively, by black and white bars, those of the oligo(dT)-cellulose-bound L-strand transcripts by hatched bars. Left and right arrows indicate the direction of H- and L-strand transcription, respectively. The vertical arrow, marked O_H, and the rightward arrow at the top indicate the location of the origin and the direction of H-strand synthesis; the arrow marked O_L indicates the origin of L-strand synthesis. (COI, COII, and COIII) Subunits I, II, and III of cytochrome *c* oxidase; (ND1, 2, 3, 4, 4L, 5, and 6) NADH dehydrogenase subunits 1, 2, 3, 4, 4L, 5, and 6; (CYTb) apocytochrome *b*; (H^+-ATPase 6 and 8) subunits 6 and 8 of H^+-ATPase (modified from Chomyn et al. 1983). The original designations of the URFs, namely URF1, URF2, URF3, URF4, URF4L, URF5, URF6, and URFA6L, have been changed to ND1, ND2, ND3, ND4, ND4L, ND5, ND6, and ATPase 8, on the basis of their functional identification described in this paper.

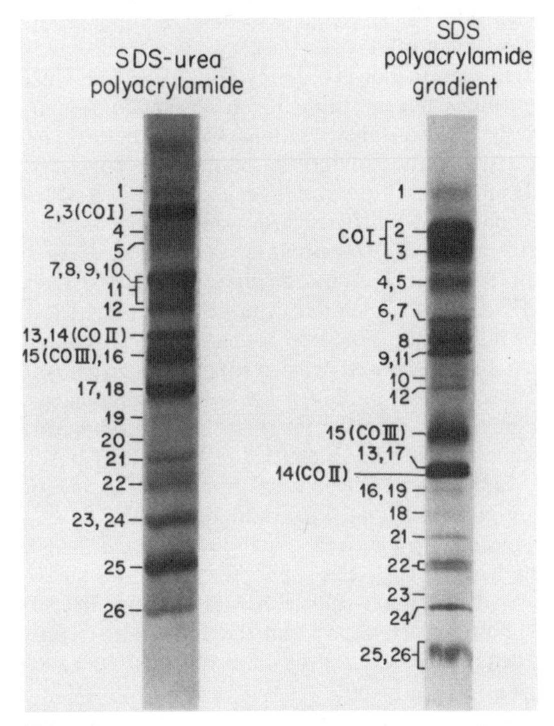

Figure 2. Fluorograms, after electrophoresis through an SDS–8M urea–15% polyacrylamide gel or an SDS–15–25% polyacrylamide gradient gel, of the proteins from the mitochondrial fraction of HeLa cells labeled for 2 hr with [^{35}S]methionine in the presence of 100 μg/ml emetine. Modified from Ching and Attardi (1982), on the basis of the results obtained with anti-peptide antibodies (Chomyn et al. 1983, in prep.; Mariottini et al. 1986b).

compared with the actual molecular weights. This is due to a variably higher degree of SDS binding, and to a correspondingly higher electrophoretic mobility, of the hydrophobic mitochondrially synthesized polypeptides relative to the hydrophilic marker proteins. Therefore, the range of actual molecular weights of the mitochondrial translation products is in agreement with the size range expected for the proteins encoded in the mtDNA reading frames (between 7900 and 66,600 [Anderson et al. 1981]).

The number of protein components resolved by bidimensional fractionation in the two gel systems described above was larger than predicted from the number of reading frames revealed by the sequence of human mtDNA (Anderson et al. 1981). This suggested that some of these components may be related to the primary translation products by processing or secondary modification, or may result from premature termination of translation or proteolytic cleavage occurring in mitochondria either under physiological conditions or under the artificial labeling conditions used in these experiments (Ching and Attardi 1982). Several of the minor components appear to be metabolically unstable, i.e., disappear after a chase (Ching and Attardi 1982; A. Chomyn, unpubl.), a finding that is consistent with their being fragments or precursors of mature products. The evidence obtained in the past few years, using peptide-specific antibodies (Chomyn et al. 1983, in prep.; Mariottini et al. 1983, 1986b), has led to the identification of some of the other minor components.

Gene Assignment of the Mitochondrially Synthesized Polypeptides

At the time the human mtDNA sequence was published (Anderson et al. 1981), only the polypeptides corresponding to COI, COII, and COIII had been identified, by analysis of the electrophoretic properties and site of synthesis of the subunits of the purified human enzyme complex (Hare et al. 1980; Ching and Attardi 1982, 1985) (Fig. 2). The other mitochondrially synthesized polypeptides shown in Figure 2 were expected to include the products of the H^+-ATPase subunit 6 and apocytochrome *b* genes that had been identified in the DNA sequence on the basis of their amino acid sequence homology to the equivalent yeast mtDNA genes, as well as the putative products of the eight URFs. To identify the polypeptides corresponding to these 10 genes we chose the powerful approach based on the use of antibodies directed against peptides predicted from the DNA sequence of the individual reading frames (Walter et al. 1980). This approach proved to be fully successful despite the fact that the hydrophobic character of almost all the mitochondrial translation products (Kyte and Doolittle 1982) reduced the opportunitites for the choice of effective immunogenic epitopes. Table 1 shows the amino acid sequence of the peptides synthesized that proved to be effective in eliciting the production of antibodies capable of reacting with the natural polypeptides. In most cases, an amino-terminal or carboxy-terminal peptide or both of each putative gene product were chosen for synthesis, the main constraint being that the peptide had a minimum degree of hydrophilic character and, therefore, solubility. The reasonable assumption was made that the amino- and carboxy-terminal segments of the protein were more likely to be exposed in the SDS-denatured or native protein and were therefore more likely to react with the antibodies. Indeed, more than 50% of the antibodies prepared against an amino- or carboxy-terminal sequence proved to be effective in reacting with the corresponding polypeptide. In a few cases, antibodies were prepared against an internal sequence of the putative gene product, chosen for its hydrophilic character from the hydropathy profile of the protein (Kyte and Doolittle 1982). Of three such types of antibodies, one proved to be effective (Table 1).

For the detection of the polypeptide(s) reacting with each type of peptide-specific antibody, advantage was taken of the possibility of selectively labeling the mitochondrial translation products in vivo. This was

Table 1. Synthetic Peptides and Modes of Attachment

Source	Sequence[a]	Attachment procedure[b]	Degree of BSA substitution[c]
COII-N (amino terminus)	Met-Ala-His-Ala-Ala-Gln-Val-Gly-Leu-Gln-(Glu)	EDC	15
COII-C (carboxy terminus)	Lys-Ile-Phe-Glu-Met-Gly-Pro-Val-Phe-Thr-Leu	GLUT	19
ATPase 6-C (carboxy terminus)	(Lys)-Val-Ser-Leu-Tyr-Leu-His-Asp-Asn-Thr	GLUT	21
URF1-C (carboxy terminus)	(Lys)-Pro-Ile-Thr-Ile-Ser-Ser-Ile-Pro-Pro-Gln-Thr	GLUT	26
URF2-N (amino terminus)	Ile-Asn-Pro-Leu-Ala-Gln-Pro-Val-(Glu)	EDC	21
URF A6L-N (amino terminus)	Met-Pro-Gln-Leu-Asn-Thr-Thr-Val-(Tyr)	BDB	33
URF A6L-C (carboxy terminus)	(Lys)-Ser-Leu-His-Ser-Leu-Pro-Pro-Gln-Ser	GLUT	5
URF3-C (carboxy terminus)	Lys-Gly-Leu-Asp-Trp-Thr-Glu	GLUT	35
URF4L-C (carboxy terminus)	(Lys)-His-Asn-Leu-Asn-Leu-Leu-Gln-X[d]	GLUT	24
URF4-C (carboxy terminus)	(Lys)-Pro-Asp-Ile-Ile-Thr-Gly-Phe-Ser-Ser	GLUT	18
URF5-N (amino terminus)	Met-Thr-Met-His-Thr-Thr-Met-(Ala-Glu)	EDC	16
URF6-I1 (internal; 14,422–14,454[e])	Ala-Ile-Glu-Glu-Tyr-Pro-Glu-Ala-Trp-Gly-Ser	GLUT	24

[a]Residues in parentheses were incorporated for attachment purposes.
[b](EDC) Water-soluble carbodiimide; (GLUT) glutaraldehyde; (BDB) bisdiazobenzidine.
[c]Number of peptide molecules/BSA molecule.
[d](X) α-Aminobutyric acid.
[e]Positions in the DNA sequence (Anderson *et al.* 1981).

achieved by inhibiting cytoplasmic protein synthesis either irreversibly with emetine, or reversibly, in the case of pulse-chase experiments, with cycloheximide. A mitochondrial fraction from HeLa cells labeled with [^{35}S]methionine under the conditions described above was lysed with SDS and allowed to react with the anti-peptide antibodies. The specific precipitate was then isolated by adsorption on formalin-fixed *Staphylococcus aureus* and analyzed by electrophoresis on polyacrylamide gels (Chomyn et al. 1983, in prep.; Mariottini et al. 1983, 1986b). This approach proved to be by far more sensitive and specific than immunoblot assays carried out on mitochondrial or total cell proteins transferred onto nitrocellulose membranes. In nearly every case, only one gene product was shown to react with each type of peptide-specific antibodies, the reaction being inhibited by the specific peptide, and always, this polypeptide had the size expected for the product of the corresponding reading frame. In a single case, where one type of antibody (i.e., anti-URFA6L-C antibodies) reacted with two different gene products (i.e., the URFA6L product and the URF1 product), this was shown to be due to a significant amino acid sequence homology between the carboxy-terminal nonapeptide of the URFA6L product and the carboxy-terminal nonapeptide of the URF1 product (Chomyn et al. 1983; Mariottini et al. 1983). Thus, three criteria, namely the site of synthesis of the polypeptide reacting with the antibodies, its specific reactivity with these antibodies, and the size of the reacting polypeptide, provided convincing evidence that the mitochondrial gene product had been identified. In some cases, minor components reacting with peptide-specific antibodies appeared to be fragments of the major gene product (Chomyn et al. 1983; Mariottini et al. 1983); in other cases, the same antibody reacted with what appeared to be two forms of the same gene product, which were electrophoretically separable under certain conditions for undetermined reasons (Mariottini et al. 1986b; Chomyn et al., in prep.).

Figure 3 shows one example of specific immunoprecipitation of mitochondrial translation products by peptide-specific antibodies. In this experiment, the mitochondrial lysate had been prepared from cells grown for 22 hr in the presence of chloramphenicol to permit the accumulation of cytoplasmically synthesized subunits of the inner membrane enzyme complexes, then labeled for 2 hr with [^{35}S]methionine in the presence of cycloheximide at 100 μg/ml, and finally chased for 2 hr in drug-free medium. One can see that each type of antibody produced a precipitate exhibiting a single specific band in an SDS-urea-polyacrylamide gel. Thus, the anti-URF2-N antibodies precipitated specifically component 11, and the anti-URF4L-C antibodies precipitated component 26; similarly, the pattern of the precipitate obtained with the anti-URF4-C antibodies exhibited a band corresponding in mobility to components 4 and 5. The electrophoretic analysis of the precipitate in these experiments was carried out using MCB SDS (DX-2490-3; MCB Manufacturing Chemists, Inc.) in the preparation and running of the gel. It was observed that the use of Sigma SDS (L-5750) or Spectrum SDS (SO180, Spectrum Chemical Mfg. Corp.) allowed the separation of components 4 and 5 over a heterogeneous background. This phenomenon may be related to the presence of a substantial proportion of C_{14} and C_{16} alkyl chains, which have higher affinity for certain sequences (Dohnal and Garvin 1979; Lacks et al. 1979; Best et al. 1981), in Sigma SDS (~32%) and in Spectrum SDS (~33%), but not in MCB SDS (<1%). Both components 4 and 5 were found to be precipitated by anti-URF4-C antibodies, suggesting that they represent two forms of the URF4 product. The nature of the difference that causes them to be separated in an SDS-urea-polyacrylamide gel prepared and run in Sigma or Spectrum SDS is not known. In Figure 3, one can see that each specific peptide competed completely with the corresponding mitochondrial translation product in the precipitation by the anti-peptide antibodies.

The identification of two URF products, namely the

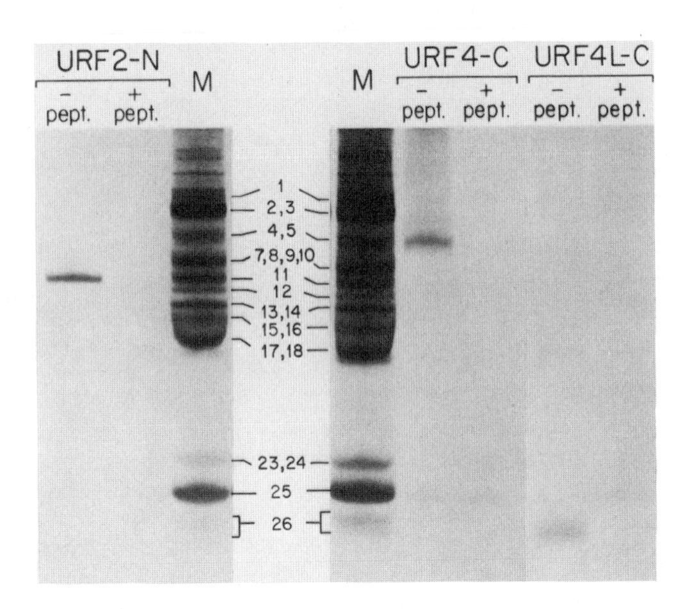

Figure 3. Specific precipitation of the products of URF2, URF4, and URF4L by peptide-specific antibodies. Samples of an SDS mitochondrial lysate (~130 μg protein) from HeLa cells grown for 22 hr in the presence of 40 μg/ml chloramphenicol, then labeled for 2 hr with [^{35}S]methionine in methionine-free medium in the presence of cycloheximide at 100 μg/ml, and finally chased for 2 hr in complete, unlabeled medium in the absence of inhibitors were incubated with anti-URF2-N, anti-URF4-C, or anti-URF4L-C γ-globulins (125 μg) in the absence (−pept.) or presence (+pept.) of 5 μg of URF2-N, URF4-C, or URF4L-C peptide, respectively. The immunoprecipitates were electrophoresed through SDS-8M urea–15% polyacrylamide gels (in MCB SDS). (Lanes *M*) Pattern of mitochondrial translation products. (Modified from Mariottini et al. 1986b.)

polypeptides encoded in URF5 and URF6, by the antipeptide antibody approach proved to be particularly difficult. Antibodies against the amino-terminal peptide of the putative URF5 product were marginally effective in precipitating one component, i.e., component 1 of the mitochondrially synthesized polypeptides (Fig. 2), whereas antibodies against the carboxy-terminal hexapeptide were totally ineffective in precipitating any mitochondrial translation product (Mariottini et al. 1986b). URF5 is indeed the only reading frame large enough to code for a protein of the size of component 1. However, it was desirable to corroborate the tentative assignment of this component as the URF5 product by another approach. This was achieved by fingerprinting analysis of a tryptic digest of component 1, in particular by the detection of the dipeptide methionyl-lysine in the digest. On the basis of the DNA sequence, this dipeptide was expected to occur not only in the URF5 product, but also in the URF2 and URFA6L products. However, the much smaller size of the latter two URF products and the lack of reactivity of anti-URF2-N and anti-URFA6L antibodies (Mariottini et al. 1983, 1986b) with component 1 clearly excluded URF2 and URFA6L as the genes encoding this component, thus supporting its identification as the URF5 product.

As to the URF6 product, of four different types of URF6-specific antipeptide antibodies tested, namely, antibodies against an amino-terminal hexapeptide or a carboxy-terminal nonapeptide or two different internal peptides of the putative URF6 product, only one of the latter two types of antibodies (I1, Table 1) proved to be effective in specifically precipitating a mitochondrial translation product, i.e., component 22 (Fig. 4). The size of this component, which is very close to that predicted from the DNA sequence, supported the conclusion that it was the URF6 product. One complication in the search for the polypeptide encoded in URF6 proved to be the anomalous electrophoretic behavior of component 22. In fact, this component was recognizable in an SDS-urea-polyacrylamide gel prepared and run in Sigma SDS (Fig. 4) or Spectrum SDS (not shown), but not one run in MCB SDS, the type routinely used in our earlier work. Here again, the phenomenon may be related to the presence in the SDS from different sources of varying amounts of contaminating alkyl chains longer than C_{12}. That, however, the presence of urea also plays a role in this phenomenon is suggested by the observation that, after electrophoresis of the mitochondrial lysate in SDS-polyacrylamide gels (Chomyn et al. 1983; Mariottini et al. 1986b), the band corresponding to component 22, which sometimes appears as a doublet (Fig. 1), is more prominent than in SDS-urea-polyacrylamide gels (Figs. 1 and 4). Furthermore, in the non-urea-containing gel systems, the intensity of component 22 is not affected by the type of SDS (not shown). It is known that urea reduces the affinity of SDS for proteins (Weber and Kuter 1971). It is conceivable that the presence of an appreciable proportion of chains longer than C_{12} in the SDS compensates in part for the effect of urea. Another factor that was recognized to affect the electrophoretic behavior of component 22 is exposure of the sample to high temperatures. In fact, heating a sample of the mitochondrial fraction, solubilized in SDS, to 100°C for 3 min, as routinely done in the earlier im-

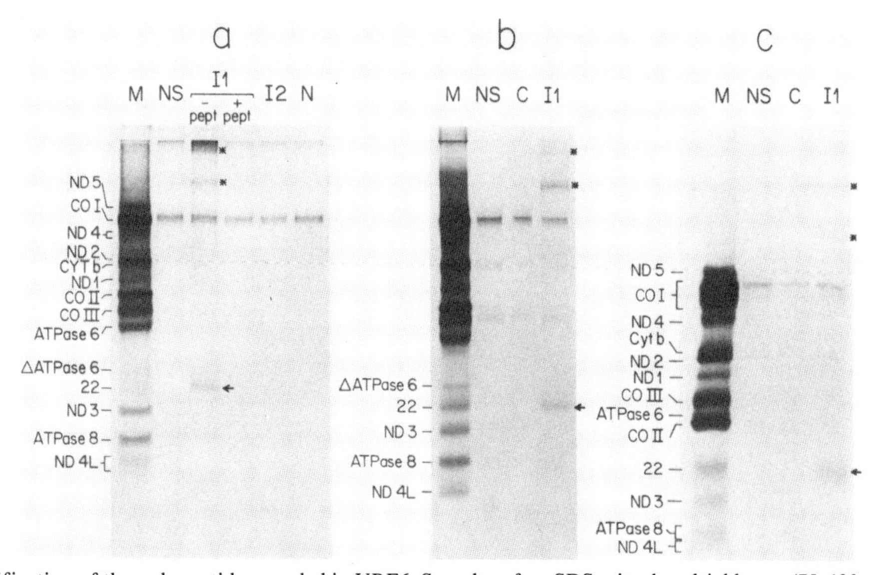

Figure 4. Identification of the polypeptide encoded in URF6. Samples of an SDS mitochondrial lysate (75–100 μg protein) from HeLa cells labeled for 2.5 hr with [^{35}S]methionine in the presence of emetine were incubated with 125 μg γ-globulins from antisera prepared against the internal undecapeptide I1 (positions 14,422–14,454 [Anderson et al. 1981]) (Table 1) or the internal decapeptide I2 (positions 14,593–14,622) or the amino-terminal hexapeptide (N) or the carboxy-terminal nonapeptide (C) of the URF6 product, or from normal serum (NS). The immunoprecipitation with anti-URF6-I1 was carried out in the absence (– pept.) or presence (+ pept.) of 10 μg of the corresponding peptide. The immunoprecipitates were run in an SDS-8M urea-15% polyacrylamide gel (*a* and *b*) or in an SDS-15–25% polyacrylamide gradient gel (*c*) (both in Sigma SDS). (Lane *M*) Pattern of HeLa cell mitochondrial translation products.

munoprecipitation experiments, caused the complete disappearance from the pattern of the component 22 band, while heating at 37°C had no effect.

The anti-peptide antibody approach was also applied successfully to the identification of the polypeptide corresponding to the H^+-ATPase subunit 6 gene (Chomyn et al. 1983) (Fig. 4). An antibody prepared against an amino-terminal decapeptide of the apocytochrome *b* gene product, though reactive with the corresponding polypeptide coupled to a carrier, proved, on the contrary, to be ineffective in precipitating any mitochondrial translation product. Identification of apocytochrome *b* among the mitochondrially synthesized polypeptides (Fig. 4) was, however, made possible by the isolation of the b-c_1 complex and analysis of the electrophoretic properties and site of synthesis of its subunits (C. Doersen and G. Attardi, in prep.).

A plot of electrophoretic mobility in an SDS–17.5% polyacrylamide gel (in Sigma SDS) versus molecular weight predicted from the DNA sequence for the HeLa cell mtDNA-coded polypeptides showed that all these polypeptides have mobilities consistent with their expected molecular weights. Only the mobility of ATPase 8 was slightly lower than predicted, presumably due to the lower degree of SDS binding by this protein (Mariottini et al. 1986b). In fact, ATPase 8 is unique among the human mitochondrial translation products for having the amino acid composition of a soluble protein (Kyte and Doolitte 1982).

Seven URF Products Are Components of the Respiratory Chain NADH Dehydrogenase

The smallest of the URFs (URFA6L), a 207-nt reading frame overlapping out of phase the amino-terminal portion of the ATPase 6 gene, specifies a protein exhibiting a significant amino acid sequence homology and other structural similarities to the product of a yeast mtDNA gene, *aap1*, which is subunit 8 of the H^+-ATPase (Macreadie et al. 1983). On the basis of this homology and of the results of immunoprecipitation experiments (Chomyn et al. 1985a), the URFA6L product has been identified as the animal equivalent of the yeast H^+-ATPase subunit 8.

Until last year, nothing was known about the nature of the remaining seven URF products, except that they are hydrophobic proteins (Kyte and Doolittle 1982) and therefore, presumably, integral proteins of the inner mitochondrial membrane. The possibility that at least some of these URF products are functionally related and, possibly, components of some kind of a complex was suggested by the observations indicating that the URFs, universally present in animal mtDNA, are either totally absent from the mtDNA of other organisms, as in yeast, or present as subsets, as in mtDNA of filamentous fungi and protozoa. To test the above possibility, conditions were investigated, using cytochrome *c* oxidase as a model system, that would allow precipitation of an entire enzyme complex from a mitochondrial lysate by antibodies directed against a peptide of one of the subunits. It was found that antibodies against a carboxy-terminal undecapeptide of COII, under appropriate conditions, could precipitate from a Triton X-100 mitochondrial lysate of HeLa cells all the 13 subunits which have been identified in mammalian cytochome *c* oxidase (Mariottini et al. 1986a). Using these conditions, we obtained the first hint that the URF products are indeed part of a complex. In fact, it was observed that antibodies against one of several URF products could precipitate from a Triton X-100 mitochondrial lysate several other URF products. This was most clearly seen with an antiserum against the carboxy-terminal heptapeptide of the URF4L product, which precipitated from a Triton X-100 mitochondrial lysate not only the URF4L product, but also the products of URF5, URF4, URF2, URF1, and URF3 (Chomyn et al. 1985b). By contrast, the same antiserum precipitated only the URF4L product from an SDS mitochondrial lysate (Fig. 3).

Of the two complexes of the oxidative-phosphorylation machinery that had not been found as yet to contain subunits encoded in mtDNA, namely, the rotenone-sensitive NADH dehydrogenase (complex I [Hatefi et al. 1985]) and the succinate dehydrogenase (complex II [Hatefi et al. 1985]), the first seemed to be a good candidate for having among its subunits at least some of the URF products. In fact, complex I is known to contain a shell of ≥ 15 hydrophobic proteins surrounding the low-molecular-weight NADH dehydrogenase or flavoprotein fragment and the iron–protein fragment (Ragan 1980; Hatefi et al. 1985). Therefore, two rabbit antisera against highly purified native bovine complex I (Heron et al. 1979), which cross-reacted with the human complex I solubilized from kidney submitochondrial particles (Chomyn et al. 1985b), were tested for their capacity to immunoprecipitate URF products. Figure 5 shows one such experiment, in which a 0.2% potassium deoxycholate (KDOC) lysate of the mitochondrial fraction from cells grown for 22 hr in the presence of 40 μg/ml chloramphenicol, then labeled for 2.5 hr with [^{35}S]methionine in the presence of 100 μg/ml cycloheximide, and finally chased for 18 hr in unlabeled medium, was incubated with one of the anti-complex I antisera. Portions of the precipitate were run in an SDS–8M urea–15% polyacrylamide gel (Fig. 5a) and in an SDS–15–25% polyacrylamide gradient gel (Fig. 5b) (both in Sigma SDS). The results were dramatic. In both types of gels, the precipitate exhibited specific bands corresponding to the URF5, URF4, URF2, URF1, URF6, URF3, and URF4L products. In the precipitate run in the SDS-urea-polyacrylamide gel, the URF6 product is considerably underrepresented, in molar terms, relative to the URF3 and URF4L products. By contrast, in the SDS-polyacrylamide gradient gel, the URF6 product appears to be present in molar amount comparable to the URF3 and URF4L products. The reason for the reduced amount of the URF6 product in the urea gel is not known, but it may be related to the effects of urea discussed above.

The results described above clearly indicate that the

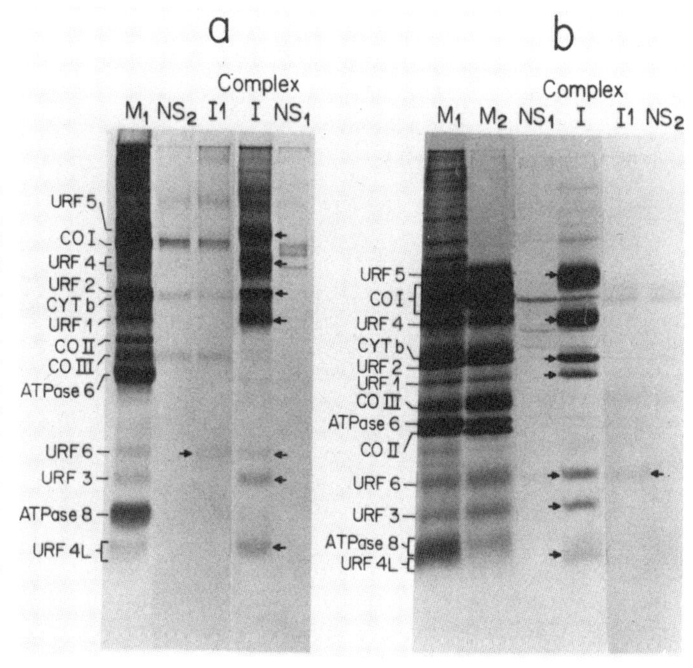

Figure 5. Precipitation of seven URF products by antibodies against native bovine respiratory chain NADH dehydrogenase. Samples of a 0.2% KDOC mitochondrial lysate from HeLa cells grown for 22 hr in the presence of 40 μg/ml chloramphenicol, then labeled for 2.5 hr with [^{35}S]methionine in the presence of 100 μg/ml cycloheximide, and finally chased for 18 hr in unlabeled medium in the absence of inhibitors were incubated with an antiserum against native bovine complex I or normal serum (NS_1). The immunoprecipitates were run in an SDS-urea-polyacrylamide gel (*a*) or in an SDS-polyacrylamide gradient gel (*b*) (both in Sigma SDS), in parallel with immunoprecipitates obtained by incubating samples of an SDS mitochondrial lysate with anti-URF6-I1 γ-globulins or normal serum γ-globulins (NS_2), as in the experiments of Fig. 4. (M_1 and M_2) Patterns of HeLa cell mitochondrial translation products from cells labeled as described above, and, respectively, from cells labeled as in the experiments of Fig. 4.

URF products are closely associated with the respiratory chain NADH dehydrogenase, and are most probably part of this complex. From the densitometric tracing of appropriate exposures of the autoradiograms and the methionine content of the URF products, one could estimate that the products of URF2, URF1, and smaller URF products are underrepresented in molar amounts in the immunoprecipitate relative to the mitochondrial fraction, after electrophoresis either in an SDS-polyacrylamide gradient gel (Table 2) or in an SDS-urea-polyacrylamide gel (not shown). It seemed likely that, after an 18-hr chase, the mtDNA-coded NADH dehydrogenase subunits labeled during a 2.5-hr [^{35}S]methionine pulse are fully incorporated into the mature complex and that, therefore, the relative molar amounts of labeled subunits in the mitochondrial fraction represent their proportions in the complex. The underrepresentation of the smaller subunits in the immunoprecipitate may be due to a loss during the precipitation procedure, possibly as a result of a destabilizing effect of the antibodies. The probably peripheral position of these subunits in the complex (see below) presumably accounts for this behavior.

The conclusion from the immunochemical evidence discussed above that the human respiratory chain NADH dehydrogenase contains polypeptides encoded in seven mtDNA URFs was strongly supported by the results of enzyme fractionation studies. In these studies, a mixture of [^{35}S]methionine-labeled HeLa cell mitochondria and human heart mitochondria was subjected to the fractionation scheme that has been used to purify NADH-cytochrome *c* reductase (complex

Table 2. Relative Molar Representation of Labeled NADH Dehydrogenase Subunits after a Pulse-chase

	Relative molar amounts[a]						
Material	ND5	ND4	ND2	ND1	ND6	ND3	ND4L
Mitochondrial fraction	1.2	1.6	5	1	1.2	1	0.7
Complex precipitated by anti-complex I antibodies	5	6	5	1	0.9	1	0.8
Complex precipitated by anti-49-kD subunit antibodies	1.2	1.4	5	1.1	1	1	0.6

HeLa cells were grown for 22 hr in the presence of 40 μg/ml chloramphenicol, then labeled for 2.5 hr with [^{35}S]methionine in the presence of 100 μg/ml cycloheximide, and finally chased in unlabeled medium in the absence of inhibitors. A 0.2% potassium deoxycholate mitochondrial lysate and a 0.5% Triton X-100 mitochondrial lysate were incubated with an antiserum against bovine native complex I and, respectively, with γ-globulins from an anti-49-kD iron-sulfur subunit antiserum. Samples of the immunoprecipitates and of an SDS or SDS-urea mitochondrial lysate were run in an SDS-polyacrylamide gradient gel and in an SDS-urea-polyacrylamide gel.

[a]Estimated from densitometric tracings of appropriate exposures of the SDS-polyacrylamide gradient gel (except for the molar amount of ND4L in the mitochondrial fraction, which was estimated relative to ND3 in a urea gel, because of the lack of resolution in an SDS-polyacrylamide gradient gel) and from the methionine content of the various subunits.

I + III of the respiratory chain) from beef heart mitochondria (Hatefi and Stigall 1978). Two fractions were isolated in which a great enrichment of URF products was specifically correlated with an enrichment in NADH-Q_1 and NADH-$K_3Fe(CN)_6$ oxidoreductase activities (Chomyn et al. 1985b).

Nature of the NADH Dehydrogenase Components Encoded in mtDNA

Considering the hydrophobic nature of the mtDNA-coded subunits of complex I, it seemed very likely that these polypeptides belong to the shell of hydrophobic proteins surrounding the catalytic subunits. This possibility has been confirmed for some of the URF products by immunoblot assays using SDS-denatured beef heart complex I and antibodies raised against human URF peptides that are well conserved in the bovine mitochondrial genome, i.e., the carboxy-terminal peptides of the URF1 (Chomyn et al. 1983), URF3 (Chomyn et al. 1983), and URF4L (Mariottini et al. 1986b) products. In fact, the anti-URF1-C antibodies reacted with a polypeptide of the hydrophobic fraction of complex I with an estimated M_r of ~33,000. This polypeptide has been clearly distinguished from the 30-kD component of the iron-protein fragment in immunoblots tested with a mixture of anti-URF1-C antiserum and anti-30-kD antiserum. If the 33-kD band corresponds to a single polypeptide, then the URF1 product can be identified as the rotenone-binding protein, and therefore must be closely involved in ubiquinone reduction (Ragan et al. 1985). This possibility is presently under investigation. The anti-URF3-C antibodies react with a 13.6-kD polypeptide of the hydrophobic fraction, which migrates a little slower than the 13-kD protein of the iron-protein fragment. The anti-URF4L-C antibodies reacted with a 10-kD protein of the hydrophobic fraction.

To test the possibility that some of the URFs may code for one or more of the iron-sulfur subunits of the respiratory chain NADH dehydrogenase, the site of synthesis of the 51-kD and 24-kD iron-sulfur proteins of the flavoprotein fragment (Ragan et al. 1982b) and of the 75-kD, 49-kD, 30-kD, and 13-kD iron-sulfur proteins of the iron-protein fragment (Ragan et al. 1982a) was investigated. For this purpose, advantage was taken of the availability of polyclonal antibodies raised against the individual electrophoretically purified iron-sulfur proteins from the beef heart complex I (Cleeter et al. 1985; Ragan et al. 1985), which cross-reacted with the homologous human subunits (Cleeter and Ragan 1985). These antibodies were used in immunoprecipitation experiments with SDS mitochondrial lysates from HeLa cells labeled for 2 hr with [^{35}S]methionine in the presence or absence of an inhibitor of mitochondrial protein synthesis (chloramphenicol) or cytoplasmic protein synthesis (cycloheximide or emetine). These experiments clearly showed that these polypeptides are synthesized in the cytoplasm, as illustrated for the 49-kD subunit in Figure 6a. The protein precipitated by the anti-49-kD antibodies exhibited four bands in an SDS-urea-polyacrylamide gel. However, on reduction and alkylation of the cysteine residues, the immunoprecipitated protein migrated as a single band (Fig. 6a), indicating that the multiplicity of bands was due to intramolecular disulfide bond formation.

Antibodies against the 49-kD Iron-Sulfur Protein Precipitate All Seven URF Products

The availability of polyclonal antibodies against individual highly purified and functionally characterized subunits of the beef heart NADH dehydrogenase, like the iron-sulfur subunits, offered the opportunity to test the identification of the mtDNA URF products as NADH dehydrogenase subunits. In fact, if one or more of these iron-sulfur subunits were exposed in the complex, it seemed likely that antibodies against any of such proteins would precipitate from a mitochondrial lysate the entire complex I, including the mitochondrially synthesized polypeptides of the hydrophobic shell. There is good evidence, from labeling studies with hydrophilic probes, that the 75-kD, 49-kD, and 30-kD iron-sulfur subunits of the iron-protein fragment of NADH dehydrogenase are in part exposed, both in the intact enzyme and in the membrane (Ohnishi et al. 1985; Ragan et al. 1985). Figure 6b shows the results of an immunoprecipitation experiment carried out with antibodies against the 49-kD iron-sulfur protein. A Triton X-100 mitochondrial lysate from HeLa cells that had been grown in the presence of chloramphenicol, then labeled for 2.5 hr with [^{35}S]methionine in the presence of cycloheximide and chased, as described above, was incubated with γ-globulins from an anti-49-kD subunit antiserum. This antiserum had been shown to react extensively with the human homologous subunit from SDS-dissociated complex I (Cleeter and Ragan 1985). Although the reactivity of these antibodies with non-SDS-dissociated complex I was weak, one can see that all seven URF products are precipitated with absolute specificity by the anti-49-kD iron-sulfur protein γ-globulins. It is interesting that the molar proportions of the URF products in this immunoprecipitate are almost identical to the proportions of the same polypeptides in the mitochondrial fraction used for immunoprecipitation (M_1 lane) (Table 2). This strongly suggests that complex I is precipitated by the anti-49-kD subunit antibodies in a more intact form than by the anti-complex I antibodies. A similar precipitation of URF products, although less efficient, was obtained with the γ-globulins from an antiserum against the 30-kD iron-sulfur protein (not shown). These results, besides confirming the accessibility to nonpermeant probes of the 49-kD and 30-kD subunits in the intact complex, provide strong additional support for the functional assignments of the URF products as NADH dehydrogenase subunits.

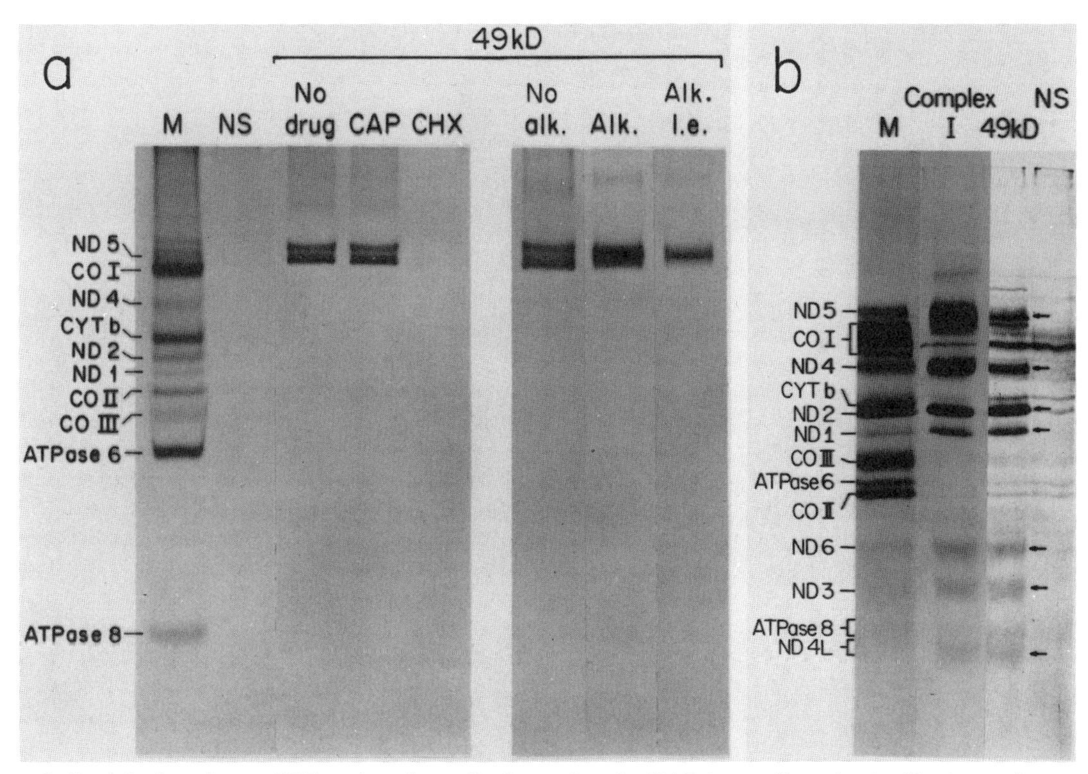

Figure 6. Precipitation of seven URF products by antibodies against the 49-kD iron-sulfur subunit of bovine respiratory chain NADH dehydrogenase. (*a*) Samples of an SDS lysate from HeLa cells labeled for 2 hr with [^{35}S]methionine in the absence of drug (ND) or in the presence of 100 μg/ml chloramphenicol (CAP) or of 100 μg/ml cycloheximide (CHX) were precipitated with γ-globulins from an antiserum against the purified 49-kD iron-sulfur protein of the bovine complex (49 kD) or from normal serum (NS). (*b*) A sample of a 0.5% Triton X-100 mitochondrial lysate from HeLa cells labeled as in Fig. 5 was precipitated with the 49-kD-specific γ-globulins or with normal serum γ-globulins (NS). A sample of a 0.2% KDOC mitochondrial lysate from the same cells was precipitated with an anti-complex I antiserum. The immunoprecipitates were run in an SDS-polyacrylamide gradient gel. (Lane *M*) Pattern of HeLa cell mitochondrial translation products labeled as in the experiments of Fig. 4; (lane *alk.*) proteins were alkylated with iodoacetamide before running on the gel; (*alk. l.e.*) lower exposure of the alkylated sample.

DISCUSSION

The results of the immunoprecipitation and enzyme fractionation studies discussed above have provided strong evidence indicating that the products of URF1, URF2, URF3, URF4, URF4L, URF5, and URF6 of human mtDNA are subunits of the respiratory chain NADH dehydrogenase. The experiments showing that antibodies against a highly purified and functionally characterized subunit of complex I, i.e., the cytoplasmically synthesized 49-kD iron-sulfur protein, precipitate with absolute specificity the seven URF products from a Triton X-100 mitochondrial lysate, substantially exclude the possibility that some protein contaminating the preparation of complex I used for immunization is responsible for the results obtained with anti-complex I antibodies. Furthermore, the molar proportions of the labeled URF products, after a 2.5-hr [^{35}S]methionine pulse and a long chase, which were precipitated by anti-49-kD iron-sulfur subunit antibodies are almost identical to the proportions of the same polypeptides in the mitochondrial fraction. This observation is fully consistent with the conclusion that the URF products are part of complex I. Further support for this conclusion has come from the recent finding that the purified rotenone-sensitive NADH dehydrogenase from *Neurospora crassa* contains several subunits synthesized within the mitochondria (Ise et al. 1985), and from the observation that the stopper mutant of *N. crassa*, whose mtDNA lacks two genes homologous to URF2 and URF3, has no functional complex I (de Vries et al. 1986). On the basis of the functional assignment discussed above, we have proposed that the designation of the mitochondrial genes encoding subunits of the NADH dehydrogenase be changed to ND (for NADH dehydrogenase) 1, 2, 3, 4, 4L, 5, and 6 (Chomyn et al. 1985b, in prep.). The functional assignment of the URF products discussed in this paper completes the work on the elucidation of the informational content of the human mitochondrial genome which was started more than 15 years ago in the Caltech laboratory (Attardi et al. 1971).

The striking conclusion of the present studies is that almost 60% of the protein-coding capacity of the human mitochondrial genome and, by extrapolation, of all animal mitochondrial genomes, is used for the synthesis of subunits of the respiratory chain NADH dehydrogenase. The physiological significance of the important role of animal mtDNA in the assembly of the first enzyme complex of the respiratory chain is not known, and has to be viewed in the context of the great variability among different organisms in the genetic control of this enzyme complex. While the mtDNA of many lower eukaryotic cells has been shown to contain one to several genes having a convincing amino acid sequence homology to the mammalian mtDNA URFs (Benne 1985; Breitenberger and RajBhandary 1985; Brown et al. 1985; Burger and Werner 1985; Vahrenholz et al. 1985; de Vries et al. 1986; Simpson 1986), such genes have, most remarkably, not been found in the mitochondrial genomes of *S. cerevisiae* (Grivell 1983) and *S. pombe* (Lang et al. 1983). The NADH dehydrogenase region of the respiratory chain exhibits functional and structural differences in *Saccharomyces* strains as compared with the mammalian counterpart, and it is conceivable that these differences are due to, or can be correlated with, the absence of some or all polypeptides homologous to the mammalian mtDNA URF products from the yeast complex. On the other hand, it is quite possible that in yeast some or all of these polypeptides are encoded in the nucleus and imported from the cytoplasm.

The differences in genetic control of the respiratory chain NADH dehydrogenase in various organisms represent another striking example of the variability in informational content that has been observed in the mitochondrial genomes from different sources. Figure 7 summarizes the present knowledge concerning the distribution of protein-coding genes in the mtDNAs of different organisms. Among these mtDNAs, only the human genome has been completely elucidated in its genetic content. Of the others, more than 85% of the *S. cerevisiae* mtDNA (de Zamoroczy and Bernardi 1985), more than 80% of the *N. crassa* mtDNA (Breitenberger and RahBhandary 1985), and almost all the transcribed portion of *Leishmania tarentolae* mtDNA (Simpson 1986) have been sequenced; all of these have a variable number of additional reading frames as yet not identified in their function. Even in this incomplete form, the diagram of Figure 7 clearly illustrates the heterogeneity in genetic content of the various mtDNAs. One can anticipate that the collection of more-extensive data on the gene content of mtDNA in different organisms will allow the construction of a genealogy based on function of the mitochondrial genomes, which will complement the phylogenetic trees based on sequence data of mtDNA and mitochondrial proteins. A detailed analysis of the functional genealogy of mtDNA should provide valuable information for reconstructing the multiple evolutionary pathways that have led to the present-day variety of mitochondrial genomes.

The URFs of animal mtDNA exhibit a higher divergence rate than the "universal" mitochondrial genes (Attardi 1985). Similarly, a comparison between the mitochondrial genes of animal cells and the homologous genes of other eukaryotic cells reveals that the URFs are in general less conserved, in both size and sequence, than the "universal" genes (Brown et al. 1983; de la Cruz et al. 1984; Burger and Werner 1985; de Vries et al. 1986). This is, on the surface, a surprising observation, since both the products of the URFs and those of the other genes are components of indispensable enzymes. A reasonable explanation for this difference may be that the URF products are components of the hydrophobic shell of the NADH dehydrogenase, which has presumably the function of creating an appropriate environment for the catalytic moieties of the enzyme complex within the inner mitochondrial membrane. This function may not depend strictly on the primary sequence. It is indeed interesting that the amino acid sequences of the URF-related mtDNA genes

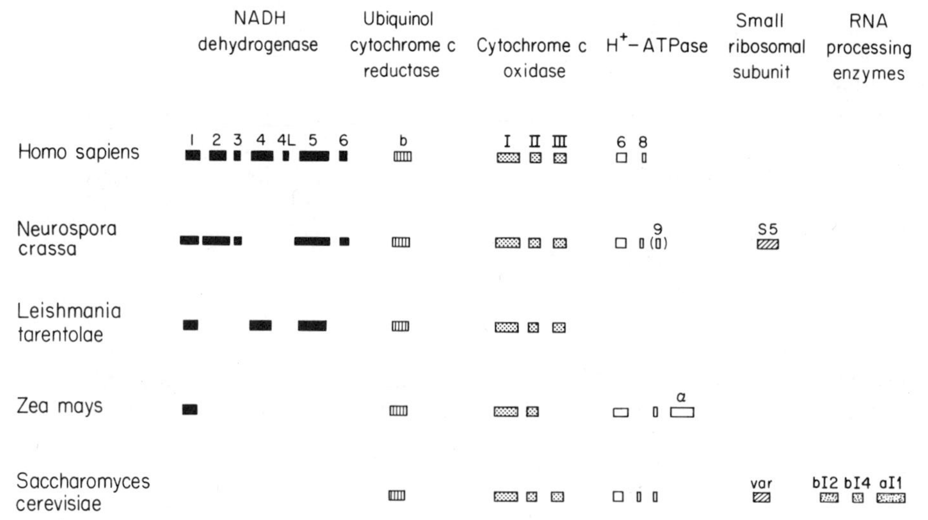

Figure 7. Distribution of currently known protein-coding genes in the mitochondrial genomes from different organisms. (bI2, bI4, aI1) "mRNA maturases" encoded in introns 2 and 4 of the apocytochrome *b* gene and in intron 1 of the COI gene of *S. cerevisiae* (Kotylak et al. 1985). The rectangle in parentheses designated 9 in *N. crassa* mtDNA represents an apparently nonfunctional H^+-ATPase subunit 9 gene (Breitenberger and RajBhandary 1985). The data on which this figure is based were derived from Anderson et al. (1981); Grivell (1983); Braun and Levings III (1985); Breitenberger and RajBhandary (1985); Chomyn et al. (1985b, in prep.); Dewey et al. (1985a,b); Kotylak et al. (1985); Leaver et al. (1985); Simpson (1986); Dawson et al. (1986); and from the work described here.

in filamentous fungi and protozoa often have only a low degree of homology to the sequences of the equivalent mammalian genes, but, on the contrary, an almost perfect correspondence in hydropathy profile over large segments (de la Cruz et al. 1984; Burger and Werner 1985; de Vries et al. 1986).

ACKNOWLEDGMENTS

These investigations were supported by grant GM-11726 to G.A., National Science Foundation grant PCM-8118172 to R.F.D., and a grant from SERC to C.I.R.

REFERENCES

Anderson, S., A.T. Bankier, B.G. Barrell, M.H.L. de Bruijn, A.R. Coulson, J. Drouin, I.C. Eperon, D.P. Nierlich, B.A. Roe, F. Sanger, P.H. Schreier, A.J.H. Smith, R. Staden, and I.G. Young. 1981. Sequence and organization of the human mitochondrial genome. *Nature* **290:** 457.

Attardi, G. 1985. Animal mitochondrial DNA: An extreme example of genetic economy. *Int. Rev. Cytol.* **93:** 93.

Attardi, G., Y. Aloni, B. Attardi, D. Ojala, L. Pica-Mattoccia, D.L. Robberson, and B. Storrie. 1971. Transcription of mitochondrial DNA in HeLa cells. *Cold Spring Harbor Symp. Quant. Biol.* **35:** 599.

Attardi, G., M. Albring, F. Amalric, R. Gelfand, J. Griffith, D. Lynch, C. Merkel, W. Murphy, and D. Ojala. 1976. Organization and expression of the mitochondrial genome in HeLa cells. In *Genetics and biogenesis of chloroplasts and mitochondria* (ed. T. Bücher et al.), p. 573. Elsevier/North-Holland, Amsterdam.

Benne, R. 1985. Mitochondrial genes in trypanosomes. *Trends Genet.* **1:** 117.

Best, D., P.J. Warr, and K. Gull. 1981. Influence of the composition of commercial sodium dodecyl sulfate preparations on the separation of α- and β-tubulin during polyacrylamide gel electrophoresis. *Anal. Biochem.* **114:** 281.

Braun, C.J. and C.S. Levings III. 1985. Nucleotide sequence of the F_1-ATPase α subunit gene from maize mitochondria. *Plant Physiol.* **79:** 571.

Breitenberger, C.A. and U.L. RajBhandary. 1985. Some highlights of mitochondrial research based on analyses of *Neurospora crassa* mitochondrial DNA. *Trends Biochem. Sci.* **10:** 478.

Brown, T.A., R.B. Waring, C. Scazzocchio, and R.W. Davies. 1985. The *Aspergillus nidulans* mitochondrial genome. *Curr. Genet.* **9:** 113.

Brown, T.A., R.W. Davies, J.A. Ray, R.B. Waring, and C. Scazzocchio. 1983. The mitochondrial genome of *Aspergillus nidulans* contains reading frames homologous to the human URFs 1 and 4. *EMBO J.* **2:** 427.

Burger, G. and S. Werner. 1985. The mitochondrial URF1 gene in *Neurospora crassa* has an intron that contains a novel type of URF. *J. Mol. Biol.* **186:** 231.

Ching, E. and G. Attardi. 1982. High resolution electrophoretic fractionation and partial characterization of the mitochondrial translation products from HeLa cells. *Biochemistry* **21:** 3188 (see correction, 1985, **24:** 7853).

Chomyn, A., P. Mariottini, M.W.J. Cleeter, C.I. Ragan, R.F. Doolittle, A. Matsuno-Yagi, Y. Hatefi, and G. Attardi. 1985a. Functional assignment of the products of the unidentified reading frames of human mitochondrial DNA. In *Achievements and perspectives of mitochondrial research,* vol. 2. *Biogenesis* (ed. E. Quagliariello et al.), p. 259. Elsevier Scientific, Amsterdam.

Chomyn, A., P. Mariottini, M.W.J. Cleeter, C.I. Ragan, A. Matsuno-Yagi, Y. Hatefi, R.F. Doolittle, and G. Attardi. 1985b. Six unidentified reading frames of human mitochondrial DNA encode components of the respiratory chain NADH dehydrogenase. *Nature* **314:** 592.

Chomyn, A., P. Mariottini, N. Gonzalez-Cadavid, G. Attardi, D.D. Strong, D. Trovato, M. Riley, and R.F. Doolittle. 1983. Identification of the polypeptides encoded in the ATPase 6 gene and in the unassigned reading frames 1 and 3 of human mtDNA. *Proc. Natl. Acad. Sci.* **80:** 5535.

Cleeter, M.W.J. and C.I. Ragan. 1985. The polypeptide composition of the mitochondrial NADH-ubiquinone reductase complex from several mammalian species. *Biochem. J.* **230:** 739.

Cleeter, M.W.J., S.H. Bannister, and C.I. Ragan. 1985. Chemical cross-linking of mitochondrial NADH dehydrogenase from bovine heart. *Biochem. J.* **227:** 467.

Costantino, P. and G. Attardi. 1975. Identification of discrete electrophoretic components among the products of mitochondrial protein synthesis in HeLa cells. *J. Mol. Biol.* **96:** 291.

Crews, S. and G. Attardi. 1980. The sequence of the small ribosomal RNA gene and the phenylalanine tRNA gene are joined end to end in human mitochondrial DNA. *Cell* **19:** 775.

Dawson, A.J., T.P. Hodge, P.G. Isaac, C.J. Leaver, and D.M. Lonsdale. 1986. Location of the genes for cytochrome oxidase subunits I and II, apocytochrome b, α-subunit of the F_1 ATPase and the ribosomal RNA genes on the mitochondrial genome of maize (*Zea mays* L.). *Curr. Genet.* **10:** 561.

de la Cruz, V.F., N. Neckelmann, and L. Simpson. 1984. Sequences of six genes and several open reading frames in the kinetoplast maxicircle DNA of *Leishmania tarentolae*. *J. Biol. Chem.* **259:** 15136.

de Vries, H., B. Alzner-DeWeerd, C.A. Breitenberger, D.D. Chang, J.C. de Jonge, and U.L. RajBhandary. 1986. The E35 stopper mutant of *Neurospora crassa*: Precise localization of endpoints in mitochondrial DNA and evidence that the deleted DNA codes for a subunit of NADH dehydrogenase. *EMBO J.* **5:** 779.

Dewey, R.E., C.S. Levings III, and D.H. Timothy. 1985a. Nucleotide sequence of ATPase subunit 6 gene of maize mitochondria. *Plant Physiol.* **79:** 914.

Dewey, R.E., A.M. Schuster, C.S. Levings III, and D.H. Timothy. 1985b. Nucleotide sequence of F_0-ATPase proteolipid (subunit 9) gene of maize mitochondria. *Proc. Natl. Acad. Sci.* **82:** 1015.

de Zamoroczy, M. and G. Bernardi. 1985. Sequence organization of the mitochondrial genome of yeast—A review. *Gene* **37:** 1.

Dohnal, J.C. and J.E. Garvin. 1979. The interaction of dodecyl and tetradecyl sulfate with proteins during polyacrylamide gel electrophoresis. *Biochim. Biophys. Acta* **576:** 393.

Gelfand, R. and G. Attardi. 1981. Synthesis and turnover of mitochondrial ribonucleic acid in HeLa cells: The mature ribosomal and messenger ribonucleic acid species are metabolically unstable. *Mol. Cell. Biol.* **1:** 497.

Grivell, L.A. 1983. Mitochondrial gene expression 1983. In *Mitochondria 1983. Nucleo-mitochondrial interactions* (ed. R.J. Schweyen et al.), p. 25. de Gruyter, Berlin.

Hare, J.F., E. Ching, and G. Attardi. 1980. Isolation, subunit composition, and site of synthesis of human cytochrome c oxidase. *Biochemistry* **19:** 2023.

Hatefi, Y. and D.L. Stigall. 1978. Preparation and properties of NADH: Cytochrome c oxidoreductase (complex I-III). *Methods Enzymol.* **53:** 5.

Hatefi, Y., C.I. Ragan, and Y.M. Galante. 1985. The enzymes and the enzyme complexes of the mitochondrial oxidative phosphorylation system. In *The enzymes of biological membranes* (ed. A. Martonosi), vol. 4, p. 1. Plenum Press, New York.

Heron, C., S. Smith, and C.I. Ragan. 1979. An analysis of the polypeptide composition of bovine heart mitochondrial NADH-ubiquinone oxidoreductase by two-dimensional

polyacrylamide-gel electrophoresis. *Biochem. J.* **181:** 435.
Ise, W., H. Haiker, and H. Weiss. 1985. Mitochondrial translation of subunits of the rotenone-sensitive NADH: Ubiquinone reductase in *Neurospora crassa*. *EMBO J.* **4:** 2075.
Kotylak, Z., J. Lazowska, D.C. Hawthorne, and P.P. Slonimski. 1985. Intron encoded proteins in mitochondria: Key elements of gene expression and genomic evolution. In *Achievements and perspectives of mitochondrial research,* vol. 2. *Biogenesis* (ed. E. Quagliariello et al.), p. 1. Elsevier Scientific, Amsterdam.
Kyte, J. and R.F. Doolittle. 1982. A simple method for displaying the hydropathic character of a protein. *J. Mol. Biol.* **157:** 105.
Lacks, S.A., S.S. Springhorn, and A.L. Rosenthal. 1979. Effect of the composition of sodium dodecyl sulfate preparations on the renaturation of enzymes after polyacrylamide gel electrophoresis. *Anal. Biochem.* **100:** 357.
Lang, B.F., F. Ahne, S. Distler, H. Trinkl, F. Kaudewitz, and K. Wolf. 1983. Sequence of the mitochondrial DNA, arrangement of genes and processing of their transcripts in *Schizosaccharomyces pombe*. In *Mitochondria 1983. Nucleo-mitochondrial interactions* (ed. R.J. Schweyen et al.), p. 313. de Gruyter, Berlin.
Leaver, C.J., P.G. Isaac, J. Bailey-Serres, I.D. Small, D.K. Hanson, and T.D. Fox. 1985. Recombination events associated with the cytochrome *c* oxidase subunit I gene in fertile and cytoplasmic male sterile maize and sorghum. In *Achievements and perspectives of mitochondrial research,* vol. 2. *Biogenesis* (ed. E. Quagliariello et al.), p. 111. Elsevier Scientific, Amsterdam.
Macreadie, I.G., C.E. Novitski, R.J. Maxwell, U. John, B.-G. Ooi, G.L. McMullen, H.B. Lukins, A.W. Linnane, and P. Nagley. 1983. Biogenesis of mitochondria: The mitochondrial gene (*aapl*) coding for mitochondrial ATPase subunit 8 in *Saccharomyces cerevisiae*. *Nucleic Acids Res.* **11:** 4435.
Mariottini, P., A. Chomyn, R.F. Doolittle, and G. Attardi. 1986a. Antibodies against the COOH-terminal undecapeptide of subunit II, but not those against the NH_2-terminal decapeptide, immunoprecipitate the whole human cytochrome *c* oxidase complex. *J. Biol. Chem.* **261:** 3355.
Mariottini, P., A. Chomyn, G. Attardi, D. Trovato, D.D. Strong, and R.F. Doolittle. 1983. Antibodies against synthetic peptides reveal that the unidentified reading frame A6L, overlapping the ATPase 6 gene, is expressed in human mitochondria. *Cell* **32:** 1269.
Mariottini, P., A. Chomyn, M. Riley, B. Cottrell, R.F. Doolittle, and G. Attardi. 1986b. Identification of the polypeptides encoded in the unassigned reading frames 2, 4, 4L and 5 of human mitochondrial DNA. *Proc. Natl. Acad. Sci.* **83:** 1563.
Ojala, D., C. Merkel, R. Gelfand, and G. Attardi. 1980. The tRNA genes punctuate the reading of genetic information in human mitochondrial DNA. *Cell* **22:** 393.
Ohnishi, T., C.I. Ragan, and Y. Hatefi. 1985. EPR studies of iron-sulfur clusters in isolated subunits and subfractions of NADH-ubiquinone oxidoreductase. *J. Biol. Chem.* **260:** 2782.
Ragan, C.I. 1980. The molecular organization of NADH dehydrogenase. *Subcell. Biochem.* **7:** 267.
Ragan, C.I., Y.M. Galante, and Y. Hatefi. 1982a. Purification of three iron-sulfur proteins from the iron-protein fragment of mitochondrial NADH-ubiquinone oxidoreductase. *Biochemistry* **21:** 2518.
Ragan, C.I., M.W.J. Cleeter, F.G.P. Earley, and S. Patel. 1985. Structural aspects of NADH dehydrogenase. In *Achievements and perspectives of mitochondrial research,* vol. 1. *Bioenergetics* (ed. E. Quagliariello et al.), p. 61. Elsevier Scientific, Amsterdam.
Ragan, C.I., Y.M. Galante, Y. Hatefi, and T. Ohnishi. 1982b. Resolution of mitochondrial NADH dehydrogenase and isolation of two iron-sulfur proteins. *Biochemistry* **21:** 590.
Simpson, L. 1986. Kinetoplast DNA in trypanosomid flagellates. *Int. Rev. Cytol.* **99:** 119.
Vahrenholz, C., E. Pratje, G. Michaelis, and B. Dujon. 1985. Mitochondrial DNA of *Chlamydomonas reinhardtii*: Sequence and arrangement of URF5 and the gene for cytochrome oxidase subunit I. *Mol. Gen. Genet.* **201:** 213.
Walter, G., K.-H. Scheidtmann, A. Carbone, A.P. Laudano, and R.F. Doolittle. 1980. Antibodies specific for the carboxy- and amino-terminal regions of simian virus 40 large tumor antigen. *Proc. Natl. Acad. Sci.* **77:** 5197.
Weber, K. and D.J. Kuter. 1971. Reversible denaturation of enzymes by sodium dodecyl sulfate. *J. Biol. Chem.* **246:** 4504.

Approaches to Physical Mapping of the Human Genome

C.L. SMITH AND C.R. CANTOR
Department of Genetics and Development, College of Physicians and Surgeons, Columbia University, New York, New York 10032

Macroscopic methods of examining the map of the human genome include cytogenetics, somatic cell genetics, and linkage analysis. Each of these methods is powerful, and all are quite complementary. In practice, however, each becomes progressively more tedious when structural data are required at finer resolution than 10,000 kb. It seems unlikely that any of these methods will be extended, in the near future, to provide routine analysis of the human genome at 1000-kb resolution.

Molecular methods of examining the map of the human genome involve cloning, restriction mapping, and sequencing DNA fragments. In practice the analysis of a 10- to 50-kb region by restriction mapping is nearly trivial. The extension of such maps to several hundred kilobases is possible by chromosome-walking techniques. However, these become quite tedious as the size of the region increases and as segments of DNA become dominated by highly repeated sequences. Although walks of 1000 kb are practical, it would be hard to approach such a task with great enthusiasm unless the particular DNA region were of compelling interest.

To bridge the molecular and macroscopic techniques, what is needed is a way to construct a physical map with 100- to 1000-kb resolution. In this article, we describe four techniques that, together, allow the construction of such a coarse restriction map. Each technique was developed or tested on simple organisms like yeast or bacteria, but each has now been shown to be feasible for comparable studies on human samples. We also outline a strategy that appears to be quite an efficient way to apply these techniques to construct complete physical maps of each human chromosome. Finally, we briefly discuss the utility of such maps.

Physical Mapping Methods

Isolation of unbroken genomic DNA. High-molecular-weight DNA samples are prepared by suspending live cells in liquid, low-gelling agarose (Schwartz and Cantor 1984). After solidification, extensive detergent, protease, and salt treatments are used to remove all cellular constituents except the DNA. This is possible because the pores of the agarose are large enough to allow rapid diffusion of proteins and other small macromolecules while genomic DNA is retained, quantitatively. We have found it convenient to use roughly cubic agarose samples, about 100 μl in volume, which we call inserts. However, it will probably be equally effective to use cells suspended in agarose microbeads (Cook 1984).

For bacterial samples, a typical insert will be made from 10 to 100 million cells resulting in 0.1–1 μg of DNA (Smith and Cantor 1986a). For mammalian samples, we typically use 1 million cells per insert, resulting in 10 μg of DNA (Smith et al. 1986b). Detailed protocols for preparing such samples are described elsewhere (Smith et al. 1986a). DNA samples made in inserts show a negligible level of double-strand breaks. They are stable indefinitely at 4°C and can be kept at room temperature for more than a month with no detectable damage. Thus, in practice, agarose inserts are transported routinely by ordinary mail.

Separation of large DNA molecules. The technique of pulsed-field gel (PFG) electrophoresis allows high-resolution size fractionation of DNA (Schwartz et al. 1983). In this technique, molecules are periodically forced to change their direction by a change in the direction of the applied electrical field. A number of variations in experimental geometry have been described in detail elsewhere (Carle and Olson 1984; Schwartz and Cantor 1984; Carle et al. 1986). The general type of apparatus that we find optimal for physical mapping studies (Smith and Cantor 1986b) contains a 1% agarose running gel 20 cm square placed at 45° between two inhomogeneous electrical fields. Two convenient sizes for the apparatus are 33 cm square, which is the smallest size that will accommodate a 20-cm gel at 45°, and 55 cm square. In practice, with typical applied electrical fields of 10 V/cm, this apparatus fractionates DNA molecules with 5% or better size resolution through an order of magnitude in size. The frequency at which the electrical fields are switched will determine the size range of the fractionation, which can vary from 5 kb for 0.1-sec pulses to 5000 kb for 1-hr pulses (C.L. Smith et al., unpubl.).

For PFG electrophoresis, the agarose sample block is inserted directly into a slot cut in the running gel and then the alternate fields are applied. Typical running times are 40 hr in the 28-cm apparatus and 70 hr in the larger apparatus. In general, resolution improves as the DNA concentration is lowered. We have found that use of halves or thirds of inserts at the sample concentrations described above represents a convenient trade-off between the resolution and sensitivity of resulting Southern blots for bacterial and mammalian DNA samples. In practice, ordinary Southern blotting and hybridization techniques suffice for the analysis of PFG electrophoretic bands. However, it is necessary to fragment the DNA prior to the Southern transfer, and our experience is that nicking with UV light has been consistently dependable, whereas acid nicking has been unreliable (Smith et al. 1986b). In our hands the sensitivity of hybridization detection of DNA blotted from PFGs is usually not as high as ordinary Southern blotting, but the reason for this is unknown.

Accurate size standards are important in assessing the performance of PFG electrophoresis and vital in using the techniques for macrorestriction mapping as described below. We have found it most convenient to use tandemly annealed oligomers of wild-type and deletion strain λ DNA as finely spaced size markers (Smith et al. 1986a). The apparent sizes of concatemers of two different monomer lengths are consistent, which supports the use of such samples as true size standards. In practice, concatemers up to the 34-mer are routinely prepared.

Specific fragmentation of genomic DNA into large pieces. PFG electrophoresis has shown excellent ability to separate small, natural linear chromosomal DNAs ranging in size from 50-kb parasite microchromosomes to multimillion-bp yeast chromosomes (Carle and Olson 1984; Schwartz and Cantor 1984; Van der Ploeg et al. 1984; Kemp et al. 1985). However, intact human chromosomes range in size from 50 million to 250 million bp (mb), too large for direct PFG separations. Bacterial chromosomes fall into the size range currently accessible by PFG electrophoresis, but their circular topology prevents their entry into gels. Thus, in practice, for the analysis of both bacterial and mammalian samples it is necessary to cut genomic DNA into large, discrete fragments. This cannot be done by normal solution techniques because of the susceptibility of large DNA to shear breakage. Instead, restriction nucleases are diffused into DNA samples inside agarose, and digestion is allowed to occur in situ.

Enzymes are chosen that recognize very infrequent sequences. *Not*I, which cleaves at GCGGCCGC, *Sfi*I, which cleaves at GGCCNNNNNGGCC, and *Mlu*I, which cleaves at ACGCGT, are particularly effective with human DNA. They yield fragments averaging in the range of 250–1000 kb. Methylase-nuclease combinations can yield even larger fragments (McClelland et al. 1984, 1985; C.L. Smith et al., unpubl.).

Protocols that provide total digestion of DNA in agarose were first developed with bacterial samples (Smith et al. 1986b). An example of the appearance of such a digest is given in Figure 1. Twenty to thirty bands are seen ranging in size from less than 50 kb to more than 400 kb. The progressive increase in staining of bands as a function of molecular weight (except for obvious multiples) and the consistency of the sum of band sizes with known genome sizes indicate that the digests are complete. This can be verified by Southern blotting and hybridization with known single-copy probes resulting in a single band.

The critical variable in preparing specific large DNA fragments was found to be the particular agarose batch. Once a suitable batch has been identified, complete digests of DNA can be obtained with 10–20 units of most enzymes per microgram of DNA. An example of a typical digest of human DNA analyzed by PFG electrophoresis is shown in Figure 2. Depending on the enzyme, the average fragment size varies from a few hundred kilobases to more than 1000 kb. In most samples, only a continuous smear of DNA is seen. This is reasonable since, from the haploid genome size of 3 billion bp, the digests should consist of 3000 to 10,000 discrete fragments. However, some enzyme digests show discrete fragments in the highest molecular-weight range examined. Elution of such material from the gels could provide relatively pure samples of 500–1000-kb regions of the genome. It remains to be determined whether any of these regions are particularly interesting ones.

Methylation is potentially a very serious complication for the production of unique large fragments of mammalian DNA. Many of the potentially most useful restriction enzymes with rare sites contain the sequence CpG in their recognition site. Indeed, it is the relative rarity of this sequence, 20% of the expected value (Bird and Taggart 1980), that makes the sites of these enzymes rare. It is estimated that more than 50% of the CpG's in mammalian genomes are methylated (Gruenbaum et al. 1981; Kunnath and Locker 1982), and many enzymes will not cut at the methylated sequence. If the pattern of methylation at each site is all or none, the result will be an apparent reduction in the number of cutting sites, but the digest will still appear to be complete. However, an intermediate methylation pattern will result in incomplete digests, which can seriously compromise some approaches for assembling the order of large fragments. Ironically, however, controlled incomplete digests are just what is desired for other macrorestriction mapping strategies.

In practice, the problem of methylation is unlikely to be as serious as it appears at first glance. The recognition sequences for the most useful restriction enzymes with rare sites appear to occur preferentially, if not exclusively, in HTF islands. These regions of the genome are predominantly single-copy DNA except for the ribosomal RNA genes (Bird et al. 1985). In HTF regions, the CpG sequences are generally not methylated. Thus, the appropriate choice of restriction nucleases may largely eliminate both the problems of methylation as well as the problems of repeated DNA in the specialized junction libraries described below.

Macrorestriction mapping. Blotting a PFG electrophorogram of digested human DNA and hybridizing with a single-copy probe identify a large DNA fragment that contains the DNA neighborhood of that probe. Where a genetic map already exists, or large

Figure 1. PFG electrophoretic analysis of the *Escherichia coli* genome after digestion with restriction nucleases. The lanes, from left to right, are: *Saccharomyces cerevisiae*; λ*vir*; digests of *E. coli* with *Not*I, *Eco*RI, *Hin*dIII, *Sfi*I, *Mlu*I, *Xho*I; λ*vir*; and *S. cerevisiae*. The electrophoresis was performed in a 55-cm apparatus at 500 V with 25-sec pulse times for 70 hr.

numbers of probes are available, direct Southern blotting of separated large fragments can reveal their order just as in conventional methods for mapping smaller DNA regions. This approach has been used on *E. coli* to provide a nearly complete physical map in less than a year of effort (C.L. Smith and C.R. Cantor, unpubl.). The same approach has provided a map of much of the human major histocompatibility complex (Lawrance et al., this volume).

Ambiguities can arise when two fragments have the same size. However, this is less serious than in ordinary restriction mapping by Southern blotting because the total number of fragments is much less. If the density of available probes is great enough, an unambiguous map can be assembled by overlapping two different enzyme digests. But often this is not the case, and then it becomes difficult to prove whether two particular fragments are actually adjacent.

In general, a more powerful and efficient approach is to screen or select probes that contain just the ends of large fragments. We call these probes junction probes (Smith et al. 1986b). There are two types of such probes: Linking probes span rare cutting sites and thus contain just the ends of two contiguous large fragments; jumping probes contain just the ends of a single large fragment. These are shown schematically in Figure 3. Linking probes can be prepared by selecting just those clones from a complete small-insert library that contain a particular rare restriction enzyme site. Probing a genomic digest of DNA generated by the same enzyme will reveal two large fragments (Smith et al. 1986a). These must be adjacent, and thus a physical map of the distances between rare cutting sites can be generated systematically by sampling all of the linking clones in the library.

Ambiguities will arise in the analysis of linking clones when two similarly sized pairs of fragments exist. For example, if two different 250-kb fragments occur, each adjacent to an 840-kb fragment, it will not be possible to tell which is next to which. This problem can be eliminated, and the general utility of the linking library approach can be enhanced by the use of a library of jumping probes. These are made by circularizing large fragments around selectable markers and then remov-

Figure 2. PFG electrophoretic analysis of human DNA digested with various restriction nucleases. The lanes, from left to right, are: *S. cerevisiae*; λ*vir*; digests of human lymphoblastoid cells with *Not*I, *Sfi*I, *Sal*I, *Pvi*I, *Xho*I, *Mlu*I, *Apa*I; λ*vir*; and *S. cerevisiae*. The electrophoresis was performed as described in the legend to Fig. 1, except that 120-sec pulses were used.

ing almost all of the genomic DNA before recycling (Collins and Weissman 1984; Poustka and Lehrach 1986). Each resulting small-insert jumping probe should identify a single large fragment in a macrorestriction digest and two unique linking probes in a linking library. Similarly, each linking probe should identify two unique jumping probes. In principle, it should be possible to walk between linking and jumping probes and thus establish their order without any direct analysis of large DNA fragments. If the jumping library is made from size-fractionated DNA, the distance covered by each step will be known.

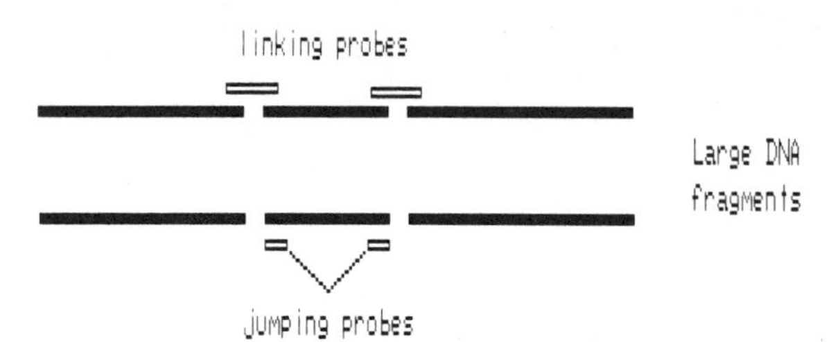

Figure 3. Schematic of two types of junction clones particularly useful in the rapid physical mapping of large DNA fragments.

Together, the two types of junction probes and the PFG electrophoretic analyses contain enough information to place the large fragments in order and determine their size. They provide sufficient redundancy to allow efficient discrimination against experimental errors. In practice, it will be desirable to analyze at least two different enzyme digests in parallel. This will provide valuable overlap information as further protection against accumulated errors. It will also help walk through any areas where junction probes are too imbedded in highly repeated DNA to be useful. In addition, as described above, the optimal choice of restriction nuclease may yield junction libraries relatively free of repeated DNA. At present, the strategies of constructing junction libraries have been tested and have been shown to work in limited cases. It will be desirable to optimize further the efficiency of creating such libraries before large-scale applications to human genome mapping are initiated.

Prospects for a Human Physical Map

It is possible to outline a simple scheme for physically mapping the entire human genome. This scheme makes use only of the existing technology outlined above and the known or presumed availability of human-rodent hybrid cell lines and libraries of small-insert, flow-sorted single human chromosomes. If one could work with single human chromosomes, appropriate digests could fragment these into 50 to a few hundred pieces. Indeed, the entire genome of *Drosophila melanogaster* is about the size of a single typical human chromosome, and a significant fraction of this genome can be visualized as discrete DNA pieces by PFG electrophoresis after digestion with a nuclease with rare sites (C.L. Smith and C.R. Cantor, unpubl.). Assembling these pieces into order would be no more difficult than assembling a macrorestriction map of a bacterial genome since the number of pieces is comparable even though the human pieces are fivefold larger in size.

One cannot currently work directly with flow-sorted human chromosomes because procedures for preparing these samples lead to too much DNA breakage. However, in practice, one can reduce the problem of working with the human genome to working with chromosomes one at a time by using hybrid cell lines containing only a single human chromosome and libraries made from flow-sorted DNA from the same chromosome. A typical human chromosome is 150 mb in size. This will be cut into 300 fragments by a single restriction nuclease with rare sites. Analysis of these will require 600 junction clones consisting of 300 linking clones and 300 jumping clones. Assuming that two nucleases are used, the number of clones needed doubles. Summaries of the numbers of clones needed for the largest and smallest human chromosomes are given in Table 1. These numbers are large but not unreasonable, especially in view of the many additional uses for the junction clones described below.

Available mapped markers and cytogenetic markers will serve as bench marks for the physical map. Wherever cell lines and flow-sorted material representing only parts of a single chromosome are available, the task will be even simpler. Such samples will allow the physical map of segments of a chromosome to be completed first. Then these can be linked up to make entire chromosomal maps. Provided that the distribution of the human material in the hybrid cell lines is known unambiguously, it will always be easier to use the principle of divide and conquer.

It is tempting to consider focusing initial mapping efforts on selected, interesting regions of the genome. The recent successes in identifying linked markers to Huntington's disease, polycystic kidney disease, cystic fibrosis, and Duchenne muscular dystrophy provide just a taste of the interesting regions likely to emerge over the next few years. To approach the mapping of such a region, one must first identify the large DNA fragment on which the linked marker resides and then clone the ends of that fragment. This will not be particularly difficult. Assume one can start with a hybrid cell line containing only a single human chromosome, 150 mb in size, on which the marker resides. A digest of that sample with a rare restriction nuclease will yield 300 human DNA fragments averaging 500 kb in size.

Current PFG electrophoretic resolution can provide a gel slice in which the desired human DNA fragment, containing the linked marker, is contaminated on average by only four other large human DNA fragments. Suppose one elutes DNA from the gel slice, fragments it with some other restriction nuclease, and clones it into a vector requiring pieces ending in the rare cutting site. Four of the twelve discrete human-DNA-containing clones will represent an end of the fragment of interest. Cell hybrids containing partial deletions or translocations of the human chromosome of interest will serve to identify which of the six clones are desired. Then physical mapping can proceed as described above. In practice, however, it will probably be more efficient to map regions of individual chromosomes without regard to particularly interesting linked markers. These markers can be used to test the correctness of the emerging physical map. Then they will provide the basis for staging the construction of a finer map of the large DNA fragments estimated to be in the actual region of the particular disease gene of interest.

As outlined above, the task of making a complete physical map of each human chromosome is arduous but now practical. Particularly for the smallest or most densely mapped chromosomes, the time seems appropriate to actually proceed with the task. However, several advances in technology seem likely to occur over the next few years that should accelerate the task still further. The size range of PFG electrophoresis has recently been extended up to 10 mb (C.L. Smith, T. Matsumoto, O. Niwa, M. Yanagida, and C.R. Cantor, unpubl.) and may be extendable still further. Potential methods for cutting human DNA into pieces this large have been described in detail. As these begin to work in practice, the possibility emerges of cutting the DNA of a chromosome into 10 to 50 pieces before fractionating these and then subcutting each into 5 to 20 smaller fragments. This would greatly facilitate physical mapping.

In all the strategies outlined above, the use of hybrid cell lines is a major limitation because one never actually has large human DNA fragments free from rodent contaminants. Problems with rodent-human cross-hybridization will inevitably arise. In addition, the need to analyze all separations by blotting rather than direct DNA visualization is quite labor-intensive. These problems would all be solved if it were possible either to (1) clone large DNA fragments directly, (2) separate a large fragment from others of the same size by virtue

Table 1. Number of Clones Required to Construct a Chromosome Map with 500-kb Resolution

Chromosome size[a]	One enzyme	Two enzymes
50 mb	100	200
150 mb	300	600
250 mb	500	1000

[a](mb) Million bases.

of some feature of its sequence, or (3) separate intact human chromosomes or chromosomal DNA without significant double-strand breakage. In view of the rate of progress over the past few years and the current level of interest in these problems, it is hard to believe that one or more of these technical breakthroughs will not be achieved long before a physical map is complete.

Applications of a Human Physical Map

The human physical map, constructed as described above, will consist of a set of cloned DNA markers spaced at accurately known positions throughout the entire genome. The average resolution of the map will be 500 kb. This is 10 times the resolution of the human genetic linkage map currently being developed. It is also 10 times the resolution of existing cytogenetic methods. This extra resolution should allow the visualization of many DNA rearrangements currently invisible cytogenetically. It will also dramatically speed the search for genes associated with inherited diseases. With even extremely tenuous evidence for genetic linkage for a particular disease, one will be able to use the map to select the appropriate DNA probes to provide a clear test of possible inheritance. If this is confirmed, then the map will serve to accelerate the search for the gene involved.

The power of the physical map is best seen by its potential for analyzing variation in DNA structure in the human population. This is illustrated in Table 2. Each linking probe allows one to examine two adjacent large fragments. Suppose that PFG electrophoresis is used to search for restriction-fragment-length polymorphism. If 25 probes could be used per lane, one could examine up to a third of the genome on a single gel at 20-kb resolution. This resolution is two orders of magnitude better than cytogenetics is currently able to achieve. Furthermore, the bands in a normal individual would always appear in the same place, so no subjective image analysis would be required and the entire process could potentially be automated.

The physical map will also serve to calibrate the genetic map. This will reveal the relationship between average recombination frequency and chromosome position. It may provide fundamental insights into the mechanisms of human meiosis. The physical map should be at high enough resolution to reveal whether the genetic linkage map is badly distorted by hot spots for DNA rearrangements. This may in turn allow the discovery of additional loci important in understanding human disease as well as human evolution.

Table 2. Potential for PFG Analysis of Polymorphism or Genome Alteration Assuming 50 Bands Probed per Gel Lane

Average fragment size	Base pairs monitored[a] per lane	Base pairs monitored[a] per gel	Resolution
500 kb	25 mb	250 mb	5 kb
1000 kb	50 mb	500 mb	10 kb
2000 kb	100 mb	1000 mb	20 kb

[a](mb) Million bases.

The initial low-resolution human physical map will also set the stage for the construction of a higher-resolution map. All the techniques needed to proceed efficiently from a 500-kb resolution map to a 50-kb map are already developed. Although it is premature to discuss higher-resolution strategies in detail, one approach seems particularly powerful and is already bearing fruit in the analysis of bacterial genomes. This is the Smith-Birnstiel strategy, which is shown schematically in Figure 4 (Smith and Birnstiel 1976). Genomic DNA is digested to completion with an enzyme with very rare sites and then digested partially with an enzyme that cuts more frequently. The digest is fractionated by length and then analyzed by hybridization with one half of a linking or jumping probe. The lengths of the resulting DNA bands reveal the positions of the more common restriction sites.

The power of the method is that many sites are mapped unambiguously on a single lane of the gel. The Smith-Birnstiel method requires accurate DNA sizing, but this requirement is well met by the high resolution of PFG electrophoresis and the availability of reliable length markers. The method also requires that controlled partial digests be carried out in agarose. The ability to do this is demonstrated by the example for *E. coli* DNA illustrated in Figure 5. The major disadvantage of the method is that partial digests inevitably require greater detection sensitivity. Current sensitivity in detecting single-copy mammalian genes on macrorestriction fragments may have to be improved before Smith-Birnstiel mapping becomes a routine approach for human DNA mapping.

The higher-resolution map is equivalent in information to an ordered set of cosmid clones spanning the entire genome. It can serve in practice to place an existing cosmid library in order. This in turn will set the stage for determining the DNA sequence of any or all regions of the genome.

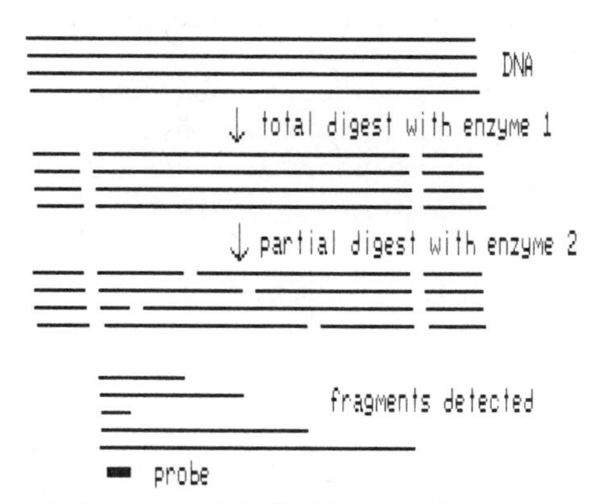

Figure 4. Schematic of the Smith-Birnstiel method of restriction mapping as applied to large DNA fragments.

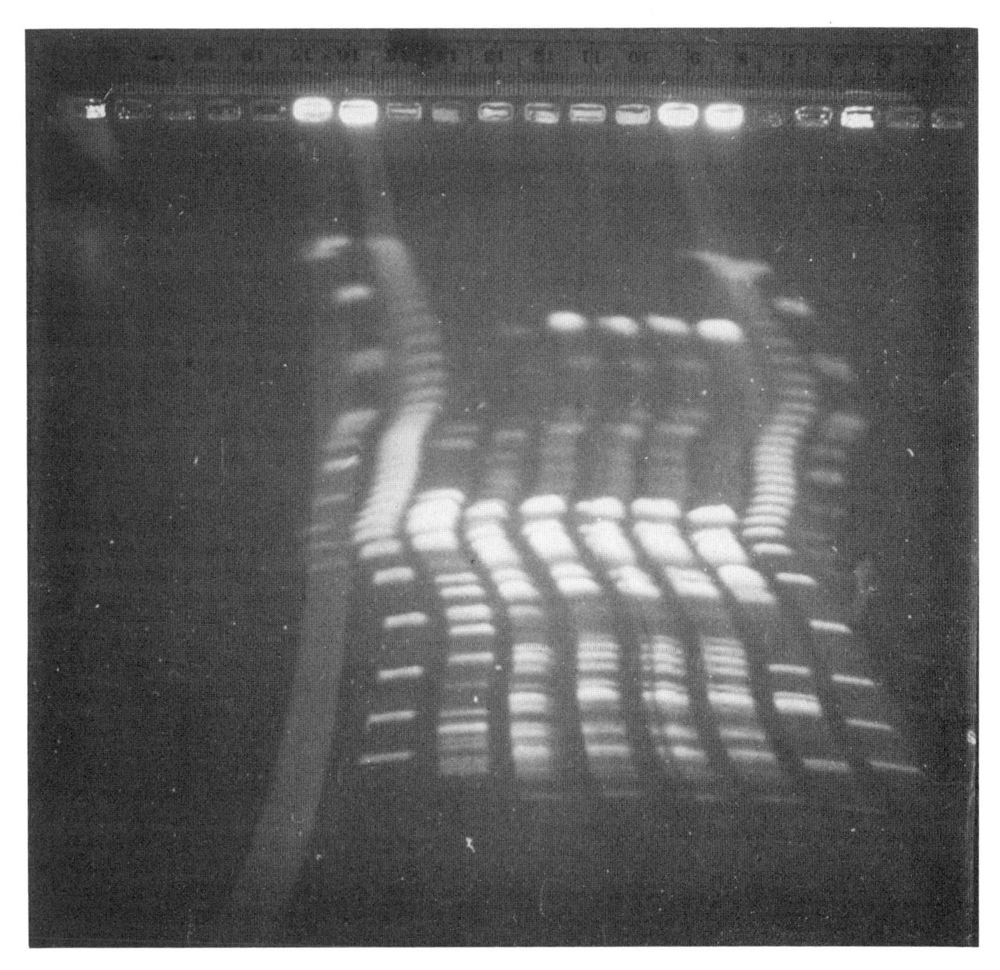

Figure 5. An example of Smith-Birnstiel mapping applied to macrorestriction fragments of *E. coli*. The outer lanes are *S. cerevisiae* and λ*vir*. The center lanes show a total *Not*I digest of *E. coli* that was subsequently treated with progressively larger amounts (from right to left) of *Sfi*I. The separation pattern of this gel is unusual because a program of two different pulse times (15 sec for 36 hr, then 120 sec for 36 hr) was used to generate high-resolution separations at both small and large molecular weights simultaneously. Otherwise, electrophoresis conditions were the same as described in the legend to Fig. 1.

ACKNOWLEDGMENTS

This work was supported by grants from the National Institutes of Health (GM 14825), the National Cancer Institute (CA 39782), the Hereditary Disease Foundation, and LKB Produkter-AB. The assistance of Jason Econome and Peter Warburton was most valuable.

REFERENCES

Bird, A.P. and M.H. Taggart. 1980. Variable patterns of total DNA and rDNA methylation in animals. *Nucleic Acids Res.* **8:** 1485.

Bird, A., M. Taggart, M. Frommer, O.J. Miller, and D. Macleod. 1985. A fraction of the mouse genome that is derived from islands of nonmethylated, CpG-rich DNA. *Cell* **40:** 91.

Carle, G.F. and M.V. Olson. 1984. Separation of chromosomal DNA molecules from yeast by orthogonal-field-alternation gel electrophoresis. *Nucleic Acids Res.* **12:** 5647.

Carle, G.F., M. Frank, and M.V. Olson. 1986. Electrophoretic separations of large DNA molecules by periodic inversion of the electric field. *Science* **232:** 65.

Collins, F.S. and S.M. Weissman. 1984. Directional cloning of DNA fragments at a large distance from an initial probe: A circularization method. *Proc. Natl. Acad. Sci.* **81:** 6912.

Cook, P.R. 1984. A general method for preparing intact nuclear DNA. *EMBO J.* **3:** 1837.

Gruenbaum, Y., R. Stein, H. Cedar, and A. Razin. 1981. Methylation of CpG sequences in eukaryotic DNA. *FEBS Lett.* **124:** 67.

Kemp, D.J., L.M. Corcoran, R.L. Coppel, H.D. Stahl, A.E. Bianco, G.V. Brown, and R.F. Anders. 1985. Size variation in chromosomes from independent cultured isolates of *Plasmodium falciparum*. *Nature* **315:** 347.

Kunnath, L. and J. Locker. 1982. Characterization of DNA methylation in the rat. *Biochim. Biophys. Acta* **699:** 264.

McClelland, M., L.G. Kessler, and M. Bittner. 1984. Site specific cleavage of DNA at 8- and 10-base-pair sequences. *Proc. Natl. Acad. Sci.* **81:** 983.

McClelland, M., M. Nelson, and C.R. Cantor. 1985. Purification of Mbo II methylase (GAAGmA) from *Moraxella bovis:* Site specific cleavage of DNA at nine and ten base pair sequences. *Nucleic Acids Res.* **13:** 7171.

Poustka, A. and H. Lehrach. 1986. Jumping libraries and linking libraries: The next generation of molecular tools in mammalian genetics. *Trends Genet.* **2:** 174.

Schwartz, D.C. and C.R. Cantor. 1984. Separation of yeast chromosome-sized DNAs by pulsed field gradient gel electrophoresis. *Cell* **37:** 67.

Schwartz, D.C., W. Saffran, J. Welsh, R. Haas, M. Goldenberg, and C.R. Cantor. 1983. New techniques for purifying large DNAs and studying their properties and packaging. *Cold Spring Harbor Symp. Quant. Biol.* **47:** 189.

Smith, C.L. and C.R. Cantor. 1986a. Purification, specific fragmentation and separation of large DNA molecules. *Methods Enzymol.* (in press).

———. 1986b. Pulsed field gel electrophoresis of large DNA molecules. *Nature* **319:** 701.

Smith, C.L., P.W. Warburton, A. Gaal, and C.R. Cantor. 1986a. Analysis of genome organization and rearrangements by pulsed field gradient gel electrophoresis. In *Genetic engineering* (ed. J.K. Setlow and A. Hollaender), vol. 8, p. 45. Plenum Press, New York.

Smith, C.L., S.K. Lawrance, G.A. Gillespie, C.R. Cantor, S.M. Weissman, and F.S. Collins. 1986b. Strategies for mapping and cloning macro-regions of mammalian genomes. *Methods Enzymol.* (in press).

Smith, H.O. and M.L. Birnstiel. 1976. A simple method for DNA restriction site mapping. *Nucleic Acids Res.* **3:** 2387.

Van der Ploeg, L.H.T., D.C. Schwartz, C.R. Cantor, and P. Borst. 1984. Antigenic variation in *Trypanosoma brucei* analyzed by electrophoretic separation of chromosome-sized DNA molecules. *Cell* **37:** 77.

Molecular Approaches to the Characterization of Megabase Regions of DNA: Applications to the Human Major Histocompatibility Complex

S.K. LAWRANCE,*† R. SRIVASTAVA,* B. RIGAS,* M.J. CHORNEY,* G.A. GILLESPIE,*
C.L. SMITH,‡ C.R. CANTOR,‡ F.S. COLLINS,§ AND S.M. WEISSMAN*

*Department of Human Genetics, Yale University School of Medicine, New Haven, Connecticut 06510;
‡Department of Genetics and Development, Columbia University, New York, New York 10032;
§Departments of Internal Medicine and Human Genetics, University of Michigan Medical School
Ann Arbor, Michigan 48109

The major histocompatibility complex (MHC) is a cluster of linked genes whose products are important in various aspects of the function of the immune system. In both man and mouse, some of these genes are among the most polymorphic known, and the propensity for various diseases often varies markedly with the particular allelic forms that an individual expresses. Because of the role of MHC products in basic immunology and pathophysiology, the gene complex has been the subject of intensive study.

The MHC consists of at least three families or classes of genes. The class I genes of man include the human leukocyte antigen genes, *HLA-A*, *-B*, and *-C*. The products of these genes are major determinants of the immunologic response in graft rejection between individuals and serve as self-recognition elements for cytolytic T cells. More than 20 other class I genes are also located within the MHC (Srivastava et al. 1985). The function and organization of these genes are unknown at present. The human MHC also contains a group of class II genes (Boss et al. 1985) that correspond structurally and functionally to the murine immune response genes. The products of these genes function in antigen presentation to and in self-recognition by T helper cells. A third type of MHC gene, the class III genes (Carroll et al. 1985b), encode some of the components of the serum complement system. Additional genes are also interspersed with, or closely linked to, the three classical types of human or murine MHC genes. These include the genes encoding steroid 21-hydroxylase (Carroll et al. 1985a), a group of small lymphocyte-specific peptides (Monaco and McDevitt 1984), the hemochromatosis allele (Simon et al. 1977), and some of the homozygous lethal genes of the murine small T/large T system (Shin et al. 1984). Very recently, the genes for tumor necrosis factor and lymphotoxin have been mapped to the same chromosomal arm that contains the MHC (Nedwin et al. 1985). In view of these reports and the extent of DNA contained within the MHC, it remains possible that yet other genes or gene families are contained within the complex.

†*Present address:* Department of Immunology IMM8, Scripps Clinic and Research Foundation, 10666 North Torrey Pines Road, La Jolla, California 92037.

Analysis of the antigenicity and amino acid sequences of the class I, class II, and class III proteins has permitted the preparation of suitable nucleic acid probes and the cloning of extensive regions of the complex. About 1.8 million bp of the murine MHC (Flavell et al. 1985; Steinmetz et al. 1986) and about 1.4 million bp of the human MHC (R. Srivastava and S. Lawrance, unpubl.) have been cloned. In spite of these extensive cloning efforts, however, there are several gaps of unknown size between the cosmid clusters derived from both the murine and human MHCs. In addition, there are extensive regions between the classically defined MHC genes in which no genes are recognizable or in which genes with no known function reside.

The difficulty in isolating and analyzing a region of DNA of this size has stimulated efforts to develop more-rapid procedures for cloning, mapping, and recognizing functional sequences within large segments of DNA. New procedures, which we are currently developing and applying to the human MHC, have the potential for expediting the construction of molecular maps and the cloning of large segments of mammalian genomes. The successful application of these techniques will complement classical genetic studies both in the resolution of questions about the MHC as well as in the full-scale analysis of the organization and variation of the human genome.

Molecular Mapping the MHC at a Resolution of 10^0–10^5 bp

We approached the molecular organization of the human MHC by cloning the class I *HLA-B* gene and the class II *HLA-DR*α gene, using probes prepared by extension of oligonucleotide primers on lymphoblast RNA (Sood et al. 1981; Das et al. 1983). Hybridization experiments with these clones as probes showed that each cross-hybridizes with an extensive group of related sequences in genomic Southern blots. For example, the class I *HLA-B* gene hybridizes with at least eight distinct *Eco*RI bands (Biro et al. 1981) while the class II *HLA-DRα* gene hybridizes with at least six distinct *Eco*RI bands (Lawrance et al. 1985), depending upon the DNA tested.

By constructing cosmid libraries of MHC hemizygous and homozygous DNA and selecting clones that hybridize with these probes, we have collected an extensive group of distinct syntenic *HLA* class-I/II-related clones. These clones include the *HLA-DRα* and *β* genes as well as clones of the class-II-related regions—*DP, DQ, DX*, and a novel class-II-related sequence that we call *DW* (Lawrance 1986). Our data and those of other groups indicate that the class II genes are organized into clusters of α and β genes. For example, we and others have shown that the *DP* subregion extends over 80 kb and consists of two α and two β genes (Lawrance et al. 1985). The combined data of several groups indicate that, in total, the class II family consists of six α genes and six to eight β genes, depending upon the DNA tested.

The class I cosmid clones include the *HLA-B* gene and the other classical transplantation antigen genes, *HLA-A* and *HLA-C*, as well as a number of additional class-I-related genes and pseudogenes. We currently estimate that the number of distinct class-I-related sequences in the DNAs we have examined is in excess of 25 (R. Srivastava, unpubl.). Structural analysis of these clones indicates that they contain both pseudogenes as well as apparently functional gene sequences. One of these, which we call RS5, encodes an expressible class-I-type protein product distinct from HLA-A, -B, and -C (R. Srivastava, in prep.).

In some cases it has been possible to orient these clones with respect to one another by cosmid-walking experiments. In most cases, however, cloning gaps of unknown extent separate the clusters of class I and class II genes (see Fig. 2). The organization of some of the genes can be inferred by reference to classical genetic studies. In the murine MHC it has been possible to do this with considerable precision (Steinmetz et al. 1986). For the most part, however, the organization of the human MHC remains poorly defined. Thus, the application of new techniques for molecular mapping is particularly appropriate.

Molecular Mapping the MHC at a Resolution of 10^4–10^7 bp

The recent development of the technique of pulsed-field gel electrophoresis (PFGE) has made it possible to resolve DNA fragments that range in size from 10,000 to well over 1 million bp in conventional agarose gels (Schwartz and Cantor 1984; Smith and Cantor 1986; Smith et al. 1986a,b). Consequently, PFGE affords a new and powerful method for examining the molecular organization of megabase expanses of DNA. Mammalian DNA can be broken into fragments within the range accessible to PFGE analysis by digestion with infrequently cutting restriction enzymes such as *Not*I, *Sfi*I, and *Mlu*I. These enzymes recognize 8-bp and/or CpG-rich restriction sites. As shown in Figure 1, the size of DNA fragments in this range can be accurately estimated by noting their migration relative to a ladder of bacteriophage λ DNA oligomers.

DNA blotting and hybridization to the large DNA fragments separated by PFGE can be performed in much the same manner as conventional Southern blotting. As shown in Figure 1, even with a DNA band as large as 1 million bp, if the DNA in the gel is fragmented by treatment with UV light before blotting, one can obtain hybridization with unique and multicopy probes that is nearly as intense as with a standard Southern blot of a fragment several hundred times smaller.

To approach the organization of the human MHC, we probed Southern blots of PFGE-separated *Not*I fragments of human DNA with cloned fragments of each of the three classes of MHC genes. To avoid problems that might arise because of heterozygosity, we examined DNA from an MHC hemizygous cell line (3.1.0) that has deleted the short arm of one chromosome 6 (Gladstone et al. 1982).

To locate class-I-related *Not*I resriction fragments, we probed with the *HLA-B* cDNA clone (Sood et al. 1981). This probe cross-hybridizes strongly with the *HLA-A* and *-C* genes, as well as with class-I-related genes, such as *RS5*. Since all of the class I genes mapped to date are clustered at the telomeric end of the MHC, the hybridization pattern of this probe approximates the extent of the class I region. Three distinct *Not*I fragments (1090, 540, and 150 kb) were de-

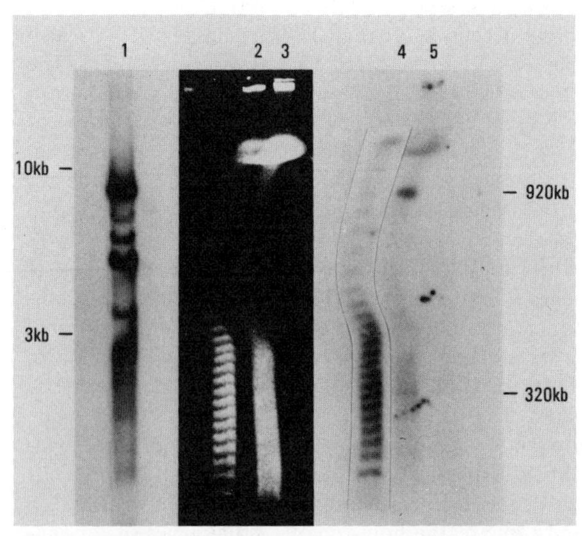

Figure 1. PFGE and Southern blotting of human DNA with an *HLA-DRβ* probe. (Lane *1*) An autoradiograph of the hybridization of the probe with an *Eco*RI digest of human (3.1.0) DNA electrophoresed in a conventional manner; the hybridizing bands include the *β* genes of the *DP, DQ*, and *DR* subregions. (Lanes *2* and *3*) Ethidium bromide stains of PFG electrophoresed concatemers of λ*vir* (42.5 kb monomer) and a *Not*I digest of human (3.1.0) DNA, respectively. These samples were run in a 55-cm PFGE apparatus for 72 hr at 500 V with 2-min pulses between 90° field reorientations. (Lanes *4* and *5*) Autoradiograms of the hybridization of lanes *3* and *4* with ^{32}P-labeled phage λ DNA and *HLA-DRβ* probe, respectively.

tected. The class II *DP, DQ, DX,* and *DR* subregions, which are clustered toward the centromeric end of the MHC, each contain β genes that cross-hybridize strongly with one another. *HLA-DRβ* and *-DQβ* cDNA clones (Bell et al. 1985) were employed to assess the extent of the class II region. Both probes hybridized with two *Not*I fragments of 920 and 320 kb. The class III region, which has been completely linked within 100 kb by cosmid walking (Carroll et al. 1985b) and located between the class I and II regions by recombination, was detected by probing with a *C4* cDNA clone (Whitehead et al. 1983). A single 920-kb *Not*I band, identical with the larger of the two class II hybridizing bands, was observed.

Combining these data with the orientations established for the MHC genes in genetic studies, a preliminary molecular map of the complex can be constructed. As shown in Figure 2, the class I region spans approximately 1800 kb, the class II and III regions span approximately 1200 kb, and the total extent of the MHC exceeds 3000 kb. Although it is possible that very weakly cross-reactive fragments or fragments not resolved by the gel may have been missed in this study, specific probes for MHC loci, which have also determined the orientation of the five *Not*I fragments, have failed to detect any additional fragments. In addition, probings of DNA digested with other infrequently cutting restriction enzymes and of DNA isolated from other cell lines have confirmed the size estimates obtained with *Not*I (S. Lawrance, in prep.).

These preliminary results indicate that the human MHC is composed minimally of five *Not*I fragments totalling approximately 3000 kb in length. Less than 50% of this distance has been isolated in cosmid clones, suggesting that large stretches of DNA exist within the *HLA* complex in which as yet undescribed genes or gene families may be found. The results also demonstrate the effectiveness of PFGE in megabase-scale molecular mapping of the human genome. A molecular map, such as that established in these studies, can be readily extended and resolved in greater detail through the use of double, triple, and partial digestion restriction mapping procedures and through the mapping of additional genes to PFGE-defined restriction fragments. These procedures are limited only by the availability of suitable probes.

General Approaches to the Molecular Mapping of Megabase Regions of DNA

The ease with which DNA blotting can be combined with PFGE suggests that techniques could be developed for rapid restriction mapping of large portions of human DNA without prior genetic information or the high density of probes that were available for the MHC. We are currently working on two sets of approaches.

Jumping and linking libraries. One approach depends on generating two types of junction fragment libraries (Smith et al. 1986b). As shown schematically in Figure 3, the linking fragment (Smith et al. 1986a) is a genomic DNA fragment that contains embedded within it a recognition site for an infrequently cutting restriction enzyme. For example, the set of *Eco*RI fragments that contain an internal *Not*I site would comprise a *Not*I linking library. The *Not*I linking library can be prepared either from total genomic DNA or from DNA isolated into flow-sorted chromosome libraries. The DNA is digested with *Not*I and recloned around a selectable marker, either a suppressor tRNA gene or an antibiotic-resistance gene flanked by *Not*I sites.

The second type of junction library, the hopping or jumping library (Collins and Weissman 1984), contains the two ends of a fragment that are generated by complete digestion with an infrequently cutting restriction

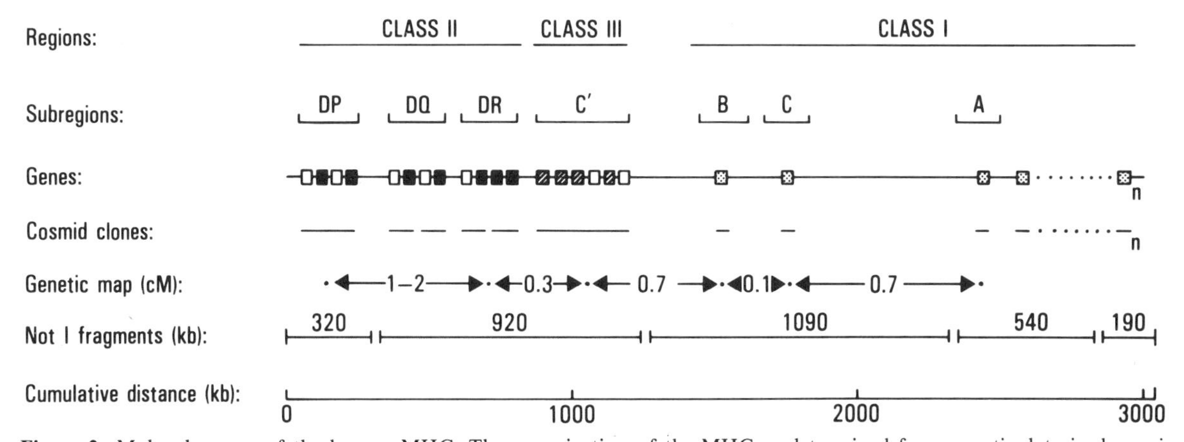

Figure 2. Molecular map of the human MHC. The organization of the MHC as determined from genetic data is shown in alignment with the results obtained in the PFGE blotting experiments described in the text. The boxes represent class II α genes (open boxes), class II β genes (closed boxes), class III genes (hatched boxes), and class I genes (stippled boxes). The open boxes in the class III cluster represent the steroid 21-hydroxylase genes. The dots in the class I region indicate the putative position of the more than 20 non-*A,B,C,* class-I-related sequences. Some of these may also be interspersed between the *A, B,* and *C* subregions. Some of the gene and subregion orientations are speculative. The *Not*I fragment orientations were determined with *HLA*-gene-specific probes (S. Lawrance, in prep.). Genetic data are taken from Robson and Lamm (1984).

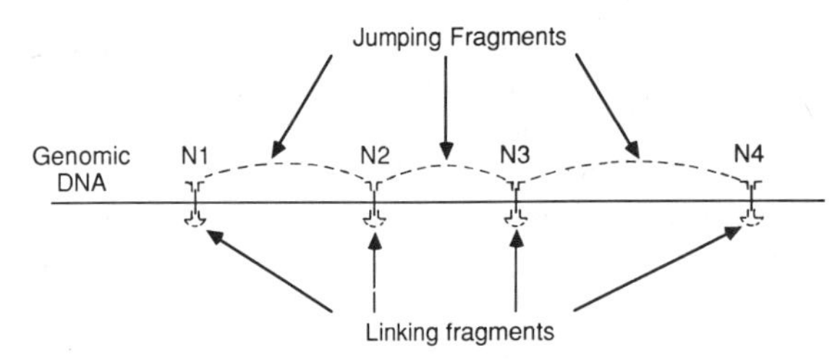

Figure 3. Schematic representation of the relationship between *Not*I linking and jumping fragments. Linking clones include the ends of two adjacent *Not*I fragments. Jumping clones join the two ends of a single *Not*I fragment. (N) *Not*I.

enzyme and that are joined together by circularization (Fig. 3). For example, a library of *Not*I jumping fragments can be constructed by digesting high-molecular-weight DNA to completion with *Not*I and circularizing around a selectable marker. Subsequent cloning of, for example, *Eco*RI-digested products of the circularization will generate a *Not*I jumping library.

The linking and jumping fragments can be overlapped in several ways. For example, each jumping fragment should overlap two linking fragments and each linking fragment should overlap two jumping fragments. Once the pattern of overlaps is obtained, the restriction map should be automatically generated, in principle without any use of gel electrophoresis.

Difficulties may be encountered in obtaining jumping fragments from very large or very small *Not*I fragments or from *Not*I sites that are blocked by methylation. For example, it may prove difficult to cross the clusters of CpG residues that are found in association with many mammalian genes (Bird 1986). Thus, it is advisable to prepare, in parallel, libraries consisting of jumping and linking fragments generated by other infrequently cutting restriction enzymes (e.g., *Mlu*I) and by enzymes that do not contain CpG in their recognition sequences (e.g., *Sfi*I). The total number of *Not*I, *Mlu*I, and *Sfi*I fragments in the human genome are estimated to be about 3,000, 3,000, and 12,000, respectively (Smith et al. 1986b). Therefore, the numbers of jumping and linking fragments required to represent the entire genome are not very great. As a final control, each linking and jumping fragment used in constructing a map should be examined to confirm that it was generated by circularization of a single fragment rather than by concatenation or some other side process. This can be done by hybridizing the jumping junction fragment to a blot of human DNA digested with *Not*I. To control for the linking fragment, one hybridizes instead to a blot of genomic DNA digested with the enzyme that created the ends of the fragments (e.g., *Eco*RI). In each case, a single band should hybridize.

A limitation of this approach is that it requires the ability to identify an infrequently cutting restriction enzyme that generates linking or jumping clones of interest. This may be difficult in regions, unlike the MHC, for which extensive cosmid clones are not available, since a probe for DNA located adjacent to the nearest rare restriction site of interest may not be available.

An alternative strategy, applicable to cases in which a probe for a rare site linked to the region of interest is not available, but in which the chromosome assignment is known, is to clone all of the *Not*I linking fragments from a library constructed from the appropriate sorted chromosome. As few as 200 such clones may be required to span an average human chromosome. Consequently, the task of identifying a linking clone that hybridizes to the same *Not*I fragment as the gene of interest would be a manageable one.

A more general approach is that of constructing a jumping library from DNA that has been completely digested with an infrequently cutting restriction enzyme such as *Not*I and partially digested with a frequently cutting restriction enzyme. This would allow one to jump from any position in the genome to the nearest *Not*I restriction site. Once a restriction site has been located near the region of interest, the combination of linking and jumping libraries enables one to rapidly determine the restriction pattern and molecular organization of extensive regions of DNA adjacent to the original probe. In principle, this region includes the entire chromosome from which the initial linking or jumping clone was isolated.

PFGE analysis of partial digests. An alternative procedure for megabase resolution mapping is even simpler in principle and has the potential for generating data rapidly. This procedure employs an experimental maneuver that we have used for some time to map cosmid clones and is designed to obtain restriction maps from partial enzymatic digests without the use of fragment-end labeling. In this procedure, as shown schematically in Figure 4, one takes a *Not*I linking fragment, for example, and separates unique probes from either side of the *Not*I site. A blot of a partial *Not*I digest of genomic DNA is then probed successively with the right-hand and left-hand probes. Some bands in the partial digest will light up with both probes and therefore represent products generated without cutting the *Not*I site in question. Other fragments will react only with the left-hand or the right-hand probe, generating series of nested fragments extending leftward and rightward from the original *Not*I site. The fragments reacting with both probes would have a length corresponding to the sum of one or more leftward and one or more rightward fragments.

This procedure has the advantage that linking fragments need not be generated for every *Not*I site in order to generate a map. The extent of the region that can be mapped in this way increases as it becomes possible to fractionate larger fragments by PFGE. There are some technical limitations that stem predominantly from the difficulty of obtaining sufficient sensitivity to detect multiple hybridizing bands in PFGE Southern blots. Thus, it is desirable to work with an enzyme that cuts relatively infrequently in order to decrease the ratio of singly to doubly hybridizing fragments, to in-

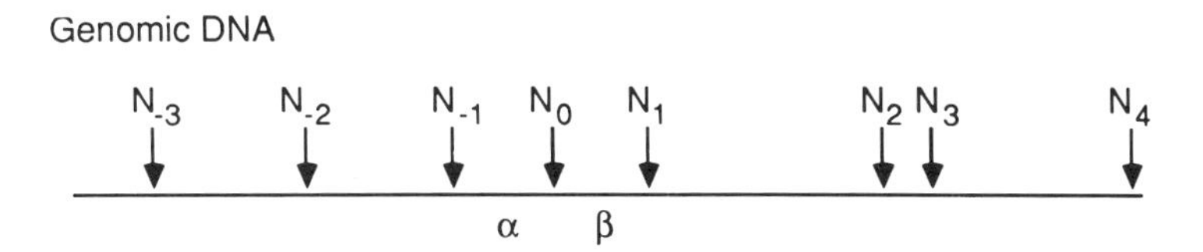

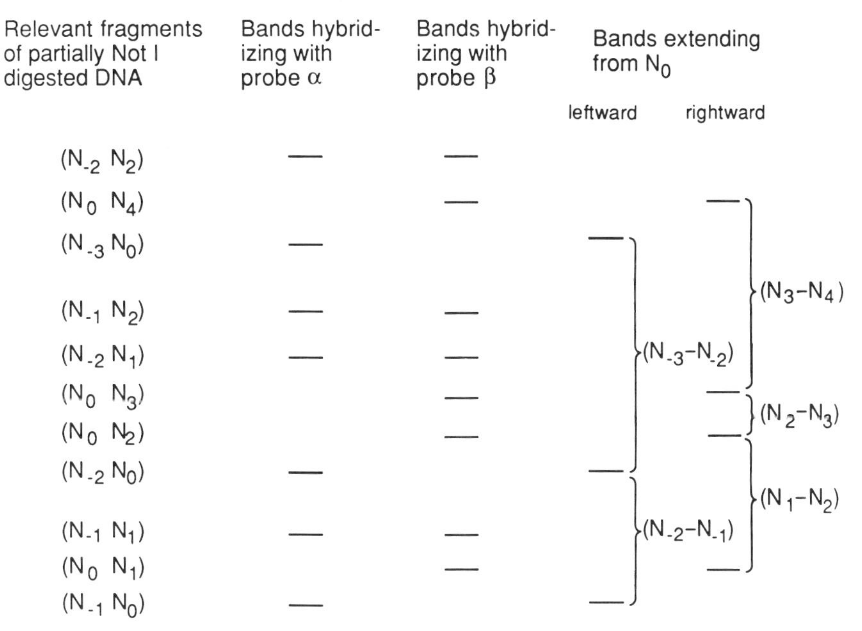

Figure 4. External mapping of large DNA fragments. Shown is a scheme for restriction mapping by partial digestion without end-labeled DNA. A *Not*I site embedded in a linking fragment is represented by N_0. To map sites proximal to N_0, a partial *Not*I digest of genomic DNA is prepared, separated by PFGE, and blotted. Probes β and α are, respectively, the left and right halves of the *Not*I linking fragment. The bands detected with these probes are designated by *Not*I sites at their two ends; e.g., $N_{-1}N_{-2}$ is the band derived by cleavage at position N_{-1} and N_{-2} during partial digestion.

crease the signal per fragment, and to decrease the frequency of overlap and unresolved regions. The development of sensitive nonradioactive detection systems would also further expedite this procedure. A range of partial digestive conditions should be explored to obtain an optimal display of bands. Multiple successive probings of each partial digest can be performed by blotting onto nylon membranes.

Hybridization to partial digests can also be used to map a probe within a *Not*I fragment. This can be done, for example, by preparing a complete digest of DNA with *Not*I and then partially digesting it with an enzyme, such as *Sfi*I, that cuts somewhat more frequently. Comparison of the bands hybridizing in a partial *Sfi*I digest to those retained in a partial *Sfi*I plus complete *Not*I digest would show the position of the probe within the fragment, because the smallest band removed by *Not*I would correspond to the distance from the further end of the *Sfi*I fragment containing probe sequences to the end of the *Not*I fragment. This approach could provide a sensitive method for detecting deletions. A deletion in a locus on one chromosome would show up as band doubling or a more extensive shift in banding pattern beyond the deletion point. In principle, one gel lane could be used to scan thousands of kilobases for a deletion.

Approaches to Cloning Megabase Regions of DNA

In addition to restriction mapping large regions of DNA, it is desirable to obtain ordered sets of clones encompassing such regions. A technique that we have used to expedite the cloning of larger regions of DNA uses the chromosome jumping procedure (Collins and Weissman 1984). One of us (F.S.C.) has prepared extensive chromosome jumping libraries, including a 100-kb jump library from the MHC hemizygous 3.1.0. cell line. This library was probed with the class I *HLA-B* cDNA probe (Sood et al. 1981) and yielded about seven positives per million plaques screened. This is in line with expectations, assuming that 1 million *Eco*RI-bounded junction fragments is about one genome's worth and that the coding regions overlap approximately the center one-half of the *Eco*RI fragment carrying a representative gene. We are currently using these jumping fragments in a second round of selection to isolate cosmids linked at 100-kb distances to class I genes.

Once a jumping library has been prepared and amplified, it can provide a resource for work with many different genes. It has become clear, however, that one of the more tedious steps in working with jumping libraries, as well as with standard genomic libraries, is the necessity of screening a number of genome equivalents of library clones in order to isolate an extensive set of junction fragments linked to a unique sequence probe. To expedite this screening, we have been working on an accelerated enrichment procedure for DNA sequences homologous to a given probe. The procedure we have used depends on *E. coli recA* protein (RecA) to promote rapid synapse formation between probe DNA and double-stranded target DNA (Cox et al. 1982; Radding 1982) and the use of probes tagged with biotinylated nucleotides (Langer et al. 1981) to promote rapid physical separation of hybridized from unhybridized target DNA molecules. In the first step of this procedure, biotinylated DNA probes are prepared by end labeling or nick translation with biotinylated dUTP. These probes are denatured, coated with RecA, and added to a mixture of homologous and nonhomologous target molecules under conditions where RecA promotes synapse formation. Under these conditions, pairing between the probe and cosmid or plasmid target molecules occurs within one minute. Probes associated with the nonhomologous DNA sequence are released, whereas the probe associated with homologous DNA sequences form D-loops. The reaction can be monitored by gel electrophoresis, because the radioactive probe stays asssociated with the plasmid DNA during the course of the electrophoresic run.

One of the major problems with use of avidin-biotin-RecA complexes is that they are so stable that it is difficult to recover the DNA in free form. This problem has been dealt with in two alternative approaches. In one, the biotin moiety of the modified nucleotide is attached to the base moiety by a disulfide linkage (Shimkus et al. 1985). In this case the probe-target complex is passed over a streptavidin column. The column is extensively washed, and the probe and target molecules are released by treatment with dithiothreitol. By use of a sufficiently long alkyl linker between the dUTP and the biotin, this procedure can be made to consistently yield near quantitative retention and release of labeled probe.

A second successful approach for recovering target DNA from biotinylated complexes is based on the binding of avidin to Cu^{++} (Porath et al. 1975). After the probe-target-RecA complex is formed, a limiting amount of avidin is added to the solution. Copper sulfate is added to saturate a chelating column, and the probe-target-RecA-avidin complex is passed over the column. After washing, the probe and target can be recovered by treatment of the column with EDTA.

We have used both of the above systems to recover plasmid DNA that can be biologically scored by bacterial transformation. A representative experiment using an artificially reconstituted mix of 200-fold excess of plasmid with a heterologous insert to plasmid with insert homologous to the probe is shown in Figure 5. Almost 10^5-fold purification was obtained in this experiment. Purifications in excess of 10,000-fold have been obtained repeatedly with either system.

A potential advantage of this system is that it could permit rapid and efficient screening of jumping and linking libraries with a given unique sequence probe. A number of such enrichments could be performed in parallel in a single experiment, although at the present time it is still necessary that the enriched plasmid preparations be rescreened by conventional Grunstein-Hogness hybridization procedures. A second potential application of these enrichment procedures would be the screening of cosmid libraries. Through a combination of jumping, linking, and cosmid libraries with the RecA-promoted screening procedure, one can envisage a strategy that would enable the direct cloning of overlapping cosmids spanning extensive chromosomal regions. An overall approach would be to initially obtain linking fragments from a segment of a chromosome (e.g., the *Not*I sites of the human MHC). Each linking fragment would be used to screen a jumping library and these in turn could be used to screen a cosmid library. Although each of the several steps may introduce a certain number of unlinked clones, a large fraction of the isolated cosmids would be derived from regions near the initial *Not*I sites. The final step would be to use a rapid cosmid fingerprinting procedure to arrange the cosmids into overlapping clusters.

CONCLUSION

In conclusion, we have applied cosmid cloning, PFGE, and chromosome jumping techniques to the analysis of the molecular organization of the human MHC. Together, these techniques enable extended genetic regions, such as the MHC, to be studied at any level of resolution between the single nucleotide and several million base pairs. We have also described several additional techniques that have potential usefulness in mapping and cloning very large regions of DNA for which little or no genetic or molecular data are available. Various combinations of jumping, linking, and cosmid libraries and the RecA-mediated enrichment procedures have considerable potential for the efficient analyses of the organization of DNA segments, in principle as large as entire human chromosomes. The refinement of these techniques should expedite the full-scale structural analysis of mammalian genomes, provide new insights into the processes of molecular evolution, and aid in the delineation of the structural variations associated with genetic disorder.

ACKNOWLEDGMENTS

The authors wish to thank Drs. Charles Radding, David Ward, and Andrew Welcher for helpful advice and discussion; Sherry Rucker, Peter Warburton, and Jason Econome for excellent technical assistance; and Ann Mulvey and Donna Uranowski for preparing the

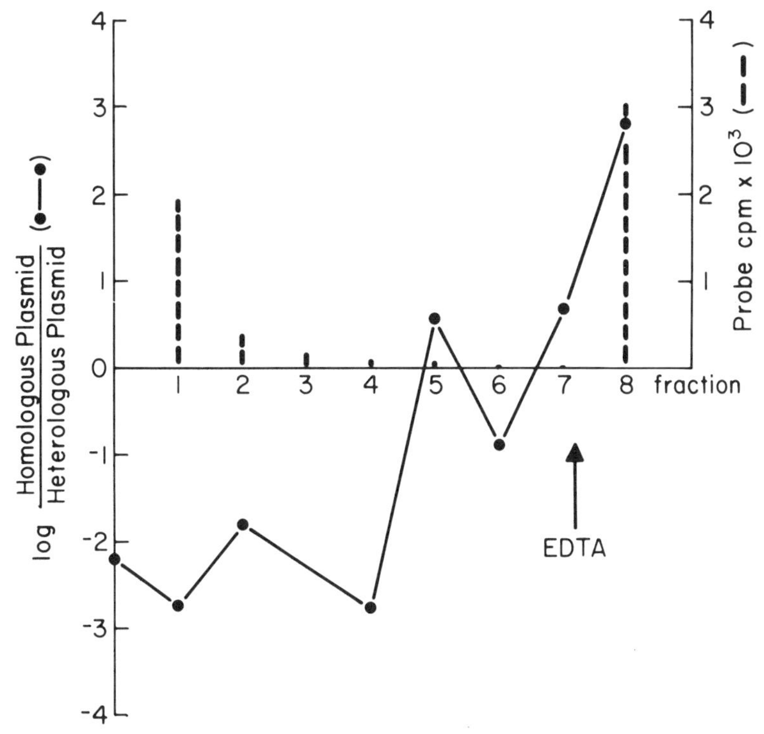

Figure 5. RecA-assisted selective plasmid isolation. In the presence of ATP-γ S, RecA was polymerized on a single-stranded probe with a biotinylated end (the 6.5-kb class I MHC pseudogene LN11A), which was then reacted with a mixture of two plasmids. The homologous plasmid (Tet^r, Kan^s) was pBR328 containing LN11A; the heterologous plasmid (Tet^s, Kan^r), present in 200-fold molar excess, was a pBR322 derivative without any sequence homology to the probe. Following synapse formation, avidin was added in the presence of high salt concentration and the mixture was loaded onto an agarose-iminodiacetic acid column saturated with $CuSO_4$ and carrier DNA. Heterologous plasmid and uncomplexed homologous plasmid were washed away. Target-probe-avidin complexes were released with EDTA. The synapse was dissociated by heating in low ionic strength. Competent *E. coli* cells were transformed with aliquots of each column fraction after desalting. Plasmids were monitored based on their different antibiotic sensitivities; the probe was monitored by using a radiolabeled nucleotide in the fill-in reaction.

manuscript. This work was supported in part by grants 5-RO1-CA30398 and R35-CA39782 from the National Institutes of Health and by the Hereditary Disease Foundation.

REFERENCES

Bell, J.I., P. Estess, T. St. John, R. Saiki, D.L. Watling, H.A. Erlich, and H.O. McDevitt. 1985. DNA sequence characterization of human class II major histocompatibility complex beta chains from the DR1 haplotype. *Proc. Natl. Acad. Sci.* **82:** 3405.

Bird, A.P. 1986. CpG-rich islands and the function of DNA methylation. *Nature* **321:** 209.

Biro, P.A., A.K. Sood, B. de Martinville, U. Francke, and S.M. Weissman. 1981. The structure of the human major histocompatibility locus. *ICN-UCLA Symp. Mol. Cell. Biol.* **20:** 315.

Boss, J.M., T. Spies, R. Sorrentino, K. Okada, and J.L. Strominger. 1985. Genetic complexity and expression of human class II histocompatibility antigens. *Immunol. Rev.* **85:** 45.

Carroll, M.C., R.D. Campbell, and R.R. Porter. 1985a. Mapping of the steroid 21-hydroxylase gene adjacent to complement component C4 genes in HLA, and major histocompatibility complex in man. *Proc. Natl. Acad. Sci.* **82:** 521.

Carroll, M.C., K.T. Belt, T.A. Palsdottir, and Y. Yung. 1985b. Molecular genetics of human complement and steroid 21-hydroxylase. *Immunol. Rev.* **87:** 39.

Collins, F.S. and S.M. Weissman. 1984. Directional cloning of DNA fragments at a large distance form an initial probe: A circularization method. *Proc. Natl. Acad. Sci.* **81:** 6812.

Cox, M.M., D.A. Soltis, Z. Zivneh, and I.R. Lehman. 1982. DNA strand exchange promoted by *recA* protein and single-stranded DNA-binding protein of *Escherichia coli. Cold Spring Harbor Symp. Quant. Biol.* **47:** 803.

Das, H.K., S.K. Lawrance, and S.M. Weissman. 1983. Structure and nucleotide sequence of the heavy chain gene of HLA-DR. *Proc. Natl. Acad. Sci.* **80:** 3543.

Flavell, R.A., H. Allen, B. Huber, C. Wake, and G. Widera. 1985. Organization of the MHC of the C57 Black/10 mouse. *Immunol. Rev.* **84:** 29.

Gladstone, P., L. Fueresz, and D. Pious. 1982. Gene dosage and gene expression in the HLA region: Evidence from deletion variants. *Proc. Natl. Acad. Sci.* **79:** 1235.

Langer, P.R., A.A. Waldrop, and D.C. Ward. 1981. Enzymatic synthesis of biotin-labeled polynucleotides: Novel nucleic acid affinity probes. *Proc. Natl. Acad. Sci.* **78:** 6633.

Lawrance, S.K., H.K. Das, J. Pan, and S.M. Weissman. 1985. The genomic organization and nucleotide sequence of the HLA-SB(DP) alpha gene. *Nucleic Acids Res.* **113:** 7515.

Lawrance, S.K. 1986. Molecular studies of the immune response (class II) genes of the human major histocompatibility complex. Ph.D. thesis, Yale University, Connecticut.

Monaco, J.J. and H.O. McDevitt. 1984. H-2 linked low molecular weight polypeptide antigens assemble into unusual macromolecular complex. *Nature* **309:** 797.

Nedwin, G.E., S.L. Naylor, A.Y. Sakaguchi, D. Smith, J.J. Nedwin, D. Pennica, D.V. Goedder, and P.W. Gray. 1985. Human lymphotoxin and tumor necrosis factor gene structure, homology and chromosomal localization. *Nucleic Acids Res.* **13:** 636.

Porath, J., J. Carlsson, T. Olsson, and G. Belfrage. 1975. Metal chelate affinity chromatography, a new approach to protein fractionation. *Nature* **258:** 598.

Radding, C. 1982. Homologous pairing and strand exchange in genetic recombination. *Annu. Rev. Genet.* **16:** 405.

Robson, E.B. and L.U. Lamm. 1984. Report of the committee on the genetic constitution of chromosome 6. Human Gene Mapping 7. *Cytogenet. Cell. Genet.* **37:** 47.

Schwartz, D.C. and C.R. Cantor. 1984. Separation of yeast chromosome sized DNAs by pulsed field gel electrophoresis. *Cell* **37:** 67.

Shimkus, M., J. Levy, and T. Herman. 1985. A chemically cleavable biotinylated nucleotide: Usefullness in the recovery of protein-DNA complexes from avidin affinity columns. *Proc. Natl. Acad. Sci.* **82:** 2593.

Shin, H.S., D. Bennet, and K. Artzt. 1984. Gene mapping within the T/t complex of the mouse. IV: The inverted MHC is intermingled with several t-lethal genes. *Cell* **39:** 573.

Simon, M., M. Bourel, B. Genetet, and R. Fauchet. 1977. Idiopathic hemochromatosis. Demonstration of recessive transmission and early detection by family HLA typing. *N. Engl. J. Med.* **297:** 1.

Smith, C.L. and C.R. Cantor. 1986. Purification, specific fragmentation, and separation of large DNA molecules. *Methods Enzymol.* (in press).

Smith, C.L., P.E. Warburton, A. Gal, and C.R. Cantor. 1986a. Analyis of genome organization and rearrangements by pulsed field gradient gel electrophoresis. In *Genetic engineering* (ed. J.K. Setlow), vol. 8, p. 45. Plenum Press, New York.

Smith, C.L., S.K. Lawrance, G.A. Gillespie, C.R. Cantor, S.M. Weissman, and F.S. Collins. 1986b. Mapping and cloning macro-regions of mammalian genomes. *Methods Enzymol.* (in press).

Sood, A.K., D. Pereira, and S.M. Weissman. 1981. Isolation and partial nucleotide sequence of a cDNA for human histocompatibility antigen HLA-B by use of an oligodeoxynucleotide primer. *Proc. Natl. Acad. Sci.* **78:** 616.

Srivastava, R.S., B.W. Duceman, P.A. Biro, A.K. Sood, and S.M. Weissman. 1985. Molecular organization of the genes of the human major histocompatibility complex. *Immunol. Rev.* **84:** 93.

Steinmetz, M., D. Stephan, and K.F. Lindahl. 1986. Gene organization and recombinational hotspots in the murine major histocompatibility complex. *Cell* **44:** 895.

Whitehead, A.A., G. Goldberger, D.W. Woods, A.F. Markham, and H.R. Colten. 1983. Use of a cDNA clone for the fourth component of human complement (C4) for analysis of a genetic deficiency of C4 in guinea pig. *Proc. Natl. Acad. Sci.* **82:** 5387.

Molecular Approaches to Mammalian Genetics

A. Poustka, T. Pohl, D.P. Barlow, G. Zehetner, A. Craig,
F. Michiels, E. Ehrich*, A.-M. Frischauf, and H. Lehrach
European Molecular Biology Laboratory, D-69 Heidelberg, Federal Republic of Germany

In the last few years a number of new genetic and molecular techniques have emerged for analyzing the mammalian, and especially the human, genome. The genetic analysis has concentrated on the new possibilities of using DNA probes as genetic markers, allowing major progress on both establishing a human linkage map (Botstein et al. 1980) and identifying and using closely linked DNA markers in the genetic analysis of genes defined by mammalian (human and mouse) mutations.

However, molecular analysis, based on the availability of cloned genes, has up to now been restricted mostly to genes identified either by their gene products (e.g., globin) or by their somatically selectable function (e.g., oncogenes), a fairly small subset of the genetic information necessary (and possibly sufficient) to encode human beings (or mice).

Many of the as yet unclonable genes, which at the moment can only be defined by the effect of alleles or mutations, are important due to their immediate effect on human health (e.g., Huntington's chorea, cystic fibrosis, and Duchenne muscular dystrophy). Other genes will be expected to be involved in more-complex phenotypes (e.g., susceptibilities to cancer or mental illness) or might provide as yet unknown gene products potentially capable of modifying the course of diseases.

Two possible routes to identify and clone these additional classes of genes can be used. (1) Although for some years mutations in, for example, *Drosophila melanogaster* have been used to define and clone genes characterized by developmental mutations progress in similar work in mammalian systems has been very difficult. This is mainly due to the difficulties of analyzing and manipulating distances, typically of millions of base pairs, which separate even genetically closely linked markers, by molecular analysis and cloning techniques, which are best suited for tens to at best hundreds of kilobase pairs.

(2) A second, still highly speculative approach, especially required for the identification of genes not defined previously by mutations, is the use of large-scale structural analysis of mammalian chromosomes. In this approach to identifying previously unknown genes and gene products one can rely on the identification of sequences conserved between man and mouse, for example, and/or on the use of rare restriction enzyme cutting sites as pointers to potential gene sequences (Bird 1986; G. Rappold et al., in prep.).

Both the identification of genes from mutations and the isolation of genes by, for instance, the identification of conserved sequences depend critically on new approaches to DNA analysis and molecular cloning. Therefore a number of new techniques potentially capable of bridging the gap between the distances defined by genetic or cytogenetic techniques in mammals and the size of DNA segments currently most easily analyzed by molecular approaches have been developed (Fig. 1) (Poustka and Lehrach 1986; Smith et al. 1986).

We outline here the current status of development of these techniques in our laboratory. The approaches considered fall into two main classes: the serial techniques logically analogous to chromosome walking, which will predominantly be used to walk between markers and mutations; and the parallel procedures, which are best suited to derive maps (or identify genes) from larger regions of the genome.

PROCEDURES

Construction of the human* Not*I*/Bam*HI chromosome jumping library. The construction of the human *Not*I/*Bam*HI jumping library will be described elsewhere (A. Poustka et al., in prep.). In short, 2 million cells of LCL127 (gift of Uta Francke), a human lymphoblast cell line carrying four X chromosomes, were suspended in 80-μl blocks of low-gelling-temperature agarose (Bethesda Research Laboratories), and DNA was prepared according to a modification of a protocol described for yeast DNA (Schwartz and Cantor 1984; D.P. Barlow, unpubl.). The DNA was cut within the blocks by *Not*I and the enzyme was removed by digestion with proteinase K. The protease was then inactivated by treatment with phenylmethylsulfonyl fluoride (PMSF), and the agarose was melted and diluted to give a final DNA concentration of approximately 0.4 μg/ml. Then the DNA was ligated to pMLS(*Not-Mlu*) (Levinson et al. 1984; Huang et al. 1986) (linearized with *Not*I [New England Biolabs] and dephosphorylated by treatment with calf intestine alkaline phosphatase) at plasmid concentrations of 0.01 and 0.1 μg/ml. The ligase was then inactivated by heating, and the DNA was cut to completion with *Bam*HI, precipitated, dephosphorylated, and ligated into

Present address: Department of Internal Medicine, Stanford University Medical Center, Stanford, California 94305.

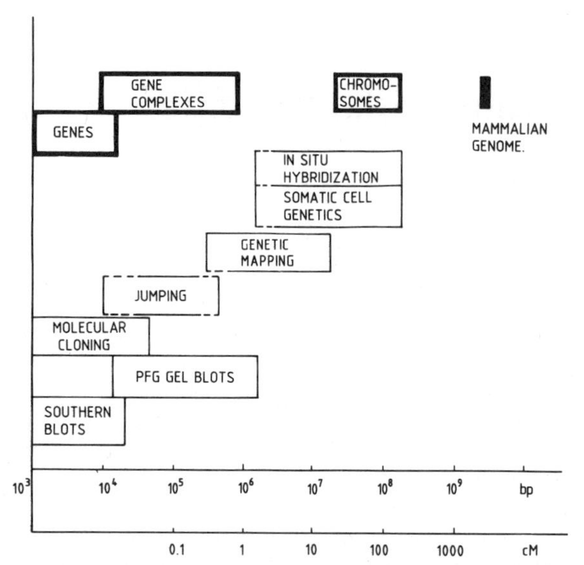

Figure 1. Comparison of the molecular (bp) and genetic (cM scale) distances covered by different techniques in the analysis of a typical mammalian genome. (Reprinted, with permission, from Poustka and Lehrach 1986.)

*Bam*HI-cut DNA from an *A*am*Bam*Sam derivative of NM1151 (Murray 1983). After in vitro packaging, the reaction mix was purified over a CsCl step gradient and plated on *E. coli* MC1061 (Casadaban and Cohen 1980) to select phages carrying the supressor plasmid sequence. To analyze the structure of the clones, random clones were picked, DNA was prepared, and the inserts were immediately subcloned as plasmids after excision by *Bam*HI and religation.

Construction of human* Not*I linking clones. The protocol for the construction of a *Not*I linking library will be described in more detail elsewhere (Frischauf et al. 1986; A.M. Frischauf, unpubl.). In short, DNA from a human-hamster cell line carrying fragments of human chromosomes 4 and 5 in a hamster background (provided by John Wasmuth) was partially cleaved with *Sau*3A to an average fragment size of 10–20 kb, circularized in the presence of *Bam*HI-cut pDSΔRI, a plasmid carrying a suppressor tRNA gene as selectable marker (Huang et al. 1986). The DNA was then cut with *Not*I, phosphatased, and ligated into *Not*I-cleaved EMBL5 (NotEMBL3A) (Frischauf et al. 1986). The ligation reaction was packaged in vitro and plated on MC1061 to select for suppressor-independent phages arising by ligation of the suppressor plasmid sequences into the phage vector carrying amber mutations. Clones hybridizing to human *Alu* repeats (Crampton et al. 1981) were analyzed further. Southern blots and hybridizations were performed as described by Herrmann et al. (1986).

RESULTS AND DISCUSSION

Chromosome Jumping

Chromosome jumping is a logical extension of chromosome walking techniques, which were designed to overcome the inherent limitation of the step size by the capacity of available cloning vectors (Poustka and Lehrach 1986). To take steps larger than the maximal size of clonable DNA, the size of large fragments is reduced by an internal deletion, leaving the ends of the fragments intact. In this approach (Fig. 2) large DNA restriction fragments are circularized in the presence of a selectable marker under conditions designed to maximize intramolecular ligation reactions. Circularized molecules are cleaved with another restriction enzyme cleaving the circle into many fragments, and the junction fragments are selectively cloned (Collins and Weissman 1984; Poutska and Lehrach 1986). As can be seen in Figure 4, chromosome jumping is inherently directional, that is, one always jumps from the first (rare-cutting) enzyme site in the direction of the second enzyme site used for library construction. We have concentrated on the construction of libraries made from DNA fragments generated by digestion with enzymes, like *Not*I, that cut only rarely in mammalian DNA. The required complexity of such libraries is determined by the number of restriction sites for the corresponding restriction enzymes in the genome. Thus the use of enzymes that recognize a small number of sites should simplify both construction and screening of these libraries by two to three orders of magnitude compared with enzymes that cut more frequently.

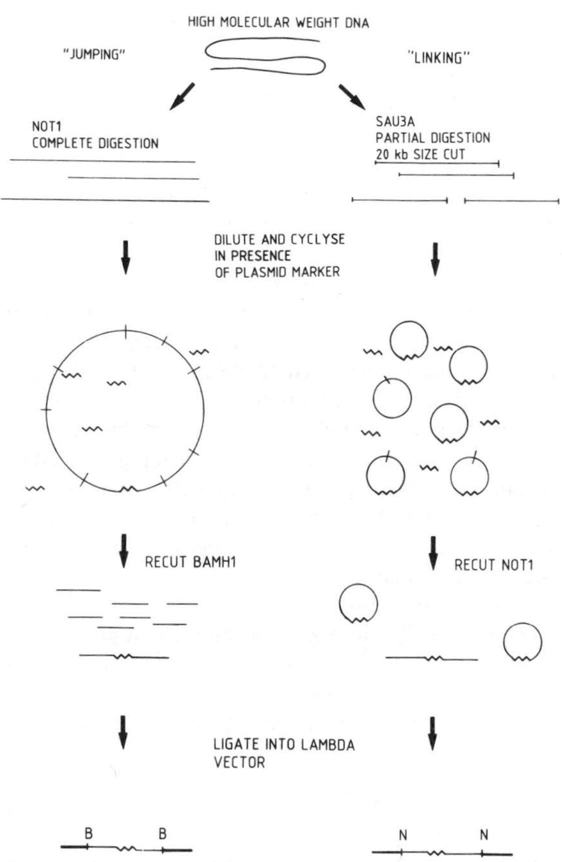

Figure 2. Schematic outline of construction of a *Not*I jumping library (*left*) and the complementary *Not*I linking library (*right*). (Reprinted, with permission, from Poustka and Lehrach 1986.)

A *Not*I chromosome jumping library of 100,000 clones (15–30 genomes, assuming an average size of 500–1000 kb for *Not*I fragments) was constructed from DNA from a human cell line with four X chromosomes. *Bam*HI was used as second enzyme to introduce the deletions. Random clones were characterized by restriction analysis and hybridization of both the jumping end-fragments to filters prepared from pulsed-field gradient gels (PFG) (Schwartz and Cantor 1984) of *Not*I-cleaved human DNA. An example of such a characterization is shown in Figure 3a. The expected structure of the jumping clone is verified here by hybridization of both presumed jumping end-points to identical *Not*I fragments of 200 kb. In general, the size of the jump will correspond to the size of the hybridizing fragment, though occasionally we jumped to a partially cut *Not*I site internal to the main hybridizing band. Many of the rare-cutting enzymes cut different sites at vastly different rates, some sites being hardly cut at all in genomic DNA. This latter phenomenon is probably due to methylation. An inherent difficulty in the use of jumping libraries is the unavoidable background of clones arising from intermolecular ligation events, reduced but not eliminated by carrying out the circularization step at very low DNA concentrations. Even then, the very large *Not*I fragments will circularize very slowly or will be lost preferentially by random breakage. Therefore, jumping clones from very large fragments will be underrepresented in the library. This problem can, however, be overcome either by using a set of libraries with different rare-cutting enzymes or by using libraries constructed from, for example, *Not*I complete/*Eco*RI partial digests (Poustka and Lehrach 1986).

Linking Libraries

Linking libraries consist of the subset of clones of a normal (phage λ or cosmid) library that contains sites for a chosen rare-cutting enzyme. A number of strategies and vectors can be used to selectively clone out such a subset of sequences (Frischauf et al. 1986; Poustka and Lehrach 1986; Smith et al. 1986). Clones of this type can be used in several applications. An obvious possibility is to use such clones as intermediates between the end of one jump and the start of the next, which is required in a series of jumping steps (Fig. 4).

In addition, linking clones can be used separately or in combination with the complementary jumping library in parallel mapping strategies, in which either the linear order of clones or the restriction map are being determined. The latter approach is being tested in collaboration with Charles Cantor, Jim Gusella, and John Wasmuth, by using linking clones from the area of the

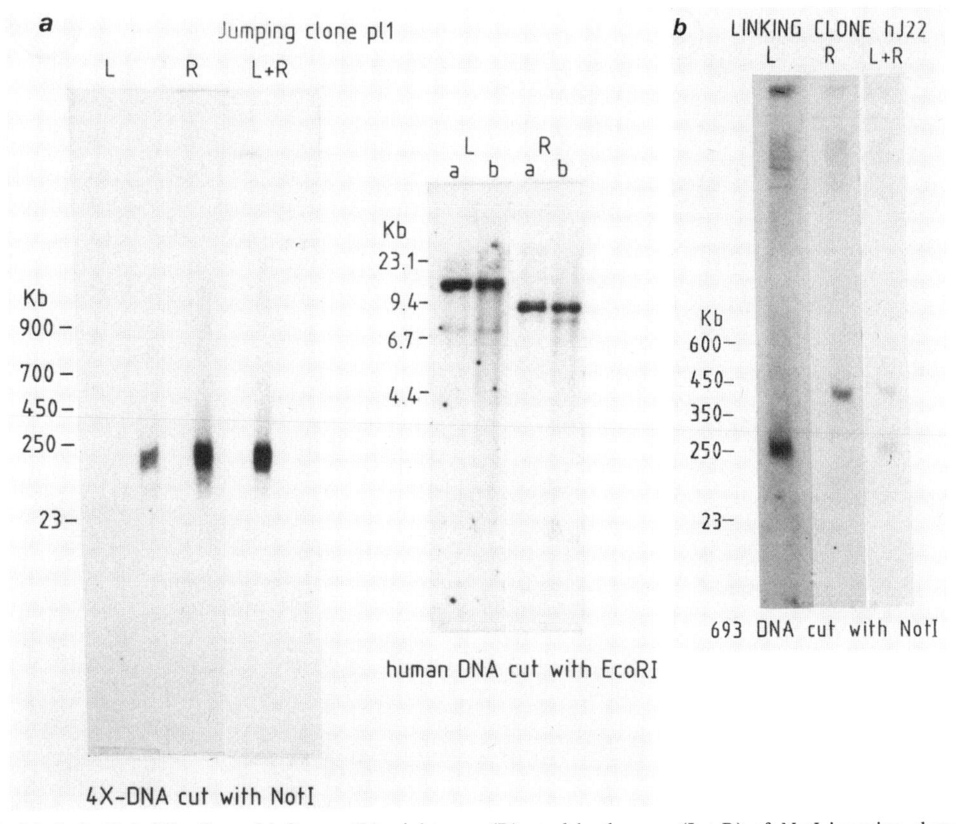

Figure 3. (*a*) *Left*: Hybridization of left arm (L), right arm (R), and both arms (L + R) of *Not*I jumping clone pl1 to *Not*I-cleaved DNA from the 4X cell line (LCL127). The DNA is separated by field inversion gel electrophoresis (FIGE; Carle et al. 1986). *Right*: Hybridization of left (L) and right (R) arms to two different human DNAs (a and b) cut with *Eco*RI. (*b*) Hybridization of left arm (L), right arm (R), and both arms (L + R) of linking clone hJ22 to *Not*I-cleaved 4X cell line DNA separated by FIGE.

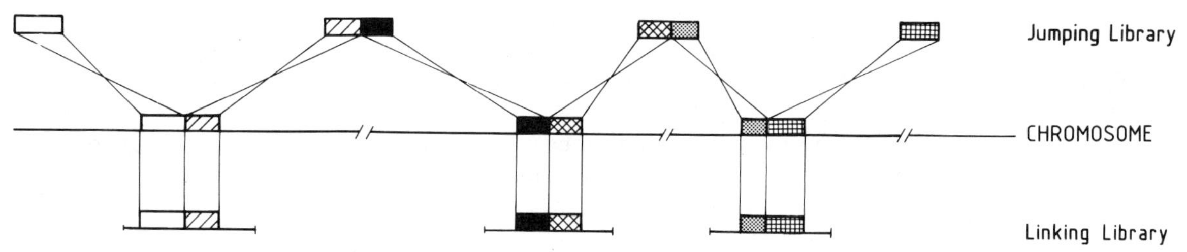

Figure 4. Schematic drawing of the relation between *Not*I jumping and linking clones on the chromosome. (Reprinted, with permission, from Poustka and Lehrach 1986.)

Huntington's chorea locus to determine a partial *Not*I restriction map of the region. Figure 3B shows the hybridization of the two arms of one of these clones to a *Not*I digest of DNA from a cell line constructed by Wasmuth containing parts of chromosomes 4 and 5 in a hamster background. Since each *Not*I linking clone spans a *Not*I site, the two sides should hybridize to different *Not*I fragments. If all linking clones of a subregion of the genome have been identified, it should be possible to at least partly deduce their linear order, since neighboring clones can be identified by hybridizing to the same *Not*I fragments.

An alternative possibility to derive similar information is the parallel identification of pairwise-overlapping jumping and linking clones by, for example, an approach similar to the one described for the cosmid linkup (see below). In this strategy, complementary *Not*I jumping and linking libraries provide overlapping subsets of the genome, essentially contracting the mammalian genome to the size of yeast or *E. coli*.

For further higher-resolution restriction mapping, arms from linking clones can be used in a partial restriction mapping approach (Smith and Birnstiel 1976; Rackwitz et al. 1984). In this technique DNA is cut

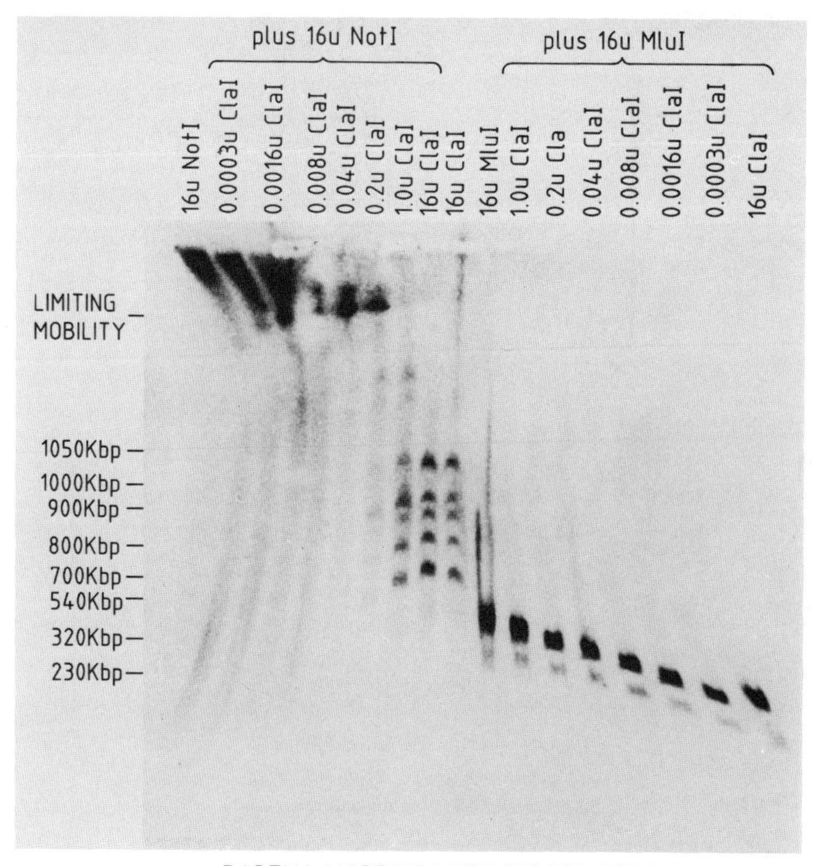

PARTIAL MAPPING WITH PULSED FIELD GELS

Figure 5. Partial restriction mapping of *Cla*I sites extending over a region of more than 1 million bp proximal to the mouse major histocompatibility complex. (*Left*) DNA from 129/129 spleen cells cut to completion with *Not*I and cut partially with increasing amounts of *Cla*I. The probe used is WII, a DNA fragment located proximally to the *H-2 K* locus (Weiss et al. 1984). (*Right*) DNA cut to completion with *Mlu*I and cut partially with *Cla*I. The same probe is used.

completely with *Not*I and partially with other enzymes. After PFG electrophoresis, blotting, and hybridization with the fragment from the linking clone, DNA fragments extending from the *Not*I site to any of the partially cleaved sites will hybridize, allowing the immediate derivation of the restriction map (Fig. 5).

Chromosome Linkup

The use of jumping and linking clones will allow the measurement of physical distances between markers and will provide markers close to genes, thus complementing the genetic analysis at a resolution of hundreds to thousands of kilobase pairs. Further analysis may require covering subregions by overlapping phage λ or cosmid clones. The highest-resolution analysis, DNA sequencing, has up to now been reserved for very short regions, covering, for example, the immediate locus of a gene. This graded increase in resolution (and effort) during the progress toward specific genes might however not be the optimal strategy for the analysis of longer DNA regions containing many genes, entire chromosomes, or the mammalian genome. Therefore, strategies to clone, analyze, and possibly sequence hu-

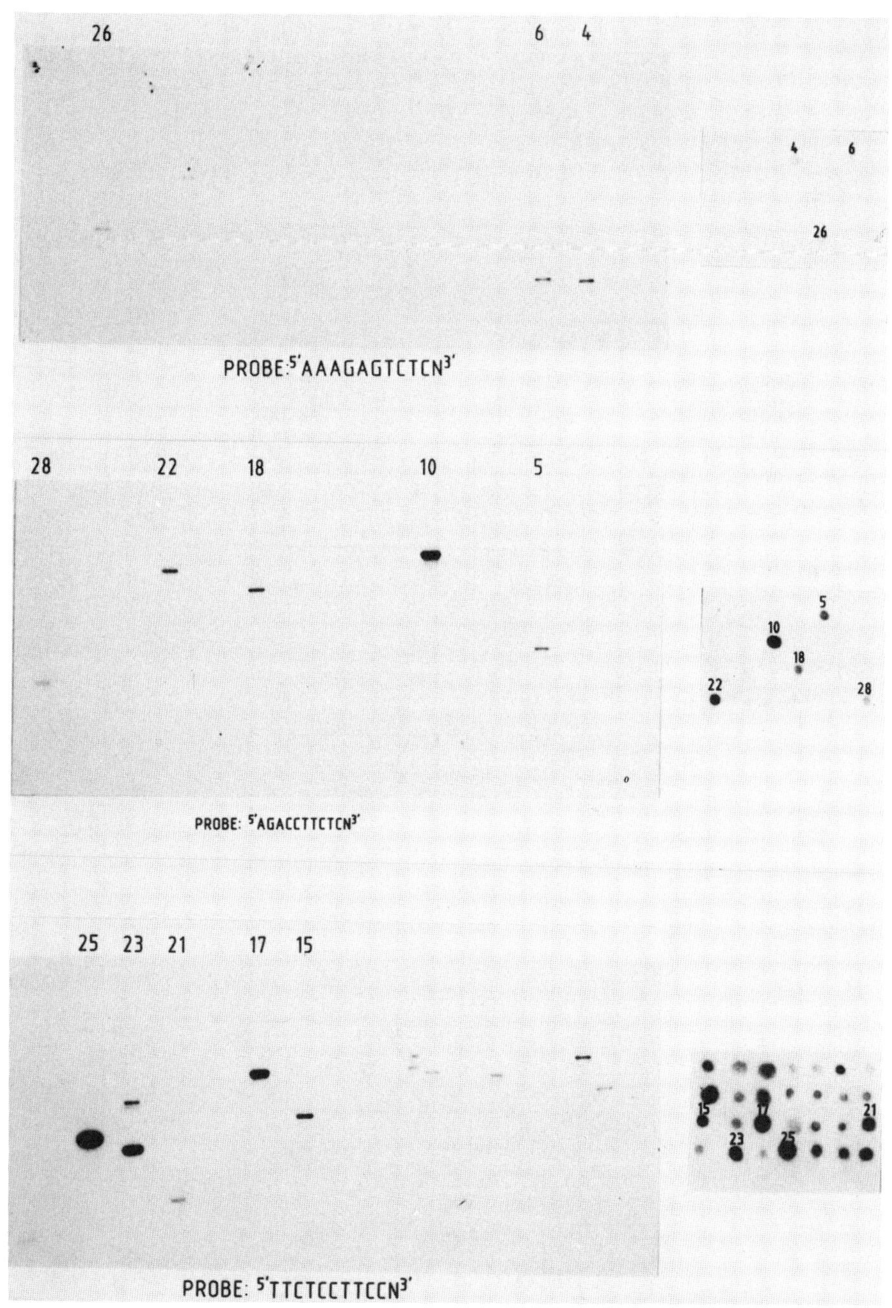

Figure 6. Hybridization of three oligonucleotide probes to Southern blots (*Eco*RI digests) and the corresponding dot-blots of DNA from 28 yeast cosmid clones.

man chromosomes or the entire human genome represented in the form of overlapping cosmid libraries might become attractive.

A number of such strategies have been considered and applied to small genomes. In the currently most efficient approach (Coulson et al. 1986), random cosmids are picked, DNA is prepared at small scale and cleaved by one enzyme (cutting approximately 10 times in the cosmid), the ends are radioactively labeled, and the DNA is recleaved with a second enzyme generating short DNA fragments. These fragments then are separated over sequencing gels, autoradiographed, and the length of the radioactive fragments is stored in the data bank. Overlapping cosmids are identified by restriction fragments of identical size.

Since the analysis of a mammalian genome by this approach would require the individual handling and analysis of at least 500,000 cosmids, we have attempted to develop alternative protocols reducing or eliminating the need to process individual cosmids. Though many technical difficulties still must be solved, we can report on prototype experiments and computer simulations.

The experiment we envisage is the following: First, DNA from individual cosmids or phages is applied to nylon membranes or generated in situ by phage or colony lysis. These filters are then hybridized with radiolabeled oligonucleotide probes chosen to hybridize to a significant fraction (10–50%) of the individual insert sequences. Conditions are designed to minimize the influence of base composition on melting temperature of the hybrid and to allow the selective scoring of perfectly matched hybrids (Wood et al. 1985). As shown later, probes expected to hybridize randomly to one in three cosmids appear to be well suited. After washing, hybridizing colonies are detected by autoradiography (Fig. 6) and the pattern of hybridization is read by computer, using an appropriate input device. Filters are then stripped of the old probe and the cycle is repeated, until 60 to 100 different probes have been

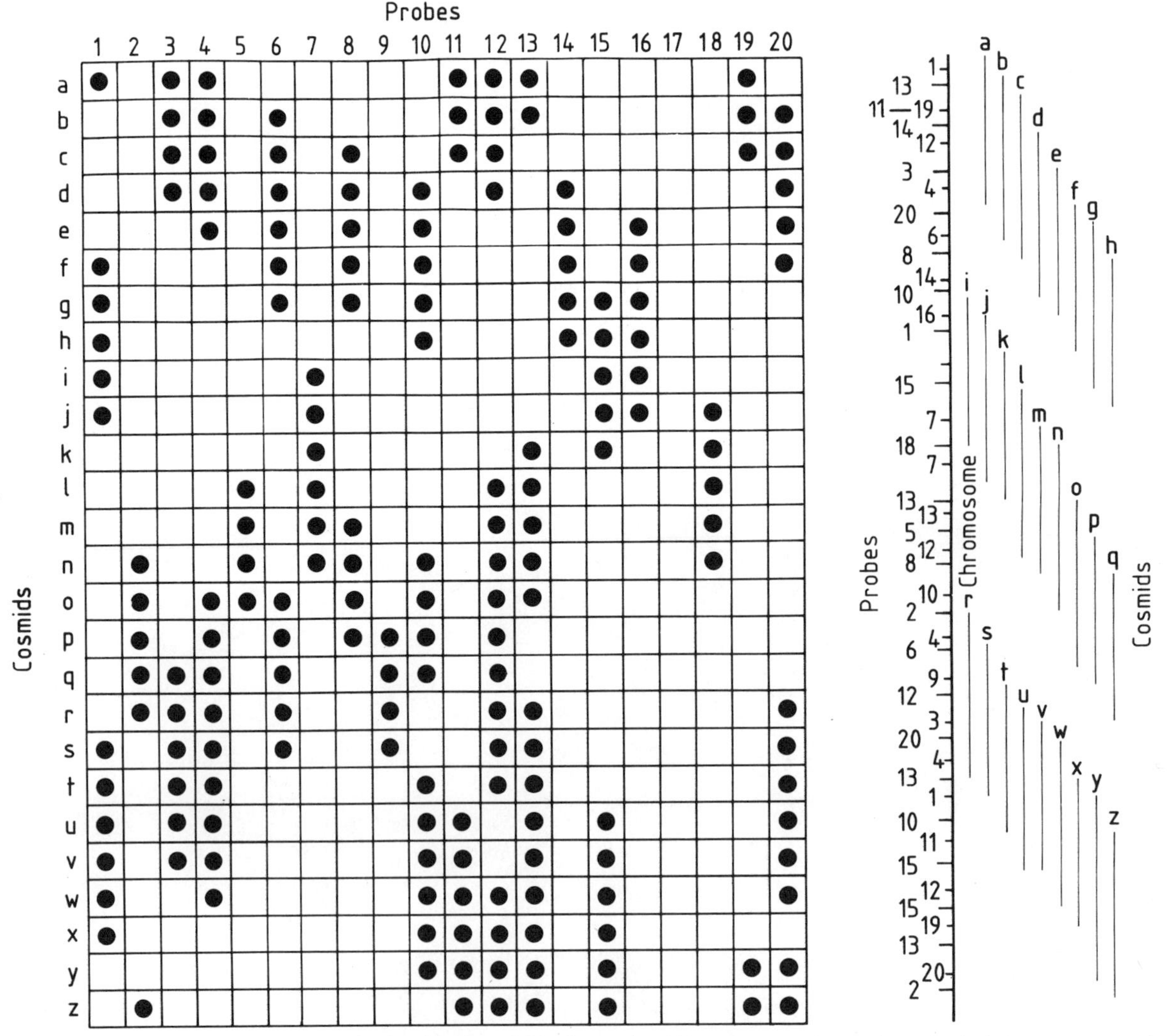

Figure 7. Schematic outline showing the localization of 20 oligonucleotide probes (1–20) on a hypothetical chromosome and the hybridization patterns of overlapping cosmids (a–z) to these oligonucleotides. The order of the cosmids can be deduced from the hybridization patterns.

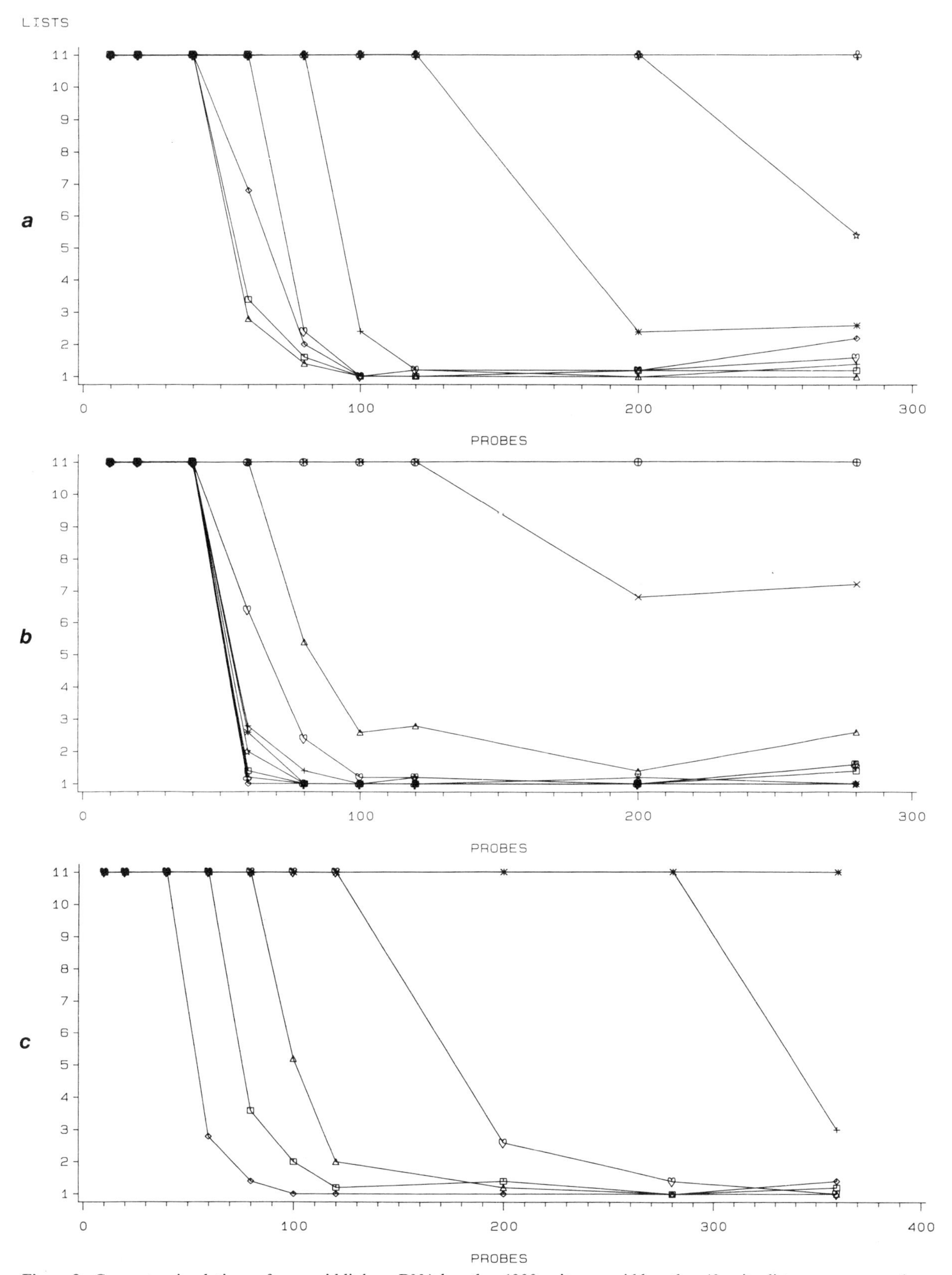

Figure 8. Computer simulations of a cosmid linkup. DNA length = 1000 units; cosmid length = 40 units; lists = average number from five simulations. (*a*) Dependence of the linkup on hybridization probabilities of oligonucleotide probes. Percentages of the average probability of hybridization of an oligonucleotide to a cosmid: (◇) 50%; (□) 40%, (△) 30%; (♡) 20%; (χ) 10%; (∗) 5%; (☆) 2%; (♧) 1%. (*b*) Dependence of the linkup on the number of cosmids. Symbols represent the number of cosmids covering the fixed genome size of 1000 kb; (⊕) 50; (×) 125; (△) 200; (♡) 275; (+) 350; (∗) 425; (☆) 500; (□) 1000; (◇) 2000. (*c*) Dependence of the linkup on the assumed error in the input data. Percentage of error: (◇) 0%; (□) 3%; (△) 5%; (♡) 10%; (+) 15%; (∗) 20%.

scored. If only binary information (hybridization vs. no hybridization) is scored, each cosmid will be characterized by a string of binary numbers in which each digest will provide information on one particular probe. Hybridization intensities might provide additional information and also reduce the sensitivity of the process to scoring errors. Overlapping cosmids will tend to have similar hybridization patterns, identified by the similarity between the binary numbers characterizing each cosmid. This has to be distinguished from the level of agreement expected randomly, a process reminiscent of identifying linked markers by similarities in their strain distribution pattern in recombinant, inbred strain mapping (Silver and Buckler 1986). If clusters of cosmids are being extended, additional information on the relative order of probes is available from their serial acquisition (or loss) by successive cosmids. This additional information can be used to further improve the identification of overlaps. The distribution of a small number of probes on a set of overlapping cosmids is illustrated schematically in Fig. 7.

To test the feasibility of this approach in a computer analysis and to determine the sensitivity of the protocol to different parameters, a number of Monte Carlo simulations were performed. In these, a number of cosmids distributed randomly over 1000-kb-long stretches of DNA were scored for hybridizing probes, their order was randomized, and, if possible, again deduced from the hybridization pattern. The success in this attempt is given as the number of clusters ("lists") identified. A score of 1 indicates perfect linkup. Scores higher than 10 are given as 11 (11 or more lists). Values represent the average of five simulations for each data point.

Figure 8a shows a family of curves, demonstrating the effect of the number of probes (*x* coordinate) on the number of cosmid clusters ("lists") for different hybridization probabilities (different curves). In this simulation the number of cosmids was kept constant at 350. It can be seen that at a hybridization probability of 30% the smallest number of probes are required for the experiment.

Figure 8b similarly shows the effect of the number of probes (*x* coordinate) on the number of lists (*y* coordinate) for different numbers of cosmids (different curves). The hybridization probability is kept constant at 0.3, the optimal value in the simulations shown before. As expected, there is a trade-off between the number of cosmids and the number of probes. At least 275 to 350 cosmids and at least 60 to 80 probes are required to give acceptable results. Linkup of larger genomes can be achieved with the same redundancy of cosmids and only 5 to 10 extra probes.

Figure 8c shows the influence of error on the simulated linkup. Again, the *x* coordinate shows the number of probes, and the *y* coordinate shows the number of resulting cosmid clusters. Different curves correspond to increasing random error introduced in scoring hybridizations. The results demonstrate that the approach should be quite insensitive to a reasonable level of error caused either by experimental difficulties or by heterozygosity in human DNA.

In our view these calculations demonstrate that even the relatively simple contingency table approach used in these programs is quite successful in analyzing these simulated data sets. We do expect that more-sophisticated programs, using Bayesian statistics, that are able to take into account additional information will allow further reduction in the amount of data required. Programs of this type are now being developed.

CONCLUSION

The technical difficulties hindering the analysis of the mammalian genome are being attacked from a number of sides, with major potential effects on our understanding of the mammalian genome and, ultimately, ourselves.

ACKNOWLEDGMENTS

We thank Cassandra Smith, Charles Cantor, and Francis Collins for discussions, Uta Francke and John Wasmuth for cell lines, and Margit Burmeister and Lisa Stubbs for contributions to the techniques described in this manuscript. This work was partially supported by Grant Le 496/2-1 from the Deutsche Forschungsgemeinschaft and by a collaborative research agreement funded by the Hereditary Disease Foundation.

REFERENCES

Bird, A.P. 1986. CpG-rich islands and the function of DNA methylation. *Nature* **321:** 209.

Botstein, D., R.L. White, M. Skolnick, and R.W. Davies. 1980. Construction of a genetic linkage map using restriction fragment length polymorphism. *Am. J. Hum. Genet.* **32:** 314.

Carle, G.F., M. Frank, and M.V. Olson. 1986. Electrophoretic separations of large DNA molecules by periodic inversions of the electric field. *Science* **232:** 65.

Casadaban, M.J. and S.N. Cohen. 1980. Analysis of gene control signals by DNA fusion and cloning in *Escherichia coli. J. Mol. Biol.* **138:** 179.

Collins, F.S. and S.M. Weissman. 1984. Directional cloning of DNA fragments at a large distance from an initial probe: A circularization method. *Proc. Natl. Acad. Sci.* **81:** 6812.

Coulson, A., J. Sulston, S. Brenner, and J. Karn. 1986. Towards a physical map of the genome of the nematode *Caenorhabditis elegans. Proc. Natl. Acad. Sci.* **83:** 7821.

Crampton, J.M., K.E. Davies, and T.F. Knapp. 1981. The occurrence of families of repetitive sequences in a library of cloned cDNA from human lymphocytes. *Nucleic Acids Res.* **9:** 3821.

Frischauf, A.-M., N.E. Murray, and H. Lehrach. 1986. Lambda phage vectors. *Methods Enzymol.* (in press).

Herrmann, B., M. Bucan, P.E. Mains, A.-M. Frischauf, L.M. Silver, and H. Lehrach. 1986. Genetic analysis of the proximal portion of the mouse t complex: Evidence for a second inversion within t haplotypes. *Cell* **44:** 469.

Huang, H.V., P.F.R. Little, and B. Seed. 1986. In *Vectors: A survey of molecular cloning vectors and their uses* (ed. R. Rodriguez). Butterworth, London. (In press.)

Levinson, A., D. Silver, and B. Seed. 1984. Minimal size plasmids containing an M13 origin for production of single stranded transducing particles. *J. Mol. Appl. Genet.* **2:** 507.

Murray, N.E. 1983. Phage lambda and molecular cloning. In *Lambda II* (ed. R.W. Hendrix et al.), p. 395. Cold Spring Harbor Laboratory, Cold Spring Harbor, New York.

Poustka, A.-M. and H. Lehrach. 1986. Jumping libraries and linking libraries: The next generation of molecular tools in mammalian genetics. *Trends Genet.* **2:** 174.

Rackwitz, H.-R., G. Zehetner, A.-M. Frischauf, and H. Lehrach. 1984. Rapid restriction mapping of DNA cloned in lambda phage vectors. *Gene* **30:** 195.

Schwartz, D. and C.R. Cantor. 1984. Separation of chromosome-sized DNAs by pulsed field gradient electrophoresis. *Cell* **37:** 67.

Silver, J. and C.E. Buckler. 1986. Statistical considerations for linkage analysis using recombinant inbred strains and backcrosses. *Proc. Natl. Acad. Sci.* **83:** 1423.

Smith, C.L., P.W. Warburton, A. Gaal, and C.R. Cantor. 1986. Analysis of genome organization and rearrangements by pulsed field gradient gel electrophoresis. In *Genetic engineering* (ed. J.K. Setlow and A. Hollaender), vol. 8, p. 45. Plenum Press, New York.

Smith, H.O. and M.L. Birnstiel. 1976. A simple method for DNA restriction site mapping. *Nucleic Acids Res.* **3:** 2387.

Weiss, E.H., L. Golden, K. Fahrner, A.L. Mellor, J.J. Devlin, H. Bullman, H. Tiddens, H. Bud, and R.A. Flavell. 1984. Organization and evolution of the class I gene family in the major histocompatibility complex of the C57B1/10 mouse. *Nature* **310:** 650.

Wood, W.I., J. Gitschier, L.A. Lasky, and R.M. Lawn. 1985. Base composition-independent hybridization in tetramethylammonium chloride: A method for oligonucleotide screening of highly complex gene libraries. *Proc. Natl. Acad. Sci.* **82:** 1585.

Flow Karyotyping and Sorting of Human Chromosomes

J.W. GRAY, J. LUCAS, D. PETERS, D. PINKEL,
B. TRASK, G. VAN DEN ENGH, AND M. VAN DILLA
Biomedical Sciences Division, Lawrence Livermore National Laboratory, Livermore, California 94550

Flow cytometry and sorting are becoming increasingly useful as tools for chromosome classification and for the detection of numerical and structural chromosome aberrations. Chromosomes of a single type can be purified with these tools to facilitate gene mapping or production of chromosome-specific recombinant DNA libraries (Gray et al. 1975, 1979a; Stubblefield et al. 1975; Carrano et al. 1979; Davies et al. 1981; Young et al. 1981; Sillar and Young 1981; Kunkel et al. 1982; Lebo 1982). For analysis of chromosomes with flow cytometry, the chromosomes are extracted from mitotic cells, stained with one or more fluorescent dyes, and classified one by one according to their dye content(s). Thus, the flow approach is fundamentally different than conventional karyotyping, where chromosomes are classified within the context of a metaphase spread. Flow sorting allows purification of chromosomes that can be distinguished by flow cytometry. We describe here the basic principles of flow cytometric chromosome classification (hereafter called flow karyotyping) and chromosome sorting, and we describe several recent applications. In addition, we present recent developments in slit-scan flow cytometry and high-speed flow sorting that promise to improve chromosome classification and purification.

Basic Principles

Chromosomes are dealt with one at a time during flow karyotyping and sorting. Therefore, the chromosomes must be extracted from mitotic cells and stained with one or more fluorescent dyes (usually DNA-specific) that allow discrimination between the various chromosome types.

Chromosome isolation is simple in concept. Typically, mitotic cells are accumulated (e.g., by treating exponentially growing monolayer cultures with mitotic blocking agents such as colcemid) for several hours. The cells are then swollen in hypotonic medium and the chromosomes are released into a stabilizing agent by mechanical shearing. A variety of stabilizing buffers have been used successfully, including magnesium sulfate (van den Engh et al. 1984, 1985), polyamines (Sillar and Young 1981), propidium iodide and other intercalating agents (Aten et al. 1980; Yu et al. 1981), and hexylene glycol (Gray et al. 1975).

Isolated chromosomes may be stained with a variety of fluorescent dyes. However, the best chromosome discrimination is achieved by using pairs of DNA-specific dyes (Gray et al. 1979a; Langlois et al. 1980) that exhibit DNA-base-composition preference, such as Hoechst 33258 (Ho; binds preferentially to AT-rich DNA) and chromomycin A_3 (CA_3; binds preferentially to GC-rich DNA). Alternately, DAPI can be used in place of Ho (Lebo 1982) and mithramycin can be used in place of CA_3 (Crissman and Tobey 1974).

Chromosomes stained with Ho and CA_3 are classified according to their stain contents by using dual-beam flow cytometry (Fig. 1). The stained chromosomes flow sequentially through two laser beams; one adjusted to the UV (351 + 363 nm) to selectively excite Ho and the other adjusted to 458 nm to selectively excite CA_3. The intensities of the two fluorescent flashes emitted as each chromosome traverses the two laser beams are recorded and added to a bivariate distribution showing the distribution of Ho and CA_3 contents among the chromosomes of the population being processed. The insert in Figure 1 shows the Ho vs. CA_3 distribution (flow karyotype) measured for chromosomes isolated from human cells. Numerous peaks are visible in the bivariate flow karyotype; most are produced by only one chromosome type. In fact, only the peaks for chromosomes 9–12 and sometimes 14 + 15 and 16 + 17 are composite. It is usual to achieve measurement precisions of about 2% while processing several hundred chromosomes per second using flow cytometry. Thus, analysis of several hundred thousand chromosomes takes only a few minutes.

Flow sorting allows purification of chromosomes that can be distinguished by flow cytometry as described in the legend to Figure 1. In commercially available sorters, it is possible to purify chromosomes of a single type at rates up to 20/sec. The purity of the sort depends on how well the desired chromosome type can be distinguished from other chromosomes or from chromosomal debris on the basis of its Ho and CA_3 contents. However, purities of up to 90% (as judged by flow karyotype analysis and in situ hybridization analysis of sorted chromosomes) are possible for well-resolved chromosomes.

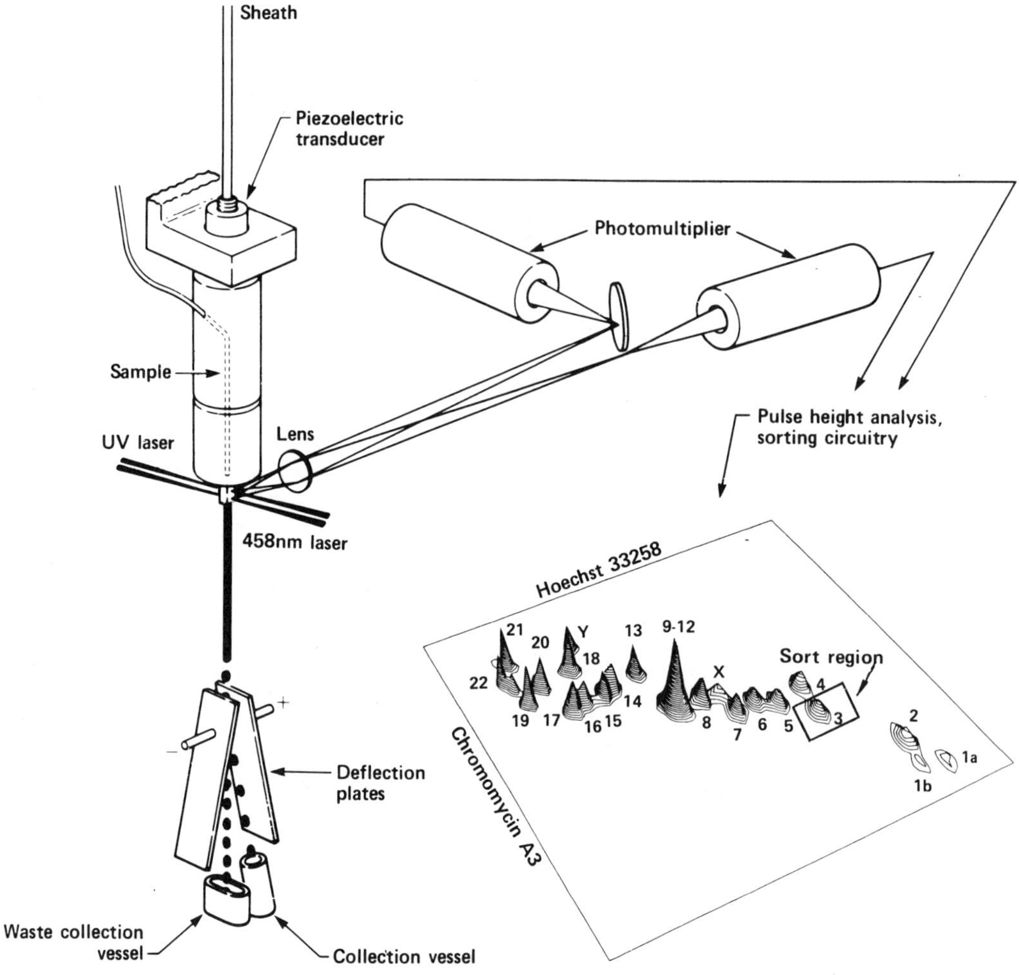

Figure 1. Diagram showing the principles of dual-beam flow cytometry and sorting. Isolated chromosomes stained with Ho and CA_3 flow one by one through two laser beams, one adjusted to UV to excite Ho and the other adjusted to 458 nm to excite CA_3. The resulting fluorescence flashes are detected photometrically and are recorded in a bivariate distribution. The insert shows a bivariate Ho vs. CA_3 distribution measured for chromosomes from karyotypically normal cells. Each peak is produced by a homogeneously staining group of chromosomes. The chromosomes producing each peak are indicated. After flow cytometric classification, the chromosomes continue down a narrow flow channel until they are ejected into air at high velocity in a thin liquid jet. The chromosomes are carried in the jet until they reach the point where they break into droplets. The droplets containing chromosomes to be sorted are charged as they separate from the jet so that the chromosomes to be sorted reside in charged droplets. The remaining droplets are uncharged and are either empty or contain chromosomes that are not wanted. The charged droplets separate from the others as they fall through an electric field and are collected for visual, biochemical, or biological analysis. Two classes of chromosomes can be sorted simultaneously by charging the droplets containing one class positively and charging the droplets containing the other class negatively.

Flow Karyotyping

Classification and purification of chromosomes from a variety of human tissues has become increasingly practical during the past few years as procedures for chromosome isolation and staining have developed. For example, we have measured flow karyotypes for a variety of human cell types, including human fibroblasts, lymphocytes, amniocytes, chorionic villus cells, colon carcinoma cells, and Burkitt's lymphoma cells (Yu et al. 1981; Langolis et al. 1982; Gray et al. 1984; van den Engh et al. 1985; J. Gray et al., in prep.). These flow karyotypes measured on one instrument for human chromosomes from karyotypically normal individuals have proved to be remarkably reproducible from day to day, from person to person, and among the various tissues. The relative peak means have been shown to vary by less than about 0.5% from day to day in distributions measured for chromosomes from the same cell type or individual (Langlois et al. 1982). The variation among individuals is somewhat larger but is sufficiently small to allow detection of alterations in chromosomal structure that cause changes in DNA content or base composition as small as 10^6 bases. Figure 2 shows 95%-tolerance ellipses describing the variation in peak means observed in flow karyotypes measured for chromosomes from 50 karyotypically normal primary human amniocyte cultures (J. Gray et al., in prep.). The ellipses do not overlap except for chromosomes 9–12 and 14 + 15. Thus, normal chromosomes

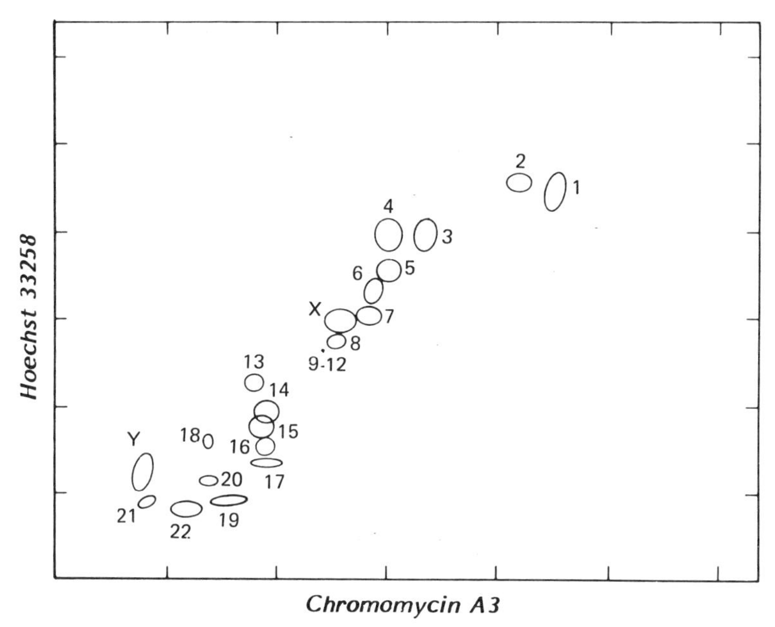

Figure 2. Regions with 95% tolerance, showing the variability in the means of the peaks in bivariate flow karyotypes. These tolerance regions were calculated using peak means determined from analysis of flow karyotypes from 50 different, karyotypically normal amniocyte cell cultures (J. Gray et al., in prep.).

can be classified according to their relative peak means, and structurally aberrant chromosomes can be recognized as producing peaks falling outside of one of the 95%-tolerance ellipses.

The flow karyotypes also provide information about the relative frequency of each chromosome group in the population. Specifically, the volume of each chromosome peak is proportional to the relative frequency of that group in the population (J. Gray et al., in prep.). Figure 3 shows the relative peak volumes calculated for the peaks in a fluorescence distribution measured for a karyotypically normal individual and normalized so that the volume of the peak for chromosomes 9–12 is 8. The volumes of the autosomal chromosomes for

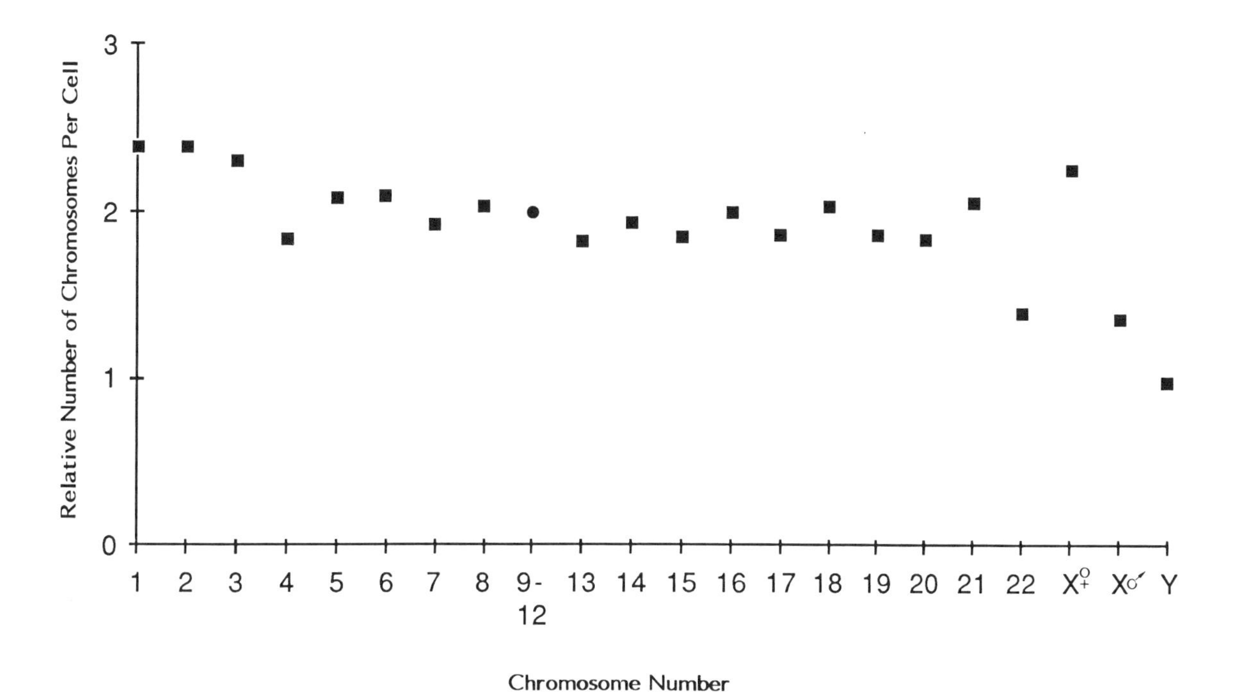

Figure 3. Relative volumes of the peaks in a flow karyotype measured for a karyotypically normal cell culture. The peak volumes are normalized so that the volume of the peak for the 9–12 chromosomes is 8.0.

three different flow karyotypes (excluding chromosomes 9–12) average 1.99, while the average volumes of the peaks for chromosomes X and Y are 1.36 and 0.98, respectively, for male cells. Only one peak (for chromosome 22) shows a volume that is not within one-half chromosome unit of its expected value (1.4 measured vs. 2.0 expected).

In summary, the peak means in bivariate flow karyotypes are sufficiently reproducible to allow classification of normal chromosomes according to their peak means. The relative frequencies of chromosomes can be determined from the relative volume of their peaks in the fluorescence distribution. This information is essentially the same as that derived from banding analysis and justifies the term "flow karyotype" to describe the Ho vs. CA_3 fluorescence distribution measured flow cytometrically.

Aberrant Chromosome Detection

Homogeneously occurring aberrant chromosomes can be detected in flow karyotypes because of the change(s) in the peak means and/or peak volumes that result from the aberrations. Flow karyotypes can be measured quickly and quantitatively and are amenable to automation. For these reasons, flow karyotyping is appealing as an adjunct to banding analysis for prenatal detection of disease-linked aberrant chromosomes and for detection of diagnostically important marker chromosomes in tumor cells.

Numerical aberration detection. Numerical chromosome aberrations (e.g., trisomies and monosomies) are detected by analysis of the relative peak volumes. A trisomy involving an autosomal chromosome should produce a peak increased in volume by 50%, while a monosomy should cause a reduction in peak volume by a factor of 2. The detection of chromosomal trisomies is illustrated in Figure 4, which shows bivariate flow karyotypes measured for three human amniocyte cultures. Figure 4a shows a normal culture. Amniocyte cultures carrying trisomies of chromosomes 21 and 18 are shown in Figure 4b and c, respectively. The peaks for the trisomic chromosomes are clearly increased in volume in both abnormal flow karyotypes. The numbers in parentheses show the volumes for the trisomic peaks and for some of the nearby peaks; all are normalized so that the volume of the peak for chromosome 9–12 is 8. The volumes for the peaks for the autosomal chromosomes are close to 2, except those for the trisomic chromosomes, which are greater than 3. This technique has provided accurate trisomy detection in more than 20 cultures carrying trisomies involving chromosomes 21, 18, and Y (J. Gray et al., in prep.). The principal limitations to the technique are the following:

1. Accurate peak volume analysis requires high-quality flow karyotypes showing a minimal contribution due to chromosomal or nuclear debris in the vicinity of the small chromosomes (often visible as a smooth continuum underlying the peaks and decreasing with increasing DNA content).
2. Flow karyotyping is poorly suited to detection of mosaic populations in which only a fraction of the cells are monosomic or trisomic.

Structural aberration detection. Structural aberrations usually result in derivative chromosomes with altered DNA content and/or altered DNA base composition. As a result, the peaks for these derivative chromosomes often fall outside of a 95%-tolerance region and the volumes of the peaks for the remaining normal chromosomes are reduced (J. Gray et al., in prep.). These concepts are illustrated in Figure 5, which shows the flow karyotypic rearrangements caused by a balanced translocation between chromosomes 14 and 15 in human amniocytes. The flow karyotype shows two new peaks, marked T1 and T2, that result from the translocation. In addition, the peaks for normal chromosomes 14 and 15 are reduced in volume by one chromosome equivalent each. Flow karyotyping is also well-suited to detection of marker chromosomes in karyotypically unstable human malignancies. Figure 5b, for example, shows a flow karyotype for a Burkitt's lymphoma cell line. Distinct peaks for two marker chromosomes (labeled $14q^+$ and 15^-) are clearly visible, in spite of the existence of considerable karyotypic instability among the cells of the culture. However, the karyotypic instability in this line precludes the use of peak volumes as a guide to the loss or gain of normal chromosomes.

Chromosome Sorting

Flow sorting allows collection of most human chromosomes with purities unequaled by other methods. As a result, chromosomal material purified by sorting has been useful in several gene mapping efforts (Lebo et al. 1979; Lebo 1982) and for production of recombinant DNA libraries (Davies et al. 1981; Krumlauf et al. 1982; Kunkel et al. 1982; Lalande et al. 1984; Van Dilla et al. 1986; Deaven et al., this volume). The main determinant of the purity with which chromosomes can be sorted is the degree to which that chromosome type can be resolved during flow karyotyping. The flow karyotype in Figure 1, for example, shows chromosomes 13, 18–22, and Y to be particularly well resolved. Chromosomes of these types can be sorted with high purity as long as the presence of DNA fragments in the immediate vicinity of these chromosome peaks is minimal. Such debris may come from disrupted interphase cells or from chromosomes destroyed by excessive shearing during chromosome isolation. The debris may be minimized by optimizing cell culture and swelling and/or by monitoring chromosome morphology during the shearing process. Peaks due to chromosomes 1–8 and X are rather close together so that purification of these chromosomes is more difficult. Purification

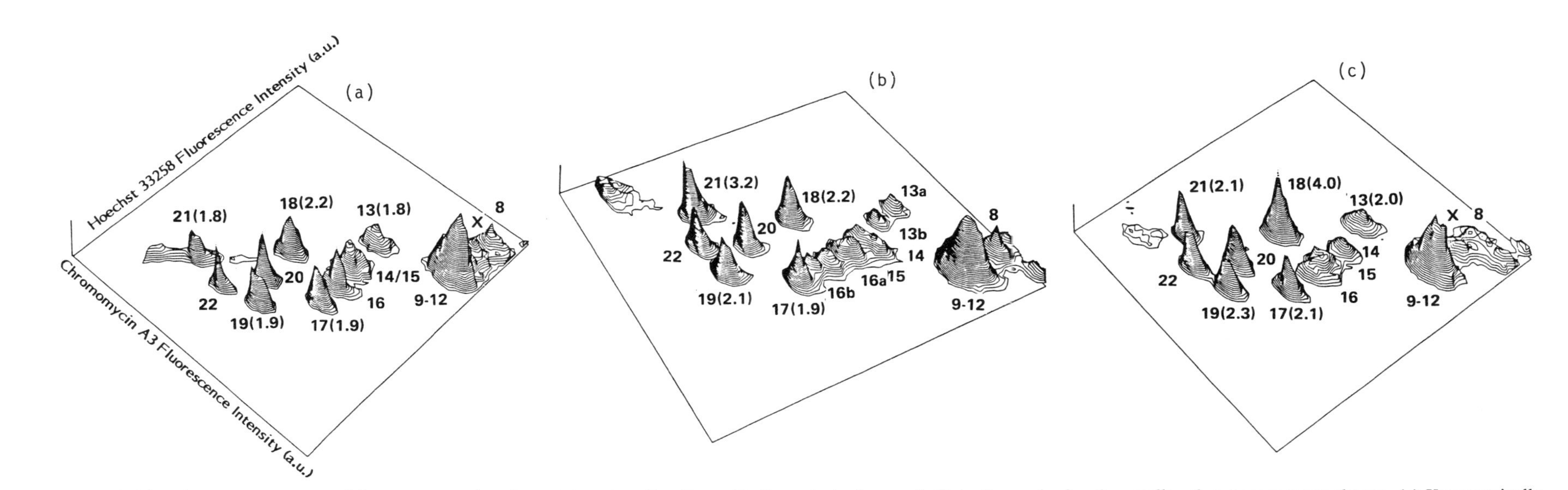

Figure 4. Flow karyotypes measured for one normal and two trisomic cell cultures (J. Gray et al., in prep.). Only the peaks for the smaller chromosomes are shown. (*a*) Karyotypically normal culture. (*b*) Trisomy 21. (*c*) Trisomy 18. The numbers in parentheses near some of the peaks indicate the relative volume of those peaks, normalized so that the volume of the peak for chromosome 9–12 is 8.

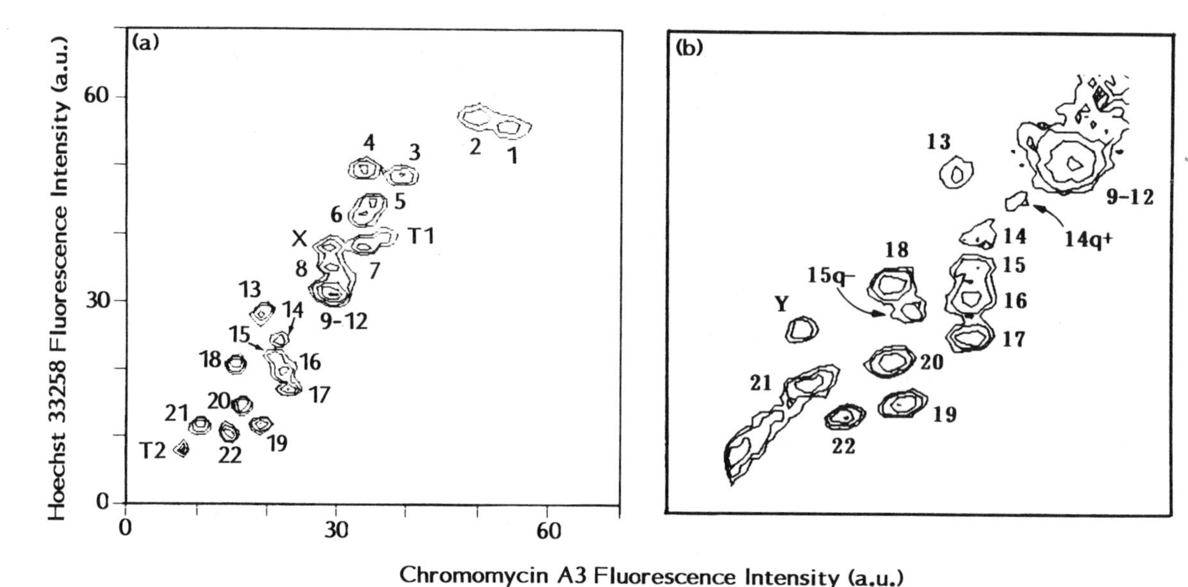

Figure 5. Flow karotypes for cells carrying structural aberrations. (*a*) Flow karyotype for a culture carrying two derivative chromosomes resulting from a translocation between chromosomes 14 and 15 (J. Gray et al., in prep.). The two derivative chromosomes are labeled T1 and T2. The numbers in parentheses show the relative volumes of several peaks; the other numbers show the chromosomes producing the various peaks. (*b*) Flow karyotype measured for a Burkitt's lymphoma (Gray and Langlois 1986). Two marker chromosomes are apparent (labeled $14q^+$ and $15q^-$).

of chromosomes 9–12 is impossible from human chromosome suspensions stained with Ho and CA_3 since they cannot be separately resolved. In these situations, chromosomes of the desired type may be purified from human/hamster hybrids containing only one or a few human chromosomes. Figure 6a, for example, shows a Ho vs. CA_3 flow karyotype measured for a Chinese hamster cell line. This figure also shows the peak locations expected for several larger human chromosomes in hybrids made using this Chinese hamster line as the parent. Most larger human chromosomes can be distinguished from the hamster chromosomes. Figure 6b shows a flow karyotype measured for a hybrid cell line containing only human chromosomes 4, 8, and 21. Chromosomes 4 and 8 are clearly resolved from all other chromosomes and can be readily purified by sorting.

Approximately 30,000 chromosomes (sufficient for gene mapping using spot-blot hybridization; Lebo et al. 1984) can be purified by sorting in about 20 min whereas 1 μg of chromosomal DNA (sufficient for production of complete digest recombinant DNA libraries; Deaven et al., this volume) can be purified from a chromosome of average size in about 50 hr by using a conventional sorter. The size of the DNA isolated from sorted chromosomes has been shown to be in excess of 50 kb (Peters et al. 1985). (Recent studies have shown the size of DNA from sorted chromosomes to be at least several hundred kilobases.) Thus, DNA from chromosomes purified by sorting is suitable for production of recombinant DNA libraries in which the insert size of the DNA to be cloned is less than about 20 kb. The principal limitations to chromosome purification by sorting come from (1) the difficulty in maintaining high resolution during long sorting runs (so that the sort purity remains high), (2) the extended sorting times required for production of microgram quantities of DNA, and (3) the possible contamination of sorted chromosomal material by DNA-containing debris fragments.

New Technical Developments

Slit-scan flow cytometry and high-speed sorting may lead to substantially improved flow karyotyping and sorting by allowing improved chromosome discrimination and by increasing the rate at which chromosomes can be purified.

Slit-scan flow cytometry. Slit-scan flow cytometry allows classification of chromosomes according to their shape as well as their DNA content (Gray et al. 1979b; Lucas et al. 1983; Lucas and Pinkel 1986). In this approach, illustrated schematically in Figure 7, the chromosomes flow lengthwise through a thin (e.g., 1.3 μm) laser beam that excites the dye with which the chromosomes have been stained (Lucas and Pinkel 1986). Fluorescence intensity is recorded as the chromosomes flow across the scanning beam. If the dye is DNA-specific, the recorded profile shows the distribution of DNA along the chromosome. Since chromosomal DNA content is reduced at the centromere, the centromere location can be determined by the location of the dip in the recorded profile. Thus, each chromosome profile contains two chromosome descriptors: the total DNA content (proportional to the area under the profile) and the chromosome centromeric index (CI = area

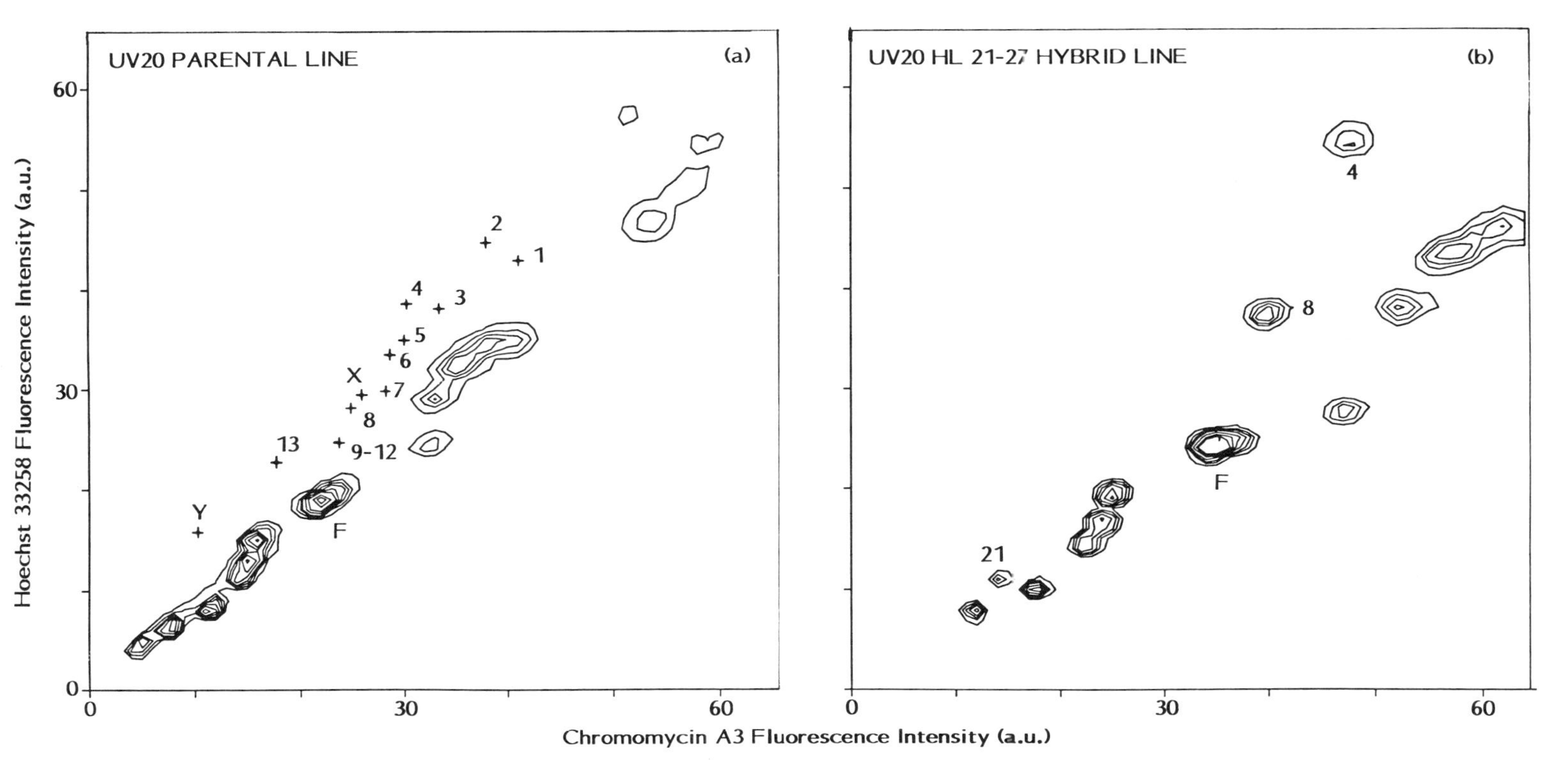

Figure 6. Flow karyotypes for human/hamster hybrid cells. (*a*) Flow karyotype for Chinese hamster ovary cells (line UV20). Also shown are the peak locations expected for human chromosomes in human/hamster hybrids made using UV20 as the hamster parent. (*b*) Flow karyotype measured for a human/hamster hybrid (line UV20HL21-27) carrying human chromosomes 4, 8, and 21. The peaks produced by the human chromosomes are indicated (Van Dilla et al. 1986).

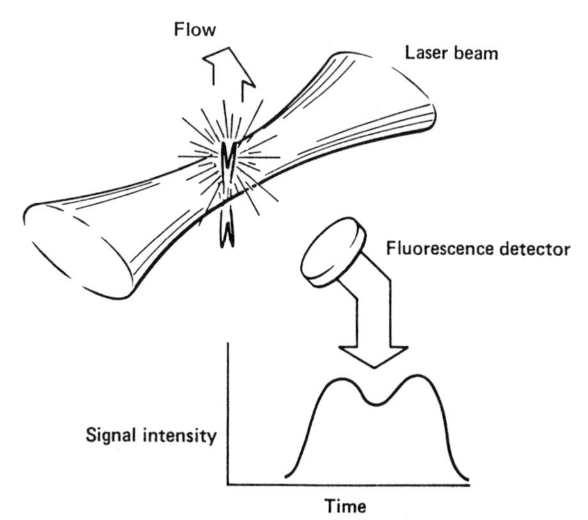

Figure 7. Schematic illustration showing the principles of slit-scan flow cytometry. Fluorescently stained chromosomes flow lengthwise through a thin (typically 1.3 μm) laser beam. Fluorescence is recorded during the passage of each chromosome. For chromosomes stained with DNA-specific dyes, the profile shows the distribution of DNA along the scanned chromosome. Such profiles typically show a dip at the centromere location due to the reduced DNA content in this region.

under long arm/total area). Figure 8 shows a bivariate distribution of DNA vs. CI measured for a karyotypically normal amniocyte cell culture (Lucas and Gray 1986). The DNA vs. CI flow karyotype resolves several of the larger human chromosomes that are not well-resolved in Ho vs. CA_3 flow karyotypes (e.g., 1–3). In addition, chromosomes 9–12 are partially resolved in the DNA vs. CI karyotype, whereas they are not resolved in the Ho vs. CA_3 flow karyotype. Since slit-scan chromosome classification does not allow discrimination of as many chromosome groups as does Ho vs. CA_3 classification, the slit-scan approach seems especially attractive if it can be combined with Ho and CA_3 flow karyotyping to allow classification of chromosomes according to Ho, CA_3, and CI.

High-speed sorting. One of the main limitations of chromosome purification by sorting is the relatively small amount of chromosomal material that can be obtained using commercially available sorters in a reasonable period of time. This limitation has been overcome to some extent by the development of a dual-beam high-speed sorter (HiSS; Peters et al. 1985) that operates approximately 1 order of magnitude faster than conventional sorters. HiSS is conceptually similar to conventional dual-beam sorters (see Fig. 1). However, compared with conventional sorters, HiSS operates at higher pressure (200 psi vs. ∼10 psi), the jet velocity is higher (50 m/sec vs. 10 m/sec), and the frequency at which droplets are made is higher (∼200,000/sec vs. ∼20,000/sec). As a result, chromosomes can be sorted as much as an order of magnitude faster by using HiSS. Our experience in the use of HiSS for purification of chromosomes stained with Ho and CA_3 has shown that chromosomes of a single type can be sorted at rates up to 200/sec (Van Dilla et al. 1986; Deaven et al., this volume). Thus, purification of microgram quantities can be accomplished in less than a day by using HiSS. This makes HiSS particularly attractive for production of DNA for production of

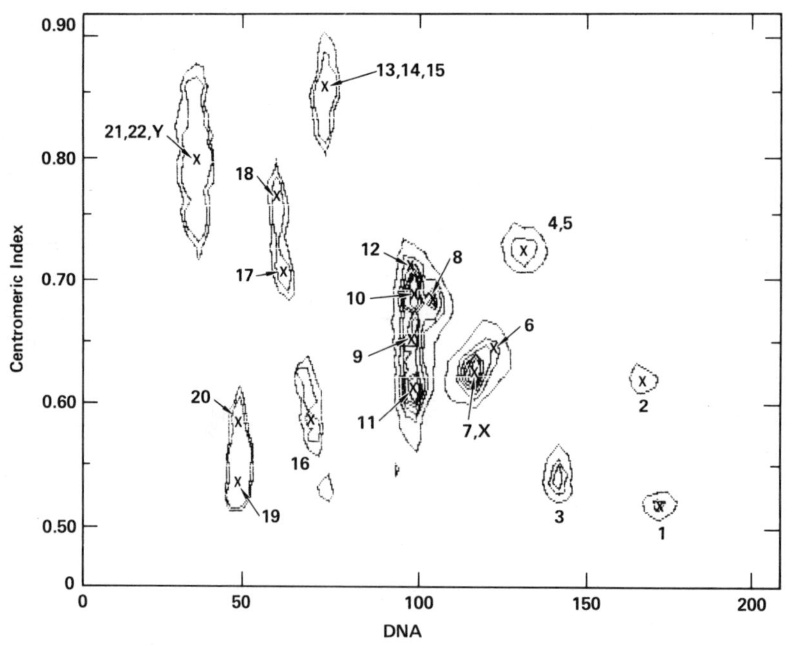

Figure 8. Flow karyotype showing DNA content vs. CI for human chromosomes measured by slit-scan flow cytometry. Each peak is produced by a group of chromosomes with similar DNA content and CI. The numbers indicate the chromosomes producing each peak (Lucas and Gray 1986).

large-insert recombinant DNA libraries since the production of these libraries is likely to require more DNA than the smaller-insert libraries produced to date.

ACKNOWLEDGMENTS

This work was performed under the auspices of the U.S. Department of Energy by the Lawrence Livermore National Laboratory under contract number W-7405-ENG-48 with support from grants HD 17665 and GM 25076 from the National Institutes of Health.

REFERENCES

Aten, J., J. Kipp, and E. Barendsen. 1980. Flow cytometric determination of damage to chromosomes from X-irradiated Chinese hamster cells. In *Flow cytometry* (ed. O. Laerum et al.), vol. 4, p. 287. Universitetsforlaget, Bergen.

Carrano, A., J. Gray, R. Langlois, K. Burkhart-Schultz, and M. Van Dilla. 1979. Measurement and purification of human chromosomes by flow cytometry and sorting. *Proc. Natl. Acad. Sci.* **76:** 1382.

Crissman, H. and R. Tobey. 1974 Cell-cycle analysis in 20 minutes. *Science* **184:** 1297.

Davies, K., B. Young, R. Elles, M. Hoill, and R. Williamson. 1981. Cloning of a representative genomic library of the human X chromosome after sorting by flow cytometry. *Nature* **293:** 364.

Gray, J. and R. Langlois. 1986. Chromosome classification and purification using flow cytometry and sorting. *Annu. Rev. Biophys. Chem.* **15:** 195.

Gray, J., J. Lucas, L. Yu, and R. Langlois. 1984. Flow cytometric detection of aberrant chromosomes. In *Biological dosimetry* (ed. W. Eisert and M. Mendelsohn), p. 25. Springer-Verlag, New York.

Gray, J., R. Langlois, A. Carrano, K. Burkart-Schultz, and M. Van Dilla. 1979a. High resolution chromosome analysis: One and two parameter flow cytometry. *Chromosoma* **73:** 9.

Gray, J., D. Peters, J. Merrill, R. Martin, and M. Van Dilla. 1979b. Slit-scan flow cytometry of mammalian chromosomes. *J. Histochem. Cytochem.* **27:** 441.

Gray, J., A. Carrano, L. Steinmetz, M. Van Dilla, D. Moore, B. Mayall, and M. Mendelsohn. 1975. Chromosome measurement and sorting by flow systems. *Proc. Natl. Acad. Sci.* **72:** 1231.

Krumlauf, R., M. Jeanpierre, and B. Young. 1982. Construction and characterization of genomic libraries from specific human chromosomes. *Proc. Natl. Acad. Sci.* **79:** 2971.

Kunkel, L., U. Tantravahi, M. Eisenhard, and S. Latt. 1982. Regional localization on the human X of DNA segments cloned from flow sorted chromosomes. *Nucleic Acids Res.* **10:** 1557.

Lalande, M., L. Kunkel, A. Flint, and S. Latt. 1984. Development and use of metaphase chromosome flow sorting methodology to obtain recombinant phage libraries enriched for parts of the human X chromosome. *Cytometry* **5:** 101.

Langlois, R., A. Carrano, J. Gray, and M. Van Dilla. 1980. Cytochemical studies of metaphase chromosomes by flow cytometry. *Chromosoma* **37:** 229.

Langlois, R., L. Yu, J. Gray, and A. Carrano. 1982. Quantitative karyotyping of human chromosomes by dual beam flow cytometry. *Proc. Natl. Acad. Sci.* **79:** 7876.

Lebo, R. 1982. Chromosome sorting and DNA sequence localization. *Cytometry* **3:** 145.

Lebo, R., A. Carrano, K. Burkhart-Schultz, A. Dozy, L. Yu, and K. Yan. 1979. Assignment of human β-, γ-, and δ-globin genes to the short arm of chromosome 11 by chromosome sorting and DNA retriction analysis. *Proc. Natl. Acad. Sci.* **76:** 5804.

Lebo, R., F. Gorin, R. Fletterick, F. Kao, M. Cheng, B. Bruce, and Y. Kan. 1984. High-resolution chromosome sorting and DNA spot-blot analysis assign McArdle's syndrome to chromosome 11. *Science* **225:** 57.

Lucas, J.N. and J.W. Gray. 1986. Centromeric index versus DNA content flow karotypes of human chromosomes measured using slit-scan flow cytometry. *Cytometry* (in press).

Lucas, J. and D. Pinkel. 1986. Orientation measurements of microsphere doublets and metaphase chromosomes in flow. *Cytometry* (in press).

Lucas, J., J. Gray, D. Peters, and M. Van Dilla. 1983. Centromeric index measurement by slit-scan flow cytometry. *Cytometry* **4:** 109.

Peters, D., E. Branscomb, P. Dean, J. Merril, D. Pinkel, M. Van Dilla, and J. Gray. 1985. The LLNL high speed sorter: Design features, operational characteristics and biological utility. *Cytometry* **6:** 290.

Sillar, R. and B. Young. 1981. A new method for preparaton of metaphase chromosomes for flow analysis. *J. Histochem. Cytochem.* **29:** 74.

Stubblefield, E., L. Cram, and L. Deaven. 1975. Flow microfluorometric analysis of isolated Chinese hamster chromosomes. *Exp. Cell Res.* **94:** 464.

van den Engh, G., B. Trask, S. Cram, and M. Bartholdi. 1984. Preparation of chromosome suspensions for flow cytometry. *Cytometry* **5:** 108.

van den Engh, G., B. Trask, J. Gray, R. Langlois, and L. Yu. 1985. Preparation and bivariate analysis of suspensions of human chromosomes. *Cytometry* **6:** 92.

Van Dilla, M., L. Deaven, K. Albright, N. Allen, M. Aubuchon, M. Bartholdi, N. Brown, E. Campbell, A. Carrano, L. Clark, L. Cram, B. Crawford, J. Fuscoe, J. Gray, C. Hildebrand, P. Jackson, J. Jett, J. Longmire, C. Lozes M. Luedemann, J. Martin, J. McNinch, L. Meincke, M. Mendelsohn, J. Meyne R. Moyzis, A. Munk, J. Perlman, D. Peters, A. Silva, and B. Trask. 1986. Human chromosome-specific DNA libraries: Construction and availability. *Biotechnology* **4:** 537.

Young, B., M. Feguson-Smith, R. Sillar, and E. Boyd. 1981. High-resolution analysis of human peripheral lymphocyte chromosomes by flow cytometry. *Proc. Natl. Acad. Sci.* **78:** 7727.

Yu, L., J. Aten, J. Gray, and A. Carrano. 1981. Human chromosome isolation from short-term lymphocyte culture for flow cytometry. *Nature* **293:** 154.

Cytogenetic Analysis by In Situ Hybridization with Fluorescently Labeled Nucleic Acid Probes

D. PINKEL,* J.W. GRAY,* B. TRASK,* G. VAN DEN ENGH,* J. FUSCOE,* AND H. VAN DEKKEN†

*Biomedical Sciences Division, Lawrence Livermore National Laboratory, Livermore, California 94550;
†Radiobiological Institute, TNO, The Netherlands

The ability to detect cytogenetic abnormalities has increased as improved methods for chromosome identification have been developed. Banding techniques (Casperson et al. 1968; Yunis 1981) now allow identification of all human chromosomes as well as detection of a broad spectrum of structural and numerical chromosome aberrations. However, banding analysis has important limitations. It requires preparation of high-quality metaphase spreads and the attendant cell culture; it is not easily automated; and it requires time-consuming work by skilled observers. These limitations affect the application of banding analysis in a variety of clinical and experimental settings. Prenatal detection of aberrations is slow and laborious because of the cell culture and scoring process. Quantitative analysis of the frequency of translocated chromosomes for biological dosimetry is also hampered by the lengthy scoring process. Tumor cytogenetic analysis is limited by the difficulty of preparing high-quality metaphase spreads that are representative of the tumorigenic cells in the population. Banding is not useful for investigating the locations of chromosomes in interphase nuclei.

A new approach to chromosome identification—in situ hybridization with chromosome-specific nucleic acid probes—promises to overcome some of the limitations of banding and offers new opportunities. Essential to this approach to chromosome staining are nucleic acid probes that are homologous with DNA sequences located predominantly on specific chromosomes or chromosomal regions. These may either be probes for repetitive sequences or consist of collections of probes for unique sequences. Repetitive probes that bind predominantly to one chromosome type have been found for many of the human chromosomes (Yang et al. 1982; Graham et al. 1984; Jabs et al. 1984; Lau and Schonberg 1984; Burk et al. 1985; Devine et al. 1985; Waye and Willard 1985; Devilee et al. 1986; R. Moyzis et al., in prep.), and new ones are being discovered continually.

The use of probes that are modified chemically rather than radioactively (Langer et al. 1981; Landegent et al. 1984; Ruth and Bryan 1984; Forster et al. 1985; Hopman et al. 1986) has advantages for chromosome labeling. Hybridization is detected with reagents that bind with high affinity to the chemically modified probe and carry various types of reporter molecules. Among the reporting systems that have proved useful are enzymes that form precipitates in the presence of substrate, fluorescent dyes, and gold particles. These allow localization of the bound probe with higher resolution than is possible using autoradiography of radioactive probes. In addition, discrimination of the hybridization of more than one probe to the same target becomes possible by using a different chemical modification and reporting molecule for each probe. We have focused on the use of fluorescent labeling because it is particularly suited for quantitative analysis through intensity measurement and high-resolution imaging. We will call this procedure fluorescence hybridization.

In this paper we demonstrate the application of fluorescence hybridization to the: (1) rapid screening of interspecies hybrid cell lines for retained chromosomes and chromosomal rearrangements; (2) use of human chromosome-specific probes to label selected chromosomes in interphase nuclei; (3) study of the three-dimensional chromosomal organization in nuclei; and (4) flow cytometric measurement of hybridization to human cell nuclei. In addition, we speculate on the utility of this technology to detection of clinically important aberrations in metaphase spreads, to detection of aneuploidy in human interphase nuclei, and to analysis of the architecture of interphase nuclei.

MATERIALS AND METHODS

Cell and chromosome preparations. Human chromosome spreads fixed with methanol-acetic acid (3:1) were prepared according to the procedure of Harper et al. (1981), using methotrexate synchronization. Similarly fixed metaphase spreads of human/hamster hybrid cells were prepared from mitotic cells collected during a 4–6 hr colcemid block. Unstimulated human lymphocytes were separated from peripheral blood with Lymphocyte Separation Medium (Litton Bionetics), fixed in methanol-acetic acid, and dropped on slides. Sorted chromosomes were cyto-centrifuged onto slides and immersed in methanol-acetic acid. Slides containing cells or chromosomes were enclosed in sealed plastic bags containing nitrogen gas and stored at −20°C until used. The slides were baked in air at 65°C for 4 hr prior to either storage or hybridization to preserve chromosome morphology during hybridization. Nuclei for suspension hybridization were isolated, fixed in ethanol, and treated with 0.1 N HCl and 0.05% Triton X-100 (B. Trask et al., in prep.).

Probe modification. DNA probes were chemically modified by nick translation with biotin-labeled deoxyuridine triphosphate (dUTP; Langer et al. 1981) or by treatment with 2-acetylaminofluorene (AAF). DNA probes were biotinylated by nick-translation according to the protocol of the supplier (Bethesda Research Laboratories) except that the amount of biotin-modified nucleotides was sometimes doubled. Approximately 10–30% of the thymidine in the probes was substituted with biotin-dUTP with this procedure. AAF modification was performed according to the procedure of Landegent (Landegent et al. 1984).

In situ hybridization. Hybridization on slides was carried out as described (Harper et al. 1981; Pinkel et al. 1986). Briefly, cells and chromosomes were treated with RNase (100 μg/ml; 37°C; 1 hr.) and denatured (70% formamide, 2×SSC; 70°C; 2 min [1× SSC is 0.15 M NaCl, 0.015 M sodium citrate]). The hybridization mix (50% formamide, 2×SSC, 10% dextran sulfate, 1 mg/ml herring sperm DNA, and probe DNA) was applied, a coverslip was added and sealed with rubber cement, and the slide was placed at 37°C. Hybridizations using genomic DNA as a probe were typically accomplished during a 2-hr incubation at a probe concentration of 1 μg/ml. The RNase treatment was usually omitted in these preparations. Chromosome-specific repetitive probes were used at 0.1–0.5 μg/ml and the hybridization was allowed to take place overnight. After hybridization, slides were washed in 50% formamide, 2×SSC (pH 7), followed by 2× SSC and finally the same buffer in which the first cytochemical detection reagent was carried. All washes were at 45°C.

Hybridization to cells in suspension was basically similar to that described by Trask et al. (1985), with the addition of 10% dextran sulfate to the hybridization mixture. Briefly, fixed nuclei were suspended in the hybridization mix as above, heated at 70°C for 10 min to denature both probe and target, and incubated overnight at 37°C. After hybridization, the nuclei were washed (50% formamide, 2× SSC) for 10 min at 45°C, followed by 10 min at room temperature in 2× SSC. Dimethylsuberimidate-fixed mouse red blood cells were added to the first wash to minimize loss of nuclei. Finally, the nuclei were suspended in the buffer appropriate for cytochemical detection.

Probe detection. Detection of the biotin-labeled probes was accomplished with fluorescently labeled avidin. The fluorescence intensity was amplified, if desired, using biotinylated goat anti-avidin, followed by an additional layer of avidin (Pinkel et al. 1986). Up to three layers of avidin were used in some studies. (Avidin and anti-avidin were obtained from Vector, Inc.) AAF-modified probes were detected with rabbit polyclonal or mouse monoclonal anti-AAF antibodies, followed by a fluorescently labeled second antibody. For microscope observation, the cells were stained for total DNA with either propidium iodide, Hoechst 33258, or DAPI. The DNA stain was carried in an antifade solution (Johnson and Aroujo 1981) for microscopy. Hoechst 33258 was used as the DNA stain in the flow measurements.

Observation and measurement. Fluorescence microscopy was performed using a Zeiss fluorescence microscope. Photographs were made with Ektachrome 400 color slide film. A SIT Vidicon TV camera interfaced to an image analysis system (Trapix 55/64, Recognition Concepts, Inc.) was used for quantitative microscope measurements. Images from cells labeled with fluorochromes that could not be simultaneously excited were obtained by digitally storing an image of each fluorochrome, assigning it a range of false color, and displaying the composite image on a color monitor. Three-dimensional nuclear reconstructions were produced by collecting a series of images at 1 μm focal intervals using a 100×, n.a. 1.3 objective (J. Mullikin et al., in prep.). The boundaries of the chromosomal domains in each image were defined by intensity thresholding. A stack of these thresholded images was made in computer memory and displayed with a contour program that allowed arbitrary image rotation and production of stereo image pairs. Additional contours were interpolated each 0.25 μm. Flow cytometric measurements were performed with a dual-beam flow cytometer. One beam was adjusted to emit in the UV range to excite the Hoechst DNA stain, and the other beam was adjusted to 488 nm to excite the fluorescein attached to the hybridized probe.

RESULTS

Species-specific Chromosome Staining in Hybrid Cell Lines

Human chromosomes in human/hamster hybrid cell lines were fluorescently stained by hybridizing with human genomic DNA (Fig. 1a) (Durnam et al. 1985; Manuelidis 1985b; Shardin et al. 1985; Pinkel et al. 1986). Specific hybridization to human chromosomes occurred because of the sequence differences between the repetitive DNA of the two species. The complete protocol could be accomplished easily in 3 hr, including a 2-hr hybridization incubation. However, chromosome-specific hybridization was visible after hybridizing for only a few minutes. Maximum intensity was reached after hybridizing for 6–8 hr. The human chromosomes were stained with high contrast both in metaphase spreads and nuclei. The existence of well-defined human chromosome domains in interphase hybrid nuclei is a general feature.

The intensity of the probe fluorescence was proportional to the amount of target sequence present in cells and chromosomes. In a hybrid line containing one copy each of human chromosomes 8 and 12, which differ by only 8% in DNA content (Mendelsohn et al. 1973), the fluorescence was proportional to the number of chro-

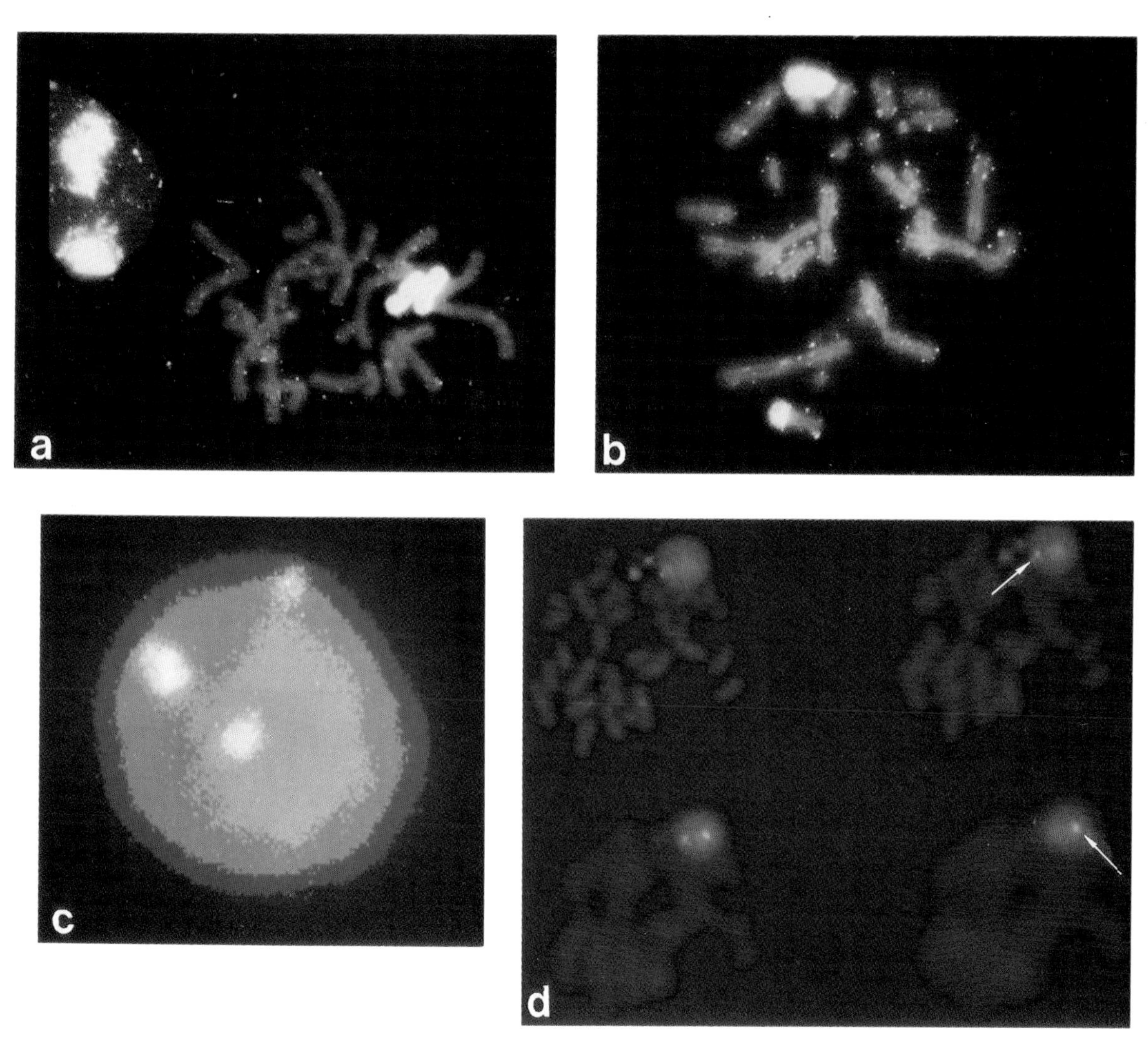

Figure 1. (*a*) Hybridization of human genomic DNA to human/hamster hybrid cells and chromosomes. Probe DNA was labeled with biotin and hybridized probe was detected with fluorescein-avidin. All of the DNA was stained with the red fluorescing dye propidium iodide. One copy each of human chromosomes 8 and 12 are visible in the metaphase spread and interphase nucleus. (*b*) Detection of an interspecies translocation. Chromosomes were treated as in *a*. The translocation is indicated by the presence of a bicolored chromosome. (*c*) Simultaneous hybridization of two chromosome-specific probes to a human cell. The binding of the AAF-labeled probe for the Y chromosome was detected with FITC, rendered white in this image. The biotinylated chromosome 9-specific probe was detected with phycoerythrin, rendered pink in this image. The DNA was counterstained with DAPI. Each fluorophore was independently imaged with an image analysis system; the three images were then redisplayed in false color. (*d*) Positions of chromosome 18 in a human nucleus. Optical sectioning of a lymphocyte nucleus hybridized with a probe specific for the pericentric region of chromosome 18 shows the relative positions of the two homologs (arrows) in three dimensions.

mosomal domains in the nuclei. Cells with zero, one, or two domains were present due to karyotypic instability of the line. Cells with no human domain had probe fluorescence below the threshold of the measurement; cells with one domain had an intensity of 0.57 ± 0.04 (mean ± S.E. of the mean); and cells with two domains had an intensity of 1.0 ± 0.1. In metaphase spreads of a hybrid line that contained human chromosomes 4, 8, and 21, the relative fluorescence intensities of these chromosomes were 0.56 ± 0.03, 0.34 ± 0.03, and 0.11 ± 0.01, respectively. The total intensity of these three chromosomes was normalized to 1.0 in each metaphase spread. The relative DNA contents of these three chromosomes are 0.50, 0.38, and 0.12, respectively (Mendelsohn et al. 1973).

The high-contrast, species-specific labeling in hybrid cells is particularly suited for rapid cytogenetic screening for translocations. For example, Figure 1b shows that interspecies translocations can be detected simply by noticing the presence of distinctive bicolored chromosomes. Visual scoring can be accomplished using low-magnification (16–40 ×) dry optics. Once a spread is found, the presence of a translocation can be established in several seconds. Translocations that are too small to be readily detectable by banding are easily seen. We have examined spreads at a rate of more than 100/hr to develop a dose-response curve for the induction of interspecies translocations by radiation (Pinkel et al. 1986).

Human Chromosome-specific Labeling

The simultaneous use of chromosome-specific repetitive probes for human chromosomes 9 and Y to label human male lymphocytes is shown in Figure 1c. The chromosome Y-specific probe pY3.4 (Burk et al. 1985) was modified with AAF and detected with a fluorescein-labeled second antibody. The chromosome 9-specific probe pHuR98 (R. Moyzis et al., in prep.) was labeled with biotin and detected with avidin conjugated to phycoerythrin. The localization of the probe fluorescence to small regions within the nucleus is a general feature following hybridization with chromosome-specific repeat sequence probes and is an indication of the localization of chromosomes in human nuclei (Zorn et al. 1979; Rappold et al. 1984; Manuelidis 1984, 1985a).

Structure of Interphase Nuclei

Hybridization to nuclei whose three-dimensional structure has been preserved allows study of the spatial relationships of chromosomes. When fluorescent techniques are used, the interphase distribution of labeled chromosomes can be explored by optical sectioning (Agard and Sedat 1983). The depth information comes from exploitation of the limited depth of field of microscope objectives with high numerical apertures. Figure 1d shows four views of a human metaphase spread and a cell nucleus to which has been hybridized a probe with predominant binding to chromosome 18 (Devilee et al. 1986). This nucleus has retained substantial thickness during preparation of the slide and the hybridization. Both copies of chromosome 18 in the metaphase spread are located near the nucleus. The chromosomes are blurry since focus has been adjusted to be at the bottom of the nucleus. In the other three views, the focal plane has been sequentially raised. This is indicated by the progressive blurring of the chromosomes. The labeled regions of the two homologs of chromosomes 18 are seen to be at different depths in the nucleus.

Figure 2 shows three views, separated by 45° rotations, of a computer reconstruction of a human/hamster hybrid cell nucleus hybridized with human genomic DNA. This cell line contains three human chromosomes. The hybridization was done in liquid suspension to preserve nuclear morphology.

Flow Cytometry

Figure 3 shows the quantitative measurement of bound probe by flow cytometry. Human cells were hybridized with AAF-labeled human genomic DNA, an AAF-labeled human chromosome 9-specific reagent,

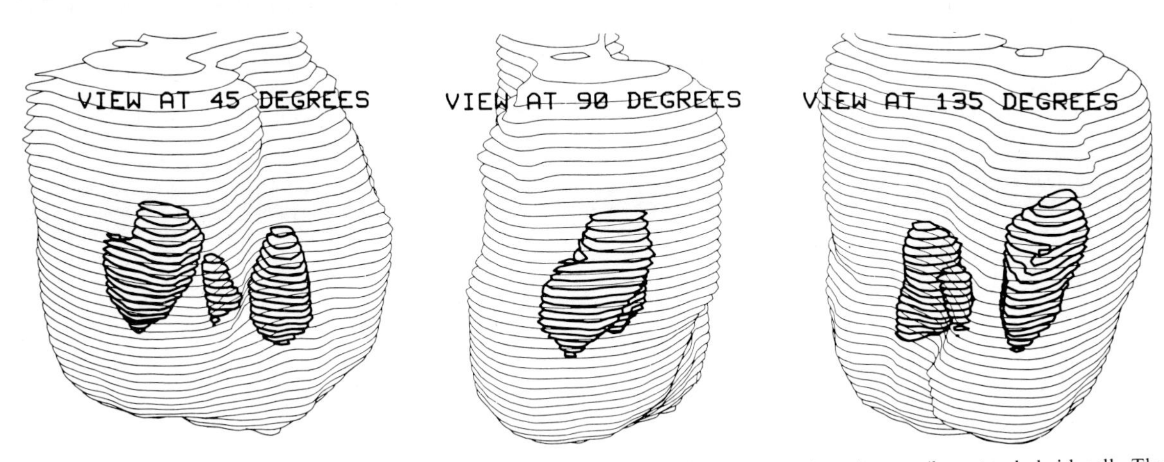

Figure 2. Three-dimensional reconstruction of the positions of human chromosomes in a human/hamster hybrid cell. The nucleus, which contains human chromosomes 4, 8, and 21, is shown from three different angles separated by 45°C.

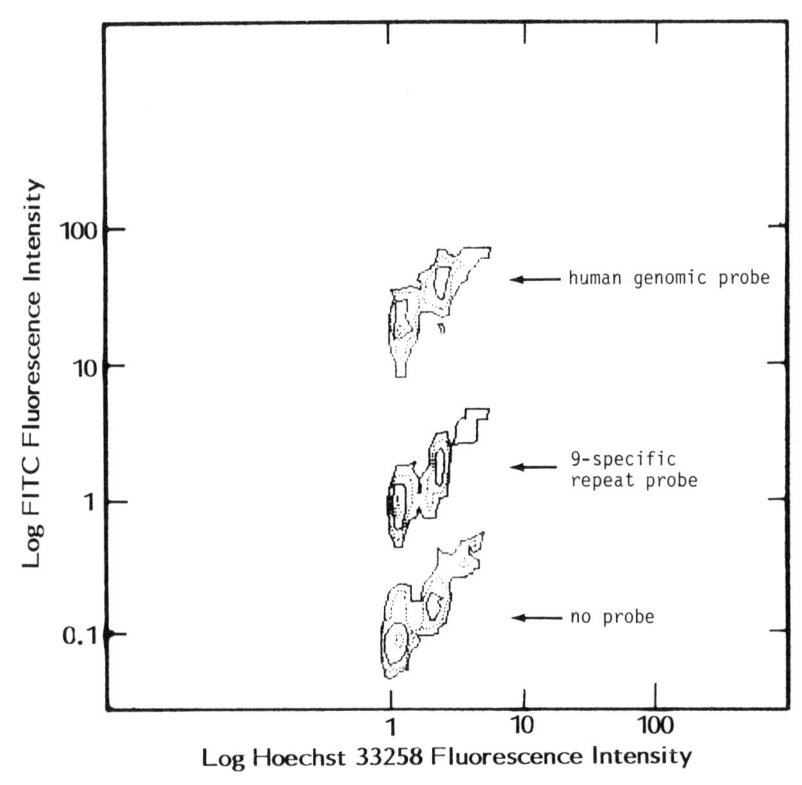

Figure 3. Flow cytometric measurement of fluorescence hybridization. Human nuclei were hybridized with an AAF-labeled probe specific for human chromosome 9, with AAF-labeled genomic human DNA, or with no probe. Bound probe was rendered fluorescent with FITC. Nuclear DNA was counterstained with Hoechst 33258. The two peaks seen in the contour plots for each population correspond to the G_1 and G_2 phases of the cell cycle. The hybridization signal increases proportionately to the nuclear DNA content.

or with no probe and counterstained with Hoechst 33258. A separate bivariate measurement of the probe-linked FITC fluorescence versus Hoechst fluorescence was made for each population, and the results were combined into one histogram. The cells labeled with the chromosome 9-specific probe are clearly brighter than the cells in the control population, which received no probe, and they are substantially less fluorescent than cells hybridized with human genomic DNA. The doubling in DNA content through the cell cycle is seen in both the intensity of the DNA stain and probe-linked stain. The relative intensity of the cells hybridized with genomic DNA and the chromosome 9-specific probe cannot be interpreted quantitatively in this measurement since insufficient genomic probe was used to assure saturation of the target sites in the nuclei.

DISCUSSION

Hybridization with genomic DNA is a very powerful and rapid approach to cytogenetic analysis of interspecies hybrid cells. High contrast, species-specific chromosome staining is possible by using protocols that take only a few hours from start to finish. The number of chromosomes of each species present in metaphase spreads and interspecies exchanges of DNA are obvious in Figure 1,a and b. Translocations too small to be detected reliably by banding are distinctly visible. The ease of the analysis makes this technique ideal for studying the induction of chromosomal abnormalities by low doses of toxic agents. It is useful for establishing the species identity of presumed human chromosomes sorted from hybrid cell lines for production of chromosome-specific libraries and for determining the purity of the sort (H. van Dekken et al., in prep.). It may be used to determine the incorporation site in interspecies transfections if enough DNA is involved. In addition, fluorescence hybridization with species-specific probes also facilitates determination of the number of chromosomes of that species that have been retained and detection of spontaneous interspecies translocations in hybrid cells. If the number of chromosomes involved is small (e.g., if one is interested in whether or not there is a single chromosome of a given species present), then hybridization to nuclei may yield sufficient information. In any case, fluorescence hybridization with species-specific probes may be preferable in many instances to banding analysis of G11 staining (Burgerhout 1975) for cytogenetic analysis of hybrid cells.

Repetitive probes specific for several human chromosomes have been discovered (Yang et al. 1982; Graham et al. 1984; Jabs et al. 1984; Lau and Schonberg 1984; Burk et al. 1985; Devilee et al. 1986; Devine et al. 1985; Waye and Willard 1985; N. Joste et al., in prep.), and it is possible that probes for all of them will eventually be found (Willard 1985). The binding distribution of these probes is usually restricted to a small portion of the chromosome, such as the long arm of chromosome Y or the pericentric region of the autosomes. The degree of chromosome specificity varies from probe to probe and depends in many cases on the stringency of the hybridization. However, these probes are useful for identifying chromosomes in metaphase spreads even if the binding is not perfectly chromosome-specific. This allows rapid screening of spreads for specific aneuploidies. The currently known repetitive probes are not particularly useful for detecting translocations because the probability that the break point will involve the labeled region of the chromosome is small, although it can happen (Lau et al. 1985).

Staining of a chromosome in a nucleus requires a probe of higher specificity than for staining in a spread since chromosome morphology cannot be used to assist with the interpretation. Many of the repetitive probes are sufficiently specific for interphase analysis, as shown in Figure 1c, where the Y and 9 chromosomes have been labeled with different colors. We anticipate that these will prove extremely valuable for detecting aneuploidy directly in nuclei. There are several approaches to accomplish this.

First, it may be sufficient to simply count the number of spots in two-dimensional images such as Figure 1c. Our experience in the analysis of lymphocytes, fibroblasts, and amniocytes with probes for chromosomes 9 and Y suggests the feasibility of this approach. The binding of these probes is tightly localized so that the number of chromosome domains to which the probe binds can be counted. In addition, the two homologs of chromosome 9 are sufficiently separated on average so that overlaps are not common. This suggests that there may be some control of the position of this chromosome in these nuclei. However, other chromosomes and/or other cell types may demonstrate different behavior. Even if the homologs of a given chromosome are regularly separated in nuclei, spot counting in two-dimensional images will not allow differentiation between a truly monosomic cell and one in which the two spots overlap due to the particular orientation of the cell on the slide. For homogeneous populations, where the abnormality is present in every cell, this will not be a major difficulty, but it will limit the ability to detect rare aneuploid cells. This may be overcome by the methods discussed below. Simultaneous hybridization with multiple probes (Fig. 1c) allows analysis of more than one chromosome at a time. In addition, the probes serve reciprocally as controls for the success of the hybridization and permit recognition of general aneuploidy of a cell. For example, when using probes for a sex chromosome and an autosome in male cells, detection of the autosomal probe should insure that there was sufficient access for the hybridization reagents to a particular nucleus. Thus, the absence of a signal from the sex chromosome could be interpreted as absence of that chromosome. Additionally, if two copies of the sex chromosome are found, the number of spots from the autosomal probe should allow differentiation between an aneuploidy of the sex chromosome and a tetraploid cell.

A second approach to interphase aneuploidy detection is quantitative measurement of the intensity of probe-linked fluorescence. This method is not affected by the possible overlap of the labeled chromosomal domains. Both quantitative microscopy and flow cytometry have sufficient precision to differentiate populations differing by 50% in target sequence. In Figure 3, the doubling of both DNA content and probe-linked fluorescence for a human chromosome 9–specific probe in human cells is clearly evident during the cell cycle. With the current measurement precision, detection of aneuploidy is limited to populations in which the proportion of abnormal cells is high.

The morphology of nuclei hybridized in suspension is well-preserved, permitting three-dimensional study of the chromosomal domains. This offers a third approach to the detection of aneuploidy, because domain overlaps in two-dimensional images are no longer a problem. Figure 2 shows the same nucleus viewed from three angles separated by 45°. Two-dimensional projections of these images would show one (center), two (right), or three (left) labeled domains.

Fluorescence hybridization with chromosome-specific probes to nuclei in suspension is a powerful approach for the study of chromosomal positions in interphase nuclei (H. van Dekken et al., in prep.). It permits the use of optical sectioning for analysis of the organization in three dimensions at the resolution level of the light microscope, which is potentially tenths of a micrometer. Figure 1d demonstrates optical sectioning of a human nucleus stained for chromosome 18, while Figure 3 shows a full reconstruction of the human chromosome domains in a hybrid cell nucleus. By using multiple probes detected with different fluorophores, it may be possible to address such questions as the systematics of chromosome order in different cell types, the cell-cycle dependence of the organization, the relation of the organization to cellular function, and mechanisms responsible for production of chromosome abnormalities (Manuelidis 1984, 1985a; Rappold et al. 1984).

The flow cytometric detection of the binding of a chromosome-specific repetitive probe demonstrates the ability to detect targets of several hundred kilobases with signals substantially above background. This represents sensitivity a thousand-fold higher than reported previously (Trask et al. 1985) and opens the possibility of detecting the distribution of the amount of specific sequences, such as amplified genes, in a cell population. Hybridization to chromosomes in suspension, currently under development, would greatly facilitate chromosome discrimination and purification by flow cytometry. This would permit detection of rare translocations, signified by bicolored chromosomes, at high rates.

ACKNOWLEDGMENTS

We thank Dr. R. Moyzis for use of the chromosome 9-specific probe pHuR 98, Dr. P. Pearson for the chromosome 18-specific repeat L1.84, and Dr. K. Smith for the Y-specific repeat pY3.4. We thank Dr. R. Baan for antibodies against AAF. This work was conducted under the auspices of the U.S. Department of Energy by the Lawrence Livermore National Laboratory under contrast number W-7405-ENG-48 with support from the U.S. Public Health Service (grant HD17655).

REFERENCES

Agard, D. and J. Sedat. 1983. Three dimensional architecture of a polytene nucleus. *Nature* **302:** 676.

Burgerhout, W. 1975. Identification of interspecific translocation chromosomes in human-Chinese hamster hybrid cells. *Humangenetik* **29:** 229.

Burk, R., P. Szabo, S. O'Brien, W. Nash, L. Yu, and K. Smith. 1985. Organization and chromosomal specificity of autosomal homologs of human Y chromosomes repeated DNA. *Chromosoma* **92:** 225.

Casperson, T., S. Farber, G. Folley, J. Kudynowski, E. Modest, E. Simonsson, U. Wash, and L. Zech. 1968. Chemical differentiation along metaphase chromosomes. *Exp. Cell. Res.* **49:** 219.

Devilee, P., T. Cremer, P. Slagboom, E. Bakker, H. Scholl, H. Hager, A. Stevenson, C. Cornelisse, and P. Pearson. 1986. Two subsets of human alphoid repetitive DNA show distinct preferential localization in the pericentric regions of chromosomes 13, 18 and 21. *Cytogenet. Cell Genet.* **41:** 193.

Devine, E., S. Nolin, E. George, E. Jenkins, and W. Brown. 1985. Chromosomal localization of several families of repetitive sequences by in situ hybridization. *Am. J. Hum. Genet.* **37:** 114.

Durnam, D., R. Gelinas, and D. Myerson. 1985. Detection of species specific chromosomes in somatic cell hybrids. *Somatic Cell Mol. Genet.* **11:** 571.

Forster, A., J. McInnes, D. Skingle, and R. Symons. 1985. Non-radioactive hybridization probes prepared by the chemical labeling of DNA and RNA with a novel reagent, photobiotin. *Nucleic Acids Res.* **13:** 745.

Graham, G., T. Hall, and M. Cummings. 1984. Isolation of repetitive DNA sequences from human chromosome 21. *Am. J. Hum. Genet.* **36:** 25.

Harper, M., A. Ullrich, and G. Saunders. 1981. Localization of the human insulin gene to the distal end of the short arm of chromosome 11. *Proc. Natl. Acad. Sci.* **78:** 4458.

Hopman, A., J. Wiegand, and P. van Duijn. 1986. A new hybridocytochemical method based on mercurated nucleic acid probes and sylfhydryl-hapten ligands. *Histochemistry* **84:** 179.

Jabs, E., S. Wolf, and B. Migeon. 1984. Characterization of a cloned DNA sequence that is present at centromeres of all human autosomes and the X chromosome and shows polymorphic variation. *Proc. Natl. Acad. Sci.* **81:** 4884.

Johnson, G. and G. Aroujo. 1981. A simple method of reducing the fading of immunofluorescence during microscopy. *J. Immunol. Methods* **43:** 349.

Landegent, J., N. Jansen In De Wal, R. Baan, J. Hoeijmakers, and M. van der Ploeg. 1984. 2-acetylaminofluorene-modified probes for the indirect hybridocytochemical detection of specific nucleic acid sequences. *Exp. Cell Res.* **153:** 61.

Langer, P., A. Waldrop, and D. Ward. 1981. Enzymatic synthesis of biotin-labeled polynucleotides: Novel nucleic acid affinity probes. *Proc. Natl. Acad. Sci.* **78:** 6633.

Lau, Y. and S. Schonberg. 1984. A male-specific DNA probe detects heterochromatin sequences in a familial Yq-chromosome. *Hum. Genet.* **36:** 1394.

Lau, Y., K. Ying, and G. Donnel. 1985. Identification of a case of Y:18 translocation using a Y-specific repetitive DNA probe. *Hum. Genet.* **69:** 102.

Manuelidis, L. 1984. Different central nervous system cell types display distinct and nonrandom arrangements of satellite DNA sequences. *Proc. Natl. Acad. Sci.* **81:** 3123.

———. 1985a. Indications of centromere movement during interphase and differentiation. *Ann. NY Acad. Sci.* **450:** 205.

———. 1985b. Individual interphase chromosome domains revealed by in situ hybridization. *Hum. Genet.* **71:** 288.

Mendelsohn, M., B. Mayall, E. Bogart, D. Moore II, and B. Perry. 1973. DNA content and DNA based centromeric index of the 24 human chromosomes. *Science* **179:** 1126.

Pinkel, D., T. Straume, and J. Gray. 1986. Cytogenetic analysis using quantitative, high sensitivity, fluorescence hybridization. *Proc. Natl. Acad. Sci.* **83:** 2934.

Rappold, G., T. Cremer, H. Hager, E. Davies, C. Muller, and T. Yang. 1984. Sex chromosome positions in human interphase nuclei as studied by in situ hybridization with chromosome specific DNA probes. *Hum. Genet.* **67:** 317.

Ruth, J. and R. Bryan. 1984. Chemical synthesis of modified oligonucleotides and their utility as non-radioactive hybridization probes. *Fed. Proc.* **43:** 2048.

Shardin, M., T. Cremer, H. Hager, and M. Lang. 1985. Specific staining of human chromosomes in Chinese hamster x human hybrid cell lines demonstrates interphase chromosome territories. *Hum. Genet.* **71:** 281.

Trask, B., G. van den Engh, J. Landegent, N. Jansen In De Wal, and M. van der Ploeg. 1985. Detection of DNA sequences in nuclei in suspension by in situ hybridization and dual beam flow cytometry. *Science* **230:** 1401.

Waye, J. and H. Willard. 1985. Chromosome-specific alpha satellite DNA: Nucleotide sequence analysis of the 2.0 kilobase pair repeat from the human X chromosome. *Nucleic Acids Res.* **13:** 2731.

Willard, H. 1985. Chromosome-specific organization of human alpha satellite DNA. *Am. J. Hum. Genet.* **37:** 524.

Yang, T., S. Hansen, K. Oishi, O. Ryder, and B. Hamkalo. 1982. Characterization of a cloned repetitive DNA sequence concentrated on the human X chromosome. *Proc. Natl. Acad. Sci.* **79:** 6593.

Yunis, J.J. 1981. Mid-prophase human chromosomes. The attainment of 2000 bands. *Hum. Genet.* **56:** 293.

Zorn, C., C. Cremer, T. Cremer, and J. Zimmer. 1979. Unscheduled DNA synthesis after partial UV irradiation of the cell nucleus. *Exp. Cell Res.* **124:** 111.

Construction of Human Chromosome-specific DNA Libraries from Flow-sorted Chromosomes

L.L. Deaven,* M.A. Van Dilla,† M.F. Bartholdi,* A.V. Carrano,† L.S. Cram,* J.C. Fuscoe,† J.W. Gray,† C.E. Hildebrand,* R.K. Moyzis,* and J. Perlman†

*Los Alamos National Laboratory, Life Sciences Division, University of California, Los Alamos, New Mexico 87545; †Lawrence Livermore National Laboratory, Biomedical Sciences Division, University of California, Livermore, California 94550

Recent developments in recombinant DNA technology have had a dramatic impact on our ability to study the properties of the human genome. It is now possible to construct libraries of cloned DNA fragments that can be used to map and to isolate each human gene. A variety of different types of DNA libraries can be constructed with distinct advantages of one type over another for a selected application. For example, for mapping purposes, a library with relatively small inserts (1–4 kb) would be ideal. Furthermore, it would be beneficial if the library were constructed from a subset of the human genome rather than from total genomic DNA. On the other hand, a library optimal for the isolation of an entire human gene or two linked genes and perhaps some flanking sequences would consist of larger DNA inserts (20–40 kb) in order to increase the probability of finding intact genes within cloned fragments. Here again, however, it would be beneficial to construct the library from a subset of the entire human genome. The availability of sets of libraries made for specific applications would greatly accelerate the rate of gene mapping and gene isolation. Because the construction of chromosome-specific subset libraries from flow-sorted chromosomes is beyond the technical means of many laboratories, we initiated the National Laboratory Gene Library Project in 1983. Our aims are to produce a series of chromosome-specific DNA libraries and to make these libraries available to interested investigators throughout the world.

The feasibility of constructing DNA libraries and of using them to isolate and map specific DNA sequences was first demonstrated with the relatively small genome of *Drosophila melanogaster* (Wensink et al. 1974). Application of these methods to larger and more complex genomes had to await the development of more-efficient methods for library construction and for the screening of large numbers of clones. These improvements were reported over the next three years and include methods for in situ plaque hybridization (Grunstein and Hogness 1975; Benton and Davis 1977), improved λ cloning vectors (Blattner et al. 1977; Leder et al. 1977), and in vitro packaging systems (Hohn and Murray 1977). Shortly thereafter, a human genomic DNA library was constructed and clones containing human γ- and β-globin genes were isolated from it (Lawn et al. 1978). Taken together, these and other developments had clearly established a new approach to mapping the human genome and to studies of gene structure and function.

Attempts to construct genomic subset libraries of chromosome-specific human DNA sequences were first described in the literature in the late 1970s and the early 1980s (Kunkel et al. 1977; Schmeckpeper et al. 1979; Gusella et al. 1980; Olsen et al. 1980). These libraries are useful but are limited in coverage because of sequence homologies between rodent and human DNA (Ohno 1974) and are difficult to construct because rodent/human hybrid cells are often characterized by chromosome instability. A more direct approach to the construction of chromosome-specific DNA sequence libraries is through the use of chromosomes purified by fluorescence-activated flow sorting (Gray et al. 1975; Shay and Cram 1985). The applicability of this approach was demonstrated with the construction of libraries enriched for sequences of the human and mouse X chromosomes (Davies et al. 1981; Disteche et al. 1982) and human chromosomes 21 and 22 (Krumlauf et al. 1982), for an abnormal human chromosome 1 (Latt et al. 1982), and for Chinese hamster chromosomes (Griffith et al. 1984).

Our work on the construction of several complete sets of human chromosome-specific DNA libraries began in 1983. At that time, we planned to construct two sets of complete-digest libraries and to make them available to all interested researchers as quickly as possible. When this work was completed, we planned to reevaluate the need for partial-digest libraries and to continue the project by constructing a set of libraries with larger inserts if this were still considered to be a significant contribution to human genetics research. Our rationale for these decisions was that, in 1983, the preeminent need for human chromosome-specific DNA libraries was to accelerate the rate of probe production for gene mapping and genetic disease diagnosis. We decided to construct two sets of libraries, one set using *Eco*RI digests and the other set using *Hin*dIII digests, in order to provide broad coverage of the DNA sequences in each chromosome.

Detailed descriptions of the project strategy, the progress, and the results have been published elsewhere

(Deaven et al. 1986; Van Dilla et al. 1986). This paper is a general summary of our work with updates of new data since our last report on the project.

METHODS

Chromosome sources. We have used diploid human fibroblast or lymphoblastoid cells as a source for the smaller human chromosomes (nos. 13–22 and Y) and Chinese hamster/human hybrid cells as a source for the larger chromosomes (nos. 1–12 and X) with only minor exceptions. The specific cell sources for individual chromosomes are listed in Table 1. The cells were cultured using conventional techniques.

Chromosome isolation and staining. We have examined the applicability of four different chromosome isolation methods for use in this project. The major advantages and disadvantages of each procedure are summarized in Table 2. In general, we utilized the hexylene glycol and the propidium iodide procedures in the early part of the project (1983 and 1984) but changed to the magnesium sulfate and polyamine procedures for our most recent work. The major reasons for making these changes were to improve flow histogram resolution and to obtain chromosomal DNA of high molecular weight. All of the methods involved a hypotonic treatment and the application of a shearing force to metaphase cells to aid in membrane disruption and chromosome dispersion. Unfortunately, a shear force large enough to disperse most of the small chromosomes also results in breakage of the large chromosomes. This depletes the population of large chromosomes and makes sorting impractical. If the shearing is reduced in an effort to retain most of the large chromosomes, clumps of small chromosomes remain undispersed and contaminate the peaks of the large chromosomes, making sort purities unacceptably low. This was one reason that we turned to hybrid cells as a source for the larger chromosomes.

Isolated chromosomes were stained with Hoechst 33258 and chromomycin A_3 for bivariate flow analysis and sorting. These two fluorescent stains differentiate chromosomes on the basis of total DNA content and also on the basis of relative binding preference for AT-rich and GC-rich DNA. In one instance, chromosomes were stained with 4′,6-diamidino-2-phenylindole (DAPI) and chromomycin A_3 and counterstained with Netropsin (Meyne et al. 1984). This stain combination enhances the fluorescence of C-band heterochromatin and enables sorting of chromosome 9 from diploid human cells. Chromosomes isolated by the Aten procedure were stained with propidium iodide as part of that method. Although this method was used in the early part of the project, we were not satisfied with the chromosome resolution obtained, and all libraries currently available were constructed with chromosomes from bivariate sorts.

Table 1. Cultured Cells Used for Chromosome Isolation

Identification number	Source	Human chromosomes used in library construction
	Human cells	
Diploid fibroblasts		
LLNL761	LLNL	13,16–18
LLNL811	LLNL	14/15,19–22,Y
HSF-7	D. Chen, LANL	9,13,16,18–22,Y
Lymphoblastoid cells		
GM130A	Inst. for Med.	17
GM131	Res. Camden, NJ	14,15
1634	T. Friedmann, UCSD	14,15
	Hamster/human hybrid cells	
UV20HL21-27	L. Thompson, LLNL	4,8
UV21HL4	L. Thompson, LLNL	9
UV20HL4	L. Thompson, LLNL	11
UV20HL15-33	L. Thompson, LLNL	6
UV24HL5	L. Thompson, LLNL	2
UV24HL10-12	L. Thompson, LLNL	1
81P5D	S. O'Brien, NIH	X,12,15
80H10	S. O'Brien, NIH	11
81H10E	S. O'Brien, NIH	–
MR3.31-6TG6	D. Ledbetter, Baylor, LANL	7
J-1	C. Jones, Denver	–
314-16	C. Jones, Denver	3
640-12	C. Jones, Denver	5
762-8A	C. Jones, Denver	10

Chromosome sorting. For the purposes of library construction, chromosome sorting must be relatively fast and it must be accomplished with a high level of purity. Library construction would not be practical if it took months of sorting in order to obtain a sufficient amount of DNA for starting material, and it would not be useful if the starting material were heavily contaminated with extraneous DNA. We will briefly discuss the characteristics of the sorters used in this project with respect to sorting rates and purity. For details of sorter design and data analysis, there are extensive descriptions published elsewhere (Van Dilla et al. 1985).

We have utilized four different instruments that sort at either conventional rates typical of commercial instruments or at speeds that are approximately an order of magnitude higher (high-speed sorters). The conventional sorters are capable of analyzing 1000–2000 chromosomes/sec and of sorting up to 50 chromosomes of a particular type each second. Optimal chromosome preparations and instrumental performance will yield 1×10^6 copies of a chromosome in 8 hr of operating time. High-speed sorters can analyze 20,000 chromosomes/sec and sort up to 200 chromosomes/sec. Under optimal operating conditions, they will yield 5×10^6 chromosomes in 8 hr of operating time. Thus, high-speed sorters have a distinct advantage over conventional sorters, especially if relatively large amounts of DNA (microgram quantities) are required as starting material for library construction.

The purity of sorted chromosomes is critical to the

Table 2. Advantages and Disadvantages of Four Chromosome Isolation Methods

Method	Advantages	Disadvantages
Hexylene glycol (Wray and Stubblefield 1970)	sorted chromosomes can be banded	DNA degradation; high debris levels
Propidium iodide (Aten et al. 1980)	low debris levels; undegraded DNA	no chromosome banding; one color stain
Polyamine (Sillar and Young 1981)	low debris levels; undegraded DNA; high flow resolution	no chromosome banding
$MgSO_4$ (van den Engh et al. 1984)	low debris levels; high flow resolution; undegraded DNA	no chromosome banding

construction of chromosome-specific DNA libraries, and a large amount of effort has gone into purity determinations of sorter outputs. The capability of sorters to differentiate and sort specific objects from a mixed population with high levels of accuracy and precision can be determined with fluorescent microspheres. In test runs, sorters used in this project can sort a single population of microspheres from mixed populations with 95–100% purity at operating speeds. Purity determinations of sorted chromosomes are more difficult and require direct examinations of sorted chromosomes when practical, as well as indirect measurements such as cytological examination of cells used for chromosome sources and library purity determinations.

Indirect examinations that we have employed include:

1. Karyotype analysis (G-banding, Q-banding, G-11 staining, DAPI-Netropsin staining) of metaphase cells from cell strains and lines used as chromosome sources: This analysis confirms that a chromosome of interest is present in the cell culture, the frequency of its presence, and whether it is normal or rearranged.
2. Isozyme analysis of hybrid lines to support karyotype analysis.
3. Measurements of chromosome peak locations and peak volumes in flow histograms and comparisons of that information with cytological karyotype analysis: This type of information on chromosome frequency (peak volume) and chromosome DNA content (peak location) is especially useful when hybrid cells are being used and human chromosome content is unstable.
4. Examinations of libraries for the presence of Chinese hamster sequences or for extraneous human DNA sequences.

Direct measurements of the purity of sorted chromosomes include:

1. Chromosome banding analysis of chromosomes sorted directly onto microscope slides: This method is extremely useful; however, it only works well with hexylene glycol isolated chromosomes.
2. DAPI-Netropsin staining of sorted chromosomes: This method is useful only for the human chromosomes with large blocks of centromeric heterochromatin, but it is valuable if isolation methods are used that do not permit G-band analysis.
3. Hybridization of human and hamster genomic DNA to sorted chromosomes or chromosomal DNA from hamster/human hybrids: This procedure can be done in situ on sorted chromosomes and used to identify hamster chromosomes or human/hamster translocations (Pinkel et al. 1986) or to DNA from sorted chromosomes mounted on nitrocellulose filters.

Library Construction

Cloning procedures and materials that were well established and reliable were selected for this project; however, the construction of chromosome-specific DNA sequence libraries from 2×10^5 to 1×10^6 flow-sorted chromosomes has presented new challenges to workers using classical cloning strategies. The major objective was to maximize cloning efficiencies in order to obtain libraries with high levels of sequence representation from minimal amounts of chromosomal DNA. The addition of a vacuum microdialysis step to the cloning procedure (Fuscoe et al. 1986) to aid in purifying the chromosomal DNA had a dramatic effect on the completeness of endonuclease digestion and on the overall efficiency of the procedure. This step is highly recommended for cloning procedures using flow-sorted chromosomes as the source of target DNA. The entire cloning procedure is outlined below.

A. Vector Preparation (Maniatis et al. 1982)
   1. Isolate vector DNA.
   2. Digest vector DNA with excess *Eco*RI or *Hin*dIII.
   3. Dephosphorylate vector arms.

B. Chromosomal DNA Preparation
   1. Chromosomes isolated by hexylene glycol, propidium iodide or $MgSO_4$ may be concentrated by centrifugation; for maximum DNA yields, omit this concentration step for polyamine-isolated chromosomes.

2. Hydrolyze chromosomal proteins with proteinase K/SDS.
3. Protein extraction with phenol and chloroform.
4. Dialyze in 0.5-ml volume against Tris-EDTA.
5. Digest with excess *Hind*III or *Eco*RI.
6. Protein extraction and microdialysis.

C. Ligation of Vector and Chromosomal DNA Fragments
1. Add arms to purified and restricted chromosomal DNA.
2. Vacuum concentration and ethanol precipitation.
3. 70% ethanol wash.
4. Ligation of DNA mixture.

D. Packaging and Amplification
1. Package recombinant phage into infectious particles.
2. Amplify by infecting *E. coli* host strain (LE392).

The frequency of nonrecombinants was estimated by taking an aliquot of vector arms through each of the cloning steps in the absence of target DNA. A small aliquot of target DNA was removed at each step of its preparation for cloning and analyzed by electrophoresis on a minigel followed by capillary transfer of DNA to a filter and hybridization with ^{32}P-labeled total human genomic DNA probe. This procedure confirms the high molecular weight of the isolated chromosomal DNA, the completeness of restriction endonuclease digestion, and the efficiency of the ligation with vector arms.

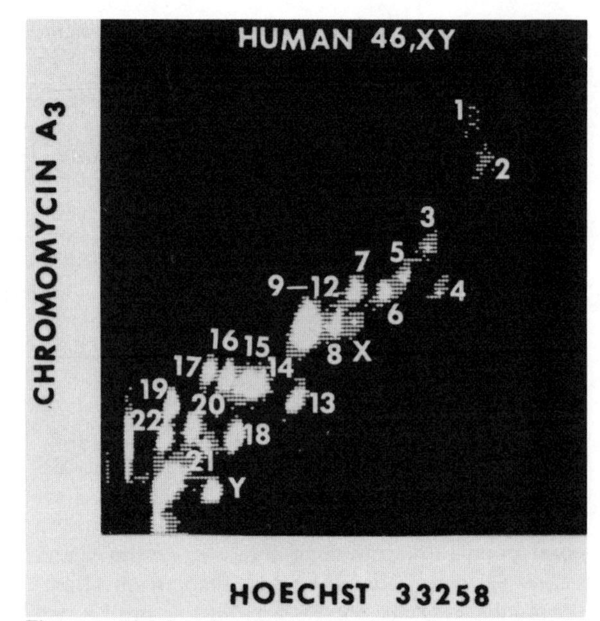

Figure 1. Bivariate flow histogram of the chromosomes in a diploid human (46, XY) fibroblast cell strain (HSF-7).

RESULTS

The most important results from our work are the libraries and the characteristics that define each library; however, chromosome-sorting data contribute significantly to the quality of the libraries, so we present in this section results from our sorting experiments as well as the available data on library construction and characterization.

Chromosome Sorting

A typical bivariate histogram of human (46, XY) chromosomes is shown in Figure 1. A histogram of the small human chromosomes at a higher gain setting is shown in Figure 2. Sorter windows of an EPICS V flow sorter are set on chromosomes 19 and Y. In some cell lines or strains, chromosomes 14 and 15 are sufficiently well resolved for sorting so all of the small chromosomes can be sorted from diploid human cells. For the larger half of the karyotype, we have utilized hamster/human hybrids with a small complement of human chromosomes. A histogram of one of these lines is shown in Figure 3. As indicated on the figure, this line (UV20HL 21-27) contains human chromosomes 4 and 8. Each of these chromosomes is sufficiently isolated from the hamster chromosomes (unlabeled peaks) for sorting. In a histogram of the base of the peaks shown in Figure 3, the human chromosomes are still well resolved in a background of debris (Fig. 4). Some human chromosomes are not as well isolated from the hamster continuum of chromosomes as those in Figure 3. An example of this is human chromosome 1, which is well

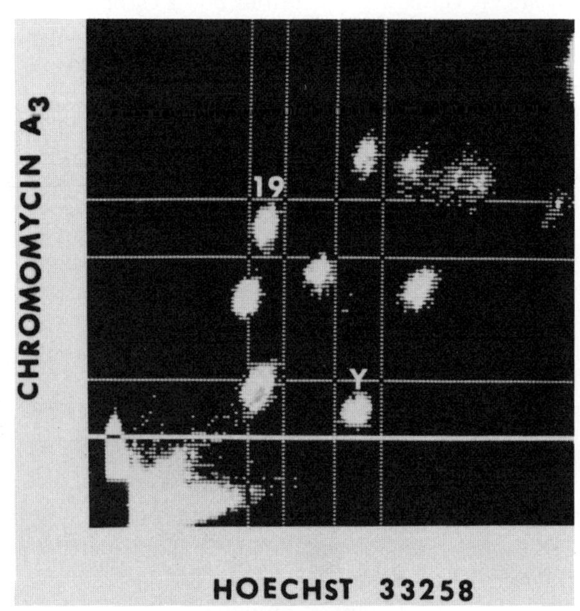

Figure 2. Bivariate flow histogram of the small human chromosomes at increased instrumental gain settings as compared with Fig. 1. Sorter windows are set on chromosomes 19 and Y.

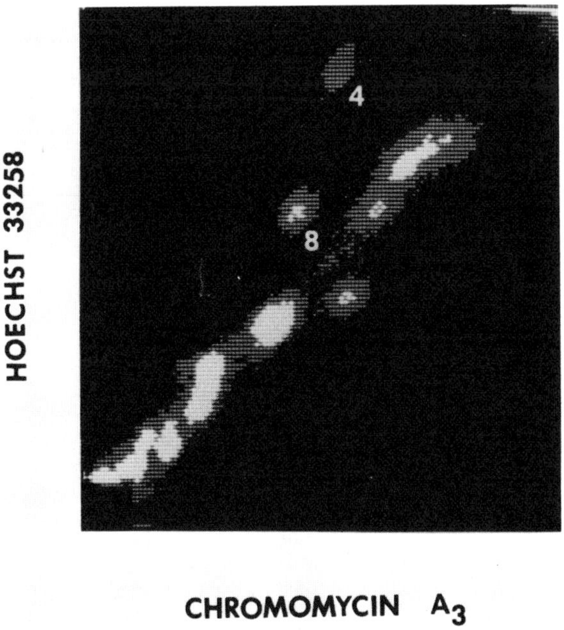

Figure 3. Bivariate flow histogram of the chromosomes in a Chinese hamster/human hybrid cell line (UV20HL21-27) that contains human chromosomes 4 and 8.

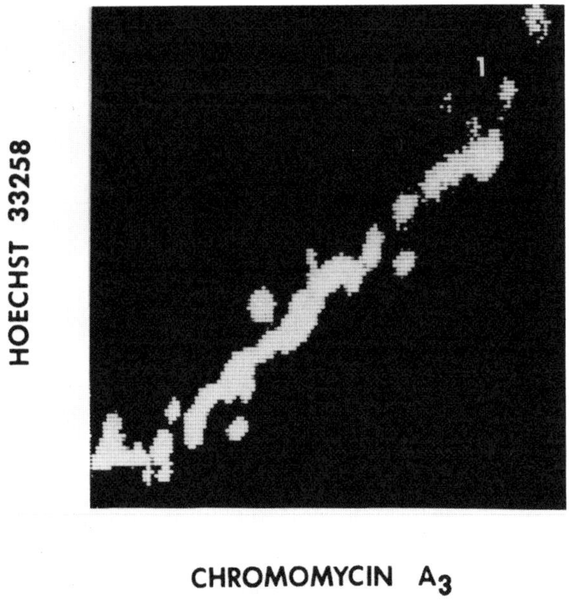

Figure 5. Bivariate flow histogram of the chromosomes in a Chinese hamster/human hybrid cell line (UV24HL10-12) that contains a human chromosome 1.

enough resolved to permit sorting, but which lies close to the hamster chromosomes (Fig. 5).

Purity of Sorted Chromosomes

In addition to the purity checks listed in the Methods section, an examination of Figures 3–5 will illustrate another method we have utilized to verify chromosome peak identities. The expected location of each human chromosome peak can be calculated, and this expected location can be compared with the location observed in a hybrid histogram. Human chromosomes with histogram positions differing by greater than 10% of the expected location are considered questionable because normal polymorphism variability of greater than 10% is rare (Langlois et al. 1982). We also compare the histograms of hamster parental lines with the histograms of hybrids when parental cells are available. This comparison aids in the identification of peaks of human origin but is more complex because hamster chromosomes may undergo rearrangements as part of the chromosomal evolution in hybrid cells.

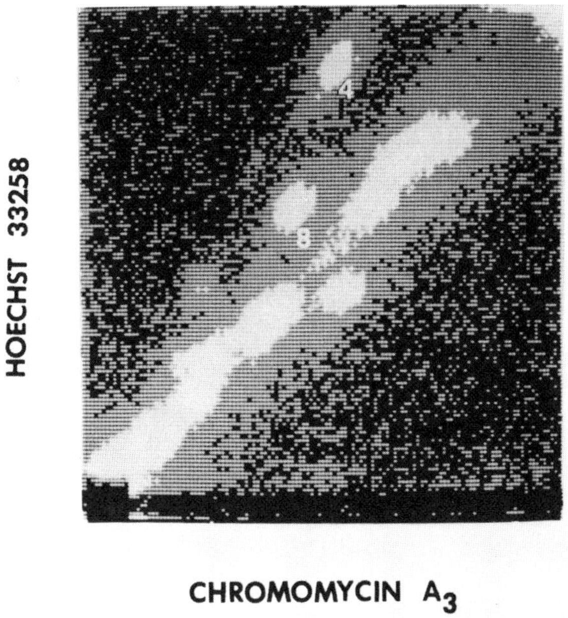

Figure 4. Bivariate flow histogram of the chromosomes in Fig. 3, showing debris levels at the base of the peaks.

Chromosome purity determinations by Q-banding of sorted chromosomes were done with hexylene glycol–isolated chromosomes. When chromosome isolation procedures were changed to methods not compatible with Q-banding, parallel isolations were made with the polyamine method and the hexylene glycol method in order to verify peak assignments and peak purity. In these examinations, identifiable human chromosomes used to construct libraries were at least 90% pure. More recently, we have used in situ hybridization to measure sort purity in sorts from hamster/human hybrids. These results are listed in Table 5.

Libraries

The libraries constructed as of May 1986 are listed in Tables 3 (*Eco*RI digests; Los Alamos) and 4 (*Hin*dIII

Table 3. Characteristics of 26 Chromosome-specific DNA Libraries Cloned into the *Eco*RI Site of Charon 21A

Chromosomal material		Packaging reaction				
library ID no.	chromosome source	no. cloned ($\times 10^6$)	no. indep. recomb. ($\times 10^5$)	no. chromosome equiv.[a]	freq. nonrecomb.	Amplified-plate lysates (total pfu $\times 10^{10}$)
LA *01* NS01	UV24HL10-12	0.5	12.7	31	<0.01	450
LA *02* NS01	UV24HL5	0.5	0.70	1.8	0.04	17
LA *03* NS01	314-16	0.4	0.27	0.8	0.04	8.4
LA *04* NS01	UV20HL21-27	0.9	0.51	1.6	<0.01	8
LA *04* NS02	UV20HL21-27	0.9	0.74	2.3	<0.01	16
LA *05* NS01	640-12	0.4	13	43	<0.01	0.18
LA *06* NS01	UV20HL15-33	1.9	0.48	1.7	0.06	14.8
LA *07* NS01	MR3.316TG6	0.1	2.39	9.2	0.06	132
LA *08* NS04	UV20HL21-27	0.5	0.36	1.5	0.10	0.10
LA *09* NS01	HSF-7	0.3	1.59	7	0.07	0.02
LA *10* NS01	762-8A	0.3	3.97	18	<0.01	0.3
LA *11* NS02	80H10	0.4	0.62	2.8	0.17	0.01
LA *12* NS01	81P5d	1.5	6	27	<0.01	97.5
LA *13* NS03	HSF-7	0.4	0.75	4.2	0.14	0.02
LA *14* NS01	1634	0.5	6.15	36	<0.01	348
LA *15* NS02	81P5d	0.3	4	20	0.06	0.61
LA *15* NS03	1634	0.4	0.68	4	0.09	43
LA *16* NS02	HSF-7	3.0	0.4	2	0.08	6.9
LA *17* NS03	GM130A	0.6	1.11	7.9	0.01	42
LA *18* NS04	HSF-7	0.5	2.51	19	0.02	112
LA *19* NS03	HSF-7	0.5	1.1	11	0.16	0.11
LA *20* NS01	HSF-7	0.5	0.16	1.5	0.24	0.10
LA *21* NS01	HSF-7	0.9	10.7	137	<0.01	450
LA *22* NS03	HSF-7	0.5	0.93	11	0.19	0.11
LA *0X* NS01	81P5d	1.2	2.14	8.5	0.02	4.23
LA *0Y* NS01	HSF-7	0.5	1.07	11.5	0.10	0.02

[a]Number of recombinants for 1 chromosome equivalent $= \frac{(3 \times 10^9)(0.65)(f)}{4100}$

where: 0.65 is the clonable fraction; *f* is the fraction of cellular DNA in particular chromosome; 4100 is average fragment size.

Table 4. Characteristics of 18 Chromosome-specific DNA Libraries Cloned into the *Hin*dIII Site of Charon 21A

Chromosomal material		Packaging reaction				
library ID no.	cell line	no. cloned ($\times 10^6$)	no. indep. recomb. ($\times 10^5$)	no. chromosome equiv.[a]	freq. nonrecomb.	Amplified plate lysates (total pfu $\times 10^{10}$)
LL *04* NS01	UV20HL21-27	1.0	0.23	0.8	0.009	0.34
LL *06* NS01	UV20HL15-33	0.4	7.6	27	0.09	6
LL *08* NS02	UV20HL21-27	0.7	22	93	0.03	22
LL *09* NS01	UV41HL4	0.4	3	13	0.095	11
LL *10* NS01	762-8A	0.5	2.4	10.6	0.009	200
LL *11* NS01	UV20HL4	0.2	1.1	4.9	0.003	6.6
LL *13* NS01	761	1.0	0.22	1.3	0.03	0.9
LL *14* NS01	GM131	0.5	23	135	0.005	730
LL *45* NS01	811	1.0	26	152	0.02	140
LL *15* NS01	GM131	1.0	0.7	4.4	0.0007	29
LL *16* NS02	761	0.7	45	300	0.02	22
LL *17* NS01	761	1.2	0.17	1.3	0.04	0.62
LL *18* NS01	761	1.0	8.9	72	0.03	42
LL *19* NS01	811	1.0	15	145	0.05	15
LL *20* NS01	811	1.0	39	354	0.01	29
LL *21* NS02	811	0.5	4.7	60	0.15	7.5
LL *22* NS01	811	0.5	6.1	71	0.01	11
LL *0Y* NS01	811	1.0	2.5	27	0.0006	2.6

[a]Number of recombinants for 1 chromosome equivalent $= \frac{(3 \times 10^9)(0.65)(f)}{4100}$

where: 0.65 is the clonable fraction; *f* is the fraction of cellular DNA in particular chromosome; 4100 is average fragment size.

digests; Livermore). The libraries are labeled with an eight-digit code. The first two digits refer to the laboratory of origin (LA = Los Alamos; LL = Lawrence Livermore); the second two digits refer to the chromosomal type; the fifth digit refers to chromosomal structure (N = normal; T = translocation; etc.); the sixth refers to insert size (S = small, complete digest; L = large, partial digest); and the final two digits represent the library construction number. In addition, the tables contain information from the cloning procedure that can help a library user determine how suitable a library may be for a particular experiment. The cell source is given for each chromosome so a comparison of Tables 3 and 4 with Table 1 will identify the chromosome content of the cell culture from which each library was made.

The most important information for library users is found in Table 5. Here we present characterization data from 24 of the libraries listed in Tables 3 and 4. Much of the information was generated at Los Alamos or Livermore, but seven different user groups have also reported characterization results to us; these are included in Table 5. The most extensive mapping data come from the libraries for chromosomes 4 and 7, reflecting interests in the search for DNA polymorphisms linked to the defective gene for Huntington's disease and cystic fibrosis, respectively. In these libraries, the human sequences map almost exclusively to the proper human chromosome; this was expected because the only other human chromosomes in the cell hybrids used for these sorts were not close in the flow histogram. The relatively high levels of hamster contamination were not expected, and they are not yet completely understood. The first report of hamster contamination (T.C. Gilliam et al., in prep.) prompted us to examine the libraries originating from hybrid cells for the presence of hamster sequences. This was done using total human and total hamster libraries as standards (Perlman and Fuscoe 1986). These results, suggesting that some libraries contain high levels of hamster sequences, were surprising because our earlier sort purity determinations from hexylene glycol preparations were in the 90% or greater range. More recently, we used in situ hybridization with total human DNA probes to determine contamination levels after sorting. A comparison of the results from the two methods (Table 5) indicates that, in general, the library impurity is higher than the sort impurity. These differences may be due to fragments of hamster DNA that cannot be detected with in situ hybridization, by the presence of recombinant phage containing multiple inserts, or by features of the organization of the mammalian genome; i.e., repetitive sequence abundance, chromosomal distribution of repetitive sequences, and fre-

Table 5. Characterization Data on 25 Libraries

Library	Mapping unique sequence clones	Avg. insert size (kb)	Estimated hamster impurity (%): by in situ hybridization of sorted chromosomes	Estimated hamster impurity (%): by plaque hybridization	Reference
LA *01* NS01			10	26	this paper
LA *02* NS01			14	39	this paper
LA *03* NS01			6	23	this paper
LL *04* NS01	47/47 #4	2.0		~20	T.C. Gilliam et al. (in prep.)
				17	Perlman and Fuscoe (1986)
LA *04* NS01			3	11	this paper
LA *05* NS01			4	25	this paper
LL *06* NS01				23	Perlman and Fuscoe (1986)
LA *06* NS01			6	55	this paper
LA *07* NS01	18/19 #7				P.J. Scambler (pers. comm.)
	53/55 #7			20–30	D.F. Barker (pers. comm.)
			26	48	this paper
LL *08* NS02				16	Perlman and Fuscoe (1986)
LA *08* NS04			20	33	this paper
LL *09* NS01				38	Perlman and Fuscoe (1986)
LA *10* NS01			12	53	this paper
LL *11* NS01				52	Perlman and Fuscoe (1986)
LA *11* NS01	6/6 #11			40–50	T. Shows (pers. comm.)
LA *11* NS02			21	27	this paper
LA *12* NS01			9	21	this paper
LL *45* NS01	3/4 #14/15	3.5			T. Van Tuinen (pers. comm.)
LA *15* NS02			2	22	this paper
LL *16* NS02	1/11 #16	3.4			V.J. Hyland and G.R. Sutherland (pers. comm.)
LA *16* NS02	1/1 #16	2.3			C.E. Hildebrand and R. Stallings (pers. comm.)
LL *17* NS01	10/20 #17	2.0			T. Van Tuinen (pers. comm.)
LL *18* NS01	3/3 #18	1.9			Perlman and Fuscoe (1986)
LL *20* NS01	5/6 #20	3.1			Perlman and Fuscoe (1986)
LA *0X* NS01			5	25	this paper

quency of *Eco*RI or *Hin*dIII restriction sites in repetitive sequence clusters. Studies are in progress to quantitatively evaluate these possibilities.

The limited data available on libraries made from chromosomes sorted from diploid human cells suggest that some are pure (*Hin*dIII 18 and 20) and some are not (*Hin*dIII 16 and 17). Libraries that are determined to have limited usefulness will be reconstructed.

Library Storage

Initially, the libraries were stored in SM buffer at 4°C as plate lysates. Unfortunately, several libraries suffered severe titer drops within 1 year under these conditions. Because titer maintenance is important in preserving library characteristics, more-stable storage methods were sought. Libraries transferred to 4 M CsCl and held at 4°C had stable titers for 1 year, and methods were later devised for holding the libraries in the vapor phase of liquid nitrogen without loss in viability (Dr. William C. Nierman, American Type Culture Collection). Currently, libraries are held in liquid nitrogen in SM buffer or in CsCl with 10% glycerol. Prior data indicate that they will remain stable for at least 5 years under these conditions.

DISCUSSION

The libraries listed in Tables 3 and 4 are currently available from the American Type Culture Collection, Rockville, Maryland 20852. Prior to transferring the libraries to the ATCC (February 1, 1986), more than 1200 library aliquots had been shipped from Los Alamos and Livermore to user groups throughout the world.

Current and future work on the project includes further attempts to improve the purity levels of sorted chromosomes without a reduction of sorter yields. The in situ hybridization procedure described in this paper provides a good means of assessing sort purity from hybrid cells. We hope to extend this method to chromosomes sorted from diploid human cells by isolating chromosome-specific probes. Three such probes are now available for chromosomes 9, 15, and 16 (R.K. Moyzis et al., pers. comm.). We are also examining the chromosomal evolution in hybrid cells that carry human chromosomes with integrated selectable markers (Athwal et al. 1985). If these cells are more stable than the hybrid lines we are currently using, a large amount of uncertainty regarding the specific chromosome constitution of a cell line at a given point in time will be eliminated. We are also investigating the use of short synthetic oligomers that encompass the *Hin*dIII or *Eco*RI sites in Charon 21A in order to detect the presence of nonrecombinant phage in the libraries. Several lines of evidence suggest that the libraries contain an appreciable number of phage with very small inserts (<300 bp). It would be useful to be able to rapidly detect and separate these phage from nonrecombinants.

In addition to finishing and improving the two sets of complete digest libraries, we are initiating studies to determine the type of partial digest library that would best meet our constraints of limited chromosomal starting material and best serve the needs of the community of library users. A prototype partial digest library of the human X chromosome has been constructed in Charon 35 (acceptance range, 17–20 kb) and work is in progress on a similar construction using cosmid vectors (acceptance range, 35–40 kb). These preliminary libraries will be characterized before a final decision is made on the vector choice for a complete set of large insert libraries.

ACKNOWLEDGMENTS

We wish to thank the following individuals who served on the Advisory Committee for the project: Paul Berg, Stanford University School of Medicine; Fred Blattner, University of Wisconsin; Thomas Caskey, Baylor College of Medicine; Marshall Edgell, University of North Carolina; Richard Gelinas, University of Washington; Samuel Latt, Harvard Medical School; Thomas Maniatis, Harvard University; Arno Motulsky, University of Washington School of Medicine; William Rutter, University of California/San Francisco; Carl Schmid, University of California/Davis; Thomas Shows, Roswell Park Memorial Institute.

We thank R. Archuleta for assistance in manuscript preparation.

This work was performed under the auspices of the U.S. Department of Energy by the Lawrence Livermore National Laboratory under contract number W-7405-ENG-48 and by the Los Alamos National Laboratory under contract number W-7405-ENG-36.

REFERENCES

Athwal, R.S., M. Smarsh, B.M. Searle, and S.S. Deo. 1985. Integration of a dominant selectable marker into human chromosomes and transfer of marked chromosomes to mouse cells by microcell fusion. *Somatic Cell Mol. Genet.* **11:** 177.

Athwal, R.S., M. Smarsh, B.M. Searle, and S.S. Deo. 1984. Integration of a dominant selectable marker into human chromosomes and transfer of marked chromosomes to mouse cells by microcell fusion. *Somatic Cell Mol. Genet.* **11:** 177.

Benton, W.D. and R.W. Davis. 1977. Screening λgt recombinant clones by hybridization to single plaques in situ. *Science* **196:** 180.

Blattner, F.R., B.G. Williams, A.E. Blechl, K.D. Thompson, H.E. Faber, L.A. Furlong, D.J. Grunwald, D.O. Kiefer, D.D. Moore, J.W. Schumm, E.L. Sheldon, and O. Smithies. 1977. Charon phages: Safer derivatives of bacteriophage lambda for DNA cloning. *Science* **196:** 161.

Davies, K.E., B.D. Young, R.G. Elles, M.E. Hill, and R. Williamson. 1981. Cloning of a representative genomic library of the human X chromosome after sorting by flow cytometry. *Nature* **293:** 374.

Deaven, L.L., C.E. Hildebrand, J.C. Fuscoe, and M.A. Van Dilla. 1986. Construction of human chromosome specific DNA libraries: The national laboratory gene library project. In *Genetic engineering* (ed. J. Setlow and A. Hollaender), vol. 8, p. 317. Plenum Press, New York.

Disteche, C.M., L.M. Kunkel, A. Lojewski, S.H. Orkin, M. Eisenhard, E. Sahar, B. Travis, and S.A. Latt. 1982. Isolation of mouse x-chromosome specific DNA from an x-enriched lamba phage library derived from flow sorted chromosomes. *Cytometry* **2:** 282.

Fuscoe, J.C., L.M. Clark, and M.A. Van Dilla. 1986. Construction of fifteen human chromosome-specific DNA libraries. *Cytogenet. Cell Genet.* (in press).

Gray, J.W., A.V. Carrano, L.L. Steinmetz, M.A. Van Dilla, D.H. Moore, B.H. Mayall, and M.L. Mendelsohn. 1975. Chromosome measurement and sorting by flow systems. *Proc. Natl. Acad. Sci.* **72:** 1231.

Griffith, J.K., L.S. Cram, B.D. Crawford, P.J. Jackson, J. Schilling, R.T. Schimke, R.A. Walters, M.E. Wilder, and J.H. Jett. 1984. Construction and analysis of DNA sequence libraries from flow-sorted chromosomes: Practical and theoretical considerations. *Nucleic Acids Res.* **12:**4019.

Grunstein, M. and D.S. Hogness. 1975. Colony hybridization: A method for the isolation of cloned DNAs that contain a specific gene. *Proc. Natl. Acad. Sci.* **72:** 3961.

Gusella, J.F., C. Keys, A. Vassanyi-Breiner, F.T. Kao, C. Jones, T.T. Puck, and D. Housman. 1980. Isolation and localization of DNA segments from specific human chromosomes. *Proc. Natl. Acad. Sci.* **77:** 2829.

Hohn, B. and K. Murray. 1977. Packaging recombinant DNA molecules into bacteriophage particles in vitro. *Proc. Natl. Acad. Sci.* **74:** 3259.

Krumlauf, R., M. Jeanpierre, and B. Young. 1982. Construction and characterization of genomic libraries from specific human chromosomes. *Proc. Natl. Acad. Sci.* **79:** 2971.

Kunkel, L.M., K.D. Smith, S.H. Boyer, D.S. Borgaonker, S.S. Wachtel, O.J. Miller, W.R. Breg, H.W. Jones, and J.M. Rary. 1977. Analysis of human y-chromosome-specific reiterated DNA in chromosome variants. *Proc. Natl. Acad. Sci.* **74:** 1245.

Langlois, R.G., L.-C. Yu, J.W. Gray, and A.V. Carrano. 1982. Quantitative karyotyping of human chromosomes by dual beam flow cytometry. *Proc. Natl. Acad. Sci.* **79:** 7876.

Latt, S.A., F.W. Alt, R.R. Schreck, N. Kanda, and D. Baltimore. 1982. The use of chromosome flow sorting and cloning to study amplified DNA sequences. In *Gene amplification* (ed. R.T. Schimke), p. 283. Cold Spring Harbor Laboratory, Cold Spring Harbor, New York.

Lawn, R.M., E.F. Fritsch, R.C. Parker, G. Blake, and T. Maniatis. 1978. The isolation and characterization of a linked δ- and β-globin gene from a cloned library of human DNA. *Cell* **15:** 1157.

Leder, P., D. Tiemeier, and L. Enquist. 1977. ED2 derivatives of bacteriophage lambda useful in the cloning of DNA from higher organisms: The λgtWES system. *Science* **196:** 175.

Maniatis, T., E.F. Fritsch, and J. Sambrook. 1982. *Molecular cloning: A Laboratory Manual*, p. 545. Cold Spring Harbor Laboratory, Cold Spring Harbor, New York.

Meyne, J., M.F. Bartholdi, G. Travis, and L.S. Cram. 1984. Counterstaining human chromosomes for flow karyology. *Cytometry* **5:** 580.

Ohno, S. 1974. Conservation of ancient linkage groups in evolution and some insight into the genetic regulatory mechanism of X-inactivation. *Cold Spring Harbor Symp. Quant. Biol.* **38:** 155.

Olsen, A.S., D.W. McBride, and M.D. Otey. 1980. Isolation of unique sequence human X chromosomal deoxyribonucleic acid. *Biochemistry* **19:** 2419.

Perlman, J. and J.C. Fuscoe. 1986. Molecular characterization of human chromosome-specific DNA libraries. *Cytogenet. Cell Genet.* (in press).

Pinkel, D., T. Straume, and J.W. Gray. 1986. Cytogenetic analysis using quantitative, high sensitivity, fluorescence hybridization. *Proc. Natl. Acad. Sci.* **83:** 2934.

Schmeckpeper, B.J., K.D. Smith, B.P. Dorman, F.H. Ruddle, and C.C. Talbot. 1979. Partial purification and characterization of DNA from the human X chromosome. *Proc. Natl. Acad. Sci.* **76:** 6525.

Shay, J.W. and L.S. Cram. 1985. Cell fusion and chromosome sorting. In *Molecular cell genetics* (ed. M. Gottesman), p. 155. Wiley, New York.

Sillar, R. and B.D. Young. 1981. A new method for the preparation of metaphase chromosomes for flow analysis. *J. Histochem. Cytochem.* **29:** 74.

Van den Engh, G., B. Trask, S. Cram, and M. Bartholdi. 1984. Preparation of chromosome suspensions for flow cytometry. *Cytometry* **5:** 108.

Van Dilla, M., L.L. Deaven, K.L. Albright, N.A. Allen, M.R. Aubuchon, M.F. Bartholdi, N.C. Brown, E.W. Campbell, A.V. Carrano, L.M. Clark, L.S. Cram, J.C. Fuscoe, J.W. Gray, C.E. Hildebrand, P.J. Jackson, J.H. Jett, J.L. Longmire, C.R. Lozes, M.L. Luedemann, J.C. Martin, J. Meyne, J.S. McNinch, L.J. Meincke, M.L. Mendelsohn, R.K. Moyzis, A.C. Munk, J. Perlman, D.C. Peters, A.J. Silva, and B.J. Trask. 1986. Human chromosome-specific DNA libraries: Construction and availability. *Biotechnology* **4:** 537.

Van Dilla, M.A., P.N. Dean, O.D. Laerum, and M.R. Melamed, eds. 1985. *Flow cytometry: Instrumentation and data analysis*. Academic Press. London.

Wensink, P.C., D.J. Finnegan, J.E. Donelson, and D.S. Hogness. 1974. A system for mapping DNA sequences in the chromosomes of *Drosophila melanogaster*. *Cell* **3:** 315.

Wray, W. and E. Stubblefield. 1970. A new method for the rapid isolation of chromosomes, mitotic apparatus, or nuclei from mammalian fibroblasts at near neutral pH. *Exp. Cell Res.* **5:** 469.

Flow-sorting Analysis of Normal and Abnormal Human Genomes

R.V. LEBO, L.A. ANDERSON, Y.-F.C. LAU, R. FLANDERMEYER, AND Y.W. KAN
Howard Hughes Medical Institute and Department of Medicine, University of California, San Francisco, California 94143

Flow analysis of human chromosomes is a powerful new approach to study the human genome. Human chromosomes can be resolved almost completely as a result of improvements in metaphase chromosome suspension preparations, new fluorescent dye combinations, and advances in sorter optical design. We have applied this technique to map genes, detect abnormal chromosome constitutions, and construct individual recombinant chromosomal DNA libraries.

The dual-laser sorter can separate all the human chromosomes except 10 and 11 (Lebo and Bruce 1986). Individual chromosomes are sorted directly onto filter paper and hybridized to radiolabeled gene probe (Lebo et al. 1984, 1985c). Chromosomes from cell lines with appropriate translocations and deletions are used to localize genes subchromosomally (Lebo et al. 1985b; Lebo 1986). With these methods, we have located 37 recently cloned genes. When homologous gene sequences such as ferritin-L (Lebo et al. 1985a) and von Willebrand factor are located on different chromosomes, Southern blot analysis of sorted chromosomal DNA directly assigns the genes to specific chromosomes.

Flow analysis can detect altered chromosome frequencies and abnormal chromosomes with a 10% change in total DNA content. We have examined 11 clinical samples with deletions, insertions, translocations, and aneuploidy and found excellent agreement with results obtained by Giemsa-banded karyotypes (Lebo et al. 1986a). These results indicate that flow cytogenetics is potentially useful for patient screening.

A chromosome-specific library facilitates the identification of many chromosome-specific DNA fragments that can be tested to locate a polymorphism close to a mapped disease locus. A library with large DNA inserts will also facilitate walking toward a disease locus from a closely linked polymorphic locus that has been identified. We have currently constructed an EMBL-4 phage library of chromosome 1 to locate a polymorphic DNA fragment close to the Charcot-Marie-Tooth disease locus (Lebo et al. 1986b). The fragments are being sublocalized by spot-blot analysis of derivative chromosomes, and those fragments found in the vicinity of the disease locus will be tested for polymorphism. Subsequent pedigree analysis will establish which fragment can be used to test the disease status of an at-risk fetus.

METHODS

Chromosome suspension preparation. Normal whole chromosomes were sorted from suspensions prepared from normal male and female permanent human lymphocyte cultures. Derivative chromosomes were sorted from fibroblast cell lines generally purchased from the Human Genetic Mutant Cell Repository (Camden, New Jersey). Mitotic cells were collected by colcemid block, suspended in hypotonic buffer, and then dissociated by homogenizing or vortexing in spermine buffer (Lebo et al. 1985b,c; Lebo and Bruce 1986). Chromosomes were then stained with the complementary dyes DIPI-chromomycin or Hoechst-chromomycin (Lebo and Bruce 1986) several days before analysis.

Chromosome sorting. Dual-laser flow analysis measures total DNA-bound fluorescent dye emission from each chromosome as it passes through two tightly focused laser beams (Gray et al. 1979). The A-T-specific dye Hoechst 33258 (Latt and Wohlleb 1975) or DIPI (Schnedl et al. 1977) emits blue fluorescence, and the G-C-specific dye chromomycin A_3 (Latt 1977) emits yellow fluorescence, which is correlated with the base-pair content of individual chromosomes (Bartholdi 1985) as well as with the Giemsa-banding patterns (Lebo et al. 1986a). We have improved the optics on the sorting bench in two phases in order to improve upon the chromosome sorting resolution, along with other system improvements that facilitate switching between cell and chromosome sorting (Lebo and Bastian 1982; Lebo et al. 1987). The most recent instrument with the highest chromosome resolution has three independently focused, high-resolution beams for use with all currently used fluorochromes (Lebo et al. 1987). All fluid lines were cleaned with sodium hypochlorite and ethanol to provide a sterile, aseptic chromosome delivery and sort. Filters were chosen to resolve further the blue Hoechst or DIPI signal from the yellow chromomycin A_3 signal, which were already separated by 0.008 inches in space and 15 μsec in time (Lebo et al. 1987). A central vacuum line was used to remove the undeflected droplets carrying unwanted chromosomes and interphase nuclei so that the two sorted fractions deflected right and left could be collected directly on a nitrocellulose filter paper under vacuum (Fig. 2) (Lebo et al. 1985c).

Spot-blot hybridization. Filters were denatured with alkali, neutralized, baked in a vacuum oven, prehybridized, and hybridized at high stringency to a radiolabeled gene probe. The excess probe was washed away under high stringency, and the filters were autoradiographed to visualize the chromosome-specific gene signal (Lebo et al. 1985c). Whenever possible, hybridization and washing conditions were chosen to assure that the primary gene sequence would give a signal that was considerably darker than homologous gene sequences (Fig. 3).

Flow cytogenetics of chromosomes. Chromosomes were analyzed on a custom dual-laser FACS sorter (Lebo and Bastian 1982). Two histograms were collected for each cell line and each strain pair: one histogram for all chromosomes and one histogram for chromosomes 7–22. The identity of peaks representing the previously known abnormal chromosomes was determined by quinacrine-banding sorted chromosomes and spot-blot analysis (Lebo et al. 1984, 1985c). The signals were recorded on a Nuclear Data 64×128 channel histogram. The histogram peak channels were determined by placing gates horizontally and then vertically on both edges of the peak and calculating the mean horizontal and vertical values from the extreme coordinates. The PDP 1103 Nuclear Data computer in the FACS console was used to place rectangular windows around each peak and calculate the number of events within the window. The number of events in each peak was compared to the number in immediately adjacent histogram peaks.

EMBL phage chromosome library construction. Five million chromosome 1's were sorted into an SW-40 centrifuge tube, pelleted by ultracentrifugation, digested with proteinase K, extracted with phenol, and precipitated with ethanol without added carrier. The chromosomal DNA was digested partially with *Mbo*I until most of the DNA was between 10 and 25 kb in length. The fragments were then cloned into the *Bam*HI site of the EMBL-4 phage. Entire phage DNAs were radiolabeled by random priming and hybridized to sorted spots of chromosome 1 versus chromosome 2 and derivative chromosomes in cell line GM4618 carrying fragments of chromosome 1 under hybridization conditions similar to those defined by Litt and White (1985).

RESULTS

Gene Mapping

Mitotic chromosome suspensions are stained with two complementary DNA-specific dyes: Hoechst 33258 or DIPI, which preferentially stain Giemsa-positive bands, and chromomycin A_3, which preferentially stains Giemsa-negative bands. These chromosomes are directed single file through two separate, independently focused laser beams (Gray et al. 1979). Each chromosome flows through the first blue-green laser beam that excites the chromomycin A_3 to fluoresce yellow. Then the chromosome flows through the second invisible ultraviolet laser beam (15 μsec later) that excites Hoechst or DIPI to fluoresce blue (Lebo et al. 1984). The two independent fluorescent signals are further separated by filters that pass only one fluorescent wavelength to each detector. A flow histogram is generated by electronically recording two signals from each of about 200,000 chromosomes in a three-dimensional histogram (Fig. 1). A high-resolution dual-laser sorter can separate and sort 20 of the 24 types of human chromosomes uniquely from a suspension of Hoechst-chromomycin-stained chromosomes. Staining suspensions with two different concentrations of DIPI-chromomycin can further resolve chromosome 9 from 10, 11, and 12 and chromosomes 9 and 12 from 10 and 11 (Lebo and Bruce 1986).

We have used these multiple chromosome separations as the basis of a rapid, direct gene-mapping method. Thirty thousand chromosomes of each separate type are sorted directly onto 1-inch circular nitrocellulose filters (Fig. 2). The chromosomal DNA is denatured and hybridized to a radiolabeled gene probe. The gene-specific signal hybridizes only to the chromosome-specific spot(s) containing homologous DNA

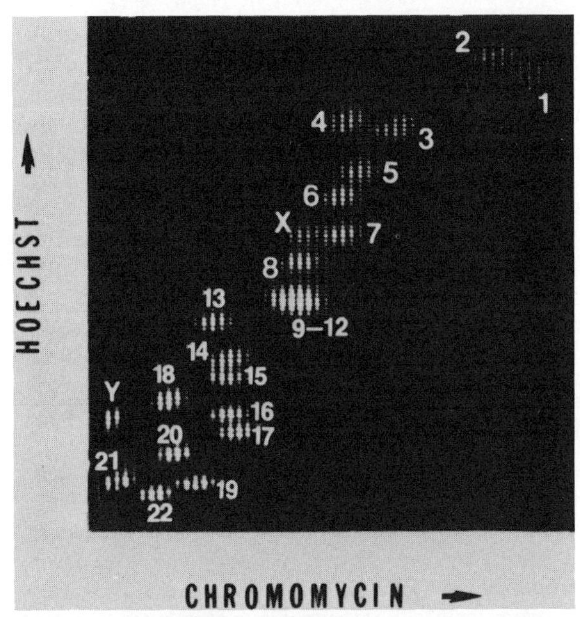

Figure 1. Flow histogram of normal male human chromosomes. A 64×128 channel three-dimensional histogram of flow-analyzed lymphocyte chromosomes viewed from above on a Nuclear Data electronic display. The two laser-excited fluorescence signals were measured separately and correspond linearly to chromomycin fluorescence from the first laser on the abscissa and to Hoechst fluorescence from the second laser on the ordinate. The light intensity for each dot in the three-dimensional histogram is linearly proportional to the number of chromosomes recorded in each channel. Note that half as many events are recorded in the chromosome-X and -Y histogram peaks as in the adjacent normal chromosome peaks. (Reprinted, with permission, from Lebo et al. 1984.)

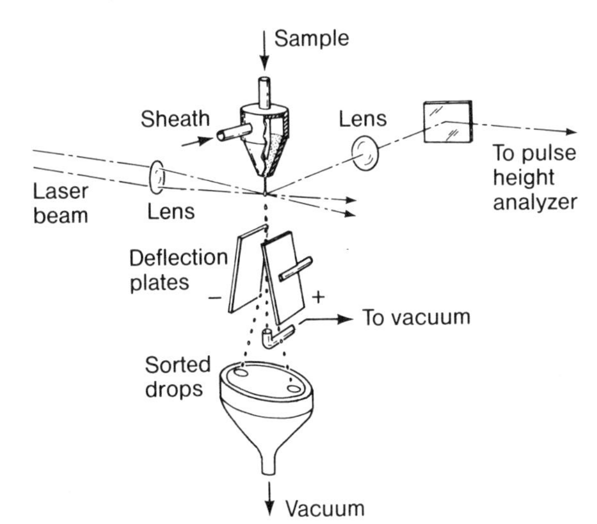

Figure 2. Spot-blot filter construction. The chromosome suspension is surrounded by sheath fluid and directed through the focused laser beam. The shorter laser-light photons are converted by the DNA-bound dye to longer wavelength fluorescent light. The scattered laser light is blocked by filters that pass the fluorescent photons onto a photomultiplier tube that converts the signal to an electrical impulse. Depending on the intensity of the fluorescent signal, the chromosome-containing droplets of interest are deflected onto one of two spots of a nitrocellulose filter applied to a scintered-glass Millipore filter holder under vacuum. The remaining chromosomes in undeflected drops are collected by the central vacuum tube. Thirty thousand chromosomes of each type were deflected to either side directly onto the nitrocellulose filter on spots delineated by the circles. An entire filter set with all chromosomes represented can be sorted in 5 hr. Once a filter set has been prepared, it may be hybridized and rehybridized about eight times to locate additional cloned genes quickly.

sequences. About half of the gene probes we mapped like insulin (Lebo et al. 1985b) hybridized to a single chromosome spot. The other probes hybridized to two or more independent chromosome spots that contain homologous DNA sequences like the aldolase-B gene probe (Fig. 3). The primary aldolase-B sequence on chromosome 9 hybridized most intensely to the isozyme-specific gene locus.

Even though some probes were hybridized to a spot-blot filter panel at high stringency, two or more different chromosome spots revealed strong signals. We then used restriction enzyme analysis of sorted chromosomal DNA to distinguish which homologous locus is located on each positive chromosome. Since the previously used standard-size Southern blots required many days of chromosome sorting to collect enough DNA for analysis, we applied a miniaturized procedure to decrease the number of sorted chromosomes required for analysis (Law et al. 1984). The loci homologous to the ferritin light-chain gene clone exhibit five *Bgl*II restriction enzyme fragments. When we sorted and tested each of several chromosomes that gave positive spot-blot hybridization, we found that the five homologous ferritin-L gene fragments were carried by three chromosomes (Fig. 4). We obtained similar results with the von Willebrand gene probe. Although the primary gene locus is on chromosome 12, a homologous locus on chromosome 22 gave a strong spot-blot signal. The unique restriction enzyme fragments on chromosome 22 were identified by miniaturized restriction enzyme analysis of sorted chromosomes.

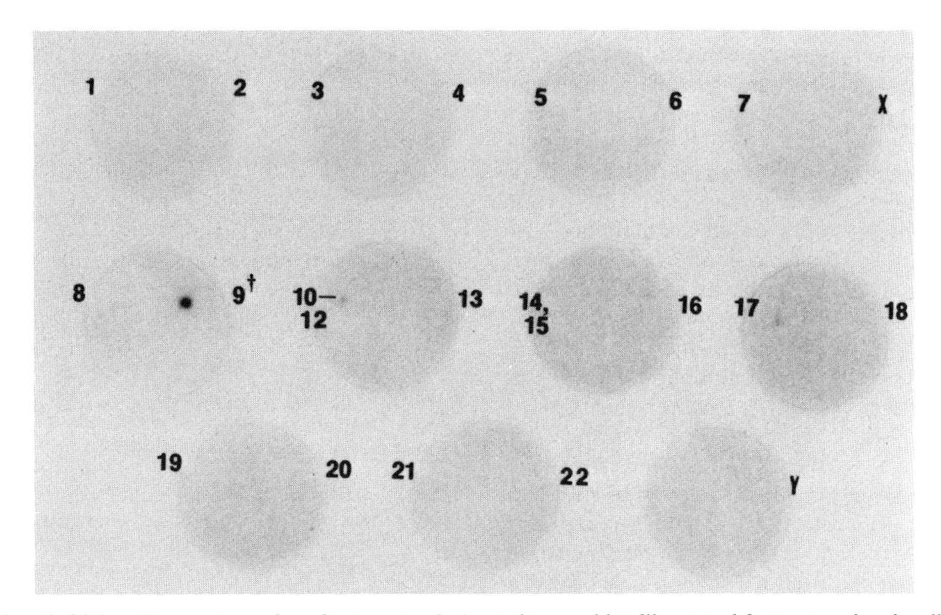

Figure 3. Cloned aldolase-B gene mapped to chromosome 9. An entire spot-blot filter panel from normal male cell line GM130 (Fig. 1) stained and sorted with DIPI-chromomycin (Lebo et al. 1984) was hybridized to the radiolabeled aldolase-B-gene clone. The chromosome-9 spot hybridized most intensely to this gene-specific probe, with lighter signal in the chromosome-17 and chromosome-10–12 spots. The chromosome-17 spot represents specific hybridization of the homologous aldolase-C gene probe (D.R. Tolan et al., in prep.), whereas the chromosome-10–12 spot may result from contamination from the neighboring chromosome-9 peak or from the aldolase-ΨA gene sequence on chromosome 10 (D.R. Tolan et al., in prep.). (Reprinted, with permission, from Lebo et al. 1985c.)

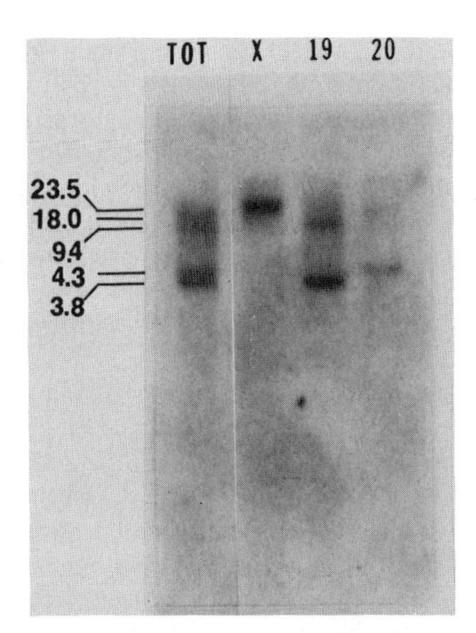

Figure 4. Miniature restriction enzyme analysis of sorted chromosomes. In the left lane, restriction enzyme analysis of total *Bgl*II-digested human DNA (TOT) reveals all five restriction fragment lengths. *Bgl*II-digested sorted X, 19, and 20 chromosomal DNA hybridized to the ferritin-L probe reveals that each of the five restriction fragment lengths are carried by unique chromosomes. (Reprinted, with permission, from Lebo et al. 1985a.)

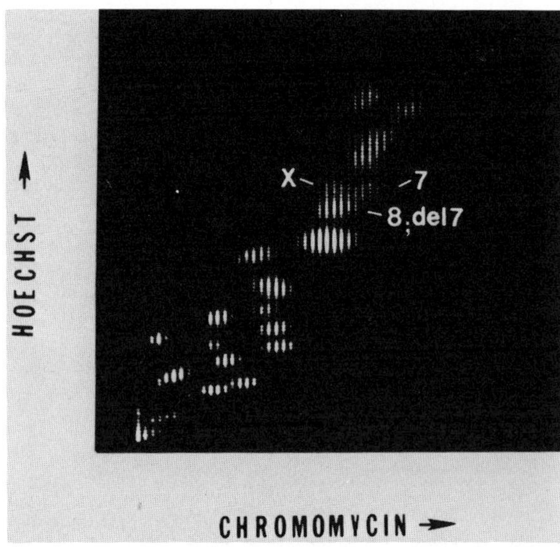

Figure 5. Flow cytogenetic analysis of a deleted chromosome-7 karyotype. This male cell line has a single chromosome X as expected. However, the events in the normal chromosome-7 and -8 peaks are altered to reveal that a deleted chromosome 7 has moved into the chromosome-8 peak (listed in Lebo et al. 1986a).

Flow Cytogenetics

Flow histograms like that in Figure 1 characterize the chromosomal DNA bundles from each cell line by the peak positions and the number of events in each peak. The normal variability of the heterochromatic regions of chromosomes 1, 9, and 16 and the different-size satellites of the acrocentric chromosomes (13, 14, 15, 21, and 22) cause minor, predictable shifts in peak positions (Lebo et al. 1986a). More major changes will result in larger predictable changes in peak position or in the number of events in a chromosome peak. Major changes resulting from monosomy or a single copy of a chromosome like X and Y (Fig. 1) will decrease the number of events by 50%, whereas trisomy will increase the number of chromosome events by 50%. A chromosome deletion will move the chromosome peak toward the origin, and an insertion will move the chromosome peak away from the origin. A reciprocal translocation between two chromosomes generally results in two new chromosome peaks (Lebo et al. 1986a).

For example, we identified a chromosome-7 deletion from the histogram in Figure 5. Since this male cell line has one copy of the X chromosome, we expect to find twice as many events in peaks 7 and 8. In fact, an equal number of events are in the chromosome-7 peak and three times as many events are in the chromosome-8 peak. Since a monosomy 7 trisomy 8 is extremely rare, we correctly designated this a deletion chromosome 7.

We also tested an Epstein-Barr-virus-transformed permanent human B-lymphocyte cell line (Fig. 6). Five new peaks appeared, and the frequency of several normal chromosome peaks also varied. Marker chromosomes half the size of chromosome 21 were clearly resolved. One abnormal chromosome moved into a normal chromosome peak from which another chromosome moved. Thus, one less event was scored by flow analysis than by Giemsa-banded karyotyping. However, the two results were entirely consistent.

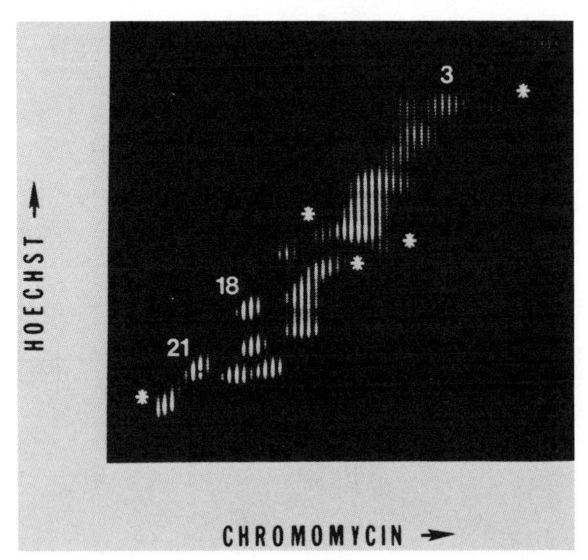

Figure 6. Flow cytogenetic analysis of transformed lymphocyte GM3639. Numerous chromosome rearrangements are indicated by five new histogram peaks (*) as well as the considerable variability in the number of events in the normal chromosomal peak positions. Note that the chromosome fragments half the size of the normal chromosome 21 are resolved uniquely. One abnormal chromosome moved into a new peak position from which another normal chromosome moved. Thus, one less chromosome rearrangement event was detected by flow cytogenetics than by Giemsa-banded karyotyping. Nevertheless, both methods gave entirely consistent results (listed in Lebo et al. 1986a). Multiple chromosome abnormalities generally occur in Epstein-Barr-transformed lymphocyte cell lines.

Chromosome-1 Library

To find a polymorphic DNA restriction enzyme fragment for prenatal diagnosis of a defined gene deficiency, the gene is cloned and the clone is used to identify polymorphic fragments. Defining a polymorphic restriction fragment in the vicinity of a genetic disease of unknown etiology requires a different strategy. First, the segregation of known polymorphic DNA or protein markers is tested in pedigrees with patients exhibiting the disease. Nonrandom segregation of a polymorphic marker in pedigrees with the disease phenotype defines a linked polymorphic locus. Subsequently, the polymorphic locus is mapped to a unique chromosome that simultaneously localizes the linked disease to a specific chromosome. Finally, the search commences for more closely linked genetic markers, with the ultimate goal being to identify and isolate the disease locus.

Several methods have been used to search for polymorphic markers of disease loci already mapped to a unique chromosome. All the cloned genes and restriction enzyme fragments mapped to the same chromosome can be tested for polymorphism. Pedigrees with disease patients are then tested further for nonrandom segregation between the disease and the polymorphism. Given enough mapped genes and DNA fragments, this method will eventually succeed. The other approach is to develop a chromosome-specific recombinant DNA library that can provide many chromosome-specific fragments quickly near a known disease locus like Huntington's disease (Lebo 1982). These are sublocalized and only the fragments in the chromosome region of interest are tested for segregation with the disease locus. We are using this approach to study the Charcot-Marie-Tooth (CMT) syndrome on chromosome 1 (Lebo et al. 1986b).

The CMT syndrome (Charcot and Marie 1886; Tooth 1886) is a human genetic disease affecting at least 1 in 50,000 people in the United States. Patients exhibit degenerative neurological symptoms that generally begin in the second or third decade. About two thirds of those affected have slower nerve conduction. Studies of pedigrees with slow nerve conduction patients show (1) autosomal dominant CMT type 1 (CMT-1) linked to the Duffy blood-group locus on chromosome 1 (Bird et al. 1982; Stebbins and Conneally 1982), (2) X-linked inheritance, and (3) autosomal inheritance unlinked to Duffy.

One method to identify unique, closely linked gene probes is to map genes randomly and to further test genes located on the same chromosome as the disease locus. Of the 37 genes we mapped recently by spot-blot analysis, the cloned cellular *src*-2 oncogene (Parker et al. 1985) and the 4.1 erythrocyte membrane protein (Conboy et al. 1985) were both mapped to chromosome 1. However, further mapping by spot-blot analysis of a translocation involving the short arm of chromosome 1 (cell line GM201; Fig. 7) sublocalized both loci to the terminal portion of the short arm (see Table 1). The Duffy blood-group locus linked to CMT-1 is on the proximal long arm. Genes on the terminal portion of the short arm are so removed physically that these loci are likely to assort randomly with the *CMT-1* locus, i.e., be at least 50 cM away. Clearly, searching for a closely linked genetic marker by testing random probes is a tedious approach to the problem. Then again, other genes mapped to the same chromosome region as the disease locus by other laboratories (Human Gene Mapping 1985) can be tested for polymorphism and segregation with *CMT-1*.

To identify a useful polymorphic DNA probe for *CMT-1*, we first tested the segregation of the previously mapped antithrombin III gene and the *CMT-1* locus in a large pedigree (Lebo et al. 1986b). Previous results indicated that the genetic recombination distance between Duffy and *CMT-1* was 9% and that between Duffy and antithrombin III was 10%. On the basis of all available data, the antithrombin gene locus was concluded to be 1 cM (1% crossover) between *CMT-1* and antithrombin III (Human Gene Mapping). Our tests of a large CMT-1 pedigree in which the *CMT-1* locus recombines with the Duffy locus 10% of the time, with a lod score of 3.11, showed that the antithrombin III locus recombined with the *CMT-1* locus at least three times in ten opportunities. This indicates that the *CMT-1* and antithrombin III loci are quite separated on chromosome 1 (R.V. Lebo et al., in prep.).

Mapping chromosome library fragments is the most rapid, direct approach to identify more closely linked library fragments. Since fragments can be isolated from chromosome libraries in much larger quantities than unique gene probes have been mapped, a chromosome region with a known disease locus can be saturated more effectively. These chromosome-specific fragments can be isolated and sublocalized. Probes in the correct chromosome region that are found to be polymorphic are then tested in affected pedigrees for segregation between the mapped polymorphic DNA marker and the disease locus. With this in mind, we established a human chromosome 1 (Lebo et al. 1986b) phage library in the EMBL-4 phage vector from 5×10^6 sorted human chromosomes of each. The chromosomal DNA was partially digested with *Mbo*I until most fragments were greater than 10 kb to decrease the number of plaques with two or more fragments. These chromosomal fragments were cloned into the *Bam*HI site of the EMBL-4 λ phage. The EMBL-4 vector has a stuffer fragment with λ early functions. When the stuffer fragment is replaced with cloned DNA, recombinants can only grow on *Escherichia coli* strains lysogenic for the temperate phage P2. Since the stuffer fragment must be removed for phage growth and lysis, only recombinants with inserts form plaques. In addition, only internal DNA fragments between 5 and 22 kb will grow. Based on this size range, the resulting 50,000-plaque chromosome library is 95% complete. Initial analysis showed that five of the five total radiolabeled phage hybridized to spot-blot filters were derived from chromosome 1, with four of five on the terminal portion of the long arm (1q23→1qter) (Fig. 7)

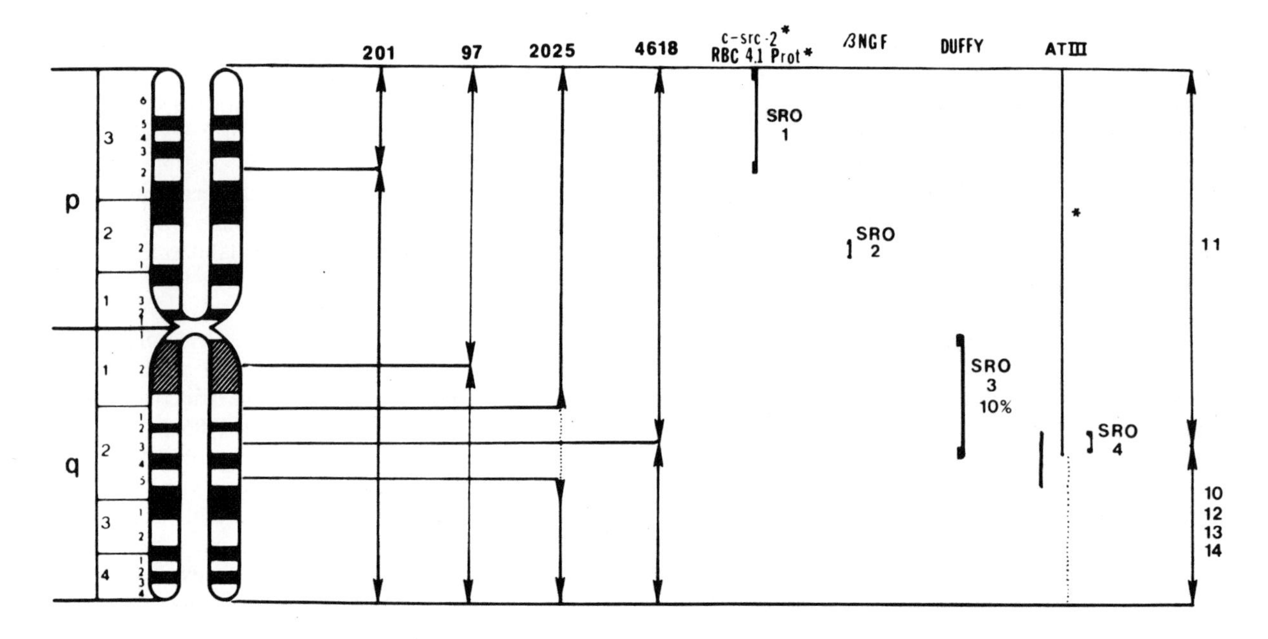

Figure 7. Summary of initial chromosome-1 studies. On the idiogram of chromosome 1 illustrated at the left are indicated the breakpoints of three translocations GM201, GM97, and GM4618 as well as the breakpoints of the deletion in GM2025. These rearrangements are being used to sublocalize further genes and DNA fragments from the chromosome-1 library. To the right are indicated the smallest regions of overlap (SRO) of gene-mapping studies with the c-*src*-2 oncogene (Lebo and Bruce 1986), the erythrocyte 4.1 protein (Conboy et al. 1986), the β-nerve growth-factor gene (Human Gene Mapping-8 1985), a working approximation of the Duffy blood-group locus based on previous linkage analysis (Human Gene Mapping-8 1985), and the antithrombin III locus (Human Gene Mapping-8 1985; and R.V. Lebo et al., in prep.). The solid line below antithrombin III (ATIII) indicates the region to which the antithrombin III locus has been mapped by spot-blot analysis of translocated chromosome from cell line GM4618. At the far right are indicated the locations of the five EMBL-4 phage clones (10–14) that have been mapped further by the GM4618 translocation. (Reprinted, with permission, from Lebo et al. 1986.)

The fifth clone was on the other portion of the chromosome. We will test more clones and further localize those in the region of the Duffy locus, using another translocation and deletion to identify fragments close to the *CMT-1* locus. This general approach can be applied to any genetic disease for which the molecular defect has not been defined.

DISCUSSION

The human genome has been mapped more slowly and with considerably more effort than the *Drosophila* (fruit fly) genome, since the generation time is longer, the number of offspring is smaller, the number of linkage groups is several times larger, and the matings are uncontrolled. More sophisticated techniques had to be developed to study the human genome with its more immediate relevance to human genetic disease. Genes have been mapped with somatic-cell hybrids (Ruddle and Cregan 1981), in situ hybridization (Harper et al. 1981), and spot-blot analysis of sorted human chromosomes. Polymorphic restriction fragments that were first used to diagnose a human genetic disease prenatally (Kan and Dozy 1978) are now available on many more chromosome sites to test linkage between gene loci (cf. Lebo et al. 1983). Human chromosome-specific recombinant DNA libraries (Disteche et al. 1982) facilitate isolating many chromosome-specific fragments to saturate the human map by linkage analysis of polymorphic restriction enzyme fragments. The classical genetics of the laboratory fruit fly has provided the basis to study the segregation of linked genes in pedigrees, but segregation in human pedigrees is now analyzed with computer programs that calculate recombination frequencies and lod scores (Ott 1974) of multiple gene loci. The human karyotype no longer consists of 46 unbanded chromosomes, but rather of a banded karyotype with up to 2000 bands per haploid genome (Francke and Oliver 1978; Yunis and Chandler 1978). With these improved methods, chromosome rearrangements can be characterized with greater precision and separated directly after identification by dual-laser flow sorting for subchromosomal gene localization. Information obtained by mapping oncogenes and immunoglobulin genes to chromosome bands involved in chromosome translocations in specific lymphomas and leukemias led to a greater understanding of molecular events in the etiology of these cancers (cf. Davis et al. 1984). Indeed, the tools to study human genetics including flow cytogenetics are undergoing continual evolution. Perhaps these tools will contribute to further understanding gene expression in health and disease.

Table 1. New Gene Assignments by Spot-blot Analysis

Gene	Chromosome location[a]	Probe source
c-*src*-2 oncogene	1p32→pter*	J.M. Bishop and R. Parker
RBC 4.1 protein	1p32→pter	J. Conboy
Fetal alkaline phosphatase	2*	W. Kam
Apolipoprotein B	2*	S. Deeb
Alcohol dehydrogenase	4	A. Yoshida
HLA-DRγ	5	H. Erlich
Glucocorticoid receptor	5,16*	R.M. Evans
Phosphoglycerokinase	6,X,19*	S. Gartler
Multiple drug resistance	7	I. Pastan
Erythropoietin	7	J. Powell
Tissue-type plasminogen activator	8*	G. Opdenakker
Aldolase B	9	D. Tolan
Aldolase ψA	10q11→qter	D. Tolan and E. Penhoet
Interleukin-2 receptor	10	W. Greene
Myophosphorylase	11q13→qter	F. Gorin and R. Fletterick
Parathyroid hormone	11p15.1→pter	P. O'Connell and R. White
ADJ-762	11p15.1→pter*	R. White
LDH-A	11p15.1→pter	G. Bruns
Calcitonin	11p15.1→pter*	M. Rosenfeld
von Willebrand factor	12*,22	F. Chehab
Procollagen (α1 type IV)	13*,8	C. Boyd and Prokoff
Her's Disease	14	R. Fletterick
β spectrin	14	J. Prchal
Aldolase A	16	D. Tolan and E. Penhoet
Aldolase C	17	D. Tolan and E. Penhoet
Homeo box	17q11→qter*	R. Tjian and G. Martin
Insulin receptor	19*	W. Rutter
Ferritin light chain	19,20,X*	J. Drysdale and D. Boyd
Growth-hormone-releasing factor	20	K.E. Mayo and R.M. Evans
c-*src*-1 oncogene	20*	J.M. Bishop and R. Parker
Prion	20	Y.-C. Liao
G-6-PD	X,17*	A. Yoshida

[a]Asterisks indicate confirmation by another method.

ACKNOWLEDGMENTS

We thank Farid Chehab for von Willebrand's data, Dean Tolan for aldolase data, and Barry Bruce for chromosome sorting. This work was supported in part by a grant from the Muscular Dystrophy Association to R.V.L. R.V.L. is an associate and Y.W.K. is an investigator of the Howard Hughes Medical Institute.

REFERENCES

Bartholdi, M.F. 1985. DNA content and base composition of human chromosomes. *J. Colloid Interface Sci.* **105:** 426.

Bird, R.D., J. Ott, and E.R. Giblett. 1982. Evidence for linkage of Charcot-Marie-Tooth neuropathy to the Duffy locus on chromosome 1. *Am. J. Hum. Genet.* **34:** 388.

Charcot, J.M. and P. Marie. 1886. Sur une forme particuliere d-atrophie musculaire progressive, souvent familiale, debutant par les pieds et les jambes et atteignant plus tard les mains. *Rev. Med.* **6:** 97.

Conboy, J.G., N. Mohandas, C. Wang, G. Tchernia, S.B. Shohet, and Y.W. Kan. 1985. Molecular cloning and characterization of the gene coding for red cell membrane skeletal protein 4.1. *Blood* **66:** 31a.

Davis, M., S. Malcom, and T.H. Rabbitts. 1984. Chromosome translocation can occur on either side of the c-myc oncogene in Burkitt lymphoma cells. *Nature* **308:** 286.

Disteche, C.M., L.M. Kunkel, A. Lojewski, S.H. Orkin, M. Eisenhard, E. Sahar, B. Travis, and S.A. Latt. 1982. Isolation of mouse X-chromosome specific DNA from an X-enriched lambda phage library derived from flow sorted chromosomes. *Cytometry* **2:** 282.

Francke, U. and N. Oliver. 1978. Quantitative analysis of high-resolution trypsin-Giemsa bands on human prometaphase chromosomes. *Hum. Genet.* **45:** 137.

Gray, J.W., R.G. Langlois, A.V. Carrano, K. Burkhart-Schultz, and M.A. Van Dilla. 1979. High resolution chromosome analysis: One and two parameter flow cytometry. *Chromosoma* **73:** 9.

Harper, M.E., A. Ullrich, and G.F. Saunders. 1981. Localization of the human insulin gene to the distal end of the short arm of chromosome 11. *Proc. Natl. Acad. Sci.* **78:** 4458.

Human Gene Mapping 8. 1985. Eighth International Workshop on Human Gene Mapping. Helsinki. *Cytogent. Cell Genet.* **40:** 1.

Kan, Y.W. and A.M. Dozy. 1978. Antenatal diagnosis of sickle cell anaemia by DNA analysis of amniotic-fluid cells. *Lancet* **II:** 910.

Latt, S.A. 1977. Fluorescent probes of chromosome structure and replication. *Can. J. Genet. Cytol.* **19:** 603.

Latt, S.A. and J.C. Wohlleb. 1975. Optical studies of the interaction of 33258 Hoechst with DNA, chromatin, and metaphase chromosomes. *Chromosoma* **52:** 297.

Law, D.J., P.M. Fossard, and D.L. Rucknagel. 1984. Highly sensitive and rapid gene mapping using miniaturized blot hybridization: Application to prenatal diagnosis. *Gene* **28:** 153.

Lebo, R.V. 1982. Chromosome sorting and DNA sequence localization: A review. *Cytometry* **3:** 145.

———. 1986. Gene mapping strategies and flow cytogenetics. In *Flow cytogenetics* (ed. J. Gray). Academic Press, New York. (In press.)

Lebo, R.V. and A.M. Bastian. 1982. Design and operation of a dual laser chromosome sorter. *Cytometry* **3:** 213.

Lebo, R.V. and B.D. Bruce. 1986. Gene mapping with sorted chromosomes. *Methods Enzymol.* (in press).

Lebo, R.V., M.S. Golbus, and M.-C. Cheung. 1986a. Detecting abnormal human chromosome constitutions by dual laser flow cytogenetics. *Am. J. Med. Genet.* (in press).

Lebo, R.V., B.D. Bruce, P.F. Dazin, and D.G. Payan. 1987. Design and application of a versatile triple-laser cell and chromosome sorter. *Cytometry* **8:** (in press).

Lebo, R.V., L.A. Anderson, Y.-F. Lau, V. Carver, and P.M. Conneally. 1986b. Linkage analysis of Charcot-Marie-Tooth syndrome. *Muscle & Nerve* **9:** 235.

Lebo, R.V., Y.W. Kan, M.C. Cheung, S.K. Jain, and J. Drysdale. 1985a. Human ferritin light chain gene sequences mapped to several sorted chromosomes. *Hum. Genet.* **71:** 325.

Lebo, R.V., M.-C. Cheung, B.D. Bruce, V.M. Riccardi, F.-T. Kao, and Y.W. Kan. 1985b. Mapping parathyroid hormone, β-globin, insulin and LDH-A genes within the human chromosome 11 short arm by spot blotting sorted chromosomes. *Hum. Genet.* **69:** 316.

Lebo, R.V., D.R. Tolan, B.D. Bruce, M.-C. Cheung, E.D. Penhoet, and Y.W. Kan. 1985c. Spot-blot analysis of sorted chromosomes assigns the aldolase B gene to chromosome 9. *Cytometry* **6:** 478.

Lebo, R.V., A. Chakravarti, K.H. Buetow, H. Cann, M.-C. Cheung, B. Cordell, and H. Goodman. 1983. Chiasma within and between the human insulin and β-globin gene loci. *Proc. Natl. Acad. Sci.* **80:** 4808.

Lebo, R.V., F. Gorin, R.J. Fletterick, F.-T. Kao, M.-C. Cheung, B.D. Bruce, and Y.W. Kan. 1984. High-resolution chromosome sorting and DNA spot-blot analysis assign McArdle's syndrome to chromosome 11. *Science* **225:** 57.

Litt, M. and R.L. White. 1985. A highly polymorphic locus in human DNA revealed by cosmid-derived probes. *Proc. Natl. Acad. Sci.* **82:** 6206.

Ott, J. 1974. Estimation of the recombination fraction in human pedigrees: Efficient computation of the likelihood for human linkage studies. *Am. J. Hum. Genet.* **26:** 588.

Parker, R.C., G. Mardon, R.V. Lebo, H.E. Varmus, and J.M. Bishop. 1985. Isolation of duplicated human c-*src* genes located on chromosomes 1 and 20. *Mol. Cell. Biol.* **5:** 831.

Ruddle, F.H. and R.P. Creagan. 1981. Parasexual approaches to the genetics of man. *Annu. Rev. Genet.* **9:** 407.

Schnedl, W., M. Breitenbach, A.V. Mikelsaar, and G. Stranzinger. 1977. Mithramycin and DIPI: A pair of fluorochromes specific for GC- and AT-rich DNA, respectively. *Hum. Genet.* **36:** 299.

Stebbins, H.B. and P.M. Conneally. 1982. Linkage of dominantly inherited Charcot' Marie Tooth neuropathy to the Duffy locus in an Indiana Family. *Am. J. Hum. Genet.* **34:** 195A.

Tooth, H.H. 1886. *The peroneal type of progressive muscular atrophy*. H.K. Lewis, London.

Yunis, J.J. and M.E. Chandler. 1978. High-resolution trypsin-Giemsa bands on human prometaphase chromosomes. *Hum. Genet.* **45:** 137.

Cloning the Gene for the Inherited Disorder Chronic Granulomatous Disease on the Basis of Its Chromosomal Location

B. ROYER-POKORA,* L.M. KUNKEL,† A.P. MONACO,† S.C. GOFF,* P.E. NEWBURGER,‡ R.L. BAEHNER,* F.S. COLE,** J.T. CURNUTTE,§ AND S.H. ORKIN,*¶

*Division of Hematology-Oncology, Childrens Hospital, Dana-Farber Cancer Institute, Department of Pediatrics; †Division of Genetics, Childrens Hospital, Department of Pediatrics and the Program in Neuroscience; **Division of Cell Biology, Childrens Hospital, Department of Pediatrics, Harvard Medical School, Boston, Massachusetts 02115; ‡Division of Pediatric Hematology Department of Pediatrics, University of Massachusetts Medical School, Worcester, Massachusetts 01605; § Division of Pediatric Hematology, Department of Pediatrics, University of Michigan Medical School, Ann Arbor, Michigan 48109; ¶ Howard Hughes Medical Institute, Childrens Hospital, Boston, Massachusetts 02115

Human inherited disorders often result from mutations in genes whose protein products are unknown. In principle, gene cloning procedures may be employed to isolate and characterize such genes without reference to protein data. Subsequently, the protein encoded by a gene of this kind may be detected using antibody reagents prepared from synthetic peptides predicted by cDNA sequences or from polypeptides expressed in prokaryotic cells. In this manner, new insights into a cell biologic system may ultimately be gained through study of human disease.

The use of genetic linkage (Botstein et al. 1980) to provide a rough location of a disease gene within the chromosome complement has been viewed as a first step in this venture. Three major diseases of uncertain etiology, Duchenne muscular dystrophy (DMD) (Davies et al. 1983; Monaco et al. 1985), Huntington's disease (Gusella et al. 1983), and cystic fibrosis (Knowlton et al. 1985; Wainwright et al. 1985; White et al. 1985), have recently been the focus of intensive analysis. It is commonly believed that genomic cloning in the appropriate chromosomal regions, coupled with searches for transcribed sequences, will ultimately provide the means to identify each locus in precise genetic terms. Here, we summarize our application of this general approach to the mapping, identification, and characterization of the gene involved in chronic granulomatous disease (CGD), a major inherited disorder of phagocytic cells (Tauber et al. 1983).

Individuals afflicted with CGD have impaired host defenses against infection with common microorganisms. Upon ingestion of bacteria or particles, phagocytes (granulocytes, monocytes, and eosinophils) of CGD patients fail to generate superoxide and other activated oxygen species. The lesion in the membrane-associated NADPH-oxidase of these cells is unknown despite considerable biochemical investigation. X-chromosome-linked and recessive varieties of the disorder have been identified. In the majority of families, the disease is X-linked. Limited linkage analyses previously suggested linkage of the disease to Xg, the most distal Xp (short-arm) marker. Biochemical studies have demonstrated a deficiency of the spectrum of an unusual *b*-type cytochrome in the vast majority of X-linked patients. In view of inherent problems in conventional biochemical analysis of the disorder, we have undertaken a genetic approach. Our goals have been to locate the *CGD* gene on the X chromosome, identify the gene, and characterize its predicted product as a means of delineating at least one critical component of the oxidase system. From the findings summarized below, we believe that CGD is the first disease for which molecular cloning has provided a specific gene without reference to protein data (Royer-Pokora et al. 1986).

RESULTS

The General Approach

Four steps are involved in the approach we have adopted to the molecular analysis of CGD: (1) positioning of the gene on the X chromosome, (2) identification of transcribed sequences derived from the appropriate chromosomal region, (3) demonstration of the relevance of the identified transcribed sequences to the disorder, and (4) characterization of the predicted protein by analysis of cDNA clones.

Mapping of the *X-CGD* Gene to Xp21.1

The position of the *X-CGD* on the X chromosome was assigned by both deletion analysis and formal linkage analysis. The existence of two patients (BB and NF) affected with both CGD and DMD and an interstitial deletion in Xp21 suggested that the initial assignment of the gene to distal Xp was incorrect (Francke et al. 1985; Baehner et al. 1986). In view of potential rearrangements of Xp in these patients that might invalidate this conclusion, we performed a linkage analysis among typical X-CGD families using cloned DNA probes derived from Xp segments. As summarized in Figure 1, evidence for linkage with CGD was obtained

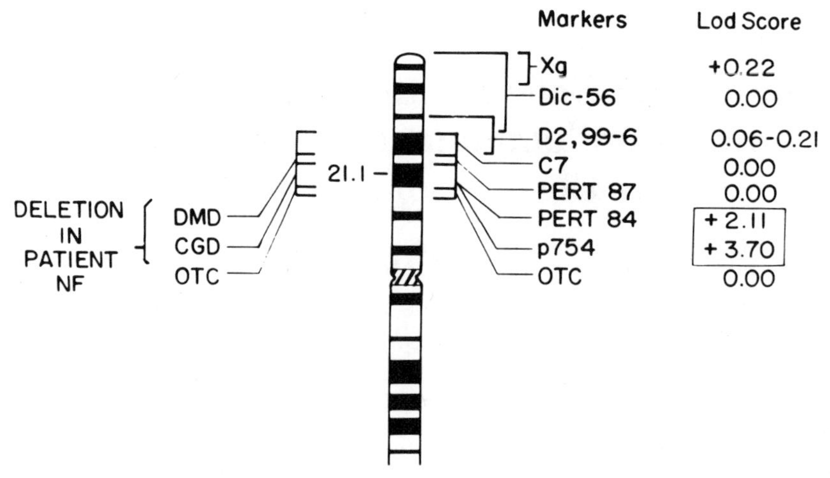

Figure 1. Localization of the *X-CGD* gene to Xp21.1. The markers utilized in the linkage analysis of X-chromosome-linked CGD are shown next to the idiogram of the short arm of the X chromosome (Xp). The calculated lod scores are given to the right for each probe. DNA markers p754 and pERT84 demonstrated significant linkage with *X-CGD*. To the left is depicted the Xp21 deletion observed in patients BB (Francke et al. 1985) and NF (Baehner et al. 1986).

for two Xp21 markers (p754 and pERT84), precisely the region deleted in patients BB and NF. On the basis of these findings, we have assigned *X-CGD* to Xp21.1 rather than to distal Xp as originally suggested (Baehner et al. 1986).

Identification of Expressed Xp21 Sequences

The deletion and linkage data position the *X-CGD* locus within about 300–500 kb of DNA of Xp. From chromosome walking studies initially aimed at characterizing the *DMD* locus, a series of bacteriophage clones derived from Xp21 and spanning at most 10% of this region was available (Kunkel et al. 1985). We searched for transcribed Xp21 sequences using the strategy illustrated in Figure 2. An enriched, radiolabeled cDNA was prepared from granulocytic human leukemic HL-60 cells (Collins et al. 1978) and employed as hybridization probe to a Southern blot filter containing restriction-enzyme-digested bacteriophage DNAs from Xp21. Granulocytic HL-60 cells were chosen as a source for preparation of cDNA, since they express the phagocyte-specific oxidase system. cDNA was enriched for granulocytic sequences by subtraction with RNA prepared from a B-cell line of patient NF, who had a deletion of Xp21 removing the *DMD* and *CGD* loci. The enriched probe used in the transcript search represented approximately 500 individual mRNAs. By the experiment shown in Figure 2, two overlapping bacteriophages, originally isolated using the pERT clone 379, were identified.

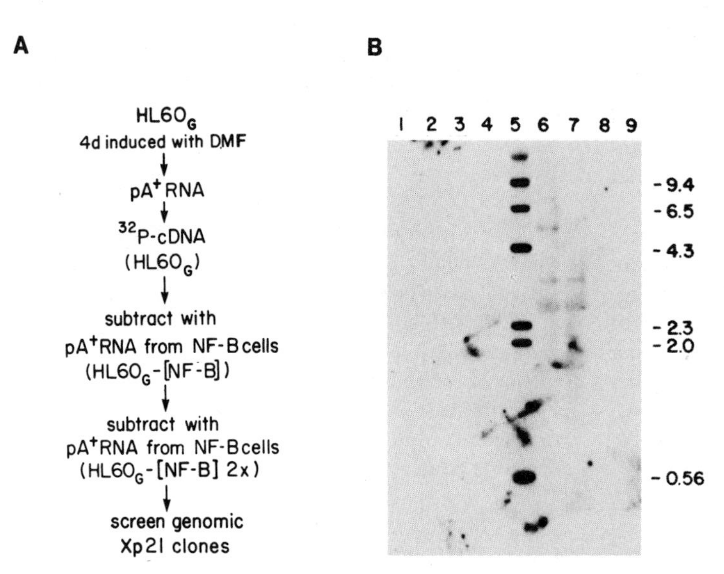

Figure 2. Detection of a transcribed region of Xp21. (*A*) Strategy employed for the preparation of an enriched, radioactive cDNA from granulocyte HL-60 cells (HL-60_G). cDNA was subjected to competitive subtractions (Davis et al. 1984) with mRNA from an EBV-transformed cell line from patient NF who had an Xp21 deletion removing the *CGD* and *DMD* loci. (*B*) Southern blot hybridization of the enriched cDNA with DNAs from bacteriophage derived from Xp21. Bacteriophage clones derived from Xp21 were isolated using pERT clones 469, 378, 379, 55, and 145 as probes (Kunkel et al. 1985; Monaco et al. 1985; L.M. Kunkel and A.P. Monaco, unpubl.). DNAs were digested with *Eco*RI and *Hin*dIII, electrophoresed in agarose, and transferred to a filter for hybridization with the radioactive enriched cDNA. Lanes *1–4* and *6–9* display independent bacteriophage isolated with pERT clones 469, 378, 378, 378, 379, 379, 55, and 145, respectively. Overlapping bacteriophages 379-A6 and 379π (lanes *6* and *7*) contained 2.5- and 3.3-kb hybridizing fragments.

Expression of the Xp21 RNA in Phagocytic Cells

Subcloned DNA fragments from the pERT379 bacteriophage were used in northern blot analysis to identify specific transcribed regions. In this manner, a segment was identified that detected a nearly 5-kb RNA transcript in granulocytic HL-60 cell RNA and was subsequently used to isolate cDNA clones from cDNA libraries.

The Xp21 transcript is relatively abundant in granulocytic HL-60 cells and virtually absent in undifferentiated cells (Fig. 3). It is abundant in monocytic HL-60 cells and in normal human monocytes but absent in RNA of fibroblastic, kidney, or liver origin (not shown). A low level of transcript was found in normal cultured B-cell lines. We estimate the transcript abundance as 0.02–0.10% in granulocytes or monocytes. The identity of the transcript as phagocyte-specific initially suggested that it might represent the product of the *X-CGD* locus.

Relevance of the Xp21 Transcript to *X-CGD*

If the transcript were derived from the relevant disease locus, we would predict quantitative and/or qualitative RNA or DNA alterations in material prepared from affected patients. RNA isolated from cultured human monocytes of normal and X-CGD origin were examined by northern blot analysis (Fig. 4). The CGD patients examined were classic in that their phagocytic cells were NBT-test negative (Tauber et al. 1983), and the spectrum of the phagocyte-specific cytochrome *b* was absent. Analysis of RNA of two patients is shown in Figure 4. In one patient (Fig. 4, lane 1), the Xp21 transcript appears to be grossly normal in abundance and size, whereas in the other patient (Fig. 4, lane 2), it is absent. The presence of phosphoglycerate kinase (PGK) RNA in all samples establishes the integrity of the RNA preparations (Fig. 4). Two additional patients studied also lacked transcripts in monocyte RNA. The three RNA-negative patients had structurally normal genes by Southern blot analysis using probes spanning the entire cDNA transcript (not shown).

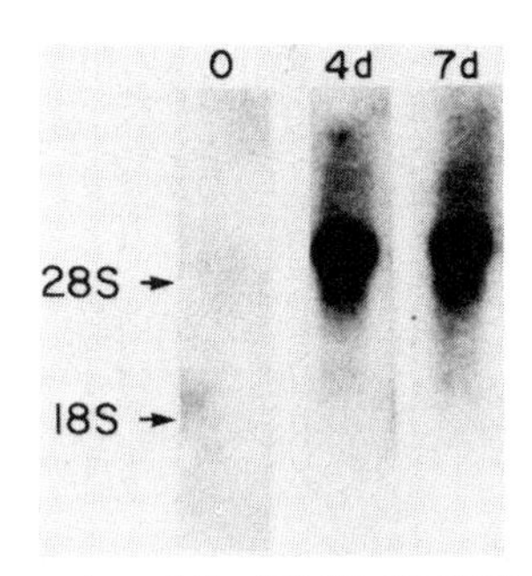

Figure 3. Expression of the 379 (Xp21) RNA in HL-60 cells. Total cellular RNA (10 μg) isolated from uninduced (0) or DMF-induced (either 4 days or 7 days) human leukemic HL-60 cells was analyzed with a fragment of 379 cDNA. Northern blot analysis was performed as described previously (Maniatis et al. 1982).

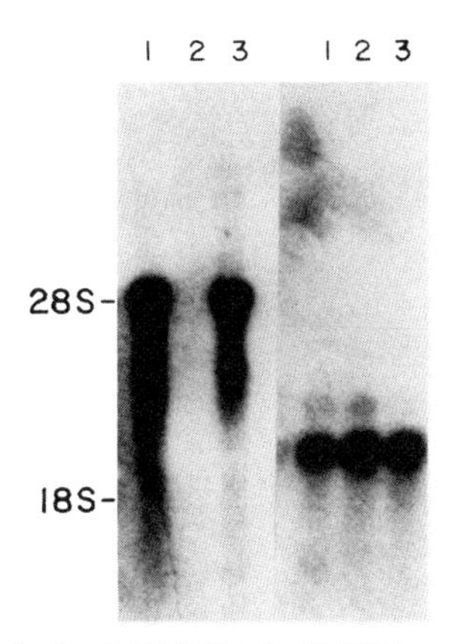

Figure 4. Analysis of 379 RNA in X-CGD monocytes. Total cellular RNA (5 μg) prepared from cultured monocytes of two X-CGD patients (lanes *1* and *2*) and from a normal individual (lane *3*) was examined for 379 sequences (*left*) or for phosphoglycerate kinase (PGK) sequences (*right*) (Michelson et al. 1983) by northern blot analysis. Peripheral blood monocytes were isolated and cultured as described previously (Strunk et al. 1985).

Although the RNA sample in lane 1 of Figure 4 appeared grossly normal, it was disrupted by an interstitial deletion, including the segment of transcribed sequences shown in Figure 5. Regions of the mRNA transcript 5′ and 3′ to the deleted segment were present

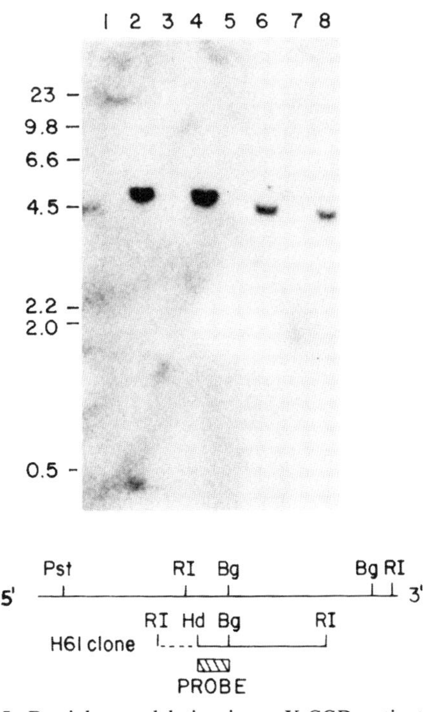

Figure 5. Partial gene deletion in an X-CGD patient. Total cellular DNA (5 μg) was digested with *Hin*dIII (*1–4*) or *Bgl*II (*5–8*) and hybridized with an 0.3-kb probed derived from the central portion of the 379 cDNA. (*1,5*) Mother of patient JW; (*2,6*) X-CGD patient JW; (*3,7*) normal; (*4,8*) Xp21 deletion patient NF. The monocyte RNA of patient JW is shown in lane 1 of Fig. 4.

```
CTTCCTCTGCCACCATCGGGGAACTGGCTGTGAATGAGGGGCTCTCCATTTTTGCTATTCTGGTTTGGCTGGGGTTGAACGTCTTCCTCTTTGTCTGGTATTACCGGGTTTATGATATT  120

CCACCTAAGTTCTTTTACACAAGAAAACTTCTTGGGTCAGCACTGGCACTGGCCAGGGCCCCTGCAGCCTCCCTGAATTTCAACTGCATGCTGATTCTCTTGCCAGTCTGTCGAAATCTG  240
                                                                                                 1                  10
CTGTCCTTCCTCAGGGGTTCCAGTGCGTGCTGCTCAACAAGAGTTCGAAGACAACTGGACAGGAATCTCACCTTTCATAAAATGGTGGCATGGATGATTGCACTTCACTCTGCGATTCAC  360
                                                                                  MetValAlaTrpMetIleAlaLeuHisSerAlaIleHis
             20                  30                            40                            50
ACCATTGCACATCTATTTAATGTGGAATGGTGTGTGAATGCCCGAGTCAATAATTCTGATCCTTATTCAGTAGCACTCTCTGAACTTGGAGACAGGCAAAATGAAAGTTATCTCAATTTT  480
ThrIleAlaHisLeuPheAsnValGluTrpCysValAsnAlaArgValAsnAsnSerAspProTyrSerValAlaLeuSerGluLeuGlyAspArgGlnAsnGluSerTyrLeuAsnPhe
             60                  70                            80                            90
GCTCGAAAGAGAATAAAGAACCCTGAAGGAGGCCTGTACCTGGCTGTGACCCTGTTGGCAGGCATCACTGGAGTTGTCATCACGCTGTGCCTCATATTAATTATCACTTCCTCCACCAAA  600
AlaArgLysArgIleLysAsnProGluGlyGlyLeuTyrLeuAlaValThrLeuLeuAlaGlyIleThrGlyValValIleThrLeuCysLeuIleLeuIleIleThrSerSerThrLys
             100                 110                           120                           130
ACCATCCGGAGGTCTTACTTTGAAGTCTTTTGGTACACACATCATCTCTTTGTGATCTTCTTCATTGGCCTTGCCATCCATGGACCTGAACGAATTGTACGTGGGCAGACCGCAGAGAGT  720
ThrIleArgArgSerTyrPheGluValPheTrpTyrThrHisHisLeuPheValIlePhePheIleGlyLeuAlaIleHisGlyAlaGluArgIleValArgGlyGlnThrAlaGluSer
             140                 150                           160                           170
TTGGCTGTGCATAATATAACAGTTTGTGAACAAAAAATCTCAGAATGGGGAAAAATAAAGGAATGCCCAATCCCTCAGTTTGCTGGAAACCCTCCTATGACTTGGAAATGGATAGTGGGT  840
LeuAlaValHisAsnIleThrValCysGluGlnLysIleSerGluTrpGlyLysIleLysGluCysProIleProGlnPheAlaGlyAsnProProMetThrTrpLysTrpIleValGly
             180                 190                           200                           210
CCCATGTTTCTGTATCTCTGTGAGAGGTTGGTGCGGTTTTGGCGATCTCAACAGAAGGTGGTCATCACCAAGGTGGTCACTCACCCTTTCAAAACCATCGAGCTACAGATGAAGAAGAAG  960
ProMetPheLeuTyrLeuCysGluArgLeuValArgPheTrpArgSerGlnGlnLysValValIleThrLysValValThrHisProPheLysThrIleGluLeuGlnMetLysLysLys
             220                 230                           240                           250
GGGTTCAAAATGGAAGTGGGACAATACATTTTTGTCAAGTGCCCAAAGGTGTCCAAGCTGGAGTGGCACCCTTTTACACTGACATCCGCCCCTGAGGAAGACTTCTTTAGTATCCATATC  1080
GlyPheLysMetGluValGlyGlnTyrIlePheValLysCysProLysValSerLysLeuGluTrpHisProPheThrLeuThrSerAlaProGluGluAspPhePheSerIleHisIle
             260                 270                           280                           290
CGCATCGTTGGGGACTGGACAGAGGGGCTGTTCAATGCTTGTGGCTGTGATAAGCAGGAGTTTCAAGATGCGTGGAAACTACCTAAGATAGCGGTTGATGGGCCCTTTGGCACTGCCAGT  1200
ArgIleValGlyAspTrpThrGluGlyLeuPheAsnAlaCysGlyCysAspLysGlnGluPheGlnAspAlaTrpLysLeuProLysIleAlaValAspGlyProPheGlyThrAlaSer
             300                 310                           320                           330
GAAGATGTGTTCAGCTATGAGGTGGTGATGTTAGTGGGAGCAGGGATTGGGGTCACACCCTTCGCATCCATTCTCAAGTCAGTCTGGTACAAATATTGCAATAACGCCACCAATCTGAAG  1320
GluAspValPheSerTyrGluValValMetLeuValGlyAlaGlyIleGlyValThrProPheAlaSerIleLeuLysSerValTrpTyrLysTyrCysAsnAsnAlaThrAsnLeuLys
             340                 350                           360                           370
CTCAAAAAGATCTACTTCTACTGGCTGTGCCGGGACACACATGCCTTTGAGTGGTTTGCAGATCTGCTGCAACTGCTGGAGAGCCAGATGCAGGAAAGGAACAATGCCGGCTTCCTCAGC  1440
LeuLysLysIleTyrPheTyrTrpLeuCysArgAspThrHisAlaPheGluTrpPheAlaAspLeuLeuGlnLeuLeuGluSerGlnMetGlnGluArgAsnAsnAlaGlyPheLeuSer
             380                 390                           400                           410
TACAACATCTACCTCACTGGCTGGGATGAGTCTCAGGCCAATCACTTTGCTGTGCACCATGATGAGGAGAAAGATGTGATCACAGGCCTGAAACAAAAGACTTTGTATGGACGGCCCAAC  1560
TyrAsnIleTyrLeuThrGlyTrpAspGluSerGlnAlaAsnHisPheAlaValHisHisAspGluGluLysAspValIleThrGlyLeuLysGlnLysThrLeuTyrGlyArgProAsn
             420                 430                           440                           450
TGGGATAATGAATTCAAGACAATTGCAAGTCAACACCCTAATACCAGAATAGGAGTTTTCCTCTGTGGACCTGAAGCCTTGGCTGAAACCCTGAGTAAACAAAGCATCTCCAACTCTGAG  1680
TrpAspAsnGluPheLysThrIleAlaSerGlnHisProAsnThrArgIleGlyValPheLeuCysGlyProGluAlaLeuAlaGluThrLeuSerLysGlnSerIleSerAsnSerGlu
             460
TCTGGCCCTCGGGGAGTGCATTTCATTTTCAACAAGGAAAACTTCTAACTTGTCTCTTCCATGAGGAAATAAATGTGGGTTGTGCTGCCAAATGCTCAAATAATGCTAATTGATAATATA  1800
SerGlyProArgGlyValHisPheIlePheAsnLysGluAsnPhe---
```

AATACCCCCTGCTTAAAAATGGACAAAAAGAAACTATAATGTAATGGTTTTCCCTTAAAGGAATGTCAAAGATTGTTTGATAGTGATAAGTTACATTTATGTGGAGCTCTATGGTTTTGA 1920

GAGCACTTTTACAAACATTATTTCATTTTTTTCCTCTCAGTAATGTCAGTGGAAGTTAGGGAAAAGATTCTTGGACTCAATTTTAGAATCAAAAGGGAAAGGATCAAAAGGTTCAGTAAC 2040

TTCCCTAAGATTATGAAACTGTGACCAGATCTAGCCCATCTTACTCCAGGTTTGATACTCTTTCCACAATACTGAGCTGCCTCAGAATCCTCAAAATCAGTTTTTATATTCCCCAAAAGA 2160

AGAAGGAAACCAAGGAGTAGCTATATATTTCTACTTTGTGTCATTTTTGCCATCATTATTATCATACTGAAGGAAATTTTCCAGATCATTAGGACATAATACATGTTGAGAGTGTCTCAA 2280

CACTTATTAGTGACAGTATTGACATCTGAGCATACTCCAGTTTACTAATACAGCAGGGTAACTGGGCCAGATGTTCTTTCTACAGAAGAATATTGGATTGATTGGAGTTAATGTAATACT 2400

CATCATTTACCACTGTGCTTGGCAGAGAGCGGATACTCAAGTAAGTTTTGTTAAATGAATGAATGAATTTAGAACCACACAATGCCAAGATAGAATTAATTTAAAGCCTTAAACAAAATT 2520

TATCTAAAGAAATAACTTCTATTACTGTCATAGACCAAAGGAATCTGATTCTCCCTAGGGTCAAGAACAGGCTAAGGATACTAACCAATAGGATTGCCTGAAGGGTTCTGCACATTCTTA 2640

TTTGAAGCATGAAAAAAGAGGGTTGGAGGTGGAGAATTAACCTCCTGCCATGACTCTGGCTCATCTAGTCCTGCTCCTTGTGCTATAAAATAAATGCAGACTAATTTCCTGCCCAAAGTG 2760

GTCTTCTCCAGCTAGCCCTTATGAATATTGAACTTAGGAATTGTGACAAATATGTATCTGATATGGTCATTTGTTTTAAATAACACCCACCCCTTATTTTCCGTAAATACACACACAAAA 2880

TGGATCGCATCTGTGTGACTAATGGTTTATTTGTATTATATCATCATCATCATCCTAAAATTAACAACCCAGAAACAAAAATCTCTATACAGAGATCAAATTCACACTCAATAGTATGTT 3000

CTGAATATATGTTCAAGAGAGAGTCTCTAAATCACTGTTAGTGTGGCCAAGAGCAGGGTTTTCTTTTTGTTCTTAGAACTGCTCCCATTTCTGGGAACTAAAACCAGTTTTATTTGCCCC 3120

ACCCCTTGGAGCCACAAATGTTTAGAACTCTTCAACTTCGGTAATGAGGAAGAAGGAGAAAGAGCTGGGGGAAGGGCAGAAGACTGGTTTAGGAGGAAAAGGAAATAAGGAGAAAAGAGA 3240

ATGGGAGAGTGAGAGAAAATAAAAAAGGCAAAAGGGAGAGAGAGGGGAAGGGGGTCTCATATTGGTCATTCCCTGCCCCAGATTTCTTAAAGTTTGATATGTATAGAATATAATTGAAGG 3360

AGGTATACACATACTGATGTTGTTTTGATTATCTATGGTATTGAATCTTTTAAAATCTGGTCACAAATTTTGATGCTGAGGGGGATTATTCAAGGGACTAGGATGAACTAAATAAGAACT 3480

CAGTTGTTCTTTGTCATACTACTATTCCTTTCGTCTCCCAGAATCCTCAGGGCACTGAGGGTAGGTCTGACAAATAAGGCCTGCTGTGCGAATATAGCCTTTCTGAAATGTACCAGGATG 3600

GTTTCTGCTTAGAGACACTTAGGTCCAGCCTGTTCACACTGCACCTCAGGTATCAATTCATCTATTCAACAGATATTTATTGTGTTATTACTATGAGTCAGGCTCTGTTTATTGTTTCAA 3720

TTCTTTACACCAAAGTATGAACTGGAGAGGGTACCTCAGTTATAAGGAGTCTGAGAATATTGGCCCTTTCTAACCTATGTGCATAATTAAAACCAGCTTCATTTGTTGCTCCGAGAGTGT 3840

TTCTCCAAGGTTTTCTATCTTCAAAACCAACTAAGTTATGAAAGTAGAGAGATCTGCCCTGTGTTATCCAGTTATGAGATAAAAAATGAATATAAGAGTGCTTGTCATTATAAAAGTTTC 3960

CTTTTTATCTCTCAAGCCACCAGCTGCCAGCCACCACGAGCCAGCTGCCAGCCTAGCTTTTTTTTTTTTTTTTTTTAGCACTTACTATTTAGCATTTATTAACAGGTACTCTAAGAAT 4080

GATGAAGCATTGTTTTTAATCTTAAGACTATGAAGGTTTTTCTTAGTTCTTCTGCTTTTGCAATTGTGTTTGTGAAATTTGAATACTTGCAGGCTTTGTATGTGAATAATTCTAGCGGGG 4200

GACCTGGGAGATAATTCTACGGGGAATTCTTAAAACTGTGCTCAACTATTAAAATGAATGAGCTTTCAAAAAAAAAA

Figure 6. Sequence of the *X-CGD* cDNA. The DNA sequence was assembled from overlapping cDNA clones described previously (Royer-Pokora et al. 1986). The first four ATGs in the sequence are underscored. The predicted initiator codon is underscored twice. The presumed processed cDNA sequence near the 3′ end of the transcript is boxed. The putative poly(A)-addition signal is overscored.

in the RNA transcript of this patient (not shown). The interstitial deletion in this patient's DNA and monocyte RNA overlaps the 3′ terminus of the large open reading frame of the cDNA described below.

In summary, the Xp21 transcript in the monocytes of four out of four X-CGD patients was abnormal. These findings provide the conclusive genetic evidence that the transcript defines the product of the *X-CGD* locus, rather than that of another, unrelated Xp21 gene. Initial studies suggest that the RNA-negative phenotype is most common among X-CGD patients.

Structure of the *X-CGD* Transcript

cDNA clones for the Xp21 transcript derived from the pERT379 region were isolated from granulocytic HL-60 libraries constructed in bacteriophage λgt10 and λgt11 (Gubler and Hoffman 1983; Young and Davis 1983; Ginsburg et al. 1985). As shown in Figure 5, cDNA clones spanning the entire mRNA were obtained. By DNA sequence analysis, the transcript measures 4.27 kb rather than nearly 5 kb estimated by the northern blot studies. Primer extension analysis suggests that the entire 5′ end of the transcript was obtained.

The 3′-untranslated region of the mRNA is 2.5 kb in length. A putative polyadenylation signal (ATTAAA) is present 14 nucleotides before a short poly(A) tract of one cDNA clone. A notable feature of the 3′-untranslated region is the presence of a $(T)_{21}$ tract followed 19 nucleotides downstream by TTTATT. On the complementary DNA strand, this organization resembles the 3′ end of a processed transcript (Nishioka et al. 1980; Sharp 1983).

Although several potential initiator ATGs are present in the 5′ segment of the mRNA, we have tentatively assigned the most favorable initiator codon (Kozak 1984) to position 322 (Fig. 6). This predicts an open reading frame extending to nucleotide 1725 and a polypeptide of 468 amino acids. The predicted protein is basic (pI = 9.5) and has four potential *N*-glycosylation sites with the canonical form Asn-X-(Thr/Ser). Analyses of the GENBANK and the protein data base revealed no significant homology of the cDNA or its predicted protein with known sequences.

DISCUSSION

Our studies represent the first successful cloning and characterization of a human disease gene without reference to protein structure, reagents, or an assayable function (Royer-Pokora et al. 1986). On the basis of linkage data, genomic cloning within a restricted region, and hybridization with an enriched cDNA probe, we have identified and analyzed the transcript that is specifically altered in CGD, a major inherited abnormality of host defense. The findings that the identified RNA transcript is derived from Xp21, expressed specifically in phagocytic cells, and deranged in X-CGD patients constitute the essential aspects of the proof that the appropriate locus has been recognized.

Although the approach we have employed is entirely general in principle, several factors may influence its applicability to other situations. Many circumstances were favorable for the successful execution of these studies, including (1) the fortuitous presence of a portion of the *X-CGD* gene in bacteriophage clones isolated with the pERT379 probe, (2) the presence of an extensive target for hybridization (the 3′-untranslated region of 2.5 kb) within the 379 bacteriophage, (3) a convenient source of phagocyte RNA (induced HL-60 cells) for preparation of an enriched probe, and (4) the availability of an Epstein-Barr virus (EBV)–transformed B-cell line from patient NF with an interstitial deletion of Xp21.

Our studies emphasize the advantages of a genetic approach to dissection of a complex cellular system. Although conventional biochemical analysis of the NADPH-oxidase system of the phagocyte has been pursued for nearly two decades, no clear consensus exists regarding its specific components and their relationship to various forms of CGD. The most likely candidate protein for the X-linked variety of CGD was the unusual *b*-type cytochrome (Segal et al. 1983; Harper et al. 1985; Segal 1985), whose spectrum is almost universally absent in affected patients. Analysis of the protein predicted by our cDNA sequence, comparison of the amino acid composition with that of purified cytochrome *b* reported by Segal and associates (Harper et al. 1985), and the apparent presence of cytochrome-*b* protein (as opposed to spectrum) in affected X-CGD patients (Segal 1985) indicate in toto that the primary genetic abnormality in X-CGD is *not* referable to the candidate cytochrome *b*. We propose that the protein predicted by our cDNA sequence is an essential component of the oxidase system and interacts with other components of unknown type and number (but possibly including the cytochrome *b* and a flavoprotein) to form an active complex in vivo. At the biochemical level, many abnormalities might be expected in X-CGD cells if the stability and integrity of other proteins rely on normal abundance and function of the protein predicted by our molecular studies. Characterization of the X-CGD protein product in vivo and examination of proteins with which it associates should provide new insights into the organization and function of the oxidase system of the phagocyte. Dissection of functional domains of the predicted protein will ultimately rely on the introduction and expression of the cDNA in phagocytic cells of X-CGD patients, an approach that may suggest the basis for genetic correction of the disorder in bone-marrow-derived cells in the future.

ACKNOWLEDGMENTS

S.H.O. is a senior investigator of the Howard Hughes Medical Institute. P.E.N. was supported by a National

Institutes of Health grant (CS-38325) and by an Established Investigatorship of the American Heart Association, with funds contributed in part by its Massachusetts affiliate. This work was supported in part by grants from the Muscular Dystrophy Association to L.M.K. and the National Institutes of Health to J.T.C. (AI-21320) and to S.H.O. (HD-18661).

REFERENCES

Baehner, R.L., L.M. Kunkel, A.P. Monaco, J.L. Haines, P.M. Conneally, C. Palmer, N. Heerema, and S.H. Orkin. 1986. DNA linkage analysis of X chromosome-linked chronic granulomatous disease. *Proc. Natl. Acad. Sci.* **83:** 3398.

Botstein, D., R.L. White, M. Skolnick, and R.W. Davis. 1980. Construction of a genetic linkage map in man using restriction fragment length polymorphism. *Am. J. Hum. Genet.* **32:** 314.

Collins, S.J., F.W. Ruscetti, R.E. Gallagher, and R.C. Gallo. 1978. Terminal differentiation of human promyelocytic leukemia cells induced by dimethyl sulfoxide and other polar compounds. *Proc. Natl. Acad. Sci.* **75:** 2458.

Davies, K.E, P.L. Pearson, P.S. Harper, M.M. Murray, T. O'Brien, M. Sarfarazi, and R. Williamson. 1983. Linkage analysis of two cloned DNA sequences flanking the Duchenne muscular dystrophy locus on the short arm of the human X chromosome. *Nucleic Acids Res.* **11:** 2303.

Davis, M., D.I. Choen, E.A. Nielsen, M. Steinmetz, W.E. Paul, and L. Hood. 1984. Cell-type-specific cDNA probes and the murine I region: The localization and orientation of A^d alpha. *Proc. Natl. Acad. Sci.* **81:** 2194.

Francke, U., H.D. Ochs, B. deMartinville, J. Giancolone, V. Lindren, C. Dieteche, R.A. Pagon, M.H. Hofker, G.J.B. van Ommen, P.L. Pearson, and R.J. Wedgewood. 1985. Minor Xp21 chromosome deletion in a male associated with expression of Duchenne muscular dystrophy, chronic granulomatous disease, retinitis pigmentosa, and McLeod syndrome. *Am. J. Hum. Genet.* **37:** 250.

Ginsburg, D., R.I. Handin, D.T. Bonthron, T.A. Donlon, G.A.P. Bruns, S.A. Latt, and S.H. Orkin. 1985. Human von Willebrand Factor (vWF): Isolation of complementary DNA (cDNA) clones and chromosomal localization. *Science* **228:** 1401.

Gubler, U. and B.J. Hoffman. 1983. A simple and very efficient method for generating cDNA libraries. *Gene* **25:** 263.

Gusella, J.F., N.S. Wexler, R.M. Conneally, S.L. Naylor, M.A. Anderson, R.E. Tanzi, P.C. Watkins, K. Ottina, M.R. Wallace, A.Y. Sakaguchi, A.B. Young, I. Shoulson, E. Bonilla, and J.B. Martin. 1983. A polymorphic DNA marker genetically linked to Huntington disease. *Nature* **306:** 234.

Harper, A.M., M.F. Chaplin, and A.W. Segal. 1985. Cytochrome b-245 from human neutrophils is a glycoprotein. *Biochem. J.* **227:** 783.

Knowlton, R.G., O. Cohen-Haguenauer, N.V. Cong, J. Frezal, V.A. Brown, D. Barker, J.C. Braman, J.W. Schumm, L.-C. Tsui, M. Buchwald, and H. Donis-Keller. 1985. A polymorphic DNA marker linked to cystic fibrosis is located on chromosome 7. *Nature* **318:** 380.

Kozak, M. 1984. Compilation and analysis of sequences upstream from the translational start site in eukaryotic mRNAs. *Nucleic Acids Res.* **12:** 857.

Kunkel, L.M., A.P. Monaco, W. Middleworth, H.D. Ochs, and S.A. Latt. 1985. Specific cloning of DNA fragments absent from the DNA of a male patient with an X chromosome deletion. *Proc. Natl. Acad. Sci.* **82:** 4778.

Maniatis, T., E. Fritsch, and J. Sambrook. 1982. *Molecular Cloning: A laboratory manual.* Cold Spring Harbor Laboratory, Cold Spring Harbor, New York.

Michelson, A.M., A.F. Markham, and S.H. Orkin. 1983. Isolation and DNA sequence of a full-length cDNA clone for human X chromosome-encoded phosphoglycerate kinase. *Proc. Natl. Acad. Sci.* **80:** 472.

Monaco, A.P., C.J. Bertelson, W. Middlesworth, C.-A. Colletti, J. Aldridge, K.H. Fischbeck, R. Bartlett, M.A. Pericak-Vance, A.D. Roses, and L.M. Kunkel. 1985. Detection of deletions spanning the Duchenne muscular dystrophy locus using a tightly linked DNA segment. *Nature* **316:** 842.

Nishioka, Y., A. Leder, and P. Leder. 1980. Unusual alpha-globin-like gene that cleanly lost both globin intervening sequences. *Proc. Natl. Acad. Sci.* **77:** 2806.

Royer-Pokora, B., L.M. Kunkel, A.P. Monaco, S.C. Goff, P.E. Newburger, R.L. Baehner, F.S. Cole, J.T. Curnutte, and S.H. Orkin. 1986. Cloning the gene for an inherited human disorder—chronic granulomatous disease—on the basis of its chromosomal location. *Nature* **322:** 32.

Segal, A.W. 1985. Variations on the theme of chronic granulomatous disease. *Lancet* **I:** 1378.

Segal, A.W., A.R. Cross, R.C. Garcia, N. Borregaard, N.H. Valerius, J.F. Soothill, and O.T.G. Jones. 1983. Absence of cytochrome b-245 in chronic granulomatous disease: A multicenter European evaluation of its incidence and relevance. *N. Engl. J. Med.* **308:** 245.

Sharp, P.A. 1983. Conversion of RNA to DNA in mammals: Alu-like elements and pseudogenes. *Nature* **301:** 471.

Strunk, R.C., A.S. Whitehead, and F.S. Cole. 1985. Pretranslational regulation of the synthesis of the third component of complement in human mononuclear phagocyte by the lipid A portion of lipopolysaccharide. *J. Clin. Invest.* **76:** 985.

Tauber, A.I., N. Borregaard, E. Simons, and J. Wright. 1983. Chronic granulomatous disease: A syndrome of phagocyte oxidase deficiencies. *Medicine* **62:** 286.

Wainwright, B.J., P.J. Scambler, J. Schmidtke, E.A. Watson, H.-Y. Law, M. Farrall, H.J. Cooke, H. Eiberg, and R. Williamson. 1985. Localization of cystic fibrosis locus to human chromosome 7cen-q22. *Nature* **318:** 384.

White, R., S. Woodward, M. Leppert, P. O'Connell, M. Hoff, J. Herbst, J.-M. Lalouel, M. Dean, and G. Vande Woude. 1985. A closely linked genetic marker for cystic fibrosis. *Nature* **318:** 382.

Young, R.A. and R.W. Davis. 1983. Efficient isolation of genes by using antibody probes. *Proc. Natl. Acad. Sci.* **80:** 1194.

Reduced Recombination Rate on Chromosomes 21 That Have Undergone Nondisjunction

S.E. Antonarakis* A. Chakravarti,† A.C. Warren,* S.A. Slaugenhaupt,† C. Wong,* S.L. Halloran,† and C. Metaxotou‡

*Genetics Unit, Department of Pediatrics, The Johns Hopkins University School of Medicine, Baltimore, Maryland 21205; †Human Genetics Program, Department of Biostatistics, University of Pittsburgh, Pittsburgh, Pennsylvania 15261; ‡Cytogenetics Unit First Department of Pediatrics, Athens University Medical School, Athens, Greece

Down syndrome (trisomy 21) is the most common known genetic cause of mental retardation, with an incidence of 1.0–1.3 per 1000 live births, or about 0.45% of all clinically recognized pregnancies (Hassold and Jacobs 1984). The cytogenetic mechanism leading to trisomy 21 is meiotic nondisjunction, which could occur either at the first or the second meiotic division (Polani 1981). Several authors have suggested that homologous chromosomes must be linked by chiasmata in diakinesis in order to segregate at the first meiotic metaphase (Darlington 1929; Dobzhansky 1933; Mather 1938). It was hypothesized that at least one chiasma per bivalent was necessary for normal segregation (Mather 1938). Experimental studies in the mouse suggest that meiotic nondisjunction due to failure of separation of homologs at anaphase is rare and is usually due to aberrant segregation of univalents (Henderson and Edwards 1968). These univalents may be produced by two processes, namely, asynapsis (failure of any chiasma formation during the zygotene-pachytene stage) or desynapsis (premature unpairing of homologs following normal chiasma formation) (Henderson 1970). It is currently not known which of these two processes causes meiotic nondisjunction involving chromosomes 21.

A direct test of the above hypothesis requires the examination of chiasmata on chromosomes that underwent nondisjunction. Since this is not possible, we chose to study genetic recombination among DNA markers on these chromosomes 21. There is considerable evidence that chiasma formation and genetic recombination are directly related (Beadle 1932; Jones 1971; Polani and Crolla 1982). Trisomy 21 due to asynapsis is equivalent to no recombination, whereas desynapsis is equivalent to normal recombination. These hypotheses pertain to nondisjunction that occurs in the first meiotic division; nondisjunction in the second meiotic division is generally assumed to be unrelated to chiasma formation and, therefore, recombination.

The purpose of this study is to specifically test the hypothesis that asynapsis has occurred on chromosomes 21 that undergo nondisjunction. We have used several DNA polymorphisms as chromosome 21 markers in families with a Down syndrome child and developed new methods for linkage analyses of these markers on chromosomes that undergo nondisjunction and on chromosomes that disjoin normally (both from control families and from Down syndrome families). Our studies show that chromosomes that have participated in nondisjunction demonstrate reduced recombination when compared with chromosomes that disjoin normally. These findings support the hypothesis that asynapsis, probably due to defective pairing of chromosomes 21 in the first meiotic division, is an etiologic factor that leads to trisomy 21.

METHODS

Subjects. Linkage analyses were performed on a total of 50 Caucasian control families from different ethnic backgrounds (two or three generations). These families were used to construct a linkage map for chromosome 21. For the estimation of map distances in chromosomes 21 that undergo nondisjunction, a total of 34 Greek families with a trisomy 21 offspring and at least one normal offspring were used.

Restriction endonuclease analysis. Nuclear DNA was isolated from leukocytes from 10–15 ml of EDTA anticoagulated blood or from cultured lymphoblastoid cells contained in a 25-cm^2 tissue culture flask as previously described (Kunkel et al. 1977). For each restriction endonuclease used, 5 μg of DNA were digested under conditions recommended by the commerical suppliers. Gel electrophoresis, transfer of DNA fragments to nitrocellulose filters, probe hybridizations, and autoradiography were performed as described elsewhere (Scott et al. 1979; Southern 1979).

Probes and DNA polymorphisms. The following cloned DNA fragments were used as probes:

1. Genomic *Eco*RI fragments D21S1 (1.5 kb; formerly named pW228C), D21S11 (1.85 kb; formerly named pW236B), D21S3 (2.1 kb; formerly named pW231C), and D21S23 (0.95 kb; formerly named pW244D); all of these were cloned in pBR328 (Watkins et al. 1985).
2. The 9-kb genomic *Eco*RI fragment D21S13 cloned in phage λ, formerly named D21K9 (Davies et al. 1984). These DNA fragments map to the long arm of chromosome 21 and are present in single copy;

more specifically, D21S1 and D21S11 map on 21q11.2-q21 (Munke et al. 1985), D21S3 maps on 21q proximal to q21.1 (Stewart et al. 1985), and D21S3 maps on 21q22.3 (Wong et al. 1986; Kazazian et al. 1985; Van Keuren et al. 1986).
3. Genomic and cDNA fragments of the superoxide dismutase gene (*SOD1*) (Levanon et al. 1985), which map to 21p22.1 (Tan et al. 1973).
4. Genomic *Pvu*II-*Sph*I fragment CW21pc (0.65 kb) cloned in M13. This fragment is present in single copy and was cloned from the junction fragment of a ring chromosome 21; the long-arm break point has been assigned to DNA fragment DS21S3 at 21q22.3, and the short-arm break point has been assigned to the pericentromeric region of chromosome 21 (Wong et al. 1986).

The following restriction endonucleases generated polymorphic DNA fragments that were detected by these probes: *Bam*HI and *Msp*I sites adjacent to probe D21S1 (Watkins et al. 1985); *Eco*RI and *Taq*I sites adjacent to D21S11 (Watkins et al. 1985); *Taq*I and *Hin*dIII sites adjacent to D21S3 (Kittur et al. 1985; Wong et al. 1986); *Msp*I site adjacent to SOD1 (Kittur et al. 1985); *Taq*I site adjacent to D21S13 (Davies et al. 1984); *Eco*RI and *Msp*I sites adjacent to D21S23 (Watkins et al. 1985); *Sst*I and *Hin*cII sites adjacent to CW21pc (Wong et al. 1986).

Linkage analysis. Tests of linkage in the control families were performed using the maximum likelihood lod score method of Morton (1955) and the computer program LIPED (Ott 1974). For each lod table, the maximum likelihood estimate of θ and the maximum lod score Z were computed using the interpolation formulae of Rao et al. (1978). The 95% confidence limits on the recombination value were computed as $\hat{\theta} \pm 1.96\sqrt{V\theta}$, where the variance of θ, $V\theta$, was calculated by the method of Buetow et al. (1985). Multilocus linkage analysis was performed using the computer program LINKAGE (Lathrop et al. 1984). For this analysis the positions of several loci were assigned on a map-distance centimorgan (cM) scale using the Haldane (1919) mapping function (no interference). By varying the location of one locus at a time and by computing the likelihood of the joint segregation of multiple markers, the relative odds for various gene orders were computed by comparing the likelihoods (location scores) directly. The maximum likelihood locations from the most likely gene order were then transformed into recombination values using Haldane's map function.

Gene mapping relative to the centromere. Gene mapping relative to the centromere can be performed whenever two or more members of a tetrad can be recovered. To produce a trisomic offspring, the parent in whom nondisjunction occurs transmits a disomic gamete (unordered half-tetrad), whereas the other parent transmits a usual monosomic gamete. To ascertain whether recombination between a marker and the centromere has occurred or not, it is first necessary to know whether the disomic gamete has arisen from a meiosis I or meiosis II error. Clearly, if no recombination occurs and the host is constitutionally heterozygous at both centromere and marker, a disomic gamete arising from a meiosis I error will be heterozygous at both centromere and marker, whereas a disomic gamete arising by a meiosis II error will be homozygous at both centromere and marker (Côté and Edwards 1975). If recombination occurs, the probability of heterozygosity of the marker is reduced if it arose as a meiosis I error and increased if it arose as a meiosis II error, the magnitude depending on the number of exchanges (chiasmata). Specifically, these probabilities can be written as a function of a new linkage parameter y as shown in Table 1 (A. Chakravarti and S. Slaugenhaupt, in prep.; N.E. Morton, pers. comm.). The parameter y can be related to both the map distance (w) and the recombination value (θ), and when there are at most two chiasmata in an interval (Morton and MacLean 1984), $y = 3\theta - w$. For complete linkage of a marker to the centromere ($\theta = 0$) $y = 0$; for no linkage to the centromere ($\theta = 1/2$) $y = 2/3$; whereas for short map distances $y = 2\theta$ (Ott et al. 1976; Morton and MacLean 1984; A. Chakravarti and S. Slaugenhaupt, in prep.). The map function used in the formula $y = 3\theta - w$, to relate w and θ, is the one suggested by *Neurospora* tetrad analysis and *Drosophila* attached-X analysis (Morton et al. 1985).

Our aim is to estimate y by the method of maximum likelihood for markers on chromosomes that have participated in nondisjunction. Using the above mentioned probabilities, one can explicitly write the likelihood function of the marker genotype of a trisomy 21 offspring given the parental genotypes. This likelihood is one of the above probabilities when the origin of nondisjunction is known. When it is not, the likelihood is calculated as a weighted average, weighted by the probabilities of each meiotic error. These probabilities, as estimated by Hassold and Jacobs (1984), are: paternal I, 0.13; paternal II, 0.06; maternal I, 0.68; maternal II, 0.13. We present our results as a lod score,

$$Z(y) = \log_{10} \frac{L(y)}{L(2/3)}$$

where $L(y)$ is the likelihood function for all families. This method has been extended to include normal off-

Table 1. Probability of Heterozygosity as a Function of Linkage Parameter y

Type of nondisjunction	Probability of heterozygosity	Probability of homozygosity
Meiosis I error	$1 - \frac{1}{2}y$	$\frac{1}{2}y$
Meiosis II error	y	$1 - y$

spring in the same family to calculate the recombination value θ on chromosomes that did not participate in nondisjunction. Then, the lod score is

$$Z(\theta, y) = \log_{10} \frac{L(\theta, y)}{L(\frac{1}{2}, \frac{2}{3})}$$

A detailed description of these new methods is provided in A. Chakravarti and S. Slaugenhaupt (in prep.). The lod score calculations are performed by the computer programs CENMAP and DSLINK, which we have developed. The maximum likelihood estimates are obtained by interpolation using the methods of Rao et al. (1978).

RESULTS

Linkage Map of Human Chromosome 21 in Control Families

Using DNA polymorphisms adjacent to the single-copy DNA fragments derived from human chromosome 21 (described in Methods), we constructed a preliminary linkage map of human chromosome 21. DNA markers D21S1 and D21S11 are closely linked to one another with no recombinants being observed in more than 135 scorable meioses (Kittur et al. 1985). Also, DNA markers D21S3 and D21S23 are closely linked with one recombinant being observed in 23 meioses (Tanzi et al. 1985; A.C. Warren et al., unpubl.) with an estimated recombination value of 4%. These markers D21S3 and D21S23 are also treated as one locus in our subsequent analyses. SOD1 shows a 7% recombination value with the D21S1/D21S11 locus ($\hat{\theta} = 0.07$; $\hat{Z} = 5.28$; 95% confidence limits (CL): 4-11%), while the DNA marker D21S13 show 17% recombination with D21S1/D21S11 ($\hat{\theta} = 0.17$; $\hat{Z} = 2.44$; 95% CL: 4-31%). The pericentromeric DNA marker CW21pc shows 14% recombination with DNA fragment D21S13 ($\hat{\theta} = 0.14$; $\hat{Z} = 2.63$; 95% CL: 5-23%) and 17% recombination with D21S1/D21S11 ($\hat{\theta} = 0.17$; $\hat{Z} = 3.80$; 95% CL: 9-25%). Our data on the map distances between D21S3 and D21S23 and the rest of the markers show no linkage and, therefore, the distance with the closest available marker is probably greater than 30 cM. Table 2 shows the recombination fraction and lod score results for our markers, using the computer program LIPED.

Next, multilocus linkage analysis was performed using the computer program LINKAGE to ascertain the order of the polymorphic loci tested (Lathrop et al. 1984). This multipoint analysis suggested that the most likely order of these loci is (A.C. Warren et al., unpubl.): CW21pc-D21S13-D21S1/D21S11-SOD1-D21S3/D21S23. Furthermore, multipoint analysis estimated the distance between SOD1 and D21S3/D21S23 to be 22 cM.

Figure 1 shows the linkage map of chromosome 21 obtained using the families examined in this study. A more detailed map with additional DNA markers and collection of data from large pedigrees has been obtained by J.F. Gusella and coworkers (pers. comm.). There are no major differences between these linkage maps for the loci that have been studied by both groups.

Table 2. Pairwise Linkage Analysis Using Computer Program LIPED

Loci	$\hat{\theta}$	$\hat{Z}$	95% confidence limits
D21S1-D21S11	0.00	6.91	–
D21S3-D21S23	0.04	5.14	0-13%[a]
CW21pc-D21S13	0.14	2.63	5-23%
CW21pc-D21S1/D21S11	0.17	3.80	9-25%
D21S13-D21S1/D21S11	0.17	2.44	4-31%
D21S1/D21S11-SOD1	0.07	5.28	4-11%

[a]We observed 1 recombinant in 5 informative meioses; all the other data were reported by Tanzi et al. (1985), where no recombinants were observed between these two markers in 18 meioses.

Using the data presented here and assuming complete interference on chromosome 21 (Laurie and Hulten 1985), the total genetic map length for chromosome 21q is approximately 60 cM.

Linkage Map of Human Chromosomes 21 That Participate in Nondisjunction

A total of 34 Greek families were studied. In each family, DNA from father, mother, trisomy 21 offspring, and at least one unaffected sibling was examined. All samples were collected in the cytogenetics division of the "Agia Sophia" Children's Hospital of Athens Medical School. Each DNA was digested with the appropriate enzyme that detects a DNA polymorphism with each particular probe. In the individual with trisomy 21, the differences in the intensity of the hybridizing allelic fragments on autoradiographs permited the identification of the three chromosome 21 alleles per locus. In this manner, the DNA polymorphisms at CW21pc, D21S13, D21S1/D21S11, and D21S3/D21S23 were analyzed (Antonarakis et al. 1985).

Any chromosome 21 marker may provide information on the parental origin of nondisjunction, whereas only a centromeric marker may give information on the meiotic stage at which nondisjunction occurs. DNA marker CW21pc, which is pericentromeric, provided information on the stage of meiosis in which the error had occurred. Using the above markers, we could identify the parental origin and/or the stage of the meiotic error in all but 9 families.

Table 3 presents the number of families in which the DNA markers were informative for the parental origin of nondisjunction. Table 4 shows the origin of nondisjunction inferred from all the marker data in the 34 families.

Next, we estimated the probabilities of paternal I, paternal II, maternal I, and maternal II errors from

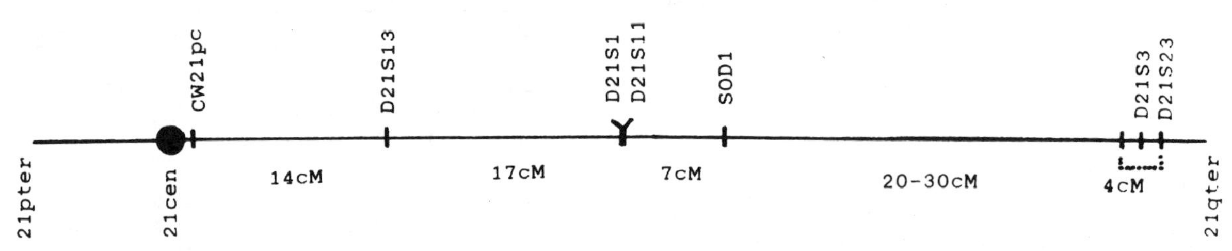

Figure 1. Linkage map of human chromosome 21. Map distances between DNA markers are shown in centimorgans. The dotted line for D21S23 indicates that the order 21CEN–D21S23–D21S3 vs. 21CEN–D21S3–D21S23 has not been determined.

our 34 families, using the method of Jacobs and Morton (1977), to be 0.10, 0.00, 0.85, and 0.05, respectively. These estimates were not significantly different ($\chi^2 = 7.52$, 5 df, $P > 0.18$) from the values in Hassold and Jacobs (1984). We employed the published figures because they are based on more than 600 families.

Subsequently, we used our methods to estimate the linkage parameter y for each marker in the 34 families. The results are presented as lod scores for linkage between a marker and the centromere, in terms of y, the lod score being calculated against the null hypothesis of no linkage ($y = 2/3$). Initially, we estimated $\hat{y} = 0$ at the maximum lod score $\hat{Z} = 1.25$ for the marker CW21pc. This represents a significant linkage ($\chi^2 = 5.76$, 1 df, $P = 0.017$) and confirms that CW21pc is a pericentromeric marker. Given this, CW21pc genotypes were used to calculate posterior probabilities for each of the four meiotic errors in each family. Next, y was estimated for each marker locus from the 34 families and from the 9 families with maternal meiosis I nondisjunction. These results, together with the maximum lod scores, are presented in Table 5.

The above results demonstrate that crossing-over is greatly reduced on chromosomes that have participated in nondisjunction, since for the markers D21S13, D21S1/D21S11, and D21S3/D21S23, we estimate $\hat{y}$ to be 0.00, 0.00, and 0.05, respectively, whereas from the normal map we would have expected these values to be approximately 0.28, 0.34, and 0.67, respectively. The latter values were calculated as $y = 2\theta$ with the θ values being obtained from Table 2.

It would, however, be desirable to estimate y and θ simultaneously from each family, so that the recombination value between the centromere and the test loci on chromosomes that participate in nondisjunction (θ_T) and on chromosomes that do not (θ_C) can be directly compared. θ_T is obtained from the y value as described in Methods. These results are presented in Table 6 and show that recombination is reduced on chromosomes that participate in nondisjunction.

DISCUSSION

In this study we tested the hypothesis that there is reduced recombination on chromosome 21 that participates in nondisjunction in trisomy 21. A total of 34 families containing one individual with trisomy 21 have been studied using DNA polymorphisms on human chromosome 21. The map distances of these loci from the centromere were estimated and compared with (1) map distances obtained by classical linkage analysis in control families, and (2) map distances obtained from the chromosomes 21 that disjoined normally in the trisomy 21 families. The data suggest that the map distances for the same sets of DNA markers were very different between "control" and "affected" chromosomes. ("Affected" chromosomes were those that underwent nondisjunction.) This implies that there is a reduced recombination rate on chromosomes 21 undergoing nondisjunction in meiosis I. There is considerable evidence from other species (*Drosophila melanogaster* and *Saccharomyces cerevisiae*) that mutations that primarily decrease recombination also cause elevated frequencies of nondisjunction. There are a

Table 3. Number of Families in Which the Following DNA Markers Were Informative for the Parental Origin of Nondisjunction

Loci	No. of informative families[a]
CW21pc	11
D21S13	6
D21S1/D21S11	12
D21S3/D21S23	16

[a]No. of families examined = 34.

Table 4. Origin of Nondisjunction in 34 Greek Families as Analyzed by Using DNA Polymorphisms

Origin unknown	9 families
Maternal	22 families[a]
Paternal	3 families

[a]Meiosis I error = 9; meiosis II error = 1.

Table 5. Map Distance Relative to the Centromere of the Following Markers on Chromosomes 21 Participating in Nondisjunction in 34 Greek Families with Trisomy 21 Using the Program CENMAP

Loci	All 34 families		9 families[a]	
	$\hat{y}$	$\hat{Z}$	$\hat{y}$	$\hat{Z}$
21 cen-D21S13	0.00	0.935	0.00	0.432
21 cen-D21S1/D21S11	0.00	1.873	0.00	0.687
21 cen-D21S3/D21S23	0.05	1.709	0.00	1.215

[a]Nondisjunction due to meiosis I error.

number of mutations in *D. melanogaster* that cause defects of recombination and/or segregation (Baker et al. 1976). In general, these recombination-defective mutants cause decreased levels of recombination and increased frequencies of nondisjunction of all chromosomal pairs. In all mutants examined, nondisjunction occurs exclusively at the first meiotic division and involves only these chromosomes lacking recombination. In *S. cerevisiae*, it seems that recombination is regulated at two levels: There is a general control of the overall frequency of crossing-over, and there are more localized controls that influence crossing-over only in particular regions. In *D. melanogaster*, aging results in reduced crossing-over on all chromosomes and increased nondisjunction (Baker et al. 1976).

In mice, chiasma frequency declines with increasing age in females (Polani 1981; Hassold and Jacobs 1984). It is not known, however, whether these mice produce abnormal gametes. In 1968 Henderson and Edwards proposed the production line hypothesis: Oocytes released later in life have undergone meiosis later with fewer chiasmata and more univalents. It has been difficult to confirm or disprove this hypothesis since there exist both supportive and nonsupportive data for this hypothesis. Hassold et al. (1980) have suggested that nondisjunction resulting from age-dependent loss of chiasmata would be most likely to involve chromosomes with the fewest chiasmata, namely, the small chromosomes. Laurie and Hulten (1985) have published the mean chiasma frequencies for each autosome in male meiosis. Chromosome 21 has the lowest mean chiasma frequency, 1.06 per bivalent.

Our results show that there is no recombination between the loci examined and the centromere, and we can speculate that no chiasmata exist in the zygotene-pachytene stage between chromosome 21 pairs that undergo meiosis I nondisjunction. Clearly, more data are needed to answer the following questions:

1. In trisomy 21 due to first meiotic division error, is reduced recombination found only on chromosome 21 or on all chromosomes?
2. Is there any association with maternal or paternal age?
3. Are there specific genes that control recombination in early meiosis I and, if so, are they chromosome specific?
4. Are the women (or men) who gave two chromosomes 21 to the Down syndrome offspring carriers of defective recombination controlling genes?
5. Is reduced recombination found in other aneuploidies, including those that involve sex chromosomes?

If reduced recombination in trisomy 21 families is found to occur only in chromosomes 21 and no other chromosomes, then the existence of a chromosome-specific gene that controls cross-over and recombination can be proposed. This hypothesis may also explain the data by Antonarakis et al. (1985) that certain DNA polymorphism haplotypes for DNA markers on chromosome 21 are more frequent in chromosomes 21 that underwent nondisjunction.

This study demonstrates the use of DNA polymorphisms in the study of the process of proper chromosomal segregation in meiosis.

ACKNOWLEDGMENTS

We gratefully acknowledge the help of Dr. H.H. Kazazian, Jr., for encouragement throughout the project, Drs. N.E. Morton and K. Buetow for helpful discussions, and Dr. S. Sherman for calculations of the probabilities of meiosis errors. We also thank Drs. J. Gusella, K. Davies, and Y. Groner for providing us with cloned DNA fragments. This study was supported by grants GM33771 and HD19491 from the National Institutes of Health to A.C. and S.E.A. and the National Foundation–March of Dimes grant to S.E.A. A.C.W. was supported by a Daland Fellowship from the American Philosophical Society and a fellowship from the John Douglas French Foundation for Alzheimer disease. The authors thank Emily Pasterfield for expert assistance in the preparation of the manuscript.

Table 6. Estimation of θ_T and θ_C Using Computer Program DSLINK

Loci	$\hat{\theta}_T$	$\hat{\theta}_C$	$\hat{Z}(\theta_T,\theta_C)$	X^2 for $\theta_T=\theta_C$ vs. $\theta_T<\theta_C$	P
CW21pc-D21S13	0.00	0.30	1.23	0.61	0.22
CW21pc-D21S1/D21S11	0.00	0.20	2.67	3.01	0.04
CW21pc-D21S3/D21S23	0.00	0.25	1.95	2.92	0.04

REFERENCES

Antonarakis, S.E., S.D. Kittur, C. Metaxotou, P.C. Watkins, and A.S. Patel. 1985. Analysis of DNA haplotypes suggests a genetic predisposition to trisomy 21 associated with DNA sequences on chromosome 21. *Proc. Natl. Acad. Sci.* **82:** 3360.

Baker, B.S., A.T.C. Carpenter, M.S. Esposito, R.E. Esposito, and L. Sandler. 1976. The genetic control of meiosis. *Annu. Rev. Genet.* **10:** 53.

Beadle, G.W. 1932. The relation of crossing-over to chromosome association in Zea-Euchlaena hybrids. *Genetics* **17:** 481.

Buetow, K.H., A. Chakravarti, R.L. Nussbaum, and R.E. Ferrell. 1985. Sampling variance and confidence limits on the recombination value: XPR and DXS7. *Cytogenet. Cell Genet.* **40:** 595 (Abstr.).

Côté, G.B. and J.H. Edwards. 1975. Centromeric linkage in autosomal trisomies. *Ann. Hum. Genet.* **39:** 51.

Darlington, C.D. 1929. Chromosome behaviour and structural hybridity in the Tradescantiae. *J. Genet.* **21:** 207.

Davies, K.E., K. Harper, D. Bonthron, R. Krumlauf, A. Polkey, M.E. Pembrey, and R. Williamson. 1984. Use of a chromosome 21 cloned DNA probe for the analysis of nondisjunction in Down syndrome. *Hum. Genet.* **66:** 54.

Dobzhansky, T. 1933. Studies on chromosome conjugation: The relation between crossing-over and disjunction of chromosomes. *Z. Indukt. Abstammungs. Vererbungsl.* **64:** 269.

Haldane, J.B.S. 1919. The combination of linkage values and the calculation of distances between the loci of linked factors. *J. Genet.* **8:** 299.

Hassold, T.J. and P.A. Jacobs. 1984. Trisomy in man. *Annu. Rev. Genet.* **18:** 69.

Hassold, T.J., P. Jacobs, J. Kline, Z. Stein, and D. Warburton. 1980. Effect of maternal age on autosomal trisomies. *Ann. Hum. Genet.* **44:** 29.

Henderson, S.A. 1970. The time and place of meiotic crossing-over. *Annu. Rev. Genet.* **4:** 295.

Henderson, S.A. and R.G. Edwards. 1968. Chiasma frequency and maternal age in mammals. *Nature* **218:** 22.

Jacobs, G.H. and N.E. Morton. 1977. Origin of human trisomies and polyploids. *Hum. Hered.* **27:** 59.

Jones, G.H. 1971. The analysis of exchanges in tritium labeled meiotic chromosomes II. Stethophyme grossum. *Chromosoma* **34:** 367.

Kazazian, H.H., Jr., S.E. Antonarakis, C. Wong, S.P. Trusko, G. Stetten, M. Oliver, M. Potter, J.F. Gusella, and P.C. Watkins. 1985. Ring chromosome 21: Characterization of DNA sequences on sites of breakage and reunion. *Ann. N.Y. Acad. Sci.* **450:** 33.

Kittur, S.D., S.E. Antonarakis, R.E. Tanzi, D.A. Meyers, A. Chakravarti, Y. Groner, J.A. Phillips, P.C. Watkins, J.F. Gusella, and H.H. Kazazian, Jr. 1985. A linkage map of three anonymous DNA fragments and SOD-1 on chromosome 21. *EMBO J.* **4:** 2257.

Kunkel, L.M., K.D. Smith, S.H. Boyer, D.S. Borgaonkar, S.F. Wechtel, O.S. Miller, W.R. Berg, H.W. Jones, Jr., and J.M.H. Rary. 1977. Analysis of human Y chromosome specific reinterated DNA in chromosome variants. *Proc. Natl. Acad. Sci.* **74:** 1245.

Lathrop, G.M., J.M. Lalouel, C. Julier, and J. Ott. 1984. Strategies for multilocus linkage analysis in humans. *Proc. Natl. Acad. Sci.* **81:** 3443.

Laurie, D.A. and M.A. Hulten. 1985. Further studies on bivalent chiasma frequency in human males with normal karyotypes. *Ann. Hum. Genet.* **49:** 189.

Levanon, D., J. Lieman-Hurwitz, N. Dafni, M. Widgerson, L. Sherman, Y. Bernstein, Z. Laver-Rudich, E. Danciger, O. Stein, and Y. Groner. 1985. Architecture and anatomy of the chromosomal locus in human chromosome 21 encoding the Cu-Zu superoxide dismutase. *EMBO J.* **4:** 77.

Mather, K. 1938. Crossing-over. *Biol. Rev. Camb. Philos. Soc.* **13:** 252.

Morton, N. 1955. Sequential tests for the detection of linkage. *Am. J. Hum. Genet.* **7:** 277.

Morton, N.E. and C.J. Maclean. 1984. Multilocus recombination frequencies. *Genetic Res.* **44:** 99.

Morton, N.E., C.J. MacLean, and R. Lew. 1985. Test of hypotheses on recombination frequencies. *Genet. Res.* **45:** 279.

Munke, M., J. Kraus, P. Watkins, R. Tanzi, J.F. Gusella, A. Mullington-Ward, M. Watson, and U. Francke. 1985. Homocystinuria gene on chromosome 21 mapped with cloned cystathione beta synthase probe and in situ hybridizaiton of other chromosome 21 probes. *Cytogenet. Cell Genet.* **40:** 706 (Abstr.).

Ott, J. 1974. Estimation of the recombination fraction in human pedigrees: Efficient computation of the likelihood for human linkage studies. *Am. J. Hum. Genet.* **26:** 588.

Ott, J., D. Linder, B.K. McCaw, E.W. Lourin, and F. Hecht. 1976. Estimating distances from the centromere by means of benign ovarian teratomas in man. *Ann. Hum. Genet.* **40:** 191.

Polani, P.E. 1981. Chiasmata, Down syndrome and nondisjunction: An overview. In *Trisomy 21 research perspectives* (ed. F.F. delaCruz and P.S. Gerald), p. 111. University Park Press, Baltimore.

Polani, P.E. and J.A. Crolla. 1982. Experiments on female mammalian meiosis. In *Genetic control of gamete production and function* (ed. P.G. Crosignani and B.L. Rubin), p. 171. Academic Press, London.

Rao, D.C., B.J.B. Keats, N.E. Morton, S. Yee, and R. Lew. 1978. Variability of human linkage data. *Am. J. Hum. Genet.* **30:** 516.

Scott, A.F., J.A. Phillips III, and B.R. Migeon. 1979. DNA restriction endonuclease analysis for localization of human beta and delta globin genes on chromosome 11. *Proc. Natl. Acad. Sci.* **76:** 4563.

Southern, E.M. 1979. Gel electrophoresis of DNA fragments. *Methods Enzymol.* **68:** 152.

Stewart, G.D., P. Harris, J. Galt, and M.A. Ferguson-Smith. 1985. Cloned DNA probes regionally mapped to human chromosome 21 and their use in determining the origin of nondisjunction. *Nucleic Acids Res.* **13:** 4125.

Tan, Y.H., J. Tischfield, and F.H. Ruddle. 1973. The linkage of genes for the human interferon-induced anti-viral protein and indophenol exidase-beta traits to chromosome G-21. *J. Exp. Med.* **137:** 317.

Tanzi, R., P. Watkins, K. Gibbons, A. Faryniarz, M.M. Wallace, R. Hallewell, P.M. Conneally, and J. Gusella. 1985. A genetic linkage map of human chromosome 21. *Cytogenet. Cell Genet.* **40:** 760.

Van Keuren, M., P. Watkins, H. Drabkin, E.W. Jabs, J.F. Gusella, and D. Patterson. 1986. Regional localization of DNA sequence on chromosome 21 using somatic cell hybrids. *Am. J. Hum. Genet.* **38:** 793.

Watkins, P.C., R. Tanzi, K. Gibbons, J. Tricoli, G. Landes, R. Eddy, T. Shows, and J.F. Gusella. 1985. The isolation of polymorphism DNA segments from human chromosome 21. *Nucleic Acids Res.* **13:** 6075.

Wong, C., H.H. Kazazian, Jr., P.C. Watkins, and S.E. Antonarakis. 1986. Ring chromosome 21: The breakage and reunion sites occurred in regions of single copy DNA in 21q and 21p. *Pediatr. Res.* **20:** 274A (Abstr.).

Genetic Recombination and Disease

M. SINISCALCO
Memorial Sloan-Kettering Cancer Center, New York, New York 10021

The reports in the following section deal primarily with genetic recombination, both legitimate and illegitimate, along the sex chromosomes. Legitimate recombination occurs between the two X chromosomes at ovogenesis or between the homologous portion of the X and Y at spermatogenesis; illegitimate recombination may occur in either female or male gametogenesis, leading respectively to X-linked mutations of the duplication/deletion type or to the formation of zygotes with an altered sex phenotype. To introduce the section, it is thus appropriate to focus on the role of genetic recombination in the generation of human disease. Apart from its biological relevance, such an issue has obvious bearing on one of the main themes of the Symposium, namely, the strategies to be devised for applying molecular biology to preventive and diagnostic medicine.

The molecular evidence demonstrating the association of a diverse group of X-linked diseases with interstitial deletions in the midportion of the X-chromosome short arm is reviewed in detail in this volume (Davies et al.; Kunkel et al.; Pearson et al.; Worton et al.). These deletions can appararently be detected in about 10% of patients with X-linked muscular dystrophies (DMD or BMD), using an array of random DNA sequences that map to the narrow region of the X-chromosome short arm referred to as Xp21. Among these sequences are the PERT probes, an acronym alluding to the phenol emulsion reassociation technique of Kohne et al. (1977). This technique was cleverly applied for the isolation of these sequences by Kunkel and colleagues (1985) through competitive reassociation of normal X-chromosomal DNA with DNA derived from a unique patient who had a detectable deletion at Xp21 and the combined phenotypes of DMD, retinitis pigmentosa, and chronic granulomatous disease (Franke et al. 1985). Contrary to expectation, restriction-fragment-length polymorphisms (RFLPs) detected with at least three of the PERT probes, and with other probes that overlap with the DMD and BMD deletions, have not been found to segregate in complete linkage association with the mutant phenotypes (Kunkel et al.; Davies et al.; Pearson et al.; all this volume). This has been explained by assuming that deletion-related mutations associated with the variable phenotypes of X-linked muscular dystrophy may occur within a very large segment of the X-chromosome short-arm DNA (Kunkel et al., this volume). However, the data thus far gathered do not exclude the possibility that such types of mutations may occur within a relatively short domain of the Xp21 region that happens to be highly prone to meiotic recombination. This latter hypothesis would be sufficient not only to explain the difficulties in finding a RFLP marker closely linked to the X-linked muscular dystrophies, but also to account for the frequent occurrence of de novo mutations, if these are regarded as deletion mutations generated through the mechanism of unequal crossing-over (Pearson et al., this volume).

The subtelomeric region of the X-chromosome long arm is another obvious candidate for a deletion-related disease. In the recent past, the locus for glucose-6-phosphate-dehydrogenase (G6PD), that for the coagulation factor VIII, and the DNA fragments homologous to probes St14 and DX13 have been mapped to band Xq28, whereas the locus for coagulation factor IX has been assigned to band Xq27 (Szabo et al. 1984; Buckle et al. 1985; Mattei et al. 1985; Purrello et al. 1985). Some of these assignments were made by in situ hybridization of the relevant molecular probes to prometaphase preparations from patients with the fragile X syndrome (the well-known X-linked type of severe mental retardation associated with an inducible fragile site at Xq27.3), thus permitting precise localization of the above-mentioned loci to the regions immediately flanking the proximal (*FIX*) or distal side (*G6PD* cluster) of the inducible fragile site (*FRAX*). This suggests that the maximum distance between the loci flanking this site must be equal or smaller than the overall length of the subtelomeric region Xq27-Xqter, which is estimated to include approximately 10 million bp. Such relatively close physical distance between the factor IX locus (*F9*) and the *G6PD* cluster was a somewhat unexpected finding, because the mutant gene for FIX deficiency (hemophilia B) was known to segregate independently from G6PD deficiency, factor VIII deficiency (hemophilia A), and all other loci of the *G6PD* cluster. However, the real surprise came after the publication of the first set of linkage data between the fragile X syndrome and two of its flanking markers, namely G6PD deficiency of Mediterranean type (Filippi et al. 1983) and the FIX-*Taq*I RFLP (Camerino et al. 1983). Since both of these loci were found to segregate in close linkage association with the *FRAX* mutation, the genetic distance between the *FIX* locus and the *G6PD* cluster turned out to be considerably different (Szabo et al. 1984; Purrello et al. 1985b). This difference was dependent upon whether it was estimated directly from pedigrees segregating at the *FIX* locus and the *G6PD* cluster or indirectly by summing up the two separate estimates of linkage obtained from fragile X pedigrees that segregated also at the *G6PD* cluster or at the *Taq*I polymorphic site detected by the *FIX* probe. As it is emphasized by Dr. Mandel's report in this volume, the question of a possible effect of the *FRAX* mutation on the frequency of crossing-overs in the subregion Xq26-Xqter is still under scrutiny since there exist other pedigrees where the occurrence of recombinants between the *FRAX* mutation and its neighboring loci is far from being rare (Choo et al. 1984; Davies et al. 1985; Warren et al. 1985; Brown et al. 1986). The situation is further complicated by the existence of pedigrees with clear instances of normal males, who are able to transmit the mutation through their unaffected daughters to their affected grandsons (Turner et al. 1986). Since these pedigrees are the same ones that have yielded the evidence favoring a close linkage between the fragile X syndrome and its neighboring loci, it has been suggested that the suppression of an otherwise high proneness of the region Xq27.3 to genetic recombination may be related to the presence of nonpenetrant males (Brown et al. 1986). Indeed, all the above findings could be explained in a unitary manner, if also the *FRAX* mutations are regarded as the result of relatively common, mistaken crossing-over events, much in the same way as the classical *bar*, *infrabar*, and *double bar* mutations of *Drosophila melanogaster* are the result of duplication/deletion mutations resulting from unequal crossing-over (Sturtevant 1925). Under such a hypothesis, a patient with the fragile X syndrome would be the deficient product of the unequal crossing-over, thus with a genome that is hemizygous for a deletion of different degree, similar to those proven to be associated with the *DMD* mutations at the region Xp21. Correspondingly, the normal transmitting males would be the redundant products (i.e., the carriers of a tandem duplication of different length). Such duplications—phenotypically undetectable in the transmitting males—might favor the occurrence of additional instances of aberrant recombination in the meioses of their heterozygous daughters with the production of deletion mutations that are then transmitted to the next generation as X-linked traits. One possible mechanism for this to happen (worth mentioning only because it can be easily proven or disproven through the analysis of relatively few informative pedigrees) is that the putative duplicated region may lead to the formation of a loop, which can be turned into a deletion by intrachromatid crossing-over without necessarily disturbing the arrangements of genes at the flanking loci, thus mimicking a complete suppression of recombination between them. It is of interest to point out in this connection the frequent occurrence of XXY children born to mothers who are obligatory heterozygotes for the *FRAX* mutation. I am aware of at least eight cases reported thus far in the literature (Barbi and Steinbach 1985) or by personal communication (W.T. Brown and P.A. Jacobs). This number seems much too high to be a chance association, since the frequency of XXY males at birth is less than 1 in 1000, and the total number of independent pedigrees with the *FRAX* mutation thus far known hardly exceeds a few hundred. In view of the postulated relationship between chiasmata formation and segregation of homologous chromosomes during the first meiosis (Darlington 1929; Mather 1938), it is tantalizing to speculate that the above finding may be yet another manifestation of the alleged crossing-over suppression in the heterozygotes for the *FRAX* mutation.

The possibility that the mammalian sex heterochromosomes might have a homologous region within which genetic recombination occurs regularly was postulated by Koller and Darlington (1934), following their observation of X/Y partial meiotic pairing of male rats. Stimulated by this report, J.B.S. Haldane (1936) predicted the existence of partial sex-linkage for genes located in the pairing region of the X and Y now known as "pseudoautosomal region" (Burgoyne 1982). The studies reported in this section by Dr. Weissenbach (Rouyer et al.), Dr. Cooke (Cooke and Smith), and Dr. Goodfellow (Darling et al.) have now unequivocally proven that this region engages in regular meiotic exchanges with a frequency that goes from obligatory recombination at the Xp telomere to the absence of it, proximally to the site Xp22.32. This site segregates the most centromeric pseudoautosomal marker (the 12E7 antigen of Goodfellow et al. 1983, 1985) from the loci for X-linked ichthyosis and for the Xg(a) blood group, which are the most telomeric X-linked markers of the human X-chromosome short arm. On the Y-chromosome side, the pseudoautosomal region is immediately distal to the gene(s) involved in the determination of maleness, so that the association between an aberrant inheritance of the aforementioned markers with an aberrant sex phenotype (XX males and XY females) has been visualized as the result of illegitimate crossing-over events (Tiepolo et al. 1980; Ferguson-Smith et al. 1982). The molecular characterization of the sex-chromosome upsets, evaluated in conjunction with the evidence for a hot spot of recombination at the Xpter/Ypter subtelomeric regions, stresses once again the role of genetic recombination in the generation of deletion-related diseases (see Page; Seboun et al.; de la Chapelle; all this volume).

Recent molecular data (White et al. 1985) suggest that the severe form of 21-hydroxylase deficiency and its unique association with the *HLA-B47* haplotype is yet another example of a deletion-related mutation occurring at a chromosomal region (6p21) whose proneness to genetic rearrangement resulting from unequal crossing-over has been postulated as the mechanism required for the evolutionary duplication of the *HLA* loci (Ceppellini 1971). Other classical examples of recurrent deletion/duplication events generated by inaccurate recombination are the well-known Lepore/anti-Lepore types of mutation known to occur within the $\beta\delta$-globin gene complex at 11p15 and within the haptoglobin gene complex at 16q22 (Smithies et al. 1962; Baglioni 1963).

From all this it seems apparent that the number of

cases of well-proven deletion-related diseases has progressively increased as detailed information on the molecular fine structure of complex gene clusters accumulated. Given the role played by genetic recombination in the generation of these gene clusters, it is natural to wonder whether the deletion-related diseases are the price that the species has to pay for its ability to upgrade its genomic variability through genetic recombination. This is to say that—as J.B.S. Haldane once remarked in one of his writings for the lay public—a significant number of the deleterious mutations known may be genetic "holes" rather than "things" and may occur with a frequency considerably higher than those resulting from single-base-pair mutations, such as transversions and transitions. It is thus legitimate to ask oneself how many of the diseases reported by Dr. McKusick (this volume) in his map of the morbid anatomy of *Homo sapiens* are the expression of such genetic holes and therefore unsuitable targets for gene hunters, who try to isolate the altered DNA sequences from cDNA libraries of the diseased tissues. Moreover, if dreadful diseases may arise from mistaken crossover events in normal individuals, it would make sense to make a special effort to map the hot spots of recombination throughout the entire genome and to study in detail the DNA sequences existing at these sites in hopes of understanding the intimate molecular mechanism of normal and aberrant recombination. Accordingly, the molecular strategies for the detection of diseased phenotypes will have to include the construction of PERT-like probes within and around each chromosomal site recognized as being prone to a high frequency of genetic recombination. The identification of such regions requires a systematic comparison of genetic vs. physical distances for all segregating loci. This endeavor is going to be greatly facilitated by the recently developed technique of pulse-field gel electrophoresis of DNA fragments of very large size that subdivide the human genome into megabases (Smith and Cantor, this volume), thus bridging the gap between the already available cytogenetical and genetical maps. It is hardly necessary to emphasize that, well beyond its medical application, a detailed molecular knowledge of the structure of our genome and of the ways that structure can be altered will contribute critical data toward understanding the basic mechanisms underlying somatic differentiation and biological evolution.

REFERENCES

Baglioni, C. 1963. Correlations between genetics and chemistry of human hemoglobins. In *Molecular genetics* (ed. J.H. Taylor), vol. 1, p. 405. Academic Press, New York.

Barbi, G. and P. Steinbach. 1985. Letter to the editor: Fragile X and Martin-Bell syndrome: New source of information. *Am. J. Med. Genet.* **22:** 415.

Brown, W.T., A.C. Chan, and E.C. Jenkins. 1986. DNA linkage studies in the fragile-X syndrome suggest genetic heterogeneity. *Am. J. Med. Genet.* **23:** 643.

Buckle, V., I.W. Craig, D. Hunter, and J.H. Edwards. 1985. Fine assignment of the coagulation factor IX gene. *Cytogenet. Cell Genet.* **40:** 593.

Burgoyne, P.S. 1982. Genetic homology and crossing over in the X and Y chromosomes of mammals. *Hum. Genet.* **61:** 85.

Camerino, G., M.G. Mattei, J.H. Mattei, M. Jaye, and J.L. Mandel. 1983. Close linkage of fragile-X mental retardation syndrome to haemophilia B and transmission through a normal male. *Nature* **306:** 701.

Ceppellini, R. 1971. Old and new facts and speculations about transplantation antigens in man. In *Progress in immunology* (ed. B. Amos), p. 973. Academic Press, New York.

Choo, K.H., D. George, G. Filby, J.L. Halliday, M. Levensha, G. Webb, and D.M. Danks. 1984. Linkage analysis of X-linked mental retardation with and without fragile-X using factor IX probe. *Lancet* **II:** 349.

Darlington, C.D. 1929. Chromosome behaviour and structural hybridity in the tradescantiae. *J. Genet.* **21:** 207.

Davies, K.E., M.G. Mattei, J.F. Mattei, H. Veenema, S. McGlade, K. Harper, N Tommerup, K.B. Nielsen, M. Mikkelsen, P. Beighton, B. Drayna, R. White, and M.E. Pembrey. 1985. Linkage studies of X-linked mental retardation: High frequency of recombination in the telomeric region of the human X-chromosome. *Hum. Genet.* **70:** 249.

Ferguson-Smith, M.A., R. Sanger, P. Tippett, D.A. Aitken, and E. Boyd. 1982. A familial t(X;Y) translocation which assigns the Xg blood group locus to the region Xp22.3→ter. *Cytogenet. Cell Genet.* **32:** 273.

Filippi, G., A. Rinaldi, N. Archidiacono, M. Rocchi, I. Balazs, and M. Siniscalco. 1983. Linkage between G6PD and fragile-X syndrome. *Am. J. Med. Genet.* **15:** 113.

Francke, U., H.D. Ochs, B. De Martinville, J. Giacalone, V. Lindgren, C. Disteche, R. Pagon, M.H. Hofker, G.J.B. Van Ommen, P. Pearson, and R.J. Wedgwood. 1985. Minor Xp21 chromosome deletion in a male associated with expression of Duchenne muscular dystrophy, chronic granulomatous disease, retinitis pigmentosa, and McLeod syndrome. *Am. J. Hum. Genet.* **37:** 250.

Goodfellow, P., S. Darling, and J. Wolfe. 1985. The human Y chromosome. *J. Med. Genet.* **22:** 329.

Goodfellow, P., G. Banting, D. Sheer, H.H. Ropers, A. Caine, M.A. Ferguson-Smith, S. Povey, and R. Voss. 1983. Genetic evidence that a Y-linked gene in man is homologous to a gene on the X-chromosome. *Nature* **302:** 346.

Haldane, J.B.S. 1936. A search for incomplete sex linkage in man. *Ann. Eugen.* **7:** 28.

Koller, P.C. and C.D. Darlington. 1934. The genetical and mechanical properties of the sex chromosomes. *J. Genet.* **29:** 159.

Kohne, D.E., S.A. Levison, and M.J. Byers. 1977. Room temperature method for increasing the rate of DNA reassociation by many thousandfold. The phenol emulsion reassociation technique. *Biochemistry* **16:** 5329.

Kunkel, L.M., A.P. Monaco, W. Middlesworth, H. Ochs, and S.A. Latt. 1985. Specific cloning of DNA fragments absent from the DNA of a male patient with an X-chromosome deletion. *Proc. Natl. Acad. Sci.* **82:** 4778.

Mather, K. 1938. Crossing over. *Biol. Rev. Camb. Philos. Soc.* **13:** 252.

Mattei, M.G., M.A. Batteman-Voelkel, R. Heilig, I. Oberle, K. Davies, J.L. Mandel, and J.F. Mattei. 1985. Localization by *in situ* hybridization of the coagulation factor IX gene and of two polymorphic DNA probes (DXS51 and DXS52) with respect to the fragile site. *Cytogenet. Cell Genet.* **40:** 692.

Purrello, M., B. Alhadeff, M. Rocchi, N. Archidiacono, D. Drayna, and M. Siniscalco. 1985a. Relative position of polymorphic DNA loci of the human X-chromosome long arm subtelomeric region with respect to the fragile-X site. *Cytogenet. Cell Genet.* **40:** 726.

Purrello, M., B. Alhadeff, D. Esposito, P. Szabo, M. Rocchi, M. Truett, F. Masiarz, and M. Siniscalco. 1985b. The human genes for hemophilia A and hemophilia B flank the X-chromosome fragile site at Xq27.3. *EMBO J.* **4:** 725.

Smithies, O., G.E. Connell, and G.H. Dixon. 1962. Chromosomal rearrangements and the evolution of haptoglobin genes. *Nature* **196:** 232.

Sturtevant, A.H. 1925. The effects of unequal crossing over at the bar locus in *Drosophilia. Genetics* **10:** 117.

Szabo, P., M. Purrello, M. Rocchi, N. Archidiacono, B. Alhadeff, G. Filippi, D. Toniolo, G. Martini, L. Luzzato, and M. Siniscalco. 1984. Cytological mapping of the human glucose-6-phosphate dehydrogenase gene distal to the fragile-X site suggests a high rate of meiotic recombination across this site. *Proc. Natl. Acad. Sci.* **81:** 7855.

Tiepolo, L., O. Zuffardi, M. Fraccaro, D. Di Natale, L. Gargantini, C.R. Muller, and H.H. Ropers. 1980. Assignment by deletion mapping of the steroid sulfatase X-linked ichtyosis locus to Xp22.3. *Hum. Genet.* **54:** 205.

Turner, G., J.M. Opitz, W.T. Brown, K.E. Davies, P.A. Jacobs, E.C. Jenkins, M. Mikkelsen, M.W. Partington, and G.R. Sutherland. 1986. Conference report: Second International Workshop on the fragile X and on X-linked mental retardation. *Am. J. Med. Genet.* **23:** 11.

Warren, S.T., T.W. Glover, R.L. Davidson, and P. Jagadeeswaran. 1985. Linkage and recombination between fragile-X-linked mental retardation and the factor IX gene. *Hum. Genet.* **69:** 44.

White, P.C., D. Grossberger, B.J. Onufer, O.D. Chaplin, M.I. New, B. Dupont, and J.L. Strominger. 1985. Two genes encoding steroid 21-hydroxylase are located near the genes encoding the fourth component of complement in man. *Proc. Natl. Acad. Sci.* **82:** 1089.

Genetic Mapping of the Human X Chromosome: Linkage Analysis of the q26-q28 Region That Includes the Fragile X Locus and Isolation of Expressed Sequences

J.L. MANDEL,* B. ARVEILER,* G. CAMERINO,† A. HANAUER,* R. HEILIG,* M. KOENIG,* AND I. OBERLÉ*

*Laboratoire de Génétique Moléculaire des Eucaryotes du CNRS, Unité 184 de Biologie Moléculaire et de Génie Génétique de l'INSERM, Faculté de Médicine, 67085 Strasbourg Cédex, France; †Dipartimento di Genetica e Microbiologia, Universita di Pavia, Pavia, Italy

The X chromosome is probably the most studied of all human chromosomes, in part because more than 115 diseases show X-linked inheritance, because of the relative ease of linkage analysis, and because of the interest in diagnostic applications, since these diseases can appear in successive or collateral generations within a family. About 70 polymorphic DNA markers have been isolated and partially characterized as of August 1985, most of them corresponding to anonymous genomic sequences, and nine genes have been cloned (Goodfellow et al. 1985). The restriction-fragment-length polymorphism (RFLP) markers are being used to construct a genetic map of the whole chromosome (Drayna and White 1985), and many disease loci have already been placed on this map (Goodfellow et al. 1985). However, in most cases the genetic localizations are still imprecise, whereas for diagnostic applications in a given disease it is mandatory (when the gene has not been cloned) to have closely linked flanking markers, which implies that a detailed map needs to be constructed.

We have concentrated our efforts on the q26-q28 region because within it several important diseases are present, including hemophilia A and B, adrenoleukodystrophy, and the fragile X mental retardation syndrome. The latter is the most common of all X-linked diseases and presents important problems both in diagnosis and in the understanding of its peculiar segregation pattern. We have characterized several informative markers in this region. In particular, we have established the structural basis for the polymorphism at the hypervariable locus *DXS52*, which is closely linked to hemophilia A. We have determined the order of seven polymorphic markers around the fragile X locus (*FRAX*). The genetic distances between markers in this region suggest that recombination is not evenly distributed. At present, the markers most closely linked to *FRAX* (*DXS52* and the coagulation factor IX gene *F9*) map at 10–15 cM from the disease locus, and there is evidence for heterogeneity in the genetic distance between *F9* and *FRAX* in different families.

As a complementary approach toward mapping of the human X chromosome, we have searched for expressed X-linked sequences since a map of such sequences might allow one to correlate a disease locus with a cloned gene located in the same region. We have partially characterized several X-linked genes and pseudogenes. We show that evolutionary considerations can be used fruitfully in this approach, to distinguish X-linked genes from X-linked pseudogenes, and to find expressed sequences within randomly cloned genomic fragments by searching for homologies with rodent genomes. The comparison of the homologous human and rodent sequences allows one to characterize putative protein-coding regions, even though the corresponding mRNA has not yet been detected. This should be useful in "genome walking" strategies designed to isolate a disease gene.

RESULTS AND DISCUSSION

The St14 (*DXS52*) Hypervariable Locus

The *DXS52* locus has been originally detected using a 9.3-kb *Eco*RI genomic fragment (St14-9) that is part of a multisequence family since it reveals five additional *Eco*RI fragments, all present in the q26-q28 region of the X chromosome. The *DXS52* locus was characterized by a highly informative *Taq*I RFLP with at least 10 different alleles (80% heterozygosity) and 3 two-allele *Msp*I RFLPs and was genetically mapped to the Xq28 region (Oberlé et al. 1985a). *DXS52* was shown to be very closely linked to the hemophilia A–coagulation factor VIII locus in two independent studies, and no recombinations were found in a total of 105 informative meioses (Oberlé et al. 1985c; Gitschier et al. 1985), suggesting that the two loci are less than 2 cM apart at a 90% confidence limit (lod-1 support). Although rare recombination events have been recently recorded (M. Goossens; H. Kazazian, both pers. comm.), the St14 probe remains an important diagnostic tool in hemophilia A families because of its very high heterozygosity, whereas the two RFLPs known within the factor VIII gene (Antonarakis et al. 1985; Gitschier et al. 1985) are informative only in less than 40% of European families (M. Goossens, pers. comm.). Combined with coagulation assays, the analysis at *DXS52* provides an accuracy for carrier diagnosis of at least 99% (Oberlé et al. 1985c). For prenatal diagnosis the risk of recombination should be taken into account, and it might be recommended to propose fetal blood analysis when the St14 segregation data suggest that the fetus is unaffected.

The high multiplicity of *Taq*I alleles suggests that they result from variations in the number of a tandemly repeated sequence unit. Jeffreys et al. (1985) have described a family of such minisatellite sequences that provide extremely useful tools for genetic analysis. It is possible that several such families might exist in the genome, and it was thus of interest to determine the structural basis of the *Taq*I RFLP at the *DXS52* locus. The original probe showed only cross-hybridization to the polymorphic region. We first cloned two additional *Eco*RI fragments homologous to St14-9: the 3.0-kb St14-l and the 3.8-kb St14-k fragments. The St14-l probe hybridized to the allelic *Taq*I fragments under very high stringency conditions, whereas all but one of the remaining fragments disappeared or gave faint signals. Restriction mapping of genomic DNA showed that the St14-l fragment was contiguous to, but did not include, the polymorphic region contained in *Bam*HI fragments of variable length. A partial genomic library was constructed after enrichment for these *Bam*HI fragments, starting from the genomic DNA of a woman heterozygous for a large (6.6-kb) and a small (3.6-kb) *Taq*I allele. Two phage clones were obtained corresponding to the smaller allele, and none for the larger one, while four clones were obtained for a constant *Bam*HI fragment that should have been less enriched during the size selection. This suggested that the polymorphic region was rather unstable under our cloning conditions in keeping with similar observations for other hypervariable loci (Wyman et al. 1985). The region responsible for the *Taq*I polymorphism was mapped to a 350-bp *Sna*BI-*Ban*I fragment, which was sequenced. This revealed the presence of a 206-bp region consisting of alternating thymidines and purines, including three repeats of a 60-bp unit (two perfect repeats and one deleted for a TA dinucleotide) (Fig. 1). This sequence has no homology with previously described hypervariable minisatellite sequences (Jeffreys et al. 1985 and references therein). However, tracks of $(TG)_n$ appear to be frequent in the human genome and

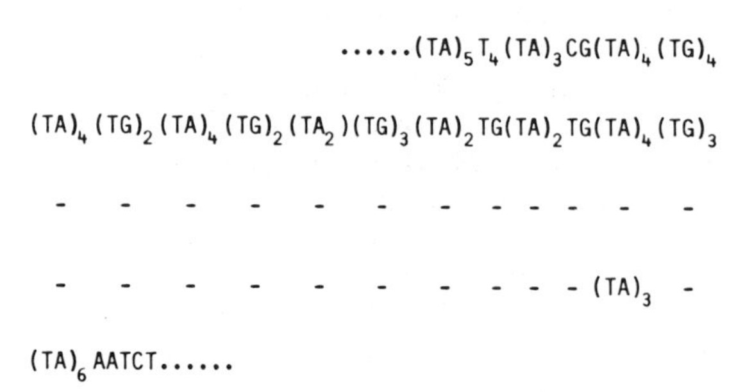

Figure 1. Sequence of the hypervariable region at the St14 locus (*DXS52*). This sequence corresponds to a 3.6-kb *Taq*I allele (designated 7A). Dashes indicate identity between the three-repeat units.

have been found within the γ-globin and cardiac α-actin genes (Slightom et al. 1980; Hamada et al. 1982). In fact, when the *Sna*BI-*Ban*I fragment is used to probe blots of human genomic DNA, it reveals, in addition to the homologous band, a smear characteristic of repetitive sequences. It is unlikely that this probe would be efficient for the cloning of additional useful hypervariable loci since polymorphisms in such simple sequences generally involve very small length variations (Slightom et al. 1980; Spritz 1981).

An interesting feature of the *DXS52* locus is the large number of additional RFLPs (Fig. 2), in contrast with the relatively low frequency of RFLPs detected with most other X-linked probes (Hofker et al. 1985; Oberlé et al. 1986a). The *Taq*I RFLP defined by fragments α and β (Fig. 2A) is relatively frequent: The β-allele is found in about 10% of the X chromosomes we analyzed, but this frequency might be subject to large ethnic-group-specific variation. This polymorphism shows strong linkage disequilibrium with the hypervariable RFLP since allele β is found specifically associated to the smallest allele (the 3.4-kb allele *8*) or the largest one (the 7.5-kb allele *1A*, frequent in an American black population; C. Schwartz, pers. comm.). Some of the other RFLPs have useful frequencies (e.g., the *Bcl*I two-allele RFLP, Fig. 2B), while others appear to be rare (Fig. 2C). This high level of variability might be related to genetic exchanges occurring within this cluster of related sequences, as was suggested for the β-globin locus (Jeffreys 1979) or for the MHC/HLA loci (Kourilsky 1983; Weiss et al. 1983). The existence of strong linkage disequilibrium between some of the two-allele RFLPs and the hypervariable region would suggest that the minisatellite sequence is not a hot spot for recombination, as has been suggested for tandemly repeated simple sequences in the major histocompatibility complex (Steinmetz et al. 1986), but detailed mapping of the variant sites is needed before a definitive conclusion can be made.

Multipoint Genetic Mapping of the Xq26-q28 Region

To map new polymorphic probes physically assigned to the Xq26-q28 region with respect to *FRAX* and to its flanking markers *F9* (the coagulation factor IX gene) and *DXS52* (St14), we have performed a segregation analysis in three large fragile X families. The seven markers analyzed (Fig. 3) include the expressed gene locus corresponding to the cDNA probe C11 (see below). All probes but one (45h) are present in two somatic hybrid cell lines (63R and GM97) and are thus distal to the corresponding Xq26 translocation breakpoints (Oberlé et al. 1986a). The probe 45h (*DXS100*) was mapped, using independent hybrids, to the q26-q27.3 region (Riddell et al. 1985), and is thus localized in q26, proximal with respect to the other probes and to the *HPRT* gene. Each of the three families was informative for at least five marker loci. Of the 55

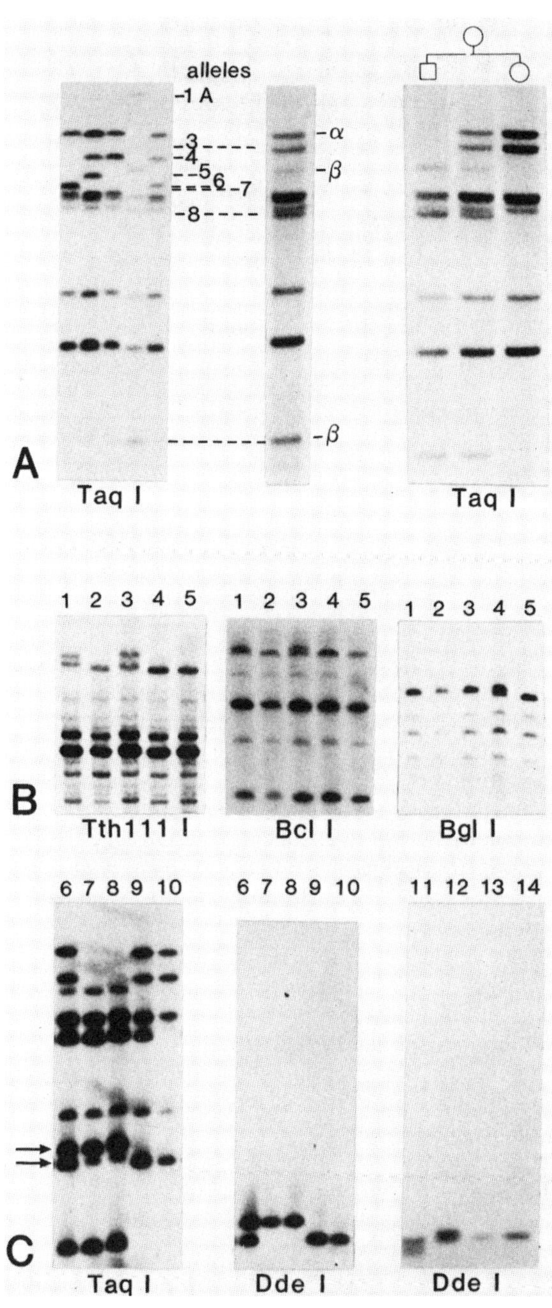

Figure 2. Polymorphisms at the *DXS52* locus. All the results presented have been obtained with probe St14-1. (*A*) *Taq*I allelic fragments corresponding to the minisatellite region are numbered 1–8 (1A is larger than the previously described allele 1). The additional two-allele RFLP is defined by the α fragment (major allele) and the two β fragments (minor allele). The β allele is shown associated to allele *1A* (*left* panel) or to allele *8* (*right* panel). (*B*) RFLPs detected in three different restriction digests performed on the same five genomic DNAs. (*C*) Multiallele RFLPs detected in *Taq*I or *Dde*I digests. These correspond probably to the same deletion-insertion polymorphism since there appears to be a correlation between the *Taq*I and *Dde*I alleles as shown in the *left* panel (allelic fragments indicated by arrows) and in the *middle* one. At least three and possibly four alleles have been seen.

meioses analyzed, 40 showed no evidence for crossover, while in the 15 remaining ones, segregation of alleles could be explained by assuming a single cross-over (Fig. 3, A–K). The loci defined by probes C11, St1, and 45h showed 3–6% recombination with 52A (*DXS51*) and 13–17% recombination with the *F9* locus (Table 1). On the other hand, recombination of the same loci with the St14 and DX13 loci was 25–33%. This suggested that C11, St1, and 45h are closer to 52A and *F9* than to the St14-DX13 cluster. The order of loci was derived more precisely from the recombinant meioses, since the close linkage between 45h and *F9* renders double recombination events unlikely. Thus, meiosis A places St1 proximal to C11 and 52A, and meiosis B places C11 between St1 and 52A. Meiosis D places 52A proximal to *F9*, in agreement with previous linkage data (Drayna and White 1985). This multipoint analysis localizes 45h proximal to 52A but does not give information as to its position relative to St1 or C11. Physical mapping has, however, suggested that 45h is proximal with respect to the latter loci. No recombinants were found between St14 and DX13 in 19 meioses, in agreement with the very tight linkage previously found (Oberlé et al. 1985a). Our results support the order centromere–45h–St1–C11–52A–*F9*–*FRAX*–(St14, DX13), and none of the new DNA markers tested map closer to *FRAX* than St14 and *F9* (see Note Added in Proof). It is interesting to note that the five loci studied here (from 45h to *F9*) as well as four others (*HPRT*, 43-15, *DXS10*, and *DXS79*) all appear to map within a 15-cM region, indicating a very close spacing of polymorphic DNA markers (Boggs and Nussbaum 1984; Drayna and White 1985; Murphy et al. 1985). An even tighter cluster (≤ 5 cM) of seven loci is present in the q28 region, including the loci for G6PD, hemophilia A, adrenoleukodystrophy, deutan and protan color blindness, and *DXS52* and *DXS15* (Szabo et al. 1984; Oberlé et al. 1985a). This close spacing of loci in both q26-q27 and q28 contrasts with the large genetic distance (25–30% recombination) between the *F9* gene and the Xq28 cluster. This supports the hypothesis that *FRAX* lies within a region of preferential recombination, as also suggested by the apparent discrepancy between physical distances (estimated from in situ hybridization data) and the genetic distances (Szabo et al. 1984; Mattei et al. 1985). The lack of RFLP markers in this region could also be due to the presence of a high proportion of repetitive sequences, which would be selected against when screening for single copy probes.

Table 1. Two-point Linkage Data for Markers in q26-q27

	45h	St1	C11	52A	F9
45h		0/16[a]	0/14[a]	1/17[a]	2/15[a]
St1	0		1/22[a]	1/27[a]	1/6[a]
C11	0	4.5		1/31[a]	3/23[a]
52A	5.9	3.7	3.2		1/16[a]
F9	13.3	16.7	13.0	6.25	

[a]Above diagonal = no. of recombinants per meioses analyzed; below diagonal = % recombination.

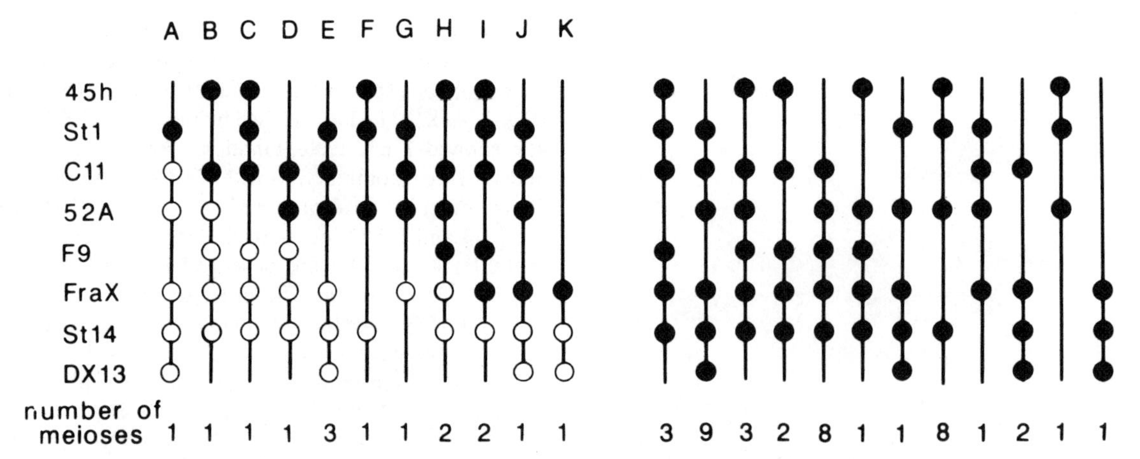

Figure 3. Representation of chromosomes derived from informative meioses. Informative loci in each meiosis are represented by a circle. Filled and empty circles correspond to loci derived from one or the other of the maternal chromosomes. Lanes *A* to *K* represent recombinant meioses; the number of meioses in each category is indicated. The official name for each loci is *DXS100* (45h), *DXS86* (St1), *DXS144* (C11), *DXS51* (52A), *F9* (coagulation factor IX), *FRAX A* (FraX), *DXS52* (St14), and *DXS15* (DX13) (Goodfellow et al. 1985).

DNA Probes and the Fragile X Mental Retardation Syndrome

The fragile X mental retardation syndrome is the most frequent of all X-linked diseases, with an incidence of about 1 in 1500 newborn males (for review, see Turner et al. 1986). The phenotypic trait associates mental retardation, some dysmorphic features, and the presence of a fragile site at Xq27.3 induced in vitro by thymidine deprivation. Its segregation departs from classical recessive X-linked inheritance since cytogenetically and/or clinically normal males can transmit the disease, while the percentage of females with clinical manifestations is much higher in other sex-linked diseases (Sherman et al. 1985). Segregation analysis suggested that the mutation rate is very high. Finally, this disease poses diagnostic problems since cytogenetic analysis detects only 55% of carrier females and may not be completely reliable for prenatal diagnosis. Segregation analysis with polymorphic DNA markers closely linked to *FRAX* could be useful in investigating such problems.

The St14 (*DXS52*) locus and the factor IX gene are at present the closest flanking markers for *FRAX* (see Note Added in Proof). From a study of 16 families we have estimated the genetic distances between *F9* and *FRAX* (12 ± 8% recombination) and between St14 and *FRAX* (10 ± 6%) (Oberlé et al. 1986b). These markers can thus effectively be used in conjunction with cytogenetic analysis to provide, in about 40% of the families, more accurate carrier and prenatal diagnosis, by decreasing the proportion of false negatives (Oberlé et al. 1985b). The segregation analysis allows one to identify families where the disease was transmitted through a normal male: These families appeared to be quite frequent in our study (about one in three), in agreement with the high frequency (20%) of phenotypically normal male carriers, which was suggested by Sherman et al. (1985).

The study of the *F9-FRAX* linkage in our laboratory as well as in others has, however, revealed a complicating factor: the apparent heterogeneity in the recombination fraction between the two loci in different families. Thus, whereas the family we originally described showed no recombination in 16 informative meioses (Camerino et al. 1983), in other families linkage appears loose and several statistical tests suggested that these differences are significant (Brown et al. 1985; Davies et al. 1985; Oberlé et al. 1986b). In no case in our data set did the segregation data suggest that *FRAX* might be outside the *F9*-St14 interval. It remains to be verified in a larger set of data that families with normal male carriers would show less recombination between *F9* and *FRAX* than the other families, as proposed by Brown et al. (1985). On the other hand, no statistically significant heterogeneity in recombination frequency has been found for the *FRAX*-St14 interval, both in our data and those of Brown et al. (1986), as well as for the St14-*F9* interval between fragile X and normal families (Oberlé et al. 1986b).

Further genetic and physical mapping should reveal whether this heterogeneity is characteristic of the *FRAX* region (and/or mutation) or whether it might be found in other regions of the genome. At least in the major histocompatibility complex region of the mouse, it has been shown that, on a smaller scale, different strains may show important variations in genetic distances (from .02 cM to ~1.5 cM) between two loci, while the physical distances are similar (Steinmetz et al. 1986). The presence of variant chromosomal arrangements (such as inversions) could also account for inherited differences in recombination over a large region, as in the case of the *t* locus in mouse (Herrmann et al. 1986).

Search for Expressed X-linked Sequences: Genes and Pseudogenes

The genetic map being built for the human X chromosome associates expressed genes that correspond to known proteins, polymorphic loci defined by anonymous probes, and disease loci for which the primary defect is unknown. For the latter category, it will be important to go from the disease location to the gene itself in order to characterize its protein product and its function. A map that would include unidentified expressed sequences characterized for their expression at the RNA level (tissue and/or developmental specificity) might be used to try to correlate a disease locus to an expressed gene present in the same region. If the disease affects a particular tissue, one might look preferentially, but not exclusively, for genes with a similar specific expression. This approach might be particularly favorable in the case of the X chromosome since on average 1 gene in 15 to 20 should correspond to a known disease (assuming that about 3000 genes are present on this chromosome). A first success (albeit in mouse) was obtained by Cohen et al. (1985), who showed that sequences related to an anonymous lymphocyte-specific cDNA and mapped to the region of the *xid* immunodeficiency are abnormally expressed in the mutant mice. Systematic identification of expressed sequences would also be an important goal in projects designed to completely clone a large chromosome segment, for instance in a walk toward a particular disease locus.

We have used three different strategies in order to evaluate the feasibility and efficiency of such an approach, and our results are summarized in Table 2. In the first method used, we tried to identify sequences common to a genomic library enriched in X chromosome sequences (Davies et al. 1981) and to pools of about 90 cDNA clones (devoid of highly repetitive sequences) isolated from a human skeletal muscle cDNA library (Hanauer and Mandel 1984). From the testing of 11 cDNA pools, genomic and cDNA clones corresponding to two different sequences were isolated and extensively characterized. These were shown to correspond to two X-linked, intronless pseudogenes. A cDNA for glyceraldehyde-3-phosphate dehydrogenase (a gene previously mapped to chromosome 12) detected 20–30 autosomal sequences in man in addition to the X chromosome sequence (Hanauer et al. 1984). A second cDNA was shown by sequence analysis to correspond to glutamate dehydrogenase, a gene that had not been cloned previously in higher eukaryotes or chromosomally mapped in man. We have shown that the X-linked genomic sequence is intronless, whereas a single additional autosomal sequence is present on chromosome 10 (Hanauer et al. 1985).

In an alternative approach cDNA clones picked at random were hybridized, individually or in groups of two or three, to Southern blots containing genomic DNA from cells with one, two, or four X chromosomes. This allowed us to demonstrate the presence of sequences related to actin cDNAs on both the X and Y chromosome, which have been subsequently isolated and sequenced and shown to correspond to intronless pseudogenes (Heilig et al. 1984; J.P. Moisan and R. Heilig, unpubl.). Of about 25 nonrepetitive random cDNA clones tested, clone C15 detected only X-coded

Table 2. Search for Expressed X-linked Sequences

Method of selection	Sequence (locus)	X localization	Additional autosomal sequences	Status of X sequence	Detection of poly(A)$^+$RNA[a]
Gene dosage	Actin (*ACTLI*)	p11-q11	~25	pseudogene (intronless)	α,β,γ actin mRNAs
	C15 (*DXS145*)	q24-q26	no	gene	+
	C11-C69 (*DXS144*)	q26-q27	1	gene	+
Cross-screening	(GAPDPI[b])	p21-p11	~25	pseudogene (intronless)	GAPDH mRNA
	(GLUDPI[c])	q24-q26	1	pseudogene? (intronless)	GDH mRNA
Homology with rodent genome	M2C (*DXS32*)	p21	no	gene (sequence) (X-linked in mouse)	(+)
	St3 (*DXS54*)	q13-q22	2–4	?	+
	St14 (*DXS52*)	q28	no	(X-linked in mouse)	–
	6 anonymous probes		no	(all X-linked in mouse)	not tested

[a]C11 and St3 detect poly(A)$^+$ RNA species of ~2.5 and 1.8 kb, respectively, which are abundant in many tissues (including liver, muscle, and placenta). C15 detects a ~4-kb RNA of low abundance in liver and placenta. Preliminary evidence has been obtained for a rare liver poly(A)$^+$ RNA corresponding to M2C.

[b]GAPDPI = Glyceraldehyde-3-phosphate-dehydrogenase pseudogene.

[c]GLUDPI = Glutamate dehydrogenase pseudogene.

sequences, while clones C11 and C69 were shown to be overlapping and detected both an X-linked and an autosomal sequence. In fact, other groups that have been working along similar lines and that had isolated putative X-linked expressed clones either by gene dosage analysis or by hybridization of uncloned cDNA to an X-enriched library have encountered the same problem: Most of the clones isolated cross-react with both X-linked and autosomal sequences (Kunkel et al. 1983; Balazs et al. 1984; Yen et al. 1984), and the identification of the expressed sequence(s) would involve the detailed characterization of both cDNA and genomic clones. (When pseudogenes are of very recent origin, hybridization stringency cannot be used to distinguish them from the expressed gene, because homology will be very high. Thus, for almost all of these sequences, the evidence that they correspond to X-linked genes is lacking.)

We have developed an alternative and less tedious solution to this problem that should allow one in many cases to differentiate between X-linked genes and pseudogenes. This is based on the so-called Ohno law of conservation of X linkage in mammals: that is, a gene that is X-linked in one species should be X-linked in all other mammalian species (due to the fact that genes on the X chromosome function at a single dose and that, in consequence, X autosome translocations are unlikely to be fixed within a species) (Ohno 1969). This hypothesis has been substantiated in all cases studied since, and the long-lived controversy about localization of the steroid sulfatase gene (*STS*), which is X-linked in man but showed apparent autosomal linkage in mouse, was solved recently when it was shown that *STS* is in the pseudoautosomal region of the mouse X and Y chromosomes (Keitges et al. 1985).

We hypothesized that human pseudogene sequences that can be recognized by cross-hybridization with an expressed sequence under conditions of moderate stringency should be of relatively recent origin (i.e., $<60 \times 10^6$ years) since they are not subject to selective pressure and will accumulate mutations at a higher rate than expressed sequences (for an estimation of the rates of DNA sequence divergence, see Britten 1986). Thus, although a truly X-linked gene in man should appear X-linked in lemur (divergence time, 50×10^6–75×10^6 years) or in mouse (divergence time, 75×10^6–110×10^6 years), X-linked pseudogenes should not show such conservation (unless X-linked pseudogenes arose independently in the species tested). We have hybridized the C11 cDNA clone and a glutamate dehydrogenase probe, both of which recognize in man an X-linked and an autosomal sequence, to a blot containing DNA from male or female brown lemurs digested with various enzymes. With probe C11, all bands show a clear male/female dosage indicative of X linkage, whereas no such dosage is obtained with the GDH probe (Fig. 4; in the latter case this is most clearly seen in the *Eco*RI digest since polymorphism obscures the interpretation for some fragments in the other digests). In mouse, C11 recognizes both an X-linked and an autosomal sequence (not shown). These results confirm that the *GDH* gene is autosomally coded and provide good evidence that the C11 probe corresponds to an X-linked expressed gene.

Identification of Expressed Sequences on the Basis of Their Conservation During Evolution

While we were mapping anonymous X-linked genomic probes by hybridization to a panel of rodent/human hybrids, we noticed that a significant proportion of them (∼25%) showed clear cross-reaction with a limited number of restriction fragments in the mouse and/or hamster genomes, under hybridization conditions of moderate stringency (i.e., 40% formamide,

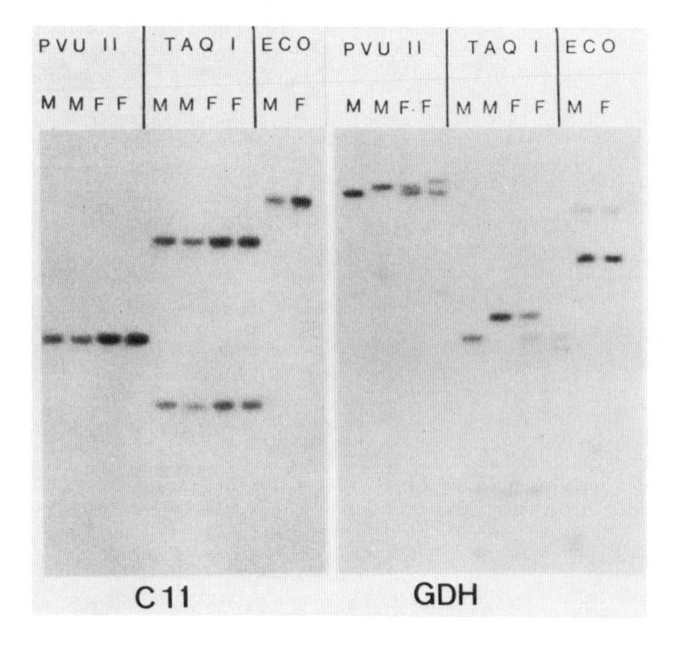

Figure 4. Sex-linked gene dosage analysis in lemurs. Equal amounts of DNA from male (M) or female (F) brown lemurs were digested with *Pvu*II, *Taq*I, or *Eco*RI, as indicated, and blotted onto diazobenzyloxymethyl paper (DBM) as described in Oberlé et al. (1986a). Hybridization was performed with the C11 cDNA probe (*DXS144*) or with a genomic probe corresponding to half of the glutamate dehydrogenase (GDH) protein-coding sequence (derived from the X-linked intronless sequence).

42°C, in 0.9 M NaCl, washing at 60°C in 0.5×SSC). This suggested to us that these human genome fragments might contain protein-coding exon sequences, which are the most likely to be conserved by selective pressure. Of the 9 probes for which we have found an indication of sequence conservation (and which include St14), we have studied probes St3 and M2C in more detail. Preliminary experiments showed that St3 (*DXS54*) detects a poly(A)$^+$ RNA expressed in several human tissues (liver, placenta, and various cell lines), whereas no signal was detected under the same conditions with either the M2C or St14 probes. However, in genomic blots of human DNA, St3 cross-hydridizes to a small number of autosomal fragments (2–4, depending on the conditions used). A similar pattern of X and autosomal linkage is seen in lemur DNA (even under high stringency conditions), whereas the probe detects about 10 fragments in mouse, which do not appear X-linked. Thus, there is little doubt that St3 recognizes an expressed gene, but this gene might not be X-linked. More species should be studied to know whether the conflicting results in lemur and mouse are due to independent occurrence of pseudogenes on the X chromosome in lemur and in man or might represent an exception to the conservation of X linkage.

The locus defined by M2C (*DXS32*) was studied in both mouse and man. Cross-hybridization to the mouse genome was sufficient to allow Dr. P. Avner and colleagues (Pasteur Institute, Paris) to map by linkage analysis the sequence homologous to M2C at about 10 cM of the mouse OTC (*Spf*) locus on the X chromosome (P. Avner, pers. comm.), a localization very similar to that found in man since OTC and M2C are present in the same p21 subband (De Martinville et al. 1985). To get more-decisive proof that such sequence homologies detectable between rodent and man in genomic Southern blot experiments represent protein-coding regions, we isolated a mouse cosmid clone that hybridized to M2C and sequenced both the human and mouse cross-hybridizing restriction fragments. We found a 500-bp region of very high homology, which starts abruptly in both mouse and human by a typical acceptor-splice-site consensus sequence (Fig. 5). From this point the two sequences show a conserved open reading frame of 57 codons (92% homology at the nucleotide level), followed by the same in-phase stop codons. This suggests that the sequence could correspond to the most 3′ protein-coding exon in the putative M2C gene. Excellent sequence conservation (~82%) is found past the stop codons for 340 bp, but with a few small deletions and insertions, which would not be expected in a protein-coding sequence. If this corresponds indeed to a 3′ untranslated region, it would appear to be under high selective pressure. Although this is not a general case, instances of similarly very high conservation of 3′ untranslated sequences between rodent and human have been described (e.g., in the α cardiac actin gene; Mayer et al. 1984).

The results we have obtained with both M2C and St3 show conclusively that sequence conservation is a very good and easily tested indication for the presence of exons (or related pseudogene sequences in some cases) within a randomly cloned fragment. This should greatly facilitate the identification of transcribed regions in a genomic walk (which might easily be missed if intron sequences are used as probes), especially since this screening test is independent of the level or tissue specificity of expression. Because of the cross-hybridization, it is then possible to analyze expression in animal tissues when the corresponding human ones are not available. Comparative sequence analysis is probably an effective way to obtain protein-coding se-

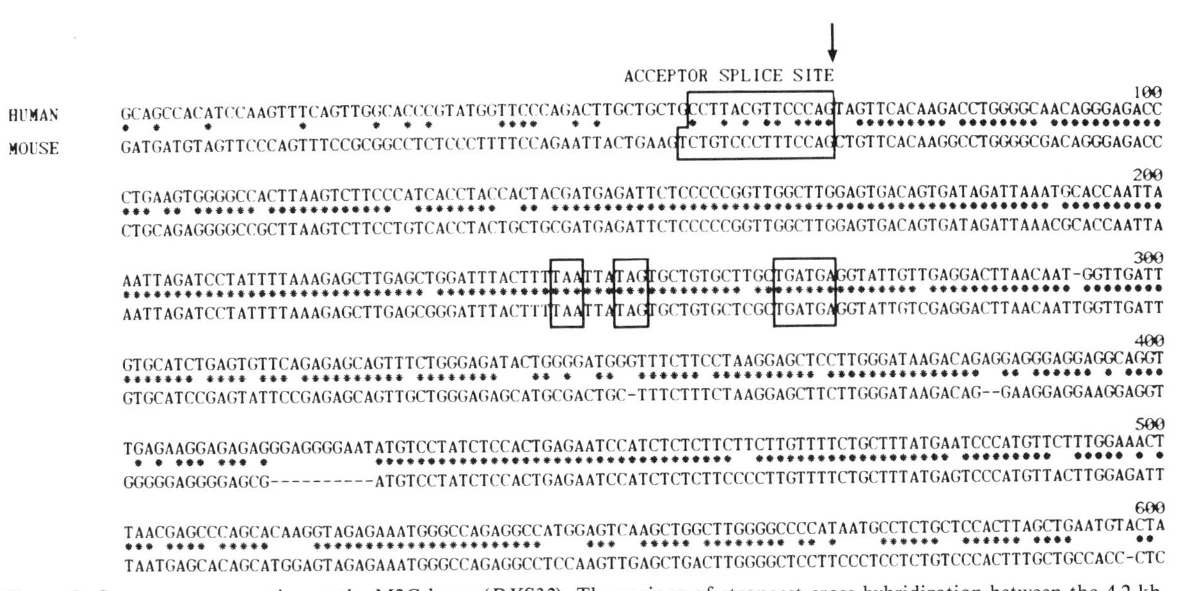

Figure 5. Sequence conservation at the M2C locus (*DXS32*). The regions of strongest cross-hybridization between the 4.2-kb M2C human fragment and a mouse cosmid clone were mapped and sequenced. The acceptor-splice-site consensus sequence and the in-phase stop codons are boxed. The arrow indicates the 5′ end of the tentative exon. Homology between the mouse and human sequences decreases rapidly past position 600.

quence information even when the corresponding mRNA sequence has not been characterized (because exons are generally small, identification of protein-coding regions within a genomic sequence is not, in general, a trivial problem). With the availability of chromosome-specific libraries, it should be relatively easy to map on each human chromosome many new gene sequences that could be partially characterized by the size of the poly(A)$^+$ RNA detected (if any), its pattern of expression, and so forth. It has been suggested recently that a way to precisely map rare, monogenic diseases is to search for deletions in many independent patients (Lange et al. 1985). Probes from expressed regions would be prime candidates for such a strategy. These probes should also be useful to establish detailed comparative linkage maps in mouse and man, thus allowing a better correlation between mouse mutants and human genetic diseases.

Note Added in Proof

Recent results obtained by multipoint linkage analysis have shown that probe 55.7 (*DXS105*) from Dr. P. Pearson (Leiden) is closer to the fragile X locus than *F9*, on the proximal side, by about 6 cM (Veenema et al., in prep.; and our unpublished results).

ACKNOWLEDGMENTS

We wish to thank Ms. C. Sengel for technical assistnace, Drs. K. Wrogemann, J.P. Moisan (Strasbourg), and Ms. E. Raimondi (Pavia) for their participation in some of the work reported here; Drs. B.N. White (Kingston, Ontario) and K. Davies (Oxford) for probes; and C. Kister and B. Boulay for the preparation of the manuscript. We are grateful to the clinicians and cytogeneticists who allowed us to study many fragile X families. This work was supported by grants from CNRS (ATP 06931) and CNAMTS (to J.L.M.) and from Progetto Finalizzato Ingegneria Genetica e Basi Molecolari Delle Malattie Ereditari (to G.C.).

REFERENCES

Antonarakis, S.E., P.G. Weber, S.D. Kittur, A.S. Patel, H.H. Kazazian, M.A. Mellis, R.B. Counts, G. Stamatoyannopoulos, E.J.W. Bowie, D.N. Fass, D.D. Pittman, J.M. Wozney, and J.J. Toole. 1985. Hemophilia A: Detection of molecular defects and of carriers by DNA analysis. *N. Engl. J. Med.* **313:** 842.

Balazs, I., M. Purrello, D.M. Kurnit, K.H. Grzeschik, and M. Siniscalco. 1984. Isolation and characterization of human random cDNA clones homologous to DNA from the X chromosome. *Somatic Cell. Mol. Genet.* **10:** 385.

Boggs, B.A. and R.L. Nussbaum. 1984. Two anonymous X-specific human sequences detecting restriction fragment length polymorphisms in region Xq26→qter. *Somatic Cell Mol. Genet.* **10:** 607.

Britten, R.J. 1986. Rates of DNA sequence evolution differ between taxonomic groups. *Science* **231:** 1393.

Brown, W.T., A.C. Gross, C.B. Chan, and E.C. Jenkins. 1985. Genetic linkage heterogeneity in the fragile X syndrome. *Hum. Genet.* 71: 11.

———. 1986. DNA linkage studies in the fragile X syndrome suggest genetic heterogeneity. *Am. J. Med. Genet.* **23:** 643.

Camerino, G., M.G. Mattei, J.F. Mattei, M. Jaye, and J.L. Mandel. 1983. Close linkage of fragile X-mental retardation syndrome to haemophilia B and transmission through a normal male. *Nature* **306:** 701.

Cohen, D.I., A.D. Steinberg, W.E. Paul, and M.M. Davis. 1985. Expression of an X-linked gene family (XLR) in late-stage B cells and its alteration by *Xid* mutation. *Nature* **314:** 372.

Davies, K.E., B. Young, R. Elles, M. Hill, and R. Williamson. 1981. Cloning of a representative genomic library of the human X chromosome after sorting and flow cytometry. *Nature* **293:** 374.

Davies, K.E., M.G. Mattei, J.F. Mattei, H. Veenema, S. McGlade, K. Harper, N. Tommerup, K.B. Nielsen, M. Mikkelsen, P. Beighton, D. Drayna, R. White, and M.E. Pembrey. 1985. Linkage studies of X-linked mental retardation: High frequency of recombination in the telomeric region of the human X chromosome. *Hum. Genet.* **70:** 249.

De Martinville, B., L.M. Kunkel, G. Bruns, F. Morle, M. Koenig, J.L. Mandel, A. Horwich, S.A. Latt, J.F. Gusella, D. Housman, and U. Francke. 1985. Localization of DNA sequences in region Xp21 of the human X chromosome: Search for molecular markers close to the Duchenne muscular dystrophy locus. *Am. J. Hum. Genet.* **37:** 235.

Drayna, D. and R. White. 1985. The genetic linkage map of the human X chromosome. *Science* **230:** 753.

Gitschier, J., D. Drayna, E.G.D. Tuddenham, R.L. White, and R.M. Lawn. 1985. Genetic mapping and diagnosis of haemophilia A achieved through a *BclI* polymorphism in the factor VIII gene. *Nature* **314:** 738.

Goodfellow, P.N., K.E. Davies, and H.H. Ropers. 1985. Report of the committee on the genetic constitution of the X and Y chromosomes. *Cytogenet. Cell Genet.* **40:** 296.

Hamada, H., M.G. Petrino, and T. Kakunaga. 1982. A novel repeated element with Z-DNA-forming potential is widely found in evolutionarily diverse eukaryotic genomes. *Proc. Natl. Acad. Sci.* **79:** 6465.

Hanauer, A. and J.L. Mandel. 1984. The glyceraldehyde-3-phosphate dehydrogenase gene family: Structure of a human cDNA and of an X chromosome linked pseudogene; amazing complexity of the gene family in mouse. *EMBO J.* **3:** 2627.

Hanauer, A., J.L. Mandel, and M.G. Mattei. 1985. X-linked and autosomal sequences corresponding to glutamate dehydrogenase (GLUD) and to an anonymous cDNA. *Cytogenet. Cell Genet.* **40:** 647.

Heilig, R., A. Hanauer, K.H. Grzeschik, M.C. Hors-Cayla, and J.L. Mandel. 1984. Actin-like sequences are present on the X and Y chromosomes. *EMBO J.* **3:** 1803.

Herrmann, B., M. Bucán, P.E. Mains, A.M. Frischauf, L.M. Silver, and H. Lehrach. 1986. Genetic analysis of the proximal portion of the mouse *t* complex: Evidence for a second inversion within *t* haplotypes. *Cell* **44:** 469.

Hofker, M.H., M.C. Wapenaar, N. Goor, E. Bakker, G.J.B. van Ommen, and P.L. Pearson. 1985. Isolation of probes detecting restriction fragment length polymorphisms from X chromosome-specific libraries: Potential use for diagnosis of Duchenne muscular dystrophy. *Hum. Genet.* **70:** 148.

Jeffreys, A.J. 1979. DNA sequence variants Gγ, Aγ-, and β-globin genes of man. *Cell* **18:** 1.

Jeffreys, A.J., V. Wilson, and S.L. Thein. 1985. Hypervariable "minisatellite" regions in human DNA. *Nature* **314:** 67.

Keitges, E., M. Rivest, M. Siniscalco, and S.M. Gartler. 1985. X-linkage of steroid sulphatase in the mouse is evidence for a functional Y-linked allele. *Nature* **321:** 226.

Kourilsky, P. 1983. Genetic exchanges between partially homologous nucleotide sequences: Possible implications for multigene families. *Biochimie* **65:** 85.

Kunkel, L.M., U. Tantravahi, D.M. Kurnit, M. Eisenhard, G.P. Bruns, and S.A. Latt. 1983. Identification and isolation of transcribed human X chromosome DNA sequences. *Nucleic Acids Res.* **11:** 7961.

Lange, K., L. Kunkel, J. Aldridge, and S.A. Latt. 1985. Accurate and superaccurate gene mapping. *Am. J. Hum. Genet.* **37:** 853.

Mattei, M.G., M.A. Baeteman, R. Heilig, I. Oberlé, K. Davies, J.L. Mandel, and J.F. Mattei. 1985. Localization by in situ hybridization of the coagulation factor IX gene and of two polymorphic DNA probes with respect to the fragile X site. *Hum. Genet.* **69:** 327.

Mayer, Y., H. Czosnek, P.E. Zeelon, D. Yaffe, and U. Nudel. 1984. Expression of the genes coding for the skeletal muscle and cardiac actins in the heart. *Nucleic Acids Res.* **12:** 1087.

Murphy, P.D., J.R. Kidd, W.R. Breg, F.H. Ruddle, and K.K. Kidd. 1985. Isolation of an anonymous polymorphic probe from distal Xq. *Cytogenet. Cell Genet.* **40:** 707.

Oberlé, I., D. Drayna, G. Camerino, R. White, and J.L. Mandel. 1985a. The telomeric region of the human X chromosomes long arm: Presence of a highly polymorphic DNA marker and analysis of recombination frequency. *Proc. Natl. Acad. Sci.* **82:** 2824.

Oberlé, I., J.L. Mandel, J. Boué, M.G. Mattei, and J.F. Mattei. 1985b. Polymorphic DNA markers in prenatal diagnosis of fragile X syndrome. *Lancet* **I:** 871.

Oberlé, I., G. Camerino, R. Heilig, L. Grunebaum, J.P. Cazenave, C. Crapanzano, P.M. Mannucci, and J.L. Mandel. 1985c. Genetic screening for hemophilia A (classic hemophilia) with a polymorphic DNA probe. *N. Engl. J. Med.* **312:** 682.

Oberlé, I., G. Camerino, C. Kloepfer, J.P. Moisan, K.H. Grzeschik, B. Hellkuhl, M.C. Hors-Cayla, N. Van Cong, D. Weil, and J.L. Mandel. 1986a. Characterization of a set of X-linked sequences and of a panel of somatic cell hybrids useful for the regional mapping of the human X chromosome. *Hum. Genet.* **72:** 43.

Oberlé, I., R. Heilig, J.P. Moisan, C. Kloepfer, M.G. Mattei, J.F. Mattei, J. Boué, U. Froster-Iskenius, P.A. Jacobs, G.M. Lathrop, J.M. Lalouel, and J.L. Mandel. 1986b. Genetic analysis of the fragile X–mental retardation syndrome with two flanking polymorphic DNA markers. *Proc. Natl. Acad. Sci.* **83:** 1016.

Ohno, S. 1969. Evolution of sex chromosomes in mammals. *Annu. Rev. Genet.* **3:** 495.

Riddell, D.C. H.S. Wang, J. Beckett, J.J.A. Holden, L. Mulligan, A. Phillips, N.E. Simpson, K. Wrogemann, B.N. White, and J.L. Hamerton. 1985. Assignment and regional localization of a series of X chromosome specific DNA probes. *Cytogenet. Cell Genet.* **40:** 733.

Sherman, S.L., P.A. Jacobs, N.E. Morton, U. Froster-Iskenius, P.N. Howard-Peebles, K.B. Nielsen, M.W. Partington, G.R. Sutherland, G. Turner, and M. Watson. 1985. Further segregation analysis of the fragile X syndrome with special reference to transmitting males. *Hum. Genet.* **69:** 289.

Slightom, J.L., A.E. Blechl, and O. Smithies. 1980. Human fetal $^{G}\gamma$- and $^{A}\gamma$-globin genes: Complete nucleotide sequences suggest that DNA can be exchanged between these duplicated genes. *Cell* **21:** 627.

Spritz, R.A. 1981. Duplication/deletion polymorphism 5′ to the human β globin gene. *Nucleic Acids Res.* **9:** 5037.

Steinmetz, M., D. Stephan, and K.F. Lindahl. 1986. Gene organization and recombinational hotspots in the murine major histocompatibility complex. *Cell* **44:** 895.

Szabo, P., M. Purrello, M. Rocchi, N. Archidiacono, B. Alhadeff, G. Filippi, D. Toniolo, G. Martini, L. Luzzatto, and M. Siniscalco. 1984. Cytological mapping of the human glucose-6-phosphate dehydrogenase gene distal to the fragile X site suggests a high rate of meiotic recombination across this site. *Proc. Natl. Acad. Sci.* **81:** 7855.

Turner, G., J.M. Opitz, W.T. Brown, K.E. Davies, P.A. Jacobs, E.C. Jenkins, M. Mikkelsen, M.W Partington, and G.R. Sutherland. 1986. Conference report: Second international workshop on the fragile X and on X-linked mental retardation. *Am. J. Med. Genet.* **23:** 11.

Weiss, E.H., A. Mellor, L. Golden, K. Fahrner, E. Simpson, J. Hurst, and R.A. Flavell. 1983. The structure of a mutant H-2 gene suggests that the generation of polymorphism in H-2 genes may occur by gene conversion-like events. *Nature* **301:** 671.

Wyman, A.R., L.B. Wolfe, and D. Botstein. 1985. Propagation of some human DNA sequences in bacteriophage λ vectors requires mutant *Escherichia coli* hosts. *Proc. Natl. Acad. Sci.* **82:** 2880.

Yen, P.H., B. Marsh, T.K. Mohandas, and L.J. Shapiro. 1984. Isolation of genomic clones homologous to transcribed sequences from human X chromosome. *Somatic Cell Mol. Genet.* **10:** 561.

Molecular Genetics of *MIC2*: A Gene Shared by the Human X and Y Chromosomes

S.M. Darling, P.J. Goodfellow, B. Pym, G.S. Banting, C. Pritchard, and P.N. Goodfellow
Laboratory of Human Molecular Genetics, Imperial Cancer Research Fund, Lincoln's Inn Fields, London, WC2A 3PX United Kingdom

Although the mammalian sex chromosomes differ extensively in morphology and genetic content (Goodfellow et al. 1985a), meiotic pairing and segregation have been taken to indicate that the sex chromosomes have sequences in common (Koller and Darlington 1934; Burgoyne 1982). This sequence homology was postulated to reside in the morphological pairing region, which in humans occurs between the tip of the X-chromosome short arm and the Y-chromosome short arm (Pearson and Bobrow 1970; Chen and Falek 1971; Chandley et al. 1984).

The human sex chromosomes share a pair of related genes that independently encode the cell-surface antigen defined by the monoclonal antibody 12E7. This antibody was produced by immunizing a BALB/c mouse with leukemic cells from a patient with T-cell acute lymphoblastic leukemia (Levy et al. 1979). The 12E7 antibody reacts with a cell-surface antigen expressed on all human cells but fails to react with the surface of rodent cells (Goodfellow 1983). Human/rodent somatic cell hybrids that independently retain the human X and Y chromosomes and express the 12E7 cell-surface antigen have been used to define the genes *MIC2X* (Goodfellow et al. 1980) and *MIC2Y* (Goodfellow et al. 1983), respectively.

In immunoblot analysis the 12E7 antibody reacts strongly with a 32-kD human-specific cell-surface molecule. The antibody also recognizes a 29-kD cytoplasmic molecule found in both mouse and human cells. The cell-surface products of the human X and Y *MIC2* loci cannot be distinguished by either size or charge (Banting et al. 1985). Expression cloning in bacterial expression vectors has been used to identify cDNA clones encoding the 12E7 epitope (Darling et al. 1986). The cDNA probes react equally well with genomic sequences on the human X and Y chromosomes, and we have been unable to detect any differences between *MIC2X* and *MIC2Y* at the DNA level by Southern blot analysis (Darling et al. 1986). In situ hybridization using the *MIC2* cDNA clone has further localized *MIC2X* to Xp22.32-pter, the terminal region of the X-chromosome short arm, and the *MIC2Y* locus has been assigned to the Y-chromosome short arm in the distal region Yp11.2-pter (Buckle et al. 1985). These locations are consistent with the postulated pairing regions.

Burgoyne (1982) has followed Darlington (1937) in proposing that recombination between the X and Y chromosomes is required for correct segregation of the sex chromosomes at meiosis. If this is the case, the X and Y chromosomes will share sequences by exchange. Depending on the position of recombination events, genes and DNA sequences present in the shared region will show only partial sex linkage if occasionally exchanged, or no sex linkage if frequently exchanged. Such behavior has been termed pseudoautosomal. Recently, several random DNA sequences have been shown to be inherited in a pseudoautosomal fashion in man (Cooke et al. 1985; Simmler et al. 1985; Rouyer et al. 1986). The assignment of the *MIC2X* and *MIC2Y* loci to the pairing region, the apparent identity of the X and Y products, and the inability to detect differences at the DNA level between *MIC2X* and *MIC2Y* suggested that they are a pair of pseudoautosomal genes. Consistent with this suggestion, the X-located *MIC2* gene escapes X inactivation, thereby maintaining genetic dosage between males and females for expressed *MIC2* genes (Goodfellow et al. 1984).

In this review we describe a partial amino acid sequence deduced from the longest open reading frame (ORF) in a cDNA clone derived from an *MIC2* gene. A monoclonal antibody has been raised against a synthetic peptide based on part of the deduced sequence. This antibody reacts with the cell-surface product of the *MIC2* genes, confirming that we have identified the correct ORF and providing additional proof that we have cloned sequences derived from *MIC2*. The cDNA clone and corresponding genomic clones have been used in family studies, and we have found a low frequency of exchange of *MIC2* genes between the sex chromosomes. We conclude that the *MIC2* gene is the first pseudoautosomal gene to be described in man. Finally, we describe experiments designed to exploit the chromosomal location of the *MIC2* gene for exploring the region of the Y chromosome responsible for sex determination.

METHODS

DNA sequence analysis. Isolation of the cDNA clone, pSG1, which encodes the 12E7 epitope has been described previously (Darling et al. 1986). A restriction map of pSG1 was compiled and fragments were subcloned into the vector M13mp18. In addition, to verify the sequence across the chosen restriction fragments, the SG1 insert was ligated and subjected to partial *Hae*III digestion. These products were also subcloned into M13mp18. Sequencing was performed using the dideoxy chain termination procedure (Sanger et al. 1977) with the modifications described by Bankier and Barrell (1983).

DNA transfectants. Full details of the 12E7 antigen-positive transfectants will be presented elsewhere. Briefly, mouse LMTK$^-$ cells (Kit et al. 1963) were transfected with DNA from MOLT-4 (Minowada et al. 1972) and the selectable plasmid pTK1 (Wilkie et al. 1979), using the calcium phosphate precipitation method (Wigler et al. 1977). After selection for incorporation of pTK1, the transfectants were sorted for 12E7 antigen expression on the fluorescence-activated cell sorter (FACS). The brightest 5% of transfectants were isolated, expanded, and resorted repeatedly until a 12E7 antigen-positive population was isolated. Continued rounds of sorting on the FACS selected a transfectant (TKM1EP13) expressing elevated levels of 12E7 antigen. DNA from a "nonamplified" primary transfectant was transfected into LMTK$^-$ cells. Selection for 12E7 expression on the FACS led to the isolation of a secondary transfectant (EP5CP4C12). pSG1 reacts with sequences present in the 12E7 transfectants and, in addition, the concentration of reacting sequences is increased in the amplified transfectant TKM1EP13 (Darling et al. 1986). DNA from the secondary transfectant fails to react with *Alu* repeat or total human DNA in Southern blot analysis.

Indirect radioimmunoassay. Indirect radioimmunoassays (IRIA) were performed as previously described (Goodfellow et al. 1980; Tsu and Herzenberg 1980).

Antibodies. The antibodies 12E7 (Levy et al. 1979), RFB-1 (Bodger et al. 1981), 013, and F21 (Dracopoli et al. 1984, 1985) have been described elsewhere. BAN-R4A5-5 was raised by immunizing a BALB/c mouse with human red blood cells prior to fusing spleen cells from the sacrificed mouse to P3X63Ag8.NS-1 cells as previously described (Kohler and Milstein 1975; Goodfellow et al. 1979). P3X63Ag8.NS-1 (P3X) is a negative-control antibody (Kohler and Milstein 1975). 11-4.1 is a monoclonal antibody that recognizes the product of the mouse *H-2K^k* locus (Oi et al. 1979).

Full details of the production of MSGB1 will appear elsewhere. Briefly, a peptide spanning amino acid residues 69–81 of the SG1 conceptual translation was constructed on an Applied Biosystems 430A peptide synthesizer, coupled to thyroglobulin or bovine serum albumin (BSA), and supplied to us by Dr. J. Rothbard (Department of Molecular Immunology, ICRF). A BALB/c mouse was immunized with the thyroglobulin-peptide conjugate, then sacrificed, and its spleen cells were fused to P3X63Ag8.NS-1 cells as described above. Tissue-culture supernatants from the resulting hybridoma colonies were initially screened against BSA and the BSA-peptide conjugate by IRIA. Those specifically recognizing the conjugate were rescreened against whole human cells (HEB7A) (Wallace et al. 1975). Hybridoma colonies producing antibodies recognizing the peptide and whole cells were subjected to two rounds of dilution cloning and rescreening. MSGB1 is one of several resulting monoclonal antibodies.

Chromosome-mediated gene transfer. The technique has been described in detail elsewhere (Pritchard and Goodfellow 1986). Briefly, metaphase chromosomes from the human cell line OX (49,XYYYY) (Sirota et al. 1981) were transferred to LMTK$^-$ mouse cells (Kit et al. 1963) by calcium phosphate coprecipitation with the plasmid pTK1 (Wilkie et al. 1979). Chromosome-mediated gene transfer (CMGT) transfectants expressing the 12E7 antigen were isolated by several rounds of selection on the FACS. The introduced human chromosome fragments are retained in mouse cells by selection with HAT medium.

RESULTS AND DISCUSSION

Sequence of pSG1 and Structural Features of the Product

SG1, a cDNA insert that encodes the 12E7 epitope (Darling et al. 1986), was subjected to nucleotide sequence analysis. The sequence is shown in Figure 1 along with the conceptual translation of the longest ORF. pSG1 is not a full-length cDNA clone for the *MIC2* gene since the size of the mRNA detected in human cells is approximately 1.3 kb. In addition, we have yet to identify the initiating AUG methionine codon. The sequence contains an ORF of 390 nucleotides and a 3′ untranslated region of approximately 670 nucleotides. There is a poly(A) tract at the 3′ end that is separated by 14 bases from the consensus sequence AAUAAA (underlined in Fig. 1), characteristic of the in vivo polyadenylation signal of eukaryotic mRNA (Proudfoot and Brownlee 1976). Approximately 80 bases upstream of this predominant signal sequence is the sequence AAUUAA (underlined in Fig. 1), which can also direct poly(A) addition at an appropriate distance (Birnstiel et al. 1985). We have recently isolated a second cDNA clone that cross-hybridizes with SG1. It is 80 bases shorter than SG1 at the 3′ end since the mRNA from which it has been derived uses the upstream poly(A) addition signal.

The predicted protein sequence contains a hydrophobic region between amino acid residues 3 and 27; this sequence is of sufficient length to span the cell-surface membrane. Hydrophobicity plots (Eisenberg 1984) are consistent with a monomeric transmembrane domain (data not shown). Two charged residues are positioned at the amino-terminal end of the transmembrane domain and, by analogy with other membrane proteins, we would predict that these lie at the cytoplasmic face as an anchor (for review, see Rapoport and Wiedmann 1985). If we have correctly identified the transmembrane region, then the carboxyl terminus

1 ... 17

CGG CGC GCT CTG GGG CCA TGG CCC GCG GGG TGC GCT GGC GTG CTG CTC TTC

Arg Arg Ala Leu Gly Pro Trp Pro Ala Gly Cys Ala Gly Val Leu Leu Phe

35

GGC CTG CTG GGT GTT CTG GTC GCC GCC CCG GAT GGT GGT TTC GAT TTA TCC GAT

Gly Leu Leu Gly Val Leu Val Ala Ala Pro Asp Gly Gly Phe Asp Leu Ser Asp

53

GCC CTT CCT GAC AAT GAA AAC AAG AAA CCC ACT GCA ATC CCC AAG AAA CCC AGT

Ala Leu Pro Asp Asn Glu Asn Lys Lys Pro Thr Ala Ile Pro Lys Lys Pro Ser

71

GCT GGG GAT GAC TTT GAC TTA GGA GAT GCT GTT GTT GAT GGA GAA AAT GAC GAC

Ala Gly Asp Asp Phe Asp Leu Gly Asp Ala Val Val Asp Gly Glu Asn Asp Asp

89

CCA CGA CCA CCG AAC CCA CCC AAA CCG ATG CCA AAT CCA AAC CCC AAC CAC CCT

Pro Arg Pro Pro Asn Pro Pro Lys Pro MET Pro Asn Pro Asn Pro Asn His Pro

107

AGT TCC TCC GGT AGC TTT TCA GAT GCT GAC CTT GCG GAT GGC GTT TCA GGT GGA

Ser Ser Ser Gly Ser Phe Ser Asp Ala Asp Leu Ala Asp Gly Val Ser Gly Gly

125

GAA GGA AAA CAG GCA GTG ATG GTG GAG GCA GCC ACA GGA AAG AAG GGG AAG AGG

Glu Gly Lys Glu Ala Val MET Val Glu Ala Ala Thr Gly Lys Lys Gly Lys Arg

130

CCG ACG CCC CAG GCG TGA TCC CCG GGA TTG TGG GGG CTG TCG TGG TCG CCG TGG

Pro Thr Pro Gln Ala ●

CTG GAG CCA TCT CTA GCT TCA TTG CTT ACC AGA AAA AGA AGC TAT GCT TCA AAG

AAA ATG AAC AAG GGG AGG TGG ACA TGG AGA GCC ACC GGA ATG CCA ACG CAG AGC

CAG CTG TTC AGC GTA CTC TTT TAG AGA AAT AGA AGA TTG TCG GCA GAA ACA GCC

CAG GCG TTG GCA GCA GGG TTA GAA CAG CTG CCT GAG GCT CCT CCC TGA AGG ACA

CCT GCC TGA GAG CAG AGA TGG AGG CCT TCT GTT CAC GGC GGA TTC TTT GTT TTA

ATC TTG CGA TGT GCT TTG CTT GTT GCT GGG CGG ATG ATG TTT ACT AAC GAT GAA

TTT TAC ATC CAA AGG GGG ATA GGC ACT TGG ACC CCC ATT CTC CAA GGC CCG GGG

GGG CGG TTT CCC ATG GGA TGT GAA AGG CTT GGC CAT TAT TAA GTC CCT GTA ACT

CAA ATG TCA ACC CCA CCG AGG CAC CCC CCG TCC CCC AGA ATC TTG GCT GTT TAC

AAA TCA CGT GTC CAT CGA GCA CGT CTG AAA CCC CTG GTA GCC CCG ACT TCT TTT

TAA TTA AAA TAA GGT AAG CCT TAA TTT GTT TCT TCA ATA TTT CTT TCA TTT GTA

1067

GGG TAT TTG TTT TCT ATA CAG ACT AAT AAA AAG AAA TTA GAA CC poly(A)tract

Figure 1. The nucleotide sequence and conceptual translation of the longest ORF of pSG1, a cDNA insert that encodes the 12E7 epitope. The amino acid residues are numbered 1–130; dibasic amino acids are boxed; a synthetic peptide was made from amino acids 69–81, which are underlined. 1067 denotes the last nucleotide prior to the poly(A) tract. Nucleotides underlined at the 3′ end show used polyadenylation signals.

of the protein must be extracellular, since the 12E7 antibody detects an epitope on the external surface of human cells.

Some other structural features of potential interest include a cysteine residue within the transmembrane domain, several pairs of basic amino acids, a proline-rich region, and a conserved repeat. In other transmembrane domains, cysteine is the site of fatty acid acylation (Schlesinger et al. 1980; Omary and Trowbridge 1981; Kaufman et al. 1984). Paired basic amino acids often form the cleavage sites for proteases (Steiner et al. 1980), and this might explain the previously noted protease sensitivity of the 12E7 antigen (Goodfellow 1983). The conserved repeats KKPTA and KKPSA occur in the middle of the extracellular domain, and a partial repeat of the sequence, KRPT, occurs at the carboxyl terminus.

Homology searches have been carried out in the EMBL and Genbank DNA data bases and the Dayhoff and Doolittle protein data bases. No significant homology has been detected with any previously characterized molecule, nor have we found any clue to the function of the amino acid repeat sequence.

The 130 amino acid residues specified by the major ORF comprise a polypeptide with a molecular mass of approximately 18 kD. Even allowing for the sequences missing from the cDNA clones, the predicted molecular mass is likely to be less than the 32 kD of the 12E7 cell-surface antigen. One possible explanation for this discrepancy is secondary modification of the polypeptide; although the external domain lacks amino-linked glycosylation sites, it contains several potential sites for oxygen-linked glycosylation. In addition, the relatively high proline content may cause aberrantly slow migration on SDS-polyacrylamide gels.

Monoclonal Antibodies That Recognize the Product of the *MIC2* Loci

To confirm that we had identified the correct ORF and to provide new reagents for functional studies, we have raised monoclonal antibodies to peptides deduced from the sequence of the ORF. A peptide spanning the unique and potentially linear stretch of amino acids 69–81 (underlined in Fig. 1) was synthesized, conjugated to thyroglobulin, and used to immunize mice. A monoclonal antibody MSGB1 was raised that reacts with the immunizing peptide, human cells, and the DNA transfectants (Table 1a). Previously we have shown that the transfectants contain pSG1-related sequences and that the amplified primary transfectant has amplified these sequences (Darling et al. 1986).

Table 1. Antibodies Recognizing Products of the *MIC2* Locus

	Antibody							
	P3X	11-4.1	12E7	MSGB1	BANR4A5-5	RFB-1	F21	013
a. Cell line/peptide								
LMTK⁻	167 (64)		215 (15)	148 (36)				
HEB7A	216 (28)		10677 (261)	4846 (64)				
EP5CP4C12	187 (39)		8685 (617)	3195 (520)				
TKM1EP13	197 (75)		17362 (355)	13650 (1601)				
BSA-peptide conjugate	75 (20)		673 (72)	16099 (-)				
BSA	80 (17)		526 (58)	350 (-)				
b. Cell line								
LMTK⁻	1584 (187)	37905 (1099)	1616 (298)		1552 (111)	1475 (368)	1422 (493)	1683 (131)
HEB7A	1438 (228)	1554 (219)	20961 (1644)		9638 (279)	11550 (1276)	18848 (2468)	22555 (94)
EP5CP4C12	1128 (420)	37826 (844)	17830 (676)		9814 (64)	15270 (271)	21115 (444)	21435 (152)
TKM1EP13	1369 (160)	37026 (1692)	31121 (807)		15062 (1707)	32191 (256)	32824 (354)	40724 (704)
c. Peptide								
BSA-peptide conjugate	200 (34)		673 (72)		14593 (2828)	642 (108)	599 (58)	694 (38)
BSA	176 (32)		526 (58)		484 (67)	509 (79)	597 (54)	606 (74)

IRIAs using cell lines, peptides, and monoclonal antibodies are described in Methods. Figures given are the means of three samples with the standard deviations shown in parentheses.

Taken together, these data provide strong evidence that the deduced amino acid sequence is derived from the correct ORF.

Rettig and colleagues have described two monoclonal antibodies that react with antigenic products of both human sex chromosomes (Dracopoli et al. 1984, 1985; Rettig et al. 1984). In addition to the 12E7 antibody, we have also identified two antibodies that react with X- and Y-encoded antigens. All five independently produced antibodies react with the 12E7-antigen-positive transfectants and show increased binding to the amplified primary transfectants (Table 1b). We conclude that these antibodies all react with the product of the *MIC2* locus. One of the monoclonal antibodies, BANR4A5-5, which was raised against human red blood cells, reacts directly with the peptide used to raise the monoclonal antibody MSGB1 (Table 1c).

Recombination at the *MIC2* Locus

pSG1 detects multiple restriction-fragment-length polymorphisms (RFLPs) in genomic human DNA. The pattern of hybridization observed in DNAs digested with *Taq*I, *Hin*dIII, *Pvu*II, and *Pst*I is complex, with several constant and polymorphic bands of varying intensity. For ease of analysis, we have isolated genomic *MIC2* sequences that give simple RFLPs. One genomic sequence, p19B, detects RFLPs in *Taq*I and *Pvu*II digestions. The p19B-defined *Taq*I polymorphism consists of two alleles of 2.5 kb and 3.2 kb with frequencies in a Caucasian population of 0.65 and 0.35, respectively. Table 2 shows the segregation of *MIC2* and the pseudoautosomal sequences *DXYS17*, *DXYS15*, and *DXYS14*, with respect to sexual phenotype as analyzed by *Taq*I RFLPs. A total of 46 informative male meioses were examined, and a single recombinant between *MIC2* and sex was observed.

The low frequency with which *MIC2* recombines with the sexual phenotype is in marked contrast to the frequencies observed for the pseudoautosomal sequences described previously (Rouyer et al. 1986). *DXYS14*, *DXYS15*, and *DXYS17* were reported to exchange between the X and Y chromosomes at frequencies of 49.5%, 31.5%, and 14%, respectively. These frequencies are similar to those reported here although the rates of recombination at *DXYS15* could not be estimated accurately as there were too few informative matings. The low rate of exchange at the *MIC2* locus suggests it is the most proximal of the pseudoautosomal loci studied so far and thus is the closest distal marker to the testis-determining region (see Fig. 2).

Table 2. Recombination between the Pseudoautosomal Loci and Sexual Phenotype at Male Meiosis

	Locus			
	MIC2	*DXYS17*	*DXYS15*	*DXYS14*
Recombined X chromosomes	1	3	0	4
Recombined Y chromosomes	0	4	3	11
Total recombined sex chromosomes	1	7	3	15
Meioses with X	20	16	1	15
Meioses with Y	26	22	6	22
Total meioses	46	38	7	37
% recombined sex chromosomes	2	18	42	41

MIC2 Expression as a Tool for Locating the Testis-determining Factor

Based on the data in the previous section and analysis of XX males and a few XY females (Goodfellow et al. 1985a; Vergnaud et al. 1986), the testis-determining factor (*TDF*) has been placed proximal to *MIC2* and distal to sequences defined by cosmid 47 (*DXYS5*) on the Y-chromosome short arm (Fig. 2). The order of loci proximal to the *TDF* assumes that XX males are generated by a single abnormal recombination event between the X and Y chromosomes and has been derived mainly from the data of Vergnaud et al. (1986). If the underlying assumptions are correct, then *MIC2* and *DXYS5* identify the presently defined boundaries of the sex-determining region.

In previous CMGT experiments (Pritchard and Goodfellow 1986) it has been shown that inclusion of a DNA plasmid encoding a selectable marker in the chromatin–calcium phosphate transfection mix, followed by selection for uptake of the marker, will enable preselection for cells that have incorporated chromosomes. The selectable gene may also integrate into donor chromosome fragments. This has enabled us to combine CMGT techniques with the use of the FACS

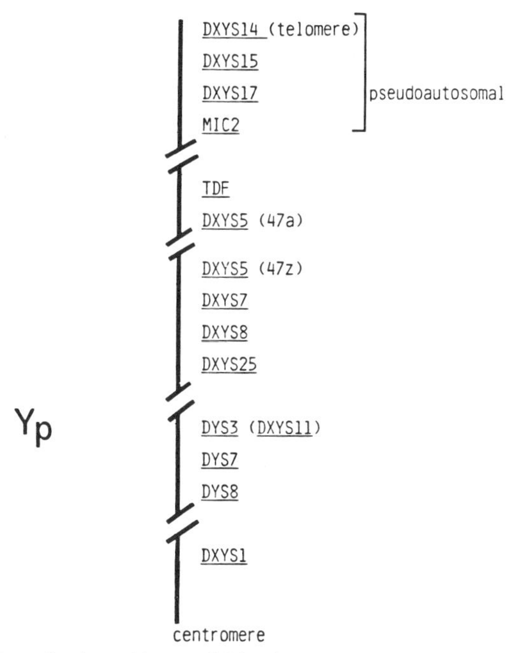

Figure 2. A working model for the Y-chromosome short arm. The model is based on data from Goodfellow et al. (1985a) and Vergnaud et al. (1986).

to isolate transfectants containing fragments of human chromosomes encoding cell-surface antigens. Metaphase chromosomes from the cell line OX (49,XYYYY) were cotransferred with the selectable plasmid pTK1 into mouse LMTK⁻ cells. After selection in HAT medium, transfectants expressing the 12E7 antigen were isolated by several rounds of selection on the FACS. Using this strategy, we have generated fragments of the human Y chromosome that contain integrated copies of the *TK* gene.

Nine primary CMGT transfectants have been analyzed for the cotransfer of loci syntenic with *MIC2* (Table 3). Variable but overlapping portions of the Y chromosome are transferred. Interstitial deletions are often generated in the CMGT process (C. Pritchard, unpubl.) and are observed in some of the Y transfectants. The cell line IP2.2 appears to have retained the region spanning *TDF* as the only Y-derived contribution; it contains *MIC2*, *DXYS5*, *DXYS17*, and none of the other Y-derived sequences that have been tested. A second round of chromosome transfer with selection for the exogenously added *TK* gene will allow the generation of a different set of breaks within this region.

CONCLUSIONS

The existence of a pseudoautosomal region was originally postulated to explain sex chromosome pairing in male meiosis (Darlington 1937). Subsequently, the existence of pseudoautosomal expressed genes was deduced from the abnormal phenotype associated with Turner's syndrome (Ferguson-Smith 1969; Burgoyne 1982). The *MIC2* gene is the first pseudoautosomal gene to be described in man and the first mammalian pseudoautosomal gene to be cloned. In the mouse, the gene encoding steroid sulfatase, *STS*, is almost certainly pseudoautosomal (Keitges et al. 1985); however, in man the presumed equivalent *STS* gene is X-linked (Shapiro et al. 1978). No other candidate pseudoautosomal genes have been described (see, however, Haldane 1936). The pseudoautosomal region in man is unlikely to be larger than 5×10^6 bp (Rouyer et al. 1986; and unpubl.); however, this is sufficient DNA to encode at least 30 or 40 genes. Hence, we cannot conclude that there is a relationship between expression of *MIC2* and Turner's syndrome.

The X-located *MIC2* gene escapes X inactivation (Goodfellow et al. 1984), and the cloned *MIC2* sequences should allow us to test current hypotheses about the mechanisms of X inactivation (Gartler and Riggs 1983). The cloned sequences will also allow us to explore at the molecular level the relationship between *MIC2*, the X-linked *XG* locus, and the Y-linked *YG* locus. The latter loci control the level of expression of the 12E7 antigen on red blood cells (Goodfellow and Tippett 1981).

There are few clues as to the biochemical function of the 12E7 antigen. The molecule is unlikely to have a role in tissue-specific differentiation or gene expression since it is found on nearly all human tissues tested (Goodfellow 1983). Unfortunately, the sequence deduced from the partial cDNA clone lacks homology with any previously described molecule, and it does not indicate any obvious structural features that might imply a particular function. We intend to explore the function of the 12E7 antigen by selecting antigen-negative variants of cells and by producing new monoclonal antibodies that may have the potential for perturbing function.

The chromosomal location of *MIC2* provides a tool for exploration of the region on the Y chromosome immediately adjacent to the pseudoautosomal region. This region is of interest for two reasons. First, sequences in the pseudoautosomal region recombine with a very high frequency in male meiosis. Presumably, at the boundary of the pseudoautosomal region a recombination barrier must occur. Second, the region below the pseudoautosomal sequences may contain the testis-determining gene(s). We intend to explore this region by utilizing the CMGT clones containing the relevant fragments of the Y chromosome. The transfectants will be used to map the region between *MIC2* and *DXYS5* and as a source for cloning sequences from this region.

ACKNOWLEDGMENTS

We would like to thank all our numerous friends and colleagues who have contributed to the study of the

Table 3. Analysis of Primary CMGT Transfectants for the Cotransfer of Loci Syntenic with *MIC2*

		CMGT transfectant								
Locus	Probe	E2P3	E1P4	IP2.1	IP2.2	IP2.6	JP5	K1P4	K2P2	K3P2
DXYS14	29Cl	+	+	–	–	–	–	–	+	+
DXYS15	113	–	–	–	–	–	–	+	+	–
DXYS17	601	+	+	–	+	+	–	+	+	–
MIC2	pSG1	+	+	+	+	+	+	+	+	+
DXYS5	47z	+	+	–	+	+	+	+	+	+
DXYS25	75/79	+	+	–	–	–	–	+	+	+
DYS7	50f2	+	–	+	–	+	+	+	+	–
DYS3	52d	+	+	–	–	+	–	+	+	–
DYS8	118	+	+	+	–	+	+	+	+	–
DXYS1	pDP34	+	–	–	–	+	–	+	–	–

(+) Positive reaction on Southern hybridization analysis; (–) negative reaction. References to the probes used can be found in Goodfellow et al. (1985b).

MIC2 gene. This manuscript could not have been prepared without the able editorial assistance of C. Middlemiss. We would also like to acknowledge M.V. Wiles for help with Figure 1 and J. Rothbard for constructing peptides and for advice on protein structure.

REFERENCES

Bankier, A.T. and B.G. Barrell. 1983. Shotgun DNA sequencing. In *Techniques in nucleic acid biochemistry* (ed. R.A. Flavell), p. 1. Elsevier Scientific, Ireland.

Banting, G.S., B. Pym, and P.N. Goodfellow. 1985. Biochemical analysis of an antigen produced by both human sex chromosomes. *EMBO J.* **4:** 1967.

Birnstiel, M.L., M. Busslinger, and K. Strub. 1985. Transcription termination and 3′ processing: The end is in site. *Cell* **41:** 349.

Bodger, M.P., G.E. Francis, D. Delia, S.M. Granger, and G. Janossy. 1981. A monoclonal antibody specific for immature human haemopoietic cells and T lineage cells. *J. Immunol.* **127:** 2269.

Buckle, V., C. Mondello, S. Darling, I.W. Craig, and P.N. Goodfellow. 1985. Homologous expressed genes in the human sex chromosome pairing region. *Nature* **317:** 739.

Burgoyne, P.S. 1982. Genetic homology and crossing over in the X and Y chromosomes of mammals. *Hum. Genet.* **61:** 85.

Chandley, A.C., P. Goetz, T.B. Hargreave, A.M. Joseph, and R.M. Speed. 1984. On the nature and extent of XY pairing at meiotic prophase in man. *Cytogenet. Cell Genet.* **38:** 241.

Chen, A.T. and A. Falek. 1971. Cytological evidence for the association of the short arms of the X and Y in the human male. *Nature* **232:** 555.

Cooke, H.J., W.R.A. Brown, and G.A. Rappold. 1985. Hypervariable telomeric sequences from the human sex chromosomes are pseudoautosomal. *Nature* **317:** 687.

Darling, S.M., G.S. Banting, B. Pym, J. Wolfe, and P.N. Goodfellow. 1986. Cloning an expressed gene shared by the human sex chromosomes. *Proc. Natl. Acad. Sci.* **83:** 135.

Darlington, C.D. 1937. *Recent advances in cytology*. J. and A. Churchill, London.

Dracopoli, N.C., W.J. Rettig, B.A. Spengler, H.F. Oettgen, J.L. Biedler, and L.J. Old. 1984. Assignment of genes determining human cell-surface antigens defined by monoclonal antibodies to chromosomes 12,X and Y. *Cytogenet. Cell Genet.* **37:** 456.

Dracopoli, N.C., W.J. Rettig, A.O. Albino, D. Esposito, N. Archidiacono, M. Rocchi, M. Siniscalco, and L.J. Old. 1985. Genes controlling gp25/30 cell-surface molecules map to chromosomes X and Y and escape X-inactivation. *Am. J. Hum. Genet.* **37:** 199.

Eisenberg, D. 1984. Three-dimensional structure of membrane and surface proteins. *Annu. Rev. Biochem.* **53:** 595.

Ferguson-Smith, M.A. 1969. Phenotypic aspects of sex chromosome aberrations. *Birth Defects Orig. Artic. Ser.* **5:** 3.

Gartler, S.M. and A.D. Riggs. 1983. Mammalian X-chromosome inactivation. *Annu. Rev. Genet.* **17:** 155.

Goodfellow, P. 1983. Expression of the 12E7 antigen is controlled independently by genes on the human X and Y chromosomes. *Differentiation* **23(S):** S35.

Goodfellow, P.N. and P. Tippett. 1981. A human quantitative polymorphism related to the Xg blood group. *Nature* **289:** 404.

Goodfellow, P., S. Darling, and J. Wolfe. 1985a. The human Y chromosome. *J. Med. Genet.* **22:** 329.

Goodfellow, P.N., K.E. Davies, and H.-H. Ropers. 1985b. Human gene mapping. *Cytogenet. Cell Genet.* **40:** 296.

Goodfellow, P.N., J.R. Levinson, V.E. Williams, and H.O. McDevitt. 1979. Monoclonal antibodies reacting with murine teratocarcinoma cells. *Proc. Natl. Acad. Sci.* **76:** 377.

Goodfellow, P., B. Pym, T. Mohandas, and L.J. Shapiro. 1984. The cell surface antigen locus, *MIC2X*, escapes X-inactivation. *Am. J. Hum. Genet.* **36:** 777.

Goodfellow, P.N., G. Banting, R. Levy, S. Povey, and A. McMichael. 1980. A human X-linked antigen defined by a monoclonal antibody. *Somatic Cell Genet.* **6:** 777.

Goodfellow, P., G. Banting, D. Sheer, H.-H. Ropers, A. Caine, M.A. Ferguson-Smith, S. Povey, and R. Voss. 1983. Genetic evidence that a Y-linked gene in man is homologous to a gene on the X chromosome. *Nature* **302:** 346.

Haldane, J.B.S. 1936. A search for incomplete sex-linkage in man. *Ann. Eugenics* **7:** 28.

Kaufman, J.F., M.S. Krangel, and J.L. Strominger. 1984. Cysteines in the transmembrane region of major histocompatibility complex antigens are fatty acylated via thioester bonds. *J. Biol. Chem.* **259:** 7230.

Keitges, E., M. Rivest, M. Siniscalco, and S.M. Gartler. 1985. X linkage of steroid sulphatase in the mouse is evidence for a functional Y-linked allele. *Nature* **315:** 226.

Kit, S., D.R. Dubbs, L.J. Piekawski, and T.C. Hsu. 1963. Deletion of thymidine kinase activity from L cells resistant to bromo-deoxyuridine. *Exp. Cell Res.* **31:** 297.

Kohler, G. and C. Milstein. 1975. Continuous cultures of fused cells secreting antibody of predefined specificity. *Nature* **256:** 495.

Koller, P.C. and C.D. Darlington. 1934. The genetical and mechanical properties of the sex chromosomes. 1. *Rattus norvegicus. J. Genet.* **29:** 159.

Levy, R., J. Dilley, R.I. Fox, and R. Warnke. 1979. A human thymus leukaemia antigen defined by hybridoma monoclonal antibodies. *Proc. Natl. Acad. Sci.* **76:** 6552.

Minowada, J., T. Ohmima, and G.E. Moore. 1972. Rosette forming human lymphoid lines. *J. Natl. Cancer Inst.* **49:** 891.

Oi, V.T., P.P. Jones, J.W. Goding, and L.A. Herzenberg. 1979. Properties of monoclonal antibodies to mouse Ig allotypes, H-2 and Ia antigens. *Curr. Top. Microbiol. Immunol.* **81:** 115.

Omary, M.B. and I.S. Trowbridge. 1981. Covalent binding of fatty acid to the transferrin receptor in cultured human cells. *J. Biol. Chem.* **256:** 4715.

Pearson, P.L. and M. Bobrow. 1970. Definitive evidence for the short arm of the Y chromosome associating with the X chromosome during meiosis in the human male. *Nature* **226:** 959.

Pritchard, C. and P.N. Goodfellow. 1986. Development of new methods in human gene mapping: Selection for fragments of the human Y chromosome after chromosome-mediated gene transfer. *EMBO J.* **5:** 979.

Proudfoot, N.J. and G.G. Brownlee. 1976. 3′ non-coding region sequences in eukaryotic mRNA. *Nature* **263:** 211.

Rapoport, T.A. and M. Wiedmann. 1985. Application of the signal hypothesis to the incorporation of integral membrane proteins. *Curr. Top. Membr. Transp.* **24:** 1.

Rettig, W.J., N.C. Dracopoli, T.A. Goetzger, B.A. Spengler, J.L. Biedler, H.F. Oettgen, and L.J. Old. 1984. Somatic cell genetic analysis of human cell-surface antigens: Chromosomal assignments and regulation of expression in rodent-human hybrid cells. *Proc. Natl. Acad. Sci.* **81:** 6437.

Rouyer, F., M.-C. Simmler, C. Johnsson, G. Vergnaud, H.J. Cooke, and J. Weissenbach. 1986. A gradient of sex linkage in the pseudoautosomal region of the human sex chromosomes. *Nature* **319:** 291.

Sanger, F., S. Nicklen, and A.R. Coulson. 1977. DNA sequencing with chain terminating inhibitors. *Proc. Natl. Acad. Sci.* **74:** 5463.

Schlesinger, M.J., A.I. Magee, and M.F.G. Schmidt. 1980. Fatty acid acylation of proteins in cultured cells. *J. Biol. Chem.* **255:** 10021.

Shapiro, L.J., R. Weiss, M.M. Buxman, J. Vidgoff, R.L. Dimond, J.A. Roller, and R.S. Wells. 1978. Enzymic basis of typical X-linked ichthyosis. *Lancet* **2:** 756.

Simmler, M.-C., F. Rouyer, G. Vergnaud, M. Nystrom-Lahti, K.Y. Ngo, A. de la Chapelle, and J. Weissenbach. 1985.

Pseudoautosomal DNA sequences in the pairing region of the human sex chromosomes. *Nature* **317:** 692.

Sirota, L., Y. Zlotogova, F. Shabtai, I. Halbrecht, and E. Eliam. 1981. 49,XYYYY. A case report. *Clin. Genet.* **19:** 87.

Steiner, D.F., P.S. Quinn, S.J. Chan, J. Marsh, and H.S. Tager. 1980. Processing mechanisms in the biosynthesis of proteins. *Ann. N.Y. Acad. Sci.* **343:** 1.

Tsu, T.T. and L.A. Herzenberg. 1980. Solid-phase radioimmune assay. In *Selected methods in cellular immunology* (ed. B.B. Mishell and S.M. Shiigi), p. 373. W.H. Freeman, San Francisco.

Vergnaud, G., D.C. Page, M.-C. Simmler, L. Brown, F. Rouyer, B. Noel, D. Botstein, A. de la Chapelle, and J. Weissenbach. 1986. A deletion map of the human Y chromosome based on DNA hybridisation. *Am. J. Hum. Genet.* **38:** 109.

Wallace, D.C., C.L. Burn, and J.M. Eisenstadt. 1975. Cytoplasmic transfer of chloramphenicol resistance in human tissue culture cells. *J. Cell Biol.* **67:** 174.

Wigler, M., S. Silverstein, L.S. Lee, A. Pellicer, Y.-C. Cheng, and R. Axel. 1977. Transfer of purified herpes virus thymidine kinase gene to cultured mouse cells. *Cell* **11:** 223.

Wilkie, N.M., J.B. Clements, W. Boll, N. Montei, D. Lonsdale, and C. Weissman. 1979. Hybrid plasmids containing an active thymidine kinase gene of Herpes simplex virus 1. *Nucleic Acids Res.* **7:** 859.

Variability at the Telomeres of the Human X/Y Pseudoautosomal Region

H.J. COOKE AND B.A. SMITH
MRC Mammalian Genome Unit, King's Buildings, Edinburgh, EH9 3JT United Kingdom

It has been evident for some time that the telomeres (ends) of eukaryotic chromosomes must have properties that fulfill a number of functional requirements. For example, unlike ends of chromosome fragments, telomeres protect the natural chromosome end against fusion with other chromosome ends. They are often associated with the nuclear envelope. They pose a replication problem for the cell since all known DNA polymerases require a primer and synthesize only in the 5'→3' direction. This implies that without a special structure, only one daughter chromosome would be completely replicated, the other being shortened, compared with the parental chromosome, by excision of a primer. This shortening would occur at each cell division and could presumably not be tolerated beyond a certain point. A number of models have been suggested that circumvent this problem (Blackburn and Szostak 1984).

The molecular biology of telomeres has until recently been confined to yeast and to those organisms that, in the development of the soma, fragment their germ-like chromosomes into multiple somatic chromosome fragments, usually with accompanying loss of DNA. Tetrahymena is probably the best studied of these organisms. Yeast genetics and molecular biology have been combined to analyze the requirements for telomere function. Since we depend on the knowledge of the properties of telomeres in these lower eukaryotes to formulate questions about human telomeres, we briefly review their structure and behavior.

The only sequences that are necessary to stabilize the ends of a linear plasmid yeast are short terminal repeats (Szostack and Blackburn 1982). These sequences—$(C_{1\text{-}3}A)_n$ in the case of yeast and $C_{1\text{-}8}(T/A)_4$ in tetrahymena—can also function as primers for the addition of tetrahymena repeat units by a terminal transferase activity present in vegetative and mating tetrahymena nuclear extracts (Greider and Blackburn 1985). This type of activity seems likely to account for the heterogeneity in the observed length of restriction fragments that terminate at telomeres in a variety of different organisms. In the case of trypanosome telomeres, synchronous addition of approximately one repeat per cell division has been observed, with decreases caused by infrequent, large deletions (Bernards et al. 1983).

In yeast the number of terminal repeats per chromosome is under genetic control. Mutations at the *CDC17* locus, a temperature-sensitive lethal mutation causing an arrest in the cell cycle, produce longer than wild-type telomeric fragments most probably due to changes in the number of terminal repeats (Carson and Hartwell 1985). Two other cell-lethal loci, *tel1* and *tel2*, are known to cause shorter than wild-type telomere lengths (Lustig and Petes 1986). It is tempting to speculate that these genes are involved in the untemplated addition of $(C_{1\text{-}3}A)$ sequences to the termini of yeast chromosomes, whereas *CDC17* affects a shortening function and that the abnormal telomeres generated are the cause of lethality at the nonpermissive temperature.

In yeast two more-complex repeat structures (Y' and X) are found immediately adjacent to the terminal repeats of the telomeres. The overall arrangement is $\text{poly}(C_{1\text{-}3}A)\text{-}(Y')_{0\text{-}4}\text{poly}(C_{1\text{-}3}A)(X)$. Y' is about 6.7 kb long, and X is a family of related repeats between 300 and 3750 bp in length (Chan and Tye 1983; Walmsley et al. 1984). These sequences are subject to frequent rearrangements (Horowitz et al. 1984) and can exist in free circular forms (Horowitz and Haber 1985). There appears to be no absolute functional requirement for these sequences.

In the case of humans, the large genome size and absence of interventionist genetics has precluded many of the approaches used to isolate telomeres in the lower eukaryotes. We have reported the isolation of a cosmid clone that terminates within 15–20 kb of the telomeres of the X- and Y-chromosome short arms (Cooke et al. 1985). Probes derived from this cosmid can be used to show that terminal heterogeneity exists at the ends of at least these human chromosomes. Sequences within 40 kb of the telomere show high levels of variability in the population, and this variability has been used to show the absence of sex linkage predicted for these regions of the mammalian sex chromosomes (Rouyer et al. 1986). We have analyzed the structure and variability found in these sex chromosome telomeric regions within and between individuals.

METHODS

DNA extractions were carried out by cell lysis into 2% SDS, 50 mM NaCl, 10 mM EDTA, 10 mM Tris (pH 8), 100 μg/ml proteinase K followed by extraction with an equal volume of chloroform/isoamyl alcohol (25/24/1). DNA was precipitated by 1 vol of isopropanol after the addition of 0.1 vol 3 mM NaOAc (pH 5.5). White cells were prepared from heparinized blood by lysis of the red cells in 155 mM NH_4Cl, 10 mM KHCO, 0.1 mM EDTA followed by centrifugation at 3000*g* for 5 min. Sperm samples were washed by centrifugation in 10 mM Tris, 1 mM EDTA (TE), incubated at 33°C overnight in TE plus 100 μg/ml proteinase K, followed by lysis into 2% SDS, 50 mM NaCl, 10 mM EDTA, 10 mM Tris (pH 8), 5% mercaptoethanol.

Enzyme reactions were carried out according to Maniatis et al. (1982). DNA digests were transferred to either nitrocellulose or nylon membranes according to the manufacturers' recommendations. Nitrocellulose filters were hybridized in 10% dextran sulfate, 5× SSC, 5× Denhardt's, 0.1% SDS, 0.1% NaPP: 100 μg/ml sonicated salmon sperm DNA at 65°C for 18 hr. Nylon filters were hybridized in 7% SDS, 0.5 M $NaPO_4$ (pH 8). Filters were washed three times for 15 min in 0.1% SDS, 0.1× SSC at 68°C. Autoradiography was carried out according to Maniatis et al. (1982).

RESULTS

Telomeric-associated Repeated Sequences

We have described the cosmid-derived probe 29C1 (Cooke et al. 1985), which detects variable numbers of *Pst* fragments in DNA from different individuals. Where analyzed in families, these fragments can be assigned to one of two alleles present either on both X chromosomes or on the X and Y chromosomes. We have not been able to determine the number of different alleles in the population, but preliminary data suggest it is in excess of 50. The nature of this variability is unclear, but it seems probable that expansion and contraction of repeats within a minisatellite (Jeffreys et al. 1984) block is not responsible. Although this would alter the size of fragments detected in digests with restriction enzymes that flank such a minisatellite block, the number of such fragments would not be predicted to vary. The subclone 29Cl detects variable numbers of fragments of different sizes in different individuals with a variety of enzymes. Restriction mapping of two 29Cl homologous sequences derived from different individuals suggests that substantial map differences occur in some regions with close conservation in adjacent regions (C.F. Inglehearn, pers. comm.). Further understanding of the nature of this variability will come from sequence analysis of a number of copies of this region from different individuals.

Although the copies of this sequence show extensive variability in the population, different tissues from the same individual are identical. We have determined the organization of these copies in one individual by a combination of restriction enzyme mapping and BAL-31 digestion. The terminal fragments in a restriction enzyme digest are those that are shortened first in a time course of BAL-31 digestion. In Figures 1 and 2 we show the results of an experiment in which high-molecular-weight DNA from a lymphoblastoid line (PES) was treated with BAL-31 and sampled at increasing times of digestion. This DNA was then digested either with *Hin*dIII or *Eco*RI and then probed with 29Cl after gel electrophoresis and transfer. Four fragments are visible in both sets of digests (Fig. 1A). In the case of the *Hin*dIII digests, the two smallest fragments are progressively shortened by BAL-31, whereas the larger fragments are unaffected. In the *Eco*RI digest, the two smallest fragments disappear abruptly rather than show a progressive shortening. This is consistent with the small fragments being close to the terminus of two chromosomes, the X and Y chromosomes in this case.

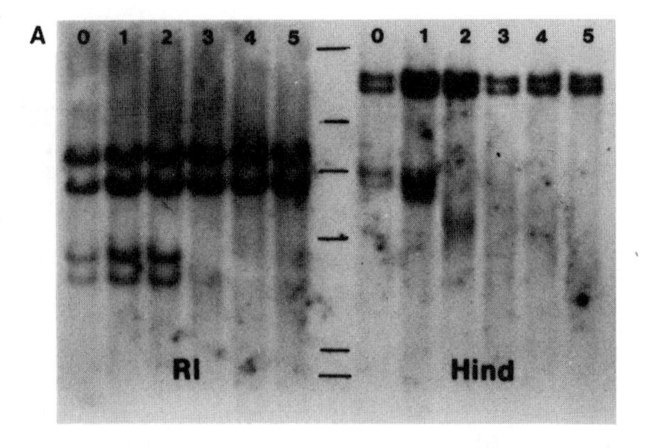

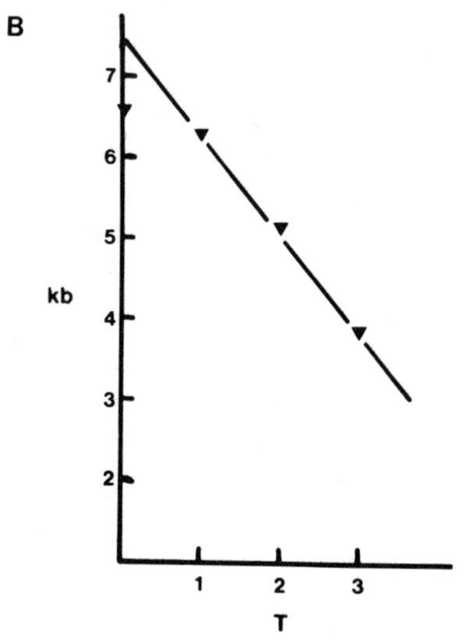

Figure 1. (*A*) Mapping of repeats on two chromosomes. DNA (100 μg) from a lymphoblastoid cell line (PES) was incubated with 15 μg of BAL-31 nuclease, and samples were withdrawn at 0, 1, 2, 3, 4, and 5 hr (lanes *0–5*). Samples were then split and digested with either *Eco*RI or *Hin*dIII, separated by gel electrophoresis, and transferred to nitrocellulose. The filter was hybridized with probe 29Cl. A marker digest of phage λ *Hin*dIII fragments is shown at the center, between lanes *5* and *0*. (*B*) Sizes of peak hybridization of the smaller pair of *Hin*dIII fragments plotted against time of digestion. The difference between the extrapolated and measured time at $t = 0$ suggests an approximate distance of 1 kb between the telomere and the first detectable *Hin*dIII site.

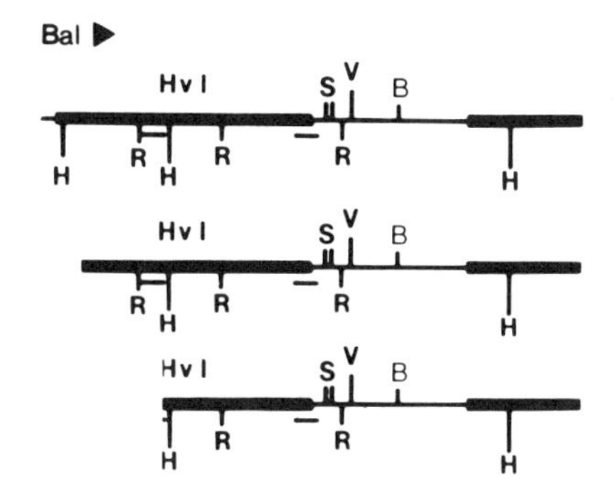

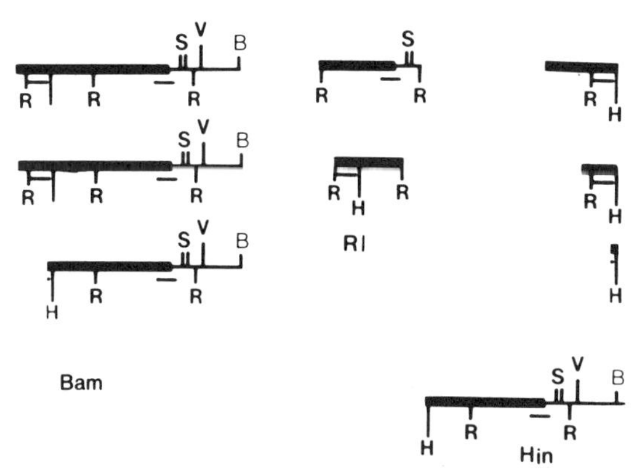

Figure 2. Restriction map of one end of one sex chromosome in lymphoblastoid cell line PES. BAL-31 digestion produces shortened molecules, depicted schematically at three stages of digestion (*top*). Different restriction-enzyme-digest products are shown at *bottom*. *Bam*HI digestion gives a single, progressively shortened fragment. *Eco*RI digestion gives two fragments, one internal and unaffected by this extent of BAL-31 digestion and one that is not detected once the BAL-31 digestion enters this region of the map since probe homology is confined to the distal part of this fragment. *Hin*dIII digestion gives one large internal fragment and one fragment that is progressively shortened. Underlined areas represent regions of homology to 29Cl.

We can estimate the distance from the distal *Hin*dIII site of this fragment to the chromosome end from the kinetics of BAL-31 digestion. If we extrapolate the size of the *Hin*dIII fragment to zero time of BAL-31 digestion, we reach a value of 7.5 kb (Fig. 1B). The *Hin*dIII fragment itself is 6.5 kb long, implying that the distal *Hin*dIII site is 1 kb from the natural end of the chromosome in this lymphoblastoid line. The behavior of the *Eco*RI fragments is consistent with the probe homology being confined to the distal end of these fragments. BAL-31 digestion past the distal *Eco*RI site removes those sequences on this restriction fragment that are homologous to the probe so that as soon as these fragments are shortened, they become undetectable (Fig. 2). This organization varies from chromosome to chromosome in the population since we have seen examples in which there appear, from pedigree analysis and studies using somatic cell hybrids, to be chromosomes with a single region of homology to 29Cl.

The original cosmid, CY29, contains another repeated sequence, unrelated to 29Cl, located at the centromere-proximal end of the cosmid. Probes from the proximal 10 kb of this cosmid all detect the same fragments in genomic digests, suggesting a tandemly repeated sequence organization. The size of this block of repeats is variable between about 15 and 40 kb in different individuals. Variability within this block can be detected in *Taq* digests of human DNA (Fig. 3B). Studies in large families show that these repeated sequences are very closely linked to the repeats detected by 29Cl (J. Weissenbach, pers. comm.), as would be predicted if all copies of these repeats were located in a single cluster 25 kb from the X- and Y-chromosome short-arm telomeres.

Since we do not know the allele numbers or frequencies for either of these systems, we cannot determine whether these loci are in linkage equilibrium. In Figure 3 we show that apparently identical copies of the repeats detected by the proximal repeat probe 29A24 are associated with different copies of 29Cl repeats, suggesting that any linkage disequilibrium is less than total.

Terminal Heterogeneity

When DNA from nucleated blood cells is digested with restriction enzymes that do not cut between the 29Cl probe and the end of the chromosome, the hybridization signal given is to a disperse size-range of fragments. Since this effect is observed in blood from single individuals (i.e., a cloned cell population), it implies that there is a cell-to-cell variation in the DNA. This size variation is also found in liver, placenta, lung, muscle, spleen, brain, ovary, and adrenals (Fig. 4).

Terminal heterogeneity within a cloned cell population is a characteristic feature of telomeres, and from the yeast model it is probably due to a variation from chromosome to chromosome in the number of short terminals repeats. In the human case we have so far been unable to analyze the nature of the chromosome end directly. Instead, we have asked questions about variation at the telomere itself indirectly, using the available probe.

Since in the fertilized male zygote there is only one X and one Y chromosome, any variation in the cells produced by division of this zygote can arise only by addition of variable amounts of DNA at the ends of the chromosome. We reasoned that the germ line provided the basic reference point from which either of these processes would start. Sperm provides a readily available source of germ-line DNA, so we have analyzed the distance from the terminal *Bam* site to the chromosome end in sperm DNA from a number of individuals. This region encompasses the hypervariable

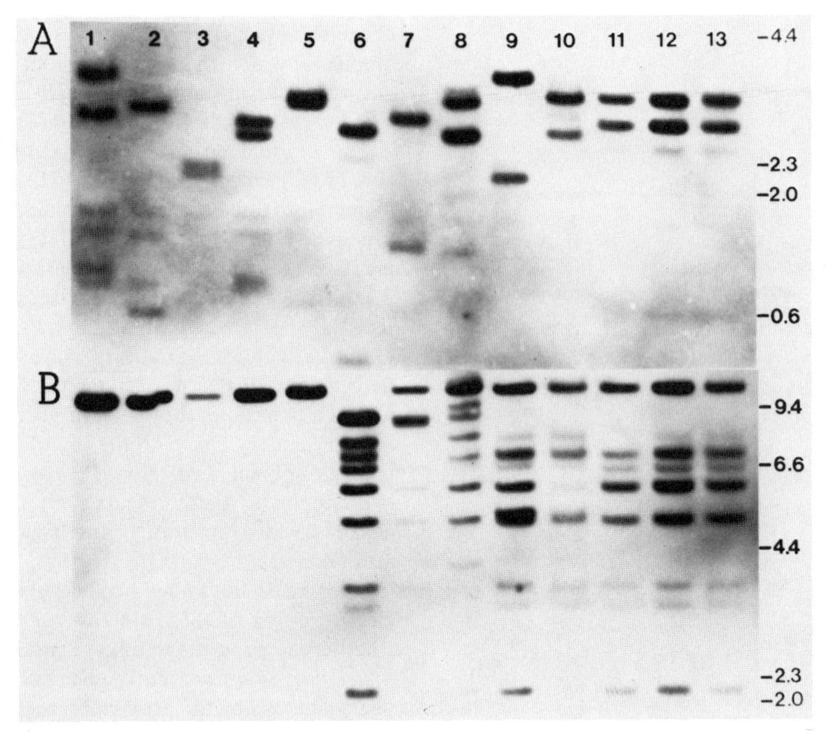

Figure 3. Individual DNA samples digested with *Taq* and probed with telomere-associated repeats. (*A*) (Lanes *1–10*) Individual blood DNA samples; (lane *11*) PES lymphoblastoid line; (lane *12*) blood DNA from the same individual; (lane *13*) sperm DNA from the same individual probed with 29Cl. (*B*) Same filter probed with the 29A24 repeat probe.

region, so we compared sperm and blood samples from the same individuals. For one individual (PES), we have shown above that sperm, blood, and a lymphoblastoid line have identical restriction fragments up to 1 kb from the telomere (Fig. 3, lanes 11,12,13); thus, variation in length cannot be caused by polymorphisms in this region.

In the individuals shown in Figure 5, the size of the terminal fragment detected in sperm DNA is larger by about 5 kb than the average size found in blood from the same individual. This is true in all six individuals

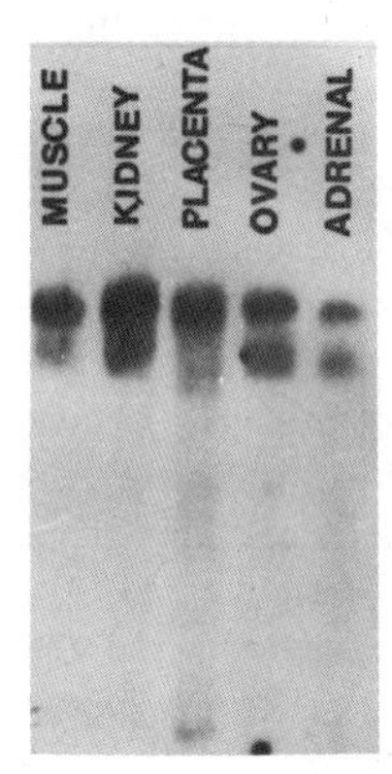

Figure 4. DNA from different tissues of the same individual. Muscle, kidney, placenta, ovary, and adrenal DNA samples were prepared from a single fetus, digested with *Bam*HI, and analyzed by blotting, using 29Cl as a probe.

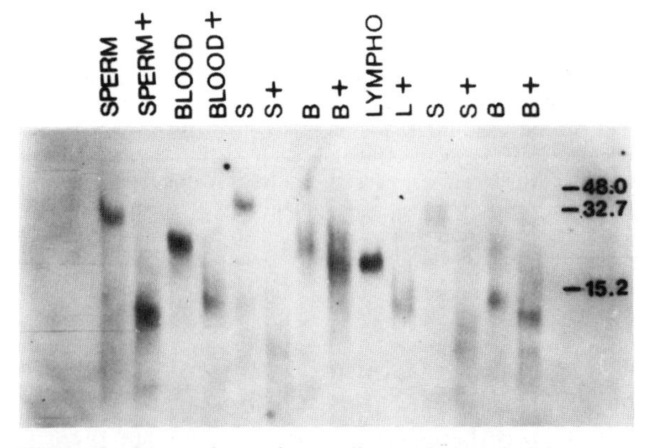

Figure 5. Comparison of germ-line and somatic DNA sample. Sperm and blood DNAs were prepared from three individuals (and in one case a lymphoblastoid line from the same individual), digested with *Bam*HI, and analyzed by blotting and hybridization with probe 29Cl. Samples were labeled and then digested with 1 μg of BAL-31 nuclease for 2 hr before *Bam*HI digestion. (*Left* to *right*) From individual 1: sperm DNA; sperm DNA digested with BAL-31; blood DNA; blood DNA digested with BAL-31. From individual 2: sperm DNA; sperm DNA digested with BAL-31; blood DNA; blood DNA digested with BAL-31; lymphoblastoid line DNA; lymphoblastoid line DNA digested with BAL-31. From individual 3: sperm DNA; sperm DNA digested with BAL-31; blood DNA; blood DNA digested with BAL-31. Each lane was loaded with 5 μg of *Bam*HI-digested DNA.

for which data are available. The fragments detected remain sensitive to BAL-31 digestion and are therefore still terminal. This eliminates the possibility that a rearrangement has occurred that converts an internal germ-line fragment to a terminal somatic one. It seems that a repeat addition process, as seen in trypanosomes, is not occurring, but rather that loss of sequence occurs at some point in the development of the soma.

One possible mechanism for this loss could be that somatic cells, but not the germ line, are deficient in the type of terminal transferase activity demonstrated in tetrahymena, resulting in a loss of DNA at each cell division. This could imply that the number of terminal repeats would limit the number of all divisions possible. It seems, however, that the situation is more complex than this. When DNA from two lymphoblastoid cell lines that had been in continuous culture for 10 years was compared with that from the same cell line grown from early stocks stored frozen over the same period, no differences were evident. A regular loss of a few base pairs per cell division would have been easily detected in this experiment (Fig. 6).

One striking feature of the patterns given by lymphoblastoid cell lines is that there is very much less terminal heterogeneity than in the blood cell population as a whole. We have examined this residual heterogeneity in a mass culture of an XO lymphoblastoid line and individual cloned lines from this mass culture. In this case we used an XO cell line to remove any contribution to heterogeneity from a second chromosome (X or Y). To control for different mobilities of DNA samples in different gel lanes, phage λ DNA digested with *Sal*I was added to all DNA samples. In this cell line the terminal *Bam* fragment comigrates with the 15.2-kb λ *Sal* fragment. It is evident that there are small but detectable differences in mobility of the *Bam* fragment between subclone DNAs that are not reflected in the mobility of the markers. This is most evident when comparing lanes i and j in Figure 7.

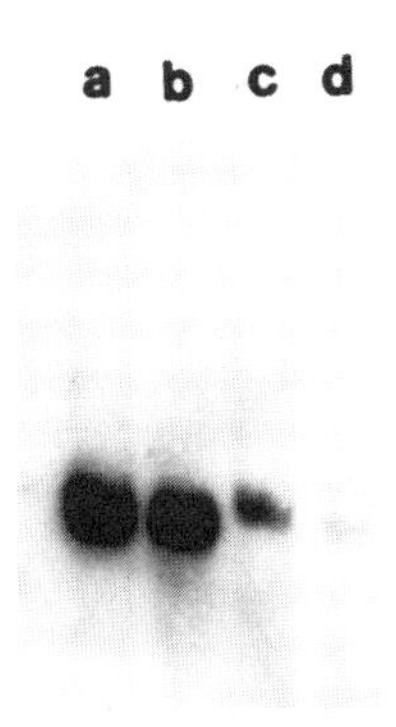

Figure 6. Comparison of early- and late-passage lymphoblastoid lines. (lane *a*) XXXXY lymphoblastoid DNA early passage; (lane *b*) XXXXY lymphoblastoid DNA late passage; (lane *c*) XY lymphoblastoid DNA early passage; (lane *d*) XY lymphoblastoid line DNA late passage. 5 μg of each DNA was digested with *Bam*HI and probed after blotting with 29Cl.

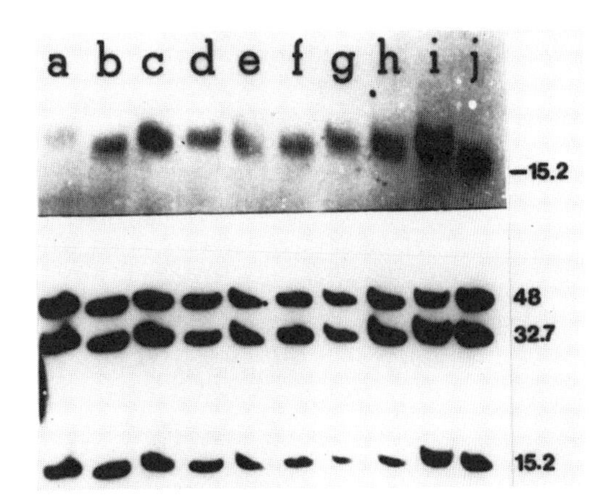

Figure 7. Analysis of cloned cell lines. DNAs prepared from a mass culture of GM3453 (an XO lymphoblastoid cell line) and from nine cloned lines from this mass culture were digested with *Bam*HI, and to 5 μg of each digest 1 ng of a *Sal*I digest of λ DNA was added. After electrophoresis and blotting, the filter was first probed with 29Cl, then with λ DNA. Lane *a* contains the mass culture DNA, and lanes *b–j* contain the nine clones derived from this mass culture.

One possible explanation of the heterogeneity seen in tissues is that there are multiple cell types present, but that there is only one in a cloned lymphoblastoid line. To assess this possibility, we have analyzed DNA from flow-sorted B and T lymphocytes and granulocytes and compared the size distribution in these populations with that found in the whole nucleated blood cell population in the same individual. The size distribution is identical in these fractionated cell populations to that found in the total blood DNA. It seems that in cloned cell populations, whatever process generates large-scale heterogeneity is no longer at work. This is not a function of the in vitro transformation process since blood DNA samples from individuals with B-cell leukemia have an almost homogeneous size of terminal restriction fragment. In these individuals most of the blood cells are clonally derived from a single precursor cell.

DISCUSSION

The study of human telomeres is in its infancy compared with our knowledge of telomeres of lower eukaryotes. Some similarities emerge, but their significance is at this stage unclear. First, both yeast and human telomeres have associated, complex repeated sequences that show extreme variation in the population (see Fig. 8 for a structural comparison). In yeast, one mechanism for generating this variation is meiotic recombination. The human XY telomeric sequences in male meiosis are in a region of the genome that exhibits a recombination rate 10 times that found in the same region in female meiosis. It seems likely that this high recombination frequency in the pseudoautosomal region will provide at least a partial explanation for the variability seen in these sequences. Other nontelomeric

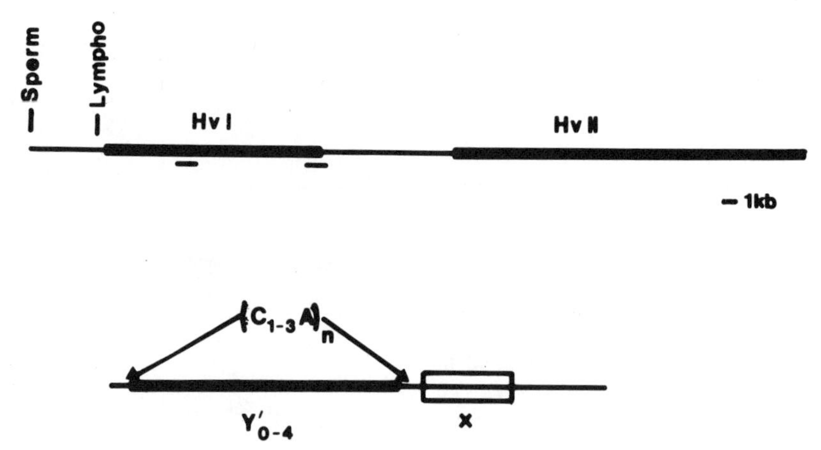

Figure 8. Comparison of human and yeast telomere-associated repeat structure. (Upper line) Human X and Y telomeres; (lower line) yeast general structure. In both cases the centromere is to the right. Heavy lines denote regions of high variability (HvI and HvII).

sequences in this region also have multiple alleles and high recombination rates. Analysis of sufficient families should lead to the detection of new variants within the telomeric region and a molecular analysis will tell us if a reciprocal recombination event is sufficient to explain this or if other mechanisms are at work.

By analogy with yeast it seems unlikely that these telomeric-associated repeats have a direct telomeric function. They are also confined to one end of one chromosome pair in humans, whereas a functionally significant sequence might reasonably be expected to be present at all telomeres. They are not present in mammals other than the primates. The only suggestion that there may be some function in this region is the presence of a nonpolymorphic segment between the two blocks of variable repeats (Fig. 8). Any functional constraints upon this DNA segment may not be related to telomere function. We would predict that sequences at the telomere would be present at all human telomeres and possibly at all mammalian telomeres. By walking toward the Xp/Yp telomere, we should be able to generate clones that will allow us to analyze other telomeric regions of the human genome and define common features. Our experience to date suggests that this may be difficult since a number of fragments that we know to have homology to the 29Cl probe have proved to be extremely difficult to clone in a wide variety of vector-host systems. Our view is that repeated sequences in this region of the human genome may lead to recombinational deletion during phage growth.

The terminus of the X and Y chromosome short arms seems likely to consist of short terminal repeats. We have no direct evidence to support this contention, but the type of heterogeneity observed is similar to that seen in those organisms known to have these structures. It is clear that there is a loss of DNA when the germ line gives rise to the soma. This phenomenon is not confined to human chromosomes. For example, the nematode *Ascaris suum* has a karyotype with 24 chromosomes, which in the germ line carry satellite DNA at their telomeres visible as heterochromatic blocks. Fragments containing this telomeric heterochromatin are cleaved off the chromosomes during mitotic condensation in presomatic blastomeres and are lost into the cytoplasm in subsequent cell divisions. About 25% of the genome is lost in this way from the soma but is retained in the germ line (Roth and Moritz 1981). In the case of the human sex chromosomes, this process is less extreme, with a loss of about 5 kb from the germ line. The sequences lost from the *Ascaris* chromosomes do not consist of the type of short repeat found at yeast and tetrahymena telomeres but are satellite sequences with repeat units of 125 and 131 bp (Streeck et al. 1982). The structures at the ends of *Ascaris* chromosomes are unknown. We have no knowledge of the sequence lost from the human sex chromosomes.

Our data suggest that either variable amounts of DNA are lost during somatic differentiation or that a constant amount is lost followed by addition of variable lengths of short terminal repeats. Whichever of these processes occurs, the size of the terminal structure is in some way stabilized after the diminution process so that in cloned cell populations all cells have similar lengths of telomeric sequence. A low level of heterogeneity remains, probably due to a continual gain and loss of a few short terminal repeats. This scheme is outlined in Figure 9.

Our intentions are to exploit the availability of telomeric-associated probes to study the molecular biology of the telomeres themselves. Once functional human telomeres are available, they provide one requirement for building a synthetic human chromosome. Such a chromosome would be of value in analyzing phenomena such as DNA replication, segregation, centromere function, and recombination. It would also have practical use as a vector for DNA sequences that would be controlled in copy number and free from problems of insertional mutagenesis.

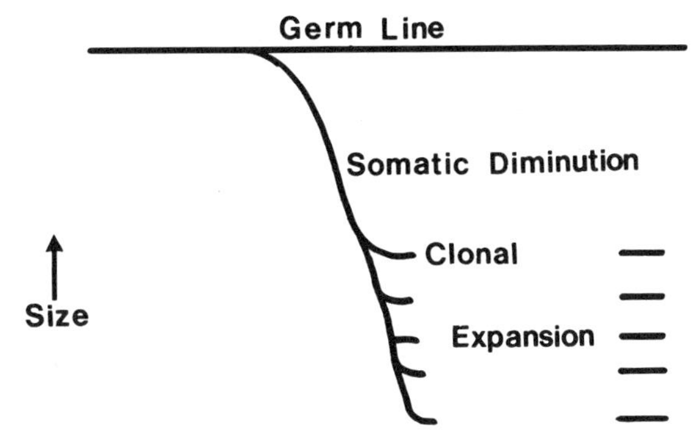

Figure 9. A possible scheme of changes occurring at the telomeres of the human X- and Y-chromosome short arms.

ACKNOWLEDGMENT

This work was supported by the Medical Research Council.

REFERENCES

Bernards, A., P.A.M. Michels, C.R. Linche, and P. Borst. 1983. Growth of chromosomal ends in multiplying trypanosomes. *Nature* **303:** 592.

Blackburn, E.H. and J.W. Szostak. 1984. The molecular structure of centromeres and telomeres. *Annu. Rev. Biochem.* **53:** 164.

Carson, M.J. and L. Hartwell. 1985. CDC 17: An essential gene that prevents telomere elongation in yeast. *Cell* **42:** 249.

Chan, C.S.M. and B.K. Tye. 1983. Organisation of DNA sequences and replication origins at yeast telomeres. *Cell* **33:** 563.

Cooke, H.J., W.R.A. Brown, and G.A. Rappold. 1985. Hypervariable telomeric sequences from the human sex chromosomes are pseudoautosomal. *Nature* **317:** 687.

Greider, C.W. and E.H. Blackburn. 1985. Identification of a specific telomere terminal transferase activity in tetrahymena. *Cell* **43:** 405.

Horowitz, H.P. and J.E. Harber. 1985. Identification of autonomously replicating circular subtelomeric Y′ telemeres in *Saccharomyces cerevisiae*. *Mol. Cell Biol.* **5:** 2369.

Horowitz, H., P. Thorburn, and J.E. Harber. 1984. Rearrangements of highly polymorphic regions near telomeres of *Saccharomyces cerevisiae*. *Mol. Cell Biol.* **4:** 2509.

Jeffreys, A.J., V. Wilson, and S.L. Thein. 1984. Hypervariable "minisatellite" regions in human DNA. *Nature* **314:** 67.

Lustig, A.J. and T.D. Petes. 1986. Identification of yeast mutants with altered telomere structure. *Proc. Natl. Acad. Sci.* **83:** 1398.

Maniatis, T., E.F. Fritsch, and J. Sambrook. 1982. *Molecular cloning: A laboratory manual.* Cold Spring Harbor Laboratory, Cold Spring Harbor, New York.

Roth, G.E. and K.B. Moritz. 1981. Restriction enzyme analysis of the germ line limited DNA of *Ascaris suum*. *Chromosoma* **83:** 169.

Rouyer, F., M.-C. Simmler, C. Johnsson, G. Vergnaud, H.J. Cooke, and J. Weissenbach. 1986. A gradient of sex linkage in the pseudoautosomal region of the human sex chromosomes. *Nature* **319:** 291.

Streeck, R.E., K.B. Moritz, and K. Beer. 1982. Chromatic diminution in *Ascaris suum*: Nucleotide sequence of the eliminated satellite DNA. *Nucleic Acids Res.* **10:** 3495.

Szostak, J.W. and E.H. Blackburn. 1982. Cloning yeast telomeres on linear plasmid vectors. *Cell* **29:** 245.

Walmsley, R.M., C.S. Chan, B.-K. Tye, and T.D. Petes. 1984. Unusual DNA sequences associated with the ends of yeast chromosomes. *Nature* **310:** 157.

The Pseudoautosomal Region of the Human Sex Chromosomes

F. ROUYER, M.-C. SIMMLER, G. VERGNAUD, C. JOHNSSON, J. LEVILLIERS,
C. PETIT, AND J. WEISSENBACH

*Unité de Recombinaison et Expression Génétique, INSERM U-163, CNRS UA-271,
Institut Pasteur, 75015 Paris, France*

Some 50 years ago, Koller and Darlington (1934) reported that synapsis between terminal parts of the sex chromosomes occurred at male meiosis in *Rattus norvegicus*. This finding, applied to other mammals, led to the proposal that the mammalian X and Y chromosomes could undergo crossing-over in their pairing region. Additional theoretical arguments and cytogenetic observations supporting this view have been presented since then (Solari 1980; Burgoyne 1982; Polani 1982; Holm and Rasmussen 1983), but until recently this issue has remained unsettled and even controversial (Ashley 1984).

Transfer of the *Sxr* mutation from an abnormal Y chromosome to the paternal X chromosome in 50% of the XX karyotyped progeny provided a first genetic clue in favor of X/Y recombination in mouse (Evans et al. 1982; Singh and Jones 1982). Evidence for recombination between normal mammalian sex chromosomes has now been obtained by enzymatic dosage in mouse and by DNA analysis in man. In mouse, the steroid sulfatase enzyme is encoded by an apparently non-sex-linked gene, although it maps on the X chromosome (Keitges et al. 1985). In man, several DNA loci have been mapped to the tip of the X- and Y-chromosome short arms (Buckle et al. 1985a; Cooke et al. 1985; Simmler et al. 1985). Alleles from such loci can be exchanged between both sex chromosomes in male meiosis (Cooke et al. 1985; Simmler et al. 1985; Rouyer et al. 1986). These results have demonstrated the earlier hypotheses postulating the existence of pseudoautosomal loci, which display either no or partial sex linkage (Burgoyne 1982).

A first analysis of three distinct pseudoautosomal DNA loci performed on about 50 meioses has shown that such loci are each partially linked to sex, according to a gradient increasing from the non-sex-linked telomere to the more proximal loci (Rouyer et al. 1986). These results are consistent with a single and obligatory crossover between the human X and Y chromosomes and indicate that this event is not uniquely localized.

We have now extended this quantitative analysis to five pseudoautosomal DNA loci in more than 100 male and female meioses. The present results totally confirm our initial findings and show that the sex-linkage gradient can vary from no linkage at all at the pseudoautosomal telomere to nearly absolute linkage for the more proximal loci. In addition, pseudoautosomal recombination is characterized by absence of double crossover and by a 10-fold to 20-fold increase of recombination frequency (RF) in male vs. female meiosis. These features add strong support to the view that the crossing-over taking place at male meiosis between the X and Y chromosomes is a single and obligatory event.

MATERIALS AND METHODS

Family analysis. DNA samples from large families have essentially been provided by the Centre d'Etude du Polymorphisme Humain (CEPH). These DNAs were digested with *Taq*I (and *Eco*RI for a few samples) and analyzed by Southern blotting. In most instances paternal phase was deduced from an analysis of the paternal grandparents. However, when DNA from grandparents was not available, paternal phase was assessed on the basis that the more proximal loci (which recombine with sexual phenotype at frequencies below 5%) cannot recombine in the majority of a sibship and that double recombination must be much less frequent than single recombination. Maternal phase was determined by assuming a linkage between the pseudoautosomal loci at female meiosis.

Pseudoautosomal DNA probes. Probes 29Cl, 113D, 601, and pSG1 have been used as originally described (Cooke et al. 1985; Simmler et al. 1985; Darling et al. 1986; Rouyer et al. 1986). Chromosome Y and Xp223-pter location was determined for the additional probes as previously described (Simmler et al. 1985).

Probe 29Cl defines telomeric locus *DXYS14* (Cooke et al. 1985) and detects hypervariable DNA fragments in any DNA digest.

Probe 362A is a tandemly repeated sequence situated between 6 kb and 30–40 kb proximal to 29Cl. It detects highly polymorphic *Hin*dIII and *Taq*I fragments.

Probe 113D defines pseudoautosomal locus *DXYS14* (Simmler et al. 1985) and detects a multiallelic *Taq*I fragment between 2.0 and 2.8 kb.

Probe 601 defines pseudoautosomal locus *DXYS17* (Rouyer et al. 1986) and detects a multiallelic *Taq*I fragment between 1.1 and 1.9 kb and a diallelic *Eco*RI fragment of 17 or 7 kb.

Probe pSG1 is a cDNA probe from locus *MIC2* (Darling et al. 1986) that detects several polymorphic *Taq*I fragments (P. Goodfellow, pers. comm.).

Probe 68B is a repeated *DXYZ2* element that detects numerous pseudoautosomal *Taq*I fragments, some of which are polymorphic (Simmler et al. 1985). The largest *Taq*I fragment is polymorphic and appears as a triallelic locus with alleles of approximately 18, 13, and 12 kb. This locus, defined by a very proximal *DXYZ2* repeat, is provisionally called *DXYZ2prox* hereafter.

The Mendelian behavior of the DNA loci analyzed in this study has been previously validated in appropriate family studies.

RESULTS

A Gradient of Sex Linkage

DNA samples from large families, provided by CEPH or from other sources, were analyzed for the restriction-fragment-length polymorphisms (RFLPs) detected by pseudoautosomal probes 29Cl, 362A, 113D, 601, pSG1, and 68B. *Taq*I or *Eco*RI Southern blots of the available samples of the paternal grandparents, parents, and children from 14 families were probed with all six pseudoautosomal sequences. As an example, segregation of alleles from four different pseudoautosomal loci in paternal meiosis has been analyzed with five different probes tested on DNA digests from a single family (Fig. 1). No recombination is observed with probe 68B (locus *DXYZ2prox*) for which all male offspring harbor the paternal grandfather's allele and similarly all female progeny possess the paternal grandmother's allele. Probes 29Cl and 362A (locus *DXYS14*) detect the grandmaternal allele in male sibs 1, 7, and 10 and the grandpaternal allele in female sibs 4, 6, and 9, which thus all result from a recombination at paternal meiosis. The recombinants detected by probes 601 (locus *DXYS17*) and 113D (locus *DXYS15*) represent only a subset of the recombinants for locus *DXYS14*. So recombination between sex and three of these four loci occurs in this family as schematized on the right-hand part of Figure 1. Segregation of *DXYS14*, *DXYS15*, *DXYS17*, *MIC2*, and *DXYZ2prox* with respect to sexual phenotype has been analyzed in the same way for the 14 studied families, and numbers of recombination events for each locus are scored in Table 1. With the exception of locus *DXYS14*, informative in any family, segregation of the other loci could only be followed in some of the 14 families. It nevertheless appears that all

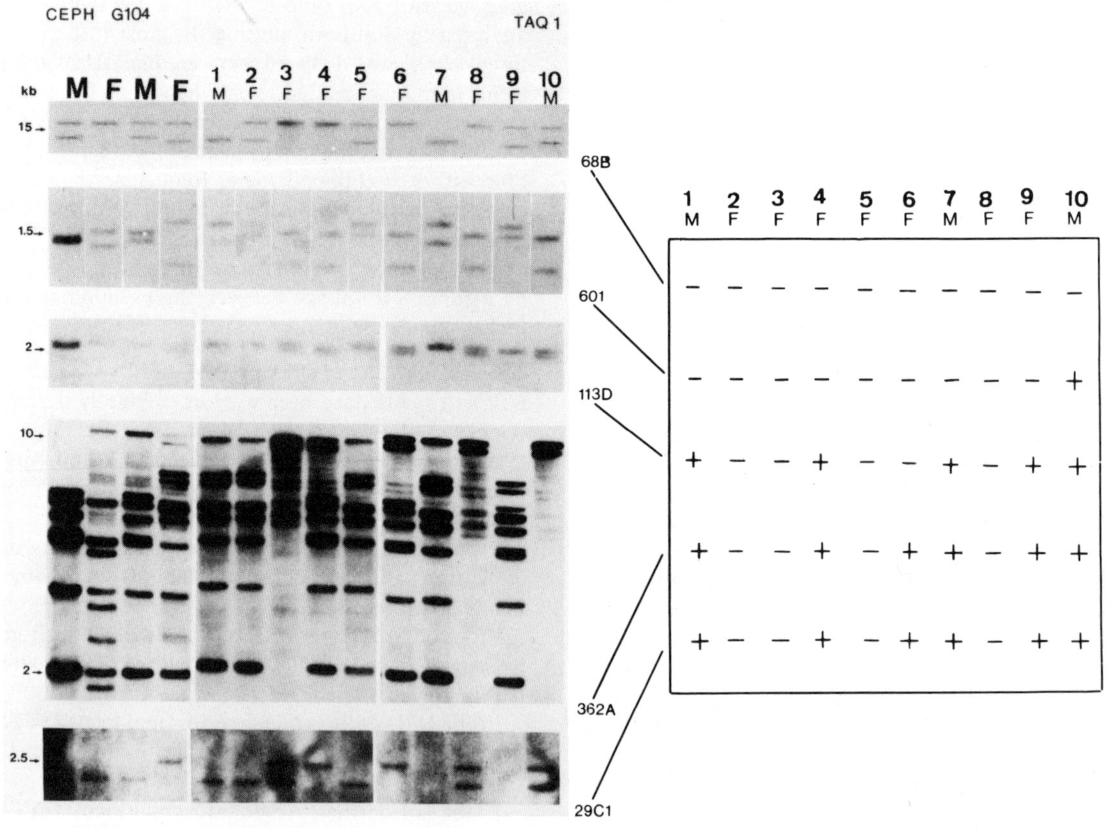

Figure 1. Hybridization patterns of pseudoautosomal probes 68B, 601, 113D, 362A, and 29Cl to Southern blots of *Taq*I restriction digests of DNAs from a single family. (M) Male sample; (F) female sample. (*Left*) Grandparental samples are the two leftmost lanes, followed by parental samples. Samples from the 10 children are marked by smaller letters designating sex and are numbered. (*Right*) Chart showing presence or absence of recombination events between sex and locus; numbers indicate the 10 children in the same order as on the blots. (−) No recombination; (+) recombination. For probes 29Cl, 113D, 601, and pSG1, hybridization has been performed as indicated elsewhere (Cooke et al. 1985; Simmler et al. 1985; Darling et al. 1986; Rouyer et al. 1986). Probe 68B is used in the same conditions as other *DXYZ2* probes 708 and 113F (Simmler et al. 1985). Blots probed with probe 362A are washed at 68°C in the presence of 2× SSC.

Table 1. Sex Linkage of Pseudoautosomal Loci

Loci (probes)	Meioses	Recombinations	Recombination frequency
DXYS14 (29C1, 362A)	114	57	50%
DXYS15 (113D)	82	29	35.5%
DXYS17 (601)	86	10	11.5%
MIC2 (pSG1)	37	1	2.5%
DXYZ2prox (68B)	46	1	2.2%

The recombinations between sexual phenotype and different pseudoautosomal loci are deduced from linkage analysis of male meiosis.

Table 3. Nonoccurrence of Double Recombination Events at Male Meiosis

	Three-point analysis		
Recombinations	type A	type B	type C
DXYS14 exclusively	11	32	/
DXYS15 exclusively	0	/	13
DXYS17 exclusively	/	0	0
Both pseudoautosomal loci	30	10	6
Nonrecombinants	41	44	35
Total meioses	82	86	54

Type A analysis: Segregation of *DXYS14, DXYS15,* and sex.
Type B analysis: Segregation of *DXYS14, DXYS17,* and sex.
Type C analysis: Segregation of *DXYS15, DXYS17,* and sex.

five of the studied loci can be transferred from one sex chromosome to the other and thus meet the pseudoautosomal criterion.

The assumption that each pseudoautosomal locus may reassort independently in male meiosis allows the prediction of a frequency for each of the different possible combinations. However, the transmission of the three most informative loci—*DXYS14*, *DXYS15*, and *DXYS17*—shows that the less frequently recombined locus never segregates independently from the locus recombining more frequently (Table 2). Similarly, recombination of locus *DXYS15* always involves the most frequently recombined locus *DXYS14* (Table 2). In addition, some of the frequencies predicted when loci are in equilibrium are in total discrepancy with the observed values: in type A crosses (Table 2) we would expect, respectively, 18% of recombinants for both loci (0.5×0.355), 32% of recombinants with *DXYS14* ($0.5 \times [1-0.355]$), 18% of recombinants with *DXYS15* ($[1-0.5] \times 0.355$), and 32% of nonrecombined paternal sex chromosomes ($[1-0.5] \times [1-0.355]$). Instead, we observe 37% of recombinants for both loci, 13% of recombinants mobilizing *DXYS14* exclusively, no recombinants with *DXYS15* exclusively, and 50% of nonrecombinants. Similar results can be observed in other three-point analyses B and C. These three-point analyses allow us to eliminate the possibility of an independent reassortment of pseudoautosomal loci in male meiosis. In addition, RFs between these three loci range from 13.5% to 36% in male meiosis (Table 3), indicating an obvious linkage between them. Cosegregation of three independent pseudoautosomal loci is not consistent with recombination via gene conversion, which is supposed to take place on short stretches of DNA and should result in an independent segregation of each locus. On the contrary, linkage of the three loci and absence of independent segregation of the less frequently recombining ones demonstrate the occurrence of crossing-over events within the pseudoautosomal regions of the X and Y chromosomes.

The telomeric locus *DXYS14* recombines with sex at a frequency of 50% and is thus not sex-linked. This complete absence of sex linkage is not observed for the other four pseudoautosomal loci analyzed, all of which display sex linkage to a certain extent. Similar observations have been made for different pseudoautosomal

Table 2. Recombination between Pseudoautosomal Loci at Female and Male Meiosis

Loci (probes)	*DXYS14* (29C1) (362A)	*DXYS15* (113D)	*DXYS17* (601)	*MIC2* (pSG1)	*DXYZ2prox* (68B)
DXYS14 (29C1) (362A)	–	0/70 0%	2/88 2.5%	1/50 2%	2/38 5.3%
DXYS15 (113D)	11/82 13.5%	–	2/62 3.2%	1/31 3.2%	0/22 0%
DXYS17 (601)	32/86 37%	13/54 24%	–	0/29 0%	1/19 5.3%
MIC2 (pSG1)	18/37 48.5%	13/37 35%	2/18 11%	–	1/36 2.8%
DXYZ2prox (68B)	25/46 54.5%	17/46 37%	2/27 7.4%	0/21 0%	–

Fractions represent no. of recombinations vs. no. of informative meioses between loci indicated above in column headings and loci indicated at the same horizontal level in the left column. Percentages represent recombination frequencies (RFs) between these loci. Values in the upper right half of the table are obtained from linkage analysis of female meiosis; values in the lower left half are obtained from analysis of male meiosis.

loci (not shown) and suggest that absence of sex linkage should be restricted to the very distal part of the human pseudoautosomal region. The four nontelomeric loci show considerable variation in their partial sex linkage, which can be intermediate (67% for *DXYS15*), high (89% for *DXYS17*), or very high (97–98% for *MIC2* and *DXYZ2prox*). This sex-linkage gradient can be represented on a linear map where the five loci can be ordered with respect to their recombination distances with the X and Y chromosomal sex-specific blocks (Fig. 2). Since only a single recombinant with sex has been observed for *MIC2* and *DXYZ2prox*, in families which were not simultaneously informative for both loci, they cannot be placed relative to each other on the sole basis of their sex linkage.

Recombination between Pseudoautosomal Loci

Recombination at male meiosis. An RF of 50% for *DXYS14* is consistent with a single and obligatory event occurring between an Xp and a Yp chromatid proximally to the telomere. It could otherwise be explained by multiple events. In this latter possibility one should observe independent segregation of the different loci or at least numerous cases of double recombinations. However, as discussed above, the pseudoautosomal loci do not segregate independently. Moreover, two widely separated loci, *DXYS14* and *DXYS17*, are 37% recombination units apart. In *DXYS17* recombinants, a second crossing-over proximal to the telomeric locus *DXYS14* is theoretically possible in such a recombination interval. No example of double recombination has been found in any of the 10 *DXYS17* recombinants analyzed (Table 3). Similarly, no second event is found in the smaller *DXYS14/DXYS15* interval (13.5%) in the 30 *DXYS15* recombinants. Since double recombination should otherwise be frequent to account for an RF of 50% at the telomere, it is concluded that the crossing-over occurs once in each X/Y bivalent at varying locations. Genetic distances between the pseudoautosomal loci are given as RFs in Table 2. According to these values, *DXYS15* is located within the *DXYS14/DXYS17* interval, and *DXYS17* lies within the *DXYS15/MIC2* interval (see Fig. 2). Strikingly, the sum of the distances between *DXYS14* and *DXYS15* (13.5%) and between *DXYS15* and *DXYS17* (24%) is almost equal to the value of the *DXYS14/DXYS17* (37%) interval. Similarly, the distance between *DXYS14* and *MIC2* (48.5%) nearly equals, on the one hand, the sum of *DXYS14/DXYS17* (37%) and *DXYS17/MIC2* (11%) intervals and, on the other hand, the sum of the *DXYS14/DXYS15* (13.5%) and *DXYS15/MIC2* (35%) intervals. Figure 2 shows a map established from RFs between pseudoautosomal loci and the specific part of the sex chromosomes along with an alternate map representing mean interloci distances as obtained from two- and three-point analyses. The maps compare remarkably to each other, and both illustrate the absence of the classical underestimation of larger genetic distances. Thus the rule of additivity of RFs, which is usually valid for the 0–15% range, appears to apply to the whole pseudoautosomal region, which stretches over a distance of 50% recombination units. This feature characterizes the human pseudoautosomal region and represents additional strong evidence in favor of a single crossing-over per male meiosis between the human sex chromosomes.

Recombination at female meiosis. Out of 100 informative female meioses, 4 result from an XX recombination within the pseudoautosomal region, whereas nearly 50% of recombination has been observed at male meiosis in the same chromosomal interval. This considerable increase in male recombination appears even more markedly in the more distal part of the region, but this tendency must still be confirmed by linkage analysis on a larger scale. However, strikingly, a crossover has been observed within the extremely small interval *MIC2/DXYZ2prox*. In this recombination, *MIC2* cosegregates with the more distal loci and allows the mapping of *DXYZ2prox* proximal to *MIC2*. The higher RF in the pseudoautosomal region at male meiosis contrasts with a significant general increase reported for the autosomal RF in female meiosis as compared with the autosomal RF in male meiosis (Cook 1965; Renwick 1968). In humans, 1% recombination (1 cM) is generally taken to represent a chromosomal distance of 1000 kb (Renwick 1969), and accordingly loci *DXYS14* and *MIC2* (or *DXYZ2prox*), which are separated by 50 cM at male meiosis, should be located some 50,000 kb apart. The actual size of this interval is possibly more than 1 order of magnitude lower as reflected by the RF observed in female meiosis (3–4%). Such an overestimation is consistent with the view that a single crossover between the X and Y chromosomes is an obligatory step in male meiosis. It has been proposed that chiasmata play a part in the disjunction of chro-

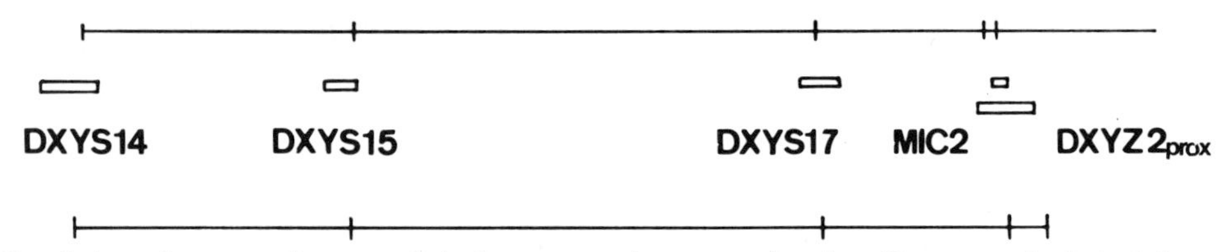

Figure 2. Correlation of two genetic maps of the human pseudoautosomal region. Upper map is derived from the RFs of pseudoautosomal loci with sexual phenotype. Lower map represents mean RFs between pseudoautosomal loci at male meiosis as calculated from two- and three-point linkage analyses. Open boxes indicate intervals within which values obtained from two-point analyses fluctuate.

mosomes (Koller and Darlington 1934) during the first meiotic division and that at least one chiasma takes place in each bivalent to ensure proper segregation. This view may apply to the pseudoautosomal region of the X/Y pair, where the apparent 10-fold to 20-fold increase in male RF is then a direct consequence of a chiasma having to be formed between each chromosomal pair in this limited region.

The *DXYZ2* Repeats

The extent of the pseudoautosomal region remains unknown. According to cytogenetic measurements, the size of the Y-chromosome short arm is generally estimated to range from 5,000 to 10,000 kb, whereas the size of the pairing region is variable and can even stretch beyond Yp (Chandley et al. 1984). Since numerous Yp-specific DNA sequences have no homologous counterpart on Xp, it follows that X/Y synapsis is partly nonhomologous, and no accurate correlation between X/Y pairing and sequence homologies can be drawn. Several tentative estimations suggest a size of 2000–4000 kb.

The *DXYZ2* repeat may be a useful tool in those estimations. Figure 3 shows a hybridization pattern of probe 68B, a *DXYZ2* element, to DNA digests from a human family. Several polymorphic elements (indicated by arrows) mapping to the X and Y chromosomes (not shown) have recombined with sex in a variable number of children, indicating that these elements are located in different areas of the pseudoautosomal region. In addition, about 7–8 *DXYZ2* elements detected under stringent hybridization conditions have been isolated in cosmid and lambda phage clones. In each of these clones, we observe the presence of one *DXYZ2* element only, confirming that most if not all of these elements are interspersed (Simmler et al. 1985). Mapping these *DXYZ2*-positive clones indicates that all are located within the pseudoautosomal region and recombine with sex at different frequencies, confirm-

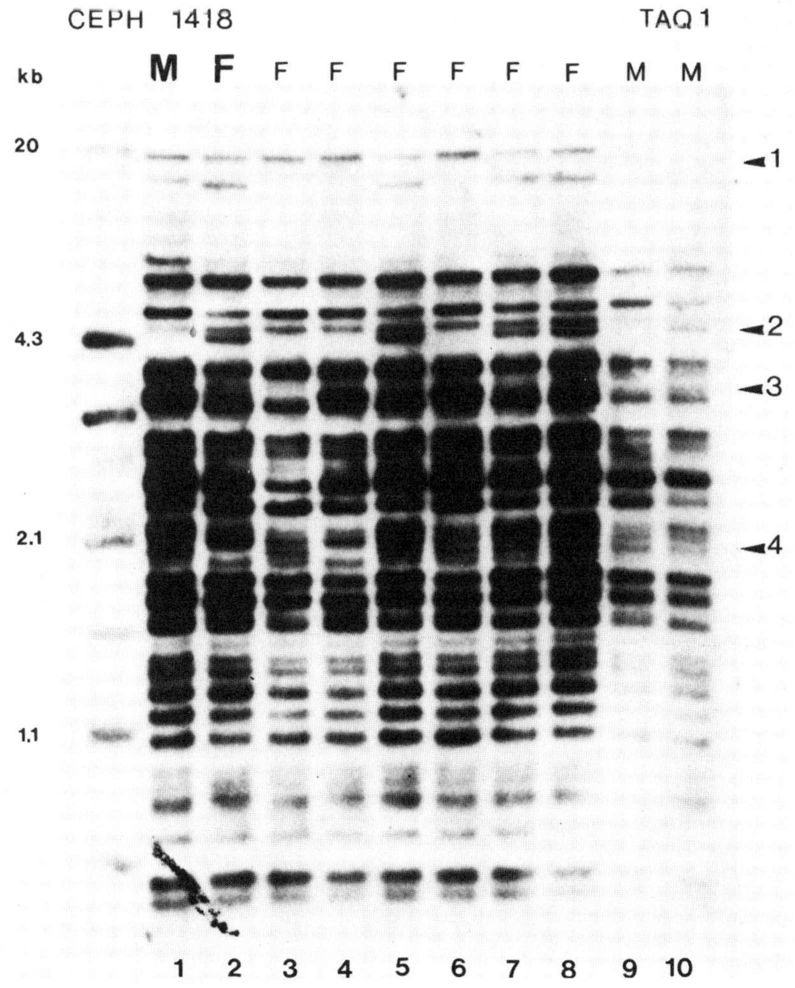

Figure 3. Pseudoautosomal *DXYZ2* repeats in a single family. *DXYZ2* repeats were analyzed with probe 68B, used in the same conditions as probes 708 or 113F (Simmler et al. 1985). Positions of hybridizing bands corresponding to polymorphic loci are indicated by numbered arrowheads. RFLP 1 corresponds to *DXYZ2prox*, which does not recombine with sex in this family. RFLP 2 is not informative in male meiosis but does not recombine with *DXYZ2prox* in female meiosis. Allelic combinations of sibs 2, 4, 9, and 10 result from paternal crossing-over between locus with RFLP 3 and sex. Similarly, linkage analysis for RFLP 4 shows that sibs 2, 9, and 10 result from recombination between this locus and sex.

ing the observations presented in Figure 3. A first element maps in close vicinity of sequence 362A, which recombines at 50% and is located near the telomere. A second element is found within the *DXYS14/DXYS15* interval (results not shown). We showed previously that 113F is part of locus *DXYS15* (Simmler et al. 1985). A fourth element described above, *DXYZ2prox*, maps to a particularly proximal location. It thus appears that *DXYZ2* elements stretch on the entire pseudoautosomal part of the sex chromosomes and represent a structural feature of this region. Summing up the size of the restriction fragments containing the *DXYZ2* repeat, provided these fragments would be both large enough and well-resolved, could thus contribute to a valuable size estimation of the human pseudoautosomal region. This could be achieved, for instance, by probing a Southern blot of DNA digested with a rare cutter restriction enzyme and separated by pulsed-field gel electrophoresis.

It has been mentioned in a previous report that sequences homologous to *DXYZ2* have been found in other mammals (e.g., bovines). The pseudoautosomal location of such sequences has been confirmed recently in great apes (W. Schemmp, pers. comm.) by in situ hybridization. It is tempting to speculate that a repeated element mapping exclusively to this very peculiar chromosomal region and conserved in other species may exert a biological function related to a specific feature of the pseudoautosomal region (i.e., noninactivation, chromosome pairing, or recombination). However, more distantly related elements (*DXYZ2*-like repeats) have been found in clones of human DNA libraries by using less stringent hybridization conditions. Some of these sequences map on autosomes, whereas others have been found in the sex-specific part of the human Y chromosome. These latter elements may represent remnants of a larger ancestral pseudoautosomal region and are presently dispersed in other chromosomal parts. This possibility is consistent with the fact that we have recently found X-specific sequences mapping to Xp22.3-Xpter and homologous to Y-specific sequences adjacent to *DXYZ2*-like repeats (F. Rouyer, unpubl.). This would imply that several important rearrangements of the Y chromosome have resulted in transposition of formerly pseudoautosomal sequences to other parts of the Y chromosome. Such rearrangements are suggested by the location of the chimpanzee pseudoautosomal region to the tip of the long arm. It can also be conceived that human *STS*, as opposed to murine *STS*, has lost its pseudoautosomal character after a rearrangement of the Y chromosome that would have displaced or removed the ancestral *STS* homolog of the Y chromosome.

The Human Pseudoautosomal Region and Human XX Maleness

It has been shown that numerous XX males (Y[+] XX males) have inherited a portion of their father's Y-chromosome short arm (Guellaën et al. 1984; Page et al. 1985; Müller et al. 1986; Vergnaud et al. 1986). In Y(+) cases studied by in situ hybridization, this portion could be mapped to the short-arm tip of one of the X chromosomes (Buckle et al. 1985b; Casanova et al. 1985; Magenis et al. 1985). Thus, most of the human XX males probably result from an abnormal paternal X/Y interchange event (Ferguson-Smith 1966; de la Chapelle et al. 1984), but several features of this process need further clarification. It is not yet established whether this interchange includes the whole terminal part of the Y chromosome or whether it originates through a complex rearrangement of the Y chromosome (Goodfellow et al. 1985; Vergnaud et al. 1986), like a transfer of an interstitial Y-chromosomal segment. Similarly, apart from cases who have lost the paternal Xg(a) allele (de la Chapelle 1981), the occurrence and extent of the deletion (if any) of the short-arm tip of the paternal X chromosome, involved in the interchange (Page and de la Chapelle 1984), is ignored. On the other hand, origin of maleness in XX males without detectable Y-specific sequences (Y[−] XX males) has not yet been defined.

These questions can now be addressed by probing DNAs from XX males with polymorphic pseudoautosomal sequences. To determine the number of copies of pseudoautosomal loci in XX males, 12 cases have been probed with different pseudoautosomal DNA sequences. As shown in Figure 4, 1 Y(+) XX male (lane 5) displays three copies of locus *DXYS15*. There is additional evidence for the presence of three copies of more-proximal loci. However, analysis of more-distal loci demonstrated the presence of two copies only (data not shown). Although parental origin of these loci could not be investigated, the overall results are consistent with the loss of the terminal part of one of the paternal pseudoautosomal regions. In the other 11 cases studied (2 shown in Fig. 4), we have no evidence so far for the presence of a third copy of pseudoautosomal

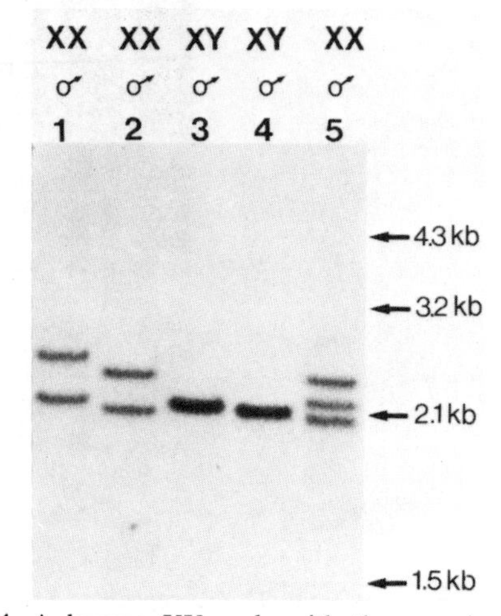

Figure 4. A human XX male with three copies of locus *DXYS15*. Probe 113D has been hybridized to *Taq*I digests of DNA samples from two normal males (lanes *3* and *4*) and three 46,XX males (lanes *1*, *2*, and *5*) as indicated previously (Simmler et al. 1985).

loci. Thus, the abnormal interchange may often result in complete loss of one of the paternal pseudoautosomal regions.

Chromosomal origin of the paternal pseudoautosomal region can be determined in rare instances by analysis of the alleles from one or both paternal grandparents. In Figure 5 we show that a Y(+) XX male has inherited the pseudoautosomal telomere from the paternal Y chromosome and conversely has lost the telomere from the paternal X chromosome. This suggests that a single Y-chromosomal fragment, including both the sex-determining region and Yp telomere, has been translocated to the X chromosome. Other experiments designed to investigate the fate of the paternal X and Y pseudoautosomal regions in both Y(+) and Y(−) XX males are currently in progress and should shed some new light on the etiology of human XX maleness.

CONCLUSIONS

Although the existence of a region common to the human X and Y chromosomes where exchange of genetic material can take place was not unexpected, it has taken 50 years to obtain experimental evidence fulfilling the prediction first made by Koller and Darlington (1934). Acquisition of these data has even been greatly facilitated by the fact that an unusually elevated proportion of the isolated pseudoautosomal probes detects multiallelic polymorphisms. Multiallelic polymorphism generally results from copy-number variation of short tandem repeats or minisatellites probably arising from unequal crossovers (Jeffreys et al. 1985). To test if the multiallelic polymorphisms of the pseudoautosomal region could be generated through similar mechanisms, we sequenced the highly variable DNA segments of loci *DXYS15* and *DXYS17*. In both instances, we could correlate the allelic variations with the presence of short tandemly repeated sequences (M.C. Simmler et al., in prep.). However, these sequences share no similarity with Jeffreys's hypervariable myoglobin minisatellite family nor with any other highly variable, known sequence. The apparent high density of minisatellites in the human pseudoautosomal region may be related to its high recombination activity at male meiosis, which could imply that male pseudoautosomal recombination has some specific features.

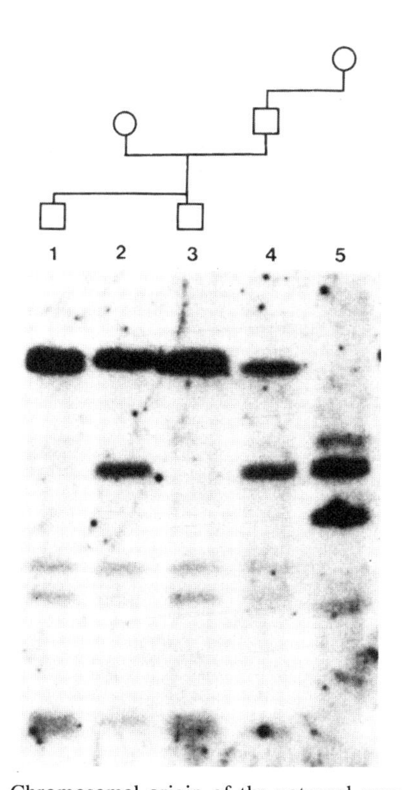

Figure 5. Chromosomal origin of the paternal pseudoautosomal region of a Y(+) human XX male. Hybridization of probe 29Cl to DNAs from several members of a Y(+) XX male family: (lane *1*) brother; (lane *2*) mother; (lane *3*) proband; (lane *4*) father; (lane *5*) paternal grandmother. The paternal allele of the proband is not originating from the paternal grandmother and is thus assigned to the father's Y chromosome.

With the magnifying effect on distance evaluation allowed by X/Y recombination analyses, extensive genetic mapping of the pseudoautosomal region can be achieved using several probes, which cannot be ordered by the more classical X-mapping studies on female meiosis. It would be of interest to reinvestigate whether some of the unassigned genetic diseases exhibit partial sex linkage and hence map close to a pseudoautosomal locus.

Consequences of the existence of pseudoautosomal genes in man have been outlined elsewhere (Ferguson-Smith 1966; Burgoyne 1982), and the question of the evolutionary conservation of such genes has also been raised (Ohno 1979; Burgoyne 1982, 1986). The genetic effect of sex chromosome monosomy (XO condition) is much less dramatic in mouse as compared with the human Turner syndrome and argues against a strong conservation. Similarly, an early replicating segment has been evidenced at the short-arm tip of the human sex chromosomes by BrdU incorporation experiments (Schemmp and Meer 1983) and can be correlated with the pseudoautosomal region (W. Schemmp, pers. comm.). No similar segment is present in the mouse, suggesting that the pseudoautosomal region has been considerably reduced in this latter species. However, an opposite direction of evolution is illustrated by murine steroid sulfatase, which is pseudoautosomal, whereas the human X-linked gene maps close to, but not within, the pseudoautosomal region and is partially inactivated on the second X chromosome. In spite of an apparently important divergence, it remains possible that some other loci require a pseudoautosomal location to fulfill their functions.

ACKNOWLEDGMENTS

We wish to thank the Centre d'Etude du Polymorphisme Humain (CEPH) for the family DNA samples;

Dr. S. Castillo Taucher for providing blood samples from a three-generation family with a sporadic case of XX maleness; Drs. B. Noel, P. Jalbert, V. Amice, M. Fellous, and D. Page for cells or blood samples from XX males; H. Cooke and P. Goodfellow for pseudoautosomal DNA probes; P. Tiollais for encouragement and support; and C. Bishop for reading this manuscript. G.V. is a member of Délégation générale de l'Armement.

REFERENCES

Ashley, T. 1984. A re-examination of the case for homology between the X and Y chromosomes of mouse and man. *Hum. Genet.* **67:** 372.

Buckle, V., C. Mondello, S. Darling, I.W. Craig, and P.N. Goodfellow. 1985a. Homologous expressed genes in the human sex chromosome pairing region. *Nature* **317:** 739.

Buckle, V., Y. Boyd, I.W. Craig, N. Fraser, P.N. Goodfellow, and J. Wolfe. 1985b. Localization of Y chromosomal sequences in normal and XX males. *Cytogenet. Cell Genet.* **40:** 593.

Burgoyne, P.S. 1982. Genetic homology and crossing over in the X and Y chromosomes of mammals. *Hum. Genet.* **61:** 85.

———. 1986. Mammalian X and Y crossover. *Nature* **319:** 258.

Casanova, M., E. Seboun, P. Leroy, C. Junien, I. Henry, E. Boucekkine, E. Magenis, and M. Fellous. 1985. Yp encoded sequences in the short arm of paternal X in human XX males and XX males with true hermaphroditism. *Cytogenet. Cell Genet.* **40:** 600.

Chandley, A.C., P. Goetz, T.B. Hargreave, A.M. Joseph, and R.M. Speed. 1984. On the nature and extent of XY pairing at meiotic prophase in man. *Cytogenet. Cell Genet.* **38:** 241.

Cook, P.J.L. 1965. Lutheran-secretor recombination fraction in man: A possible sex difference. *Am. J. Hum. Genet.* **28:** 393.

Cooke, H.J., W.A.R. Brown, and G. Rappold. 1985. Hypervariable telomeric sequences from the human sex chromosomes are pseudoautosomal. *Nature* **317:** 688.

Darling, S.M., G.S. Banting, B. Pym, J. Wolfe, and P.N. Goodfellow. 1986. Cloning an expressed gene shared by the human sex chromosomes. *Proc. Natl. Acad. Sci.* **83:** 135.

de la Chapelle, A. 1981. The etiology of maleness in XX men. *Hum. Genet.* **58:** 105.

de la Chapelle, A., P. Tippett, G. Wetterstrand, and D. Page. 1984. Genetic evidence of X-Y interchange in a human XX male. *Nature* **307:** 170.

Evans, E.P., M.D. Burtenshaw, and B.M. Cattanach. 1982. Meiotic crossing over between the X and Y chromosomes of male mice carrying the sex-reversing (Sxr) factor. *Nature* **300:** 443.

Ferguson-Smith, M.A. 1966. X-Y chromosomal interchange in the aetiology of true hermaphroditism and of XX Klinefelter's syndrome. *Lancet* **II:** 475.

Goodfellow, P.N., K.E. Davies, and H.H. Ropers. 1985. Report of the committee on the genetic constitution of the X and Y chromosomes. *Cytogenet. Cell Genet.* **40:** 296.

Guellaën, G., M. Casanova, C. Bishop, D. Geldwerth, G. André, M. Fellous, and J. Weissenbach. 1984. Human XX males with Y single-copy DNA fragments. *Nature* **307:** 172.

Holm, P.B. and S.W. Rasmussen. 1983. Human meiosis. VI. Crossing-over in human spermatocytes. *Carlsberg Res. Commun.* **48:** 385.

Jeffreys, A.J., V. Wilson, and S.L. Thein. 1985. Hypervariable "minisatellite" regions in human DNA. *Nature* **314:** 67.

Keitges, E., M. Rivest, M. Siniscalco, and S.M. Gartler. 1985. X-linkage of steroid sulphatase in the mouse is evidence for a functional Y-linked allele. *Nature* **315:** 226.

Koller, P.C. and C.D. Darlington. 1934. The genetical and mechanical properties of the sex chromosomes. 1. *Rattus norvegicus. J. Genet.* **29:** 159.

Magenis, R.E., R. Sheehy, S. Olson, M.G. Brown, M. Casanova, and M. Fellous. 1985. Genes for maleness (TDF) localized to distal one-half of Y chromosome short arm: Evidence from in situ hybridization of a Y-derived single copy DNA probe. *Cytogenet. Cell Genet.* **40:** 686.

Müller, U., M. Lalande, T. Donlon, and S.A. Latt. 1986. Moderately repeated DNA sequences specific for the short arm of the human Y chromosome are present in XX males and reduced in copy number in an XY female. *Nucleic Acids Res.* **14:** 1225.

Ohno, S., ed. 1979. *Major sex determining genes*. Springer, Berlin.

Page, D.C. and A. de la Chapelle. 1984. The parental origin of X chromosomes in XX males determined using restriction fragment length polymorphisms. *Am J. Hum. Genet.* **36:** 565.

Page, D.C., A. de la Chapelle, and J. Weissenbach. 1985. Chromosome Y-specific DNA in related human XX males. *Nature* **315:** 224.

Polani, P.E. 1982. Pairing of X and Y chromosomes, non inactivation of X-linked genes, and the maleness factor. *Hum. Genet.* **60:** 207.

Renwick, J.H. 1968. Ratios of female to male recombination in man. *Bull. Eur. Soc. Hum. Genet.* **2:** 7.

———. 1969. Progress in mapping human autosomes. *Br. Med. Bull.* **25:** 65.

Rouyer, F., M.C. Simmler, C. Johnsson, G. Vergnaud, H. Cooke, and J. Weissenbach. 1986. A gradient of sex linkage in the pseudoautosomal region of the human sex chromosomes. *Nature* **319:** 291.

Schempp, W. and B. Meer. 1983. Cytologic evidence for three human X-chromosomal segments escaping inactivation. *Hum. Genet.* **63:** 171.

Simmler, M.C., F. Rouyer, G. Vergnaud, M. Nyström-Lahti, K.Y. Ngo, A. de la Chapelle, and J. Weissenbach. 1985. Pseudoautosomal DNA sequences in the pairing region of the human sex chromosomes. *Nature* **317:** 692.

Singh, L. and K.W. Jones. 1982. Sex reversal in the mouse (*Mus musculus*) is caused by a recurrent nonreciprocal crossover involving the X and the aberrant Y chromosome. *Cell* **28:** 205.

Solari, A.J. 1980. Synaptonemal complexes and associated structures in microspread human spermatocytes. *Chromosoma* **81:** 315.

Vergnaud, G., D.C. Page, M.C. Simmler, L. Brown, F. Rouyer, B. Noel, D. Botstein, A. de la Chapelle, and J. Weissenbach. 1986. A deletion map of the human Y chromosome based on DNA hybridization. *Am. J. Hum. Genet.* **38:** 109.

Sex Reversal: Deletion Mapping the Male-determining Function of the Human Y Chromosome

D.C. Page
Whitehead Institute for Biomedical Research, Cambridge, Massachusetts 02142

The Developmental Genetics of Sex Differentiation

Mammalian molecular geneticists face the challenge of understanding how a fertilized egg develops into a mature organism, with all of its complex organ systems. Much has been learned about the molecular correlates of differentiation in, for example, hematopoietic cell lineages, but the molecular mechanisms of the development of mammalian organ systems remain virtually unexplored.

In mammals, the reproductive tract will undoubtedly prove to be among the organ systems most amenable to developmental genetic studies. Consider the invertebrates, where developmental genetics has been pursued with greatest success in *Drosophila melanogaster* (fruit fly) and *Caenorhabditis elegans* (nematode). It is no mere coincidence that, in both fruit fly and nematode, the best understood developmental pathway is sex differentiation (Belote et al. 1985; Hodgkin et al. 1985; Maine et al. 1985; Nothiger and Steinmann-Zwicky 1985). The reasons are twofold and they apply to mammals as well as to invertebrates. First, although mutations in the development of most other organ systems (e.g., circulatory system, respiratory tract, liver) will often be lethal, the result of mutations in the development of the reproductive tract is generally sex reversal and infertility. Because they are not lethal, a great number and variety of mutations affecting sex differentiation can be observed. At least 19 distinct Mendelian mutations are known to perturb the pathway of sex differentiation in the human (Wilson and Goldstein 1975). A number of inherited disorders of sex differentiation are also known to occur in the mouse (e.g., Eicher 1982).

A second advantage in the analysis of sex differentiation is that the primary sex-determining signal can often be pinpointed. In both fruit fly and nematode, for instance, the sex-determining signal is the ratio of X chromosomes to autosomes. In most developmental pathways other than sex differentiation, one cannot make such inferences as to the initiating signal.

In mammals, the primary sex-determining signal is the Y chromosome. Positioned at the head of the sex differentiation pathway, the presence or absence of the Y chromosome determines the fate of the indifferent gonad in embryogenesis. XY, XXY, XXXY, and XXXXY embryos develop testes, while X, XX, XXX, and XXXX embryos develop ovaries. In turn, the embryonic testes or ovaries establish, respectively, a male or female hormonal environment. That embryonic hormonal environment determines the remainder of the sex phenotype, including the sex of the internal accessory organs and external genitalia (Jost 1970). The entire sex phenotype—male or female—hinges upon the function of a gene or genes on the Y chromosome.

This Y-borne gene or gene complex is referred to as the testis-determining factor, or *TDF*. Much debate in the field of mammalian sex determination has focused on the nature of this master regulatory gene. A model that has enjoyed widespread acceptance proposes that TDF and the H-Y antigen are synonymous (Wachtel et al. 1975). Another model assumes that the *TDF* gene will be found among a family of evolutionarily conserved, heterogametic-sex-specific DNA sequences (Epplen et al. 1983; Singh et al. 1984). Recent studies cast doubt on both of these hypotheses (McLaren et al. 1984; Kiel-Metzger et al. 1985).

In reality, we have no meaningful biochemical or cell biological insights into the nature or mode of action of TDF. It seems likely that the function of TDF will come to be understood only through cloning of the gene or gene complex. Despite our ignorance as to the biochemical and cell biological events set in motion by TDF, and despite the lack of a selection for TDF function in cell culture, it may well prove possible to clone the *TDF* gene purely on the basis of its genetic or chromosomal map position. In this sense, the search for the *TDF* gene is analogous to the searches for the genes underlying X-linked chronic granulomatous disease, Duchenne muscular dystrophy, Huntington's disease, and cystic fibrosis (all described elsewhere in this volume).

The remainder of this article focuses on two intertwined themes. The first of these is the chromosomal mapping of *TDF*. The second is the chromosomal basis of gonadal sex reversal (e.g., XX males and XY females).

A Deletion Map of the Y Chromosome

How then does one go about determining the precise genetic map position of *TDF* on the Y chromosome? Most of the Y, the only haploid human chromosome, does not participate in meiotic recombination. (This is not true of the small portion of the Y that exhibits "pseudoautosomal" inheritance; Rouyer et al.; Cooke and Smith; Darling et al.; all this volume; D. Page, unpubl.) It is therefore not possible to construct a genetic linkage map of the Y chromosome from recombination frequencies among markers. However, the natural occurrence of a wide variety of structural abnormalities of the Y chromosome suggests the possibility of constructing a deletion map. Indeed, attempts were made to infer the regional location of *TDF* on the human Y chromosome by karyotype-phenotype correlation (Buhler 1980; Davis 1981). Unfortunately, descriptions of structurally abnormal Y chromosomes from chromosome-banding studies are usually of limited precision and accuracy. Such studies left unresolved the debate as to whether *TDF* maps to the short arm (Yp), centromeric region, or long arm (Yq), or whether in fact multiple *TDF* genes might map to both Yp and Yq.

Hybridization with Y-DNA probes – in conjunction with chromosome-banding studies – is a superior method for characterizing Y-chromosome anomalies and hence for constructing a deletion map of the Y chromosome. We have tested more than 80 individuals for the presence of as many as 140 Y-DNA loci by hybridization to Southern (1975) transfers of restriction-digested genomic DNAs. The majority of the persons tested are XX males or XY females (to be described later) or have, as judged by cytogenetic analysis, a structurally abnormal Y chromosome. DNA studies showed that 50 of the individuals tested carry part but not all of the Y chromosome (Page et al. 1985; Disteche et al. 1986b; Vergnaud et al. 1986; D. Page, unpubl.). That is, each of these 50 individuals has some but not all of the Y-specific restriction fragments invariably present in normal (XY) males. Although only 19 of these 50 Y deletions had been detected by chromosome-banding studies, every Y deletion detected by chromosome banding was also revealed by DNA hybridization. The various patterns of Y-DNA loci present in these 50 individuals carrying part but not all of the Y chromosome are, as a group, most simply explained by the 8-interval deletion map shown in Figure 1. (The intervals are numbered according to the 7-interval map described by Vergnaud et al. [1986]. The subdivision of interval 4 into 4A and 4B is based on the additional findings of Disteche et al. [1986b]. Earlier ambiguities regarding the ordering of intervals on Yp [Vergnaud et al. 1986] have largely been resolved [D. Page, unpubl.].) With the exception of only one of these 50 individuals, this map accounts for each case on the basis of a single Y breakpoint; that is, this deletion map reconciles our hybridization data with the presence of a single, contiguous portion of the Y chromosome in all but 1 of these 50 individuals. The strength of the map lies in its internal consistency. There is little or no basis for ordering the 140 Y-DNA loci apart from this hybridization analysis of deleted-Y individuals. Though a few of these DNA loci have been regionally mapped on the Y chromosome by in situ hybridization – the results of which are consistent with the deletion map – the resolving power of deletion mapping is greater.

The 8 deletion intervals shown in Figure 1 have been ordered with respect to each other without reference to cytogenetic findings. However, by correlating the results of the DNA hybridization studies with cytogenetic findings, one can orient this otherwise abstract map with respect to the long and short arms of the Y chromosome. Several males with microscopically detectable deletions of distal portions of Yq and an apparently intact Yp (see XYq^- males in Fig. 1) are among the deleted-Y patients studied (Vergnaud et al. 1986; D. Page, unpubl.). Conversely, as judged by staining of extended (prometaphase) chromosomes, some of the XY females studied (Fig. 1) have deletions of minute portions of Yp and an apparently intact Yq (Disteche et al. 1986b). The DNA hybridization results obtained with these XYq^- males and XYp^- females allow us to assign intervals 1–4A to Yp and 5–7 to Yq. Evidence of two sorts indicates that interval 4B contains the centromere. First, 4B is the only interval present in all deleted but independently segregating Y chromosomes (as in XYq^- males and XYp^- females). Second, Wolfe et al. (1985) localized an alphoid repeated sequence to the Y centromere by in situ hybridization, and deletion analysis maps that repeated sequence to interval 4B (D. Page, unpubl.).

The map shown in Figure 1 provides a simple ac-

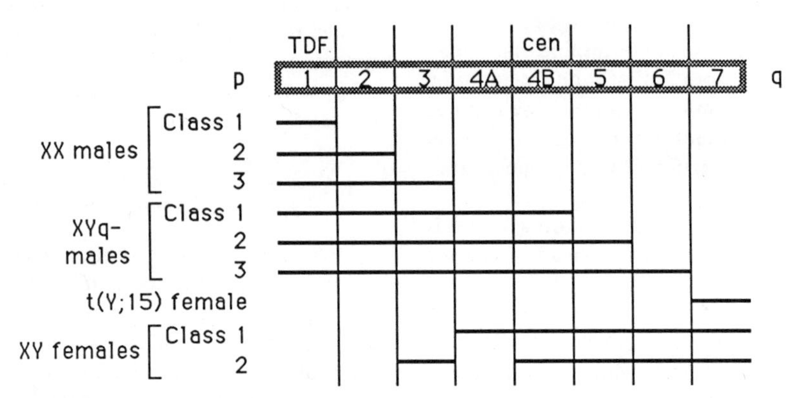

Figure 1. Eight intervals in a deletion map of the Y chromosome, based on DNA hybridization. This map is of the portion of the Y chromosome showing strictly sex-linked (as opposed to pseudoautosomal) inheritance. (p) Short arm; (cen) centromere; (q) long arm. (*TDF*) Testis-determining factor.

counting of all 50 Y deletions except that found in the class 2 XY female, who would appear to carry two noncontiguous portions of the Y chromosome. The class 2 XY female can be more easily accommodated if an implicit assumption underlying the map is relaxed. It has been assumed that the order of intervals on the Y chromosome is invariant among normal males, including the fathers of the patients studied. Suppose, however, that among the population, there exist two different Y chromosomes that differ by an inversion of intervals 3 and 4A. Perhaps such an inversion polymorphism would be easily tolerated on the Y, most of which does not participate in meiotic exchange. Then one might explain the class 3 XX males and class 1 XY females as arising from a Y chromosome of the 2-3-4A-4B form (as shown in Fig. 1) and the class 2 XY female as arising from a Y chromosome of the 2-4A-3-4B variety (with intervals 3 and 4A inverted). This would allow us to account for each of the cases via a single breakpoint on the Y. This dimorphism would also rationalize the finding that, on the 2-3-4A-4B chromosome, *DXYS1*-like sequences are found in two noncontiguous regions (Vergnaud et al. 1986); on the 2-4A-3-4B chromosome, the *DXYS1*-like sequences would occur in a single block (D. Page, unpubl.), consistent with their having transposed from the X as a unit (Page et al. 1984).

A working model of the Y chromosome is shown in Figure 2. Little is known about the physical size of the 8 deletion invervals, except that interval 7, composed largely of Y-specific tandemly repeated sequences (Cooke 1976; Kunkel et al. 1976), accounts for nearly half the chromosome. Intervals 1–4A are on the short arm, the centromere is in interval 4B, and intervals 5–7 are on the long arm. The pseudoautosomal domain is probably distal to all Y-specific sequences assigned to the short arm by deletion mapping (Rouyer et al.; Cooke and Smith; Darling et al.; all this volume; D. Page, unpubl.).

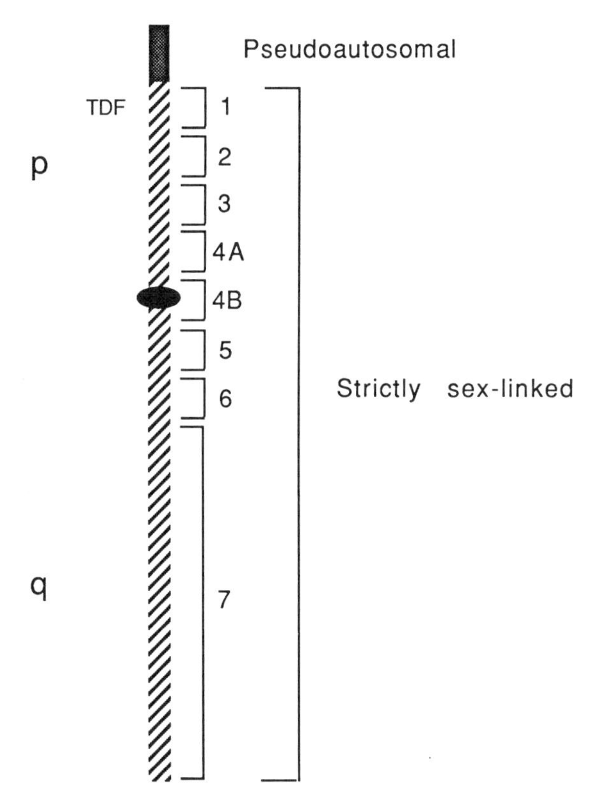

Figure 2. Cartoon of the Y chromosome.

How detailed a map of the Y can one derive from such DNA hybridization studies of naturally occurring deletions? Flow-sorted (Deaven, this volume) and other Y-enriched libraries provide an effectively endless supply of new Y-DNA hybridization probes. To be useful in deletion mapping, probes need not be absolutely Y-specific in their pattern of hybridization, so long as they detect one or more Y-specific restriction fragments on Southern transfers. Accordingly, one is ultimately limited by the distribution of endpoints among spontaneously deleted Y chromosomes – and by one's ability to screen for such deletions among the population. Many such deletions are ascertained because of resulting sterility or other abnormalities of sexual development. Deletions without marked phenotypic consequence will go unnoticed unless they occur at sufficient frequency to be detected in surveys of normal populations. Y deletions can also be produced in tissue culture (Darling et al., this volume), but then one does not have the opportunity to examine the organismal phenotype of the deletion. Recent studies suggest that characterization of an expanded set of deleted-Y patients with a growing number of Y-DNA probes will yield a substantially refined map. Intervals 1, 3, and 6, for example, have recently been subdivided (D. Page, unpubl.).

Mapping *TDF* and Other Genes

Since it is based on the results of hybridization with cloned Y-chromosomal DNA, this deletion map is quite revealing with respect to the organization of Y-specific repeated sequences and of X-Y homologous sequences within the Y chromosome (Vergnaud et al. 1986). Nonetheless, the principal reason for constructing such a map is to facilitate our understanding the biological roles of Y-borne genes. Foremost among these is the male-determining function of the Y.

Is the male-determining function of the Y chromosome the responsibility of a single gene or of several? If there are multiple male-determining genes on the Y, are they redundant, such that any one alone can initiate testis differentiation, or do they act in concert? Our results to date are consistent with there being a single *TDF* gene or gene complex, located on Yp (the short arm). The DNA hybridization results obtained with XX males (Fig. 1) suggest that interval 1 of the Y chromosome is *sufficient* to induce testis differentiation. Indeed, among the patients in whom we have detected part but not all of the Y chromosome, all those with testes carry interval 1 of the Y chromosome. If more than one Y-borne gene is required to initiate testis differentiation, then those genes must all be located within interval 1. Conversely, the DNA hybridization results obtained with XY females (Fig. 1) suggest that interval 1 of the Y chromosome is *necessary* to induce testis differentiation. That is, among the patients in whom we have detected part but not all of the Y chromosome, all those lacking testicular tissue also lack interval 1. (Some XY females have recently been found to have deletions smaller than those shown in Fig. 1; D. Page, unpubl.) The male-determining function of the human Y chromosome, then, appears to reside entirely within a gene or gene complex found in interval 1, on Yp (Figs. 1 and 2).

This deletion map may also be of use in resolving other long-standing controversies regarding the Y chromosome. In particular, what is the function of the H-Y antigen? H-Y is a male-specific, minor histocompatibility antigen, first identified by graft rejection (Eichwald and Silmser 1955). Although it has been proposed that TDF and H-Y antigen are synonymous (Wachtel et al. 1975), H-Y does not appear to be required for testicular differentiation in mice (McLaren et al. 1984). A reliable method for H-Y typing of human B-cell lines using a cytotoxic-T-cell assay has been developed (Goulmy et al. 1983; Simpson 1986). H-Y typing of B-cell lines from patients with well-characterized Y deletions may allow one to unambiguously assign a gene responsible for H-Y antigen expression to a particular deletion interval. If H-Y antigen and TDF are one and the same, then this *H-Y* gene must map to interval 1 of the Y chromosome, where *TDF* resides (Fig. 2). If the *H-Y* gene maps outside interval 1, then it is not TDF. If H-Y antigen is required for spermatogenesis (Burgoyne et al. 1986), then one might expect it to map to one of the intervals on Yq, portions of which may be necessary for male fertility (Tiepolo and Zuffardi 1976).

The presence of a Y chromosome in a gonadal female appears to strongly predispose her to gonadal neoplasia, particularly gonadoblastoma and dysgerminoma (Manuel et al. 1976). If this predisposition is the result of a single Y-borne gene or gene complex, then Y-DNA studies of females with these neoplasms and deleted Y chromosomes should allow one to map the locus. The presence of gonadoblastoma in a class 2 XY female (case 2 described by Disteche et al. 1986b; Fig. 1) suggests that this "gonadoblastoma locus," if it exists, is somewhere in intervals 3, 4B, 5, 6, or 7. It appears, then, that *TDF* itself is not the gonadoblastoma gene.

XX Males and XY Females: The X-Y Interchange Model

Two human "sex reversal" syndromes have been mentioned but not properly introduced. "XX males" are sterile but otherwise phenotypically male individuals whose karyotype is 46,XX (de la Chapelle 1981). They have testicular but no ovarian tissue. As judged by standard cytogenetic methods, they have the chromosomes of a normal female. "XY females" are sterile but otherwise phenotypically female individuals whose karyotype is 46,XY. They have "streak" ovaries devoid of follicles and no testicular tissue. The internal accessory structures are female. Secondary sexual characteristics are female but variably developed, and the somatic features of Turner syndrome (a phenotype classically associated with a 45,X karyotype) are present in some. As judged by routine cytogenetic analysis, XY females have the chromosomes of a normal male, with no evidence of mosaicism for a 45,X cell line. Thus, in both XX males and XY females, there is a discordancy between the gonadal sex and the chromosomal sex—as least as judged by chromosome banding. Given the Y-DNA hybridization findings already described, XX males and many XY females no longer represent exceptions to the rule that the Y chromosome is male-determining. Rather, they provide the basis for a stronger, refined rule: Interval 1 of the Y chromosome is male-determining.

Ferguson-Smith (1966) proposed that human XX males are the result of aberrant X-Y interchanges occuring during paternal meiosis, when terminal portions of the X and Y chromosomes pair (Fig. 3). According to this X-Y interchange hypothesis, an XX male inherits from his father an X chromosome whose terminus has been replaced by a *TDF*-bearing portion of the Y chromosome. Ferguson-Smith suggested that XY females might also result from aberrant X-Y interchanges. I will argue that this X-Y interchange hypothesis continues to provide a good working model on which to base investigations of XX males and XY fe-

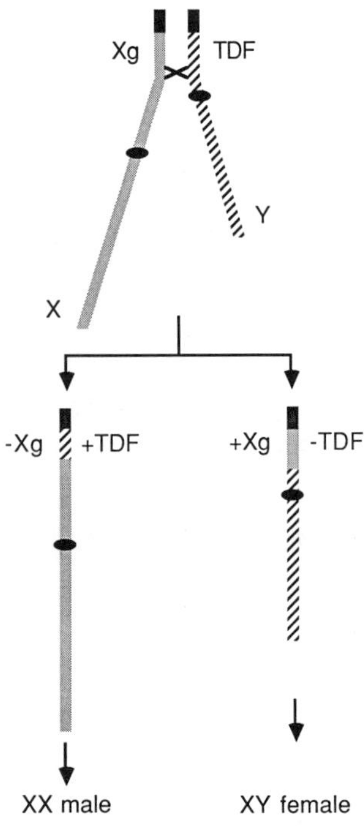

Figure 3. The X-Y interchange hypothesis: XX males and XY females receive reciprocal products of similar aberrant X;Y translocations occurring during or prior to paternal meiosis.

males. Specifically, our recent findings suggest that XX males and XY females may carry reciprocal products of similar X;Y translocations (Fig. 3). I will also outline additional predictions of the model that have not yet been tested.

The X-Y interchange hypothesis was motivated in large part by the anomalous inheritance in XX males of *Xg*, a dominant, X-linked antigenic marker. Many an XX male does not express his father's allele for *Xg* (de la Chapelle 1981), as if he has inherited two X chromosomes from his mother and none from his father. According to the X-Y interchange model, however, such an XX male has received a paternal X chromosome or, more precisely, a paternal X;Y translocation product that lacks *Xg* but carries *TDF* (Fig. 3).

To determine the parental origin of the X chromosomes in XX males, the inheritance of several X-linked restriction-fragment-length polymorphisms (RFLPs) was followed in 10 XX males and their immediate relatives. While *Xg* maps to distal Xp, the X-linked RFLPs used to establish the parental origin of the X chromosomes map to Xq or to more proximal portions of Xp. In each of these 10 families, the XX male was found to inherit a paternal and a maternal X-linked RFLP allele (Page and de la Chapelle 1984 and unpubl.). Among these XX males were 2 who were known not to express their father's alleles for *Xg*. Thus, most if not all XX males inherit a paternal X and a maternal X, as normal 46,XX females do.

If XY females received the reciprocal product of similar X;Y translocations, then many an XY female should express her father's allele for *Xg*—despite having a single, maternally derived X chromosome. The only informative matings will be those in which the XY female's father is phenotypically Xg(a$^+$) and the mother Xg(a$^-$). Given the frequencies of the *Xg* alleles, less than 10% of matings will be of this sort. Unfortunately but not surprisingly, then, the two XY females shown to have deletions of *TDF* were products of matings uninformative for *Xg* (Disteche et al. 1986b).

The X-Y interchange model predicts, of course, the presence of Y-chromosomal material in the genomes of XX males. DNA and antigenic marker studies provided direct evidence of such Y material in XX males. Guellaen et al. (1984) detected certain Y-specific DNA sequences in three XX males by Southern (1975) blotting. We described an XX male who expressed his father's allele for 12E7, a Y-linked antigenic marker; he did not express his father's allele for Xg (de la Chapelle et al. 1984).

It is the distal short arms of the X and Y chromosomes that pair in male meiosis (Fig. 3; Chandley et al. 1984). According to the X-Y interchange model, then, the Y material present in XX males should originate from distal Yp. As judged by DNA hybridization, XX males do appear to carry terminal portions of Yp, albeit of varying size (Fig. 1). Conversely, the model would predict the absence of terminal portions of Yp in XY females. Class 1 XY females do appear to lack terminal portions of Yp (Fig. 1). Indeed, the deletion in class 1 XY females corresponds to the portion of the Y present in class 3 XX males, suggesting that class 1 XY females and class 3 XX males carry reciprocal products of similar recombination events. If, as earlier postulated, the class 2 XY female's father has a Y chromosome in which intervals 3 and 4A are inverted, then she too lacks a terminal portion of Yp.

The X-Y interchange model predicts that the Yp DNA found in XX males would be transferred to the distal short arm of an X chromosome. This prediction was confirmed by the results of in situ hybridization with a probe detecting Y-specific repeated sequences present in class 3 XX males. In all three class 3 XX males tested, this Y-specific probe hybridized unambiguously to the most distal portion of Xp (Andersson et al. 1986).

According to the X-Y interchange model, XX males should be hemizygous for some strictly X-linked (as opposed to pseudoautosomal) DNA sequences on distal Xp (Fig. 3). At such a locus, the single copy present in an XX male should be of maternal origin. We have not found any X-specific sequences for which XX males are hemizygous, but these negative results do not disprove their existence. The simplest explanation for many XX males not expressing their fathers' alleles for *Xg* is hemizygosity for the *Xg* locus. However, in the absence of a DNA probe for *Xg*, it remains possible that, in many XX males, the paternal *Xg* locus is present but is not expressed because of the position effect of a nearby translocation breakpoint.

Conversely, the model predicts, that, despite having only one X chromosome, many an XY female should have two copies of some strictly X-linked DNA sequences from distal Xp. At such a locus, one copy should be of maternal origin and the other of paternal origin. Further, the paternally derived copy should be transferred to distal Yp.

If each XX male is the result of a single Xp-Yp crossover proximal to the pseudoautosomal region (Fig. 3), then XX males should carry pseudoautosomal sequences derived from their fathers' Y chromosomes but not from their fathers' X chromosomes. Conversely, XY females should carry pseudoautosomal sequences derived from their fathers' X chromosomes but not from their fathers' Y chromosomes. Highly informative pseudoautosomal RFLPs (Rouyer et al.; Cooke and Smith; Darling et al.; all this volume; D. Page, unpubl.) should allow these predictions of the model to be tested.

The *Sxr* mutation in mice results in sex reversal, and it provides an interesting contrast to the phenomenon of human XX males. The *Sxr* strain is characterized by a high frequency of sterile XX males. The trait is transmitted by carrier males, in whom a duplicate of the testis-determining locus is found in the pseudoautosomal region of the Y chromosome. Meiotic recombination regularly transfers this pseudoautosomal copy of the testis-determining gene from the abnormal Y to the X, resulting in XX males (Burgoyne 1982; Evans et al. 1982; Singh and Jones 1982). Thus, sex reversal in *Sxr* mice is due to frequent recombination between an aberrant Y chromosome and a normal X chromosome.

In contrast, human sex reversal appears to be due to aberrant recombination between normal X and Y chromosomes. We have studied the parents and siblings of many XX males by Southern hybridization using Y-DNA probes (Page et al. 1985; Vergnaud et al. 1986; D. Page, unpubl.). No Y DNA was detected in mothers or sisters, and the Y chromosomes of fathers and unaffected brothers appeared to be normal. In particular, fathers do not carry a duplication of the male-determining region that they transmit to their XX male sons. These results were particularly notable in the case of a family with three XX males (Page et al. 1985), in which one might have suspected a mechanism more like that seen in *Sxr* mice. Similarly, no Y DNA has been detected in the mothers of XY females with Y deletions, and their fathers appear to have normal Y chromosomes as judged by DNA hybridization (Disteche et al. 1986b; D. Page unpubl.). Thus, each such human XX male and XY female appears to be the result of a new mutation. This is particularly remarkable given that one in 20,000 males is an XX male (de la Chapelle 1981); this constitutional translocation is generated anew at a high frequency.

Unexplained Sex Reversal

XX males and XY females are not the only examples of apparent inconsistency between chromosomal sex and gonadal sex. X males have testes but are sterile, and they have a 45,X karyotype. As judged by Y-DNA studies, one X male is mosaic for a Y-bearing cell line (de la Chapelle et al. 1986), while another has a cytogenetically undetected translocation of a *TDF*-bearing portion of the Y to chromosome 15 (Disteche et al. 1986a). Not all cases of gonadal sex reversal, however, are due to anomalies of the Y chromosome. For example, several families have been described in which there are multiple XY females with streak ovaries. (The XY females in whom we detected Y deletions are all sporadic cases.) In most of these families, sex reversal is inherited in an X-linked or autosomal manner (e.g., see Espiner et al. 1970), suggesting mutations in genes that function downstream of *TDF* in the pathway of gonadal differentiation; the Y chromosome is probably intact. To be sure, autosomal mutation can cause gonadal sex reversal in XY mice (Washburn and Eicher 1983). In addition, there is at least one X male and several XX males and XX hermaphrodites (testicular and ovarian tissue) in whom no Y DNA has been detected (de la Chapelle et al. 1986; Vergnaud et al. 1986; D. Page, unpubl.). Some of these individuals may yet prove to have very small, *TDF*-bearing portions of the Y chromosome. On the other hand, perhaps some of these individuals do not carry *TDF* but have testicular tissue because of a gain-of-function mutation in a gene that acts downstream of *TDF*.

SUMMARY

An 8-interval deletion map of the human Y chromosome has been constructed using DNA hybridization to characterize naturally occurring structural abnormalities. The map is oriented with respect to the short and long arms, and the position of the centromere is known. The deletion map may be dimorphic; that is, within the normal population there may occur two Y chromosomes that differ by an inversion within the short arm.

Y-DNA studies have reconciled many cases of gonadal "sex reversal" with the rule that the human Y chromosome is male-determining. Gonadal sex is determined according to the presence or absence of interval 1 which is on the short arm of the Y chromosome. XX males are the result of X;Y translocations that occur during or prior to meiosis in the fathers. Some XY females may carry the reciprocal product of similar X;Y translocations. Not all cases of gonadal sex reversal can be explained on the basis of Y-chromosome aberrations, and these provide evidence of genes that function downstream of *TDF* in the pathway of gonadal differentiation.

ACKNOWLEDGMENTS

I am indebted to my co-workers, and in particular to David Botstein, Laura Brown, Albert de la Chapelle, Christine Disteche, and Jean Weissenbach. This work was supported in part by grant HD20059 from the National Institutes of Health.

REFERENCES

Andersson, M., D.C. Page, and A. de la Chapelle. 1986. Chromosome Y-specific DNA is transferred to the short arm of X chromosome in human XX males. *Science* **33:** 786.

Belote, J.M., M.B. McKeown, D.J. Andrew, T.N. Scott, M.F. Wolfner, and B.S. Baker. 1985. Control of sexual differentiation in *Drosophila melanogaster. Cold Spring Harbor Symp. Quant. Biol.* **50:** 605.

Buhler, E.M. 1980. A synopsis of the human Y chromosome. *Hum. Genet.* **55:** 145.

Burgoyne, P.S. 1982. Genetic homology and crossing over in the X and Y chromosomes of mammals. *Hum. Genet.* **61:** 85.

Burgoyne, P., E.R. Levy, and A. McLaren. 1986. Spermatogenic failure in mice lacking H-Y antigen. *Nature* **320:** 170.

Chandley, A.C., P. Goetz, T.B. Hargreave, A.M. Joseph, and R.M. Speed. 1984. On the nature and extent of XY pairing at meiotic prophase in man. *Cytogenet. Cell Genet.* **38:** 241.

Cooke, H. 1976. Repeated sequence specific to human males. *Nature* **262:** 182.

Davis, R.M. 1981. Localization of male determining factors in man: A thorough review of structural anomalies of the Y chromosome. *J. Med. Genet.* **18:** 161.

de la Chapelle, A. 1981. The etiology of maleness in XX men. *Hum. Genet.* **58:** 105.

de la Chapelle, A., P.A. Tippett, G. Wetterstrand, and D. Page. 1984. Genetic evidence of X-Y interchange in a human XX male. *Nature* **307:** 170.

de la Chapelle, A., D.C. Page, L. Brown, U. Kaski, T. Parvinen, and P.A. Tippett. 1986. The origin of 45,X males. *Am. J. Hum. Genet.* **38:** 109.

Disteche, C.M., L. Brown, H. Saal, C. Friedman, H.C. Thuline, D.I. Hoar, R.A. Pagon, and D.C. Page. 1986a. Molecular detection of translocation (Y;15) in a 45,X male. *Hum. Genet.* (in press).

Disteche, C.M., M. Casanova, H. Saal, C. Friedman, V. Sybert, J. Graham, H. Thuline, D.C. Page, and M. Fellous. 1986b. Small deletions of the short arm of the Y chromosome in 46,XY females. *Proc. Natl. Acad. Sci.* **83:** 7841.

Eicher, E.M. 1982. Primary sex determining genes in mice. In *Prospects for sexing mammalian sperm* (ed. R.P. Amann and G.E. Seidel, Jr.), p. 121. Colorado Associated University Press, Boulder.

Eichwald, E.M. and C.R. Silmser. 1955. Communication. *Transplant. Bull.* **2:** 148.

Epplen, J.T., A. Cellini, S. Romero, and S. Ohno. 1983. An attempt to approach the molecular mechanisms of primary sex determination: W- and Y-chromosomal conserved simple repetitive DNA sequences and their differential expression in mRNA. *J. Exp. Zool.* **228:** 305.

Espiner, E.A., A.M.O. Veale, V.E. Sands, and P.H. Fitzgerald. 1970. Familial syndrome of streak gonads and normal male karyotype in five phenotypic females. *N. Engl. J. Med.* **283:** 6.

Evans, E.P., M.D. Burtenshaw, and B.M. Cattanach. 1982. Meiotic crossing-over between the X and Y chromosomes of male mice carrying the sex-reversing (Sxr) factor. *Nature* **300:** 443.

Ferguson-Smith, M.A. 1966. X-Y chromosomal interchange in the aetiology of true hermaphroditism and of XX Klinefelter's syndrome. *Lancet* **II:** 475.

Goulmy, E., A. van Leeuwen, E. Blokland, E.S. Sachs, and J.P.M. Geraedts. 1983. The recognition of abnormal sex chromosome constitution by HLA-restricted anti-H-Y cytotoxic T cells and antibody. *Immunogenetics* **17:** 523.

Guellaen, G., M. Casanova, C. Bishop, D. Geldwerth, E. Andre, M. Fellous, and J. Weissenbach. 1984. Human XX males with Y single-copy DNA fragments. *Nature* **307:** 172.

Hodgkin, J., T. Doniach, and M. Shen. 1985. The sex determination pathway in the nematode *Caenorhabditis elegans*: Variations on a theme. *Cold Spring Harbor Symp. Quant. Biol.* **50:** 585.

Jost, A. 1970. Hormonal factors in the sex differentiation of the mammalian foetus. *Philos. Trans. R. Soc. Lond. B* **259:** 119.

Kiel-Metzger, K., G. Warren, G.N. Wilson, and R.P. Erickson. 1985. Evidence that the human Y chromosome does not contain clustered DNA sequences (Bkm) associated with heterogametic sex determination in other vertebrates. *N. Engl. J. Med.* **313:** 242.

Kunkel, L.M., K.D. Smith, and S.H. Boyer. 1976. Human Y-chromosome-specific reiterated DNA. *Science* **191:** 1189.

Maine, E.M., H.K. Salz, P. Schedl, and T.W. Cline. 1985. Sex-lethal, a link between sex determination and sexual differentiation in *Drosophila melanogaster. Cold Spring Harbor Symp. Quant. Biol.* **50:** 595.

Manuel, M., P.K. Katayama, and H.W. Jones. 1976. The age of occurrence of gonadal tumors in intersex patients with a Y chromosome. *Am. J. Obstet. Gynecol.* **124:** 293.

McLaren, A., E. Simpson, K. Tomonari, P. Chandler, and H. Hogg. 1984. Male sexual differentiation in mice lacking H-Y antigen. *Nature* **312:** 552.

Nothiger, R. and M. Steinmann-Zwicky. 1985. A single principle for sex determination in insects. *Cold Spring Harbor Symp. Quant. Biol.* **50:** 615.

Page, D.C. and A. de la Chapelle. 1984. The parental origin of X chromosomes in XX males determined using restriction fragment length polymorphisms. *Am. J. Hum. Genet.* **36:** 565.

Page, D.C., A. de la Chapelle, and J. Weissenbach. 1985. Chromosome Y-specific DNA in related human XX males. *Nature* **315:** 224.

Page, D.C., M.E. Harper, J. Love, and D. Botstein. 1984. Occurrence of a transposition from the X-chromosome long arm to the Y-chromosome short arm during human evolution. *Nature* **311:** 119.

Simpson, E. 1986. The H-Y antigen and sex reversal. *Cell* **44:** 813.

Singh, L. and K.W. Jones. 1982. Sex reversal in the mouse (*Mus musculus*) is caused by a recurrent nonreciprocal crossover involving the X and an aberrant Y chromosome. *Cell* **28:** 205.

Singh, L., C. Phillips and K.W. Jones. 1984. The conserved nucleotide sequences of Bkm, which define *Sxr* in the mouse, are transcribed. *Cell* **36:** 111.

Southern, E.M. 1975. Detection of specific sequences among DNA fragments separated by gel electrophoresis. *J. Mol. Biol.* **98:** 503.

Tiepolo, L. and O. Zuffardi. 1976. Localization of factors controlling spermatogenesis in the non-fluorescent portion of the human Y chromosome long arm. *Hum. Genet.* **34:** 119.

Vergnaud, G., D.C. Page, M.-C. Simmler, L. Brown, F. Rouyer, B. Noel, D. Botstein, A. de la Chapelle, and J. Weissenbach. 1986. A deletion map of the human Y chromosome based on DNA hybridization. *Am. J. Hum. Genet.* **38:** 330.

Wachtel, S.S., S. Ohno, G.C. Koo, and E.A. Boyse. 1975. Possible role for H-Y antigen in the primary determination of sex. *Nature* **257:** 235.

Washburn, L.L. and E.M. Eicher. 1983. Sex reversal in XY mice caused by dominant mutation on chromosome 17. *Nature* **303:** 338.

Wilson, J.D. and J.L. Goldstein. 1975. Classification of hereditary disorders of sexual development. *Birth Defects Orig. Artic. Ser.* **11:** 1.

Wolfe, J., S.M. Darling, R.P. Erickson, I.W. Craig, V.J. Buckle, P.W.J. Rigby, H.F. Willard, and P.N. Goodfellow. 1985. Isolation and characterization of an alphoid centromeric repeat family from the human Y chromosome. *J. Mol. Biol.* **182:** 477.

A Molecular Approach to the Study of the Human Y Chromosome and Anomalies of Sex Determination in Man

E. SEBOUN,* P. LEROY,* M. CASANOVA,* E. MAGENIS,† C. BOUCEKKINE,‡ C. DISTECHE,§ C. BISHOP,* AND M. FELLOUS*

*Unité d'Expression des Gènes du Complexe Majeur d'Histocompatibilité, INSERM U. 276, Institut Pasteur 75724 Paris Cédex, France; †Department of Medical Genetics, Crippled Children's Division, University Hospitals Cytogenetics Laboratory, Oregon Health Sciences University, Portland, Oregon 97201; ‡Faculté de Médecine d'Alger, Centre Hospitalier et Universitaire Bologhine, Bains Romains, Alger, Algérie; §Department of Pathology, University of Washington, Seattle, Washington 98195

The primary event of sex determination is the differentiation of the indifferent fetal gonad into a testis or ovary. The simplest hypothesis is that a dominant Y-linked gene(s)—the testis-determining gene (*TDY*)—encodes a single factor, the testis-determining factor (TDF), which induces testicular development. Of course, more-complex models can be envisaged, and the postulated *TDY* has not been accurately mapped or cloned. For this purpose, the construction of genomic libraries from the human or mouse Y chromosome has been achieved by several groups (Bishop et al. 1984, 1985; Wolf et al. 1985). Such libraries have allowed a rapid advance in understanding the structure of the human Y chromosome.

The use of random probes derived from these libraries to analyze two anomalies of human sex determination—XX males and XY females—has been particularly informative in this respect (Bühler 1980; de la Chapelle 1981). Several hypotheses have been proposed to explain the occurrence of XX males (de la Chapelle et al. 1964; Ferguson-Smith 1966; Evans et al. 1979). These include undetected low-level mosaicism, mutation of autosomal or X-linked genes involved in sex determination, and the presence of cryptic Y-chromosomal material in the otherwise XX genome. The existence of some familial cases could argue in favor of the autosomal mutation hypothesis. However, direct detection of Y-specific DNA sequences in XX males demonstrates that one of the primary causes of XX maleness is the transfer of primary sex-determining gene(s) into the female genome (Guellaen et al. 1984). It has been postulated that this transfer is due to X-Y interchange during male meiosis (Ferguson-Smith 1966). The work presented here defines the precise origin of the Y-DNA sequences detected in XX males, allowing the mapping of the *TDY* locus on the Y chromosome. We also present evidence that this Y-chromosomal material is localized to the tip of the paternal X chromosome, which is consistent with X-Y interchange during male meiosis.

The use of DNA probes derived from the human or mouse Y-chromosome library is also a powerful way to study genes expressed from the Y chromosome.

To date, approximately 3000 genetic loci have been identified in the human genome (McKusick 1983). If we consider that the number of genes is proportional to the DNA content and that the euchromatic part of the Y represents less than 1% of the haploid genome, about 30 genes should be on the human Y chromosome.

In fact, no more than 5 loci have been so far assigned to the Y. Among them, 3 are Y specific: The *TDY*, the *H-Y* histocompatibility gene(s), and the gene(s) controlling spermatogenesis. Two other characterized genes (Goodfellow et al. 1985) are shared by the X and Y chromosomes: *MIC2*, encoding the 12E7 cell-surface antigen, and *Yg*, a regulatory gene from *Xg* and the 12E7 antigen. Studies of Y-chromosome deletions in man suggest that this *TDY* gene(s) is located on the pericentric region of the short arm of the human Y chromosome. Similar studies have localized gene(s) necessary for normal spermatogenesis on the long arm of the euchromatic region. However, in the mouse, it has been shown that autosomal gene(s) are also involved in primary sex determination (Eicher et al. 1982, 1983) and function in a coordinate fashion with Y sex-determining genes.

We also present several recent findings concerning the use of Y-chromosome probes to detect transcribed sequences.

EXPERIMENTAL PROCEDURES

Human cell lines. Peripheral blood or lymphoblastoid cell lines were from 46,XX males, 46 XYp⁻ females with features of Turner syndrome (Magenis et al. 1984a,b); one 46,X psu dic (Y) (q11,22) male (Magenis 1982); and 46,XYq⁻ male patients with a nonfluorescent Y chromosome broken at different regions within the distal part of the Yq euchromatin (Bühler 1980; Magenis and Donlon 1982); as well as from 49, XYYYY and 49,XXXXY cell lines taken as controls (Bishop et al. 1983). 3E7 is a human/mouse somatic cell hybrid containing a modal number of 4 human Y chromosomes (Marcus et al. 1976). Tera I and Tera II are undifferentiated human male teratocarinoma cell lines, and PA1 is an ovarian carcinoma cell line (Avner et al. 1981). Human testes were kindly provided by Prof.

Cukier (Hôpital Necker, Paris) from prostatic cancer patients. Human ovaries were obtained from a 46-year-old patient with uterine fibroma. These human tissues were frozen in liquid nitrogen immediately after surgical resection.

Northern and Southern blots. RNAs from cells and tissues were prepared by the LiCl/urea precipitation method (Auffray and Rougeon 1980). Poly(A)$^+$ RNAs were selected on a poly(U) Sephadex column. Poly(A)$^+$ RNA (5 μg) was fractionated through a vertical 1.5% agarose gel containing 10 mM methyl mercuric hydroxide and subsequently transferred to a nitrocellulose membrane (Bailey and Davison 1976). Hybridization was as for Southern blots (see below).

DNA was prepared from lymphocytes or cell lines by standard procedures; 15 μg was then digested to completion with the appropriate restriction enzyme, separated on 0.8% agarose gels, and transferred to nylon membranes. DNA probes (purified insert DNA) were labeled with ^{32}P by nick translation to a specific activity of about 2×10^8 cpm/μg. Hybridization was at 42°C for 16 hr in 50% formamide, 5× SSC, 5× Denhardt's, 50 mM sodium phosphate (pH 7.0), 0.5% SDS, and 100 μg/ml salmon sperm DNA. Approximately 1×10^6 cpm/ml of probe was used. Washing was performed three times for 20 min in 0.1× SSC, 0.1% SDS at 50°C for northern blots. For Southern blots, three washes were done for 20 min in 2× SSC, 0.1% SDS at 68°C (low stringency) followed by three washes for 20 min in 0.1× SSC, 0.1% SDS at 68°C (high stringency). Filters were then exposed to Kodak XAR-5 film at −70°C with intensifying screens.

Construction of a human testis cDNA library. A cDNA library was constructed from 5 μg of poly(A)$^+$ RNA extracted for human testis. First-strand synthesis was performed with AMV reverse transcriptase. RNA-DNA hybrids were then converted to double-stranded cDNA with RNase H and DNA polymerase I (Gubler and Hoffman 1983). The cDNA was methylated with *Eco*RI methylase, the ends were repaired with DNA T4 polymerase, and *Eco*RI linkers were added. cDNA was ligated to the *Eco*RI-cut arms of the bacteriophage NM1149 (Murray 1983). After in vitro packaging, recombinant clones were selected on *E. coli* NM514 (*hflr$^-$m$^+$*). 5×10^5 clones were screened with p12F3, and positive clones were plaque-purified.

In situ hybridization. In situ hybridization was performed using the technique of Harper and Saunders (1981). The chromosomes were treated for 1 hr with RNase A at a concentration of 200 μg/ml in 2× SSC at 37°C, denatured in 70% formamide (MC/B, Norwood, Ohio), 2× SSC at 70°C for 2 min, then immersed in 70%, 80%, 95%, and 100% ethanol for 1 min each. The probe was diluted to a final concentration of 0.1 μg/ml in pH 7 hybridization buffer containing 50% formamide/2× SSC, 10% dextran sulfate, and 0.1 mg/ml sonicated, denatured salmon sperm DNA. Chromosomes were hybridized for 12 hr at 37°C; excess probe was then removed with three changes of 50% formamide/2× SSC for 3 min each, followed by five changes of 2× SSC for 2 min each, all at 39°C. Chromosome preparations were again dehydrated with an alcohol series, and Kodak autoradiographic NTB-2 liquid emulsion was applied to the slides for a 4- or 6-day exposure.

After the slides were developed, they were either G-banded (Yunis and Chandler 1977) or R-banded by using a modification of the technique of Schweizer (1980). This modification consisted of staining the preparations with a 0.5 mg/ml solution of chromomycin A (Sigma) for 15 min followed by a 5-min incubation in a 0.1 mg/ml solution of distamycin. Both antibiotics were dissolved in 50% McIlvaine's buffer (pH 7.0) containing a 5 mM concentration of $MgCl_2$. The slide preparations were rinsed, air-dried, and mounted in 100% glycerol, using a clean coverslip. They were examined and photographed using a Zeiss photoscope III with a KP 490 excitor filter, an FT 510 dichroic mirror, and an LP 520 barrier filter. Silver grains were analyzed over fluorescent R-banded preparations by a double-illumination system similar to that described by Sawin et al. (1978). For illustrative purposes, some of these preparations were destained in methanol and stained with Wright's stain (Yunis and Chandler 1977).

Somatic cell hybrid. Peripheral blood lymphocytes from two XX males were fused with HPRT$^-$ Chinese hamster fibroblasts (CHO), using polyethylene glycol. Several independent somatic cell hybrid clones were isolated, using HAT selective medium (Littlefield 1964). They were karyotyped with the help of R. Berger, A. Bernheim, and A. Abadie by using the technique of R-banding (Dutrillaux and Lejune 1971). G6PD was analyzed by following the method of Meera-Khan (1971). Genomic DNA was extracted and analyzed as described above.

RESULTS

Molecular Analysis of Human XX Males

Detection of Y-specific DNA sequences among a panel of XX males. DNAs originating from the blood of 40 nonrelated XX males were digested with *Eco*RI, *Taq*I, or *Hin*dIII, fractionated on agarose gels, and blotted onto Zetapore membranes as described. The filters were successively hybridized to 10 different genomic probes detecting Y-specific DNA sequences and usually washed after each hybridization under low-stringency conditions, allowing identification of DNA fragments not strictly homologous to the probes. Out of 40 DNA samples tested, 65% reacted at least with derived probe 47b + c Y (Table 1).

Among the patients positive with Y-DNA probes, four groups were distinguishable, depending on the number of Y probes giving a positive signal. Group e consisted of 16 patients in which no Y-specific sequences could be detected. The other groups (a,b,c, and d) showed a variable amount of Y-derived DNA fragments, demonstrating a genetic heterogeneity among human XX males. Except for the group (e) pa-

Table 1. Male-specific DNA Sequences in the Genome of Human XX Males

Loci	Y-DNA probes	Normal male	XX Males a (3)	b (6)	c (7)	d (8)	e (16)
DXYS5	p47b + c	2(T)	2	2	2	2	–
DYS7	p50f2	4(T)	2	2	2	–	–
DYS3	p52d	3(T)	2	2	2	–	–
DYS8	p118	8(T)	4	4	4	–	–
DYS6	p48	1(H)	1	1	–	–	–
DYS5	p27	1(T)	1	–	–	–	–
	p64a7	1(E)	–	–	–	–	–
	p37c	1(E)	–	–	–	–	–
DYS1	p49f	3(T)	–	–	–	–	–
DYS11	p12f3	1(T)	–	–	–	–	–

The results refer to *Taq*I (T), *Hin*dIII (H), or *Eco*RI (E) genomic blots after hybridization with different human Y-genomic probes and low-stringency washing conditions, except for probes p64a7, p37c, and p49f. In each vertical column, numbers indicate the male-specific bands detected with each probe. Numbers in parentheses correspond to the number of patients belonging to each group.

tients, no obvious phenotypic differences could be seen among the patients from groups a, b, c, or d.

In these positive groups the presence of Y-DNA is found to be partial at two levels: First, a limited amount of Y-DNA sequences are detected, since more than half of the positive patients react only with 5 probes (p47b + c, p50f2, p118, p52d, and p48); second, probes detecting multiple Y-specific bands on normal male DNA revealed only some of these Y bands when hybridized to XX male DNAs. For example (Fig. 1A),

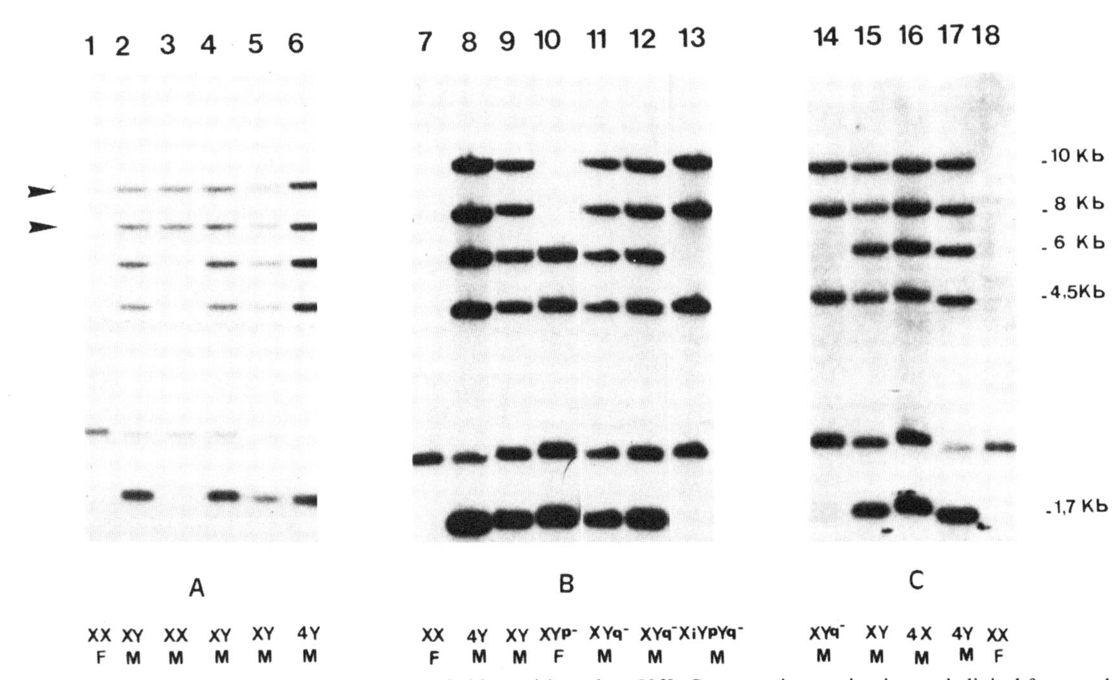

Figure 1. Hybridization patterns of *Eco*RI genomic blots with probe p50f2. Cytogenetic examination and clinical features have been performed by Dr. E. Magenis for patients AM and CC, Dr. S. Gilgenkrantz for patient LL226, and Dr. E. Buhler for patients BB and BU8. (*A*) Lane *1*: 46,XX female (mother of patient BS); lanes *2* and *4*: 46,XY males (brothers of patient BS); lane *3*: 46,XX male BS; lane *5*: 46,XY male; lane *6*: 49,XYYYY male. (*B*) Lane *7*: 46,XX female; lane *8*: 49,XYYYY male; lane *9*: 46,XY male (father of patient AM); lane *10*: 46,XYp⁻ female AM; lane *11*: 46,XYq⁻ male BU8; lane *12*: 46,XYq⁻ BB; lane *13*: 46,X iso Yp del Yq male CC. (*C*) Lane *14*: 46,XYq⁻ male LL226; lane *15*: 46,XY male; lane *16*: 49,XXXXY male; lane *17*: 49,XYYYY male; lane *18*: 46,XX female.

probe 50f2 detects five Y-specific bands in normal males but only detects two Y-specific bands in the XX males analyzed. Similar results are observed with p52d and p118 (Table 1). All the Y-specific bands detected are the same in all positive patients analyzed and appear to be the same size as compared with normal male DNAs (within the limits of the three restriction enzymes tested).

Finally, the Y-DNA sequences observed in XX male DNA are not randomly distributed. The polarization of the Y-derived sequences as shown in Table 1 shows the presence of overlapping Y-chromosomal fragments in the genome of 65% of XX males tested.

To further analyze this inclusive overlapping presence of Y sequences in XX males and to define the Yp or Yq origin of these sequences, comparative analysis of DNAs from XX males and patients with various deletions of the Y chromosome has been performed.

Origin of Y-specific DNA sequences found in XX males, using genomic DNA from a panel of human Yp and Yq deletions. DNA from XX males and from patients with deletions of the short or long arm of the Y chromosome was digested with *Eco*RI and hybridized with five probes detecting Y-specific DNA sequences. Probe p47b + c detects only one Y-specific band in 46,XY normal males DNAs as compared with normal females. This band can be assigned to the short arm of the Y chromosome, since it is missing in DNA from three 46,XYp⁻ females tested (data not shown). Probe p50f2 detects five Y-specific bands in normal male DNA (Fig. 1A). Two of these Y-specific sequences that are detected in XX males are absent from 46,XYp⁻ females but present in 46,XYq⁻ males (Fig. 1B,C). Similar results have been obtained with p52d, p118, and p48. These data demonstrate that the Y-specific sequences detected by p47b + c and p50f2, p52d, p118, and p48 in the genome of XX males originate from the distal short arm of the Y because they are missing from the genome of 46,XYp⁻ females.

Chromosomal localization of Y-specific DNA sequences in XX males. Chromosome preparations from nine subjects—a normal 46,XY male; a 47,XYY male; two 46,X nonfluorescent Y males; two 46,XX males; an XO male; a 46,Xdel (Yp) female; and a normal XX female—were hybridized with Y-derived DNA probe p50f (Bishop et al. 1983, 1985). Chromosomes were prepared from peripheral blood and lymphoblast cultures by using methotrexate synchronization to increase the yield of early metaphase chromosomes.

The human Y-derived DNA sequence p50f (*DYS7*), which has been found in the genome of some XX males, was radiolabeled by nick translation to a specific activity of 1.63×10^7 cpm/μg, and hybridized to the metaphase preparations according to the methods of Harper and Saunders (1981). Slides were then developed and R-banded. Nineteen of 200 cells from the XY male had silver grains on the Y short arm, indicating significant molecular hybridization of the probe to this region; 16% of XYY cells and 12% of cells from the males with 46,Xiso (Yp) showed hybridization to that same region. The X chromosomes of these males showed no significant hybridization, whereas there was a clustering of grains to the X short arm in preparations examined from XX males: 13% and 7% of cells had grains.

Hybridization of the probe to normal 46,XX female preparations showed 5 of 100 cells with labeling on the X but no grains on the short arm. In preparations from the 46,Xdel (Yp) female, there were only two silver grains found in 200 cells examined on the short arm of the Y, and one additional grain on the long arm. These results indicate that Y-short-arm material is present on the short arm of the X chromosome in the XX males analyzed, and that this same material is absent from the female missing the distal Y short arm.

Somatic cell hybrid analysis shows that the Y-specific sequences are carried by the paternally derived X chromosome in XX males. Somatic cell hybrids were obtained from fusion between HPRT⁻ CHO hamster fibroblasts and peripheral blood lymphocytes from two different XX males belonging, respectively, to groups c and d. Somatic cell hybrids obtained with lymphocytes from an XX male belonging to group c are described in Table 2. Four independent somatic cell hybrids selected in HAT medium were analyzed. Clone T2T was derived from clone HA2 by back-selection with thioguanine in order to eliminate the human X chromosome. The four Y DNA sequences present in the genome of the XX male used are detected in HA1, HA2, and HA3 clones but not in clone T2T and N4. They cosegregate as a unit. Their detection correlates with the presence of a human X chromosome. However, the paternal or maternal origin of the X chromosome could not be characterized since both the isoenzyme (G6PD) and RFLP (factor IX) were not informative in this family.

A second set of somatic cell hybrids were obtained between CHO fibroblasts and lymphocytes from an XX male belonging to group d (positive only with p47b + c DNA probe; see Table 3). Seven somatic cell hybrids were analyzed. The Y-DNA fragments cosegregate with a human X chromosome. A RFLP defined by a factor IX DNA probe was informative in this family and showed that the paternal X chromosome was present in the Y-positive hybrids. Since all XX males so far analyzed carry a paternal and a maternal X chromosome, we can conclude that in the somatic cell hybrid studied, the Y-DNA sequence is correlated with the presence of the human X chromosome of paternal origin. These results allow us to conclude that the Y-DNA sequences present in the genome of XX males are carried by the paternal X chromosome.

Clinical characteristics of XX male patients where no Y-DNA sequences could be detected. Group e is composed of 16 patients, half of which are clinically indistinguishable from patients of groups a, b, c, and

Table 2. Analysis of XX Male/CHO Somatic Cell Hybrids: Segregation of Y-DNA Sequences

	Blood				Somatic cell hybrid clones[a]				
	father	mother	XX male[b]	CHO[c]	ClHA1[d]	ClHA2[d]	ClHA3[d]	ClT2T[e]	ClN4[f]
Y-specific probes									
DXYS5 (p47b + c)	+	–	+	–	+	+	+	–	–
DYS7 (p50F2)	+	–	+	–	+	+	+	–	–
DYS3 (p52d)	+	–	+	–	+	+	+	–	–
DYS8 (p118)	+	–	+	–	+	+	+	–	–
Human chromosomes[g]	46,XY	46,XX	46,XX	0	*X*,1,6,7,9 13,19,21	*X*,3,6,7 8,22	*X*,3,5,6 7,10,15 18,21	3,6,7 8	5,6 10,15 18,21
G6PD (human)	A	A	A	0	A	A	A	0	0

[a]Hybrids were selected in HAT medium.
[b]XX male is a group c patient.
[c]Peripheral blood lymphocytes were fused with HPRT$^-$ Chinese hamster fibroblasts (CHO).
[d]HAT medium.
[e]Back selection against the X with thioguanine.
[f]Grown in normal medium.
[g]Human chromosomes were analyzed using the R-banding technique.

d: They are sterile adults with an XX karyotype, yet they are phenotypically male and have normal genitalia with small testes. The other 8 patients studied were 1 month to 14 years old. They were detected by pediatricians because they have external genital ambiguity such as anterior or posterior hypospadias, bilateral cryptorchism, or vaginal pouches. In these latter cases, we analyzed three familial cases where two brothers carry such ambiguities. They too were negative for the Y probes tested. These results demonstrate a genetic heterogeneity of the XX male syndrome. They strongly suggest that different mechanisms might be involved in the etiology of such anomalies.

Y Chromosome and Transcribed Sequences

In an attempt to identify gene(s) expressed from the Y chromosomes, we used pools of DNA probes derived from our Y genomic library to identify transcripts. Four pools of probes were prepared, each containing 12 probes. Pool 1 was made up of Y-specific probes, some of which were localized to the short arm. Pools 2–4 contained a mixture of X-Y, X-Y-A, and Y-autosomal probes. Each pool was hybridized to a northern blot of testis and liver poly(A)$^+$ RNA.

Two probes, p12F and p49F, identified transcripts in the testis RNA.

Analysis of transcripts detected by p12F. p12F is a 5.2-kb *Eco*RI fragment that detects multiple hybridizing bands located on the Y and X chromosomes and on autosomes (Figs. 2 and 3). It can be subdivided by *Bgl*II digestion into three contiguous fragments, 12F1, 12F2, and 12F3. Probing northern blots with all three subclones showed that it was exclusively 12F3 (2 kb), which detected the transcription. Although 12F3 shows a less complex hybridization pattern than 12F, several bands are detected both on the Y chromosome and autosomes. Using the panel of Y deletions, it can be shown that Y sequences map to the euchromatic portion of Yq. When 12F3 (Fig. 4) is used to probe poly(A)$^+$ RNA from testis, ovary, and several male and female teratoma cell lines, a faintly hybridizing band of approximately 5.3 kb could be seen in all tissues, and a very strongly hybridizing band of 1.8 kb was detected exclusively in the testis.

A testis cDNA library was constructed in the phage NM1149, and we have isolated the cDNA corresponding to the 5.3-kb RNA. This cDNA is partial since its size is 4.8 kb. When hybridized to an *Eco*RI-digested Southern blot, the cDNA detects only autosomal sequences (Fig. 5). This indicates that this cDNA is transcribed from an autosomal loci. Analysis of cDNA clones corresponding to the testis-specific RNA detected by 12F3 are in progress.

Analysis of transcripts detecting p49F. p49F is a 2.8-kb *Eco*RI restriction fragment. On *Eco*RI-restricted genomic blots (Fig. 6) under stringent conditions, this probe hybridizes only to the Y chromosome, whereas under relaxed conditions it also detects an autosomally located band. Interestingly, when it is used to probe *Taq*I-digested blots under nonstringent conditions, it detects several Y-located polymorphisms, allowing the distinction of 16 Y haplotypes (Ngo et al. 1986).

When used to probe northern blots, a 4-kb hybridizing band can be seen in testis RNA (Fig. 7). This band is absent from human ovary and several male and female teratoma-derived cell lines. In addition, it cannot be detected in the somatic cell hybrid 3E7, which contains 4 Y chromosomes on a mouse background. We have isolated an apparently full length cDNA clone from our testis library. When this cDNA clone 49HT8

Table 3. Analysis of XX Male/CHO Somatic Cell Hybrids: Segregation of Y-DNA Sequences

	Blood				Somatic cell hybrid clones[a]						
	father	mother	XX male[b]	CHO[c]	ClHA1[d]	ClHA2[d]	ClHA3[d]	ClHA4[d]	Cl1T[e]	Cl27[e]	Cl3T[e]
Y-specific probes											
DXYS5 (p47b + c)	+	–	+	–	+	+	+	–	–	–	–
DYS7 (p50F2)	+	–	–	–	–	–	–	–	–	–	–
DYS11 (p52d)	+	–	–	–	–	–	–	–	–	–	–
Human chromosomes[f]	46,XY	46,XX	46,XX	0	*X*,9,21	*X*,9	*X*	*X*	9	9,21	0
X polymorphic probe factor IX	A	A,B	A,B	–	A	A	A	B	0	0	0

[a]Hybrids were selected in HAT medium.
[b]XX male is a group d patient.
[c]Peripheral blood lymphocytes were fused with $HPRT^-$ Chinese hamster fibroblasts (CHO).
[d]HAT medium.
[e]Back selection against the X with thioguanine.
[f]Human chromosomes were analyzed using the R-banding technique.

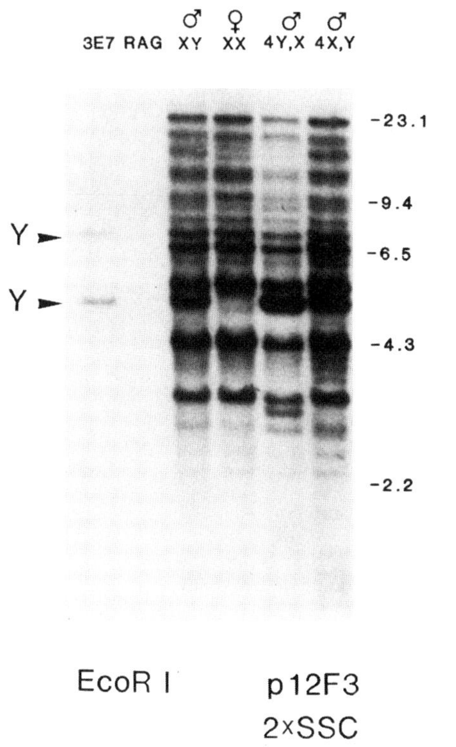

Figure 2. Southern blot analysis of the genomic probe p12F3 in nonstringent conditions of hybridization. DNA (15 μg) from the somatic cell hybrid 3E7, the mouse cell line RAG, the human cell line with 49,XYYYY chromosomes (X,4Y), the human cell line with 49,XXXXY chromosomes (4X,Y), and lymphocytes of normal male (XY) and normal female (XX) were digested wtih *Eco*RI. Restriction fragments were separated by electrophoresis on an 0.8% agarose gel and transferred onto a nylon membrane. The filter was hybridized to the nick-translated probe; 12F3 detects several autosomal bands and two Y bands.

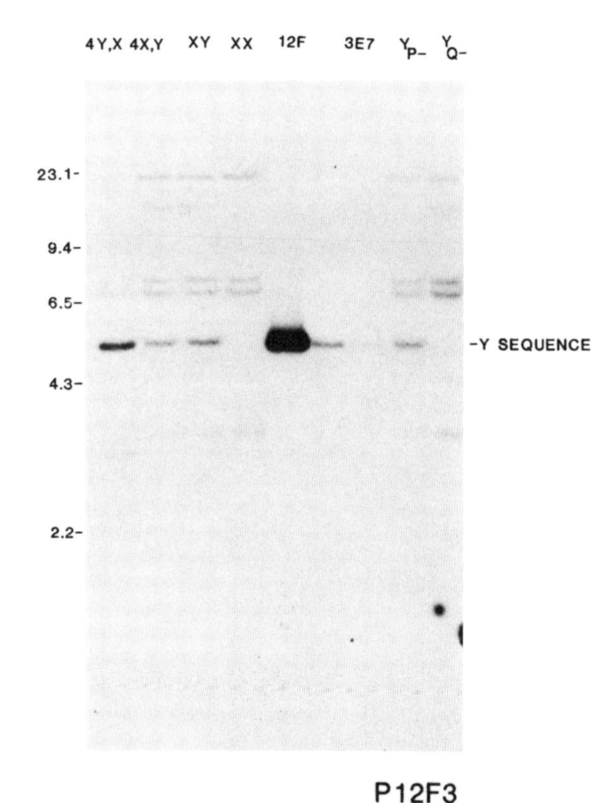

Figure 3. Southern blot analysis of the genomic probe 12F3 in stringent conditions of hybridization. DNAs from 3E7, 4X,Y, 4Y,X, XX, XY, cell lines with Y-short-arm deletion (Yp^-) and Y-long-arm deletion (Yq^-), and the plasmid 12F were digested with *Eco*RI. Restriction fragments separated on an 0.8% agarose gel were transferred onto a nylon membrane. After hybridization, 12F3 detects five autosomal sequences and the 12F fragment corresponding to the Y sequence. The genomic probe is located on the Y long arm.

is used to probe *Eco*RI-digested Southern blots under nonstringent conditions, both Y-specific and autosomal bands were detected (Fig. 8). Unexpectedly, when washed under stringent conditions, the Y bands are lost and only the autosomal sequences remain (Fig. 9). This would strongly suggest that the cDNA is transcribed from an autosome, the genomic probe p49F representing perhaps a Y-located pseudogene.

Preliminary data using somatic cell hybrids suggest that the gene itself maps to chromosome 3. Furthermore, it is highly conserved, being detectable in primates, mice, and even snakes. Using in situ hybridization to testis tissue sections, we have been able to demonstrate that it is detectably transcribed only in the germ line at around meiosis II, probably in spermatid cells.

DISCUSSION

Two hypotheses have been suggested to explain the occurrence of sterile males with an apparent XX karyotype (de la Chapelle 1981): a mutation in a postulated gene(s) not localized on the Y chromosome and required for primary sex determination, or inheritance of the paternal Y-chromosome sex-determining gene(s) as the result of an abnormal exchange between X and Y. Our results suggest that both hypotheses are not mutually exclusive and both may be correct. About two thirds of the 40 XX males studied contain Y-derived sequences. These Y-DNA sequences detected in XX males originate from Yp in the vicinity of the X-Y pairing region. Genes for maleness must therefore be localized on the distal part of the Y-chromosome short arm. Using both in situ hybridization and somatic cell genetics, we have shown that Y-DNA material can be detected on the distal part of the short arm of the paternal X chromosome in XX males, again in the proximity of the X-Y pairing region. These results favor an anomaly of X-Y exchange during male meiosis as a mechanism. They show similarities to the mouse *Sxr* (sex-reversed) model in which the *Sxr* factor is trans-

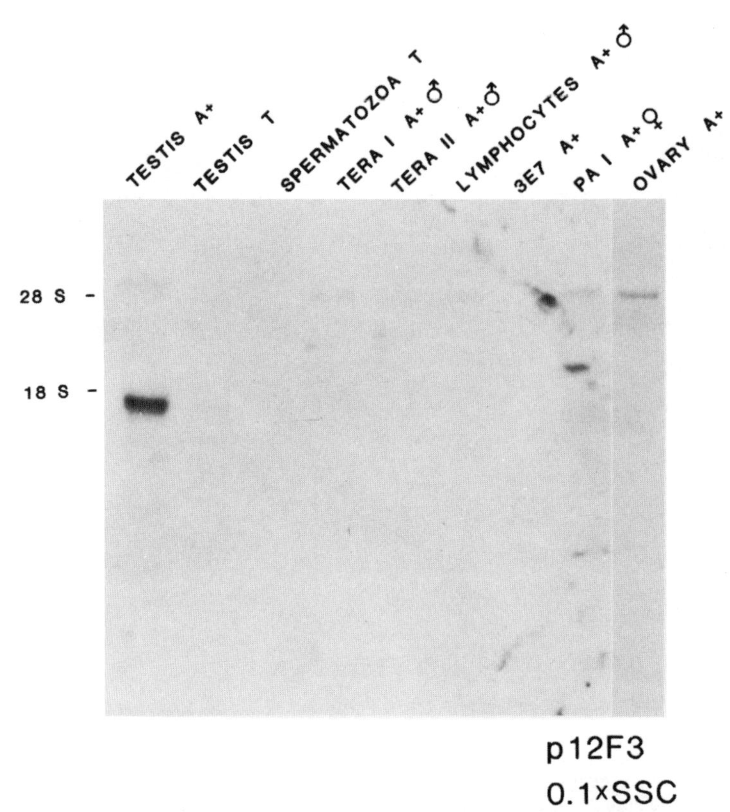

Figure 4. Northern blot analysis of the genomic probe p12F3 in stringent conditions of hybridization. Five μg of poly(A)$^+$ RNA (labeled A$^+$) from testis and ovary, male teratoma cell lines tera I and tera II, male lymphocytes, female teratoma cell line PA1, the somatic cell hybrid 3E7, and 10 μg of total RNA (T) from testis and spermatozoa were separated on a 1.5% denaturing agarose gel and transferred to a nitrocellulose membrane. 12F3 detects a 5.3-kb poly(A)$^+$ RNA common to all tissues or cell lines and a 1.7-kb poly(A)$^+$ RNA transcribed in testis.

ferred from the mouse Y chromosome to the X chromosome by what appears to be an obligate crossover (Evans et al. 1982; Singh and Jones 1982). These results demonstrate that in man the term sex reversal is not justified since XX males carry part of the Y chromosome and therefore cannot be defined as true XX individuals. On the contrary, one third of XX males analyzed so far do not contain Y-DNA sequences. In this group we favor a different mechanism, such as a mutation of an autosomal or an X-linked gene required for primary sex determination. Such non-Y-linked genes are known to exist in the mouse (Eicher 1982; 1983). The fact that XX males without detectable Y-DNA sequences are clinically distinct from classical XX males, with frequently ambiguous sexual genitalia, is consistent with this hypothesis. We recently analyzed three familial cases of XX maleness. In each case, two XX brothers belong to this Y-DNA-negative group. A high incidence of XX males without so-far-detectable Y-DNA sequences exists in North Africa (Algeria and Tunisia). In these areas consanguinity has been found to be high (C. Boucekkine and A. Spira, pers. comm.). We are presently trying to map the proposed mutations in several of such XX male families. A fascinating hypothesis would localize this gene to the short arm of the X chromosome in an area considered as homologous with Yp (Burgoyne 1982).

A counterpart of XX males, the 46,XY female syndrome, has been shown in these studies to result from a small deletion of the short arm of the Y chromosome. Several Y-specific DNA probes present in the majority of XX males (p47b + c) were found to be deleted in the two 46,XY females with Turner stigmata. DNA analysis showed that the two deletions were different but included a common overlapping region likely to be essential for male determination (Fig. 10).

Finally, Y-DNA sequences are not randomly present in XX males but can be present as an overlapping system. Thus, the first group reacts only with probe p47b + c; the second group reacts with probes p47b + c, p50f2, p52d, and p118; the third group reacts with all the above probes plus probe p48; and the fourth group reacts additionally with probe p27. If we assumed that XX males are caused by variable transfer of Yp-located

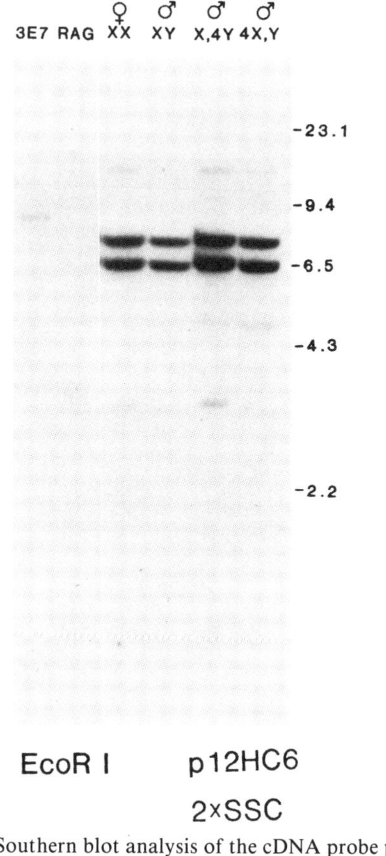

Figure 5. Southern blot analysis of the cDNA probe p12HC6 in nonstringent conditions of hybridization. Fifteen μg of DNA of 3E7, RAG, XX, XY, 4X,Y, and X,4Y were cleaved with *Eco*RI. Restriction fragments were separated on an 0.8% agarose gel and transferred onto a nylon filter. p12HC6 is a 4.8-kb cDNA corresponding to the large poly(A)$^+$ RNA of 5.3 kb common to all cell types. This cDNA recognized only autosomal bands. Some bands are common with those detected by 12F3. No Y sequences are seen.

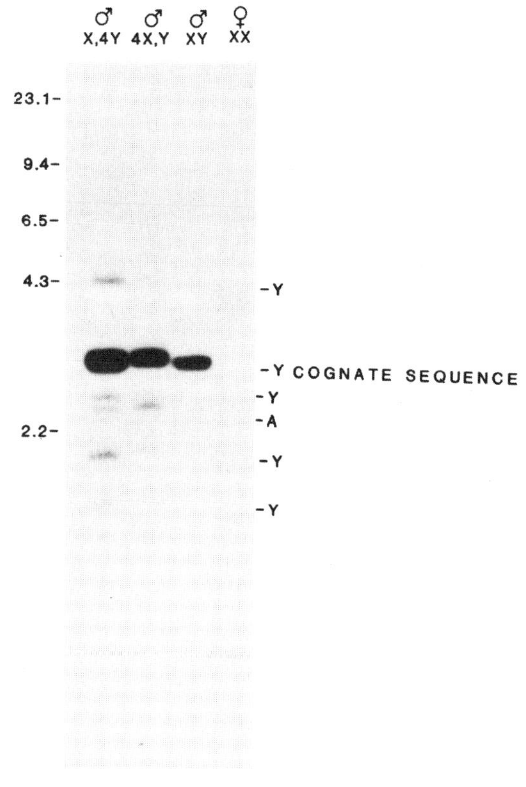

Figure 6. Southern blot analysis of the genomic probe p49F in nonstringent conditions of hybridization. DNAs from XY, XX, X,4Y, and 4X,Y were digested with *Eco*RI. Restriction fragments were separated on an 0.8% agarose gel and transferred onto a nylon membrane. p49F detects five Y sequences (Y) and one autosomal band (A). Among these five Y sequences, p49F recognized the cognate sequence of 2.8 kb.

TDY genes and the transfer involved a single chromosomal event, then the model present in Figure 10 could represent the first tentative map of the Yp chromosome in man. Such a tentative map has also been proposed by Vergnaud et al. (1986).

In an effort to identify Y-encoded genes, we have identified two Y-derived genomic probes that detect RNA transcribed in human testis. These two probes have common features. They both detect RFLPs, they are located on the long arm of the Y chromosome, and both are Y-autosomal probes. 12F detects an *Eco*RI or *Taq*I polymorphism probably generated by an insertion/deletion mechanism (Casanova 1985). The polymorphism visualized by 49F with *Taq*I is due to point mutations within the restriction sites. The CpG dinucleotide of the *Taq*I site is considered as a hot spot for point mutations (Barner et al. 1984). This is because if the C is methylated, it can frequently deaminate to a T. Although p49F is undoubtedly Y-derived, the mRNA it detects is transcribed from human chromosome 3.

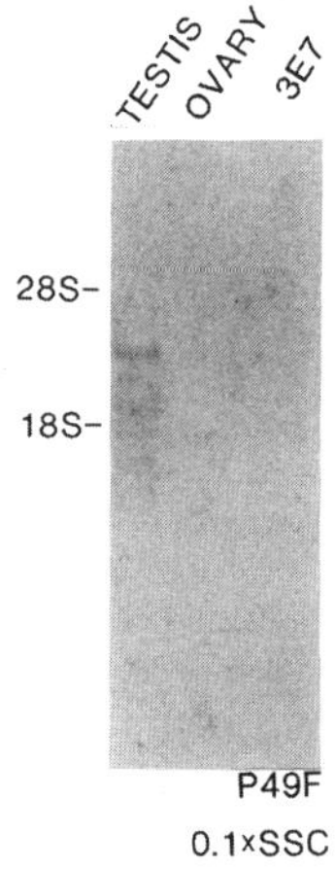

Figure 7. Northern blot analysis of the genomic probe p49F in stringent conditions of hybridization. Five μg of poly(A)$^+$ RNA from testis, ovary, and the somatic cell hybrid 3E7 were separated on a 1.5% agarose gel and transferred onto a nitrocellulose sheet. The filter was hybridized with nick-translated probe p49F. A 4-kb poly(A)$^+$ RNA is detected in testis.

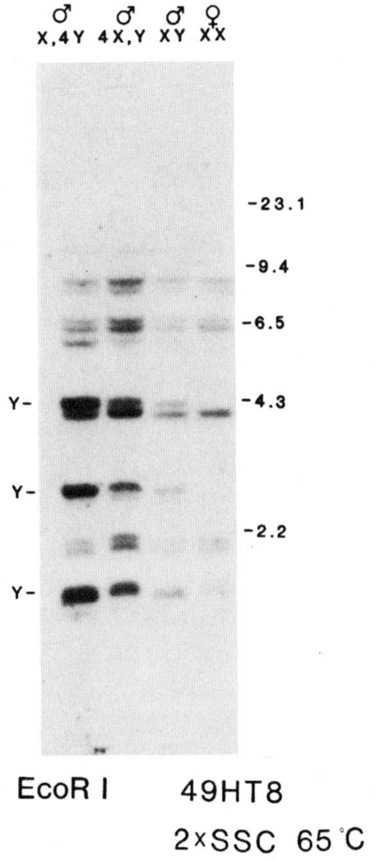

Figure 8. Southern 49HT8, 2× SSC. The cDNA corresponding to the 4-kb poly(A)$^+$ RNA detected by the genomic probe 49F has been cloned. This cDNA, named 49HT8, when hybridized to a Southern blot in nonstringent conditions, recognized autosomal sequences and three Y sequences present in male, amplified in 4Y,X, and absent from female.

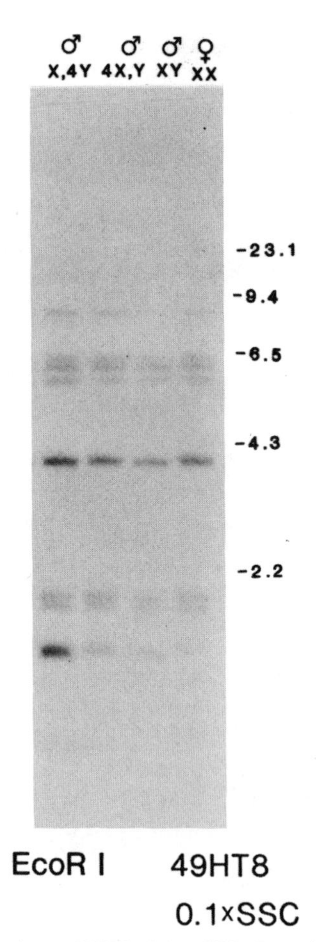

Figure 9. Southern 49HT8, 0.1× SSC. In stringent conditions, the cDNA 49HT8 recognized only autosomal sequences. All the Y sequences are lost in these conditions.

Hence it is possible that p49F identifies the remains of a functional region of a Y-linked gene or that it is a pseudogene of a chromosome-3-located gene. The presence of argininosuccinate synthetase and actin-like sequences localized on the Y chromosome have been reported (Daiger et al. 1982; Heilig et al. 1984). We have tested other Y-autosomal probes, but none of them identifies transcript polymorphisms or RFLPs. At present, it is not known if the polymorphism detected by these probes is a key point.

All the Yp probes tested, even those present in XX males, fail to detect a transcript. Either the RNA they could identify is expressed at an early stage of development or these probes are only markers of Y regions. Identification of transcripts during gonadal differentiation is difficult in man, and we are now using the mouse models to facilitate this.

The Y chromosome plays a fundamental role in sex determination and in spermatogenesis, but genes localized on autosomes are also involved in these two functions. In the mouse, several genes for spermatogenesis map in the *t* locus (Willison and Dudley 1985); genes for testis differentiation are also located on autosomes (Eicher et al. 1982); and specific β tubulins localized on chromosome 2 are expressed only in the testis (Gerhard et al. 1985).

The presence of several pseudogenes on the Y chromosome raise the question of whether this chromosome has lost functional genes during evolution: The described transcribed sequences could be the remains of this evolution. The other possibility is that pseudogenes occurred on the Y as on autosomes by duplication or by retrotranscription. The haploid state of the Y chromosome has probably favored accumulation of pseudogenes.

ACKNOWLEDGMENTS

This work is dedicated to Dr. E. Fellous. We are very grateful to Drs. R. Berger, A. Bernheim, J.L. Mandel, and M. Koenig for their invaluable help. The work was supported by grants from INSERM and ANVAR.

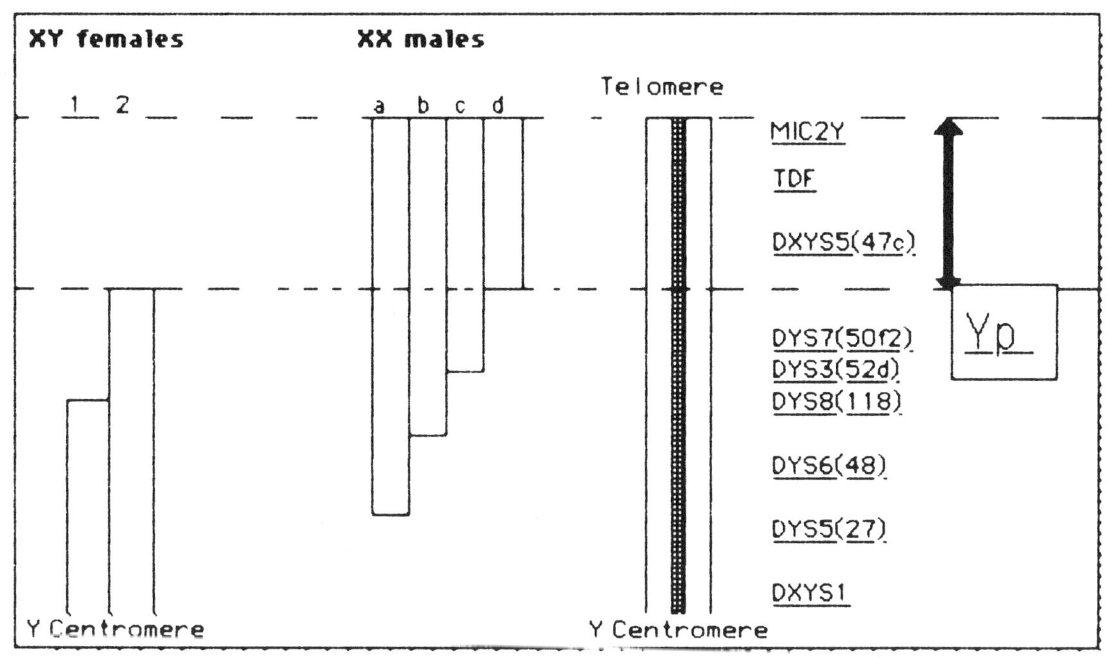

Figure 10. Y-DNA fragments present in different groups of XX males, two cases of XYp females, and the working model for the Y-chromosome short arm.

REFERENCES

Auffray, C. and F. Rougeon. 1980. A simple procedure to obtain a high yield of intact mRNA from animal tissues. *Eur. J. Biochem.* **107:** 303.

Avner, P., R. Bono, R. Berger, and M. Fellous. 1981. Characterization of human teratoma cell lines for their in vitro developmental properties. *J. Immunogenet.* **8:** 151.

Bailey, J.M. and N. Davidson. 1976. Methylmercure as a reversible denaturing agent for agarose gel electrophoresis. *Anal. Biochem.* **70:** 75.

Barner, D., M. Schafer, and R. White. 1984. Restriction sites containing CpG show a higher frequency of polymorphism in human DNA. *Cell* **36:** 131.

Bishop, C.E., P. Boursot, B. Baron, P. Bonhomme, and D. Hatat. 1985. Most classical *M.m.domesticus* laboratory mouse strains carry a *M.m. musculus* Y chromosome. *Nature* **315:** 70.

Bishop, C., G. Guellaen, D. Geldwerth, M. Fellous, and J. Weissenbach. 1984. Extensive sequence homologies between Y and other human chromosomes. *J. Mol. Biol.* **173:** 403.

Bishop, C.E., G. Guellaen, D. Geldwerth, R. Voss, M. Fellous, and J. Weissenbach. 1983. Single copy DNA sequences specific to the human Y chromosome. *Nature* **303:** 832.

Bühler, E.M. 1980. A synopsis of the human Y chromosome. *Hum. Genet.* **55:** 145.

Burgoyne, P.O.S. 1982. Genetic homology and crossing over in the X and Y chromosomes in mammals. *Hum. Genet.* **61:** 85.

Casanova, M., P. Leroy, C. Boucekkine, J. Weissenbach, C. Bishop, and M. Fellous. 1985. A human Y linked DNA polymorphism and its potential for estimating genetic and evolutionary distance. *Science* **230:** 1403.

Daiger, S.P., R.S. Wildin, and T.-S. Su. 1982. Sequences on the human Y chromosome homologous to the autosomal gene for argininosuccinate synthetase. *Nature* **298:** 682.

de la Chapelle, A. 1981. The etiology of maleness in XX men. *Hum. Genet.* **58:** 105.

de la Chapelle, A., H. Hortling, M. Niemi, and J. Wennstrom. 1964. XX sex chromosomes in a human male: First case. *Acta Med. Scand.* (suppl.) **412:** 25.

Dutrillaux, B. and J. Lejeune. 1971. Sur une nouvelle technique d'analyse du caryotype humain. *C.R. Acad. Sci.* **272:** 2638.

Eicher, E.M., S.J. Phillips, and L.L. Washburn. 1983. The use of molecular probes and chromosomal rearrangements to partition the mouse Y chromosome into functional regions. In *Recombinant DNA and medical genetics* (ed. A. Messer and I.H. Porter), p. 57. Academic Press, New York.

Eicher, E.M., L.L. Washburn, J. Barry Whitney III, and K.E. Morrow. 1982. Mus poschiavinus Y chromosome in the C57BL/6J murine genome causes sex reversal. *Science* **217:** 535.

Evans, E.P., M. Burtenshaw, and B.M. Cattenach. 1982. Cytological evidence for meiotic crossing over between X and Y chromosomes of male mice carrying the sex reveral (Sxr) factor. *Nature* **300:** 443.

Evans, H.J., K.E. Buckton, G. Spowart, and A.D. Carothers. 1979. Heteromorphic X chromosomes in 46,XX males: Evidence for the involvement of X-Y interchange. *Hum. Genet.* **49:** 11.

Ferguson-Smith, M.A. 1966. X-Y chromosomal interchange in the aetiology of true hermaphroditism and of XX Klinefelter's syndrome. *Lancet* **II:** 475.

Gerhard, D.S., P.R. Dobner, and G. Bruns. 1985. Testis specific α-tubulin is on chromosome 2q. *Cytogenet. Cell Genet.* **40:** 639.

Goodfellow, P., S. Darling, and J. Wolfe. 1985. The human Y chromosome. *J. Med. Genet.* **22:** 329.

Gubler, U. and B.J. Hoffman. 1983. A single copy and very efficient method for generating cDNA libraries. *Gene* **25:** 263.

Guellaen, G., M. Casanova, C. Bishop, D. Geldwerth, G. Andre, M. Fellous, and J. Weissenbach. 1984. Human XX males with Y single-copy DNA fragments. *Nature* **307:** 172.

Harper, M.E. and G.F. Saunders. 1981. Localization of single copy DNA sequences of G-banded chromosomes by in situ hybridization. *Chromosoma* **83:** 431.

Heilig, R., A. Hanauer, K.H. Grzeschik, M.C. Hors-Cayla, and J.L. Mandel. 1984. Actin-like sequences are present on human X and Y chromosomes. *EMBO J.* **3:** 1803.

Littlefield, J.W. 1964. Selection of hybrids from mating of fibroblast in vitro and their presumed recombinant. *Science* **145:** 109.

Magenis, R.E. and T. Donlon. 1982. Non-fluorescent Y chromosomes. Cytologic evidence of origin. *Hum. Genet.* **60:** 133.

Magenis, R.E., M.L. Tochen, K.P. Holahan, T. Carey, L. Allen, and M.G. Brown. 1984a. Turner syndrome resulting from partial deletion of Y chromosome short arm: Localization of male determinants. *J. Pediatr.* **105:** 916.

Magenis, R.E., M.J. Webb, R.S. McKean, D. Tomar, L.J. Allen, H. Kammer, D.L. Van Dyke, and E. Lowrien. 1982. Translocation (X;Y)(p22.33;p11.2) in XX males: Etiology of male phenotype. *Hum. Genet.* **62:** 271.

Magenis, R.E., D. Tomar, M.G. Brown, L. Allen, H. Kammer, D.L. Van Dyke, M.J. Webb, and M. Tochen. 1984b. Localisation of male determining factors to the Y short arm, bands p11.2→pter. *Cytogenet. Cell Genet.* **37:** 529.

Marcus, M., R. Trantravali, J.G. Dev, D.A. Miller, and O.J. Miller. 1976. Human-mouse cell hybrid within human multiple Y chromosomes. *Nature* **262:** 63.

McKusick, U. 1983. *Mendelian inheritance in man: Catalogs of autosomal dominant, autosome recessive and X linked phenotypes*, 6th edition. Johns Hopkins Press, Baltimore, Maryland.

Meera-Khan, P. 1971. Enzyme electrophoresis on cellulose acetate gel: Zymogram patterns in man-mouse and man Chinese hamster somatic cell hybrids. *Arch. Biochem. Biophys.* **145:** 470.

Murray, N.E. 1983. Phage lambda and molecular cloning. In *Lambda II* (ed. J.W. Roberts et al.), p. 395. Cold Spring Harbor Laboratory, Cold Spring Harbor, New York.

Ngo, K.Y., G. Vergnaud, C. Johnsson, G. Lucotte, and J. Weissenbach. 1986. A DNA probe detecting multiple haplotypes of the human Y chromosome. *Am. J. Hum. Genet.* **38:** 407.

Sawin, U.L., A.M. Stalka, and W. Wray. 1978. Simultaneous observation of quinacrine bands and silver grains on radiolabeled metaphase chromosomes. *Histochemistry* **59:** 1.

Schweitzer, D. 1980. Simultaneous fluorescent staining of R-banding and specific heterochromatic regions (DA-DAPI bands) in human chromosomes. *Cytogenet. Cell Genet.* **27:** 190.

Singh, L. and K. Jones. 1982. Sex reversal in the mouse (*Mus musculus*) is caused by a recurrent nonreciprocal crossover involving the X and an aberrant Y chromosome. *Cell* **28:** 205.

Vergnaud, G., D.C. Page, M.C. Simmler, L. Brown, R. Rouyer, B. Noel, D. Botstein, A. de la Chapelle, and J. Weissenbach. 1986. Deletion map of the human Y chromosome based on DNA hybridization. *Am. J. Hum. Genet.* **38:** 109.

Willison, K. and K. Dudley. 1985. Haploid gene expression and the mouse t-complex. Genetic manipulation of the early mammalian embryo. *Banbury Rep.* **20:** 57.

Wolf, J., S.M. Darling, R.P. Erickson, I.W. Craig, V.J. Buckle, P.W.J. Rigby, H.F. Willard, and P.N. Goodfellow. 1985. Isolation and characterization of an alphoid centromeric repeat family from the human Y chromosome. *J. Mol. Biol.* **182:** 477.

Yunis, J.J. and M.E. Chandler. 1977. High resolution chromosome analysis in clinical medicine. *Prog. Clin. Pathol.* **7:** 267.

Genetic and Molecular Studies on 46,XX and 45,X Males

A. DE LA CHAPELLE
Department of Medical Genetics, University of Helsinki, 00290 Helsinki, Finland

The gene termed testis-determining factor, or *TDF* (Shows et al. 1984), has not yet been cloned or characterized, nor is anything known about its product. However, solid evidence indicates that it is the primary male-determining gene, and that it is located on the Y chromosome (Davis 1981; Goodfellow et al. 1985a). Much of the recent progress in our understanding of the organization of the human Y chromosome became possible when a Y-chromosome-specific DNA library was constructed (Bishop et al. 1983). When it was shown that the genomes of XX males contain DNA sequences that normally only occur in the Y chromosome (Guellaen et al. 1984; Vergnaud et al. 1986), the enigma of maleness in the apparent absence of a Y chromosome was, on the whole, solved. It is reasonable to assume that the presence in the genomes of XX males of arbitrary Y-derived DNA sequences marks the presence of *TDF* in these individuals, as well, and so explains their maleness. There are now several lines of evidence to show that a mechanism by which XX males can acquire Y-chromosomal sequences is by interchange between the Y and X in the meiosis of the father. Some of the evidence is reviewed below, and implications for the gene order in the *TDF* region are discussed. Assuming that the Y-X interchange hypothesis is correct, then it remains to determine whether all maleness without a Y chromosome arises in this way.

Other hypotheses that have been proposed to explain maleness without a Y chromosome are considered as alternative explanations. Mosaicism involving a cell line containing a Y chromosome is reviewed. Finally, the existence of a gene not normally involved in sex determination but that might turn into one as a result of mutation is considered.

Evidence in Favor of Y-X Interchange

Sequence homology between the X and Y chromosomes. The existence of DNA sequence homology between the human X and Y chromosomes has been amply demonstrated. After the description of pDP34, the first single-copy probe reported that hybridizes with sequences both on the Y and the X (Page et al. 1982), a considerable number of such probes has been reported. At the 8th International Human Gene Mapping Workshop, which was held in 1985, a total of 22 single-copy and 2 repetitious X-Y homologous probes were listed (Goodfellow et al. 1985b). Of the single-copy probes, 9 were described as polymorphic. Mapping data show that contrary to what might have been expected, these homologies are by no means restricted to the pairing region (Burgoyne 1982; Polani 1982) of the X and Y chromosomes. Some of the most widely used X-Y homologous probes, such as pDP34, which characterizes locus *DXYS1*, map to the short arm of Y and the proximal part of the long arm of the X (Page et al. 1984). However, other probes recognize homologous sequences with different localizations, such as Yp-Xp, Yq-Xq, and so forth. A complicated crisscross pattern of homology (for review, see Koenig et al. 1984, 1985) has emerged, indicating that complex rearrangements have characterized the evolution of the X-Y chromosome pair.

The pseudoautosomal region. The existence of a so-called pseudoautosomal segment on the human sex chromosomes, postulated earlier (Burgoyne 1982; Polani 1982), was recently established. Three groups of workers described probes that recognize sequences on the distal ends of Xp and Yp and that recombine in male meiosis (Buckle et al. 1985; Cooke et al. 1985; Simmler et al. 1985). How extensive the male recombination is became apparent when the segregation of three highly polymorphic probes from the pseudoautosomal region was studied in large families (Rouyer et al. 1986). The observed 50% recombination rate indeed corroborated the previous suggestion (Burgoyne 1982) about an "obligatory" crossover in this region (Rouyer et al. 1986). Interestingly, in female meiosis the recombination is much less, perhaps only 10% of that in the male. The finding of the pseudoautosomal region lends strong support to the Y-X interchange hypothesis in the etiology of XX maleness in that it proves that crossovers take place normally. Below we briefly consider some aspects of the mechanism by which the presumptive abnormal interchange takes place.

Abnormal Crossing-over between X and Y in the Etiology of 46,XX Maleness

The recombination frequency in the pseudoautosomal region is 50% in the male (Rouyer et al. 1986). In the female, genetic distances between known polymorphic loci proximal of the pseudoautosomal region on the short arm of the X (Drayna and White 1985) indicate that Xp may be approximately 100 cM long. Assuming only 10% recombination in the pseudoautosomal region and an average of 2 crossovers elsewhere in Xp, then the pseudoautosomal region in the X may be physically short. It must be very short indeed on the Y if it occupies only part of its tiny short arm. The most distal probes hitherto studied that show regular female recombination (e.g., p782 and M1A) have been assigned to the terminal band Xp223. Hence, the pseudoautosomal region may comprise only part of band Xp223. Let us further assume that homology in the pseudoautosomal region is almost 100% (Rouyer et al. 1986) and contiguous. Near the pairing region the degree of homology may either gradually decrease or consist of stretches of sequences with extensive homology interspersed between stretches with little or no homology. These assumptions form the basis of the hypothesis that, under normal circumstances, crossing-over in the male is confined to the pseudoautosomal region but may under abnormal circumstances occur elsewhere in the pairing region. Strong additional support for the Y-X interchange model comes from in situ hybridization experiments performed by us in which a Y-chromosome-specific DNA sequence was shown to hybridize to the tip of the short arm of one X chromosome in three XX males (Andersson et al. 1986). As shown by cytogenetic methods, pairing between the X and Y chromosomes can cover a length corresponding to up to 60% of the length of the entire Y chromosome and may be variable from cell to cell (Chandley et al. 1984). Pairing is probably a prerequisite for the postulated abnormal crossovers. The way in which these crossovers occur has implications for the order of the genes in the region. Below we examine these implications separately under two assumptions. The first assumption is that these crossovers are equal and terminal; the second is that unequal and/or interstitial crossing-over can also occur.

Gene order if abnormal crossovers are equal and terminal. The first XX male described (de la Chapelle et al. 1964) was Xg(a −), while his father was Xg(a +) (Fig. 1). We showed later with the aid of X-chromosomal restriction-fragment-length polymorphisms (RFLPs) that one of his X chromosomes was indeed paternal (Page and de la Chapelle 1984). Parenthetically, the RFLP study of seven father–mother–XX male families gave an informative result as to the parental origin of the X chromosomes in five, and all five showed that the proband had received one X from each parent (Page and de la Chapelle 1984), so we can assume that most XX males have indeed a paternal X chromosome. The failure of an XX male to express his father's *Xg* allele has been observed at least 12 times (for review, see de la Chapelle 1981 and unpubl.). This indicates that the crossover that leads to the transfer of *TDF* onto the paternal X also often leads to the loss of the *Xg*; this would place the crossover proximal of both *Xg* and *TDF* (breakpoint 1, Fig. 2). An exceptionally informative XX male provided evidence of a crossover involving *Xg* and its putative counterpart on the Y chromosome, the postulated *Yg* (de la Chapelle et al. 1984a). This somewhat hypothetical locus determines the level of expression of the 12E7 antigen (Goodfellow and Tippett 1981; Goodfellow et al. 1983; Tippett et al. 1986). It could be shown that the XX male proband had (1) one X from the father (D. Page and A. de la Chapelle, unpubl.), (2) did not express his father's *Xg* allele, but (3) expressed the *Yg* allele present on his father's Y chromosome (de la Chapelle et al. 1984a). These data taken together indicate that the point of crossover in this patient was proximal of both *TDF* and *Yg* (breakpoint 2, Fig. 2) corresponding to breakpoint 1 on the X chromosome.

We have studied an Xg(a +) XX male whose father is Xg(a +) and mother Xg(a −) (de la Chapelle et al. 1984b) (Fig. 3). The same constellation has been seen in at least four other XX males (for review, see de la Chapelle 1981). Assuming equal exchange, the breakpoint on the X chromosome must be distal of *Xg* (breakpoint 3, Fig. 2), but since *TDF* is presumably involved, it is proximal of *TDF* on the Y (breakpoint 4, Fig. 2). This would place *TDF* (on the Y) distal of *Xg* (on the X).

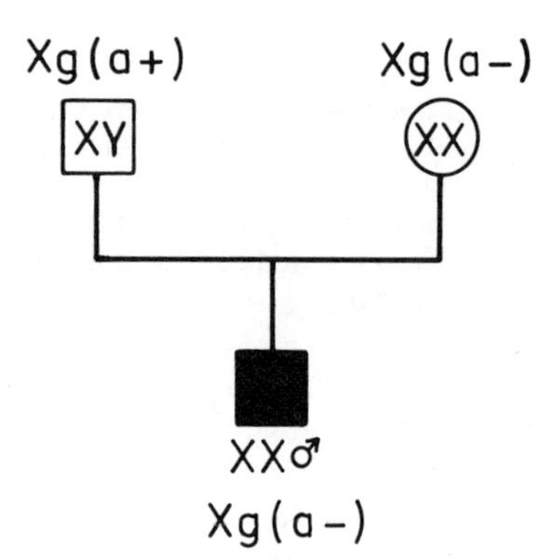

Figure 1. Xg blood group phenotypes in a father, mother, and XX male son. The XX male does not express his father's allele. This pattern has been observed at least 12 times.

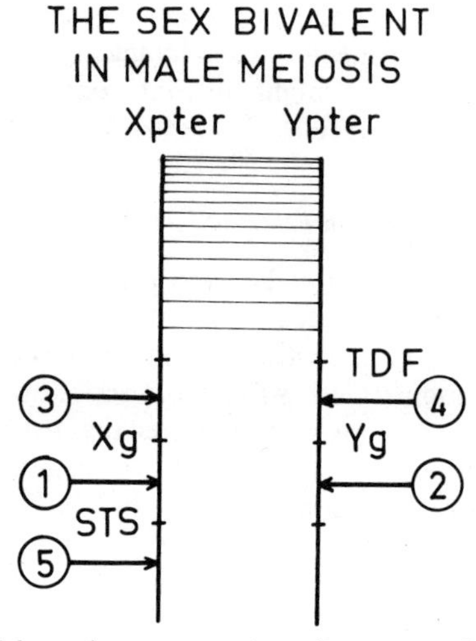

Figure 2. Schematic representation of genes on the human sex bivalent in male meiosis. Numbers in circles refer to breakpoints discussed in the text. (Xg) Blood group Xg; (STS) steroid sulfatase; (TDF) testis-determining factor; (Yg) locus on Y chromosome controlling the expression of 12E7 antigen.

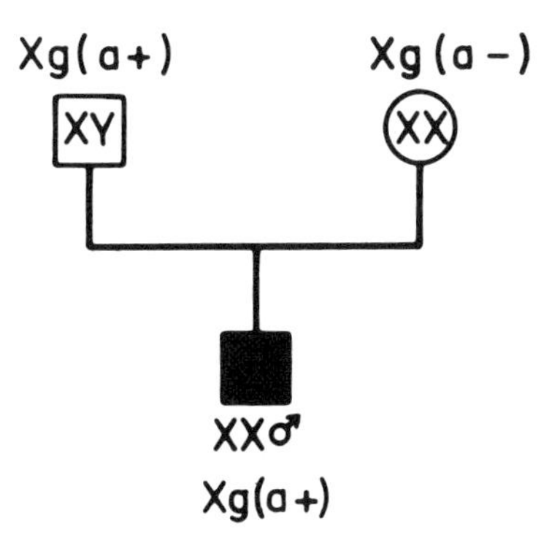

Figure 3. Xg blood group phenotypes in a father, mother, and XX male son. The XX male expresses his father's allele. This pattern has been observed at least five times.

Steroid sulfatase gene. Further evidence on gene order comes from the study of the steroid sulfatase gene, *STS*. Such studies have been hampered by the fact that the *STS* gene has not been cloned, that it undergoes partial inactivation in females (Migeon et al. 1982), and that the determination of STS activity requires fresh specimens and high technical skill. It was shown several years ago that XX males fall into two classes with respect to STS expression: viz., "female" expressers (presumably disomic for the *STS* region of the X) and "male" expressers (presumably monosomic) (Pierella et al. 1981; Ropers et al. 1981). Experiments using somatic cell hybrids containing one or the other X chromosome from such an XX male later indicated that indeed one X had apparently lost the *STS* locus (Wieacker et al. 1983). This places the breakpoint proximal of *STS* on the X (breakpoint 5, Fig. 2).

Evidence from X;Y translocations. An X;Y translocation presumably involving the terminal band Xp22.3 and band Yq11 produces an abnormal derivative chromosome der(X),t(X;Y)(p22.3;q11) (Khudr et al. 1973), which has furnished some important information on the gene order in Xp. Women who have a normal X chromosome and a der(X) are phenotypically almost normal and fertile. When an ovum containing the der(X) is fertilized by a Y-carrying sperm, a boy is born who is nullisomic for the distal part of Xp (and disomic for parts of Yq). Boys with der(X) and Y are phenotypically abnormal with hypogonadism, mental retardation, and the type of generalized ichthyosis that is caused by STS deficiency (Tiepolo et al. 1980). Consequently, the breakpoint is proximal of *STS* on the X (breakpoint 5, Fig. 2). Interestingly, in two such patients (Geller et al. 1986; H.H. Ropers et al., pers. comm.) the break separated *MIC2* and *STS* and so probably resided in region 1 (Fig. 2). The translocation data show that regions in which abnormal crossing-over can take place in the male sex bivalent may also be prone to form translocations with a region on the proximal long arm of the Y chromosome. These findings are consistent with the hypothesis that abnormal exchange takes place in hot spots, but confirmation of the hypothesis will have to await data on the DNA sequences rather than their products.

Gene order if abnormal exchange is not equal and terminal. The gene order shown in Figure 2 may be unlikely because it places *TDF* distal of *Xg-Yg*. Present evidence is consistent with the hypothesis that the *MIC2* gene is X-Y homologous and behaves in a pseudoautosomal fashion (Buckle et al. 1985; Darling et al. 1986). Circumstantial evidence links *MIC2* to *Xg-Yg* so that these phenotypes may in fact be derived from DNA sequences belonging to the *MIC2* locus. Given that *MIC2* is pseudoautosomal and *TDF* is distal of it, *TDF* might also be pseudoautosomal. If *TDF* were pseudoautosomal and subject to normal crossovers, then the well-known chromosomal sex determination mechanism would break down. Alternatively, mechanisms preventing most crossover products from being included in functional sperm or other selective mechanisms would have to be involved.

A simpler assumption is that *TDF* is proximal of *Xg-Yg*. At least five XX males exist who express their paternal *Xg* allele (Fig. 3). If *TDF* occurs in their genomes, it could have been transferred from Yp to Xp through an unequal crossing-over with one break and reunion or as a result of an interstitial exchange involving two breaks and reunions. In addition, two different explanations should be considered: (1) *TDF* might have been transferred to an autosome by regular translocation (cf. below); (2) *TDF* is not involved in testis determination in these XX males. The present research activities concerning the Y chromosome will probably soon result in more DNA probes being available to test these various possibilities. It may be anticipated that the gene order around *TDF* will soon be more decisively determined.

Frequency of Different Crossover Points

It has been suggested that the frequency of abnormal crossovers decreases with increasing distance from the telomere (Polani 1982). Only scanty data are presently available to test this hypothesis. If the simple assumption is made that most abnormal crossovers are equal and terminal, and if the gene order is as shown in Figure 2, then XX maleness should occur more frequently than the loss of *Xg* from an X chromosome, and the loss of *STS* should be even rarer. Moreover, it also follows that any XX individual whose paternal X chromosome has lost *Xg* (or *STS*) by crossover should be a male. The population frequency of XX males is on the order of 1 in 20,000 males. Population studies of the Xg blood group phenotypes have disclosed four Xg(a−) daughters of Xg(a+) fathers among 2066 daughters tested (Sanger et al. 1971; Race and Sanger 1975). These females could have a paternal X chromosome that had lost its *Xg* allele through abnormal crossing-over and would refute the hypothesis that *TDF* must be acquired by the X in such circumstances. However, as chromosome studies have not been done on these females, other interpretations are possible, such as the karyotypes 46,XY or 45,X. If Xg(a−) daughters of Xg(a+) fathers were found to have the karyotype 46,XX, it would still be necessary to prove that the apparent nonexpression of the paternal *Xg* could not be due to other causes. In fact, all the four Xg(a−) daughters of Xg(a+) fathers were exceptional (Race and Sanger 1975), and as far as I know these questions have not been further pursued.

Conclusions Regarding Y-X Interchange

In conclusion, the gene order at the end of Xp is Xpter–pseudoautosomal-*MIC2X,Xg–STS*. Assuming that this region undergoes homologous pairing with the Y chromosome at male meiosis and that all exchanges are equal and terminal, then *TDF* maps between the pseudoautosomal region and *Yg* on the Y in the male sex bivalent. If unequal or interstitial exchanges sometimes occur, then *TDF* probably maps proximal of *Yg*. Present evidence indicates that abnormal crossing-over and Xp;Yq translocation can take place in at least three different intervals in the *Xg-STS* region of Xp. Present data on the frequency of XX males and females who do not express their paternal *Xg* and *STS* alleles are not sufficient to prove or disprove the hypothesis that the frequency with which abnormal crossovers occur proximal of the pseudoautosomal region decreases with increasing distance. The existence of hot spots at which abnormal crossing-over is more likely to occur than elsewhere must also be considered. Such hot spots should probably show high recombination in female meiosis. Both hypotheses should be testable when suitable polymorphic probes become available. These will also make it possible to test whether or not most abnormal crossovers are equal and terminal.

Mosaicism in the Etiology of Maleness without Y

Cytogenetic and clinical aspects of this subject have been previously reviewed (de la Chapelle 1972; de la Chapelle et al. 1977). Extremely low grade or circumscribed mosaicism involving a cell line with a Y chromosome has been found in a small number of male patients who were predominantly XX. In others, inconclusive cytogenetic data have been invoked to suggest mosaicism. The earlier conclusion that mosaicism is unlikely to be of etiologic significance in the great majority of XX males (de la Chapelle 1981) has been confirmed by molecular studies using Y-specific repetitive DNA sequences that mostly occur in Yq. In more than 30 XX males and XX true hermaphrodites in whom DNA mostly from blood cells and fibroblasts has been screened, no evidence for the presence of these sequences has been obtained (Vergnaud et al. 1986; D. Page and A. de la Chapelle, pers. comm.). The sensitivity of these experiments is such that a 1–5% admixture of cells with a Y chromosome is detectable (de la Chapelle et al. 1986). These data confirm cytogenetic results showing absence of the Y chromosome in mitoses from the same tissues (de la Chapelle 1981).

45,X males. The situation may be different in the rare 45,X male. As far as I know, maleness in association with a 45,X karyotype has been described in less than 10 patients (for review, see de la Chapelle et al. 1986). By definition, no Y chromosome can be detected in these individuals. The phenotype is variable and differs markedly from that of XX males. Many X males are quite short and have other features often seen in 45,X Turner's syndrome in females.

In a detailed cytogenetic and molecular study of two 45,X males (de la Chapelle et al. 1986), one was found to be a remarkably low grade mosaic. Lymphocyte mitoses studied on numerous occasions showed no evidence of any other karyotype than 45,X, and no Y-chromosome-specific DNA was detected in blood. However, on several occasions 1–5% of fibroblast mitoses were 46,XY with a structurally normal Y chromosome identical with that of the father. In DNA extracted from parallel fibroblast cultures, Southern blotting showed the presence of Y-specific repeated DNA sequences (DYZ1 and DYZ2). The Y DNA was present in greatly reduced amount compared with the father's DNA, however. In fact, there was good correlation between the cytogenetically determined degree of mosaicism (1–5%) and the results of Y-DNA hybridization (3%). These findings indicate that mosaicism involving a normal 46,XY cell line can account for "45,X" maleness. The paucity and tissue-limited distribution of the 46,XY cell line emphasize the need for very thorough investigations in cases of suspected mosaicism.

In the second 45,X male patient studied by us, evidence for mosaicism could be found neither cytogenetically nor by DNA analysis. The findings do not exclude the possibility that the patient has or has had mosaicism of a very circumscribed nature, but this cannot be proven. Other mechanisms leading to maleness must be considered. However, X-Y interchange involving *TDF*, which is a common etiologic mechanism in XX males (see above), is unlikely because the single X of many 45,X males is of maternal origin. This was shown with the aid of the blood group Xg and with several X-chromosome-specific RFLPs in the two patients studied by us. Likewise, the only X was maternally derived in the patient studied by Schempp et al. (1985). In the absence of a paternally derived X chromosome, maleness cannot be due to X-Y interchange. However, if Yp-derived DNA sequences were found in the genomes of such 45,X males, Y-to-autosome translocation of *TDF* would be a possibility.

Y;Autosome Translocation

As mentioned above, Schempp et al. (1985) described a 45,X male in whose genome Yp-derived single-copy DNA was present. The patient had a cytogenetic abnormality of one 15p, and the authors suggested that Y-DNA (and *TDF*) might be present on 15p. In the absence of direct evidence in favor of the translocation, the case is unproven. However, both *Xg* and X-chromosomal RFLPs showed the patient's only X to be maternally derived. This raises the possibility that Y-DNA is indeed present on an autosome in this 45,X male patient. It remains to be shown what autosome is involved. Further Y;autosome translocations reported in "45,X" males have been described by Subrt and Blehova (1974), Koo et al. (1977), and Turleau et al. (1980). Even though in each case the authors suggested translocation of euchromatic Y-chromosomal material onto an autosome to explain the patient's maleness, firm evidence is presently lacking. The question whether Y-DNA might exist on an autosome rather than an X chromosome not only in some 45,X males but also in some XX males requires further study.

Gene Mutations Causing Maleness

There is not evidence to suggest that genes exist in man that, if mutated, become testis-determining. The *Sxr* "mutation" in mice was long thought to be an autosomal one until it could be shown that *Sxr* is a transposition and duplication of the testis-determining gene on the Y associated with its abnormal meiotic transfer to the X (Cattanach et al. 1982; Evans et al. 1982; Singh and Jones 1982). XX males or hermaphrodites occur in other mammals, such as pigs (Sittman et al. 1980) and dogs (Selden et al. 1984), but the mechanisms have not been clarified.

One argument against the existence of a gene mutation causing maleness in man is the paucity of familial XX or X males, notably among brothers. The few familial cases reported have been previously reviewed (de la Chapelle 1981). Among these, the two XX male brothers born to unrelated parents (Minowada et al. 1979) might yield important information if studied by molecular methods. A well-studied case of related XX males (de la Chapelle et al. 1977) is briefly discussed here. The pedigree shows two male second cousins who were both XX males (Fig. 4). These two individuals were distantly related to a third XX male. In the case of the second cousins, male-to-male transmission excludes X-chromosomal inheritance (Fig. 4). The transmission through females to the more distantly related third XX male (not shown) precludes autosomal dominant inheritance. Hence the possibility of autosomal

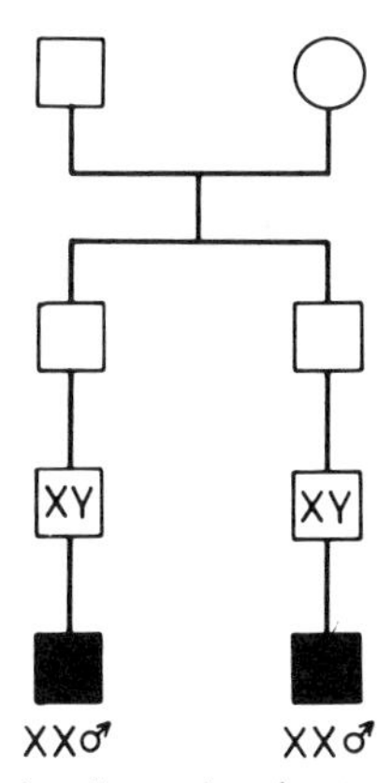

Figure 4. Simplified pedigree showing two XX males who are second cousins.

recessive inheritance was raised (de la Chapelle et al. 1978; Burgoyne 1984).

Molecular studies using DNA probes derived from the short arm of the Y chromosome showed that all three related XX males did indeed have Y-DNA. The fragments detected with three probes were identical in the two second cousins but different in the distantly related third XX male (Page et al. 1985). These findings indicate that the molecular event causing maleness in the distantly related individual was different from that in the second cousins and so the relationship to the third patient may be fortuitous. On the other hand, the molecular event appears to have been the same in both second cousins and so might have been genetically transmitted from their common ancestors, that is, the paternal grandparents of the two fathers. However, as indicated earlier, male-to-male transmission excludes X-chromosomal inheritance. It remains to be shown, for example, by in situ hybridization, in which chromosome the Y-DNA resides. If it is found on an autosome, it probably was genetically transmitted through three generations. If it is found on an X chromosome, an identical abnormal X-Y crossover must have occurred in the meiosis of both fathers. To account for this rare event occurring with the same breakpoints in two first cousins (the fathers of the XX males), the existence of a common gene predisposing to abnormal crossovers has been suggested (Page et al. 1985). Alternatively, a structural peculiarity of the DNA in the critical region of Yp (which the fathers share) and of Xp (which they may share) might predispose to meiotic breakage and reunion in exactly the same places in both. As more probes become available, this problem may eventually be elucidated.

In conclusion, familial occurrence of maleness without a Y chromosome is very rare. No evidence exists to suggest that a mutation in a hitherto undescribed gene can make it into a testis-determining one. Presently it is not possible to distinguish between several hypotheses as to the mechanisms involved in the rare instances of familial XX maleness.

ACKNOWLEDGMENTS

This work was supported by grants from the Sigrid Juselius Foundation, the Finska Läkaresällskapet, and the Academy of Finland. Part of the research was performed at the Folkhälsan Institute of Genetics, Helsinki.

REFERENCES

Andersson, M., D.C. Page, and A. de la Chapelle. 1986. Chromosome Y-specific DNA is transferred to the short arm of the X-chromosome in human XX males. *Science* **233:** 386.

Bishop, C.E., G. Guellaen, D. Geldwerth, R. Voss, M. Fellous, and J. Weissenbach. 1983. Single-copy DNA sequences specific for the human Y chromosome. *Nature* **303:** 831.

Buckle, V., C. Mondello, S. Darling, I.W. Craig, and P.N. Goodfellow. 1985. Homologous expressed genes in the human sex chromosome pairing region. *Nature* **317:** 739.

Burgoyne, P.S. 1982. Genetic homology and crossing over in the X and Y chromosomes of mammals. *Hum. Genet.* **61:** 85.

———. 1984. The origins of men with two X chromosomes. *Nature* **307:** 109.

Cattanach, B.M., E.P. Evans, M. Burtenshaw, and J. Barlow. 1982. Male and female development in mice of identical chromosome constitution. *Nature* **300:** 445.

Chandley, A.C., P. Goetz, T.B. Hargreave, A.M. Joseph, and R.M. Speed. 1984. On the nature and extent of XY pairing at meiotic prophase in man. *Cytogenet. Cell Genet.* **38:** 241.

Cooke, H.J., W.R.A. Brown, and G.A. Rappold. 1985. Hypervariable telomeric sequences from the human sex chromosomes are pseudoautosomal. *Nature* **317:** 687.

de la Chapelle, A. 1972. Nature and origin of males with XX sex chromosomes. *Am. J. Hum. Genet.* **24:** 71.

———. 1981. The etiology of maleness in XX men. *Hum. Genet.* **58:** 105.

de la Chapelle, A., G.C. Koo, and S.S. Wachtel. 1978. Recessive sex-determining genes in human XX male syndrome. *Cell* **15:** 837.

de la Chapelle, A., H. Hortling, M. Niemi, and J. Wennström. 1964. XX sex chromosomes in a human male. First case. *Acta Med. Scand. Suppl.* **412:** 25.

de la Chapelle, A., J. Schröder, J. Murros, and G. Tallqvist. 1977. Two XX males in one family and additional observations bearing on the etiology of XX males. *Clin. Genet.* **11:** 91.

de la Chapelle, A., P.A. Tippett, G. Wetterstrand, and D. Page. 1984a. Genetic evidence of X-Y interchange in a human XX male. *Nature* **307:** 170.

de la Chapelle, A., D.C. Page, L. Brown, U. Kaski, T. Parvinen, and P.A. Tippett. 1986. The origin of 45,X males. *Am. J. Hum. Genet.* **38:** 330.

de la Chapelle, A., H. Savikurki, R. Herva, P.A. Tippett, F. Knutar, P. Gröhn, H. Siponen, K. Huovinen, and T. Korhonen. 1984b. Aetiological studies in males with the karyotype 46,XX. In *Aspects of human genetics with special reference to X-linked disorders* (ed. C. San Roman and A. McDermott), p. 125. Karger, Basel.

Darling, S.M., G.S. Banting, B. Pym, J. Wolfe, and P.N. Goodfellow. 1986. Cloning an expressed gene shared by the human sex chromosomes. *Proc. Natl. Acad. Sci.* **83:** 135.

Davis, R.M. 1981. Localization of male determining factors in man: A thorough review of structural anomalies of the Y chromosome. *J. Med. Genet.* **18:** 161.

Drayna, D. and R. White. 1985. The genetic map of the human X-chromosome. *Science* **230:** 753.

Evans, E.P., M. Burtenshaw, and B.M. Cattanach. 1982. Cytological evidence for meiotic crossing over between X and Y chromosomes of male mice carrying the sex reversing (Sxr) factor. *Nature* **300:** 445.

Geller, R.L., L.J. Shapiro, and T.K. Mohandas. 1986. Fine mapping of the distal short arm of the human X chromosome using X/Y translocations. *Am. J. Hum. Genet.* **38:** 884.

Goodfellow, P.N. and P. Tippett. 1981. A human quantitative polymorphism related to Xg blood groups. *Nature* **289:** 404.

Goodfellow, P.N., S. Darling, and J. Wolfe. 1985a. The human Y chromosome. *J. Med. Genet.* **22:** 329.

Goodfellow, P.N., K.E. Davies, and H.H. Ropers. 1985b. Human Gene Mapping 8. Report of the committee on the genetic constitution of the X and Y chromosomes. *Cytogenet. Cell Genet.* **40:** 295.

Goodfellow, P., G. Banting, D. Sheer, H.H. Ropers, A. Caine, M.A. Ferguson-Smith, S. Povey, and R. Voss. 1983. Genetic evidence that a Y-linked gene in man is homologous to a gene on the X chromosome. *Nature* **302:** 346.

Guellaen, G., M. Casanova, C. Bishop, D. Geldwerth, G. Andre, M. Fellous, and J. Weissenbach. 1984. Human XX males with Y single-copy fragments. *Nature* **307:** 172.

Khudr, G., K. Benirschke, H.L. Judd, and J. Strauss. 1973. Y to X translocation in a woman with reproductive failure. *J. Am. Med. Assoc.* **226:** 544.

Koenig, M., G. Camerino, R. Heilig, and J.-L. Mandel. 1984. A DNA fragment from the human X chromosome short arm which detects a partially homologous sequence on the Y chromosome long arm. *Nucleic Acids Res.* **12:** 4097.

Koenig, M., J.P. Moisan, R. Heilig, and J.-L. Mandel. 1985. Homologies between X and Y chromosomes detected by DNA probes: Localization and evolution. *Nucleic Acids Res.* **13:** 5485.

Koo, G.C., S.S. Wachtel, K. Krupen-Brown, L.R. Mittl, W.R. Breg, M. Genel, I.M. Rosenthal, D.S. Borgaonkar, A.D. Miller, R. Tantravahi, R.R. Scheck, B.F. Erlanger, and O.J. Miller. 1977. Mapping the locus of the H-Y gene on the human Y chromosome. *Science* **198:** 940.

Migeon, B.R., L.J. Shapiro, R.A. Norum, T. Mohandas, J. Axelman, and R.L. Dabora. 1982. Differential expression of steroid sulphatase levels in XX males, including observations on two affected cousins. *Hum. Genet.* **59:** 87.

Minowada, S., K. Kobayashi, K. Isurugi, K. Fukutani, H. Ikeuchi, T. Hasegawa, and K. Yamada. 1979. Two XX male brothers. *Clin. Genet.* **15:** 399.

Page, D.C. and A. de la Chapelle. 1984. The parental origin of X chromosomes in XX males determined using restriction fragment length polymorphisms. *Am. J. Hum. Genet.* **36:** 565.

Page, D.C., A. de la Chapelle, and J. Weissenbach. 1985. Chromosome Y-specific DNA in related human XX males. *Nature* **315:** 224.

Page, D., M.E. Harper, J. Love, and D. Botstein. 1984. Occurrence of a transposition from the X chromosome long arm to the Y-chromosome short arm during human evolution. *Nature* **311:** 119.

Page, D., B. de Martinville, D. Barker, A. Wyman, R. White, U. Francke, and D. Botstein. 1982. Single copy sequence hybridizes to polymorphic and homologous loci on human X and Y chromosomes. *Proc. Natl. Acad. Sci.* **79:** 5352.

Pierella, P., I. Craig, M. Bobrow, and A. de la Chapelle. 1981. Steroid sulphatase levels in XX males including observations on two affected cousins. *Hum. Genet.* **59:** 87.

Polani, P.E. 1982. Pairing of X and Y chromosomes, non-inactivation of X-linked genes, and the maleness factor. *Hum. Genet.* **60:** 207.

Race, R.R and R. Sanger. 1975. *Blood groups in man*, 6th edition. Blackwell Scientific, Oxford.

Ropers, H.H., B. Migl, J. Zimmer, and C.R. Müller. 1981. Steroid sulfatase activity in cultured fibroblasts of XX males. *Cytogenet. Cell Genet.* **30:** 168.

Rouyer, F., M.-C. Simmler, C. Johnsson, G. Vergnaud, H.J. Cooke, and J. Weissenbach. 1986. A gradient of sex linkage in the pseudoautosomal region of the human sex chromosomes. *Nature* **319:** 291.

Sanger, R., P. Tippett, and J. Gavin. 1971. The X-linked blood group system Xg: Tests on unrelated people and families of Northern European ancestry. *J. Med. Genet.* **8:** 427.

Schempp, W., B. Weber, A. Serra, G. Neri, A. Gal, and U. Wolf. 1985. A 45,X male with evidence of a translocation of Y euchromatin onto chromosome 15. *Hum. Genet.* **71:** 150.

Selden, J.R., P.S. Moorhead, G.C. Koo, S.S. Wachtel, M.E. Haskins, and D.F. Patterson. 1984. Inherited XX sex reversal in the cocker spaniel dog. *Hum. Genet.* **67:** 62.

Shows, T.B., P.J. McAlpine, and R.L. Miller. 1984. Human Gene Mapping 7. The 1983 catalogue of mapped human genetic markers and report of the nomenclature committee. *Cytogenet. Cell Genet.* **37:** 340.

Simmler, M.-C., F. Rouyer, G. Vergnaud, M. Nyström-Lahti, K.Y. Ngo, A. de la Chapelle, and J. Weissenbach. 1985. Pseudoautosomal DNA sequences in the pairing region of the human sex chromosomes. *Nature* **317:** 692.

Singh, L. and K. Jones. 1982. Sex reversal in the mouse (*Mus musculus*) is caused by a recurrent non-reciprocal cross over involving the X and aberrant Y chromosome. *Cell* **28:** 205.

Sittman, K., A.J. Breeuwsma, and J.H.A. TeBrake. 1980. On the inheritance of intersexuality in swine. *Can. J. Genet. Cytol.* **22:** 507.

Subrt, I. and B. Blehova. 1974. Robertsonian translocation between the chromosomes Y and 15. *Humangenetik* **23:** 305.

Tiepolo, L., O. Zuffardi, M. Fraccaro, B. di Natale, L. Gargantini, C.R. Müller, and H.H. Ropers. 1980. Assignment by deletion mapping of the steroid sulfatase X-linked ichthyosis locus to Xp223. *Hum. Genet.* **54:** 205.

Tippett, P., M.-A. Shaw, C.A. Green, and G.L. Daniels. 1986. The 12E7 red cell quantitative polymorphism: Control by the Y-borne locus, Yg. *Ann. Hum. Genet.* **50:** (in press).

Turleau, C., F. Chavin-Colin, and J. de Grouchy. 1980. A 45,X male with translocation of euchromatic Y chromosome material. *Hum. Genet.* **53:** 299.

Vergnaud, G., D.C. Page, M.-C. Simmler, L. Brown, F. Rouyer, B. Noel, D. Botstein, A. de la Chapelle, and J. Weissenbach. 1986. A deletion map of the human Y chromosome based on DNA hybridization. *Am. J. Hum. Genet.* **38:** 109.

Wieacker, P., J. Voiculescu, C.R. Müller, and H.H. Ropers. 1983. An XX male with a single STS gene dose. *Cytogenet. Cell Genet.* **35:** 72.

Application of Synthetic DNA Probes to the Analysis of DNA Sequence Variants in Man

R.B. WALLACE,* L.D. PETZ,† AND P.Y. YAM†
*Department of Molecular Genetics, Beckman Research Institute of the City of Hope, Duarte, California 91010; †Department of Clinical and Experimental Immunology, City of Hope Medical Center, Duarte, California 91010

It has been estimated that the human genome contains between 10^7 (Cooper et al. 1985) and 3×10^7 (Jeffreys 1979) polymorphic DNA sequence variants. Some of this sequence variation can be detected by the analysis of the gain or loss of specific restriction endonuclease cleavage sites, resulting in restriction-fragment-length polymorphism (RFLP) (Botstein et al. 1980; White et al. 1985). Such analysis, however, only permits the sampling of a small fraction (perhaps 1%) of the vast pool of genetic variability (Jeffreys 1979). The reasons for this limitation are clear. (1) Only a small fraction of nucleotide sequences are known to be restriction enzyme cleavage sites. For example, of the 4096 possible hexanucleotide sequences, only 64 (1.5%) have the palindromic symmetry typical of most class II restriction enzyme cleavage sites. Of these 64 sequences, only 41 have been identified as known restriction enzyme cleavage sites (Kessler et al. 1985). (2) Some of the enzymes recognizing known cleavage sites are not practical to use because the enzymes are not commercially available, are expensive, or produce unacceptably small or large fragments. (3) The resolution of the RFLP approach is such that small differences in fragment length can be overlooked. (4) Sequence variants present in the repetitive DNA (30% of all sequences) are not easily analyzed by conventional probes. Due to these limitations, identification of sufficient RFLP loci to produce a complete map of the human genome is far from being complete. Indeed, RFLP of a locus is often difficult to demonstrate. For example, in a recent study of the human factor VIII gene, only 1 out of 35 enzymes tested was found to reveal polymorphism (Gitschier et al. 1985).

In recent studies, Barker et al. (1984) have shown that RFLP loci are more often seen using *Taq*I and *Msp*I restriction enzymes than with others. These authors proposed that there is a greater frequency of polymorphism at these two restriction sites because of the presence of the CpG dinucleotide in the recognition sequence. The CpG dinucleotide is underrepresented in the mammalian genome (Singer et al. 1979), and the C is usually methylated (Riggs and Jones 1983). By analogy with bacteria, where 5meC residues are hot spots for transition mutations (Miller 1983), it has been proposed that the 5meC sequence in the CpG dinucleotide is a mutational hot spot in mammalian cells (i.e., CpG→TpG or CpA). In bacteria, 5meC deaminates to T and therefore escapes recognition by repair enzymes (C, on the other hand, deaminates to U and is recognized and repaired). Thus, *Msp*I (CCGG) and *Taq*I (TCGA) sites will be lost at a higher rate than other restriction enzyme sites. Barker et al. (1984) estimated that a CpG dinucleotide is about 10 times more likely to be polymorphic than any other sequence. This phenomenon has resulted in a more frequent identification of RFLP loci using *Msp*I and *Taq*I, but these enzymes still only allow the sampling of less than 1% of the possible tetranucleotide sequences (2/256).

Another approach to the sampling of DNA sequence variation is through the use of synthetic oligonucleotide probes. Under appropriate conditions, synthetic oligonucleotide hybridization probes display essentially absolute hybridization specificity; i.e., every nucleotide must form a Watson-Crick base pair so that the probe forms a stable duplex. All of the non-Watson-Crick base pairs, including G-T, have a destabilizing effect (for review, see Itakura et al. 1984). Thus, it is possible to choose stringent conditions of hybridization such that, although a perfectly matched duplex between an oligonucleotide and complementary DNA will form, duplexes mismatched at one or more positions will not. Sickle cell anemia, a human genetic disease found predominantly in the black population, is the result of a single-base-pair change in the β-globin gene (GAG→GTG, Glu→Val at codon 6). Because of their absolute hybridization specificity under stringent conditions, oligonucleotide hybridization probes complementary to the normal β-globin gene, β^A, or to the sickle cell allele, β^S, in the region of the point mutation can distinguish between these two alleles (Conner et al. 1983). DNA from individuals homozygous for the normal gene will hybridize to the normal gene probe but not to the sickle cell gene probe, and DNA from those homozygous for the sickle cell gene will hybridize to the sickle cell gene probe but not to the normal gene probe. Only DNA from heterozygotes will hybridize to both probes. Examples of mutant genes (alleles) that have been distinguished from their normal counterparts by this approach are $\beta^{39(UAG)}$ thalassemia (Pirastu et al. 1983), $\beta^{IVS\text{-}1}$ thalassemia (Orkin et al. 1983; Thein et al. 1985b), α_1-antitrypsin (Kidd et al. 1983), β^C (Studencki et al. 1984), and β^E (Thein et al. 1985a).

Clearly, oligonucleotide hybridization allows the detection of base substitutions between two alleles that differ by as little as a single base pair. In the case where the sequence of only one allele is known, it is possible to detect the presence of all other alleles with a different sequence in the probe region by the absence of hybridization. Oligonucleotides allow the sampling of more nucleotides per site than a restriction enzyme does. More importantly, there is no constraint on which sequences are chosen to sample as there is in the case of RFLPs.

Probe Design

There are several ways in which oligonucleotide probes could be used to reveal polymorphism in the human genome. One way would be to use these probes to detect polymorphisms due to single-base substitutions that do not affect restriction enzyme cleavage sites and thus do not produce RFLPs. The problem is to determine which sequences are polymorphic. Since it has been estimated that between 1 in 100 and 1 in 300 bp in the human genome are polymorphic, one approach would be to synthesize oligonucleotide probes complementary to known sequences and test each one on a panel of DNA from unrelated individuals. For oligonucleotide probes 20 bases long, between 1 in 5 and 1 in 15 would be expected to reveal polymorphism.

An alternative approach is to use oligonucleotide probes complementary to specific repetitive DNA sequences. Highly repeated DNA sequences might be expected to be highly polymorphic due to processes such as unequal crossing over or gene conversion. Cloned members of such repeat families hybridize to many related sequences that are interspersed in the genome, producing complex hybridization patterns. These sequences are thus avoided as probes for polymorphism analysis (White et al. 1985). Synthetic DNA sequences complementary to a sequence present in the repeat family will hybridize to those members of the family containing that sequence but not to those containing a related but not identical one. Thus, it is expected that the synthetic DNA probe will hybridize to a subset of the repeat family, producing a less complex pattern and revealing any RFLPs present in that subset. This latter approach was used in the present study. The sequence chosen is complementary to the highly repeated, evolutionarily conserved simple repeat $(GACA)_n$ (Epplen et al. 1982; Singh et al. 1984).

Mixed Hematopoietic Chimerism

Various genetic markers have been used to document the engraftment of donor marrow after bone-marrow transplantation (BMT). These include the use of red blood cell antigens, chromosome analysis, immunoglobulin allotypes, red and white blood cell enzymes, and white blood cell antigens (Yam et al. 1986). Each of these methods has its limitations. For example, red blood cell antigens and immunoglobulin allotypes cannot be used in the early posttransplant period due to the clinical necessity of blood transfusions, making determination of the patient's phenotype impossible for 3–6 months following BMT. Red and white cell enzymes and white cell antigens are not highly informative genetic markers, and they only give information about a single cell type when they are useful.

Recently, several groups have used RFLPs as genetic markers to document the origin of hematopoietic cells in patients following BMT (Schubach et al. 1982; Blazar et al. 1985; Ginsburg et al. 1985; Minden et al. 1985; Witherspoon et al. 1985; Petz et al. 1986; Yam et al. 1986). We have used highly polymorphic DNA probes to document engraftment in the early posttransplant period, to demonstrate both transient and chronic stable mixed hematopoietic chimerism and to study the origin of leukemic cells in patients with recurrent disease (Petz et al. 1986; Yam et al. 1986). RFLPs represent powerful genetic markers for this purpose. In order for there to be a high probability that there exists a difference in the donor and recipient restriction fragment patterns, one needs a small number of RFLP probes with a high polymorphism information content (PIC). Unfortunately, at this moment, one needs to use as many as five probes and each probe may use a different enzyme. What would be better would be a single probe using a single enzyme. To this end, we began investigating probes to highly repeated DNA sequences.

EXPERIMENTAL PROCEDURES

Oligonucleotide synthesis and labeling. The oligodeoxyribonucleotide 5′-GACAGACAGACAGACA, abbreviated $(GACA)_4$, was synthesized, using a Systec Microsyn 1450 automated DNA synthesizer, and purified by polyacrylamide gel electrophoresis and/or high-performance liquid chromatography (HPLC), using a PRP-1 reversed-phase column (Hamilton). Oligonucleotide (20 pmoles) was labeled at the 5′-terminal hydroxyl group using [γ-^{32}P]ATP (40 pmoles; New England Nuclear, 7000 Ci/mmole) and T_4 polynucleotide kinase (5–9 units; New England Nuclear) in a 10-μl reaction containing 67 mM Tris-HCl (pH 8.0), 10 mM $Mg(OAc)_2$, and 10 mM dithiothreitol (DTT). The kinase reactions were performed at 37°C for 30 minutes. The labeled oligonucleotide was then separated from the [γ-^{32}P]ATP by chromatography over DE-52 cellulose (Whatman). After the kinase reaction, the labeled oligonucleotide was diluted tenfold in 10 mM Tris-HCl and 1 mM EDTA (TE) (pH 8.0). The sample was then applied to a small column of DE-52 (bed volume, ~0.2 ml) previously equilibrated with TE. The column was then washed with 5 bed volumes of TE followed by 5 bed volumes of 0.2 M NaCl in TE, which removes the unincorporated label. The labeled oligonucleotide was then eluted with 0.5 ml of 1 M NaCl in TE and filtered through a 0.45-μm pore size Acrodisc disposable filter (Gelman) prior to use.

Restriction enzyme digestion and hybridization of genomic DNAs. Human DNAs were prepared from peripheral blood samples as described previously (Conner et al. 1983). Genomic DNAs (5 μg) were digested in

100-μl reactions with 30 units *Mbo*I in 100 mM NaCl, 10 mM Tris-HCl (pH 7.5), 10 mM $MgCl_2$, and 1 mM DTT for 16 hours at 16°C. After digestion, samples were recovered by ethanol precipitation. Samples (1 μg) in 10 μl of 20 mM Tris-HCl (pH 8.2), 2 mM EDTA, 10% glycerol, and 0.1% bromocresol green were loaded onto a horizontal 0.6% agarose gel (Seakem, Maine) in a buffer containing 0.04 M Tris-acetate (pH 8.0) and 2 mM EDTA and subjected to electrophoresis at 2 V/cm for 16 hours. After running, the gel was stained with ethidium bromide and photographed under UV light. The gel was then prepared for direct hybridization (Conner et al. 1983; Tsao et al. 1983). The DNA was denatured in situ with 200 ml of 0.4 N NaOH and 0.8 M NaCl at room temperature for 30 minutes and neutralized with 200 ml of 0.5 M Tris-HCl (pH 7.2) and 0.15 M NaCl at 0°C for 30 minutes. The gel was then dried onto Whatman 3MM paper using a Bio-Rad gel dryer. The gel membrane was removed from its paper backing by wetting the paper with distilled water.

The gel was directly hybridized with the labeled oligonucleotide in 5× SSPE (1× SSPE is 10 mM sodium phosphate [pH 7.0], 0.18 M NaCl, 1 mM EDTA), 0.1% SDS, 10 μg/ml sonicated, denatured *Escherichia coli* DNA, and 1×10^6 cpm/ml labeled probe at 42°C for 16 hours. After the hybridization, the gel was washed at room temperature two times for 15 minutes, followed by a third wash for at least 2 hours in 6× SSC (1× SSC is 0.15 M NaCl, 0.015 M sodium citrate). The gel was then washed at 42°C in 6× SSC for 1 minute. The washed gel was then exposed to X-ray film with two du Pont Lightning-Plus intensifying screens at −70°C for 4–16 hours.

RESULTS

Repeated DNA Probes

Recently, Epplen and co-workers (pers. comm.) demonstrated that oligonucleotide probes complementary to the evolutionarily conserved simple repeated sequences $(GACA)_n$ and $(GATA)_n$ (Epplen et al. 1982; Singh et al. 1984) hybridize to restriction endonuclease fragments of varying sizes in different individuals. Family studies were consistent with these RFLPs being inherited as Mendelian traits. We were interested in exploiting the GACA RFLPs as markers to trace the donor and recipient cells in bone-marrow-transplant patients. Since the frequency of polymorphism of these RFLPs seems to be rather high, it was hoped that a large proportion of donor/recipient pairs would show different patterns. In preliminary studies, it was found that in six out of nine pairs, the patterns were different in the recipient and donor. Two donor/recipient pairs, UPN 301 and UPN 307, were chosen for further study (UPN refers to unique patient number).

Hybridization to Human Genomic DNA

DNA from peripheral lymphocytes was isolated from both the donors and the recipients prior to the transplant. Following the BMT, DNA was isolated from the recipients on day 30, 60, and 93 in the case of UPN 301 and on day 30, 40, and 60 in the case of UPN 307. The DNA was digested with the restriction enzyme *Mbo*I and subjected to electrophoresis on a 0.6% agarose gel. The DNA was denatured, and the gel was neutralized and then dried. The DNA was then hybridized directly in the gel with the $(GACA)_4$ probe. The gel was then washed in 6× SSC and exposed to X-ray film. The resulting films are shown in Figures 1 and 2. In the case of UPN 301 (Fig. 1), there is a unique hybridizing band present in the donor (~6 kb) and in the recipient (~ 4.4 kb). The 30- and 60-day posttransplant samples show a complete donor pattern, indicating engraftment of the donor marrow. On day 93, however, the reappearance of the recipient-specific band and the reduced intensity of the donor-specific band indicate the recurrence of the leukemia. These results are in complete agreement with those previously reported for this patient using other genetic markers, including RFLPs (Yam et al. 1986). In the case of UPN 307 (Fig. 2), there is only a recipient-specific band (~4.6 kb). The posttransplant samples all show a donor-specific pattern, indicating engraftment of the donor marrow without a detectable level of mixed hematopoietic chimerism.

DISCUSSION

Oligonucleotide probes provide a tool to discriminate between any two alleles on the basis of hybridiza-

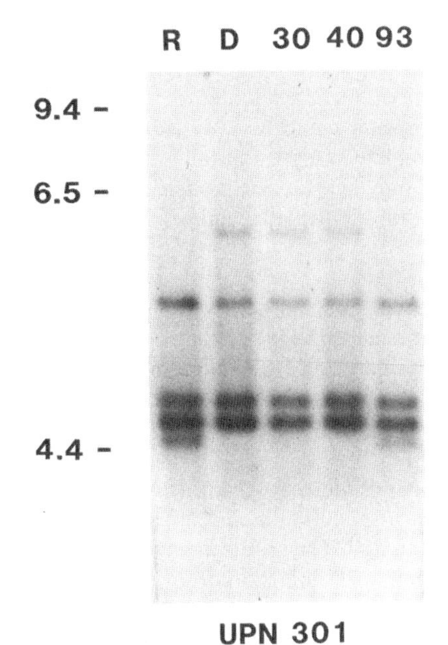

Figure 1. Autoradiogram of genomic DNA samples isolated from the peripheral lymphocytes of the donor (D) and the recipient UPN 301. R is the pretransplant recipient sample. Also shown are the 30-, 60-, and 93-day posttransplant samples. The DNA was digested with *Mbo*I and electrophoresed on a 0.6% agarose gel. The DNA was denatured, and the gel was neutralized and dried. The dried gel was then hybridized with the $(GACA)_n$ probe as described in Experimenal Procedures. The sizes of the marker DNA bands (λ DNA digested with *Hin*dIII) are given in kilobases.

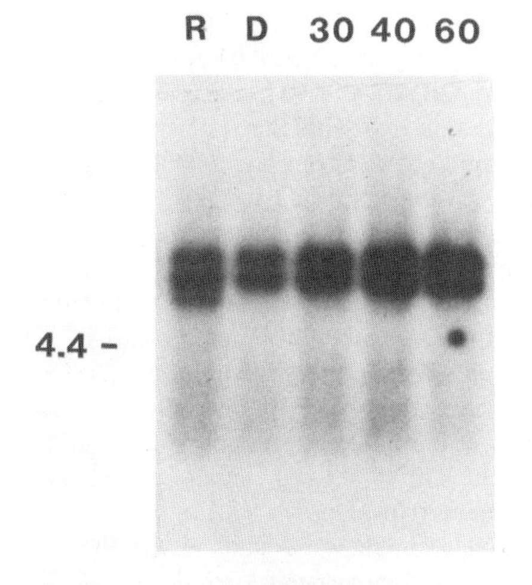

Figure 2. Autoradiogram of genomic DNA samples isolated from the peripheral lymphocytes of the donor (D) and the recipient UPN 307. R is the pretransplant recipient sample. Also shown are the 30-, 40- and 60-day posttransplant samples. The DNA was electrophoresed and hybridized with the $(GACA)_n$ probe as described in Fig. 1.

tion. Since the single-copy DNA sequence heterozygosity of the human genome is at least 0.0037 (Jeffreys 1979; Copper et al. 1985), random sampling of the genome with different oligonucleotide probes should reveal polymorphism in a certain percentage of the cases. In the hope of identifying polymorphic regions more efficiently, we chose to take advantage of the proposed hypermutability of repeated DNA sequences and the specificity of oligonucleotide hybridization. Since, under appropriate conditions, oligonucleotide probes require complete base pairing for hybridization to occur, they will only hybridize to a subset of the members of a repeat family when all members of the family are not identical.

The results presented here suggest that oligonucleotide hybridization can be used to extend the genomic sequences that can be tested for the presence of RFLPs. This expands the tools available to human genetics. In addition, the results suggest that repeated DNA sequences are indeed more polymorphic than single-copy sequences.

The analysis of marrow engraftment would be greatly improved with the availability of a single RFLP probe with a high PIC. The results presented here are encouraging, since 67% of the donor/recipient pairs were found to have informative differences in their pattern of hybridization to the $(GACA)_4$ probe. This degree of polymorphism is still not sufficient, however. As seen in the case of UPN 307, there only existed a recipient-specific band, making it impossible to follow the fate of the donor cells. We are continuing to investigate other probes complementary to repeated DNA sequences, such as the $(GATA)_n$ sequence, with the aim of finding a probe for which there is a greater than 95% probability of finding informative differences between the donor and recipient DNA.

ACKNOWLEDGMENTS

We thank Drs. K. Blume, J. Epplen, and J. Schmidtke for helpful discussions. The work described here was supported by National Institutes of Health grant RO1-GM-31261 and National Science Foundation grant DCB-8515365 to R.B.W. and National Institutes of Health grant PO1-CA-30206. R.B.W. is a member of the cancer center of the City of Hope (NIH-CA-33572).

REFERENCES

Barker, D., M. Schafer, and R. White. 1984. Restriction sites containing CpG show a higher frequency of polymorphism in human DNA. *Cell* **36:** 131.

Blazar, B.R., H.T. Orr, D.C. Arthur, J.H. Kersey, and A.H. Filipovich. 1985. Restriction fragment length polymorphisms as markers of engraftment in allogeneic marrow transplantation. *Blood* **66:** 1436.

Botstein, D., R.L. White, M. Skolnick, and R.W. Davis. 1980. Construction of a genetic linkage map in man using restriction fragment length polymorphisms. *Am. J. Hum. Genet.* **32:** 314.

Connor, B.J., A.A. Reyes, C. Morin, K. Itakura, R.L. Teplitz, and R.B. Wallace. 1983. Detection of sickle cell β^S-globin allele by hybridization with synthetic oligonucleotides. *Proc. Natl. Acad. Sci.* **80:** 278.

Cooper, D.N., B.A. Smith, H.J. Cooke, S. Niemann, and J. Schmidtke. 1985. An estimate of unique DNA sequence heterozygosity in the human genome. *Hum. Genet.* **69:** 201.

Epplen, J.T., J.R. McCarrey, S. Sutou, and S. Ohno. 1982. Base sequence of a cloned snake W-chromosome DNA fragment and identification of a male-specific putative mRNA in the mouse. *Proc. Natl. Acad. Sci.* **79:** 3798.

Ginsburg, D., J.H. Antin, B.R. Smith, S.H. Orkin, and J.M. Rappeport. 1985. Origin of cell populations after bone marrow transplantation. Analysis using DNA sequence polymorphisms. *J. Clin. Invest.* **75:** 596.

Gitschier, J., D. Drayna, E.G.D. Tuddenheim, R.L. White, and R.M. Lawn. 1985. Genetic mapping and diagnosis of haemophilia A achieved through a *BclI* polymorphism in the factor VIII gene. *Nature* **314:** 738.

Itakura, K., J.J. Rossi, and R.B. Wallace. 1984. Synthesis and use of synthetic oligonucleotides. *Annu. Rev. Biochem.* 53: 323.

Jeffreys, A.J. 1979. DNA sequence variants in the $^G\gamma$-, $^A\gamma$-, δ- and β-globin genes of man. *Cell* **18:** 1.

Kessler, C., P.S. Neumaier, and W. Wolf. 1985. Recognition sequences of restriction endonucleases and methylases—A review. *Gene* **33:** 1.

Kidd, V.J., R.B. Wallace, K. Itakura, and S.L.C. Woo. 1983. α_1 antitrypsin deficiency detection by direct analysis of the mutation in the gene. *Nature* **304:** 230.

Miller, J.H. 1983. Mutational specificity in bacteria. *Annu. Rev. Genet.* **17:** 215.

Minden, M.D., H.A. Messner, and A. Belch. 1985. Origin of leukemic relapse after bone marrow transplantation detected by restriction fragment length polymorphisms. *J. Clin. Invest.* **75:** 91.

Orkin, S.H., A.F. Markham, and H.H. Kazazian. 1983. Direct detection of the common Mediterranean β-thalassemia gene with synthetic DNA probes. *J. Clin. Invest.* **71:** 775.

Petz, L.D., P. Yam, R.B. Wallace, A.D. Stock, G. de Lange, R.G. Knowlton, V.A. Brown, H. Donis-Keller, and K.G. Blume. 1986. Mixed hematopoietic chimerism following bone marrow transplantation for hematologic malignancies: Incidence, characterization, and implications of marrow transplantation. *UCLA Symp. Mol. Cell. Biol. New Ser.* **53:** (in press).

Pirastu, M., Y.W. Kan, A. Cao, B.J. Conner, R.L. Teplitz, and R.B. Wallace. 1983. Prenatal diagnosis of β thalassemia: Direct detection of a single nucleotide mutation in DNA. *N. Eng. J. Med.* **309:** 284.

Riggs, A.D. and P.A. Jones. 1983. 5-Methylcytosine, gene regulation and cancer. *Adv. Cancer Res.* **40:** 1.

Schubach, W.H., R. Hackman, P.E. Neiman, G. Miller, and E.D. Thomas. 1982. A monoclonal immunoblastic sarcoma in donor cells bearing Epstein-Barr virus genomes following allogeneic marrow grafting for acute lymphoblastic leukemia. *Blood* **60:** 180.

Singer, J., J. Roberts-Ems, and A.D. Riggs. 1979. Methylation of mouse liver DNA studied by means of the restriction enzymes *Msp*I and *Hpa*II. *Science* **203:** 1019.

Singh, L., C. Phillips, and K.W. Jones. 1984. The conserved nucleotide sequences of Bkm, which define Sxr in the mouse, are transcribed. *Cell* **36:** 111.

Studencki, A.B., B.J. Conner, C.C. Impraim, R.L. Teplitz, and R.B. Wallace. 1984. Discrimination among the human β^A, β^S and β^C-globin genes using allele specific oligonucleotide hybridization probes. *Am. J. Hum. Genet.* **37:** 42.

Thein, S.L., J.R. Lynch, D.J. Weatherall, and R.B. Wallace. 1985a. Direct detection of haemoglobin E with synthetic oligonucleotides. *Lancet* **I:** 93.

Thein, S.L., J.S. Wainscoat, J.M. Old, M. Sampietro, G. Fiorelli, R.B. Wallace, and D.J. Weatherall. 1985b. Feasibility of prenatal diagnosis of β-thalassaemia with synthetic DNA probes in two Mediterranean populations. *Lancet* **II:** 345.

Tsao, S.G.S., C.F. Brunk, and R.E. Perlman. 1983. Hybridization of nucleic acids directly in agarose gels. *Anal. Biochem.* **131:** 365.

White, R., M. Leppert, D.T. Bishop, D. Barker, J. Berkowitz, C. Brown, P. Callahan, T. Holm, and L. Jerominski. 1985. Construction of linkage maps with DNA markers for human chromosomes. *Nature* **313:** 101.

Witherspoon, R.P., W. Schubach, P. Neiman, P. Martin, and E.D. Thomas. 1985. Donor cell leukemia developing six years after marrow grafting for acute leukemia. *Blood* **65:** 1172.

Yam, P.Y., L.D. Petz, R.G. Knowlton, R.B. Wallace, A.D. Stock, G. de Lange, V.A. Brown, H. Donis-Keller, and K.G. Blume. 1986. Use of DNA restriction fragment length polymorphisms to document marrow engraftment and mixed hematopoietic chimerism following bone marrow transplantation. *Transplantation* (in press).

Specific Enzymatic Amplification of DNA In Vitro: The Polymerase Chain Reaction

K. MULLIS, F. FALOONA, S. SCHARF, R. SAIKI, G. HORN, AND H. ERLICH
Cetus Corporation, Department of Human Genetics, Emeryville, California 94608

The discovery of specific restriction endonucleases (Smith and Wilcox 1970) made possible the isolation of discrete molecular fragments of naturally occurring DNA for the first time. This capability was crucial to the development of molecular cloning (Cohen et al. 1973); and the combination of molecular cloning and endonuclease restriction allowed the synthesis and isolation of any naturally occurring DNA sequence that could be cloned into a useful vector and, on the basis of flanking restriction sites, excised from it. The availability of a large variety of restriction enzymes (Roberts 1985) has significantly extended the utility of these methods.

The de novo organic synthesis of oligonucleotides and the development of methods for their assembly into long double-stranded DNA molecules (Davies and Gassen 1983) have removed, at least theoretically, the minor limitations imposed by the availability of natural sequences with fortuitously unique flanking restriction sites. However, de novo synthesis, even with automated equipment, is not easy; it is often fraught with peril due to the inevitable indelicacy of chemical reagents (Urdea et al. 1985; Watt et al. 1985; Mullenbach et al. 1986), and it is not capable of producing, intentionally, a sequence that is not yet fully known.

We have been exploring an alternative method for the synthesis of specific DNA sequences (Fig. 1). It involves the reciprocal interaction of two oligonucleotides and the DNA polymerase extension products whose synthesis they prime, when they are hybridized to different strands of a DNA template in a relative orientation such that their extension products overlap. The method consists of repetitive cycles of denaturation, hybridization, and polymerase extension and seems not a little boring until the realization occurs that this procedure is catalyzing a doubling with each cycle in the amount of the fragment defined by the positions of the 5′ ends of the two primers on the template DNA, that this fragment is therefore increasing in concentration exponentially, and that the process can be continued for many cycles and is inherently very specific.

The original template DNA molecule could have been a relatively small amount of the sequence to be synthesized (in a pure form and as a discrete molecule) or it could have been the same sequence embedded in a much larger molecule in a complex mixture as in the case of a fragment of a single-copy gene in whole human DNA. It could also have been a single-stranded DNA molecule or, with a minor modification in the technique, it could have been an RNA molecule. In any case, the product of the reaction will be a discrete double-stranded DNA molecule with termini corresponding to the 5′ ends of the oligonucleotides employed.

We have called this process polymerase chain reaction or (inevitably) PCR. Several embodiments have been devised that enable one not only to extract a specific sequence from a complex template and amplify it, but also to increase the inherent specificity of this process by using nested primer sets, or to append sequence information to one or both ends of the sequence as it is being amplified, or to construct a sequence entirely from synthetic fragments.

MATERIALS AND METHODS

PCR amplification from genomic DNA. Human DNA (1 μg) was dissolved in 100 μl of a polymerase buffer containing 50 mM NaCl, 10 mM Tris-Cl (pH 7.6), and 10 mM $MgCl_2$. The reaction mixture was adjusted to 1.5 mM in each of the four deoxynucleoside triphosphates and 1 μM in each of two oligonucleotide primers. A single cycle of the polymerase chain reaction was performed by heating the reaction to 95°C for 2 minutes, cooling to 30°C for 2 minutes, and adding 1 unit of the Klenow fragment of *Escherichia coli* DNA polymerase I in 2 μl of the buffer described above containing about 0.1 μl of glycerol (Klenow was obtained from U.S. Biochemicals in a 50% glycerol solution containing 5 U/μl). The extension reaction was allowed to proceed for 2 minutes at 30°C. The cycle was terminated and a new cycle was initiated by returning the reaction to 95°C for 2 minutes. In the amplifications of human DNA reported here, the number of cycles performed ranged from 20 to 27.

Genotype analysis of PCR-amplified genomic DNA using ASO probes. DNA (1 μg) from various cell lines was subjected to 25 cycles of PCR amplification. Aliquots representing one thirtieth of the amplification mixture (33 ng of initial DNA) were made 0.4 N in NaOH, 25 mM in EDTA in a volume of 200 μl and applied to a Genatran-45 nylon filter with a Bio-Dot spotting apparatus. Three replicate filters were prepared. ASO probes (Table 1) were 5′-phosphorylated with [λ-^{32}P]ATP and polynucleotide kinase and purified by spin dialysis. The specific activities of the probes were between 3.5 and 4.5 μCi/pmole. Each filter

  0-87969-052-6/86 $1.00

was prehybridized individually in 8 ml of 5× SSPE, 5× DET, and 0.5% SDS for 30 minutes at 55°C. The probe (1 pmole) was then added, and hybridization was continued at 55°C for 1 hour. The filters were rinsed twice in 2× SSPE and 0.1% SDS at room temperature, followed by a high stringency wash in 5× SSPE and 0.1% SDS for 10 minutes at 55°C (for 19C) or 60°C (for 19A and 19S) and autoradiographed for 2.5 hours at −80°C with a single intensification screen.

Cloning from PCR-amplified genomic DNA using linker primers. An entire PCR reaction was digested at 37°C with *Pst*I (20 units) and *Hin*dIII (20 units) for 90 minutes (for β-globin) or *Bam*HI (24 units) and *Pst*I (20 units) for 60 minutes (for HLA-DQα). After phenol extraction, the DNA was dialyzed to remove low-molecular-weight inhibitors of ligation (presumably the dNTPs used in PCR) and concentrated by ethanol precipitation. All (β-globin) or one tenth (DQα) of the material was ligated to 0.5 μg of the cut M13 vector under standard conditions and transformed into approximately 6×10^9 freshly prepared competent JM103 cells in a total volume of 200 μl. These cells (10–30 μl) were mixed with 150 μl of JM103 culture, plated on IPTG/X-Gal agar plates, and incubated overnight. The plates were scored for blue (parental) plaques and lifted onto BioDyne A filters. These filters were hybridized either with one of the labeled PCR oligonucleotide primers to visualize all of the clones containing PCR-amplified DNA (primer plaques) or with a β-globin oligonucleotide probe (RS06) or an HLA-DQα cDNA probe to visualize specifically the clones containing target sequences. Ten β-globin clones from this latter category were sequenced by using the dideoxy extension method. Nine were identical to the expected β-globin target sequence, and one was identical to the homologous region of the human δ-globin gene.

PCR construction of a 374-bp DNA fragment from synthetic oligodeoxyribonucleotides. 100 pmoles of TN10, d(CCTCGTCTACTCCCAGGTCCTCTTCAA-GGGCCAAGGCTGCCCCGACTATGTGCTCCTCAC-CCACACCGTCAGCC), and TN11, d(GGCAGGGGC-TCTTGACGGCAGAGAGGAGGTTGACCTTCTCCT-GGTAGGAGATGGCGAAGCGGCTGACGGTGTGG), designed so as to overlap by 14 complementary bases on their 3′ ends, were dissolved in 100 μl of buffer containing 30 mM Tris-acetate (pH 7.9), 60 mM sodium acetate, 10 mM magnesium acetate, 2.5 mM dithiothreitol, and 2 mM each dNTP. The solution was heated to 100°C for 1 minute and cooled in air at about 23°C for 1 minute; 1 μl containing 5 units Klenow fragment of *E. coli* DNA polymerase I was added, and the polymerization reaction was allowed to proceed for 2 minutes. Gel electrophoresis on 4% NuSieve agarose in the presence of 0.5 μg/ml ethidium bromide indicated that eight repetitions of this procedure were required before the mutual extension of the two primers on each other was complete (Fig. 5, lane I). A 2-μl aliquot of this reaction without purification was added to 100 μl of a second-stage reaction mixture identical to the one above except that 300 pmoles each of oligonucleotides LL09, d(CCTGGCCAATGGCATGGATCTGAAAGATAACC-AGCTGGTGGTGCCAGCAGATGGCCTGTACCTCG-TCTACTCCC), and LL12, d(CTCCCTGATAGATGG-GCTCATACCAGGGCTTGAGCTCAGCCCCCTCTG-GGGTGTCCTTCGGGCAGGGGCTCTTG), were substituted for TN10 and TN11. LL09 and LL12 were designed so that their 3′ ends would overlap with 14 complementary bases on the 3′ ends of the single-stranded fragments released when the 135-bp product of the previous reaction was denatured (see Fig. 4, no. 2). The cycle of heating, cooling, and adding Klenow fragment was repeated 15 times in order to produce the 254-bp fragment in lane II of Figure 5; 2 μl of this reaction mixture without purification was diluted into a third-stage reaction mixture as above but containing 300 pmoles of TN09, d(TGTAGCAAACCATCAAGTT-GAGGAGCAGCTCGAGTGGCTGAGCCAGCGGGC-CAATGCCCTCCTGGCCAATGGCA), and TN13, d(GATACTTGGGCAGATTGACCTCAGCGCTGAGT-TGGTCACCCTTCTCCAGCTGGAAGACCCCTCCC-TGATAGATG). After 15 cycles of PCR, the 374-bp product in lane III of Figure 5 was evident on gel electrophoresis. After gel purification, restriction analysis of the 374-bp fragment with several enzymes resulted in the expected fragments (data not shown).

PCR amplification with oligonucleotide linker primers. DNA was amplified by mixing 1 μg of genomic DNA in the buffer described above with 100 pmoles of each primer. Samples were subjected to 20 cycles of PCR, each consisting of 2 minutes of denaturation at 95°C, 2 minutes of cooling to 37°C, and 2 minutes of polymerization with 1 unit of Klenow DNA polymerase. After amplification, the DNA was concentrated by enthanol precipitation, and half of the total reaction was electrophoresed on a gel of 4% NuSieve agarose in TBE buffer. The ethidium-bromide-stained gel was photographed (see Fig. 6A), and the DNA was transferred to Genatran nylon membrane and hybridized to a labeled probe (RSO6) specific for the target sequence (Fig. 6B). The blot was then washed and autoradiographed. For the amplification of β-globin, the starting DNA was either from the Molt-4 cell line or from the globin deletion mutant GM2064 For lane 5, the starting material was 11 pg of the β-globin recombinant plasmid pBR328::β^A, the molar equivalent of 5 μg of genomic DNA. For lane 6, the reaction was performed as in lane 1 except that no enzyme was added. To increase the efficiency of amplification of the longer HLA-DQα segment, DMSO was added to 10% (v/v), and the polymerization was carried out at 37°C for 27 cycles. The starting DNA for the amplification of DQα was either from the consanguineous HLA typing cell line LG-2 or from the HLA class II deletion mutant LCL721.180.

DISCUSSION

Extraction and Amplification

Figure 1 describes the basic PCR process that results in the extraction and amplification of a nucleic acid sequence. "Extraction" is used here in the sense that the sequence, although contained within a larger molecule that may in fact be a heterogeneous population of larger molecules, as in the case of a chromosomal DNA preparation, will be amplified as a single discrete molecular entity, and thus extracted from its source. This feature of the chain reaction can be, in some cases, as important as the amplification itself. The source DNA is denatured and allowed to hybridize to an excess of primers that correspond to the extremities of the fragment to be amplified (Fig. 1A). The oligonucleotide primers are employed in micromolar concentrations, and thus the hybridization is rapid and complete but not particularly stringent. A DNA polymerase is added to the reaction, which already contains the four deoxynucleoside triphosphates, and the primers are extended (Fig. 1B,C). After a short time, the reaction is stopped, and the DNA is denatured again by heating. On cooling, the excess of primers again hybridizes rapidly, and this time there are twice as many sites for hybridization on sequences representing the target. Sequences not representing the target, which may have been copied in the first polymerization reaction by virtue of an adventitious interaction with one of the primers, will only very rarely have generated an additional site for either of the primers. The unique property of the targeted sequence for regenerating new primer sites with each cycle is intrinsic to the support of a chain reaction, and the improbability of its happening by chance accounts for the observed specificity of the overall amplification.

In Figure 1D, the discrete nature of the final product becomes evident. Whereas initial extension of the primers on templates with indefinite termini results in products with definite 5′ ends but indefinite 3′ ends (Fig. 1C), their extension on a template that is itself an extension product from a previous cycle results in a DNA strand that has both ends defined (Fig. 1D). It is the number of these DNA fragments, with both ends defined by primers, that increases exponentially during subsequent cycles. All other products of the reaction increase in a linear fashion. Thus, the reaction effectively amplifies only that DNA sequence that has been targeted by the primers.

Nested Primer Sets

Despite its intrinsic specificity, the amplification of fragments of single-copy genes from whole human DNA sometimes produces a molecule exclusively representative of the intended target, and sometimes it does not. Given the complexity of the human genome and the tendency for many genes to have some sequences in common with other related genes, this lack of complete specificity is not surprising. In many applications, extreme specificity is not required of the amplification process because a second level of specificity is invoked by the process of detecting the amplified products, e.g., the use of a labeled hybridization probe to detect a PCR-amplified fragment in a Southern blot. However, in other applications, such as attempting to visualize an amplified fragment of the β-globin gene by the ethidium bromide staining of an agarose gel, further specificity than that obtained by a simple amplification protocol was required. We achieved this by doing the amplification in two stages (Fig. 2). The first stage amplified a 110-bp fragment; the second stage employed two oligomers that primed within the sequence of this fragment to produce a subfragment of 58 bp (Table 1). By thus employing the specificity inherent in the requirement of four independent but coordinated priming events, we were able to amplify exclusively a fragment of the human β-globin gene approximately 2,000,000-fold (Mullis and Faloona 1986).

Addition and Amplification

Figure 3 depicts a PCR in which one of the oligonucleotides employed, primer B, has a 5′ sequence that is not homologous to the target sequence. During cycle 1 of the amplification, priming occurs on the basis of the homologous 3′ end of the oligomer, but the nonhomologous bases become a part of the extension product. In later cycles, when this extension product is copied by virtue of the fact that it contains a site for primer A, the entire complement to primer B is incorporated in this copy; thereafter, no further primer-template interactions involving nonhomologous bases are required, and the product that increases exponentially in the amplification contains sequence information on one end that was not present in the original target.

A similar situation results if both primers carry an additional 5′ sequence into the reaction, except that now both ends of the product will have been appended in the course of amplification. We have employed this embodiment of the PCR to insert restriction site linkers onto amplified fragments of human genomic sequences to facilitate their cloning; it also underlies the strategy described below for construction of DNA from oligonucleotides alone.

If one or both primers are designed so as to include an internal mismatch with the target sequence, specific in vitro mutations can be accomplished. As in the case above, the amplified product will contain the sequence of the primers rather than that of the target.

Construction and Amplification

The process whereby extrinsic sequences are appended to a fragment during the PCR amplification can be utilized in the stepwise construction of a totally synthetic sequence from oligonucleotides. The advan-

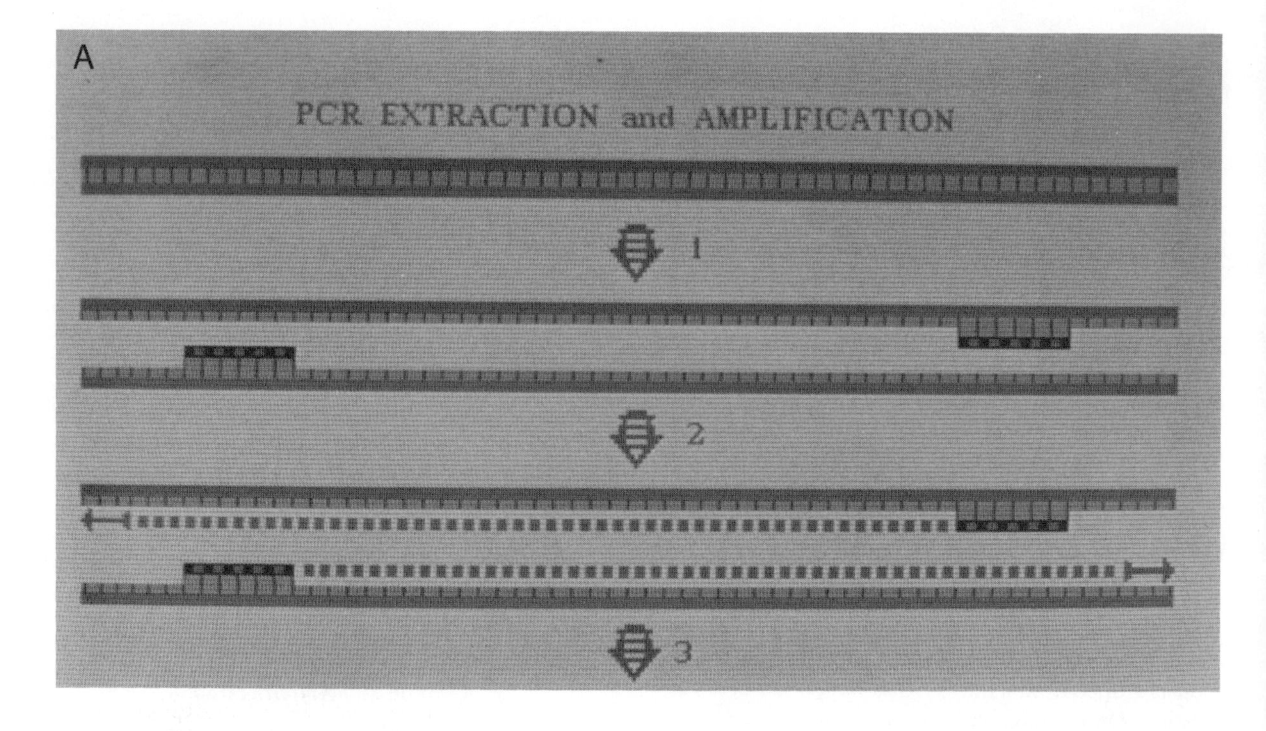
A
PCR EXTRACTION and AMPLIFICATION
1
2
3

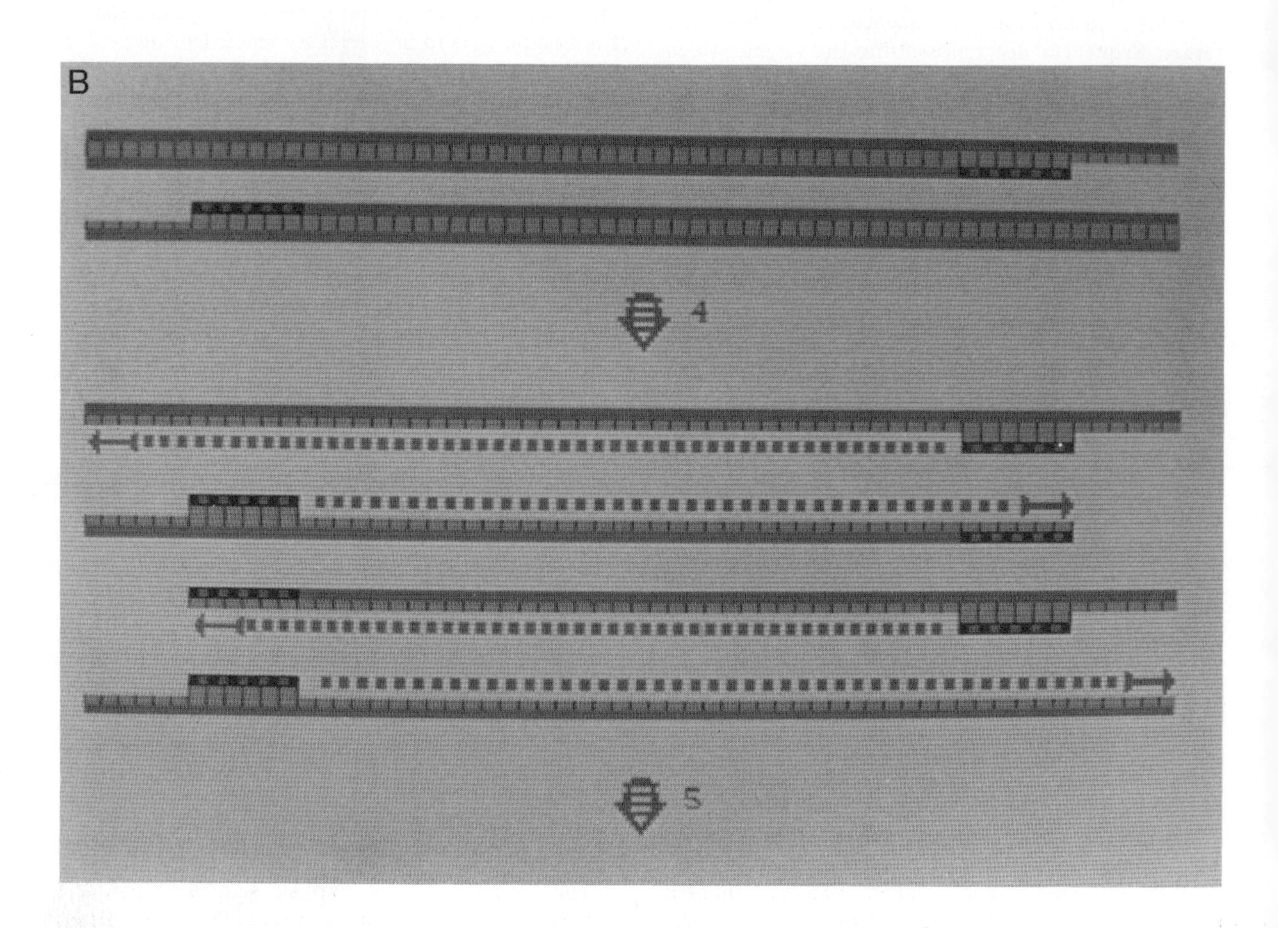
B
4
5

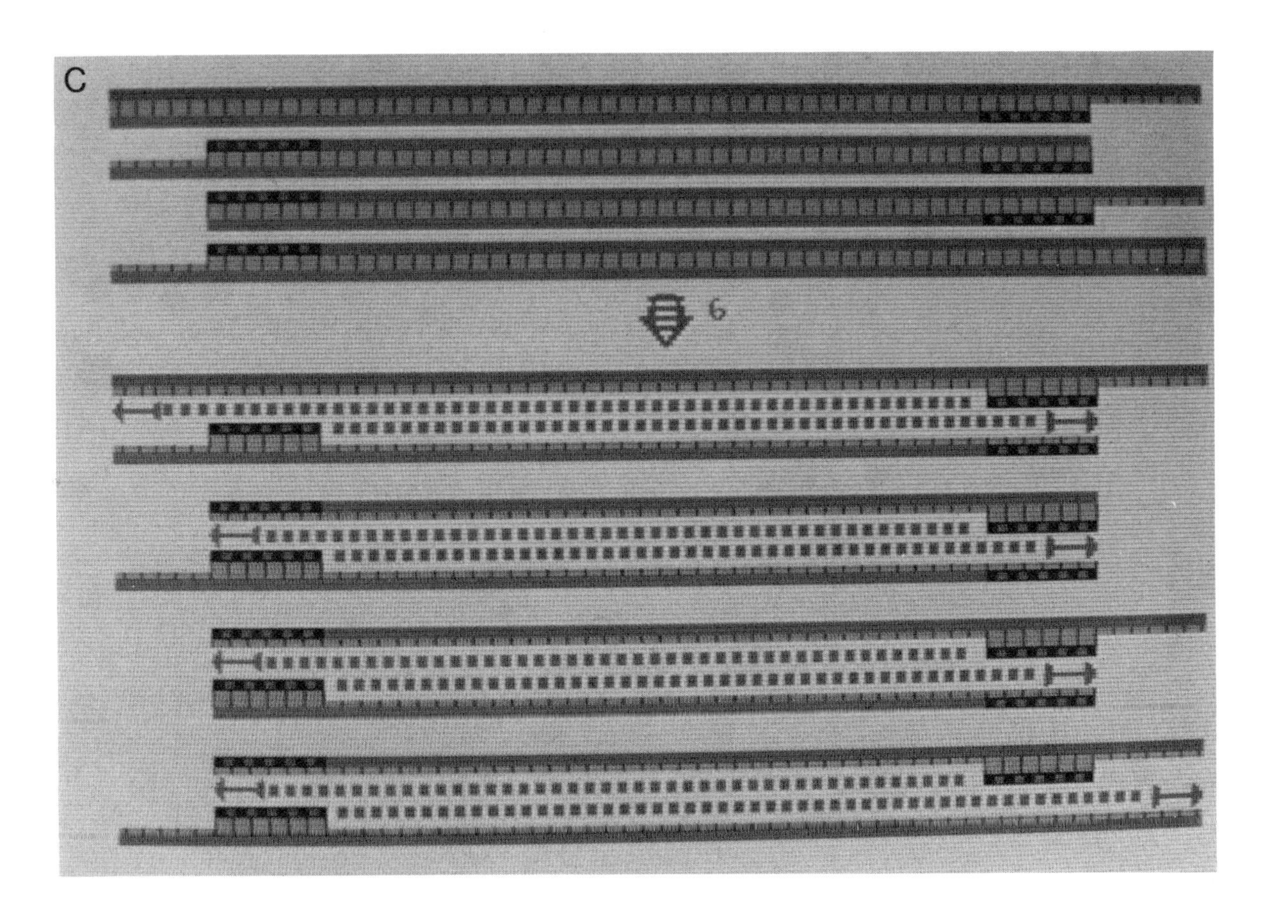

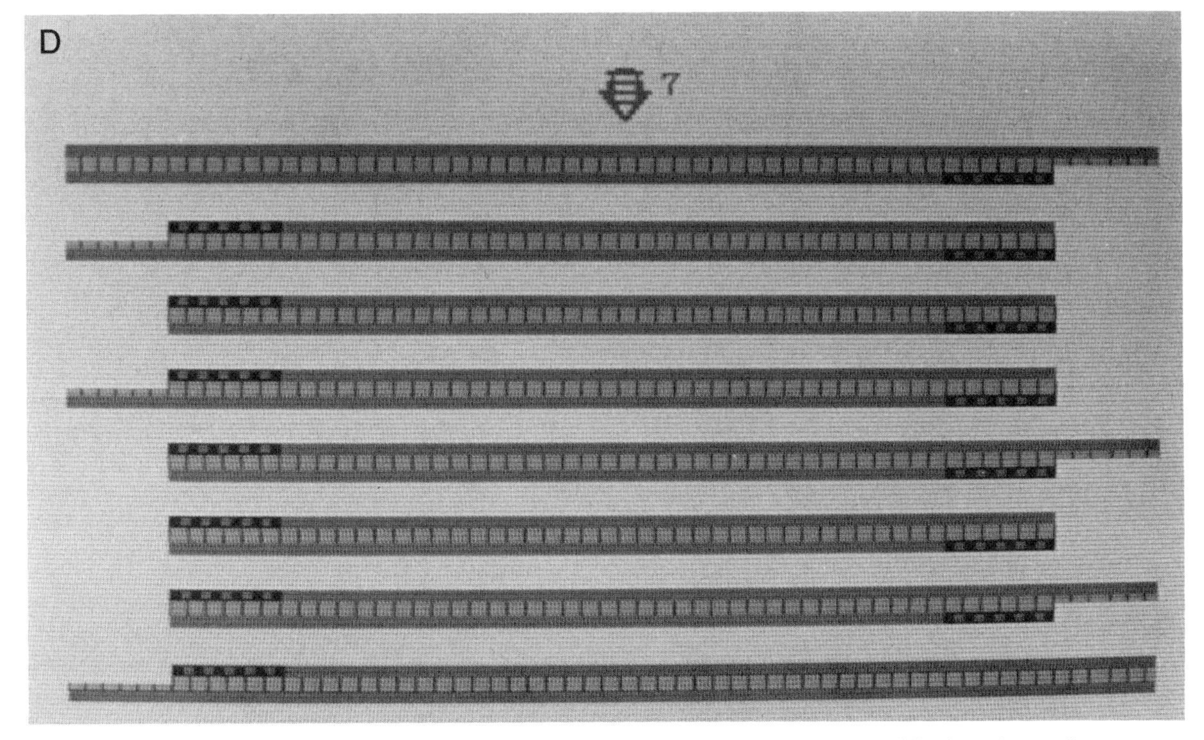

Figure 1. Three complete cycles of the polymerase chain reaction resulting in the eightfold amplification of a template sequence defined by the 5′ ends of two primers hybridized to different strands of the template are depicted above. The first cycle is shown in detail as reactions 1, 2, and 3. The second cycle is depicted in less detail as reactions 4 and 5, and the third cycle, as reactions 6 and 7. By cycle 3, a double-stranded DNA fragment is produced that has discrete termini, and thus the targeted sequence has been extracted from its source as well as amplified.

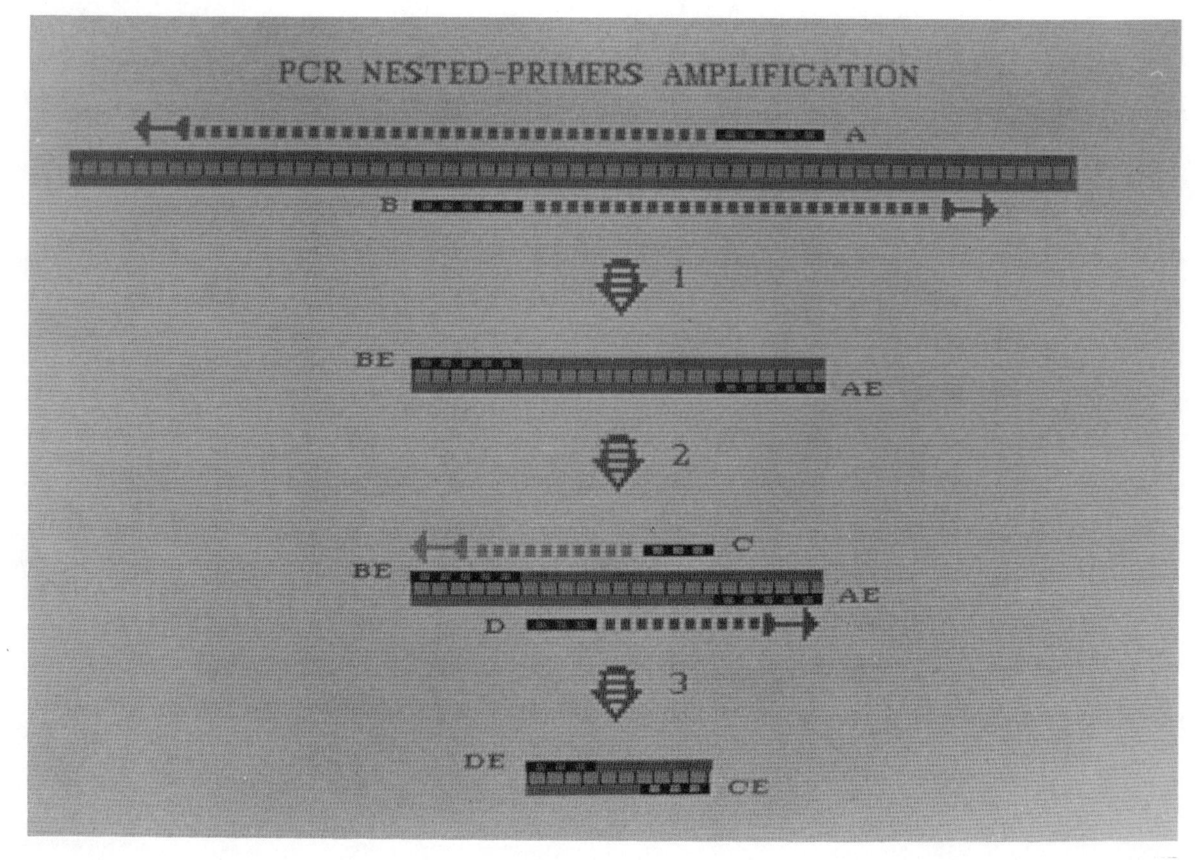

Figure 2. A PCR amplification employing oligonucleotides A and B results after a number of cycles in the fragment depicted as AE:BE. A subfragment, CE:DE, of this product can be extracted and further amplified by use of oligonucleotides C and D. This overall process requires four independent and coordinated primer-template interactions and is thereby potentially more specific than a single-stage PCR amplification.

tages of this approach (Figs. 4 and 5) over methods currently in use are that the PCR method does not require phosphorylation or ligation of the oligonucleotides. Like the method employed by Rossi et al. (1982), which involves the mutual extension of pairs of oligonucleotides on each other by polymerase (Fig. 4, no. 1), the PCR method does not require organic synthesis of both strands of the final product. Unlike this method, however, the PCR method is completely general in that no particular restriction enzyme recognition sequences need be built into the product for purposes of accomplishing the synthesis. Furthermore, the PCR method offers the convenience of enabling the final product, or any of the several intermediates, to be amplified during the synthesis or afterwards to produce whatever amounts of these molecules are required.

Table 1. Oligodeoxyribonucleotides

	Sequence	Use
PC03	ACACAACTGTGTTCACTAGC	produce a 110-bp fragment
PC04	CAACTTCATCCACGTTCACC	from β-globin 1st exon
PC07	CAGACACCATGGTGCACCTGACTCCTG	produce a 58-bp subfragment
PC08	CCCCACAGGGCAGTAACGGCAGACTTCTCC	of the 110-bp fragment above
GH18	CTTCTGCAGCAACTGTGTTCACTAGC	as PC03 and PC04, but with *Pst*I
GH19	CACAAGCTTCATCCACGTTCACC	and *Hin*dIII linkers added
GH26	GTGCTGCAGGTGTAAACTTGTACCAG	242-bp from HLA-DQα
GH27	CACGGATCCGGTAGCAGCGGTAGAGTTG	with *Pst*I and *Bam*HI linkers
RS06	CTGACTCCTGAGGAGAAGTCTGCCGTT ACTGCCCTGTGGG	probe to central region of 110-bp fragment produced by PC03 and PC04
19A	CTCCTGAGGAGAAGTCTGC	ASO probe to β^{A}-globin
19S	CTCCTGTGGAGAAGTCTGC	ASO probe to β^{S}-globin
19C	CTCCTAAGGAGAAGTCTGC	ASO probe to β^{C}-globin

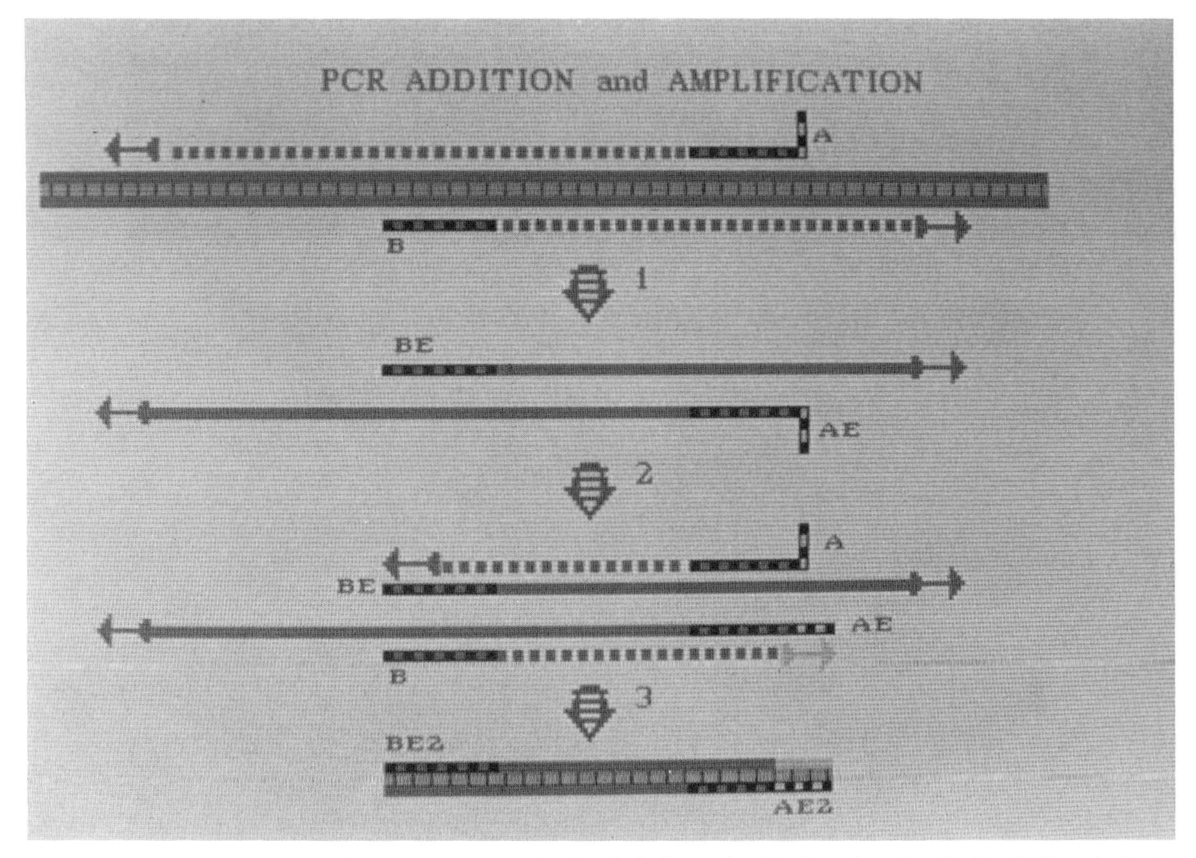

Figure 3. Addition of the nonhomologous sequence on the 5′ end of oligonucleotide A results, after the first PCR cycle, in the incorporation of this nonhomologous sequence into extension product AE, which, in a later cycle as template for B, directs the synthesis of BE2 that together with AE2 comprises the double-stranded DNA fragment AE2:BE2. This fragment will accumulate exponentially with further cycles.

Genetic Analysis

In addition to its use as an in vitro method for enzymatically synthesizing a specific DNA fragment, the PCR, followed by hybridization with specific probes, can serve as a powerful tool in the analysis of genomic sequence variation. Understanding the molecular basis of genetic disease or of complex genetic polymorphisms such as those in the HLA region requires detailed nucleotide sequence information from a variety of individuals to localize relevant variations. Currently, the analysis of each allelic variant requires a substantial effort in library construction, screening, mapping, subcloning, and sequencing. Here, using PCR primers modified near their 5′ ends to produce restriction sites, we describe a method for amplification of specific segments of genomic DNA and their direct cloning into M13 vectors for sequence analysis. Moreover, the cloning and sequencing of PCR-amplified DNA represents a powerful analytical tool for the study of the specificity and fidelity of this newly developed technique, as well as a rapid method of genomic sequencing. In addition, allelic variation has been analyzed using PCR amplification prior to hybridization with allele-specific oligonucleotide probes in a dot-blot format. This is a simple, general, and rapid method for genetic analysis and recently has been demonstrated in crude cell lysates, eliminating the need for DNA purification.

PCR Cloning and Direct Sequence Analysis

To develop a rapid method for genomic sequence determination and to analyze the individual products of PCR amplification, we chose the oligonucleotide primers and probes previously described for the diagnosis of sickle cell anemia (Saiki et al. 1985). These primers amplify a 110-bp segment of the human β-hemoglobin gene containing the Hb-S mutation and were modified near their 5′ ends (as described above in Fig. 3) to produce convenient restriction sites (linkers) for cloning directly into the M13mp10 sequencing vector. These modifications did not affect the efficiency of PCR amplification of the specific β-globin segment. After amplification, the PCR products were cleaved with the appropriate restriction enzymes, ligated into the M13 vector, and transformed into the JM103 host, and the resulting plaques were screened by hybridization with a labeled oligonucleotide probe to detect the β-globin clones. The plaques were also screened with the labeled PCR oligonucleotide primers to identify all of the clones containing amplified DNA (Table 2). In-

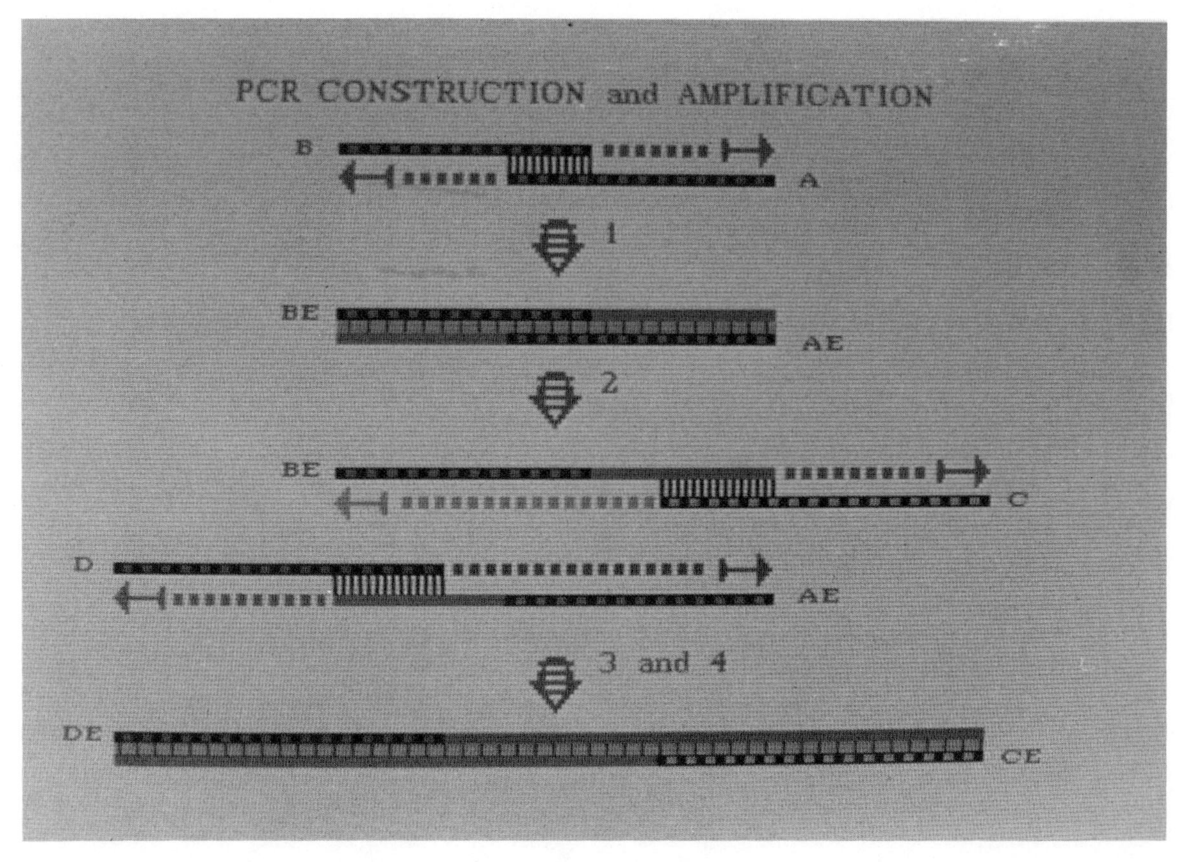

Figure 4. PCR construction and amplification. Oligonucleotides with complementary overlapping 3′ ends are extended on each other using DNA polymerase. The double-stranded DNA product of this reaction is extended on both ends by a second stage of several PCR cycles depicted by reactions 2, 3, and 4.

dividual clones were then sequenced directly by using the dideoxy primer-extension method.

Over 80% of the clones contained DNA inserts with the PCR primer sequences, but only about 1% of the clones hybridized to the internal β-globin probe. These nonglobin fragments presumably represent amplifications of other segments of the genome. This observation is consistent with the gel and Southern blot analysis of the PCR-amplified DNA from a β-globin deletion mutant and a normal cell line (Fig. 6). The similarity of the observed gel profiles reveals that most of the amplified genomic DNA fragments arise from nonglobin templates. Sequence analysis of two of these nontarget clones showed that the segments between the PCR primer sequences were unrelated to the β-globin gene and contained an abundance of dinucleotide repeats, similar to some genomic intergenic spacer sequences.

When ten of the clones that hybridized to the β-globin probe were sequenced, nine proved to be identical to the β-globin gene and one contained five nucleotide differences but was identical to the δ-globin gene. Each β-globin PCR primer has two mismatches with the δ-globin sequence. Each of these ten sequenced clones contains a segment of 70 bp originally synthesized from the genomic DNA template during the PCR amplification process. Since no sequence alterations were seen in these clones, the frequency of nucleotide misincorporation during 20 cycles of PCR amplification is less than 1 in 700.

To analyze the molecular basis of genetic polymorphism and disease susceptibility in the HLA class II loci, this approach has been extended to the amplification and cloning of a 242-bp fragment from the second exon, which exhibits localized allelic variability, of the HLA-DQα locus. In this case, the primer sequences (based on conserved regions of this exon) contain 5′-terminal restriction sites that have no homology with the DQα sequence. The specificity of amplification achieved by using these primers is greater than that achieved with the β-globin primers, since gel electrophoresis of the PCR products reveals a discrete band at 240 bp, which is absent from an HLA deletion mutant (Fig. 6). In addition, hybridization screening of the M13 clones from this amplification indicates that about 20% are homologous to the DQα cDNA probe (data not shown), an increase of 20-fold over the β-globin amplification. At this time, the basis for the difference in the specificity of amplification, defined here as the ratio of target to nontarget clones, is not clear, but may reflect the primer sequences and their genomic distribution. As described above, in some cases, the specificity of the PCR amplification can be significantly enhanced by using nested sets of PCR primers (Fig. 2).

Three HLA-DQα PCR clones derived from the

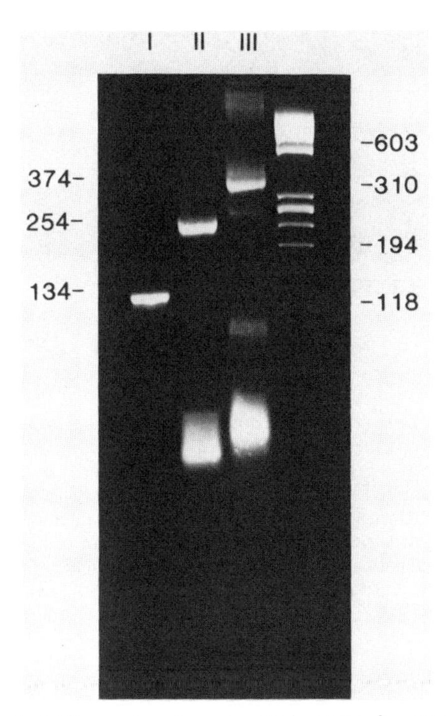

Figure 5. PCR construction of a 374-bp DNA fragment from synthetic oligodeoxynucleotides. (Lane I) 134-bp mutual extension product of TN10 and TN11; (lane II) 254-bp fragment produced by polymerase chain reaction of LL09 and LL12 with product in lane I; (lane III) 374-bp fragment produced by polymerase chain reaction of TN08 and TN13 with product in lane II; extreme right-hand lane is molecular-weight markers.

homozygous typing cell LG2 were subjected to sequence analysis. Two clones were identical to an HLA-DQα cDNA clone from the same cell line. One differed by a single nucleotide, indicating an error rate of approximately 1/600, assuming the substitution occurred during the 27 cycles of amplification. We are also currently using this procedure to analyze sequences from polymorphic regions of the HLA-DQα and DRβ loci. These preliminary studies suggest that the error rate over many cycles of amplification appears to be sufficiently low so that reliable genomic sequences can be determined directly from PCR amplification and cloning. This approach greatly reduces the number of cloned DNA fragments to be screened, circumvents the need for full genomic libraries, and allows cloning from nanogram quantities of genomic DNA.

Table 2. Cloning β-Globin Sequences from PCR-amplified DNA

Category	Number	Frequency (%)
Total plaques	1496	100
White plaques	1338	89
Primer plaques	1206	81
Globin plaques	15	1

Analysis of PCR-amplified DNA with Allele-specific Oligonucleotide Probes

Allelic sequence variation has been analyzed by oligonucleotide hybridization probes capable of detecting single-base substitutions in human genomic DNA that has been digested by restriction enzymes and resolved by gel electrophoresis (Conner et al. 1983). The basis of this specificity is that, under appropriate hybridization conditions, an allele-specific oligonucleotide (ASO) will anneal only to those sequences to which it is perfectly matched, a single-base-pair mismatch being

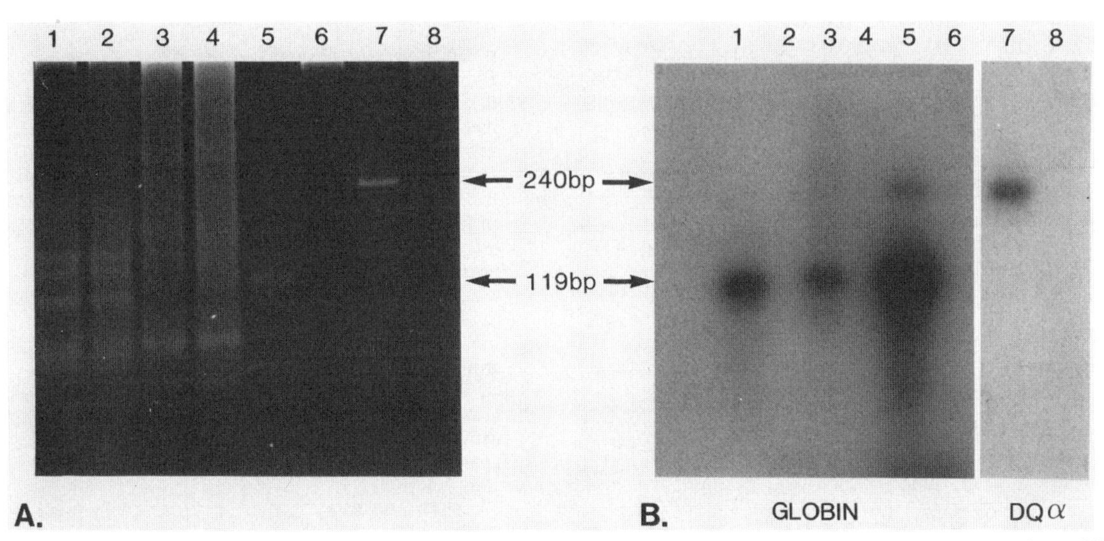

Figure 6. PCR amplification with oligonucleotide linker primers. (*A*) Ethidium-bromide-stained gel showing total amplified products. (*1*) Primers PC03 and PC04 on Molt-4 DNA; (*2*) PC03 and PC04 on GM2064 DNA; (*3*) GH18 and GH19 on Molt-4; (*4*) GH18 and GH19 on GM2064; (*5*) GH18 and GH19 on pBR328::β^A DNA; (*6*) PC03 and PC04 on Molt-4, no enzyme; (*7*) primers GH26 and GH27 on LG-2 DNA; (*8*) GH26 and GH27 on LCL721.180 DNA. (*B*) Southern blots showing specific amplified products. Lanes are numbered as in *A*. Lanes *1* through *6* were hybridized to the labeled RS06 oligonucleotide probe. Lanes *7* and *8* were hybridized to a cloned DBα cDNA probe labeled by nick translation.

sufficiently destabilizing to prevent hybridization. To improve the sensitivity, specificity, and simplicity of this approach, we have used the PCR procedure to amplify enzymatically a specific β-globin sequence in human genomic DNA prior to hybridization with ASOs. The PCR amplification, which produces a greater than 10^5-fold increase in the amount of target sequence, permits the analysis of allelic variation with as little as 1 ng of genomic DNA and the use of a simple "dot-blot" for probe hybridization. As a further simplification, PCR amplification has been performed directly on crude cell lysates, eliminating the need for DNA purification.

To develop a simple and sensitive method, we have chosen the sickle cell anemia and hemoglobin C mutations in the sixth codon of the β-globin gene as a model system for genetic diagnosis. We have used ASOs specific for the normal (β^A), sickle cell (β^S), and hemoglobin C (β^C) sequences as probes to detect these alleles in PCR-amplified genomic samples. The sequences of the 19-base ASO probes used here are identical to those described previously (Studenski et al. 1985).

DNA was extracted from six blood samples from individuals whose β-globin genotypes comprise each possible diploid combination of the β^A, β^S, and β^C alleles and from the cell line GM2064, which has a homozygous deletion of the β-globin gene. Aliquots (1 μg) of each sample were subjected to 25 cycles of PCR amplification, and one thirtieth of the reaction product (33 ng) was applied to a nylon filter as a dot-blot. Three replicate filters were prepared and each was hybridized with one of the three ^{32}P-labeled ASOs under stringent conditions. The resulting autoradiogram (Fig. 7) clearly indicates that each ASO annealed only to those DNA samples containing at least one copy of the β-globin allele to which the probe was perfectly matched and not at all to the GM2064 deletion mutant. The frequency of the specific β-globin target to PCR-amplified DNA has been estimated by the analysis of cloned amplification products to be approximately 1% (Table 2), an enrichment of greater than 10^5 over unamplified genomic DNA. It is this substantial reduction of complexity that allows the application of 19-base probes in a dot-blot format and the use of shorter oligonucleotide probes, capable of allelic discrimination using less stringent conditions. In addition, this approach has been applied to the analysis of genetic polymorphism in the HLA-DQα locus by using four different ASO probes and is being extended to the HLA-DQβ and DRβ loci.

PCR amplification has been used to detect β-globin genotypes in as little as 0.5 ng of genomic DNA and, recently, in the crude lysate of 75 cells. This ability to rapidly analyze genetic variation of minute amounts of purified DNA or in cell lysates has important implications for a wide variety of clinical genetic analyses. With the use of nonisotopic probes and PCR automation, this procedure combining in vitro target amplification and ASO probes in a dot-blot format promises to be a general and simple method for the detection of allelic variation.

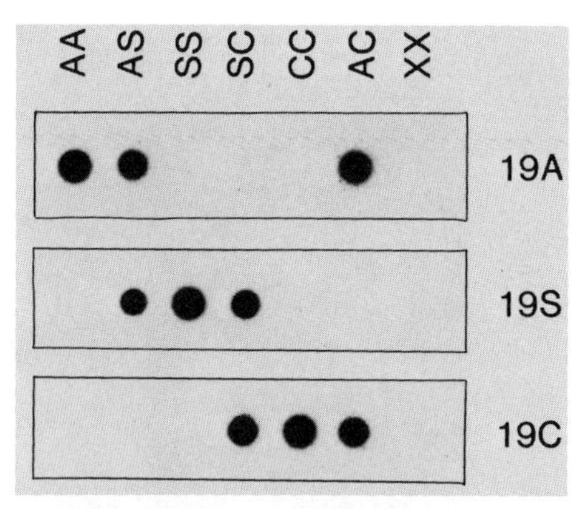

Figure 7. Genotype analysis of PCR-amplified genomic DNA using ASO probes. DNA extracted from the blood of individuals of known β-globin genotype and from the β-globin deletion mutant cell line GM2064 were subjected to PCR amplification. Aliquots of PCR reactions equivalent to 33 ng starting DNA were applied to replicate filters in a dot-blot format and probed with allele-specific oligonucleotides. (AA) $\beta^A\beta^A$; (AS) $\beta^A\beta^S$; (SS) $\beta^S\beta^S$; (SC) $\beta^S\beta^C$; (CC) $\beta^C\beta^C$; (AC) $\beta^A\beta^C$; (XX) GM2064.

ACKNOWLEDGMENTS

We thank Corey Levenson, Lauri Goda, and Dragan Spasic for the synthesis of oligonucleotides, Eric Ladner for graphic support, and Kathy Levenson for preparation of the manuscript.

REFERENCES

Cohen, S., A. Chang, H. Boyer, and R. Helling. 1973. Construction of biologically functional bacterial plasmids *in vitro*. *Proc. Natl. Acad. Sci.* **70:** 3240.

Conner, B.J., A. Reyes, C. Morin, K. Itakura, R. Teplitz, and R.B. Wallace. 1983. Detection of sickle cell β-S-globin allele by hybridization with synthetic oligonucleotides. *Proc. Natl. Acad. Sci.* **80:** 272.

Davies, J.E. and H.G. Gaessen. 1983. Synthetic gene fragments in genetic engineering—The renaissance of chemistry in molecular biology. *Angew Chem. Int. Chem. Ed. Engl.* **22:** 13.

Mullenbach, G.T., A. Tabrizi, R.W. Blacher, and K.S. Steimer. 1986. Chemical synthesis and expression in yeast of a gene encoding connective tissue activating peptide-III. *J. Biol. Chem.* **261:** 719.

Mullis, K.B. and F. Faloona. 1986. Specific synthesis of DNA *in vitro* via a polymerase catalyzed chain reaction. *Methods Enzymol.* (in press).

Roberts, R.J. 1985. Restriction and modification enzymes and their recognition sequences. *Nucleic Acids Res.* **9:** 75.

Rossi, J.J., R. Kierzek, T. Huang, P.A. Walker, and K. Itakura. 1982. An alternate method for synthesis of double-stranded DNA segments. *J. Biol. Chem.* **257:** 9226.

Saiki, R.K., S. Scharf, F. Faloona, K.B. Mullis, G. Horn, H.A. Erlich, and N. Arnheim. 1985. Enzymatic amplification of β-globin genomic sequences and restriction site analysis for diagnosis of sickle cell anemia. *Science* **230:** 1350.

Smith, H.O. and K.W. Wilcox. 1970. A restriction enzyme from *Hemophilus influenzae*. *J. Mol. Biol.* **51:** 379.

Studencki, A.B., B. Conner, C. Imprain, R. Teplitz, and R.B. Wallace. 1985. Discrimination among the human β-a, β-s, and β-c-globin genes using allele-specific oligonucleotide hybridization probes. *Am. J. Hum. Genet.* **37:** 42.

Urdea, M.S., L. Ku, T. Horn, Y.G. Gee, and B.D. Warner. 1985. Base modification and cloning efficiency of oligodeoxynucleotides synthesized by phosphoramidite method; methyl versus cyanoethyl phosphorous protection. *Nucleic Acids Res.* **16:** 257.

Watt, V.M., C.J. Ingles, M.S. Urdea, and W.J. Rutter. 1985. Homology requirements for recombination in *Escherichia coli. Proc. Natl. Acad. Sci.* **82:** 4768.

Recent Advances in the Development of Methods for Detecting Single-base Substitutions Associated with Human Genetic Diseases

R.M. MYERS* AND T. MANIATIS†

*Department of Physiology S-762, University of California, School of Medicine, San Francisco, California 94143;
†Department of Biochemistry and Molecular Biology, Harvard University, Cambridge, Massachusetts 02138

The ability to detect single-base mutations in human genomic DNA is of fundamental importance in the diagnosis of genetic diseases. This was first accomplished by using restriction endonucleases and the Southern blotting technique to detect the loss or gain of restriction enzyme recognition sites (Flavell et al. 1978; Geever et al. 1981). Unfortunately, this approach is limited by the fact that most single-base substitutions do not alter restriction endonuclease cleavage sites. In fact, sickle cell anemia is the only genetic disease that has been directly diagnosed by this method in the clinic on a routine basis (Chang and Kan 1982; Orkin et al. 1982). However, many genetic diseases can be diagnosed by this method by virtue of their linkage to restriction site polymorphisms (see, e.g., Kan and Dozy 1978; Gusella et al. 1982; Orkin et al. 1982). Base substitutions that do not alter restriction enzyme cleavage sites can be detected by differential hybridization of synthetic oligonucleotide probes if the DNA sequences of the mutation and the surrounding nucleotides are known (Kidd et al. 1983; Pirastu et al. 1983). This method does not provide a practical approach to screen for new mutations or to detect multiallelic genetic diseases.

Despite considerable effort, no general method capable of detecting all possible single-base mutations in human genomic DNA is currently available. However, two recently developed methods significantly increase the number of single-base mutations that can be detected. One method is based on the melting behavior of DNA and the use of denaturing gradient gel electrophoresis (Lerman 1986; Myers et al. 1985a,b,c,d, 1986), and the other method is based on the susceptibility of single-base mismatches in RNA:DNA duplexes to ribonuclease cleavage (Myers et al. 1985a). These methods should not only be useful for diagnosing the mutations directly responsible for specific genetic diseases, but also be valuable in identifying new single-base polymorphisms for genetic linkage studies. In this paper, we describe the development and application of these methods and discuss the advantages and limitations of each approach.

RESULTS

Denaturing Gradient Gel Electrophoresis

DNA fragments differing by single-base substitutions can be separated from each other by electrophoresis in polyacrylamide gels containing an ascending gradient of the DNA denaturants urea and formamide (Fischer and Lerman 1983; Lerman et al. 1984; Myers et al. 1985c,d,e). DNA fragments of identical size, but differing by a single-base change, will initially move through the polyacrylamide gel at a constant rate. As they migrate into a critical concentration of denaturant, specific regions or "domains" within the fragment melt to produce partially denatured DNA. Melting of a domain is accompanied by an abrupt decrease in mobility, which is a consequence of the entanglement of branched DNA molecules in the acrylamide matrix. The position in the denaturing gradient gel at which the decrease in mobility is observed corresponds to the melting temperature (T_m) of the domain. Since a single-base substitution within the melting domain results in a T_m difference, partial denaturation of the mutant and wild-type DNA fragments will occur at different positions in the gel. DNA molecules can therefore be separated on the basis of very small differences in the T_m values of their melting domains.

A typical DNA fragment of about 500 bp will have more than one melting domain. Although a decrease in the mobility of the fragment will occur when the domain with the lowest T_m melts, the molecules continue to move slowly into higher concentrations of denaturant, where the other melting domains undergo strand separation. Since single-base mutations in any of these domains will alter their melting properties, these mutations will lead to differences in the patterns of electrophoresis. However, when the highest-temperature domain melts, the fragment undergoes complete strand dissociation, and the resolving power of the gel is lost (Myers et al. 1985c,d). Thus, mutations in the melting domain with the highest T_m cannot be separated from the wild-type DNA fragment. In cloned DNA molecules, this limitation has been overcome by attaching a GC-rich DNA sequence, designated a "GC-clamp," next to the DNA fragment of interest (Myers et al. 1985b,c,d). This GC-clamp makes it possible to detect single-base substitutions in all melting domains of an attached DNA fragment by preventing complete strand dissociation as it proceeds through the denaturing gradient gel.

An example of the analysis of DNA fragments containing single-base mutations by denaturing gradient gel electrophoresis is shown in Figure 1. All of the frag-

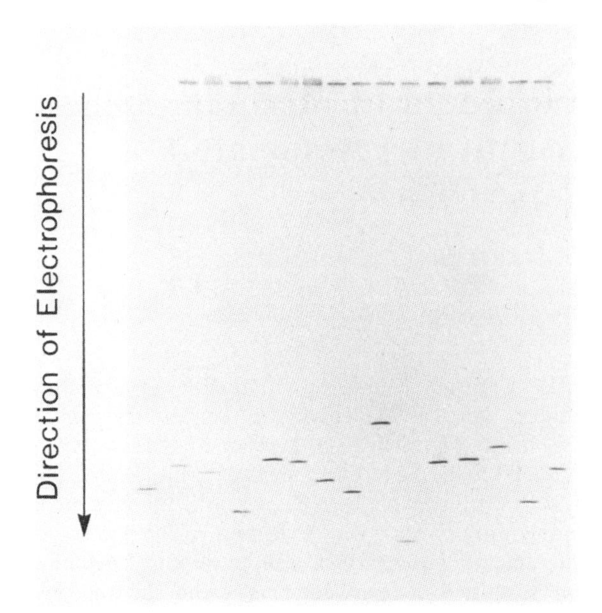

Figure 1. Separation of DNA fragments differing by a single-base substitution by denaturing gradient gel electrophoresis. A negative image of an ethidium-stained denaturing gradient gel containing DNA fragments 440 bp in length, which includes a 300-bp GC-clamp. The left lane contains the wild-type DNA fragment, and the remaining lanes contain examples of mutant DNA fragments carrying different single-base substitutions. The fragments shown in this gel are all homoduplexes.

ments are 440 bp in length and consist of 140 bp of DNA containing the mouse β-globin promoter and a 300-bp GC-clamp. Approximately 90% of all single-base substitutions in the promoter region of this fragment are expected to resolve, based on theoretical calculations and the analysis of over 130 different single-base mutations (Myers et al. 1985b,c,d). The separation between normal and mutant DNA fragments can be as great as 14 mm (Fig. 1). An important point is that most single-base substitutions that do not alter the base composition (e.g., A:T to T:A transversions) can be readily separated on the gel. Thus, the changes in T_m due to differences in base stacking alone are sufficient to allow discrimination.

A significant increase in resolution can be achieved by examining heteroduplexes between wild-type and mutant DNA molecules containing a single-base mismatch (R.M. Myers et al., in prep.). This is presumably a result of the instability caused by the disruption of the helix at the position of the mismatch. Examples of the effect of forming heteroduplexes on resolution in denaturing gradient gels are shown in Figure 2. As shown in Figure 2A, heteroduplexes are formed by mixing equal amounts of the wild-type and mutant DNA fragments, denaturing, and reannealing. Four distinct species are generated by this reassortment: wild-type homoduplexes, mutant homoduplexes, and two different heteroduplexes. Remarkably, all four of these DNA species are resolved on the gel, and in every case, the resolution of the heteroduplexes from the wild-type homoduplex is significantly greater than the corresponding resolution of the mutant homoduplexes (Fig. 2B).

This increase in resolution was effectively exploited to detect single-base mutations in total human genomic DNA from individuals with β-thalassemia (Myers et al. 1985e). As shown in Figure 3A, a radio-

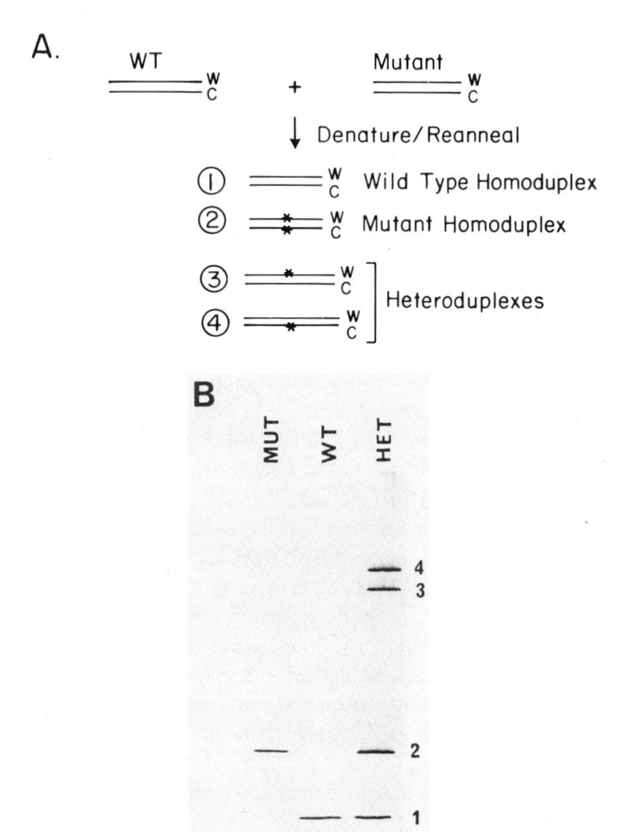

Figure 2. Analysis of mismatched DNA:DNA heteroduplexes by denaturing gradient gel electrophoresis. (*A*) Diagram showing the formation of heteroduplexes between cloned wild-type and mutant DNA fragments. Wild-type and mutant fragments were mixed in roughly equal proportions, heated to denature the strands, and then annealed to allow a reassortment of the single-stranded fragments with their complements. Four different DNA species are expected from such a reassortment: wild-type homoduplexes, mutant homoduplexes, and two different heteroduplexes composed of one wild-type strand and one mutant strand (each heteroduplex carries a single-base mismatch). (*B*) Denaturing gradient gel electrophoresis of heteroduplex fragments. Shown is a negative image of an ethidium-stained denaturing gradient gel comparing the gel positions of homo- and heteroduplex species. DNA fragments 440 bp in length are composed of a 300-bp GC-clamp and a 140-bp sequence containing the mouse β-major globin promoter. Heteroduplexes were prepared as indicated in *A* and then subjected to electrophoresis in a 6.5% polyacrylamide gel containing a gradient of 40–80% denaturants. The gel was prepared as described by Myers et al. (1986). The lanes labeled MUT, WT, and HET contain, respectively, mutant homoduplexes, wild-type homoduplexes, and the mixture of homo- and heteroduplexes formed by reassortment of the mutant and wild-type DNA strands after denaturation and reannealing. The bands labeled *1* through *4* in the HET lane correspond to the four DNA species indicated in *A*.

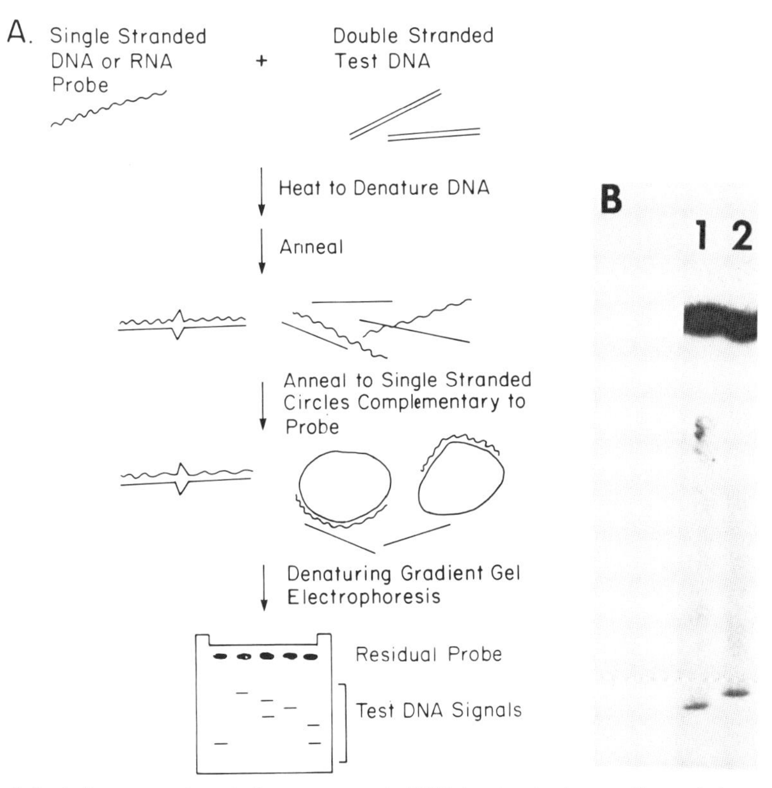

Figure 3. Detection of single-base mutations in human genomic DNA by denaturing gradient gel electrophoresis. (*A*) Diagram showing the experimental procedure used to detect mutations in genomic DNA. The initial hybridization step was carried out in the presence of 10–20-fold molar excess of DNA or RNA probe to drive the reaction to completion. The excess probe was removed by hybridization to single-stranded circular DNA, followed by electrophoresis. This probe remains at the top of the gel because of the large size of the single-stranded circular DNA to which it is annealed. (*B*) Autoradiogram showing the detection of a single-base mutation in the human β-globin gene that results in β-thalassemia. In this experiment, 5 μg of total genomic DNA from an individual with normal β-globin genes and from an individual homozygous for the G-to-A mutation at codon 39 was used. Genomic DNA was annealed to a radioactively labeled single-stranded DNA probe (272 nucleotides) composed of a wild-type sequence covering the codon 39 region. After annealing, nonlabeled single-stranded, circular M13 DNA containing sequences complementary to the probe was added to hybridize to residual probe, and the mixture was subjected to electrophoresis on a 14% acrylamide gel containing a 30–60% gradient of denaturants. Electrophoresis and autoradiography were carried out as described by Myers et al. (1985e). (Lane *1*) DNA from an individual with normal β-globin genes; (lane *2*) DNA from a thalassemic individual homozygous for the codon 39 mutation. The large amount of radioactive material near the top of the gel is the residual probe hybridized to the M13 circular DNA.

actively labeled DNA or RNA probe derived from the normal gene is mixed in 10–20-fold molar excess with double-stranded genomic DNA, and the mixture is heated to separate the DNA strands. After incubating the mixture to allow the probe to anneal to its complementary strand in the genomic DNA, a large single-stranded, circular molecule containing sequences complementary to the probe is added and allowed to hybridize to the residual excess probe. This step, developed by M. Collins (pers. comm.), circumvents the need to digest the reaction mix with S1 nuclease to remove unreacted probe, a step that was used in the original protocol that sometimes resulted in background radioactivity on the autoradiograms (Myers et al. 1985e). The products following this second annealing reaction are then electrophoresed on a denaturing gradient gel. Visualization of the products by autoradiography reveals radioactive signals in the middle portion of the gel corresponding to duplexes between the probe and genomic DNA test sequences and a large amount of radioactivity at the top portion of the gel corresponding to the residual probe bound to the single-stranded, circular DNA.

The results of an analysis of total genomic DNA from a thalassemic individual homozygous for a G-to-A mutation at codon 39 in the human β-globin gene are presented in Figure 3B. A separation of 2 mm was observed between the homoduplexes formed by hybridizing the probe to DNA from an individual with normal β-globin genes (Fig. 3B, lane 1) and the heteroduplexes formed between the probe and the DNA from the thalassemic individual (Fig. 3B, lane 2).

This genomic heteroduplex method has several advantages over procedures that rely on the use of restriction endonucleases and Southern blotting. The solution hybridization step is more efficient and convenient than hybridization on filters; in addition, the Southern transfer step is eliminated, since the reaction products are examined by direct autoradiography of the gel. Finally, a large fraction of all possible mutations can be detected by the denaturing gradient gel method; on the basis of theoretical and experimental analyses, approximately 50–75% of all mutations could be detected (Myers et al. 1985c,d,e; Lerman et al., this volume). The failure of certain single-base substitutions to separate from wild-type DNA is a consequence of the melting behavior of DNA. If the substitution is located in the highest-temperature melting domain, or if the fragment melts as a single domain, resolution from the wild-type DNA fragment is not achieved. Although this limitation has been overcome in cloned DNA fragments by attaching a GC-rich "clamp" sequence to cloned test DNA fragments (Myers et al. 1985b,c,d), this approach has not yet been extended to the analysis of total genomic DNA.

Another limitation of the denaturing gradient gel electrophoresis system at present is that only relatively short DNA fragments (100–1000 bp) can be used in the analysis. The mobility of larger fragments is too low to achieve resolution of mutant DNA fragments; this problem is compounded by the fact that the melting of each domain leads to even further decreases in mobility. Therefore, only a small fraction of a large DNA fragment would be accessible to analysis. In an effort to overcome this difficulty, we have attempted to analyze a number of small DNA fragments, which together encompass an entire gene, on a single gel. To accomplish this objective, we utilized a gel in which the gradient of denaturants is perpendicular, rather than parallel, to the direction of electrophoresis (Fischer and Lerman 1983). As shown in Figure 4A, the DNA sample is layered across the top of the gel and then subjected to electrophoresis. DNA fragments on the side of the gel containing no denaturant migrate with the rapid linear mobility expected for duplex DNA. In contrast, DNA fragments on the side of the gel containing high concentrations of denaturants melt immediately upon entering the gel and remain near the top of the gel. At intermediate concentrations of denaturant, transitions from fully duplex DNA to partially denatured DNA occur at positions in the gel that are distinct for each fragment. The cooperativity of melting of each domain results in a sharp transition of mobility, reflected in the sigmoidal shape of the DNA pattern on the gel. If a number of DNA fragments that differ in size are electrophoresed on the same gel, a series of melting transitions corresponding to each fragment can be visualized (Lerman et al. 1984), as shown in the diagram of Figure 4A. In principle, if a mixture of wild-type and mutant DNA fragments are run on such a gel, a single melting curve, consisting of one or more transitions, will be seen for each wild-type DNA fragment. In contrast, the melting curve for the mutant DNA fragment will differ in one transition corresponding to the melting domain containing the mutation.

As in the case of analyzing single fragments on parallel gels, the formation of a heteroduplex between wild-type and mutant DNAs should lead to the generation of four DNA fragments corresponding to the duplexes shown in the diagram of Figure 2. To determine the feasibility of this multifragment approach, we have analyzed a 4000-bp DNA fragment containing the 1500-bp human β-globin gene and 2500 bp of flanking sequences. As shown in Figure 4A, this was accomplished by mixing equal amounts of cloned wild-type

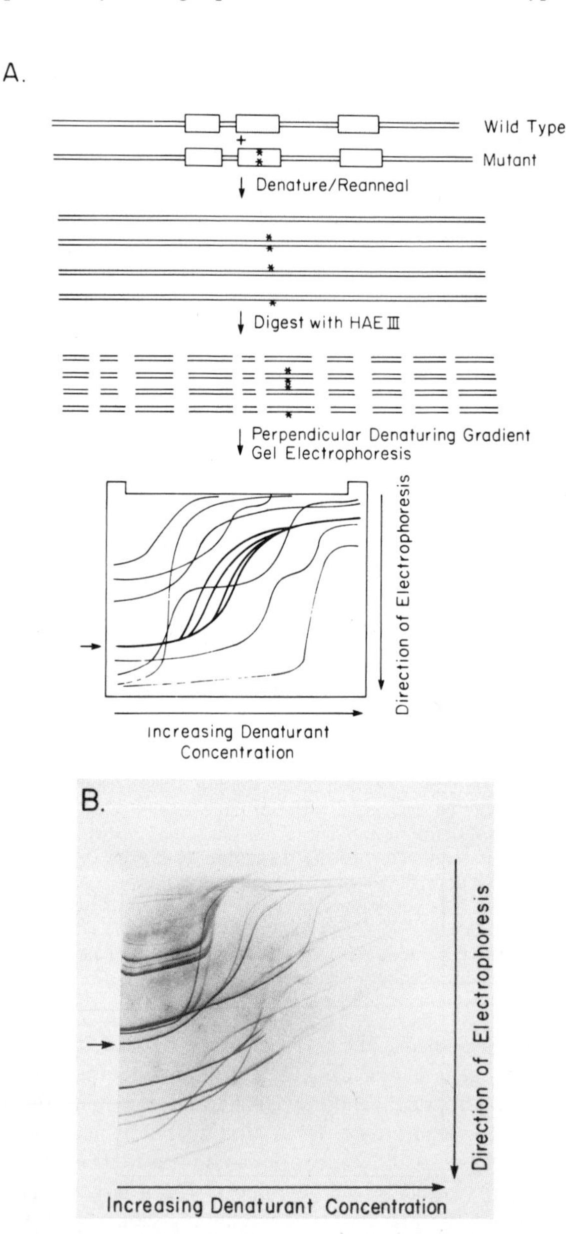

Figure 4. Analysis of multiple DNA fragments by electrophoresis on a perpendicular denaturing gradient gel. (*A*) Diagram showing the experimental approach used to screen the entire human β-globin gene and flanking sequences for single-base mutations. Cloned DNA fragments containing the wild-type and mutant globin genes were denatured and reannealed and then digested with the restriction enzyme *Hae*III. The entire collection of homo- and heteroduplexes was then layered across the top of a perpendicular denaturing gradient gel and subjected to electrophoresis. Asterisks indicate the site of the mutation. The drawing at the bottom of the figure shows the behavior of each fragment on the gel. Arrow indicates the fragment containing the mutation, as evidenced by the split in the melting curve. (*B*) Negative image of an ethidium-stained gel showing the separation of DNA fragments from mutant and wild-type human β-globin genes. Arrow indicates the fragment containing the β-thalassemia mutation.

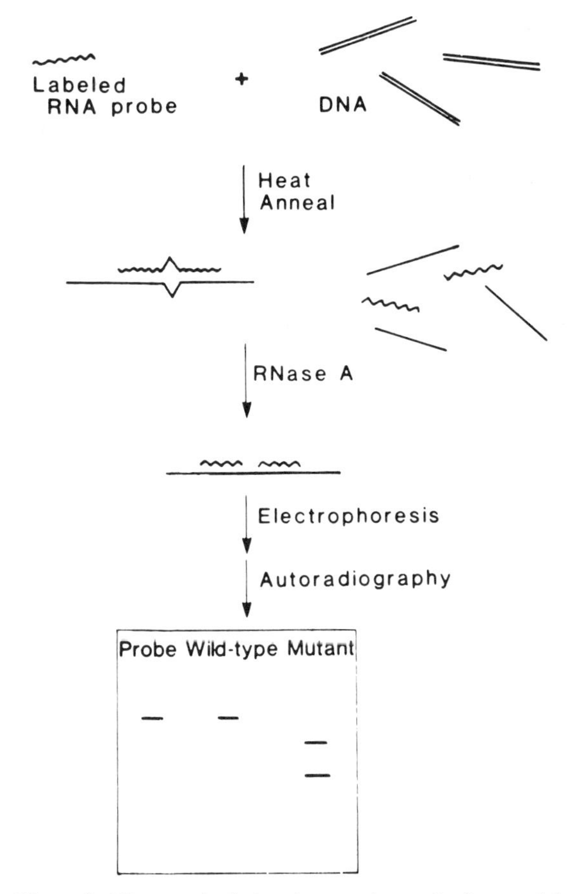

Figure 5. Diagram depicting the experimental scheme of the RNase cleavage procedure. A radioactively labeled single-stranded RNA probe was synthesized in vitro by using a bacteriophage promoter transcription system and annealed to cloned or genomic DNA. The products of the annealing reaction were then treated with RNase A under conditions that result in cleavage at some mismatches but do not affect the perfectly paired regions of the duplex. The RNase-resistant products were then size-fractionated by denaturing gel electrophoresis and detected by autoradiography. In the absence of a mutation, the full-length probe fragment is observed. If the test DNA contains a single-base mutation, cleavage at the resulting mismatch generates two RNA fragments whose total lengths are equal to that of the probe.

and β-thalassemia DNAs, denaturing, reannealing, and then digesting with the restriction enzyme *Hae*III, which generates approximately 20 DNA fragments ranging in size from 50 bp to several hundred base pairs in length. The analysis of these DNA fragments on a perpendicular denaturing gradient gel is shown in Figure 4B. For all of the DNA fragments except one, a single melting curve can be traced starting from the left side of the gel where there is no denaturant. However, the melting curve from the fragment indicated by the arrow splits into three curves; the lower curve corresponds to the wild-type homoduplexes, the middle curve corresponds to the mutant homoduplex, and the upper curve corresponds to the two heteroduplexes. The size of this DNA fragment, as determined by its mobility on the left side of the gel, corresponds to that of the DNA fragment known to contain the single-base thalassemic mutation. Thus, by using this general approach, it is possible to scan an entire cloned gene for single-base mutations. Of course, since the DNA fragments used in this analysis do not contain a GC-clamp, not all possible single-base substitutions in the gene and flanking regions are detectable.

It should also be possible to use this approach to analyze total genomic DNA. In this case, it would be necessary to prepare a labeled single-stranded DNA probe that spans the entire gene and flanking sequences and then anneal this DNA to genomic DNA prior to digestion with *Hae*III. By including small amounts of labeled wild-type homoduplex tracer DNA, equivalent to one-fifth to one-tenth the molar amount of the test genomic DNA, it should be possible to detect differences between mutant and wild-type DNA fragments present in the test genomic DNA. The presence of a mutant DNA fragment would be revealed by additional melting curves emanating from the curve corresponding to the tracer DNA. By adjusting the ratio of tracer to test DNA appropriately, it should be possible to distinguish between homozygous wild-type, homozygous mutant, and simple or compound heterozygous genomes. Experiments to test this possibility are in progress. This approach should also be useful in the search for new neutral polymorphisms in DNA, since much longer stretches of DNA can be scanned in a single gel.

Ribonuclease Cleavage at Mismatches in RNA:DNA Duplexes

We have developed an alternative method for detecting single-base substitutions in cloned and genomic DNAs that involves the enzymatic cleavage of RNA at a single-base mismatch in an RNA-DNA hybrid. This approach was made possible by the development of methods for synthesizing RNA probes (Green et al. 1983; Zinn et al. 1983; Melton et al. 1984) and on the observation that many ribonucleases are specific for single-stranded RNA under appropriate reaction conditions (Brownlee 1972). A systematic analysis of a large number of single-base mutations revealed that many mismatches can be cleaved specifically by ribonuclease A (RNase A; Myers et al. 1985a). In addition, previous studies indicated that presumptive single-base mismatches in RNA:RNA heteroduplexes formed between vesicular stomatitis virus mRNA and virion RNA can be cleaved by ribonucleases (Freeman and Huang 1981). More recently, known single-base mutations in human c-Ki-*ras* genes were detected in hybrids between a single-stranded RNA probe and c-Ki-*ras* mRNA (Winter et al. 1985).

The steps in the RNase cleavage procedure developed in our laboratory are outlined in Figure 5. A ^{32}P-labeled RNA probe is synthesized from a wild-type DNA template using a bacteriophage transcription system. The RNA probe is hybridized to denatured test DNA in solution, and the resulting RNA-DNA hybrid is treated with RNase A. The RNA products are then analyzed by electrophoresis in a denaturing gel. If the test DNA is identical to wild-type DNA, a single band is observed in the autoradiogram of the gel, since the RNA-DNA hybrid is not cleaved by RNase. However, if the test DNA contains a single-base substitution, and if the mismatch is recognized by RNase A, two new RNA fragments will be detected. The total size of these fragments should equal the size of the single RNA fragment observed with wild-type DNA. Thus, the mutation can be localized relative to the ends of the RNA probe by determining the sizes of the cleavage products. The end of the RNA probe mapping nearest to the substitution can be determined when the experiment is performed with DNA digested with an additional restriction enzyme, thus localizing the substitution unambiguously.

To establish optimal conditions for recognizing single-base mismatches, and to determine which types of mismatches can be cleaved by RNase, we examined a large number of single-base substitutions in the mouse β-major globin promoter region and a number of cloned and genomic human β-globin DNA fragments carrying β-thalassemia mutations. With this collection, it was possible to examine all 12 types of mismatches possible in RNA-DNA hybrids in several different sequence contexts. We found that many, but not all, mismatches could be cleaved by RNase A by using this method. Some mismatches were not cleaved to completion under the conditions of the assay, and some types of mismatches that were cleaved in one sequence context were not cleaved in another context. A compilation of the results of analyzing over 100 different mismatches is summarized in Figure 6; the numbers represent the fractions of each type of mismatch that were cleaved to 50% or greater in the assay.

We found that 3 out of the 12 possible types of mismatches (C:A, C:C, and C:T) are recognized efficiently by RNase A in all sequence contexts tested. Thus, approximately 25% of all possible single-base substitutions can be detected using an RNA probe homologous to one strand of the test DNA. This number can be doubled by using a second RNA probe, homologous to the opposite strand of the test DNA. For example, a G:T mismatch formed between an RNA probe and its complementary strand in the test DNA may not be cleaved by RNase A. However, when the other DNA strand is hybridized to its homologous RNA probe, the C:A mismatch at that same position will be cleaved by RNase. Thus, approximately 50% of all possible single-base substitutions should be detected. This is clearly a minimum estimate, since we have observed cleavage at eight of the remaining nine possible types of mismatches in some sequence contexts.

RNA	DNA	Fraction	RNA	DNA	Fraction
C	A	22/22	G	T	0/14
C	C	4/4	G	G	0/6
C	T	9/9	G	A	1/7
U	G	3/11	A	C	4/12
U	C	0/3	A	G	2/3
U	T	1/4	A	A	0/6

Figure 6. Summary of the analysis of a large number of single-base mismatches by the RNase cleavage method (Myers et al. 1985a). In each case, the upper strand is the RNA probe and the lower strand, the complementary test DNA. The number next to each type of mismatch is the fraction of different mismatches of that type that was cleaved to greater than 50% completion by RNase A. For a complete compilation of these data, see Myers et al. (1985a).

The data summarized in Figure 6 were obtained by performing the RNase reactions for a fixed length of time (30 min). In a time-course experiment, we found that many mismatches that are only partially cleaved in 30 minutes can be cleaved almost to completion in 90 minutes under the same conditions with only a slight increase in background. As shown in Figure 7, digestion of an RNA:DNA duplex formed between wild-type DNA and wild-type RNA probes generates a single full-length protected fragment when treated with RNase for 30, 60, or 90 minutes. The duplex is insensitive to RNase cleavage at all time points, since only a small amount of RNA shorter than the full-length probe is observed (Fig. 7, lanes 1–3). Similarly, an RNA:DNA duplex in which the DNA contains a single-base mutation that is cleaved to completion in 30 minutes is not further degraded after 60 and 90 minutes in incubation (Fig. 7, lanes 4–6). Other mutations that show partial cleavage at 30 minutes are cleaved to completion when incubated for 90 minutes (Fig. 7, lanes 7–9 and 10–12). As with the wild-type RNA-DNA hybrid, very little nonspecific degradation is observed with these mutations. Although the reasons for differences in susceptibility of different mismatches to RNase cleavage are not understood, it may be possible to achieve complete cleavage of most susceptible RNA:DNA mismatches by carrying out the RNase reaction for 90 minutes.

It may be possible to use a variation of the RNase cleavage procedure to screen rapidly the exon sequences of very large genes for single-base mutations. A case in point is the factor VIII gene that is 180,000 bp in length but encodes an mRNA of only 9000 nucleotides (Gitschier et al. 1984; Toole et a. 1984). Although it would be impractical to screen the entire factor VIII gene for mutations by any method currently available, it should be possible to screen exon sequences using an RNA probe corresponding to the full-length cDNA. To test this possibility, we have attempted to detect mutations in the exons of the human β-globin gene using the approach outlined in the schematic diagram of Figure 8A. A labeled SP6 RNA transcript of a nearly full-length β-globin cDNA clone is synthesized and annealed to the cloned β-globin gene.

Figure 7. Time course of RNase cleavage of single-base mismatches. RNA:DNA duplexes formed as shown in Fig. 5 were treated for 30, 60, or 90 min with RNase A and then size-fractionated on a 8% polyacrylamide, 7 M urea gel. Lanes *1–3* contain wild-type duplexes, so that a single band corresponding to full-length protection of the probe is observed. In the remaining lanes, mutant DNA fragments were annealed to the probe, so that the two bands correspond to the two fragments generated by cleavage at the mismatched bases.

In the resulting hybrid, the exon sequences of the gene will base pair with the RNA probe, whereas the single-stranded intron sequences will be looped out. In principle, digestion of a hybrid containing the wild-type gene and the wild-type probe will produce three protected RNA fragments corresponding to the sizes of the three exons. If the genomic DNA contains a single-base mutation, one of the three fragments should be digested at the site of the mismatch. In practice, we have found that the RNA sequences at the exon junctions of the probe are not readily accessible to RNase cleavage, suggesting that the RNA across from the looped-out introns of the gene cannot be cleaved. This could be due to the lack of accessible pyrimidines at or near the exon junction.

An alternative approach is to digest the genomic DNA prior to hybridization with restriction enzymes that cleave within the introns (Fig. 8B). If the probe is in excess over genomic DNA, each genomic DNA fragment will hybridize to a different RNA probe molecule. When the hybridization mix is then digested with RNase, a series of RNA fragments of predictable size will be protected from cleavage. If the genomic DNA contains a single-base mutation, one of the fragments will be cleaved at the site of the mismatch. As shown in Figure 8C, single-base mutations in exons can be detected by this approach. An RNA probe corresponding to the cDNA of the human β-globin gene was annealed to *Hin*fI-digested cloned human β-globin genes containing both introns. In lane 1, the RNA probe was annealed to the normal human β-globin gene, and in lane 2, the probe was annealed to a human β-globin gene carrying a single-base mutation at codon 39 in exon 2. The patterns of RNase protection are identical in the two lanes except for the absence of one band and the appearance of two new bands in lane 2. The size of the RNA fragment that is missing in lane 2 corresponds to that expected for the fragment containing the single-base mutation. The sizes of the two smaller fragments in lane 2 are consistent with being the products of digestion at the single-base mismatch at codon 39. On the basis of these results, it should be possible to use this approach to detect and localize mutations within exons of much larger genes.

DISCUSSION

The ability to clone human genes has made it possible to identify and localize single-base mutations that cause specific genetic diseases. These mutations were initially identified by comparing the DNA sequences of these genes isolated from normal and affected individuals. The availability of the normal and mutant genes has provided the opportunity to establish general approaches for detecting mutations associated with a variety of genetic diseases. The normal genes can be used as hybridization probes to detect single-base mutations that alter restriction endonuclease cleavage sites. However, a relatively small fraction of all possible mutations can be detected by this approach even when a large battery of restriction endonucleases are tested. The two approaches for detecting single-base mutations described here involve cloning fragments of the normal gene into plasmid vectors that facilitate the synthesis of labeled single-stranded DNA or RNA probes. The specific activity of these probes can be made sufficiently high to detect conveniently single-copy genes with a few micrograms of human genomic DNA. The sensitivity of the methods could be dramatically increased by amplifying the genomic DNA sequences of interest by multiple rounds of primer-dependent DNA synthesis (Saiki et al. 1985; Mullis et al., this volume).

Comparison of Methods for Detecting Single-base Mutations in Genomic DNA

The two methods described here are each capable of detecting over 50% of all possible single-base substitutions in DNA fragments of 100–1000 bp in length.

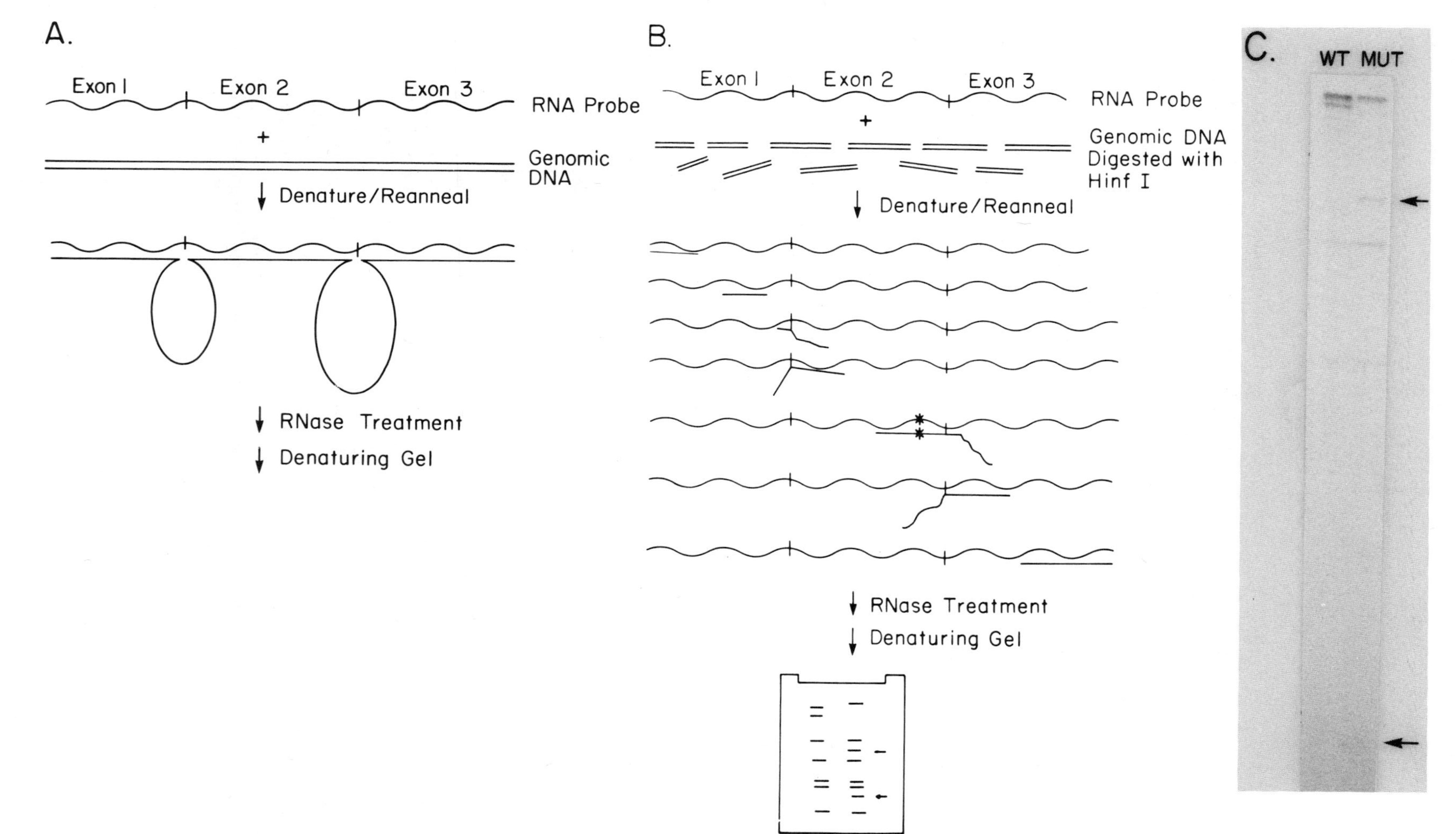

Figure 8. Detection of mutations in exons by RNase cleavage. (*A*) Diagram showing a scheme in which an RNA probe synthesized from a cDNA fragment containing exons 1 through 3 of the human β-globin gene was annealed to genomic DNA and then digested with RNase. In principle, three fragments will be generated, corresponding to the three exon sequences. For a discussion of attempts to use this approach, see text. (*B*) Diagram showing a scheme in which the genomic DNA was digested with a restriction enzyme that cleaves within the introns prior to hybridization to the RNA probe containing exons 1–3. Because the RNA probe was in excess, probe molecules were never annealed to more than one DNA fragment. Digestion of this mixture of RNA probe and DNA fragments generated a series of protected RNA fragments of predicted size. If a single-base mismatch is present in any of the fragments, a fragment protected in the wild-type DNA sample will be lost, and it will be replaced by two smaller fragments generated by cleavage at the mismatch. (*C*) Autoradiogram showing the RNA fragments protected from RNase cleavage after annealing to the wild-type (lane *1*) and mutant (lane *2*) human β-globin gene following the scheme presented in *B*. Arrows indicate the two fragments generated by cleavage at the mismatch.

Both methods involve solution hybridization followed by gel electrophoresis and autoradiography, and thus transfer of the DNA onto filters is unnecessary. The primary limitation of the denaturing gradient gel electrophoresis method is that substitutions in the highest melting domain of a fragment cannot be detected in genomic DNA. Efforts are in progress to attach a GC-clamp to specific fragments in total genomic DNA. On the basis of theoretical calculations and previous studies of the effect of a GC-clamp on the detection of known mutations in cloned DNA, the use of this method in the analysis of genomic DNA should make virtually all possible single-base substitutions accessible to detection.

Another limitation of the denaturing gradient gel system is that only small DNA fragments can be analyzed. Thus, it would be necessary to use prohibitively large numbers of probes to screen large eukaryotic genes for mutations. However, in this paper, we have demonstrated the feasibility of screening many small DNA fragments, which together span over 4000 bp of DNA. In this method, only one probe is required in a single solution hybridization reaction, and multiple fragments can be analyzed on a single gel.

Although all of the analyses with denaturing gradient gels described here involve the use of single stranded DNA probes, we and others (Smith et al. 1986; S. Wolf, pers. comm.) have demonstrated the feasibility of using labeled single-stranded RNA probes with this method. Similar results are obtained with RNA and DNA probes, although the melting behavior of RNA-DNA hybrids is different from the corresponding DNA-DNA hybrids. It has also been demonstrated that mismatches in RNA-RNA hybrids can be detected by the gel system (Smith et al. 1986). The primary advantage of the analysis of RNA:DNA duplexes is the ease of preparation of labeled single-stranded RNA probes using cloning vectors with bacteriophage promoters.

The advantages of the RNase cleavage procedure are that the analysis is carried out by conventional gel electrophoresis and that the method not only detects, but also localizes the mutations. In addition, the procedure used to prepare the duplex molecules containing the single-base mismatches is somewhat simpler than that used for the denaturing gradient gel procedure. One limitation of the RNase cleavage procedure is that, thus far, relatively small probes have been routinely used (up to 1500 nucleotides). Thus, screening large genes for mutations by this method may require the use of multiple probes. Although we and others have had success in synthesizing very large RNA probes (over 7000 nucleotides), in preliminary experiments, we find that nonspecific degradation of the RNA:DNA duplex produces a background that makes the detection of cleavage at the mismatch difficult.

As with the denaturing gradient gel electrophoresis method, the RNase cleavage procedure does not allow the detection of all possible single-base mutations. Although there is some overlap in the mutations that can be detected by the two methods, it is possible to detect some mutations with one procedure that are not detected by the other. Thus, it should be possible to detect a very large fraction of all possible mutations by combining the two methods. A convenient way of accomplishing this would be to form an RNA:DNA heteroduplex and then analyze one aliquot by denaturing gradient gel electrophoresis and another aliquot by RNase cleavage. A systematic evaluation of this approach with a large number of different single-base substitutions is in progress.

In addition to the detection of base substitutions that cause genetic diseases, these methods should be useful in identifying and utilizing single-base neutral polymorphisms in the establishment of human genetic linkage maps and in the diagnosis of genetic defects linked to the polymorphisms. In fact, the denaturing gradient gel electrophoresis method has been used to identify neutral polymorphisms within the human factor VIII gene (M. Collins and E. Fritsch, pers. comm.). The advantage of detecting single-base polymorphisms by the methods discussed here compared to the restriction-fragment-length polymorphism (RFLP) approach are clear. Rather than screening a large number of restriction enzymes for single-base polymorphisms, a high percentage of all polymorphisms within a region of 1000 bp can be detected on a single gel lane.

Although the methods described in this paper do not allow the detection of all possible single-base mutations in total genomic DNA, the fraction that can be detected is large. Because the denaturing gradient gel electrophoresis method can be used to detect more than 90% of all possible base substitutions in cloned DNA linked to a GC-clamp, it seems likely that similar approaches can be developed to increase the fraction of base substitutions that can be detected in genomic DNA. Similarly, by investigating the susceptibility of single-base mismatches in RNA-DNA hybrids to a variety of ribonucleases with different specificities, and systematically examining digestion conditions, it may be possible to increase significantly the number of different mismatches that can be recognized with the RNase cleavage procedure. In any case, these methods provide valuable tools for identifying mutations and diagnosing genetic diseases. In addition, they have proven to be useful in genetic studies of gene regulation and development.

ACKNOWLEDGMENTS

We wish to acknowledge that the application of the denaturing gradient gel electrophoresis system to the detection of mutations associated with human genetic diseases was carried out in close collaboration with Leonard Lerman, who along with Stuart Fischer developed the gel system and the theoretical basis upon which strategies for its application were designed. In addition, we thank Zoia Larin who contributed to the development of the RNase cleavage method. We also thank Mary Collins and Ed Fritsch, and Stan Wolf for permission to cite their results prior to publication and for useful discussions. This work was supported by a grant from the National Institutes of Health to T.M. and by grants from the Wills Foundation to T.M. and R.M. Part of this work was supported by a fellowship from the Leukemia Society of American, Inc., to R.M.

REFERENCES

Brownlee, G.G. 1972. Determination of sequences in RNA. In *Laboratory techinques in biochemistry and molecular biology* (ed. T.S. Work and E. Work), p. 67. American Elsevier, New York.

Chang, J.C. and Y.W. Kan. 1982. A sensitive new prenatal test for sickle cell anemia. *N. Engl. J. Med.* **307:** 30.

Fischer, S.G. and L.S. Lerman. 1983. DNA fragments differing by single base pair substitutions are separated in denaturing gradient gels: Correspondence with melting theory. *Proc. Natl. Acad. Sci.* **80:** 1579.

Flavell, R.A., J.M. Kooter, E. DeBoer, P.F.R. Little, and R. Williamson. 1978. Analysis of the beta-delta-globin gene loci in normal and Hb lepore DNA: Direct determination of gene linkage and intergene distance. *Cell* **15:** 25.

Freeman, G.J. and A.S. Huang. 1981. Mapping temperature-sensitive mutants of vesicular stomatitis virus by RNA heteroduplex formation. *J. Gen. Virol.* **57:** 103.

Geever, R.F., L.B. Wilson, F.S. Nallaseth, P.F. Milner, M. Bittner, and J.T. Wilson. 1981. Direct identification of sickle cell anemia by blot hybridization. *Proc. Natl. Acad. Sci.* **78:** 5081.

Gitschier, J., W.I. Wood, T.M. Goralka, K.L. Wion, E.Y. Chen, D.H. Eaton, G.A. Vehar, D.J. Capon, and R.M. Lawn. 1984. Characterization of the human factor VIII gene. *Nature* **312:** 326.

Green, M.R., T. Maniatis, and D.A. Melton. 1983. Human beta-globin pre-mRNA synthesized in vitro is accurately spliced in *Xenopus* oocyte nuclei. *Cell* **32:** 681.

Gusella, J.F., N.S. Wexler, P.M. Conneally, S.L. Naylor, M.A. Anderson, R.E. Tanzi, P.C. Watkins, K. Ottina, M.R. Wallace, A.Y. Sakaguchi, A.B. Young, I. Shoulson, E. Bonilla, and J.B. Martin. 1982. A polymorphic marker genetically linked to Huntington's disease. *Nature* **306:** 234.

Kan, Y.W. and A.M. Dozy. 1978. Polymorphism of DNA sequence adjacent to human beta-globin structural gene: Relationship to sickle mutation. *Proc. Natl. Acad. Sci.* **75:** 5631.

Kidd, V.J., R.B. Wallace, I. Itakura, and S.L.C. Woo. 1983. Alpha-antitrypsin deficiency detection by direct analysis of the mutation in the gene. *Nature* **304:** 230.

Lerman, L.S. 1986. Electrophoresis of DNA in denaturing gradient gels. In *Genetic engineering, principles and methods* (ed. J.K. Setlow and A. Hollaender), p. 221. Plenum Press, New York.

Lerman, L.S., S.G. Fischer, I. Hurley, K. Silverstein, and N. Lumelsky. 1984. Sequence-determined DNA separation. *Annu. Rev. Biophys. Bioeng.* **13:** 399.

Melton, D.A., P.A. Krieg, M.R. Rebagliati, T. Maniatis, K. Zinn, and M.R. Green. 1984. Efficient in vitro synthesis of biologically active RNA and RNA hybridization probes from plasmids containing a bacteriophage SP6 promoter. *Nucleic Acids Res.* **12:** 7035.

Myers, R.M., Z. Larin, and T. Maniatis. 1985a. Detection of single base substitutions by ribonuclease cleavage at mismatches in RNA:DNA duplexes. *Science* **230:** 1242.

Myers, R.M., L.S. Lerman, and T. Maniatis. 1985b. A general method for saturation mutagenesis of cloned DNA fragments. *Science* **229:** 242.

Myers, R.M., T. Maniatis, and L.S. Lerman. 1986. Detection and localization of single base changes by denaturing gradient gel electrophoresis. *Methods Enzymol.* (in press).

Myers, R.M., S.G. Fischer, L.S. Lerman, and T. Maniatis. 1985c. Nearly all single base substitutions in DNA fragments attached to a GC-clamp can be detected by denaturing gradient gel electrophoresis. *Nucleic Acids Res.* **13:** 3131.

Myers, R.M., S.G. Fischer, T. Maniatis, and L.S. Lerman. 1985d. Modificiation of the melting properties of duplex DNA by attachment of a GC-rich DNA sequence as determined by denaturing gradient gel electrophoresis. *Nucleic Acids Res.* **13:** 3111.

Myers, R.M., N. Lumelsky, L.S. Lerman, and T. Maniatis. 1985e. Detection of single base substitutions in total genomic DNA. *Nature* **313:** 495.

Orkin, S.H., P.F.R. Little, H.H. Kazazian, and C.D. Boehm. 1982. Improved detection of the sickle mutation by DNA analysis. *N. Engl. J. Med.* **307:** 32.

Pirastu, M., Y.W. Kan, A. Cao, B.J. Connor, R.L. Teplitz, and R.B. Wallace. 1983. Prenatal diagnosis of beta-thalassemia: Detection of a single nucleotide mutation in DNA. *N. Engl. J. Med.* **309:** 284.

Saiki, R.K., S. Scharf, F. Faloona, K.B. Mullis, G.T. Horn, H.A. Erlich, and N. Arnheim. 1985. Enzymatic amplification of beta-globin genomic sequences and restriction site analysis for diagnosis of sickle cell anemia. *Science* **230:** 1350.

Smith, F.I., J.D. Parvin, and P. Palese. 1986. Detection of single base substitutions in influenza virus RNA molecules by denaturing gradient gel electrophoresis of RNA-RNA or DNA-RNA heteroduplexes. *Virology* **150:** 55.

Toole, J.J., J.L. Knopf, J.M. Wozney, L.A. Sultzman, J.L. Bueker, D.D. Pittman, R.J. Kaufman, E. Brown, C. Shoemaker, E.C. Orr, G.W. Amphlett, W.B. Foster, M.L. Coe, G.J. Knutson, D.N. Fass, and R.M. Hewick. 1984. Molecular cloning of a cDNA encoding human antihaemophilic factor. *Nature* **312:** 342.

Winter, E., E. Yamamoto, C. Almoguera, and M. Perucho. 1985. A method to detect and characterize point mutations in transcribed genes: Amplification and overexpression of the mutant c-Ki-*ras* allele in human tumor cells. *Proc. Natl. Acad. Sci.* **82:** 7575.

Zinn, K., D. DiMaio, and T. Maniatis. 1983. Identification of two distinct regulatory regions adjacent to the human beta-interferon gene. *Cell* **34:** 865.

Searching for Gene Defects by Denaturing Gradient Gel Electrophoresis

L.S. Lerman, K. Silverstein, and E. Grinfeld
Genetics Institute, Cambridge, Massachusetts 02140

In attempting to ascertain the presence of particular alleles of critical genes in a human fetus, it would be most desirable to know precisely the base sequence of its genome. It is not clear how soon direct sequencing techniques will become fast and easy enough to permit scrutiny of 50 kb within the limits of Medicare reimbursement set by the diagnostic related groupings (DRGs), but for the present, other paths to detection of departures from normal sequence seem to be necessary.

The presence of a dominant or sex-linked allele can be traced through a family by means of linkage to an appropriate sequence marker. Current practice depends mostly on restriction-fragment-length polymorphisms (RFLPs) as markers, but simple hypervariable repeats are also promising. The estimate of the probability that the fetus is affected depends on which and how many other family members are available for testing and on the closeness of the linkage between the gene and marker. For recessive genes where there is no biochemical test for a gene product, linkage to nearby markers must be traced through two families. Linkage analysis is of limited help where a male carrying a sex-linked defect is unlikely to reproduce; one third of all defects represent new mutations. The need to test the DNA of only one individual, rather than a family, would radically alter the outlook for testing and for a secure conclusion.

The technical problem is difficult because the precise site of departure from normal sequence is generally not known in advance. The type of sequence difference responsible for sickle cell anemia, a particular base substitution at a particular site, is exceptional rather than typical for human inherited disease. β-Thalassemia represents a more generally applicable model. At least 36 different sequence aberrations have been reported as a basis for defects in human β-globin synthesis, and it is likely that many more sites that are less frequently involved will be discerned (Orkin and Kazazian 1984). Some of these show phenotypic differences, particularly whether β-globin synthesis is partially or entirely abolished.

The same difficulty arises with respect to other questions. An effort to determine the rate of spontaneous or induced mutations in man might either examine the incidence of changes at a particular recognizable locus in a very large number of persons or examine a very large sample of sequences for changes anywhere within a much smaller number of genomes. The evaluation of genetically determined risk factors for particular disorders may be more troublesome than prenatal detection of thalassemia because of the subtlety or late onset of the defects and the lower likelihood that the DNA of the parents and grandparents will be available for linkage analysis.

A number of means are available for recognition of departures from a prototype sequence in genomic DNA without requiring full sequence determination. Single-base changes may remove or add restriction sites. Binding of an oligodeoxyribonucleotide bracketing the putative altered site may be severely reduced. If an RNA transcript is available as a probe, a mismatch formed in hybridization with a genomic strand carrying a base substitution may be vulnerable to attack by ribonuclease A. A mismatch with a DNA probe may be detected by the reaction of the unpaired parent with carbodiimide (Novack et al. 1986). Mismatches between the genomic strand and a prototype DNA probe alter the melting properties of the heteroduplex relative to either homoduplex and are readily detectable by denaturing gradient gel electrophoresis. Each of these has its own limitations at the present state of development. The oligonucleotide technique discussed by Wallace et al. (this volume) scrutinizes a very small site that must be specified in advance and requires that the prototype base sequence surrounding that site be known. A small number of bases are scrutinized by the test for restriction site changes and, indeed, few of the substitutions responsible for β-thalassemia can be detected (Orkin and Kazazian 1984). The applicability of the recent RNase method is discussed by Myers and Maniatis (this volume). Denaturing gradient gel electrophoresis (Fischer and Lerman 1979a; Lerman et al. 1984; Myers et al. 1985e) can examine a large section of most sequences but does not approach comprehensive detections of all changes unless the genomic sample to be tested is cloned in a special way (Myers et al. 1985c).

A Small Sample of the Genome

The present discussion is concerned with estimating the likelihood that a random base change in an unknown sequence will be detected following the simplest scheme for applying denaturing gradient gel electrophoresis. The estimate will be based on a set of theoretical calculations that simulate experimental results in detail and have been shown to agree closely with experiments where they have been tested. We will take a large contiguous portion (about 35.4 kb) of thc human β-globin gene cluster as a sample of the genome; allow it to be broken into fragments by each of two 4-base-specific restriction endonucleases; hybridize the digest with similarily fragmented, guaranteed wild-type DNA; and apply the resulting mixture to a denaturing gradient gel. The effect of a possible mismatch due to a base substitution, an insertion, or a deletion at every tenth base will be systematically determined. The total fraction of the full 35.4 kb in which changes might be discerned will be compiled from a relatively conservative criterion for a distinct signal.

Principles of the Method

The experimental procedure, previously described in detail (Fischer and Lerman 1979b; Myers et al. 1986), separates double-helical DNA molecules according to the nature of their sequence—unlike conventional gel methods that are sensitive mainly to length. This procedure is sensitive enough to respond to single-base substitutions (even though perfect pairing is maintained) or single-base insertions or deletions, but it encompasses a broad enough spectrum of sensitivity to effect separations depending on differences of hundreds of base pairs. Although it is simple to execute with apparatus commonly found in molecular biology laboratories, the modified electrophoresis unit we have described permits a more convenient and reliable operation. The end results in most experiments, separation of bands in a lane of a gel, look like conventional gel electrophoresis but with two important differences: (1) The bands are considerably sharper (they are focused by the mechanics of the process) and (2) they define a position in the gradient, not an electrophoretic mobility. Where the DNA molecule, as a heteroduplex between the genomic strand and the probe carrying the prototype sequence, contains a non-Watson-Crick base pair or looped-out base, the band will be substantially displaced from the position of the parental homoduplexes, sometimes as much as 6 cm.

The appearance of bands in a denaturing gradient gel (ethidium staining) and the shifts due to sequence changes and matching errors are shown in the right-hand part of Figure 1, one lane of a gel in which the urea-formamide concentration increases from the top to bottom, parallel to the direction of electrophoretic migration. The 445-bp molecule is a recombinant containing about 140 bp of the mouse β-major globin promoter, previously described by Myers et al. (1985d). The lowest band is that of the prototype sequence, fully base paired. The second band from the bottom is a mutant in which a GC pair is deleted at 33 bp from the end of the promoter. The two uppermost bands represent reassortment of the mutant and prototype strands such that each molecule has an extra base, looped out of the helix. The external G forms a slightly more stable structure than an external C, giving a slightly lower band. Although the gradient clearly resolves the fully paired mutant, the heteroduplexes provide a much stronger separation.

The gel pattern on the left is derived from the same mixture applied as a continuous line across the top of a gel with a gradient of denaturant increasing from right to left, perpendicular to the direction of the electric field. Each molecule travels through a particular, unchanging concentration of denaturant. The final pattern represents the mobility of that molecule as a function of denaturant concentration and provides a direct indication of the concentration (or equivalent temperature) at which each domain melts, effecting a drop in mobility. From right to left, the curves correspond to the heteroduplex with C looped out, the heteroduplex with G looped out, the homoduplex with a GC deletion, and the prototype sequence. The pattern in the parallel gradient at the right can be seen to re-

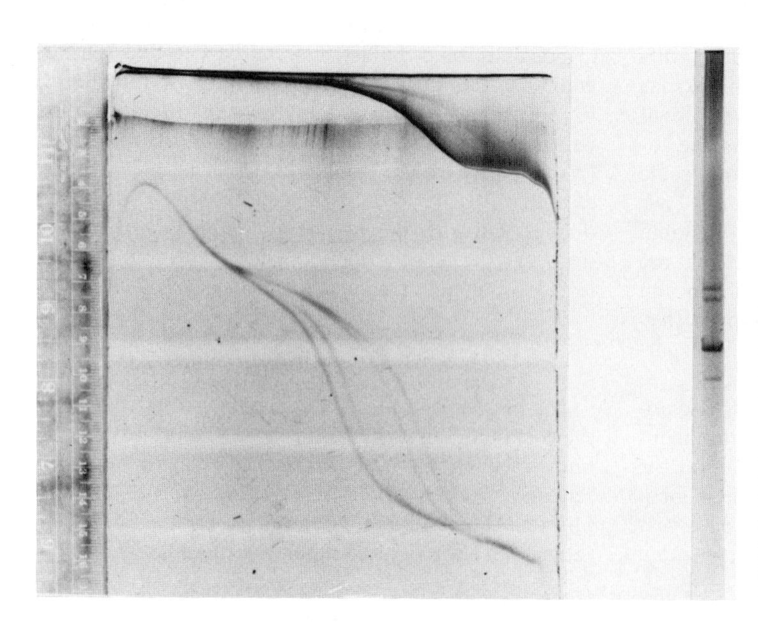

Figure 1. Perpendicular (*left*) and parallel (*right*) denaturing gradient patterns. The gels were loaded with a mixture of a 435-bp composite molecule containing the mouse β-major globin promoter and a mutant with a GC deletion; these had been heated to effect strand dissociation and reassociation. The gel samples contained the two parental types and two different heteroduplexes. The sequence and origin of the molecule are given in Myers et al. (1985b,d). The site of the GC deletion is -10 with respect to the cap site in the mouse promoter. The denaturant concentration increases from right to left in the perpendicular pattern and from top to bottom in the parallel pattern.

flect the slowing down of each molecule as it reaches its characteristic depth.

Although theoretical simulation of the experimental results has had only limited challenge up to now, it has shown a useful level of agreement with the data from λ DNA (Fischer and Lerman 1983), pBR322 (K. Silverstein and L.S. Lerman, unpubl.), human mitochondria (Lerman et al. 1983), and human (Lumelsky 1984) and mouse β-globin (Myers et al. 1985c,d). The theory has been able to account quantitatively for the major experimental observations: the changes in mobility that various molecules undergo in the gradient; the denaturant concentrations required to effect those changes for different sequences; the detection of base substitutions (without mismatches) in some regions of the molecule and not in others; the amount of separation in the gradient given by the various permutations of base substitution; the need for different running time in the electrophoretic procedure for optimum separation according to the position of the base change in the molecule; and other features of the experiments.

The theoretical foundation of the calculation has two components. The first is a relation between the electrophoretic mobility of a DNA molecule in a gel as a function of the number of base pairs that are melted. The second is the calculation of how much and what parts of any molecule will be melted at any temperature, given its sequence. We have relied on the statistical-mechanical principles and algorithms developed by Poland (1974) and Fixman and Friere (1977), together with the use of the nearest-neighbor base-pair doublet parameters introduced by Gotoh and Tagashira (1981), and some additional slight modifications. The melting properties can be predicted using a digital computer for any sequence. The melting calculation at any fixed temperature gives the probability that each base pair is in either the melted or helical state. Repetition of the melting probability calculation at a series of increasing temperatures permits estimation of the temperature for each base pair at which it has equal probability of either the helical or melted state. A plot of these values along the sequence is termed a melting map. It indicates the extended regions, termed domains, within which the melting of contiguous bases is very closely coupled; these appear as plateaus or flat valleys in the melting map. Computer programs for carrying out these calculations are described in more detail elsewhere (Lerman and Silverstein 1986).

Figure 2 shows the melting map of a portion of the human β-globin gene extending from the last part of the first coding sequence, through the first intron, and into the second coding sequence. The schematic diagram of a molecule on the right shows the progression of melting as the temperature of its environment, or the concentration of denaturant, rises, assuming that the rate of temperature rise is sufficiently slow to maintain equilibrium between the helical and melted structures. The abrupt changes in transition temperature along the sequence indicated by the boundaries between domains mark off the portions within which single-base changes can be detected, the lowest melting domains of a molecule or, in which they are imperceptible, the higher domains of the molecule. The melting map itself provides no indication of the swing of the equilibrium toward complete dissociation of the two strands. In general, this equilibrium will be driven to dissociation at a temperature below the top plateaus of the melting map. Since separation in the denaturing gradient depends on perturbations that take effect during partial melting, base substitutions or mismatches that pertain to regions of the molecule where the contour of the melting map lies in the region of strong dissociation will not be detectable.

Strand dissociation becomes relevant even at small values of the equilibrium constant for dissociation. A band moving through a gel is not a closed system, and any substantial difference between the mobilities of the helix and the separated strands will drive the reaction

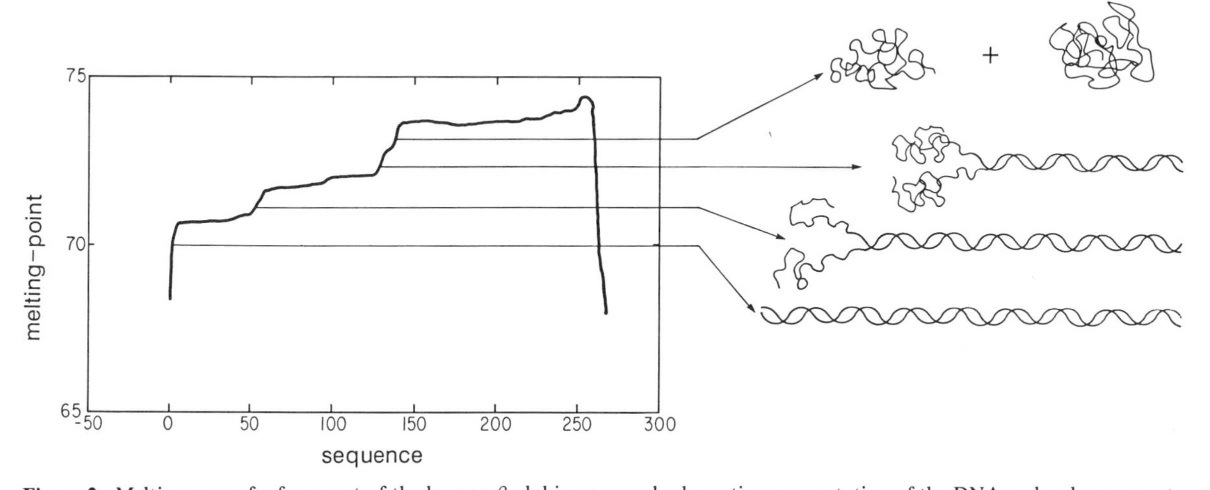

Figure 2. Melting map of a fragment of the human β-globin gene and schematic representation of the DNA molecule representing the extent of melting at several temperature levels. The fragment includes the last 10 bp of the first coding sequence and extends 262 bp into the first intervening sequence. The map indicates the calculated portion of the sequence that will retain a stable helical conformation at each temperature, disregarding the possibility of dissociation. Equilibrium favors complete strand dissociation at temperatures below the highest part of the curve at ordinary DNA concentrations.

toward dissociation. The temperature at which the rate of strand dissociation becomes significant sets a practical upper limit as to how far any molecule can usefully be drawn into the gradient and how far into upper domain levels changes will be scored.

The presence of mismatches due to non-Watson-Crick pairs, or insertions or deletions, alters the melting map; this effect must be taken into account in attempting to estimate the portion of a molecule in which changes can result in good separations. The features of the melting map contributed by any particular short component sequence are also significantly affected by the position of that portion relative to the ends of the molecule. The same sequence may establish a lower-lying domain when it is near an end than when it is central. For this reason, the melting map of a fragment can differ substantially from the corresponding portion of the melting map of the intact longer molecule. To approach the answer that we are after, i.e., the fraction of a long gene in which sequence alterations can be detected, reliable conclusions can be derived only from calculations based on the details of each specific fragment, including any perturbation due to a helix defect.

The calculation gives us the probability that any base pair is melted at each temperature; these probabilities are summed to give, in effect, the total number of pairs melted at each temperature. An argument based on polymer dynamics is used to convert the melted length into a prediction of the electrophoretic mobility of the molecule. Figure 3 shows a comparison of the theoretical outcome, including both calculated melting and calculated mobility, with actual results in a gel for a heteroduplex molecule in which two bases are looped out of the fully base-paired reference molecule. The defective heteroduplex was prepared by reassociating strands from two different deletion mutants isolated by R. Myers. The solid line represents a digitized tracing of the curve from a gel and the dashed line represents the theoretical counterpart, allowing adjustment of one parameter—the thermodynamic effect of the mismatch—for a best fit.

As in Figure 1, the homoduplex falls to the right and the defective helix to the left, corresponding to melting of the defective domains at a lower equivalent temperature. The discrepancy between the experimental and calculated curves at very high denaturant concentration (the upper right) reflects the beginning of strand dissociation, which sets the minimum mobility at that of separated single strands. This process was not included in the mobility calculation. The agreement between experiment and theory is not always this good, but the usual level of discrepancy is insignificant relative to the effect of base changes.

Knowing the mobility as a function of temperature, it is easy to calculate the level a molecule reaches in a parallel gradient. The sharp drop in mobility at a critical level in a parallel gradient may bring the molecule to nearly a dead stop. This difference in level between wild type and the defective helix is the most useful final result.

Simulating the Search

The problem at hand is to estimate the utility of denaturing gradient gel electrophoretic separations in de-

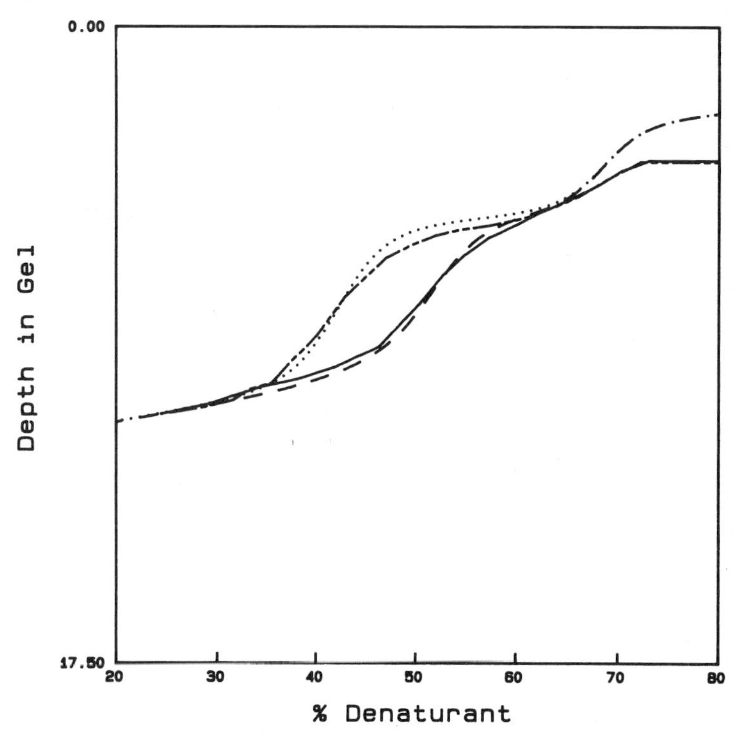

Figure 3. Comparison of experimental and calculated relations between electrophoretic mobility and denaturant concentration for a fully paired homoduplex. The solid line (———) and the long dash- dot-dot (— · · —) line were manually digitized from a photograph of ethidium fluorescence, as in Fig. 1; they correspond to the same prototype homoduplex shown in Fig. 1 and to a heteroduplex between a CG deletion and a TA deletion in which the two looped-out bases are separated by 25 bp. The theoretical curves for the homoduplex (– – –) and heteroduplex (· · · ·) were based on the sequence-defined calculation with a small number of empirical parameters (see Lerman and Silverstein 1986). The only difference in the calculation between the homo- and heteroduplexes is the replacement of the nearest-neighbor stability values bracketing the sites of each of the two loop-outs by smaller values, each adjusted to fit the mobility vs. denaturant curves of the corresponding heteroduplex with a single loop-out (E. Grinfeld et al., in prep.).

tecting all possible single-base changes in a presumably representative part of the human genome. The procedure will be limited to simulation of the simplest possible operations: merely fragmenting the genome with an arbitrarily chosen restriction endonuclease, hybridizing the digest with a series of probes representing the corresponding fragments of the wild-type sequence, and applying the annealed heteroduplexes to a parallel denaturing gradient. Initially, we will disregard the possibility of making recombinants of all of the fragments obtained from the genome to be tested, but as will be seen later, this procedure, where it is practical, changes the outlook substantially. The prospect of detecting single-base changes can be estimated with sufficient reliability by means of computer simulation, based on the theory we have sketched. To minimize a priori bias in the type of sequence to which this test is applied, we have picked a very long, continuous sequence available in GenBank[R] (Release 40.4, 1986; a registered trademark of the U.S. Dept. of Health and Human Services, distributed by BBN Laboratories) and compiled the results that might be expected in a denaturing gradient from the total collection of restriction fragments. The calculation inserts the thermodynamic perturbation of a single mismatch at one end of each restriction fragment and determines the amount of separation in the gradient that would be expected between the mismatch molecule and a wild-type fully matched helix. The calculation is repeated, moving the mismatch 10 bp further down the molecule with each repetition. Where the mismatch generates an expected gel separation of 0.1°C (in equivalent temperature) or more, the test is scored as a displacement detectable in ordinary gel experiments. In positions along the sequence for which the expected displacement is calculated to be less than 0.1°C, the test is scored as negative. The 0.1°C criterion is conservative; with an appropriate gradient, it can correspond to a 3-mm separation of mutant and wild-type bands.

We have carried through a calculation using 35.4 kb of the complete continuous sequence of human β-globin cluster of chromosome 11, as compiled by Collins and Weissman (1984). The region includes both the γ and δ genes, the pseudogene, and β-globin, together with four *Alu* sequences. Sampling only every tenth base may allow a slight underestimate of the total number of detectable alterations in each fragment, but the 1:10 sampling was necessary to restrict the computer time required. Almost 100 hours was needed on a 7 MIPS computer.

We have also calculated a melting map of 67 kb of the β-globin gene cluster to get some idea of the character of the genome with respect to the local variations in helix stability on which denaturing gradient separations depend. A number of physical features of the map can be identified with biological function, but there is a wealth of detail of unknown (if any) significance. The map is included as an appendix to this paper. There are striking similarities in the patterns of the various members of the β-globin gene family but clear individual differences.

To calculate the fraction of sequence changes that would be detectable, it was necessary to introduce some more or less arbitrary assumptions. We chose a single, uniform, more or less typical value for the enthalpic destabilization of each mismatch, a small value compared to the largest we have observed to prevent an unrealistic overestimate of yield. All fragments shorter than 50 bp were discarded because they can be expected either to dissociate without partial melting or to undergo an ineffectively small change in mobility if a small portion melts. The least certain aspect of the calculation, and one which needs to be examined experimentally, is in the specification of where electrophoresis must be terminated because of the onset of complete strand dissociation. There is some arbitrariness in deciding which regions of the sequence that melt above the lowest melting domain will retain stable branched structures and respond to base changes and which regions will be uninformative because strand dissociation accompanies their melting. For the present calculation, we have assumed that if the calculated equilibrium dissociation constant is relatively small, the dissociation rate will also be small relative to the rate of migration and will permit clear separations. The calculation was carried through for several values of the equilibrium dissociation constant serving as cut-off limits; the results were found to be relatively insensitive to the choice. Only separations that occur at temperatures below that at which the dissociation constant reaches one of the critical values are scored as detectable. In every case, the dissociation constant was calculated for the molecule including the mismatch, not the fully matched prototype helix.

RESULTS

Let us suppose that all of the *Alu*I fragments of this part of the β-globin cluster are hybridized with the *Alu*I fragments of a genome containing precisely 1 base substitution or a single base insertion or deletion in the β-globin region. The mixture is then applied to a denaturing gradient gel in a way appropriate (discussed later) for discriminating 100 or so bands. Genomic errors at some positions in each fragment will result in a shift away from the wild-type position, or a new band with that shift will appear in the heterozygote. The fraction of sequence positions in which mismatches are detectable is shown as a function of the length of each fragment in Figure 4. Detection is cut off for each fragment at the temperature that drives the strand dissociation constant to 10^{-8} M. There are a few fragments not shown that are longer than 1 kb. The scatter reflects the diversity of melting patterns among genomic fragments. Points representing a number of relatively short fragments are found at zero yield, indicating dissociation without useful retardation from any mismatch.

We find that a large fraction (49.4% of all possible

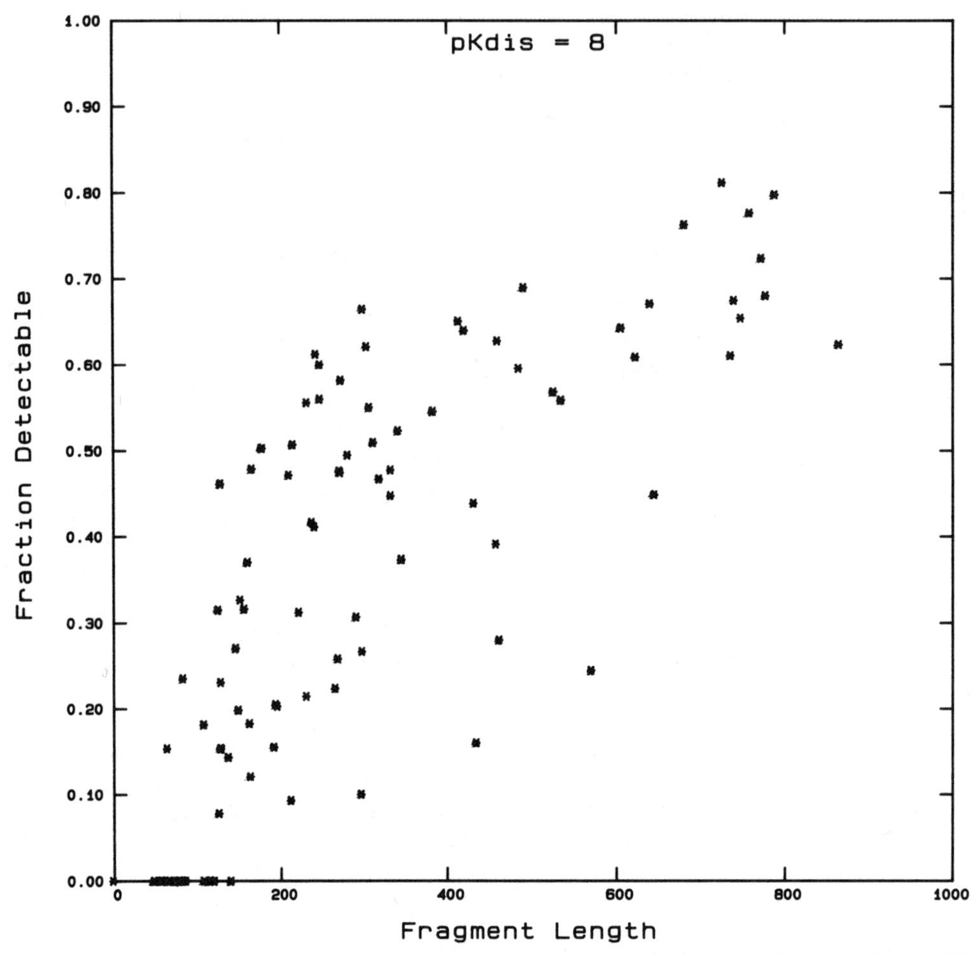

Figure 4. Distribution of gradient-separable mismatch sites. Each point represents one *Alu*I fragment of a 35.4-kb portion of the human β-globin gene cluster. The thermodynamic effect of a typical mismatch in a heteroduplex between a probe and a genomic strand is introduced at each position along the sequence by subtracting 50° from the stability values of the nearest neighbors that include the mismatch. The calculated position of a band of the modified sequence in a parallel gradient is compared with the calculated position of a perfectly matched molecule (as in the gel on the right in Fig. 1). If the calculated gradient separation is sufficiently large to be seen easily and if it occurs at a denaturant concentration less than that which brings about complete strand dissociation, that site is scored as detectable with respect to base substitutions and single-base insertions or deletions. The plot shows the fraction of sites in each fragment in which changes might be detected as a function of the length of the fragment. The distribution is generally similar if the calculation is based on *Hae*III cleavage, and it changes relatively little if a less stringent limit for strand dissociation is substituted.

base changes) should be detectable within the set of all *Alu*I fragments, and a large fraction (52.6%) can also be detected with the set of all *Hae*III fragments, even with a restrictive choice of the strand-dissociation constant. However, some sites of base change that cannot be seen in the *Alu* set are visible in *Hae*III and vice versa. Adding a set from the second enzyme increases the yield about 26%, and the overall fraction of sites that should be screenable is roughly 66.5%, if both sets of fragments are tested. The extension may be attributable mainly to a different distribution of the very small fragments that do not contribute (or do not contribute significantly) to the yield from either pattern of restriction alone. If the strand-dissociation constant determining the cut-off is raised to 10^{-6} M, the overall yield is increased only slightly, to 68.9%. This conclusion from the calculation does not conflict with the results of Myers et al. (1985e) and Lumelsky (1984), who found that of the 12 base changes associated with β-thalassemia tested, 11 could clearly be demonstrated in denaturing gradient gels in the laboratory. The high yield in the simulation depends on separations that the calculation shows should be expected after some retardation in the gradient has occurred due to the melting of the lowest domain. Although we have demonstrated clear separations of second domain mutants (Myers 1985c) and have examined their theoretical properties with respect to gradient parameters and time (Lerman and Silverstein 1986), we have not yet studied corresponding genomic molecules. Technical modifications suggested by the calculations should simplify the experiments.

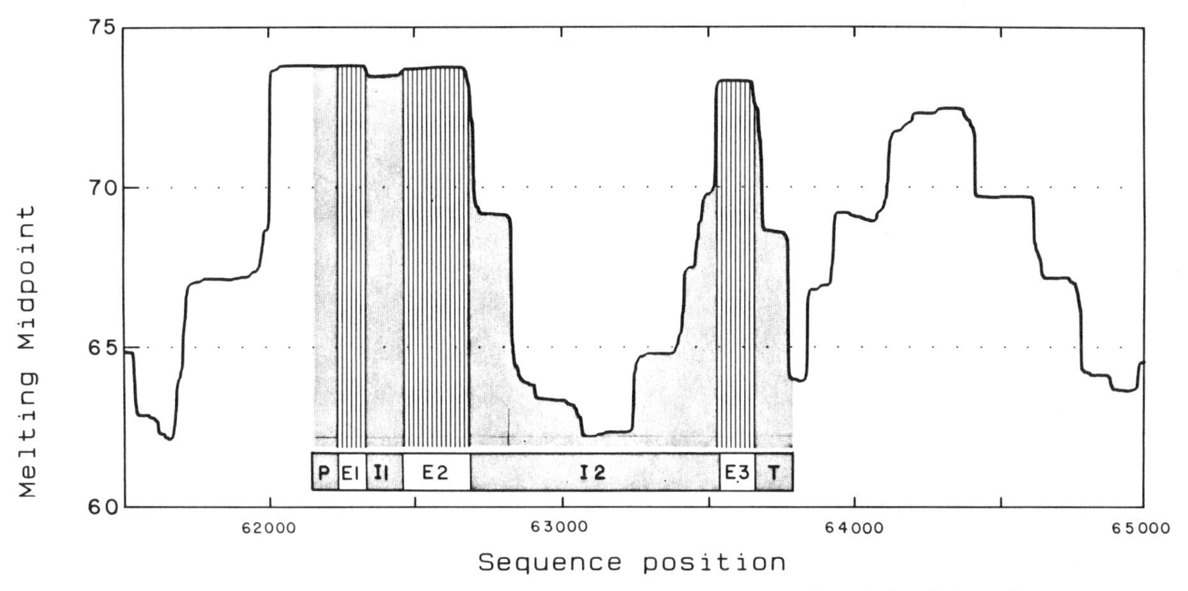

Figure 5. Melting map of the local region containing the human β-globin gene. The boundaries of the coding sequences are indicated. The map is part of a calculation based on a much longer sequence rather than a fragment with the sequence limits shown. The fragment mapped in Fig. 2 would fall between 62321 and 62592 here. Note that the conspicuous difference in domain levels at the boundary between the first intron and the second coding sequence in the Fig. 2 fragment is much smaller here. Melting properties are strongly influenced by neighboring sequences.

The Possibility of Bias in Sensitivity

There are, however, some problems. The melting map of β-globin (Fig. 5) shows that the 3′ flank of the second exon and both flanks of the third exon abut domains that melt at lower temperatures. Although they are not identical, melting maps of the other genes in the cluster have similar features (see Appendix). Will coding sequences generally lie in the high-melting part of any fragment and thereby be invisible in the gradient?

To the extent that coding sequences are more GC-dense than adjacent sequences, there will be a bias against sensitivity in the portion of a coding sequence that lies in a restriction fragment that straddles a boundary and carries a significant length of noncoding adjacent sequence. As pointed out by Bird (1985), a large number of genes are accompanied by a GC-dense noncoding sequence, usually on the upstream flank; human α-globin and HPRT are examples. Bird has termed these regions HTF islands. Their contribution to a restriction fragment will render the amino-terminal coding sequences the sensitive part of the fragment.

Restriction fragments lying entirely within a coding sequence or other high plateaus on the melting map, such as the β-globin fragment presented in Figure 2, will not be affected by the properties of the surrounding sequences and can be expected to show typical yields of gradient-responsive sites. Fragmentation usually generates useful end domains, and the presence of a mismatch can convert a portion of a flat melting map into a useful plateau.

Nevertheless, coding sequences, especially those in short exons, are likely to be underrepresented. Comparison of the base composition of cDNA with the average composition of the long genomic segments carrying the corresponding gene shows about a 10% higher GC-density in the exons (Bernardi et al. 1985). Of several we have examined, only the preproglucagon cDNA is lower in GC than the mean of the human genome.

Up to now, we have considered only unselected, shotgun fragmentation. Restriction sites can be selected more judiciously and not necessarily with the same specificity at both ends. Probes prepared from restriction sites chosen on the basis of melting calculations can provide fragments in which the interesting portion of the sequence is favorably situated in the domain structure.

Ensuring Comprehensive Detection

For those parts of the sequences that do not fall into low-melting domains of any naturally defined restriction fragments and appear to be undetectable by calculation, a composite probe can be designed to guarantee separation. Any portion of the genome can be recombined with an arbitrary sequence of high melting temperature (a "clamp") by cloning in an appropriate vector. With Myers and Maniatis, we have shown (Myers et al. 1985c) that the attachment of a high-melting sequence derived from the 5′-flanking region of human α-globin to a 140-bp section of the mouse β-major globin promoter renders all base changes in the promoter sequence detectable. All of a smaller number that have actually been tested were found to give significant, clear separations. The promoter fragment alone would show separations due to changes at only 76 sites. More extreme comparisons can be expected, depending on the nature of any particular sequence. To be sure, the need to clone the DNA of every fetus to be subjected to prenatal diagnosis is impractical, but simpler means to an equivalent result can be proposed.

Detection of sequence changes in the blind spots, the higher melting portion of any fragment of any natural (unclamped) sequence, is under study in our laboratory and presumably elsewhere. There is some hope that by changing the conditions of denaturation, the melting map might be inverted or at least radically revised, such that the regions of high GC density might be rendered lower melting than the remainder.

Effect of Sequence Repetitions

A small part of the genome within repeated sequences or just adjacent to repetitions is likely to be lost and invisible using a procedure that depends on hybridizing genomic fragments to similar labeled probes. Depending on the repetition frequency, the degree of homology among repetitions, and the ratio of probe to genomic molecules, hybridization among repetitions may compete effectively with the formation of useful heteroduplexes. The loss may be most severe among unique sequences immediately flanking repetitions in fragment populations produced by an endonuclease other than that characterizing the repeated unit.

Changes in Large Segments

Where the sequences of the wild-type and aberrant forms differ by a substantial insertion, deletion, or inversion, rather than a change in one or a few bases, there will be again a definitive alteration in the denaturing gradient pattern given by appropriate probes. Obviously, if a sequence is deleted, some bands will be missing. The fragments at the boundaries of any of these extended changes will carry a region of mismatch between probes and genomic fragments. A single mismatch at the end will be seen as a displacement as usual, and a longer mismatch may prevent the fragment from moving beyond the top boundary of the gel. A sample calculation on the introduction of a modestly destabilizing mismatch at the clamped end of a region dense in GC shows that only one mismatch would generate a distinct displacement; any mismatch in the terminal five bases would be detectable, despite the GC density. Only if both ends of an insertion, deletion, inversion, or translocation should precisely coincide with restriction sites would that sequence difference fail to generate a signal in the gel.

Practical Considerations

The analysis of *Alu*I or *Hae*III fragments of 35.4 kb of the β-globin cluster has presumed that the laboratory is capable of examining the response to 105 different probes in denaturing gradient gels. The seriousness of the questions asked in prenatal diagnosis can justify relatively great effort if decisive and reliable results can be expected. Nevertheless, if the labor required is indeed exorbitant, no procedure will have widespread application. The number of probes in the globin cluster calculation may represent a middling size gene, but the gene for factor VIII spans about 187 kb and HPRT spans about 58 kb. As noted previously, it is possible that the entire sequence does not need to be examined but that experience will reveal critical regions affecting promotion, initiation, splicing, unequal recombination, and insertion sites, and other parts of the sequence may be safely disregarded. Even so, the number of probes needed may remain very large.

This project can be accommodated within present techniques and within reasonable effort in at least two ways if fully stable (nonradioactive) probes are available with adequately sensitive labeling. Hybridization can be carried out with single-stranded probes (Myers et al. 1986) in mixtures containing ten or more different probes, selected such that they will be distributed broadly over the gradient at the end of the run in the absence of sequence aberrations. A single denaturing gradient gel with ten or more lanes could easily accommodate 100 probes, and the presence of a new band would be discernible without difficulty. Quantitative measurements should also permit reliable detection of bands containing half the expected diploid amount of DNA. Alternatively, the fragmented genomic DNA might hybridize simultaneously to a large number of probes and the results sorted on a two-dimensional gel, where the fragments are first separated by length using conventional gel electrophoresis and then separated in the perpendicular direction in the denaturing gradient. We have reported the resolution of a mixture of at least 300 fragments into discrete spots in a two-dimensional gel (Fischer and Lerman 1979a), although under slightly different conditions. It might be preferable to use double-stranded probes, if sensitive labeling permits, to provide a redundant indicator of an altered fragment. Positions along the length coordinate would also signal some of the information provided by RFLPs.

Interpretation of Results

The identification of sequence aberrations with abnormal gene function tends to be complicated by the presence of sequence changes that have no known phenotypic consequences. These are termed polymorphisms when the frequency of the less common allele is at least 1%. At least 17 such positions where a base substitution alters a restriction site have been identified in the β-globin gene cluster as well as a few others in β-globin that do not alter restriction sites (Orkin and Kazazian 1984). Where the sites and type of polymorphic base substitutions are adequately known, denaturing gradient detection of significant changes will generally remain comprehensive and definitive because of the additional information derived from the gradient. The extent of separation from the position of the wild-type fragment in the gradient, as we have noted before, depends on both the particular base change and its position in the sequence. It is possible that in rare instances, a polymorphism might mimic the separation given by a significant substitution, but these effects can be predicted from knowledge of the sequence of the gene and circumvented by the use of probes of the opposite polarity, which give different separations, or by a somewhat different sequence selection. The possibility of both a polymorphic substitution and a significant base change in the same fragment does not necessarily introduce ambiguity. We find that the separations attributable to double mismatches in the same domain are either additive if the sites are not too close or more than additive if they are quite nearby.

CONCLUSIONS

The comprehensive calculation has provided some surprises and a new guidance in the design of experiments. The overall yield is higher than previous guesses, but it depends on implied modifications of techniques that will have to be tested. In particular, we will examine the utility of driving molecules deeper into the gradient, beyond the initial retardation that has concluded previous runs. It appears that prospects are reasonably favorable toward the detection of the majority of phenotypically significant sequence aberrations in any individual human gene using the denaturing gradient technique, and the detection of 100% does not seem out of the question in the forseeable future. Augmenting the denaturing gradient by other means for determining mismatches, such as cleavage by ribonuclease in DNA-RNA hybrids containing a CA mismatch (Myers et al. 1985a) might raise the probability of detection within immediately available techniques closer to 95%.

It is clear that this approach to direct detection of genetic defects from DNA of a fetus or the proband alone can be applied only after the relevant gene or gene region has been cloned to provide probes, but complete sequencing is not required. A body of experience would be needed to distinguish phenotypically neutral polymorphisms from significant sequence defects. Until the DNA of any locus of interest is isolated and identified, linkage analysis will remain useful. For some applications, the nondestructive aspect of denaturing gradient separations might be useful. Fragments of interest can be eluted from the gel and recovered by cloning (Myers et al. 1985b).

ACKNOWLEDGMENTS

We are grateful to Robert Anderson for helpful discussions, to Rick Myers for providing the mutants, and to Karen Pichel for preparation of the manuscript. This work was supported by grant RO1-GM-35095 from the National Institutes of Health.

APPENDIX

A Melting Map of the Region of the Human β-Globin Gene Cluster

K. Silverstein, R. Andersen, J. Slack, and L.S. Lerman

Genetics Institute, Cambridge, Massachusetts 02140

The plot shows the calculated temperature at which each base pair has an equal probability of adopting a helical, base-paired conformation or an unpaired, random-chain conformation. The base sequence was taken directly from the GenBank file, humhbb, version 40.0. The calculation was carried out as described by Lerman and Silverstein (1986) using the Poland-Fixman Friere algorithms and Gotoh-Tagashira nearest-neighbor stability values for 0.02 M sodium. A detailed discussion of the theory of DNA melting and its correspondence with experimental measurements has been presented by Wartell and Benight (1985). It should be noted that the map describes the helix stability only near the melting temperature or under denaturing conditions equivalent to elevated temperature, not at physiological temperatures. A melting map for φX174 has been presented by Gotoh (1983) in a review on melting theory, together with comments on relations between the calculated thermal stability of the helix and some biological functions, particularly at low temperatures. Correlations between domain structure and biologically functional boundaries are reviewed by Wada and Suyama (1986).

Regions of biological relevance are indicated on the map. Coding regions for the β-like globin proteins are represented by solid blocks. The map shows the peptide-coding sequence of ε-globin from 19559 to 20979, $^{G}\gamma$-globin from 34549 to 36000, $^{A}\gamma$-globin from 39485 to 39576, the pseudogene from 45741 to 47157, δ-globin from 54842 to 56311, and β-globin from 62239 to 63662. Sets of exons are bounded on either side by vertical dash/dot lines. The first of these represents the beginning of the promoter, the second indicates the mRNA cap site, and the third represents the mRNA poly(A) site. *Kpn* and *Alu* repetitive sequences are represented by hatched and cross-hatched blocks, respectively. Arrows (base of map) indicate open reading frames that are 150 or more bases long. All six frames are represented; left-to-right arrows are forward frames. These frames were identified using the MAP software (initiation: ATG; termination: TAA, TGA, TAG) found within the Seqanal package, University of Wisconsin.

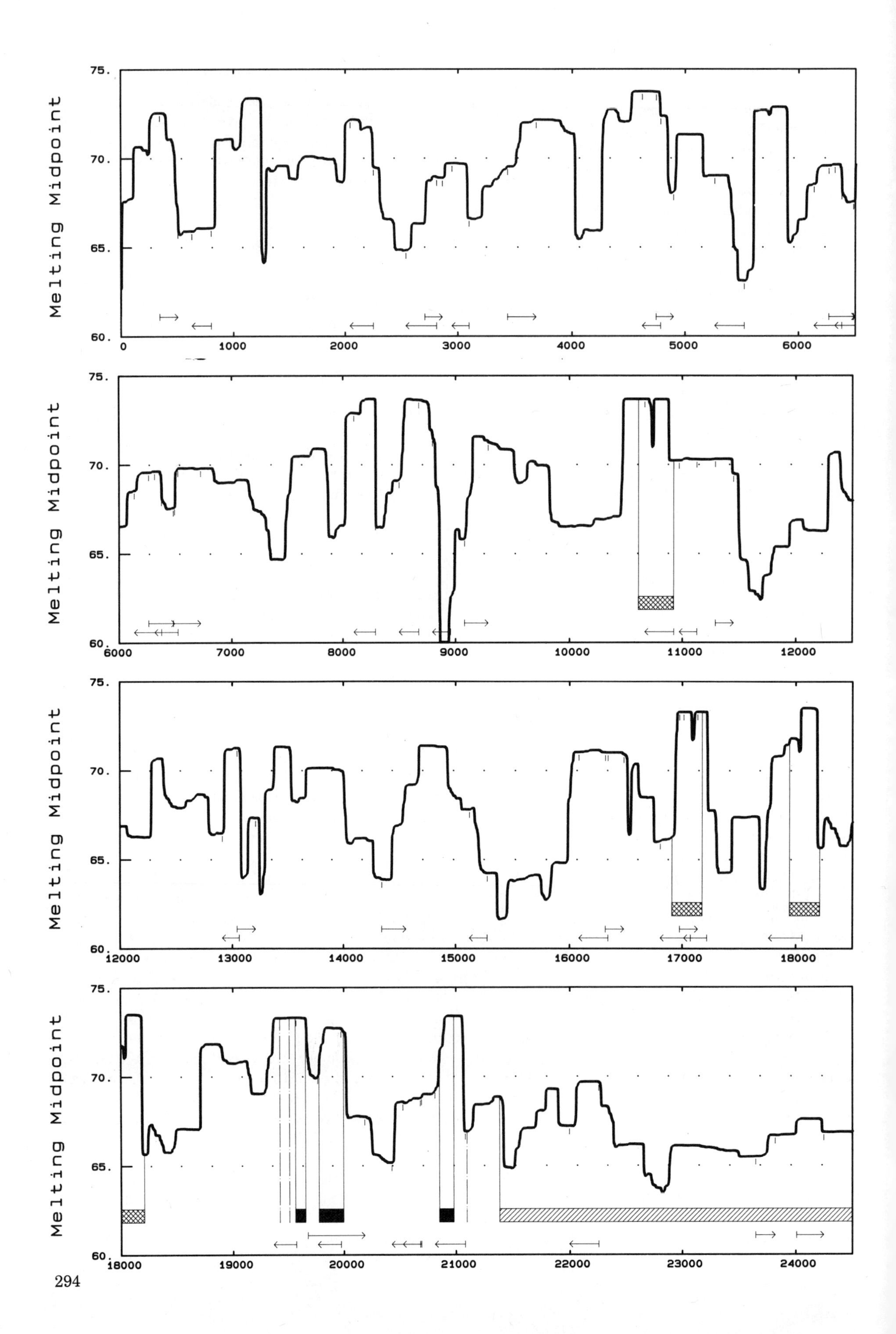

Melting Midpoint
75.
70.
65.
60.
0
1000
2000
3000
4000
5000
6000
Melting Midpoint
75.
70.
65.
60.
6000
7000
8000
9000
10000
11000
12000
Melting Midpoint
75.
70.
65.
60.
12000
13000
14000
15000
16000
17000
18000
Melting Midpoint
75.
70.
65.
60.
18000
19000
20000
21000
22000
23000
24000

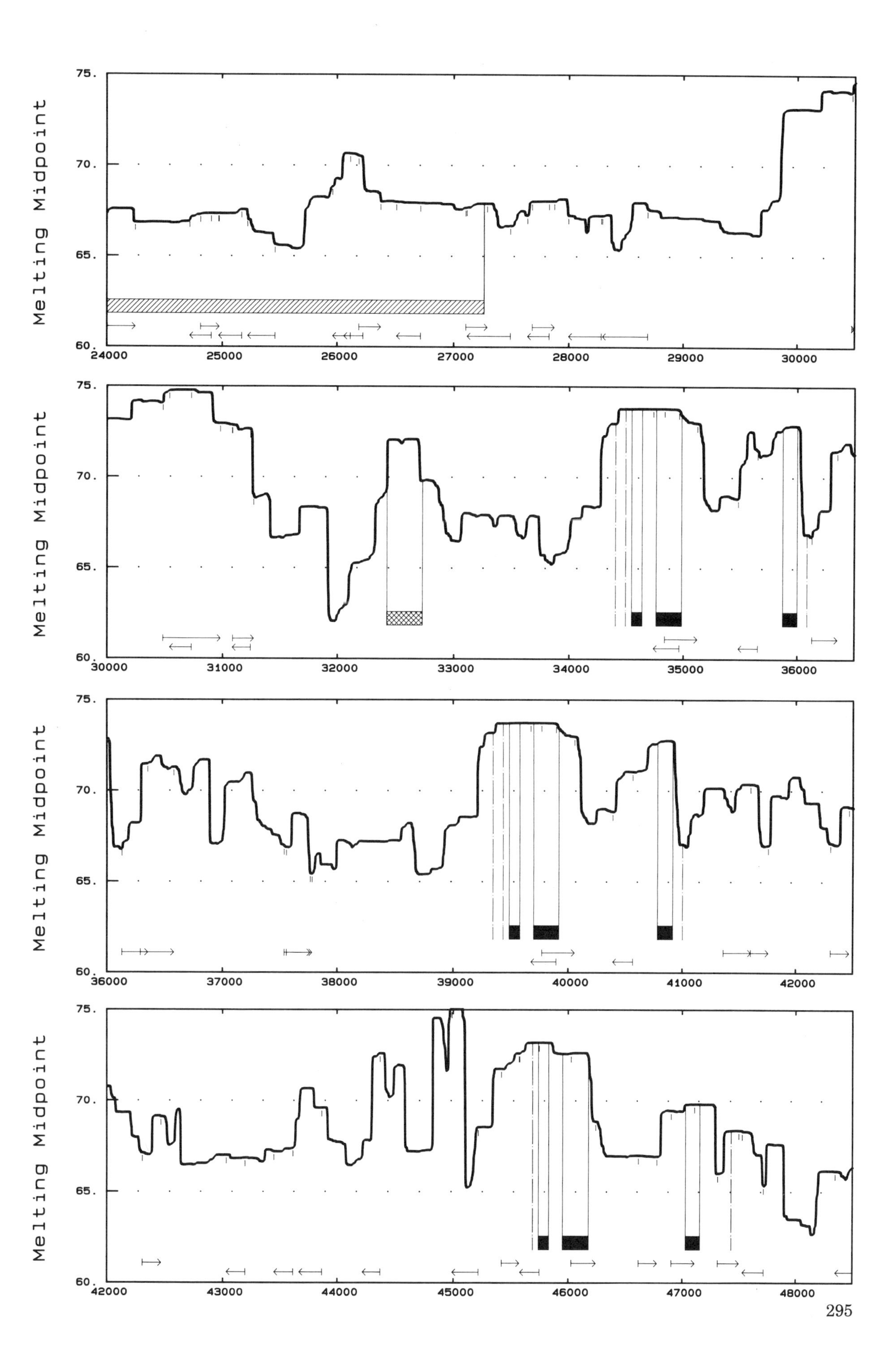

Melting Midpoint
75.
70.
65.
60.
24000
25000
26000
27000
28000
29000
30000
Melting Midpoint
75.
70.
65.
60.
30000
31000
32000
33000
34000
35000
36000
Melting Midpoint
75.
70.
65.
60.
36000
37000
38000
39000
40000
41000
42000
Melting Midpoint
75.
70.
65.
60.
42000
43000
44000
45000
46000
47000
48000

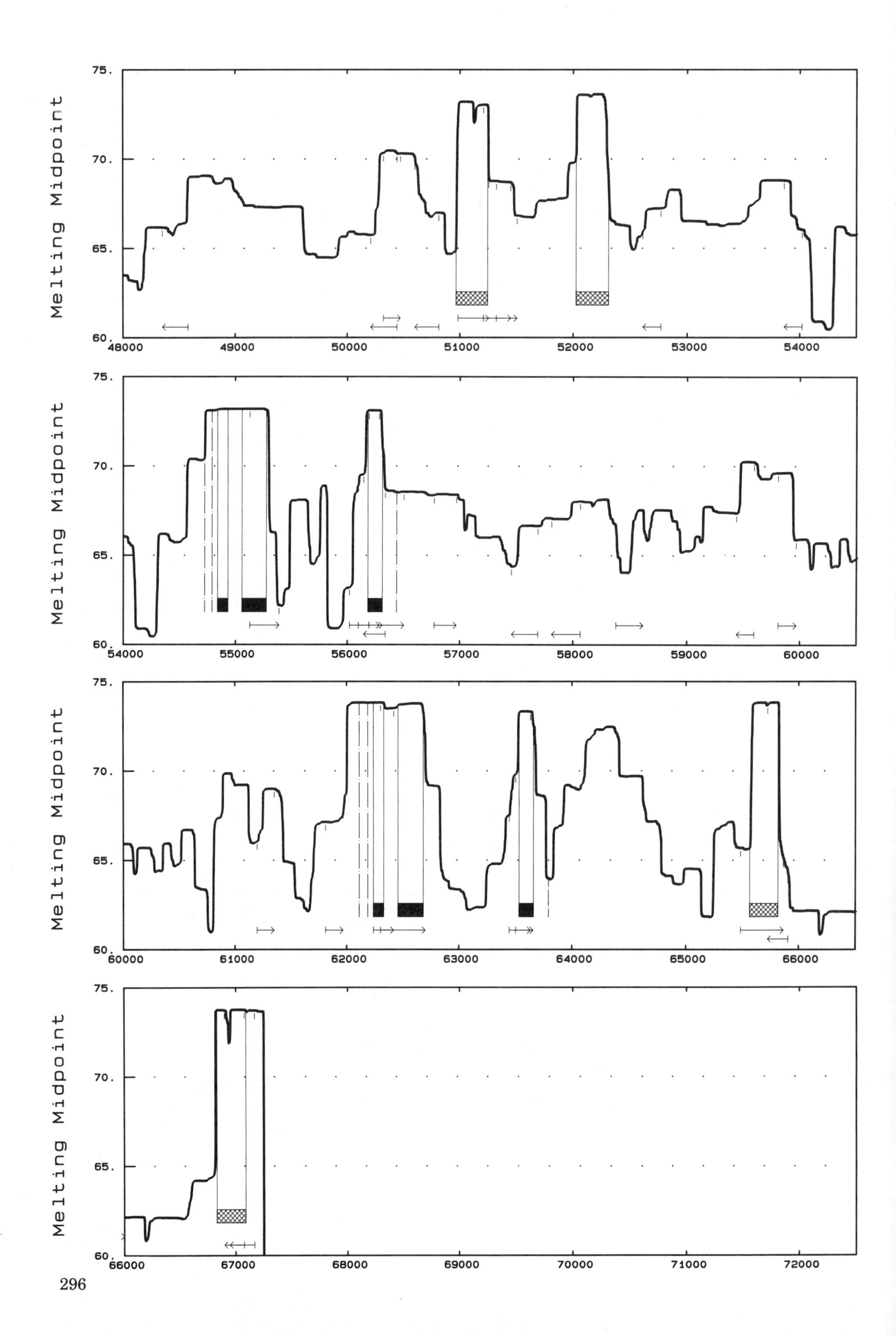

Melting Midpoint
75.
70.
65.
60.
48000
49000
50000
51000
52000
53000
54000
Melting Midpoint
75.
70.
65.
60.
54000
55000
56000
57000
58000
59000
60000
Melting Midpoint
75.
70.
65.
60.
60000
61000
62000
63000
64000
65000
66000
Melting Midpoint
75.
70.
65.
60.
66000
67000
68000
69000
70000
71000
72000

REFERENCES

Bernardi, G., B. Olofsson, J. Filipski, M. Zerial, J. Salinas, G. Cuny, M. Meunier-Rotival, and F. Rodier. 1985. The mosaic genome of warm-blooded vertebrates. *Science* **228:** 953.

Bird, A., M. Taggart, M. Frommer, O.J. Miller, and D. Macleod. 1985. A fraction of the mouse genome is derived from islands of nonmethylated, CpG-rich DNA. *Cell* **40:** 91.

Collins, F.S. and S.M. Weissman. 1984. The molecular genetics of human hemoglobin. *Prog. Nucleic Acid Res.* **31:** 315.

Fischer, S.G. and L.S. Lerman. 1979a. Length independent separation of DNA restriction fragments in two-dimensional gel electrophoresis. *Cell* **16:** 191.

———. 1979b. Two dimensional electrophoretic separation of restriction enzyme fragments of DNA. *Methods Enzymol.* **68:** 183.

———. 1983. DNA fragments differing by single base-pair substitutions are separated in denaturing gradient gels: Correspondence with melting theory. *Proc. Natl. Acad. Sci.* **80:** 1579.

Fixman, M. and J.J. Friere. 1977. Theory of DNA melting curves. *Biopolymers* **16:** 2693.

Gotoh, O. 1983. Prediction of melting profiles and local helix stability for sequenced DNA. *Adv. Biophys.* **16:** 1.

Gotoh, O. and Y. Tagashira. 1981. Stabilities of nearest neighbor doublets in double helical DNA determined by fitting calculated melting profiles to observed profiles. *Biopolymers* **20:** 1033.

Lerman, L.S. and K. Silverstein. 1986. Computational simulation of DNA melting and its application to denaturing gradient gel electrophoresis. *Methods Enzymol.* (in press).

Lerman, L.S., S.G. Fischer, and N. Lumelsky. 1983. Inferring variations in genotype from properties of DNA. In *Recombinant DNA and medical genetics* (ed. A. Messer and I. Porter), vol. 13, p. 157. Academic Press, New York.

Lerman, L.S., S.G. Fischer, I. Hurley, K. Silverstein, and N. Lumelsky. 1984. Sequence determined DNA separations. *Annu. Rev. Biophys. Bioeng.* **13:** 399.

Lumelsky, N. 1984. "Application of denaturing gel electrophoresis to the detection of single base changes in the human genome." Ph.D. thesis, State University of New York at Albany.

Myers, R.M., Z. Larin, and T. Maniatis. 1985a. Detection of single base substitutions by ribonuclease cleavage at mismatches in RNA:DNA duplexes. *Science* **230:** 1242.

Myers, R.M., L.S. Lerman, and T. Maniatis. 1985b. A general method for saturation mutagenesis of cloned DNA fragments. *Science* **229:** 242.

Myers, R.M., T. Maniatis, and L.S. Lerman. 1986. Detection and localization of single base changes by denaturing gradient gel electrophoresis. *Methods Enzymol.* (in press).

Myers, R.M., S.G. Fischer, L.S. Lerman, and T. Maniatis. 1985c. Nearly all single base substitutions in DNA fragments joined to a GC-clamp can be detected by denaturing gradient gel electrophoresis. *Nucleic Acids Res.* **13:** 3131.

Myers, R.M., S.G. Fischer, T. Maniatis, and L.S. Lerman. 1985d. Modification of the melting properties of duplex DNA by attachment of a GC-rich DNA sequence as determined by denaturing gradient gel electrophoresis. *Nucleic Acids Res.* **13:** 3111.

Myers, R.M., N. Lumelsky, L.S. Lerman, and T. Maniatis. 1985e. Detection of single base substitutions in total genomic DNA. *Nature* **313:** 495.

Novack, D.F., N.J. Casna, S.G. Fischer, and J.P. Ford. 1986. Detection of single base-pair mismatches in DNA by chemical modification followed by electrophoresis in 15% polyacrylamide gel. *Proc. Natl. Acad. Sci.* **83:** 586.

Orkin, S.H. and H.H. Kazazian. 1984. The mutation and polymorphism of the human beta globin gene and its surrounding DNA. *Annu. Rev. Genet.* **18:** 131.

Poland, D. 1974. Recursion generation of probability profiles for specific-sequence macromolecules with long-range correlations. *Biopolymers* **13:** 1859.

Wada, A. and A. Suyama. 1986. Local stability of DNA and RNA secondary structure and its relation to biological functions. *Prog. Biophys. Mol. Biol.* **47:** 113.

Wartell, R.M. and A.S. Benight. 1985. Thermal denaturation of DNA molecules: A comparison of theory with experiment. *Phys. Rep.* **126:** 67.

DNA-based Detection of Chromosome Deletion and Amplification: Diagnostic and Mechanistic Significance

S.A. LATT,*§ M. LALANDE,*¶ T. DONLON,* A. WYMAN,** E. ROSE,* Y. SHILOH,*//
B. KORF,* U. MÜLLER,* K. SAKAI,*†† N. KANDA,*# J. KANG,* H. STROH,*
P. HARRIS,* G. BRUNS,* R. WHARTON,‡ AND L. KAPLAN*†

*Division of Genetics, †Developmental Evaluation Clinic, and ‡Division of Ambulatory Pediatrics, Children's Hospital, and Departments of *†‡Pediatrics and §Genetics, Harvard Medical School, Boston, Massachusetts 02115; **Department of Biology, M.I.T., Cambridge, Massachusetts 02138

Recombinant DNA libraries constructed from highly purified flow-sorted metaphase chromosomes (for review, see Young 1984) provide convenient starting points for detailed analysis of regions of specific chromosomes. These include the X (Davies et al. 1981; Kunkel et al. 1982; Lalande et al. 1984a), no. 7 (Scambler et al. 1986), no. 13 (Lalande et al. 1984b), the proximal part of no. 15 (Lalande et al. 1985; Donlon et al. 1986), and Y (Müller et al. 1986b) chromosomes, and a no. 1 chromosome with a large homogeneously staining region from a human neuroblastoma (Kanda et al. 1983). In addition to providing molecular-linkage markers for human diseases, such as Duchenne muscular dystrophy (Davies et al. 1983; Aldridge et al. 1984; Hofker et al. 1985) and cystic fibrosis (Scambler et al. 1986), such libraries can yield probes detecting microscopic or even submicroscopic structural changes within human chromosomes. The latter use of chromosome-specific DNA segments can serve not only to complement cytological analysis of metaphase chromosome abnormalities, but also to guide studies examining the nature of the molecular processes responsible for the chromosomal abnormalities observed.

METHODS

Cell lines and fibroblast cultures. The lymphoblastoid cell lines used were established from peripheral blood samples (Forget et al. 1976). The neuroblastoma cell line IMR-32 (Tumilowicz et al. 1970) was obtained from F. Alt; NB-9, NB-16, NB-19, and NB-69 (Gilbert et al. 1982) were obtained from F. Gilbert, and LA-N-5 was obtained from R. Seeger. Fibroblasts from 46,XX males, a 46,XX true hermaphrodite, and 46,XY gonadal dysgenesis patients were obtained as described by Müller et al. (1986b).

DNA panels, probe labeling, and hybridization conditions. DNA was extracted essentially as described by Aldridge et al. (1984) (leukocytes) or Kunkel et al. (1982) (other cells). DNA for human-rodent somatic-cell hybrid panels was provided by G. Bruns (Bruns et al. 1979). DNA was cleaved with the restriction endonuclease *Hin*dIII under conditions recommended by the manufacturer. DNA samples quantitated fluorometrically (Brunk et al. 1979) were then separated on agarose gels and blotted (Southern 1975). DNA probe hybridization and blot washing were carried out as described previously (Tantravahi et al. 1983; Müller et al. 1986b). DNA probes were labeled with ^{32}P using either T4 DNA polymerase end-labeling (O'Farrel 1981), as modified by Kunkel et al. (1982), or the random primer method of Feinberg and Vogelstein (1983).

Construction of chromosome-enriched recombinant libraries. Charon 21A chromosome-enriched libraries were created from *Hin*dIII-digested DNA from IMR-32 no. 1 + HSR (Kanda et al. 1983), no. 13 (Lalande et al. 1984b), inv dup(15) (Lalande et al. 1985; Donlon et al. 1986), or Y chromosomes (Müller et al. 1986a,b) by methodologies detailed in these references. All but the very early library utilized the metaphase chromosome isolation technique of Sillar and Young (1981). Unidimensional flow histograms were obtained on a Becton-Dickinson FACSII, using Hoechst 33258 as a stain (see, e.g., Lalande et al. 1984b), whereas bivariate flow histograms were obtained on a Becton-Dickinson FACSIV, using the dye-pair Hoechst 33258 and Chromomycin A_3 (see, e.g., Sahar and Latt 1978; Langlois et al. 1982; Lalande et al. 1985; Donlon et al. 1986; Müller et al. 1986b). Fractions containing the desired chromosomes were isolated, and the DNA was isolated, cleaved, and cloned in phage Charon 21A, as detailed by Lalande et al. (1984a). Typically, phage were propagated on LE392 bacteria (Maniatis et al. 1982). A proportion of the inv dup(15) library was also propagated on bacterial strain DB1257 (Donlon et al. 1986). Strain DB1257 is related to strain DB1170 (Wyman et al. 1985), which has the genotype *recBC*$^-$, *sbcB*$^-$.

Present addresses: ¶National Research Council, Montreal, Canada; //Department of Genetics, University of Tel Aviv, Tel Aviv, Israel; ††Department of Radiation Biophysics, University of Tokyo, Tokyo, Japan; #Department of Anatomy, Tokyo Women's Medical College, Tokyo, Japan.

Screening of libraries. Most phages were subjected to multiple rounds of Benton-Davis (1977) screening, using radiolabeled human DNA as a probe, and inserts lacking chromosome nonspecific repeated sequences were selected for further analysis using Southern (1975) blot hybridization (see, e.g., Kunkel et al. 1982). Some phages from the DB1257-grown library (see Results) were processed directly, and others were picked onto lawns of both LE392 and DB1257; phage growing poorly or not at all on LE392 were selected for further analysis. Phage inserts were employed either directly or after subcloning into plasmids, e.g., pBR322 (Boliver et al. 1977).

Chromosomal mapping. Initial mapping of single-copy autosomal inserts was accomplished using DNA from rodent-human somatic-cell hybrid panels (Lalande et al. 1984b; Donlon et al. 1986). Inserts homologous to Y DNA could be identified by selective hybridization to normal male DNA (Müller et al. 1986b), whereas amplified DNA sequences were detected by their increased hybridization intensity in DNA blots (see, e.g., Kanda et al. 1983). Scanning densitometry for estimation of DNA sequence copy number was performed essentially as described by Tantravahi et al. (1983). In situ hybridization, utilizing tritiated probes, was performed according to the method of Harper et al. (1981), as modified by Donlon et al. (1983).

DNA sequencing. DNA sequencing utilized M13 (Messing et al. 1977) dideoxynucleotide (Sanger et al. 1977) methodology.

RESULTS

Molecular Detection of Chromosome-13 Deletions in Retinoblastoma

Acquisition of a flow-sorted human chromosome-13-enriched library utilized staining with a single dye, Hoechst 33258, and employed 80 ng of chromosomal DNA to construct a library of 1.5×10^5 pfu (Lalande et al. 1984b). Many of the single-copy inserts from this library thus far examined localize, using somatic-cell hybrids (for a description, see Lalande et al. 1984b), to the region between 13q12 and 13q22. Two probes (H1-2 [D13S29] and H2-10 [D13S26]) hybridize with one-copy intensity (i.e., detect a deletion) to the DNA of one of three patients with retinoblastoma and detect deletions ranging from 13q13 to 13q21.2 (Fig. 1). Two other probes (H3-8 [D13S30] and H2-42 [D13S25]) detect the deletions in all three patients. A fifth probe, H2-26 (D13S28), maps more proximal on 13q and detected no deletions.

In an attempt to localize these probes more precisely, we combined information from hybridization of probes to DNA from patients with 13q deletions and that from autoradiographic in situ hybridization. Both techniques have limited resolution. The intention was to determine a region of overlap from the best estimates of probe position from deletion mapping and in situ

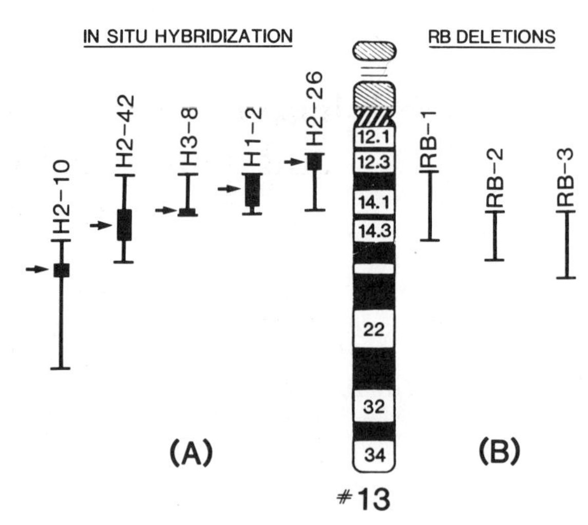

Figure 1. Use of in situ hybridization data on DNA probes and DNA dosage blot data with different human chromosome-13 probes and deletions for probe localization. (*A*) Range of possible location of probes on chromosome 13 by in situ hybridization; (*B*) approximate extents of chromosome deletions in RB-1, RB-2, and RB-3 based on cytology, used to obtain dosage blot data, with one copy of probes H3-8 and H2-42 deleted in all three patients, H2-10 in RB-3, and H1-2 in RB-1. Arrows indicate the regions of overlap between probe locations obtained by in situ hybridization and DNA dosage blot data. These constitute the best estimates for location of these probes. (Data from Lalande et al. 1986.)

hybridization (Lalande et al. 1986). The net result was an ordering of the five probes mentioned above (see Fig. 1).

One probe, H3-8, seems particularly close to the region of 13q14, whose deletion is most involved in sporadic retinoblastomas (Sparkes 1985), and it may hence be very close to the locus responsible for retinoblastoma. The probes identified may, along with others from this region, prove useful for a high-resolution delineation of chromosome-13-associated deletions that is less open to the subjective interpretation of these deletions by cytological methodology.

Molecular Analysis of Chromosome 15 Deleted in the Prader-Willi Syndrome

The Prader-Willi syndrome is a condition associated with congenital hypotonia, developmental delay, hypogonadism, mental retardation, dysmorphic features, and hyperphagia (Butler et al. 1986; L. Kaplan et al., in prep.; R. Wharton, pers. comm.). Ledbetter et al. (1981, 1982) observed that the Prader-Willi syndrome can be associated with a very small deletion involving band 15q11.2 and perhaps 15q12 (Fig. 2). Such deletions, which can be very difficult to detect cytologically, are spontaneous, and yet they virtually always involve the paternal chromosome 15 (Butler and Palmer 1983; Butler et al. 1986). As described by Butler et al. (1986), the Prader-Willi syndrome can also involve translocations or even duplications of proximal 15q, and both deletions (L. Kaplan et al., unpubl.) and du-

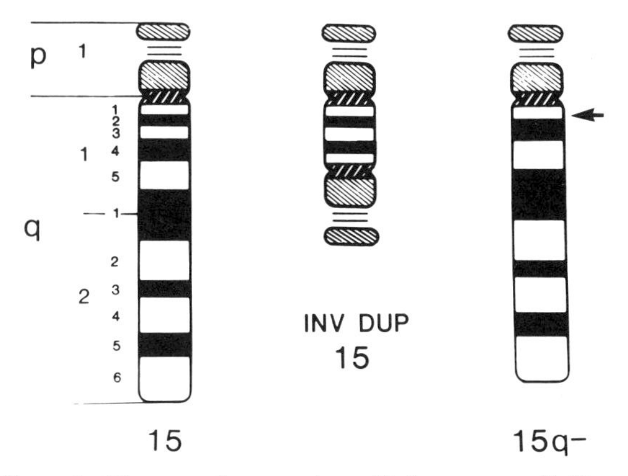

Figure 2. Diagram of a normal no. 15 chromosome (*left*), an inverted duplication no. 15 chromosome with a breakpoint at band 15q13 (*center*), and a no. 15 chromosome with a proximal long-arm deletion (arrow), such as often seen in the Prader-Willi syndrome (*right*).

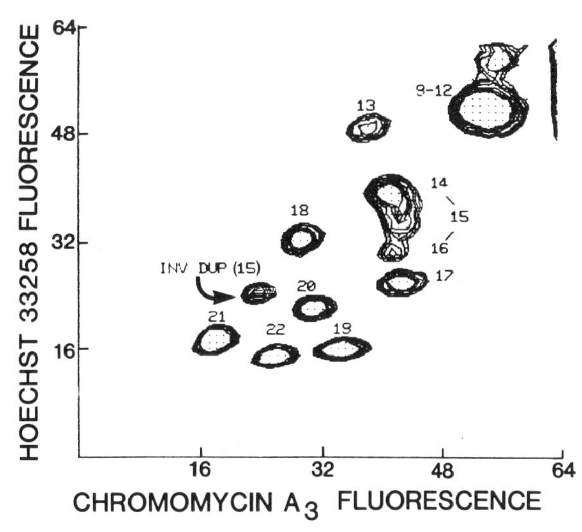

Figure 3. Bivariate flow histogram of cell line ALD-24 showing the distribution of smaller metaphase chromosomes. The chromosomes were isolated and stained with a combination of Hoechst 33258 (ordinate) and Chromomycin A_3 (abscissa). Contour lines correspond to the relative numbers of chromosomes. The peak marked inv dup(15), as judged from previous analyses (Lalande et al. 1985), was used for chromosome sorting. (Reprinted, with permission, from Donlon et al. 1986.)

plications (e.g., inv dup[15]) chromosomes (Fig. 2) (Schreck et al. 1977; Stetten et al. 1981) of proximal 15q can be associated with other clinical entities. The inv dup(15) chromosomes have the additional interesting property of being derived from two different parental chromosomes (Schreck et al. 1977). Hence, for the Prader-Willi syndrome and other abnormalities associated with aberrations localized to proximal 15q, one is confronted with the need for sensitive molecular detection of deletions and specific delineation of clinical heterogeneity. Of interest as well is the basis for the apparent structural lability of this region, including the mechanisms responsible for proximal 15q deletions, rearrangements, and duplications.

Isolation of DNA segments from a region as small as 15q11.2 was achieved by chromosome sorting, which was greatly facilitated by starting with a structurally abnormal chromosome (an inv dup[15]) with a breakpoint at 15q13, itself enriched for this region (Fig. 2). This chromosome could be reliably resolved over background only by use of a bivariate flow histogram (Lalande et al. 1985; Donlon et al. 1986) utilizing two dyes, Hoechst 33258 and Chromomycin A_3 (Fig. 3). In the 47,XX + inv dup(15) lymphoblast chromosomes shown, the inv dup(15) could be easily resolved and sorted. DNA (40 ng) from sorted chromosomes yielded 40,000 pfu on LE392, whereas another 10 ng of DNA was packaged and plated out on the *recBC*$^-$, *sbcB*$^-$ bacterial strain DB1257 (Donlon et al. 1986). Thus far, these studies have yielded a few phages with inserts exhibiting homology with the proximal long arm of human chromosome 15. One of these, probe pTD3-21, is shown in Figure 4 to blot with single-dose intensity to DNA from the Prader-Willi-syndrome-derived lymphoblast line DON-10 but not to DNA from DON-5, which has a smaller deletion of 15q11.2.

Of equal interest is that the structural nature of the probes isolated from 15q11→15q13 may shed some light on the structural lability of this region. There is at least suggestive evidence that phages containing DNA from this region are isolated in greater yield when grown on DB1257, a strain with a phenotype that permits the propagation of phages containing inserts with inverted repetitions (Wyman et al. 1985), than when grown on

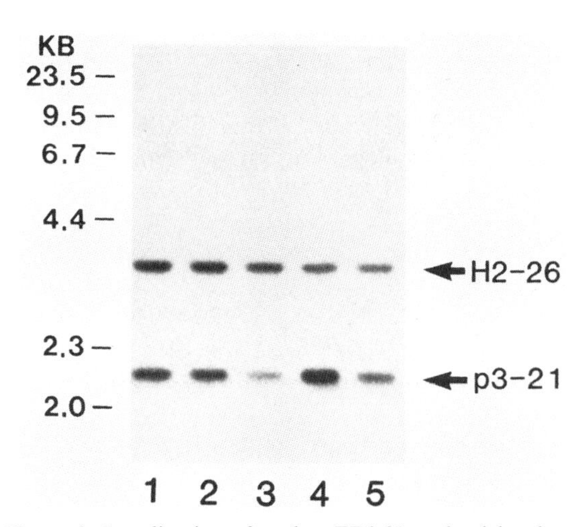

Figure 4. Localization of probe pTD3-21 to the deleted segment of 15q11.2 in DNA from cell line DON-10 (Prader-Willi syndrome [PWS] patient) using quantitative DNA blotting. Each lane contains 1.5 μg of DNA, digested with *Hin*dIII endonuclease. DNA in the lanes shown was obtained from 46,XY (*1*); DON-5 (PWS; very small 15q11.2 deletion) (*2*); DON-10 (PWS; somewhat larger 15q11.2 deletion) (*3*); ALD-24 inv dup(15) (breakpoint at 15q13) (*4*); and ALD-6, inv dup(15) (breakpoint near 15q11) cells (*5*). Reduced hybridization of the 2.2-kb band in lane *3* (DON-10) indicates that it is a deletion from one homolog in this cell line. Probe H2-26, a 3.5-kb insert from chromosome 13, was used as an internal standard for hybridization intensity. (Reprinted, with permission, from Donlon et al. 1986.)

LE392, a strain not conducive to the propagation of such phage. Consistent with this, both phages that could be propagated on DB1257 but not LE392 and that exhibit homology with DNA in band 15q11.2 contain, by heteroduplex mapping, inverted repeats (Donlon et al. 1986). Either inverted or direct repeats could labilize a chromosome band such as 15q11.2 to deletion or rearrangement.

Molecular Analysis of Submicroscopic Insertion or Deletion of Y Chromosomal DNA

Microscopic or submicroscopic chromosome insertions of human Y DNA can be present in the chromosomes of 46,XX males (see, e.g., Ferguson-Smith 1966; de la Chapelle et al. 1984; Guellaen et al. 1984; Page et al. 1985; Müller et al. 1986b; Vergnaud et al. 1986; U. Muller et al., unpubl.). Typically, these DNA fragments map to or near the short arm of the human Y. Models for the organization of these sequences on Yp, based on the frequency of their occurrence in 46,XX males, are beginning to emerge (Vergnaud et al. 1986; U. Müller et al., in prep.). Conversely, cytological deletions of Yp (Magenis et al. 1984; Disteche et al. 1986) or a deletion of Yp sequences in the absence of detectable cytological changes (Müller et al. 1986b) can be associated with abnormal female sexual differentiation. Implicit in this work is that one or more male sex determinants are located on Yp (Zuffardi et al. 1982; Goodfellow et al. 1985).

DNA segments to study 46,XX males and 46,XY females were obtained from a Y-enriched chromosome fraction isolated using two-dimensional flow sorting (Müller et al. 1986a,b); 40 ng of DNA yielded approximately 50,000 pfu (Müller et al. 1986b). More than half of the mappable inserts from this library have at least some Y homology; thus far, 30 such probes have been identified (Müller et al. 1986a,b and unpubl.). Consistent with other studies (see, e.g., Page et al. 1982; Bishop et al. 1984; Goodfellow et al. 1985; Vergnaud et al. 1986), many of the probes from this library are homologous not only to Y DNA, but also to DNA from other chromosomes. However, several of these probes are purely Y-specific, hybridizing to male DNA as either single-copy or Y-specific repeated sequences (Müller et al. 1986b and unpubl.).

One Y-chromosome repeat probe (Y-156) (Müller et al. 1986b) embodies many of the interesting properties of the Y-specific probes isolated from this library. Y-156 maps by in situ hybridization to Yp (Fig. 5). When hybridized against normal male DNA but not normal female DNA, it identifies an approximately 4.5-kb *Hin*dIII restriction fragment with an approximately 10 copy number intensity, plus a few bands of weaker intensity (Fig. 6). Importantly, Y-156 is present at normal intensity in several 46,XX males and at reduced intensity in one of two 46,XY females (Fig. 6) (Müller et al. 1986b). Probes such as Y-156 are of potential diagnostic use and may be useful in a molecular search for putative male sex determinants thought to exist on Yp.

Amplified DNA Sequences in Neuroblastoma Tissue

An alternative and cytologically dramatic example of DNA insertion into a chromosome other than its normal location occurs in homogeneously staining regions (HSRs) of neuroblastomas. These sequences map to the short arm of chromosome 2 (Kanda et al. 1983), most often (Schwab et al. 1984), but not exclusively (Shiloh et al. 1985), to or very near band 2p24. Yet, they are typically amplified at a distance, either as HSRs, e.g., in IMR-32, or as extrachromosomal double-minute (DM) bodies, e.g., in NB-9 cells. DNA probes isolated from flow sorting the no. 1 + HSR

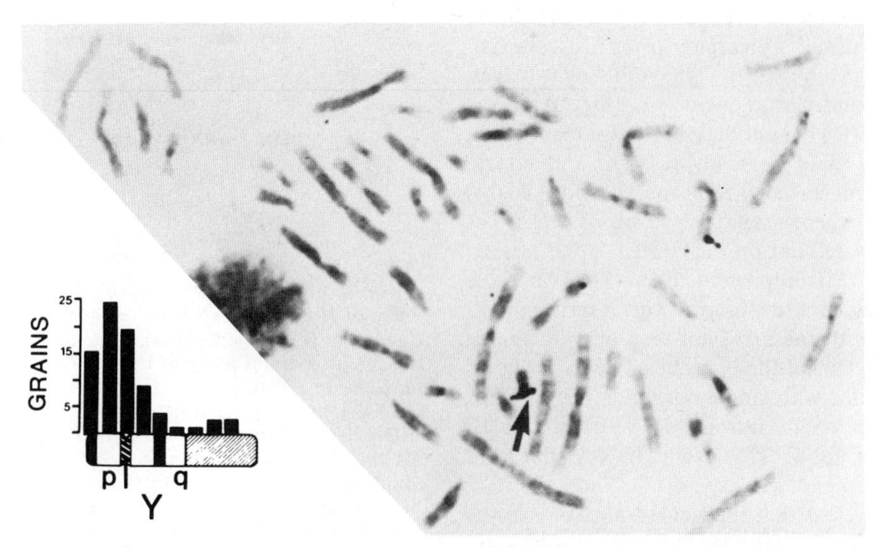

Figure 5. In situ hybridization of Y-156 to a 46,XY derived metaphase. Of all metaphases examined, 55% showed at least one grain over Yp, as shown by the arrow. (*Inset*) Grain distribution obtained from 50 cells analyzed. (Reprinted, with permission, from Müller et al. 1986b.)

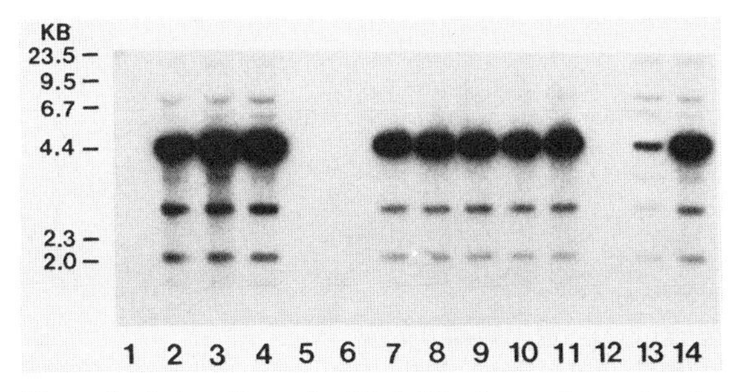

Figure 6. Autoradiograph of hybridization pattern of probe Y-156 with *Hin*dIII-digested DNA from normal 46,XX females (lanes *1,12*), normal 46,XY males (lanes *2-4*), a 46,XX true hermaphrodite (lane *5*), 46,XX males (lanes *6-11*), and 46,XY females (lanes *13,14*). Identical amounts of DNA were applied in each lane. Note the strikingly reduced hybridization signal of the approximately 4.5-kb fragment in the 46,XY gonadal dysgenesis patient (lane *13*) and its presence in many of the 46,XX males. (Reprinted, with permission, from Müller et al. 1986b.)

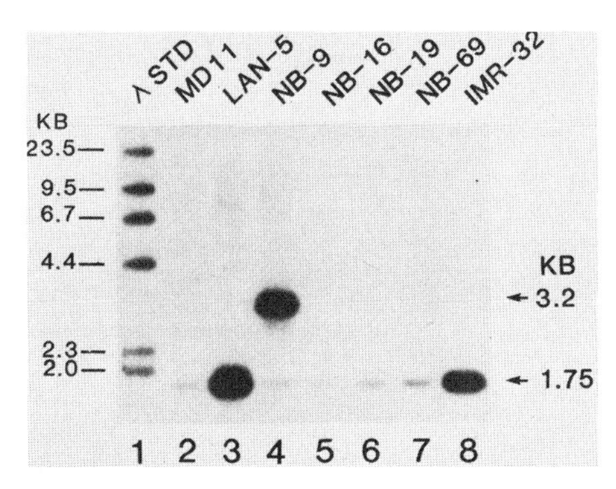

Figure 7. Hybridization of an IMR-32 HSR-derived 1.75-kb *Hin*dIII fragment with equal amounts of *Hin*dIII-digested DNA from different human neuroblastoma cell lines. Note the presence of a band at 1.75 kb with at least single-copy intensity in lanes *2-8*. (Reprinted, with permission, from Latt et al. 1984.)

chromosomes of IMR-32 (Kanda et al. 1983) have provided one means of analyzing interesting phenomena associated with amplification of DNA sequences in both neuroblastoma cell lines and primary neuroblastomas.

Properties of amplified neuroblastoma sequences have been described in detail previously and are only briefly summarized here. First, some of the IMR-32-derived probes exhibit amplification in other neuroblastoma cell lines, which were subsequently screened with known oncogenes, including c-*myc*, leading to the identification of a new oncogene, N-*myc* (Kohl et al. 1983). N-*myc* was independently discovered in systematic screens of neuroblastoma cells with oncogene panels (Montgomery et al. 1983; Schwab et al. 1983). N-*myc* amplification has been observed not only in neuroblastoma cell lines with HSRs or DM bodies, but also in advanced-stage primary neuroblastomas (Brodeur et al. 1984), in some retinoblastoma cell lines (Lee et al. 1984; Sakai et al. 1985), and in some small-cell lung cancers (Nau et al. 1984). Rearrangements associated with amplification of IMR-32 probes, in particular the eighth largest (probe no. 8; 1.75-kb *Hin*dIII fragment), were observed fairly early (Fig. 7) (Kohl et al. 1983; Latt et al. 1984). Subsequent IMR-32-derived and other DNA sequences isolated in related studies demonstrated that this amplification-associated rearrangement was relatively common in neuroblastoma tissue (Latt et al. 1986b; Y. Shiloh et al., in prep.).

DNA rearrangement associated with neuroblastoma displays at least three interesting properties. First, restriction mapping of the 1.75-kb probe 8 and the 3.2-kb amplified sequence homologous to it in NB-9 (Fig. 7) showed that, at one end of the two *Hin*dIII segments, there was complete homology (Kohl et al. 1983), whereas at the other end, no homology exists. This initially suggested that a simple DNA interchange was associated with amplification of the 3.2-kb fragment in NB-9. A more complicated phenomenon associated with DNA amplification and relocation made evident with the IMR-32-derived DNA fragments was a long-range DNA splicing (Shiloh et al. 1985). Specifically, DNA segments mapping to the terminal, mid, or proximal short arm of chromosome 2 were shown to be organized into the same 3000-kb amplified unit in IMR-32. Finally, a careful search in the vicinity of the junction between probe-8 homologous and nonhomologous sequences in the 3.2-kb NB-9 *Hin*dIII fragment (for a summary, see Latt et al. 1986a,b; K. Sakai, unpubl.) detected a hint of inserted DNA that was verified upon DNA sequencing (Fig. 8). Hence, either the rearrangement leading to the NB-9 3.2-kb *Hin*dIII fragment was complex (involving multiple splicing events) or it involved a direct insertion of a 155-bp segment, whose origin and properties are now being studied. In addition to the likely biological importance of the N-*myc* oncogene discovered in the neuroblastoma gene-amplification experiments, these studies have revealed a remarkable fluidity to the human genome whose cause-and-effect relationship to neuroblastoma development remains to be determined.

DISCUSSION

One of the immediate goals of applying molecular biology to human cytogenetics has been the acquisition of a deeper understanding of the nature and genesis of chromosome structural abnormalities. Related to this is attainment of greater insight into the dynamics of structural changes in human chromosomes. Such studies can involve cytologically visible chromosome changes or those too small to be detected readily by light microscopy. In this regard, it should be remembered that even the most extended pictures of human chromosomes (e.g., 2000-band karyotypes; Yunis 1981) resolve the human genome into segments with an average size of 1.5 million base pairs. Within such a region there can be cytologically undetected a great deal of deletions and insertions. Also, rearrangements such as sister chromatid exchanges (Latt 1981) appear recip-

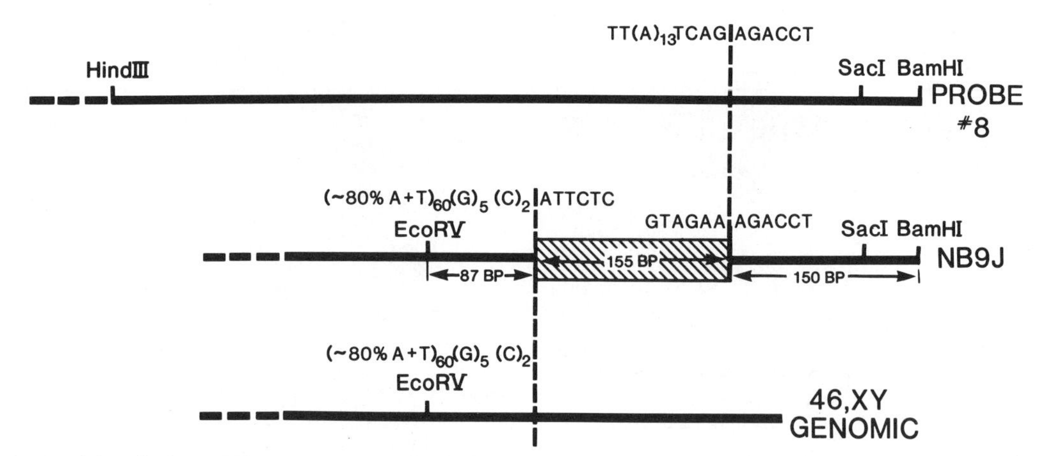

Figure 8. Partial preliminary DNA sequence revealing 155-bp insert making NB-9 rearrangement detected by IMR-32 probe 8 at least a three-way splice. A description is also given of the DNA sequence bordering the sites at which the DNA flanking this insert are spliced, as well as the first few base pairs inside of the inserted segment (K. Sakai, unpubl.). (Reprinted, with permission, from Latt et al. 1986a,b.)

rocal by light microscopy but are of unknown status at a molecular level. A combination of cytological and molecular methodologies presents a new way to approach these phenomena. When it is apparent to which region of the genome one's attentions are turned, metaphase chromosome sorting, followed by the construction of chromosome-enriched recombinant libraries, can then provide many molecular probes for a highly detailed examination of the region in question. Techniques for expanding the region explorable around the small DNA fragments isolated from these libraries (see, e.g., Bender et al. 1983; Collins and Weissman 1984; Schwartz and Cantor 1984; Carle et al. 1986) can then extend the range of these small molecular probes, with at least the feasible goal of reaching the cytological level, providing, in principle, a coherent molecular-cytological view of human chromosomes.

This paper has indicated a few of the many possible examples in which a molecular cytogenetic approach can ultimately lead to a new, important understanding about the statics and dynamics of human chromosome structure. In the case of retinoblastoma, cytological observations of deletions (Knudson et al. 1976) and linkage analysis (Sparkes et al. 1983) have positioned the "retinoblastoma" locus to band 13q14. This locus is grossly deleted in some spontaneous tumors and converted, for example, by somatic recombination or gene conversion (Cavenee et al. 1983) to homozygosity in others. It is still necessary to locate more precisely and characterize the nature of the retinoblastoma locus, as well as the basis for the heterogeneity in deletions removing one copy of this locus. One is left with the possibility that those deletions that may be observed cytologically reflect but "the tip of the iceberg" of deletions; detection of others may require molecular probes. A related question is the nature of the DNA sequences at the deletion boundaries and the role they play in promoting these deletions.

In the case of deletions in the proximal long arm of human chromosome 15, such as the 15q11.2 deletion(s) of the Prader-Willi syndrome, one is left with the same questions about detection of deletion heterogeneity and boundaries, plus one other, i.e., the molecular basis for the vulnerability of this chromosome region not only to deletion, but also to duplication and exchange. Perhaps the propensity for sequences isolated from this region to propagate on a recombination-altered host plus their internal repeat structure contains a hint about this instability. High-resolution analysis of genetic recombination within this region might shed further light about its fluidity.

Insertions of Y chromosomal sequences into the DNA of 46,XX males present several new challenges. Foremost among these at a functional level is probably the isolation of those DNA sequences responsible for male sexual differentiation. It is also important to locate the sites of insertion of these sequences and, once again, the molecular nature of the regions bordering these insertions. Similarly, the regions surrounding Yp deletions in 46,XY females are worthy of detailed study.

The neuroblastoma gene amplification is, on the one hand, a recapitulation of the phenomena outlined above and, on the other, one example of the chromosomal remodeling that occurs in human cancer. One has DNA-amplification, associated with relocation of sequences, copies of which might well be retained at the original loci. This necessitates both an extrachromosomal intermediate plus sequences predisposing to insertion of these intermediates at a distance. In addition, there is dramatic DNA sequence reorganization, ranging from long-range DNA splicing to amplification-associated rearrangements that appear simple at the restriction map level but very complicated at the DNA sequence level. Again, junctional sequences become of interest, as does the organization (e.g., head-head or head-tail) of the amplified units. Of equal interest is the suggestion, as illustrated by the IMR-32 probe 8 NB-9 rearrangement case (Fig. 8), that specific, e.g., A-T-rich, sequences predispose for such rearrangements or inserts. Roles for specific sequence repeats (Singer 1982), topoisomerases (Wang 1985), and even A-T-recognizing proteins (Solomon et al. 1986) might all be hypothesized as underlying these events. In the case of neoplasia, one has the additional complication that a single rearrangement might promote neoplastic behavior, while this behavior itself may further promote additional chromosome rearrangement.

When one adds to these observations the many known examples of chromosome translocation, somatic recombination as recognized by restriction fragment polymorphism (see, e.g., Cavenee et al. 1983), quadriradial figures (German 1964), or sister chromatid exchanges (Latt 1981), as well as the additional detection of chromosomal fragile sites, one is tempted to speculate about possible unifying underlying mechanisms. As with the case of sister chromatid exchanges, this could reside in the DNA sequences themselves or in changes in these sequences, e.g., due to environmental insults or to special proteins necessary for the evolution of chromosomal structure. Unlike the case of sister chromatid exchanges, however, which have thus far defied definitive attempts to go from the cytological observations to molecular explanations, one has, in the chromosomal changes illustrated above, valuable molecular probes with which to begin definitive molecular studies. Success in such studies may be distant, but its achievement would provide a new and more coherent picture of the remarkable fluidity of the human genome.

ACKNOWLEDGMENTS

This work was sponsored by grants from the National Institutes of Health, the American Cancer Society, and the National Foundation March of Dimes. U.M. is the recipient of a Heisenberg fellowship from the Deutsche Forschungsgemeinschaft. We thank Drs. Umadevi Tantravahi and Louis Kunkel for advice, and numerous previous colleagues, in particular Dr. Rhona Schreck, for their participation in the very early stages of some of the work presented.

REFERENCES

Aldridge, J., L. Kunkel, G. Bruns, U. Tantravahi, M. Lalande, T. Brewster, E. Moreau, M. Wilson, W. Bromley, T. Roderick, and S.A. Latt. 1984. A strategy to reveal high frequency RFLPs along the human X chromosome. *Am. J. Hum. Genet.* **36:** 546.

Bender, W., M. Arkham, F. Karch, P.A. Beachy, M. Pfeifer, P. Spierer, E.B. Lewis, and D. Hogness. 1983. Molecular genetics of the bithorax complex in *Drosophila melanogaster*. *Science* **221:** 23.

Benton, W.D. and R.W. Davis. 1977. Screening λgt recombinant clones by hybridization to single plaques in situ. *Science* **196:** 180.

Bishop, C.E., G. Guellen, D. Geldwerth, M. Fellous, and J. Weissenbach. 1984. Extensive sequence homologies between Y and other human chromosomes. *J. Mol. Biol.* **173:** 403.

Bolivar, I., R.C. Rodriquez, P.J. Green, M.D. Betlach, H.L. Heyneker, H.W. Boyer, J.H. Cross, and S. Falkow. 1977. Construction and characterization of new cloning vehicles. II. A multipurpose cloning system. *Gene* **2:** 95.

Brodeur, G.M., R.C. Seeger, M. Schwab, H.F. Varmus, and J.M. Bishop. 1984. Amplification of N-*myc* in untreated human neuroblastomas correlates with advanced disease stage. *Science* **224:** 1121.

Brunk, C.F., C.K. Jones, and T.W. James. 1979. Assay for nanogram quantities of DNA in cellular homogenates. *Anal. Biochem.* **92:** 497.

Bruns, G.A., B.J. Mintz, A.C. Leary, V.M. Regina, and P.S. Gerald. 1979. Human lysosomal genes: Arylsulfatase A and β-galactosidase. *Biochem. Genet.* **17:** 1031.

Butler, M.G. and C.G. Palmer. 1983. Parental origin of chromosome 15 deletions in Prader-Willi syndrome. *Lancet* **I:** 1285.

Butler, M.G., F.J. Meaney, and C. Palmer. 1986. Clinical and cytogenetic survey of 39 individuals with Prader-Labhart-Willi syndrome. *Am. J. Med. Genet.* **23:** 793.

Carle, G.F., M. Frank, and M.V. Olson. 1986. Electrophoretic separations of large DNA molecules by periodic inversion of the electric field. *Science* **232:** 65.

Cavenee, W.K., T.P. Dryja, R.A. Phillips, W.F. Benedict, R. Godbout, B.L. Gallie, A.L. Murphee, L.C. Strong, and R.L. White. 1983. Expression of recessive alleles by chromosomal mechanisms in retinoblastoma. *Nature* **305:** 779.

Collins, F. and S. Weissman. 1984. Directional cloning of DNA fragments at a large distance from an initial probe. A circularization method. *Proc. Natl. Acad. Sci.* **81:** 6812.

Davies, K.E., B.D. Young, R.G. Elles, M.E. Hill, and R. Williamson. 1981. Cloning of a representative genomic library of the human X chromosome after sorting by flow cytometry. *Nature* **293:** 374.

Davies, K.E., P.L. Pearson, P.S. Harper, J.M. Murray, T. O'Brien, M. Sarfaraazi, and R. Williamson. 1983. Linkage analysis of two cloned DNA sequences flanking the Duchenne muscular dystrophy locus on the short arm of the human X chromosome. *Nucleic Acids Res.* **11:** 2303.

de la Chapelle, A., P.A. Tippett, G. Wetterstrand, and D. Page. 1984. Genetic evidence of X-Y interchange in a human XX male. *Nature* **307:** 170.

Disteche, C.M., M. Casanova, H. Saal, C. Friedman, V. Sybert, J. Graham, H. Thuline, D. Page, and M. Fellous. 1986. Small deletions of the short arm of the Y chromosome in 46,XY females. *Proc. Natl. Acad. Sci.* (in press).

Donlon, T.A., M. Litt, S.R. Newcom, and R.E. Magenis. 1983. Localization of the restriction fragment length polymorphism D14S1 (pAW101) to chromosome 14q32.1→32.2. *Am. J. Hum. Genet.* **35:** 1097.

Donlon, T.A., M. Lalande, A. Wyman, G. Bruns, and S.A. Latt. 1986. Isolation of molecular probes associated with the chromosome 15 instability in the Prader-Willi syndrome. *Proc. Natl. Acad. Sci.* **83:** 4408.

Feinberg, A.P. and B. Vogelstein. 1983. A technique for radiolabeling DNA restriction endonuclease fragments to high specific activity. *Anal. Biochem.* **132:** 6.

Ferguson-Smith, M.A. 1966. X-Y chromosome interchange in the aetiology of true hermaphroditism and of XX Klinefelter's syndrome. *Lancet* **II:** 475.

Forget, B.C., D.G. Hillman, H. Lazarus, E.E. Barell, E.L. Benz, Jr., T.H.J. Huisman, W.A. Schroder, and D. Housman. 1976. Absence of messenger RNA and gene DNA for β-globin chains in hereditary persistance of fetal hemoglobin. *Cell* **7:** 323.

German, J. 1964. Cytological evidence for crossing-over in vitro in human lymphoid cells. *Science* **144:** 298.

Gilbert, F., G. Balaban, P. Moorhead, D. Bianchi, and H. Schlesinger. 1982. Abnormalities of chromosome lp in human neuroblastoma tumors and cell lines. *Cancer Genet. Cytogenet.* **7:** 33.

Goodfellow, P., S. Darling, and J. Wolfe. 1985. The human Y chromosome. *J. Med. Genet.* **22:** 329.

Guellaen, G., M. Casanova, C. Bishop, D. Geldwerth, G. Andre, M. Fellous, and J. Weissenbach. 1984. Human XX males with Y single-copy DNA fragments. *Nature* **307:** 172.

Harper, M.E., A. Ullrich, and G.F. Saunders. 1981. Localization of the human insulin gene to the short arm of chromosome 11. *Proc. Natl. Acad. Sci.* **78:** 4458.

Hofker, M., M. Wapenaar, N. Goor, E. Bakker, O. Van Ommen, and P. Pearson. 1985. Isolation of probes detecting restriction fragment length polymorphisms from X chromosome-specific libraries: Potential use for diagnosis of Duchenne muscular dystrophy. *Hum. Genet.* **70:** 148.

Kanda, N., R. Schreck, F. Alt, G. Bruns, D. Baltimore, and S.A. Latt. 1983. Isolation of amplified DNA sequences from IMR-32 human neuroblastoma cells: Facilitation by fluorescence-activated flow sorting of metaphase chromosomes. *Proc. Natl. Acad. Sci.* **80:** 4069.

Knudson, A.G., A.T. Meadows, W.W. Nichols, and R. Hill. 1976. Chromosomal deletion and retinoblastoma. *N. Eng. J. Med.* **295:** 1120.

Kohl, N.E., N. Kanda, R.R. Schreck, G. Bruns, S.A. Latt, F. Gilbert, and F.W. Alt. 1983. Transposition and amplification of oncogene-related sequences in human neuroblastomas. *Cell* **35:** 359.

Kunkel, L.M., U. Tantravahi, M. Eisenhard, and S.A. Latt. 1982. Regional localization on the human X of DNA segments cloned from flow sorted chromosomes. *Nucleic Acids Res.* **10:** 1557.

Lalande, M., L. Kunkel, A. Flint, and S.A. Latt. 1984a. Development and use of metaphase chromosome flow sorting methodology to obtain recombinant phage libraries enriched for parts of the human X chromosome. *Cytometry* **5:** 101.

Lalande, M., R.R. Schreck, R. Hoffman, and S.A. Latt. 1985. Identification of inverted duplicated no. 15 chromosomes using bivariate flow cytometric analysis. *Cytometry* **6:** 1.

Lalande, M., T. Donlon, R.A. Petersen, R. Liberfarb, S. Manter, and S.A. Latt. 1986. Molecular detection and differentiation of deletions in band 13q14 in human retinoblastoma. *Cancer Genet. Cytogenet.* **23:** 151.

Lalande, M., T.P. Dryja, R.R. Schreck, J. Shipley, A. Flint, and S.A. Latt. 1984b. Isolation of human chromosome 13-specific DNA sequences cloned from flow sorted chromosomes and potentially linked to the retinoblastoma locus. *Cancer Genet. Cytogenet.* **13:** 283.

Langlois, R.S., L.C. Yu, J.W. Gray, and A.V. Carrano. 1982. Quantitative karyotyping of human chromosomes by dual beam flow cytometry. *Proc. Natl. Acad. Sci.* **79:** 7876.

Latt, S.A. 1981. Sister chromatid exchange formation. *Annu. Rev. Genet.* **15:** 11.

Latt, S.A., N. Kanda, L. Kunkel, M. Lalande, F. Alt, N. Kohl, G. Bruns, J. Aldridge, R. Schreck, and U. Tantravahi. 1984. Sorting, cloning and analysis of specific human chromosomes. *Chromosomes Today* **8:** 15.

Latt, S.A., Y. Shiloh, K. Sakai, G. Brodeur, T. Donlon, B. Korf, J. Shipley, G. Bruns, M. Heartlein, N. Kanda, N. Kohl, F. Alt, and R. Seeger. 1986a. Novel DNA rearrangement phenomena associated with DNA amplification in human neuroblastoma cell lines. In *Genetic toxicology of environmental chemicals. A. Basic principles and mechanisms of action* (ed. C. Ramel et al.), p. 601. A.R. Liss, New York.

Latt, S.A., Y. Shiloh, K. Sakai, E. Rose, G. Brodeur, T. Donlon, B. Korf, N. Kanda, M. Heartlein, J. Kang, H. Stroh, P. Harris, G. Bruns, and R. Seeger. 1986b. DNA rearrangement, relocation and amplification in neuroblastoma cell lines and primary tumors. *ICN-UCLA Symp. Mol. Cell Biol.* (in press).

Ledbetter, D.H., J.T. Mascarello, V.M. Riccardi, V.D. Harper, S.D. Airhart, and R.J. Strobel. 1982. Chromosome 15 abnormalities and the Prader-Willi syndrome: A follow-up report of 40 cases. *Am. J. Hum. Genet.* **34:** 278.

Ledbetter, D.H., V.M. Riccardi, S.D. Airhart, R.J. Strobel, S.B. Keenan, and J.D. Crawford. 1981. Deletions of chromosome no. 15 as a cause of the Prader-Willi syndrome. *N. Engl. J. Med.* **304:** 325.

Lee, W.H., A.L. Murphee, and W.F. Benedict. 1984. Expression and amplification of the N-*myc* gene in primary retinoblastoma. *Nature* **309:** 458.

Magenis, R.E., M.L. Tochen, K.P. Holahan, T. Carey, L. Allen, and M.G. Brown. 1984. Turner syndrome resulting from partial deletion of Y chromosome short arm: Localization of male determinants. *J. Pediatr.* **105:** 916.

Maniatis, T., E.F. Fritsch, and J. Sambrook. 1982. *Molecular cloning: A laboratory manual*, p. 504. Cold Spring Harbor Laboratory, Cold Spring Harbor, New York.

Messing, J., B. Gronenborn, B. Muller-Hill, and P.H. Hofschneider. 1977. Filamentous coliphage M13 as a cloning vehicle: Insertion of a *Hin*dII fragment of the *lac* regulatory region in M13 replicating form in vitro. *Proc. Natl. Acad. Sci.* **74:** 3642.

Montgomery, K.T., J.L. Biedler, B.A. Spengler, and P.W. Mera. 1983. Specific DNA sequence amplification in human neuroblastoma cells. *Proc. Natl. Acad. Sci.* **80:** 5724.

Müller, U., M. Lalande, C.M. Disteche, and S.A. Latt. 1986a. Construction, analysis, and application of 46,XY gonadal dysgenesis of a recombinant phage DNA library from flow sorted human Y chromosomes. *Cytometry* (in press).

Müller, U., M. Lalande, T. Donlon, and S.A. Latt. 1986b. Moderately repeated DNA sequences specific for the short arm of the human Y chromosome are present in XX males and reduced in copy number in an XY female. *Nucleic Acids Res.* **14:** 1325.

Nau, M.M., D.N. Carney, J.B. Johnson, C. Little, A. Gozdar, and J.D. Mina. 1984. Amplification, expression, and rearrangement of c-*myc* and N-*myc* oncogenes in human lung cancer. *Curr. Top. Microbiol. Immunol.* **113:** 172.

O'Farrel, P. 1981. Replacement synthesis method of labeling DNA fragments. *Focus* **3:** 1.

Page, D.C., A. de la Chapelle, and J. Weissenbach. 1985. Chromosome Y specific DNA in related human XX males. *Nature* **315:** 224.

Page, D.C., B. de Martinville, D. Barker, A. Wyman, R. White, U. Francke, and D. Botstein. 1982. Single-copy sequence hybridizes to polymorphic and homologous loci on human X and Y chromosomes. *Proc. Natl. Acad. Sci.* **79:** 5352.

Sahar, E. and S.A. Latt. 1978. Enhancement of banding patterns in human metaphase chromosomes by energy transfer. *Proc. Natl. Acad. Sci.* **75:** 5650.

Sakai, K., N. Kanda, Y. Shiloh, T. Donlon, R. Schreck, J. Shipley, T. Dryja, E. Chaum, R.S.K. Chaganti, and S. Latt. 1985. Molecular and cytological analysis of DNA amplification in retinoblastoma. *Cancer Genet. Cytogenet.* **17:** 95.

Sanger, F., S. Nicklen, and A.R. Coulson. 1977. DNA sequencing with chain terminating inhibitors. *Proc. Natl. Acad. Sci.* **74:** 5463.

Scambler, P.J., B.J. Wainwright, E. Watson, G. Bates, G. Bell, R. Williamson, and M. Farral. 1986. Isolation of a further anonymous informative DNA sequence from chromosome seven closely linked to cystic fibrosis. *Nucleic Acids Res.* **14:** 1951.

Schreck R.R., W.R. Breg, B.F. Erlanger, and O.J. Miller. 1977. Preferential derivation of abnormal human G-group-like chromosomes from chromosome 15. *Human Genet.* **36:** 1.

Schwab, M., H.E. Varmus, J.M. Bishop, K.H. Grzeschik, S.L. Naylor, A.V. Sakaguchi, G. Brodeur, and J. Trent. 1984. Chromosome localization in normal human cells and neuroblastomas of a gene related to c-*myc*. *Nature* **308:** 288.

Schwab, M., K. Alitalo, K.H. Klempaneur, H. Varmus, J.M. Bishop, R. Gilbert, G. Brodeur, M. Goldstein, and J. Trent. 1983. Amplified DNA with limited homology to *myc* cellular oncogene is shared by human neuroblastoma cell lines and a neuroblastoma tumor. *Nature* **305:** 245.

Schwartz, D.C. and C.R. Cantor. 1984. Separation of yeast chromosome-size DNAs by pulsed-field gradient gel electrophoresis. *Cell* **37:** 67.

Shiloh, Y., J. Shipley, G.M. Brodeur, G. Bruns, B. Korf, T. Donlon, R.R. Schreck, R. Seeger, K. Sakai, and S.A. Latt.

1985. Differential amplification, assembly, and relocation of multiple DNA sequences in human neuroblastomas and neuroblastoma cell lines. *Proc. Natl. Acad. Sci.* **82:** 3761.

Sillar, R. and B.D. Young. 1981. A new method for the preparation of metaphase chromosomes for flow analysis. *J. Hisochem. Cytochem.* **29:** 74.

Singer, M.F. 1982. Highly repeated sequences in mammalian genomes. *Int. Rev. Cytol.* **76:** 67.

Solomon, M.J., F. Strauss, and A. Varshavsky. 1986. A mammalian high mobility group protein recognizes any stretch of six A-T base pair duplex DNA. *Proc. Natl. Acad. Sci.* **83:** 1276.

Southern, E.M. 1975. Detection of specific sequences among DNA fragments separated by gel electrophoresis. *J. Mol. Biol.* **48:** 503.

Sparkes, R.S. 1985. The genetics of retinoblastoma. *Biochim. Biophys. Acta* **780:** 95.

Sparkes, R.S., A.L. Murphree, R.W. Lingua, M.C. Sparkes, L.L. Field, S.J. Funderburk, and W.F. Benedict. 1983. Gene for hereditary retinoblastoma assigned to chromosome 13 by linkage to esterase D. *Science* **219:** 971.

Stetten, G., B. Sroka-Zaczek, and L. Corson. 1981. Prenatal detection of an accessory chromosome identified as an inversion duplication (15). *Human Genet.* **57:** 357.

Tantravahi, U., D.A. Kirschner, L. Beauregard, L. Page, L. Kunkel, and S.A. Latt. 1983. Cytological and molecular analysis of 46,XXq-cells to identify a DNA probe for a putative human X chromosome inactivation center. *Human Genet.* **64:** 33.

Tumilowicz, J.J., W.W. Nichols, J.J. Cholon, and A.E. Greene. 1970. Definition of a human cell line derived from neuroblastoma. *Cancer Res.* **30:** 2110.

Vergnaud, G., D.C. Page, M.-C. Simmler, L. Brown, F. Rouyer, B. Moel, D. Botstein, A. de la Chapelle, and J. Weissenbach. 1986. A deletion map of the human Y chromosome based on DNA hybridization. *Am. J. Hum. Genet.* **38:** 109.

Wang, J.C. 1985. DNA topoisomerases. *Annu. Rev. Biochem.* **54:** 665.

Wyman, A.L., L. Wolfe, and D. Botstein. 1985. Propagation of some human vectors requires mutant *Escherichia coli* hosts. *Proc. Natl. Acad. Sci.* **82:** 2880.

Young, B.D. 1984. Chromosome analysis by flow cytometry: A review. *Basic Appl. Histochem.* **28:** 9.

Yunis, J.J. 1981. Mid-prophase human chromosomes. The attainment of 2000 bands. *Human Genet.* **56:** 293.

Zuffardi, O., P. Maraschio, F.L. Curto, U. Muller, A. Giarola, and L. Perotti. 1982. The role of Yp in sex determination: New evidence from X/Y translocations. *Am. J. Med. Genet.* **12:** 175.

Molecular Genetics and the Basic Defect Causing Cystic Fibrosis

R. WILLIAMSON, G. BELL, J. BELL, G. BATES, K.A. DAVIES, X. ESTIVILL, M. FARRALL, H. KRUYER, H.Y. LAW, N. LENCH, P. SCAMBLER, P. STANIER, B. WAINWRIGHT, E. WATSON, AND C. WORRALL

Cystic Fibrosis Genetics Research Group, Department of Biochemistry, St. Mary's Hospital Medical School, University of London, London W2 1PG, England

Cystic fibrosis (CF) is the most common inherited disease affecting North Europeans. The carrier frequency in Britain is approximately 1 in 20, and the incidence of the disease is 1 in 1600. The mode of inheritance is autosomal recessive, and the disease has a relatively constant clinical course. A proportion of infants are born with compacted meconium, which may have to be corrected surgically, but most cases present as "failure to thrive" at between 6 months and 5 years of age. The osmolarity of the sweat is approximately two to three times the normal range; elevated sweat sodium and chloride are diagnostic for the disease (Goodchild and Dodge 1985).

The major clinical symptoms are due to insufficient pancreatic secretion (which has to be corrected by adding pancreatic enzyme extracts to the diet) and to viscous mucus that leads to lung infection. The lung involvement can be reduced by continuous and vigorous physiotherapy throughout life. This in turn improves resistance to infection, but ultimately (usually between the ages of 15 and 30, depending on the aggressiveness of clinical care and on individual variations that are not understood) antibiotic-resistant strains of bacteria colonize the lungs, often *Pseudomonas* spp. Few CF patients survive beyond the age of 30, even in countries such as Australia where clinical care is excellent and patients are grouped in CF centers. Carriers of the CF mutation show no known symptoms, nor known heterozygous advantage; there is at present no carrier test that can be applied to the general population.

In this paper, we describe three general approaches that apply molecular biology techniques to the study of a disease for which the biochemical defect is unknown (Williamson 1985). Although they will be described separately, their usefulness is obviously increased when they are used in a complementary fashion.

Candidate Genes

We have defined a candidate gene for any inherited disease as one where there is a presumption that a mutation of the DNA sequence may cause or (in the case of polygenetic or multifactorial disease) contribute to the genetic component of the clinical phenotype (Davies et al. 1983a). Over the years, there have been a large number of suggestions as to which genes might be mutated and so cause CF. Among these are superoxide dismutase, calmodulin, albumin, and complement 3 (see, e.g., Wilson and Bahm 1980). The availability of a gene probe for any gene that has been proposed as a candidate for the causal mutation makes it relatively simple to test the hypothesis that the gene is involved if the gene reveals a DNA polymorphism (RFLP) and families with multiple affected sibs are available.

Let us consider complement 3 or albumin. In each case, the gene has been cloned and polymorphisms have been identified (Whitehead et al. 1982; Murray et al. 1983). For any family with multiple affected sibs, each sib *must* inherit the same allele from a heterozygote parent, since recombination is extremely unlikely within a coding gene (the chance is of the order of 0.001% for each meiosis). In the case of complement 3, we found one family with three affected sibs, each of whom had a different genotype. Assuming that CF is a single-locus disease, this single experiment with one family excludes complement 3 as the mutated gene (Davies et al. 1983a). The approach is powerful because it is not statistical but unequivocal. The same approach was applied to albumin (Scambler et al. 1985) and several other candidate genes in our laboratory.

Candidate Regions of Chromosomes

It is also possible to treat candidate chromosomal regions in a similar fashion. We first made use of a "candidate region of a chromosome" in our attempts to isolate DNA markers linked to Duchenne muscular dystrophy (DMD). For DMD, there were several female cases of this sex-linked disease that had been described in which the female had a translocation near to the middle of the short arm of the X chromosome. We used this as the starting point in our search for linked probes from our human X-chromosome library and found a recombinant mapping to Xp that showed linkage (Murray et al. 1982; Davies et al. 1983b). Once a mutation for a disease has been localized, it in turn becomes a candidate region for any genetic disease that might be a related variant, as shown for Becker muscular dystrophy, which is located very close to DMD (Kingston et al. 1983).

Two such regions have been proposed for CF: chromosome 4 and the tip of the long arm of chromosome 13. The data for chromosome 4 depended on the expression of a ciliary dyskinesis factor in medium in which cells containing various different human chromosomes were growing. Since Bowman and her colleagues had shown that ciliary dyskinesis is a feature of epithelial cells from CF patients, there was a possibility that chromosomes elaborating factors causing dyskinesis were the location of the mutation causing CF (Mayo et al. 1980). Although the hypothesis seemed unlikely (as many factors can cause dyskinesis, a point made by the original group reporting the data), we still started with chromosome 4 in our chromosome-by-chromosome exclusion, on the grounds that any clue, however weak, is better than none at all. However, a careful linkage analysis with several probes shows that CF is not on chromosome 4 (Scambler et al. 1985).

In the case of chromosome 13, involvement was suggested by a family in which the mother had a balanced translocation between the tip of the long arm of chromosome 13 and chromosome 6 (Edwards et al. 1984). Both parents were, of course, phenotypically normal; they have two children, both mentally retarded and one with CF. The child who has CF is monosomic for 13q34-ter; the other child is trisomic for this region.

Edwards and his colleagues argued that 13q34-ter is a candidate for the locus of the mutation, since monosomy at the region of the locus would result in CF in 1 case in 20, rather than 1 in 1600. Although this argument is ingenious and has been applied fruitfully in other cases (such as the association of DMD, glycerol kinase deficiency, OTC deficiency, adrenal hypoplasia, and chronic granulomatous disease with a set of deletions on the short arm of the X chromosome, often associated with mental retardation), in this case, it was not correct. We showed that there is no linkage between CF and probes that are located at 13q34-ter (Scambler et al. 1986a), and we have excluded this chromosomal region from involvement in CF.

In conclusion, the use of candidate genes or chromosomal regions, although relatively simple in terms of effort, did not result in progress in the search for the molecular defect causing CF.

Comparison of Libraries

Most human cell types express approximately 10,000–20,000 different genes as poly(A)$^+$ mRNA and presumably as protein. In the case of autosomal recessive diseases, most known biochemical defects are either mutant proteins or the absence of synthesis of a protein. Therefore, it is at least a reasonable assumption that the defect in CF will be biochemically apparent as a missing or altered protein in affected tissues.

Comparison of libraries of expressed genes (cDNA libraries) poses at least four major problems. The first is that many of the sequences expressed as mRNA are present in very low concentrations, perhaps only one or two copies per cell. To study these by hybridization comparison techniques, very long annealing times are required. Second, the cells being compared must be isogenic in expression in all ways apart from the inherited disease. This is unrealistic, since persons differ in many ways (most nonpathological) apart from an inherited disease such as CF. Third, the techniques are much more suitable when a gene is not expressed than for a point mutation, and we have no way of knowing into which of these two classes CF will fall. Finally, a tissue that expresses the disease must be available. In the case of CF, these tissues are sweat glands, the pancreatic and salivary ducts, and lung epithelium – none of them standard tissues easy to obtain.

Our research group spent several years constructing cDNA libraries from human lymphocytes from CF patients and normal controls, and we developed the techniques that permitted us to maintain and analyze them (Woods et al. 1980; Crampton et al. 1981). However, we were never able to detect differences that were characteristic of CF, as opposed to those due to clinical treatment or associated with sex. In cases where this technique has worked, such as comparison of untransformed and SV40-transformed cell lines, the isogenic expression criterion is rigorously met (Scott et al. 1983). We do not know, even now, whether CF is expressed in lymphocytes or whether it is due to a point mutation or a gene deletion.

In principle, it is also possible to compare genomic libraries. There are several problems in such an exercise – the presence of interspersed repetitive elements (of varying degrees of repetitiveness), the sheer complexity of the genome, and (perhaps most important) the high frequency of random differences between normal genomes. It has been estimated that such single-base changes may occur as often as once every 100 or 200 base pairs (Jeffreys 1979). The frequency of small deletions and inversions in the normal genome is also unknown. Therefore, comparison of total genomic libraries must be regarded as unfeasible at this time, although it is now possible to probe a genomic library directly for a coding sequence using a long and carefully constructed oligonucleotide probe (Berent et al. 1985).

Total Human Gene Mapping

The fact that markers that are close chromosomal neighbors cosegregate has been known since Sturtevant's classical studies of *Drosophila* genetics (1913). However, the first DNA linkage demonstrated was between the hemoglobinopathies and markers in the globin gene region by Kan and Dozy (1978) for sickle cell disease (1978), followed by our work on β-thalassemia (Little et al. 1980). After Kan and Dozy's paper appeared, Solomon and Bodmer (1979) immediately pointed out in a short note that this pointed toward the possibility of creating a complete human gene linkage map with a small number of selectively neutral informative DNA markers, each of which exhibits a restriction-fragment-length polymorphism (RFLP).

Shortly after, Botstein et al. (1980) published a detailed analysis of the construction of a human linkage map based on RFLPs. They showed that as few as 200 fully informative DNA probes, equally spaced among and along the human chromosome complement, would allow the mapping of any gene to a chromosomal locus using a small number of families. The difficulty of the exercise depends on the structure of the families, the mode of inheritance, and the precise spacing of the probes. At the last Human Gene Mapping 8 conference (HGM8), approximately 1000 markers were reported, the majority of them anonymous probes, each localized chromosomally.

Although there are enough probes, in principle, to map the entire genome, it is still necessary to construct a linkage map chromosome by chromosome by studying probe/probe segregation in large pedigrees. At this time, chromosomes 4, 7q, 11, 13, 16, 19, 20, 21, 22, and X have more or less complete linkage maps, using DNA probes and protein markers, and it is probable that sets of linked markers spanning each human chromosome will be available within the next year.

Linkage to CF

During the Human Gene Mapping 8 conference, the various groups that are attempting to isolate the gene mutated in CF were able to integrate their exclusion data (Fig. 1) (the groups involved were those working in Berkeley/Stanford, Copenhagen, Integrated Genetics Boston, St. Mary's London, Salt Lake City, and Toronto). However, they also were able to assess data from the Copenhagen group, working in collaboration with our group and several others, showing linkage between CF and paraoxonase (PON) (see Eiberg et al. 1985a,b).

PON is a hydrolase found in human serum, which hydrolyzes the insecticide parathion. When its activity is assayed using the *p*-nitrophenol ester of parathion, it is found that persons have either a high or low serum enzyme activity and that low activity is transmitted in families as a Mendelian recessive. (The genetics are slightly more complex than this—there is also a null allele in a very small number of cases, and it is not clear whether homozygous high metabolizers have the same activity as heterozygotes, but analysis as a simple Mendelian trait is possible.) When PON activity was run against CF in families with multiple affected sibs from Denmark, London, and Toronto, linkage at approximately 15 map units was found, with a lod score of approximately 3.

Although the linkage to PON was extremely exciting, it was of limited immediate use to molecular biologists, since the protein is not well characterized (there are no sequence data) and had not been chromosomally located. However, Eiberg's results also gave a great deal of marker/marker linkage data, since all polymorphic protein markers had been run through the same set of CF families. The linkage of PON to each of the other markers studied could be deduced, to generate a rough exclusion map for the enzyme. It was possible to put this map together with the DNA probe exclusion map, at least in a tentative way.

With these data in hand, each of the groups immediately focused attention on those chromosomal regions that had not been excluded either from linkage to CF or from linkage to PON. In our laboratory, we had already decided to focus on probes for chromosomes 7 and 8 earlier in the year and therefore were fortunate enough to have the probes in hand to commence study of chromosome 7, which was one of those with an obvious linkage "gap."

Almost simultaneously, three groups obtained linkage for probes on chromosome 7: the Toronto group was first with a Collaborative Research probe, *917*, followed within days by the Salt Lake City group with the probe *met* and our group with probes *TCRB, COL1A2,* and *pJ3-11* (Scambler et al. 1985; Tsui et al. 1985; Wainwright et al. 1985; White et al. 1985). Of these probes, both *pJ3-11* and *met* are sufficiently close to be used in diagnosis. The present map of human chromosome 7, with the position of the probes relative to *CF,* is shown in Figure 2.

Other markers have since been isolated. We have reported the characterization of an anonymous probe *7C22*, which is approximately 4 map units from *CF* (Scambler et al. 1986b). At least four other probes have been found to be linked to the mutation and are useful in family studies, but their map positions have not yet been reported. However, none of the probes described to date show allele disequilibrium with *CF*. This suggests either that *CF* is a very old mutation, and has reached equilibrium with closely neighboring probes, or that the *CF* mutation has occurred many times during evolution.

It now is possible to use these informative markers in linkage studies of families in which CF occurs to determine whether unaffected siblings are carriers or homozygous normals (Farrall et al. 1986a). This is the first time that accurate assessment of carrier status has been possible, although the error rate is still approximately 2%. The proportion of carriers to homozygous normals is approximately 2:1, as predicted.

First trimester prenatal diagnosis for CF is also possible for informative families using linked probes by DNA analysis of fetal tissue (Williamson et al. 1981). We have carried out approximately 15 such tests (Farrall et al. 1986b) using fetal DNA prepared from chorionic villi taken transcervically and using *pJ3-11* and *met* as probes. Risk calculations show that the expected false positive and negative rates are approximately 6% and 2%, respectively, for informative, typical nuclear families with one affected living child. Existing probes are sufficiently informative to allow full diagnosis in approximately two thirds of couples presenting; in half the remainder, diagnosis of the inheritance of one parental mutant chromosome can be made (Farrall et al. 1986b). In rare cases, it is even possible to carry out prenatal diagnosis in the absence of living affected persons with CF if there is an extended family history (H.Y. Law et al., in prep.).

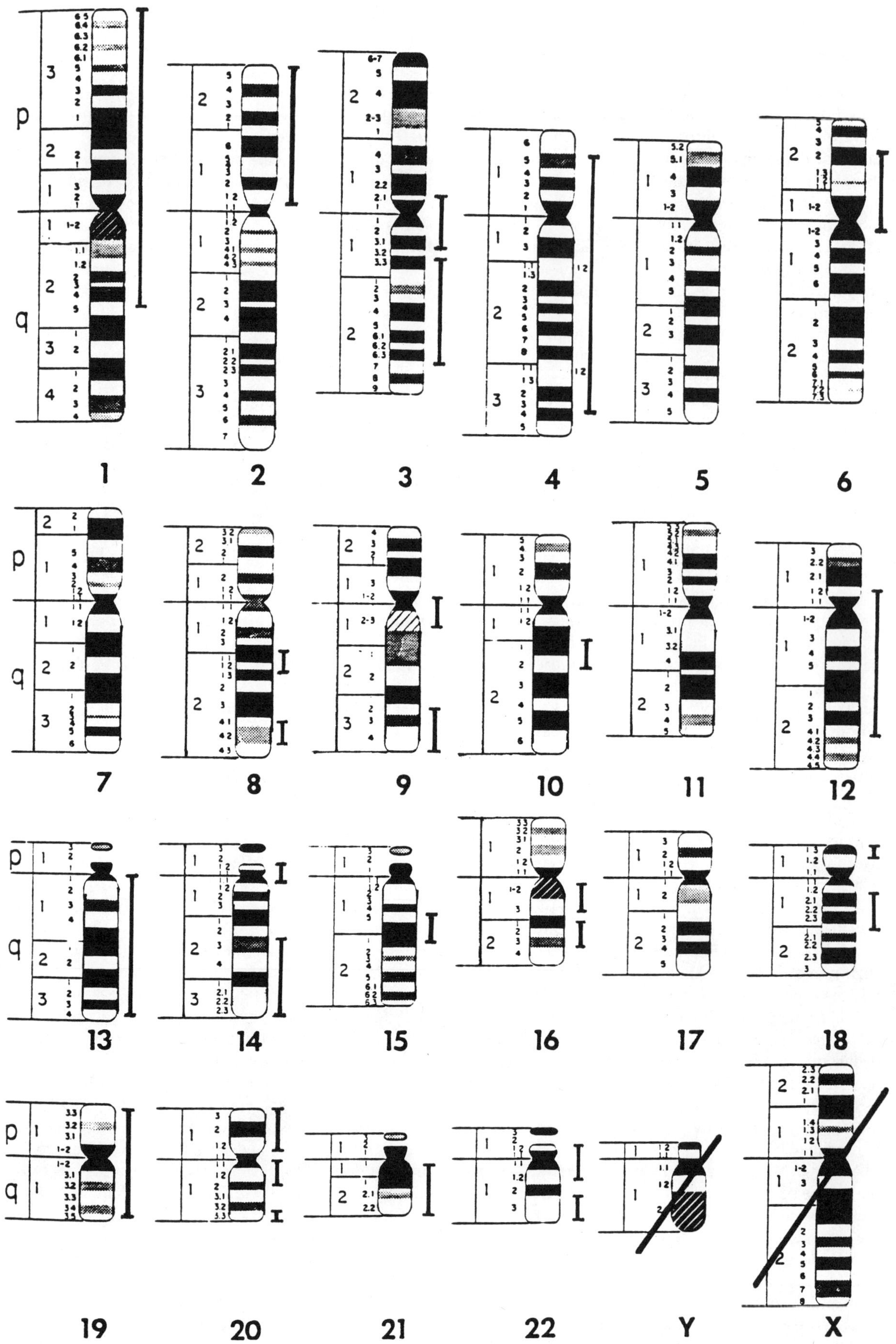

Figure 1. Representation of the human karyotype showing (vertical lines to the right of the chromosomal region) the approximate exclusions obtained using gene probes and linkage analysis at the time of HGM8, August 1985.

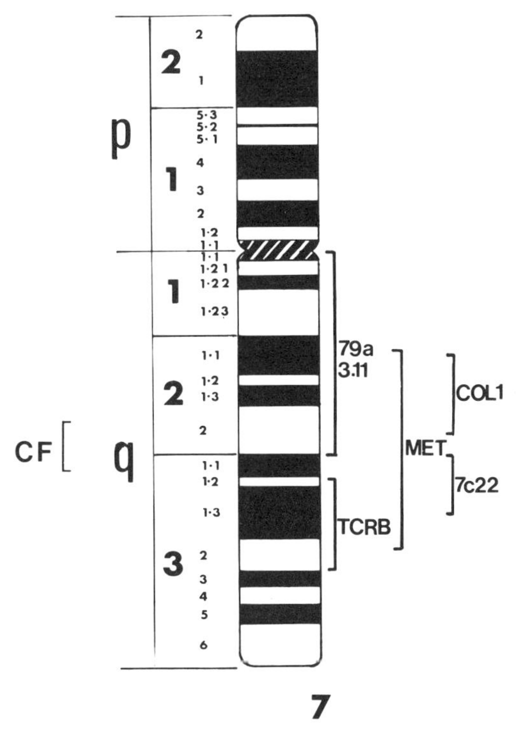

Figure 2. Present map of human chromosome 7 showing the approximate position of *CF* and its linked probes.

There are, however, several reasons why antenatal diagnosis using linkage in phase-known families is far from adequate. First, there will be a proportion of families that are not informative (in this case, approximately 15%) or are only partially informative (again, 15%). Of those families that are informative, a proportion will be unwilling to make use of antenatal diagnosis, although, in our experience, this proportion is very small once a family has an affected child (Kaback et al. 1984). Linkage is of no benefit to patients who have CF. Finally, there will be little reduction in the total number of CF cases, since it is only after a CF child has been diagnosed that antenatal analysis is available, since linkage does not permit carrier detection in the population in the absence of disequilibrium.

It is therefore necessary to attempt to isolate the gene that is mutated so as to cause CF, to permit carrier detection in the general population and in the hope this will allow new forms of treatment to be introduced that will put therapy on a rational basis.

From Linkage to Gene

Here, we only outline the approaches that we have considered to move from linkage to the *CF* gene, since all are under intensive study in a number of laboratories and new developments for each should occur during the next year.

It is possible to saturate the region with closely linked markers. These can be isolated from flow-sorted chromosome libraries (Davies et al. 1981), such as those provided by the Lawrence-Livermore Laboratory project. Alternatively, it is possible to microdissect human chromosomes under phase contrast and to microclone the region of interest, as has been done for mouse chromosomes by Fisher et al. (1985). We have begun such experiments, but chromosome 7 is rather difficult to identify under phase contrast, which limits the usefulness of the technique.

Linked probes are difficult to order once they are within a few million base pairs of each other. Therefore, it is attractive to begin to construct a set of cosmid clones spanning the entire region including the *CF* mutation. To do this, the probe order (including *CF*) is required to ensure that any "walk" has a start and an end. Cosmid clones containing existing probes such as *pJ3-11* and *met* are obvious starting points, particularly if vectors that permit directional walks (Little and Cross 1985) are used. In addition, it is relatively simple to saturate a region of the genome with cosmids that can then be analyzed for overlaps; a region 2 map units in length in general corresponds to approximately 2000 kb, or 50 cosmids end to end.

In some ways, the easiest solution to the problem is to start with a chromosomal fragment that includes the *CF* mutation and closest flanking markers, perhaps 2000 kb in length, in a rodent-human hybrid cell for which it is the only human component. This can be done using chromosome-mediated gene transfer, with one of several selection procedures. The simplest is to select with an endogenous gene (as was done for fragments of the X chromosome with irradiated cells using HPRT by Goss and Harris [1977], among others); in this case, *met* can be used for selection because it transforms NIH-3T3 cells and confers a selectable phenotype. Other possible selection techniques include the use of protein markers on the cell surface or added drug resistance genes that are cotransfected with the human chromosome preparation and used to identify cells competent for uptake of exogenous material.

With these techniques, it is common to obtain transfer of between 1 and 10 million base pairs of DNA, as a human chromosomal fragment in a mouse background (Porteous et al. 1986). This is precisely the genome size that is most amenable to analysis using pulse-field gel electrophoresis (Schwartz and Cantor 1984; Carle et al. 1986).

Suppose that a cell line is obtained that contains 5 megabases of the human chromosomal region 7q22, introduced by chromosome-mediated gene transfer followed by selection for *met*. This region would contain sequences for *met* by definition that could be recognized by blotting, as well as sequences for other probes (such as *pJ3-11* or *7C22*) within the cotransferred chromosomal sequence. Since there are restriction enzymes, such as *Not*I, *Sfi*I, and *Mlu*I, that recognize sites that occur only rarely in the genome, it should be possible to map the human chromosome-7 fragment with respect to both restriction sites and known linked markers for CF.

Such a map, in which the *CF* mutation can be located by its relation to known markers, permits walking to commence. The objective will be to use each restriction site for a rare cutter, such as *Not*I, as the starting point for a cosmid walk that must give a set of overlapping cosmids, including the gene that is mutated so as to cause CF. Since it is very difficult to map using recombination in families to less than 1 cM (~1 million base pairs; Hartley et al. 1984), we assume that this strategy will yield a set of 20 cosmids spanning approximately one megabase and including the *CF* mutation with one linked marker to each side.

It may be possible to define the mutation by one of three methods: searching for small deletions in this region (as for DMD, PKU, or hemophilia A), attempting to find a progressive trend toward disequilibrium between RFLP alleles and the *CF* mutation as one moves toward the defect, or looking for variant mRNA sequences in tissues such as lung epithelium or sweat gland, which are known to be affected by CF pathology.

Fortunately, many of these approaches have been developed in order to study diseases such as Duchenne and Becker muscular dystrophies and chronic granulomatous disease on the well-mapped X chromosome, and therefore it will be possible to apply them with some confidence to the study of CF. Further developments for CF will in turn apply to other diseases where an "unknown" coding gene is to be isolated from a defined small region of a chromosome: Huntington's disease on chromosome 4 and myotonic dystrophy on chromosome 19.

To those of us who have been working on the molecular biology of CF for many years, it is both remarkable and satisfying that there has been so much movement during the past few months, after such a long period of relative lack of progress in spite of considerable effort. There seems little doubt that one of the groups studying CF will isolate the gene within the next few months and it will then be possible to understand both the molecular and clinical genetics of this common disease.

SUMMARY

Recombinant DNA sequences are now available that allow the mapping of the entire human genome, but the first linkage to CF came using classical protein polymorphisms. The enzyme paraoxonase was shown to be loosely linked to CF by the Copenhagen group, to be followed quickly by six cloned DNA sequences: *pJ3-11*, *7C22*, *COL1A2*, and *TCRB* (St. Mary's), *917* (Toronto), and *met* (Salt Lake City). Both *pJ3-11* and *met* are very close genetically to the *CF* mutation and can be used for carrier detection and prenatal diagnosis in many informative families with a CF child. There is no evidence for heterogeneity of the *CF* locus. The collection of informative markers surrounding the *CF* locus is now sufficient to permit attempts to be made to isolate the defective gene using a combination of chromosome-mediated gene transfer, pulse-field gel electrophoresis, cosmid mapping, and chromosome walking techniques, although the difficulty of obtaining tissue in which the defect is known to be expressed remains a problem.

ACKNOWLEDGMENTS

During the past 6 years, the Cystic Fibrosis Genetics Research Group has been funded generously by the Cystic Fibrosis Research Trust (UK) and by the Medical Research Council. We wish to thank all of our colleagues who have helped in this group effort, the clinicians and scientists in other Departments with whom we have collaborated, and the CF families for their understanding and patience when faced with repeated requests for samples. Dr. X. Estivill is a visiting research fellow from the Hospital de Sant Pau, Autonomous University of Barcelona and Dr. H.Y. Law is a visiting research fellow from the Institute of Cellular and Molecular Biology, University of Singapore.

REFERENCES

Berent, S.L., M. Mahmoudi, R.M. Torczynski, P.W. Bragg, and A.P. Bollon. 1985. Comparison of oligonucleotide and long DNA fragments as probes in DNA and RNA dot, Southern, Northern, colony and plaque hybridizations. *Biotechniques* **3:** 208.

Botstein, D., R.L. White, M. Skolnick, and R.W. Davies. 1980. Construction of a genetic linkage map using restriction fragment length polymorphisms. *Am. J. Hum. Genet.* **32:** 314.

Carle, G., M. Frank, and M. Olsen. 1986. Electrophoretic separations of large DNA molecules by periodic inversion of the electric field. *Science* **232:** 65.

Crampton, J.M., K.E. Davies, and R. Williamson. 1981. The occurrence of families of repetitive sequences in a library of cloned cDNA from human lymphocytes. *Nucleic Acids Res.* **9:** 3821.

Davies, K.E., C. Gilliam, and R. Williamson. 1983a. Cystic fibrosis is not caused by a defect in the gene coding for human complement three. *Mol. Biol. Med.* **1:** 185.

Davies, K.E., B.D. Young, R.G. Elles, M.E. Hill, and R. Williamson. 1981. Cloning of a representative genomic library of the human X chromosome after sorting by flow cytometry. *Nature* **293:** 374.

Davies, K.E., P.L. Pearson, P.S. Harper, J.M. Murray, T. O'Brien, M. Sarfarazi, and R. Williamson. 1983b. Linkage analysis of two cloned DNA sequences flanking the Duchenne muscular dystrophy locus on the short arm of the human X chromosome. *Nucleic Acids Res.* **11:** 2303.

Edwards, J.H., J.A. Johasson, and N.L. Blackwell. 1984. Locus for cystic fibrosis. *Lancet* **I:** 1020.

Eiberg, H., J. Mohr, K. Schmiegelow, L.S. Nielsen, and R. Williamson. 1985a. Linkage relationships of paraoxonase (PON) with other markers: Indication of PON-cystic fibrosis synteny. *Clin. Genet.* **28:** 265.

Eiberg, H., K. Schmiegelow, L.-C. Tsui, M. Buchwald, E. Niebuhr, P.D. Phelan, R. Williamson, W. Warwick, C. Koch, and J. Mohr. 1985b. Cystic fibrosis, linkage with PON. *Cytogenet. Cell Genet.* **40:** 623.

Farrall, M., P.J. Scambler, K.W. Klinger, K.A. Davies, C. Worrall, R. Williamson, and B.J. Wainwright. 1986a. Cystic fibrosis carrier detection using a linked gene probe. *J. Med. Genet.* **23:** 295.

Farrall, M., H.-Y. Law, C.H. Rodeck, R. Warren, P. Stanier, M. Super, W. Lissens, P. Scambler, E. Watson, B. Wainwright, and R. Williamson. 1986b. First trimester prenatal diagnosis of cystic fibrosis with linked DNA probes. *Lancet* **I:** 1402.

Fisher, E.M.C., J.S. Cavanna, and S.D.M. Brown. 1985. Microdissection and microcloning of the mouse X chromosome. *Proc. Natl. Acad. Sci.* **82:** 5846.

Goodchild, M.C. and J.A. Dodge, eds. 1985. *Cystic fibrosis*, 2nd ed. Bailliere Tindall, Eastbourne, England.

Goss, S.J. and H. Harris. 1977. Gene transfer by means of cell fusion. *J. Cell Sci.* **25:** 17.

Hartley, D.A., K.E. Davies, D. Drayna, R.L. White, and R. Williamson. 1984. A cytological map of the human X chromosome—Evidence for non-random recombination. *Nucleic Acids Res.* **12:** 5277.

Jeffreys, A.J. 1979. DNA sequence variants in $^{G}\gamma$-, $^{A}\gamma$-, δ- and β-globin genes of man. *Cell* **18:** 1.

Kaback, M., D. Zippin, P. Boyd, and R. Canton. 1984. Attitude toward prenatal diagnosis of cystic fibrosis among parents of affected children. In *Cystic fibrosis: Horizons* (ed. D. Lawson), p. 15. J. Wiley, New York.

Kan, Y.W. and A.M. Dozy. 1978. Antenatal diagnosis of sickle cell anaemia by DNA analysis of amniotic fluid cells. *Lancet* **II:** 910.

Kingston, H.M., N.S.T. Thomas, P.L. Pearson, M. Sarfarazi, and P.S. Harper. 1983. Genetic linkage between Becker muscular dystrophy and a polymorphic DNA sequence on the short arm of the X chromosome. *J. Med. Genet.* **20:** 255.

Little, P.F.R. and S.H. Cross. 1985. A cosmid vector that facilitates restriction enzyme mapping. *Proc. Natl. Acad. Sci.* **82:** 3159.

Little, P.F.R., G. Annison, S. Darling, R. Williamson, L. Camba, and B. Modell. 1980. Model for antenatal diagnosis of β-thalassaemia and other monogenic disorders by molecular analysis of linked DNA polymorphisms. *Nature* **285:** 144.

Mayo, B.J., R.J. Klebe, D.R. Barnett, B.J. Lankford, and B.H. Bowman. 1980. Somatic cell genetic studies of the cystic fibrosis mucociliary inhibitor. *Clin. Genet.* **18:** 379.

Murray, J.C., C.M. Demopulos, R.M. Lawn, and A.G. Motulsky. 1983. Molecular genetics of serum albumin. *Proc. Natl. Acad. Sci.* **80:** 5951.

Murray, J.M., K.E. Davies, P.S. Harper, L. Meredith, C.R. Mueller, and R. Williamson. 1982. Linkage relationship of a cloned DNA sequence on the short arm of the X chromosome to Duchenne muscular dystrophy. *Nature* **300:** 69.

Porteous, D.J., J.E.N. Morten, G. Cranston, J.M. Fletcher, A. Mitchell, V. van Heyningen, J.A. Fantes, P.A. Boyd, and N.D. Hastie. 1986. Molecular and physical arrangements of human DNA in H*ras* 1-selected chromosome mediated transfectants. *Mol. Cell. Biol.* **6:** 2223.

Scambler, P.J., B.J. Wainwright, R. MacGillivray, M.R. Fung, and R. Williamson. 1986a. Exclusion of human chromosome 13q34 as the site of the cystic fibrosis mutation. *Am. J. Hum. Genet.* **38:** 567.

Scambler, P.J., B.J. Wainwright, E. Watson, G. Bates, R. Williamson, and M. Farrall. 1986b. Isolation of a further anonymous informative DNA sequence from chromosome seven closely linked to cystic fibrosis. *Nucleic Acids Res.* **14:** 1951.

Scambler, P., M. Farrall, P. Stanier, G. Bell, K. Ramirez, B. Wainwright, J. Bell, N.J. Lynch, K. Kruyer, and R. Williamson. 1985. Linkage of COL1A2 collagen gene to cystic fibrosis and its clinical implications. *Lancet* **II:** 1241.

Schwartz, D. and C. Cantor. 1984. Separation of yeast chromosome-sized DNAs by pulsed field gel electrophoresis. *Cell* **34:** 67.

Scott, M.R.D., K.H. Westphal, and P.W.J. Rigby. 1983. Activation of mouse genes in transformed cells. *Cell* **34:** 557.

Solomon, E. and W.F. Bodmer. 1979. Evolution of a sickle variant gene. *Lancet* **I 28:** 923.

Sturtevant, A.H. 1913. The linear arrangement of six sex linked factors in *Drosophila*, as shown by their mode of association. *J. Exp. Zool.* **14:** 43.

Tsui, L.-C., M. Buchwald, D. Barker, J.C. Braman, R. Knowlton, J.W. Schumm, H. Eiberg, J. Mohr, D. Kennedy, N. Plasvsic, D. Markiewicz, G. Akots, V. Brown, C. Helms, T. Gravius, C. Parker, K. Rediker, and H. Donis-Keller. 1985. Cystic fibrosis locus defined by a genetically linked polymorphic DNA marker. *Science* **230:** 1054.

Wainwright, B.J., P.J. Scambler, J. Schmidtke, E.A. Watson, H.-Y. Law, M. Farrall, H.J. Cooke, H. Eiberg, and R. Williamson. 1985. Localization of cystic fibrosis locus to human chromosome 7cen-q22. *Nature* **318:** 384.

White, R., S. Woodward, M. Leppert, P. O'Connell, M. Hoff, J. Herbst, J.-M. Lalouel, M. Dean, and G. Vande Woude. 1985. A closely linked genetic marker for cystic fibrosis. *Nature* **318:** 382.

Whitehead, A.S., E. Solomon, S. Chambers, W.F. Bodmer, S. Povey, and G. Fey. 1982. Assignment of the structural gene for the third component of human complement to chromosome 19. *Proc. Natl. Acad. Sci.* **79:** 5021.

Williamson, R. 1985. Towards a total human gene map. *Royal Institution Proceedings* **57:** 45.

Williamson, R., J. Eskdale, D.V. Coleman, M. Niazi, F.E. Loeffler, and B.M. Modell. 1981. Direct gene analysis of chorionic villi. *Lancet* **II:** 1126.

Wilson, G.B. and V.J. Bahm. 1980. Synthesis and secretion of cystic fibrosis ciliary dyskinesia substances by purified subpopulations of lymphocytes. *J. Clin. Invest.* **66:** 1010.

Woods, D., J. Crampton, B. Clarke, and R. Williamson. 1980. The construction of a recombinant cDNA library representative of the poly(A)$^{+}$ mRNA population from normal human lymphocytes. *Nucleic Acids Res.* **8:** 5157.

Highly Polymorphic RFLP Probes as Diagnostic Tools

H. DONIS-KELLER, D.F. BARKER, R.G. KNOWLTON, J.W. SCHUMM, J.C. BRAMAN, AND P. GREEN

Department of Human Genetics, Collaborative Research, Inc., Lexington, Massachusetts 02173

Diagnosis in advance of symptoms is very useful for disease prevention and as a guide for instituting effective therapeutic regimens. In principle, linkage methods should become applicable to virtually any inherited disorder, given an adequate supply of genetic markers. Among such genetic markers are restriction-fragment-length polymorphisms (RFLPs) (Botstein et al. 1980), regions of variation of DNA sequence among individuals, which are commonly detected using radioactively labeled cloned human DNA probes on gel transfers of human genomic DNA cleaved with restriction enzymes. These RFLP loci behave as codominant Mendelian alleles and thus can be used in family studies to trace the inheritance of traits that cosegregate with a RFLP locus, even though almost nothing else may be known about the trait. For example, RFLP probes are currently used in the presymptomatic diagnosis of disorders such as Duchenne muscular dystrophy (Kunkel et al. 1985), hemophilia A and B (Giannelli et al. 1984; Gitschier et al. 1985), sickle cell anemia (Orkin et al. 1982), and thalassemia (Boehm et al. 1983). In addition, RFLP loci linked to the loci for cystic fibrosis (CF) (Tsui et al. 1985; Wainwright et al. 1985; White et al. 1985), Huntington's disease (Gusella et al. 1983), and adult polycystic kidney disease (Reeders et al. 1985) have been found, opening the way to diagnostic testing and possibly to the identification and characterization of the genes that cause these disorders.

To follow the inheritance of the disease genes in a family, it is necessary that the linked polymorphic loci be sufficiently informative to allow all four parental chromosomes to be distinguished. RFLPs with a large number of alleles evenly distributed in the population are ideal for this purpose. Two allele markers with low polymorphism information content (PIC; Botstein et al. 1980) are acceptable, if inefficient, for general mapping purposes but are of limited diagnostic use because they are often not informative in small families; however, several such markers closely linked to the disease locus may be informative as a single highly polymorphic marker. Even with a marker locus located close to the disease locus, errors in diagnosis may occur due to meiotic recombination events between the two loci. Such events can be detected by the use of genetic markers that flank the disease gene.

In this paper, we describe the identification of highly polymorphic RFLP loci and their application to genotyping in humans and to mapping the *CF* gene to chromosome 7. We also report the construction of a high-resolution genetic map of chromosome 7 and summarize progress toward the development of a presymptomatic diagnostic test for CF that should be useful in virtually every case.

RESULTS

Identification of Highly Polymorphic RFLP Loci

Although more than 300 polymorphisms have been identified and localized to chromosomes (Willard et al. 1985), the vast majority are simple two-allele systems that are not particularly informative in linkage studies. Although such probes can be used to construct a genetic map, a more productive approach (especially when one considers eventual diagnostic applications) is to identify highly polymorphic RFLP probes randomly distributed throughout the genome. About 150 fully informative RFLP loci evenly spaced at 20-cM intervals will span the human genome (Botstein et al. 1980). With this number as a minimum estimate of highly informative RFLP loci required to build a useful genetic map, we initiated a large-scale search for new RFLPs. A rapid and efficient screening procedure was developed that preferentially selects highly polymorphic RFLPs. Linkage relationships among the various loci were determined by following their inheritance in three-generation families with large sibships, it being much more efficient to localize just one member of a linkage group to a chromosome by physical methods than to determine the chromosomal origin of every probe in advance of testing for polymorphism or linkage.

We utilized a bacteriophage λ human genomic library (Lawn et al. 1978) as a source of cloned DNA to test for polymorphism because the insert size was relatively large (10–20 kb) and because these clones could be used directly (i.e., without needing to purify the insert or subclone) in tests for polymorphism. Clones that failed to hybridize with nick-translated total genomic DNA in plaque-filter hybridization experiments were selected as candidate single-copy sequences. These constitute about 1% of the library (Wyman and White 1980). We were able to use clones in the remaining 99% of the library by prehybridizing them to total genomic DNA, which "blocks" the repeat sequences but leaves the unique sequence DNA available to be tested for polymorphism (Litt and White 1985; J.W. Schumm et al., in prep.). This permits virtually all unique sequence regions within the library to be screened. To diminish the chance of picking up rare polymorphisms, our standard screening blot consisted of only five unrelated individuals, digested with each of six restriction enzymes. Nylon membranes were chosen as the solid support for Southern hybridization, since they could be reused as many as ten times.

Over 500 polymorphisms were identified from about 1700 clones tested in the study. To date, about 200 have been characterized and have PIC values above 0.3. As Figure 1 shows, 94 RFLPs have PIC values between 0.3 and 0.5, 38 are in the range of 0.5–0.7, and 29 have PIC values greater than 0.7 with a single enzyme. Thus, about one third of the clones surveyed were found to reveal polymorphism, and of these, one in seven was found to have a PIC value of greater than 0.5.

Application of Highly Polymorphic RFLP Probes to Genotyping

Until quite recently, the best method for distinguishing individuals on the basis of inherited characteristics was the use of histocompatibility cell-surface antigens (HLA) and red blood cell antigens. The HLA system is highly polymorphic but of limited use in distinguishing between siblings because its genes lie in a single region of chromosome 6. Highly polymorphic DNA probes offer an even better method for distinguishing even closely related individuals because loci from a variety of chromosomes can be examined. Jeffreys et al. (1985) have also used DNA probes for genotyping; their approach differs in that they simultaneously look at multiple loci by using probes that hybridize to families of tandem repeat sequences. We have applied our most highly polymorphic RFLP probes to the problem of determining the origin of hematopoietic cells in recipients of bone-marrow transplants.

In most cases, candidates for bone-marrow transplantation are matched as closely as possible to donors (usually siblings) who share the same HLA haplotype, which makes it difficult to determine the donor or recipient origin of hematopoietic cells following transplantation. It is important to make this distinction to monitor the course of engraftment. We find that using our most highly polymorphic probes, we can distinguish donor from recipient cells in better than 99% cases and can routinely detect as little as 1–5% of one cell population mixed with the other (Knowlton et al. 1986). Prior to transplantation, we test a panel of probes against genomic DNA purified from blood of the donor and recipient in gel-transfer experiments. An RFLP probe–enzyme combination is then selected that clearly separates one or more donor-specific fragments from one or more recipient-specific restriction fragments, and this probe is then used in posttransplant follow ups. Figure 2 shows a case in which both donor and recipient cells are present following engraftment. Initially, the patient showed the presence of only donor cells; however, at 1 year posttransplant, approximately 50% of cells were recipient and 50% donor. This example shows how RFLP probes simplify and quanti-

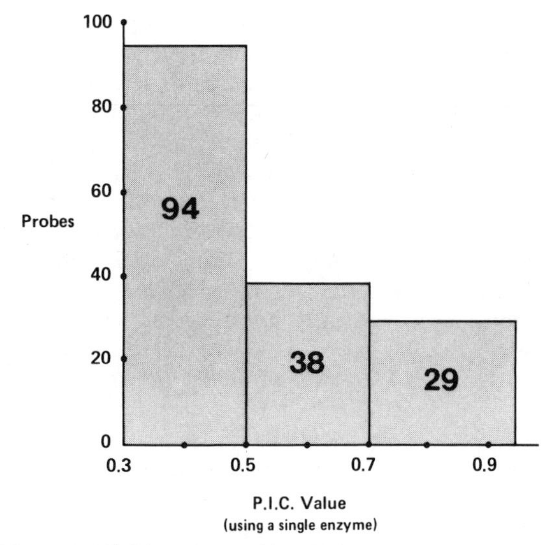

Figure 1. Highly polymorphic RFLP probes. RFLP probes characterized thus far from screening random genomic clones for polymorphism are indicated (see Table 1). PIC (Botstein et al. 1980) values are based on a single restriction endonuclease.

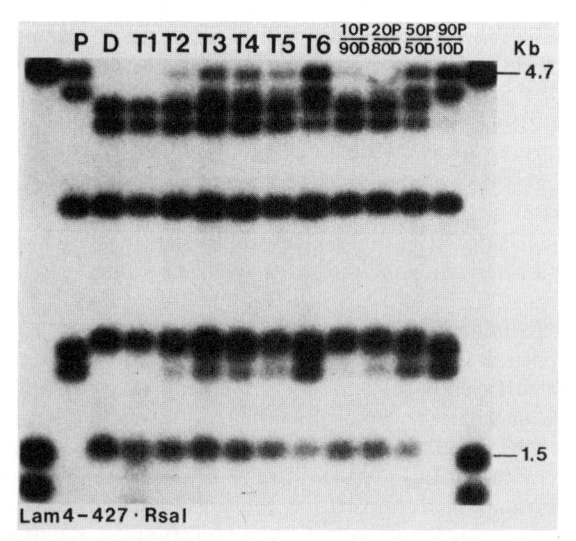

Figure 2. Monitoring bone-marrow engraftment with RFLP probes: Lympho-hematopoietic chimerism. Engraftment of donor marrow in the patient is followed by reappearance of cells of patient origin. Probe Lam4-427 reveals polymorphism with the restriction endonuclease *Rsa*I. (P) Pretransplant patient DNA; (D) donor DNA; (T1) cultured peripheral blood, 3 months after bone-marrow transplant; (T2–T5) peripheral blood at 4, 7, 9, and 12 months after bone-marrow transplant; (T6) bone marrow at 12 months after bone-marrow transplant. All lanes contain 4 μg of DNA. Proportions of cells of patient and donor genotype in samples showing mixed chimerism are estimated by comparison with mixtures of known composition in adjacent lanes. (Reprinted, with permission, from Knowlton et al. 1986.)

tate the genotyping of blood-forming tissue. The use of single-copy probes allows straightforward quantitation at the minor cost of having to screen (before transplantation) several probes in place of just one. These RFLP probes provide a useful tool for studying the biology of leukemia and engraftment. In particular, they should aid greatly the further study of the role of chimerism in graft versus host disease and in survival after bone-marrow transplantation.

The probes can be used in other genotyping applications, e.g., paternity exclusion (Kazazian et al. 1986) and even inclusion to a nearly unlimited degree of certainty. Figure 3 demonstrates the use of RFLP probes to confirm paternity in a case of apparent spontaneous mutation to thalassemia. Maternal-specific and paternal-specific restriction fragments were identified, and allele frequencies in the general population were used to determine the probability that the pattern seen could have resulted from paternity by a random individual in the population. The probability in this particular case was calculated to be 2×10^{-9}. Table 1 shows 17 RFLP probes that were used for this paternity determination.

Mapping the *CF* Locus to Chromosome 7

We became interested in mapping the *CF* locus by genetic linkage for the following reasons. First, the autosomal recessive mode of inheritance had been well established, although differences in the severity of the disease among individuals suggested the possibility that CF could be caused by any one of several loci (Thompson 1980). Second, CF is very common in Caucasian populations (1 in 2000 births; Talamo et al. 1983) and is, in general, a lethal disorder, although improved therapies have extended life expectancy into early adulthood for many. Third, families with two or more CF-affected individuals were available (Tsui et al., this volume) for use in linkage studies using DNA from cell lines or directly purified from blood samples. Finally, we calculated a greater than 85% chance of detecting an RFLP linked to the *CF* locus by testing each random probe in our collection against the CF family collection.

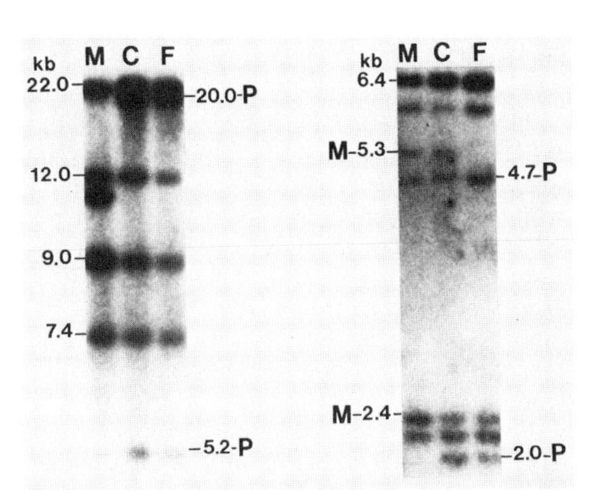

Figure 3. Confirmation of paternity by genotypic analysis with RFLP probes. DNA samples from the mother (M), affected child (C), and father (F) were digested and tested with each of the probes listed in Table 1. (*Left*) *Bgl*II-digested DNAs hybridized to Lam4-1214. (*Right*) *Pst*I-digested samples probed with Lam4-159. The child inherited the 4.7- and 2.0-kb fragments from her father and the 5.3- and 2.4-kb fragments from her mother. (M) Known maternal fragments; (P) known paternal fragments. (Reprinted, with permission, from Kazazian et al. 1986.)

Genetic Linkage of the First DNA Marker to the *CF* Locus

A number of CF families with two or more CF-affected individuals collected by L.C. Tsui and M. Buchwald (Tsui et al., this volume) were used in a collaborative study to find the first linkage to the *CF* gene (Tsui et al. 1985). This first linked RFLP (called D0CRI-917) was found to be about 15 cM from the *CF* locus and about 5 cM from the paraoxonase (*PON*) locus, a non-DNA marker that had previously been found to be linked to CF. Immediately thereafter, we showed that D0CRI-917 is on human chromosome 7 (Knowlton et al. 1985).

Mapping Chromosome 7 and High-resolution Mapping of the *CF* Locus with RFLPs

It was clear that the *917* locus was too far away from the *CF* locus to be useful in a diagnostic test, and, in any case, we wanted to find markers on both sides of the gene. Three DNA probe resources were available to us: (1) probes for polymorphisms already known to originate from chromosome 7, (2) our large collection

Table 1. Polymorphic DNA Probes for Genotypic Analysis of a Child and Her Parents

Probe	Enzyme	PIC	Frequency of potential paternal alleles
Lam4-45	*Msp*I	0.60	0.34
Lam4-159	*Pst*I	0.90	0.06
Lam4-355	*Bgl*II	0.90	0.04
Lam4-368	*Hin*dIII	0.58	0.23
Lam4-281	*Msp*I	0.70	0.20
Lam4-1020	*Taq*I	0.80	0.37
Lam4-1214	*Bgl*II	0.94	0.17
RL4-265	*Bgl*II	0.75	0.25
RL4-392	*Bgl*II	0.65	0.43
Lam4-123	*Rsa*I	0.81	0.69
Lam4-171	*Msp*I	0.70	0.11
Lam4-427	*Rsa*I	0.95	0.69
Lam4-605	*Msp*I	0.70	0.53
Lam4-966	*Rsa*I	0.88	0.65
Lam4-1065	*Rsa*I	0.75	0.71
Lam4-1209	*Taq*I	0.70	0.93
Lam4-962	*Msp*I	0.70	0.66

For each of the polymorphic loci, those alleles that could account for the restriction fragment patterns seen in the affected child (given the mother's genotype) were classified as potential paternal alleles. The sum of the frequencies of all the potential paternal alleles is then the probability that an allele from a random individual would give the same pattern as the one observed, and the probability that alleles at all 17 loci would be consistent with this pattern is the product of these values, 2×10^{-9}. Data from Kazazian et al. (1986).

of random polymorphisms, and (3) a chromosome-7 library from the collection of libraries established by the researchers at the Los Alamos laboratory (Van Dilla et al. 1986). We and a number of other researchers began testing the markers known to reside on chromosome 7, and results were quickly obtained that showed four RFLPs linked to the *CF* locus: *TCRB* (T-cell receptor β-chain), *COL1A*, (α1 collagen), *met* (met oncogene), and *pJ3-11*. The latter two were found to be very close to the locus (probably within 5 cM or 5 million base pairs and probably on the same side) (Buchwald et al. 1986; Wainwright et al. 1985; White et al. 1985; Tsui et al., this volume).

Since chromosome 7 constitutes about 5% of the human genome, we estimated that about 25 probes or 5% of our random collection should map there. The overall strategy that we first applied to the random RFLP probe collection was first to check for linkage to *917*, *met*, and *pJ3-11*, using the CEPH (Centre d'Etude du Polymorphisme Humain) families, then to confirm the chromosome-7 assignment of linked probes, using somatic-cell hybrid panels, and finally to characterize further probes that had positive lod scores on CF families. This strategy was repeated with polymorphic probes isolated from the chromosome-7 library. In this way, we expected to construct a high-resolution map of the *CF* locus and a fairly detailed map of chromosome 7 that would be of general utility.

Because the chromosome-7 library contained mainly small inserts (less than 10 kb), we fractionated the library by a CsCl gradient and selected the fraction with the largest clones (95% were >4 kb) for further study. The clones were first sorted to distinguish those with human single-copy sequences from those with hamster DNA, by plaque-filter hybridization with total human genomic DNA and total hamster genomic DNA (a hamster-human cell hybrid was used in making the library). Candidate single-copy and repeat-containing clones were then tested, as previously described, for their ability to reveal polymorphism. Once candidate polymorphisms had been identified, we confirmed the chromosome-7 assignment and tested for linkage relationships using the CEPH families. All possible pairwise comparisons of our set of chromosome-7 markers were performed.

Our map-making progress to date is shown in Figure 4. A total of 40 markers have been placed on this preliminary genetic map of chromosome 7; 36 are new markers discovered in our search for CF-linked markers. The map consists of 64 pairwise scores with lod scores greater than 5.0 for all but three linkages. Approximately 15 markers are thought to lie within 15 cM of the *CF* locus. The map is continuous except for one linkage group (S-161–S-194–L4-966–L4-281) that maps to chromosome 7 by hybrid panel analysis (not shown) but is as yet genetically unlinked to other markers on chromosome 7. The total genetic distance is estimated to be about 130 cM (the markers from S-003 to S-166 constitute a distance of about 120 cM and the unlinked group, S-161–L4-281, is about 10 cM in length). Linkage analysis results using two-point comparisons with CF-affected families are consistent with respect to map order and spacing between the genetic markers with the data from the panel of three-generation reference families from the CEPH collection as shown in Figure 4. Since chromosome 7 is estimated to be about 136 cM (Morton et al. 1982), we believe that the chromosome is fairly well covered. Eight probes from the random collection were found to be linked to other markers on chromosome 7; one marker, Lam4-1033, shows no recombinants in CF families with other markers known to be tightly linked to CF. The remaining 28 markers came from the chromosome-7 library.[1]

Several ambiguities regarding probe order and genetic distances remain. In addition, approximately ten more markers that are candidates for linkage on chromosome 7 remain to be fully characterized. We are currently working to resolve the mapping ambiguities in three ways. First, we are characterizing the markers on further CEPH and CF families to generate more overlapping linkage data. Second, we are working on increasing the informativeness of those markers near the *CF* gene by chromosome walking or "locus-expansion" techniques. Cosmid libraries have been constructed that will be checked for clones homologous to the *CF* locus markers. Third, new methods for analyzing data that can handle multipoint linkage questions (e.g., as many as 50 markers need to be ordered on the chromosome-7 map) are required. In collaboration with E. Lander and his colleagues (Massachusetts Institute of Technology), we are developing such methods. Thus far, we are able to compare all possible orders of a set of eight to ten phase-known markers. The development of these new analytical techniques will be described by E. Lander and P. Green (in prep.) and in this volume (see Lander and Botstein). One feature of the chromosome-7 map of particular importance to diagnosis of CF is that several markers are near *CF*, but further to the left of the closest CF-linked markers (*met* and *pJ3-11*) than *CF* is likely to be. This means that some (if not all) of these flank *CF*. As described below, these near-flanking markers could become, when expanded to become more informative, important components of a truly precise diagnostic test.

In summary, we have identified more than 40 RFLP markers that originate from chromosome 7 and have developed a preliminary map using two-point analysis. It is clear that we have a number of candidate RFLPs that not only flank the *CF* gene, but are apparently very close to it, perhaps as close as if not closer than *pJ3-11* and *met*. In fact, one or more of these markers may be part of the locus itself. Current plans are to construct a physical map of the region surrounding the *CF*

[1]Interestingly, one of the most highly polymorphic markers we have ever identified was found in the chromosome-7 library, although it was later determined to be homologous to the X and Y chromosomes (R.G. Knowlton et al., in prep.) and did not originate from chromosome 7. The hybrid cell line from which the chromosome-7 library was constructed was known to contain the X chromosome in earlier passages of the line.

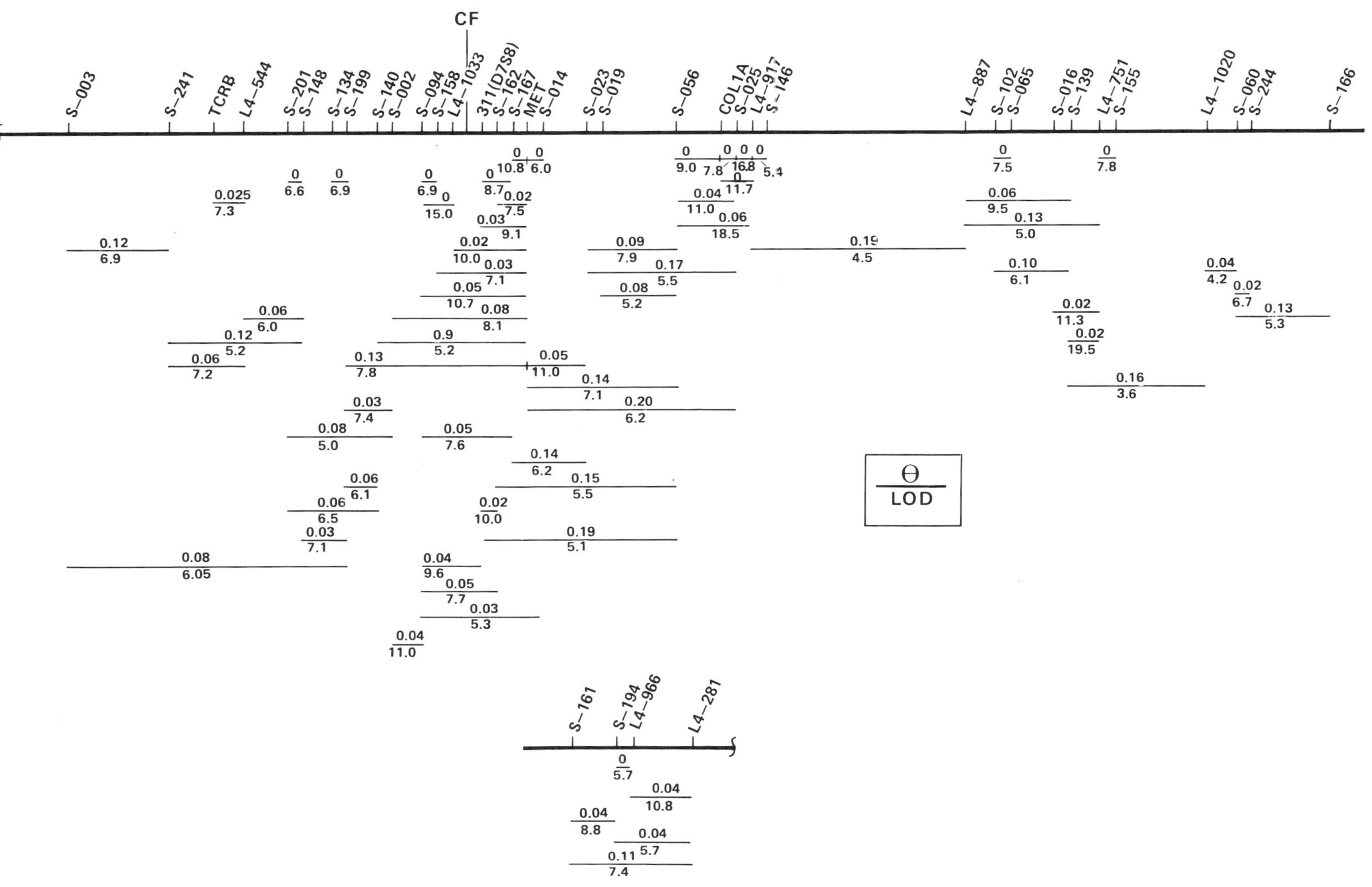

Figure 4. Genetic linkage map of chromosome 7. Two-point linkage calculations were performed for 40 genetic markers. For the 64-pair scores, all lods are greater than 5.0. The total genetic distance is estimated to be about 130 cM. Recombination fraction, Θ, is shown above each lod score for the pairwise comparison. A four-member linkage group (probes S-161 to L4-281), although confirmed as originating from chromosome 7, is not yet linked to other markers on the chromosome-7 map.

locus that can be compared with the genetic map of the same region. This may help to identify the *CF* gene or at least to delimit its extent more precisely. The chromosome-7 map can also be used to exclude other simply inherited disorders such as neurofibromatosis in genetic mapping studies.

Presymptomatic Diagnostic Testing for CF with RFLPs

Criteria for a useful linkage test. The usefulness of an RFLP-linkage test for presymptomatic diagnostic testing depends on (1) the informativeness of the RFLP loci, i.e., the probability that all four parental chromosomes can be distinguished in any offspring, and (2) the genetic distance between the RFLP locus and the disease locus (the greater the distance between the two, the more likely a recombination event [or crossover] will occur during meiosis, thereby uncoupling the disease allele from the RFLP allele present in the parent as the chromosomes are passed on to the children). In addition, the error rate of such tests can be dramatically reduced when genetic markers that bracket the disease locus are used to track inheritance of disease alleles.

Informativeness of an RFLP locus is measured as the probability that in a random mating, the parental origin of the RFLP alleles can be determined in the offspring. This is quantified as the PIC value (Botstein et al. 1980); this differs from the heterozygosity estimate, which is just the frequency of heterozygotes in a population. For example, a two-allele system with each allele present at a frequency of 0.5 in the population has a heterozygosity of 50% and a PIC value of 0.37. Figure 5 shows what is generally meant by informativeness with a two-allele RFLP marker: The locus is either fully informative, partially informative, or uninformative with respect to unequivocal identification of the genotypes. In making a diagnosis, the partially informative case is often of no utility in determining the status of the individual at risk when the carrier and affected genotypes cannot be distinguished (i.e., when the genotype of the affected individual cannot be distinguished from the carrier or normal genotypes).

In the case of diagnostic testing for CF using the RFLP loci *met* and *pJ3-11*, the prediction of the number of fully informative cases is less than one half, even when *met* and *pJ3-11* are used as a single locus with no linkage disequilibrium. With a combination of RFLP markers tightly linked to the *CF*, *met*, and *pJ3-11* loci, the number of fully informative cases increases to greater than 90%. Table 2 shows the increase in informativeness possible as RFLP markers are added to *met* and *pJ3-11*. The calculation assumes no linkage disequilibrium between the markers.

Single-locus RFLP test versus flanking-marker RFLP test. A single-locus RFLP-linkage test relies on the identity or nonidentity of the at-risk genotype compared with that of the affected individual. Crossovers

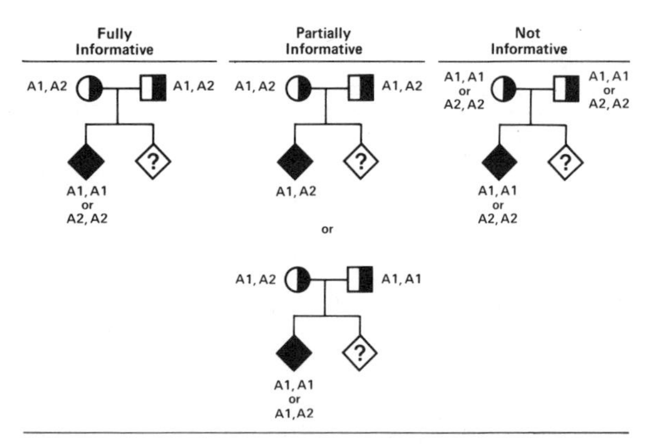

Figure 5. Informativeness with a two-allele RFLP. Three types of informativeness encountered in diagnostic testing are shown for locus *A*, alleles 1 and 2. (○) Carrier mother; (□) carrier father; (♦) CF-affected child; (?) individual at risk.

cannot be detected, and the same error rate prevails each time the test is performed. A flanking-marker test (in which the disease locus is bracketed by RFLP loci) is intrinsically better because in almost all cases, a crossover or recombination event can be detected. Even in the case of a fully informative RFLP that is very close to the disease locus, the error rate of this single-marker test is much higher than with a flanking-marker test. Figures 6 and 7 illustrate the improvement of the accuracy of diagnosis using flanking markers compared with that of a single fully informative marker 1 cM (1% recombination) from a disease locus. In this example, RFLP marker *A* is 1 cM from the disease locus. As the alleles are passed on to the offspring during meiosis, there is a 1% chance of a crossover between locus *A* and the disease locus for the two chromosomes that are passed on to the affected individual and a 1% chance of a crossover in each of the two chromosomes passed on to the individual at risk. Therefore, each time a diagnosis is done, there is a 4% chance of error. However, with flanking markers, the error rate can be substantially reduced because crossover events can be

Table 2. Informativeness of RFLP Markers in Diagnosis

Probe	Allele frequencies	Fully informative (%)	Partially informative (%)	Total (%)
Met	0.35, 0.49, 0.16	29.7	47.5	77.2
+3-11	0.40, 0.60	48.8	41.6	90.3
+S-094	0.48, 0.52	63.7	32.0	95.7
+S-002	0.64, 0.36	73.7	24.3	97.9
+L-1033	0.46, 0.54	81.0	18.0	99.0
+S-162	0.31, 0.69	85.9	13.6	99.5
+S-158	0.19, 0.81	89.0	10.7	99.7
+S-167	0.81, 0.19	91.4	8.48	99.8
+S-014	0.90, 0.10	92.6	7.25	99.9

Table entries give the expected percentage of CF families in which all four parental chromosome 7s (fully informative) or only the two chromosomes of one parent (partially informative) can be distinguished using the indicated probe together with all probes above it. It is assumed that all marker loci are in linkage equilibrium with each other and with the *CF* locus.

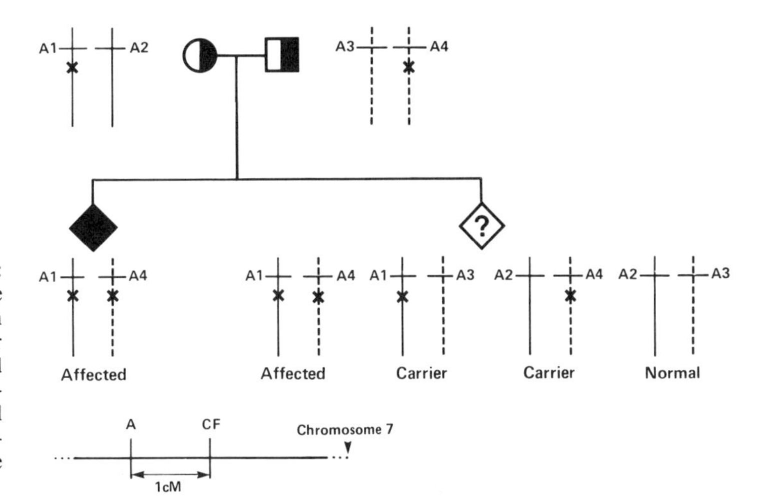

Figure 6. Example of single-marker diagnosis: Marker fully informative. In this example, the RFLP marker is 1 cM (1% recombination) from the *CF* locus. The probability of a crossover event is 1% for each chromosome considered (the affected and at-risk individuals). Therefore, in 4% of the cases, a crossover event will occur (1% × 4 chromosomes), and it will be undetected by RFLP analysis; thus, the error rate is 4%.

detected. In the example shown, the flanking marker is 2 cM on the other side of the disease locus. Therefore, there is a 3% chance of a crossover for each of the four chromosomes under consideration, i.e., a 12% probability of a crossover occurring during meiosis. For 88% of cases in which no crossover is observed, the error rate is reduced to the probability of a double-recombination event occurring between the two markers on a chromosome ([2% × 1%] × 4]) = 0.08%; the test is 99.92% accurate. For the roughly 11% of cases in which a crossover is detected in one parent only, the genotype at the disease locus on the chromosome derived from the other parent can still be determined with 99.96% accuracy. Thus, a diagnosis that this allele is normal (indicating that the individual is unaffected but possibly a carrier) will be possible in roughly half such cases or 5% of the total; in the other half, the allele will be determined (with 99.96% accuracy) to be the disease allele and no diagnosis will be possible. In 1% of cases, crossovers in both parents will be detected.

We are currently completing the development of such a flanking-marker test for the presymptomatic diagnosis of CF. As described above, a number of candidate flanking markers very close to the *CF* locus have been identified. Each of the markers identified is not fully informative at this time. The informativeness of the loci can be improved by expanding the loci by chromosome walking. Phage or cosmid genomic libraries are screened for overlapping clones and each of these clones is tested for polymorphism with a variety of restriction enzymes. In this way, locus haplotypes are identified.

CONCLUSION

We have shown above that highly polymorphic RFLP probes have many uses. It is worth stressing that for mapping purposes, markers with very high informativeness are essential if small families are to be used efficiently. This problem of informativeness is even more crucial when diagnostic tests based on linkage are contemplated. Although it is true that several poorly informative markers can, in combination, substitute for a single highly informative one, the practical advantages continue to make the search for individual probes displaying high informativeness worthwhile.

The experience with chromosome 7 and the *CF* locus illustrates the problem. On the one hand, we have found a number of RFLPs in the CF region and, indeed, they must flank the *CF* locus. However, at the

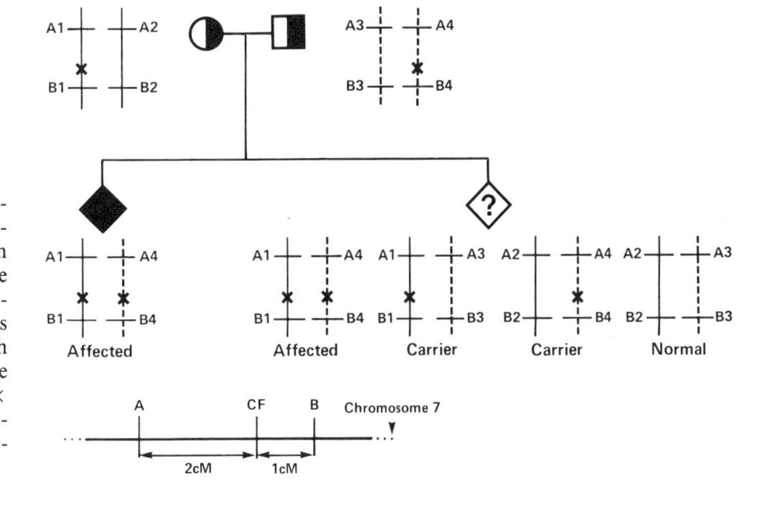

Figure 7. Example of flanking-marker diagnosis: Markers fully informative. In this example, locus *A* is 2 cM and locus *B* is 1 cM from the *CF* gene. Therefore, there is a 3% chance of a crossover between *A* and *B* for each chromosome (3% × 4 chromosomes) = 12% of cases will be crossovers. For the 88% of cases with no crossovers, the error rate is 0.08% (i.e., the probability of a double crossover is 0.02 × 0.01 = 0.02% for each chromosome × 4 chromosomes = 0.08%). Thus, the test is 99.92% accurate.

present time, only the aggregate of all the RFLPs taken together really suffices to make a *fully* informative diagnostic test applicable to virtually every family. Considerable further effort is being put into increasing the informativeness of the individual markers in the CF region so as to allow more precise resolution of their positions relative to the *CF* locus and to make them individually useful in a *fully* informative diagnostic test using flanking markers.

ACKNOWLEDGMENTS

We are grateful for the continuing collaboration of Manuel Buchwald, Jean Frezal, Haig Kazazian, Lawrence Petz, Jerome Ritz, Thomas Shows, Lap-Chee Tsui, Huntington Willard, and their groups. We acknowledge the invaluable assistance of CEPH in providing DNAs of three-generation reference families. We thank the scientists in our group (Gita Akots, Valerie Brown, Tom Gravius, Cindy Helms, Chris Nelson, Carol Parker, Joan Thurston, and Ken Rediker) for their enthusiastic and skillful assistance. We also thank David Botstein for his comments on the manuscript and Ronald Davis and Eric Lander for discussions.

REFERENCES

Boehm, C.D., S.E. Antonarakis, J.A. Phillips, G. Stetten, and H.H. Kazazian, Jr. 1983. Prenatal diagnosis using DNA polymorphisms. *N. Engl. J. Med.* **308:** 1054.

Botstein, D., R. White, M. Skolnick, and R. Davis. 1980. Construction of a genetic linkage map in man using restriction fragment length polymorphisms. *Am. J. Hum. Genet.* **32:** 314.

Buchwald, M., M. Zsiga, D. Markiewicz, N. Plavsic, D. Kennedy, S. Zengerling, H.F. Willard, P. Tsipouras, K. Schmiegelow, M. Schwartz, H. Eibert, J. Mohr, D. Barker, H. Donis-Keller, and L.-C. Tsui. 1986. Linkage of cystic fibrosis to the pro-α2(I) collagen gene, COL1A2, on chromosome 7. *Cytogenet. Cell Genet.* **41:** 234.

Eiberg, H., J. Mohr, K. Schmiegelow, L.S. Nielsen, and R. Williamson. 1985. Linkage relationships of paraoxonase (PON) with other markers: Indication of PON-cystic fibrosis synteny. *Clin. Genet.* **28:** 265.

Giannelli, F., K.H. Choo, P.R. Winship, C.R. Rizza, D.S. Anson, D.J.G. Rees, N. Ferrari, and G.G. Brownlee. 1984. Characterization and use of an intragenic polymorphic marker for detection of carriers of haemophilia B (Factor IX deficiency). *Lancet* **I:** 241.

Gitschier, J., D.E.G. Drayna, D. Tuddenham, R.L. White, and R.M. Lawn. 1985. Genetic mapping and diagnosis of haemophilia A achieved through a *BclI* polymorphism in the factor VIII gene. *Nature* **314:** 738.

Gusella, J.F., N.S. Wexler, P.M. Conneally, S.L. Naylor, M.A. Anderson, R.E. Tanzi, P.C. Watkins, K. Ottina, M.R. Wallace, A.Y. Sakaguchi, A.B. Young, I. Whoulson, E. Bonilla, and J.B. Martin. 1983. A polymorphic DNA marker genetically linked to Huntington's disease. *Nature* **306:** 234.

Jeffreys, A.J., V. Wilson, and S.L. Thein. 1985. Hypervariable "minisatellite" regions in human DNA. *Nature* **314:** 67.

Kazazian, H.H., Jr., S.H. Orkin, C.D. Boehm, S.C. Goff, C. Wong, C.E. Dowling, P.E. Newburger, R.G. Knowlton, V. Brown, and H. Donis-Keller. 1986. Characterization of a spontaneous mutation to a B-thalassemia allele. *Am. J. Hum. Genet.* **38:** 860.

Knowlton, R.G., V.A. Brown, J.C. Braman, D. Barker, J.W. Schumm, C. Murray, T. Takvorian, J. Ritz, and H. Donis-Keller. 1986. Use of highly polymorphic DNA probes for genotypic analysis following bone marrow transplantation. *Blood* **68:** 378.

Knowlton, R.G., O. Cohen-Haguenauer, N. Van Cong, J. Frezal, V.A. Brown, D. Barker, J.C. Braman, J.W. Schumm, L.-C. Tsui, M. Buchwald, and H. Donis-Keller. 1985. A polymorphic DNA marker linked to cystic fibrosis is located on chromosome 7. *Nature* **318:** 380.

Kunkel, L.M., A.P. Monaco, W. Middlesworth, H.D. Ochs, and S.A. Latt. 1985. Specific cloning of DNA fragments absent from the DNA of a male patient with an X-chromosome deletion. *Proc. Natl. Acad. Sci.* **82:** 4778.

Lawn, R., E. Fritsch, R. Parker, G. Blake, and T. Maniatis. 1978. The isolation and characterization of linked δ and β globin genes from a cloned library of human DNA. *Cell* **15:** 1157.

Litt, M. and R. White. 1985. A highly polymorphic locus in human DNA revealed by cosmid derived probes. *Proc. Natl. Acad. Sci.* **82:** 6206.

Morton, N.E., J. Lindsten, L. Isellus, and S. Yee. 1982. Data and theory for a revised chiasma map of man. *Hum. Genet.* **62:** 266.

Orkin, S.H., P.F.R. Little, H.H. Kazazian, Jr., and C.D. Boehm. 1982. Improved detection of the sickle mutation by DNA analysis. *N. Engl. J. Med.* **307:** 32.

Reeders, S.T., M.H. Bruening, K.E. Davies, R.D. Nicholls, A.P. Jarman, D.R. Higgs, P.L. Pearson, and D.J. Weatherall. 1985. A highly polymorphic DNA marker linked to adult polycystic kidney disease on chromosome 16. *Nature* **317:** 542.

Talamo, R.C., B.J. Rosenstern, and R.W. Berninger. 1983. Cystic fibrosis. In *The metabolic basis of inherited disease* (ed. J.B. Stanbury et al.), p. 1889. McGraw-Hill, New York.

Thompson, M.W. 1980. Genetics of cystic fibrosis. In *Perspectives in cystic fibrosis* (ed. J.M. Sturgess), p. 281. Canadian Cystic Fibrosis Foundation, Toronto.

Tsui, L.-C., M. Buchwald, D.F. Barker, J.C. Braman, R.G. Knowlton, J.W. Schumm, H. Eiberg, J. Mohr, D. Kennedy, N. Plavsic, M. Zsiga, D. Markiewicz, G. Akots, V. Brown, C. Helms, T. Gravius, C. Parker, K. Rediker, and H. Donis-Keller. 1985. Cystic fibrosis locus defined by genetically linked polymorphic DNA marker. *Science* **230:** 1054.

Van Dilla, M.A., L.L. Deaven, K.L. Albright, N.A. Allen, M.R. Aubuchon, M.F. Bartholdi, N.C. Brown, E.W. Campbell, A.V. Carrano, L.M. Clark, L.S. Cram, B.D. Crawford, J.C. Fuscoe, J.W. Gray, C.E. Hildebrand, P.J. Jackson, J.H. Jett, J.L. Longmire, C.R. Lozes, M.L. Luedemann, J.C. Martin, J.S. McNinch, L.J. Meincke, M.L. Mendelsohn, J. Meyne, R.K. Moyzis, A.C. Munk, J. Perlman, D.C. Peters, A.J. Silva, and B.J. Trask. 1986. Human chromosome-specific DNA libraries: Construction and availability. *Biotechnology* **4:** 537.

Wainwright, B., P. Scambler, J. Schmidtke, E. Watson, H.-Y. Law, M. Farral, H. Cooke, H. Eiberg, and R. Williamson. 1985. Localization of cystic fibrosis locus to human chromosome 7cen-q22. *Nature* **318:** 384.

White, R., S. Woodward, M. Leppert, P. O'Connell, M. Huff, J. Herbst, J.M. Lalouel, M. Dean, and G. Vande Woude. 1985. A closely linked genetic marker for cystic fibrosis. *Nature* **318:** 382.

Willard, H.F., M.H. Skolnick, P.L. Pearson, and J.L. Mandel. 1985. Report of the committee on human gene mapping by recombinant DNA techniques. *Cytogenet. Cell Genet.* **40:** 360.

Wyman, A.R. and R. White. 1980. A highly polymorphic locus in human DNA. *Proc. Natl. Acad. Sci.* **77:** 6754.

Mapping of the Cystic Fibrosis Locus on Chromosome 7

L.-C. TSUI, S. ZENGERLING, H.F. WILLARD, AND M. BUCHWALD
Department of Genetics, Research Institute, The Hospital for Sick Children, Toronto, Ontario, Canada M5G 1X8; Departments of Medical Genetics and Medical Biophysics, University of Toronto, Ontario, Canada

Cystic fibrosis (CF) is the most common autosomal recessive disorder affecting approximately 1 out of 2000 live births in the Caucasian population (Talamo et al. 1983). The major clinical manifestations of CF include chronic obstructive pulmonary disease, pancreatic enzyme insufficiency, and elevated sweat electrolyte levels, all of which are suggestive of an inborn error of metabolism involving exocrine glands. If untreated, CF individuals usually die at an early age due to severe lung infection, but, because of the advances in disease treatment, many patients now survive into adulthood. On the other hand, despite extensive research efforts, the basic defect of this disease remains unknown. Although a defective regulation of chloride ion transport seems to be the most consistent measurable parameter in CF (Knowles et al. 1983; Quinton 1983; Quinton and Bijman 1983; Sato 1984; Widdicombe et al. 1985; Yankaskas et al. 1985), it is still not clear whether this abnormality is the primary cause of the disease.

The results of extensive studies based on first-cousin marriages, however, suggested that CF is due to mutation(s) in a single gene (Danks et al. 1984; Romeo et al. 1985). The simple autosomal recessive mode of inheritance of CF further submits that it is possible to map the disease locus (*CF*) by genetic linkage analysis. Although several early attempts using conventional protein markers were not successful (Steinberg et al. 1956; Steinberg and Morton 1956; Goodchild et al. 1976), the feasibility of this approach was first demonstrated by the discovery of a linkage between *CF* and *PON,* a genetic determinant for serum paraoxonase (PON) activity (Eiberg et al. 1985; Schmiegelow et al. 1986).

As the first step in our approach to the basic defect in CF, we have initiated a CF linkage study using DNA markers defined by restriction-fragment-length polymorphisms (RFLPs) (Botstein et al. 1980). This was based on the assumption that a marker linked to CF would not only be useful in obtaining information on the chromosome location of *CF* and as a means for genetic diagnosis, but also serve as a reference point for searching the primary genetic lesion(s).

After screening a large number of markers, covering a significant portion of the human genome (Tsui et al. 1985a,b), we have recently discovered a linkage between *CF* and a randomly isolated DNA marker, *D7S15* (formerly *DOCRI-917*) (Tsui et al. 1985c). Subsequent localization of this DNA marker to chromosome 7 has established that *CF* is also on chromosome 7 (Knowlton et al. 1985). This assignment has since been confirmed by the demonstration of a linkage between *CF* and a number of other DNA markers known to be on chromosome 7, including the met oncogene (*met*) (White et al. 1985), DNA marker *D7S8* (Wainwright et al. 1985), the proα2(I) collagen gene (*COL1A2*) (Scambler et al. 1985; Buchwald et al. 1986), the T-cell receptor β gene (*TCRB*) (Wainwright et al. 1985), and DNA marker *7C22* (Scambler et al. 1986).

This paper describes the experiments that first allowed us to demonstrate the linkage between *CF* and *D7S15* and the subsequent analysis that provided further information on the location of *CF* on chromosome 7. Part of the data presented here has appeared in previous publications (Tsui et al. 1985c; Buchwald et al. 1986).

MATERIALS AND METHODS

Family collection. Blood samples from CF families with two or more affected children were collected through the cooperation of seven clinics in Canada. Informed consent was obtained from each of the participating individuals. In addition to the family history, the diagnosis of CF was confirmed by the presence of elevated sweat chloride levels (>60 meq/liter) in conjunction with chronic obstructive lung disease and/or pancreatic insufficiency. The clinical data obtained from these patients are consistent with the classical description of CF. The number of affected males (60) was slightly higher than that of affected females (46), but the difference was not statistically significant. Similar observations were made in previous studies by Warwick et al. (1975) and Gurwitz et al. (1979), who attributed the difference to the increased mortality of CF females.

Forty-seven Canadian CF families were used in the present study, and their structure is shown in Figure 1. Of these, 39 were families with two affected children and eight were families with three or more affected children. Lymphoblast cultures of all of the 106 CF patients, together with the 94 parents and 44 unaffected sibs, were established using the Epstein-Barr virus transformation procedure of Anderson and Gusella (1984). Additional families used in this study were obtained from the Human Genetic Mutant Cell Repository, Camden, New Jersey, and CF research laboratories as indicated.

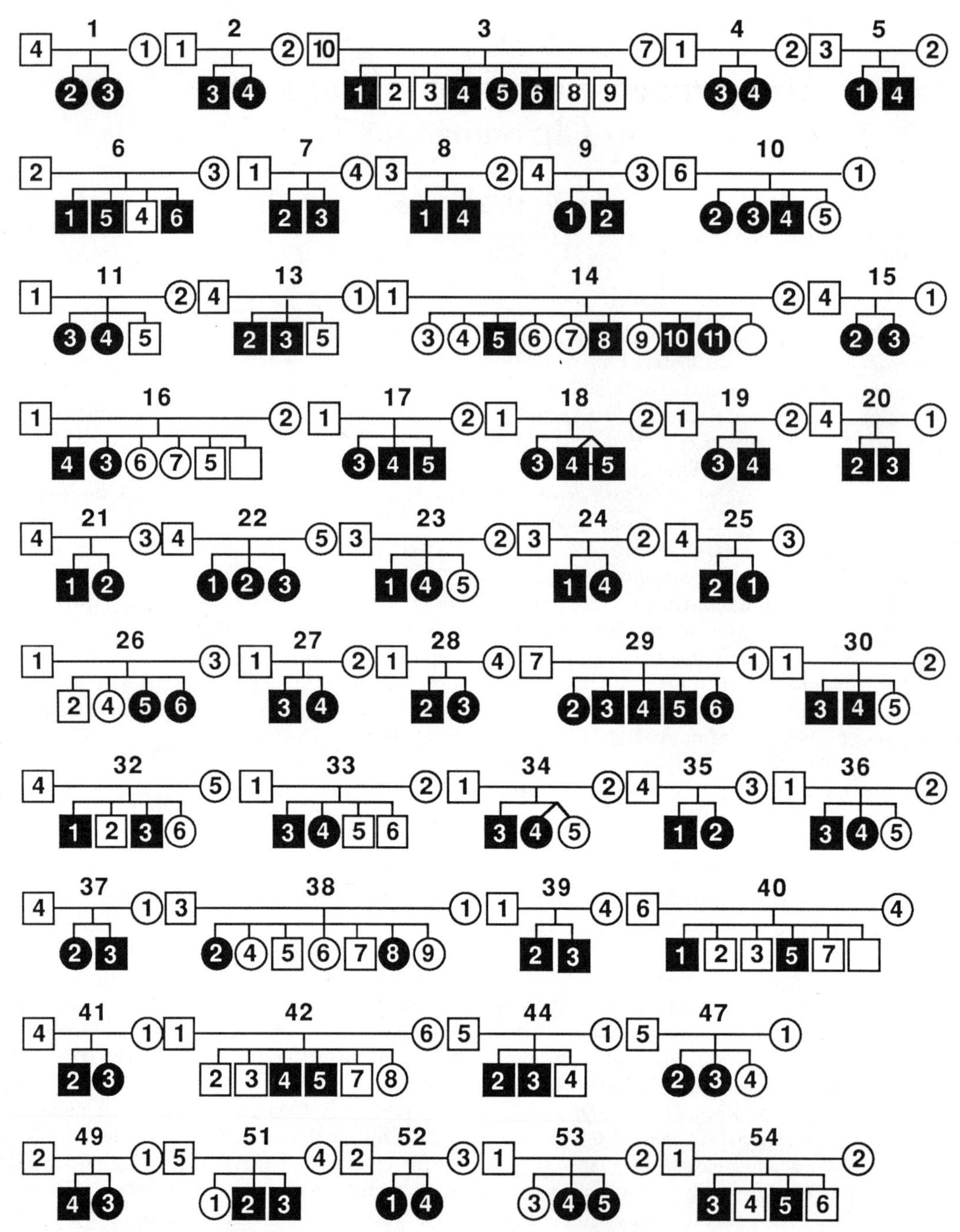

Figure 1. Family pedigrees used in the present study. The CF status of each individual is indicated by open (unaffected) or closed (affected) symbols.

DNA probes. *D7S15* is defined by a human genomic clone Lam4-917 isolated and provided as a collaborative study by Collaborative Research, Inc. A detailed description of Lam4-917 has been presented previously (Tsui et al. 1985c). It detects two RFLPs, *Hin*dIII and *Hin*cII, as shown in Figure 2. Both of these RFLPs are two-allele systems, with respective frequencies of 0.7 and 0.3 for the major (6.3 kb) and minor (5.3 + 1.0 kb) *Hin*dIII RFLP alleles and approximately equal distribution of the two (4.3 and 2.3 + 1.8 kb) *Hin*cII alleles. In addition, the *Hin*dIII and *Hin*cII polymorphic sites were found to be close to random association, i.e., in linkage equilibrium, in the population.

The three DNA probes for *COL1A2*, NJ-3, NJ-1 (Tsipouras et al. 1984), and Hf-32 (Myers et al. 1981), used in this study were provided by F. Ramirez and P. Tsipouras. These probes detect three different RFLPs, *Eco*RI, *Msp*I, and *Rsa*I, respectively.

The *Bgl*II RFLP associated with *TCRB* was detected with a cDNA probe generously provided by T. Mak (Ontario Cancer Institute).

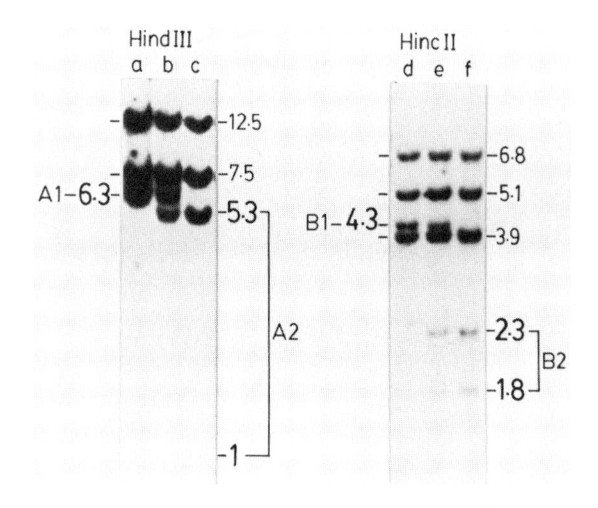

Figure 2. Blot hybridization analysis of RFLPs associated with the *D7S15* locus. Autoradiograms show the Lam4-917 hybridization patterns of DNA samples (six random individuals) digested witn *Hin*dIII (*a–c*) and *Hin*cII (*d–f*), electrophoresed on a 0.7% agarose gel, and transferred to Zetabind membranes. Numbers indicate fragment sizes in kilobases.

The *metH* and *metD* probes for the *met* locus were gifts from R. White (Howard Hughes Medical Institute, Salt Lake City). The *metH* probe detects a *Taq*I and a *Msp*I RFLP; the *metD* probe detects a *Taq*I and a *Ban*I RFLP (R. White; A. Beaudet; both pers. comm.). The *pJ3-11* probe that detected the *Msp*I and *Taq*I RFLPs associated with *D7S8* were obtained from J. Schmidtke (Institut für Humangenetik der Universität, Göttingen).

DNA analysis. DNA samples of individual family members were prepared either directly from peripheral blood samples or from lymphoblast cultures. The procedures for DNA analysis were carried out essentially as described previously (Tsui et al. 1985c). Restriction endonuclease digestions were performed according to conditions specified by suppliers (New England BioLabs, Boehringer-Mannheim, and GIBCO-BRL). The digested samples were fractionated by electrophoresis on agarose gels and transferred to Zetabind membranes (AMF Cuno, manufacturer) following the procedure of Southern (1975). Radioactive ^{32}P-labeled probes were prepared using the random primer procedure of Feinberg and Vogelstein (1983).

Linkage analysis. The maximal likelihood estimate of the recombination fraction (θ) between genetic markers was obtained using the lod score analysis method of Morton (1955). The calculation of lod (z) score was performed with the LIPED computer program (Ott 1974). The disease status of each individual was coded by their phenotype based on the recessive mode of inheritance, and the RFLPs were coded as codominant. If more than one RFLP were informative for a marker loci, haplotypes were determined by inspection when possible. The unaffected sibs were also included in the analysis, although their CF status was undetermined; the algorithm considered them to be either homozygous normal or heterozygous carrier.

The confidence interval for the maximal likelihood estimate was estimated by taking the θ values that corresponded to the lod score of ($\hat{z} - 1$) according to Conneally et al. (1985). The order of genetic loci was examined using the LINKAGE computer program (version 3.5) (Lathrop et al. 1984).

Paraoxonase assay. The serum PON assay was performed in the laboratory of H. Eiberg as described previously (Eiberg et al. 1985; Schmiegelow et al. 1986). The "high" and "low" alleles were scored as codominant in the present study.

Somatic-cell hybrid lines. Chromosomal localization of *D7S15* was accomplished by hybridization analysis using DNA prepared from human-rodent somatic-cell hybrids (Knowlton et al. 1985). Additional hybrid lines used in the study were established as described previously (Willard and Riordan 1985; Willard et al. 1985).

RESULTS AND DISCUSSION

Linkage between *CF* and *D7S15*

To investigate the linkage relationship between *D7S15* and *CF*, we examined a total of 43 families, of which 39 were found to be informative for the analysis, since one or both of the parents in these were heterozygous for either the *Hin*dIII or the *Hin*cII RFLP, or both. A standard lod (z) score analysis (Morton 1985) was performed using the LIPED computer program based on the segregation patterns of the RFLPs and the disease in each of the informative families. Assuming recombination frequencies between male and female being equal, the maximal likelihood estimate for recombination fraction (θ) between *D7S15* and *CF* was found to be 0.14 with a lod (z) score of 3.96 (odds ratio of 9100:1) (Table 1). Thus, the lod score was well above 3, the value generally accepted for proof of linkage. The estimated value for θ also agreed well with that derived from counting apparent recombinants in CF children among informative meioses (13/101). The confidence interval for θ derived from the lod score analysis was between 0.07 and 0.25. Subsequent to our previous report (Tsui et al. 1985c), we examined additional families collected by ourselves and other CF research laboratories. As a result, 28 more families were found to be informative for linkage analysis between *D7S15* and *CF*. The revised maximal likelihood estimate for θ was 0.17 with a lod score of 6.68 (Table 1).

PON

Since a previous report indicated that *CF* and *PON* were genetically linked (Eiberg et al. 1985) and since PON typing data for our families were available (Schmiegelow et al. 1986), we next examined the linkage relationship between *D7S15* and *PON*. As shown

Table 1. Linkage Relationships between CF and Adjacent Markers on Chromosome 7

Marker loci	Number of informative families	Lod (z) scores at recombination fraction (θ) of									θ_{max}	z_{max}	Confidence intervals
		0.01	0.05	0.10	0.15	0.20	0.25	0.30	0.35	0.40			
CF-D7S15	39[a]	−5.88	1.67	3.63	3.95	3.62	2.97	2.18	1.38	0.67	0.14	3.96	0.07–0.25
CF-D7S15	67[b]	−21.52	−0.77	5.09	6.61	6.43	5.43	4.05	2.58	1.26	0.17	6.68	0.11–0.24
CF-PON	31[c]	−1.83	2.02	2.85	2.80	2.42	1.90	1.34	0.82	0.39	0.12	2.88	
CF-COL1A2	51	−15.78	−1.50	2.62	3.77	3.77	3.21	2.4	1.52	0.74	0.17	3.87	0.11–0.27
CF-TCRB	31	−14.80	−4.73	−1.32	0.08	0.66	0.81	0.73	0.52	0.27	0.25	0.81	
CF-met	44	19.28	18.21	15.74	13.07	10.04	7.81	5.4	3.26	1.54	0.01	19.28	0.001–0.05
CF-D7S8	46	19.58	18.41	15.82	13.07	10.34	7.73	5.32	3.20	1.51	0.01	19.58	0.001–0.05
D7S15-PON	20	5.36	6.48	6.15	5.45	4.59	3.66	2.71	1.78	0.94	0.06	6.47	0.01–0.20
D7S15-COL1A2	30	11.02	13.25	12.58	11.14	9.38	7.48	5.51	3.58	1.83	0.05	13.25	0.02–0.11
D7S15-TCRB	18	−8.15	−1.93	0.1	0.87	1.12	1.1	0.91	0.63	0.33	0.22	1.14	
D7S15-met	28	−11.74	−0.86	2.41	3.42	3.51	3.11	2.44	1.63	0.84	0.18	3.55	0.10–0.29
D7S15-D7S8	21	−14.19	−2.51	1.2	2.52	2.89	2.73	2.27	1.61	0.89	0.21	2.89	
COL1A2-PON	21	−1.90	1.47	2.25	2.29	2.05	1.67	1.23	0.79	0.39	0.13	2.33	
COL1A2-TCRB	22	−14.80	−4.68	−1.20	0.24	0.86	1.03	0.92	0.66	0.35	0.25	1.03	
COL1A2-met	29	−8.35	0.10	2.61	3.34	3.34	2.94	2.31	1.56	0.81	0.17	3.40	0.09–0.29
COL1A2-D7S8	32	−15.03	−3.80	−0.12	1.3	1.78	1.77	1.47	1.02	0.54	0.22	1.82	
TCRB-PON	8	−7.08	−3.15	−1.66	−0.84	−0.52	−0.28	−0.13	−0.06	−0.02	0.50	0.00	
TCRB-met	12	−3.11	0.49	1.54	1.81	1.76	1.53	1.19	0.8	0.41	0.16	1.82	
TCRB-D7S8	17	−2.46	1.64	2.72	2.9	2.69	2.28	1.74	1.16	0.6	0.14	2.90	
met-PON	16	−7.02	−1.39	0.48	1.19	1.43	1.39	1.18	0.86	0.49	0.22	1.44	
met-D7S8	27	16.63	16.06	14.17	12.02	9.79	7.54	5.35	3.31	1.59	0.02	16.75	
D7S8-PON	15	−9.24	−2.91	−0.73	0.18	0.57	0.67	0.6	0.43	0.23	0.25	0.67	

[a]These include 36 Canadian and 3 HGCMR CF families.
[b]These include the families in footnote a and 9 from Copenhagen, 4 from London, 5 from Salt Lake City, and 9 from Stanford-Berkeley.
[c]These include 22 Canadian and 9 Danish families.

in Table 1, 20 families were found to be informative for this marker-marker linkage analysis. The maximal likelihood estimate for θ was 0.06, with a lod score of 6.47 and confidence interval for θ between 0.01 and 0.20. This result thus confirmed the linkage relationship between *CF* and *D7S15*.

On the basis of the reported distance between *CF* and *PON*, which was approximately 10 cM (Eiberg et al. 1985; Schmiegelow et al. 1986), and our linkage estimates between *CF* and *D7S15* and between *D7S15* and *PON*, it was apparent that the order for these three genetic loci was *D7S15–PON–CF*. However, since the confidence limits for these estimates were wide, we examined the probabilities of each of the three possible orders using the LINKAGE computer program (Lathrop et al. 1984). By fixing the genetic distance between *D7S15* and *PON* as 5 cM, the most likely location for *CF* was calculated based on data from 11 CF families that were informative for both marker loci. As shown in Figure 3a, although the result of this analysis did not greatly favor the order *D7S15–PON–CF* over *CF–D7S15–PON*, it seemed highly unlikely that *CF* was in between *D7S15* and *PON*.

Localization of *D7S15* to the Long Arm of Chromosome 7

The chromosomal location for the RFLP locus *D7S15* and by inference, *CF*, was subsequently determined by DNA hybridization analysis using a panel of human-rodent somatic-cell hybrid lines containing different human chromosome complements (Knowlton et al. 1985). Comparison of the Lam4-917 probe hybridization results with the chromosome content of 25 dif-

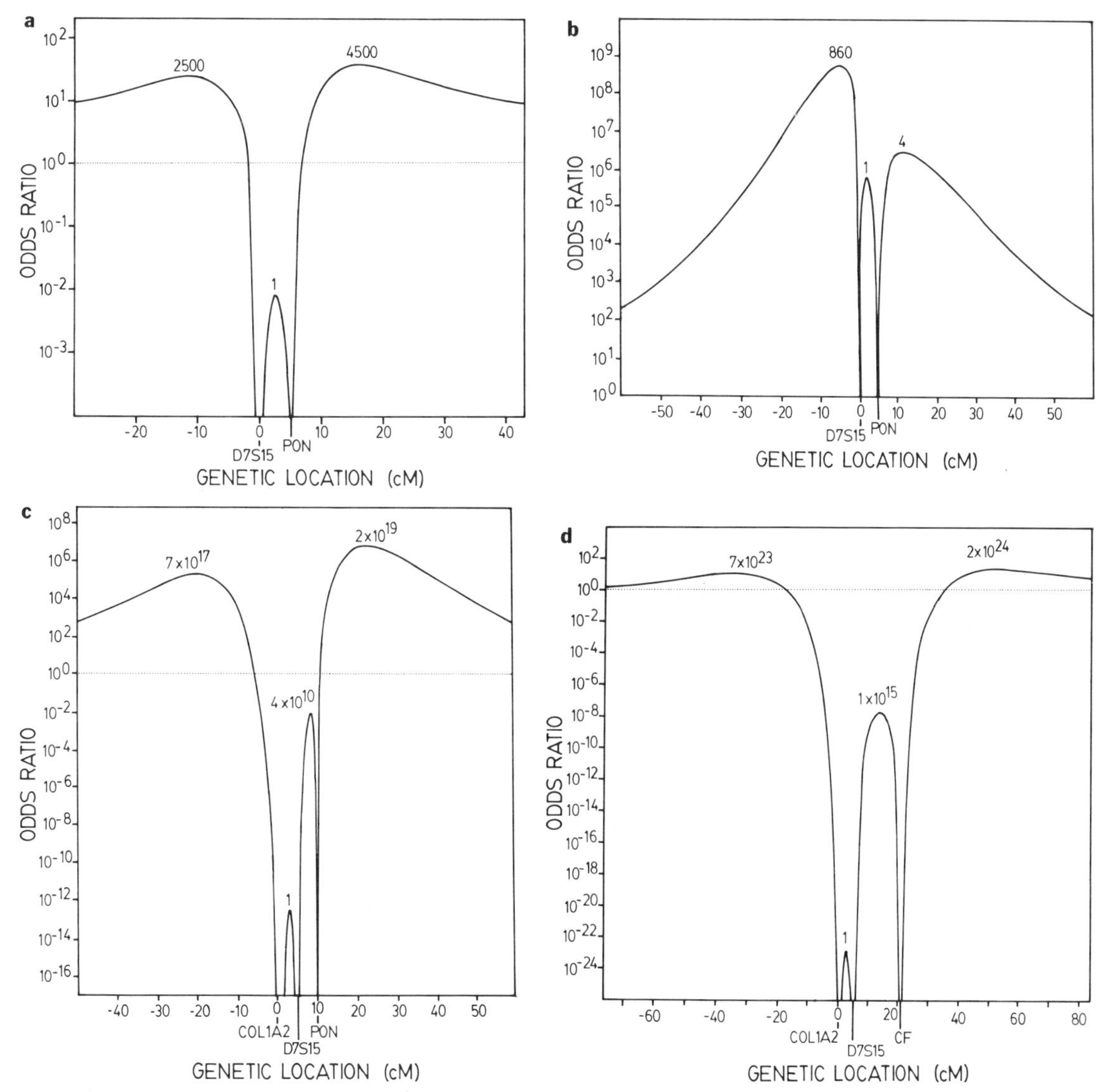

Figure 3. Likelihood of the map locations of *CF, COL1A2,* and *TCRB* with respect to other linked loci: (*a*) *CF* vs *D7S15* and *PON*; (*b*) *COL1A2* vs *D7S15* and *PON*; (*c*) *CF* vs *COL1A2, D7S15,* and *PON*; (*d*) *TCRB* vs *COL1A2, D7S15,* and *CF.* Horizontal axis: genetic distance from the marker locus set at 0; vertical axis: odds ratio for location of the test locus at given distance from 0 vs at infinite distance (i.e., no linkage). The relative odds for the test marker being at various test locations are indicated above each peak. (*a, b,* and *c,* redrawn, with permission, from Tsui et al. 1985a,c.)

ferent hybrid cell lines indicated that *D7S15* was on chromosome 7. This assignment was confirmed by similar hybridization analysis using another 23 independently established human-rodent hybrids (Willard and Riordan 1985; Willard et al. 1985). The result of the latter analysis is shown in Table 2. In 13 hybrid cell lines where human chromosome-7 material was present, a Lam4-917 hybridizing sequence was also present, whereas in 10 hybrids, the absence of chromosome 7 coincided with the absence of a Lam4-917 hybridization signal, showing 100% concordance (23/23).

Positive hybridization signals in two hybrid lines containing portions of chromosome 7 revealed further information on the location of *D7S15*. The retention of the Lam4-917 hybridizing sequence in cell line t60-14 (del 7 [7pter→q31]) suggests that *D7S15* is located either on the short arm or on the proximal half of the long arm (Table 2). In addition, the presence of the Lam4-917 hybridizing sequence in cell line CH13 (see Knowlton et al. 1985), which contained only a part of human chromosome 7, lacking *GUSB* (cent→q22) but retaining *TCRB* (q35), was consistent with *D7S15* being on the long arm of chromosome 7. Therefore, these results suggested the most probable location for *D7S15* to be 7cent→q31.

COL1A2

In our attempts to define the location of CF further, we next examined its linkage relationship with *COL1A2*, a well-characterized marker that had been localized to 7q21→q22 (Junien et al. 1982) and, more recently, to 7q21.3→q22.1 (Retief et al. 1985). Fifty-one families were found to be informative for the analysis (Table 1). The maximal likelihood estimate for the recombination fraction between *CF* and *COL1A2* was obtained at $\theta = 0.17$, with a lod score of 3.87. A strong linkage was also detected between *COL1A2* and *D7S15* at θ of 0.05 with a lod score of 13.25.

From the available two-point linkage data (Table 1), it was apparent that *COL1A2, D7S15*, and *PON* were all on the same side of *CF*. However, to examine the order of these genetic loci further, the multipoint LINKAGE program was employed. First, the proper order for *COL1A2, D7S15,* and *PON* was investigated using data from 22 CF families in which all of these markers were found to be informative. As shown in Figure 3b, the most probable order for these three markers was *COL1A2–D7S15–PON*. Then, by fixing the distance between *COL1A2* and *D7S15* and between *D7S15* and *PON*, the most likely location for *CF* was calculated. The latter analysis was based on marker information from 50 CF families, each of which showed segregation of at least two of the marker loci. The result in Figure 3c suggested that the most likely order for these genetic loci was *COL1A2–D7S15–PON–CF*. On the basis of the linkage distance from *COL1A2*, it seemed probable that *CF* was also on the long arm of chromosome 7, either close to the centromere or near the middle of the long arm.

TCRB

We then examined the relationship between *CF* and *TCRB,* another well-studied marker that had been regionally assigned to band q32 or q31 (Barker et al. 1984; Collins et al. 1984; Le Beau et al. 1984; Morton et al. 1985). Thirty-one families were informative for the analysis, but no significant linkage between these two genetic loci was detectable, with the highest lod score being 0.81 at θ of 0.25 (Table 1). This apparent lack of close linkage is consistent with the number of obligatory crossovers (15/70) observed by direct counting of informative meioses. Therefore, this result is slightly different, but not in disagreement with that reported by Wainwright et al. (1985, 1986).

To investigate the linkage relationship between *CF* and *TCRB* further, we attempted to determine the location for *TCRB* with respect to *COL1A2–D7S15–CF* using the LINKAGE analysis program. As shown in Figure 3d, although *TCRB* was highly unlikely to be located between *COL1A2* and *D7S15* or between *D7S15* and *CF*, the order *COL1A2–D7S15–CF–TCRB* was only slightly favored over *TCRB–COL1A2–D7S15–CF.* Additional analysis using *met* and *D7S8* in place of *CF* also yielded ambiguous results (data not shown). However, on the basis of the chromosome location of *COL1A2* and *TCRB*, it seems probable that *CF* is flanked by these two markers and, therefore, in the middle of the long arm. In support of this view, a significant linkage was detected between *CF* and *TCRB* in a different set of CF families, although no linkage relationship could be observed between *COL1A2* and *TCRB* (Wainwright et al. 1986).

met and *D7S8*

Further information on the regional localization of *CF* derived from the discovery of close linkage between *CF* and *met* (White et al. 1985) and that between *CF* and *D7S8* (Wainwright et al. 1985). Although our family data also indicated a close linkage between *CF* and *met* and between *CF* and *D7S8* (Table 1), several recombination events were noted. By direct counting of the informative chromosomes in CF children, we detected 2 apparent crossovers between *met* and *CF* out of 123 meioses and 2 between *CF* and *D7S8* out of 120. We also detected two crossovers between *met* and *D7S8*, but since they were only observed in unaffected children, no useful information could be derived for determining the order of these three closely linked genetic loci.

However, based on the map location of *met* and its close linkage with *CF* (Dean et al. 1985), it has been suggested that *CF* also resides within region q21→q31 (White et al. 1985). Consistent with this assignment, we noticed by hybridization intensity difference that both *met* and *D7S8* were missing in GM1059, a human mutant cell line that had deleted band q31 [46 XX, del 7 (pter→q22::q32→qter)] (L.-C. Tsui et al., unpubl.).

Table 2. Segregation of D7S15 Sequences with Human Chromosomes in Somatic-cell Hybrids

Hybrids	Chromosomes																								Probe
	1	2	3	4	5	6	7	8	9	10	11	12	13	14	15	15	17	18	19	20	21	22	X	Y	917
A60-12		+	+	+			+	+				+	+	+	+	+	+	+	+	+	+		+		+
A60-2A		+		+	+		+										+				+	+	+		+
A60-9				+	+		+	+		+				+			+			+			+		+
t60-8	+		+	+	+		+	+		+	+	+	+	+	+	+	+	+	+	+	+		+		+
t60-14				+			+	+			+		+	+				+			+	+	+		+
t11-4A-7B	+			+			+	+				+		+				+	+		+		+		+
A48-1GAZ43					+		+			+	/	+						+		+	+		/		+
A60-1A					+		+							+						+	+	+	+		+
LT23-4C	+	+		+	+	+	+			+	+		+	+	+		+	+	+	+	+		+		+
A23-1A	+		+				+	+		+	+	+		+	+	+			+	+			+		+
A48-1G					+		+			+	+							+		+		+	+		+
L23-1B	+		+		+	+	+					+		+		+	+			+	+				+
W4-3A				+			+							+					+				+		+
A54-8Acl1	+		+															+	+		+		+		−
A60-11					+					+		+											+		−
A60-7			+	+		+					+			+	+	+	+			+		+			−
t60-11	+		+	+	+			+			+	+	+	+	+	+	+	+	+	+	+	+	+		−
A48-1GAZ44					+						/							+		+	+		/		−
L23-4B					+						+						+	+		+	+		+		−
A23-2E1		+	+										+	+			+			+			+		−
y15/1																								+	−
W44-14A	+		+	+					+		+			+		+			+				+		−
W53-5Bcl5			+		+	+					+	+	+	/	+	+	+				+	+	/		−
Concordant																									
+/+	5	3	4	8	8	2	13	6	0	6	5	6	4	10	4	4	6	7	6	9	9	4	11	0	
−/−	7	9	4	7	5	8	10	9	9	9	4	7	7	5	7	6	5	6	7	5	5	7	1	9	
Discordant																									
+/−	3	1	6	3	5	2	0	1	1	1	5	3	3	4	3	4	5	4	3	5	5	3	7	1	
−/+	8	10	9	5	5	11	0	7	13	7	7	7	9	3	9	9	7	6	7	4	4	9	1	13	
No score											2			1									3		
Total discordant ($n = 23$)	11	11	15	8	10	13	0	8	14	8	12	10	12	7	12	13	12	10	10	9	9	12	8	14	

+/+ = chromosome present, *D7S15* sequence present; +/− = chromosome present, *D7S15* sequence absent; −/+ = chromosome absent, *D7S15* sequence present; −/− = chromosome absent, *D7S15* absent; / = not scored, due to translocation in parental human cells.

Location of *CF*

Taking together the linkage relationship with *D7S15, COL1A2, TCRB, met,* and *D7S8*, and the chromosomal assignments for these markers, evidence is highly suggestive that *CF* is located within the middle third of the long arm of chromosome 7 (Fig. 4). Furthermore, assuming a direct correlation between genetic and physical distances, it is probable that *CF* maps within band q31. A more detailed analysis on the regional localization of *met* and *D7S8* should prove this point. In the mean time, this provisional assignment of *CF* should be useful for the isolation of additional markers for CF linkage studies, especially by cloning of microdissected chromosomal DNA (Röhme et al. 1984). It should also be useful for the identification of chromosomal abnormalities involving the *CF* locus. In this regard, we are in the process of segregating the two chromosome 7s of GM1059 in hybrid cell lines generated by fusion with mouse A9 cells. The resulting cell line containing the chromosome 7 with the deletion would be particularly useful for rapid screening of new markers randomly isolated from a chromosome-7-specific genomic library or for testing of candidate gene probes for their relationship to CF.

Genetic Diagnosis Using Linked DNA Markers

Another immediate application of the CF-linked DNA markers is their use in genetic diagnosis for families at risk, i.e., families who already have individuals with CF. Based on the markers inherited by the affected individuals, predictions can be made for other members of the family if their CF status is not already known. However, the usefulness of these markers is dictated by three major factors, namely, the distance from *CF*, the informativeness of the probe, and the degree of genetic heterogeneity if it exists. Since both *met* and *D7S8* are approximately 1 cM away from *CF*, they should be adequate for this purpose. A combined use of all six RFLPs associated with these markers would predict an accuracy of diagnosis approaching 100% in some cases. However, the usefulness of these markers is presently limited by their informativeness due to a high degree of allele association (linkage disequilibrium) between these two loci (R. White; B. Williamson; both pers. comm.; L.-C. Tsui et al., unpubl.).

Since it is difficult to estimate the polymorphic information content (Botstein et al. 1980) for the multiple RFLPs associated with *met* and *D7S8*, we have performed a retrospective study based on our existing family data. Using information derived from the first affected child in each family to set the phase of the parental chromosomes, we counted the number of families (matings) under each of the three possible categories: fully informative, partially informative, and noninformative (see Table 3). Interestingly, although separate use of probes for *met* and *D7S8* would permit only 20–25% of fully informative diagnosis, the combined use of the two would allow over 50%. If the test

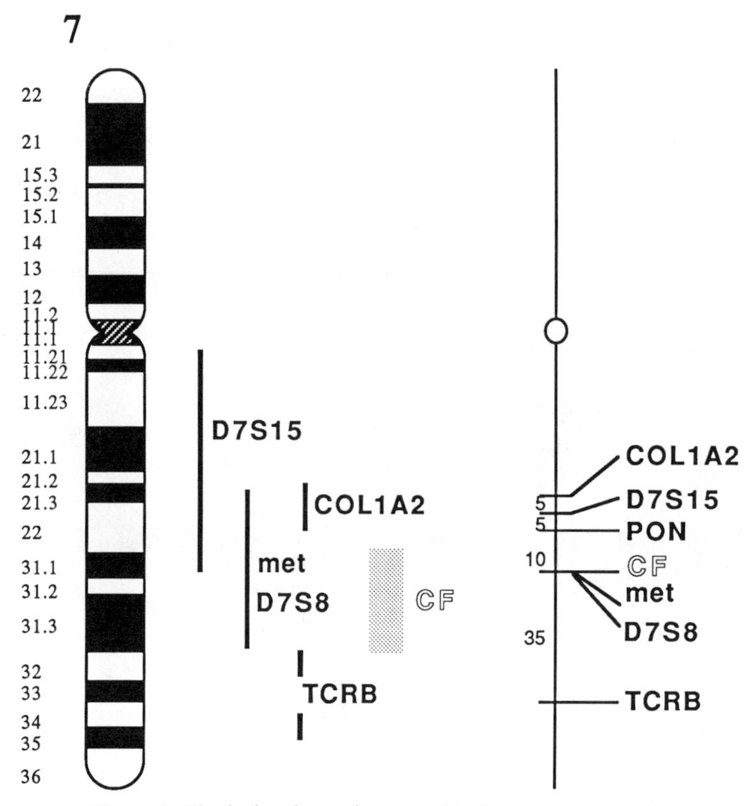

Figure 4. Physical and genetic maps of loci on chromosome 7.

Table 3. Informativeness of the *met* and *D7S8* RFLPs in Genetic Diagnosis Based on Data from Retrospective Studies

Probes	I. Fully informative[a]	II. Partially informative		III. Non-informative	Total
		a	b		
met H + D[b]	14(24)[c]	3(5)	28(48)	13(22)	58
pJ3-11 (*D7S8*)	10(19)	10(19)	27(51)	6(11)	53
met H + D and *pJ3-11*	27(53)	2(4)	22(43)	0(0)	51

[a]The informativeness of genetic diagnosis was determined by inheritance pattern of the test marker. Example for I: AB × AB and the first CF child being AA or BB; IIa: AB × AB and the CF child being AB; IIb: AB × AA and the CF child being AA or AB; III: both parents being homozygous for the test marker.

[b]All RFLPs were scored for the *met* and *D7S8* loci. Haplotypes were used when possible.

[c]Number of families in each category with percentage of total in parentheses.

were intended for detecting carrier status in unaffected individuals or identifying CF children in utero, the fraction of satisfactory diagnosis would even increase to over 75%. Furthermore, no family would be noninformative if all RFLPs were examined.

Genetic Heterogeneity

Genetic heterogeneity in CF has been considered as one of the possible explanations for the high frequency of the disease (Thompson 1980). On the basis of linkage data from this and other studies (Wainwright et al. 1985; White et al. 1985), there is no strong evidence for a second *CF* locus, thus consistent with the previous assumption that CF is due to defects in a single gene (Danks et al. 1984; Romeo et al. 1985). However, although linkage analysis indicates that the majority of the *CF* mutations in the population map within the same locus, it is possible that there is a small proportion of CF families segregating mutations at another unlinked locus. A critical evaluation of this problem awaits identification and molecular characterization of the *CF* gene itself.

Linkage Disequilibrium

During the course of our study, we also detected a strong disequilibrium between *CF* and *met* and to a lesser extent between *CF* and *D7S8* (Table 4). This information was derived from the chromosomes of parents of CF children, since the association between the normal (N) and disease (CF) alleles at the *CF* locus and the various *met* or *D7S8* alleles or haplotypes could easily be determined by inspecting the affected children in each family. A total of 92 parents or 184 chromosomes were thus studied. As shown in Table 4, there is an apparent nonrandom association of alleles. For example, we detected twice as many *CF* alleles associated with one haplotype (A) at the *met* locus than the normal allele and three times as many normal alleles associated with another haplotype (B) than the *CF* allele. Although this nonrandom association of alleles might be expected for closely linked genetic markers, its significance remains to be investigated.

ACKNOWLEDGMENTS

We thank Collaborative Research, Inc. for providing the probe Lam4-917 for the present study; Helen Donis-Keller, David Barker, Ken Buetow, and Aravinda Chakravarti for helpful discussions; Dara Kennedy, Natasa Plavsic, Danuta Markiewicz, and Marta Zsiga for expert technical assistance; and the CF clinics at the Hospital for Sick Children and other hospitals throughout Canada for their assistance in collecting the family blood samples. We also thank the Canadian Cystic Fibrosis Foundation (CCFF) and the patients and their family members who participated in this study. This work was supported by grants from CCFF, the National Institutes of Health, the Medical Research Council of Canada, the North Dakota Cystic Fibrosis Foundation, and the Sellers Fund from HSC. L.-C.T. is a Research Scholar of CCFF.

Table 4. Linkage Disequilibrium between *CF* and *met* and *D7S8*

Haplotype[a]		N	CF	
met	A	29[b]	58	
	B	19	6	$\chi^2 = 20$, d.f. = 2; $p<0.01$
	C	44	28	
D7S8	A	18	28	
	B	3	8	$\chi^2 = 7.7$, d.f. = 2; $p<0.05$
	C	42	27	

[a]*met* haplotype A = *metH TaqI* allele 1 + *metD MspI* allele 1; haplotype B = *metH TaqI* allele 1 + *metD MspI* allele 2; haplotype C = *metH TaqI* allele 2 + *metD MspI* allele 1; *D7S8* haplotype A = *MspI* allele 1 + *TaqI* allele 1; haplotype B = *MspI* allele 1 + *TaqI* allele 2; haplotype C = *MspI* allele 2 + *TaqI* allele 1; haplotype D for both *met* and *D7S8* are missing.

[b]Number of chromosomes carrying the normal (N) or the CF allele.

REFERENCES

Anderson, M.A. and J.F Gusella. 1984. Use of cyclosporin A in establishing Epstein-Barr virus-transformed human lymphoblastoid cell lines. *In vitro* **20:** 856.

Barker, P.E., F.H. Ruddle, H.D. Royer, O. Acuto, and E.L. Reinherz. 1984. Chromosomal location of human T-cell receptor gene TiBeta. *Science* **23:** 348.

Botstein, D., R.L. White, M. Skolnick, and R.W. Davis. 1980. Construction of a genetic linkage map in man using restriction fragment length polymorphisms. *Am. J. Hum. Genet.* **32:** 314.

Buchwald, M., M. Zsiga, D. Markiewicz, N. Plavsic, D. Kennedy, S. Zengerling, H.F. Willard, P. Tsipouras, K. Schmiegelow, M. Schwartz, H. Eiberg, J. Mohr, D. Barker, H. Donis-Keller, and L.-C. Tsui. 1986. Linkage of cystic fibrosis to the proα2(1) collagen gene, *COL1A2*, on chromosome 7. *Cytogenet. Cell Genet.* **41:** 234.

Collins, M.K.L., P.N. Goodfellow, M.J. Dunne, N.K. Spurr, E. Soloman, and M.J. Owens. 1984. A human T-cell antigen receptor beta chain gene maps to chromosome 7. *EMBO J.* **3:** 2347.

Conneally, P.M., J.H. Edwards, K.K. Kidd, J.-M. Lalouel, N.E. Morton, J. Ott, and R. White. 1985. Report of the committee on methods of linkage analysis and reporting. Eighth International Workshop on Human Gene Mapping. Helsinki. *Cytogenet. Cell Genet.* **40:** 356.

Danks, D.M., P.D. Phelan, and C. Chapman. 1984. Retraction: No evidence for more than one locus in cystic fibrosis. *Am. J. Hum. Genet.* **36:** 1398.

Dean, M., M. Park, M.M. Le Beau, T.S. Robbins, M.O. Diaz, J.D. Rowley, D.G. Blair, and G.F. Vande Woude. 1985. The human *met* oncogene is related to the tyrosine kinase oncogenes. *Nature* **318:** 385.

Eiberg, H., J. Mohr, K. Schmiegelow, L.S. Nielsen, and R. Williamson. 1985. Linkage relationships of paraoxonase (PON) with other markers: Indication of PON-cystic fibrosis synteny. *Clin. Genet.* **28:** 265.

Feinberg, A.P. and B. Vogelstein. 1983. A technique for radiolabeling DNA restriction endonuclease fragments to high specific activity. *Anal. Biochem.* **132:** 6.

Goodchild, M.C., J.H. Edwards, K.P. Glenn, C. Grindey, R. Harris, P. Mackintosh, and J. Wentzel. 1976. A search for linkage in cystic fibrosis. *J. Med. Genet.* **7:** 417.

Gurwitz, D., M. Corey, P.W.J. Francis, D. Crozier, and H. Levison. 1979. Perspectives in cystic fibrosis. *Pediatr. Clin. N. Am.* **26:** 603.

Junien, D., D. Weil, J.C. Meyers, V.C. Nguyen, M.-L. Chu, C. Foubert, M.S. Gross, D.J. Prockop, J.C. Kaplan, and F. Ramirez. 1982. Assignment of the human proα2(I) collagen structural gene (*COL1A2*) to chromosome 7 by molecular hybridization. *Am. J. Hum. Genet.* **34:** 381.

Knowles, M.R., M.J. Stutts, A. Spock, N. Fischer, J.T. Gatzy, and R.C. Boucher. 1983. Abnormal ion permeation through cystic fibrosis respiratory epithelium. *Science* **221:** 1067.

Knowlton, R.G., O. Cohen-Haguenauer, V.C. Nguyen, J. Frézal, V. Brown, D. Barker, J.C. Braman, J.W. Schumm, L.-C. Tsui, M. Buchwald, and H. Donis-Keller. 1985. A polymorphic DNA marker linked to cystic fibrosis is located on chromosome 7. *Nature* **318:** 380.

Lathrop, G.M., J.M. Lalouel, C. Julier, and J. Ott. 1984. Strategies for multilocus linkage analysis in humans. *Proc. Natl. Acad. Sci.* **81:** 3443.

Le Beau, W.A., M.O. Diaz, J.D. Rowley, and T.W. Mak. 1984. Chromosome 7 localization of the human T-cell receptor beta chain genes. *Cell* **41:** 435.

Meyers, J.C., M.-L. Chu, S.H. Faro, W.J. Clark, D.J. Prockop, and F. Ramirez. 1981. Cloning a cDNA for the proα2 chain of human type 1 collagen. *Proc. Natl. Acad. Sci.* **78:** 3516.

Morton, C.C., A.D. Duby, R.L. Eddy, T. Shows, and J.G. Seidman. 1985. Genes for the beta chain of human T-cell antigen receptor map to regions of chromosomal rearrangement in T cells. *Science* **228:** 582.

Morton, N.E. 1955. Sequential tests for the detection of linkage. *Am. J. Hum. Genet.* **7:** 277.

Ott, J. 1974. Estimation of the recombinant fraction in human pedigrees: Efficient computation of the likelihood for human linkage studies. *Am. J. Hum. Genet.* **26:** 588.

Quinton, P.M. 1983. Chloride impermeability in cystic fibrosis. *Nature* **301:** 421.

Quinton, P.M. and J. Bijman. 1983. Higher bioelectric potentials due to decreased chloride absorption in the sweat glands of patients with cystic fibrosis. *N. Engl. J. Med.* **308:** 1185.

Retief, E., M.I. Parker, and A.E. Retief. 1985. Regional chromosome mapping of human collagen genes alpha 2(I) and alpha (I) (*COL1A2* and *COL1A1*). *Hum. Genet.* **69:** 304.

Röhme, D., H. Fox, B. Herrmann, A.-F. Fritchauf, J.-E. Edström, P. Mains, L.M. Silver, and H. Lehrach. 1984. Molecular clones of the mouse *t* complex derived from microdissected metaphase chromosomes. *Cell* **36:** 783.

Romeo, G., M. Bianco, M. Devoto, P. Menozzi, G. Mastella, A.M. Giunta, C. Micalizzi, M. Antonelli, A. Battistini, F. Santamaria, D. Castello, M. Marianelli, A.G. Marchi, A. Manca, and A. Miano. 1985. Incidence in Italy, genetic heterogeneity, and segregation analysis of cystic fibrosis. *Am. J. Hum. Genet.* **37:** 338.

Sato, K. 1984. Differing luminal potential difference of cystic fibrosis and control sweat secretory coils *in vitro*. *Am. J. Physiol.* **247:** R646.

Scambler, P.J., B.J. Wainwright, E. Watson, G. Bates, G. Bell, R. Williamson, and M. Farrall. 1986. Isolation of a further anonymous informative DNA sequence from chromosome seven closely linked to cystic fibrosis. *Nucleic Acids Res.* **14:** 1951.

Scambler, P.J., B.J. Wainwright, M. Farrall, J. Bell, P. Stanier, N.J. Lench, G. Bell, H. Kruyer, F. Ramirez, and R. Williamson. 1985. Linkage of *COL1A2* collagen gene to cystic fibrosis and its clinical implications. *Lancet* **II:** 1241.

Schmiegelow, K., H. Eiberg, L.-C. Tsui, M. Buchwald, P.D. Phelan, R. Williamson, W. Warwick, E. Niebuhr, J. Mohr, M. Schwartz, and C. Koch. 1986. Linkage between the loci for cystic fibrosis and paraoxonase. *Clin. Genet.* **29:** 374.

Southern, E.M. 1975. Detection of specific sequences among DNA fragments separated by gel electrophoresis. *J. Mol. Biol.* **98:** 503.

Steinberg, A.G. and N.E. Morton. 1956. Sequential test for linkage between cystic fibrosis of the pancreas and the MNS locus. *Am. J. Hum. Genet.* **8:** 177.

Steinberg, A.G., H. Schwachman, F.H. Allen, and R.R. Dooley. 1956. Linkage studies with cystic fibrosis of the pancreas. *Am. J. Hum. Genet.* **8:**162.

Talamo, R.C., B.J. Rosenstein, and R.W. Berninger. 1983. Cystic fibrosis. In *The metabolic basis of inherited disease* (ed. J.B. Stanbury et al.), p. 1889. McGraw-Hill, New York.

Thompson, M.W. 1980. Genetics of cystic fibrosis. In *Perspectives in cystic fibrosis* (ed. J.M. Sturgess). p. 281. Canadian Cystic Fibrosis Foundation, Toronto.

Tsipouras, P., A.L. Borresen, L.A. Dickson, K. Berg, D.J. Prockop, and F. Ramirez. 1984. Molecular heterogeneity in the mild autosomal dominant forms of osteogenesis imperfecta. *Am. J. Hum. Genet.* **36:** 1172.

Tsui, L.-C., D.W. Cox, P.A. McAlpine, and M. Buchwald. 1985a. Cystic fibrosis: Analysis of linkage of the disease locus to red cell and plasma protein markers. *Cytogenet. Cell Genet.* **39:** 238.

Tsui, L.-C., M. Zsiga, D. Kennedy, N. Plavsic, D. Markiewicz, and M. Buchwald. 1985b. Cystic fibrosis: Progress in mapping the disease loci using polymorphic DNA markers. I. *Cytogenet. Cell Genet.* **39:** 299.

Tsui, L.-C., M. Buchwald, D. Barker, J.C. Braman, R. Knowlton, J.W. Schumm, H. Eiberg, J. Mohr, D. Kennedy, N. Plavsic, M. Zsiga, D. Markiewicz, G. Akots, V.

Brown, C. Helms, T. Gravius, C. Parker, K. Rediker, and H. Donis-Keller. 1985c. Cystic fibrosis locus defined by a genetically linked polymorphic DNA marker. *Science* **230:** 1054.

Wainwright, B.J., P. Scambler, M. Farrall, M. Swartz, and R. Williamson. 1986. Linkage between the cystic fibrosis locus and markers on chromosome 7q. *Cytogenet. Cell Genet.* **41:** 191.

Wainwright, B.J., P.J. Scambler, J. Schmidtke, E.A. Watson, H.-Y. Law, M. Farrall, H.J. Cooke, H. Eiberg, and R. Williamson. 1985. Localization of cystic fibrosis locus to human chromosome 7cen-q22. *Nature* **318:** 382.

Warwick, W.J., R.E. Pogue, H.U. Gerber, and C.J. Nesbitt. 1975. Survival patterns in cystic fibrosis. *Am. J. Hum. Genet.* **36:** 1399.

White, R., S. Woodward, M. Leppert, P. O'Connell, Y. Nakamura, M. Hoff, J. Herbst, J.-M. Lalouel, M. Dean, and G. Vande Woude. 1985. A closely linked genetic marker for cystic fibrosis. *Nature* **318:** 382.

Widdicombe, J.H., M.J. Welsh, and W.E. Finkbeiner. 1985. Cystic fibrosis decreases the apical membrane chloride permeability of monolayers cultured from cells of tracheal epithelium. *Proc. Natl. Acad. Sci.* **82:** 6167.

Willard, H.F. and J.R. Riordan. 1985. Assignment of the gene for myelin proteolipid protein to the X-chromosome in man and mouse: Implications for X-linked inherited disorders of myelin. *Science* **230:** 940.

Willard, H.F., S. Meakin, L.-C. Tsui, and M.L. Breitman. 1985. Assignment of the human γ-crystallin multigene family to chromosome 2. *Somat. Cell Mol. Genet.* **11:** 511.

Yankaskas, J.R., M.R. Knowles, J.T. Gatzy, and R.C. Boucher. 1985. Persistence of abnormal chloride ion permeability in cystic fibrosis nasal epithelial cells in heterologous culture. *Lancet* **I:** 954.

Molecular Analysis of Human X-linked Diseases

K.E. Davies, S.P. Ball, H.R. Dorkins, S.M. Forrest, S.J. Kenwrick, A.W. King, I.J.D. Lavenir, S.A. McGlade, M.N. Patterson, T.J. Smith, L. Wilson, K. Paulsen,* A. Speer,† and C. Coutelle†

*Nuffield Department of Clinical Medicine, John Radcliffe Hospital, Oxford OX3 9DU; *Institüt für Humagenetik und Anthropologie, Freiburg, Federal Republic of Germany; †Institute of Molecular Biology, Academy of Sciences of the German Democratic Republic, 1115 Berlin*

Many genes and phenotypes have been mapped to the human X chromosome because of their characteristic sex-linked mode of inheritance in families (McKusick 1983). Thus, some 100 disease loci and at least 100 DNA markers have been localized to the human X chromosome in recent years (Goodfellow et al. 1985). Until a few years ago, only two main linkage groups were established, one localized at Xpter and the other at Xqter (McKusick and Ruddle 1977). With the advent of DNA recombinant technology and the application of restriction-fragment-length polymorphisms (RFLPs) as genetic markers, it has been possible to construct a map spanning the entire human X chromosome (Botstein et al. 1980; Drayna and White 1985). Markers are sufficiently well distributed along the chromosome to enable the mapping of almost any common sex-linked disorder. In particular, some of these have been used for the development of markers for the carrier detection and antenatal diagnosis of Duchenne muscular dystrophy and hemophilia A, and studies are under way to find closely linked markers for the fragile X syndrome (for review, see Davies 1985). These linkage studies have provided the first step in the identification of the basic biochemical defects in several X-linked disorders. Various strategies are now being employed to proceed from these linked markers to the identification of the gene sequences involved in the development of the phenotype.

Genetic Mapping of the X Chromosome

Using eight RFLPs and conventional markers such as the *XG* antigen, we constructed a map of the human X chromosome where we estimated the length to be about 200 cM (Drayna et al. 1984). These studies demonstrated an increase in frequency of recombination toward the telomeres, similar to that observed previously for autosomes (Laurie et al. 1981). In particular, *factor IX* (localized at Xq27) and *DX13* (localized within Xq28) were found to be only loosely linked despite the relatively short distance between them (Drayna et al. 1984). Similarly, on the short arm of the human X chromosome, *RC8* (localized at Xp22.3) is not closely linked to *XG* (Sarfarazi et al. 1983) and only shows loose linkage with the steroid sulphatase locus (Wieacker et al. 1983a) and the retinoschisis locus (Wieacker et al. 1983b). It is therefore apparent that the ease with which a genetic disease can be mapped depends very much on its localization on the chromosome. More markers will be needed to localize sequences near the telomeres compared with those that lie more centromeric.

The great limitation of the present X-chromosome linkage map is that the vast majority of the markers are dimorphic, and therefore several families segregating for a particular genetic disease are required in order to establish linkage. This also severely limits the use of some of these probes in antenatal diagnosis and carrier detection of X-linked disorders. We, and others, have also noted that random X-chromosome DNA probes appear to be less polymorphic than random probes isolated from autosomes (Oberle et al. 1986). Unfortunately, screening our X library with the hypervariable sequences isolated by Jeffreys et al. (1985) suggested that this type of sequence does not exist on the human X chromosome. We are currently using other hypervariable markers to screen the human X library in order to identify loci that are highly polymorphic.

Duchenne Muscular Dystrophy

Linkage studies and diagnosis. Duchenne muscular dystrophy (DMD) and Becker muscular dystrophy (BMD) are X-linked recessive disorders, with DMD being the most common and more severe, with a frequency of about 1 in 2500 newborn males (Moser 1984). The biochemical defects responsible for these diseases are unknown, and until recently, the methods of carrier detection were not totally reliable. Over the past 3 years, however, it has become possible to approach the study of DMD and BMD through gene mapping using RFLPs (Murray et al. 1982; Davies et al. 1983). These studies have localized the DMD and BMD mutations to the same region of the human X chromosome within Xp21. This is the region where breakpoints occur in female patients with balanced X;autosome translocations (for review, see Elejalde and Elejalde 1983). These females are affected because the translocated X chromosome is active, whereas the normal X chromosome is inactive. The break at Xp21 may modify gene expression in this region either by directly interrupting gene expression or by indirectly exerting a position effect on a gene complex.

The first RFLP found to be linked to the *DMD* locus was *RC8*, a DNA sequence localized at Xp22.3 proximal to *XG*. This locus was found to be linked at a recombination fraction of 0.17 to the *DMD* locus (Murray et al. 1982). There are at least 11 RFLPs bridging the *DMD* and *BMD* loci for use in carrier status determinations (Aldridge et al. 1984; Bakker et al. 1985; Davies 1985). More recently, sequences have been isolated that are deleted in a small percentage of DMD patients (*pERT87* probes) (Monaco et al. 1985), and a probe has been isolated that hybridizes to the breakpoint in the female with an X;21 translocation (Ray et al. 1985) (*XJ1.1*). Surprisingly, these recombine with the mutation in at least 5% of the meioses studied. Both of these probes are polymorphic and useful for antenatal diagnosis. However, because of the uncertainty of the position of the mutation in DMD and BMD families, it is not possible to use these markers in conjunction with any other as bracketing the mutation. In addition, some of the RFLPs are in linkage disequilibrium, which limits their use. Figure 1 shows the use of the *Bst*XI RFLP of *pERT87-8* for carrier status determination in a family segregating for a deletion. Both of the affected males are deleted for the *pERT87-8* locus. The female (Fig. 1, proband II-1) has inherited the 2.2-kb band paternally and does not show any maternal allele. She must therefore have inherited a deleted X chromosome maternally and be at high risk of being a carrier.

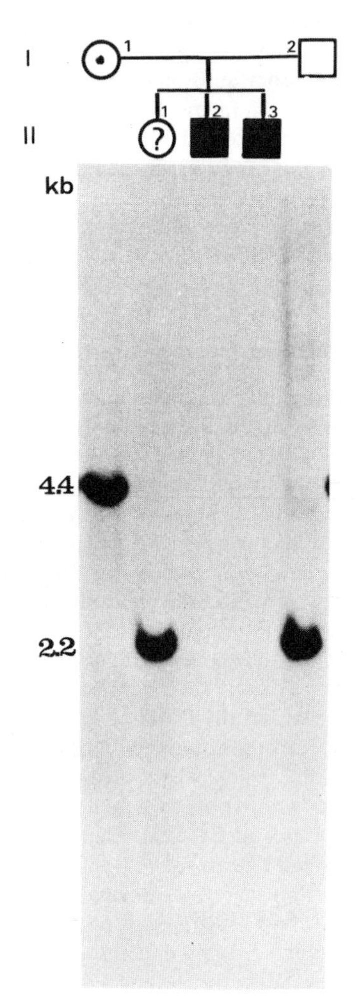

Figure 1. Southern blot analysis with *pERT87-8* of a family affected by DMD.

One of the most surprising results has been that one of the probes, *754*, which lies within the deletion of 6000 kbp in a patient suffering from DMD, chronic granulomatous disease, and retinitis pigmentosa, is only linked at a recombination fraction of 0.10 (Davies et al. 1985; Francke et al. 1985; Goodfellow et al. 1985). This suggests that there might be a high rate of recombination in this region of the human X chromosome. It is not yet clear whether this is a consequence of the *DMD* mutation or whether the same linkage relationships are observed in normal families. We have investigated this by comparing recombination rates in DMD families with those in "normal" families. The latter included families that are affected by an X-linked disease but should be normal with respect to Xp21. Tables 1 and 2 present the linkage data that we have obtained between markers that bracket the *DMD* and *BMD* loci. The figures given include previously published data (Davies et al. 1985; Dorkins et al. 1985). Note that in most cases, data are only available for the DMD/BMD families. However, where possible, the two family types were compared using the A test (Ott 1983). Two hypotheses were used: In the first, all families are linked at a single θ value, and in the second, although both are linked, the recombination rate differs in DMD and normal families. Recombination rates in the intervals *754-p99-6* and *pERT87-p99-6* are higher in our DMD families. The A test suggests that this represents a real difference, the posterior probability of the DMD and normal families being of the same type is 0.5. For the other comparison, there was no evidence of genetic heterogeneity. Further work is under way to validate this conclusion.

Isolation of expressed sequences. The identification and characterization of sequences on the X chromosome that are expressed, or related to expressed genes, are important for an understanding of the role of the X chromosome in development, sex determination, and X-linked genetic disease. We previously isolated a series of X-coded genes from a human X-chromosome recombinant library that recognize a single or a few mRNA species (Benham et al. 1984). Assuming that there are 50,000 genes in the human genome and that only 10,000 of these are expressed in each tissue type, then the X chromosome would encode approximately 400 muscle proteins. Any one of these would be a candidate gene sequence for an involvement in the etiology of DMD and BMD. Some of these sequences can be isolated from an X-chromosome-enriched library by screening it with mRNA molecules isolated from fetal or adult muscle cells. Alternatively, a cDNA library can be screened with X-library DNA that has been enriched for single-copy sequences. Sequences identified by either of these approaches can then be

Table 1. Lod Scores between Marker Loci in DMD/BMD Families

θ	= 0.00	0.05	0.10	0.20	0.30	0.40
DMD-L1.28	– ∞	– 1.31	– 0.82	– 0.40	– 0.20	– 0.07
DMD-OTC	– ∞	2.11	2.60	2.42	1.69	0.81
DMD-754	– ∞	6.16	7.09	6.39	4.50	2.16
DMD-pERT87	3.82	3.48	3.09	2.24	1.35	0.56
DMD-C7	– ∞	7.01	6.98	5.66	3.76	1.72
DMD-p99-6	0.68	0.62	0.55	0.40	0.25	0.11
DMD-pD2	– ∞	– 0.39	– 0.15	0.02	0.05	0.04
DMD-RC8	– 0.18	– 0.16	– 0.14	– 0.10	– 0.06	– 0.03
L1.28-OTC	0.24	0.20	0.17	0.11	0.07	0.03
L1.28-754	– ∞	0.60	0.72	0.63	0.38	0.12
L1.28-pERT87	– ∞	– 0.79	– 0.50	– 0.22	– 0.09	– 0.02
L1.28-p99-6	– ∞	– 1.32	– 0.78	– 0.33	– 0.12	– 0.03
L1.28-pD2	– ∞	– 0.68	– 0.41	– 0.17	– 0.07	– 0.02
OTC-754	– ∞	0.72	0.80	0.68	0.45	0.21
OTC-pERT87	– ∞	– 0.25	– 0.03	0.12	0.12	0.08
OTC-C7	0.30	0.28	0.26	0.20	0.15	0.08
OTC-pD2	– ∞	– 0.46	– 0.23	– 0.06	– 0.01	0.00
OTC-RC8	0.30	0.28	0.26	0.20	0.15	0.08
754-pERT87	– ∞	0.61	1.17	1.31	1.03	0.56
754-C7	– ∞	3.21	3.56	3.11	2.08	0.85
754-p99-6	– ∞	– 1.93	– 1.15	– 0.48	– 0.20	– 0.06
754-pD2	– ∞	– 0.66	– 0.19	0.17	0.20	0.14
754-RC8	0.30	0.28	0.26	0.20	0.15	0.08
pERT87-C7	– ∞	0.39	0.53	0.50	0.35	0.18
pERT87-p99-6	0.43	0.38	0.33	0.22	0.11	0.02
pERT87-pD2	– ∞	0.07	0.24	0.28	0.20	0.10
C7-RC8	0.30	0.28	0.26	0.20	0.15	0.08

θ = recombination fraction. Lod scores were calculated using the program MLINK (Lathrop and Lalouel 1984).

localized to the Xp21 region using somatic-cell hybrids or in situ hybridization.

We have isolated several clones that lie within or very close to the DMD region using this method (K. Paulsen et al., in prep.). The X library was screened with mRNA in the presence of sonicated human DNA as competitor, and the putative expressed sequences were subsequently localized using a hybrid panel and a series of patients carrying deletions. Figure 2 shows a schematic diagram for the localization of two of these clones (*2bA3* and *1aE3*), together with other sequences within Xp21 isolated in our laboratory and elsewhere (Dorkins et al. 1985; Hofker et al. 1985; Ingle et al. 1985; Monaco et al. 1985; Ray et al. 1985; Oberle et al. 1986). The clones have been mapped by their absence or presence in DNA isolated from patients with different deletions of Xp21. Some of these patients suffer from glycerol kinase deficiency (GK) and adrenal hypoplasia (AHC), in addition to DMD (Wieringa et al. 1985; Dunger et al. 1986). Interestingly, one of the patients has a deletion extending 6000 kb, as assayed by flow cytometry (Wilcox et al. 1986). This patient is deleted for *2bA3* but not *1aE3*.

Characterization of deletion patients. Much evidence is now accumulating to suggest that there may be great heterogeneity in the genetic lesions that give rise to the DMD phenotype. Also, in view of the recombination rate between the *pERT87* and *XJ1.1* loci and the *DMD* locus, it is likely that the DMD region is very large. This may mean that the gene in this disease (if there is a gene involved) is at least a megabase long or that there are several genes that can be mutated to cause the disease. Several patients with complex phenotypes including DMD have now been analyzed. We have made a detailed study of two patients suffering

Table 2. Lod Scores between Marker Loci in Normal Families

θ	= 0.00	0.05	0.10	0.20	0.30	0.40
L1.28-754	0.27	0.29	0.30	0.25	0.16	0.06
L1.28-pD2	0.06	0.06	0.06	0.04	0.03	0.02
OTC-754	– 0.05	– 0.04	– 0.03	– 0.02	– 0.01	0.00
OTC-RC8	– 0.05	– 0.04	– 0.03	– 0.02	– 0.01	0.00
754-pERT87	1.22	1.12	1.02	0.82	0.59	0.32
754-p99-6	0.90	0.84	0.77	0.61	0.44	0.24
754-pD2	0.38	0.32	0.25	0.13	0.04	– 0.01
754-RC8	– ∞	0.48	0.84	0.92	0.69	0.32
pERT87-p99-6	– ∞	– 0.79	– 0.10	0.37	0.42	0.28

θ = recombination fraction. Lod scores were calculated using the program MLINK (Lathrop and Lalouel 1984).

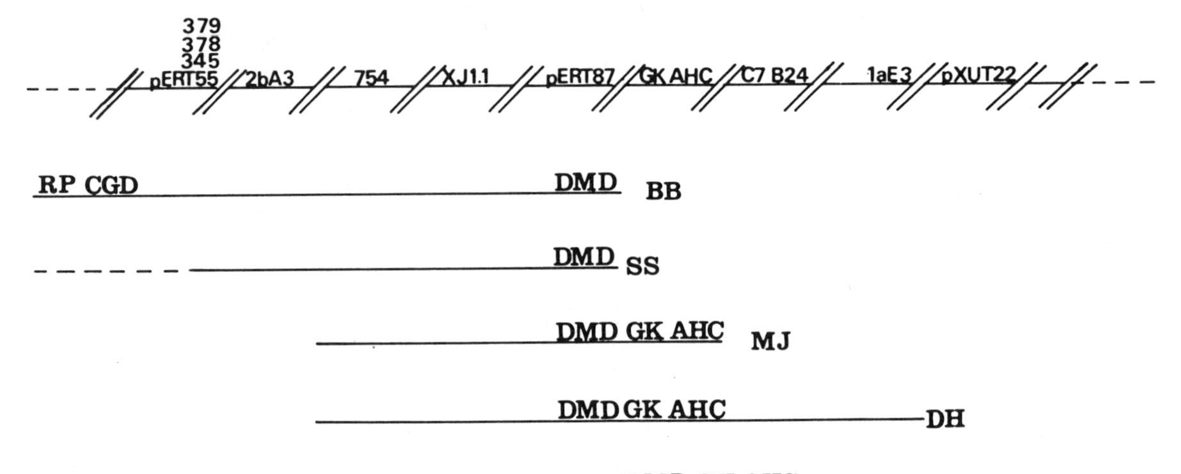

Figure 2. Schematic presentation of the mapping of probes across Xp21 within the deletions of unrelated patients displaying different clinical phenotypes.

from GK, AHC, and DMD (Dunger et al. 1986). These patients do not show cytologically visible deletions within Xp21, and one of them is not deleted for either the *pERT87* or the *XJ1.1* loci. This patient must therefore have a mutation giving rise to DMD that lies distal to the *pERT87* region if this clinical phenotype is the result of a simple deletion. Neither of these patients suffers from chronic granulomatous disease (CGD). However, the patient characterized by Francke et al. (1985), and shown to have a deletion of Xp21.2, suffers from CGD, DMD, and retinitis pigmentosa. This suggests that the *CGD* locus must lie proximal to the *DMD* mutation and that the *GK* and *AHC* loci lie distal to the *DMD* mutation. However, we have also characterized one patient suffering from DMD and CGD who is clearly not deleted for any of the probes so far isolated.

One of the clones isolated by screening the X library with fetal muscle mRNA is *2bA3*, which maps in the BB deletion. This clone does not map in the deletions of the two patients who are *GK*$^-$, *AHC,* and *DMD* mentioned above. Therefore, *2bA3* must lie between the breakpoints of these patients and the BB breakpoint. We are investigating the extent of this region using pulsed-field gradient (PFG) electrophoresis.

PFG electrophoresis is a technique that can be used for resolving large fragments of DNA of 10–1000 kb. Unsheared high-molecular-weight DNA is isolated from growing cells that have been encapsulated in agarose beads. The pores in the beads are of a size that permit the free exchange of quite large molecules such as protein and RNA, but not chromosomal DNA. The cells can be lysed and the protein and RNA extracted, leaving DNA in the form of essentially native chromatin in the beads. In this state, the DNA can be digested with infrequently cutting restriction enzymes such as *Not*I, *Sfi*I, and *Sma*I. These enzymes tend to cleave the DNA into large fragments because they all have the sequence CpG in their recognition sites, which is underrepresented in the mammalian genome (Bird et al. 1985). The DNA fragments are then resolved by PFG electrophoresis, blotted, and hybridized to probes using standard methods (Carle and Olson 1984; Schwartz and Cantor 1984). Figure 3 shows an example of one of the X short-arm probes hybridized to human DNA cut with the enzyme *Sfi*I.

The mapping of sequences relative to the deletion breakpoints in patients should give us some information about the extent of the region that is involved in the mutation in DMD.

X-linked Mental Retardation

The most common form of X-linked mental retardation (the Martin Bell syndrome) is associated with a fragile site at Xq27.3 (for review, see Turner and Jacobs 1984). The fragile site can be induced in vitro by growing lymphocytes in folate-deficient medium (Suther-

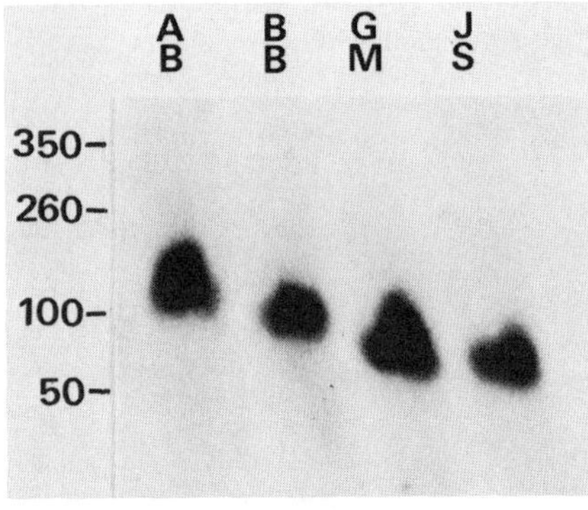

Figure 3. Analysis of DNA from four patients by pulsed-field electrophoresis after digestion with *Sfi*I. Hybridization with the short arm probe pSB1.8.

land 1983). It is not always observed in affected individuals nor is it always seen in obligate carriers of the disorder. The disorder was thought originally to follow classical X-linked inheritance, but recently it has become apparent that phenotypically normal males can also transmit the disease (for review, see Sherman et al. 1984, 1985). Normal-transmitting males have affected sibs, and their daughters are almost never mentally retarded. Approximately 30% of the daughters of carrier females are mentally retarded. These puzzling observations make the analysis of this condition by conventional methods difficult and severely limit the antenatal diagnosis of the fragile X by cytogenetic techniques. The RFLP markers that have been studied fall into two main linkage groups: those that lie distal to the fragile site and those that lie proximal in the region Xq27. *Factor IX (F9)*, which lies proximal to the fragile site, demonstrates linkage to the fragile X syndrome with a peak lod score of 4.88 at a recombination fraction of 0.25. Linkage analysis for probes distal to the fragile site in the cluster close to *hemophilia A, DX13, St14* indicate that these markers are linked at a recombination frequency of approximately 0.20. Brown et al. (1985) have noted differences in the frequency of recombination between families where segregation occurred through a normal-transmitting male and those families where segregation was through an obligate carrier female. In our families, there are certainly instances of male transmission where recombination occurs very frequently across the fragile X locus. We therefore suggest that although there may be heterogeneity, the basis for this is not determined by the class of heterozygote (those that inherit a mutation from their phenotypically normal fathers and those that inherit a mutation from their obligatory carrier mothers). We are attempting to identify new markers in the region of the fragile site and are currently investigating a number of these for RFLPs and ordering them relative to existing markers. We are also studying this region by PFG electrophoresis in order to determine whether there is any chromosomal rearrangement involved in the development of the mutation.

We have suggested the existence of a premutation to account for the segregation patterns observed in the families (Pembrey et al. 1985, 1986). Such a premutation might be converted into the full syndrome by a chromosomal structural change occurring during female meiosis. Alternative theories have been proposed by Friedman and Howard-Peebles (1986) and Hoegerman and Rary (1986), where they have suggested the involvement of a transposable genetic element. More recently, a paper was presented that suggests the involvement of an autosomal suppressor gene (Steinbach 1986). To explain the data, the suppressor would have to be recessive in males carrying the mutant X-linked locus but dominant in hemizygous females. This could be the result of Lyonization. However, this is not the explanation in all cases as shown from the twin study of Tuckerman et al. (1985). The investigation of the underlying biochemical mechanism causing this syndrome awaits the isolation of more DNA probes in the Xq27/28 region.

Mapping of Rare X-linked Disorders

The linkage map established for RFLP markers in normal families can greatly assist in the mapping of rare genetic disorders. The information available on the linkage distances between the RFLP markers themselves contained in normal families can be used to do multilocus linkage analysis of the segregation of the disease locus, together with the informative markers in a particular family. An example of the usefulness of this approach is demonstrated by a mapping of X-linked spastic paraplegia (Kenwrick et al. 1986). We analyzed one large kindred from Iowa City with several X-linked markers. The lod scores were found to be negative for all of the markers except *DX13* and *St14*. From this family alone, information on the pairwise linkage distances between these markers and the spastic paraplegia locus (calculated with MLINK) showed linkage to *DX13* and *St14*, with a lod score of 3.18 for *DX13* and a lod score of 3.14 for *St14*. By using the linkage program LINKMAP, we determined a location score of 14.3 for the spastic paraplegia locus being at Xq27/28. Other families will be needed in order to establish whether the mutation lies distal or proximal to these loci.

ACKNOWLEDGMENTS

We thank the Muscular Dystrophy Group of Great Britian, The Muscular Dystrophy Association of America, and the Medical Research Council for their generous support. Part of this work was supported by a Wellcome Trust grant to St. Mary's Hospital Medical School allowing a 3-month working visit for C.C. at its Biochemistry Department. We are also grateful to Rachel Kitt for careful typing of the manuscript; for gifts of probes from Professor Brownlee (Oxford), Drs. Kunkel (Boston), Mandel (Strasbourg), Pearson (Leiden), Willard (Toronto), and Worton (Toronto); and for cell lines from Drs. Francke (Yale) and Ferguson-Smith (Glasgow).

REFERENCES

Aldridge, J., L. Kunkel, G. Bruns, M. Lalande, U. Tantravahi, and S. Latt. 1984. A strategy for construction of highly polymorphic DNA haplotypes in specific human chromosomal regions: Application to linkage analysis of X chromosome specific diseases such as Duchenne muscular dystrophy. *Ital. J. Neurol. Sci.* **3:** 39.

Bakker, E., M.H. Hofker, N. Goor, J.L. Mandel, K. Wrogemann, K.E. Davies, L.M. Kunkel, H.F. Willard, W.A. Fenton, L. Sandkuyl, D. Majoor-Krakauer, A.J. Van Essen, M.G.J. Jahoda, E.S. Sachs, G.J.B. Van Ommen, and P.L. Pearson. 1985. Prenatal diagnosis and carrier detection of Duchenne muscular dystrophy with closely linked RFLPs. *Lancet* **I:** 655.

Benham, F.J., S. Hodgkinson, and K.E. Davies. 1984. A glyceraldehyde-3-phosphate dehydrogenase pseudogene on the short arm of the human X chromosome defines a multigene family. *EMBO J.* **3:** 2635.

Bird, A., M. Taggart, M. Frommer, O.J. Miller, and D. Macleod. 1985. A fraction of the mouse genome that is derived from islands of nonmethylated, CpG-rich DNA. *Cell* **40:** 91.

Botstein, D., R.L. White, M.H. Scolnick, and R.W. Davis. 1980. Construction of a genetic linkage map in man using restriction fragment length polymorphisms. *Am. J. Hum. Genet.* **32:** 314.

Brown, W.T., A.C. Gross, C.B. Chan, and E.C. Jenkins. 1985. Genetic linkage heterogeneity in the fragile X syndrome. *Hum. Genet.* **71:** 11.

Carle, G.F. and M.V. Olson. 1984. Separation of chromosomal DNA molecules from yeast by orthogonal-field alternation gel electrophoresis. *Nucleic Acids Res.* **12:** 5647.

Davies, K.E. 1985. Molecular genetics of the human X chromosome. *J. Med. Genet.* **22:** 243.

Davies, K.E., P.L. Pearson, P.S. Harper, J.M. Murray, T. O'Brien, M. Sarfarazi, and R. Williamson. 1983. Linkage analysis of the two cloned DNA sequences flanking the Duchenne muscular dystrophy locus on the short arm of the human X chromosome. *Nucleic Acids Res.* **11:** 2303.

Davies, K.E., A. Speer, F. Herrmann, A.W.J. Spiegler, S. McGlade, M.H. Hofker, P. Briand, R. Hanke, M. Schwartz, V. Steinbicker, R. Szibor, H. Korner, D. Sommer, P.L. Pearson, and C. Coutelle. 1985. Human X chromosome markers and Duchenne muscular dystrophy. *Nucleic Acids Res.* **13:** 3419.

Dorkins, H.R., C. Junien, J.-L. Mandel, K. Wrogemann, J.P. Moisan, M. Martinez, J.M. Old, S. Bundey, M. Schwartz, N.J. Carpenter, D. Hill, M. Lindlof, A. de la Chapelle, P.L. Pearson, and K.E. Davies. 1985. Segregation analysis of a marker localised Xp21.2-Xp21.3 in Duchenne and Becker muscular dystrophy families. *Hum. Genet.* **71:** 103.

Drayna, D. and R. White. 1985. The genetic linkage map of the human X chromosome. *Science* **230:** 753.

Drayna, D., K.E. Davies, D.A. Hartley, J.-L. Mandel, G. Camerino, R. Williamson, and R. White. 1984. Genetic mapping of the human X chromosome by using restriction fragment length polymorphisms. *Proc. Natl. Acad. Sci.* **81:** 2836.

Dunger, D.B., K.E. Davies, M. Pembrey, B. Lake, P. Pearson, D. Williams, A. Whitfield, and M.J.D. Dillon. 1986. Deletion on the X chromosome detected by direct DNA analysis in one of two unrelated boys with glycerol kinase deficiency, adrenal hypoplasia and Duchenne muscular dystrophy. *Lancet* **I:** 585.

Elejalde, B.R. and M.M. Elejalde. 1983. Phenotypic manifestations of X-autosome translocations. In *Cytogenetics of the X chromosome* (ed. A.A. Sandberg), p. 225. A.R. Liss, New York.

Francke, U., H.D. Ochs, B. De Martinville, J. Giacalone, V. Lindgren, C. Disteche, R.A. Pagon, M.H. Hofker, G.J.B. Van Ommen, P.L. Pearson, and R.J. Wedgewood. 1985. Minor Xp21 chromosome deletion in a male associated with expression of Duchenne muscular dystrophy, chronic granulomatous disease, retinitis pigmentosa, and McLeod syndrome. *Am. J. Hum. Genet.* **37:** 250.

Friedman, J.M. and P.N. Howard-Peebles. 1986. Inheritance of fragile X syndrome: An hypothesis. *Am. J. Med. Genet.* **23:** 701.

Goodfellow, P., K.E. Davies, and H.-H. Ropers. 1985. Report of the committee on the genetic constitution of the X and Y chromosome. In *Human gene mapping conference VIII,* p. 296. March of Dimes Birth Defects Foundation, White Plains, New York.

Hoegerman, S.F. and J.M. Rary. 1986. Speculation on the role of transposable elements in human genetic disease with particular attention to achondroplasia and the fragile X syndrome. *Am. J. Med. Genet.* **23:** 685.

Hofker, M.H., M.C. Wapenaar, N. Goor, E. Bakker, G.-J.B. Van Ommen, and P.L. Pearson. 1985. Isolation of probes detecting restriction fragment length polymorphisms from X chromosome-specific libraries: Potential use for diagnosis of Duchenne muscular dystrophy. *Hum. Genet.* **70:** 148.

Ingle, C., R. Williamson, A. de la Chapelle, R.R Herva, K. Haapala, G. Bates, H.F. Willard, P.L. Pearson, and K.E. Davies. 1985. Mapping DNA sequences in a human X chromosome deletion which extends across the region of the Duchenne muscular dystrophy mutation. *Am. J. Hum. Genet.* **37:** 451.

Jeffreys, A.J., V. Wilson, and S.L. Thein. 1985. Hypervariable "minisatellite" regions in human DNA. *Nature* **314:** 67.

Kenwrick, S.J., V. Ionasescu, G. Ionasescu, C. Searby, A. King, M. Dubowitz, and K.E. Davies. 1986. Linkage studies of X-linked recessive spastic paraplegia using DNA probes. *Hum. Genet.* **73:** 264.

Lathrop, G.M. and J.M. Lalouel. 1984. Easy calculations of lod scores and genetic risks on small computers. *Am. J. Hum. Genet.* **36:** 460.

Laurie, D.A., M. Hulten, and G.H. Jones. 1981. Chiasma frequency and distribution in a sample of human males. *Cytogenet. Cell Genet.* **31:** 153.

McKusick, V.A. 1983. *Mendelian inheritance in man.* 6th edition. Johns Hopkins Press, Baltimore.

McKusick, V.A. and F.H. Ruddle. 1977. The status of the gene map of the human chromosomes. *Science* **196:** 390.

Monaco, A.P., C.J. Bertelson, W. Middelsworth, C.-A. Colletti, J. Aldridge, K.H. Fischbeck, R. Bartlett, M.A. Pericak-Vance, A.D. Roses, and L.M. Kunkel. 1985. Detection of deletions spanning the Duchenne muscular dystrophy locus using a tightly linked DNA segment. *Nature* **316:** 842.

Moser, H. 1984. Duchenne muscular dystrophy: Pathogenic aspects and genetic prevention. *Hum. Genet.* **66:** 17.

Murray, J.M., K.E. Davies, P.S. Harper, L. Meredith, C.R. Mueller, and R. Williamson. 1982. Linkage relationship of a cloned DNA sequence on the short arm of the X chromosome to Duchenne muscular dystrophy. *Nature* **300:** 69.

Oberle, I., G. Camerino, C. Kloepfer, J.P. Moisan, K.H. Grzeschik, B. Hellkuhl, M.C. Hors-Cayla, N. van Cong, D. Weil, and J.-L. Mandel. 1986. Characterization of a set of X-linked sequences and a panel of somatic cell hybrids useful for the regional mapping of the human X chromosome. *Hum. Genet.* **72:** 43.

Ott, J. 1983. Linkage analysis and family classification under heterogeneity. *Ann. Hum. Genet.* **47:** 311.

Pembrey, M.E., R. Winter, and K.E. Davies. 1985. A premutation that generates a defect at crossing-over explains the inheritance of fragile (X) mental retardation. *Am. J. Med. Genet.* **21:** 709.

———. 1986. Fragile X mental retardation: Current controversies. *TINS Trends Neurosci.* **9:** 58.

Ray, P.N., B. Belfall, C. Duff, C. Logan, V. Kean, M.W. Thompson, J.E. Sylvester, J.L. Gorski, R.D. Schmickel, and R.G. Worton. 1985. Cloning of the breakpoint of an X;21 translocation associated with Duchenne muscular dystrophy. *Nature* **318:** 672.

Sarfarazi, M., P.S. Harper, H.M. Kingston, J.M. Murray, T. O'Brien, K.E. Davies, R. Williamson, P. Tippett, and R. Sanger. 1983. Genetic linkage relationships between the Xg blood group system and two X chromosome DNA polymorphisms in families with Duchenne and Becker muscular dystrophy. *Hum. Genet.* **65:** 169.

Schwartz, D.A. and C.R. Cantor. 1984. Separation of yeast chromosome-sized DNAs by pulsed field gradient gel electrophoresis. *Cell* **37:** 67.

Sherman, S.L., N.E. Morton, P.A. Jacobs, and G. Turner. 1984. The marker (X) syndrome: A cytogenetic and genetic analysis. *Ann. Hum. Genet.* **48:** 21.

Sherman, S.L., P.A. Jacobs, N.E. Morton, U. Froster-Iskenius, P.N. Howard-Peebles, K.B. Nielsen, M.W. Partington, G.R. Sutherland, G. Turner, and M. Watson. 1985. Further segregation analysis of the fragile X syndrome with special reference to transmitting males. *Hum. Genet.* **69:** 289.

Sutherland, G.R. 1983. The fragile X chromosome. *Int. Rev. Cytol.* **81:** 107.

Steinbach, P. 1986. Mental impairment in Martin-Bell syndrome is probably determined by interaction of several genes: Simple explanation of phenotypic differences between unaffected males with the same X chromosome. *Hum. Genet.* **72:** 248.

Tuckerman, E., T. Webb, and S.E. Bundey. 1985. Frequency and replication status of the fragile X, fra (X)(q27-28), in a pair of monozygotic twins and markedly differing intelligence. *J. Med. Genet.* **22:** 85.

Turner, G. and P.A. Jacobs. 1984. Marker (X) linked mental retardation. *Adv. Hum. Genet.* **13:** 83.

Wieacker, P., K.E. Davies, B. Bevorah, and H.-H. Ropers. 1983a. Linkage studies in a family with X-linked recessive ichthyosis employing a cloned DNA sequence from the distal short arm of the X chromosome. *Hum. Genet.* **63:** 113.

Wieacker, P., T.F. Wienker, B. Dallapiccola, K. Bender, K.E. Davies, and H.-H. Ropers. 1983b. Linkage relationships between retinoschisis, Xg and a cloned DNA sequence from the distal short arm of the X chromosome. *Hum. Genet.* **64:** 143.

Wieringa, B., T. Hustinx, J. Scheres, M. Hofker, H.H. Ropers, and B. Ter Haar. 1985. Glycerol kinse deficiency syndrome explained as X chromosomal deletion. *Cytogenet. Cell Genet.* **40:** 777.

Wilcox, D.E., A. Cooke, G. Colgan, E. Boyd, D.A. Aitkin, L. Sinclair, L. Glasgow, J.B.P. Stephenson, and M.A. Ferguson-Smith. 1986. Duchenne muscular dystrophy due to familial Xp21 deletion detectable by DNA analysis and flow cytometry. *Hum. Genet.* **73:** 175.

Analysis of an X-autosome Translocation Responsible for X-linked Muscular Dystrophy

R.G. WORTON, P.N. RAY, S. BODRUG, AND M.W. THOMPSON

Genetics Department and Research Institute, The Hospital for Sick Children, Toronto, Ontario, Canada M5G 1X8; Department of Medical Genetics, University of Toronto, Toronto, Canada

Duchenne muscular dystrophy (DMD) is an X-linked recessive disorder affecting approximately 1 in 3300 males, making it the most common of the neuromuscular dystrophies. Affected children show first signs of muscular weakness at about age 3, are confined to a wheelchair by age 10, and rarely live beyond age 20 (Moser 1984). Becker muscular dystrophy (BMD) is a similar disease, less frequent and less severe, but also X-linked. Evidence suggests that the genes responsible for the two disorders map close together on the X-chromosome short arm at band Xp21 (Kingston et al. 1983). Indeed, the two diseases may be caused by different mutations in the same gene locus.

Mapping of the *DMD* and *BMD* loci at band Xp21 comes from three lines of evidence. The first indication was the observation of several females with DMD or BMD and X-autosome translocations, the exchange point in the X chromosome occurring consistently at Xp21. Nonrandom inactivation of the normal X chromosome in all or most somatic cells of these females resulted in complete or near-complete expression of the disease phenotype. Because the translocation and the disease appeared de novo in the affected females, it has been taken as evidence that the translocation event has disrupted the activity of the *DMD* or *BMD* gene to cause the disease (Verellen-Dumoulin et al. 1984; Boyd and Buckle 1986).

The second line of evidence consisted of family studies showing segregation of the *DMD* and *BMD* gene loci with restriction-fragment-length polymorphisms (RFLPs) mapping on both sides of Xp21 (Davies et al. 1983; Davies 1985). Finally, cytologically visible deletions of the Xp21 region were found in patients ascertained with complex phenotypes, including DMD together with chronic granulomatous disease, glycerol kinase deficiency, and/or congenital adrenal hypoplasia (Francke et al. 1985; Kunkel et al., this volume).

Despite the fact that a single locus at band Xp21 seems to be responsible for DMD and BMD, recent data suggest that the locus may be very large. The earliest indication of this came from Boyd and Buckle (1986), who reexamined with high-resolution banding the chromosomes from several of the translocation females. They found that although many of the exchange points occurred in subband Xp21.2, two exchange points appeared to occur on the proximal side in subband Xp21.1 and two others appeared more distal at Xp21.3. This suggested that the exchange points, although close together on a cytogenetic scale, may be scattered over several thousand kilobases on a molecular scale.

In our studies, we have isolated by molecular cloning sequences from the junction of one of the X-autosome translocations. From this junction, we have walked along the X chromosome to isolate 60 kb of flanking sequence, have characterized the translocation in detail, and have isolated DNA fragments revealing RFLPs that segregate in families with the *DMD* gene. Our results, summarized here, are also consistent with a very large gene, or perhaps a set of related genes, at the *DMD/BMD* locus.

RESULTS

Isolation of an X:21 Junction Fragment

One of the first translocation females recognized was a Belgian girl identified by C. Verellen-Dumoulin to have a mild form of DMD, or in retrospect, BMD. The translocation, t(X;21) (p21;p12), is between the X chromosome and chromosome 21, with exchange points at band p21 in the X chromosome and at band p12 in chromosome 21 (Verellen-Dumoulin et al. 1984). Band 21 p12 is the site of 30–40 tandemly repeated copies of the gene encoding 18S and 28S rRNA, and the translocation has split the block of "ribosomal DNA" (rDNA) into two smaller blocks on each of the two translocation chromosomes (Worton et al. 1984). As depicted schematically in Figure 1, most of the rDNA repeat units have remained on the derivative 21 chromosome, whereas three-to-five repeat units were translocated onto the derivative X chromosome (Worton et

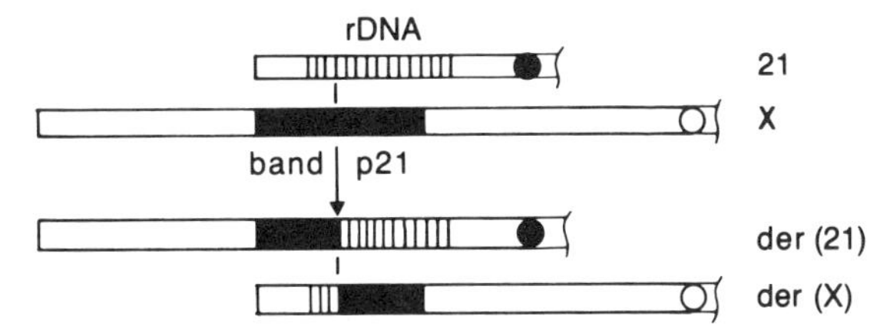

Figure 1. Schematic of t(X;21) translocation. Only short arms of chromosomes are shown. The translocation exchange points are in a block of ribosomal RNA genes in chromosome 21 and in band p21 of the X chromosome.

al. 1984). Our approach has been to separate the two translocation-derived chromosomes from other human chromosomes (numbers 13, 14, 15, 21, and 22) known to contain blocks of rDNA, by segregation in somatic-cell hybrids, and then to use existing rDNA probes to identify DNA fragments spanning the X:21 translocation junction.

Following fusion of patient's fibroblasts to mouse A9 cells, two segregated hybrids were available for this study. Hybrid A2 carries the der(X) chromosome as its only human chromosome, whereas hybrid C2-T10 carries the der(21) chromosome (Worton et al. 1984). Southern blot analysis of hybrid A2 DNA with a set of four rDNA probes revealed a unique junction fragment near the 3′ end of the 28S gene (Worton et al. 1986). Having mapped the junction site, we isolated a small human-specific segment (Fig. 2, 100-3) of the 28S gene, which was used as a hybridization probe to detect a 12-kb *Bam*HI fragment containing a short rDNA sequence at the 5′ end and about 11 kb of X chromosome at the 3′ end. To clone the junction fragment, DNA from the A2 hybrid line was cleaved with *Bam*HI, and size-selected fragments of 11–15 kb were cloned into λ Charon 35 (Ray et al. 1985). One of the X-junction clones (*XJ1*) isolated by plaque hybridization with probe 100-3 is depicted in Figure 2. The clone contains 620 bp of rDNA from the 28S gene (Fig. 2). The adjacent non-rDNA sequences were found to map to the short arm of the X chromosome, verifying that clone *XJ1* is the X:21 junction fragment (Ray et al. 1985). A subclone of *XJ1* (*XJ1.1*) contains a 1-kb *Nsi*I fragment (Fig. 2) that reveals an RFLP that segregates with the *DMD* gene in families and whose sequence is deleted in a proportion of male DMD patients (Ray et al. 1985). Thus, the *XJ1* clone contains sequences at or near the *DMD* locus.

In more recent studies (P.N. Ray et al., unpubl.), we have used the *Nsi*I fragment of clone *XJ1.1* as a probe to isolate clone *XJ2* from a 4X human DNA library. Clone *XJ2* extends leftward (Fig. 2) toward the telomere and crosses over the translocation site. Sequences from the left (5′) end of *XJ2* were used to isolate *XJ4*. In the other direction, sequences from the 3′ end of *XJ1* allowed the isolation of *XJ3*. In subsequent chromosome walking, over 60 kb of the X chromosome have been isolated in a series of six overlapping λ clones. Two new subclones (*XJ1.2* and *XJ2.3*) have been found to reveal RFLPs that also segregate with the *DMD* locus. Further discussion of the family studies is given below.

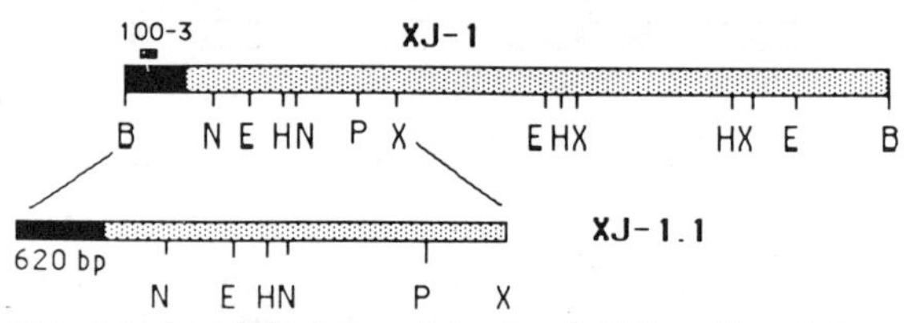

Figure 2. Restriction map of the clone λ *XJ1* and its subclone *XJ1.1*. The rDNA probe 100-3 was used to isolate the clone that contains 620 bp of rDNA at the 5′ end. (B) *Bam*HI; (E) *Eco*RI; (H) *Hind*III; (N) *Nsi*I; (P) *Pst*I; (X) *Xba*I.

Paternal Inheritance of Translocation Chromosomes

In hypothesizing that the translocations in affected females act by disrupting the *DMD/BMD* gene activity, the underlying assumption is that the defective gene is not inherited independently from a carrier mother. To test this for our case, a series of X-chromosome probes were used to examine RFLP markers from the patient and her parents. Two markers, 782 and L1.28, proved to be heterozygous in the patient and homozygous in her mother. Southern blot analysis (data not shown) of DNA from the hybrids A2 and C2-T10 carrying, respectively, the der(X) and der(21) chromosomes revealed that the translocation chromosomes both carry paternally derived markers (V.M. Kean et al., in prep.). Thus, it is the maternal gene on the normal X chromosome that is silenced by nonrandom inactivation, and the paternally derived de novo translocation chromosomes must carry a de novo mutation at the *DMD/BMD* gene. This strengthens the argument that the translocation is a mutational event.

Reciprocal Nature of Translocation

One possibility in this and other patients is that a major deletion has occurred at the translocation junction, and the disease results not from the translocation itself but from the deletion of a gene several hundred kilobases from the sequences contained in clone *XJ1*. To test this, clone *XJ2* was isolated as part of the chromosome walk described above and extends several kilobases in the 5′ (telomeric) direction from the rDNA:X junction of clone *XJ1*. A subclone from *XJ2* was found to hybridize with sequences on a Southern blot of DNA from hybrid C2-T10 carrying the der(21) chromosome. Further analysis (data not shown) with this probe revealed that the restriction maps of the der(X) and der(21) chromosomes are essentially colinear with the maps of the rDNA on 21 and the normal X chromosome, with a reciprocal exchange and no detectable deletion (S. Bodrug et al., unpubl.). This rules out the possibility of a significant deletion and suggests that the reciprocal translocation has disrupted the gene itself, separated it from a *cis*-acting regulatory sequence, or altered the chromatin configuration in such a way as to alter its activity.

Mapping of Other Translocation Exchange Points

Boyd and Buckle (1986) have used high-resolution chromosome banding to map cytogenetically the translocation exchange points in several DMD/BMD females and found them to be located in bands Xp21.1, Xp21.2, and Xp21.3. The most proximal exchange points were separated from the most distal ones by perhaps as much as several million base pairs.

Y. Boyd and her colleagues, in collaboration with our laboratory, have utilized the *XJ* probes to analyze DNA from somatic-cell hybrids carrying the derivative X or the derivative autosome from several of these X-autosome translocations. Also utilized in this study were the *pERT87* probes developed by Kunkel et al. (1985) and shown to reveal deletions in a proportion of male patients (Monaco et al. 1985; Kunkel et al., this volume). As illustrated in Figure 3, *pERT87* maps distal to the X:21 junction that we have cloned (Monaco et al. 1985). The *pERT87* and *XJ* regions have been expanded by chromosome walking to cover about 140 kb (Kunkel et al., this volume) and 60 kb, respectively. As the two regions appear so far to be nonoverlapping, the *pERT87/XJ* region spans a minimum of 200 kb. It is flanked on the proximal and distal sides by the randomly generated probes *754* and *C7* (Goodfellow et al. 1985), respectively.

Mapping of other translocation exchange points involved selection of somatic-cell hybrids carrying either the der(X) or the der(autosome) and Southern blot analysis of these hybrids with a series of *pERT87* and *XJ* probes (Y. Boyd et al., in prep.). Figure 3 shows that two exchange points map distal to the *pERT87/XJ* region and two others map proximal to this region. Thus, the translocation exchange points, although clustered in Xp21, are separated from one another by a minimum of 200 kb or, if cytogenetic estimates are accurate, by as much as 3000–4000 kb. If the translocations act by disrupting the *DMD/BMD* locus, it is indeed a very large locus.

Family Studies with *XJ* Probes

The finding of three polymorphic markers in clones *XJ1* and *XJ2* has allowed the segregation of the RFLP markers to be followed in Duchenne and Becker families. The three probes and their associated RFLP markers are listed in Table 1. Although our initial expectation was that the *XJ* region might segregate 100% with the *DMD/BMD* gene, the heterogeneity of translocation exchange points suggested that *XJ1* might be several hundred to a few thousand kilobases from the gene.

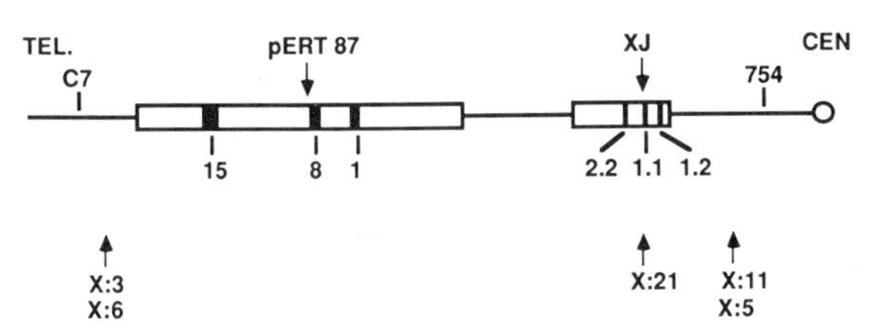

Figure 3. Schematic of the *pERT/XJ* region. Clones that detect RFLP markers are designated as *pERT87-15, 8,* and *1*, and *XJ1.1*, *2.2*, and *1.2*. The positions of random markers *C7* and *754* are indicated as are the exchange points of other translocations. The figure shows probe *XJ2.2*, whereas the table and text refer to probe *XJ2.3*. These are adjacent segments of the X chromosome and reveal the same polymorphism.

Table 1. *XJ* Allele Frequencies

Marker	Enzyme	Allele sizes	Allele frequencies
XJ1.1	*Taq*I	3.1, 3.8	.72/.28
XJ1.2	*Bcl*I	1.7, 2.0	.70/.30
XJ2.3	*Taq*I	6.4, 7.8	.70/.30

To date, 31 Toronto families have been informative for the *XJ* and other closely linked probes. The recombination fraction is 2 of 40 meioses for recombination between *XJ* and *DMD* and 2 of 46 meioses for *pERT* and *DMD*. One of the latter is recombined for both *pERT* and *XJ* and is shown in Figure 4. Of interest in this family is the fact that the flanking markers *D-2* and *754* are nonrecombinant with the *DMD* gene. Thus, the mutation could be on either side of the *pERT/XJ* region, and one of the males is a double recombinant.

One family with BMD was analyzed, and of five meioses, one recombinant between *pERT* and *BMD* was observed; *XJ* was not informative.

Clearly, a great deal more information is required from family studies to obtain an accurate recombination fraction and to assess the position of the mutations relative to the *pERT/XJ* region.

DISCUSSION

With close to 20 females now identified with DMD or BMD, and in each case the finding of X-autosome translocation involving band Xp21, there can be little doubt that disruption of the Xp21 region leads to muscular dystrophy (Verellen-Dumoulin et al. 1984). On the other hand, careful cytogenetic analysis has suggested that the translocation exchange points may be widely separated on a molecular scale with a few million base pairs between the most distal and the most proximal points so far observed (Boyd and Buckle 1986).

Our cloning of one of the t(X;21) translocation exchange points has allowed detailed analysis of this clone and mapping of several other exchange points relative to it. The finding that the de novo translocation in the t(X;21) patient is paternally derived argues

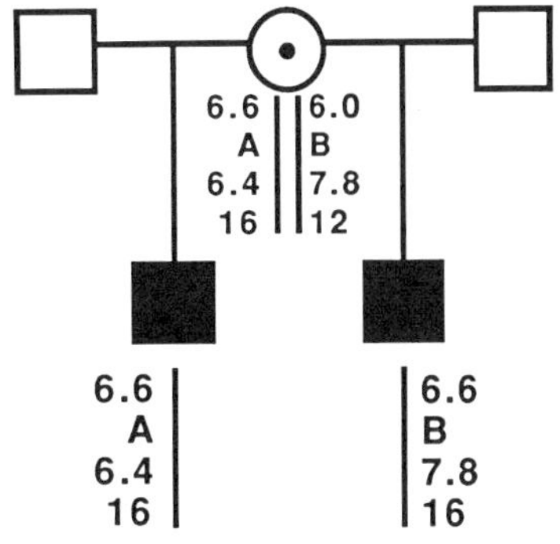

Figure 4. Pedigree showing double recombination. The second affected male is nonrecombinant for *D-2* (alleles 6.0, 6.6) and *754* (alleles 12, 16) but recombinant for *pERT87* (marker *A/B*, where *A* and *B* are haplotypes related to *pERT87-1* and *87-8*) and for *XJ1.2* (alleles 6.4, 7.8).

strongly that the maternal chromosome is the inactive one, and the mutation must be carried on one of the translocation chromosomes. This, in turn, suggests that the translocation is responsible for the mutation. The finding that there is no substantial deletion at the translocation point argues that the translocation must have disrupted the gene itself, separated it from a *cis*-acting regulator, or disrupted the chromatin configuration so as to disrupt the gene activity.

If the gene itself has been split by the translocation, then an examination of nearby sequences for open reading frames with intron-exon splice junctions or of sequences that hybridize to muscle or other mRNAs on a Northern blot may lead to the identification of the DMD/BMD-coding region and ultimately to the complete gene and its product. Such studies are in progress.

One puzzling fact in these studies is the unexpectedly high frequency of recombination between the *pERT/XJ* region and the *DMD/BMD* gene complex. Averaged over the human genome, meiotic recombination events occur about one per million base pairs. A recombination fraction of nearly 5% would suggest that there might be 5 million base pairs separating the *pERT/XJ* region from the *DMD/BMD* complex. Alternatively, the separation distance may be considerably smaller, and the gene might be in a region of frequent recombination. Indeed, one is tempted to speculate at this point that in some cases it may be a recombination event that deletes one or more of a set of required genes as proposed by Winter and Pembry (1982). Whatever the truth turns out to be, a genetic locus with widely scattered translocations, frequent megabase deletions, and high recombination frequency is bound to be of great biological interest. The fact that this locus is responsible for the most frequent and most devastating form of muscular dystrophy makes study of it all the more interesting and exciting.

ACKNOWLEDGMENT

The collaboration of R. Schmickel, J. Sylvester, and G. Gorski in providing rDNA probes that allowed isolation of the translocation junction is gratefully acknowledged.

REFERENCES

Boyd, Y. and V.J. Buckle. 1986. Cytogenetic heterogeneity of translocations associated with Duchenne muscular dystrophy. *Clin. Genet.* **29:** 108.

Davies, K.E. 1985. Molecular genetics of the human X chromosome. *J. Med. Genet.* **22:** 243.

Davies, K.E., P.L. Pearson, P.S. Harper, J.M. Murray, T.O. O'Brien, M. Sarfarazi, and R. Williamson. 1983. Linkage analysis of two cloned DNA sequences flanking the Duchenne muscular dystrophy locus on the short arm of the human X chromosome. *Nucleic Acids Res.* **8:** 2302.

Francke, U., H.D. Ochs, B. de Martinville, J. Giacalone, V. Lindgren, C. Disteche, R.A. Pagon, M.H. Hofker, G.-J.B. van Ommen, P.L. Pearson, and R.J. Wedgwood. 1985. Minor Xp21 chromosome deletion in a male associated with expression of Duchenne muscular dystrophy, chronic granulomatous disease, retinitis pigmentosa, and McLeod syndrome. *Am. J. Hum. Genet.* **37:** 250.

Goodfellow, P.N., K.E. Davies, and H.H. Ropers. 1985. Report of the committee on the genetic constitution of the X and Y chromosomes. *Cytogenet. Cell Genet.* **40:** 296.

Kingston, H.M., N.S.T. Thomas, P.L. Pearson, M. Sarfarazi, and P.S. Harper. 1983. Genetic linkage between Becker muscular dystrophy and a polymorphic DNA sequence on the short arm of the X chromosome. *J. Med. Genet.* **20:** 255.

Kunkel, L.M., A.P. Monaco, W. Middlesworth, H.D. Ochs, and S.A. Latt. 1985. Specific cloning of DNA fragments absent from the DNA of a male patient with an X chromosome deletion. *Proc. Natl. Acad. Sci.* **82:** 4778.

Monaco, A.P., C.J. Bertelson, W. Middlesworth, C.-A. Colletti, J. Aldridge, K.H. Fischbeck, R. Bartlett, M.A. Pericak-Vance, A.D. Roses, and L.M. Kunkel. 1985. Detection of deletions spanning the Duchenne muscular dystrophy locus using a tightly linked DNA segment. *Nature* **316:** 842.

Moser, H. 1984. Duchenne muscular dystrophy: Pathogenetic aspects and genetic prevention. *Hum. Genet.* **66:** 17.

Ray, P.N., B. Belfall, C. Duff, C. Logan, V. Kean, M.W. Thompson, J.E. Sylvester, J.L. Gorski, R.D. Schmickel, and R.G. Worton. 1985. Cloning of the breakpoint of an X;21 translocation associated with Duchenne muscular dystrophy. *Nature* **318:** 672.

Verellen-Dumoulin, C., M. Freund, R. DeMeyer, C. Laterre, J. Frederic, M.W. Thompson, V.D. Markovic, and R.G. Worton. 1984. Expression of an X-linked muscular dystrophy in a female due to translocation involving Xp21 and non-random inactivation of the normal X chromosome. *Hum. Genet.* **67:** 115.

Winter, R.M. and M.E. Pembry. 1982. Does unequal crossing over contribute to the mutation rate in Duchenne muscular dystrophy? *Am. J. Med. Genet.* **12:** 437.

Worton, R.G., C. Duff, J.E. Sylvester, R.D. Schmickel, and H.F. Willard. 1984. Duchenne muscular dystrophy involving translocation of the *dmd* gene next to ribosomal RNA genes. *Science* **224:** 1447.

Worton, R.G., C. Duff, C. Logan, P.N. Ray, V. Kean, M.W. Thompson, J.E. Sylvester, and R.D. Schmickel. 1986. Approaching the Duchenne muscular dystrophy gene through a translocation involving ribosomal RNA genes. In *Molecular biology of muscle development* (ed. C. Emerson et al.), p. 887. A.R. Liss, New York.

Molecular Genetics of Duchenne Muscular Dystrophy

L.M. Kunkel,*† A.P. Monaco,*‡ C.J. Bertelson,* and C.A. Colletti*

*Division of Genetics, Mental Retardation Center, The Children's Hospital, Boston, Massachusetts 02115; †Department of Pediatrics, Harvard Medical School, Boston Massachusetts 02115; ‡Program in Neuroscience, Harvard University, Cambridge, Massachusetts 02138

Duchenne muscular dystrophy (DMD) is a human X-linked disorder in which the affected males show progressive muscle weakness that usually results in death during the third decade. Despite extensive research, the defective product of the *DMD* locus has yet to be discovered. Over the last 10 years, nearly a dozen females have been described that exhibit phenotypes similar to DMD (for review, see Boyd and Buckle 1986). In each case, the females were heterozygous for an X;autosome translocation where the break on the X occurred in Xp21, with no apparent specificity to the autosomal break. The coincidence of breaks in Xp21 in these females and the apparent DMD phenotype was presumed evidence that the *DMD* locus might reside in or nearby Xp21. Human-X-chromosome-enriched recombinant libraries have served as sources of random DNA fragments derived from the human X (Davies et al. 1981; Kunkel et al. 1982). Those fragments that recognized restriction-fragment-length polymorphisms (RFLPs; Botstein et al. 1980) have been used as linkage markers in families segregating the *DMD* mutation (Davies et al. 1983; Human Gene Mapping 8 1985). Those DNA loci that were mapped physically within or nearby Xp21 exhibited linkage to the *DMD* mutation, further substantiating the Xp21 location of the *DMD* mutation (de Martinville et al. 1985).

Recently, a male patient has been described who exhibited the phenotype of DMD associated with the additional X-linked phenotypes of chronic granulomatous disease (CGD), retinitis pigmentosa (RP), McCleod red cell phenotype, and a small interstitial deletion within Xp21 (Francke et al. 1985). DNA isolated from a cell line established from the patient's lymphocytes was used along with 49,XXXXY DNA in a subtractive reassociation experiment (Kunkel et al. 1985). DNA fragments derived from the 49,XXXXY DNA sample that found no complement in the 46,Ydel(X) DNA sample were separated from the surrounding DNA fragments by molecular cloning (Palmer and Lamar 1984). Southern blot hybridization of the Xp21-enriched DNA fragments confirmed their location and absence from the original 46,Ydel(X) DNA sample. Eight DNA segments were obtained in this manner and confirmed the original observation that the patient's DNA contained a deletion that was indicated by the absence of the cloned segment *DXS84* (*754*) (Francke et al. 1985). To subdivide the nine cloned segments further, the DNA of males who exhibited just the phenotype of DMD were tested for normal hybridization at each of these DNA loci. One of the nine segments (*pERT87*) was found to be absent from the DNA of approximately 9% of DMD males; the remainder hybridized to normal-sized restriction fragments and with normal intensity (Monaco et al. 1985). No unaffected males showed absence of the *pERT87* clone. Chromosome "walking" (Bender et al. 1983) was initiated from the original 200-bp *pERT87* clone, and unique sequence subclones spanning a 137-kb segment of contiguous DNA were further tested for deletion in DMD males. In an international cooperative effort, 1346 DNA samples isolated from unrelated DMD and Becker muscular dystrophy (BMD) males (the less severe X-linked muscular dystrophy) were studied. Eighty-eight DNA samples exhibited complete or partial absence of subcloned DNA fragments from the *DXS164* (*pERT87*) locus (Kunkel et al. 1986). Among the deletion DNA samples, two were found that were isolated from boys with the less-severe BMD phenotype. The coincidence of *DXS164* deletions in both DMD and BMD males provides evidence that these two phenotypes are derived from two very closely spaced loci or are the result of different alleles at the same locus, consistent with previous linkage results (Kingston et al. 1984; Human Gene Mapping 8 1985). There was also a nearly twofold increase in the incidence of deletions between DNA samples isolated from DMD males with a clear family history of the disease when compared to those DMD males with no previous family history. Theoretically, the greater incidence of deletion mutation in familial cases of DMD might be a reflection of the deletion mutation frequency in male and female meioses. Any truly isolated case of DMD must result from female meiotic mutation, whereas familial cases would have a contribution of both male and female meiotic events. The observation that the incidence of deletions is different is consistent with the conclusion that female and male deletion mutation frequencies are nearly equal (Kunkel et al. 1986).

The *DXS164* subclones, *pERT87-1, 8,* and *15*, each recognize useful RFLP alleles and have been utilized in linkage analysis in families segregating the *DMD* mutation. These studies indicate that the *DXS164* locus is indeed close to or within the *DMD* locus, as indicated by the ability of *DXS164* subclones to detect deletions in DNA isolated from DMD and BMD males. The RFLP-detecting subclones from the *DXS164* locus have been made available to numerous investigators throughout the world for genotype prediction in DMD and BMD families. The *DXS164* locus has been observed to recombine with both BMD and DMD in approximately 5% of informative meioses (Kunkel et al. 1986). Such recombinants indicate that the *DXS164* locus is best used for genotype prediction in conjunction with other Xp21 RFLP-detecting loci to increase the accuracy of prediction by the combined segregation results.

One of the major objectives of the present study has been the identification of DNA regions within the *DXS164* locus that might be important in the expression of the DMD and BMD phenotype. The deletion breakpoints within the *DXS164* locus appear to spread over the entire length and extend in both directions. The only possible region of importance indicated by the deletion analysis is a relatively small (10-kb) region of deletion overlap, which is commonly deleted in 90% of deletions analyzed. The size of this region and that of the entire *DXS164* locus limits the use of practical DNA-sequencing strategies to uncover possible coding sequences within the *DXS164* locus. An attempt was made to identify sequences that might be represented among cellular RNA transcripts. Poly(A)$^+$ RNA and total cellular RNA were isolated from a variety of human tissues. Northern blots were prepared and unique sequence-containing subclones from the *DXS164* locus were radiolabeled and hybridized to nitrocellulose filters in sets of pooled hybridization probes. No transcript was identified reproducibly among the various RNA sample/probe combinations tested.

In an effort to increase the accuracy of identification and facilitate the analysis, an alternative procedure was attempted. DNA was isolated from a number of different mammalian species and the chicken. The DNA was cleaved with *Hin*dIII and size separated by agarose gel electrophoresis. Following transfer to nitrocellulose membranes, the immobilized DNA was hybridized with various unique and moderately repeated DNA segments from the *DXS164* locus. An example of the analysis is given in Figure 1. In Figure 1B the subclone *pERT87-1* only hybridizes to the human X chromosome in a rodent-human hybrid (lane 1) and the new world primate (lane 6) DNA samples. This sequence was judged not to be extensively conserved between mammalian species. In contrast, the hybridization results presented in Figure 1A indicate a human *DXS164* subclone *(pERT87-25)* that hybridized to a single *Hin*dIII fragment in all DNA samples tested including chicken DNA. These results are consistent with the *DXS164* subclone *pERT87-25* maintaining nucleotide sequence conservation among all DNA samples. Among the nearly 50 subclones tested, only two subclones exhibited any appreciable degree of sequence conservation. The two DNA fragments have been subcloned as smaller fragments (each originally was a little over 1 kb), and each smaller segment was tested separately for sequence conservation. Both nucleotide sequence conserved regions were delimited to fragments 150–300 bp in size. Nucleotide sequencing has been initiated for each fragment. Each region has also been utilized as an hybridization probe against a total genomic library constructed from size-selected partial *Mbo*I digests of mouse DNA in *Bam*HI-cleaved EMBL-4 (Frischauf et al. 1983). Hybridizing phage have been plaque-purified and DNA prepared from each phage. Subcloning is currently under way, and DNA sequence analysis will be used on the conserved segments of mouse DNA as well as the human. It is anticipated that the nucleotide sequence of both the human and the mouse segments will reveal the nature or reason for the extensive nucleotide sequence conservation.

The use of random fragments of DNA from a defined region of the human X chromosome has led to the identification of a region of DNA that exhibits deletion in patients with DMD. The same region is tightly linked to the disorder in families segregating the disease. A systematic search of the region has failed to detect reproducible transcription in any of the tissues tested, including both human adult and fetal tissue. A possibly more simple and straightforward approach is to utilize nucleotide sequence conservation between species to identify presumably important segments. Two conserved DNA fragments were found, and DNA

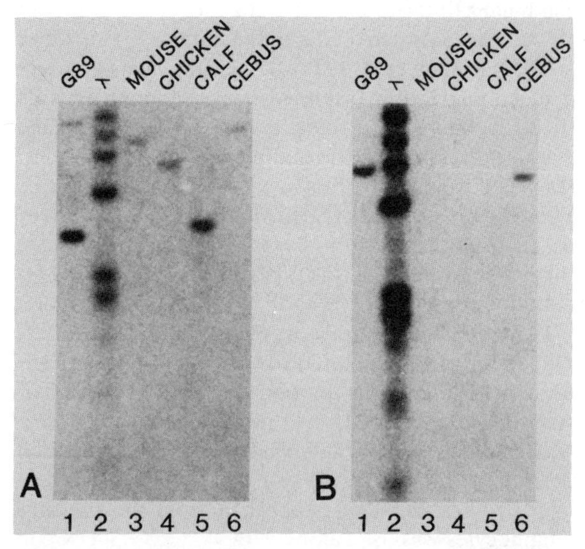

Figure 1. Hybridization of two *DXS164* subclones to DNA samples isolated from different species. Autoradiographs are presented for two identical nitrocellulose filters. The filter in panel *A* had the subclone *pERT87-25* as a hybridization probe, and the filter in panel *B* had *pERT87-1* as a hybridization probe. The *Hin*dIII-cleaved genomic DNA samples (1.5 μg) were separated by electrophoresis and subsequently transferred to a nitrocellulose membrane. (Lane *1*) G89, a hamster-human hybrid cell line DNA sample with only human X chromosome representation among rodent chromosomes. In panel *A*, the human *Hin*dIII fragment is at approximately 3 kb, and the hamster hybridizing fragment is at approximately 18 kb. (Lane *2*) ^{32}P-end-labeled *Hin*dIII-cleaved bacteriophage λ DNA; (lane *3*) DNA isolated from a mouse; (lane *4*) DNA isolated from chicken livers; (lane *5*) DNA prepared from calf thymus; (lane *6*) DNA isolated from the new world primate, Cebus (provided by A. Scott of Johns Hopkins University). Hybridization and washing were at standard stringency as described previously (Aldridge et al. 1984).

sequence analysis of both the mouse and human segments should soon determine if indeed these regions are important to the normal expression of the *DMD* gene.

ACKNOWLEDGMENTS

We thank Dr. Alan Scott for sending us new world primate DNA samples and Drs. L. Villa-Komaroff and B. Wentworth for providing the mouse genomic library. This work was supported by the Muscular Dystrophy Association of America.

REFERENCES

Aldridge, J.F., L.M. Kunkel, G. Bruns, U. Tantravahi, M. Lalande, T. Brewster, E. Moreau, M. Wilson, W. Bromley, T. Roderick, and S. Latt. 1984. A strategy to reveal high frequency RFLPs along the human X chromosome. *Am. J. Hum. Genet.* **36:** 546.

Bender, W., M. Arkam, F. Karch, P.A. Beachy, M. Pfeifer, P. Spierer, E.B. Lewis, and D.S. Hogness. 1983. Molecular genetics of the bithorax complex in *Drosophila melanogaster. Science* **196:** 180.

Botstein, D., R.L. White, M. Skolnick, and R.W. Davis. 1980. Construction of genetic linkage map in man using restriction fragment length polymorphisms. *Am. J. Hum. Genet.* **32:** 314.

Boyd, Y. and V.J. Buckle. 1986. Cytogenetic heterogeneity of translocations associated with Duchenne muscular dystrophy. *Clin. Genet.* **29:** 108.

Davies, K.E., B.D. Young, R.G. Elles, M.E. Hill, and R. Williamson. 1981. Cloning of a representative genomic library of the human X chromosome after sorting by flow cytometry. *Nature* **293:** 374.

Davies, K.E., P.L. Pearson, P.S. Harper, J.M. Murray, T. O'Brien, M. Sarfarazi, and R. Williamson. 1983. Linkage analysis of two cloned DNA sequences flanking the Duchenne muscular dystrophy locus on the short arm of the human X-chromosome. *Nucleic Acids Res.* **11:** 2303.

de Martinville, B., L.M. Kunkel, G. Bruns, F. Morle, M. Koenig, J.L. Mandel, A. Horwich, S.A. Latt, J.F. Gusella, D. Housman, and U. Francke. 1985. Localization of DNA sequences in the region Xp21 of the human X chromosome: Search for molecular markers close to the Duchenne muscular dystrophy locus. *Am. J. Hum. Genet.* **37:** 235.

Francke, U., H.D. Ochs, B. de Martinville, J. Giacalone, V. Lindgren, C. Disteche, R.A. Pagon, M.H. Hofker, G.K.B. van Ommen, P.L. Pearson, and R.J. Wedgwood. 1985. Minor Xp21 chromosome deletion in a male associated with expression of Duchenne muscular dystrophy, chronic granulomatous disease, retinitis pigmentosa, and McLeod syndrome. *Am. J. Hum. Genet.* **37:** 250.

Frischauf, A.M., H. Lehrach, A. Poustka, and N. Murray. 1983. Lambda replacement vectors carrying polylinker sequences. *J. Mol. Biol.* **170:** 827.

Human Gene Mapping 8. 1985. Eighth International Workshop on Human Gene Mapping, Helsinki. *Cytogenet. Cell Genet.* **40:** 1.

Kingston, H.M., M. Sarfarazi, N.S.T. Thomas, and P.S. Harper. 1984. Localisation of the Becker muscular dystrophy gene on the short arm of the X chromosome by linkage to cloned DNA sequences. *Hum. Genet.* **67:** 6.

Kunkel, L.M., U. Tantravahi, M. Eisenhard, and S.A. Latt. 1982. Regional localization on the human X of DNA segments cloned from flow sorted chromosomes. *Nucleic Acids Res.* **10:** 1557.

Kunkel, L.M., A.P. Monaco, W. Middlesworth, H. Ochs, and S.A. Latt. 1985. Specific cloning of DNA fragments absent from the DNA of a male patient with an X chromosome deletion. *Proc. Natl. Acad. Sci.* **82:** 4778.

Kunkel, L.M. et al. (including 73 contributors from 25 different centers). 1986. Analysis of deletions in the DNA from patients with Becker and Duchenne muscular dystrophy. *Nature* **322:** 73.

Monaco, A.P., C.J. Bertelson, W. Middlesworth, C.-A. Colletti, J. Aldridge, K.H. Fischbeck, R. Bartlett, M.A. Pericak-Vance, A.D. Roses, and L.M. Kunkel. 1985. Detection of deletions spanning the Duchenne muscular dystrophy locus using a tightly linked DNA segment. *Nature* **316:** 842.

Palmer, E. and E.E. Lamar. 1984. Y-encoded, species-specific DNA in mice: Evidence that the Y chromosome exists in two polymorphic forms in inbred strains. *Cell* **37:** 171.

Carrier Detection and Gene Analysis of Duchenne Muscular Dystrophy

P.L. Pearson, G.J.B. van Ommen, and E. Bakker
Department of Human Genetics, Sylvius Laboratories, 2333 AL Leiden, Netherlands

Duchenne muscular dystrophy (DMD) is an inherited disorder demonstrating the classical features of an X-linked recessive disease, namely, that of female transmission and male expression. The most important clinical feature is progressive muscle degeneration inevitably leading to wheelchair confinement in the early teens and death around age 20. The patients never transmit the disease gene themselves to the next generation, and the disease must be regarded as a genetic lethal. This, combined with the observation of a frequency of DMD patients varying between 1 in 3000 to 4000 newborn males (Moser 1984) in all western communities examined to date, suggests an extremely high mutation rate unmatched by other human disease loci (Vogel 1977).

Until recently, prediction of DMD carrier status was entirely dependent on a combination of pedigree analysis and creatine phosphokinase determinations in serum. Unfortunately, the latter are notoriously variable (Bullock et al. 1979) and fail to make a distinction with normal female values in approximately 30% of cases. The association of X-autosome translocations with breakpoints in band Xp21, with rare cases of females demonstrating the Duchenne phenotype, suggested that the *DMD* locus may be located at this chromosomal position. Linkage studies confirmed this (Davies et al. 1983). The introduction of restriction-fragment-length polymorphism (RFLP) analysis, based on random probes obtained from X-chromosome-specific banks (Hofker et al. 1985), now permits a more reliable carrier prediction and prenatal diagnosis to be carried out in many situations (Wieacker et al. 1983; Bakker et al. 1985 and in prep.). Furthermore, the occurrence of minor interstitial deletions causing DMD and various other syndromes (Francke et al. 1985; Patil et al. 1985; Wieringa et al. 1985; Dunger et al. 1986) and submicroscopic deletions in 6–8% of all male DMD patients (Monaco et al. 1985; Kunkel et al. 1986) presents us with a powerful research tool. Together with the X-autosomal DMD translocations, these anomalies allow the concise delineation and detailed study of the genetically vital segments of the Xp21 region by the physical mapping of the extent and the boundaries of the rearrangements. This can be done by the recently developed method of pulsed-field gradient electrophoresis (Schwartz and Cantor 1984; Carle and Olson 1984; von Ommen and Verkerk 1986). In this paper, we summarize the Leiden experience on the use of DNA probe analysis for DMD post- and prenatal diagnoses, comment on the frequency and possible significance of meiotic recombination in the region of the *DMD* locus, and finally, present and discuss recent Leiden data in the construction of a megabase map of the DMD region.

Postnatal and Prenatal DMD Screening

We are currently using the probes shown in Figure 1 for family RFLP screening. The list is given according to their order along the short arm of the X chromosome as determined primarily by three-point cross-information (Goodfellow et al. 1985; own data) and secondarily by physical mapping information (see below, Megabase Map of the DMD Region). From recent pulsed-field data, we conclude that the order of the loci *C7* and *B24* should be the reverse of that shown in Figure 1, with *B24* being more distal than *C7*. The loci *cX5.7* and *754* are defined by random probes isolated in Leiden (Goodfellow et al. 1985; Hofker et al. 1985) from various X-chromosome-specific banks and localized to the DMD region using somatic-cell hybrid and deletion cell lines. The isolation and characterization of the locus *pERT87* have been described by Monaco et al. (1985) and *XJ1.1* by Ray et al. (1985). All four of these polymorphic loci are distinguished by their presence within a deletion region detected in a patient exhibiting combined chronic granulomatous disease (CGD) and DMD (Francke et al. 1985). Accordingly, they must be regarded as being physically close to the presumptive site of the *DMD* locus. The evidence from other smaller deletion mutations suggests that *pERT87* and *XJ1.1* may be located within the *DMD* locus itself (L. Kunkel et al.; R. Worton; both pers. comm.). However, family studies from our own group, and elsewhere, demonstrate the occurrence of meiotic recombination between the *DMD* locus and these marker loci and place the *DMD* locus distal to *pERT87*. It may very well be that these studies demonstrate occurrence of intragenic recombination and that the recombination is distal to *pERT87* and proximal to a preferred mutation site within a locus stretching down below *XJ1.1*. In four Leiden families involving recombination between the marker *754* and the *DMD* locus, two also involved a recombination with *pERT87*, confirming the general results of the field, which show that the *pERT* locus has a recombination with *DMD* that is ap-

```
Telomere

     RC-8          B     TaqI
     782           D     EcoRI
     XUT23         H     BglII
     D2            E     PvuII
21   99.6          F     PstI
     C7            J     EcoRV
     B24           I     MspI

   ?DMD, BMD

     PERT 87.1   (K1-n) MspI(M), EcoRV(P), XmnI(R), BstXI(S)
          87.8   ( ,, ) TaqI(K), BstXI(Q)
          87.15  ( ,, ) TaqI(L), XmnI(O)
     XJ1.1          U   TaqI

   ?DMD, BMD

     754           C     PstI
     754.11        T     EcoRI (BglII=C)
     cX5.7         N     MspI
21   OTC           G     MspI
     L1.28         A     TaqI

Centromere
```

Figure 1. Probes used for RFLP screening.

proximately half that of *754*, i.e., 5% (Kunkel et al. 1986). The possibility of recombination dictates that the markers are always used in flanking pairs. In practice, *D2* and *OTC* are currently the most distant markers used, and these have recombination frequencies with *DMD* of approximately 15% and 13%, respectively (Goodfellow et al. 1985). Fortunately, the polymorphism frequency of the middle loci immediately adjacent to the *DMD* locus is such that they can be employed in the majority of situations and, in the case of a combination of *pERT87* with *C7*, provides a reliability of greater than 99.6%. Time is frequently of the essence in carrying out family studies, and we have adopted the restriction and hybridization regime outlined in Figure 2. Each DNA sample is restricted with the five enzymes shown (Fig. 2) and then hybridized with the respective probes shown at the top. In the event that a particular probe is uninformative, the next probe in the same column is taken.

Carrier detection. The probes have been used for carrier detection in 61 pedigrees, of which 30 were familial with 2 or more patients and 31 were sporadic. Sporadic is defined here as occurrence of a single patient and takes no cognizance of the number of carriers. Of the 136 females tested, 68 were shown to be noncarriers with more than 95% reliability and 36 were carriers with the same certainty. No decision could be taken in a further 32 females. However, prenatal diagnosis would be possible in 65 out of the total of 68 proven carriers plus "unknowns." In the latter case, this is based on detection of a chromosome in the fetus proven to be normal elsewhere in the pedigree, i.e., the grandpaternal chromosome.

Prenatal diagnosis. Prenatal diagnoses have been performed to date in 23 male fetuses; 13 were demonstrated to be not affected since they carried the presumptively normal chromosome. To date, seven have come to term and all have been shown to have normal CPK values. In retrospect, ten of the analyses were unnecessary, since it subsequently appeared from the family study that the mother was not a carrier or not informative for markers (two cases). This underlines the importance of completing the family investigation before the prenatal study is carried out. However, we use chorionic villus samples taken at the tenth or eleventh week of pregnancy for our DNA analyses. This implies that the family studies should be carried out either before pregnancy takes place (planned pregnancy) or very early in the pregnancy to provide sufficient time for all necessary investigations.

Eight of the pregnancies proved to be potentially high-risk cases involving the following situations: three had the affected chromosome, four demonstrated a meiotic recombination, and one turned out to have two

Design of family blots

Blot		EcoRV	MspI	PstI	PvuII	TaqI
Probe	1	C7+87.1	OTC	754+99.6	D2	87.15(+XJ1.1)
	2		87.1			87.8
	3		cX5.7			L1.28+RC8
	4		B24			

Figure 2. Restriction/hybridization regime.

X chromosomes and a Y, although cytogenetically it had been diagnosed as a normal XY male. In the latter case, the decision was taken to terminate the pregnancy on the basis of the discrepancy between the cytogenetic and DNA analyses and the fact that the fetus carried, besides a normal X chromosome, the affected chromosome.

Our tentative conclusion is that prenatal diagnosis of DMD families is proving to be an acceptable method of permitting high-risk carriers to have normal male children without undue fear or favor, but that problems arise in the timing of the procedure. It normally takes between 3 and 4 weeks to carry out a family investigation adequately; however, if the investigation is commenced after the seventh week of pregnancy, this overlaps with the timing of the chorionic biopsy, resulting in unnecessary risk to the fetus in women subsequently shown to be noncarriers. We have noted that we can rarely carry out a family investigation without having to approach the family on multiple occasions, which implies that it is vital to maintain close patient contact via the genetic counselors and social workers and that the laboratory work must be carried out in a totally integrated setting. This can rarely or never be achieved by commercial laboratories operating outside fully staffed genetic centers.

Origin of *DMD* Mutations

We have accumulated information on the origin of the mutation in 12 cases of sporadic DMD using flanking RFLPs (E. Bakker and P.L. Pearson, in prep.). In four cases, the grand-paternal chromosome was affected, and in eight cases, either the maternal or grandmaternal chromosome was affected. Interestingly, meiotic recombination had occurred between *pERT87* and the closest distal flanking markers in four of the eight female origins, suggesting that meiotic mutational events such as unequal crossing-over contribute to *DMD* mutation (Winter and Pembrey 1982) and that there may be a preferential mutation site distal to *pERT87* (Fig. 3). This is supported by the appearance of a deletion mutation in the same meiosis as the recombination in pedigree D. The sample size is too small to draw firm conclusions on whether the male and female contribution to mutation differs. However, the observed 2-to-1 ratio of female to male is approximately what one would expect with equal mutation rates in male- and female-derived X chromosomes (Vogel 1977; Winter 1980).

Megabase Map of the DMD Region

The discovery that deletion mutations associated with DMD appear to extend over large lengths of the genome indicates that the *DMD* locus itself may be extremely large. By making use of the central DNA markers depicted in Figure 1 and various deletion cell lines, we have constructed a megabase map based on pulsed-field electrophoresis (Carle and Olson 1984; Schwarz and Cantor 1984). The techniques used were described elsewhere by van Ommen and Verkerk (1986). The results will not be described in detail, since they will be published elsewhere (G.J.B. van Ommen et al., in prep.). Here, we will give only the most important conclusions (for most of the salient details, see Fig. 4). Figure 4 shows the presumptive positions of three disease loci, namely, *CGD, DMD,* and *GK* (glycerol kinase deficiency). Below this, we have positioned the locations of DNA loci, with arrows indicating those that have been used in hybridization experiments. Following is a pulsed-field gradient (PFG) electrophoresis map based on *Sfi*I fragments. None of the probes used map into the same *Sfi*I fragment. The total length of the map is currently about 4.5 megabases. This must be regarded as a minimum estimate, since it is not known how many *Sfi*I fragments occur between those observed. This will only be resolved by the use of other probes and enzymes. However, other infrequent cutters such as *Not*I or *Mlu*I result in hybridization patterns that are difficult to interpret, probably through methylation. We feel that this region of the X chromosome is particularly prone to methylation, since the same filters frequently provide clear single bands when used for hybridization with autosomal probes. Further below in Figure 4, we see the situation depicted for a series of deletion patients. The names of the deletions are given at the right. The ends of the breakpoints have been drawn relative to the position of the DNA loci shown above, and the lengths of the *Sfi*I end fragments are indicated. Those ends adjacent to a known *Sfi*I site have been indicated with an arrow. The bottom horizontal line in Figure 4 gives the position of the breakpoint in the X;21 female Duchenne patient (Ray et al. 1985). It appears to be minimally 110 kb proximal to the distal *Sfi*I site in the 860-kb fragment seen with the *XJ1.1* locus. Finally, the approximate positions of breakpoints in other X-autosome translocations are shown at the bottom. In collaboration with Y. Boyd at Oxford, work is currently proceeding to specify the position of these breakpoints more accurately.

CONCLUSIONS

We conclude that the genetic region involved with the *DMD* locus is extremely large. It is likely that part of the predisposition to high mutation at the *DMD* locus is a reflection of the large target size. However, many aspects of the problem remain unresolved and further in-depth studies will be required to determine, among other things, why meiotic recombination apparently occurs at such a high frequency. The ultimate goal will be structure and regulation studies of the gene or gene cluster itself, which, considering the complexities already encountered, is likely to be a mammoth task requiring the involvement of many research groups. However, the stakes in terms of human suffering are high, and neither the scientific nor the funding communities should shirk their responsibilities.

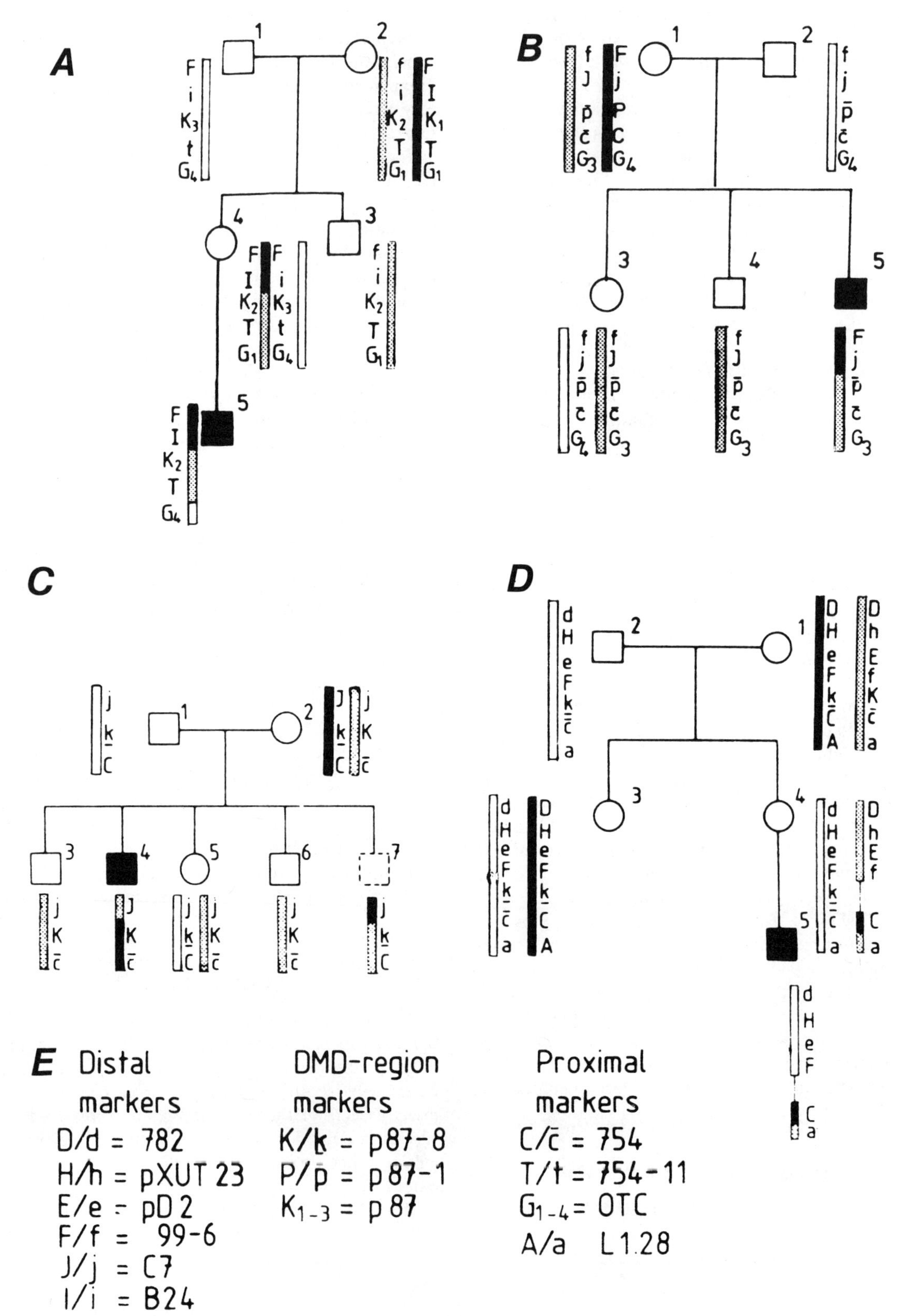

Figure 3. Origin of DMD mutations.

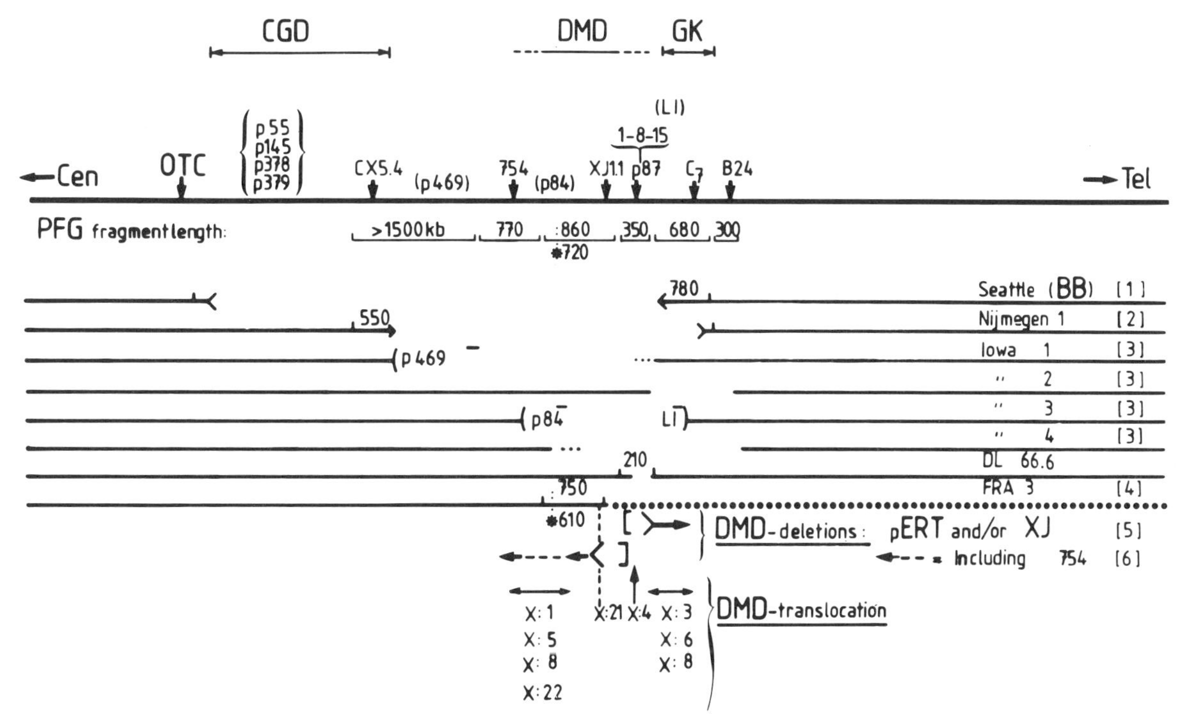

Figure 4. Megabase map of the DMD region.

ACKNOWLEDGMENTS

We thank the Netherlands Foundation for Medical Research, the British Muscular Dystrophy Group, and the Netherlands Disease Prevention Funds.

REFERENCES

Bakker, E., N. Goor, K. Wrogeman, L.M. Kunkel, W.A. Fenton, D. Major-Krakauer, M.G.J. Jahoda, G.J.B. van Ommen, M.H. Hofker, J.L. Mandel, K.E. Davies, H.F. Willard, L. Sandkuyl, A.J. van Essen, E.S. Sachs, and P.L. Pearson. 1985. Prenatal diagnosis and carrier detection of Duchenne muscular dystrophy with closely linked RFLPs. *Lancet* **I:** 656.

Bullocks, D.G., F.M. McSweeny, T.P. Whitehead, and J.H. Edwards. 1979. Serum creatine kinase activity and carrier status for Duchenne muscular dystrophy. *Lancet* **II:** 1151.

Carle, G.F. and M.V. Olson. 1984. Separation of chromosomal DNA molecules from yeast by orthogonal field alternation gel electrophoresis. *Nucleic Acids Res.* **12:** 5647.

Davies, K.E., P.L. Pearson, P.S. Harper, J.M. Murray, T. O'Brien, M. Sarfaraz, and R. Williamson. 1983. Linkage analysis of two cloned DNA sequences flanking the Duchenne muscular dystrophy locus on the short arm of the human X-chromosome. *Nucleic Acids Res.* **11:** 2303.

Dunger, D.B., K.E. Davies, M. Pembrey, B. Lake, P.L. Pearson, D. Williams, A. Whitfield, and M.S.D. Dillon. 1986. Deletion of the X chromosome detected by direct DNA analysis in one of two unrelated boys with glycerol kinase deficiency, adrenal hypoplasia, and Duchenne muscular dystrophy. *Lancet* **I:** 585.

Francke, U., H.D. Ochs, B. de Martinville, J. Giacalone, V. Lindgren, C. Disteche, R.A. Pagon, M.H. Hofker, G.J.B. van Ommen, P.L. Pearson, and R. Wedgewood. 1985. X chromosome deletion in a male associated with the expression of several X-linked recessive disorders. *Am. J. Hum. Genet.* **37:** 250.

Goodfellow, P., K.E. Davies, and H.H. Ropers. 1985. Report of the committee on the genetic constitution of the X and Y chromosomes. *Cytogenet. Cell Genet.* **40:** 296.

Hofker, M.H., M.C. Wapenaar, N. Goor, E. Bakker, G.J.B. van Ommen, and P.L. Pearson. 1985. Isolation of probes detecting restriction fragment length polymorphisms from X-chromosome specific libraries: Potential use for diagnosis of Duchenne muscular dystrophy. *Hum. Genet.* **70:** 148.

Kunkel, L.M. et al. (including 73 contributors from 25 different centers). 1986. Analysis of deletions in the DNA from patients with Becker and Duchenne muscular dystrophy. *Nature* **322:** 73.

Monaco, P.A., C.J. Bertelson, W. Middlesworth, C.A. Colletti, J. Aldridge, K.H. Fishbeck, R. Bartlett, M.A. Pericak-Vancell, A.D. Roses, and L. Kunkel. 1985. Detection of deletions spanning the Duchenne muscular dystrophy locus using a tightly linked DNA segment. *Nature* **316:** 842.

Moser, H. 1984. Review of studies on the proportion and origin of new mutants in Duchenne muscular dystrophy. In *Research into the origin and treatment of muscular dystrophy* (ed. L. Kate et al.). Excerpta Medica, Amsterdam.

Patil, S.R., J.A. Bartley, J.C. Murray, V.V. Iona Sescu, and P.L. Pearson. 1985. X-linked glycose kinase, adrenal hypoplasia and myopathy maps at Xp21. *Cytogenet. Cell Genet.* **40:** 7200.

Ray, P.N., B. Belfall, C. Duff, C. Logan, V. Kean, M.W. Thompson, J.E. Sylvester, J.L. Gorski, R.D. Schmikel, and R.G. Worton. 1985. Cloning of the breakpoint of an X;21 translocation associated with Duchenne muscular dystrophy. *Nature* **318:** 672.

Schwarz, D.C. and C.R. Cantor. 1984. Separation of yeast chromosome-sized DNAs by pulsed field gradient electrophoresis. *Cell* **37:** 67.

van Ommen, G.J.B. and J.M.H. Verkerk. 1986. Restriction analysis of chromosomal DNA in a size range up to 2 million base pairs by pulsed field gradient electrophoresis. In *Human genetic disease. A practical approach* (ed. K. Davies), p. 113. IRL Press, Oxford.

Wieacker, P., K.E. Davies, P.L. Pearson, and H.H. Ropers. 1983. Carrier detection in Duchenne muscular dystrophy by use of cloned DNA sequences. *Lancet* **II:** 1325.

Wieringa, B., T. Hustinx, J. Scheres, W. Reiner, M. Hofker, and B. ter Haar. 1985. Complex glycerol kinase deficiency syndrome explained as X-chromosome deletion. *Clin. Genet.* **27:** 522.

Winter, R.M. 1980. Estimation of male to female ratio of mutation rates of Duchenne muscular dystrophy. *J. Med. Genet.* **17:** 106.

Winter, R.M. and M.E. Pembrey. 1982. Does unequal crossing-over contribute to the mutation rate in Duchenne muscular dystrophy? *Am. J. Hum. Genet.* **12:** 437.

Vogel, F. 1977. A probable sex difference in some mutation rates. *Am. J. Hum. Genet.* **29:** 312.

Molecular Genetics of Huntington's Disease

J.F. GUSELLA,* T.C. GILLIAM,* R.E. TANZI,* M.E. MACDONALD,* S.V. CHENG,* M. WALLACE,†
J. HAINES,† P.M. CONNEALLY,† AND N.S. WEXLER‡
*Neurogenetics Laboratory, Massachusetts General Hospital and Department of Genetics,
Harvard University, Boston, Massachusetts 02114; †Department of Medical Genetics, Indiana University,
Indianapolis, Indiana 46223; ‡President, Hereditary Disease Foundation, Santa Monica, California 90401

Huntington's disease (HD) is an autosomal dominant neurodegenerative disorder characterized by premature localized nerve-cell death (Hayden 1981; Martin 1984). In most cases, signs of the disease are not apparent until middle age. The onset is insidious, involving minor adventitious movements, but it is followed by inexorable progression to severe chorea. The movement disorder, which is eventually totally debilitating, is paralleled by progressive dementia. In some cases, psychiatric symptoms such as impulsive or aggressive behavior or chronic depression accompany or even precede the onset of motor symptoms. The risk of suicide is also increased in this disease. There is no effective treatment for preventing or delaying the onset and progression of HD. Affected individuals, most of whom have already had children before their symptoms appear, become increasingly debilitated over a 15–20-year span until death ensues, usually as a result of heart disease due to the incessant choreic movements or from pneumonia secondary to aspiration.

In the century since the disorder was first described in detail by George Huntington (Huntington 1872), a Long Island physician, extensive neuroanatomical and neurochemical investigations have provided a clear description of the neuronal loss in HD without elucidating its underlying cause (Martin 1984). Cell loss is most prominent in the striatum but also affects many other regions of the brain. Neurons that are lost represent several different cell types carrying different neurotransmitters. Postmortem neurochemical determinations have demonstrated decreases in striatal levels of most neurotransmitters, although somatostatin-containing cells appear to be selectively spared (Ferrante et al. 1985). Recent neuroanatomical studies have suggested that the medium spiny neurons of the caudate may be the first cells to die in HD, but the biochemical basis for the cell death is not known (Graveland et al. 1985).

In 1980, we set out to investigate HD using the increasing power of recombinant DNA technology. The straightforward approach applied to other genetic diseases of isolating a cDNA clone encoding the defective protein was not applicable, since no abnormal protein had been identified in HD. The task of identifying such a cDNA blindly, based on a difference between HD and normal mRNA, is further complicated by the difficulty of obtaining the appropriate tissue from HD patients. Although it seems likely that the primary defect is expressed in the cells of the striatum, there is no guarantee that this is the case. Furthermore, striatal tissue can only be obtained postmortem, and by the time a patient with a firm diagnosis of HD expires, the cells of interest have already disappeared. Finally, this approach ultimately depends on the assumption that the defect is expressed through transcription into RNA. We chose instead to attack HD with a strategy that was independent of any assumptions concerning the nature of the gene defect. The clear-cut autosomal dominant mode of transmission and high penetrance of the disorder made it an ideal candidate for the application of genetic linkage analysis to discover the map location of the defect. In 1983, we discovered a polymorphic DNA marker genetically linked to HD and thereby assigned the defect to chromosome 4 (Gusella et al. 1983). The subsequent characterization of this marker has laid the groundwork for its use in presymptomatic detection and prenatal diagnosis of HD. Of greater importance, it has set the stage for cloning the HD gene on the basis of its map position and elucidating the nature of this genetic defect. In this paper, we summarize current knowledge concerning the linked marker and chromosomal position of the HD gene.

MATERIALS AND METHODS

DNA from members of HD pedigrees. We have established a permanent lymphoblastoid cell line to act as a permanent source of DNA for every member of the HD pedigrees we are studying (Anderson and Gusella 1984). DNA was prepared from these lines and from hybrid cell lines for mapping purposes as described previously (Gusella et al. 1983).

Mapping and typing DNA markers. Digestion of genomic DNA, gel electrophoresis, transfer to a nylon membrane, and hybridization to labeled probes were performed as described by Gusella et al. (1983). Every putative RFLP, detected as a difference in the pattern of hybridization of a probe to DNA from unrelated individuals, was traced through several pedigrees to determine that it displayed Mendelian segregation. PIC values were calculated as described by Botstein et al. (1980) from allele frequencies estimated by counting alleles from 50 unrelated North Americans.

Analysis of linkage data. Since all markers were codominant systems, the phenotypes obtained in Southern transfer and hybridization experiments as outlined above were directly indicative of individual genotypes, except where the coupling phase of the various alleles could not be determined unambiguously and there was consequent uncertainty in assigning haplotypes. The data were analyzed for linkage of the restriction-fragment-length polymorphisms (RFLPs) to each other and to HD, using the computer program LIPED (Ott 1974).

RESULTS AND DISCUSSION

G8 (*D4S10*) Marker

The HD gene was localized to chromosome 4 by linkage to RFLP detected by the probe G8, a 17.6-kb segment of single-copy DNA cloned in bacteriophage λ Charon 4A (Gusella et al. 1983). The marker locus detected by G8 was assigned the symbol *D4S10* at Human Gene Mapping Workshop 7 (Scolnick et al. 1984). Hybridization of G8 to genomic DNA digested with the enzyme *Hin*dIII reveals two independent RFLPs. Approximately 60% of normal North American individuals are heterozygous for one or both of these sites, giving the marker a polymorphism information content (PIC; Botstein et al. 1980) of 0.56.

In view of the difficulty of identifying and sampling large HD families from around the world, our immediate goal was to expand the informativeness of the marker to make optimal use of scarce family material. We hybridized G8 and subclones of the G8 insert to restriction enzyme digests of genomic DNA from five unrelated individuals in order to identify additional RFLPs at the locus. We also isolated R7, an overlapping genomic clone, and included it in the screen for new RFLPs. Figure 1 shows the probes used to detect RFLPs at the *D4S10* locus. A total of 11 RFLPs have been identified with these probes to date (Gusella et al. 1984a; R.L. Heft et al., in prep.). These are listed in Table 1 with their individual PIC values. The original two *Hin*dIII RFLPs are in complete equilibrium with each other, but the other RFLPs display varying degrees of disequilibrium. We have distinguished at least 30 different haplotypes. The overall level of heterozygosity of the *D4S10* locus that can be achieved using these RFLPs exceeds 90% in the general population, making it a highly informative genetic marker.

Table 1. RFLPs at the *D4S10* Locus

Restriction enzyme site	Probe	PIC
*Hin*dIII site 1	G8, pK082	0.32
*Hin*dIII site 2	G8, pK082	0.28
*Bgl*I site 1	R7	0.37
*Bgl*I site 2	pK082	0.21
*Eco*RI site 1	pK083	0.37
*Eco*RI site 2	G8	0.07
*Nci*I	pK082	0.28
*Pst*I	pK082	0.15
*Taq*I	pK082	0.33
*Pvu*II	R7	0.37
*Sac*I	R7	0.37

PIC (polymorphism information content) for each single-site RFLP was calculated as described previously (Botstein et al. 1980). The probe used to detect each RFLP is shown in Fig. 1.

Linkage of the *D4S10* Locus to HD

Positive scores for linkage of HD to the *D4S10* marker locus were first obtained using only two large HD kindreds, an American family of Germanic ancestry and a very large pedigree from Venezuela (Gusella et al. 1983). In the initial analysis, no obligate recombination events were detected between the marker and the disease locus, but the confidence interval on this recombination estimate remained quite large.

We have attempted to identify and type additional large HD kindreds from various parts of the world for two reasons. First, we would like to obtain as accurate an estimate as possible for the rate of recombination between these two loci and in doing so identify recombination events that can act as landmarks for our efforts to home in on and isolate the gene defect. Second, we must obtain firm scientific evidence that there is no nonallelic heterogeneity in HD if the discovery of linkage is to lead to a practical presymptomatic test for the disorder.

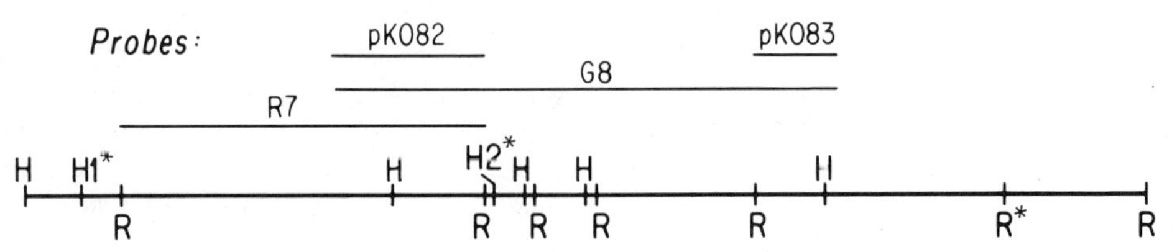

Figure 1. Probes from the *D4S10* locus. A map of the *Hin*dIII (H) and *Eco*RI (R) sites of the *D4S10* is shown. Asterisks indicate the polymorphic sites. The probes used to detect the various polymorphic sites listed in Table 1 are shown above the map. The G8 insert is cloned in Charon 4A and bordered by artificial *Eco*RI sites generated by the cloning process. The R7 insert is an overlapping *Eco*RI fragment also cloned in Charon 4A. The inserts of pK082 and pK083 are *Eco*RI fragments from G8 cloned in pBR328. The polymorphic *Eco*RI site shown in the map is referred to in Table 1 as *Eco*RI site 1. A much less frequent polymorphic site not shown, *Eco*RI site 2, reduces the size of the largest *Eco*RI fragment of the G8 insert and can be detected readily by hybridization with G8.

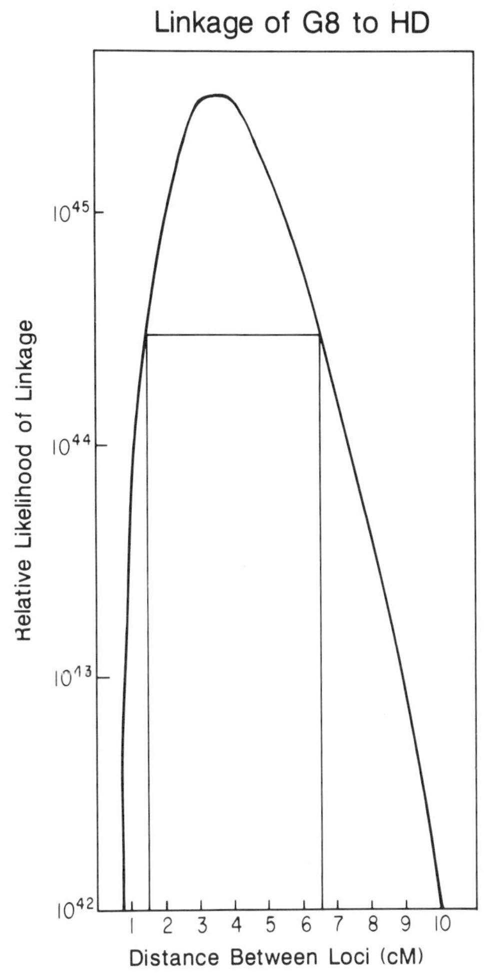

Figure 2. Linkage of the G8 Marker (*D4S10*) to HD. Linkage data for the families described in the text were analyzed by the computer program LIPED (Ott 1974) to determine the likelihood of linkage between *D4S10* and HD at different recombinational distances relative to nonlinkage. The peak of the curve represents the best estimate for the genetic distance separating the two loci. The horizontal line drawn one order of magnitude below the peak approximates a 95% confidence interval for this estimate.

Figure 2 summarizes the published linkage data for those families tested to date and indicates overwhelming support for linkage of the *D4S10* marker to HD. The data derive from the two families outlined above (Gusella et al. 1983, 1984b), several families from Wales (Harper et al. 1985), one American black family, and a second American family of Germanic ancestry (Folstein et al. 1985). The latter two families are particularly interesting because they represent opposite ends of the spectrum of symptoms seen in HD. The disease in the former family displays a relatively early age of onset (late 20s), with primarily motor disturbance, whereas in the latter family, onset is relatively late (early 60s) and accompanied by prominent psychiatric symptoms. All of the families outlined above show positive linkage to the *D4S10* marker, suggesting that they possess mutations at the same locus. A number of recombination events were detected in this analysis, resulting in an estimate of 4% recombination between the two loci. The 1-lod unit confidence interval for this estimate (which approximates a 95% confidence interval) extends from 1.5% to 6.5% recombination.

The number of families included in this analysis is not sufficient to rule out the possibility that a very significant proportion of families with HD actually have a defect at a second locus not linked to the G8 marker on chromosome 4. Until recently, there has been insufficient data to assess this possibility of nonallelic heterogeneity adequately. Consequently, we did not immediately embark on the use of the G8 marker for presymptomatic diagnosis. In view of the nature of HD, its effect on cognitive function, the increased risk of suicide, and other potentially devastating effects of presymptomatic diagnosis in the absence of viable treatment (Wexler et al. 1985), we decided to first develop a firm scientific basis for clinical use of the marker. Many additional families have now been typed with the marker by ourselves and others. These as yet unpublished data suggest that if a second locus is involved in some HD families, the proportion of these does not exceed 10%. Thus, it may be possible in the near future to proceed with clinical use of the probe on an experimental basis to determine the nature and extent of counseling and resources needed to protect and support adequately the "at risk" individual who requests presymptomatic diagnosis.

Precise Localization of the HD Defect

Our ultimate goal in applying recombinant DNA methods to HD is to clone and characterize the defect based on its map position. Although the initial discovery of a linkage marker and subsequent identification of recombination events between the marker and the disease gene are significant steps in this direction, it is also essential to map precisely the physical location of the disease gene on the chromosome. The marker locus has now been mapped in a number of ways to the terminal band of the short arm of chromosome 4. We have demonstrated that the *D4S10* locus is hemizygous in patients with Wolf-Hirschhorn syndrome, a birth defect involving heterozygous deletion of band 4p16 (Gusella et al. 1985). Several studies using in situ hybridization with both normal and abnormal chromosomes have confirmed this result and suggest that the *D4S10* locus resides approximately in the middle of the 4p16 band, perhaps 5,000,000 bp below the telomere (Magenis et al. 1985; Wang et al. 1985; Zabel et al. 1985). Unfortunately, the mapping of the G8 segment on chromosome 4 gives only an approximate location for the HD gene because of the uncertain relationship between the frequency of recombination and physical distance in this region of the genome. Using an average estimate for the whole genome (1 cM = 1,000,000 bp), the HD defect should be approximately 4,000,000 bp above or below the G8 segment.

The Search for a Flanking Marker

A critical step in more precisely defining the location of the HD gene is the isolation of additional DNA markers linked to the defect. In particular, a flanking marker on the opposite side of the HD locus from the *D4S10* would permit us to bracket the defect to a segment of DNA bordered by the nearest detectable recombination event on either side. We estimate that the pedigrees available provide enough meiotic events to reduce such a segment to a region of less than 1% recombination. In addition, a flanking marker would dramatically increase the potential accuracy of presymptomatic diagnosis by allowing detection of recombination events that cause an inherent error rate in a predictive linkage test with a single marker locus.

In conjunction with a project aimed at construction of a linkage map of human chromosome 4, we have isolated and physically mapped over 200 single-copy clones from this autosome (Gilliam et al. 1985 and in prep.). The distribution of these, which were chosen at random from a library of DNA from flow-sorted chromosome 4, is shown in Figure 3. The majority of the 53 clones mapping in the terminal region of the short arm have yielded RFLPs that have been tested for genetic linkage to *D4S10*. Unfortunately, all of these markers display greater than 15% recombination with *D4S10* and none flanks the HD gene. The fact that these markers do not show close linkage to HD suggests that the defect maps *above D4S10* close to the telomere of the short arm. If this is the case, it may be very difficult to identify a flanking marker, since the region between the defect and the telomere could be quite small. On the other hand, it has been suggested that recombination may be increased in the region of telomeres. Thus, the 4% recombination between HD and *D4S10* could correspond to considerably less than 4,000,000 bp.

The location of the HD gene near the telomere of chromosome 4 has a number of implications for further work. As indicated, attempts to isolate a flanking marker by random selection of clones mapping to the terminal portion of the short arm may continue to be unsuccessful if the region distal to the defect is quite small. In this case, however, the telomere itself would form a natural border that would still permit the bracketing of the defect to a very small segment of DNA without resorting to recombination events as landmarks on the distal side. Until the HD gene is identified, however, a flanking marker would remain very important for clinical use. It might be obtained by employing specialized cloning strategies to isolate a fragment representing the telomere sequence.

The failure to obtain additional clones detecting RFLPs closely linked to *D4S10* also suggests that this marker may be located closer to the telomere than previously estimated. If this is case, the HD gene may already be localized to a DNA segment as short as 2,000,000 bp between *D4S10* and the telomere. It should be straightforward to analyze a region of this size by fine-structure physical mapping, given the recent development of pulsed-field gel electrophoresis (Schwartz and Cantor 1984) and inversion-field gel electrophoresis (Carle et al. 1986). Both methods are capable of resolving DNA fragments of greater than 1,000,000 bp in size. The region containing the HD gene might, for example, be contained in only a few *Not*I restriction fragments. Thus, the preparative use of these gel electrophoretic techniques to isolate large fragments for redigestion and cloning might well be the most efficient for obtaining additional markers near the defect.

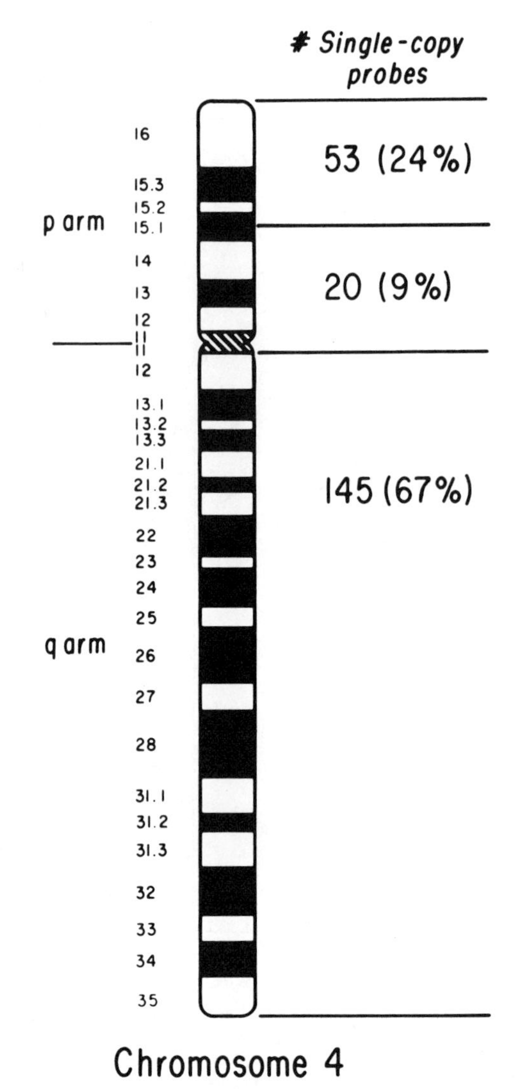

Figure 3. Anonymous DNA probes mapped to chromosome 4. Anonymous single-copy DNA probes were chosen at random from a chromosome library and mapped using somatic-cell hybrid lines containing chromosome-4 translocations with the breakpoints shown (Bocian et al. 1979; J.J. Wasmuth et al., in prep.).

Identification of the HD Gene

The precise genetic and physical mapping of the HD gene, combined with the constant improvement of recombinant DNA techniques, has placed us at the brink of cloning a small section of chromosome 4 that must contain the defect. The initial size of this segment will be determined by the recombination event nearest to the HD gene on the *D4S10* side and its distance from the telomere, but it is unlikely to exceed 1,000,000 bp. It should be possible to narrow the region further by detecting recombination events distal to the HD gene once the entire segment has been cloned. The difficult task of identifying the actual defect among the cloned fragments of DNA will then remain.

One of the great strengths of the genetic approach to HD is that it makes no assumptions concerning the nature of the defect. Therefore, unless an obvious structural alteration such as an insertion or deletion underlies the HD defect, it may be wise to continue to narrow the minimum region containing the locus by genetic methods, rather than relying on intensive analysis of all genes in this part of the chromosome. Having exhausted all meiotic events in existing pedigrees, it might still be possible to use historical recombination events to home in on the defect. The strategy for this approach would be to examine RFLP markers in the region for linkage disequilibrium with the HD gene. If all HD gene carriers in the world carry the same primary defect because the mutation occurred once in a common ancestor, then any apparently unrelated HD patients could be used for this analysis. The occurrence of a single HD mutation once in human history remains a formal possibility, since there is currently no conclusive proof of a new mutation to HD. In this instance, the RFLPs closest to the defect may be in complete linkage disequilibrium with HD. The size of the region of complete disequilibrium will depend on how long ago the original mutation occurred. Even if more than one HD mutation has contributed to the pool of patients in the world today, it might still be possible to apply the linkage disequilibrium approach to patients from a particular ethnic group or geographical area if HD had been introduced only once into this select population.

Ultimately, when genetic techniques have delineated the smallest possible stretch of chromosome 4 that could contain the HD gene, it may be necessary to make certain assumptions concerning the nature of the defect in order to prioritize candidate sequences for intensive anaysis. The first assumption that would seem reasonable is that the HD mutation qualitatively or quantitatively alters expression of a protein-coding sequence. A second more tentative assumption might be that the primary defect is expressed in this striatum.

When the time comes to prioritize sequences for analysis, the assumptions to be made will be determined by the full extent of knowledge then available concerning the possible nature of the defect. Of particular interest in this regard is recent evidence we have obtained concerning the phenotype of individuals homozygous for the HD gene (N.S. Wexler et al., in prep.). The manifestation of HD in these individuals does not differ in any detectable way from that in typical heterozygotes, indicating that HD has a completely dominant phenotype. The defect shows no dosage effect, and the presence of a normal allele in heterozygotes does not alter the onset or progression of the disorder. This suggests that the HD mutation results in a gain of function rather than a loss. The defect could confer a new property on a protein, such as altering a receptor to bind a new ligand, or it could cause increased or inappropriate expression of an otherwise normal protein. In any event, HD is unlikely to result in a loss of function in an enzyme or structural protein unless it is one whose level must be maintained above a strict threshold.

Identification of the HD defect would be considerably easier if there were a biochemical assay for the gene function. The availability of a linked marker permits us to begin to correlate the genotype of "at risk" individuals with various biochemical parameters that can be monitored in peripheral tissues and have been reported to be associated with HD. If any of these effects can be unequivocally related to the presence of the HD gene, they may provide the needed assay. Ultimately, however, it may be necessary to introduce the putative HD gene into animal embryos in an attempt to produce adult animals with the neuropathology typical of HD. The characterization of such animals would not only permit firm identification of the HD gene, but also provide a model system for delineating the biochemical defect and testing possible therapeutic interventions.

SUMMARY

The discovery of a DNA marker linked to the HD gene has provided new avenues into the investigation of this devastating disorder. Genetic investigations have determined that in most and possibly all HD families, the disease is caused by a defect that maps near the telomere on the short arm of chromosome 4. DNA markers will soon provide presymptomatic diagnosis for this disorder, but this increased capability may be a mixed blessing in the absence of effective treatment. The most hopeful route to developing such treatment lies in cloning and characterization of the primary defect. Precise genetic and physical mapping using DNA markers and improvements in techniques for analyzing large segments of DNA have set the stage for cloning of the disease gene in the near future. It will undoubtedly reveal an interesting mechanism for complete phenotypic dominance in man for comparison with completely dominant mutations in other species, particularly *Drosophila*. The nature of the defect may provide new insights into the functional organization of the central nervous system. For the sake of the many individuals who are afflicted by HD or who are asymptomatic gene carriers, it is to be hoped that cloning and characterizing the disease gene will also yield the necessary information to develop an effective therapy.

ACKNOWLEDGMENTS

This work was supported by NINCDS grants NS-16367 (Huntington's Disease Center Without Walls),

NS-20012, and NS-22031, and by grants from the McKnight Foundation and Hereditary Disease Foundation. J.F.C. is a Searle Scholar of the Chicago Community Trust. T.C.G. and M.E.M. were supported by the Anna Mitchell Fellowship and Grace E. Neuman Fellowship, respectively, from the Hereditary Disease Foundation. S.V.C. received a fellowship from the National Huntington Disease Association. We thank Drs. M. Smith and J. Wasmuth for providing hybrid cell lines and Dr. M. VanDilla for providing the chromosome 4 library. The chromosome-4-specific library used in this work was constructed at the Biomedical Sciences Division, Lawrence Livermore National Laboratory, Livermore, California, under the auspices of the National Laboratory Gene Library Project which is sponsored by the U.S. Department of Energy.

REFERENCES

Anderson, M.A. and J.F. Gusella. 1984. The use of cyclosporin A in establishing human EBV-transformed lymphoblastoid cell lines. *In vitro* **20:** 856.

Bocian, M., T. Mohandas, R.S. Sparkes, and S. Funderbunk. 1979. Peptidase S (*PEPS*) and phosphoglucomutase 2 (*PGM*) loci on the short arm of human chromosome 4. *Cytogenet. Cell Genet.* **25:** 138.

Botstein, D., R.L. White, M. Skolnick, and R.W. Davies. 1980. Construction of a genetic linkage map in man using restriction fragment length polymorphism. *Am. J. Hum. Genet.* **32:** 314.

Carle, G.F., M. Frank, and M.V. Olson. 1986. Electrophoretic separations of large DNA molecules by periodic inversion of the electric field. *Science* **232:** 65.

Ferrante, R.J., N.W. Kowall, M.F. Beal, E.P. Richardson, Jr., E.D. Bird, and J.B. Martin. 1985. Selective sparing of a class of striatal neurons in Huntington's disease. *Science* **230:** 561.

Folstein, S., J. Phillips, D. Myers, G. Chase, M. Abbott, M. Franz, P. Waber, H. Kazazian, P.M. Conneally, W. Hobbs, R.E. Tanzi, K. Gibbons, and J.F. Gusella. 1985. Huntington's disease: Two families with differing clinical features show linkage to the G8 probe. *Science* **229:** 776.

Gilliam, T.C., S. Healey, R. Tanzi, G. Stewart, and J. Gusella. 1985. Polymorphic DNA probes from a human chromosome 4 library. *Cytogenet. Cell Genet.* **40:** 641.

Graveland, G.A., R.S. Williams, and M. DiFiglia. 1985. Evidence for degenerative and regenerative changes in neostriatal spiny neurons in Huntington's disease. *Science* **227:** 770.

Gusella, J.F., R.E. Tanzi, P.I. Bader, M.C. Phelan, R. Stevenson, M.R. Hayden, K.J. Hofman, A.G. Faryniarz, and K. Gibbons. 1985. Deletion of the Huntington's disease linked G8 (*D4S10*) locus in Wolf-Hirschhorn syndrome. *Nature* **318:** 75.

Gusella, J.F., K. Gibbons, W. Hobbs, R. Heft, M. Anderson, R. Rashtchian, S. Folstein, M. Wallace, P.M. Conneally, and R. Tanzi. 1984a. The G8 locus linked to HD. *Am. J. Hum. Genet.* **36:** 1395.

Gusella, J.F., R.E. Tanzi, M.A. Anderson, W. Hobbs, K. Gibbons, R. Raschtchian, T.C. Gilliam, M.R. Wallace, N.S. Wexler, and P.M. Conneally. 1984b. DNA markers for nervous system disorders. *Science* **225:** 1320.

Gusella, J.F., N.S. Wexler, P.M. Conneally, S. Naylor, M.A. Anderson, R.E. Tanzi, P.C. Watkins, K. Ottina, M. Wallace, A. Sakaguchi, A. Young, I. Shoulson, E. Bonilla, and J.B. Martin. 1983. A polymorphic DNA marker genetically linked to Huntington's disease. *Nature* **306:** 234.

Harper, P.S., S. Youngman, M.A. Anderson, M. Sarafarazi, O. Quarrell, R. Tanzi, D. Shaw, P. Wallace, P.M. Conneally, and J.F. Gusella. 1985. Genetic linkage between Huntington's disease and the DNA polymorphism G8 in South Wales families. *J. Med. Genet.* **22:** 447.

Hayden, M.R. 1981. *Huntington's chorea.* Springer-Verlag, New York.

Huntington, G. 1872. On chorea. *Med. Surg. Reporter* **26:** 317.

Magenis, E., J. Gusella, K. Weliky, G. Haight, and B. Sheehy. 1985. Huntington disease-linked (*HD*) restriction fragment length polymorphism localized to band p16 of chromosome 4 by in situ hybridization. *Cytogenet. Cell Genet.* **40:** 685.

Martin, J.B. 1984. Huntington's disease: New approaches to an old problem. *Neurology* **34:** 1059.

Ott, J. 1974. Estimation of the recombination fraction in human pedigrees: Efficient computation of the likelihood for human linkage studies. *Am. J. Hum. Genet.* **26:** 588.

Schwartz, D.C. and C.R. Cantor. 1984. Separation of yeast chromosome-sized DNAs by pulsed field gradient gel electrophoresis. *Cell* **37:** 67.

Scolnick, M.H., H.F. Willard, and L.A. Menlove. 1984. Report of the committee on human gene mapping by recombinant DNA techniques. *Cytogenet. Cell Genet.* **37:** 210.

Wang, H.S., C.R. Greenberg, D. Kalousek, J. Gusella, D. Horsman, and M.R. Hayden. 1985. Subregional assignment of the linked marker (*D4S10*) for Huntington disease by in situ hybridization. *Cytogenet. Cell Genet.* **40:** 772.

Wexler, N.S., P.M. Conneally, D. Housman, and J.F. Gusella. 1985. A DNA polymorphism for Huntington's disease markers for the future. *Arch. Neurol.* **42:** 20.

Zabel, B.U., S.L. Naylor, A.Y. Sakaguchi, and J.F. Gusella. 1985. Regional localization of a DNA polymorphism (*D4S10*) linked to Huntington's disease. *Cytogenet. Cell Genet.* **40:** 787.

Cloned Factor VIII and the Molecular Genetics of Hemophilia

R.M. LAWN, W.I. WOOD, J. GITSCHIER,* K.L. WION, D. EATON,
G.A. VEHAR, AND E.G.D. TUDDENHAM†

Department of Molecular Biology and Biochemistry, Genentech, Inc., South San Francisco, California 94080; †Haemophilia Centre and Academic Department of Haematology, Royal Free Hospital, London NW3 2QG England

Hemophilia is the most common inherited bleeding disorder in man. This disease, which primarily affects males, has been described since Biblical times, but it is only in the past several decades that the proteins responsible for hemophilia have been identified. The recent contributions of recombinant DNA research to this field include the elucidation of the gene and primary structures of the proteins, the availability of DNA-based prenatal diagnosis and carrier detection, and the possible means to provide improved treatment of the disease.

A stable blood clot results from the catalytic amplification of an initial stimulus by a series of interacting proteins that comprise the clotting cascade. In essential form, the intrinsic clotting pathway consists of a series of inactive proteases (zymogens) and their appropriate cofactors, which are serially activated to finally convert the soluble protein fibrinogen to its insoluble form, fibrin, which forms a filamentous network and stabilizes the platelet plug. (The hemostatic "pathway" is actually replete with feedback loops and inputs to and from other protein interactions.) Located in the middle of this cascade are the serine protease factor IX and its cofactor, VIII. The complex of activated factor IX, factor VIII, calcium ions, and the phospholipid surface activates factor X by proteolytic cleavage. The two common forms of hemophilia result from a deficiency of either factor VIII (hemophilia A; comprising ~85% of all cases) or factor IX (hemophilia B or Christmas disease). The frequent involvement of these particular clotting factors in bleeding disease is largely due to the location of the genes of both factor VIII and factor IX on the X chromosome, where a single defective allele will cause hemophilia in males.

Without blood-replacement treatments, which were not available until the middle of the twentieth century, hemophiliacs led a restricted life and usually died at an early age from bleeding complications. In the past 30 years, hemophilia has been treated with whole plasma and more recently with concentrates of the appropriate clotting factor. Although dramatically improving the prospects for hemophiliacs, these preparations are still quite impure, and since they are often prepared from large pools of blood donors, they carry the risk of spreading hepatitis and AIDS. The factor VIII protein itself remained largely uncharacterized, save for its definition as the clotting activity lacking in hemophilia A patients. Thus, both scientific curiosity and the desire to produce an improved pharmaceutical product spurred the efforts to clone and express factor VIII.

*Present address: Howard Hughes Medical Institute, University of California, San Francisco, California 94143.

The Human Factor VIII Gene and Protein

Three considerations influenced the strategy that led to the successful cloning and expression of the factor VIII gene: the extremely low abundance of circulating factor VIII (about one-millionth the molar concentration of albumin, for example), its large size and lability (reported molecular weights from 100,000 to 360,000), and uncertainty about its site of synthesis. Both ourselves (Gitschier et al. 1984; Wood et al. 1984) and Toole et al. (1984) used fragmentary protein sequence data to design synthetic oligonucleotide probes for the initial isolation of factor VIII clones from recombinant genomes. Only after partial genomic clones had been characterized could larger and more specific restriction fragment probes be used with confidence to identify suitable sources of factor VIII mRNA and to commence cDNA cloning. (A recent survey of a variety of cell types using hybridization to a factor VIII gene probe [Wion et al. 1985] shows that the mRNA can be detected in liver hepatocytes and a number of other tissues, including spleen, lymph node, pancreas, kidney, muscle, and placenta.) By obtaining overlapping cDNA and genomic clones, we isolated both the 9000 bp of factor VIII cDNA and the 186,000 bp of the gene. When connected to viral promoters and transfected into mammalian tissue-culture cells, the cDNA directs the synthesis and secretion of biologically active factor VIII, which can correct the clotting time of hemophilic plasma.

Restriction mapping and DNA sequencing of the cloned factor VIII gene revealed several interesting features. The overall structure of the gene reflects the large size of the protein it encodes. The 9 kb of coding DNA is divided into 26 exons, which span 186 kb, or roughly 0.1% of the entire X chromosome (Fig. 1). The gene includes several unusually large exons (up to 3.1 kb) and introns (up to 32 kb). Since almost no direct pro-

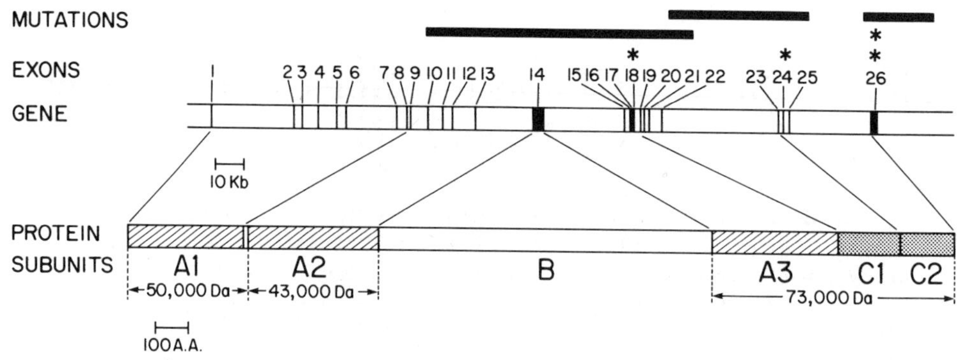

Figure 1. The human factor VIII gene and its encoded protein and known mutations. The human factor VIII gene, spanning 186,000 nucleotides, is shown with exons (■) numbered from the 5′ to 3′ end. Above the gene are shown the location of mutations occurring in the gene of hemophiliacs (Gitschier et al. 1985c, 1986; Antonarakis et al. 1985). Asterisks denote point mutations, and the bars represent deletions. The three deletions and three nonsense mutations occur in the gene of severe hemophiliacs, whereas a missense mutation occurs in the gene of a mild hemophiliac. Below the gene is shown the single chain primary translation product (minus the prepeptide), divided into the A, B, and C regions discussed in the text. The entire intermediate B domain, which is eliminated during activation, is encoded by the 3-kb exon 14. The locations of the 50,000-, 43,000- and 73,000-dalton subunits of the thrombin-activated protein (see Fig. 3) are shown at the bottom.

tein sequence of factor VIII had been published, the primary structure of the protein was first determined by DNA sequence analysis. The sequence of the cDNA revealed an open reading frame coding for a 19-amino-acid hydrophobic signal peptide and a 2332-amino-acid mature protein that, accounting for glycosylation, is large enough to be the M_r ~300,000 single-chain form of the molecule.

Purification of the single-chain form of factor VIII has been reported only once (Rotblat 1985). Due to its susceptibility to proteolysis, factor VIII purified from plasma or concentrates generally consists of multiple polypeptides with relative molecular weights ranging from 80,000 to 210,000 (Fig. 2). Analysis of these proteins by amino-terminal sequence analysis (and the subsequent alignment of the obtained sequences to the factor VIII DNA sequence) allowed their assignment to positions in the single-chain precursor (Vehar et al. 1984; Eaton et al. 1986). The M_r 210,000 and 80,000 proteins represent the amino- and carboxy-terminal regions of factor VIII. Further proteolysis of the M_r 210,000 protein yields a series of proteins with M_r 90,000–180,000. The M_r 80,000 protein is thought to form a metal-linked (perhaps Ca^{++}) complex with each of the M_r 90,000–210,000 proteins (Fig. 2). Once separated, neither the M_r 80,000 protein nor the M_r 90,000–210,000 proteins exhibit coagulant activity (Eaton et al. 1986).

Both the single chain (M_r ~300,000) and multiple polypeptide (M_r 80,000–210,000) forms of factor VIII appear to be less active or inactive precursors that are activated through specific proteolysis by thrombin. Factor VIII activated by thrombin consists of three subunits, with M_r 50,000 and 43,000 fragments derived from contiguous regions of the amino terminus and an M_r 73,000 fragment derived from the carboxyl terminus of the single-chain precursor (Fig. 3). All three subunits are required for activity and are probably bound in a metal-dependent complex. Thrombin is also involved in the eventual proteolytic inactivation of factor VIII via the activation of protein C by the thrombin-thrombomodulin complex. Activated protein C inactivates factor VIII by a single cleavage of the M_r 50,000 polypeptide. These sites of proteolytic processing of factor VIII as determined by Eaton et al. (1986) are shown in Figure 3.

Even before these details of processing were elucidated, the translated DNA sequence revealed some interesting features of the factor VIII protein. The sequence contains two distinct types of internal repetition. A domain of about 350 amino acids is repeated three times in the molecule, with about 30% amino acid homology (Fig. 1, A), and a distinct 150-amino-acid region (Fig. 1, C) is repeated twice with about 35% homology. These regions are separated by the nearly 1000-amino-acid intermediate B domain, which is heavily glycosylated and is eliminated during proteolytic activation of factor VIII.

Unexpectedly, the triplicated A domains bear 30% amino acid homology with what was presumed to be an unrelated protein, ceruloplasmin. Ceruloplasmin is a copper-binding plasma protein (M_r 130,000) containing little else than three copies of the A-like domain (Takahashi et al. 1984). Small blue copper-binding proteins have probably existed for 3 billion years, as attested by the homologous copper-binding bacterial azurins and the plant plastocyanins (Ryden 1984). An ancestral oxidase may have been a precursor of the large multicopper oxidases, fungal laccase and ceruloplasmin, about 1.5 billion years ago. A doubling of a 170-residue plastocyanin-like precursor might have prefigured the 350-residue units of the recent blue oxidases (Ryden 1984). More recently, factor VIII and the closely related factor V have presumably emerged from this gene family. It is noteworthy that the entire B domain, which separates two of the ceruloplasmin-like A domains of factor VIII and is eliminated during activation, is wholly encoded by a 3106-bp exon of the fac-

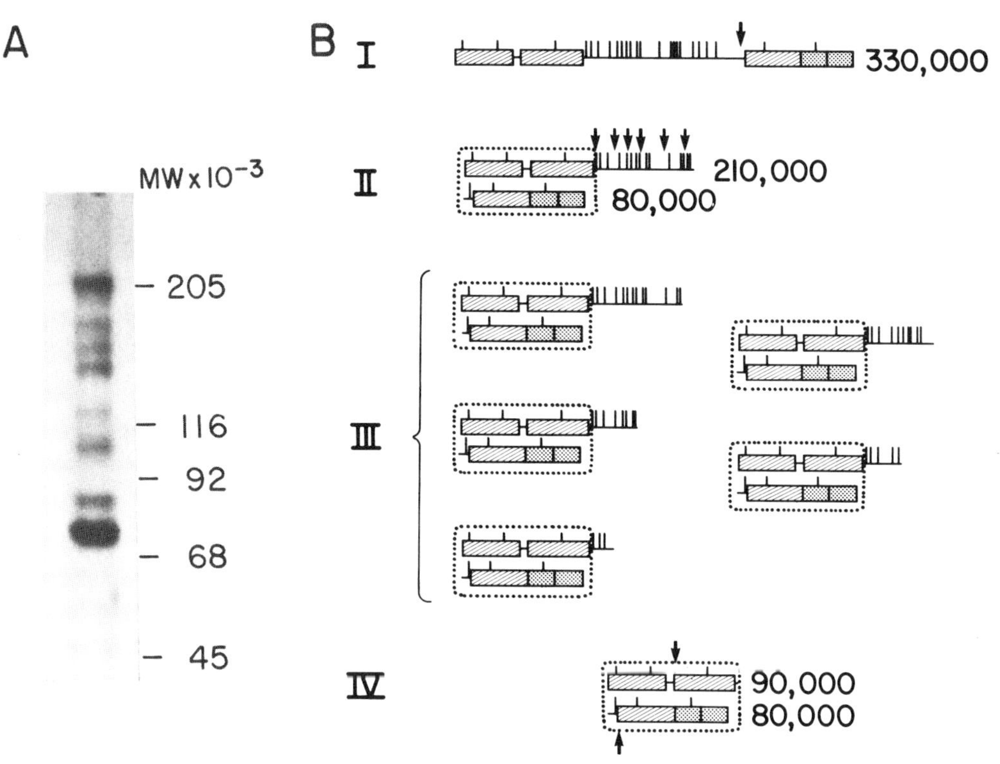

Figure 2. Factor VIII structure. (*A*) Purified human factor VIII (3 μg) was electrophoresed in a 5–10% SDS-polyacrylamide gradient gel and silver-stained. All of the bands seen in this stained gel react with factor VIII monoclonal antibodies (Eaton et al. 1986). (*B*) Diagrammatic representation of the fragmentation of human factor VIII. (Line I) The single chain, 2332-amino-acid form of factor VIII of M_r ~330,000 (not seen in the gel) is cleaved to form the M_r ~210,000 and 80,000 subunits shown in line II. The triplicated A domains are indicated by the hatched boxes, and the C domains are indicated by the dotted boxes. The vertical lines represent the position of potential asparagine-linked glycosylation sites and arrows show the approximate locations of proteolytic cleavage. The dotted box depicts the noncovalent association between the subunits. The five forms depicted in group III derive from differential processing of the carboxyl terminus of the 210,000 subunit and account for gel bands between the M_r 205,000 and 92,000 markers. Molecule IV is the bound form of the 90,000 and 80,000 subunits; arrows depict the two points of thrombin cleavage to maximally activate the protein.

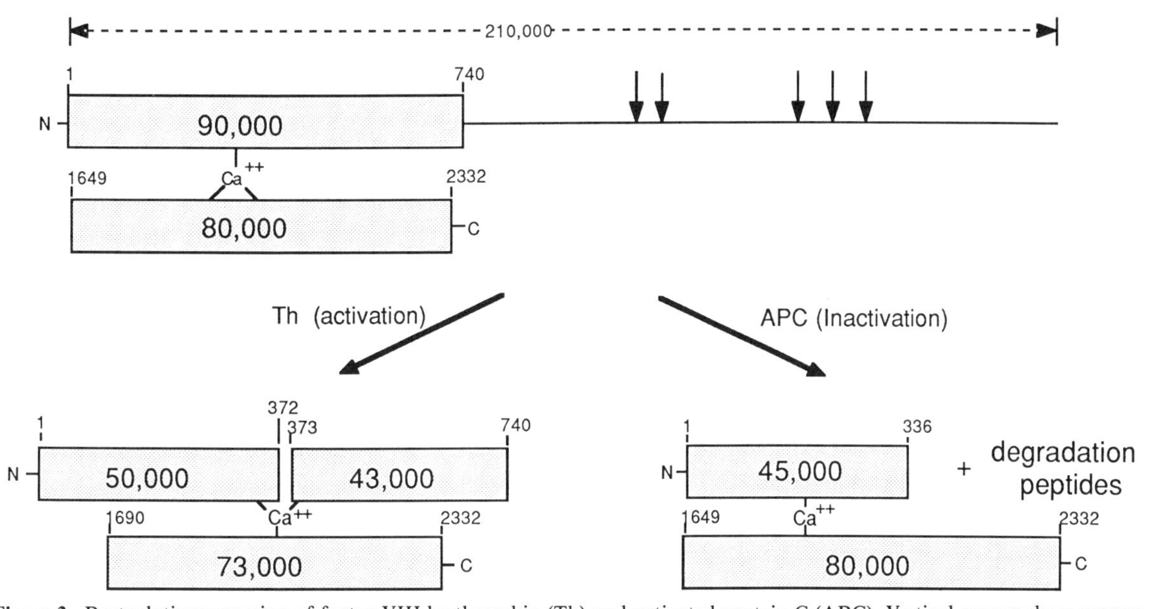

Figure 3. Proteolytic processing of factor VIII by thrombin (Th) and activated protein C (APC). Vertical arrows show approximate sites of cleavage of the M_r 210,000 subunit. Numbers above the protein subunits represent terminal amino acids (Vehar et al. 1984; Eaton et al. 1986). After APC inactivation, the M_r 43,000 subunit probably remains bound to the complex of 45,000 and 80,000 subunits.

tor VIII gene. This far exceeds the size of most eukaryotic exons, which usually range from 60 to 200 bp. Perhaps during evolution, a copy of a processed gene was inserted between two of the triplicated domains of the ancestral factor VIII gene.

Factor VIII Mutations and Hemophilia

Hemophilia A results from a heterogeneous collection of genetic lesions. Until the treatment available in the past 30 years, hemophilia was a fatal disease. As Haldane predicted in 1935, a supply of de novo mutations are necessary to account for the persistence in populations of an X-linked "lethal" disease (Haldane 1935). Until recently, a particular hemophilic gene would persist in a family for only a few generations. The mutation arising with Queen Victoria is a noteworthy example. Hemophilia A also displays a wide range of clinical severity. In addition, about 10% of hemophilia victims generate high levels of neutralizing antibodies to exogenously supplied factor VIII. These "inhibitor" patients present a particular challenge to clinicians. With the cloning of the normal factor VIII gene, it has become possible to define some hemophilia-causing mutations at the DNA level, with the eventual hope of correlating some classes of defects or genetic haplotypes with mild, severe, or inhibitor forms of the disease.

The factor VIII genes of over 100 hemophilia-A patients have now been examined. Both deletion and single-base (nonsense) mutations were found in severe hemophiliacs (with less than 1% normal activity, these included some inhibitor and noninhibitor patients), and a missense mutation has been located in the factor VIII gene of a mild hemophiliac (with 6% of normal activity). Such surveys are biased toward detecting deletion mutations. Deletions of greater than about 100 bp can be readily detected by Southern blot analysis of genomic DNA, whereas single-base changes can only occasionally be discovered. However, restriction enzymes such as *Taq*I that contain CpG in their recognition sequence are especially useful for detecting nonsense mutations by blotting. CpG is the major site of methylation in human DNA and possibly a mutation hot spot due to methyl-cytosine deamination, which may lead to C→T transitions (Barker et al. 1984). Five *Taq*I sites in the factor VIII gene could convert to in-frame termination codons by such a mechanism. We screened 92 hemophilic DNA samples with *Taq*I blots and found four case-specific *Taq*I variations in the hybridization patterns of four severe hemophiliacs who have no detectable factor VIII activity (Gitschier et al. 1985c). Mapping, cloning, and DNA sequencing revealed that two *Taq*I site alterations did indeed result from C→T transitions in exons that changed arginine codons to termination signals. (In the other two cases, the restriction site changes occurred in introns and do not appear to disrupt or to create splice sites by DNA sequence analysis.) The nonsense mutations appeared in exon 24 of an inhibitor patient and exon 26 of a severe hemophiliac without neutralizing antibodies. The inhibitor case was especially interesting because it showed the appearance of a new hemophilia mutation. This patient is the first hemophiliac in the family, and only his DNA, but not that of his normal mother and siblings, has lost the *Taq*I restriction site by blot analysis. Antonarakis et al. (1985) have now reported a third distinct nonsense mutation detected by a *Taq*I site loss in exon 18 of an inhibitor patient.

More recently, Gitschier et al. (1986) detected in a mild hemophiliac the absence of the same *Taq*I site in exon 26 that was altered in one of the severe hemophiliacs described above. In the severe hemophiliac, the CpG to TpG transition converted an arginine codon to a stop codon. We reasoned that a likely occurrence was a similar methyl-cytosine-induced transition in the complementary strand. This would lead to an arginine to glutamine substitution, which might be consistent with the mild hemophilia phenotype. Synthetic 18-base oligonucleotides based on normal and suspected mutant gene sequences were employed as hybridization probes to digested genomic DNAs to confirm this prediction.

In addition to these point mutations, two deletions were detected in the survey of 92 hemophiliacs (Gitschier et al. 1985c). In an inhibitor patient, a deletion removes exons 23–25, whereas a deletion in a noninhibitor patient removes the final exon, 26, of the factor VIII gene. Antonarakis et al. (1985) have characterized one additional deletion in the seven factor VIII inhibitor patients they examined. The location of reported factor VIII mutations is shown in Figure 1. Since only about 2 out of 21 inhibitor patients analyzed contained a detectable gene deletion or rearrangement, it appears unlikely that there is a correlation between gross alterations in the factor VIII gene and the presence of antibodies, as had been noted in a survey of factor IX (hemophilia B) inhibitor patients (Giannelli et al. 1983). One conclusion that can be drawn from this preliminary study of defective factor VIII genes is that hemophilia A is indeed the result of many different genetic lesions. None of these first-reported mutations have been found in more than the family from which it was isolated. However, a report of further hemophilia-causing mutations appearing in this volume (Kazazian et al.) adds some interesting examples. Kazazian and colleagues have now found examples of the same point mutation in the factor VIII gene in unrelated families and can point to the individual in which the mutation independently arose. Furthermore, they describe five different deletions in the factor VIII gene, none of which occurs in a hemophiliac with inhibitor antibodies.

DNA-based Diagnosis

DNA restriction-fragment-length polymorphisms (RFLPs) have become valuable tools for prenatal diagnosis and carrier detection of genetic diseases. Fortunately, RFLPs need not involve the base changes that lead to defective proteins and can thus be useful in diseases like hemophilia, which are caused by many independent mutations. Diagnostic RFLPs can be detected either by probes comprising the cloned gene itself or by probes shown to be tightly linked to the disease locus. We initially reported an RFLP in the factor VIII gene caused by the presence or absence of a *Bcl*I site within intron 18 of the gene (Gitschier et al. 1985b). The two alleles occur at frequencies of about 70% and 30%, so that the predicted frequency of informative females heterozygous for this polymorphism is about 42%. The advantages of DNA-based diagnosis for hemophilia are that fetal DNA can be obtained (without incision) at 8 weeks of gestation by chorion villus biopsy, in contrast to the technique of fetal blood sampling for clotting analysis, which is performed at 20–22 weeks. In addition, current methods of carrier detection in women are often equivocal due to the wide range of clotting activity in normal and heterozygous females. Examples of both prenatal diagnosis and carrier detection of hemophilia A by Southern blot analysis are shown in Gitschier et al. (1985a).

Several additional RFLPs are now available to increase the percentage of informative families and provide checks on the interpretation of the data. These include two "random" RFLPs that have been mapped to within 5% recombination of the factor VIII locus (Gitschier et al. 1985b) and two further RFLPs within the gene (Antonarakis et al. 1985; Wion et al. 1986). One of these RFLPs entails a polymorphic *Xba*I site in intron 22 of the factor VIII gene. DNA probes from this region hybridize to a second site in the human genome even under stringent conditions. However, mapping of the cloned gene and analysis of DNA from a hemophiliac with intron 22 deleted corroborate that it is the *Xba*I site in the gene that is polymorphic. This RFLP has a favorable allele frequency (~59% and 41%) and does not appear in a fixed relationship with the *Bcl*I polymorphism in intron 18 (Wion et al. 1986). Hence, it provides a useful addition to the preexisting markers.

CONCLUSIONS

In the past 5 years, there has been an enormous increase in the understanding of factor VIII and hemophilia. The cloning of the human factor VIII gene and subsequent studies have provided a clearer picture of the protein and its processing. Further studies, aided by the recombinant DNA-driven expression of altered factor VIII molecules, will be needed to define its mode of action precisely. Cloned DNA probes have revealed the molecular basis of some cases of hemophilia, have clarified the uncertainty about the site of factor VIII synthesis, and have provided DNA-based carrier detection and prenatal diagnosis of hemophilia A. Certainly, in addition to these contributions to the molecular biology of homo sapiens, we hope that the cloning of the factor VIII gene will lead to improved treatment of this genetic disease.

REFERENCES

Antonarakis, S., P. Waber, S. Kittur, A. Patel, H. Kazazian, M. Mellis, R. Counts, G. Stamatoyannopoulos, W. Bowie, D. Fass, D. Pittman, J. Wozney, and J. Toole. 1985. Hemophilia A, detection of molecular defects and of carriers by DNA analysis. *N. Engl. J. Med.* **313:** 842.

Barker, D., M. Schafer, and R. White. 1984. Restriction sites containing CpG show a higher frequency of polymorphism in human DNA. *Cell* **36:** 131.

Eaton, D., H. Rodriquez, and G.A. Vehar. 1986. Proteolytic processing of human factor VIII. *Biochemistry* **25:** 505.

Giannelli, F., K. Choo, D. Rees, Y. Boyd, C. Rizza, and G. Brownlee. 1983. Gene deletions in patients with haemophilia B and anti-factor IX antibodies. *Nature* **303:** 181.

Gitschier, J., W.I. Wood, M. Shuman, and R.M. Lawn. 1986. Identification of a missense mutation in the factor VIII gene of a mild hemophiliac. *Science* **232:** 1415.

Gitschier, J., D. Drayna, E. Tuddenham, R. White, and R. Lawn. 1985b. Genetic mapping and diagnosis of haemophilia A achieved through a *Bcl*I polymorphism in the factor VIII gene. *Nature* **314:** 6013.

Gitschier, J., R. Lawn, F. Rotblat, E. Goldman, and E. Tuddenham. 1985a. Antenatal diagnosis and carrier detection of haemophilia A using factor VIII gene probe. *Lancet* **I:** 1093.

Gitschier, J., W.I. Wood, E. Tuddenham, M. Shuman, T. Goralka, E. Chen, and R. Lawn. 1985c. Detection and sequence of mutations in the factor VIII gene of haemophiliacs. *Nature* **315:** 427.

Gitschier, J., W.I. Wood, T. Goralka, K. Wion, E. Chen, D. Eaton, G. Vehar, D. Capon, and R. Lawn. 1984. Characterization of the human factor VIII gene. *Nature* **312:** 326.

Haldane, J.B.S. 1935. The rate of spontaneous mutation of a human gene. *J. Genet.* **31:** 317.

Rotblat, F., D. O'Brien, F. O'Brien, A. Goodall, and E. Tuddenham. 1985. Purification of human factor VIII:C and its characterization by Western blotting using monoclonal antibodies. *Biochemistry* **24:** 4294.

Ryden, L. 1984. Structure and evolution of the small blue proteins. In *Copper proteins and copper enzymes* (ed. R. Lontie), vol. 1, p. 157. CRC Press, Boca Raton, Florida.

Takahashi, N., T. Ortel, and F. Putnam. 1984. Single-chain structure of human ceruloplasmin. *Proc. Natl. Acad. Sci.* **81:** 390.

Toole, J., J. Knopf, J. Wozney, L. Sultzman, J. Buecker, D. Pittman, R. Kaufman, E. Brown, C. Shoemaker, E. Orr, G. Amphlett, B. Foster, M. Coe, G. Knutson, D. Fass, and R. Hewick. 1984. Molecular cloning of a cDNA encoding human antihaemophilic factor. *Nature* **312:** 342.

Vehar, G., B. Keyt, D. Eaton, H. Rodriquez, D. O'Brien, F. Rotblat, H. Oppermann, R. Keck, W.I. Wood, R. Harkins, E. Tuddenham, R. Lawn, and D. Capon. 1984. Structure of human factor VIII. *Nature* **312:** 337.

Wion, K., E. Tuddenham, and R. Lawn. 1986. A new polymorphism in the factor VIII gene for prenatal diagnosis of hemophilia A. *Nucleic Acids Res.* **14:** 4535.

Wion, K., D. Kelley, J. Summerfield, E. Tuddenham, and R. Lawn. 1985. Distribution of factor VIII mRNA and antigen in human liver and other tissues. *Nature* **317:** 726.

Wood, W.I., D. Capon, C. Simonsen, D. Eaton, J. Gitschier, B. Keyt, P. Seeburg, D. Smith, P. Hollingshead, K. Wion, E. Delwart, E. Tuddenham, G. Vehar, and R. Lawn. 1984. Expression of active human factor VIII from recombinant DNA clones. *Nature* **312:** 330.

Comparison of Deficiency Alleles of the β-Globin and Factor VIII:C Genes: New Lessons from a Giant Gene

H.H. KAZAZIAN, JR., S.E. ANTONARAKIS, H. YOUSSOUFIAN, C.E. DOWLING, D.G. PHILLIPS, C. WONG, AND C.D. BOEHM

Genetics Unit, Department of Pediatrics, The Johns Hopkins Hospital, Baltimore, Maryland 21205

Certain variables strongly influence the type of deficiency alleles observed at a locus, e.g., the size of the relevant gene, its chromosomal location, and whether deficiency alleles are under the influence of positive selection. In the case of the β-globin locus, the gene is small, the locus is autosomal in location, and β-thalassemia alleles are under positive selection in regions endemic for malaria that are populated by 50% of mankind. Thus, not all deficiency alleles of β-globin may be seen because (1) individuals carry two copies of the gene and (2) selection may result in the oldest deficiency mutations attaining the highest gene frequency.

In contrast, the factor VIII:C gene is quite large, and, because deficiency alleles are essentially X-linked lethals (until recently the reproductive fitness of hemophilia A patients was low), the mutations have a life of two to four generations in a stationary population (Haldane 1935) and nearly all deficiency alleles are observed. Moreover, deficiency alleles are not under positive selection and are observed at the same frequency in all ethnic groups. Thus, the frequency of the various types of mutant alleles observed should be relatively unbiased for this locus or any large, X-linked locus at which deficiency alleles are genetic lethals.

EXPERIMENTAL PROCEDURES

Most of the methods used in this work are in general use. These include (1) DNA isolation from blood and tissues, (2) Southern blotting, (3) oligonucleotide hybridization to dried gels, (4) DNA cloning in λ phage, and (5) subcloning into M13 vectors for DNA sequencing by the dideoxy method using ^{35}S-labeled deoxynucleotide triphosphates. Methods for 1, 2, 4, and 5 are all well established and do not require extensive description.

The method of oligonucleotide hybridization in dried gels is modified from that of Studencki and Wallace (1984). Oligonucleotides are end-labeled using T_4 polynucleotide kinase and [^{32}P]ATP. After labeling, the specific activity of the oligonucleotide is maximized by separating ^{32}P-labeled oligomer from unlabeled oligomer by electrophoresis on a 20% polyacrylamide sequencing gel. The gel region containing labeled oligomer is excised and the oligomer is extracted by the crush and soak method. Hybridization of labeled oligomer with the dried gel is carried out as described at T_m −4°C for 3 hours. The gel is washed in 2× SSPE/0.1% SDS, and the final wash in 2× SSPE is 1–2° below the calculated T_m for 2–3 minutes (1× SSPE is 10 mmoles/liter sodium phosphate, 0.18 mole/liter sodium chloride, 1 mmole/liter EDTA, and pH 7.4). Gels are then autoradiographed as described in Studencki and Wallace (1984). Gels hybridized to probe in this manner can be reused with new probe (5–8 reuses of the same gel are standard in the lab) after they are stripped by soaking in denaturing buffer for 30 minutes and neutralizing buffer for 30 minutes.

RESULTS AND DISCUSSION

β-Thalassemia Alleles

The thalassemias are inherited hemolytic, hypochromic anemias that are secondary to defects in the synthesis of one or the other globin chain (α or β) of adult hemoglobin (Weatherall and Clegg 1981). For example, when defects affecting synthesis are present in both of the genes for β-globin, an excess of α-chains is synthesized; these α-chains precipitate within the erythroid precursor cell and premature cell lysis results (Fessas 1963; Nathan et al. 1969). In addition, surviving cells have a deficiency of hemoglobin tetramers, even though some compensation occurs through production of γ-globin chains of fetal hemoglobin. Mutations affecting α-globin production lead to α-thalassemia, whereas those altering β-globin synthesis lead to β-thalassemia.

Deletion as a cause of β-thalassemia is quite uncommon, presumably because the general incidence of deletion is much less than that of a point mutation. The Lepore-type deletions (fusion of δ-globin and β-globin genes following mispairing and unequal crossing-over) account for about 1% of β-thalassemia genes in the Mediterranean basin, and a particular 619-bp deletion of obscure origin that eliminates all of the third exon of the β-globin gene accounts for about 30% of β-thalassemia genes in India (Spritz and Orkin 1982). Point mutations known to cause β-thalassemia are presented in Table 1.

Transcription mutations. Four different mutations have been found in the TATA box, a region located about 30 nucleotides 5′ to the cap site (−30) which is thought to be important for high-level transcription of overly expressed specialized genes, such as β-globin

Table 1. Mutations in β-Thalassemia (Total number = 37, March, 1986)

Mutant class	Type (0 = β^0, + = β^+)	Origin	Direct detection
I. Nonfunctional mRNA			
a. Nonsense mutants			
1. codon 17 (A-T)	0	Chinese	oligonucleotide
2. codon 39 (C-T)	0	Mediterranean	oligonucleotide
3. codon 15 (G-A)	0	Asian Indian	oligonucleotide
4. codon 121 (A-T)	0	Polish	*Eco*RI
5. codon 37 (G-A)	0	Saudi Arabian	*Ava*II
b. Frameshift mutants			
6. −2 codon 8	0	Turkish	oligonucleotide
7. −1 codon 16	0	Asian Indian	oligonucleotide
8. −1 codon 44	0	Kurdish	oligonucleotide
9. +1 codons 8/9	0	Asian Indian	oligonucleotide
10. −4 codons 41/42	0	Asian Indian	oligonucleotide
11. −1 codon 6	0	Mediterranean	*Mst*II
12. +1 codons 71/72	0	Chinese	oligonucleotide
II. RNA Processing mutants			
a. Splice junction changes			
1. IVS-1 position 1 (G-A)	0	Mediterranean	oligonucleotide
2. IVS-1 position 1 (G-T)	0	Asian Indian	oligonucleotide
3. IVS-2 position 1 (G-A)	0	Mediterranean	*Hph*I
4. IVS-1 3′ end −17 bp	0	Kuwaiti	*Fnu*4H
5. IVS-1 3′ end −25 bp	0	Asian Indian	*Fnu*4H
6. IVS-2 3′ end (A-G)	0	American Black	oligonucleotide
b. Consensus changes			
7. IVS-1 position 5 (G-C)	+	Indian	oligonucleotide
8. IVS-1 position 5 (G-T)	+	Mediterranean	oligonucleotide
9. IVS-1 position 6 (T-C)	+	Mediterranean	oligonucleotide
10. IVS-2 3′-end CAG-AAG	+	Iranian	oligonucleotide
c. Internal IVS changes			
11. IVS-1 position 110 (G-A)	+	Mediterranean	oligonucleotide
12. IVS-1 position 116 (T-G)	?	Mediterranean	*Mae*I
13. IVS-2 position 705 (T-G)	+	Mediterranean	oligonucleotide
14. IVS-2 position 745 (C-G)	+	Mediterranean	*Rsa*I
15. IVS-2 position 654 (C-T)	0	Chinese	oligonucleotide
d. Coding regions substitutions affecting processing			
16. codon 26 (G-A)	E	Southeast Asian	HbE
17. codon 24 (T-A)	+	American Black	oligonucleotide
18. codon 27 (G-T)	Knossos	Mediterranean	Hb Knossos
III. Transcriptional mutants			
1. −88 C-T	+	American Black	oligonucleotide
2. −87 C-G	+	Mediterranean	*Avr*II
3. −31 A-G	+	Japanese	oligonucleotide
4. −29 A-G	+	American Black	oligonucleotide
5. −28 A-C	+	Kurdish	oligonucleotide
6. −28 A-G	+	Chinese	oligonucleotide
IV. RNA cleavage + polyadenylation mutants			
1. AATAAA - AACAAA	+	American Black	oligonucleotide

(Orkin and Kazazian 1984). Another region thought to be important in transcription of genes such as β-globin is the CCAAT homology at −70 nucleotides. No mutation in this region has yet been found in β-thalassemia. At −85 to −90 is a six-base sequence, ACACCC, which has also been implicated through site-specific mutagenesis experiments as a crucial promoter element for transcription of β-globin genes (Dierks et al. 1983). Two mutations have been observed in this distal promoter element in β-thalassemia genes. They are in the two C residues at the 3′ end of the sequence, at positions −87 and −88 (Orkin and Kazazian 1984). All six mutations in this region, the two at −87 and −88 and the four in the TATA box homology, are relatively mild, reducing the RNA output of the mutant gene by 75–80%, but not eliminating it (Orkin and Kazazian

1984; Bunn and Forget 1986). These mutations are termed β^+-thalassemia alleles because significant though reduced numbers of β-chains result from these mutant genes. β^0-Thalassemia genes contain mutations that completely eliminate β-globin production.

RNA cleavage and 3′ polyadenylation mutation. A single mutation has been observed that greatly affects cleavage of the β-globin RNA transcript (Orkin et al. 1985). It lies in the six-nucleotide sequence homology, AATAAA, which is present about 15 nucleotides 5′ to the RNA cleavage site in nearly all protein-encoding genes and is thought to signal an endonuclease cleavage site of the growing RNA transcript. The β-thalassemia mutation in question substitutes a C for the T in the AATAAA signal sequence. This substitution greatly inhibits endonucleolytic cleavage of the growing RNA. Indeed, an RNA species containing 900 additional nucleotides is found in vivo. This RNA is the product of failed cleavage at the AACAAA sequence followed by cleavage 10–15 nucleotides 3′ to the next AATAAA sequence downstream. How this mutation actually produces β-thalassemia is not clear. Theoretically, the elongated mRNA should be normally translated. However, its low concentration is reticulocytes in vivo would suggest that it is unstable.

RNA splicing. Of the total of 37 different mutations presently known that produce β-thalassemia, 18 are mutations affecting RNA splicing (Kazazian 1985). These mutations alter RNA splicing either by changing normal splicing sites or by creating new sites in introns or exons. These new sites may never otherwise be used for splicing, even in the presence of mutations in normal splice sites, or they may be cryptic sites that are not used under normal circumstances but are used when mutation affects the normal site.

Mutations that alter the GT or AG at the splice junctions lead to complete elimination of normal splicing and are β^0-thalassemia genes (Orkin and Kazazian 1984) (see Table 1 for examples). Mutations in consensus sequences at positions 5 and 6 of the donor site of intervening sequence (IVS)-1 and at position −3 in the acceptor site of IVS-2 do not eliminate use of the normal splice site and are β^+-thalassemia genes (Treisman et al. 1983). These mutations greatly slow normal splicing and often with these mutations splicing occurs at cryptic splice sites.

Single nucleotide substitutions that produce new donor sites in IVS-2 at positions 654, 705, and 745 (Orkin and Kazazian 1984) alter normal splicing by causing aberrant splices from the site of the mutation to the normal acceptor site. When this occurs, the normal acceptor site has been used and the normal donor site then undergoes splicing with the cryptic acceptor at IVS-2 nucleotide 579 (Orkin and Kazazian 1984).

Mutations at IVS-1 nucleotide 110 and IVS-1 nucleotide 116 affect sequences that are similar to normal acceptor sites, thereby turning them into excellent acceptor sites (Fukamski et al. 1982; Bunn and Forget 1986). These mutations result in small segments of IVS-1 remaining in many of the β-mRNA molecules, and these mRNA molecules cannot be properly translated into β-globin.

One cryptic donor site in exon 1 has been subject to three different β-thalassemia mutations, including the common β-globin variant, β^E. These mutations have merely altered the sequence in this region of the gene to make it conform more closely to the consensus sequence for donor splice sites. They lead to use of the cryptic donor site that is never used in normal RNA processing.

Mutations affecting translation. These mutations are of two types, chain terminator nonsense mutations and frameshift mutations. Five different nonsense mutations are known, all of which are single nucleotide substitutions that alter triplet codons for an amino acid to produce one of the termination codons in mRNA, i.e., UAG, UAA, and UGA (Orkin and Kazazian 1984). Although 29 single nucleotide substitutions could produce nonsense codons in the coding region of the β-globin gene, only five nonsense mutations have been observed.

Frameshift mutations are deletions or additions of one, two, or four nucleotides in the coding region of the gene. These mutations lead to a completely altered amino acid sequence following the mutation and a premature termination of the protein chain. Seven frameshift mutations have been found in β-thalassemia genes (Orkin and Kazazian 1984).

Origins and Spread of β-Thalassemia Genes

Since the large majority of β-thalassemia genes are found only in a single ethnic or racial group, it is apparent that the mutant alleles observed in modern times have arisen and expanded in frequency after the divergence of the human races. Among the major population groups affected, 13 different alleles have been observed in the Mediterranean basin, eight in Asiatic Indians, seven in Chinese, and six in American Blacks. Analysis of the normal polymorphisms in the β-globin gene suggests that a small number of normal β-globin gene types, or frameworks, were present in man well before the divergence of the races. Since the four common β-gene frameworks are present in all racial groups, this suggests that the population size at the time of racial divergence was, in all likelihood, in the range of thousands, not hundreds or tens. This makes it likely that β-thalassemia mutations present prior to racial divergence would have survived the moderate population bottleneck attendant with racial separation, and ancient mutations under positive selection would be observed in all racial groups. This is not the case. Since mutational events to alleles affecting β-globin expression must have occurred since the dawn of man, the ethnic distribution of mutant alleles seen today suggests that the onset of the positive selection force, which may have been falciparum malaria, occurred after racial divergence.

Yet a growing number of β-thalassemia genes (seven at last count) have been found in more than one ethnic group and in very different chromosome backgrounds (see below). The best explanation for the occurrence of these alleles is two separate independent origins of the same mutation. For example, the −29 A-G allele is present in Blacks and Chinese (Huang et al. 1986), and the −88 A-G allele is present in Blacks and Asiatic Indians (Wong et al. 1986). Since these mutations have occurred in different β-globin gene backgrounds in these groups, mutant gene migration is highly improbable. Interestingly, the common Mediterranean β-thalassemia allele, nonsense codon 39, has been observed as a new mutation in a European (Chehab et al. 1986).

In a handful of instances, the same mutation is observed within an ethnic group on different β-globin genes as determined by intragenic polymorphisms (Orkin and Kazazian 1984). These observations are difficult to explain by independent origins of the mutation since the mutations have only been found in one ethnic group. The best explanation for this type of spread of a mutation from one genetic background to another is interallelic gene conversion. This is the process, observed in yeast and *Drosophila*, by which short stretches of DNA sequence are passed from one allele at a locus to another (Jackson and Fink 1981). If such events have occurred in the β-globin gene to produce the observations cited, the converted regions would have to be 500 nucleotides or less. Another curious observation is that, even though a large number of known β-globin mutations are present in introns, none of these have participated in a presumed interallelic gene conversion event. In contrast, there are five examples of such events affecting mutations in exons or coding regions. This observation makes it tempting to speculate that these events affecting the β-globin gene are mediated by an RNA transcript. However, traces of β-globin transcripts have not been found in human germ cells, the cells in which such an event would have to take place.

Mutations of the Factor VIII:C Gene: New Lessons from a Giant Gene

Hemophilia A (classic hemophilia) is the most common inherited disease of blood coagulation. The disorder, which is caused by a deficiency or abnormality of clotting factor VIII:C, is inherited as a X-linked trait. The locus for factor VIII:C has been assigned to the long arm of the human X chromosome at Xq28 (Filipi et al. 1983). About 50% of patients have factor VIII:C levels that are less than 5% of normal, and the remainder have levels that are between 5% and 20% of normal. Both forms that are positive and forms that are negative for cross-reacting material have been described. Specific antibodies against factor VIII:C, called factor VIII:C inhibitors, develop in approximately 6% of patients with hemophilia A (Brinkhous et al. 1972). Inhibitors usually appear in such patients only after transfusion therapy. Hemophilic patients with the lowest factor VIII:C levels (less than 1 unit per deciliter) are the most severely affected and the most prone to the development of circulating factor VIII:C inhibitors.

Over the past year we at Hopkins and Gitschier et al. at Genentech, Inc. have made an effort to determine the molecular basis of classical hemophilia (hemophilia A). In 1984, Gitschier et al. and Toole et al. reported the cloning of the factor VIII:C gene, the gene for the clotting factor which, when defective, produces hemophilia A (Gitschier et al. 1984; Toole et al. 1984). In contrast to the small β-globin gene (1.5 kb), the factor VIII:C gene is a giant gene (186 kb). It is composed of 9 kb of exon sequences in 26 separate exons and 177 kb of introns.

In the study of 120 males with hemophilia A, Gitschier et al. found five different mutant alleles, two deletions and three point mutations (Gitschier et al. 1985). The deletions were all present toward the 3′ end of the gene, and the point mutations were CGA-TGA mutations producing nonsense codons in exons 24 and 26 and at the same codon in exon 26, a CGA to CAA missense mutation in a mildly affected male.

Antonarakis et al. screened 10 hemophilia A individuals initially and discovered a large 70-kb deletion in the middle of the factor VIII:C gene, a CGA-TGA nonsense mutation in exon 18, and a probable point mutation which creates a new *Hin*dIII site in exon 19 (Antonarakis et al. 1985). Recently, H. Youssoufian et al. (in prep.) have extended the characterization of mutant factor VIII:C alleles from 70 patients whose families were referred to us for prenatal diagnosis and/or carrier detection. Nearly all of these 70 patients have been completely studied by the following screening procedure. Three endonucleases are used, *Sst*I, *Eco*RI, and *Taq*I, and two cDNA probes, probe A, which spans exons 1–12, and probe B, which includes exons 14–26 (Fig. 1). Thus, each gene is studied under six different conditions of enzyme and probe. Eight mutations among the 70 patients have been characterized and the characterization of five more mutant alleles is still incomplete. Certain alleles will require cloning and sequencing of critical gene regions in order to complete their characterization. The recent results of H. Youssoufian (in prep.) are similar to those of Antonarakis et al., who had a 3 in 10 yield of mutant alleles, and differ from those of Gitschier et al., who found five mutant alleles among 120 patients studied.

Deletions in the Factor VIII:C Gene

The gene defects recently discovered include five partial gene deletions. These are (1) a 7-kb deletion that eliminates exon 6; (2) a 2.5-kb deletion that partially eliminates exon 14; (3) a 6-kb deletion that eliminates exon 22; (4) a deletion of at least 3.4 kb that removes exons 24 and 25; and (5) a deletion of at least 9.5 kb

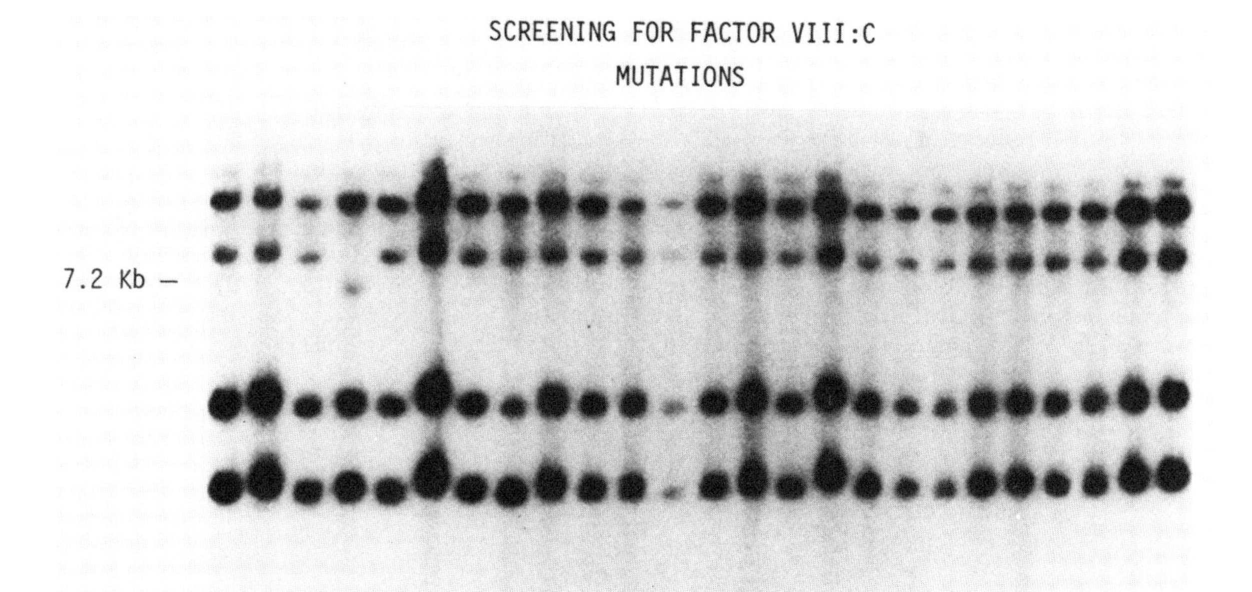

Figure 1. Southern blot of hemophilia A DNA samples digested with *SstI* and hybridized to probe B. An abnormal fragment at 7.2 kb is observed in lane 4 from the left and was shown to result from deletion of exons 24 and 25.

that eliminates exons 23–25 (Fig. 2). None of these deletions is associated with inhibitor production by the affected individuals, and all but the exon 22 deletion produce severe hemophilia A. This latter deletion removes 52 amino acids "in frame" and the affected male has roughly 5% of normal factor VIII:C activity. The lack of inhibitor production in these patients is interesting in light of the production of high-titer factor VIII:C inhibitors by patients with two other deletions, one of which also removes exons 23–25 (Gitschier et al. 1985) and another deletion which removes exons 10–22 (Antonarakis et al. 1985). Correlation of the structure of the factor VIII:C protein that results from various gene deletions with the function of the molecule and the production of inhibitors in affected patients will require further investigation.

Point Mutations in the Factor VIII:C Gene

The point mutations observed by H. Youssoufian et al. (in prep.) have been of even greater interest. It has long been known that there exists a relative deficiency of CpG dinucleotides in mammalian DNA. Nussinov has surveyed a number of sequenced mammalian genes and found that CpG dinucleotides are 37% of expected in exons (Nussinov 1981). Barker et al. (1984) have postulated that CpGs are relatively mutation sensitive, since a high fraction of DNA polymorphisms are seen

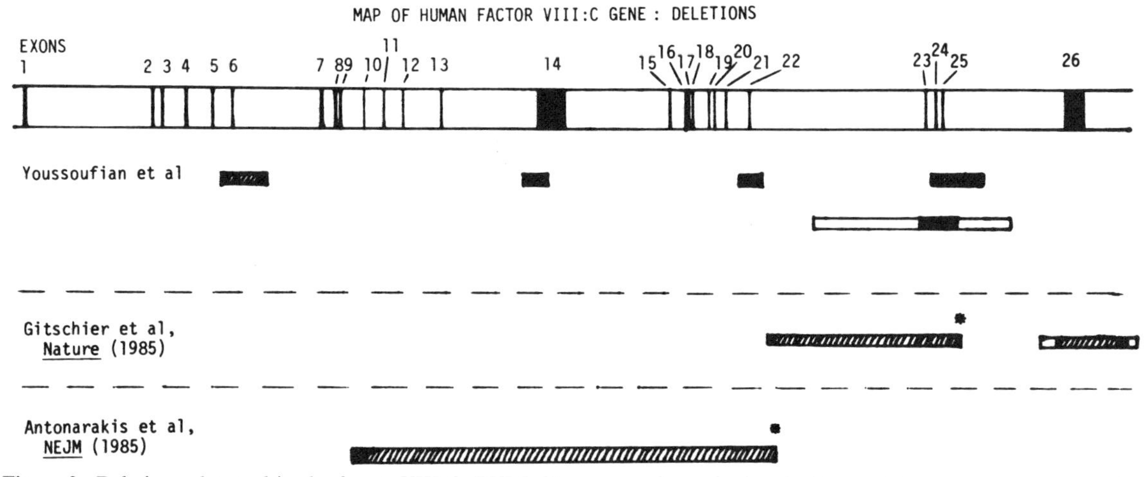

Figure 2. Deletions observed in the factor VIII:C (FVIII:C) gene are shown by hatched boxes. The open areas at the ends of two deletions indicate their possible extent. Asterisks indicate that two deletions out of eight are associated with inhibitor production in the affected patients.

in *Taq*I and *Msp*I sites, which have CpG as part of their recognition sequences. Cytosine 5′ to guanosine is preferentially methylated and the 5-methyl cytosine produced can be spontaneously deaminated to thymine. Thus, CG-TG or CG-CA changes are predicted by this mechanism, the latter when the C-T substitution occurs on the antisense strand of DNA. As stated earlier, Gitschier et al. found two CG-TG mutations and one CG-CA change among 120 hemophilia A genes surveyed. In addition, Nussbaum et al. (1986) found two mutations affecting *Taq*I sites among only six deficiency alleles of the ornithine transcarbamylase gene. Even though these mutant alleles have not yet been characterized, there is a good chance that they are nonsense mutations of the CGA-TGA variety. Among only five point mutations described in the factor IX gene producing hemophilia B, two are of the CG-CA variety in exons (Noyes et al. 1983; Bentley et al. 1986). It now appears that CG-TG substitutions are much more common than expected in hemophilia A genes.

A recurrence of the CGA-TGA nonsense mutation at codon 1960 of the factor VIII:C gene has been found. This was first discovered by Southern blotting with *Taq*I and the 3′ cDNA probe (probe B). A new 5-kb

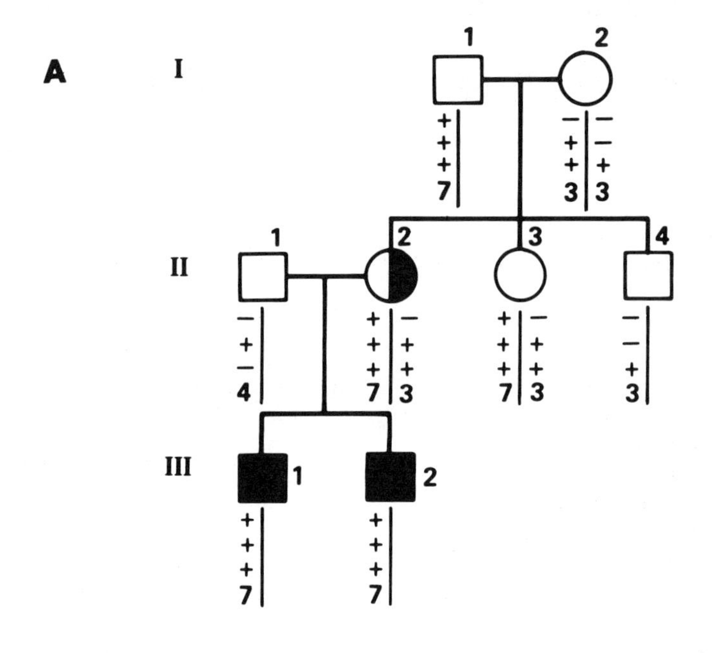

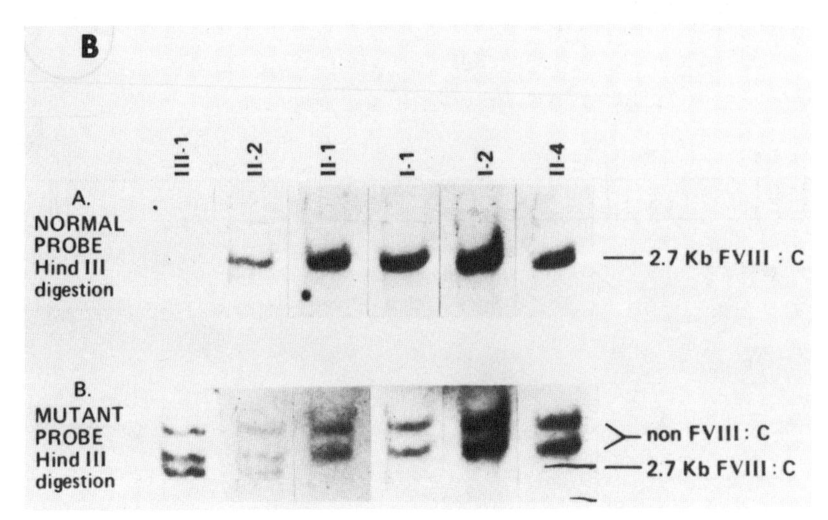

Figure 3. (*A*) Pedigree of family. (*B*) Detection of the cytosine to thymine nonsense mutation in exon 18 of the factor VIII:C (FVIII:C) gene by oligonucleotide hybridization. Individual III-1 shows lack of hybridization with normal probe and hybridization with mutant probe, and his mother, individual II-2, shows hybridization with both normal and mutant probes. Note that both grandparents I-1 and I-2 lack the mutation. DNA polymorphism analysis shown by the haplotypes in A next to the family members indicates that the affected FVIII:C gene in the proband is inherited as a new mutation from the maternal grandfather.

fragment and absence of a normal 2.5-kb fragment was observed in the patient, and this pattern was identical to that observed in a previous patient from another unrelated family. Oligonucleotide hybridization using probes specific for the normal sequence in the region of codon 1960 and the proposed mutant sequence in this region of the gene provided definitive evidence of the C-T mutation in both hemophilia genes (Fig. 3).

Additionally, another altered *Taq*I pattern (a new fragment at 13 kb and absence of a 5.8-kb fragment) was seen in two other unrelated patients. A *Taq*I site in exon 22, codon 2135, was altered by both of these mutations. We synthesized oligonucleotides corresponding to the normal and predicted mutant sequence in exon 22 around codon 2135 and the oligonucleotide hybridization conclusively showed that a CGA-TGA nonsense mutation was present in both families (Fig. 4). Thus, two separate mutations have been shown to recur and account for four of the mutant alleles among 80 screened by our group. One can calculate that, if the frequency of these alleles is similar in the world to that observed in these studies, over 1000 separate recurrences of each of these mutations are present in individuals alive today. One can also estimate that these alleles may have recurred over 10,000 times in human history. Thus, these CpG dinucleotides may be up to 50–100 times more sensitive to mutation than other dinucleotide pairs. Obviously, more work is needed to quantify the in vivo sensitivity of CpG to mutation. The estimates of mutation recurrence rates presented above are derived from (1) the frequency of the disease (1 in 10,000 males), (2) the lack of ethnic variation in disease incidence, and (3) the apparent lack of positive selection in females heterozygous for deficiency alleles.

It is noteworthly that three CG-TG or CA mutations have been observed by Gitschier et al. in exons 24 and 26. Thus, combined data indicate that seven CG-TG or CA mutant alleles have been discovered, even though only five *Taq*I sites (which contain a CGA coding for arginine) in exons have been screened (Fig. 5). The seven mutations affect four of the five *Taq*I sites. Seven other CGA arginine codons exist in the factor VIII:C gene, but because they are preceded by C, G, or A and not T, they are not *Taq*I sites and have not yet been screened for mutations producing hemophilia A.

As mentioned earlier, five other defects have been observed by Youssoufian in his series, but they have not yet been fully characterized. One is a *Taq*I site change in IVS-4. Four others appear to be *Taq*I site mutations in exons, three of which may affect the *Taq*I site in exon 24. These mutations are characterized by a new *Taq*I fragment of 4.2 kb, the same size as the new *Taq*I fragment observed by Gitschier et al. in DNA of the

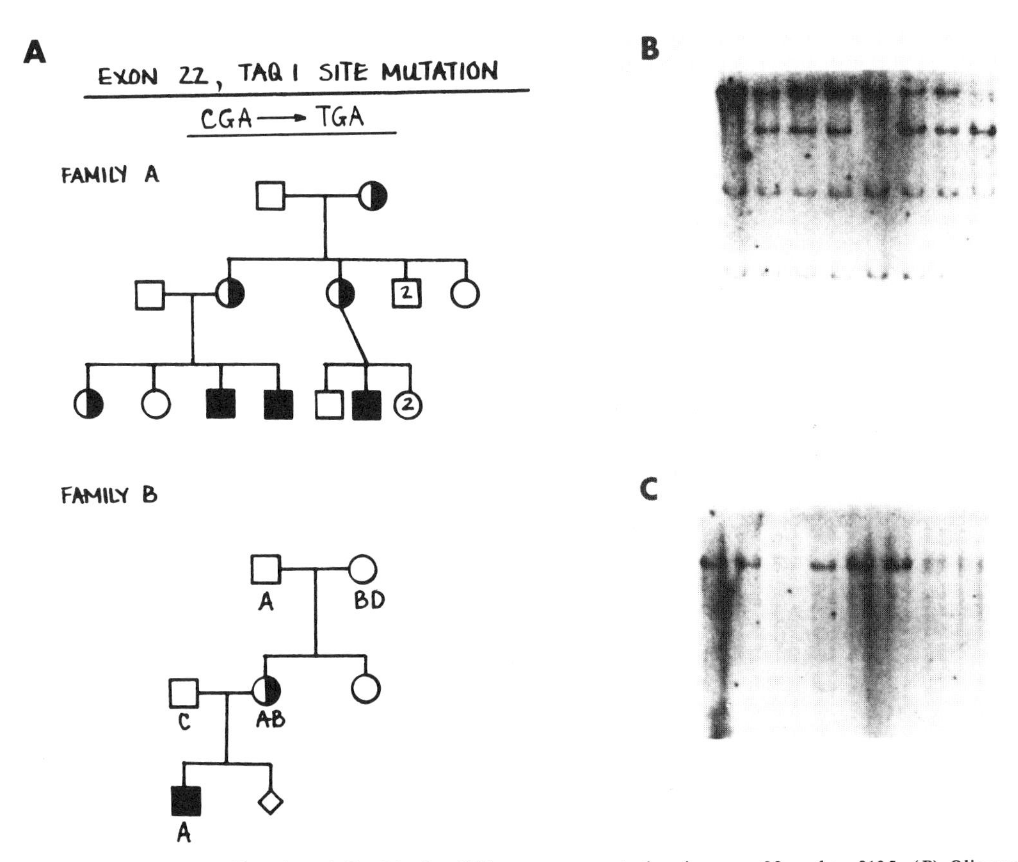

Figure 4. (*A*) Pedigrees of families A and B with the C-T nonsense mutation in exon 22 codon 2135. (*B*) Oligonucleotide hybridization with normal (*top*) and mutant probes specific for either CGA or TGA as codon 2135. Note that one patient from each family and two carriers from family A and one carrier from family B (lanes *1, 2, 4, 5,* and *6*) are positive for hybridization with mutant probe, while the patients in families A and B are negative with normal probe.

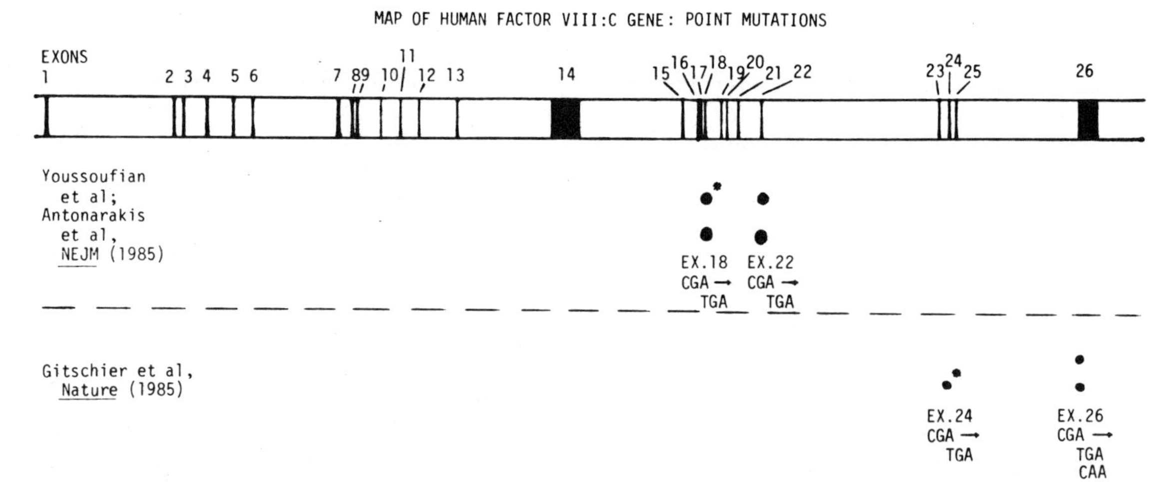

Figure 5. Summary of point mutations reported to date in the factor VIII:C gene. Asterisks refer to inhibitor production by the affected individuals with the mutation. Note that two families are known with the exon 18 CGA-TGA mutation and the affected individuals in one family have inhibitors, whereas those in the other do not even though they are 15 and 16 years old.

patient with the exon 24 mutation (Gitschier et al. 1985). The fourth *Taq*I site change in an exon appears to eliminate the site in exon 23, the fifth *Taq*I site in exons. These five *Taq*I site mutations still require further characterization.

The observation of detectable gene defects in 20% (16 of 80) of hemophilia A patients has led to a marvelous opportunity to study Haldane's predictions of 50 years ago regarding X-linked disorders. These include the incidence of de novo mutations among hemophilia A patients and the parental age at which new mutations occur in males and females. To date, our studies have documented three new mutations that were present in the mother of the affected male. Two point mutations were shown by DNA analysis of intragenic polymorphisms, the polymorphic *Bcl*I and *Bgl*I sites, to have arisen in germ cells of grandfathers when they were ages 33 and 32. A single deletion arose in grandmaternal germ cells in a woman aged 27.

To prove new mutations in paternal germ cells, one must document the stated paternity in each case. Paternity documentation is crucial because the real father could be affected with hemophilia A, thereby rendering the conclusion of new mutations false. In the two cases mentioned above, we have used the multiple allele polymorphisms developed by Donis-Keller and her colleagues to document paternity at the 10^5–10^6 level of certainty (Schumm et al. 1985).

COMMENT

There are a number of reasons that we discussed in the introduction to explain why our view of mutation events producing genetic disease may have been skewed by attempts to generalize from the studies of β-globin. The β-globin gene is small, it is autosomal, and mutations affecting its expression are under positive selection in malarial endemic regions. Thus, any mutation occurring in malarious regions that produces a deficiency of β-globin gene expression will attain a high frequency. This means that chance occurrence of a particular mutational event, whether the event is generally a common or a rare one, will play a large role in the present observed frequency of the allele. In addition, the β-globin gene is very small, leading to underrepresentation of mutations of uncommon sequences, such as CpG dinucleotides.

We appear to observe a different distribution of mutant alleles in the factor VIII:C gene from that observed in the β-globin gene. Since (1) no apparent selection exists for alleles producing hemophilia A genes, (2) these alleles turn over relatively rapidly in the population, and (3) the factor VIII:C gene involved is very large, the observed alleles should provide a relatively unbiased frequency of occurrence of various deleterious mutations. In addition, the observation of a high frequency of CG-TG or CG-CA changes in the factor VIII:C gene is quite striking and suggests that these sequences are highly sensitive to specific mutation.

ACKNOWLEDGMENTS

This work was supported in part by grants from the National Institutes of Health, National Foundation-March of Dimes, the Cooley's Anemia Foundation of Maryland, the Genetics Institute, Inc., and an institutional grant from The Johns Hopkins University School of Medicine. We thank Emily Pasterfield for expert assistance in the preparation of this manuscript.

REFERENCES

Antonarakis, S.E., P.G. Waber, S.D. Kittur, A.S. Patel, H.H. Kazazian, M.A. Mellis, R.B. Counts, G. Stamatoyannopoulos, E.J.W. Bowie, D.N. Fass, D.D. Pittman, J.M. Wozney, and J.J. Toole. 1985. Hemophilia A: Detection of molecular defects and of carriers by DNA analysis. *N. Engl. J. Med.* **313:** 842.

Barker, D., M. Schafer, and R. White. 1984. Restriction sites containing CpG show a high frequency of polymorphism in human DNA. *Cell* **36:** 131.

Bentley, A.K., D.J.G. Rees, C. Rizza, and G.G. Brownlee. 1986. Defective propeptide processing of blood clotting factor IX caused by mutation of arginine −4 to glutamine. *Cell* **45:** 343.

Brinkhous, K.M., H.R. Roberts, and A.E. Weiss. 1972. Prevalance of inhibitors in hemophilia A and B. *Thromb. Diath. Haemorrh.* (suppl.) **41:** 315.

Bunn, H.F. and B.G. Forget, eds. 1986. *Hemoglobin: Molecular, genetic and clinical aspects*. W.B. Saunders, Philadelphia.

Chehab, F.F., G.R. Honig, and Y.W. Kan. 1986. Spontaneous mutation in β-thalassemia producing the same nucleotide substitution as that in a common hereditary form. *Lancet* **I:** 3.

Dierks, P., A.V. Ooyen, M.D. Cochran, C. Dobkin, J. Reiser, and C. Weissman. 1983. Three regions upstream from the cap site are required for efficient and accurate transcription of the rabbit β-globin gene in mouse 3T3 cells. *Cell* **32:** 695.

Fessas, P. 1963. Inclusion of hemoglobin in erythroblasts and erythrocytes of thalassemia. *Blood* **21:** 21.

Filipi, G., A. Rinaldi, N. Archidiacono, M. Ricchi, I. Balazs, and M. Siniscalco. 1983. Linkage between G6PD and fragile X syndrome. *Am. J. Hum. Genet.* **15:** 113.

Fukamaki, Y., P.K. Ghosh, E.J. Benz, Jr., D.B. Reddy, P. Lebowitz, B.G. Forget, and S.M. Weissman. 1982. Abnormally spliced messenger RNA in erythroid cells from patients with β^+-thalassemia and monkey kidney cells expressing a cloned β^+-thalassemia gene. *Cell* **28:** 585.

Gitschier, J., W.I. Wood, E.G.D. Tuddenham, M.A. Shuman, T.M. Goralka, E.Y. Chen, and R.M. Lawn. 1985. Detection and sequence of mutations in the factor VIII gene of hemophiliac. *Nature* **315:** 427.

Gitschier, J., W.I. Wood, T.M. Goralka, K.L. Wion, E.Y. Chen, D.H. Eaton, G.A. Vehar, D.J. Capon, and R.M. Lawn. 1984. Characterization of the human factor VIII gene. *Nature* **312:** 327.

Haldane, J.B.S. 1935. The rate of spontaneous mutation of a human gene. *J. Genet.* **31:** 317.

Huang, S., C. Wong, S.E. Antonarakis, W.H.Y. Lo, and H.H. Kazazian, Jr. 1986. The same TATA box β-thalassemia mutation in Chinese and U.S. Blacks: Another example of independent origins of mutation. *Hum. Genet.* (in press).

Jackson, J.A. and G.R. Fink. 1981. Gene conversion between duplicated genetic elements in yeast. *Nature* **292:** 306.

Kazazian, H.H., Jr. 1985. The nature of mutation. *Hosp. Pract.* **21:** 55.

Nathan, D.G., T.B. Stossel, R.B. Gunn, H.S. Zarkowsky, and M.T. Laforet. 1969. Influence of hemoglobin precipitation on erythrocyte metabolism in α and β-thalassemia. *J. Clin. Invest.* **48:** 33.

Noyes, C.M., M.J. Griffith, J.R. Roberts, and R.L. Lundblad. 1983. Identification of the molecular defect in factor IX Chapel Hill: Substitution of histidine for arginine at position 145. *Proc. Natl. Acad. Sci.* **80:** 4200.

Nussbaum, R.L., B.A. Boggs, A.L. Beaudet, S. Doyle, J.L. Potter, and W.E. O'Brien. 1986. New mutation and prenatal diagnosis in ornithin transcarbamylase deficiency. *Am. J. Hum. Genet.* **38:** 149.

Nussinov, R. 1981. Eukaryotic and dinucleotide preference rules and their implications for degenerate codon usage. *J. Mol. Biol.* **149:** 125.

Orkin, S.H. and H.H. Kazazian, Jr. 1984. The mutation and polymorphisms of the human β-globin gene and its surrounding DNA. *Annu. Rev. Genet.* **18:** 131.

Orkin, S.H., T.-C. Cheng, S.E. Antonarakis, and H.H. Kazazian, Jr. 1985. Thalassemia due to a mutation in the cleavage-polyadenylation signal of the human β-globin gene. *EMBO J.* **4:** 453.

Schumm, J., R. Knowlton, J. Braman, D. Barker, G. Vovis, G. Akots, V. Brown, T. Gravius, C. Helms, K. Hsiao, K. Rediker, J. Thurston, D. Botstein, and H. Donis-Keller. 1985. Detection of more than 500 single-copy RFLPs by random screening. *Human Gene Mapping* **8** (21): 739 (Abstr.)

Spritz, R.A. and S.H. Orkin. 1982. Duplication followed by deletion accounts for the structure of an Indian deletion β-thalassemia gene. *Nucleic Acids Res.* **19:** 8025.

Studencki, A.B. and R.B. Wallace. 1984. Allele-specific hybridization using oligonucleotide probes of very high specific activity: Discrimination of the human B^A- and B^S-globin genes. *DNA* **3:** 7.

Toole, J.J., J.L. Knoph, J.M. Wozney, L.A. Sultzman, J.L. Buecker, D.D. Pittman, R.J. Kaufman, E. Brown, C. Shoemaker, E.C. Orr, G.W. Amphlett, W.B. Foster, M.O. Coe, G.J. Knutson, D.N. Fass, and R.M. Hewick. 1984. Molecular cloning of a cDNA encoding human antihemophilic factor. *Nature* **312:** 342.

Treisman, R., S.H. Orkin, and T. Maniatis. 1983. Specific transcription and RNA splicing defects in five cloned RNA thalassemia genes. *Nature* **302:** 591.

Weatherall, D.J. and J.B. Clegg, eds. 1981. *The thalassemia syndromes*. Blackwell, Oxford, England.

Wong, C., S.E. Antonarakis, S.C. Goff, S.H. Orkin, C.D. Boehm, and H.H. Kazazian, Jr. 1986. On the origin and spread of β-thalassemia: Recurrent observation of four mutations in different ethnic groups. *Proc. Natl. Acad. Sci.* **83:** 6529.

Molecular Genetics of Down's Syndrome: Overexpression of Transfected Human Cu/Zn-Superoxide Dismutase Gene and the Consequent Physiological Changes

Y. GRONER, O. ELROY-STEIN, Y. BERNSTEIN, N. DAFNI, D. LEVANON, E. DANCIGER, AND A. NEER

Department of Virology, The Weizmann Institute of Science, Rehovot 76100, Israel

Down's syndrome (DS), first described as "congenital anomalies and mental retardation syndrome" in 1846 by Edward Seguin (Zellweger 1981) and later on in 1866 by John Langdon Down, is still the most common genetic abnormality, occurring once per 1000 live births (Hook 1981). It is characterized by severe mental retardation as well as a wide variety of physiological defects, such as reduced viability, morphogenetic abnormalities, increased incidence of leukemia, high susceptibility to infections, and some signs of premature aging (for review, see Smith and Berg 1976; Burgio et al. 1981; de la Cruz and Gerald 1981). In 1959, J. Lejeune demonstrated that patients with DS have an extra copy of chromosome 21 in their cells, a condition called trisomy 21 (Lejeune et al. 1959). The causative factor for most of the trisomy is an ovarian nondisjunction at meiosis, although nondisjunction of paternal origin occurring at the first or second meiotic division accounts for approximately 20% of DS (Bond and Chandley 1983; Hassold and Jacob 1984). The risk of a child being born with trisomy 21 increases sharply with maternal age, and since nowadays many couples are postponing parenthood until the fourth decade of life, the incidence of DS babies is expected to increase. Furthermore, because of the continuous improvement in all aspects of clinical treatment, life expectancy of DS patients has tripled over the last two decades. Therefore, despite the instituted prenatal screening by amniocentesis, the prevalence of DS individuals in our society will not significantly decrease in the near future.

Patients with DS usually have a karyotype with 47 chromosomes (46 plus one additional chromosome 21). However, cases of DS have been identified in which only a portion of chromosome 21 is present in triplicate, usually translocated to another chromosome. This finding has permitted the localization of the region "responsible" for the syndrome on segment 21q22—the distal portion of the long arm of chromosome 21 (for review, see Summit 1981). Of the 12 genes assigned, so far, to chromosome 21 (see Fig. 1), 8 have been mapped to segment 21q22: the cell-surface receptor for α- and β-interferon (IFRC) (Tan et al. 1973; Raziuddin et al. 1984); cytoplasmic cuper zinc superoxide dismutase (CuZnSOD) (Tan et al. 1973); two enzymes of the purine biosynthetic pathway, glycineamide phosphoribonucleotide synthetase (GARS) (Moore et al. 1977) and aminoimidazole ribonucleotide synthetase (AIRS) (Patterson et al. 1981); the liver-type subunit of 6-phosphofructokinase (PFKL) (Vora and Francke 1981; Cox et al. 1984; Van Keuren et al. 1986); cystathionine β-synthase (CBS) (Skovby et al. 1984; Chadefaux et al. 1985; Kraus et al. 1986); the human ets-2 (ETS2) (Watson et al. 1985); and an estrogen-inducible gene (BCEI) (Moisan et al. 1985).

Although the syndrome was described more than 100 years ago and the relationship between trisomy 21 and DS has been known for more than 25 years, there is as yet no effective treatment and very little is known about the way in which the additional chromosomal segment (21q22) causes the disease (Smith and Warren 1985). It is generally assumed that the extra chromosome or segment codes for normal products and that the abnormalities found in the syndrome are produced by an excess of some of these proteins (see de la Cruz and Gerald 1981; Epstein et al. 1981, 1982; Patterson et al. 1982; Scoggin and Patterson 1982).

The gene encoding the "housekeeping" enzyme Cu/ZnSOD (E.C.1.15.1.1) resides at the segment known to be involved in the syndrome (Fig. 1) (for review, see Epstein and Epstein 1981; Francke 1981; Sinet 1982). It may therefore be considered a suspect for involvement in the etiology of DS. CuZnSOD is a protective enzyme responsible for maintaining lower levels of superoxide radicals within the cells (for review, see Fridovich 1978, 1979). In all eukaryotic cells, which utilize oxygen for their metabolism, unstable highly reactive oxygen radicals are formed (Malmstrom 1982). During the process, starting from molecular oxygen and generating H_2O on the other end, four electrons, one at a time, are acquired by the oxygen, and as a result, a series of reactive intermediates are generated (Fig. 2). The sequence is superoxide-$O_2^{\overline{\cdot}}$, hydrogen peroxide-H_2O_2, and hydroxyl radical-$HO^{\cdot}$. The most reactive and noxious species is the hydroxyl radical. Its reactivity is so great that when it is formed in living systems, it will react immediately with whatever biological molecule is in its vicinity (Badwey and Karnovsky 1980). One of the major targets is the polyunsaturated fatty acids of membrane phospholipids. If RH in Figure 2 is a phospholipid, the attack initiates a chain reaction that eventuates in the oxidative degradation of phospho-

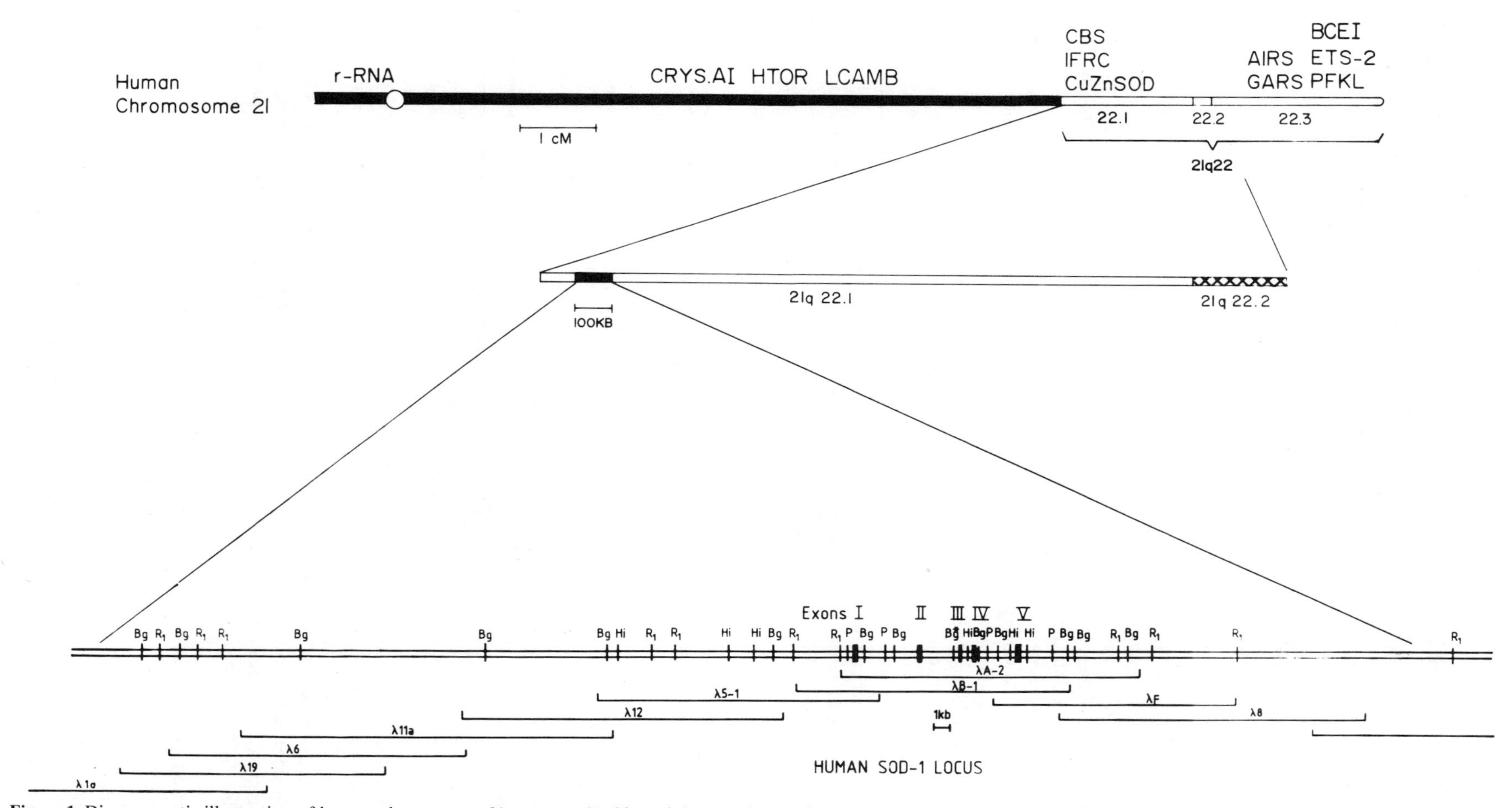

Figure 1 Diagrammatic illustration of human chromosome 21, segment 21q22, and the gene locus of CuZnSOD. Locations of the various genes mapped to this chromosome are shown (see text for explanation).

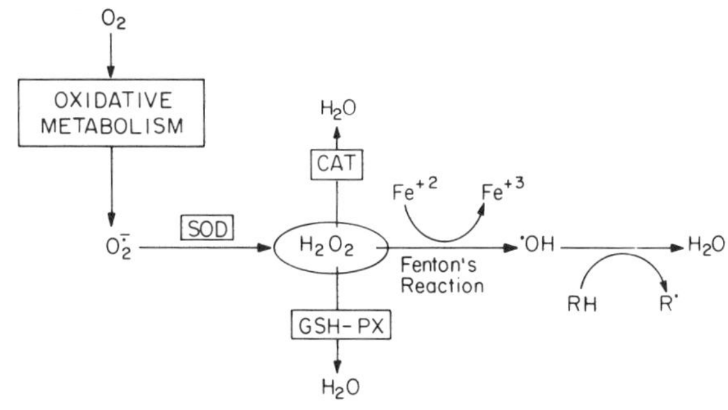

Figure 2. Schematic diagram showing the steps of oxygen reduction in biological systems leading to the formation of reactive oxygen species.

lipids, a process known as lipid peroxidation (Halliwell and Gutteridge 1985). The mechanism of formation of hydroxyl radical in biological systems was disputed for a long time, but it is now generally accepted that HO$^{\cdot}$ is produced by interaction of H_2O_2 with transition metals, mainly iron, in what is called Fenton's reaction (Fig. 2) (Aust et al. 1985; Halliwell and Gutteridge 1985). The enzymes catalase (CAT) and glutathione peroxidase (GSH-Px) metabolize H_2O_2 and thus limit the rate of Fenton's reaction within cells. However, when the level of SOD rises without a concomitant increase in CAT or GSH-Px, there is danger of an increased steady-state level of H_2O_2 and generation of HO$^{\cdot}$ (Fig. 2). A gene-dosage response of CuZnSOD in trisomy 21 has been shown in blood cells and in cultured fibroblasts (Francke 1981; Epstein and Epstein 1981; Sinet 1982; Sherman et al. 1983). More recently, we have quantitated the level of CuZnSOD RNA in various tissues of DS abortuses and have found it higher than in age-matched controls (A. Neer et al., unpubl.). Brooksbank and Balazs (1984) have reported that in addition to increased activity of CuZnSOD, in vitro lipoperoxidation is also enhanced in cerebral cortex homogenates of DS fetuses.

To investigate a possible involvement of CuZnSOD overproduction in the etiology of DS, we first cloned its cDNA (Lieman-Hurwitz et al. 1982), isolated the gene, and studied its molecular structure and expression in normal individuals versus DS patients (Sherman 1983, 1984; Groner et al. 1985; Levanon et al. 1985; Danciger et al. 1986). We then introduced (as described here) both the gene and the cDNA into mouse and human cells, obtained transformants overproducing CuZnSOD, and examined them for consequent physiological changes.

EXPERIMENTAL PROCEDURES

Construction of Recombinant Plasmids Expressing Human CuZnSOD

pSV2-cSOD-SVneo. A cDNA clone containing the entire coding region of human (h)-CuZnSOD was utilized (Lieman-Hurwitz et al. 1982; Sherman et al. 1983). An *Nde*I site (CATATG) containing the authentic ATG was created at the 5′ terminus of the cDNA clone (J.R. Hartman et al., unpubl.). This clone was digested with *Nde*I, and the 5′-protruding terminus was filled in and ligated to *Hin*dIII linkers. The plasmid was then digested with *Alu*I (which removes 45 nucleotides of the 3′-untranslated region) plus *Hin*dIII, and the resulting 520-bp fragment was inserted at the proper orientation into a pSV2 derivative (Subramani et al. 1981) to create the pSV2-cSOD (Fig. 3A). A 2.7-kb *Bam*HI fragment encompassing the *neo* transcriptional unit was removed from pSV2-*neo* (Southern and Berg 1982) and inserted into the unique *Bam*HI site of pSV2-cSOD 3′-proximal to the cDNA (Fig. 3A). Expression of both the h-CuZnSOD cDNA and the *neo* gene was controlled by SV40 regulatory elements (early promoter, enhancer, termination, and polyadenylation site). The plasmid used in the construction of pSV2-cSOD was derived from a modified pSV2-DHFR (Canaani and Berg 1982), which contains the *Eco*RI-*Sal*I fragment of pML (Lusky and Botchan 1981), by removing the DHFR cDNA in three consecutive steps: (1) digestion with *Bgl*II, (2) repair to blunt ends, and (3) digestion with *Hin*dIII, followed by isolation of a 4.3-kb pSV2 vector.

pGSOD-SVneo. The gene coding for h-CuZnSOD was previously isolated from human genomic libraries (Levanon et al. 1985). A 13-kb *Eco*RI fragment from λB-1 containing the entire gene as well as 0.55 kb and 5.0 kb from the 5′- and 3′-flanking sequences, respectively, was inserted into the *Eco*RI site of pHG165 (a pBR322 derivative containing a polylinker (Fig. 3B). This plasmid was digested with *Bam*HI, which cuts twice, in the polylinker and in the 3′ region of the gene, removing a 1.5-kb 3′-flanking fragment. A 2.7-kb *Bam*HI fragment containing the *neo* transcriptional unit described above was then ligated to the p-GSOD construct to create p-GSOD-SV*neo* (Fig. 3B).

Growth and Transfection of Cells

Human HeLa and mouse L cells were maintained in Dulbecco's modified Eagle's medium (DMEM) containing penicillin, streptomycin, and 10% FCS. The antibiotic G-418 (GIBCO) was stored in DMEM (8 mg/ml) and diluted into culture medium as needed. Supercoiled plasmid DNA was introduced into tissue-culture cells (10 μg for ~5×10^6 cells) as a calcium phosphate precipitate (Graham and Van der Eb 1973; Wigler et al. 1978), followed by a glycerol shock after 4 hours. About 48 hours later, the cells were trypsinized and replated at a 1:10 dilution. Selection for G-418 at 800 μg/ml was instituted 16 hours later. G-418-resistant sublines were maintained in medium containing 400 μg/ml antibiotic and screened for the expression of CuZnSOD. Selected clones were grown to mass culture for biochemical analysis.

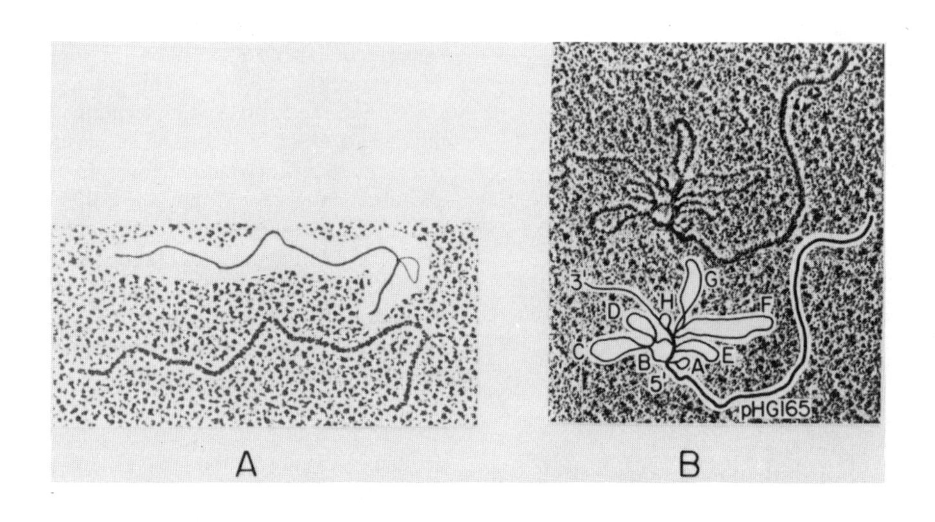

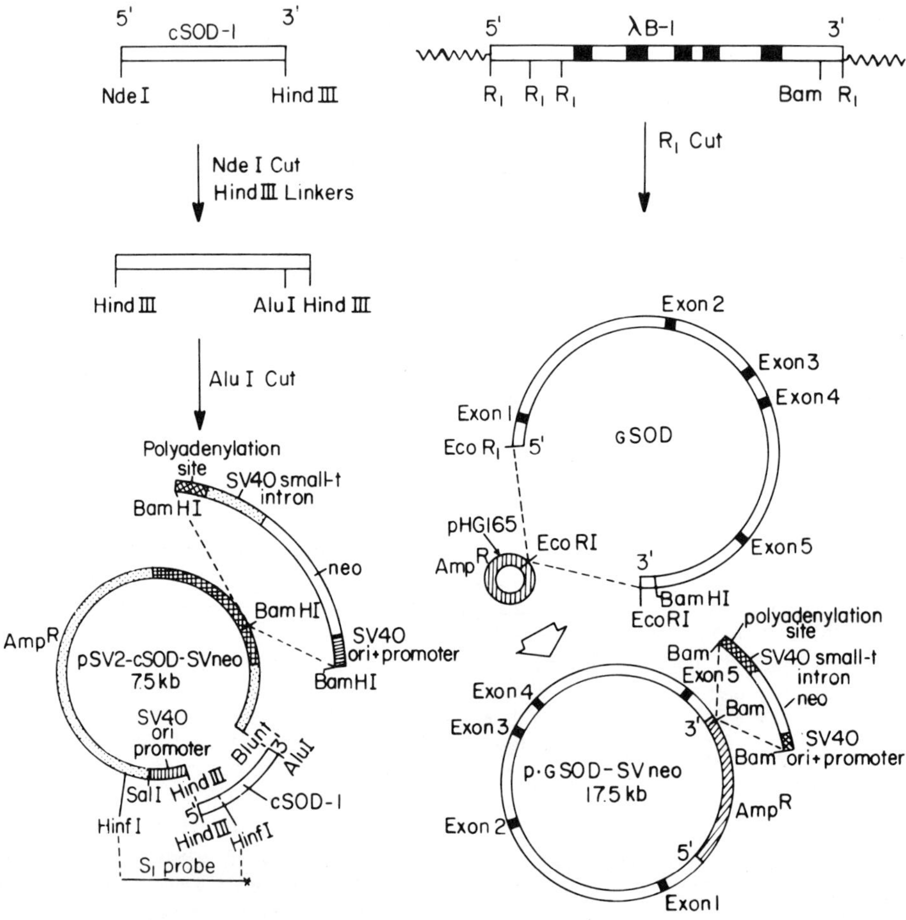

Figure 3. Structure of CuZnSOD recombinant plasmids. Details of the construction are given in Experimental Procedures. (*A*) Shuttle vector expressing the CuZnSOD cDNA (cSOD). Stippled segment represents the pBR322 DNA that contains the pBR322 origin of DNA replication and the β-lactamase gene. Open segments represent either the *neo* gene or the CuZnSOD cDNA. Hatched segment contains the SV40 origin of DNA replication, the early promoter, and enhancer. Stippled and cross-hatched segments contain the SV40 small-T-antigen intron and the site at which the SV40 early transcript is polyadenylated. (*B*) Shuttle vector expressing the CuZnSOD gene (GSOD). Hatched segments represent the pHG165 DNA that contains the pBR322 origin of DNA replication, the β-lactamase gene, and the polylinker. Open segment represents the *neo* gene, and the cross-hatched segment represents the SV40 regulatory elements indicated above. Closed segments indicate the exons of the CuZnSOD gene. (*Top*) Electron micrographs and tracing of heteroduplex molecules between the CuZnSOD cDNA plasmid and pBR322 (*A*) or genomic fragment (*B*) used in pGSOD-SV*neo*.

Enzymatic Assays of CuZnSOD

Gel electrophoresis assay (adapted from Beauchamp and Fridovich 1971) was performed in 10% polyacrylamide slab gels containing 8 M urea. Gels and running buffer did not contain SDS, and samples were not heated prior to electrophoresis. SOD was localized by soaking the gels in 2.45×10^{-3} M nitroblue tetrazolium for 20 minutes, followed by immersion for 15 minutes in a solution containing 0.028 M tetramethylethylenediamine, 2.8×10^{-5} M riboflavin, and 0.036 M potassium phosphate at pH 7.8. The gels were then placed in glass trays and illuminated for 5–15 minutes. During illumination, the gels became uniformly blue except at positions containing SOD. Illumination was discontinued when maximum contrast between the achromatic zone and the general blue color had been achieved. The gels were then photographed.

Inhibition of nitrite formation was determined by a spectrophotometric method (adapted from Elstner and Heupel 1976), and nitrite formation from hydroxylammonium chloride was determined under the following conditions: The reaction mixture (0.5 ml) contained 250 μl of SOD standard or cell extract in 65 mM phosphate buffer, 25 μl xanthine (1.142 mg/ml), 25 μl hydroxylammonium chloride (0.69 mg/ml), 125 μl KCN (4 mM) or H_2O, and 75 μl xanthine oxidase (0.1 U/ml). The mixture was incubated at 25°C for 20 minutes, followed by addition of 0.5 ml of α-naphthylamine (1 ng/ml) and 0.5 ml of sulfonic acid (3.3 mg/ml). Incubation was continued at room temperature for 20 minutes and the optical density at 530 nm was determined.

Assay of Paraquat Cytotoxicity

G-418-resistant clones and parental cell lines were grown in DMEM supplemented with 10% FCS; 24 hours before application of paraquat, cells were seeded in 9-mm microwells at 4×10^4 cells per 100 μl per well. Paraquat, at various concentrations, was applied in triplicate, and the cells were further incubated for 48 hours. Viability of the cells was determined by a modification of the procedure described by Finter (1969). Cells were incubated with neutral red for 2 hours, excess dye was washed away, and the neutral red that was taken up by the cells was extracted with Sorensen's citrate buffer–ethanol mixture and quantitated colorimetrically at 570 nm with a MicroELISA Autoreader (Dynatech, Alexander, Virginia).

Assay of Lipid Peroxidation

Cells, 5×10^5 per 9-cm plate, were maintained for 48 hours in DMEM containing penicillin, streptomycin, 4% FCS, and 50 μM linoleic acid plus arachidonic acid. DMEM containing 10% FCS was then substituted, and the cells were maintained for a further 24 hours. Cell monolayers were washed once with PBS plus 1% BSA and three times with plain PBS. Cells were then scraped off, quantitatively transferred to an Eppendorf tube, washed with 50 mM Tris (pH 7.0), resuspended in 400 μl of the same buffer, and lysed using a Dounce homogenizer. The lipid peroxidation assay was adapted from Yagi (1976) and Boehme et al. (1977). To determine the extent of in vivo lipid peroxidation, aliquots (75–150 μl) of cell extracts were immediately removed into an equal volume of 10% trichloroacetic acid (TCA). Following centrifugation at 12,000*g* for 15 minutes, two volumes of 0.67% thiobarbituric acid (TBA) were added to the supernatants, and the reaction mixtures were heated for 15 minutes in boiling water. The TBA-positive material was determined by measuring the relative intensity at 553 nm using a Perkin-Elmer MPF-44A fluorometer with excitation at 515 nm. Blanks without cell extracts were similarly processed. For determination of in vitro lipid peroxidation, aliquots (75–150 μl) were removed into glass tubes containing 10 μM $FeSO_4$ and 250 μM ascorbic acid in a final volume of 300 μl, and the tubes were incubated uncovered by shaking at 37°C for 2 hours. Reactions were stopped by TCA, and TBA-positive material was determined as before, except that the absorbance at 532 nm was measured by a spectrophotometer (Beckman DU-5). A standard curve of 0–10 nmoles of malondialdehyde (MDA) was prepared from 1,1,3,3-tetraethoxypropane (Sigma) and read either by the fluorometer or by the spectrophotometer; the results are expressed as pmoles MDA per microgram of protein.

Other procedures used in this work have been described previously. Labeling, extraction, and immunoprecipitation of CuZnSOD from human and mouse cells were carried out as described by Lieman-Hurwitz et al. (1982). DNA-blot hybridization and S1-nuclease analysis were done as described by Sherman et al. (1983, 1984) and Levanon et al. (1985).

RESULTS

Transfection of Human and Mouse Cells with CuZnSOD Recombinant Plasmids

To examine the consequences of CuZnSOD gene dosage, apart from the gene-dosage effects contributed by other genes residing on chromosome 21, we decided to generate clones of cells overproducing CuZnSOD. Recombinant plasmids containing the cDNA were introduced into homologous recipient–HeLa cells, and extra copies of the gene were introduced into heterologous host–mouse L cells. A cDNA clone that contains the entire coding region (Sherman et al. 1983) was trimmed at the 3′-untranslated region and inserted into the pSV2 vector (see Mulligan and Berg 1980) between the SV40 promoter and small T intron (Fig. 3A). A 2.7-kb *Bam*HI fragment containing the *neo* transcriptional unit (Southern and Berg 1982) was then inserted into the unique *Bam*HI site 3′-proximal to the CuZnSOD cDNA. In this construct (pSV2-cSOD-*neo*), both the CuZnSOD cDNA and the *neo* gene are under the control of SV40 regulatory elements. Expression of the *neo* gene confers upon cells resistance to the toxic aminoglycoside G-418 (Southern and Berg 1982). A second plasmid, p-GSOD-SV*neo* (Fig. 3B), contained in addition to the *neo* transcriptional unit a 12-kb *Eco*RI-*Bam*HI fragment encompassing the CuZnSOD gene previously isolated and characterized (Levanon et al. 1985). Cells were transfected by the calcium phosphate precipitation method (Graham and Van der Eb 1973; Wigler et al. 1978) and selected for G-418 resistance. Colonies that grew in the presence of the antibiotic were collected, and a few dozen of them were expanded into cell clones and examined for CuZnSOD activity and for the presence of integrated CuZnSOD and *neo* sequences in their chromosomal DNA. Several CuZnSOD producers, as well as nonproducers, were selected for further study. High-molecular-weight DNA from HeLa-derived G-418-resistant clones (originally transfected with pSV-cSOD-*neo*) was digested with *Bam*HI and analyzed by blot hybridization (Fig. 4). This enzyme cuts the plasmid twice, separating the *neo* and CuZnSOD transcription units (Fig. 3A), and is therefore diagnostic of an intact CuZnSOD fragment. The seven clones analyzed contained one to five copies of plasmid in their genomes (Fig. 4). In clones A-2, A-11, and A-14, the CuZnSOD-containing sequences occurred mostly as a 4.8-kb band corresponding to the larger *Bam*HI fragment of the transfecting plasmid, indicating a head-to-tail arrangement of the integrated copies. As indicated below, these clones expressed the integrated plasmids as enzymatically active CuZnSOD, whereas B-6, B-11, and B-19, which apparently lack the intact 4.8 kb, did not. Comparable genomic blot analysis of the L-cell-derived transformants generated by transfecting the gene-containing plasmid p-GSOD-SV*neo* also revealed a pattern indicating integration of two to ten tandemly arranged copies (data not shown).

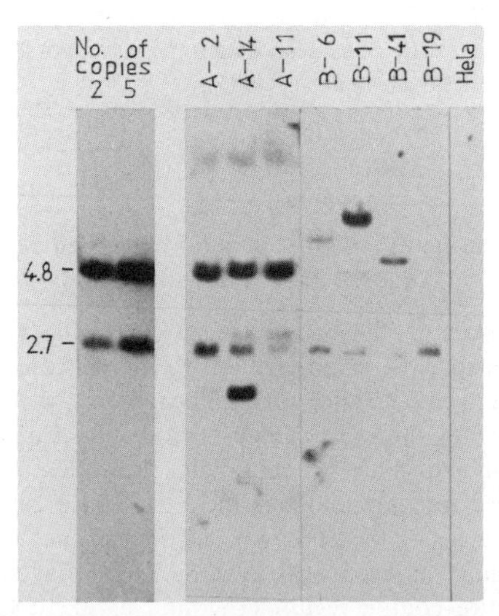

Figure 4. DNA sequences derived from the transfected plasmid pSV2-cSOD-SV*neo* in high-molecular-weight DNA extracted from G-418-resistant HeLa cells. DNA (20 μg) from selected clones was digested with *Bam*HI, electrophoresed in 1% agarose gel, and transferred to nitrocellulose filters. The blots were hybridized with ^{32}P-labeled pSV2-*neo* DNA. The 4.8-kb fragment contains the CuZnSOD cDNA and the pBR322 sequences (see Fig. 3A), and the 2.7-kb fragment encompasses the *neo* transcription unit. To estimate the number of integrated plasmids, DNA equal to 2 and 5 copies of pSV2-cSOD-SV*neo* was added to the HeLa cells prior to digestion with *Bam*HI (*left* panel).

Production of Enzymatically Active h-CuZnSOD by the Integrated Plasmids

The synthesis of h-CuZnSOD mRNA and protein by the human and mouse transformed clones was analyzed in a variety of ways. In human cells, the unique functional gene (located on chromosome 21) expresses two mRNAs of 0.7 and 0.9 kb. They differ in length at their 3′-untranslated regions, and the 0.7-kb mRNA is four times more abundant than the 0.9-kb mRNA (Sherman et al. 1983, 1984; Levanon et al. 1985; Danciger et al. 1986). To characterize and quantitate the CuZnSOD RNA made in the mouse L-cell-derived clones, a 636-bp *Ava*II-*Hin*dIII genomic fragment that spans the 3′ exon sequences and extends 250 bp further downstream was labeled at the 3′ end of the *Ava*II site and used for S1-nuclease analysis (Fig. 5C) (Berk and Sharp 1977; Weaver and Weissmann 1979). Analysis of RNA extracted from the parental line and from one of the G-418-resistant CuZnSOD-negative clones (G-51) is shown (Fig. 5), along with several clones that produced different amounts of CuZnSOD transcripts. All the mouse clones expressing the gene produced the two mRNA species at a ratio similar to that found in human cells (Fig. 5A,B). RNA extracted from the HeLa-derived clones was similarly analyzed by S1 nuclease using as a probe the 660-bp *Hin*fI-*Hin*fI fragment encompassing the SV40 promoter and the 5′ region of the CuZnSOD cDNA (see Fig. 3). Of the seven G-418-resistant clones examined (as listed in Fig. 4), A-2, A-11, and A-14 had substantial amounts of vector-derived RNA, whereas B-6, B-11, B-14, and B-19 did not (data not shown).

The ability of the L clones to translate the CuZnSOD RNA was assayed by immunoprecipitation. Transformants were labeled with [^{35}S]cysteine, and extracts were examined by *Staphylococcus*-protein A immunoprecipitation with rabbit anti-h-CuZnSOD. A protein with an approximate molecular weight of 16,000 comigrating with authentic h-CuZnSOD was precipitated from all the clones except no. 51 (Fig. 6). In addition to the h-CuZnSOD, a mouse protein with a molecular weight of 14,000 was also detected, since the anti-h-CuZnSOD serum cross-reacts with the mouse SOD (Fig. 6). The relative intensity of the two bands can serve as an indicator for the amount of the human enzyme produced by the various clones, and it is clear that clones 60 and 99-C expressed between four and five times more human CuZnSOD than did clones 73 or 81.

After gel electrophoresis to separate the human and mouse enzymes, the enzymatic activity of CuZnSOD in the transfected cells was determined using the nitroblue tetrazolium assay (Fig. 7), as well as by spectrophotometric assay with the inhibition of nitrite for-

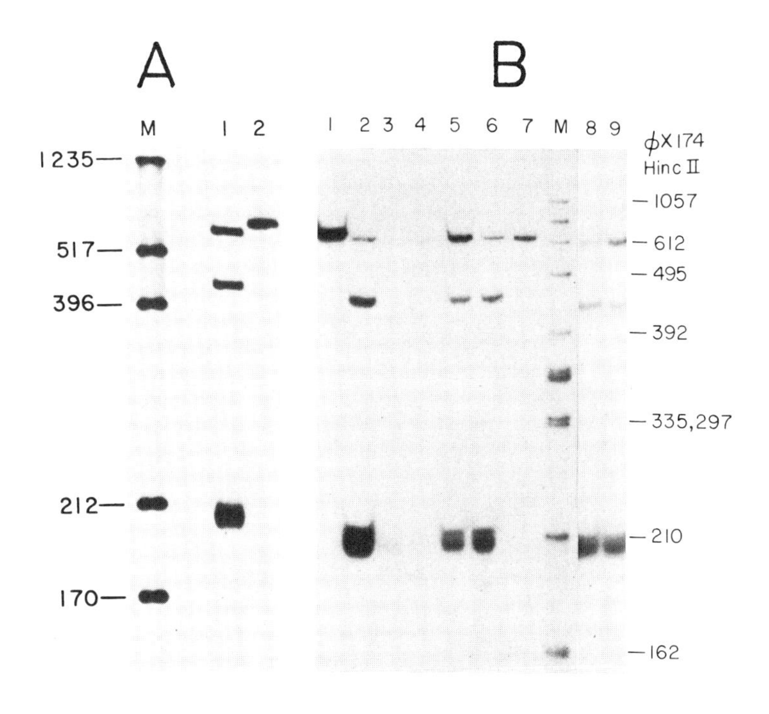

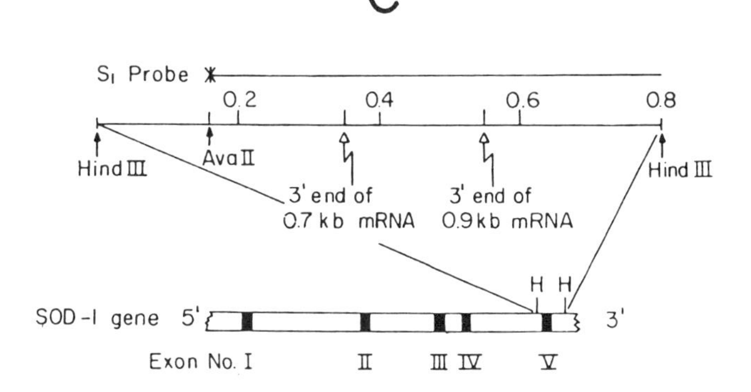

Figure 5. S1-nuclease analysis of h-CuZnSOD-specific mRNAs. The *Hin*dIII-*Ava*II fragment shown in *C* was labeled at the 3′ end of the *Ava*II site, strand-separated, and hybridized to 10 μg of poly(A)$^+$ RNA from human placenta (*A*, lane *1*), and control, minus RNA (*A*, lane *2*). (*B*) Total cytoplasmic RNA (20 μg) from the various L-cell-derived CuZnSOD clones. (Lanes *1–6*) Clones 51, 99C, 73, 81, 88, and 60, respectively; (lane *7*) RNA from the parental L cells; (lanes *8,9*) longer exposure (4 days rather than 1) of lanes *3* and *4*.

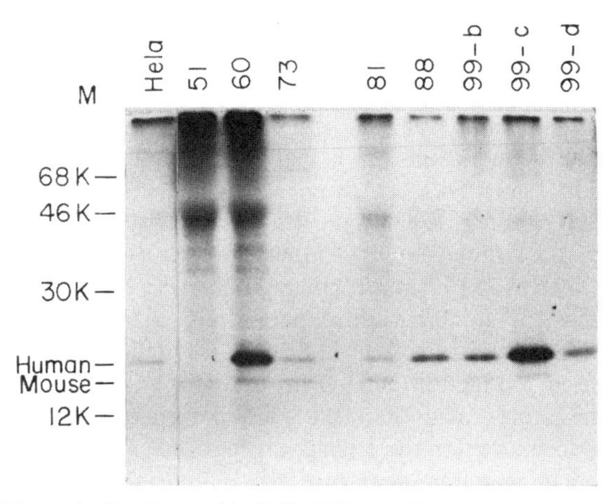

Figure 6. Synthesis of h-CuZnSOD protein detected in mouse L-cell transformants. Cells were labeled with [^{35}S]cysteine for 4 hr, and total cell lysates were prepared, subjected to immunoprecipitation with rabbit anti-human CuZnSOD serum, and analyzed by SDS-gel electrophoresis. Positions of human and mouse CuZnSOD are indicated; 51 through 99 are the various L-cell-derived clones containing the h-CuZnSOD gene.

mation method (Table 1). Of the HeLa-derived transformants, B-6 had, as expected, a level of CuZnSOD activity very similar to that of the parental line, whereas A-11 and A-14 contained more activity (Fig. 7A). Among the L-cell derivatives, clone 51 was similar to the parental line, whereas the other clones produced between two and five times more enzymatically active h-CuZnSOD (Fig. 7B and Table 1). In summary, the collection of G-418-resistant transformants contained clones expressing various levels of CuZnSOD RNA and protein. The recombinant plasmids encoding the CuZnSOD were integrated into the cellular DNA and the protein made was enzymatically active.

Physiological Effects Resulting from Overproduction of h-CuZnSOD

To understand at the molecular level how an extra set of what seems to be normal genes produces DS, the genes involved in the pathology associated with the syndrome should be identified by demonstrating a con-

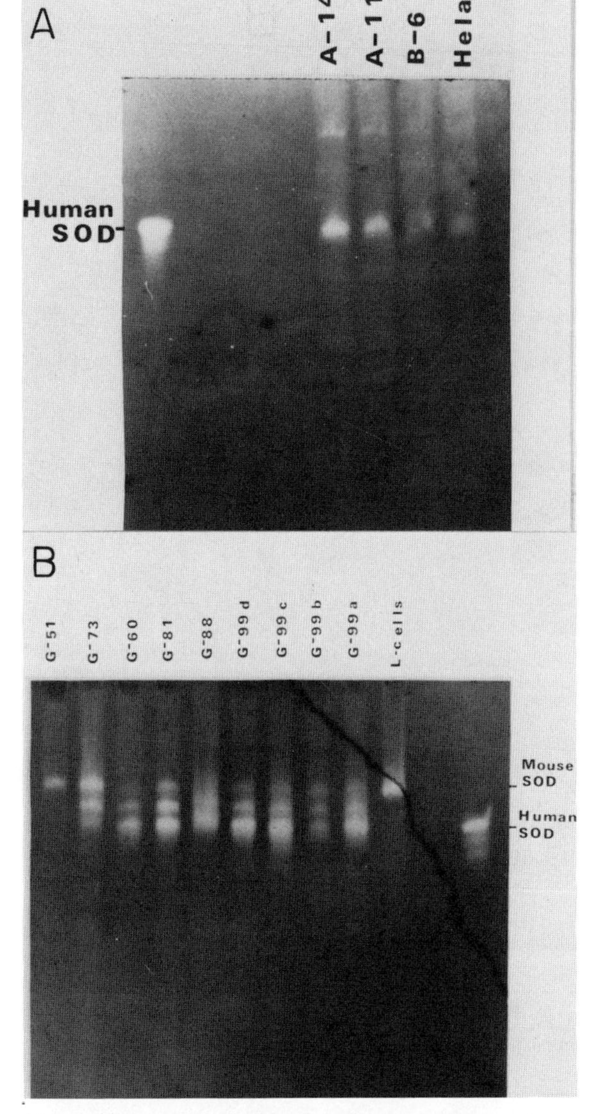

Figure 7. Enzymatic activity of CuZnSOD in HeLa-cell- and L-cell-derived transformants. In each case, an NP-40 extract of 1.0×10^6 cells was made and loaded on 10% acrylamide-urea gel, and electrophoresis was performed for 2 hr at 150 V. For measurements of CuZnSOD activity, the gels were treated as described in Experimental Procedures. (*A*) HeLa-cell-derived clones; (*B*) L-cell-derived clones. Positions of mouse and human CuZnSOD are indicated. The middle band present in the L-cell-derived clones expressing the h-CuZnSOD is probably a hybrid dimer of human subunit plus mouse subunit.

nection between overproduction of their gene products and clinical symptoms of DS. We therefore examined the physiological consequences of CuZnSOD overproduction. We first analyzed the ability of the human and mouse transformants overproducing CuZnSOD to withstand increasing concentrations of paraquat. The herbicide paraquat (1,1'dimethyl-4,4'-bipyridinium dichloride) is a nonselective weed killer, highly cytotoxic and lethal to animals (for review, see Autor 1977). It is known to increase the production of superoxide

Table 1. Specific Activity of CuZnSOD in L-cell Transformants

Clone	Specific activity (U/mg protein)	Specific activity relative to L cells
L cells	4.9	1.0
51	4.9	1.0
73	5.0	1.0
99-a	16.6	3.4
60	17.7	3.6
99-c	28.5	5.8

Cell extracts (0.5% NP-40) were prepared, and specific activity of CuZnSOD (U/mg protein) was determined as described in Experimental Procedures. Units of SOD-1 represent the KCN-sensitive activity in the extracts and was deduced from a standard curve obtained with purified h-CuZnSOD. Each number represents the average of specific activity measurements on three different extracts.

radicals ($O_2^{\cdot-}$) by a mechanism that involves reduction of paraquat by NADPH-diaphorase to a relatively stable paraquat radical ($PQ^{\cdot+}$) that reacts rapidly with oxygen to produce the superoxide radical (Farrington et al. 1973; Hassan and Fridovich 1979; Moody and Hassan 1982). In animals, paraquat toxicity mainly affects the lungs; it was proposed that the pulmonary lesions are related to peroxidation of membrane lipids (Bus et al. 1977). It was reported that paraquat stimulates SOD activity in *Escherichia coli* (Hassan and Fridovich 1977) and in *Salmonella typhimurium* (Moody and Hassan 1982) and that due to the increase in SOD activity, the cells become more resistant to paraquat toxicity (Moody and Hassan 1982).

The various clones were treated for 48 hours with different concentrations of paraquat, and the extent of survival was determined (Fig. 8). The HeLa-derived overproducer clone A-11 was totally resistant to paraquat concentrations that killed all the cells of B-6, a G-418-resistant clone that expresses only the native CuZnSOD. Clone A-2, which overexpresses CuZnSOD to a lesser extent than A-11, was less resistant but not as sensitive as B-6. Surprisingly, A-14, which produced higher levels of CuZnSOD than A-11, was less resistant (Fig. 8A). Similar results were obtained with the L-cell-derived clones. G-51, an h-CuZnSOD-negative clone, was sensitive to even 0.2 mM paraquat, a concentration not affecting clone 60, a high CuZnSOD producer. Again, clone 99-C, which possessed higher enzymatic activity than clone 60, was less resistant to the drug (Fig. 8B). Based on the knowledge that H_2O_2 is the product of $O_2^{\cdot-}$ dismutation, our tentative conclusion is that the combination of paraquat and higher CuZnSOD activity in A-14 and 99-C resulted in an increased production of H_2O_2 to the point where it became more toxic than the paraquat-generated $O_2^{\cdot-}$. Moreover, even in the other overproducers, the protection against paraquat cytotoxicity is limited, and A-11, G-60, and G-99-C will die upon longer (>48 hr) exposure to paraquat concentrations as low as 0.05 mM. We assume that this cell death is due to the accumulation of lesions caused by elevated levels of hydrogen peroxide and the generation of hydroxyl radicals in the clones possessing high CuZnSOD activity (see Fig. 2). The

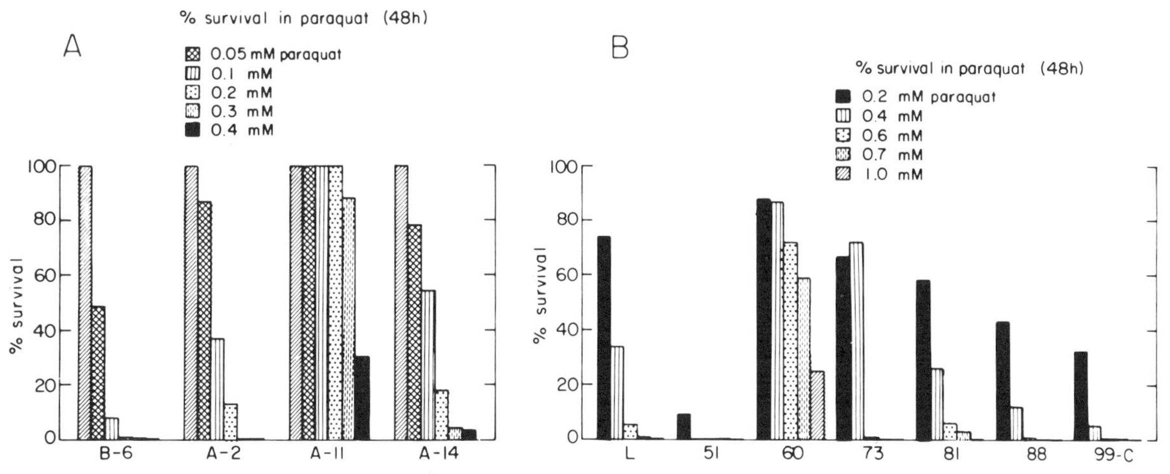

Figure 8. Increased resistance to paraquat of the transfected clones overexpressing CuZnSOD. Paraquat cytotoxicity was measured as described in Experimental Procedures. (*A*) HeLa-cell-derived clones; (*B*) L-cell-derived clones.

knowledge that perturbation in the delicate balance between the rates of hydrogen peroxide formation and its removal can cause lipid peroxidation (Yagi 1982; Kanner and Harel 1985), as well as the more directly related observations of enhanced lipid peroxidation, possibly linked to increased CuZnSOD activity, in homogenates of DS brains (Brooksbank and Balazs 1984), prompted us to examine the degree of lipid peroxidation in the overproducing clones. Tappel (1973) has defined lipid peroxidation as "the reaction of oxidative deterioration of polyunsaturated lipids." Although the exact mechanism of this chain reaction has not been elucidated, it was shown that lipid hydroperoxides undergo decomposition through several possible intermediates to produce MDA. This compound will react with TBA to yield a pink-colored TBA pigment with characteristic absorption and fluorescence spectra (Dahle et al. 1982; Pryor et al. 1976; Yagi 1976). The extent of lipid peroxidation in two of the L-derived overproducers (clones 60 and 99-C) was determined and compared with the level found in the parental L cells (Table 2). The abovementioned TBA reaction (Niehaus and Samuelsson 1968; Brooksbank and Balazs 1984; Kanner and Harel 1985) was employed. Both in vivo and in vitro lipid peroxidation were significantly enhanced in clones 60 and 99-C (Table 2). The process was augmented by paraquat, and as before (Fig. 8B), its effect was more pronounced in clone 99-C than in clone 60. These results support the contention that the increased activity of CuZnSOD created within the cells an oxidative stress leading to enhanced lipid peroxidation.

DISCUSSION

Human and Mouse Cell Clones Overexpressing the h-CuZnSOD

The physiological and mental abnormalities associated with DS are complex and probably involve more than an extra dose of a single gene (Epstein et al. 1981; Smith 1985). A gene implicated in the syndrome would most likely (1) reside at the 21q22 segment of chromosome 21, (2) be overexpressed in all or certain tissues of DS fetuses, and (3) be connected by virtue of its overproduction or mistiming of expression to clinical symptoms of DS. The CuZnSOD gene fulfills the first two requirements, but do higher levels of the enzyme produce phenotypic changes? In an attempt to answer this question and to analyze how an excess of a normal gene product affects cell physiology, a series of cultured cells overproducing the human CuZnSOD was generated and examined for consequent metabolic

Table 2. Lipid Peroxidation in L-cell Transformants

Cells	pmoles MDA per microgram of protein			
	in vivo	in vivo (+ PQ)	in vitro	in vitro (+ PQ)
Parental L cells	0.074	0.483	45.45	108.00
Clone 60	0.109(147)	0.371	82.25(180)	126.45
Clone 99-c	0.139(188)	0.632	87.20(191)	131.80

MDA formation in vivo and in vitro was determined as described in Experimental Procedures. Paraquat at 0.4 mM was added for 24 hr prior to harvesting the cells. Values are representative of 11 separate determinations, with different batches of clones 60 and 99-c, all of which showed a significant increase (compared with the parental L cells) of MDA formation both in vivo and in vitro. These extracts were also assayed for h-CuZnSOD activity as described in Table 1 and Fig. 7 to verify the consistency of the increase in enzymatic activity. Numbers in parenthesis represent the percentage of the values in L cells.

changes. Plasmids containing the selectable marker *neo* and either the CuZnSOD cDNA or the gene were transfected into human and mouse cells, respectively, and G-418-resistant clones overproducing various amounts (up to sixfold) of enzymatically active CuZnSOD protein were isolated (Figs. 6 and 7 and Table 1). These CuZnSOD overproducers enabled us to study the cellular response to elevated levels of the enzyme in a more defined background, distinct from cellular effects caused by the gene dosage of other chromosome-21-encoded genes.

Overproduction of CuZnSOD Extenuates Paraquat-mediated Cytotoxicity

When the transfected cells were exposed for 48 hours to paraquat, a herbicide known to act as an in vivo generator of superoxide radicals, the clones overproducing CuZnSOD were substantially more resistant than control cells (Fig. 8). In previous reports, it was shown that *E. coli* (Hassan and Fridovich 1977) and *S. typhimurium* (Moody and Hassan 1982) cells become more resistant to paraquat toxicity by inducing SOD activity (~fourfold). Preventing this induction greatly augmented paraquat toxicity. The results presented here are therefore in good correlation with the data obtained for *E. coli* and *S. typhimurium,* although the human CuZnSOD and the bacterial Mn-SOD are different proteins (see Oberley 1982). As for the correlation between CuZnSOD overexpression and resistance to paraquat, it was interesting to note that although all the transformants were totally protected against the toxic G-418 due to the expression of the neo^R gene, the clones that possess the highest CuZnSOD activity (A-14 and 99-C) were less resistant to paraquat, suggesting the existence of additional processes in which elevated activity of CuZnSOD is not advantageous. Moreover, even the most resistant clones (A-14 and A-60) were gradually deteriorating in the presence of relatively low concentrations (0.05 mM) of paraquat, indicating that the elevated levels of CuZnSOD did not protect these cells from the cumulative damage caused by paraquat by-products. Based on the knowledge that hydrogen peroxide is produced by the dismutation of superoxide, our conclusion is that a combination of paraquat and the higher CuZnSOD of A-14 and 99-C resulted in increased production of H_2O_2 and other active forms of oxygen ($OH^\cdot$, $^1O_2$) to the point where they became more toxic than just the $O_2^{\cdot -}$ generated by paraquat. Accordingly, even in the absence of paraquat, higher levels of CuZnSOD may interfere with the normal metabolism of reduced oxygen species and create an oxidative stress. In this context, it should be mentioned that so far, we have failed to isolate a transfected clone possessing greater than sixfold CuZnSOD activity, suggesting that higher levels may be noxious to the cell. Gene-transfer experiments in which the CuZnSOD expression is controlled by an inducible promoter should clarify this point.

CuZnSOD Overproduction: Enhancement of Lipid Peroxidation and DS

Under normal conditions, a delicate balance exists between the rate of hydrogen peroxide formation via dismutation of superoxides and its removal by GSH-Px and CAT. Increased activity of CuZnSOD, as in the transfected clones, could lead to an increase in the steady-state levels of hydrogen peroxides, resulting in formation of hydroxyl radical ($OH^\cdot$) and singlet oxygen ($^1O_2$), which may enhance lipid peroxidation (see Fig. 2) (Yagi 1982; Halliwell and Gutteridge 1985). Lipid peroxidation is defined as a nonenzymatic breakdown of unsaturated fatty acids, giving rise to peroxy radicals and ultimately to MDA. The physiological consequences of increased lipid peroxidation result from the damage caused to cellular membranes and organelles and their associated enzymes. The overall mechanism of the process is not yet completely understood, but the involvement of hydrogen peroxides, hydroxyl radicals, and singlet oxygen is firmly established (Yagi 1982; Halliwell and Gutteridge 1985). For example, Kanner and Harel (1985) have recently shown that even small amounts of H_2O_2 could interact with metmyoglobin and generate active radicals that initiate membranal lipid peroxidation. The data presented in Table 2 clearly show a connection between the higher level of CuZnSOD activity and lipid peroxidation (both in vivo and in vitro). Furthermore, the lipid peroxidation was enhanced by paraquat (0.4 mM) even in those clones having high CuZnSOD, indicating that H_2O_2 or other oxygen derivatives are involved in the process. The transformants in Table 2 overexpressed 3.6 and 5.8 times more CuZnSOD than the parental L cells, whereas in DS, the documented CuZnSOD activity amounted to only 1.5 times the normal level. Nevertheless, the clear correlation in Table 2 between elevated CuZnSOD activity and lipid peroxidation supports the idea of Brooksbank and Balazs (1984) (see below) that in brains of DS fetuses, there is a higher potential for lipid peroxidation due to the higher activity of CuZnSOD. It should be emphasized that during embryonal development, the levels of CuZnSOD in certain regions of the DS brain may increase above the conventional gene dosage.

In DS erythrocytes, GSH-Px activity is significantly increased along with that of CuZnSOD (Sinet et al. 1975; Frischer et al. 1981; Kedziora et al. 1982), whereas CAT levels are normal (Pantelakis et al. 1972). This is believed to be an adaptive response to the elevated levels of hydrogen peroxides produced by the augmented CuZnSOD activity, rather than a gene-dosage effect, since the gene coding for GSH-Px is located on human chromosome 3 (Wijnen et al. 1978). However, as was reported by Brooksbank and Balazs (1984), this adaptive mechanism does not function in the brain of DS fetuses in which CuZnSOD activity was raised, but GSH-Px activity was not. Furthermore, in vitro lipid peroxidation was substantially elevated in those brain specimens (Brooksbank and Balazs 1984). It was also found that the level of polyunsaturated fatty acids in cell membranes of DS patients is significantly reduced (Shah 1979) and that during brain development of DS fetuses, a perturbation in the metabolism of essential fatty acids occurs, distorting their incorporation into phosphoglycerides (Brooksbank et al. 1985). Thus, also in brain of DS fetuses, there seems to be a correlation between potential increase of lipid peroxidation and elevated CuZnSOD activity. Since the molecular structure of the lipid component has a profound effect on the properties of the membranes, enhanced lipid peroxidation may be related to the reported functional abnormalities in DS membranes, e.g., decrease in Na^+/K^+ ATPase and serotonin uptake in DS platelets (McCoy and Enns 1978), increased adhesiveness of DS fetal fibroblasts in vitro (Wright et al. 1984), and abnormal electric membrane properties of DS neurons in cell culture (Scott et al. 1982). It will be interesting to determine whether neuronal cells programmed to express elevated levels of CuZnSOD show any of the abovementioned defects.

Although various other genes residing in the 21q22 segment are undoubtedly involved in the etiology of the Down's phenotype, the data presented here are consistent with the possibility that the gene dosage of CuZnSOD leads to increased lipid peroxidation and thus contributes to some of the clinical symptoms associated with Down's syndrome.

ACKNOWLEDGMENTS

We thank Dan Canaani and Paul Berg for providing the pSV2 vectors, Jacob Hartman and Marian Gorecki for the SOD-1 plasmid containing the *Nde*I site, and Jonathan Kuhn for the plasmid pHG165. We also thank Ephraim Yavin for his help and comments concerning the analysis of fatty acids and lipid peroxidation and Meir Shinizky for helpful discussions about lipid peroxidation and membrane functions. This work was supported by a basic research grant (no. 1-906) from the March of Dimes Birth Defects Foundation, by a research grant from BioTechnology General Corp. (Israel), by the Minerva Foundation (Munich, Germany), and by the Weizmann Institute's Leo and Julia Forchheimer Center of Molecular Genetics.

REFERENCES

Aust, S.D., L.A. Morehouse, and C.E. Thomas. 1985. Role of metals in oxygen radical reactions. *J. Free Radicals Biol. Med.* **1:** 3.

Autor, A.P., ed. 1977. *Biomedical mechanisms of paraquat toxicity*. Academic Press, New York.

Badwey, J.A. and L.M. Karnovsky. 1980. Active oxygen species and the functions of phagocytic leukocytes. *Annu. Rev. Biochem.* **49:** 695.

Beauchamp, C. and I. Fridovich. 1971. Superoxide dismutase: Improved assays and an assay applicable to acrylamide gels. *Anal. Biochem.* **44:** 276.

Berk, A.J. and P.A. Sharp. 1977. Sizing and mapping of early adenovirus mRNAs by gel electrophoresis of S1 nuclease digested hybrids. *Cell* **12:** 721.

Boehme, D.H., R. Kosecki, S. Carson, F. Stern, and N. Marks. 1977. Lipoperoxidation in human and rat brain tissue: Developmental and regional studies. *Brain Res.* **136:** 11.

Bond, D.J. and A.C. Chandley, ed. 1983. *Aneuploidy*. Oxford University Press, London.

Brooksbank, B.W.L. and R. Balazs. 1984. Superoxide dismutase, glutathione peroxidase and lipoperoxidation in Down's syndrome fetal brain. *Dev. Brain Res.* **16:** 37.

Brooksbank, B.W.L., M. Martinez, and F. Balazs. 1985. Altered composition of polyunsaturated fatty acyl groups in phosphoglycerides of Down's syndrome fetal brain. *J. Neurochem.* **44:** 869.

Burgio, G.R., M. Fraccaro, L. Tiepolo, and U. Wolf, eds. 1981. *Trisomy 21*. Springer-Verlag, Berlin.

Bus, J.S., S.D. Aust, and J.E. Gibson. 1977. In *Biochemical mechanisms of paraquat toxicity* (ed. A.P. Autor), p. 157. Academic Press, New York.

Canaani, D. and P. Berg. 1982. Regulated expression of human interferon β1 gene after transduction into cultured mouse and rabbit cells. *Proc. Natl. Acad. Sci.* **79:** 5166.

Chadefaux, B., M.O. Rethore, O. Raol, I. Cebollos, M. Poissonnier, S. Gilgenkranz, and D. Allard. 1985. Cystathionine β-synthase: Gene dosage effect in trisomy 21. *Biochem. Biophys. Res. Commun.* **128:** 40.

Cox, D.R., H. Kawashima, S. Vora, and C.J. Epstein. 1984. Regional mapping of SOD-1, PRGS, and PFK-L on human chromosome 21. *Cytogenet. Cell Genet.* **37:** 441.

Dahle, L.K., E.G. Hill, and R.T. Holman. 1962. Thiobarbituric acid reaction and the autoxidation of polyunsaturated fatty acid methyl esters. *Arch. Biochem. Biophys.* **98:** 253.

Danciger, E., N. Dafni, Y. Bernstein, Z. Laver-Rudich, A. Neer, and Y. Groner. 1986. Human Cu/Zn superoxide dismutase gene family: Molecular structure and characterization of four Cu/Zn superoxide dismutase-related pseudogenes. *Proc. Natl. Acad. Sci.* **83:** 3619.

de La Cruz, F.F. and P.S. Gerald, eds. 1981. *Trisomy 21 (Down's syndrome). Research perspectives*. University Park Press, Baltimore.

Elstner, E.F. and A. Heupel. 1976. Inhibition of nitrit formation from hydroxylammonium-chloride: A simple assay for superoxide dismutase. *Anal. Biochem.* **70:** 616.

Epstein, C.J. and L.B. Epstein. 1981. Gene dosage effects in trisomy 21. In *Trisomy 21 (Down's syndrome). Research perspectives* (ed. F.F. de la Cruz and P.S. Gerald), p. 253. University Park Press, Baltimore.

Epstein, C.J., L.B. Epstein, D.R. Cox, and J. Weil. 1981. Functional implication of gene dosage effect in trisomy 21. In *Trisomy 21* (ed. A.R. Burgio et al.), p. 155. Springer-Verlag, New York.

Epstein, C.J., L.B. Epstein, J. Weil, and D.R. Cox. 1982. Trisomy 21: Mechanisms and models. *Ann. N.Y. Acad. Sci.* **396:** 107.

Farrington, J.A., M. Ebert, E.J. Land, and K. Fletcher. 1973. Pulse radiolysis studies of the reaction of paraquat radical with oxygen. Implications for the mode of action of bipyridyl herbicides. *Biochim. Biophys. Acta* **314:** 372.

Finter, N.B. 1969. Dye uptake methods for assessing virol cytopathogenicity and their application to interferon assay. *J. Gen. Virol.* **5:** 419.

Francke, U. 1981. Gene dosage studies on Down's syndrome: A review. In *Trisomy 21 (Down's syndrome). Research perspectives* (ed. F.F. de la Cruz and P.S. Gerald), p. 237. University Park Press, Baltimore.

Fridovich, I. 1978. The biology of oxygen radicals. *Science* **201:** 875.

———. 1979. Superoxide and superoxide dismutases. In *Advances in inorganic biochemistry* (ed. G.L. Eichhorn and L.G. Marzilli), p. 67. Elsevier/North-Holland, New York.

Frischer, H., L.K. Cher, T. Ahmad, P. Justice, and G.F. Smith. 1981. Superoxide dismutase and glutathione peroxidase abnormalities in erythrocytes and lymphoid cells in Down's syndrome. *Prog. Clin. Biol. Res.* **55:** 269.

Graham, F.L. and A.J. Van der Eb. 1973. A new technique for the assay of infectivity of human adenovirus 5 DNA. *Virology* **52:** 456.

Groner, Y., J. Lieman-Hurwitz, N. Dafni, L. Sherman, D. Levanon, Y. Bernstein, E. Danciger, and O. Elroy-Stein. 1985. Molecular structure and expression of the gene locus on chromosome 21 encoding the Cu/Zn-superoxide dismutase and its relevance to Down's syndrome. *Ann. N.Y. Acad. Sci.* **450:** 133.

Halliwell, B. and J.M.C. Gutteridge. 1985. *Free radicals in biology and medicine.* Clarendon Press, Oxford.

Hassan, H.M. and I. Fridovich. 1977. Regulation of the synthesis of superoxide dismutase in *Escherichia coli. J. Biol. Chem.* **252:** 7667.

———. 1979. Intracellular production of superoxide radical and hydrogen peroxide by redox active compounds. *Arch. Biochem. Biophys.* **196:** 385.

Hassold, T.J. and P.A. Jacobs. 1984. Trisomy in man. *Annu. Rev. Genet.* **18:** 69.

Hook, E.B. 1981. Down's syndrome: Its frequency in human populations and some factors pertinent to variation in rats. In *Trisomy 21 (Down's syndrome). Research perspectives* (ed. F.F. de la Cruz and P.S. Gerald), p. 3. University Park Press, Baltimore.

Kanner, J. and S. Harel. 1985. Initiation of membranal lipid peroxidation by activated metmyoglobin and methemoglobin. *Arch. Biochem. Biophys.* **237:** 314.

Kedziora, J., R. Lukaszewicz, M. Koter, G. Bartosz, B. Pawlowska, and D. Aitkin. 1982. Red blood cell glutathione peroxidase in simple trisomy 21 and translocation 21/22. *Experientia* **38:** 543.

Kraus, J.P., C.L. Williamson, F.A. Firgaira, T.L. Yang-Feng, M. Munke, U. Francke, and L.E. Rosenberg. 1986. Cloning and screening with nanogram amounts of immunopurified mRNAs: cDNA cloning and chromosomal mapping of cystathionine β-synthase and the β subunit of propionyl-CoA carboxylase. *Proc. Natl. Acad. Sci.* **83:** 2047.

Lejeune, J.M., M. Gautier, and R. Turpin. 1959. Etude des chromosomes somatique de neuf enfants mongoliens. *C.R. Acad. Sci.* **248:** 1721.

Levanon, D., J. Lieman-Hurwitz, N. Dafni, M. Wigderson, L. Sherman, Y. Bernstein, Z. Laver-Rudich, E. Danciger, O. Stein, and Y. Groner. 1985. Architecture and anatomy of the chromosomal locus in human chromosome 21 encoding the Cu/Zn superoxide dismutase. *EMBO J.* **4:** 77.

Lieman-Hurwitz, J., N. Dafni, V. Lavie, and Y. Groner. 1982. Human cytoplasmic superoxide dismutase cDNA clone: A probe for studying the molecular biology of Down's syndrome. *Proc. Natl. Acad. Sci.* **79:** 2808.

Lusky, M. and M. Botchan. 1981. Inhibition of SV40 replication in simian cells by specific pBR322 DNA sequences. *Nature* **293:** 79.

Malmstrom, B.G. 1982. Enzymology of oxygen. *Annu. Rev. Biochem.* **51:** 21.

McCoy, E.E. and L. Enns. 1978. Sodium transport, oubain binding and Na^+/K^+ ATPase activity in Down's syndrome platelets. *Pediatr. Res.* **12:** 685.

Moisan, J.P., M.G. Mattei, M.A. Baeteman-Volkel, J.F. Mattei, A.M.C. Brown, J.M. Garnier, J.M. Jeltsch, P. Mesiakowsky, M. Roberts, and J.L. Mandel. 1985. A gene expressed in human mammary tumor cell under estrogen control is located in 21g223 and defines an RFLP. *Cytogenet. Cell Genet.* **40:** 701 (Abstr.).

Moody, C.S. and H.M. Hassan. 1982. Mutagenicity of oxygen free radicals. *Proc. Natl. Acad. Sci.* **79:** 2855.

Moore, E.E., C. Jones, E.-T. Kao, and D.C. Oates. 1977. Synteny between glycinamide ribonucleotide synthetase and superoxide dismutase (soluble). *Am. J. Hum. Genet.* **29:** 389.

Mulligan, R.C. and P. Berg. 1980. Expression of a bacterial gene in mammalian cells. *Science* **209:** 1422.

Niehuas, W.G., Jr. and B. Samuelsson. 1968. Formation of malonaldehyde from phospholipid arachidonate during microsomal lipid peroxidation. *Eur. J. Biochem.* **6:** 126.

Oberley, L.W. ed. 1982. *Superoxide dismutase.* CRC Press, Boca Raton, Florida.

Pantelakis, S.N., A.G. Karaklis, D. Alexion, E. Verdas, and T. Valaes. 1972. Red cell enzymes in trisomy 21. *Am. J. Hum. Genet.* **22:** 184.

Patterson, D., S. Graw, and C. Jones. 1981. Demonstration, by somatic cell genetics, of coordinate regulation of genes for two enzymes of purine synthesis assigned to chromosome 21. *Proc. Natl. Acad. Sci.* **78:** 405.

Patterson, D., C. Jones, C. Scoggin, Y.E. Miller, and S. Graw. 1982. Somatic cell genetic approaches to Down's syndrome. *Ann. N.Y. Acad. Sci.* **396:** 69.

Pryor, W.A., J.P. Stanley, and E. Blair. 1976. Autoxidation of polyunsaturated fatty acids: 11. A suggested mechanism for the formation of TBA-reactive materials from prostaglandin-like endoperoxides. *Lipids* **11:** 370.

Raziuddin, A., F.H. Sarkar, R. Dutkowski, L. Schulman, F.H. Ruddle, and S.L. Gupta. 1984. Receptors for human α and β-interferon but not γ interferon are specified by human chromosome 21. *Proc. Natl. Acad. Sci.* **81:** 5504.

Scoggin, C.H. and D. Patterson. 1982. Down's syndrome as a model disease. *Arch. Intern. Med.* **142:** 462.

Scott, B.S., T.L. Petit, L.E. Becker, and B.A.V. Edwards. 1982. Abnormal electric membrane properties of Down's syndrome DRG neurons in cell culture. *Dev. Brain Res.* **2:** 257.

Shah, S.N. 1979. Fatty acid composition of lipids of human brain myelin and synaptosomes: Changes in phenylketonuria and Down's syndrome. *Int. J. Biochem.* **10:** 477.

Sherman, L., N. Dafni, J. Lieman-Hurwitz, and Y. Groner. 1983. Nucleotide sequence and expression of human chromosome 21-encoded superoxide dismutase mRNA. *Proc. Natl. Acad. Sci.* **80:** 5465.

Sherman, L., D. Levanon, J. Lieman-Hurwitz, N. Dafni, and Y. Groner. 1984. Human Cu/Zn superoxide dismutase gene: Molecular characterization of its two mRNA species. *Nucleic Acids Res.* **12:** 9349.

Sinet, P.M. 1982. Metabolism of oxygen derivatives in Down's syndrome. *Ann. N.Y. Acad. Sci.* **396:** 83.

Sinet, P.M., A.M. Michelson, A. Bazin, J. Lejeune, and H. Jerome. 1975. Increase in glutathione peroxidase activity in erythrocytes from trisomy 21 subjects. *Biochem. Biophys. Res. Commun.* **67:** 910.

Skovby, F., N. Krassikoff, and U. Francke. 1984. Assignment of the gene for cystathionine β-synthase to human chromosome 21 in somatic cell hybrids. *Hum. Genet.* **65:** 291.

Smith, G.F., ed. 1985. Molecular structure of the number 21 chromosome and Down's syndrome. *Ann. N.Y. Acad. Sci.* **450:**

Smith, G.F. and J.M. Berg. 1976. *Down's anomaly*, 2nd edition. Churchill Ltd., London.

Smith, G.F. and S.T. Warren. 1985. The biology of Down's syndrome. *Ann. N.Y. Acad. Sci.* **450:** 1.

Southern, P.J. and P. Berg. 1982. Transformation of mammalian cells to antibiotic resistance with bacterial gene under control of the SV40 early gene promoter. *J. Mol. App. Genet.* **1:** 327.

Subramani, S., R. Mulligan, and P. Berg. 1981. Expression of the mouse dihydrofolate reductase complementary deoxyribonucleic acid in simian virus 40 vectors. *Mol. Cell. Biol.* **1:** 854.

Summitt, R.L. 1981. Chromosome 21: Specific segments that cause the phenotype of Down's syndrome. In *Trisomy 21 (Down's syndrome), Research perspectives* (ed. F.F. de la Cruz and P.S. Gerald), p. 225. University Park Press, Baltimore.

Tan, Y.H., J. Tischfield, and F.H. Ruddle. 1973. The linkage of genes for the human interferon-induced antiviral protein and indophenol oxidase-B traits to human chromosome 6-21. *J. Exp. Med.* **137:** 317.

Tappel, A.L. 1973. Lipid peroxidation damage to cell components. *Fed. Proc.* **32:** 1870.

Van Keuren, M., H. Drabkin, I. Hart, D. Harker, D. Patterson, and S. Vora. 1986. Regional assignment of human liver-type 6-phosphofructokinase to chromosome 21g22.3 by using somatic cell hybrids and monoclonal anti-L antibody. *Hum. Genet.* **74:** 34.

Vora, S., and U. Francke. 1981. Assignment of the human gene for liver-type 6-phosphofructokinase isozyme (PFKL) to chromosome 21 by using somatic cell hybrids and monoclonal antibody. *Proc. Natl. Acad. Sci.* **78:** 3738.

Watson, D.K., M.J. Smith, C. Kozak, R. Reeves, J. Gearhart, M.F. Nunn, W. Nash, J.R. Fowle, P. Duesberg, T.S. Papas, and S.J. O'Brien. 1985. Conserved chromosomal positions of dual domains of the *ets* proto-oncogene in cats, mice and man. *Proc. Natl. Acad. Sci.* **83:** 1792.

Weaver, R.F. and C. Weissman. 1979. Mapping of RNA by a modification of the Berk-Sharp procedure: The 5′ termini of 15S β-globin mRNA precursor and mature 10S β-globin mRNA have identical map coordinates. *Nucleic Acids Res.* **7:** 1175.

Wigler, M., A. Pellicer, S. Silverstein, and R. Axel. 1978. Biochemical transfer of single-copy eukaryotic genes using total cellular DNA as donor. *Cell* **14:** 725.

Wijnen, L.M.M., M. Monteba-Ven Heusel, P.L. Pearson, and P.M. Khan. 1978. Assignment of a gene for glutathione peroxidase (GPX1) to human chromosome 3. *Cytogenet. Cell Genet.* **22:** 232.

Wright, T.C., M. Destrempes, R. Orkin, and D.M. Kurnit. 1984. Increased adhesiveness of Down's syndrome fetal fibroblasts *in vitro*. *Proc. Natl. Acad. Sci.* **81:** 2426.

Yagi, K. 1976. A simple fluorometric assay for lipoperoxide in blood plasma. *Biochem. Med.* **15:** 212.

———. 1982. *Lipid peroxidation in biology and medicine.* Academic Press, New York.

Zellweger, H. 1981. The story of Down's syndrome which preceded Langdon Down. *Down's syndrome.* **4:** 1.

Molecular Basis of Phenylketonuria and Potential Somatic Gene Therapy

S.L.C. Woo, A.G. DiLella, J. Marvit, and F.D. Ledley
Howard Hughes Medical Institute, Department of Cell Biology and Institute of Molecular Genetics, Baylor College of Medicine, Houston, Texas 77030

Classical phenylketonuria (PKU) is an autosomal recessive human genetic disorder caused by a deficiency in hepatic phenylalanine hydroxylase (PAH; phenylalanine 4-mono-oxygenase, E.C. 1.14.16.1). The enzyme is a mixed-function oxidase that catalyzes hydroxylation of phenylalanine to tyrosine, which is the rate-limiting step in phenylalanine catabolism. A deficiency of this enzyme causes an accumulation of phenylalanine, resulting in heperphenylalaninemia and abnormalities in the metabolism of many compounds derived from the aromatic amino acids. If left untreated, these metabolic abnormalities cause postnatal brain damage and severe mental retardation. PKU is the most common inborn error in amino acid metabolism, with an average incidence of about 1 in 10,000 Caucasians. The mutant gene frequency is such that 1 in 50 individuals is a carrier of the disease trait (for review, see Scriver and Clow 1980a,b,c).

Purified PAH from rat (Kaufman and Fisher 1970), monkey (Cotton and Grattan 1975), or human liver (Friedman and Kaufman 1973; Woo et al. 1974) has a molecular weight of about 100,000 and is a dimer composed of 49,000–50,000-dalton subunits. We have recently reported the isolation of a full-length human PAH cDNA and have determined its sequence (Kwok et al. 1985). The amino acid sequence deduced from the cDNA clone predicts a peptide of 451 amino acids and a molecular weight of 51,672. DNA-mediated gene-transfer experiments demonstrated that the cDNA clone contains all the genetic information necessary to code for a functional enzyme (Ledley et al. 1985), providing direct evidence that the enzyme is a homodimer and encoded by a single mRNA.

Southern hybridization analysis of DNA isolated from PKU individuals demonstrated that the disorder is not caused by any obvious deletions, insertions, or rearrangements in the *PAH* gene (Woo et al. 1983). Using the full-length cDNA clone as the hybridization probe, Lidsky et al. (1985b) have determined a total of eight restriction-fragment-length polymorphisms (RFLPs) in the human *PAH* locus. Application of RFLP analyses to trace the transmission of *PAH* genes in PKU families has demonstrated conclusively that the RFLPs segregated concordantly with the disease state in various PKU kindreds (Woo et al. 1983; Daiger et al. 1986) and resulted in the possibility for prenatal diagnosis of PKU (Lidsky et al. 1985a). The polymorphism information content (PIC) is such that almost 90% of PKU families are informative for analysis (Chakraborty et al. 1986).

The genetic lesions causing the loss of enzyme activity in PKU are not known. The liver-specific expression of the *PAH* gene and the scarcity of human liver biopsy tissue have precluded the examination of mutant *PAH* proteins or mRNA for characterizing the molecular basis of PKU. Alternatively, PKU mutations can be identified at the genomic DNA level by cloning and sequencing the mutant alleles, followed by gene-transfer and expression experiments. These studies are based on knowledge of the normal chromosomal human *PAH* gene structure that has recently been established. The gene contains 13 exons and is about 90 kb in length to code for a mature mRNA of about 2400 nucleotides (DiLella et al. 1986a).

To select appropriate mutant alleles for molecular analysis, we first performed extensive RFLP haplotype analysis of the *PAH* locus. In PKU kindreds among Caucasians of Northern European ancestry, haplotype analysis showed that mutant alleles of four different RFLP haplotypes constitute 90% of PKU chromosomes in the population.

In this paper, we discuss the molecular characterization of one of these mutant alleles and its tight association with a particular RFLP haplotype due to linkage disequilibrium, as well as potential therapy of PKU by somatic gene transfer.

RESULTS

RFLPs at the Human *PAH* Locus

Figure 1 shows the map locations of the eight RFLP sites on the *PAH* locus. Except for the *Hin*dIII site, all RFLPs seem to be substitutional nucleotide changes arising by point mutations. The *Hin*dIII polymorphism has three alleles, apparently the result of insertion/deletion of a 0.2-kb unit (A.G. DiLella, unpubl.). From the map distances of the RFLP sites, it is evident that these eight polymorphic sites are not uniformly distributed over the entire *PAH* gene. Three sites, *Bgl*II, *Pvu*II(a), and *Pvu*II(b), are within 22 kb of each other, whereas five other sites, *Eco*RI, *Xmn*I, *Msp*I, *Hin*dIII, and *Eco*RV, are within 31 kb of each other. Between these two clusters, there is a 43-kb gap in which no RFLP site was detected.

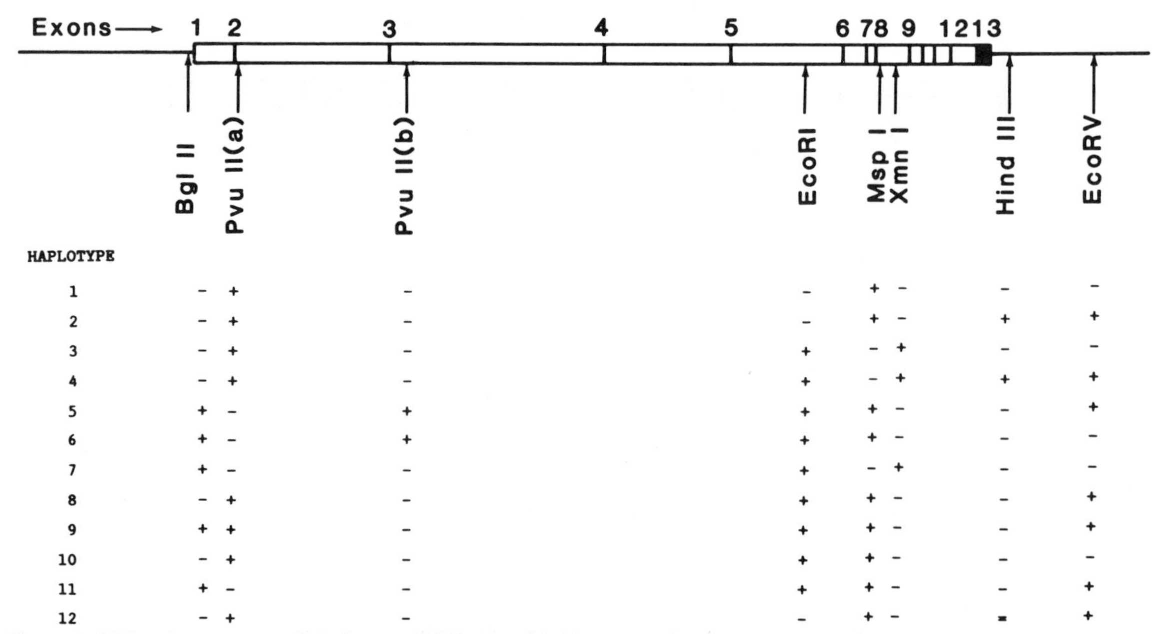

Figure 1. Molecular structure of the human *PAH* gene with 13 exons and 12 introns. Arrows indicate the map positions of the polymorphic restriction sites. RFLP haplotypes of 66 normal and 66 mutant *PAH* alleles were determined in 33 Danish PKU families, and a total of 12 haplotypes were observed. The presence or absence of each of the polymorphic restriction sites for each of the haplotypes is indicated by the plus (+) and minus (−) symbols below the respective restriction sites.

RFLP Haplotypes and Their Frequencies

The RFLP haplotypes of 66 normal and 66 mutant alleles in 33 Danish kindreds were determined, and a total of 12 RFLP haplotypes was observed (Fig. 1). The haplotype frequencies defined by the eight RFLPs determined from 132 parental chromosomes are shown in Table 1. Of the 12 haplotypes detected in the sample, 6 occur on both normal and PKU chromosomes, 3 on non-PKU chromosomes only, and the remaining 3 on PKU chromosomes only. Thus, even though the kindred analysis indicated that PKU is tightly linked (with a maximum lod score of 9.96) with the RFLPs in the *PAH* locus (Daiger et al. 1986), no apparent association between PKU and a particular RFLP haplotype of the *PAH* gene is evident at the population level.

Table 1. RFLP Haplotype Distribution of the *PAH* Alleles in Denmark

	Normal alleles		Mutant alleles	
Haplotypes	no.[a]	%	no.[a]	%
1	23	34.8	12	18.2
2	3	4.6	13	19.7
3	2	3.0	25	37.9
4	21	31.8	9	13.6
5	7	10.6	0	0
6	0	0	2	3.0
7	7	10.6	1	1.5
8	1	1.5	0	0
9	0	0	1	1.5
10	1	1.5	0	0
11	1	1.5	1	1.5
12	0	0	2	3.0

[a]Number of alleles observed in a total of 66 chromosomes analyzed.

Nevertheless, it is important to note that although nine different haplotypes have been observed for the PKU chromosomes, about 90% of all mutant alleles are associated with only four haplotypes (haplotypes 1–4).

Isolation and Sequence Characterization of a Mutant *PAH* Gene

Since 38% of the mutant alleles have a single haplotype (haplotype 3), which is relatively rare among the normal gene pool (3%) (Table 1), a cosmid genomic DNA library was constructed from leukocyte DNA isolated from a PKU individual who is homozygous for haplotype 3. The library was screened with a full-length human *PAH* cDNA probe, and several overlapping clones spanning about 135 kb of contiguous genomic DNA were obtained. To identify the mutation that causes PKU, exon-containing regions of the gene were subcloned from the cosmid into M13mp18 for sequence analysis. The mutant gene sequence is identical to that of the normal gene, with the exception of a G→A transition at the 5′ splice donor site of intron 12, altering the obligatory donor dinucleotide GT to AT (Fig. 2A; DiLella et al. 1986b).

The Mutant Allele Causes Aberrant RNA Splicing

Gene-transfer and expression experiments demonstrated that this mutation results in abnormal *PAH* mRNA processing and the absence of *PAH* protein in the cell. In the human liver, this mutation causes the skipping of exon 12 in the mature mRNA, as evidenced by cloning and sequence characterization of a correspondingly deleted *PAH* cDNA clone from the liver of a heterozygote bearing this mutant allele (J. Marvit et al., in prep.).

An analogous mutation has been observed previously in the human β-globin locus causing β-thalassemia. Here, a mutation at the 5′ end of intron 2 of the human β-globin gene causes skipping of exon 2 in mature β-globin mRNA in the patient's bone-marrow cells (T. Maniatis, pers. comm.). Since there are only three exons in the β-globin gene, there can be no alternative exons for splicing after skipping exon 2. Such is not the case in the human *PAH* gene. Since there are 13 exons in the gene, exon 13 could have been spliced together with any of the first 11 exons. The fact that it only skipped exon 12 would suggest that the entire exon is disregarded by the splicing mechanism as a result of the GT→AT transition at the splice donor site of intron 12. Thus, we propose that there may be cooperative interactions between the small ribonucleoprotein particles in the intron/exon boundary, such that not only introns, but also exons are looped out in a splicesome (Fig. 2B).

Detection of Haplotype-3 Mutant Alleles in PKU Kindreds by Oligonucleotide Hybridization

Since the GT→AT transition in the allele does not alter the sequence at that region to either create or destroy a restriction enzyme cleavage site, two specific oligonucleotides, one specific for the mutant allele and the other for the normal allele, were synthesized. Under appropriate hybrdization conditions, the synthetic oligonucleotide probes were used to identify the mutant alleles in genomic DNA and to analyze their segregation in PKU kindreds. The analysis of a representative kindred is shown in Figure 3. The father is a haplotype-3 homozygote and an obligate carrier of the PKU trait. He bears a mutant haplotype-3 allele that hybridized to the mutant probe (Fig. 3, lane 1) and a normal haplotype-3 allele that hybridized to the normal probe (Fig. 3, lane 7). The results demonstrated conclusively that the splicing mutation can be readily detected in the genomic DNA of a heterozygote and that the splicing mutation is not a constitutive part of the normal haplotype-3 allele. In contrast, the mother bears one each of a normal and a mutant haplotype-1 allele. The mutant probe did not hybridize to DNA isolated from this individual (Fig. 3, lane 2), whereas the normal probe did (Fig. 3, lane 8). The data indicated that the PKU mutation associated with the haplotype-1 allele in this family is not the same as that identified in the mutant haplotype-3 allele. The proband in this family, having inherited a mutant haplotype-3 allele from the father and a mutant haplotype-1 allele from the mother, hybridized to both the mutant and normal probes as expected (Fig. 3, lanes 3 and 9, respectively).

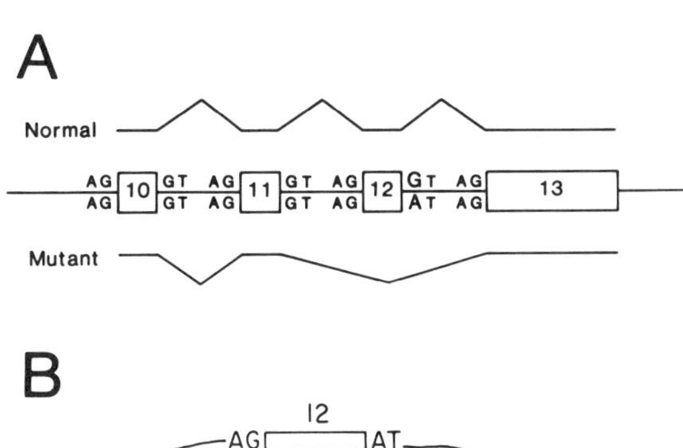

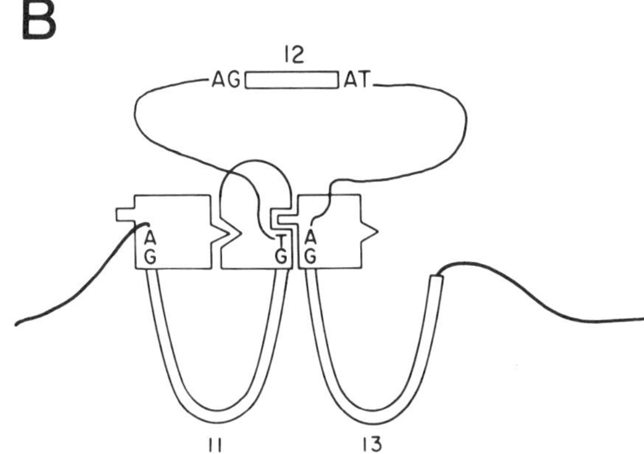

Figure 2. (*A*) Schematic representation for products of aberrant RNA splicing due to a GT→AT transition at the 5′ donor splicing site of intron 12. (*B*) Model for RNA splicing in a splicesome.

Mutation Frequency and Haplotype Association

We next determined the frequency of the splicing mutation and the degree of its association with haplotype 3 in the Danish population. RFLP haplotype analysis of 33 informative PKU families (66 normal and 66 mutant PAH chromosomes) provided a repertoire of mutant alleles for such an analysis (Table 1). Genomic DNA samples isolated from PKU individuals of defined RFLP haplotypes were analyzed by hybridization using the normal and mutant oligonucleotide probes for the splice donor site mutation sequence. A total of 91% (60 out of 66) of the available PKU chromosomes and 27% (18 out of 66) of the available normal chromosomes were analyzed. The population genetic data are summarized in Table 2, which shows that all mutant haplotype-3 alleles hybridized with the mutant probe. The remaining mutant alleles of other haplotypes and all of the normal alleles hybridized with the normal probe. The results demonstrated unambiguously that the splicing mutation is specifically associated with the mutant haplotype-3 alleles, and the association is both inclusive and exclusive within the Danish PKU population.

Retrovirus-mediated Gene Transfer of Human *PAH*

As a first step directed toward the ultimate goal of somatic gene therapy for PKU, we introduced the full-length *PAH* cDNA sequence into pzip-*neo*SV(X),

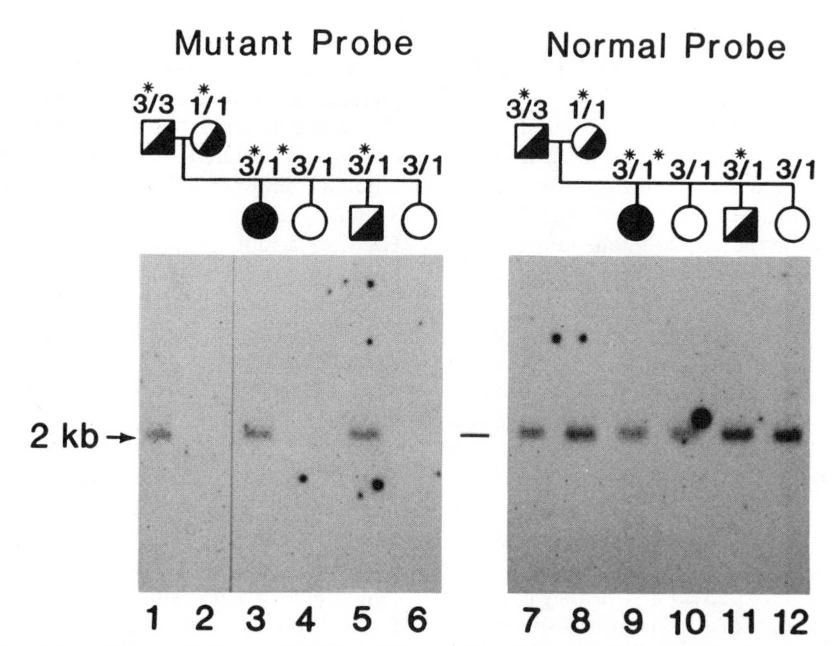

Figure 3. Detection of the splicing mutation in a PKU family by specific oligonucleotide hybridization. DNA was isolated from lymphocytes of family members, digested with *Pvu*II, and analyzed by dried-gel hybridization using mutant (lanes *1–6*) and normal (lanes *7–12*) specific oligonucleotide probes. Exon 12 plus flanking intron sequences are contained in a 2-kb *Pvu*II fragment. The segregation of the PKU alleles (*) and the appropriate RFLP haplotypes are shown at the top of each autoradiograph. High specific activity probes (i.e., 10^{10} cpm/μg) were generated by the primer extension method (Studencki and Wallace 1984). The normal probe was synthesized on a 21-base template (5′-TCCATTAACAGTAAGTAATTT-3′) by extension of a hybridized 9-base primer (3′TTCATTAAA-5′), and the mutant probe was synthesized by extension of a 9-base primer (5′-TCCATTAAC) hybridized to a 21-base template (3′-AGGTAATTGTTATTCATTAAA-5′). Dried gels were hybridized overnight at 37°C in 6× NET (0.9 mM NaCl, 6 mM EDTA, 0.5% SDS, and 0.09 M Tris at pH 7.5) containing 0.2 mg of salmon sperm DNA and 2×10^6 cpm of probe per milliliter of hybridization solution. The gel membranes were then washed twice at 0°C for 30 min in TMA (3 M tetramethylammonium chloride [Aldrich], 2 mM EDTA, and 50 mM Tris at pH 8.0), once each at 23°C (30 min) and 60°C (7 min) in TMA containing 0.2% SDS, and at 23°C for 30 min in TMA. The gel strips were then autoradiographed between two Quanta III intensifier screens (du Pont) at −80°C.

which is a retroviral vector containing the bacterial *neo* gene. The recombinant was transfected into ψ2 cells, which provide synthesis of the retroviral capsid (Mann et al. 1983; Cepko et al. 1984). Recombinant virus was detected in the culture medium of the transfected ψ2 cells, which is capable of transmitting the human *PAH* cDNA into mouse NIH-3T3 cells by infection, leading to stable incorporation of the recombinant provirus. The infected NIH-3T3 cells expressed *PAH* mRNA, immunoreactive *PAH* protein, and pterin-dependent *PAH* activity (Ledley et al. 1986).

Human *PAH* Activity Is Expressed in Mouse Hepatoma Cells after Infection with the Recombinant Virus

Mouse hepa1-a hepatoma cells were infected with vZPAH(+) as a model for introduction of the recombinant virus into hepatic tissue. The hepa1-a cell line is $HPRT^-$ and preserves many differentiated properties of hepatocytes (Darlington et al. 1980, 1982), but it has no detectable *PAH* mRNA, protein, or enzymatic activity. hepa1-a cells were infected by cocultivation with either the recombinant or vector-virus-producing cell lines. Virus-infected mouse hepatoma cells were selected for G418 and 6-thioguanine resistance. Approximately 10^3 and 10^4 resistant colonies were obtained by

Table 2. Association of Splicing Mutation with Mutant Haplotype-3 Alleles in Denmark

	Mutant probe analysis[a]		Normal probe analysis[a]	
Haplotype	normal	PKU	normal	PKU
1	0/5	0/10	5/5	10/10
2	n.d.	0/12	n.d.	12/12
3	0/2	23/23	2/2	0/23
4	0/6	0/8	6/6	8/8
5	0/1	0	1/1	0
6	0	0/2	0	2/2
7	0/2	0/1	2/2	1/1
8	0/1	0	1/1	0
9	0	0/1	0	1/1
10	0/1	0	1/1	0
11	n.d.	0/1	n.d.	1/1
12	0	0/2	0	2/2

[a]Number of hybridizing alleles/number of genes analyzed. n.d. indicates not determined due to lack of DNA samples.

cocultivation with the recombinant and vector-virus-producing cell lines, respectively. The drug-resistant cell colonies had the morphological characteristics and karyotype of the hepa1-a cell line. Mixed cultures representing 10^3–10^4 colonies were propagated, and cellular extracts were assayed for *PAH* activity. *PAH* activity was present in extracts of the recombinant virus-infected hepatoma cells but not cells infected with the vector virus (Ledley et al. 1986).

Phenylalanine hydroxylation is a complex oxidation-reduction reaction that requires molecular oxygen and a pterin cofactor (tetrahydrobiopterin). The cofactor is cooxidized with the phenylalanine substrate and must be regenerated by reduction carried out by another enzyme (dihydropteridine reductase). To determine whether the phenylalanine hydroxylation system was successfully reconstituted in the infected mouse hepatoma cells, the cells were cultured in tyrosine-free media for several passages. It is apparent that the cells infected by the vector virus failed to survive, whereas the cells infected by the recombinant virus grew to confluency (Fig. 4).

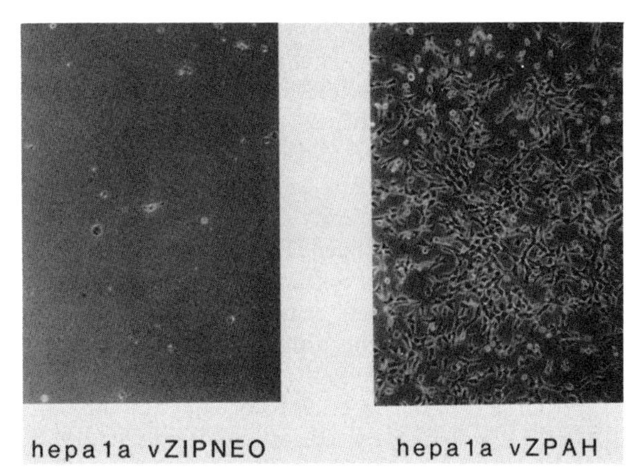

Figure 4. Culture of hepa1-a cells in tyrosine-free medium after infection with vector virus (*left*) and recombinant virus containing human *PAH* cDNA (*right*). hepa1-a cells (NEO^-, $HPRT^-$, PAH^-) were cocultivated with ψ2 cells producing an ecotrophic retrovirus containing the *neo* resistance gene and the human *PAH* cDNA (NEO^+, $HPRT^+$, PAH^+). After 48 hr cultivation in the 2 μg/ml polybrine, the mixed cultures were subjected to selection with 6-thioguanine and G418. $HPRT^-$ cells transformed to G418 resistance by infection with the recombinant retrovirus also exhibited PAH activity as evidenced by in vitro assays and the ability to grow in tyrosine-free media (NEO^+, $HPRT^-$, PAH^+).

DISCUSSION

Although PKU appears to be a heterogeneous disorder at the clinical level, whether this heterogeneity results from multiple mutations in the *PAH* gene of various degrees of severity has been a controversial topic in the literature (for review, see Scriver and Clow 1980a,b,c). We recently reported that the cause of PKU must be heterogeneous, since *PAH* mRNA was detected in some, but not all, PKU liver biopsy specimens (DiLella et al. 1985). Southern hybridization analysis of DNA isolated from a large number of PKU individuals demonstrated that, in general, the disorder is not caused by any obvious deletions, insertions, or rearrangements in the *PAH* gene. Liver-specific expression of the *PAH* gene in man and the general lack of hepatic tissues of sufficient quantity and quality have precluded studying the molecular basis of PKU at the mRNA and protein levels. Having established the normal *PAH* gene structure, we can identify PKU mutations at the genomic DNA level by characterization of the mutant genes.

Extensive RFLP haplotype analysis of the *PAH* locus in the Danish population provided the theoretical basis for selection of mutant alleles for molecular characterization. Ninety percent of the PKU genes in the Danish population are confined to four haplotypes, two of which are relatively uncommon in the total normal gene pool. Forty percent of the PKU alleles are associated with haplotypes 1 and 4, which are also prevalent among the normal alleles (75%), suggesting that the corresponding PKU mutations arose on chromosomes of common background haplotypes. The majority of the PKU alleles (60%), however, are associated with haplotypes 2 and 3, which are rare among the normal alleles. We reported the molecular cloning of a mutant *PAH* gene associated with haplotype 3, which comprises 38% of all mutant alleles in the Danish population. Sequence analysis of this gene demonstrated a single-base substitution at the 5′ splice donor site of intron 12. DNA-hybridization analysis using an oligonucleotide probe specific for the genetic lesion demonstrated an absolute association of this mutation with the mutant haplotype-3 alleles in the Danish population. This observation has recently been extended to various regions throughout Europe. The results would strongly suggest that the GT→AT mutation occurred relatively recently on a normal haplotype-3 gene, which then spread in the Caucasian population by a founder effect. In addition, the specific mutation is not present in mutant alleles of other RFLP haplotypes, providing unambiguous evidence that there are multiple and distinct mutations in the *PAH* gene that are responsible for PKU.

The future characterization of PKU mutations arising on the other prevalent haplotypes (i.e., haplotypes 1, 2, and 4) will clarify whether PKU is caused by a limited number of mutations that spread throughout the Caucasian race by founder effect or whether multiple PKU mutations arose independently in various population backgrounds. Such a study will provide insight into the evolution and molecular origin of mutations in the *PAH* gene that cause PKU. If our hypothesis regarding the linkage of specific mutations and RFLP haplotypes in the *PAH* locus can be verified by these studies, it would be possible to design a cassette of oligonucleotide probes to detect 90% of the mutation chromosomes in the Caucasian population and provide a potential molecular means for carrier detection among random individuals without a prior family history of PKU.

Finally, we demonstrated that human *PAH* cDNA can be transmitted efficiently into NIH-3T3 cells and a mouse hepatoma cell line. The transduced cDNA is expressed in the infected cells into mature mRNA, immunoreactive protein, and enzymatic activity. In particular, the ability to reconstitute the phenylalanine hydroxylation system in cells of hepatic origin is important in considering gene-replacement therapy for PKU. *PAH* activity requires not only the intact *PAH* apoenzyme, but also a reduced pterin cofactor (tetrahydrobiopterin), which is oxidized to a quinonoid dihydropterin in conjunction with the hydroxylation of phenylalanine. The reduced tetrahydrobiopterin must then be replenished by the enzyme dihydropteridine reductase (Kaufman 1976). The liver contains the enzymes required for pterin biosynthesis and reduction and is thus able to provide the necessary constituents for the hydroxylation system. NIH-3T3 cells do not provide all of these enzymatic functions. Thus, despite the fact that a functional *PAH* enzyme can be introduced into these cells by infection with the recombinant virus, no activity can be detectable in the living cells in situ (Ledley et al. 1986). In contrast, *PAH* activity is present in recombinant virus-infected hepa1-a cells in situ. This indicates that it is possible to reconstitute complete *PAH* holoenzyme activity in cells that have selective deficiency of the *PAH* apoenzyme. This is the situation in the livers of PKU individuals who are deficient only in the *PAH* apoenzyme and are able to synthesize and reduce the pterin cofactor normally.

It should be emphasized that the present study only constitutes a first step toward the ultimate goal of somatic genetic therapy of PKU by demonstrating the feasibility of using recombinant viruses containing the human *PAH* gene for genetic transfer into cells normally lacking *PAH* activity. Amphotropic variants of Moloney murine leukemia virus exist that produce viruses with a broader host range capable of infecting cells from many species, including man. Thus, it is conceivable that recombinant viruses such as the one described here might be useful to introduce a functioning *PAH* gene into liver cells, enabling an individual with PKU to metabolize excess phenylalanine to tyrosine and thus ameliorating the biochemical abnormalities and clinical symptoms of *PAH* deficiency.

ACKNOWLEDGMENTS

This work was partially supported by National Institutes of Health grant HD-17711. S.L.C.W. is an investigator and F.D.L. is an assistant investigator of the Howard Hughes Medical Institute. A.G.D. is the recipient of a National Institutes of Health postdoctoral fellowship (HD-06495). We thank Ms. Kelly Porter for typing the manuscript.

REFERENCES

Cepko, C.L., B.E. Roberts, and R.C. Mulligan. 1984. Constructions and applications of a highly transmissible murine retrovirus shuttle vector. *Cell* **37:** 1053.

Chakraborty, R., A.S. Lidsky, S.P. Daiger, F. Güttler, S. Sullivan, A.G. DiLella, and S.L.C. Woo. 1986. Polymorphic DNA haplotypes at the human phenylalanine hydroxylase locus and their relationship with phenylketonuria. *Proc. Natl. Acad. Sci.* (in press).

Cotton, R.G.H. and P.J. Grattan. 1975. Phenylalanine hydroxylase of *Macaca irus:* A simple purification by affinity chromatography. *Eur. J. Biochem.* **60:** 427

Daiger, S.P., R. Chakraborty, F. Güttler, A.S. Lidsky, R. Koch, and S.L.C. Woo. 1986. Effective use of polymorphic DNA haplotypes at the phenylalanine hydroxylase locus in prenatal diagnosis of phenylketonuria. *Lancet* **I:** 229.

Darlington, G.J., H.P. Bernhard, R.A. Miller, and F.H. Ruddle. 1980. TI expression of liver phenotypes in cultured mouse hepatoma cells. *J. Natl. Cancer Inst.* **64:** 809.

Darlington, G.J., J. Papaconstanatinou, D.W. Sammons, C.C. Brown, E.Y. Wong, A.L. Esterman, and J. Kang. 1982. Generation and characterization of variants of mouse hepatoma cells with defects in hepato-specific gene expression. *Somatic Cell Genet.* **8:** 451.

DiLella, A.G., S.C.M. Kwok, F.D. Ledley, J. Marvit, and S.L.C. Woo. 1986a. Molecular structure and polymorphic map of the human phenylalanine hydroxylase gene. *Biochemistry* **25:** 743.

DiLella, A.G., F.D. Ledley, F. Rey, A. Munnich, and S.L.C. Woo. 1985. Detection of phenylalanine hydroxylase messenger RNA in PKU liver biopsies. *Lancet* **I:** 160.

DiLella, A.G., J. Marvit, A.S. Lidsky, F. Güttler, and S.L.C. Woo. 1986b. A splicing mutation in phenylketonuria is tightly linked to a specific RFLP haplotype of the human phenylalanine hydroxylase gene. *Nature* (in press).

Friedman, P.A. and S. Kaufman. 1973. Some characteristics of partially purified human liver phenylalanine hydroxylase. *Biochim. Biophys. Acta* **293:** 56.

Kaufman, S. 1976. Phenylketonuria, biochemical mechanisms. *Adv. Neurochem.* **2:** 1.

Kaufman, S. and D.B. Fisher. 1970. Purification and some physical properties of phenylalanine hydroxylase from rat liver. *J. Biol. Chem.* **245:** 4745.

Kwok, S., F.D. Ledley, A.G. DiLella, K.J.H. Robson, and S.L.C. Woo. 1985. Nucleotide sequence of a full-length complementary DNA clone and amino acid sequence of human phenylalanine hydroxylase. *Biochemistry* **24:** 556.

Ledley, F.D., H.E. Grenett, M. McGinnis-Shelnutt, and S.L.C. Woo. 1986. Retroviral-mediated gene transfer of human phenylalanine hydroxylase into NIH 3T3 and hepatoma cells. *Proc. Natl. Acad. Sci.* **83:** 409.

Ledley, F.D., H.E. Grenett, A.G. DiLella, S.C.M. Kwok, and S.L.C. Woo. 1985. Gene transfer and expression of human phenylalanine hydroxylase. *Science* **228:** 77.

Lidsky, A.S., F. Güttler, and S.L.C. Woo. 1985a. Prenatal diagnosis of classic phenylketonuria by DNA analysis. *Lancet* **I:** 549.

Lidsky, A.S., F.D. Ledley, A.G. DiLella, S.C.M. Kwok, S.P. Daiger, K.J.H. Robson, and S.L.C. Woo. 1985b. Extensive restriction site polymorphism at the human phenylalanine hydroxylase locus and application in prenatal diagnosis of phenylketonuria. *Am. J. Hum. Genet.* **37:** 619.

Mann, R., R.C. Mulligan, and D.B. Baltimore. 1983. Con-

struction of a retrovirus packaging mutant and its use to produce helper-free defective retrovirus. *Cell* **33:** 153.
Scriver, C.R. and C.L. Clow. 1980a. Phenylketonuria and other phenylalanine hydroxylation mutants in man. *Annu. Rev. Genet.* **4:** 179.
———. 1980b. Phenylketonuria—Epitome of human biochemical genetics. *N. Engl. J. Med.* **303:** 1336.
———. 1980c. Phenylketonuria—Epitome of human biochemical genetics. II. A review. *N. Engl. J. Med.* **303:** 1394.
Studencki, A.B. and R.B. Wallace. 1984. Allele-specific hybridization using oligonucleotide probes of very high specific activity: Discrimination of the human β^{A}- and β^{S}-globin genes. *DNA* **3:** 7.
Woo, S.L.C., S.S. Gillam, and S.I. Woolf. 1974. Properties of phenylalanine hydroxylase from human fetal liver. *Biochem. J.* **139:** 741.
Woo, S.L.C., A.S. Lidsky, F. Güttler, T. Chandra, and K.J.H. Robson. 1983. Cloned human phenylalanine hydroxylase gene allows prenatal diagnosis and carrier detection of classical phenylketonuria. *Nature* **306:** 151.

Molecular Genetics of Apolipoproteins and Coronary Heart Disease

S. DEEB, A. FAILOR, B.G. BROWN, J.D. BRUNZELL, J.J. ALBERS, AND A.G. MOTULSKY
Departments of Medicine and Genetics, University of Washington, Seattle, Washington 98195

The principal impact of genetics on medicine has been on two broad classes of genetic disorders: Mendelian traits transmitted by single-gene inheritance (monogenic diseases) and various chromosomal aberrations. The application of concepts and techniques from molecular biology to the monogenic diseases has been spectacular. The various papers in this volume testify to the many exciting conceptual and practical advances in this area.

The full implications of genetics for medicine, however, are not exhausted with study of the relatively rare monogenic diseases and chromosomal disorders. Genetic factors play a significant role in many common diseases (Motulsky 1982; Vogel and Motulsky 1986). These disorders include various birth defects (such as congenital heart disease, cleft lip and palate, and neural tube defects), common diseases of middle life (such as coronary heart disease [CHD], diabetes, allergies, and autoimmune diseases), and the common major psychiatric disorders (schizophrenia and manic-depressive disease). In all of these disorders, familial aggregation is observed, yet the frequencies are lower than those expected with Mendelian segregation. Identical twin pairs are more commonly affected than dizygous twin pairs, but 100% concordance for the disease is never observed among identical twins with such disorders. These data and others based on adoption designs (Schulsinger 1985) suggest that environmental factors shared by families cannot account for the various findings and that genetic factors contribute significantly to the pathogenesis of these diseases. One genetic approach has been to utilize the presence or absence of a given medical diagnosis (i.e., diabetes and schizophrenia) as a criterion by which the data are tested against various models of genetic and cultural transmission. The availability of computer programs that calculate the likelihood of the best-fitting transmission model allows large-scale testing of the family data against all conceivable modes of inheritance. The results can be useful when the data are homogeneous, and valid conclusions regarding genetic transmission can often be made. However, the number and nature of the genes involved remain unknown. Quite frequently, current medical diagnostic categories include cases with different etiological causes, and thus a formal genetic analysis of this sort will give confusing results. It is already apparent that conditions such as hypertension, diabetes, and CHD may be produced by a variety of different mechanisms. The results of a genetic analysis of a heterogeneous disease therefore can be misleading.

More analytic attempts to arrive at an understanding of the genetic factors underlying these diseases proceeds with a consideration of the various biochemical, pathophysiological, and immunological factors involved in causation. The geneticist attempts to search for genetic variation in the pathways that have been uncovered as important in the etiology of a given disease. This approach is aimed at uncovering those genes that have a major influence on disease susceptibility (Motulsky 1984). Classical genetic theory often assumes that a large number of genes (polygenes), each of small effect, interact additively to cause the disease. In many diseases, however, it is likely that one or a few genes are the major determinants, whereas the many genes that influence the probability of the occurrence of the disease and its outcome may be considered as the genetic background, which is quite variable in our outbred species. Finding that a given gene is related to a disease does not imply that this gene plays an important role in its etiology. An identifiable gene associated with a disease may only contribute slightly to susceptibility, i.e., blood group O in peptic ulcer.

The long-term goal is the identification of a susceptible subpopulation that is at high risk for a given disease. Since, in most such conditions, the genetically determined susceptibility requires interaction with specified environmental factors, this approach aims at disease prevention by using drugs, diets, or other specific measures directed at interfering with a unique gene-environment interaction that causes a specific disease phenotype.

This approach of identifying a genetically susceptible subpopulation for environmental manipulation presupposes that the environmental measures that are to be applied are onerous, difficult, unpopular, potentially hazardous, or expensive. If environmental measures are simple and inexpensive, they can be instituted to the benefit of the entire population without any attempts at identifying genetically susceptible individuals. Thus, even though there is genetic variability in dental caries susceptibility, we add fluoride to the water supply of the entire population. Similarly, we vaccinate all individuals against poliomyelitis regardless of genetic differences that affect the development of paralysis.

Approaches to Coronary Heart Disease

CHD is a major public health problem in the Western world. More males than females are affected, and there is familial aggregation in that first-degree relatives of affected patients have a two- to sixfold risk of the disease. Identical twins are more frequently concordant than nonidentical twins. Environmental factors are important, since mortality for CHD in the United States has declined by at least one third within the last 15 years (Rao et al. 1984). Furthermore, genetically uniform populations in different environments have different frequencies of CHD; e.g., Japanese in Japan have a lower frequency, Japanese in Hawaii have an intermediate frequency, and Japanese in the United States have a higher frequency of the disease (Robertson et al. 1977a,b).

Other evidence for environmental factors comes from our studies which show that the family members of unaffected female spouses of men with myocardial infarcts had almost as high a frequency of CHD as did the family members of the coronary patients themselves, whereas families of control subjects had the expected lower frequency of CHD (ten Kate et al. 1984). These data were interpreted as evidence for assortative mating for the life style that predisposes to CHD, i.e., smoking, diet, etc. It is postulated that a man who will develop CHD is more likely to marry a woman who comes from a family where the various environmental risk factors for CHD aggregate.

All of these findings suggest that CHD is caused by interaction of both genetic and environmental factors. In attempting to study genetic factors in this disease, consideration of the genetic contribution to the various risk factors that have been established by epidemiologic studies is the next step in a genetic approach. Hyperlipidemia, hypertension, and diabetes are risk factors that have a definite genetic basis. Study of the genetics of CHD can therefore be advanced by investigations of the genetics of these risk factors. However, the familial aggregation of CHD cannot be fully accounted for by the genetics of hyperlipidemia, hypertension, and diabetes alone, since we have shown that familial aggregation of CHD remains even after the effects of cholesterol and triglycerides, blood pressure, and high blood sugar have been discarded (ten Kate et al. 1982). This residue of familial aggregation is likely to be the result of yet unidentified genetic factors and/or common environmental factors that are shared by families.

The hyperlipidemias are important risk factors and offer an opportunity to study genetic susceptibility to CHD using a variety of biochemical and molecular markers, since understanding of the biochemistry, pathophysiology, and genetics of the hyperlipidemias has progressed rapidly. In hypertension and diabetes, the pathway between genotypes and phenotypes is less well defined, although progress is also being made.

Hyperlipidemia

What is known about hyperlipidemias? Elevated cholesterol levels clearly predispose to CHD. In a few cases, hypercholesterolemia can be caused by a single gene defect affecting the binding of low-density lipoprotein (LDL) to its receptor. Most individuals do not owe their hypercholesterolemia to receptor abnormalities, although minor inherited alterations in receptor function have been suggested to affect cholesterol levels (Maartmann-Moe et al. 1981; Magnus et al. 1981). Genetic variation at the apolipoprotein E (*apo*ϵ) locus has significant effects on cholesterol levels. Of the three *apoE* alleles, $\epsilon 2$ reduces LDL cholesterol levels by 12.5%, whereas the $\epsilon 4$ allele raises LDL cholesterol levels by 6.4%. Monogenic variation at this locus contributes 8.3% of the total variation of LDL cholesterol levels (Sing and Davignon 1985).

"Combined hyperlipidemia" or "multiple-type hyperlipidemia" can cause cholesterol elevations and appears to be transmitted as a monogenic trait (Brunzell and Motulsky 1984). Affected persons may exhibit cholesterol elevations alone, triglyceride elevations alone, or elevations of both cholesterol and triglyceride levels. The fundamental defect remains unknown, and the existence of this entity has been contested by statistically oriented investigators as an artifact of case selection. However, some large pedigrees fit the criteria of this condition and show Mendelian segregation. Recent work suggests that affected subjects have elevations of apolipoprotein B (Brunzell et al. 1983). In this regard, elevated apoB levels have been frequently reported among patients with CHD in the absence of other lipid elevations (Brunzell et al. 1984). The proportion of such individuals who owe their apoB elevation to combined hyperlipidemia remains unclear. Individuals with variants of Lp(a) have an increased frequency of coronary abnormalities (Berg 1983). The Lp(a) variants may possibly relate to apoB abnormalities of unspecified nature.

Reduced high-density lipoprotein (HDL) levels have been consistently reported in CHD. Apolipoprotein AI (*apoAI*) is an important constituent of HDL. A rare genetic variant resulting from a DNA inversion at the *apoAI-CIII* locus has led to very low HDL levels and CHD (Karathanasis et al. 1983). Reduced apoAI levels have also been reported in large populations with "garden variety" CHD (Avogaro et al. 1980; Maciejko et al. 1983; Brunzell et al. 1984). Elevated triglyceride levels by themselves have not been clearly related to coronary atherosclerosis.

The existence of DNA variants at the apolipoprotein loci allows assessment of the possible role of these genes in hyperlipidemias and CHD (see Discussion). In London, Rees et al. (1983) showed that the less common allele at a *Sst*I polymorphism adjacent to the *apoCIII* gene is severalfold more frequent in a hypertriglyceridemic group as compared with a control population. The same group showed an increase in frequency of the same allele in postmyocardial infarction patients (Ferns et al. 1985). Other investigators, however, could not find such a difference in allelic distribution at this locus between normal and hypertriglyceridemic groups from London, Norway, Scotland, and Japan (Kessling et al. 1985, 1986; Morris and Price 1985; Rees et al. 1986).

An association of the rare allele of an *Msp*I polymorphism adjacent to the *apoAII* locus (chromosome 1) with high-plasma apoAII levels and altered HDL composition (lower AI/AII ratio) has been reported (Scott et al. 1985). Recently, Ordovas et al. (1986) observed that the rare allele of the *Pst*I polymorphism flanking the *apoAI* gene occurs at a much higher frequency (42%) in patients with severe CHD than in controls (4.1%). A high proportion (58%) of the patients studied had HDL cholesterol levels below the tenth percentile of the normal level. Furthermore, this rare *Pst*I allele was also found in 8 of 12 index cases of kindreds with familial hypoalphalipoproteinemia.

In this study, we have used several restriction enzyme polymorphisms at the apolipoprotein loci (Table 1; Fig. 1) to investigate the possible association of particular alleles with CHD or altered levels of lipids and lipoproteins. The frequency of DNA variants was compared in patients with angiocardiographically proven heart disease and in appropriate ethnic controls. The distribution, averages, and highest and lowest quintiles of apoB, HDL-C, cholesterol, and triglyceride levels were determined among carriers of the polymorphic apoprotein variants and compared with the same measurements among those carrying the common alleles.

Table 1. Apolipoprotein RFLPs

Gene	Enzyme	Fragment size (kb)	Frequency[a]
AI	*Pst*I	2.2	0.9
		3.3	0.1
AI/CIII	*Sst*I	3.2	0.06
		4.2	0.94
AI	*Msp*I	1.0	0.94
		1.7	0.06
AII	*Msp*I	3.0	0.87
		3.7	0.13
B	*Taq*I	4.8	0.92
		5.2	0.08
B	*Pvu*II	5.5	0.06
		7.8	0.94
B	*Eco*RI	10.5	0.86
		12.5	0.14
B	*Xba*I	5.0	0.50
		8.5	0.50
CII	*Taq*I	3.5	0.35
		3.8	0.65

[a]In a Caucasoid Seattle population.

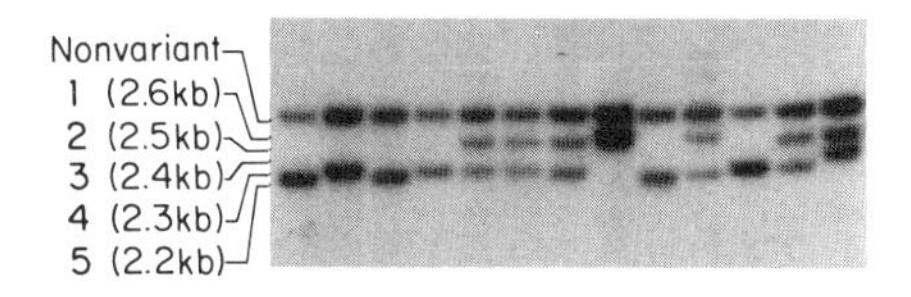

Allele frequency : 1 : 0.036
2 : 0.140
3 : 0.027
4 : 0.710
5 : 0.080

Figure 1. Minisatellite polymorphism at the 3′ end of the *apoB* gene. (*Top*) Autoradiograph of a Southern blot showing the five bands corresponding to the alleles at this polymorphic site. The fragments differ in size by approximately 100 bp. The polymorphism is due to variation in the number of repeat sequences. (*Bottom*) Frequencies of various alleles in a Caucasoid population in Seattle.

MATERIALS AND METHODS

Restriction endonuclease analysis. DNA was prepared (Ponz et al. 1983) from 10–20 ml of peripheral blood. Aliquots (10 μg) were digested to completion with the appropriate restriction enzyme and subjected to electrophoresis on 0.8–1.2% agarose gels. Southern blotting and hybridization to ^{32}P-labeled cDNA were performed as described elsewhere (Maniatis et al. 1982).

cDNA and genomic probes. ApoAI-113 and AB1 were gifts from J. Breslow (Breslow et al. 1982) and B. Levy-Wilson (Knott et al. 1985), respectively. Probes for *apoAII*, *apoB8*, and *apoCII* were isolated in our laboratory.

Populations. The populations used in this study were Caucasoids from the Seattle area. The CHD population sample was composed of consecutive male patients less than 65 years of age (Cardiac Catherization Laboratory, University of Washington). They were all angiocardiographically proven cases with CHD (NV_{50} $\geq$ 1.0). As an example, an NV_{50} of 2 refers to the fact that two coronary arteries had 50% or more stenosis. The "normal" population was composed of ostensibly healthy students, staff, and faculty from the University of Washington and spouses from familial investigations. Volunteers for a lipid screening study from an industrial firm in the Seattle area were used also for the study assessing the effects of the DNA apolipoprotein variants on lipid and lipoprotein levels.

Lipid analysis. Plasma cholesterol, triglycerides, and HDL cholesterol were determined in all patients and control groups at the Northwest Lipid Research Center (LRC) by standard technologies of the LRC program (1974). ApoB was measured by radioimmunoassay (Albers et al. 1975).

Statistical analysis. Comparisons of restriction-fragment-length polymorphism (RFLP) allele frequencies between normals and CHD patients as well as between the higher and lower quintiles of the various lipid and lipoprotein parameters were done by Chi-square analysis. One-way analysis of variance was used to compare averages of lipid levels in the three genotypes at various apolipoprotein loci.

RESULTS

Frequency of RFLP Alleles for Apolipoproteins in Coronary Heart Disease and Normals

The only statistically significant differences observed ($p = 0.05$) were in the allelic distribution at the *apoAI-CIII SstI* and *apoB* minisatellite (allele 5) polymorphic sites (Table 2). The *Sst*I site is in the ApoAI-CIII-AIV cluster of genes on the long arm of chromosome 11 (Fig. 2). ApoAI is the major protein of HDL, the plasma level of which is characteristically low in individuals with CHD. In both cases, the frequency of the rare allelic variant was twice or more among patients with CHD as compared with that of controls.

Table 2. Frequency of Apolipoprotein RFLP Alleles in CHD and Normals

RFLP	CHD population ($NV_{50} \geq 1.0$)[a]	Random "normals"
*AI-Pst*I	0.07 ($n = 140$)	0.10 ($n = 114$)
*AI/CIII-Sst*I	0.12* ($n = 140$)	0.06 ($n = 101$)
*AII-Msp*I	0.17 ($n = 107$)	0.13($n = 56$)
*B-Taq*I	0.07 ($n = 140$)	0.08 ($n = 87$)
*B-Pvu*II	0.09 ($n = 139$)	0.06 ($n = 78$)
*B-Eco*RI	0.19 ($n = 123$)	0.14 ($n = 108$)
*B-Xba*I	0.50 ($n = 117$)	0.54 ($n = 102$)
B-minisatellite		
allele 2	0.13 ($n = 103$)	0.20 ($n = 62$)
allele 4	0.66 ($n = 103$)	0.66 ($n = 62$)
allele 5	0.15* ($n = 103$)	0.06 ($n = 62$)
*CII-Taq*I	0.44 ($n = 110$)	0.36 ($n = 103$)

[a] NV_{50} indicates the number of coronary arteries with 50% or more stenosis. *n* denotes the number of individuals. Asterisks indicate significance at the 0.05 level.

Frequency of Apolipoprotein RFLP Alleles as a Function of Plasma Lipid and Lipoprotein Concentrations

A group of 300–400 individuals made up of both normal controls and CHD patients was examined for possible association of alleles at apolipoprotein polymorphic sites with either the upper or lower quintiles of plasma lipid and lipoprotein levels. No statistically significant differences in allelic frequencies were observed among individuals with high and low levels of *apoB-100*, total cholesterol, HDL-C, and triglycerides for the following apolipoprotein RFLPs: *AI-Pst*I, *AI-CIII-Sst*I, *AII-Msp*I, *B-Eco*RI, and *CII-Taq*I.

Significantly higher frequencies ($p = 0.05$) of the less common allele of *apoB-Taq*I and the 8.5-kb allele of *Xba*I were observed in individuals with elevated levels (upper 20%) of total cholesterol and triglycerides, respectively. The frequency of the rare allele of *apoB-Pvu*II was significantly higher ($p = 0.05$) in individuals with low plasma HDL levels (lower 20%).

Plasma Lipid and Lipoprotein Concentrations as a Function of Genotype at Apolipoprotein Loci

The three genotypes at each of the dimorphic apolipoprotein RFLPs (Table 1) were examined for differences in mean plasma cholesterol, HDL-C, triglycerides, and *apoB-100* levels in a population composed of normal controls and individuals with CHD. The RFLPs for which significant differences in mean lipid values were observed are shown in Table 3.

Higher mean levels of total cholesterol were associated with the less common alleles of the *apoCII Taq*I and *apoB Taq*I polymorphisms. Lower HDL levels are associated with genotypes that include the common allele of *apoAII Msp*I (significant at the 0.01 level) and the rare allele of *apoB Pvu*II. *apoAI* is a component of the HDL particle. The common allele of *apoAII Msp*I is also associated with high levels of triglycerides. No significant associations of genotypes at any of the apolipoprotein loci were observed with plasma *apoB* levels.

Linkage Equilibrium Relationships at the *apoAI-CIII* Locus

The *Msp*I and *Sst*I variants at this locus (Fig. 2) were in strong linkage disequilibrium, using several standard statistical tests. However, using the same tests, the *Pst* locus located between the *Msp* and *Sst* sites appeared to be in linkage equilibrium with both the *Msp* and *Sst* sites. The distance involved is no more than 4–5 kb. Biologically, one would expect the *Pst* site also to be in linkage disequilibrium. However, several other observations of such anomalous linkage equilibrium

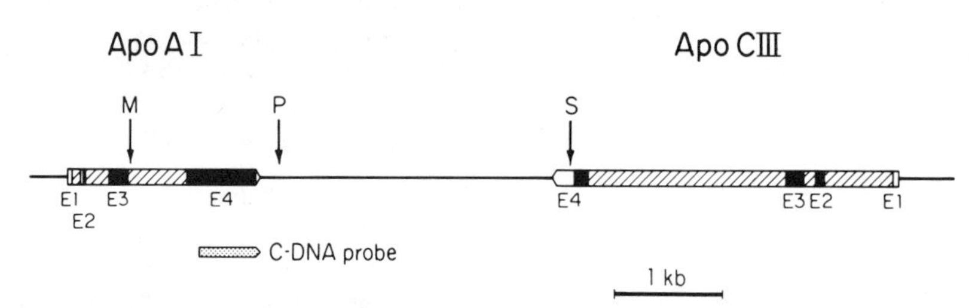

Figure 2. Map of the *apoAI-CIII* gene complex. RFLPs shown are *Msp*I (M), *Pst*I (P), and *Sst*I (S). Each gene is composed of four exons (E1–E4). Solid bars denote translated sequences; open bars, untranslated sequences of mRNA; hatched bars, introns; and single lines, DNA flanking the genes. The genes are transcribed in opposite directions (Protter et al. 1984).

Table 3. Serum Lipid and Lipoprotein Concentrations in Three Genotypes at Apolipoprotein Loci

RFLP	Genotype								
	FF[a]			FS			SS		
	n	mean	S.D.	*n*	mean	S.D.	*n*	mean	S.D.
Total cholesterol									
*CII-Taq*I[b]	24	218	45	76	216	58	36	245	67
*B-Taq*I[c]	336	215	49	49	227	45	2	256	95
HDL									
*AII-Msp*I[d]	94	37	9	35	39	13	3	55	10
*B-Pvu*II[b]	0	–	–	59	39	13	326	44	15
Triglycerides									
*AII-Msp*I[b]	94	179	97	35	146	59	3	80	53

[a]*n* indicates number of individuals. S.D. is the standard deviation.
[b]Significant at the 0.05 level.
[c]Significant at the 0.06 level.
[d]Significant at the 0.01 level.

relationships have been reported previously (Barker et al. 1984; Chakravarti et al. 1984a,b). We investigated this problem further with the help of E. Thompson, an expert genetical statistician. Her studies (which will be reported elsewhere) clearly show that the sample sizes required to show linkage disequilibrium of the *Pst* site are very large and would rarely be reached in most human studies. Thus, it is unlikely when considering distances of a few kilobases that unexpected results of linkage equilibrium are biologically meaningful when more remote markers at the same locus are in linkage disequilibrium.

DISCUSSION

The rationale for examining the apolipoprotein loci for a possible association with the hyperlipidemias and CHD is straightforward. It is possible that structural or regulatory mutations of the apolipoproteins may be associated with hyperlipidemia and with CHD. A simplified method for assessing possible apolipoprotein mutations screens for the frequency of DNA variants (RFLPs and minisatellites) at a given apolipoprotein locus. It would be expected that the frequency of a DNA variant tightly linked to a given structural or regulatory apolipoprotein mutation would be altered as compared with the frequency of that variant in the control population. Furthermore, any association of lipoprotein DNA variants with CHD is likely to be mediated via quantitative lipid parameters, since abnormal lipid levels are the intervening variables predisposing to CHD. Thus, carriers of a given apolipoprotein DNA variant might have higher mean lipid levels (i.e., ApoB, cholesterol, etc.) or lower lipid levels (i.e., HDL, AI). We therefore examined the upper and lower quintiles of various lipid and lipoprotein levels to ascertain whether carriers of the lipoprotein DNA variants were underrepresented or overrepresented among these segments of the lipid distribution. Similarly, we examined mean lipid levels among the three genetic classes ("normal" homozygotes and heterozygotes and "abnormal" homozygotes) for the various apolipoprotein variants. The statistically significant associations are given in Results and are summarized in Table 4.

The biological significance of these associations cannot yet be fully assessed. All but one association are statistically significant only at the 5% level, and the variants that show association with CHD (*apoAI-CIII Sst*I and *B* minisatellite 5) do not exhibit any changes in lipid values. However, other workers (Ordovas et al. 1986) have shown lower HDL levels in carriers of the rare allele of *apoAI-CIII Pst*I, as well as an increased frequency of CHD among carriers of the

Table 4. Statistically Significant Associations between Apolipoprotein DNA Variants, Hyperlipidemia, and CHD (This Study)

DNA apolipoprotein variant[a]	Associations
*AI-CIII Sst*I	higher frequency of coronary heart disease ($p = 0.05$)
*AII-Msp*I (common)	lower HDL levels ($p = 0.01$)
*AIIMsp*I (common)	higher triglyceride levels ($p = 0.05$)
*B-Taq*I	higher cholesterol levels ($p = 0.06$)
*B-Pvu*II	lower HDL levels ($p = 0.05$)
B-minisatellite 5	higher frequency of CHD ($p = 0.05$)
*B-Xba*I (8.5 kb)	higher triglyceride levels ($p = 0.05$)
*CII-Taq*I (common)	higher cholesterol levels ($p = 0.05$)

[a]Rare variant unless noted.

same allele. We are now exploring the effect of various haplotypes on hyperlipidemias and CHD. Before effects of haplotypes can be fully evaluated, we need to assess further the linkage disequilibrium problem (see Results). Ultimately, the interaction between various apolipoprotein markers at different loci needs to be considered in the elaboration of "marker profiles." Sufficiently encouraging results have been obtained to accumulate additional data in order to work out the genotypic contributions of the apolipoprotein loci to CHD susceptibility.

SUMMARY

A variety of DNA markers for apolipoprotein genes were examined among patients with angiocardiographically proven heart disease and among a variety of normal individuals with various lipid values. An increased frequency of an *apoAI-CIII SstI* RFLP and an *apoB* minisatellite (allele 5) was found among patients with CHD. Higher levels of cholesterol were found among carriers of the rare *apoB TaqI* and the common *apoCII TaqI* variants, whereas higher levels of triglycerides were found in carriers of the common *apoAII MspI* and the rare *apoB XbaI* variants. Lower levels of HDL were found among carriers of the common *apoAII MspI* and the rare *apoB PvuII* variants.

The biological significance of these results and those of other investigators for the pathogenesis of CHD and hyperlipidemia is suggestive but not yet fully clarified. Additional genetic epidemiologic studies and family investigations will be required.

Currently used statistical methodology may lead to false inferences regarding the genetic equilibrium or disequilibrium status of closely linked DNA variants. Conclusions regarding the presence of genetic equilibrium if closely linked flanking markers are in disequilibrium may be faulty.

ACKNOWLEDGMENTS

We wish to thank Elaine Loomis for her excellent technical assistance in Southern blot analysis. This work was supported by U.S. Public Health Service grants GM-15253 and HL30086.

REFERENCES

Albers, J.J., V.G. Cabana, and W.R. Hazzard. 1975. Immunoassay of human plasma apolipoprotein B. *Metabolism* **24:** 1399.

Avogaro, P., B.G. Bittolo, G. Cazzolato, and E. Rorai. 1980. Relationship between apolipoproteins and chemical components of lipoproteins in survivors of myocardial infarction. *Atherosclerosis* **37:** 69.

Barker, D., T. Holm, and R. White. 1984. A locus on chromosome 11p with multiple restriction site polymorphisms. *Am. J. Hum. Genet.* **36:** 1159.

Berg, K. 1983. Genetics of coronary heart disease. *Prog. Med. Genet.* **5:** 35.

Breslow, J.L., D. Ross, J. McPherson, H. Williams, D. Kurnit, A.L. Nussbaum, S.K. Karathanasis, and V.I. Zannis. 1982. Isolation and characterization of cDNA clones for human apolipoprotein A-I. *Proc. Natl. Acad. Sci.* **79:** 6861.

Brunzell, J.D. and A.G. Motulsky. 1984. Status of "familial combined hyperlipidemia." In *Genetic epidemiology of coronary heart disease: Past, present, and future* (ed. C.D. Rao et al.), p. 541. A.R. Liss, New York.

Brunzell, J.D., A.D. Sniderman, J.J. Albers, and P.O. Kwiterovitch, Jr. 1984. Apoproteins B and A-I and coronary artery disease in humans. *Arteriosclerosis* **4:** 79.

Brunzell, J.D., J.J. Albers, A. Chait, S.M. Grundy, E. Groszek, and G.B. McDonald. 1983. Plasma lipoproteins in familial combined hyperlipidemia and monogenic familial hypertriglyceridemia. *J. Lipid Res.* **24:** 147.

Chakravarti, A., J.A. Phillips III, K.H. Mellits, K.H. Buetow, and P.H. Seeburg. 1984a. Patterns of polymorphism and linkage disequilibrium suggest independent origins of the human growth hormone gene cluster. *Proc. Natl. Acad. Sci.* **81:** 6085.

Chakravarti, A., K.H. Buetow, S.E. Antonarakis, P.G. Waber, C.D. Boehm, and H.H. Kazazian, Jr. 1984b. Nonuniform recombination within the human β-globin gene cluster. *Am. J. Hum. Genet.* **36:** 1239.

Ferns, G.A.A., C. Ritchie, J. Stocks, and D.J. Galton. 1985. Genetic polymorphisms of apolipoprotein C-III and insulin in survivors of myocardial infarction. *Lancet* **II:** 300.

Karathanasis, S.K., V.I. Zannis, and J.L. Breslow. 1983. A DNA insertion has occurred in the apolipoprotein AI gene of patients with premature atherosclerosis. *Nature* **305:** 823.

Kessling, A.M., B. Horsthemke, and S.E. Humphries. 1985. A study of DNA polymorphisms around the human apolipoprotein AI gene in hypertriglyceridemic and normal individuals. *Clin. Genet.* **28:** 296.

Kessling, A.M., K. Berg, E. Mockleby, and S.E. Humphries. 1986. DNA polymorphisms around the apo AI gene in normal and hyperlipidaemic individuals selected for a twin study. *Clin. Genet.* **29:** 485.

Knott, T.J., S.C. Rall, Jr., T.L. Innerarity, S.F. Jacobson, M.S. Urdea, B. Levy-Wilson, L.M. Powell, R.J. Pease, R. Eddy, H. Nakai, M. Byers, L.M. Priestly, E. Robertson, L.B. Rall, C. Betsholtz, T.B. Shows, R.W. Mahly, and J. Scott. 1985. Human apoprotein B: Structure of carboxyl-terminal domains, sites of gene expression, and chromosomal localization. *Science* **230:** 37.

Lipid Research Clinic's Manual of Laboratory Operations. 1974. U.S. Department of Health, Education, and Welfare, Washington, D.C.

Maartmann-Moe, K., P. Magnus, W. Golden, and K. Berg. 1981. Genetics of low density lipoprotein receptor. III. Evidence for multiple normal alleles at the low density lipoprotein receptor locus. *Clin. Genet.* **20:** 113.

Maciejko, J.J., D.R. Holmes, B.A. Kottke, Z.R. Zinsmeister, D.M. Dinh, and S.J.T. Mao. 1983. Apoprotein AI as a marker of angiographically assessed coronary artery disease. *N. Engl. J. Med.* **309:** 385.

Magnus, P., K. Maartmann-Moe, W. Golden, W.E. Nance, and K. Berg. 1981. Genetics of low density lipoprotein receptor. II. Genetic control of variation in cell membrane low density lipoprotein receptor activity in cultured fibroblasts. *Clin. Genet.* **20:** 104.

Maniatis, T., E.F. Fritsch, and J. Sambrook. 1982. *Molecular cloning: A laboratory manual.* Cold Spring Harbor Laboratory, Cold Spring Harbor, New York.

Morris, S.W. and W.H. Price. 1985. DNA sequence polymorphisms with apolipoprotein A-I/C-III gene cluster. *Lancet* **II:** 1127.

Motulsky, A.G. 1982. Genetic approaches to common diseases. In *Human genetics*, part B: *Medical aspects* (ed. B. Bonne-Tamir), p. 89. A.R. Liss, New York.

———. 1984. Genetic research in coronary heart disease. In *Genetic epidemiology of coronary heart disease: Past, present, and future* (ed. C.D. Rao et al.), p. 541. A.R. Liss, New York.

Ordovas, J.M., E.J. Schaefer, D. Salem, R.H. Ward, C.J. Glueck, C. Vergani, P.W.F. Wilson, and S.K. Karathanasis. 1986. Apolipoprotein A-I gene polymorphism associated with premature coronary artery disease and familial hypoalphalipoproteinemia. *N. Engl. J. Med.* **314:** 671.

Ponz, M., D. Solowiejcky, B. Harpel, Y. Mory, E. Swartz, and S. Surrey. 1983. Construction of human gene libraries from small amounts of peripheral blood: Analysis of β-like globin genes. *Hemoglobin* **6:** 27.

Protter, A.A., B. Levy-Wilson, J. Miller, G. Beneen, T. White, and J.J. Seilhamer. 1984. Isolation and sequence analysis of the human apolipoprotein CIII gene and the intergenic region between the apo AI and apo CIII genes. *DNA* **3:** 449.

Rao, C.D., R.C. Elston, L.H. Kuller, M. Feinleib, C. Carter, and R. Havlik, eds. 1984. *Genetic epidemiology of coronary heart disease: Past, present, and future.* A.R. Liss, New York.

Rees, A., J. Stocks, H. Paul, Y. Ohuchi, and D. Galton. 1986. Haplotypes identified by DNA polymorphisms at the apolipoprotein A-I and C-III loci and hypertriglyceridaemia. A study in a Japanese population. *Hum. Genet.* **72:** 168.

Rees, A., J. Stocks, C.C. Shoulders, D.J. Galton, and F.E. Baralle. 1983. DNA polymorphism adjacent to human apoprotein AI gene: Relation to hypertriglyceridemia. *Lancet* **I:** 444.

Robertson, T.L., H. Kata, T. Gordon, A. Kagan, G.G. Rhoads, C.E. Land, R.M. Worth, J.C. Belsky, D.S. Dock, M. Miyanishi, and S. Kawamoto. 1977a. Epidemiologic studies of coronary heart disease and stroke in Japanese men living in Japan, Hawaii, and California. Coronary heart disease risk factors in Japan and Hawaii. *Am. J. Cardiol.* **39:** 244.

Robertson, T.L., H. Kato, G.G. Rhoads, A. Kagan, M. Marmot, S.L. Syne, T. Gordon, R.M. Worth, J.C. Belsky, D.S. Dock, M. Miyanishi, and S. Kawamoto. 1977b. Epidemiologic studies of coronary heart disease and stroke in Japanese men living in Japan, Hawaii, and California: Incidence of myocardial infarction and death from coronary heart disease. *Am. J. Cardiol.* **39:** 239.

Schulsinger, F. 1985. The experience from the adoption method in genetic research. *Prog. Clin. Biol. Res.* **177:** 351.

Scott, J., L.M. Priestly, T.J. Knott, M.E. Robertson, and D.V. Mann. 1985. High-density lipoprotein composition is altered by a common DNA polymorphism adjacent to apoprotein AII gene in man. *Lancet* **I:** 771.

Sing, C.F. and J. Davignon. 1985. Role of the apolipoprotein E polymorphism in determining normal plasma lipid and lipoprotein variation. *Am. J. Hum. Genet.* **37:** 265.

ten Kate, L.P., H. Boman, S.P. Daiger and A.G. Motulsky. 1982. Familial aggregation of coronary heart disease and its relation to known genetic risk factors. *Am. J. Cardiol.* **50:** 945.

———. 1984. Increased frequency of coronary heart disease in relatives of wives of myocardial infarct survivors: Assortative mating for lifestyle and risk factors? *Am. J. Cardiol.* **53:** 399.

Vogel, F. and A.G. Motulsky. 1986. *Human genetics. Problems and approaches*, 2nd edition. Springer-Verlag, New York.

DNA Markers and Genetic Variation in the Human Species

L.L. CAVALLI-SFORZA,* J.R. KIDD,† K.K. KIDD,† C. BUCCI,* A.M. BOWCOCK,*
B.S. HEWLETT,‡ AND J.S. FRIEDLAENDER§
*Department of Genetics, Stanford University Medical Center, Stanford, California 04305;
†Department of Human Genetics, Yale University School of Medicine, New Haven, Connecticut 06510;
‡Department of Sociology and Anthropology, Southern Oregon State College, Ashland, Oregon 97520;
§Department of Anthropology, Temple University, Philadelphia, Pennsylvania 19122

The first attempt at using genetic markers for anthropological purposes goes back to 1918, when Hirszfeld and Hirszfeld (1918–1919) showed that ABO blood groups varied with the ethnic origin of population samples. Using ABO, MN, and Rh blood group data, Boyd (1950) showed one could assign some populations to their ethnic group of origin. Only rarely is an individual locus sufficiently informative that conclusions about ethnic origins can be drawn. Nevertheless, evolutionary theory predicts that by accumulating information from many genes one may be able to reconstruct the evolutionary history of a group of populations. With this in mind, Cavalli-Sforza and Edwards (1964) attempted to reconstruct a tree of descent of human ethnic groups using five loci for which information was then available for 15 populations. A new method specifically designed for reconstruction of phylogenies and a classical statistical method of reduction of dimensions (the analysis of principal components) both showed that all populations clustered in a geographically sensible way. A figure showing similar clusterings was published in the 1964 Cold Spring Harbor Symposium volume which, like the present Symposium, was dedicated to humans (Cavalli-Sforza et al. 1965, Fig. 2).

Tree structures can depict two kinds of information—the existing genetic relationships among the populations (both qualitative and quantitative) and the evolutionary history represented by the relationships of the populations to the origin, or root, of the tree. The two methods that were developed in the early 1960s (least-squares fitting of an additive tree, and minimum path) did not generate a root. The root in those early trees was placed about midway between the two most remote populations: Bantu-speaking Africans and Australian aborigines in the first tree. This divided the world in half: Europe and Africa on one side, and Asia, America, and Australia on the other. America was most probably settled from northeastern Asia and Australia from southeastern Asia; therefore they can be considered as part of "Greater Asia." Thus, the suggestion was made that the first separation in the genesis of human groups was between "Eurafrica" and (greater) Asia. Addition of other genes did not change this conclusion, e.g., haptoglobin (*HP*) and transferrin (*TF*) in Kidd (1973) and *HLA* in Piazza et al. (1975), but tended to show Caucasians on very short branches connected to the tree close to the root and Africans clustered at the tip of a very long branch (Cavalli-Sforza 1972).

When the comparison was extended to many new enzyme markers that were later developed, and restricted to three representative populations or to population groups of Europe, East Asia, and Africa, different results were obtained than with blood groups (Nei 1978; Nei and Roychoudhury 1982). While the blood group data were in favor of the first split leaving Eurafrica on one side and Greater Asia on the other, enzyme data favored a first split between Africans on one side and, on the other side, all other populations (Eurasia, we might say, where Asia is still meant in the extended form).

While placing the root is of great interest, the assumptions required (e.g, constant evolutionary rates and no admixture) make such an estimate, based on current methods and available data, problematic (Cavalli-Sforza and Edwards 1967; Kidd and Sgaramella-Zonta 1971; Kidd and Cavalli-Sforza 1974; Astolfi et al. 1979). It may be that only with a substantial increase in the amount of data will the position of the root become unassailable.

Until DNA markers became available, the number of marker loci that could be studied was small, and the most important ones had already been exploited. Since the standard errors of distances among populations are primarily a function of the number of informative genes studied, there was little hope of further clarification. Today these limits have been removed: There are already probably more than 1000 polymorphic probes available (Willard et al. 1985). Similarly, establishing permanent cell lines, for instance with Epstein-Barr virus (EBV) transformation of lymphocytes, makes it possible to obtain potentially any amount of the DNA of a subject from a single sampling of a very modest amount of peripheral blood. Individual samples can be preserved and expanded practically indefinitely; thus, with suitable organization one can avoid the incompleteness of data that makes the analysis of information existing today for non-DNA polymorphisms highly unsatisfactory. Although expense precludes testing as many individuals as was customary for classical markers, one can compensate by increasing the number of loci tested. We are therefore re-embarking on studies of human population genetics with much more powerful techniques.

Results for two types of DNA polymorphisms have been described so far in several human ethnic groups: mtDNA variants and genomic restriction fragment length polymorphisms (RFLPs) in the β-globin region. Similar work on the Y chromosome is in progress in several laboratories. Both major studies of mtDNA variation (Johnson et al. 1983; Wilson et al. 1985) have shown very long branches for Africans compared with simple expectations. One set of data on polymorphisms in the β-globin region (Wainscoat et al. 1986) also shows a longer branch for Africans, again in agreement with an early Africa/Eurasia split. One might expect, however, that any single probe or short segment of DNA will have peculiarities that limit the general validity of relationships inferred. For example, the strong selection associated with hemoglobin S and the thalassemias may have obscured the effect of random genetic drift. Only by accumulating data over a large number of independent systems will the accurate reconstruction of phylogenies be possible.

To make such data collection possible, we have established permanent lymphoblastoid cell lines on samples of populations of importance. We shall describe ideal criteria for sampling and the difficulties we have encountered, and then present some interesting preliminary results.

METHODS

Selection of populations. There are two types of populations of interest: traditional and modern. Among traditional ones, clearly the choice should be populations that have little, or no, recent admixture. For American natives this may limit the choice considerably. If the admixture is ancient, it can be neglected; at some level, all populations have had some admixture. Modern populations are probably best represented in rural areas where there is likely to be less admixture than in urban areas. Unfortunately, in collecting our Oriental samples we have been, as yet, unable to apply this rule, as will be indicated later. Storing the samples, extracting DNA, and testing the probes are so costly that procedures will have to be developed for optimal results even if the ideal sampling scheme is especially difficult to carry out in practice.

Sample sizes and nature of each population sample. We have chosen to sample small family units rather than unrelated individuals because we wish to take advantage of the additional power in haplotype data. We recognize that family units will represent fewer independent chromosomes but the segregation data present in families will be essential for identifying the haplotypes actually present at many multisite systems. The current technology for typing RFLPS imposes practical limits to the number of individuals that can be studied. Our initial theoretical studies indicate that a sample of at least 100 independent chromosomes obtained from 25 families of two unrelated parents and two children will allow identification of most haplotypes from segregation data and be large enough to yield acceptably accurate estimates of their frequencies. As noted below, we are finding it difficult, at the moment, to meet this objective for such remote populations as the Pygmies and Melanesians.

Markers to be studied. The choice of RFLP marker systems is of considerable importance. Basically, there are two major categories of probes available: clones of known genes and anonymous clones, i.e., DNA segments of unknown function. They are both of extreme interest for different reasons. The first may show greater, and the second smaller, effects of natural selection (on average). The difference, however, will not usually be sharp. A restriction marker will not necessarily be associated with a mutation under selection, even if the marker is located within a segment that codes for a specific protein. Similarly, a random segment may actually be a coding region or may be associated with a selectively important, closely linked gene. The choice of pseudogenes as probes will help by guaranteeing that no coding region under natural selection is part of the probe, but cannot guarantee that neighboring sequences and loci are not. Special methods will have to be devised to distinguish between natural selection for the segment under study and for a nearby marker that hitchhikes with a selected marker.

Interest in distinguishing potentially selected polymorphisms from unselected, neutral ones is twofold. We may have important cues about selection, say under climatic or other environmental conditions, by comparing populations for specific genes. So far we have only had limited luck in understanding selective differentials in gene frequencies, but the functions of most of the classical markers investigated, from blood groups to almost all enzymes, are sufficiently poorly understood that they offer very little help for understanding possible selective effects. While markers under selection tell about the ecological history, truly neutral markers tell about the history of separation of populations. It is only with selectively neutral markers that, on average, genetic divergence will increase with the time of separation. There will, of course, be many sources of distortions. Even for neutral changes one can expect substantial variations in local evolutionary rates because of different rates of drift in populations of different sizes, or because of admixture, gene flow, or hitchhiking. Selected markers may be useful for reconstruction of the history of separations but will require many more caveats.

PRELIMINARY RESULTS

Populations Studied

Samples of three aboriginal populations have so far been transformed as lymphoblastoid cell lines: (1) Biaka Pygmies from the village of Bagandu in the Southwest corner of the Central African Republic; (2) Mbuti Pygmies from the Ituri Forest in northeastern Zaire; and (3) Nasioi Melanesians from Bougainville in the Solomon Islands. The Biaka Pygmies are the same general group earlier referred to as Babinga in various studies (e.g., Cavalli-Sforza et al. 1969). These and the Mbuti are described and discussed in more detail in Cavalli-Sforza (1986). The Nasioi are Melanesians who speak a non-Austronesian language and represent the population that was indigenous to the Solomon Islands prior to the wave of Austronesian-speaking peoples (including the Polynesians) who spread through the area about 2000 years ago. As such, the Nasioi should show closer affinities to the New Guinea highlanders and Australian Aboriginals than to Polynesians. The Nasioi and other populations of Bougainville are described in detail by Friedlaender (1986).

For all three of these aboriginal populations, blood samples collected in the field were brought to the laboratories as quickly as possible after collection. However, in the time necessary for travel (just over 3 days), there was a high mortality of the cells. Transformations were done in duplicate at both Stanford and Yale, and the percentage of success was higher in the laboratory that was closer to the area of collection. Although about 25 families with one to three children were collected in each case, the percentage of successfully EBV-transformed cultures was only about 30%. As most cell lines did not cluster in families, we have only samples of 50–60 random chromosomes for each population. Further sampling of these populations will be required to obtain the necessary family material. However, while not ideal, these samples are giving interesting and useful data.

Caucasian samples are easily available, and data on polymorphisms given in the literature are almost exclusively on Caucasians. The Stanford and Yale laboratories each has a collection of biologically unrelated Caucasians that are members (parents and married-in spouses) of large families. In both labs, the ancestry of the individuals can only be described as mixed European, predominantly western European. To provide an Asian sample, individuals of Chinese origin, born in mainland China and living in the San Francisco Bay area, have been sampled. At this juncture this is an important reference population although it too fails to meet our anthropological ideal.

New Genetic Variation in Isolated Populations

We have identified a new pattern of genetic variation in Pygmies for the *D1721* locus. This "locus" described by Waye and Willard (1986) and Willard et al. (1986) is an alpha satellite DNA at the centromere of chromosome 17 showing two orders of tandem repeats. The common variation seen in Caucasians in length and copy number of the higher-order repeat, though not allelic in the classical sense, has been shown to segregate in a Mendelian fashion. The polymorphic variation found in Caucasians and most other ethnic groups studied (J.S. Waye and H.F. Willard, pers. comm.) consists of four different chromosomal types using *Eco*RI and a two-allele polymorphism with *Pvu*II. The *Pvu*II pattern in Caucasians is shown in the first two lanes of Figure 1 and the variation is in the presence or absence of the band at about 2.2 kb and is the only pattern previously seen in a survey of several hundred individuals from several ethnic groups (including 50 Chinese and 20 American Blacks) (H.F. Willard, pers. comm.).

The polymorphic variation shown at this locus among Pygmies is shown in lanes 3, 4, 5, 6, and 7 of Figure 1. Most Pygmies seem to have the 2.2-kb band (absent in many Caucasians) plus a band of variable intensity at about 2.0 kb and another at about 1.9 kb, neither of which has been observed in any other samples.

If we assume that these multiple copies of homologous DNA of different lengths are on chromosome 17, at least two new "alleles" are required to account for these four patterns. While new "alleles" at *D1721* are the most parsimonious explanation of these patterns, at present we lack the necessary family data to show the formal segregation of new allelic forms. In addition, due to the lack of family material, we cannot formally demonstrate by linkage that these new bands represent DNA on chromosome 17 at the *D1721* locus. Additional molecular studies using, for example, *Eco*RI digests of the Pygmy samples, isolation of an individual chromosome 17 in somatic cell hybrids, and in situ hybridization will help resolve these questions.

We find these new phenotypic patterns in both Pygmy groups but do not believe that we understand the patterns well enough to estimate allele frequencies.

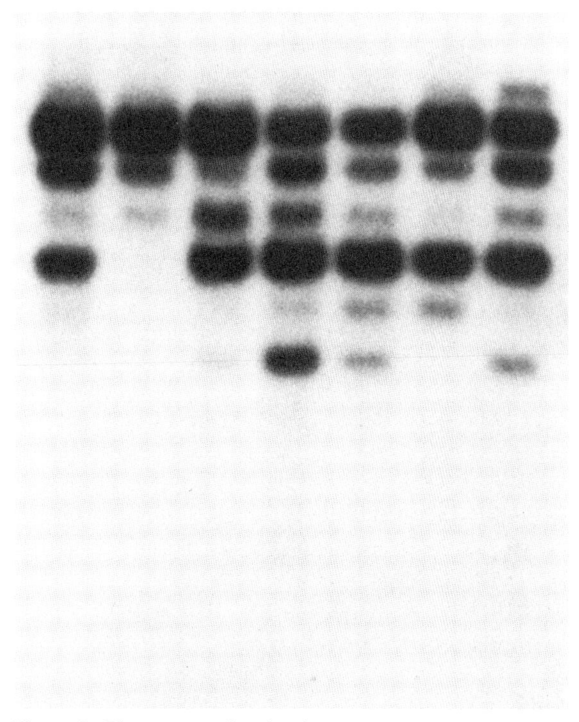

Figure 1. The patterns for the chromosome 17 DNA revealed by *Pvu*II. Each lane is a *Pvu*II digest of genomic DNA of a different individual. The two Caucasians in the two lanes on the left show the two common patterns found in most populations. The Biaka Pygmies in the other six lanes show additional patterns not previously found.

Table 1. RFLPs Typed at Stanford

Locus	Map position	Source of probe	
CD8(Leu2)	2p12	R. Margolskee	Department of Biochemistry, Stanford University
		P. Kavathas	Department of Laboratory Medicine, Yale University Medical School
POMC	2p23	S. Cohen	Department of Genetics, Stanford University
PLAT	8p12	T.J.R. Harris	Celltech Limited
COL1A2	7q21.3–q22.1	J. C. Myers	Department of Medicine, University of Pennsylvania
COL4A1	13qter	C. Boyd	Department of Surgery, Rutgers Medical School
NGFB	1p22.1	A. Ullrich	Genentech
A2M	12	G. Bell	Chiron Corporation
TG	8q24	G.J.B van Ommen	Department of Human Genetics, University of Leiden
D13S10	13q14	M. Leppert	Howard Hughes Medical Institute, University of Utah
IGF1	12q22–q24.1	A. Ullrich	Genentech
pL19.2	?	L.L. Cavalli-Sforza	
pL18.1	14	L.L. Cavalli-Sforza	

The genetic markers used in the study.

The fact that these patterns have not been found in a moderate sample of American Blacks suggests that these may be Pygmy-specific variants. Confirmation of that will require study of several additional African groups. Although we could speculate on what alternative interpretations could be given depending upon what is found, it is clear that the pattern of variation for these new forms will tell us much about the evolutionary relationships of African and non-African groups.

Allele Frequency Variation among Populations

Although this study has just begun, we shall summarize the results to date of our efforts to repeat with DNA markers the analysis of major human ethnic groups. The loci already studied are listed in Tables 1 and 2. Both known genes and anonymous loci are included; selection of these loci was based primarily on availability of the probes and their current use in the laboratories for other studies, though efforts were made to insure that several different chromosomes were represented. In Tables 3, 4, and 5 we present several allele frequencies and statistics that summarize the patterns of allele frequency variation we have found. The number of chromosomes studied for each locus in Table 3 averaged 51 for the Biaka Pygmies, 65 for the Mbuti Pygmies, 39 for Melanesians, 35 for Chinese, and 92 for the lab panel of Caucasians. For each locus studied in Table 4, the number of chromosomes studied averaged 49 for the Biaka Pygmies, 42 for the Mbuti Pygmies, 50 for the lab panel of Caucasians, and 97 for the Caucasian data from the original investigators who found each RFLP (tabulated in Willard et al. 1985 and pers. comm.).

In Table 3 the chi square testing of the heterogeneity of the five populations has four degrees of freedom; only two of the 20 chi squares are not significant. F_{ST}, the standardized measure of variation of gene frequencies, is calculated by the formula suggested by Reynolds et al. (1983), which incidentally gives results similar to the simple procedure, suggested in Cavalli-Sforza and Bodmer (1971), of dividing the chi square shown in the second column by its degrees of freedom, subtract-

Table 2. RFLPs Typed at Yale

Locus	Map position	Source of probe	
ADH3	4q21–q25	M. Smith	Department of Pediatrics, University of California at Irvine
AMY1	1q21	M. Meisler	Department of Human Genetics, University of Michigan Medical School
APOB	2pter–p23	J. Scott	Molecular Medicine Research Group, Clinical Research Center, MRC, Harrow, England
AT3	1q23–q25	S. Orkin	Department of Pediatrics, Harvard Medical School
MT2P1	4p11–q21	L.-C. Tsui	Research Institute, The Hospital for Sick Children, Toronto
SST	3q28	G. Bell	Chiron Corporation
D3S5	3q21–qter	A. Driesel	Dechema
D4S14	4q26–qter	R. Williamson	Department of Biochemistry, St. Mary's Hospital Medical School, London
D4S35	4pter–q12	J. Gusella	Massachusetts General Hospital, Harvard Medical School
D13S2	13q22	W. Cavenee	College of Medicine, University of Cincinnati
D17Z1	17cen	H. Willard	Department of Medical Genetics, University of Toronto

The genetic markers used in the study.

Table 3. Stanford Allele Frequency Data

Locus/site	Allele tallied (kb)	Pygmies		Melanesian	Chinese	Caucasian	$\overline{H}$ ± S.E.	Chi-square between populations ($df = 4$)	F_{ST}
		Biaka	Mbuti						
*CD8/Dra*I (1st)	3.0	0.42	0.16	0.91	0.77	0.66	0.345 ± 0.060	85.5	0.332
*CD8/Dra*I (2nd)	3.3	0.13	0.0	0.0	0.0	0.0	0.057 ± 0.047	32.8	0.104
*POMC/Rsa*I	0.8	1	1	1	0.63	0.66	0.184 ± 0.112	72.4	0.266
*POMC/Sst*I	15	0.31	0.58	0.31	0.21	0.28	0.414 ± 0.026	23.1	0.076
*PLAT/Eco*RI	2.9	0.30	0.29	0.05	0.73	0.50	0.363 ± 0.070	44.7	0.175
*COL1A2/Rsa*I	2.5	1	1	1	0.45	0.66	0.189 ± 0.116	75.0	0.401
*COL4A1/Xmn*I	4.7	0.02	0.11	0.02	0.29	0.06	0.159 ± 0.069	30.1	0.100
*COL4A1/Hin*dIII	2.1	0.46	0.38	0.62	0.38	0.53	0.481 ± 0.007	8.5	0.013
*NGFB/Bgl*II	5.6	0.04	0.23	0.46	0.41	0.22	0.352 ± 0.076	22.8	0.092
*NGFB/Taq*I	6.0	0.50	0.24	0.09	0.10	0.17	0.299 ± 0.062	33.3	0.110
*A2M/Bgl*II	6.5	0.29	0.23	0.07	0.04	0.29	0.276 ± 0.072	21.0	0.059
*A2M/Eco*RV	9.0	0.22	0.07	0.07	0.02	0.29	0.211 ± 0.071	29.4	0.097
*A2M/Hae*III	2.0	0.52	0.46	0.07	0.03	0.25	0.310 ± 0.094	49.0	0.159
*A2M/Hin*fI	0.9	0.56	0.46	0.50	0.13	0.49	0.444 ± 0.054	19.4	0.054
*A2M/Pvu*II	8.5	0.17	0.06	0.03	0.02	0.25	0.175 ± 0.067	24.1	0.083
*TG/Eco*RV	14	1	1	1	0.75	0.85	0.126 ± 0.079	30.1	0.169
*D13S10/Dra*I	0.9	0.32	0.17	0.13	0.40	0.35	0.372 ± 0.083	5.9	0.004
*IGF1/Xmn*I	6.3	0.01	0.0	0.14	0.23	0.21	0.187 ± 0.052	37.4	0.131
*PL19.2/Sst*I	3.2	0.0	0.05	0.82	0.08	0.63	0.202 ± 0.075	112.3	0.555
*PL18.1/Bgl*II	2.6	0.13	0.22	0.35	0.50	0.57	0.401 ± 0.053	12.8	0.101

Summary of genetic variation for 20 polymorphic restriction sites studied at Stanford in samples of five populations: Biaka Pygmies (Central African Republic), Mbuti Pygmies (Zaire), Nasioi Melanesians (Bougainville), Chinese, and Caucasians. For each site the table gives, in order, the kilobase size of the allele used to define a two-allele system, the frequency of that allele, the unweighted average heterozygosity of that system across the five populations, the chi-square measuring significance of the allele frequency variation among populations, and the F_{ST} (standardized Wahlund's variance) across the five populations.

ing one, and dividing by the average number of chromosomes per population. It is clear that the genes showing the highest variation are *CD8*, a lymphocyte antigen; *POMC* or proopiomelanocortin; the α2-chain of type I collagen; and a random probe pL19.2, which has not yet been assigned to a chromosome. In all cases, the genes that provide greater discrimination separate Africans from Eurasians; some genes, however, group Melanesians with Pygmies while others group them with Eurasians, where they would be expected to fall.

The average heterozygosities in both Tables 3 and 4 are generally quite large. The singular exception is for the second *Dra*I site for *CD8*, a polymorphism discovered in the Biaka population. Of course, all other sites were studied because they were known to be polymorphic, with high heterozygosities, in Caucasians. The large average heterozygosities for sites, such as *POMC/Sst*I, that have low F_{ST} values show that the lack of discrimination is not for want of polymorphism—both alleles are frequent in all populations.

Taking each locus as an independent datum, the average heterozygosities for each population are given in Table 5. These data show that heterozygosities are greatest in Caucasians, probably reflecting the fact that these loci were selected for polymorphism in Caucasians. Of more significance than the relative values is the fact that the primitive populations still have appreciable levels of heterozygosity. Most polymorphisms

Table 4. Yale Allele Frequency Data

Locus/site	Allele tallied (kb)	Pygmies		Lab panel	Published Caucasians	$\overline{H}$ ± S.E.
		Biaka	Mbuti			
*AMY1/Pst*I	8	0.0	0.0	0.12	0.15	0.117 ± 0.073
ADH3/Msp	12	0.98	0.95	0.71	0.50	0.261 ± 0.114
	14					
*APOB/Eco*RI		0.13	0.10	0.09	0.20	0.225 ± 0.106
*AT3/Pst*I	10.5	0.07	0.0	0.62	0.50	0.276 ± 0.123
*MT2P1/Eco*RI	4.8	0.88	0.96	0.49	0.50	0.321 ± 0.078
*SST/Eco*RI	12	0.65	0.80	0.69	0.91	0.341 ± 0.037
*D3S5/Msp*I	3.0	0.05	0.09	0.21	0.33	0.260 ± 0.085
*D4S14/Msp*I	"larger"	0.86	0.93	0.50	0.65	0.332 ± 0.062
*D4S35/Msp*I	5.8	0.14	0.29	0.16	0.58	0.324 ± 0.041
*D13S2/Msp*I	15	0.88	0.83	0.73	0.58	0.341 ± 0.067

Summary of genetic variation for 10 polymorphic loci studied at Yale in samples of three populations: Biaka Pygmies (Central African Republic), Mbuti Pygmies (Zaire), and two Caucasian samples. Information is organized as in Table 3 except that chi-square and F_{ST} were not calculated because fewer populations have been studied so far. All allele frequencies were calculated by gene counting except for AMY1, the only dominant system, which assumed Hardy-Weinberg.

Table 5. Average Heterozygosities for Populations Studied

Population	*H*
Populations studied for the 20 sites in Table 3	
Biaka Pygmies	0.269 ± 0.046
Mbuti Pygmies	0.243 ± 0.042
Nasioi Melanesians	0.200 ± 0.040
Chinese	0.294 ± 0.090
Caucasians	0.378 ± 0.030
Populations studied for the 10 sites in Table 4	
Biaka Pygmies	0.185 ± 0.040
Mbuti Pygmies	0.166 ± 0.043
Caucasians (lab panel)	0.369 ± 0.037
Caucasians (published/original data)	0.401 ± 0.037

Table 7. Matrix of Genetic Distances Based on Data Collected at Yale

1. Biaka Pygmy–Mbuti Pygmy	0.019 ± 0.008
2. Pygmy (both)–Caucasian (both)	0.196 ± 0.043
3. Lab panel–literature data	0.042 ± 0.019

found in Caucasians exist in such diverse populations as Pygmies, Melanesians, and Chinese. Only four of the 29 sites polymorphic in Caucasians were monomorphic in Pygmies and, where tested, Melanesians: *POMC/Rsa*I, *COIA2/Rsa*I, *TG/ECO*RV, and *AMY1/Pst*I.

Distances calculated by the Reynolds et al. (1983) formula and averaged over all loci in Table 3 are given in Table 6. Clearly the two Pygmy groups show the shortest distance, as expected because they have considerable similarity. The next shortest distance is between Chinese and Caucasians, and this distance (0.086) is shorter than that between Caucasians and Pygmies (on average, 0.136). Within the limits of this new sample, therefore, Eurasia versus Africa appears a better split than Eurafrica versus Asia. The sample is larger in terms of genes (but not of populations) than the original one on which this work was first started (Cavalli-Sforza and Edwards 1964), but it would be premature to consider the problem closed, given that, based on this sample alone, the difference between the two distances is not significant. It does point, however, in the direction of splitting Africa from the rest of the world. Today some consensus is developing that remains of modern *Homo sapiens* from South Africa at the Border Caves, Natal, and at Klasies River Mouth near the Cape (Rightmire 1984) are older than any remains found so far in Eurasia. This is sometimes explained by assuming an earlier split between African and all Eurasian human groups.

The same distances are given in Table 7 for selected pairs based on the data for the loci in Table 4. To be noted is a significant difference between the lab panel and the literature data, though only at $p < 5\%$. This difference may indicate that considerable heterogeneity among Caucasians will be documented when ethnically more homogeneous groups are studied. The comparison between Pygmies and Caucasians is in agreement with the distance in Table 6. Among the 10 loci studied, those that show the greatest difference between Africans and Caucasians are alcohol dehydrogenase 3 (*ADH3*), antithrombin III (*AT3*), and the anonymous locus *D4S14*.

An analysis testing for differences between random probes, pseudogenes, and probes coding for known genes will be possible when a larger number of probes have been tested. It is also worth noting that all probes tested so far have been selected for polymorphism in Caucasians. Indeed, in our data (Tables 5 and 6) Caucasians have the highest average heterozygosities.

The conclusion with respect to the problem briefly described at the beginning of this article is that the data accumulated so far favor an early African/Eurasian split, but are not sufficient, in isolation, to reach statistical significance. The difference observed, however, suggests that with three or four times as many probes as those studied here it may be possible to reach a statistically significant solution. In any case it will also be necessary to extend the range of populations considered. For Melanesians, where some genes resemble Africans more than one might expect, further studies may make it possible to choose between a hypothesis of selective convergence, given the similarity of the habitats, and one of residual similarity between two ancient lineages.

One finding of clear relevance to understanding of human evolution is the ubiquity of these polymorphisms. It is premature to reach specific conclusions on the evolutionary factors involved, just as it is not yet possible to evaluate the significance of our finding of new forms of the chromosome 17 alphoid DNA. It is clear, in any case, that we can expect completely new vistas, in addition to the solution of a number of old problems, as additional data accumulate.

ACKNOWLEDGMENTS

This work was supported in part by U.S. Public Health Service grants GM20467 (to L.C.-S) and CA32066 and MH39239 (to K.K.K.); National Science Foundation Anthropology Section, grant BNS8311854 (to J.S.F.); a grant from the Lucille P. Markey Foun-

Table 6. Matrix of Genetic Distances Based on Data Collected at Stanford

	Biaka Pygmy	Mbuti Pygmy	Melanesian	Chinese
1. Biaka Pygmy	–			
2. Mbuti Pygmy	0.047 ± 0.013	–		
3. Melanesian	0.165 ± 0.047	0.131 ± 0.050	–	
4. Chinese	0.248 ± 0.038	0.205 ± 0.045	0.171 ± 0.053	–
5. Caucasian	0.130 ± 0.034	0.143 ± 0.034	0.103 ± 0.023	0.086 ± 0.024

dation (to L.C.-S.) for funding trips to Africa to collect samples; and grants from the John D. and Catherine T. MacArthur Foundation (to K.K.K.)

REFERENCES

Astolfi, P., A. Piazza, and K.K. Kidd. 1979. Testing of evolutionary independence in stimulated phylogenetic trees. *Syst. Zool.* **27:** 391.

Boyd, W.C. 1950. *Genetics and races of man.* Little Brown, Boston.

Cavalli-Sforza, L.L. 1972. Origin and differentiation of human races. In *Proceedings of the Royal Anthropological Institute*, p. 15.

———. 1986. *African Pygmies.* Academic Press, Orlando, Florida.

Cavalli-Sforza, L.L. and W.F. Bodmer. 1971. *The genetics of human populations.* W.H. Freeman, San Francisco.

Cavalli-Sforza, L.L. and A.W.F. Edwards. 1964. Analysis of human evolution. Genetics today. *Proc. Int. Congr. Genet.* **2:** 923.

———. 1967. Phylogenetic analysis: Models and estimation procedures. *Am. J. Hum. Genet.* **9:** 234.

Cavalli-Sforza, L.L., I. Barrai, and A.W.F. Edwards. 1965. Analysis of human evolution under random genetic drift. *Cold Spring Harbor Symp. Quant. Biol.* **29:** 9.

Cavalli-Sforza, L.L., L.A. Zonta, F. Nuzzo, L. Mernini, W.W.W. DeJong, P. Meera Khan, A.K. Ray, L.N. Went, M. Siniscalco, L.E. Nijenhuis, E. van Loghem, and G. Modiano. 1969. Studies on African Pygmies. I. A pilot investigation of Babinga Pygmies in the Central African Republic (with an analysis of genetic distance). *Am. J. Hum. Genet.* **21:** 252.

Friedlaender, J.S., ed. 1986. *The Solomon Islands project. A Long-term study of health, human biology, and culture change.* Oxford University Press, England. (In press.)

Hirszfeld, L. and H. Hirszfeld. 1918–1919. Essai d'applicatio methodes serologiques au probleme des races. *Anthropologie* **29:** 505.

Johnson, M.J., D.C. Wallace, S.D. Ferris, M.C. Rattazzi, and L.L. Cavalli-Sforza. 1983. Radiation of human mitochondria DNA types analyzed by restriction endonuclease cleavage patterns. *J. Mol. Evol.* **19:** 255.

Kidd, K.K. 1973. Genetic approaches to human evolution. In *Quaderno No. 182, Atti del Colloqui Internazionale sul Tema: L'Origine Dell'Uoma.*, p. 149. Academia Nazionale dei Lincei, Rome, Italy.

Kidd, K.K. and L.L. Cavalli-Sforza. 1974. The role of genetic drift in the differentiation of Icelandic and Norwegian cattle. *Evolution* **28:** 381.

Kidd, K.K. and L.A. Sgaramella-Zonta. 1971. Phylogenetic analysis: Concepts and methods. *Am. J. Hum. Genet.* **23:** 235.

Nei, M. 1978. The theory of genetic distance and evolution of human races. *Jpn. J. Hum. Genet.* **23:** 341.

Nei, M. and A.K. Roychoudhury. 1982. Genetic relationship and evolution of human races. *Evol. Biol.* **14:** 1.

Piazza, A., L. Sgaramella-Zonta, P.D. Gluckman, and L.L. Cavalli-Sforza. 1975. The Fifth Histocompatibility Workshop. Gene frequency data: A phylogenetic analysis. *Tissue Antigens* **5:** 445.

Reynolds, J., B.S. Weir, and C.C. Cockerham. 1983. Estimation of the coancestry coefficient: Basis for a short-term genetic distance. *Genetics* **105:** 767.

Rightmire, G.P. 1984. *Homo sapiens* in subsaharan Africa. In *The origin of modern humans* (ed. F.H. Smith and F. Spencer), p. 295. A.R. Liss, New York.

Wainscoat, J.S., A.V.S. Hill, A.L. Boyce, J. Flint, M. Hernandez, S.L. Thein, J.M. Old, J.R. Lynch, A.G. Falusi, D.J. Weatherall, and J.B. Clegg. 1986. Evolutionary relationships of human populations from an analysis of nuclear DNA polymorphisms. *Nature* **319:** 491.

Waye, J.S. and H.F. Willard. 1986. Structure, organization, and sequence of alpha satellite DNA from human chromosome 17: Evidence for evolution by unequal crossing-over and an ancestral pentamer repeat shared with the human X chromosome. *Mol. Cell. Biol.* **6:** 3156.

Willard, H.F., M.H. Skolnick, P.L. Pearson, and J.-L. Mandel. 1985. Report of the committee on human gene mapping by recombinant DNA techniques. *Cytogenet. Cell Genet.* **40:** 356.

Willard, H.F., J.S. Waye, M.H. Skolnick, C.E. Schwartz, V.E. Powers, and S.B. England. 1986. Detection of fragment length polymorphisms at the centromeres of human chromosomes by using chromosome-specific α satellite DNA probes: Implications for development of centromere-based genetic linkage maps. *Proc. Natl. Acad. Sci.* **83:** 5611.

Wilson, A.C., R.L. Cann, S.M. Carr, M. George, B. Gyllensten, K. Helm-Bychowski, R.G. Higuchi, S.R. Palumbi, E.M. Prager, R.D. Sage, and M. Stoneking. 1985. Mitochondrial DNA and two perspectives on evolutionary genetics. *Biol. J. Linn. Soc.* **26:** 375.

Fossil Evidence on Human Origins and Dispersal

P. ANDREWS
Department of Paleontology, British Museum (Natural History), London SW7 5BD, United Kingdom

The close relationship of humans with apes has been recognized for many years. Darwin (1871) was very specific in his inferences about the common ancestor between humans and apes. He recognized the subdivision of the Catarrhini from other primates, and divided it into the monkeys and the anthropomorphous apes, with humans part of the latter group; and he further states that as "chimpanzees and gorillas...are now man's nearest allies, it is somewhat more probable that our early progenitors lived on the African continent than elsewhere." Darwin also had some more speculative things to say about early human ancestors, for instance, inferring that they had frugivorous diets (based on analogy of morphological form of teeth), that they were restricted to hot climates (loss of hair), and that they had diverged from the apes by the middle Miocene (because of the known existence of the fossil ape *Dryopithecus* in the middle Miocene deposits in Europe).

The evidence on which Darwin based the above conclusions was a mixture of comparative and fossil evidence, with the former predominating because of the extreme shortage of any fossil evidence. His comparative evidence, and that of Huxley (1863), provided a sound basis on which the mounting fossil evidence could subsequently be examined, and it is a tragedy of scientific misdirection that during the middle part of this century the evidence from fossils gained an ascendancy over the comparative method in interpreting hominoid phylogeny. It has taken an entirely new source of comparative data to redress the balance, that from molecular studies of proteins and DNA, but there is a danger now that in correcting and redirecting work on hominoid phylogeny some workers would like to eliminate fossils altogether. I think both sources of evidence have much to offer, however, and what is important is not to see them as competing with each other but as complementary disciplines that have much to offer each other.

The nature of fossil evidence is very different from that provided by comparative data. In the 1950 Cold Spring Harbor Symposium, Simpson stressed that the evidence of relationship provided by the comparative morphology of living animals gives only the end results of clearly historical processes, but tells us nothing of the processes themselves. This is true equally of gross morphology and of molecular similarity, and to translate the similarity-based sets of relationships into phylogenies requires an additional set of inferences about how phylogenetic change is presumed to have occurred. These inferences are sometimes integrated into the similarity matrices in a way that blurs the distinction between the observable fact; for instance, similarity based on shared characters, and inference about evolutionary processes.

By contrast, fossils provide first-hand historical data (Simpson 1951), but the problem here is the incompleteness of the fossil record. The fossil remains are incomplete themselves, succession through time is broken by many gaps, and the spatial distribution of fossil faunas is incomplete, so that all we have to work on is isolated fragments of evidence. This greatly reduces the historical value of the fossil record. Simpson's view in 1950, as that of the majority of paleontologists of that period, was that despite these deficiences the fossil record was still the primary source of historical data. This may be true of some groups of animals, but in my view the fossil evidence for hominoid evolution is too incomplete to base any general statement of hominoid phylogeny on it alone.

For the purposes of this paper I will accept the possibility of two sets of relationships for the Hominoidea (Fig. 1). These are identical except for the relationship between humans, chimpanzees, and gorillas: One set puts humans and chimpanzees more closely related, and this is based on the evidence from DNA–DNA hybridization (Sibley and Ahlquist 1984) and also supported by some evidence from DNA sequencing (Ferris et al. 1981a,b; Bishop and Friday 1985; Hasegawa et al. 1985), amino acid sequencing (Goodman et al. 1983, 1984; Koop et al. 1986), and some chromosomal evidence (Yunis and Prakash 1982). The other set of relationships groups chimpanzees and gorillas together in an African ape clade to which humans are related, and this is supported by morphological evidence (Andrews 1986b), some DNA sequencing (Brown et al. 1982; Templeton 1983, 1985), and some chromosomal evidence (Marks 1983). These references are not exhaustive (see Andrews 1986a,b for a full list of references), but they illustrate the lack of uniformity of the evidence and its interpretation. Any resolution of this discordance must await additional data, but in the meantime in this paper the fossil evidence for hominoid evolution will be discussed with these two possible sets of relationships in mind.

Three things have to be considered when using fossil evidence for dating evolutionary events. One is that any branching point always has two sets of descendants, and evidence for the timing of the branching point may come from either. For example, the divergence of the catarrhine primates into monkeys and apes (Cercopithecoidea and Hominoidea) could be dated by the first appearance of either in the fossil record, so that the

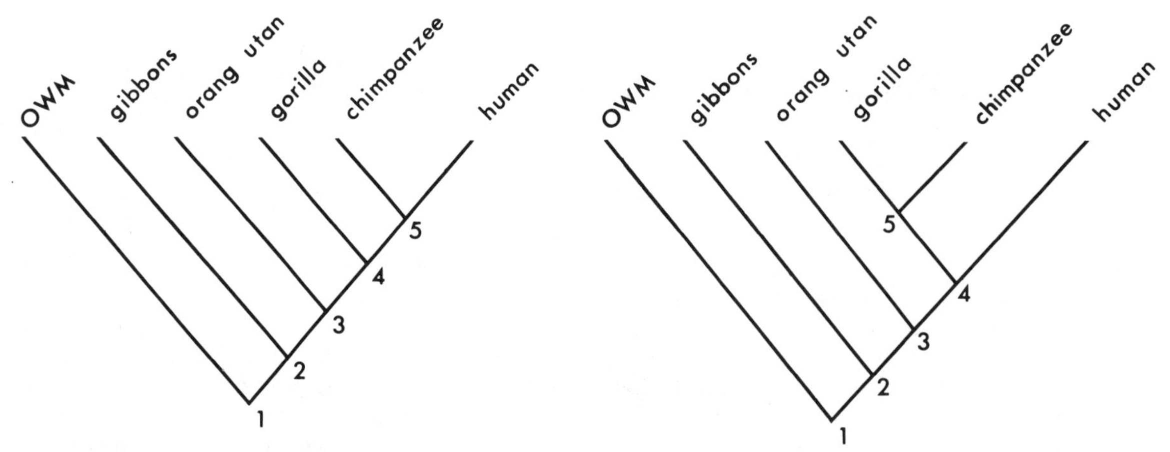

Figure 1. Relationships of the Hominoidea. On the left is shown one possibility with the chimpanzee most closely related to humans, and on the right is shown the alternative of chimpanzees and gorillas more closely related.

date of origin of the hominoids could be indicated by the first appearance of cercopithecoids, if that were earlier than the earliest hominoid fossils. Stemming from this is the second point, that fossils only provide a minimum date of divergence, never a maximum, so that they indicate a time by which a lineage has originated, not the actual time. Finally, the way by which fossil taxa are identified is important, and in this paper the assignment of a fossil is only accepted when the presence of synapomorphies with particular lineages can be demonstrated.

Hominoid Diversity

The hominoids have traditionally been divided into the lesser apes and the great apes and humans. The lesser apes include the six species of gibbon and siamang; and the great apes include the orangutan from Borneo and Sumatra (recognized as two very distinct subspecies or perhaps two species), two species of chimpanzees from west and central Africa, and one species of gorilla with three relatively distinct subspecies having more limited distribution than chimpanzees.

The six species of gibbon and siamang have allopatric distributions and morphological variations indicative of recent speciation. The African great apes, on the other hand, have sympatric distributions and contracting ranges suggestive of earlier speciation and present relic status. The orangutan range is also apparently contracting, being formerly known from the mainland of southeast Asia, and its distribution is also suggestive of relic status.

Rates of divergence in the mitochondrial DNA molecule within and between species adds an interesting perspective to these general statements about extant apes. Within any one of the hominoid species the greatest percentage divergence in the mtDNA molecule is seen in the orangutan (Wilson et al. 1985). The sequence divergence within this species is 3.65%, compared with a slightly greater figure for the divergence of the two species of chimpanzee and much lower figures within any of the other hominoid species (0.36–1.33%). The human figure is the lowest within the hominoids, and this will be commented on below, but all of the hominoids except the orangutan appear, on this evidence, to have had comparatively recent origins.

The fossil evidence for the separate lineages leading to extant hominoids will be discussed below, but there is some fossil evidence for the present relic status of most or all of the extant hominoids. Hominoids are first known from the fossil record about 22 million years ago (Myr), and between then and about 18 Myr they reach the highest diversity ever achieved, when 10 species are known within a very restricted area of east Africa (Andrews 1981; updated by including recent fossil discoveries). Between 15 and 18 Myr the number of fossil ape species falls to seven, but fewer sites are known from this time period, and so this fall may not be real. By the late Miocene, however, only two fossil ape species are known from Africa, so that within this continent, and between 22 and 8 Myr, there has been a dramatic fall in the diversity of fossil hominoids. This is shown in Figure 2 compared with total catarrhine diversity in the same area of east and north Africa. By calculating the regression of species numbers against time, it has been shown that the projected decline in numbers of hominoid species during the Miocene could have led to their extinction in Africa about 3 Myr (Andrews 1981). The number of species actually present at 3 Myr could not have been less than three (proto-chimpanzee, -gorilla, and -human), and probably it was no more than four, and these species may well have been represented by populations no larger than those of extant African ape species.

The fossil evidence for decrease in hominoid diversity, to the extent that they can be considered to have been bordering on extinction 3 Myr, is entirely consistent with their present relic distributions. The relatively high sequence divergence of mtDNA within the small populations of living great apes suggests they are rem-

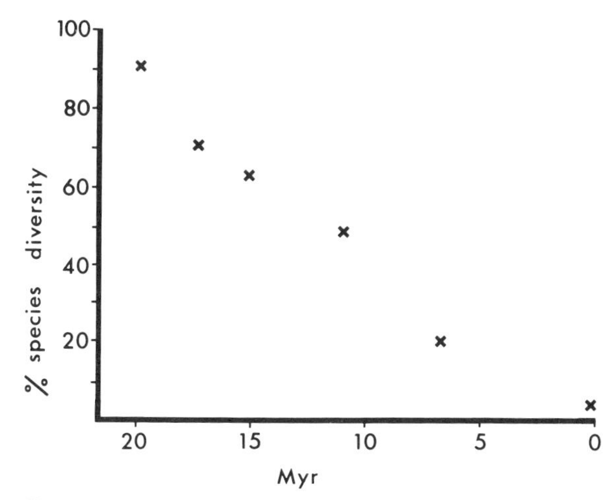

Figure 2. Hominoid species diversity. (*Vertical axis*) Species diversity of hominoids as a percentage of catarrhines in Africa; (*horizontal axis*) time in millions of years.

nants of formerly larger populations (Ferris et al. 1981b). They may have become subdivided into many populations in the past, and with contraction of their ranges the high mtDNA divergence resulting from their former variability could have been retained in the smaller relic population.

Hominoid Origins

In the preceding section, attention on hominoid diversity changes was focused on Africa, because the early evolutionary history of hominoids is entirely restricted to Africa, covering a period of approximately 9 Myr. The earliest known fossil that can be identified as hominoid is from 22-Myr deposits in Kenya (Andrews et al. 1981), and it is not until approximately 13 Myr that hominoids are known outside Africa (Andrews and Tobien 1977; Raza et al. 1983; Barry et al. 1985).

Two consequences of importance emerge from this observation. One is that the probable place of origin of the Hominoidea is somewhere in Africa. There is no shortage of fossil localities in Europe and Asia during the early part of the Miocene and late Oligocene, the periods covering the 15- to 30-Myr time interval, and the total absence of hominoids from these faunas can be considered significant in view of their number and range of habitats sampled.

The other consequence of the early hominoid fossil record in Africa is that the origin of the group can be put at earlier than 22 Myr. How much earlier it is difficult to be sure, but it has been shown in the preceding section that it was close to the beginning of this time that the hominoids reached their greatest species diversity, and this suggests that they were already well established by then and must have originated several million years prior to 22 Myr. There is no fossil record in Africa for the period 22–30 Myr, and therefore no direct evidence, but on the basis of the nature and extent of the early hominoid radiation the origin of the group probably occurred at least 3–5 Myr earlier than this, in other words between 25 and 28 Myr.

The sister group of the Hominoidea is the Cercopithecoidea, the Old World monkeys (Fig. 1). Fossil evidence for this group might also be expected to provide some information on the time and place of origin of the hominoids, but in fact little additional evidence is available. The earliest fossil monkey is from early Miocene deposits in Africa which also yield early hominoids. The monkeys during the first half of the Miocene are rather rare (Delson 1975), and the great radiation leading to the many species living today did not take place until late in the Miocene when hominoid diversity was dropping quickly. It has been suggested that the two events may be causally linked (Andrews 1981). Like the hominoids, the early fossil record of monkeys is restricted to Africa, and in this case they are not found in Eurasian fossil deposits until late in the Miocene about 8 Myr ago. Therefore, the fossil evidence for monkeys gives added support to Africa as the center for catarrhine diversification but little additional information about the time of origin of either group.

For the timing of hominoid origins there is evidence available from comparative biology through the various estimates from molecular clocks. There are a number of problems with the interpretation of molecular clocks, many of which have been reviewed elsewhere (Read and Lestrel 1970; Read 1975; Corruccini et al. 1980; Goodman et al. 1983; Gingerich 1984; Wilson et al. 1985; Andrews 1986a,b; Koop et al. 1986). These have demonstrated fluctuations in rates of molecular change both within and between lineages that make the extrapolation of constant rates doubtful in many cases. One exception to this is the averaged rate demonstrated by Sibley and Ahlquist (1984), who were able to minimize the effects of homoplasy and functional constraints. Constant rates of change assume that all change is divergent and nonfunctional, and if either of these is incorrect the rates of change can be expected to fall over longer periods of time. By analyzing the whole of the DNA molecule, and restricting similarity to multiple units of DNA rather than to single-base changes (see below), the effects of function have been minimized and the probability of homoplasy greatly reduced by the hybridization process.

The molecular clocks provide a relative time scale that has to be calibrated by some independent means, usually the fossil record. This unfortunately introduces an element of circularity into the timing of evolutionary events. In the case of DNA-DNA hybridization, the calibration point used was the divergence of the orangutan (see below) from the other apes, assuming a divergence date of 13–16 Myr from the fossil record. On the basis of this the divergence of hominoids and cercopithecoids was calculated to be between 27 and 33 Myr. This calibration from the orangutan is probably accurate, and my results suggest further that orangutan divergence was closer to 13 than to 16 Myr, so that the more likely range of dates for the divergence of the hominoids is between the lower estimate, 27 Myr, and about 30 Myr. This calibration will be discussed in more detail in the section on orangutan divergence, but it can be noted here that the range of dates provided by DNA-DNA hybridization of 27-30 Myr and the suggested minimum dates from the fossils of 25-28 Myr (Fig. 3) overlap to a certain extent.

Work on nuclear DNA sequencing used as calibration places the Anthropoidea divergence at 35-45 Myr (Koop et al. 1986). By this is probably meant the divergence of New World monkeys (Platyrrhines) and Old World monkeys and apes (Catarrhines). Again, this seems to be a reasonable calibration (Gingerich [1984] gives a date of 40 Myr for this event), and based on this Koop et al. (1986) obtain an age range of 22-28 Myr for the divergence of cercopithecoids and hominoids. This also overlaps with the fossil evidence, and, since the younger dates can be shown to be less likely by fossil evidence, it seems possible to conclude that a realistic date is between 25 and 28 Myr for the separation of cercopithecoids and hominoids (Fig. 3).

Gibbon Divergence

There is little information from the fossil record about the time and place of gibbon origins. No fossil species that are clearly part of the gibbon clade have yet been recognized, and all we have to go on are fragmentary fossil hominoids from Kenya and Saudi Arabia that are considered to belong to the great ape and human clade, that is the sister group to the gibbons (Andrews et al. 1978). One specimen is from a locality called Ad Dabtiyah in Saudi Arabia, and it is dated by its faunal remains to about 16-17 Myr. Saudi Arabia geologically is on the African plate and so must be counted as part of Africa during the early part of the Miocene. The other specimens are from Buluk in northern Kenya (Leakey and Walker 1985), and have been radiometrically dated to older than 17 Myr. Both hominoids appear to be related to the later fossil genus *Kenyapithecus*, and, on the basis of this relationship and the characters they share with later and more complete hominoids, it can be concluded that both belong to the great ape and human clade, after, that is, the divergence of the gibbons. The gibbons, therefore, must have already been separate by this time, even if totally unknown as fossils, so this gives a minimum age of 17 Myr for gibbon origins.

There are several dates for gibbon divergence provided by molecular clocks. The DNA-DNA hybridization data give an age range of 18-22 Myr based on the calibration range accepted for orangutans of 13-16 Myr (Sibley and Ahlquist 1984). It was indicated in the previous section that the younger of these two dates is likely to be more realistic for the orangutan (see below), and in consequence the time of gibbon divergence is probably closer to 18 than to 22 Myr. Divergence data in nuclear DNA sequences are unfortunately not provided for the gibbon branching point (Koop et al. 1986), but extrapolating between the two adjacent branching points it would clearly have a similar if slightly younger age.

Sequencing of mtDNA has also been used to calculate divergence dates for the gibbons (Hasegawa et al. 1985). This is calibrated by assuming divergence of primates and ungulates at 65 Myr. This is a reasonable estimate based on available fossil evidence, but it appears likely that it is a branching point too remote from the hominoid branching points being considered here,

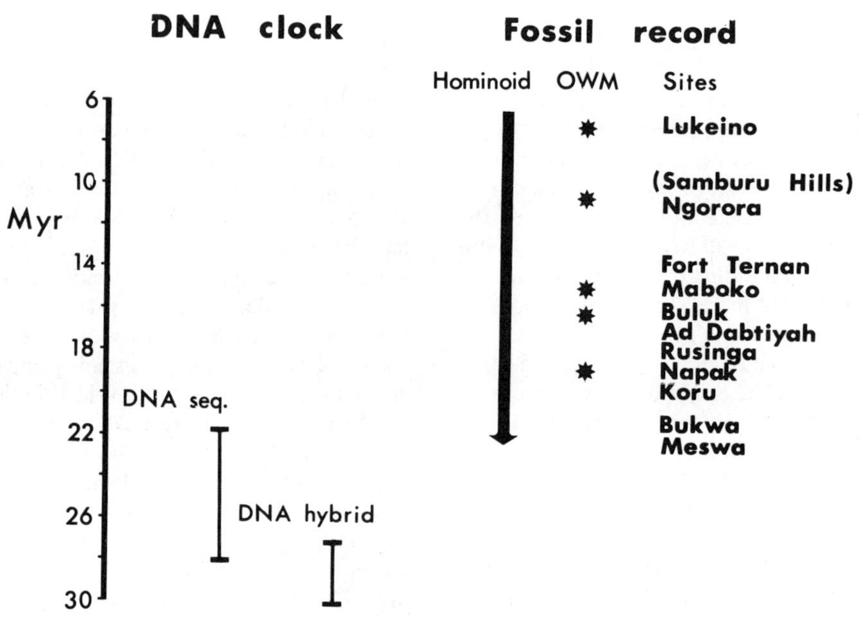

Figure 3. Time ranges for hominoid origins provided by the molecular clock on the left and fossil evidence on the right. The hominoid fossil record extends back to just over 22 Myr, while that of cercopithecoids (OWM) extends to 19 Myr.

so that with only a moderate amount of slowing down of sequence change the dates obtained would all be too young. The first date given for the gibbon branching point is 13.3 ± 1.5 Myr, and this is considerably younger than dates from other sources; succeeding branching points diverge still more from other sources. These workers now acknowledge (M. Hasegawa, pers. comm.) that their dates are in fact too young.

The dating of the gibbon branching point by the two nuclear DNA clocks gives a range of around 17–22 Myr for this event. By narrowing down the calibration range from the fossil record (based on other branching points), it is possible to give a more precise estimate at 17–20 Myr. This corresponds with the published evidence so far available from the fossil record, which indicates an age of greater than 17 Myr for the gibbon branching point.

Orangutan Divergence

A recent spate of papers have documented previously unrecognized members of the orangutan clade in the fossil record (Andrews and Tekkaya 1980; Andrews and Cronin 1982; Pilbeam 1982; Ward and Pilbeam 1983; Andrews 1986b). These group the fossil genera *Sivapithecus* and *Ramapithecus* in the orangutan clade on the basis of a number of shared derived characters of the palate and face (Andrews and Cronin 1982). All of the fossil specimens that can be attributed to these genera with any degree of certainty (that is based on the shared presence of derived characters with the orangutan) come from the Miocene deposits in India, Pakistan, and Turkey. They have an age range of 8–11.8 Myr in Pakistan, a similar age in India, and an age of approximately 11 Myr in Turkey (Andrews and Tekkaya 1980; Raza et al. 1983; Barry et al. 1985). The earliest secure date is the one for the Chinji Formation in Pakistan at 11.8 Myr, and this indicates an age of at least 12 Myr or more for the branching point of the orangutan.

A number of other fossil discoveries have been attributed to *Sivapithecus* and *Ramapithecus*, and these must be examined briefly here. A number of discoveries from China, Greece, and Hungary extend the range of these genera, if correctly identified, but add little more information, for instance, in terms of dating. The discovery of large numbers of hominoid teeth at Pasalar in Turkey, some of which have been described by Andrews and Tobien (1977), appears to confirm the presence of *Sivapithecus* in these middle Miocene deposits, which are dated faunally to between 13 and 14 Myr. This may extend the temporal range of this genus to 14 Myr, and hence also the branching point of the orangutan. The Pasalar hominoids are very incomplete, however, and do not preserve most of the relevant body parts to identify orangutan synapomorphies, but their incisor proportions and the conformation of the enamel dentine junction (Martin 1985) of their molars are shared with later specimens of *Sivapithecus* and the orangutan. It is considered possible, therefore, that they should be referred to *Sivapithecus* and considered members of the orangutan clade. If this is correct, the time of divergence of the orangutan is placed at or before 14 Myr; if it is incorrect, the earliest date is about 12 Myr.

Historically, a number of specimens from Africa have been attributed to *Ramapithecus* and *Sivapithecus*. This was based on the shared presence of thick molar tooth enamel in these genera, but since this has been shown to be an ancestral great ape and human character (Martin 1985), and since the postcranial distinctions of the African material have been clarified (Harrison 1982), it has become evident that these African specimens from the sites of Fort Ternan and Maboko Island are cladistically distinct. They have therefore been returned to the genus originally described for them by Louis Leakey (1962), *Kenyapithecus*. This would be placed between nodes 2 and 3 of Figure 1, in contrast to *Sivapithecus* which is placed within node 3.

Finally, one additional discovery, also from Africa, must be commented on briefly. This is the hominoid from Buluk in northern Kenya, consisting of extremely fragmentary remains attributed to *Sivapithecus* by Brown et al. (1985). There are some similarities with this genus to be seen in the Buluk specimens, but the character of thick molar enamel, which is now being used to define the great ape and human clade including *Sivapithecus*, (Fig. 4), and which is also present in *Kenyapithecus*, appears not to be present in the Buluk material. Lacking this and other derived characters of *Kenyapithecus* and *Sivapithecus*, the Buluk material appears to represent a more primitive, perhaps unnamed, genus.

The conclusion from the fossils, therefore, is that the branching point of the orangutan took place at least 12 Myr and perhaps more than 14 Myr. All of the fossil discoveries that can be linked with the orangutan with any degree of certainty come from European and Asian deposits, and in fact the fossils from Pasalar appear to represent the earliest hominoid of any type yet known from outside Africa. It is likely that the *Sivapithecus* remains from the 13 to 14 Myr deposits at Pasalar represent the earliest emigration of hominoids out of Africa, and, although other types of hominoid are known from later deposits, the orangutan lineage was established in Eurasia by this time.

The early branching of the orangutan lineage and its formerly more widespread distribution are consistent with the results from comparative biology. In this instance, the DNA–DNA hybridization results cannot be used to date the branching point, because this was the one used to calibrate the DNA clock in the first place. The nuclear DNA sequencing results give a date of 12.7–16.4 Myr for this branching point (Koop et al. 1986), which is compatible with the fossil record (Andrews and Cronin 1982), but the mtDNA clock of Hasegawa et al. (1985) gives a younger date of 9.7–12.1 Myr. As mentioned earlier, this date is probably too young.

It can be concluded that a combination of fossil and molecular dating is consistent in indicating an age of

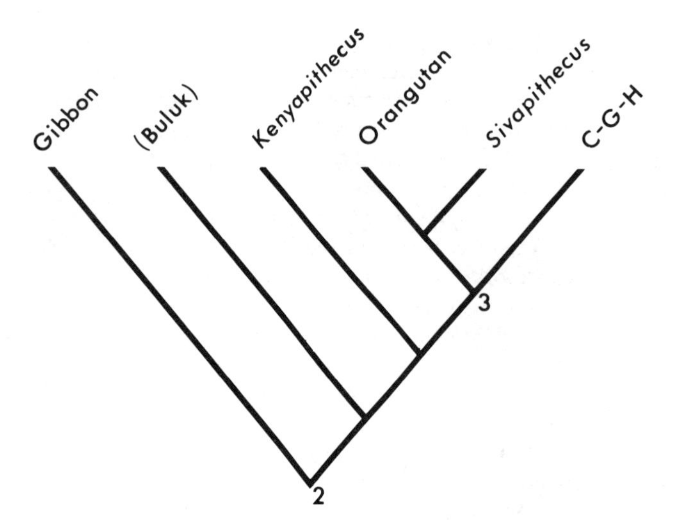

Figure 4. Cladogram showing the relationships of *Sivapithecus*, *Kenyapithecus*, and the new Buluk fossil hominoids. C-G-H signifies chimpanzees, gorillas, and humans.

about 13–14 Myr for the branching point of the orangutan. The lineage is likely to have had an African origin, but this is unknown at present.

African Ape and Human Divergence

On this subject, full of uncertainties, there is little that I can add to my recent review (Andrews 1986b). The problem has already been laid out in the introduction and Figure 1, and as the fossil record provides no help in this instance I will just contrast two of the extremes of evidence and interpretation from comparative biology.

One of the strengths claimed for DNA–DNA hybridization (Sibley and Ahlquist 1984) is that reassociation of the DNA strands to form hybrid pairs depends not just on single-base equivalence of the DNA chains but on identity of multiple groups of bases. Sibley argues that whereas single-base changes could be duplicated by homoplasy, and hence distort the significance of DNA similarity, the multiple groups are complex and unlikely to be duplicated independently and so must be homologous. This is the basis for his claim that what is being measured by the hybridization technique is only, or at least mainly, homology, and it appears to be a convincing argument. This provides strong evidence in support of the relationship of humans and chimpanzees, with gorillas as the outgroup. This relationship is also supported by at least two uniquely shared nuclear DNA substitutions at the *n* locus (Koop et al. 1986) and at least one amino acid substitution (Goodman et al. 1983).

In a recent paper I have tried to formulate a functional argument for assessing morphological change that in fact is essentially the same as Sibley's argument for DNA hybridization. Simply stated this says that if a number of characters are functionally correlated so as to relate to a single function, the more complex the function, and the greater the number of characters relating to it, the less likely it is to be achieved independently and the more likely is the similarity to be the result of homology. The individual characters contributing to the complex cannot be treated separately but simply as part of a single multiple character, and for homology to be indicated all the parts of this multiple character must be identical. Shared presence of even a single highly complex multiple character is taken as strong evidence of homology and therefore of relationship. Two such complexes were provided linking chimpanzees and gorillas (Andrews 1986b), namely the characters of shoulder, elbow, wrist, and hand (all related to knuckle-walking); and the enamel prism morphology, developmental type, proportions, and growth rate (all relating to molar enamel thickness). These provide evidence for relationship between chimpanzees and gorillas that is at least as strong as that linking humans and chimpanzees.

One fossil that may be relevant to this branching point is the recently described upper jaw from the Samburu Hills in the Kenya rift valley (Ishida et al. 1984). This is faunally dated at between 7 and 9 Myr, and it is of particular interest because it is probably part of the African ape and human clade. Which part, however, is difficult to tell on the evidence available. It has the expanded premolar crowns with reduced sectoriality characteristic of the clade. It appears to have thick enamel that is considered primitive for the clade. In its molar proportions, however, with a particularly elongated third molar, and in the discrete nature of the molar cusps, lacking well-defined ridges connecting the cusps, there are similarities with the gorilla that are otherwise unique to this great ape. With only one specimen to go on, it is not possible to be sure of the significance of these characters, but it is possible that they are gorilla synapomorphies and that this fossil represents an early member of the gorilla lineage. Its date of 7–9 Myr is interesting in this respect, but it does not resolve the issue about order of splitting within the clade, although perhaps it lends some circumstantial support for the initial divergence of the gorilla.

This single fossil from the Samburu Hills in Kenya also provides some support for an African origin for the African ape and human clade (node 4 of Fig. 1). This is likely anyway on a vicariance model, with Africa the only continent common to all three living members of the clade, and it is further supported by the early human fossil record. It will be shown in the next section that the early history of the human lineage is confined to Africa so that while now the human species has a world-wide distribution it was originally an African group with a distribution probably very like that of the extant African apes.

The dates provided by the DNA clocks for the African ape and human branching point vary considerably (Fig. 5). The results from DNA–DNA hybridization give dates 8–10 Myr for gorilla divergence and 6.3–7.7 Myr for chimpanzee divergence (Sibley and Ahlquist 1984). The combined threeway split dated by Koop et al. (1986) is put at 6.3–8.1 Myr for nuclear DNA sequencing; and very young dates of 2.7–3.7 Myr are provided by Hasegawa et al. (1985) from mtDNA sequencing, dates that are acknowledged to be too young now by Hasegawa (pers. comm.). These dates are compatible with that from the Samburu Hills hominoid, but lack of additional fossils relating to this most interesting of branching points makes it difficult to say more at this stage.

Human Origins

The last stage of human evolution for which there is both fossil evidence and evidence for comparative biology is the emergence of modern humans, *Homo sapiens*. In addition, a number of intermediate stages are documented only by the fossil record covering evolution before the emergence of modern humans.

The earliest record of human fossils is in Africa at between 4 and 5 Myr. The remains are too fragmentary to do more than to document their existence at this time (Patterson et al. 1970; Hill 1985), but between 3 and 4 Myr much more complete fossil remains are known (White 1977; Johanson et al. 1982). These are all generally attributed to *Australopithecus*, and together with later australopithecines from South and East Africa (Howell 1978) they appear to form a clade, including both gracile and robust species (Rak 1983). Species belonging to this clade appear to be limited to Africa, as is the earliest known species attributed to the genus *Homo*, *H. habilis*. There is thus every indication from the fossil record that for at least 3 Myr of recorded history the human lineage was confined to Africa, that there were at least two major subdivisions of the human lineage, and that the genus *Homo* also originated in Africa.

Early members of the genus *Homo* are well documented in East African deposits at Olduvai and Koobi Fora (Leakey and Walker 1976, 1985; Howell 1978; Leakey and Leakey 1978; Walker and Leakey 1978). They range from relatively small-brained *H. habilis* to larger-brained individuals, which have been referred to *H. erectus* but which may belong to the species named by Groves and Mazak (1975), *H. ergaster*. As originally defined, the species *H. erectus* was based on material with southeast Asia and China, and examination of these fossils and comparison with the African specimens attributed to the same species shows that the type material has a considerable number of derived characters not present in the African specimens or in later specimens of the genus *Homo*, including *H. sapiens* (Andrews 1984; Stringer 1984; Wood 1984). This raises the possibility that the Asian material, which probably dates to within the last million years, and which represents a derived version of the African fossils, was an emigrant offshoot from Africa that evolved in isolation in eastern Asia and did not contribute to the ancestry of modern humans.

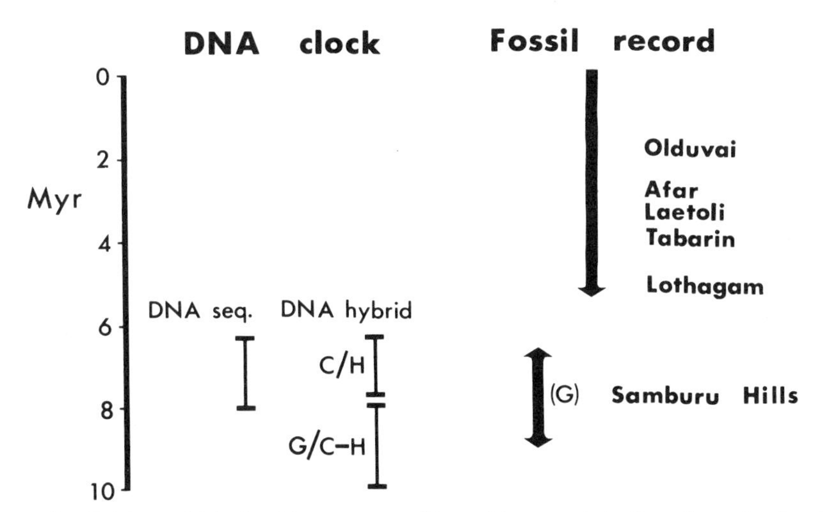

Figure 5. Time ranges for divisions within the chimpanzee, gorilla, and human clade, from the molecular clock on the left and fossil evidence on the right. The data from DNA sequencing is for the three-way split between them, while DNA–DNA hybridization provides two dates, one for the initial divergence of gorillas from humans and chimpanzees, and a second for the divergence of the two latter species (see Fig. 1). The fossil record shows the time range of human fossils back to 5 Myr and the probable time range of the fossil from the Samburu Hills that may be related to the gorilla.

The same contention would hold even if the African early Pleistocene fossils were referred to *Homo erectus*. There are taxonomic difficulties with such an arrangement, however, for there are no characters that define the species without making part of the species more closely related to *H. sapiens* than to the rest of *H. erectus*. This is clearly an unstable taxonomic arrangement, and it seems more reasonable to recognize the Asian material as an isolated emigrant population from an earlier African population and to distinguish them with different specific names.

Whatever taxonomic scheme is preferred, the major issue in question here is the nature of the transition from the early Middle Pleistocene species of *Homo* to *Homo sapiens*. I cannot review this complex subject adequately here, but some comment can be made on the collections of fossil remains from the three geographical regions relevant to the discussion.

The Far East. Specimens of *Homo erectus* come from deposits spanning in excess of half a million years, and the latest middle Pleistocene specimens of this species are contemporary with more advanced forms both in the same localities and elsewhere in the world. The undescribed Yingkou partial skeleton has clear affinities with modern humans, lacking the *H. erectus* synapomorphies but including many *H. sapiens* ones (Wu Ru-Kang, pers. comm.). Since this skeleton apparently occurs at levels contemporary with *H. erectus*, the strong possibility must be raised that it represents part of an immigrant population rather than having evolved in place in the Far East. Similarly, the Dali and Maba crania show modern human characters not present in *H. erectus* and could be later derivatives of this immigrant population.

Europe and the Middle East. Between 300 and 35 thousand years ago (Kyr) the Neanderthals emerged in this area (Cook et al. 1982; Hublin 1982; Stringer 1985, 1986; Vandermeersch 1985). These authors identify pre-Neanderthal characters (synapomorphies with later classic Neanderthals) in such early fossils as Petralona and Arago, which probably date back to between 300 and 400 Kyr. Additional Neanderthal synapomorphies are developed on later fossils, for instance in the sequence Swanscombe, Steinheim, Biache-Saint-Vaast, Saccopastore, Ehringsdorf, Krapina. These span approximately 200 Kyr, culminating in the classic Neanderthals of the last glaciation.

This evidence indicates the existence of a Neanderthal clade becoming progressively more differentiated through time (Vandermeersch 1985). It is replaced relatively abruptly at about 30–35 Kyr by modern humans in Western Europe, although evidence from Eastern Europe is claimed to show intermediate stages between Neanderthals and modern humans.

Africa. The two early human populations from the Far East and Europe, which are represented by the Yinkow and Dali specimens in the former and the Neanderthal lineage in the latter, are often collectively known as archaic *Homo sapiens*. This is essentially a grade concept meaning large-brained humans retaining many primitive characters. The same grade is also seen in Africa where it is exemplified by such specimens as those from Broken Hill and Saldanha. The dating of these sites is problematic, but they appear to be older than 200 Kyr (Stringer 1986). Similar aged and older fossils from North Africa take the fossil record back approximately 500 Kyr (e.g., at Ternifine), but the relationships of these fragmentary fossils with both earlier and later human fossils taxa are uncertain, so that no connection can presently be traced between the early Pleistocene so-called *Homo erectus* (= *Homo ergaster*?; Groves and Mazak 1975) from East Africa and these later forms.

After about 200 Kyr, a group of specimens is recognized spanning Africa from north to south. These show a mixture of archaic retentions with some modern human synapomorphies, as seen in, for instance, Florisbad, Djebel Irhoud 1, Ngaloba, and Omo 2. Later still, between 50 and 100 Kyr, is seen a population with modern human characteristics, and again it is pan-African. This includes such skulls as those from Omo 1, Dar-es-Soltan, Djebel Irhoud 2, and Klasies (and perhaps Border Cave). The earliest of these is a single specimen from Klasies, which is immediately above a raised beach with a date of 116–127 Kyr, and below a bed with two uranium series dates of 98 and 110 Kyr. This single specimen has uncertain affinities, and most of the specimens from Klasies River Mouth that have modern morphology are above this bed. These are older than 70 Kyr and may be as much as 100 Kyr or slightly younger. These, then, are the oldest recorded fossil of *Homo sapiens*, and they occur as part of a well-documented sequence going back in time to 200 Kyr or more. In Africa, therefore, the transition to *Homo sapiens* occurred at or before 100 Kyr and later than 200 Myr.

Interregional migrations. Combining the fossil evidence from these three regions shows that for at least four stages of human evolution the pattern has been: first appearance in Africa, sometimes followed by subsequent appearance and further development in other parts of the world. These four stages can be summarized as follows:

1. The human fossil record is limited to Africa for the first 3 Myr of recorded history.
2. The first appearance of *Homo* is also African, and at least two species are known, *H. habilis*, and a more advanced species that has been called either *H. erectus* or *H. ergaster*; these are dated at 1.6–2 Myr, and they are followed by the later appearance of the first non-African fossil humans in eastern Asia with the type and referred material of *H. erectus*. These are fairly clearly descended from the earlier African species.
3. The first appearance of what is generally called "archaic *Homo sapiens*" also takes place in Africa at about half a million years ago, and again there are

later derivatives of this in Europe about 100 Kyr later and in China about 200 Kyr later. The Mauer jaw from Germany may be as old as the earliest African specimens, although its affinities are uncertain. In this case, therefore, the inference that this form evolved in Africa first is less well supported.

The taxomonic classification of these archaic humans is uncertain, but in view of their distinctiveness from *Homo sapiens*, itself a markedly polytypic species, it would seem better not to classify them together. There is less agreement about this interpretation in the literature, and some authorities suggest that these earliest representatives of *H. sapiens* evolved independently from separate populations of *H. erectus* in Europe, Asia, and Africa. This would imply the branching point for the human species at least 500 Kyr ago.

4. The earliest appearance of humans morphologically indistinguishable from modern humans occurs also in Africa at about 100 Kyr. This is followed by records in Europe and Asia between 30 and 40 Kyr. Again the inference is that modern humans evolved in Africa and subsequently dispersed into the rest of the world.

Biochemical evidence. Several papers have recently put forward evidence for recent origin and population bottlenecks in the evolution of modern humans. Since this evidence is being discussed elsewhere in this volume, I will only mention some of the results here. Brown (1980) described low mtDNA sequence diversity for a small number of human samples, and he interpreted that as indicating a severe bottleneck in human evolution 180–360 Kyr ago. Avise et al. (1984) challenged this view with a model showing the possibility of stochastic extinction of mtDNA lineages in nonexpanding populations. Nei (1982) used 62 protein and 23 blood group loci to calculate proportions of polymorphic loci and average heterozygosity per locus for three races of humans, and he found that 90% of genetic diversity occurred within rather than between the populations, and that the negroid population was more distinct from the mongoloid and caucasoid groups than either was from the other. On the basis of protein evidence alone, he suggested dates of 26–56 Kyr for the separation of the caucasoid and mongoloid populations, and 79–150 Kyr for the separation of the negroids from them.

The evidence indicating an initial split between African populations and the rest (Nei 1982) has been greatly augmented by recent DNA sequencing. Five linked polymorphic restriction enzyme sites in the β-globin gene cluster have been examined for eight diverse human populations (Wainscoat et al. 1986) and these show a basic split between African and non-African populations, with the non-African populations losing the distinctive African haplotype through founder effect. No date is given for either the initial branching or for the bottleneck, but the higher genetic diversity of African populations may indicate a greater age for them.

A date is given by Cann et al. (1986) when they demonstrate similar results from mtDNA sequencing. They give a date of 140–290 Kyr for the origin of *Homo sapiens* based on 134 types of mtDNA and a mean rate of DNA divergence of 2–4% per million years. The initial split is again between African and non-African populations. The oldest cluster of human samples with no African members has an age range of 90–180 Kyr, and the inference here is that the ancestor must have left Africa by then (or left no African descendants). These dates from protein and DNA, and the evidence of African origin for the species, are entirely consistent with the model based on fossil evidence just presented and are strongly at variance with other interpretations of fossil evidence which take the origin of *Homo sapiens* back to 500 Kyr or more.

ACKNOWLEDGMENTS

I benefitted greatly from discussion with Allan Wilson and Chris Stringer while preparing this paper, and I am also grateful to Chris Stringer and Lawrence Martin for comments on the manuscript.

REFERENCES

Andrews, P. 1981. Species diversity and diet in monkeys and apes during the Miocene. In *Aspects of human evolution* (ed. C.B. Stringer), p. 25. Taylor and Francis, London.

———. 1984. An alternative interpretation of the characters used to define *Homo erectus*. *Cour. Forschungsinst. Senckenb.* **69:** 167.

———. 1986a. Molecular evidence for catarrhine evolution. In *Major topics in primate and human evolution* (ed. B. Wood et al.), p. 107. Cambridge University Press, Cambridge, England.

———. 1986b. Aspects of hominoid phylogeny. In *Molecules and morphology in evolution—Conflict or compromise* (ed. C. Patterson). Cambridge University Press, Cambridge, England. (In press.)

Andrews, P. and J. Cronin. 1982. The relationships of *Sivapithecus* and *Ramapithecus* and the evolution of the orang utan. *Nature* **297:** 541.

Andrews, P. and I. Tekkaya. 1980. A revision of the Turkish Miocene hominoid *Sivapithecus meteai*. *Palaeontology* **23:** 85.

Andrews, P. and H. Tobien. 1977. New Miocene locality in Turkey with evidence on the origins of *Ramapithecus* and *Sivapithecus*. *Nature* **268:** 699.

Andrews, P., W.R. Hamilton, and R.J. Whybrow. 1978. Dry-pithecines from the Miocene of Saudi Arabia. *Nature* **274:** 249.

Andrews, P., T. Harrison, L. Martin, and M. Pickford. 1981. Hominoid primates from a new Miocene locality named Meswa Bridge in Kenya. *J. Hum. Evol.* **19:** 123.

Avise, J.C., J.E. Neigle, and J. Arnold. 1984. Demographic influences on mtDNA lineage survivorship in animal populations. *J. Mol. Evol.* **20:** 99.

Barry, J.C., N.M. Johnson, S.M. Raza, and L. Jacobs. 1985. Neogene mammalian faunal change in southern Asia: Correlations with climatic, tectonic and eustatic events. *Geology* **13:** 637.

Bishop, M.J. and A.E. Friday. 1985. Molecular sequences and hominoid phylogeny. In *Major topics in primate and human evolution* (ed. B. Wood et al.), p. 150. Cambridge University Press, Cambridge, England.

Brown, F., J. Harris, R.E. Leakey, and A.C. Walker. 1985.

Early *Homo erectus* skeleton from west Lake Turkana, Kenya. *Nature* **316:** 788.
Brown, W.M. 1980. Polymorphism in mtDNA of humans revealed by restriction endonuclease analysis. *Proc. Natl. Acad. Sci.* **77:** 3605.
Brown, W.M., E.M. Prager, A. Wang, and A.C. Wilson. 1982. Mitochondrial DNA sequences of primates: Tempo and mode of evolution. *J. Mol. Evol.* **18:** 225.
Cann, R.L., M. Stoneking, and A.C. Wilson. 1986. Mitochondrial DNA and human evolution. *Nature* (in press).
Cook, J., C.B. Stringer, A.P. Currant, H.P. Schwarcz, and A.G. Wintle. 1982. A review of the chronology of the European Middle Pleistocene hominid record. *Yearb. Phys. Anthropol.* **25:** 19.
Corruccini, R.S., M. Baba, M. Goodman, R. Ciochon, and J.E. Cronin. 1980. Non-linear macromolecular evolution and the molecular clock. *Evolution* **34:** 1216.
Darwin, C. 1871. *The descent of man and selection in relation to sex.* Murray, London.
Delson, E. 1975. Evolutionary history of the Cercopithecidae. In *Approaches to primate paleobiology* (ed. F.S. Szalay), p. 167. Karger, Basel.
Ferris, S.D., A.C. Wilson, and W.M. Brown. 1981a. Evolutionary tree for apes and humans based on cleavage maps of mtDNA. *Proc. Natl. Acad. Sci.* **78:** 2432.
Ferris, S.D., W.M. Brown, W.S. Davidson, and A.C. Wilson. 1981b. Extensive polymorphism in the mtDNA of apes. *Proc. Natl. Acad. Sci.* **78:** 6319.
Gingerich, P.D. 1984. Primate evolution: Evidence from the fossil record, comparative morphology and molecular biology. *Yearb. Phys. Anthropol.* **27:** 57.
Goodman, M., G. Braunitzer, A. Stangl, and B. Schrank. 1983. Evidence on human origins from haemoglobins of African apes. *Nature* **303:** 546.
Goodman, M., B.F. Koop, J. Czelusniak, M.L. Weiss, and J.L. Slightom. 1984. The η globin gene, its long evolutionary history in the beta globin gene family of mammals. *J. Mol. Evol.* **180:** 803.
Groves, C.P. and V. Mazak. 1975. An approach to the taxonomy of the Hominidae: Gracile Villafranchian hominids of Africa. *Cas. Min. Geol.* **20:** 225.
Harrison, T. 1982. "Small bodied apes from the Miocene of East Africa." Ph.D. thesis, University College, London.
Hasegawa, M., H. Krishino, and T. Yano. 1985. Dating of the human-ape splitting by a molecular clock of DNA. *J. Mol. Evol.* **22:** 160.
Hill, A. 1985. Early hominid from Baringo, Kenya. *Nature* **315:** 222.
Howell, F.C. 1978. Hominidae. In *Evolution of African mammals* (ed. V.C. Maglio and H.B.S. Cooke), p. 154. Harvard University Press, Cambridge, Massachusetts.
Hublin, J.J. 1982. Les Anteneandertaliens: Presapiens ou preneandertaliens. *Geobios Mem. Spec.* **6:** 345.
Huxley, T.H. 1893. *Evidence as to man's place in nature.* Williams and Norgate, London.
Ishida, H., M. Pickford, H. Nakaya, and Y. Nakaya. 1984. Fossil anthropoids from Nachola and Samburu Hills, Samburu District, Kenya. *Afr. Study Monogr.* **2:** 73.
Johanson, D.C., M. Taieb, and Y. Coppens. 1982. Pliocene hominids from the Hadar Formation, Ethiopia. *Am. J. Phys. Anthropol.* **57:** 373.
Koop, B.F., M. Goodman, P. Xu, K. Chan, and J.L. Slightom. 1986. Primate η-globin DNA sequences and man's place among the great apes. *Nature* **319:** 234.
Leakey, L.S.B. 1962. A new lower Pliocene fossil primate from Kenya. *Ann. Mag. Nat. Hist.* **4:** 689.
Leakey, M.G. and R.E. Leakey. 1978. *The Koobi Fora research project: The fossil hominids and an introduction to their context, 1968--1974.* Oxford University Press, Oxford.
Leakey, R.E. and A.C. Walker. 1976. *Australopithecus, Homo erectus* and the single species hypothesis. *Nature* **261:** 572.
———. 1985. New higher primates from the early Miocene of Buluk, Kenya. *Nature* **318:** 173.
Marks, J. 1983. Hominoid cytogenetics and evolution. *Yearb. Phys. Anthropol.* **26:** 131.
Martin, L. 1985. Significance of enamel thickness in hominoid evolution. *Nature* **314:** 260.
Nei, M. 1982. Evolution of human races at the gene level. In *Human genetics, Part A: The unfolding genome* (ed. B. Bonné-Tamis et al.), p. 167. A.R. Liss, New York.
Patterson, B., A.K. Behrensmeyer, and W.D. Gill. 1970. Geology and fauna of a new Pliocene locality in Northwestern Kenya. *Nature* **226:** 918.
Pilbeam, D.R. 1982. New hominoid skull material from the Miocene of Pakistan. *Nature* **295:** 232.
Raza, S.M., J.C. Barry, D.R. Pilbeam, M.D. Rose, S.M.I. Shah, and S. Ward. 1983. New hominoid primates from the middle Miocene Chinji Formation, Potwar Plateau, Pakistan. *Nature* **306:** 52.
Rak, Y. 1983. *The Australopithecine face.* Harvard University Press, Cambridge, Massachusetts.
Read, D.W. 1975. Primate phylogeny, neutral mutations and "molecular clocks." *Syst. Zool.* **24:** 209.
Read, D.W. and P.E. Lestrel. 1970. Hominid phylogeny and immunology: A critical appraisal. *Science* **168:** 578.
Sibley, C.G. and J.E. Ahlquist. 1984. The phylogeny of the hominoid primates, as indicated by DNA–DNA hybridization. *J. Mol. Evol.* **20:** 2.
Simpson, G.G. 1951. Some principles of historical biology bearing on human origins. *Cold Spring Harbor Symp. Quant. Biol.* **15:** 55.
Stringer, C.B. 1984. The definition of *Homo erectus* and the existence of the species in Africa and Europe. *Cour. Forschungsinst. Senckenb.* **69:** 131.
———. 1985. Middle Pleistocene variability and the origin of late Pleistocene humans. In *Ancestors: The hard evidence* (ed. E. Delson), p. 289. A.R. Liss, New York.
———. 1986. Documenting the origin of modern humans. In *The origins of modern humans* (ed. E. Trinkhaus). Cambridge University Press, Cambridge, England. (In press.)
Templeton, A.R. 1983. Phylogenetic inference from restriction endonuclease cleavage site maps with particular reference to the evolution of humans and apes. *Evolution* **37:** 221.
———. 1985. The phylogeny of the hominoid primates: A statistical analysis of the DNA–DNA hybridization data. *Mol. Biol. Evol.* **2:** 420.
Vandermeersch, B. 1985. The origin of the Neanderthals. In *Ancestors: The hard evidence* (ed. E. Delson), p. 306. A.R. Liss, New York.
Wainscoat, J.S., A.V.S. Hill, A.L. Boyce, J. Flint, M. Hernandez, S.L. Theim, J.M. Old, J.R. Lynch, A.G. Falusi, D.J. Weatherall, and J.B. Clegg. 1986. Evolutionary relationships of human populations from an analysis of nuclear DNA polymorphisms. *Nature* **319:** 491.
Walker, A.C. and R.E. Leakey. 1978. The hominids of East Turkana. *Sci. Am.* **239:** 44.
Ward, S. and D.R. Pilbeam. 1983. Maxillofacial morphology of Miocene hominoids from Africa and Indo-Pakistan. In *New interpretations of ape and human ancestry* (ed. R.L. Ciochon and R.S. Corruccini), p. 211. Plenum Press, New York.
White, T.D. 1977. New fossil hominids from Laetolil, Tanzania. *Am. J. Phys. Anthropol.* **46:** 197.
Wilson, A.C., R.L. Cann, S.M. Carr, M. George, U.B. Gyllensten, K.M. Helm-Bychowsky, R.G. Higuchi, S.R. Palumbi, E.M. Prager, R.D. Sage, and M. Stoneking. 1985. Mitochondrial DNA and two perspectives on evolutionary genetics. *Biol. J. Linn. Soc.* **26:** 375.
Wood, B.A. 1984. The origin of *Homo erectus. Cour. Forschungsinst. Senckenb.* **69:** 99.
Yunis, J.J. and O. Prakash. 1982. The origin of man: A chromosomal pictorial legacy. *Science* **215:** 1525.

Hominoid Evolution Based on the Structures of Immunoglobulin Epsilon and Alpha Genes

S. UEDA,* Y. WATANABE,* H. HAYASHIDA,† T. MIYATA,† F. MATSUDA,‡ AND T. HONJO‡

*Department of Anthropology, Faculty of Science, The University of Tokyo, Tokyo 113, Japan; †Department of Biology, Faculty of Science, Kyushu University, Fukuoka 812, Japan; and ‡Department of Medical Chemistry, Kyoto University Faculty of Medicine, Kyoto 606, Japan

The organization of the heavy-chain constant (C_H) region genes of the human immunoglobulin (Ig) is 5'-C_μ-C_δ-$C_{\gamma3}$-$C_{\gamma1}$-$C_{\epsilon2}$-$C_{\alpha1}$....ΨC_γ....$C_{\gamma2}$-$C_{\gamma4}$-$C_{\epsilon1}$-$C_{\alpha2}$-3', located on chromosome 14 (Ellison and Hood 1982; Flanagan and Rabbitts 1982; Krawinkel and Rabbitts 1982; Lefranc et al. 1982; Max et al. 1982; Nishida et al. 1982; Takahashi et al. 1982; Bech-Hansen et al. 1983; Hisajima et al. 1983; Flanagan et al. 1984; Migone et al. 1984). In addition to this cluster, the human genome contains a processed C_ϵ pseudogene ($C_{\epsilon3}$) on chromosome 9 (Battey et al. 1982; Ueda et al. 1982). The human Ig C_ϵ gene family thus consists of lists of three members: the $C_{\epsilon1}$ gene (active), the $C_{\epsilon2}$ gene (truncated pseudogene), and the $C_{\epsilon3}$ gene (processed pseudogene). Since mouse contains only one C_ϵ gene (Shimizu et al. 1982), creation of two C_ϵ pseudogenes in the human genome seems to have taken place after mammalian radiation. Using the human C_ϵ probe specific to each family member, we have recently found that Old World monkeys have two C_ϵ genes, the $C_{\epsilon1}$ and $C_{\epsilon3}$ genes. Among the hominoids only the gorilla and human genomes contain the three C_ϵ genes, whereas other hominoids including the chimpanzee have the $C_{\epsilon1}$ and $C_{\epsilon3}$ genes but not the $C_{\epsilon2}$ gene (Ueda et al. 1985). The results indicate two alternative possibilities: (1) gorilla is more closely related to man than chimpanzee is, or (2) chimpanzee has lost the $C_{\epsilon2}$ gene after divergence of this species.

To distinguish these possibilities we have isolated DNA fragments containing the $C_{\epsilon1}$, $C_{\epsilon2}$, $C_{\alpha1}$, and $C_{\alpha2}$ genes from gorilla and chimpanzee DNAs. Characterization of these DNA segments by nucleotide sequence determination demonstrated that chimpanzee had deleted the entire $C_{\epsilon2}$ gene after its divergence. In addition, comparison of nucleotide sequences of the $C_{\epsilon3}$ pseudogenes of man, gorilla, and chimpanzee allowed us to obtain statistically different branching nodes of gorilla and chimpanzee from the human lineage. Such studies indicate that man is more closely related to chimpanzee than to gorilla.

Only the Human and the Gorilla Genomes Contain the Three C_ϵ Genes

The organization of the C_ϵ genes in DNAs from various species of primates was studied by the Southern hybridization method using the human C_ϵ gene fragments as probes. The species of nonhuman primates examined include 13 species of Cercopithecoidea (Old World monkeys) and five species of Hominoidea (hominoids). As summarized in Table 1, two interesting results were obtained: (1) the processed C_ϵ gene is present in all the catarrhines (Old World monkeys and hominoids) examined and (2) among the hominoids only the gorilla and the human genomes contained the three C_ϵ genes whereas other hominoids, including chimpanzee, had the $C_{\epsilon1}$ and the $C_{\epsilon3}$ genes but not the $C_{\epsilon2}$ gene (Ueda et al. 1985). There was no variation in Southern hybridization patterns among 11 individual chimpanzee DNAs examined.

The first result suggests that the processed C_ϵ gene evolved before the divergence between Old World monkeys and hominoids. The second result appears to be inconsistent with the currently popular hominoid phylogenetic trees: (1) chimpanzee is more closely related to man than gorilla is (Sibley and Ahlquist 1984; Koop et al. 1986), (2) the African apes (chimpanzee and gorilla) are equally distant from man and closer to each other (Ferris et al. 1981; Brown et al. 1982), or (3) the three species are equally distant (Benveniste and Todaro 1976; Andrews 1982). If one assumes these evolutionary trees, the $C_{\epsilon2}$ gene in the chimpanzee genome must have been deleted after the divergence of its ancestor containing three C_ϵ genes.

Alternatively, our findings suggest that gorilla is more closely related to man than chimpanzee is. In this case duplication of the C_ϵ gene followed by truncation

Table 1. Organization of the C_ϵ Genes in Primates

Species	$C_{\epsilon1}$ gene (active)[a]	$C_{\epsilon2}$ gene (truncated)[a]	$C_{\epsilon3}$ gene (processed)[a]
Man	2.7	5.9	8.0
Gorilla	2.7	6.9	15
Chimpanzee	2.7	–	8.0
Orangutan	2.7	–	7.0
Gibbons	2.7	–	5.6
Macaques	2.7	–	7.1
Baboons	2.7	–	7.1
Gelada	2.7	–	7.1
Patas monkey	7.4	–	7.1
Green monkey	7.6	–	7.1

[a]Numbers indicate sizes of *Bam*HI fragments in kilobases.

in one ($C_{\epsilon2}$ gene) of the duplicated C_ϵ genes would have occurred during a relatively short period of time between divergence of chimpanzee and gorilla from the human lineage. It is difficult to explain the presence of the truncated $C_{\epsilon2}$ gene in gorilla by polymorphism in this species because this involves two genetic events, duplication and truncation. The third alternative is the independent evolution of the truncated C_ϵ gene in both man and gorilla. We think that this possibility is the least likely among the three because of the same reason, described above, excluding polymorphism in gorilla.

Chimpanzee Has Lost the $C_{\epsilon2}$ Gene Completely

To distinguish three possibilities described above, we have isolated DNA fragments containing the $C_{\epsilon1}$, $C_{\epsilon2}$, $C_{\alpha1}$, and $C_{\alpha2}$ genes from gorilla and chimpanzee DNAs. Characterization of these DNA segments by nucleotide sequence determination demonstrated that the coding sequence of the $C_{\epsilon2}$ gene was completely deleted from the chimpanzee genome. The gorilla $C_{\epsilon2}$ gene had the 3′ half of the CH2 exon, and the CH3 and CH4 exons. The deletion of the $C_{\epsilon2}$ began approximately 2.1 kb upstream of the CH1 exon in both species, and ended approximately 1 kb downstream to the CH4 exon in the chimpanzee $C_{\epsilon2}$ gene and at the 64th nucleotide of the CH2 exon in the gorilla $C_{\epsilon2}$ gene. In the human $C_{\epsilon2}$ gene the deletion begins at a similar position to the above two species but ends at the second nucleotide of the CH3 exon (Max et al. 1982; Hisajima et al. 1983). The nucleotide sequences upstream to the deletion points of the chimpanzee and the gorilla $C_{\epsilon2}$ genes matched well with those of the corresponding region of the human $C_{\epsilon2}$ gene. In summary, the starting points of the deletion are roughly identical but the ending points of the deletion are different among the three species. In the genomes of all the three species, the $C_{\epsilon1}$ and $C_{\epsilon2}$ genes had been linked to the $C_{\alpha2}$ and $C_{\alpha1}$ genes, respectively, although the $C_{\epsilon2}$ exons were completely deleted from the chimpanzee genome. These results indicated that multiple recombinational events took place in the Ig C_ϵ and C_α genes during hominoid evolution.

The hinge region of the human $C_{\alpha1}$ gene consists of two tandem repeats of a 30-bp unit with 6-bp overlap (Flanagan et al. 1984). The 30-bp unit can be divided into two tandem 15-bp units. The hinge region of the human $C_{\alpha2}$ gene consists of the 15-bp unit. The $C_{\alpha2}$ genes of man, chimpanzee, and gorilla have the homologous hinge regions. On the other hand, the hinge regions of the human and the chimpanzee $C_{\alpha1}$ genes are almost identical to each other but different in length from that of the gorilla $C_{\alpha1}$ gene.

Although the branching patterns of the lineages and the datings of the divergence nodes among man and the African apes are still in dispute, the structural comparison of the $C_{\alpha1}$ hinge regions of the three species suggests that chimpanzee is more closely related to man than gorilla. Assuming that it is the case, the most simple evolutionary development of the Ig C_ϵ and C_α genes is shown in Figure 1. Organization of the human heavy-chain C_H genes suggests that duplication of DNA containing the C_γ, C_ϵ, and C_α genes took place after divergence from mouse in the common ancestor of the three species. Then, deletion of a part of the hinge region (15-bp unit) took place in the $C_{\alpha2}$ gene. After divergence of gorilla, there was duplication of the hinge-coding region (30-bp unit) in the $C_{\alpha1}$ gene of the common ancestor of man and chimpanzee. Moreover, deletion of $C_{\epsilon2}$ gene exons occurred independently in the individual lineages. The present study has shown that DNA rearrangement is a good marker to trace evolutionary lineage, as we have proposed previously (Ueda et al. 1985).

Comparison of Nucleotide Sequences of the $C_{\epsilon3}$ Genes

In spite of a large amount of sequence information for primate DNA, phylogenetic relationships among these three species (man, chimpanzee, and gorilla) and their divergence dates are still in dispute. This is because the three species are so closely related that the divergence of the gene sequences is too small to be quantitated accurately. The degrees of the sequence differences with the statistical standard errors so far estimated overlap each other (Koop et al. 1986). Because pseudogenes diverge faster than active genes, we determined the complete nucleotide sequences of the processed C_ϵ pseudogenes of chimpanzee and gorilla and compared them with that of the human $C_{\epsilon3}$ gene. The basic structures of these $C_{\epsilon3}$ genes are identical. The $C_{\epsilon3}$ genes deleted the three introns precisely at the splicing signals and had adenine-rich sequences 16 bp 3′ to the putative poly(A) additional signal. Moreover, the 100-bp segment located in the 5′-flanking region of the $C_{\epsilon1}$ gene, which is interrupted by a spacer region in the $C_{\epsilon1}$ gene, was translocated immediately 5′ to the pseudo-CH1 exons of the $C_{\epsilon3}$ genes.

In calculating the nucleotide difference (K), we treated all regions as noncoding regions. The nucleotide difference K was corrected for superimposed substitutions by the formula: $K^C = -(3/4)\ln[1-(4/3)K]$ (Kumura and Ohta 1972). The standard errors were evaluated as $\sqrt{K^C/N}$, where N was the number of sites compared. The K^C value for human versus chimpanzee (0.0178 ± 0.0027) was the smallest among the three combinations. On the other hand, that for human versus gorilla (0.0266 ± 0.0033) was nearly equal to that for chimpanzee versus gorilla (0.0258 ± 0.0032). These values were 1.5 times greater than that for human versus chimpanzee. Of 75 substitutions in total, 39 were common between man and chimpanzee, 20 between chimpanzee and gorilla, and 16 between man and gorilla, respectively. From these results, we constructed the phylogenetic tree as shown in Figure 2. Our results show that chimpanzee is more closely related to man than gorilla without overlapping the ranges of the K^C values. Using the evolutionary divergence rate estimated for the primate η-globin pseudogene (1.55×10^{-9}

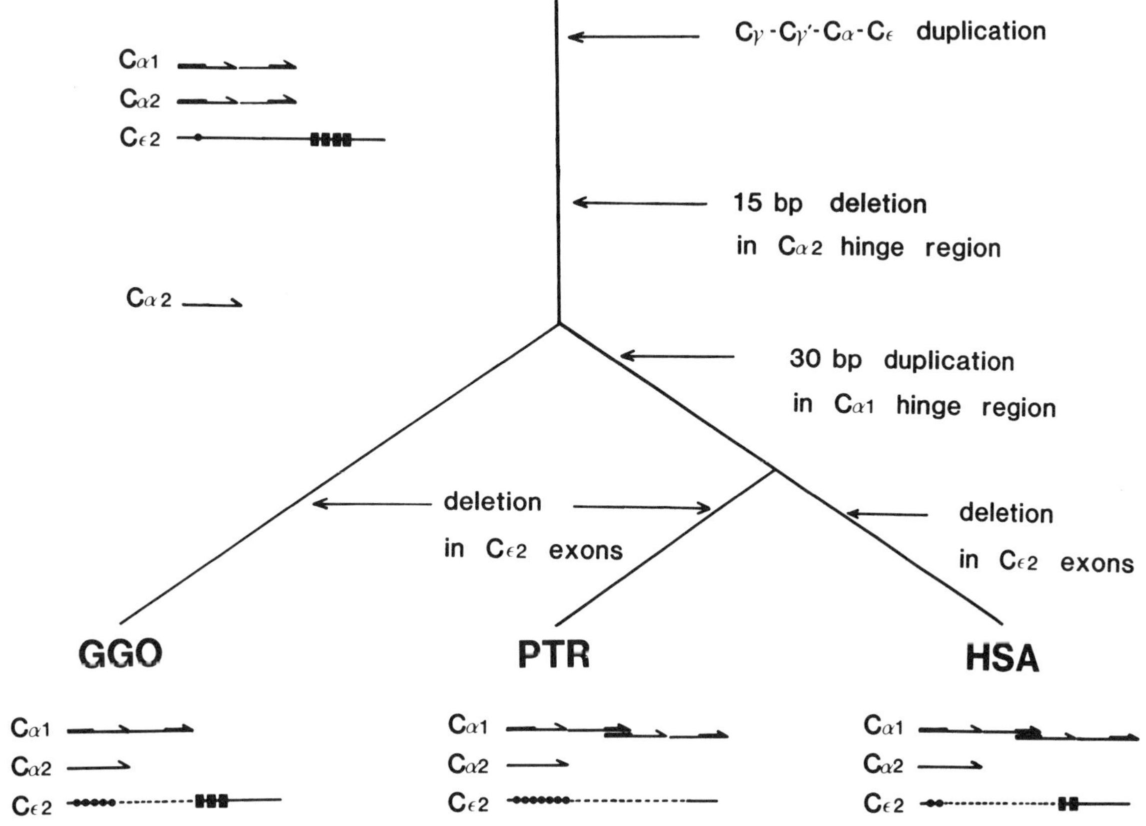

Figure 1. Evolutionary order of rearrangements in the C_ϵ and C_α genes of man, gorilla, and chimpanzee. Genetic events in the C_ϵ and C_α genes are ordered according to the phylogenetic tree of man, gorilla, and chimpanzee as described in Fig. 2. Schematic structures of the hinge regions of the C_α genes are shown by horizontal arrows representing 15-bp units. Conserved 6-bp were indicated by thicker lines. Structures of the $C_{\epsilon 2}$ genes are shown by horizontal lines with closed rectangles and circles which indicate exons and 35-bp repeating units, respectively. Broken lines show deletions in the $C_{\epsilon 2}$ genes and their flanking regions. (HSA) Man; (PTR) chimpanzee; and (GGO) gorilla.

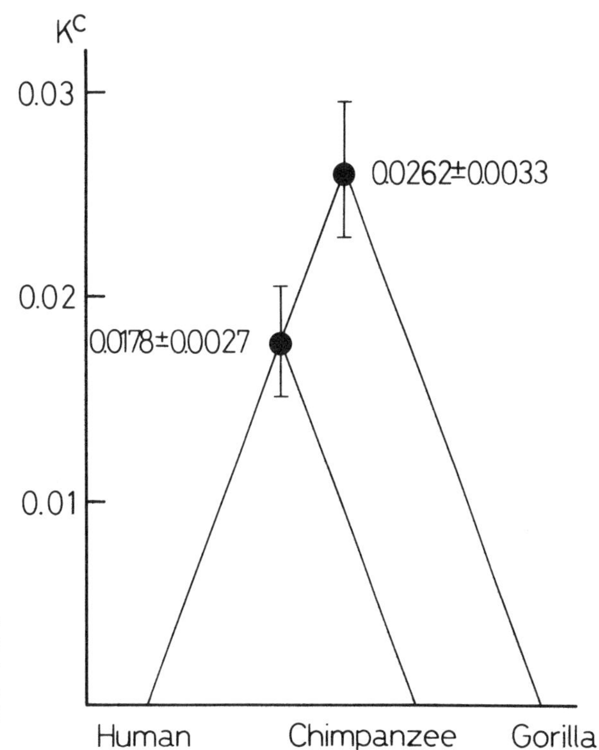

Figure 2. Phylogenetic relationship among man, chimpanzee, and gorilla. The numbers show the nucleotide differences corrected for superimposed substitutions (K^C values) and their standard errors. The K^C value for human–chimpanzee versus gorilla was the average of that between man and gorilla and that between chimpanzee and gorilla.

nucleotide difference per site per year), the divergence dates of chimpanzee and gorilla from human lineage were estimated to be 5.9 ± 0.9 and 8.5 ± 1.1 million years ago, respectively.

ACKNOWLEDGMENT

We are grateful to Prof. K. Omoto, the University of Tokyo, and Prof. O. Takenaka, Kyoto University, for their continuous encouragement. This study was supported in part by grants from the Ministry of Education, Science, and Culture and by a cooperative research grant from Primate Research Institute, Kyoto University.

REFERENCES

Andrews, P. 1982. Hominoid evolution. *Nature* **295:** 185.

Battey, J., E.E. Max, W.O. McBride, D. Swan, and P. Leder. 1982. A processed human immunoglobulin ϵ gene has moved to chromosome 9. *Proc. Natl. Acad. Sci.* **79:** 5956.

Bech-Hansen, N.T., P.S. Linsley, and D.W. Cox. 1983. Restriction fragment length polymorphisms associated with immunoglobulin C_γ genes reveal linkage disequilibrium and genomic organization. *Proc. Natl. Acad. Sci.* **80:** 6952.

Benveniste, R.E. and G.J. Todaro. 1976. Evolution of type C viral genes: Evidence for an Asian origin of man. *Nature* **261:** 101.

Brown, W.M., E.M. Prager, A. Wang, and A.C. Wilson. 1982. Mitochondrial DNA sequences of primates: Tempo and mode of evolution. *J. Mol. Evol.* **18:** 225.

Ellison, J. and L. Hood. 1982. Linkage and sequence homology of two human immunoglobulin γ heavy chain constant region genes. *Proc. Natl. Acad. Sci.* **79:** 1984.

Flanagan, J.G. and T.H. Rabbitts. 1982. Arrangement of human immunoglobulin heavy chain constant region genes implies evolutionary duplication of a segment containing γ, ϵ, and α genes. *Nature* **300:** 709.

Flanagan, J.G., M. Lefranc, and T.H. Rabbitts. 1984. Mechanisms of divergence and convergence of the human immunoglobulin $\alpha 1$ and $\alpha 2$ constant region gene sequences. *Cell* **36:** 681.

Ferris, S.D., A.C. Wilson, and W.M. Brown. 1981. Evolutionary tree for apes and humans based on cleavage maps of mitochondrial DNA. *Proc. Natl. Acad. Sci.* **78:** 2432.

Hisajima, H., Y. Nishida, S. Nakai, N. Takahasi, S. Ueda, and T. Honjo. 1983. Structure of the human immunoglobulin $C_{\epsilon 2}$ gene, a truncated pseudogene; implications for its evolutionary origin. *Proc. Natl. Acad. Sci.* **80:** 2995.

Kimura, M. and T. Ohta. 1972. On the stochastic model for estimation of mutational distance between homologous proteins. *J. Mol. Evol.* **2:** 87.

Koop, B.F., M. Goodman, P. Zu, K. Chan, and J.L. Slightom. 1986. Primate η-globin DNA sequences and man's place among the great apes. *Nature* **319:** 234.

Krawinkel, U. and T.H. Rabbitts. 1982. Comparison of the hinge-coding segments in human immunoglobulin gamma heavy chain genes and the linkage of the gamma 2 and gamma 4 subclass genes. *EMBO J.* **1:** 403.

Lefranc, M., G. Lefranc, and T.H. Rabbitts. 1982. Inherited deletion of immunoglobulin heavy chain constant region genes in normal human individuals. *Nature* **300:** 760.

Max, E.E., J. Battey, R. Ney, I.R. Kirsh, and P. Leder. 1982. Duplication and deletion in human immunoglobulin ϵ genes. *Cell* **29:** 691.

Migone, N., S. Oliviero, G. Lange, D.L. Delacroix, D. Boschis, F. Altruda, L. Silengo, M. DeMarchi, and A.O. Carbonara. 1984. Multiple gene deletions within the human immunoglobulin heavy-chain cluster. *Proc. Natl. Acad. Sci.* **81:** 5811.

Nishida, Y., T. Miki, H. Hisajima, and T. Honjo. 1982. Cloning of human immunoglobulin ϵ chain genes: Evidence for multiple C_ϵ genes. *Proc. Natl. Acad. Sci.* **79:** 3833.

Shimizu, A., N. Takahashi, Y. Yaoita, and T. Honjo. 1982. Organization of the constant-region gene family of the mouse immunoglobulin heavy chain. *Cell* **28:** 499.

Sibley, C.G. and J.E. Ahlquist. 1984. The phylogeny of the hominoid primates, as indicated by DNA-DNA hybridization. *J. Mol. Evol.* **20:** 2.

Takahashi, N., S. Ueda, M. Obata, T. Nikaido, S. Nakai, and T. Honjo. 1982. Structure of human immunoglobulin gamma genes: Implications for evolution of a gene family. *Cell* **29:** 671.

Ueda, S., O. Takenaka, and T. Honjo. 1985. A truncated immunoglobulin ϵ pseudogene is found in gorilla and man but not in chimpanzee. *Proc. Natl. Acad. Sci.* **82:** 3712.

Ueda, S., S. Nakai, Y. Nishida, H. Hisajima, and T. Honjo. 1982. Long terminal repeat-like elements flank a human immunoglobulin epsilon pseudogene that lacks introns. *EMBO J.* **1:** 1539.

Rate of Sequence Divergence Estimated from Restriction Maps of Mitochondrial DNAs from Papua New Guinea

M. STONEKING,* K. BHATIA,† AND A.C. WILSON*
*Department of Biochemistry, University of California, Berkeley, California 94720;
†Papua New Guinea Institute of Medical Research, Goroka, Papua New Guinea

mtDNA, by virtue of its strictly maternal mode of inheritance and rapid rate of evolution (Wilson et al. 1985), is a source of new perspectives on the origin and history of our species (Brown 1980; Johnson et al. 1983; Cann et al. 1984, 1986; Horai et al. 1984; Cann 1985; Wallace et al. 1985; Bonné-Tamir et al. 1986; Horai and Matsunaga 1986; Stoneking et al. 1986). Among the most provocative findings to arise from these studies is the suggestion by Cann et al. (1986) that all of the mtDNA diversity in modern human populations can be traced back to a common female ancestor who lived in Africa 150,000–300,000 years ago.

An estimate of the absolute rate of human mtDNA sequence divergence was required to arrive at this date. Cann et al. (1986) used an estimate of 2–4% per million years, which came in part from interspecies comparisons of a variety of mammals, including humans (Brown et al. 1979; Wilson et al. 1985). However, the applicability of this rate to human mtDNA evolution can be questioned on the basis of claims that the rate of DNA sequence evolution has slowed in hominoids and particularly in hominids, for both nuclear DNA (Britten 1986; Koop et al. 1986) and mtDNA (Templeton 1983, 1985; but see Nei and Tajima 1985 for rebuttal). Cann et al. (1986) interpreted the origin and spread of mtDNA diversity out of Africa as possibly reflecting an African origin for *Homo sapiens*, followed by later spread from Africa. If, however, the rate were lower, the mtDNA results might actually correspond to the spread of *Homo erectus* from Africa, beginning approximately 1 million years ago (Pope 1983; Stringer 1984). A calibration of the rate of mtDNA evolution, specifically for the human species, is obviously needed. Fortunately, human populations probably afford the best opportunity to arrive at an intraspecies calibration, since more historical information is available for human populations than for populations of any other species, due to the efforts of archaeologists and physical anthropologists.

To provide such a calibration, it is necessary to study mtDNA variation in a human population that colonized a specific region at a defined time and remained in relative isolation following colonization. It is particularly important that migration be unidirectional, i.e., that migration back toward the source population would not have occurred to any significant extent. Thus, mtDNA types that are specific to a particular geographic region will reflect the primary colonization of that area. Although no human population will satisfy these requirements exactly, aboriginal populations of Papua New Guinea provide a reasonable approximation.

This paper presents a calibration of the rate of human mtDNA evolution based on mtDNA variation in Papua New Guinea, the first such estimate based entirely on within-species comparisons. Published data on mtDNA restriction site polymorphisms are available for 26 individuals, largely from the Eastern Highlands of Papua New Guinea (Cann et al. 1986; Stoneking et al. 1986). The present study adds 29 more individuals from coastal and island areas of New Guinea. Phylogenetic analysis of the mtDNAs allows us to estimate the rate of human mtDNA evolution in two ways, (1) by considering the average divergence from the common ancestor within each New Guinea-specific cluster of mtDNA types and (2) by considering the average divergence between each New Guinea-specific cluster and the most closely related types outside New Guinea. We also present estimates derived in a similar fashion for Australian mtDNA (Cann 1985; Cann et al. 1986) and American Indian mtDNA (Wallace et al. 1985). These results all agree with the broadly defined rate of mtDNA evolution of 2–4% per million years, derived from comparisons of diverse vertebrate species (Wilson et al. 1985).

MATERIALS AND METHODS

mtDNA was purified to homogeneity from placentas of 55 aboriginal Papua New Guineans by differential ultracentrifigation through cesium chloride density gradients, as described by Brown et al. (1979) and Cann (1982). Twenty-six of these people were largely from the Eastern Highlands of Papua New Guinea (Stoneking et al. 1986); the additional 29 were from the Gulf Province (7), the Central Province (18), and the island of New Britain (4).

Purified mtDNAs were mapped by the sequence comparison method with the 12 restriction enzymes that we have employed before (Cann et al. 1984, 1986; Stoneking et al. 1986). High-resolution mapping of polymorphic restriction sites by this method is described in detail for New Guinea mtDNAs by Stoneking et al. (1986). Briefly, the precise location of a polymorphic restriction site was determined by comparing the variant restriction fragment pattern to the pattern predicted for the published reference sequence (Anderson et al. 1981).

Phylogenetic trees relating mtDNA types were built with the help of the computer program PAUP (obtained from D. Swofford). These are trees of minimal length, based on the parsimony principle, that account for the evolution of all of the mtDNA types from a common ancestor with the fewest changes (point mutations). Maximum likelihood estimates of sequence divergence between mtDNA types were made by the method of Nei and Tajima (1983).

RESULTS

Restriction Site Variation

The 12 restriction enzymes used recognized an average of 370 restriction sites per mtDNA genome. Thirty-five polymorphic restriction sites were identified among the 26 mtDNAs analyzed previously (Cann et al. 1986; Stoneking et al. 1986); an additional 15 polymorphisms were found among the 29 new samples. These 50 polymorphisms defined 30 mtDNA types, 17 of which occurred in the first 26 individuals; 13 new types were found in the 29 additional individuals. These 30 types were distributed among the 55 individuals as follows: 23 types occurred once, two types occurred twice, one type three times, one type five times, two types occurred six times each, and the last type was found in eight individuals. No individual possessed more than one mtDNA type, and none of these 30 types is known to occur outside New Guinea (Cann et al. 1986; Horai and Matsunaga 1986).

Phylogenetic Analysis

A phylogenetic tree of minimal length relating the 30 mtDNA types to 117 other types of human mtDNA from Africa, Asia, Europe, and Australia (Cann et al. 1986) is depicted in Figure 1A. This tree, constructed by a maximum parsimony algorithm, required 315 changes at 97 phylogenetically informative sites to account for the observed mtDNA diversity. Branch lengths reflect the averge sequence divergence (Nei and Tajima 1983) between mtDNA types. The common ancestor (root) was placed at the midpoint of the longest path connecting two mtDNA types, and thus assumes a constant rate of mtDNA sequence divergence. In interpreting this tree, we note that mtDNA is strictly maternally inherited and does not recombine. Therefore, the phylogenetic tree of mtDNA types represents a genealogy of maternal lineages in modern human populations.

This "new" tree in Figure 1A contains 14 clusters of mtDNA types specific to New Guinea. For comparative purposes we have included in Figure 1B the "old" tree relating the 17 mtDNA types in the original 26 New Guineans to the 117 mtDNA types from around the world (Cann et al. 1986). The old tree contains seven clusters of New Guinea-specific mtDNA types; the 13 additional New Guinea types fall into two of these clusters and define seven new clusters (Fig. 1).

Origin of New Guinea mtDNA Lineages

In all 14 instances the nearest relative of a New Guinea-specific cluster is either exclusively Asian or contains Asian mtDNA types. Australia, Africa, and Europe occur, together with Asia, as the nearest relative of a New Guinea-specific cluster nine, five, and ten times, respectively. Furthermore, of the four instances in which the nearest relative of a New Guinea-specific cluster consists of mtDNA types from a single region, all are from Asia. Since most of the Asian types come from Southeast Asia (Cann 1982), this area is the likely source of the New Guinea mtDNA types. There is no obvious association between New Guinea mtDNA types and Australian mtDNA types, implying that Australia was not colonized via New Guinea or vice versa (cf. Kirk 1982; Nei 1985).

Rate of Human mtDNA Evolution

The degree of divergence of each cluster of New Guinea mtDNA types from the nearest non-New Guinea mtDNA type, and the divergence of mtDNA types within each New Guinea-specific cluster, provide two estimates of the rate of sequence divergence for human mtDNA. Figure 2 illustrates the rationale behind this procedure. Node A represents the common ancestor of both a hypothetical New Guinea-specific cluster and a non-New Guinea mtDNA type, while node B is the immediate common ancestor of all of the mtDNA types within this New Guinea-specific cluster. The time corresponding to the colonizaton of New Guinea by this lineage is presumably after time t_2 (corresponding to node A) and before time t_1 (corresponding to node B). The amount of pairwise sequence divergence through node A (averaged across all individuals, not types, within a cluster) thus provides a maximal estimate of the extent of human mtDNA sequence divergence since the colonization of New Guinea. Likewise, the average sequence divergence through node B is a minimal estimate.

Table 1 presents the average pairwise sequence divergence through nodes A and B for each of the 14 New Guinea-specific clusters (Fig. 1A). Note that for the seven clusters represented by a single individual, the divergence through node B (divergence within a New Guinea-specific cluster) cannot be estimated. Either we have not discovered other mtDNA types that would be included in one of these seven clusters, or these lineages colonized New Guinea so recently that diversification within a lineage has not yet occurred. The estimate for average divergence within a New Guinea-specific cluster is 0.09% and that for average divergence from a non-New Guinea ancestor is 0.28% (Table 1).

Fossil and archaeological evidence for the presence of humans in New Guinea is limited; the earliest evi-

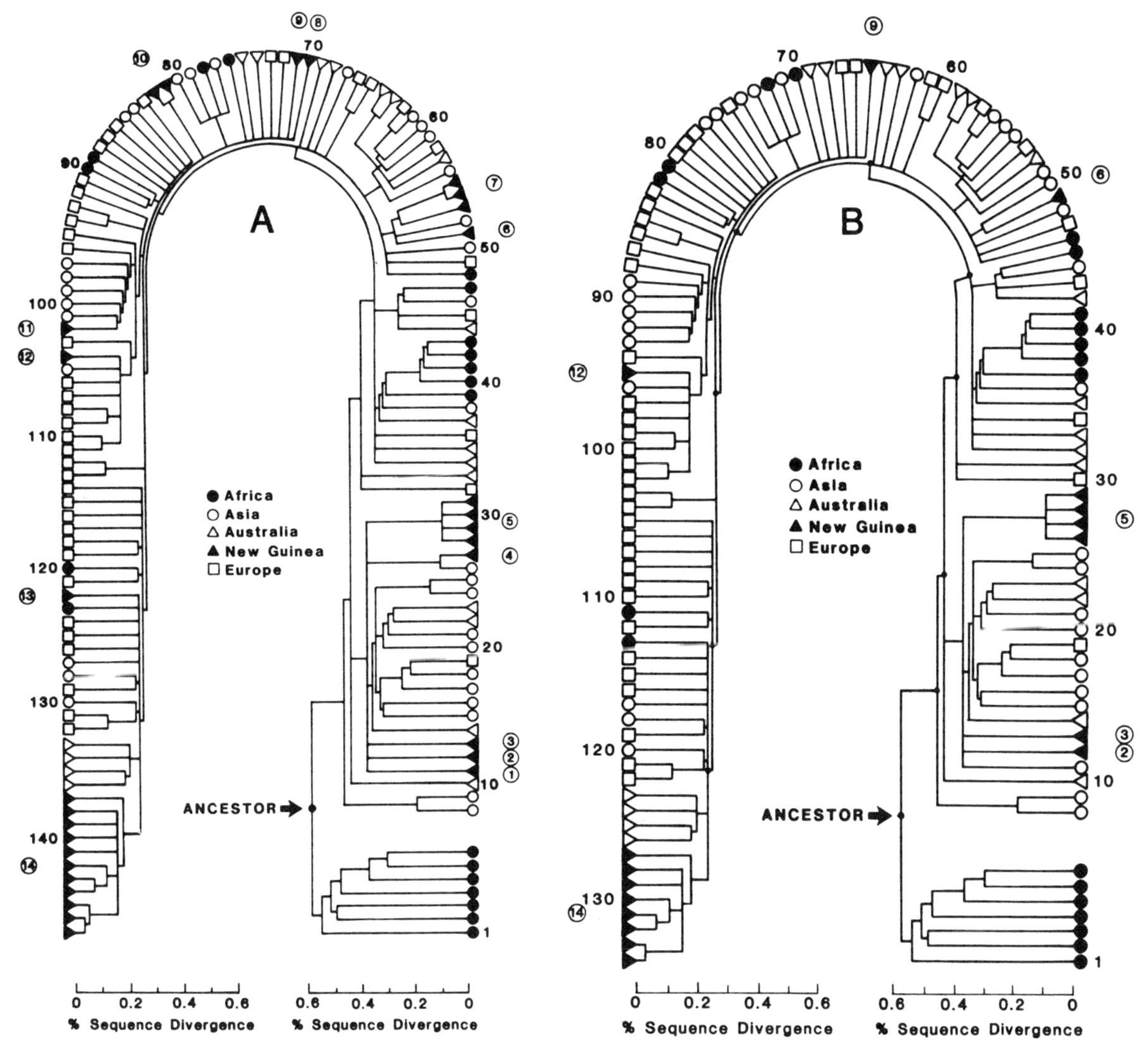

Figure 1. (*A*) Phylogenetic tree of minimal length relating 147 mtDNA types from five geographic regions. Circled numbers identify the 14 New Guinea-specific clusters. (*B*) Tree published by Cann et al. (1986) for 134 mtDNA types, including seven New Guinea-specific clusters identified as in Fig. 1A. These trees were produced by maximum parsimony analysis of the mtDNA restriction site data. Cann et al. (1986) provide a list of the polymorphic sites defining each of the original 134 mtDNA types; a similar list for the additional 13 mtDNA types (types 11, 27, 53–55, 70, 81–82, 102, 122, 138, 141, and 145) follows. Sites are numbered according to Anderson et al. (1981), with restriction enzymes indicated by the following single-letter code: (a) *Alu*I, (b) *Ava*II, (c) *Dde*I, (d) *Fnud*II, (e) *Hae*III, (f) *Hha*I, (g) *Hin*fI, (h) *Hpa*I, (i) *Hpa*II, (j) *Mbo*I, (k) *Rsa*I, (l) *Taq*I. Displacement loop—*207h* 138 141 145; 12S rRNA—*1043c* 55, *1403a* 11 27 53–55; 16S rRNA—1715*c* 138, *2390j* 53; cytochrome oxidase (CO) I—5983*g*/*5985k* 145, 6904*j* 81 82, 7103*c* 138; COII—7598*f* 53–55; ATPase 8—*8466a* 53; ATPase 6—8729*j* 53, 9053*f* 102; COIII—9553*e* 102; NADH dehydrogenase (ND) 3—*10394c* 11 27 53–55; ND4—10830*g* 54 55; ND5—*12345k* (*12350,12528*) 141 145, 12406*h* 102, 13065*c* 11; cytochrome *b*—15047*e* 138, *15606a* 138 141 145; displacement loop—16049*k* 53, 16096*k* 53, 16208*k* 27 141, 16303*k* 102 141, 16310*k* 11 27 70 102, *16389g*/16390*b* 53–55, 16517*e* 27 53 122 138 141 145. Italicized sites were present in the indicated mtDNA types; nonitalicized sites were absent in the indicated types; parentheses indicate alternative placements of inferred sites; and sites separated by a slash indicate polymorphisms for two different restriction enzymes caused by a single inferred nucleotide substitution.

dence dates back about 30,000 years (White and O'Connell 1982; Hope et al. 1983). However, since humans were known to be in southern Australia by 40,000 years ago (White and O'Connell 1982; Wolpoff et al. 1984), it has been suggested that colonization of New Guinea began some 50,000 years ago. With 30,000 years as a value for t_1 and t_2 and the averages from Table 1, we estimate that human mtDNA sequence divergence proceeds at a rate of 3.0–9.3% per million years. If 50,000 years is used instead, the estimated rate is 1.8–5.6% per million years.

Rate Estimated from Australian mtDNA

Similarly, one can estimate the rate of human mtDNA evolution from Australian-specific clusters. The trees in Figure 1 contain 15 clusters, consisting of 20 mtDNA types (from 21 Australian mtDNAs ana-

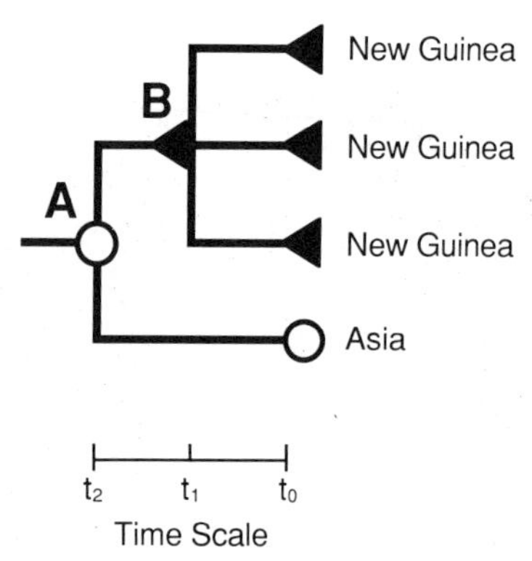

Figure 2. New Guinea-specific cluster of three mtDNA types and a related type from Asia. The common ancestor of the New Guinea cluster (B) lived at time t_1, and the common ancestor of all four types (A) lived at time t_2. For the purpose of calculating the rate of mtDNA evolution, it is assumed that New Guinea was colonized by an Asian lineage between t_2 and t_1 (see text).

lyzed). Table 2 gives the average divergence within each cluster and between each cluster and its nearest non-Australian mtDNA type. Using 40,000 years as an estimate for the colonization of Australia results in a rate of human mtDNA sequence divergence of 4.0–7.2% per million years.

Rate Estimated from American Indian mtDNA

Published data on mtDNA variation in American Indians (Wallace et al. 1985) can also be used to estimate the rate of human mtDNA evolution. Wallace and co-workers surveyed an average of 50 restriction sites per individual and found eight mtDNA types in 74 American Indians. In comparing these results to the New Guinean and Australian data, we note that sequence differences less than about 0.25% would not have been detected by Wallace et al. (1985), because of the small number of sites surveyed. By contrast, our study of approximately 370 sites per individual allowed us to detect sequence differences of 0.03%.

Table 1. Average Sequence Divergence between Each New Guinea-specific Cluster and Its Nearest Non-New Guinea Relative (A), and within Each New Guinea-specific Cluster (B)

Cluster	Number of types	Number of individuals	Divergence A (%)	Divergence B (%)
1	1	1	0.39	–
2	1	1	0.37	–
3	1	1	0.37	–
4	1	2	0.18	0.00
5	4	8	0.37	0.09
6	1	1	0.25	–
7	3	4	0.25	0.23
8	1	8	0.28	0.00
9	1	3	0.28	0.00
10	2	7	0.28	0.01
11	1	1	0.16	–
12	1	1	0.18	–
13	1	1	0.25	–
14	11	16	0.25	0.16
Average[a]			0.28	0.09

[a]Weighted by the number of individuals within a cluster.

Table 2. Average Sequence Divergence between Each Australia-specific Cluster and Its Nearest non-Australian Relative (A), and within Each Australia-specific Cluster (B)

Cluster	Number of types	Number of individuals	Divergence A (%)	Divergence B (%)
1–2	1	1	0.39	–
3	2	2	0.29	0.27
4–9	1	1	0.31	–
10	2	3	0.21	0.02
11–14	1	1	0.29	–
15	4	4	0.25	0.20
Average[a]			0.29	0.16

[a]Weighted by the number of individuals within a cluster.

Figure 3 shows a parsimony tree relating the eight types of American Indian mtDNA, with branch lengths drawn proportional to average sequence divergence. Four of the eight American Indian mtDNA types are identical to Asian types, which presumably reflects the lower resolving power of the method and the recent colonization of the Americas from Asia. The average divergence between the American Indian types and the nearest non-American Indian type (e.g., average divergence through node A) therefore cannot be estimated in the usual fashion. However, if we assume that the two clusters of four types each in Figure 3 represent two separate mtDNA lineages that colonized the Americas, then the average divergence between individ-

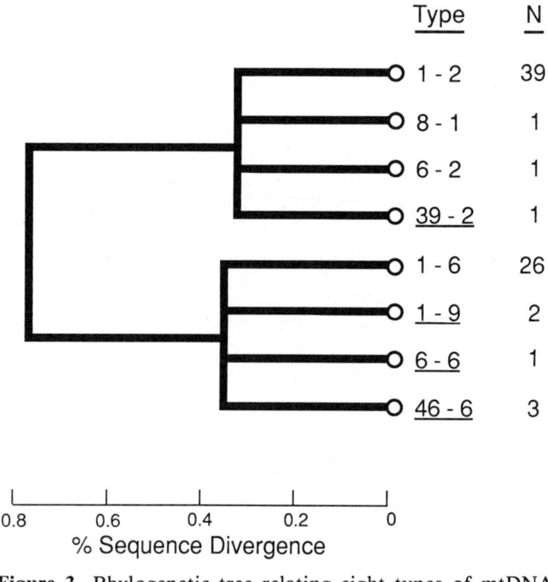

Figure 3. Phylogenetic tree relating eight types of mtDNA found in American Indians by Wallace et al. (1985). mtDNA types are designated by their nomenclature; N refers to the number of individuals of each type. Underlined types were unique to American Indians, whereas the remaining types occur in Asia as well.

uals within the two clusters (e.g., average divergence through node B) can still be estimated.

For the first cluster, consisting of 42 individuals, the average divergence is 0.03%; for the second cluster, consisting of 32 individuals, it is 0.08%. The overall average divergence within a cluster is thus 0.05%. Human occupation of the Americas is usually assumed to have occurred 12,000 years ago (Owen 1984), which leads to an estimated rate of 4.2% per million years. Recent evidence, however, has added support to the contention that the first human occupation was as early as 20,000 years ago (Dorn et al. 1986), which would lead to an estimated rate of 2.5% per million years.

Rate Estimates and a Statistical Consideration

Table 3 presents a summary of the estimates of the rate of sequence divergence derived from New Guinean, Australian, and American Indian mtDNAs. These values are in general agreement with an overall rate of 2–4% per million years, and therefore seem to be no lower than in other mammals (Wilson et al. 1985).

A statistical test of the hypothesis that the average rate is lower in humans would be desirable. Unfortunately, standard errors associated with the estimates in Table 3 cannot be calculated (see Discussion). Instead, Table 4 presents the total number of mutations observed within region-specific clusters of more than one individual, and the number of mutations expected for various values of the rate of sequence divergence. The observed number of mutations is small for American Indians, because of the low resolving power of the method. It is also small for Australians, because only three clusters of more than one individual (involving only nine of the 21 Australians) occurred. Fortunately, a substantial number of mutations (66) were observed within New Guinea-specific clusters, many more than would be expected if the true rate of human mtDNA evolution were much less than 3% per million years. In this case, it seems unlikely that the rate could be as low as 1% per million years.

DISCUSSION

The mapping of 29 additional New Guinea mtDNAs raises the total number of high-resolution restriction maps from this geographic region to 55. Despite the large increase in the number of New Guinea-specific clusters (from 7 to 14), it is likely that most of the major mtDNA lineages that colonized New Guinea have now been discovered. Preliminary results from an additional 64 New Guinea mtDNAs from the southern highland and northern coastal regions reveal only a few new clusters of New Guinea-specific lineages, none consisting of more than two individuals (M. Stoneking et al., unpubl.).

Table 3. Summary of Estimates of the Rate of Human mtDNA Evolution Derived from New Guineans, Australians, and American Indians

Population	Age in years	Percent divergence per million years
New Guinean	30,000	3.0–9.3
	50,000	1.8–5.6
Australian	40,000	4.0–7.2
American Indian	12,000	4.2
	20,000	2.5

Our results on New Guinea-specific and Australia-specific mtDNA clusters, and published data on American Indian mtDNA types, were used to estimate the mean rate of human mtDNA sequence divergence. These estimates, summarized in Table 3, are the first based specifically on comparisons within a species, rather than between species. They therefore represent the best available values for the rate of human mtDNA evolution, although there are reasons for viewing them with caution, as we now discuss.

There are several potential sources of error in these estimates. The analysis assumes that the most closely related mtDNA type from outside each population has been correctly identified; if additional types either break up a region-specific cluster or are more closely related to a cluster, the estimate of the rate will be reduced. A wider survey of Asian (in particular, Indonesian) mtDNA types is necessary to assess this for the New Guinean and Australian data. However, for the estimate derived from New Guinea-specific clusters to be reduced very much, many new mtDNA types would have to enter the two main clusters of closely related types (clusters 5 and 14, Fig. 1A), which does not seem likely.

A second assumption of the analysis is that migration to these areas was essentially in one direction, with

Table 4. Observed and Expected Numbers of Mutations within Region-specific Clusters

Geographic region	Number of individuals[a]	Observed mutations[b]	Age[c]	Expected mutations[d] 3%	1%	0.3%
New Guinea	48/55	66	30	66.0	22.0	6.6
Australia	9/21	20	40	15.0	5.0	1.5
America	74/74	9	12	6.4	2.1	0.6

[a]The number of individuals occurring in clusters of more than one individual, followed by the number of individuals surveyed.

[b]Within region-specific clusters of more than one individual.

[c]In thousands of years.

[d]For three rates of divergence, ranging from 0.3% to 3% per million years.

no appreciable back-migration. A large amount of migration from any of these areas back toward the source populations would result in similar mtDNA types being found outside these areas, and consequently lead to an underestimate of the rate of human mtDNA sequence divergence. Although difficult to assess, there is no reason to doubt the validity of the assumption of unidirectional migration for Australia, New Guinea, and the Americas.

The procedure used to calculate the rate requires an estimate of the time of colonization of an area. Inability to estimate this time accurately from fossil, archaeological, or biogeographical information can lead to incorrect rate estimates. We have used the minimum and maximum plausible dates for the first occupation of an area, to see how this affects the rate estimates (Table 3). A more accurate assessment of the time depth will lead to a more accurate calibration of the rate.

Another potential source of error would be multiple colonization events occurring at various times. This probably does not apply to New Guinea, Australia, or the Americas, for the following reason: A likely scenario for the colonization of these areas is rapid population expansion following the first entry of humans; later invasions would thus encounter established populations. The most probable outcome, mixing or hybridization between the invading and resident populations, mediated chiefly by invading males and resident females, would result in the retention of mtDNA types of the resident population. If, however, the mtDNA types of later, invading populations did become predominant, then our approach will result in an underestimate of the rate.

A final consideration is stochastic variation in the estimates of pairwise sequence divergence. There are two sources of such variation: sampling error and stochastic (and systematic) variation in the mutational process itself (Engels 1981; Ewens 1983; Tajima 1983). The sample sizes for the New Guinea and American Indian estimates are adequate (Tajima 1983), although the sample of Australians is smaller than desirable. The error associated with stochastic variation in the accumulation of pairwise sequence divergence by the occurrence and fixation of mtDNA mutations is difficult to estimate (M. Nei, pers. comm.), but is undoubtedly large. Standard errors of the branching points of a distance tree for a subset of these data also indicate a large stochastic variation (Nei 1985).

The stochastic error is dependent on the number of restriction sites surveyed per mtDNA genome; since the New Guinea and Australian estimates are based on an average of about 370 restriction sites per genome, while the American Indian estimate is based on an average of about 50 sites per genome (Wallace et al. 1985), we have more confidence in the first two estimates. The large number of mutations that have occurred within New Guinea-specific clusters (Table 4) further indicates a smaller stochastic error for the associated rate estimate. Increasing the number of restriction sites examined, or direct sequence analysis, would be required to reduce the stochastic error.

Interspecies comparisons of mtDNA from a wide variety of mammals, including humans, have led to the conclusion that the rate of mtDNA evolution is about 2–4% per million years (Wilson et al. 1985). The rate estimates presented here for human mtDNA, derived from intraspecies comparisons, do not contradict this view. In particular, there is no evidence for a decrease in the rate of human mtDNA evolution; the most likely sources of error in these data would lead us to underestimate the rate. An important consequence of this analysis is that it further substantiates the hypothesis of Cann et al. (1986) that the common ancestor of all existing mtDNA types in modern human populations lived some 150,000–300,000 years ago.

ACKNOWLEDGMENTS

We thank R. Cann, U. Gyllensten, E. Prager, and V. Sarich for helpful discussions. This work was supported by grants from the National Institutes of Health and the National Science Foundation.

REFERENCES

Anderson, S., A.T. Bankier, B.G. Barrell, M.H.L. de Bruijn, A.R. Coulson, J. Drouin, I.C. Eperon, D.P. Nierlich, B.A. Roe, F. Sanger, P.H. Schreier, A.J.H. Smith, R. Staden, and I.G. Young. 1981. Sequence and organization of the human mitochondrial genome. *Nature* **290:** 457.

Bonné-Tamir, B., M.J. Johnson, A. Natali, D.C. Wallace, and L.L. Cavalli-Sforza. 1986. Human mitochondrial DNA types in two Israeli populations—A comparative study at the DNA level. *Am J. Hum. Genet.* **38:** 341.

Britten, R.J. 1986. Rates of DNA sequence evolution differ between taxonomic groups. *Science* **231:** 1393.

Brown, W.M. 1980. Polymorphism in mitochondrial DNA of humans as revealed by restriction endonuclease analysis. *Proc. Natl. Acad. Sci.* **77:** 3605.

Brown, W.M., M. George, and A.C. Wilson. 1979. Rapid evolution of animal mitochondrial DNA. *Proc. Natl. Acad. Sci.* **76:** 1967.

Cann, R.L. 1982. "The evolution of human mitochondrial DNA." Ph.D. thesis, University of California, Berkeley.

———. 1985. Mitochondrial DNA variation and the spread of modern populations. In *Out of Asia—Peopling the Americas and the Pacific* (ed. R. Kirk and E. Szathmary), p. 113. Journal of Pacific History, Australian National University, Canberra.

Cann, R.L., W.M. Brown, and A.C. Wilson. 1984. Polymorphic sites and the mechanism of evolution in human mitochondrial DNA. *Genetics* **106:** 479.

Cann, R.L., M. Stoneking, and A.C. Wilson. 1986. Mitochondrial DNA and human evolution. *Nature* (in press).

Dorn, R.I., D.B. Banforth, T.A. Cahill, J.C. Dohrenwend, B.D. Turrin, D.J. Donahue, A.J.T. Jull, A. Long, M.E. Macko, E.B. Weil, D.S. Whitley, and T.H. Zabel. 1986. Cation-ratio and accelerator radiocarbon dating of rock varnish on Mojave artifacts and landforms. *Science* **231:** 830.

Engels, W.R. 1981. Estimating genetic divergence and genetic variability with restriction endonucleases. *Proc. Natl. Acad. Sci.* **78:** 6329.

Ewens, W.J. 1983. The role of models in the analysis of molecular genetic data, with particular reference to restriction fragment data. In *Statistical analysis of DNA sequence data* (ed. B.S. Weir), p. 45. Marcel Dekker, New York.

Hope, G.S., J. Golson, and J. Allen. 1983. Paleoecology and prehistory in New Guinea. *J. Hum. Evol.* **12:** 37.

Horai, S. and E. Matsunaga. 1986. Mitochondrial DNA polymorphism in Japanese. II. Analysis with restriction enzymes of four or five base pair recognition. *Hum. Genet.* **72:** 105.

Horai, S., T. Gojobori, and E. Matsunaga. 1984. Mitochondrial DNA polymorphism in Japanese. *Hum. Genet.* **68:** 324.

Johnson, M.J., D.C. Wallace, S.D. Ferris, M.C. Rattazzi, and L.L. Cavalli-Sforza. 1983. Radiation of human mitochondria DNA types analyzed by restriction endonuclease cleavage patterns. *J. Mol. Evol.* **19:** 255.

Kirk, R.L. 1982. Microevolution and migration in the Pacific. In *Human genetics*, part A: *The unfolding genome* (ed. B. Bonné-Tamir et al.), p. 215. A.R. Liss, New York.

Koop, B.F., M. Goodman, P. Xu, K. Chan, and J.L. Slightom. 1986. Primate η-globin DNA sequences and man's place among the great apes. *Nature* **319:** 234.

Nei, M. 1985. Human evolution at the molecular level. In *Population genetics and molecular evolution* (ed. K. Aoki and T. Ohta), p. 41. Springer-Verlag, Berlin.

Nei, M. and F. Tajima. 1983. Maximum likelihood estimation of the number of nucleotide substitutions from restriction site data. *Genetics* **105:** 207.

———. 1985. Evolutionary change of restriction cleavage sites and phylogenetic inference for man and apes. *Mol. Biol. Evol.* **2:** 189.

Owen, R.C. 1984. The Americas: The case against an ice-age human population. In *The origins of modern humans: A world survey of the fossil evidence* (ed. F.H. Smith and F. Spencer), p. 517. A.R. Liss, New York.

Pope, G.G. 1983. Evidence on the age of the Asian Hominidae. *Proc. Natl. Acad. Sci.* **80:** 4988.

Stoneking, M., K. Bhatia, and A.C. Wilson. 1986. Mitochondrial DNA variation in Eastern Highlanders of Papua New Guinea. In *Genetic variation and its maintenance in tropical populations* (ed. D.F. Roberts and G. DeStefano). Cambridge University Press, England. (In press.)

Stringer, C.B. 1984. The definition of *Homo erectus* and the existence of the species in Africa and Europe. *Cour. Forschungsinst. Senckenb.* **69:** 131.

Tajima, F. 1983. Evolutionary relationship of DNA sequences in finite populations. *Genetics* **105:** 437.

Templeton, A.R. 1983. Phylogenetic inference from restriction endonuclease cleavage site maps with particular reference to the evolution of humans and the apes. *Evolution* **37:** 221.

———. 1985. The phylogeny of the hominoid primates: A statistical analysis of the DNA-DNA hybridization data. *Mol. Biol. Evol.* **2:** 420.

Wallace, D.C., K. Garrison, and W.C. Knowler. 1985. Dramatic founder effects in Amerindian mitochondrial DNAs. *Am. J. Phys. Anthropol.* **68:** 149.

White, J.P. and J.F. O'Connell. 1982. *A prehistory of Australia, New Guinea, and Sahul.* Academic Press, New York.

Wilson, A.C., R.L. Cann, S.M. Carr, M. George, U.B. Gyllensten, K.M. Helm-Bychowski, R.G. Higuchi, S.R. Palumbi, E.M. Prager, R.D. Sage, and M. Stoneking. 1985. Mitochondrial DNA and two perspectives on evolutionary genetics. *Biol. J. Linn. Soc.* **26:** 375.

Wolpoff, M.H., W.X. Zhi, and A.G. Thorne. 1984. Modern *Homo sapiens* origins: A general theory of hominid evolution involving the fossil evidence from East Asia. In *The origins of modern humans: A world survey of the fossil evidence* (ed. F.H. Smith and F. Spencer), p. 411. A.R. Liss, New York.

Molecular Genetic Investigations of Ancient Human Remains

S. PÄÄBO

Department of Cell Research, The Wallenberg Laboratory, University of Uppsala, S-75122 Uppsala, Sweden; Institute of Egyptology, Gustavianum, University of Uppsala, S-75120 Uppsala, Sweden

Molecular evolution is a diachronic process of genetic change, taking place as a result of stochastic events, as well as natural selection. However, most of the evidence available for the study of molecular evolution is indirect and circumstantial, involving deduction from the distribution of polymorphism within and between modern species. A fundamentally new approach to evolutionary biology would be opened if it could be proved possible to investigate directly the fate of genes in a species or a population over substantial time periods. The present communication summarizes some preliminary attempts to achieve this by studying the preservation of DNA in ancient human remains.

Human soft tissue remains have been preserved in various locations around the world, generally as a result of desiccation (for review, see Cockburn and Cockburn 1980; Table 1). Two geographical areas stand out due to the large number of preserved individuals as well as the time span that those individuals represent. These areas are Egypt and the central Andean region, where the remains of several thousand individuals have been retrieved and additional material continues to be found by archaeologists every year. At least in theory, one could therefore hope to obtain representative samples from ancient populations in these areas. For this reason, the attempts to establish methods for the study of ancient DNA have been concentrated on archaeological remains from Egypt and Peru.

EXPERIMENTAL PROCEDURES

Archaeological Samples

Fourteen mummy samples were analyzed, the first four being Peruvian and the remainder being Egyptian. From the Institute for Anthropology and Human Genetics, University of Munich, Munich, FRG, were: (a) Cahuachi 341 (new number, 579); (b) Pacatnamu z20 (new number, 305); (c) Las Trancas 120 (new number, 642); (d) Las Trancas 83 (new number, 606); (e) big package, right side of the chest of ÄS73b; (f) big package, left side of the chest of ÄS73b; (g) package containing Horus figure of ÄS12d. See Wick et al. (1980) for the Peruvian mummies and Ziegelmayer (1985), for the Egyptian mummies. From the Manchester Museum, Manchester, England, was: (h) liver from a canopic jar of mummy 21470 (see David 1979). From the British Museum, London, England, were: (i) a loose fragment of mummy 32752; (j) superficial tissue from the left lower leg of 32753; (k) tissue from right lower leg of 32753; (l) loose fragment of 32754, probably from the hand; (m) tissue from a fragment of the knee of 32755; (n) skin from a loose fragment of 57353 (see Dawson and Gray 1968).

Table 1. Overview of Mummified Human Remains

Provenance	Approximate time period	Approximate number of individuals
Egypt	3000 BC–present	10^4–10^5
Andean area	4000 BC–1700 AD	10^2–10^3
Southwestern USA	500 AD–1400 AD	10^2
Aleutian Islands and Alaska	600 AD–1700 AD	10^1
Australia and Melanesia	19th century AD	10^1
Japan	1100 AD–1900 AD	10^1

Histology

Tissue samples for microscopy were rehydrated in an aqueous solution of 1% Na_2CO_3 (wt/vol), 0.5% formalin, and 28.5% ethanol (Sandison 1955), containing 0.85% NaCl (Turner and Holton 1981), for 1–2 days. Paraffin embedding and hematoxylin and eosin staining were performed using routine protocols.

DNA Extraction

DNA extractions were performed using a modification of the procedure described by Blin and Stafford (1976). Briefly, approximately 30 mg of dried tissue was carefully minced and suspended in 400 μl of 10 mM Tris (pH 8.0), 2 mM EDTA, and 10 mM NaCl. A 100-μl aliquot of the same solution containing 5% SDS, 30 mg/ml DTT, and 2 mg/ml proteinase K was then added. The samples were incubated at 37°C with gentle agitation for 4 hours. An additional 100 μl of buffer containing SDS, DTT, and proteinase K was added and the incubation was continued overnight. Insoluble material was removed by centrifugation and the brownish supernatant was extracted once with phenol and once with chloroform/isoamylalcohol (4:1) before being precipitated with ethanol. The precipitates were dissolved in 50 μl of distilled water.

Enzymatic Procedures and Cloning

Ten microliters of the extracts were incubated with T4 DNA polymerase (Pharmacia P-L Biochemicals) in the presence of [α-^{32}P]dATP, according to the instructions of the manufacturer. After terminating the reaction by the addition of EDTA, 20% of the material was run on a 1.8% agarose gel and visualized by autoradiography of the dried gel. DNA polymerase I and mung bean nuclease were used according to manufacturer's instructions (Pharmacia P-L Biochemicals).

For cloning, DNA was prepared from 260 mg of tissue from sample d (Las Trancas 83) and the extraction was scaled up to a total volume of 10 ml. Cloning in λgt10 was performed according to Huynh et al. (1984), with preannealing with λ*c*I857 Sam7 DNA (Higuchi et al. 1984).

Other Procedures

DNA extracted from the Windover brain was analyzed by agarose gel electrophoresis in a 1.5% gel and transferred to a nitrocellulose filter for hybridization (Southern 1975) with a nick-translated (Rigby et al. 1977) *Bgl*II-*Sph*I restriction fragment (Larhammar et al. 1985), which contains an *Alu* repeat (Schmid and Jelinek 1982).

RESULTS

Histological Investigations

Since the beginning of this century, numerous histological investigations on rehydrated mummified tissues have been performed and the preservation of cell nuclei has been noted (Ruffer 1911; Sandison 1955; Daniels et al. 1970; Tapp 1979; Chapel et al. 1981). In all cases except one (Tapp 1979), the visualized nuclei have been found in the epidermis. By the use of DNA-specific dyes, such as ethidium bromide, the presence of DNA in these cell nuclei can be demonstrated (Pääbo 1984). Most of these investigations have been performed on Egyptian mummies prepared during Pharaonic times with crystalline salts as a dehydrating agent (Lucas 1932). However, numerous so-called natural mummies also exist from various regions of the world where arid environmental conditions have provided for their postmortem dehydration. Histological examination of a number of such mummies has demonstrated that they often are equally well or better preserved than the artificially mummified bodies from Pharaonic Egypt. Figure 1 shows the epidermis of one such 5000-year-old natural mummy from Egypt, where cell nuclei can be clearly seen.

DNA Extraction

Ancient human nucleic acids previously have been extracted from the liver of a 2000-year-old Chinese corpse (Wang and Lu 1981) and from two approximately 2400-year-old Egyptian mummies. In one case, the DNA contained modified pyrimidines and could not be cloned in bacteria (Pääbo 1985a), whereas in another case it could be cloned, using a plasmid vector (Pääbo 1985b).

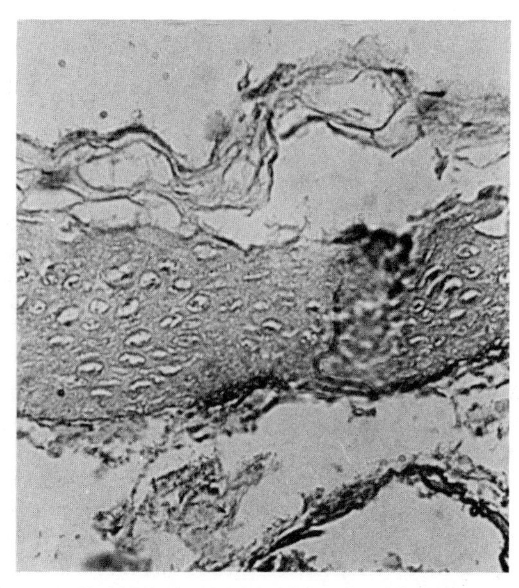

Figure 1. Tissue section of the epidermis of a 5000-year-old Egyptian mummy (British Museum, 32753). Shrunken cell nuclei are visible in the epidermis. Hematoxylin and eosin staining; initial magnification, 180×.

To elucidate how various environmental conditions and mummification methods have affected the preservation of DNA in ancient organic remains, DNA was extracted from approximately 30 mg of tissue removed from a number of Peruvian and Egyptian mummies. The purified DNA samples were analyzed by the incorporation of radioactive nucleotides, using T4 DNA polymerase and agarose gel electrophoresis, followed by autoradiography of the dried gel. Figure 2 summarizes the results. Skin and subcutaneous tissue from four mummies, originating from the coastal parts of Peru and displaying no signs of artificial preservation, are represented in Figure 2 (lanes a–d). They have been dated to the 5th–6th century AD, the 11th–15th century AD, the 7th–8th century AD, and the 4th–5th century AD, respectively (Wick et al. 1980).

Internal organs from ancient Egyptian mummies are represented in Figure 2 (lanes e–h). Viscera were often removed from the body cavities during the mummification process and dried separately before being put back into the body or placed in so-called canopic jars in the tomb. Organ packages found in the chest cavity of Djehuti-irdis, a boy who died in his teens during the 3rd century BC (Ziegelmayer 1985), are represented in Figure 2 (lanes e,f). The samples are liver and intestine, respectively. Lane g is the extract from an unidentified internal organ found in a package in the chest cavity of a man from the New Kingdom (approximately 1550–1100 BC) (Ziegelmayer 1985). Lane h is the extract of a liver sample from a canopic jar found in the tomb of a man named Nekht-ankh who lived during

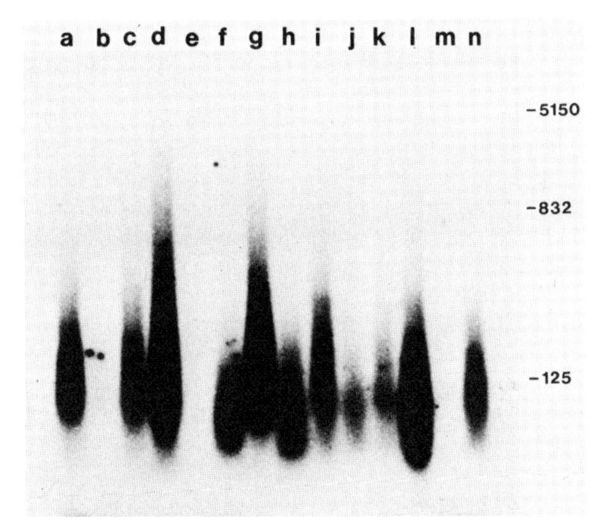

Figure 2. Agarose gel electrophoresis of radioactively labeled DNA extracted from ancient human soft tissues. (Lanes *a–d*) Samples from Peruvian mummies; (lanes *e–h*) samples from artificially mummified organs of Egyptian mummies; (lanes *i–n*) samples from predynastic natural mummies from Egypt. See Experimental Procedures for details. Molecular size markers are indicated in number of base pairs.

the 20th or 19th century BC (David 1979). Lanes i–n represent extracts of skin and subcutaneous tissues from a number of late predynastic bodies found at Gebelein in Egypt, dated to approximately 3000 BC (Dawson and Gray 1968). They have been preserved due to the dehydrating effect of the desert sand in which they were buried.

It is evident from Figure 2 that three out of four Peruvian mummies and three out of the four artificially mummified organs from Pharaonic Egypt, as well as five out of the six samples of the naturally mummified, predynastic Egyptian bodies, contain DNA into which nucleotides can be incorporated by DNA polymerase. The amounts of DNA present in the tissues vary considerably among the individual mummies but are not consistently higher in any of the groups. In all cases, the DNA is degraded to an average molecular size of between 100 and 400 bp. Only small amounts of higher-molecular-mass DNA are present and can be visualized by a longer exposure of the autoradiogram (not shown).

Molecular Cloning

The DNA extract from the 4th century Peruvian mummy, illustrated in Figure 2 (lane d) was used for molecular cloning, using the bacteriophage λ vector λgt10 (Huynh et al. 1984). When 100 ng of the DNA was made blunt-ended by T4 polymerase before the addition of a synthetic *Eco*RI linker and ligation to the vector arms, only about 500 clones were obtained. Treatment of the DNA with DNA polymerase I resulted in an equally small number of clones, whereas digestion of the DNA by mung bean nuclease yielded an approximately fourfold increase in the number of clones obtained. Intriguingly, the sequential treatment of the DNA by DNA polymerase I and mung bean nuclease resulted in an increase in cloning efficiency that was between one and two orders of magnitude. The detailed analysis of these clones is currently under way.

An 8000-year-old Human Brain

A human brain, excavated from peat below the bottom of a pond at the Windover site in Florida and radiocarbon-dated to approximately 8000–8200 BP (B. Peck, pers. comm.), was investigated. In spite of the fact that the histological examination of the tissue showed only occasionally preserved cells (not shown), DNA could be extracted and was found to be nearly 10,000 bp in size. That this DNA is indeed of human origin was ascertained by Southern hybridization (Southern 1975) using a probe containing an *Alu* repeat (Schmid and Jelinek 1982), which under the hybridization conditions used is specific for human DNA (Fig. 3). Others have demonstrated the presence of full-length mitochondrial DNA in extracts from this brain (G.H. Doran et al., pers. comm.).

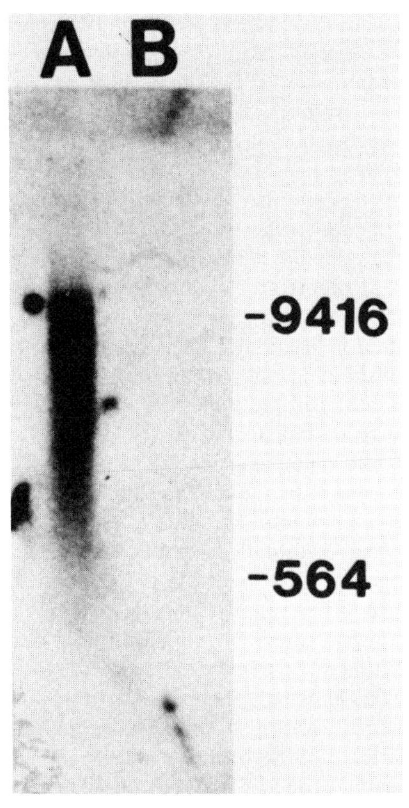

Figure 3. Southern blot of DNA extracted from an 8000-year-old human brain from Florida. Approximately 50 ng of DNA (lane *A*) was electrophoresed in a 1.5% agarose gel along with an equal amount of *E. coli* DNA (lane *B*). The gel was blotted to a nitrocellulose filter and hybridized to a probe containing an *Alu* repeat sequence. Molecular size markers are indicated in numbers of base pairs.

DISCUSSION

The histological examinations establish that superficial and peripheral parts of ancient mummified bodies, in general, are the best preserved (see, e.g., Fig. 1). In addition, they show that the antiquity of an individual mummy per se need not determine its state of preservation. Instead, the amount of time that has elapsed between the death and the dehydration of the tissues can be speculated to determine the amount of autolytic and microbial degradation of the body. Internal organs are well preserved only when it can be assumed that they have been removed soon after death and rapidly dried.

The results presented in Figure 2 demonstrate that if superficial parts of mummified bodies or organs are investigated, the retrieval of DNA is not a rare exception but instead a frequent finding. However, the bulk of the DNA is always degraded to a low average molecular mass. It is not known if this should be attributed primarily to some time-dependent process, such as background radiation, or if it is due to autolytic processes taking place soon after death (Rebrov et al. 1983). It has recently been shown that high-molecular-mass DNA can be isolated from up to 4-year-old, air-dried blood (Gill et al. 1985). Future work involving DNA extraction from parts of ancient bodies that can be expected to have dried exceptionally fast after death or to have been protected from autolytic processes will answer the question if high-molecular-mass DNA can be obtained from remote time periods.

A low efficiency of cloning seems to be a general phenomenon when ancient DNA samples are investigated. In one case, the base composition of an ancient Egyptian DNA has been analyzed by acidic hydrolysis and HPLC (Pääbo 1985a). A relative loss of pyrimidines was demonstrated and a number of unidentified compounds appeared early in the chromatogram. This is consistent with oxidative damage to the DNA, which affects pyrimidines more than purines (Scholer et al. 1960). That macromolecules have indeed been subjected to oxidation in the studied mummy sample is shown by the amino acid analysis of proteins extracted from the same mummy, which demonstrated oxidation products of methionine as well as cysteine (data not shown). That high-molecular-mass DNA can be isolated from the Windover brain (Fig. 3), which in contrast to the other samples investigated has been preserved in highly anaerobic conditions, further stresses the importance of oxidative damage to DNA over long time periods and suggests that oxidation might be involved also in causing strand breakage, e.g., through the oxidation of deoxyribose residues.

The treatment of the DNA with mung bean nuclease, a single-strand-specific nuclease, increases the clonability of the DNA. This indicates that the extracted DNA might contain single-stranded gaps. However, treatment of the DNA with polymerase I, which should repair such gaps, does not increase its clonability. This could be due to other types of damage to the DNA, which may constitute complete blocks to DNA replication. For example, DNA-DNA cross-links have been observed when DNA is irradiated in the dried state (Smith 1968). The sequential treatment of the ancient DNA by polymerase I and mung bean nuclease yields a severalfold increase in the number of clones obtained. This may be explained if the ancient DNA contains modifications that represent blocks to its replication. Polymerase I would then replicate the DNA up to the modified sites and stop at these points, where it would leave one or a few-nucleotide-long, single-stranded gaps, presumably accessible to a single-strand-specific endonuclease. Digestion of the DNA by mung bean nuclease after the replication of the DNA by polymerase I might thus yield DNA fragments free of modifications that inhibit their replication and thus their cloning. Further investigations will have to address the generality of this finding as well as the question of how faithfully the ancient DNA sequences are reproduced by the clones isolated.

The study of ancient DNA bears the potential of following the molecular evolution of genes over time. The recent results involving human DNA as well as the recent cloning of 140-year-old quagga DNA (Higuchi et al. 1984) and the extraction of DNA from up to 44,600-year-old plant remains (Rogers and Bendich 1985) raises the hope of retrieving even older DNA samples from fossil remains. However, by molecular genetic studies of a few thousand-year-old human populations, some evolutionary questions of general importance can be addressed already.

One such question concerns the rate and mode of evolution of human parasites, which by their short generation times or mode of DNA replication can be speculated to evolve rapidly. The feasibility of such investigations is suggested by the demonstration of papilloma virus in a Chilean mummy (Horne and Kawasaki 1984), as well as the presence on Egyptian mummies of skin eruptions that may represent smallpox (Sandison 1980; and Fig. 4).

Gene transfers between species involving retroviral sequences have been demonstrated in a number of cases (for review, see Benveniste 1985). It can be speculated that the close cohabitation of man and various animals, introduced with the domestication of animals, increased the frequency of such transfer events. This could be studied directly by molecular investigations of human as well as animal remains in Egypt.

Furthermore, the impact of abrupt changes in selective pressures could be studied in human populations. One attractive example would be the study of immune-response gene frequencies in the Andean population before and after the catastrophic epidemics that followed the Spanish Conquest, and the introduction of a number of infectious diseases from Europe. These epidemics killed a majority of the indigenous population and could thus prove to be a unique possibility for gaining further insight into how the genetic constitution of a population determines its susceptibility to infectious diseases. Finally, ancient population movements could be studied, as well as some more controversial theories involving genetic changes taking

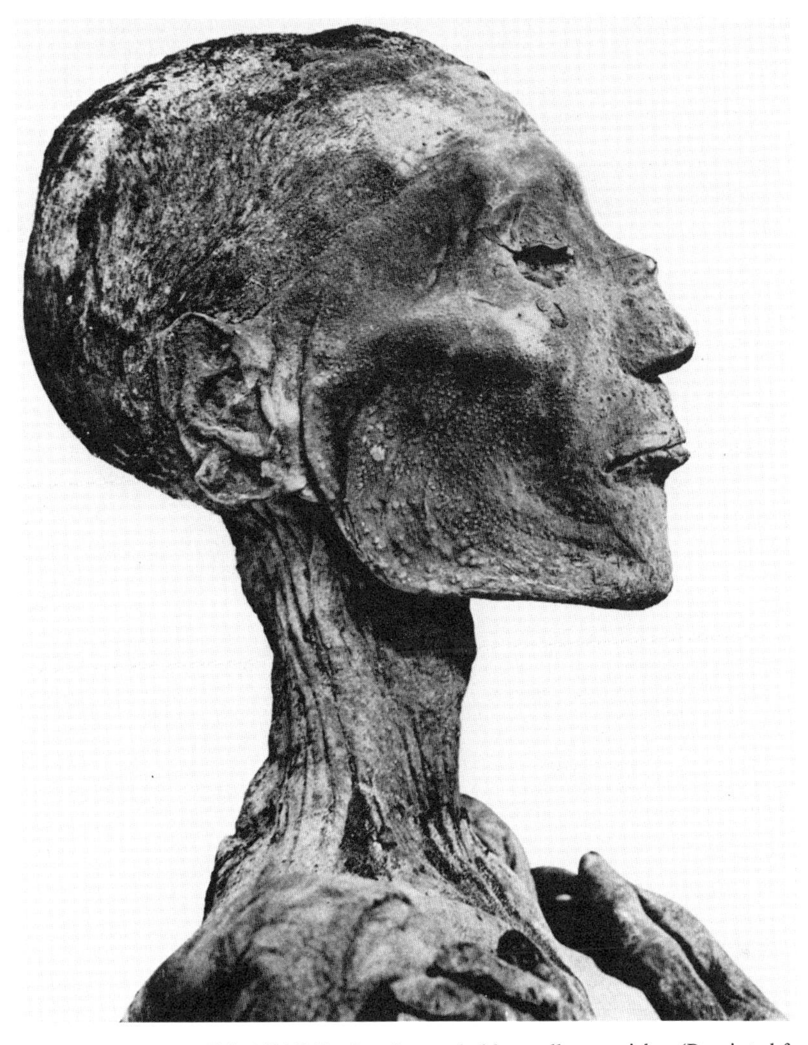

Figure 4. Head of Pharaoh Ramses V (~1100 BC), showing probable smallpox vesicles. (Reprinted from Smith 1912.)

place over historic time (Lumsden and Wilson 1981; Wilson 1985).

ACKNOWLEDGMENTS

I am greatly indebted to Professor Per A. Peterson for providing excellent working facilities; to Ms. Ann-Charlotte Runn for help with the histological work; to Mr. Peter Wenkler for the synthesis of oligonucleotides; to Drs. Franz Parsche, Rosalie David, Vivian Davies, and Bruff Peck for generously giving me access to archaeological material; to Djehuti-irdis, Nekht-ankh, and numerous anonymous men and woman who actually contributed their tissues for the analyses; to Professor Rostislav Holthoer for continuous encouragement and advice; and to the Faculty of Medicine of the University of Uppsala for a Lennander grant.

REFERENCES

Benveniste, R.E. 1985. The contribution of retroviruses to the study of mammalian evolution. In *Molecular evolutionary genetics* (ed. R.J. MacIntyre), p. 354. Plenum Press, New York.

Blin, N. and D.W. Stafford. 1976. A general method for isolation of high molecular weight DNA from eukaryotes. *Nucleic Acids Res.* **3:** 2303.

Chapel, T.A., A.H. Mehregan, and T.A. Reyman. 1981. Histologic findings in mummified skin. *J. Am. Acad. Dermatol.* **4:** 27.

Cockburn, A. and E. Cockburn, eds. 1980. *Mummies, disease and ancient culture.* Cambridge University Press, England.

Daniels, F., Jr. and P.W. Post. 1970. The histology and histochemistry of prehistoric mummy skin. *Adv. Biol. Skin* **10:** 279.

David, A.R. 1979. A catalogue of Egyptian human and animal mummified remains. In *Manchester Museum mummy project* (ed. A.R. David), p. 1. Manchester University Press, Manchester, England.

Dawson, W.R. and P.H.K. Gray. 1968. *Catalogue of Egyptian antiquities in the British Museum.* I. *Mummies and human remains.* British Museum, London.

Gill, P., A.J. Jeffreys, and D.J. Werrett. 1985. Forensic applications of DNA "fingerprints." *Nature* **318:** 577.

Higuchi, R., B. Bowman, M. Freiberger, O.A. Ryder, and A.C. Wilson. 1984. DNA sequences from the quagga, an extinct member of the horse family. *Nature* **312:** 282.

Horne, P.R.T. and S.Q. Kawasaki. 1984. The prince of el Plomo: A paleopathological study. *Bull. N.Y. Acad. Med.* **60:** 925.

Huynh, T.V., R.A. Young, and R.W. Davis. 1984. Construction and screening cDNA libraries in λgt10 and λgt11. In *DNA cloning techniques: A practical approach* (ed. D.M. Glover), p. 49. IRL, Oxford, England.

Larhammar, D., B. Servenius, L. Rask, and P.A. Peterson. 1985. Characterization of an HLA DR_β pseudogene. *Proc. Natl. Acad. Sci.* **82:** 1475.

Lucas, A. 1932. The use of natron in mummification. *J. Egypt. Archaeol.* **18:** 125.

Lumsden, C.J. and E.O. Wilson. 1981. *Genes, mind and culture*. Harvard University Press, Cambridge, Massachusetts.

Pääbo, S. 1984. Uber den Nachweiss von DNA in altägyptischen Mumien. *Das Altertum* **30:** 213.

———. 1985a. Preservation of DNA in ancient Egyptian mummies. *J. Archaeol. Sci.* **12:** 411.

———. 1985b. Molecular cloning of ancient Egyptian mummy DNA. *Nature* **314:** 644.

Rebrov, L.B., V.L. Kozeltdev, S.S. Shishkin, and S.S. Debov. 1983. Nekotorie enzimaticheskie aspekty posmertnogo autoliza. *Vestn. Akad. Med. Nauk SSSR* **10:** 82.

Rigby, P.W.J., M. Diekman, C. Rhodes, and P. Berg. 1977. Labeling deoxyribonucleic acid to high specific activity in vitro by nick translation with DNA polymerase I. *J. Mol. Biol.* **133:** 237.

Rogers, S.O. and A.J. Bendich. 1985. Extraction of DNA from milligram amounts of fresh, herbarium and mummified plant tissue. *Plant Mol. Biol.* **5:** 69.

Ruffer, M.A. 1911. Histological studies on Egyptian mummies. *Mémoires sur l'Egypte: Institute d'Egypte* **6**(3): 1.

Sandison, A.T. 1955. The histological examination of mummified material. *Stain Technol.* **30:** 277.

———. 1980. Diseases in ancient Egypt. In *Mummies, disease and ancient culture* (ed. A. Cockburn and E. Cockburn), p. 29. Cambridge University Press, England.

Schmid, C.W. and W.R. Jelinek. 1982. The Alu family of dispersed repetitive sequences. *Science* **216:** 1065.

Scholer, G., J.F. Ward, and J.J. Weiss. 1960. Mechanisms of the radiation-induced degradation of nucleic acids. *J. Mol. Biol.* **2:** 379.

Smith, G.E. 1912. Catalogue générale des antiquités égyptiennes du Musée du Caire. *The Royal mummies*, plate LVI. Imrimerie de l'Institute Francais d'Archéologie Orientale, Cairo, Egypt.

Smith, K.C. 1968. The biological importance of UV-induced DNA-protein cross-linking in vivo and its probable chemical mechanism. *Photochem. Photobiol.* **7:** 651.

Southern, E.M. 1975. Detection of specific sequences among DNA fragments separated by gel electrophoresis. *J. Mol. Biol.* **98:** 504.

Tapp, E. 1979. Disease in the Manchester mummies. In *Manchester Museum mummy project* (ed. A.R. David), p. 95. Manchester University Press, Manchester, England.

Turner, P.J. and D.B. Holton. 1981. The use of fabric softener in the reconstitution of mummified tissue prior to paraffin wax sectioning for light microscopical examination. *Stain Technol.* **56:** 35.

Wang, G.H. and C.C. Lu. 1981. Isolation and identification of nucleic acids of the liver from a corpse from the Changsha Han tomb. *Sheng Wu Hua Hsueh Yu Sheng Wu Li Chin Chan* (in Chinese) **39:** 70.

Wick, G., M. Haller, R. Timpl, H. Cleve, and G. Ziegelmayer. 1980. Mummies from Peru. Demonstration of antigenic determinants of collagen in the skin. *Int. Arch. Allergy Appl. Immunol.* **62:** 76.

Wilson, A.C. 1985. The molecular basis of evolution. *Sci. Am.* **253:** 148.

Ziegelmayer, G. 1985. *Münchener Mumien. Lipp GmbtH,* München.

Relationships of Human Protein Sequences to Those of Other Organisms

R.F. DOOLITTLE, D.F. FENG, M.S. JOHNSON, AND M.A. MCCLURE
Department of Chemistry, University of California, San Diego, La Jolla, California 92093

Presently more than 4000 amino acid sequences are known, including more than 200 representing human proteins. Many of the sequences, which are being reported at an ever-increasing rate, have been inferred from nucleic acid sequences, and often the protein itself is not necessarily in hand. In such cases there is much that can be learned by comparison with sequences from other species or with other related proteins. Furthermore, the entire course of human history lies in our genes and can be read through the sequences of our proteins. In this article we attempt to chart some of the major divergences that have led to human proteins over the past few billion years. In this regard we have categorized the sequenced proteins into several arbitrary groups, beginning with the most ancient and continuing through to the modern.

Inasmuch as some of the most recent proteins in the human repertory appear to be the result of "exon shuffling," we have taken advantage of this forum to express some quarrelsome opinions about the invention of introns during evolution. Although some of these views run contrary to the current dogma, they are rendered here in the spirit of calling attention to unanswered questions rather than with the notion of proclaiming some alternate scheme of how introns came into being.

MATERIALS AND METHODS

The amino acid sequences examined here were all taken from either the NEWAT sequence bank (Doolittle 1981) or Release 6.0 of the National Biomedical Research Foundation Protein Identification Resource (Barker et al. 1985). For the most part, the programs used for searching and alignment have been described elsewhere (Feng et al. 1985).

About molecular clocks. It is important to understand that molecular clocks involving sequence data are ultimately set by recourse to the fossil record. In this regard, one is usually on quite firm ground when discussing the divergence times of vertebrate animals, and we often use sequence comparisons between existing vertebrates to gauge how fast a given protein is changing. The divergence times for vertebrates and invertebrates, on the one hand, and plants and animals, on the other, are also reasonably firm dates. When we make comparisons among the prokaryotes, however, we must be more cautious. Some approximate divergence times are listed in Table 1.

Some reminders. It must be recalled that the "evolutionary distance," usually expressed as a time since some divergence point, is related to sequence resemblance in the form of a negative exponential (Fig. 1). It must also be understood that two sequences can never diverge to the point of zero resemblance, if only because all proteins are made of the same 20 amino acids. Further, because deletions and insertions are constantly occurring during the evolution of proteins, and because alignment schemes must take account of these events, two unrelated or random sequences will have an apparent resemblance equal to 10–15% identity after a typical computer alignment that has inserted appropriate gaps.

These considerations can be reduced to a few simple rules-of-thumb about resemblance expressed in terms of percent identities after optimum alignment. First, any two sequences of reasonable length—a hundred residues or more—that are more than 25% identical are almost certainly homologous, which is to say that they have descended from a common ancestral sequence. If, on the other hand, two sequences are less than 15% identical after optimum alignment, it is unlikely that a convincing case for common ancestry can be made on the basis of the two sequences alone.

In between lies the "twilight zone" (Fig. 1). Keep in mind that there is just as much evolutionary distance between 25% and 15% identity as there is in all the span from the point of divergence (100% identity) to 25% identity.

Different proteins change at different rates. It is well known that proteins change at different but characteristic rates during their evolution. Doubtless a number of factors are involved, but the peculiar vul-

Table 1. Some Important Dates in History

Origin of the universe	– 15[a]	± 4
Formation of the solar system	– 4.6	
First self-replicating system	– 3.5	± 0.5
Prokaryotic–eukaryotic divergence	– 1.8	± 0.3
Plant–animal divergence	– 1.0	
Invertebrate–vertebrate divergence	– 0.5	
Mammalian radiation beginning	– 0.1	

[a]Billion years.

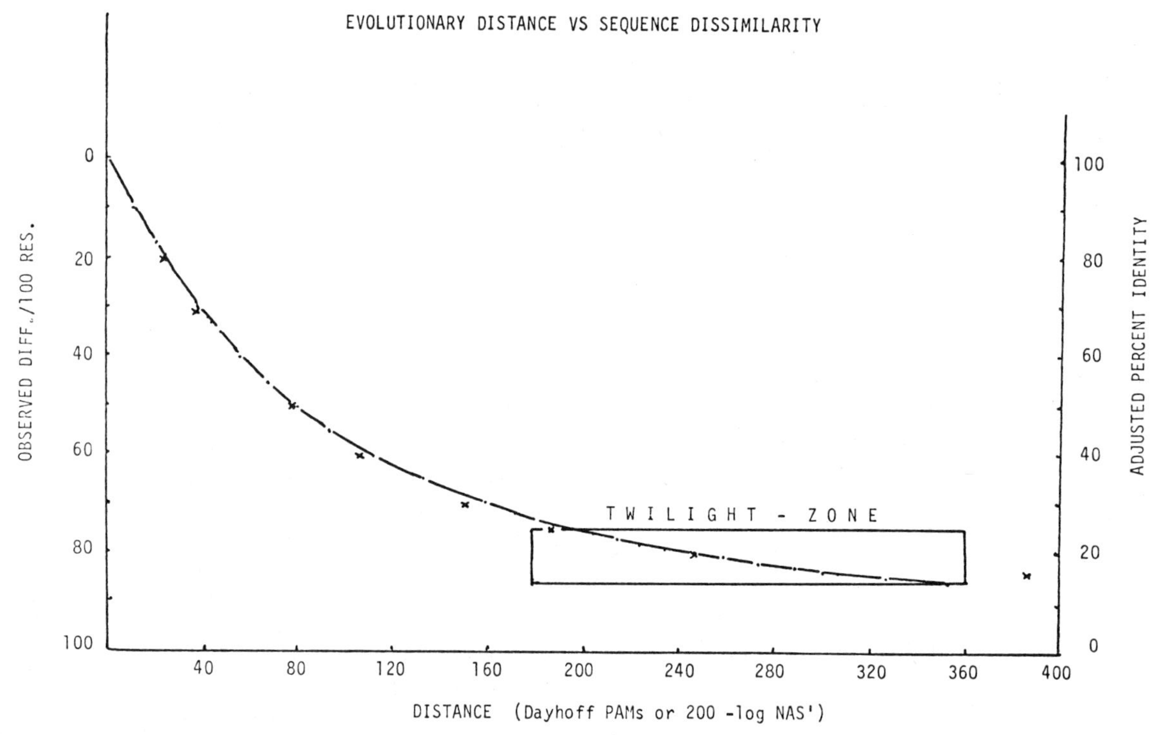

Figure 1. The relationship between the percent difference and the actual number of changes (evolutionary distance) as expressed in accepted point mutations (PAM; Dayhoff 1978) or as a function of the alignment score normalized to 100 residues (NAS; Doolittle 1981).

nerability of a given protein's structure to disruption by amino acid replacement seems most important. Not unexpectedly also, different parts of a given protein may change at different rates. For example, the exterior portions of a protein generally change more rapidly than the interior ones, and binding sites and catalytic assemblages are more or less conserved wherever they may be. In proteins with disulfide bonds, the cysteine residues are among the slowest changing of residues.

Natural selection and rejection. There must be a time in a protein's natural history, during its infancy, when natural selection plays a role in shaping structure in line with function. In most mature, functional proteins of the sort we ordinarily encounter, however, most of the replacements that survive are effectively neutral (for a review of various points of view, see Doolittle 1979). In either case the vast majority of amino acid replacements inflicted by mutation must be disruptive and are weeded out by natural rejection.

There are exceptions, of course. In the case of parasitic organisms, whether they be viruses, bacteria, protozoa, or worms, the evasion of immunologic surveillance by rapid change is likely to be due, in large measure, to selection.

Lookback times. How far back can we look? In theory, it is possible to identify remnants of the very first coded proteins that appeared in the most primitive forms of life on the Earth 3.5 billion years ago. In this paper, however, we will limit ourselves to the last 2 billion years or so. Naturally, the slower a protein is changing, the more useful it will be for connecting very distantly related species or other related proteins. Indeed, if we know how fast a protein is changing, as the result of comparing its sequence from two divergent species, then we can estimate how far back we can follow that sequence and still recognize it, presuming a constant rate of change (Table 2).

Some of the most ancient proteins are changing very slowly. As has been remarked upon by others (Alberts et al. 1983), this is at least partly because they have had so long to attain perfection. Many are so good at their

Table 2. Different Rates of Change for Different Proteins

Protein	PAMs[a]/100 residues 10^8 years	Theoretical lookback time[b]
Pseudogenes	400	45[c]
Fibrinopeptides	90	200[c]
Lactalbumins	27	670[c]
Lysozymes	24	750[c]
Ribonucleases	21	850[c]
Hemoglobins	12	1.5[d]
Acid proteases	8	2.3[d]
Triosephosphate isomerase	3	6[d]
Phosphoglyceraldehyde dehydrogenase	2	9[d]
Glutamate dehydrogenase	1	18[d]

[a]PAMs, Accepted point mutations.
[b]Useful lookback time = 360 PAMs.
[c]Million years.
[d]Billion years.

job that they are under "diffusion control." In this regard, metabolic enzymes from humans and prokaryotes are usually in the neighborhood of about 40–50% identical. The relationship of the human enzyme 3-phosphoglyceraldehyde dehydrogenase to those from a range of organisms is shown in Table 3.

New proteins from old. Most "new" proteins are derived from the blueprints of "old," or at least previously existing, proteins as the result of gene duplications. Very often the rate of change of the descendants remains the same as that of the ancestral protein, although occasionally a speed-up occurs if one of the descendants assumes a less demanding role. An oft-quoted example of the latter involves lactalbumin, an accessory protein found in milk, that is descended from lysozyme, but is changing somewhat faster than enzyme (Table 2).

Not all proteins need be the result of gene duplications of previously existing proteins. Highly repetitive proteins may be the result of serial tandem duplications of very short stretches of DNA and, as such, may be regarded as having de novo origins (Ycas 1972). Other proteins have been fashioned by the genetic rearrangement of portions of preexisting proteins (vide infra).

Table 4. Classification of Human Proteins by Invention Period

I. Ancient proteins
   A. First editions. Direct-line descendancy to human and contemporary prokaryotes. Mostly mainstream metabolism enzymes. Example: triosephosphate isomerase (46% identical).
   B. Second edition. Homologous sequences in human and prokaryotic proteins, but apparently different functions or special circumstances. Example: glutathione reductase (human red blood cells) and mercury reductase (*Pseudomonas*)(27% identical).

II. Middle-age proteins. Proteins found in most eukaryotes but prokaryotic counterparts are as yet unknown. Example: actin.

III. Modern proteins
   A. Recent vintage. Proteins found in animals or plants but not both. Not found in prokaryotes. Example: collagen.
   B. Very recent inventions. Proteins found in vertebrate animals but not elsewhere. Example: plasma albumin.
   C. Recent mosaics. Modern proteins clearly the result of exon shuffling. Example: LDL receptor.

DISCUSSION

Categorizing the Known Human Protein Sequences

We have grouped the known sequences of human proteins by the time of their likely invention (Table 4). As noted above, most mainstream enzymes are remarkedly similar to their prokaryotic counterparts. We will refer to these ancient proteins as "first editions," by which we imply a straight-line descendancy without change of function since the time when prokaryotes and eukaryotes last shared a common ancestry about 2 billion years ago.

There is another group of ancient proteins, also mainly enzymes, that we call "second editions," if only because they are the result of gene duplications involving other known proteins. The distinction is a subtle one and does not imply that a closer prokaryotic relative does not exist: Merely that the sequences we presently have for comparison are "cousins" rather than straight-line descendants. An example is the human enzyme hypoxanthine-guanine phosphoribosyl transferase (HGPRT) which is homologous with the prokaryotic enzyme glutamine phosphoribosyltransferase (Table 5).

"Middle-Age" proteins in this classification scheme are those that are found more or less ubiquitously in eukaryotes but, so far, have not been found in prokaryotes. Actin, for example, appears to have been invented about the time of the major eukaryotic divergence. Although a number of actin sequences have been determined for fungi, plants, and animals, and although the actin sequence is very conservative and slow-changing, we can find nothing in the existing sequence collections that remotely resembles it among the prokaryotes.

Modern proteins. Modern proteins, by this designation, are those that occur in either plants or animals, but not both, and are not found in prokaryotes. For example, collagen is found in both invertebrates and vertebrates. Of even more recent vintage are those proteins that appear to be exclusive to vertebrates. Plasma albumin is a protein with no invertebrate counterpart and whose invention remains mysterious.

Table 3. Percent Identities and Relative Evolutionary Distances for the Sequence of 3-Phosphoglyceraldehyde Dehydrogenase

	Human	Pig	Lobster	Yeast	*B. stearothermophilus*
				Percent identities	
Human	–	90	73	65	54
Pig	11	–	72	66	52
Lobster	34	34	–	67	52
Yeast	46	45	44	–	54
B. stearothermophilus	73	76	76	73	–
	Relative evolutionary distances				

Table 5. Some Ancient Human Proteins with Prokaryotic Counterparts

A. First edition type

Human protein	Prokaryotic comparison	Resemblance (%)
Triosephosphate isomerase	*E. coli*	46
Phosphoglyceraldehyde dehydrogenase	*B. stearothermophilus*	52
Alkaline phosphatase	*E. coli*	31
Dihydrofolate reductase	*E. coli*	30
Superoxide dismutase (Cu-Zn)	*P. leiognathi*	26

B. Second edition type

Human protein	Prokaryotic protein	Resemblance (%)
Glutathione reductase	Mercuric reductase, *Pseudomonas*	27
Glutamate dehydrogenase (NAD)	Glutamate dehydrogenase (NADP), *E. coli*	26
Ornithine transcarbamylase	Aspartate transcarbamylase, *E. coli*	26
Hypoxanthine-guanine phosphoribosyl transferase	Glutamine phosphoribosyl-PP_i transferase, *E. coli*	19

Finally, there are a number of vertebrate proteins that have been identified recently as being the result of some degree of exon shuffling. Some of these are enzymes of ancient stock that have had appendages of one sort or other attached (Fig. 2). Tissue plasminogen activator, which is a typical serine protease, has a collection of added units at its amino terminus that clearly have been acquired from other proteins, including fibronectin and epidermal growth factor-type proteins (Banyai et al. 1983). The blood clotting factors IX and X and Protein C have a similar history, being serine proteases that have acquired regulatory devices by some kind of genetic exchange with the epidermal growth factor precursor (Scott et al. 1983; Doolittle et al. 1984). The low-density lipoprotein (LDL)-receptor is another protein that has combined preexisting units (Yamamoto et al. 1984; Sudhof et al. 1985), including contributions from both EGF-like domains and another unit of about the same size that is found in the complement protein C9 (Stanley et al. 1985). More recently, both these segment types, which are about 40 residues long and contain three disulfide bonds (Fig. 3), have been found within two invertebrates proteins, the "notch" protein from *Drosophila* (Wharton et al. 1985) and the *lin-12* protein from a nematode worm (Greenwald 1985).

The distribution of introns has been studied in several of these systems (Ny et al. 1984; Sudhof et al. 1985), and it is quite obvious that exon shuffling resulting from recombination within introns lies at the heart of the invention of these mosaic proteins.

The Intron Controversy: Early or Late Invention

Introns were discovered not very long after the notion was introduced that certain modular units in proteins—specifically nucleotide-binding domains—may have been shuffled about during the earliest stages of life on Earth. The basis for this thinking derived from structural comparisons of NAD-binding segments from a series of proteins whose X-ray structures were known (Rossmann et al. 1974). It was postulated that a prototype NAD-binding domain could have combined with various peptide segments that bound different substrates.

Introns seemed a simple way of encouraging such combinations and recombinations (Blake 1978; Gilbert 1978), and the possible importance to protein evolution was widely heralded. Still, the evidence for such a role was, for the most part, only circumstantial, and a number of energetic if indirect efforts had to be made

Table 6. Some Animal Proteins That Have Sequence Segments in Common

A. EGF-type
- Epidermal growth factor precursor
- Tumor growth factors
- Low-density lipoprotein receptor
- Factor IX
- Factor X
- Protein C
- Tissue plasminogen activator
- Urokinase
- Complement C9
- Notch protein (*Drosophila*)
- *lin-12* (nematode)

B. C9-type
- Complement C9
- Low-density lipoprotein receptor
- (Fibronectin?)
- Notch (*Drosophila*)
- *lin-12* (nematode)

C. Fibronectin "finger"
- Fibronectin
- Tissue plasminogen activator

D. Proprotease "Kringle"
- Plasminogen
- Tissue plasminogen activator
- Urokinase
- Prothrombin

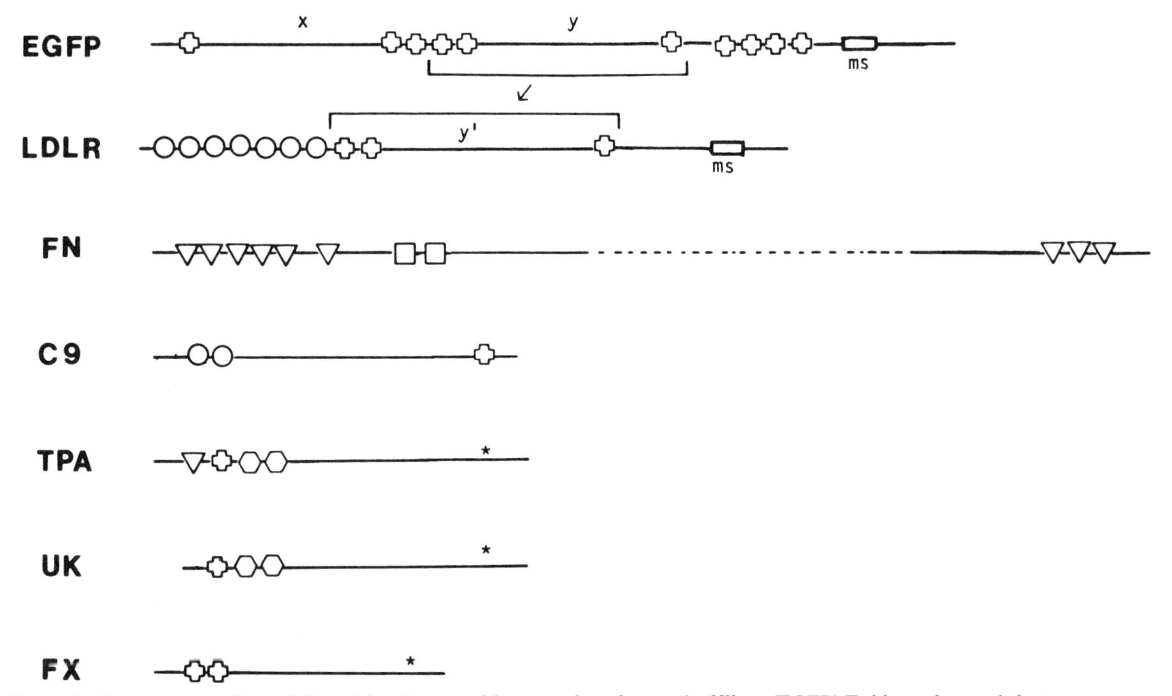

Figure 2. Some modern "mosaic" proteins that provide examples of exon shuffling. (EGFP) Epidermal growth factor precursor; (LDLR) low-density lipoprotein receptor; (FN) fibronectin; (C9) complement component C9; (TPA) tissue plasminogen activator; (UK) urokinase; (FX) blood coagulation Factor X; (MS) membrane spanning units; (*) active site of serine proteases. The sections labeled x, y, and y′ have homologous sequences, over and beyond the five modular units described in Fig. 3. (Reprinted, with permission, from Doolittle 1985.)

Symbol	Typical Occurrence	Unit Length	Disulfides	Disulfide Arrangement
○	Complement C9	40 ± 2	3	
✥	Epidermal Growth Factor	40 ± 2	3	
▽	Fibronectin, Type I	45 ± 2	2	
□	Fibronectin, Type II	60 ± 2	2	
⬡	Proprotease "Kringle"	80 ± 2	3	

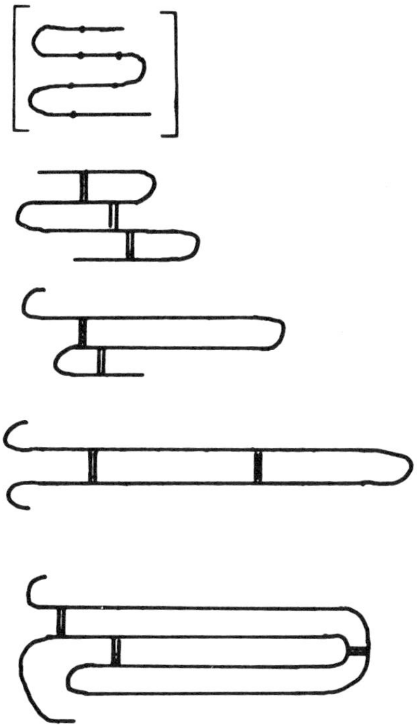

Figure 3. Modular units frequently shuffled among vertebrate proteins. Note the placement of the various units in the mosaic proteins illustrated in Fig. 2. The disulfide arrangement in the C9-type unit is not known. (Reprinted, with permission, from Doolittle 1985.)

to buttress the hypothesis. Among these were many attempts to correlate the locations of introns with the boundaries of structurally independent segments, or domains. Unfortunately, it was mostly a subjective attack, with arrows boldly proclaiming the occasional success. In the cases of many proteins even the most ardent proclaimers conceded a lack of correlation (Blake 1985).

There was another flaw in the grand plan; the genes for prokaryotic proteins did not contain introns. To explain this and save the exonic playground, Ford Doolittle (1978) proposed that prokaryotes had simply discarded all their introns in the interests of a more effective and rapid replication of their circular genomes.

The discovery of rearranged units in the vertebrate proteins noted above (Fig. 2) strengthened the general hypothesis, however, and in short order the notion that introns were present from the very earliest living systems was on its way to becoming dogma. Moreover, the discovery of an intron in a bacteriophage protein (Chu et al. 1984) lent credence to the notion that prokaryotic genes actually contained intervening sequences in the past. Finally, the discovery that splicing could be achieved autocatalytically and that it follows a common theme in all systems (Cech 1986) seemed to settle the matter.

There are still a few facts that the skeptic might like explained, however. In this regard, most students of

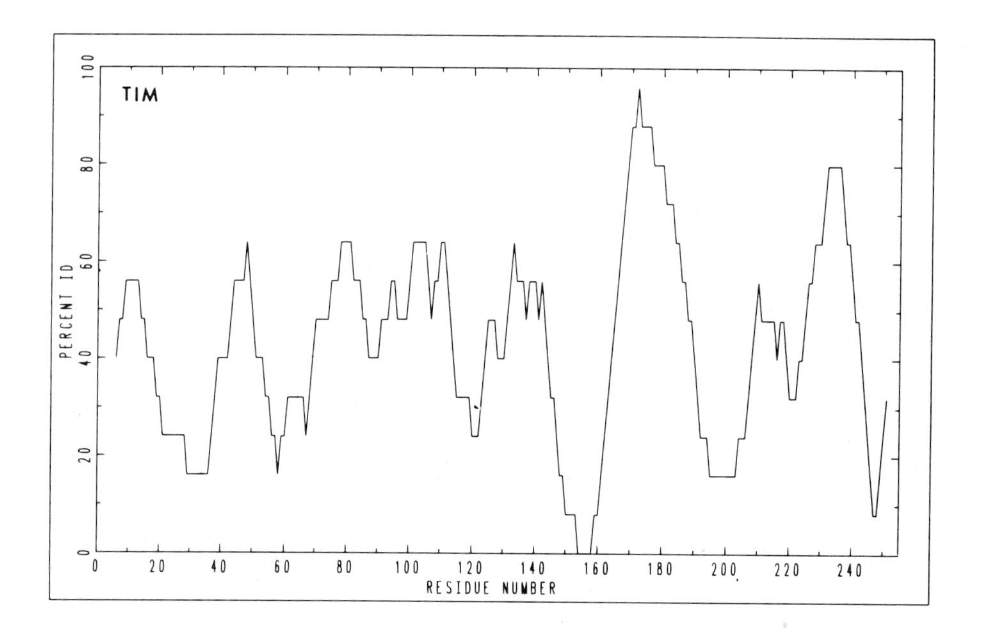

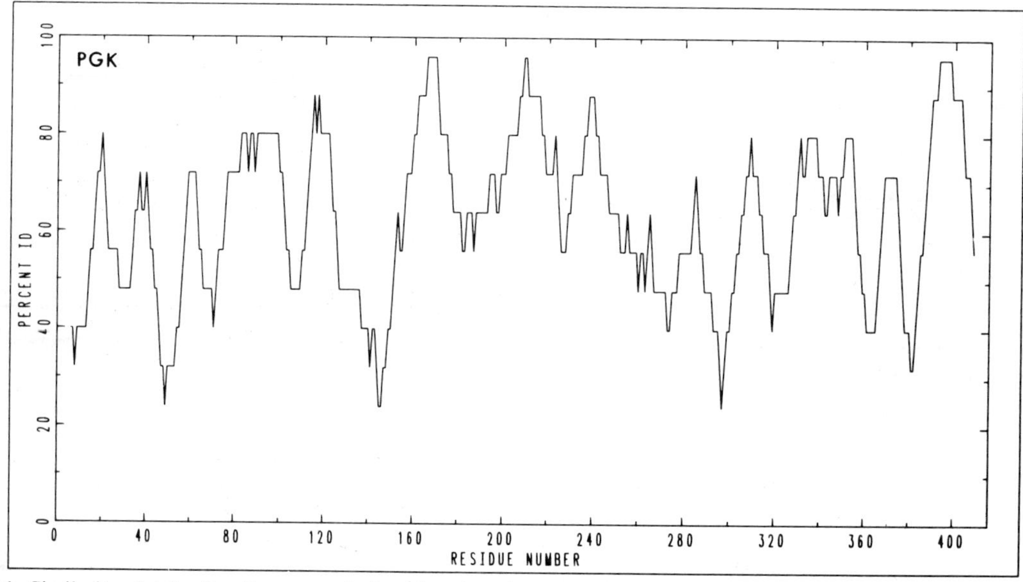

Figure 4. Similarity plot for the alignment of triosephosphate isomerase (TIM) from human and *E. coli* (*upper panel*) and for the alignment of phosphoglyceraldehyde kinase (PGK) from human and *E. coli* (*lower panel*). The tracing reflects the percent identity averaged over 20 residues after optimal alignment.

evolution still regard prokaryotes as being ancestral to eukaryotes. This being the case, then the major endosymbiotic events (the "big gulps") must have occurred before prokaryotes lost their introns. Thus, after a 2-billion-year history of life with intervening sequences (from −3.5 to −1.5 billion years ago), prokaryotes then shed their introns during the next 1.5–2 billion years, while eukaryotes, in contrast, tenaciously retained them.

Interestingly, none of the ancient protein descendants leading to the human line has engaged in exon shuffling during that 2 billion years, even though the enhanced recombinatorial wherewithal has been available. We can say this with certainty because these mainstream enzymes have maintained resemblance to prokaryotic stock from the amino terminus to the carboxyl terminus (Fig. 4). The sequences do not exhibit the sharp discontinuities and hybrid qualities that are seen in the shuffling of units in modern proteins, as is so apparent in the rearranged EGF or complement C9 units. Moreover, the exons in these humans (or other mammalian) sequences do not correlate at all with the regions of greatest conservation (Fig. 5).

A CONCLUDING COMMENT

Despite the large size of today's sequence collections, only a small fraction of contemporary proteins is rep-

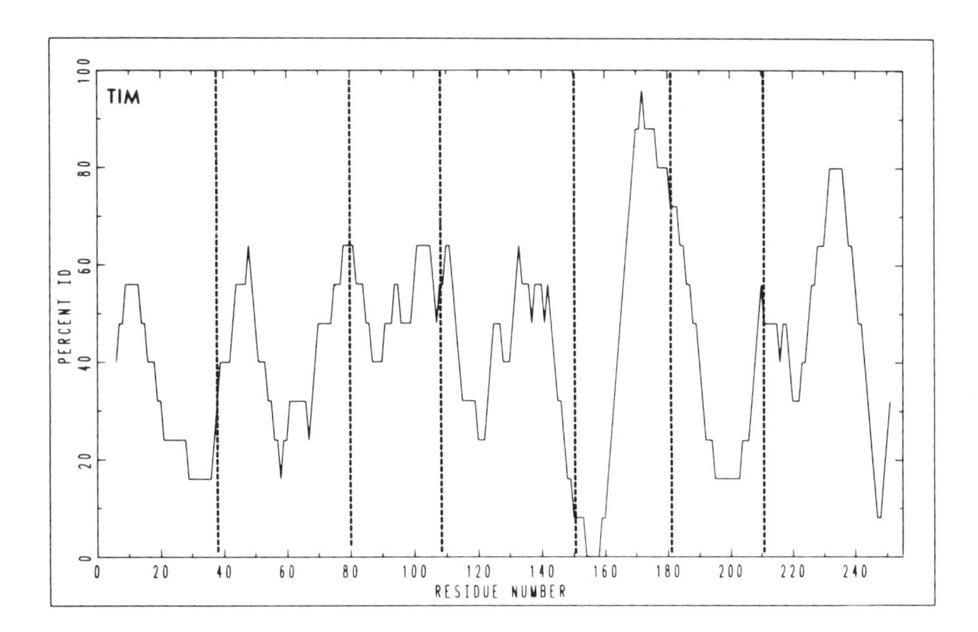

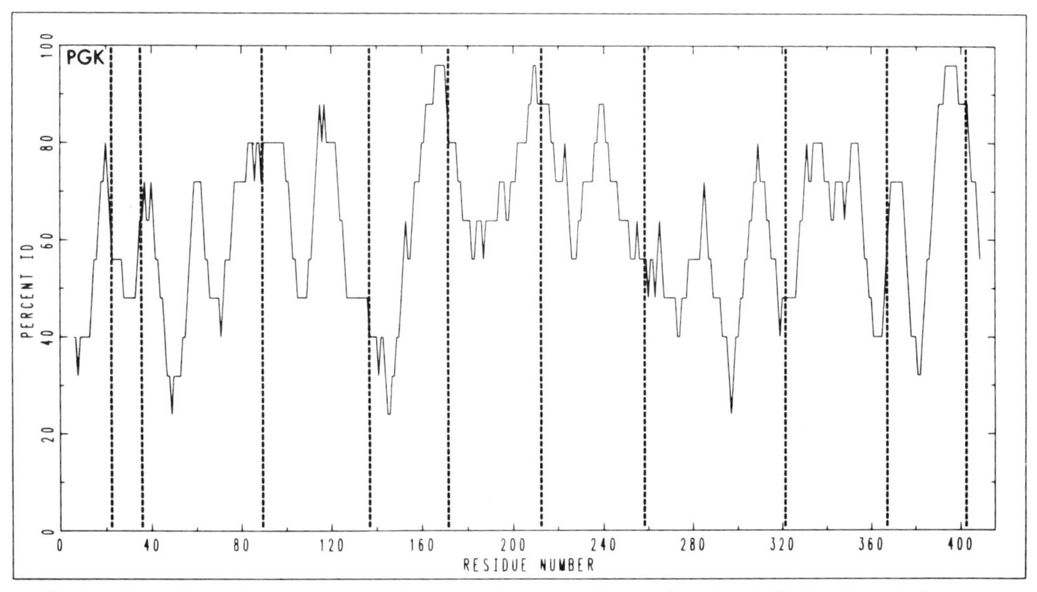

Figure 5. The locations of exon-intron-exon junctions (- - - -) are superimposed on the similarity plots of alignment depicted in Fig. 4. (TIM) Triosephosphate isomerase; (PGK) phosphoglyceraldehyde kinase. Note that the intron locations are not associated with regions of low sequence homology but are randomly distributed. A similar study (but with different conclusions) has been reported for glyceraldehyde 3-phosphate dehydrogenase by Stone et al. (1985).

```
Human   T E C L T T V K S V N K T D S Q T L L T T F G S L E Q L I A
Coli    T S S L E T I E G V G P K R R Q M L L K Y M G G L Q G L R N

Human   A S R E D L A L C P G L G P Q K A R R L F D V L H E
Coli    A S V E E I A K V P G I S Q G L A E K I F W S L K H
```

Figure 6. Alignment of a carboxy-terminal fragment from the human DNA repair enzymes (residues 236–291; van Duin et al. 1986) showing homology to a carboxy-terminal portion of the UV repair enzyme subunit UvrC from *E. coli* (residues 533–588; Sancar and Rupp 1983). Nineteen of the 56 aligned residues are identical. The significance of the authentic match score is more than 10 standard deviations higher than the average scores of randomized comparisons.

resented. During the next few years, newly reported sequences will likely clarify a still fuzzy picture of early events. Perhaps some ancient proteins will be found that do in fact exhibit evidence of rearranged segments of the sort observed in modern proteins. Recently, in a collaboration with Aziz Sancar and his co-workers (Doolittle et al. 1986), we used the computer to search and characterize the *E. coli* UvrABC excision nuclease, a DNA repair enzyme composed of three subunits. One of the interesting features to emerge from our searching was a resemblance between the subunit UvrC and a recently reported sequence for the human excision nuclease *ERCC-1* (Fig. 6). The resemblances, while statistically very sound, occur in very different parts of the polypeptide chains. Whether this could possibly be a vestige of some antiquitous exon shuffle remains to be seen.

In the meantime, the unanswered questions still seem to be the following:

1. Are the introns we see today direct-line descendants of prebiotic junk sequences?
2. Did prokaryotes really rid themselves of introns, or did they never have them?
3. Are introns still being introduced in some situations?
4. How did introns get into modern proteins like actin, or collagen, or plasma albumin?

The answers to these questions lie hidden in our genes and those of all other extant creatures. When all the data are in, the dispute will almost certainly have disappeared.

ACKNOWLEDGMENTS

We thank Karen Anderson for assistance in maintaining the NEWAT sequence collection and for considerable help in assembling this manuscript. This study was supported by National Institutes of Health grant GM 32833 and a grant from the American Cancer Society.

REFERENCES

Alberts, B., D. Bray, J. Lewis, M. Raff, K. Roberts, and J.D. Watson. 1983. *Molecular biology of the cell*. Garland, New York.

Barker, W.C., L.T. Hunt, D.G. George, and B.C. Orcutt. 1985. A resource for protein identification. In *The role of data in scientific progress* (ed. P.S. Glaser), p. 127. Elsevier Scientific, North-Holland, Amsterdam.

Blake, C.C.F. 1978. Do genes-in-pieces imply protein-in-pieces? *Nature* **273:** 267.

———. 1985. Exons and the evolution of proteins. *Int. Rev. Cytol.* **93:** 149.

Banyai, L., A. Varadi, and L. Patthy. 1983. Common evolutionary origin of the fibrin-binding structures of fibronectin and tissue-type plasminogen activator. *FEBS Lett.* **163:** 37.

Cech, T.R. 1986. The generality of self-splicing RNA: Relationship to nuclear mRNA splicing. *Cell* **44:** 207.

Chu, F.K., G.F. Maley, F. Maley, and M. Belfort. 1984. Intervening sequence in the thymidylate synthase gene of bacteriophage T4. *Proc. Natl. Acad. Sci.* **81:** 3049.

Dayhoff, M.O., ed. 1978. A model of evolutionary change in proteins. Matrices for detecting distant relationships. In *Atlas of protein sequence and structure* (ed. M.O. Dayhoff), vol. 5, p. 345. National Biomedical Foundation, Washington, D.C.

Doolittle, R.F. 1979. Protein evolution. In *The proteins* (ed. H. Neurath and R.L. Hill), vol. 4, p. 1. Academic Press, New York.

———. 1981. Similar amino acid sequences: Chance or common ancestry? *Science* **214:** 149.

———. 1985. The genealogy of some recently evolved vertebrate proteins. *Trends Biochem. Sci.* **10:** 233.

Doolittle, R.F., D.-F. Feng, and M.S. Johnson. 1984. Computer-based characterization of epidermal growth factor precursor. *Nature* **307:** 558.

Doolittle, R.F., M.S. Johnson, I. Hussain, B. van Houten, D.C. Thomas, and A. Sancar. 1986. Domainal evolution of a prokaryotic DNA-repair and its relationship to active transport proteins. *Nature* (in press).

Doolittle, W.F. 1978. Genes in pieces: Were they ever together? *Nature* **272:** 581.

Feng, D.F., M.S. Johnson, and R.F. Doolittle. 1985. Aligning amino acid sequences: Comparison of commonly used methods. *J. Mol. Evol.* **21:** 112.

Gilbert, W. 1978. Why genes in pieces? *Nature* **271:** 501.

Greenwald, I. 1985. *lin-12*, a nematode homeotic gene, is homologous to a set of mammalian proteins that includes epidermal growth factor. *Cell* **43:** 583.

Ny, T., F. Elgh, and B. Lund. 1984. The structure of the human tissue-type plasminogen activator gene: Correlation of intron and exon structures to functional and structural domains. *Proc. Natl. Acad. Sci.* **81:** 5355.

Rossmann, M.G., D. Moras, and K.W. Olsen. 1974. Chemical and biological evolution of a nucleotide-binding protein. *Nature* **250:** 194.

Sancar, A. and W.D. Rupp. 1983. A novel repair enzyme: UVRABC excision nuclease of *Escherichia coli* cuts a DNA strand on both sides of the damaged region. *Cell* **33:** 249.

Scott, J., M. Urdea, M. Quiroga, R. Sanchez-Pescador, N. Fong, M. Selby, W.J. Rutter, and G.I. Bell. 1983. Structure of a mouse submaxillary messenger RNA encoding epidermal growth factor and seven related proteins. *Science* **221:** 236.

Stanley, K.K., H.-P. Kocher, J.P. Luzio, P. Jackson, and J. Tschopp. 1985. The sequence and topology of human complement component C9. *EMBO J.* **4:** 375.

Stone, E.M., K.N. Rothblum, and R.J. Schwartz. 1985. Intron-dependent evolution of chicken glyceraldehyde phosphate dehydrogenase gene. *Nature* **313:** 498.

Sudhof, T.C., D.W. Russell, J.L. Goldstein, M.S. Brown, R. Sanchez-Pescador, and G.I. Bell. 1985. Cassette of eight exons shared by genes for LDL receptor and EGF precursor. *Science* **228:** 893.

van Duin, M., J. de Wit, H. Odjik, A. Westerveld, A. Yasui, M.H.M. Koken, J.H.J. Hoeijmakers, and D. Bootsma. 1986. Molecular characterization of the human excision repair gene *ERCC-1*: cDNA cloning and amino acid homology with the yeast DNA repair gene. *RAD10*. *Cell* **44:** 913.

Wharton, K.A., K.M. Johansen, T. Xu, and S. Artavanis-Tsakonas. 1985. Nucleotide sequence from the neurogenic locus notch implies a gene product that shares homology with proteins containing EGF-like repeats. *Cell* **43:** 567.

Yamamoto, T., C.G. Davis, M.S. Brown, W.J. Schneider, M.L. Casey, J.L. Goldstein, and D.W. Russell. 1984. The human LDL receptor: A cysteine-rich protein with multiple Alu sequences in its mRNA. *Cell* **39:** 27.

Ycas, M. 1972. *De novo* origin of periodic proteins. *J. Mol. Evol.* **2:** 17.

The Abundant LINE-1 Family of Repeated DNA Sequences in Mammals: Genes and Pseudogenes

J. SKOWRONSKI* AND M.F. SINGER

Laboratory of Biochemistry, National Cancer Institute, National Institutes of Health, Bethesda, Maryland 20892

Mammalian genomes are peppered with the members of several families of dispersed, highly repeated sequences (for reviews, see Jelinek and Schmid 1982; Rogers 1985; Singer and Skowronski 1985; Deininger and Daniels 1986). For present purposes, the definition of "highly repeated" is more than 10^4 copies per haploid genome. About 8% of human and other Old World primate DNA is accounted for by the two most abundant families, *Alu* and LINE-1 (L1). The 300-bp long *Alu* family members are classified as SINES, and their counterparts in other mammalian species are a panoply of short interspersed repetitive families. In general, SINES are species or order specific and are homologous, at least in part, to class III genes such as those for 7SL RNA (e.g., primate *Alu* and rodent B1) (Ullu and Tschudi 1984) and tRNA (the B2 rodent family) (Deininger and Daniels 1986). Several different SINE families occur in many species, although in the Old World primates one, the *Alu* family, predominates. Two families of LINES (for long interspersed repeated sequences) have been described in humans; the LINE-1 (or L1) family and the THE-1 family (Paulson et al. 1985). Families homologous to L1 occur in all placental mammals and marsupials examined to date (Burton et al. 1986; T. Fannig, pers. comm.).

A general schematic representation of a human L1 segment is shown in Figure 1. The longest known family members are about 6.1 kbp and appear to define a unit length (Hattori et al. 1985). Such unit-length family members occur about 4×10^3 times in primate genomes (Adams et al. 1980; Grimaldi et al. 1984). As many as 10^5 truncated L1s are also found and these almost always lack sequences from the 5′ end of the unit length (Kole et al. 1983; Grimaldi et al. 1984; Hwu et al. 1986). In addition, many L1 elements contain internal deletions or inversions. There are no direct or inverted repeats at the termini of unit-length L1s; the 3′ ends are typically rich in A residues. Many, although not all, L1 units are, however, flanked by short direct repeats. The flanking repeats differ in length and sequence from one L1 to another and are generally assumed to be target site duplications, as was demonstrated directly in two cases (Thayer and Singer 1983; Wakasugi et al. 1985). The homologous L1 families in rodents (Soares et al. 1985; D'Ambrosio et al. 1986;

*Present address: Cold Spring Harbor Laboratory, Cold Spring Harbor, New York 11724.

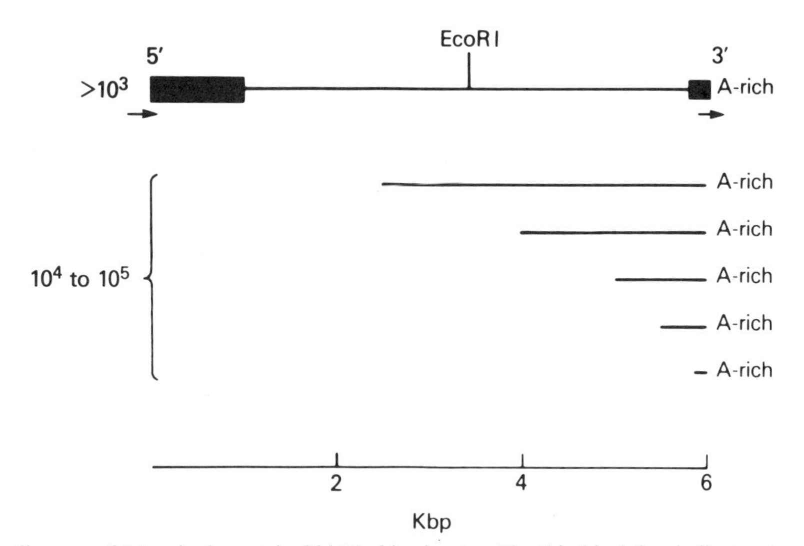

Figure 1. Schematic diagram of L1 unit element in Old World primates. The thin black bar indicates the region that contains extensive open reading frames and is highly conserved in other mammals. The heavy bars at the 3′ and 5′ ends indicate the regions that are less conserved. The small arrows indicate the flanking direct repeats found around many but not all L1 units. The diagrams below the unit element show schematically the truncated forms of most L1 elements; on the left is summarized the relative abundance of unit-length and truncated elements.

Loeb et al. 1986), lagomorphs (Demers et al. 1986), and carnivores (Katzir et al. 1985; T. Fanning, pers. comm.) appear to have similar structural properties, although the details differ.

An internal portion of the unit length L1, ~4-5 kbp, is conserved in all mammals; nucleotide sequence homology between mouse and primate varies from 50% or less near the 5′ end to about 70% near the 3′ end of the conserved region. In contrast, the sequences at the 5′ and 3′ ends of L1 units are not highly conserved; they vary in length and consensus nucleotide sequence. The variation is extensive between L1s from different mammalian orders, but is less marked when L1s from species within an order are compared.

The sequences of randomly cloned genomic L1 segments indicate the presence of open reading frames (in the 5′ to 3′ direction of the strand shown in Fig. 1) covering most of the 4-5 kbp of conserved sequence (Manuelidis 1982; Martin et al. 1984; Potter 1984; Loeb et al. 1986). However, no genomic family member with an unbroken 5 kbp open reading frame has been reported from any species. In several instances, a small number of base changes could join the individual open reading frames into a 5-kbp-long open frame; the interruptions do not have the properties of introns. The stop codons that terminate the 3′-most open reading frame mark, in each case, the end of the interspecies homologies and the start of the nonconserved 3′ ends. The interspecies homologies between the nucleotide sequences yield corresponding homologies among the predicted amino acid sequences for at least 3900 bp, and the ratio of silent to replacement amino acid changes calculated from the nucleotides coding for these amino acids is that expected for a conserved protein coding sequence (Demers et al. 1986; Loeb et al. 1986).

The structures of human and other mammalian L1 units resemble processed pseudogenes in several ways: (1) a conserved potential protein coding region incapacitated by single-base-pair changes, small deletions, and insertions; (2) nonconserved regions that could represent the 5′ leader and 3′ trailer segments of an mRNA; (3) A-rich 3′ terminal segments; and (4) flanking direct repeats that represent target site duplications. The possibility that many L1 units are processed pseudogenes suggests that one or more family members in each species may be functional genes encoding homologous proteins. Moreover, this possibility also suggests models that might account for the dispersal and amplification of L1 units (DiGiovanni et al. 1983; Lerman et al. 1983; Martin et al. 1984; Voliva et al. 1984). This paper will summarize and report data that are relevant to these suggestions.

MATERIALS AND METHODS

Unless otherwise indicated all materials and methods were as previously described (Grimaldi et al. 1984; Skowronski and Singer 1985). The cDNA library was prepared using cytoplasmic, polyadenylated RNA from NTera2D1 cells with the embryonal carcinoma morphology. *Eco*RI linkers were attached to the double-stranded cDNA which was then size-selected to enrich for molecules longer than 1.5 kbp, ligated to λgt10 arms, and packaged into phage particles in vitro (J. Skowronski, in prep.). The resulting cDNA library contained about 1.5×10^6 independent recombinants and the mean size of the inserts was 2-3 kbp. After selection with L1 probes (see Table 2, below), portions of the isolated cDNA clones were subcloned in pUC18. DNA sequencing was carried out using denatured, double-stranded plasmid DNA as template, selected oligonucleotide primers, and the dideoxynucleotide technique (Hattori et al. 1986a).

The probes used to isolate the NTera2D1 cDNA clones (see Table 2, below) were all isolated from monkey L1 elements and have been described previously. The four probes cover the following regions of the 6-kbp element described in Figure 1: p2, 0.22-0.8 kbp; p3, 1.0-1.8 kbp; p4, 1.8-3.4 kbp (Grimaldi et al. 1984); p600, 5.0-5.5 kbp (Lerman et al. 1983).

Consensus sequences represent the most abundant nucleotide at each residue within a set of homologous sequences. The cDNA consensus sequence was calculated from the data summarized in Figure 4A. The consensus of randomly cloned human genomic L1 elements was calculated using sequences reported in the literature, several of them being found by searching the Genetic Sequence Data Bank, Los Alamos National Library. The references for the sequences are as follows and the numbers correspond to the identifying numbers on Figure 4B (below): 1, Potter and Jones 1983; 2, Potter 1984; 3, Hattori et al. 1985; 4, Yoshitake et al. 1985; 5, Kelly and Trowsdale 1985; 6 and 7, Sun et al. 1984; 8, 10, and 12, DiGiovanni et al. 1983; 9, Larhammar et al. 1985; 11, Deininger et al. 1981; 13, Li et al. 1985; 14, Ullrich et al. 1982; 15, 18, 19, and 20, Collins and Weissman 1984; 16, Wakasugi et al. 1985; 17, Comb et al. 1983.

RESULTS

The Transcription of Human L1 Sequences

Nuclear RNA from various cell types is rich in L1 sequences (Kole et al. 1983; Lerman et al. 1983; Shafit-Zagardo et al. 1983; Schmeckpeper et al. 1984; Skowronski and Singer 1985). These RNA polymerase II transcripts (Shafit-Zagardo et al. 1983) range from about 0.5 to over 14 kb in length and are homologous to both L1 strands. The relative abundance of the 5′- and 3′-end sequences mirrors their genomic frequency (Skowronski and Singer 1985); thus, 3′-end sequences occur about 10 times more often than the 5′-end sequences in HeLa cell nuclear RNA. Most of these transcripts are neither transported to the cytoplasm nor polyadenylated and many are likely to represent L1 sequences that are included in unrelated transcription units (DiGiovanni et al. 1983; Kole et al. 1983; Skowronski and Singer 1985); two out of five clones isolated from human fibroblast cDNA libraries by selection with L1 probes contain L1 sequences joined to unrelated DNA segments (DiGiovanni et al. 1983; S. Contente and M.F. Singer, unpubl.). Cytoplasmic RNAs of several sizes have been reported in the lymphoblastoid cell line, Jurkat (Kole et al. 1983).

A polyadenylated, cytoplasmic L1 RNA that contains sequences spanning the entire 6 kbp of the unit-length human L1 has been reported in the human teratocarcinoma cell line, NTera2D1 (Skowronski and Singer 1985). This RNA specifically anneals to the L1 strand that is complementary to that containing the long open reading frames. It is most abundant in NTera2D1 cells having the embryonal carcinoma morphology, is barely detectable in sparsely growing cells, and is undetectable after NTera2D1 cells are induced to differentiate with retinoic acid. Primer extension experiments that are summarized in Figure 2 have recently demonstrated that the bulk of this RNA starts with the same nucleotide as was previously identified as the 5′-most residue common to the unit-length L1s (Grimaldi et al. 1984). Altogether, these results suggest that the NTera2D1 cytoplasm contains RNA that could represent mRNA produced by transcription from one or more genomic L1 units. To examine this possibility further, we analyzed clones selected with L1 DNA probes from a cDNA library constructed with NTera-2D1 RNA.

Characterization of L1 cDNA Clones

The cDNA library was constructed as described in Materials and Methods. Quadruplicate screening of phage plaques was carried out with four cloned probes that together encompass more than 5 kbp of the unit-length L1 element (Table 1). About 40 clones that annealed to all four probes (group I) or to all but the 5′-most probe (group II) were found and those that contained the single *Eco*RI site commonly found in genomic L1s (see Fig. 1) were analyzed further. Several of the inserts in group I are close to 6 kbp long. The *Eco*RI site divides the inserts into 3′ portions (about 2.5 kbp) and 5′ portions and these were subcloned; their properties, as far as is known, are summarized in Table 1. The complete nucleotide sequence of the 3′ portion of eight of the cDNAs has been determined and partial

Table 1. Summary of the Properties of the cDNA Clones Isolated Using NTera2D1 Polyadenylated, Cytoplasmic RNA

cDNA[a]	Approximate[b] length	Open/closed[c]	Percent divergence from cDNA consensus[d] ORF (2385 bp)
2 II	5.3	c	2.6
3 II	4.3	c	1.0
4 II	5.5	c	2.2
5 II	4.5	o	0.6
6 II	4.1	c	2.2
9 II	4.6	c	2.1
10 II	4.2	c	3.9
11 I	5.7	o	0.7
12 I	5.7		1.0
13 II	4.3		1.8
14/15 I[e]	6.2	o	1.6
16 I	6.2	c	0.4
21 II	4.6	o	0.6
23 I	5.4	c	0.8
25 II	4.6	o	1.0
26 I	4.0	o	1.7
27 II	4.4	c	1.2
28 II	4.6	c	5.0
29 II	6.1	c	2.9

[a]The λgt10 clones in group I annealed with probes p2, p3, p4, and p600 whereas those in group II annealed with p3, p4, and p600 but not p2.

[b]All cDNA clones listed have the *Eco*RI site at approximately 3.4 kbp from the 5′ end as shown in Fig. 1. Lengths are the sum of the sizes of *Eco*RI fragments released from the λgt10 phage.

[c]Refers to the region (2385 bp) from the *Eco*RI site to the TGA stop codon 210 bp from the 3′ end of the typical L1 (Fig. 1); characterization is based on primary sequence data. cDNAs known to have 2385-bp open reading frames are marked o for open; those known to have in-frame stop codons are marked c for closed. cD12 and cD13 are open as far as is known but the data are incomplete.

[d]Where the sequence data are incomplete, divergence from the consensus was calculated from the available sequence information.

[e]Two clones, cD14 and 15, were identical and are treated as one.

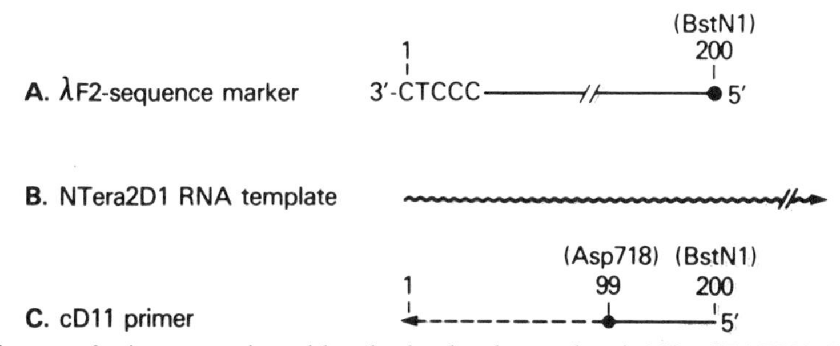

Figure 2. Schematic diagram of primer extension with polyadenylated, cytoplasmic NTera2D1 RNA (J. Skowronski, in prep.). (*A*) The cloned monkey genomic L1 sequence called λF2 showing the GC base pair previously identified as residue 1 in the unit length L1 element by alignment with other monkey and human L1 units (Grimaldi et al. 1984). The opposite strand from that in Fig. 1 is shown. The 200-bp segment from residue 1 through 200 was labeled on one strand (black dot) and Maxam-Gilbert sequencing reactions provided markers for the primer extension. (*B*) Schematic of NTera2D1 L1 RNA. (*C*) The primer (full line) was prepared by cleavage of cD11 (see text) with restriction endonucleases, labeling by filling in at the Asp-718 restriction endonuclease site using [α-^{32}P]dATP, and cleavage with endonuclease *Bst*N1. After primer extension with reverse transcriptase, the length of the extended primer (— — — —) was determined by gel electrophoresis.

sequence information in this portion was obtained for the others. Because the junction between the 3′-most part of each insert and the vector was, with one exception, unique, we concluded that each of the cDNAs came from an independent cloning event. The only exception was a pair of clones, cD14 and cD15, which were identical throughout and must have represented two selections of the same clone.

The structure of all the cDNA clones is consistent with their being derived from RNAs presenting the L1 strand with the long open reading frames. Within the limits of the available data (from the *Eco*RI site to the 3′ terminus), the cDNAs have the following properties. With one exception (cD28), all the cDNAs contain only sequences that had previously been identified as L1 sequences. Several extend about 6 kbp, ending 30–40 bp 3′ of the 5′ end predicted from the primer extension experiments; others terminate about 1300 bp 3′ of that end. No two of the cDNAs have the same sequence. Individual clones diverge from a consensus sequence of all the cDNAs an average of 1.7% and 3.6% in the potential coding and 3′-trailer regions, respectively (Table 1 and Fig. 4A). Notably, six of the cDNAs (cD5, 11, 14, 21, 25, and 26) contain a single unbroken reading frame stretching 795 codons (2385 bases) from the *Eco*RI site to the TGA stop codon that marks the beginning of the 3′-trailer region. All the other cDNAs contain one or more deletions, insertions, or single-base changes that introduce stop codons in the long open reading frame between the *Eco*RI site and the TGA stop codon. Additional preliminary and incomplete sequence data in the region 5′ to the *Eco*RI site indicates that the reading frame in cD26 is closed while the reading frames in cD11 and 14 extend at least 380 bp on the 5′ side of the site.

Several conclusions can be drawn from these data. First, because each of the cDNAs has a different sequence, multiple genomic L1 units must be transcribed in NTera2D1 cells. It is likely that each of these genomic units is associated with common regulatory signals that account for the specificity of transcription and transport to the cytoplasm in NTera2D1 cells with the embryonal carcinoma phenotype. Second, at most only a minority of the cytoplasmic, polyadenylated RNAs in NTera2D1 cells can be translated into a polypeptide of the length predicted by the unit length L1s. Until the sequence analysis of the 5′ portions of the inserts is complete, we will not know the full coding potential of the remaining cDNA.

The 3′ Trailer Region of the cDNAs

Figure 3 compares the consensus sequence for the 3′ trailer region of the cDNA clones with the consensus sequence for the same region as derived from random human genomic L1 segments. The 3′-trailer region is defined as the 210 bp starting with the TGA stop codon at the end of the reading frame and continuing to the 3′ end just preceding the A-rich region. The two differ at 16 positions. Figure 4A shows the divergence of each of the cDNAs from the cDNA consensus sequence and Figure 4B the divergence of each of the genomic sequences from the genomic consensus. The randomly selected genomic segments diverge much more from their own consensus than do the cDNA sequences from the cDNA consensus. Three of the 20 randomly selected human genomic segments used to construct the consensus sequence are more similar to the cDNA consensus than to the genomic consensus (sequences numbered 1, 2, and 12). In contrast, in only one of the cDNA clones does a majority of the 16 positions have the sequence typical of the genomic rather than cDNA consensus (cD6). We conclude from these data that a subset of genomic L1 segments, called here subset T, gives rise to the bulk of the polyadenylated cytoplasmic transcripts in NTera2D1 cells; subset T is defined by the special consensus sequence in the 3′-trailer region and is probably also associated with common regulatory elements, as suggested above. To distinguish subset T from other unit-length L1 units, we shall refer to those that resemble the genomic consensus sequence in the 3′-trailer region as subset U.

Within subset T, four of the cDNAs, 5, 21, 23, and 27, define a distinctive group (subset Ta). Each of these (shown on the top four lines of Fig. 4A) differs from the cDNA consensus in identical ways at four residues (the affected residues are underlined in Fig. 3). Within subset Ta, two of the cDNAs have open reading frames as far as our data extend.

The Coding Region of the cDNA Clones

In the region for which extensive sequence data are available, namely, from the *Eco*RI site to the TGA stop

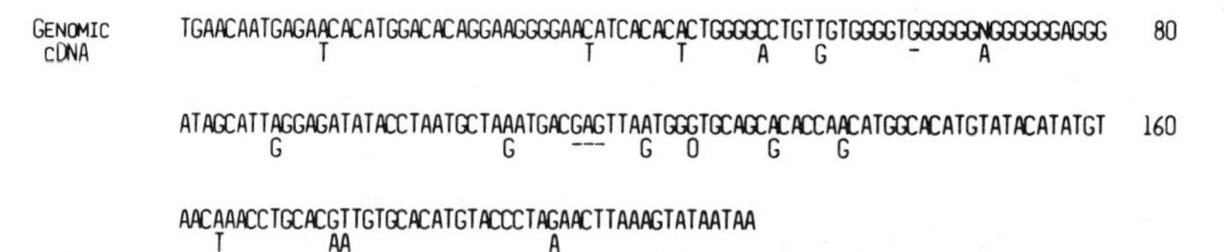

Figure 3. Consensus sequences of 20 randomly selected genomic L1s and the L1 cDNAs in the 3′-trailer region (210 bp starting with the TGA stop codon that ends the long open reading frame). The consensus represents the most frequent base at every residue. The cDNA consensus was determined from our data. The genomic consensus was constructed as described in the section on Materials and Methods. (The details are available from our laboratory upon request.) Only those residues in the cDNA consensus that differ from the genomic consensus are shown; the 0 means that the base pair is deleted. The bars under residues 64, 115, 116, and 117 indicate the base pairs that differ in subset Ta, where they are, respectively, C, A, C, and A. The strand is shown 5′ to 3′ and is the strand with the open reading frames.

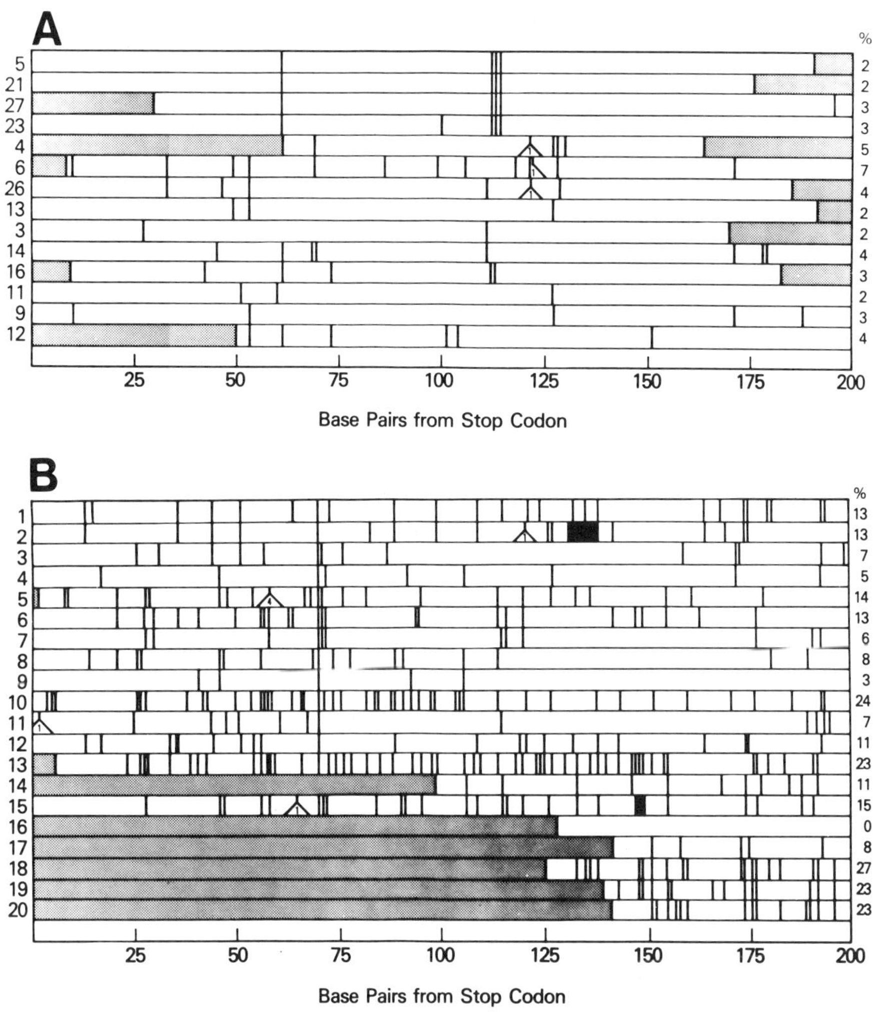

Figure 4. Divergence of individual cDNA (*A*) and human genomic (*B*) sequences from the respective consensus sequences in the 3′-trailer region. The consensus sequences are given in Fig. 3. Each line represents an individual sequence. The bars show residues where the sequence differs from the consensus. The triangles with numbers indicate the number of extra nucleotides inserted at that point. In *A*, the numbers on the left side refer to the number assigned each cDNA clone; in *B*, they refer to the sequences as described in Materials and Methods. The numbers on the right side indicate the percent that each sequence diverges from the consensus sequence.

codon, the open reading frame regions of the cDNAs diverge an average of 1.7% from the cDNA consensus. This is markedly less than the 5% divergence from their own consensus of randomly selected human genomic L1 segments (Soares et al. 1985) and is thus consistent with the occurrence of a special transcribed subset of genomic units, subset T (and subset Ta), as proposed above.

The nucleotide sequences of randomly selected genomic L1 units from human (Hattori et al. 1985), mouse (Loeb et al. 1986), and rat (D'Ambrosio et al. 1986) have recently been published. The mouse unit has a completely open reading frame in the region homologous to the sequenced portions of our cDNA clones. Comparison of the polypeptide predicted by the mouse sequence with the 795-amino-acid-long sequence predicted from the cDNAs indicates an overall homology of 65%. Thus, the cDNA structures confirm the earlier conclusions regarding the highly conserved nature of the L1 coding sequences. A more detailed comparison of the human cDNA consensus with the mouse sequences indicates that the degree of homology varies from one region to another within the 795-amino-acid-long predicted polypeptide. Taking groups of 20 amino acids at a time, the homology ranges from 35 to 95% considering only identical amino acids.

The Mobility of L1 Elements

The dispersed positions of L1 elements between genes, in introns, and interrupting tandem arrays of species-specific satellites (for review, see Soares et al. 1985), as well as the target site duplications associated with many L1s, are hallmarks of movable elements. It is reasonable to conclude that L1s were movable in the past. How recently can mobility be documented? The discovery in dogs, rats, and mice of polymorphic allelic positions that differ by the presence or absence of L1

Table 2. Mammalian Polymorphic Alleles That Differ by the Presence and Absence of L1 Units

Species	Locus	Approximate size of L1 insert (kbp)	Reference
Dog	*myc*	1.3	Katzir et al. (1985)
Rat	*Igh*	6.5	Economou-Pachnis et al. (1985)
Rat	*mlvi*-2	6.5	Economou-Pachnis et al. (1985)
Rat	insulin-1	2.7	Lakshmikumaran et al. (1985)
Mouse	β-globin	>6	Burton et al. (1985) Shyman and Weaver (1985)

units suggests that transposition occurred in the relatively recent past (Table 2). Moreover, an L1 unit has been found inserted in a *myc* allele in DNA from a human breast adenocarcinoma (C.C. Morse and P. Rothberg, pers. comm.). These data, together with the absence of any reason to argue otherwise, suggest that L1s are still capable of being inserted into new genomic loci. In this connection, it is of interest to point out the structural similarity between L1 units and two *Drosophila melanogaster* transposable elements, F (DiNocera et al. 1983) and I factors (D.J. Finnegan 1985, pers. comm.), both of which also lack long terminal repeats, end in A-rich regions, and are associated with variable-length target site duplications.

DISCUSSION

Subsets of Human L1s

The data reported here allow us to distinguish several subsets of human L1 units; subset U, subset T, and subset Ta. Earlier data had already made it clear that a variety of L1 units occur in human DNA. Within subset U there are several variant types: Those lacking variable numbers of base pairs from the 5′ end or from the 3′ end (Miyake et al. 1983) and those having inversions and/or internal deletions of portions of the typical L1 as well as those that are both truncated and rearranged. Human L1 units are also polymorphic regarding the presence of several of the typical restriction endonuclease sites. Approximately 50% of human L1s that extend to the 5′-end contain 131 additional base pairs inserted after residue 766 in the unit shown in Figure 1 (Hattori et al. 1985); it is not known if any of the cDNAs contain this insertion. In another group of L1s, the 5′-end depicted in Figure 1 appears to be replaced by a different sequence called A36, and a few thousand such units occur (Collins and Weissman 1984; P.K. Rogan and S.M. Weissman, pers. comm.). L1s with and without apparent target site duplications can also be distinguished.

Models for the Amplification and Dispersal of L1 Units

Most published speculations regarding possible mechanisms for amplification and dispersal of L1 are based on the formal structural analogy between L1s and processed pseudogenes (e.g., Lerman et al. 1983; Martin et al. 1984; Loeb et al. 1986). The widely accepted and reasonable model for the origin of processed pseudogenes is that they arise through reverse transcription of mRNAs (for review, see Rogers 1985; Vanin 1985), but this has not yet been demonstrated experimentally. The extension of the model to the L1 family has been problematic because of the lack of known corresponding functional genes or unit-length transcripts. The discovery in NTera2D1 cytoplasm of approximately 6.5-kb-long RNAs representing the putative coding strand of L1 units (Skowronski and Singer 1985) and the finding, reported here, that the 5′ end of the bulk of these RNAs corresponds to the 5′ end of the longest known L1 units indicate that an appropriate template for reverse transcription may indeed be available. Further, as pointed out previously, the occurrence of these RNAs in pluripotent embryonic cells, such as NTera2D1, lends support to the model because heritable duplications and transpositions of L1 sequences presumably take place in germ line cells or their progenitors (Skowronski and Singer 1985).

By itself, however, the reverse transcription model does not explain the extraordinary abundance of L1 elements—between 10^4 and 10^5 copies. Actin mRNA, for example, is much more abundant in NTera2D1 cytoplasm than is L1 RNA, yet an estimated total of only about 30 actin genes and pseudogenes occur in human DNA (Soriano et al. 1982). There are several possible ways in which reverse transcription of L1 RNA might be favored, including an abundance of a specific primer or a special efficiency as a template. Another possibility that has been discussed in the field for some years (see, for example Martin et al. 1984; Loeb et al. 1986) is that the functional L1 units may encode reverse transcriptase. According to this model, when the translation of an L1 mRNA is completed, the proximity of template and enzyme might favor the reverse transcription of the L1 mRNA itself. Analysis of rodent (Loeb et al. 1986) and human (Y. Sakaki et al., pers. comm.) genomic L1 sequences as well as of the human cDNAs described here indeed indicates that the 2300 bp of open reading frame at the 3′ side of the long L1 unit encodes a polypeptide with homology to the conserved peptides in a variety of retrovirus and retrotransposon reverse transcriptases (Patarca and Haseltinc 1984). Moreover, those regions of L1 that have homology to reverse transcriptase (Hattori et al. 1986b; Loeb et al. 1986) overlap with the portions of human and mouse L1s that are most highly conserved between the species. It will be of great interest to learn whether an active reverse transcriptase can be generated by translation of L1 RNA. In considering these models we point out that all known reverse transcriptions appear to occur within ribonucleoprotein particles. This is true of retroviruses, of retrotransposons, of cauliflower mosaic virus, and probably also of the hepadnaviruses (see, for example, Garfinkel et al. 1985; Seeger et al. 1986). No such particles involving L1 transcripts have been described, although this would be one way of assuring the proximity of L1 RNA and reverse transcriptase. Terminal inverted and direct repeats are another feature of the templates and products of reverse transcription that is unknown to L1 sequences. As pointed out above, the *Drosophila* F and I factor transposable elements also lack terminal repeats.

If the L1 transcription units that give rise to polyadenylated, cytoplasmic RNA in NTera2D1 cells are like most class II transcription units, they may be associated with external regulatory signals. Each of the transcribed units may then be surrounded, on one or both ends, by homologous sequences that are not found in subset U. Amplification by the reverse transcription model, as described above, would not preserve any surrounding regulatory sequences. This suggests that L1 subsets T and Ta may arise through a mechanism that is different from that operating with subset U. One possibility is that as yet unidentified transcripts longer than unit length mediate the amplification of subsets T and Ta. Another is that the transcriptional regulatory signals reside within the transcribed unit rather than the flanking DNA. In this case, all L1s might arise by replicative transposition through reverse transcription, with the distinctive subsets being separately homogenized. While it might have been expected that all subsets would belong to one homogenization network, concerted evolution, operating, for example, through gene conversion, may tend to homogenize more efficiently within a subset than between them. Yet another model is that subsets T and Ta arose by reiteration of sequences at the DNA level by one or more mechanisms, such as unequal crossing-over or reiterative replication or recombination with extrachromosomal DNA circles containing L1 sequences. Such circles have been characterized (Jones and Potter 1985; Schindler and Rush 1985). The existence of subsets of repeated sequences that are distinguishable on functional grounds as well as mechanism of duplication would not be unique. An analogous situation occurs with another family of class II genes, the human U1 genes, pseudogenes, and processed pseudogenes (Bernstein et al. 1985).

Does the L1 Family Contain Functional Genes?

In conclusion, we review the current status of the proposal that the L1 family includes functional genes. The interspecies homology between the coding regions of L1 elements in both nucleotide sequence and predicted amino acid sequence, and the ratio of silent to replacement-type mutations, are all similar to that seen in a variety of conserved genes. Similarly, the divergence between different species within the apparently noncoding 5′ leader and 3′ trailer regions is typical of many conserved genes. At least in the human teratocarcinoma cell line NTera2D1, polyadenylated, cytoplasmic RNAs that may prove to be functional mRNAs occur. These transcripts diverge less from one another than do the randomly selected genomic units. Thus, the evidence that selective pressure is operating to maintain the reading frames in humans is now substantial.

ACKNOWLEDGMENTS

We are grateful to Michael Brownstein for the preparation of oligonucleotides, to Tom Fanning for discussion and help, and to Marshall Edgell, Oliver Smithies, Sherman Weissman, and Yoshiyuki Sakaki for giving us information prior to its publication.

REFERENCES

Adams, J.W., R.E. Kaufman, P.J. Kretschmer, M. Harrison, and A.W. Nienhuis. 1980. A family of long reiterated DNA sequences, one copy of which is next to the human beta globin gene. *Nucleic Acids Res.* **8:** 6113.

Bernstein, L.B., T. Manser, and A.M. Weiner. 1985. Human U1 small nuclear RNA genes: Extensive conservation of flanking sequences suggests cycles of gene amplification and transposition. *Mol. Cell. Biol.* **5:** 2159.

Burton, F.H., D.D. Loeb, S.F. Chao, C.A. Hutchinson III, and M.H. Edgell. 1985. Transposition of a long member of the L1 major interspersed DNA family into the mouse beta globin locus. *Nucleic Acids Res.* **13:** 5071.

Burton, F.H., D. Loeb, C.F. Voliva, S.L. Martin, M.H. Edgell, and C.A. Hutchinson III. 1986. Conservation throughout mammalian and extensive protein coding capacity of the highly repeated DNA long interspersed sequences 1. *J. Mol. Biol.* **187:** 291.

Collins, F.S. and S.M. Weissman. 1984. The molecular genetics of human hemoglobin. *Prog. Nucleic Acid Res. Mol. Biol.* **31:** 317.

Comb, M., H. Rosen, P. Seeburg, J. Adelman, and E. Herbert. 1983. Primary structure of the human proenkephalin gene. *DNA* **2:** 213.

D'Ambrosio, E., S.D. Waitzkin, R.R. Witney, A. Salemme, and A.V. Furano. 1986. Structure of the highly repeated long interspersed DNA family (LINE or L1Rn) of the rat. *Mol. Cell. Biol.* **6:** 411.

Deininger, P.L. and G.R. Daniels. 1986. The recent evolution of mammalian repetitive DNA elements. *Trends Genet.* **2:** 76.

Deininger, P.L., D.J. Jolly, D.M. Ruben, T. Friedman, and C.W. Schmid. 1981. Base sequence studies of 300 nt renatured repeated human DNA clones. *J. Mol. Biol.* **151:** 17.

Demers, G.W., K. Brech, and R.C. Hardison. 1986. Long interspersed L1 repeats in rabbit DNA are homologous to L1 repeats of mouse and human in an open reading frame. *Mol. Biol. Evol.* **3:** 179.

DiGiovanni, L., S.R. Haynes, C. Misra, and W.R. Jelinek. 1983. *Kpn*I family of long-dispersed repeated DNA sequences of man: Evidence for entry into genomic DNA of DNA copies of polyA-terminated *Kpn*I RNAs. *Proc. Natl. Acad. Sci.* **80:** 6533.

DiNocera, P.P., M.E. Digan, and I.B. Dawid. 1983. A family of oligo-adenylate-terminated transposable elements in *Drosophila melagnogaster. J. Mol. Biol.* **168:** 1983.

Economou-Pachnis, A., M.A. Lohse, A.V. Furano, and P.N. Tsichlis. 1985. Insertion of long interspersed repeated elements at the *Igh* (immunoglobulin heavy chain and *Mlvi-2* (Moloney leukemia virus integration 2) loci of rats. *Proc. Natl. Acad. Sci.* **82:** 2857.

Finnegan, D.J. 1985. Transposable elements in eukaryotes. *Int. Rev. Cytol.* **93:** 281.

Garfinkel, G.J., J.D. Boeke, and G.R. Fink. 1985. Ty element transposition: Reverse transcriptase and virus-like particles. *Cell* **42:** 507.

Grimaldi, G., J. Skowronski, and M.F. Singer. 1984. Defining the beginning and end of *Kpn*I family segments. *EMBO J.* **3:** 1753.

Hattori, M., S. Hidaka, and Y. Sakaki. 1985. Sequence analysis of a *Kpn*I family member near the 3′-end of the human β-globin gene. *Nucleic Acids Res.* **13:** 7813.

Hattori, M., H. Okamoto, and Y. Sakaki. 1986a. An improved rapid dideoxy sequencing method using denatured plasmid templates. *Anal. Biochem.* **152:** 232.

Hattori, M., S. Kuhara, O. Takenaka, and Y. Sakaki. 1986b. L1 family of repetitive DNA sequences in primates may be derived from a sequence encoding a reverse trancriptase-related protein. *Nature* **321:** 625.

Hwu, H.R., J.W. Roberts, E.H. Davidson, and R.J. Britten. 1986. Insertion and/or deletion of many repeated DNA sequences in human and higher ape evolution. *Proc. Natl. Acad. Sci.* **83:** 3875.

Jelinek, W.R. and C.W. Schmid. 1982. Repetitive sequences in eukaryotic DNA and their expression. *Annu. Rev. Biochem.* **51:** 813.

Jones, R.S. and S.S. Potter. 1985. L1 sequences in HeLa extrachromosomal circular DNA: Evidence for circularization by homologous recombination. *Proc. Natl. Acad. Sci.* **82:** 1989.

Katzir, N., G. Rechavi, J.B. Cohen, T. Unger, F. Simoni, S. Segal, D. Cohen, and D. Givol. 1985. "Retroposon" insertion into the cellular oncogene c-*myc* in canine transmissible venereal tumor. *Proc. Natl. Acad. Sci.* **82:** 1054.

Kelly, A. and J. Trowsdale. 1985. A 'Bam5' element in the first intron of the HLA-DPβ1 gene. *Nucleic Acids Res.* **13:** 1607.

Kole, L.B., S.R. Haynes, and W.R. Jelinek. 1983. Discrete and heterogeneous high molecular weight RNAs complementary to a long dispersed repeat family (a possible transposon) of human DNA. *J. Mol. Biol.* **165:** 257.

Lakshmikumaran, M.S., E. D'Ambrosio, L.A. Laimins, D.T. Lin, and A.V. Furano. 1985. Long interspersed repeated DNA (LINE) causes polymorphism at the insulin 1 locus. *Mol. Cell. Biol.* **5:** 2197.

Larhammar, D., B. Servenius, L. Rask, and P.A. Peterson. 1985. Characterization of an HLA DRβ psuedogene. *Proc. Natl. Acad. Sci.* **82:** 1475.

Lerman, M.I., R.E. Thayer, and M.F. Singer. 1983. *Kpn*I family of long interspersed repeated DNA sequences in primates: Polymorphism of family members and evidence for transcription. *Proc. Natl. Acad. Sci.* **80:** 3966.

Li, Q., P.W. Powers, and O. Smithies. 1985. Nucleotide sequence of 16-kilobase pairs of DNA 5′ to the human ϵ-globin gene. *J. Biol. Chem.* **260:** 14901.

Loeb, D.D., R.W. Padgett, S.C. Hardies, W.R. Shehee, M.B. Comer, M.H. Edgell, and C.A. Hutchinson III. 1986. The sequence of large L1Md element reveals a tandemly repeated 5′-end and several features found in retrotransposons. *Mol. Cell. Biol.* **6:** 168.

Manuelidis, L. 1982. Nucleotide sequence definition of a major human repeated DNA, the *Hin*dIII 1.9 kb family. *Nucleic Acids Res.* **10:** 3211.

Martin, S.L., C.F. Voliva, F.H. Burton, M.H. Edgell, and C.A. Hutchinson III. 1984. A large interspersed repeat found in mouse DNA contains a long open reading frame that evolves as if it encodes a protein. *Proc. Natl. Acad. Sci.* **81:** 2308.

Miyake, T., K. Migita, and Y. Sakaki. 1983. Some *Kpn*I family members are associated with the Alu family in the human genome. *Nucleic Acids Res.* **11:** 6837.

Patarca, R. and W.A. Haseltine. 1984. Sequence similarity among retroviruses–Erratum. *Nature* **309:** 728.

Paulson, K.E., N. Deka, C.W. Schmid, R. Misra, C.W. Schindler, M.G. Rush, L. Kadyk, and L. Leinwand. 1985. A transposon-like element in human DNA. *Nature* **316:** 359.

Potter, S.S. 1984. Rearranged sequences of a human *Kpn*I element. *Proc. Natl. Acad. Sci.* **81:** 1012.

Potter, S.S. and R.S. Jones. 1983. Unusual domains of human alphoid satellite DNA with contiguous non-satellite sequences: Sequence analysis of a junction region. *Nucleic Acids Res.* **11:** 3137.

Rogers, J.H. 1985. The origin and evolution of retroposons. *Int. Rev. Cytol.* **93:** 188.

Schindler, C.W. and M.G. Rush. 1985. The *Kpn*I family of interspersed nucleotide sequences is present on discrete sizes of circular DNA in monkey (BSC-1) cells. *J. Mol. Biol.* **181:** 161.

Schmeckpeper, B.J., A.F. Scott, and K.D. Smith. 1984. Transcripts homologous to a long repeated DNA element in the human genome. *J. Biol. Chem.* **259:** 1218.

Seeger, C., D. Ganem, and H.E. Varmus. 1986. Biochemical and genetic evidence for the hepatitis B virus replication strategy. *Science* **232:** 477.

Shafit-Zagardo, B., F.L. Brown, P.J. Zavodny, and J.J. Maio. 1983. Transcription of the *Kpn*I families of long interspersed DNAs in human cells. *Nature* **304:** 277.

Shyman, S. and S. Weaver. 1985. Chromosomal rearrangements associated with LINE elements in the mouse genome. *Nucleic Acids Res.* **13:** 5085.

Singer, M.F. and J. Skowronski. 1985. Making sense out of LINES: Long interspersed repeat sequences in mammalian genomes. *Trends Biochem. Sci.* **10:** 119.

Skowronski, J. and M.F. Singer. 1985. Expression of a cytoplasmic LINE-1 transcript is regulated in a human teratocarcinoma cell line. *Proc. Natl. Acad. Sci.* **82:** 6050.

Soares, M.B., E. Schon, and A. Efstratiadis. 1985. Rat Line-1: The origin and evolution of a family of long interspersed middle repetitive DNA elements. *J. Mol. Evol.* **22:** 117.

Soriano, P., P. Szabo, and G. Bernardi. 1982. The scattered distribution of actin genes in the mouse and human genomes. *EMBO J.* **1:** 579.

Sun, L., K.E. Paulson, C.W. Schmid, L. Kadyk, and L. Leinwand. 1984. Non-Alu family interspersed repeats in human DNA and their transcriptional activity. *Nucleic Acids Res.* **12:** 2669.

Thayer, R.E. and M.F. Singer. 1983. Interruption of an α-satellite array by a short member of the *Kpn*I family of interspersed highly repeated monkey DNA sequences. *Mol. Cell. Biol.* **3:** 967.

Ullrich, A., A. Gray, D.V. Goeddel, and T.J. Dull. 1982. Nucleotide sequence of a portion of human chromosome 9 containing a leukocyte interferon gene cluster. *J. Mol. Biol.* **156:** 467.

Ullu, E. and C. Tschudi. 1984. Alu sequences are processed 7SL RNA genes. *Nature* **312:** 171.

Vanin, E.F. 1985. Processed pseudogenes: Characteristics and evolution. *Annu. Rev. Genet.* **19:** 253.

Voliva, C.F., S.L. Martin, C.A. Huthinson III, and M.H. Edgell. 1984. The dispersal process associated with the L1 family of interspersed repetitive sequences. *J. Mol. Biol.* **178:** 795.

Wakasugi, S., H. Nomiyama, M. Fukuda, T. Tsuzuki, and K. Shimida. 1985. Insertion of a long *Kpn*I family member within a mitochondrial-DNA-like sequence present in the human nuclear genome. *Gene* **36:** 281.

Yoshitake, S., B.G. Schach, D.C. Foster, E.W. Davie, and K. Kurachi. 1985. Nucleotide sequence of the gene for human factor IX (antihemophilic factor B). *Biochemistry* **24:** 3736.

The LINE-1 Family of Primates May Encode a Reverse Transcriptase-like Protein

Y. SAKAKI,* M. HATTORI,* A. FUJITA,* K. YOSHIOKA,* S. KUHARA,† AND O. TAKENAKA‡
*Research Laboratory for Genetic Information, Kyushu University, Fukuoka 812, Japan;
†Department of Agricultural Chemistry, Faculty of Agriculture, Kyushu University, Fukuoka 812, Japan;
‡Primate Institute, Kyoto University, Inuyama, Aichi Prefecture, Japan

Primate and mouse LINE-1 (L1) family sequences usually have an A-rich stretch at the 3′ end and, in most instances, are truncated at the 5′ end; they are often flanked by short direct repeats. These features of the L1 family are believed to be due to reverse transcription beginning at the 3′ end of L1 family transcripts and terminating prematurely, and to the target site duplication caused by the insertion of the cDNA, as in the case of a processed pseudogene (for review, see Singer and Skowronski 1985). This reverse transcription model is further supported by the recent findings of Skowronski and Singer (1985), who showed the pres-

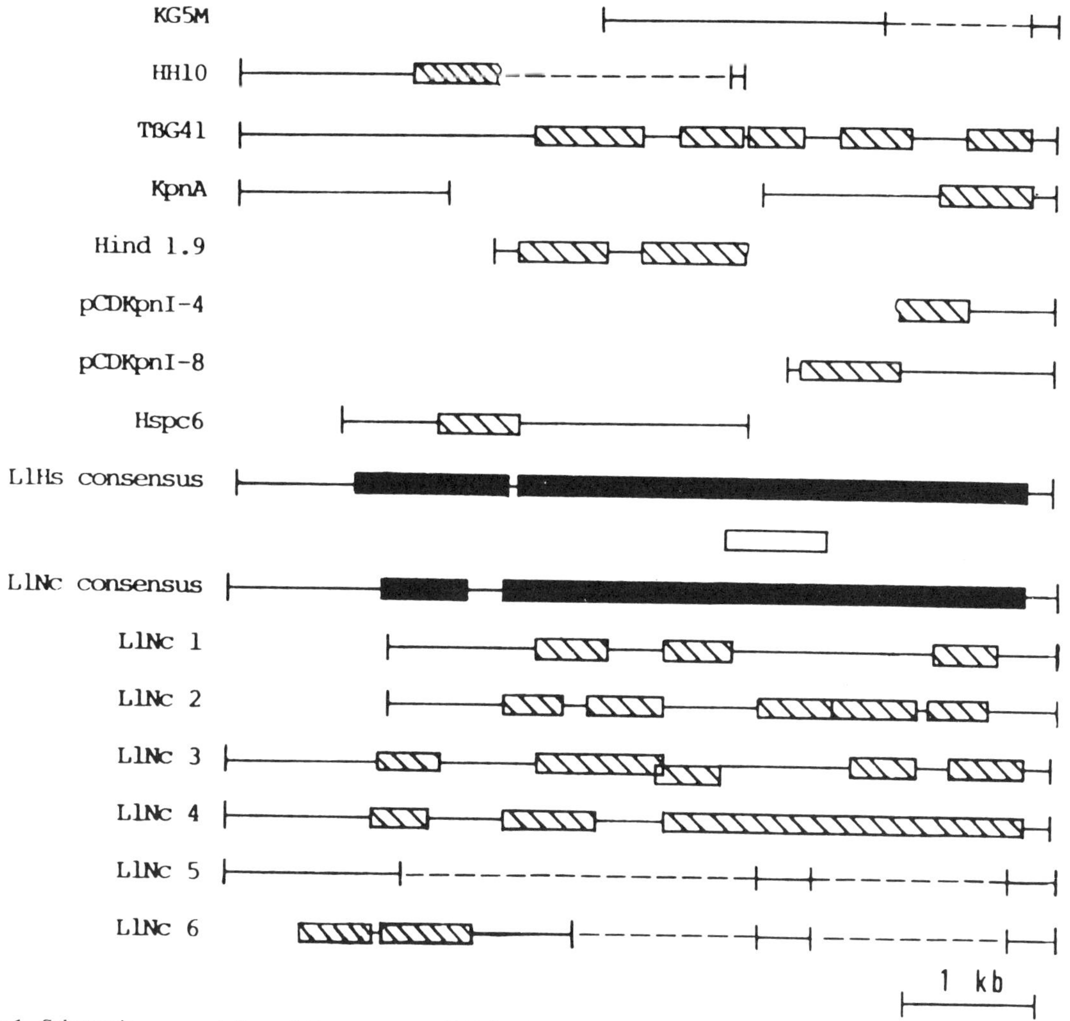

Figure 1. Schematic representation of the structure of L1 family members and the consensus sequences. Sequences of TβG41 (Hattori et al. 1985), KpnA (Potter 1984), Hind1.9 (Manuelidis 1982), pCDKpnI-4, and pCDKpnI-8 (DiGiovanni et al. 1983) were taken from published data. HH10 and Hspc6 have been cloned by Miyake et al. (1983) and Kunisada and Yamagishi (1984), respectively. Filled bars in the consensus sequences and hatched bars in each member show positions of ORFs over 420 bp long. An open bar between two consensus sequences indicates the region homologous to reverse transcriptase sequences (see text). Broken lines show the region where the sequences is undetermined.

Figure 2. *(See facing page for legend.)*

ence of a polyadenylated L1 family transcript in human pluripotential embryonic cells (see also, Skowronski and Singer, this volume). However, it has not been clear why the L1 family could produce an extraordinarily large number of copies (more than 10^4 per haploid genome) during evolution. It seems likely that the progenitor of the L1 family by itself carries (or carried) functions or signals that promote the active dispersion of the L1 family sequence, and that such function is conserved during evolution. Therefore, we compared L1 family sequences from evolutionarily distant species to find out functions and signals that may be important for the active dispersion of the L1 family.

METHODS

DNA cloning and sequencing. *Nycticebus coucang* liver DNA was partially digested with *Eco*RI, and 15- to 20-kb fragments were introduced into Charon 4A by in vitro packaging. The phage library was screened with ^{32}P-labeled fragments of a human L1 family TβG41 (Adams et al. 1980; Hattori et al. 1985). Positive phage clones were analyzed further by Southern blotting, and selected DNA fragments were subcloned into pUC13 of pUC18 and subjected to sequence determination by the method described by Hattori and Sakaki (1986).

Computer analysis. DNA and amino acid sequences were analyzed by the GENAS system (Kuhara et al. 1984; at Kyushu University Computer Center), which enables us to retrieve sequence data from DNA and protein data bases and to analyze them by application programs. The search for the L1 family-homologous sequences was carried out with the program of Wilbur and Lipman (1983) in GENAS.

RESULTS AND DISCUSSION

Consensus L1 Sequence

For comparative analysis, humans and prosimians, which diverged from each other about 70 million years ago, were chosen. Several L1 family members were isolated from genomic DNA libraries of the slow loris (*N. coucang*) and human, and their nucleotide sequences were determined (Hattori et al. 1986). As shown in Figure 1, a large open reading frame (ORF) of 927 amino acid residues was found in L1Nc4; all the large ORFs in other members were found to use the same reading frame as the large ORF of LINc4, and they overlapped each other. From these sequence data, together with some published data, consensus nucleotide sequences for the prosimian and human L1 families were deduced (available on request) and were found to be capable of encoding a very large ORF of more than 1200 amino acid residues (Fig. 1). Human and prosimian large ORFs showed a homology of about 65% of each other, but the non-ORF regions showed very low homologies (40% or less). This homology profile suggests, from an evolutionary viewpoint, a structural resemblance of the L1 family to eukaryotic mRNA; that is, evolutionarily conserved protein-coding region, evolutionarily variable 5′ end and 3′ noncoding regions, and a poly(A) tail at the 3′ end. These results provide additional support for the idea that the L1 family is derived from sequences encoding proteins (Martin et al. 1984).

Sequence Homology Analysis

We then asked the question, what kind of protein could be encoded by the ORFs? Upon searching the Protein Sequence Database of the National Biochemical Research Foundation for amino acid sequences homologous to ORFs in the human L1 family sequence TβG41, a relatively high homology was found with transferrins (Hattori et al. 1985). Using the same approach, significant homology was also found between consensus ORFs and some retroviral reverse transcriptases. Alignment of the amino acid sequences of reverse transcriptases (RNA-dependent DNA polymerases) of various origins was then carried out together with those of L1 family ORFs (Hattori et al. 1986). As shown in Figure 2, the L1 family ORF is evidently homologous to RNA-dependent DNA polymerases. Particularly, the ORF is highly homologous to the ORFs of yeast mitochondrial class II introns (Oxi3-1 and Oxi3-2), which show a high sequence homology to RNA-dependent DNA polymerases (Michel and Lang 1985). Thirty-seven out of 102 amino acid residues were identical between the consensus L1 ORFs and Oxi3-1 ORF. Using the mutation matrix of Schwartz and Dayhoff (1972), the significance of each homologous sequence block was estimated by a computer algorithm. The statistical test showed that homologies of most sequence blocks between L1 ORFs and yeast intron-encoded ORFs are significant with the probability of less than 1.3×10^{-3} (more than 3 σ).

Toh et al. (1983) described 10 invariant amino acid residues in five viral RNA-dependent DNA polymerases. Eight out of the 10 invariant amino acid residues were found to be conserved in human and prosimian consensus ORFs (indicated by filled circle in Fig. 2). Toh and his colleagues (1985) also pointed out 27 po-

Figure 2. Homology of amino acid sequences between L1 consensus ORFs and RNA-dependent DNA polymerases of various origins. Amino acids that are common to at least five polymerase sequences and either L1 consensus ORF, and amino acids in polymerase sequences identical to those of either L1 consensus ORF, are boxed. Ten invariant amino acid residues (Toh et al. 1983) are indicated by circles, among which the residues conserved in L1 ORFs are shown by filled circles. The 27 identical and chemically similar amino acids residues by Toh et al. (1985) are indicated by triangles, among which residues conserved in L1 ORFs are shown by filled triangles (5 out of 27 amino acids residues are not included in the region shown in this figure). Identical amino acid residues between prosimian and mouse (L1Md-A2, Loeb et al. 1986) L1 ORFs are indicated by filled squares. (See Hattori et al. [1986] for references for RNA-dependent polymerases.)

sitions at which identical or chemically similar amino acid residues were found in eight RNA-dependent DNA polymerases. At 20 out of the 27 positions, the consensus ORFs were found to possess amino acid residues common to the polymerases (indicated by filled triangles in Fig. 2). A highly conserved sequence, YXDD, flanked by three hydrophobic residues, was described by Toh et al. (1983, 1985). L1 consensus ORFs use F (phenylalanine) instead of Y (tyrosine) in the sequence, but it should be noted that the RNA-dependent DNA polymerase of yeast Ty transposable element (Clare and Farabaugh 1985; Hauber et al. 1985) uses F instead of Y, just as in the case of L1 ORFs. Homology between a mouse L1 family ORF and RNA-dependent polymerases was also reported by Loeb et al. (1986). As indicated by the filled squares in Figure 2, the region homologous to RNA-dependent polymerases is highly conserved between mouse and primates. Thus, it seems very likely that the protein potentially encoded by the L1 family or its progenitor is closely related to RNA-dependent DNA polymerases. If the protein is actually a reverse transcriptase, it is conceivable that the enzyme binds to its own mRNA immediately after translation to start cDNA synthesis. This supports a preferential and active dispersion model of L1 family sequences.

The human L1 family was also found to have some homology with chick *Blym-1* at the nucleotide sequence level (Hattori et al. 1985). Recently, Rogers (1986) and Cooper et al. (1986) pointed out a striking homology between *Blym-1* and the mouse L1 family. These results are summarized schematically in Figure 3.

Conserved Sequences at the 5′ Terminal Region

Although the non-ORF regions showed low homology between prosimian and human as a whole, a highly conserved sequence was identified at the 5′ terminal region. It was located 10–15 nucleotides downstream from the 5′ end of the L1 family sequence and had a sequence of AAGATGGCNNNNNAGNAACAGCT. This sequence was conserved in all human and prosimian L1 family members tested (four human and four prosimian). Considering the very low homology of the non-ORF regions between these two species, this highly

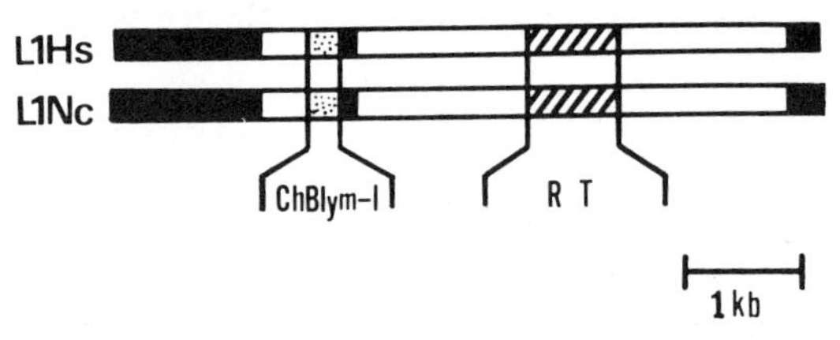

Figure 3. Schematic representation of L1 family structure. The ORF and the non-ORF regions are shown by an open bar and filled bar, respectively.

conserved sequence appears to have some significance. Skowronski and Singer (1985) have identified transcript of the L1 family specific for the undifferentiated stage of embryonic teratocarcinoma cell. A certain signal may be required for such stage-specific transcription starting at the specific site.

In conclusion, some evolutionarily conserved functions and signals were identified in the L1 family, making it more likely that the L1 family is a kind of retrotransposon.

ACKNOWLEDGMENT

We thank Drs. Maxine Singer and Jacek Skowronski for valuable discussion. This work was supported by grants from the Ministry of Education, Science and Culture, Japan to Y.S.

REFERENCES

Adams, J.W., R.E. Kaufman, P.J. Kretschmer, M. Harrison, and A.W. Nienhuis. 1980. A family of long reiterated DNA sequences, one copy of which is next to the human beta globin gene. *Nucleic Acids Res.* **8:** 6113.

Clare, J. and P. Farabaugh. 1985. Nucleotide sequence of a yeast Ty element: Evidence for an unusual mechanism of gene expression. *Proc. Natl. Acad. Sci.* **82:** 2829.

Cooper, G.M., G. Goubin, A. Diamond, and P. Neiman. 1986. Relationship of Blym genes to repeated sequences. *Nature* **320:** 579.

DiGiovanni, L., S.R. Haynes, R. Misra, and W.R. Jelinek. 1983. KpnI family of long-dispersed repeated DNA sequences of man: Evidence for entry into genomic DNA of DNA copies of poly(A)-terminated KpnI RNAs. *Proc. Natl. Acad. Sci.* **80:** 6533.

Hattori, M. and Y. Sakaki. 1986. Dideoxy sequencing method using denatured plasmid templates. *Anal. Biochem.* **152:** 232.

Hattori, M., S. Hidaka, and Y. Sakaki. 1985. Sequence analysis of a KpnI family member near the 3′ end of human β-globin gene. *Nucleic Acids Res.* **13:** 7813.

Hattori, M., S. Kuhara, O. Takenaka, and Y. Sakaki. 1986. L1 family of repetitive DNA sequences in primates may be derived from a sequence encoding a reverse transcriptase-related protein. *Nature* **321:** 625.

Hauber, J., P. Nelböck-Hochstetter, and H. Feldmann. 1985. Nucleotide sequence and characterization of a Ty element from yeast. *Nucleic Acids Res.* **13:** 2745.

Kuhara, S., F. Matsuo, S. Futamura, A. Fujita, T. Shinohara, T. Takakgi, and Y. Sakaki. 1984. GENAS: A database system for nucleic acid sequence analysis. *Nucleic Acids Res.* **12:** 89.

Kunisada, T. and H. Yamagishi. 1984. Sequence repetition and genomic distribution of small polydisperse circular DNA purified from HeLa cells. *Gene* **31:** 213.

Loeb, D.D., R.W. Padgett, S.C. Hardies, W.R. Shehee, M.B. Comer, M.H. Edgell, and C.A. Hutchinson III. 1986. The sequences of a large L1Md element reveals a tandemly repeated 5′ end and several features found in retrotransposons. *Mol. Cell. Biol.* **6:** 168.

Manuelidis, L. 1982. Nucleotide sequence definition of a major human repeated DNA, the *Hin*dIII 1.9 kb family. *Nucleic Acids Res.* **10:** 3211.

Martin, S.L., C.F. Voliva, F.H. Burton, M.H. Edgell, and C.A. Hutchinson III. 1984. A large interspersed repeat found in mouse DNA contains a large open reading frame that evolves as if it encodes a protein. *Proc. Natl. Acad. Sci.* **81:** 2308.

Michel, F. and B.F. Lang. 1985. Mitochondrial class II introns encode proteins related to the reverse transcriptases of retroviruses. *Nature* **316:** 641.

Miyake, T., K. Migita, and Y. Sakaki. 1983. Some KpnI family members are associated with the Alu family in human genome. *Nucleic Acids Res.* **11:** 6837.

Potter, S.S. 1984. Rearranged sequences of a human KpnI element. *Proc. Natl. Acad. Sci.* **81:** 1012.

Rogers, J. 1986. Relationship of Blym genes to repeated sequences. *Nature* **320:** 579.

Schwartz, R.M. and M.O. Dayhoff. 1972. Matrices for detecting distant relationships. In *Atlas of protein sequence and structure* (ed. M.O. Dayhoff), vol. 5, p. 353. National Biomedical Research Foundation, Washington, D.C.

Singer, M.F. and J. Skowronski. 1985. Making sense out of LINES: Long interspersed repeat sequences in mammalian genomes. *Trends Biochem. Sci.* **10:** 119.

Skowronski, J. and M.F. Singer. 1985. Expression of a cytoplasmic LINE-1 transcript is regulated in a human teratocarcinoma cell line. *Proc. Natl. Acad. Sci.* **82:** 6050.

Toh, H., H. Hayashida, and T. Miyata. 1983. Sequence homology between retroviral reverse transcriptase and putative polymerases of hepatitis B virus and cauliflower mosaic virus. *Nature* **305:** 827.

Toh, H., R. Kikuno, H. Hayashida, T. Miyata, W. Kugimiya, S. Inouye, S. Yuki, and K. Saigo. 1985. Close structural resemblance between putative polymerases of a *Drosophila* transposable genetic element 17.6 and pol gene product of Molony murine leukemia virus. *EMBO J.* **4:** 1267.

Wilbur, W.J. and D.J. Lipman. 1983. Rapid similarity searches of nucleic acid and protein data banks. *Proc. Natl. Acad. Sci.* **80:** 726.

Repetitive Human DNA Sequences

I. Evolution of the Primate α-Globin Gene Cluster and Interspersed *Alu* Repeats

I. SAWADA AND C.W. SCHMID
Department of Chemistry, University of California at Davis, Davis, California 95616

II. Properties of a Transposon-like Human Element

N. DEKA, K.E. PAULSON, C. WILLARD, AND C.W. SCHMID
Department of Chemistry, University of California at Davis, Davis, California 95616

I. EVOLUTION OF THE PRIMATE α-GLOBIN GENE CLUSTER AND INTERSPERSED *ALU* REPEATS

The human *Alu* family consists of 0.5×10^6 members that share a 300-bp consensus sequence and are broadly distributed throughout the genome (Schmid and Jelinek 1982; Schmid and Shen 1985). The consensus sequences of the chimpanzee and New World monkey *Alu* families are indistinguishable from the human consensus sequences (Daniels et al. 1983; Sawada et al. 1985). The *Alu* families in these higher primates probably resulted from common founder sequences and as an extreme may consist of orthologous members. Human and chimpanzee share orthologous *Alu* sequences at each of seven loci investigated to date (Maeda et al. 1983; Sawada et al. 1985).

There are at least two distinct families of *Alu*-like sequences in galago, a prosimian primate (Daniels et al. 1983; Daniels and Deininger 1983). One galago family is ancestrally related to the human family but shows species-specific differences from the human *Alu* family consensus sequence (Daniels et al. 1983). The other galago *Alu*-like family can be regarded as an entirely different structure that lacks a closely related human counterpart (Daniels and Deininger 1983).

The *Alu*-like sequences in galago and higher primates (monkey, ape, and human) have been homogenized to the extent that the species-specific differences described above are apparent upon comparing randomly selected member sequences from the two lineages. A long-standing question is whether amplification or conversion maintains this homogeneity in repetitive sequence families (reviewed by Weiner and Denison 1983). As an experimental approach toward answering this question, we compare the structures of the α-globin gene clusters and interspersed repeats in human, chimpanzee, monkey, and galago.

Galago α-Globin Gene Cluster

The galago α-globin gene cluster consists of three genes labeled 5′AαBαCα3′ (Fig. 1; I. Sawada and C.W. Schmid, in prep.). Blot hybridization studies do not detect additional galago α-globin genes, so that at least one of these three genes is functional. The Aα and Bα genes are closely related sequences reflecting a recent duplication or conversion, whereas the Cα gene differs significantly from Aα and Bα (Fig. 1). Either the Aα or Bα gene might code for a product, and the predicted product of both genes resembles galago α-globin (I. Sawada and C.W. Schmid, in prep.). The Cα gene is almost certainly inactive.

The organization of galago genes is very similar to the organization of α-globin genes in rabbit, horse, and goat (Clegg et al. 1984; Lingrel et al. 1985; Hardison and Gelinas 1986; Fig. 1). A pseudogene, ψα, is positioned on the 3′ end of both the horse and rabbit gene clusters (Fig. 1). The unique sequences flanking the horse α1 gene are homologous to the unique sequences flanking the galago Bα gene (Fig. 1). It is therefore quite probable that the galago Cα and horse ψα gene are orthologous. The functional goat Iα and IIα genes are nearly identical sequences that, in analogy to the similar galago Aα and Bα sequences, result from a very recent duplication or correction (Fig. 1).

The unique sequences flanking the goat Iα and IIα genes are homologous to the corresponding regions flanking the galago Aα and Bα genes (Fig. 1). Goat and galago therefore share an orthologous pair of recently corrected α-globin genes. There are other structural organizations of mammalian α-globin genes. For example, in rabbit there have been several duplications of the entire α-globin cluster (Cheng et al. 1986; Hardison and Gelinas 1986). In summary, galago and some other mammalian species have a similar organization of α-globin genes consisting of two nearly identical

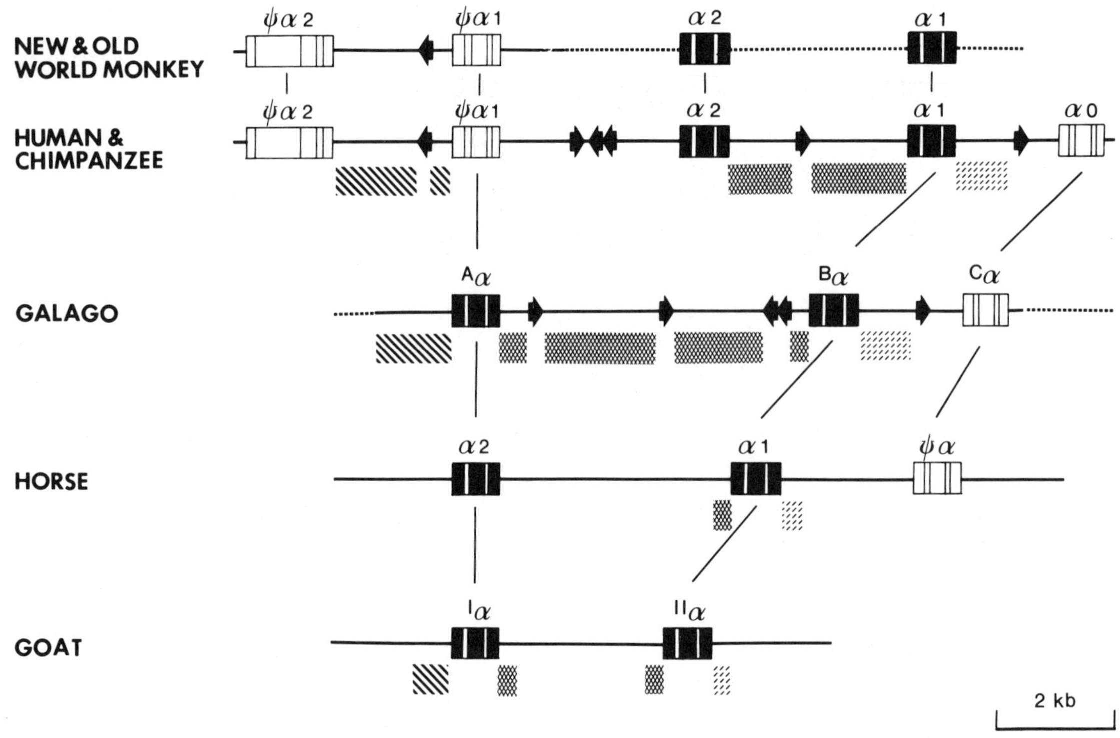

Figure 1. Schematic diagram of orthologous relationships between the α-globin gene clusters of several species. Base sequence determinations of human, chimpanzee, horse, and goat α-globin clusters are reviewed by Hardison et al. (1986), Sawada et al. (1985), and Hardison and Gelinas (1986). Marks et al. (1986) report the α0 gene. Hardison and Gelinas (1986) previously reported orthologous alignments between most of these species; here we extend their alignments to galago and monkey based on sequence and mapping data (I. Sawada and C.W. Schmid, in prep.). The order of the ψα1, α2, and α1 genes in human and in New World monkey (red howler) and in Old World monkey is identical. However, the sizes of the intergenic regions in three *Macaca* species of Old World monkeys are somewhat different than in human. Active and converted α-globin genes are designated by filled-in exons, whereas pseudogenes and unconverted genes are designated by open boxes. Various cross-hatched symbols designate homologous sequences to identify the orthologous alignments schematically depicted. Arrows depict sequenced *Alu*-like repeats. All human, chimpanzee, and monkey (*Macaca fasciculans*) (i.e., higher primate) *Alu* repeats depicted are orthologous, whereas all galago *Alu*-like repeats are distinct sequences inserted at the distinct sites compared to those of higher primates.

genes and a divergent pseudogene on the 3′ end of the cluster.

Homologous flanking sequences demonstrate that the human ψα1 and galago Aα genes are orthologous (Fig. 1). The orthologous ψα1 gene is also present in both Old and New World monkeys (Fig. 1; I. Sawada and C.W. Schmid, in prep.). These phylogenetic comparisons show that the human ψα1 gene was inactivated following the divergence of human and galago but before the divergence of human and monkey. In agreement with this finding, Proudfoot and Maniatis (1980) estimated that the human ψα1 gene diverged from the active α2/α1 genes 60 million years ago (Myr) and was inactivated 45 Myr. A date of 45 Myr would precede the divergence of the monkey and ape lineages but would follow the divergence of the prosimian and higher primate lineages (Goodman et al. 1982).

Human α-globin contains two nearly identical functional genes, α2 and α1, which have very similar 5′-flanking sequences (Lauer et al. 1980; Hess et al. 1983, 1984; Michelson and Orkin 1983). Unique sequences flanking the human α1 gene, but not the α2 gene, are homologous to sequences flanking the galago Bα gene. The human α1 and galago Bα genes are ancestrally related and may be regarded as orthologous (Fig. 1). The close homology between the unique 3′-flanking regions of the human α1 and galago Bα genes confirms this assignment (Fig. 1).

According to this assignment the human α2 gene and its 5′-flanking region should be regarded as the most recent duplication in the human α-globin cluster. As described above, the organization of the galago Aα, Bα, and Cα genes probably resembles the gene organization in the ancestral primate lineage. Simple unequal crossing over between the Aα and Bα genes in a presumptive primate ancestor would give rise to a 5′...Aα(B/A)α Bα Cα...3′ recombinant correctly predicting the present-day correspondence between the human and galago α-globin clusters (Fig. 1).

Alu Family Evolution

Alu-like sequences are interspersed throughout the primate α-globin gene cluster (Fig. 1). Human and chimpanzee share orthologous *Alu* repeats positioned 5′ to the α2, ψα1 (Fig. 1), and α1 genes (Sawada et al.

1985). The divergence of these orthologous chimpanzee/human pairs of *Alu* repeats approximates the value expected for nonselected DNA sequences.

The orthology of human and monkey *Alu* repeats has not yet been extensively investigated. However, human and monkey share orthologous *Alu* repeats positioned 5′ to their respective $\psi\alpha 1$ genes (Fig. 1). The human and monkey *Alu* sequences in this orthologous pair have diverged by 9.4%, which is the extent of divergence expected between nonselected human and monkey DNAs. The very similar restriction maps of the human and monkey α-globin clusters makes it seem likely that human and monkey will share other orthologous *Alu* pairs interspersed within the α-globin cluster (I. Sawada and C.W. Schmid, in prep.).

None of the galago *Alu*-like repeats are present in the human gene cluster and none of the human *Alu* sequences are present in the galago cluster (Fig. 1). In each case, the flanking direct repeats surrounding an *Alu*-like sequence in one species can be recognized as the unoccupied target site in the other species. The *Alu*-like sequences in the human and galago α-globin clusters are independently generated by different founder sequences.

In conclusion, most human *Alu* repeats were inserted following the divergence of human and galago, but prior to the divergence of human and chimpanzee and possibly prior to the divergence of the ape and monkey lineages. Subsequent to their insertion, *Alu* repeats acquire mutations at the rate expected for nonselected sequences. These results confirm the classic saltatory replication model advanced by Britten and Kohne (1968) to account for the homogenization of species-specific differences in repeat sequence families (Weiner and Denison 1983).

II. PROPERTIES OF A TRANSPOSON-LIKE HUMAN ELEMENT: THE-1 REPEATS

Interspersed repeats such as *Alu* probably transpose via an RNA intermediate (Jagadeeswaran et al. 1981; Van Ardsell et al. 1981; Schmid and Shen 1985). New members of such repetitive "retroposon" families have abandoned necessary upstream transcriptional control elements and like processed RNA pseudogenes would normally be transcriptionally inactive (Ullu and Weiner 1985). *Alu* repeats are not transcribed in HeLa cells by RNA polymerase III, although they are transcribed by polymerase III in vitro (K.E. Paulson and C.W. Schmid, in prep.).

Because of their flanking long terminal repeats (LTRs), transposons such as copia in *Drosophila*, Ty in yeast, and mammalian proretrovirus can retain transcriptional activity upon insertion at a new site. The LTRs contain promoter and enhancer elements, and 3′-end processing signals (Temin 1982; Varmus 1983). Given these transcriptional control signals, it is not surprising that the insertion of a transposon often disrupts the expression of neighboring genes (Roeder and Fink 1983; Varmus 1983).

Properties of the THE-1 Repeat Family

The human THE-1 family of repeats consists of 10,000 members that share a 2.3-kb consensus sequence (Fig. 2; Paulson et al. 1985). Individual members are flanked by 350-bp sequences that resemble LTRs. THE-1 sequences, as well as solitary LTRs, are flanked by short direct repeats that result from a duplication of the genomic entry site (Fig. 3).

The genomic entry sites of most transposon families

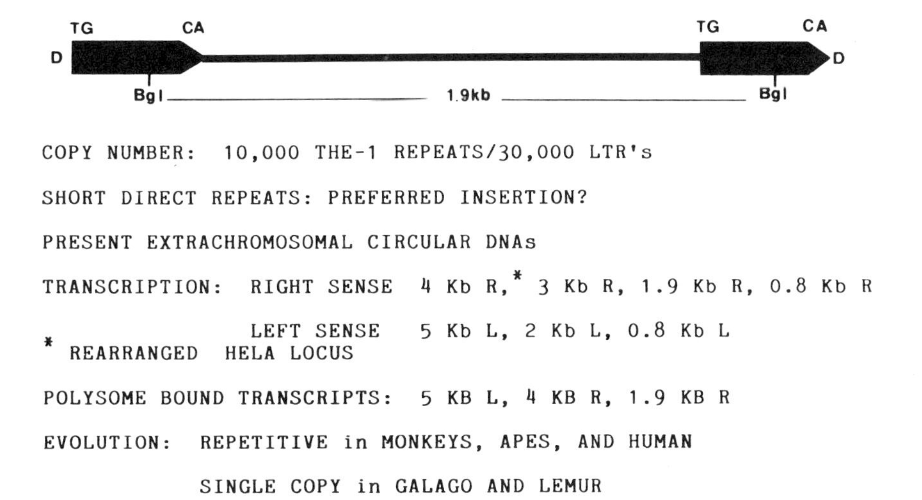

Figure 2. Summary of properties of the THE-1 family repeats (Paulson et al. 1985, in prep.). The entries for transcription and translation identify both the length and orientation (R or L defined by Figs. 4–6) of the RNA molecules.

5'	CAGATACtgat	THE 1A	tacaGATAC	3'
5'	CATACtgat	THE 1B	deleted 3' LTR	
5'	GATTACtgac	SOL 0_5	tacaGTTAC	3'
5'	TAGGCtgat	THE 1J	tgcaCAGGCT	3'
5'	TCATTtgat	SOL 0_4	tacaTCATT	3'

Figure 3. Short direct repeats surround THE-1 elements and solitary LTRs written in capital letters are indicated by underlining arrows. Small letters indicate the 5′ and 3′ ends of LTRs which usually begin with 5′ TG... and end with ...CA 3′.

and proretrovirus are usually thought to be randomly selected sequences (Inouye et al. 1984). However this may not be the case for THE-1 family members, as the flanking direct repeats in several cases resemble variants of the sequence...gATAc...(Fig. 3). A *Drosophila* copia-like element inserts into variants of...TATA..., giving precedence for transposition directed to a preferred target site (Inouye et al. 1984). Yeast Ty elements also have a preferred AT-rich landing site that effectively targets them toward promoter elements (Eibel and Philippsen 1984).

Transposition is thought to proceed by an RNA intermediate, which is converted into a extrachromosomal circular DNA prior to insertion (Flavell 1984; Baltimore 1985). THE-1 sequences are present on discrete-length extrachromosomal circular DNAs (Paulson et al. 1985).

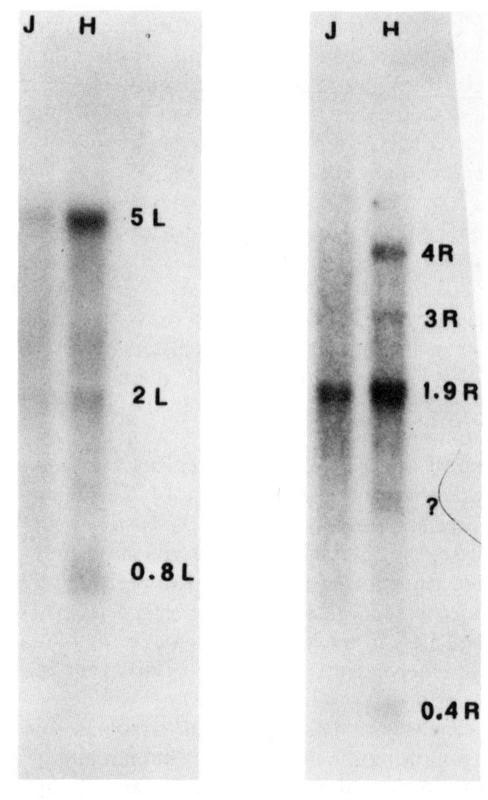

Figure 4. Northern blot hybridization of poly(A)$^+$ RNAs from HeLa (lane *H*) and Jurkat (lane *J*) cells with SP6-transcribed probes representing two orientations of THE-1. The "left" strand (L) corresponds to the same strand reported by Paulson et al. (1985); right RNA strands (R) complement the sequence reported by K.E. Paulson et al. (in prep.). The lengths of the RNAs are reported in kilobase units.

THE-1 Transcription

The transcriptional activity of the human THE-1 family is complex (Paulson et al. 1985). Both transcriptional orientations are represented on discrete-length poly(A)$^+$ RNA, including molecules that exceed the 2.3-kb THE-1 consensus length (Fig. 4). The insertion of THE-1 into other transcription units accounts for these longer RNA lengths.

One cDNA clone, cDNA 7, selected from the cDNA library constructed from an SV40-transformed fibroblast cell line (Okayama and Berg 1983), results from a promoter upstream to the THE-1 element (Fig. 5; K.E. Paulson et al., in prep.). The transcript was polyadenylated within the 3′ LTR of THE-1 element (Fig. 5). Sequences in THE-1 located near this polyadenylation site include a variant polyadenylation signal, a consensus poly(A) addition site, CA, and downstream elements that may terminate transcripts (Fig. 5.; Birnstiel et al. 1985). In this case a THE-1 element processes the transcript resulting from an upstream promoter. The functional polyadenylation signals of this transcriptional orientation suggest that it is the sense strand for any THE-1 gene products. This cDNA 7 sequence corresponds to the left strand transcript, as defined in Figure 4.

A 3.3-kb cDNA clone, cDNA 5, isolated from the Okayama-Berg library represents the right transcriptional orientation of THE-1 (Fig. 6). Primer extension experiments show that the major right-strand transcript in HeLa cells results from a promoter located 600 bp upstream from the 3′ end of the left LTR. Assuming that this transcriptional start site gives rise to the 3.3-kb cDNA clone, then cDNA 5 is an incomplete cDNA corresponding to 4-kb right-strand transcript (Figs. 4 and 6). In agreement with this prediction, a unique sequence from the 3′ end of cDNA 5 hybridizes to a 4-kb RNA from HeLa cells (Fig. 6).

The 4-kb R transcript is not found in thymus, spleen, neuroblastoma, and Jurkat cells (K.E. Paulson et al., in prep.). It is present in HeLa RNA and presumably the cell line RNA used to construct the Berg-Okayama cDNA library. One possibility is that this transcriptional activity reflects a DNA sequence rearrangement involving a THE-1 element. In support of this possibility, the unique 3′-flanking region of cDNA 5 hybridizes to an additional restriction fragment in HeLa DNA that is not present in placental DNA (Fig. 6). The exact structure of this HeLa locus remains to be determined but restriction mapping is consistent with the insertion of a THE-1 element next to the unique sequence corresponding to the 4-kb R transcript. In addition, a human genomic λ clone isolated to have homology with the unique 3′-flanking region of cDNA 5 contains a THE-1 element next to the unique region.

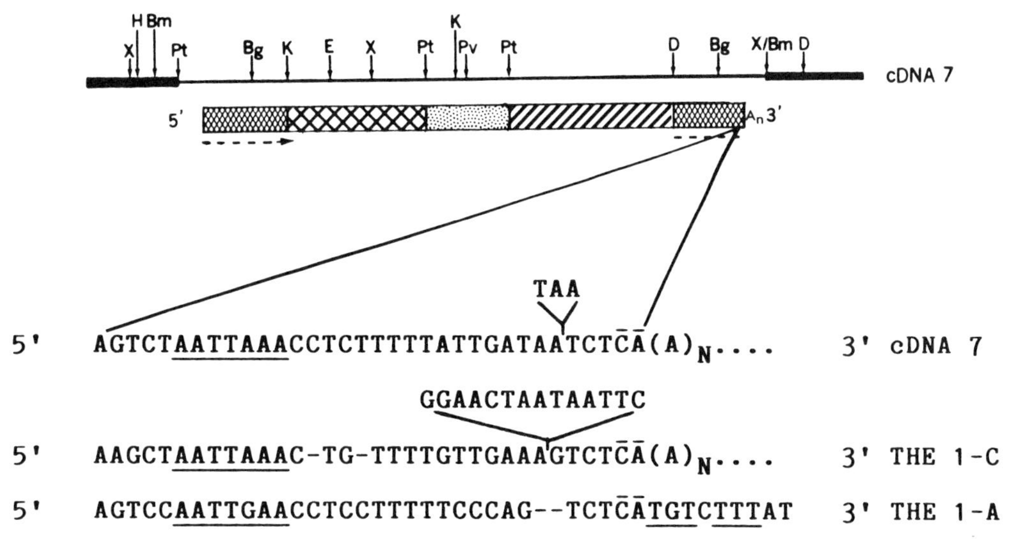

Figure 5. cDNA 7 isolated from the Okayama and Berg (1983) library is a left-strand transcript as defined in Fig. 3. The 5′ end of cDNA 7 lies upstream from its left LTR so that the cDNA 7 RNA resulted from an upstream promoter. Polyadenylation occurs as indicated in the 3′ LTR of cDNA 7. THE-1A is a genomic clone of the THE-1 member having a variant polyadenylation signal, poly(A) addition site, and downstream TGT and TTT signals as indicated by over- and underlining. THE-1C is a genomic clone of a THE-1 member having truncated 5′ and 3′ LTRs. The 3′ LTR is polyadenylated at the same site as cDNA 7.

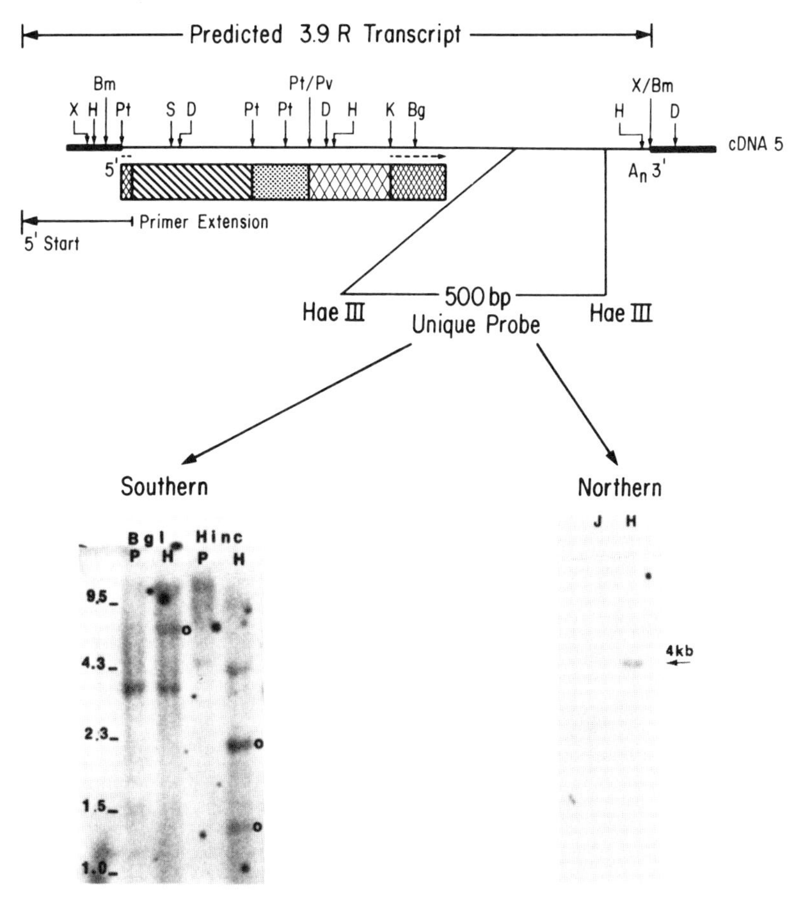

Figure 6. cDNA 5 isolated from the Okayama-Berg (1983) library has the complementary transcriptional orientation to cDNA 7 (Fig. 5). Primer extension indicates that the major rightward transcript initiates from an upstream promoter positioned 670 at 5′ to the 3′ end of the left LTR. In the case of cDNA 5 this start site would correspond to a 4-kb transcript. In agreement with this prediction, a unique probe derived from the 3′-flanking region of cDNA 5 hybridizes to a 4-kb RNA from Hela cells. This 4-kb RNA is not present in Jurkat cells (Fig. 4) or other human cells investigated (K.E. Paulson et al., in prep.). In Southern blot hybridization, this unique probe hybridizes to additional bands in HeLa DNA (Lane *H*) that are not observed in human placental DNA (lane *P*).

There are approximately 10,000 THE-1 repeats as well as an additional 10,000 solitary THE-1 LTRs within the human genome. In analogy to other systems we expect the transcriptional control elements within THE-1 LTRs to have a variety of effects on neighboring genes. This expectation is confirmed by the present results demonstrating that a THE-1 element processes the 3′ end of an upstream transcription unit and that the transcriptionally active 4-kb *R* locus in HeLa results from a DNA rearrangement involving a THE-1 repeat. In the experiments reported here, we would only observe transcripts containing THE-1 repeats. It is probable that THE-1 elements affect the expression of other transcription units, resulting in transcripts that do not contain THE-1 sequences.

Evidence for a Prototransposon

The abundant THE-1 repeat sequence families in human, chimpanzee, and Old and New World monkey DNAs share similar consensus restriction maps (Fig. 7). THE-1 sequences have not been amplified as repetitive sequences in prosimian, but instead THE-1 sequences hybridize to a small number of single-copy restriction fragments in galago and lemur DNAs (Fig. 7). Presumably these prosimian single-copy sequences are ancestrally related to the THE-1 element, which successfully amplified within the higher primate genome. The human THE-1 LTR does not hybridize to galago DNA (data not shown).

We have isolated two galago genomic clones, galago 6 and galago 7, which hybridize to human THE-1 hybridization probes under reduced stringency. Neither galago clone hybridizes to the human THE-1 LTR. Restriction fragments from each of these galago clones hybridize under reduced stringency to a repetitive 1.9-kb *Bgl*II restriction fragment in human DNA that is diagnostic for the human THE-1 repeat sequence family (Fig. 7). However, each of these galago restriction fragments hybridizes to distinct single-copy loci in galago DNA (Fig. 7). Neither of the two THE-1 homologs in galago exists as a repeat sequence family. Galago 6 and galago 7 probes each hybridize to distinct restriction fragments in galago DNA (Fig. 7). Presumably these

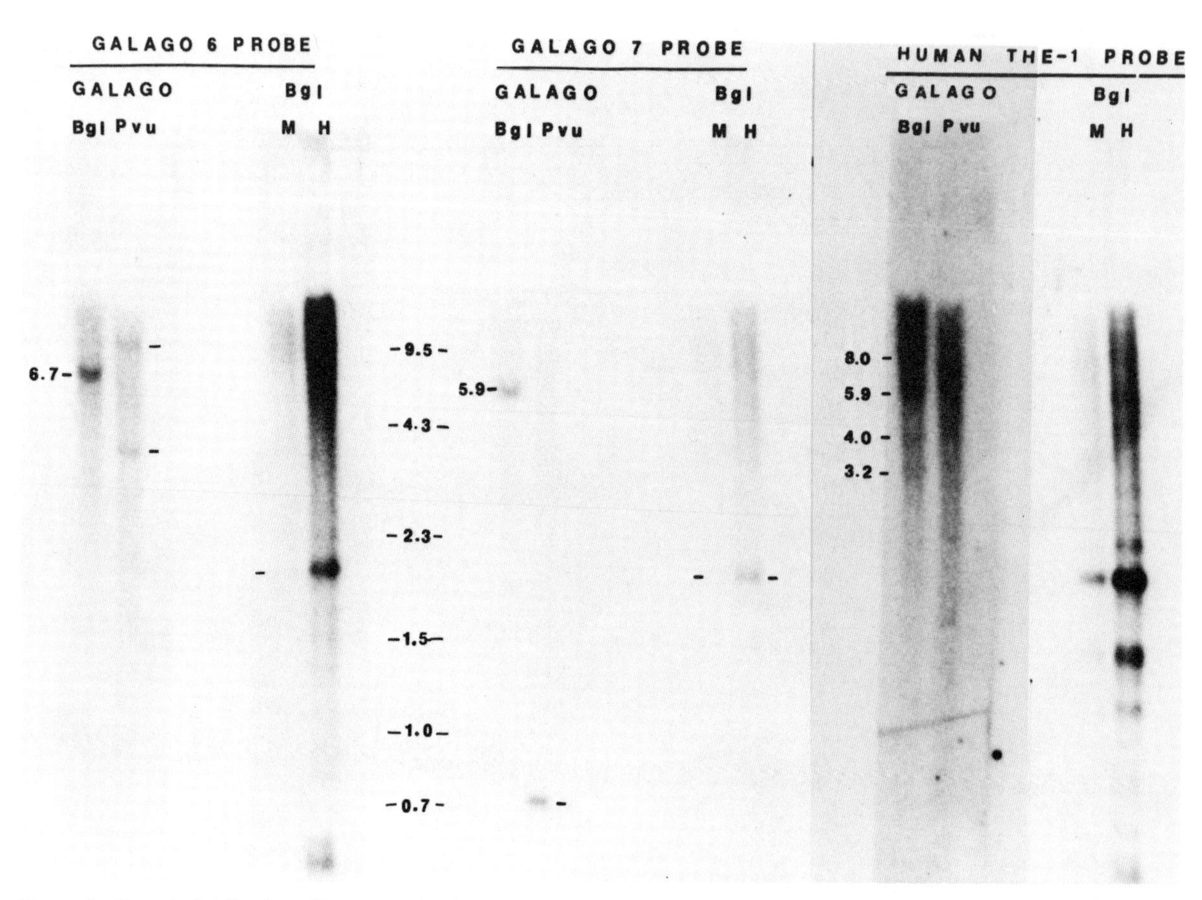

Figure 7. Cross-hybridization of human and galago THE-1 homologous sequences. PUC 0_4, a human THE-1 subclone, hybridizes to a repetitive 1.9-kb *Bgl*II restriction fragment in human and monkey DNA and to several single-copy bands in galago DNA. A restriction fragment derived from the genomic clone, galago 6, hybridizes to the repetitive 1.9-kb *Bgl*II restriction fragments in human and monkey and to a 6.7-kb single-copy *Bgl*II band in galago DNA. A restriction fragment derived from the genomic clone, galago 7, hybridizes to the repetitive 1.9-kb *Bgl*II restriction fragment in human and monkey and to a 5.9-kb single-copy *Bgl*II band in galago DNA. Washing is performed at 3× SSC (60°C) when cross-hybridizing galago and higher primate DNAs and is performed at 1× SSC (60°C) when hybridizing galago to galago or human to higher primate DNAs.

distinct THE-1 homologs in galago DNA are ancestrally related to different elements of the human THE-1 family. In analogy to the protovirus hypothesis (Temin 1971), these galago THE-1 homologs might be examples of a prototransposon. Conceivably, sequence elements that are ancestrally related to those contained within the galago 6 and 7 clones rearranged to form the sequence that subsequently successfully amplified as the human THE-1 family of repeated sequences. It is also conceivable that the human genome contains single-copy sequences coding for the same function as galago 6 and 7. The hypothetical human single-copy sequences would be obscured by the abundant THE-1 family.

ACKNOWLEDGMENT

This research is supported by U.S. Public Health Service grant GM21346.

REFERENCES

Baltimore, D. 1985. Retroviruses and retrotransposons. *Cell* **40:** 481.

Birnstiel, M.L., M. Busslinger, and K. Strub. 1985. Transcription termination and 3′ processing: The end is in site *Cell* **41:** 349.

Britten, R.J. and D.E. Kohne. 1968. Repeated sequences in DNA. *Science* **61:** 529.

Chen, J.-F., L. Raid, and R.C. Hardison. 1986. Isolation and nucleotide sequence of the rabbit globin gene cluster $\psi\zeta$-α1-$\psi\alpha$: Absence of a pair of α globin genes evolving in concert. *J. Biol. Chem.* **216:** 834.

Clegg, J.B., S.E.Y. Goodbourn, and M. Braend. 1984. Genetic organization of the polymorphic equine α-globin locus and the sequence of the BII α1 gene. *Nucleic Acids Res.* **12:** 7847.

Daniels, G.R. and P.L. Deininger. 1983. A second major class of Alu family repeated DNA sequences in a primate genome. *Nucleic Acids Res.* **11:** 7595.

Daniels G.R., G.M. Fox, D. Loewensteiner, C.W. Schmid, and P.L. Deininger. 1983. Species-specific homogeneity of the primate Alu family of repeated DNA sequences. *Nucleic Acids Res.* **11:** 7579.

Eibel, H. and P. Philippsen. 1984. Preferential integration of yeast transposable element Ty into a promoter region. *Nature* **307:** 386.

Flavell, A.J. 1984. Role of reverse transcription in the generation of extrachromosomal *copia* mobile genetic elements. *Nature* **310:** 514.

Goodman, R., A.E. Romero-Herrera, H. Dene, J. Czelusniak, and R.E. Tashian. 1982. In *Macromolecular sequences in systematic and evolutionary biology* (ed. M. Goodman), p. 115. Plenum Press, New York.

Hardison, R.C. and R.E. Gelinas. 1986. Assignment of orthologous relationships among mammalian alpha-globin genes by examining flanking regions reveals a rapid rate of evolution. *Mol. Biol. Evol.* **3:** 243.

Hardison, R.C., I. Sawada, J.-F. Chen, C.-K.J. Shen, and C.W. Schmid. 1986. A previously undetected pseudogene in the human alpha globin gene cluster. *Nucleic Acids Res.* **14:** 1903.

Hess, J.F., C.W. Schmid, and C.-K.J. Shen. 1984. A gradient of sequence divergence in the human adult α-globin duplication units. *Science* **226:** 67.

Hess, J.F., M. Fox, C.W. Schmid, and C.-K.J. Shen. 1983. Molecular evolution of the human adult α-globin like region. *Proc. Natl. Acad. Sci.* **80:** 5970.

Inouye, S., S. Yuki, and K. Saigo. 1984. Sequence-specific insertion of the *Drosophila* transposable element 17.6. *Nature* **310:** 332.

Jagadeeswaran, P., B.G. Forget, and S.M. Weissman. 1981. Short interspersed repetitive DNA elements in eukaryotes. *Cell* **26:** 141.

Lauer, J., C.-K.J. Shen, and T. Maniatis. 1980. The chromosomal arrangement of human α-like globin genes. *Cell* **20:** 119.

Lingrel, J.B., T.M. Townes, S.G. Shapiro, S.M. Wernke, P.A. Liberator, and A.G. Menon. 1985. Structural organization of the α and β globin loci of the goat. In *Experimental approaches for the study of hemoglobin switching* (ed. G. Stamatoyannopoulos and A.W. Nienhus), p. 67. A.R. Liss, New York.

Maeda, N., J.B. Fliska, and O. Smithies. 1983. Recombination and balanced chromosome polymorphism suggested by DNA sequences 5′ to the human β-globin gene. *Proc. Natl. Acad. Sci.* **80:** 5012.

Marks, J., J.P. Shaw, and J.C.-K. Shen. 1986 The orangutan α-globin gene locus. *Proc. Natl. Acad. Sci.* **83:** 1413.

Michelson, A.M. and S.H. Orkin. 1983. Boundaries of gene conversion within duplicated human α-globin genes. *J. Biol. Chem.* **258:** 15245.

Okayama, H. and P. Berg. 1983. A cDNA cloning vector that permits expression of cDNA inserts in mammalian cells. *Mol. Cell. Biol.* **3:** 280.

Paulson, K.E., N. Deka, C.W. Schmid, R. Misra, C.W. Schindler, M.G. Rush, L. Kadyk, and L. Leinwand. 1985. A transposon-like element in human DNA. *Nature* **316:** 359.

Proudfoot, N. and T. Maniatis. 1980. The structure of a human α-globin pseudogene and its relationship to α-globin gene duplication. *Cell* **21:** 537.

Roeder, G.S. and G.R. Fink. 1983. Transposable elements in yeast. In *Mobile genetic elements* (ed. J.A. Shapiro), p. 299. Academic Press, Orlando, Florida.

Sawada, I., C. Willard, C.-K.J. Shen, B. Chapman, A.C. Wilson, and C.W. Schmid. 1985. Evolution of Alu family repeats since the divergence of human and chimpanzee. *J. Mol. Evol.* **22:** 316.

Schmid, C.W. and W.R. Jelinek. 1982. The Alu family of dispersed repetitive sequences. *Science* **216:** 1065.

Schmid, C.W. and C.-K.J. Shen. 1985. The evolution of interspersed repetitive DNA sequences in mammals and other vertebrates. In *Molecular evolutionary genetics* (ed. R.J. MacIntyre). Plenum Press, New York.

Temin, H.M. 1971. The protovirus hypothesis. *J. Natl. Cancer Inst.* **46:** 3.

———. 1982. Function of the retrovirus long terminal repeat. *Cell* **28:** 3.

Ullu, E. and A.H. Weiner. 1985. Upstream sequences modulate the internal promoter of the human 7S L RNA gene. *Nature* **318:** 371.

Van Ardsell, S.W., R.A. Denison, L.B. Bernstein, and A.M. Weiner. 1981. Direct repeats flank three small nuclear RNA pseudogenes in the human genome. *Cell* **26:** 11.

Varmus, H.E. 1983. Retroviruses. In *Mobile genetic elements* (ed. J.A. Shapiro), p. 411. Academic Press, Orlando, Florida.

Weiner, A.M. and R.A. Denison. 1983. Either gene amplification or gene conversion may maintain the homogeneity of the multigene family encoding human U1 small nuclear RNA. *Cold Spring Harbor Symp. Quant. Biol.* **47:** 1141.

The Human Genome and Its Evolutionary Context

G. Bernardi and G. Bernardi
Laboratoire de Génétique Moléculaire, Institut Jacques Monod, 75005 Paris, France

Our knowledge of the structure, chromosomal location, and transcription of human genes (as well as of eukaryotic genes in general) has made impressive advances in recent years, as witnessed by this Symposium. In contrast, the wider issue of genome organization has largely remained a *terra incognita*. Indeed, the only widely known general picture of the human genome is still based on the reassociation kinetics of short DNA fragments. This information, essentially concerning the relative amounts of sequence classes endowed with different reassociation rates and the interspersion pattern of repetitive and nonrepetitive DNA (Schmid and Deininger 1975), not only is very limited in itself, but also can barely be expanded any further. On the other hand, our advances regarding genes, however important otherwise, concern a minute fraction of the mammalian genome. They are, therefore, unlikely to lead to a breakthrough in our understanding of the overall organization and evolution of the genome. A more hopeful source of information is represented by work on specific families of interspersed repeats (Singer 1982; Singer and Skowronski 1985), but this is a difficult area of research still in a pioneering stage. These points stress the fact that real progress on the issue under consideration here can only come through new approaches, such as those developed in our laboratory over the past years. Here we will review briefly some of our recent results as an introduction, and then present a number of new data on the organization and evolution of the vertebrate genome as well as some general conclusions.

The Compositional Compartmentalization of the Genome of Warm-blooded Vertebrates

Several years ago, we realized that density-gradient centrifugation in the presence of certain DNA ligands could not only provide the expected separation of satellite DNAs, but also a fractionation of the so-called "main-band" DNA from warm-blooded vertebrates into a small number of "light" (GC-poor) and "heavy" (GC-rich) "major components," the modal buoyant densities of which corresponded to maxima of DNA distribution (Fig. 1). The resolution of major components not only was much better in Cs_2SO_4/3,6-bis(acetatomercurimethyl)dioxane (BAMD) or Cs_2SO_4/Ag^+ than in CsCl, but also improved (1) with increasing GC, because GC-rich fractions were less abundant in the genome and less overlapping with other fractions, and (2) with increasing DNA size, because of a lower band-spreading by diffusion. In fact, the families of DNA fragments forming the major components derive by preparative breakage from very long (>200 kb) DNA segments, which are fairly homogeneous in base composition, belong to a small number of major classes distinguished by different GC levels, and probably correspond to the DNA stretches present in Giemsa and reverse chromosomal bands. These DNA segments were called isochores, for "similar regions."

The strong compositional compartmentalization of the genomes of warm-blooded vertebrates contrasts with the weak one shown by those of cold-blooded vertebrates; the latter basically either lack or have scarce GC-rich isochores. This difference is paralleled by chromosomal banding, which is strong in warm-blooded vertebrates and weak or absent in cold-blooded vertebrates.

The separation of major components from the genomes of warm-blooded vertebrates allowed us to study the distribution of any sequence that could be probed, and led to a number of conclusions, which can be summarized as follows (see Bernardi et al. 1985, and references 1–22 therein).

1. Single-copy or clustered genes are usually found in just one of the major components of Figure 1; their distribution appears to be largely conserved within each class of warm-blooded vertebrates (mammals, birds). In contrast, scattered genes and pseudogenes belonging to the same family (like the actin family) may be located on different major components.
2. Genes present in a given major component may be located on different chromosomes; conversely, genes present in different major components may be located on the same chromosome.
3. The distribution of genes (or gene clusters) is highly nonuniform; the majority of those localized so far were found in the heaviest components, which only represent a few percent of total DNA. Families of interspersed repeats are concentrated in given major components. As a consequence, renaturation kinetics of different major components of human DNA are different; these kinetics are also different from those of isopycnic components of mouse DNA. Integrated viral sequences are also mainly located in a given major component.
4. The GC levels of genes, introns, exons, codon third position, specific families of interspersed repeats, and integrated viral sequences, as well as the level of the CpG doublet, are linearly correlated with the

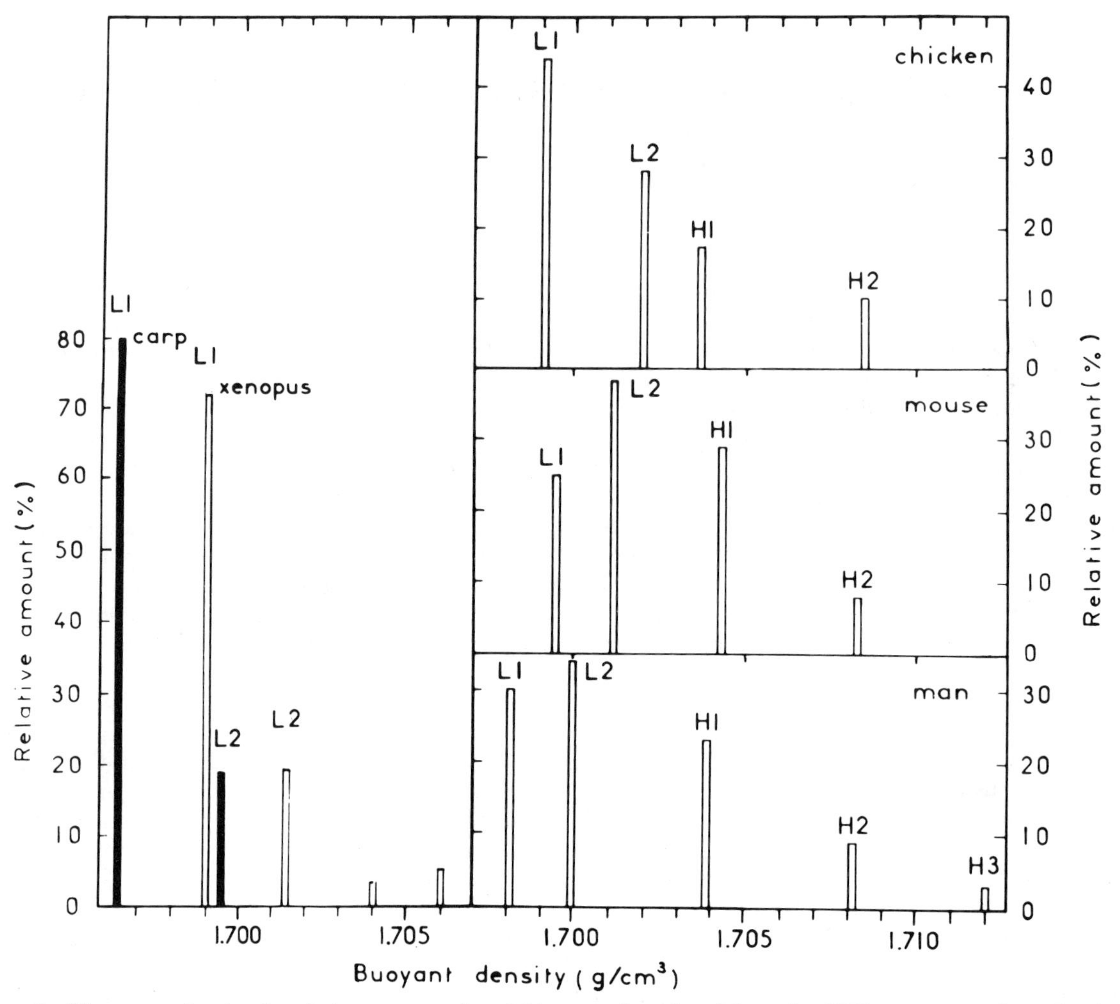

Figure 1. Histograms showing the relative amounts of modal buoyant densities of the major DNA components from *Cyprinus carpio, Xenopus laevis* (*left panel*), chicken, mouse, and man (*right panel*), as estimated by analytical CsCl density gradient centrifugation of fractions obtained from preparative Cs_2SO_4 centrifugation in the presence of Ag^+ or BAMD. Satellite and minor components (namely, components representing each less than 3% of DNA) are not shown. Carp and *Xenopus* genomes represent extreme cases of low and high heterogeneity among cold-blooded vertebrates. Notice that even in *Xenopus*, DNA having a density higher than 1.704 represents less than 10% of the genome, as compared with 30–40% for warm-blooded vertebrates. (Modified, from Bernardi et al. 1985.)

GC levels of the major components harboring the sequences under consideration.

5. The evolutionary origin of heavy components is mainly associated with regional GC increases in ancestral sequences from cold-blooded vertebrates, and, occasionally, also with the amplification of preexisting repeated heavy sequences (like the *Alu* sequences). Incidentally, the latter process implies a "targeted" integration of mobile repeats, like that found for viral sequences, and seems, at least in part, to be responsible for the formation of the H3 component in the human genome.
6. The identification of isochores with the DNA segments present in G and R bands rests not only on the points already mentioned (parallelism of compositional compartmentalization and chromosomal banding in vertebrates, presence of different isochores on the same chromosome), but also on the presence in early-replicating DNA (namely in R bands) of early-replicating genes located in the heavy components and in late-replicating DNA (namely in G bands), of late-replicating genes located in light components as well as on other points (Bernardi et al. 1985).

The compositional compartmentalization of the genome of warm-blooded vertebrates indicates that strong compositional constraints are operative on both coding and noncoding sequences. Here we report on (1) the general features of compositional constraints as studied on over 300 genes derived from more than 50 genomes spanning a very wide phylogenetic range (Table 1); (2) the effects of compositional constraints on the structure and function of DNA, RNA, and proteins; (3) the causes of compositional compartmentalization and, more specifically, of the formation of "heavy" isochores; and (4) some more general issues.

Table 1. List of Genomes Examined and Numbers of Genes Analyzed

Genomes	Number of genes	Genomes	Number of genes
Prokaryotes		Viruses	
1A Lambda (left arm)	23	31 Abelson murine leukemia virus	1
1B Lambda (right arm)	39	32 Adenovirus type 12	5
02 *Agrobacterium tumefaciens*	26	33 AKV murine leukemia virus	2
03 *Anacystis nidulans*	1	34 Avian sarcoma virus Y73	1
04 *Bacillus licheniformis*	2	35 Hepatitis B virus	2
05 *Bacillus megaterium*	1	36 Herpes simplex virus type 1	2
06 *Bacillus pumilus*	1	37 Human adult T-cell leukemia virus	3
07 *Bacillus stearothermophilus*	1	38 Human papilloma virus	4
08 *Bacillus subtilis*	7	39 Human papovavirus BK	7
09 *Erwinia amylovora*	2	40 Mouse hepatitis virus	2
10 *Escherichia coli*	43	41 Polyoma virus	6
11 *Haemophilus haemolyticus*	1	42 Tobacco mosaic virus	6
12 *Klebsiella pneumoniae*	4	51* Adenovirus 2 associated virus	
13 *Pseudomonas aeruginosa*	1	52* Adenovirus type 3	
14 *Pseudomonas putida*	1	53* Adenovirus type 25	
15 *Rhizobium* sp.	3	54* Adenovirus type 28	
16 *Salmonella typhimurium*	6	55* ERP	
17 *Shigella dysenteriae*	3	56* Herpes simplex virus type 2	
18 *Streptomyces fradiae*	1	57* Pseudorabies	
19 *Thermus thermophilus*	1	58* Simian adenovirus SA7	
20 *Vibrio cholerae*	4	59* Shope	
		60* Frog virus type 3	
Vertebrates		Vertebrates	
81 *Cyprinus carpio*	1	91 Man	20
82 *Lophius americanus*	1	92 Chicken (L2)	4
83 *Xenopus laevis*	2	93 (H2)	5
84 Chicken	2	94 Mouse (L1)	1
85 Mouse	7	95 (L2)	5
86 Hamster	1	96 (H2)	3
87 Rat	2	97 Rabbit (L2)	1
88 Dog	1	98 (H2)	1
89 Calf	3	99 Man (L2)	5
90 Ape	2	100 (H1)	2
		101 (H3)	6

The nucleotide sequence data for the ~300 genes analyzed in the present study were obtained from the GenBank Genetic Sequence Data Bank (Bilofsky et al. 1986), Release 38.0 (November, 1985). In selecting protein coding sequences we relied mainly on the "Features" tables of the GenBank Data base, and only complete genes, starting with an initiation codon and ending with one of the stop codons, were used in the present work. Numbers with asterisks refer to viral genomes for which only the CpG/GpC ratio was available (Russell 1974). In the case of warm-blooded vertebrates, the table lists the isochore classes or genome compartments (L1, L2, H1, H2, H3) in which several genes were localized (Bernardi et al. 1985). These genes were different from those considered under items 84, 85, and 91. A list of all the genes used in this analysis is available upon request.

Compositional Constraints Affect both Coding and Noncoding Sequences

If the GC levels found at each codon position of genes from different genomes are plotted against the GC levels of the coding sequences from the same genomes, linear relationships characterized by roughly the same slopes and the same intercepts are found for prokaryotes, viruses, and vertebrates (Fig. 2; the only exception concerns second positions of viruses). The slopes generally increase from second-, to first-, and to third-position plots.

In the case of prokaryotic and viral genomes, which are almost exclusively formed by coding sequences, the relationships described for coding sequence GC obviously coincide with those for gene GC and genome GC. In contrast, in the case of vertebrate genomes, where noncoding sequences are abundant, the relationships generally show slight, but significant, changes when plots are made against gene GC or genome GC. The slopes of "gene plots" (not shown) are slightly lower than those of exon plots, due to the influence of intron GC, whereas the linear relationships of genome plots are shifted to the left; this indicates a lower GC level in intergenic sequences compared with exons or genes (Fig. 2).

A special compositional constraint present in both coding and noncoding sequences of vertebrate genomes leads to a shortage of the doublet CpG (Swartz et al. 1962). On the basis of the different CpG levels exhibited by the genomes of small and large vertebrate viruses, it was suggested (1) that CpG shortage is associated with constraints from the translation machinery (Subak-Sharpe et al. 1967; Russell 1974) and (2) that the bulk of vertebrate DNA derives from, and maintains the CpG shortage of, polypeptide-specifying DNA (Russell 1974; Russell et al. 1976). Both suggestions (for a more detailed discussion, see Bernardi 1985)

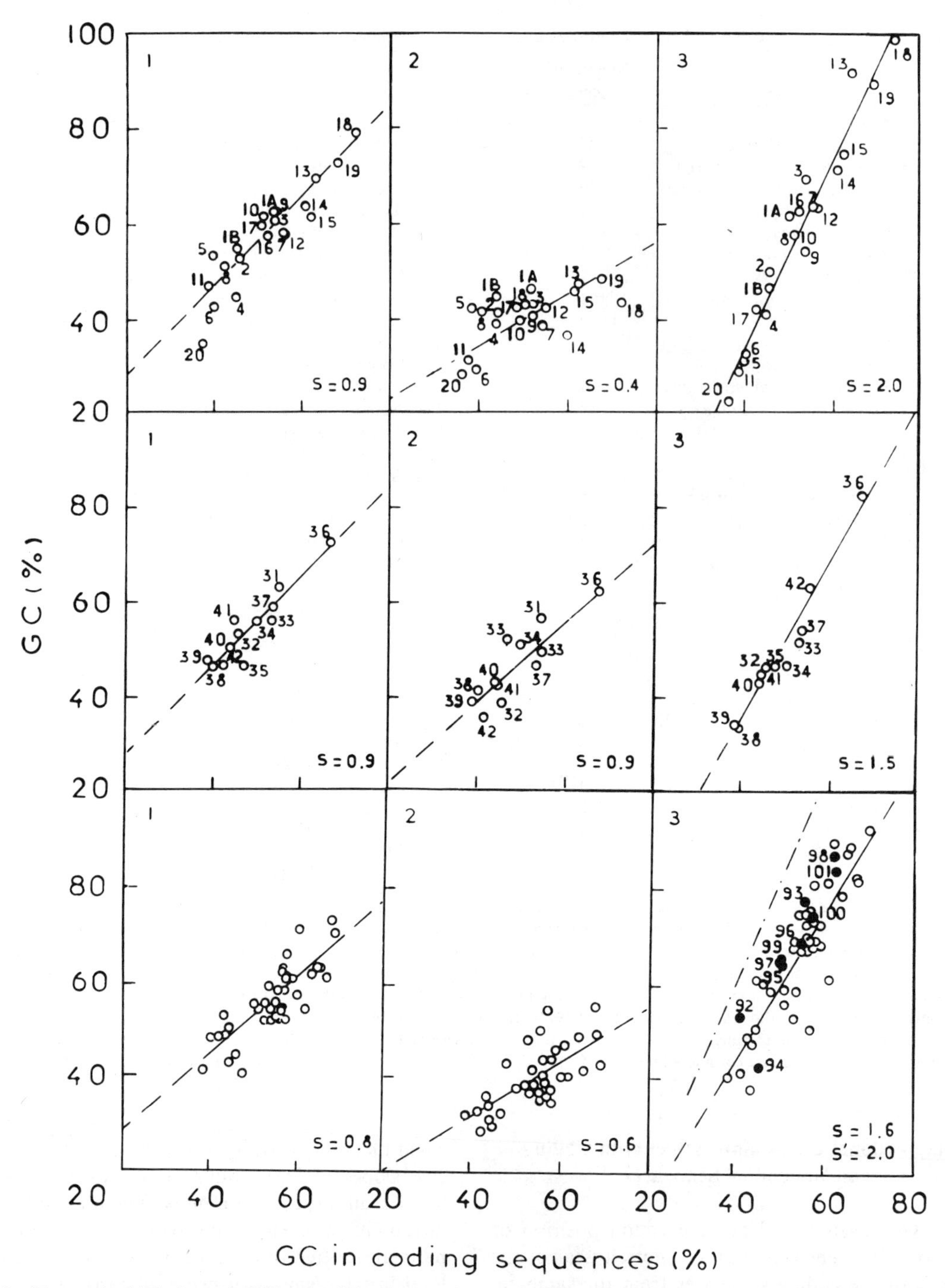

Figure 2. (*Top and middle frames*) GC levels of the three codon positions (1,2,3) of prokaryotic (*top frames*) and viral (*middle frames*) genes are plotted against the GC levels of the corresponding genomes. The scatter of points belonging to different genes from the same genome was small and average values, weighted for gene size, were therefore used. Numbers refer to the genomes listed in Table 1. Lines were drawn using the least-squares method; the slopes, *s*, are indicated; correlation coefficients were 0.91, 0.58, and 0.97 for prokaryotic genomes, and 0.91, 0.88 and 0.95 for viral genomes, respectively. (*Bottom frames*) (○) GC levels of codon positions of individual genes from vertebrates are plotted against the GC levels of the corresponding exons; (●) average values for third position GC of genes belonging to the same compartment of a given genome (see Table 1) are plotted against the GC levels of coding sequences of the genome compartment; correlation coefficients were 0.81, 0.67, and 0.88, respectively; the dash-and-point line is the third position plot against GC of genome compartments; its slope is given by s′.

are contradicted, however, by the findings that the CpG level of the genomes of vertebrate viruses is not related to the small or large size of the viral genome (and its supposed dependence or independence upon the translation machinery of the host cell), but simply upon the GC levels of the viral genomes (the shortage disappearing at high GC levels; Fig. 3), and that the relationship for viral genomes is very close to that previously seen

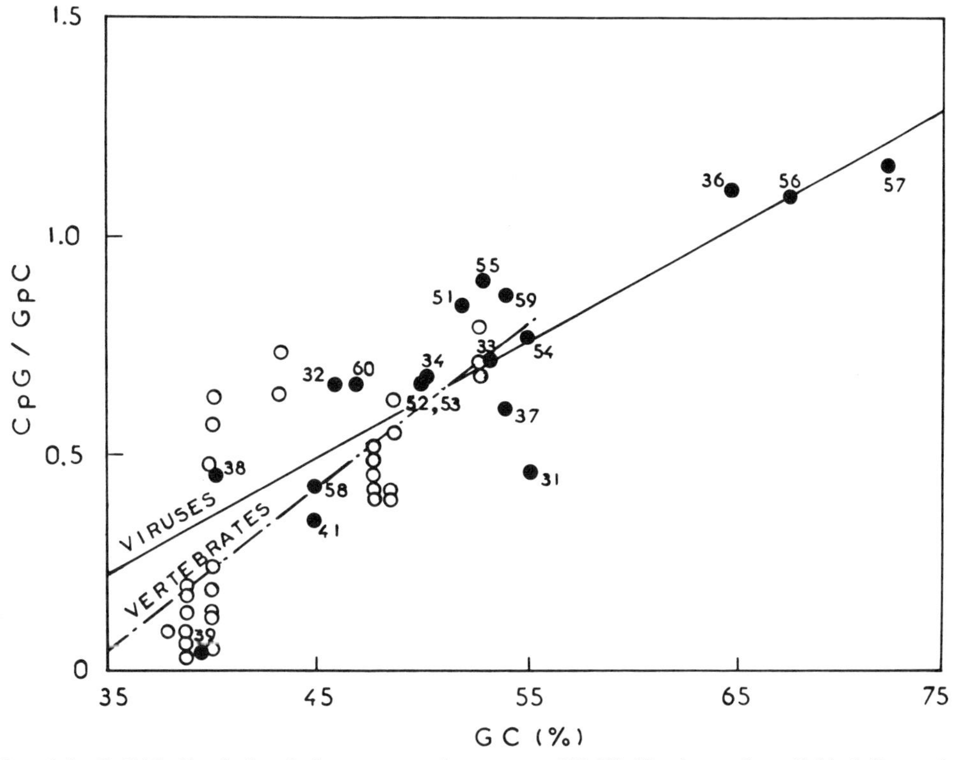

Figure 3. Plot of the CpG/GpC ratio for viral genomes against genome GC (●). Numbers refer to Table 1. Except for frog virus 3 (point 60), all viruses were from warm-blooded vertebrates. The dash-and-point line and the open symbols correspond to a similar plot, as obtained for vertebrate genes or exons (Bernardi et al. 1985).

(Bernardi et al. 1985) for vertebrate genes or exons (Fig. 3).

The general conclusion to be drawn from the above results is that essentially the same compositional constraints affect coding and noncoding sequences (which represent more than 90% of DNA in mammalian genomes). In coding sequences, these constraints concern not only GC levels, but also the levels of individual bases in different codon positions (not shown), and lead to specific relationships for different bases; obviously, such relationships are not the result of a random increase in G and C in different codon positions. Constraints on introns and intergenic sequences are indicated by the linear relationships found between the GC levels of codon positions and those of genes or genomes (Fig. 2; see also Bernardi et al. 1985); the slopes and intercepts of these relationships are close to those exhibited by coding sequences.

GC Increases in Coding Sequences Affect mRNA and Protein Stability

All GC changes in second-codon positions entail changes in the amino acid composition of proteins; so also do most first-position changes and a few third-position changes. An analysis of the amino acid replacements that accompany the GC increases in codon positions has revealed that they comprise those that lead to thermodynamically more stable proteins (Argos et al. 1979; Zuber 1981) (Table 2). Indeed, the amino acids (alanine and arginine) that are most frequently acquired in thermophiles and that most contribute to an increased stability increase, whereas those (serine and lysine) that are correspondingly lost and that diminish stability decrease, with increasing exon GC (Fig. 4). In the case of compartmentalized genomes, these changes may take place within the same genome; for instance, in the human genome the (Ala + Arg)/(Ser + Lys) molar ratio varies by a factor of four between the AT-richest and GC-richest genes (Fig. 4).

In conclusion, the compositional changes that make

Table 2. Amino Acid Exchanges Observed in Thermophiles and the Accompanying GC Changes in Their Codons

Exchanges			Codon
mesophiles		thermophiles	GC
*Gly	→	Ala	0
*Ser	→	Ala	+
Ser	→	Thr	0
Lys	→	Arg	+
Asp	→	Glu	±
Ser	→	Gly	+
*Lys	→	Ala	+

Exchanges are given in order of decreasing frequency. Asterisks indicate the largest expected increase in stability (Argos et al. 1979). Only the Lys → Ala exchange requires more than one base change per codon. The right column indicates increases (+), decreases (−), and no changes (0) in codon GC levels.

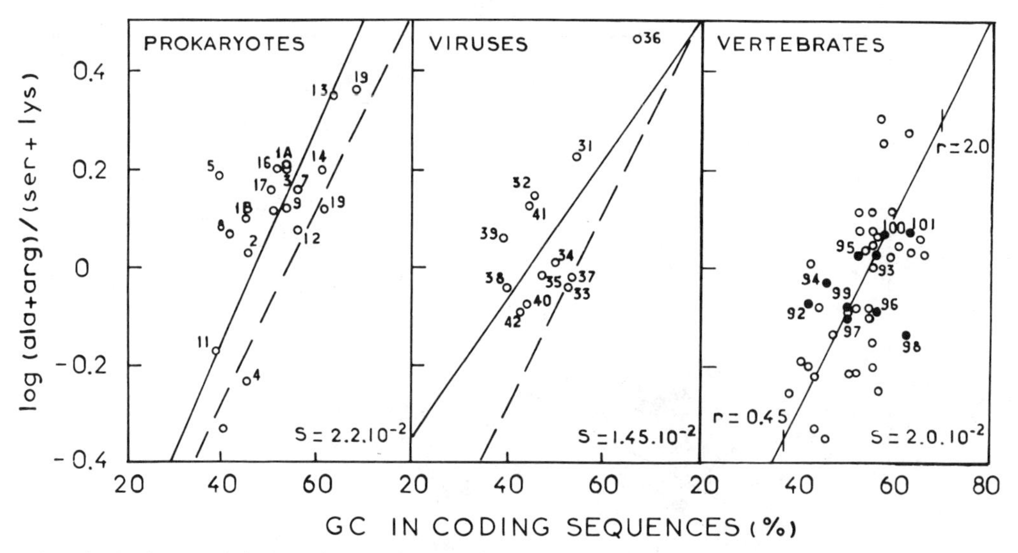

Figure 4. Plot of (alanine + arginine)/(serine + lysine) molar ratio against GC levels of coding sequences from prokaryotes, viruses, and vertebrates (see Table 1). In the vertebrate plot: (○) individual genes (for man, chimpanzee, mouse, hamster, rat, calf, and chicken); (●) average values of genes belonging to the same compartment of a given genome (see Table 1); the two *r* values correspond to the lowest and highest ratios found for human genes. Slopes are indicated by *s*; the vertebrate line is also shown in the other two diagrams (*broken lines*), for the sake of comparison.

DNA thermodynamically more stable also increase the thermodynamic stability of the encoded proteins. The same changes obviously also lead to higher GC levels in mRNAs, a factor known (Hasegawa et al. 1979) to increase their base-pairing and stability. The limited data available so far suggest that similar changes may occur in rRNAs and tRNAs.

These points suggest that the formation of GC-rich isochores may have begun in cold-blooded vertebrates exposed to higher temperatures, by a preferential fixation of A/T→G/C changes in the genes of proteins that could be most affected in their function by temperature. This process was apparently accompanied by a GC enrichment in the neighboring noncoding sequences and led to a limited formation of heavy isochores (experimentally found in a number of cold-blooded vertebrates; see Bernardi et al. 1985). The same process went much farther in warm-blooded vertebrates, where it led to an extensive formation of GC-rich isochores. The suggestion that genes were the initial nuclei for the regional increases in GC would account, at least in part, for the high gene concentration in GC-rich isochores (Bernardi et al. 1985). It also raises the possibility that a particular set of genes was preferentially located in heavy components. This set might correspond to housekeeping genes because these genes appear to be only present in R bands (Goldman et al. 1984), namely in the GC-rich components of the genome, whereas tissue-specific genes are preferentially located in GC-poor G bands. It is clear, however, that a number of the latter genes were later translocated into GC-rich components, where they subsequently underwent a GC increase insuring a better protection against DNA breathing and mutability (L. Orgel, pers. comm.).

Codon Usage Is Largely Determined by Compositional Constraints

Since nonrandomness of codon usage was first discovered, several not mutually exclusive explanations have been provided for this phenomenon. These comprise: (1) the optimization of codon–anticodon interaction energy (Grosjean et al. 1978) and the consequent optimization of translation efficiency in highly expressed genes (Grantham et al. 1981; Bennetzen and Hall 1982); (2) the fulfillment of requirements for mRNA secondary structure and stability (Hasegawa et al. 1979); and (3) an adaptation of codons to the actual populations of isoaccepting tRNAs (Ikemura 1985). These explanations essentially rest on intraspecific differences in the usage of all synonymous codons.

In contrast, our results concern interspecific and intercompartmental differences in the usage of synonymous codons characterized by different GC levels in third positions; this subset of codons corresponds to two-thirds of all synonymous codons. Our results lead to the following conclusions.

1. Interspecific and intercompartmental differences in codon usage largely depend upon the compositional constraints affecting the genome, or the genome compartments. This provides, to a large extent, a rationale for the "genome strategy of codon usage" (Grantham et al. 1980), which comprises several "compartmental strategies" in compartmentalized genomes.
2. The proposals that mRNA structure (Hasegawa et al. 1979) or the abundance of synonymous tRNAs (Ikemura 1985) are the causes and not the effects of codon usage should be reversed. Since third posi-

tions are under essentially the same compositional constraints as noncoding sequences, the primary phenomenon is at the DNA level and the effects are at the mRNA or tRNA levels; this point was already well demonstrated by the changes in tRNA distributions that occur in the silk gland of *Bombyx mori* in connection with the expression of the fibroin and sericin genes (Chevallier and Garel 1979). As already mentioned, our results do not bear on the intraspecific differences in codon usage that have been shown in unicellular organisms, like *Escherichia coli* and *Saccharomyces cerevisiae* (Ikemura 1985). These differences should probably be visualized as due to the selection of a subset of codons for highly expressed genes on the basis of optimization of codon–anticodon interaction energies (Grosjean et al. 1978).

3. The results shown in Figure 2 account for the so-called "contextual constraints" previously seen in different codon positions (Nussinov 1980, 1981; Lipman and Wilbur 1983), for the frequency of pyrimidine-purine doublets in third- and first-codon positions, and for the 3-bp periodicity detected in coding sequences (Trifonov and Sussman 1980; Shepherd 1981).

Mutations Are Fixed Mainly Through Positive Darwinian Selection

The evolution of living organisms is primarily caused by mutations that may subsequently be eliminated or become fixed in the genome. It is generally agreed that elimination affects deleterious mutations and occurs by negative selection. In contrast, fixation has been visualized as due either (1) to positive Darwinian selection acting on advantageous mutations or (2) to random genetic drift acting on selectively neutral (i.e., selectively equivalent) mutations. Our findings have a direct bearing on the mechanism by which mutations become fixed.

Intragenomic changes. (1) Changes in any codon position appear to have been fixed, on the average, under the influence of compositional constraints, in conformity with the base composition of the genome, or the genome compartment, in which genes are located. For instance, the base changes that occurred over evolutionary time in the genomes of warm-blooded vertebrates preferentially kept low and high GC levels in different codon positions of GC-poor and GC-rich genes, respectively. Moreover, the scatter in the base compositions of individual codon positions of different genes belonging to the same genome, or the same genome compartment, is small (a fact justifying our averaging values in Fig. 2). These findings indicate that, on the average, changes are conservative. By analogy with the "genome strategy of codon usage" (Grantham et al. 1980), one should, therefore, take into consideration a more general "compositional strategy of coding sequences," which also concerns nonsilent changes and may comprise several compartmental strategies in compartmentalized genomes. (2) Changes in noncoding sequences of eukaryotes conform with the same general rules as changes in coding sequences. In eukaryotes, the compositional strategy of coding sequences is therefore part of a general compositional strategy which also affects noncoding sequences. (3) The CpG level (which corresponds to the level of potential methylation sites) in both coding and noncoding sequences of vertebrates also appears to be subject to the same compositional constraints as the base changes just discussed; indeed, the CpG shortage is different in different genome compartments and is related to their GC levels (Fig. 3).

To sum up, intragenomic GC changes clearly indicate that most mutations are fixed, in both coding and noncoding sequences, not at random but under the influence of compositional constraints, in compliance with a general compositional strategy. Random fixation of neutral mutations (Kimura 1968, 1983, 1986; King and Jukes 1969) certainly also occurs, but only to such an extent that the general compositional strategy is not blurred.

Intergenomic changes. (1) The decreasing extents of GC changes from the third to the first and to the second positions appear to be correlated with the corresponding increasing impacts on amino acid composition of proteins. In other words, the slopes of Figure 2 are correlated with the different fixation rates that have been detected in different codon positions of a number of genes (Zuckerkandl 1976; Kimura 1983, 1986; Li et al. 1985). While the latter results concern differences at specific positions of homologous genes from different organisms, the data of Figure 2 compare average values for codon positions of genes from given organisms with those from other organisms. (2) A clear directionality is shown by the amino acid substitutions, the silent base changes, the changes in noncoding seqeunces, and the CpG changes that accompanied the transition from cold-blooded to warm-blooded vertebrates, to lead to the formation of GC-rich genes and GC-rich isochores in the genomes of the latter.

In conclusion, the compositional constraints within a genome (or within genome compartments) and the directional changes just mentioned can only be explained by a positive Darwinian selection acting on mutations that confer selective advantages in relationship with environmental pressures. Some of these advantages have been identified. For instance, in the transition from cold-blooded to warm-blooded vertebrates, silent changes appear to lead to an optimization of structure and function at the level of both DNA and RNA; nonsilent changes lead, in addition, to an optimization of structure and function at the protein level.

Obviously, our conclusions reverse the proposals of the "neutral mutation–random drift hypothesis" (1) "that the great majority of evolutionary changes at the molecular level are caused not by Darwinian selection acting on advantageous mutations, but by random fixation of selectively neutral and nearly neutral mutants" and (2) "that only a minute fraction of DNA changes are adaptive in nature" (Kimura 1986). Both proposals rest, in fact, on the premise (see next section) that the phenotype of an organism only corresponds to its "gene products"; as a logical consequence, silent mutations and changes in noncoding sequences were visualized as drifting at random and having no evolutionary impact.

CONCLUSIONS

The main finding reported here concerns the demonstration of compositional constraints that affect the genomes of living organisms and are more general and stronger than other genome constraints, namely (1) fixation rate constraints (previously called functional constraints [Kimura 1983]; this terminology should be abandoned since all constraints have functional consequences) that increasingly limit the fixation of mutations taking place in third-, first-, and second-codon positions; (2) constraints associated with the CpG (and potential methylation) levels of vertebrate genomes; and (3) constraints associated with the codon usage of highly expressed genes from unicellular organisms.

Indeed, compositional constraints affect both coding and noncoding sequences from all organisms explored, whereas other constraints are more limited in the range of sequences and/or organisms concerned, are evident in all codon positions independently of differences in fixation rates (Fig. 2), lead to the disappearance of the CpG shortage in GC-rich genomes (Fig. 3), and are not hidden by the preferential use of a subset of codons of highly expressed genes in unicellular organisms.

Compositional constraints indicate that most mutations are not fixed at random, but are fixed in relationship to a general compositional strategy of the genome. This appears to be the result of positive Darwinian selection of mutations that is advantageous as far as environmental pressures are concerned. Neutral mutations obviously also exist, but their fixation is small enough so as not to distort the general compositional strategy. These conclusions lead to two general ideas: (1) that the environment can shape up the genome through selection; and (2) that genome evolution depends more on natural selection than on random events.

Compositional constraints largely affect the structure and stability of the genome (at its different DNA, chromatin, and chromosome levels), of the transcripts, and even of proteins (as exemplified by the higher stability accompanying GC increases), as well as codon usage. At the same time, they also conceivably touch on a number of basic functions, such as replication, recombination, transcription, and translation, that are sensitive to the compositional/structural features just mentioned.

In eukaryotes, both coding and noncoding sequences appear to be under essentially the same compositional constraints, and therefore under the same selection pressures. This finding leads to a number of conclusions. First of all, it stresses the fundamental unity of the genome, already suggested by the genome strategy of codon usage (Grantham et al. 1980), and contradicts what has been called the "bean bag" view of the genes within the genome (Mayr 1976). Second, it points to the genome as the unit upon which natural selection acts. Third, it does not support the view that noncoding sequences can be equated with functionless "junk DNA" (Ohno 1972). In contrast, it suggests that noncoding sequences largely play a physiological role, which may have to do with the modulation of basic genome functions.

As far as the latter suggestion is concerned, it was already pointed out that the fixation of advantageous mutations concerns the majority, but not the totality, of mutations. Likewise, the functional role of noncoding sequences is not an all-or-none issue, and should be visualized as associated with the majority, and not the totality, of noncoding sequences. Moreover, this suggestion, although not a new one (Britten and Davidson 1969; Davidson and Britten 1979), does not rest anymore on "adaptive stories" (Gould and Lewontin 1979), which can be criticized (Doolittle and Sapienza 1980; Orgel and Crick 1980), but on the newly demonstrated existence of compositional constraints. Interestingly, similar conclusions have been reached on the basis of independent evidence for the noncoding sequences of mitochondrial genome of yeast (Bernardi 1982, 1983; de Zamaroczy and Bernardi 1985, 1986, in prep.).

Finally, compositional constraints identify a new component in the organismal phenotype, which may be called the "genome phenotype." This component adds to the other classical component of the phenotype, which is formed by "gene products," and is defined by nonsilent mutations in the genes and by mutations in regulatory signals.

ACKNOWLEDGMENTS

The senior author thanks the Fogarty International Center for Advanced Study in the Health Sciences, National Institutes of Health, Bethesda, Maryland, for a scholarship. Sequence data treatments were performed using computer facilities at CITI2 in Paris on a PDP8 computer with the help of the French Ministere de la Recherche et la Technologie (Programme Mobilisateur "Essor des Biotechnologies").

REFERENCES

Argos, P., M.G. Rossmann, U.M. Grau, A. Zuber, G. Franck, and J.D. Tratschin. 1979. Thermal stability and protein structure. *Biochemistry* **18:** 5698.

Bennetzen, J.L. and B.D. Hall. 1982. Codon selection in yeast. *J. Biol. Chem.* **257:** 3026.

Bernardi, G. 1982. The evolutionary origin and the biological role of non-coding sequences in the mitochondrial genome of yeast. In *Mitochondrial genes* (ed. G. Attardi et

al.), p. 269. Cold Spring Harbor Laboratory, Cold Spring Harbor, New York.
———. 1983. Genome instability and the selfish DNA issue. *Folia Biol.* **29:** 82.
———. 1985. The organization of the vertebrate genome and the problem of the CpG shortage. In *Biochemistry and biology of DNA methylation* (ed. G.L. Cantoni and A. Razin), p. 3. A.R. Liss, New York.
Bernardi, G., B. Olofsson, J. Filipski, M. Zerial, J. Salinas, G. Cuny, M. Meunier-Rotival, and F. Rodier. 1985. The mosaic genome of warm-blooded vertebrates. *Science* **228:** 953.
Bilofsky, H.S., C. Burks, J.W. Fickett, W.B. Gould, F.L. Lewitter, W.P. Rindone, C.D. Swindell, and C.S. Tung. 1986. The GenBank genetic sequence data bank. *Nucleic Acids Res.* **14:** 1.
Britten, R.J. and E.H. Davidson. 1969. Gene regulation for higher cells: A theory. *Science* **165:** 349.
Chevallier, A. and J.R. Garel. 1979. Studies on tRNA adaptation, tRNA turnover, precursor tRNA and tRNA gene distribution in *Bombyx mori* by using two-dimensional polyacrylamide gel electrophoresis. *Biochimie* **61:** 245.
Davidson, E.H. and R.J. Britten. 1979. Regulation of gene expression: Possible role of repetitive sequences. *Science* **204:** 1052.
Doolittle, W.F. and C. Sapienza. 1980. Selfish genes, the phenotype paradigm and genome evolution. *Nature* **284:** 601.
Goldman, M.A., G.P. Holmquist, M.C. Gray, L.A. Caston, and A. Nag. 1984. Replication timing of genes and middle repetitive sequences. *Science* **224:** 686.
Gould, S.J. and R.C. Lewontin. 1979. The spandrels of San Marco and the Panglossian paradigm: A critique of the adaptionist program. *Proc. Roy. Soc. Lond.* **B205:** 581.
Grantham, R., C. Gautier, M. Gouy, M. Jacobzone, and R. Mercier. 1981. Codon catalog usage is a genome strategy modulated for gene expressivity. *Nucleic Acids Res.* **9:** r43.
Grantham, R., C. Gautier, M. Gouy, R. Mercier, and A. Paré. 1980. Catalog usage and the genome hypothesis. *Nucleic Acids Res.* **8:** r49.
Grosjean, H., D. Sankoff, W. Min Jou, W. Fiers, and R.J. Cedergren. 1978. Bacteriophage MS 2RNA: A correlation between the stability of the codon:anticodon interaction and the choice of code words. *J. Mol. Evol.* **12:** 113.
Hasegawa, H., T. Yasunaga, and T. Miyata. 1979. Secondary structure of MS2 phage RNA and bias in code word usage. *Nucleic Acids Res.* **7:** 2073.
Ikemura, T. 1985. Codon usage and tRNA content in unicellular and multicellular organisms. *Mol. Biol. Evol.* **2:** 13.
Kimura, M. 1968. Evolutionary rate at the molecular level. *Nature* **217:** 624.
———. 1983. *The neutral theory of molecular evolution.* Cambridge University Press, England.
———. 1986. DNA and the neutral theory. *Phil. Trans. R. Soc. Lond.* **B312:** 343.
King, J.L. and T.H. Jukes. 1969. Non-darwinian evolution. *Science* **164:** 788.
Li, W.-H., C.-C. Luo, and C.-I. Wu. 1985. Evolution of DNA sequences. In *Molecular evolutionary genetics* (ed. R.J. Mac Intyre), p. 1. Plenum Press, New York.
Lipman, D.J. and W.J. Wilbur. 1983. Contextual constraints on synonymous codon choice. *J. Mol. Biol.* **163:** 363.
Mayr, E. 1976. *Evolution and the diversity of life.* Harvard University Press, Cambridge, Massachusetts.
Nussinov, R. 1980. Some rules in the ordering of nucleotides in the DNA. *Nucleic Acids Res.* **8:** 4545.
———. 1981. The universal dinucleotide asymmetry rules and the amino acid codon choice. *J. Mol. Evol.* **17:** 237.
Ohno, S. 1972. An argument for the genetic simplicity of man and other mammals. *J. Hum. Evol.* **1:** 651.
Orgel, L.E. and F.H.C. Crick. 1980. Selfish DNA: The ultimate parasite. *Nature* **284:** 604.
Russell, G.J. 1974. "Characterization of deoxyribonucleic acids by doublet frequency analysis." Ph.D. Thesis, University of Glasgow, Scotland.
Russell, G.J., P.M.B. Walker, R.A. Elton, and J.H. Subak-Sharpe. 1976. Doublet frequency analysis of fractionated vertebrate nuclear DNA. *J. Mol. Biol.* **108:** 1.
Schmid, C.W. and P.L. Deininger. 1975. Sequence organization of the human genome. *Cell* **6:** 345.
Shepherd, J. 1981. Method to determine the reading frame of a protein from the purine/pyrimidine genome sequence and its possible evolutionary justification. *Proc. Natl. Acad. Sci.* **78:** 1596.
Singer, M.F. 1982. SINES and LINES: Highly repeated short and long interspersed sequences in mammalian genomes. *Cell* **28:** 433.
Singer, M.F. and J. Skowronski. 1985. Making sense out of LINES: Long interspersed repeats in mammalian genomes. *Trends Biochem. Sci.* **10:** 119.
Subak-Sharpe, H., R.R. Burk, L.V. Crawford, J.M. Morrison, J. Hay, and H.M. Keir. 1967. An approach to evolutionary relationships of mammalian DNA viruses through analysis of the pattern of nearest neighbor base sequences. *Cold Spring Harbor Symp. Quant. Biol.* **31:**737.
Swartz, M.N., T.A. Trautner, and A. Kornberg. 1962. Enzymatic synthesis of deoxyribonucleic acid. XI. Further studies on nearest neighbor base sequences in deoxyribonucleic acids. *J. Biol. Chem.* **237:** 1961.
Trifonov, E.N. and J.L. Sussman. 1980. The pitch of chromatin is reflected in its nucleotide sequence. *Proc. Natl. Acad. Sci.* **77:** 3816.
de Zamaroczy, M. and G. Bernardi. 1985. Sequence organization of the mitochondrial genome of yeast—A review. *Gene* **37:** 1.
———. 1986. The GC clusters of the mitochondrial genome of yeast and their evolutionary origin. *Gene* **41:** 1.
Zuber, H. 1981. Structure and function of thermophilic enzymes. In *Structural and functional aspects of enzyme catalysis* (ed. H. Eggerer and R. Huber), p. 114. Springer Verlag, Berlin.
Zuckerkandl, E. 1976. Evolutionary processes and evolutionary noise at the molecular level. II. A selectionist model for random fixation in proteins. *J. Mol. Evol.* **7:** 269.

The Population Genetics of α-Thalassemia and the Malaria Hypothesis

A.V.S. HILL
M.R.C. Molecular Haematology Unit, Nuffield Department of Clinical Medicine, University of Oxford, John Radcliffe Hospital, Oxford OX3 9DU, United Kingdom

Since the advent of rapid DNA analysis techniques, surveys of numerous tropical and subtropical regions have revealed remarkably high frequencies of α-thalassemia. Indeed, it now appears that α^+-thalassemia, which leads to decreased α-globin production by an affected chromosome, is the commonest single gene disorder in man (Higgs and Weatherall 1983). The finding of β-thalassemia in areas that were or had been malarious prompted Haldane (1949) to suggest that thalassemia heterozygotes might enjoy increased fitness in the presence of this parasitic disease. Epidemiological and in vitro studies on sickle cell carriers subsequently provided convincing support for the malaria hypothesis as applied to the hemoglobinopathy Hb S (for review, see Allison 1965; Pasvol and Wilson 1982). However, for β-thalassemia there is only limited evidence from the geographical distribution of this condition to support the hypothesis. In the case of the common mild forms of α^+-thalassemia there is no evidence that malarial selection is responsible. This is mainly due to the more complex genetics of α-thalassemia that have made it impossible to determine genotypes unequivocally on the basis of parameters such as red cell morphology, reduced hemoglobin content, and α-globin synthesis (Higgs and Weatherall 1983). Hence, before the relatively recent introduction of DNA analysis, it has been extremely difficult to study the population genetics of α-thalassemia.

There are several reasons why evaluation of the malaria hypothesis as applied to α-thalassemia is of general interest and importance and not just an elaboration of a principle that has already been established for Hb S. First, α^+-thalassemia is clinically a far milder disorder than either sickle cell anemia or β-thalassemia and is often hematologically undetectable in the heterozygous state. Hence this disorder, for which the term "disease" is hardly warranted, is far more typical of the type of variant on which natural selection usually acts. New variants, such as the Hb S gene, which have selection coefficients of the order of 0.15, must be extremely rare in human evolution (Cavalli-Sforza and Bodmer 1971). Second, also a consequence of this mild phenotype, it seemed quite plausible that factors other than malaria might be responsible for this polymorphism. The reason that *Plasmodium falciparum* was the outstanding candidate for maintaining high frequencies of the Hb S gene was that only a common lethal agent could produce the large difference in survival observed between Hb AA and Hb AS individuals. Since homozygosity for α^+-thalassemia is a very mild condition, other less severe environmental factors might be implicated. Alternatively, milder forms of malaria, such as *P. vivax,* might be sufficient to produce such selection. In addition, it has been suggested that at least part of the selective pressure for α^+-thalassemia alleles in Africa might be generated by its ameliorating effect on the clinical course of sickle cell anemia (Pagnier et al. 1984). Another point of interest is whether α^+-thalassemia ($-\alpha$) alleles represent a balanced or a transient polymorphism. Since the $-\alpha/-\alpha$ hematologic phenotype is relatively mild, a significant protective advantage should lead to the fixation of the $-\alpha$ allele. That this is not known to have occurred in any human population must mean either that insufficient time has elapsed for this to happen or that the $-\alpha$ allele represents a balanced polymorphism.

Although the existing data on the global distribution of α-thalassemia suggested that malaria might be a selective agent, such macroepidemiological studies can probably never be considered conclusive because of the complicating effects of possible racial differences and the covariation of many other factors with the distribution of malaria. It seemed important, therefore, to attempt to test the malaria hypothesis for α-thalassemia by a large-scale microepidemiological study. The island populations of the Southwest Pacific presented the most suitable location for this project for several reasons. Malaria endemicity in Melanesia (Figs. 1 and 2) varies in two dimensions, with altitude and with latitude (Lambert 1949; Black 1955): the high interior of New Guinea was almost devoid of malaria before the arrival of Europeans toward the end of the last century. Whereas malaria is hyper- to holoendemic in north coastal Papua New Guinea (based on spleen rate surveys, Fig. 2), it decreases as one moves southeast through Island Melanesia and is absent in New Caledonia and Fiji. This provides the opportunity to compare the frequencies of α-thalassemia in Melanesians living in areas with different malaria endemicities. Furthermore, the β-globin chain variants Hb S and Hb E, which are probably under stronger selection and might complicate the analysis of any relationship between the distribution of malaria and α-thalassemia, are not found in Melanesia (Livingstone 1985). Similarly, β-thalassemia, which has a more severe phenotype, is absent on most of the islands and α^0-thalassemia, caused

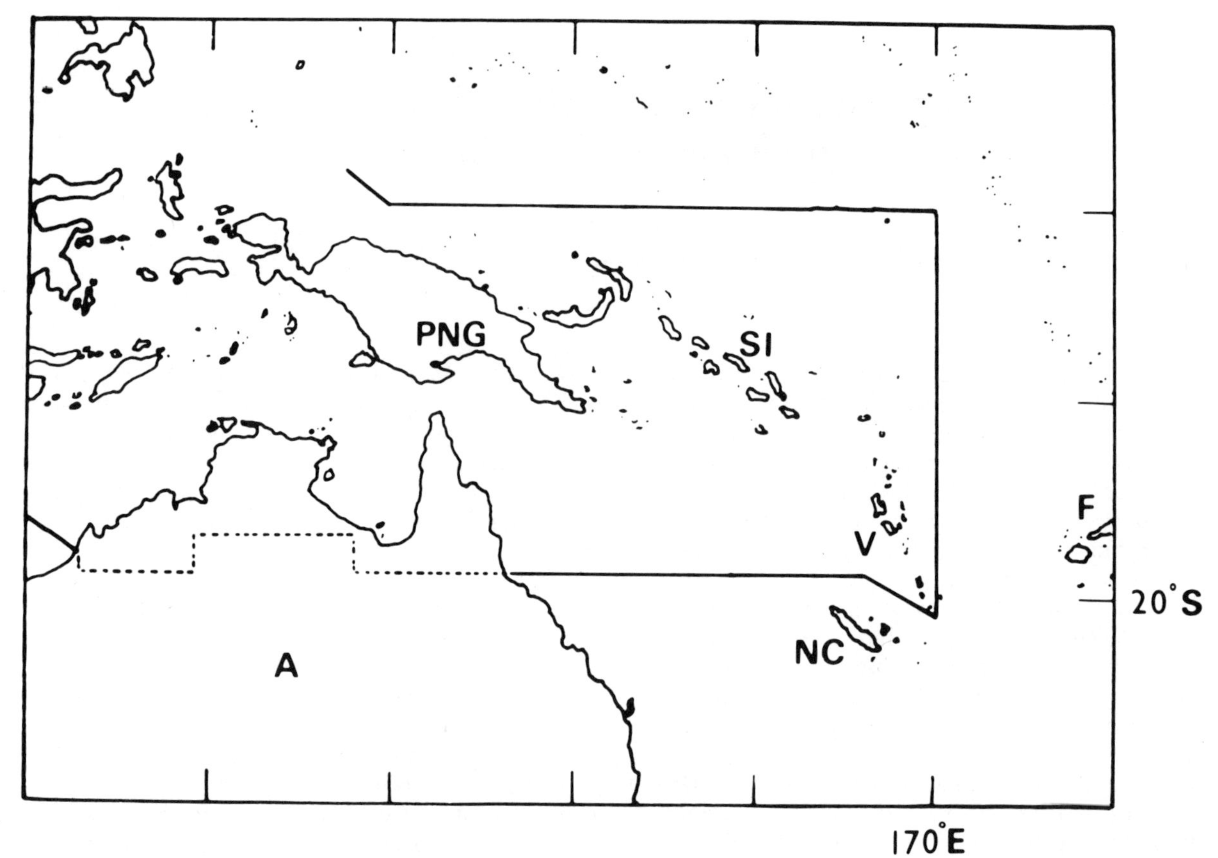

Figure 1. Distribution of malaria in the Southwest Pacific. Malaria is found west of 170°E and north of 20°S in Papua New Guinea (PNG) (except above 1800 meters), the Solomon Islands (SI), and most of Vanuatu (V), but not in New Caledonia (NC) or Fiji (F). (A) Australia. The islands in Melanesia, excluding the mainland of New Guinea, are referred to as Island Melanesia.

by a deletion of both linked α genes, is found only sporadically in Melanesia (Bowden et al. 1985). Because of the suggested competitive interaction in a population between α^+- and α^0-thalassemia alleles (Wills and Londo 1981), a significant prevalence of the latter might well have obscured any frequency correlation between the common single gene deletion forms of α^+-thalassemia and malaria. Finally, because the present-day endemicities of malaria in the region have been altered by control programs, it is fortunate that numerous malaria surveys were performed before such intervention began.

In this study comparison of the α^+-thalassemia frequencies of individuals from different parts of Melanesia with historical preintervention endemicities of malaria showed a strong correlation, whether the latter varied with latitude or altitude. DNA analysis of other unlinked polymorphisms failed to show significant population differences for these markers. Moreover, subtyping and restriction enzyme haplotype analysis of $-\alpha$ chromosomes from Melanesians argue for multiple local origins for these deletions. It is proposed that these data strongly support the hypothesis of selection by malaria.

METHODS

The Papua New Guinea samples were collected in the larger towns from schools and hospitals attended by villagers from across the island. The North Solomons samples were collected in Kieta, Bougainville, and the samples from Vanuatu from throughout the archipelago. The New Caledonian samples were collected from healthy Melanesians on that island. The Filipinos were from Manila, the Icelanders from Reykjavik, and the British individuals from Oxford. DNA extraction and determination of α-globin, γ-globin, and haptoglobin genotypes by the Southern blot procedure were as described previously (Old and Higgs 1983; Hill et al. 1985b, 1986a,b). α-Globin haplotypes were determined as described in Higgs et al. (1986). Analysis of *Bgl*II site polymorphisms with the probes 2.30 and 3.6 and a *Sst*I polymorphism by a C3 probe were as described (Davies et al. 1983; Gilliam et al. 1984; Povey et al. 1985). Four hypervariable region (HVR) allele size classes were detected using the insulin gene probe *phins 214* and an *Rsa*I digest (Bell et al. 1984): class 1, 4.7 kb; class 2, 3.9 kb; class 3, 2.6 kb; class 4, 2.1–2.4 kb. Six allele size classes were detected

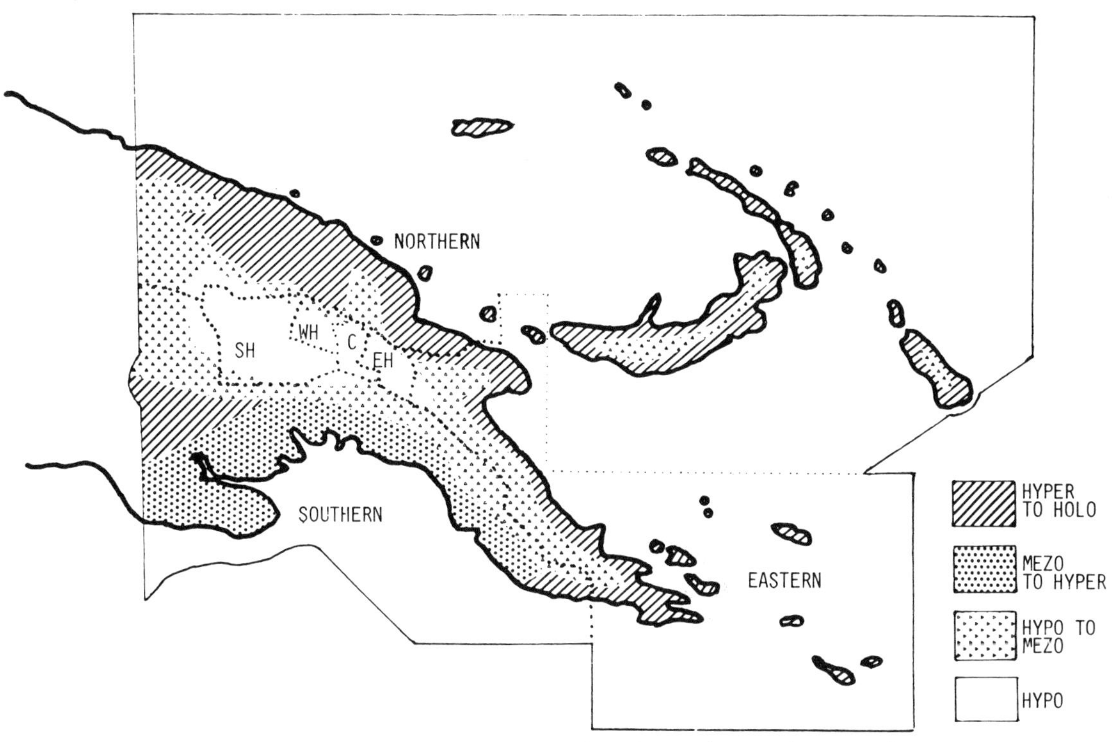

Figure 2. Map of Papua New Guinea showing the three coastal regions from which samples were collected (northern, southern, and eastern), and the four highlands regions (Southern Highlands and Enga Provinces; [SH]; Western Highlands Province [WH]; Chimbu Province [C]; Eastern Highlands Province [EH]) superimposed on estimated malarial endemicity calculated from spleen rate surveys on children aged 2–9 years: holoendemic >75% splenomegaly; hyperendemic 50–75%; mesoendemic 10–50%; hypoendemic <10%.

with *Hin*fI using a *Bam*HI-*Sst*I fragment from pBL 116-4 containing the *HRAS1* HVR (Capon et al. 1983): class 1, 4.3 kb; class 2, 3.8 kb; class 3, 3.0 kb; class 4, 2.8 kb; class 5, 2.7 kb; class 6, 2.0 kb.

RESULTS

The Variety of Globin Gene Variants in Melanesia

In contrast to neighboring southeast Asia where Hb E, Hb Constant Spring, and Southeast Asian α^0-thalassemia deletion are common, these variants are absent in Melanesia where the predominant hemoglobinopathy is α-thalassemia of the single gene deletion type ($-\alpha$) (Bowden et al. 1985; Livingstone 1985). Although a variety of γ- and ζ-globin gene variants are also present at significant frequencies in Melanesians, these appear to have negligible phenotypic effects (Hill et al. 1985a, 1986a). Molecular analysis of β-thalassemia alleles in this region suggest that these also have had local origins (Hill 1986). There are two common deletion forms of α^+-thalassemia (Embury et al. 1980). One involves a loss of 3.7 kb from between the two α-globin genes or removes the $\alpha 1$ gene ($-\alpha^{3.7}$); the other deletion is 4.2 kb long and removes the entire $\alpha 2$ gene ($-\alpha^{4.2}$). Both are present in Melanesians. The commoner $-\alpha^{3.7}$ deletion may be divided into three subtypes according to the precise position of the crossover that produces the deletion (Higgs et al. 1984). The predominant subtype in Melanesia has been shown to be the $-\alpha^{3.7}$ III defect that is found only in Melanesians and Polynesians (Hill et al. 1985b).

α-Thalassemia and Its Correlation with Malaria in Melanesia

Table 1 shows the results of α-globin genotype determination in Melanesians. Clearly α-thalassemia is uncommon in the highlands of Papua New Guinea but very common in both coastal regions, confirming an earlier pilot study (Oppenheimer et al. 1984). Furthermore there is a geographical cline in the frequency of the $-\alpha$ deletion from the north coast of Papua New Guinea through Vanuatu to New Caledonia. Both $-\alpha^{3.7}$ and $-\alpha^{4.2}$ deletions contribute to these $-\alpha$ frequencies, and the former were mainly of the $-\alpha^{3.7}$ III subtype (87/99 chromosomes studied). Nondeletion forms of α^+-thalassemia would not have been detected by the screening method used, but, as in all other populations investigated to date, in Melanesia these are much less common than the deletion forms (Hill 1986).

Extensive surveys of malaria in the Southwest Pacific have been reported (for review, see Lambert 1949; Black 1955; Hill 1986). In Papua New Guinea malaria is hyper- to holoendemic along the north coast but, in

Table 1. α-Thalassemia in Melanesia

	Number	$\alpha\alpha/\alpha\alpha$	$-\alpha/\alpha\alpha$	$-\alpha/-\alpha$	Percent $-\alpha^{3.7}$	Percent $-\alpha^{4.2}$	Percent $-\alpha$
Highlands PNG							
Madang (Bundi)	82	75	6	1	0	5	5
Chimbu	89	85	4	0	0	2	2
E. Highlands	157	143	14	0	2	3	5
S. Highlands	190	179	11	0	1	2	3
W. Highlands	21	20	1	0	0	2	2
Total	539	502	36	1	1	3	4
Coastal PNG							
Northern	63	8	24	31	8	60	68
Southern	105	68	28	9	13	9	22
Eastern	20	8	9	3	18	20	38
Total	188	84	61	43	12	27	39
North Solomons	30	9	15	6	23	22	45
Vanuatu							
Espiritu Santo	178	72	75	31	28	10	38
Maewo	169	80	67	22	28	5	33
Pentecost	149	80	56	13	26	1	27
Malekula	21	9	10	2	19	14	33
Paama	25	15	9	1	16	6	22
Emae	85	46	33	6	10	16	26
Makura-Mataso	28	23	5	0	3	6	9
Efate	34	20	11	3	17	8	25
Tanna	224	149	69	6	11	7	18
Futuna	56	51	4	1	3	3	6
Aneityum	66	56	10	0	7	1	8
New Caledonia	84	74	10	0	6	1	6

The number of individuals with different α-globin genotypes and the percentage gene frequencies of α-globin gene deletions in Melanesia. The countries, and the islands within Vanuatu, are listed from northwest to southeast (Fig. 1). In Papua New Guinea coastal samples are from below 1500 m; highland samples are from above 1500 m.

general, less prevalent in south coastal provinces. However, before European contact at the end of the nineteenth century, the high interior of the island appears to have been free of the disease. The first malaria survey in Island Melanesia was performed by Buxton (1926) in Vanuatu (formerly the New Hebrides) followed by several others before the introduction of any control measures (Black 1954). As in New Guinea both *P. falciparum* and *P. vivax* were found to be common but *P. malariae* was rare. In Vanuatu malaria is seasonal with substantial local variation perhaps mainly due to differences in the amount of surface water. However, there was and still is, overall, more intense malaria transmission in the northern islands than in the south. The only island east of 170°E in Vanuatu, Futuna, is the only island free of malaria. Although the most southerly island in Vanuatu, Aneityum, now has malaria, historically it too may have been free of the disease or have harbored only *P. malariae* (Laird 1954). New Caledonia, the most southerly island in Melanesia and Fiji, like all of Polynesia to the east, has always been malaria free. So, in general, as one moves south and east from north coastal Papua New Guinea, the endemicity of malaria declines until it is absent in New Caledonia and Fiji.

As shown in Table 2 there is a strong correlation between the earliest recorded endemicities of malaria and the frequencies of α-thalassemia in parts of Island Melanesia. From north coastal Papua New Guinea through the Solomon Islands and Vanuatu to New Caledonia, there is a cline in the frequency of α-thalassemia that is paralleled by a decrease in malaria endemicity. New Caledonia and Futuna (in Vanuatu) are the only islands free of malaria and these have the lowest α-thalassemia frequencies. Fiji, to the east, is also malaria free and there α-thalassemia is rare in Melanesians (gene frequency ~0.01; D.K. Bowden, pers. comm.). The rank correlation between $-\alpha$ frequencies and malaria endemicity is highly significant ($\tau = 0.876$; $p < 0.001$).

Multiple α-Thalassemia Deletions on Melanesian Haplotypes

Analysis of polymorphic restriction enzyme sites in the globin gene complexes provides information on the origin and evolution of variant alleles. Haplotypes thus defined have been used to argue for single or multiple origins for particular globin variants (Antonarakis et al. 1985; Hill and Wainscoat 1986). Eight such polymorphisms in the α gene complex of $-\alpha$-deletion chromosomes from different parts of Melanesia were analyzed to investigate the origin of $-\alpha$ deletions in this region (Table 3). It has been reported previously

Table 2. Distribution of Malaria and DNA Variants in Island Melanesia

	Percent $-\gamma$	Percent $\gamma\gamma\gamma$	Percent Hp[1]	Percent $-\alpha$	Malaria endemicity
North coastal PNG	4.6	0.8	64	68	hyper- to holo-
South coastal PNG	0.9	0.9	71	24	meso- to hyper-
North Solomons	3.3	1.7	62	45	hyper-
Espiritu Santo	11.0	3.0	73	38	hyper-
Maewo	9.8	8.9	76	33	hyper-
Pentecost	2.2	4.0	68	27	hyper-
Emae	11.8	2.2	63	26	
Efate	2.6	3.1	61	25	hypo- to meso-
Tanna	2.1	1.1	71	18	hypo-
Futuna	12.5	0.0	69	6	absent
Aneityum	9.8	0.9	61	8	? absent
New Caledonia	1.7	8.3	59	6	absent

Correlation of malaria endemicity with $-\alpha$, $-\gamma$, $\gamma\gamma\gamma$, and Hp[1] gene frequencies in Island Melanesia. Samples from both southern and eastern regions of Papua New Guinea (Table 1) are included here in the south coastal group. Malaria endemicity is from the earliest available surveys (detailed references in Hill 1986), excluding areas where malaria control measures have been initiated. (No early data are available for Emae.)

that the predominant $-\alpha^{3.7}$ deletion in Melanesians is of the $-\alpha^{3.7}$ III subtype not found in other parts of the world, and this is associated with a single restriction enzyme haplotype, strongly supporting a single origin (Hill et al. 1985b). $-\alpha^{4.2}$ deletions in Melanesia are associated with at least three different haplotypes, none of which is the same as that associated with six southeast Asian $-\alpha^{4.2}$ alleles. At least two different haplotypes are associated with $-\alpha^{3.7}$ I chromosomes in Island Melanesia. Hence, it would appear from subtyping and haplotype analysis that there have probably been at least six origins for $-\alpha$ deletions in Melanesia. It is also of relevance that these deletion chromosomes have the same haplotypes as the common normal ($\alpha\alpha$) haplotypes in Melanesia. These Melanesian $\alpha\alpha$ haplotypes are unusual in being relatively uncommon in other world populations that have been surveyed (Higgs et al. 1986). This suggests that the $-\alpha$ deletions now prevalent in Melanesia arose locally on the $\alpha\alpha$ haplotypes common in this region and were not imported from other malarious areas. Hence it would appear that some local mechanism is responsible for their elevation to the high frequencies observed today.

Other DNA Variants Do Not Correlate with Malaria Endemicity

If the cline in α-thalassemia frequency is related to differing degrees of selection by malaria rather than to population migration, such a cline should not be ob-

Table 3. Haplotypes of $-\alpha$ and $\alpha\alpha$ Chromosomes

Population		Number	Haplotype							
			X	S	B	H	Z	A	R	P
Vanuatu	$-\alpha^{3.7}$III	19	–	–	+	M	Z	–	–	del
	$-\alpha^{4.2}$	2	–	–	+	M	Z	–	del	–
	$-\alpha^{4.2}$	2	+	–	–	S	PZ	+	del	+
	$\alpha\alpha$	13	–	–	+	M	Z	–	–	–
	$\alpha\alpha$	4	+	–	+	M	Z	–	–	–
	$\alpha\alpha$	16	+	–	–	S	PZ	+	–	+
	$\alpha\alpha$	3	+	–	–	S	PZ	+	–	–
	$\alpha\alpha$	1	–	–	–	S	PZ	+	–	+
	$\alpha\alpha$	1	+	+	–	M	PZ	+	+	–
Papua New Guinea	$-\alpha^{3.7}$III	2	–	–	+	M	Z	–	–	del
	$-\alpha^{4.2}$	29	–	–	+	M	Z	–	del	–
	$-\alpha^{4.2}$	7	–	–	+	M	Z	–	del	+
	$\alpha\alpha$	14	–	–	+	M	Z	–	–	–
	$\alpha\alpha$	4	+	–	+	M	Z	–	–	–
	$\alpha\alpha$	2	+	–	–	S	PZ	+	–	+
	$\alpha\alpha$	2	+	–	–	S	PZ	+	–	–
	$\alpha\alpha$	1	–	–	–	S	PZ	+	–	+
	$\alpha\alpha$	1	–	–	–	S	PZ	+	–	–
Southeast Asia	$-\alpha^{4.2}$	6	+	+	–	M	PZ	+	del	–

Restriction enzyme haplotype analysis of $-\alpha$ and $\alpha\alpha$ chromosomes from Melanesia and, for comparison, six $-\alpha^{4.2}$ deletions from southeast Asia. The eight polymorphic sites in the α complex are as described in Higgs et al. (1986), except that the 3′ *Pst*I site is not included here. "del" indicates that the single α gene deletion removes that polymorphic site. M and S indicate medium- and small-length alleles of the interzeta hypervariable region (H). PZ/Z indicates the $\psi\zeta1/\zeta1$ polymorphism.

served in genetic markers unaffected by malaria. Because there is very little population genetic data available for Island Melanesia, the frequencies of two genetic polymorphisms readily detectable by DNA analysis were analyzed in the same samples used for the α-thalassemia study. These are the haptoglobin polymorphism, with alleles Hp^1 and Hp^2, and the frequencies of the variant γ-globin genotypes, $-\gamma$ and $\gamma\gamma\gamma$ (Hill et al. 1986a,b). Unlike the frequencies of the $-\alpha$ deletion, the Hp^1, $-\gamma$, and $\gamma\gamma\gamma$ gene frequencies showed no geographical trend (Table 2) and no significant association with malaria endemicity (Fig. 3).

Unlike Island Melanesia, a large amount of data on blood groups and protein polymorphisms have been collected on the inhabitants of Papua New Guinea, and no nonglobin locus so far examined has the degree of divergence between highland and coastal populations discovered for α^+-thalassemia (Simmons and Booth 1971). Although the classical microepidemiological study of the relationship between β-thalassemia and malaria is that of Siniscalco et al. (1966) in Sardinia, little attention has been paid to the data available for New Guinea. Review of the published studies, which include over 3500 individuals, reveals a striking difference in carrier rates between the highland (0.2%) and coastal (5.0%) populations (Hill 1986). To confirm that the previous observations on nonglobin loci applied to the samples studied here, allele frequencies for three unlinked restriction enzyme site polymorphisms and two highly informative hypervariable regions were determined (Table 4). The results indicate only small frequency differences between the highland and coastal populations. A partition of genetic variation (Smouse and Ward 1978) showed a highly significant difference between the highland and coastal regions for $-\alpha$ gene

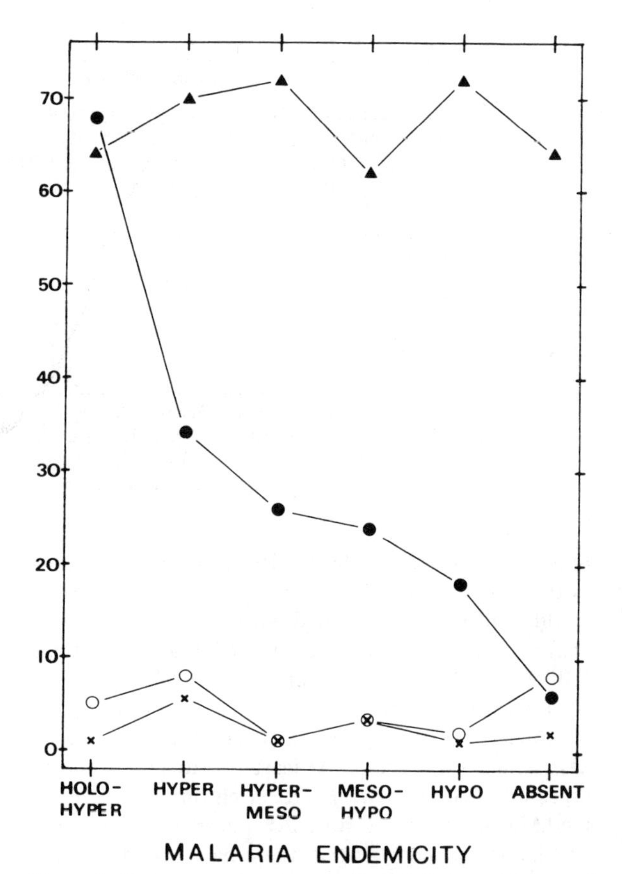

Figure 3. Percentage gene frequencies of the $-\alpha$, Hp^1, $-\gamma$, and $\gamma\gamma\gamma$ alleles in regions of various malarial endemicities in Melanesia. (●) $-\alpha$; (▲) Hp^1; (○) $-\gamma$; (×) $\gamma\gamma\gamma$.

Table 4. Distribution of DNA Variants in Papua New Guinea

	Papua New Guinea highlands						Papua New Guinea coast			
Probe	Madang (Bundi)	Chimbu	eastern highland	southern highland	western highland	highlands average	northern region	southern region	eastern region	coastal average
C3	43 (56)	42 (78)	53 (18)	58 (83)	50 (10)	48	48 (33)	52 (62)	50 (8)	51
3.6	10 (80)	21 (82)	10 (111)	5 (148)	0 (13)	10	9 (53)	7 (97)	11 (14)	8
2.30	24 (55)	28 (38)	46 (37)	22 (154)	42 (13)	27	19 (48)	26 (89)	5 (10)	22
Insulin HVR class										
1	1	2	4	4	0	3	3	0	0	2
2	0	0	0	0	0	0	5	4	0	4
3	0	1	0	4	0	1	0	0	0	0
4	99	97	96	92	100	96	92	96	100	94
	(57)	(42)	(52)	(79)	(18)		(48)	(26)	(7)	
HRAS1 HVR class										
1	21	7	7	21	8	14	24	12		18
2	29	31	26	16	17	23	26	29		28
3	5	2	0	0	0	1	1	1		1
4	0	5	6	8	8	6	7	4		5
5	45	55	61	55	50	55	35	48		42
6	0	0	0	0	0	0	3	3		3
Variants	0	0	0	0	17	1	4	3		3
	(19)	(21)	(52)	(50)	(6)		(40)	(38)		

Percentage gene frequencies of the presence of three site polymorphisms, detected by the unlinked C3, 3.6, and 2.30 probes, and the percentage frequencies of different length allele classes detected by two HVR probes, in different parts of Papua New Guinea. The numbers of individuals studied is shown in brackets.

frequencies ($F_{[1,6]} = 22.04$, $p < 0.001$) but not for other loci. It is clear that none of the alleles studied exhibits an altitude- or latitude-dependent variation comparable to that found for α-thalassemia.

The α-Thalassemia Cline: Migration, Genetic Drift, or Selection?

Although the most plausible explanation for the observed geographic distribution of α-thalassemia in Melanesia is that malarial selection is responsible, other factors such as genetic drift and population migration should be considered. Genetic drift and, in particular, founder effects are likely to be of importance in small island populations. Examples of the clustering of particular genetic variants on single islands in Vanuatu (Bowden et al. 1985) suggest that, despite their proximity, historically these islands have remained relatively isolated, thus increasing the potential for genetic drift. However, there are several difficulties in ascribing the observed cline in α-thalassemia frequency simply to this mechanism. The gene frequencies of the $-\alpha$ deletion, particularly in the north of Papua New Guinea, seem too high to attribute to random fluctuations, unless the founding population was extremely small. Furthermore, genetic drift should produce uneven variation from island to island as seen for the $-\gamma$ and $\gamma\gamma\gamma$ frequencies rather than a frequency cline. Moreover, assuming that the $-\alpha$ chromosome in most of the founding populations of these islands was less common than the normal $\alpha\alpha$ chromosome, in the absence of selection, the deletion chromosome should have been lost completely on some islands through genetic drift. The observation that this has not occurred suggests that selection is involved.

Migration of two populations with different genetic compositions into a single area can readily produce a gene frequency cline. For example, had Melanesia been populated from the southeast by a population with little α-thalassemia and from the northeast by a people in whom it was very prevalent, a cline somewhat similar to that observed might have been generated. However, with the exception of some recent backmigration from Polynesia, all known migrations into Melanesia have come from the northwest (Bellwood 1978). The islands of Futuna and Emae in Vanuatu have ethnically Polynesian populations and, culturally, there is probably more Polynesian influence in the south than the north of Vanuatu. Similarly, relatively recent admixture of Polynesian genes into the south coastal Papua New Guinea population may, in part, explain the lower overall prevalence of α-thalassemia there than on the north coast. However, such backmigration has almost certainly been too limited to explain more than a minor part of the observed α-thalassemia cline. For example, the "Polynesian" population of Emae has a much higher α-thalassemia frequency than anywhere in Polynesia. Similarly, although a late migration from the northwest by a very large population with a high frequency of α^+-thalassemia could have produced a $-\alpha$ frequency cline, this is inconsistent with the haplotype data which indicate that most of the $-\alpha$ chromosomes found in Melanesia have not been imported from southeast Asia. Furthermore, no simple migration model could explain the generation of a cline made up of different proportions of the various $-\alpha$ deletions in different locations.

The DNA polymorphism and other population data from Papua New Guinea suggest that here also drift and migration are insufficient to account for the $-\alpha$ frequency difference observed. Unless by chance this and previous studies have only looked at unaffected loci, drift acting to divide the population may be excluded as an explanation of the observed α-thalassemia distribution because such drift should have affected other gene frequencies to a similar extent. Similarly, admixture of α-thalassemia only into coastal populations leaving existing frequencies unchanged would only have been possible if the invaders were genetically very similar to the resident population. Furthermore, evidence from other disciplines (Bellwood 1978) agrees with previous gene frequency data in indicating a definite but small Austronesian admixture into the coastal Papuan stock. Hence, even if $-\alpha$ deletions were imported they would have produced only a low overall frequency on the coast.

α^+-Thalassemia in Other Malarious Regions

In view of the remarkably high prevalences of single α-gene deletions found in Melanesia, it is at first sight puzzling that in other populations with more intense malaria these deletions are often much less common. For example $-\alpha$ allele frequencies in African Blacks are of the order of 0.10–0.30 (Pagnier et al. 1984) and in southeast Asians appear to be lower (0.02–0.10) (Wasi 1983). Recent DNA analysis of 306 chromosomes from Filipinos (A.V.S. Hill and G. Watt, unpubl.) revealed only seven $-\alpha$ deletions (gene frequency, 0.02). Part of the explanation appears to be that other hemoglobinopathies such as Hb S or Hb E, which are probably more strongly selected in the presence of malaria, are present in these populations (Livingstone 1983). In Filipinos at least two types of α^0-thalassemia deletions are present and these double α-gene deletions probably compete with $-\alpha$ chromosomes. Livingstone (1983) has shown that α^0-thalassemia will inhibit the increase of $-\alpha$ frequencies if it attains polymorphic frequencies first (through loss of the double heterozygotes), and this may have happened in southeast Asia. Another factor to be considered is that fitness values for the various genotypes may differ between regions because of differing intensities of malaria. Thus, although α^0-thalassemia heterozygotes appear to be selected for in southeast Asia, it cannot be assumed that in Melanesia individuals with the phenotypically similar $-\alpha/-\alpha$ genotype are fitter than normals. Finally, it has recently been possible to demonstrate a small difference in phenotypic severity between $-\alpha^{4.2}$ and (milder) $-\alpha^{3.7}$ III deletions (Hill 1986). Although this is much less than the difference between the α^{+}- and α^{0}-thalassemias, it is another factor to consider in modeling variations in allele frequencies between populations.

α^{+}-Thalassemia in Nonmalarious Regions

Whereas α^{+}-thalassemia is prevalent in all countries of the "malaria-belt" that have been studied, there has been no formal testing of the other implication of the malaria hypothesis, i.e., that in populations historically free of malaria α-thalassemia should be rare. Because of the mild phenotype of $-\alpha$ deletions, it was possible that these alleles might be present at polymorphic frequencies in nonmalarious regions but remain undetected. The recent discovery of a cluster of cases of α^{0}-thalassemia in British people (Higgs et al. 1985) made investigation of this problem appear quite pertinent. Therefore 500 chromosomes from Icelanders and Britons were analyzed and no $-\alpha$ deletion was found (Flint et al. 1986). Recently published data on Japanese α-thalassemia (Shimizu et al. 1986) are also included in Table 5 for comparison, and in this nonmalarious region α-thalassemia is again rare. Malaria has probably always been absent in Iceland and only a mild form of malaria, probably *P. vivax,* has been present episodically in parts of Britain (Dobson 1980). Hence these data documenting the rarity of α^{+}-thalassemia deletions in populations historically free of malaria are entirely in keeping with the malaria hypothesis. The one apparent exception is the presence of α^{+}-thalassemia at gene frequencies of up to 0.12 in parts of Polynesia, all of which is free of malaria. However, in these islands the distribution of α-thalassemia is patchy, and detailed molecular analysis of the $-\alpha$ deletion found there shows it to be identical to the $-\alpha^{3.7}$ III variant prevalent in Island Melanesia. It seems most likely, therefore, that α-thalassemia alleles were brought into Polynesia by the original migrants from malarious Melanesia (Hill et al. 1985b).

DISCUSSION

At least six approaches might be taken to evaluate the malaria hypothesis as applied to α-thalassemia. All have been attempted in the case of Hb S (for review, see Livingstone 1983) and together the results amount to convincing evidence in the case of that variant. First, $-\alpha$ frequencies might be measured amongst those with fatal cerebral malaria and compared with the total population. This was the approach that provided the most persuasive early data on malaria and the Hb S gene. However, because of the probable much smaller selective advantage of the $-\alpha$ allele, large studies in areas with high attack rates would be required and these are the regions where many other complicating hemoglobinopathies are found. Second, comparison of the α genotypes in different age groups might provide evidence for differential viability and indicate the age group on which the selective agent acted. Indeed the limited available data from Melanesia do suggest an excess of $-\alpha/-\alpha$ individuals in adulthood (Hill 1986). Here again, however, a relatively small selective advantage might be easily missed if, for example, significant migration in and out of the area under study occurred. A third approach would be to compare malarial parasite rates and densities in individuals of different α genotypes. The assumption here would be that, as with Hb S, protection of carriers might manifest itself in lower parasite rates or densities. However, the very limited data available for α-thalassemia suggest the opposite, that the $-\alpha$ deletion might be associated with increased parasitemia (Oppenheimer et al. 1986). Although the meaning of these observations is at present unclear, they do suggest that parasite rate comparisons might not be the most straightforward approach. A fourth strategy employed in the case of Hb S was to inject volunteers of various genotypes with parasites and observe their responses. In α^{+}-thalassemia, this would probably have to be done on large numbers of nonimmune individuals to hope to detect any difference, making such a study impracticable and ethically unacceptable. The fifth approach, and a potentially very informative one, is to compare in vitro the behavior of parasites in erythrocytes of different α genotypes. However, such studies have failed to demonstrate any abnormality of invasion or growth of *P. falciparum* in the red cells of α^{+}-thalassemia heterozygotes or homozygotes, unless the cells are maintained under conditions of unusual oxidant stress (Pasvol and Wilson 1982).

Given the difficulties with the above approaches, the sixth strategy of epidemiological comparison of α-thalassemia distribution with historical malarial endemicities seemed most appropriate. The available macroepidemiological information was suggestive and the microepidemiological data from Melanesia reported here strongly implicate malaria as the selective agent responsible for the high prevalences of α-thalassemia in many tropical and subtropical regions. Because both *P. falciparum* and *P. vivax* are found throughout the malarious parts of Melanesia, the data do not indicate which species is involved. The limited data documenting the rarity of α-thalassemia in populations historically free of malaria provide further support for the hypothesis.

Evidence that malaria is involved in the generation of high frequencies of α^{+}-thalassemia does not exclude the involvement of other factors. However, in human studies, the problem has generally been to produce any evidence of the action of natural selection, not the choice between the candidates. Thus, for example, al-

Table 5. α^{+}-Thalassemia in Nonmalarious Regions

	Number	$\alpha\alpha\alpha$	$\alpha\alpha$	$-\alpha$	Percent $-\alpha$
Britain	280	3	277	0	0
Iceland	220	0	220	0	0
Japan	464	7	452	5	1

Number of chromosomes with different α-globin gene arrangements in British, Icelandic, and Japanese individuals (Flint et al. 1986; Shimizu et al. 1986).

though a role for its alleviation of sickle cell disease in elevating α^+-thalassemia frequencies cannot be excluded, there is no convincing evidence to support this proposal. Furthermore, because all Hb SS individuals in traditional African societies appear to die before reproductive age, it is difficult to see how α-thalassemia alleles could be selected in this way.

While the mechanism of the apparent selective advantage conferred by α-thalassemia remains obscure, the epidemiological data do provide some indication of its probable magnitude. It is unclear whether the observed gene frequencies in Melanesia are at equilibrium or whether the single α-gene deletion is still increasing in frequency. If the latter, because Island Melanesia has been inhabited for over 3500 years and Papua New Guinea for much longer (Bellwood 1978), the observation that the $-\alpha$ chromosome has not yet reached its equilibrium frequency would suggest that its selective advantage is relatively small. On the other hand, if the present frequencies are at or near equilibrium, the selective advantage in the presence of malaria must be balanced by some disadvantage associated with homozygosity for α^+-thalassemia. The mildness of the latter condition, which contrasts markedly with homozygosity for β-thalassemia or Hb S, would again suggest that the balancing selective advantage of α^+-thalassemia against malaria is small.

An alternative approach to an estimate of the possible fitness of the $-\alpha$ heterozygote may be made if one assumes that there is some rough proportionality between the extent of red cell changes in thalassemia and protection against malaria, and that the highest recorded frequencies of β-thalassemia, ~0.10, are at or near equilibrium. To maintain this β-thalassemia frequency, a selection coefficient of 0.11 is required (Cavalli-Sforza and Bodmer 1971). Homozygotes for α^+-thalassemia have somewhat milder red cell changes than β-thalassemia heterozygotes and $-\alpha$ heterozygotes may be indistinguishable from normal (Weatherall and Clegg 1981), suggesting that the latter enjoy an increased fitness of perhaps only a few percent.

Theoretical considerations show that such a weak selection pressure is sufficient to maintain the cline observed in Island Melanesia, despite significant population movement (gene flow). Given the 2000-mile ($= w$) cline (Fig. 1), selection can maintain this despite gene flow if the selection pressure is greater than $(2x/w)^2$, where x is the dispersal range (standard deviation of distance between parent and offspring) (Barton and Hewitt 1985). So, even if x were as great as 100 miles, a selection pressure as low as $(2 \times 100/2000)^2 = 0.01$ would suffice. This low value may explain the failure to detect differences between the $-\alpha$ and $\alpha\alpha$ haplotypes in the in vitro studies.

It would thus appear that these data on α-thalassemia in Melanesia provide a remarkable example of how a small selective advantage can lead to large variations in gene frequencies between human populations. It remains to identify the mechanism of protection and to obtain direct estimates of the selection pressure. Although the former should be feasible, both time and places to measure the latter are limited. It is perhaps ironic that practical applications of the same DNA technology that made this study feasible (Miller et al. 1984) may soon reduce further the possibilities for studying the action of malarial selection.

ACKNOWLEDGMENTS

I am grateful to D.K. Bowden, S.J. Oppenheimer, P.R. Sill, S.W. Serjeantson, J. Bana-Koiri, K. Bhatia, G. Watt, and J.-P. Moreau for providing samples; to J. Flint for extensive analysis of the New Guinea samples; to A.J. Boyce for statistical analysis; to G. Pasvol for discussions; to Linda Roberts for preparing the manuscript; to the Rockefeller Foundation for financial support; and to D.R. Higgs, D.J. Weatherall, and J.B. Clegg for advice and encouragement. During this work the author held an MRC Training Fellowship and the Staines Medical Research Fellowship at Exeter College, Oxford.

REFERENCES

Allison, A.C. 1965. Polymorphism and natural selection in human populations. *Cold Spring Harbor Symp. Quant. Biol.* **29:** 137.

Antonarakis, S.E., H.H. Kazazian, and S.H. Orkin. 1985. DNA polymorphism and the molecular pathology of the human globin gene clusters. *Hum. Genet.* **69:** 1.

Barton, N.H. and G.M. Hewitt. 1985. Analysis of hybrid zones. *Annu. Rev. Ecol. Syst.* **16:** 113.

Bell, G.I., S. Horita, and J.H. Karam. 1984. A polymorphic locus near the human insulin gene is associated with insulin dependent diabetes mellitus. *Diabetes* **33:** 76.

Bellwood, P. 1978. *Man's conquest of the Pacific*. Collins, London.

Black, R.H. 1954. Some aspects of malaria in the New Hebrides. *South Pacific Commission Technical Paper* No. 60.

———. 1955. Malaria in the Southwest Pacific. *South Pacific Commission Technical Paper* No. 81.

Bowden, D.K., A.V.S. Hill, D.R. Higgs, D.J. Weatherall, and J.B. Clegg. 1985. The relative roles of genetic factors, dietary deficiency and infection in anaemia in Vanuatu, Southwest Pacific. *Lancet* **II:** 1025.

Buxton, P.A. 1926. The depopulation of the New Hebrides and other parts of Melanesia. *Trans. R. Soc. Trop. Med. Hyg.* **19:** 420.

Capon, D.J., E.Y. Chen, A.D. Levinson, P.A. Seeburg, and D.V. Goeddel. 1983. Complete nucleotide sequence of the T24 human bladder carcinoma oncogene and its normal homologue. *Nature* **302:** 33.

Cavalli-Sforza, L.L. and W.F. Bodmer. 1971. *The genetics of human populations*. W.H. Freeman, San Francisco.

Davies, K.E., J. Jackson, R. Williamson, P.S. Harper, S. Ball, T. O'Brien, L. Meredith, and G. Fey. 1983. Linkage analysis of myotonic dystrophy and sequences on chromosome 19 using a cloned complement 3 gene probe. *J. Med. Genet.* **20:** 259.

Dobson, M. 1980. "Marsh fever"—The geography of malaria in England. *J. Hist. Geog.* **4:** 357.

Embury, S.H., J.A. Miller, A.M. Dozy, Y.W. Kan, V. Chan, and D. Todd. 1980. Two different molecular organisations for the single α-globin gene of the α-thalassemia-2 genotype. *J. Clin. Invest.* **66:** 1319.

Flint, J., A.V.S. Hill, J.B. Clegg, D.J. Weatherall, and D.R. Higgs. 1986. Alpha globin genotypes in two North European populations. *Br. J. Haematol.* **63:** 796.

Gilliam, T.C., P. Scrambler, T. Robbins, C. Ingle, R. Williamson, and K.E. Davies. 1984. The positions of three restriction fragment length polymorphisms on chromosome 4 relative to known genetic markers. *Hum. Genet.* **68:** 154.

Haldane, J.B.S. 1949. The rate of mutation of human genes. *Proc. Int. Congr. Genet.* **8:** 267.

Higgs, D.R. and D.J. Weatherall. 1983. Alpha thalassemia. *Curr. Top. Hematol.* **4:** 37.

Higgs, D.R., A.V.S. Hill, D.K. Bowden, D.J. Weatherall, and J.B. Clegg. 1984. Independent recombination events between the duplicated human α globin genes: Implications for their concerted evolution. *Nucleic Acids Res.* **12:** 6965.

Higgs, D.R., H. Ayyub, J.B. Clegg, A.V.S. Hill, R.D. Nicholls, H. Teal, J.S. Wainscoat, and D.J. Weatherall. 1985. α Thalassemia in British people. *Br. Med. J.* **290:** 1303.

Higgs, D.R., J.S. Wainscoat, J. Flint, A.V.S. Hill, S.L. Thein, R.D. Nicholls, H. Teal, H. Ayyub, T.E.A. Peto, A.G. Falusi, A.P. Jarman, J.B. Clegg, and D.J. Weatherall. 1986. Analysis of the human α-globin gene cluster reveals a highly informative genetic locus. *Proc. Natl. Acad. Sci.* **83:** 5165.

Hill, A.V.S. 1986. "The distribution and molecular basis of thalassemia in Oceania." Ph.D. thesis, University of Oxford, England.

Hill, A.V.S. and J.S. Wainscoat. 1986. The evolution of the α- and β-globin gene clusters in human populations. *Hum. Genet.* **74:** 16.

Hill, A.V.S., D.K. Bowden, D.J. Weatherall, and J.B. Clegg. 1986a. Chromosomes with one, two, three or four fetal globin genes: Molecular and hematologic analysis. *Blood* **67:** 1611.

Hill, A.V.S., R.D. Nicholls, S.L. Thein, and D.R. Higgs. 1985a. Recombination within the human embryonic ζ globin locus: A common ζ-ζ chromosome produced by gene conversion of the $\psi\zeta$ gene. *Cell* **42:** 809.

Hill, A.V.S., D.K. Bowden, J. Flint, D.B. Whitehouse, D.A. Hopkinson, S.J. Oppenheimer, S.W. Serjeantson, and J.B. Clegg. 1986b. A population genetic survey of the haptoglobin polymorphism in Melanesians by DNA analysis. *Am. J. Hum. Genet.* **38:** 382.

Hill, A.V.S., D.K. Bowden, R.J. Trent, D.R. Higgs, S.J. Oppenheimer, S.L. Thein, K.N.P. Mickelson, D.J. Weatherall, and J.B. Clegg. 1985b. Melanesians and Polynesians share a unique α-thalassemia mutation. *Am. J. Hum. Genet.* **37:** 571.

Laird, M. 1954. *Anopheles* and malaria at Aneityum, New Hebrides. *Bull. Entomol. Res.* **45:** 279.

Lambert, S.M. 1949. Malaria incidence in Australia and the South Pacific. In *Malariology* (ed. M.F. Boyd), p. 820. W.B. Saunders, Philadelphia.

Livingstone, F.B. 1983. The malaria hypothesis. In *Distribution and evolution of the hemoglobin and globin loci* (ed. J. Bowman), p. 15. Elsevier/North-Holland, New York.

———. 1985. *Frequencies of hemoglobin variants.* Oxford University Press, England.

Miller, L.H., P.H. David, and T.J. Hadley. 1984. Perspectives for malaria vaccination. *Philos. Trans. R. Soc. Lond B* **307:** 99.

Old, J.M. and D.R. Higgs. 1983. Gene analysis. *Methods Haematol.* **6:** 74.

Oppenheimer, S.J., D.R. Higgs, D.J. Weatherall, J. Barker, and R.A. Spark. 1984. α Thalassemia in Papua New Guinea. *Lancet* **I:** 424.

Oppenheimer, S.J., A.V.S. Hill, F.D. Gibson, S.B. MacFarlane, and J.B. Moody. 1986. The interaction of alpha thalassemia with malaria. *Trans R. Soc. Trop. Med. Hyg.* (in press).

Pagnier, J., O. Dunda-Belkhodja, I. Zohoun, J. Teyssier, H. Baya, G. Taeger, R.L. Nagel, and D. Labie. 1984. α Thalassemia among sickle cell anemia patients in various African populations. *Hum. Genet.* **68:** 318.

Pasvol, G. and R.J.M. Wilson. 1982. The interaction of malaria parasites with red blood cells. *Br. Med. Bull.* **38:** 133.

Povey, S., N.E. Martin, and S.L. Sherman. 1985. Report of the committee on the genetic constitutions of chromosomes 1 and 2. *Cytogenet. Cell Genet* **40:** 67.

Shimizu, K., T. Harano, K. Harano, S. Miwa, Y. Amenomori, Y. Ohba, F. Kutlar, and T.H.J. Huisman. 1986. Abnormal arrangements in the α- and γ-globin gene clusters in a relatively large group of Japanese newborns. *Am. J. Hum. Genet.* **38:** 45.

Simmons, R.T. and P.B. Booth. 1971. *A compendium of Melanesian genetic data I-IV.* Commonwealth Serum Laboratories, Victoria, Australia.

Siniscalco, M., L. Bernini, G. Filippi, B. Latte, P. Meera Khan, S. Piomelli, and M. Rattazzi. 1966. Population genetics of haemoglobin variants, thalassemia and glucose-6-phosphate dehydrogenase deficiency, with particular reference to the malaria hypothesis. *Bull. WHO* **34:** 379.

Smouse, P.E. and R.H. Ward. 1978. A comparison of the genetic infrastructure of the Ye'cuana and Yanomama: A likelihood analysis of genotypic variation among populations. *Genetics* **88:** 611.

Wasi, P. 1983. Hemoglobinopathies in southeast Asia. In *Distribution and evolution of the hemoglobin and globin loci.* (ed. J. Bowman). Elsevier/North-Holland, New York.

Weatherall, D.J. and J.B. Clegg. 1981. *The thalassemia syndromes*, 3rd edition. Blackwell, Oxford, England.

Wills, C. and D.R. Londo. 1981. Is the doubly deleted α-thalassemia gene a fugitive allele. *Am. J. Hum. Genet.* **33:** 215.

The Primate α-Globin Gene Family: A Paradigm of the Fluid Genome

J. MARKS, J.-P. SHAW, C. PEREZ-STABLE, W.-S. HU,
T.M. AYRES, C. SHEN, AND C.-K.J. SHEN
Department of Genetics, University of California, Davis, California 95616

The human α-globin gene family, like the β-globin family, is excellent for the study of the mechanisms of gene regulation and molecular evolution. First, the family consists of genes that are differentially and coordinately expressed during development (Bunn et al. 1977; Maniatis et al. 1980; Collins and Weissman 1984). Second, the human α-globin-like cluster has been cloned, and the bulk of its nucleotide sequence has been determined (see references below). Third, there exist well-documented clinical data and molecular studies of both quantitative and qualitative variations in the gene products from this region (Higgs et al. 1985; Kan 1985; Livingstone 1985). Fourth, the various DNA rearrangements that have occurred and continue to be found in human populations (Goossens et al. 1980; Higgs et al. 1981, 1985; Goodbourn et al. 1983; Hess et al. 1983, 1984; Kan 1985) make it a useful system in which to study both homologous and illegitimate recombination processes in mammalian cells. Finally, of the two blood globin clusters, that of the α-globins is of considerably greater compactness and simplicity than the β-globins, and is consequently the focus of our investigations.

In this paper we will describe several recent developments from this laboratory on the primate α-globin gene family. This research and related discussion fall into four broad categories: the use of molecular cloning and comparative phylogenetic data for the determination of molecular evolutionary rates; the use of comparative phylogenetic data to illuminate aspects of the structure, function, and recent history of the human α-globin cluster; the discovery of a new subfamily of the α-globin-like gene family, including an apparently functional, α-globin-like gene linked to the adult α1 gene; and our efforts in setting up in vitro and in vivo systems for the study of molecular mechanisms of DNA recombination.

Structure of the α-Globin Gene Cluster

The linkage map of the human α-globin gene cluster, which is located on chromosome 16 (Deisseroth et al. 1977), is shown in Figure 1. It contains a functional embryonic gene, ζ; a nonfunctional embryonic pseudogene, ψζ (Proudfoot et al. 1982), which is occasionally functional in humans (Hill et al. 1985), and functional in the chimpanzee as well (Willard et al. 1985); an adjacent pseudogene of remote ancestry, ψα2 (Hardison et al. 1986); the ψα1 pseudogene (Proudfoot and Maniatis 1980); the duplicated adult α-globin genes, α2 and α1 (Liebhaber et al. 1980, 1981; Michelson and Orkin 1980, 1983); and a related gene of unknown function, θ (Marks et al. 1986b). All vertebrate globin genes contain three coding sequences separated by two introns; and the total length of the α- (or ζ)-globin protein is 141 amino acids, slightly shorter than the β-like globins and myoglobins.

The adult α-globin genes are contained within two tandemly arranged, duplicated regions spanning approximately 4 kb each (Lauer et al. 1980; Hess et al.

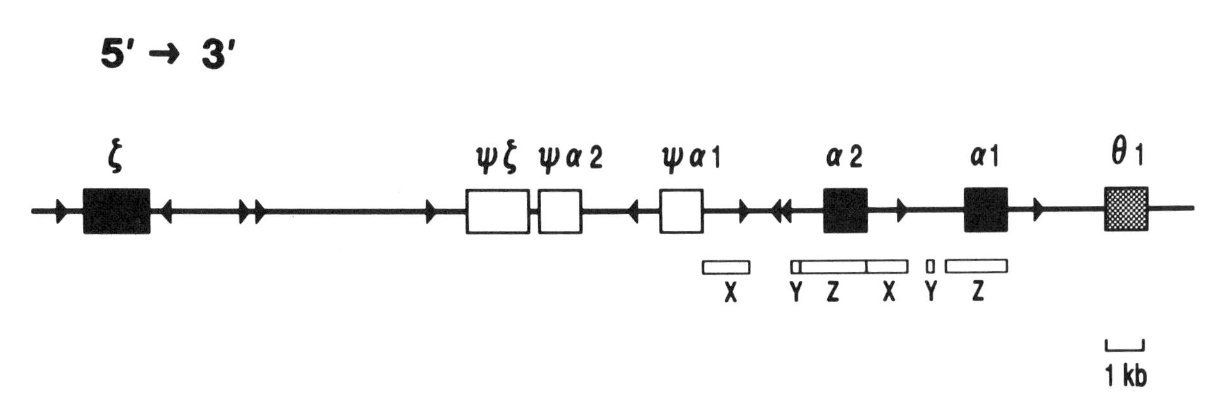

Figure 1. Linkage map of the human α-globin gene cluster on chromosome 16. Filled boxes represent functional genes; empty boxes represent pseudogenes. θ1, which is inferentially functional, is given as a stippled box. Arrows indicate the locations and orientations of the *Alu* family repeats. The X, Y, and Z homology blocks of the adult α-globin duplication units are shown below the linkage map.

1983, 1984; Michelson and Orkin 1983). These duplication units are subdivided into the X, Y, and Z blocks given in Figure 1, in which the gene is located at the 3′ end of the Z block.

Gene correction is detectable between these duplication units. For example, the $\alpha 2$ and $\alpha 1$ genes are more similar within any given mammalian species than across species, no matter how closely related the species. In humans, these cryptic processes keep the Z blocks containing the genes virtually identical: up to the termination codon, $\alpha 2$ and $\alpha 1$ differ by but two point mutations and a 7-bp length difference in intron 2. Downstream from the termination codon, the genes diverge by 16–17%.

More than ten different *Alu* family repeats are scattered throughout the α-globin cluster (Shen and Maniatis 1982; Hess et al. 1983; Sawada et al. 1983, 1985; Higgs et al. 1985). Variability exists in human populations for intergenic sequences upstream from $\psi\zeta$ and downstream from θ (Higgs et al. 1981; Goodbourn et al. 1983). These have been termed "hypervariable regions," and some have been used as genetic markers for adult polycystic kidney disease (Reeders et al. 1985).

The α-Globins as a Paradigmatic Gene System

Three types of comparisons can be made on molecular data derived from the α-globin gene region: (1) differences among similar sequences in the same genome (paralogous sequences, such as the Z blocks of the human adult α-globin duplication units); (2) differences among genomes within a species (polymorphic or polytypic variation); and (3) differences among related species at a single locus (orthologous sequences). Thus, the α-globin gene family of humans and their zoological relatives provide a model system for the study of three fundamental genetic problems: (1) The generation of genetic diversity (mutation); (2) the persistence or alteration of genetic structures through time and across lineages (evolution); and (3) the mechanism of functional control of the genetic material (regulation).

Mutation. The α-globins allow us to analyze the two modes of mutation in gene clusters. First and foremost are point mutations, or single-nucleotide changes. These are readily tabulated and can be manipulated statistically, and are therefore particularly useful when summed as evolutionary distances. Establishment of rates of sequence divergence of the α-globin genes and their flanking DNA across lineages may also aid in understanding the phylogeny and evolution of the primates, and the nature of the molecular clock (Wilson et al. 1977).

The second mode of genomic variation involves recombination events, which can be usefully subdivided into homologous and nonhomologous recombinations. Homologous recombination involves strand pairing and genetic exchange with a region of high similarity. If these regions are adjacent to one another, the result is a duplication in one chromosome and a deletion in the other. In the adult α-globin region, such an event occurring within the X blocks has resulted in the $-\alpha^{4.2}$ thalassemia genotype; whereas a similar event within the two Z blocks has resulted in the $-\alpha^{3.7}$ thalassemia genotype prevalent in many malarious zones of the Old World (Orkin et al. 1979; Embury et al. 1979; Higgs and Weatherall 1983). The Z block recombination actually results in a single hybrid gene ($\alpha 2/\alpha 1$) on one chromosome, and one reciprocal hybrid and two intact genes on the other. Both haplotypes are found in human populations, but perhaps owing to a selective advantage against malaria, the one-gene haplotype is much more prevalent than the three-gene haplotype (Higgs et al. 1985; Kan 1985). Alternatively, the higher frequency of the one-gene state may be the result of intrachromosomal, rather than interchromosomal recombination. This distribution, however, seems to be reversed in chimpanzees (Zimmer et al. 1980).

To study the molecular mechanisms of homologous recombination, we have reconstructed the two major α-thalassemia genotypes in COS-7 cells (Fig. 2). A major portion of the human adult α-globin gene region was cloned into an SV40 *ori*-containing vector psvod (Mellon et al. 1981). The resulting plasmid psvA2A1 contains the entire $\alpha 2$ duplication unit and most of the $\alpha 1$ duplication unit (Fig. 2A). In this plasmid, the target sizes are approximately 1.3 kb for both the X block and Z block recombination units. After introducing psvA2A1 into T antigen-expressing COS-7 cells (Gluzman 1981) by the DEAE-dextran transfection technique, low-molecular-weight DNAs (Hirt 1967) were isolated at different times posttransfection. Southern blot analyses of these samples indicate that homologous recombination has occurred in COS-7 cells at high frequencies within the X blocks and Z blocks, respectively (Fig. 2B). Since very few, if any, recombinant plasmids are observed to contain three duplication units, the detectable X and Z block recombinations are most likely the result of intrachromosomal recombination. Furthermore, the frequency of X block recombination is approximately twofold higher than that of the Z block recombination, despite the similarity in target sizes. This is observed throughout the time course after transfection (Fig. 2C). For example, at 72 hours posttransfection, the proportion of the X block recombinants is 24%, while that of the Z block recombinants is 13%.

These and other unpublished data suggest that the host cell–episome system we have constructed is feasible for studying the homologous recombination processes that occur in human populations. It is possible that these recombination events are modulated by a *cis*-acting regulatory element (W.-S. Hu and C.-K.J. Shen, in prep.).

Nonhomologous or illegitimate recombinations involve quantitative alterations in the structure of the genome, and occur in the absence of significant pairing between similar regions. This would include major insertions and deletions, and most importantly the interpolation of repetitive sequences (Schmid and Shen

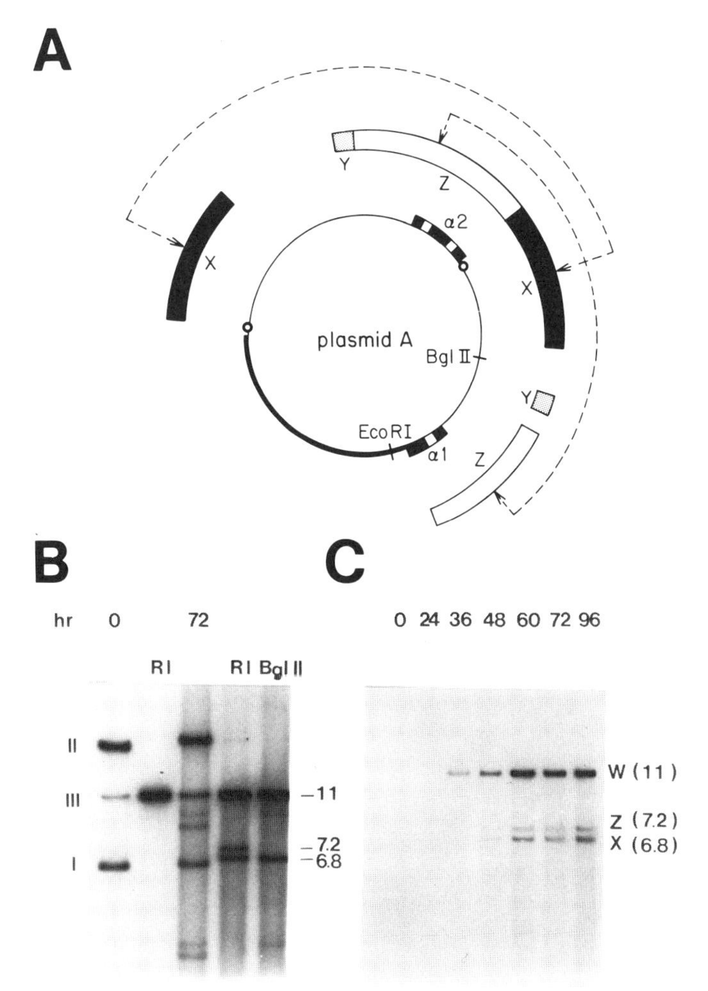

Figure 2. (*A*) The structure of plasmid psvA2A1 (or plasmid A), containing two adult α-globin duplication units cloned into shuttle vector psvod. This plasmid contains the complete α2 duplication unit and all of the α1 duplication unit except for 400 bp of its 3′ portion. The details of the cloning procedure will be described elsewhere (W.-S. Hu and C.-K.J. Shen, in prep.). (*B*) Southern blot analysis of the transfection experiment. Hirt DNA samples isolated at 0 hr (lanes *1, 2*) and 72 hr (lanes *3–5*) posttransfection from COS-7 cells transfected with psvA2A1 were electrophoresed with and without prior enzyme digestion, blotted, and hybridized with ^{32}P-labeled probe. (Lanes *1* and *3*) No digestion; (lanes *2* and *5*) digestion with *Bgl*II; (lane *4*) digestion with *Eco*RI. When homologous recombination occurs between the two X blocks, 4.2. kb of DNA are deleted and yield a 6.8-kb plasmid, the X recombinant. This recombinant can be linearized by either *Eco*RI or *Bgl*II. On the other hand, when homologous recombination occurs between the two Z blocks, 3.8 kb of DNA is deleted. This results in a 7.2-kb plasmid, the Z recombinant. This recombinant can be cleaved once by *Eco*RI but not by *Bgl*II (see also Fig. 2A). W, X, and Z indicate the linearized plasmids corresponding to the wild-type psvA2A1 (W), the X recombinants, and the Z recombinants, respectively. (*C*) Southern blot analysis of Hirt DNA isolated at different times posttransfection. Lanes *1–7* are samples collected at 0, 24, 36, 48, 60, 72, and 96 hr posttransfection. All thc samples were double-digested with *Eco*RI, which linearizes the plasmids, and with *Dpn*I, which eliminates the signals from input plasmid DNA.

1985). The insertion of *Alu* family repeats in the human α-globin cluster and their possible effect on gene correction have been described previously (Hess et al. 1983). Since topoisomerase I has been suggested to be involved in illegitimate recombination (Edwards et al. 1982; Bullock et al. 1985), we have initiated in vitro studies of the interaction between this enzyme and the *Alu* repeats in the human α-globin gene cluster.

Topoisomerase I is an enzyme capable of changing the topological state of circular DNA by breaking and resealing single strands (Wang and Liu 1979; Wang 1985). The antitumor drug camptothecin strongly inhibits the activity of this topoisomerase I by trapping an enzyme–DNA intermediate and preventing the resealing interaction (Hsiang et al. 1985). End-labeled, double-stranded DNA probes containing different *Alu* repeats and their flanking sequences were used to compile a new consensus topoisomerase I nicking site on human DNA (C. Perez-Stable et al., in prep.), as follows:

$$5'\ A/T - A/T - Pu - A/T - T{\downarrow}Pu - Pu - A/T\ 3'$$

Figure 3 is an example of DNA nicking by HeLa topoisomerase I (Edwards et al. 1982; Liu 1983; Been et al. 1984) within and flanking the 3′ α1 *Alu* repeat (Hess et al. 1985). Strong nicking sites were located at the

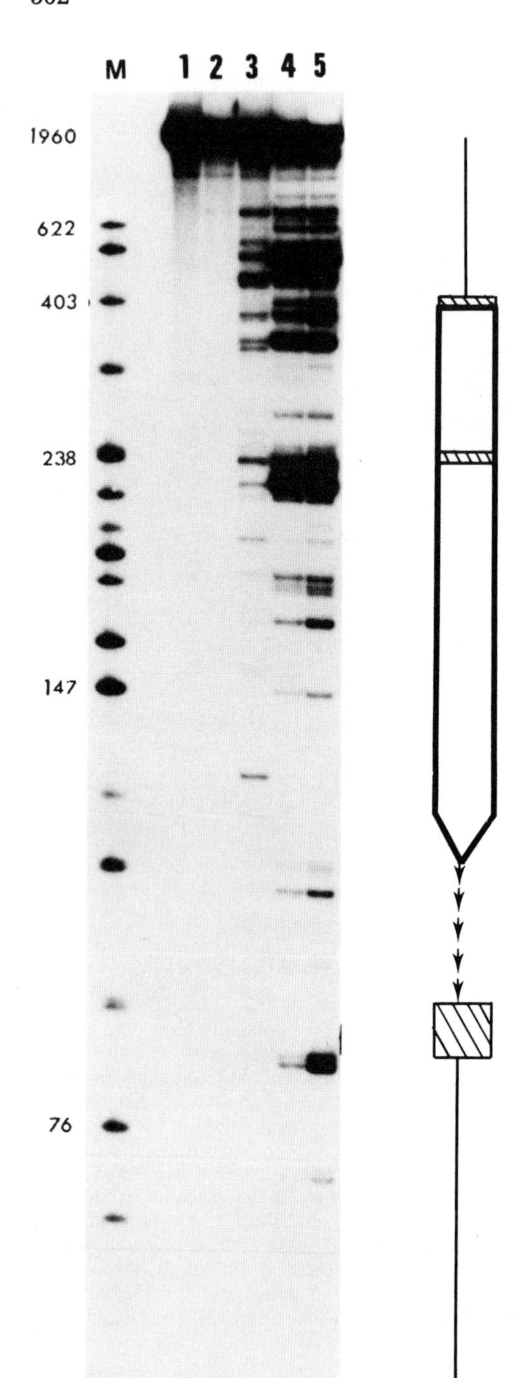

Figure 3. HeLa topoisomerase I nicking of a human *Alu* family repeat. A 1.96-kb *Eco*RI-*Hin*dIII DNA fragment containing the 3′ α1 *Alu* repeat (Lauer et al. 1980; Shen and Maniatis 1982) was 3′ end-labeled at the *Eco*RI site with ^{32}P and used as a substrate for HeLa topoisomerase I nicking according to Edwards et al. (1982). Samples were denatured, electrophoresed on a 6% polyacrylamide–7 M urea gel, and analyzed by autoradiography. The *Alu* repeat is represented as the open arrow in the direction of its transcription, alongside the autoradiogram. The direct repeats flanking and internal to the *Alu* repeat (Hess et al. 1985) are shown as stippled boxes. The five small arrows represent the repetition of the 5′-TTTAA-3′ oligonucleotide near the 3′ end of the *Alu* repeat. Thin lines represent flanking human DNA sequence. (Lane *1*) No enzyme; (lane *2*) 50 ng enzyme, no SDS; (lane *3*) enzyme plus SDS treatment; (lane *4*) enzyme plus 1 μM camptothecin, and treated with SDS; (lane *5*) enzyme plus 25 μM camptothecin, and treated with SDS. Markers (m) were denatured, end-labeled, *Hpa*II-cut pBR322. Numbers to the left of the panel represent fragment sizes in nucleotides.

direct repeat flanking this *Alu* sequence. This may reflect a role of topoisomerase I in an insertion (i.e., nonhomologous recombination) process. Strong topoisomerase I nicking sites were also located within and flanking several other *Alu* repeats. The 3′ ends of many *Alu* repeats contain the repeated oligonucleotide sequences $(TA_n)_m$ (where n = 2–4 and m = 4–7), just prior to the direct repeat. This sequence serves as a strong topoisomerase I nicking site in vitro (C. Perez-Stable et al., in prep.). Three classes of topoisomerase I-induced nicking sites were observed and classified according to their response to camptothecin (Fig. 3). This suggests that different conformations of the enzyme–DNA complex may exist. We are curently investigating whether topoisomerase I binds specifically with *Alu* repeats in vivo.

Evolution. The mutational processes directly translate into evolutionary products. Accumulation of point mutations results in sequence divergence in evolutionary time. The homologous recombinations are involved in gene duplication and correction, whereas nonhomologous recombinations generate insertions and deletions. All of these evolutionary patterns are detectable in the α-globin gene cluster. Examples of the utility of the α-globin gene family for the study of molecular evolution in the primates will be presented below.

Regulation. The α-globin family affords a fairly simple system in which to study the processes of genetic control or regulation. Genetic regulation is now believed potentially to be able to answer many, if not most, of the significant questions of interest to evolutionary biology (King and Wilson 1975). Insofar as the primary structure of DNA may affect gene expression, the study of point mutations of the α-globins may illuminate aspects of gene structure and function. On a gross level, the coordinated and differential expression of the globin genes seems to correlate with location of the genes within the cluster (Maniatis et al. 1980), the state of methylation of the cluster (see references in Shen 1984), and the chromatin configuration (see references in Yagi et al. 1986). All of these factors can be modulated by changes in nucleotide sequence and DNA arrangement within the cluster.

We may even use phylogenetic data to derive inferences about the structure and function of genes. As shown below, certain hypotheses about the human globin genes can be tested, and certain other predictions can be made, based on the structure of the genes in other Old World higher primates.

Studies on Nonhuman Primate α-Globin Genes

Length polymorphism of the globin intergenic DNA. Genomic maps of the adult α-globin gene regions of human, orangutan (*Pongo pygmaeus),* rhesus macaque (*Macaca mulatta*), and olive baboon (*Papio anubis*) are compared in Figure 4A. Zimmer et al. (1980) have previously compared this region among the apes (chimpanzee, gorilla, orangutan, and gibbon), and found it to be highly conserved for intergenic length and restriction fragment sizes. Only a 0.9-kb insertion downstream from α2 in the gibbon and a 0.3-kb insertion downstream from α1 in the orangutan were necessary to make the maps of all the apes essentially identical for this region (Zimmer 1980; Zimmer et al. 1980).

Our data, as shown in Figure 4A, indicate that the lengths of the intergenic regions vary among the primates. There are two DNA deletions, each of approximately 0.3 kb, detected in the region upstream from the α2 globin gene of the rhesus macaque and olive baboon examined. The α2–α1 intergenic DNA of this baboon is also 0.9 kb longer than the human, while that of the macaque is 1.8 kb longer. Interestingly, another macaque is only 0.9 kb longer than the human in this intergenic region, rather than 1.8 kb. (J.-P. Shaw, unpubl.). The locations of the 0.3-kb deletions mentioned above have been mapped (J.-P. Shaw and C.-K.J. Shen, in prep.), but the specific locations of the 0.9-kb and 1.8-kb length differences are still unknown. Nucleotide sequence analysis of this DNA will provide greater insight into the nature of these length differences. It should be noted that among other mammals, the intergenic region varies in size from 3.6 kb in sheep (Rando et al. 1986) to 5.0 kb in horses (Clegg et al. 1984), and 13 kb in the mouse (Leder et al. 1981).

Sequence divergence and gene correction. The most widespread use of molecular data in an evolutionary context involves the rates of divergence of genes in various lineages (Efstradiatis et al. 1980; Perler et al. 1980; Li et al. 1985; Britten 1986). We find, for example, that the rate of divergence of the human from orangutan α-globin genes (Marks et al. 1986a) approximates very closely the rate calculated by previous workers on globins in other organisms. This rate has been calculated as a 1% change in silent sites per 1.2 million years (unit evolutionary period [UEP] = 1.2), and a 1% change in replacement sites per 10 million years (UEP = 10; Efstradiatis et al. 1980; Perler et al. 1980). The human and chimpanzee α-globin genes, however, are more similar to one another than this rate predicts, by either molecule-derived or morphology-derived divergence dates (Liebhaber and Begley 1983; Marks et al. 1986a). We interpret this as evidence in favor of a slowdown in the rate of molecular evolution in the African apes (Goodman 1961; Goodman et al. 1982), although the possibility that the paucity of differences between human and chimpanzee is the result of sampling error should be noted.

Beyond the molecular clock calculations, the phylogenic data also allow us to test hypotheses about the evolution of the human α-globin genes. For example, sequence analysis of the cloned orangutan adult α-globin genes shows that, as in the human, gene correction has been operating on the two duplication units (Marks et al. 1986a). The 7-bp insertion in intron 2 of the human α1 globin gene is not present in either orangutan gene, indicating that this insertion is not causally related to the sequence divergence in the 3′ untranslated regions (UTR) of the two genes. Interestingly, preliminary data on the α2 3′UTR of the rhesus macaque suggest that the domain of gene correction may extend further downstream in this monkey than in the hominoid apes (J.-P. Shaw et al., unpubl.).

Another testable hypothesis should be mentioned here, although we have not yet collected sufficient data to deal with it adequately. Hess et al. (1984) found that the X blocks of the human adult α-globin duplication units diverge from one another more significantly downstream than they do upstream. They therefore suggested that perhaps nonhomologous recombination events, such as those that generated nonhomology blocks I and II (Hess et al. 1983), are adversely affecting the correction process of the two X blocks. Sequence analysis of the X blocks of the nonhuman primates will undoubtedly provide further clarification of the molecular mechanism for this genetic process.

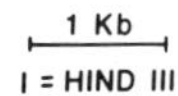

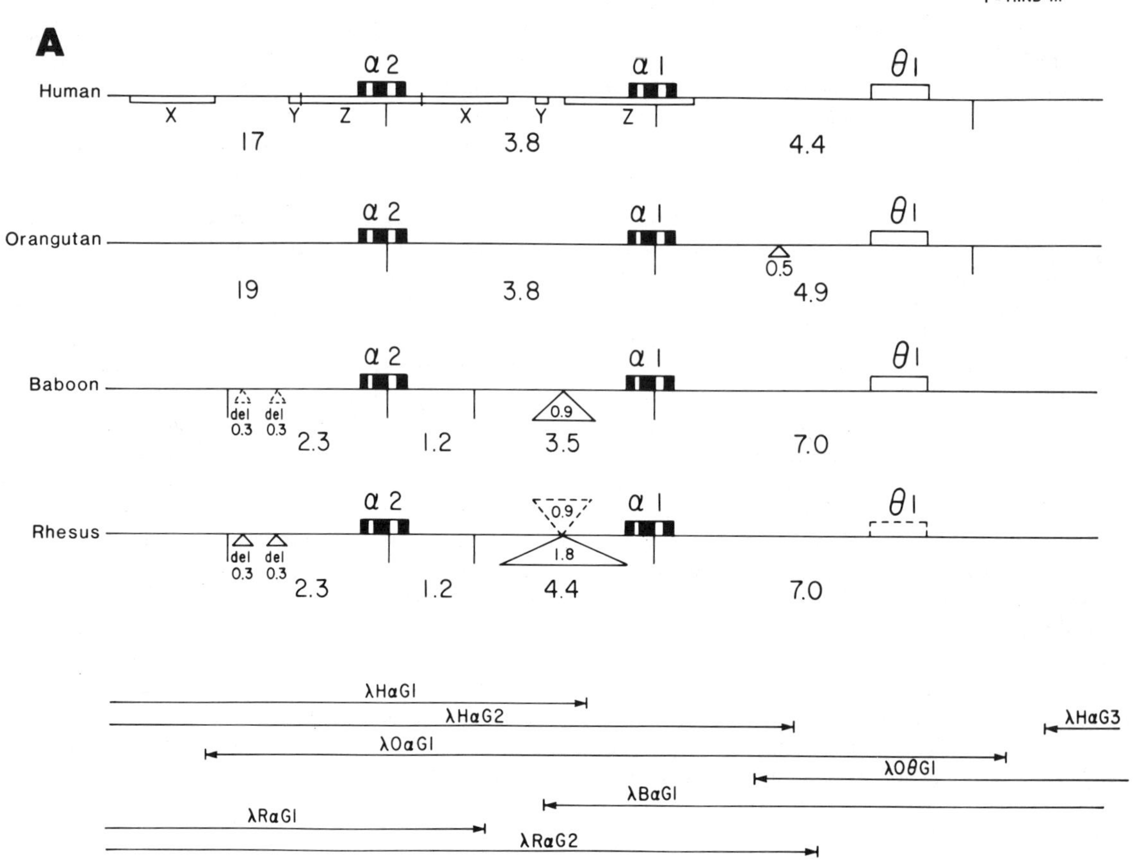

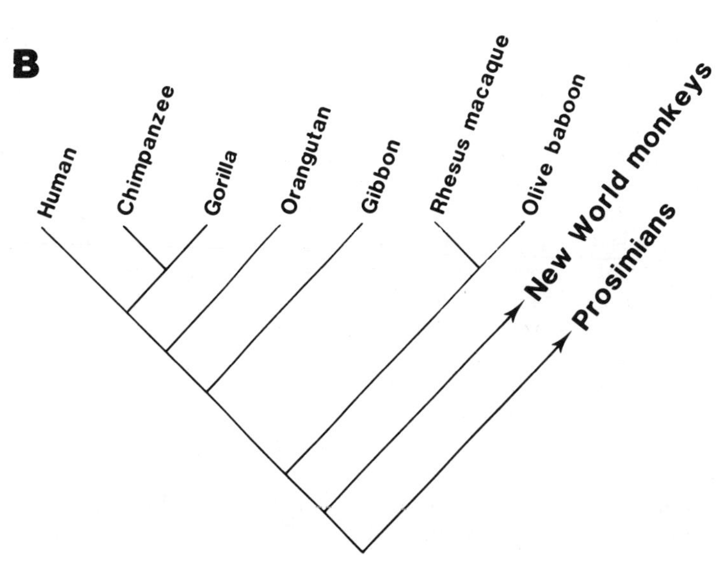

Figure 4. (*A*) (*Above*) Alignment of the adult α-globin gene regions of the human and three nonhuman primates: orangutan, olive baboon, and rhesus macaque. Solid-line triangles denote length differences detected by both genomic blot analysis and molecular cloning; broken-line triangles denote length polymorphisms based only on genomic blotting data. These length differences among human, macaque, and baboon are discussed in the text. (*Below*) Phage clones containing the α-globin gene regions analyzed. These are: human λHαG1, λHαG2, and λHαG3; orangutan, λOαG1, and λOθG1; baboon, λBαG1; rhesus, λRαG1, and λRαG2. (*B*) Relationships of the primates, including those discussed above. Chimpanzee, gorilla, and orangutan are usually grouped as the "great apes," although this does not accurately reflect their phylogeny. Together with the gibbons ("lesser apes") and humans, these are the hominoids (i.e., Superfamily Hominoidea). The olive baboon and rhesus macaque are but two representatives of a very diverse superfamily (Cercopithecoidea) of Old World monkeys (cf. Szalay and Delson 1979).

The θ1-Globin Gene

An unexpected finding that emerged from our work on the α-globin genes of the orangutan was the discovery of an α-globin-like gene located downstream from the α1 gene (Marks et al. 1986a). Multiple copies of this new α-like globin gene, designated as θ1, exist in human, orangutan, macaque, and baboon. Determination of the complete primary structure of the θ1 gene revealed that it is indeed an α-globin-like gene (Marks et al. 1986b).

We give in Figure 5 the primary sequence and translation of the orangutan θ1-globin gene. It possesses initiation and termination codons, although the termination codon is TGA, instead of TAA, as in α1. The upstream promoters CCAAT and ATA are present, though about 150 bp further upstream from the initiation codon than in α1. The polyadenylation signal is AGTAAA instead of AATAAA. Though no θ-globin protein has yet been detected, θ1 is not an obvious pseudogene. It may well have a stage-specific or tissue-specific pattern of expression. Indeed, we interpret its sequence differences from α1 upstream and downstream from the coding region as being indicative of a different pattern of regulation for the two genes.

In addition to the inference of function from the structural integrity of θ1, a second indirect evidence that θ1 may be functional derives from a comparison

```
ccaattttg tgtttttagt agagactaaa aaccatatgg tgaacaccta agacgggggg

ccttggatcc agggcaattc agagggcccc ggtcggagct gtcggagatg gagcgcgcgc

gctcccggga tcccggacga ggccctggac cccagggcgg cgaggctgca gcgcggcgcc

ccctggaggc cgcgggaccc ctagccggtc cgcgcaggcg cggcggggac gcagggcgcg

                         INI Ala Leu Ser Ala Glu Asp Arg Ala Leu Val A
gcgggttcca gcgcgggg ATG GCG CTG TCC GCG GAG GAC CGG GCG CTG GTG C

rg Ala Leu Trp Lys Lys Leu Gly Ser Asn Val Gly Val Tyr Thr Thr Gl
GC GCC CTG TGG AAG AAG CTG GGC AGC AAC GTC GGC GTC TAC ACG ACA GA

u Ala Leu Glu Ar
G GCC CTG GAG AG gtgcggcgag gctgggcgcc cccgcccccca ggggccctcc ctcc

                                           g Thr Phe Leu Ala P
ccaagc cccccggact cgcctcaccc acgttcctct cgcag G ACC TTC CTG GCC T

he Pro Ala Thr Lys Thr Tyr Phe Ser His Leu Asp Leu Ser Pro Gly Se
TC CCC GCA ACG AAG ACC TAC TTC TCC CAC CTG GAC CTG AGC CCC GGC TC

r Ser Gln Val Arg Ala His Gly Gln Lys Val Ala Asp Ala Leu Ser Leu
C TCA CAG GTC AGA GCC CAC GGC CAG AAG GTG GCG GAC GCG CTG AGC CTC

Ala Val Glu Arg Leu Asp Asp Leu Pro His Ala Leu Ser Ala Leu Ser H
GCC GTG GAG CGC CTG GAC GAC CTA CCC CAC GCG CTG TCC GCG CTG AGC C

is Leu His Ala Cys Gln Leu Arg Val Asp Pro Ala Ser Phe Gln
AC CTG CAC GCG TGC CAG CTG CGA GTG GAC CCG GCC AGC TTC CAG gtgagc

ggct gccgtgctgg gcccctgtcc ccgggagggc cccggcgggg cgggtgcggg gggcg

                                                Leu Leu Gly His
tgcac ggcgggtgca ggcgagtgag ccttgagcgc tcgccgcag CTC CTG GGC CAC

Cys Leu Leu Val Thr Leu Ala Arg His Tyr Pro Gly Asp Phe Ser Pro A
TGC CTG CTG GTA ACC CTC GCC CGG CAC TAC CCC GGA GAC TTC AGC CCC G

la Leu Gln Ala Ser Leu Asp Lys Phe Leu Ser His Val Ile Ser Ala Le
CG CTG CAG GCG TCG CTG GAC AAG TTC CTG AGC CAC GTG ATC TCG GCG CT

u Ala Ser Glu Tyr Arg TER
G GCT TCC GAG TAC CGC TGA actgtgggtg ggtggccgcg ggacccccac gcgact

ttcc ccgtatttga gtaaagtctc tccaaggagc agccttcttg ccgtgctctc tccag

ggcag gacgcgagag gaaggcgccg cccctcccca aggaaaggcg agggcctggg gcgc

accccc agtgcccgga tcc
```

Figure 5. Nucleotide sequence of the orangutan θ1 gene. Noncoding regions are given in small letters; presumptive coding regions in capitals. Above the coding regions is given the amino acid translation of the presumptive θ1 protein. The two promoter sequences (CCAAT and ATA) are underlined, as are the apparent polyadenylation signal and site. Sequence data are taken from Marks et al. (1986b).

of the orangutan $\theta 1$ sequence with that of the olive baboon (J.-P. Shaw et al., in prep.). If the gene is functional, and thereby constrained by natural selection, we would anticipate finding fewer differences in the coding than in noncoding sequences, and a preponderance of silent changes over substitution mutations in the coding sequence. This is indeed the case (Fig. 6). The coding sequences diverge by 6.3%, whereas noncoding regions diverge by over 9%. And of the 27 differences in the coding sequence, 22 are silent third-position changes. We interpret these data as strongly in support of a functional role for $\theta 1$ and for strong constraints on its evolution.

The evolutionary history of the primate $\theta 1$ gene can be estimated from the percentage of replacement-site substitutions when compared with other α-like globin genes. Assuming that the globin genes tend to accumulate amino acid substitution mutations at a rate of 1% change per 10 million years, we calculate that the $\theta 1$ gene began to diverge from the adult α-globin gene approximately 260 million years ago (Fig. 7A). The relationships of the organisms in which an apparent θ-globin homolog has been located are shown in Figure 7B.

We anticipate that the $\theta 1$ globin gene isolated from the human will turn out to be functional, as its orthologous counterparts in the orangutan and baboon seem to be. An interesting problem is why the primary structure of the galago (or bushbaby, a prosimian primate) and the rabbit seems to preclude the possibility of expression in these organisms (Cheng et al. 1986; I. Sawada and C. Schmid, in prep.). One possibility is that the novel form of placentation evolved by the higher primates (Luckett 1975) placed demands for globin synthesis on the genetic system to which the retention of a functional θ-globin gene is an adaptive response. Another possibility is that the gene exists in multiple copies in the bushbaby and rabbit, and while the ortholog of $\theta 1$ may be silenced, another, yet undetected copy, may be functional. It is of interest to note here that one of the $\theta 1$-homologous sequences in the human genome is a solitary pseudogene (J.-P. Shaw, et al., in prep.).

SUMMARY AND CONCLUSIONS

The α-globins are an extraordinarily useful model for analyzing the dynamic processes that produced the human genome and that still occur within it, most of which could hardly have been imagined 20 years ago. We are approaching problems of genomic turnover from a structural, biophysical, and cellular perspective, in the hopes of deepening our knowledge of the mechanisms involved in these recombinations. Nevertheless, it is always possible, in the highly focused realm of molecular studies, to lose sight of the larger spectacle of phylogeny and evolution. We appreciate that our work interfaces with other disciplines, such as systematics, primatology, and biological anthropology, and that interchange can be mutually illuminating.

The $\theta 1$ gene was discovered first in the orangutan because the human α-cluster had proven difficult to clone downstream from $\alpha 1$, and all the known globin proteins were accounted for genomically and there seemed no reason to expect anything of interest to exist there. By focusing our attention not only on genomic evolutionary processes, but on their evolutionary products in other closely related species as well, we have been able to frame and test hypotheses about the structure of the human genome, and reciprocally to make inferences about the processes themselves.

It is now obvious that the genome is not the stable entity it was once considered to be. On the contrary, it is in a fairly constant state of flux. The 30-kb stretch comprising the human α-globin cluster affords us a window through which we may glimpse the genomic turnover processes operating, model them, and, in the end, gain a deeper understanding of the cryptic factors that govern the genetic evolution of our species.

ACKNOWLEDGMENTS

This paper is in memory of Dr. Che-Wen James Shen. We thank Ben Koop, Morris Goodman, Francisco Ayala, and Allan Wilson for their encouragement and interest in our work. Leroy Liu, Tom Maniatis, and Bob Tjian have provided invaluable materials. The communication of unpublished results from Ross Hardison and Carl Schmid is greatly appreciated. We also thank Yerkes Regional Primate Center and the California Primate Research Center for primate tissues; and Kimbery Strauch for typing the manuscript. This research was supported by grants from the National Institutes of Health and the American Cancer Society, and by a Research Career Development Award to C.-K.J. Shen.

Orangutan θ / Baboon θ

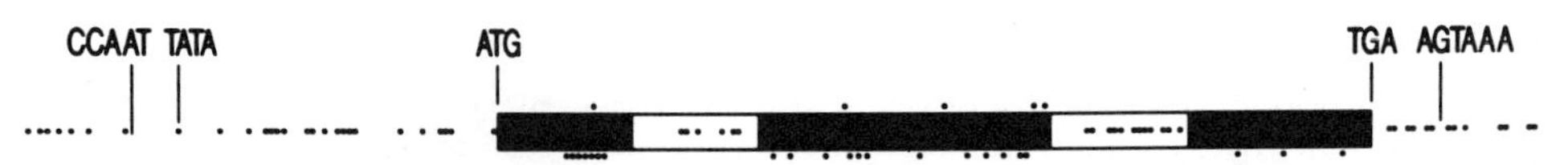

Figure 6. Distribution of nucleotide substitutions, as denoted by dots, between the $\theta 1$ homologs of orangutan and olive baboon. The gene is given with presumptive exons filled in. Above the gene are the locations of amino acid replacement mutations; below the gene are the locations of silent mutations. Mutations of noncoding regions are given on the line.

A

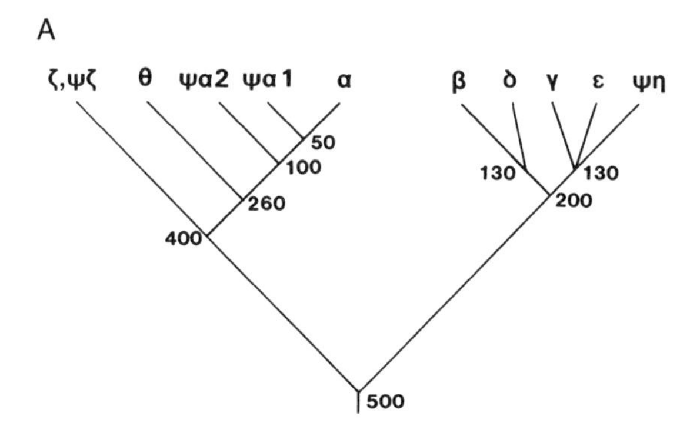

B

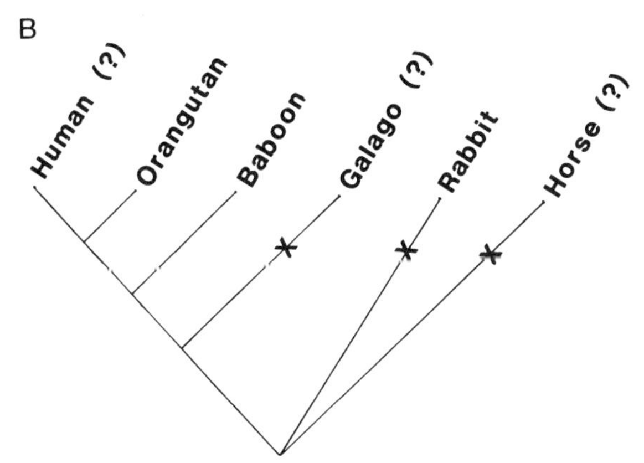

Figure 7. (*A*) Evolutionary relationships of the human blood globin gene family. Dates for the β-cluster are derived from analyses by Efstradiatis et al. (1980), Hardison (1984), Harris et al. (1984), Goodman et al. (1984), and Collins and Weissman (1984). Dates for the α-cluster are derived from analyses by Proudfoot and Maniatis (1980), Proudfoot et al. (1982), Czelusniak et al. (1982), and this study. Calculations of α-globin divergences are based upon the replacement site changes within the coding sequences, and therefore reflect the time at which the coding regions began to diverge, and not necessarily the time of actual physical duplication. Distances between the branch points are not in proportion to the time scale. (*B*) Relationships of the organisms in which an apparent θ1 homolog has been detected. Branches marked with an X indicate those lineages in which the θ1 homolog appears to be a pseudogene. A question mark indicates some ambiguity in the conclusions derived. For the human, no sequence data are yet available. The galago C gene is a pseudogene and may be an ortholog of θ1 (I. Sawada and C. Schmid, pers. comm.). The horse gene is reported to be a pseudogene, although no sequence data have been published (Clegg et al. 1984).

REFERENCES

Been, M.D., R.R. Burgess, and J.J. Champoux. 1984. Nucleotide sequence preference at rat liver and wheat germ type I DNA topoisomerase breakage sites in duplex SV40. *Nucleic Acids Res.* **12:** 3097.

Britten, R.J. 1986. Rates of DNA sequence evolution differ between taxonomic groups. *Science* **231:** 1393.

Bullock, P., J. Champoux, and M. Botchan. 1985. Association of crossover points with topoisomerase I cleavage sites: A model for nonhomologous recombination. *Science* **230:** 954.

Bunn, H.F., B.G. Forget, and H.M. Ranney. 1977. *Human hemoglobins.* W.B. Saunders, Philadelphia.

Cheng, J.-F., L. Raid, and R.C. Hardison. 1986. Isolation and nucleotide sequence of the rabbit globin gene cluster ψζ-α1-ψα. *J. Biol. Chem.* **261:** 839.

Clegg, J.G., S.E.Y. Goodbourn, and M. Braend. 1984. Genetic organization of the polymorphic equine α globin locus and sequence of the BII α1 gene. *Nucleic Acids Res.* **12:** 7847.

Collins, F.S. and S.M. Weissman. 1984. The molecular genetics of human hemoglobin. *Prog. Nucleic Acid Res. Mol. Biol.* **34:** 315.

Czelusniak, J., M. Goodman, D. Hewett-Emmett, M.L. Weiss, P.J. Venta, and R.E. Tashian. 1982. Phylogenetic origins and adaptive evolution of avian and mammalian haemoglobin genes. *Nature* **298:** 297.

Deisseroth, A., P. Nienhuis, P. Turner, R. Velez, W.F. Anderson, F. Ruddle, J. Lawrence, R. Creagen, and R. Kuchlerlapati. 1977. Localization of the human alpha globin gene to chromosome 16 in somatic cell hybrids by molecular hybridization assay. *Cell* **12:** 205.

Edwards, K.A., B.D. Halligan, J.L. Davis, N.L. Nivera, and L.F. Liu. 1982. Recognition sites of eukaryotic DNA topoisomerase I: DNA nucleotide sequencing analysis of Topo I cleavage sites on SV40 DNA. *Nucleic Acids Res.* **10:** 2565.

Efstradiatis, A., J.W. Posakony, T. Maniatis, R.M. Lawn, C. O'Connell, R.A. Spritz, J.K. DeRiel, B.G. Forget, S.M. Weissman, J.L. Slightom, A.E. Blechl, O. Smithies, F.E. Baralle, C.C. Shoulders, and N.J. Proudfoot. 1980. The structure and evolution of the human β-globin gene family. *Cell* **21:** 653.

Embury, S.H., S. Lebo, A.M. Dozy, and Y.W. Kan. 1979. Organization of the α globin genes in the Chinese α thalassemia syndromes. *J. Clin. Invest.* **63:** 1307.

Gluzman, Y. 1981. SV40-transformed simian cells support the replication of early SV40 mutants. *Cell* **23:** 175.

Goodbourn, S.E.Y., D.R. Higgs, J.B. Clegg, and D.J. Weatherall. 1983. Molecular basis of length polymorphism in the human ζ-globin gene complex. *Proc. Natl. Acad. Sci.* **80:** 5022.

Goodman, M. 1961. The role of immunochemical differences in the phyletic development of human behavior. *Hum. Biol.* **33:** 131.

Goodman, J., M.L. Weiss, and J. Czelusniak. 1982. Molecular evolution above the species level: Branching pattern, rates, and mechanisms. *Syst. Zool.* **31:** 376.

Goodman, M., B.F. Koop, J. Czelusniak, M.L. Weiss, and J.L. Slightom. 1984. The η-globin gene: Its long evolutionary history in the β-globin gene family of mammals. *J. Mol. Biol.* **180:** 803.

Goossens, M., A.M. Dozy, S.H. Embury, Z. Zachariades, M.G. Hadjiminas, G. Stamatoyannopoulos, and Y.W. Kan. 1980. Triplicated α-globin loci in humans. *Proc. Natl. Acad. Sci.* **77:** 518.

Hardison, R. 1984. Comparison of the β-like globin gene families of rabbits and humans indicates that the gene cluster 5′-ε-γ-δ-β-3′ predates the mammalian radiation. *Mol. Biol. Evol.* **1:** 390.

Hardison, R., I. Sawada, J.-F. Cheng, C.-K. Shen, and C.W. Schmid. 1986. A previously undetected pseudogene in the human alpha globin gene cluster. *Nucleic Acids Res.* **14:** 1903.

Harris, S., P.A. Barrie, M.L. Weiss, and A.J. Jeffreys. 1984. The primate ψβ1 gene: An ancient β-globin pseudogene. *J. Mol. Biol.* **180:** 785.

Hess, J.F., C.W. Schmid, and C.-K.J. Shen. 1984. A gradient of sequence divergence in the human adult α-globin duplication units. *Science* **226:** 67.

Hess, J., F.M. Fox, C. Schmid, and C.-K.J. Shen. 1983. Molecular evolution of the human adult α-globin-like gene region: Insertion and deletion of *Alu* family repeats and non-*Alu* DNA sequences. *Proc. Natl. Acad. Sci.* **80:** 5970.

Hess, J., C. Perez-Stable, G.J. Wu, B. Weir, I. Tinoco, and C.-K.J. Shen. 1985. End-to-end transcription of an Alu family repeat: A new type of polymerase-III-dependent terminator and its evolutionary implications. *J. Mol. Biol.* **184:** 7.

Higgs, D.R. and D.J. Weatherall. 1985. Alpha-thalassemia. *Curr. Top. Hematol.* **4:** 37.

Higgs, D.R., S.E.Y. Goodbourn, J.S. Wainscoat, J.B. Clegg, and D.J. Weatherall. 1981. Highly variable regions of DNA flank the human globin genes. *Nucleic Acids Res.* **9:** 4213.

Higgs, D.R., A.V.S. Hill, R. Nicholls, S.E.Y. Goodbourn, H. Ayyub, J.B. Clegg, and D.J. Weatherall. 1985. Molecular rearrangements of the human α-globin gene cluster. *Ann. N.Y. Acad. Sci.* **445:** 45.

Hill, A.V.S., R.D. Nicholls, S.L. Thein, and D.R. Higgs. 1985. Recombination within the human embryonic ζ-globin locus: A common ζ-ζ chromosome produced by gene conversion of the ψζ gene. *Cell* **42:** 809.

Hirt, G. 1967. Selective extraction of polyoma DNA from infected mouse cell cultures. *J. Mol. Biol.* **26:** 365.

Hsiang, Y.-H., R. Hertzberg, S. Hecht, and L.F. Liu. 1985. Camptothecin induces protein-linked DNA breaks via mammalian DNA topoisomerase I. *J. Biol. Chem.* **260:** 14871.

Kan, Y.W. 1985. Molecular pathology of α-thalassemia. *Ann. N.Y. Acad. Sci.* **445:** 28.

King, M.-C. and A.C. Wilson. 1975. Evolution at two levels in humans and chimpanzees. *Science* **188:** 107.

Lauer, J., C.-K.J. Shen, and T. Maniatis. 1980. The chromosomal arrangement of human α-like globin genes: Sequence homology and α-globin gene deletions. *Cell* **20:** 199.

Leder, A., D. Swan, F. Ruddle, P. D'Eustachio, and P. Leder. 1981. Dispersion of α-like globin genes of the mouse to three different chromosomes. *Nature* **293:** 196.

Li, W.-H., C.-C. Luo, and C.-I Wu. 1985. Evolution of DNA sequences. In *Molecular evolutionary genetics* (ed. R.J. MacIntyre), p. 1. Plenum Press, New York.

Liebhaber, S.A. and K.A. Begley. 1983. Structural and evolutionary analysis of the two chimpanzee α-globin mRNAs. *Nucleic Acids Res.* **11:** 8915.

Liebhaber, S.A., M. Goossens, and Y.W. Kan. 1981. Homology and concerted evolution at the α1 and α2 loci of human α-globin. *Nature* **290:** 26.

Liebhaber, S.A., M. Goossens, R. Poon, and Y.W. Kan. 1980. The primary structure of the α-globin gene cloned from normal human DNA. *Proc. Natl. Acad. Sci.* **77:** 7054.

Liu, L.F. 1983. HeLa topoisomerase I. *Methods Enzymol.* **100:** 133.

Livingstone, F.B. 1985. *Frequencies of hemoglobin variants.* Oxford University Press, New York.

Luckett, W.P. 1975. Ontogeny of the fetal membranes and placenta: Their bearing on primate phylogeny. In *Phylogeny of the primates* (ed. W.P. Luckett and F.S. Szalay), p. 157. Plenum Press, New York.

Maniatis, T., E. Fritsch, J. Lauer, and R. Lawn. 1980. The molecular genetics of human hemoglobins. *Annu. Rev. Genet.* **14:** 145.

Marks, J., J.-P. Shaw, and C.-K.J. Shen. 1986a. The orangutan adult α-globin gene locus: Duplicated functional genes and a newly detected member of the primate α-globin gene family. *Proc. Natl. Acad. Sci.* **83:** 1413.

———. 1986b. Sequence organization and genomic complexity of the primate θ1 globin gene, a novel α-globin-like gene. *Nature* **321:** 785.

Mellon, P., V. Parker, Y. Gluzman, and T. Maniatis. 1981. Identification of DNA sequences required for transcription of the human α1-globin gene in a new SV40 host-vector system. *Cell* **27:** 279.

Michelson, A.M. and S.H. Orkin. 1980. The 3′ untranslated regions of the duplicated human α-globin genes are unexpectedly divergent. *Cell* **22:** 371.

———. 1983. Boundaries of gene conversion within the duplicated human α-globin genes. *J. Biol. Chem.* **258:** 15245.

Orkin, S.H., J. Old, H. Lazarus, C. Altay, A. Gurgey, D.J. Weatherall, and D.G. Nathan. 1979. The molecular basis of α-thalassemia: Frequent occurrence of dysfunctional α loci among non-Asians with Hb H disease. *Cell* **17:** 33.

Perler, F., A. Efstradiatis, P. Lomedico, W. Gilbert, A. Kolodner, and J. Dodgson. 1980. The evolution of genes: The chicken preproinsulin gene. *Cell* **20:** 555.

Proudfoot, N.J. and T. Maniatis. 1980. The structure of a human α-globin pseudogene and its relationship to α-globin gene duplication. *Cell* **21:** 537.

Proudfoot, N.J., A. Gil, and T. Maniatis. 1982. The structure of the human zeta-globin gene and a closely linked, nearly identical pseudogene. *Cell* **31:** 553.

Rando, A., L. Ramunno, and P. Masina. 1986. Variation in the number of α-globin loci in sheep. *Mol. Biol. Evol.* **3:** 168.

Reeders, S.T., M.H. Breuning, K.E. Davies, R.D. Nicholls, A.P. Jarman, D.R. Higgs, P.L. Pearson, and D.J. Weatherall. 1985. A highly polymorphic DNA marker linked to adult polycystic kidney disease on chromosome 16. *Nature* **317:** 542.

Sawada I., M.P. Beal, C.-K.J. Shen, B. Chapman, A.C. Wilson, and C. Schmid. 1983. Intergenic DNA sequences flanking the pseudo alpha globin genes of human and chimpanzee. *Nucleic Acids Res.* **11:** 8087.

Sawada, I., C. Willard, C.-K.J. Shen, B. Chapman, A.C. Wilson, and C.W. Schmid. 1985. Evolution of Alu family repeats since the divergence of human chimpanzee. *J. Mol. Evol.* **22:** 316.

Schmid, C.W. and C.-K.J. Shen. 1985. The evolution of interspersed repetitive DNA sequences in mammals and other vertebrates. In *Molecular evolutionary genetics* (ed. R.J. MacIntyre), p. 323. Plenum Press, New York.

Shen, C.-K.J. 1984. DNA methylation. In *DNA methylation: Biochemistry and biological significance* (ed. A. Razin et al.), p. 249. Springer-Verlag, New York.

Shen, C.-K.J. and T. Maniatis. 1982. The organization, structure and *in vitro* transcription of Alu family RNA polymerase III transcription units in the human α-like globin gene cluster: Precipitation of *in vitro* transcripts by lupus anti-La antibodies. *J. Mol. Appl. Genet.* **1:** 343.

Szalay, F. and E. Delson. 1979. *Evolutionary history of the primates*. Academic Press, New York.

Wang, J. 1985. DNA topoisomerases. *Annu. Rev. Biochem.* **54:** 665.

Wang, J. and L. Liu. 1979. DNA topoisomerase: Enzymes that catalyze the concerted breaking and rejoining of DNA backbone bonds. In *Molecular Genetics* (ed. J.H. Taylor), p. 65. Academic Press, New York.

Willard, C., E. Wong, J.F. Hess, C.-K.J. Shen, B. Chapman, A.C. Wilson, and C.W. Schmid. 1985. Comparison of human and chimpanzee ζ1 globin genes. *J. Mol. Evol.* **22:** 309.

Wilson, A.C., S.S. Carlson, and T.J. White. 1977. Biochemical evolution. *Annu. Rev. Biochem.* **46:** 573.

Yagi, M., R. Gelinas, J.T. Elder, M. Peretz, T. Papayannopoulou, G. Stamatoyannopoulos, and M. Groudine. 1986. Chromatin structure and developmental expression of the human α-globin cluster. *Mol. Cell. Biol.* **6:** 1108.

Zimmer, E.A. 1980. "Evolution of primate globin genes." Ph.D. thesis, University of California, Berkeley.

Zimmer, E.A., S.L. Martin, S.M. Beverley, Y.W. Kan, and A.C. Wilson. 1980. Rapid duplication and loss of genes coding for the α chains of hemoglobin. *Proc. Natl. Acad. Sci.* **77:** 2158.

Structural Features of the Proteins Participating in Blood Coagulation and Fibrinolysis

E.W. DAVIE, A. ICHINOSE, AND S.P. LEYTUS
Department of Biochemistry, University of Washington, Seattle, Washington 98195

In recent years, a great deal of new information has emerged regarding the structure and function of the various proteins involved in blood coagulation, fibrinolysis, complement activation, and other physiological events that occur in blood. The structural data for these proteins have been obtained by a combination of amino acid sequence analysis of purified preparations and cDNA cloning. This information has made it clear that there are many common structures or domains that are shared by the plasma proteins and other proteins from different tissues and organs. For example, the carboxy-terminal 250–300 amino acids that constitute the catalytic domain in the serine proteases of plasma show a high degree of amino acid sequence homology with pancreatic trypsin (Katayama et al. 1979). The function of this domain in the plasma serine proteases is to catalyze the hydrolysis of arginine-containing peptide bonds. It contains the triad of serine, histidine, and aspartic acid that is involved in catalysis, and these residues are located at positions analogous to those in trypsin. Also, the catalytic domain contains an aspartic acid residue six amino acids prior to the active site serine. In trypsin, this aspartic acid is located at the bottom of the substrate binding pocket and is responsible for its substrate specificity toward basic amino acids.

In the case of the blood serine proteases, the substrate specificity is much more restrictive in that enzymes, such as thrombin; factors VIIa, IXa, Xa, XIa, and XIIa; plasma kallikrein; activated protein C; tissue plasminogen activator; and urokinase, hydrolyze arginine-containing peptide bonds in only a few proteins.

Domains in the Noncatalytic Chain of the Serine Proteases

A number of different domains or structures that are in the amino-terminal region of the serine proteases have also been identified. These domains often play a specific role in the binding or interaction of the serine proteases with their cofactors, phospholipid membranes, surfaces, or small molecules. These structures include (1) the kringle domains (~85 amino acids) that are present in prothrombin (Magnusson et al. 1975), factor XII (McMullen and Fujikawa 1985), plasminogen (Magnusson et al. 1976), tissue plasminogen activator (Pennica et al. 1983), and urokinase (Gunzler et al. 1982); (2) the γ-carboxyglutamic acid domains (~55 amino acids) that are present in the vitamin-K-dependent proteins, including prothrombin (Magnusson et al. 1975), factor VII (Hagen et al. 1986), factor IX (Katayama et al. 1979), factor X (Titani et al. 1975), and protein C (Stenflo and Fernlund 1982); (3) the epidermal growth-factor domains (Savage et al. 1972; Gray et al. 1983; Scott et al. 1983) (~45 amino acids) that are homologous to structures present in factor VII (Hagen et al. 1986), factor IX (Yoshitake et al. 1985), factor X (Young et al. 1978), factor XII (McMullen and Fujikawa 1985), protein C (Foster and Davie 1984), tissue plasminogen activator (Ny et al. 1984), and urokinase (Gunzler et al. 1982); (4) the type I and type II domains of fibronectin (Petersen et al. 1983; Skorstengaard et al. 1984) that are homologous to structures in factor XII (McMullen and Fujikawa 1985) and tissue plasminogen activator (Ny et al. 1984); and (5) the tandem repeats (~90 amino acids) that have been found in factor XI (Fujikawa et al. 1986) and plasma kallikrein (Chung et al. 1986) (Fig. 1). Since all of these proteins are secreted into plasma, they are also synthesized with a hydrophobic leader sequence.

The evolution of these proteins apparently occurred through a series of translocation events during the last 500 million years or so (Patthy 1985; Yoshitake et al. 1985). This would involve the transfer of segments of DNA coding for the various domains from one gene to another. Eventually, new serine proteases are generated with different biological functions. In some cases, the introns in these genes may be involved in this shuffling process (Blake 1978, 1979; Gilbert 1978), since the genes for these proteins often have similar organization and structure.

Domains in Other Plasma Proteins

Other plasma proteins that contain common domains include factor V and factor VIII (Fass et al. 1985). These two proteins participate in blood coagulation as cofactors rather than serine proteases. Each contains type-A domains (~350 amino acids), and these domains are homologous to ceruloplasmin (Church et al. 1984; Toole et al. 1984; Vehar et al. 1984; Kane and Davie 1986). In addition, factor V and factor VIII contain C domains (~150 amino acids) that are homologous to each other. The central or connecting region of factor V contains, in addition, 20 or more

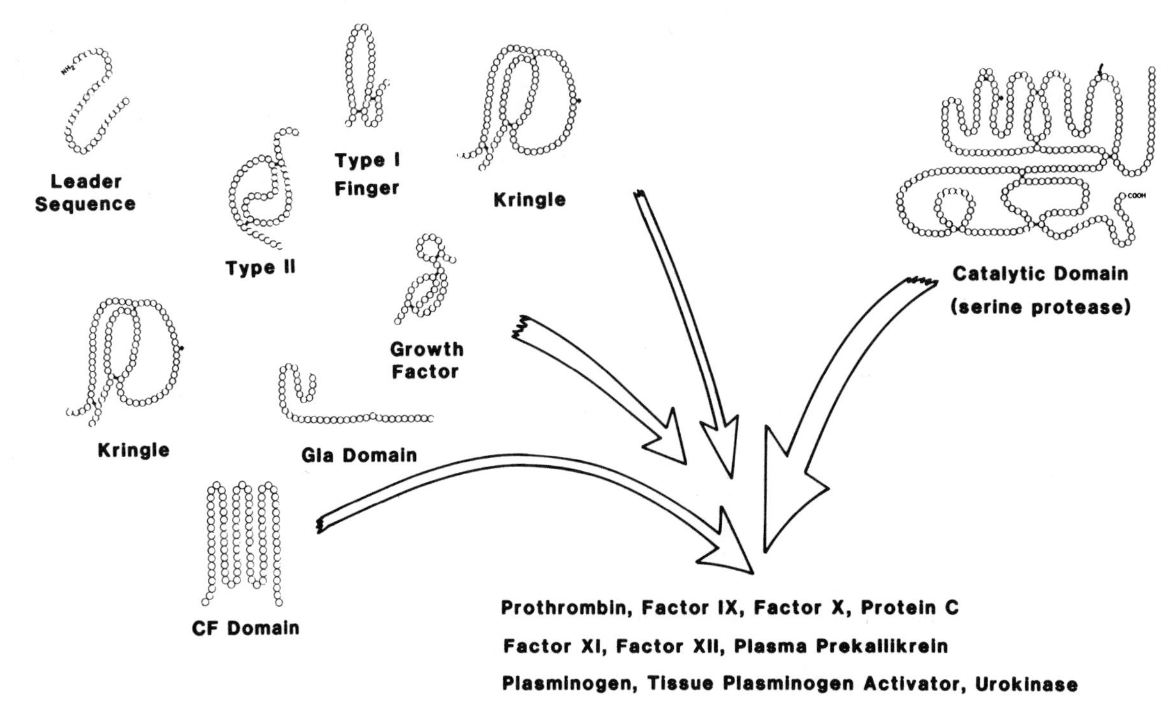

Figure 1. Assembly of various domains into plasma serine proteases. Each of the serine proteases is synthesized with a leader sequence that is removed during biosynthesis. Type I (finger) and type II domains were originally described by Petersen et al. (1983) in fibronectin. The contact factor (CF) domains are present in factor XI and plasma prekallikrein (Chung et al. 1986; Fujikawa et al. 1986), whereas the Gla domain (γ-carboxyglutamic acid domain) is present in the vitamin-K-dependent proteins. The kringle domains were first described in prothrombin by Magnusson et al. (1975), and the growth-factor structure was initially reported by Savage et al. (1972). (Reprinted, with permission, from Davie 1986.)

tandem repeats of nine amino acids that appear to be unique to this protein (Kane and Davie 1986).

von Willebrand factor is another of the large multidomain proteins present in plasma. It is involved in platelet adhesion to subendothelium and contains five different repeats that occur as duplications or triplications (Sadler et al. 1985; Shelton-Inloes et al. 1986; Titani et al. 1986). These five unrelated repeats are approximately 220, 20, 115, 330, and 45 amino acids in length for repeats A, B, C, D, and E, respectively. The A domains of von Willebrand factor are also homologous to one of the domains present in complement factor B.

GP-I Domains

Recently, we have completed the amino acid sequence of human factor XIII by a combination of protein sequencing and cDNA cloning (Ichinose et al. 1986a,b). This protein circulates in blood as a tetramer consisting of two a subunits and two b subunits. In the final stages of blood coagulation, factor XIII is converted by thrombin to factor XIIIa, an enzyme with transglutaminase activity. In the presence of calcium ions, factor XIIIa catalyzes the cross-linking of fibrin monomers by the formation of intermolecular ϵ-(γ-glutamyl)lysine bonds. This results in the formation of a fibrin polymer with increased mechanical strength and resistance to degradation by plasmin. The a subunit (731 amino acids) contains the catalytic site of the enzyme and an active site sequence of Tyr-Gly-Gln-Cys-Trp. This sequence is homologous to that of tissue transglutaminase. The b subunit (641 amino acids) is composed of ten repetitive, homologous sequences of about 60 amino acids. These repeats, which have been called GP-I repeats, are homologous to regions in a number of other plasma and membrane proteins, including β_2-glycoprotein I (with five; Lozier et al. 1984), complement factor B (with three; Mole et al. 1984), complement receptor type I (CR I, complement C3b/C4b receptor) (with at least three; Wong et al. 1985), the A chain of complement Clr (with two; Leytus et al. 1986), complement Cls (with two; Spycher et al. 1986), the receptor for interleukin-2 (with one; Leonard et al. 1984; Nikaido et al. 1984), and haptoglobin (with one or two; Kurosky et al. 1980) (Fig. 2). The internal repeats in these proteins contain four half-cysteine residues that are highly conserved. In β_2-glycoprotein I, the first and third and the second and fourth half-cysteine residues are linked by disulfide bonds (Lozier et al. 1984). On the basis of the amino acid sequence homology with the GP-I repeats, it seems likely that most, if not all, of the tandem repeats present in the b subunit of factor XIII have disulfide bonds in similar positions (Fig. 3). The alignment scores for several of these proteins containing the GP-I domains are shown in Table 1. For most of these domains, the alignment scores are higher than five. Recently, Kristensen and

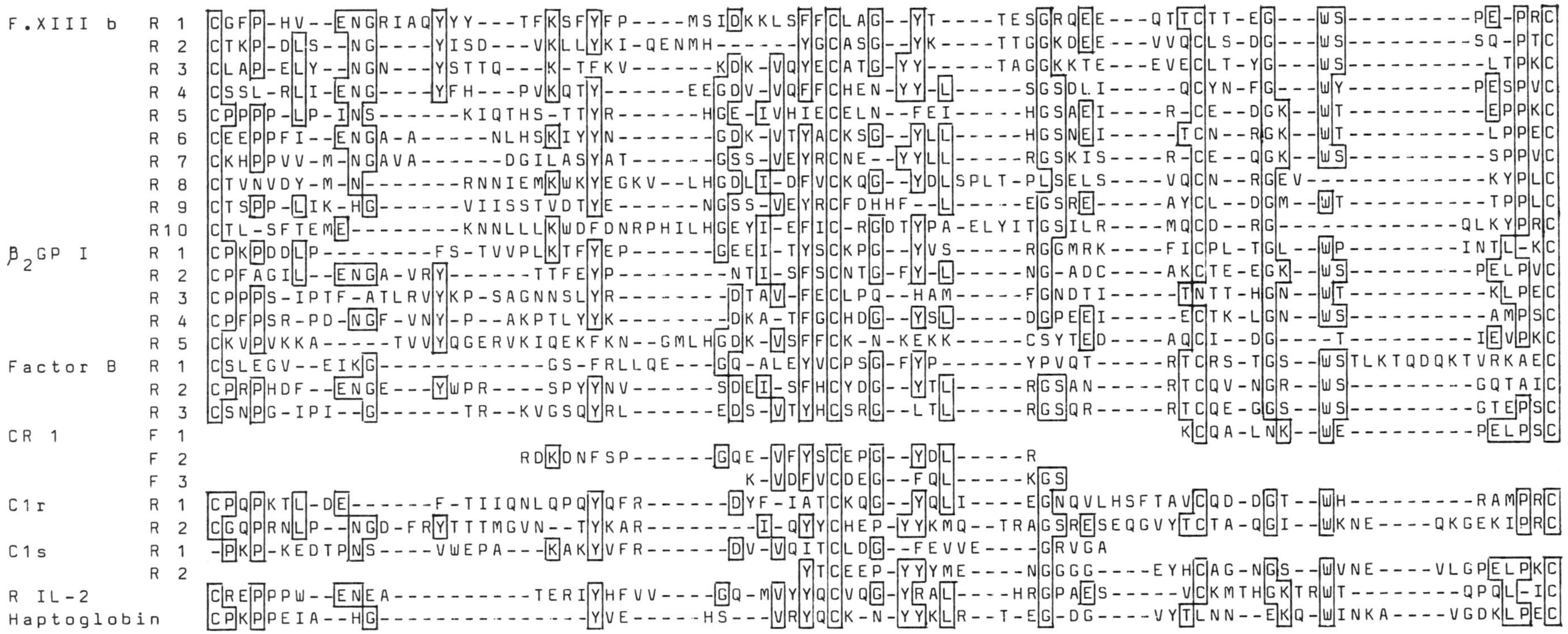

```
F.XIII b    R 1   CGFP-HV--ENGRIAQYYY---TFKSFYFP----MSIDKKLSFFCLAG--YT----TESGRQEE---QTTCTT-EG---WS---------PE-PRC
            R 2   CTKP-DLS--NG----YISD---VKLLYKI-QENMH------YGCASG--YK----TTGGKDEE---VVQCLS-DG---WS---------SQ-PTC
            R 3   CLAP-ELY--NGN---YSTTQ---K-TFKV------KDK-VQYECATG-YY-----TAGGKKTE---EVECLT-YG---WS----------LTPKC
            R 4   CSSL-RLI-ENG----YFH---PVKQTY------EEGDV-VQFFCHEN-YY-L-----SGSDLI-----QCYN-FG---WY---------PESPVC
            R 5   CPPPP-LP-INS------KIQTHS-TTYR------HGE-IVHIECELN--FEI-----HGSAEI----R-CE--DGK--WT----------EPPKC
            R 6   CEEPPFI--ENGA-A-----NLHSKIYYN-------GDK-VTYACKSG--YLL-----HGSNEI-----TCN--RGK--WT----------LPPEC
            R 7   CKHPPVV-M-NGAVA------DGILASYAT------GSS-VEYRCNE--YYLL-----RGSKIS----R-CE--QGK--WS----------SPPVC
            R 8   CTVNVDY-M-N--------RNNIEMKWKYEGKV--LHGDLI-DFVCKQG--YDLSPLT-PLSELS----VQCN--RGEV--------------KYPLC
            R 9   CTSPP-LIK-HG------VIISSTVDTYE------NGSS-VEYRCFDHHF--L-----EGSRE-----AYCL--DGM--WT-----------TPPLC
            R10   CTL-SFTEME--------KNNLLLKWDFDNRPHILHGEYI-EFIC-RGDTYPA-ELYITGSILR----MQCD--RG---------------QLKYPRC
β2GP I      R 1   CPKPDDLP--------FS-TVVPLKTFYEP------GEEI-TYSCKPG--YVS-----RGGMRK----FICPL-TGL--WP---------INTL-KC
            R 2   CPFAGIL--ENGA-VRY------TTFEYP--------NTI-SFSCNTG-FY-L-----NG-ADC----AKCTE-EGK--WS---------PELPVC
            R 3   CPPPS-IPTF-ATLRVYKP-SAGNNSLYR--------DTAV-FECLPQ--HAM-----FGNDTI-----TNTT-HGN--WT-----------KLPEC
            R 4   CPFPSR-PD-NGF-VNY-P--AKPTLYYK--------DKA-TFGCHDG--YSL-----DGPEEI-----ECTK-LGN--WS-----------AMPSC
            R 5   CKVPVKKA-----TVVYQGERVKIQEKFKN--GMLHGDK-VSFFCK-N-KEKK------CSYTED----AQCI--DG----T----------IEVPKC
Factor B    R 1   CSLEGV--EIKG-------------GS-FRLLQE---GQ-ALEYVCPSG-FYP-------YPVQT-----RTCRS-TGS--WSTLKTQDQKTVRKAEC
            R 2   CPRPHDF--ENGE---YWPR----SPYYNV------SDEI-SFHCYDG--YTL-----RGSAN-----RTCQV-NGR--WS----------GQTAIC
            R 3   CSNPG-IPI--G------TR--KVGSQYRL------EDS-VTYHCSRG--LTL-----RGSQR-----RTCQE-GGS--WS----------GTEPSC
CR 1        F 1                                                                     KCQA-LNK--WE----------PELPSC
            F 2                      RDKDNFSP------GQE-VFYSCEPG--YDL-----R
            F 3                                     K-VDFVCDEG--FQL-----KGS
C1r         R 1   CPQPKTL-DE------F-TIIQNLQPQYQFR------DYF-IATCKQG--YQLI----EGNQVLHSFTAVCQD-DGT--WH----------RAMPRC
            R 2   CGQPRNLP--NGD-FRYTTTMGVN--TYKAR---------I-QYYCHEP-YYKMQ--TRAGSRESEQGVYTCTA-QGI--WKNE----QKGEKIPRC
C1s         R 1   -PKP-KEDTPNS----VWEPA---KAKYVFR------DV-VQITCLDG--FEVVE----GRVGA
            R 2                                                   YTCEEP-YYYME----NGGGG----EYHCAG-NGS--WVNE----VLGPELPKC
R IL-2            CREPPPW--ENEA-----------TERIYHFVV----GQ-MVYYQCVQG-YRAL----HRGPAES------VCKMTHGKTRWT---------QPQL-IC
Haptoglobin       CPKPPEIA--HG---------------YVE-----HS---VRYQCK-N-YYKLR--T-EG-DG----VYTLNN--EKQ-WINKA----VGDKLPEC
```

Figure 2. Alignment of the amino acid sequences of GP-I sructures in the b subunit of human factor XIII (F.XIII b), β_2-glycoprotein I (β_2-GP I), complement factor B (Factor B), the type I receptor for complement C3b/C4b (CR 1), complement Clr, complement Cls, the receptor for interleukin-2 (R IL-2), and haptoglobin. R refers to the repeat in each protein and F refers to a fragment of a repeat. Seven or more identical residues in the same position are boxed. Gaps were inserted to obtain optimal alignment.

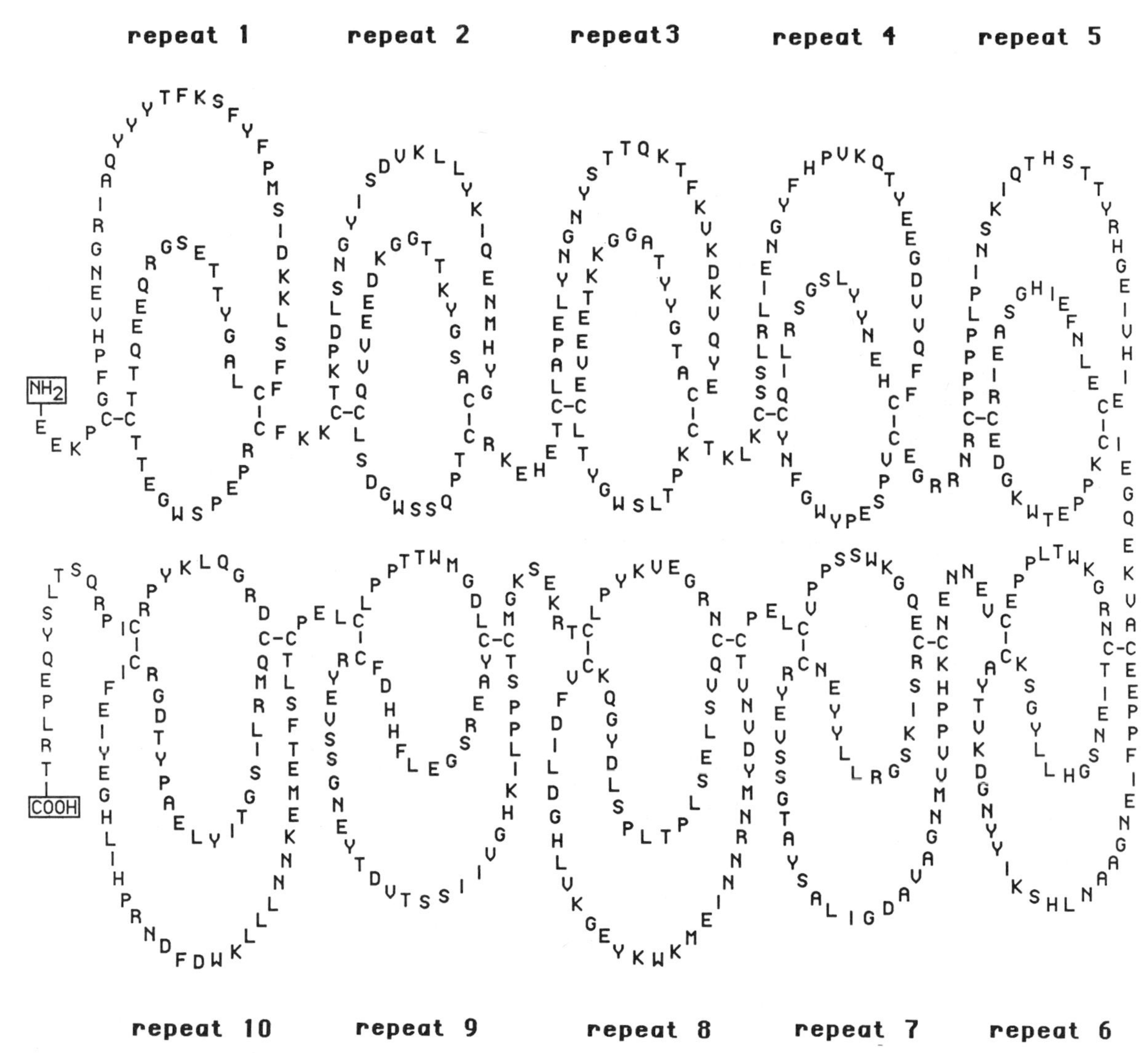

Figure 3. Amino acid sequence and proposed structure for the b subunit of human coagulation factor XIII.

Table 1. Alignment Scores for Comparisons between GP-I Segments of Various Proteins

First segment (residues)		Second segment (residues)		Alignment score (S.D. units)
XIIIb	(193–247)	β_2-GP-I	(186–241)	9.8
XIIIb	(193–247)	F.B	(78–133)	8.4
XIIIb	(316–369)	R IL-2	(125–184)	7.1
XIIIb	(376–430)	Clr	(309–371)	6.7
XIIIb	(376–430)	HP	(15–68)	4.6
β_2-GP-I	(4–60)	F.B	(78–133)	10.8
β_2-GP-I	(65–118)	Clr	(376–447)	7.6
β_2-GP-I	(65–118)	R IL-2	(125–184)	6.0
β_2-GP-I	(186–241)	HP	(15–68)	5.6
F.B	(78–133)	R IL-2	(125–184)	7.6
F.B	(78–133)	HP	(15–68)	4.3
F.B	(140–193)	Clr	(309–371)	7.3
Clr	(309–371)	R IL-2	(125–184)	5.4
Clr	(376–447)	HP	(15–68)	8.5
R IL-2	(125–184)	HP	(15–68)	3.7

Abbreviations: (XIIIb) The b subunit of factor XIII; (β_2-GP-I) β_2-glycoprotein I; (F.B) complement factor B; (R IL-2) the receptor for interleukin-2; (HP) haptoglobin.

Tack (1986) have reported the presence of 20 of the GP-I domains as tandem repeats in complement protein H. The repetitive units are also present in complement C2 (Bentley and Campbell 1985) and complement C4 binding protein (Chung et al. 1985). Thus, 11 different proteins have been identified that contain the GP-I repeats. Accordingly, it is evident that these proteins are members of another distinct superfamily.

It is noteworthy that these proteins share a similar protein-binding function, in addition to their amino acid sequence homology. Furthermore, the GP-I repeat in haptoglobin is homologous to the kringles of plasminogen (Kurosky et al. 1980), and the kringles share homology with the type II structure of fibronectin (Patthy et al. 1984). Also, the kringle and type II structures participate as protein-binding modules. In comparing the GP-I, kringle, and type II structures, it is evident that the GP-I structure lacks the lower part of the kringle region, whereas the type II structure has a

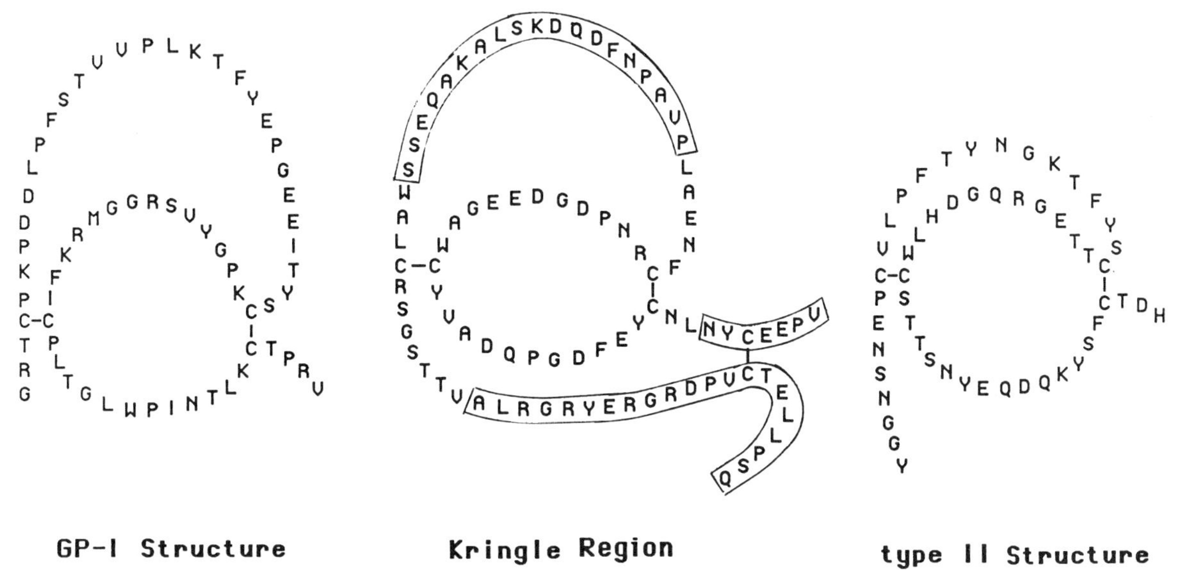

Figure 4. Structures of the GP-I domain from β_2-glycoprotein I (Lozier et al. 1984), a kringle region from prothrombin (Magnusson et al. 1975), and the type II homology from fibronectin (Petersen et al. 1983; Skorstengaard et al. 1984). The upper and lower boxed regions in the kringle are absent in the type II structure, whereas the lower boxed region is absent in the GP-I structure.

deletion of the upper and lower parts of the kringle structure. These deleted regions have been boxed in the kringle structure shown in Figure 4. Accordingly, GP-I, kringle, and type II structures may have evolved from a common ancestral protein-binding module.

Complement Clr is a plasma protein that is closely related to the proteins participating in blood coagulation. It is also a multidomain protein containing five domains linked to a serine protease domain (Leytus et al. 1986). The first and third domains are duplications of approximately 110 amino acids, and these sequences are similar to the amino-terminal region of complement Cls. The fourth and fifth domains are homologous to GP-I, and the second domain is homologous to epidermal growth factor. Accordingly, it represents another of the plasma proteins from the complement family that has evolved from a series of translocation events involving segments of DNA coding for specific regions or domains.

ACKNOWLEDGMENTS

We thank Dr. K. Walsh for his help in the computer search and Drs. K. Fujikawa, D. Chung, W. Kane, D. Foster, and K. Kurachi for helpful discussions. We also thank Dr. J. Herriott and B. McMullen for their assistance in the operation of the Vax computer. This study was supported by research grant HL-16919 from the National Institutes of Health.

REFERENCES

Bentley, D.R. and R.D. Campbell. 1985. Primary structure of C2 and relationship to other components of the complement system. *Complement* 2: 9.

Blake, C.C.F. 1978. Do genes-in-pieces imply proteins-in-pieces? *Nature* **273:** 267.

———. 1979. Exons encode protein functional units. *Nature* **277:** 598.

Chung, D.W., K. Fujikawa, B.A. McMullen, and E.W. Davie. 1986. Human plasma prekallikrein, a zymogen to a serine protease that contains four tandem repeats. *Biochemistry* **25:** 2410.

Chung, L.P., D.R. Bentley, and K.B.M. Reid. 1985. Molecular cloning and characterization of the cDNA coding for C4b-binding protein, a regulatory protein of the classic pathway of the human complement system. *Biochem. J.* **230:** 133.

Church, W.R., R.L. Jernigan, J. Toole, R.M. Hewick, J. Knopf, G.J. Knutson, M.E. Nesheim, K.G. Mann, and D.N. Fass. 1984. Coagulation factors V and VIII and ceruloplasmin constitute a family of structurally related proteins. *Proc. Natl. Acad. Sci.* **81:** 6934.

Davie, E.W. 1986. Introduction of the blood coagulation cascade and cloning of the blood coagulation factors. *J. Protein Chem.* (in press).

Fass, D.N., R.M. Hewick, G.J. Knutson, M.E. Nesheim, and K.G. Mann. 1985. Internal duplication and sequence homology in factor V and VIII. *Proc. Natl. Acad. Sci.* **82:** 1688.

Foster, D. and E.W. Davie. 1984. Characterization of a cDNA coding for human protein C. *Proc. Natl. Acad. Sci.* **81:** 4766.

Fujikawa, K., D.W. Chung, L.E. Hendrickson, and E.W. Davie. 1986. Amino acid sequence of human factor XI, a blood coagulation factor with four tandem repeats that are highly homologous with plasma prekallikrein. *Biochemistry* **25:** 2417.

Gilbert, W. 1978. Why genes in pieces? *Nature* **271:** 501.

Gray, A., T.J. Dull, and A. Ullrich. 1983. Nucleotide sequence of epidermal growth factor cDNA predicts a 128,000-molecular weight protein precursor. *Nature* **303:** 722.

Gunzler, W.A., G.J. Steffens, F. Otting, S.-M.A. Kim, E. Frankus, and L. Flohe. 1982. The primary structure of high molecular mass urokinase from human urine. The complete amino acid sequence of the A chain. *Hoppe-Seyler's Z. Physiol. Chem.* **363:** 1155.

Hagen, F.S., C.L. Gray, P. O'Hara, F.J. Grant, G.C. Saari, R.G. Woodbury, C.E. Hart, M. Insley, W. Kisiel, K. Kurachi, and E.W. Davie. 1986. Characterization of a cDNA coding for human factor VII. *Proc. Natl. Acad. Sci.* **83:** 2412.

Ichinose, A., B.A. McMullen, K. Fujikawa, and E.W. Davie. 1986a. Amino acid sequence of the b subunit of human factor XIII, a protein composed of ten repetitive segments. *Biochemistry* **25:** 4633.

Ichinose, A., L.E. Hendrickson, K. Fujikawa, and E.W. Davie. 1986b. The amino acid sequence of the a subunit of human factor XIII. *Biochemistry* **25:** (in press).

Kane, W.H. and E.W. Davie. 1986. Cloning of a cDNA coding for human factor V, a blood coagulation factor homologous to factor VIII and ceruloplasmin. *Proc. Natl. Acad. Sci.* **83:** 6800.

Katayama, K., L.H. Ericsson, D.L. Enfield, K.A. Walsh, H. Neurath, E.W. Davie, and K. Titani. 1979. Comparison of amino acid sequence of bovine coagulation factor IX (Christmas Factor) with that of other vitamin K-dependent plasma proteins. *Proc. Natl. Acad. Sci.* **76:** 4990.

Kristensen, T. and B.F. Tack. 1986. Murine protein H is composed of 20 repeating units, 61 amino acids in length. *Proc. Natl. Acad. Sci.* **83:** 3963.

Kurosky, A., D.R. Barnett, T.-H. Lee, B. Touchstone, R.E. Hay, M.S. Arnott, B.H. Bowman, and W.M. Fitch. 1980. Covalent structure of human haptoglobin: A serine protease homolog. *Proc. Natl. Acad. Sci.* **77:** 3388.

Leonard, W.J., J.M. Depper, G.R. Crabtree, S. Rudikoff, J. Pumphrey, R.J. Robb, M. Kronke, P.B. Svetlik, N.J. Peffer, T.A. Waldmann, and W.C. Greene. 1984. Molecular cloning and expression of cDNAs for the human interleukin-2 receptor. *Nature* **311:** 626.

Leytus, S.P., K. Kurachi, K.S. Sakariassen, and E.W. Davie. 1986. Nucleotide sequence of the cDNA coding for human complement Clr. *Biochemistry* **25:** 4855.

Lozier, J., N. Takahashi, and F.W. Putnam. 1984. Complete amino acid sequence of human plasma β_2-glycoprotein I. *Proc. Natl. Acad. Sci.* **81:** 3640.

Magnusson, S., T.E. Petersen, L. Sottrup-Jensen, and H. Claeys. 1975. Complete primary structure of prothrombin: Isolation, structure and reactivity of ten carboxylated glutamic acid residues and regulation of prothrombin activation by thrombin. *Cold Spring Harbor Conf. Cell Proliferation* **2:** 123.

Magnusson, S., L. Sottrup-Jensen, T.E. Petersen, G. Dudek-Wojciechowska, and H. Claeys. 1976. Homologous "kringle" structures common to plasminogen and prothrombin. Substrate specificity of enzymes activating prothrombin and plasminogen. In *Proteolysis and physiological regulation* (ed. D.W. Ribbons and K. Brew), p. 203. Academic Press, New York.

McMullen, B.A. and K. Fujikawa. 1985. Amino acid sequence of the heavy chain of human α-factor XIIa (activated Hageman factor). *J. Biol. Chem.* **260:** 5328.

Mole, J.E., J.K. Anderson, E.A. Davison, and D.E. Woods. 1984. Complete primary structure for the zymogen of human complement factor B. *J. Biol. Chem.* **259:** 3407.

Nikaido, T., A. Shimizu, N. Ishida, H. Sabe, K. Teshigawara, M. Maeda, T. Uchiyama, J. Yodoi, and T. Honjo. 1984. Molecular cloning of cDNA encoding human interleukin-2 receptor. *Nature* **311:** 631.

Ny, T., F. Elgh, and B. Lund. 1984. The structure of the human tissue-type plasminogen activator gene: Correlation of intron and exon structures to functional and structural domains. *Proc. Natl. Acad. Sci.* **81:** 5355.

Patthy, L. 1985. Evolution of the proteases of blood coagulation and fibrinolysis by assembly from modules. *Cell* **41:** 657.

Patthy, L., M. Trexler, Z. Vali, L. Banyai, and A. Varadi. 1984. Kringles: Modules specialized for protein binding. Homology of the gelatin-binding region of fibronectin with the kringle structures of proteases. *FEBS Lett.* **171:** 131.

Pennica, D., W.E. Holmes, W.J. Kohr, R.N. Harkins, G.A. Vehar, C.A. Ward, W.F. Bennett, E. Yelverton, P.H. Seeburg, H.L. Heyneker, D.V. Goeddel, and D. Collen. 1983. Cloning and expression of human tissue-type plasminogen activator cDNA in *E. coli. Nature* **301:** 214.

Petersen, T.E., H.C. Thogersen, K. Skorstengaard, K. Vibe-Pedersen, P. Sahl, L. Sottrup-Jensen, and S. Magnusson. 1983. Partial primary structure of bovine plasma fibronectin: Three types of internal homology. *Proc. Natl. Acad. Sci.* **80:** 137.

Sadler, J.E., B.B. Shelton-Inloes, J.M. Sorace, J.M. Harlan, K. Titani, and E.W. Davie. 1985. Cloning and characterization of two cDNAs coding for human von Willebrand factor. *Proc. Natl. Acad. Sci.* **82:** 6394.

Savage, C.R., T. Inagami, and S. Cohen. 1972. The primary structure of epidermal growth factor. *J. Biol. Chem.* **247:** 7612.

Scott, J., M. Urdea, M. Quiroga, R. Sanchez-Pescador, N. Fong, M. Selby, W.J. Rutter, and G.I. Bell. 1983. Structure of mouse submaxillary messenger RNA encoding epidermal growth factor and seven related proteins. *Science* **221:** 236.

Shelton-Inloes, B.B., K. Titani, and E. Sadler. 1986. cDNA sequences for human von Willebrand factor reveal five types of repeated domains and five possible protein sequence polymorphisms. *Biochemistry* **25:** 3164.

Skorstengaard, K., H.C. Thogersen, and T.E. Petersen. 1984. Complete primary structure of the collagen-binding domain of bovine fibronectin. *Eur. J. Biochem.* **140:** 235.

Spycher, S.E., H. Nick, and E.E. Rickli. 1986. Human complement Cls. Partial sequence determination of the heavy chain and identification of the peptide bond cleaved during activation. *Eur. J. Biochem.* **156:** 49.

Stenflo, J. and P. Fernlund. 1982. Amino acid sequence of the heavy chain of bovine protein C. *J. Biol. Chem.* **257:** 12180.

Titani, K., K. Fujikawa, D.L. Enfield, L.H. Ericsson, K.A. Walsh, and H. Neurath. 1975. Bovine factor X_1 (Stuart factor): Amino acid sequence of heavy chain. *Proc. Natl. Acad. Sci.* **72:** 3082.

Titani, K., S. Kumar, K. Takio, L.H. Ericksson, R.D. Wade, K. Ashida, K.A. Walsh, M.W. Chopek, J.E. Sadler, and K. Fujikawa. 1986. Amino acid sequence of human von Willebrand factor. *Biochemistry* **25:** 3171.

Toole, J.J., J.L. Knopf, J.M. Wozney, L.A. Sultzman, J.L. Buecker, D.D. Pittman, R.J. Kaufman, E. Brown, C. Shoemaker, E.C. Orr, G.W. Amphlett, W.B. Foster, M.L. Coe, G.J. Knutson, D.N. Fass, and R.M. Hewick. 1984. Molecular cloning of a cDNA encoding human antihaemophilic factor. *Nature* **312:** 342.

Vehar, G.A., B. Keyt, D. Eaton, H. Rodriguez, D.P. O'Brien, F. Rotblat, H. Oppermann, R. Keck, W.I. Wood, R.N. Harkins, E.G.D. Tuddenham, R.M. Lawn, and D.J. Capon. 1984. Structure of human factor VIII. *Nature* **312:** 337.

Wong, W.W., L.B. Klickstein, J.A. Smith, J.H. Weis, and D.T. Fearon. 1985. Identification of a partial cDNA clone for the human receptor for complement fragments C3b/C4b. *Proc. Natl. Acad. Sci.* **82:** 7711.

Yoshitake, S., B.G. Schach, D.C. Foster, E.W. Davie, and K. Kurachi. 1985. Nucleotide sequence of the gene for human factor IX (antihemophilic factor B). *Biochemistry* **24:** 3736.

Young, C.L., W.C. Barker, C.M. Tomaselli, and M.O. Dayhoff. 1978. Serine proteases. In *Atlas of protein sequence and structure* (ed. M.O. Dayhoff), vol. 5, suppl. 3, p. 73. The National Biological Research Foundation, Washington, D.C.

Cloning of cDNA and Genomic DNA for Human von Willebrand Factor

J.E. Sadler,* B.B. Shelton-Inloes,* J.M. Sorace,* and K. Titani†
*Howard Hughes Medical Institute, Departments of Medicine and Biological Chemistry, Washington University, St. Louis, Missouri 63110; †Department of Biochemistry, University of Washington, Seattle, Washington 98195

Human von Willebrand factor (vWF) is an essential hemostatic glycoprotein that is found in platelet α-granules, in subendothelial connective tissue, and in plasma at a concentration of 10 μg/ml. It is synthesized by endothelial cells (Jaffe et al. 1973, 1974) and also by megakaryocytes (Nachman et al. 1977; Sporn et al. 1985). The biosynthesis of vWF is quite complicated. The primary translation product appears to be a polypeptide of over 300,000 daltons that rapidly dimerizes in the endoplasmic reticulum by disulfide bond formation. The dimers undergo further polymerization into a series of homologous multimers ranging from the dimer up to species of over 10,000,000 daltons. Multimer formation is associated with the removal of an amino-terminal 75–100-kD peptide (Lynch et al. 1983; Wagner and Marder 1983; Fay et al. 1986), the incorporation of inorganic sulfate (Browning et al. 1983), glycosylation (Wagner and Marder 1983, 1984), and additional disulfide bond formation (Wagner et al. 1985, 1986). In endothelial cells, mature vWF is stored in unique organelles called Weibel-Palade bodies (Weibel and Palade 1964; Wagner et al. 1982). Its concentration in blood increases in response to a variety of stimuli, including estrogens, adrenergic agents, and vasopressin analogs (Bloom 1979).

vWF is not an enzyme, but participates in hemostasis through several binding interactions. It is required for normal platelet adhesion to areas of damage to the vascular endothelium, forming a bridge between platelet receptors and components of the subendothelial connective tissue. In the presence of ristocetin (an antibiotic), vWF binds to glycoprotein Ib on the platelet plasma membrane (Jenkins et al. 1976). Inherited deficiency of glycoprotein Ib, called Bernard-Soulier syndrome, causes severe bleeding that is probably due to poor platelet adhesion. Platelets that have been activated with thrombin or other platelet agonists can bind vWF through a second receptor, the glycoprotein IIb/IIIa complex. Fibrinogen and fibronectin compete with vWF for these latter sites, and the physiological significance of the vWF-glycoprotein IIb/IIIa interaction is not completely understood (Pietu et al. 1984; Haverstick et al. 1985; Plow et al. 1985). The component of the subendothelium to which vWF binds is probably collagen (Santoro 1981), although binding to collagen-free extracellular matrix has also been reported (Wagner et al. 1984). Finally, vWF binds to factor VIII, which constitutes approximately 1–2% of the mass of circulating vWF. This interaction is necessary for normal factor VIII survival (Tuddenham et al. 1982).

These binding properties account for the symptoms of inherited vWF deficiency. Individuals with severe von Willebrand disease suffer from mucocutaneous and gastrointestinal bleeding that mimics platelet dysfunction. However, if the level of vWF is sufficiently low, there may be a secondary deficiency of factor VIII, and such patients can also have soft tissue bleeding and hemarthroses that are characteristic of classical hemophilia A. This relationship between factor VIII and vWF gave rise to a confusing nomenclature in which "factor VIII" referred to either protein and also to the proposal that vWF was a precursor of factor VIII (for review, see Sadler and Davie 1986).

von Willebrand disease is one of the most common inherited bleeding disorders worldwide, with a prevalence of at least 125 per million. However, most of those affected have mild symptoms, and severe disease affects only 0.5–3 per million. In most cases, the disorder shows autosomal dominant inheritance, but in some families, only homozygous or doubly heterozygous individuals have symptoms. Pedigrees that appear to exhibit both dominant and recessive inheritance have been described, suggesting a complex interaction between the abnormal vWF genes and the genetic environment in which they occur. This genetic variability is reflected in the phenotypic heterogeneity of von Willebrand disease. Most patients appear to have a quantitative deficiency of vWF, but perhaps one fourth of the patients synthesize a qualitatively abnormal protein. This latter group can be further subdivided according to whether the vWF exhibits decreased polymerization into multimers, abnormal spacing between multimers upon gel electrophoresis, and decreased or increased sensitivity to ristocetin in platelet aggregation assays. Thus, many different genetic lesions can give rise to von Willebrand disease (for review, see Sadler and Davie 1986).

To better understand the complex biology and pathophysiology of vWF, we have recently characterized cDNA and genomic DNA clones for this protein. These studies provide a foundation for determining the genetic basis of von Willebrand disease.

EXPERIMENTAL PROCEDURES

The methods employed for endothelial cell culture, cDNA library construction and screening, DNA sequencing, RNA preparation, Northern blot analysis, and the computer analysis of DNA and protein sequences are presented in Sadler et al. (1985) and Shelton-Inloes et al. (1986). Conditions for the screening of a human genomic DNA bacteriophage λ library and characterization of positive isolates were carried out as described by Malinowski et al. (1984).

RESULTS AND DISCUSSION

cDNA Cloning

Human umbilical vein endothelial cells synthesize vWF in culture (Jaffe et al. 1973, 1974). Using this tissue as the source of poly(A)$^+$RNA, cDNA was synthesized by conventional methods, employing Klenow and reverse transcriptase for second-strand synthesis, and S1 nuclease for hairpin loop cleavage. The cDNA was used to prepare a λgt11 library, which was screened with an affinity-purified rabbit antibody to human vWF. Among the 2.5 million recombinants screened, two positives were plaque-purified and characterized. One, λHvWF1 (Fig. 1), contained a 404-bp insert, with one continuous open reading frame. By comparison with the amino acid sequence derived from native vWF and from an amino-terminal cyanogen bromide peptide, this isolate was shown to encode the amino-terminal 110 residues of the mature vWF subunit and 24 residues of precursor peptide. The second isolate, λHvWF3, contained a 4.9-kb insert (Fig. 1), with an open reading frame of 4576 nucleotides, a stop codon, a 3′-noncoding region of 134 nucleotides, and a poly(A) tail of over 150 nucleotides. The translated amino acid sequence of this isolate contained that determined independently for several cyanogen bromide peptides of vWF (Sadler et al. 1985; Titani et al. 1986).

The inserts from these initial isolates were employed to rescreen the cDNA library by hybridization, and several additional positives were identified. Two of these, λHvWF2 and λHvWF4, have been sequenced (Fig. 1). The 556-bp cDNA insert of λHvWF2 encoded amino acids 22–206 of the vWF subunit. Similarly, the 1.9-kb cDNA insert of λHvWF4 specified amino acids 18–661 (Shelton-Inloes et al. 1986). A second human umbilical vein endothelial λgt11 cDNA library, constructed using a different method of cDNA synthesis (Gubler and Hoffman 1983), has been screened with the cDNA insert of λHvWF4. Several isolates have now been partially sequenced that completely span the remaining sequence encoding the vWF precursor and at least 240 nucleotides of the 5′-noncoding region. One of these, λHvWF5, is shown in Figure 1 (B.B. Shelton-Inloes and J.E. Sadler, in prep.). In summary, overlapping cDNA inserts have been characterized that encompass approximately 9.0 kb of vWF cDNA sequence. This correlates well with the size of the mRNA estimated by Northern blotting of 8.5 kb (Fig. 2) and is consistent with the results of cDNA cloning and Northern blotting reported by other investigators (Ginsburg et al. 1985; Lynch et al. 1985; Verweij et al. 1985).

DNA and Amino Acid Sequence Comparisons

Analysis of the nucleotide and translated amino acid sequences of vWF has revealed several remarkable structural features of the protein and also has permitted some predictions concerning its biosynthetic processing. The primary translation product appears to have a typical cleavable signal peptide of 22 amino acids (Fig. 3), and the subsequent predicted sequence exactly matches that determined by amino acid sequencing for the amino-terminal 31 residues of von Willebrand antigen II (vWAgII) (B.B Shelton-Inloes et al., unpubl.). Similar findings have recently been reported by Fay et al. (1986). vWAgII is a 75–100-kD plasma protein of unknown function that is deficient or absent in von Willebrand disease (Montgomery and Zimmerman 1978). Like vWF, vWAgII is synthesized by endothelial cells and is found in Weibel-Palade bodies as well as platelets. However, vWF and vWAgII are immuno-

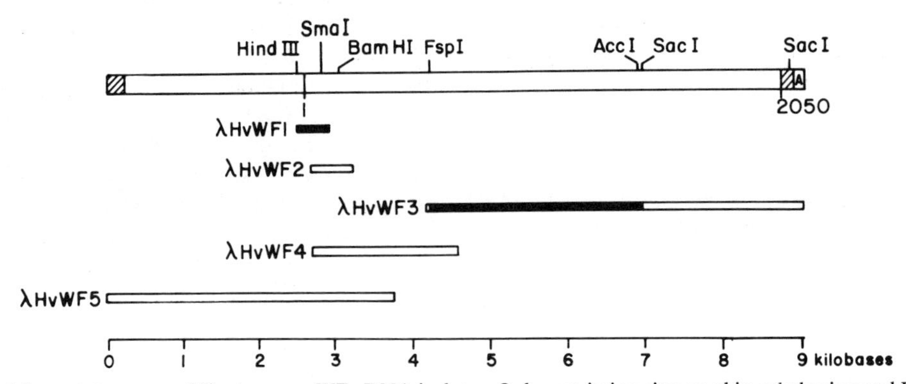

Figure 1. Partial restriction map of five human vWF cDNA isolates. Only restriction sites used in subcloning and DNA sequencing are shown. In the summary map at the top of the figure, the hatched regions represent 5′- and 3′-noncoding sequences, and the segment at the 3′ terminus labeled *A* represents the poly(A) tail. The black regions of λHvWF1 and λHvWF3 represent fragments employed in cDNA screening and chromosome localization, as described in the text. The scale is in kilobases.

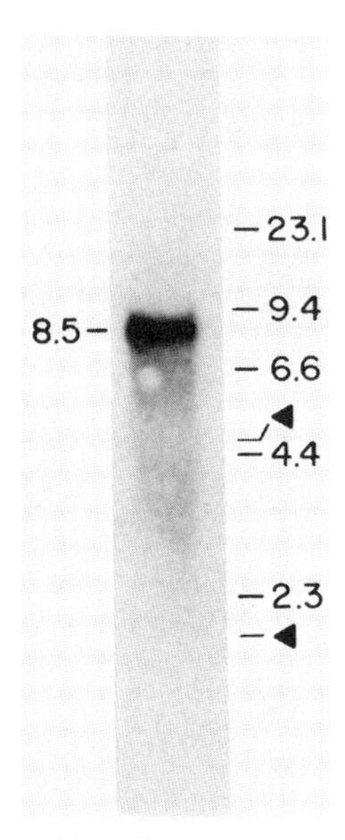

Figure 2. Northern blot of RNA from cultured human umbilical vein endothelial cells probed with human vWF cDNA. The probe was derived from isolate λHvWF1. Arrowheads indicate the mobility of bovine 28S and 18S ribosomal RNA, and the size of denatured DNA markers is shown in kilobases. (Reprinted, with permission, from Shelton-Inloes et al. 1986.)

chemically distinct in plasma (McCarroll et al. 1985). The cDNA sequence clarifies the relationship between these two proteins. The vWAgII polypeptide is apparently cleaved from the mature vWF subunit in the Golgi apparatus (Wagner and Marder 1984). Whether this processing is necessary for the assembly of functional multimers is not known. Perhaps 1% of all vWF subunits secreted by cultured endothelial cells still contain the vWAgII polypeptide (Wagner and Marder 1984). Because this fraction is similar to the factor VIII content of the circulating factor VIII–vWF complex, it is tempting to speculate that vWAgII might bind to factor VIII.

Proteolytic processing at other sites can be excluded by comparing the amino acid sequence predicted from the cDNA with that determined by protein sequencing. In particular, the carboxy-terminal cyanogen bromide fragment has been identified as the only peptide lacking homoserine lactone at its carboxyl terminus (Titani et al. 1986), and the nucleotide sequence encoding this peptide is followed by a stop codon in the cDNA (Fig. 3). Thus, all proteolytic processing during the biosynthesis of vWF must occur solely at the amino-terminal end of the precursor molecule.

The amino acid sequence of the vWF precursor around the amino-terminal serine residue of the mature subunit is His-Arg-Ser-Lys-Arg-*Ser*-Leu (Fig. 3). Thus, the vWAgII propeptide is separated from vWF by cleavage after a Lys-Arg dipeptide. Cleavage after a Lys-Arg or Arg-Arg dipeptide is characteristic of the posttranslational processing for many protein precursors (Table 1). The amino acid at position −4 is frequently a basic residue, and a mutation of arginine to glutamic acid at this site in the factor IX precursor appears to prevent processing and to cause hemophilia B

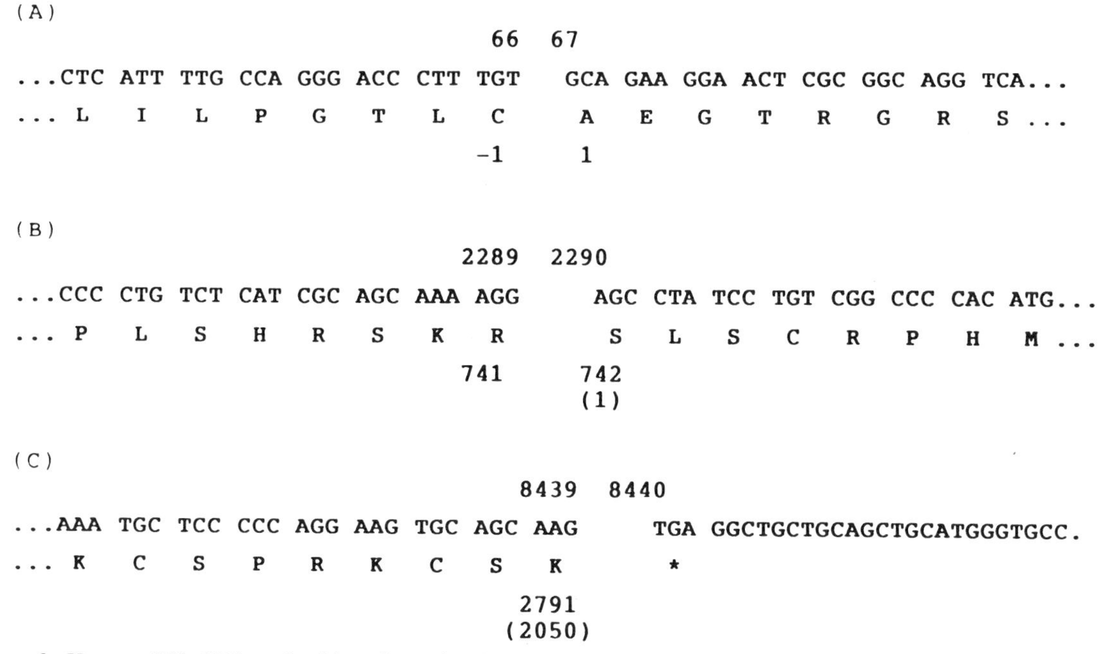

Figure 3. Human vWF cDNA nucleotide and translated amino acid sequence. The partial sequences shown encode the junction between the signal peptide and vWAgII (*A*), the amino terminus of the mature subunit, at the site of proteolytic processing that removes the vWAgII propeptide (*B*), and the carboxyl terminus of the subunit (*C*). Nucleotide numbering begins with the first nucleotide of the initiator methionine. Amino acid numbering begins with the first residue of the pro-vWF polypeptide so that signal peptide residues have negative numbers. Numbers in parentheses refer to the position relative to the amino-terminal serine of the mature vWF subunit.

Table 1. Protein Precursors That Are Posttranslationally Processed after Paired Basic Residues

Protein precursor		Sequence at cleavage site		
		−4 −3 −2 −1	+1	
Human von Willebrand factor	−4	Arg-Ser-Lys-Arg	Ser	+1
Human serum albumin	−4	Val-Phe-Arg-Arg	Asp	+1
Human apolipoprotein A-II	−4	Leu-Val-Arg-Arg	Gln	+1
Human factor IX	−4	Arg-Pro-Lys-Arg	Tyr	+1
Human prothrombin	−4	Arg-Val-Arg-Arg	Ala	+1
Human protein C				
leader site	−4	Ile-Ser-Lys-Arg	Ala	+1
interchain site	+154	His-Leu-Lys-Arg	Asp	+158
Human proinsulin				
A–C chain site	+53	Lys-Thr-Arg-Arg	Glu	+57
C–B chain site	+86	Leu-Gln-Lys-Arg	Gly	+90
Human parathyroid hormone	+3	Val-Lys-Lys-Arg	Ser	+7
Human β-nerve growth factor	+163	Arg-Ser-Lys-Arg	Ser	+167
Rat calcitonin	+81	Arg-Ser-Lys-Arg	Cys	+85
Sindbis virus polyprotein (protein E2 site)	+325	Arg-Ser-Lys-Arg	Ser	+329
Aspergillus awamori glucoamylase	−4	Ile-Ser-Lys-Arg	Ala	+1
Yeast mating factor-α polyprotein	+82	Leu-Asp-Lys-Arg	Glu	+86
(four sites total, two shown)	+101	Met-Tyr-Lys-Arg	Glu	+105

Sequences were obtained from the NBRF Protein Sequence Database, except for human apolipoprotein A-II (Gordon et al. 1983), human prothrombin (Degen et al. 1983), human protein C (Foster and Davie 1984; Beckmann et al. 1985), *A. awamori* glucoamylase (Innis et al. 1985), and human vWF (Sadler et al. 1985). The position of amino acids relative to the site of cleavage is indicated by the headings, −4 to +1.

(Bentley et al. 1986). The list in Table 1 is not exhaustive; in particular, many hormone precursors are not included. Nevertheless, the similarities among these sites from distantly related eukaryotes suggest that a conserved mechanism may be employed to generate the mature secreted products. A mutation in yeast, *kex2*, has been characterized that abolishes the processing of the mating factor-α precursor (Julius et al. 1984). To date, no homologous protease activity has been identified with certainty in higher eukaryotes.

As determined by both protein and cDNA sequencing, the mature vWF subunit consists of 2050 amino acids. Comparisons among the coding regions of all the currently reported cDNA sequences (Ginsburg et al. 1985; Lynch et al. 1985; Sadler et al. 1985; Verweij et al. 1985; Shelton-Inloes et al. 1986) show a total of eight single nucleotide differences, of which four alter the translated protein sequence. These are 26-Ala/Thr, 89-Arg/Gln, 94-Asn/Asp, 618-Ala/Thr, with the predominant amino acid found by protein sequencing given first (Titani et al. 1986). Another potential polymorphism is suggested by the discrepancy between the translated cDNA sequence (histidine) and the protein sequence (proline) at residue 7. Polymorphism at residue 26 has been confirmed by the identification of both alanine and threonine in a ratio of approximately 4:1 at that position by protein sequencing (Titani et al. 1986). Whether the remaining discrepancies represent polymorphisms or artifacts of cDNA cloning is unknown.

There are a total of 13 potential N-linked glycosylation sites with the sequence Asn-X-Thr/Ser in the vWF subunit, of which 11 are glycosylated. The glycosylation site at Asn-94 would be abolished by the potential polymorphism at that residue discussed above. One additional asparagine, in the unusual sequence Asn(384)-Ser-Cys, is also glycosylated (Titani et al. 1986). The location of ten O-linked glycosylation sites, eight on threonine and two on serine residues, has also been determined (Titani et al. 1986).

Repeated Domains in the vWF Protein Sequence

There are five unrelated types of repeated domains (A–E) in the vWF subunit, and the distribution of these sequence elements is shown schematically in Figure 4. The A domains are composed of 193–220 amino acids and are present in three imperfect tandem copies between residues 497 and 1111. The B domains are composed of 25–35 amino acids and are triplicated between residues 1533 and 1636. The C domains contain 116–119 amino acids and are present in two copies between residues 1637 and 1899. In contrast to these tandemly arranged repeated sequences, the two copies of domain D, consisting of 270–289 amino acids, are separated by 804 residues that contain all three A domains. Similarly, the duplicated E domains each contain 46 amino acids and are separated by 1383 residues that include the A and D domains. The degree of sequence identity between pairs of repeated segments varies from 23% to

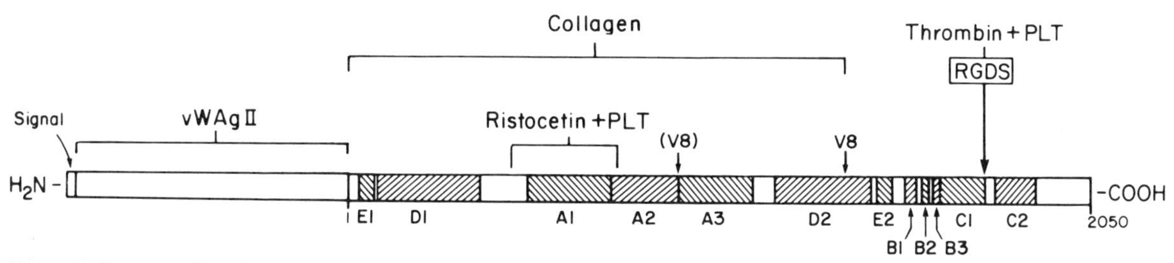

Figure 4. Structure-function relationships for the human vWF precursor. The position of the signal peptide and von Willebrand antigen II propeptide (vWAgII) relative to the mature subunit is shown. Numbers below the schematic subunit indicate the amino-terminal serine (1) and the carboxy-terminal lysine (2050). Two principal sites of staphylococcal V8-protease cleavage are indicated by arrows, with the minor site in parentheses. The hatched areas show the extent of internal sequence duplications, or domains, discussed in the text. The locations of peptide fragments that bind to platelet glycoprotein Ib (*Ristocetin + PLT*), collagen (*Collagen*), and platelet glycoprotein IIb/IIIa (*Thrombin + PLT*) are shown, as is the position of the Arg-Gly-Asp-Ser (RGDS) tetrapeptide that may mediate the latter binding activity.

43%. The A domains are notably poor in cysteine, containing a total of 6 (1%) compared to 169 in the entire subunit (8%). The remaining domains (B–E) are quite rich in cysteine. Altogether, repeated domains account for over 75% of the sequence of mature vWF, indicating that the evolution of this protein has required many separate gene segment duplications (Sadler et al. 1985; Shelton-Inloes et al. 1986).

Possible Homology between vWF and Complement Factor B

Comparison of the vWF cDNA sequence with entries in the Genbank Genetic Sequence Data Bank has not uncovered any similar sequences. However, examination of the National Biomedical Research Foundation Protein Sequence Database has revealed one protein that may be homologous to vWF. A 225-residue segment of complement factor B can be aligned with each of the three A domains of vWF, as shown in Figure 5. Complement factor B is a serine protease zymogen in the alternate complement pathway. It is activated by factor D, which cleaves between Arg-234 and Lys-235 to generate the factor Ba and Bb peptides. The carboxy-terminal Bb peptide contains the serine protease domain, and participates in the C3 and C5 convertase complexes of the alternate pathway. The segment of factor B that is similar to vWF includes five amino acids of the Ba peptide and 220 amino acids from the amino terminus of the Bb peptide prior to the protease domain. This segment is particularly variable among serine proteases, and the vWF A domains do not show significant similarity to the corresponding regions of other related proteases. The genomic organization of vWF shows some resemblance to that of factor B (Fig. 5). The first two A domains of vWF are encoded mostly by one exon. However, the third A domain and the aligned segment of factor B are encoded by five exons, and except for one site in the factor B gene, the intron-exon boundaries occur in roughly corresponding positions (Campbell et al. 1984; Sorace et al. 1986).

The function of complement component C2 in the classical pathway is analogous to that of factor B in the alternate pathway, and these two proteases are homologous. Twenty-five residues of the component C2a sequence that follow the activation cleavage site have been reported (Parkes et al. 1983), and that sequence also appears to align well with the vWF A domains. However, this segment of component C2 is too short to permit a strong assertion of homology to vWF.

Efforts to demonstrate functional similarity between complement factor B and vWF have been unsuccessful. Neither factor B nor the Bb peptide inhibits the ristocetin-induced binding of vWF to platelets. Conversely, vWF inhibits neither the activation of factor B by factor D nor the factor-B-dependent generation of C3/C5 convertase activity (E.J. Brown and J.E. Sadler, unpubl.). The apparently significant alignment of these protein sequences and their possibly conserved genomic organization may reflect descent from a common ancestor. If so, the biological significance of this relationship is unknown.

Structure-Function Relationships for vWF

Determination of the primary structure of vWF, by a combination of protein sequencing and cDNA cloning, allows the assignment of some binding activities to specific amino acid sequences. These structure-function correlations are summarized in Figure 4.

Digestion of the intact protein with staphylococcal V8 protease yields two major fragments, an amino-terminal homodimer of 170-kD peptides and a carboxy-terminal homodimer of 100-kD peptides. The amino-terminal fragment retains the ability to bind to both glycoprotein Ib of platelets (Girma et al. 1986a) and to collagen (Fressinaud et al. 1985). The segment that interacts with glycoprotein Ib has been further localized to a tryptic fragment containing residues 449–729 (Fujimura et al. 1986; Titani et al. 1986). This fragment also can be isolated as a homodimer. Thus, at least one of the seven cysteine residues in this sequence must form an intersubunit disulfide bond (Titani et al. 1986). Most of this segment is encoded by a single exon that specifies amino acids 463–921 (J.M. Sorace and J.E.

```
           1       5         10        15        20        25        30        35        40        45        50        55        60        65
A1   497   E D I S E - P P L H D F Y - C S R L L D L V F L L D G S S R L S E A E F E V L K A F V V D M M E R L R I S Q K W V R V A V V E Y H
A2   717   P G L L G V S T L G P K R N - S M V L D V A F V L E G S D K I G E A D F N R S K E F M E E V I Q R M D V G Q D S I H V T V L Q Y -
A3   910   E G L - Q I P T L S P A P D C S Q P L D V I L L L D G S S S F P A S Y F D E M K S F A K A F I S K A N I G P R L T Q V S V L Q Y -
FB   228 PGE Q Q K R K I V L D P - - S G S M - - N I Y L V L D G S D S I G A S N F T G A K K C L V N L I E K V A S Y G V K P R Y G L V T Y -
           E               L   P         S M           L V L D G S D S I G A S   F       K       V       I E K                 R     V   Y

           66      70        75        80        85        90        95        100       105       110       115       120       125       130
A1   560   D G S - - - - H A Y I G L K D R K R - P S E - L R R I A S Q V K Y A G S Q V - - - A S T S E V L K Y T L - - - - F Q I F S K I D -
A2   780   - - S Y M V T V E Y - P F S E A Q S - K G D I L Q R V - R E I R Y Q G G N R - - - T N T G L A L R Y - L S D H S F L V - S Q G D -
A3   973   - G S - - I T T I D V P W N V V P E - K A H L L S L V D V M Q R - E G G P S Q I G D A L G F A V R Y - L T - - - - - - - S E M H G
FB   290   - A T Y - - P K I W V K V S E A D S S N A D W V T K Q L N E I N Y E D H K L K S G T N T K K A L Q A V Y S M M S W P D D V P P E G
                 Y         I   V     S E A   S     A D               E I   Y E             G T N T     A L         S     S                 G

           131     135       140       145       150       155       160       165       170       175       180       185       190       195
A1   611   - R P E A S R - I A L L L M A - - - - - - - - - - - S Q E P Q R M S R N F V R Y V Q G L K K K K V I V I P V G I G P H A N L K Q I
A2   834   - R E Q A P N - L V Y M V T G - - - - - - - - N P A S D E I K R L P G - - - - - - - - - - - - D I Q V V P I G V G P N A N V Q E L
A3  1025   A R P G A S K A V V I L V T D V S V D S V D A A A D A A - - - - - - - - - - - - - - - - R S N R V T V F P I G I G D R Y D A A Q L
FB   352   W N R T - - R H V I I L M T D G L H N M G G D P I T V I D E I R D L L Y I G K D R K N P R E D Y L D V Y V F G V G P L V N Q V N I
                       R   V   I L   T D                           D E I R   L                     R           V       G V G P     N       I

           196     200       205       210       215       220       225       230       235       240       245       250
A1   663   R L I E - - K Q A P E N K A F V L S S V D E L E Q Q R D E I V S Y L C D L A P E A P P P T L P P D M A Q V T V G
A2   877   E R I - G W P N A P - - - - - I L - - I Q D F E - - - - - - - - - - - T L P R E A P - - D L V - - L Q R C C S G
A3  1074   - R I L A G P A G D S N - - - V V K - L Q R I E - - - - - - - - - - - D L P T M V T L G N S F - - L H K L C S G
FB   415   N A - L A S K K D N E Q H V F K V K D M E N L E - - - - - - - - - - - - - - D V F Y Q M I D E S Q S L S L C - G
                 L A   K       E       F   V K         L E                                                         L C   G
```

Figure 5. Alignment of human vWF A domains with complement factor B. Amino acid numbering of the three A domains (A1, A2, A3) is relative to the amino-terminal serine of the plasma vWF subunit. Numbering of the complement factor B peptide (FB) was done according to Campbell et al. (1984). Residues that are identical between any pair of sequences are shown in boldface, and conserved residues are underlined. The fifth line shows residues that are common to factor B and at least one A domain in boldface. Arrows indicate the positions of intron-exon boundaries. Methods employed for this analysis are described in Shelton-Inloes et al. (1986).

Sadler, unpubl.). The collagen-binding site has not yet been further localized.

The carboxy-teminal V8 protease fragment binds to the glycoprotein IIb/IIIa complex of thrombin-activated platelets (Girma et al. 1986b). Both fibronectin and fibrinogen compete with vWF for this site, and the binding activity of fibronectin has been localized to a tetrapeptide with the sequence Arg-Gly-Asp-Ser. The same sequence mediates the cell attachment of fibronectin and several other adhesive glycoproteins (for review, see Ruoslahti and Pierschbacher 1986). Small peptides containing Arg-Gly-Asp-Ser inhibit the binding of all three proteins to activated platelets (Gartner and Bennett 1985; Haverstick et al. 1985; Plow et al. 1985), and the same tetrapeptide is found between residues 1744 and 1747 of vWF, at the carboxyl terminus of the first C domain (Fig. 4) (Sadler et al. 1985). Thus, this segment is probably responsible for the binding of vWF to activated platelets. Fibronectin, fibrinogen, and vWF do not appear to be homologous, and if this activity is physiologically important, the presence of the Arg-Gly-Asp-Ser sequence in these unrelated proteins would represent convergent evolution. Factor VIII binding activity has not yet been reported for any fragment of vWF.

Characterization of the vWF Gene

Probes derived from λHvWF3 (Fig. 1) have been employed to screen a human genomic DNA library in the Charon 4A bacteriophage λ (Lawn et al. 1978). Among 600,000 recombinants, 8 unique positive isolates have been characterized by restriction mapping, Southern blotting, and DNA sequencing. These isolates represent approximately 60 kb of the vWF gene, corresponding to about 3.3 kb of the cDNA insert of λHvWF3. Ten intron-exon boundaries have been placed within the cDNA sequence. Eight of these define six exons that encode amino acid residues 463–922, 922–961, 961–1008, 1008–1055, 1055–1111, and 1867–1899. The two remaining intron-exon boundaries occur at residues 1185 and 1942/1943. Southern blotting of human genomic DNA suggests at least eight additional exons must exist (Sorace et al. 1986). There is a limited degree of correspondence between the position of the known intron-exon boundaries and the apparent repeated domain structure of the vWF subunit. Most of domains A1 and A2 are encoded in a single 1.4-kb exon, whereas domain A3 is encoded by five exons (Fig. 5). Splice junctions occur 13 amino acids from the domain A2–A3 boundary, 15 amino acids from the amino-terminal end of domain D2, at the carboxy-terminal end of domain A3 (residue 1111), and also at the carboxy-terminal end of domain C2 (residue 1899) (Sorace et al. 1986).

The identification of exons by Southern blotting of genomic DNA is complicated by the presence of at least two loci in the human genome that hybridize to one segment of the cDNA. Both the cDNA insert of λHvWF1 and the *Sac*I-*Sac*I fragment from the 3′ end of λHvWF3 hybridize only to chromosome 12 (F. Chehab and J.E. Sadler, unpubl.). This agrees with the localization of the vWF gene to chromosome 12 reported by other investigators (Ginsburg et al. 1985; Lynch et al. 1985; Verweij et al. 1985). However, the *Fsp*I-*Sac*I fragment from the 5′ end of λHvWF3 hybridizes both to chromosome 12 and to chromosome 22 under conditions of very high stringency (F. Chehab and J.E. Sadler, unpubl.). The related sequence on chromosome 22 has not yet been further characterized, but it is not the complement component C2 or factor B genes, which are on chromosome 6 (Campbell et al. 1984).

Molecular Defects in von Willebrand Disease

Ginsburg et al. (1985) examined DNA from two patients with severe von Willebrand disease by Southern blotting. Using a nearly full-length cDNA probe, the observed *Bam*HI hybridization patterns were indistinguishable from those of a normal control. In collaboration with A. Federici and P.M. Manucci (University of Milan, Italy), we have examined leukocyte DNA from 17 patients with typical autosomal-dominant von Willebrand disease (type I) and 19 patients with severe von Willebrand disease (type III). Preliminary results show that two unrelated patients with type II disease have large gene deletions corresponding to at least the 6.5 kb of cDNA contained in λHvWF1, λHvWF3, and λHvWF4. In each case, the bands due to related sequences on chromosome 22 appear normal. Interestingly, these are the only patients studied that have alloantibodies to transfused vWF. All of the remaining patients appear to have normal *Eco*RI hybridization patterns with all probes (B.B. Shelton-Inloes et al., in prep.).

SUMMARY

The recent isolation of cDNA and genomic DNA clones for human vWF by ourselves and others has finally laid to rest the historical notion that factor VIII and vWF might have a precursor-product or other complex relationship. These two hemostatic activities are clearly present on two distinct proteins, each encoded by a separate gene. Whether there is coordinate regulation of the factor VIII and vWF genes is still unknown. The structure and structure-function relationships of the vWF protein have been elucidated by many investigators, only some of whom can be cited in this short paper. Together, these molecular biology and protein chemistry studies have shown that vWAgII is the amino-terminal propeptide of vWF, explaining the proportional deficiency of these two plasma proteins in von Willebrand disease (type I). In addition, sites of proteolytic processing, glycosylation, and disulfide bond formation have been defined, and some of the binding functions of mature vWF have been identified with specific amino acid sequences. The protein has a highly repeated structure, suggesting a complex evolutionary history. The A domains of vWF appear to be homologous to complement factor B, and perhaps to component C2, although the biological meaning of this similarity is unknown. Genomic DNA clones have been isolated corresponding to approximately one third of the vWF gene on chromosome 12, and a fragment of the cDNA also hybridizes to uncharacterized sequences on chromosome 22. Preliminary studies in von Willebrand disease have revealed two patients with very large deletions in the vWF gene. Continuing studies will yield an increasingly detailed picture of the biosynthesis, structure-function relationships, and genetic organization and evolution of vWF, as well as a better understanding of the genetic basis of von Willebrand disease.

ACKNOWLEDGMENTS

We thank Earl W. Davie for his interest and encouragement and George J. Broze for determining the amino acid sequences of vWAgII. This work was supported in part by research grant HL-29595 (to K.T.) from the National Institutes of Health. K.T. is also a visiting professor of the Fujita-Gakuen Health University, Nagoya, Japan.

REFERENCES

Beckmann, R.J., R.J. Schmidt, R.F. Santerre, J. Plutzky, G.R. Crabtree, and G.L. Long. 1985. The structure and evolution of a 461 amino acid human protein C precursor and its messenger RNA, based upon the DNA sequence of cloned human liver cDNAs. *Nucleic Acids Res.* **13:** 5233.

Bentley, A.K., D.J.G. Rees, C. Rizza, and G.G. Brownlee. 1986. Defective propeptide processing of blood clotting factor IX caused by mutation of arginine to glutamine at position −4. *Cell* **45:** 343.

Bloom, A.L. 1979. The biosynthesis of factor VIII. *Clin. Haematol.* **8:** 53.

Browning, P.J., E.H. Ling, T.S. Zimmerman, and D.C. Lynch. 1983. Sulfation of von Willebrand factor by human umbilical vein endothelial cells. *Blood* **65:** 218a.

Campbell, R.D., D.R. Bentley, and B.J. Morley. 1984. The factor B and C2 genes. *Philos. Trans. R. Soc. Lond. B* **306:** 367.

Degen, S.J.F., R.T.A. MacGillivray, and E.W. Davie. 1983. Characterization of the complementary deoxyribonucleic acid and gene coding for human prothrombin. *Biochemistry* **22:** 2087.

Fay, P.J., Y. Kawai, D.D. Wagner, D. Ginsburg, D. Bonthron, B.M. Ohlsson-Wilhelm, S.I. Chavin, G.N. Abraham, R.I. Handin, S.H. Orkin, R.R. Montgomery, and V.J. Marder. 1986. Propolypeptide of von Willebrand factor circulates in blood and is identical to von Willebrand antigen II. *Science* **232:** 995.

Foster, D. and E.W. Davie. 1984. Characterization of a cDNA coding for human protein C. *Proc. Natl. Acad. Sci.* **81:** 4766.

Fujimura, Y., K. Titani, L.Z. Holland, S.R. Russell, J.R. Roberts, J.H. Elder, Z.M. Ruggeri, and T.S. Zimmerman. 1986. Von Willebrand factor. A reduced and alkylated 52/48-kDa fragment beginning at amino acid residue 449 contains the domain interacting with platelet glycoprotein Ib. *J. Biol. Chem.* **261:** 381.

Gartner, T.K. and J.S. Bennett. 1985. The tetrapeptide analogue of the cell attachment site of fibronectin inhibits platelet aggregation and fibrinogen binding to activated platelets. *J. Biol. Chem.* **260:** 11891.

Ginsburg, D., R.I. Handin, D.T. Bonthron, T.A. Donlon, G.A.P. Bruns, S.A. Latt, and S.H. Orkin. 1985. Human von Willebrand factor (vWF): Isolation of complementary DNA (cDNA) clones and chromosomal localization. *Science* **228:** 1401.

Girma, J.P., M.W. Chopek, K. Titani, and E.W. Davie. 1986a. Limited proteolysis of human von Willebrand factor by *Staphylococcus aureus* V-8 protease: Isolation and partial characterization of a platelet-binding domain. *Biochemistry* **25:** 3156.

Girma, J.P., M. Kalafatis, G. Pietu, J.-M. Lavergne, M.W. Chopek, T.S. Edgington, and D. Meyer. 1986b. Mapping of distinct von Willebrand factor domains interacting with platelet GPIb and GPIIb/IIIa and with collagen using monoclonal antibodies. *Blood* **67:** 1356.

Gordon, J.I., K.A. Budelier, H.F. Sims, C. Edelstein, A.M. Scanu, and A.W. Strauss. 1983. Biosynthesis of human preproapolipoprotein A-II. *J. Biol. Chem.* **258:** 14054.

Gubler, U. and B.J. Hoffman. 1983. A simple and very efficient method for generating cDNA libraries. *Gene* **25:** 263.

Haverstick, D.M., J.F. Cowan, K.M. Yamada, and S.A. Santoro. 1985. Inhibition of platelet adhesion to fibronectin, fibrinogen, and von Willebrand factor substrates by a synthetic tetrapeptide derived from the cell binding domain of fibronectin. *Blood* **66:** 946.

Innis, M.A., M.J. Holland, P.C. McCabe, G.E. Cole, V.P. Wittman, R. Tal, K.W.K. Watt, D.H. Gelfand, J.P. Holland, and J.H. Meade. 1985. Expression, glycosylation, and secretion of an *Aspergillus* glucoamylase by *Saccharomyces cerevisiae*. *Science* **228:** 21.

Jaffe, E.A., L.W. Hoyer, and R.L. Nachman. 1973. Synthesis of antihemophilic factor antigen by cultured human endothelial cells. *J. Clin. Invest.* **52:** 2757.

———. 1974. Synthesis of von Willebrand factor by cultured human endothelial cells. *Proc. Natl. Acad. Sci.* **71:** 1906.

Jenkins, C.S.P., D.R. Phillips, K.J. Clemetson, D. Meyer, M.-J. Larrieu, and E.F. Luscher. 1976. Platelet membrane glycoproteins implicated in ristocetin-induced aggregation. *J. Clin. Invest.* **57:** 112.

Julius, D., A. Brake, L. Blair, R. Kunisawa, and J. Thorner. 1984. Isolation of the putative structural gene for the lysine-arginine-cleaving endopeptidase required for processing of yeast prepro-α-factor. *Cell* **37:** 1075.

Lawn, R.M., E.F. Fritsch, R.C. Parker, G. Blake, and T. Maniatis. 1978. The isolation and characterization of a linked δ- and β-globin gene from a cloned library of human DNA. *Cell* **15:** 1157.

Lynch, D.C., R. Williams, T.S. Zimmerman, E.P. Kirby, and D.M. Livingston. 1983. Biosynthesis of the subunits of factor VIIIR by bovine aortic endothelial cells. *Proc. Natl. Acad. Sci.* **80:** 2738.

Lynch, D.C., T.S. Zimmerman, C.J. Collins, M. Brown, M.J. Morin, E.H. Ling, and D.M. Livingston. 1985. Molecular cloning of cDNA for human von Willebrand factor: Authentication by a new method. *Cell* **41:** 49.

McCarroll, D.R., E.G. Levin, and R.R. Montgomery. 1985. Endothelial cell synthesis of von Willebrand antigen II, von Willebrand factor, and von Willebrand factor/von Willebrand antigen II complex. *J. Clin. Invest.* **75:** 1089.

Malinowski, D.P., J.E. Sadler, and E.W. Davie. 1984. Characterization of a complementary deoxyribonucleic acid coding for human and bovine plasminogen. *Biochemistry* **23:** 4243.

Montgomery, R.R. and T.S. Zimmerman. 1978. von Willebrand's disease antigen II: A new plasma and platelet antigen deficient in severe von Willebrand's disease. *J. Clin. Invest.* **61:** 1498.

Nachman, R., R. Levine, and E.A. Jaffe. 1977. Synthesis of factor VIII antigen by cultured guinea pig megakaryocytes. *J. Clin. Invest.* **60:** 914.

Parkes, C., J. Gagnon, and M.A. Kerr. 1983. The reaction of iodine and thiol-blocking reagents with human complement components C2 and factor B. *Biochem. J.* **213:** 201.

Pietu, G., G. Cherel, G. Marguerie, and D. Meyer. 1984. Inhibition of von Willebrand factor-platelet interaction by fibrinogen. *Nature* **308:** 648.

Plow, E.F., M.D. Pierschbacher, E. Ruoslahti, G.A. Marguerie, and M.H. Ginsburg. 1985. The effect of Arg-Gly-Asp-containing peptides on fibrinogen and von Willebrand factor binding to platelets. *Proc. Natl. Acad. Sci.* **82:** 8057.

Ruoslahti, E. and M.D. Pierschbacher. 1986. Arg-Gly-Asp: A versatile cell recognition signal. *Cell* **44:** 517.

Sadler, J.E. and E.W. Davie. 1986. Hemophilia A, hemophilia B, and von Willebrand disease. In *Molecular basis of blood diseases* (ed. G. Stamatoyannopoulos et al.). Saunders, Philadelphia. (In press.)

Sadler, J.E., B.B. Shelton-Inloes, J.M. Sorace, J.M. Harlan, K. Titani, and E.W. Davie. 1985. Cloning and characterization of two cDNAs coding for two human von Willebrand factor. *Proc. Natl. Acad. Sci.* **82:** 6394.

Sakariassen, K.S., E. Fressinaud, J.-P. Girma, H.R. Baumgartner, and D. Meyer. 1986. Mediation of platelet adhesion to fibrillar collagen in flowing blood by a proteolytic fragment of human von Willebrand factor. *Blood* **67:** 1515.

Santoro, S.A. 1981. Absorption of von Willebrand factor/factor VIII by the genetically distinct interstitial collagens. *Thromb. Res.* **21:** 689.

Shelton-Inloes, B.B., K. Titani, and J.E. Sadler. 1986. cDNA sequences for human von Willebrand factor reveal five types of repeated domains and five possible protein sequence polymorphisms. *Biochemistry* **25:** 3164.

Sorace, J.M., B.B. Shelton-Inloes, and J.E. Sadler. 1986. Isolation and characterization of genomic clones for human von Willebrand factor. *Fed. Proc.* **45:** 1639.

Sporn, L.A., S.I. Chavin, V.J. Marder, and D.D. Wagner. 1985. Biosynthesis of von Willebrand protein by human megakaryocytes. *J. Clin. Invest.* **76:** 1102.

Titani, K., S. Kumar, K. Takio, L.H. Ericsson, R.D. Wade, K. Ashida, K.A. Walsh, M.W. Chopek, J.E. Sadler, and K. Fujikawa. 1986. Amino acid sequence of human von Willebrand factor. *Biochemistry* **25:** 3171.

Tuddenham, E.G.D., R.S. Lane, F. Rotblat, A.J. Johnson, T.J. Snape, S. Middleton, and P.B.A. Kernoff. 1982. Response to infusions of polyelectrolyte fractionated human factor VIII concentrate in human hemophilia A and von Willebrand's disease. *Br. J. Haematol.* **52:** 259.

Verweij, C.L., C.J.M. de Vries, B. Distel, A.-J. van Zonneveld, A.G. von Kessel, J.A. van Mourik, and H. Pannekoek. 1985. Construction of cDNA coding for human von Willebrand factor using antibody probes for colony-screening and mapping of the chromosomal gene. *Nucleic Acids Res.* **13:** 4699.

Wagner, D.D. and V.J. Marder. 1983. Biosynthesis of von Willebrand protein by human endothelial cells. Identification of a large precursor polypeptide chain. *J. Biol. Chem.* **258:** 2065.

———. 1984. Biosynthesis of von Willebrand protein by human endothelial cells: Processing steps and their intracellular localization. *J. Cell Biol.* **99:** 2123.

Wagner, D.D., T. Mayadas, and V.J. Marder. 1986. Initial glycosylation and acidic pH in the Golgi apparatus are required for multimerization of von Willebrand factor. *J. Cell Biol.* **102:** 1320.

Wagner, D.D., J.B. Olmsted, and V.J. Marder. 1982. Immunolocalization of von Willebrand protein in Weibel-Palade bodies of human endothelial cells. *J. Cell Biol.* **95:** 355.

Wagner, D.D., M. Urban-Pickering, and V.J. Marder. 1984. von Willebrand protein binds to extracellular matrices independently of collagen. *Proc. Natl. Acad. Sci.* **81:** 471.

Wagner, D.D., T. Mayadas, M. Urban-Pickering, B.H. Lewis, and V.J. Marder. 1985. Inhibition of disulfide bonding of von Willebrand protein by monensin results in small, functionally defective multimers. *J. Cell Biol.* **101:** 112.

Weibel, E.R. and G.E. Palade. 1964. New cytoplasmic components in arterial endothelia. *J. Cell Biol.* **23:** 101.

Structure and Evolution of the Human Genes Encoding Protein C and Coagulation Factors VII, IX, and X

G.L. LONG*

Division of Molecular and Cell Biology, Lilly Research Laboratories, Indianapolis, Indiana 46285

The blood-clotting cascade in animals contains several vitamin-K-dependent serine proteases, including factors VII, IX, and X, prothrombin, and protein C. Figure 1 shows schematically the position of these proteins in the intrinsic coagulation pathway. Factor VII (not shown in Fig. 1) affects clotting through the extrinsic pathway by the activation of factor X.

Protein C is involved in the regulation of the clotting cascade (Stenflo 1984). At the site of clot formation, the circulating inactive protein-C zymogen precursor is converted to an active serine protease by limited thrombin cleavage. Activated protein C specifically degrades two nonenzymatic protein cofactors in the coagulation cascade, Va and VIIIa, thereby serving as an on-demand, feedback down-regulator of both the intrinsic and extrinsic pathways of coagulation. Human protein C has been purified and partially characterized (Kisiel 1979).

Recently, the entire amino acid sequence of the human protein-C precursor has been reported (Beckmann et al. 1985), based on the cDNA sequence. The precursor consists of five distinct domains (Fig. 2): leader peptide (amino acids -42 to -1), a γ-carboxyglutamate (Gla) segment (amino acids 1–45), two epidermal growth-factor domains (amino acids 46–91 and 92–137), and a classical trypsin-like serine protease region (amino acids 158–419).

The human protein-C precursor undergoes extensive posttranslational modification, including vitamin-K-dependent carboxylation of nine glutamyl residues, hydroxylation of an aspartyl residue, glycosylation (23% weight), disulfide bond formation, and limited proteolysis. Circulating inactive protein C consists predominantly of a "light" ($M_r = 21,000$) chain and a "heavy" ($M_r = 41,000$) chain joined by disulfide bridging. The circulating two-chain molecule is generated by cleavage of a linking Lys-Arg dipeptide from the internal portion of the precursor single-chain protein (Beckmann et al. 1985).

*Present address: Department of Biochemistry, College of Medicine, University of Vermont, Burlington, Vermont 05405.

Comparison of Protein C with Vitamin-K-dependent Coagulation Factors

Protein C shares sequence homology with other vitamin-K-dependent coagulation factors, including factors VII, IX, and X and prothrombin (Fernlund and Stenflo 1982; Stenflo and Fernlund 1982). For example, human protein C has sequence homology throughout (34% identical amino acids overall) when compared with human coagulation factor IX (Beckmann

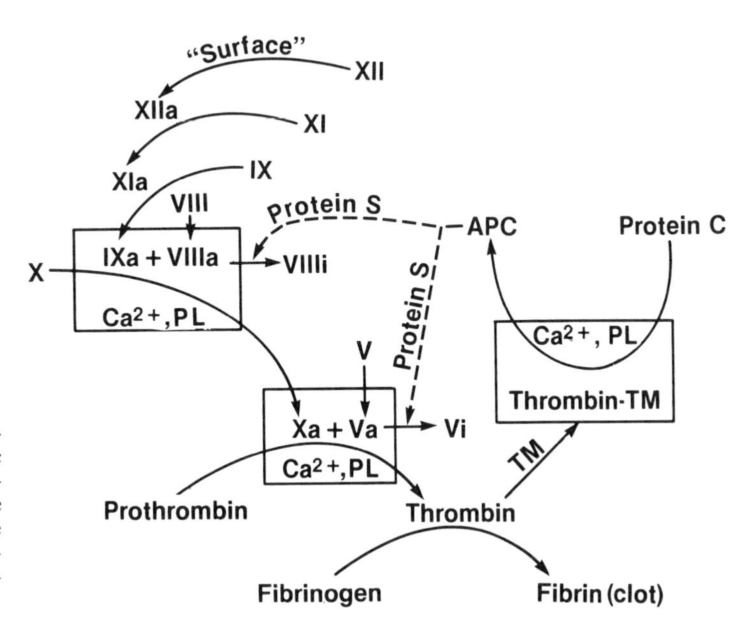

Figure 1. Intrinsic pathway for blood coagulation. Boxes enclose multicomponent enzymatic complexes on the vascular cell surface. The extrinsic pathway (not shown) enters the above path at the point of factor X activation in the presence of tissue factor, Ca^{++}, and phospholipid (PL), i.e., altered cell surface. (APC) Activated protein C; (TM) thrombomodulin.

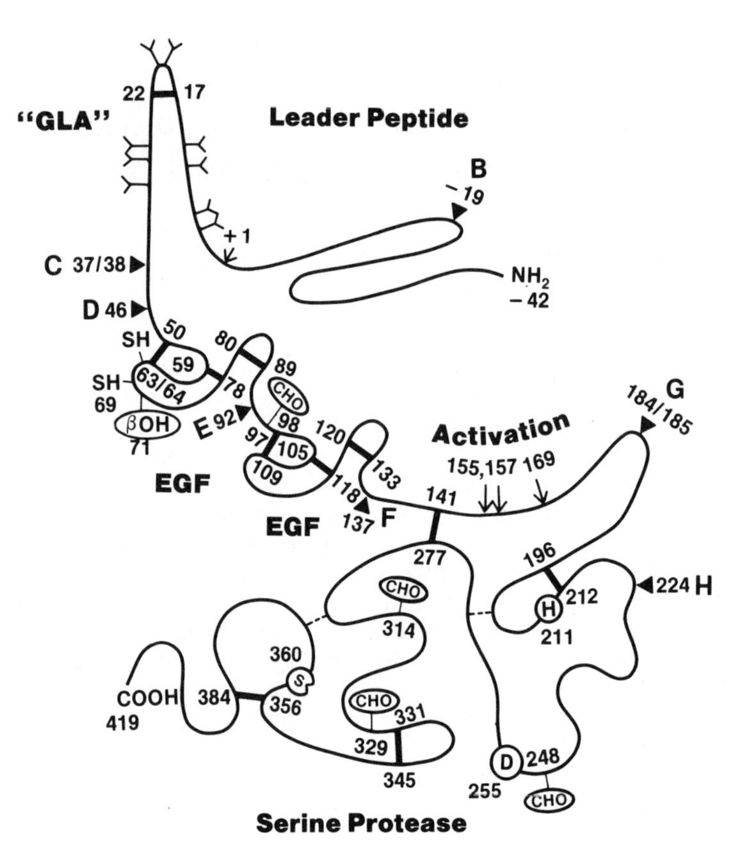

Figure 2. Schematic representation of protein C structure. The single-chain precursor is represented by the thin curving line. Numbers refer to amino acid positions in the precursor protein. For a description of the noted protein domains, see Beckmann et al. (1985) and the text. Proposed positions of the disulfide bridges (thick bars), β-hydroxyaspartate (β-OH), γ-carboxyglutamate (Y), and carbohydrate attachment (CHO) in the mature protein are also shown and are based on homology with related proteins as discussed elsewhere (Beckmann et al. 1985). The catalytic triad (Ser, His, Asp) of the serine protease domain are represented by the circled letters S, H, and D, respectively. Known proteolytic cleavage sites resulting in two-chain, activated protein C are shown with arrows. Corresponding positions of introns B–H in the gene are portrayed by closed triangles (▲) (for a detailed description, see Plutzky et al. 1986).

et al. 1985), and consequently they share the same protein domains. Because of the structural closeness of the two proteins, a comparison of the genes for human protein C and factor IX was performed (Plutzky et al. 1986). Both genes contain seven intervening segments (introns) in the protein-coding portions of the mRNA precursors, and *all* seven of the introns occur at identical positions in the two genes. Figure 2 shows the position of the introns relative to the protein sequence. The position of the introns correlates well with the protein's proposed structural domain boundaries. These results are consistent with the hypothesis put forth by Gilbert (1978) several years ago that introns may bound coding segments for distinct protein structural domains and may participate in the genetic reshuffling of common domains among many different proteins. In contrast to their positional identity, the introns for protein C and factor IX show no similarity in size, sequence, or base composition (Long 1986). All of the protein-C introns are approximately 20% richer in guanine (G) and cytosine (C) bases than the factor IX counterparts (∼60% vs ∼40%). Furthermore, the G + C content of the coding segments of the protein-C gene is 21% greater than that for factor IX (Long 1986). A comparison of the coding regions has been extended to factors VII and X, as shown in Table 1. The base composition for factor IX is distinctly different from that for the other three homologous proteins. The codon usage patterns for protein C, factor VII, and factor X are also very similar to one another but different from that for factor IX, as shown in Table 2. There is a general cytosine to thymine (T) and guanine to adenine (A) transition when comparing protein-C or factors VII and X codons with those of factor IX. As expected from a consideration of the genetic code redundancy, the greatest difference in base frequency

Table 1. Base Composition of Coding Regions

Base	Protein C	Factor VII	Factor X	Factor IX	Factor VIII	HPRT
Thymine	240	242	229	383	1807	190
Cytosine	418	405	408	249	1547	109
Adenine	287	261	358	431	2218	197
Guanine	441	427	432	323	1484	159
Total	1386	1335	1427	1386	7056	655
%G + C	62.0	62.3	58.9	41.3	43.0	41.0

cDNA sequences were taken from the following sources: protein C (Beckmann et al. 1985); factor VII (Hagen et al. 1986); factor X (Fung et al. 1985); factor IX (Anson et al. 1984); factor VIII (Wood et al. 1984); hypoxanthine-phosphoribosyltransferase (HPRT) (Jolly et al. 1983).

Table 2. Human Codon Usage for Protein C and Factors VII, IX, and X

	PC	VII	X	IX		PC	VII	X	IX		PC	VII	X	IX		PC	VII	X	IX
TTT:Phe	0	2	2	12	TCT:Ser	1	4	2	6	TAT:Tyr	1	2	2	11	TGT:Cys	7	9	11	19
TTC:Phe	15	12	19	9	TCC:Ser	7	11	8	5	TAC:Tyr	7	10	12	5	TGC:Cys	17	16	13	5
TTA:Leu	0	0	3	6	TCA:Ser	2	4	2	6	TAA:Stop	0	0	0	1	TGA:Stop	0	0	1	0
TTG:Leu	3	4	1	3	TCG:Ser	2	1	0	0	TAG:Stop	1	1	0	0	TGG:Trp	15	8	6	7
CTT:Leu	5	3	2	9	CCT:Pro	4	2	4	4	CAT:His	2	2	1	6	CGT:Arg	3	1	2	1
CTC:Leu	16	13	14	5	CCC:Pro	11	9	10	3	CAC:His	16	12	7	4	CGC:Arg	10	7	7	1
CTA:Leu	1	1	1	2	CCA:Pro	1	8	2	8	CAA:Gln	3	2	1	7	CGA:Arg	2	2	0	6
CTG:Leu	24	26	16	3	CCG:Pro	4	2	2	0	CAG:Gln	12	21	18	7	CGG:Arg	8	12	4	3
ATT:Ile	4	3	3	17	ACT:Thr	0	1	1	12	AAT:Asn	5	4	4	15	AGT:Ser	2	4	2	7
ATC:Ile	12	13	14	7	ACC:Thr	10	10	21	7	AAC:Asn	8	9	17	17	AGC:Ser	22	7	15	3
ATA:Ile	2	1	1	1	ACA:Thr	5	2	5	10	AAA:Lys	3	5	8	12	AGA:Arg	2	1	2	8
ATG:Met	8	5	7	6	ACG:Thr	3	10	9	1	AAG:Lys	21	12	25	16	AGG:Arg	3	6	11	1
GTT:Val	1	1	0	22	GCT:Ala	1	2	2	10	GAT:Asp	5	4	3	12	GGT:Gly	2	0	3	8
GTC:Val	11	12	7	3	GCC:Ala	15	14	21	5	GAC:Asp	25	15	24	7	GGC:Gly	20	23	15	8
GTA:Val	1	1	1	5	GCA:Ala	5	3	2	8	GAA:Glu	4	4	14	33	GGA:Gly	3	5	5	15
GTG:Val	16	17	4	7	GCG:Ala	3	6	3	1	GAG:Glu	30	26	30	10	GGG:Gly	10	12	15	4

Sources of data and abbreviations are the same as those noted in Table 1.

between factor IX and the other homologous genes is in the third position of codons and the smallest difference is in the second position. The largest change is a C→T transition in the third position. Possibly related to this is the fact that the most abundant base for all four genes in the first position of the codons is guanine.

DISCUSSION

A comparison of the intron positions for human protein-C, factor X (R.T.A. MacGillivray, unpubl.), and factor IX genes shows that they are identical, suggesting that the genes have a close evolutionary relationship. However, a comparison of intron size and gene base composition for protein-C and factor IX genes shows that they are significantly divergent. This disparity for factor IX also exists in coding regions when compared with protein C and factors VII and X (Tables 1 and 2).

No clear explanation for the compositional divergence of the factor IX genes exists. One possibility is based on observed high mutation rates in bacteriophage (Baltz et al. 1976), bacteria (Coulondre et al. 1978; Duncan and Weiss 1982), *Neurospora* (Selker and Stevens 1985), and eukaryotes (Taylor 1984) of cytosine to thymine (and indirectly G→A on the opposite strand) via methylation and subsequent deamination. It has been proposed (Coulondre et al. 1978; Duncan and Weiss 1982) that a polarized C→T transition results from the inability of DNA-uracil glycosylase to excise and repair mismatched thymine bases resulting from cytosine methylation and deamination (Fig. 3). In animals, the predominant (90–95%) site of methylation is at the CpG dinucleotide (Doscocil and Sorm 1962). The report on the mutation of *Neurospora* 5S RNA genes (Selker and Stevens 1985) may be particularly significant in regard to the divergence of the factor IX gene. Two of the genes, ζ and η, lie close together and are apparently the result of a tandem duplication. Within the duplicated 794-nucleotide regions, there are 113 base differences, all of which are C-T or G-A differences. Comparison with other *Neurospora* 5S RNA genes also shows that the mutations were polarized (C→T, G→A) and that the ζ and η regions are heavily methylated. The fact that within the repeated regions there are no transversion mutations (G-C, G-T, A-C, or A-T) suggests that the duplication was a recent evolutionary event, followed by the accumulation of transition mutations via methylation and mutation. In the case of the factor IX gene, methylation of cytosine residues followed by mutation to thymine could result in the compositional differences reported here. Consistent with this hypothesis are the observations that the most common transition (C-T) occurs in the third codon position, which is the most permissive position (i.e., no amino acid change), and is most frequently followed by a guanine base (first position). This situation reflects the predominant site of eukaryotic DNA methylation: CpG, resulting in the mutation product TpG (Doscosil and Sorm 1962). One consequence of this hypothesis is that the genes containing the greater amount of G-C bases (protein C, factor VII, and factor X) can be considered as more closely representing the gene duplication precursor. The similarity of composition presented above for protein-C and factors VII and X coding regions relative to that of factor IX also argues for considering the factor IX gene as being more divergent than the others from a common ancestral gene.

One additional observation that may be related to the above hypothesis is the chromosomal location of the genes for protein C and factors X, IX, and VII.

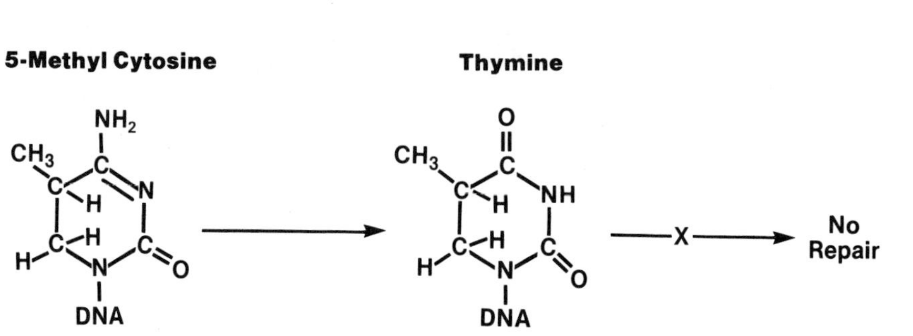

Figure 3. Mutational effect of cytosine methylation.

The human protein-C gene is located on chromosome 2 (Povey et al. 1985; S.S. Naylor and G.L. Long, unpubl.), and the factor X and factor VII genes are located on chromosome 13 (Cox and Gedde-Dahl 1985), whereas factor IX resides on the X chromosome (Camerino et al. 1984). DNA methylation appears to play an important role in X-chromosome inactivation (Gartler and Riggs 1983), and the area where factor IX resides may have undergone hypermethylation. Recently, the compositionally mosaic nature of genomic DNA in higher animals has been described (Bernardi et al. 1985; this volume). The marked differences in gene structure for highly homologous proteins reported here is an example of this phenomenon at the molecular level. In this regard, coding regions for two genes located on the X chromosome near factor IX (Drayna and White 1985) were examined (included in Table 1). Coding regions for human factor VIII (Wood et al. 1984) and hypoxanthine-phosphoribosyltransferase (Jolly et al. 1983) both have G + C contents of approximately 40%, similar to that for factor IX, and may represent parts of a common genomic compositional unit (isochore).

Although the above hypothesis may help to explain the process by which the factor IX gene has diverged, the question of why the proposed precursor form of the genes and the present form of the protein-C, factor VII, and factor X genes are so G + C-rich remains unanswered.

ACKNOWLEDGMENT

The author thanks Ms. Sherry Pike for excellent secretarial assistance.

REFERENCES

Anson, D.S., K.H. Choo, D.J.G. Rees, F. Giannelli, K. Gould, J.A. Huddleston, and G.G. Brownlee. 1984. The gene structure of human anti-haemophilic factor IX. *EMBO J.* **3:** 1053.

Baltz, R.H., P.M. Bingham, and J.W. Drake. 1976. Heat mutagenesis in bacteriophage T4: The transition pathway. *Proc. Natl. Acad. Sci.* **73:** 1269.

Beckman, R.J., R.J. Schmidt, R.F. Santerre, G. Plutzky, G.R. Crabtree, and G.L. Long. 1985. The structure and evolution of a 461 amino acid human protein C precursor and its messenger RNA, based upon the DNA sequence of cloned human liver cDNAs. *Nucleic Acids Res.* **13:** 5233.

Bernardi, G., B. Olofsson, J. Filipski, M. Zerial, J. Salinas, G. Cuny, M. Meunier-Rotival, and F. Rodier. 1985. The mosaic genome of warm-blooded vertebrates. *Science* **228:** 953.

Camerino, G., K.H. Grzeschik, M. Jaye, H. De La Salle, P. Tolstoshev, J.P. Lecocq, R. Heilig, and J.L. Mankel. 1984. Regional localization on the human X chromosome and polymorphism of the coagulation factor IX gene (hemophila B locus). *Proc. Natl. Acad. Sci.* **81:** 498.

Coulondre, C., J.H. Miller, P.J. Farabauch, and W. Gilbert. 1978. Molecular basis of base substitution hot spots in *Escherichia coli. Nature* **274:** 755.

Cox, D.R. and T. Gedde-Dahl. 1985. Human gene mapping 8: Report of the committee on the genetic constitution of chromosomes 13, 14, 15, and 16. *Cytogenet. Cell Genet.* **40:** 206.

Doscocil, J. and F. Sorm. 1962. Distribution of 5-methylcytosine in pyrimidine sequences of deoxyribonucleic acid. *Biochim. Biophys. Acta* **55:** 953.

Drayna, D. and R. White. 1985. The genetic linkage map of the human X chromosome. *Science* **230:** 753.

Duncan, B.K. and B. Weiss. 1982. Specific mutator effects of *ung* (uracil-DNA glycosylase) mutation in *Escherichia coli. J. Bacteriol.* **151:** 750.

Fung, M.R., C.W. Hay, and R.T.A. MacGillivray. 1985. Characterization of an almost full-length cDNA coding for human blood coagulation factor X. *Proc. Natl. Acad. Sci.* **82:** 3591.

Furnlund, P. and J. Stenflo. 1982. Amino acid sequence of the light chain of bovine protein C. *J. Biol. Chem.* **257:** 12170.

Gartler, S.M. and A.D. Riggs. 1983. Mammalian X-chromosome inactivation. *Annu. Rev. Genetics* **17:** 155.

Gilbert, W. 1978. Why genes in pieces? *Nature* **271:** 501.

Hagen, F.S., C.L. Gray, P. O'Hara, F.J. Grant, G.C. Saari, R.G. Woodbury, C.E. Hart, M. Insley, W. Kisiel, K. Kurachi, and E.W. Davie. 1986. Characterization of a cDNA coding for human factor VII. *Proc. Natl. Acad. Sci.* **83:** 2412.

Jolly, D.J., H. Okayama, P. Berg, A.C. Esty, D. Filpula, P. Bohlen, G.G. Johnson, J.E. Shivley, T. Hunkpillar, and T. Friedman. 1983. Isolation and characterization of a full-length expressible cDNA for human hypoxanthine phosphoribosyltransferase. *Proc. Natl. Acad. Sci.* **80:** 177.

Kisiel, W. 1979. Human plasma protein C: Isolation, characterization, and mechanism of activation by α-thrombin. *J. Clin. Invest.* **64:** 761.

Long, G.L. 1986. Structure and evolution of the human genes encoding protein C and coagulation factor IX. *J. Cell. Biochem.* (in press).

Plutzky, J., J. Hoskins, G.L. Long, and G.R. Crabtree. 1986. Evolution and organization of the human protein C gene. *Proc. Natl. Acad. Sci.* **83:** 546.

Povey, S., N.E. Morton, and S.L. Sherman. 1985. Human gene mapping 8: Report of the committee on the genetic constitution of chromosomes 1 and 2. *Cytogenet. Cell Genet.* **40:** 67.

Selker, E.U. and J.N. Stevens. 1985. DNA methylation at asymmetric sites is associated with numerous transition mutations. *Proc. Natl. Acad. Sci.* **82:** 8114.

Stenflo, J. 1984. Structure and function of protein C. *Semin. Thromb. Hemostasis* **10:** 109.

Stenflo, J. and P. Fernlund. 1982. Amino acid sequence of the heavy chain of bovine protein C. *J. Biol. Chem.* **257:** 12180.

Taylor, J.H. 1984. *DNA methylation and cellular differentiation.* Springer-Verlag, New York.

Wood, W.I., D.J. Capon, C.C. Simonsen, D.L. Eaton, J. Gitschier, B. Keyt, P.H. Seeburg, D.H. Smith, P. Hallingshead, K.L. Wion, E. Delwart, E.G.D. Tuddenham, G.A. Vehar, and R.M. Lawn. 1984. Expression of active human factor VIII from recombinant DNA clones. *Nature* **312:** 330.

Isolation and Expression of cDNAs Encoding Human Factor VII

K. BERKNER,* S. BUSBY,* E. DAVIE,† C. HART,* M. INSLEY,* W. KISIEL,‡ A. KUMAR,* M. MURRAY,* P. O'HARA,* R. WOODBURY,* AND F. HAGEN*

*ZymoGenetics, Inc., Seattle, Washington 98103; †Department of Biochemistry, University of Washington, Seattle, Washington 98195; ‡Department of Pathology, University of New Mexico, Albuquerque, New Mexico 87131

Two alternative pathways exist for the activation of factor X, a step required to convert prothrombin to thrombin in the later stages of the coagulation cascade. In the intrinsic pathway of blood coagulation, a series of protease-mediated activations ultimately generates factor IXa (see Fig. 1). This protein, coupled with factor VIIIa, produces factor Xa from factor X by cleavage of a glycopeptide from the heavy-chain amino terminus (Davie et al. 1979). An identical proteolysis is effected by factor VIIa and tissue factor in the extrinsic pathway of blood coagulation. Tissue factor, a membrane-bound protein, normally does not circulate in plasma. Upon vessel disruption, however, it can complex with factor VIIa to catalyze the factor X activation in a Ca^{++}- and phospholipid-dependent manner (Østerud and Rapaport 1977).

Factor VII, a 56K protein, normally circulates in the plasma as a single-chain zymogen at a concentration of approximately 100–500 ng/ml (Kisiel and Davie 1975; Radcliffe and Nemerson 1975; Broze and Majerus 1980; Bajaj et al. 1981). It can be activated to a two-chain form by factor Xa (Radcliffe and Nemerson 1975, 1976), factor XIIa (Kisiel et al. 1977; Radcliffe et al. 1977; Seligsohn et al. 1979; Broze and Majerus 1980), factor IXa (Seligsohn et al. 1979), or thrombin (Radcliffe and Nemerson 1975), although the physiological significance of these different activities is not clear. Like several other plasma proteins involved in hemostasis (e.g., prothrombin, factor IX, factor X, protein C, protein S, and protein Z), it is dependent on vitamin K for its activity. This cofactor is required for a modification of factor VII, i.e., the γ-carboxylation of multiple glutamic acid residues that are clustered in the amino terminus of this protein. These γ-carboxylated glutamic acids (termed glas) are required for the metal-associated interaction of factor VII with phospholipids (Jackson and Nemerson 1980; Suttie 1980).

The ability to express recombinant factor VII DNA would be of obvious value in studying its role and regulation in the coagulation cascade. Detailed analyses of the mechanism by which the factors interact in the coagulation cascade have sometimes been hampered by the presence of trace amounts of contaminating factors in highly purified preparations. The ability to gen-

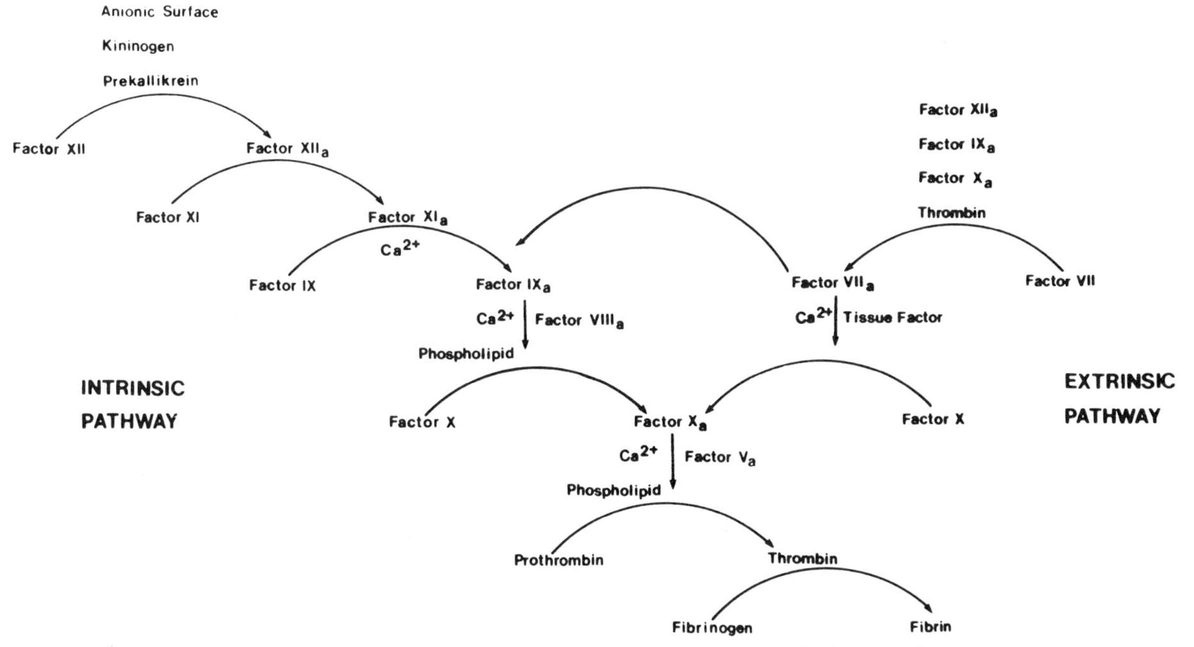

Figure 1. The Intrinsic and extrinsic pathways in the coagulation cascade.

erate homogeneous factor VII via recombinant techniques, along with the increasing availability of other recombinant plasma proteins, e.g., factor IX (Kurachi and Davie 1982; Jaye et al. 1983), factor X (Fung et al. 1984; Leytus et al. 1984), and protein C (Foster and Davie 1984; Long et al. 1984), should eventually obviate this problem. In addition, one of the most powerful applications for the recombinant factor VII gene may be in its therapeutic potential for treating factor-VII-deficient patients and possibly in treating type-A (factor-VIII-deficient) and type-B (factor-IX-deficient) hemophiliacs. The ability of factor VIIa to bypass the intrinsic pathway leading to the generation of thrombin has been tested by Hedner and Kisiel (1983). After administering multiple doses of factor VIIa to two type-A hemophiliacs who had developed alloantibodies against factor VIII, these authors observed a rapid return to normal hemostasis with both patients. Production of what is normally a trace plasma protein by recombinant means would also provide a protein source free of viral agents, such as those causing AIDS or hepatitis. For these reasons, we have attempted the isolation and expression of a cDNA encoding human factor VII, and these studies are described here.

EXPERIMENTAL PROCEDURES

Culturing and transfection of mammalian cell lines. BHK cells and 293 cells (Graham et al. 1977) were grown in Dulbecco's medium containing 10% FCS. The factor IX–BHK cells were supplemented with 0.5 mg/ml G-418 and 5 μg/ml vitamin K (phytonadione; Merck). Human hepatoma cells (ATCC HTB-52 and HTB-8065) were cultured in modified Eagle's medium containing 10% FCS, 2 mM glutamine, 0.1 mM nonessential amino acids, and 1 mM sodium pyruvate. Transient transfections were performed using either calcium phosphate precipitation (Graham and van der Eb 1973) or DEAE-dextran (Danna and Sompayrac 1982). To establish permanently expressing factor VII cell lines, recombinant factor VII was cotransfected either with a plasmid encoding the methotrexate (MTX)-resistant form of dihydrofolate reductase ($DHFR^R$) (Simonsen and Levinson 1983) or with neomycin. Two days posttransfection, cells were split at an approximate 1:10 ratio into media containing 0.1–1 μM MTX (Sigma; for cells containing the $DHFR^R$ plasmid) or 500 μg/ml G-418 (GIBCO; for cells transfected with the neomycin gene). Vitamin K (5 μg/ml) was included in the media immediately posttransfection, and colonies were isolated 10–14 days after selection was applied. To generate non-γ-carboxylated recombinant protein, cells were passaged for at least 20 generations in the absence of vitamin K.

Protein analysis. Recombinant factor IX and factor VII harvested from transiently or stably transfected cells were quantitated using an enzyme-linked immunosorbent assay (ELISA). A mouse monoclonal anti-factor VII or IX antibody was used as the catching antibody, and a rabbit polyclonal anti-factor VII or IX antibody was used for detection. An alkaline phosphatase conjugate of a goat anti-rabbit IgG fraction was used for subsequent quantitation of the antigen-antibody complex. Purified plasma factor VII or factor IX was used as a reference standard. The biological activity of factor IX, factor VII, or the fusion products (factor IX/VII-1 or 2), was determined from the ability of a sample to correct the prolonged clotting time observed with congenitally deficient factor IX or factor VII plasma (George King Biomedicals). The level of activity was based on normal plasma levels of factor VII (500 ng/ml) or factor IX (3 μg/ml).

Analysis of protein samples following gel electrophoresis (Laemmli 1970) was performed either by staining the gels directly or by immunoblotting (Towbin et al. 1979) using polyclonal rabbit antibodies. Protein samples were also analyzed following barium citrate precipitation (Bajaj et al. 1981). The amino-terminal amino acid sequence of reduced and carboxymethylated (Crestfield et al. 1963) factor VII was achieved after high-performance liquid chromatography (HPLC) fractionation and Edman degradation using a gas-phase protein sequenator.

cDNA cloning and plasmid constructions. The purification and protein sequence analysis of plasma factor VII, as well as the preparation of λgt11 libraries containing factor VII plasmids and their subsequent screening, have been described in detail (Hagen et al. 1986). The isolation and characterization of the factor VII gene will be presented elsewhere (P. O'Hara et al., in prep.). The screening described in these studies was performed primarily using dideoxy-chain termination (Sanger et al. 1977) and, in some cases, using the Maxam-Gilbert method (1980). Oligonucleotides were prepared using an Applied Biosystems DNA synthesizer, followed by gel purification.

Factor VII and factor VII derivatives (described below) were expressed in mammalian cells, using an expression vector (pDX) derived from one previously used to express factor IX (Busby et al. 1985). This vector, which will be described in detail elsewhere (K.L. Berkner et al., in prep.), contains a unique *Eco*RI insertion site that positions cDNAs behind the adenovirus-2 major late promoter and tripartite leader sequences. Two different factor VII cDNAs were inserted into this *Eco*RI site either as an *Eco*RI fragment (from λHVII 2463) or by reassembling a full-length factor VII cDNA by combining the 5′ *Eco*RI-*Bgl*II fragment from λHVII 565 with the 3′ *Bgl*II-*Eco*RI fragment from λHVII 2463 (Fig. 2).

Factor VII and factor IX hybrids, whose constructions will be described in detail elsewhere (A. Kumar et al., in prep.), were prepared using oligonucleotide-directed loop-out mutagenesis (Zoller and Smith 1984). One of these proteins, factor IX/VII-1, was constructed by adjoining the gla domain of factor IX with the serine protease portion of factor VII (Fig. 3). Factor IX/VII-1 contained the leader and the first 38

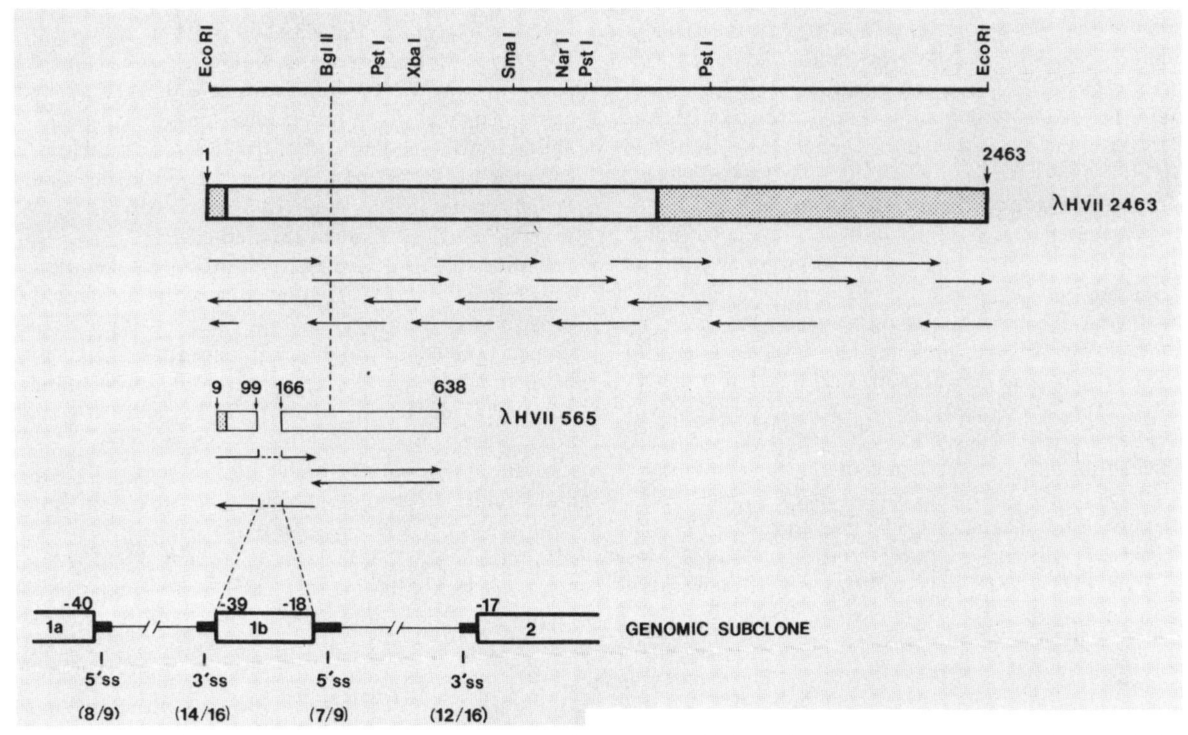

Figure 2. Genomic and cDNA sequences encoding factor VII. The two cDNAs used in these studies were derived from λHVII 2463 and λHVII 565. Open areas indicate the coding sequences and the stippled areas indicate the noncoding sequences. Each cDNA was sequenced in both directions by the dideoxy method, as indicated by the arrows. The sequences between 99 bp and 166 bp, shown above, were also found in a genomic subclone. The match of flanking sequences (indicated by the solid bars) to a 5′ or 3′ consensus splice site is indicated in parentheses for this putative exon ("1b"), as well as for exons 1a and 2. Numbers above the genomic subclone exons refer to the amino acid positions.

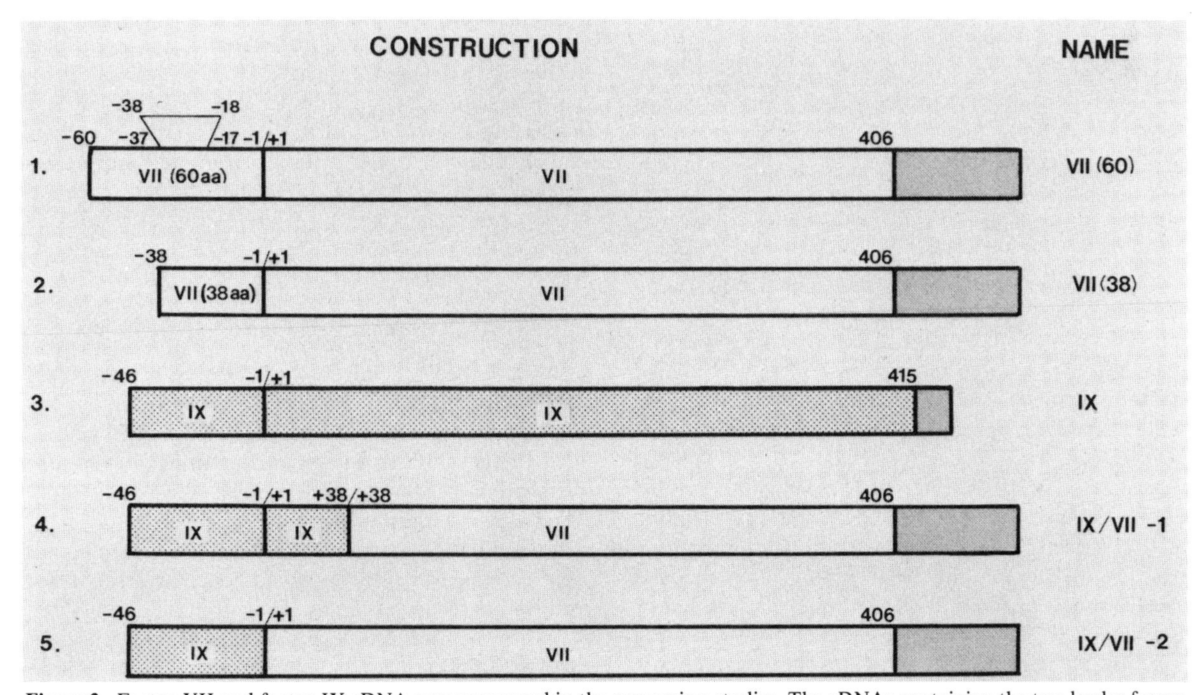

Figure 3. Factor VII and factor IX cDNA sequences used in the expression studies. The cDNAs containing the two leader forms of factor VII (*1*,*2*) or factor IX (*3*) or fusions of factor IX (stippled box) and factor VII (open box) (*4*,*5*) were each inserted into similar expression vectors, as indicated in Experimental Procedures. The amino acid identification is indicated above each construction. The constructions in *1*, *2*, *4*, and *5* all contain 3′-untranslated sequences from factor VII, and those in *3* contain factor IX 3′-untranslated sequences.

amino acids of factor IX, and all of factor VII from amino acids 38 to 406. Another factor IX/VII hybrid, factor IX/VII-2, contained the factor IX leader adjoined to the mature factor VII sequences (Fig. 3). The first 28 amino acids (i.e., up to the *Bgl*II site, indicated in Fig. 2) were encoded by a synthetic DNA sequence. The 84 bp encoding these amino acids were not identical to authentic factor VII DNA but were constructed to conform with optimal codon usage among several related blood-coagulation proteins in this highly homologous region. The remaining sequences encoding amino acids 29–406 of factor IX/VII-2 were derived from λHVII 2463 DNA. Both the IX/VII-1 and IX/VII-2 cDNAs were inserted into an expression vector similar to pDX. Factor IX/VII-2 contained an optimized translational signal immediately upstream of the first methionine in the factor IX leader (Kozak 1984). Another plasmid that was identical to factor IX/VII-2, except that it had a poor translational signal, was also constructed. This plasmid gave results identical to those described below for factor IX/VII-2.

RESULTS AND DISCUSSION

Isolation of Factor-VII-encoding cDNAs

A monoclonal antibody prepared against highly purified plasma factor VII was used to screen a λgt11 cDNA library to obtain a factor VII cDNA clone. This clone was then used as a probe to obtain additional cDNAs. The details for the protein purification, amino acid sequence analysis, and cDNA library preparation and isolation have been described previously (Hagen et al. 1986). Briefly, a λgt11 cDNA library prepared from human liver RNA (provided by S. Woo) was screened using an α-factor VII monoclonal antibody. Several factor VII cDNAs were isolated from this screening, and the longest of these lacked only a small amount of the 5′-terminal sequences (i.e., ending at nucleotide 321, using the numbering system shown in Fig. 2). A second λgt11 cDNA library, prepared from HepG2 mRNA, using a modification of the method of Gubler and Hoffman (1983), was subsequently probed using the cDNA derived from the first library. Among the set isolated from this screening, two cDNAs were found that contained factor VII 5′-terminal coding sequences; one of these cDNAs was full length (λHVII 2463) and one was truncated (λHVII 565) (Fig. 2). The partial λHVII 565 cDNA was most likely generated by oligo(dT) priming at an AT-rich region just downstream from nucleotide 638 (Fig. 2) (Hagen et al. 1986). Both cDNAs encoded a leader sequence and amino-terminal mature factor VII sequences. The identity of these cDNAs as factor-VII-encoding was based on the criteria of a comparison with amino acid sequence data compiled for the light and heavy chains of factor VII, by identity with genomic clones that were isolated and sequenced (P. O'Hara et al., in prep.), and by homology comparisons with other blood-coagulation factors.

Surprisingly, λHVII 2463 and λHVII 565 did not encode identical leader sequences (Fig. 2). The leader encoded by λHVII 2463 was 60 amino acids in length and contained an extra 22 amino acids, positioned between amino acids −17 and −18 of the 38-amino-acid leader encoded by λHVII 565 (Fig. 3) (Hagen et al. 1986). When genomic clones detected using factor VII cDNA were isolated and sequenced, a 66-bp fragment corresponding to this 22-amino-acid insert was identified. This sequence was flanked by introns and appeared to constitute an exon by homology matches of bordering sequences to consensus 3′ and 5′ splice sites (Fig. 2) and of upstream sequences to a consensus splice branch point (P. O'Hara et al., in prep.). Thus, factor VII may be encoded by two distinct mRNAs. Although there are several examples of alternative splicing events that generate multiple protein products (Early et al. 1980; Crabtree and Kant 1982; King and Piatigorsky 1983; Kitamura et al. 1983; Schwarzbauer et al. 1983; Transy et al. 1984; Breitbart et al. 1985), the situation with factor VII differs in that both mRNAs ultimately encode the same mature protein. When both factor VII leader sequences were compared with those of other vitamin-K-dependent plasma proteins, the 38-amino-acid leader demonstrated greater similarities. For example, only the 38-amino-acid leader hydrophobicity profile resembled that of the other factors, and only the 38-amino-acid leader contained a predicted peptidase cleavage site (Perlman and Halvorson 1983; von Heijne 1983) similar to the other factors.

Because the observation of multiple factor VII mRNA species was made in HepG2 cells rather than in liver, the normal source of factor VII, we analyzed liver mRNA for possible production of both factor VII leader forms (Fig. 4). Using a riboprobe that distinguished sequences encoding the 60- or 38-amino-acid leader, we only observed the sequence that generates the 38-amino-acid leader form. From the limits of detection, it could be concluded that mRNA encoding the 38-amino-acid leader represented at least 90% of the liver factor VII mRNA.

Structure of Factor VII

As described in detail elswhere, the full-length λHVII 2463 factor VII cDNA encodes a mature protein of 406 amino acids (Hagen et al. 1986). The 38- and 60-amino-acid leaders both end with four arginines, which immediately precedes the sequence Ala-Asn-Ala-Phe, the known amino terminus for plasma factor VII (Fig. 5). Since a dibasic sequence is known to be a poor signal peptidase cleavage site (Perlman and Halvorson 1983; von Heijne 1983), it is likely that a prepro-leader factor VII form is produced. The predicted signal peptidase cleavage site (Perlman and Halvorson 1983; von Heijne 1983) is most likely after amino acid −19 for the 38-amino-acid leader or after amino acid −33 for the 60-amino-acid leader. Generation of the mature form would presumably occur after processing by at least one additional protease, analogous to the processing

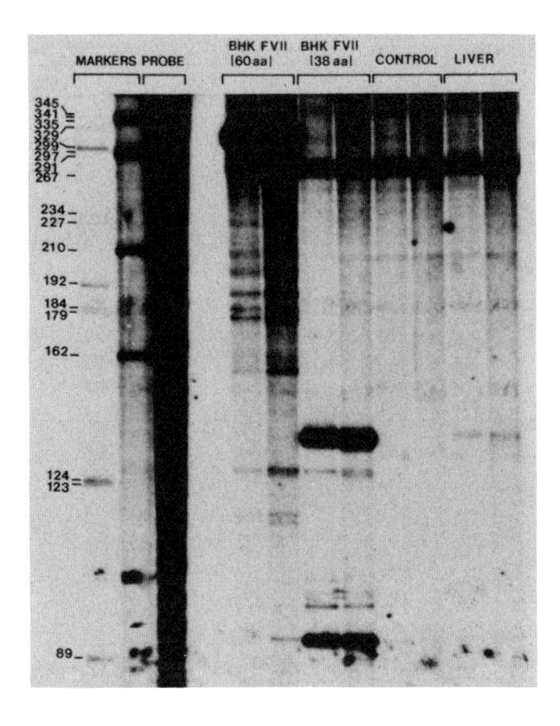

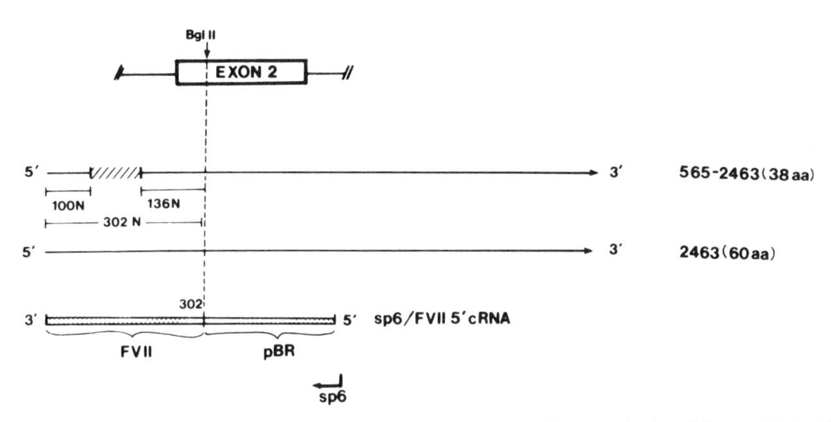

Figure 4. Detection of factor VII mRNA isolated from liver. Total cytoplasmic RNA isolated from BHK cells (CONTROL) or from BHK cells stably transfected with factor VII (BHK FVII) containing either the 60- or 38-amino-acid form was analyzed by ribonuclease mapping (Melton et al. 1984). The riboprobe used in these studies was generated by subcloning amino-terminal factor VII sequences into an sp6-based vector and transcribing off the sp6 promoter, as indicated above. The mRNA encoding the 60-amino-acid leader should give rise to a 302-nucleotide band, whereas that encoding the 38-amino-acid leader should yield bands of 100 and 136 nucleotides. Details of the probe and conditions used in analyzing the RNA will be presented elsewhere. For the analysis of liver RNA, poly(A)$^+$ mRNA was mixed with total cytoplasmic BHK RNA prior to nuclease digestion.

events suggested for other vitamin-K-dependent blood-coagulation factors (Friezner Degen et al. 1983).

The site of cleavage effecting activation of factor VII must be between Arg-152 and Ile-153, since this generates a new amino terminus with the appropriate known sequence for the heavy chain (Kisiel and McMullen 1981). The light (152 amino acids) chain of factor VII contains 10 glutamic acids clustered within the first 35 amino acids of the amino terminus. This region most likely constitutes the gla domain and is followed by two potential growth-factor domains of unknown function. The heavy chain (254 amino acids) contains the serine protease domain. The three principal catalytic residues, histidine, aspartic acid, and serine, are at positions 41, 90, and 192 of the heavy chain, respectively. There are two potential N-linked glycosylation sites, one each in the light (amino acid 145) and heavy (amino acid 170) chains. By analogy with the structure of other coagulation factors, the heavy chain is most likely covalently bound to the light chain via a single disulfide bond.

Factor VII shares a high degree of homology with other known vitamin-K-dependent plasma proteins. Of the 24 cysteines, for example, the positions of 22 are identical with those of bovine and human factor IX, factor X, protein C, and prothrombin. The homology

	-17	-16	-15	-14	-13	-12	-11	-10	-9	-8	-7	-6	-5	-4	-3	-2	-1
Factor VII																	
2463	V	F	V	T	Q	E	E	A	H	G	V	L	H	R	R	R	R
565	V	F	V	T	Q	E	E	A	H	G	V	L	H	R	R	R	R
Factor IX	V	F	L	D	H	E	N	A	N	K	I	L	N	R	P	K	R
Factor X	V	F	L	P	R	D	Q	A	H	R	V	L	Q	R	A	R	R
Prothrombin	V	F	L	A	P	Q	Q	A	R	S	L	L	Q	R	V	R	R
Protein C	V	F	S	S	S	E	R	A	H	Q	V	L	R	I	R	K	R
Bone Gla Prot	A	P	M	S	K	Q	E	G	S	K	V	V	N	R	L	R	R

	1	2	3	4	5	6	7	8	9	10	11	12	13	14	15	16	17	18	19	20	21
Factor VII	A	N	A	-	F	L	γ	γ	L	R	P	G	S	L	-	γ	R	γ	C	K	γ
Factor IX	Y	N	S	G	K	L	γ	γ	F	V	-	Q	G	N	L	γ	R	γ	C	M	γ
Factor X	A	N	S	-	F	L	γ	γ	-	V	K	Q	G	N	L	γ	R	γ	C	L	γ
Prothrombin	A	N	T	-	F	L	γ	γ	-	V	R	K	G	N	L	γ	R	γ	C	V	γ
Protein C	A	N	S	-	F	L	γ	γ	-	L	R	H	S	S	L	γ	R	γ	C	T	γ

	22	23	24	25	26	27	28	29	30	31	32	33	34	35	36	37	38	39	40	41	TOTAL GLAS
Factor VII	γ	Q	C	S	F	γ	γ	A	R	γ	I	F	K	D	A	γ	R	T	K	L	10
Factor IX	γ	K	C	S	F	γ	γ	A	R	γ	V	F	γ	N	T	γ	R	T	T	γ	12
Factor X	γ	A	C	S	L	γ	γ	A	R	γ	V	F	γ	D	A	γ	Q	T	D	γ	12
Prothrombin	γ	T	C	S	Y	γ	γ	A	F	γ	A	L	γ	S	S	T	A	T	D	V	10
Protein C	γ	I	C	D	F	γ	γ	A	K	γ	I	F	Q	N	V	D	D	T	L	A	9

Figure 5. Sequences of gla-containing proteins. The sequences for factor VII (Hagen et al. 1986), factor IX (Kurachi and Davie 1982; Jaye et al. 1983), prothrombin (Friezner Degen et al. 1983), and protein C (Foster et al. 1985) are from a human source. The factor X sequence (Fung et al. 1984) is of bovine origin, and the bone gla protein sequence (Pan and Price 1985) is from rat.

is very strong within the pro-leader sequence (i.e., amino acids −1 to −17) and the first 40 amino acids (i.e., the gla domain) (Fig. 5). The pro-leader sequence is also homologous to that of the bone gla protein, a protein of unknown function that contains three gla residues in its mature (6 kD) form (Price et al. 1981). Factor VII also shares homology with other coagulation factors in the growth-factor domains and in the serine protease domain, especially in those amino acids immediately surrounding the three principal active site residues. The overall homology of factor VII with factors IX and X, prothrombin, and protein C supports the concept of their evolution from a common ancestral gene (Katayama et al. 1979).

Biologically Active Recombinant Factor VII Is Produced in a Variety of Mammalian Cells

Factor VII, like several other blood-coagulation proteins, requires γ-carboxylation of multiple glutamic acid residues clustered at the amino terminus of the mature protein, i.e., the gla domain, for its biological activity. The effect of γ-carboxylation on factor VII activity is calcium-mediated and involves phospholipid interaction. This requirement of γ-carboxylation for biological activity introduces an unusual complexity in expressing recombinant factor VII in mammalian cells, since the capacity of cultured cells for γ-carboxylation has not been well characterized. The expression of factor IX, another plasma protein that requires γ-carboxylation, has been reported by a number of groups (Anson et al. 1985; Busby et al. 1985; de la Salle et al. 1985). Several different fibroblast cell lines were found to express factor IX that was, at best, only partially biologically active. Expression of recombinant factor IX in hepatoma cells, attempted since most of the coagulation plasma proteins are synthesized in the liver, appeared to produce protein with increased biological activity (Anson et al. 1985).

To determine whether cultured mammalian cells could express active recombinant factor VII, cells either were transfected with plasmids encoding factor VII and then assayed transiently 2–5 days later or were cotransfected with plasmids encoding factor VII and a dominant selectable marker. Colonies surviving selection were isolated and assayed. Representative results are shown in Table 1. As seen with transient analysis in 293 cells, both leader forms (i.e., 38 or 60 amino acids) of recombinant factor VII were expressed at approximately equal levels. The range in expression among isolated BHK cell colonies transfected with either leader form was also found to be the same (four of

Table 1. Expression of Factors IX and VII

(a) *Transient (293 cells)*

Protein	Sample	ELISA (μg)	Percent secreted	Clotting assay (μg)	Percent active
Factor VII/	media	158	98.3	243	154
38-amino-acid leader	pellet	2.7		n.d.	
Factor VII/	media	156	96.7	175	112
60-amino-acid leader	pellet	5.2		n.d.	
Control	media	<2			
	pellet	n.d.			
Factor IX	media	400	99.7	153	34
	pellet	1		n.d.	
Factor IX	media	250	99.7	150	53
	pellet	0.7		n.d.	
Control	media	<0.2		17	
	pellet	<0.5		n.d.	

(b) *Stable Lines*

Sample	ELISA (μg)	Clotting assay (μg)	Percent active
BHK/FVII/ 38-amino-acid leader	800	910	114
BHK/FVII/ 38-amino-acid leader	553	570	103
BHK/FVII/ 60-amino-acid leader	197	216	110
BHK/FVII/ 60-amino-acid leader	760	620	82
BHK/FIX	1500	648	44
BHK/FIX	1375	678	49
Human hepatoma/FVII/ 60-amino-acid leader	360	350	97
Human hepatoma/FVII/ 60-amino-acid leader	280	313	112

these are indicated in Table 1). A large variety of cell lines have been tested, and no difference in expression has been observed between the 38- and 60-amino-acid leader forms of recombinant factor VII.

Most of factor VII was secreted from 293 cells (Table 1a), and this was the case with several other cell lines examined (not shown). When clotting activities were compared with ELISA values, the biological activity of factor VII appeared to be around 100%. This was observed with transiently expressing 293 cells or with stably transfected BHK or human hepatoma cells (Table 1), as well as with other cell lines (data not shown). The biological activity of mammalian cells expressing factor VII was dependent on the presence of vitamin K, a cofactor required in the γ-carboxylation reaction. When cells were grown in media depleted of vitamin K, ELISA-positive secreted protein was detected but was found to be inactive.

The high levels of recombinant factor-VII-specific biological activity observed in several different cell lines contrasted with results previously reported for factor IX (Anson et al. 1985; Busby et al. 1985; de la Salle et al. 1985). In the studies described here, additional cell lines (e.g., 293 cells and HepG2 cells) were examined for their ability to express factor IX, and these were shown to also produce partially (30–50%) active factor IX (e.g., the activity in 293 cells is shown in Table 1a). The difference in activity between factors IX and VII was not vector-dependent; e.g., in the 293 cell transfection, the identical vector was used. Neither was the discrepancy between factor-VII- and factor-IX-specific biological activities due to differences in expression levels, since the levels of each protein were comparable in the different cell line studies. Moreover, the partial biological activity of factor IX did not appear to be correlated with expression levels: Approximately the same extent of factor IX biological activity (30–50%) was observed over at least a 20-fold range.

The activity of the activated form of factor VII (i.e., factor VIIa) is 25-fold higher than that of single-chain factor VII, when measured in the one-stage clotting assay (Broze and Majerus 1982). To determine whether generation of two-chain factor VIIa was causing an artifically high level of biological activity, the physical form of secreted factor VII was analyzed. Proteins labeled in vivo with [^{35}S]methionine were immunoprecipitated and subjected to gel electrophoresis. As shown in Figure 6A, only single-chain factor VII was observed. The high level of factor VII clotting activity did not therefore appear to be due to secretion of factor VIIa from stably transfected cells.

Another criterion used in examining the extent of γ-carboxylation of recombinantly produced factor IX and factor VII was their precipitability using barium citrate. This approach represents a rough assessment of the protein structure, since only partial γ-carboxylation is required for barium citrate precipitability. Moreover, the minimal number of glas required and the

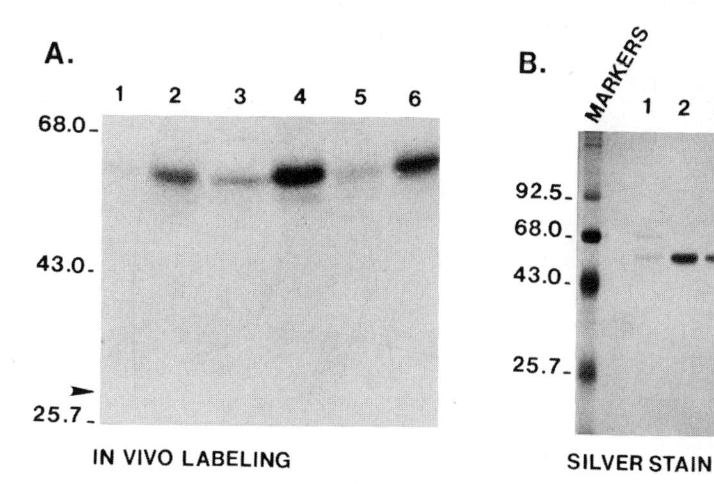

Figure 6. Protein analysis of recombinant factors VII and IX/VII-1. (*A*) Protein labeled in vivo with [^{35}S]methionine was immunoprecipitated from BHK cells using antisera against factor VII and subsequently subjected to SDS-PAGE. The BHK cells were grown in the absence (odd numbers) or presence (even numbers) of 5 μg/ml vitamin K, and the samples were factor IX/VII-1 (*1,2*), factor VII (38-amino-acid form) (*3,4*), and factor VII (60-amino-acid form) (*5,6*). The position that the factor VII heavy chain form would migrate at is indicated by an arrowhead. The light band migrating just ahead of the factor VII band in several lanes was also observed with control BHK cells. (*B*) Factor VII carried through different stages of purification was analyzed by gel electrophoresis and subsequent silver staining. Immunoblotting of this gel using antisera against factor VII identified the major band as factor VII (data not shown).

effect of the position of these glas on adsorption are not well understood. Nevertheless, when the factor VII and factor IX proteins secreted from BHK cells were compared, marked differences in barium citrate adsorption were observed, presumably reflecting differences in the extent or pattern of γ-carboxylation (Table 2). Although virtually the entire ELISA-positive factor VII in the medium was precipitated, only 50% of the secreted factor IX protein was adsorbed. In both cases, essentially all of the biologically active protein was present in the precipitated fraction, as expected since activity depends on more extensive γ-carboxylation than does barium citrate adsorption.

Characterization of Purified Recombinant Factor VII

Factor VII was prepared from the spent media of stably transfected BHK cells using immunochemical and ion-exchange fractionation. Purified material analyzed by gel electrophoresis was shown to be homogeneous by silver staining (Fig. 6B). Subsequent Western analysis using an α-factor VII polyclonal antibody confirmed its identify as factor VII (not shown). The recombinant product comigrated with plasma factor VII as a 56K protein, indicating that it was glycosylated by the BHK cells. Virtually all of the purified protein was present in single-chain form.

Purified recombinant factor VII was subjected to Edman degradation using a gas-phase protein sequenator. Analysis of the first ten amino acids was consistent with a recombinant factor VII protein preparation composed totally of processed, mature protein. No evidence for amino-terminal sequence stuttering that could be generated by multiple processing events within the four arginines preceding the first mature amino acid (i.e., alanine) (Fig. 5) was detected. In addition, no glutamic acids were observed at amino acid positions 6 and 7 (Fig. 5), indicating indirectly that these positions were fully γ-carboxylated in the purified protein.

Construction of Factor IX–Factor VII Fusion Proteins

As discussed above, factor VII exhibits a high degree of homology with other blood-coagulation factors, and

Table 2. Barium Citrate Precipitation of Factors IX, VII, and IX/VII-1 from Stably Transfected BHK Cells

Protein	Sample	ELISA (μg)	Activity (μg) [a]
Factor VII	media	3.88	5.26
	supernatant fraction after precipitation and dialysis	0.11	0.01
	redissolved precipitate	3.42	1.84
Factor IX	media	5.29	3.19
	supernatant fraction after precipitation and dialysis	2.64	0.01
	redissolved precipitate	1.50	1.94
Factor IX/VII-1	media	16.60	1.85
	supernatant fraction after precipitation and dialysis	2.07	0.01
	redissolved precipitate	13.65	0.88

[a]The recovery of activity after precipitation has never been 100% and there is presently no explanation for this.

this is especially pronounced within the gla domain (Fig. 5). This homology prompted us to test the possibility of exchanging the gla domain of one factor with that of another. DNA sequences encoding the leader and gla domains of factor IX were therefore adjoined to sequences encoding the growth-factor and serine protease domains of factor VII, using the two-primer method of site-directed mutagenesis (Zoller and Smith 1984). A resultant cDNA (factor IX/VII-1; Fig. 5, no. 4), whose construction will be described in detail elsewhere, contained the 46-amino-acid leader and first 38 amino acids of factor IX fused to factor VII amino acids 38–406. The genomic structures for both factor IX (Anson et al. 1984; Yoshitake et al. 1985) and factor VII (P. O'Hara et al., in prep.) have been determined. The fusion described here corresponds almost precisely to the interchange of two factor IX exons (corresponding to amino acid −46 to −17 of the leader and amino acid −17 to 38 of the leader and gla domain) with two (or three) factor VII exons (amino acid −38 or −60 to −17 and amino acid −17 to 37).

The factor IX/VII-1 cDNA was inserted into an expression vector (see Experimental Procedures), and BHK cells were cotransfected with this plasmid and one encoding resistance to G-418. To detect stably transfected cells expressing factor IX/VII-1, cell supernatants were analyzed by ELISAs using α-factor VII antisera. Several cell lines were obtained, and one was chosen for further study. When the clotting activity of secreted factor IX/VII-1 was assayed, the fusion protein was found to be biologically active (Table 2). This activity was dependent on the presence of vitamin K in the media, and it could be specifically blocked using α-factor VII polyclonal antibody (data not shown). The factor IX/VII-1 and factor VII biological activities, however, were not identical. The level of fusion protein activity was found by repeated measurements to be only about 24% that of recombinant factor VII.

The γ-carboxylation of factor IX/VII-1 was also examined using barium citrate precipitation (Table 2). When BHK cell spent media expressing factor IX/VII-1 were treated with barium citrate, 82% of the protein was precipitated, indicating that most of the protein molecules contain glas. As observed with factors IX and VII, greater than 99% of the active protein was contained in the precipitate. When the adsorbed material was resuspended, dialyzed, and subjected to gel electrophoresis, a 56K protein, the size expected for the single-chain form, was detected by Western analysis using α-factor VII monoclonal antibody (data not shown). Thus, a factor IX/VII fusion protein containing the factor IX leader and gla domain is capable of being correctly processed, and the mature protein must be γ-carboxylated, since this modification is required for both biological activity and barium citrate adsorption. Moreover, the factor IX gla domain in the fusion protein can function in the activation of the factor VII serine protease activity. At present, it is not known why the specific activity of the fusion protein is less than that of factor VII. Differences in γ-carboxylation, in the level of cleavage resulting in activation of the serine protease, or in distinct conformations possibly inherent in a hybrid molecule versus a native protein are all possibilities that are presently being addressed.

The biological activity observed with the factor IX/VII-1 fusion protein, in which both the factor IX leader and gla domain were substituted for those of factor VII, contrasted with results obtained when a substitution of only the leader sequence was made. In this construction, factor IX/VII-2 (Fig. 3, no. 5), the 46-amino-acid factor IX leader was adjoined to the 406-amino-acid mature sequences of factor VII. The amino-terminal 84 bp of factor VII were derived from a synthetic fragment containing codons optimized for translation of blood-coagulation factors, which encoded an authentic factor VII amino acid sequence. This synthetic sequence did not differ significantly from the factor VII cDNA sequences with regard to optimal factor VII codon usage. When the factor IX/VII-2 fusion cDNA was inserted into an expression vector identical with that used for factor VII and transfected into BHK cells or 293 cells, neither factor VII activity nor protein was observed. In a parallel transfection experiment, normal recombinant factor VII expression was detected. In addition, the product of a gene cotransfected with factor IX/VII-2 (i.e., the chloramphenicol *trans*-acetylase gene) was observed, indicating normal transfection efficiencies. The inability to detect functional protein did not appear to be due to an inadvertent mutation acquired during cloning, since the two fragments used to construct factor IX/VII-2 were sequenced in their entirety in both directions. Furthermore, trace amounts of factor IX/VII-2 were observed when immunofluorescence using α-factor VII antibody was performed on transfected cells.

The contrasting results obtained with the two different factor IX/VII fusion genes are not understood but very likely reflect the complex set of events involved in the posttranslational modification and proteolytic processing of the vitamin-K-dependent plasma proteins. It has been suggested that the propeptide region of the vitamin-K-dependent proteins contains a signal for γ-carboxylation (Pan and Price 1985). This suggestion is supported by the observation of homology between the propeptides of these proteins and that of the bone gla protein, a vitamin-K-dependent protein that is functionally unrelated to the plasma proteins and shares homology only in the propeptide. Thus, the gla domain may actually comprise both a propeptide gla signal (between amino acid −1 and −17) and the sequences to be γ-carboxylated (between amino acid 1 and 38). The fusion of the factor IX leader with the factor VII mature protein in the factor IX/VII-2 construction may have generated a nonfunctional, hybrid gla domain. In the biologically active factor IX/VII-1 fusion protein, both the putative signal and γ-carboxylated sequences were derived from factor IX. An alternative explanation for the lack of factor IX/VII-2 activity may come from the uniqueness of factor IX among the vitamin-K-dependent coagulation factors in the first amino acid; factor IX contains a tyrosine and the rest contain an alanine (Fig. 5). The substitution of the factor VII mature amino terminus in the factor IX/VII-2 fusion protein may have interfered with normal processing. With either explanation, however, it is difficult to account for the ability to detect only trace levels of the factor IX/VII-2 protein. Additional permutations in the factor IX/VII fusion proteins are currently being constructed. Clearly, these kinds of experiments, complemented by the analysis of available phenotypic variants such as that recently reported for a factor IX hemophiliac (Bentley et al. 1986), should be of value in determining the mechanism by which these proteins are processed and rendered biologically active.

ACKNOWLEDGMENTS

We thank the following people for their valued contributions to these studies: Cheryl Bradley, Frank Grant, Charles Gray, Betty Haldeman, Laurel Halfpap, Michael Joseph, Donna Prunkard, Gena Saari, Geoff Tupper, and Susan Yarnold. We also acknowledge the helpful comments by Don Foster and the careful preparation of the manuscript by Margo Rogers.

REFERENCES

Anson, D.S., D.E.G. Austen, and G.G. Brownlee. 1985. Expression of active human clotting factor IX from recombinant DNA clones in mammalian cells. *Nature* **315:** 683.

Anson, D.S., K.H. Choo, D.J.G. Rees, F. Giannelli, K. Gould, J.A. Huddleston, and G.G. Brownlee. 1984. Gene structure of human anti-haemophilic factor IX. *EMBO J.* **3:** 1053.

Bajaj, S.P., S.I. Rapaport, and S.F. Brown. 1981. Isolation and characterization of human factor VII. *J. Biol. Chem.* **256:** 253.

Bentley, A.K., D.J.G. Rees, C. Rizza, and G.G. Brownlee. 1986. Defective propeptide processing of blood clotting factor IX caused by mutation of arginine to glutamine at position −4. *Cell* **45:** 343.

Breitbart, R.E., H.T. Nguyen, R.M. Medford, A.T. Destree, V. Mahdavi, and B. Nadal-Ginard. 1985. Intricate combinatorial patterns of exon splicing generate multiple regulated troponin T isoforms from a single gene. *Cell* **41:** 67.

Broze, G.J. and P.W. Majerus. 1980. Purification and characterization of human coagulation factor VII. *J. Biol. Chem.* **255:** 1242.

———. 1982. Human factor VII. *Methods Enzymol.* **80:** 228.

Busby, S., A. Kumar, M. Joseph, L. Halfpap, M. Insley, K. Berkner, K. Kurachi, and R. Woodbury. 1985. Expression of active human factor IX in transfected cells. *Nature* **316:** 271.

Crabtree, G.R. and J.A. Kant. 1982. Organization of the rat γ-fibrinogen gene: Alternative mRNA splice patterns produce the γA and γB (γ′) chains of fibrinogen. *Cell* **31:** 159.

Crestfield, A.M., S. Moore, and W.H. Stein. 1963. Preparation of enzymatic hydrolysis of reduced and *S*-carboxymethylated proteins. *J. Biol. Chem.* **238:** 622.

Danna, K.J. and L.M. Sompayrac. 1982. Efficient infection of monkey cells with SV40 DNA. II. Use of low-molecular-weight DEAE-dextran for large-scale experiments. *J. Virol. Methods* **5:** 335.

Davie, E.W., K. Fujikawa, K. Kurachi, and W. Kisiel. 1979. The role of serine proteases in the blood coagulation cascade. *Adv. Enzymol.* **48:** 277.

de la Salle, H., W. Altenburger, R. Elkaim, K. Dott, A. Dieterlé, R. Drillien, J.P. Cazenave, P. Tolstoshev, and J.P. Lecocq. 1985. Active γ-carboxylated human factor IX expressed using recombinant DNA techniques. *Nature* **316:** 268.

Early, P., J. Rogers, M. Davis, K. Calame, M. Bond, R. Wall, and L. Hood. 1980. Two mRNAs can be produced from a single immunoglobulin μ gene by alternative RNA processing pathways. *Cell* **20:** 313.

Foster, D. and E.W. Davie. 1984. Characterization of a cDNA coding for human protein C. *Proc. Natl. Acad. Sci.* **81:** 4766.

Foster, D.C., S. Yoshitake, and E.W. Davie. 1985. The nucleotide sequence of the gene for human protein C. *Proc. Natl. Acad. Sci.* **82:** 4673.

Friezner Degen, S.J., R.T.A. MacGillivray, and E.W. Davie. 1983. Characterization of the complementary deoxyribonucleic acid and gene coding for human prothrombin. *Biochemistry* **22:** 2087.

Fung, M.R., R.M. Campbell, and R.T.A. MacGillivray. 1984. Blood coagulation factor X mRNA encodes a single polypeptide chain containing a prepro leader sequence. *Nucleic Acids Res.* **12:** 4481.

Graham, F.L. and A.J. van der Eb. 1973. A new technique for the assay of infectivity of human adenovirus 5 DNA. *Virology* **52:** 456.

Graham, F.L., J. Smiley, W.C. Russell, and R. Nairn. 1977. Characteristics of a human cell line transformed by DNA from human adenovirus type 5. *J. Gen. Virol.* **36:** 59.

Gubler, U. and B.J. Hoffman. 1983. A simple and very efficient method for generating cDNA libraries. *Gene* **25:** 263.

Hagen, F.S., C.L. Gray, P. O'Hara, F.J. Grant, G.C. Saari, R.G. Woodbury, C.E. Hart, M. Insley, W. Kisiel, K. Kurachi, and E.W. Davie. 1986. Characterization of a cDNA coding for human factor VII. *Proc. Natl. Acad. Sci.* **83:** 2412.

Hedner, U. and W. Kisiel. 1983. Use of human factor VIIa in the treatment of two hemophilia A patients with high-titer inhibitors. *J. Clin. Invest.* **71:** 1836.

Jackson, C.M. and Y. Nemerson. 1980. Blood coagulation. *Annu. Rev. Biochem.* **49:** 765.

Jaye, M., H. de la Salle, F. Schamber, A. Balland, V. Kohli, A. Findeli, P. Tolstoshev, and J.P. Lecocq. 1983. Isolation of a human anti-haemophilic factor IX cDNA clone using a unique 52-base synthetic oligonucleotide probe deduced from the amino acid sequence of bovine factor IX. *Nucleic Acids Res.* **11:** 2325.

Katayama, K., L.H. Ericsson, D.L. Enfield, K.A. Walsh, H. Neurath, E.W. Davie, and K. Titani. 1979. Comparison of the amino acid sequence of bovine coagulation factor IX (Christmas factor) with other vitamin K-dependent plasma proteins. *Proc. Natl. Acad. Sci.* **76:** 4990.

King, C.R. and J. Piatigorsky. 1983. Alternative RNA splicing of the murine αA-crystallin gene: Protein-coding information within an intron. *Cell* **32:** 707.

Kisiel, W. and E.W. Davie. 1975. Isolation and characterization of bovine factor VII. *Biochemistry* **14:** 4928.

Kisiel, W. and B.A. McMullen. 1981. Isolation and characterization of human factor VII_a. *Thromb. Res.* **22:** 375.

Kisiel, W., K. Kujikawa, and E.W. Davie. 1977. Activation of bovine factor VII (proconvertin) by factor XII_a (activated Hageman factor). *Biochemistry* **16:** 4189.

Kitamura, N., Y. Takagaki, S. Furuto, T. Tanaka, H. Nawa, and S. Nakanishi. 1983. A single gene for bovine high molecular weight and low molecular weight kininogens. *Nature* **305:** 545.

Kozak, M. 1984. Compilation and analysis of sequences upstream from the translational start site in eukaryotic mRNAs. *Nucleic Acids Res.* **12:** 857.

Kurachi, K. and E.W. Davie. 1982. Isolation and characterization of a cDNA coding for human factor IX. *Proc. Natl. Acad. Sci.* **79:** 6461.

Laemmli, U.K. 1970. Cleavage of structural proteins during the assembly of the head of bacteriophage T4. *Nature* **227:** 680.

Leytus, S.P., D.W. Chung, W. Kisiel, K. Kurachi, and E.W. Davie. 1984. Characterization of a cDNA coding for human factor X. *Proc. Natl. Acad. Sci.* **81:** 3699.

Long, G.L., R.M. Belagaje, and R.T.A. MacGillivray. 1984. Cloning and sequencing of liver cDNA coding for bovine protein C. *Proc. Natl. Acad. Sci.* **81:** 5653.

Maxam, A.M. and W. Gilbert. 1980. Sequencing end-labeled DNA with base-specific chemical cleavages. *Methods Enzymol.* **65:** 499.

Melton, D.A., P.A. Krieg, M.R. Rebagliati, T. Maniatis, K. Zinn, and M.R. Green. 1984. Efficient *in vitro* synthesis of biologically active RNA and RNA hybridization probes from plasmids containing a bacteriophage SP6 promoter. *Nucleic Acids Res.* **12:** 7035.

Østerud, B. and S.I. Rapaport. 1977. Activation of factor IX by the reaction product of tissue factor and factor VII: Additional pathway for initiating blood coagulation. *Proc. Natl. Acad. Sci.* **74:** 5260.

Pan, L.C. and P.A. Price. 1985. The propeptide of rat bone γ-carboxyglutamic acid protein shares homology with other vitamin K-dependent protein precursors. *Proc. Natl. Acad. Sci.* **82:** 6109.

Perlman, D. and H.O. Halvorson. 1983. A putative signal peptidase recognition site and sequence in eukaryotic and prokaryotic signal peptides. *J. Mol. Biol.* **167:** 391.

Price, P.A., M.K. Williamson, and J.W. Lothringer. 1981. Origin of the vitamin K-dependent bone protein found in plasma and its clearance by kidney and bone. *J. Biol. Chem.* **256:** 12760.

Radcliffe, R. and Y. Nemerson. 1975. Activation and control of factor VII by activated factor X and thrombin. *J. Biol. Chem.* **250:** 388.

———. 1976. Mechanism of activation of bovine factor VII. *J. Biol. Chem.* **251:** 4797.

Radcliffe, R.A., A. Bagdasarian, R. Colman, and Y. Nemerson. 1977. Activation of bovine factor VII by Hageman factor fragments. *Blood* **50:** 611.

Sanger, F., S. Nicklen, and A.R. Coulson. 1977. DNA sequencing with chain-terminating inhibitors. *Proc. Natl. Acad. Sci.* **74:** 5463.

Schwarzbauer, J.E., J.W. Tamkun, I.R. Lemischka, and R.O. Hynes. 1983. Three different fibronectin mRNAs arise by alternative splicing within the coding region. *Cell* **35:** 421.

Seligsohn, U., B. Østerud, S.F. Brown, J.H. Griffin, and S.I. Rapaport. 1979. Activation of human factor VII in plasma and in purified systems. Roles of activated factor IX, kallikrein, and activated factor XII. *J. Clin. Invest.* **64:** 1056.

Simonsen, C.C. and A.D. Levinson. 1983. Isolation and expression of an altered mouse dihydrofolate reductase cDNA. *Proc. Natl. Acad. Sci.* **80:** 2495.

Suttie, J.W. 1980. Mechanism of action of vitamin K: Synthesis of γ-carboxyglutamic acid. *CRC Crit. Rev. Biochem.* **8:** 191.

Towbin, H., T. Staehelin, and J. Gordon. 1979. Electrophoretic transfer of proteins from polyacrylamide gels to nitrocellulose sheets: Procedure and some applications. *Proc. Natl. Acad. Sci.* **76:** 4350.

Transy C., J.L. Lalanne, and P. Kourilsky. 1984. Alternative splicing in the 5′ moiety of the H-2K^d gene transcript. *EMBO J.* **3:** 2383.

von Heijne, G. 1983. Patterns of amino acids near signal-sequence cleavage sites. *Eur. J. Biochem.* **133:** 17.

Yoshitake, S., B.G. Schach, D.C. Foster, E.W. Davie, and K. Kurachi. 1985. Nucleotide sequence of the gene for human factor IX (anti-haemophilic factor B). *Biochemistry* **24:** 3736.

Zoller, M.J. and M. Smith. 1984. Oligonucleotide-directed mutagenesis: A simple method using two oligonucleotide primers and a single-stranded DNA template. *DNA* **3:** 479.

Exploration of Structure-Function Relationships in Human Factor VIII by Site-directed Mutagenesis

J.J. TOOLE,* D. PITTMAN, P. MURTHA, L.C. WASLEY, J. WANG, G. AMPHLETT, R. HEWICK, W.B. FOSTER, R. KAMEN, AND R.J. KAUFMAN

Genetics Institute, Inc., Cambridge, Massachusetts 02140

Hemophilia A is an X-linked bleeding disorder (which occurs in ~10–20 males in every 100,000) caused by deficiency or abnormality of a particular clotting protein, factor VIII. Afflicted individuals suffer episodes of uncontrolled bleeding and are currently treated with concentrates rich in factor VIII derived from human plasma. The available therapy, although reasonably effective, is very costly and is associated with a finite risk of infection.

Factor VIII functions in the blood-clotting cascade as the cofactor for factor IXa proteolytic activation of factor X. The blood-clotting pathway in which factor VIII participates eventually results in the proteolytic cleavage of fibrinogen to form insoluble fibrin polymers. In vivo, fibrin deposition in conjunction with platelet aggregation act to curtail blood loss from a damaged vessel.

The cofactor activity of factor VIII acts to increase the V_{max} of the factor-IXa-dependent activation of factor X by at least four orders of magnitude (van Dieijen et al. 1981). Factor VIII does not function proteolytically in this reaction but can itself be proteolytically activated by other coagulation enzymes such as factor X or thrombin. Neither the mechanism of factor VIII activation nor the nature of its cofactor activity is well understood.

The entire human factor VIII gene, spanning over 185 kb of the X chromosome, and full-length cDNA have been cloned (Gitschier et al. 1984; Toole et al. 1984; Wood et al. 1984). The DNA sequence of the cDNA and amino-terminal amino acid sequence analysis of the plasma protein show factor VIII to be synthesized as a single-chain precursor of 2351 amino acids, from which a 19-amino-acid secretory peptide is cleaved during translation. The factor VIII molecule is especially susceptible to proteolytic cleavage and is generally purified as a two-chain molecule, with a "heavy chain" of about 200 kD or various proteolytic fragments derived therefrom (see Fig. 2A) and a carboxy-terminal "light-chain" doublet of about 80 kD (Fulcher et al. 1983; Rotblat et al. 1985).

The cDNA sequence showed factor VIII to contain 25 potential N-linked glycosylation sites. Interestingly, most of these sites are situated in the middle third of the molecule as shown schematically in Figure 1A. A computer-aided search for factor VIII intramolecular homologies revealed three distinct structural domains, including a triplicated region (A domain) of approximately 330 amino acids, a duplicated region (C domain) of approximately 150 amino acids, and a unique region (B domain) of approximately 980 amino acids. These domains are arranged in the order A1-A2-B-A3-C1-C2 (see Fig. 1B) as described by Vehar et al. (1984). The B domain is essentially delimited by residues 740 and 1689, which are sites cleaved during proteolytic activation of the molecule, and therefore appears like an unusually large activation peptide. The A domains exhibit about 35% homology with the triplicated domains of ceruloplasmin, a plasma copper-binding protein that is not involved, even peripherally, in blood coagulation. Ceruloplasmin, in contrast to factor VIII, does not contain any regions other than the A domains. In addition to the human clones, we isolated a porcine factor VIII genomic clone homologous to the human factor VIII recombinant phage containing exon 14, a 3106-bp exon encoding the entire B domain. We subsequently walked 3′ in the porcine genomic library and isolated a clone homologous to human exon 15. DNA sequencing of both porcine exons was completed, and the porcine factor VIII amino acid sequence was deduced. A comparison of the corresponding porcine and human factor VIII amino acid sequences revealed that, remarkably, there exist several substantial deletions totaling over 200 amino acids in the porcine B domain relative to the human domain (Toole et al. 1986). In addition, there is a marked diminution of homology within the B domain compared either with that observed in the A2-A3 regions deduced from the DNA sequence or with the limited regions derived from amino-terminal amino acid sequencing of porcine factor VIII (Toole et al. 1984). Specifically, there is approximately 50% homology, excluding the deletions, in the B domain, whereas the known amino acid sequence elsewhere in the factor VIII molecule exhibits about 80–85% homology.

Considering this striking divergence and since porcine factor VIII corrects the hemophilia A coagulation deficiency in vivo (Kernoff et al. 1984) and in vitro, we postulated that the B domain is not involved in procoagulant activity per se. Also, blood-clotting factor V, which is structurally related to factor VIII (Church et al. 1984), is synthesized as a single-chain precursor of 330,000 daltons and is proteolytically activated by

*Present address: Department of Medicine, Stanford University School of Medicine, Palo Alto, California 94307.

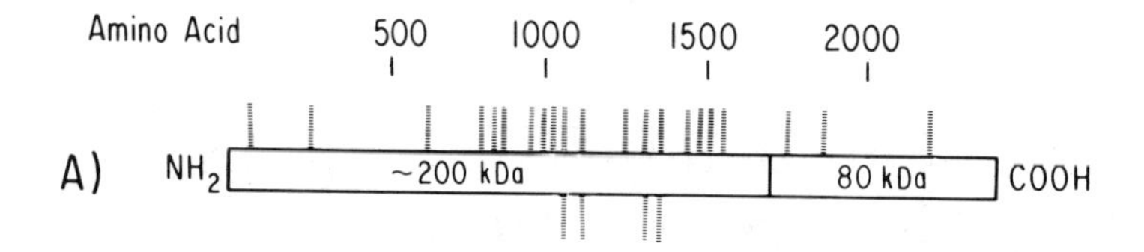

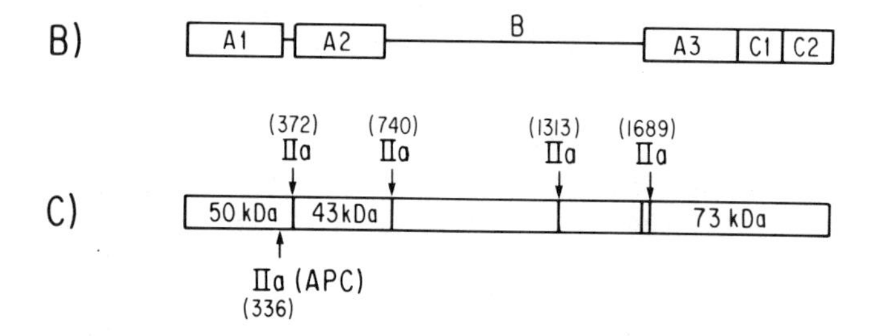

Figure 1. Structural features of factor VIII. (*A*) Relative positions of the two-chain form, with the vertical bars representing sites of potential N-linked glycosylation. (*B*) Relative positions of the domains comprising factor VIII. (*C*) Position of thrombin (IIa) and activated protein C (APC) cleavage sites.

thrombin or factor Xa to yield factor Va (Nesheim et al. 1979; Foster et al. 1983). Significant levels of factor Va activity are obtained when the amino-terminal (94,000 daltons) and carboxy-terminal (74,000 daltons) thrombin cleavage fragments are reconstituted in the presence of divalent cations (Esmon 1979).

In this paper, we describe in vitro mutagenesis of factor VIII, which demonstrates that most of the B domain is dispensable for in vitro procoagulant activity. In addition, we describe the effects of thrombin cleavage-site mutants, produced in vitro, on factor VIII activity.

METHODS

Plasmid constructions. To construct pDGR-2, 10 μg of the plasmid pACE, a pSP64 derivative (Promega Biotec, Madison, Wisconsin), containing nucleotides 562–7269 of human factor VIII cDNA (nucleotide 1 is the A of the ATG initiator methionine codon) was subjected to partial *Bam*HI digestion in 100 μl, containing 50 mM Tris-HCl (pH 8.0), 50 mM $MgCl_2$, and 2.4 units *Bam*HI (New England BioLabs) for 30 minutes at 37°C. The reaction was terminated by the addition of EDTA to 10 mM and then extracted once with phenol and once with chloroform, ethanol-precipitated, and pelleted by centrifugation. The DNA was redissolved, cleaved to completion in 50 μl, using 40 units *Sac*I for 1.5 hours at 37°C. DNA was then electrophoresed through a buffered 0.6% agarose gel. An 8.1-kb fragment corresponding to the partial *Bam*HI-*Sac*I fragment of pACE, lacking only the sequence corresponding to nucleotides 2992–4774 of the factor VIII sequence, was purified from the gel using the glass-powder technique (Vogelstein and Gillespie 1979). Purified DNA was ligated with 100 pmoles of the following double-stranded oligonucleotide.

5′ – P – C A T G G A C C G – 3′
3′ – T C G A G T A C C T G G C C T A G – 5′

DNA was then used to transform competent *Escherichia coli* bacteria, and DNA from several ampicillin-resistant transformants was analyzed by restriction mapping to identify a plasmid harboring the desired *Sac*I-*Bam*HI deletion mutant. DNA from this plasmid was digested to completion with *Kpn*I, which cleaves the plasmid uniquely at nucleotide 1816 of the factor-VIII-coding sequence. This DNA was ligated with a *Kpn*I DNA fragment containing nucleotides 1–1815 of factor VIII DNA and a synthetic *Sal*I site at nucleotides – 11 to – 5 and then used to transform competent *E. coli* bacteria.

Plasmid DNA was isolated and oriented by restriction mapping to identify a plasmid containing the correct 5′→3′ orientation of the *Kpn*I insert. *Sal*I digestion, which excises the entire polypeptide-coding region from the plasmid, was performed, and the DNA was electrophoresed through a buffered 0.6% agarose gel. The 5.3-kb *Sal*I fragment was purified from the gel as described above. This DNA fragment was ligated with *Xho*I-cut pXMT2 DNA to give rise to plasmid pDGR-2. pXMT2 is a plasmid capable of expressing heterologous genes when introduced into mammalian cells, such as the COS-1 African green monkey kidney cell line, and is a derivative of the expression vector p91023(B) (Kaufman 1985; Wong et al. 1985). The unique *Xho*I site of pXMT2 allows for expression of inserted cDNA from the upstream promoter. Restriction mapping of transformants identified a plasmid, pDGR-2, containing the correct 5′→3′ orientation of the polypeptide-coding sequence relative to the direction of transcription.

pLA-2 was created using the loop-out mutagenesis technique described by Morinaga et al. (1984). Briefly, 1 μg each of pDGR-2 and pXMT2 was linearized by restriction endonuclease digestion with *Eco*RV and *Xho*I, respectively, denatured in 20 μl, containing 0.2 N NaOH for 10 minutes at room temperature, and then neutralized by addition of 180 μl 0.02 N HCl and 0.1 M Tris-HCl (pH 8.0). The DNA was then reannealed for 90 minutes at 68°C. The sample (40 μl) was then slowly cooled to room temperature after addition of an excess of the kinased 40-mer oligonucleotide 5'AAAAGCAA-TTTAATGCCACCCCACCAGTCTTGAAACGCCA. The sample was brought to 50 μl, containing 0.1 M Tris-HCl (pH 8.0), 20 mM NaCl, 2 mM $MgCl_2$, 500 μM ATP, 100 μM deoxynucleotide triphosphate, 4 U/μl Klenow fragment of *E. coli* DNA polymerase, and 400 U/μl T4 ligase. The reaction was incubated for 16 hours at 15°C and then terminated by phenol-chloroform extraction, and the nucleic acids were ethanol-precipitated. DNA was then used to transform competent *E. coli* bacteria; ampicillin-resistant colonies were screened using the ^{32}P-labeled 14-mer 5'TGCCACCCCACCAG. Restriction mapping identified a plasmid pLA-2, containing the desired deletion.

The cleavage-site mutations were produced using 40-mer oligonucleotides encoding the mutation and the procedures described above. For convenience, the mutations were introduced into plasmid pDGR-2 encoding an abridged factor VIII molecule. All mutations were verified by DNA sequencing (Sanger et al. 1979).

COS cell transfections and analysis of procoagulant activity. COS cells (Gluzman 1981) were transfected with 8 μg of each plasmid per 10-cm dish in 4 ml of DEAE-dextran in Dulbecco's modified Eagle's conditioned medium as described previously (Kaufman 1985). Forty-eight hours posttransfection, 10 ml of fresh media containing 10% fetal bovine serum was applied, and samples were taken 24 hours later for factor VIII activity assay as described by Toole et al. (1984).

Immunoprecipitation analysis. Cells were transfected and, 48 hours posttransfection, labeled with 0.5 mCi/ml of [^{35}S]methionine (sp. act. >800 Ci/mmole; New England Nuclear) in the presence of serum-free media. Six hours later, equal aliquots of conditioned media were taken for immunoprecipitation with a monoclonal antibody (Hybritech) that recognizes the light chain (80 kD) for factor VIII, using rabbit anti-mouse IgG as the precipitating antibody. Immunoprecipitates were analyzed by SDS-reducing-gel electrophoresis as described by Kaufman and Sharp (Kaufman and Sharp 1982).

Thrombin analysis of factor VIII. Recombinant human factor VIII was purified from conditioned medium using monoclonal antibody affinity chromatography and ion-exchange chromatography. The polypeptide content of the purified factor VIII preparations was monitored by SDS-PAGE and high-performance liquid chromatography (HPLC) before and after digestion with thrombin. Samples were digested and analyzed as follows: After ion-exchange chromatography, fractions were digested for 0–15 minutes at 2 U/ml thrombin (Sigma) at 37°C in 4 mM 4-morpholine ethanesulfonic acid (MES) (pH 6.25), 0.02% Tween 80, and 80 mM NaCl. At various times, aliquots were subjected to SDS-PAGE (8% acrylamide) according to the method of Laemmli (1970), and bands were visualized after silver staining according to the method of Merril et al. (1981). Similar aliquots were subjected to HPLC, and thrombin cleavage products were detected by their adsorbance at 280 nm. Various HPLC fractions were subjected to amino-terminal sequence analysis, using an Applied Biosystems gas-phase sequenator according to the method of Hewick et al. (1981).

RESULTS

We recently reported the expression of low levels of biologically active factor VIII from mammalian cells (Toole et al. 1984). We have now obtained greatly increased levels of expression that allows the purification of recombinant factor VIII. To compare the recombinant molecules with the plasma-derived molecule, we purified factor VIII from both sources. The purification from plasma or conditioned media was done using conventional chromatographic techniques in addition to an immunoaffinity step employing an anti-factor VIII monoclonal antibody. Figure 2 shows a silver-stained SDS gel of plasma (Fig. 2A) and recombinant (Fig. 2B) factor VIII before and after exposure to thrombin.

We consistently observe that recombinant factor VIII is of much greater integrity than what we and others obtained from plasma (Fulcher et al. 1983; Rotblat et al. 1985; Eaton et al. 1986). Both recombinant and plasma factor VIII were cleaved with thrombin to yield similar patterns as observed on SDS-PAGE. The various thrombin fragments were isolated and subjected to microsequence analysis as described in Methods. The thrombin cleavage sites of the recombinant factor VIII were identical to those reported by Eaton et al. (1986), except that we found a cleavage site at residue 1331 and,

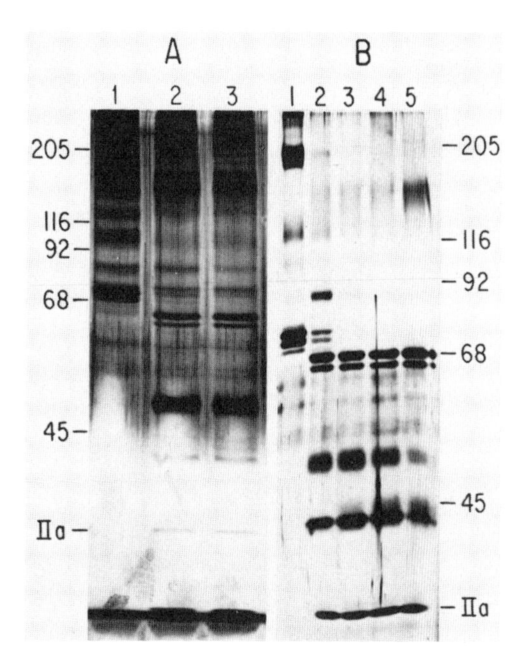

Figure 2. SDS-PAGE analysis of thrombin-cleaved human factor VIII. (*A*) Plasma-derived factor VIII before (lane *1*) and after exposure to 2 μ/ml of thrombin for 1 min (lane *2*) and 5 min (lane *3*). (*B*) Recombinant factor VIII before (lane *1*) and after exposure to 2 μ/ml of thrombin for 1, 5, 10, and 30 min (lanes *2–5*, respectively).

with prolonged exposure to thrombin, cleavage at residue 336. This latter site is the inactivating cleavage site of APC and factor Xa (Eaton et al. 1986). A map of the thrombin cleavage sites is shown in Figure 1C.

As thrombin cleavage appears to be effecting the excision of the B domain and because this region shows striking porcine-human sequence divergence, we asked whether segments of polypeptide could be removed without a loss of biological activity. We used either restriction enzymes or site-directed loop-out mutagenesis to remove a segment of human factor VIII cDNA encoding the B domain and tested the protein encoded by the new constructs for procoagulant activity. Restriction enzymes *Sac*I and *Bam*HI and an oligonucleotide linker allowed us to remove cDNA, encoding 581 amino acids corresponding to residues 1001 through 1581, while maintaining the translational reading frame. We also utilized a 40-mer oligonucleotide to loop out cDNA encoding 880 amino acids, following the procedure described by Morinaga et al. (1984). This larger deletion begins 19 amino acids after the thrombin cleavage site at residue 740 and ends 50 amino acids before the thrombin cleavage site at residue 1689. The positions of the deletions in the factor VIII molecule are shown schematically in Figure 3. The smaller deletion removes 13 potential asparagine-linked glycosylation sites; the larger deletion removes 18 of the 20 total sites in the B domain. Factor VIII molecules missing 581 or 880 amino acids are encoded by plasmids pDGR-2 and pLA-2, respectively.

The factor VIII cDNAs deficient in B-domain polypeptide were inserted into the mammalian cell expression vector pXMT-2, a derivative of p91023(B) (Kaufman 1985; Wong et al. 1985). We have previously demonstrated that a similar vector (pCVSVL) directs the synthesis of biologically active factor VIII when the DNA is introduced by transfection into the African green monkey cell line COS-1 (Gluzman 1981).

The expression plasmids containing the modified factor VIII cDNAs and the full-length cDNA, pXMT-VIII, were introduced into COS-1 cells via the DEAE-dextran transfection protocol (Sompayrac and Dana 1981). Conditioned media were harvested 72 hours posttransfection and assayed for factor VIII activity as described previously (Toole et al. 1984). The results of the experiment are summarized in Table 1. Surprisingly, both plasmids containing the deleted factor VIII cDNAs yielded procoagulant activity, and, moreover, the activity was greater than that obtained from wild-type cDNA. One possible explanation for this increased activity level is that these novel molecules are in a conformation that mimics to some extent the proteolytically activated form of the protein. To examine this, we determined the ability of thrombin to activate, as determined by a factor-VIII-deficient plasma assay, the modified factor VIII molecules in the conditioned media. If the abridged factor VIII molecules were in a partially or fully activated conformation, then exposure to catalytic amounts of thrombin should result in a decreased enhancement of activity. We found that the novel forms of factor VIII had their activity enhanced to the same extent (~20–30-fold) as wild type after exposure to thrombin. We conclude from these data that removal of up to 880 amino acids (~95,000 daltons) in a defined domain of human factor VIII does not destroy cofactor activity. Furthermore, these abridged procoagulant proteins retain their ability to be activated by thrombin.

We next wanted to determine the reason for the increased activity observed in the conditioned medium of pDGR-2 and pLA-2 transfectants. To do this, cells were labeled with [^{35}S]methionine at 48 hours posttransfection, and conditioned medium from the cells was immunoprecipitated using a monoclonal antibody specific for the 80-kD light chain of human factor VIII. The precipitate was then subjected to SDS-PAGE autoradiography. As shown in Figure 4, considerably more of the 80-kD polypeptide was precipitated from the media of cells transfected with pDGR-2 compared with pXMT-VIII (wild type). The increase in immunopre-

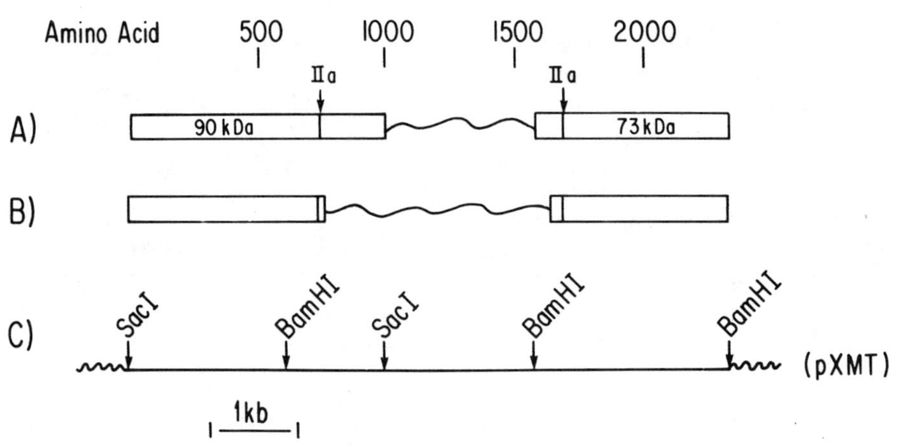

Figure 3. Abridged factor VIII molecules. (*A,B*) Wavy lines depict the regions deleted from human factor. The molecules diagramed in *A* and *B* are encoded by plasmids pDGR-2 and pLA-2, respectively. (*C*) Restriction map showing the relative location of *Bam*HI and *Sac*I sites in human factor VIII cDNA. Also shown in *A* are the thrombin (IIa) sites that yield the 90-kD and 73-kD polypeptides.

Table 1. Expression of Abridged Factor VIII Molecules

	Number of amino acids deleted	Chromogenic activity (mU/ml^{-1})	Clotek activity	
			− IIa	+ IIa (fold)
No DNA	–	0		
pXMT-VIII	–	15.1	–	450
pDGR-2	581	114	250	5750 (23 ×)
pLA-2	880	162	330	9240 (28 ×)

The plasmids indicated were transfected into COS cells, and 48 hr post-transfection, the conditioned media were changed and taken for assay by the Kabi Coatest factor VIII:C method (chromogenic activity) and by the one-stage-activated partial thromboplastin time (APTT) coagulation assay (Clotek activity), using factor VIII-C-deficient plasma. For thrombin activation, samples were pretreated 1–10 min with 0.2 U/ml thrombin (IIa) at room temperature. Activation coefficients are provided in parentheses. Activity from media from the wild-type (pXMT-VIII) transfection was too low to measure directly Clotek activity before thrombin activation. From other experiments where the wild-type factor VIII activity was concentrated, it was demonstrated to be approximately 30-fold activatable. Also, separate cotransfection experiments have demonstrated that the differences in activities are not due to differences in transfection efficiencies.

cipitable radiolabeled factor VIII approximately reflects the differences in activities observed in the cultures (~tenfold). Thus, the removal of polypeptide from the B domain of human factor VIII results in procoagulant proteins having a specific activity similar to that of the natural protein. However, these abridged forms of factor VIII accumulate to higher levels, compared with wild type, in media conditioned by the transfected cells. Preliminary data indicate that the increased accumulation is not due to a change in half-life but to an increased synthetic and/or secretion rate.

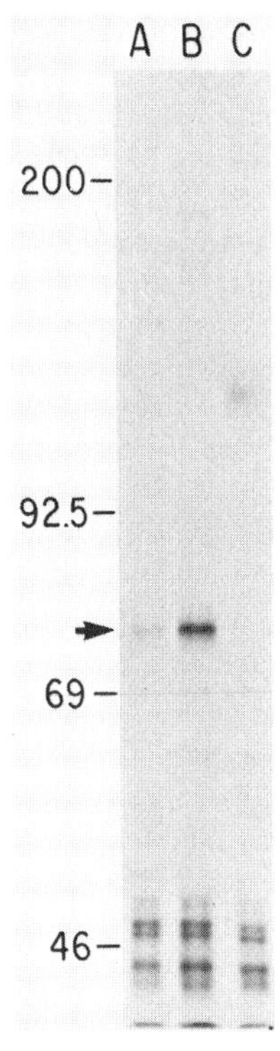

Figure 4. Factor VIII synthesis and secretion in transfected COS cells. COS cells were transfected and labeled with [^{35}S]methionine at 48-54 hr posttransfection. [^{35}S]Methionine-labeled factor VIII was detected by immunoprecipitation and autoradiography as described in Methods. Results with transfection of pXMT-VIII (*A*), pDGR-2 (*B*), and mock (no DNA) (*C*) are presented. The arrow indicates the predicted 80-kD species reactive with the anti-factor VIII monoclonal antibody. The heavy chain of factor VIII is not detected in this analysis.

We next wanted to determine which of the known cleavage sites are critical for the activation of factor VIII. To examine this, we used site-directed mutagenesis to change the arginine codon to either an isoleucine codon or a lysine codon. The arginine is the only amino-terminal residue known to be cleaved by thrombin, APC, or factor Xa. We again employed the loop-out mutagenesis procedure of Morinaga et al. (1984), using pDGR-2 (described above) as the parental plasmid. After the mutants were identified and isolated and the DNA sequence was verified, the plasmids were transfected onto COS cells, and the conditioned media assayed as described above. The results are summarized in Table 2. All mutant plasmids, with the exception of pCSM-372, directed the synthesis of biologically active factor VIII at levels comparable to that of pDGR-2. pCSM-372 mutates the cleavage site that produces the 50-kD and 43-kD polypeptides from the 90-kD precursor. These results support those of Eaton et al. (1986), who demonstrated that, after exposure to thrombin, maximal factor VIII activity coincided with the generation of the 50-kD and 43-kD polypeptides.

Table 2. Mutational Analysis of Cleavage Sites in Factor VIII

Plasmid	Amino acid change	Activity mU/ml/day	IIa activity
pDGR	581 a.a.Δ	288	+
pCSM-336	R_{336} ——K	431	+
pCSM-372	R_{372} ——I	0	–
pCSM-740	R_{740} ——I	114	+
pCSM-1648	R_{1648} ——I	246	+

Expression of factor VIII molecules containing thrombin cleavage-site mutations. The plasmids indicated were transfected into COS cells, and activity was assayed as described in Table 1. The one letter code is used for amino acids and the subscript refers to the corresponding residue number of mature factor VIII.

Although production of the 90-kD polypeptide generally precedes the appearance of the 50-kD species, our results suggest that this cleavage is not required for the cleavage at Arg-372, since pCSM-740 directed the synthesis of factor VIII. We cannot, however, rule out that alternative sites proximal to residue 740 are being cleaved, which allows the 50-kD and 43-kD polypeptides to be generated. We are presently isolating [^{35}S]methionine-labeled factor VIII from the various conditioned media in order to examine the thrombin-cleavage patterns produced from the mutant proteins. We are also determining whether pCSM-336 actually yields a protein resistant to APC inactivation.

DISCUSSION

We have purified recombinant factor VIII from mammalian cell cultures and have determined the fragmentation of the molecule after exposure to thrombin. The kinetics and pattern of cleavage observed were similar to those obtained from plasma-derived factor VIII. Microsequence analysis of the recombinant factor VIII fragments revealed two cleavage sites previously unidentified. One site, within the B domain at residue 1313, is probably inconsequential with regard to activation, since this region is dispensable for activity. The other site at residue 336 has previously been reported to be the site of factor Xa and APC cleavage (Eaton et al. 1986). Thus, thrombin is analogous to factor Xa in that both enzymes have the specificity that allows for both activation and inactivation of the factor VIII molecule.

Site-directed mutagenesis of several cleavage sites demonstrated that cleavage at residue 372, which produces the 50-kD and 43-kD fragments from a 90-kD precursor, is critical for the activation of factor VIII. It also implies that the uncleaved form of factor VIII has little, if any, procoagulant activity. Alternatively, the lack of activity exhibited by the Arg-372 mutation could be that arginine per se, not necessarily cleavage, is critical. Thrombin availability, therefore, probably does not measure the enhancement of activity but instead is an indication of the amount of factor VIII already cleaved to yield an activated form. If the cleavages at residues 336 and 372 are inactivating and activating, respectively, it implies a functional significance of those 36 amino acids. This region, like the 41-amino-acid peptide bounded by the 80-kD cleavage sites, is highly acidic. It is possible that one or both of these regions must be intact for the subunits to remain associated, possibly by coordination of Ca^{++} ion. Alternatively, these regions of factor VIII could be important for interaction with its substrates, factors IX and X.

We also utilized mutagenesis to demonstrate that up to 880 amino acids from the B domain of factor VIII, a region that exhibits an unusually high degree of porcine-human amino acid sequence divergence (Toole et al. 1986), can be removed without a concomitant loss of procoagulant activity as measured in vitro. Recent work suggests that the factor VIII molecule lacking the 880-amino-acid segment is also active in hemophilic dogs (G. Amphlett and A. Giles, pers. comm.). The fact that these abridged molecules are active is highly reminiscent of factor V, another blood-clotting factor that is structurally related and acts analogously to factor VIII (Church et al. 1984). Factor Va activity can be reconstituted from thrombin cleavage fragments that lack a large internal region of the molecule (Esmon 1979).

What role can be ascribed to the B domain? As an activation peptide, it is an unusually large fraction of the molecule. The activation peptides for the serine proteases of the blood-clotting cascade are generally much smaller. Specifically, the activation peptide of human factor X is 52 residues out of a 448-amino-acid zymogen. Interestingly, the bovine-human amino acid sequence homology is a mere 14% within the activation peptide, although much higher overall (Fung et al. 1985).

It is possible that the B domain of factor VIII is important for interaction with the von Willebrand factor (vWF), a large plasma glycoprotein that noncovalently associates with factor VIII in plasma. This interaction is important, since patients lacking von Willebrand factor also have severely reduced levels of plasma factor VIII. We are now initiating studies to determine whether the abridged forms of factor VIII bind to vWF-Sepharose beads. Another interesting possibility for the function of this region is that it may be involved in intracellular processing or storage in the cells that normally process factor VIII in vivo or that the B domain (or proteolytic products derived therefrom) has procoagulant, anticoagulant, or vasoactive properties heretofore unknown. For example, the vasodilator bradykinin is a nonapeptide proteolytically released from high-molecular-weight kininogen that, like factor VIII, participates in the intrinsic pathway of blood coagulation.

ACKNOWLEDGMENTS

We acknowledge the technical assistance of James Booth for performing thrombin activation, Dr. Simon Jones for oligonucleotides, John Leger for chromogenic assays, Katherine Smith for coordination and project management, Gabriel Schmergel for his continued support and encouragement during this project, and Marybeth Erker for typing the manuscript. The characterization of recombinant wild-type and plasma-derived factor VIII was supported by Baxter Travenol, Inc. Part of the remaining work was supported by a Small Business Innovation Research grant (1R43HL-35946-01) from the Department of Health and Human Services.

REFERENCES

Church, W.R., R.L. Jernigan, J.J. Toole, R.M. Hewick, J. Knopf, G.J. Knutson, M.E. Nesheim, K.G. Mann, and

D.N. Fass. 1984. Coagulation factors V and VIII and ceruloplasmin constitute a family of structurally related proteins. *Proc. Natl. Acad. Sci.* **81:** 6934.

Eaton, D., H. Rodriquez, and G.A. Vehar. 1986. Proteolytic processing of human factor VIII. Correlation of specific cleavages by thrombin, factor Xa, and activated protein C with activation and inactivation of factor VIII coagulant activity. *Biochemistry* **25:** 505.

Esmon, C.T. 1979. The subunit structure of thrombin activated factor V. Isolation of activated factor V, separation of subunits, and reconstitution of biological activity. *J. Biol. Chem.* **254:** 964.

Foster, W.B., M.E. Nesheim, and K.G. Mann. 1983. The factor Xa-catalyzed activation of factor V. *J. Biol. Chem.* **259:** 3187.

Fulcher, C., J.R. Roberts, and T.S. Zimmerman. 1983. Thrombin proteolysis of purified factor VIII procoagulant protein: Correlation with generation of a specific polypeptide. *Blood* **61:** 807.

Fung, M.R., C.W. Hay, and R.T.A. MacGillivray. 1985. Characterization of an almost full-length cDNA coding for human blood coagulation factor X. *Proc. Natl. Acad. Sci.* **82:** 3591.

Gitschier, J., W.I. Wood, T.M. Goralka, K.L. Wion, E.Y. Chen, D.H. Eaton, G.A. Vehar, D.J. Capon, and R.M. Lawn. 1984. Characterization of the human factor VIII gene. *Nature* **312:** 326.

Gluzman, Y. 1981. SV40-transformed simian cells support the replication of early SV40 mutants. *Cell* **23:** 175.

Hewick, R.M., M.W. Hunkapiller, L.E. Hood, and W.J. Dreyer. 1981. A gas-liquid solid phase peptide and protein sequenator. *J. Biol. Chem.* **256:** 7990.

Kaufman, R. 1985. Identification of the components necessary for adenovirus translational control and their utilization in cDNA expression vectors. *Proc. Natl. Acad. Sci.* **82:** 689.

Kaufman, R.J. and P.A. Sharp. 1982. Amplification and expression of sequences cotransfected with a modular dihydrofolate reductase complementary DNA gene. *J. Mol. Biol.* **159:** 601.

Kernoff, P.B.A., N.D. Thomas, P.A. Lilley, K.B. Mathews, E. Goldman, and E.G.D. Tuddenham. 1984. Clinical experience with polyelectrolyte-fractionated porcine factor VIII concentrate in the treatment of hemophiliacs with antibodies to factor VIII. *Blood* **63:** 31.

Laemmli, U.K. 1970. Cleavage of structural proteins during the assembly of the head of bacteriophage T_4. *Nature* **227:** 680.

Merril, C.R., D. Goldman, S.A. Sedman, and M.H. Ebert. 1981. Ultrasensitive strain for proteins in polyacrylamide gels shows regional variation in cerebrospinal fluid proteins. *Science* **211:** 1437.

Morinaga, Y., T. Franceschini, S. Inonye, and M. Inonye. 1984. Improvement of oligonucleotide-directed site-specific mutagenesis using double-stranded plasmid DNA. *Biotechnology* **84:** 636.

Nesheim, M.E., K.H. Myrmel, L. Hibbard, and K.G. Mann. 1979. Isolation and characterization of single chain bovine factor V. *J. Biol. Chem.* **254:** 508.

Rotblat, R., D.P. O'Brien, F.J. O'Brien, A.H. Goodall, and E.G.D. Tuddenham. 1985. Purification of human factor VIII:C and its characterization by western blotting using monoclonal antibodies. *Biochemistry* **24:** 4294.

Sanger, F., S. Nicklen, and A.R. Coulson. 1977. DNA sequencing with chain-terminating inhibitors. *Proc. Natl. Acad. Sci.* **74:** 5463.

Sompayrac, L.M. and K.J. Dana. 1981. Efficient infection of monkey cells with DNA of simian virus 40. *Proc. Natl. Acad. Sci.* **78:** 7575.

Toole, J.J., J.L. Knopf, J.M. Wozney, L.A. Sultzman, J.L. Buecker, D.D. Pittman, R.J. Kaufman, E. Brown, C. Shoemaker, E.C. Orr, G.N. Amphlett, W.B. Foster, M.L. Coe, G.S. Knutson, D.N. Fass, and R.M. Hewick. 1984. Molecular cloning of a cDNA encoding human antihaemophilic factor. *Nature* **312:** 342.

Toole, J.J., D.D. Pittman, E.C. Orr, P. Murtha, L.C. Wasley, and R.J. Kaufman. 1986. A large region (∼95kDa) of human factor VIII is dispensable for *in vitro* procoagulant activity. *Proc. Natl. Acad. Sci.* **83:** 5939.

van Dieijen, G., G. Tans, J. Rosing, and H.C. Hemker. 1981. The role of phospholipid and factor VIII in the activation of bovine factor X. *J. Biol. Chem.* **256:** 3433.

Vehar, G.A., B. Keyt, D. Eaton, H. Rodriguez, D.P.O. O'Brien, F. Rotblat, H. Oppermann, R. Keck, W.I. Wood, R.W. Harkins, E.G.D. Tuddenham, R.M. Lawn, and D.J. Capon. 1984. Structure of human factor VIII. *Nature* **312:** 337.

Vogelstein, B. and D. Gillespie. 1979. Preparative and analytical purification of DNA from agarose. *Proc. Natl. Acad. Sci.* **76:** 615.

Wong, G.C., J.S. Witek, P.A. Temple, K.M. Wilkens, A.C. Leary, D.P. Luxenberg, S.S. Jones, E.L. Brown, R.M. Kay, E.C. Orr, C.S. Shoemaker, D.W. Golde, R.J. Kaufman, R.M. Hewick, E.A. Wang, and S.C. Clark. 1985. Human GM-CSF: Molecular cloning of the complimentary DNA and purification of the natural and recombinant proteins. *Science* **228:** 810.

Wood, W.I., D.J. Capon, C.C. Simonsen, D.L. Eaton, J. Gitschier, B. Keyt, P.H. Seeburg, D.H. Smith, P. Hollingshead, K.L. Wion, E. Delwart, E.G.D. Tuddenham, G.A. Vehar, and R.M. Lawn. 1984. Expression of active human factor VIII from recombinant DNA cloned. *Nature* **312:** 330.

Characterization Studies of Human Tissue-type Plasminogen Activator Produced by Recombinant DNA Technology

G.A. Vehar, M.W. Spellman, B.A. Keyt, C.K. Ferguson, R.G. Keck, R.C. Chloupek, R. Harris, W.F. Bennett, S.E. Builder, and W.S. Hancock
Genentech, Inc., South San Francisco, California 94080

A crucial aspect of hemostasis in mammals is the control of the formation and dissolution of a fibrin matrix. The enzymatic system in plasma for fibrin formation consists of more than a dozen proteases and cofactors that act in series (Jackson and Nemerson 1980). Localization of these reactions is achieved by the requirement for the phospholipid surface, which is supplied by activated platelets at the site of trauma. Regulation of fibrin formation is achieved through control of protease activation, the requirement for a phospholipid surface at certain steps of the cascade, and feedback activation and inactivation reactions involving coagulation factors, other plasma proteins, and cell-surface proteins. The resulting fibrin matrix serves to prevent unwanted blood loss at the site of an injury.

Equal in importance to fibrin formation, fibrin dissolution is accomplished by a much simpler series of enzymatic reactions. The removal of fibrin is accomplished through the proteolytic cleavage of the insoluble fibrin strands into soluble fragments by the action of the serine protease plasmin (for a recent review, see Danø et al. 1985). Control of fibrin dissolution is achieved by the extent and localization of the activation of the glycoprotein plasminogen, the circulating zymogen form of plasmin. The best-known mammalian plasminogen activators are urokinase and tissue plasminogen activator (t-PA) (Danø et al. 1985). Through selective release and subsequent localization of urokinase, t-PA, and plasmin(ogen), selective and localized dissolution of fibrin is accomplished.

Although the body can effectively remove fibrin in the process of hemostasis, a number of human disease states result from the presence of unwanted blood clots. The desirability of a means to enhance the endogenous rate of clot lysis has long been recognized as essential to the effective treatment of such disease states as myocardial infarction, pulmonary embolism, deep vein thrombosis, and stroke (Verstraete 1980). This has led to the development of therapeutics (urokinase and streptokinase) that systemically activate plasminogen in the hope that sufficient plasmin will adsorb to the clot to enhance the rate of clot lysis. Although streptokinase and urokinase have both been used for such thrombolytic therapy, their usefulness in the management of thromboembolic disease remains to be established (Brogden et al. 1973; Paoletti and Sherry 1977). Unfortunately, plasmin is a relatively nonspecific protein (Robbins 1978; Castellino and Powell 1981; Robbins et al. 1981) that, as it circulates, will degrade fibrinogen as well as inactivate the coagulation factors V and VIII. This unwanted systemic proteolysis can lead to a serious bleeding state that can be as serious as the complication necessitating treatment.

For these reasons, a more effective and safe fibrinolytic agent is required. The ideal activator of plasminogen would selectively activate plasminogen in the vicinity of a clot while effecting minimal systemic activation. t-PA possesses these properties. This plasminogen activator specifically binds fibrin (Thorsen et al. 1972; Rijken and Collen 1981), and its ability to activate plasminogen is markedly increased in the presence of fibrin (Danø et al. 1985). These two properties (fibrin binding and activity enhancement) would be expected to give t-PA the specificity required of an effective and specific fibrinolytic agent. Increasing the level of t-PA in plasma was therefore expected to accelerate clot lysis while having minimal effect on circulating plasminogen. Although preliminary studies on t-PA produced by a melanoma cell line were promising (Weimar et al. 1981; Van de Werf et al. 1984), it was not clear whether sufficient material could be produced to make a sufficient fibrinolytic product at a cost-effective price using the natural sources (Collen et al. 1982). The application of recombinant DNA techniques to solve this problem led initially to the expression of the protein in bacteria (Pennica et al. 1983). We present here a report on subsequent work directed toward the production and characterization of recombinant tissue plasminogen activator (rt-PA) expressed in Chinese hamster ovary (CHO) cells, and we describe some of the physical properties of the resulting protein.

EXPERIMENTAL PROCEDURES

Protein. Recombinant t-PA (Activase[R]) is produced by Genentech, Inc. (South San Francisco, California). The protein was purified from the supernatants of CHO cells that have been transfected with a t-PA expression plasmid (Goeddel et al. 1983).

Amino acid sequence analysis. The protein samples were analyzed on the Beckman 890C protein sequencer for 11 cycles using 0.1 M Quadrol. The first cycle was performed without the addition of the coupling agent phenylisothiocyanate (PITC) to remove free amino acids that may coelute with the protein sample during the purification process. The next ten cycles were run with the addition of PITC and analyzed for PTH amino acids using a Waters M6000 high-performance liquid chromatography (HPLC) system with a 5-μm C_8-reversed-phase Microsorb column as described by Rodriquez et al. (1984). The amino acids were identified by comparison of retention times to a standard mixture of PTH amino acids.

Electrophoresis. SDS-PAGE was performed on 10% polyacrylamide gels according to the procedure of Laemmli (1970). Protein was visualized by staining with Coomassie blue R-250.

Plasmin activation of t-PA. rt-PA was converted to the two-chain form with human plasmin (Helena Laboratories). Plasmin (50 casein units) was covalently coupled to cyanogen bromide preactivated Sepharose (Pharmacia). rt-PA was applied to the immobilized plasmin column (5 ml) at a flow rate of 10 ml/hr. Conversion of rt-PA to the two-chain form was greater than 95% as determined by SDS-gel electrophoresis.

Separation of kringle and protease domains. Plasmin-treated rt-PA was reduced and carboxymethylated according to the procedure of Crestfield et al. (1963). The reduced, carboxymethylated two-chain t-PA (3 mg/2 ml) was applied to a Sephadex G-75 superfine column (1.5 × 100 cm) equilibrated in 0.1 M ammonium bicarbonate at pH 8.3. The column was developed under gravity flow, and the effluent was monitored for absorbance at 280 nm.

Tryptic digestion. Recombinant t-PA was reduced and carboxymethylated (RCM) according to the method of Crestfield et al. (1963). The product was digested in 0.1 M ammonium bicarbonate at 25°C with an addition of L-1-*p*-tosylamino-2-phenylethyl chloromethyl ketone (TPCK)-treated trypsin (Cooper Biomedical) at an enzyme-to-substrate ratio of 1:100 (w/w) followed by a second addition after 8 hours. The digestions were stopped after 24 hours by freezing at −20°C.

HPLC tryptic mapping. The trifluoroacetic acid (TFA)-based tryptic map of rt-PA was achieved on a 5-μm Nova Pak-C18 column. Mobile phase A consisted of 0.1% TFA, and mobile phase B consisted of acetonitrile with 0.08% TFA. The linear gradient was carried out at a rate of 0.5% per minute for 25 minutes, followed by 1.0% per minute for 35 minutes. The sodium-phosphate-based tryptic map of rt-PA was performed under the same conditions except that 0.05 M sodium phosphate was the ionic modifier. The samples were loaded in 0.1 M ammonium bicarbonate and monitored at 214 and 280 nm. The separation was achieved on a Waters gradient liquid chromatograph that included two 610 pumps, a 720 controller, and a WISP injector.

Sialic acid determination. Sialic acid content was determined by the thiobarbituric acid (TBA) procedure of Warren (1959).

Glycosyl residue composition. Neutral and amino sugars were identified and quantitated by gas-liquid chromatography after hydrolysis in 4 N TFA (Nesser and Schweizer 1984) and conversion to alditol acetates (Albersheim et al. 1967).

Glycosyl-linkage analysis. Glycosyl-linkage composition was determined by methylation analysis (Lindberg 1972) as modified by Waeghe et al. (1983). Methylation was carried out with methyl iodide and methylsulfinylcarbanion in dimethyl sulfoxide. Methylated oligosaccharides were recovered by chromatography on C-18 Sep-Pak cartridges (Waters, Inc.) and then hydrolyzed, reduced, and acetylated. The resulting partially methylated alditol acetates were resolved and identified by combined gas-liquid chromatography/mass spectrometry.

RESULTS AND DISCUSSION

Structure of t-PA

The identification of cDNA clones that included the coding region of human t-PA resulted in the first complete picture of the primary structure of the protein (Pennica et al. 1983). The cDNA revealed a leader sequence of 35 amino acids preceding the amino-terminal serine residue of the secreted protein. The first 20–23 amino acids of the leader sequence are hydrophobic, typical of a signal sequence. This is followed by 12–15 hydrophilic amino acids of unknown function that do not appear on the secreted form of the protein. The function, if any, of this pre-pro-leader structure is unknown at present. Following processing of the pre-pro t-PA, the secreted form of the protein can be obtained as a single polypeptide chain containing 527 amino acids, 35 cysteine residues, and 4 potential N-linked glycosylation sites (Pennica et al. 1983; Pohl et al. 1984; Vehar et al. 1984). The protein sequences revealed by clones obtained by several independent laboratories are in agreement on the amino acid sequence of the protein, although some nucleic acid sequence differences have been noted (Ny et al. 1984; Fisher et al. 1985; Kaufman et al. 1985). Only one gene has been detected in the human genome (Pennica et al. 1983; Ny et al. 1984).

On the basis of sequence homology with other proteins, t-PA is believed to be a multidomain protein (Fig. 1). Starting from the amino-terminal end, it has been proposed to contain a potential type I finger domain (Banyai et al. 1983), a growth-factor domain (Banyai et al. 1983), two kringle domains (Pennica et al. 1983), and a serine protease or catalytic domain (Pennica et al. 1983).

A proteolytic cleavage site at Arg-275 (Pennica et al. 1983) divides the protein into two approximately equally sized polypeptide chains. The chain arising

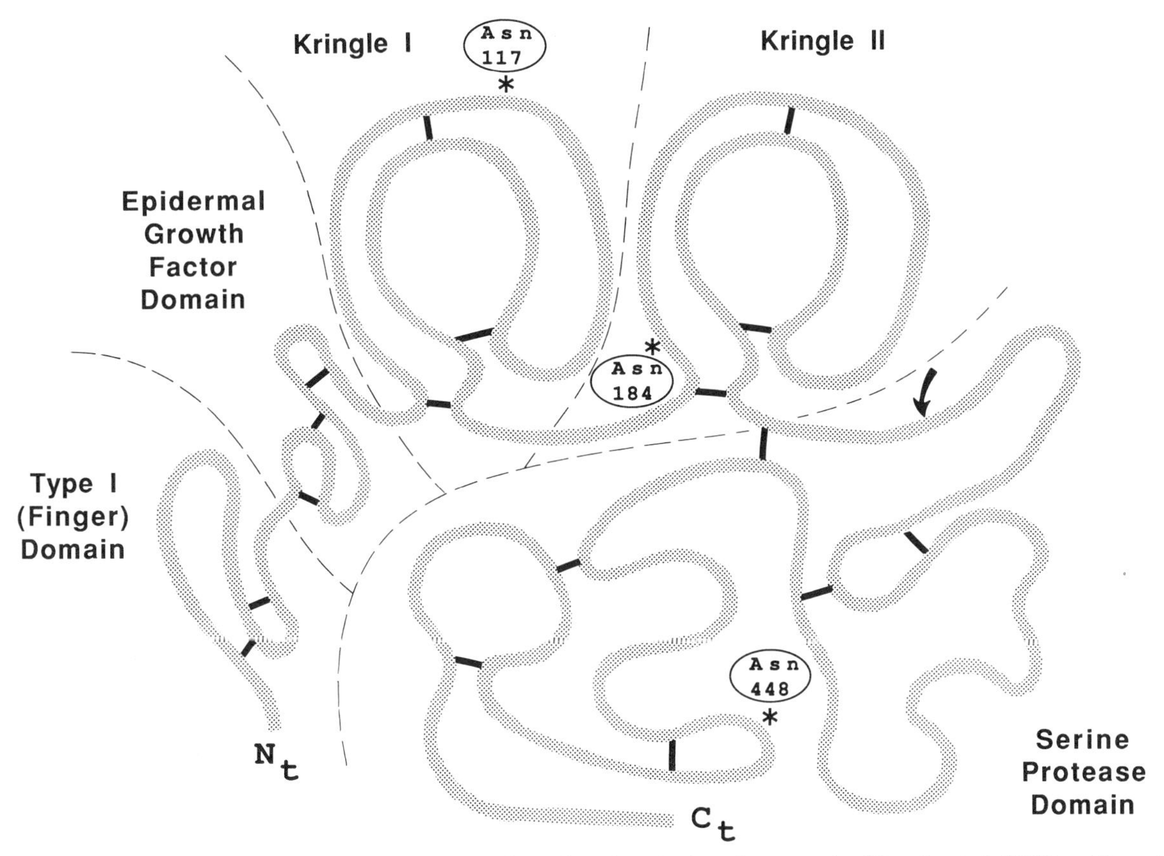

Figure 1. Schematic diagram of the domain structure of t-PA based on the domain proposals of Pennica et al. (1983) and Banyai et al. (1983). Stippled line represents the polypeptide chain. Solid bars represent the disulfide bonds. Sites of glycosylation are identified by asterisks, with the glycosylated asparagine residue numbers identified in the circle. Arrow indicates the position of the cleavage site at Asn-275. The amino(N_t)- and carboxy(C_t)-terminal ends are labeled. Dashed lines delineate the borders of the different domains.

from the carboxy-terminal portion of t-PA is the serine protease domain and shares considerable amino acid sequence homology with other members of this family of proteins (Strassburger et al. 1983). By analogy with other serine proteases, five intradomain disulfide bonds have been proposed for 10 of the 11 cysteine residues of this region of t-PA (Pennica et al. 1983). The extra cysteine residue of the protease domain has been proposed to connect the protease to the polypeptide chain derived from the amino terminus of the protein (Pennica et al. 1983). In this manner, the catalytic domain of t-PA remains covalently attached to the remaining domains following the proteolytic cleavage at position 275.

The amino-terminal 275 amino acids of t-PA contain the several proposed domains. The first 43 amino acids share limited sequence homology with type I finger structures (containing two disulfide bonds) (Banyai et al. 1983) that were originally identified in fibronectin (Peterson et al. 1983). These structures occur in a fragment of fibronectin known to be responsible for the fibrin affinity of the protein (Sekiguchi et al. 1981; Seidl and Hoermann 1983). By analogy with fibronectin, it was originally proposed that the type I structure was involved with the affinity of t-PA for fibrin (Banyai et al. 1983). However, it has subsequently been reported that expression of a form of t-PA lacking the finger domain results in a protein that will still bind fibrin (Kagitani et al. 1985). Further work will be required to clarify the function of this structure in t-PA.

Six amino acids separate the type I structure from a 34-amino-acid, three-disulfide-bond-containing region that shares sequence homology with growth factors (Banyai et al. 1983). Analogous growth-factor-like domains have been proposed to occur in factor X, factor IX, factor XII, protein C, and urokinase (Banyai et al. 1983; McMullen and Fujikawa 1985). The vitamin-K-dependent proteases (factor IX, factor X, protein C) each contains two growth-factor domains, of which one domain contains a β-hydroxyaspartic acid residue (Fernlund and Stenflo 1983; McMullen et al. 1983). An aspartic acid does not occur in the corresponding position of t-PA, such that this growth-factor domain would not be expected to contain this posttranscriptional modification. The function, if any, of the growth-factor domains in these proteases is unknown.

The growth-factor domain is separated by seven amino acids from the first of two proposed kringle structures (Pennica et al. 1983). A six-amino-acid segment separates the two t-PA kringle structures. Kringles were first identified in prothrombin (Magnusson et al. 1975) and have subsequently been found in plasminogen (Claeys et al. 1976), urokinase (Günzler et al. 1982), and factor XII (McMullen and Fujikawa 1985). The first of the five plasminogen kringle structures has been shown to be responsible for the interaction of this protein with fibrin (Lerch et al. 1980; Thorsen et al. 1981). It is not known at present whether the t-PA kringles play a comparable role in interactions with fibrin.

Two amino acids follow the second kringle domain before the cysteine involved in the disulfide link to the protease is positioned (Pennica et al. 1983). The eleventh amino acid following this cysteine is Arg-275, which immediately precedes the serine protease domain. t-PA is structured such that very short polypeptide segments separate the various domains. This gives the protein considerable resistance to proteolysis in the native state, possibly reflecting the fact that the protein must function in the presence of the relatively nonspecific protease plasmin.

Host Selection

Escherichia coli was the expression system of choice for the early recombinant products (e.g., somatostatin, insulin, and the interferons). A number of factors were involved with this selection, including high expression levels, ease of growth, well-known genetics, suitable plasmids, and efficient promoters. The successful production of such products as human growth hormone eased concerns over the possibility of improper disulfide formation due to the intracellular reducing environment of *E. coli*. The production of fully active γ-interferon demonstrated that the lack of a glycosylation system in *E. coli* would not necessarily affect the product quality. *E. coli* was therefore the host of choice for expression of t-PA. Successful expression was achieved as evidenced by the ability of *E. coli* extracts to cause the dissolution of fibrin matrix (Pennica et al. 1983). More efficient expression was, however, achieved through the application of mammalian cell expression systems, wherein a properly processed and glycosylated molecule could be obtained from the cell-conditioned media (Collen et al. 1984; Zamarron et al. 1984).

Recombinant pharmaceutical products derived from the mammalian cell lines present unique safety concerns (Petricciani 1985). These can be addressed by careful validation of both the master working cell bank (Lubiniecki and May 1984; Martin 1985; Lubiniecki 1986; Palladino et al. 1986) and the process used to purify the products (Jones and O'Connor 1983).

The purification process for the production of rt-PA results in a highly purified product as shown in Figure 2. Nonreduced, a closely spaced doublet is observed that exhibits an apparent molecular weight of approximately 50,000. Upon reduction, the majority of the protein migrates with an apparent molecular weight of approximately 63,000. In addition, a faint band is visible with an apparent molecular weight of 35,000, indicating that a small amount of material has been

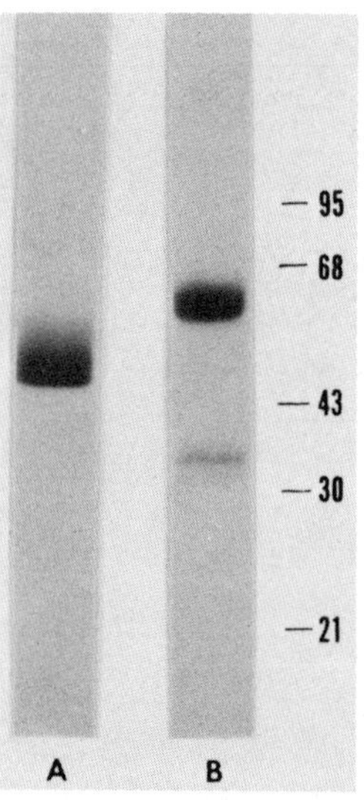

Figure 2. SDS-PAGE analysis of rt-PA. (*A*) Sample prepared without reduction; (*B*) sample reduced with mercaptoethanol prior to electrophoresis. The mobilities of molecular-weight standards are indicated at the right of the figure, with the molecular weight indicated ($\times$ 1000).

cleaved to the two-chain form. Amino acid sequence analysis of the material confirms that this cleavage has occurred at Arg-275.

Plasmin-catalyzed Cleavage of t-PA

The large size of rt-PA complicates detailed analysis of the protein. Advantage was taken of the fact that limited proteolysis of rt-PA results in the generation of two polypeptide chains of approximately equal size. A single interdomain disulfide bond (Cys-264–Cys-395) has been proposed to link the kringle-containing domain (1–275) with the serine protease region (276–527). The action of plasmin on t-PA results in a single-peptide-bond cleavage that yields a two-chain, disulfide-bonded form of the protein (Wallen et al. 1983; Pohl et al. 1984). To confirm the site of plasmin cleavage within the t-PA molecule, the individual plasmin fragments of t-PA were prepared, isolated, and sequenced. rt-PA was fully converted to its two-chain form with immobilized plasmin and then dialyzed into 8 M urea, 0.5 M Tris-Cl at pH 8.3 as a modification of the procedure of Crestfield et al. (1963). Plasmin-treated t-PA was reduced with 10 mM dithiothreitol at 37°C for 30 minutes and carboxymethylated with 25 mM iodoacetate for 20 minutes at 25°C. The reduced and carboxymethylated kringle-containing region (RCM 1–275) and protease domain (RCM 276–527) were separated by gel filtration on a column of superfine Sephadex G-75 as seen in the elution profile (Fig. 3, top). RCM 276–527 and RCM 1–275 eluted in separate peaks as identified by SDS-gel electrophoresis of the collected fractions (Fig. 3, bottom). The chromatographic elution profile of this column was characterized by nonideal behavior of the separated components. RCM 276–527 and RCM 1–275

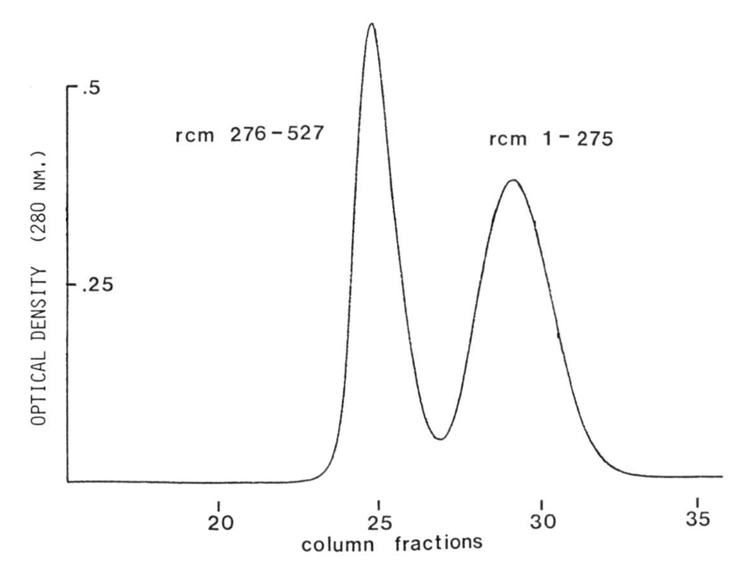

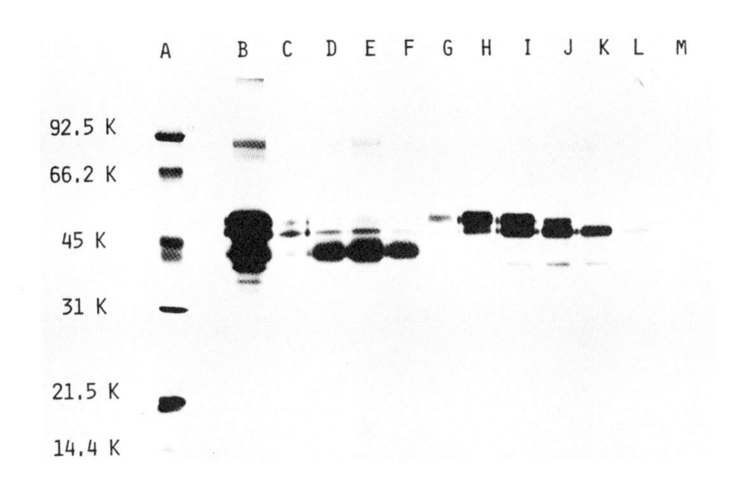

Figure 3. (*Top*) Gel filtration of RCM 1-275 and RCM 276-527. Elution profile of Sephadex G-75 superfine column (1.5 × 100 cm) in 0.1 M ammonium bicarbonate at pH 8.3. An aliquot (3 mg) of reduced and carboxymethylated t-PA (two-chain) was loaded onto the column. Fractions (5 ml) were monitored for optical density at 280 nm. The early and late eluting peaks were identified by amino acid sequencing and SDS-gel electrophoresis (Fig. 3) as protease (RCM 276-527) and kringle (RCM 1-275), respectively. (*Bottom*) Silver stain (Oakley) of G-75 profile. The elution profile of the Sephadex G-75 separation of two-chain RCM rt-PA (Fig. 2, top) was analyzed by SDS-PAGE. Clear separations of protease (lanes *D-F*) and kringle (lanes *H-K*) can be seen. Type I and type II kringles are also partially resolved by gel filtration (lanes *H* vs. *K*). (*A*) Molecular-weight standards; (*B*) RCM rt-PA, column load; (*C*) fraction 23; (*D*) fraction 24; (*E*) fraction 25; (*F*) fraction 26; (*G*) fraction 27; (*H*) fraction 28; (*I*) fraction 29; (*J*) fraction 30; (*K*) fraction 31; (*L*) fraction 32; (*M*) fraction 33.

were not expected to be resolved by gel filtration as they have similar molecular weights (28,600 and 31,900, respectively). Both components of the RCM mixture eluted with high molecular weights as determined by comparison with gel-filtration standards (data not shown). These separated fractions of rt-PA were useful for further characterization of this complex glycoprotein. Amino acid sequencing of RCM 1-275 and RCM 276-527 confirmed the first ten cycles for the amino terminus of rt-PA (serine 1) and for the site of plasmin cleavage after Arg-275 (Fig. 4, top and bottom). The presence of cysteine at position six of the amino terminus of the kringle-containing region was established by the yield of carboxymethyl cysteine. The purified fractions of rt-PA, RCM 1-275 and RCM 276-527, are well resolved and unequivocally identified by SDS-gel electrophoresis. The apparent size heterogeneity of the kringle-containing region (1-275, type I and type II) is clearly resolved by reduction and carboxymethylation of the sample prior to SDS-gel electrophoresis (Fig. 3, bottom). Further characterization of the structure and sequence of rt-PA was accomplished by examination of tryptic fragments.

Peptide Analysis of rt-PA

Reversed-phase HPLC has been demonstrated to be a powerful analytical technique suitable for the analysis of low-molecular-weight pharmaceuticals (U.S. Pharmacopeia 1985). It was hoped that this procedure would prove useful for the characterization of different production lots of rt-PA.

In 1978, it was shown that reversed-phase HPLC could be used to separate efficiently mixtures of peptides prepared by tryptic digestion of low-molecular-weight proteins (Hancock et al. 1978). Subsequently, this technique was extensively used for sequence determination of proteins, in which the tryptic peptides were isolated for characterization (Olieman and Voskamp 1984). In this application, the addition of TFA to the mobile phase as a volatile ionic modifier was particularly successful, as the purified peptides could be read-

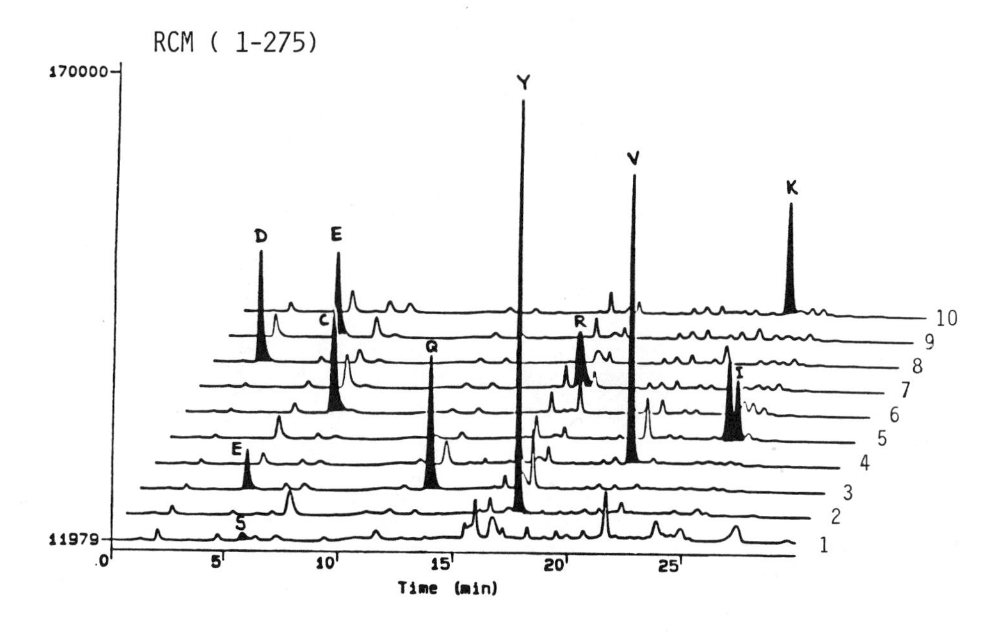

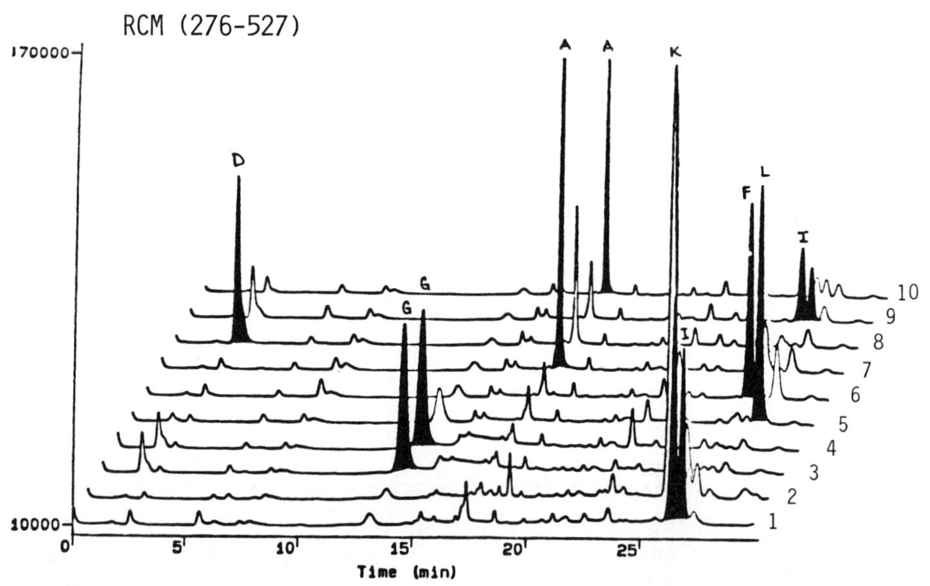

Figure 4. (*Top*) HPLC profiles obtained from ten cycles of automated amino acid sequencing RCM 1–275, i.e., kringle domain of rt-PA. (*Bottom*) HPLC profiles obtained from ten cycles of automated amino acid sequencing RCM 276–527, i.e., protease domain of rt-PA.

ily isolated in a salt-free form by lyophilization (Hearn and Hancock 1979; Olieman and Voskamp 1984). However, a number of studies have demonstrated that the addition of phosphate salts instead of TFA to the mobile phase gave improved peak shapes, better recoveries, and improved column lifetimes (Hancock and Harding 1984). The use of involatile phosphate severely hinders characterization of the purified peptides. Therefore, elution profiles of the reversed-phase tryptic map were performed for both the TFA and phosphate-containing mobile phases.

On the basis of the primary structure, tryptic digestion of rt-PA would be expected to yield 50 peptides. The elution profiles for the tryptic maps shown in Figures 5 and 6 demonstrate that reversed-phase HPLC can indeed resolve so complex a mixture. Approximately 44 and 51 peptides are resolved with the TFA and phosphate mobile phases, respectively. The higher number of peptides resolved in the phosphate-based system can be related to the increased resolving power of reversed-phase HPLC with this mobile phase due to mixed-mode chromatography, where the peptides in-

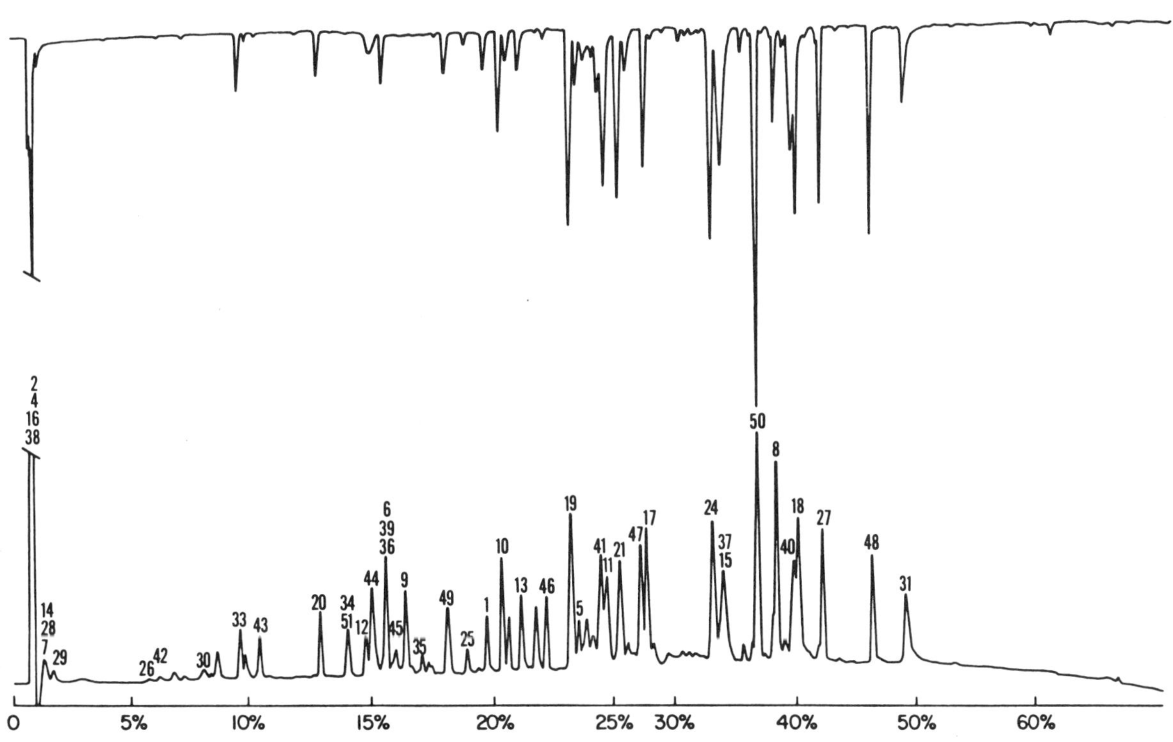

Figure 5. TFA-based tryptic map of rt-PA. A tryptic digest of rt-PA (200 μg in 0.2 ml) was applied to the column and analyzed as described in Experimental Procedures.

teract with the stationary phase by both polar and nonpolar interactions (Hancock and Sparrow 1981).

The presence of peptides in addition to the expected tryptic fragments can be related to heterogeneity in the protein caused by side reactions such as deamidation (particularly under the conditions of the tryptic digestion). Also, chymotryptic-like cleavages can result in additional peptide fragments. However, all peptide fragments were found to be consistent with the proposed structure for rt-PA. The average recovery of each peptide was approximately 50% of the 18 nmoles injected on the reversed-phase column.

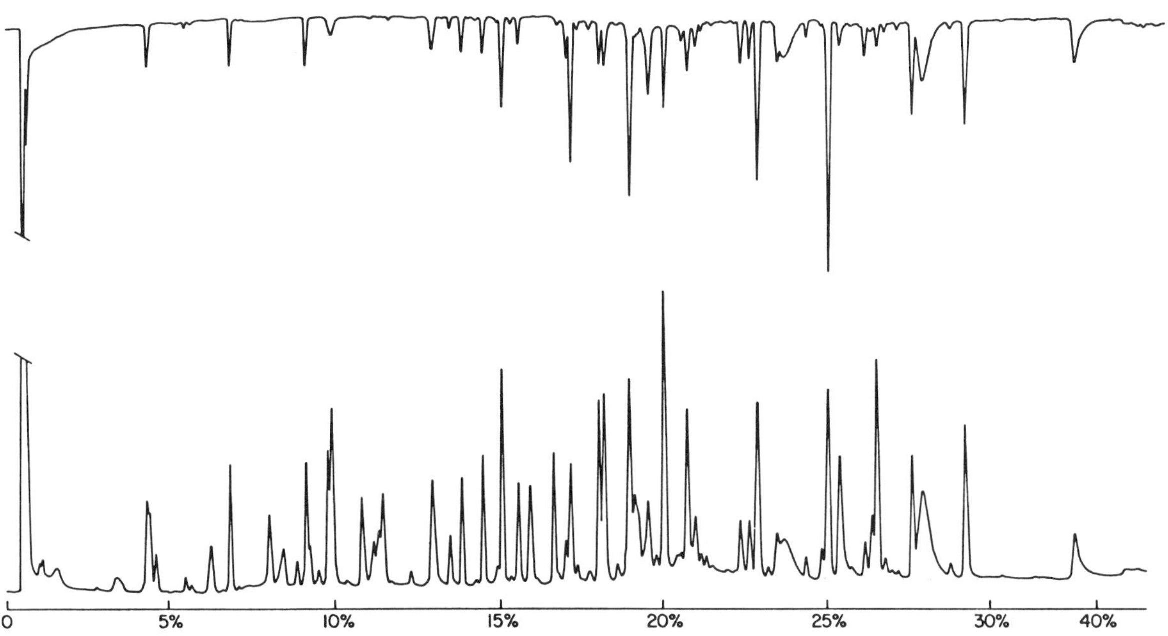

Figure 6. Sodium-phosphate-based tryptic map of rt-PA. A tryptic digest of rt-PA (70 μg in 0.1 ml) was applied to the column and analyzed as described in Experimental Procedures.

Figure 7 shows the primary structure proposed for rt-PA based on the cDNA sequence. Also shown is the location of the tryptic peptides reported by Pohl ct al. (1984). All of the expected tryptic peptides were demonstrated to be present in the reversed-phase tryptic map. Each eluted peptide was identified by a combination of amino acid analysis and sequencing techniques. The identification of the tryptic peptides in the reversed-phase HPLC separation is shown in Figure 7. The order of elution follows approximately predicted retention times based on polarity measurements of the side-chain functional groups (Sasagawa and Teller 1984).

Carbohydrate Characterization of rt-PA

Oligosaccharide side chains of mammalian glycoproteins can be linked via the hydroxyl group of a serine or threonine residue (O-linked) or via the amide nitrogen of an asparagine residue (N-linked). O-linked oligosaccharides usually contain *N*-acetylgalactosamine and may also contain galactose and sialic acid. N-linked oligosaccharides are divided into two major categories: high-mannose oligosaccharides, composed of mannose and *N*-acetylglucosamine, and complex oligosaccharides, composed of mannose, *N*-acetylglucosamine, galactose, fucose, and sialic acid (Kornfeld and Kornfeld 1980).

The amino acid sequence of t-PA includes four potential N-linked glycosylation sites (Asn-X-Ser/Thr) (Marshall 1972): asparagine residues 117, 184, 218, and 448 (Pennica et al. 1983). Position 218, however, is not glycosylated in melanoma t-PA (Pohl et al. 1984; Vehar et al. 1984). The amino acid sequence at this site is Asn-Pro-Ser. The presence of proline in the middle position has been shown to interfere with glycosylation in other systems (Marshall 1974; Scheffer and Beintema 1974) and presumably is the cause for the absence of carbohydrate on Asn-218.

Melanoma-derived t-PA has been shown to exist in two variants (types I and II) that differ in carbohydrate content (Bennett 1983; Pohl et al. 1985). Melanoma-derived type I t-PA was found to contain N-linked glycans at positions 117, 184, and 448, whereas the type II t-PA lacks the glycan at position 184 (Bennett 1983; Pohl et al. 1985). The carbohydrate composition of melanoma-derived t-PA has been further investigated by Rijken et al. (1985). The protein was found to contain sialic acid, fucose, mannose, galactose, *N*-acetylglucosamine, and *N*-acetylgalactosamine. The type I and II forms were found to contain a high proportion of mannose, suggesting the presence of high-mannose

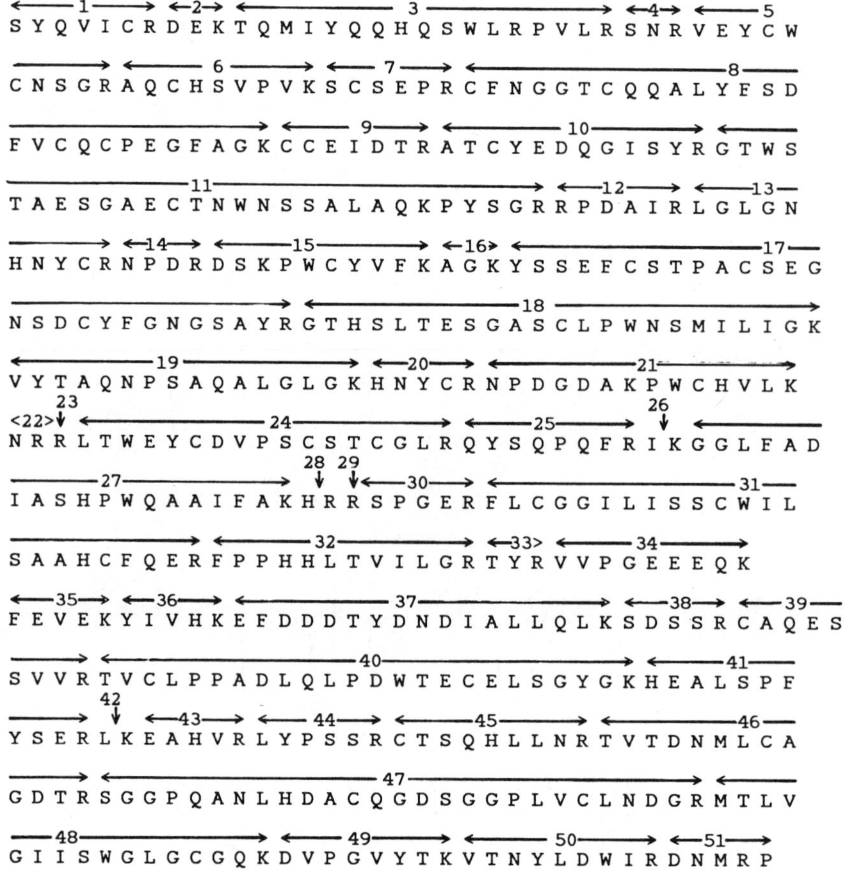

Figure 7. Linear sequence of rt-PA with tryptic peptides. Arrows define each peptide, and the number refers to the identity of the peptide listed in Fig. 4.

oligosaccharides. It has also been reported that both forms of melanoma-derived t-PA contain a high-mannose oligosaccharide at position 117 (Pohl et al. 1985).

The availability of glycosylated rt-PA allowed the characterization of the carbohydrate structures. The total carbohydrate content was determined using the thiobarbituric acid assay for sialic acid and the alditol acetate method for neutral and amino sugars (Table 1). The results indicate that rt-PA contains approximately 7% carbohydrate by weight. The presence of sialic acid, mannose, galactose, *N*-acetylglucosamine, and fucose was consistent with N-linked glycans of the fucosylated complex type, but the mannose content was higher than would be expected if the protein contained only complex oligosaccharides. The results suggest that rt-PA contains N-linked oligosaccharides of both the high-mannose and complex types.

The glycosyl linkage composition was determined by methylation analysis of pronase-digested rt-PA (Table 2). The linkage composition of rt-PA was consistent with a mixture of high-mannose and fucosylated complex oligosaccharides. The comparatively large amount of terminal mannose indicates the presence of high-mannose oligosaccharides. The linkage pattern observed with galactose indicates that sialic acid is attached to the 3-position of this residue in the complex oligosaccharides. Only a small amount of terminal galactose was observed, indicating that sialylation is nearly complete. Biantennary complex oligosaccharides contain 3,6-linked mannose as branching point; the presence of 2,4-linked mannose and 2,6-linked mannose indicates that rt-PA also contains some tri- and/or tetraantennary structures.

The attachment positions of the N-linked oligosaccharides were determined by characterizing tryptic glycopeptides. rt-PA was reduced, carboxymethylated, and digested with trypsin. Tryptic glycopeptides were isolated by lectin chromatography on concanavalin A–Sepharose followed by reversed-phase HPLC and were identified by amino acid analysis (Table 3). The major peptide containing residue 117 resulted from "chymotryptic" cleavage at tyrosine residue 126, whereas the peptides containing residues 184 and 448 resulted from the expected tryptic cleavage.

The neutral and amino sugar compositions of the isolated glycopeptides were determined by the alditol acetate method (Table 3). This analysis indicates that position 117 contains an oligosaccharide of the high-mannose type (probably $Man_5GlcNAc_2$ or $Man_6GlcNAc_2$). Positions 184 and 448 were found to contain fucosylated complex oligosaccharides.

Table 1. Carbohydrate Composition of Intact rt-PA

Residue[a] (mole/mole rt-PA)				
Fuc	Man	Gal	GlcNAc	NeuAc[b]
2.5	9.8	3.7	7.4	2.7

[a]Abbreviations: Fuc, fucose; Man, mannose; Gal, galactose; GlcNAc, *N*-acetylglucosamine; NeuAc, *N*-acetylneuraminic acid (sialic acid).

[b]Determined by thiobarbituric acid assay.

Table 2. Glycosyl Linkage Composition of rt-PA

Residue	Positions of O-methyl groups[a]	Deduced linkage	Area[b] (%)
Fuc	2,3,4	terminal	8
Man	2,3,4,6	terminal	13
Man	3,4,6	2-linked	11
Man	3,6	2,4-linked	2
Man	3,4	2,6-linked	trace
Man	2,4	3,6-linked	17
Gal	2,3,4,6	terminal	4
Gal	2,4,6	3-linked	16
GlcNAc	2,3,6	4-linked	27
GlcNAc	2,3	4,6-linked	3

[a]Determined from electron-impact mass spectra.

[b]From gas-liquid chromatography using flame-ionization detection.

In summary, rt-PA contains a high-mannose oligosaccharide at residue 117 and complex oligosaccharides at residues 184 and 448. This N-linked glycosylation pattern is similar to that reported for melanoma-derived t-PA (Pohl et al. 1985).

CONCLUSIONS

The ability to safely increase the rate of lysis of unwanted fibrin clots in the vasculature has been a goal for many years. Current therapeutics have unwanted side effects that complicate safe and effective therapy. The efficacy of melanoma-cell-derived t-PA for such use was first demonstrated in two patients with venous thrombosis (Weimer et al. 1981). Subsequent studies of acute myocardial infarction documented t-PA-induced thrombolysis without significant fibrinogenolysis (Van de Werf et al. 1984). These studies indicated that t-PA had the properties of an ideal fibrinolytic agent. Unfortunately, the t-PA production level of these cell lines was sufficiently low that it was not clear whether sufficient material could be produced to treat effectively the more than one million potential patients that could benefit from the availability of a superior thrombolytic agent.

The generation of CHO cells capable of producing human t-PA has necessitated the development of large-scale tissue-culture fermentation and purification procedures (Builder and Grossbard 1986). The resulting product is highly purified and for the most part consists of the single chain form. The rt-PA has full catalytic and biologic activity (Collen et al. 1984; Zamarron et al. 1984). Analysis of the protein and its fragments has proven the fidelity of the CHO cells synthesizing a protein identical in amino acid sequence to the t-PA produced by human melanoma cells. The only differences due to expression in a nonhuman cell line would arise in the nature of the carbohydrate moieties. The availability of large amounts of rt-PA has allowed the preliminary analysis of these structures.

Table 3. Characterization of Tryptic Glycopeptides

Peptide[a]	Attachment residue no.	Molar ratio Fuc	Man	Gal	GlcNAc	Glycosylation type
GTWSTAESGAECT-NWNSSALAQKPY	117	0.3	5.8	0.4	2.0	high mannose[b]
YSSEFCSTPACSE-GNSDCYFGNGSAYR	184	0.8	3.0	1.9	2.0	complex[c]
CTSQHLLNR	448	0.9	3.0	1.5	2.6	complex[c]

[a]Identification based on amino acid analysis.
[b]Normalized to 2 *N*-acetylglucosamine.
[c]Normalized to 3 mannose.

Human clinical trials have shown rt-PA (Activase[R]) to be an effective thrombolytic agent (Topol et al. 1985; Gold et al. 1986; Graor et al. 1986; Risius et al. 1986). Continued clinical testing will allow the evaluation of the potential benefit of rt-PA infusions in patients presenting with myocardial infarction, pulmonary embolism, deep vein thrombosis, and peripheral arterial occlusion. The success to date with rt-PA provides another example of the tremendous potential of recombinant DNA technology.

ACKNOWLEDGMENT

We acknowledge the invaluable contributions of Dr. Andrew Jones to the characterization of rt-PA.

REFERENCES

Albersheim, P., D.J. Nevins, P.D. English, and A. Karr. 1967. A method for the analysis of sugars in plant cell-wall polysaccharides by gas-liquid chromatography. *Carbohydr. Res.* **5:** 340.

Banyai, L., A. Varadi, and L. Patthy. 1983. Common evolutionary origin of the fibrin-binding structures of fibronectin and tissue-type plasminogen activator. *FEBS Lett.* **163:** 37.

Bennett, W.F. 1983. Two forms of tissue-type plasminogen activator (tPA) differ at a single specific glycosylation site. *Thromb. Res.* **50:** 106.

Brogden, R.N., T.M. Speight, and G.S. Avery. 1973. Streptokinase: A review of its clinical pharmacology, mechanism of action and therapeutic uses. *Drugs* **5:** 357.

Builder, S.E. and E. Grossbard. 1986. Laboratory and clinical experience with recombinant plasminogen activator. In *Transfusion medicine: Recent technological advances* (ed. K. Murawski and F. Peetoom), p. 303. A.R. Liss, New York.

Castellino, F.J. and J.R. Powell. 1981. Human plasminogen. *Methods Enzymol.* **80:** 365.

Claeys, H., L. Sottrup-Jensen, M. Zajdel, T.E. Peterson, and S. Magnusson. 1976. Multiple gene duplication in the activation of plasminogen. Five regions of structural homology with the two internally homologous structures in prothrombin. *FEBS Lett.* **61:** 20.

Collen, D., D.C. Rijken, J. Van Damme, and A. Billiau. 1982. Purification of human tissue-type plasminogen activator in centigram quantities from human melanoma cell culture fluid and its conditioning for use in vivo. *Thromb. Haemostasis.* **48:** 294.

Collen, D., J.M. Stassen, B.F. Marafino, Jr., S. Builder, F. De Cock, J. Ogez, D. Tajiri, D. Pennica, W.F. Bennett, J. Salwa, and C.F. Hoyng. 1984. Biological properties of human tissue-type plasminogen activator obtained by expression of recombinant DNA in mammalian cells. *J. Pharmacol. Exp. Ther.* **231:** 146.

Crestifeld, A.M., S. Moore, and W.H. Stein. 1963. The preparation and enzymatic hydrolysis of reduced and *S*-carboxymethylated proteins. *J. Biol. Chem.* **238:** 622.

Danø, K., P.A. Andreasen, J. Grondahl-Hansen, P. Kristensen, L.S. Nielsen, and L. Skriver. 1985. T1 Plasminogen activators, tissue degradation, and cancer. *Anticancer Res.* **5:** 605.

Fernlund, P. and J. Stenflo. 1983. β-Hydroxyaspartic acid in vitamin K-dependent proteins. *J. Biol. Chem.* **258:** 12509.

Fisher, R., D.K. Waller, G. Grossi, D. Thompson, R. Tizard, and W.-D. Schleuning. 1985. Isolation and characterization of the human tissue-type plasminogen activator structural gene including its 5′ flanking region. *J. Biol. Chem.* **260:** 11223.

Goeddel, D.V., W.J. Kohr, D. Pennica, and G.A. Vehar. 1983. Human tissue plasminogen activator, pharmaceutical compositions containing it, processes for making it, and DNA and transformed intermediates therefor. *United Kingdom* Patent No. 2119804.

Gold, H.K., R.C. Leinbach, H.D. Garabedian, T. Yasuda, J.A. Johns, E.B. Grossbard, I. Palacios, and D. Collen. 1986. Acute coronary reocclusion after thrombolysis with recombinant human tissue-type plasminogen activator: Prevention by a maintenance infusion. *Circulation* **73:** 347.

Graor, R.A., B. Risius, J.R. Young, K. Denny, E.G. Beven, M.A. Geisinger, N.R. Hertzer, L.P. Krajewski, F.V. Lucas, P.J. O'Hara, W.F. Ruschhaupt, S. Winton, M.G. Zelch, and E.B. Grossbard. 1986. Peripheral artery and bypass graft thrombolysis with recombinant tissue-type plasminogen activator. *J. Vasc. Surg.* **3:** 115.

Günzler, W.A., G.J. Steffens, F. Otting, S.A. Kim, E. Frankus, and L. Flohe. 1982. The primary structure of high molecular mass urokinase from human urine: The complete amino acid sequence of the A chain. *Hoppe-Seyler's Z. Physiol. Chem.* **363:** 1155.

Hancock, W.S. and D.R.K. Harding. 1984. Review of separation conditions for peptides. In *CRC handbook of HPLC for the separation of amino acids, peptides, and proteins* (ed. W.S. Hancock), vol. II, p. 3. CRC Press, Boca Raton, Florida.

Hancock, W.S. and J.T. Sparrow. 1981. The use of mixed mode HPLC for the separation of peptide and protein mixtures. *J. Chromatogr.* **206:** 71.

Hancock, W.S., C.A. Bishop, J.E. Battersby, D.R.K. Harding, and M.T.W. Hearn. 1978. Reversed phase HPLC for peptide mapping of proteins. *Anal. Biochem.* **89:** 203.

Hearn, M.T.W. and W.S. Hancock. 1979. Ion-pair partition reversed phase HPLC: A new method for the rapid analysis and isolation of underivatised amino acids, peptides, and protein. *Trends Biochem. Sci.* **4:** N58.

Jackson, C.M. and Y. Nemerson. 1980. Blood coagulation. *Annu. Rev. Biochem.* **49:** 765.

Jones, A.J.S. And J.V. O'Connor. 1985. Control of recombinant DNA produced pharmaceuticals by a combination of process validation and final product specifications. *Dev. Biol. Stand.* **59:** 175.

Kagitani, H., M. Tagawa, K. Hatanaka, T. Ikari, A. Saito, H. Bando, K. Okada, and O. Matsuo. 1985. Expression in *E. coli* of finger-domain lacking tissue-type plasminogen activator with high fibrin affinity. *FEBS Lett.* **189:** 145.

Kaufman, R.J., L.D. Wasley, A.J. Spiliotes, S.D. Gossels, S.A. Latt, G.R. Larsen, and R.M. Kay. 1985. Coamplification and coexpression of human tissue-type plasminogen activator and murine dihydrofolate reductase sequences in Chinese hamster ovary cells. *Mol. Cell. Biol.* **5:** 1750.

Kornfeld, R. and S. Kornfeld. 1980. Structure of glycoproteins and their oligosaccharide units. In *The biochemistry of glycoproteins and proteoglycans* (ed. W.J. Lennarz), p. 1. Plenum Press, New York.

Laemmli, U.K. 1970. Cleavage of structural proteins during the assembly of the head of bacteriophage T4. *Nature* **227:** 680.

Lerch, P.G., E.E. Rickli, W. Lergier, and D. Gillessen. 1980. Localization of individual lysine-binding regions in human plasminogen and investigations on their complex-forming properties. *Eur. J. Biochem.* **107:** 7.

Levinson, A.D., L.P. Svedersky, and M.A. Palladino, Jr. 1985. Tumorigenic potential of DNA derived from mammalian cell lines. *In Vitro Monogr.* **6:** 161.

Lindberg, B. 1972. Methylation analysis of polysaccharides. *Methods. Enzymol.* **28B:** 178.

Lubiniecki, A. 1986. Safety considerations for cell culture-derived biologicals. In *Large scale cell culture technology.* (ed. K. Lydersen). Hanser GmbH, Munich. (In press.)

Lubiniecki, A.L. and L.H. May. 1984. Cell bank characterization for recombinant DNA mammalian cell lines. *Dev. Biol. Stand.* **60:** 141.

Martin, D.W., Jr. 1985. An overview of risk association with viruses endogenous to cell substrates. *In Vitro Monogr.* **6:** 29.

Magnusson, S., T.E. Peterson, L. Sottrup-Jensen, and H. Claeys. 1975. The complete primary structure of prothrombin: Isolation, structure, and reactivity of ten carboxylated glutamic acid residues and regulation of prothrombin activation by thrombin. *Cold Spring Harbor Conf. Cell Proliferation* **2:** 123.

Marshall, R.D. 1972. Glycoproteins. *Annu. Rev. Biochem.* **41:** 673.

———. 1974. Nature and metabolism of the carbohydrate/peptide linkages of glycoproteins. *Biochem. Soc. Symp.* **40:** 17.

McMullen, B.A. and K. Fujikawa. 1985. Amino acid sequence of the heavy chain of human factor XIIa (activated Hageman factor). *J. Biol. Chem.* **260:** 5328.

McMullen, B.A., K. Fujikawa, and W. Kisiel. 1983. The occurrence of β-hydroxyaspartic acid in the vitamin K-dependent blood coagulation zymogens. *Biochem. Biophys. Res. Commun.* **115:** 8.

Nesser, J.R. and J.F. Schweizer. 1984. A quantitative determination by capillary gas-liquid chromatography of neutral and amino sugars (as *O*-methyloxime acetates), and a study on hydrolytic conditions for glycoproteins and polysaccharides in order to increase sugar recoveries. *Anal. Biochem.* **142:** 58.

Ny, T., F. Elgh, and B. Lund. 1984. The structure of the human tissue-type plasminogen activator gene: Correlation of intron and exon structures to functional and structural domains. *Proc. Natl. Acad. Sci.* **81:** 5355.

Olieman, C. and D. Voskamp. 1984. Perfluoroalkanoic acids. Mobile phase modifiers. In *CRC handbook of HPLC for the separation of amino acids, peptides, and proteins* (ed. W.S. Hancock), vol. I, p. 161. CRC Press, Boca Raton, Florida.

Palladino, M.A., A.D. Levinson, L.P. Svedersky, and J.F. Obijeski. 1986. Safety issues related to the use of recombinant DNA derived cell culture products. I. Cellular components. In *International association of biological standardization.* (In press.)

Paoletti, R. and S. Sherry, eds. 1977. *Thrombosis and urokinase.* Academic Press, London.

Pennica, D., W.E. Holmes, W.J. Kohr, R.N. Harkins, G.A. Vehar, C.A. Ward, W.F. Bennett, E. Yelverton, P.H. Seeburg, H.L. Heyneker, D.V. Goeddel, and D. Collen. 1983. Cloning and expression of human tissue-type plasminogen activator cDNA in *E. coli. Nature* **301:** 214.

Peterson, T.E., H.C. Thogersen, K. Skorstengaard, K. Vibe-Pedersen, P. Sahl, L. Sottrup-Jensen, and S. Magnusson. 1983. Partial primary structure of bovine plasma fibronectin: Three types of internal homology. *Proc. Natl. Acad. Sci.* **80:** 137.

Petricciani, J.C. 1985. Regulatory considerations for products derived from the new biotechnology. *Pharm. Manuf.* **2(3):** 31.

Pohl, G., M. Einarsson, B. Nilsson, and S. Svensson. 1985. The size heterogeneity in melanoma tissue plasminogen activator is caused by carbohydrate differences. *Thromb. Res.* **50:** 163.

Pohl, G., M. Kollstrom, N. Bergsdorf, P. Wallen, and H. Jornvall. 1984. Tissue plasminogen activator: Peptide analyses confirm an indirectly derived amino acid sequence, identify the active site serine residue, establish glycosylation sites, and localize variant differences. *Biochemistry* **23:** 3701.

Rijken, D.C. and D. Collen. 1981. Purification and characterization of the plasminogen activator secreted by human melanoma cells in culture. *J. Biol. Chem.* **256:** 7035.

Rijken, D.C., J.J. Emeis, and G.J. Gerwig. 1985. On the composition and function of the carbohydrate moiety of tissue-type plasminogen activator from human melanoma cells. *Thromb. Haemostasis* **54:** 788.

Risius, B., R.A. Graor, M. Geisingeer, M. Zelch, F.V. Lucas, J.R. Young, and E. Grossbard. 1986. Recombinant human tissue-type plasminogen activator for thrombolysis in peripheral arteries and bypass grafts. *Radiology* (in press).

Robbins, K.C. 1978. Plasmin. *Fibrinolytics Antifibrinolytics* **46:** 317.

Robbins, K.C., L. Summaria, and R.C. Wohl. 1981. Human plasma. *Methods Enzymol.* **80:** 379.

Rodriguez, H., W.J. Kohr, and R.N. Harkins. 1984. Design and operation of a completely automated Beckman microsequencer. *Anal. Biochem.* **140:** 538.

Sasagawa, T. and D.C. Teller. 1984. Prediction of peptide retention times in reversed phase HPLC. In *CRC handbook of HPLC for the separation of amino acids, peptides, and proteins* (ed. W.S. Hancock), vol. II, p. 53. CRC Press, Boca Raton, Florida.

Scheffer, A.J. and J.J. Beintema. 1974. Horse pancreatic ribonuclease. *Eur. J. Biochem.* **46:** 221.

Sekiguchi, K., M. Fukuda, and S.I. Hakamori. 1981. Domain structure of hamster plasma fibronectin. Isolation and characterization of four functionally distinct domains and their unequal distributions between two subunit polypeptides. *J. Biol. Chem.* **256:** 6452.

Seidl, M. and H. Hoermann. 1983. Affinity chromatography on immobilized fibrin monomer. IV. Two fibrin-binding peptides of a chymotryptic digest of human plasma fibronectin. *Hoppe-Seyler's Z. Physiol. Chem.* **364:** 83.

Strassburger, W., A. Wollmer, J.E. Pitts, I.D. Glover, I.J. Tickle, T.L. Blundell, G.J. Steffens, W.A. Günzler, F. Ötting, and L. Flohé. 1983. Adaptation of plasminogen activator sequences to known protease structures. *FEBS Lett.* **157:** 219.

Thorsen, S., P. Glas-Greenwalt, and T. Astrup. 1972. Differences in the binding to fibrin of urokinase and tissue plasminogen activator. *Thromb. Diath. Haemorrh.* **28:** 65.

Thorsen, S., I. Clemmensen, L. Sottrup-Jensen, and S. Magnusson. 1981. Adsorption to fibrin of native fragments of known primary structure from human plasminogen. *Biochim. Biophys. Acta* **668:** 377.

Topol, E.J., A.A. Ciuffo, T.A. Pearson, J. Dillman, S. Builder, E. Grossbard, M.L. Weisfeldt, and B.H. Bulkley. 1985. Thrombolysis with recombinant tissue plasminogen activator in atherosclerotic thrombotic occlusion. *J. Am. Coll. Cardiol.* **5:** 85.

U.S. Pharmacopeia. 1985. No. 21, Rockville, Maryland.

Van de Werf, F., P.A. Ludbrook, S. Bergmann, A.J. Teifenbrunn, K.A.A. Fox, H. De Geest, M. Verstraete, and D. Collen. 1984. Clot selective coronary thrombolysis with tissue-type plasminogen activator in patients with evolving myocardial infarction. *N. Engl. J. Med.* **310:** 609.

Vehar, G.A., W.J. Kohr, W.F. Bennett, D. Pennica, C.A. Ward, R.N. Harkins, and D. Collen. 1984. Characterization studies on human melanoma cell tissue plasminogen activator. *Bio/tech.* **2:** 1051.

Verstraete, M. 1980. A far-reaching program: Rapid, safe and predictable thrombolysis in man. In *Fibrinolysis* (ed. D.L. Kline and K.N.N. Reddy), p. 129. CRC Press, Cleveland.

Waeghe, T.J., A.G. Darvill, M. McNeil, and P. Albersheim. 1983. Determination, by methylation analysis, of the glycosyl-linkage compositions of microgram quantities of complex carbohydrates. *Carbohydr. Res.* **123:** 281.

Wallen, P., G. Pohl, N. Bergsdorf, M. Ranby, T. Ny, and H. Jörnvall. 1983. Purification and characterization of a melanoma cell plasminogen activator. *Eur. J. Biochem.* **132:** 681.

Warren, L. 1959. The thiobarbituric acid assay of sialic acids. *J. Biol. Chem.* **234:** 1971.

Weimar, W., J. Stibbe, A.J. Van Seyen, A. Billiau, P. De Somer, and D. Collen. 1981. Specific lysis of an iliofemoral thrombus by administration of extrinsic (tissue-type) plasminogen activator. *Lancet* **II:** 1018.

Zamarron, C., H.R. Lijnen, and D. Collen. 1984. Kinetics of the activation of plasminogen by natural and recombinant tissue-type plasminogen activator. *J. Biol. Chem.* **259:** 2080.

Biochemical and Biological Properties of Single-chain Urokinase-type Plasminogen Activator

D.C. STUMP,* H.R. LIJNEN, AND D. COLLEN
Center for Thrombosis and Vascular Research, University of Leuven, Leuven, Belgium

Conversion of the circulating proenzyme plasminogen to the active serine protease plasmin is the key step that must occur for physiological fibrinolysis to take place. Plasmin generated in this way is a very efficient enzyme for the lysis of a fibrin clot. However, if it circulates freely in plasma, it can also degrade other normal plasma proteins such as fibrinogen, factor V, and factor VIII. Absence of these key coagulation factors will ultimately lead to an anticoagulated state with potential predisposition to life-threatening bleeding. Under physiological conditions, the free circulation of active plasmin in blood is prevented by the presence of a rapid inhibitor of plasmin, α_2-antiplasmin. At the fibrin surface, however, inhibition of plasmin by α_2-antiplasmin proceeds only very slowly. As a result, most efficient plasminogen activation occurs near the fibrin clot, where the enzymatic action of plasmin is effected both without inhibition by α_2-antiplasmin and with increased specificity for fibrin as opposed to normal plasma coagulation factors (Collen 1980). The physiological importance of this process is realized with the safe maintenance of vascular patency and the resulting avoidance of tissue damage from either arterial or venous thrombosis.

The enzymes responsible for the conversion of plasminogen to plasmin are known as plasminogen activators. These serine proteases are generally classified into two immunologically distinct types. The first, tissue-type plasminogen activator (t-PA), was first purified from human uterus (Rijken et al. 1979) and later from cultured human melanoma cells (Rijken and Collen 1981). It binds directly to fibrin (Rijken and Collen 1981) where it causes more efficient plasminogen activation as a result of a fibrin-dependent increased affinity (low K_m) for its natural substrate plasminogen (Hoylaerts et al. 1982). The recognition of the potential of t-PA to induce fibrin-specific clot lysis has led to intense interest in its production by recombinant DNA methodology (Pennica et al. 1983) and its successful therapeutic application in patients with thrombotic disorders (Collen 1985). The second class of plasminogen activator is found in human urine and is designated urokinase-type plasminogen activator (u-PA). It has been purified for large-scale commercial use from both urine (Plough and Kjeldgaard 1957; White et al. 1966) and fetal kidney cell cultures (Barlow and Lazer 1972). As a two-chain polypeptide (M_r 54,000), it activates plasminogen equally well in the presence or absence of fibrin. Thus, in contrast to t-PA, u-PA causes fibrinolysis only with the generation of freely circulating plasmin and its associated hemostatic breakdown. As a result, its use in patients has not been widespread.

In 1973, it was observed that latent urokinase-like activity was present in conditioned media of some human cells (Bernik 1973), and this phenomenon was later confirmed (Nolan et al. 1977). Shortly thereafter, this latent form of urokinase was purified from urine (Husain et al. 1983), plasma (Wun et al. 1982b), and conditioned cell-culture media (Nielsen et al. 1982; Sumi et al. 1982a; Wun et al. 1982a). It was characterized physically by single-chain structure on SDS-PAGE in the presence of reductants and by immunological identity to two-chain u-PA. Functionally, this material displayed plasminogen-dependent activity on fibrin films, was inactive toward low-molecular-weight chromogenic substrates, but could be converted to active urokinase by treatment with plasmin. Because of this proenzyme-like behavior, the name "pro-urokinase" was adopted.

Pro-urokinase was found to have specific high affinity for fibrin in some studies (Sumi et al. 1982b; Husain et al. 1983; Kohno et al. 1984; Kasai et al. 1985) but not in others (Eaton et al. 1984). Nevertheless, initial studies with pro-urokinase did show potential for a considerable degree of fibrin-specific clot lysis (Sumi et al. 1982b; Collen et al. 1984b; Gurewich et al. 1984; Zamarron et al. 1984). Identification of enriched sources and development of straightforward methods for the purification of human single-chain u-PA (scu-PA) (Stump et al. 1986a,b), in addition to its cloning and expression in *Escherichia coli* (Holmes et al. 1985), have now made highly purified scu-PA more readily available for study. These materials have been utilized to characterize further the unique biochemical and biological properties of scu-PA as presented in this paper.

EXPERIMENTAL PROCEDURES

Materials

Natural scu-PA was purified from human urine and conditioned medium, from cultured Calu-3 (American Type Culture Collection, HTB-55) human lung aden-

*Permanent address: Department of Internal Medicine, University of Vermont, College of Medicine, Burlington, Vermont 05405.

ocarcinoma cells by a combination of chromatography on zinc chelate-Sepharose, SP-Sephadex C-50, and Sephadex G-100 with (urine) or without (cells) additional immunoadsorption on murine anti-human urokinase monoclonal antibody-Sepharose (Stump et al. 1986a,b). In all cases, contaminating two-chain u-PA was removed (<1%) by chromatography on benzamidine-Sepharose. Natural scu-PA was also obtained from the Mochida Chemical Company (Tokyo, Japan) as purified from a human renal adenocarcinoma cell line (American Type Culture Collection 1611). Natural two-chain u-PA (urokinase) was a gift from E. Murano (Bureau of Biologics, Bethesda, Maryland). Recombinant scu-PA (rscu-PA) was obtained after expression in *E. coli* (Holmes et al. 1985) from Genentech Inc. (South San Francisco, California) or from Grunenthal GmbH (Aachen, W. Germany).

Methods

Kinetics of plasminogen activation. Activation of plasminogen by scu-PA was measured by determination of plasmin generation in the presence of excess (1 mM) chromogenic substrate D-Val-Leu-Lys-*p*-nitroanilide (S-2251), and conversion of scu-PA to u-PA was measured with chromogenic substrate Pyroglu-Gly-Arg-*p*-nitroanilide (S-2444) (Collen et al. 1986).

Mechanism of action. Activation of plasminogen by scu-PA in plasma was determined in a similar manner after the addition of varying amounts of human plasma depleted of both plasminogen and α_2-antiplasmin (Lijnen et al. 1986). Binding of scu-PA to fibrin was measured by quantitation of scu-PA remaining in the supernatant after thrombin clotting of human fibrinogen or recalcification of human plasma.

In vitro clot lysis. Relative fibrinolytic and fibrinogenolytic properties of scu-PA were determined with ^{125}I-labeled human plasma clots immersed in citrated human plasma (Matsuo et al. 1981). Clots were aged for 1 hour prior to addition of scu-PA, and lysis was measured by quantitation of soluble radioactivity at timed intervals. Plasma fibrinogen and α_2-antiplasmin levels were measured as described previously (Vermylen et al. 1963; Edy et al. 1976).

In vitro thrombolysis in animals. Thrombolysis with natural scu-PA was measured after its intravenous infusion into rabbits with jugular vein thrombosis (Collen et al. 1983) or into dogs with copper-coil-induced coronary thrombosis (Van de Werf et al. 1984). Thrombolysis with rscu-PA was measured in the same rabbit jugular vein thrombosis model and also in baboons with coronary artery thrombosis (Flameng et al. 1985).

Pharmacokinetics of scu-PA. Levels of scu-PA in plasma were measured by enzyme-linked immunosorbent assay (Darras et al. 1986) at timed intervals after bolus intravenous injection into rabbits or squirrel monkeys. Organ clearance was determined by quantitation of organ radioactivity in animals sacrificed at timed intervals after bolus intravenous injection of ^{125}I-labeled scu-PA. Postinfusion clearance of scu-PA was measured in all of the animal models of thrombosis described above in addition to patients receiving scu-PA for coronary artery thrombolysis.

Coronary thrombolysis in humans. Patients with chest pain typical of myocardial ischemia for more than 30 minutes and less than 6 hours duration and who gave informed consent to the protocol as approved by the Human Studies Committee of the University of Leuven were studied. Those with angiographically demonstrable occlusion of the infarct-related coronary artery were given a continuous intravenous infusion of scu-PA over 1 hour. Reperfusion was documented by angiography done at 15-minute intervals or done with the occurrence of clinical signs of reperfusion. Plasma fibrinogen and α_2-antiplasmin levels were performed both during and for 3 hours after the infusion by the methods described above.

RESULTS

Biochemical Characterization

Physical properties. scu-PA from human urine is a molecule of apparent M_r 54,000 on both gel-filtration chromatography and SDS-PAGE. Unlike two-chain u-PA, it migrates as a single-chain polypeptide when analyzed under reducing conditions and does not adsorb to benzamidine-Sepharose. It is immunologically identical to two-chain u-PA but distinct from t-PA, and its amino acid composition is similar to that of two-chain u-PA. It is relatively inactive toward the urokinase chromogenic substrate S-2444 but can be converted to fully active two-chain u-PA by limited treatment with plasmin. The specific fibrinolytic activity of scu-PA is plasminogen-dependent, about 50,000 IU/mg on bovine fibrin films as compared to the International Reference Preparation for urokinase (66/46) (D.C. Stump et al., in prep.). scu-PA purified from Calu-3 cells (Stump et al. 1986a), or obtained by recombinant DNA techniques (Holmes et al 1985; Lijnen et al. 1986), has properties very similar to the properties of urinary scu-PA, except for a higher specific fibrinolytic activity of 110,000 IU/mg for rscu-PA.

Kinetic properties. The activation of plasminogen by scu-PA occurs as summarized in Table 1 via three reactions, all of which obey Michaelis-Menten kinetics. In reaction I, scu-PA activates plasminogen directly with unusually high affinity ($K_m < 1$ μM) but low turnover number ($k_{cat} = 0.0009$ s^{-1}). The generation of plasmin at this step is not abolished by the addition of the plasmin inhibitors aprotinin or α_2-antiplasmin, confirming the intrinsic plasminogen activation potential of scu-PA. Furthermore, in the presence of excess chromogenic substrate S-2251, no conversion of scu-PA to two-chain u-PA occurs, suggesting that the initial

Table 1. Kinetic Constants for the Consecutive Reactions Occurring during the Conversion of Plasminogen to Plasmin by scu-PA

	$U + P \overset{K_m}{\rightleftarrows} U.P$	$\overset{k_{cat}}{\rightarrow} U + p$	$p + U \overset{K_m}{\rightleftarrows} p.U$	$\overset{k_{cat}}{\rightarrow} p + u$	$u + P \overset{K_m}{\rightleftarrows} u.P$	$\overset{k_{cat}}{\rightarrow} u + p$
	K_m (μM)	k_{cat} (s^{-1})	K_m (μM)	k_{cat} (s^{-1})	K_m (μM)	k_{cat} (s^{-1})
Natural scu-PA	0.8	0.0009	5	0.23	50	5.3
Recombinant scu-PA	0.23	0.017	5	1.2	30	5.0

(U) scu-PA; (u) two-chain u-PA; (P) plasminogen; (p) plasmin.

generation of plasmin is not a result of reaction III (Lijnen et al. 1986).

In reaction II, plasmin generated from reaction I converts scu-PA to two-chain u-PA with relatively low affinity ($K_m = 5$ μM) but higher turnover number ($k_{cat} = 0.23$ s^{-1}). The two-chain u-PA generated from reaction II causes further activation by plasminogen in reaction III. This occurs with very low affinity ($K_m = 50$ μM) but rapid turnover rate ($k_{cat} = 5.3$ s^{-1}). rscu-PA displays very similar kinetic properties except for a 19-fold higher k_{cat} in reaction I. The uniquely high substrate affinity for reaction I, however, is comparable with $K_m = 0.23$ μM.

Mechanism of action. A theoretical mechanism of action of scu-PA in plasma is shown in Figure 1. Despite the potential of scu-PA to activate plasminogen very efficiently at physiologic plasma plasminogen levels (1–2 μM), very little plasmin is generated by the addition of purified scu-PA to plasma. Moreover, plasma depleted of plasminogen and α_2-antiplasmin inhibits the activation of plasminogen by scu-PA in a concentration-dependent manner. This inhibitory effect is reversible by the addition of cyanogen-bromide-digested fibrinogen, a valid substitute for fibrin, but not by purified undigested fibrinogen (Lijnen et al. 1986). However, when fibrin is formed by thrombin-induced clotting of purified fibrinogen in the presence of scu-PA, nearly all of the scu-PA remains in the supernatant, suggesting very little affinity of scu-PA for fibrin. Similar results are found in plasma clotted either with the addition of thrombin or by recalcification (Lijnen et al. 1986; Stump et al. 1986a,b). In contrast, t-PA in the same system is quantitatively absorbed to the fibrin clot (Rijken and Collen 1981). Thus, fibrin exerts its stimulatory effect on scu-PA-mediated plasminogen activation by reversing the inhibitory effect of plasma in the absence of direct binding of scu-PA to fibrin.

Biological Characterization

In vitro thrombolytic properties. In human plasma, scu-PA caused a time- and dose-dependent lysis of an immersed ^{125}I-labeled plasma clot. Significant lysis occurred within 5 hours at concentrations above 100 IU/ml for natural scu-PA and above 50 IU/ml for rscu-PA. This lysis occurred without associated fibrinogen degradation at concentrations below 200 IU/ml and 100 IU/ml, respectively. At higher concentrations of both, lysis was associated with both α_2-antiplasmin consumption and decreased plasma fibrinogen levels. In addition, the fibrinolytic potential of scu-PA was stable in plasma even after preincubation for up to 48 hours (Zamarron et al. 1984; Stump and Collen 1986). These properties contrast with those of two-chain u-PA, which caused fibrinolysis only with concomitant fibrinogen degradation and was inhibited after only a few hours of preincubation.

In vitro thrombolytic properties in animal models of thrombosis. Intravenous infusion of scu-PA into animals with various types of experimental thrombosis yielded the results summarized in Table 2. Both natural (Stump and Collen 1986) scu-PA and recombinant (Collen et al. 1984b) scu-PA have been applied to the same model of rabbit jugular vein thrombosis in similar doses. For both, lysis levels were significant over similar ranges, each in the absence of fibrinogen degradation. Both forms of scu-PA have also been studied in animal models of coronary artery thrombosis. Natural scu-PA given intravenously to dogs with a 1-

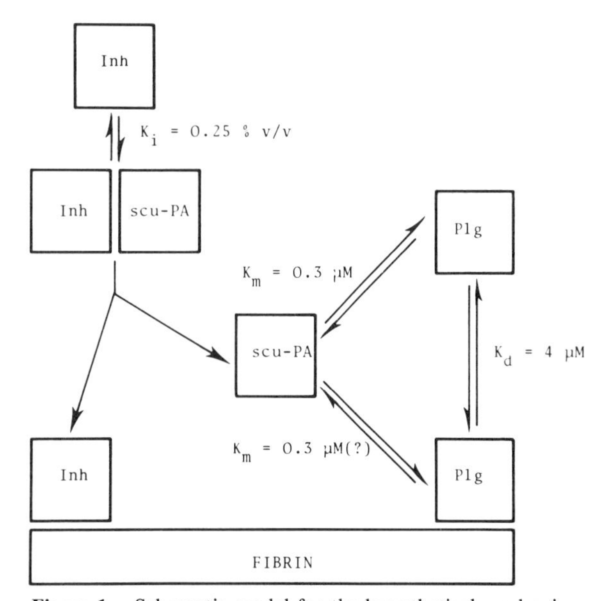

Figure 1. Schematic model for the hypothetical mechanism of action of scu-PA in plasma. scu-PA is normally associated with an inhibitory factor(s) (Inh) that is dissociated in the presence of fibrin, permitting direct activation of plasminogen (Plg) by scu-PA.

Table 2. Thrombolytic Properties of scu-PA in Animal Models of Thrombosis

Animal model	Source of scu-PA	Number	Administration	Results	Fibrinogen
Rabbits with jugular vein thrombosis	natural	13	8,700–35,000 IU/kg IV over 4 hr	13–34% lysis	normal
Dogs with coronary artery thrombosis	natural	6	20 μg/kg/min IV for 30 min	reperfusion in 23 ± 2 min	normal
		2	10 μg/kg/min IV for 30 min	no reperfusion	normal
Rabbits with jugular vein thrombosis	recombinant	9	7,500–30,000 IU/kg IV over 4 hr	14–60% lysis	normal
Baboons with coronary artery thrombosis	recombinant	6	20 μg/kg/min for 60 min	reperfusion in 21 ± 4 min	normal

hour-old left anterior descending coronary artery thrombosis caused reperfusion after 23 ± 2 minutes when infused continously at 20 μg/kg/min but not within 30 minutes at 10 μg/kg/min. Neither dose was associated with degradation of fibrinogen (Collen et al. 1985). Recombinant scu-PA given to open-chest baboons with a 45-minute-old thrombosis of the same coronary artery caused reperfusion similarly after 21 ± 4 minutes, likewise with normal fibrinogen levels (Flameng et al. 1986). In all four of these studies, the use of two-chain u-PA in control animals also caused significant thrombolysis but only in association with marked systemic fibrinolytic activation characterized by plasma fibrinogen degradation and frequent wound hemorrhage.

Pharmacokinetic properties. The turnover and clearance of both natural and recombinant scu-PA were studied in rabbits and squirrel monkeys as shown in Table 3. The disappearance of scu-PA antigen after intravenous injection could adequately be described by a sum of two exponential terms representing an initial rapid half-life of 3–6 minutes and a second more delayed half-life of 18–20 minutes. The disappearance of plasma euglobulin fibrinolytic activity was similarly rapid, suggesting that the main mechanism of turnover was via removal of the molecule from blood. The major organ of clearance as determined by uptake of ^{125}I-labeled scu-PA was the liver, although in rabbits some clearance was also mediated by the kidney. This was confirmed by an experimental hepatectomy that prolonged the initial half-life to 20–30 minutes, whereas experimental nephrectomy had no effect (Collen et al. 1984a; Stump and Collen 1986). Steady-state plasma levels were proportional to the doses infused while postinfusion half-lives were comparably short (4–7 min) when measured following experimental in vivo thrombolysis in the above animal models of thrombosis. Similarly, rapid postinfusion half-lives were also found in patients undergoing thrombolytic therapy with scu-PA, suggesting both that the clearance mechanism is nonsaturable and that continuous intravenous infusion will likely be required for the maintenance of therapeutic levels of scu-PA, which can reach 5–10 μg/ml plasma.

Thrombolytic properties in man. To date, six patients have been treated with natural scu-PA (Van de Werf et al. 1986). All were diagnosed as having an acute myocardial infarction of less than 5 hours duration and had angiographically confirmed coronary artery occlusion. As shown in Table 4, intravenous infusion of 40 mg of scu-PA over 60 minutes resulted in reperfusion of the infarct-related vessel in four of the six patients, with a fifth patient being reperfused after a 30-minute intracoronary infusion of 20 mg. No evidence of systemic fibrinogen degradation was observed in five

Table 3. Pharmacokinetic Properties of scu-PA

Species	Source of scu-PA	Administration	Postinjection half-life early (min)	Postinjection half-life late (min)	Clearance site
Rabbit	natural	IV bolus	3	18	liver (primary) kidney (secondary)
Squirrel monkey	natural	IV bolus	3	20	liver
Rabbit	natural	4-hr infusion	6	–	–
Dog	natural	30-min infusion	7	–	–
Rabbit	recombinant	IV bolus	3–6	–	liver
Squirrel monkey	recombinant	IV bolus	3.5	–	liver
Baboon	recombinant	60-min infusion	5	–	–
Man	natural	60-min infusion	4	–	–
Man	recombinant	60-min infusion	8	–	–

Table 4. Thrombolytic Properties of scu-PA in Man

Thrombus localization	Source of scu-PA	Number	Administration	Results	Fibrinogen (% baseline)	Side effects
Acute myocardial infarction; <5-hr duration	natural	6	40 mg IV over 60 min	reperfusion in 4	normal in 5	none
			± IC over 30 min	reperfusion in 1	30% in 1	
Acute myocardial infarction; <6-hr duration	recombinant	6	40 mg IV over 60 min	reperfusion in 2; partial reperfusion in 3	86%	none
			70 mg IV over 60 min	reperfusion in 4	62%	none

of the six patients, but one patient experienced a drop in fibrinogen level to 25% of baseline. In three of the patients with normal fibrinogen levels, some evidence of systemic fibrinolytic activation was evident, however, from significant consumption of α_2-antiplasmin to 40% of baseline levels.

Twelve patients, also with acute myocardial infarction and coronary occlusion, have now been treated with recombinant scu-PA (Van de Werf et al. 1986b). Six received 40 mg over 1 hour, leading to reperfusion in five at 34 ± 10 minutes. Three of the five, however, had evidence of residual thrombosis despite reperfusion. Plasma fibrinogen levels fell slightly to 86%, whereas α_2-antiplasmin levels decreased to 51% of baseline. Infusion of 70 mg over 1 hour led to reperfusion in four of six additional patients within 30 ± 7 minutes, with residual thrombosis evident in two of the four. Evidence of systemic fibrinolytic activation was more pronounced at the higher dose, with fibrinogen and α_2-antiplasmin levels of 62% and 29% of baseline, respectively, at the end of the 1-hour infusion. No significant bleeding or other toxicity was observed in any of the patients receiving either the natural or recombinant form of scu-PA.

DISCUSSION

scu-PA was first isolated as a proenzyme form of urokinase with high affinity for fibrin. The preliminary observation, in fact, of its fibrin-specific thrombolytic properties resulted in the conclusion that its mechanism of action was probably very similar to that of t-PA, which had previously been shown to have high affinity for fibrin. The current availability of a straightforward purification procedure for scu-PA from natural biological sources has facilitated further studies detailed here that have given further insight into the properties of this unique plasminogen activator.

scu-PA is clearly physically similar to urokinase, both being translation products of the same mRNA. In contrast to the well-characterized two-chain urokinase, however, scu-PA is a single-chain polypeptide structure. It has low specific activity toward low-molecular-weight chromogenic substrates but can be converted to the fully active two-chain form by treatment with plasmin. However, scu-PA can activate its natural substrate plasminogen directly via a reaction characterized by unique high affinity (low K_m). This property, not typical of a proenzyme, has led to the adoption of the nomenclature "single-chain urokinase-type plasminogen activator" (Subcommittee Report 1985), or scu-PA, as a reflection of this intrinsic plasminogen activation potential.

The mechanism of action of scu-PA is distinct from that of t-PA. Plasminogen activation by t-PA, which is slow in the absence of fibrin, is markedly enhanced in the presence of fibrin as a result of the generation of a higher affinity of the enzyme for its substrate after direct binding of the enzyme to fibrin. In contrast, scu-PA can activate plasminogen very efficiently with high affinity even in the absence of fibrin in purified systems but is inhibited in plasma. This inhibition is reversible (even after several hours of preincubation) by the presence of fibrin, an effect that occurs without direct binding of scu-PA to fibrin itself. The factors responsible for this unusual plasma stability of scu-PA remain to be defined.

The specific fibrinolytic properties of scu-PA are now more clear. In an in vitro plasma clot system, in animal models of thrombosis, and in patients with coronary artery occlusion, scu-PA causes clot-selective fibrinolysis with a relative lack of systemic fibrinolytic activation, as reflected by normal plasma fibrinogen and α_2-antiplasmin levels, although in some but not all patients, evidence of systemic fibrinolysis is observed. Pharmacokinetically, scu-PA is removed from the circulation by the liver with an initial rapid half-life of only a few minutes and a later more delayed half-life of 20–30 minutes. This means that in its natural form, continuous intravenous infusion will most likely be required for its clinical use. This also indicates that within a short time postinfusion, most patients will have restoration of normal hemostatic function, which may be of great value in those requiring surgery or other invasive procedures.

Finally, although natural sources of scu-PA presently exist, its production is expensive and is not yet applicable for large-scale clinical needs. Now, scu-PA produced by recombinant DNA techniques provides a more readily available alternative. As shown here, rscu-PA is biologically identical to natural scu-PA in terms of both pharmacokinetic and thrombolytic properties. Results in preliminary clinical studies have now shown that it can induce fibrin-specific thrombolysis in patients with acute myocardial infarction and coronary artery occlusion. Still, its use is associated with an unavoidable degree of systemic fibrinolytic activation marked by varying degrees of fibrinogen degradation in some patients.

Thus, the challenge of the future will be to apply the advances in biotechnology toward the design and production of new and improved plasminogen activators with even greater efficacy and fibrin specificity. Only then will thrombolytic therapy offer realistic hope for the reduction of morbidity and mortality that result from thromboembolic disease.

ACKNOWLEDGMENTS

These studies were supported by grants from the Geconcerteerde Onderzoeksacties and the Fonds voor Geneeskundig Wetenshappelijk Onderzoek. D.C.S. is the recipient of Clinical Investigator Award (KO8-01489) from the National Heart, Lung, and Blood Institute of the National Institutes of Health.

REFERENCES

Barlow, G.H. and L. Lazer. 1972. Characterization of the plasminogen activator isolated from human embryo kidney cells: Comparison with urokinase. *Thromb. Res.* **1:** 201.

Bernik, M.B. 1973. Increased plasminogen activator (urokinase) in tissue culture after fibrin deposition. *J. Clin. Invest.* **52:** 823.

Collen, D. 1980. On the regulation and control of fibrinolysis. *Thromb. Haemostasis* **43:** 77.

———. 1985. Human tissue-type plasminogen activator from the laboratory to the bedside. *Circulation* **72:** 18.

Collen, D., F. DeCock, and H.R. Lijnen. 1984a. Biological and thrombolytic properties of proenzyme and active forms of human urokinase. II. Turnover of natural and recombinant urokinase in rabbits and squirrel monkeys. *Thromb. Haemostasis* **52:** 24.

Collen, D., J.M. Stassen, and M. Verstraete. 1983. Thrombolysis with human extrinsic (tissue-type) plasminogen activator in rabbits with experimental jugular vein thrombosis. Effect of molecular form and dose of activator, age of the thrombus, and route of administration. *J. Clin. Invest.* **71:** 368.

Collen, D., C. Zamarron, H.R. Lijnen, and M. Hoylaerts. 1986. Activation of plasminogen by pro-urokinase. II. Kinetics. *J. Biol. Chem.* **261:** 1259.

Collen, D., J.M. Stassen, M. Blaber, M. Winkler, and M. Verstraete. 1984b. Biological and thrombolytic properties of proenzyme and active forms of human urokinase. III. Thrombolytic properties of natural and recombinant urokinase in rabbits with experimental jugular vein thrombosis. *Thromb. Haemostasis* **52:** 27.

Collen, D., D. Stump, F. Van de Werf, I.K. Jang, M. Nobuhara, and H.R. Lijnen. 1985. Coronary thrombolysis in dogs with intravenously administered human pro-urokinase. *Circulation* **72:** 384.

Darras, V., M. Thienpont, D.C. Stump, and D. Collen. 1986. Measurement of urokinase-type plasminogen activator (u-PA) with an enzyme-linked immunoadsorbent assay (ELISA) based on three murine monoclonal antibodies. *Thromb. Haemostasis* (in press).

Eaton, D.L., R.W. Scott, and J.B. Baker. 1984. Purification of human fibroblast urokinase proenzyme and analysis of its regulation by proteases and protease nexin. *J. Biol. Chem.* **259:** 6241.

Edy, J., F. DeCock, and D. Collen. 1976. Inhibition of plasmin by normal and antiplasmin-depleted plasma. *Thromb. Res.* **8:** 513.

Flameng, W., F. Van de Werf, J. Vanhaecke, M. Verstraete, and D. Collen. 1985. Coronary thrombolysis and infarct size reduction after intravenous infusion of recombinant tissue-type plasminogen activator in nonhuman primates. *J. Clin. Invest.* **75:** 84.

Flameng, W., J. Vanhaecke, D.C. Stump, F. Van de Werf, W. Holmes, W. Gunzler, L. Flohé, and D. Collen. 1986. Coronary thrombolysis by intravenous infusion of recombinant single chain urokinase-type plasminogen activator or recombinant urokinase in baboons. *J. Am. Coll. Cardiol.* **8:** 118.

Gurewich, V., R. Pannell, S. Louie, P. Kelley, R.L. Suddith, and R. Greenlee. 1984. Effective and fibrin-specific clot lysis by a zymogen precursor form of urokinase (pro-urokinase). A study in vitro and in two animal species. *J. Clin. Invest.* **73:** 1731.

Holmes, W.E., D. Pennica, M. Blaber, M.W. Rey, W.A. Gunzler, G.J. Steffens, and H.L. Heyneker. 1985. Cloning and expression of the gene for pro-urokinase in *Escherichia coli*. *Biotechnology* **3:** 923.

Hoylaerts, M., D.C. Rijken, H.R. Lijnen, and D. Collen. 1982. Kinetics of the activation of plasminogen by human tissue plasminogen activator. *J. Biol. Chem.* **257:** 2912.

Husain, S., V. Gurewich, and B. Lipinski. 1983. Purification and partial characterization of a single-chain high-molecular-weight form of urokinase from human urine. *Arch. Biochem. Biophys.* **220:** 31.

Kasai, S., H. Arimura, M. Nishida, and T. Suyama. 1985. Proteolytic cleavage of single-chain pro-urokinase induces conformational change which follows activation of the zymogen and reduction of its high affinity for fibrin. *J. Biol. Chem.* **260:** 12377.

Kohno, T., P. Hopper, J.S. Lillquist, R.L. Suddith, R. Greenlee, and D.T. Moir. 1984. Kidney plasminogen activator: A precursor form of human urokinase with high fibrin affinity. *Biotechnology* **2:** 628.

Lijnen, H.R., C. Zamarron, M. Blaber, M.E. Winkler, and D. Collen. 1986. Activation of plasminogen by pro-urokinase. I. Mechanism. *J. Biol. Chem.* **261:** 1253.

Matsuo, O., D.C. Rijken, and D. Collen. 1981. Comparison of the relative fibrinogenolytic, fibrinolytic and thrombolytic properties of tissue plasminogen activator and urokinase in vitro. *Thromb. Haemostasis* **45:** 225.

Nielsen, L.S., J.G. Hansen, L. Skriver, E.L. Wilson, K. Kaltoft, J. Zeuthen, and K. Dano. 1982. Purification of zymogen to plasminogen activator from human glioblastoma cells by affinity chromatography with monoclonal antibody. *Biochemistry* **21:** 6410.

Nolan, C., L. Hall, G. Barlow, and I.I.E. Tribby. 1977. Plasminogen activator from human embryonic kidney cell cultures. Evidence for a proactivator. *Biochim. Biophys. Acta* **496:** 384.

Pennica, D., W.E. Holmes, W.J. Kohr, R.N. Harkins, G.A. Vehar, C.A. Ward, W.F. Bennett, E. Yelverton, P.H. Seeburg, H.L. Heyneker, D.V. Goeddel, and D. Collen. 1983. Cloning and expression of human tissue-type plasminogen activator cDNA in *E. coli*. *Nature* **301:** 214.

Plough, J. and K.O. Kjeldgaard. 1957. Urokinase: An activator of plasminogen from human urine. I. Isolation and properties. *Biochim. Biophys. Acta* **24:** 278.

Rijken, D.L. and D. Collen. 1981. Purification and characterization of the plasminogen activator secreted by human melanoma cells in culture. *J. Biol. Chem.* **256:** 7035.

Rijken, D.C., G. Wijngaards, M. Zaal-de Jong, and J. Welbergen. 1979. Purification and partial characterization of plasminogen activator from human uterine tissue. *Biochim. Biophys. Acta* **580:** 140.

Stump, D.C. and D. Collen. 1986. Comparative biological properties of single chain forms of urokinase-type plasminogen activator. *Circulation* (in press).

Stump, D.C., H.R. Lijnen, and D. Collen. 1986a. Purification and characterization of single chain urokinase-type plasminogen activator (scu-PA) from human cell cultures. *J. Biol. Chem.* **261:** 1274.

Stump, D.C., M. Thienpont, and D. Collen. 1986b. Urokinase-related proteins in human urine. Isolation and characterization of single chain urokinase (prourokinase) and urokinase-inhibitor complex. *J. Biol. Chem.* **261:** 1267.

Subcommittee Report. 1985. Report of the Subcommittee on fibrinolysis of the International Society of Thrombosis and Haemostasis, San Diego. *Thromb. Haemostasis* **54:** 893.

Sumi, H., T. Kosugi, O. Matsuo, and H. Mihara. 1982a. Physiochemical properties of highly purified kidney cultured plasminogen activator (single polypeptide chain-urokinase). *Acta Haematol. Jpn.* **45:** 119.

Sumi, H., M. Maruyama, O. Matsuo, H. Mihara, and N. Toki. 1982b. Higher fibrin-binding and thrombolytic properties of single polypeptide chain-high molecular weight urokinase. *Thromb. Haemostasis* **47:** 297.

Van de Werf, F., M. Nobuhara, and D. Collen. 1986a. Coronary thrombolysis with human single chain urokinase-type plasminogen activator (scu-PA) in patients with acute myocardial infarction. *Ann. Int. Med.* **104:** 345.

Van de Werf, F., J. Vanhaecke, H. DeGeest, M. Verstraete, and D. Collen. 1986b. Coronary thrombolysis with recombinant single chain urokinase-type plasminogen activator (rscu-PA) in patients with acute myocardial infarction. *Circulation* (in press).

Van de Werf, F., S.R. Bergmann, K.A.A. Fox, H. DeGeest, C.F. Hoyng, B.E. Sobel, and D. Collen. 1984. Coronary thrombolysis with intravenously administered human tissue-type plasminogen activator produced by recombinant DNA technology. *Circulation* **69:** 605.

Vermylen, C., R. DeVreker, and M. Verstraete. 1963. A rapid enzymatic method for assay of fibrinogen fibrin polymerization time (FPT-test). *Clin. Chim. Acta* **8:** 418.

White, W.F., G.H. Barlow, and M.M. Mozen. 1966. The isolation and characterization of plasminogen activators (urokinase) from human urine. *Biochemistry* **5:** 2160.

Wun, T.C., L. Ossowski, and E. Reich. 1982a. A proenzyme form of human urokinase. *J. Biol. Chem.* **257:** 7262.

Wun, T.C., W.D. Schleuning, and E. Reich. 1982b. Isolation and characterization of urokinase from human plasma. *J. Biol. Chem.* **257:** 3276.

Zamarron, C., H.R. Lijnen, B. Van Hoef, and D. Collen. 1984. Biological and thrombolytic properties of proenzyme and active forms of urokinase. I. Fibrinolytic and fibrinogenolytic properties in human plasma in vitro of urokinases obtained from human urine or by recombinant DNA technology. *Thromb. Haemostasis* **52:** 19.

Interferon Production from Human Cell Cultures

N.B. Finter, G.D. Ball, K.H. Fantes, M.D. Johnston, and W.G. Lewis
Wellcome Biotech, Langley Court, Beckenham, Kent BR3 3BS, England

Human proteins required for medical purposes, e.g., insulin, interferons, tissue plasminogen activator, and growth hormone, can now often be produced by recombinant DNA procedures. They can also be obtained by extraction from the appropriate human organ (e.g., growth hormone from pituitaries) or by stimulating human cells grown in culture to make the protein of interest. We pioneered production by the latter approach because we needed to make enough human interferon for large-scale clinical trials; we continue to use this route today, because we think it has certain advantages. To understand our reasons, some points from the history of interferons must be considered.

Discovery and Development of Interferons

Fifty years ago, Hoskins (1935) and Magrassi (1935) independently discovered the interference phenomenon: A virus replicating in an animal's organ, e.g., the brain, prevented that organ during the next few days from superinfection with any of a number of serologically unrelated viruses. Although the phenomenon was intensively studied (Henle and Henle 1984), the underlying mechanism was not identified until Isaacs and Lindenmann (1957), working in the laboratories of the Medical Research Council in London, discovered "the interferon." This was a protein released from chicken cells exposed to irradiated influenza virus, which could interact with fresh cells so that these became resistant to virus infection. It seemed that interferons might be broad-spectrum antiviral substances with medical applications, and accordingly, they attracted considerable interest. This interest increased when interferons were found to be active against tumor viruses in culture and in animals and also, and at first surprisingly, against tumors induced by chemicals or arising spontaneously (for review, see Oxman 1973). However, clinical evaluation proved extraordinarily difficult for several reasons. One was that because interferons from the cells of different animals are not identical in their properties (Tyrrell 1959), interferon from human cells was needed for such studies. This raised a major technical problem: How could the enormous numbers of cells needed to make worthwhile amounts of interferon be obtained? For a time, this problem seemed almost insuperable.

Leukocyte Interferon

The first procedure for making human interferon in quantity was devised by Strander and Cantell (1966). They used citrated blood donated for transfusion as the source, collecting the layer of leukocytes that forms at the interface between the red cells and the serum when the blood is centrifuged. These "buffy-coat" cells were induced to form interferon by adding a mouse parainfluenza virus, Sendai. The crude interferon present in the medium after overnight incubation was separated and partially purified. The final product was by today's standards of only low purity (~ 0.5–1% pure), but it proved acceptable for clinical use. There were some side effects associated with its use, which were at the time attributed to the presence of impurities, but it is now clear that most are due to the interferons themselves.

Through the cooperation of the Finnish Red Cross Blood Transfusion Service, which provided the large numbers of buffy coats, leukocyte interferon was made over the years in increasing amounts from hundreds of thousands of blood donations. This material was used in almost all clinical studies until 1979, and thus played a vital role in the development of interferons for medical use. However, it could not be made in amounts sufficient to meet all demands, and it is inherently expensive to make.

Wellcome and Interferons

Our involvement with interferon began in 1959 when some of our scientists joined the Scientific Committee on Interferon, set up by the British Medical Research Council, to expedite the development of this seemingly promising material; work on interferons has continued in our laboratories ever since. In the early 1970s, we followed with keen interest the early clinical studies with leukocyte interferon in Scandinavia and in California, which gave hints of benefit in patients with virus infections or with cancer. It was obvious that much would need to be done to prove or disprove the results and that for the additional studies, very large amounts of interferon would be needed. We wanted to help to provide these, but felt that some source of interferon other than buffy-coat cells must be found. There was at the time only one possible alternative, namely, to grow the large number of human cells required in culture and use them to make interferon. From experience with the production of human virus vaccines, we knew that to culture the necessary large numbers of anchorage-dependent cells would involve formidable technical problems. We also had access to considerable expertise in the technology of growing mammalian cells in suspension cultures: Colleagues were making large amounts of foot-and-mouth disease virus vaccine for veterinary use from cells of the BHK line of baby hamster kidney grown in tanks of up to 3000-liter capacity. We were therefore attracted to the idea of making interferon from suspension cultures of transformed human cells, which was an almost heretical idea at the time. Many scientists felt strongly that such cells should not be used to make a substance intended for medicinal use, arguing that the product might be contaminated with some oncogenic substance derived from the cells. However, there seemed to be no firm experimental evidence to support their views, and since interferons seemed to be beneficial in some patients, including in some with cancer, we decided to discount these concerns. In 1974, we started the task of trying to make safe and acceptable preparations of interferon from transformed cells.

Namalwa Cells

Our first task was to screen a number of human cells known to grow well in suspension culture for their ability to produce interferon. Unfortunately, most produced very small amounts. However, in a publication that appeared most opportunely, Strander et al. (1975) reported that cells from 21 human lymphoblastoid lines produced appreciable quantities of interferon, with cells of the Namalwa line making particularly large amounts. We ourselves tested 120 lymphoblastoid cell lines (Christofinis et al. 1981) but also found Namalwa cells to be best producers. As these also grew rapidly in culture and to a high cell concentration, we chose them as our source of interferon.

Namalwa was an Ugandan girl who developed a Burkitt tumor, and the cell line with her name was derived by Professor George Klein in Stockholm from a biopsy of tumor tissue. Thus, in choosing to make clinically acceptable interferon from these cells, we knew that we had set ourselves a considerable challenge.

The Namalwa cells we received were contaminated with *Mycoplasma orale*. Having cleared this infection, we stored 227 samples of 5 ml from a single lot of cells in liquid nitrogen to serve as our master cell bank.

Production of Interferon

The procedure for making lymphoblastoid interferon from Namalwa cells is straightforward. Cells thawed from the master bank are grown at 35–36°C in cultures of increasing size, first in stationary flasks, then in magnetically stirred flasks, and, finally, in the smallest steel fermenter tank in a series. The medium used is RPMI 1640, which is supplemented with γ-irradiated (3–5 MegaRads) adult bovine serum at a maximum concentration of up to 3.5%. The cells usually adapt quickly to growing in tank culture, where for the first time the environment can be closely controlled in terms of pH, redox, temperature, and agitation. Probably as a result, they start to grow more rapidly than before and reach a higher final maximum cell concentration. When the cell count reaches a level of about 1–2 million cells/ml, fresh growth medium is added, and this process is repeated until the tank is full. Some of the contents are then transferred to the next tank in the series, medium is added to both tanks to dilute the cells to about 0.5 million/ml, and the process is repeated. From the second tank, cells are moved in due course to the next larger tank, and so on until the largest vessels are seeded. When these production tanks are half full with cells at a high concentration, their contents are diluted with an equal volume of serum-free medium, sodium butyrate is added to a final concentration of 2 mM, and incubation is continued for a further 2 days. This procedure, developed in our laboratory by Johnston (1979, 1980), leads to a much more consistent and higher yield of interferon from one batch to another than is obtained otherwise. Formation of interferon is induced by adding the predetermined optimum amount of Sendai virus to the butyrate-treated cells, and incubation is continued for 14–16 hours, by which time most of the interferon has been released into the medium. The medium containing the crude interferon is separated from the cells by centrifugation or filtration, or both, and transferred to a holding tank. From here, it passes through a four-stage purification procedure, which reduces the volume about 2000 times and eliminates all but minute traces of impurity proteins. The final product is at least 95% pure interferon protein: This represents a purification of some 10,000 times.

We developed this process on a pilot plant, which was completed in 1978 and housed a single 1000-liter production tank. Subsequently, we have constructed three plants for the production of lymphoblastoid interferon with progressively larger production tanks. Those commissioned in 1983 and 1985 contained 8000-liter tanks, and the one enlarged this year has 10,000-liter tanks. Since essentially the same numbers of staff are needed to operate tanks irrespective of their size and to carry out testing and quality control procedures whatever the batch size, there are very great economies as the scale of production is increased.

Safety Considerations

We first had to assess the magnitude of the problem, if any, associated with the crude product. We therefore exhaustively checked Namalwa cells from our master bank for the presence of active or latent infection with microorganisms and viruses: We detected nothing except that all the cells contain the genome of Epstein-Barr virus (EBV), as evidenced by staining specifically for EBV nuclear antigen. However, EBV is an extremely common infection of man, and on the basis of serological evidence, most people in all countries have been infected by the time they are adult. Thus, the presence of this virus in Namalwa cells (as in cells from the great majority of B lymphoblastoid cell lines) did not seem a matter of serious concern. In any event, the virus in Namalwa cells does not enter into a replicative cycle, even when cells are treated with inducing chemicals, such as bromodeoxyuridine: EBV early antigens are never expressed and no infectious virus is ever formed.

The crude product must inevitably contain some DNA derived from disintegrating cells, but this can be detected only in a minority of batches and at most in an extremely low concentration. For example, in 15 consecutive harvests of crude interferon, DNA was detected by a dot hybridization method in only 5 and at concentrations of 10–20 ng/ml. The highest concentration was found in a batch in which, due to an instrument failure, many cells died during the period of interferon induction, so that there must have been considerable lysis. Negligible amounts of interferon were formed and the batch was discarded.

It can be calculated that the amount of EBV DNA in the crude product can only be a small fraction of that of cellular DNA. It was thus not surprising that viral DNA could not be detected even by a sensitive hybridization test, i.e., less than 12 pg/ml was present.

In view of the concerns expressed, the onus was on us to prove that our final product was safe. Therefore, when choosing a method for purifying our crude interferon preparation, we finally adopted a four-step purification sequence that led to a very pure product and, equally importantly, would make it acceptable for clinical use even if there were indeed major problems associated with the crude product. To validate this point, we deliberately added (one at a time) various marker substances, representing every type of contaminant actually or potentially present in the starting material, to experimental batches of crude interferon, which were then taken through the purification procedure. These studies have been reviewed previously (Finter and Fantes 1980; Finter et al. 1985), but, in short, large amounts of every virus tested were totally inactivated by the end of the third stage in the procedure, and most by the end of the first stage. Large amounts of the sheep pathogen, scrapie, used as a representative slow virus, were totally eliminated. The radioactivity of heavily radiolabeled samples of DNA was reduced to very low levels or to background, whether added to crude interferon or to the intermediate product from the first stage of purification. The same result was obtained with high-molecular-weight DNA from Namalwa cells and with cleaved fragments corresponding in size to that of a small oncogene. From our studies, it could be calculated that the amounts of residual EBV DNA in our final product would be less than 0.09 molecules per million units of interferon, even if every batch of crude were to contain as much as 20 μg/ml of total DNA, the highest value we have ever measured (Finter et al. 1985). However, for further reassurance, each batch is tested directly for its DNA content by dot hybridization and shown to contain less than 10 pg per million units of interferon.

Composition of Wellcome's Lymphoblastoid Interferon, Wellferon

Allen and Fantes (1980) showed that our interferon preparation, Wellferon, contains at least seven different subtypes of α-interferon, distinguished by their physicochemical properties and chemical composition. Subsequently, Zoon and co-workers (1985, 1986), using sequential immunoabsorbent chromatography on columns made from four different anti-α-interferon monoclonal antibodies, followed by high-pressure liquid chromatography, separated 16 α-interferons from lymphoblastoid interferon which we supplied. Weck et al. (1981) showed that each individual α-interferon subtype made by recombinant DNA procedures from bacteria had biological properties that were unique and different from those of a preparation of leukocyte interferon. This latter material contains a different mixture of α-interferon subtypes. Our final product also has its own spectrum of biological properties distinct from those of a preparation that, for example, consists only of subtype α_2-interferon.

Analytical Procedures

The purity and the uniformity of composition of Wellferon have been monitored in a number of ways, of which the most useful has proved to be SDS-polyacrylamide electrophoresis applied to samples of the final purified product under rigidly controlled nonreducing conditions; the technique is not useful for the crude and intermediate products (Fig. 1). When the gels are stained with Coomassie blue and examined by densitometry, two large and two small peaks are observed, together with a much smaller peak of higher-molecular-weight material (Fig. 2): All of these stain when Western blotting is carried out with monoclonal antibodies to α-interferon. The two larger peaks account for most of the total interferon protein, and these and the larger of the remaining peaks are quite constant in their proportions. The smallest peak is more variable, but to a large extent, this probably reflects the difficulty of establishing where to draw the base line.

In some of our preparations, there is a further small peak that stains with an antiserum raised against Sendai virus grown in fertile hens' eggs; it is probable that the antigen detected is egg protein. The amount of this protein never exceeds 2% of the total measured on the gel and represents less than 50 ng of impurity protein per million units of interferon. The other contaminants that might be carried over from the crude product are Namalwa cell and calf serum proteins, but no reaction has been detected in Western blotting with appropriate antisera. The amounts of residual bovine albumin measured in the final product by radioimmu-

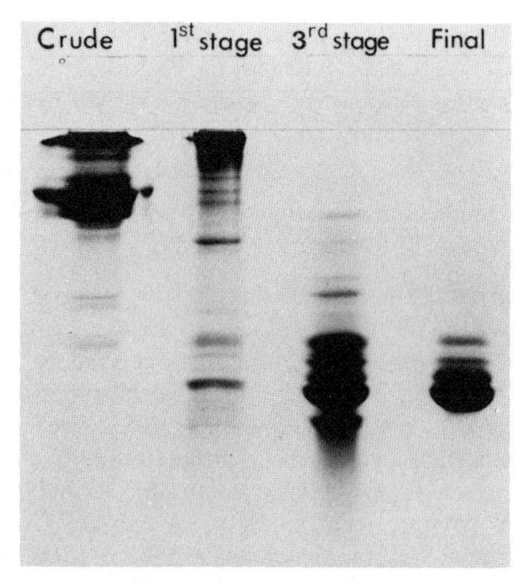

Figure 1. Elimination of impurity proteins during the routine purification of Wellferon, as monitored by SDS-PAGE. The gels were loaded with: samples of CRUDE interferon-containing harvest; 1st STAGE and 2nd STAGE intermediate products from the four-stage purification process; and FINAL purified product (before addition of human serum albumin as a stabilizer). All the protein bands in the Final gel are interferons; they can also be seen in the intermediate products but not in the Crude harvest.

noassay are very low; an upper limit of 50 ng per million units of interferon is set for control purposes. The purity of each blend can be estimated on the basis of the SDS-PAGE analysis described above. The mean

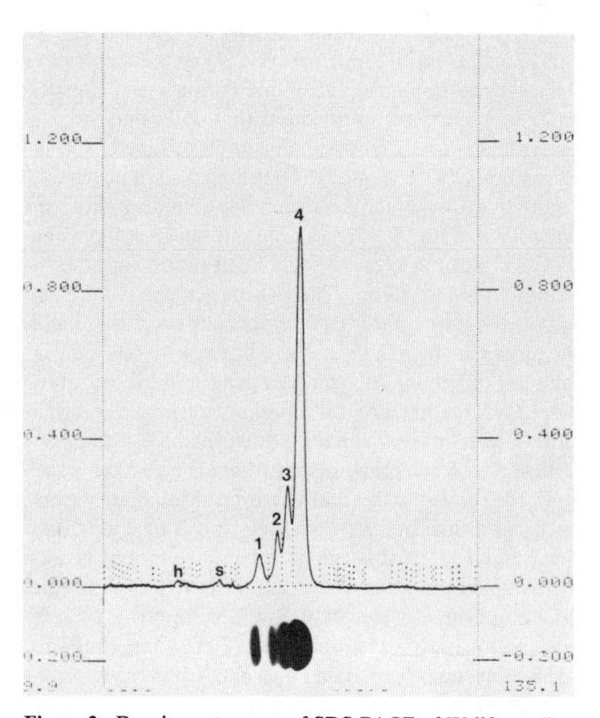

Figure 2. Densitometer scan of SDS-PAGE of Wellferon (before addition of stabilizer). Peaks *1–4* and *h* react with specific antisera to α-interferons in Western blots; peak *h* is complexed interferons. Peak *s* is Sendai virus/egg protein.

value is 97.3%, and we require a minimum of 95% purity.

It is likely that the physiological state of the Namalwa cells in the production tanks varies to some extent from one occasion to another. We therefore assume that not only will the amount of interferon formed vary to some extent from one crude harvest to another, as is observed, but also some differences in the particular mix of α-interferon subtypes might be present. To overcome this potential problem, the final product issued for clinical use is made by blending the purified mixtures of α-interferon subtypes obtained from at least 20 (and often many more) separate production batches. On analysis, such blends have a composition that is uniform within narrow limits.

DISCUSSION

Our program was started when there seemed to be no alternative way of making interferons in quantity, and some years before the development of recombinant DNA procedures for making individual α-interferon subtypes. When these preparations became available, we were at first concerned that our mammalian cell-production system might become outmoded. We no longer think that this is likely. As already pointed out, our polyclonal product differs in its composition and properties from any single interferon of recombinant DNA origin. How important this will be in terms of clinical utility remains to be established. In theory, there should occasionally be a situation where a particular α-interferon subtype is the one most appropriate to the medical situation and should ideally be used. However, since at present there is no way of knowing this, it may prove more generally beneficial to make use of the broader spectrum of a natural mixture if other factors such as general availability and price have little bearing on the choice. In this connection, the various plants built to our design specifications within the Wellcome group and by our associates overseas can easily make all the α-interferon needed for currently established medical purposes worldwide. It is true that the manufacturing cost per million units of interferon is likely to be slightly more for our Namalwa cell product than for one made from bacteria. However, as with any pharmaceutical preparation, the bulk of the final price reflects the costs of testing, packaging, and distribution.

Wellferon has been used in clinical trials in England since 1979 and in the United States and many other countries since 1981. It has already shown promise in a number of medical conditions and has proved to be extremely effective in the treatment of hairy cell leukemia (Worman et al. 1985): It was the first α-interferon preparation to be licensed in England for use in this condition.

In the long term, probably the major significance of our interferon program is that we have shown that products acceptable for clinical use can be made from large suspension cultures of human or animal cells.

This route to manufacture is likely to become of increasing commercial importance in relation to the production by recombinant DNA procedures of proteins that are too large or complex to obtain from transfected bacteria or that for other reasons, such as the particular glycosylation required, cannot be satisfactorily obtained from other hosts such as yeast or insect cells. We now think it unlikely that any one type of expression system will serve as a universal source of proteins and that the one most appropriate for the purpose in hand should be chosen. We have shown that it is no longer necessary to leave mammalian cells out of consideration for purely technical reasons.

REFERENCES

Allen, G. and K.H. Fantes. 1980. Family of structural genes for human lymphoblastoid (leukocyte-type) interferon. *Nature* **287:** 408.

Christofinis, G.J., C.M. Steel, and N.B. Finter. 1981. Interferon production by human lymphoblastoid cells of different origins. *J. Gen. Virol.* **52:** 169.

Finter, N.B. and K.H. Fantes. 1980. The purity and safety of interferons prepared for clinical use. The case for lymphoblastoid interferon. In *Interferon 1980* (ed. I. Gresser), vol. 2, p. 65. Academic Press, New York.

Finter, N.B., K.H. Fantes, M.J. Lockyer, W.G. Lewis, and G.D. Ball. 1985. The DNA content of crude and purified human lymphoblastoid (Namalwa cell) interferon prepared by Wellcome Biotechnology Limited. In *Abnormal cells, new products, and risk* (ed. H.E. Hopps and J.C. Petricciani), monogr. 6, p. 125. Tissue Culture Association, Gaithersburg, Maryland.

Henle, W. and G. Henle. 1984. The road to interferon: Interference by inactivated influenza virus. In *Interferon: General and applied aspects* (ed. N.B. Finter), vol. 1, p. 3. Elsevier/North-Holland, Amsterdam.

Hoskins, M. 1935. A protective action of neurotropic against viscerotropic yellow fever virus in *Macacus rhesus*. *Am. J. Trop. Med. Hyg.* **15:** 675.

Isaacs, A. and J. Lindenmann. 1957. Virus interference. I. The interferon. *Proc. R. Soc. Lond. B.* **147:** 258.

Johnston, M.D. 1979. Improvements in or relating to a process for producing interferon. *European Patent No.* EP 0 000 520 A.

———. 1980. Enhanced production of interferon from human lymphoblastoid (Namalwa) cells pretreated with sodium butyrate. *J. Gen. Virol.* **50:** 191.

Magrassi, F. 1935. Studii sull'infezione e sull'immunita da virus ereptico. III. Rapporti tra infezione e superinfezione di fronte ai processi immunitari: Sulla possibilita di profondamente modificare il decorso e gli esiti del processo infettivo gia in atto. *Z. Hyg. Infektionskr.* **117:** 573.

Oxman, M.N. 1973. Interferon, tumors and tumor viruses. In *Interferons and interferon inducers* (ed. N.B. Finter), p. 391. Elsevier/North-Holland, Amsterdam.

Strander, H. and K. Cantell. 1966. Production of interferon by human leukocytes in vitro. *Ann. Med. Exp. Biol. Fenn.* **44:** 265.

Strander, H., K.E. Mogensen, and K. Cantell. 1975. Production of human lymphoblastoid interferon. *J. Clin. Microbiol.* **1:** 116.

Tyrrell, D.A.J. 1959. Interferon produced by cultures of calf kidney cells. *Nature* **184:** 452.

Weck, P.K., S. Apperson, L. May, and N. Stebbing. 1981. Comparison of the antiviral activities of various cloned human interferon-alpha subtypes in mammalian cell cultures. *J. Gen. Virol.* **57:** 233.

Worman, C.P., D. Catovsky, P.C. Bevan, L. Camba, M. Joyner, P.J. Green, H.J.H. Williams, J.M. Bottomley, and E.C. Gordon-Smith. 1985. Interferon is effective in hairy-cell leukaemia. *Br. J. Haematol.* **60:** 759.

Zoon, K.C., R.-Q. Hu, D. zur Nedden, and N.Y. Nguyen. 1985. The purification and characterization of multiple species of human lymphoblastoid interferon alpha. In *The biology of the interferon system, 1984* (ed. H. Kirchner and H. Schellekens), p. 61. Elsevier/North-Holland, Amsterdam.

Zoon, K.C., R.-Q. Hu, D. zur Nedden, T.L. Gerrard, J.C. Enterline, R.A. Boykins, and N.Y. Nguyen. 1986. Further studies on the purification and partial characterization of human lymphoblastoid interferon alphas. In *The biology of the interferon system, 1985* (ed. W.E. Stewart II and H. Schellekens), p. 55. Elsevier/North-Holland, Amsterdam.

Interleukin-2 and Its Receptor: Structure and Functional Expression of the Genes

T. TANIGUCHI,*† T. FUJITA,*† M. HATAKEYAMA,* H. MORI,* H. MATSUI,†‡ T. SATO,*‡
J. HAMURO,‡ S. MINAMOTO,* G. YAMADA,*† AND H. SHIBUYA*

*Institute for Molecular and Cellular Biology, Osaka University, Suita-shi, Osaka 565 Japan;
†Cancer Institute, Japanese Foundation for Cancer Research, Toshima-ku, Tokyo 170 Japan;
‡Central Research Laboratories, Ajinomoto Co. Inc., Totsuka-ku, Yokohama 244 Japan

The immune system requires mechanisms that ensure the selective, clonal expansion of antigen-stimulated lymphocytes. Interleukin-2 (IL-2), also referred to as T-cell growth factor, is a lymphokine that plays a major role in the proliferation of T lymphocytes by virtue of its interaction with a specific cell-surface receptor (IL-2R) (Morgan et al. 1976; Smith 1984; Taniguchi et al. 1986). Expression of both IL-2 and IL-2R primarily requires T-cell activation by the interaction of antigen/major histocompatibility complex (MHC) and T-cell receptor complex (T3/Ti complex). Hence, clonal growth of antigen-specific T cells appears to occur via a process of signal transduction wherein the exquisitely specific antigen/MHC:T3/Ti interaction is converted into a pathway of cellular growth mediated by the IL-2 system. More recently, Reinherz and his colleagues reported the presence of a novel lymphokine that bypasses the antigen stimulation for IL-2-mediated T-cell growth (Milanese et al. 1986).

In addition, much evidence has been accumulated for the multiple biological functions of IL-2, including B-cell growth and differentiation (Kishi et al. 1985; Mond et al. 1985), generation of lymphokine-activated killer (LAK) cells (Grimm et al. 1982), augmentation of natural killer (NK) cell activity (Henny et al. 1981), and induction of cytotoxic T-cell reactivity (Gillis et al. 1979). Furthermore, we demonstrate in this paper that IL-2 is also a growth-inhibitory factor in certain neoplastic T cells (Hatakeyama et al. 1985; Sugamura et al. 1985). These studies strongly support the notion that the expression of IL-2 in vivo is one of the most crucial events in the control of the entire immune reaction (Smith 1980, 1984; Taniguchi et al. 1986).

Given the various biological properties of IL-2, attention has been focused on the clinical potential of this lymphokine, particularly in cancer therapy (Rosenberg et al. 1985). However, the limited amount of IL-2 produced by human T cells has been an impediment for its basic and clinical application.

Elucidation of the structure and control mechanism of IL-2 and its receptor genes would give insight into the regulatory function of the IL-2 system in lymphocyte growth and differentiation. Furthermore, the use of a cloned gene for the human IL-2 should make it possible to provide essentially unlimited amounts of recombinant IL-2. In this paper, we describe our studies on the structure and expression of IL-2 and its receptor genes.

EXPERIMENTAL PROCEDURES

All plasmid constructions were made essentially following generally established procedures (Maniatis et al. 1982). Cloning procedures of the human and mouse cDNA and chromosomal genes have been described previously (Fujita et al. 1983; Taniguchi et al. 1983; Fuse et al. 1984; Kashima et al. 1985). Construction of the IL-2:chloramphenicol acetyltransferase (IL-2:CAT) (Gorman et al. 1982) fusion genes (the 5′ deletion mutants, pIL2-551cat, pIL2-380cat, pIL2-319cat, pIL2-264cat, and pIL2-222cat) has been described elsewhere (Fujita et al. 1986). To delimitate the 3′ boundary of the IL-2 gene 5′-flanking region, pIL2(Sal)-319 was first cut by *Hin*dIII, whose single cleavage site is located at the boundary of IL-2 and CAT sequences (Fujita et al. 1986), and subsequently treated by BAL-31. Fragments of DNA each containing the IL-2 5′-flanking sequence with various 3′ termini were then excised by *Sal*I (the site was converted from the *Xba*I site of pIL2-319) and cloned into the upstream site of the human IFN-β promoter sequence (-66 to $+19$, relative to the cap site; Ohno and Taniguchi 1980) that was previously inserted into pSV0cat. Thus, pIL2(319–98)IFNcat, pIL2(319–115)IFNcat, pIL2(319–127)IFNcat, pIL2(319–145)IFNcat, and pIL2(319–169)IFNcat were obtained.

Procedures for DNA transfection, induction of EL-4 cells by 12-*O*-tetradecanoylphorbol 13 acetate (TPA), and CAT enzyme assay have been described elsewhere (Hatakeyama et al. 1985; Fujita et al. 1986).

FACS analysis and ^{125}I-labeled IL-2-binding assay were performed as described previously (Hatakeyama et al. 1985). For the IL-2-mediated growth-inhibition assay, human IL-2R-bearing EL-4 cells (ELT-5) were cultured in the presence of various concentrations of either recombinant human or mouse IL-2 (provided by Ajinomoto, Inc.), and [^{3}H]thymidine ([^{3}H]TdR) incorporation was monitored exactly as described by Hatakeyama et al. (1985).

RNA blotting analysis was performed as described by Thomas (1980). Various oncogene probes were kindly provided by T. Sekiya (National Cancer Center Institute, Tokyo) and K. Shimizu (Kyushu University).

RESULTS

Human and Mouse IL-2 cDNA: Structure and Expression

We prepared a cDNA library by using as template the IL-2 mRNA-enriched poly(A) RNA isolated from a human T-cell line (Jurkat-111) and subsequently fractionated by sucrose gradient centrifugation (Taniguchi et al. 1983). We then screened for IL-2-specific clones by means of the hybridization-translation assay (Harpold et al. 1978), which was successfully used for the initial human INF-β gene cloning (Taniguchi et al. 1979, 1980). By screening 432 clones, we identified one clone whose cDNA specifically hybridized to the human IL-2 mRNA. Using this cDNA insert as probe, we subsequently obtained a clone, pIL2-50A, that contains the full-length cDNA. The cDNA was shown to direct the production and secretion of biologically active human IL-2 when it is expressed in monkey COS cells under the control of SV40 promoter-enhancer sequences (Taniguchi et al. 1983). The human IL-2 cDNA was used to screen for the mouse IL-2 cDNA clone, since they cross-hybridized to each other (Kashima et al. 1985). Nucleotide sequence analysis of the cloned cDNAs allowed us to uncover the hitherto unknown primary structure of human and mouse IL-2 (Fig. 1). In fact, human IL-2 is synthesized as a precursor consisting of 153 amino acids, and, upon secretion, its signal peptide (20 amino acids) is removed to form the mature IL-2. The predicted primary structure of the mature IL-2 was confirmed by analysis of the affinity-purified IL-2 (Robb et al. 1984b). Further biochemical and functional studies revealed that cysteine residues at positions 58 and 105 (Fig. 1) are involved in an intramolecular disulfide bridge essential for IL-2 in its active conformation (Wang et al. 1984; Robb et al. 1984b). The mouse IL-2 shows approximately 63% amino acid sequence homology with the human IL-2 and it consists of 149 amino acids. The relatively large size of mouse IL-2 compared with human IL-2 is mostly attributable to the presence of a stretch of unusual 12 consecutive glutamine residues that are encoded by unique repeats of a CAG sequence (between positions 15 and 26) (Fig. 1). The "glutamine repeats" seem to be absent in the bovine IL-2 (Cerretti et al. 1986; Reeves et al. 1986).

To assess further the functional role of IL-2 in vivo, we recently generated a series of retroviruses, each of which gives rise to high production of either human or mouse IL-2 molecules in a variety of infected host cells (further details will be given elsewhere by G. Yamada and T. Taniguchi).

Production of Recombinant Human IL-2 in *Escherichia coli*

To provide human IL-2 in sufficient quantity for basic as well as clinical application, we attempted to express the cloned IL-2 cDNA in *Escherichia coli*. It has been reported that the human IL-2 polypeptide undergoes very minor, if any, posttranslational modifications (Robb et al. 1984b). With regard to glycosylation, there is no N-linked glycosylation site in the mature IL-2, and a threonine residue at position 3 seems to be *O*-glycosylated occasionally (Robb et al. 1984b). Thus, it was expected that the IL-2 produced in *E. coli* would manifest the same biological properties as IL-2 made in human T cells.

As illustrated in Figure 2, we constructed a plasmid vector, pT9-11, in which the human IL-2 cDNA sequence is preceded by *E. coli trp* promoter-operator sequences and followed by a synthetic transcription terminator. When introduced into an *E. coli* host, this expression vector should give rise to the synthesis of an IL-2 protein that would start with fMet-Ala-Pro-Thr... at its amino terminus. When the *E. coli* harboring pT9-11 was induced with 3-indole acrylic acid (IAA) for the *trp* promoter, production and accumulation of a protein with a molecular weight of 15,000 became readily detectable upon electrophoretic analysis of the whole-cell extracts in SDS-PAGE. In fact, the IL-2 protein represented 20–30% of the total cellular protein, and the IL-2 activity recovered was consistently higher than 10^7 units/ml of bacterial culture. Furthermore, molecular analysis of the recombinant IL-2 revealed that, like most of the bacterial proteins, the fMet residue is effectively removed from the IL-2 polypeptide, providing evidence that the recombinant IL-2 is essentially identical in structure to that produced by human T cells (H. Matsui and J. Hamuro, unpubl.).

Structural Organization and Regulation of the IL-2 Gene

It has been well documented that production of IL-2 is under strict control; with some exceptions, only mature (mainly helper) T cells produce IL-2 upon stimulation by antigen or mitogen such as concanavalin A (Con A) (Smith 1980; Taniguchi et al. 1983). A number of studies on the IL-2 mRNA synthesis indicate that it is controlled primarily at the transcription level (Efrat et al. 1982; Taniguchi et al. 1983).

To elucidate the molecular mechanism operating in IL-2 gene regulation, we cloned the chromosomal genes encoding human and mouse IL-2 (Fujita et al. 1983; Fuse et al. 1984). Both genes are present in a single copy, and the human gene has been assigned to chromosome 4 (Shows et al. 1984). As depicted in Figure 3, the human IL-2 gene is divided into four blocks. The structural organization of the human and mouse IL-2 genes is very similar (Fuse et al. 1984). Within these genes, we noted the presence of highly conserved DNA sequences in their 5′-flanking region. To assess the functional role of such sequences, a DNA segment containing the 5′-flanking region of the human IL-2 gene (−584 to +51 relative to the cap site) was fused to the CAT structural gene. Using the transient induction-expression system, we found that the DNA segment indeed mediates distal CAT gene expression in a manner specific to activated T-cell clones (Jurkat and

```
                                                         50
Human  ATCACTCTCTTTAATCACTACTCACAGTAACCTCAACTCCTGCCACA ATG TAC AGG ATG CAA CTC CTG TCT TGC ATT GCA CTA AGT
                                                       Met Tyr Arg Met Gln Leu Leu Ser Cys Ile Ala Leu Ser
                                                       -21                                         -10

                                                       -21                                         -10
                                                       Met Tyr Ser Met Gln Leu Ala Ser Cys Val Thr Leu Thr
Mouse  ATCACCCTTGCTAATCACTCCTCACAGTGACCTCAAGTCCTGCAGGC ATG TAC AGC ATG CAG CTC GCA TCC TGT GTC ACA TTG ACA
                                                       50

        100
CTT GCA CTT GTC ACA AAC AGT GCA CCT ACT TCA AGT TCT ACA AAG AAA ... ACA ... ... ... CAG CTA CAA CTG
Leu Ala Leu Val Thr Asn Ser Ala Pro Thr Ser Ser Ser Thr Lys Lys ... Thr ... ... ... Gln Leu Gln Leu
                        1                                           10

                        1                                       10
Leu Val Leu Leu Val Asn Ser Ala Pro Thr Ser Ser Ser Thr Ser Ser Ser Thr Ala Glu Ala Gln Gln Gln Gln
CTT GTG CTC CTT GTC AAC AGC GCA CCC ACT TCA AGC TCC ACT TCA AGC TCT ACA GCG GAA GCA CAG CAG CAG CAG
    100                                                                         150

                                150
... ... ... ... ... ... ... ... GAG CAT ... ... ... TTA CTG CTG GAT TTA CAG ATG ATT TTC AAT GGA ATT AAT
... ... ... ... ... ... ... ... Glu His ... ... ... Leu Leu Leu Asp Leu Gln Met Ile Leu Asn Gly Ile Asn
                                                                20

    20                                  30                                      40
Gln Gln Gln Gln Gln Gln Gln Gln His Leu Glu Gln Leu Leu Met Asp Leu Gln Glu Leu Leu Ser Arg Met Glu
CAG CAG CAG CAG CAG CAG CAG CAG CAC CTG GAG CAG CTG TTG ATG GAC CTA CAG GAG CTC CTG AGC AGG ATG GAG
                                                    200

200                                                                     250
AAT TAC AAG AAT CCC AAA CTC ACC AGG ATG CTC ACA TTT AAG TTT TAC ATG CCC AAG AAG GCC ACA GAA CTG AAA
Asn Tyr Lys Asn Pro Lys Leu Thr Arg Met Leu Thr Phe Lys Phe Tyr Met Pro Lys Lys Ala Thr Glu Leu Lys
30                                  40                                          50

                        50                                      60
Asn Tyr Arg Asn Leu Lys Leu Pro Arg Met Leu Thr Phe Lys Phe Tyr Leu Pro Lys Gln Ala Thr Glu Leu Lys
AAT TAC AGG AAC CTG AAA CTC CCC AGG ATG CTC ACC TTC AAA TTT TAC TTG CCC AAG CAG GCC ACA GAA TTG AAA
        250                                                                     300

                                300
CAT CTT CAG TGT CTA GAA GAA GAA CTC AAA CCT CTG GAG GAA GTG CTA AAT TTA GCT CAA AGC AAA AAC TTT CAC
His Leu Gln Cys Leu Glu Glu Glu Leu Lys Pro Leu Glu Glu Val Leu Asn Leu Ala Gln Ser Lys Asn Phe His
                    60                                  70

    70                                  80                                  90
Asp Leu Gln Cys Leu Glu Asp Glu Leu Gly Pro Leu Arg His Val Leu Asp Leu Thr Gln Ser Lys Ser Phe Gln
GAT CTT CAG TGC CTA GAA GAT GAA CTT GGA CCT CTG CGG CAT GTT CTG GAT TTG ACT CAA AGC AAA AGC TTT CAA
                                                350

        350                                                             400
TTA ... AGA CCC AGG GAC TTA ATC AGC AAT ATC AAC GTA ATA GTT CTG GAA CTA AAG GGA TCT GAA ACA ACA TTC
Leu ... Arg Pro Arg Asp Leu Ile Ser Asn Ile Asn Val Ile Val Leu Glu Leu Lys Gly Ser Glu Thr Thr Phe
80                                          90                                  100

                        100                                     110
Leu Glu Asp Ala Glu Asn Phe Ile Ser Asn Ile Arg Val Thr Val Val Lys Leu Lys Gly Ser Asp Asn Thr Phe
TTG GAA GAT GCT GAG AAT TTC ATC AGC AAT ATC AGA GTA ACT GTT GTA AAA CTA AAG GGC TCT GAC AAC ACA TTT
            400                                                                 450

                                        450
ATG TGT GAA TAT GCT GAT GAG ACA GCA ACC ATT GTA GAA TTT CTG AAC AGA TGG ATT ACC TTT TGT CAA AGC ATC
Met Cys Glu Tyr Ala Asp Glu Thr Ala Thr Ile Val Glu Phe Leu Asn Arg Trp Ile Thr Phe Cys Gln Ser Ile
                        110                                     120

    120                                 130                                     140
Glu Cys Gln Phe Asp Asp Glu Ser Ala Thr Val Val Asp Phe Leu Arg Arg Trp Ile Ala Phe Cys Gln Ser Ile
GAG TGC CAA TTC GAT GAT GAG TCA GCA ACT GTG GTG GAC TTT CTG AGG AGA TGG ATA GCC TTC TGT CAA AGC ATC
                                                500

    500                                                       550
ATC TCA ACA CTA ACT TGA TAA TTA AGTGCTTCCCACTTAAAACATATCAGGCCTTCTATTTATTTAAATATTTAAATTTTATATTTATTGT
Ile Ser Thr Leu Thr
    130

Ile Ser Thr Ser Pro Gln
ATC TCA ACA AGC CCT CAA TAA CTA TGTACCTCCTGCTTACAACACATAAGGCTCTCTATTTATTTAAATATTTAAC.TTTA.ATTTATTTT
            550                                                              600
            600                                               650
TGAATGTATGGTTTGCTACCTATTGTAACTATTATTCTTAATCTTAAAACTATAAATATGGATCTTTTATGATTCTTTTTGTAAGCCCTAGGGGCTCTA

TGGATGTATTGTTTACTATCTTTTGTAACTACTAGTCTT...CAGATG...ATAAATATGGATCTTTAAAGATTCTTTTTGTAAGCCCCAAGGGCTCAA
                                650                                                    700
             700                                                  750
AAATGGTTTCACTTATTTATCCCAAAATATTTATTATTATGTTGAAT.GTTAAATATAGTATCTATGTAGATTGGTTAGTAAAACTATTTAATAAATTT

AAATGTTTTAAACTATTTATCTGAAATTATTTATTAT.A..TTGAATTGTTAAATATCATGTGTAGGTAGACTCATTAATAAAAGTATTT.AGA....T
                                       750                                              800
                 800
GAT..AAATATAAAAAAAAAAAAACpolyA

GATTCAAATATAAATAAGCTCAGATGTCTGTCATTTTTAGGACAGCACAAAGTAAGCGCTAAAATAACTTCTCAGTTATTCCTGTGAACTCTATGTTAA
                                     850                                                   900

TCAGTGTTTTCAAGAAATAAAGCTCTCCTCTpolyA
```

Figure 1. Structure of the cDNA and of the deduced primary sequences for human and mouse IL-2. Gaps were introduced to align the sequences for maximal homology. Coincident amino acids are framed. (Data from Taniguchi et al. [1983] and Kashima et al. [1985].)

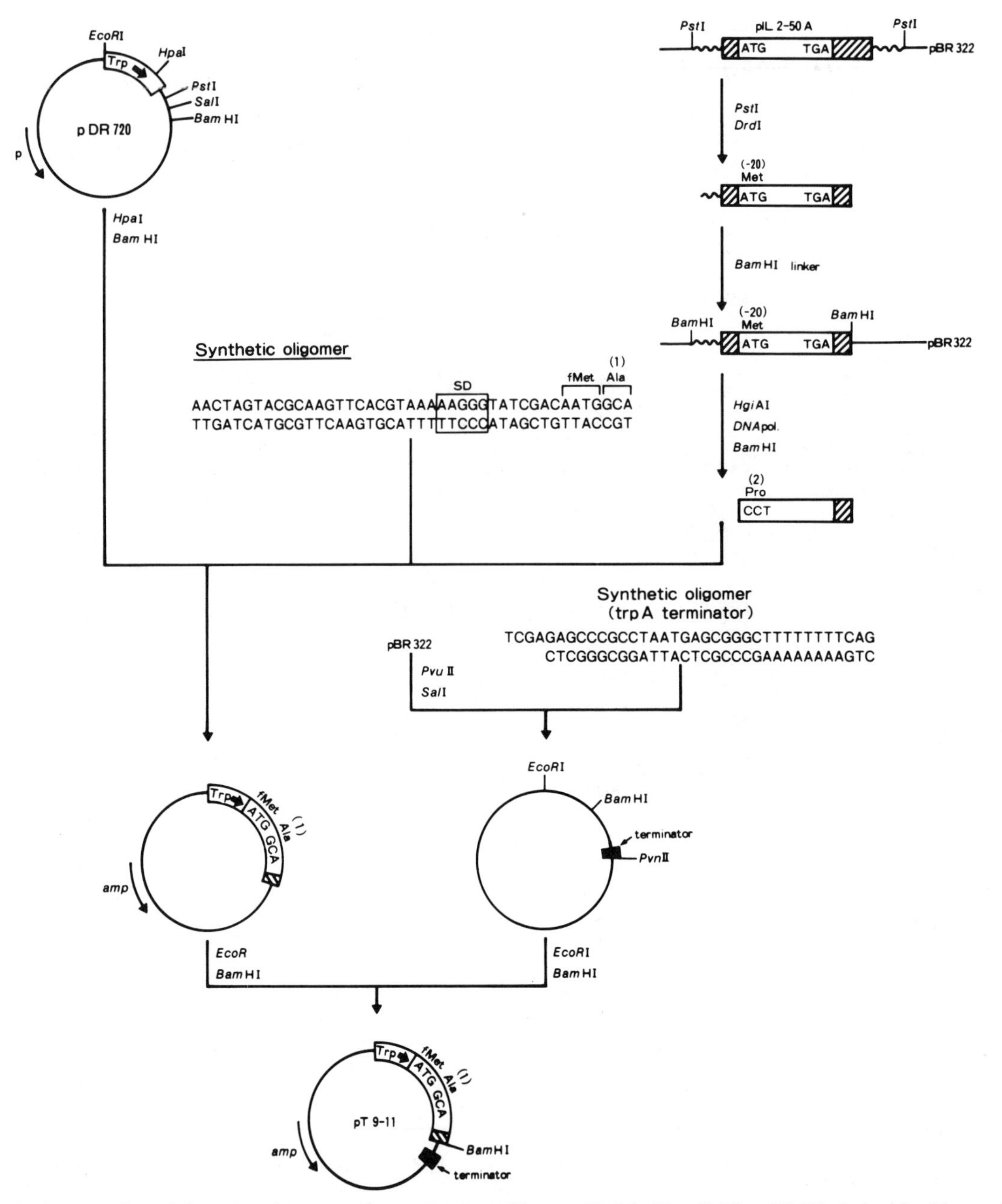

Figure 2. Construction of the expression vector for production of human IL-2 in *E. coli*. The pBR322-derived backbone piece of DNA containing the *trp* promoter was ligated with the IL-2 cDNA sequences and a fragment of DNA that possesses a transcription terminator. The resulting plasmid pT9-11 gives rise to about 5×10^7 units of IL-2 per milliliter of bacterial culture.

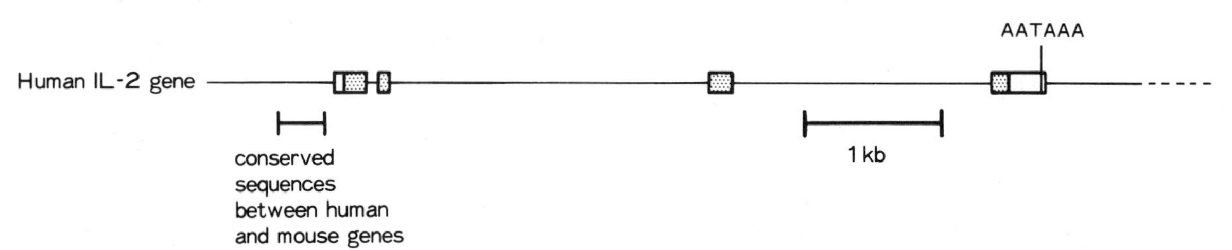

Figure 3. Organization of the human IL-2 gene. The gene is divided into four exons (for details, see Fujita et al. 1983).

EL-4). The initiation of the mRNA transcript from the correct site in this expression system was also confirmed by S1-mapping analysis (Fujita et al. 1986). We then generated a series of 5′ and 3′ deletion mutants to delimitate the upstream and downstream boundaries of the functional DNA sequences. The results are summarized in Figure 4. The upstream boundary appears to lie between −319 and −264 (Fujita et al. 1986). In this figure, the 3′ deletion mutants (in all the mutants, the 5′ boundary is −319) were each linked to a promoter sequence (−66 to +19) of the human IFN-β gene. Unless activated by an additional promoter or enhancer sequence, the promoter itself is inert in essentially all cell types (Fujita et al. 1985). Therefore, the functional properties of the IL-2-specific DNA sequences in activated T cells can be readily monitored by measuring the expression of the CAT gene, whose transcription is initiated from the IFN-β gene cap site in the activated T-cell clone (EL-4). As shown in Figure 4 (right panel), the 3′ boundary of the functional DNA sequences appears to lie between −127 and −145. Essentially the same results were obtained by generating internal deletions in this region of the IL-2 gene (Fujita et al. 1986). In both sets of 3′ deletion mutant genes, the DNA segment spanning −98 to −319 is always less effective than the DNA segments of shorter length (i.e., −115 to −319 and −127 to −319). One could envisage the presence of an inhibitory DNA sequence(s) between −98 and −115. The DNA segment consisting of about 200 bp (−319 to −127) thus appears to mediate the controlled IL-2 gene expression in T lymphocytes. We also obtained essentially the same results by using a human T cell line, Jurkat; the same DNA sequences are required for the Con-A-induced Jurkat cells (Fujita et al. 1986). In addition, it has been shown to function in an orientation-independent manner and contains regions that have sequence homology with long terminal repeat (LTR) sequences of HIV (or HTLV-III/LAV) and the 5′ upstream region of the IL-2 receptor and IFN-γ genes (Fujita et al. 1986). The promoter sequence of the human IL-2 gene (−81 to +51) appears to function in nonlymphoid cells under certain circumstances (Fujita et al. 1986).

Functional Expression of the IL-2 Receptor Gene

A number of studies demonstrated that IL-2 delivers its growth signal, and perhaps other signals as well, by virtue of its interaction with a specific cell-surface receptor (IL-2R) (Robb et al. 1981; Smith 1984). The molecular nature of the human IL-2R became apparent with the availability of a monoclonal antibody, anti-Tac, which recognizes an epitope of the IL-2R (Uchiyama et al. 1981). The putative IL-2R, the Tac antigen, was shown to be a glycoprotein (m.w. = 55,000) that is present on the surface of activated (but not resting) T cells (Uchiyama et al. 1981; Leonard et al. 1983). The Tac antigen was purified and, based on its partial primary sequence, the corresponding cDNAs were cloned (Leonard et al. 1984; Nikaido et al. 1984; Cosman et al. 1984).

IL-2-binding studies also demonstrated the presence of two classes of IL-2R (Robb et al. 1984a). The finding that IL-2-mediated T-cell growth correlates well with the expression of the high-affinity receptor indicated that only the high-affinity IL-2R is functional in signal transduction. When the cloned cDNA for the Tac antigen was expressed in mouse L cells, only low-

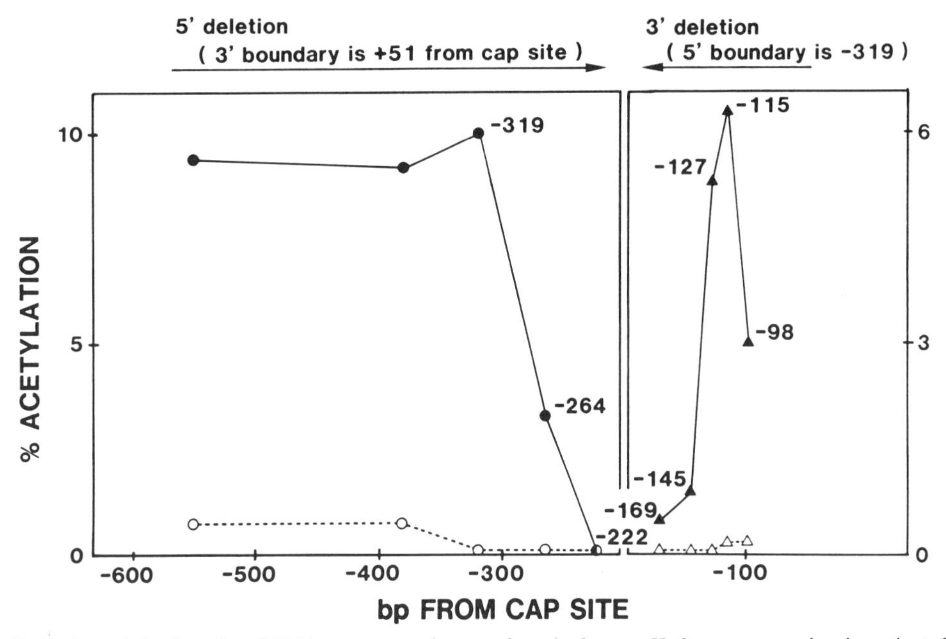

Figure 4. Delimitation of the functional DNA sequences that regulate the human IL-2 gene expression in activated T lymphocytes. Function of the IL-2 5′-flanking sequences was monitored by the distal CAT gene expression in TPA-activated EL-4 cells. The levels of CAT expression are represented at the ordinate. The results using pIL2cat series were also verified by S1 analysis (Fujita et al. 1986). (●,▲) With TPA stimulation; (○,△) without TPA stimulation.

affinity IL-2R molecules became detectable on the cell surface (Greene et al. 1985). In view of this finding and the unique structure of the Tac-antigen molecule, which contains a very short intracytoplasmic tail (consisting of 13 amino acids) as revealed from the cDNA sequence analysis, it was important or even essential to clarify whether or not the molecule itself was functional as IL-2R, i.e., whether the molecule can under any circumstances exhibit high affinity to IL-2 and mediate signal transduction.

We therefore attempted to reconstitute biologically functional IL-2R molecules by expressing the cloned gene in various host cells. We cloned a cDNA encoding the Tac antigen from activated normal human T cells. The cDNA, which encodes a protein with a sequence identical to that from HTLV-I-infected human T cells (Cosman et al. 1984; Leonard et al. 1984; Nikaido et al. 1984), was linked to the LTR sequence of Rous sarcoma virus. The resulting chimeric gene, inserted into pSV2-gpt (Mulligan and Berg 1981), was then introduced into EL-4 (a mouse line with Thyl$^+$, Lyl$^+$, Lyt2$^-$, and L3T4$^+$ phenotype) and L929 cells (Hatakeyama et al. 1985). Transformants were grown in a selective medium containing mycophenolic acid, and Tac-positive cells were cloned. EL-4 is advantageous for studying the heterologous IL-2R expression: These T lymphoid cells do not express autologous mouse IL-2R, which would otherwise function in response to human IL-2 and thus obscure the interpretation of the results.

We next determined whether the expressed molecule manifests high or low affinity for human IL-2 by radiolabeled IL-2-binding assays. As shown in Figure 5, Scatchard analysis of an EL-4 transformant, but not an L929 transformant, showed curvilinear features, demonstrating the presence of two distinct receptors with different affinities for the recombinant human IL-2. Thus, by expressing the cDNA for Tac antigen, both high-affinity IL-2R and low-affinity IL-2R can be generated specifically in the T lymphoid cells. Very interestingly, the human IL-2R expressed in EL-4 cells responded to IL-2 and mediated a "reversed signal." Growth of the EL-4 cells expressing the human IL-2R was inhibited specifically by human, but not mouse, recombinant IL-2 (Fig. 6). The signal transduction occurred at IL-2 concentrations (1.3–13 pmoles) in which only high-affinity IL-2R can interact with the ligand, and it was completely abrogated by monoclonal antibodies against Tac antigen (Hatakeyama et al. 1985; M. Hatakeyama and T. Taniguchi, unpubl.). In addition, we obtained a number of EL-4 transformants that expressed different densities of IL-2R on their surfaces. Since most of the transformants contain a few copies of the gene, the difference of the receptor density is likely to be caused by the different sites of gene integration that would affect transcription efficiencies

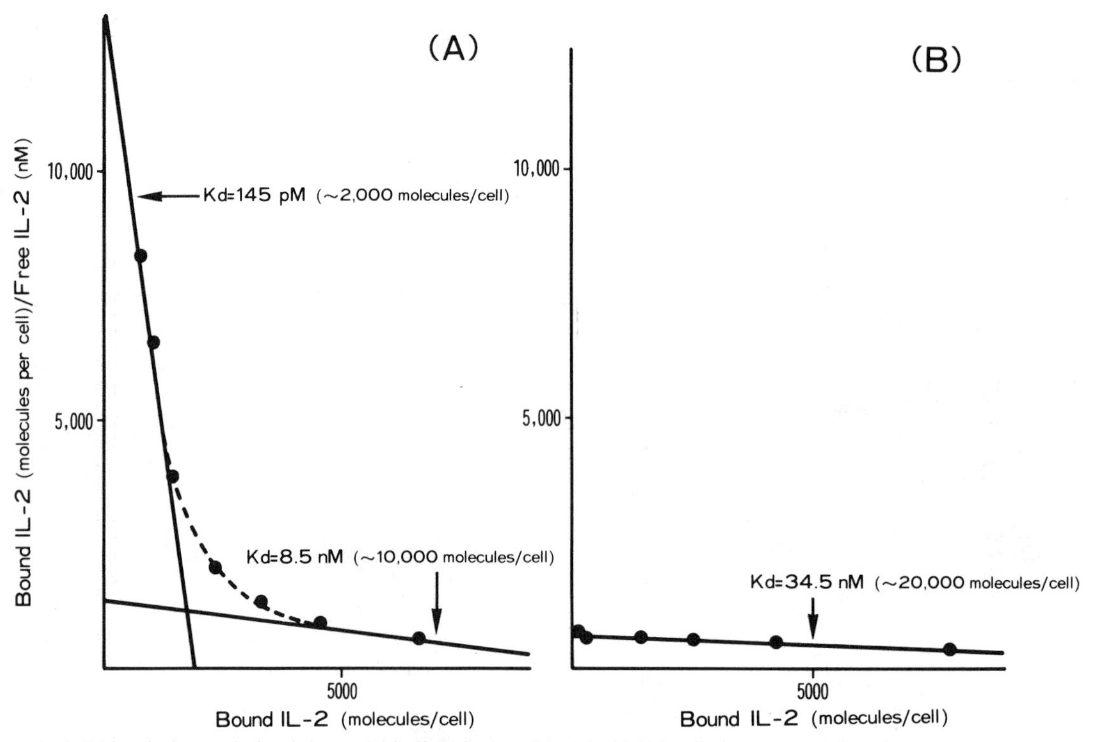

Figure 5. Scatchard plot analysis of the IL-2-binding data on EL-4 and L929 cell clones expressing the Tac antigen. An expression plasmid pSVIL2R-3 (Hatakeyama et al. 1985) in which the Tac-antigen cDNA was linked to the LTR sequence of Rous sarcoma virus was transfected into EL-4 and L929 cells. After cloning, the EL-4 and L cell transformants (ELT-19 and LCT-4, respectively) were subjected to the ^{125}I-labeled IL-2-binding assay (Robb et al. 1984b; Hatakeyama et al. 1985). (*A*) ELT-19 cells; (*B*) LCT-4 cells.

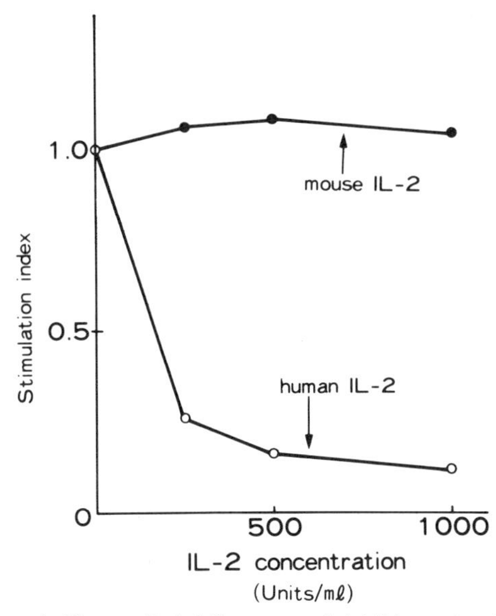

Figure 6. Human IL-2 delivers growth inhibitory signal(s) via the reconstituted receptor on EL-4. A human IL-2R-expressing clone (ELT-5) was cultured with various concentrations of either recombinant human or mouse IL-2 for 72 hr. The [^{3}H]TdR incorporation was then measured as described by Hatakeyama et al. (1985).

in the host chromosomes. As shown in Table 1, amplitude of the signal transduction (as represented by the stimulation index, S.I.) correlates well with the density of high-affinity IL-2R.

Table 1. Receptor Number and Magnitude of Cellular Response

Cell line	Number of high-affinity IL-2 receptors	Cellular response to IL-2 (S.I.)
ELT-5	3000	0.12
ELT-13	4500	0.23
ELT-19	1950	0.09
ELT-21	1800	0.35
ELT-22	1700	0.57
ELT-23	1000	0.74
ELT-24	400	1.03
ELT-25	300	0.93
EL-4	0	1.00

The number of high-affinity IL-2 receptors was determined for each clone by the ^{125}I-labeled IL-2-binding assay. Cells were cultured in the presence of 1000 U/ml of human IL-2 for 72 hr and S.I. was determined as previously described (Hatakeyama et al. 1985).

Effect of IL-2-specific Signaling in the Cellular Oncogene Expression in EL-4 Transformants

A number of reports demonstrate that extracellular signaling by growth factors or mitogens leads to the activation of cellular oncogenes (Kelly et al. 1983; Greenberg and Ziff 1984; Reed et al. 1985a,b). Regulation of cellular oncogenes certainly plays a key role in cell growth and differentiation. The prior accumulation of c-*myc* mRNA caused by IL-2 seems to correlate with the growth of T-cell clones (Reed et al. 1985a,b).

We addressed the question of whether or not the expression of any cellular oncogene is affected by IL-2 via the reconstituted human IL-2R in mouse EL-4 cells. An EL-4 transformant, ELT-5, which expresses about 3000 molecules per cell of the high-affinity IL-2R, was incubated with recombinant human IL-2 (1000 units/ml) at various intervals. Poly(A) RNA or total RNA was then prepared and subjected to RNA blotting analysis (Thomas 1980). As shown in Figure 7A, the parental EL-4 and its transformant, ELT-5, constitutively express a number of oncogenes (their nature is not yet characterized). The mRNA level of those genes was not affected by IL-2, except for c-*myc*-specific mRNA, whose level significantly decreased after IL-2 treatment. The decrease of c-*myc* mRNA correlated well with the inhibition of cellular growth by IL-2 (Fig. 7B) (Hatakeyama et al. 1985). Thus, taken together with the reports of Reed et al. (1985a,b), the expression level of the c-*myc* gene and the IL-2-mediated cellular responses appears to be closely related. Of interest was the finding that IL-2 also mediated a rapid and transient accumulation of c-*fos* mRNA in the IL-2R-bearing EL-4 transformant (Fig. 7B). No change was detectable in the c-*myc* mRNA level soon after IL-2 treatment (result not shown).

DISCUSSION

The work described here focuses primarily on (1) the molecular nature of T-cell growth control, which is mediated by the IL-2 system, and (2) the production of IL-2 by means of genetic engineering for basic and therapeutic investigations.

We first isolated and expressed cDNAs encoding the human and mouse IL-2. Similar work has also been done by other groups (Devos et al. 1983; Clark et al. 1984; Holbrook et al. 1984; Yokota et al. 1985). When the deduced primary structures of both IL-2 molecules were compared, approximately 63% sequence homology was observed. It has been well documented that human IL-2 is functional on mouse T cells, whereas mouse IL-2 is nonfunctional on human T cells (Gillis et al. 1978; Smith 1980). Accordingly, mouse IL-2 is totally nonfunctional on mouse cells that express heterologous human IL-2R but not the autologous receptor (Fig. 6). The structural moiety of the IL-2 molecule primarily responsible for the specific binding to the receptors is not known at present. Removal of several amino acids from either the amino- or carboxy-terminal region of the human IL-2 seems to abolish its activity (H. Matsui, unpubl.; H. Miyaji and S. Ito, pers. comm.). Generation of chimeric IL-2 molecules by means of the recombination procedure for homologous genes (Weber and Weissmann 1983) may provide a clue for the understanding of the specific ligand-receptor interactions.

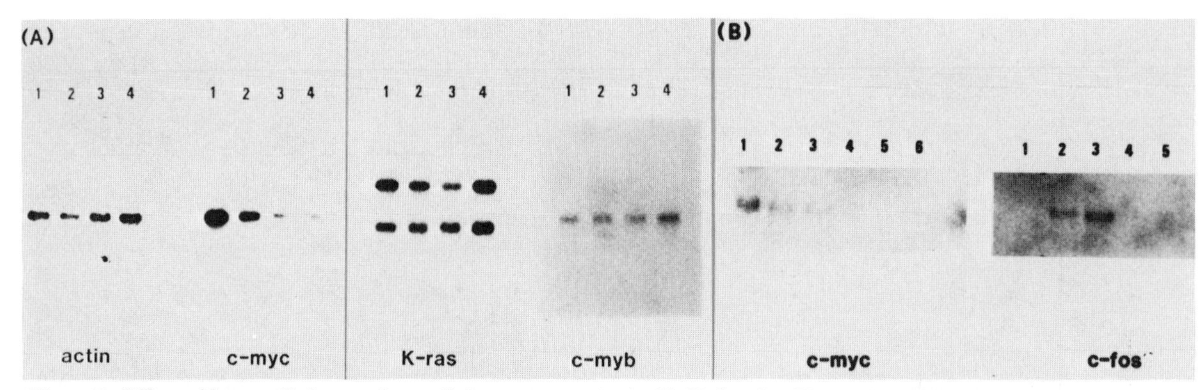

Figure 7. Effect of human IL-2 on various cellular oncogenes in the IL-2R-bearing EL-4 cells. (*A*) Poly(A) RNAs were isolated from 10^7 cells of parental EL-4 (lane *1*) and ELT-5 after culturing in the presence of human IL-2 (1000 U/ml) for 0 (lane *2*), 6 (lane *3*), and 24 hr (lane *4*). They were then subjected to the RNA blotting analysis (Thomas et al. 1980). (*B*) Total RNAs were isolated from 10^6 cells at each point and then analyzed directly. For the c-*myc* gene, cells were cultured with the IL-2 (1000 U/ml) for 0 (lane *1*), 6 (lane *2*), 12 (lane *3*), 24 (lane *4*), 48 (lane *5*), and 72 hr (lane *6*). For c-*fos* gene expression, RNA was extracted shortly after the IL-2 treatment (lane *1*, zero time; lane *2*, 15 min; lane *3*, 30 min; lane *4*, 60 min; lane *5*, 120 min).

Production of the recombinant human IL-2 in *E. coli* has made it possible to investigate further the biological role of this lymphokine and to test it for clinical potentials. So far, the recombinant IL-2 seems to be indistinguishable from the natural IL-2. Rosenberg et al. (1985) first reported the successful application of recombinant IL-2 for adoptive cancer immunotherapy. It remains to be seen whether IL-2 will be effective also in the treatment of other immune deficiencies.

To elucidate the molecular mechanism operating in IL-2 gene regulation, we cloned the chromosomal genes encoding human and mouse IL-2. Functional DNA sequences were identified that mediate the regulated gene expression in T lymphocytes. The sequences, located in the 5′ upstream region of the gene, function in an orientation-independent manner (Fujita et al. 1986) and also activate a heterologous promoter (Fig. 4). So far, we have not detected any other functional DNA sequences for the induction of IL-2 gene transcription within the cloned DNA (7 kb). It is possible, however, that within the DNA segment, sequences are present that are involved in the posttranscriptional control of gene expression (Efrat and Kaempfer 1984). Requirements of about 200 bp of DNA for the regulated gene expression may be interpreted to mean that multiple factors are involved in the gene control.

IL-2 delivers the extracellular signal through its own receptor. As demonstrated in Figure 5, the cDNA encoding Tac antigen can give rise to the expression of high-affinity IL-2R specifically in a T lymphoid cell line. Our results indicate strongly that Tac antigen per se is insufficient in functioning as IL-2R. Indeed, it manifests only low affinity to IL-2 when expressed on the L-cell surface (Fig. 5; Greene et al. 1985; Hatakeyama et al. 1985). Thus, we envisage the formation of a functional "IL-2 receptor complex" when the Tac antigen associates with other membrane determinant(s) whose presence is restricted in certain sets of lymphocytes.

We also found that the human IL-2R expressed in EL-4 cells responds to human IL-2 and delivers growth inhibitory signals. The extent of this signaling appears to correlate with the density of the high-affinity IL-2R (Table 1). This phenomenon may be related to the efficiency of receptor clustering that is likely to occur on the cell surface after the ligand binding. Analysis of cellular oncogene mRNAs in the IL-2R-bearing EL-4 transformant revealed that a specific decrease of the c-*myc* mRNA correlates with cell-growth inhibition by IL-2. Thus, the IL-2-mediated growth and inhibition of T cells appear to be accompanied by the respective stimulation (Reed et al. 1985a,b) and suppression (Fig. 7) of the c-*myc* mRNA level, respectively. Further work will be required to appreciate the significance of this correlation. However, such bidirectional c-*myc* gene regulation may be of some physiological importance.

Is the suppressive mechanism for cell growth by IL-2 operative in vivo? The IL-2-mediated cell-growth inhibition has also been observed in some HTLV-I-infected human T cells (Sugamura et al. 1985). The observation that the subset of normal, human $T4^+$ cells becomes refractory to IL-2 earlier than $T8^+$ cells (Gullberg and Smith 1986) may also relate to our findings. A tempting idea would be the function of IL-2 per se as a feedback regulator to control the overgrowth under certain processes of the clonal expansion of lymphocytes. Alternatively, as we have shown the c-*fos* gene induction by IL-2 (Fig. 7), the phenomenon described here may be viewed as mimicking the process of lymphocyte differentiation mediated by IL-2.

Our results, together with those of other investigators (for review, see Sporn and Roberts 1985), draw attention to the multiple functions of cytokines and other mediators whose effect is controlled by the complex mechanism operating in a given responder cell.

ACKNOWLEDGMENTS

We thank Drs. T. Sekiya and K. Shimizu for oncogene probes and Drs. T. Kishimoto and T. Uchiyama for valuable comments. We also thank Dr. E.L. Barsoumian for critical reading of the manuscript and Ms. M. Nagatsuka for her excellent assistance. This work was supported in part by a grant-in-aid for Special Project Research, Cancer-Bioscience from the Ministry of Education, Science, and Culture of Japan and by grant-in-aid from The Mochida Memorial Foundation for Medical and Pharmaceutical Research.

REFERENCES

Cerretti, D.P., K. McKereghan, A. Larsen, M.A. Cantrell, D. Anderson, S. Gillis, D. Cosman, and P.E. Baker. 1986. Cloning, sequence, and expression of bovine interleukin 2. *Proc. Natl. Acad. Sci.* **83:** 3223.

Clark, S.C., S.K. Arya, F. Wong-Staal, M. Matsumoto-Kobayashi, R.M. Kay, R.J. Kaufman, E.L. Brown, C. Schoemaker, T. Copeland, S. Droszlan, K. Smith, M.G. Sarngadharam, S.G. Linder, and R. Gallo. 1984. Human T-cell growth factor: Partial amino acid sequence cDNA cloning, and organization and expression in normal and leukemic cells. *Proc. Natl. Acad. Sci.* **81:** 2543.

Cosman, D., D.P. Cerretti, A. Larsen, L. Park, C. March, S. Dower, S. Gillis, and D. Urdall. 1984. Cloning, sequence and expression of human interleukin-2 receptor. *Nature* **312:** 768.

Devos, R., G. Plaetinck, H. Cheroutre, G. Simons, W. Degrave, J. Tavernier, E. Remaut, and W. Fiers. 1983. Molecular cloning of human interleukin 2 cDNA and its expression in *E. coli. Nucleic Acids Res.* **11:** 4307.

Efrat, S. and R. Kaempfer. 1984. Control of biologically active interleukin 2 messenger RNA formation in induced human lymphocytes. *Proc. Natl. Acad. Sci.* **81:** 2601.

Efrat, S., S. Pilo, and R. Kaempfer. 1982. Kinetics of induction and molecular size of mRNAs encoding human interleukin-2 and γ-interferon. *Nature* **297:** 236.

Fujita, T., S. Ohno, H. Yasumitsu, and T. Taniguchi. 1985. Delimitation and properties of DNA sequences required for the regulated expression of human interferon-β gene. *Cell* **41:** 489.

Fujita, T., C. Takaoka, H. Matsui, and T. Taniguchi. 1983. Structure of the human interleukin 2 gene. *Proc. Natl. Acad. Sci.* **80:** 7437.

Fujita, T., H. Shibuya, T. Ohashi, K. Yamanishi, and T. Taniguchi. 1986. Regulation of human interleukin-2 gene: Functional DNA sequences in the 5′ flanking region for the gene expression in activated T lymphocytes. *Cell* **46:** 401.

Fuse, A., T. Fujita, H. Yasumitsu, N. Kashima, K. Hasegawa, and T. Taniguchi. 1984. Organization and structure of the mouse interleukin-2 gene. *Nucleic Acids Res.* **12:** 9323.

Gillis, S., P.E. Baker, N.A. Union, and K.A. Smith. 1979. The *in vitro* generation and sustained culture of nude cytotoxic T-lymphocytes. *J. Exp. Med.* **149:** 1460.

Gillis, S., M.M. Ferm, W. Ou, and K.A. Smith. 1978. T-cell growth factor: Parameters of production and a quantitative microassay for activity. *J. Immunol.* **120:** 2027.

Gorman, C.M., L.F. Moffat, and B.H. Howard. 1982. Recombinant genomes which express chloramphenicol acetyltransferase in mammalian cell. *Mol. Cell. Biol.* **2:** 1044.

Greenberg, M.E. and E.B. Ziff. 1984. Stimulation of 3T3 cells induces transcription of the c-*fos* proto-oncogene. *Nature* **311:** 433.

Greene, W., R.J. Robb, P.B. Svetlik, G.M. Rusk, J.M. Depper, and W.J. Leonard. 1985. Stable expression of cDNA encoding the human interleukin 2 receptor in eukaryotic cells. *J. Exp. Med.* **162:** 363.

Grimm, E.A., A. Mazumder, H.Z. Zharg, and S.A. Rosenberg. 1982 Lymphokine-activated killer cell phenomenon: Lysis of natural killer-resistant fresh solid tumor cells by interleukin-2-activated autologous human peripheral blood lymphocytes. *J. Exp. Med.* **155:** 1823.

Gullberg, M. and D.A. Smith. 1986. Regulation of T cell autocrine growth, T4+ cells become refractory to interleukin 2. *J. Exp. Med.* **163:** 270.

Harpold, M.M., R.R. Dobner, R.M. Evans, and F.C. Bancroft. 1978. Construction and identification of positive hybridization-translation of a bacterial plasmid containing a rat growth hormone structural gene sequence. *Nucleic Acids Res.* **5:** 2039.

Hatakeyama, M., S. Minamoto, T. Uchiyama, R.R. Hardy, G. Yamada, and T. Taniguchi. 1985. Reconstitution of functional receptor for human interleukin-2 in mouse cells. *Nature* **318:** 467.

Henny, C.S., K. Kuribayashi, D.E. Kern, and S. Gillis. 1981. Interleukin-2 augments natural killer cell activity. *Nature* **291:** 335.

Holbrook, N.J., K.A. Smith, A.J. Fornace, Jr., C.M. Comeau, R.L. Wilskocil, and G. Crabtree. 1984. T-cell growth factor: Complete nucleotide sequence and organization of the gene in normal and malignant cells. *Proc. Natl. Acad. Sci.* **81:** 1634.

Kashima, N., C. Nishi-Takaoka, T. Fujita, S. Taki, G. Yamada, J. Hamuro, and T. Taniguchi. 1985. Unique structure of murine interleukin-2 as deduced from cloned cDNAs. *Nature* **313:** 402.

Kelly, K., B.H. Cochran, C.D. Stiles, and P. Leder. 1983. Cell-specific-regulation of the c-*myc* gene by lymphocyte mitogens and platelet-derived growth factor. *Cell* **35:** 603.

Kishi, H., S. Inui, A. Muraguchi, T. Hirano, Y. Yamamura, and T. Kishimoto. 1985. Induction of IgG secretion in a human B cell clone with recombinant IL2. *J. Immunol.* **134:** 3104.

Leonard, W.J., J.M. Depper, R.J. Robb, T.A. Waldmann, and W.C. Greene. 1983. Characterization of the human receptor for T-cell growth factor. *Proc. Natl. Acad. Sci.* **80:** 6957.

Leonard, W.J., J.M. Depper, G.R. Crabtree, S. Rudikoff, J. Pumphrey, R.J. Robb, M. Kronke, P.B. Svetlik, N.J. Peffer, T.A. Waldmann, and W.C. Greene. 1984. Molecular cloning and expression of cDNAs for the human interleukin-2 receptor. *Nature* **311:** 626.

Maniatis, T., E.F. Fritsch, and J. Sambrook. 1982. *Molecular cloning: A laboratory manual.* Cold Spring Harbor Laboratory, Cold Spring Harbor, New York.

Milanese, C., N.E. Richardson, and E.L. Reinherz. 1986. Identification of a T helper cell-derived lymphokine that activates resting T lymphocytes. *Science* **231:** 1118.

Mond, J.J., C.T. Thompson, F.D. Finkelman, J. Farrar, M. Schaefer, and R.J. Robb. 1985. Affinity-purified interleukin-2 induces proliferation of large but not small B cells. *Proc. Natl. Acad. Sci.* **82:** 1518.

Morgan, D.A., F.W. Ruscetti, and R. Gallo. 1976. Selective *in vitro* growth of T-lymphocytes from normal human bone marrows. *Science* **193:** 1007.

Mulligan, R.C. and P. Berg. 1981. Selection for animal cells that express the *Escherichia coli* gene coding for xanthine-guanine phosphoribosyl transferase. *Proc. Natl. Acad. Sci.* **78:** 2072.

Nagata, S., H. Taira, A. Hall, L. Johnsrud, M. Streuli, J. Ecsodi, W. Boll, K. Cantell, and C. Weissmann. 1980. Synthesis in *E. coli* of a polypeptide with human leukocyte interferon activity. *Nature* **284:** 316

Nikaido, T., A. Shimizu, N. Ishida, H. Sabe, K. Teshigawara, M. Maeda, T. Uchiyama, J. Yodoi, and T. Honjo. 1984. Molecular cloning of cDNA encoding human interleukin-2 receptor. *Nature* **311:** 631.

Ohno, S. and T. Taniguchi. 1980. Structure of a chromosomal gene for human interferon-β. *Proc. Natl. Acad. Sci.* **78:** 5305.

Reed, J.C., P.C. Nowell, and R.G. Hoover. 1985a. Regulation of c-*myc* mRNA levels in normal human lymphocytes by modulators of cell proliferation. *Proc. Natl. Acad. Sci.* **82:** 4221.

Reed, J.C., D.E. Sabath, R.G. Hoover, and M.B. Prystowsky. 1985b. Recombinant interleukin-2 regulates levels of c-*myc* mRNA in a cloned murine T lymphocyte. *Mol. Cell. Biol.* **5:** 3361.

Reeves, R., A.G. Spies, M.S. Nissen, C.D. Buck, A.D. Weinberg, P.J. Barr, N.S. Magnuson, and J.A. Magnuson. 1986. Molecular cloning of a functional bovine interleukin 2 cDNA. *Proc. Natl. Acad. Sci.* **83:** 3228.

Robb, R.J., W.C. Greene, and C.M. Rusk. 1984a. Low and high affinity cellular receptors for interleukin 2. Implications for the level of Tac antigen. *J. Exp. Med.* **160:** 1126.

Robb, R.J., A. Munck, and K.A. Smith. 1981. T cell growth factor receptors. Quantitation, specificity, and biological relevance. *J. Exp. Med.* **154:** 1455.

Robb, R.J., R.M. Kutny, M. Panico, H.R. Morris, and V. Chowdhry. 1984b. Amino acid sequence and post-translational modifications of human interleukin 2. *Proc. Natl. Acad. Sci.* **81:** 6486.

Rosenberg, S.A., M.T. Lotze, L.M. Muul, S. Leitman, A.E. Chang, S.E. Ettinghausen, Y.L. Matory, J.M. Skibber, E. Shiloni, J.T. Vetto, C.A. Seipp, C. Simpson, and C.M. Reichert. 1985. Observations on the systematic administration of autologous lymphokine-activated killer cells and recombinant interleukin-2 to patients with metostatic cancer. *N. Engl. J. Med.* **313:** 1485.

Shows, T., R. Eddy, L. Haley, M. Byers, M. Henry, T. Fujita, T. Matsui, and T. Taniguchi. 1984. Interleukin 2 (IL2) is assigned to human chromosome 4. *Somatic Cell Mol. Genet.* **10:** 315.

Smith, K.A. 1980. T-cell growth factor. *Immunol. Rev.* **51:** 337.

———. 1984. Interleukin 2. *Annu. Rev. Immunol.* **2:** 319.

Sporn, M.B. and A.B. Roberts. 1985. Autocrine growth factors and cancer. *Nature* **313:** 745.

Sugamura, K., S. Nakai, M. Fujii, and Y. Hinuma. 1985. Interleukin 2 inhibits in vitro growth of human T cell lines carrying retrovirus. *J. Exp. Med.* **161:** 1243.

Taniguchi, T., Y. Fujii-Kuriyama, and M. Muramatsu. 1980. Molecular cloning of human interferon cDNA. *Proc. Natl. Acad. Sci.* **77:** 4003.

Taniguchi, T., M. Sakai, Y. Fujii-Kuriyama, M. Muramatsu, S. Kobayashi, and T. Sudo. 1979. Construction and identification of a bacterial plasmid containing the human fibroblast interferon gene sequence. *Proc. Jpn. Acad.* **55:** 464.

Taniguchi, T., H. Matsui, T. Fujita, C. Takaoka, N. Kashima, R. Yoshimoto, and J. Hamuro. 1983. Structure and expression of a cloned cDNA for human interleukin-2. *Nature* **302:** 305.

Taniguchi, T., H. Matsui, T. Fujita, M. Hatakeyama, N. Kashima, A. Fuse, J. Hamuro, C. Nishi-Takaoka, and G. Yamada. 1986. Molecular analysis of the interleukin-2 system. *Immunol. Rev.* **92:** 121.

Thomas, P.S. 1980. Hybridization of denatured RNA and small DNA fragments transferred to nitrocellulose. *Proc. Natl. Acad. Sci.* **77:** 5201.

Uchiyama, T., D.L. Nelson, T.A. Fleisher, and T. Waldmann. 1981. A monoclonal antibody (anti-Tac) reactive with activated and functionally mature human T cells. I. Production of anti-Tac monoclonal antibody and distribution of Tac(+) cells. *J. Immunol.* **126:** 1393.

Wang, A., S.-D. Lu, and D.F. Mark. 1984. Site-specific mutagenesis of the human interleukin-2 gene: Structure-function analysis of the cysteine residues. *Science* **224:** 1431.

Weber, H. and C. Weissmann. 1983. Formation of genes coding for hybrid proteins by recombination between related, cloned genes in *E. coli. Nucleic Acids Res.* **11:** 5661.

Yokota, T., N. Arai, F. Lee, D. Rennick, T. Mosman, and K. Arai. 1985. Use of a cDNA expression vector for isolation of mouse interleukin 2 cDNA clones: Expression of T-cell growth-factor activity after transfection of monkey cells. *Proc. Natl. Acad. Sci.* **82:** 68.

Lymphokines and Monokines in Anti-cancer Therapy

W. Fiers, P. Brouckaert, R. Devos, L. Fransen,* G. Leroux-Roels,
E. Remaut, P. Suffys, J. Tavernier,* J. Van der Heyden,* and F. Van Roy
Laboratory of Molecular Biology, State University of Ghent, 9000 Ghent, Belgium
**Biogent (Subsidiary of Biogen S.A.), 9000 Ghent, Belgium*

About 10 years ago, it became evident that many types of cancer cells were causally related to irreversible changes in their genetic makeup. Hence, elimination of these malignant cells was the only possible therapy. Except for surgery, classical treatments like irradiation and chemotherapy are fairly unselective and have unpleasant side effects. Can molecular biology provide drugs and treatments that are more selective and effective while being more biocompatible and less toxic?

Only 5 years after its discovery, interferon was shown to have an antiproliferative effect on transformed cells (Paucker et al. 1962). Nevertheless, it took an awfully long time before the clinical potential of interferon was extensively tested, and the reasons for this slow progress were more complex than simple availability (Cantell 1979). But the emerging genetic engineering methodology brought the promise to produce human interferon cheaply in unlimited amounts. So we and others started out to clone the corresponding genes and to express them in bacteria. The first results were obtained with the type I interferons, leukocyte or IFN-α and fibroblast or IFN-β (Derynck et al. 1980a,b; Nagata et al. 1980; Taniguchi et al. 1980a,b). Because of the restricted species specificity of the interferons, it is usually necessary to isolate and develop not only the human gene, but also the homologous mouse system so that extensive biological and preclinical studies can be carried out.

Immune Interferon or IFN-γ

Type I interferons are naturally produced as a result of a viral challenge. IFN-γ, on the other hand, is formed following an immune cell interaction, possibly by way of interleukin-2 (IL-2) acting as a direct inducer (Vilcek et al. 1985). Therefore, IFN-γ can be more properly considered as an immunomodulator, rather than as an antiviral effector. Moreover, IFN-γ seemed to have a more pronounced antiproliferative activity on some transformed cells than type I interferons (Crane et al. 1978; Blalock et al. 1980), although, with hindsight, there is some doubt as to whether these IFN-γ preparations were really free from contamination with lymphotoxin.

The human IFN-γ was first cloned and characterized by Gray and colleagues (1982) and independently by our group (Devos et al. 1982). The mature protein can be very efficiently synthesized in *Escherichia coli*, and expression levels of about 25% of total protein were achieved (Simons et al. 1984). The scale-up and purification process was developed by Biogen, and the first clinical trials with this IFN-γ under the trade name Immuneron began in 1983.

Natural human IFN-γ is a glycoprotein, and the primary amino acid sequence deduced from the nucleotide sequence data reveals two potential *N*-glycosylation sites. On denaturing gel electrophoresis, the protein gives a 21-kD band and a 25-kD band (Yip et al. 1982), whereas the mature, unglycosylated polypeptide corresponds to 17 kD. It was shown by Rinderknecht and colleagues (1984) that the 21-kD chain was glycosylated at the first site, whereas the slower moving 25-kD chain was glycosylated at both positions. We also have expressed the human IFN-γ gene in Chinese hamster ovary (CHO) cells, and after methotrexate amplification, expression levels of about 5–10 mg/liter were achieved (Scahill et al. 1983). This product was purified to homogeneity (Devos et al. 1984). Surprisingly—considering the different cell types making it—the ratio of mono- to diglycosylated polypeptide chains (i.e., 21 kD and 25 kD) was about the same for the CHO material as for the natural IFN-γ. In collaboration with J. Mutsaers, J. Kamerling, and J. Vliegenthart, we have established the chemical structure of the glycosyl group (Mutsaers et al. 1986). It is a classical biantenna structure exhibiting microheterogeneity as to the terminal sialic acid and the core α(1–6)-fucosyl group.

The availability of both the unglycosylated bacterial IFN-γ and its glycosylated CHO-derived counterpart provided a unique opportunity to evaluate the effect of glycosylation on biological function. The half-life in circulation is two- to threefold longer for the glycosylated molecule. But otherwise, besides some sporadic observations, no clear-cut difference in biological activity between the glycosylated and its unglycosylated counterpart has emerged.

Since 1985, Immuneron has been in phase III clinical trials for renal cell carcinoma. Some beneficial effects have also been observed with rheumatoid arthritis. But possibly the most important use of human IFN-γ in cancer therapy will be based on its synergism with tumor necrosis factor.

Tumor Necrosis Factor: Background and Molecular Biology

It was reported as early as the second half of the previous century, first by P. Bruns and later by W.B. Coley, that bacterial infections could lead to spontaneous regressions of certain tumors in some cancer patients. A major turning point, which opened the eyes of at least some molecular biologists, came in 1975 with the publication by Carswell et al. (1975). These authors showed that mice, which had been treated for 2–3 weeks with Bacillus Calmette-Guerin (BCG), followed by treatment with endotoxin for a few hours, contained in their serum a substance that caused in vivo necrosis of Meth A sarcomas and therefore was called tumor necrosis factor (TNF). This TNF can also be released by macrophages and by monocyte-derived cell lines in tissue culture (Männel et al. 1980). The background of TNF has been well reviewed and discussed (Ruff and Gifford 1981; Old 1985; Beutler and Cerami 1986).

Human TNF cDNA has been cloned and sequenced by a number of groups (Pennica et al. 1984; Marmenout et al. 1985; Wang et al. 1985). In our case, we first purified TNF from serum of rabbits treated with BCG and endotoxin. We then determined a partial amino acid sequence, synthesized oligonucleotide probes, and finally could pick up a mouse TNF cDNA from a library made with mRNA from lipopolysaccharidc-induced PU 5.1.8. cells (Fransen et al. 1985). Using the mouse cDNA as a probe, the human TNF cDNA clone was then isolated from a library made with mRNA from the monocytic cell line U937 induced with phorbol ester and retinoic acid (Marmenout et al. 1985). The human genome contains a single gene for TNF that is interrupted by three introns (Marmenout et al. 1985; Nedwin et al. 1985; Shirai et al. 1985); it has been located on chromosome 6 (Nedwin et al. 1985).

The human TNF cDNA codes in an open reading frame for 233 amino acids. The amino-terminal sequence of mature TNF purified from the supernatant of induced U937 cells was determined, and on this basis, we could deduce that 76 amino acids have to be processed away from the amino terminus of the precursor (Marmenout et al. 1985). This is unusually long for a signal sequence and in fact the hydrophobic domain only extends from about amino acid 30 to amino acid 58 of the precursor. Such a long, atypical presequence has also been observed in interleukin 1 (IL-1-α and IL-1-β), likewise products of induced macrophages (March et al. 1985). Nevertheless, other cells, such as

	(1–10)
HUMAN LT	LEU PRO GLY VAL GLY LEU THR PRO SER ALA ALA GLN THR ALA ARG GLN HIS PRO LYS MET HIS LEU ALA HIS SER THR LEU
HUMAN TNF	VAL ARG SER SER SER ARG THR PRO SER ASP
MOUSE TNF	LEU ARG SER SER SER GLN ASN SER SER ASP
RABBIT TNF	LEU ARG SER ALA SER ARG ALA LEU SER ASP (GLY)

	(11–40; 20, 30, 40)
HUMAN LT	LYS PRO ALA ALA HIS LEU ILE GLY ASP PRO SER LYS GLN ASN SER LEU LEU TRP ARG ALA ASN THR ASP ARG ALA PHE LEU GLN ASP GLY
HUMAN TNF	LYS PRO VAL ALA HIS VAL VAL ALA ASN PRO GLN ALA GLU GLY GLN LEU GLN TRP LEU ASN ARG ARG ALA ASN ALA LEU LEU ALA ASN GLY
MOUSE TNF	LYS PRO VAL ALA HIS VAL VAL ALA ASN HIS GLN VAL GLU GLU GLN LEU GLU TRP LEU SER GLN ARG ALA ASN ALA LEU LEU ALA ASN GLY
RABBIT TNF	LYS PRO LEU ALA HIS VAL VAL ALA ASN PRO GLN VAL GLU GLY GLN LEU GLN TRP LEU SER GLN ARG ALA ASN ALA LEU LEU ALA ASN GLY

	(41–70; 50, 60, 70)
HUMAN LT	PHE SER LEU SER ASN ASN SER LEU LEU VAL PRO THR SER GLY ILE TYR PHE VAL TYR SER GLN VAL VAL PHE SER GLY LYS ALA TYR SER
HUMAN TNF	VAL GLU LEU ARG ASP ASN GLN LEU VAL VAL PRO SER GLU GLY LEU TYR LEU ILE TYR SER GLN VAL LEU PHE LYS GLY GLN GLY CYS PRO
MOUSE TNF	MET ASP LEU LYS ASP ASN GLN LEU VAL VAL PRO ALA ASP GLY LEU TYR LEU VAL TYR SER GLN VAL LEU PHE LYS GLY GLN GLY CYS PRO
RABBIT TNF	MET LYS LEU THR ASP ASN GLN LEU VAL VAL PRO ALA ASP GLY LEU TYR LEU ILE TYR SER GLN VAL LEU PHE SER GLY GLN GLY CYS ARG

	(71–100; 80, 90, 100)
HUMAN LT	PRO LYS ALA THR SER SER PRO LEU TYR LEU ALA HIS GLU VAL GLN LEU PHE SER SER GLN TYR PRO PHE HIS VAL PRO LEU LEU SER SER
HUMAN TNF	SER THR HIS VAL LEU LEU THR HIS THR ILE SER ARG ILE ALA VAL SER TYR GLN THR LYS VAL ASN LEU LEU SER ALA ILE LYS SER PRO
MOUSE TNF	ASP TYR – VAL LEU LEU THR HIS THR VAL SER ARG PHE ALA ILE SER TYR GLN GLU LYS VAL ASN LEU LEU SER ALA VAL LYS SER PRO
RABBIT TNF	SER TYR – VAL LEU LEU THR HIS THR VAL SER ARG PHE ALA VAL SER TYR PRO ASN LYS VAL ASN LEU LEU SER ALA ILE LYS SER PRO

	(101–130; 110, 120, 130)
HUMAN LT	GLN LYS MET VAL TYR PRO GLY LEU GLN GLU – – PRO TRP LEU HIS SER MET TYR HIS GLY ALA ALA PHE GLN LEU THR GLN GLY ASP
HUMAN TNF	CYS GLN ARG GLU THR PRO GLU GLY ALA GLU ALA LYS PRO TRP TYR GLU PRO ILE TYR LEU GLY GLY VAL PHE GLN LEU GLU LYS GLY ASP
MOUSE TNF	CYS PRO LYS ASP THR PRO GLU GLY ALA GLU LEU LYS PRO TRP TYR GLU PRO ILE TYR LEU GLY GLY VAL PHE GLN LEU GLU LYS GLY ASP
RABBIT TNF	CYS HIS ARG GLU THR PRO GLU GLU ALA GLU PRO MET ALA TRP TYR GLU PRO ILE TYR LEU GLY GLY VAL PHE GLN LEU GLU LYS GLY ASP

	(131–157; 140, 150, 157)
HUMAN LT	GLN LEU SER THR HIS THR ASP GLY ILE PRO HIS LEU VAL LEU SER PRO SER THR – VAL PHE PHE GLY ALA PHE ALA LEU
HUMAN TNF	ARG LEU SER ALA GLU ILE ASN ARG PRO ASP TYR LEU ASP PHE ALA GLU SER GLY GLN VAL TYR PHE GLY ILE ILE ALA LEU
MOUSE TNF	GLN LEU SER ALA GLU VAL ASN LEU PRO LYS TYR LEU ASP PHE ALA GLU SER GLY GLN VAL TYR PHE GLY VAL ILE ALA LEU
RABBIT TNF	ARG LEU SER THR GLU VAL ASN GLN PRO GLU TYR LEU ASP LEU ALA GLU SER GLY GLN VAL TYR PHE GLY ILE ILE ALA LEU

Figure 1. Amino acid sequences of human lymphotoxin (Gray et al. 1984), human TNF (Marmenout et al. 1985), mouse TNF (Fransen et al. 1985; Pennica et al. 1985), and rabbit TNF (Itoh 1985). Homologies are boxed, heavy lines indicating homology between the four species and dashes showing the cysteine residues. Particularly conserved domains are underscored. The antennae represent potential *N*-glycosylation sites.

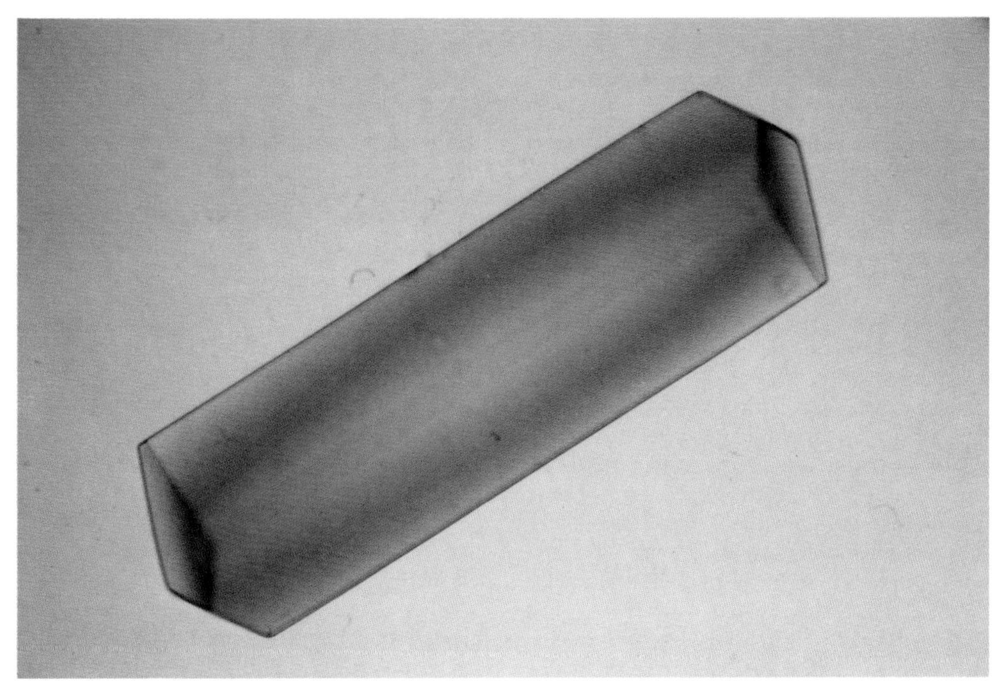

Figure 2. Crystal of r-hTNF (J. Rosa and J. Tavernier, unpubl.).

Xenopus laevis oocytes and CHO, can also correctly process the TNF precursor and secrete the mature product, although a fraction of the molecules are extended at the amino terminus for 15–30 residues (reminiscent of lymphotoxin, vide infra) (Müller et al. 1986).

The amino acid sequence of human TNF is shown in Figure 1 and is compared with those of mouse TNF, rabbit TNF, and human lymphotoxin (sometimes called TNF-β). Clearly, TNFs from the three species are closely related; the polypeptides of mature human and mouse TNFs show 79% homology and, remarkably, the preregions (76 residues for human; 79 for mouse) are even 86% homologous. As discussed by Gray et al. (1984) and Pennica et al. (1984), lymphotoxin, a product of T lymphocytes, is about 30% homologous to TNF (i.e., similar to the relatedness between IFN-α and IFN-β). The TNF polypeptide, but not lymphotoxin, contains two cysteine residues linked by a disulfide bridge; it is precisely this region, spanning the two cysteines, that seems to be completely unrelated to lymphotoxin.

There is no evidence that natural human TNF is glycosylated or otherwise modified. Also, the amino acid sequence does not contain a potential *N*-glycosylation site. The first two amino acids (Val-Arg) can easily be cleaved off during purification without a drop in biological activity (Marmenout et al. 1985). The mature gene has been appropriately engineered and can be very efficiently expressed in *E. coli*. Starting from such cells, human TNF has been purified, without involvement of denaturing solvents, to over 99% homogeneity and with a nearly undetectable endotoxin level. Depending on the conditions and the host strain used, a fraction of the polypeptides can still retain an amino-terminal methionine, but these molecules can be purified away. The pure material is in all measurable characteristics identical with its natural counterpart and has now also been crystallized (Fig. 2). Recombinant human TNF (r-hTNF) moves on a gel filtration column with a molecular weight of about 46,000, while the subunit molecular weight is 17,356. Cross-linking studies, however, indicate a trimer structure (expected m.w. = 52,000).

Natural mouse TNF is a glycoprotein, and indeed its amino acid sequence reveals an *N*-glycosylation site (Fig. 1). Nevertheless, the mature gene can be efficiently expressed in *E. coli*, and the product was again completely purified and characterized.

TNF Action on Malignant Cells and Its Synergism with IFN-γ

It is known, since the early work of Carswell et al. (1975) and others, that TNF is selectively toxic to some tumor cell lines. Moreover, inhibition of transcription or protein synthesis with drugs like actinomycin D or cycloheximide dramatically enhances the sensitivity to TNF. One could therefore assume the existence of a repair system that counteracts the toxic effect of TNF; this repair system requires transcription-translation and may be induced following TNF action or may be metabolically labile. TNF is often assayed in the presence of actinomycin D on sensitive L929 cells (Ruff and Gifford 1981). We find a specific activity of about 2.5×10^7 U/mg for r-hTNF and 7.5×10^7 U/mg for r-m(urine)TNF; obviously, these values are relative and depend, among other things, on the particular L929 subline used. An increase in temperature also dramatically increases the sensitivity to TNF (an indication for combination treatment with thermotherapy!).

In addition to using toxic drugs like actinomycin, cycloheximide, or mitomycin C to render cells more sensitive to TNF, one can also use interferon (Williamson et al. 1983). IFN-γ is especially effective.

Taking all this into consideration, we have tested r-hTNF on a large number of malignant and untransformed cell lines (Fransen et al. 1986b). Basically, three types of assays were carried out: The first assay was done in the presence of actinomycin D (1 μg/ml); here one starts with a confluent monolayer, and the cells that are still present after 18 hours can be scored. In the second assay, one starts with a subconfluent culture, and the scoring occurs after 3 days of treatment with TNF. The third assay is the same, except that IFN-γ (in a series of concentrations) is also present. Some representative results are shown in Table 1 and Figure 3. A number of conclusions can be drawn: (1) Some transformed cell lines are sensitive to TNF and others become sensitive when also treated with IFN-γ, but primary, untransformed cells are not affected, even in the presence of IFN-γ. (2) Sensitivity to TNF is not correlated with a particular tumor type; e.g., some breast carcinoma lines are sensitive while others are not, and some hepatoma lines are sensitive, while others are not. (3) Increased malignancy, as determined by other parameters, is sometimes, but certainly not always, correlated with increased sensitivity to TNF. (4) Synergism with actinomycin D may occur independently of synergism with IFN-γ and vice versa.

r-hTNF and r-mTNF were also tested on a number of human and mouse cell lines in the presence and absence of the species-specific IFN-γ (Fransen et al. 1986a). It is generally assumed that TNF is not species-specific, but in these detailed studies, a small species bias was nevertheless observed. Furthermore, the action of TNF in the presence of IFN-γ was evaluated in an organotypic culture system as described by Mareel et al. (1979). Embryonic heart fragments were confronted with melanoma cells. When untreated, the cancer cells invade the healthy tissue, which necrotizes; after 7 days in the presence of r-mTNF + mIFN-γ, however, the melanoma cells were completely eliminated while the heart tissue remained perfectly healthy. These results on the very selective toxicity of TNF (often in combination with IFN-γ) for many types of malignant cells offer much hope for the clinical potential of TNF.

TNF Activities on Normal Cells

The selective toxicity of TNF for transformed cells might perhaps suggest that only these cells carry the corresponding receptor. However, this is very far from the truth. Early on, we realized that r-hTNF does in fact act on normal fibroblasts not by inhibiting their growth, but by promoting it (Fiers et al. 1986). Similar results have been reported by Sugarman et al. (1985) and Vilcek et al. (1986). Cerami and colleagues have carried out extensive studies on cachexia and shock as a consequence of invasive stimuli such as infection or cancer. They identified the mediator, which they called cachectin, and found that cachectin is in fact identical to TNF (Beutler and Cerami 1986). On endothelial cells, r-hTNF induces procoagulant activity (Bevilacqua et al. 1986), stimulates adhesiveness (Gamble et al. 1985), and induces the antigen H4/18 (Pober et al. 1986), all of which are phenomena that can equally be induced by IL-1. Nevertheless, r-hTNF and IL-1 act on different receptors (Pober et al. 1986) (but the secondary messenger may be similar or identical). On synovial cells, r-hTNF acts like IL-1, namely, it induces release of prostaglandins and collagenase (Dayer et al. 1985). r-hTNF as well as IL-1 can stimulate osteoclasts. Another system that we have studied is called cytotoxic-inducing activity. This involves a precursor T-cell line that can be differentiated to a cytotoxic T cell by the combination of two factors, IL-2 and IL-1 (Erard et al. 1984). Again, IL-1 can be replaced by TNF (G. Plaetinck et al., in prep.). Nevertheless, in the classical assay system for IL-1 (the induction of IL-2 release by thymocytes), rTNF is not active, possibly because these

Table 1. Sensitivity of Transformed and Primary Cell Lines to r-hTNF in Function of the hIFN-γ Concentration

	Cell line (U/ml TNF)										
hIFN-γ (IU/ml)	HeLa D98	HeLa H21	BT-20	MCF-7	ME-180	HT-29	SK-BR-3	SK-CO-1	SK-OV-3	McG (30–80)6	FS4 WI38 E_1SM
0	100	3300	30	80	370	>3300	>33,300	>33,300	>33,300	>3300	>3300
0.9	100	2200	30	80	370	>3300	>33,300	>33,300	>33,300	>3300	>3300
2.7	100	3300	30	80	370	>3300	>33,300	>33,300	>33,300	>3300	>3300
8	30	3300	30	80	370	>2200	>33,300	140	>33,300	2220	>3300
25	10	1110	30	80	2	35	>33,300	140	>33,300	1110	>3300
75	12	750	20	80	2	9	270	—	>33,300	—	>3300
220	10	250	9	60	2	5	—	100	—	30	>3300
670	5	125	5	30	2	3	100	—	11,100	10	>3300
2000	7	80	5	30	10	9	270	15	11,100	10	>3300
6000	5	60	5	30	10	9	400	—	—	30	>3300
18000	7	35	3	30	10	5	400	15	1,250	30	>3300

Cells were tested after 3 days of treatment (cf. Fig. 3). Sensitivity is expressed as units per milliliter r-hTNF required to produce a decrease of 50% in the number of cells, as compared with the culture treated with hIFN-γ alone. FS4, WI38, and E_1SM are untransformed, nontumorigenic, primary lines. (Results adapted from Fransen et al. 1986b.)

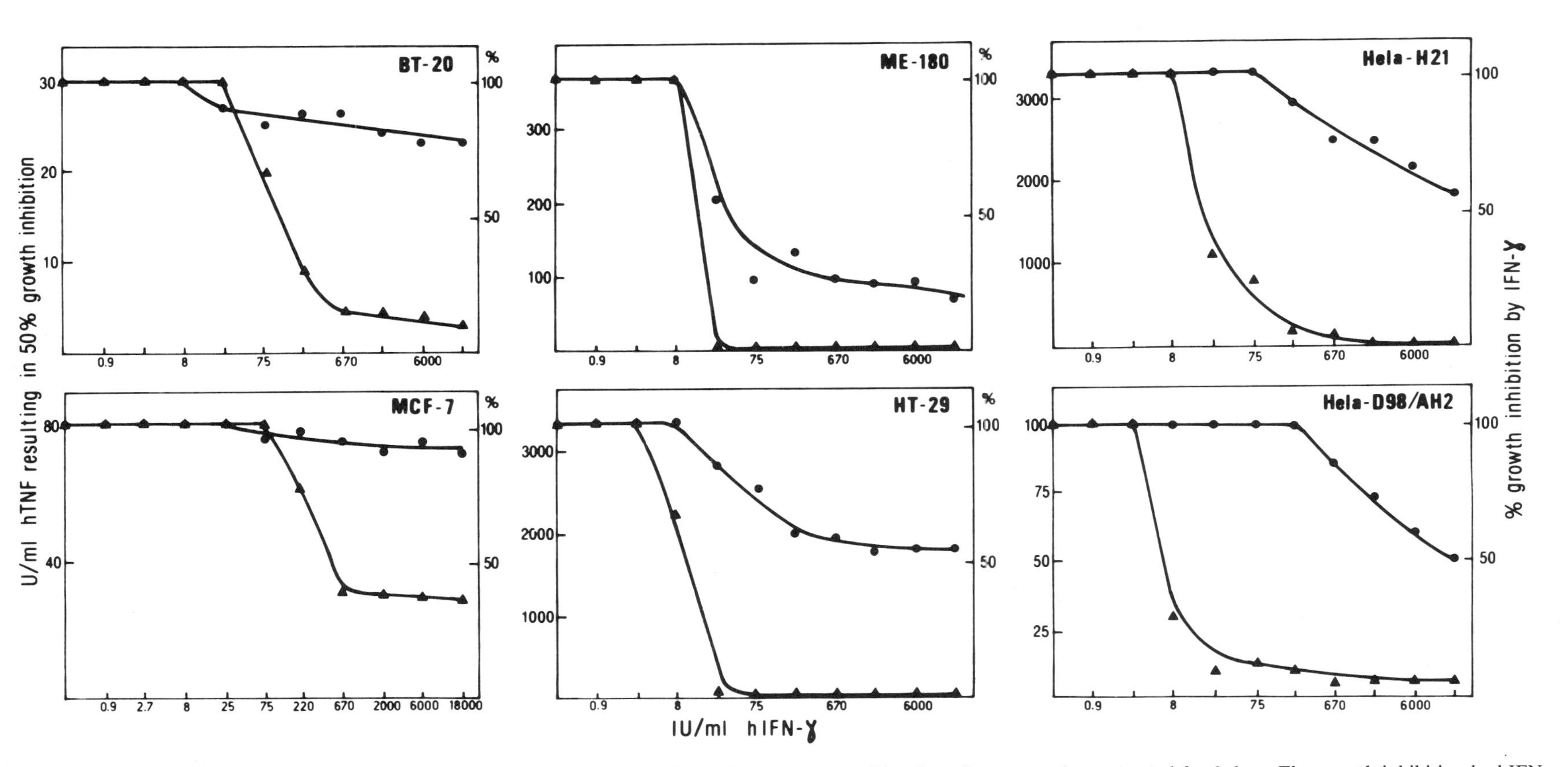

Figure 3. Synergism of r-hTNF and hIFN-γ on different human malignant cell lines. The assay starts with subconfluent monolayers treated for 3 days. The growth inhibition by hIFN-γ alone (●, right ordinate) is expressed as a percentage of the untreated control. TNF sensitivity (▲) is expressed as units per milliliter r-hTNF (left ordinate) required to obtain a 50% inhibition in cell number in the presence of the corresponding hIFN-γ concentration. (Reprinted, with permission, from Fransen et al. 1986b.)

cells do not carry TNF receptors. The various activities of r-TNF are summarized in Table 2.

TNF receptors have been characterized using labeled ligand. The results vary somewhat according to cell type and to laboratory, and a summary is given in Table 3. Not unexpectedly, it turns out that normal cells also carry TNF receptors, and the difference with transformed cells is certainly not at this level. Only some transformed B-cell lines, untransformed lymphocytes, and erythrocytes lack receptors. The number of receptors is often in the range of 10^3 to 10^4 per cell and the binding constant is about 2×10^{-10} M. Several groups have observed that the number of receptors per cell can increase two- to threefold upon rIFN-γ treatment; however, this modest enhancement is presumably not sufficient to explain the dramatic increase in sensitivity by treatment with IFN-γ. After binding to the receptor, r-hTNF is internalized in the cell via the endosome pathway and becomes rapidly degraded (Baglioni et al. 1985).

TNF: In Vivo Antitumor Effects

Natural TNF has a clear antitumor effect in vivo (see, e.g., Haranaka et al. 1984). With the unlimited availability of recombinant TNF, much more extensive in vivo studies become possible. Experiments with the Meth A sarcoma are not very informative, since this tumor is rather immunogenic and easy to cure. We have preferred to work with the B16B16 melanoma in syngeneic mice as a model system (Brouckaert et al. 1986). In vitro, these tumor cells are only sensitive to TNF in the presence of IFN-γ. In vivo, however, we could obtain complete regression with r-mTNF, even without IFN-γ treatment. Preliminary data suggest that IFN-γ might be required to obtain complete cures. With r-hTNF, the injection had to be given daily at the site of the tumor, whereas with r-mTNF, success was also achieved after intraperitoneal injection.

Balkwill et al. (1986) have studied the effect of intratumorally administered r-hTNF on nude mice carrying human xenografts. Total regression was seen in three out of six groups, whereas stasis occurred in another two groups (i.e., a very significant response in five out of six xenograft lines). When r-hTNF was injected intraperitoneally, a good response was observed only when IFN-α or IFN-γ was also given. Considering the species specificity, this is a clear indication that direct effects of the biological drugs on the tumor cells are involved.

I. Gresser and colleagues (in prep.), on the other hand, observed very extensive tumor regression, although the cell line used, a Friend erythroleukemia clone, is resistant to rTNF and to IFN-α/β–IFN-γ in tissue culture. Clearly, indirect effects are involved in this in vivo system.

The main limitation to the more extensive use of rTNF, however, is its toxicity. The therapeutic index is

Table 2. Biological Activities of Recombinant Tumor Necrosis Factor

Cell type	Biological action	References
IN VITRO		
Transformed cells	cytocidal (growth inhibition)	Sugarman et al. (1985); Fransen et al. (1986b)
Fibroblast cells	growth factor	Fiers et al. (1986); Vilcek et al. (1986); Sugarman et al. (1985)
	induction of class I MHC Ag;	Collins et al. (1986)
Synovial cells and dermal fibroblasts	stimulation of collagenase and PGE2 production	Dayer et al. (1985)
Hybridoma mouse CTL × rat thymoma	induction of cytotoxicity	G. Plaetinck et al. (in prep.)
Adipocytes (3T3-L1)	suppression of LPL activity; suppression of activity of lipogenic enzymes	Beutler et al. (1985b); Torti et al. (1985)
Endothelial cells	induction of class I MHC Ag; procoagulant activity; stimulation of adhesiveness; induction of H4/18 Ag; reorganization	Collins et al. (1986); Bevilacqua et al. (1986); Gamble et al. (1985); Pober et al. (1986); Stolpen et al. (1986)
Neutrophils	activation	Shalaby et al. (1985)
Eosinophils	activation	Silberstein and David (1986)
Osteoclasts	stimulation	Bertolini et al. (1986)
IN VIVO		
Tumor necrosis		
neoplastic cells		Brouckaert et al. (1986)
xenographs		Balkwill et al. (1986)
Cachexia		Beutler and Cerami (1986)
Reduction of parasitemia		J. Taverne et al. (in prep.)
Endotoxin shock		Beutler et al. (1985a)

Table 3. TNF Receptors

	Cell line	No. of Receptors	Binding constant (M)	Increase of no. by IFN-γ	Reference
human rTNF	HeLa S2	6,000	2×10^{-10}		Baglioni et al. (1985)
	Daudi; Raji	n.d.			
	Jurkat	1,100	2.5×10^{-10}		
	HT-29	800	2×10^{-10}	2–3 ×	Ruggiero et al. (1986)
	HeLa D98/AH2	2,400	2×10^{-10}	2–3 ×	
	SK-MEL-109	9,000	2×10^{-10}		
	L929	2,200	6.1×10^{-10}		Tsujimoto et al. (1986a)
	FS4	7,500	3.2×10^{-10}		
	HeLa	15,000	1.6×10^{-10}		Tsujimoto et al. (1986b)
	HT-29	4,500	2.0×10^{-10}		
	WISH	6,800	2.4×10^{-10}	1.5–2 ×	
	A 673	5,700	1.9×10^{-10}		
	RPMI 7272	18,000	2.3×10^{-10}		
	WI38	2,000	2×10^{-10}		Sugarman et al. (1985)
	T24	1,500	2.5×10^{-10}		
	ME-180	2,000	2×10^{-10}	2–3 ×	Aggarwal et al. (1985)
human LuKII	L(M)				Rubin et al. (1985)
	HeLa	200–300	1×10^{-10}		
murine J774.1	L(M)	1,100	3×10^{-12}		Kull et al. (1985)
	L929	770	3×10^{-12}		
	3T3A31	1,100	2.5×10^{-12}		
	J774.1	1,100	3×10^{-12}		
	CPAE (bovine)	1,500	3×10^{-12}		
murine RAW264.7	3T3 L1	10,000	3×10^{-9}		Beutler et al. (1985b)
	C2	10,000	3×10^{-9}		
	erythrocytes	n.d.			
	lymphocytes	n.d.			

The type of cytotoxic factor used is indicated at the left: r-human TNF; human LuKII is related or identical to human lymphotoxin; murine J774.1 and RAW 264.7 are related or identical to mouse natural TNF. n.d. indicates not detectable.

very narrow. This toxicity is in fact not surprising, considering the many effects of TNF on normal cells (Table 2). The toxicity is very similar to that caused by endotoxin, and in fact, Beutler et al. (1985a) have shown that TNF is an essential component in the series of events leading to toxic shock. Nevertheless, with the knowledge gained regarding the action of TNF on various cells, it is not unreasonable to hope that ways will be found to alleviate the toxic effects while preserving the antitumor activity.

TNF: How Does It Act?

The exact mechanism of action of TNF is not yet known. Nevertheless, a number of road signs are beginning to appear. We propose that TNF action on target cells will lead to mobilization of polyunsaturated fatty acids, such as arachidonic acid, perhaps by the activation of phospholipase A_2. Peroxydation of the polyunsaturated fatty acids then leads to intracellular reactive oxygen (superoxide, hydrogen peroxide, hydroxyl radical, or singlet oxygen), which in turn causes cell damage, such as lipoxygenation and DNA breakage (Fig. 4). Some evidence that supports this hypothesis: Dexamethasone, which blocks phospholipase A_2, possibly by induction or activation of lipocortin, also inhibits TNF action. We have seen above (section on TNF Activities on Normal Cells) that TNF action leads to release of prostaglandins (Table 2), which are derived from arachidonic acid by cyclooxygenase. TNF action on neutrophils and eosinophils leads to leukotriene synthesis (A. Dessein, pers. comm.). These leukotrienes are likewise derived from arachidonic acid by lipooxygenase action. Some compounds like methionine and ethanol, which can act as reactive oxygen scavengers, protect the cell to a certain extent from

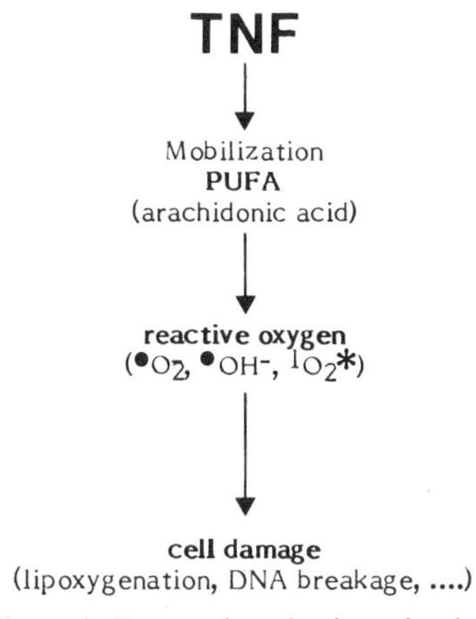

Figure 4. Proposed mechanism of action.

TNF-induced damage. Finally, both lipid cell membrane damage and DNA breakage have been observed after TNF treatment and are the likely outcome of intracellular reactive oxygen havoc.

Obviously, many other explanations are possible for the above observations, but at least some key elements of the proposed action scheme can be tested. Even if the proposed mechanism is correct, one must still explain why "normal" cells are not damaged and why many transformed cells need IFN treatment to become TNF-sensitive. In short, we are back to the perennial question: What is a cancer cell?

ACKNOWLEDGMENTS

We thank our colleagues who allowed us to refer to their still unpublished results. F.V.R. is a senior research associate of the NFWO; P.S. thanks the IWONL for a fellowship. Work in the senior author's laboratory has been supported by grants from the Fonds voor Geneeskundig Wetenschappelijk Onderzoek (FGWO), the Algemene Spaar- en Lijfrentekas (ASLK) and Biogen S.A.

REFERENCES

Aggarwal, B.B., T.E. Eessalu, and P.E. Hass. 1985. Characterization of receptors for human tumour necrosis factor and their regulation by γ-interferon. *Nature* **318:** 665.

Baglioni, C., S. McCandless, J. Tavernier, and W. Fiers. 1985. Binding of human tumor necrosis factor to high-affinity receptors on HeLa cells and lymphoblastoid cells sensitive to growth inhibition. *J. Biol. Chem.* **260:** 13395.

Balkwill, F.R., A. Lee, G. Aldam, E. Moodie, J.A. Thomas, J. Tavernier, and W. Fiers. 1986. Human tumor xenografts treated with recombinant human tumor necrosis factor alone or in combination with interferons. *Cancer Res.* **46:** 3990.

Bertolini, D.R., G.E. Nedwin, T.S. Bringman, D.D. Smith, and G.R. Mundy. 1986. Stimulation of bone resorption and inhibition of bone formation *in vitro* by human tumour necrosis factors. *Nature* **319:** 516.

Beutler, B. and A. Cerami. 1986. Cachectin and tumor necrosis factor as two sides of the same biological coin. *Nature* **320:** 584.

Beutler, B., I.W. Milsark, and A.C. Cerami. 1985a. Passive immunization against cachectin/tumor necrosis factor protects mice from lethal effect of endotoxin. *Science* **229:** 869.

Beutler, B., J. Mahoney, N. Le Trang, P. Pekala, and A. Cerami. 1985b. Purification of cachectin, a lipoprotein lipase-suppressing hormone secreted by endotoxin-induced raw 264.7 cells. *J. Exp. Med.* **161:** 984.

Bevilacqua, M.P., J.S. Pober, G.R. Majeau, W. Fiers, R.S. Cotran, and M.A. Gimbrone, Jr. 1986. Recombinant tumor necrosis factor induces pro-coagulant activity in cultured human vascular endothelium: Characterization and comparison with the actions of interleukin 1. *Proc. Natl. Acad. Sci.* **83:** 4533.

Blalock, J.E., J.A. Georgiades, M.P. Langford, and H.M. Johnson. 1980. Purified human immune interferon has more potent anticellular activity than fibroblast or leukocyte interferon. *Cell. Immunol.* **49:** 390.

Brouckaert, P.G.G., G. Leroux-Roels, Y. Guisez, J. Tavernier, and W. Fiers. 1986. In vivo antitumour activity of recombinant human and murine TNF, alone and in combination with murine IFN-gamma, on a syngeneic murine melanoma. *Int. J. Cancer* (in press).

Cantell, K. 1979. Why is interferon not in clinical use today? In *Interferon 1* (ed. I. Gresser), p. 1. Academic Press, New York.

Carswell, E.A., L.J. Old, R.L. Kassel, S. Green, N. Fiore, and B. Williamson. 1975. An endotoxin-induced serum factor that causes necrosis of tumors. *Proc. Natl. Acad. Sci.* **72:** 3666.

Collins, T., L.A. Lapierre, W. Fiers, J.L. Strominger, and J.S. Pober. 1986. Recombinant human tumor necrosis factor increases mRNA levels and surface expression of HLA-A,B antigens in vascular endothelial cells and dermal fibroblasts *in vitro*. *Proc. Natl. Acad. Sci.* **83:** 446.

Crane, J.L., L.A. Glasgow, E.R. Kern, and J.S. Youngner. 1978. Inhibition of murine osteogenic sarcomas by treatment with type I or type II interferon. *J. Natl. Cancer Inst.* **61:** 871.

Dayer, J.-M., B. Beutler, and A. Cerami. 1985. Cachectin/tumor necrosis factor stimulates collagenase and prostaglandin E_2 production by human synovial cells and dermal fibroblasts. *J. Exp. Med.* **162:** 2163.

Derynck, R., J. Content, E. Declercq, G. Volckaert, J. Tavernier, R. Devos, and W. Fiers. 1980a. Isolation and structure of a human fibroblast interferon gene. *Nature* **285:** 542.

Derynck, R., E. Remaut, E. Saman, P. Stanssens, E. DeClercq, J. Content, and W. Fiers. 1980b. Expression of human fibroblast interferon gene in *Escherichia coli*. *Nature* **287:** 193.

Devos, R., H. Cheroutre, Y. Taya, and W. Fiers. 1982. Molecular cloning of human immune interferon cDNA and its expression in eukaryotic cells. *Nucleic Acids Res.* **10:** 2487.

Devos, R., C. Opsomer, S.J. Scahill, J. Van der Heyden, and W. Fiers. 1984. Purification of recombinant glycosylated human IFN-gamma expressed in transformed Chinese hamster ovary cells. *J. Interferon Res.* **4:** 461.

Erard, F., P. Corthesy, K.A. Smith, W. Fiers, A. Conzelmann, and M. Nabholz. 1984. Characterization of soluble factors that induce the cytolytic activity and the expression of T cell growth factor receptors of a T cell hybrid. *J. Exp. Med.* **160:** 584.

Fiers, W., P. Brouckaert, Y. Guisez, E. Remaut, F. Van Roy, R. Devos, L. Fransen, G. Leroux-Roels, A. Marmenout, J. Tavernier, and J. Van der Heyden. 1986. Recombinant interferon gamma and its synergism with tumor necrosis factor in the human and mouse systems. In *The biology of the interferon system,* 1985 (ed. W.E. Stewart and H. Schellekens), p. 241. Elsevier Scientific, Amsterdam.

Fransen, L., M.R. Ruysschaert, J. Van der Heyden, and W. Fiers. 1986a. Recombinant tumor necrosis factor: Species specificity for a variety of human and murine transformed cell lines. *Cell. Immunol.* **100:** 260.

Fransen, L., J. Van der Heyden, M.R. Ruysschaert, and W. Fiers. 1986b. Recombinant tumor necrosis factor: Its effect and its synergism with interferon gamma on a variety of normal and transformed human cell lines. *Eur. J. Cancer & Clin. Oncol.* **22:** 419.

Fransen, L., R. Muller, A. Marmenout, J. Tavernier, J. Van der Heyden, E. Kawashima, A. Chollet, R. Tizard, H. Van Heuverswyn, A. Van Vliet, M.-R. Ruysschaert, and W. Fiers. 1985. Molecular cloning of mouse tumour necrosis factor cDNA and its eukaryotic expression. *Nucleic Acids Res.* **13:** 4417.

Gamble, J.R., J.M. Harlan, S.J. Klebanoff, and M.A. Vadas. 1985. Stimulation of the adherence of neutrophils to umbilical vein endothelium by human recombinant tumor necrosis factor. *Proc. Natl. Acad. Sci.* **82:** 8667.

Gray, P.W., B.B. Aggarwal, C.V. Benton, T.S. Bringman, W.J. Henzel, J.A. Jarrett, D.W. Leung, B. Moffat, P. Ng, L.P. Svedersky, M.A. Palladino, and G.E. Nedwin. 1984. Cloning and expression of cDNA for human lymphotoxin, a lymphokine with tumour necrosis activity. *Nature* **312:** 721.

Gray, P.W., D.W. Leung, D. Pennica, E. Yelverton, R. Najarian, C.C. Simonsen, R. Derynck, P.J. Sherwood, D.M.

Wallace, S.L. Berger, A.D. Levinson, and D.V. Goeddel. 1982. Expression of human immune interferon cDNA in *E. coli* and monkey cells. *Nature* **295:** 503.

Haranaka, K., N. Satomi, and A. Sakurai. 1984. Antitumor activity of murine tumour necrosis factor (TNF) against transplanted murine tumors and heterotransplanted human tumors in nude mice. *Int. J. Cancer* **43:** 264.

Itoh, H. 1985. *European Patent Application* No. 84105149.3.

Männel, D.N., R.N. Moore, and S.E. Mergenhagen. 1980. Macrophages as a source of tumoricidal activity (tumor-necrotizing factor). *Infect. Immun.* **30:** 523.

March, C.J., B. Mosley, A. Larsen, D.P. Cerretti, G. Braedt, V. Price, S. Gillis, C.S. Henney, S.R. Kronheim, K. Grabstein, P.J. Conlon, T.P. Hopp, and D. Cosman. 1985. Cloning, sequence and expression of two distant human interleukin-1 complementary DNAs. *Nature* **315:** 641.

Mareel, M., J. Kint, and C. Meyvisch. 1979. Methods of study of the invasion of malignant C3H mouse fibroblasts into embryonic chick heart in vitro. *Virchows Arch. B Cell Pathol.* **30:** 95.

Marmenout, A., L. Fransen, J. Tavernier, J. Van der Heyden, R. Tizard, E. Kawashima, A. Shaw, M.J. Johnson, D. Semon, and R. Muller. 1985. Molecular cloning and expression of human tumor necrosis factor and comparison with mouse tumor necrosis factor. *Eur. J. Biochem.* **152:** 515.

Müller, R., A. Marmenout, and W. Fiers. 1986. Synthesis and maturation of recombinant human tumor necrosis factor in eukaryotic systems. *FEBS Lett.* **197:** 99.

Mutsaers, J.H.G.M., J.P. Kamerling, R. Devos, Y. Guisez, W. Fiers, and J.F.G. Vliegenthart. 1986. Structural studies of the carbohydrate chains of human γ-interferon. *Eur. J. Biochem.* **156:** 651.

Nagata, S., H. Taira, A. Hall, L. Johnsrud, M. Streuli, J. Ecsödi, W. Boll, K. Cantell, and C. Weissmann. 1980. Synthesis in *E. coli* of a polypeptide with human leukocyte interferon activity. *Nature* **284:** 316.

Nedwin, G.E., S.L. Naylor, A.Y. Sakaguchi, D. Smith, J. Jarret-Nedwin, D. Pennica, D.V. Goeddel, and P.W. Gray. 1985. Human lymphotoxin and tumor necrosis factor genes: Structure, homology and chromosomal localization. *Nucleic Acids Res.* **13:** 6361.

Old, L.J. 1985. Tumor necrosis factor (TNF). *Science* **230:** 630.

Paucker, K., K. Cantell, and W. Heule. 1962. Quantitative studies on viral interference in suspended L-cells. III. Effect of interfering viruses and interferon on the growth rate of cells. *Virology* **17:** 324.

Pennica, D., J.S. Hayflick, T.S. Bringman, M.A. Palladino, and D.V. Goeddel. 1985. Cloning and expression in *Escherichia coli* of the cDNA for murine tumor necrosis factor. *Proc. Natl. Acad. Sci.* **82:** 6060.

Pennica, D., G.E. Nedwin, J.S. Hayflick, P.H. Seeburg, R. Derynck, M.A. Palladino, W.J. Kohr, B.B. Aggarwal, and D.V. Goeddel. 1984. Human tumour necrosis factor: Precursor structure, expression and homology to lymphotoxin. *Nature* **312:** 724.

Pober, J.S., M.P. Bevilaqua, D.L. Mendrick, L.A. Lapierre, W. Fiers, and M.A. Gimbrone, Jr. 1986. Two distinct monokines, interleukin 1 and tumor necrosis factor, each independently induce biosynthesis and transient expression of the same antigen on the surface of culture human vascular endothelial cells. *J. Immunol.* **136:** 1680.

Rinderknecht, E., B.H. O'Connor, and H. Rodriguez. 1984. Natural human interferon-γ. *J. Biol. Chem.* **259:** 6790.

Rubin, B.Y., S.L. Anderson, S.A. Sullivan, B.D. Williamson, E.A. Carswell, and L.J. Old. 1985. High affinity binding of ^{125}I-labeled human tumor necrosis factor (LuKII) to specific surface receptors. *J. Exp. Med.* **162:** 1099.

Ruff, M.R. and G.E. Gifford. 1981. Tumor necrosis factor. In *Lymphokines* (ed. E. Pick), vol. 2, p. 235. Academic Press, New York.

Ruggiero, V., J. Tavernier, W. Fiers, and C. Baglioni. 1986. Induction of the synthesis of tumor necrosis factor receptors by interferon-γ. *J. Immunol.* **136:** 2445.

Scahill, S.J., R. Devos, J. Van der Heyden, and W. Fiers. 1983. Expression and characterization of the product of a human interferon cDNA gene in Chinese hamster ovary cells. *Proc. Natl. Acad. Sci.* **80:** 4654.

Shalaby, M.R., B.B. Aggarwal, E. Rinderknecht, L.P. Svedersky, B.S. Finkle, and M.A. Palladino, Jr. 1985. Activation of human polymorphonuclear neutrophil functions by interferon-γ and tumor necrosis factors. *J. Immunol.* **135:** 2069.

Shirai, T., H. Yamaguchi, H. Ito, C.W. Todd, and R.B. Wallace. 1985. Cloning and expression in *Escherichia coli* of the gene for human tumour necrosis factor. *Nature* **313:** 803.

Silberstein, D.S. and J.R. David. 1986. Tumor necrosis factor enhances eosinophil toxicity to *Schistosoma mansoni* larvae. *Proc. Natl. Acad. Sci.* (in press).

Simons, G., E. Remaut, B. Allet, R. Devos, and W. Fiers. 1984. High-level expression of human interferon gamma in *E. coli* under control of the P_L-promoter of bacteriophage lambda. *Gene* **28:** 55.

Stolpen, A.H., E.C. Guinan, W. Fiers, and J.S. Pober. 1986. Recombinant tumor necrosis factor and immune interferon act singly and in combination to reorganize human vascular endothelial cell monolayers. *Am. J. Pathol.* **123:** 16.

Sugarman, B.J., B.B. Aggarwal, P.E. Hass, I.S. Figari, M.A. Palladino, and H.M. Shepard. 1985. Recombinant human tumor necrosis factor-α: Effects on proliferation of normal and transformed cells in vitro. *Science* **230:** 943.

Taniguchi, T., Y. Fujii-Kuriyama, and M. Muramatsu. 1980a. Molecular cloning of human interferon cDNA. *Proc. Natl. Acad. Sci.* **77:** 4003.

Taniguchi, T., S. Ohno, Y. Fujii-Kuriyama, and M. Muramatsu. 1980b. The nucleotide sequence of human fibroblast interferon cDNA. *Gene* **10:** 11.

Torti, F.M., B. Dieckman, B. Beutler, A. Cerami, and G.M. Ringold. 1985. A macrophage factor inhibits adipocyte gene expression: An in vitro model of cachexia. *Science* **229:** 867.

Tsujimoto, M., Y.K. Yip, and J. Vilcek. 1986a. Tumor necrosis factor: Specific binding and internalization in sensitive and resistant cells. *Proc. Natl. Acad. Sci.* **82:** 7626.

———. 1986b. Interferon-gamma enhances expression of cellular receptors for tumor necrosis factor. *J. Immunol.* **136:** 2441.

Vilcek, J., D. Henriksen-Destephano, R.J. Robb, and J. Le. 1985. Interleukin-2 as the inducing signal for interferon gamma in peripheral blood leukocytes stimulated with mitogen or antigen. In *The biology of the interferon system 1984* (ed. H. Kichner and H. Schellekens), p. 385. Elsevier Scientific, Amsterdam.

Vilcek, J., V.J. Palombella, D. Henriksen-Destefano, C. Swenson, R. Feinman, M. Hirai, and M. Tsujimoto. 1986. Fibroblast growth enhancing activity of tumor necrosis factor and its relationship to other polypeptide growth factors. *J. Exp. Med.* **163:** 632.

Wang, A.M., A.A. Creasy, M.B. Ladner, L.S. Lin, J. Strickler, J.N. Van Arsdell, R. Yamamoto, and D.F. Mark. 1985. Molecular cloning of the complementary DNA for human tumor necrosis factor. *Science* **228:** 149.

Williamson, B.D., E.A. Carswell, B.Y. Rubin, J.S. Prendergast, and L.J. Old. 1983. Human tumor necrosis factor produced by human B-cell lines: Synergistic cytotoxic interaction with human interferon. *Proc. Natl. Acad. Sci.* **80:** 5397.

Yip, Y.K., B.S. Barrowclough, C. Urban, and J. Vilcek. 1982. Purification of 2 subspecies of human γ (immune) interferon. *Proc. Natl. Acad. Sci.* **79:** 1820.

Tumor Necrosis Factors: Gene Structure and Biological Activities

D.V. GOEDDEL, B.B. AGGARWAL, P.W. GRAY, D.W. LEUNG, G.E. NEDWIN, M.A. PALLADINO,* J.S. PATTON,* D. PENNICA, H.M. SHEPARD,* B.J. SUGARMAN,* AND G.H.W. WONG
*Departments of Molecular Biology and *Pharmacological Sciences, Genentech, Inc., South San Francisco, California 94080*

Tumor necrosis factor (TNF) is the name given to a serum-derived factor that is cytotoxic for many transformed cell lines in vitro and causes the necrosis of certain tumors in vivo (Carswell et al. 1975). The name lymphotoxin was proposed in 1968 for a factor with similar biological properties that is synthesized by mitogen-stimulated lymphocytes (Granger and Kolb 1968; Ruddle and Waksman 1968). Both of these activities are now known to correspond to distinct proteins. Many other proteins with cytotoxic activities have been described and given a variety of names; however, whether any of these activities can be attributed to cytokines distinct from TNF and lymphotoxin is still not clear. On the basis of structural homology and similarity in biological function, the names TNF-α and TNF-β have been given to TNF and lymphotoxin, respectively. Reviews on the current status of research on TNF-α (Old 1985; Pennica and Goeddel 1986) and TNF-β (Gray 1986) have appeared recently.

TNF Protein and cDNA Structure

Both TNF-α and TNF-β are assayed by measuring cytotoxic activity on actinomycin-D-treated L-M cells, a sensitive clone of murine L929 fibroblasts (Kramer and Carver 1986). The biochemical characterization of the TNFs was greatly aided by the identification of cell lines capable of producing these cytokines after exposure to phorbol myristate acetate (PMA). TNF-α and TNF-β were found to be produced by the human promyelocytic cell line HL-60 (Pennica et al. 1984) and the human lymphoblastoid cell line RPMI 1788 (Aggarwal et al. 1984), respectively. Both proteins were purified to homogeneity by Aggarwal et al. (1984, 1985b,c). Amino acid sequence analysis of the purified cytokines permitted the design of synthetic DNA hybridization probes that were used to screen the appropriate cDNA libraries. Cloned cDNAs corresponding to human TNF-α (Pennica et al. 1984) and TNF-β (Gray et al. 1984) were isolated and characterized.

On the basis of sequence analysis of cloned cDNAs, human TNF-α mRNA was shown to encode a precursor protein of 233 amino acids (Pennica et al. 1984). Amino-terminal sequence analysis of natural TNF-α (Aggarwal et al. 1985c) demonstrated that the mature protein of 157 amino acids is preceded by a 76-amino-acid signal sequence involved in protein secretion. The calculated monomeric molecular weight of 17,356 agrees with the value determined experimentally by SDS-PAGE under reducing conditions. The two cysteine residues of the mature TNF-α are linked by a disulfide bridge that is essential for cytotoxic activity (Aggarwal et al. 1985c).

TNF-β mRNA contains an open reading frame of 205 codons, the first 34 of which constitute a secretion signal sequence. Native TNF-β isolated from the RPMI 1788 cell line is a glycoprotein that exists in two forms (20 kD, 148 amino acids, and 25 kD, 171 amino acids) differing by 23 amino acids at their amino termini (Aggarwal et al. 1985b). Unlike TNF-α, TNF-β does not contain any cysteine residues.

A comparison of TNF-α and TNF-β reveals a high degree of amino acid sequence homology. If two gaps are introduced, the sequences can be aligned so that 44 amino acids (28%) occur in identical positions (Fig. 1). The introduction of two additional gaps permits the alignment of nine more amino acids (34% overall homology; Aggarwal et al. 1985c). It is likely that the two highly conserved regions (amino acids 35–66 and 110–133; TNF-α numbering) are important for the similar cytotoxic activities of the two molecules and/or the recognition of the TNF receptor. Interesting differences between the two proteins are found in the amino-terminal portion and in the region from amino acids 67 to 109 (TNF-α numbering), where there are only two identical residues.

Expression plasmids were constructed that direct the synthesis in *Escherichia coli* of mature recombinant human TNF-α (Pennica et al. 1984) and TNF-β (Gray et al. 1984). Both recombinant products have been purified to homogeneity free of contaminating lipopolysaccharide (LPS or endotoxin). Specific activities of approximately 10^8 U/mg were determined for both TNFs in the L-M cell in vitro cytotoxic assay.

The amino acid sequences of TNF-α (murine, rabbit, and bovine) and TNF-β (murine and bovine) from other species have been deduced by cDNA and genomic DNA sequencing. These amino acid sequences are compared in Figures 2 and 3. The high degree of amino acid con-

```
                                                                               1
TNF-α                                                                         val arg ser
TNF-β     leu pro gly val gly leu thr pro ser ala ala gln thr ala arg gln his pro lys met
           1                                  10                                      20

                                  10                                      20
TNF-α     ser ser arg thr pro ser asp lys pro val ala his val val ala asn pro gln ala glu
                                       *   *       *   *                   *
TNF-β     his leu ala his ser thr leu lys pro ala ala his leu ile gly asp pro ser lys gln
                                              30                                      40

                                  30                                      40
TNF-α     gly gln leu gln trp leu asn arg arg ala asn ala leu leu ala asn gly val glu leu
                   *       *                           *       *           *           *
TNF-β     asn ser leu leu trp arg ala asn thr asp arg ala phe leu gln asp gly phe ser leu
                                              50                                      60

                                  50                                      60
TNF-α     arg asp asn gln leu val val pro ser glu gly leu tyr leu ile tyr ser gln val leu
                   *       *       *   *           *       *           *   *   *   *
TNF-β     ser asn asn ser leu leu val pro thr ser gly ile tyr phe val tyr ser gln val val
                                              70                                      80

                                  70                                      80
TNF-α     phe lys gly gln gly cys pro ser thr his val leu leu thr his thr ile ser arg ile
           *       *
TNF-β     phe ser gly lys ala tyr ser pro lys ala thr ser ser pro leu tyr leu ala his glu
                                              90                                      100

                                  90                                      100
TNF-α     ala val ser tyr gln thr lys val asn leu leu ser ala ile lys ser pro cys gln arg
                                                                       *
TNF-β     val gln leu phe ser ser gln tyr pro phe his val pro leu leu ser ser gln lys met
                                              110                                     120

                                  110                                     120
TNF-α     glu thr pro glu gly ala glu ala lys pro trp tyr glu pro ile tyr leu gly gly val
                   *               *           *   *                   *       *
TNF-β     val tyr pro gly leu gln glu --- --- pro trp leu his ser met tyr his gly ala ala
                                                      130

                                  130                                     140
TNF-α     phe gln leu glu lys gly asp arg leu ser ala glu ile asn arg pro asp tyr leu asp
           *   *   *           *   *       *   *                                   *
TNF-β     phe gln leu thr gln gly asp gln leu ser thr his thr asp gly ile pro his leu val
              140                                     150

                                  150
TNF-α     phe ala glu ser gly gln val tyr phe gly ile ile ala leu
                       *           *       *   *           *   *
TNF-β     leu ser pro ser thr --- val phe phe gly ala phe ala leu
              160                                         170
```

Figure 1. Comparison of the amino acid sequences of TNF-α and TNF-β. Two gaps have been introduced into the TNF-β sequence to increase the homology. Asterisks indicate identical amino acids.

servation among the various TNF-α and TNF-β sequences is not surprising in view of the cross-species activities observed for these proteins (Pennica et al. 1985).

TNF Gene Structure

TNF-α and TNF-β are homologous cytokines that share many biological properties, yet are produced by distinct cell types. To understand more precisely their differential regulation, we compared the structure and organization of the human TNF genes (Nedwin et al. 1985). Human TNF-α and TNF-β were found to each be encoded by a single gene. Furthermore, both genes are about 3 kbp in size and are split by three introns. However, only the third intron is located in a homologous position in the two genes (preceding amino acids 18 and 35 of TNF-α and TNF-β, respectively; see Fig. 1). Only the fourth exons, which encode the majority

```
                    1
Human       -76  MSTESMIRDVELAEEALPKKTGGPQGSRRCLFLSLFSFLIVAGATTLFCLLHFGVIGPQR
Bovine      -77  MSTKSMIRDVELAEEVLSEKAGGPQGSRSCLCLSLFSFLLVAGATTLFCLLHFGVIGPQR
Rabbit      -79  MSTESMIRDVELAEGPLPKKAGGPQGSKRCLCLSLFSFLLVAGATTLFCLLHFRVIGPQE
Mouse       -79  MSTESMIRDVELAEEALPQKMGGFQNSRRCLCLSLFSFLLVAGATTLFCLLNFGVIGPQR
Consensus        MST SMIRDVELAE  L  K GG Q S  CL LSLFSFL VAGATTLFCLL F VIGPQ

                                   1
Human       -16  EEF-PRDLSLISPLAQA--VRSSSRTPSDKPVAHVVANPQAEGQLQWLNRRANALLANGV
Bovine      -17  EEQVPSGPSINSPLVQ--TLRSSSQASSNKPVAHVVADINSPGQLRWWDSYANALMANGV
Rabbit      -19  EEQSPNNLHLVNPVAQMVTLRSASRALSDKPLAHVVANPQVEGQLQWLSQRANALLANGM
Mouse       -19  DEKFPNGLPLISSMAQTLTLRSSSQNSSDKPVAHVVANHQVEEQLEWLSQRANALLANGM
Consensus         E  P          Q    RS S   S KP AHVVA      QL W    ANAL ANG

Human        42  ELRDNQLVVPSEGLYLIYSQVLFKGQGCPSTHVLLTHTISRIAVSYQTKVNLLSAIKSPC
Bovine       42  KLEDNQLVVPAEGLYLIYSQVLFRGQGCP-PPPVLTHTISRIAVSYQTKVNILSAIKSPC
Rabbit       42  KLTDNQLVVPADGLYLIYSQVLFSGQGCR-SYVLLTHTVSRFAVSYPNKVNLLSAIKSPC
Mouse        42  DLKDNQLVVPADGLYLVYSQVLFKGQGCP-DYVLLTHTVSRFAISYQEKVNLLSAVKSPC
Consensus         L DNQLVVP  GLYL YSQVLF GQGC      LTHT SR A SY  KVN LSA KSPC

Human       102  QRETPEGAEAKPWYEPIYLGGVFQLEKGDRLSAEINRPDYLDFAESGQVYFGIIAL    157
Bovine      101  HRETPEWAEAKPWYEPIYQGGVFQLEKGDRLSAEINLPDYLDYAESGQVYFGIIAL    156
Rabbit      101  HRETPEEAEPMAWYEPIYLGGVFQLEKGDRLSTEVNQPEYLDLAESGQVYFGIIAL    156
Mouse       101  PKDTPEGAELKPWYEPIYLGGVFQLEKGDQLSAEVNLPKYLDFAESGQVYFGVIAL    156
Consensus           TPE AE   WYEPIY GGVFQLEKGD LS E N P YLD AESGQVYPG IAL
```

Figure 2. Amino acid sequences of human (Pennica et al. 1984), bovine (D. Goeddel et al., unpubl.), rabbit (Itoh and Wallace 1985), and murine (Pennica et al. 1985) TNF-α. (-79 to -1) Amino acids of the signal sequences. "Consensus" indicates residues that are identical in all four sequences.

of the mature proteins, have significant DNA sequence homology (56%).

The human TNF genes were localized to the p23→q12 region of chromosome 6 through Southern blot analysis of human-murine somatic-cell hybrids (Nedwin et al. 1985). We have recently shown that the two genes are very closely linked; the polyadenylation site of the TNF-β gene is separated from the transcription-initiation site of the TNF-α gene by only 1221 bp. Both have the same orientation with respect to the direction of transcription. The regions immediately flanking the two genes are extremely homologous in the human,

```
                                                   1
Human       -34  MTPPERLFLPRVCGTTLHLLLLGLLLVLLPGAQGLPGVGLTPSAAQTARQHPKMHLAHST
Bovine      -33  MTPPGRLYLLRVCSTPP-LLLLGLLLALPLEAQGLRGIGLTPSAAQPAHQQLPTPFTRGT
Mouse       -33  MTLLGRLHLLRVLGTPP-VFLLGLLLALPLGAQGLSGVRF--SAARTAHPLPQKHLTHGI
Consensus        MT   RL L RV  T     LLGLLL L   AQGL G     SAA  A

Human        27  LKPAAHLIGDPSKQNSLLWRANTDRAFLQDGFSLSNNSLLVPTSGIYFVYSQVVFSGKAY
Bovine       27  LKPAAHLVGDPSTQDSLRWRANTDRAFLRHGFSLSNNSLLVPTSGLYFVYSQVVFSGRGC
Mouse        25  LKPAAHLVGYPSKQNSLLWRASTDRAFLRHGFSLSNNSLLIPTSGLYFVYSQVVFSGESC
Consensus        LKPAAHL G PS Q SL WRA TDRAFL  GFSLSNNSLL PTSG YFVYSQVVFSG

Human        87  SPKATSSPLYLAHEVQLFSSQYPFHVPLLSSQKMVYPGLQEPWLHSMYHGAAFQLTQGDQ
Bovine       87  FPRATPTPLYLAHEVQLFSPQYPFHVPLLSAQKSVCPGPQGPWVRSVYQGAVFLLTRGDQ
Mouse        85  SPRAIPTPIYLAHEVQLFSSQYPFHVPLLSAQKSVYPGLQGPWVRSMYQGAVFLLSKGDQ
Consensus         P A   P YLAHEVQLFS QYPFHVPLLS QK V PG Q PW  S Y GA F L  GDQ

Human       147  LSTHTDGIPHLVLSPSTVFFGAFAL    171
Bovine      147  LSTHTDGISHLLLSPSSVFFGAFAL    171
Mouse       145  LSTHTDGISHLHFSPSSVFFGAFAL    169
Consensus        LSTHTDGI HL  SPS VFFGAFAL
```

Figure 3. Amino acid sequences of human (Gray et al. 1984), bovine (D. Goeddel et al., unpubl.), and murine (P.W. Gray, unpubl.) TNF-β. (-34 to -1) Amino acids of the signal sequences. "Consensus" indicates the residues that are identical in all three sequences.

murine, and bovine genomes (D.V. Goeddel et al., unpubl.).

Regulation of TNF-α and TNF-β Synthesis

TNF-α and TNF-β are produced by activated macrophages (Carswell et al. 1975) and lymphocytes (Granger and Kolb 1968; Ruddle and Waksman 1968), respectively (Stone-Wolff et al. 1984). We have examined the inducibility of both TNF-α and TNF-β by mitogens on 30 different cell lines representing many cell types. As TNF-α and TNF-β behave similarly in the L-M cell cytotoxic assay, they were distinguished by neutralization of activity with the corresponding monoclonal antibodies. Northern blot analysis using TNF-α and TNF-β cDNA probes was also performed to determine which TNF was expressed.

Although normal T lymphocytes have been reported to be the cellular source of TNF-β (Ruddle et al. 1983), we have not detected any TNF activity after exposure to the mitogen PMA using two human T-cell lines (Molt-4 and Jurkat). However, all six B lymphoblastoid cell lines examined produced high levels of TNF-β. Similar results were seen by Williamson et al. (1983), but they did not distinguish TNF-α and TNF-β activities.

The majority of the human and murine macrophage cell lines tested (including HL-60, U937, PU5-1.8, RAW 264, P388D$_1$, J774, and WRI-7) produce TNF-α but not TNF-β when stimulated with mitogens. This result was confirmed by Northern blot hybridization with the TNF-α and TNF-β cDNA probes. Other cell types, such as normal endothelial cells, rat glial cells (C6), IL-3-dependent normal bone-marrow-derived mast cells, and a mast cell line (A1) are capable of producing TNF-α but not TNF-β. Most nonlymphoid cell lines (including HeLa, A549, T24, WI38, K562, ME-180, HT-29, MCF, SKCO-1, A-431, HT1080, 7860, NRK, Rat-1, and C127) do not produce TNFs in the presence or absence of the mitogen.

The expression of both TNF-α and TNF-β mRNAs is transient even in the presence of continuous mitogenic stimuli. TNF-α mRNA in murine macrophage PU5-1.8 cells is detectable 1 hour after induction with PMA, is highest at approximately 2–4 hours, and becomes undetectable by 12 hours. In contrast, TNF-β mRNA in the B lymphoblastoid cell line 1788 begins to increase at 4 hours, reaches its maximal level at 12–24 hours, and decreases to the basal level after 48 hours. Thus, induction of TNF-α is faster than induction of TNF-β.

The effect of physiological mediators or cytokines on the expression of TNF was also examined. Glucocorticoids, known to be immunosuppressive agents (Claman 1972; Vischer 1972), have been shown to inhibit Ia antigen expression (Wong et al. 1984) and lymphokine production (Arya et al. 1984; Culpepper and Lee 1985). We found that the glucocorticoid dexamethasone inhibited TNF-α mRNA expression in HL-60 cells (Fig. 4). Similar results were obtained by Beutler

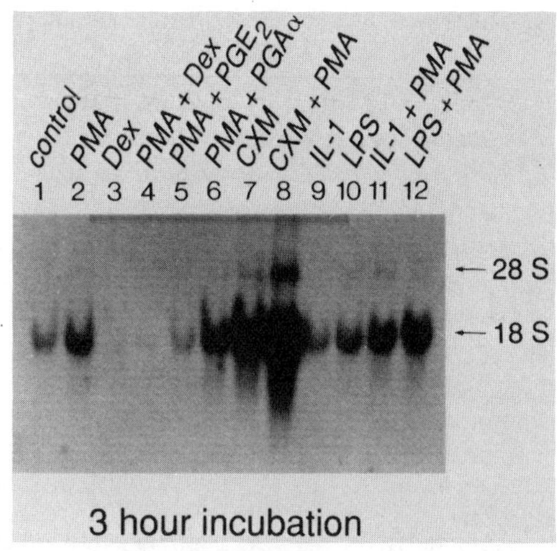

Figure 4. Regulation of TNF-α mRNA levels in HL-60 cells. HL-60 cells (5×10^6 cells/ml) were treated with PMA (50 ng/ml) in the presence or absence of dexamethasone (Dex) (5×10^{-7} M), PGE$_2$ (50 ng/ml), PGAα (50 ng/ml), cycloheximide (1 μg/ml), interleukin-1 (100 U/ml), and LPS (10 μg/ml). After 3 hr, total RNA was extracted, poly(A) RNA was prepared, and Northern hybridization was performed as described previously (Thomas 1980). Each lane contained about 1 μg of poly(A) RNA. Hybridization was performed using a ^{32}P-labeled TNF-α cDNA probe.

et al. (1986) using murine macrophages. Other glucocorticoids (prenisolone, hydrocortisone, and corticosterone), but not sex steroids (testosterone, estrogen, and progesterone), inhibited the production of TNF-α in macrophage cell lines (HL-60, U937, PU5-1.8) and TNF-β in B lymphoid cells (RPMI 1788, IM9).

Prostaglandin E (PGE), which is an important modulator of inflammation and cellular immune responses (Goodwin and Webb 1980), also down-regulates the expression of TNF-α mRNA in HL-60 cells (Fig. 4), whereas prostaglandin Aα or Fα had no inhibitory effect. Similar inhibitory effects of PGE$_1$ and PGE$_2$ on the expression of TNF-α by other macrophage cell lines (U937, PU5-1.8, RAW 264, P388D$_1$, and J774) were observed. Indomethacin, a cyclooxygenase inhibitor (Sheen and Winter 1977), enhances the production of TNFs in both macrophages and B lymphoid cells. Surprisingly, PGE$_1$ or PGE$_2$ does not suppress the production of TNF-β mRNA or biological activity in two B lymphoid cell lines (RPMI 1788 and IM9). It will be of interest to examine how the differential expression of TNF-α and TNF-β is regulated by PGEs.

The protein synthesis inhibitor cycloheximide further enhances the accumulation of both TNF-α (Fig. 4) and TNF-β mRNAs induced by PMA. The production of TNFs by macrophages and lymphoid cells can also be enhanced by the lymphokine interferon-γ (IFN-γ).

Normally, LPS alone does not induce TNF-β in 1788 or IM9 cells. However, LPS in combination with PMA gave at least a tenfold greater increase in both TNF-β mRNA and biological activity than did PMA alone. Similarly, the induction of TNF-α mRNA in HL-60 cells by the combination is greater than by either induction alone (Fig. 4). These results demonstrate that two different mitogens can synergistically induce the expression of both TNF-α and TNF-β genes. The regulation of the TNF genes by glucocorticoids, prostaglandins, cycloheximide, IFN-γ, and different mitogens observed in cell lines was similarly observed in normal human peripheral blood leukocytes. In addition to bacterial LPS and the mitogen PMA, both viruses and poly(I):poly(C) can induce the synthesis of TNF-α and TNF-β in TNF-producing cell lines and normal human leukocytes (Aderka et al. 1986; G. Wong and D. Goeddel, in prep.).

Antitumor Properties of TNFs

TNF is a term that was initially used to describe an activity present in the serum of bacillus Calmette-Guerin-infected endotoxin-treated mice that induced the hemorrhagic necrosis of certain transplantable tumors in inbred mice (Carswell et al. 1975). We have utilized this classic assay to compare further the ability of TNF-α and TNF-β to induce necrosis of an established Meth-A sarcoma intradermal implant in (BALB/c × C57BL/6)F_1 mice. Intravenous injection of 1–50 μg of either recombinant human TNF-α (r-hTNF-α) or r-hTNF-β induced significant degrees of hemorrhagic necrosis 24 hours after injection (Table 1). We have previously reported similar antitumor activity following intraperitoneal, intralesional, and intramuscular injections of TNF-α and TNF-β (Gray et al. 1984; Pennica et al. 1984).

Additional experiments were performed to characterize the in vivo antitumor activities of TNF-α against two subcutaneously implanted, chemically induced sarcomas of BALB/c origin (WEHI-164 and Meth A). Intravenous treatment with TNF-α significantly inhibited the growth of both sarcomas (Table 2). However, there were no complete remissions in the WEHI-164 group and only one of ten animals in the Meth-A group completely rejected the tumor after TNF-α treatment. The site of tumor implantation appears to be a critical factor for the demonstration of antitumor activities of TNF-α against transplantable sarcomas, since Meth A and M5076 (a spontaneously arising ovarian sarcoma of C57BL/6 origin) implanted intraperitoneally are almost completely refractory to the antitumor effects of TNF-α (M.A. Palladino, unpubl.). In contrast, treatment of the Meth-A tumor implanted intradermally results in a cure rate of almost 100%.

We have previously reported the ability of both TNF-α and TNF-β to augment superoxide radical production, antibody-dependent cellular cytotoxicity, and phagocytosis by neutrophils (Shalaby et al. 1985). The role of neutrophils in host defense responses is well established, and they have been shown to release activated oxygen intermediates (Babior 1978) and interleukin-1 (Tiku et al. 1986). Other studies have shown that neutrophils can mediate endothelial damage in vitro and that TNF-α can directly stimulate endothelial cells to produce procoagulant activity (Harlan et al. 1981; Nawroth and Stern 1986). It is therefore quite possible that the association between changes in neutrophil functions and endothelial cell homeostatic properties mediates the antitumor activities of TNF in vivo (Pennica et al. 1986). This hypothesis is supported by the fact that intraperitoneally implanted Meth A is refractory to the antitumor effects of TNF-α, whereas intradermally implanted Meth A is highly sensitive. Studies directed at addressing this hypothesis further are in progress and may give greater insight into the in vivo antitumor mechanisms of TNF.

In Vitro Growth Activities of TNFs

TNF-α and TNF-β have been shown to have comparable cytostatic and cytolytic properties in vitro (Ag-

Table 1. Necrosis of Meth-A Sarcoma after Intravenous Injection of r-hTNF-α or r-hTNF-β

Intravenous treatment	Dose (μg)	Hemorrhagic necrosis score: + + +	+ +	+	−	Percentage of mice with >25% necrosis
PBS		0	0	2	18	0
r-hTNF-α	50	15	3	1	0	95
	15	13	6	1	0	95
	5	13	5	0	1	95
	1	3	7	7	3	50
r-hTNF-β	50	14	1	4	1	75
	15	15	2	2	1	85
	5	11	5	2	2	80
	1	4	6	6	4	50

(BALB/c × C57BL/6)F_1 female mice were injected intradermally with 5×10^5 Meth-A sarcoma cells. After 7 days (average tumor diameter 0.75 cm), TNF was injected intravenously in 0.1 ml PBS; 24 hr later, the tumors were excised, sectioned, and scored as described previously (Carswell et al. 1975; Pennica et al. 1984). + + + represents between 50% and 75% of tumor mass necrotic; + + represents between 25% and 50% of tumor mass necrotic; + represents less than 25% of tumor mass necrotic; − represents no visible necrosis.

Table 2. Antitumor Effects of r-hTNF-α against Subcutaneously Implanted Sarcomas

Sarcoma	Intravenous treatment	Tumor size (mm) on days 7	14	21	*p* Value
Meth A	PBS	2.6	7.1	15.1	
	r-hTNF-α	4.3	4.3	7.6	<0.01
WEHI-164	PBS	12.4	25.2	34.9	
	r-hTNF-α	9.3	14.3	22.9	<0.01

r-hTNF-α was administered daily from day 7 to day 18 for Meth A and from day 5 to day 12 for WEHI-164 at a dose of 25 μg/day and 15 μg/day, respectively. There were 10 animals per group.

garwal et al. 1984, 1985b,c). In addition, IFN-γ can potentiate the antiproliferative effects of both cytokines on certain tumor cell lines (Williams and Bellanti 1983; Williamson et al. 1983; Lee et al. 1984; Stone-Wolff et al. 1984; Sugarman et al. 1985). Subsequent experiments using recombinant TNF-α and TNF-β demonstrated that they compete for the same binding sites on tumor cells, which probably accounts for their similar effects on tumor cell growth in vitro (Aggarwal et al. 1985a).

A number of studies characterizing the effects of TNF-α or TNF-β on the proliferation of various cell lines have been reported (Williamson et al. 1983; Lee et al. 1984; Kull et al. 1985; Sugarman et al. 1985; Ruggiero et al. 1986). Initial results indicated that cell lines can be subdivided into three categories on the basis of their response to TNF-α: (1) a cytostatic or cytolytic effect, (2) little or no antiproliferative effect, and (3) enhanced growth (Table 3). Furthermore, addition of exogenous growth factors can either inhibit the antiproliferative response or enhance the growth-promoting effect of TNF-α (Vilcek et al. 1986; B.J. Sugarman et al., unpubl.). More recent results have shown that tumor cells sensitive to the cytotoxic effects of TNF-α are actually growth stimulated at relatively low concentrations of TNF-α (G. Lewis et al., unpubl.).

Approximately 40% of the established tumor cell lines are sensitive to the antiproliferative effects of TNF-α (Sugarman et al. 1985). Cells such as L929, WEHI 164, and UV1591-RE are extremely susceptible to TNF-α-mediated cytotoxicity (i.e., <10 U/ml reduces cell viability by 50%), whereas a similar decrease in the viability of the most sensitive human cell lines (e.g., ME-180 and BT-20) requires a tenfold greater concentration of TNF-α. IFN-γ can enhance the cytotoxic effects of TNF-α on some TNF-sensitive cell types (e.g., ME-180 and BT-20) as well as cell lines that are insensitive to its antiproliferative effects (e.g., A549, B16F10, Saos-2, and WI38 VA13) (Sugarman et al. 1985, unpubl.). However, treatment of tumor cells with both IFN-γ and TNF-α does not always result in an enhanced cytotoxic response; no synergistic antiproliferative response is seen on T24 bladder carcinoma, Calu-3 lung carcinoma, or RPMI 7272 melanoma cell lines, which are all refractory to the cytotoxic effects of TNF-α alone (Sugarman et al. 1985; Tsujimoto et al. 1986).

The response of normal fibroblasts to treatment with TNF-α in vitro is completely different from that of other cell lines. Their growth is stimulated in the presence of picomolar concentrations of TNF-α (Sugarman et al. 1985; Vilcek et al. 1986). An antibody that neutralizes the cytotoxic effects of TNF-α on tumor cells abrogates TNF-α-induced proliferation of normal fibroblasts (Sugarman et al. 1985). IFN-γ interferes with TNF-α-induced growth of normal fibroblasts in a dose-dependent manner (Sugarman et al. 1985; Vilcek et al. 1986). This TNF-α-induced fibroblast proliferation can also be augmented by exogenously added factors. Insulin and platelet-derived growth factor enhance the proliferation of NRK-49F fibroblasts (G. Lewis et al., unpubl.).

It is not clear how TNF-α can stimulate the growth of certain cell lines while inhibiting the growth of others. Variations in the proliferative responses induced by TNF-α are not necessarily due to differences in the number of binding sites per cell or their affinity for TNF-α (Kull et al. 1985; Sugarman et al. 1985; Tsujimoto et al. 1986). The ability of TNFs to both stimulate and interfere with cell proliferation is similar to transforming growth factor-β, a distinct growth factor/inhibitor (Tucker et al. 1984; Roberts et al. 1985), and may be characteristic of proteins involved in homeostatic regulatory mechanisms.

Antiviral Properties of TNFs

In response to viral infection in vivo, interferons (IFNs) are secreted and induce a state of viral resistance in noninfected cells (Stewart 1981). TNF-α and TNF-β also have antiviral activity on some cells and antiviral enhancing activity on most cells tested.

TNF-α and TNF-β inhibit the cytopathic effects of vesicular stomatitis virus (VSV) in murine epithelial cells (C127) and rat fibroblasts (Rat-1). In addition to inhibition of the VSV-mediated cytopathic effect, TNF-α and TNF-β also dramatically decrease the VSV yield in these cells (Fig. 5). This antiviral activity is dose-dependent and is neutralized by respective monoclonal antibodies against human TNF-α and TNF-β. The antiviral activity of TNFs is not caused by the induction of IFNs from these cells, since the antiviral activity is not abolished by polyclonal or monoclonal antibodies against IFN-α, -β, or -γ. The antiviral effects of TNF-α and TNF-β are also observed with encephalomyocar-

Table 3. Response of Various Cell Lines to TNF-α or TNF-β In Vitro

Cell line	Reference
Growth Enhancement	
CCD-18Co (normal human colon)	Sugarman et al. (1985)
Detroit 551 (normal human fetal skin)	Sugarman et al. (1985)
FS-4 (normal human foreskin)	Vilcek et al. (1986)
FS-48 (normal human foreskin)	Kohase et al. (1986)
LL24 (normal human lung)	Sugarman et al. (1985)
NRK-49F (normal rat kidney)	G.D. Lewis et al. (in prep.)
Osteoclasts	Bertolini et al. (1986)
WI38 (normal human fetal lung)	Lee et al. (1984); Sugarman et al. (1985)
WI-1003 (normal human lung)	Sugarman et al. (1985)
Null response	
A549 (human lung carcinoma)	Sugarman et al. (1985)
B16 (murine melanoma)	Lee et al. (1984)
B16F10 (murine melanoma)	Sugarman et al. (1985)
Calu-3 (human lung carcinoma)	Sugarman et al. (1985)
CMT-93 (murine rectal carcinoma)	Sugarman et al. (1985)
G-361 (human melanoma)	Sugarman et al. (1985)
HeLa (human cervical carcinoma)	Sugarman et al. (1985)
HeLa D98 (human cervical carcinoma)	Ruggiero et al. (1986)
HT-29 (human colon carcinoma)	Ruggiero et al. (1986); Tsujimoto et al. (1986)
HT1080 (human fibrosarcoma)	Sugarman et al. (1985)
KB (human oral epidermoid carcinoma)	Sugarman et al. (1985)
LS174T (human colon carcinoma)	Sugarman et al. (1985)
RD (human rhabdosarcoma)	Sugarman et al. (1985)
Saos 2 (human osteogenic sarcoma)	Sugarman et al. (1985)
SK-CO-1 (human colon carcinoma)	Sugarman et al. (1985)
SK-LU-1 (human lung carcinoma)	Sugarman et al. (1985)
SK-OV-3 (human ovarian carcinoma)	Sugarman et al. (1985)
SK-UT-1 (human uterine carcinoma)	Sugarman et al. (1985)
S49 (murine lymphoma)	Sugarman et al. (1985)
T24 (human bladder carcinoma)	Sugarman et al. (1985)
WI38 VA13 (human transformed WI38)	Lee et al. (1984); Sugarman et al. (1985)
Antiproliferative response	
BT-20 (human breast carcinoma)	Sugarman et al. (1985)
BT-475 (human breast carcinoma)	Sugarman et al. (1985)
B6MS2 (murine sarcoma)	Sugarman et al. (1985)
B6MS5 (murine sarcoma)	Sugarman et al. (1985)
CMS4 (murine sarcoma)	Sugarman et al. (1985)
CMS16 (murine sarcoma)	Sugarman et al. (1985)
L929 (murine fibroblast)	Sugarman et al. (1985)
MCF7 (human breast carcinoma)	Sugarman et al. (1985)
ME-180 (human cervical carcinoma)	Sugarman et al. (1985)
Meth A (murine sarcoma)	Sugarman et al. (1985)
MMT (murine breast carcinoma)	Sugarman et al. (1985)
SAC (Moloney-transformed murine 3T3)	Sugarman et al. (1985)
SK-MEL-109 (human melanoma)	Sugarman et al. (1985)
SK-OV-4 (human ovarian carcinoma)	Sugarman et al. (1985)
UV1591-RE (murine fibrosarcoma)	Urban et al. (1986)
WEHI-164 (murine sarcoma)	Sugarman et al. (1985)
WiDr (human colon carcinoma)	Sugarman et al. (1985)

ditis (EMCV) and herpes simplex virus types 1 and 2 (HSV-1 and HSV-2). Coincubation with the RNA synthesis inhibitor actinomycin D (1 μg/ml) or the protein synthesis inhibitor cycloheximide (1 μg/ml) abolishes the antiviral activity of TNFs. A 24-hour preincubation with TNF provides complete protection against VSV in C127 cells. The antiviral activity of TNF is also observed in human renal carcinoma 7860, lymphoid RPMI 8226, and murine macrophage RAW 264 cell lines. Human TNFs exhibit antiviral activity on some murine cell lines, and murine TNF-α also protects some human cells from viral infection. This suggests that the antiviral activity of TNF-α and TNF-β is not species-specific.

TNF-α or TNF-β alone protects against viral infection in some cell lines but is inactive in most cells. However, they enhance the antiviral activity of IFN-α, -β, and -γ on a variety of cell types tested. An example of their synergistic antiviral action using EMCV and the human lung carcinoma A549 cell line is shown in Figure 6. Similarly, the activity of bovine IFN-γ against VSV in bovine MDBK cells is also enhanced by human TNF-α or TNF-β.

Although IFN-γ is active in protecting most cells from VSV infection, 1 μg/ml (10^5 units in the EMCV assay) of human IFN-γ has no detectable activity against VSV in the A549 cell line. However, IFN-γ is effective in protecting against VSV infection in the

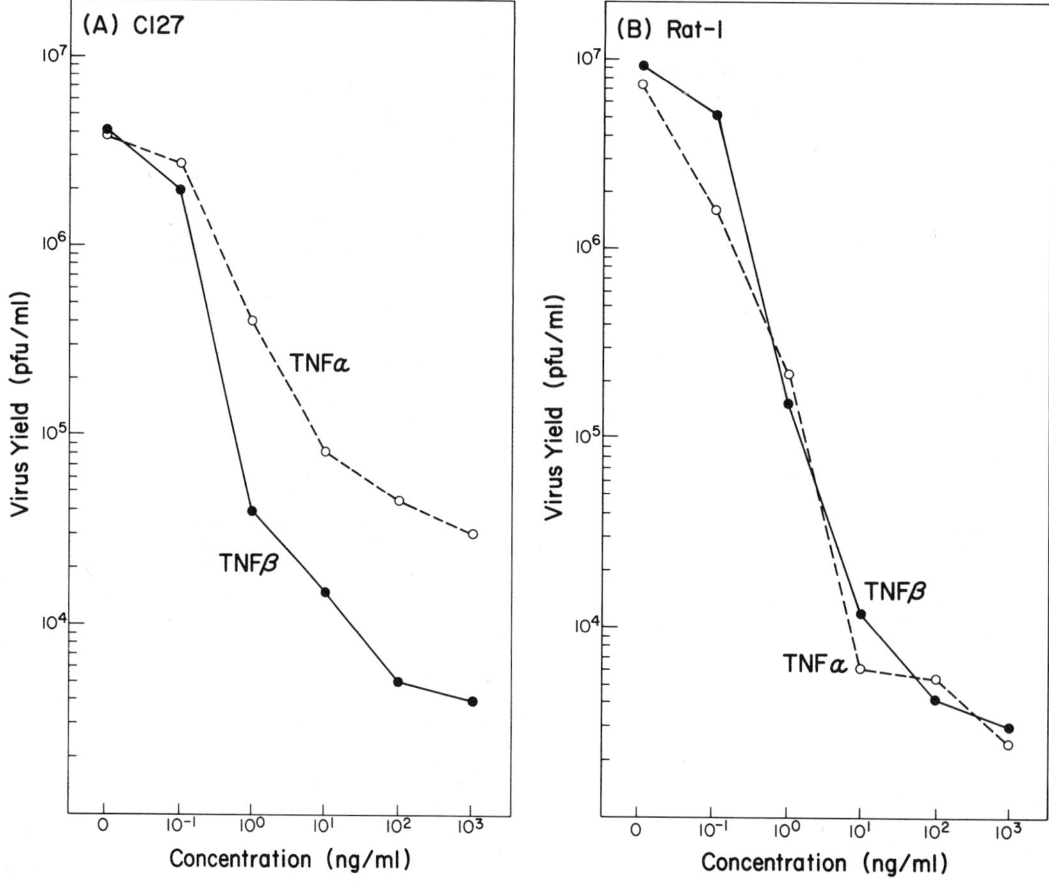

Figure 5. TNF-α and TNF-β inhibit VSV yield in murine C127 (*A*) and Rat-1 (*B*) cells. The cells were grown to confluency in 24-well plates and then treated wtih the indicated concentrations of TNF-α or TNF-β for 24 hr. The medium was removed before challenging with VSV at a multiplicity of infection (m.o.i.) of 10. Two hours later, the supernatants were aspirated to remove excess virus. After 24 hr, the cultures were assayed for virus yield in terms of plaque-forming units per milliliter on A549 cells (Rager-Zisman and Merigan 1973).

presence of TNF-α or TNF-β, although IFNs alone have no activity in these cells. As little as 1 ng/ml of IFN-γ and either TNF-α or TNF-β gave complete protection. Furthermore, this combination strongly inhibits the replication in A549 cells of the DNA viruses adenovirus-2, HSV-1, and HSV-2. IFN-γ alone is relatively ineffective. TNFs also enhance the antiviral activity of IFN-α and IFN-β, although to a lesser extent than IFN-γ.

TNF-α or TNF-β also enhances the antiviral activity of IFN-γ on a variety of transformed and normal cell lines. These include transformed cell lines of human (HeLa cervical carcinoma, HT1080 fibrosarcoma, 7860 renal carcinoma, T24 bladder carcinoma, HT-29 colon carcinoma, ST-486 Burkitt lymphoma, RPMI 8226 myeloma, and U87MG glioblastoma), rat (C6 glialoma), and murine (C127 epthelialoma and RAW 264 macrophage) origin, and three normal fibroblast cell lines (murine 3T3 and rat NRK and Rat-1). Thus, the antiviral potentiating activity of TNF-α or TNF-β is not virus-, cell-type-, or species-specific.

In addition to its preventive role against virus infection, TNF-α and TNF-β selectively kill virus-infected cells under certain conditions. This killing is enhanced by IFN-α, -β, or -γ. Thus, one major function of TNFs may be to broaden and extend the antiviral activity of IFNs by inducing cellular resistance in uninfected cells and by selectively destroying virus-infected cells. These results have implications for the therapy of the many medically important viral diseases caused by DNA and RNA viruses.

Effect of TNFs on HLA-B7 and HLA-DR Expression

The expression of the major histocompatibility complex (MHC) is essential for the initiation and regulation of the immune response. TNFs, like IFNs, can regulate the expression of both class I (HLA-B7) and class II (HLA-DR) MHC genes. Human bladder carcinoma T24 cells express detectable levels of HLA-B7 but not HLA-DRβ mRNA. Incubation with TNF-α (0.1 μg/ml) or IFN-γ (0.01 μg/ml) for 24 hours results in a fivefold increase of HLA-B7 but not HLA-DR mRNA. The induction is greater for the combination of IFN-γ and TNF-α than for either cytokine alone. Induction of HLA-B7 by IFN-γ is direct because an increase in

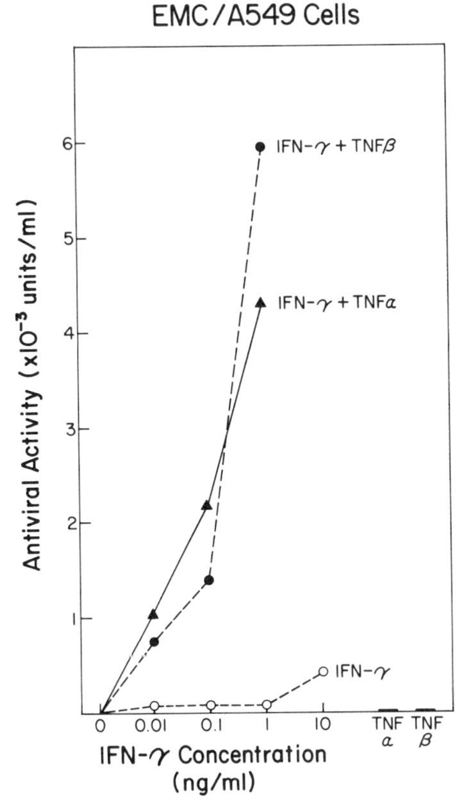

Figure 6. TNF-α and TNF-β enhance the antiviral activity of IFN-γ in A549 cells. A549 cells were grown to confluency, treated with human TNF-α (0.1 μg/ml), TNF-β (0.1 μg/ml), the indicated concentrations of IFN-γ, or the combinations for 24 hr before challenge with EMCV at a m.o.i. of 1. After 24 hr, the cytopathic effect was determined by staining the cells with crystal violet, and the titer was quantitatively monitored using a micro-ELISA autoreader. The assay was standardized to the International Reference Sample of the National Institutes of Health human IFN-γ (Gg 23-901-530).

HLA-B7 mRNA by IFN-γ can be detected after 4 hours of treatment in the presence of cycloheximide. In contrast, the induction of HLA-B7 by TNF-α can be blocked by cycloheximide, and the induction can only be detected after 12 hours of treatment. These results suggest that the induction of HLA-B7 mRNA by TNF-α is indirect. It is possible that TNF-α induces the synthesis of IFNs that cause the increase in HLA-B7 mRNA. However, the HLA-B7-inducing activity of TNF-α in T24 cells cannot be blocked by antibodies that neutralize IFN-α, -β, or -γ.

Although TNF-α and IFN-γ do not induce HLA-DR in most cell lines, they do induce HLA-DRβ mRNA in two human monocyte-like cell lines: the promyelocytic leukemia HL-60 and the histocytic lymphoma U937. The class-II-MHC-inducing activity of TNF-α is less potent than IFN-γ but is detectable at 24 hours. The combination of TNF-α and IFN-γ induces approximately tenfold higher levels of the HLA-DRβ mRNA in these cells than TNF-α or IFN-γ alone. TNF-β also enhances the expression of both HLA-B7 and HLA-DR mRNAs in HL-60 and U937 cells.

The expression of class I and class II MHC antigens has emerged as an essential component for antigen presentation in an immune response and in the control of tumor growth in vivo (Tanaka et al. 1985). Thus, TNFs and IFN-γ may potentiate the immune response through the synergistic amplification of MHC gene expression on a variety of cell types.

Catabolic Effects of TNFs

TNF-α is thought by some to be the agent that causes cachexia or wasting in parasite-infected animals (Beutler and Cerami 1986). However, there is no direct evidence to support this hypothesis. In cultured adipocytes, TNF can inhibit the transcription of genes encoding enzymes involved in fatty acid uptake and lipid synthesis (Torti et al. 1985). This fact, coupled with the observation that the serum of wasting animals can have elevated lipid levels, has led to the TNF-cachexia theory (Beutler and Cerami 1986).

When the ability of TNF-α to inhibit lipid anabolism in adipocytes is compared with that of other cytokines, we found that this property is not unique to TNF-α but is instead a general property of many cytokines (Patton et al. 1986). In addition to TNF-α and TNF-β, IFN-α, IFN-β, IFN-γ, and interleukin-1 appear to inhibit lipid anabolism (Keay and Grossberg 1980; Beutler and Cerami 1985; Patton et al. 1986). Table 4 shows the effect of a variety of cytokines on ^{3}H-labeled acetate uptake into lipid by 3T3 L1 fibroblasts and adipocytes. Both human TNF-α and TNF-β act across species boundaries to inhibit acetate uptake in mouse adipocytes. Murine IFN-γ exhibits 90% inhibition against mouse adipocytes, in contrast to human IFN-γ, which shows no effect. However, ^{3}H-labeled acetate uptake is inhibited by human IFN-α (hybrid IFN-α2/αl [*Bgl*II]), which does exhibit cross-species antiviral activity. Uptake of ^{3}H-labeled acetate by undifferentiated adipocytes (3T3 L1 fibroblasts) is also inhibited by the cytokines, although not as markedly (35–40%) as in the differentiated cells. Furthermore, these same cytokines inhibit lipoprotein lipase activity (the enzyme responsible for removing lipid from the serum) and stimulate the release of fatty acids into the medium (Patton et al. 1986). Thus, the net effect of cytokine action on a fat cell is catabolic; fat uptake and synthesis are inhibited and fat mobilization is stimulated (a situation that arises during starvation).

The short-term in vivo catabolic effect of cytokines is to mobilize fat into the circulation for utilization by the immune system (Beutler and Cerami 1986). If the infection becomes chronic, wasting ensues. Although exudates from endotoxin-stimulated macrophages have been reported to cause dose-dependent wasting in mice when injected over a period of days (Cerami et al. 1985), we have been unable to induce wasting in rats with repeated injections of highly purified recombinant human TNF-α. In our experiment, the initial injections cause a loss of appetite and an increase in blood lipids that lasts for about 24 hours. However,

Table 4. Effect of Cytokines on ^{3}H-labeled Acetate Uptake in 3T3 L1 Adipocytes and Fibroblasts

Cell type	Cytokine	pmoles acetate uptake/hr/mg protein
3T3 L1 Murine Adipocytes	Control	53.6 ± 9.7[a]
	r-hTNF-α	4.6 ± 0.6[b]
	r-hTNF-α + r-hTNF-α antibody	57.6 ± 8.9
	r-hTNF-β	4.0 ± 0.4[b]
	r-mIFN-γ	5.3 ± 0.4[b]
	r-hIFN-γ	53.7 ± 7.7
	r-hIFN-α2/α1(*Bgl*II)	4.9 ± 0.4[b]
3T3 L1 Murine Fibroblasts	Control	8.0 ± 0.5
	r-hTNF-α	5.2 ± 1.0[c]
	r-hTNF-β	4.5 ± 0.6[d]
	r-mIFN-γ	5.2 ± 0.8[c]

Cells were treated with cytokines (~1.5 nM) for 24 hr and then given ^{3}H-labeled acetate for 1 hr.

[a]Mean ± S.D. ($n = 3$).

[b]$p < 0.0001$ relative to control value.

[c]$p < 0.005$ relative to control value.

[d]$p < 0.025$ relative to control value.

appetite returned and tolerance developed by the second day (J. Patton et al., unpubl.). Comparisons between these two experiments are difficult. The fact that rapid tolerance to TNF did develop is intriguing and we are investigating this further. Another question that merits further work is whether or not the cytotoxic effect of cytokines on certain cells is simply an expression of the same catabolic effects that are seen on fat cells.

The observation that a variety of cytokines can be catabolic suggests that release of individual cytokines in the animal may be separated in location and time. Perhaps during the multiple phases of host response to infection there is a succession of cytokines appearing and disappearing in the circulation or at the site of infection. Alternatively, different infections (e.g., viral, microbial, and parasitic) or different tissues subject to the same infection may induce a different set of cytokines. Thus, if energy mobilization is to occur in many types of infection or in many sites within the body, it is critical that several cytokines with overlapping biological function share this host defense activity.

TNF Receptor Binding Studies

Both TNF-α and TNF-β interact with cells via specific cell-surface receptors (Aggarwal et al. 1985a; Hass et al. 1985; Sugarman et al. 1985). The binding is both time- and temperature-dependent, reaching a maximum within 1 hour at 37°C. The binding requires 4–6 hours to plateau at 4°C. For both cytokines, a single class of high-affinity receptors with a K_d of approximately 10^{-10} M has been identified on a variety of tumor cell lines. In a few cases (e.g., 3T3 L1 adipocytes; Patton et al. 1986), both low- and high-affinity receptors for TNF-α were observed. Most of the tumor cells examined have 1000–5000 receptors per cell.

Whether full receptor occupancy is needed for the biological response of TNF-α or TNF-β is not yet clear. The concentration of TNF-β required for 50% killing of murine L929 cells is the same as that required to displace 50% of maximum binding, suggesting that full receptor occupancy is essential (Hass et al. 1985). However, the cytotoxic activity of TNF-α and TNF-β can be observed at severalfold lower concentrations with actinomycin-D-treated or mitomycin-C-treated cells. These metabolic inhibitors have no effect on the K_d of ligand binding, suggesting that under these conditions a very small fraction of the total receptors must be occupied for the cytotoxic response. The binding of ^{125}I-labeled TNF-α to cells can be effectively competed by unlabeled TNF-β, suggesting that both molecules are recognized by the same cell-surface receptor. This is probably a result of the structural similarity of TNF-α and TNF-β and explains their many common biological properties.

The receptor for TNF is probably a protein, since the binding of TNFs to cells can be abolished by pretreating the cells with proteolytic enzymes. After protease removal, binding of ligand can be completely restored in 24 hours (B.B. Aggarwal and T. Eessalu, unpubl.). The binding of some peptide hormones to cell surfaces can be abolished by gangliosides (Van Heyningen 1974). However, gangliosides were found to have no effect on the binding of TNFs to the TNF receptor.

The binding of TNFs to the TNF receptor is not directly correlated with effects on cell proliferation (Sugarman et al. 1985). Cell lines on which treatment with TNF-α or TNF-β had no effect on growth (e.g., T24 bladder carcinoma) bound these ligands with the same affinity as cell lines that were highly sensitive (Fig. 7). A cell line whose growth was inhibited by TNF (ME-180 cervical carcinoma) bound to the ligand with the same affinity as cells that were growth stimulated by TNF (WI38 lung fibroblasts). A similar number of receptors per cell were determined on all three cell types. The lack of correlation between binding and biological response of a given hormone has been observed previously for other proteins (Marchand-Brustel et al. 1985). This could be due to a cascade of events involved in the ultimate response of the cell to a given ligand. It is conceivable that nonresponsive cells may be defective for any one or more of these events.

We have reported that the cytotoxic activity of both TNF-α and TNF-β can be potentiated synergistically by IFN-γ (Lee et al. 1984; Sugarman et al. 1985). Although the mechanism of this synergy is not known, preexposure of cells to IFN-γ increases the total number of TNF receptors without affecting the affinity of the receptor-ligand interaction (Aggarwal et al. 1985a). A typical enhancement in TNF binding to B16 melanoma cells after exposure to IFN-γ is shown in Figure 8. Treatment of these cells for 16 hours with IFN-γ results in an approximately twofold increase in receptor number. Since this increase in TNF-α binding requires protein synthesis, it can be proposed that IFN-γ induces the de novo synthesis of TNF receptors. Whether the increase in receptor number is sufficient to explain synergistic action of TNF with IFN-γ is uncertain. Receptor induction may be just one part of the total mechanism of synergy. Since inhibitors of protein and RNA synthesis also potentiate the cytotoxic activity of TNFs, it is possible that IFN-γ also suppresses the synthesis of certain proteins that antagonize the actions of TNFs.

TNF-α and TNF-β display little species specificity. Radiolabeled human TNF-α and TNF-β bind to both human and murine cell lines, and this binding can be competed with both unlabeled murine and human TNFs. This is likely due to the close homology of human and murine TNFs (>80% amino acid sequence identity).

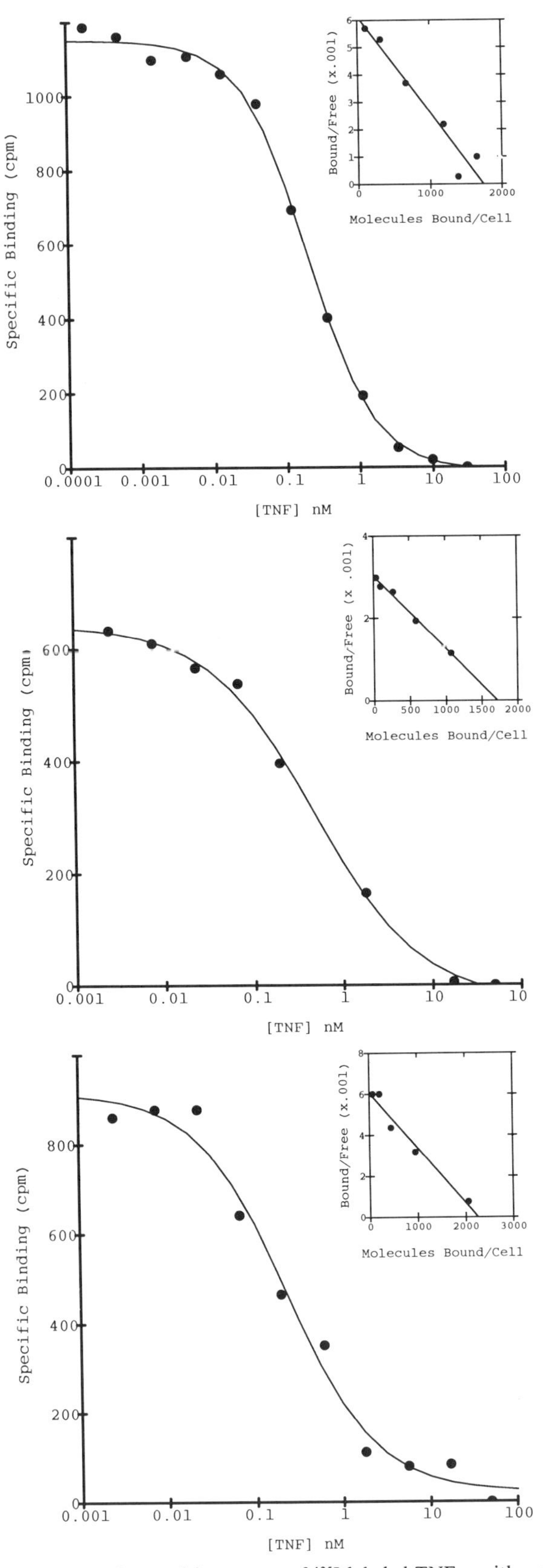

Figure 7. Competition curves of ^{125}I-labeled TNF-α with unlabeled TNF-α for binding to various cell lines. (*Top*) ME-180 cervical carcinoma; (*middle*) T24 bladder carcinoma; (*bottom*) WI38 lung fibroblast. (*Insets*) Scatchard analysis of the binding data. For experimental details, see Aggarwal et al. (1985a) and Hass et al. (1985).

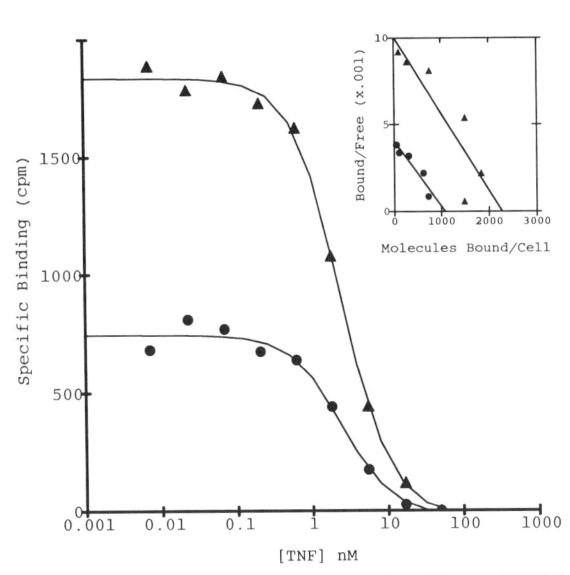

Figure 8. Induction of TNF-α receptors by IFN-γ on B16F10 melanoma cells. Competitive curves for untreated cells (●) and for cells treated overnight with IFN-γ (▲) are indicated. (*Inset*) Scatchard analysis of the binding data.

CONCLUSION

Our understanding of the TNF system has advanced tremendously over the past few years. The elucidation of the sequences of TNF-α and TNF-β by direct biochemical analysis and cDNA cloning has revealed that these cytokines are structurally related. Furthermore, the TNF genes are closely linked, their expression is tightly regulated, and the encoded gene products share many important biological properties. The biological activities of TNF-α and TNF-β are mediated through interaction with a common cell-surface receptor. The availability of cloned TNF genes and highly purified TNF preparations from recombinant *E. coli* should make it possible to address the many questions that remain unanswered concerning the regulation of TNF expression and the mechanisms of TNF action.

ACKNOWLEDGMENTS

We thank Drs. W. Wood and E. Chen for their help with bovine TNF-α and TNF-β sequence determination and Jeanne Arch for preparation of the manuscript.

REFERENCES

Aderka, D., H. Hoffmann, L. Toker, T. Hahn, and D. Wallach. 1986. Tumor necrosis factor induction by Sendai virus. *J. Immunol.* **136:** 2938.

Aggarwal, B.B., T.E. Eessalu, and P.E. Hass. 1985a. Characterization of receptors for human necrosis factor and their regulation by γ-interferon. *Nature* **318:** 665.

Aggarwal, B.B., B. Moffat, and R.N. Harkins. 1984. Human lymphotoxin. *J. Biol. Chem.* **259:** 686.

Aggarwal, B.B., W.J. Henzel, B. Moffat, W.J. Kohr, and R.N. Harkins. 1985b. Primary structure of human lymphotoxin derived from 1788 lymphoblastoid cell line. *J. Biol. Chem.* **260:** 2334.

Aggarwal, B.B., W.J. Kohr, P.E. Hass, B. Moffat, S.A. Spencer, W.J. Henzel, T.S. Bringman, G.E. Nedwin, D.V. Goeddel, and R.N. Harkins. 1985c. Human tumor necrosis factor: Production, purification and characterization. *J. Biol. Chem.* **260:** 2345.

Arya, S.K., F. Wong-Staal, and R.C. Gallo. 1984. Dexamethasone-mediated inhibition of human T cell growth factor and γ-interferon messenger RNA. *J. Immunol.* **133:** 273.

Babior, B.M. 1978. Oxygen-dependent microbial killing by phagocytes. *N. Engl. J. Med.* **298:** 659.

Bertolini, D.R., G.E. Nedwin, T.S. Bringman, D.D. Smith, and G.R. Mundy. 1986. Stimulation of bone resorption and inhibition of bone formation *in vitro* by human tumor necrosis factors. *Nature* **319:** 516.

Beutler, B.A. and A. Cerami. 1985. Recombinant interleukin 1 suppresses lipoprotein lipase activity in 3T3 L1 cells. *J. Immunol.* **135:** 3969.

———. 1986. Cachectin and tumour necrosis factor as two sides of the same biological coin. *Nature* **320:** 584.

Beutler, B., N. Krochin, I.W. Milsark, C. Luedke, and A. Cerami. 1986. Control of cachectin (tumor necrosis factor) synthesis: Mechanisms of endotoxin resistance. *Science* **232:** 977.

Carswell, E.A., L.J. Old, R.L. Cassel, S. Green, N. Fiore, and B. Williamson. 1975. An endotoxin-induced serum factor that causes necrosis of tumors. *Proc. Natl. Acad. Sci.* **72:** 3666.

Cerami, A., Y. Ikeda, N. Le Trang, P.J. Hotez, and B. Beutler. 1985. Weight loss associated with an endotoxin-induced mediator from peritoneal macrophages: The role of cachectin (tumor necrosis factor). *Immunol. Lett.* **11:** 173.

Claman, H.N. 1972. Corticosteroids and lymphoid cells. *N. Engl. J. Med.* **287:** 388.

Culpepper, J.A. and F. Lee. 1985. Regulation of IL-3 expression by glucocorticoids in cloned murine T lymphocytes. *J. Immunol.* **135:** 3191.

Goodwin, J.S. and D.R. Webb. 1980. Regulation of immune response by prostaglandin. *Clin. Immunol. Immunopathol.* **15:** 106.

Granger, G.A. and W.P. Kolb. 1968. Lymphocyte *in vitro* cytotoxicity: Mechanisms of immune and non-immune small lymphocyte mediated target L-cell destruction. *J. Immunol.* **101:** 111.

Gray, P.W. 1986. Molecular characterization of human lymphotoxin. In *Gene cloning in lymphokine research* (ed. D. Webb and D.V. Goeddel). Academic Press, New York. (In press.)

Gray, P.W., B.B. Aggarwal, C.V. Benton, T.S. Bringman, W.J. Henzel, J.A. Jarrett, D.W. Leung, B. Moffatt, P. Ng, L.P. Svedersky, M.A. Palladino, and G.E. Nedwin. 1984. Cloning and expression of cDNA for human lymphotoxin, a lymphokine with tumor necrosis activity. *Nature* **312:** 721.

Harlan, J.M., P.D. Killen, L.A. Harker, G.E. Stricker, and D.G. Wright. 1981. Neutrophil mediated endothelial injury *in vitro*: Mechanisms of cell detachment. *J. Clin. Invest.* **68:** 1394.

Hass, P.E., A. Hotchkiss, M. Mohler, and B.B. Aggarwal. 1985. Characterization of specific high affinity receptors for human tumor necrosis factor on mouse fibroblasts. *J. Biol. Chem.* **260:** 12214.

Itoh, H. and R.B. Wallace. 1985. A novel human physiologically active polypeptide. *European Patent Application* No. 84105149.3.

Keay, S. and S.E. Grossberg. 1980. Interferon inhibits the conversion of 3T3 L1 fibroblasts into adipocytes. *Proc. Natl. Acad. Sci.* **77:** 4099.

Kohase, M., D. Henriksen-DeStefano, L.T. May, J. Vilcek, and P.B. Sehgal. 1986. Induction of β_2-interferon by tumor necrosis factor: A homeostatic mechanism in the control of cell proliferation. *Cell* **45:** 659.

Kramer, S.M. and M.E. Carver. 1986. Serum-free *in vitro* bioassay for the detection of tumor necrosis factor. *J. Immunol. Methods* (in press).

Kull, F.C., S. Jacobs, and P. Cuatrecasas. 1985. Cellular receptor for ^{125}I-labeled tumor necrosis factor: Specific binding, affinity labeling, and relationship to sensitivity. *Proc. Natl. Acad. Sci.* **92:** 5756.

Lee, S.H., B.B. Aggarwal, E. Rinderknecht, F. Assisi, and H. Chiu. 1984. The synergistic and anti-proliferative effect of γ-interferon and human lymphotoxin. *J. Immunol.* **133:** 1083.

Marchand-Brustel, Y.L., T. Gremeaux, P. Ballotti, and E. Van Obberghen. 1985. Insulin receptor tyrosine kinase is defective in skeletal muscle of insulin resistant obese mice. *Nature* **315:** 676.

Nawroth, P.P. and D.M. Stern. 1986. Modulation of endothelial cell homeostatic properties by tumor necrosis factor. *J. Exp. Med.* **163:** 740.

Nedwin, G.E., S.L. Naylor, A.Y. Sakaguchi, D. Smith, J. Jarrett-Nedwin, D. Pennica, D.V. Goeddel, and P.W. Gray. 1985. Human lymphotoxin and tumor necrosis factor genes: Structure, homology and chromosomal localization. *Nucleic Acids Res.* **13:** 6361.

Old, L.J. 1985. Tumor necrosis factor (TNF). *Science* **230:** 630.

Patton, J.S., H.M. Shepard, H. Wilking, G.D. Lewis, B.B. Aggarwal, T.E. Eessalu, L.A. Gavin, and C. Grunfeld. 1986. Interferons and tumor necrosis factors have similar catabolic effects on 3T3 L1 cells. *Proc. Natl. Acad. Sci.* (in press).

Pennica, D. and D.V. Goeddel. 1986. Cloning and characterization of the genes for human and murine tumor necrosis factors. In *Gene cloning in lymphokine research* (ed. D. Webb and D.V. Goeddel). Academic Press, New York. (In press.)

Pennica, D., M.R. Shalaby, and M.A. Palladino. 1986. Tumor necrosis factors alpha and beta. In *Recombinant lymphokines and their receptors* (ed. S. Gillis). Marcel Dekker, New York. (In press.)

Pennica, D., J.S. Hayflick, T.S. Bringman, M.A. Palladino, and D.V. Goeddel. 1985. Cloning and expression in *Escherichia coli* of the cDNA for murine tumor necrosis factor. *Proc. Natl. Acad. Sci.* **82:** 6060.

Pennica, D., G.E. Nedwin, J.S. Hayflick, P.H. Seeburg, R. Derynck, M.A. Palladino, W.J. Kohr, B.B. Aggarwal, and D.V. Goeddel. 1984. Human tumor necrosis factor: Precursor structure, expression, and homology to lymphotoxin. *Nature* **312:** 724.

Rager-Zisman, B. and T.C. Merigan. 1973. A useful quantitative semimicromethod for viral plaque assay (37202). *Proc. Soc. Exp. Biol. Med.* **142:** 1174.

Roberts, A.B., M.. Anzano, L.M. Wakefield, N.S. Roche, D.F. Stern, and M.B. Sporn. 1985. Type β transforming growth factor: A bifunctional regulator of cellular growth. *Proc. Natl. Acad. Sci.* **82:** 119.

Ruddle, N.H. and B.H. Waksman. 1968. Cytotoxicity mediated by soluble antigens and lymphocytes in delayed hypersensitivity. *Science* **157:** 1060.

Ruddle, N.H., M.B. Powell, and B.S. Conta. 1983. Lymphotoxin, a biologically relevant model lymphokine. *Lymphokine Res.* **2:** 23.

Ruggiero, V., J. Tavernier, W. Fiers, and C. Baglioni. 1986. Induction of the synthesis of tumor necrosis factor receptors by interferon-γ. *J. Immunol.* **136:** 2445.

Shalaby, M.R., B.B. Aggarwal, E. Rinderknecht, L.P. Svedersky, B.S. Finkle, and M.A. Palladino. 1985. Activation of human polymorphonuclear neutrophil functions by gamma interferon and tumor necrosis factors. *J. Immunol.* **135:** 2069.

Shen, T. and C.A. Winter. 1977. Chemical and biological studies on indomethacin, sulindac and their analogs. *Adv. Drug. Res.* **12:** 89.

Stewart, W.E., II. 1981. *The interferon system*. Springer-Verlag, Vienna.

Stone-Wolff, D.S., Y.K. Yip, H.C. Kelker, J. Le, D. Henriksen-DeStefano, B.Y. Rubin, E. Rinderknecht, B.B. Aggarwal, and J. Vilcek. 1984. Interrelationship of human interferon-gamma with lymphotoxin and monocyte cytotoxin. *J. Exp. Med.* **159:** 828.

Sugarman, B.J., B.B. Aggarwal, P.E. Hass, I.S. Figari, M.A. Palladino, and H.M. Shepard. 1985. Recombinant human tumor necrosis factor-α: Effects on proliferation of normal and transformed cells *in vitro*. *Science* **230:** 943.

Tanaka, K., K.J. Isselbacher, G. Khoury, and G. Jay. 1985. Reversal of oncogensis by the expression of a major histocompatibility complex class I gene. *Science* **228:** 26.

Thomas, P.S. 1980. Hybridization of denatured RNA and small DNA fragments transferred to nitrocellulose. *Proc. Natl. Acad. Sci.* **77:** 5201.

Tiku, K., M.L. Tiku, and J.L. Skosey. 1986. Interleukin 1 production by human polymorphonuclear neutrophils. *J. Immunol.* **136:** 3677.

Torti, F.M., B. Dieckmann, B. Beutler, A. Cerami, and G.M. Ringold. 1985. A macrophage factor inhibits adipocyte gene expression: An *in vitro* model of cachexia. *Science* **229:** 867.

Tsujimoto, M., Y.K. Yip, and J. Vilcek. 1986. Inteferon-γ enhances expression of cellular receptors for tumor necrosis factor. *J. Immunol.* **136:** 2441.

Tucker, R.F., G.D. Shipley, H.L. Moses, and R.W. Holley. 1984. Growth inhibitor from BSC-1 cells closely related to platlet type β transforming growth factor. *Science* **226:** 705.

Urban, J.L., H.M. Shepard, J.L. Rothstein, B.J. Sugarman, and H. Schreiber. 1986. Tumor necrosis factor: A potent effector molecule for tumor cell killing by activated macrophages. *Proc. Natl. Acad. Sci.* **83:** 5233.

Van Heyningen, W.E. 1974. Gangliosides as membrane receptors for tetanus toxin, cholera toxin and seratonin. *Nature* **249:** 415.

Vilcek, J., V.J. Palombella, D. Henricksen-DeStefano, C. Swenson, R. Feinman, M. Hirai, and M. Tsujimoto. 1986. Fibroblast growth enhancing activity of tumor necrosis factor and its relationship to other polypeptide growth factors. *J. Exp. Med.* **163:** 632.

Vischer, T.L. 1972. Effect of hydrocortisone on the reactivity of thymus and spleen cells of mice to *in vitro* stimulation. *Immunology* **23:** 77.

Williams, T.W. and J.A. Bellanti. 1983. *In vitro* synergism between human interferons and human lymphotoxins: Enhancement of lymphotoxin-induced target cell killing. *J. Immunol.* **130:** 518.

Williamson, B.D., E.A. Carswell, B.Y. Rubin, J.S. Prendergast, and L. Old. 1983. Human tumor necrosis factor produced by human B-cell lines: Synergistic cytotoxic interaction with human interferon. *Proc. Natl. Acad. Sci.* **80:** 5397.

Wong, G.H.W., I. Clark-Lewis, J.A. Hamilton, and J.W. Schrader. 1984. P cell stimulating factor and glucocorticoids oppose the action of interferon-γ in inducing Ia antigens on T-dependent mast cells (P cells). *J. Immunol.* **133:** 2043.

Tandem Arrangement of Genes Coding for Tumor Necrosis Factor (TNF-α) and Lymphotoxin (TNF-β) in the Human Genome

S.A. Nedospasov,* A.N. Shakhov,* R.L. Turetskaya,* V.A. Mett.* M.M. Azizov,* G.P. Georgiev,* V.G. Korobko,† V.N. Dobrynin,† S.A. Filippov,† N.S. Bystrov,† E.F. Boldyreva,† S.A. Chuvpilo,† A.M. Chumakov,† L.N. Shingarova,† and Y. A. Ovchinnikov†

*Institute of Molecular Biology and † Shemyakin Institute of Bioorganic Chemistry, USSR Academy of Sciences, Moscow, USSR

Tumor necrosis factor and lymphotoxin, also termed TNF-α and TNF-β (Nedwin et al. 1985a), are polypeptides secreted by peripheral blood leukocytes (PBLs). Their production is induced by mitogens, phorbol esters, or interleukin 2 (Haranaka et al. 1984; Chroboczek Kelker et al. 1985; Nedwin et al. 1985a). Both factors are highly cytotoxic to a wide range of tumor cells in vitro and in vivo (for review, see Ruff and Gifford 1981; Old 1985; Sugarman et al. 1985). It appears that TNF-α and TNF-β are either the only or the major cytotoxic proteins present in the crude lymphokine preparations from stimulated PBLs (Chroboczek Kelker et al. 1985). Apparently, both proteins bind to the same receptor molecules on the surfaces of normal and tumor cells (Aggarwal et al. 1985a). Although they originate from different types of cells, TNF-α and TNF-β share 30% amino acid homology and show similar cytostatic and cytolytic activities in vitro and in vivo (Aggarwal et al. 1985b,c). TNF-α, first found in sera of mice infected with *Mycobacterium bovis* (strain Bacillus Calmette-Guerin) and treated with bacterial endotoxin (Carswell et al. 1975), is produced by activated macrophages (Ruff and Gifford et al. 1981; Nedwin et al. 1985a), whereas TNF-β (lymphotoxin) is produced by stimulated lymphocytes (Aggarwal et al. 1984, 1985b; Chroboczek Kelker et al. 1985).

Tumor cells differ in their sensitivity to the cytotoxic action of TNFs; these variations are apparently not due to differences in the number of binding sites per cell and/or different affinities to TNF (Sugarman et al. 1985; Tsujimoto et al. 1985). Antitumor activity of both TNF-α and TNF-β is enhanced by γ-interferon (Williamson et al. 1983; Lee et al. 1984; Sugarman et al. 1985). Normal cells are resistant to the cytotoxic action of TNFs; moreover, several normal cell lines have shown enhancement of growth in response to TNF-α (Sugarman et al. 1985).

Recently, TNF-α has been shown to be identical to the peptide hormone cachectin (Beutler et al. 1985; Beutler and Cerami 1986). Biochemical studies showed that TNF/cachectin inhibits an anabolic enzyme lipoprotein lipase in adipocytes and also activates production of catabolic substances—collagenase and prostaglandin E_2 in synovial cells and dermal fibroblasts (Dayer et al. 1985; Torti et al. 1985). In addition, TNF-α is known to inhibit transcription of adipocyte-specific genes (Torti et al. 1985) as well as to enhance dramatically expression of the class I major histocompatibility complex (MHC) genes (Collins et al. 1986) and of another surface antigen in cultured endothelial cells (Pober et al. 1986). TNF-α/cachectin is now considered a multipotent mediator of inflammatory reactions (Beutler and Cerami 1986). The molecular mechanism of TNF necrotic action remains obscure.

cDNAs for both TNF-β and TNF-α have been cloned (Gray et al. 1984; Pennica et al. 1984; Marmenout et al. 1985; Wang et al. 1985) and the corresponding human genes have been characterized (Nedospasov et al. 1985; Nedwin et al. 1985b; Shirai et al. 1985). As a result of these studies, expression vectors for TNF production have been constructed, and biologically active recombinant proteins are now available.

In this paper, we present data on the close proximity of human TNF-α and TNF-β genes in regard to the study of molecular organization of human genome. Biological experiments with engineered TNF, as well as their potential clinical applications, are discussed in detail in this volume (Fiers et al.; Goeddel et al.).

EXPERIMENTAL PROCEDURES

Oligonucleotide synthesis. Both phosphate and phosphite synthetic approaches for formation of internucleotide linkages have been employed. Oligonucleotides designed as hybridization probes have been synthesized using the modified solid-phase phosphotriester procedure (Dobrynin et al. 1983). Nucleotide chain elongations were carried out by mono- and dinucleotide-3′-chlorophenyl-phosphates, activated by arylsulfonyl chloride in the presence of *N*-methyl-imidazole. Internucleotide condensations, subsequent capping, and 5′-deblocking steps were performed in anhydrous dichloroethane. The same solvent was used for washing the polymer support after each reaction step. Oligonucleotides designed for assembling double-stranded DNA coding for amino-terminal parts of mature proteins (TNF-α and TNF-β) have been synthesized by means of the phosphoramidite method (Barone et al. 1984), which we have modified to reduce side reactions involving deoxyguanosine residues (Filippov et al. 1987). Deprotected oligonucleotides were isolated by reversed-phase chromatography as trityl derivatives. The subsequent purification of oligonucleotides was performed after detritylation by high-performance liquid chromatography (HPLC) or by denaturing gel electrophoresis.

Preparation of labeled oligonucleotide probes. Pairs of partially complementary oligonucleotides (17–20 mers) were labeled after annealing by filling in with the Klenow fragment of *Escherichia coli* DNA polymerase (kindly donated by I. Zaitzev). In a typical reaction, 5 pmoles of two oligonucleotides were annealed by cooling from 45°C to 15°C in 5 μl of nick-translation buffer (Maniatis et al. 1982). Labeled precursors (two or all four dNTPs at high specific activity) and cold dNTPs (if necessary) were then added in 2–3 molar excess (according to nucleotide composition of oligonucleotides to be labeled), and the filling reaction was carried out with 2–5 units of Klenow fragment in 20 μl at 12°C for 30 minutes, followed by a cold chase in the presence of 100 μM of each dNTP. The specific activity of the probe was usually between 5×10^7 and 10^8 cpm/pmole, which enabled us to detect single-copy sequences on genomic blots and to identify positive clones by plaque hybridization with genomic libraries.

Genomic libraries. The human genomic library for initial cloning of the TNF-α gene was constructed from a mixture of DNAs from several human placentas. DNA was digested with *Eco*RI and analyzed by genomic blot-hybridization with TNF-specific probes. A single band 2.8 kb long was identified. Milligram quantities of *Eco*RI-digested DNA were then fractionated by gel electrophoresis in low-melting-temperature agarose. DNA from different zones was excised, purified, and used for the construction of enriched genomic libraries. In particular, to clone the TNF-α gene, λgt*WES*λB (Leder et al. 1977) was used. For characterization of the genomic locus surrounding the TNF-α gene and for the search of TNF homologous sequences, the human genomic library in Charon 4A phage described by Lawn et al. (1978) was used (courtesy of T. Maniatis). Screening with oligonucleotide probes was performed on amplified replicas prepared in duplicate according to the method of Woo (1979).

DNA sequencing. In most cases, the solid-phase modification of the Maxam-Gilbert procedure was employed. As first described by Chuvpilo and Kravchenko (1983, 1985), this technique is less laborious and less time-consuming. Base-specific reactions of the Maxam-Gilbert procedure, modified as described previously by Dobrynin et al. (1980), were performed on small disks of DEAE paper. After piperidine hydrolysis, the DNA was eluted and analyzed by denaturing gel electrophoresis (Maxam and Gilbert 1977). Certain fragments cloned in M13 have been sequenced by the chain-termination method (Sanger et al. 1977), using TNF-specific synthetic oligonucleotides as sequencing primers.

***Expression of recombinant TNF in* E. coli.** To construct the expression vector, we used a genomic *Xho*I/*Hin*dIII fragment (see Fig. 4), which contained a part of the fourth exon of TNF-α gene. This fragment was cloned into *Sal*I/*Hin*dIII sites of the pUR278 plasmid (Ruther and Muller-Hill 1983). The resultant construction, pTNF1, induced by IPTG, can express β-galactosidase fused to a protein that contains 140 carboxy-terminal residues of mature TNF (see Pennica et al. 1984). This hybrid protein was purified and used for raising polyclonal antibodies.

To obtain a complete sequence coding for mature TNF, we synthesized overlapping oligonucleotides and then assembled double-stranded DNA (98 bp long) containing the ATG translation-initiation sequence, followed by codons for 30 amino-terminal amino acids of mature TNF (according to Shirai et al. 1985). Codon usage for three serine residues was chosen in such a way that a convenient *Xho*I site was introduced in this sequence. In addition, the 98-bp synthetic sequence was designed to contain cohesive *Sal*I and *Msp*I ends. Expression vector pTNF2 was assembled by ligating a large *Sal*I/*Hin*dIII fragment from pUR291 (Ruther and Muller-Hill 1983) with the *Msp*I/*Hin*dIII 505-bp-long genomic TNF fragment from pTNF1 (see above) in the presence of an excess of nonphosphorylated synthetic 98-bp DNA. As a result, the mature TNF sequence was fused to β-galactosidase through the methionine residue. This construction was used for preparation of the hymeric protein, which was helpful in the preparation of monoclonal and polyclonal antibodies against TNF.

For direct expression, the TNF gene was introduced into a plasmid derived from pBR325 (Prentki et al. 1981) under the control of the *trp* promoter. Analysis of crude bacterial lysates by SDS-PAGE clearly showed the presence of a new 17-kD polypeptide (data not shown). This protein was then purified to near homogeneity by ion-exchange chromatography.

Cytotoxicity assay. Cytotoxic activities of crude lysates and purified preparations of recombinant TNF were monitored by standard assay on transformed murine fibroblasts (L929) in the presence of actinomycin D (Fransen et al. 1985). The specific activity of the protein was 1×10^7 to 2×10^7 units/mg. This activity could be specifically blocked by antisera raised against recombinant TNF protein.

RESULTS

Cloning of the Human TNF-α Gene

To isolate the human gene coding for TNF-α, we have screened several human genomic libraries with an oligonucleotide probe designed according to the pub-

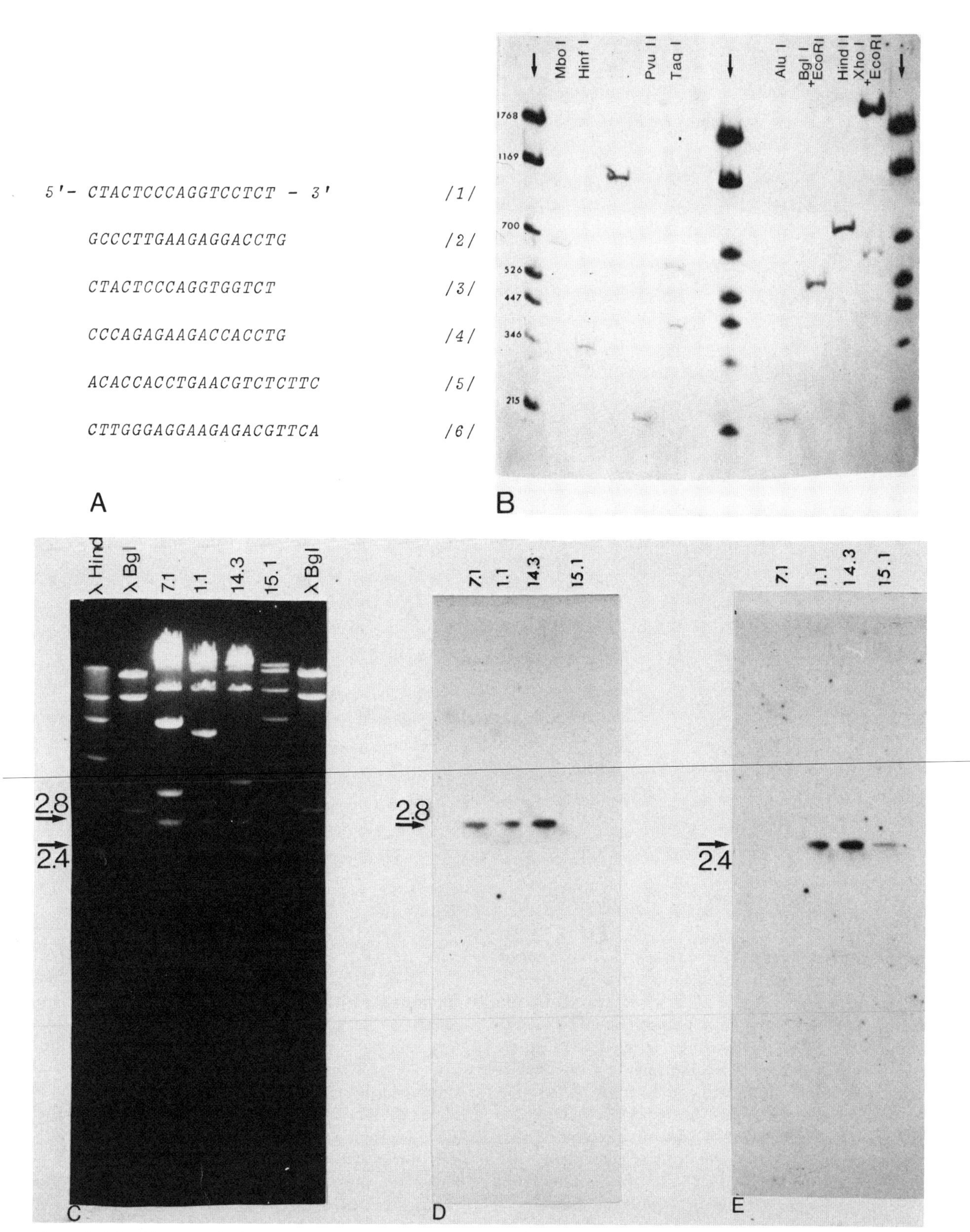

Figure 1. Identification of TNF-α- and TNF-β-specific clones by Southern analysis of recombinant phage DNA. (*A*) Pairs of partially complementary oligonucleotides used to screen the human genomic library. (*1,2*) TNF-α (exon 4); (*3,4*) TNF-β (exon 4); (*5,6*) TNF-β (exon 2). (*B*) Example of Southern hybridization of recombinant phage 7.1 DNA with TNF-α-specific probe 1,2. Phage DNA has been digested with different restriction enzymes. Marker is labeled SV40-M DNA (see Nedospasov et al. 1984) digested with *Hin*dIII + *Msp*I. (*C*) *Eco*RI digestion of DNAs of selected clones and Southern analysis with TNF-α TNF-β probes. Ethidium-bromide-stained gel with *Eco*RI digest of four DNAs: 7.1 (α-positive, β-negative); 1.1. and 14.3 (double positive); 15.2 (α-negative, β-positive). Markers are *Bgl*I and *Hin*dIII digests of λ DNA. (*D*) Same as *C*, but after hybridization with TNF-α probe 1,2. (*E*) Same as *C*, but after hybridization with TNF-β probe 5,6 or 3,4.

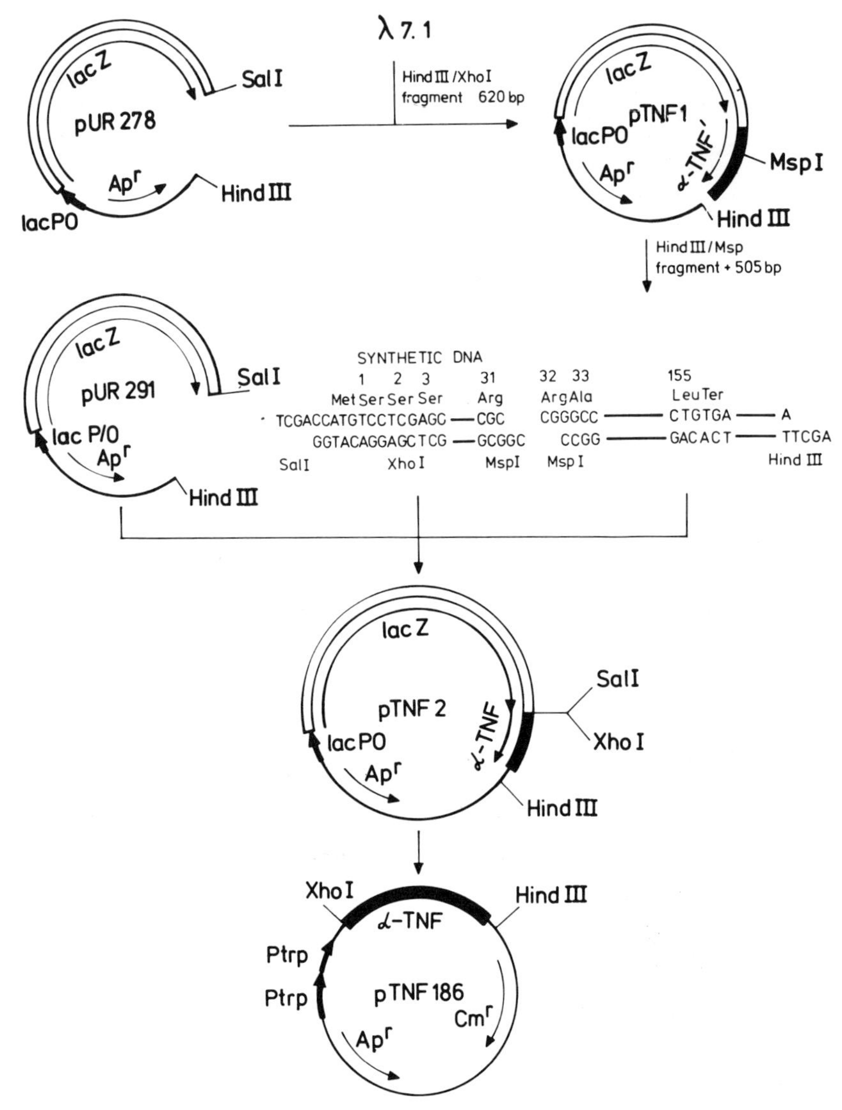

Figure 2. Construction of plasmids for expression of TNF-α in bacteria (See Experimental Procedures).

lished cDNA sequence (Pennica et al. 1984). Hoping to obtain a probe that is not interrupted by intron-exon junctions, we have chosen a short sequence from the region of highest homology between the TNF-α and lymphotoxin (TNF-β) genes (Gray et al. 1984; Pennica et al. 1984). Two partially complementary oligonucleotides have been synthesized and used for preparation of a radioactive probe with high specific activity (see Experimental Procedures). This probe corresponded to nucleotide 554–578 from the TNF-α cDNA sequence described by Pennica et al. (1984) (nucleotides 5770–5794 of the sequence shown in Fig. 5; see also Figs. 1 and 4). A number of positive recombinant clones have been purified from several libraries used throughout this study. Here, we describe in detail only those TNF clones that have been isolated from one particular genomic library (Lawn et al. 1978).

We analyzed 1.5×10^6 plaques and isolated ten clones that unambiguously hybridized to the TNF-α oligonucleotide probe. DNAs from these recombinant phages were purified and subjected to Southern analysis (Southern 1975). At least seven of them looked different after *Eco*RI digestion, although they shared several restriction fragments. One of these common fragments, 2.8 kb long, hybridized with the oligonucleotide probe specific for TNF-α (Fig. 1). This fragment was subcloned into the pUC12/13 plasmid (Vieira and Messing 1982) and M13 vector WB238/239 (Barnes et al. 1983; Sahli et al. 1985) and subjected to further restriction enzyme analysis and sequencing, in particular, using the same oligonucleotides as sequencing primers (see above). We concluded that this 2.8-kb fragment corresponded to a single-copy TNF-α gene described by Shirai et al. (1985).

Cloning of the Human TNF-β Gene

The lymphotoxin (TNF-β) gene has been cloned unexpectedly from the same library in the course of our search for TNF-α-related sequences. We have noted

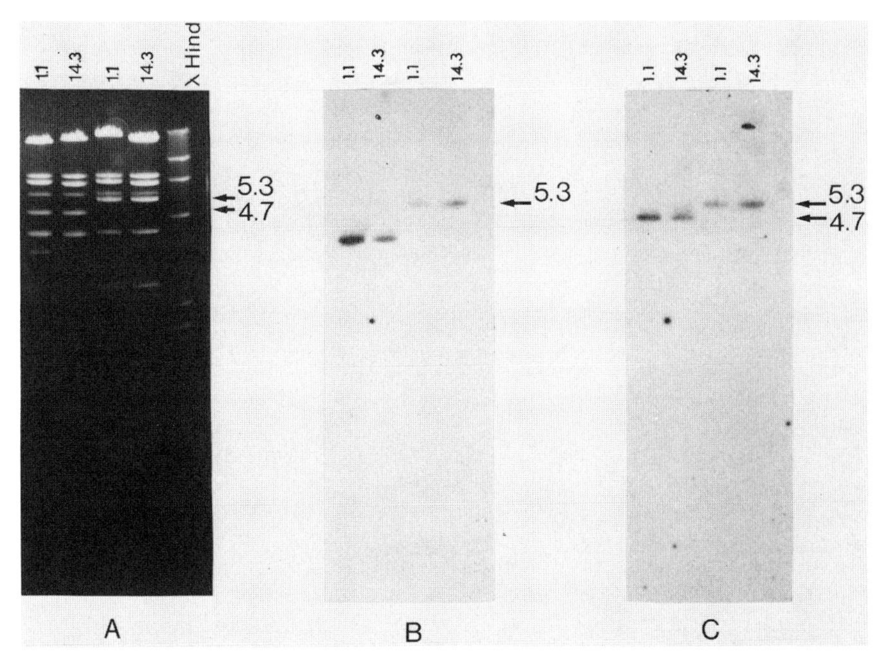

Figure 3. Close proximity and mutual orientation of TNF genes in the human genome. (*A*) Ethidium-bromide-stained gel with *Bam*HI/*Xho*I (*left*) and *Bam*HI/*Hin*dIII (*right*) double digestions of two different phage clones containing both TNF genes (1.1 and 14.3; see Fig. 6). (*B*) Same as *A* but after Southern hybridization with the probe prepared from oligonucleotides 1,2 (see Fig. 1*A*). (*C*) Same as *A* but after hybridization with the probe prepared from oligonucleotides 3,4.

that most of the TNF-α-positive clones on the same master plates also hybridized with TNF-β-specific probes, corresponding to nucleotides 83–112 and 406–431 from the cDNA clone described by Gray et al. (1984). Southern analysis showed that a 2.4-kb *Eco*RI fragment contained most of the sequences coding for TNF-β (see Fig. 1). This fragment was recloned into pUC12/13, characterized with restriction enzymes, and sequenced.

Construction of Bacterial Vectors for TNF-α and TNF-β Expression

Sequences necessary for the production of mature TNF proteins in *E. coli* have been constructed by combining large parts of the last exons for both TNF genes and synthetic parts (Fig. 2) according to TNF-α and TNF-β nucleotide and amino acid sequences (Gray et al. 1984; Pennica et al. 1984; Aggarwal et al. 1985a,b). An analogous approach has been published by Shirai et al. (1985). Strains of bacteria producing large quantities of TNF-α have been obtained, and highly purified biologically active protein has been used for a number of biological experiments. Similar design and constructions have also been made for TNF-β.

Structure of the 7-kb Human DNA Locus Containing Complete Genomic Copies of TNF-α and TNF-β

The map was constructed from restriction nuclease analysis of several recombinant phage clones containing both TNF genes. *Eco*RI fragments 2.8 kb and 2.4 kb long have been subcloned into plasmid and single-stranded phages and analyzed by sequencing. The precise distance between these *Eco*RI fragments was determined from double digestions of recombinant phage DNAs with *Bam*HI/*Hin*dIII and *Bam*HI/*Xho*I. As known from restriction enzymes and sequencing analysis, the 2.4-kb *Eco*RI fragments (TNF-β) contained two *Bam*HI sites and no *Hin*dIII or *Xho*I sites, whereas the 2.8-kb *Eco*RI fragment (TNF-α) contained a single internal *Hin*dIII and *Xho*I cleavage site but no *Bam*HI sites. We have therefore been able to find both the mutual orientation of the two genes and the distance between them. After hybridization with TNF-β probes, the *Bam*HI/*Xho*I fragment appeared to be shorter than the *Bam*HI/*Hin*dIII fragment (Fig. 3). Since the direction of transcription of the two genes was known from the sequencing data of large *Eco*RI fragments, we concluded that the TNF-β gene is located upstream of TNF-α gene in the same orientation (see Fig. 4). This conclusion was confirmed after sequencing of the whole 7-kb fragment.

Both TNF-β and TNF-α probes hybridized to the same 5.3-kb *Bam*HI/*Hin*dIII fragment (see Fig. 3). This fragment was subcloned into pUC13 and analyzed in detail by means of restriction enzymes. The interval 2.2-kb *Pvu*II fragment overlapping the two genes was purified and used for sequencing.

In the course of our study of TNF genes, several groups published results of sequence analysis of genomic copies for TNF-α (Marmenout et al. 1985; Nedwin et al. 1985b; Shirai et al. 1985) and TNF-β (Nedwin et al. 1985b). Therefore, we compared our data with

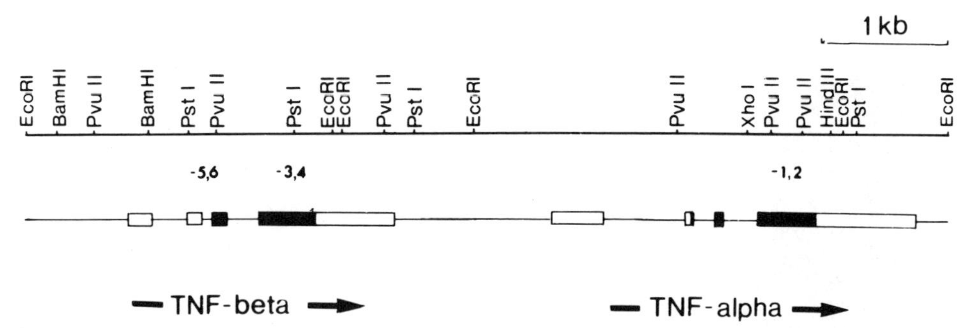

Figure 4. Restriction map of the 7-kb segment of the human genome containing TNF genes (see Fig. 5 for nucleotide sequence). Diagram shows a schematic representation of TNF-β and TNF-α exon-intron structures. Regions coding for mature proteins are represented by solid bars. Positions of the sequences homologous to oligonucleotides 1-6 (Fig. 1*A*) are indicated. Arrows show direction of transcription.

those presented by others as well as with cDNA sequences (Gray et al. 1984; Pennica et al. 1984). The most important is an A-to-C change between the genomic sequence of Nedwin et al. (1985b) (nucleotide 1540) and our data (nucleotide 1540 in Fig. 5), which confirms that amino acid 26 of mature lymphotoxin (TNF-β) is threonine, in agreement with the data on protein and cDNA structure (see Gray et al. 1984; Aggarwal et al. 1985b). Other differences are listed in Table 1.

The whole sequence of TNF locus was constructed from those of the 2.8-kb and 2.4-kb *Eco*RI and 2.2-kb *Pvu*II fragments overlapping the two genes (see Fig. 4). In some cases where we still do not have unambiguous data (e.g., the 0.8-kb *Eco*RI fragment in 3′-nontranslated region of TNF-α and the *Eco*RI-*Pvu*II fragment in 3′-nontranslated region of TNF-β), we have included certain parts of the sequences determined by other groups of investigators (Marmenout et al. 1985; Nedwin et al. 1985b; Shirai et al. 1985) (Fig. 5).

Mapping of the 30-kb Genetic Locus Surrounding TNF Genes

Recombinant phage DNAs positive for TNF-α and/or TNF-β probes have been subjected to restriction enzyme analysis. Restriction maps of overlapping human DNA inserts for *Eco*RI, *Hin*dIII, *Kpn*I, *Bam*HI, and *Xho*I have been constructed and aligned to give a map of a human DNA fragment approximately 30 kb long (see Fig. 6). The restriction map of the central part containing both TNF genes has been verified by Southern hybridization (see above). After digestion of different DNAs from human placentas with *Bam*HI/*Hin*dIII and Southern hybridization with TNF-α- and TNF-β-specific probes, we observed the same band with the expected length of 5.3 kb (Fig. 7). In these experiments, we also observed an additional band corresponding to a *Hin*dIII fragment 3.1 kb long that is not compatible with the map shown in Figure 4. This band was absent on the initial genomic blots with DNA from human lymphocyte cell culture (data not shown). Thus, the significance of this observation is not clear at the present time.

Two recombinant phage DNAs containing only the TNF-β gene and extended about 10 kb upstream of the TNF locus had different structures, which can be interpreted as a deletion of 4 kb in one of them (Fig. 6). To decide which variant corresponded to the structure of human genomic locus, *Hin*dIII digests of different DNAs were analyzed by Southern hybridization with the TNF-β probe. As shown in Figure 7, in all DNAs, the large *Hin*dIII fragment that contains the TNF sequences is 17 kb long. This value was compatible with the longer clone (15.1) (Fig. 6) of the two recombinant DNAs. The deletion observed in the second clone (11.1) may result from rearrangements during cloning procedures or, alternatively, may correspond to a rare allelic variant of the TNF locus.

We have also started to map positions of repetitive

Figure 5. Nucleotide sequence of the tandem of TNF genes. The complete sequence (7112 nucleotides) has been compiled from sequences of DNA fragments formed by restriction nucleases (see Fig. 4). Most of the sequencing was done by solid-phase modification of the Maxam-Gilbert procedure (see Experimental Procedures). The *Hin*dIII-*Xho*I fragment and both ends of the *Eco*RI fragments of 2.8 kb and 2.4 kb (see Fig. 4) were sequenced by the chain-termination method. Internal *Pvu*II fragments have been sequenced by the following strategy: DNA was digested separately by *Mva*I or *Ava*II (both having alternating A/T in recognition sites). Digests were labeled by the filling-in reaction in the presence of either [α-^{32}P]dATP or [α-^{32}P]-TTP and run on the native gel in parallel. Autoradiography revealed different patterns, depending on the presence of an A or T at the ends of the fragments. Those that could be labeled with both dATP and TTP had different sequences at both ends and therefore contained a single label in each case. These fragments were excised from the gel and sequenced without strand separation. Obviously, those fragments that were labeled in only one of the two reactions contained label on both ends; therefore, sequencing was done after strand separation. Overlapping sequences of *Mva*I and *Ava*II subfragments (both strands) were used for reconstruction of the nucleotide sequence of initial *Pvu*II fragments. Sequences between nucleotides 516–706 and 4992–5520 were compiled from Nedwin et al. (1985b). A number of alterations as compared to the sequence published by Nedwin et al. (1985b) are listed in Table 1.

```
    / Eco RI
   1 GAATTCTCGA AACTTCCTTT GTAGAAAACT TTGGAAGGTG TCTGCCACAT   50
  51 TGATCCTGGA ATGTGTGTTT ATTTGGGGTT ATATAAATCT GTTCTGTGGA  100
 101 AGCCACCTGA AGTCAGGAAG AGATGGAGGG CATCCTTCAG GAGTGAGATG  150
 151 AGACCTCATC ATACTTGACT GTCCAGCATC ATCTCTGAGT GAGGGGACCA  200
                                                  / Bam HI
 201 AAAAATTTAT CTTCCAAACT AGGACACTTT CAAGAGTGGA AGGGGGATCC  250
 251 ATTAATATTT TCACCTGGAC AAGAGGCAAA CACCAGAATG TCCCCGATGA  300
 301 AGGGGATATA TAATGGACCT TCTTGATGTG AAACCTGCCA GATGGGCTGG  350
 351 AAAGTCCGTA TACTGGGACA AGTATGATTT GAGTTGTTTG GGACAAGGAC  400
 401 AGGGGTACAA GAGAAGGAAA TGGGCAAAGA GAGAAGCCTG TACTCAGCCA  450
 451 AGGGTGCAGA GATGTTATAT ATGATTGCTC TTCAGGGAAC CGGGCCTCCA  500
                / Pvu II
 501 GCTCACACCC CAGCTGCTCA ACCACCTCCT CTCTGAATTG ACTGTCCCTT  550
 551 CTTTGGAACT CTAGGCCTGA CCCCACTCCC TGGCCCTCCC AGCCCACGAT  600
 601 TCCCCTGACC CGACTCCCTT TCCCAGAACT CAGTCGCCTG AACCCCCAGC  650
 651 CTGTGGTTCT CTCCTAGGCC TCAGCCTTTC CTGCCTTTGA CTGAAACAGC  700
 701 AGTATCTTCT AAGCCCTGGG GGCTTCCCCG GGCCCCAGCC CCGACCTAGA  750
                                   TATA box TNF-beta
 751 ACCCGCCCGC TGCCTGCCAC GCTGCCACTG CCGCTTCCTC TATAAAGGGA  800
 801 CCTGAGCGTC CGGGCCCAGG GGCTCCGCAC AGCAGGTGAG GCTCTCCTGC  850
 851 CCCATCTCCT TGGGCTGCCC GTGCTTCGTG CTTTGGACTA CCGCCCAGCA  900
                                             Bam HI /
 901 GTGTCCTGCC CTCTGCCTGG GCCTCGGTCC CTCCTGCACC TGCTGCCTGG  950
 951 ATCCCCGGCC TGCCTGGGCC TGGGCCTTGG TGGGTTTGGT TTTGGTTTCC 1000
1001 TTCTCTGTCT CTGACTCTCC ATCTGTCAGT CTCATTGTCT CTGTCACACA 1050
1051 TTCTCTGTTT CTGCCATGAT TCCTCTCTGT TCCCTTCCTG TCTCTCTCTG 1100
1101 TCTCCCTCTG CTCACCTTGG GGTTTCTCTG ACTGCATCTT GTCCCCTTCT 1150
1151 CTGTCGATCT CTCTCTCGGG GGTCGGGGGG TGCTGTCTCC CAGGGCGGGA 1200
1201 GGTCTGTCTT CCGCCGCGTG CCCCGCCCCG CTCACTGTCT CTCTCTCTCT 1250
               / Pst I          *
1251 CTCTCTTTCT CTGCAGGTTC TCCCCATGAC ACCACCTGAA CGTCTCTTCC 1300
1301 TCCCAAGGGT GTGTGGCACC ACCCTACACC TCCTCCTTCT GGGGCTGCTG 1350
1351 CTGGTTCTGC TGCCTGGGGC CCAGGTGAGG CAGCAGGAGA ATGGGGGCTG 1400
1401 CTGGGGTGGC TCAGCCAAAC CTTGAGCCCT AGAGCCCCCC TCAACTCTGT 1450
                                           / Pvu II
1451 TCTCCCCTAG GGGCTCCCTG GTGTTGGCCT CACACCTTCA GCTGCCCAGA 1500
1501 CTGCCCGTCA GCACCCCAAG ATGCATCTTG CCCACAGCAC CCTCAAACCT 1550
1551 GCTGCTCACC TCATTGGTAA ACATCCACCT GACCTCCCAG ACATGTCCCC 1600
1601 ACCAGCTCTC CTCCTACCCC TGCCTCAGGA ACCCAAGCAT CCACCCCTCT 1650
1651 CCCCCAACTT CCCCCACGCT AAAAAAAACA GAGGGAGCCC ACTCCTATGC 1700
1701 CTCCCCCTGC CATCCCCCAG GAACTCAGTT GTTCAGTGCC CACTTCCTCA 1750
1751 GGGATTGAGA CCTCTGATCC AGACCCCTGA TCTCCCACCC CCATCCCCTA 1800
1801 TGGCTCTTCC TAGGAGACCC CAGCAAGCAG AACTCACTGC TCTGGAGAGC 1850
1851 AAACACGGAC CGTGCCTTCC TCCAGGATGG TTTCTCCTTG AGCAACAATT 1900
1901 CTCTCCTGGT CCCCACCAGT GGCATCTACT TCGTCTACTC CCAGGTGGTC 1950
1951 TTCTCTGGGA AAGCCTACTC TCCCAAGGCC ACCTCCTCCC CACTCTACCT 2000
```

Figure 5. (*Sequence printout continued on next page. See facing page for legend.*)

```
2001 GGCCCATGAG GTCCAGCTCT TCTCCTCCCA GTACCCCTTC CATGTGCCTC  2050
                                            / Pst I
2051 TCCTCAGCTC CCAGAAGATG GTGTATCCAG GGCTGCAGGA ACCCTGGCTG  2100
2101 CACTCGATGT ACCACGGGGC TGCGTTCCAG CTCACCCAGG GAGACCAGCT  2150
2151 ATCCACCCAC ACAGATGGCA TCCCCCACCT AGTCCTCAGC CCTAGTACTG  2200
                             **
2201 TCTTCTTTGG AGCCTTCGCT CTGTAGAACT TGGAAAAATC CAGAAAGAAA  2250
2251 AAATAATTGA TTTCAAGACC TTCTCCCCAT TCTGCCTCCA TTCTGACCAT  2300
2301 TTCAGGGGTC GTCACCACCT CTCCTTTGGC CATTCCAACA GCTCAAGTCT  2350
                                       / Eco RI
2351 TCCCTGATCA AGTCACCGGA GCTTTCAAAG AAGGAATTCT AGGCATCCCA  2400
2401 GGGGACCACA CCTCCCTGAA CCATCCCTGA TGTCTGTCTG GCTGAGGATT  2450
                   / Eco RI
2451 TCAAGCCTGC CTAGGAATTC CCAGCCCAAA GCTGTTGGTC TTGTCCACCA  2500
2501 GCTAGGTGGG GCCTAGATCC ACACACAGAG GAAGAGCAGG CACATGGAGG  2550
2551 AGCTTGGGGG ATGACTAGAG GCAGGGAGGG GACTATTTAT GAAGGCAAAA  2600
2601 AAATTAAATT ATTTATTTAT GGAGGATGGA GAGAGGGGAA TAATAGAAGA  2650
2651 ACATCCAAGG AGAAACAGAG ACAGGCCCAA GAGATGAAGA GTGAGAGGGC  2700
2701 ATGCGCACAA GGCTGACCAA GAGAGAAAGA AGTAGGCATG AGGGATCACA  2750
                                        / Pvu II
2751 GGGCCCCAGA AGGCAGGGAA AGGCTCTGAA AGCCAGCTGC CGACCAGAGC  2800
                                        poly(A)
2801 CCCACACGGA GGCATCTGCA CCCTCGATGA AGCCCAATAA ACCTCTTTTC  2850
2851 TCTGAAATGC TGTCTGCTTG TGTGTGTGTG TCTGGGAGTG AGAACTTCCC  2900
2901 AGTCTATCTA AGGAATGGAG GGAGGGACAG AGGGCTCAAA GGGAGCAAGA  2950
2951 GCTGTGGGGA GAACAAAAGG ATAAGGGCTC AGAGAGCTTC AGGGATATGT  3000
                                      / Pst I
3001 GATGGACTCA CCAGGTGAGG CCGCCAGACT GCTGCAGGGG AAGCAAAGGA  3050
3051 GAAGCTGAGA AGATGAAGGA AAAGTCAGGG TCTGGAGGGG CGGGGGTCAG  3100
3101 GGAGCTCCTG GGAGATATGG CCACATGTAG CGGCTCTGAG GAATGGGTTA  3150
3151 CAGGAGACCT CTGGGGAGAT GTGACCACAG CAATGGGTAG GAGAATGTCC  3200
3201 AGGGCTATGA AAGTCGAGTA TGGGGACCCC CCCTTAACGA AGACAGGGCC  3250
3251 ATGTAGAGGG CCCCAGGGAG TGAAAGAGCC TCCAGGACCT CCAGGTATGG  3300
3301 AATACAGGGG ACGTTTAAGA AGATATGGCC ACACACTGGG GCCCTGAGAA  3350
3351 GTGAGAGCTT CATGAAAAAA ATCAGGGACC CCAGAGTTCC TTGGAAGCCA  3400
3401 AGACTGAAAC CAGCATTATG AGTCTCCGGG TCAGAATGAA AGAAGAGGGC  3450
                        / Eco RI
3451 CTGCCCCAGT GGGGTCTGTG AATTCCCGGG GGTGATTTCA CTCCCCGGGG  3500
3501 CTGTCCCAGG CTTGTCCCTG CTACCCGCAC CCAGCCTTTC CTGAGGCCTC  3550
3551 AAGCCTGCCA CCAAGCCCCC AGCTCCTTCT CCCCGCAGGG CCCAAACACA  3600
3601 GGCCTCAGGA CTCAACACAG CTTTTCCCTC CAACCCCGTT TTCTCTCCCT  3650
3651 CAACGGACTC AGCTTTCTGA AGCCCCTCCC AGTTCTAGTT CTATCTTTTT  3700
3701 CCTGCATCCT GTCTGGAAGT TAGAAGGAAA CAGACCACAG ACCTGGTCCC  3750
3751 CAAAAGAAAT GGAGGCAATA GGTTTTGAGG GGCATGGGGA CGGGGTTCAG  3800
3801 CCTCCAGGGT CCTACACACA AATCAGTCAG TGGCCCAGAA GACCCCCCTC  3850
3851 GGAATCGGAG CAGGGAGGAT GGGGAGTGTG AGGGGTATCC TTGATGCTTG  3900
3901 TGTGTCCCCA ACTTTCCAAA TCCCCGCCCC CGCGATGGAG AAGAAACCGA  3950
3951 GACAGAAGGT GCAGGGCCCA CTACCGCTTC CTCCAGATGA GCTCATGGGT  4000
```

Figure 5. (*Continued on next page.*)

```
4001 TTCTCCACCA AGGAAGTTTT CCGCTGGTTG AATGATTCTT TCCCCGCCCT  4050
                          TATA box TNF-alpha
4051 CCTCTCGCCC CAGGGACATA TAAAGGCAGT TGTTGGCACA CCCAGCCAGC  4100
4101 AGACGCTCCC TCAGCAAGGA CAGCAGAGGA CCAGCTAAGA GGGAGAGAAG  4150
4151 CAACTACAGA CCCCCCCTGA AAACAACCCT CAGACGCCAC ATCCCCTGAC  4200
4201 AAGCTGCCAG GCAGGTTCTC TTCCTCTCAC ATACTGACCC ACGGCTTCAC  4250
                               *
4251 CCTCTCTCCC CTGGAAAGGA CACCATGAGC ACTGAAAGCA TGATCCGGGA  4300
4301 CGTGGAGCTG GCCGAGGAGG CGCTCCCCAA GAAGACAGGG GGGCCCCAGG  4350
4351 GCTCCAGGCG GTGCTTGTTC CTCAGCCTCT TCTCCTTCCT GATCGTGGCA  4400
4401 GGCGCCACCA CGCTCTTCTG CCTGCTGCAC TTTGGAGTGA TCGGCCCCCA  4450
4451 GAGGGAAGAG GTGAGTGCCT GGCCAGCCTT CATCCACTCT CCCACCCAAG  4500
4501 GGGAAATGGA GACGCAAGAG AGGGAGAGAG ATGGGATGGG TGAAAGATGT  4550
4551 GCGCTGATAG GGAGGGATGG AGAGAAAAAA ACGTGGAGAA AGACGGGGAT  4600
4601 GCAGAAAGAG ATGTGGCAAG AGATGGGGAA GAGAGAGAGA GAAAGATGGA  4650
4651 GAGACAGGAT GTCTGGCACA TGGAAGGTGC TCACTAAGTG TGTATGGAGT  4700
4701 GAATGAATGA ATGAATGAAT GAACAAGCAG ATATATAAAT AAGATATGGA  4750
4751 GACAGATGTG GGGTGTGAGA AGAGAGATGG GGGAAGAAAC AAGTGATATG  4800
4801 AATAAAGATG GTGAGACAGA AAGAGCGGGA AATATGACAG CTAAGGAGAG  4850
4851 AGATGGGGGA GATAAGGAGA GAAGAAGATA GGGTGTCTGG CACACAGAAG  4900
4901 ACACTCAGGG AAAGAGCTGT TGAATGCCTG GAAGGTGAAT ACACAGATGA  4950
4951 ATGGAGAGAG AAAACCAGAC ACCTCAGGGC TAAGAGCGCA GGCCAGACAG  5000
             / Pvu II
5001 GCAGCCAGCT GTTCCTCCTT TAAGGGTGAC TCCCTCGATG TTAACCATTC  5050
5051 TCCTTCTCCC CAACAGTTCC CCAGGGACCT CTCTCTAATC AGCCCTCTGG  5100
5101 CCCAGGCAGT CAGTAAGTGT CTCCAAACCT CTTTCCTAAT TCTGGGTTTG  5150
5151 GGTTTGGGGG TAGGGTTAGT ACCGGTATGG AAGCAGTGGG GGAAATTTAA  5200
5201 AGTTTTGGTC TTGGGGGAGG ATGGATGGAG GTGAAAGTAG GGGGTATTT   5250
5251 TCTAGGAAGT TTAAGGGTCT CAGCTTTTTC TTTTCTCTCT CCTCTTCAGG  5300
5301 ATCATCTTCT CGAACCCCGA GTGACAAGCC TGTAGCCCAT GTTGTAGGTA  5350
5351 AGAGCTCTGA GGATGTGTCT TGGAACTTGG AGGGCTAGGA TTTGGGGATT  5400
5401 GAAGCCCGGC TGATGGTAGG CAGAACTTGG AGACAATGTG AGAAGGACTC  5450
5451 GCTGAGCTCA AGGGAAGGGT GGAGGAACAG CACAGGCCTT AGTGGGATAC  5500
5501 TCAGAACGTC ATGGCCAGGT GGGATGTGGG ATGACAGACA GAGAGGACAG  5550
                              / Xho I
5551 GAACCGGATG TGGGGTGGGC AGAGCTCGAG GGCCAGGATG TGGAGAGTGA  5600
5601 ACCGACATGG CCACACTGAC TCTCCTCTCC CTCTCTCCCT CCCTCCAGCA  5650
5651 AACCCTCAAG CTGAGGGGCA GCTCCAGTGG CTGAACCGCC GGGCCAATGC  5700
                                          / Pvu II
5701 CCTCCTGGCC AATGGCGTGG AGCTGAGAGA TAACCAGCTG GTGGTGCCAT  5750
5751 CAGAGGGCCT GTACCTCATC TACTCCCAGG TCCTCTTCAA GGGCCAAGGC  5800
5801 TGCCCCTCCA CCCATGTGCT CCTCACCCAC ACCATCAGCC GCATCGCCGT  5850
5851 CTCCTACCAG ACCAAGGTCA ACCTCCTCTC TGCCATCAAG AGCCCCTGCC  5900
5901 AGAGGGAGAC CCCAGAGGGG GCTGAGGCCA AGCCCTGGTA TGAGCCCATC  5950
                        / Pvu II
5951 TATCTGGGAG GGGTCTTCCA GCTGGAGAAG GGTGACCGAC TCAGCGCTGA  6000
```

Figure 5. (*Continued on next page.*)

```
6001 GATCAATCGG CCCGACTATC TCGACTTTGC CGAGTCTGGG CAGGTCTACT   6050
                     **
6051 TTGGGATCAT TGCCCTGTGA GGAGGACGAA CATCCAACCT TCCCAAACGC   6100
6101 CTCCCCTGCC CCAATCCCTT TATTACCCCC TCCTTCAGAC ACCCTCAACC   6150
                                                / Hind III
6151 TCTTCTGGCT CAAAAAGAGA ATTGGGGGCT TAGGGTCGGA ACCCAAGCTT   6200
6201 AGAACTTTAA GCAACAAGAC CACCACTTCG AAACCTGGGA TTCAGGAATG   6250
                                      / Eco RI
6251 TGTGGCCTGC ACAGTGAAGT GCTGGCAACC ACTAAGAATT CAAACTGGGG   6300
6301 CCTCCAGAAC TCACTGGGGC CTACAGCTTT GATCCCTGAC ATCTGGAATC   6350
                                             / Pst I
6351 TGGAGACCAG GGAGCCTTTG GTTCTGGCCA GAATGCTGCA GGACTTGAGA   6400
6401 AGACCTCACC TAGAAATTGA CACAAGTGGA CCTTAGGCCT TCCTCTCTCC   6450
6451 AGATGTTTCC AGACTTCCTT GAGACACGGA GCCCAGCCCT CCCCATGGAG   6500
6501 CCAGCTCCCT CTATTTATGT TTGCACTTGT GATTATTTAT TATTTATTTA   6550
6551 TTATTTATTT ATTTACAGAT GAATGTATTT ATTTGGGAGA CCGGGGTATC   6600
6601 CTGGGGGACC CAATGTAGGA GCTGCCTTGG CTCAGACATG TTTTCCGTGA   6650
6651 AAACGGAGCT GAACAATAGG CTGTTCCCAT GTAGCCCCCT GGCCTCTGTG   6700
6701 CCTTCTTTTG ATTATGTTTT TTAAAATATT TATCTGATTA AGTTGTCTAA   6750
6751 ACAATGCTGA TTTGGTGACC AACTGTCACT CATTGCTGAG CCTCTGCTCC   6800
                                                 poly(A)
6801 CCAGGGGAGT TGTGTCTGTA ATCGCCCTAC TATTCAGTGG CGAGAAATAA   6850
6851 AGTTTGCTTA GAAAAGAAAC ATGGTCTCCT TCTTGGAATT AATTCTGCAT   6900
6901 CTGCCTCTTC TTGTGGGTGG GAAGAAGCTC CCTAAGTCCT CTCTCCACAG   6950
6951 GCTTTAAGAT CCCTCGGACC CAGTCCCATC CTTAGACTCC TAGGGCCCTG   7000
7001 GAGACCCTAC ATAAACAAAG CCCAACAGAA TATTCCCCAT CCCCCAGGAA   7050
7051 ACAAGAGCCT GAACCTAATT ACCTCTCCCT CAGGGCATGG GAATTTCCAA   7100
            / Eco RI
7101 CTCTGGGAAT TC                                             7150
```

sequences on the 30-kb chromosomal segment shown in Figure 6. Southern hybridization with nick-translated total human DNA and subcloned *Alu* repeats (BLUR 8; Deininger et al. 1981) demonstrated that highly repetitive sequences surrounded the TNF locus shown in Figure 4. *Alu* repeats were mapped to the 4-kb *Hin*dIII fragment downstream from the sequenced TNF locus and to the 1.9-kb *Eco*RI fragment just upstream of the TNF-β gene (Fig. 6). Hybridization with total human DNA failed to reveal any repeats in the 7-kb segment shown in Figure 4.

Finally, we have hybridized certain subcloned internal fragments of the TNF locus to genomic blots with digested human DNA. Our data (Fig. 7) are consistent with the conclusion that sequences in *Eco*RI fragments 2.8, 2.4, and 1.0 kb long (TNF-α, TNF-β, and intergenic sequence, respectively; see Fig. 4) are unique in the human genome. On the other hand, we have preliminary evidence that certain parts of the TNF-α gene cross-hybridize to a not-yet-identified single-copy sequence in the human genome.

DISCUSSION

We have presented data on close linkage and tandem arrangement of the TNF-α/cachectin and TNF-β/lymphotoxin genes in the human genome. The close proximity between TNF genes first became evident in the course of screening of genomic libraries. Of the seven different recombinant clones we have characterized in detail, four turned out to hybridize to both TNF-α- and TNF-β-specific probes. These clones shared at least four *Eco*RI fragments (including those 2.8, 2.4, 1.0, and 0.8 kb long). Southern analysis revealed that two larger internal fragments corresponded to TNF-α and TNF-β genes, respectively (see Fig. 4). We have also found TNF-α-positive and TNF-β-negative clones, and vice versa (see Fig. 6). Restriction enzyme analysis confirmed that these recombinant clones contained overlapping inserts of human DNA each approximately 15 kb long.

If a set of randomly overlapping clones with the length of insertion L hybridize to two short genomic

Table 1. Alterations in Nucleotide Sequences between Figure 5 and Nedwin et al. (1985b)

No.	Nucleotide number in Fig. 5	Nucleotide	
		Fig. 5	Nedwin et al. (1985b)
1	1069	A	G
2	1540	C (Thr)	A (Asn)
3	3477–3479	CGG	–
4	3496–3498	CGG	–
5	3507	C	–
6	3518	C	–
7	3527	G	–
8	3590	G	A
9	3637	C	–
10	4508/4509	–	A
11	4569/4570	–	A
12	4583	G	A
13	4927/4928	–	C

sequences separated by distance l on the same chromosome, then one can easily calculate that the ratio of double-positive clones to the total number of positive clones N is given by the simple formula $N = L - l/L + l$. Therefore, if N is known from screening of a large representative genomic library with good statistics, then one can estimate the unknown distance between two sequences $l = L\ (1 - N)/1 + N$. Since we analyzed in detail seven recombinant clones positive for at least one probe, from the genomic library with $L = 15$ kb, our value $N = 4/7$ is only a rough estimation; however, one can still calculate that the fourth exon of TNF-α and the second exon of TNF-β are separated by approximately 4 kb. Note that sequences of oligonucleotides used as hybridization probes for TNF-α and TNF-β genes came from internal regions of *Eco*RI fragments 2.8 kb and 2.4 kb long (see above and Fig. 4). Therefore, these *Eco*RI fragments are separated from each other on the human genomic map by a distance even smaller than 4 kb. All of these estimations proved to be correct when the map of the human TNF locus was constructed. The actual distance between the TNF-β probe (oligonucleotides 5 and 6; Fig. 4) and the TNF-α probe (oligonucleotides 1 and 2; Fig. 4) is 4.5 kb.

We have shown that the TNF-α/cachectin gene and the TNF-β/lymphotoxin gene are tandemly arranged in a short segment of the human genome. As discussed above, most of recombinant λ bacteriophages from standard human genomic libraries that contain any of two TNF genes contain both of them. Therefore, it seems surprising that the close proximity of the two genes has been overlooked in recent studies. The similar structure, close linkage, and tandem arrangement of the TNF genes reinforce the fact that they are functionally related and probably evolved after duplication from a common ancestor. The structure of both genes and their mutual arrangement seem to be conserved in evolution. Recent data show that the map of the TNF locus in mice closely resembles that of its human counterpart, with almost the same distance between TNF-α and TNF-β (S.A. Nedospasov and C.V. Jongeneel, unpubl.). Sequencing analysis of the mouse TNF locus is under way.

TNF-α/cachectin and TNF-β/lymphotoxin genes are shown to be expressed in distinct cell populations of the immune system. If this differential expression is maintained at the level of transcription, then it would be interesting to find regulatory factors common to both genes and those specific for each of them and to

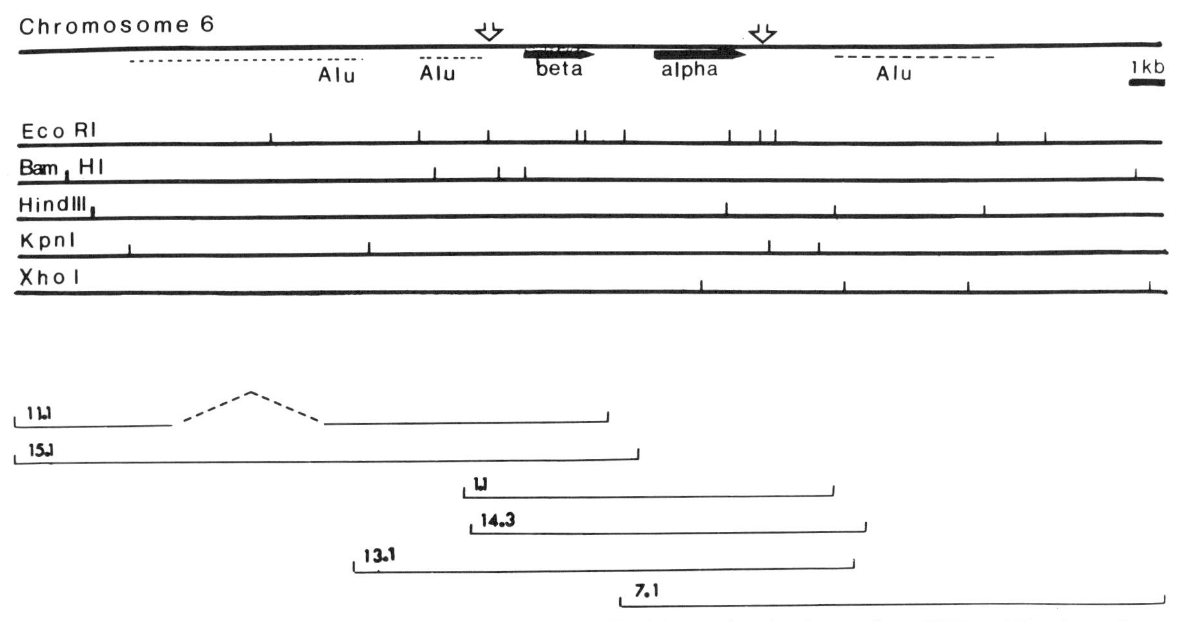

Figure 6. Map of the 30-kb locus of human chromosome 6 defined by overlapping inserts from TNF-positive phage clones. Horizontal arrows correspond to sequences coding mRNAs for TNF-β and TNF-α genes. Vertical arrows indicate the 7-kb TNF locus with known sequence (see Figs. 4 and 5). Dashed lines indicate restriction fragments that contain highly repetitive *Alu* sequences.

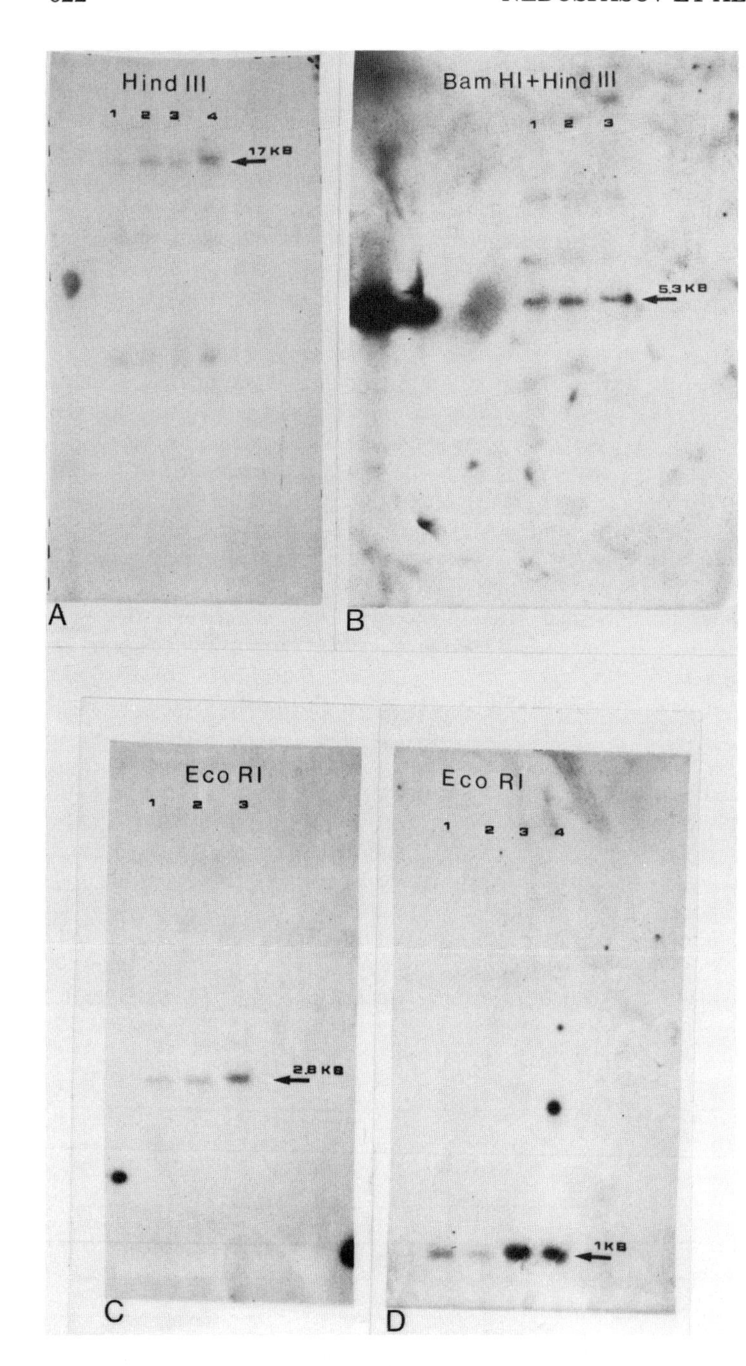

Figure 7. Examples of Southern hybridization with restriction-enzyme-digested genomic DNA. (*A*) *Hin*dIII digests of four DNAs from four human placentas, probed by nick-translated TNF-β probe (*Eco*RI fragment, 2.4 kb long). Arrow indicates position of the 17-kb band that is compatible with the map of the insert from clone 15.1 (see Fig. 6). (*B*) *Bam*HI/*Hin*dIII double digest of three placental DNAs probed with the TNF-α *Eco*RI fragment (2.8 kb). Two slots at the left correspond to different dilutions of recombinant phage DNA (clone 14.3) digested with *Bam*HI/*Hin*dIII. Position of the 5.3-kb fragment is indicated. Upper bands result from partial digestion with *Bam*HI. (*C*) *Eco*RI digest of human DNAs shown in *A* probed with the nick-translated *Xho*I/*Eco*RI 0.7-kb fragment, containing the fourth exon of TNF-α. (*D*) The same, but probed with the *Eco*RI 1.0-kb fragment containing intergenic sequences (Fig. 4).

localize their target sequences inside the TNF locus. In this regard, a detectable homology has been noted between putative control regions of these genes.

Possible linkage of TNF genes to the MHC locus on human chromosome 6 (Kaufman et al. 1984) might also be of functional importance to coordinate regulation of the immune response. More data on the structural organization of corresponding human loci are needed to prove or disprove this intriguing possibility.

ACKNOWLEDGMENT

We thank our many colleagues for advice, sharing materials, and helpful discussions.

REFERENCES

Aggarwal, B.B., T.E. Eessalu, and P.E. Hass. 1985a. Characterization of receptors for human tumor necrosis factor and their regulation by γ-interferon. *Nature* **318:** 665.

Aggarwal, B.B., B. Moffat, and R.N. Harkins. 1984. Human lymphotoxin. *J. Biol. Chem.* **259:** 689.

Aggarwal, B.B., W.J. Henzel, B. Moffat, W.J. Kohr, and R.N. Harkins. 1985b. Primary structure of human lymphotoxin derived from 1788 lymphoblastoid cell line. *J. Biol. Chem.* **260:** 2334.

Aggarwal, B.B., W.J. Kohr, P.E. Hass, B. Moffat, S.A. Spencer, W.J. Henzel, T.S. Bringman, G.E. Nedwin, D.V. Goeddel, and R.N. Harkins. 1985c. Human tumor necrosis factor. Production, purification, and characterization. *J. Biol. Chem.* **260:** 2345.

Barnes, W.M., M. Bevan, and P.H. Son. 1983. Kilo-sequencing: Creation of an ordered nest of asymmetric deletions across a large target sequence carried on phage M13. *Methods Enzymol.* **101:** 98.

Barone, A.D., J.-Y. Tang, and M.H. Caruthers. 1984. In situ activitation of bis-dialkyl-aminophosphines—A new method for synthesizing deoxyoligonucleotides on polymer supports. *Nucleic Acids Res.* **12:** 4051.

Buetler, B. and A. Cerami. 1986. Cachectin and tumor necrosis factor as two sides of the same biological coin. *Nature* **320:** 584.

Beutler, B., D. Greenwald, J.D. Hulmes, M. Chang, Y.-C. Pan, J. Mathison, R. Ulevitch, and A. Cerami. 1985. Identity of tumor necrosis factor and the macrophage-secreted factor cachectin. *Nature* **316:** 552.

Carswell, E.A., L.J. Old, R.J. Kassel, S. Green, N. Fiore, and B. Williamson. 1975. An endotoxin-induced serum factor that causes necrosis of tumors. *Proc. Natl. Acad. Sci.* **72:** 3666.

Chroboczek Kelker, H.C., J.D. Oppenheim, D. Stone-Wolff, D. Henriksen-de Stefano, B.B. Aggarwal, H.C. Stevenson, and J. Vilcek. 1985. Characterization of human tumor necrosis factor produced by peripheral blood monocytes and its separation from lymphotoxin. *Int. J. Cancer* **36:** 69.

Chuvpilo, S.A. and V.V. Kravchenko. 1983. A solid-phase method for sequencing DNA. *Bioorg. Khim.* **9:** 1634.

———. 1985. A simple and rapid method for sequencing DNA. *FEBS Lett.* **179:** 34.

Collins, T., L.A. Lapierre, W. Fiers, J.L. Strominger, and S. Pober. 1986. Recombinant human tumor necrosis factor increases mRNA levels and surface expression of HLA-A,B antigens in vascular endothelial cells and dermal fibroblasts in vitro. *Proc. Natl. Acad. Sci.* **83:** 446.

Dayer, J.-M., B. Buetler, and A. Cerami. 1985. Cachectin/tumor necrosis factor stimulates collagenase and prostaglandin E_2 production by human synovial cells and dermal fibroblasts. *J. Exp. Med.* **162:** 2163.

Deininger, P.L., D.J. Jolli, C.M. Rubin, T. Friedman, and C.M. Schmid. 1981. Base sequence studies of 300 nucleotide renatured repeat human DNA clones. *J. Mol. Biol.* **151:** 17.

Dobrynin, V.N., S.A. Filippov, N.S. Bystrov, I.V. Severtsova, and M.N. Kolosov. 1983. Methylimidazole catalyzed rapid phosphotriester synthesis of oligonucleotides on a silica gel support in dichloroethane. *Bioorg. Khim.* **9:** 706.

Dobrynin, V.N., V.G. Korobko, N.S. Bystrov, I.V. Severtsova, S.A. Chuvpilo, and M.N. Kolosov. 1980. Synthesis of model promoter for gene expression in *Escherichia coli*. *Nucleic Acids Res. Symp. Ser.* **7:** 365.

Filippov, S.A., S.V. Kalinichenko, N.S. Bystrov, and V.N. Dobrynin. 1987. Phosphoramidite synthesis of oligonucleotides with rich content of dG residues. *Bioorg. Khim.* (in press).

Fransen, L., P. Müller, A. Marmenout, J. Tavernier, J. Van der Heyden, E. Kawashima, A. Cholett, R. Tizard, H. Van Heuverswin, A. Van Vliet, M.-R. Ruysschaert, and W. Fiers. 1985. Molecular cloning of mouse tumor necrosis factor cDNA and its eukaryotic expression. *Nucleic Acids Res.* **13:** 4417.

Gray, P.W., B.B. Aggarwal, C.V. Benton, T.S. Bringman, W.C. Henzel, J.A. Jarett, D.W. Leung, B. Moffat, P. Ng, L.P. Svedersky, M.A. Palladino, and G.E. Nedwin. 1984. Cloning and expression of cDNA for human lymphotoxin, a lymphokine with tumor necrosis activity. *Nature* **312:** 721.

Haranaka, K., N. Satomi, and A. Sakurai. 1984. Antitumor activity of murine tumor necrosis factor (TNF) against transplanted murine tumors and heterotransplanted human tumors in nude mice. *Int. J. Cancer* **34:** 263.

Kaufman, J.F., C. Auffray, A.J. Korman, D.A. Shackelford, and J. Strominger. 1984. The class II molecules of the human and murine major histocompatibility complex. *Cell* **36:** 1.

Lawn, R.M., E.F. Fritsch, R.C. Parker, G. Blake, and T. Maniatis. 1978. The isolation and characterization of a linked δ- and β-globin gene from a cloned library of human DNA. *Cell* **15:** 1157.

Leder, P., D. Tiemeier, and L. Enquist. 1977. EK 2 derivatives of bacteriophage lambda useful in the cloning of DNA from other organisms: The λ gtWES system. *Science* **196:** 175.

Lee, S.H., B.B. Aggarwal, E. Rinderknecht, F.A. Assisi, and H. Chiu. 1984. The synergistic anti-proliferative effect of γ-interferon and human lymphotoxin. *J. Immunol.* **133:** 1.

Maniatis, T., E.F. Fritsch, and J. Sambrook. 1982. *Molecular cloning: A laboratory manual*. Cold Spring Harbor Laboratory, Cold Spring Harbor, New York.

Marmenout, A.L., L. Fransen, J. Tavernier, J. Van der Heyden, R. Tizard, E. Kawashima, A. Shaw, M.-J. Johnson, D. Semon, R. Müller, N.-R. Ruysshaert, A. Van Vliet, and W. Fiers. 1985. Molecular cloning and expression of human tumor necrosis factor and comparison with mouse tumor necrosis factor. *Eur. J. Biochem.* **152:** 515.

Maxam, A.M. and W. Gilbert. 1977. A new method for sequencing DNA. *Proc. Natl. Acad. Sci.* **74:** 560.

Nedospasov, S.A., A.N. Shakhov, and R. Sahli. 1984. Structural and functional organization of SV40 chromosome. II. Characterization of virus genome containing deletion in regulatory region. *Mol. Biol.* **18:** 889.

Nedospasov, S.A., A.N. Shakhov, R.L. Turetskaya, V.A. Mett, G.P. Georgiev, V.N. Dobrynin, and V.G. Korobko. 1985. Molecular cloning of human tumor necrosis factor gene: Tandem arrangement of TNF-alpha and TNF-beta genes in a short segment of human genome. *Dokl. Akad. Nauk. SSSR* **285:** 1487.

Nedwin, G.E., L.P. Svedersky, T.S. Bringman, M.A. Palladino, Jr., and D.V. Goeddel. 1985a. Effect of interleukin 2, interferon-γ, and mitogens on the production of tumor necrosis factors α and β. *J. Immunol.* **135:** 2492.

Nedwin, G.E., S.L. Naylor, A.J. Sakagucho, D. Smith, J. Jarett-Nedwin, D. Pennica, D.V. Goeddel, and P.W. Gray. 1985b. Human lymphotoxin and tumor necrosis factor genes: Structure, homology and chromosomal localization. *Nucleic Acids Res.* **13:** 1487.

Old, L.J. 1985. Tumor necrosis factor (TNF). *Science* **230:** 630.

Pennica, D., G.E. Nedwin, J.S. Hayflick, P.H. Seeburg, R. Derynck, M.A. Palladino, W.J. Kohr, B.B. Aggarwal, and D.V. Goeddel. 1984. Human tumor necrosis factor: Precursor structure, expression and homology to lymphotoxin. *Nature* **312:** 724.

Pober, J.S., M.P. Bevilacqua, D.L. Mendrick, L.A. Lapierre, W. Fiers, and M.A. Gimbrone, Jr. 1986. Two distinct monokines, interleukin 1 and tumor necrosis factor, each independently induces biosynthesis and transient expression of the same antigen on the surface of cultured human vascular endothelial cells. *J. Immunol.* **136:** 1680.

Prentki, O., F. Karch, S. Iida, and J. Meyer. 1981. The plasmid cloning vector pBR325 contains a 482 base-pair-long inverted duplication. *Gene* **14:** 289.

Ruff, M.R. and G.E. Gifford. 1981. Tumor necrosis factor. In *Lymphokines* (ed. E. Rick), vol. 2, p. 235. Academic Press, New York.

Ruther, U. and B. Muller-Hill. 1983. Easy identification of cDNA clones. *EMBO J.* **2:** 1791.

Sahli, R., G.K. McMaster, and B. Hirt. 1985. DNA sequence comparison between two tissue-specific variants of the autonomous parvovirus, minute virus of mice. *Nucleic Acids Res.* **13:** 3617.

Sanger, F., S. Nicklen, and A.R. Coulson. 1977. DNA sequencing with chain-terminating inhibitors. *Proc. Natl. Acad. Sci.* **74:** 5463.

Shirai, T., H. Yamaguchi, H. Ito, C.V. Todd, and R.B. Wallace. 1985. Cloning and expression in *Escherichia coli* of the gene for human tumor necrosis factor. *Nature* **313:** 803.

Southern, E. 1975. Detection of specific sequences among

DNA fragments separated by gel electrophoresis. *J. Mol. Biol.* **98:** 503.

Sugarman, B.J., B.B. Aggarwal, P.E. Hass, J.S. Figari, M.A. Paladino, Jr., and H.M. Shephard. 1985. Recombinant human tumor necrosis factor-α: Effect on proliferation of normal and transformed cells in vitro. *Science* **230:** 943.

Torti, F.M., B. Dieckmann, B. Beutler, A. Cerami, and G.M. Ringold. 1985. A macrophage factor inhibits adipocyte gene expression: An in vitro model of cachexia. *Science* **229:** 867.

Tsujimoto, M., Y.K. Yip, and J. Vilcek. 1985. Tumor necrosis factor: Specific binding and internalization in sensitive and resistant cells. *Proc. Natl. Acad. Sci.* **82:** 7626.

Vieira, J. and J. Messing. 1982. The pUC plasmids, and M13mp7-derived system for insertion mutagenesis and sequencing with synthetic universal primers. *Gene* **19:** 268.

Wang, A.M., A.A. Creasey, M.B. Ladner, L.S. Lin, J. Stickler, J.N. Van Arsdell, R. Yamamoto, and D. Mark. 1985. Molecular cloning of the complementary DNA for human tumor necrosis factor. *Science* **228:** 5397.

Williamson, B.D., E.A. Carswell, B.Y. Rubin, J.S. Prendergast, and L.J. Old. 1983. Human tumor necrosis factor produced by human B-cell lines: Synergistic cytotoxic interaction with human interferon. *Proc. Natl. Acad. Sci.* **80:** 5397.

Woo, S.L.C. 1979. A sensitive and rapid method for recombinant phage screening. *Methods Enzymol.* **68:** 389.

Cachectin: The Dark Side of Tumor Necrosis Factor

A. CERAMI AND B. BEUTLER*
Laboratory of Medical Biochemistry, The Rockefeller University, New York, New York 10021

Animals or humans infected with various bacteria, parasites, or viruses develop a catabolic state that, if unremitting, can advance to severe cachexia (wasting), shock, and death. Over the years, various explanations for this commonly observed phenomenon have been offered. Prevalent among these was the concept that the organisms were usurping the energy supply of the host or producing a toxic metabolite. Several years ago, we began to question these explanations, since cattle and rabbits infected with parasites of the Trypanosomatidae family undergo a profound wasting diathesis while exhibiting a relatively low parasite burden. Studies of the metabolic changes evoked by *Trypanosoma brucei* in rabbits revealed a marked lipemia in the last few weeks of this terminal disease. This was quite surprising since the animals had complete anorexia and resorption of body fat stores. Investigation into the mechanism of this paradoxical hypertriglyceridemia revealed a deficiency of the enzyme lipoprotein lipase (LPL), which is essential for clearing triglyceride from the blood (Rouzer and Cerami 1980). In the rabbit, very low density lipoprotein (VLDL) synthesis by the liver continues unabated, leading to very high VLDL levels in the plasma in the cachectic state. Other species (e.g., the mouse and rat) apparently compensate by reducing the production of VLDL and do not achieve the degree of lipemia observed in rabbits.

The first indication that an endogenous mediator caused the loss of LPL was obtained through the use of two closely related strains of mice, C3He/HeN and C3H/HeJ, which are sensitive and resistant, respectively, to the administration of endotoxin (lipopolysaccharide [LPS]). When endotoxin-sensitive mice (C3He/HeN) are injected with nonlethal doses of endotoxin, they develop a systemic LPL deficiency and a mild hypertriglyceridemia, whereas endotoxin-resistant mice (C3H/HeJ) do not (Kawakami and Cerami 1981). On the other hand, a loss of LPL activity could be elicited in C3H/HeJ mice if they were injected with serum from endotoxin-o-sensitive animals that had been treated 2 hours previously with endotoxin (Kawakami and Cerami 1981). Moreover, a factor capable of suppressing LPL in the C3H/HeJ mice could be produced in vitro by endotoxin treatment of macrophages obtained from sensitive mice (Kawakami et al. 1982).

This macrophage factor could also suppress LPL activity on the surface and within cells of the adipocyte line 3T3-L1 (Kawakami et al. 1982). In fact, this activity served as the basis for the subsequent isolation of the monokine, cachectin (Beutler et al. 1985b). In addition to endotoxin, products of parasites (e.g., *T. brucei*, *Plasmodium berghei*) were able to elicit expression of the LPL-suppressing factor by macrophages (Hotez et al. 1984). Other key lipogenic enzymes including acetyl-CoA carboxylase and fatty acid synthetase were suppressed by the endotoxin-induced monokine (Pekala et al. 1983). Radiolabeling and specific immunoprecipitation experiments implicated an inhibition of the synthesis of these key anabolic protcins. Total protein synthesis continued without reduction in the presence of the monokine, although the synthesis of these specific proteins diminished rapidly.

Subsequent studies with specific cDNA clones pointed to the ability of this monokine to suppress biosynthesis of a number of differentiation-specific mRNA molecules (e.g., glycerol phosphate dehydrogenase and the fatty acid binding protein) present in the adipocyte (Torti et al. 1985). This suppression was specific since transcription rates of "household" mRNA species, e.g., actin, were not affected. In addition, evidence was obtained indicating that the monokine had the ability to activate selectively synthesis of specific gene products.

The mechanism for this switching on and off of particular genes is not known. Apparently, a repertoire of genes is altered that switches the metabolism of the cell from an anabolic state to a catabolic state. Addition of the monokine to 3T3-L1 preadipocytes also prevents the morphological differentiation to form adipocytes, whereas the addition to mature adipocytes laden with lipid prompts a loss of lipid droplets and a form of "in vitro cachexia" (Torti et al. 1985). The monokine-prompted change in cellular metabolism is reversible; when removed, these cells switch back to an anabolic mode of biosynthetic machinery and become laden with lipid again. The ability of the monokine to selectively switch mRNA biosynthesis in the fat cell presumably reflects a more generalized action of this monokine on cells. Further work is needed to define more accurately the manner in which this monokine can achieve these effects on differentiation and host metabolism.

To understand the monokine better, it was isolated from the culture medium of the murine macrophage line RAW 264.7, following stimulation of the cells by endotoxin (Beutler et al. 1985b). The protein was purified to homogeneity and was found to have a molec-

*Present address: The Howard Hughes Medical Institute, The University of Texas Health Science Center at Dallas, 5323 Harry Hines Boulevard, Dallas, Texas 75235.

ular weight of 17,000 and a pI of 4.7. We named the protein cachectin because of its presumed role in cachexia. Cachectin is one of the major secretory proteins produced by the macrophage in response to endotoxin, accounting for 1–2% of the total secretory protein.

Purified cachectin was radiolabeled with ^{125}I using the iodogen method without significant loss of biological activity (Beutler et al. 1985b). Using this radiolabeled material, we were able to define approximately 10^4 high-affinity receptors on 3T3-L1 adipocytes and C-2 myoblasts with an affinity estimated at 3×10^9 M^{-1}. The similarity in molecular weight and pI of cachectin and another monokine, IL-1, suggested a possible relationship. This was ruled out by showing that IL-1 could not displace radiolabeled cachectin from its receptor and that cachectin had no lymphocyte-activating factor (LAF) activity.

A strong similarity was observed, however, when the amino acid sequence of murine cachectin was compared with that of the monokine tumor necrosis factor (TNF), derived from human cells (Beutler et al. 1985c); 14 of the first 19 residues were identical. Subsequent work has confirmed the identity of these two monokines. Several groups (Fransen et al. 1985; Pennica et al. 1985; Caput et al. 1986) have reported the complete DNA sequence of mouse TNF cDNA, which exactly predicts the amino-terminal sequence of mouse cachectin (Beutler et al. 1985c).

The primary structure of cachectin/TNF has been reported for three species (Pennica et al. 1984, 1985; Fransen et al. 1985; Shirai et al. 1985; Wang et al. 1985; Caput et al. 1986; Ito et al. 1986a,b). The mature human protein contains 157 residues, the mouse protein contains 156 residues, and the rabbit protein contains 154 residues. (Approximately 80% identity is noted on comparison of the primary structure of human and mouse or human and rabbit proteins.) An even stronger homology (86%) is noted when comparing the residues of the rather long propeptide sequence attached to the amino terminus of the mature protein. At present, it is not known whether this propeptide or fragments derived from it have biological activity. A highly conserved region between residues 114 and 130 of the mature protein is also highly conserved in the lymphokine, lymphotoxin. This region may contribute to the receptor binding site, since lymphotoxin has been noted to bind to the same receptor as cachectin/TNF (Aggarwal et al. 1985).

In addition to the conservation noted between cachectin proteins derived from different species, a conserved sequence was noted in the 3′-untranslated region downstream from the termination codon of the cachectin/TNF gene (Caput et al. 1986). The sequence consisted of tandem and overlapping repeats of the octamer TTATTTAT. This conserved sequence is uncommon among mammalian genes sequenced to date but was found in repeats of varying lengths in the 3′-untranslated segment of mRNAs specifying several inflammatory mediators. These include the mRNAs encoding lymphotoxin, all of IL-1 monokines, granulocyte-macrophage colony-stimulating factor, and most of the α, β, and γ interferons analyzed to date. We have postulated that this sequence may play a regulatory role, in which it is perhaps involved in the translational control of these mediators (Beutler et al. 1986a; Caput et al. 1986).

Biological Activities

The availability of recombinant cachectin (TNF) has permitted a number of investigations into its biological activities. In fact, it is fair to anticipate that most of the activities are yet to be determined. From the data reported at present, it appears that cachectin affects many cell types in the body.

One of the interesting aspects of the biological activities associated with cachectin is the overlap with those described previously for IL-1. These two monokines are not structurally related and do not bind to the same receptor. Yet, cachectin can function as an endogenous pyrogen (Dinarello et al. 1986). When injected into rabbits, cachectin prompts a biphasic fever. The first is the result of a direct effect on hypothalamic neurons, and the second peak reflects the cachectin-induced release of IL-1.

Cachectin also shares with IL-1 the ability to stimulate the release of collagenase and prostaglandin E_2 (PGE_2) by human rheumatoid synovial cells and dermal fibroblasts (Dayer et al. 1985). These effects are believed to be important in the pathogenesis of inflammatory diseases of the joints and skin (Krane et al. 1982). In addition, IL-1 and cachectin are capable of stimulating release of Ca^{++} from osteocytes in vitro and account for the osteoclast-activating factor (OAF) activity (Bertolini et al. 1986).

Both IL-1 and cachectin can alter the hemostatic properties of endothelial cells. Stern and colleagues have shown that these monokines can induce procoagulant activity and decrease the expression of thrombomodulin on cultured endothelial cells (Stern and Nawroth 1986). This converts the endothelial cell from an anticoagulating surface to a coagulation-inducing surface by stimulating the intrinsic coagulation pathway and by decreasing the rate of thrombomodulin-activated protein C formation.

A number of recent papers point to a major effect of cachectin/TNF on other leukocytes. The ability of neutrophils to phagocytize latex beads and lyse antibody-coated chicken red cells is significantly augmented by cachectin. Cachectin also promotes the adherence of neutrophils to endothelial cells (Gamble et al. 1985; Shalaby et al. 1985). Analysis of these phenomena revealed effects on both the neutrophil and endothelial cells. The effect on the neutrophil was rapid (occurring in less than 5 min) and did not require protein or RNA synthesis. The effect on the endothelial cells was maximal only after 4 hours following exposure to cachectin and required RNA and protein synthesis. As described below, administration of cachectin to animals leads to neutrophil margination and transudation. Cachectin has also been shown to activate eosinophils, enhancing their ability to kill schistosomula in vitro (Silberstein and David 1986). Monocytes and macrophages are themselves sensitive to cachectin, producing IL-1 in vitro and in vivo (Dinarello et al. 1986).

Recent studies (Degliantoni et al. 1985) suggest that cachectin/TNF is produced in small amounts by natural killer (NK) cells and is responsible for the myeloid colony-inhibiting activity (NK-CIA) produced by NK cells.

The wasting of muscle mass associated with infection points to a possible effect of cachectin on muscle cells. Although 10^4 high-affinity receptors per cell are present on myotubules grown in culture (Beutler et al. 1985b), little is known of the biochemical changes that are prompted by cachectin. Recently, Tracey et al. (1986a) reported that cachectin can lower the transmembrane potential of muscle cells, consistent with the observation of a decreased muscle-cell membrane potential in septic individuals. It is not known whether this effect of cachectin results from increased membrane permeability or from inefficiency of the Na^+/K^+-dependent ATPase pump. It will be of interest to see if this effect on membrane potential is shared by other cells as well.

Pathogenesis of Septic Shock

One of the most important of the biological activities of cachectin is its apparent ability to mediate the lethal effects of endotoxin. It has been known for many years that an injection of endotoxin (the LPS portion of gram-negative bacteria) can evoke a lethal effect in mammals. The infusion of microgram quantities of LPS into mice, rabbits, dogs, and certain primates results in fever, leukopenia, coagulation abnormalities, end-organ injury, shock, and death. The occurrence of these findings in patients or animals with septicemia suggests that endotoxin is the agent responsible for the death of infected individuals. Although many investigators originally believed that LPS was directly toxic to mammalian tissues, this is no longer a widely held idea. Recent evidence has pointed to an important role of the hematopoietic system.

Insight into the nature of endotoxin toxicity has been gained from studies of endotoxin-sensitive (C3H/HeN) and -resistant (C3H/HeJ) mice. The resistant mouse differs from the sensitive mouse by a single codominant allele (*lps*^{d}) that imparts an ability to withstand the effects of several milligrams of injected endotoxin. Endotoxin sensitivity can be restored to C3H/HeJ mice if the mice are reconstituted hematopoietically after irradiation by infusion of bone marrow from C3H/HeN donors. Similarly irradiated C3H/HeN mice reconstituted with C3H/HeJ marrow are resistant to endotoxin. The macrophage has been strongly implicated as the cell mediating endotoxin sensitivity. Hyperplasia of the reticuloendothelial system induced by Bacillus Calmette-Guerin (BCG) and *Mycobacterium lepraemurium* imparts a greatly increased sensitivity to the lethal effects of endotoxin (Vogel et al. 1980; Ha et al. 1983). Indication of a macrophage mediator was obtained recently when it was shown that peritoneal macrophages from sensitive mice could produce a soluble factor in response to endotoxin, which could kill the endotoxin-resistant mice (Cerami et al. 1985). An obvious candidate in defining the molecular identity of this lethal mediator was cachectin.

Indications for the role of cachectin were first obtained by studies using rabbit antiserum to mouse cachectin. Passive immunization of endotoxin-sensitive mice rendered these mice resistant to the effects of endotoxin (Beutler et al. 1985a). Maximum protection was afforded to those animals that received the antiserum prior to administration of endotoxin. The rapid induction of a lethal effect is consistent with the rapid kinetics of cachectin production and clearance. Although several hours pass before an animal succumbs to endotoxin, the lethal injury is instituted within a short time after cachectin production.

The availability of recombinant human cachectin has allowed us to confirm that this monokine acts as a major mediator of endotoxic shock (Tracey et al. 1986b). Injection of recombinant human cachectin (600 μg/kg) into rats produces fever, hypotension, and a relentless metabolic acidosis coupled with a respiratory alkalosis. Several hours after the infusion, the animals succumb to respiratory arrest. In the period prior to death the animals also display transient hyperglycemia (with glucose concentrations exceeding 500 mg/dl in some instances) and hemoconcentration. At lower doses of cachectin (200 μg/kg) the rats display tachypnea and respiratory alkalosis without the severe metabolic acidosis.

At necropsy, the animals receiving cachectin have inflammatory lesions of the lungs, kidneys, and gastrointestinal tract. The large bowel is often ischemic or infarcted, and punctate hemorrhages are observed in the lungs. Microscopically, these lesions are marked by massive infiltration of polymorphonuclear cells in the lungs and bowel and by acute tubular necrosis in the kidney. The pathologic findings induced by cachectin reproduce the effects noted in animals receiving endotoxin.

When cachectin is repeatedly administered in sublethal doses over a period of several days, weight loss and anorexia are generally observed. It has not as yet been possible to measure the production of cachectin in vivo in chronic wasting illnesses, perhaps because of the brief half-life of the hormone in the circulation. However, primates treated with *Escherichia coli* generate lethal quantities of the hormone during the early stages of sepsis (A. Cerami and B. Beutler, unpubl.). Studies on septic and otherwise debilitated human subjects are currently in progress.

Control of Cachectin Biosynthesis

The availability of specific antibodies to cachectin, and a cDNA probe capable of recognizing cachectin mRNA, has allowed a study of the mechanisms of endotoxin resistance, as it occurs in the C3H/HeJ mouse and in glucocorticoid-treated animals (Beutler et al. 1986a). Macrophages obtained from the C3H/HeJ mice fail to exhibit normal cachectin gene transcription in response to LPS, until high concentrations of LPS (1 μg/ml) are utilized. However, even under these conditions, when considerable amounts of cachectin mRNA are present within the cells, none of the mature protein is formed. Thus, the endotoxin resistance of these animals is the result of a dual lesion involving both cachectin gene transcription and cachectin mRNA translation.

Interestingly, it is possible to circumvent partially the effects of the C3H/HeJ defects by pretreating in vitro the macrophages with interferon-γ (Beutler et al. 1986b). This lymphokine, which itself cannot induce cachectin production, augments the LPS-induced biosynthesis of cachectin mRNA and relieves the translational blockade that precludes cachectin biosynthesis. The mechanism by which interferon-γ can circumvent the lesion imposed by the *lps*^{d} mutation is not presently understood.

Pretreatment of macrophages from endotoxin-sensitive mice with the glucocorticoid, dexamethasone, reduces cachectin transcription and completely prevents the translation of any cachectin mRNA that escapes the transcriptional blockade (Beutler et al. 1986a). The inhibiting effect on translation is only effective if dexamethasone is added prior to LPS activation. Once activated, macrophages are indifferent to the addition of dexamethasone and express cachectin in normal amounts. This finding is consistent with the clinical observation that steroids are most effective if administered early in the course of sepsis.

The cachectin gene has been conserved throughout mammalian evolution, suggesting that it must confer some selective advantage upon the individual or group. Yet, from the foregoing discussion, it is apparent that cachectin is lethal when produced in large quantities (as occurs, for example, in endotoxemic states). It has previously been suggested that at low doses, cachectin may help to mobilize energy reserves for use by cells of the immune system (Kawakami and Cerami 1981). Moreover, neutrophil (Gamble et al. 1985; Shalaby et al. 1985) and eosinophil (Silberstein and David 1986) activation may serve a useful purpose. Further studies are needed before it will be possible to understand how this monokine can benefit the mammalian host.

Pharmacotherapeutic Applications

Oncologists have shown considerable interest in the use of cachectin/TNF as an antineoplastic agent. Indeed, the isolation of TNF was predicated on the assumption that the molecule would prove to be devoid of endotoxin activities while retaining the tumorolytic activity evoked by endotoxin. It is now quite clear, however, that in addition to its role as a mediator of tumorolysis, cachectin is one of the principal endogenous mediators of endotoxic shock (Tracey et al. 1986b).

It remains to be seen whether a tumor-necrotizing effect can ever be achieved without inducing the deleterious effects evoked by LPS. Indeed, the sensitivity of some tumors to the effects of cachectin/TNF in vivo most likely reflects their more tenuous vascular supply and, perhaps, their greater susceptibility to the thrombotic effects of the hormone. However, a precise understanding of the mechanism by which cachectin/TNF can selectively destroy tumor cells in vitro may ultimately be helpful in the design of novel antineoplastic drugs.

On the other hand, control of cachectin production and inhibition of its effects at the level of the receptor and beyond may offer a new approach to the management of septic shock and other inflammatory states. Current studies of the mechanism by which cachectin biosynthesis is controlled, and of the events that follow hormone-receptor interaction, are sure to provide valuable insight into these important clinical problems.

ACKNOWLEDGMENTS

We acknowledge the material support provided by Chiron Corporation of Emeryville, California. This work was also funded, in part, by National Institutes of Health grants AM-01314 and AI-21359.

REFERENCES

Aggarwal, B.B., T.E. Eessalu, and P.E. Hass. 1985. Characterization of receptors for human tumour necrosis factor and their regulation by gamma-interferon. *Nature* **318:** 665.

Bertolini, D.R., G. Nedwin, T. Bringman, D. Smith, and G.R. Mundy. 1986. Stimulation of bone resorption and inhibition of bone formation in vitro by human tumour necrosis factor. *Nature* **319:** 516.

Beutler, B., I.W. Milsark, and A. Cerami. 1985a. Passive immunization against cachectin/tumor necrosis factor (TNF) protects mice from the lethal effect of endotoxin. *Science* **229:** 869.

Beutler, B., N. Krochin, I.W. Milsark, C. Luedke, and A. Cerami. 1986a. Control of cachectin (tumor necrosis factor) synthesis: Mechanisms of endotoxin resistance. *Science* **232:** 977.

Beutler, B., J. Mahoney, N. Le Trang, P. Pekala, and A. Cerami. 1985b. Purification of cachectin, a lipoprotein lipase-suppressing hormone secreted by endotoxin-induced RAW 264.7 cells. *J. Exp. Med.* **161:** 984.

Beutler, B., V. Tkacenko, J.W. Milsark, N. Krochin, and A. Cerami. 1986b. The effect of interferon-gamma on cachectin expression by mononuclear phagocytes: Reversal of the *lps*[d] (endotoxin resistance) phenotype. *J. Exp. Med.* (in press).

Beutler, B., D. Greenwald, J.D. Hulmes, M. Chang, Y.-C.E. Pan, J. Mathison, R. Ulevitch, and A. Cerami. 1985c. Identity of tumour necrosis factor and the macrophage-secreted factor cachectin. *Nature* **316:** 552.

Caput, D., B. Beutler, K. Hartog, S. Brown-Shimer, and A. Cerami. 1986. Identification of a common nucleotide sequence in the 3′-untranslated region of mRNA molecules specifying inflammatory mediators. *Proc. Natl. Acad. Sci.* **83:** 1670.

Cerami, A., Y. Ikeda, N. Le Trang, P.J. Hotez, and B. Beutler. 1985. Weight loss associated with an endotoxin-induced mediator from peritoneal macrophages: The role of cachectin (tumor necrosis factor). *Immunol. Lett.* **11:** 173.

Dayer, J.-M., B. Beutler, and A. Cerami. 1985. Cachectin/tumor necrosis factor (TNF) stimulates collagenase and PGE2 production by human synovial cells and dermal fibroblasts. *J. Exp. Med.* **162:** 2163.

Degliantoni, G., M. Murphy, M. Kobayashi, M.K. Francis, B. Perussia, and G. Trinchieri. 1985. Natural killer (NK) cell-derived hematopoietic colony-inhibiting activity and NK cytotoxic factor. Relationship with tumor necrosis factor and synergism with immune interferon. *J. Exp. Med.* **162:** 1512.

Dinarello, C.A., J.G. Cannon, S.M. Wolff, H.A. Bernheim, B. Beutler, A. Cerami, M.A. Palladino, and J.V. O'Connor. 1986. Tumor necrosis factor (cachectin) is an endogenous pyrogen and induces production of interleukin-1. *J. Exp. Med.* **163:** 1433.

Fransen, L., R. Muller, A. Marmenout, J. Tavernier, J. Van der Heyden, E. Kawashima, A. Chollet, R. Tizard, H. Van Heuverswyn, A. Van Vliet, M.-R. Ruysschaert, and W. Fiers. 1985. Molecular cloning of mouse tumour necrosis factor cDNA and its eukaryotic expression. *Nucleic Acids Res.* **13:** 4417.

Gamble, J.R., J.M. Harlan, S.J. Klebanoff, A.F. Lopez, and M.A. Vadas. 1985. Stimulation of the adherence of neutrophils to umbilical vein endothelium by human recombinant tumor necrosis factor. *Proc. Natl. Acad. Sci.* **82:** 8667.

Ha, D.K., I.D. Gardner, and J.W. Lawton. 1983. Characterization of macrophage function in *Mycobacterium lepraemurium*-infected mice: Sensitivity of mice to endotoxin and release of mediators and lysosomal enzymes after endotoxin treatment. *Parasite Immunol.* **5:** 513.

Hotez, P.J., N. Le Trang, A.H. Fairlamb, and A. Cerami. 1984. Lipoprotein lipase suppression in 3T3-L1 cells by a haematoprotozoan-induced mediator from peritoneal exudate cells. *Parasite Immunol.* **6:** 203.

Ito, H., T. Shirai, S. Yamamoto, M. Akira, S. Kawahara, C.W. Todd, and R.B. Wallace. 1986a. Molecular cloning of the gene encoding rabbit tumor necrosis factor. *DNA* **5:** 157.

Ito, H., S. Yamamoto, S. Kuroda, H. Sakamoto, J. Kajihara, T. Kiyota, H. Hayashi, M. Kato, and M. Seko. 1986b. Molecular cloning and expression in *Escherichia coli* of the cDNA coding for rabbit tumor necrosis factor. *DNA* **5:** 149.

Kawakami, M. and A. Cerami. 1981. Studies of endotoxin-induced decrease in lipoprotein lipase activity. *J. Exp. Med.* **154:** 631.

Kawakami, M., P.H. Pekala, M.D. Lane, and A. Cerami. 1982. Lipoprotein lipase suppression in 3T3-L1 cells by an endotoxin-induced mediator from exudate cells. *Proc. Natl. Acad. Sci.* **79:** 912.

Krane, S.M., S.R. Goldring, and J.-M. Dayer. 1982. Interactions among lymphocytes, monocytes and other synovial cells in the rheumatoid synovium. In *Lymphokines* (ed. E. Pick), vol. 7, p. 75. Academic Press, New York.

Pekala, P.H., M. Kawakami, C.W. Angus, M.D. Lane, and A. Cerami. 1983. Selective inhibition of synthesis of enzymes for de novo fatty acid biosynthesis by an endotoxin-induced mediator from exudate cells. *Proc. Natl. Acad. Sci.* **80:** 2743.

Pennica, D., J.S. Hayflick, T.S. Bringman, M.A. Palladino, and D.V. Goeddel. 1985. Cloning and expression in *Escherichia coli* of the cDNA for murine tumor necrosis factor. *Proc. Natl. Acad. Sci.* **82:** 6060.

Pennica, D., G.E. Nedwin, J.S. Hayflick, P.H. Seeburg, R. Derynck, M.A. Palladino, W.J. Kohr, B.B. Aggarwal, and D.V. Goeddel. 1984. Human tumor necrosis factor: Precursor structure, expression and homology to lymphotoxin. *Nature* **312:** 724.

Rouzer, C.A. and A. Cerami. 1980. Hypertriglyceridemia associated with *Trypanosoma brucei brucei* infection in rabbits: Role of defective triglyceride removal. *Mol. Biochem. Parasitol.* **2:** 31.

Shalaby, M.R., B.B. Aggarwal, E. Rinderknecht, L.P. Svedersky, B.S. Finkle, and M.A. Palladino, Jr. 1985. Activation of human polymorphonuclear neutrophil functions by interferon-gamma and tumor necrosis factors. *J. Immunol.* **135:** 2069.

Shirai, T., H. Yamaguchi, H. Ito, C.W. Todd, and R.B. Wallace. 1985. Cloning and expression in *Escherichia coli* of the gene for human tumour necrosis factor. *Nature* **313:** 803.

Silberstein, D.S. and J.R. David. 1986. Tumor necrosis factor enhances eosinophil toxicity to *Schistosoma mansoni* larvae. *Proc. Natl. Acad. Sci.* **83:** 1055.

Stern, D.M. and P.P. Nawroth. 1986. Modulation of endothelial hemostatic properties by tumor necrosis factor. *J. Exp. Med.* **163:** 740.

Torti, F.M., B. Dieckmann, B. Beutler, A. Cerami, and G.M. Ringold. 1985. A macrophage factor inhibits adipocyte gene expression: An in vitro model of cachexia. *Science* **229:** 867.

Tracey, K.J., S.F. Lowry, B. Beutler, A. Cerami, J.D. Albert, and G.T. Shires. 1986a. Cachectin/tumor necrosis factor mediates changes of skeletal muscle plasma membrane potential. *J. Exp. Med.* **164:** 1368.

Tracey, K.J., B. Beutler, S.F. Lowry, J. Merryweather, S. Wolpe, I.W. Milsark, R.J. Hariri, T.J. Fahey III, A. Zentella, J.D. Albert, G.T. Shires, and A. Cerami. 1986b. Shock and tissue injury induced by recombinant human cachectin. *Science* **234:** 470.

Vogel, S.N., R.N. Moore, J.D. Sipe, and D.L. Rosenstreich. 1980. BCG-induced enhancement of endotoxin sensitivity in C3H/HeJ mice. I. In vivo studies. *J. Immunol.* **124:** 2004.

Wang, A.M., A.A. Creasey, M.B. Ladner, L.S. Lin, J. Strickler, J.N. Van Arsdell, R. Yamamoto, and D.F. Mark. 1985. Molecular cloning of the complementary DNA for human tumor necrosis factor. *Science* **228:** 149.

Molecular Biology of Interleukin-1

P.T. Lomedico, P.L. Kilian, U. Gubler, A.S. Stern, and R. Chizzonite
Departments of Immunopharmacology, Molecular Genetics, and Protein Biochemistry, Hoffmann-La Roche Inc., Roche Research Center, Nutley, New Jersey 07110

Interleukin-1 (IL-1) was first described more than a decade ago as lymphocyte-activating factor (LAF) for its ability to costimulate mitogen-activated mouse thymocyte proliferation (Gery et al. 1971). Since this initial observation, it has become clear that IL-1 has a number of other biological activities and that there are at least two major types of IL-1 proteins produced by activated macrophages and by other cell types. Some of the activities ascribed to IL-1 include the ability to stimulate B-cell function, fibroblast proliferation, muscle proteolysis, prostaglandin E_2 (PGE_2) and collagenase production by rheumatoid synovial cells, hepatic acute-phase protein synthesis, and to induce fever (for review, see Dinarello 1984). Although its activities are numerous and diverse, it has been suggested that IL-1 plays an important role in stimulating immune and inflammatory responses in the body's defense against infection and other forms of trauma. IL-1 is also believed to contribute to the pathology observed in certain chronic inflammatory conditions (e.g., rheumatoid arthritis). In the past few years, the cloning and expression of the genes coding for the IL-1 proteins have greatly expanded our knowledge of IL-1 structure and function.

IL-1 Protein Structure and Gene Evolution

IL-1 preparations from most species are heterogeneous with respect to polypeptide size (molecular weights between 12,000 and 19,000) and charge (multiple forms between pI 5 and pI 7). Because of the difficulty in preparing sufficient amounts of IL-1 protein for structural studies, the relationship between the different forms has not been clear until recently. We now know from mammalian gene-cloning experiments that there are a family of nonallelic genes that code for IL-1 precursor proteins. Expression of the different IL-1 genes within the macrophage, plus imprecise proteolytic processing of the precursors, probably explains IL-1 heterogeneity.

Our studies began as a collaboration with S. Mizel to understand IL-1 biosynthesis in the murine macrophage-like tumor cell line $P388D_1$. A highly specific goat antibody to murine IL-1 (Mizel et al. 1983) was used to demonstrate that mRNA prepared from stimulated $P388D_1$ cells programmed the cell-free synthesis of a 33,000-molecular-weight polypeptide (Lomedico et al. 1984; Giri et al. 1985). Pulse-labeling experiments indicated that the 33,000-molecular-weight polypeptide is synthesized by stimulated $P388D_1$ cells and normal peritoneal macrophages and is the precursor to the heterogeneous collection of low-molecular-weight IL-1 polypeptides found in the culture supernatant of stimulated cells (Giri et al. 1985). A cDNA copy of the mRNA coding for the mouse IL-1 precursor was cloned (Lomedico et al. 1984) using an mRNA hybrid selection/translational analysis (Parnes et al. 1981). The nucleotide sequence of this cDNA predicts a protein of 270 amino acids (Fig. 1) with a calculated molecular weight of 31,026. Partial amino acid sequencing of $P388D_1$-derived IL-1 confirmed that the protein sequence predicted by nucleotide sequencing was correct and indicated that a subset of the processed IL-1 polypeptides possesses an amino terminus that begins at position 115 of the 270-amino-acid precursor. This observation suggested that IL-1 bioactivity maps to the carboxy-terminal 156 amino acids of the precursor. To prove this point, amino acids 115–270 were expressed in *Escherichia coli*, yielding a protein of the expected size that reacted with the goat anti-IL-1 antibodies and was active in the murine thymocyte proliferation assay (half-maximal activity between 10^{-11} and 10^{-12} M).

The murine IL-1 cDNA was used as a probe to isolate the human homolog from cDNA generated to mRNA isolated from lipopolysaccharide (LPS)-stimulated human leukocytes (Gubler et al. 1986). The nucleotide sequence of this cDNA predicts a protein of 271 amino acids (Fig. 1), which shows ≈61% homology with its murine counterpart. This same cDNA has also been cloned by March et al. (1985) and Furutani et al. (1985). Expression of the carboxy-terminal 154 amino acids of this precursor in *E. coli* yields a biologically active IL-1 protein of the expected size (Gubler et al. 1986). Both recombinant murine and human IL-1 proteins exhibit a pI of ≈5 (Dukovich et al. 1986; A. Stern, unpubl.).

Recently, Furutani et al. (1985) described the cloning of rabbit IL-1 cDNA from a library prepared from stimulated alveolar macrophages. The nucleotide sequence of this cDNA predicts a protein of 267 amino acids (Fig. 1), which is highly related (between 61% and 65% homologous) to the previously described murine and human IL-1 gene products. We assume that these three proteins are the products of a homologous locus, termed the IL-1α or pI 5 gene.

The first human IL-1 cDNA cloning and sequencing was accomplished by Auron et al. (1984). IL-1 preparations from stimulated human monocytes exhibit charge heterogeneity, with the major species possessing a pI of ≈7 and several minor species between pI 5 and

```
        10        20        30        40        50        60
MAKVPDLFEDLKNCFSENEEYSSAIDHLSLNQK-SFYDASYEPLHEDCMNKVVSLSTSET     RABBIT α
:::::: ::::::: ::::: :: ::::::::: :::  :: :::: ::   :::: :::
MAKVPDMFEDLKNCYSENEEDSSSIDHLSLNQK-SFYHVSYGPLHEGCMDQSVSLSISET     HUMAN α
:::::: ::::::::::::  :: ::::::::: :::  ::: ::: : :: :::  :::
MAKVPDLFEDLKNCYSENEDYSSAIDHLSLNQK-SFYDASYGSLHETCTDQFVSLRTSET     MOUSE α
:: :: :       :: :::            : :: :     :     :    :: :
MAEVPKLASEMMAYYSGNEDDLFFEADGPKQMKCSFQDLDLCPL-----DGGIQLRISDH     HUMAN β
:: :: :  ::    :  : ::::: :::  ::  ::  ::        :  ::: ::
MATVPELNCEMPPFDSD-ENDLFFEVDGPQKMKGCFQTFDLGCP-----DESIQLQISQQ     MOUSE β
** **          *  *             *  *                  *  *

         70           80        90         100       110
SVSPNLTFQENVVAVTA---SGKILKKRRLSLNQPITDVDL--ETNVSDPEEGIIKPRSV     RABBIT α
     ::: :  : :      :: :::::::: : ::: ::  :    : :: ::::::
SKTSKLTFKESMVVVAT---NGKVLKKRRLSLSQSITDDDL--EAIANDSEEEIIKPRSA     HUMAN α
 :  :  ::::: : :     ::: ::::::: :   : :::    :  : :: :  ::::
SKMSNFTFKESRVTVSATSSNGKILKKRRLSFSETFTEDDL--QSITHDLEETIQ-PRSA     MOUSE α
     :    : :             :       :: : ::         :: :
HYSKGFRQAASVVVAM------DKLRKMLVPCPQTFQENDLSTFFPFIFEEEPIFFDTWD     HUMAN β
:  : :::: :  ::        ::    :  : :::  : :::: ::::::::  : ::
HINKSFRQAVSLIVAV------EKLWQLPVSFPWTFQDEDMSTFFSFIFEEEPILCDSWD     MOUSE β
                                       *          ** *

   120       130       140        150         160       170
PYTFQRNMRYKYLRIIKQEFTLNDALNQSLVRDTSDQYLRAAPL--QNLGDAVKFDMGVY     RABBIT α
:  :  :  :   :::: :: ::::::::  :   ::::  :: :   ::  ::::::: :
PFSFLSNVKYNFMRIIKYEFILNDALNQSIIRAN-DQYLTAAAL--HNLDEAVKFDMGAY     HUMAN α
:    :   :  :      :  :: ::: :       ::    :    :   ::::: ::
PYTYQSDLRYKLMKLVRQKFVMNDSLNQTIYQDVDKHYLSTTWL--NDLQQEVKFDMYAY     MOUSE α
               ::          :          :    :   :  : : : :
NEAY------VHDAPVRSLNCTLRDSQQKSLVMSGPYELKALHLQGQDMEQQVVFSMSFV     HUMAN β
          : : : : :   ::: :::::: : ::::::::: ::   ::: ::::::
DDDNLL----VCDVPIRQLHYRLRDEQQKSLVLSDPYELKALHLNGQNINQQVIFSMSFV     MOUSE β
                           *          *    *        * * *

      180       190       200       210         220
MTSEDSILP-VTLRISQTPLFVSAQNEDEPVLLKEMPETPRIITDS--ESDILFFWETQG     RABBIT α
  : :     : :::: : : : :: :: ::::::::: :  :: :  :   :::::: :
KSSKDDAKITVILRISKTQLYVTAQDEDQPVLLKEMPEIPKTITGS--ETNLLFFWETHG     HUMAN α
 :  :: :  : : ::  :: : :: ::::::::: :: :: ::::  :: : :::
SSGGDDSKYPVTLKISDSQLFVSAQGEDQPVLLKELPETPKLITGS--ETDLIFFWKSIN     MOUSE α
       : :: :      :  :    :    :      ::       :    :    ::
QGEESNDKIPVALGLKEKNLYLSCVLKDDKPTLQLESVDPKNYPKKKMEKRFVFNKIEIN     HUMAN β
::: :::::::::::: :::::::: ::  ::::::::::: ::::::::::::::::
QGEPSNDKIPVALGLKGKNLYLSCVMKDGTPTLQLESVDPKQYPKKKMEKRFVFNKIEVK     MOUSE β
          * *      *       *    *      *        *    *

        240         250         260
NKNYFKSAANPQLFIATK--PEHLVHMAR--GLPSMTDFQIS                       RABBIT α
 :::: : : : ::::::      :  :   : :: :::::
TKNYFTSVAHPNLFIATK--QDYWVCLAG--GPPSITDFQILENQA                   HUMAN α
 :::::: : : ::::::      : ::   : :: :::::
SKNYFTSAAYPELFIATK--EQSRVHLAR--GLPSMTDFQIS                       MOUSE α
 :  : ::  :   : :   :   : :    :    :::
NKLEFESAQFPNWYISTSQAENMPVFLGGTKGGQDITDFTMQFVSS                   HUMAN β
 : ::::: :::::::::::: ::::::   : ::: ::::  :::
SKVEFESAEFPNWYISTSQAEHKPVFLGNNSG-QDIIDFTMESVSS                   MOUSE β
 *  * *   *   * *       *      *     **
```

Figure 1. Amino acid sequence alignment for the known IL-1 precursors. The sequences of rabbit IL-1α (Furutani et al. 1985), human IL-1α (Furutani et al. 1985; March et al. 1985; Gubler et al. 1986), mouse IL-1α (Lomedico et al. 1984), human IL-1β (Auron et al. 1984), and mouse IL-1β (Gray et al. 1986) were aligned using the Wilbur and Lipman (1983) algorithm. Homologies between sequences are marked with a double dot; homologies extending to all the sequences shown are marked with an asterisk.

pI 6. Using a cDNA library prepared from human peripheral blood monocytes stimulated with endotoxin, a rabbit antibody that mostly recognizes the pI≈7 IL-1 species, differential hybridization with stimulated and unstimulated monocyte probes, and mRNA-directed IL-1 expression in *Xenopus laevis* oocytes, Auron et al. were able to isolate a human IL-1 cDNA clone that codes for a protein of 269 amino acids (Fig. 1). This protein exhibits a low, but statistically significant, level of homology (between 27% and 33%) with the different members of the IL-1α gene family. This sequence homology is primarily restricted to the carboxy-terminal portions of the proteins (Fig. 1). Hence, it is likely that the Auron et al. IL-1 gene (termed IL-1β or pI 7) is evolutionarily related to the IL-1α gene. The human pI 7 IL-1 protein was purified and sequenced, first by Van Damme et al. (1985) and subsequently by March et al. (1985) and Cameron et al. (1985); the amino-terminal sequence analyses showed some heterogeneity, but the major species matched the sequence predicted by Auron et al. beginning at amino acid position 117. Hence, the pI 7 IL-1 protein probably represents the carboxy-terminal 153 amino acids (117–269) of the 269-amino-acid IL-1β precursor. Expression of this portion of the precursor in *E. coli* yields biologically active IL-1 protein (Dinarello et al. 1986b) that exhibits a pI of ≈7 (U. Gubler and A. Stern, unpubl.).

Gray et al. (1986) recently used the human IL-1β cDNA as a hybridization probe to isolate a mouse IL-1β cDNA clone from a cDNA library prepared from the macrophage-like cell line PU5-1.8. The nucleotide sequence of this cDNA predicts a protein of 269 amino acids (Fig. 1) that is ≈67% homologous to the human IL-1β precursor and ≈30% homologous to the different IL-1α precursors.

The primary structures of the IL-1 precursors, both α and β, predicted by cDNA cloning have several common features: (1) They are approximately the same size (267–271 amino acids); (2) they contain no sizeable hydrophobic region that could serve as a classical signal sequence, contrary to what would be expected for a conventional secretory protein precursor; and (3) IL-1 bioactivity maps to the carboxy-terminal half of the molecule. Although evaluation of recombinant IL-1 is continuing (see below), the gene cloning and expression results suggest the existence of a family of at least two independently evolving IL-1 genes, whose dissimilar protein products have similar biological activities. The members of the IL-1α subfamily (e.g., rabbit α, mouse α, and human α) are all between 61% and 65% homologous to one another, but only ≈30% homologous to the members of the IL-1β subfamily (e.g., mouse β and human β), which are themselves ≈67% homologous. It is likely that there was a gene duplication before or during vertebrate evolution that created two genes that began to diverge independently of each other to create the IL-1α and IL-1β subfamilies. The great difference between the α and β subfamily members represents divergence since the initial gene duplication, whereas the lesser differences among the mammalian members of either subfamily represent divergence during the past 90 million years since the start of the mammalian radiation.

Biological Properties of Recombinant IL-1

The diversity of biological effects ascribed to IL-1 has certainly challenged the view that a single molecule was responsible for all the activities. We have expressed mouse IL-1α (amino acids 115–270), human IL-1α (amino acids 118–271), and human IL-1β (amino acids 117–269) in *E. coli* and purified all three proteins to homogeneity (Lomedico et al. 1984; Dukovich et al. 1986; Gubler et al. 1986; Kilian et al. 1986). Biological studies of these recombinant proteins in vivo and in vitro have confirmed many "IL-1 activities." Specifically, recombinant murine IL-1α has been shown to (1) stimulate mouse thymocyte activation (Lomedico et al. 1984); (2) stimulate human dermal fibroblast proliferation (Dukovich et al. 1986; Gubler et al. 1986); (3) stimulate interleukin-2 production by EL-4 thymoma cells (Kilian et al. 1986); (4) stimulate PGE_2 and collagenase production by human rheumatoid synovial cells and dermal fibroblasts (Dukovich et al. 1986; Gubler et al. 1986); (5) stimulate arachidonic acid metabolism in rat liver and squirrel monkey smooth-muscle cells (Levine and Xiao 1985); (6) stimulate metallothionein gene expression in human hepatoma cells (Karin et al. 1985); (7) stimulate synthesis of certain hepatic acute-phase proteins in vivo and in vitro (Bauer et al. 1985; Ramadori et al. 1985; Perlmutter et al. 1986; Westmacott et al. 1986); (8) stimulate bone resorption in vitro (Gowen and Mundy 1986); (9) stimulate ACTH production from a mouse pituitary tumor line (Woloski et al. 1985); (10) act like cachectin (tumor necrosis factor) to suppress lipoprotein lipase activity in 3T3 L1 adipocytes (Beutler et al. 1985a); (11) have activity as a B-cell growth and differentiation factor (Pike and Nossal 1985); and (12) stimulate platelet-activating factor production in cultured endothelial cells (Bussolino et al. 1986). In vivo, recombinant murine IL-1α has been shown to (1) be pyrogenic (McCarthy et al. 1985; Ikejima et al. 1986; Tocco-Bradley et al. 1986); (2) promote leukocytosis and hypozincemia (Tocco-Bradley et al. 1986) and hypoferremia (Westmacott et al. 1986); (3) induce a transient suppression of food intake (McCarthy et al. 1985); (4) stimulate accumulation of procoagulant activity in endothelial cells (Nawroth et al. 1986); (5) protect mice from lethal doses of ionizing radiation (Neta et al. 1986); and (6) induce a local inflammatory response in the skin (Granstein et al. 1986). Comparison of these results to those of experiments conducted by Dinarello and colleagues with recombinant human IL-1β (Dinarello 1985; Dayer et al. 1986; Dinarello et al. 1986a,b) suggests that IL-1α and IL-1β have a wide but identical spectrum of activities. To date, no one has observed a biological response with either protein that cannot be qualitatively duplicated by the other. Preliminary experience so far suggests that human IL-1α and IL-1β exhibit comparable specific activities in the different biological assays.

IL-1 Precursor Processing

The mechanism by which macrophages process the biologically active and secreted form of IL-1 from its precursor is not clear. Since there is no evidence for processed IL-1 molecules inside macrophages (Giri et al. 1985), it is possible that IL-1 precursor processing occurs extracellularly or at the cell membrane. How the IL-1 precursor is translocated within the cell without using a hydrophobic signal sequence is nevertheless a mystery. Possibly the extreme amino terminus of the IL-1 precursor plays some role in cellular compartmentalization and/or secretion, since this region is well conserved, especially among the IL-1α subfamily members. The suggestion has been made that IL-1 is stored within the macrophage and only released upon significant cellular insult (Gery and Lepe-Zuniga 1984).

We have proposed a model (Lomedico et al. 1984) which predicts that the IL-1 precursor is secreted and then enzymatically converted to lower-molecular-weight forms by proteases concomitantly released by the activated macrophage. The primary processing site may be a highly charged region (KXLKKRR) that is conserved among the IL-1α subfamily members around amino acid 80. Cleavage at this site would release a ≈180-amino-acid carboxy-terminal fragment that possesses (A. Stern, unpubl.) a protease-resistant core. Subsequent proteolytic attack at the newly exposed amino terminus would generate a heterogeneous population of molecules with "ragged" amino termini. Amino-terminal-end heterogeneity, as revealed by sequencing studies (Lomedico et al. 1984; March et al. 1985; Van Damme et al. 1985), supports this model.

To test this model directly and to define the minimum structural requirements for IL-1 biological activity, we have initiated studies using site-directed mutagenesis technology to explore murine IL-1 structure-function relationships (DeChiara et al. 1986). The results suggest that (1) IL-1 polypeptides must be more than 127 amino acids in length in order to exhibit efficient biological activity; (2) the integrity of the precursor's carboxyl terminus is important for biological activity; and (3) different amino termini can be utilized to generate molecules with equivalent activities. We have generated single amino acid substitutions throughout the 127-amino-acid core that result in biologically inactive proteins (T. DeChiara, unpubl.), further confirming that this entire structure is necessary for activity.

IL-1 Synthesis by Non-macrophage Cells

There have been several reports that non-macrophage cell lines are capable of producing molecules with IL-1-like activities (Dinarello 1984). Except for keratinocyte ETAF (epidermal-cell-derived thymocyte-activating factor) synthesis (see Granstein et al. 1986), there are no data relating the structures of these IL-1-like molecules to monocyte/macrophage-derived IL-1. We have used the cDNA clones for both human IL-1α and IL-1β mRNAs to set up very sensitive S1-nuclease protection assays to measure relative levels of IL-1α and IL-1β mRNAs in total RNA preparations from a variety of human cell lines and tissue sources. In LPS-stimulated adherent monocytes (Fig. 2), in LPS-stimulated leukocyte suspension cultures (Figs. 2–4), and in the placenta (U. Gubler, unpubl.), there is approximately ten times more IL-1β mRNA than IL-1α mRNA; in the absence of LPS, there is no detectable IL-1 mRNA in adherent monocytes (Fig. 2). Human foreskin keratinocytes constitutively synthesize both IL-1 mRNAs, but the ratio appears shifted in favor of the IL-1α species (Fig. 3); these data are in agreement with biochemical and immunological studies of the ETAF/IL-1 released by these cells (T. Kupper, in prep.). Human umbilical vein endothelial cells produce mostly IL-1α mRNA following LPS stimulation (Fig. 4); these cells produce secreted IL-1 under these conditions (Stern et al. 1985). Surprisingly, certain human fibroblast lines (e.g., GM2504) constitutively synthesize both IL-1 mRNAs in a ratio similar to that observed in monocytes (Fig. 4). In addition, certain epithelial cell lines and certain Epstein-Barr virus (EBV)-transformed B cells contain both IL-1 mRNAs (U. Gubler, unpubl.), whereas many other cells (e.g., T-cell lines and myeloma cell lines ± LPS) do not produce either mRNA.

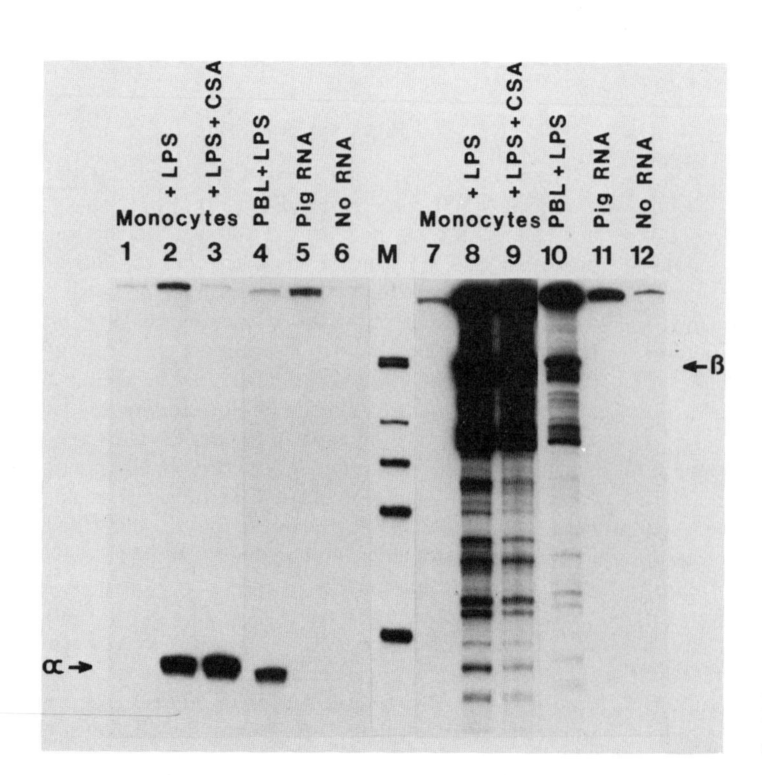

Figure 2. Measurement of IL-1 mRNA levels in human monocytes. Total RNA was isolated from cells and hybridized to end-labeled IL-1α or IL-1β probes. Following digestion with S1 nuclease, the products were fractionated on a sequencing gel and autoradiographed. The expected sizes for the protected fragments (arrows) are 210 nucleotides for IL-1α and 525 nucleotides for IL-1β. Human peripheral blood monocytes were selected following a 1-hr adherence and incubated with no stimulation (*1,7*), with 20 μg/ml LPS (*2,8*), or with LPS + 1 μg/ml cyclosporin (*3,9*) for 8 hr. Human peripheral blood leukocytes were cultured in suspension with 10 μg/ml LPS (Gubler et al. 1986) for 12 hr (*4,10*). Controls include no RNA (*6,12*) and pig liver RNA (*5,11*). (*1–6*) RNA samples hybridized to the IL-1α probe; (*7–12*) RNA samples hybridized to the IL-1β probe; (M) molecular weight markers (pBR322 cut with *Hin*fI). A. Piperno and R. Steinman (The Rockefeller University) suggested and performed the adherent monocyte RNA isolation.

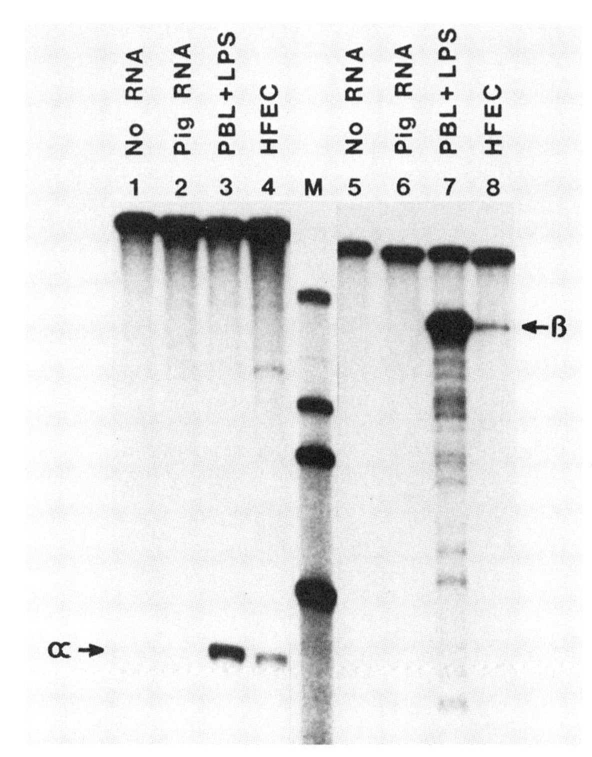

Figure 3. Measurement of IL-1 mRNA levels in human keratinocytes. The S1-nuclease protection assay was peformed as described in Fig. 2. (*1–4*) RNA samples hybridized to the IL-1α probe; (*5–8*) RNA samples hybridized to the IL-1β probe. (*1,5*) No RNA control; (*2,6*) pig liver RNA control; (*3,7*) RNA from peripheral blood leukocytes stimulated with LPS (see Fig. 2); (*4,8*) RNA isolated from human foreskin epidermal keratinocytes. T. Kupper (Yale University School of Medicine) suggested and performed the keratinocyte RNA isolation.

IL-1 Receptors

Characterization of the cell membrane receptors that bind IL-1 and mediate its action is in progress. High-affinity-binding sites for IL-1 on lymphoid and non-lymphoid cell lines have been detected using radiolabeled natural human IL-1β (Dower et al. 1985, 1986) and recombinant murine and human IL-1α (Kilian et al. 1986; S.B. Mizel et al., in prep.). Cross-linking studies suggest that the IL-1 receptor has a molecular weight of approximately 80,000 as determined by SDS-PAGE analysis (Dower et al. 1985). The events following IL-1 binding to its receptor are not known, but it appears that IL-1 is internalized but not degraded following the binding interaction and that IL-1 receptors are down-regulated by exposure of cells to IL-1 (S.B. Mizel et al., in prep.).

In view of the limited homology in amino acid sequence between the α and β forms of IL-1, the question arises as to whether the different IL-1 proteins bind to the same receptor. One initial study aimed at addressing this issue suggested that there is a common receptor for both IL-1α and IL-1β (Kilian et al. 1986). This study showed that the binding of radioiodinated recombinant murine IL-1α to murine EL-4 thymoma cells is inhibited by murine and human IL-1α, as well as by human IL-1β. This observation has now been extended in two ways. First, recombinant human IL-1α was radiolabeled with ^{125}I such that it maintains biological activity and binds to EL-4 cells in a specific and saturable manner (Fig. 5). Scatchard plot analysis of this binding interaction indicates that there are approximately 1500 binding sites per cell, in good agreement with values obtained with radioiodinated murine IL-1α (Kilian et

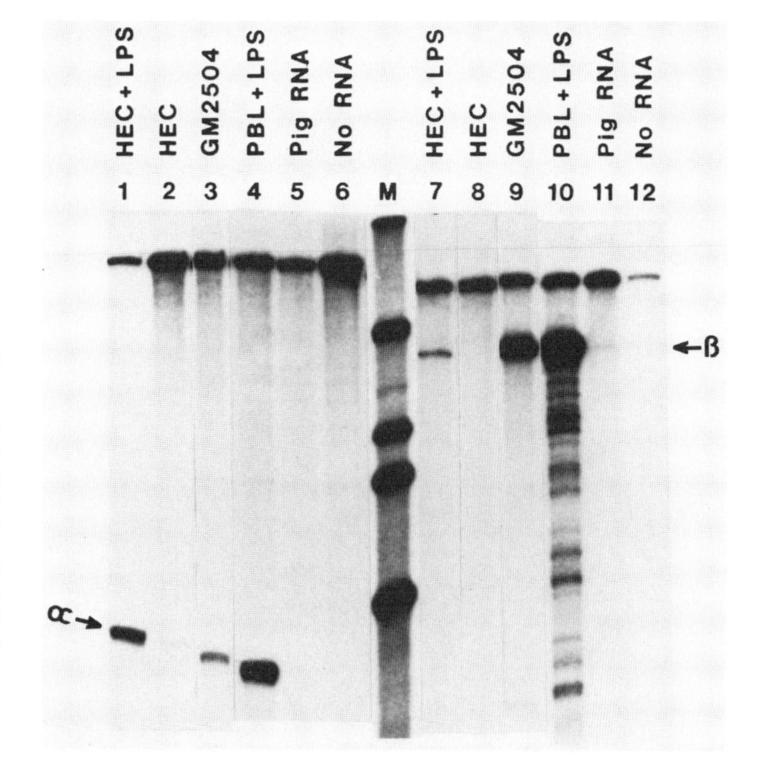

Figure 4. Measurement of IL-1 mRNA levels in human endothelial cells and fibroblasts. The S1-nuclease protection assay was performed as described in Fig. 2. (*1–6*) RNA samples hybridized to the IL-1α probe; (*7–12*) RNA samples hybridized to the IL-1β probe. Human umbilical vein endothelial cells were incubated for 24 hr with no stimulation (*2,8*) or with 1 μg/ml LPS (*1,7*) before RNA isolation. RNA was isolated from confluent cultures of GM2504, a human dermal fibroblast cell line (*3,9*). (*4,10*) RNA from peripheral blood leukocytes stimulated with LPS (see Fig. 2); (*5,10*) pig liver RNA control; (*6,12*) no RNA control. D. Stern and P. Nawroth (Oklahoma Medical Research Foundation) suggested and performed the endothelial cell RNA isolation.

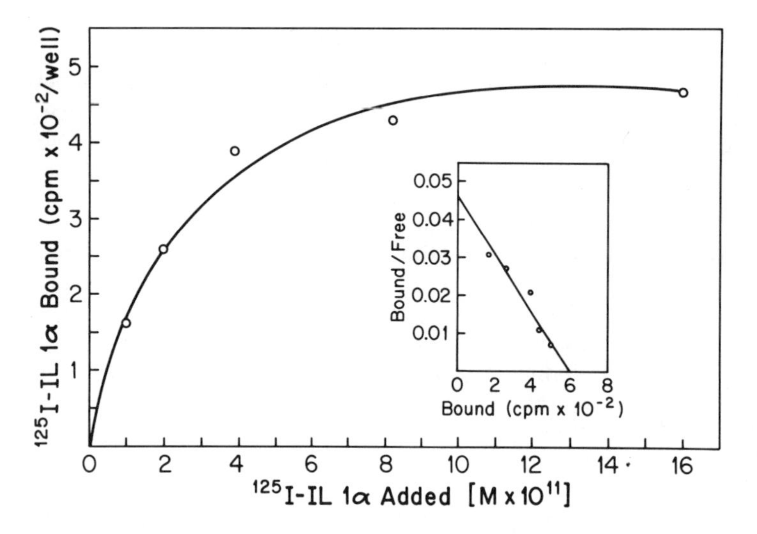

Figure 5. Specific binding of human ^{125}I-labeled IL-1α to EL-4 cells. ^{125}I-labeled IL-1α was prepared and the binding to EL-4 cells was determined as described previously (Kilian et al. 1986; S.B. Mizel et al., in prep.). (*Inset*) Analysis of binding data according to the method of Scatchard.

al. 1986). The human IL-1α binds with a dissociation constant (K_D) of approximately 2×10^{-11} to 5×10^{-11} M. The ability of recombinant human IL-1α and IL-1β to inhibit binding of the human ^{125}I-labeled IL-1α to EL-4 cells, as well as to murine Swiss 3T3 and human dermal CRL-1445 fibroblasts, was examined (Fig. 6A,B). It is clear that both IL-1 proteins inhibit this binding similarly on all three cell lines. Thus, our initial observation that IL-1α and IL-1β cross-react on the same receptors on T cells has now been extended to non-T-cell lines.

Second, the ability of recombinant human IL-1β to bind to its receptor was measured directly by radiolabeling the IL-1β with ^{125}I in a manner such that it maintains biological activity. This ^{125}I-labeled IL-1β binds to EL-4 thymoma cells with high affinity (Fig. 7). Scatchard plot analysis indicates a single type of high-affinity-binding site ($K_D \approx 0.3$ nM) and that a similar number of binding sites are detected with the ^{125}I-labeled IL-1β as with ^{125}I-labeled IL-1α. This rules out the presence of cryptic sites for IL-1β not recognized by IL-1α. Competitive inhibition experiments to examine the ability of IL-1α and IL-1β to inhibit binding of ^{125}I-labeled IL-1β to EL-4 cells were also performed (Fig. 8). It is clear that both IL-1α and IL-1β inhibit binding to these cells. The poorer affinity of IL-1β binding to EL-4 cells correlates with the approximately tenfold lower specific activity of this protein relative to IL-1α

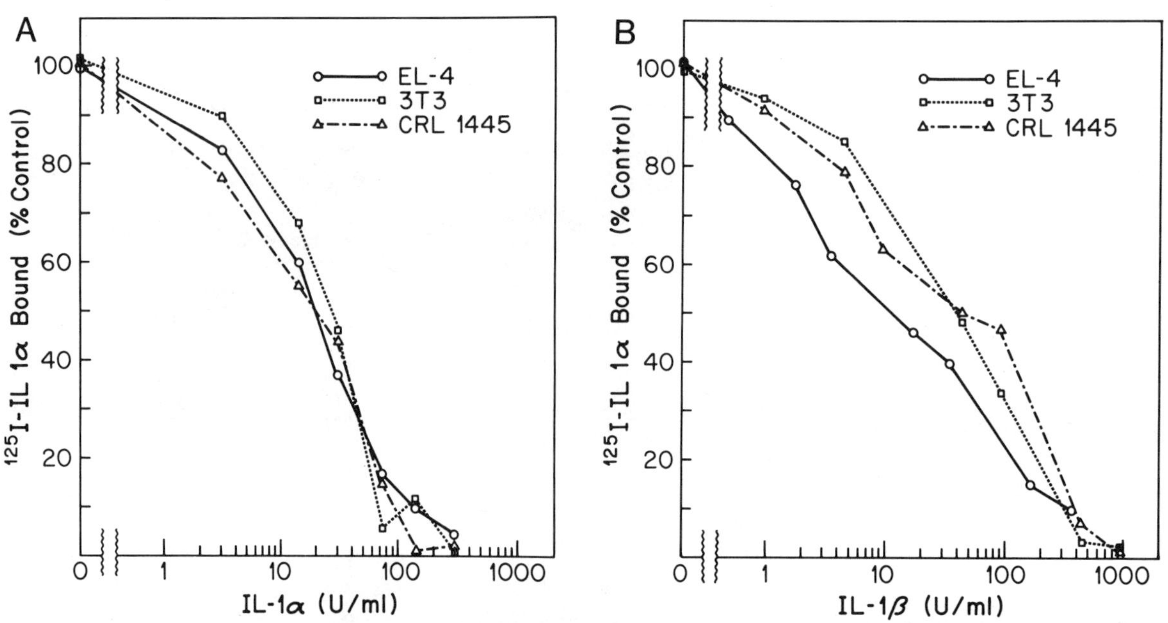

Figure 6. Effect of recombinant human IL-1α (*A*) and IL-1β (*B*) on binding of human ^{125}I-labeled IL-1α to various cell lines. The EL-4, Swiss 3T3, and CRL-1445 cells were incubated with approximately 50 pM ^{125}I-labeled IL-1α and purified recombinant human IL-1α or IL-1β at the indicated concentrations (expressed in U/ml, as determined in the LAF mouse thymocyte proliferation assay). Data represent binding as percentage of that observed in the absence of any competing ligand.

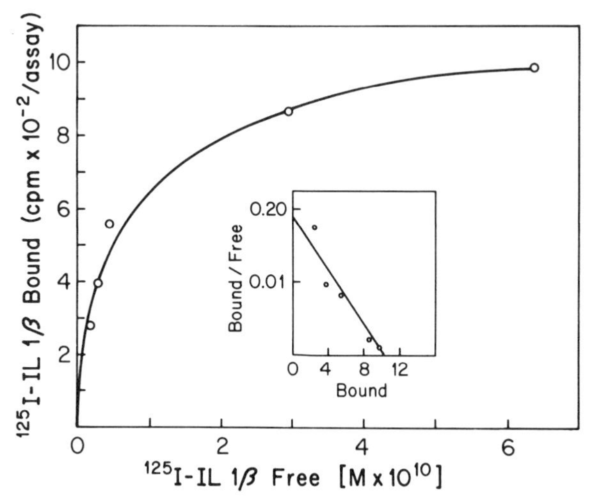

Figure 7. Specific binding of human ^{125}I-labeled IL-1β to EL-4 cells. ^{125}I-labeled IL-1β was prepared according to the method reported for IL-1α (Kilian et al. 1986). (*Inset*) Analysis of binding data according to the method of Scatchard.

in the LAF assay. In summary, results obtained to date indicate that there is only one type of IL-1 receptor to which both IL-1α and IL-1β bind. The ability of two proteins, despite major differences in amino acid sequence, to bind to the same receptor is surprising but not unique: Interferon-α and interferon-β (Branca and Baglioni 1981), epidermal growth factor and transforming growth factor-α (Marquardt et al. 1983), and tumor necrosis factor and lymphotoxin (Aggarwal et al. 1985) are all similar examples.

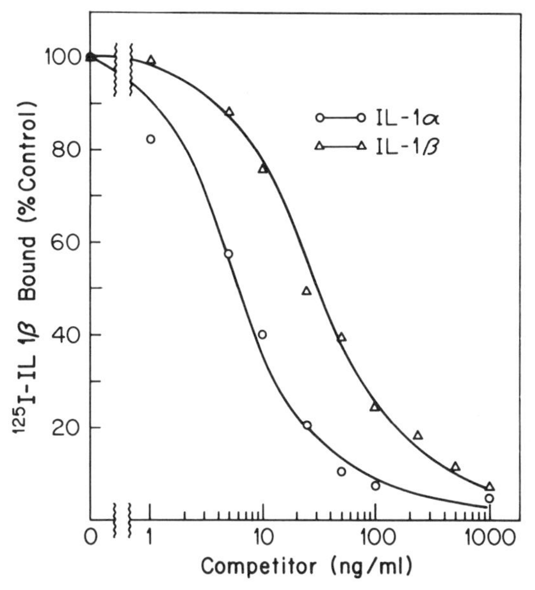

Figure 8. Effect of recombinant human IL-1α and IL-1β on binding of human ^{125}I-labeled IL-1β to EL-4 cells. Cells were incubated with aproximately 0.3 nM ^{125}I-labeled IL-1β and purified recombinant IL-1α or IL-1β at the indicated concentrations. Data represent binding as percentage of that observed in the absence of any competing ligand.

DISCUSSION

The cloning and expression of the genes coding for the different IL-1 proteins have greatly expanded our knowledge of this polypeptide hormone at the molecular level. We have learned that IL-1 is synthesized as a much larger precursor protein, which is processed in an unknown fashion to release the biologically active moiety from the carboxyl terminus. We know that a family of evolutionarily related genes code for at least two structurally dissimilar IL-1 proteins, which interact with a common cell-surface receptor and thereby exhibit an identical spectrum of biological activities. Studies of recombinant IL-1 proteins indicate that many of the diverse biological activities previously ascribed only to impure preparations of IL-1 are indeed associated with a single molecule. Finally, studies of mRNA expression in non-macrophage cells demonstrate convincingly that a wide variety of cells are able to synthesize IL-1 precursor molecules that are identical to those produced by macrophages.

While our knowledge of IL-1 biology increased, many new questions were generated that remain unanswered. Since the IL-1 precursor proteins do not resemble conventional secretory proteins, how do macrophages and other cells process the biologically active and secreted form of IL-1 from this precursor? Is it possible that the IL-1 precursor proteins have other functions *inside* cells that are distinct from the hormonal activities of the proteolytically processed and secreted lower-molecular-weight forms? Do non-macrophage cells respond in a similar fashion to the signals that stimulate IL-1 synthesis and secretion in the macrophage? Many of the activities (e.g., stimulation of fibroblast proliferation and release of PGE_2 and collagenase and stimulation of bone resorption in vitro) exhibited by IL-1 partially overlap with activities of other proteins (e.g., tumor necrosis factor, platelet-derived growth factor, fibroblast growth factor, and epidermal growth factor) that are also synthesized by activated macrophages. What are the relative contributions of IL-1 and these other proteins to normal and pathological (e.g., chronic inflammation and shock) biological processes? In this regard, the recent proof of identity of tumor necrosis factor and cachectin (Beutler et al. 1985b), and the demonstration that this protein is pyrogenic (Dinarello et al. 1986a) and shares many other activities with IL-1 (Beutler and Cerami 1986), is remarkable. And, finally, for certain activities previously ascribed to IL-1, e.g., stimulation of skeletal-muscle protein degradation (L.L. Moldawer et al., in prep.) and stimulation of the synthesis of certain hepatic acute-phase proteins (G. Crabtree, pers. comm.), there is evidence that other macrophage-derived proteins are really responsible for the observed effects.

Although it is clear that the human genome contains at least two nonallelic IL-1 genes, the question remains: Is there a physiologic basis for the existence of *both* human IL-1α and IL-1β? The evidence to date indicates that these two structurally dissimilar proteins exhibit an identical spectrum of biological activities by virtue of their ability to interact with a common cell-surface receptor. Macrophages and other cells synthesize different amounts of the two proteins in a manner that cannot be rationalized at present. Data from other organisms suggest that the monocyte IL-1α/IL-1β synthesis ratio, which is 10–20 in humans, is not conserved (e.g., stimulated pig buffy-coat leukocytes appear to synthesize comparable amounts of IL-1α and IL-1β [Saklatvala et al. 1985]). Hence, it is possible that the existence of two IL-1 genes is simply an accident of evolution and that there is no physiological reason for an organism to have to produce both gene products. Gene duplications and rearrangements (i.e., exon shuffling) are random events that provide the raw material for the evolution of new proteins with different functions. As discussed above, the gene duplication that gave rise to the IL-1α and IL-1β subfamilies was an ancient one. Since the initial duplication, the two genes began to diverge independently of one another, resulting in two gene products that share only modest sequence homology. Until we discover an activity that is not shared by the two proteins, we have to assume that the sequence divergence has not created a new function that provides a protective value to the organism. Hence, since there is no selection pressure to eliminate a gene with a nondeleterious function, it is likely that both IL-1 genes are maintained by default.

ACKNOWLEDGMENTS

We thank Steve Mizel and our colleagues at Hoffmann-La Roche for many helpful discussions. We also thank A. Piperno, R. Steinman, T. Kupper, D. Stern, and P. Nawroth for suggesting and collaborating on IL-1 mRNA measurements in different human cells. Juli Farruggia provided her usual superb secretarial assistance.

REFERENCES

Aggarwal, B.B., T.E. Eessalu, and P.E. Hass. 1985. Characterization of receptors for human tumour necrosis factor and their regulation by γ-interferon. *Nature* **318:** 665.

Auron, P.E., A.C. Webb, L.J. Rosenwasser, S.F. Mucci, A. Rich, S.M. Wolff, and C.A. Dinarello. 1984. Nucleotide sequence of human monocyte interleukin 1 precursor cDNA. *Proc. Natl. Acad. Sci.* **81:** 7907.

Bauer, J., W. Weber, T.-A. Tran-Thi, G.-H. Northoff, K. Decker, W. Gerok, and P.C. Heinrich. 1985. Murine interleukin-1 stimulates α_2-macroglobulin synthesis in rat hepatocyte primary cultures. *FEBS Lett.* **190:** 271.

Beutler, B. and A. Cerami. 1986. Cachectin and tumour necrosis factor as two sides of the same biological coin. *Nature* **320:** 584.

Beutler, B., J. Mahoney, N. Le Trang, P. Pekala, and A. Cerami. 1985a. Purification of cachectin, a lipoprotein lipase-suppressing hormone secreted by endotoxin-induced RAW 264.7 cells. *J. Exp. Med.* **161:** 984.

Beutler, B., D. Greenwald, J.D. Hulmes, M. Chang, Y.-C.E. Pan, J. Mathison, R. Ulevitch, and A. Cerami. 1985b. Identity of tumour necrosis factor and the macrophage-secreted factor cachectin. *Nature* **316:** 552.

Branca, A.A. and C. Baglioni. 1981. Evidence that types I and II interferons have different receptors. *Nature* **294:** 768.

Bussolino, F., F. Breviario, C. Tetta, M. Aglietta, A. Mantovani, and E. Dejana. 1986. Interleukin-1 stimulates platelet activating factor production in cultured human endothelial cells. *J. Clin. Invest.* **77:** 2027.

Cameron, P., G. Limjuco, J. Rodkey, C. Bennett, and J.A. Schmidt. 1985. Amino acid sequence analysis of human interleukin-1 (IL-1). Evidence of biochemically distinct forms of IL-1. *J. Exp. Med.* **162:** 790.

Dayer, J.-M., B. de Rochemonteix, B. Burrus, S. Demczuk, and C.A. Dinarello. 1986. Human recombinant interleukin-1 stimulates collagenase and prostaglandin E_2 production by human synovial cells. *J. Clin. Invest.* **77:** 645.

DeChiara, T., D. Young, R. Semionow, A.S. Stern, C. Batula-Bernardo, C. Fiedler-Nagy, K. Kaffka, P.L. Kilian, S. Yamazaki, S.B. Mizel, and P.T. Lomedico. 1986. Structure-function analysis of murine interleukin-1: Biologically active polypeptides are at least 127 amino acids long and are derived from the carboxyterminus of a 270 amino acid precursor. *Proc. Natl. Acad. Sci.* (in press).

Dinarello, C.A. 1984. Interleukin-1. *Rev. Infect. Dis.* **6:** 52.

———. 1985. An update on human Interleukin-1: From molecular biology to clinical relevance. *J. Clin. Immunol.* **5:** 287.

Dinarello, C.A., J.G. Cannon, S.M. Wolff, H.A. Bernheim, B. Beutler, A. Cerami, I.S. Figari, M.A. Palladino, and J.V. O'Connor. 1986a. Tumor necrosis factor (Cachectin) is an endogenous pyrogen and induces production of interleukin-1. *J. Exp. Med.* **163:** 1433.

Dinarello, C.A., J.G. Cannon, J.W. Mier, H.A. Bernheim, G. LoPreste, D.L. Lynn, R.N. Love, A.C. Webb, P.E. Auron, R.C. Reuben, A. Rich, S.M. Wolff, and S.D. Putney. 1986b. Multiple biological activities of human recombinant interleukin-1. *J. Clin. Invest.* **77:** 1734.

Dower, S.K., S.M. Call, S. Gillis, and D.L. Urdal. 1986. Similarity between the interleukin-1 receptors on a murine T-lymphoma cell line and on a murine fibroblast cell line. *Proc. Natl. Acad. Sci.* **83:** 1060.

Dower, S.K., S.R. Kronheim, C.J. March, P.J. Conlon, T.P. Hopp, S. Gillis, and D.L. Urdal. 1985. Detection and characterization of high affinity plasma membrane receptors for human interleukin-1. *J. Exp. Med.* **162:** 501.

Dukovich, M., J.M. Severin, S.J. White, S. Yamazaki, and S.B. Mizel. 1986. Stimulation of fibroblast proliferation and prostaglandin production by purified recombinant murine interleukin-1. *Clin. Immunol. Immunpathol.* **38:** 381.

Furutani, Y., M. Notake, M. Yamayoshi, J.-I. Yamagishi, H. Nomura, M. Ohue, R. Furuta, T. Fukui, M. Yamada, and S. Nakamura. 1985. Cloning and characterization of the cDNA for human and rabbit interleukin-1 precursor. *Nucleic Acids Res.* **13:** 5869.

Gery, I. and J.L. Lepe-Zuniga. 1984. Interleukin-1: Uniqueness of its production and spectrum of activities. *Lymphokines* **9:** 109.

Gery, I., R.K. Gershon, and B.H. Waksman. 1971. Potentiation of cultured mouse thymocyte responses by factors released by peripheral leucocytes. *J. Immunol.* **107:** 1778.

Giri, J.G., P.T. Lomedico, and S.B. Mizel. 1985. Studies on the synthesis and secretion of interleukin-1. I. A 33,000 molecular weight precursor for interleukin-1. *J. Immunol.* **134:** 343.

Gowen, M. and G.R. Mundy. 1986. Actions of recombinant interleukin-1, interleukin-2, and interferon-γ on bone resorption *in vitro*. *J. Immunol.* **136:** 2478.

Granstein, R.D., R. Margolis, S.B. Mizel, and D.N. Sauder. 1986. In vivo inflammatory activity of epidermal cell derived thymocyte activating factor and recombinant interleukin 1 in the mouse. *J. Clin. Invest.* **77:** 1020.

Gray, P.W., D. Glaister, E. Chen, D.V. Goeddel, and D. Pennica. 1986. Two interleukin-1 genes in the mouse: Cloning and expression of the cDNA for mouse interleukin-1β. *J. Immun.* (in press).

Gubler, U., A.O. Chua, A.S. Stern, C.P. Hellmann, M.P. Vitek, T.M. DeChiara, W.R. Benjamin, K.J. Collier, M. Dukovich, P.C. Familletti, C. Fiedler-Nagy, J. Jenson, K. Kaffka, P.L. Kilian, D. Stremlo, B.H. Wittreich, D. Woehle, S.B. Mizel, and P.T. Lomedico. 1986. Recombinant human interleukin-1α: Purification and biological characterization. *J. Immunol.* **136:** 2492.

Ikejima, T., M. Minami, D.M. Gill, and C.A. Dinarello. 1986. Human, rabbit and murine interleukin-1 production in response to toxic shock syndrome toxin-1. *J. Immunol.* (in press).

Karin, M., R.J. Imbra, A. Heguy, and G. Wong. 1985. Interleukin-1 regulates human metallothionein gene expression. *Mol. Cell. Biol.* **5:** 2866.

Kilian, P.L., K.L. Kaffka, A.S. Stern, D. Woehle, W.R. Benjamin, T.M. DeChiara, U. Gubler, J.J. Farrar, S.B. Mizel, and P.T. Lomedico. 1986. Interleukin-1α and interleukin-1β bind to the same receptor on T cells. *J. Immunol.* **136:** 4509.

Levine, L. and D.-M. Xiao. 1985. The stimulations of arachidonic acid metabolism by recombinant murine interleukin 1 and tumor promoters or 1-oleoyl-2-acetylglycerol are synergistic. *J. Immunol.* **135:** 3430.

Lomedico, P.T., U. Gubler, C.P. Hellmann, M. Dukovich, J.G. Giri, Y.-C.E. Pan, K. Collier, R. Semionow, A.O. Chua, and S.B. Mizel. 1984. Cloning and expression of murine interleukin-1 cDNA in *Escherichia coli*. *Nature* **312:** 458.

March, C.J., B. Mosley, A. Larsen, D.P. Cerretti, G. Braedt, V. Price, S. Gillis, C.S. Henney, S.R. Kronheim, K. Grabstein, P.J. Conlon, T.P. Hopp, and D. Cosman. 1985. Cloning, sequence and expression of two distinct human interleukin-1 complementary DNAs. *Nature* **315:** 641.

Marquardt, H., M.W. Hunkapiller, L.E. Hood, D.R. Twardzik, J.E. DeLarco, J.R. Stephenson, and G.J. Todaro. 1983. Transforming growth factors produced by retrovirus-transformed rodent fibroblasts and human melanoma cells: Amino acid sequence homology with epidermal growth factor. *Proc. Natl. Acad. Sci.* **80:** 4684.

McCarthy, D.O., M.J. Kluger, and A.J. Vander. 1985. Suppression of food intake during infection: Is interleukin-1 involved? *Am. J. Clin. Nutr.* **42:** 1179.

Mizel, S.B., M. Dukovich, and J. Rothstein. 1983. Preparation of goat antibodies against interleukin-1: Use of an immunoadsorbent to purify interleukin-1. *J. Immunol.* **131:** 1834.

Nawroth, P.P., D.A. Handley, C.T. Esmon, and D.M. Stern. 1986. Interleukin-1 induces endothelial cell procoagulant while suppressing cell-surface anticoagulant activity. *Proc. Natl. Acad. Sci.* **83:** 3460.

Neta, R., S. Douches, and J.J. Oppenheim. 1986. Interleukin-1 is a radioprotector. *J. Immunol.* **136:** 2483.

Parnes, J.R., B. Velan, A. Felsenfeld, L. Ramanathan, U. Ferrini, E. Appella, and J.G. Seidman. 1981. Mouse β2-microglobulin cDNA clones: A screening procedure for cDNA clones corresponding to rare mRNAs. *Proc. Natl. Acad. Sci.* **78:** 2253.

Perlmutter, D.H., G. Goldberger, C.A. Dinarello, S.B. Mizel, and H.R. Colten. 1986. Regulation of class III major histocompatibility complex gene products of interleukin-1. *Science* **232:** 850.

Pike, B.L. and G.J.V. Nossal. 1985. Interleukin-1 can act as a B-cell growth and differentiation factor. *Proc. Natl. Acad. Sci.* **82:** 8153.

Ramadori, G., J.D. Sipe, C.A. Dinarello, S.B. Mizel, and H.R. Colten. 1985. Pretranslational modulation of acute phase hepatic protein synthesis by murine recombinant interleukin-1 (IL-1) and purified human IL-1. *J. Exp. Med.* **162:** 930.

Saklatvala, J., S.J. Sarsfield, and J. Townshend. 1985. Pig interleukin 1. Purification of two immunologically different leukocyte proteins that cause cartilage resorption, lymphocyte activation, and fever. *J. Exp. Med.* **162:** 1208.

Stern, D.M., I. Bank. P.P. Nawroth, J. Cassimeris, W. Kisiel, J.W. Fenton II, C. Dinarello, L. Chess, and E.A. Jaffe. 1985. Self-regulation of procoagulant events on the endothelial cell surface. *J. Exp. Med.* **162:** 1223.

Tocco-Bradley, R., L.L. Moldawer, C.T. Jones, B. Gerson, G.L. Blackburn, and B.R. Bistrian. 1986. The biological activity *in vivo* of recombinant murine interleukin-1 in the rat. *Proc. Soc. Exp. Biol. Med.* **182:** 263.

Van Damme, J., M. De Ley, G. Opdenakker, A. Billiau, P. De Somer, and J. Van Beeumen. 1985. Homogeneous interferon-inducing 22K factor is related to endogenous pyrogen and interleukin-1. *Nature* **314:** 266.

Westmacott, D., J.E. Hawkes, R.P. Hill, L.E. Clarke, and D.P. Bloxham. 1986. Comparison of the effects of recombinant murine and human interleukin-1 *in vitro* and *in vivo*. *Lymphokine Res.* **5:** 87.

Wilbur, W.J. and D.J. Lipman. 1983. Rapid similarity searches of nucleic acid and protein data banks. *Proc. Natl. Acad. Sci.* **80:** 726.

Woloski, B.M.R.N.J., E.M. Smith, W.J. Meyer, G.M. Fuller, and J.E. Blalock. 1985. Corticotropin-releasing activity of monokines. *Science* **230:** 1035.

Development of Mullerian Inhibiting Substance as an Anti-cancer Drug

R.L. Cate,* E.G. Ninfa,* D.J. Pratt,* D.T. MacLaughlin,† and P.K. Donahoe†
*Biogen Research Corporation, Cambridge, Massachusetts 02142; †Pediatric Surgical Research Laboratory, Department of Surgery, Massachusetts General Hospital and Harvard Medical School, Boston, Massachusetts 02114

Normal animal cell proliferation is a precisely controlled process, which when disrupted leads to malignant transformation and uncontrolled growth. The concept of autocrine stimulation of cellular proliferation has been proposed to explain malignant transformation: Normal cells require exogenous growth factors, whereas transformed cells have gained autonomy due to the endogenous production of these factors (Todaro et al. 1977; Sporn and Todaro 1980). Many examples can be cited in support of this hypothesis (De Larco and Todaro 1978; Kaplan et al. 1982; Waterfield et al. 1983; Lang et al. 1985). Recently, the autocrine hypothesis has been extended to include growth inhibitors: Transformation may also result from failure to express or respond to specific growth-inhibitory substances that may be released by cells to regulate their orderly growth (Sporn and Roberts 1985). Consistent with this, transforming growth factor (TGF-β) can suppress the growth of the monkey kidney cells that produce this peptide (Tucker et al. 1984).

Can we use this situation given our current level of knowledge to control the growth of cancer cells? Sporn and Todaro (1980) have suggested the development of antagonists of positive autocrine growth factors as potential therapeutics in treating cancer. Alternatively, it may be possible to use negative growth factors to restore growth control of cancer cells that have reexpressed receptors for that negative growth factor. This approach has been the basis for an investigation started in 1979 to determine whether Mullerian inhibiting substance (MIS), a testicular glycoprotein that inhibits the growth of cells during male embryonic development, can also inhibit the growth of tumor cells (Donahoe et al. 1979).

Early in the sexual development of the male embryo, both the Wolffian and Mullerian ducts begin to develop in the urogenital ridges. Under the influence of testosterone, the Wolffian duct continues to develop into the epididymas, vas deferens, and seminal vesicles. The Mullerian duct, which develops into the uterus, vagina, and fallopian tubes in the normal female embryo, regresses due to MIS. Regression of the Mullerian duct is characterized by a breakdown of the basement membrane of Mullerian duct epithelial cells, followed by their dissociation and transformation to mesenchyme. Programmed cell death also plays an important role in this developmental process (Donahoe et al. 1982).

The hypothesis that MIS might be an effective anticancer agent was first tested in 1979. In this study (Donahoe et al. 1979), it was demonstrated that testicular extracts were cytotoxic to a human ovarian serous cystadenocarcinoma cell line (HOC-21). In later experiments (Fuller et al. 1982, 1985), purified bovine MIS was shown to inhibit colony growth of HOC-21 and primary tumors in soft agar. Also, bovine MIS inhibited the growth of tumors in nude mice when incubated with HOC-21 or HEC-1 cells prior to implantation (Donahoe et al. 1981; Fuller et al. 1984). In these experiments, MIS did not affect the tumorigenicity of a colon tumor cell line, indicating that it was acting in a specific manner. A summary of the effects of MIS on inhibiting tumor growth is shown in Table 1.

These experiments have provided a basis for developing MIS as a biologic therapy for cancer. However, extensive testing of MIS in animals has been difficult due to its scarcity. To solve this problem, we have isolated the human gene and expressed it in Chinese hamster ovary (CHO) cells (Cate et al. 1986). The protein is glycosylated by the CHO cells and is active in the organ culture assay.

METHODS

Transformation of CHO cells with the human MIS gene and assays. The transformation of CHO cells lacking the *dhfr* gene (Chasin and Urlaub 1980) with plasmid pBG311.hmis and plasmid SV2DHFR (Subramani et al. 1981) has been described previously (Cate et al. 1986). Following transfection, the cells were grown in selective medium (α-minimal essential medium lacking ribonucleosides and deoxyribonucleosides [GIBCO; No. 410-2000]) containing 10% fetal bovine serum (FBS). Twenty-five colonies were picked and screened for expression of human MIS mRNA by S1-nuclease analysis.

Total RNA was isolated from each cell line (10^7 cells), and S1 mapping was performed by the procedure of Weaver and Weissman (1979) using 20 μg of RNA and the 5′-end-labeled single-stranded probe shown in Figure 2B. The 340-bp *Pst*I-*Bam*HI fragment was isolated and 5′-end-labeled using [γ-^{32}P]ATP and polynucleotide kinase, and the antisense strand was purified on a strand-separation gel. Protected fragments were resolved on a 6% acrylamide/7 M urea gel.

Table 1. Summary of MIS Effects on Cell Lines and Primary Tumors

(a) *Effects on cell lines*

	Cell lines					
Assay	HOC-21 (ovarian)	HEC-1 (endometrial)	A431 (vaginal)	SW-48 (colon)	fibroblast	glioblastoma
Monolayer	+	n.d.	n.d.	n.d.	–	–
Clonogenic	+	n.d.	+	–	n.d.	n.d.
Nude mouse	+	+	n.d.	–	n.d.	n.d.

(b) *Effects on primary tumors*

	Primary tumors		
Assay	ovarian	endometrial	fallopian
Clonogenic	18/20	4/5	1/1

+ indicates that MIS inhibited growth of the cell line. For the primary tumors, the first number indicates tumors that were inhibited by MIS out of the total number attempted in the clonogenic assay.
n.d. indicates not determined.

To test for biologically active MIS in the conditioned medium, cells were grown to confluence and switched to selective medium containing horse serum (in place of FBS) for 48 hours. Benzamidine was added at 0.1% to the conditioned medium, and the protein was precipitated with ammonium sulfate. The protein was resuspended in 1/20th volume of 10 mM sodium phosphate (pH 8), 50 mM NaCl, 1 mM EDTA, and 0.03% NaN_3, dialyzed, and tested in the organ culture assay as described previously (Cate et al. 1986).

Partial purification of MIS from conditioned medium. The conditioned medium was first passed over a protein-A-Sepharose CL-4B column (Pharmacia) to remove IgG. The column was equilibrated with 20 mM sodium phosphate (pH 7.1) and 150 mM NaCl, and then 100 ml of conditioned medium from CHO cell line 311-22 or G2 was applied at a flow rate of 7.2 ml/hr. Fractions 2–12 (75 ml) were pooled (after confirming by SDS-PAGE that IgG had been removed) and then dialyzed againt 25 mM Tris-HCl (pH 7.5) and 0.5% deoxycholate. The pooled and dialyzed fractions were then applied at a flow rate of 14.4 ml/hr to a lentil-lectin–Sepharose 4B column (Pharmacia) that had been equilibrated with the same buffer. The column was washed extensively with this buffer, and the bound glycoproteins were then eluted with 0.2 M methyl-α-D-mannoside, 25 mM Tris-HCl (pH 7.5), and 0.5% deoxycholate. Fractions 1–11 (11 ml) were pooled and concentrated to a volume of 0.8 ml by ultrafiltration using a PM-30 Amicon membrane. Aliquots (20 μl) were subjected to SDS-PAGE under reducing conditions and analyzed by Western blot analysis (Towbin et al. 1979), using two different polyclonal antibodies: One antibody was raised against denatured gel-purified bovine MIS and another was raised against a peptide of human MIS containing amino acids 503–518 of the mature sequence.

Metabolic labeling of MIS in CHO cells. Cells were grown to 50% confluence in a 6-well plate, washed twice with Hank's balanced salts solution, and then incubated at 37°C in 1 ml of RPMI 1640 medium containing 0.01% glucose, 10% dialyzed FBS, and 100 μCi [^{3}H]glucosamine (30–60 Ci/mmole; New England Nuclear). Uptake of labeled glucosamine from the medium was monitored by removing 5-μl aliquots at intervals, and the radioactivity was measured in Aquasol using a Beckman LS3801 scintillation counter. After 6 hours, 1 μl of 10% glucose was added to the cells. The incubation was continued for 16 hours at 37°C, at which time 30% of the label had been taken up by the cells.

The supernatant was transferred to a 1.5-ml Eppendorf tube, and the cells were removed by centrifugation in a microfuge for 5 minutes at 4°C. Lentil-lectin–Sepharose 4B (50 μl) equilibrated with phosphate-buffered saline (PBS) (10 mM NaH_2PO_4, 1.5 mM K_2HPO_4, 27 mM KCl, 137 mM NaCl at pH 7.0) was added to the supernatants and rocked for 3 hours at 4°C. The lentil-lectin–Sepharose 4B was recovered by centrifuging for 20 seconds, washed twice with PBS, and then dispersed in 25 μl of PBS containing 0.5% SDS. After heating the samples at 60°C for 15 minutes, the samples were centrifuged, and the supernatants were recovered. A 5-μl aliquot was removed and added to protein sample buffer for SDS-PAGE analysis.

A 17-μl aliquot of the eluate from the lentil-lectin–Sepharose 4B was immunoprecipitated in a volume of 242 μl, containing 20 mM Tris-HCl (pH 7.2), 50 mM NaCl, 0.5% NP-40, 0.5% deoxycholate, and 200 μg/ml bovine serum albumin (BSA), and 2 μl of antisera raised against denatured bovine MIS. Protein-A–Sepharose CL-4B (25 μl) was added, and the samples were rocked for 16 hours at 4°C. The protein-A–Sepharose was recovered by centrifuging for 20 seconds, washed twice in the above buffer without BSA, and dispersed in 40 μl of 2× protein sample buffer containing β-mercaptoethanol. After heating at 60°C, the samples were centrifuged, and the supernatants were recovered for SDS-PAGE analysis.

RESULTS

Isolation of the Bovine and Human MIS Genes

MIS has been purified from Sertoli cells and newborn bovine testis (Josso et al. 1977; Donahoe et al. 1982), using an organ culture assay (Picon 1969). Current knowledge suggests that it is a disulfide-linked dimer with a molecular weight of about 140,000 (Picard and Josso 1984; Josso et al. 1981; Budzik et al. 1983). The molecular weight of the subunit is 70,000–74,000 (Josso et al. 1981; Budzik et al. 1985).

Our strategy for isolating the human MIS gene was to first isolate the bovine MIS cDNA and use it as a hybridization probe for screening a human cosmid library (Cate et al. 1986). Protein sequence information obtained from tryptic peptides of bovine MIS was used to design degenerate oliognucleotide probes. After the degeneracy of the probes was reduced, they were used to isolate a bovine MIS cDNA clone from a newborn bovine testis cDNA library. The bovine cDNA clone was subsequently used to isolate the bovine and human genomic clones from cosmid libraries. Using the protein and DNA sequence obtained from these clones and the results of a primer-extension analysis of bovine MIS mRNA, we have composed the schematic picture of human MIS that is shown in Figure 1.

The human MIS gene has five exons that code for a protein of 560 amino acids. The first 25 amino acids are a leader sequence, which can be divided into a 17-amino-acid presequence and a 8-amino-acid prosequence. Mature human MIS has a molecular weight of 57,000 deduced from the amino acid sequence, whereas mature bovine MIS has a predicted molecular weight of 58,000. Since mature bovine MIS migrates with a molecular weight of around 74,000 after reduction, it must undergo a substantial amount of posttranslational modification. This is probably also true for human MIS, but biochemical analysis of the human protein has been impossible up until now. Two potential N-linked glycosylation sites are indicated in Figure 1; the level of glycosylation of bovine MIS has been determined by Donahoe et al. (1982) to be 8.3%, and Picard et al. (1986) have recently measured the carbohydrate content at 13.5%.

Expression of the Human MIS Gene in CHO Cells

To express the human MIS gene in animal cells, we inserted a 4.5-kb *Afl*II fragment containing the entire gene into an expression vector carrying the SV40 early promoter (see Cate et al. 1986). This plasmid, pBG311.hmis, was introduced along with plasmid

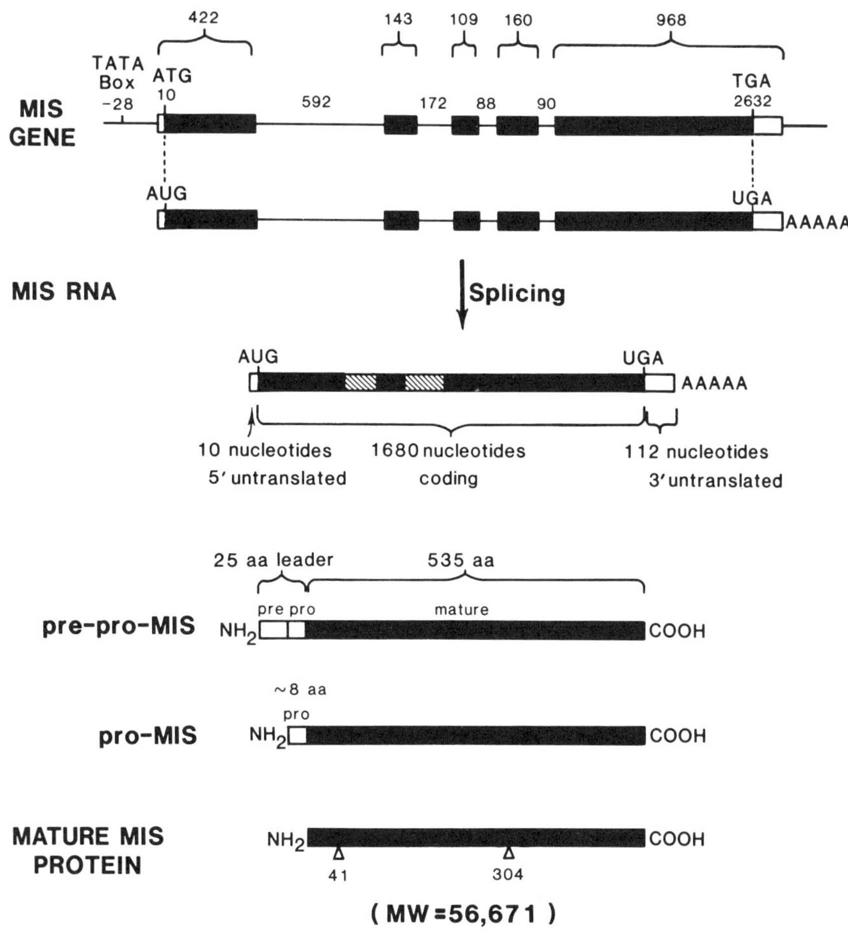

Figure 1. Schematic diagram showing the human MIS gene, RNA, and protein.

pSV2DHFR (Subramani et al. 1981) into CHO cells deficient in dihydrofolate reductase (Chasin and Urlaub 1980). Twenty-five clones that grew in medium lacking nucleosides were grown up in T75 flasks. Total RNA was isolated from these clones and analyzed for the presence of human MIS mRNA by the S1 assay shown in Figure 2B. The results of this analysis with 18 clones are shown in Figure 2A. The clones vary significantly in their expression levels of MIS mRNA. However, the ratio of protected fragment to undigested probe in those clones expressing MIS mRNA indicates that the majority of the RNA is spliced.

Do the CHO cells translate the mRNA and secrete active MIS into the medium? To answer this question, we collected conditioned medium after 48 hours, concentrated it by ammonium sulfate precipitation, and tested for biologically active MIS with the organ culture assay (Picon 1969; Donahoe et al. 1977). The results are shown in Figure 3. The conditioned medium from a cell line positive for MIS mRNA, 311-22, produced grade-3–4 regression of the Mullerian duct (Fig. 3A), whereas conditioned medium from a control cell line G2 did not cause regression (Fig. 3B). Grading is based on the reduction in size of the Mullerian duct (M) relative to the Wolffian duct (W) and the degree of basement membrane breakdown and mesenchymal condensation around the Mullerian duct.

MIS was purified from the conditioned medium of cell line 311-22 using lentil-lectin chromatography and analyzed on Western blots with two different antibodies (Fig. 4). One antibody was raised against denatured bovine MIS and the other was raised against a peptide of human MIS. In both cases, the antibodies recognize a protein in the conditioned medium of 311-22 with a molecular weight of 70,000–74,000. There is no detectable protein in the conditioned medium of CHO cell line G2. Thus, human MIS made in CHO cells is glycosylated to the same or approximately the same level as bovine MIS isolated from newborn testis.

Figure 5 shows the results of a metabolic-labeling experiment with a cell line (311-2A9) that was selected in 10 nM methotrexate. The cells were grown for 24 hours in the presence of [^{3}H]glucosamine, after which the glycoproteins were batch-purified from the condi-

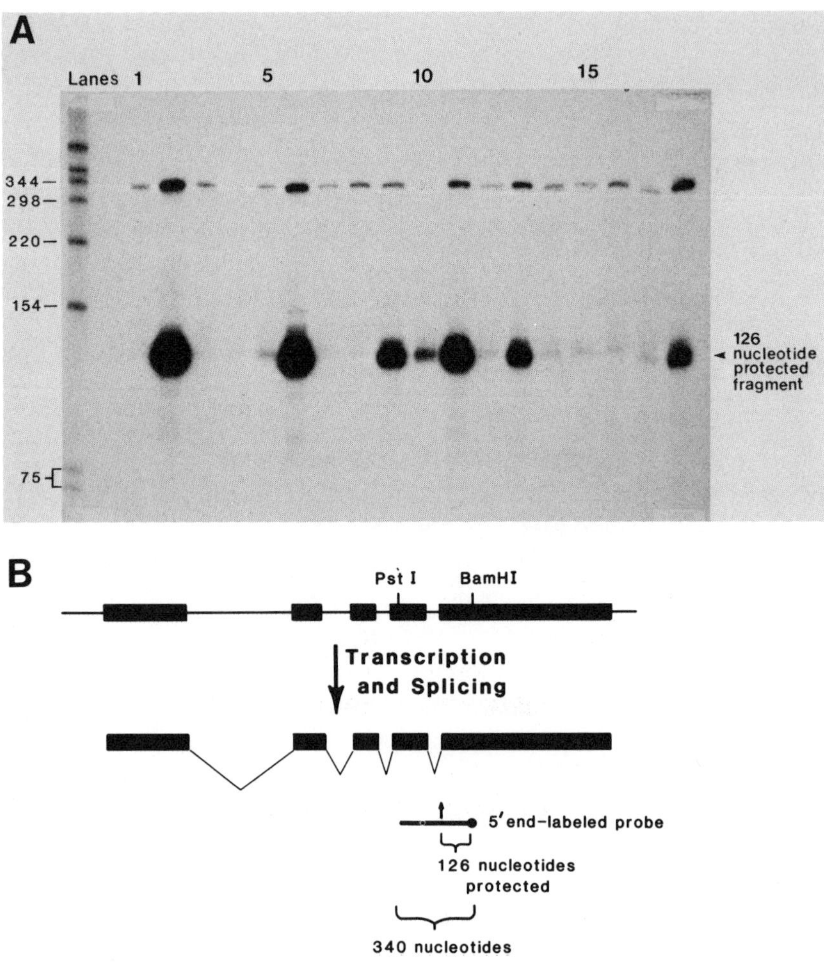

Figure 2. Analysis of CHO clones for human MIS mRNA. (*A*) S1 analysis of RNA isolated from 18 CHO cell lines transfected with the human MIS gene (pBG311.hmis) and the DHFR cDNA (SV2DHFR) with the probe shown in *B*. Markers are shown at the left. (*B*) The 340-nucleotide probe is shown relative to the fourth intron of the human MIS gene.

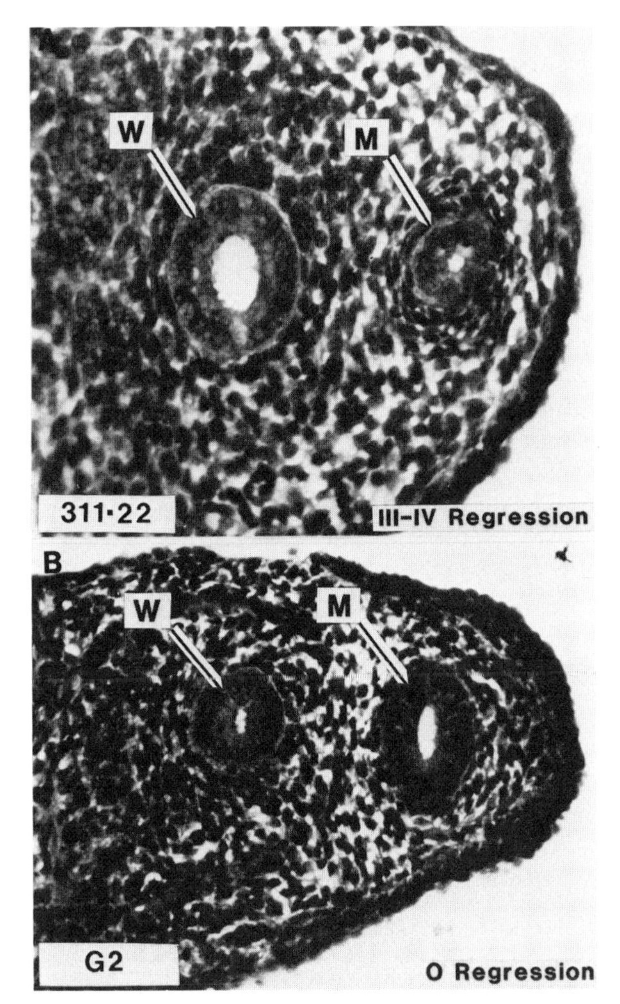

Figure 3. Regression of the rat Mullerian duct by human MIS secreted by CHO cells. Conditioned medium from cell line 311-22 (*A*) and control cell line G2 (*B*) was concentrated by ammonium sulfate precipitation and added to the organ culture assay. (M) Mullerian duct; (W) Wolffian duct.

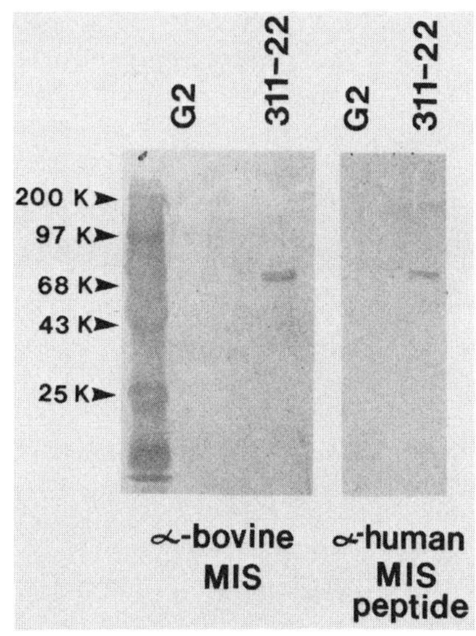

Figure 4. Western blot analysis of partially purified human MIS from the conditioned medium of CHO cell line 311-22, with two antibodies. The conditioned medium of cell line G2 (negative for MIS) was treated in a similar fashion.

tioned medium with lentil-lectin–Sepharose. MIS was then immunoprecipitated with the antibody against denatured bovine MIS. Cell line G2 again served as the negative control. MIS is the highest labeled glycoprotein secreted by cell line 311-2A9 (Fig. 5). The level of MIS produced by this line is about ten times the level produced by 311-22 (data not shown). Cell line 311-2A9 is currently being amplified in 30 nM methotrexate in order to increase the expression level still further.

DISCUSSION

The production of human MIS in animal cells is an important milestone in the development of MIS as a potential anti-cancer drug. The scarcity of MIS has hindered these investigations up until the present time. Animal cells producing MIS in the range of 1 mg/liter of conditioned medium will be able to supply sufficient quantities for animal oncology testing and for investigating the mechanism of action of MIS. We have demonstrated that human MIS in CHO cells is glycosylated to approximately the same level as bovine MIS isolated from newborn testis and that this MIS is active in regressing the rat Mullerian duct in vitro. Further experiments are needed to determine if glycosylation is necessary for biological activity and whether the recombinant protein is identical to the natural protein.

What is the prognosis for MIS as a potential anticancer drug? The results summarized in Table 1 document that MIS can inhibit tumor growth in vitro and in vivo. A number of clinical correlations carried out by other investigators (Alberts et al. 1980; Von Hoff and Weisenthal 1980; Hamburger 1981) indicate that inhibition of growth of clonogenic cells in vitro predicts clinical response in vivo. It has also been established that MIS can be injected into the tail vein of a mouse and cause regression of a rat Mullerian duct implanted in the subrenal capsule of the mouse kidney (Donahoe et al. 1984). Together, these results suggest that MIS will have a half-life sufficiently long for it to exert its inhibitory effects on implanted tumors.

Support for the effectiveness of MIS as an anti-cancer treatment is also provided by the research of Sporn and Roberts and their co-workers on TGF-β. MIS shares distinct homology with TGF-β at its carboxy-terminal end (Cate et al. 1986). Recently, Roberts et al. (1985) have demonstrated that TGF-β under certain conditions is a potent inhibitor of cellular proliferation at very low concentrations (pM). These similarities between MIS and TGF-β suggest that they may have similar mechanisms of action.

This may also be indicated by the effects that MIS and TGF-β have on the extracellular matrix. It has been reported that TGF-β can stimulate cells to synthesize

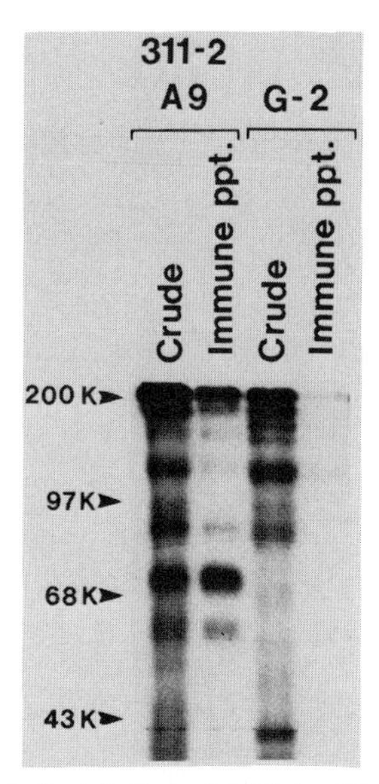

Figure 5. Metabolic labeling of MIS in CHO cell line 311-2A9. CHO cell lines 311-2A9 and G2 were incubated in [^{3}H]glucosamine, and the glycoproteins were isolated from the conditioned medium with lentil-lectin–Sepharose. Aliquots of the purified labeled glycoproteins were analyzed by SDS-PAGE before (crude) and after immunoprecipitation with an antibody against denatured bovine MIS.

extracellular matrix components (Sporn et al. 1983; Ignotz and Massague 1986; Roberts et al. 1986). Ignotz and Massague (1986) have proposed that the ability of TGF-β to potentiate the growth of cells is through an increase in the levels of fibronectin and collagen. In contrast, breakdown of the basement membrane is one of the earliest events in MIS-induced regression of the Mullerian duct. This is characterized by a loss of fibronectin and type IV collagen (Ikawa et al. 1984) and rapid turnover of hyaluronate and sulfated glycosaminoglycans (Hayashi et al. 1982). Whether this correlation between the growth effects mediated by TGF-β and MIS and the composition of the extracellular matrix is important in a causal sense remains to be determined.

Understanding the mechanism of action of MIS will be a fundamental part of developing it as an anti-cancer agent. It has been demonstrated that MIS can inhibit the autophosphorylation of the EGF receptor (Hutson et al. 1984; J.P. Coughlin et al., in prep.). Since MIS does not compete with EGF for receptor sites, this is most likely a secondary event that occurs after MIS binds to its receptor and starts the series of events that lead to regression of the Mullerian duct or the cessation of cellular proliferation. Thus, MIS may inhibit growth by antagonizing the effects of a growth factor. The knowledge that is obtained from studying the mechanism of action of MIS may provide insight into additional therapies for cancer.

ACKNOWLEDGMENTS

We thank G. Torres for expert technical assistance with tissue culture, G. Deegan for assistance in generating antibodies, J. Leban for synthesizing the human MIS peptide, and R. Ragin and T. Manganaro for performing the organ culture assays. We also thank J.L. Browning, R.J. Mattaliano, R.B. Pepinsky, J. Smart, and V. Sato for valuable discussions, B. Wallner for critical reading of the manuscript, and N. Ostrom for help in preparing this manuscript. This research was supported by Biogen Research Corporation, by U.S. Public Health Service grant 1-R43-CA-42340-01, awarded by the National Cancer Institute, DHHS, to the Biogen Research Corporation, and by National Institutes of Health grant CA-17393 and American Cancer Society grant PDT221a, awarded to P.D.K., who is also a recipient of a Johnson and Johnson Directed Giving Award.

REFERENCES

Alberts, D.S., H.S.G. Chen, B. Soehnlen, S.E. Salmon, E.A. Surwit, L. Young, and T.E. Moon. 1980. *In vitro* clonogenic assay for predicting response of ovarian cancer to chemotherapy. *Lancet* **II:** 340.

Budzik, G.P., P.K. Donahoe, and J.M. Hutson. 1985. Possible purification of Mullerian inhibiting substance and a model for its mechanism of action. In *Developmental mechanisms: Normal and abnormal*, p. 207. A.R. Liss, New York.

Budzik, G.P., S.M. Powell, S. Kamagata, and P.K. Donahoe. 1983. Mullerian inhibiting substance fractionation by dye affinity chromatography. *Cell* **34:** 307.

Cate, R.L., R.J. Mattaliano, C. Hession, R. Tizard, N.M. Farber, A. Cheung, E.G. Ninfa, A.Z. Frey, D.J. Gash, E.P. Chow, R.A. Fisher, J.M. Bertonis, G. Torres, B.P. Wallner, K.L. Ramachandran, R.C. Ragin, T.F. Manganaro, D.T. MacLaughlin, and P.K. Donahoe. 1986. Isolation of the bovine and human genes for Mullerian inhibiting substance and expression of the human gene in animal cells. *Cell* **45:** 685.

Chasin, L. and G. Urlaub. 1980. Isolation of Chinese hamster cell mutants deficient in dihydrofolate reductase activity. *Proc. Natl. Acad. Sci.* **77:** 4216.

DeLarco, J.E. and G.J. Todaro. 1978. Growth factors from murine sarcoma virus-transformed cells. *Proc. Natl. Acad. Sci.* **75:** 4001.

Donahoe, P.K., Y. Ito, S. Marfatia, and W.H. Hendren. 1977. A graded organ culture assay for the detection of Mullerian inhibiting substance. *J. Surg. Res.* **23:** 141.

Donahoe, P.K., D.A. Swann, A. Hayashi, and M.D. Sullivan. 1979. Mullerian duct regression in the embryo correlated with cytotoxic activity against human ovarian cancer. *Science* **205:** 913.

Donahoe, P.K., A.F. Fuller, Jr., R.E. Scully, S.R. Guy, and G.P. Budzik. 1981. Mullerian inhibiting substance inhibits growth of a human ovarian cancer in nude mice. *Ann. Surg.* **194:** 472.

Donahoe, P.K., I. Krane, A.E. Bogden, S. Kamagata, and G.P. Budzik. 1984. Subrenal capsule assay to test the viability of parentally delivered Mullerian inhibiting substance. *J. Pediatr. Surg.* **19:** 863.

Donahoe, P.K., G.P. Budzik, R.L. Trelstad, M. Mudgett-Hunter, A.F. Fuller, Jr., J.M. Hutson, H. Ikawa, A. Hayashi, and D.T. MacLaughlin. 1982. Mullerian inhibiting substance: An update. *Recent Prog. Horm. Res.* **38:** 279.

Fuller, A.F., Jr., G.P. Budzik, I.M. Krane, and P.K. Donahoe. 1984. Mullerian inhibiting substance inhibition of a human endometrical carcinoma cell line xenografted in nude mice. *Gynecol. Oncol.* **17:** 124.

Fuller, A.F., Jr., S.R. Guy, G.P. Budzik, and P.K. Donahoe. 1982. Mullerian inhibiting substance inhibits colony growth of a human ovarian carcinoma cell line. *J. Clin. Endocrinol. Metab.* **54:** 1051.

Fuller, A.F., Jr., I.M. Krane, G.P. Budzik, and P.K. Donahoe. 1985. Mullerian inhibiting substance reduction of colony growth of human gynecologic cancers in a stem cell assay. *Gynecol. Oncol.* **22:** 135.

Hamburger, A.W. 1981. Use of *in vitro* tests in predictive cancer chemotherapy. *J. Nat. Cancer Inst.* **66:** 981.

Hayashi, A., P.K. Donahoe, G.P. Budzik, and R.L. Trelstad. 1982. Periductal and matrix glycosaminoglycans in rat Mullerian duct development and regression. *Dev. Biol.* **92:** 16.

Hutson, J.M., M.E. Fallat, S. Kamagata, P.K. Donahoe, and G.P. Budzik. 1984. Phosphorylation events during Mullerian duct regression. *Science* **223:** 586.

Ignotz, R.A. and J. Massague. 1986. Transforming growth factor-β stimulates the expression of fibronectin and collagen and their incorporation into the extracellular matrix. *J. Biol. Chem.* **261:** 4337.

Ikawa, H., R.L. Trelstad, J.M. Hutson, T.F. Manganaro, and P.K. Donahoe. 1984. Changing patterns of fibronectin, laminin, type IV collagen, and a basement membrane proteoglycan during rat Mullerian duct regression. *Dev. Biol.* **102:** 260.

Josso, N., J.Y. Picard, and D. Tran. 1977. The anti-Mullerian hormone. *Recent Prog. Horm. Res.* **33:** 117.

Josso, N., J.Y. Picard, and B. Vigier. 1981. Purification de l'hormone anti-Mullerian bovine a l'aide d'un anticorps monoclonal. *C.R. Acad. Sci.* **293:** 447.

Kaplan, P.L., M. Anderson, and B. Ozanne. 1982. Transforming growth factor(s) production enables cells to grow in the absence of serum: An autocrine system. *Proc. Natl. Acad. Sci.* **79:** 485.

Lang, R.A., D. Metcalf, N.M. Gough, A.R. Dunn, and T.J. Gonda. 1985. Expression of a hemopoietic growth factor cDNA in a factor-dependent cell line results in autonomous growth and tumorigenicity. *Cell* **43:** 531.

Picard, J.Y. and N. Josso. 1984. Purification of testicular anti-Müllerian hormone allowing direct visualization of the pure glycoprotein and determination of yield and purification factor. *Mol. Cell. Endrocrinol.* **34:** 23.

Picard, J.-Y., C. Goulut, R. Bourrillon, and N. Josso. 1986. Biochemical analysis of bovine testicular anti-Mullerian hormone. *FEBS. Lett.* **195:** 73.

Picon, R. 1969. Action du testicule foetal sur le development *in vitro* des canaux de Muller chez le rat. *Arch. Anat. Microsc. Morphol. Exp.* **58:** 1.

Roberts, A.B., M.A. Anzano, L.M. Wakefield, N.S. Roche, D.F. Stern, and M.B. Sporn. 1985. Type B transforming growth factor: A bifunctional regulator of cellular growth. *Proc. Natl. Acad. Sci.* **82:** 119.

Roberts, A.S., M.B. Sporn, R.K. Assoian, J.M. Smith, N.S. Roche, L.M. Wakefield, U.I. Heine, L.A. Liotta, V. Falanga, J.H. Kehrl, and A.S. Fauci. 1986. Transforming growth factor-β: Rapid induction of fibrosis and angiogenesis *in vivo* and stimulation of collagen formation *in vitro*. *Proc. Natl. Acad. Sci.* **83:** 4167.

Sporn, M.B. and A.B. Roberts. 1985. Autocrine growth factors and cancer. *Nature* **313:** 745.

Sporn, M.B. and G.J. Todaro. 1980. Autocrine secretion and malignant transformation of cells. *N. Eng. J. Med.* **303:** 878.

Sporn, M.B., A.B. Roberts, J.H. Shull, J.M. Smith, J.M. Ward, and J. Sodek. 1983. Polypeptide transforming growth factors isolated from bovine sources and used for wound healing *in vivo*. *Science* **219:** 1329.

Subramani, S., R. Mulligan, and P. Berg. 1981. Expression of the mouse dihydrofolate reductase complementary deoxyribonucleic acid in simian virus 40 vectors. *Mol. Cell. Biol.* **1:** 854.

Todaro, G.J., J.E. DeLarco, S.P. Nissley, and M.M. Rechler. 1977. MSA and EGF receptors on sarcoma virus transformed cells and human fibrosarcoma cells in culture. *Nature* **267:** 526.

Towbin, H., T. Staehelin, and J. Gordon. 1979. Electrophoretic transfer of proteins from polyacrylamide gels to nitrocellulose sheets: Procedure and some applications. *Proc. Natl. Acad. Sci.* **76:** 4350.

Tucker, R.F., G.D. Shipley, and H.L. Moses. 1984. Growth inhibitor from BSC-1 cells closely related to platelet type β transforming growth factor. *Science* **226:** 705.

Von Hoff, D.D. and L. Weisenthal. 1980. *In vitro* methods to predict for patient response to chemotherapy. *Adv. Pharmacol. Chemother.* **17:** 133.

Waterfield, M.D., G.T. Scrace, N. Whittle, P. Stroobant, A. Johnsson, A. Wasteson, B. Westermark, C.-H. Heldin, J.S. Huang, and T.F. Deuel. 1983. Platelet-derived growth factor is structurally related to the putative transforming protein $p28^{sis}$ of simian sarcoma virus. *Nature* **304:** 35.

Weaver, R.F. and C. Weissmann. 1979. Mapping of RNA by a modification of the Berk-Sharp procedure: The 5′ termini of 15 S β-globin mRNA precursor and mature 10 S β-globin mRNA have identical map coordinates. *Nucleic Acids Res.* **7:** 1175.

Endogeneous and Heterologous Expression of Transforming Growth Factor-α in Mammalian Cells

R. Derynck, A. Rosenthal, P.B. Lindquist, T.S. Bringman, and D.V. Goeddel
Department of Molecular Biology, Genentech, Inc., South San Francisco, California 94080

Two distinct polypeptides have been termed transforming growth factors (TGFs): TGF-α is related to epidermal growth factor (EGF) and binds to the EGF receptor, and TGF-β is a structurally unrelated protein that recognizes a different receptor. TGF-α was originally isolated from murine-sarcoma-virus-transformed fibroblast cultures and was therefore initially called sarcoma growth factor (De Larco and Todaro 1978). Subsequent examination of a variety of cell sources showed that this factor was made by many more transformed cell lines but not by adult normal cells in culture (Ozanne et al. 1980; Roberts et al. 1980; Todaro et al. 1980, 1985). The fact that this factor was able to convert normal rat kidney (NRK) cells into phenotypically transformed cells, and its synthesis by various transformed cells, led to the name transforming growth factor. Initially, it was assumed that sarcoma or transforming growth factor was a single peptide (De Larco and Todaro 1978). Extensive biochemical purification and characterization showed later that the preparations consisted of the structurally unrelated peptides TGF-α and TGF-β and that the profound morphological changes induced in NRK fibroblasts are due to their cooperative effect (Anzano et al. 1983). TGF-β by itself will not induce any colony formation of NRK cells in soft agar, whereas pure TGF-α can elicit the formation of a few relatively small colonies. The need for both growth factors to promote phenotypic transformation is dependent on the cell system used. Both TGF-α and TGF-β are needed in the NRK system, whereas in many other cell systems, they do not have a cooperative effect on cell proliferation or transformation.

Structure of TGF-α and Its Precursor

TGF-α isolated from the media of several rodent or human tumor cell lines are heterogeneous in molecular weight as determined by gel filtration. The smallest TGF-α species (~6 kD) has been purified to homogeneity from several cell sources. This has led to the establishment of the complete amino acid sequence of the 50-amino-acid-long rat TGF-α (Marquardt et al. 1984). cDNA analysis (Derynck et al. 1984; Lee et al. 1985b) has now indicated that the 50-amino-acid TGF-α is initially translated as an internal part of a 160-amino-acid precursor from which it is derived after proteolytic cleavage (Fig. 1). A short hydrophobic sequence between positions 8 and 18 following the amino terminus of the precursor suggests the presence of a signal peptide of 20–24 residues long. The 50 amino acids of the fully processed TGF-α peptide are found at residues 40–89 of the precursor. An extremely hydrophobic domain begins nine residues downstream from the carboxyl terminus of the 50-amino-acid TGF-α. The 23-amino-acid length of this segment and the basic character of the flanking amino acids are characteristic of transmembrane domains of membrane proteins. This would thus suggest that the TGF-α precursor, following the removal of the signal peptide, would be inserted into the membrane (Fig. 1). Subsequent proteolytic cleavage at the external segment of the precursor could then release the 50-amino-acid TGF-α. If the TGF-α precursor indeed constitutes a transmembrane protein, then the 37-amino-acid carboxy-terminal portion that contains seven cysteines would remain at the cytoplasmic side of the membrane and may not undergo any disulfide bond formation.

Endogenous TGF-α Secretion by Tumor Cells

The reported secretion of TGF-α by retrovirally transformed fibroblasts and the apparent lack of its synthesis by normal fibroblasts in culture (Todaro et al. 1985) led us to examine a large variety of human tumors for the presence of TGF-α mRNA. We initially examined a variety of human cell lines derived from different tumor types for the presence of TGF-α mRNA. No TGF-α mRNA was detectable in any cell line of hematopoietic origin. However, a TGF-α mRNA species of about 4.5–4.8 kb was clearly present in many other tumor cell lines of carcinoma or sarcoma origin (Fig. 2). Examination of all the hybridization data (data not shown) indicates that TGF-α mRNA occurs most frequently in carcinoma cell lines, although some sarcoma cell lines such as HT1080 also contain relatively high levels of this mRNA. TGF-α mRNA was not detectable in the two normal fibroblast lines examined.

To determine the clinical relevance of this finding, we also examined solid tumors that had been surgically removed. RNAs from many different types of carcinomas and sarcomas were screened by Northern hybridization (data not shown). TGF-α mRNA was detected in many of the samples and its size was identical in all cases (Fig. 2). Due to the limitations of the detection method, very low levels of TGF-α mRNA may have remained undetected. Furthermore, the concentrations of TGF-α mRNA may be diluted due to the presence of nontumor tissue in the tumor specimens. TGF-α

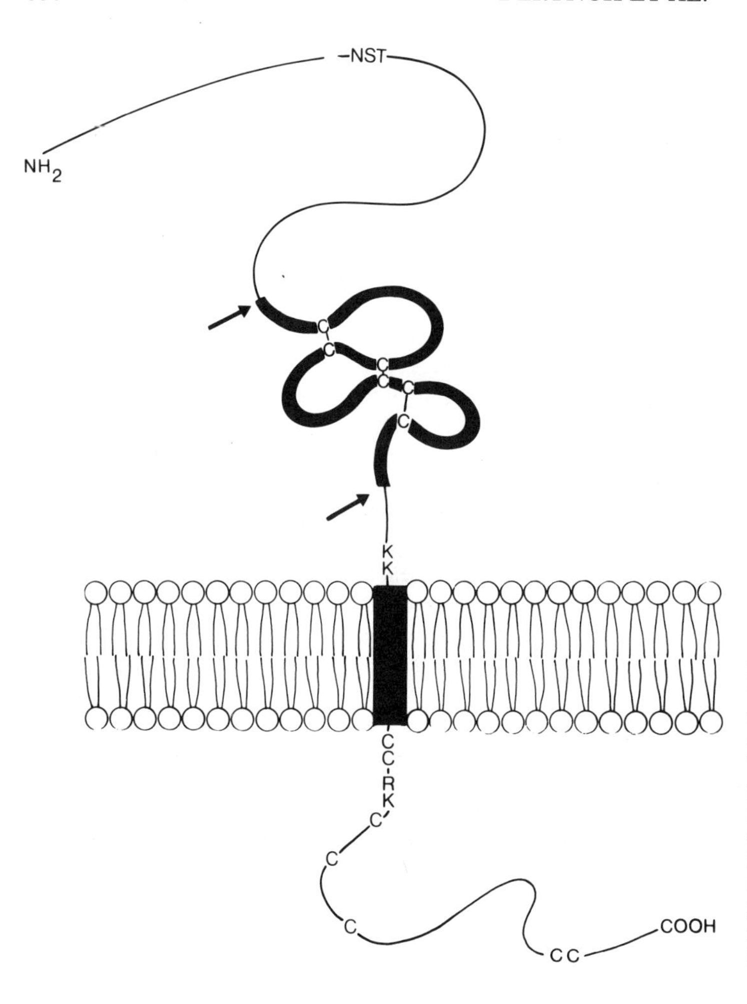

Figure 1. Depiction of a hypothetical model of the TGF-α precursor as a transmembrane protein. The amino-terminal signal sequence is shown as already cleaved from the precursor. The 50-amino-acid TGF-α with its three proposed cysteine (C)-disulfide bridges is shown as a heavy line, flanked by the proteolytic cleavage sites (arrows). The boxed transmembrane region is flanked at each side by two basic amino acids (KK and RK). The carboxy-terminal cytoplasmic domain shown below the membrane is rich in cysteines (C). NST indicates the potential *N*-glycosylation sequence Asn-Ser-Thr.

mRNA could not be detected in the normal tissues examined. The data obtained with cell lines and tissues led to the conclusion that TGF-α is made by a variety of solid tumors but may be absent in hematopoietic tumors. TGF-α mRNA levels were preferentially high in renal carcinomas and in squamous carcinomas, in-

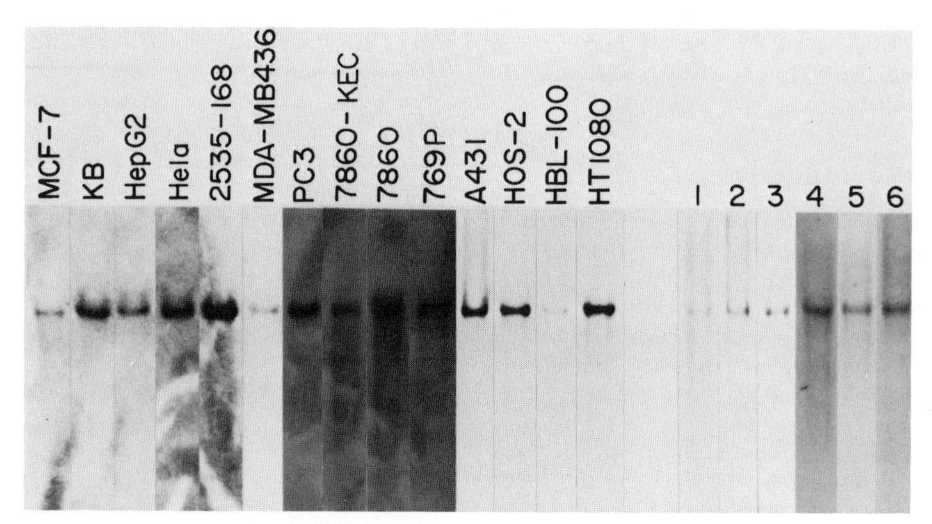

Figure 2. Some examples of Northern hybridizations for TGF-α mRNA, using RNA derived from various tumor cell lines and tumor specimens. The tumor cell lines are designated with their name above each lane, and the tumor specimens are numbered *1* through *6*. (*1*) Adenocarcinoma in endometrium of uterus (Dix); (*2*) metastatic adenocarcinoma in stomach (Glo); (*3*) squamous carcinoma of head and neck (Pit); (*4*) large-cell carcinoma in lung; (*5*) nephroma; (*6*) apudoma.

dependent of their localization. In addition, TGF-α mRNA can also be unambiguously detected in many tumors of neuronal origin and in many mammary carcinomas.

The cell line A-431, which has a 10–50-fold higher level of EGF receptors than most other cell types (Wrann and Fox 1979), is particularly interesting since the effects in these cells of exogenously added EGF have been very well studied. From our results, it is clear that A-431 cells synthesize TGF-α mRNA (Fig. 2) and also release low levels of TGF-α in the medium (data not shown). This suggests that the endogenous production of TGF-α and its interaction with EGF receptors in an autocrine fashion may have to be considered when evaluating the effects of endogenous EGF on this cell line.

Our hybridization data indicate that TGF-α mRNA is not synthesized by any of the hematopoietic tumor cell lines examined. Cells of the hematopoietic lineage are known to lack EGF receptors (Carpenter and Cohen 1979). In contrast, many tumor cells derived from solid tumors synthesize TGF-α, and all these cells also synthesize EGF receptor mRNA (data not shown). This suggests that TGF-α synthesis could act in an autocrine fashion in these solid tumors. Although TGF-α mRNA can be found in tumors belonging to many types of carcinomas and sarcomas, the overall highest and most consistent synthesis can be found in four different tumor types. Squamous carcinomas and renal carcinomas synthesize relatively high levels of TGF-α mRNA, but TGF-α mRNA can also be found in mammary carcinomas and in tumors derived from neuroectodermal origin, such as melanomas. TGF-α synthesis may also be relatively common in hepatomas, since it has been demonstrated in two hepatoma cell lines (Luetteke and Michalopoulos 1985), and TGF-α mRNA is also present in the HepG2 hepatoma cell line. The lack of a sufficient number of tumor samples of particular tumor types precludes making other generalizations.

It is tempting to speculate in the context of an autocrine mechanism that there may be a relationship between the level of TGF-α mRNA and the level of EGF receptor mRNA, since many TGF-α-mRNA-containing tumors also have relatively high levels of EGF receptor mRNA. However, an absolute relationship does not exist, since the HT1080 fibrosarcoma cell line, which contains relatively high TGF-α mRNA levels, has only low levels of EGF receptor mRNA. Also, we could not detect any TGF-α mRNA in the A172 glioblastoma cell line, which contains high levels of EGF receptor mRNA.

Malignancy-associated hypercalcemia occurs relatively frequently in patients with renal carcinoma, squamous carcinomas, breast carcinoma, and melanoma (Mundy et al. 1985). Tumors of these types are most consistent for the presence of TGF-α mRNA. In vitro studies have indicated that TGF-α is a very potent inducer of Ca^{++} release in mouse calvariae and in rat fetal long bones (Stern et al. 1985; Ibbotson et al. 1986). It can thus be speculated that the TGF-α synthesized by these various tumor types triggers or contributes to malignancy-induced hypercalcemia. In this context, it may be relevant that the HOS-2 cell line derived from an osteosarcoma contains highly elevated TGF-α mRNA levels.

Previous studies have indicated that TGF-α is a potent inducer of angiogenesis in an in vivo model (Schreiber et al. 1986). Hematopoietic tumors that do not require neovascularization for their development do not contain TGF-α mRNA. However, the endogenous synthesis of TGF-α in various types of solid tumors may not only contribute to an autocrine growth stimulation of these tumor cells, but also play a paracrine role in the tumor-induced angiogenesis, probably in concert with other angiogenic factors.

TGF-α is thus synthesized by many tumor cells but cannot be detected in medium from normal cells in vitro and is not known to be made in normal fully developed tissues. However, only a very limited number of observations have been reported, and as yet, it cannot be excluded that TGF-α could play a role in the normal physiology of the adult organism. Recent evidence from specific antibody-based detection and Northern hybridizations have indicated that TGF-α is synthesized during early fetal development (Lee et al. 1985a; Twardzik 1985), which indicates that TGF-α may function as a normal embryonic version of a family of EGF-related growth factors. The expression of the gene may be reinitiated during the process of malignant transformation and tumor development, indicating that TGF-α is an oncodevelopmental antigen.

Expression of an Introduced TGF-α cDNA in Rat Fibroblasts

Much attention has been focused during the last few years on whether the secretion of a growth factor can induce malignant transformation via an autocrine mechanism. The fact that many squamous carcinomas and renal carcinomas produce both TGF-α mRNA and relatively high levels of EGF receptor mRNA would agree with the hypothesis that TGF-α overproduction could be involved in the transformation process. We therefore tested whether expression of a human TGF-α cDNA would induce malignant transformation of rat fibroblasts in an autocrine fashion.

An expression vector was constructed in which the cDNA sequence encoding the human TGF-α precursor protein (Derynck et al. 1984) was placed downstream from the SV40 early promoter. The proximal part of the natural TGF-α 3′-untranslated region is followed by a segment of the 3′-untranslated region of the hepatitis-B surface antigen (HBsAg) gene (Crowley et al. 1983), which provides the transcription-termination and polyadenylation signal sequences. This plasmid also contains the sequences coding for aminoglycosyl phosphotransferase II as a selectable marker (Southern and Berg 1982) and for dihydrofolate reductase, thus enabling the plasmid to be amplified in mammalian cells (Alt et al. 1978). The latter two genes are also preceded by separate SV40 early promoter elements and followed by the polyadenylation signal sequences of the HBsAg gene. This plasmid, pMTE4E (Fig. 3), was introduced into Rat-1 fibroblasts using the calcium phosphate transfection method (Wigler et al. 1979), and clones resistant to 400 μg/ml of the antibiotic Geneticin G-418 were picked, expanded, and analyzed for their ability to secrete human TGF-α into the culture medium. As judged from a sensitive enzyme-linked immunosorbent assay (ELISA), using recombinant human TGF-α as a standard, approximately 40 out of the 100 clones analyzed secreted human TGF-α at concentrations ranging between 0.1 and 1 ng/ml per 10^5 cells per 24 hours (Rosenthal et al. 1986).

The acquisition of anchorage independence is generally considered a property concomitant with the transformed phenotype. The transfected Rat-1 fibroblasts were therefore examined for their growth characteristics. The parental Rat-1 cells grew on a solid support in a monolayer and displayed contact inhibition. Three clones that expressed more than 0.7 ng of TGF-α per 24 hours grew in culture plates to a higher density than the parental Rat-1 cells and displayed a tendency to pile up. Clones synthesizing TGF-α at levels of 0.4 ng/ml or higher also acquired the ability to form colonies in soft agar (Table 1). The number of colonies correlates roughly with the level of TGF-α secreted and corresponds to the number of colonies obtained with Rat-1 cells in the presence of similar concentrations of exogenously added TGF-α. Clones expressing low levels of TGF-α as well as the parental Rat-1 cells do not show any anchorage-independent growth. The number of colonies formed in soft agar increased approximately twofold when human TGF-β was added at a concentration of 2.6 ng/ml, indicating that TGF-β can potentiate the effect of endogenously produced TGF-α (Rosenthal et al. 1986).

The hypothesis that autocrine growth factors are first secreted and subsequently act through cell-surface receptors suggests that the growth of tumor cells might be controlled by extracellular antagonists for either the autocrine peptides or their receptors. To test this possibility, we analyzed the ability of antibodies raised against human TGF-α to prevent colony formation in soft agar by rat fibroblasts expressing human TGF-α. A monoclonal antibody to TGF-α does not inhibit the growth of TGF-α-expressing clones in monolayer (data not shown). This monoclonal antibody does prevent colony formation by normal Rat-1 cells in response to exogenously added human TGF-α and by the TGF-α-expressing clones (Table 1). In both cases, colony formation was not inhibited by an anti-human growth hormone (hGH) monoclonal antibody. Thus, an autocrine growth stimulation system has been experimentally generated in which TGF-α is first secreted and subsequently exerts its effects through receptors on the cell surface. Colony formation was inhibited to a large extent by anti-human TGF-α antibodies, implying an extracellular interaction of TGF-α with the EGF receptor (Rosenthal et al. 1986).

To assess the tumorigenicity of these cells in vivo, cells from two clones that express high levels of TGF-α were injected subcutaneously into nude mice, and the resulting tumors were scored. Mice injected with Rat-1 fibroblasts transformed by the TGF-α cDNA expression vector displayed a much higher frequency of tumor formation as compared with mice injected with nontransfected Rat-1 cells. All animals injected with the TGF-α-expressing Rat-1 cells had developed palpa-

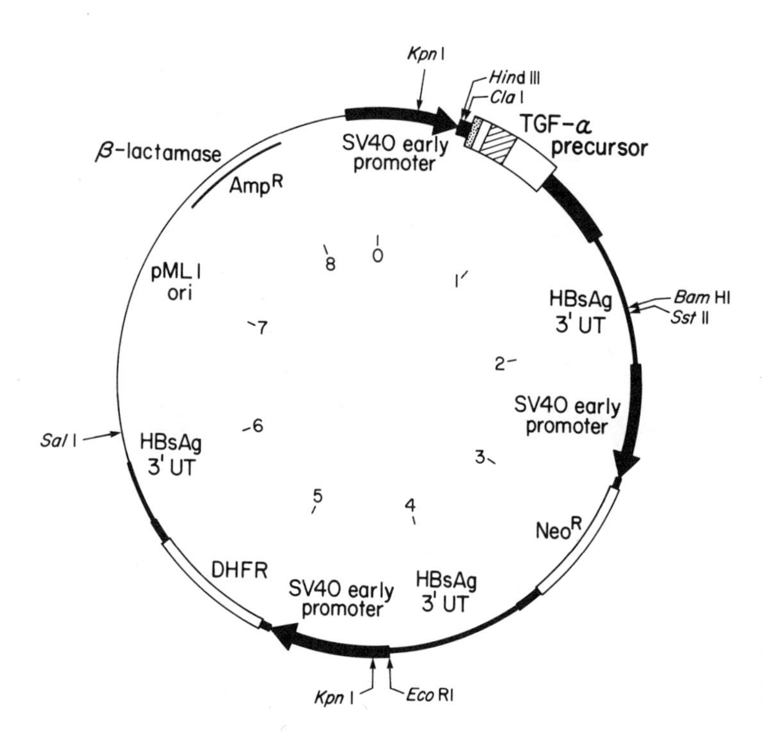

Figure 3. Schematic diagram of the pMTE4E TGF-α expression vector. The SV40 promoter segments with the direction of early transcription are indicated with black arrows. The coding sequences transcribed from the SV40 promoter elements are boxed. Within the TGF-α precursor-coding sequence, the dotted box represents the signal sequence and the dashed area corresponds to the 50-amino-acid TGF-α. Some marker restriction sites are shown. The distances in kilobase pairs are marked in the center. (HBsAg 3'-UT) Hepatitis-B surface antigen gene 3'-untranslated sequence; (DHFR) dihydrofolate reductase.

Table 1. Inhibition of Colony Formation in Soft Agar of TGF-α-expressing Clones by Anti-human TGF-α Monoclonal Antibodies

	Number of colonies in the presence of	
Clone number	100 μg/ml control monoclonal antibody	100 μg/ml anti-TGF-α monoclonal antibody
Rat-1 (+ 10 ng/ml TGF-α)	800	18
Clone 16	290	20
Clone 42	195	20
T24 c-Ha-*ras*	~5000	~5000

Anti-human growth hormone (hGH) antibodies were used as negative controls.

ble tumors after 4 weeks. But as judged from the latency period and tumor size, the c-Ha-*ras*-transformed Rat-1 cells are much more tumorigenic than the TGF-α-producing Rat-1 cells in this in vivo system (Rosenthal et al. 1986).

CONCLUSION

Unrestricted growth that leads to malignant transformation might begin with a cell acquiring the ability to produce and respond to growth peptides (Todaro et al. 1977; Sporn and Todaro 1980; Sporn and Roberts 1985). If true, this hypothesis implies that expression of an introduced growth-factor-coding sequence in cells that bear its receptors could lead to or contribute to their malignant transformation. It also implies that an extracellular antagonist for the autocrine peptide or its receptor could reverse this growth-factor-mediated phenotypic transformation. We tested this hypothesis directly by introducing and expressing a human TGF-α cDNA in normal rat fibroblasts, thus experimentally generating a TGF-α-mediated autocrine stimulation system. As predicted by the autocrine hypothesis, synthesis and secretion of human TGF-α result in transformation of Rat-1 cells. The cells show anchorage-independent growth and efficient tumor formation in nude mice. The transformed phenotype in culture could be reversed by anti-TGF-α monoclonal antibodies, indicating that TGF-α is secreted and then exerts its effects through receptors on the cell surface.

Experimental reconstitution of autocrine systems has been reported previously for two other growth factors: platelet-derived growth factor (PDGF) and granulocyte-macrophage colony-stimulating factor (GM-CSF). Expression of the normal PDGF β-chain gene (Gazit et al. 1984) or of its acquired viral counterpart, the p28$^{v\text{-}sis}$ gene of simian sarcoma virus (SSV) (Huang et al. 1984; Johnsson et al. 1985), in nontumorigenic cells results in malignant transformation. PDGF antiserum partially inhibits the growth of some, but not all, SSV-transformed fibroblast cell lines (Huang et al. 1984; Johnsson et al. 1985). The fact that only a small fraction of the PDGF produced by SSV-transformed cells is secreted (Robbins et al. 1985), combined with very limited growth inhibition of these cells by antisera to PDGF, led to the proposal that the growth factor may be able to interact and activate its receptor in an intracellular compartment. Similarly, nontumorigenic hematopoietic cells can be converted into transformed leukemic cells following expression of an introduced human GM-CSF gene. The inability of GM-CSF antiserum to prevent transformation implies that at least part of the GM-CSF receptor interaction is intracellular (Lang et al. 1985). Two separate instances of inhibition of autocrine growth stimulation by extracellular antagonists have been reported. The growth of a human small-cell lung carcinoma, which produces the growth factor bombesin, can be inhibited by anti-bombesin monoclonal antibodies (Cuttitta et al. 1985). Also, polyclonal antibodies against a macrophage growth factor (MGF) inhibited the growth of virally transformed chicken hematopoietic cells that produce this polypeptide (Adkins et al. 1984). The difference in the ability to control autocrine effects of growth factors with extracellular antagonists in these four examples may reflect differences either in the mechanism of hormone-receptor interactions, in the cell dependence on a particular growth factor for maintenance of the transformed phenotype, or in the affinity and neutralizing ability of the antibodies used.

Our results with the TGF-α system indicate that expression of an introduced growth factor gene in established cell lines can result in the acquisition of the transformed phenotype via an autocrine mechanism. The observations that the TGF-α-expressing Rat-1 cells and many other transformed cells need exogenous growth factors in order to grow in soft agar might indicate that in addition to TGF-α, other growth factors are required for the induction and maintenance of the observed transformed phenotype. This is supported by the finding that mitotic stimulation of fibroblasts by IGF-II is required for phenotypic transformation by TGF-α (Massagué et al. 1985). Possible synergistic actions between several growth factors might also explain the efficient anchorage-independent growth of nontransformed rat fibroblasts in the presence of EGF, TGF-β, and PDGF (Assoian et al. 1984).

Constitutive synthesis of TGF-α may be of significance for the development of malignancies in vivo. As discussed above, TGF-α mRNA was found in many types of solid tumors, especially squamous, renal, and mammary carcinomas, and tumors of neuroectodermal origin such as melanomas. All of the tumors that contained TGF-α mRNA also contained mRNA for the EGF receptor, supporting the possibility that autocrine action of TGF-α may play a physiological role in tumor development.

Both TGF-α and EGF mediate their activity through the same receptor and are equally effective at inducing mitosis and anchorage-independent growth of NRK fibroblasts. However, TGF-α synthesis is far more prevalent than EGF synthesis in tumors. EGF expression has been demonstrated in only a single salivary gland adenocarcinoma cell line (Sato et al. 1985). It is conceivable that synthesis and proper processing of the EGF precursor could trigger transformation in vitro in a manner analogous to that observed with TGF-α. However, no experimental data are available to support this proposal. The differential expression in vivo of these related growth factors could reflect differences in their mechanism of gene regulation whereby the TGF-α gene can be more easily activated during the transformation process. The expression of TGF-α in naturally occurring tumors may also be related to significant differences in biological activities between TGF-α and EGF. TGF-α has indeed been shown to be superior to EGF in the induction of bone resorption in vitro (Stern et al. 1985; Ibbotson et al. 1986) and of angiogenesis (Schreiber et al. 1986), activities that may play key roles in the development of malignancies in vivo. The studies described here support the notion that TGF-α synthesis could result in tumor formation under physiological conditions. Furthermore, the susceptibility of the TGF-α-expressing Rat-1 cells to the anti-TGF-α antibodies raises the intriguing possibility of inhibiting the growth of TGF-α-expressing human tumors with extracellular antagonists.

REFERENCES

Adkins, B., A. Leutz, and T. Graf. 1984. Autocrine growth induced by *src*-related oncogenes in transformed chicken myeloid cells. *Cell* **39:** 439.

Alt, F.W., R.E. Kellems, J.R. Bertino, and R.T. Schimke. 1978. Selective amplification of dihydrofolate reductase genes in methotrexate-resistant variants of cultured murine cells. *J. Biol. Chem.* **253:** 1357.

Anzano, M.A., A.B. Roberts, J.M. Smith, M.B. Sporn, and J.E. De Larco. 1983. Sarcoma growth factor from conditioned medium of virally transformed cells is composed of both type α and type β transforming growth factors. *Proc. Natl. Acad. Sci.* **80:** 6264.

Assoian, R.K., G.R. Grotendorst, D.M. Miller, and M.B. Sporn. 1984. Cellular transformation by coordinated action of three peptide growth factors from human platelets. *Nature* **309:** 804.

Carpenter, G. and S. Cohen. 1979. Epidermal growth factor. *Annu. Rev. Biochem.* **48:** 193.

Crowley, C.W., C.-C. Liu, and A.D. Levinson. 1983. Plasmid directed synthesis of hepatitis B surface antigen in monkey cells. *Mol. Cell. Biol.* **3:** 44.

Cuttitta, F., D.N. Carney, J. Mulshire, T.W. Moody, J. Fedorko, A. Fischler, and J.D. Minna. 1985. Bombesin-like peptides can function as autocrine growth factors in human small-cell lung cancer. *Nature* **316:** 823.

De Larco, J. and G.J. Todaro. 1978. Growth factors from murine sarcoma virus-transformed cells. *Proc. Natl. Acad. Sci.* **75:** 4001.

Derynck, R., A.B. Roberts, M.E. Winkler, E.Y. Chen, and D.V. Goeddel. 1984. Human transforming growth factor-α: Precursor structure and expression in *E. coli*. *Cell* **38:** 287.

Gazit, A., H. Igarashi, I.M. Chiu, A. Srinivasan, A. Yaniv, S.R. Tronick, K.C. Robbins, and S.A. Aaronson. 1984. Expression of the normal human sis/PDGF-2 coding sequence induces cellular transformation. *Cell* **39:** 89.

Huang, J.S., S.S. Huang, and T.F. Deuel. 1984. Transforming protein of simian sarcoma virus stimulates autocrine growth of SSV-transformed cells through PDGF cell surface receptors. *Cell* **39:** 79.

Ibbotson, K.J., J. Harrod, M. Gowen, S. D'Souza, M.E. Winkler, R. Derynck, and G.R. Mundy. 1986. The effects of recombinant human transforming growth factor-α on bone resorption *in vitro*. *Proc. Natl. Acad. Sci.* **83:** 2228.

Johnsson, A., C. Betsholtz, C.H. Heldin, and B. Westermark. 1985. Antibodies against platelet-derived growth factor inhibit acute transformation by simian sarcoma virus. *Nature* **317:** 438.

Lang, R.A., D. Metcalf, N.M. Gough, A.R. Dunn, and T.J. Gonda. 1985. Expression of a hemopoietic growth factor cDNA in a factor-dependent cell line results in autonomous growth and tumorigenicity. *Cell* **43:** 531.

Lee, D.C., R.M. Rochford, G.J. Todaro, and L.P. Villareal. 1985a. Developmental expression of rat transforming growth factor-α mRNA. *Mol. Cell. Biol.* **5:** 3644.

Lee, D.C., T.M. Rose, N.R. Webb, and G.J. Todaro. 1985b. Cloning and sequence analysis of a cDNA for rat transforming growth factor-α. *Nature* **313:** 489.

Luetteke, D.C. and G.K. Michalopoulos. 1985. Partial purification and characterization of a hepatocyte growth factor produced by rat hepatocellular carcinoma cells. *Cancer Res.* **45:** 6331.

Marquardt, H., M.W. Hunkapiller, L.E. Hood, and G.J. Todaro. 1984. Rat transforming growth factor type 1: Structure and relation to epidermal growth factor. *Science* **223:** 1079.

Massagué, J., B. Kelly, and C. Mottola. 1985. Stimulation by insulin-like growth factors is required for cellular transformation by type β transforming growth factor. *J. Biol. Chem.* **260:** 4551.

Mundy, G.R., K.J. Ibbotson, and S.M. D'Souza. 1985. Tumor products and the hypercalcemia of malignancy. *J. Clin. Invest.* **76:** 391.

Ozanne, B., R.J. Fulton, and P.L. Kaplan. 1980. Kirsten murine sarcoma virus transformed cell lines and a spontaneously transformed rat cell line produce transforming factors. *J. Cell. Physiol.* **105:** 163.

Robbins, K.C., F. Seal, J.H. Pierce, and S.A. Aaronson. 1985. The v-sis/PDGF-2 transforming gene product localizes to cell membranes but is not a secretory protein. *EMBO J.* **4:** 1783.

Roberts, A.B., L.C. Lamb, D.L. Newton, M.B. Sporn, J.E. De Larco, and G.J. Todaro. 1980. Transforming growth factors: Isolation of peptides from virally and chemically transformed cells by acid/ethanol extraction. *Proc. Natl. Acad. Sci.* **77:** 3494.

Rosenthal, A., P.B. Lindquist, T.S. Bringman, D.V. Goeddel, and R. Derynck. 1986. Expression in rat fibroblasts of a human transforming growth factor-α cDNA probe. *Cell* **46:** 301.

Sato, M., H. Yoshida, Y. Hayashi, K. Miyakami, T. Bando, T. Yangawa, Y. Yura, M. Azuma, and A. Ueno. 1985. Expression of epidermal growth factor and transforming growth factor-β in a human salivary gland adenocarcinoma cell line. *Cancer Res.* **45:** 6160.

Schreiber, A.B., M.E. Winkler, and R. Derynck. 1986. Transforming growth factor-α is a more potent angiogenesis mediator than epidermal growth factor. *Science* **232:** 1250.

Southern, P.J. and P. Berg. 1982. Transformation of mammalian cells to antibiotic resistance with a bacterial gene under control of the SV40 early region promoter. *J. Mol. Appl. Genet.* **1:** 327.

Sporn, M.B. and A.B. Roberts. 1985. Autocrine growth factors and cancer. *Nature* **313:** 745.

Sporn, M.B. and G.J. Todaro. 1980. Autocrine secretion and malignant transformation of cells. *N. Engl. J. Med.* **303:** 878.

Stern, P.H., N.S. Krieger, R.A. Nissenson, R.D. Williams, M.E. Winkler, R. Derynck, and G.J. Strewler. 1985. Human transforming growth factor-α stimulates bone resorption *in vitro*. *J. Clin. Invest.* **76:** 2016.

Todaro, G.J., C. Fryling, and J.E. De Larco. 1980. Transforming growth factors produced by certain human tumours: Polypeptides that interact with epidermal growth factor receptors. *Proc. Natl. Acad. Sci.* **77:** 5258.

Todaro, G.J., J.E. De Larco, S.P. Nissley, and M.M. Rechler. 1977. MSA and EGF receptors on sarcoma virus-transformed cells and human fibrosarcoma cells in culture. *Nature* **267:** 526.

Todaro, G.J., D.C. Lee, N.R. Webb, T.M. Rose, and J.P. Brown. 1985. Rat type-α transforming growth factor: Structure and possible function as a membrane receptor. *Cancer Cells* **3:** 51.

Twardzik, D.R. 1985. Differential expression of transforming growth factor-α during prenatal development of the mouse. *Cancer Res.* **45:** 5413.

Wigler, M., R. Sweet, G.K. Sim, B. Wold, A. Pellicer, E. Lacy, T. Maniatis, S. Silverstein, and R. Axel. 1979. Transformation of mammalian cells with genes from procaryotes and eucaryotes. *Cell* **16:** 777.

Wrann, M.M. and C.F. Fox. 1979. Identification of epidermal growth factor receptors in a hyperproducing human epidermoid carcinoma cell line. *J. Biol. Chem.* **254:** 8083.

Human Basic Fibroblast Growth Factor: Nucleotide Sequence, Genomic Organization, and Expression in Mammalian Cells

J.A. Abraham, J.L. Whang, A. Tumolo, A. Mergia, and J.C. Fiddes
California Biotechnology, Inc., Mountain View, California 94043

Angiogenesis, the process of new capillary growth, involves a series of steps, including the proliferation and migration of vascular endothelial cells. At the onset of angiogenesis, existing capillaries located close to the site of an angiogenic stimulus are induced to form new sprouts. These sprouts contain dividing endothelial cells, with migrating endothelial cells at the sprout tips. Eventually, the sprouts unite and form new capillary tubes (for review, see Folkman 1985).

Angiogenesis only takes place in a limited number of circumstances; endothelial cells in capillaries and large vessels are not usually dividing. Examples of normal physiological conditions under which angiogenesis occurs include the development of the corpus luteum at the time of ovulation, the growth of the placenta and fetus during gestation, and the repair of tissue after wounding. Angiogenesis is also associated with certain disease states, including the neovascularization that surrounds a solid tumor and supports its expansion beyond a diameter of 1–2 mm and the neovascularization of the retina that often leads to blindness in diabetic retinopathy.

The involvement of endothelial cell proliferation in the process of angiogenesis implicated endothelial cell growth factors as possible candidates for the signal that initiates capillary growth (Gimbrone et al. 1974; Folkman et al. 1979). A variety of endothelial cell mitogens have been reported during the past decade, some of which have indeed been shown to stimulate new capillary growth in angiogenic assays. One of the best characterized of these proteins, basic fibroblast growth factor (FGF), is not only a potent mitogen for endothelial cells, but serves as a chemoattractant for these cells as well (for review, see Gospodarowicz 1985, 1986). Basic FGF stimulates angiogenesis in both the rabbit corneal assay (Gospodarowicz et al. 1979) and the chick chorioallantoic membrane assay (Gospodarowicz et al. 1984; Esch et al. 1985b).

Basic FGF has been isolated from a wide range of tissues, including pituitary, hypothalamus, brain, kidney, placenta, corpus luteum, and adrenal gland (Gospodarowicz 1986). The protein has a pI of 9.6 (Gospodarowicz et al. 1982; Gospodarowicz 1986) and binds strongly to heparin-Sepharose (Gospodarowicz et al. 1984; Shing et al. 1984), eluting from the heparin at 1.5–2.0 M NaCl. Basic FGF purified from bovine pituitary was shown by protein sequence analysis to contain 146 amino acids (Esch et al. 1985b), although a biologically active form lacking the amino-terminal 15 amino acids has also been identified (Gospodarowicz 1986).

Basic FGF shares 55% amino acid sequence homology with a second endothelial cell mitogen, acidic FGF (also known as endothelial cell growth factor, ECGF), which has a pI of approximately 5.0–5.9 (Thomas et al. 1984; Esch et al. 1985a; Gimenez-Gallego et al. 1985; Burgess et al. 1986). As with basic FGF, acidic FGF has been shown to be angiogenic (Lobb et al. 1985; Thomas et al. 1985); unlike basic FGF, however, it has so far only been detected in neural tissues (Thomas and Gimenez-Gallego 1986). Acidic FGF also binds to heparin-Sepharose but elutes at a lower salt concentration than basic FGF (0.9–1.1 M NaCl; Maciag et al. 1984). Acidic FGF from bovine brain has been completely sequenced and shown to contain 140 amino acids (Esch et al. 1985a; Gimenez-Gallego et al. 1985). Slightly shorter and longer forms of acidic FGF have also been identified (Burgess et al. 1985, 1986; Esch et al. 1985a).

Less well-characterized endothelial cell mitogens have been reported from an array of sources, including tumor tissue (Shing et al. 1984) and tumor cell lines (Klagsbrun et al. 1986; Lobb et al. 1986). Where tested, all of these mitogens have been found to bind strongly to heparin-Sepharose, eluting at the same salt concentration as either acidic or basic FGF (Lobb and Fett 1984; Lobb et al. 1986). These results, combined in some cases with determinations of molecular weight, pI, retention time on high-performance liquid chromatography, and/or amino acid composition, suggest that all of the various factors could be essentially identical to either basic FGF or acidic FGF (Lobb and Fett 1984; Lobb et al. 1986).

Presented below is an overview of results we have recently obtained through the cloning of human and bovine basic FGF. DNA sequence analysis of the clones has shown that human basic FGF is remarkably similar to the bovine protein, differing by only two amino acids. Both the human and bovine proteins appear to be synthesized initially with an amino-terminal extension of nine residues, but without an obvious signal peptide. Southern blot analysis demonstrates that basic FGF is encoded by a single gene, confirming that the closely related, basic heparin-binding endothelial cell mitogens are in fact all products of a single gene. Finally, in keeping with the apparent lack of a signal sequence, we find that human basic FGF is not released into the medium when expressed in a vaccinia-based transient assay system.

RESULTS

Characterization of a cDNA Clone Encoding Bovine Basic FGF

To obtain clones encoding bovine basic FGF, a λgt10 cDNA library of approximately 10^6 independent recombinants was constructed from bovine pituitary RNA. This library was screened with a 40-base synthetic oligonucleotide probe to amino acids 18–31 of basic FGF (Esch et al. 1985b), and a single hybridizing recombinant was detected (Abraham et al. 1986b). Sequence analysis of this recombinant (λBB2) demonstrated that the 2.12-kb cDNA insert in the phage contained an open reading frame encoding the entire 146 amino acids of the sequenced form of basic FGF.

The coding region of the cDNA sequence in λBB2 is shown in Figure 1. Upstream of the "mature" basic FGF coding sequence, the open reading frame extends for nine codons to an in-frame ATG. The open reading frame then continues uninterrupted to the 5′ end of the cDNA; however, the sequence upstream of the ATG is extremely GC-rich (88% G + C) and is reminiscent of the GC-rich 5′-untranslated regions of the growth factors transforming growth factor-β (Derynck et al. 1985), insulin-like growth factor-II (Dull et al. 1984), and the β-chain of platelet-derived growth factor (Collins et al. 1985). It therefore seems likely that translation of the basic FGF-coding region initiates with the ATG codon at position −9. This conclusion is supported by the fact that the sequence surrounding the ATG (GGGCC*ATG*G) is in good agreement with the consensus sequence (CC$^{A}_{G}$CC*ATG*G) proposed by Kozak (1984) for the start of eukaryotic translation. Accordingly, we have adopted a 155-residue numbering system for basic FGF (rather than 146 residues), beginning with the ATG labeled −9 in Figure 1. The 155-residue system is used in the remaining figures in this paper.

Although the stretch of amino acids from positions −8 to 4 is hydrophobic (Fig. 1), the signal-peptidase cleavage sites predicted relative to this potential hydrophobic core region (von Heijne 1983) would lie carboxy-terminal to the proline at position +1 and several residues amino-terminal to the histidine at position 16 (the two amino termini of the sequenced forms of basic FGF). Thus, if the assignment proposed above for the initiating ATG is correct, basic FGF is synthesized without a conventional secretion signal sequence.

The coding sequence for mature basic FGF is followed immediately by a translation-termination codon (TGA), indicating that basic FGF has no carboxy-terminally extended precursor. The TGA codon is followed in the cDNA by 1551 bp of 3′-untranslated sequence; this sequence probably represents only a partial copy of the full 3′-untranslated region of the basic FGF mRNA, since the sequence contains neither a polyadenylation addition signal (AATAAA; Proudfoot and Brownlee 1976) nor a polyadenylate tail (Abraham et al. 1986b).

cDNA and Genomic Clones Encoding Human Basic FGF

To isolate clones encoding human basic FGF, a 1.4-kb *Eco*RI fragment spanning the bovine basic FGF-coding region was purified from λBB2 and used as a probe to screen human cDNA libraries (Abraham et al. 1986a). A total of seven cDNA clones was obtained from five different cDNA libraries: two clones (λKB2 and λKB7) from an adult kidney library of 2 × 10^6 recombinants, one (λPB2) from a term placenta library of 0.7 × 10^6 recombinants, two (λET1 and λET2) from a library of 0.35 × 10^6 clones made from a breast carcinoma, one (λHFL1) from a fetal liver library of 0.5 × 10^6 recombinants, and one (λHFH1) from a fetal heart library of 0.2 × 10^6 recombinants.

The structures of six of these clones, as determined by sequence analysis, are diagramed in Figure 2; the seventh clone (λHFH1) appeared from preliminary characterization to be derived solely from the 3′-untranslated region and was not sequenced. None of the cDNA clones contain the entire coding sequence for basic FGF, but a composite structure for the mRNA (Fig. 2, top) could be deduced from a comparison of the six clones. Three of the clones (λKB7, λKB2, and λET2) appeared to represent cDNA copies of unspliced messages. This assumption was confirmed when DNA sequence information became available for genomic clones of human basic FGF (see below). Since the cDNA libraries were generated from a mixture of nuclear and cytoplasmic polyadenylated RNA, the unspliced cDNAs presumably arose by oligo(dT) priming from A-rich regions in precursor RNA. The high proportion of cDNAs containing intron sequences suggests that the fully processed mRNA for basic FGF is unstable.

The λBB2-derived probe was also used to screen for genomic copies of the human basic FGF sequence. Two clones (λMG4 and λMG10) were obtained from a fetal liver genomic library in Charon 4A (Lawn et al. 1978); sequence analysis demonstrated that λMG4 contains the coding sequence for the amino-terminal portion of basic FGF, whereas λMG10 encodes the carboxy-terminal portion of the molecule (Fig. 2). The Charon 4A library did not appear to contain the coding sequence for 35 amino acids in the middle of basic FGF, but this sequence was present on the clone λHT1, isolated by screening a Charon 28 library (a gift from E. Fritsch) made from a chemically transformed fibroblast cell line.

Restriction maps of the genomic DNA inserts in λMG4, λHT1, and λMG10 are shown in Figure 3. Although all of the inserts span at least 12 kb, each insert contains only a single coding exon (see also Fig. 2). Since none of the inserts overlap in sequence, the full extent of the human basic FGF gene cannot be deter-

```
1          CCGGGGCCGC GCCGCGGAGC GCGTCGGAGG CCGGGGCCGG GGCGCGGCGG CTCCCCGCGC GGCTCCAGGG GCTCGGGGAC CCCGCCAGGG CCTTGGTGGG GCC

      -9                              -1   1                                  10                                      20
     met ala ala gly ser ile thr thr leu PRO ALA LEU PRO GLU ASP GLY GLY SER GLY ALA PHE PRO PRO GLY HIS PHE LYS ASP PRO LYS
104  ATG GCC GCC GGG AGC ATC ACC ACG CTG CCA GCC CTG CCG GAG GAC GGC GGC AGC GGC GCT TTC CCG CCG GGC CAC TTC AAG GAC CCC AAG

                                      30                                      40                                      50
     ARG LEU TYR CYS LYS ASN GLY GLY PHE PHE LEU ARG ILE HIS PRO ASP GLY ARG VAL ASP GLY VAL ARG GLU LYS SER ASP PRO HIS ILE
194  CGG CTG TAC TGC AAG AAC GGG GGC TTC TTC CTG CGC ATC CAC CCC GAC GGC CGA GTG GAC GGG GTC CGC GAG AAG AGC GAC CCA CAC ATC

                                      60                                      70                                      80
     LYS LEU GLN LEU GLN ALA GLU GLU ARG GLY VAL VAL SER ILE LYS GLY VAL CYS ALA ASN ARG TYR LEU ALA MET LYS GLU ASP GLY ARG
284  AAA CTA CAA CTT CAA GCA GAA GAG AGA GGG GTT GTG TCT ATC AAA GGA GTG TGT GCA AAC CGT TAC CTT GCT ATG AAA GAA GAT GGA AGA

                                      90                                      100                                     110
     LEU LEU ALA SER LYS CYS VAL THR ASP GLU CYS PHE PHE PHE GLU ARG LEU GLU SER ASN ASN TYR ASN THR TYR ARG SER ARG LYS TYR
374  TTA CTA GCT TCT AAA TGT GTT ACA GAC GAG TGT TTC TTT TTT GAA CGA TTG GAG TCT AAT AAC TAC AAT ACT TAC CGG TCA AGG AAA TAC

                                      120                                     130                                     140
     SER SER TRP TYR VAL ALA LEU LYS ARG THR GLY GLN TYR LYS LEU GLY PRO LYS THR GLY PRO GLY GLN LYS ALA ILE LEU PHE LEU PRO
464  TCC AGT TGG TAT GTG GCA CTG AAA CGA ACT GGG CAG TAT AAA CTT GGA CCC AAA ACA GGA CCT GGG CAG AAA GCT ATA CTT TTT CTT CCA

     MET SER ALA LYS SER  *
554  ATG TCT GCT AAG AGC TGA TCTTAATGGC AGCATCTGAT CTCATTTTAC ATGAAGAGGT ATATTTCAGA AATGTGTTAA TGAAAAAAGA AAAATGTGTA CAGTGAG

659  CTG CTCAGTTTGG GTAACTGTTC AGATAACCGT TTATCTAAGA GTAAAATATT TAACCATTGC
```

Figure 1. Partial nucleotide sequence of the cDNA insert in the bovine basic FGF clone λBB2. The coding sequence for the 146 amino acids of pituitary-derived basic FGF is shown; the deduced amino acid sequence agrees exactly with the amino acid sequence determined by Esch et al. (1985b). The nine-residue amino-terminal extension of the proposed basic FGF primary translation product is shown in lowercase letters (codons −1 to −9). Numbers at the left indicate nucleotide positions.

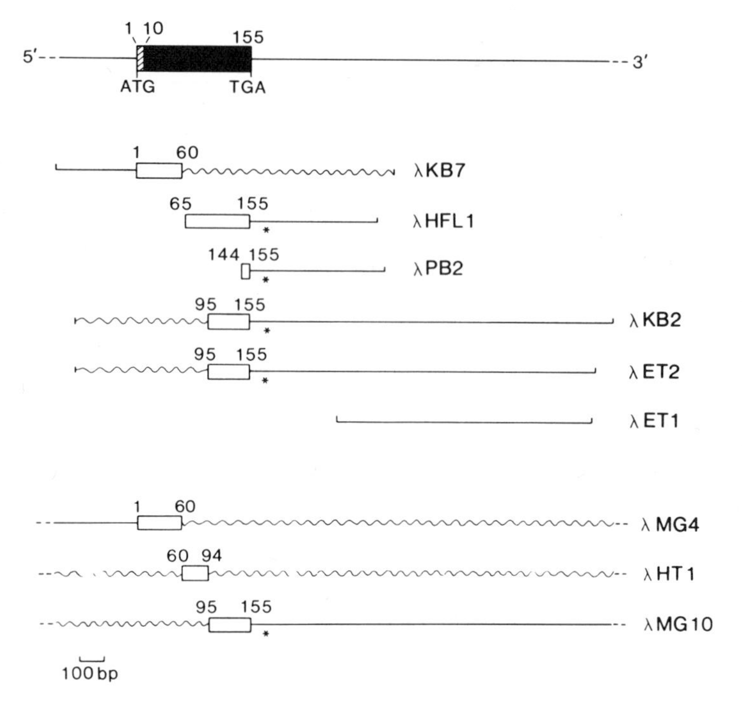

Figure 2. Structures of the human basic FGF cDNAs and of the exon-containing regions of the human basic FGF genomic clones. Numbers above each structure indicate amino acid positions within the basic FGF protein; the amino acids have been numbered 1–155, initiating with the methionine encoded nine codons upstream of the amino terminus of pituitary-derived basic FGF (see Fig. 4). The structure at the top of the figure represents a composite derived from a comparison of the various cDNA and genomic clones. The structures of six cDNAs (λKB7 through λET1) are diagramed immediately below the composite; portions of the genomic clones λMG4, λHT1, and λMG10 are shown at the bottom of the figure. Coding sequences are represented as open boxes; untranslated and intron sequences are shown as straight and wavy lines, respectively. Asterisk indicates the site of a potential polyadenylation addition signal (AATAAA).

mined from the data shown in Figure 3, but it must span at least 34 kb.

The nucleotide sequence of the coding region of human basic FGF, derived from both the cDNAs and the genomic clones, is given in Figure 4. The amino-terminal 41 amino acids of human brain basic FGF have been determined by Gimenez-Gallego et al. (1986); this sequence is in agreement with the deduced amino acid sequence shown in Figure 4 (residues 10–50). Upstream of this sequence, the human basic FGF gene encodes the same nine amino acids as are found in the bovine basic FGF extension (Fig. 1). As with the bovine gene, the sequence surrounding the ATG at the end of the human coding region for the extension (GGAC*CATG*G) agrees well with the translation-initiation consensus sequence of Kozak (1984); in addition, the sequence immediately preceding the ATG is extremely GC-rich. These observations support the proposal that this ATG encodes the initiating methionine for basic FGF.

The two introns in the human basic FGF-coding region lie within codon 60 and between codons 94 and 95. If the basic FGF protein sequence is aligned with acidic FGF to give the greatest degree of homology (see Esch et al. 1985a), the location of the basic FGF intron in codon 60 is identical to the location of an intron found within the bovine and human acidic FGF genomic sequences (Abraham et al. 1986b; A. Mergia et al., unpubl.). Basic and acidic FGFs thus share gene structure homology as well as amino acid sequence homology.

The translation-termination codon at the end of the 155-amino-acid coding region for human basic FGF is followed by at least 1.5 kb of 3′-untranslated sequence in λKB2 (Fig. 2). None of the cDNA clones isolated contain polyadenylate tails. A potential polyadenylation signal, AATAAA, lies 79 bp downstream from the termination codon (Fig. 4), but this signal was apparently not used in the formation of the RNA molecules that were copied to make λHFL1, λPB2, λKB2, λET1, and λET2 (see Fig. 2).

The amino acid sequence for human basic FGF, as deduced from the DNA sequence, is very similar to the sequence of the bovine protein (Fig. 5). The proteins differ by only two amino acids: a conservative change at amino acid 121, from serine in the bovine sequence to threonine in human basic FGF, and a nonconservative proline-to-serine change at position 137. As mentioned above, the homology between the human and bovine basic FGFs extends into the proposed amino-terminal extension region.

Basic FGF Gene Copy Number

To determine the number of genes for basic FGF in the human genome, the 1.4-kb *Eco*RI fragment from λBB2 was used to probe Southern blots of genomic DNA digested with a variety of restriction enzymes (Abraham et al. 1986a). Figure 6 shows some of the results obtained. Since the probe fragment spans all three coding exons, and the exons are known to be widely spaced in the genome (Fig. 3), we expected to see at least three hybridizing fragments in each lane of the blot. The sizes of many of these fragments could be predicted from the restriction maps of the human genomic clones shown in Figure 3. In most cases, however, only the predicted fragments spanning the middle and carboxy-terminal exons were detected (Fig. 6). The identities of these fragments were confirmed by rehybridizing the blots with synthetic oligonucleotide

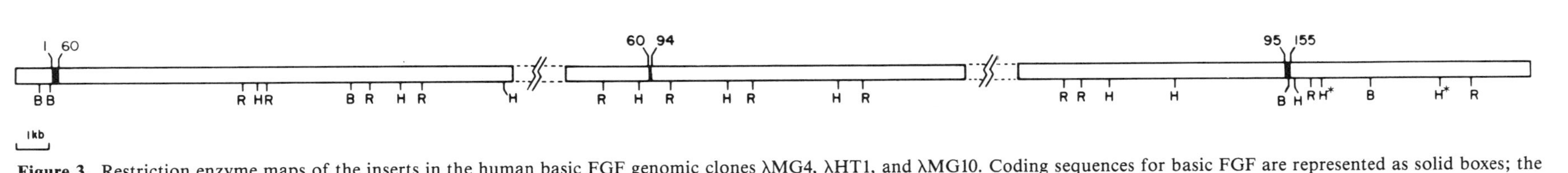

Figure 3. Restriction enzyme maps of the inserts in the human basic FGF genomic clones λMG4, λHT1, and λMG10. Coding sequences for basic FGF are represented as solid boxes; the numbers above each box indicate the amino acids encoded by each exon. Restriction sites are shown for the enzymes *Bam*HI (B), *Eco*RI (R), and *Hin*dIII (H). There are five additional unmapped *Hin*dIII sites in λMG10 in the region bracketed by asterisks, and there is an extra 0.4-kb *Eco*RI fragment in λHT1 that has not been localized.

```
                                        AAGTTGAGTC ACGGCTGGTT GCGCACGAAA AGCCCCGCAG TGTGGAGAAA GCCTAAACGT GGTTTGGGTG
                                                            ***
GTGCGGGGGT TGGGCGGGGG TGACTTTTGG GGGATAAGGG GCGGTGGAGC CCAGGGAATG CCAAAGCCCT GCCGCGGCCT CCGACGCGCG CCCCCCGCCC CTCGCCTCTC
                                        ** *
CCCCGCCCCC GACTGAGGCC GGGCTCCCCG CCGGACTGAT GTCGCGCGCT TGCGTGTTGT GGCCGAAGCC GCCGAACTCA GAGGCCGGCC CCAGAAAACC CGAGCGAGTA
                                                                 ***        ***
GGGGGCGGCG CGCAGGAGGG AGGAGAACTG GGGGCGCGGG AGGCTGGTGG GTGTGGGGGG TGGAGATGTA GAAGATGTGA CGCCGCGGCC CGGCGGGTGC CAGATTAGCG

GACGGCTGCC CGCGGTTGCA ACGGGATCCC GGGCGCTGCA GCTTGGGAGG CGGCTCTCCC CAGGCGGCGT CCGCGGAGAC ACCCATCTGT GAACCCCAGG TCCCGGGCCG
      ↑
CCGGCTCGCC GCGCACCAGG GGCCGGCGGA CAGAAGAGCG GCCGAGCGGC TCGAGGCTGG GGGACCGCGG GCGCGGCCGC GCGCTGCCGG GCGGGAGGCT GGGGGGCCGG

GGCCGGGGCC GTGCCCGGAG CGGGTCGGAG GCCGGGGCCG GGGCCGGGGG ACGGCGGCTC CCCGCGCGGC TCCAGCGGCT CGGGGATCCC GGCCGGGCCC CGCAGGGACC

ATG GCA GCC GGG AGC ATC ACC ACG CTG CCC GCC TTG CCC GAG GAT GGC GGC AGC GGC GCC TTC CCG CCC GGC CAC TTC AAG GAC CCC AAG
Met Ala Ala Gly Ser Ile Thr Thr Leu Pro Ala Leu Pro Glu Asp Gly Gly Ser Gly Ala Phe Pro Pro Gly His Phe Lys Asp Pro Lys
 1                                  10                                      20                                      30

CGG CTG TAC TGC AAA AAC GGG GGC TTC TTC CTG CGC ATC CAC CCC GAC GGC CGA GTT GAC GGG GTC CGG GAG AAG AGC GAC CCT CAC A gt
Arg Leu Tyr Cys Lys Asn Gly Gly Phe Phe Leu Arg Ile His Pro Asp Gly Arg Val Asp Gly Val Arg Glu Lys Ser Asp Pro His I
                                    40                                      50

gagtgccgacccgctctctccgcctcatttccatttcgtgggttctcg.......................aaggctctttcctctgtggtgctacaaagataatttttttcccgtt

acag TC AAG CTA CAA CTT CAA GCA GAA GAG AGA GGA GTT GTG TCT ATC AAA GGA GTG TGT GCT AAC CGT TAC CTG GCT ATG AAG GAA GAT
      le Lys Leu Gln Leu Gln Ala Glu Glu Arg Gly Val Val Ser Ile Lys Gly Val Cys Ala Asn Arg Tyr Leu Ala Met Lys Glu Asp
                                            70                                      80

GGA AGA TTA CTG GCT TCT gtaagcatactttctgttttcacacgtttttgttagcttttattgctgt.........................taataataataatgataat
Gly Arg Leu Leu Ala Ser
     90

aataacaggtaattcttcctttatttttcag AAA TGT GTT ACG GAT GAG TGT TTC TTT TTT GAA CGA TTG GAA TCT AAT AAC TAC AAT ACT TAC CGG
                                Lys Cys Val Thr Asp Glu Cys Phe Phe Phe Glu Arg Leu Gly Ser Asn Asn Tyr Asn Thr Tyr Arg
                                                    100                                     110

TCA AGG AAA TAC ACC AGT TGG TAT GTG GCA TTG AAA CGA ACT GGG CAG TAT AAA CTT GGA TCC AAA ACA GGA CCT GGG CAG AAA GCT ATA
Ser Arg Lys Tyr Thr Ser Trp Tyr Val Ala Leu Lys Arg Thr Gly Gln Tyr Lys Leu Gly Ser Lys Thr Gly Pro Gly Gln Lys Ala Ile
            120                                 130                                     140

CTT TTT CTT CCA ATG TCT GCT AAG AGC TGA TTTTAATGGC CACATCTAAT CTCATTTCAC ATGAAAGAAG AAGTATATTT TAGAAATTTG TTAATGAGAG TA
Leu Phe Leu Pro Met Ser Ala Lys Ser Ter
            150

AAAGAAAA TAAATGTGTA TAGCTCAGTT TGGATAATTG GTCAAACAAT TTTTTATCCA GTAGTA
```

Figure 4. Partial nucleotide sequence of the human basic FGF gene. The encoded amino acids are numbered 1–155, assuming that translation initiates with the methionine labeled *1*. The first and last 50 bases of each intron are shown in lowercase letters. Possible binding sites for the transcription factor Sp1 in the 5′-flanking sequence (see Discussion) are indicated by a bold underline. Asterisks indicate possible translation-initiation codons in this upstream region. Arrow indicates the 5′ end of cDNA clone λKB7; the sequence upstream of this point was derived from λMG4. A possible polyadenylation addition signal in the 3′-untranslated sequence is boxed.

Figure 5. Comparison of the human and bovine nucleotide sequences and encoded amino acid sequences. The nucleotide sequence of the coding region for human basic FGF is shown, along with the encoded amino acids. The bovine nucleotide and amino acid sequences are given above the human sequences only at the locations where bovine and human differ. Arrowheads indicate the amino termini of two isolated forms of basic FGF, which comprise amino acids 10–155 (also referred to as the 146-residue form) and 25–155 (Gospodarowicz 1986). (Reprinted, with permission, from Abraham et al. 1986a.)

probes designed to be specific for each of these two exons. The amino-terminal exon was generally not detected either with the probe fragment or with an exon-specific oligonucleotide probe. This lack of hybridization may be due to the formation of interfering secondary structure in the highly GC-rich sequences in and around the 5′ end of the FGF-coding region (Fig. 4).

All of the hybridizing fragments generated by digestion with *Bam*HI, *Hin*dIII, or *Eco*RI (Fig. 6) could be accounted for on the genomic restriction maps, even when lower stringency conditions were employed (not shown). Human basic FGF is therefore encoded by a single gene.

RNA Transcripts for Human and Bovine Basic FGF

Basic FGF transcripts were not seen in Northern blot analyses of the polyadenylated RNAs from which the cDNA libraries described above were constructed (Abraham et al. 1986a), consistent with the observation that basic FGF clones were only found in the libraries at an average frequency of 1 in 500,000. However, a single species of bovine basic FGF RNA of approximately 7.0 kb was detected in bovine hypothalamus RNA (Fig. 7, lane 1). Basic FGF RNAs were also found in RNA from a human hepatoma cell line, SK-HEP-1 (Fig. 7, lane 2), known to make an endothelial cell mitogen very similar in characteristics to basic FGF (Klagsbrun et al. 1986; Lobb et al. 1986). In this case, two transcripts were detected: a 7.0-kb RNA comigrating with the bovine hypothalamus basic FGF RNA, and a shorter RNA of approximately 3.7 kb. Initially, it appeared that the smaller RNA from the human cells could be explained by use of the AATAAA polyadenylation signal present 79 bp downstream from the translation stop codon in the human (but not bovine) 3′-

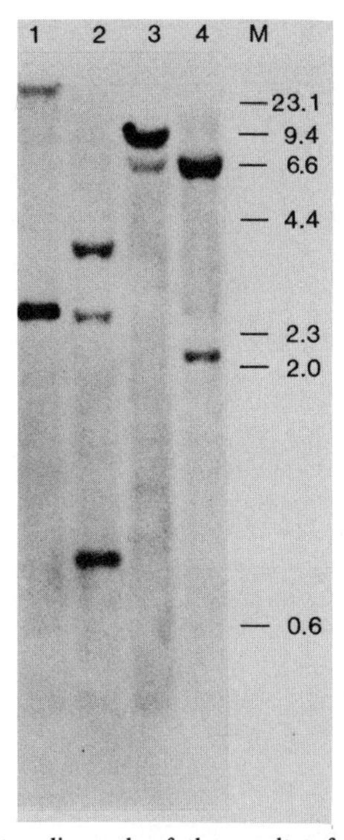

Figure 6. Autoradiograph of the results of Southern blot analysis of human genomic DNA digested with *Bam*HI (lane *1*), *Hin*dIII (lane *2*), *Pst*I (lane *3*), or *Eco*RI (lane *4*). The DNA from each digestion was fractionated on a 0.8% agarose gel, transferred to nitrocellulose, and probed with a 1.4-kb *Eco*RI fragment derived from the bovine clone, λBB2. Size markers (M) are in kilobases. (Reprinted, with permission, from Abraham et al. 1986a.)

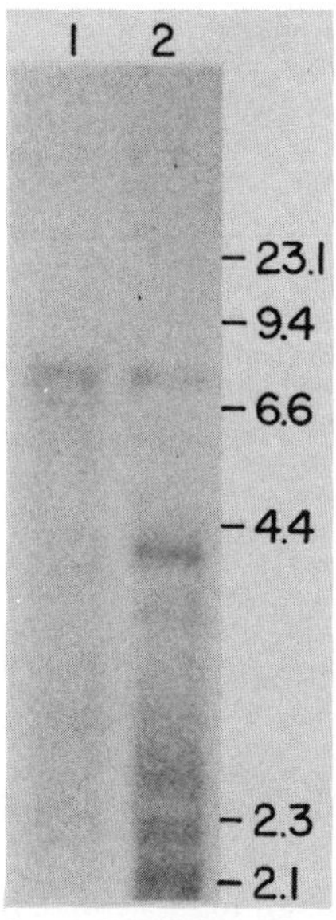

Figure 7. Autoradiograph of the results of Northern blot analysis of RNAs from bovine hypothalamus (lane *1*) and the human hepatoma cell line SK-HEP-1 (lane *2*). Extracted RNAs were fractionated by oligo(dT)-cellulose chromatography, and polyadenylated RNA was then electrophoresed on a 1.2% agarose-formaldehyde gel, transferred to nitrocellulose, and hybridized with the 1.4-kb fragment from λBB2. Locations of DNA size markers are given at the right in kilobases. (Reprinted, with permission, from Abraham et al. 1986b [copyright by the AAAS].)

untranslated sequence (Figs. 1 and 4). However, L. Schweigerer et al. (in prep.) have recently shown that RNA extracted from cultured bovine endothelial cells also contains the two basic FGF RNAs of 7.0 and 3.7 kb. The relationship of the two RNA species thus remains unclear.

Expression of Human Basic FGF in a Vaccinia-based Transient Assay System

Basic FGF has been shown to bind to membrane-associated, cell-surface receptors (Neufeld and Gospodarowicz 1985, 1986), but the proposed primary translation product for basic FGF contains no classical secretion signal sequence (Figs. 1 and 4). Two possible explanations for this apparent contradiction might be (1) that basic FGF utilizes an unclipped, non-amino-terminal secretion signal sequence or (2) that basic FGF is released into the medium by an alternative pathway (e.g., exocytosis or cell lysis).

To examine this question, we have used a transient assay expression system for human basic FGF (J.A. Abraham et al., unpubl.). A cDNA sequence encoding the human form of basic FGF was first constructed by in vitro mutagenesis of the bovine cDNA, λBB2. In a first round of mutagenesis, the two bovine-specific positions (serine codon at position 121 and proline codon at position 137; see Fig. 5) were altered to encode the amino acids found in the human protein at these positions (threonine and serine, respectively). A second round of mutagenesis then created a *Hin*dIII restriction site 34 bp downstream from the translation-termination codon. Since the 5′ end of the basic FGF-coding sequence coincides with an *Nco*I site (spanning the ATG at position 1), the human basic FGF-encoding sequence could be conveniently excised as an approximately 500-bp *Nco*I-*Hin*dIII fragment. The 500-bp fragment was then ligated into the expression vector pGS20 (Mackett et al. 1984) to form the vector pJV1 (Fig. 8A). In this vector, the FGF sequence lies immediately downstream from the promoter sequence for the vaccinia early gene 7.5K.

For transient expression, the pJV1 plasmid was transfected into vaccinia-infected CV-1 cells as described by Cochran et al. (1985) to give vaccinia-directed, cytoplasmic transcription of the basic FGF cDNA from the 7.5K promoter. After approximately 24 hours, conditioned media and cell lysates were analyzed by heparin-Sepharose chromatography, and with an in vitro growth assay using bovine brain capillary endothelial cells, for the presence of basic FGF (J.A. Abraham et al., unpubl.). No FGF-like protein was detected in the heparin-Sepharose-treated conditioned medium, but a new protein of approximately 18 kD was detected in lysates from cells transfected with pJV1, as compared with lysates from pGS20-transfected cells (Fig. 8B). This new protein is slightly larger than the 16.4-kD (146-residue) basic FGF marker, indicating that an extended form of basic FGF is being synthesized. The material produced in the pJV1-containing CV-1

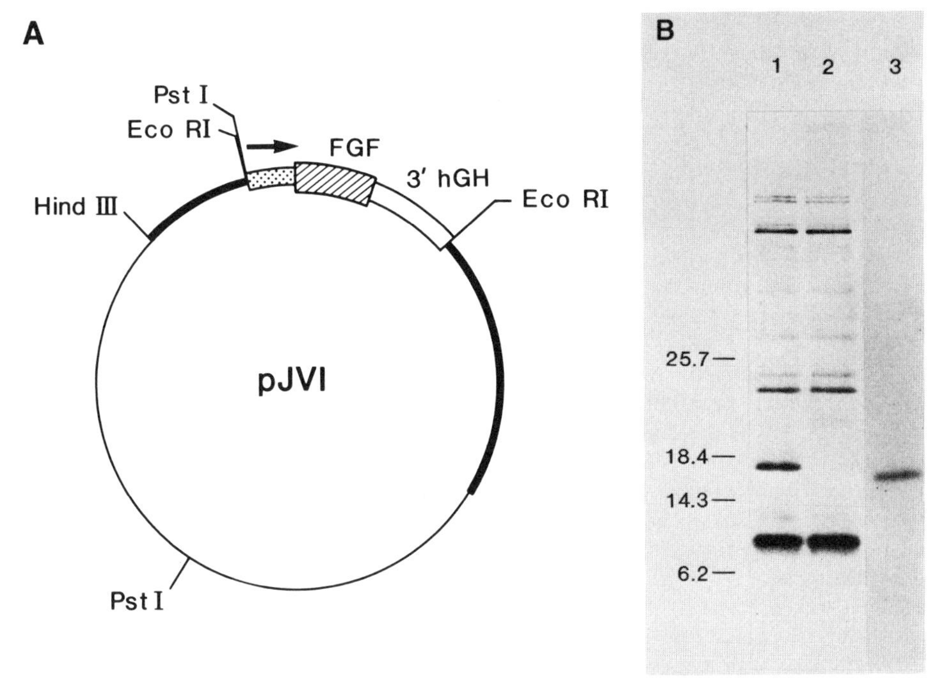

Figure 8. Transient expression of human basic FGF in vaccinia-infected CV-1 cells. (*A*) Structure of the expression vector pJV1, derived from pGS20 (Mackett et al. 1984). (Stippled box) Vaccinia 7.5K promoter; (cross-hatched box) human basic FGF-encoding cDNA sequence; (open box) 3′-untranslated sequences and polyadenylation addition signal from human growth hormone (DeNoto et al. 1981); (thick line) vaccinia thymidine kinase gene; (thin line) pBR328 sequences. The growth hormone sequences were originally added to the basic FGF sequence for use in another expression system, and probably have little effect in the system described here. (*B*) Polyacrylamide gel analysis of fractionated cell lysates from the transient assay. Transfected, vaccinia-infected cells were labeled with [^{35}S]methionine for 4 hr, lysed, and bound to heparin-Sepharose. Proteins eluting with 2 M NaCl (after a column wash with 1.1 M NaCl) were precipitated with trichloroacetic acid and electrophoresed on a 17.5% acrylamide/SDS gel. An autoradiograph of the gel is shown. (Lane *1*) pJV1-transfected cells; (lane *2*) pGS20-transfected cells; (lane *3*) ^{125}I-labeled basic FGF (146-amino-acid form). Size markers are in kilodaltons.

cells also stimulated the growth of endothelial cells; no such activity was found in the pGS20-containing cells. Biologically active basic FGF is thus produced in the transient assay, but it remains cell-associated. Further experiments will be required to determine if the recombinant basic FGF is truly intracellular or if the protein is actually secreted but subsequently binds to the membrane or extracellular matrix.

DISCUSSION

Using an oligonucleotide probe, we have isolated a pituitary cDNA clone encoding bovine basic FGF; cDNA and genomic clones encoding human basic FGF were then isolated by homology with this bovine clone. DNA sequence analysis of the human clones allowed us to deduce the amino acid sequence of human basic FGF (Fig. 4), which was found to be surprisingly homologous to its bovine counterpart, differing at only two amino acids (Fig. 5). The high degree of protein sequence conservation suggests a very strong selection pressure for conservation of function and structure.

Analysis of the DNA sequence of the bovine and human basic FGF clones also indicated that, in both species, the 146-residue form of basic FGF is initially synthesized with an amino-terminal extension of nine amino acids (Figs. 1 and 4). This proposal is supported by the fact that the related growth factor, acidic FGF, also appears to be synthesized with an amino-terminal extension in both human and bovine cells (Abraham et al. 1986b; Jaye et al. 1986; A. Mergia et al., unpubl.). Alignment of the basic and acidic FGF amino acid sequences reveals that the homology between these two proteins extends into the added residues in the proposed extensions (Fig. 9). In both the human and bovine acidic FGF genes, an in-frame translation stop codon is found two codons 5′ to the ATG at the amino terminus of the extension.

Although the best-characterized forms of basic and acidic FGF contain 146 and 140 amino acids, respectively, longer forms of these two molecules have also been reported. Klagsbrun and colleagues (Klagsbrun et al. 1986; Lobb et al. 1986) isolated an amino-terminally extended form of basic FGF from the SK-HEP-1 human hepatoma cell line; this approximately 18.5-kD protein is blocked at the amino terminus but was shown to contain threonine and leucine as the carboxy-terminal residues of the extension, in agreement with the DNA sequence data presented here. Similarly, Burgess et al. (1985, 1986) have purified an amino-terminally extended form of acidic FGF from human brain. This protein, like the hepatoma-derived factor, is blocked at its amino terminus, but mass spectrometry analysis established that the 14-residue extension is identical in

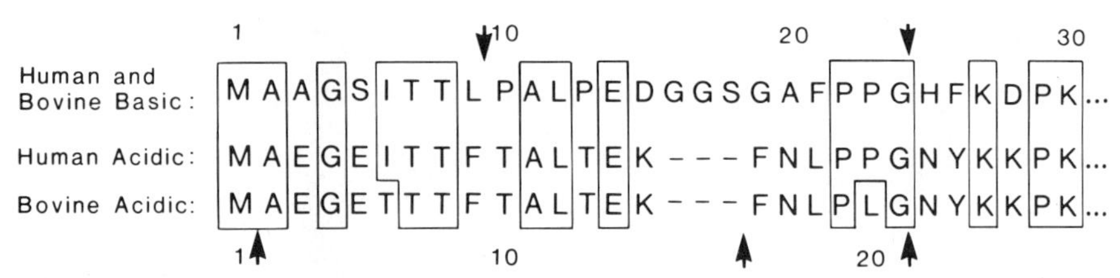

Figure 9. Comparison of the amino-terminal amino acid sequences of basic and acidic FGF. The amino termini of the proposed primary translation products of basic and acidic FGF from human and bovine cells are aligned to give the maximum degree of homology. Shared amino acids between the basic and acidic FGF sequences are boxed. Arrows represent cleavages that must occur to generate the known "mature" forms of the basic and acidic proteins.

sequence to the extension predicted from the DNA sequence (minus the initiating methionine). The existence of these "long" forms of basic and acidic FGF raises the question of what constitutes the actual "mature" form(s) of these two proteins. Specifically, it will be interesting to determine if any of the shorter forms that have been isolated are generated in a tissue-specific manner or as artifacts of purification.

Neither basic FGF (Figs. 1 and 4) nor acidic FGF (Abraham et al. 1986b; Jaye et al. 1986; A. Mergia et al., unpubl.) is synthesized with a conventional signal peptide. These two growth factors must be released from cells by some mechanism, however, since both have been shown to bind specifically to cell-surface receptors (Neufeld and Gospodarowicz 1985, 1986; Schreiber et al. 1985). One possible mechanism might be that the FGFs are somehow stored intracellularly and then released by cell lysis or specific transport in response to a signal (e.g., at the site of a wound or growing tumor). This type of mechanism would be consistent with the finding that basic FGF is not released into the medium when expressed in vaccinia-infected CV-1 cells (J.A. Abraham et al., unpubl.). It will be interesting to determine if the CV-1 cells (or other cell types either transfected with an FGF expression plasmid or naturally producing basic FGF) can be induced to release the synthesized FGF by, for example, treatment with wound fluid or coculturing with transformed cells.

In the human cell line SK-HEP-1 (Fig. 7) and in bovine endothelial cells (L. Schweigerer et al., in prep.), the basic FGF gene encodes two polyadenylated RNAs of approximately 7.0 and 3.7 kb. The promoters for these two RNAs have not yet been identified. The human genomic DNA sequences so far determined upstream of the 5′-most human cDNA (λKB7; see Fig. 4) contain no obvious TATA-like sequences, but they do contain several repeats of the binding site for the transcription factor Sp1 (CCGCCC, underlined in Fig. 4) (Dynan and Tjian 1985). This upstream region also contains several ATG codons (Fig. 4, asterisks) that, if contained within the mRNA, would have to be passed over by ribosomes for translation to initiate at the ATG marked as position 1 in Figure 4. It is also possible that additional RNA splicing events occur in the 5′-untranslated sequence to remove this region completely from the mature mRNA.

Several different groups have isolated basic endothelial cell mitogens that, like basic FGF, bind strongly to heparin-Sepharose, elute from the heparin at between 1.5 and 2.0 M NaCl, and have molecular weights in the range of 16,000–19,000 (e.g., see Lobb et al. 1986). Amino acid composition and/or antibody cross-reactivity data on some of these factors have indicated that they are very similar to basic FGF. These results suggested that either there exists a family of closely related, basic FGF-like genes or all of the factors of similar amino acid composition or cross-reactivity are derived from the same gene, differing only by posttranslational processing or modification. Southern blot analysis of human genomic DNA established that the second alternative is correct: There is only one basic FGF-like gene (Fig. 6). Thus, we conclude that basic FGF (Esch et al. 1985b), cationic hypothalamus-derived growth factor (cHDGF; Klagsbrun and Shing 1985; Lobb et al. 1986), heparin-binding growth factor β (HGF-β; Lobb and Fett 1984), cartilage-derived growth factor (CDGF; Sullivan and Klagsbrun 1985; Lobb et al. 1986), β retina-derived growth factor (β-RDGF; Baird et al. 1985), eye-derived growth factor I (EDGF-I; Courty et al. 1985), chondrosarcoma-derived growth factor (Shing et al. 1984; Lobb et al. 1986) and hepatoma-derived growth factor (HDGF; Klagsbrun et al. 1986; Lobb et al. 1986) are all encoded by the same gene.

ACKNOWLEDGMENTS

We thank D. Gospodarowicz for advice and help with the biological activity assays, J. Friedman for Southern and Northern analyses, and K. Talmadge and M. Snyder for helpful discussions and comments on the manuscript. This work was supported by the Biotechnology Research Partners and by Small Business Innovation grant GM-36762-01 from the National Institute of General Medical Sciences.

REFERENCES

Abraham, J.A., J.L. Whang, A. Tumolo, A. Mergia, J. Friedman, D. Gospodarowicz, and J.C. Fiddes. 1986a. Human basic fibroblast growth factor: Nucleotide sequence and genomic organization. *EMBO. J.* (in press).

Abraham, J.A., A. Mergia, J.L. J.L. Whang, A. Tumolo, J. Friedman, K.A. Hjerrild, D. Gospodarowicz, and J.C. Fiddes. 1986b. Nucleotide sequence of a bovine clone encoding the angiogenic protein, basic fibroblast growth factor. *Science* **233:** 545.

Baird, A., F. Esch, D. Gospodarowicz, and R. Guillemin.

1985. Retina- and eye-derived endothelial cell growth factors: Partial molecular characterization and identity with acidic and basic fibroblast growth factors. *Biochemistry* **24:** 7855.

Burgess, W.H., T. Mehlman, R. Friesel, W.V. Johnson, and T. Maciag. 1985. Multiple forms of endothelial cell growth factor: Rapid isolation and biological and chemical characterization. *J. Biol. Chem.* **260:** 11389.

Burgess, W.H., T. Mehlman, D.R. Marshak, B.A. Fraser, and T. Maciag. 1986. Structural evidence that β-endothelial cell growth factor is the precursor of both α-endothelial cell growth factor and acidic-fibroblast growth factor. *Proc. Natl. Acad. Sci.* (in press).

Cochran, M.A., M. Mackett, and B. Moss. 1985. Eukaryotic transient expression system dependent on transcription factors and regulatory DNA sequences of vaccinia virus. *Proc. Natl. Acad. Sci.* **82:** 19.

Collins, T., D. Ginsburg, J.M. Boss, S.H. Orkin, and J.S. Pober. 1985. Cultured human endothelial cells express platelet-derived growth factor B chain: cDNA cloning and structural analysis. *Nature* **316:** 748.

Courty, J., C. Loret, M. Moenner, B. Chevallier, O. Lagente, Y. Courtois, and D. Barritault. 1985. Bovine retina contains three growth factor activities with different affinity to heparin: Eye derived growth factor I, II, III. *Biochimie* **67:** 265.

DeNoto, F.M., D.D. Moore, and H.M. Goodman. 1981. Human growth hormone DNA sequence and mRNA structure: Possible alternative splicing. *Nucleic Acids Res.* **9:** 3719.

Derynck, R., J.A. Jarrett, E.Y. Chen, D.H. Eaton, J.R. Bell, R.K. Assoian, A.B. Roberts, M.B. Sporn, and D.V. Goeddel. 1985. Human transforming growth factor-β complementary DNA sequence and expression in normal and transformed cells. *Nature* **316:** 701.

Dull, T.J., A. Gray, J.S. Hayflick, and A. Ullrich. 1984. Insulin-like growth factor II precursor gene organization in relation to insulin gene family. *Nature* **310:** 777.

Dynan, W.S. and R. Tjian. 1985. Control of eukaryotic messenger RNA synthesis by sequence-specific DNA-binding proteins. *Nature* **316:** 774.

Esch, F., N. Ueno, A. Baird, F. Hill, L. Denoroy, N. Ling, D. Gospodarowicz, and R. Guillemin. 1985a. Primary structure of bovine brain acidic fibroblast growth factor (FGF). *Biochem. Biophys. Res. Commun.* **133:** 554.

Esch, F., A. Baird, N. Ling, N. Ueno, F. Hill, L. Denoroy, R. Klepper, D. Gospodarowicz, P. Böhlen, and R. Guillemin. 1985b. Primary structure of bovine pituitary basic fibroblast growth factor (FGF) and comparison with the amino-terminal sequence of bovine brain acidic FGF. *Proc. Natl. Acad. Sci.* **82:** 6507.

Folkman, J. 1985. Angiogenesis and its inhibitors. In *Important advances in oncology 1985* (ed. R. Winters), p. 42. J.B. Lippincott, Philadelphia.

Folkman, J., C.C. Haudenschild, and B.R. Zetter. 1979. Long-term culture of capillary endothelial cells. *Proc. Natl. Acad. Sci.* **76:** 5217.

Gimbrone, M.A., R.S. Cotran, and J. Folkman. 1974. Human vascular endothelial cells in culture: Growth and DNA synthesis. *J. Cell Biol.* **60:** 673.

Gimenez-Gallego, G., G. Conn, V.B. Hatcher, and K.A. Thomas. 1986. Human brain-derived acidic and basic fibroblast growth factors: Amino terminal sequences and specific mitogenic activities. *Biochem. Biophys. Res. Commun.* **135:** 541.

Gimenez-Gallego, G., J. Rodkey, C. Bennett, M. Rios-Candelore, J. DiSalvo, and K. Thomas. 1985. Brain-derived acidic fibroblast growth factor: Complete amino acid sequence and homologies. *Science* **230:** 1385.

Gospodarowicz, D. 1985. Biological activity in vivo and in vitro of pituitary and brain fibroblast growth factor. In *Mediators in cell growth and differentiation* (ed. R.J. Ford and A.L. Maizel), p. 109. Raven Press, New York.

———. 1986. Isolation and characterization of acidic and basic fibroblast growth factor. *Methods Enzymol.* (in press).

Gospodarowicz, D., H. Bialecki, and T.K. Thakral. 1979. The angiogenic activity of the fibroblast and epidermal growth factor. *Exp. Eye Res.* **28:** 501.

Gospodarowicz, D., G.-M. Lui, and J. Cheng. 1982. Purification in high yield of brain fibroblast growth factor by preparative isoelectric focusing at pH 9.6. *J. Biol. Chem.* **257:** 12266.

Gospodarowicz, D., J. Cheng, G.-M. Lui, A. Baird, and P. Böhlen. 1984. Isolation of brain fibroblast growth factor by heparin-Sepharose affinity chromatography: Identity with pituitary fibroblast growth factor. *Proc. Natl. Acad. Sci.* **81:** 6963.

Jaye, M., R. Howk, W. Burgess, G.A. Ricca, I.-M. Chiu, M. Ravera, S.J. O'Brien, W.S. Modi, T. Maciag, and W.N. Drohan. 1986. Human endothelial cell growth factor: Cloning, nucleotide sequence, and chromosome localization. *Science* **233:** 541.

Klagsbrun, M. and Y. Shing. 1985. Heparin affinity of anionic and cationic capillary endothelial cell growth factors: Analysis of hypothalamus-derived growth factors and fibroblast growth factors. *Proc. Natl. Acad. Sci.* **82:** 805.

Klagsbrun, M., J. Sasse, R. Sullivan, and J.A. Smith. 1986. Human tumor cells synthesize an endothelial cell growth factor that is structurally related to basic fibroblast growth factor. *Proc. Natl. Acad. Sci.* **83:** 2448.

Kozak, M. 1984. Compilation and analysis of sequences upstream from the translational start site in eukaryotic mRNAs. *Nucleic Acids Res.* **12:** 857.

Lawn, R.M., E.F. Fritsch, R.C. Parker, G. Blake, and T. Maniatis. 1978. The isolation and characterization of linked δ and β-globin genes from a cloned library of human DNA. *Cell* **15:** 1157.

Lobb, R.R. and J.W. Fett. 1984. Purification of two distinct growth factors from bovine neural tissue by heparin affinity chromatography. *Biochemistry* **23:** 6295.

Lobb, R.R., E.M. Alderman, and J.W. Fett. 1985. Induction of angiogenesis by bovine brain derived class 1 heparin-binding growth factor. *Biochemistry* **24:** 4969.

Lobb, R., J. Sasse, R. Sullivan, Y. Shing, P. D'Amore, J. Jacobs, and M. Klagsbrun. 1986. Purification and characterization of heparin-binding endothelial cell growth factors. *J. Biol. Chem.* **261:** 1924.

Maciag, T., T. Mehlman, R. Friesel, and A.B. Schreiber. 1984. Heparin binds endothelial cell growth factor, the principal endothelial cell mitogen in bovine brain. *Science* **225:** 932.

Mackett, M., G.L. Smith, and B. Moss. 1984. General method for production and selection of infectious vaccinia virus recombinants expressing foreign genes. *J. Virol.* **49:** 857.

Neufeld, G. and D. Gospodarowicz. 1985. The identification and partial characterization of the fibroblast growth factor receptor of baby hamster kidney cells. *J. Biol. Chem.* **260:** 13860.

———. 1986. Basic and acidic fibroblast growth factors interact with the same cell surface receptors. *J. Biol. Chem.* **261:** 5631.

Proudfoot, N.H. and G.G. Brownlee. 1976. 3′ Non-coding region sequences in eukaryotic messenger RNA. *Nature* **263:** 211.

Schreiber, A.B., J. Kenney, J. Kowalski, K.A. Thomas, G. Gimenez-Gallego, M. Rios-Candelore, J. DiSalvo, D. Barritault, J. Courty, Y. Courtois, M. Moenner, C. Loret, W.H. Burgess, T. Mehlman, R. Friesel, W. Johnson, and T. Maciag. 1985. A unique family of endothelial cell polypeptide mitogens: The antigenic and receptor cross-reactivity of bovine endothelial cell growth factor, brain-derived acidic fibroblast growth factor, and eye-derived growth factor II. *J. Cell Biol.* **101:** 1623.

Shing, Y., J. Folkman, R. Sullivan, C. Butterfield, J. Murray, and M. Klagsbrun. 1984. Heparin affinity: Purification of a tumor-derived capillary endothelial cell growth factor. *Science* **223:** 1296.

Sullivan, R. and M. Klagsbrun. 1985. Purification of cartilage-derived growth factor by heparin affinity chromatography. *J. Biol. Chem.* **260:** 2399.

Thomas, K.A. and G. Gimenez-Gallego. 1986. Fibroblast growth factors: Broad spectrum mitogens with potent angiogenic activity. *Trends Biochem. Sci.* **11:** 81.

Thomas, K.A., M. Rios-Candelore, and S. Fitzpatrick. 1984. Purification and characterization of acidic fibroblast growth factor from bovine brain. *Proc. Natl. Acad. Sci.* **81:** 357.

Thomas, K.A., M. Rios-Candelore, G. Gimenez-Gallego, J. DiSalvo, C. Bennett, J. Rodkey, and S. Fitzpatrick. 1985. Pure brain-derived acidic fibroblast growth factor is a potent angiogenic vascular endothelial cell mitogen with sequence homology to interleukin 1. *Proc. Natl. Acad. Sci.* **82:** 6409.

von Heijne, G. 1983. Patterns of amino acids near signal-sequence cleavage sites. *Eur. J. Biochem.* **133:** 17.

Human Growth Hormone: From Clone to Clinic

P.H. SEEBURG
Genentech, Inc., South San Francisco, California 94080

Pituitary growth hormone was isolated only half a century ago, and until quite recently, nobody ever expected that one day microbes would be engineered to produce this important hormone. The first indications that the pituitary gland was involved in regulating linear growth came from an understanding of acromegaly. This disorder was first described in 1886 by Pierre Marie as a "striking noncongenital hypertrophy of the extremities" (Marie 1886). O. Minkowski noted in 1887 that an enlargement of the hypophysis had occurred in all carefully examined cases of acromegaly. That such pituitary dysfunction resulted from glandular hypersecretion was first specified by Harvey Cushing in 1909 (Cushing 1909).

More than 10 years later, Herbert Evans provided experimental evidence for pituitary gland hyperactivity in acromegaly by showing that extracts from bovine pituitary anterior lobes produced excessive growth in rats (Evans 1924). With proof of the existence of a growth-promoting pituitary hormone, the search began for a preparation that would prove effective in humans.

Another 20 years elapsed before the first report on the purification of bovine growth hormone by Evans and Li (1944). There was general disappointment when clinical trials with this material in humans showed no effect (Bennett et al. 1950). Eventually, it became well established that humans respond only to growth hormone of human or closely related primate origin (Li and Papkoff 1956), in stark contrast to the situation with insulin.

A new chapter was opened in this field of clinical research with the isolation of human growth hormone (hGH) by C.H. Li and colleagues (Li and Papkoff 1956). The first human trials using this material demonstrated anabolic (Li 1957) and growth-promoting activity (Raben 1958; Hutchings et al. 1959). The efficacy of hGH in short children generated a high demand for this hormone by clinical endocrinologists and pediatricians. hGH was made available by C.H. Li in the United States in 1957. In 1963, the National Pituitary Agency (NPA) was formed to collect human pituitary glands within the United States, to organize the extraction of hGH (and other pituitary hormones), and to distribute them for hypopituitary children and for other clinical studies (Raiti 1973). By the late 1970s, this agency was collecting 60,000 pituitaries per year to treat 1600 hypopituitary dwarfs in the United States.

These achievements in providing clinical-grade hGH for the treatment of idiopathic hypopituitarism were of signal importance. However, despite an enviable record of efficacy and safety, certain urgent considerations prompted developing an alternate source for this valuable hormone. Although there seemed to be no hGH shortage for the treatment of short children, the hormone was not available in sufficient quantities to conduct large-scale clinical studies for its somatotropic and anabolic efficacy under a variety of conditions (for details and further references, see Raiti 1980). Because of its species specificity, hGH could only be obtained from human cadavers. This deplorable situation was made worse because in some parts of the world, money could be gained from collecting cadaver pituitaries.

Contamination of hGH with other pituitary hormones was a recognized problem. Less recognized was the fact that human tissues are a potential source of infectious disease. With pituitaries, there was the risk of transmission of degenerative neurologic disease agents, and no screening had been established to ban patients with such diseases from collection. Pituitaries were even used from deceased patients from mental hospitals. Transmission could potentially be caused from contamination of tissue from the neurohypophysis and adjacent hypothalamic nuclei, which are known to be involved in neurodegenerative processes (Beck et al. 1969). Despite a few earlier warnings, and mainly due to the slow nature of certain neuropathological disease processes, this danger is only fully realized today after several young adults who had received cadaver hGH died from Creutzfeldt-Jakob disease (Brown et al. 1985). The need to develop a new source of hGH coincided with the advent of recombinant DNA technology. In 1975, when I began the series of experiments that would lead to recombinant hGH, this technology was very much in its infancy; the quest for an alternate source of hGH became a test case for new development in this field. This paper traces the molecular biology of hGH from clone to clinic.

MATERIALS AND METHODS

In vitro translation. Aliquots (1 μg) of polyadenylated RNA from placenta or rat pituitary cells that had been induced by 1 μM dexamethasone and 10 nM L-triiodothyronine were translated in a cell-free translation system derived from wheat germ in the presence of [^{35}S]methionine. Radioactive protein products were separated on 12.5% polyacrylamide gels containing SDS for 4 hours at 20 mA (for details, see Seeburg et al. 1977a,b).

cDNA cleavage and cloning. For analytical reactions, 1 μg of polyadenylated RNA was transcribed into cDNA using avian myeloblastosis virus reverse transcriptase in the presence of [α-^{32}P]dCTP (final sp. act. 10^5 cpm/pmole). Aliquots (10^5 cpm) of the cDNA were cleaved with restriction endonucleases *Hae*III or *Hha*I or combinations of both. Products were separated by polyacrylamide gel electrophoresis and visualized by autoradiography. For preparative reactions, 10 μg of RNA was reverse-transcribed at a specific activity of 100 cpm/pmole dCTP and cleaved, and the gel-isolated fragments were phosphorylated using [γ-^{32}P]ATP and polynucleotide kinase for DNA sequencing (Seeburg 1977b).

Immunological detection. Bacterial colonies containing the β-lactamase–pre-rGH expression vector were assayed for the presence of rat growth hormone (rGH) sequences using a solid-phase immunological screening method. The wash buffer was phosphate-buffered saline (0.1 M NaCl, 0.025 M potassium phosphate at pH 7.4) containing gelatin (10 mg/ml) and NP-40 (0.1%). Bacterial colonies were grown on L-broth agar in petri dishes. Colonies were treated for 20 minutes with chloroform vapor and then exposed to polyvinyl strips coated with antiserum to rGH. Strips were soaked in wash buffer, incubated overnight at 4°C with affinity-column-purified ^{125}I anti-rGH IgG (sp. act. 5 μCi/μg), washed, and exposed to X-ray film.

Assembly of the hGH expression gene. The chemically synthesized DNA encoding the first 24 amino acids of hGH was assembled by kinasing and ligating 12 separate fragments (Goeddel et al. 1979) and cloned via protruding *Eco*RI and *Hin*dIII termini into plasmid pBR322. The hGH cDNA *Hae*III fragment was given dC tails using terminal transferase and cloned in the dG-tailed *Pst*I site of pBR322. The cloned synthetic DNA was recovered by *Eco*RI-*Hae*III cleavage, and the cDNA fragment was recovered by *Hae*III and *Xma*I cleavage. Both of these parts of the hybrid gene were ligated as a mixture, cleaved to unit length by *Eco*RI and *Sma*I digestion, and cloned unidirectionally into a plasmid containing the *lac* promoter.

RESULTS AND DISCUSSION

Analysis of Growth-hormone mRNA and cDNA

Significantly for its molecular biology, the hGH polypeptide had been well characterized. It is a single-chain protein of 191 amino acid residues. The primary structure of hGH was first published by Li et al. (1966). This sequence was corrected later in two important aspects by Niall (1971) and finally revised by Li (1972). Because human pituitaries containing undamaged RNA were difficult to obtain, it was of high practical importance that the human placenta produce high amounts of a polypeptide termed human chorionic somatomammotropin (hCS) with more than 85% homology with hGH (Niall et al. 1971; Li et al. 1973). The high degree of homology suggested that these hormones had evolved from a common ancestral gene (Niall et al. 1971). Since fresh placental tissue was readily available, it was used to obtain information on hCS mRNA and, by analogy, hGH mRNA.

Another system was studied in parallel, whose idiosyncrasies added additional impetus to the study of the molecular biology of growth hormones. This system was a stable line of rat pituitary tumor cells (Kohler et al. 1969) in which growth-hormone mRNA levels were shown by us to be regulated by thyroid and glucocorticoid hormones (Martial et al. 1977a,b). This finding was considerably useful in defining conditions to obtain high growth-hormone mRNA levels.

mRNA levels were initially measured through analysis of the protein products generated by in vitro translation. Figure 1 shows the in vitro translation of placental and rat pituitary cell mRNAs and demonstrates the abundance of mRNA encoding the related peptide hormones. Figure 1 also demonstrates that the primary translation product of growth-hormone mRNA is larger than the mature hormone, a fact previously noted by Sussman et al. (1976). This longer sequence includes an amino-terminal signal sequence involved in protein secretion, in accordance with the model proposed by Blobel and Dobberstein (1975).

Although the mRNA levels for rGH and for the human placental homolog were quite abundant, cDNA cloning and especially sequence-screening methods were not available in 1975. Thus, it was of the utmost importance to visualize the sequence of interest, either by purifying the mRNA or otherwise. Using human placental mRNA, a key method was developed to identify individual cDNA sequences in cDNA mixtures. The method made use of the susceptibility of single-stranded cDNA to the activity of certain restriction endonucleases (e.g., *Hae*III and *Hha*I) and was applicable to cDNAs that represented approximately 2% or more of complex cDNA mixtures (Seeburg et al. 1977b). Endonuclease digestion of such cDNA will generate abundant DNA fragments that can be detected as bands on gels. After recovery from the gels, these fragments can be subjected to end labeling and direct DNA sequencing (see Fig. 2), using the Maxam-Gilbert procedure (1977). By this method, we obtained the sequence of part of the cDNA for the placental hormone. On the basis of these results and the known protein sequence for the placental hormone, we obtained a crude restriction map of the complete cDNA. In particular, a *Hae*III-generated DNA fragment of approximately 550 nucleotides could be assigned as encoding hCS residues 23–191 and containing, in addition, 50 nucleotides of 3′-untranslated sequence. A DNA fragment of the same size was also seen in *Hae*III digests of cDNA derived from human pituitary tumors, attesting to the high sequence homology between placental and pituitary growth-hormone mRNAs (see Fig. 3). The pituitary cDNA-derived *Hae*III fragment was later used to obtain recombinant hGH (see below).

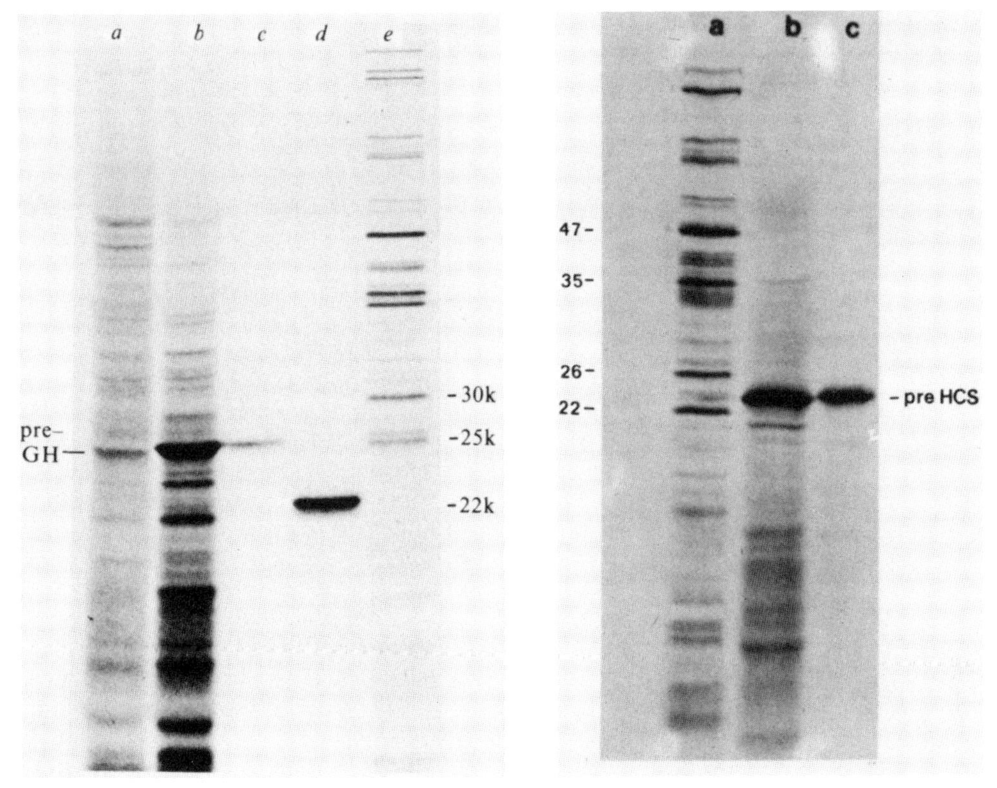

Figure 1. Products of in vitro translation of mRNA from rat pituitary cells (*left*) and human placenta (*right*) in a wheat-germ cell-free system. (*Left*) (*a*) mRNA from cells grown in absence of dexamethasone and thyroid hormone; (*b*) mRNA from hormone-induced cells; (*c*) immunoprecipitate of *b*; (*d*) immunoprecipitate of rGH secreted by cells into medium; (*e*) size standards. (*Right*) (*a*) size standards; (*b*) translation products of human placental mRNA; (*c*) immunoprecipitate of *b*.

Cloning of Growth-hormone cDNA

To develop cDNA procedures and to learn about the stability of growth-hormone-encoding DNA sequences in *Escherichia coli,* we constructed recombinant DNA using full-length rGH cDNA and the *Hae*III fragment of hCS cDNA. Cloning of cDNA was greatly aided by the construction of new cloning vectors such as pBR322 and pMB9 by F. Bolivar and H. Boyer (see Bolivar et al. 1977). These vectors were used by us for cloning growth-hormone cDNA sequences. Since we had knowledge of the restriction map of the cDNAs of interest, clones could be identified by restriction enzyme analysis. Purification of cDNA sequences prior to cloning reduced the number of clones that needed analyzing. In the case of the hCS cDNA, J. Shine developed an elegant procedure that allowed purification to homogeneity of the *Hae*III-derived cDNA fragment before ligating it into the cloning vector (Shine et al. 1977). Extensive purification was necessitated by the regulatory restrictions on the cloning of human DNA that were in effect at that time.

Sequence analysis of the cloned DNAs showed for the first time that growth-hormone cDNA sequences could be faithfully amplified in *E. coli*. The cloned rGH sequence represented the first full-length sequence for any somatotropin mRNA and allowed us to predict the protein sequence of rGH complete with its signal sequence (Seeburg et al. 1977a). This pioneered the rapid determination of protein sequences from DNA sequencing of cloned cDNAs. The derived rGH protein sequence showed high homology with that of hGH and hCS (see Fig. 4). Thus, the cloned rGH cDNA constituted an excellent model to evaluate the possibility of bacterial growth-hormone synthesis.

Synthesis of hGH by Bacteria

The cloned growth-hormone cDNA sequences were used to address the issue of whether *E. coli* could synthesize somatotropin. An expression vector was constructed that programmed the synthesis of a protein in which the amino-terminal 180 amino acids of the bacterial enzyme β-lactamase are fused to nearly the whole rat pre-growth-hormone molecule (Fig. 5). Construction was facilitated by the presence of a unique *Pst*I site in both pBR322 DNA and the cloned rGH cDNA with the identical reading frame across this restriction site in both sequences (Sutcliffe 1978). Results of this experiment showed unequivocally that *E. coli* can be genetically programmed to produce growth-hormone protein sequences (Seeburg et al. 1978). A protein of the right size (M_r = 46,000) was seen only in bacteria transformed with the expression vector, and these bacteria reacted with anti-rGH antibodies (Fig. 5).

Although this finding represented a major advance in our assessment of the feasibility of bacterial hGH

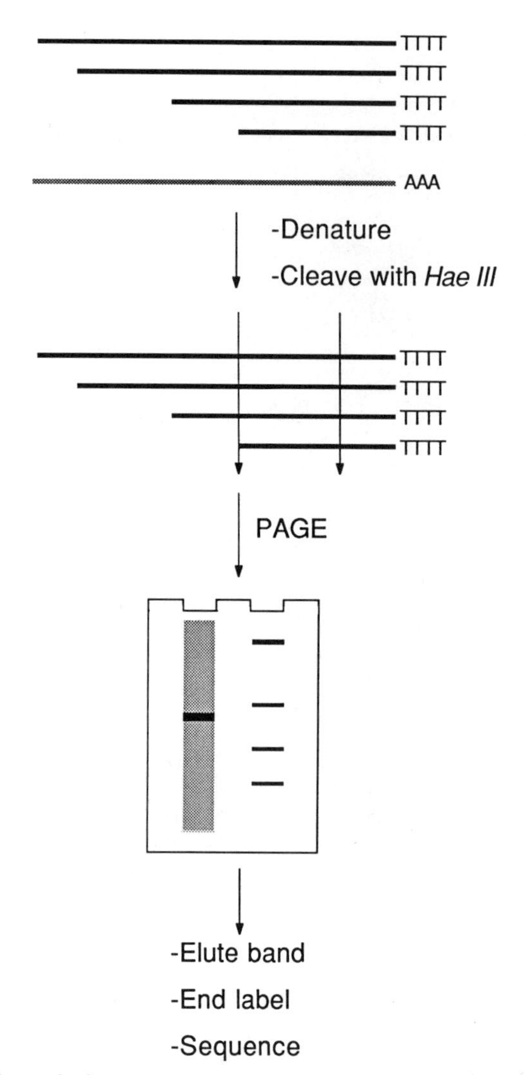

Figure 2. Schematic representation of restriction endonuclease cleavage of predominant cDNA. cDNA fragments generated from abundant mRNAs show as bands on polyacrylamide gels and can be sequence-analyzed (Seeburg et al. 1977a).

production, a bacterial fusion protein could not be expected to produce bioactive growth hormone, nor expression of the hormone in its precursor form. We therefore constructed a novel gene that would direct the synthesis of mature hGH (Goeddel et al. 1979). This gene was assembled from cloned hGH cDNA sequences and synthetic DNA (Fig. 6). The bulk of the coding sequence for hGH (residues 24–191) was contained on the 551-bp *Hae*III fragment described in Figure 3. The DNA encoding the amino-terminal 23 residues was assembled from 12 oligodeoxynucleotides, 6 for each strand. The synthetic DNA fragment was cloned and fused to the cloned cDNA fragment, resulting in a complete coding sequence for mature hGH. Since the methionine-initiation codon for protein synthesis in natural hGH mRNA precedes the signal sequence, a new initiation codon was provided by the synthetic DNA and directly preceded the codon for phenylalanine, the first amino acid in mature hGH. The use of synthetic DNA to construct this synthetic-natural hybrid gene provided additional degrees of freedom in sequence engineering, not possible through use of cDNA alone. Not only was the initiation codon for protein synthesis replaced, but some of the hGH codons were altered to preferred *E. coli* codons. At the same time, for reasons of translational efficacy, the synthetic sequence was chosen to minimize strong secondary structure formation between the bacterial Shine-Dalgarno sequence and the coding sequence for the hGH amino terminus.

For expression in *E. coli*, a plasmid was constructed that put this gene under *lac* operon control. Bacterial extracts from transfected *E. coli* showed the presence of a protein that comigrated with pituitary hGH and that could be immunoprecipitated using antisera. This constituted the first time that a human polypeptide had been directly expressed in *E. coli* in a nonprecursor form.

The synthetic-natural hybrid gene codes for hGH with an amino-terminal methionine residue. The fact that most bacterial proteins do not contain amino-terminal methionine residues suggested that it might be efficiently removed from hGH. We learned later, however, that with elevated expression levels of hGH, the methionine residue stays. Importantly, the biosynthetic methionyl-hGH was found to be biologically active (Hintz et al. 1982) in humans.

We have learned since that hGH can also be secreted by *Pseudomonas* (Gray et al. 1984) and by *E. coli* (Gray et al. 1985) using a bacterial or the natural signal sequence. In these systems, hGH is correctly processed and has no methionine at its amino terminus.

Scale-up, Purification, and Clinical Studies

Initially, bacterial hGH levels were low and there were indications of hGH instability in the microorganism. A study using different operon systems showed that the use of the inducible *E. coli trp* promoter helped increase hGH expression levels (de Boer et al. 1982). This construct is in use for our large-scale hGH production. Similarly, a study of bacterial growth conditions helped to optimize the stability of the hormone. High-density fermentation has been optimized to yield several grams of hGH from each liter of bacterial medium.

The purification of methionyl-hGH from bacteria was a considerable task (Jones and O'Connor 1982). The process was modified several times to obtain the hormone in extremely pure form, not only essentially free of endotoxin, but also free of hGH dimers and modified forms that generate anti-hGH antibodies. Initial hGH batches, although efficacious in growth-promoting properties (hypopituitary children grew on the average of 10.4 cm ± 2.3 cm/year over 3.5 ± 1.4 cm before treatment), generated a higher than usual incidence of antibodies to hGH. Additional purification

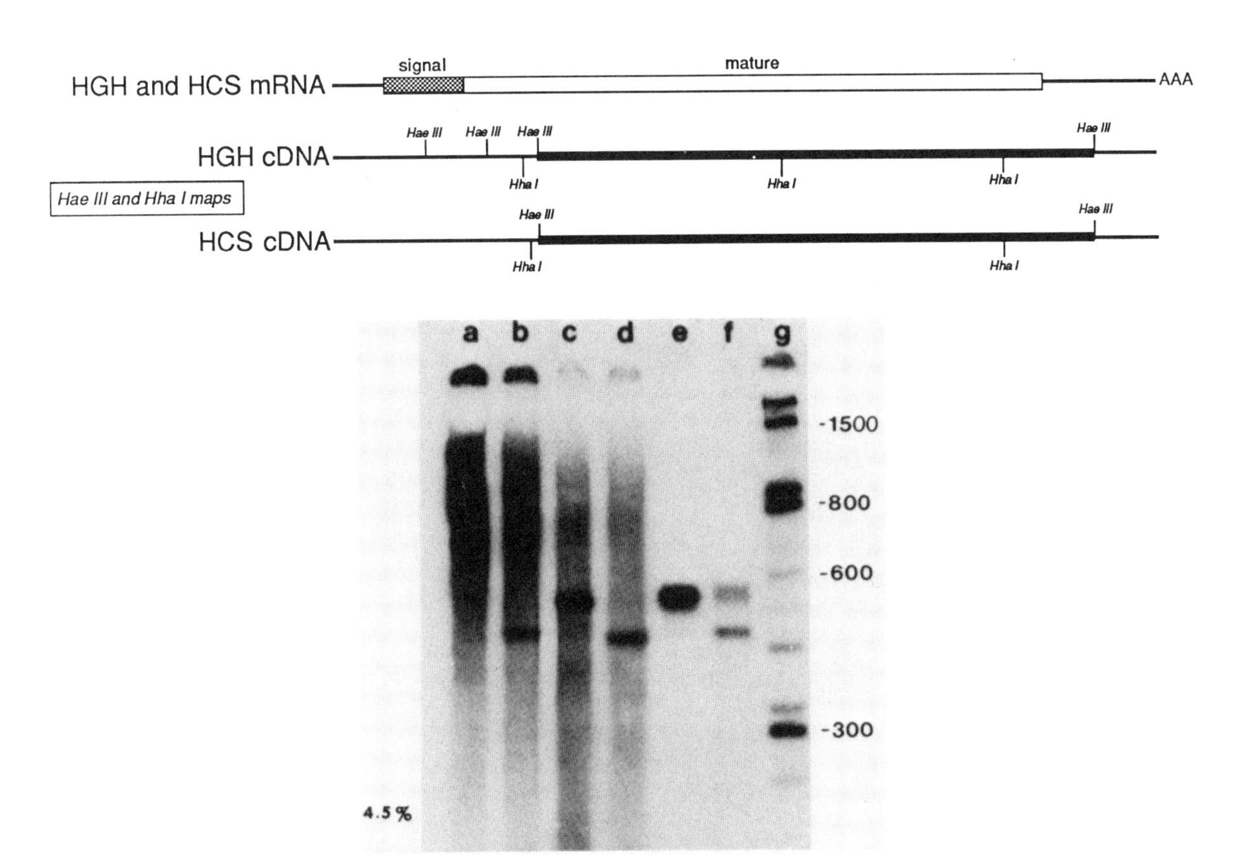

Figure 3. Schematic representation of hGH and hCS mRNAs and cleavage patterns of the respective cDNAs. The mRNAs for hGH and hCS are approximately 800 nucleotides long without the poly(A) tail. They contain a 60-nucleotide-long 5′-untranslated region, a 78-base region encoding a signal peptide of 26 amino acid residues, a 573-nucleotide-long region coding for the 191-residue-long mature polypeptide hormone, and a 3′-untranslated region of 110 bases. Since both mRNAs are abundant in their tissues of origin (pituitary and placenta), cDNA fragments generated by restriction endonucleases and corresponding in sequence to hormone mRNAs can be visualized on gels and used for DNA sequencing and/or generating restriction maps. Such maps are shown for hGH and hCS cDNAs (Seeburg et al. 1977a). The 550-bp *Hae*III fragment that contains coding sequences for amino acids 23–191 of hGH and hCS is represented by a black bar. hCS cDNA fragments generated by endonucleases *Hae*III and *Hha*I are shown in the autoradiograph below. (*a*) Uncleaved placental cDNA; (*b*) same as *a* but cleaved with *Hha*I; (*c*) cleaved with *Hae*III (note the 550-bp prominent cDNA fragment); (*d*) cleaved with both *Hae*III and *Hha*I; (*e*) the gel-isolated *Hae*III fragment; (*f*) same as *e* but *Hha*I-cleaved; (*g*) single-stranded DNA size markers.

```
       -26                          1
                                                                      ___
HGH   : MATGSRTSLL LAFGLLCLPW LQEGSAFPTI PLSRLFDNAM LRAHRLHQLA FDTYQEFEEA
HCS   :   P           A          AG VQ V        H    Q   A     I        T
RGH   :   AD Q PW   T S    L  P  AG   AM    S  A  V    QH      A   K   R

                       50
        __________ ___
HGH   : YIPKEQKYSF LQNPQTSLCF SESIPTPSNR EETQQKSNLE LLRISLLLIQ SWLEPVQFLR
HCS   :     D       HDS        D       M                     E       R
RGH   :    EG R  - I  A AAF     T  A TGK   A  RTDM     F          G     S

            100                                                150
HGH   : SVFANSLVYG ASDSNVYDLL KDLEEGIQTL MGRLEDGSPR TGQIFKQTYS KFDTNSHNDD
HCS   :  M   N   D T   DD H                      R      L              H
RGH   : RI T   MF  T  R-  EK          A   QE        I   L    D    A MRS

                                              191
HGH   : ALLKNYGLLY CFRKDMDKVE TFLRIVQCRS VEGSCGF
HCS   :                           M
RGH   :          S   K  LH A   Y  VMK  RFA S  A
```

Figure 4. Sequence comparison of the precursor proteins of hGH, hCS, and rGH. The sequences are presented in single-letter notation. For hCS and rGH, only residues that differ from those of hGH are shown. Sequences show the high degree of homology between hGH and hCS (85%) and the conserved nature of rGH (65% homology with hGH). Numbers denote residues in hGH and hCS, since rGH has two single-amino-acid deletions and one addition relative to hGH. Negative numbers are assigned to the residues in the signal sequence. The protein sequences shown are translations from previously determined nucleotide and gene sequences (Seeburg et al. 1977a,b; Goeddel et al. 1979; Seeburg 1982). The amino acid residues overscored in hGH are missing in the 20,000-dalton variant of somatotropin (Lewis et al. 1980).

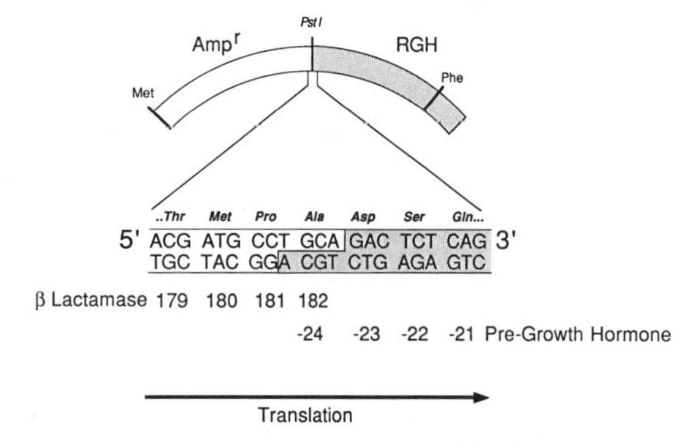

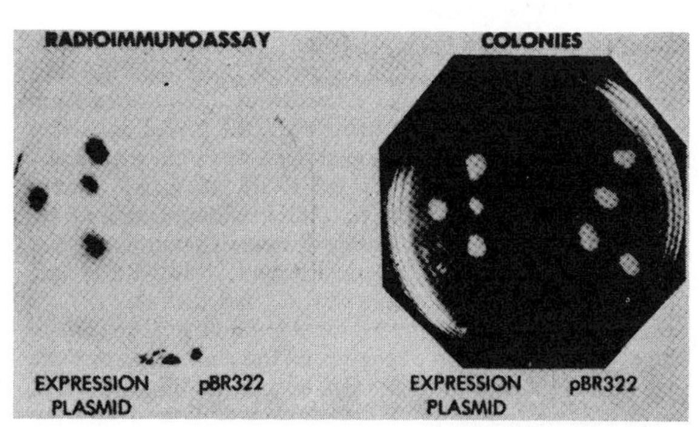

Figure 5. Expression of growth-hormone sequences in *E. coli*. (*Top*) Point of fusion between bacterial (positively numbered amino acids) β-lactamase and pre-rGH (negatively numbered amino acids) coding sequences across a *Pst*I site (CTGCAG) to generate an expression vector for bacterial synthesis of a growth-hormone fusion. (*Bottom*) Immunological detection of growth hormone in bacteria. Only bacterial colonies containing the expression vector are seen to react with iodinated anti-rGH antiserum (Seeburg et al. 1978).

steps (Fig. 7) resulted in a product that has a comparably low degree of antigenicity to pure pituitary hGH. As a result of its purity and efficacy, hGH was approved by the FDA in late 1985 and represents the second genetically engineered pharmaceutical.

The hGH Locus

This report on the molecular biology of hGH would not be complete without a description of the hGH gene locus (Fig. 8). A total of five genes are found clustered on chromosome 17 (George et al. 1981) within a region of approximately 50–60 kb (Seeburg 1982; Barsh et al. 1983; Seeburg 1985). The arrangement of these genes is 5′-*hGH-N, hCS-L, hCS-A, hGH-V, hCS-B*-3′ (see Barsh et al. 1983). Suggesting the recent evolutionary origin of this locus, all five genes are highly homologous (>85%) throughout exons, introns, and promoter regions (Seeburg 1982). Only primates and humans seem to carry such a complex growth-hormone locus. Each gene consists of five exons with introns in exactly the same relative positions in all genes.

To understand the mechanisms of tissue-specific expression of these genes, we have recently determined the complete nucleotide sequence of the hGH locus and have used specific synthetic oligonucleotides to probe for expression of each gene in human pituitary and placental tissue (E.Y. Chen et al., in prep.). In summary, the *hGH-N* gene is expressed only in the pituitary, giving rise to two products: the 191-amino-acid-long hGH and the 20K hGH (a shorter version) that is lacking residues 32–46 due to an internal splicing event removing 45 nucleotides from the third exon of hGH mRNA (De Noto et al. 1981; Seeburg 1982). No expression of *hCS-L* gene sequences was found. This seems to be due to a mutation in a splice site, giving the *L* gene a pseudogene character (H.A. Barrera-Saldana and P.H. Seeburg, unpubl.). Both the *hCS-A* and *hCS-B* genes are expressed in placental tissue (Barrera-Saldana et al. 1983) only.

The *hGH-V* gene, which is sandwiched between the *hCS-A* and *hCS-B* genes, encodes a new polypeptide hormone (Seeburg 1982) closely related to hGH but with several notable differences. There are 13 amino acid differences between the mature polypeptides. Although most of these changes are conservative in nature, the *hGH-V* protein has two tryptophan residues (at positions 85 and 126; hGH has one at position 85) and contains a possible N-linked glycosylation site at residue 140. We have recently isolated cDNA for this gene from a term placental library (P.H. Seeburg et al., in prep.), generating proof for the expression of this gene. No expression was detected in the pituitary. The placental *hGH-V* gene expression as measured by the number of cDNA clones was found to be approxi-

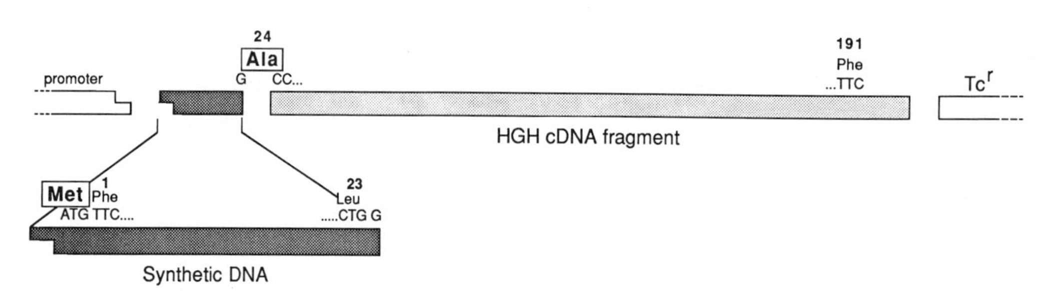

Figure 6. Assembly of the bacterial synthetic-natural hybrid gene directing hGH synthesis in *E. coli*. Synthetic DNA is highlighted and contains the methionine codon for initiation of protein synthesis directly in front of the codon for the first amino acid of mature hGH. Codon 24 is generated by fusion of the cloned synthetic DNA and the cloned 550-bp hGH cDNA fragment (Goeddel et al. 1979). This fragment contains the remainder of the hGH-coding region and, in addition, part of the 3′-untranslated region of hGH cDNA.

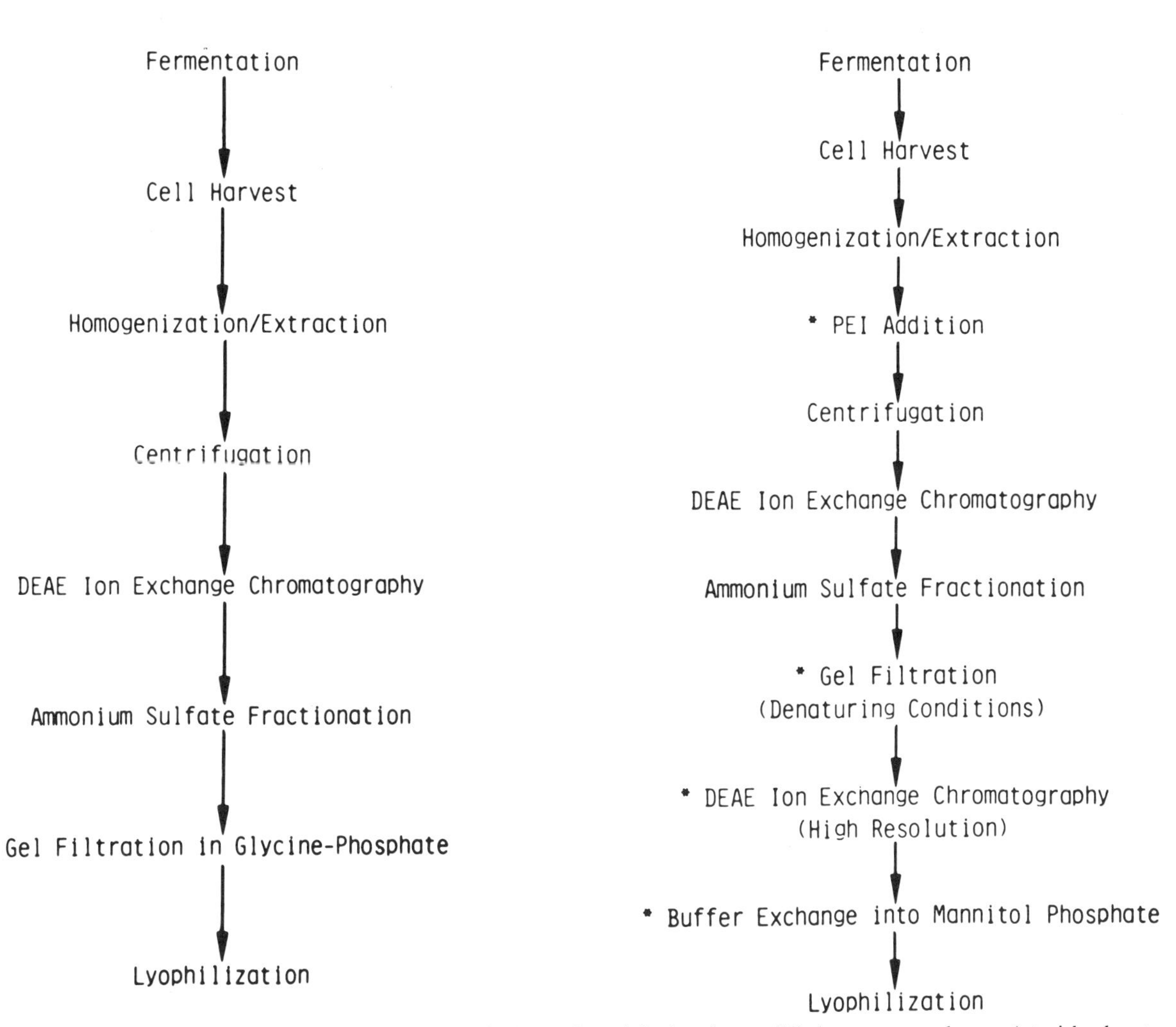

Figure 7. Purification steps for bacterially produced hGH. The original and a modified process are shown. Asterisks denote new steps in the modified process to obtain higher purity hGH.

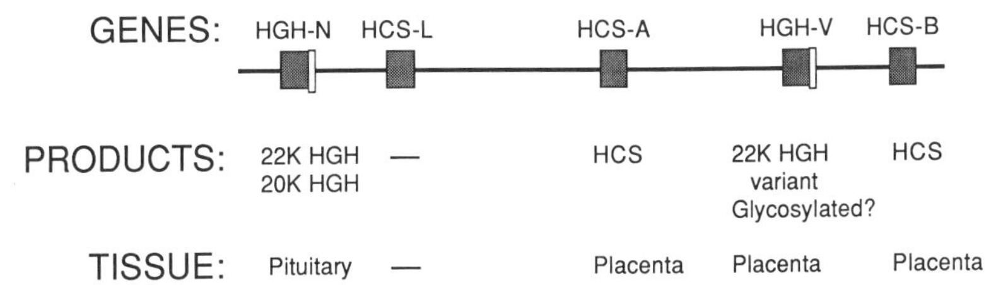

Figure 8. Chromosomal hGH locus. The arrangement of the five genes in this locus on human chromosome 17 is shown. Each gene is approximately 2 kb in length and consists of five exons. The complete locus is about 50 kb. The open bars at the end of the *hGH-N* and *hGH-V* genes are repetitive Alu elements not found in the immediate region 3′ of the polyadenylation site in the three hCS-type genes. The known products of the genes as well as the tissues in which these genes are expressed are shown below the schematic map.

mately three orders of magnitude lower than that of the *hCS-A* and *hCS-B* genes. The cloned *hGH-V* cDNAs contained the coding region for residues 32–46, thus encoding the complete 191-amino-acid protein.

The function of the *hGH-V* gene product is presently unknown. Interestingly, a protein from placenta was recently characterized (Hennen et al. 1985) with properties very similar to those expected of *hGH-V* (Seeburg et al. 1985) and the product of this gene expressed in heterologous cells (Pavlakis et al. 1981). This protein cross-reacts with anti-hGH antibodies, is placenta-specific, and has a higher isoelectric point than pituitary hGH, as predicted from the sequence of the *hGH-V* gene (Seeburg 1982).

CONCLUSION

The programming of *E. coli* with a synthetic-natural hybrid gene for hGH created an alternate source for this important hormone. This source helps to generate the equivalent amount of hGH in each liter of fermentation medium that would have to be isolated from many human cadavers. It is also free of contamination with other pituitary hormones or infectious disease agents. Due to the unlimited supply of hGH, it is now possible to construct large-scale clinical studies to assess its benefits in conditions such as familial short stature, Turner syndrome, achondroplasia, aging, postoperative repair and wasting states, nonhealing bone fractures, stress ulcer, skin burns, and hypercholesterolemia for coronary-prone subjects.

Interest in hGH and its molecular biology has also generated an understanding of the hGH gene and its chromosomal environment. This has helped in understanding that the cause of familial isolated growth-hormone deficiency type A is a deletion of this gene from chromosome 17 of afflicted individuals (Phillips et al. 1981). The structure of the hGH locus is of high evolutionary interest, since recent gene duplications have generated a total of five highly homologous genes, whose expression is tissue-specific. The existence of a new growth-hormone-related polypeptide hormone was predicted from studying this locus (Seeburg 1982), and the expression of the corresponding gene in placenta has recently been detected (P.H. Seeburg et al., in prep.). The search for the function of this new hormone and for the genetic mechanisms that control the tissue specificity of gene expression in the hGH locus represents an exciting direction for future studies concerning the molecular biology of hGH.

ACKNOWLEDGMENTS

I am indebted to my many colleagues at Genentech, Inc., in particular, Drs. David Goeddel, Herbert Heyneker, Ellson Chen, Herman de Boer, Andy Jones, Barry Morgan, and Ken Olsen, who helped to turn an idea into a product from which many will benefit. I was privileged to collaborate with John Shine, John Phillips, and Richard Gelinas, whose creative input helped elucidate the molecular biology of hGH and its related genes. I thank Dr. Ann Johanson and Dr. Hugh Niall for discussions, Carol Morita for art work, and Jeanne Arch for help with this manuscript.

REFERENCES

Barrera-Saldana, H.A., P.H. Seeburg, and G.F. Saunders. 1983. Two structurally different genes produce the same secreted human placental lactogen hormone. *J. Biol. Chem.* **258:** 3787.

Barsh, G.S., P.H. Seeburg, and R.E. Gelinas. 1983. The human growth hormone gene family: Structure and evolution of the chromosomal locus. *Nucleic Acids Res.* **11:** 3939.

Beck, E., P.M. Daniel, and W.B. Matthews. 1969. Creutzfeldt-Jakob disease: The neuropathology of a transmission experiment. *Brain* **92:** 699.

Bennett, L.L. et al. 1950. Failure of hypophyseal growth hormone to produce nitrogen storage in girl with hypophyseal dwarfism. *J. Clin. Endocrinol.* **10:** 492.

Blobel, G. and B. Dobberstein. 1975. Transfer of proteins across membranes. Part I. Presence of proteolytically processed and unprocessed nascent immunoglobulin light chains on membrane-bound ribosomes of murine myeloma cells. *J. Cell. Biol.* **67:** 835.

Bolivar, F., R. Rodriguez, M.C. Betlach, H.L. Heyneker, and H.W. Boyer. 1977. Construction and characterization of new cloning vehicles. II. A multipurpose cloning system. *Gene* **2:** 95.

Brown, P., D.C. Gajdusek, C.J. Gibbs, Jr., and D.M. Asher. 1985. Potential epidemic of Creutzfeldt-Jacob disease from human growth hormone therapy. *N. Engl. J. Med.* **313:** 728.

Cushing, H. 1909. *J. Am. Med. Assoc.* **53:** 249.

de Boer, H.A, L.J. Comstock, D.G. Yansura, and H.L. Heyneker. 1982. Construction of a tandem *trp-lac* promoter for efficient and controlled expression of human growth hormone gene in *Escherichia coli.* In *Promoters: Structure and function* (ed. R.L. Rodriguez and M.J. Chamberlin), p. 462. Praeger, New York.

De Noto, F.M., D.D. Moore, and H. Goodman. 1981. Human growth hormone DNA sequence and mRNA structure: Possible alternative splicing. *Nucleic Acids Res.* **9:** 3719.

Evans, H.M. 1924. *Harvey Lect.* **19:** 212.

Evans, H.M. and C.H. Li. 1944. Human pituitary growth hormone. *Science* **99:** 183.

George, D.L., J.A. Phillips III, U. Francke, and P.H. Seeburg. 1981. The genes for growth hormone and chorionic somatomammotropin are on the long arm of chromosome 17 in region q21-qter. *Hum. Genet.* **57:** 138.

Goeddel, D.V., H.L. Heyneker, T. Hozumi, R. Arentzen, K. Itakura, D.G. Yansura, M.J. Ross, G. Miozzari, R. Crea, and P.H. Seeburg. 1979. Direct expression in *Escherichia coli* of a DNA sequence coding for human growth hormone. *Nature* **281:** 544.

Gray, S.L., J.S. Baldridge, K.A. McKeown, H.L. Heyneker, and C.N. Chang. 1985. Periplasmic production of correctly processed human growth hormone in *Escherichia coli*: Natural and bacterial signal sequences are interchangeable. *Gene* **39:** 247.

Gray, G.L., K.A. McKeown, A.J.S. Jones, P.H. Seeburg, and H.L. Heyneker. 1984. *Pseudomonas aeruginosa* secretes and correctly processes human growth hormone. *Biotechnology* **2:** 161.

Hennen, G., F. Frankenne, J. Closset, F. Gomez, G. Pirens, and N. El Khayat. 1985. A human placental GH: Increasing levels during second half of pregnancy with pituitary GH suppression as revealed by monoclonal antibody radioimmunoassays. *Int. J. Fertil.* **30:** 27.

Hintz, R.L., D.M. Wilson, J. Finno, R.G. Rosenfeld, A. Bennett, B. McClellan, and R. Swift. 1982. Biosynthetic methionyl human growth hormone is biologically active in adult man. *Lancet* **8284:** 1276.

Hutchings, J.J., R.F. Escamilla, W.C. Deamer, and C.H. Li. 1959. Metabolic changes produced by human growth hormone (Li) in a pituitary dwarf. *J. Clin. Endocrinol. Metab.* **19:** 759.

Jones, A.J.S. and J.V. O'Connor. 1982. Chemical characterization of methionyl human growth hormone. In *Hormone drugs, proceedings of the FDA-USP workshop on drug reference standards for insulins, somatotropins and thyroid-axis hormones* (ed. J.L. Gueriguian), p. 335. U.S. Pharmacopeial Convention, Bethesda, Maryland.

Kohler, P.O., L.A. Frohman, W.E. Briolson, T. Vanha-Perttula, and J.M. Hammond. 1969. Cortisol induction of growth hormone synthesis in a cloned line of rat pituitary cells in culture. *Science* **166:** 633.

Lewis, U.J., L.F. Bonewald, and L.J. Lewis. 1980. The 20,000 dalton variant of human growth hormone: Location of the amino acid deletion. *Biochem. Biophys. Res. Commun.* **92:** 511.

Li, C.H. 1957. Properties of and structural investigations on growth hormones isolated from bovine, monkey and human pituitary glands. *Fed. Proc.* **16:** 775.

———. 1972. Hormones of the adenohypophysis. *Proc. Am. Philos. Soc.* **116:** 365.

Li, C.H. and H. Papkoff. 1956. Preparation and properties of growth hormone from human and monkey pituitary glands. *Science* **124:** 1293.

Li, C.H., J.S. Dixon, and D. Chung. 1973. Amino acid sequence of human chorionic somatomammotropin. *Arch. Biochem. Biophys.* **155:** 95.

Li, C.H., W.K. Liu, and J.S. Dixon. 1966. Human pituitary growth hormone. XII. The amino acid sequence of the hormone. *J. Am. Chem. Soc.* **88:** 2050.

Marie, P. 1886. *Rev. Med. (Paris)* **6:** 297.

Martial, J.A., J.D. Baxter, H.M. Goodman, and P.H. Seeburg. 1977a. Regulation of growth hormone messenger RNA by thyroid and glucocorticoid hormones. *Proc. Natl. Acad. Sci.* **74:** 1816.

Martial, J.A., P.H. Seeburg, D. Guenzi, H.M. Goodman, and J.D. Baxter. 1977b. Regulation of growth hormone gene expression: Synergistic effects of thyroid and glucocorticoid hormones. *Proc. Natl. Acad. Sci.* **74:** 4293.

Maxam, A. and W. Gilbert. 1977. A new method for sequencing DNA. *Proc. Natl. Acad. Sci.* **74:** 560.

Niall, H.D. 1971. Human growth hormone—Revised primary structure. *Nat. New Biol.* **230:** 90.

Niall, H.D., M.L. Hogan, R. Sauer, I.Y. Rosenblum, and F.C. Greenwood. 1971. Sequences of pituitary and placental lactogenic and growth hormones: Evolution from a primordial peptide by gene reduplication. *Proc. Natl. Acad. Sci.* **68:** 866.

Pavlakis, G.N., N. Hizuka, P. Gorden, P.H. Seeburg, and D.H. Hamer. 1981. Expression of two human growth hormone genes in monkey cells infected by simian virus 40 recombinants. *Proc. Natl. Acad. Sci.* **78:** 7398.

Phillips, J.A., III, B.L. Hjelle, P.H. Seeburg, and M. Zachmann. 1981. Molecular basis for familial isolated growth hormone deficiency. *Proc. Natl. Acad. Sci.* **78:** 6372.

Raben, M.S. 1958. Treatment of pituitary dwarf with human growth hormone. *J. Clin. Endocrinol. Metab.* **18:** 901.

Raiti, S. 1973. The National Pituitary Agency. In *Advances in human growth hormone research* (DHEW-NIH74-612) (ed. S. Raiti), p. 11. Government Printing Office, Washington, D.C.

———. 1980. Clinical studies with human somatotropin. In *Polypeptide hormones.* Twelfth Miles International Symposium (ed. R.F. Beers, Jr. and E.G. Bassett), p. 309. Raven Press, New York.

Seeburg, P.H. 1982. The human growth hormone gene family: Nucleotide sequences show recent divergence and predict a new polypeptide hormone. *DNA* **1:** 239.

———. 1985. The human growth hormone locus: The genes and their products. In *Biogenesis of neurohormonal peptides* (ed. R. Hakanson and J. Thorell), p. 83. Academic Press, New York.

Seeburg P.H., J. Shine, J.A. Martial, J.D. Baxter, and H.M. Goodman. 1977a. Nucleotide sequence and amplification in bacteria of structural gene for rat growth hormone. *Nature* **270:** 486.

Seeburg, P., J. Shine, J.A. Martial, A. Ullrich, J.D. Baxter, and H.M. Goodman. 1977b. Nucleotide sequence of part of the gene for human chorionic somatomammotropin: Purification of DNA complementary to predominant mRNA species. *Cell* **12:** 157.

Seeburg, P.H., J. Shine, J.A. Martial, R.D. Ivarie, J.A. Morris, A. Ullrich, J.D. Baxter, and H.M. Goodman. 1978. Synthesis of growth hormone by bacteria. *Nature* **276:** 795.

Shine, J., P.H. Seeburg, J.A. Martial, J.D. Baxter, and H.M. Goodman. 1977. Construction and analysis of recombinant DNA for human chorionic somatomammotropin. *Nature* **270:** 494.

Sussman, P.M., R.J. Tushinski, and F.C. Bancroft. 1976. Pregrowth hormone: Product of the translation *in vitro* of messenger RNA coding for growth hormone. *Proc. Natl. Acad. Sci.* **73:** 29.

Sutcliffe, J.G. 1978. Nucleotide sequence of the ampicillin resistance gene of *Escherichia coli* plasmid pBR322. *Proc. Natl. Acad. Sci.* **75:** 3737.

Molecular and Biological Properties of Human Macrophage Growth Factor, CSF-1

P. RALPH,* M.K. WARREN,* M.B. LADNER,† E.S. KAWASAKI,† A. BOOSMAN,‡ AND T.J. WHITE**

*Departments of *Cell Biology, †Molecular Biology, and ‡Analytical Development and **Vice President of Research, Cetus Corporation, Emeryville, California 94608*

CSF-1 belongs to a family of colony-stimulating factors (CSF) that regulate the production of the blood cells (Metcalf 1984). CSF-1 is a specific growth and differentiation factor for bone-marrow progenitor cells of the mononuclear phagocyte lineage and also promotes the proliferation of mature macrophages via specific receptors on the responding cells (Das et al. 1981; Das and Stanley 1982). CSF-1 also has a variety of stimulatory effects on the function of macrophages and monocytes. In this paper, we summarize the cloning of the cDNA and the genomic structure of human CSF-1, and we describe properties of the macrophage growth factor that may make it a useful drug in several clinical settings.

RESULTS

Genomic and cDNA Structure

Genomic clones for human CSF-1 were identified using DNA probes based on amino-terminal sequence data of the human urinary protein. Using a genomic probe, we obtained a cDNA clone from the human MIA PaCa cell line that codes for bioactive CSF-1 upon transfection of the primate COS cell line (Kawasaki et al. 1985). CSF-1 appears to be encoded by a single-copy gene, which is about 18 kb in length and contains nine exons (Kawasaki et al. 1985, 1986) (see Fig. 1).

The cDNA specifies a 32-amino-acid leader peptide followed by a 224-residue polypeptide. There are two potential N-linked glycosylation sites. At residue 59 of the mature protein, the cDNA codes for tyrosine, whereas the genomic codon is aspartic acid. This could be due to a natural polymorphism or to a reverse transcriptase error when making the cDNA library. The cDNA predicts an unusual structure for a secreted protein, namely, a very hydrophobic region of 23 amino acids (residues 166–188 of the mature protein) followed by Arg-Trp-Arg-Arg-Arg. This is typical of membrane proteins that have a transmembrane hydrophobic domain followed by three positively charged residues act-

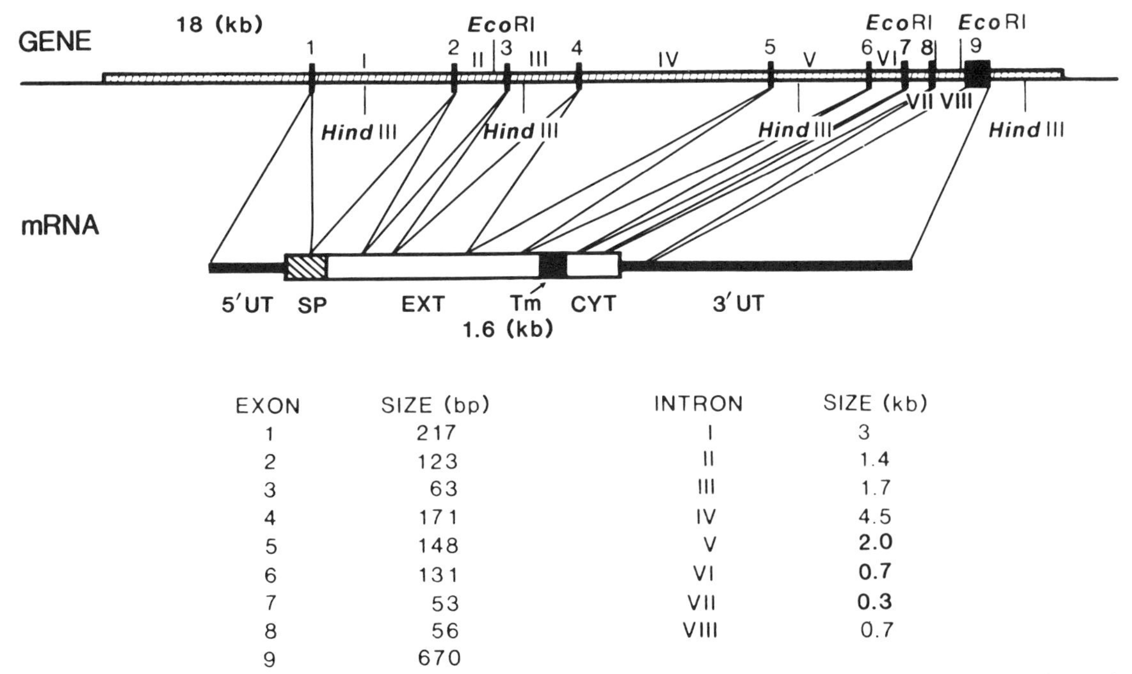

EXON	SIZE (bp)
1	217
2	123
3	63
4	171
5	148
6	131
7	53
8	56
9	670

INTRON	SIZE (kb)
I	3
II	1.4
III	1.7
IV	4.5
V	2.0
VI	0.7
VII	0.3
VIII	0.7

Figure 1. Structure of the human CSF-1 gene and mRNA. The gene is shown with some of its restriction enzyme sites and the exons. The functional units of the mRNA are 5′- and 3′-untranslated regions (UT), signal peptide (SP), extracellular domain (EXT), and putative transmembrane (Tm) and cytoplasmic (CYT) domains.

Table 1. Homology between Human and Murine CSF-1 Proteins

Source	Position 1									11										21										
Human MIA PaCa							*																							
Human cDNA	E	E	V	S	E	Y	C	S	H	M	I	G	S	G	H	L	Q	S	L	Q	R	L	I	D	S	Q	M	E	T	S
Murine L cell	×					×							×				×	×			×									

Source	31									41			
MIA PaCa	*												
cDNA	C	Q	I	T	F	E	F	V	D	Q	E	Q	L
L cell	×			×			*			*	*	*	*

Source	65						71										81							88
MIA PaCa	( )	( )		( )							( )		( )	( )	( )				( )		( )	( )	*	*
cDNA	M	R	F	R	D	N	T	P	N	A	I	A	I	V	Q	L	Q	E	L	S	L	R	L	K
L cell	( )	( )		×							(×)		×	×	(×)				×		(×)	(×)	*	(×)

Murine protein-to-human cDNA homology: 48/65 = 74% identity. Human MIA PaCa and murine L-cell CSF-1 proteins were purified to homogeneity and partially sequenced (A. Boosman et al., in prep.). Amino acid residues indentical to the translation of the cDNA (shown from the amino terminus to position 88 in single-letter code) are shown as blanks. Residues not determined (*) and residues different from the cDNA (×) are also shown. Residues from position 44 to 64 in the native protein have not been determined. Empty parentheses indicate a tentative identification agreeing with the cDNA. Parentheses containing an × indicate a tentative identification different from the cDNA.

ing as an anchor on the cytoplasmic side (Sabatini et al. 1982).

Protein Structure and Amino Acid Homology with Murine CSF-1

CSF-1 molecules from murine L929 cells (Ben-Avram 1985; Ben-Avram et al. 1985; Kawasaki et al. 1985), human urine (Kawaskaki et al. 1985), and the MIA PaCa cell line (Csejtey and Boosman 1986; A. Boosman et al., in prep.) have been purified to homogeneity and partially sequenced. The sequence data to date show that the human molecules are identical to each other and to the protein predicted from the cDNA (Table 1). The human and murine molecules show 74% amino acid identity over the 65 residues of the regions that have been sequenced and are thus highly homologous.

Both native human CSF-1 and murine CSF-1 are heavily glycosylated dimer proteins with a molecular weight of 45,000–70,000. The unglycosylated subunits of both murine and human CSF-1 are reported to have a molecular weight of 14.5K (Das and Stanley 1982), whereas the human cDNA predicts a polypeptide of 26K. Thus, the larger translated product may have another function as a membrane-bound molecule, with intracellular protein processing or perhaps a differently spliced mRNA used to produce the secreted CSF-1. There is evidence for a cell-surface-bound form of CSF-1 (Stanley et al. 1976).

Biological Properties of Native CSF-1

CSF-1 has direct stimulating effects on the mature monocyte and macrophage, in addition to being a growth and differentiation factor for bone-marrow

Table 2. Stimulation of Mature Macrophage and Monocyte Functions by CSF-1

	References	
	mouse	human
Plasminogen activator production	Lin and Gordon (1979)	
PGE production	Ralph (1984)	
Ferritin production	Broxmeyer et al. (1985a)	
IL-1 production	Moore et al. (1980)	
Myeloid CSF production	Metcalf and Nicola (1985)	Warren and Ralph (1986)
IFN production	Ralph (1984)	Warren and Ralph (1986)
Oxygen metabolites	Wing et al. (1985)	
Intracellular killing of Candida	Ralph (1984)	
Tumor cytostatic activity	Wing et al. (1982)	
Tumor cytotoxin		Warren and Ralph (1986)
Tumor cytotoxic activity	P. Ralph and I. Nakoinz (in prep.)	
Resist viral infection	M.K. Warren and M.T. Lee (in prep.)	

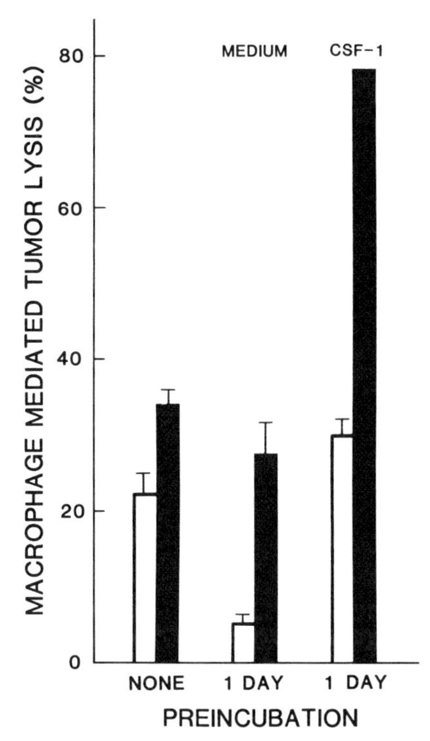

Figure 2. Tumoricidal activity of 1-day-cultured peritoneal exudate macrophages is enhanced by L-CM containing CSF-1. Adherent peritoneal exudate cells from proteose peptone-injected C3H-HeN mice were tested without preincubation or after a 1-day preincubation with medium or CSF-1 (300 U/ml in conditioned medium of L929 cells [L-CM]) for killing [^{3}H]thymidine-labeled TU5 sarcoma cells at 40:1 in a 48-hr assay (Ralph et al. 1982). At the time of tumor lysis assay, replicate wells received 10% v/v lymphokine (■; 2-day concanavalin-A spleen supernate). Background radiolabel release of 7% was subtracted.

precursors (Table 2). It stimulates the production of prostaglandins and interferon (IFN), the intracellular killing of Candida, and the production of plasminogen activator, interleukin-1, oxygen metabolites, ferritin, and a G-CSF (for review, see Ralph et al. 1986). Figure 2 shows that pretreatment of murine macrophages for 1 day with L-cell-conditioned medium (L-CM) containing CSF-1 stimulated the spontaneous killing and greatly augmented the lymphokine (LK)-induced killing of TU5 sarcoma targets. One or two days of pretreatment of macrophages with CSF-1 were optimal for the stimulation of spontaneous and LK-induced killing. Pretreatment of macrophages with 300 U/ml CSF-1 (or more) augmented tumor lysis induced with LK or LK plus CSF-1 in the killing assay, whereas 1200 U/ml CSF-1 was required for a large increase in spontaneous killing. The stimulatory activity of L-CM copurified with CSF-1 on a monoclonal immunoabsorbent column. The activity was not due to lipopolysaccharide (LPS) because the preparations had low LPS content, were active on LPS-nonresponder macrophages, and were not inhibited by LPS-neutralizing polymyxin B (P. Ralph and I. Nakoinz, in prep.). In fact, incubation of macrophages with LPS decreased their cytotoxic activity. We have also observed CSF-1 to protect macrophages from lytic infection by vesicular stomatitis virus (M.K. Warren and M.T. Lee, in prep.).

The effects of CSF-1 on human mononuclear phagocytes are just being discovered (Warren and Ralph 1986). Figure 3 shows that CSF-1 treatment of human monocytes stimulates their production of IFN in response to poly(I-C) and the production of tumor necrosis factor and a myeloid CSF in response to LPS and phorbol myristic acetate (PMA).

Pharmacologic Effects of CSF-1

A factor possibly identical to CSF-1 has been isolated from human urine. It induced in human monocytes the production of a granulocyte (G)-CSF that promotes the growth of neutrophil colonies from human bone-marrow precursors (Motoyoshi et al. 1982). Treatment of normal human donors with the urinary material revealed an increased production of G-CSF by peripheral blood monocytes and increased numbers of blood neutrophils and bone-marrow myeloid precursors (Ishizaka et al. 1985). Studies with murine CSF-1 also demonstrated in vivo stimulation of monocyte and neutrophil as well as of early pleuripotent progenitors (granulocyte/macrophage/erythrocyte/megakaryocyte colony-forming units) in mice (see Broxmeyer et al. 1985b). These clinical and preclinical results showing CSF-1 stimulation of myelopoietic events outside the mononuclear phagocyte lineage are presumably due to an indirect effect of promoting the endogenous production of G-CSF and pleuripoietins in the body. The cloning of the gene for human CSF-1 will make large amounts of this protein available for further studies.

We therefore anticipate (Table 3) that CSF-1 may find clinical utility in restoring white and red blood cell numbers that have been reduced by myelosuppressive chemotherapy or γ-irradiation for cancer treatment or bone-marrow transplantation and in naturally occurring leukopenias. CSF-1 may also have direct activating effects on mononuclear phagocytes that will improve the body's resistance to infectious diseases (viral, bacterial, and fungal) and that will stimulate the macrophages within or near tumors to destroy the neoplastic cells.

ACKNOWLEDGMENTS

We thank I. Nakoinz, M.-T. Lee, L. Brindley, K. Defay, M. Coyne, R. Tal, M. Nikoloff, J. Van Arsdell, and J. Csejtey for experimental assistance; E. Ladner, S. Nilson, and T. Culp for the graphics; and R. Bengelsdorf for typing the manuscript.

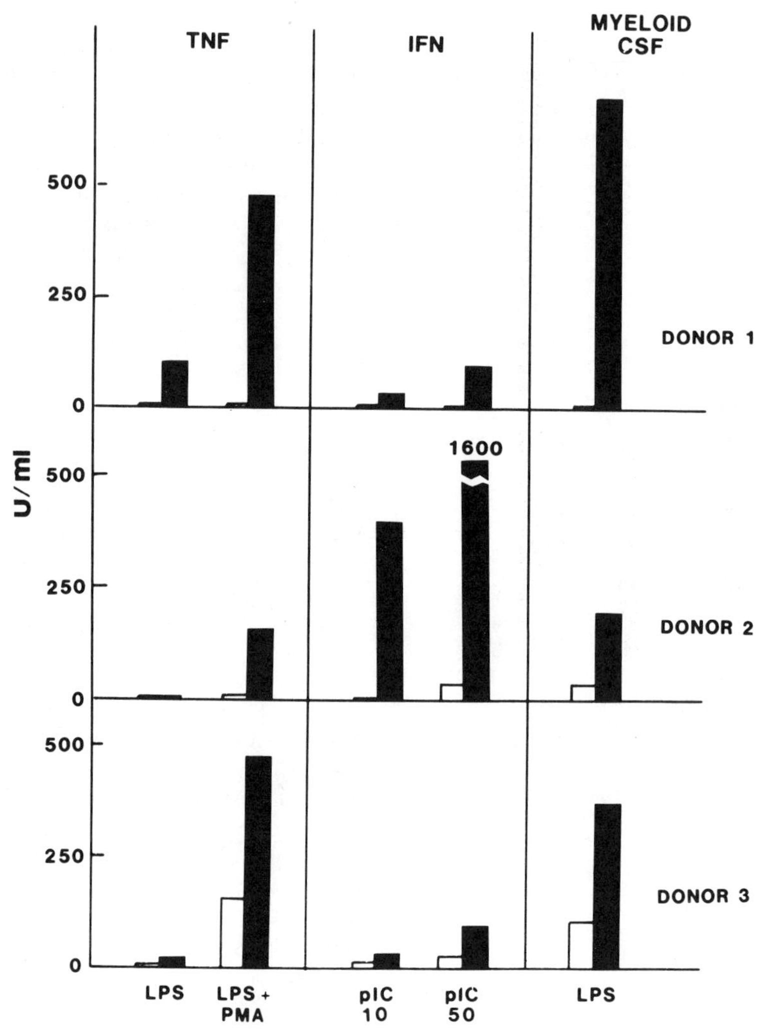

Figure 3. Stimulation of human monocyte secretion of TNF, IFN, and myeloid CSF by CSF-1. Adherent peripheral blood mononuclear cells were incubated 3 days in DME medium containing fetal calf serum in the presence (■) or absence (□) of 1000 U/ml human CSF-1. CSF-1 was purified from the MIA PaCa cell line (Ralph et al. 1986) to a specific activity of approximately 2.5×10^7 U/mg, >50% pure, <0.2 ng LPS/1000 U. The cells were washed, and 2.5×10^5 cells in 0.5 ml were recultured for 2 days with 1 μg/ml LPS (*Salmonella typhimurium*; Sigma), LPS + 20 ng/ml phorbol myristic acetate (PMA; Sigma), or 10 or 50 μg/ml polyinosinic polycytidylic acid (pIC; Sigma). Supernatants were assayed for tumor necrosis factor (TNF), interferon (IFN), and a myeloid growth factor (myeloid CSF) as described by Warren and Ralph (1986). The activity of CSF-1 was not blocked by LPS-neutralizing polymyxin B, and 0.1–1 ng/ml LPS (the maximum possible contamination in the CSF-1) did not stimulate monocyte production of the three factors.

Table 3. Expected Pharmacologic Effect of CSF-1 In Vivo

Restore monocyte numbers (and indirectly neutrophils and other blood cells) reduced

- by myelosuppressive chemotherapy and γ-irradiation for cancer and bone-marrow transplantation
- in naturally occurring anemias

Improve resistance to infection in patients at risk

- cancer, bone-marrow transplantation
- immunodeficiencies and leukopenias
- elderly
- during major surgery

Anti-cancer therapy via direct stimulation of macrophages

REFERENCES

Ben-Avram, C.H. 1985. Correction. *Proc. Natl. Acad. Sci.* **82:** 7801.

Ben-Avram, C.H., J.E. Shively, R.K. Shadduck, A. Waheed, T. Rajavashisth, and A.J. Lusis. 1985. Amino-terminal amino acid sequence of murine colony-stimulating factor 1. *Proc. Natl. Acad. Sci.* **82:** 4486.

Broxmeyer, H.E., J. Juliano, A. Waheed, and R.K. Shadduck. 1985a. Release from mouse macrophages of acidic isoferritins that suppress hematopoietic progenitor cells is induced by purified L cell colony stimulating factor and suppressed by human lacroferrin. *J. Immunol.* **136:** 3224.

Broxmeyer, H.E., D.E. Williams, S. Cooper, R.K. Shadduck, S. Gillis, A. Waheed, and D. Urdal. 1985b. The effects *in vivo* of pure murine CSF-1 and recombinant IL-3. *Blood* (suppl. 1) **66:** 146a.

Csejtey, J. and A. Boosman. 1986. Purification of human macrophage colony stimulating factor (CSF-1) from medium conditioned by pancreatic carcinoma cells. *Biochem. Biophys. Res. Commun.* **138:** 238.

Das, S.K. and E.R. Stanley. 1982. Structure-function studies of a colony-stimulating factor (CSF-1). *J. Biol. Chem.* **257:** 13679.

Das, S.K., E.R. Stanley, L.J. Guilbert, and L.W. Forman. 1981. Human colony stimulating factor (CSF-1) radioimmunoassay: Resolution of three subclasses of human colony-stimulating factors. *Blood* **58:** 630.

Ishizaka, Y., K. Motoyoski, M. Saito, Y. Miura, and F. Takaku. 1985. Mode of action of human urinary colony-stimulating factor in granulopoiesis *in vivo* and *in vitro*. *J. Leukocyte Biol.* **38:** 168.

Kawasaki, E.S., M.B. Ladner, A.M. Wang, J. Van Arsdell, M.K. Warren, M.Y. Coyne, V.L. Schweickart, M.-T. Lee, K.J. Wilson, A. Boosman, E.R. Stanley, P. Ralph, and D.F. Mark. 1985. Molecular cloning of a complementary DNA encoding human macrophage-specific colony stimulating factor (CSF-1). *Science* **230:** 291.

Kawasaki, E.S., M.B. Ladner, A.M. Wang, J. Van Arsdell, M.K. Warren, M.Y. Coyne, V.L. Schweickart, M.-T. Lee, D.M. Nikoloff, R. Tal, J.F. Weaver, L. Brindley, K.J. Wilson, A. Boosman, M.A. Innis, E.R. Stanley, P. Ralph, T.J. White, and D. Mark. 1986. Human macrophage colony stimulating factor (CSF-1): Isolation of genomic and complementary DNA clones. *UCLA Symp. Mol. Cell. Biol.* **41:** (in press).

Lin, H.S. and S. Gordon. 1979. Secretion of plasminogen activator by bone marrow-derived mononuclear phagocytes and its enhancement by colony stimulating factor. *J. Exp. Med.* **150:** 231.

Metcalf, D. 1984. *The hemopoietic colony stimulating factors.* Elsevier, Amsterdam.

Metcalf, D. and N.A. Nicola. 1985. Synthesis by mouse peritoneal macrophages of G-CSF, the differentiation inducer for myeloid leukemia cells: Stimulation by endotoxin, M-CSF and multi-CSF. *Leuk. Res.* **9:** 35.

Moore, R.J., J.J. Oppenheim, J.J. Farrar, C.S. Carter, Jr., J.A. Waheed, and R.K. Shadduck. 1980. Production of lymphocyte-activating factors (interleukin 1) by macrophages activated with colony-stimulating factors. *J. Immunol.* **125:** 1302.

Motoyoshi, K., T. Suda, K. Kusumoto, F. Takaku, and Y. Miura. 1982. Granulocyte-macrophage colony-stimulating and binding activities of purified human urinary colony-stimulating factor to murine and human bone marrow cells. *Blood* **60:** 1378.

Ralph, P. 1984. Activating factors for nonspecific and antibody-dependent cytotoxicity by human and murine mononuclear phagocytes. *Lymphokine Res.* **3:** 153.

Ralph, P., N. Williams, I. Nakoinz, H. Jackson, and J.S. Watson. 1982. Distinct signals for antibody-dependent and nonspecific killing of tumor target mediated by macrophages. *J. Immunol.* **129:** 427.

Ralph, P., M.K. Warren, M.-T. Lee, J. Csejtey, J.F. Weaver, H.E. Broxmeyer, D.E. Williams, E.R. Stanley, and E.S. Kawasaki. 1986. Inducible production of human macrophage growth factor, CSF-1. *Blood* **68:** 633.

Sabatini, D.D., G. Kreibich, T. Morimoto, and M. Adesnik. 1982. Mechanisms for the incorporation of proteins in membranes and organelles. *J. Cell Biol.* **91:** 1.

Stanley, E.R., M. Cifone, P.M. Heard, and V. Defendi. 1976. Factors regulating macrophage production and growth: Identity of colony-stimulating factor and macrophage growth factor. *J. Exp. Med.* **143:** 35.

Warren, M.K. and P. Ralph. 1986. Macrophage growth factor CSF-1 stimulates human monocyte production of interferon, tumor necrosis factor, and myeloid CSF. *J. Immunol.* **137:** 2281.

Wing, E.J., N.M. Ampel, A. Waheed, and R.K. Shadduck. 1985. Macrophage colony-stimulating factor (M-CSF) enhances the capacity of murine macrophages to secrete oxygen reduction products. *J. Immunol.* **135:** 2052.

Wing, E.J., A. Waheed, R.K. Shadduck, L.S. Nagle, and K. Stephenson. 1982. Effect of colony-stimulating factor of murine macrophages: Induction of anti-tumor activity. *J. Clin. Invest.* **69:** 270.

Effects of N-linked Carbohydrate on the In Vivo Properties of Human GM-CSF

R.E. DONAHUE, E.A. WANG, R.J. KAUFMAN, L. FOUTCH, A.C. LEARY, J.S. WITEK-GIANNETTI, M. METZGER, R.M. HEWICK, D.R. STEINBRINK, G. SHAW, R. KAMEN, AND S.C. CLARK
Genetics Institute, Inc., Cambridge, Massachusetts 02140

The hematopoietic system provides a unique opportunity for the development of useful protein pharmaceuticals. In this system, hematopoietic progenitor cells found in the bone marrow continuously divide and differentiate, ultimately resulting in the release of mature blood cells into the periphery. A number of different polypeptide growth factors have been identified that are capable, at least in vitro, of regulating the production of mature blood cells (for review, see Metcalf 1984). In principle, any of these growth factors could prove valuable in the treatment of cytopenias, both naturally arising and those induced by chemotherapy or irradiation therapy for cancer (Gasson et al. 1983). These proteins are particularly attractive as therapeutic agents because their target cells in the bone marrow are readily accessible to soluble drugs via the circulatory system.

The hematopoietic growth factors (also known as the colony-stimulating factors) were originally identified by their ability to stimulate the clonal expansion of bone-marrow progenitor cells in semisolid medium (Metcalf 1984). Traditionally, the colony-stimulating factors are classified on the basis of the types of blood cells found in the colonies grown in vitro. Thus, granulocyte-macrophage colony-stimulating factor or GM-CSF was identified as a hematopoietic growth factor that is capable of stimulating the formation of colonies containing both granulocytes and macrophages (for review, see Metcalf 1985). Both the murine and human GM-CSF cDNAs have been cloned and used to produce the respective recombinant growth factors (Gough et al. 1984; Wong et al. 1985b). With these developments, it became possible for the first time to perform detailed studies of the physical and biological properties of these proteins (Wong et al. 1985a; Metcalf et al. 1986; Tomonage et al. 1986). These studies have greatly extended the range of the known activities of GM-CSF. Thus, GM-CSF has proved to be a potent activator of mature neutrophils, eosinophils, and monocytes (Weisbart et al. 1985; Grabstein et al. 1986; S.M. Hammer et al.; A.F. Lopez et al.; both in prep.). In addition, Nathan and co-workers have found that the human GM-CSF in the presence of erythropoietin stimulates normal human progenitors to form colonies containing erythroid cells, suggesting that this hematopoietin may interact with earlier progenitors than originally thought (Donahue et al. 1985; Emerson et al. 1985; Seiff et al. 1985).

The large-scale production of recombinant human GM-CSF has permitted us to begin to study the biological properties of this molecule in vivo (Donahue et al. 1986). Previously, we have shown that the human GM-CSF is a potent stimulator of hematopoiesis in a primate model. Other investigators have used a murine model to show that another hematopoietin, interleukin-3 (IL-3), can stimulate blood cell production in vivo (Kindler et al. 1986). In preparation for initiating clinical studies with the recombinant human GM-CSF, we have extended our studies of the in vivo properties of this protein. Here, we describe the effects of glycosylation on the kinetics of clearance of GM-CSF using a rat model and present results of a study of the dose responsiveness of GM-CSF in our primate model.

METHODS

Cell culture. The growth of the Mo T-cell line has been described previously (Wong et al. 1985b). The C10-MJ2 T-cell line was grown as described by Arya et al. (1984). These cells were induced to make GM-CSF by incubation at 5×10^5 cells/ml for 24 hours in the presence of 0.3% phytohemagglutinin (PHA) and 5 ng/ml phorbol myrystic acetate (PMA) in the presence of 10% fetal calf serum. Peripheral blood lymphocyte (PBL)-conditioned medium was prepared similarly only using Ficoll-separated PBLs at a final density of 1×10^6 cells/ml. Monkey COS-1 cell transfections were performed using the DEAE-dextran protocol with chloroquin treatment as described previously (Wong et al. 1985b). The cells were pulse-labeled by incubation with 0.5 mCi of [^{35}S]methionine in 0.5 ml (per 10-cm dish) of Dulbecco's modified Eagle's medium (DMEM) for 4 hours, 48 hours after the chloroquin treatment. In one transfection, 10 μg/ml tunicamycin (Sigma) was added 30 minutes prior to labeling to inhibit the addition of N-linked carbohydrate. Plasmid pCSF-1 isolated by expression cloning was introduced into Chinese hamster ovary (CHO) cells using standard methods (Kaufman and Sharp 1982), and the CSF sequences in the resulting cells were amplified to approximately 200 copies/cell by step-wise selection with methotrexate to yield a cell line (CHO-D2) that expresses high levels of human GM-CSF (R.J. Kaufman, in prep.). Confluent dishes of CHO-D2 cells were pulse-labeled for 4 hours with 0.5 mCi of [^{35}S]methionine in 0.5 ml of DMEM.

Analysis of natural GM-CSFs. Samples (10 ml) of conditioned medium from Mo cells, C10-MJ2 cells, lectin-stimulated PBLs, and primary blasts from a patient with acute myeloblastic leukemia (kindly provided by J. Griffin, Dana-Farber Cancer Institute, Boston, Massachusetts) were passed over a 3-ml antibody column prepared from an antiserum raised in sheep by immunization with recombinant human GM-CSF (E. Wang, unpubl.). The bound GM-CSF was eluted with 0.1 M glycine (pH 2.8) and concentrated by centrifugation through a Centricon 10 concentrator (Amicon). Aliquots of these samples and an aliquot of CHO-D2-conditioned medium were fractionated by electrophoresis through a 12% polyacrylamide gel, and the positions of the GM-CSF species in each sample were visualized by transferring the protein to nitrocellulose (Bio-Rad Transblot System) and incubating the resulting filters with a rabbit anti-GM-CSF antibody. The antibody bound to the filters was visualized by treatment with ^{125}I-labeled Staph-A protein (New England Nuclear [NEN]), followed by autoradiography.

Site-specific mutagenesis of GM-CSF. Specific mutations in the GM-CSF sequence that eliminated either the first (site 1) or second (site 2) potential sites of N-linked carbohydrate addition were generated using the gapped heteroduplex site-directed mutagenesis as described by Morinaga et al. (1984). This was accomplished using two synthetic 21 mers (no. 1585 having the sequence d [TCTACTCAGCTGCAGGAGACG] and no. 1590 having the sequence d[TACTGTTTCCTGCATCTCAGC]) that spanned the DNA sequence for either site 1 or site 2 such that the codons for either Asn-27 (no. 1585) or Asn-37 (no. 1590) were positioned in the center of the oligonucleotide and were converted to glutamine codons. pCSF-1 DNA was linearized by treatment with *Sal*I, which cleaves this plasmid at a unique site in the tetracycline resistance gene, and p91023(B) was linearized at its unique cloning site by cleavage with *Eco*RI. The linearized plasmid DNAs were mixed in equimolar amounts, denatured by base treatment, and allowed to reanneal to form heteroduplexes. These heteroduplexes were annealed with either oligonucleotide 1585 or 1590 and then repaired by treatment with the large fragment of DNA polymerase I (Klenow fragment) in the presence of all four deoxynucleotide triphosphates, ATP, and T4 DNA ligase. The products of these reactions were used to transform *Escherichia coli* MC 1061, and the desired mutants were identified by colony hybridization with the appropriate oligonucleotide. The double mutant in which both sites were eliminated was made beginning with the DNA from the site-1 single mutant by an identical strategy. All of the mutations were confirmed by DNA sequence analysis.

Purification of [^{35}S]methionine-labeled GM-CSF. Conditioned medium from metabolically labeled CHO-D2 cells was diluted 1:1 with 0.02 M Tris-HCl (pH 7.4), 1.0 mM EDTA, and 0.1% Tween-80 and passed over a 1-ml column of DEAE-Trisacryl M (LKB). The column was rinsed in the same buffer containing 50 mM NaCl, and the CSF was eluted with 250 mM NaCl in the same buffer. To isolate the CSF having either one (1-N) or both (2-N) N-linked carbohydrate sites occupied, the CSF eluted from the DEAE column was passed over a Vydac C4 analytical reversed-phase column as described previously (Wong et al. 1985a). The 2-N GM-CSF, which is less hydrophobic, eluted from this column first (about 41% acetonitrile), four fractions before the 1-N (about 48% acetonitrile). To isolate GM-CSF lacking N-linked carbohydrate, the DEAE-purified protein was passed over a 1-ml column of hydroxyapatite (Bio-Rad; DNA grade). The column was washed with 10 mM sodium phosphate (pH 7.0) to elute the 1-N and 2-N proteins, and the 0-N CSF (lacking N-linked carbohydrate) was eluted with 25 mM sodium phosphate (pH 7.0). To isolate all of the forms of GM-CSF at once, the labeled CHO-D2 medium was passed over the anti-GM-CSF antibody column (3 ml) (as described for the conditioned medium samples above). The extent of sialation of the labeled CSF was assessed by binding to the lectin Ricinus Communis Agglutinin I (RCA 120, agarose-bound, from Vector Laboratories) (Debray et al. 1983). To do this, the CSF was diluted 1:1 with 0.01 M Tris-Cl (pH 7.4), 0.1 M NaCl, and 0.01% Tween-80 and passed over a 0.2-ml column of RCA 120. The column was washed with 10 volumes of the same buffer, and the bound protein was eluted with the same buffer containing 0.5 M galactose. In general, none of the labeled GM-CSF could be bound to the RCA 120, indicating that it was completely sialated. To desialate the CSF, 5×10^6 cpm were incubated in 0.3 ml of 0.25 M sodium acetate (pH 4.5) for 30 minutes at 37°C in the presence of 0.1 unit of bacterial neuraminidase (Sigma N2876). The sample was brought to neutral pH by addition of 2 M Tris base, and the desialated CSF was isolated by binding to RCA 120. Essentially all of the 1-N and 2-N CSFs specifically bound to the RCA 120 after neuraminidase treatment, indicating that at least one sialic acid residue had been removed from each molecule of CSF having N-linked carbohydrate. Little if any of the 0-N CSF was bound to RCA 120, suggesting either that the sialic acid on the O-linked carbohydrate is resistant to neuraminidase treatment or that the desialated 0-N CSF does not bind to RCA 120.

Rat clearance studies. Each sample of ^{35}S-labeled GM-CSF was concentrated by centrifugation in a Centricon 10 microconcentrator (Amicon) to 0.2 ml in phosphate-buffered saline (PBS). These samples (10^6 cpm) were injected rapidly into the tail vein of a 200–300-g male Sprague-Dawley rat that had previously been anesthetized with sodium pentobarbital (15 mg/250 g). At various times after the injection, 0.2–0.3-ml aliquots of blood were collected from the end of the tail into tubes containing 40 units of heparin. After the final sample, the rat was exsanguinated, and the major body organs (lung, kidney, and liver) were collected and weighed. Aliquots (0.05 ml) of plasma were spotted onto glass-giber filters and washed gently with cold 5% trichloracetic acid and then cold methanol and were finally dried and coupled in anhydrous scintillation fluid (NEN). Radioactivity in each organ was determined by homogenizing 200–300-mg samples in 1 ml of 1% SDS, heating to 100°C, and counting 0.1 ml in 5 ml of Aquasol (NEN).

Administration of GM-CSF to monkeys. GM-CSF was administered to three different monkeys (*M. fasicularis*) by continuous infusion through catheters surgically implanted in the jugular veins as described previously (Donahue et al. 1986). The GM-CSF was purified by the Genetics Institute Pilot Development Laboratory using lentil-lectin affinity chromatography and reversed-phase high-performance liquid chromatography (HPLC) essentially as described by Gasson et al. (1984) (E. Wang, unpubl.). This CSF had a specific activity of 1×10^6 to 2×10^6 U/ml in our CML proliferation assay as discussed elsewhere (Donahue et al. 1986). The protein was a mixture of about 30% 1-N and 70% 2-N GM-CSFs (R. Hewick and R. Steinbrink, unpubl.).

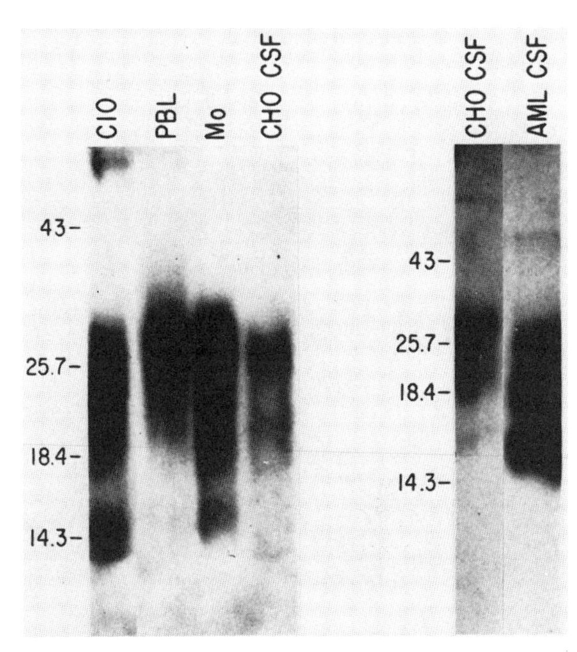

Figure 1. Size heterogeneity of GM-CSF. Preparations of GM-CSF (see Methods) from C10 cells, PBLs, Mo cells, AML cells, and crude conditioned medium from a human GM-CSF-producing CHO cell line were analyzed by SDS-PAGE and immunoblotting using anti-GM-CSF antibody as probe. The positions of low-molecular-weight protein size markers (Pharmacia) are indicated.

RESULTS

Carbohydrate Structure of GM-CSF

From the earliest attempts to purify human GM-CSF, it was recognized that this hematopoietin is heterogeneously glycosylated (Lusis et al. 1981; Wong et al. 1985a,b). However, the extent of this heterogeneity (Fig. 1) proved to be even greater than was originally thought. The GM-CSF derived from different sources, including lectin-stimulated PBLs, several T-cell lines, and primary leukemic blast cells from a patient with acute myelogenous leukemia, ranges in apparent molecular mass from 14.5 to 30 kD. As the molecular mass predicted from the primary sequence is 14.5 kD, some forms of GM-CSF comprise more than 50% carbohydrate.

Although the size range of the GM-CSF from different sources was found to be similar, the distribution of sizes within that range can be extremely variable. For example, as shown in Figure 1, GM-CSF produced by normal PBLs consisted largely of molecules at the high end of the range (22–30 kD), whereas the AML blast-cell-derived protein on the average was much less glycosylated. The distribution of species of recombinant GM-CSF produced by our engineered Chinese hamster ovary cell line (CHO-D2) was similar to the size distribution observed for the CSF from normal PBLs.

The distribution of the different species of GM-CSF from all of the different sources was observed to fall into three general size classes, most clearly evident in Figure 1 in the case of the AML-derived protein. Because the primary sequence of GM-CSF contains two consensus sequences (Asn-X-Thr/Ser) (Winzler 1973) for asparagine-linked carbohydrate (site 1 is Asn-27 and site 2 is Asn-37), we predicted that these size classes were generated by the state of occupancy of these two sites. We expected that molecules with both site 1 and site 2 modified would fall in the largest size class (22–30 kD), molecules with either site 1 or site 2 occupied would be in the intermediate size class (18–22 kD), and molecules in which neither site is occupied would fall in the smallest size class (14–18 kD). To test this prediction, we used oligonucleotide-primed site-directed mutagenesis to generate mutations in the CSF sequence that eliminated either one or both of the consensus N-linked carbohydrate addition sequences by converting the asparagine codons to codons for glutamine. These mutant DNAs, each constructed in the expression vector p91023(B) (Wong et al. 1985b), were used to transfect monkey COS-1 cells. The expressed proteins were visualized by SDS-PAGE analysis of conditioned medium from the transfected cells that had been pulse-labeled with $[^{35}S]$methionine. As shown in Figure 2, alteration of the carbohydrate addition sites resulted in the predicted shift in the size distribution of the GM-CSF. Elimination of either site 1 or site 2 prevented the synthesis of any molecules in the largest size class, whereas the simultaneous alteration of both site 1 and site 2 prevented the expression of CSF in either the large or intermediate size classes. The double mutant directed the expression of GM-CSF with a size distribution similar in range to that observed when COS-1 cells transfected with the wild-type CSF sequence were treated with tunicamycin, a drug that inhibits the addition of asparagine-linked carbohydrate (Duksin and Mahoney 1982). These results demonstrate that the

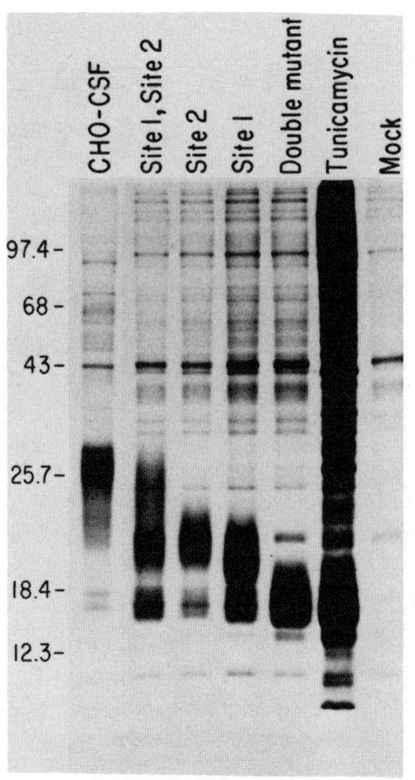

Figure 2. Analysis of the effects of N-linked carbohydrate on the size distribution of GM-CSF. Pulse-labeled conditioned medium from the GM-CSF-expressing CHO cell line (CHO); from COS-1 cells transfected with a plasmid expressing the wild-type GM-CSF sequence (Site 1, Site 2); with the mutant sequence having only site 2 remaining (Site 2); with the mutant sequence having only site 1 remaining (Site 1); with the double mutant in which both sites have been eliminated (Double mutant); with the wild-type GM-CSF sequence in the presence of tunicamycin (Tunicamycin); or mock-transfected (Mock) was fractionated by SDS-PAGE, and the patterns of protein were visualized by fluorography of the gel, which was dried after treatment with Enhance (NEN). The mobilities of the high-molecular-weight protein size markers (Pharmacia) are indicated.

major size heterogeneity of GM-CSF results from different degrees of occupancy of the two sites of N-linked carbohydrate.

Role of Carbohydrate in Clearance of GM-CSF

The function of the carbohydrate modification of GM-CSF is poorly understood. The activity of GM-CSF measured in vitro has been found to decrease with increasing carbohydrate content (E. Wang and K. Turner, unpubl.), yet the natural protein from PBLs is highly glycosylated (Fig. 1). To study the effects of the carbohydrate modifications on the clearance of the protein from the bloodstream, we have used a rat model system to follow the fate of metabolically labeled GM-CSF after intravenous injection. In this study, we used conventional protein fractionation to resolve the three distinct sizes of GM-CSF expressed by our CHO-D2 cell line. As has been observed with erythropoietin (Steinberg et al. 1986), the clearance of GM-CSF from the bloodstream of a rat follows biphasic kinetics (Fig. 3). The initial phase (called the α phase) of the clearance largely reflects distribution of the protein into all extracellular fluid of the animal, whereas true plasma clearance is represented by the second (or β phase) portion of the decay curve (Shargel and Yu 1985). As shown in Figure 3, the extent of modification of the GM-CSF via N-linked carbohydrate had a significant effect on the α but not the β portion of the curve. In the case of GM-CSF having no N-linked carbohydrate, more than 60% of the GM-CSF was lost from the bloodstream with an apparent half-life of about 2.5 minutes. The remaining 40% was cleared with an apparent half-life of 16 minutes. Approximately 60% of GM-CSF having one N-linked carbohydrate residue was lost from circulation with an apparent half-life of 4 minutes, and the remainder was cleared during the β phase with a 15-minute half-life. In contrast, for the GM-CSF in which both N-linked glycosylation sites were occupied, only 10% of the CSF was lost from circulation during the α phase with a half-life of 4.5 minutes, and 90% of the injection protein was cleared during the β phase with an apparent half-life of 15 minutes. Further work is required to determine if these differences reflect differences in the rate of distribution of the GM-CSF in the extracellular body fluids or if some mechanism of clearance also contributes to the altered kinetics seen in Figure 3.

In analyzing the distribution of the labeled GM-CSF in the various organs of the rat 30 minutes postinjection, it was clear that the predominant site of clearance for this protein is the kidney (Table 1). Because the major organs of the rat are highly vascular, their blood content would be expected to contribute significantly to the labeled CSF found in each organ. If we assume, for example, that there is no significant clearance of GM-CSF in the lungs, then it is apparent that the total level of radioactivity in the liver (which has five to seven times the mass of the lungs) resulted largely from blood contamination and not specific clearance of the glycoprotein in this organ. The high specific activity of CSF found in the kidney, however, clearly demonstrated that this organ is an important site of clearance for this protein.

In a separate study, we used the enzyme neuraminidase to remove the terminal sialic acid residues of metabolically labeled GM-CSF. The desialated CSF was isolated by binding to a column to which the lectin RCA 120 had been covalently attached. This lectin has been shown to specifically bind terminal galactose residues that are exposed upon enzymatic removal of sialic acid from glycoproteins (Debray et al. 1983). Prior to the treatment with neuraminidase, none of the labeled GM-CSF was found by the lectin column. Posttreatment, all of the labeled GM-CSF having N-linked carbohydrate was bound to the column and specifically eluted by competition with 0.5 M galactose. A comparison of the clearance of the sialated and desialated CSFs is presented in Table 1. In this experiment, the

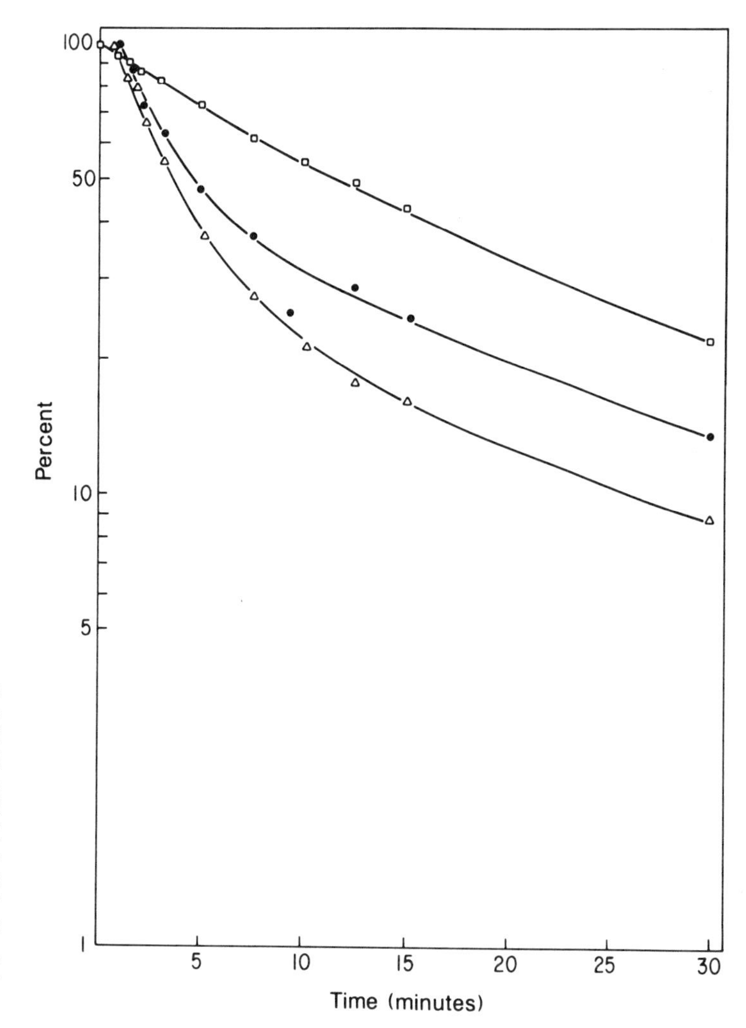

Figure 3. Effects of N-linked carbohydrate on the clearance of GM-CSF in the rat. Preparations of labeled GM-CSF having both N-linked sites (□), one N-linked site (●), or no N-linked sites (△) modified by carbohydrate addition were injected into the tail vein of a rat, blood samples were collected, and the plasma levels of labeled GM-CSF were determined. The percentage of the CSF remaining in circulation was plotted assigning the highest level of GM-CSF found in circulation (usually between 30 and 60 sec postinjection) a value of 100%. Typically, this value represented 60–100% of the CSF injected.

rats were sacrificed 15 minutes after injection, and the distribution of CSF in the various organs was determined. From these data, it is clear that the exposure of

Table 1. Organ Clearance of Different Preparations of GM-CSF after Injection into Rats

	Lung (%)	Liver (%)	Kidney (%)	Plasma (%)
GM-CSF preparation				
2-N	1.0	7.5	22.0	25.0
1-N	2.0	10.0	23.0	16.0
0-N	0.5	4.5	8.5	11.5
Sialated GM-CSF	1.0	3.0	34.0	47.0
Desialated GM-CSF	1.0	52.0	12.0	28.0

Each preparation was injected into a rat, and the kinetics of clearance were followed by sampling the blood at different times as described in Methods. Each rat was sacrificed at the end of the kinetic study, and the distribution of labeled GM-CSF (as a percentage of the circulating plasma GM-CSF at the start of the kinetic study) was determined in the indicated organs. The amount of GM-CSF remaining in the plasma was determined from the final time point of the kinetic study. Shown is a comparison of the effect of the extent of N-linked glycosylation on organ distribution (30 min postinjection): 2-N, 1-N, and 0-N CSFs have two, one, or no N-linked sites occupied. Also shown is a comparison of the organ clearance of fully sialated and desialated CSF (15 min postinjection).

the terminal galactose residues by removal of sialic acid dramatically altered the organ clearance of GM-CSF: Desialated GM-CSF was cleared in the liver and the sialated protein was cleared by the kidney. Because both of the samples represented mixtures of different size classes of CSF, the kinetics of clearance were difficult to analyze. However, the levels of circulating CSF 15 minutes after injection demonstrated that the desialated GM-CSF was cleared more rapidly than the fully sialated protein.

Dose Responsiveness of GM-CSF in Primates

The effect of the carbohydrate on the clearance of GM-CSF in the rat might be expected to influence the efficacy of the protein in vivo. Previously, we have shown that the CHO-D2-derived GM-CSF is a potent stimulator of hematopoiesis in *Macaca fasicularis* (macaque) and *Macaca mulatta* (rhesus) (Donahue et al. 1986). In preparation for human clinical trials, we have undertaken a study of the dose-response relationships of GM-CSF administration in monkeys. In early experiments, we found that desialated GM-CSF (as assessed by 100% binding to an RCA 120 column) was ineffective at stimulating primate hematopoiesis (data not shown). In contrast, GM-CSF with a much higher sialic acid content (10 moles/mole of GM-CSF) has been found to elicit a rapid leukocytosis when administered at a rate of 10 μg/kg/day as illustrated in Figure 4. The observed response in the levels of circulating blood cells is dose-dependent; when GM-CSF was administered at a rate of 5 μg/kg/day, a minimal response was observed. The dose dependency is also evident (Fig. 5) in the rate of increase in the white cell count, which increased with increasing rates of administration of the protein.

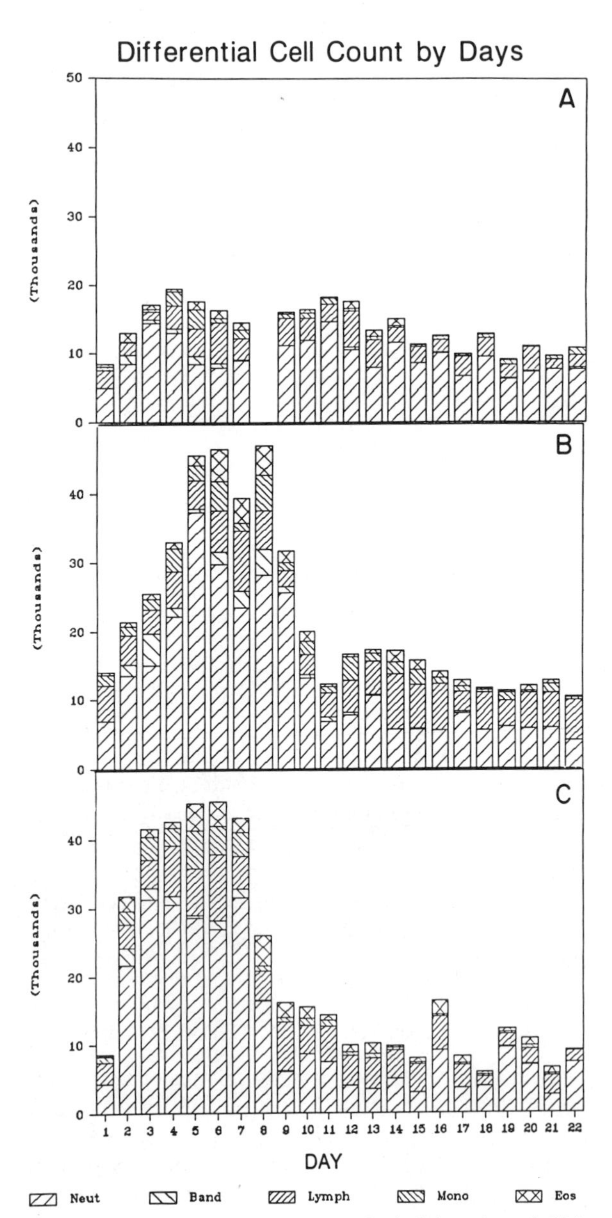

Figure 4. Effect of different rates of administration of GM-CSF on the levels of circulating blood cells in monkeys. Three different monkeys were continuously administered GM-CSF for 7 days at three different rates (5, 10, and 20 μg/kg/day for *A*, *B*, and *C*, respectively) and the differential cell counts were followed daily throughout the administration and for 2 weeks following the infusion.

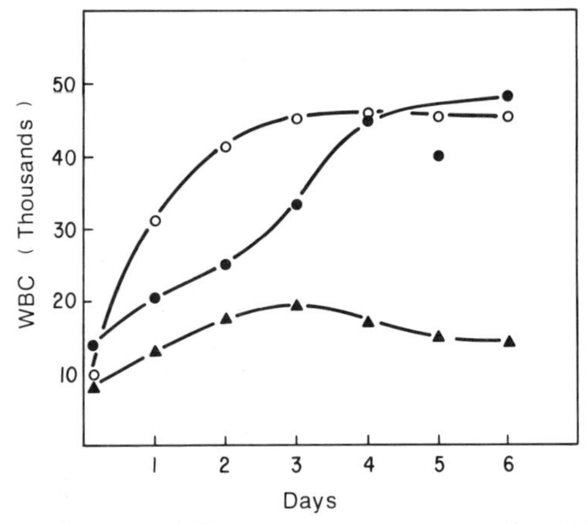

Figure 5. Effect of different rates of administration of GM-CSF on the rate of increase in the levels of circulating white blood cells in monkeys. The total white blood cell counts from Fig. 4 were plotted as a function of time of administration of GM-CSF at 5 (▲), 10 (●), and 20 (○) μg/kg/min.

DISCUSSION

Several of the natural hematopoietins, including GM-CSF, erythropoietin (Goldwasser et al. 1974), and CSF-1 (for review, see Stanley 1985), are extensively modified by the addition of a asparagine-linked carbohydrate. The modification of GM-CSF in this fashion is particularly heterogeneous and varies greatly between the different sources of the protein. The generation of mutant forms of GM-CSF plus detailed structural analysis of the native sequence (R. Hewick and R. Steinbrink, unpubl.) have revealed that this heterogeneity results in large part from different states of occupancy of the two sites for addition of N-linked sugar. The recombinant GM-CSF made in either CHO cells or monkey COS-1 cells is also further modified by the addition of O-linked sugar at several different sites in the molecule, generating further heterogeneity in the structure of the protein (R. Steinbrink and R. Hewick, unpubl.).

The function of the carbohydrate modifications of the hematopoietins is unclear. The specific activity of GM-CSF measured in vitro is significantly depressed in the largest, most fully glycosylated forms of the protein relative to the smaller, less heavily glycosylated molecules. For this reason, current efforts have focused on the effects of the carbohydrate structures on the clearance of the hormones from the bloodstream. Here, we have shown that the effective half-life of GM-CSF in the bloodstream of a rat following a single intravenous bolus injection is significantly increased by the addition of N-linked carbohydrate. The clearance of GM-CSF in the rat follows biphasic kinetics, and it is the first or α phase that is prolonged by the carbohydrate modification. In this study, we have not distinguished between simple distribution of the hormone throughout the extracellular fluid of the animal and specific organ-mediated clearance. It will be of interest to determine if the effect is on the retention of the molecule within the circulatory system of the animal or if there is a site of rapid clearance of GM-CSF that is blocked by the carbohydrate modification of the protein. Because GM-CSF is largely cleared in the kidney, this second site of clearance most likely would also reside in this organ. These questions are important in practical terms because of the possibility of altering the molecule to prolong its half-life in the circulation.

The clearance of asialoglycoproteins from the bloodstream is well documented (Neufeld and Ashwell 1980). The receptor for the penultimate galactose residues of glycoproteins that are exposed when complex carbohydrate is desialated is found in the liver and has been extensively studied (Schwarz et al. 1981). As expected, desialated GM-CSF is rapidly cleared from the circulation in the liver. This rapid clearance appears to reduce the effectiveness of the molecule in elevating the white count in monkeys. The stimulation in the levels of circulating blood cells achieved with extensively sialated GM-CSF was much more dramatic than that observed using the desialated protein. The stimulation with the sialated GM-CSF was dose-dependent as the higher doses resulted in a more rapid increase in the numbers of circulating blood cells.

We have used a mammalian cell expression system to produce large quantities of glycosylated GM-CSF that structurally closely resembles the natural molecule. This material is a potent stimulator of hematopoiesis in monkeys and has no deleterious effects on the animals. These studies support the hypothesis that recombinant GM-CSF will be effective in treating cytopenias in humans that frequently result from chemical or radiation therapy for cancer. We further hope that this molecule will prove effective in speeding the recovery of patients undergoing bone-marrow transplantation. If this proves to be the case, the high mortality from this treatment might be greatly reduced. The availability of the recombinant human GM-CSF will soon permit us to address these questions in a clinical setting.

ACKNOWLEDGMENTS

We thank D. Vanstone, S. Jone, and J. Smith for the preparation of synthetic oligonucleotides; R. Kriz for the DNA sequence analysis of the mutant DNAs; the Pilot Development Laboratory of Genetics Institute for the GM-CSF that was administered to the monkeys; H. Bader, D. Luxenberg, C. Norton, and B. Foster for providing immunological reagents; P. Sehgal of the New England Regional Primate Center for the implantation of the catheters in the monkeys; A. Schwarz for assistance with the rat clearance studies; G. Larsen for helpful discussions; and U. Roy-Chowdhury for assistance in the preparation of the manuscript.

REFERENCES

Arya, S.K., F. Wong-Staal, and R.C. Gallo. 1984. T-cell growth factor genes: Lack of expression in human T-cell leukemia-lymphoma virus-infected cells. *Science* **223:** 1086.

Debray, H., A. Pierce-Cretal, G. Spik, and J. Montreuil. 1983. Affinity of ten insolubilized lectins toward various glycopeptides with the N-glycosylamine linkage and related oligosaccharides. In *Lectins: Biology, biochemistry, clinical biochemistry* (ed. T.C. Bog-Hansen and G.A. Spengler), vol. 3, p. 335. Walter de Gruyter, Berlin.

Donahue, R.E., S.G. Emerson, E.A. Wang, G.G. Wong, S.C. Clark, and D.G. Nathan. 1985. Demonstration of burst-promoting activity of recombinant human GM-CSF on circulating erythroid progenitors using an assay involving the delayed addition of erythropoietin. *Blood* **66:** 1479.

Donahue, R.E., E.A. Wang, D.K. Stone, R. Kamen, G.G. Wong, P.K. Sehgal, D.G. Nathan, and S.C. Clark. 1986. Stimulation of haematopoiesis in primates by continuous infusion of recombinant human GM-CSF. *Nature* **321:** 872.

Duksin, D. and W.C. Mahoney. 1982. Relationship of the structure and biological activity of the natural homologues of tunicamycin. *J. Biol. Chem.* **257:** 3105.

Emerson, S.G., C.A. Sieff, E.A. Wang, G.G. Wong, S.C. Clark, and D.G. Nathan. 1985. Purification of fetal hematopoietic progenitors and demonstration of recombinant multipotential colony-stimulating activity. *J. Clin. Invest.* **76:** 1286.

Gasson, J.C., I.S.Y. Chen, C.A. Westrook, and D. Golde. 1983. Lymphokines and hematopoiesis. In *Normal and neoplastic hematopoiesis* (ed. D. Golde and P.A. Marks), p. 129. A.R. Liss, New York.

Gasson, J.C., R.H. Weisbart, S.E. Kaufman, S.C. Clark, R.M. Hewick, G.G. Wong, and D.W. Golde. 1984. Purified human granulocyte-macrophage colony stimulating factor: Direct action on neutrophils. *Science* **226:** 1339.

Goldwasser, E., C.K.H. Kung, and J. Eliason. 1974. On the mechanism of erythropoietin-induced differentiation. *J. Biol. Chem.* **249:** 4202.

Gough, N.M., J. Gough, D. Metcalf, A. Kelso, D. Grail, N.A. Nicola, and A.W. Dunn. 1984. Molecular cloning of cDNA encoding a murine haemopoietic growth regulator, granulocyte-macrophage colony stimulating factor. *Nature* **309:** 763.

Grabstein, K.H., D.L. Urda, R.J. Tushinski, D.Y. Mochizuki, V.L. Price, M.A. Cantrell, S. Gillis, and P.J. Conlon. 1986. Induction of macrophage tumoricidal activity by granulocyte-macrophage colony-stimulating factor. *Science* **232:** 506.

Kaufman, R.J. and P.A. Sharp. 1982. Amplification and expression of sequences cotransfected with a modular dihydrofolate reductase complementary DNA gene. *J. Mol. Biol.* **159:** 601.

Kindler, V., B. Thorens, S. deKossodo, B. Allet, J.F. Eliason, D. Thatcher, N. Farber, and P. Vassalli. 1986. Simulation of hematopoiesis *in vivo* by recombinant bacterial murine interleukin 3. *Proc. Natl. Acad. Sci.* **83:** 1001.

Lusis, A.J., D.H. Quon, and D.W. Golde. 1981. Purification and characterization of a human T-lymphocyte-derived granulocyte-macrophage colony-stimulating factor. *Blood* **57:** 13.

Metcalf, D. 1984. *The hemopoietic colony stimulating factors*. Elsevier, Amsterdam.

———. 1985. The granulocyte-macrophage colony-stimulating factors. *Science* **229:** 16.

Metcalf, D., G.G. Begley, G.R. Johnson, N.A. Nicola, M.A. Vada, A.F. Lopez, D.J. Williamson, G.G Wong, S.C. Clark, and E.A. Wang. 1986. Biologic properties *in vitro* of a recombinant human granulocyte-macrophage colony-stimulating factor. *Blood* **67:** 34.

Moringa, Y., T. Franceschin, S. Inouye, and M. Inouye. 1984. Improvements of oligonucleotide-directed site-specific mutagenesis using double-stranded plasmid DNA. *Biotechnology* **2:** 636.

Neufeld, E. and G. Ashwell. 1980. Carbohydrate recognition systems for receptor mediated pinocytosis. In *The biochemistry of glycoproteins and proteoglycans* (ed. W. Lennarz), p. 241. Plenum Press, New York.

Schwarz, A.L., S.F. Fridovich, B.B. Knowles, and H.F. Lodish. 1981. Characterization of the asialoglycoprotein receptor in a continuous rat hepatoma cell line. *J. Biol. Chem.* **256:** 8878.

Sieff, C.A., S.G. Emerson, R.E. Donahue, D.G. Nathan, E.A. Wang, G.G. Wong, and S.C. Clark. 1985. Human recombinant granulocyte-macrophage colony-stimulating factor: A multilineage hematopoietin. *Science* **230:** 1171.

Shargel, L. and A.B.C. Yu. 1985. *Applied biopharmaceuticals and pharmacokinetics* (ed. G. Di Sabato et al.). Appleton-Century-Crofts, Connecticut.

Stanley, E.R. 1985. The macrophage colony-stimulating factor, CSF-1. *Methods Enzymol.* **116:** 564.

Steinberg, S.E., J.F. Garcia, G.R. Metzke, and J. Mladenovic. 1986. Erythropoietin kinetics in rats: Generation and clearance. *Blood* **67:** 646.

Tomonage, M., D.W. Golde, and J.G. Gasson. 1986. Biosynthetic (recombinant) human granulocyte-macrophage colony-stimulating factor: Effect on normal bone marrow and leukemia cell lines. *Blood* **67:** 31.

Weisbart, R.H., D.W. Golde, S.C. Clark, G.G. Wong, and J.C. Gasson. 1985. Human granulocyte-macrophage colony-stimulating factor is a neutrophil activator. *Nature* **341:** 361.

Winzler, R.J. 1973. The chemistry of glycoproteins. In *Hormonal proteins and peptides* (ed. C.H. Li), vol. 1, p. 1. Academic Press, New York.

Wong, G.G., J. Witek, P.A. Temple, K.M. Wilkens, A.C. Leary, D.P. Luxenberg, S.S. Jones, E.L. Brown, R.M. Kay, E.C. Orr, C. Shoemaker, D.W. Golde, R.J. Kaufman, R.M. Hewick, S.C. Clark, and E.A. Wang. 1985a. Molecular cloning of human and gibbon T-cell-derived GM-CSF cDNAs and purification of the natural and recombinant proteins. *Cancer Cells* **3:** 235.

Wong, G.G., J. Witek, P.A. Temple, K.M. Wilkens, A.C. Leary, D.P. Luxenberg, S.S. Jones, E.C. Brown, R.M. Kay, E.C. Orr, C. Shoemaker, D.W. Golde, R.J. Kaufman, R.M. Hewick, E.A. Wang, and S.C. Clark. 1985b. Human GM-CSF: Molecular cloning of the complementary DNA and purification of the natural and recombinant proteins. *Science* **228:** 819.

Erythropoietin: Gene Cloning, Protein Structure, and Biological Properties

J.K. Browne, A.M. Cohen, J.C. Egrie, P.H. Lai, F.-K. Lin, T. Strickland, E. Watson, and N. Stebbing
Amgen, Thousand Oaks, California 91320

The successful cloning of the gene for human erythropoietin (EPO) (Jacobs et al. 1985; Lin et al. 1985) has yielded information on the genetic organization and protein structure of this hormone and has allowed assessment of its biological properties. Biological studies have clearly indicated the clinical potential for this hormone in treatment of various anemias, and initial clinical studies of recombinant-DNA-produced human EPO (r-hEPO)[1] are now under way.

EPO, a sialylglycoprotein hormone, is responsible for regulating the rate of red blood cell formation and for maintaining the red blood cell mass (Krantz and Jacobson 1970; Graber and Krantz 1978; Spivak and Graber 1980). EPO is produced primarily by the kidney in adults and by the liver during fetal life and is secreted into the circulation (Jacobsen et al. 1957; Fried 1972; Zanjani et al. 1981). Circulating levels of EPO are approximately 20 mU/ml (Koeffler and Goldwasser 1981; Cotes 1982; Garcia et al. 1982). Serum levels of EPO increase under conditions of tissue hypoxia and decrease under conditions of hyperoxia. The kidney responds to anemia by increasing the rate of EPO production, resulting in an increase of as much as 100-fold in serum EPO levels (Eschbach and Adamson 1985). Damage to the kidney, as found in chronic renal failure, results in anemia primarily due to a deficiency in EPO production (Brown 1965; Naets 1975; Erslev et al. 1980). Although postulated at the turn of the century (Carnat and Defandre 1906), EPO was first partially purified in 1971 from anemic sheep plasma (Goldwasser and Kung 1971). Human EPO was first purified to homogeneity in 1977 from the urine of aplastic anemia patients (Miyake et al. 1977). Purified human urinary EPO has an apparent molecular weight of about 34,000 and can be separated into two forms, termed α and β, which differ in their carbohydrate content (Dordal et al. 1985). A specific activity of 70,000 U/mg has been reported for purified human urinary EPO (Miyake et al. 1977).

Isolation of a gene for human EPO proved to be particularly problematical because there was no known source of mRNA. No cell lines had been characterized that produced significant amounts of EPO that could provide enriched sources of mRNA. Moreover, although the kidney had been identified as the probable site of synthesis, there remained considerable uncertainty as to whether induction of EPO represented de novo synthesis, release, or activation of an inactive precursor (Fyhrquist et al. 1984). In addition, there was no reliable, convenient assay that could be used to screen cell lines or clones rapidly in an expression system. Thus, the overall cloning strategy was based on obtaining some amino acid sequence information from the very limited amount of purified human urinary EPO that was available. Mixed DNA probes, based on all possible coding sequences, could then be used to screen a human genomic library, and candidate DNA would be expressed in mammalian cells. A cDNA clone, isolated separately, would also be required to unambiguously assign intron/exon boundaries.

Due to the lack of a suitable source of human tissue mRNA, a heterologous approach was employed in which mixed oligonucleotide probes, based on human EPO amino acid sequence information, were used to identify EPO mRNA isolated from various tissues from normal animals or animals made anemic experimentally. This approach required that the amino acid sequence be essentially identical over the region for which the oligonucleotide probes were constructed. For this reason, primates, namely, cynomolgus monkeys, were used because of their relatively close relationship to man.

Gas-phase microsequencing of a small amount of purified human urinary EPO yielded some definitive amino acid assignments but also some uncertainties in the first amino-terminal 23 positions. Due to the redundancy in the genetic code and uncertainties in the amino acid sequence, oligonucleotide probes corresponding to this region proved not to be useful for isolating the EPO gene. Amino acid sequences from internal peptide regions with lower genetic code redundancy were obtained by trypsin digestion of EPO. The amino-terminal sequence did allow, however, preparation of a monoclonal antibody directed against a synthetic peptide from this region (Egrie 1983), and this material, together with antibodies against intact EPO, proved useful in the overall cloning strategy. Because of the heterogeneous nature of urinary EPO preparations, it seemed that either there was considerable secondary

[1]r-hEPO is being jointly developed by Amgen (Thousand Oaks, California), Cilag (Schaffhausen, Switzerland), Kirin Brewery Co., Ltd. (Tokyo, Japan), and Ortho Phamaceutical Corp. (Raritan, New Jersey).

  0-87969-052-6/86 $1.00

modification of the protein or the material was heterogeneous in amino acid sequence and was perhaps coded by a gene family. Thus, considerable effort was directed toward characterization of urinary EPO with the realization that, at the very least, it could confirm primary structure features predictable from the coding gene sequence(s). As it transpired, almost complete characterization of the primary structure of human urinary EPO was achieved by direct analysis within the time taken to characterize the genes (Lai et al. 1986). The direct structural analyses proved important in resolving various structural features of this hormone. EPO proved to be a posttranslationally modified product of a single-copy gene that is highly conserved within the mammals.

EXPERIMENTAL PROCEDURES

Isolation of EPO. Human EPO was purified from the urine of aplastic anemia patients as described previously (Miyake et al. 1977). r-hEPO was produced by Chinese hamster ovary (CHO) cells stably transformed with the human gene (Lin et al. 1985) and purified to homogeneity from conditioned culture media.

Isolation and analyses of gene clones. Cloning of the human and cynomolgus monkey EPO genes, RNA isolation, cDNA cloning, and Northern and Southern blot analyses have been described recently (Lin et al. 1985, 1986). Nucleic acid sequence analysis was carried out primarily by the dideoxy method (Sanger et al. 1977), with a few regions sequenced by the chemical cleavage method (Maxam and Gilbert 1980).

DNA-mediated gene transfer. DNA-mediated gene transfer was carried out using the calcium phosphate microprecipitation method (Graham and van der Eb 1973) as modified by Wigler et al. (1978). For gene transfer into 293 cells (Graham et al. 1977), cells were treated for 4 hours with the calcium-DNA precipitate. COS-1 cells (Gluzman 1981) (ATCC no. CRL 1650) were incubated for 16 hours with the precipitate. Media were sampled for EPO 3–7 days posttransfection.

Amino acid sequence analysis. The sequencing of human EPO was described by Lai et al. (1986).

EPO assays. A radioimmunoassay (RIA) for EPO using purified urinary ^{125}I-labeled EPO and rabbit sera raised against a partially purified preparation of urinary EPO was described recently by Egrie et al. (1986). RIAs using recombinant reagents were performed as described by J.C. Egrie et al. (in prep.). The in vivo exhypoxic polycythemic mouse bioassay for EPO (Cotes and Bangham 1961) and the in vitro rat bone-marrow assay (Goldwasser et al. 1975) were used to assess the biological activity of various EPO preparations. Western blot analysis of EPO, which used a mouse monoclonal antibody raised to a synthetic peptide corresponding to the amino terminus of EPO (Egrie 1983), was described previously (Egrie et al. 1985, 1986).

Physical analyses of EPO. The carbohydrate composition of r-hEPO was determined by methanolysis, trifluoroacetylation, and separation by gas chromatography (Zanetta et al. 1972). Endoglycosidase F (New England Nuclear), neuraminidase, and *O*-glycanase (Genzyme) digestions were performed according to the manufacturers procedures. Analytical ultracentrifuge analysis, dry-weight determination, and measurement of the partial specific volume of EPO were performed as described by J. Davis et al. (in prep.).

Treatment of dogs with r-hEPO. Young adult male and female beagles (six of each sex per group) were dosed intravenously via the cephalic vein for 3 weeks, three times per week, with purified r-hEPO. The animals received either excipient control or a dose of 280 or 2800 U/kg EPO. Blood samples were taken 1 week prior to the study and weekly thereafter.

Pharmacokinetic studies. Pharmacokinetic analysis of r-hEPO in rats will be described in detail separately (A.C. Cohen et al., in prep.). CD rats (Charles Rivers Breeding Laboratories) had a polyethylene cannula implanted into the left carotid artery and a silastic cannula implanted into the right jugular vein 3 days prior to the study. To determine the effect of renal failure on the pharmacokinetics of EPO, both renal pedicles were ligated in one group of animals on the day of the study. r-hEPO was metabolically labeled with [^{35}S]methionine and cysteine and purified. ^{35}S-labeled r-hEPO was administered via the jugular cannula at 1 μCi (370 units r-hEPO) per kilogram of body weight. Blood samples were collected periodically via the carotid cannula, and ethanol precipitable counts were determined. The best-fit lines were determined by a Guass-Newton curve-fitting algorithm, and pharmacokinetic parameters were determined by standard methods.

Subtotal nephrectomized animal study. The system developed by Anagnostou et al. (1977) was used to study r-hEPO in an animal model of renal failure. One group of rats were subjected to a two-step surgical removal of all of one kidney and one half of the second kidney. Sham-operated controls were subjected to concurrent laparotomy. Starting 1 week after surgery, the two groups of animals were injected intramuscularly with 0.1 ml of saline control solution or 10 units of r-hEPO five times a week for 2 weeks. Body weight, hematocrit, and plasma urea nitrogen were determined at the beginning and end of the study (A.C. Cohen, in prep.).

Inflammatory disease animal model. The anemia of adjuvant inflammation was induced according to the procedure of Lukens et al. (1967). Rats were given a single injection of Freund's complete adjuvant in the left hind footpad. Ten days later, treatments were begun wih 0.1-ml intraperitoneal injections of either saline (control solution) or 40 units of r-hEPO. After 10 days of treatment, blood samples were obtained for determination of hematocrit, red cell mass, hemoglobin concentration, plasma iron, and iron-binding capacity. Similar determinations were made on nonadjuvant treated rats concurrently given injections of either saline or r-hEPO.

RESULTS

Cloning of the Human EPO Gene

The strategy employed to isolate a human EPO genomic gene clone and a cynomolgus monkey cDNA is outlined in Figure 1. A human EPO gene clone was isolated from a λ bacteriophage-borne human genomic library using oligonucleotide probes (Lin et al. 1985). Two sets of mixed oligonucleotide probes were used, each containing a pool of 128 sequences: One pool was a mixture of 20-nucleotide-long oligonucleotides containing all possible coding sequences for an internal hexapeptide, and the other pool was a mixture of 17-nucleotide-long oligonucleotides directed against the coding sequence for a heptapeptide. Using probes directed to two nonoverlapping regions of the EPO gene allowed rapid confirmation of putative clones, eliminating a great number of the false positives obtained with either probe alone.

DNA sequence analysis demonstrated that three of four independent genomic clones had sequences corresponding to the known EPO tryptic peptide amino acid sequences. Proof that a clone contained a complete functional gene encoding human EPO was obtained by its expression in mammalian cells, to produce a gene product with the immunological and biological properties of EPO. A 5.4-kb *Bam*HI-*Hin*dIII restriction fragment from clone λHE1 was identified as potentially carrying the entire EPO gene by Southern blot analysis (Southern 1975) using mixed nucleotide probes. This fragment was subcloned into the plasmid pUC8 (Vieira and Messing 1982) and transiently transfected into 293 cells. These cells are a human embryonic kidney line stably transformed with adenovirus type 5 (Ad5). 293 cells constitutively express the Ad5 EIA gene products that act as *trans*-acting enhancers of expression of DNA introduced into these cells by transfection.

RIA analysis demonstrated the presence of EPO in culture medium samples from transfected cells; the expressed EPO produced a dose-response curve identical to that of urinary EPO. Medium samples from cultures transfected with the plasmid lacking the EPO gene insert were uniformly negative in the RIA. The *Bam*HI-*Hin*dIII fragment was inserted into a shuttle vector, containing the SV40 origin of replication, and transiently transfected into COS-1 cells. RIA analysis again demonstrated that EPO was secreted into the culture media; this material was also shown to be biologically active as determined by the in vitro rat bone-marrow assay and the in vivo exhypoxic polycythemic mouse bioassay. Control cultures were uniformly negative in

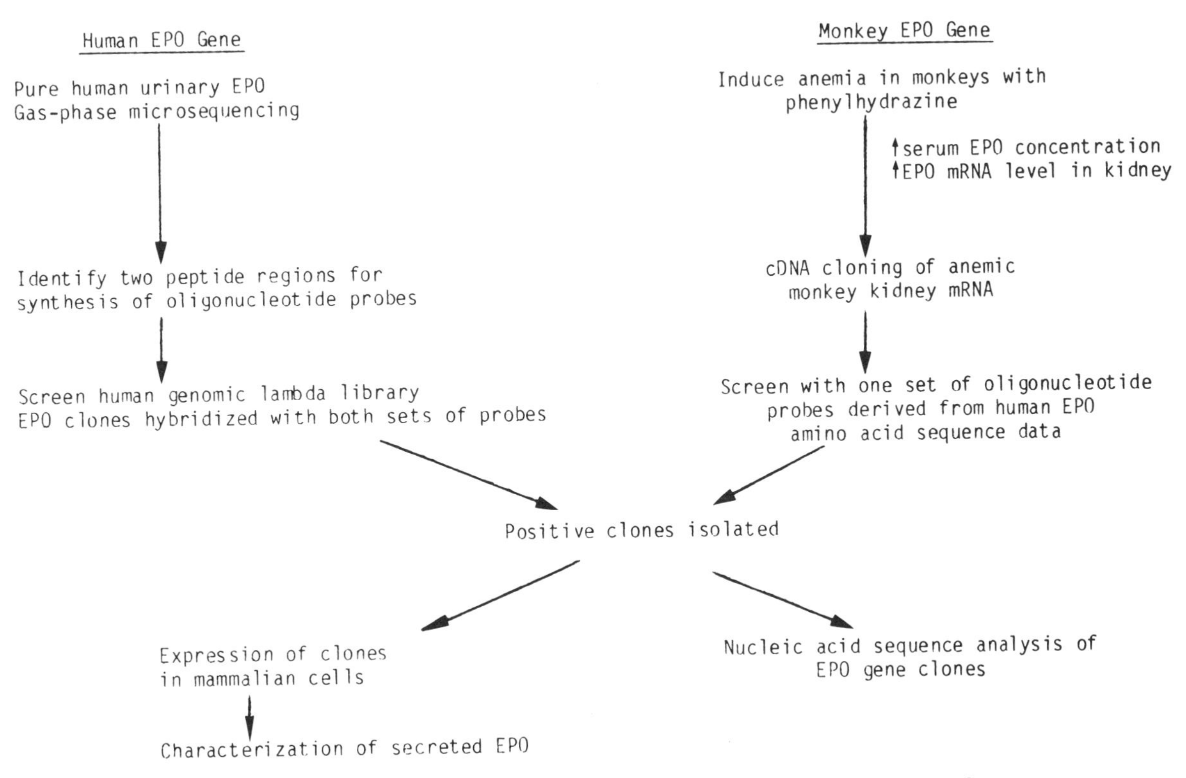

Figure 1. Strategy employed to clone, express, and characterize the human and monkey EPO genes.

these assays. The r-hEPO produced in COS-1 cells is also indistinguishable from urinary EPO by Western blot analysis (Egrie et al. 1985). Polyadenylated RNA isolated from COS-1 cells 48 hours after transfection with this construction was used to prepare a cDNA library from which human EPO cDNA clones were isolated.

The amino acid sequence of the 193-amino-acid primary translation product was deduced from the nucleic acid sequence as shown in Figure 2. The mature hormone is 166 amino acids in length (calculated M_r of 18,399). There are three potential N-linked glycosylation sites, as indicated. The first 27 amino acids predicted by the DNA coding sequence are consistent with this being a hydrophobic leader peptide. The coding portion of the gene is divided by four intervening sequences. The amino terminus of the mature protein was assigned directly from amino-terminal amino acid sequencing of urinary and r-hEPO. Intron/exon junction assignments, which all conform to consensus splice rules (Mount 1982), were made by comparison of a monkey cDNA clone, a human EPO cDNA clone prepared from mRNA isolated from COS-1 cells transfected with the genomic gene clone, and, ultimately, complete amino acid sequence analysis of human urinary EPO (Lai et al. 1986) and recombinant EPO produced in cell culture (P.H. Lai, unpubl.).

Southern blot analysis (Southern 1975) was used to analyze the human EPO gene. The restriction fragment pattern of human DNA probes with a human EPO cDNA clone and the results of low-stringency hybridizations with this probe demonstrate that there is a single copy of the human EPO gene and that there are no apparent closely related genes or pseudogenes (Lin et al. 1985). Computer searches of protein and nucleic acid databases failed to reveal a significant homology with any published sequence. In particular, there is no homology with angiotensinogen, which has been suggested as a possible precursor of EPO (Fyhrquist et al. 1984). In addition, the structure of the human EPO gene and cDNA clones and the amino acid sequence results are consistent with EPO being secreted as an active hormone rather than an inactive precursor.

Cloning of a Monkey EPO cDNA

A cynomolgus monkey EPO cDNA was isolated using the strategy outlined in Figure 1 (Lin et al. 1986). Anemia was induced in monkeys by treatment with phenylhydrazine. As a result of the anemia, serum EPO levels were found to increase in anemic monkeys as measured by RIA and Western blot analyses (Egrie et al. 1985).

A Northern blot analysis of polyadenylated mRNA isolated from normal and anemic monkey kidneys is shown in Figure 3. One of the pools of mixed oligonu-

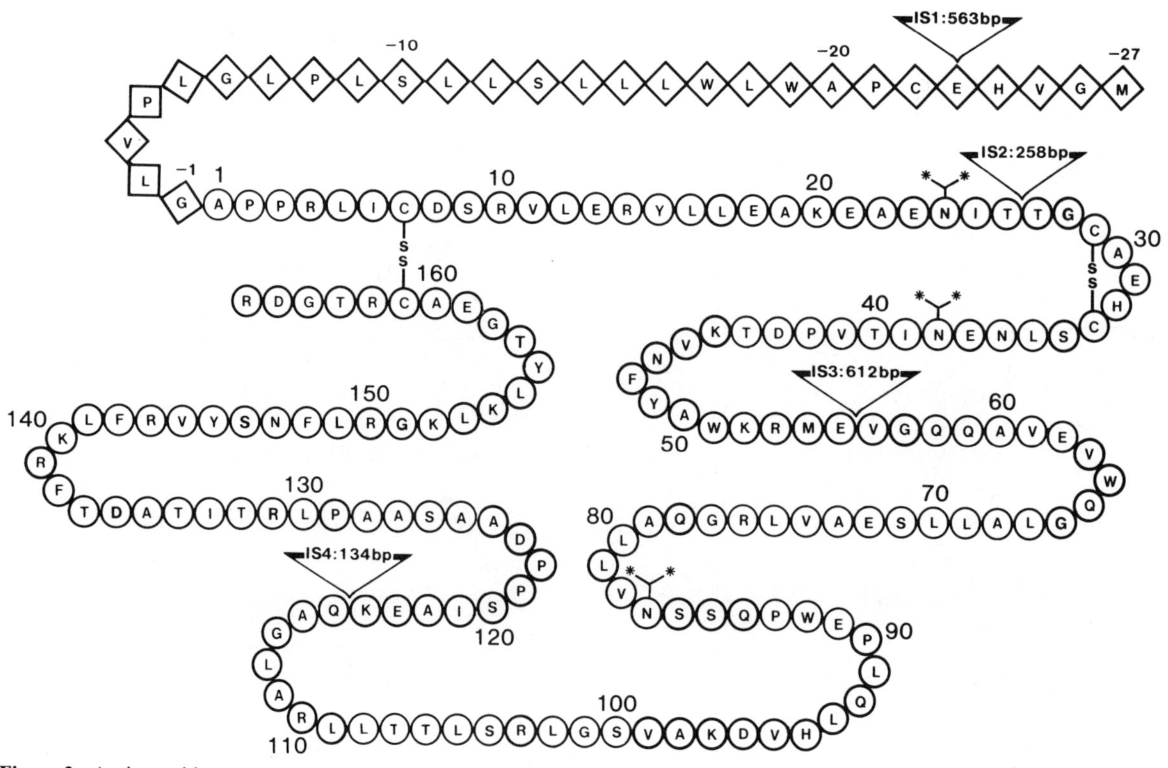

Figure 2. Amino acid sequence of the 193-amino-acid primary translation product of the human EPO gene is presented in one-letter code. The 27 amino-terminal amino acids of the putative signal peptide are boxed, and the residues in the 166-amino-acid mature hormone are circled. Positions of the two disulfide bands (S-S) and the three N-linked glycosidation sites (Y) are noted. Positions at which the four intervening sequences (IS) interrupt the coding position of the gene are given, along with the length of each IS sequence.

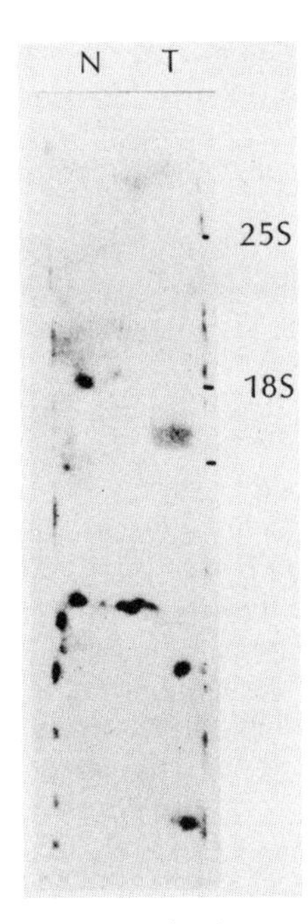

Figure 3. Northern blot analysis of cynomolgus monkey kidney poly(A) containing RNA. ^{32}P-end-labeled mixed oligonucleotide probes to EPO were used to develop the blot. (N) RNA isolated from normal monkey kidneys; (T) RNA isolated from the kidneys of monkeys made anemic by phenylhydrazine treatment. Positions of the ribosomal RNA markers are indicated.

cleotides corresponding to the human EPO amino acid sequence was used as the probe. Positive results were obtained only with mRNA from phenylhydrazine-treated animals (Fig 3, lane T). There was no detectable EPO-specific mRNA in normal monkey kidneys (Fig. 3, lane N). The size of the EPO-specific mRNA is approximately 1600 nucleotides in length. These Northern blot analysis results demonstrate that the steady-state level of EPO mRNA in the kidney is dramatically increased upon induction of anemia. The rise in serum EPO levels is therefore apparently due to increased EPO synthesis mediated by an increased transcription rate of the EPO gene and/or an increase in EPO mRNA stability as opposed to release of previously synthesized and sequestered EPO.

A monkey EPO cDNA clone was isolated from a cDNA library prepared from mRNA from anemic monkey kidneys. Only one of the two pools of mixed oligonucleotide probes proved useful for Northern blot analysis or cDNA cloning. It was subsequently determined that the second probe mixture failed to hybridize with monkey EPO due to a single-amino-acid difference between human and monkey EPO in the region corresponding to the probe. The cDNA clone sequence codes for a 192-amino-acid protein, differing at 15 residues from the human EPO gene sequence. Monkey EPO lacks lysine at position 116, when aligned with the human EPO sequence. Lysine at position 116 in the human gene is the first amino acid in the fourth exon of the human gene. The monkey gene presumably uses an alternative splice junction for the fourth exon, and the DNA sequence is consistent with this. The amino terminus of mature-cell-culture-produced recombinant monkey EPO is at position −3 relative to human EPO. Thus, monkey EPO has a 24-amino-acid signal peptide, and the mature hormone is 168 amino acids in length. A substitution of a proline for a leucine at position −2 (relative to the human gene) in monkey EPO may be the cause of the difference in the signal peptide cleavage site.

Structural Features of EPO

To verify the primary amino acid sequence determined by gene cloning and to assess the nature of secondary modifications, purified human urinary EPO was characterized beyond the initial sequencing involved in designing gene probes. A total of 565 μg of urinary EPO was used to determine the primary structure of the molecule (Lai et al. 1986). About 30 μg was used to determine the amino-terminal amino acid sequence and this allowed assignments at 42 of 50 cycles of degradation. Cyanogen bromide cleavage fragments (100 μg) allowed assignment of 45 additional positions from a total of 77 cycles. About 200 μg was then utilized for sequencing tryptic digests, and about the same amount was used for V8-protease digests. The only residues not assigned directly were the asparagines at positions 24, 38, and 83. The absence of signals for these three residues during direct sequencing is consistent with the presence of carbohydrate moieties at these positions.

The positions of the disulfides were determined by the copurification of peptides predicted from the primary amino acid sequence to be separate after protease treatments (see Fig. 2). A PTH-cysteine was detected in the seventh step of Edman degradation, and sequencing of a reduced preparation of the peptide indicated the presence of the 7-161 disulfide. Assignment of the 29-33 disulfide was indirect and based on the following observations: (1) by Ellman's reaction and attempted labeling with [^{3}H]iodoacetic acid, EPO was found to contain no free thiol residues; (2) sequencing of EPO showed no residues at positions 29 and 33, but performic-acid-oxidized material showed cysteic acid at these positions; and (3) the Glu-31, His-32 bond in oxidized r-hEPO was not hydrolyzed by V8 protease, indicating some conformational abnormality in the structure around these residues.

The molecular weight of purified r-hEPO, produced in CHO cells stably transfected with the human EPO gene inserted in an expression vector, was determined by analytical ultracentrifuge analysis. To carry out this analysis, the extinction coefficient of r-hEPO was measured by dry-weight determination, and the partical-specific volume was determined by the mechanical oscillation technique. This analysis yielded a result of 29,900 ± 400 daltons for the mass of r-hEPO (J. Davis et al., in prep.). The carbohydrate portion of the molecule, assuming a molecular weight of 18,399 for the protein portion of r-hEPO, comprises 38 ± 1% of the total weight.

Human urinary EPO and CHO-cell-derived r-hEPO migrate identically in SDS-polyacrylamide gels, indicating that both molecules are glycosylated to a similar extent. The carbohydrate composition of r-hEPO was determined as described by Zanetta et al. (1972) and compared to literature values for human urinary EPO (Dordal et al. 1985). The carbohydrate composition of r-hEPO was essentially the same as that of urinary EPO (T.W. Strickland et al., in prep.). Trace amounts of *N*-acetylgalactosamine were found in r-hEPO, indicating the presence of O-linked glycosylation.

Shown in Figure 4 are the results of a deglycosylation experiment that indicates that both r-hEPO and urinary EPO contain both N-linked and O-linked carbohydrates in similar amounts. Both r-hEPO and urinary EPO were analyzed by Western blot analysis after sequential glycosylase digestion. Figure 4 (lanes 1 and 5) shows urinary EPO and r-hEPO, respectively, prior to treatment. After treatment with endoglycosidase F, which removes N-linked carbohydrate, the apparent molecular weight of both r-hEPO and urinary EPO is shifted to approximately 19,500 with a minor band at about 18,400 (lanes 2 and 6). Following further treatment, first with sialidase (lanes 3 and 7) and then by *O*-glycanase (lanes 4 and 8), which remove O-linked carbohydrate, both r-hEPO and urinary EPO migrated as a single band with an apparent molecular weight of 18,400. Although the presence of *N*-acetylgalactosamine had not been detected previously (Dordal et al. 1985), these results demonstrate that urinary EPO, as well as r-hEPO, contains O-linked carbohydrate. In addition, direct carbohydrate analysis of endoglycosidase-F-treated r-hEPO yields galactose, sialic acid, and

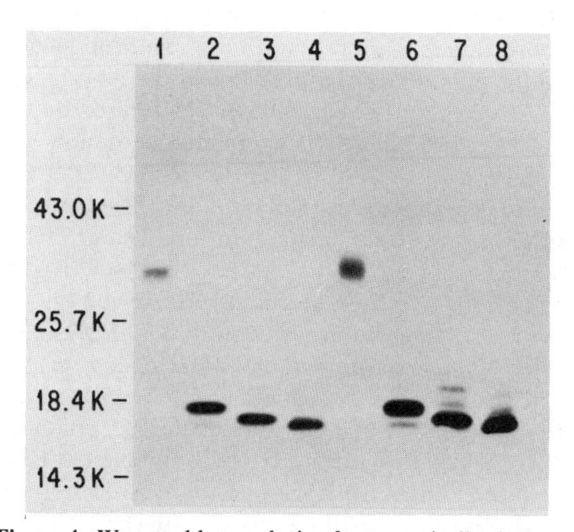

Figure 4. Western blot analysis of enzymatically deglycosylated urinary EPO (lanes *1–4*) and r-hEPO (lanes *5–8*). Samples were digested sequentially with glycosidases, and the products were separated on a 12.5% SDS-polyacrylamide gel under reducing conditions. An EPO-specific monoclonal antibody was used to visualize EPO after transfer to a nitrocellulose membrane. (*1,5*) No treatment; (*2,6*) endoglycosidase F; (*3,7*) endoglycosidase F plus neuraminidase; (*4,8*) endoglycosidase F, neuraminidase plus *O*-glycanase.

N-acetyl galactosamine, confirming the presence of O-linked carbohydrate (T.W. Strickland et al., in prep.). As shown in Figure 4, the proportion of EPO containing O-linked carbohydrate is comparable in urinary EPO and r-hEPO.

Samples of r-hEPO and urinary EPO taken over the course of endoglycosidase-F digestion were analyzed by Western blot analysis. Two clear partial digestion products, in addition to the final product, were revealed, indicating that all three potential N-linked glycosylation sites are utilized (data not shown).

The immunoreactivity of r-hEPO and urinary EPO were evaluated by comparing the dose-response curves of each preparation in a series of RIAs. The first RIA compared the ability of r-hEPO or urinary EPO to compete the binding of urinary ^{125}I-labeled EPO by a rabbit polyclonal antibody raised against a 1% pure preparation of urinary EPO. In this assay, identical dose-response curves were obtained with the recombinant and natural hormones (Egrie et al. 1985, 1986). Identical dose-response curves were also obtained for each source of hormone when ^{125}I-labeled r-hEPO was used as the tracer or when rabbit polyclonal sera to r-hEPO was used in combination with either tracer (data not shown; J.C. Egrie et al., in prep.). These experiments demonstrate that there are no epitopes present on one hormone preparation that are not present on the other.

Biological Effects of EPO

The biological activities of r-hEPO and urinary EPO were also indistinguishable, as measured by the dose-response curve of each preparation in both in vitro and in vivo biological assays (Egrie et al. 1986). The exhypoxic polycythemic mouse bioassay (Cotes and Bangham 1961) measures the incorporation of ^{59}Fe into red blood cells in mice made polycythemic by exposure to low-oxygen conditions. Due to the polycythemia, these animals have a reduced rate of red blood cell synthesis after return to normal atmospheric conditions and therefore have a low background rate of incorporation of iron into red blood cells. However, these animals will respond to exogenous administration of EPO, in a dose-dependent manner, by increasing the rate of red blood cell synthesis and incorporation of ^{59}Fe. The biological activity of EPO was also measured by the incorporation of ^{59}Fe into heme in primary cultures of rat bone-marrow cells. In each of these assays (Egrie et al. 1986), as well as in mouse BFU-e and CFU-e assays (S.B. Krantz, pers. comm.), r-hEPO and urinary EPO have indistinguishable dose-response curves. These results indicate that r-hEPO produced in CHO cells has all the biological and immunological properties of natural EPO that can be measured by these assays. In addition, r-hEPO has the same activity in the RIAs and in vitro and in vivo bioassays, indicating that it is fully biologically active (Egrie et al. 1986).

The biological and immunological activities of multiple lots of purified r-hEPO have been measured.

These assays are standardized with respect to the second international reference preparation of EPO (IRP#2) (Annable et al. 1972). r-hEPO has a constant specific activity of 174,000 units/A_{280} ± 5%, which is approximately 2.5 times the reported specific activity of urinary EPO (Miyake et al. 1977). Perhaps the harsh conditions required for purification of EPO from concentrated urine are responsible for the difference in specific activity.

Because of earlier notions that EPO may not be the only factor involved in production of red blood cell mass, we determined whether EPO alone could stimulate red blood cell production in normal mice, rats, and dogs. In all three species, there was a dose-dependent increase in reticulocyte counts and an increase in red blood cells. The results in Figure 5 show the effect on reticulocyte and red blood cell counts in dogs treated three times per week for 3 weeks. No effects were observed in excipient-treated controls, but doses of 280 U/kg and 2800 U/kg gave dose-dependent increases in both reticulocyte and red blood cell counts. It is noteworthy that the reticulocyte counts began to drop before the end of treatment and then dropped rapidly after treatment ceased, whereas the red blood cell counts continued to rise and then showed a gradual decrease. This particular study involved very high doses of EPO and thus allowed assessment of potential toxicity effects of EPO; however, no adverse effects of any note were observed. Overall, EPO was well tolerated and showed no histological abnormalities. Demonstration of an increase in red blood cells with highly purified EPO in normal, intact experimental animals indicates, for the first time, that EPO alone is capable of this effect and that other factors are not limiting in stimulating red blood cell production.

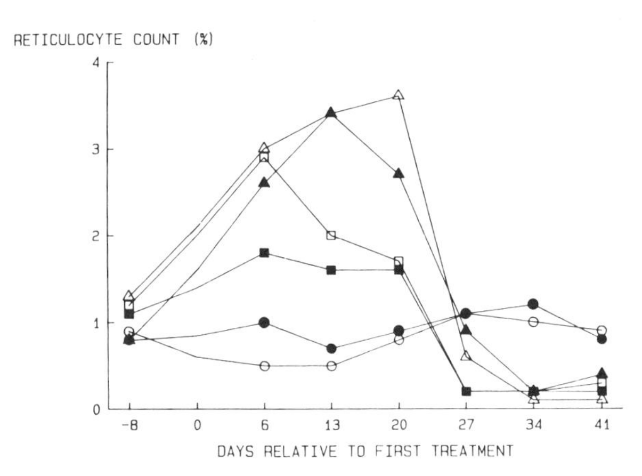

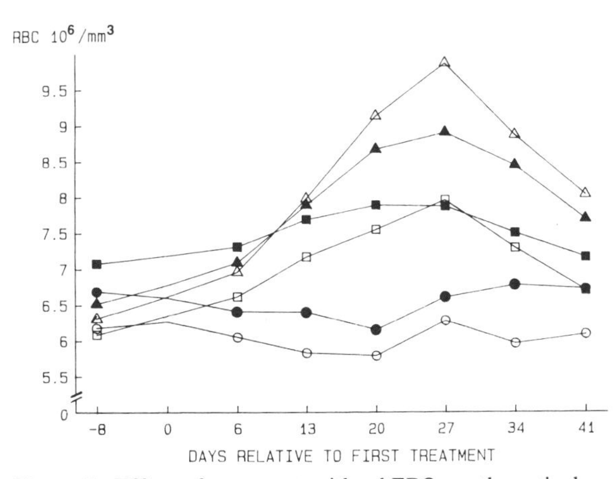

Figure 5. Effect of treatment with r-hEPO on the reticulocyte and red blood cell count of beagles. Young adult male (○, □, △) and female (●, ■ ▲) beagles were injected intravenously three times a week for 3 weeks with either a buffered saline control solution (○, ●) or a dose of r-hEPO at 280.0 U/kg (□, ■) or 2800 U/kg (△, ▲). The last treatment with r-hEPO was on day 20.

Previous studies of the pharmacokinetic features of EPO in experimental animals have involved use of biologically inactive, iodinated materials (Emmanouel et al. 1984). Production of EPO in a defined and controllable system has allowed labeling with ^{35}S-labeled amino acids during manufacture. This material has been found to be fully biologically active and its clearance is comparable to that of iodinated EPO, in terms of the rate of distribution and elimination. There was a modest reduction in clearance characteristics in functionally nephrectomized rats (A.C. Cohen et al., in prep.). These observations are encouraging in terms of the potential clinical use of EPO in patients with end-stage renal disease.

In view of the potential utility of EPO in patients with end-stage renal disease, we examined the responsiveness of nephrectomized rats to r-hEPO. Control animals were subjected to laparotomy without ligation of the kidneys. These animals showed no changes in blood urea nitrogen or hematocrit. Subtotal nephrectomized rats showed, as expected, increased blood urea nitrogen. Frank anemia was not observed in these animals presumably because insufficient renal mass had been removed. Both control (sham operated) and subtotal nephrectomized rats were responsive to EPO treatment in terms of increased hematocrits, as indicated by the results in Table 1. Because this occurred without significant changes in body weight, we conclude that the hematocrit increase is due to an increase in red blood cells and not simply contraction of intravascular fluid space.

Anemias associated with inflammatory diseases, such as rheumatoid arthritis, provide other possible indications for an agent such as EPO. In this case, abnormalities of iron turnover and compartmentalization might override the effects of EPO. Thus, we examined the effect of EPO on the anemia of adjuvant-linked inflammation in rats. Data (Table 2) show that this experimental anemia is corrected by treatment with EPO. Specifically, this type of anemia is a consequence of impaired recirculation of iron from the liver to the marrow. This was evidenced by the decrease in plasma iron and hemoglobin concentration in the adjuvant-treated rats. EPO corrected this situation by stimulating the production of erythrocytes. In the process, plasma iron was decreased further as it diverted into hemoglobin synthesis, as evidenced by a return of

Table 1. Treatment of Subtotally Nephrectomized Rats with r-hEP0

	Pretreatment		Postreatment			
	sham	nephx	sham + 0 EP0	sham + 10 units EP0	nephx + 0 EP0	nephx + 10 units EP0
Body weight (g)	210 ± 15	199 ± 16	234 ± 20	236 ± 11	218 ± 23	229 ± 12
Hematocrit	42.4 ± 2.2	40.8 ± 2.0	39.7 ± 1.1	46.8 ± 4.4[a]	41.9 ± 3.8	48.2 ± 3.2[a]
Blood urea nitrogen (mg/dl)	24.1 ± 4.6	40.1 ± 8.0[b]	29.5 ± 4.0	26.0 ± 5.4	45.5 ± 2.9[b]	47.0 ± 0.4[b]

[a]$p < 0.05$ vs. respective treatment control (0 EP0).
[b]$p < 0.05$ vs. sham control.

hemoglobin concentrations to normal values, with a concomitant decrease in the percentage of saturation. Total iron-binding capacity (TIBC) was unaffected by EPO.

DISCUSSION

Paucity of structural information, tissue of origin, and specific assays rendered the cloning of human EPO particularly difficult and required refinement of methods to allow identification of a single-gene copy in the genome using multiple DNA probes. The structural heterogeneity of urinary EPO, as exemplified by its disperse nature on SDS-PAGE, was ambiguous in terms of whether there was a family of related proteins and/or variable secondary modifications of the primary protein structure. The successful cloning and expression of the protein, as well as characterization of natural urinary EPO, has indicated that there is a single unique gene for this protein. The cloning approach used was risky in that it relied on considerable homology between human EPO and monkey EPO in the sequences used to construct probes. An alternative strategy involving use only of a human fetal cDNA library (Jacobs et al. 1985) was confounded by the possibility of a distinct fetal EPO gene. Current data indicate that there is only one gene for human EPO, and if there is also a distinct fetal gene, it is only distantly related in structure to the EPO gene described here. Although it is probable that EPO can be secreted by tissues other than the kidney (anepheric individuals retain the ability to produce low amounts of EPO), the results of the screening of monkey tissue mRNA clearly indicate that the kidney responds to anemia by increasing the rate of synthesis of EPO.

The EPO gene seems to be highly conserved in mammals. The amino acid sequences of the primary translation products in human and cynomolgus monkey EPO differ at 15 positions and the monkey protein is one residue shorter: The lysine at position 116 in human EPO is missing. Mature monkey EPO and human EPO also differ at the site of cleavage of the signal peptide. There are 33 positions different between mouse EPO and human EPO (McDonald et al. 1986). Mouse EPO has only three cysteine residues (position 33 is not a cysteine), so that the small loop maintained by a disulfide in human EPO (see Fig. 3) would not seem to be essential for activity. Estimates of evolution rates have been made from the three known mammalian EPO sequences and these indicate 1.3×10^{-9} amino acid substitutions per site per year. This is comparable (although slightly lower) to the rate of evolution of rapidly evolving proteins, such as the fibrinopeptides (9×10^{-9} substitutions/site/year), and much more rapid than the most slowly evolving proteins, such as histones (0.006×10^{-9} substitutions/site/year) (McDonald et al. 1986).

It is noteworthy that the N-linked glycosylation sites are conserved in human, monkey, and mouse EPO sequences. This is distinct from cases in which glycosylation sites in closely related proteins, even in the same species, are not conserved. For example, some subtypes of the human interferon-α family of proteins are glycosylated, but the sites are nonhomologous and distinct also from the glycosylation sites in human interferon-β, which is closely related in amino acid sequence to interferon-α (Stebbing 1986). A remarkable feature of human EPO produced in CHO cells is the similarity in the carbohydrate modifications that occur. The same broad spread of material on SDS-PAGE occurs with urinary EPO and CHO cell-derived EPO, and deglycosylation experiments yielded the same ratio of the N- and O-linked species. Furthermore, the car-

Table 2. Effect of r-hEP0 on the Anemia of Adjuvant-induced Inflammation in Rats

	Untreated		Adjuvant-treated	
	saline	EP0	saline	EP0
PCV (%)	43.6 ± 0.8	54.9 ± 1.3	43.2 ± 0.6	52.9 ± 0.8
RBCV (ml/100 g)	2.02 ± 0.02	2.75 ± 0.06	2.10 ± 0.07	2.72 ± 0.04
Hemoglobin (g/dl)	13.89 ± 0.36	14.87 ± 0.69	11.62 ± 0.31	13.60 ± 1.05
Plasma iron (μg/dl)	145.9 ± 12.8	50.7 ± 26.7	124.4 ± 9.5	31.0 ± 8.3
TIBC (μg/dl)	505.6 ± 18.1	476.3 ± 16.1	491.4 ± 23.3	522.0 ± 29.4
% Saturation	28.9 ± 2.3	10.4 ± 5.4	25.3 ± 1.6	5.8 ± 1.4

bohydrate composition of r-hEPO and urinary EPO is very similar. The mechanisms controlling these secondary modifications remain unclear. Recombinant EPO produced in *Escherichia coli*, and therefore lacking glycosylation, or r-hEPO deglycosylated enzymatically has greatly decreased in vivo activity, although in vitro, its biological activity is preserved. The reasons why glycosylation is important for biological activity are unclear.

A clinical use of EPO in end-stage renal disease is obvious from the biology reviewed here. Treatment with EPO in end-stage renal disease constitutes replacement therapy. As such, clinical schedules and doses should prove easier to establish than for other recombinant-DNA-derived materials, such as the various lymphokines now in clinical trials. The extent to which EPO may be useful in other forms of anemia remains to be established, but the preclinical studies carried out so far are promising with regard to anemia of cancers and inflammatory diseases.

ACKNOWLEDGMENTS

We thank Dr. Peter Dukes for performing the in vivo mouse exhypoxic polycythemic EPO bioassay, Joan Bennett and Sandi Olson for preparation of the manuscript, Jeanne Fitzgerald and Julie Heuston for the artwork, and Daniel Vapnek for critical reading of the manuscript.

REFERENCES

Anagnostou, A., J. Barone, A. Kedo, and W. Fried. 1977. Effect of erythropoietin therapy on the red cell volume of uraemic and non-uraemic rats. *Br J. Hematol.* **37:** 85.

Annable, L., M. Cotes, and M.V. Musset. 1972. The second international reference preparation of erythropoietin, human urinary, for bioassay. *Bull. W.H.O.* **47:** 99.

Brown, R. 1965. Plasma erythropoietin in chronic uremia. *Br. Med. J.* **2:** 1036.

Carnot, P. and C. Deflandre. 1906. Sur l'activite hematopoietique des differents organes au cours de la regeneration du sang. *C.R. Acad. Sci. D.* **143:** 432.

Cotes, P.M. 1982. Immunoreactive erythropoietin in serum. *Br. J. Hematol.* **50:** 427.

Cotes, P.M. and D.R. Bangham. 1961. Bioassay of erythropoietin in mice made polycythemic by exposure to air at reduced pressure. *Nature* **191:** 1065.

Dordal, M.S., F.F. Wang, and E. Goldwasser. 1985. The role of carbohydrate in erythropoietin action. *Endocrinology* **116:** 2293.

Egrie, J.C. 1983. Monoclonal antibodies to the N-terminal region of human erythropoietin. *Hybridoma* **2:** 136.

Egrie, J.C., J.K. Browne, P. Lai, and F.-K. Lin. 1985. Characterization of recombinant monkey and human erythropoietin. In *Experimental approaches for the study of hemoglobin switching* (ed. G. Stamatoyanopoulos and A.W. Nienhuis), p. 339. A.R. Liss, New York.

Egrie, J.C., T.W. Strickland, J. Lane, K. Aoki, A.M. Cohen, R. Smalling, G. Trail, F.-K. Lin, J.K. Browne, and D.K. Hines. 1986. Characterization and biological effects of recombinant human erythropoietin. *Immunobiology* (in press).

Emmanouel, D.S., E. Goldwasser, and A.I. Katz. 1984. Metabolism of pure human erythropoietin in the rat. *Am. J. Physiol.* **247:** F168.

Erslev, A.J., J. Caro, E. Kansu, and R. Silver. 1980. Renal and extrarenal erythropoietin production in anemic rats. *Br. J. Hematol.* **45:** 65.

Eschbach, J.W. and J.W. Adamson. 1985. Anemia of end stage renal disease (ESRD). *Kidney Int.* **28:** 1.

Fried, W. 1972. The liver as a source of extrarenal erythropoietin. *Blood* **40:** 671.

Fyhrquist, F., K. Rosenlof, C. Gronhagen-Riska, L. Hortling, and I. Tikkanen. 1984. Is renin substrate an erythropoietin precursor? *Nature* **308:** 649.

Garcia, J.F., S.N. Ebbe, L. Hollander, H.O. Cutting, M.E. Miller, and E.P. Cronkite. 1982. Radioimmunoassay of erythropoietin: Circulating levels in normal and polycythemic human beings. *J. Lab. Clin. Med.* **99:** 624.

Gluzman, Y. 1981. SV40-transformed simian cells support the replication of early SV40 mutants. *Cell* **23:** 175.

Goldwasser, E. and C.K.H. Kung. 1971. Purification of erythropoietin. *Proc. Natl. Acad. Sci.* **68:** 697.

Goldwasser, E., J.F. Eliason, and D. Sikkema. 1975. An assay for erythropoietin *in vitro* at the milliunit level. *Endocrinology* **97:** 315.

Graber, S.E. and S.B. Krantz. 1978. Erythropoietin and the control of red cell production. *Annu. Rev. Med.* **29:** 51.

Graham, F.L. and A.J. van der Eb. 1973. A new technique for the assay of infectivity of adenovirus 5 DNA. *Virology* **52:** 456.

Graham, F.L., J. Smiley, W.C. Russel, and R. Nairn. 1977. Characteristics of a human cell line transformed by DNA from adenovirus type 5. *J. Gen. Virol.* **36:** 59.

Jacobs, K., C. Shoemaker, R. Rurensdorf, S.D. Neill, R.J. Kaufman, H. Mufson, J. Seehra, S.S. Jones, R. Hewick, E.F. Fritsch, M. Kawakita, T. Shimizu, and T. Miyake. 1985. Isolation and characterization of genomic and cDNA clones of human erythropoietin. *Nature* **313:** 806.

Jacobson, L.O., E. Goldwasser, W. Fried, and L. Plzak. 1957. Studies on erythropoiesis. VII. The role of the kidney in the production of erythropoietin. *Trans. Assoc. Am. Physicians* **70:** 305.

Koeffler, H.P. and E. Goldwasser. 1981. Erythropoietin radioimmunoassay in evaluating patients with polycythemia. *Ann. Intern. Med.* **94:** 44.

Krantz, S.B. and L.O. Jacobsen. 1970. *Erythropoietin and the regulation of erythropoiesis*, p. 325. University of Chicago Press, Chicago, Illinois.

Lai, P.-H., R. Everett, F.F. Wang, T. Arakawa, and E. Goldwasser. 1986. The primary structure of human erythropoietin. *J. Biol. Chem.* **261:** 3116.

Lin, F.-K., C.H. Lin, P.-H. Lai, J.K. Browne, J.C. Egrie, R. Smalling, G.M. Fox, K.K. Chen, M. Castro, and S. Suggs. 1986. Monkey erythropoietin gene: Cloning, expression and comparison with the human erythropoietin gene. *Gene* **44:** 201.

Lin, F.-K., S. Suggs, C.-H. Lin, J.K. Browne, R. Smalling, J.C. Egrie, K.K. Chen, G.M. Fox, F. Martin, Z. Stabinsky, S.M. Badrawi, P.-H. Lai, and E. Goldwasser. 1985. Cloning and expression of the human erythropoietin gene. *Proc. Natl. Acad. Sci.* **82:** 7580.

Lukens, J.N., G.E. Cartwright, and M.M. Wintrobe. 1967. Anemia of adjuvant-induced inflammation in rats. *Proc. Soc. Exp. Biol. Med.* **126:** 346.

Maxam, A.M. and W. Gilbert. 1980. Sequencing end-labeled DNA with base-specific chemical cleavages. *Methods Enzymol.* **65:** 499.

McDonald, J.D., F.-K. Lin, and E. Goldwasser. 1986. Cloning, sequencing, and evolutionary analysis of the mouse erythropoietin gene. *Mol. Cell. Biol.* **6:** 842.

Miyake, T., C.K.H. Kung, and E. Goldwasser. 1977. Purification of human erythropoietin. *J. Biol. Chem.* **252:** 5558.

Mount, S.M. 1982. A catalog of splice junction sequences. *Nucleic Acids Res.* **10:** 459.

Naets, J.P. 1975. Hematological disorders in renal failure. *Nephron* **14:** 181.

Sanger, F., S. Nicklen, and A.R. Coulson. 1977. DNA sequencing with chain-terminating inhibitors. *Proc. Natl. Acad. Sci.* **74:** 5463.

Southern, E. 1975. Detection of specific sequences among DNA fragments separated by gel electrophoresis. *J. Mol. Biol.* **98:** 503.

Spivak, J.L. and S.E. Graber. 1980. Erythropoietin and the regulation of erythropoiesis. *Johns Hopkins Med. J.* **146:** 311.

Stebbing, N. 1986. Mechanisms of action of interferons: Evidence from studies with recombinant DNA derived subtypes and analogs. In *Mechanisms of interferon actions* (ed. L.M. Pfeffer). CRC Press, Boca Raton, Florida. (In press.)

Vieira, J. and J. Messing. 1982. The pUC plasmids, and M13mp7-derived system for insertion mutagenesis and sequencing with synthetic universal primers. *Gene* **19:** 259.

Wigler, M., A. Pellicer, S. Silverstein, and R. Axel. 1978. Biochemical transfer of single-copy eucaryotic genes using total cellular DNA as donor. *Cell* **14:** 725.

Zanetta, J.P., W.C. Breckenridge, and G. Vincendon. 1972. Analysis of monosaccharides by gas liquid chromatography of *O*-methyl glucosides as trifluoracetate derivatives. Application to glycoproteins and glycolipids. *J. Chromatogr.* **69:** 291.

Zanjani, E.D., J.L. Ascensao, P.B. McGlave, M. Banisadre, and R.C. Ash. 1981. Studies on the liver to kidney switch of erythropoietin production. *J. Clin. Invest.* **67:** 1183.